# Quantitative Chemical Analysis

["The Experiment" by Sempe © C. Charillon, Paris.]

# Quantitative Chemical Analysis

## SEVENTH EDITION

## Daniel C. Harris

*Michelson Laboratory*
*China Lake, California*

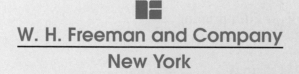

W. H. Freeman and Company

New York

*Publisher:* Craig Bleyer
*Senior Acquisitions Editor:* Jessica Fiorillo
*Marketing Manager:* Anthony Palmiotto
*Media Editor:* Victoria Anderson
*Associate Editor:* Amy Thorne
*Photo Editors:* Cecilia Varas/Donna Ranieri
*Design Manager:* Diana Blume
*Cover Designer:* Trina Donini
*Text Designer:* Rae Grant
*Text Layout:* Jerry Wilke
*Senior Project Editor:* Mary Louise Byrd
*Illustrations:* Fine Line Illustrations
*Illustration Coordinators:* Shawn Churchman/Susan Timmins
*Production Coordinator:* Paul W. Rohloff
*Composition:* TechBooks/GTS Companies, York, PA
*Printing and Binding:* RR Donnelley

Library of Congress Control Number: 2006922923

ISBN-13: 978-0-7167-7041-1
ISBN-10: 0-7167-7041-5

Printed in the United States of America

Fifth printing

W. H. Freeman and Company
41 Madison Avenue
New York, NY 10010
Houndmills, Basingstoke
RG21 6XS, England
www.whfreeman.com

# Brief Contents

# Contents

## Experiments

Experiments are found at the Web site
**www.whfreeman.com/qca7e**

1. Calibration of Volumetric Glassware
2. Gravimetric Determination of Calcium as
    $CaC_2O_4 \cdot H_2O$

## Spreadsheet Topics

Felicia

Abraham

Arthur

*My grandchildren assure me that the future is bright.*

# Preface

One of our most pressing problems is the need for sources of energy to replace oil. The chart at the right shows that world production of oil per capita has probably already peaked. Oil will play a decreasing role as an energy source and should be more valuable as a raw material than as a fuel. There is also strong pressure to minimize the burning of fuels that produce carbon dioxide, which could be altering Earth's climate.

It is my hope that some of you reading this book will become scientists, engineers, and enlightened policy makers who will find efficient, sustainable ways to harness energy from sunlight, wind, waves, biomass, and nuclear fission and fusion. Nuclear fission is far less polluting than burning oil, but difficult problems of waste containment are unsolved. Much coal remains, but coal creates carbon dioxide and more air pollution than any major energy source. There is a public misconception that hydrogen is a source of energy. Hydrogen requires energy to make and is only a means of storing energy. There are also serious questions about whether ethanol provides more energy than is required for its production. More efficient use of energy will play a major role in reducing demand. No source of energy is sufficient if our population continues to grow.

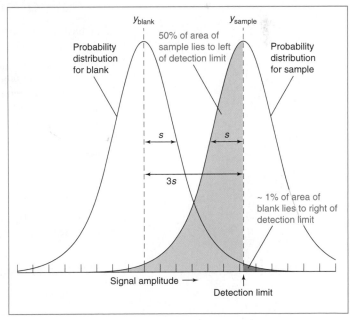

Per capita production of oil peaked in the 1970s and is expected to decrease in coming decades.*

## Goals of This Book

My goals are to provide a sound physical understanding of the principles of analytical chemistry and to show how these principles are applied in chemistry and related disciplines—especially in life sciences and environmental science. I have attempted to present the subject in a rigorous, readable, and interesting manner that will appeal to students whether or not their primary interest is chemistry. I intend the material to be lucid enough for nonchemistry majors yet to contain the depth required by advanced undergraduates. This book grew out of an introductory analytical chemistry course that I taught mainly for nonmajors at the University of California at Davis and from a course for third-year chemistry students at Franklin and Marshall College in Lancaster, Pennsylvania.

## What's New?

In the seventh edition, quality assurance was moved from the back of the book into Chapter 5 to emphasize the increasing importance attached to this subject and to link it closely to statistics and calibration. Two chapters on activity coefficients and the systematic treatment of equilibrium from the sixth edition were condensed into Chapter 8. A new, advanced treatment of equilibrium appears in Chapter 13. This chapter, which requires spreadsheets, is going to be skipped in introductory courses but should be of value for advanced undergraduate or graduate work. New topics in the rest of this book include the acidity of metal ions in Chapter 6, a revised discussion of ion sizes and an example of experimental design in Chapter 8, pH of zero charge for colloids

Quality assurance applies concepts from statistics.

*Oil production data can be found at http://bp.com/worldenergy. See also D. Goodstein, *Out of Gas* (New York: W. W. Norton, 2004); K. S. Deffeyes, *Beyond Oil: The View from Hubbert's Peak* (New York: Farrar, Straus and Giroux, 2005); and R. C. Duncan, "World Energy Production, Population Growth, and the Road to the Olduvai Gorge," *Population and Environment* **2001**, *22, 503* (or HubbertPeak.com/Duncan/Olduvai2000.htm).

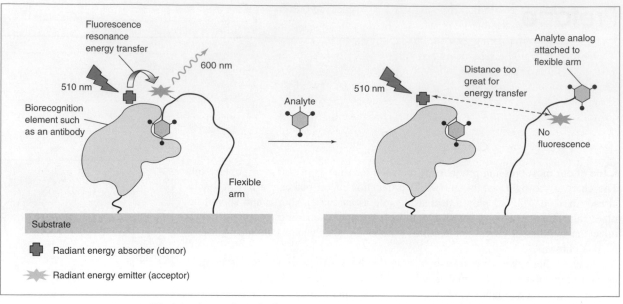

in Chapter 10, monoclonal antibodies in Chapter 12, more on microelectrodes and the Karl Fischer titration in Chapter 17, self-absorption in fluorescence in Chapter 18, surface plasmon resonance and intracellular oxygen sensing in Chapter 20, ion mobility spectrometry for airport explosive sniffers in Chapter 22, a microscopic description of chromatography in Chapter 23, illustrations of the effects of column parameters on separations in gas chromatography in Chapter 24, advances in liquid chromatography stationary phases and more detail on gradient separations in Chapter 25, automation of ion chromatography in Chapter 26, and sample concentration by sweeping in electrophoresis in Chapter 26. Updates to many existing topics are found throughout the book. Chapter 27 on gravimetric analysis now includes an example taken from the Ph.D. thesis of Marie Curie from 1903 and a description of how 20-year-old Arthur Holmes measured the geologic time scale in 1910.

## Applications

A basic tenet of this book is to introduce and illustrate topics with concrete, interesting examples. In addition to their pedagogic value, Chapter Openers, Boxes, Demonstrations, and Color Plates are intended to help lighten the load of a very dense subject. I hope you will find these features interesting and informative. **Chapter Openers** show the relevance of analytical chemistry to the real world and to other disciplines of science. I can't come to your classroom to present **Chemical Demonstrations,** but I can tell you about some of my favorites and show you color photos of how they look. **Color Plates** are located near the center of the book. **Boxes** discuss interesting topics related to what you are studying or they amplify points in the text.

New boxed applications include an arsenic biosensor (Chapter 0), microcantilevers to measure attograms of mass (Chapter 2), molecular wire (Chapter 14), a fluorescence resonance energy transfer biosensor (Chapter 19), cavity ring-down spectroscopy for ulcer diagnosis (Chapter 20), and environmental mercury analysis by atomic fluorescence (Chapter 21).

## Problem Solving

Nobody can do your learning for you. The two most important ways to master this course are to work problems and to gain experience in the laboratory. Worked **Examples** are a principal pedagogic tool designed to teach problem solving and to illustrate how to apply what you have just read. There are Exercises and Problems at the end of each chapter. **Exercises** are the minimum set of problems that apply most major concepts of each chapter. Please struggle mightily with an Exercise before consulting the solution at the back of the book. **Problems** cover the entire content of the book. **Short answers** to numerical problems are at the back of the book and complete solutions appear in the **Solutions Manual**.

**Spreadsheets** are indispensable tools for science and engineering. You can cover this book without using spreadsheets, but you will never regret taking the time to learn to use them. The text explains how to use spreadsheets and some problems ask you to apply them. If you are comfortable with spreadsheets, you will use them even when the problem does not ask you to. A few of the powerful built-in features of Microsoft Excel are described as they are needed. These features include graphing in Chapter 2, statistical functions and regression in Chapter 4, multiple regression for experimental design in Chapter 7, solving equations with GOAL SEEK in Chapters 6, 8, and 9, SOLVER in Chapters 13 and 19, and matrix operations in Chapter 19.

|  | A | B | C | D | E | F | G |
|---|---|---|---|---|---|---|---|
| 1 | x | y |  |  | Output from LINEST |  |  |
| 2 | 1 | 2 |  |  | Slope | Intercept |  |
| 3 | 3 | 3 |  | Parameter | 0.61538 | 1.34615 |  |
| 4 | 4 | 4 |  | Std Dev | 0.05439 | 0.21414 |  |
| 5 | 6 | 5 |  | R^2 | 0.98462 | 0.19612 | Std Dev (y) |
| 6 |  |  |  |  |  |  |  |
| 7 | Highlight cells E3:F5 |  |  |  |  |  |  |
| 8 | Type "= LINEST(B2:B5,A2:A5,TRUE,TRUE)" |  |  |  |  |  |  |
| 9 | Press CTRL + SHIFT + ENTER (on PC) |  |  |  |  |  |  |
| 10 | Press COMMAND + RETURN (on Mac) |  |  |  |  |  |  |

Spreadsheets are indispensable tools.

# Other Features of This Book

**Terms to Understand** Essential vocabulary, highlighted in **boldface** in the text or, sometimes, in color in the margin, is collected at the end of the chapter. Other unfamiliar or new terms are *italic* in the text, but are not listed at the end of the chapter.

**Glossary** All boldface vocabulary terms and many of the italic terms are defined in the glossary at the back of the book.

**Appendixes** Tables of solubility products, acid dissociation constants (updated to 2001 values), redox potentials, and formation constants appear at the back of the book. You will also find discussions of logarithms and exponents, equations of a straight line, propagation of error, balancing redox equations, normality, and analytical standards.

**Notes and References** Citations in the chapters appear at the end of the book.

**Inside Cover** Here are your trusty periodic table, physical constants, and other useful information.

# Supplements

## NEW! eBook

This online version of *Quantitative Chemical Analysis, Seventh Edition* combines the text and all existing student media resources, along with additional eBook features. The eBook includes

- **Intuitive navigation** to any section or subsection, as well as any printed book page number.
- In-text links to all glossary **term definitions.**
- **Bookmarking, Highlighting,** and **Notes** features, with all activity automatically saved, allow students or instructors to add notes to any page.
- A full **glossary** and **index** and **full-text search.**

**For instructors,** the eBook offers unparalleled flexibility and customization options, including

- **Custom chapter selection:** students will access only chapters the instructor selects.
- **Instructor notes:** Instructors can incorporate notes used for their course into the eBook. Students will automatically get the customized version. Notes can include text, Web links, and even images.

The *Solutions Manual for Quantitative Chemical Analysis* contains complete solutions to all problems.

The **Student Web Site,** www.whfreeman.com/qca7e, has **directions for experiments** that may be reproduced for your use. At this Web site, you will also find **lists of experiments** from the *Journal of Chemical Education,* a few **downloadable Excel spreadsheets,** and a few **Living Graph Java applets** that allow students to manipulate graphs by altering data points and variables. **Supplementary topics** at the Web site include spreadsheets for precipitation titrations, microequilibrium constants, spreadsheets for redox titration curves, and analysis of variance.

The **Instructors' Web Site,** www.whfreeman.com/qca7e, has all illustrations and tables from the book in preformatted PowerPoint slides.

## The People

A book of this size and complexity is the work of many people. At W. H. Freeman and Company, Jessica Fiorillo provided guidance and feedback and was especially helpful in ferreting out the opinions of instructors. Mary Louise Byrd shepherded the manuscript through production with her magic wand and is most responsible for creating the physical appearance of this book. Patty Zimmerman edited the copy with great care. The design was created by Diana Blume. Pages were laid out by Jerry Wilke and proofread by Karen Osborne. Photo editing and research was done by Cecilia Varas and Donna Ranieri. Paul Rohloff had overall responsibility for production.

Julian Roberts of the University of Redlands twisted my arm until I created the new Chapter 13, and he provided considerable content and critique. My consultants at Michelson Laboratory, Mike Seltzer and Eric Erickson, were helpful, as always. Solutions to problems and exercises were checked by Samantha Hawkins at Michelson Lab and Teh Yun Ling in Singapore.

My wife, Sally, worked on every aspect of this book and the *Solutions Manual*. She contributes mightily to whatever clarity and accuracy we have achieved.

## In Closing

This book is dedicated to the students who use it, who occasionally smile when they read it, who gain new insight, and who feel satisfaction after struggling to solve a problem. I have been successful if this book helps you develop critical, independent reasoning that you can apply to new problems. I truly relish your comments, criticisms, suggestions, and corrections. Please address correspondence to me at the Chemistry Division (Mail Stop 6303), Research Department, Michelson Laboratory, China Lake, CA 93555.

Dan Harris

## Acknowledgments

I am indebted to users of the sixth edition who offered corrections and suggestions and to the many people who reviewed parts of the current manuscript. John Haberman at NASA provided a great deal of help in creating the back cover of this book. Bill Schinzer (Pfizer, Inc.) offered comments and information about the Karl Fischer titration. Athula Attygalle (Stevens Institute of Technology) pointed out my misinterpretation of Kielland's "ion sizes," which led to a revision of Chapter 8. Krishnan Rajeshwar (University of Texas, Arlington) had many helpful suggestions, especially for electrochemistry. Carl E. Moore (Emeritus Professor, Loyola University, Chicago) educated me on the history of the pH electrode and the pH meter. Herb Hill (Washington State University) and G. A. Eiceman (New Mexico State University) were most gracious in providing comments and information on ion mobility spectrometry. Nebojsa Avdalovic (Dionex Corporation) provided key information on automation of ion chromatography. Shigeru Terabe (University of Hyogo, Japan) and Robert Weinberger helped with electrophoresis. Other corrections, suggestions, and helpful comments were provided by James Gordon (Central Methodist University, Fayette, Missouri), Dick Zare (Stanford University), D. Bax (Utrecht University, The Netherlands), Keith Kuwata (Macalester College), David Green (Albion College), Joe Foley (Drexel University), Frank Dalton (Pine Instrument Company), David Riese (Purdue School of Pharmacy), Igor Kaltashov (University of Massachusetts, Amherst), Suzanne Pearce (Kwantlen University College, British Columbia), Patrick Burton (Socorro, New Mexico), Bing Xu (Hong Kong), and Stuart Larsen (New Zealand).

People who reviewed parts of the seventh-edition manuscript or who reviewed the sixth edition to make suggestions for the seventh edition included David E. Alonso (Andrews University), Dean Atkinson (Portland State University), James Boiani (State University of New York, Geneseo), Mark Bryant, (Manchester College), Houston Byrd (University of Montevallo), Donald Castillo (Wofford College), Nikolay Dimitrov (State University of New York, Binghamton), John Ejnik (Northern Michigan University), Facundo Fernandez (Georgia Institute of Technology), Augustus Fountain (U.S. Military Academy), Andreas Gebauer (California State University, Bakersfield), Jennifer Ropp Goodnough (University of Minnesota, Morris), David W. Green (Albion College), C. Alton Hassell (Baylor University), Dale Hawley (Kansas State University), John Hedstrom (Luther College, Decorah, Iowa), Dan Heglund (South Dakota School of Mines and Technology), David Henderson (Trinity College, Hartford), Kenneth Hess (Franklin and Marshall College), Shauna Hiley (Missouri

Western State University), Elizabeth Jensen (Aquinas College, Grand Rapids), Mark Krahling (University of Southern Indiana), Barbara Kramer (Truman State University), Brian Lamp (Truman State University), Lisa B. Lewis (Albion College), Sharon McCarthy (Chicago State University), David McCurdy (Truman State University), Mysore Mohan (Texas A&M University), Kenneth Mopper (Old Dominion University), Richard Peterson (Northern State University, Aberdeen, South Dakota), David Rahni (Pace University, Pleasantville/Briarcliff), Gary Rayson (New Mexico State University), Steve Reid (University of Saskatchewan), Tracey Simmons-Willis (Texas Southern University), Julianne Smist (Springfield College, Massachusetts), Touradj Solouki (University of Maine), Thomas M. Spudich (Mercyhurst College), Craig Taylor (Oakland University), Sheryl A. Tucker (University of Missouri, Columbia), Amy Witter (Dickinson College), and Kris Varazo (Francis-Marion University).

# 0 | The Analytical Process

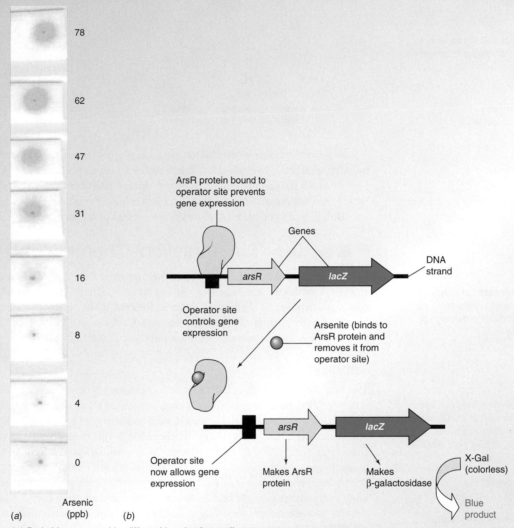

(a) Test strips exposed to different levels of arsenite. *[Courtesy J. R. van der Meer, Université de Lausanne, Switzerland.]*
(b) How the genetically engineered DNA works.

In Bangladesh, 15–25% of the population is exposed to unsafe levels of arsenic in drinking water from aquifers in contact with arsenic-containing minerals. The analytical problem is to reliably and cheaply identify wells in which arsenic is above 50 parts per billion (ppb). Arsenic at this level causes vascular and skin diseases and cancer.

Panel (*a*) shows 8 test strips impregnated with genetically engineered *E. coli* bacteria whose genes are turned on by arsenite ($HAsO_3^{2-}$). When the strips are exposed to drinking water, a blue spot develops whose size increases with the concentration of arsenite in the water. By comparing the spot with a set of standards, we can estimate whether arsenic is above or below 50 ppb. We call the test strip a *biosensor*, because it uses biological components in its operation.

Panel (*b*) shows how the assay works. Genetically engineered DNA in *E. coli* contains the gene *arsR*, which encodes the regulatory protein ArsR, and the gene *lacZ*, which encodes the protein β-galactosidase. ArsR binds to regulatory sites on the gene to prevent DNA transcription. Arsenite causes ArsR to dissociate from the gene and the cell proceeds to manufacture both ArsR and β-galactosidase. Then β-galactosidase transforms a synthetic, colorless substance called X-Gal in the test strip into a blue product. The more arsenite, the more intense the color.

Chocolate is great to eat, but not so easy to analyze. [W. H. Freeman photo by K. Bendo.]

A *diuretic* makes you urinate.
A *vasodilator* enlarges blood vessels.

Notes and references are listed at the back of the book.

*Chemical Abstracts* is the most comprehensive source for locating articles published in chemistry journals. *Scifinder* is software that accesses *Chemical Abstracts*.

**Bold** terms should be learned. They are listed at the end of the chapter and in the Glossary at the back of the book. *Italicized* words are less important, but many of their definitions are also found in the Glossary.

*Homogeneous:* same throughout

*Heterogeneous:* differs from region to region

Pestle

Mortar

*Figure 0-1* Ceramic mortar and pestle used to grind solids into fine powders.

Chocolate[3] has been the savior of many a student on the long night before a major assignment was due. My favorite chocolate bar, jammed with 33% fat and 47% sugar, propels me over mountains in California's Sierra Nevada. In addition to its high energy content, chocolate packs an extra punch with the stimulant caffeine and its biochemical precursor, theobromine.

Theobromine
A diuretic, smooth muscle relaxant, cardiac stimulant, and vasodilator

Caffeine
A central nervous system stimulant

Too much caffeine is harmful for many people, and even small amounts cannot be tolerated by some unlucky individuals. How much caffeine is in a chocolate bar? How does that amount compare with the quantity in coffee or soft drinks? At Bates College in Maine, Professor Tom Wenzel teaches his students chemical problem solving through questions such as these.[4]

But, how *do* you measure the caffeine content of a chocolate bar?

## ■ ■ ■ ■  0-1  The Analytical Chemist's Job

Two students, Denby and Scott, began their quest at the library with a computer search for analytical methods. Searching with the key words "caffeine" and "chocolate," they uncovered numerous articles in chemistry journals. Reports titled "High Pressure Liquid Chromatographic Determination of Theobromine and Caffeine in Cocoa and Chocolate Products"[5] described a procedure suitable for the equipment in their laboratory.[6]

### Sampling

The first step in any chemical analysis is procuring a representative sample to measure—a process called **sampling.** Is all chocolate the same? Of course not. Denby and Scott bought one chocolate bar in the neighborhood store and analyzed pieces of it. If you wanted to make broad statements about "caffeine in chocolate," you would need to analyze a variety of chocolates from different manufacturers. You would also need to measure multiple samples of each type to determine the range of caffeine in each kind of chocolate.

A pure chocolate bar is fairly **homogeneous,** which means that its composition is the same everywhere. It might be safe to assume that a piece from one end has the same caffeine content as a piece from the other end. Chocolate with a macadamia nut in the middle is an example of a **heterogeneous** material—one whose composition differs from place to place. The nut is different from the chocolate. To sample a heterogeneous material, you need to use a strategy different from that used to sample a homogeneous material. You would need to know the average mass of chocolate and the average mass of nuts in many candies. You would need to know the average caffeine content of the chocolate and of the macadamia nut (if it has any caffeine). Only then could you make a statement about the average caffeine content of macadamia chocolate.

### Sample Preparation

The first step in the procedure calls for weighing out some chocolate and extracting fat from it by dissolving the fat in a hydrocarbon solvent. Fat needs to be removed because it would interfere with chromatography later in the analysis. Unfortunately, if you just shake a chunk of chocolate with solvent, extraction is not very effective, because the solvent has no access to the inside of the chocolate. So, our resourceful students sliced the chocolate into small bits and placed the pieces into a mortar and pestle (Figure 0-1), thinking they would grind the solid into small particles.

Imagine trying to grind chocolate! The solid is too soft to be ground. So Denby and Scott froze the mortar and pestle with its load of sliced chocolate. Once the chocolate

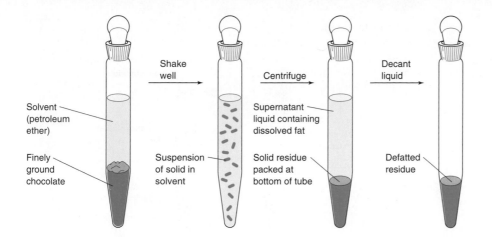

Figure 0-2 Extracting fat from chocolate to leave defatted solid residue for analysis.

**Solvent (petroleum ether)**

**Shake well**

**Finely ground chocolate**

**Suspension of solid in solvent**

**Centrifuge**

**Supernatant liquid containing dissolved fat**

**Solid residue packed at bottom of tube**

**Decant liquid**

**Defatted residue**

was cold, it was brittle enough to grind. Then small pieces were placed in a preweighed 15-milliliter (mL) centrifuge tube, and their mass was noted.

Figure 0-2 shows the next part of the procedure. A 10-mL portion of the solvent, petroleum ether, was added to the tube, and the top was capped with a stopper. The tube was shaken vigorously to dissolve fat from the solid chocolate into the solvent. Caffeine and theobromine are insoluble in this solvent. The mixture of liquid and fine particles was then spun in a centrifuge to pack the chocolate at the bottom of the tube. The clear liquid, containing dissolved fat, could now be **decanted** (poured off) and discarded. Extraction with fresh portions of solvent was repeated twice more to ensure complete removal of fat from the chocolate. Residual solvent in the chocolate was finally removed by heating the centrifuge tube in a beaker of boiling water. The mass of chocolate residue could be calculated by weighing the centrifuge tube plus its content of defatted chocolate residue and subtracting the known mass of the empty tube.

Substances being measured—caffeine and theobromine in this case—are called **analytes.** The next step in the sample preparation procedure was to make a **quantitative transfer** (a complete transfer) of the fat-free chocolate residue to an Erlenmeyer flask and to dissolve the analytes in water for the chemical analysis. If any residue were not transferred from the tube to the flask, then the final analysis would be in error because not all of the analyte would be present. To perform the quantitative transfer, Denby and Scott added a few milliliters of pure water to the centrifuge tube and used stirring and heating to dissolve or suspend as much of the chocolate as possible. Then they poured the **slurry** (a suspension of solid in a liquid) into a 50-mL flask. They repeated the procedure several times with fresh portions of water to ensure that every bit of chocolate was transferred from the centrifuge tube to the flask.

To complete the dissolution of analytes, Denby and Scott added water to bring the volume up to about 30 mL. They heated the flask in a boiling water bath to extract all the caffeine and theobromine from the chocolate into the water. To compute the quantity of analyte later, the total mass of solvent (water) must be accurately known. Denby and Scott knew the mass of chocolate residue in the centrifuge tube and they knew the mass of the empty Erlenmeyer flask. So they put the flask on a balance and added water drop by drop until there were exactly 33.3 g of water in the flask. Later, they would compare known solutions of pure analyte in water with the unknown solution containing 33.3 g of water.

Before Denby and Scott could inject the unknown solution into a chromatograph for the chemical analysis, they had to clean up the unknown even further (Figure 0-3). The slurry of chocolate residue in water contained tiny solid particles that would surely clog their expensive chromatography column and ruin it. So they transferred a portion of the slurry to a centrifuge tube and centrifuged the mixture to pack as much of the solid as possible at the bottom of the tube. The cloudy, tan **supernatant liquid** (liquid above the packed solid) was then filtered in a further attempt to remove tiny particles of solid from the liquid.

It is critical to avoid injecting solids into a chromatography column, but the tan liquid still looked cloudy. So Denby and Scott took turns between classes to repeat the centrifugation and filtration five times. After each cycle in which the supernatant liquid was filtered and centrifuged, it became a little cleaner. But the liquid was never completely clear. Given enough time, more solid always seemed to precipitate from the filtered solution.

The tedious procedure described so far is called **sample preparation**—transforming a sample into a state that is suitable for analysis. In this case, fat had to be removed from the

A solution of anything in water is called an aqueous solution.

Real-life samples rarely cooperate with you!

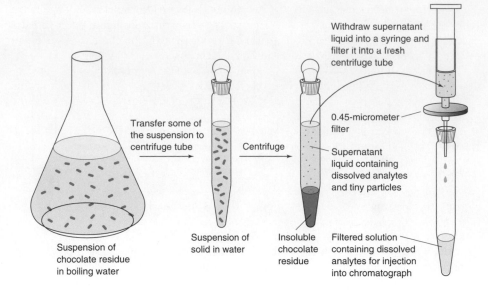

**Figure 0-3** Centrifugation and filtration are used to separate undesired solid residue from the aqueous solution of analytes.

Withdraw supernatant liquid into a syringe and filter it into a fresh centrifuge tube

Transfer some of the suspension to centrifuge tube

Centrifuge

0.45-micrometer filter

Supernatant liquid containing dissolved analytes and tiny particles

Suspension of chocolate residue in boiling water

Suspension of solid in water

Insoluble chocolate residue

Filtered solution containing dissolved analytes for injection into chromatograph

chocolate, analytes had to be extracted into water, and residual solid had to be separated from the water.

## The Chemical Analysis (At Last!)

Denby and Scott finally decided that the solution of analytes was as clean as they could make it in the time available. The next step was to inject solution into a *chromatography* column, which would separate the analytes and measure the quantity of each. The column in Figure 0-4a is packed with tiny particles of silica ($SiO_2$) to which are attached long hydrocarbon molecules. Twenty microliters ($20.0 \times 10^{-6}$ liters) of the chocolate extract were injected into the column and washed through with a solvent made by mixing 79 mL of pure water, 20 mL of methanol, and 1 mL of acetic acid. Caffeine is more soluble than theobromine in the hydrocarbon on the silica surface. Therefore, caffeine "sticks" to the coated silica particles in the column more strongly than theobromine does. When both analytes are flushed through the column by solvent, theobromine reaches the outlet before caffeine (Figure 0-4b).

Chromatography solvent is selected by a systematic trial-and-error process described in Chapter 25. The function of the acetic acid is to react with negatively charged oxygen atoms that lie on the silica surface and, when not neutralized, tightly bind a small fraction of caffeine and theobromine.

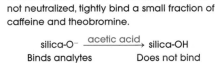

silica-O⁻ $\xrightarrow{\text{acetic acid}}$ silica-OH
Binds analytes very tightly          Does not bind analytes strongly

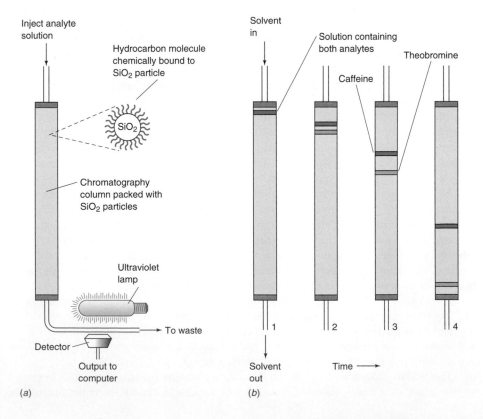

**Figure 0-4** Principle of liquid chromatography. (*a*) Chromatography apparatus with an ultraviolet absorbance monitor to detect analytes at the column outlet. (*b*) Separation of caffeine and theobromine by chromatography. Caffeine is more soluble than theobromine in the hydrocarbon layer on the particles in the column. Therefore, caffeine is retained more strongly and moves through the column more slowly than theobromine.

Inject analyte solution

Hydrocarbon molecule chemically bound to $SiO_2$ particle

Chromatography column packed with $SiO_2$ particles

Ultraviolet lamp

To waste

Detector

Output to computer

(a)

Solvent in

Solution containing both analytes

Theobromine

Caffeine

1    2    3    4

Solvent out

Time ⟶

(b)

Analytes are detected at the outlet by their ability to absorb ultraviolet radiation from the lamp in Figure 0-4a. The graph of detector response versus time in Figure 0-5 is called a *chromatogram.* Theobromine and caffeine are the major peaks in the chromatogram. Small peaks arise from other substances extracted from the chocolate.

The chromatogram alone does not tell us what compounds are present. One way to identify individual peaks is to measure spectral characteristics of each one as it emerges from the column. Another way is to add an authentic sample of either caffeine or theobromine to the unknown and see whether one of the peaks grows in magnitude.

Identifying *what* is in an unknown is called **qualitative analysis.** Identifying *how much* is present is called **quantitative analysis.** The vast majority of this book deals with quantitative analysis.

In Figure 0-5, the *area* under each peak is proportional to the quantity of compound passing through the detector. The best way to measure area is with a computer that receives output from the chromatography detector. Denby and Scott did not have a computer linked to their chromatograph, so they measured the *height* of each peak instead.

## Calibration Curves

In general, analytes with equal concentrations give different detector responses. Therefore, the response must be measured for known concentrations of each analyte. A graph of detector response as a function of analyte concentration is called a **calibration curve** or a *standard curve.* To construct such a curve, **standard solutions** containing known concentrations of pure theobromine or caffeine were prepared and injected into the column, and the resulting peak heights were measured. Figure 0-6 is a chromatogram of one of the standard solutions, and Figure 0-7 shows calibration curves made by injecting solutions containing 10.0, 25.0, 50.0, or 100.0 micrograms of each analyte per gram of solution.

Straight lines drawn through the calibration points could then be used to find the concentrations of theobromine and caffeine in an unknown. From the equation of the theobromine line in Figure 0-7, we can say that if the observed peak height of theobromine from an unknown solution is 15.0 cm, then the concentration is 76.9 micrograms per gram of solution.

## Interpreting the Results

Knowing how much analyte is in the aqueous extract of the chocolate, Denby and Scott could calculate how much theobromine and caffeine were in the original chocolate. Results

Only substances that absorb ultraviolet radiation at a wavelength of 254 nanometers are observed in Figure 0-5. By far, the major components in the aqueous extract are sugars, but they are not detected in this experiment.

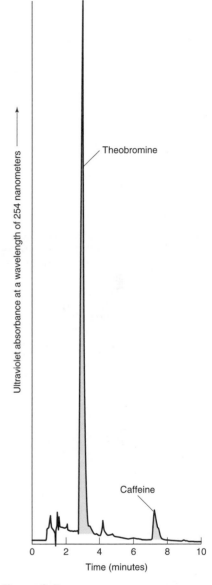

**Figure 0-5** Chromatogram of 20.0 microliters of dark chocolate extract. A 4.6-mm-diameter × 150-mm-long column, packed with 5-micrometer particles of Hypersil ODS, was eluted (washed) with water:methanol:acetic acid (79:20:1 by volume) at a rate of 1.0 mL per minute.

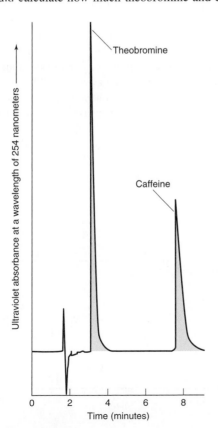

**Figure 0-6** Chromatogram of 20.0 microliters of a standard solution containing 50.0 micrograms of theobromine and 50.0 micrograms of caffeine per gram of solution.

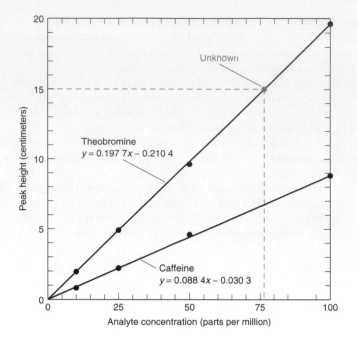

**Figure 0-7** Calibration curves, showing observed peak heights for known concentrations of pure compounds. One part per million is one microgram of analyte per gram of solution. Equations of the straight lines drawn through the experimental data points were determined by the *method of least squares*, described in Chapter 4.

for dark and white chocolates are shown in Table 0-1. The quantities found in white chocolate are only about 2% as great as the quantities in dark chocolate.

**Table 0-1** Analyses of dark and white chocolate

| Analyte | Grams of analyte per 100 grams of chocolate | |
|---|---|---|
| | Dark chocolate | White chocolate |
| Theobromine | $0.392 \pm 0.002$ | $0.010 \pm 0.007$ |
| Caffeine | $0.050 \pm 0.003$ | $0.000\ 9 \pm 0.001\ 4$ |

*Uncertainties are the standard deviation of three replicate injections of each extract.*

The table also reports the *standard deviation* of three replicate measurements for each sample. Standard deviation, discussed in Chapter 4, is a measure of the reproducibility of the results. If three samples were to give identical results, the standard deviation would be 0. If results are not very reproducible, then the standard deviation is large. For theobromine in dark chocolate, the standard deviation (0.002) is less than 1% of the average (0.392), so we say the measurement is reproducible. For theobromine in white chocolate, the standard deviation (0.007) is nearly as great as the average (0.010), so the measurement is poorly reproducible.

The purpose of an analysis is to reach some conclusion. The questions posed at the beginning of this chapter were "How much caffeine is in a chocolate bar?" and "How does it compare with the quantity in coffee or soft drinks?" After all this work, Denby and Scott dis-

**Table 0-2** Caffeine content of beverages and foods

| Source | Caffeine (milligrams per serving) | Serving size[a] (ounces) |
|---|---|---|
| Regular coffee | 106–164 | 5 |
| Decaffeinated coffee | 2–5 | 5 |
| Tea | 21–50 | 5 |
| Cocoa beverage | 2–8 | 6 |
| Baking chocolate | 35 | 1 |
| Sweet chocolate | 20 | 1 |
| Milk chocolate | 6 | 1 |
| Caffeinated soft drinks | 36–57 | 12 |

*a. 1 ounce = 28.35 grams.*

*SOURCE: Tea Association (http://www.chinamist.com/caffeine.htm).*

covered how much caffeine is in the *one* particular chocolate bar that they analyzed. It would take a great deal more work to sample many chocolate bars of the same type and many different types of chocolate to gain a more universal view. Table 0-2 compares results from analyses of different sources of caffeine. A can of soft drink or a cup of tea contains less than one-half of the caffeine in a small cup of coffee. Chocolate contains even less caffeine, but a hungry backpacker eating enough baking chocolate can get a pretty good jolt!

## ▮ ▮ ▮ 0-2 General Steps in a Chemical Analysis

The analytical process often begins with a question that is not phrased in terms of a chemical analysis. The question could be "Is this water safe to drink?" or "Does emission testing of automobiles reduce air pollution?" A scientist translates such questions into the need for particular measurements. An analytical chemist then chooses or invents a procedure to carry out those measurements.

When the analysis is complete, the analyst must translate the results into terms that can be understood by others—preferably by the general public. A most important feature of any result is its limitations. What is the statistical uncertainty in reported results? If you took samples in a different manner, would you obtain the same results? Is a tiny amount (a *trace*) of analyte found in a sample really there or is it contamination? Only after we understand the results and their limitations can we draw conclusions.

We can now summarize general steps in the analytical process:

| | |
|---|---|
| Formulating the question | Translate general questions into specific questions to be answered through chemical measurements. |
| Selecting analytical procedures | Search the chemical literature to find appropriate procedures or, if necessary, devise new procedures to make the required measurements. |
| Sampling | *Sampling* is the process of selecting representative material to analyze. Box 0-1 provides some ideas on how to do so. If you begin with a poorly chosen sample or if the sample changes between the time it is collected and the time it is analyzed, the results are meaningless. "Garbage in, garbage out!" |

### Box 0-1 Constructing a Representative Sample

In a **random heterogeneous material,** differences in composition occur randomly and on a fine scale. When you collect a portion of the material for analysis, you obtain some of each of the different compositions. To construct a representative sample from a heterogeneous material, you can first visually divide the material into segments. A **random sample** is collected by taking portions from the desired number of segments chosen at random. If you want to measure the magnesium content of the grass in the 10-meter × 20-meter field in panel (*a*), you could divide the field into 20 000 small patches that are 10 centimeters on a side. After assigning a number to each small patch, you could use a computer program to pick 100 numbers at random from 1 to 20 000. Then

harvest and combine the grass from each of these 100 patches to construct a representative bulk sample for analysis.

For a **segregated heterogeneous material** (in which large regions have obviously different compositions), a representative **composite sample** must be constructed. For example, the field in panel (*b*) has three different types of grass segregated into regions A, B, and C. You could draw a map of the field on graph paper and measure the area in each region. In this case, 66% of the area lies in region A, 14% lies in region B, and 20% lies in region C. To construct a representative bulk sample from this segregated material, take 66 of the small patches from region A, 14 from region B, and 20 from region C. You could do so by drawing random numbers from 1 to 20 000 to select patches until you have the desired number from each region.

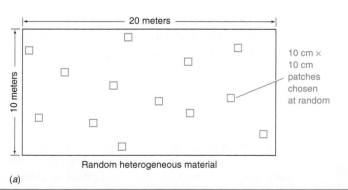

(*a*) Random heterogeneous material

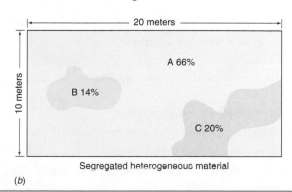

(*b*) Segregated heterogeneous material

| | |
|---|---|
| Sample preparation | *Sample preparation* is the process of converting a representative sample into a form suitable for chemical analysis, which usually means dissolving the sample. Samples with a low concentration of analyte may need to be concentrated prior to analysis. It may be necessary to remove or *mask* species that interfere with the chemical analysis. For a chocolate bar, sample preparation consisted of removing fat and dissolving the desired analytes. The reason for removing fat was that it would interfere with chromatography. |
| Analysis | Measure the concentration of analyte in several identical **aliquots** (portions). The purpose of *replicate measurements* (repeated measurements) is to assess the variability (uncertainty) in the analysis and to guard against a gross error in the analysis of a single aliquot. *The uncertainty of a measurement is as important as the measurement itself,* because it tells us how reliable the measurement is. If necessary, use different analytical methods on similar samples to make sure that all methods give the same result and that the choice of analytical method is not biasing the result. You may also wish to construct and analyze several different bulk samples to see what variations arise from your sampling procedure. |
| Reporting and interpretation | Deliver a clearly written, complete report of your results, highlighting any limitations that you attach to them. Your report might be written to be read only by a specialist (such as your instructor) or it might be written for a general audience (perhaps your mother). Be sure the report is appropriate for its intended audience. |
| Drawing conclusions | Once a report is written, the analyst might not be involved in what is done with the information, such as modifying the raw material supply for a factory or creating new laws to regulate food additives. The more clearly a report is written, the less likely it is to be misinterpreted by those who use it. |

Chemists use the term **species** to refer to any chemical of interest. Species is both singular and plural. **Interference** occurs when a species other than analyte increases or decreases the response of the analytical method and makes it appear that there is more or less analyte than is actually present. **Masking** is the transformation of an interfering species into a form that is not detected. For example, $Ca^{2+}$ in lake water can be measured with a reagent called EDTA. $Al^{3+}$ interferes with this analysis, because it also reacts with EDTA. $Al^{3+}$ can be masked by treating the sample with excess $F^-$ to form $AlF_6^{3-}$, which does not react with EDTA.

Most of this book deals with measuring chemical concentrations in homogeneous aliquots of an unknown. The analysis is meaningless unless you have collected the sample properly, you have taken measures to ensure the reliability of the analytical method, and you communicate your results clearly and completely. The chemical analysis is only the middle portion of a process that begins with a question and ends with a conclusion.

## Terms to Understand

Terms are introduced in **bold** type in the chapter and are also defined in the Glossary.

| | | | |
|---|---|---|---|
| aliquot | heterogeneous | quantitative transfer | segregated heterogeneous |
| analyte | homogeneous | random heterogeneous | material |
| aqueous | interference | material | slurry |
| calibration curve | masking | random sample | species |
| composite sample | qualitative analysis | sample preparation | standard solution |
| decant | quantitative analysis | sampling | supernatant liquid |

## Problems

Complete solutions to Problems can be found in the *Solutions Manual*. Short answers to numerical problems are at the back of the book.

**0-1.** What is the difference between *qualitative* and *quantitative* analysis?

**0-2.** List the steps in a chemical analysis.

**0-3.** What does it mean to *mask* an interfering species?

**0-4.** What is the purpose of a calibration curve?

**0-5. (a)** What is the difference between a homogeneous material and a heterogeneous material?

**(b)** After reading Box 0-1, state the difference between a segregated heterogeneous material and a random heterogeneous material.

**(c)** How would you construct a representative sample from each type of material?

**0-6.** The iodide $(I^-)$ content of a commercial mineral water was measured by two methods that produced wildy different results.[7] Method A found 0.23 milligrams of $I^-$ per liter (mg/L) and method B found 0.009 mg/L. When $Mn^{2+}$ was added to the water, the $I^-$ content found by method A increased each time more $Mn^{2+}$ was added, but results from method B were unchanged. Which of the *Terms to Understand* describes what is occurring in these measurements?

# 1 Measurements

One of the ways we will learn to express quantities in Chapter 1 is by using prefixes such as *mega* for million ($10^6$), *micro* for one-millionth ($10^{-6}$), and atto for $10^{-18}$. The illustration shows a signal due to light absorption by just *60 atoms* of rubidium in the cross-sectional area of a laser beam. There are $6.02 \times 10^{23}$ atoms in a mole, so 60 atoms amount to $1.0 \times 10^{-22}$ moles. With prefixes from Table 1-3, we will express this number as 100 *yoctomoles* (ymol) or 0.1 *zeptomole* (zmol). The prefix *yocto* stands for $10^{-24}$ and *zepto* stands for $10^{-21}$. As chemists learn to measure fewer and fewer atoms or molecules, these strange-sounding prefixes become more and more common in the chemical literature.

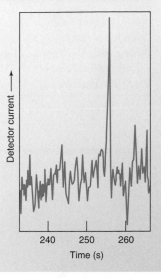

Atomic absorption signal from 60 gaseous rubidium atoms observed by laser wave mixing. A 10-microliter ($10 \times 10^{-6}$ L) sample containing 1 attogram ($1 \times 10^{-18}$ g) of Rb⁺ was injected into a graphite furnace to create the atomic vapor. We will study atomic absorption spectroscopy in Chapter 21. [F. K. Mickadeit, S. Berniolles, H. R. Kemp, and W. G. Tong, Anal. Chem. **2004**, 76, 1788.]

**P**rimed by an overview of the analytical process in Chapter 0, we are ready to discuss subjects required to get started in the lab. Topics include units of measurement, chemical concentrations, preparation of solutions, and the stoichiometry of chemical reactions.

## ■ ■ ■ 1-1 SI Units

**SI units** of measurement, used by scientists around the world, derive their name from the French *Système International d'Unités. Fundamental units* (base units) from which all others are derived are defined in Table 1-1. Standards of length, mass, and time are the *meter* (m), *kilogram* (kg), and *second* (s), respectively. Temperature is measured in *kelvins* (K), amount of substance in *moles* (mol), and electric current in *amperes* (A).

For readability, we insert a space after every third digit on either side of the decimal point. Commas are not used because in some parts of the world a comma has the same meaning as a decimal point. Two examples:

*speed of light:* **299 792 458 m/s**
*Avogadro's number:* **6.022 141 5 × 10²³ mol⁻¹**

**Table 1-1**  Fundamental SI units

| Quantity | Unit (symbol) | Definition |
|---|---|---|
| Length | meter (m) | One meter is the distance light travels in a vacuum during $\frac{1}{299\ 792\ 458}$ of a second. |
| Mass | kilogram (kg) | One kilogram is the mass of the prototype kilogram kept at Sèvres, France. |
| Time | second (s) | One second is the duration of 9 192 631 770 periods of the radiation corresponding to a certain atomic transition of $^{133}$Cs. |
| Electric current | ampere (A) | One ampere of current produces a force of $2 \times 10^{-7}$ newtons per meter of length when maintained in two straight, parallel conductors of infinite length and negligible cross section, separated by 1 meter in a vacuum. |
| Temperature | kelvin (K) | Temperature is defined such that the triple point of water (at which solid, liquid, and gaseous water are in equilibrium) is 273.16 K, and the temperature of absolute zero is 0 K. |
| Luminous intensity | candela (cd) | Candela is a measure of luminous intensity visible to the human eye. |
| Amount of substance | mole (mol) | One mole is the number of particles equal to the number of atoms in exactly 0.012 kg of $^{12}$C (approximately $6.022\ 141\ 5 \times 10^{23}$). |
| Plane angle | radian (rad) | There are $2\pi$ radians in a circle. |
| Solid angle | steradian (sr) | There are $4\pi$ steradians in a sphere. |

**Table 1-2** SI-derived units with special names

| Quantity | Unit | Symbol | Expression in terms of other units | Expression in terms of SI base units |
|---|---|---|---|---|
| Frequency | hertz | Hz | | 1/s |
| Force | newton | N | | $m \cdot kg/s^2$ |
| Pressure | pascal | Pa | $N/m^2$ | $kg/(m \cdot s^2)$ |
| Energy, work, quantity of heat | joule | J | $N \cdot m$ | $m^2 \cdot kg/s^2$ |
| Power, radiant flux | watt | W | $J/s$ | $m^2 \cdot kg/s^3$ |
| Quantity of electricity, electric charge | coulomb | C | | $s \cdot A$ |
| Electric potential, potential difference, electromotive force | volt | V | $W/A$ | $m^2 \cdot kg/(s^3 \cdot A)$ |
| Electric resistance | ohm | Ω | $V/A$ | $m^2 \cdot kg/(s^3 \cdot A^2)$ |
| Electric capacitance | farad | F | $C/V$ | $s^4 \cdot A^2/(m^2 \cdot kg)$ |

Table 1-2 lists some quantities that are defined in terms of the fundamental quantities. For example, force is measured in *newtons* (N), pressure is measured in *pascals* (Pa), and energy is measured in *joules* (J), each of which can be expressed in terms of the more fundamental units of length, time, and mass.

Pressure is force per unit area: 1 pascal (Pa) = 1 N/m². The pressure of the atmosphere is approximately 100 000 Pa.

## Using Prefixes as Multipliers

Rather than using exponential notation, we often use prefixes from Table 1-3 to express large or small quantities. As an example, consider the pressure of ozone ($O_3$) in the upper atmosphere (Figure 1-1). Ozone is important because it absorbs ultraviolet radiation from the sun that damages many organisms and causes skin cancer. Each spring, a great deal of ozone disappears from the Antarctic stratosphere, thereby creating what is called an ozone "hole." The opening of Chapter 18 discusses the chemistry behind this process.

At an altitude of $1.7 \times 10^4$ meters above the earth's surface, the pressure of ozone over Antarctica reaches a peak of 0.019 Pa. Let's express these numbers with prefixes from Table 1-3. We customarily use prefixes for every third power of ten ($10^{-9}$, $10^{-6}$, $10^{-3}$, $10^3$, $10^6$, $10^9$, and so on). The number $1.7 \times 10^4$ m is more than $10^3$ m and less than $10^6$ m, so we use a multiple of $10^3$ m (= kilometers, km):

$$1.7 \times 10^4 \text{ m} \times \frac{1 \text{ km}}{10^3 \text{ m}} = 1.7 \times 10^1 \text{ km} = 17 \text{ km}$$

Of course you recall that $10^0 = 1$.

The number 0.019 Pa is more than $10^{-3}$ Pa and less than $10^0$ Pa, so we use a multiple of $10^{-3}$ Pa (= millipascals, mPa):

$$0.019 \text{ Pa} \times \frac{1 \text{ mPa}}{10^{-3} \text{ Pa}} = 1.9 \times 10^1 \text{ mPa} = 19 \text{ mPa}$$

Figure 1-1 is labeled with km on the *y*-axis and mPa on the *x*-axis. The *y*-axis of any graph is called the **ordinate** and the *x*-axis is called the **abscissa.**

It is a fabulous idea to write units beside each number in a calculation and to cancel identical units in the numerator and denominator. This practice ensures that you know the

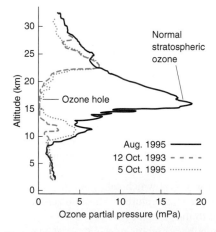

**Figure 1-1** An ozone "hole" forms each year in the stratosphere over the South Pole at the beginning of spring in October. The graph compares ozone pressure in August, when there is no hole, with the pressure in October, when the hole is deepest. Less severe ozone loss is observed at the North Pole. *[Data from National Oceanic and Atmospheric Administration.]*

**Table 1-3** Prefixes

| Prefix | Symbol | Factor | Prefix | Symbol | Factor |
|---|---|---|---|---|---|
| yotta | Y | $10^{24}$ | deci | d | $10^{-1}$ |
| zetta | Z | $10^{21}$ | centi | c | $10^{-2}$ |
| exa | E | $10^{18}$ | milli | m | $10^{-3}$ |
| peta | P | $10^{15}$ | micro | μ | $10^{-6}$ |
| tera | T | $10^{12}$ | nano | n | $10^{-9}$ |
| giga | G | $10^9$ | pico | p | $10^{-12}$ |
| mega | M | $10^6$ | femto | f | $10^{-15}$ |
| kilo | k | $10^3$ | atto | a | $10^{-18}$ |
| hecto | h | $10^2$ | zepto | z | $10^{-21}$ |
| deca | da | $10^1$ | yocto | y | $10^{-24}$ |

units for your answer. If you intend to calculate pressure and your answer comes out with units other than pascals (or some other unit of pressure), then you know you have made a mistake.

## Converting Between Units

Although SI is the internationally accepted system of measurement in science, other units are encountered. Useful conversion factors are found in Table 1-4. For example, common non-SI units for energy are the *calorie* (cal) and the *Calorie* (with a capital C, which stands for 1 000 calories, or 1 kcal). Table 1-4 states that 1 cal is exactly 4.184 J (joules).

Your *basal metabolism* requires approximately 46 Calories per hour (h) per 100 pounds (lb) of body mass to carry out basic functions required for life, apart from doing any kind of exercise. A person walking at 2 miles per hour on a level path requires approximately 45 Calories per hour per 100 pounds of body mass beyond basal metabolism. The same person swimming at 2 miles per hour consumes 360 Calories per hour per 100 pounds beyond basal metabolism.

> **Example** Unit Conversions
>
> Express the rate of energy used by a person walking 2 miles per hour (46 + 45 = 91 Calories per hour per 100 pounds of body mass) in kilojoules per hour per kilogram of body mass.
>
> **Solution**   We will convert each non-SI unit separately. First, note that 91 Calories equals 91 kcal. Table 1-4 states that 1 cal = 4.184 J; so 1 kcal = 4.184 kJ, and
>
> $$91 \text{ kcal} \times 4.184 \frac{\text{kJ}}{\text{kcal}} = 3.8 \times 10^2 \text{ kJ}$$
>
> Table 1-4 also says that 1 lb is 0.453 6 kg; so 100 lb = 45.36 kg. The rate of energy consumption is therefore
>
> $$\frac{91 \text{ kcal/h}}{100 \text{ lb}} = \frac{3.8 \times 10^2 \text{ kJ/h}}{45.36 \text{ kg}} = 8.4 \frac{\text{kJ/h}}{\text{kg}}$$
>
> We could have written this as one long calculation:
>
> $$\text{Rate} = \frac{91 \text{ kcal/h}}{100 \text{ lb}} \times 4.184 \frac{\text{kJ}}{\text{kcal}} \times \frac{1 \text{ lb}}{0.453 \text{ 6 kg}} = 8.4 \frac{\text{kJ/h}}{\text{kg}}$$

**Table 1-4**   Conversion factors

| Quantity | Unit | Symbol | SI equivalent[a] |
|---|---|---|---|
| Volume | liter | L | $*10^{-3} \text{ m}^3$ |
| | milliliter | mL | $*10^{-6} \text{ m}^3$ |
| Length | angstrom | Å | $*10^{-10} \text{ m}$ |
| | inch | in. | *0.025 4 m |
| Mass | pound | lb | *0.453 592 37 kg |
| | metric ton | | *1 000 kg |
| Force | dyne | dyn | $*10^{-5} \text{ N}$ |
| Pressure | bar | bar | $*10^5 \text{ Pa}$ |
| | atmosphere | atm | *101 325 Pa |
| | torr (= 1 mm Hg) | Torr | 133.322 Pa |
| | pound/in.$^2$ | psi | 6 894.76 Pa |
| Energy | erg | erg | $*10^{-7} \text{ J}$ |
| | electron volt | eV | $1.602 \ 176 \ 53 \times 10^{-19} \text{ J}$ |
| | calorie, thermochemical | cal | *4.184 J |
| | Calorie (with a capital C) | Cal | *1 000 cal = 4.184 kJ |
| | British thermal unit | Btu | 1 055.06 J |
| Power | horsepower | | 745.700 W |
| Temperature | centigrade (= Celsius) | °C | *K − 273.15 |
| | Fahrenheit | °F | *1.8(K − 273.15) + 32 |

*a. An asterisk (*) indicates that the conversion is exact (by definition).*

One *calorie* is the energy required to heat 1 gram of water from 14.5° to 15.5°C.
One *joule* is the energy expended when a force of 1 newton acts over a distance of 1 meter. This much energy can raise 102 g (about $\frac{1}{4}$ pound) by 1 meter.

I cal = 4.184 J

1 pound (mass) ≈ 0.453 6 kg

1 mile ≈ 1.609 km

The symbol ≈ is read "is approximately equal to."

Significant figures are discussed in Chapter 3. For multiplication and division, the number with the fewest digits determines how many digits should be in the answer. The number 91 kcal at the beginning of this problem limits the answer to 2 digits.

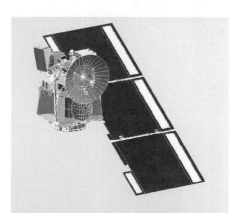

*Write the units:* In 1999, the $125 million *Mars Climate Orbiter* spacecraft was lost when it entered the Martian atmosphere 100 km lower than planned. *The navigation error would have been avoided if people had labeled their units of measurement.* Engineers who built the spacecraft calculated thrust in the English unit, pounds of force. Jet Propulsion Laboratory engineers thought they were receiving the information in the metric unit, newtons. Nobody caught the error.

## 1-2 Chemical Concentrations

A *solution* is a *homogeneous* mixture of two or more substances. A minor species in a solution is called **solute** and the major species is the **solvent.** In this book, most discussions concern *aqueous solutions,* in which the solvent is water. **Concentration** states how much solute is contained in a given volume or mass of solution or solvent.

Homogeneous means that the mixture has the same composition everywhere. When sugar dissolves in water, the mixture is homogeneous. A mixture that is not the same everywhere (such as orange juice, which has suspended solids) is *heterogeneous.*

### Molarity and Molality

A **mole** (mol) is *Avogadro's number* of particles (atoms, molecules, ions, or anything else). **Molarity** (M) is the number of moles of a substance per liter of solution. A **liter** (L) is the volume of a cube that is 10 cm on each edge. Because 10 cm = 0.1 m, 1 L = $(0.1 \text{ m})^3$ = $10^{-3} \text{ m}^3$. Chemical concentrations, denoted with square brackets, are usually expressed in moles per liter (M). Thus "$[H^+]$" means "the concentration of $H^+$."

Avogadro's number =
  number of atoms in 12 g of $^{12}C$

$$\text{Molarity (M)} = \frac{\text{moles of solute}}{\text{liters of solution}}$$

The **atomic mass** of an element is the number of grams containing Avogadro's number of atoms.[1] The **molecular mass** of a compound is the sum of atomic masses of the atoms in the molecule. It is the number of grams containing Avogadro's number of molecules.

Atomic masses are shown in the periodic table inside the cover of this book. Physical constants such as Avogadro's number are also listed inside the cover.

> **Example** Molarity of Salts in the Sea
>
> **(a)** Typical seawater contains 2.7 g of salt (sodium chloride, NaCl) per 100 mL (= 100 × $10^{-3}$ L). What is the molarity of NaCl in the ocean? **(b)** $MgCl_2$ has a concentration of 0.054 M in the ocean. How many grams of $MgCl_2$ are present in 25 mL of seawater?
>
> **Solution** **(a)** The molecular mass of NaCl is 22.99 g/mol (Na) + 35.45 g/mol (Cl) = 58.44 g/mol. The moles of salt in 2.7 g are $(2.7 \text{ g})/(58.44 \text{ g/mol})$ = 0.046 mol, so the molarity is
>
> $$\text{Molarity of NaCl} = \frac{\text{mol NaCl}}{\text{L of seawater}} = \frac{0.046 \text{ mol}}{100 \times 10^{-3} \text{ L}} = 0.46 \text{ M}$$
>
> **(b)** The molecular mass of $MgCl_2$ is 24.30 g/mol (Mg) + 2 × 35.45 g/mol (Cl) = 95.20 g/mol. The number of grams in 25 mL is
>
> $$\text{Grams of } MgCl_2 = \left(0.054 \frac{\text{mol}}{\text{L}}\right)\left(95.20 \frac{\text{g}}{\text{mol}}\right)(25 \times 10^{-3} \text{ L}) = 0.13 \text{ g}$$

An **electrolyte** is a substance that dissociates into ions in solution. In general, electrolytes are more dissociated in water than in other solvents. We refer to a compound that is mostly dissociated into ions as a *strong electrolyte.* One that is partially dissociated is called a *weak electrolyte.*

*Strong electrolyte:* mostly dissociated into ions in solution
*Weak electrolyte:* partially dissociated into ions in solution

Magnesium chloride is a strong electrolyte. In 0.44 M $MgCl_2$ solution, 70% of the magnesium is free $Mg^{2+}$ and 30% is $MgCl^+$.[2] The concentration of $MgCl_2$ molecules is close to 0. Sometimes the molarity of a strong electrolyte is called the **formal concentration** (F), to emphasize that the substance is really converted into other species in solution. When we say that the "concentration" of $MgCl_2$ is 0.054 M in seawater, we are really referring to its formal concentration (0.054 F). The "molecular mass" of a strong electrolyte is called the **formula mass** (FM), because it is the sum of atomic masses of atoms in the formula, even though there are very few molecules with that formula. *We are going to use the abbreviation FM for both formula mass and molecular mass.*

For a *weak electrolyte* such as acetic acid, $CH_3CO_2H$, some of the molecules dissociate into ions in solution:

| | Formal concentration | Percent dissociated |
|---|---|---|
| | 0.10 F | 1.3% |
| | 0.010 F | 4.1% |
| | 0.001 0 F | 12% |

**Molality** (*m*) is concentration expressed as moles of substance per kilogram of solvent (not total solution). Molality is independent of temperature. Molarity changes with temperature because the volume of a solution usually increases when it is heated.

Confusing abbreviations:

mol = moles

$$M = \text{molarity} = \frac{\text{mol solute}}{\text{L solution}}$$

$$m = \text{molality} = \frac{\text{mol solute}}{\text{kg solvent}}$$

## Percent Composition

The percentage of a component in a mixture or solution is usually expressed as a **weight percent** (wt%):

$$\text{Weight percent} = \frac{\text{mass of solute}}{\text{mass of total solution or mixture}} \times 100 \qquad (1\text{-}1)$$

A common form of ethanol ($CH_3CH_2OH$) is 95 wt%; this expression means 95 g of ethanol per 100 g of total solution. The remainder is water. **Volume percent** (vol%) is defined as

$$\text{Volume percent} = \frac{\text{volume of solute}}{\text{volume of total solution}} \times 100 \qquad (1\text{-}2)$$

Although units of mass or volume should always be expressed to avoid ambiguity, mass is usually implied when units are absent.

---

**Example** Converting Weight Percent into Molarity and Molality

Find the molarity and molality of 37.0 wt% HCl. The **density** of a substance is the mass per unit volume. The table inside the back cover of this book tells us that the density of the reagent is 1.19 g/mL.

**Solution**   For *molarity,* we need to find the moles of HCl per liter of solution. The mass of a liter of solution is $(1.19 \text{ g/mL})(1\,000 \text{ mL}) = 1.19 \times 10^3$ g. The mass of HCl in a liter is

$$\text{Mass of HCl per liter} = \left(1.19 \times 10^3 \frac{\text{g solution}}{\text{L}}\right)\left(0.370 \frac{\text{g HCl}}{\text{g solution}}\right) = 4.40 \times 10^2 \frac{\text{g HCl}}{\text{L}}$$

This is what
37.0 wt% means

The molecular mass of HCl is 36.46 g/mol, so the molarity is

$$\text{Molarity} = \frac{\text{mol HCl}}{\text{L solution}} = \frac{4.40 \times 10^2 \text{ g HCl/L}}{36.46 \text{ g HCl/mol}} = 12.1 \frac{\text{mol}}{\text{L}} = 12.1 \text{ M}$$

For *molality,* we need to find the moles of HCl per kilogram of solvent (which is $H_2O$). The solution is 37.0 wt% HCl, so we know that 100.0 g of solution contains 37.0 g of HCl and $100.0 - 37.0 = 63.0$ g of $H_2O$ ($= 0.063\,0$ kg). But 37.0 g of HCl contains $37.0 \text{ g}/(36.46 \text{ g/mol}) = 1.01$ mol. The molality is therefore

$$\text{Molality} = \frac{\text{mol HCl}}{\text{kg of solvent}} = \frac{1.01 \text{ mol HCl}}{0.063\,0 \text{ kg } H_2O} = 16.1 \ m$$

Figure 1-2 illustrates a weight percent measurement in the application of analytical chemistry to archaeology. Gold and silver are found together in nature. Dots in Figure 1-2 show the weight percent of gold in more than 1 300 silver coins minted over a 500-year period. Prior to A.D. 500, it was rare for the gold content to be below 0.3 wt%. By A.D. 600, people had developed techniques for removing more gold from the silver, so some coins had as little as 0.02 wt% gold. Colored squares in Figure 1-2 represent known, modern forgeries made from silver whose gold content is always less than the prevailing gold content in the years A.D. 200 to 500. Chemical analysis makes it easy to detect the forgeries.

## Parts per Million and Parts per Billion

Sometimes composition is expressed as **parts per million** (ppm) or **parts per billion** (ppb), which mean grams of substance per million or billion grams of total solution or mixture. Because the density of a dilute aqueous solution is close to 1.00 g/mL, *we frequently equate 1 g of water with 1 mL of water,* although this equivalence is only approximate. Therefore, 1 ppm corresponds to 1 μg/mL ($= 1$ mg/L) and 1 ppb is 1 ng/mL ($= 1$ μg/L). For gases, ppm usually refers to volume rather than mass. Atmospheric $CO_2$ has a concentration near 380 ppm, which means 380 μL $CO_2$ per liter of air. It is best to label units to avoid confusion.

---

*Margin notes:*

$$Density = \frac{\text{mass}}{\text{volume}} = \frac{\text{g}}{\text{mL}}$$

A closely related dimensionless quantity is

$$Specific \ gravity = \frac{\text{density of a substance}}{\text{density of water at 4°C}}$$

Because the density of water at 4°C is very close to 1 g/mL, specific gravity is nearly the same as density.

If you divide 1.01/0.063 0, you get 16.0. Dan got 16.1 because he kept all the digits in his calculator and did not round off until the end. The number 1.01 was really 1.014 8 and (1.014 8)/(0.063 0) = 16.1.

$$\text{ppm} = \frac{\text{mass of substance}}{\text{mass of sample}} \times 10^6$$

$$\text{ppb} = \frac{\text{mass of substance}}{\text{mass of sample}} \times 10^9$$

*Question* What does one part per thousand mean?

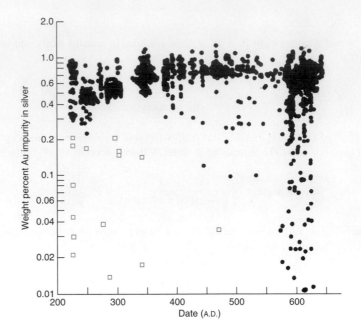

Figure 1-2 Weight percent of gold impurity in silver coins from Persia. Colored squares are known, modern forgeries. Note that the ordinate scale is logarithmic. [A. A. Gordus and J. P. Gordus, Archaeological Chemistry, Adv. Chem. No. 138, American Chemical Society, Washington, DC, 1974, pp. 124–147.]

### Example  Converting Parts per Billion into Molarity

Normal alkanes are hydrocarbons with the formula $C_nH_{2n+2}$. Plants selectively synthesize alkanes with an odd number of carbon atoms. The concentration of $C_{29}H_{60}$ in summer rainwater collected in Hannover, Germany, is 34 ppb. Find the molarity of $C_{29}H_{60}$ and express the answer with a prefix from Table 1-3.

**Solution**   A concentration of 34 ppb means there are 34 ng of $C_{29}H_{60}$ per gram of rainwater, a value that we equate to 34 ng/mL. Multiplying nanograms and milliliters by 1 000 gives 34 μg of $C_{29}H_{60}$ per liter of rainwater. Because the molecular mass of $C_{29}H_{60}$ is 408.8 g/mol, the molarity is

$$\text{Molarity of } C_{29}H_{60} \text{ in rainwater} = \frac{34 \times 10^{-6} \text{ g/L}}{408.8 \text{ g/mol}} = 8.3 \times 10^{-8} \text{ M}$$

An appropriate prefix from Table 1-3 would be nano (n), which is a multiple of $10^{-9}$:

$$8.3 \times 10^{-8} \text{ M} \left( \frac{1 \text{ nM}}{10^{-9} \text{ M}} \right) = 83 \text{ nM}$$

## ▰▰▰ 1-3  Preparing Solutions

To prepare a solution with a desired molarity from a pure solid or liquid, we weigh out the correct mass of reagent and dissolve it in the desired volume in a *volumetric flask* (Figure 1-3).

### Example  Preparing a Solution with a Desired Molarity

Copper(II) sulfate pentahydrate, $CuSO_4 \cdot 5H_2O$, has 5 moles of $H_2O$ for each mole of $CuSO_4$ in the solid crystal. The formula mass of $CuSO_4 \cdot 5H_2O$ ($= CuSO_9H_{10}$) is 249.69 g/mol. (Copper(II) sulfate without water in the crystal has the formula $CuSO_4$ and is said to be **anhydrous.**) How many grams of $CuSO_4 \cdot 5H_2O$ should be dissolved in a volume of 500.0 mL to make 8.00 mM $Cu^{2+}$?

**Solution**   An 8.00 mM solution contains $8.00 \times 10^{-3}$ mol/L. We need

$$8.00 \times 10^{-3} \frac{\text{mol}}{L} \times 0.500\,0 \; L = 4.00 \times 10^{-3} \text{ mol } CuSO_4 \cdot 5H_2O$$

The mass of reagent is $(4.00 \times 10^{-3} \text{ mol}) \times \left( 249.69 \frac{\text{g}}{\text{mol}} \right) = 0.999 \text{ g}$.

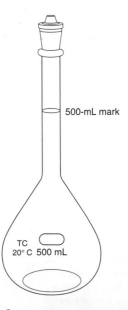

500-mL mark

TC
20° C  500 mL

Figure 1-3 A volumetric flask contains a specified volume when the liquid level is adjusted to the middle of the mark in the thin neck of the flask. Use of this flask is described in Section 2-5.

*Using a volumetric flask:* The procedure is to place 0.999 g of solid $CuSO_4 \cdot 5H_2O$ into a 500-mL volumetric flask, add about 400 mL of distilled water, and swirl to dissolve the reagent. Then dilute with distilled water up to the 500-mL mark and invert the flask several times to ensure complete mixing.

## Dilution

Dilute solutions can be prepared from concentrated solutions. A volume of the concentrated solution is transferred to a fresh vessel and diluted to the desired final volume. The number of moles of reagent in $V$ liters containing M moles per liter is the product $M \cdot V = mol/\cancel{L} \cdot \cancel{L}$, so we equate the number of moles in the concentrated (conc) and dilute (dil) solutions:

*Dilution formula:*
$$M_{conc} \cdot V_{conc} = M_{dil} \cdot V_{dil} \qquad (1\text{-}3)$$

Moles taken from concentrated solution    Moles placed in dilute solution

You may use any units for concentration and volume in this equation, as long as you use the same units on both sides. We frequently use mL for volume.

### Example  Preparing 0.100 M HCl

The molarity of "concentrated" HCl purchased for laboratory use is approximately 12.1 M. How many milliliters of this reagent should be diluted to 1.000 L to make 0.100 M HCl?

**Solution**  The dilution formula handles this problem directly:

$$M_{conc} \cdot V_{conc} = M_{dil} \cdot V_{dil}$$

$$(12.1 \text{ M}) \cdot (x \text{ mL}) = (0.100 \text{ M}) \cdot (1\,000 \text{ mL}) \Rightarrow x = 8.26 \text{ mL}$$

The symbol $\Rightarrow$ is read "implies that."

To make 0.100 M HCl, we would dilute 8.26 mL of concentrated HCl up to 1.000 L. The concentration will not be exactly 0.100 M, because the reagent is not exactly 12.1 M. A table inside the cover of this book gives volumes of common reagents required to make 1.0 M solutions.

### Example  A More Complicated Dilution Calculation

A solution of ammonia in water is called "ammonium hydroxide" because of the equilibrium

$$NH_3 + H_2O \rightleftharpoons NH_4^+ + OH^- \qquad (1\text{-}4)$$

Ammonia                    Ammonium    Hydroxide

In a chemical reaction, species on the left side are called **reactants** and species on the right are called **products**. $NH_3$ is a reactant and $NH_4^+$ is a product in Reaction 1-4.

The density of concentrated ammonium hydroxide, which contains 28.0 wt% $NH_3$, is 0.899 g/mL. What volume of this reagent should be diluted to 500.0 mL to make 0.250 M $NH_3$?

**Solution**  To use Equation 1-3, we need to know the molarity of the concentrated reagent. The solution contains 0.899 g of solution per milliliter and there is 0.280 g of $NH_3$ per gram of solution (28.0 wt%), so we can write

$$\text{Molarity of } NH_3 = \frac{899 \; \frac{\cancel{\text{g solution}}}{L} \times 0.280 \; \frac{\text{g } NH_3}{\cancel{\text{g solution}}}}{17.03 \; \frac{\text{g } NH_3}{\text{mol } NH_3}} = 14.8 \text{ M}$$

Now we find the volume of 14.8 M $NH_3$ required to prepare 500.0 mL of 0.250 M $NH_3$:

$$M_{conc} \cdot V_{conc} = M_{dil} \cdot V_{dil}$$

$$14.8 \text{ M} \times V_{conc} = 0.250 \text{ M} \times 500.0 \text{ mL} \Rightarrow V_{conc} = 8.45 \text{ mL}$$

The procedure is to place 8.45 mL of concentrated reagent in a 500-mL volumetric flask, add about 400 mL of water, and swirl to mix. Then dilute to exactly 500 mL with water and invert the flask many times to mix well.

# ■ ■ ■ ■ 1-4 Stoichiometry Calculations

*Stoichiometry* is the calculation of quantities of substances involved in a chemical reaction. It is derived from the Greek *stoicheion* (simplest component) and *metiri* (to measure).

$$^-O_2C \qquad H$$
$$\diagdown C=C \diagup$$
$$H \diagup \qquad \diagdown CO_2^-$$

Fumarate anion, $C_4H_2O_4^{2-}$

The units of formula mass (FM) are g/mol.

The symbol ~ is read "approximately."

$$mol = \frac{grams}{grams\ per\ mol} = \frac{grams}{formula\ mass}$$

The atomic mass of Fe, 55.845 g/mol, is in the periodic table inside the cover.

$$Moles = \frac{grams}{formula\ mass} = \frac{g}{g/mol}$$

You should be able to use this relationship in your sleep.

Let's apply concepts from preceding sections to a chemical analysis. Iron from a dietary supplement tablet can be measured by dissolving it and then converting the iron into solid $Fe_2O_3$. From the mass of $Fe_2O_3$, we can calculate the mass of iron in the original tablet. Chemical analysis based on weighing a final product is called *gravimetric analysis*.

Here are the steps in the procedure:

**Step 1**  Tablets containing iron(II) fumarate ($Fe^{2+}C_4H_2O_4^{2-}$) and inert binder are mixed with 150 mL of 0.100 M HCl to dissolve the $Fe^{2+}$. The solution is filtered to remove insoluble binder.

**Step 2**  Iron(II) in the clear liquid is oxidized to iron(III) with excess hydrogen peroxide:

$$2Fe^{2+} \quad + \quad H_2O_2 \quad + \quad 2H^+ \quad \rightarrow \quad 2Fe^{3+} \quad + \quad 2H_2O \qquad (1\text{-}5)$$
$$\text{Iron(II)} \qquad \text{Hydrogen peroxide} \qquad \qquad \text{Iron(III)}$$
$$\text{(ferrous ion)} \qquad \text{FM 34.01} \qquad \qquad \text{(ferric ion)}$$

**Step 3**  Ammonium hydroxide is added to precipitate hydrous iron(III) oxide, which is a gel. The gel is filtered and heated in a furnace to convert it into pure solid $Fe_2O_3$.

$$Fe^{3+} + 3OH^- + (x-1)H_2O \rightarrow FeOOH \cdot xH_2O(s) \xrightarrow{900°C} Fe_2O_3(s) \qquad (1\text{-}6)$$
$$\text{Hydroxide} \qquad\qquad \text{Hydrous iron(III) oxide} \qquad \text{Iron(III) oxide}$$
$$\text{FM 159.69}$$

We now work through some practical laboratory calculations for this analysis.

> **Example**  How Many Tablets Should We Analyze?
>
> In a gravimetric analysis, we need enough product to weigh accurately. Each tablet provides ~15 mg of iron. How many tablets should we analyze to provide 0.25 g of $Fe_2O_3$ product?

**Solution**  We can answer the question if we know how many grams of iron are in 0.25 g of $Fe_2O_3$. The formula mass of $Fe_2O_3$ is 159.69 g/mol, so 0.25 g is equal to

$$mol\ Fe_2O_3 = \frac{0.25\ g}{159.69\ g/mol} = 1.6 \times 10^{-3}\ mol$$

Each mol of $Fe_2O_3$ has 2 mol of Fe, so 0.25 g of $Fe_2O_3$ contains

$$1.6 \times 10^{-3}\ \cancel{mol\ Fe_2O_3} \times \frac{2\ mol\ Fe}{1\ \cancel{mol\ Fe_2O_3}} = 3.2 \times 10^{-3}\ mol\ Fe$$

The mass of Fe is

$$3.2 \times 10^{-3}\ \cancel{mol\ Fe} \times \frac{55.845\ g\ Fe}{\cancel{mol\ Fe}} = 0.18\ g\ Fe$$

If each tablet contains 15 mg Fe, the number of tablets required is

$$Number\ of\ tablets = \frac{0.18\ \cancel{g\ Fe}}{0.015\ \cancel{g\ Fe}/tablet} = 12\ tablets$$

> **Example**  How Much $H_2O_2$ Is Required?
>
> What mass of 3.0 wt% $H_2O_2$ solution is required to provide a 50% excess of reagent for Reaction 1-5 with 12 dietary iron tablets?

**Solution**  Twelve tablets provide 12 $\cancel{tablets} \times (0.015\ g/\cancel{tablet}) = 0.18$ g of $Fe^{2+}$, or $(0.18\ \cancel{g\ Fe^{2+}})/(55.845\ \cancel{g\ Fe^{2+}}/mol\ Fe^{2+}) = 3.2 \times 10^{-3}\ mol\ Fe^{2+}$. Reaction 1-5 requires 1 mol of $H_2O_2$ for every 2 mol of $Fe^{2+}$. Therefore $3.2 \times 10^{-3}\ mol\ Fe^{2+}$ requires $(3.2 \times 10^{-3}\ \cancel{mol\ Fe^{2+}})(1\ mol\ H_2O_2/2\ \cancel{mol\ Fe^{2+}}) = 1.6 \times 10^{-3}\ mol\ H_2O_2$. A 50% excess means that we want to use 1.50 times the stoichiometric quantity: $(1.50)(1.6 \times 10^{-3}\ mol\ H_2O_2) = 2.4 \times 10^{-3}\ mol\ H_2O_2$. The formula mass of $H_2O_2$ is 34.01 g/mol, so the required mass of pure $H_2O_2$ is $(2.4 \times 10^{-3}\ \cancel{mol})(34.01\ g/\cancel{mol}) = 0.082$ g. But hydrogen peroxide is available as a 3.0 wt% solution, so the required mass of solution is

$$Mass\ of\ H_2O_2\ solution = \frac{0.082\ \cancel{g\ H_2O_2}}{0.030\ \cancel{g\ H_2O_2}/g\ solution} = 2.7\ g\ solution$$

## Example  The Gravimetric Calculation

The final mass of $Fe_2O_3$ isolated at the end of the experiment was 0.277 g. What is the average mass of iron per dietary tablet?

**Solution**  The moles of isolated $Fe_2O_3$ are $(0.277 \text{ g})/(159.69 \text{ g/mol}) = 1.73 \times 10^{-3}$ mol. There are 2 mol Fe per formula unit, so the moles of Fe in the product are

$$(1.73 \times 10^{-3} \text{ mol } Fe_2O_3)\left(\frac{2 \text{ mol Fe}}{1 \text{ mol } Fe_2O_3}\right) = 3.47 \times 10^{-3} \text{ mol Fe}$$

The mass of Fe is $(3.47 \times 10^{-3} \text{ mol Fe})(55.845 \text{ g Fe/mol Fe}) = 0.194$ g Fe. Each of the 12 tablets therefore contains an average of $(0.194 \text{ g Fe})/12 = 0.016\ 1 \text{ g} = 16.1$ mg.

Retain all the digits in your calculator during a series of calculations. The product $1.73 \times 2$ is not 3.47; but, with the extra digits in the calculator, the correct answer is 3.47 because 1.73 was really 1.734 6.

---

## Terms to Understand

Terms are introduced in **bold** type in the chapter and are also defined in the Glossary.

| | | | |
|---|---|---|---|
| abscissa | formal concentration | molecular mass | SI units |
| anhydrous | formula mass | ordinate | solute |
| atomic mass | liter | ppb (parts per billion) | solvent |
| concentration | molality | ppm (parts per million) | volume percent |
| density | molarity | product | weight percent |
| electrolyte | mole | reactant | |

---

## Summary

SI base units include the meter (m), kilogram (kg), second (s), ampere (A), kelvin (K), and mole (mol). Derived quantities such as force (newton, N), pressure (pascal, Pa), and energy (joule, J) can be expressed in terms of base units. In calculations, units should be carried along with the numbers. Prefixes such as kilo- and milli- are used to denote multiples of units. Common expressions of concentration are molarity (moles of solute per liter of solution), molality (moles of solute per kilogram of solvent), formal concentration (formula units per liter), percent composition, and parts per million. To calculate quantities of reagents needed to prepare solutions, the relation $M_{conc} \cdot V_{conc} = M_{dil} \cdot V_{dil}$ is useful because it equates the moles of reagent removed from a stock solution to the moles delivered into a new solution. You should be able to use stoichiometry relations to calculate required masses or volumes of reagents for chemical reactions. From the mass of the product of a reaction, you should be able to compute how much reactant was consumed.

---

## Exercises

The difference between *Exercises* and *Problems* is that complete solutions to *Exercises* are provided at the back of the book, whereas only numerical answers to *Problems* are provided. Complete solutions to *Problems* are in the *Solutions Manual*. Exercises usually cover most of the major ideas in each chapter in the minimum number of questions.

**1-A.** A solution with a final volume of 500.0 mL was prepared by dissolving 25.00 mL of methanol ($CH_3OH$, density = 0.791 4 g/mL) in chloroform.
(a) Calculate the *molarity* of methanol in the solution.
(b) The solution has a density of 1.454 g/mL. Find the *molality* of methanol.

**1-B.** A 48.0 wt% solution of HBr in water has a density of 1.50 g/mL.
(a) Find the formal concentration of HBr.
(b) What mass of solution contains 36.0 g of HBr?
(c) What volume of solution contains 233 mmol of HBr?
(d) How much solution is required to prepare 0.250 L of 0.160 M HBr?

**1-C.** A solution contains 12.6 ppm of dissolved $Ca(NO_3)_2$ (which dissociates into $Ca^{2+} + 2NO_3^-$). Find the concentration of $NO_3^-$ in parts per million.

---

## Problems

### Units and Conversions

**1-1. (a)** List the SI units of length, mass, time, electric current, temperature, and amount of substance; write the abbreviation for each.
**(b)** Write the units and symbols for frequency, force, pressure, energy, and power.

**1-2.** Write the names and abbreviations for each of the prefixes from $10^{-24}$ to $10^{24}$. Which abbreviations are capitalized?

**1-3.** Write the name and number represented by each symbol. For example, for kW you should write kW = kilowatt = $10^3$ watts.
(a) mW
(b) pm
(c) kΩ
(d) μF
(e) TJ
(f) ns
(g) fg
(h) dPa

**1-4.** Express the following quantities with abbreviations for units and prefixes from Tables 1-1 through 1-3:
(a) $10^{-13}$ joules
(b) $4.317\ 28 \times 10^{-8}$ farads
(c) $2.997\ 9 \times 10^{14}$ hertz
(d) $10^{-10}$ meters
(e) $2.1 \times 10^{13}$ watts
(f) $48.3 \times 10^{-20}$ moles

**1-5.** During the 1980s, the average emission of carbon from burning fossil fuels on Earth was 5.4 petagrams (Pg) of carbon per year in the form of $CO_2$.[3]
(a) How many kg of C were placed in the atmosphere each year?
(b) How many kg of $CO_2$ were placed in the atmosphere each year?
(c) A metric ton is 1 000 kg. How many metric tons of $CO_2$ were placed in the atmosphere each year? If there were 5 billion people on Earth, how many tons of $CO_2$ were produced for each person?

**1-6.** How many joules per second and how many calories per hour are produced by a 100.0-horsepower engine?

**1-7.** A 120-pound woman working in an office consumes about $2.2 \times 10^3$ kcal/day, whereas the same woman climbing a mountain needs $3.4 \times 10^3$ kcal/day.
(a) Express these numbers in terms of joules per second per kilogram of body mass (= watts per kilogram).
(b) Which consumes more power (watts), the office worker or a 100-W light bulb?

**1-8.** (a) Refer to Table 1-4 and find how many meters are in 1 inch. How many inches are in 1 m?
(b) A mile contains 5 280 feet and a foot contains 12 inches. The speed of sound in the atmosphere at sea level is 345 m/s. Express the speed of sound in miles per second and miles per hour.
(c) There is a delay between lightning and thunder in a storm, because light reaches us almost instantaneously, but sound is slower. How many meters, kilometers, and miles away is a lightning bolt if the sound reaches you 3.00 s after the light?

**1-9.** How many joules per second (J/s) are used by a device that requires $5.00 \times 10^3$ British thermal units per hour (Btu/h)? How many watts (W) does this device use?

**1-10.** Newton's law states that force = mass $\times$ acceleration. You also know that energy = force $\times$ distance and pressure = force/area. From these relations, derive the dimensions of newtons, joules, and pascals in terms of the fundamental SI units in Table 1-1. Check your answers in Table 1-2.

**1-11.** Dust falls on Chicago at a rate of 65 mg m$^{-2}$ day$^{-1}$. Major metallic elements in the dust include Al, Mg, Cu, Zn, Mn, and Pb.[4] Pb accumulates at a rate of 0.03 mg m$^{-2}$ day$^{-1}$. How many metric tons (1 metric ton = 1 000 kg) of Pb fall on the 535 square kilometers of Chicago in 1 year?

### Chemical Concentrations

**1-12.** Define the following terms:
(a) molarity
(b) molality
(c) density
(d) weight percent
(e) volume percent
(f) parts per million
(g) parts per billion
(h) formal concentration

**1-13.** Why is it more accurate to say that the concentration of a solution of acetic acid is 0.01 F rather than 0.01 M? (Despite this distinction, we will usually write 0.01 M.)

**1-14.** What is the formal concentration (expressed as mol/L = M) of NaCl when 32.0 g are dissolved in water and diluted to 0.500 L?

**1-15.** How many grams of methanol ($CH_3OH$, FM 32.04) are contained in 0.100 L of 1.71 M aqueous methanol (that is, 1.71 mol $CH_3OH$/L solution)?

**1-16.** The concentration of a gas is related to its pressure by the *ideal gas law:*

$$\text{Concentration}\left(\frac{\text{mol}}{\text{L}}\right) = \frac{n}{V} = \frac{P}{RT}$$

$$R = \text{gas constant} = 0.083\ 14\ \frac{\text{L} \cdot \text{bar}}{\text{mol} \cdot \text{K}}$$

where $n$ is the number of moles, $V$ is volume (L), $P$ is pressure (bar), and $T$ is temperature (K).
(a) The maximum pressure of ozone in the Antarctic stratosphere in Figure 1-1 is 19 mPa. Convert this pressure into bars.
(b) Find the molar concentration of ozone in part (a) if the temperature is $-70°$ C.

**1-17.** Any dilute aqueous solution has a density near 1.00 g/mL. Suppose the solution contains 1 ppm of solute; express the concentration of solute in g/L, $\mu$g/L, $\mu$g/mL, and mg/L.

**1-18.** The concentration of the alkane $C_{20}H_{42}$ (FM 282.55) in a particular sample of rainwater is 0.2 ppb. Assume that the density of rainwater is close to 1.00 g/mL and find the molar concentration of $C_{20}H_{42}$.

**1-19.** How many grams of perchloric acid, $HClO_4$, are contained in 37.6 g of 70.5 wt% aqueous perchloric acid? How many grams of water are in the same solution?

**1-20.** The density of 70.5 wt% aqueous perchloric acid is 1.67 g/mL. Recall that grams refers to grams of *solution* (= g $HClO_4$ + g $H_2O$).
(a) How many grams of solution are in 1.000 L?
(b) How many grams of $HClO_4$ are in 1.000 L?
(c) How many moles of $HClO_4$ are in 1.000 L?

**1-21.** An aqueous solution containing 20.0 wt% KI has a density of 1.168 g/mL. Find the molality ($m$, not M) of the KI solution.

**1-22.** A cell in your adrenal gland has about $2.5 \times 10^4$ tiny compartments called *vesicles* that contain the hormone epinephrine (also called adrenaline).
(a) An entire cell has about 150 fmol of epinephrine. How many attomoles (amol) of epinephrine are in each vesicle?
(b) How many molecules of epinephrine are in each vesicle?
(c) The volume of a sphere of radius $r$ is $\frac{4}{3}\pi r^3$. Find the volume of a spherical vesicle of radius 200 nm. Express your answer in cubic meters (m$^3$) and liters, remembering that $1\ \text{L} = 10^{-3}\ \text{m}^3$.
(d) Find the molar concentration of epinephrine in the vesicle if it contains 10 amol of epinephrine.

**1-23.** The concentration of sugar (glucose, $C_6H_{12}O_6$) in human blood ranges from about 80 mg/100 mL before meals to 120 mg/100 mL after eating. Find the molarity of glucose in blood before and after eating.

**1-24.** An aqueous solution of antifreeze contains 6.067 M ethylene glycol ($HOCH_2CH_2OH$, FM 62.07) and has a density of 1.046 g/mL.
(a) Find the mass of 1.000 L of this solution and the number of grams of ethylene glycol per liter.
(b) Find the molality of ethylene glycol in this solution.

**1-25.** Protein and carbohydrates provide 4.0 Cal/g, whereas fat gives 9.0 Cal/g. (Remember that 1 Calorie, with a capital C, is really 1 kcal.) The weight percents of these components in some foods are

| Food | Wt% protein | Wt% carbohydrate | Wt% fat |
|---|---|---|---|
| Shredded wheat | 9.9 | 79.9 | — |
| Doughnut | 4.6 | 51.4 | 18.6 |
| Hamburger (cooked) | 24.2 | — | 20.3 |
| Apple | — | 12.0 | — |

Calculate the number of calories per gram and calories per ounce in each of these foods. (Use Table 1-4 to convert grams into ounces, remembering that there are 16 ounces in 1 pound.)

**1-26.** It is recommended that drinking water contain 1.6 ppm fluoride ($F^-$) for prevention of tooth decay. Consider a reservoir with a diameter of $4.50 \times 10^2$ m and a depth of 10.0 m. (The volume is $\pi r^2 h$, where $r$ is the radius and $h$ is the height.) How many grams of $F^-$ should be added to give 1.6 ppm? How many grams of sodium fluoride, NaF, contain this much fluoride?

**1-27.** Noble gases (Group 18 in the periodic table) have the following volume concentrations in dry air: He, 5.24 ppm; Ne, 18.2 ppm; Ar, 0.934%; Kr, 1.14 ppm; Xe, 87 ppb.
(a) A concentration of 5.24 ppm He means 5.24 μL of He per liter of air. Using the ideal gas law in Problem 1-16, find how many moles of He are contained in 5.24 μL at 25.00°C (298.15 K) and 1.000 bar. This number is the molarity of He in the air.
(b) Find the molar concentrations of Ar, Kr, and Xe in air at 25°C and 1 bar.

**Preparing Solutions**

**1-28.** How many grams of boric acid, $B(OH)_3$ (FM 61.83), should be used to make 2.00 L of 0.050 0 M solution? What kind of flask is used to prepare this solution?

**1-29.** Describe how you would prepare approximately 2 L of 0.050 0 *m* boric acid, $B(OH)_3$.

**1-30.** What is the maximum volume of 0.25 M sodium hypochlorite solution (NaOCl, laundry bleach) that can be prepared by dilution of 1.00 L of 0.80 M NaOCl?

**1-31.** How many grams of 50 wt% NaOH (FM 40.00) should be diluted to 1.00 L to make 0.10 M NaOH? (Answer with two digits.)

**1-32.** A bottle of concentrated aqueous sulfuric acid, labeled 98.0 wt% $H_2SO_4$, has a concentration of 18.0 M.
(a) How many milliliters of reagent should be diluted to 1.000 L to give 1.00 M $H_2SO_4$?
(b) Calculate the density of 98.0 wt% $H_2SO_4$.

**1-33.** What is the density of 53.4 wt% aqueous NaOH (FM 40.00) if 16.7 mL of the solution diluted to 2.00 L gives 0.169 M NaOH?

**Stoichiometry Calculations**

**1-34.** How many milliliters of 3.00 M $H_2SO_4$ are required to react with 4.35 g of solid containing 23.2 wt% $Ba(NO_3)_2$ if the reaction is $Ba^{2+} + SO_4^{2-} \rightarrow BaSO_4(s)$?

**1-35.** How many grams of 0.491 wt% aqueous HF are required to provide a 50% excess to react with 25.0 mL of 0.023 6 M $Th^{4+}$ by the reaction $Th^{4+} + 4F^- \rightarrow ThF_4(s)$?

# 2 Tools of the Trade

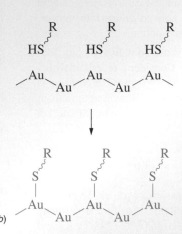

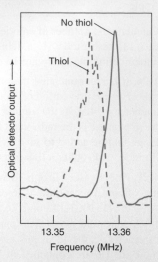

(a) Silicon cantilever with gold dot deposited on the surface, (b) Organic compound with thiol (—SH) group at the end binds to gold surface. (c) Resonant vibrational frequency of cantilever changes when thiol compound binds to gold dot. [B. Ilic, H. G. Craighead, S. Krylov, W. Senaratne, C. Ober, and P. Neuzil, "Attogram Detection Using Nanoelectromechanical Oscillators," J. Appl. Phys. **2004,** 95, 3694.]

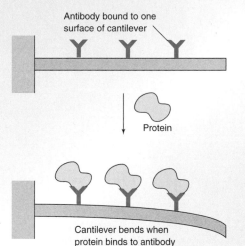

Antibody bound to one surface of cantilever

Protein

Cantilever bends when protein binds to antibody

Binding of molecules to one side creates surface stress that bends the cantilever.

Scientists can fabricate *microelectromechanical devices* such as the *cantilever* above, which is a beam of silicon anchored at one end. The beam has a resonant vibrational frequency near $13 \times 10^6$ hertz (13 MHz) when stimulated with a *piezoelectric* vibrator. (A *piezoelectric* crystal, such as quartz, is one whose dimensions change in response to an electric field.) When 93 attograms ($93 \times 10^{-18}$ g) of an organic compound bind to the gold dot near the end of the cantilever, the vibrational frequency decreases by 3.5 kHz because of the extra mass on the beam. The minimum mass that can be detected is estimated as 0.4 attogram.

Microcantilevers can be coated with DNA or antibodies to respond to biological molecules or even a single virus.[1,2,3] Bound material can be detected by the change in resonant frequency, as above, or by measuring nanometer-scale static bending, shown at the left, caused by stress on the surface of the cantilever when molecules bind.

Analytical chemistry extends from simple "wet" chemical procedures to elaborate instrumental methods. This chapter describes basic laboratory apparatus and manipulations associated with chemical measurements. We also introduce spreadsheets, which have become essential to everyone who manipulates quantitative data.

## ■ ■ ■ 2-1 Safe, Ethical Handling of Chemicals and Waste

Chemical experimentation, like driving a car or operating a household, creates hazards. *The primary safety rule is to familiarize yourself with the hazards and then to do nothing that you*

If carelessly discarded, many laboratory and household chemicals and products are harmful to plants, animals, and people.[4] For each experiment, your instructor should establish procedures for waste disposal. Options include (1) pouring solutions down the drain and diluting with tap water, (2) saving the waste for disposal in an approved landfill, (3) treating waste to decrease the hazard and then pouring it down the drain or saving it for a landfill, and (4) recycling. *Chemically incompatible wastes should never be mixed with each other, and each waste container must be labeled to indicate the quantity and identity of its contents.* Waste containers must indicate whether the contents are flammable, toxic, corrosive, or reactive, or have other dangerous properties.

A few examples illustrate different approaches to managing lab waste.[5] Dichromate ($Cr_2O_7^{2-}$) is reduced to $Cr^{3+}$ with sodium hydrogen sulfite ($NaHSO_3$), treated with hydroxide to make insoluble $Cr(OH)_3$, and evaporated to dryness for disposal in a landfill. Waste acid is mixed with waste base until nearly neutral (as determined with pH paper) and then poured down the drain. Waste iodate ($IO_3^-$) is reduced to $I^-$ with $NaHSO_3$, neutralized with base, and poured down the drain. Waste $Pb^{2+}$ solution is treated with sodium metasilicate ($Na_2SiO_3$) solution to precipitate insoluble $PbSiO_3$ that can be packaged for a landfill. Waste silver or gold is treated to recover the metal.[6] Toxic gases used in a fume hood are bubbled through a chemical trap or burned to prevent escape from the hood.

---

*(or your instructor or supervisor) consider to be dangerous.* If you believe that an operation is hazardous, discuss it first and do not proceed until sensible precautions are in place.

Preservation of a habitable planet demands that we minimize waste production and responsibly dispose of waste that is generated (Box 2-1). Recycling of chemicals is practiced in industry for economic as well as ethical reasons; it should be an important component of pollution control in your lab.

Before working, familiarize yourself with safety features of your laboratory. You should wear goggles or safety glasses with side shields (Figure 2-1) at all times in the lab to protect your eyes from liquids and glass, which fly around when least expected. Contact lenses are not recommended in the lab, because vapors can be trapped between the lens and your eye. You can protect your skin from spills and flames by wearing a flame-resistant lab coat. Use rubber gloves when pouring concentrated acids. Do not eat or drink in the lab.

Organic solvents, concentrated acids, and concentrated ammonia should be handled in a fume hood. Air flowing into the hood keeps fumes out of the lab and dilutes the fumes before expelling them from the roof. Never generate large quantities of toxic fumes that are allowed to escape through the hood. Wear a respirator when handling fine powders, which could produce a cloud of dust that might be inhaled.

*Limitations of gloves:* In 1997, popular Dartmouth College chemistry professor Karen Wetterhahn, age 48, died from a drop of dimethylmercury absorbed through the latex rubber gloves she was wearing. Many organic compounds readily penetrate rubber. Wetterhahn was an expert in the biochemistry of metals and the first female professor of chemistry at Dartmouth. She was a mother of two children and played a major role in bringing more women into science and engineering.

*Figure 2-1*  Goggles or safety glasses with side shields should be worn at all times in every lab. *[Stockdisk.]*

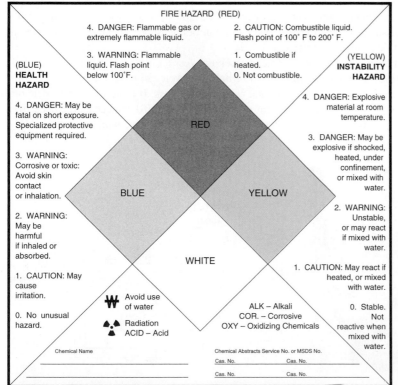

FIRE HAZARD  (RED)

4. DANGER: Flammable gas or extremely flammable liquid.

3. WARNING: Flammable liquid. Flash point below 100˚F.

(BLUE)
**HEALTH HAZARD**

4. DANGER: May be fatal on short exposure. Specialized protective equipment required.

3. WARNING: Corrosive or toxic: Avoid skin contact or inhalation.

2. WARNING: May be harmful if inhaled or absorbed.

1. CAUTION: May cause irritation.

0. No unusual hazard.

2. CAUTION: Combustible liquid. Flash point of 100˚ F to 200˚ F.

1. Combustible if heated.
0. Not combustible.

(YELLOW)
**INSTABILITY HAZARD**

4. DANGER: Explosive material at room temperature.

3. DANGER: May be explosive if shocked, heated, under confinement, or mixed with water.

2. WARNING: Unstable, or may react if mixed with water.

1. CAUTION: May react if heated, or mixed with water.

0. Stable. Not reactive when mixed with water.

RED

BLUE

YELLOW

WHITE

Avoid use of water

Radiation
ACID – Acid

ALK – Alkali
COR. – Corrosive
OXY – Oxidizing Chemicals

Chemical Name

Chemical Abstracts Service No. or MSDS No.
Cas. No.        Cas. No.

Cas. No.        Cas. No.

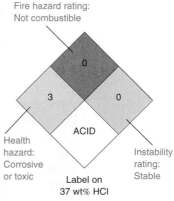

Fire hazard rating: Not combustible

0

3        0

ACID

Health hazard: Corrosive or toxic

Label on 37 wt% HCl

Instability rating: Stable

*Figure 2-2*  Chemical hazards label used by the National Fire Protection Association.

Clean up spills immediately to prevent accidental contact by the next person who comes along. Treat spills on your skin first by flooding with water. In anticipation of splashes on your body or in your eyes, know where to find and how to operate the emergency shower and eyewash. If the sink is closer than an eyewash, use the sink first for splashes in your eyes. Know how to operate the fire extinguisher and how to use an emergency blanket to extinguish burning clothing. A first aid kit should be available, and you should know how and where to seek emergency medical assistance.

*Label all vessels to indicate what they contain.* An unlabeled bottle left and forgotten in a refrigerator or cabinet presents an expensive disposal problem, because the contents must be analyzed before they can be legally discarded. National Fire Protection Association labels shown in Figure 2-2 identify hazards associated with chemical reagents. A Material Safety Data Sheet provided with each chemical sold in the United States lists hazards and safety precautions for that chemical. It gives first aid procedures and instructions for handling spills.

## ■■■■ 2-2 The Lab Notebook

The lab notebook must
1. State what was done
2. State what was observed
3. Be understandable to someone else

The critical functions of your lab notebook are to state *what you did* and *what you observed,* and it should be *understandable by a stranger.* The greatest error, made even by experienced scientists, is writing incomplete or unintelligible notebooks. Using *complete sentences* is an excellent way to prevent incomplete descriptions.

Beginning students often find it useful to write a complete description of an experiment, with sections dealing with purpose, methods, results, and conclusions. Arranging a notebook to accept numerical data prior to coming to the lab is an excellent way to prepare for an experiment. It is good practice to write a balanced chemical equation for every reaction you use. This practice helps you understand what you are doing and may point out what you do not understand about what you are doing.

The measure of scientific "truth" is the ability of different people to reproduce an experiment. A good lab notebook will state everything that was done and what you observed and will allow you or anyone else to repeat the experiment.

One fine future day, you or one of your classmates will make an important discovery and will seek a patent. The lab notebook is your legal record of your discovery. For this purpose, each page in your notebook should be signed and dated. Anything of potential importance should also be signed and dated by a second person.

Record in your notebook the names of computer files where programs and data are stored. Paste hard copies of important data into your notebook. The lifetime of a printed page is an order of magnitude (or more) greater than the lifetime of a computer disk.

## ■■■■ 2-3 Analytical Balance

An *electronic balance* uses an electromagnet to balance the load on the pan. Figure 2-3a shows a typical analytical balance with a capacity of 100–200 g and a sensitivity of 0.01–0.1 mg. *Sensitivity* is the smallest increment of mass that can be measured. A *microbalance* weighs milligram quantities with a sensitivity of 0.1 μg.

To weigh a chemical, first place a clean receiving vessel on the balance pan.[7] The mass of the empty vessel is called the **tare.** On most balances, you can press a button to reset the tare to 0. Add the chemical to the vessel and read its new mass. If there is no automatic tare operation, subtract the tare mass from that of the filled vessel. To protect the balance from corrosion, *chemicals should never be placed directly on the weighing pan.*

*Figure 2-3* (*a*) Electronic analytical balance measures mass down to 0.1 mg. *[Courtesy Fisher Scientific, Pittsburgh, PA.]* (*b*) Displacement of the balance pan generates a correction current. The electromagnet then restores the pan to its initial position. N and S are the north and south poles of the permanent magnet. *[R. M. Schoonover, "A Look at the Electronic Analytical Balance," Anal. Chem. 1982, 54, 973A. See also B. B. Johnson and J. D. Wells, "Cautions Concerning Electronic Analytical Balances," J. Chem. Ed. 1986, 63, 86.]*

(*a*)

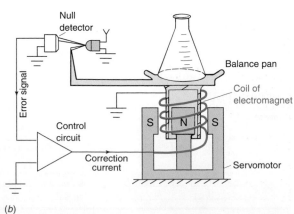

(*b*)

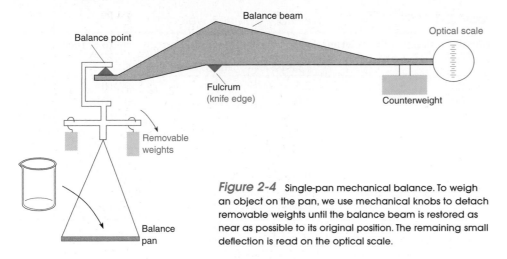

Balance beam

Balance point

Optical scale

Fulcrum
(knife edge)

Counterweight

Removable
weights

Balance
pan

*Figure 2-4* Single-pan mechanical balance. To weigh an object on the pan, we use mechanical knobs to detach removable weights until the balance beam is restored as near as possible to its original position. The remaining small deflection is read on the optical scale.

An alternate procedure, called "weighing by difference," is necessary for **hygroscopic** reagents, which rapidly absorb moisture from the air. First weigh a capped bottle containing dry reagent. Then quickly pour some reagent from that weighing bottle into a receiver. Cap the weighing bottle and weigh it again. The difference is the mass of reagent delivered from the weighing bottle. With an electronic balance, set the initial mass of the weighing bottle to zero with the tare button. Then deliver reagent from the bottle and reweigh the bottle. The negative reading on the balance is the mass of reagent delivered from the bottle.[8]

Weighing by difference

## Principle of Operation

An object placed on the pan of the balance in Figure 2-3b pushes the pan down with a force equal to $m \times g$, where $m$ is the mass of the object and $g$ is the acceleration of gravity. The null detector senses the displacement and sends an error signal to the circuit that generates a correction current. This current flows through the coil beneath the pan, thereby creating a magnetic field that is repelled by a permanent magnet under the pan. As the deflection decreases, the output of the null detector decreases. The current required to restore the pan to its initial position is proportional to the mass on the pan.

The older single-pan *mechanical balance* shown in Figure 2-4 uses standard masses and a balance beam suspended on a sharp *knife edge* to measure the mass of the object on the balance pan. The mass of the pan hanging from the balance point (another knife edge) at the left is balanced by a counterweight at the right. An object placed on the pan pushes it down. We rotate knobs to remove weights from a bar that is above the pan and hidden inside the balance. The balance beam is restored almost to its original position when the masses removed are nearly equal to the mass of the object on the pan. The slight difference from the original position is shown on an optical scale, whose reading is added to that of the knobs.

*A mechanical balance should be in its arrested position when you load or unload the pan and in the half-arrested position when you are dialing weights.* This practice minimizes wear on the knife edges, which degrades sensitivity.

## Preventing Weighing Errors

Use a paper towel or tissue to handle the vessel you are weighing, because fingerprints will change its mass. Samples should be at *ambient temperature* (the temperature of the surroundings) to prevent errors due to air currents. A sample that has been dried in an oven takes about 30 min to cool to room temperature. Place the sample in a desiccator during cooling to prevent accumulation of moisture. Close the glass doors of the balance in Figure 2-3a to prevent drafts from affecting the reading. Many top-loading balances have a plastic fence around the pan to protect it from drafts. Sensitive balances should be located on a heavy table, such as a marble slab, to minimize vibrations. The balance has adjustable feet and a bubble meter that allow you to keep it level. Avoid spilling chemicals into the gap between the coil and the permanent magnet of the servomotor.

Analytical balances calibrate themselves automatically by placing a standard mass on the load-bearing structure and measuring the current required to balance the weight. Less expensive electronic balances are calibrated at the factory, where the force of gravity may not be the same as the force of gravity in your lab. (Gravitational acceleration varies by $\approx 0.1\%$ among different locations in the United States.) Magnetic materials or electromagnetic fields

**Table 2-1** Tolerances for laboratory balance weights[a]

| Denomination | Tolerance (mg) | | Denomination | Tolerance (mg) | |
|---|---|---|---|---|---|
| Grams | Class 1 | Class 2 | Milligrams | Class 1 | Class 2 |
| 500 | 1.2 | 2.5 | 500 | 0.010 | 0.025 |
| 200 | 0.50 | 1.0 | 200 | 0.010 | 0.025 |
| 100 | 0.25 | 0.50 | 100 | 0.010 | 0.025 |
| 50 | 0.12 | 0.25 | 50 | 0.010 | 0.014 |
| 20 | 0.074 | 0.10 | 20 | 0.010 | 0.014 |
| 10 | 0.050 | 0.074 | 10 | 0.010 | 0.014 |
| 5 | 0.034 | 0.054 | 5 | 0.010 | 0.014 |
| 2 | 0.034 | 0.054 | 2 | 0.010 | 0.014 |
| 1 | 0.034 | 0.054 | 1 | 0.010 | 0.014 |

a. Tolerances are defined in ASTM (American Society for Testing and Materials) Standard E 617. Classes 1 and 2 are the most accurate. Larger tolerances exist for Classes 3–6, which are not given in this table.

from neighboring instruments can affect the balance reading. Periodically check your balance by weighing a standard mass. *Tolerances* (allowable deviations) for standard masses are listed in Table 2-1.

## Buoyancy

Your weight when swimming is nearly zero, which is why people can float. **Buoyancy** is the upward force exerted on an object in a liquid or gaseous fluid.[9] An object weighed in air appears lighter than its actual mass by an amount equal to the mass of air that it displaces. True mass is the mass measured in vacuum. A standard mass in a balance is also affected by buoyancy, so it weighs less in air than in vacuum. A buoyancy error occurs whenever the density of the object being weighed is not equal to the density of the standard mass.

If mass $m'$ is read from a balance, the true mass $m$ of the object weighed in vacuum is given by[10]

*Buoyancy equation:*
$$m = \frac{m'\left(1 - \dfrac{d_a}{d_w}\right)}{\left(1 - \dfrac{d_a}{d}\right)}$$
(2-1)

where $d_a$ is the density of air (0.001 2 g/mL near 1 bar and 25°C),[11] $d_w$ is the density of the calibration weights (typically 8.0 g/mL), and $d$ is the density of the object being weighed.

## Example Buoyancy Correction

A pure compound called "tris" is used as a *primary standard* to measure concentrations of acids. The volume of acid required to react with a known mass of tris tells us the concentration of the acid. Find the true mass of tris (density = 1.33 g/mL) if the apparent mass weighed in air is 100.00 g.

**Solution** Assuming that the balance weights have a density of 8.0 g/mL and the density of air is 0.001 2 g/mL, we find the true mass by using Equation 2-1:

$$m = \frac{100.00 \text{ g}\left(1 - \dfrac{0.001\ 2 \text{ g/mL}}{8.0 \text{ g/mL}}\right)}{1 - \dfrac{0.001\ 2 \text{ g/mL}}{1.33 \text{ g/mL}}} = 100.08 \text{ g}$$

Unless we correct for buoyancy, we would think that the mass of tris is 0.08% less than the actual mass and we would think that the molarity of acid reacting with the tris is 0.08% less than the actual molarity.

Figure 2-5 shows buoyancy corrections for several substances. When you weigh water with a density of 1.00 g/mL, the true mass is 1.001 1 g when the balance reads 1.000 0 g. The

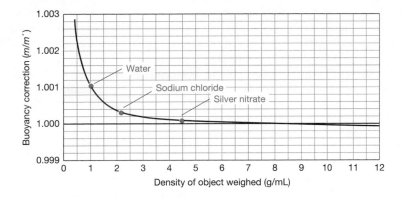

Figure 2-5 Buoyancy correction, assuming $d_a = 0.001\ 2$ g/mL and $d_w = 8.0$ g/mL. The apparent mass measured in air (1.000 0 g) is multiplied by the buoyancy correction to find the true mass.

error is 0.11%. For NaCl with a density of 2.16 g/mL, the error is 0.04%; and for $AgNO_3$ with a density of 4.45 g/mL, the error is only 0.01%.

## ▮▮▮▮ 2-4 Burets

The **buret** in Figure 2-6a is a precisely manufactured glass tube with graduations enabling you to measure the volume of liquid delivered through the stopcock (the valve) at the bottom. The 0-mL mark is near the top. If the initial liquid level is 0.83 mL and the final level is 27.16 mL, then you have delivered 27.16 − 0.83 = 26.33 mL. Class A burets (the most accurate grade) are certified to meet the tolerances in Table 2-2. If the reading of a 50-mL buret is 27.16 mL, the true volume can be anywhere in the range 27.21 to 27.11 mL and still be within the tolerance of ±0.05 mL.

When reading the liquid level in a buret, your eye should be at the same height as the top of the liquid. If your eye is too high, the liquid seems to be higher than it really is. If your eye is too low, the liquid appears too low. The error that occurs when your eye is not at the same height as the liquid is called **parallax.**

The surface of most liquids forms a concave **meniscus** like that shown in Figure 2-7.[12] It is helpful to use black tape on a white card as a background for locating the precise position of the meniscus. Move the black strip up the buret to approach the meniscus. The bottom of the meniscus turns dark as the black strip approaches, thus making the meniscus more easily

**Table 2-2** Tolerances of Class A burets

| Buret volume (mL) | Smallest graduation (mL) | Tolerance (mL) |
|---|---|---|
| 5 | 0.01 | ±0.01 |
| 10 | 0.05 *or* 0.02 | ±0.02 |
| 25 | 0.1 | ±0.03 |
| 50 | 0.1 | ±0.05 |
| 100 | 0.2 | ±0.10 |

Figure 2-6 (*a*) Glass buret. [*Courtesy A. H. Thomas Co., Philadelphia, PA.*] (*b*) Digital titrator with plastic cartridge containing reagent solution is used for analyses in the field. [*Courtesy Hach Co., Loveland, CO.*] (*c*) Battery-operated electronic buret with digital readout delivers 0.01-mL increments from a reagent bottle. This device can be used for accurate titrations in the field. [*Courtesy Cole-Parmer Co., Niles, IL.*]

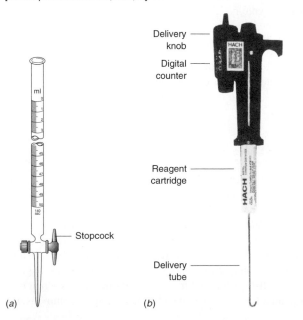

Delivery knob

Digital counter

Reagent cartridge

Delivery tube

Stopcock

(a)　　(b)　　(c)

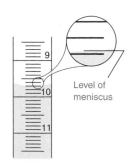

Level of meniscus

Figure 2-7 Buret with the meniscus at 9.68 mL. Estimate the reading of any scale to the nearest tenth of a division. This buret has 0.1-mL divisions, so we estimate the reading to the nearest 0.01 mL.

readable. Highly colored solutions may appear to have two meniscuses; either one may be used. Because volumes are determined by subtracting one reading from another, the important point is to read the position of the meniscus reproducibly. Always estimate the reading to the nearest tenth of a division between marks.

The thickness of the markings on a 50-mL buret corresponds to about 0.02 mL. For best accuracy, select one portion of the marking to be called zero. For example, you can say that the liquid level is *at* the mark when the bottom of the meniscus just touches the top of the mark. When the meniscus is at the *bottom* of the same mark, the reading is 0.02 mL greater.

For precise location of the end of a titration, we deliver less than one drop at a time from the buret near the end point. (A drop from a 50-mL buret is about 0.05 mL.) To deliver a fraction of a drop, carefully open the stopcock until part of a drop is hanging from the buret tip. (Some people prefer to rotate the stopcock rapidly through the open position to expel part of a drop.) Then touch the inside glass wall of the receiving flask to the buret tip to transfer the droplet to the wall of the flask. Carefully tip the flask so that the main body of liquid washes over the newly added droplet. Swirl the flask to mix the contents. Near the end of a titration, tip and rotate the flask often to ensure that droplets on the wall containing unreacted analyte contact the bulk solution.

Liquid should drain evenly down the wall of a buret. The tendency of liquid to stick to glass is reduced by draining the buret slowly (<20 mL/min). If many droplets stick to the wall, then clean the buret with detergent and a buret brush. If this cleaning is insufficient, soak the buret in peroxydisulfate–sulfuric acid cleaning solution,[13] which eats clothing and people, as well as grease in the buret. Never soak volumetric glassware in alkaline solutions, which attack glass. A 5 wt% NaOH solution at 95°C dissolves Pyrex glass at a rate of 9 μm/h.

Error can be caused by failure to expel the bubble of air often found directly beneath the stopcock (Figure 2-8). If the bubble becomes filled with liquid during the titration, then some volume that drained from the graduated portion of the buret did not reach the titration vessel. The bubble can be dislodged by draining the buret for a second or two with the stopcock wide open. You can expel a tenacious bubble by abruptly shaking the buret while draining it into a sink.

When you fill a buret with fresh solution, it is a wonderful idea to rinse the buret several times with small portions of the new solution, discarding each wash. It is not necessary to fill the buret with wash solution. Simply tilt the buret to allow all surfaces to contact the wash liquid. This same technique should be used with any vessel (such as a spectrophotometer cuvet or a pipet) that is reused without drying.

The *digital titrator* in Figure 2-6b is convenient for use in the field where samples are collected. The counter tells how much reagent has been dispensed. The precision of 1% is 10 times poorer than that of a glass buret, but many measurements do not require higher precision. The battery-operated *electronic buret* in Figure 2-6c fits on a reagent bottle and delivers up to 99.99 mL in 0.01-mL increments. For titrations requiring the very highest precision, measure the *mass* of reagent, instead of the volume, delivered from a buret or syringe.[14] Mass can be measured more precisely than can volume.

## Microscale Titrations

"Microscale" student experiments reduce costs by decreasing consumption of reagents and generation of waste. An inexpensive student buret can be constructed from a 2-mL pipet graduated in 0.01-mL intervals.[15] Volume can be read to 0.001 mL and titrations can be carried out with a precision of 1%.

## ■ ■ ■ 2-5 Volumetric Flasks

A **volumetric flask** is calibrated to contain a particular volume of solution at 20°C when the bottom of the meniscus is adjusted to the center of the mark on the neck of the flask (Figure 2-9, Table 2-3). Most flasks bear the label "TC 20°C," which means *to contain* at 20°C. (Pipets and burets are calibrated *to deliver*, "TD," their indicated volume.) The temperature of the container is relevant because both liquid and glass expand when heated.

To use a volumetric flask, dissolve the desired mass of reagent in the flask by swirling with *less* than the final volume of liquid. Then add more liquid and swirl the solution again. Adjust the final volume with as much well-mixed liquid in the flask as possible. (When two different liquids are mixed, there is generally a small volume change. The total volume is *not*

Operating a buret:
- Wash buret with new solution
- Eliminate air bubble before use
- Drain liquid slowly
- Deliver fraction of a drop near end point
- Read bottom of concave meniscus
- Estimate reading to 1/10 of a division
- Avoid parallax
- Account for graduation thickness in readings

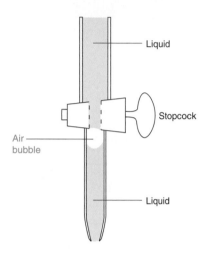

*Figure 2-8* An air bubble trapped beneath the stopcock should be expelled before you use the buret.

*Precision* refers to the reproducibility of replicate deliveries.
*Accuracy* refers to the difference between the stated volume and the actual volume delivered.

Thermal expansion of water and glass is discussed in Section 2-9. In contrast to older types of glass, volumetric glassware made of Pyrex, Kimax, or other low-expansion glass can be safely dried in an oven heated to at least 320°C without harm,[16] although there is rarely reason to go above 150°C.

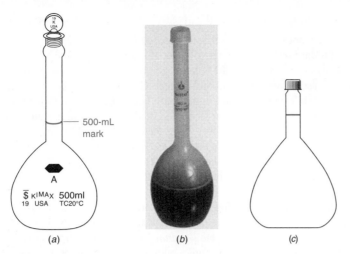

| Table 2-3 | Tolerances of Class A volumetric flasks |
|---|---|

| Flask capacity (mL) | Tolerance (mL) |
|---|---|
| 1 | ±0.02 |
| 2 | ±0.02 |
| 5 | ±0.02 |
| 10 | ±0.02 |
| 25 | ±0.03 |
| 50 | ±0.05 |
| 100 | ±0.08 |
| 200 | ±0.10 |
| 250 | ±0.12 |
| 500 | ±0.20 |
| 1 000 | ±0.30 |
| 2 000 | ±0.50 |

*Figure 2-9* (*a*) Class A glass volumetric flask. *[Courtesy A. H. Thomas Co., Philadelphia, PA.]* (*b*) Class B polypropylene plastic volumetric flask for trace analysis. *[Courtesy Fisher Scientific, Pittsburgh, PA.]* Class A flasks meet tolerances of Table 2-3. Class B tolerances are twice as big as Class A tolerances. (*c*) Short-form volumetric flask with Teflon-lined screw cap fits in the analytical balance in Figure 2-3a. Teflon protects the cap from chemical attack.

the sum of the two volumes that were mixed. By swirling the liquid in a nearly full volumetric flask before the liquid reaches the thin neck, you minimize the change in volume when the last liquid is added.) For good control, add the final drops of liquid with a pipet, *not a squirt bottle.* After adjusting the liquid to the correct level, hold the cap firmly in place and invert the flask several times to complete mixing. Before the liquid is homogeneous, we observe streaks (called *schliera*) arising from regions that refract light differently. After the schliera are gone, invert the flask a few more times to ensure complete mixing.

Figure 2-10 shows how liquid appears when it is at the *center* of the mark of a volumetric flask or a pipet. Adjust the liquid level while viewing the flask from above or below the level of the mark. The front and back of the mark describe an ellipse with the meniscus at the center.

Glass is notorious for *adsorbing* traces of chemicals—especially cations. **Adsorption** is the process in which a substance sticks to a surface. (In contrast, **absorption** is the process in which a substance is taken inside another, as water is taken into a sponge.) For critical work, you should **acid wash** glassware to replace low concentrations of cations on the surface with $H^+$. To do this, soak already thoroughly cleaned glassware in 3–6 M HCl (in a fume hood) for >1 h. Then rinse it well with distilled water and, finally, soak it in distilled water. Acid can be reused many times, as long as it is only used for clean glassware. Acid washing is *especially* appropriate for new glassware, which you should always assume is not clean. The polypropylene plastic volumetric flask in Figure 2-9b is designed for trace analysis (parts per billion concentrations) in which analyte might be lost by adsorption on the walls of a glass flask.

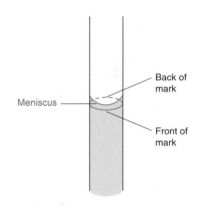

*Figure 2-10* Proper position of the meniscus—at the center of the ellipse formed by the front and back of the calibration mark when viewed from above or below. Volumetric flasks and transfer pipets are calibrated to this position.

### ▮▮▮▮ 2-6 Pipets and Syringes

**Pipets** deliver known volumes of liquid. The *transfer pipet* in Figure 2-11a is calibrated to deliver one fixed volume. The last drop does not drain out of the pipet and *should not be blown out.* The *measuring pipet* in Figure 2-11b is calibrated like a buret. It is used to deliver a variable volume, such as 5.6 mL, by starting delivery at the 1.0-mL mark and terminating at the 6.6-mL mark. The transfer pipet is more accurate, with tolerances listed in Table 2-4.

| Table 2-4 | Tolerances of Class A transfer pipets |
|---|---|

| Volume (mL) | Tolerance (mL) |
|---|---|
| 0.5 | ±0.006 |
| 1 | ±0.006 |
| 2 | ±0.006 |
| 3 | ±0.01 |
| 4 | ±0.01 |
| 5 | ±0.01 |
| 10 | ±0.02 |
| 15 | ±0.03 |
| 20 | ±0.03 |
| 25 | ±0.03 |
| 50 | ±0.05 |
| 100 | ±0.08 |

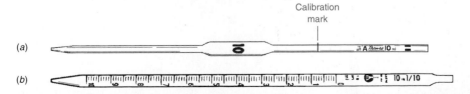

*Figure 2-11* (*a*) Transfer pipet and (*b*) measuring (Mohr) pipet. *[Courtesy A. H. Thomas Co., Philadelphia, PA.]*

## Using a Transfer Pipet

Using a rubber bulb or other pipet suction device, *not your mouth,* suck liquid up past the calibration mark. Discard one or two pipet volumes of liquid to rinse traces of previous reagents from the pipet. After taking up a third volume past the calibration mark, quickly replace the bulb with your index finger at the end of the pipet. Gently pressing the pipet against the bottom of the vessel while removing the rubber bulb helps prevent liquid from draining below the mark while you put your finger in place. (Alternatively, you can use an automatic suction device that remains attached to the pipet.) Wipe the excess liquid off the outside of the pipet with a clean tissue. *Touch the tip of the pipet to the side of a beaker* and drain the liquid until the bottom of the meniscus just reaches the center of the mark, as shown in Figure 2-10. Touching the beaker draws liquid from the pipet without leaving part of a drop hanging when the liquid reaches the calibration mark.

Do not blow the last drop out of a transfer pipet.

Transfer the pipet to a receiving vessel and drain it by gravity *while holding the tip against the wall of the vessel.* After the liquid stops, hold the pipet to the wall for a few more seconds to complete draining. *Do not blow out the last drop.* The pipet should be nearly vertical at the end of delivery. When you finish with a pipet, you should rinse it with distilled water or soak it until you are ready to clean it. Solutions should never be allowed to dry inside a pipet because removing internal deposits is very difficult.

## Micropipets

Micropipets (Figure 2-12) deliver volumes of 1 to 1 000 $\mu$L (1 $\mu$L = $10^{-6}$ L). Liquid is contained in the disposable polypropylene tip, which is stable to most aqueous solutions and many organic solvents except chloroform ($CHCl_3$). The tip is not resistant to concentrated nitric or sulfuric acids. To prevent aerosols from entering the pipet shaft, tips are available with polyethylene filters. Aerosols can corrode mechanical parts of the pipet or cross-contaminate biological experiments.

To use a micropipet, place a fresh tip tightly on the barrel. Keep tips in their package or dispenser so you do not contaminate the tips with your fingers. Set the desired volume with the knob at the top of the pipet. Depress the plunger to the first stop, which corresponds to the selected volume. Hold the pipet *vertically,* dip it 3–5 mm into the reagent solution, and

**Figure 2-12** (*a*) Microliter pipet with disposable plastic tip (*b*) Enlarged view of disposable tip containing polyethylene filter to prevent aerosol from contaminating the shaft of the pipet. (*c*) Volume selection dial set to 150 $\mu$L. [Courtesy Rainin Instrument Co., Emeryville, CA.]

Piston

Shaft

Filter

Aerosol

Disposable polypropylene tip

(*a*)       (*b*)        (*c*)

CHAPTER 2 Tools of the Trade

**Table 2-5** Manufacturer's tolerances for micropipets

| Pipet volume (μL) | At 10% of pipet volume | | At 100% of pipet volume | |
|---|---|---|---|---|
| | Accuracy (%) | Precision (%) | Accuracy (%) | Precision (%) |
| *Adjustable Pipets* | | | | |
| 0.2–2 | ±8 | ±4 | ±1.2 | ±0.6 |
| 1–10 | ±2.5 | ±1.2 | ±0.8 | ±0.4 |
| 2.5–25 | ±4.5 | ±1.5 | ±0.8 | ±0.2 |
| 10–100 | ±1.8 | ±0.7 | ±0.6 | ±0.15 |
| 30–300 | ±1.2 | ±0.4 | ±0.4 | ±0.15 |
| 100–1 000 | ±1.6 | ±0.5 | ±0.3 | ±0.12 |
| *Fixed Pipets* | | | | |
| 10 | | | ±0.8 | ±0.4 |
| 25 | | | ±0.8 | ±0.3 |
| 100 | | | ±0.5 | ±0.2 |
| 500 | | | ±0.4 | ±0.18 |
| 1 000 | | | ±0.3 | ±0.12 |

SOURCE: Data from Hamilton Co., Reno. NV.

*slowly* release the plunger to suck up liquid. The volume of liquid taken into the tip depends on the angle at which the pipet is held and how far beneath the liquid surface the tip is held during uptake. Withdraw the tip from the liquid by sliding it along the wall of the vessel to remove liquid from the outside of the tip. To dispense liquid, touch the tip to the wall of the receiver and gently depress the plunger to the first stop. Wait a few seconds to allow liquid to drain down the tip, and then depress the plunger further to squirt out the last liquid. Clean and wet a fresh tip by taking up and discarding two or three squirts of reagent. The tip can be discarded or rinsed well with a squirt bottle and reused. A tip with a filter (Figure 2-12b) cannot be cleaned for reuse.

Table 2-5 lists tolerances for micropipets from one manufacturer. As internal parts wear out, both precision and accuracy can decline by an order of magnitude. In a study[17] of 54 micropipets in use at a biomedical lab, 12 were accurate and precise to ≤1%. Five of 54 had errors ≥10%. When 54 quality control technicians at four pharmaceutical companies used a properly functioning micropipet, 10 people were accurate and precise to ≤1%. Six were inaccurate by ≥10%. Micropipets require periodic calibration and maintenance (cleaning, seal replacement, and lubrication) and operators require certification.[18]

## Syringes

Microliter *syringes,* such as that in Figure 2-13, come in sizes from 1 to 500 μL and have an accuracy and precision near 1%. When using a syringe, take up and discard several volumes of liquid to wash the glass walls and to remove air bubbles from the barrel. The steel needle is attacked by strong acid and will contaminate strongly acidic solutions with iron.

Needle          Barrel     Plunger

**Figure 2-13** Hamilton syringe with a volume of 1 μL and divisions of 0.02 μL on the glass barrel. [Courtesy Hamilton Co., Reno, NV.]

## ▪▪▪▪ 2-7 Filtration

In *gravimetric analysis,* the mass of product from a reaction is measured to determine how much unknown was present. Precipitates from gravimetric analyses are collected by filtration, washed, and then dried. Most precipitates are collected in a *fritted-glass funnel* (also

**Figure 2-14** Filtration with a Gooch crucible that has a porous (*fritted*) glass disk through which liquid can pass. Suction can be applied by the house vacuum system or by an *aspirator* that uses flowing water to create vacuum. The trap prevents backup of tap water from the aspirator into the suction flask. Alternatively, the trap prevents liquid in your suction flask from being accidentally sucked into the house vacuum system. Using a trap is always a good idea, no matter what your source of vacuum.

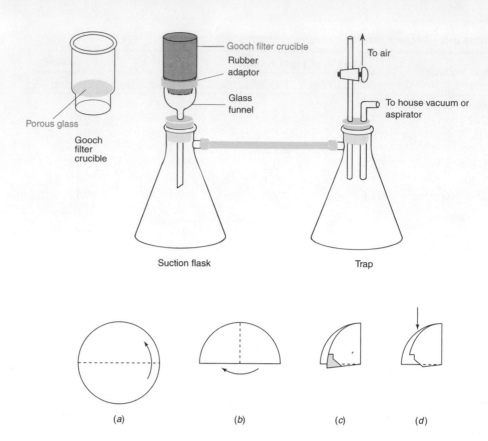

Porous glass

Gooch filter crucible

Gooch filter crucible

Rubber adaptor

Glass funnel

To air

To house vacuum or aspirator

Suction flask

Trap

**Figure 2-15** Folding filter paper for a conical funnel, (*a*) Fold the paper in half. (*b*) Then fold it in half again. (*c*) Tear off a corner to allow better seating of the paper in the funnel. (*d*) Open the side that was not torn when fitting the paper in the funnel.

(a)    (b)    (c)    (d)

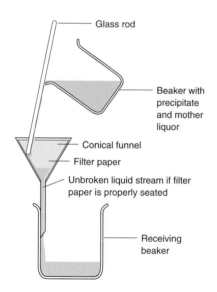

Glass rod

Beaker with precipitate and mother liquor

Conical funnel

Filter paper

Unbroken liquid stream if filter paper is properly seated

Receiving beaker

**Figure 2-16** Filtering a precipitate. The conical funnel is supported by a metal ring attached to a ring stand, neither of which is shown.

Dust is a source of contamination in all experiments, so . . .

Cover all vessels whenever possible.

called a Gooch filter crucible) with suction applied to speed filtration (Figure 2-14). The porous glass plate in the funnel allows liquid to pass but retains solids. The empty funnel is first dried at 110°C and weighed. After collecting solid and drying again, the funnel and its contents are weighed a second time to determine the mass of collected solid. Liquid from which a substance precipitates or crystallizes is called the **mother liquor.** Liquid that passes through the filter is called **filtrate.**

In some gravimetric procedures, **ignition** (heating at high temperature over a burner or in a furnace) is used to convert a precipitate into a known, constant composition. For example, $Fe^{3+}$ precipitates as hydrous ferric oxide, $FeOOH \cdot xH_2O$, with variable composition. Ignition converts it into pure $Fe_2O_3$ prior to weighing. When a precipitate is to be ignited, it is collected in **ashless filter paper,** which leaves little residue when burned.

To use filter paper with a conical glass funnel, fold the paper into quarters, tear off one corner (to allow a firm fit into the funnel), and place the paper in the funnel (Figure 2-15). The filter paper should fit snugly and be seated with some distilled water. When liquid is poured in, an unbroken stream of liquid should fill the stem of the funnel (Figure 2-16). The weight of liquid in the stem helps speed filtration.

For filtration, pour the slurry of precipitate down a glass rod to prevent splattering (Figure 2-16). (A **slurry** is a suspension of solid in liquid.) Particles adhering to the beaker or rod can be dislodged with a *rubber policeman,* which is a flattened piece of rubber at the end of a glass rod. Use a jet of appropriate wash liquid from a squirt bottle to transfer particles from the rubber and glassware to the filter. If the precipitate is going to be ignited, particles remaining in the beaker should be wiped onto a small piece of moist filter paper. Add that paper to the filter to be ignited.

## 2-8 Drying

Reagents, precipitates, and glassware are conveniently dried in an oven at 110°C. (Some chemicals require other temperatures.) *Anything that you put in the oven should be labeled.* Use a beaker and watchglass (Figure 2-17) to minimize contamination by dust during drying. It is good practice to cover all vessels on the benchtop to prevent dust contamination.

The mass of a gravimetric precipitate is measured by weighing a dry, empty filter crucible before the procedure and reweighing the same crucible filled with dry product after the procedure. To weigh the empty crucible, first bring it to "constant mass" by drying it in the oven for 1 h or longer and then cooling it for 30 min in a desiccator. Weigh the crucible and then heat it again for about 30 min. Cool it and reweigh it. When successive weighings agree to ±0.3 mg, the filter has reached "constant mass." You can use a microwave oven instead of an electric oven for drying reagents and crucibles. Try an initial heating time of 4 min, with subsequent 2-min heatings. Use a 15-min cooldown before weighing.

A **desiccator** (Figure 2-18) is a closed chamber containing a drying agent called a **desiccant** (Table 2-6). The lid is greased to make an airtight seal and desiccant is placed in the bottom beneath the perforated disk. Another useful desiccant that is not in the table is 98% sulfuric acid. After placing a hot object in the desiccator, leave the lid cracked open for a minute until the object has cooled slightly. This practice prevents the lid from popping open when the air inside warms up. To open a desiccator, slide the lid sideways rather than trying to pull it straight up.

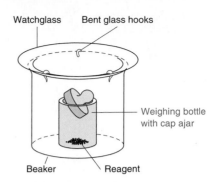

**Figure 2-17** Use a watchglass as a dust cover while drying reagents or crucibles in the oven.

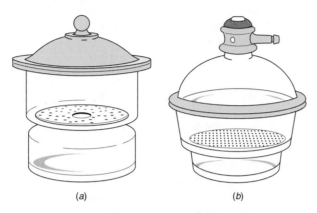

(a)                           (b)

*Figure 2-18* (a) Ordinary desiccator. (b) Vacuum desiccator that can be evacuated through the side arm at the top and then sealed by rotating the joint containing the side arm. Drying is more efficient at low pressure. [Courtesy A. H. Thomas Co., Philadelphia, PA.]

## ■ ■ ■ ■  2-9  Calibration of Volumetric Glassware

Each instrument that we use has a scale of some sort to measure a quantity such as mass, volume, force, or electric current. Manufacturers usually certify that the indicated quantity lies within a certain *tolerance* from the true quantity. For example, a Class A transfer pipet is certified to deliver 10.00 ± 0.02 mL when you use it properly. Your individual pipet might always deliver 10.016 ± 0.004 mL in a series of trials. That is, your pipet delivers an average of 0.016 mL more than the indicated volume in repeated trials. **Calibration** is the process of measuring the actual quantity of mass, volume, force, electric current, and so on, that corresponds to an indicated quantity on the scale of an instrument.

**Table 2-6** Efficiencies of drying agents

| Agent | Formula | Water left in atmosphere ($\mu g \ H_2O/L$)[a] |
|---|---|---|
| Magnesium perchlorate, anhydrous | $Mg(ClO_4)_2$ | 0.2 |
| "Anhydrone" | $Mg(ClO_4)_2 \cdot 1\text{–}1.5H_2O$ | 1.5 |
| Barium oxide | BaO | 2.8 |
| Alumina | $Al_2O_3$ | 2.9 |
| Phosphorus pentoxide | $P_4O_{10}$ | 3.6 |
| Calcium sulfate (Drierite)[b] | $CaSO_4$ | 67 |
| Silica gel | $SiO_2$ | 70 |

a. Moist nitrogen was passed over each desiccant, and the water remaining in the gas was condensed and weighed. [A. I. Vogel, A Textbook of Quantitative Inorganic Analysis, 3rd ed. (New York: Wiley. 1961), p. 178.] For drying gases, the gas can be passed through a 60-cm-long Nafion tube. At 25°C, the residual moisture is 10 µg/L. If the drier is held at 0°C, the residual moisture is 0.8 µg/L. [K. J. Leckrone and J. M. Hayes. "Efficiency and Temperature Dependence of Water Removal by Membrane Dryers." Anal. Chem. **1997**, 69, 911.]

b. Used Drierite can be regenerated by irradiating 1.5-kg batches in a 100 × 190 mm Pyrex crystallizing dish in a microwave oven for 15 min. Stir the solid, heat a second time for 15 min. and place the hot, dry material back in its original container. Use small glass spacers between the crystallizing dish and the glass tray of the oven to protect the tray. [J. A. Green and R. W. Goetz, "Recycling Drierite," J. Chem. Ed. **1991**, 68, 429.]

**Table 2-7**  Density of water

| Temperature (°C) | Density (g/mL) | Volume of 1 g of water (mL) | |
| | | At temperature shown[a] | Corrected to 20°C[b] |
| --- | --- | --- | --- |
| 10 | 0.999 702 6 | 1.001 4 | 1.001 5 |
| 11 | 0.999 608 4 | 1.001 5 | 1.001 6 |
| 12 | 0.999 500 4 | 1.001 6 | 1.001 7 |
| 13 | 0.999 380 1 | 1.001 7 | 1.001 8 |
| 14 | 0.999 247 4 | 1.001 8 | 1.001 9 |
| 15 | 0.999 102 6 | 1.002 0 | 1.002 0 |
| 16 | 0.998 946 0 | 1.002 1 | 1.002 1 |
| 17 | 0.998 777 9 | 1.002 3 | 1.002 3 |
| 18 | 0.998 598 6 | 1.002 5 | 1.002 5 |
| 19 | 0.998 408 2 | 1.002 7 | 1.002 7 |
| 20 | 0.998 207 1 | 1.002 9 | 1.002 9 |
| 21 | 0.997 995 5 | 1.003 1 | 1.003 1 |
| 22 | 0.997 773 5 | 1.003 3 | 1.003 3 |
| 23 | 0.997 541 5 | 1.003 5 | 1.003 5 |
| 24 | 0.997 299 5 | 1.003 8 | 1.003 8 |
| 25 | 0.997 047 9 | 1.004 0 | 1.004 0 |
| 26 | 0.996 786 7 | 1.004 3 | 1.004 2 |
| 27 | 0.996 516 2 | 1.004 6 | 1.004 5 |
| 28 | 0.996 236 5 | 1.004 8 | 1.004 7 |
| 29 | 0.995 947 8 | 1.005 1 | 1.005 0 |
| 30 | 0.995 650 2 | 1.005 4 | 1.005 3 |

a. Corrected for buoyancy with Equation 2-1.

b. Corrected for buoyancy and expansion of borosilicate glass (0.001 0% K$^{-1}$).

Page 38 gives a detailed procedure for calibrating a buret.

For greatest accuracy, we calibrate volumetric glassware to measure the volume actually contained in or delivered by a particular piece of equipment. We do this by measuring the mass of water contained or delivered by the vessel and using the density of water to convert mass into volume.

In the most careful work, it is necessary to account for thermal expansion of solutions and glassware with changing temperature. For this purpose, you should know the lab temperature when a solution was prepared and when it is used. Table 2-7 shows that water expands 0.02% per degree near 20°C. Because the concentration of a solution is proportional to its density, we can write

Correction for thermal expansion:
$$\frac{c'}{d'} = \frac{c}{d} \tag{2-2}$$

The concentration of the solution decreases when the temperature increases.

where $c'$ and $d'$ are the concentration and density at temperature $T'$, and $c$ and $d$ apply at temperature $T$.

**Example**  Effect of Temperature on Solution Concentration

A 0.031 46 M aqueous solution was prepared in winter when the lab temperature was 17°C. What is the molarity of the solution on a warm day when the temperature is 25°C?

**Solution**  We assume that the thermal expansion of a dilute solution is equal to the thermal expansion of pure water. Then, using Equation 2-2 and densities from Table 2-7, we write

$$\frac{c' \text{ at } 25°}{0.997\ 05 \text{ g/mL}} = \frac{0.031\ 46 \text{ M}}{0.998\ 78 \text{ g/mL}} \Rightarrow c' = 0.031\ 41 \text{ M}$$

The concentration has decreased by 0.16% on the warm day.

Pyrex and other borosilicate glasses expand by 0.001 0% per degree near room temperature. If the temperature increases by 10°C, the volume of a piece of glassware increases by (10)(0.001 0%) = 0.010%. For most work, this expansion is insignificant.

To calibrate a 25-mL transfer pipet, you should first weigh an empty weighing bottle like the one in Figure 2-17. Then fill the pipet to the mark with distilled water, drain it into the weighing bottle, and cap the bottle to prevent evaporation. Weigh the bottle again to find the mass of water delivered from the pipet. Finally, use Equation 2-3 to convert mass into volume.

$$\text{True volume} = (\text{grams of water}) \times (\text{volume of 1 g of } H_2O \text{ in Table 2-7}) \qquad (2\text{-}3)$$

Small or odd-shaped vessels can be calibrated with Hg, which is easier than water to pour out of glass and is 13.6 times denser than water. This procedure is for researchers, not students.

**Example** Calibration of a Pipet

An empty weighing bottle had a mass of 10.313 g. After the addition of water from a 25-mL pipet, the mass was 35.225 g. If the lab temperature was 27°C, find the volume of water delivered by the pipet.

**Solution** The mass of water is 35.225 − 10.313 = 24.912 g. From Equation 2-3 and the next-to-last column of Table 2-7, the volume of water is (24.912 g)(1.004 6 mL/g) = 25.027 mL at 27°C. The last column in Table 2-7 tells us what the volume would be if the pipet were at 20°C. This pipet would deliver (24.912 g)(1.004 5 mL/g) = 25.024 mL at 20°C.

The pipet delivers less volume at 20°C than at 27°C because glass contracts slightly as the temperature is lowered. Volumetric glassware is usually calibrated at 20°C.

## 2-10 Introduction to Microsoft Excel

If you already use a spreadsheet, you can skip this section. The computer spreadsheet is an essential tool for manipulating quantitative information. In analytical chemistry, spreadsheets can help us with calibration curves, statistical analysis, titration curves, and equilibrium problems. Spreadsheets allow us to conduct "what if" experiments such as investigating the effect of a stronger acid or a different ionic strength on a titration curve. We use Microsoft Excel in this book as a tool for solving problems in analytical chemistry. Although you can skip over spreadsheets with no loss of continuity, spreadsheets will enrich your understanding of chemistry and provide a valuable tool for use outside this course.

This section introduces a few basic features of Excel 2000 for a PC computer. Other versions of Excel and other spreadsheets are not very different from what we describe. Excellent books are available if you want to learn much more about this software.[19]

### Getting Started: Calculating the Density of Water

Let's prepare a spreadsheet to compute the density of water from the equation

$$\text{Density(g/mL)} = a_0 + a_1*T + a_2*T^2 + a_3*T^3 \qquad (2\text{-}4)$$

This equation is accurate to five decimal places over the range 4° to 40°C.

where $T$ is temperature (°C) and $a_0 = 0.999\,89$, $a_1 = 5.332\,2 \times 10^{-5}$, $a_2 = -7.589\,9 \times 10^{-6}$, and $a_3 = 3.671\,9 \times 10^{-8}$.

The blank spreadsheet in Figure 2-19a has columns labeled A, B, C and rows numbered 1, 2, 3, . . . , 12. The box in column B, row 4 is called *cell* B4.

Begin each spreadsheet with a title to help make the spreadsheet more readable. In Figure 2-19b, we click in cell A1 and type "Calculating Density of $H_2O$ with Equation 2-4". Then we click in cell A2 and write "(from the delightful book by Dan Harris)". The computer automatically spreads the text into adjoining cells.

We adopt a convention in this book in which constants are collected in column A. Type "Constants:" in cell A4. Then select cell A5 and type "a0=". Now select cell A6 and type the number 0.99989 (without extra spaces). In cells A7 to A12, enter the remaining constants. Powers of 10 are written, for example, as E-5 for $10^{-5}$. Your spreadsheet should now look like Figure 2-19b.

In cell B4, write the heading "Temp (°C)". Then enter temperatures from 5 through 40 in cells B5 through B12. This is our *input* to the spreadsheet. The *output* will be computed values of density in column C.

Oops! I forgot one little trick. If you need more room in a column for the number of digits, you can grab the vertical line at the top of the column with your mouse and resize the column.

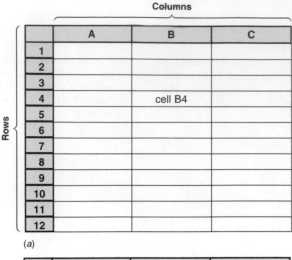

**(a)**

| | A | B | C |
|---|---|---|---|
| 1 | | | |
| 2 | | | |
| 3 | | | |
| 4 | | cell B4 | |
| 5 | | | |
| 6 | | | |
| 7 | | | |
| 8 | | | |
| 9 | | | |
| 10 | | | |
| 11 | | | |
| 12 | | | |

**(b)**

| | A | B | C |
|---|---|---|---|
| 1 | Calculating Density of H2O with Equation 2-4 | | |
| 2 | (from the delightful book by Dan Harris) | | |
| 3 | | | |
| 4 | Constants: | | |
| 5 | a0 = | | |
| 6 | 0.99989 | | |
| 7 | a1 = | | |
| 8 | 5.3322E-05 | | |
| 9 | a2 = | | |
| 10 | -7.5899E-06 | | |
| 11 | a3 = | | |
| 12 | 3.6719E-08 | | |

**(c)**

| | A | B | C |
|---|---|---|---|
| 1 | Calculating Density of H2O with Equation 2-4 | | |
| 2 | (from the delightful book by Dan Harris) | | |
| 3 | | | |
| 4 | Constants: | Temp (°C) | Density (g/mL) |
| 5 | a0 = | 5 | 0.99997 |
| 6 | 0.99989 | 10 | |
| 7 | a1 = | 15 | |
| 8 | 5.3322E-05 | 20 | |
| 9 | a2 = | 25 | |
| 10 | -7.5899E-06 | 30 | |
| 11 | a3 = | 35 | |
| 12 | 3.6719E-08 | 40 | |

Figure 2-19 Evolution of a spreadsheet for computing the density of water.

**(d)**

| | A | B | C |
|---|---|---|---|
| 1 | Calculating Density of H2O with Equation 2-4 | | |
| 2 | (from the delightful book by Dan Harris) | | |
| 3 | | | |
| 4 | Constants: | Temp (°C) | Density (g/mL) |
| 5 | a0 = | 5 | 0.99997 |
| 6 | 0.99989 | 10 | 0.99970 |
| 7 | a1 = | 15 | 0.99911 |
| 8 | 5.3322E-05 | 20 | 0.99821 |
| 9 | a2 = | 25 | 0.99705 |
| 10 | -7.5899E-06 | 30 | 0.99565 |
| 11 | a3 = | 35 | 0.99403 |
| 12 | 3.6719E-08 | 40 | 0.99223 |
| 13 | | | |
| 14 | Formula: | | |
| 15 | C5 = $A$6+$A$8*B5+$A$10*B5^2+$A$12*B5^3 | | |

Formulas begin with an equal sign. Arithmetic operations in a spreadsheet are

| | |
|---|---|
| + | addition |
| − | subtraction |
| * | multiplication |
| / | division |
| ^ | exponentiation |

Three kinds of entries:

| | |
|---|---|
| label | A3 = |
| number | 4.4E-05 |
| formula | = $A$8*B5 |

In cell C4, enter the heading "Density (g/mL)". Cell C5 is the most important one in the table. In this one, you will write the *formula*

$$= \$A\$6 + \$A\$8*B5 + \$A\$10*B5\text{\textasciicircum}2 + \$A\$12*B5\text{\textasciicircum}3$$

(It doesn't matter whether or not you use spaces around the arithmetic operators.) When you hit RETURN, the number 0.99997 appears in cell C5. The formula above is the spreadsheet translation of Equation 2-4. $A$6 refers to the constant in cell A6. (We will explain the dollar signs shortly.) B5 refers to the temperature in cell B5. The times sign is * and the exponentiation sign is ^. For example, the term "$A$12*B5^3" means "(contents of cell A12) × (contents of cell B5)$^3$."

Now comes the most magical property of a spreadsheet. Highlight cell C5 and the empty cells below it from C6 to C12. Then select the FILL DOWN command from the EDIT menu. This procedure copies the formula from C5 into the cells below it and evaluates the numbers in each of the selected cells. The density of water at each temperature now appears in column C in Figure 2-19d.

In this example, we made three types of entries. *Labels* such as "a0=" were typed in as text. An entry that does not begin with a digit or an equal sign is treated as text. *Numbers,* such as 25, were typed in some cells. The spreadsheet treats a number differently from text. In cell C5, we entered a *formula* that necessarily begins with an equal sign.

## Arithmetic Operations and Functions

Addition, subtraction, multiplication, division, and exponentiation have the symbols +, −, *, /, and ^. *Functions* such as Exp(·) can be typed by you or can be selected from the INSERT menu by choosing FUNCTION. Exp(·) raises e to the power in parentheses. Other functions such as $Ln(\cdot)$, $Log(\cdot)$, $Sin(\cdot)$, and $Cos(\cdot)$ are also available. In some spreadsheets, functions are written in all capital letters.

The order of arithmetic operations in formulas is negation first, followed by ^, followed by * and / (evaluated in order from left to right as they appear), finally followed by + and − (also evaluated from left to right). Make liberal use of parentheses to be sure that the computer does what you intend. The contents of parentheses are evaluated first, before carrying out operations outside the parentheses. Here are some examples:

$$9/5*100+32 = (9/5)*100+32 = (1.8)*100+32 = (1.8*100)+32 = (180)+32 = 212$$
$$9/5*(100+32) = 9/5*(132) = (1.8)*(132) = 237.6$$
$$9+5*100/32 = 9+(5*100)/32 = 9+(500)/32 = 9+(500/32) = 9+(15.625) = 24.625$$
$$9/5\char94 2+32 = 9/(5\char94 2)+32 = (9/25)+32 = (0.36)+32 = 32.36$$
$$-2\char94 2 = 4 \quad \text{but} \quad -(2\char94 2) = -4$$

When in doubt about how an expression will be evaluated, use parentheses to force what you intend.

## Documentation and Readability

The first important *documentation* in the spreadsheet is the name of the file. A name such as "Expt 9 Gran Plot" is much more meaningful than "Dan's Lab". The next important feature is a title at the top of the spreadsheet, which tells its purpose. To remind ourselves what formulas were used in the spreadsheet, we added text (labels) at the bottom. In cell A14, write "Formula:" and in cell A15 write "C5 = \$A\$6+\$A\$8*B5+\$A\$10*B5^2+\$A\$12*B5^3". This documentation tells us how numbers in column C were calculated.

For additional readability in Figure 2-19, column C was set to display just five decimal places, even though the computer retains many more for its calculations. It does not throw away the digits that are not displayed. You can control the way numbers are displayed in a cell by going to the FORMAT menu, selecting CELLS, and going to the Number tab.

## Absolute and Relative References

The formula "= \$A\$8*B5" refers to cells A8 and B5 in different manners. \$A\$8 is an *absolute reference* to the contents of cell A8. No matter where cell \$A\$8 is called from in the spreadsheet, the computer goes to cell A8 to look for a number. "B5" is a *relative reference* in the formula in cell C5. When called from cell C5, the computer goes to cell B5 to find a number. When called from cell C6, the computer goes to cell B6 to look for a number. If called from cell C19, the computer would look in cell B19. This is why the cell written without dollar signs is called a relative reference. If you want the computer to always look only in cell B5, then you should write "\$B\$5".

## 2-11 Graphing with Microsoft Excel

*Graphs are critical to understanding quantitative relations.* Depending on which version of Excel you have, there may be some variation from what is described here.

To make a graph from the spreadsheet in Figure 2-19d, go to the INSERT menu and select CHART. A window appears with a variety of options. The one you will almost always want is XY (Scatter). Highlight XY (Scatter) and several options appear. Select the one that shows data points connected by a smooth curve. Click Next to move to the next window.

Now you are asked which cells contain the data to be plotted. Identify the *x* data by writing B5:B12 next to Data Range. Then write a comma and identify the *y* data by writing C5:C12. The input for Data Range now looks like B5:B12,C5:C12. Click the button to show that data are in columns, not rows. Click Next.

Now a small graph of your data appears. If it does not look as expected, make sure you selected the correct data, with *x* before *y*. The new window asks you for axis labels and an optional title for the graph. For the title, write "Density of Water" (without quotation marks). For the *x*-axis, enter "Temperature (°C)" and for the *y*-axis write "Density (g/mL)". Click Next.

Now you are given the option of drawing the graph on a new sheet or on the same sheet that is already open. For this case, select "As object in Sheet 1". Click Finish and the chart

Order of operations:
1. Negation (a minus sign before a term)
2. Exponentiation
3. Multiplication and division (in order from left to right)
4. Addition and subtraction (in order from left to right)

Operations within parentheses are evaluated first, from the innermost set.

*Documentation* means labeling. If your spreadsheet cannot be read by another person without your help, it needs better documentation. (The same is true of your lab notebook!)

Absolute reference: \$A\$8
Relative reference: B5

Save your files frequently while you are working and make a backup file of anything that you don't want to lose.

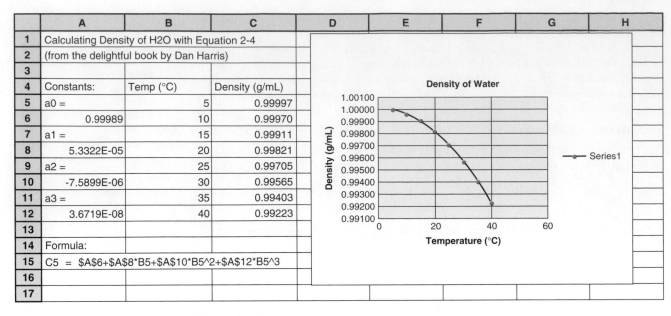

| | A | B | C | D | E | F | G | H |
|---|---|---|---|---|---|---|---|---|
| 1 | Calculating Density of H2O with Equation 2-4 | | | | | | | |
| 2 | (from the delightful book by Dan Harris) | | | | | | | |
| 3 | | | | | | | | |
| 4 | Constants: | Temp (°C) | Density (g/mL) | | | | | |
| 5 | a0 = | 5 | 0.99997 | | | | | |
| 6 | 0.99989 | 10 | 0.99970 | | | | | |
| 7 | a1 = | 15 | 0.99911 | | | | | |
| 8 | 5.3322E-05 | 20 | 0.99821 | | | | | |
| 9 | a2 = | 25 | 0.99705 | | | | | |
| 10 | -7.5899E-06 | 30 | 0.99565 | | | | | |
| 11 | a3 = | 35 | 0.99403 | | | | | |
| 12 | 3.6719E-08 | 40 | 0.99223 | | | | | |
| 13 | | | | | | | | |
| 14 | Formula: | | | | | | | |
| 15 | C5 = $A$6+$A$8*B5+$A$10*B5^2+$A$12*B5^3 | | | | | | | |
| 16 | | | | | | | | |
| 17 | | | | | | | | |

*Figure 2-20*  Initial density chart drawn by Excel.

will appear on your spreadsheet. Grab the chart with your mouse and move it to the right of the data in your spreadsheet, as shown in Figure 2-20.

You have just drawn your first graph! Excel gives us many options for changing features of the graph. Here are a few, but you should experiment with the graph to discover other formatting options. Double click on the $y$-axis and a window appears. Select the Patterns tab. Change Minor tic mark type from None to Outside and click OK. You will see new tic marks appear on the $y$-axis. Double click on the $y$-axis again and select the Number tab. Change the decimal places to 3 and click OK. Double click on the $y$-axis again and select the Scale tab. Set the minimum to 0.992 and the maximum to 1.000 and click OK.

Double click on the $x$-axis and select Patterns. Change the Minor tic mark type from None to Outside. Select Scale and set the maximum to 40, the major unit to 10, the minor unit to 5, and click OK.

Double click on the gray area of the graph and a window called Patterns appears. Select Automatic for the Border and None for the Area. This removes the gray background and gives a solid line around the graph. To add vertical lines at the major tic marks, select the graph with the mouse. Then go to the CHART menu and select CHART OPTIONS. In the window that appears, select Gridlines. For the Value (X) axis, check Major gridlines. Then select the tab for Legend and remove the check mark from Show Legend. The legend will disappear. Click OK. You should be getting the idea that you can format virtually any part of the chart.

Click on the outer border of the chart and handles appear. Grab the one on the right and resize the chart so it does not extend past column F of the spreadsheet. Grab the handle at the bottom and resize the chart so it does not extend below row 15. When you resized the chart, letters and numbers shrank. Double click on each set of numbers and change the font to 8 points. Double click on the labels and change the letters to 9 points. Your chart should now look like the one in Figure 2-21.

If you want to write on the chart, go to the VIEW menu and select TOOLBARS and DRAWING. Select the Text Box tool from the Drawing toolbar, click inside your chart, and you can begin typing words. You can move the words around and change their format. You can draw arrows on the chart with the Arrow tool. If you double click on a data point on the chart, a box appears that allows you to change the plotting symbols.

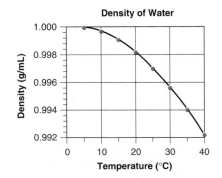

*Figure 2-21*  Density chart after reformatting.

## Terms to Understand

| | | | |
|---|---|---|---|
| absorption | buret | hygroscopic | pipet |
| acid wash | calibration | ignition | slurry |
| adsorption | desiccant | meniscus | tare |
| ashless filter paper | desiccator | mother liquor | volumetric flask |
| buoyancy | filtrate | parallax | |

## Summary

Safety requires you to think in advance about what you will do; never do anything that seems dangerous. Know how to use safety equipment such as goggles, fume hood, lab coat, gloves, emergency shower, eyewash, and fire extinguisher. Chemicals should be stored and used in a manner that minimizes contact of solids, liquids, and vapors with people. Environmentally acceptable disposal procedures should be established in advance for every chemical that you use. Your lab notebook tells what you did and what you observed; it should be understandable to other people. It also should allow you to repeat an experiment in the same manner in the future. You should understand the principles of operation of electronic and mechanical balances and treat them as delicate equipment. Buoyancy corrections are required in accurate work. Burets should be read in a reproducible manner and drained slowly for best results. Always interpolate between markings to obtain accuracy one deci-mal place beyond the graduations. Volumetric flasks are used to prepare solutions with known volume. Transfer pipets deliver fixed volumes; less accurate measuring pipets deliver variable volumes. Do not be lulled into complacency by the nice digital reading on a micropipet. Unless your pipet has been calibrated recently and your personal technique tested, micropipets can have gross errors. Filtration and collection of precipitates require careful technique, as does the drying of reagents, precipitates, and glassware in ovens and desiccators. Volumetric glassware is calibrated by weighing water contained in or delivered by the vessel. In the most careful work, solution concentrations and volumes of vessels should be corrected for changes in temperature.

If you plan to use spreadsheets in this course, you should know how to enter formulas in a spreadsheet and how to draw a graph of data in a spreadsheet.

## Exercises

**2-A.** What is the true mass of water if the measured mass in the atmosphere is 5.397 4 g? When you look up the density of water, assume that the lab temperature is **(a)** 15°C and **(b)** 25°C. Take the density of air to be 0.001 2 g/mL and the density of balance weights to be 8.0 g/mL.

**2-B.** A sample of ferric oxide ($Fe_2O_3$, density = 5.24 g/mL) obtained from ignition of a gravimetric precipitate weighed 0.296 1 g in the atmosphere. What is the true mass in vacuum?

**2-C.** A solution of potassium permanganate ($KMnO_4$) was found by titration to be 0.051 38 M at 24°C. What was the molarity when the lab temperature dropped to 16°C?

**2-D.** Water was drained from a buret between the 0.12- and 15.78-mL marks. The apparent volume delivered was $15.78 - 0.12 = 15.66$ mL. Measured in the air at 22°C, the mass of water delivered was 15.569 g. What was the true volume?

**2-E.** 🖳 To familiarize yourself with your spreadsheet and graphing software, reproduce the spreadsheet in Figure 2-19 and the graph in Figure 2-21.

## Problems

**Safety and Lab Notebook**

**2-1.** What is the primary safety rule and what is your implied responsibility to make it work?

**2-2.** After the safety features and procedures in your laboratory have been explained to you, make a list of them.

**2-3.** In Box 2-1, why is $Pb^{2+}$ converted into $PbSiO_3$ before disposal in an approved landfill?

**2-4.** Explain what each of the three numbered hazard ratings means for 37 wt% HC1 in Figure 2-2.

**2-5.** State three essential attributes of a lab notebook.

**Analytical Balance**

**2-6.** Explain the principles of operation of electronic and mechanical balances.

**2-7.** Why is the buoyancy correction equal to 1 in Figure 2-5 when the density of the object being weighed is 8.0 g/mL?

**2-8.** Pentane ($C_5H_{12}$) is a liquid with a density of 0.626 g/mL near 25°C. Use Equation 2-1 to find the true mass of pentane when the mass weighed in air is 14.82 g. Assume that the air density is 0.001 2 g/mL and the balance weight density is 8.0 g/mL.

**2-9.** The densities (g/mL) of several substances are: acetic acid, 1.05; $CCl_4$, 1.59; sulfur, 2.07; lithium, 0.53; mercury, 13.5; $PbO_2$, 9.4; lead, 11.4; iridium, 22.5. From Figure 2-5, predict which substance will have the smallest percentage of buoyancy correction and which will have the greatest.

**2-10.** Potassium hydrogen phthalate is a primary standard used to measure the concentration of NaOH solutions. Find the true mass of potassium hydrogen phthalate (density = 1.636 g/mL) if the apparent mass weighed in air is 4.236 6 g. If you did not correct the mass for buoyancy, would the calculated molarity of NaOH be too high or too low? By what percentage?

**2-11. (a)** Use the ideal gas law (Problem 1-16) to calculate the density (g/mL) of helium at 20°C and 1.00 bar.

**(b)** Find the true mass of sodium metal (density = 0.97 g/mL) weighed in a glove box with a helium atmosphere, if the apparent mass was 0.823 g and the balance weight density is 8.0 g/mL.

**2-12. (a)** What is the vapor pressure of water in the air at 20°C if the relative humidity is 42%? The vapor pressure of water at 20°C at equilibrium is 2 330 Pa. (*Relative humidity* is the percentage of the equilibrium water vapor pressure in the air.)

**(b)** Use note 11 for Chapter 2 at the end of the book to find the air density (g/mL, not g/L) under the conditions of part **(a)** if the barometric pressure is 94.0 kPa.

**(c)** What is the true mass of water in part **(b)** if the balance indicates that 1.000 0 g is present (balance weight density = 8.0 g/mL)?

**2-13.** *Effect of altitude on electronic balance.* If an object weighs $m_a$ grams at distance $r_a$ from the center of the earth, it will weigh $m_b = m_a(r_a^2/r_b^2)$ when raised to $r_b$. An object weighs 100.000 0 g on the 1st floor of a building at $r_a = 6\ 370$ km. How much will it weigh on the 10th floor, which is 30 m higher?

**Glassware and Thermal Expansion**

**2-14.** What do the symbols "TD" and "TC" mean on volumetric glassware?

**2-15.** Describe how to prepare 250.0 mL of 0.150 0 M $K_2SO_4$ with a volumetric flask.

**2-16.** When is it preferable to use a plastic volumetric flask instead of a more accurate glass flask?

**2-17.** Describe how to deliver 5.00 mL of liquid by using a transfer pipet.

**2-18.** Which is more accurate, a transfer pipet or a measuring pipet?

**2-19.** What is the purpose of the trap in Figure 2-14 and the watch-glass in Figure 2-17?

**2-20.** Which drying agent is more efficient, Drierite or phosphorus pentoxide?

**2-21.** An empty 10-mL volumetric flask weighs 10.263 4 g. When the flask is filled to the mark with distilled water and weighed again in the air at 20°C, the mass is 20.214 4 g. What is the true volume of the flask at 20°C?

**2-22.** By what percentage does a dilute aqueous solution expand when heated from 15° to 25°C? If a 0.500 0 M solution is prepared at 15°C, what would its molarity be at 25°C?

**2-23.** The true volume of a 50-mL volumetric flask is 50.037 mL at 20°C. What mass of water measured **(a)** in vacuum and **(b)** in air at 20°C would be contained in the flask?

**2-24.** You want to prepare 500.0 mL of 1.000 M $KNO_3$ at 20°C, but the lab (and water) temperature is 24°C at the time of preparation. How many grams of solid $KNO_3$ (density = 2.109 g/mL) should be dissolved in a volume of 500.0 mL at 24°C to give a concentration of 1.000 M at 20°C? What apparent mass of $KNO_3$ weighed in air is required?

**2-25.** Glass is a notorious source of metal ion contamination. Three glass bottles were crushed and sieved to collect 1-mm pieces.[20] To see how much $Al^{3+}$ could be extracted, 200 mL of a 0.05 M solution of the metal-binding compound EDTA was stirred with 0.50 g of ~1-mm glass particles in a polyethylene flask. The Al content of the solution after 2 months was 5.2 μM. The total Al content of the glass, measured after completely dissolving some glass in 48 wt% HF with microwave heating, was 0.80 wt%. What fraction of the Al was extracted from glass by EDTA?

**2-26.** The efficiency of a gas chromatography column is measured by a parameter called plate height ($H$, mm), which is related to the gas flow rate ($u$, mL/min) by the van Deemter equation: $H = A + B/u + Cu$, where $A$, $B$, and $C$ are constants. Prepare a spreadsheet with a graph showing values of $H$ as a function of $u$ for $u = 4, 6, 8, 10, 20, 30, 40, 50, 60, 70, 80, 90,$ and 100 mL/min. Use the values $A = 1.65$ mm, $B = 25.8$ mm · mL/min, and $C = 0.023$ 6 mm · min/mL.

---

## Reference Procedure: Calibrating a 50-mL Buret

This procedure tells how to construct a graph such as Figure 3-3 to convert the measured volume delivered by a buret to the true volume delivered at 20°C.

1.  Fill the buret with distilled water and force any air bubbles out the tip. See whether the buret drains without leaving drops on the walls. If drops are left, clean the buret with soap and water or soak it with cleaning solution.[13] Adjust the meniscus to be at or slightly below 0.00 mL, and touch the buret tip to a beaker to remove the suspended drop of water. Allow the buret to stand for 5 min while you weigh a 125-mL flask fitted with a rubber stopper. (Hold the flask with a tissue or paper towel, not with your hands, to prevent fingerprint residue from changing its mass.) If the level of the liquid in the buret has changed, tighten the stopcock and repeat the procedure. Record the level of the liquid.
2.  Drain approximately 10 mL of water at a rate < 20 mL/min into the weighed flask, and cap it tightly to prevent evaporation. Allow about 30 s for the film of liquid on the walls to descend before you read the buret. Estimate all readings to the nearest 0.01 mL. Weigh the flask again to determine the mass of water delivered.
3.  Now drain the buret from 10 to 20 mL, and measure the mass of water delivered. Repeat the procedure for 30, 40, and 50 mL. Then do the entire procedure (10, 20, 30, 40, 50 mL) a second time.
4.  Use Table 2-7 to convert the mass of water into the volume delivered. Repeat any set of duplicate buret corrections that do not agree to within 0.04 mL. Prepare a calibration graph like that in Figure 3-3, showing the correction factor at each 10-mL interval.

### Example  Buret Calibration

When draining the buret at 24°C, you observe the following values:

| | | |
|---|---|---|
| Final reading | 10.01 | 10.08 mL |
| Initial reading | 0.03 | 0.04 |
| Difference | 9.98 | 10.04 mL |
| Mass | 9.984 | 10.056 g |
| Actual volume delivered | 10.02 | 10.09 mL |
| Correction | +0.04 | +0.05 mL |
| Average correction | | +0.045 mL |

To calculate the actual volume delivered when 9.984 g of water is delivered at 24°C, look at the column of Table 2-7 headed "Corrected to 20°C." In the row for 24°C, you find that 1.000 0 g of water occupies 1.003 8 mL. Therefore, 9.984 g occupies (9.984 g)(1.003 8 mL/g) = 10.02 mL. The average correction for both sets of data is +0.045 mL.

To obtain the correction for a volume greater than 10 mL, add successive masses of water collected in the flask. Suppose that the following masses were measured:

| Volume interval (mL) | Mass delivered (g) |
|---|---|
| 0.03–10.01 | 9.984 |
| 10.01–19.90 | 9.835 |
| 19.90–30.06 | 10.071 |
| Sum  30.03 mL | 29.890 g |

The total volume of water delivered is (29.890 g)(1.003 8 mL/g) = 30.00 mL. Because the indicated volume is 30.03 mL, the buret correction at 30 mL is −0.03 mL.

*What does this mean?* Suppose that Figure 3-3 applies to your buret. If you begin a titration at 0.04 mL and end at 29.00 mL, you would deliver 28.96 mL if the buret were perfect. Figure 3-3 tells you that the buret delivers 0.03 mL less than the indicated amount, so only 28.93 mL were actually delivered. To use the calibration curve, either begin all titrations near 0.00 mL or correct both the initial and the final readings. Use the calibration curve whenever you use your buret.

# 3 Experimental Error

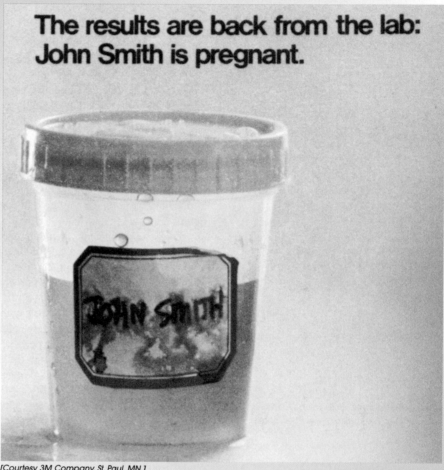

The results are back from the lab:
John Smith is pregnant.

[Courtesy 3M Company, St. Paul, MN.]

Some laboratory errors are more obvious than others, but there is error associated with every measurement. There is no way to measure the "true value" of anything. The best we can do in a chemical analysis is to carefully apply a technique that experience tells us is reliable. Repetition of one method of measurement several times tells us the *precision* (reproducibility) of the measurement. If the results of measuring the same quantity by different methods agree with one another, then we become confident that the results are *accurate*, which means they are near the "true" value.

$S$uppose that you determine the density of a mineral by measuring its mass (4.635 ± 0.002 g) and volume (1.13 ± 0.05 mL). Density is mass per unit volume: 4.635 g/1.13 mL = 4.101 8 g/mL. The uncertainties in measured mass and volume are ±0.002 g and ±0.05 mL, but what is the uncertainty in the computed density? And how many significant figures should be used for the density? This chapter discusses the propagation of uncertainty in lab calculations.

## 3-1 Significant Figures

*Significant figures:* minimum number of digits required to express a value in scientific notation without loss of accuracy

The number of **significant figures** is the minimum number of digits needed to write a given value in scientific notation without loss of accuracy. The number 142.7 has four significant figures, because it can be written $1.427 \times 10^2$. If you write $1.427\ 0 \times 10^2$, you imply that

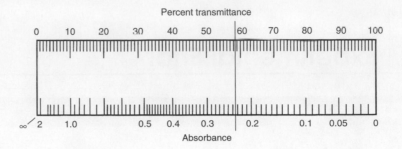

**Figure 3-1** Scale of a Bausch and Lomb Spectronic 20 spectrophotometer. Percent transmittance is a linear scale and absorbance is a logarithmic scale.

you know the value of the digit after 7, which is not the case for the number 142.7. The number $1.427\,0 \times 10^2$ has five significant figures.

The number $6.302 \times 10^{-6}$ has four significant figures, because all four digits are necessary. You could write the same number as 0.000 006 302, which also has just *four* significant figures. The zeros to the left of the 6 are merely holding decimal places. The number 92 500 is ambiguous. It could mean any of the following:

| | |
|---|---|
| $9.25 \times 10^4$ | 3 significant figures |
| $9.250 \times 10^4$ | 4 significant figures |
| $9.250\,0 \times 10^4$ | 5 significant figures |

You should write one of the three numbers above, instead of 92 500, to indicate how many figures are actually known.

Zeros are significant when they occur (1) in the middle of a number or (2) at the end of a number on the right-hand side of a decimal point.

**Significant zeros below are bold:**

106    0.010 **6**    0.106    0.106 **0**

The last significant digit (farthest to the right) in a measured quantity always has some associated uncertainty. The minimum uncertainty is ±1 in the last digit. The scale of a Spectronic 20 spectrophotometer is drawn in Figure 3-1. The needle in the figure appears to be at an absorbance value of 0.234. We say that this number has three significant figures because the numbers 2 and 3 are completely certain and the number 4 is an estimate. The value might be read 0.233 or 0.235 by other people. The percent transmittance is near 58.3. Because the transmittance scale is smaller than the absorbance scale at this point, there is more uncertainty in the last digit of transmittance. A reasonable estimate of uncertainty might be 58.3 ± 0.2. There are three significant figures in the number 58.3.

*Interpolation:* Estimate all readings to the nearest tenth of the distance between scale divisions.

When reading the scale of any apparatus, try to estimate to the nearest tenth of a division. On a 50-mL buret, which is graduated to 0.1 mL, read the level to the nearest 0.01 mL. For a ruler calibrated in millimeters, estimate distances to the nearest 0.1 mm.

There is uncertainty in any *measured* quantity, even if the measuring instrument has a digital readout that does not fluctuate. When a digital pH meter indicates a pH of 3.51, there is uncertainty in the digit 1 (and maybe even in the digit 5). By contrast, some numbers are exact—with an infinite number of unwritten significant digits. To calculate the average height of four people, you would divide the sum of heights (which is a measured quantity with some uncertainty) by the integer 4. There are exactly 4 people, not 4.000 ± 0.002 people!

## ■■■ 3-2 Significant Figures in Arithmetic

We now consider how many digits to retain in the answer after you have performed arithmetic operations with your data. Rounding should only be done on the *final answer* (not intermediate results), to avoid accumulating round-off errors.

### Addition and Subtraction

If the numbers to be added or subtracted have equal numbers of digits, the answer goes to the *same decimal place* as in any of the individual numbers:

$$\begin{array}{r} 1.362 \times 10^{-4} \\ + 3.111 \times 10^{-4} \\ \hline 4.473 \times 10^{-4} \end{array}$$

The number of significant figures in the answer may exceed or be less than that in the original data.

$$
\begin{array}{cc}
5.345 & 7.26 \times 10^{14} \\
+\ 6.728 & -\ 6.69 \times 10^{14} \\
\hline
12.073 & 0.57 \times 10^{14}
\end{array}
$$

If the numbers being added do not have the same number of significant figures, we are limited by the least-certain one. For example, the molecular mass of $KrF_2$ is known only to the third decimal place, because we only know the atomic mass of Kr to three decimal places:

$$
\begin{array}{ll}
18.998\ 403\ 2 & \text{(F)} \\
+\ 18.998\ 403\ 2 & \text{(F)} \\
+\ 83.798 & \text{(Kr)} \\
\hline
121.794\ \underbrace{806\ 4}_{\text{Not significant}}
\end{array}
$$

The number 121.794 806 4 should be rounded to 121.795 as the final answer.

When rounding off, look at *all* the digits *beyond* the last place desired. In the preceding example, the digits 806 4 lie beyond the last significant decimal place. Because this number is more than halfway to the next higher digit, we round the 4 up to 5 (that is, we round up to 121.795 instead of down to 121.794). If the insignificant figures were less than halfway, we would round down. For example, 121.794 3 is rounded to 121.794.

In the special case where the number is exactly halfway, round to the nearest *even* digit. Thus, 43.55 is rounded to 43.6, if we can only have three significant figures. If we are retaining only three figures, $1.425 \times 10^{-9}$ becomes $1.42 \times 10^{-9}$. The number $1.425\ 01 \times 10^{-9}$ would become $1.43 \times 10^{-9}$, because 501 is more than halfway to the next digit. The rationale for rounding to an even digit is to avoid systematically increasing or decreasing results through successive round-off errors. Half the round-offs will be up and half down.

In the addition or subtraction of numbers expressed in scientific notation, all numbers should first be expressed with the same exponent:

$$
\begin{array}{ll}
1.632 \times 10^5 & \\
+\ 4.107 \times 10^3 & \rightarrow \\
+\ 0.984 \times 10^6 &
\end{array}
\qquad
\begin{array}{ll}
1.632 & \times 10^5 \\
+\ 0.041\ 07 & \times 10^5 \\
+\ 9.84 & \times 10^5 \\
\hline
11.51 & \times 10^5
\end{array}
$$

The sum $11.513\ 07 \times 10^5$ is rounded to $11.51 \times 10^5$ because the number $9.84 \times 10^5$ limits us to two decimal places when all numbers are expressed as multiples of $10^5$.

## Multiplication and Division

In multiplication and division, we are normally limited to the number of digits contained in the number with the fewest significant figures:

$$
\begin{array}{ccc}
3.26 \times 10^{-5} & 4.317\ 9 \times 10^{12} & 34.60 \\
\times\ 1.78 & \times\ 3.6\ \ \times 10^{-19} & \div\ 2.462\ 87 \\
\hline
5.80 \times 10^{-5} & 1.6\ \ \ \times 10^{-6} & 14.05
\end{array}
$$

The power of 10 has no influence on the number of figures that should be retained.

## Logarithms and Antilogarithms

The base 10 **logarithm** of $n$ is the number $a$, whose value is such that $n = 10^a$:

Logarithm of n: $\qquad n = 10^a \quad \text{means that} \quad \log n = a \qquad$ (3-1)

For example, 2 is the logarithm of 100 because $100 = 10^2$. The logarithm of 0.001 is $-3$ because $0.001 = 10^{-3}$. To find the logarithm of a number with your calculator, enter the number and press the *log* function.

In Equation 3-1, the number $n$ is said to be the **antilogarithm** of $a$. That is, the antilogarithm of 2 is 100 because $10^2 = 100$, and the antilogarithm of $-3$ is 0.001 because $10^{-3} = 0.001$. Your calculator has either a *$10^x$* key or an *antilog* key. To find the antilogarithm of a number, enter it in your calculator and press *$10^x$* (or *antilog*).

Inspect the legend of the periodic table inside the cover of this book. Be sure you can interpret uncertainties in atomic masses. For F and Kr, the atomic masses are

F: $18.998\ 403\ 2 \pm 0.000\ 000\ 5$
Kr: $83.798 \pm 0.002$

Rules for rounding off numbers

Addition and subtraction: Express all numbers with the same exponent and align all numbers with respect to the decimal point. Round off the answer according to the number of decimal places in the number with the fewest decimal places.

$10^{-3} = \dfrac{1}{10^3} = \dfrac{1}{1\ 000} = 0.001$

A logarithm is composed of a **characteristic** and a **mantissa.** The characteristic is the integer part and the mantissa is the decimal part:

$$\log 339 = 2.\underbrace{530}_{} \qquad \log 3.39 \times 10^{-5} = -\underbrace{4}.\underbrace{470}$$

$$\begin{array}{cc} \text{Characteristic} & \text{Mantissa} \\ = 2 & = 0.530 \end{array} \qquad \begin{array}{cc} \text{Characteristic} & \text{Mantissa} \\ = -4 & = 0.470 \end{array}$$

The number 339 can be written $3.39 \times 10^2$. *The number of digits in the mantissa of log 339 should equal the number of significant figures in 339.* The logarithm of 339 is properly expressed as 2.530. The *characteristic,* 2, corresponds to the exponent in $3.39 \times 10^2$.

To see that the third decimal place is the last significant place, consider the following results:

$$10^{2.531} = 340 \; (339.6)$$
$$10^{2.530} = 339 \; (338.8)$$
$$10^{2.529} = 338 \; (338.1)$$

The numbers in parentheses are the results prior to rounding to three figures. Changing the exponent in the third decimal place changes the answer in the third place of 339.

In the conversion of a logarithm into its antilogarithm, *the number of significant figures in the antilogarithm should equal the number of digits in the mantissa.* Thus,

$$\text{antilog} \; (-3.\underbrace{42}) = 10^{-3.\underbrace{42}} = \underbrace{3.8} \times 10^{-4}$$

$$\underbrace{\phantom{x}}_{2 \text{ digits}} \qquad \underbrace{\phantom{x}}_{2 \text{ digits}} \qquad \underbrace{\phantom{x}}_{2 \text{ digits}}$$

Here are several examples showing the proper use of significant figures:

$$\log 0.001\,237 = -2.907\,6 \qquad \text{antilog } 4.37 = 2.3 \times 10^4$$
$$\log 1\,237 = 3.092\,4 \qquad\qquad 10^{4.37} = 2.3 \times 10^4$$
$$\log 3.2 = 0.51 \qquad\qquad 10^{-2.600} = 2.51 \times 10^{-3}$$

## Significant Figures and Graphs

When drawing a graph on a computer, consider whether the graph is meant to display qualitative behavior of the data (Figure 3-2) or precise values. If someone will use the graph (such as Figure 3-3) to read points, it should at least have tic marks on both sides of the horizontal and vertical scales. Better still is a fine grid superimposed on the graph.

Problem 3-8 shows you how to control gridlines in an Excel graph.

Number of digits in *mantissa* of log $x$ = number of significant figures in $x$:

$$\log(5.403 \times 10^{-8}) = -7.267\,4$$

$$\underbrace{\phantom{xx}}_{4 \text{ digits}} \qquad\qquad \underbrace{\phantom{xx}}_{4 \text{ digits}}$$

Number of digits in antilog $x \; (=10^x)$ = number of significant figures in *mantissa* of $x$:

$$10^{6.142} = 1.39 \times 10^6$$

$$\underbrace{\phantom{x}}_{3 \text{ digits}} \quad \underbrace{\phantom{x}}_{3 \text{ digits}}$$

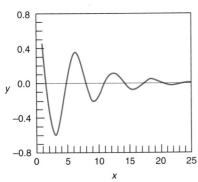

**Figure 3-2** Example of a graph intended to show the qualitative behavior of the function $y = e^{-x/6} \cos x$. You are not expected to be able to read coordinates accurately on this graph.

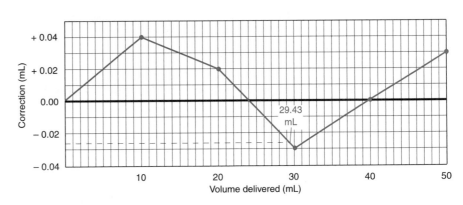

**Figure 3-3** Calibration curve for a 50-mL buret. The volume delivered can be read to the nearest 0.1 mL. If your buret reading is 29.43 mL, you can find the correction factor accurately enough by locating 29.4 mL on the graph. The correction factor on the ordinate ($y$-axis) for 29.4 mL on the abscissa ($x$-axis) is −0.03 mL (to the nearest 0.01 mL).

## ■ ■ ■ 3-3 Types of Error

*Every* measurement has some uncertainty, which is called *experimental error.* Conclusions can be expressed with a high or a low degree of confidence, but never with complete certainty. Experimental error is classified as either *systematic* or *random.*

### Systematic Error

**Systematic error,** also called **determinate error,** arises from a flaw in equipment or the design of an experiment. If you conduct the experiment again in exactly the same manner,

*Systematic error* is a consistent error that can be detected and corrected. Box 3-1 describes Standard Reference Materials designed to reduce systematic errors.

the error is reproducible. In principle, systematic error can be discovered and corrected, although this may not be easy.

For example, a pH meter that has been standardized incorrectly produces a systematic error. Suppose you think that the pH of the buffer used to standardize the meter is 7.00, but it is really 7.08. Then all your pH readings will be 0.08 pH unit too low. When you read a pH of 5.60, the actual pH of the sample is 5.68. This systematic error could be discovered by using a second buffer of known pH to test the meter.

Another systematic error arises from an uncalibrated buret. The manufacturer's tolerance for a Class A 50-mL buret is ±0.05 mL. When you think you have delivered 29.43 mL, the real volume could be anywhere from 29.38 to 29.48 mL and still be within tolerance. One way to correct for an error of this type is to construct a calibration curve, such as that in Figure 3-3, by the procedure on page 38. To do this, deliver distilled water from the buret into a flask and weigh it. Determine the volume of water from its mass by using Table 2-7. Figure 3-3 tells us to apply a correction factor of −0.03 mL to the measured value of 29.43 mL. The actual volume delivered is 29.43 − 0.03 = 29.40 mL.

A key feature of systematic error is that it is reproducible. For the buret just discussed, the error is always −0.03 mL when the buret reading is 29.43 mL. Systematic error may always be positive in some regions and always negative in others. With care and cleverness, you can detect and correct a systematic error.

Ways to detect systematic error:
1. Analyze a known sample, such as a Standard Reference Material. Your method should reproduce the known answer. (See Box 15-1 for an example.)
2. Analyze "blank" samples containing none of the analyte being sought. If you observe a nonzero result, your method responds to more than you intend. Section 5-1 discusses different kinds of blanks.
3. Use different analytical methods to measure the same quantity. If results do not agree, there is error in one (or more) of the methods.
4. *Round robin* experiment: Different people in several laboratories analyze identical samples by the same or different methods. Disagreement beyond the estimated random error is systematic error.

## Random Error

**Random error,** also called **indeterminate error,** arises from the effects of uncontrolled (and maybe uncontrollable) variables in the measurement. Random error has an equal chance of being positive or negative. It is always present and cannot be corrected. There is random error associated with reading a scale. Different people reading the scale in Figure 3-1 report a range of values representing their subjective interpolation between the markings. One person reading the same instrument several times might report several different readings. Another random error results from electrical noise in an instrument. Positive and negative fluctuations occur with approximately equal frequency and cannot be completely eliminated.

*Random error* cannot be eliminated, but it might be reduced by a better experiment.

## Precision and Accuracy

**Precision** describes the reproducibility of a result. If you measure a quantity several times and the values agree closely with one another, your measurement is precise. If the values vary widely, your measurement is not precise. **Accuracy** describes how close a measured value is to the "true" value. If a known standard is available (such as a Standard Reference Material described in Box 3-1), accuracy is how close your value is to the known value.

*Precision:* reproducibility
*Accuracy:* nearness to the "truth"

---

### Box 3-1 Standard Reference Materials

Inaccurate laboratory measurements can mean wrong medical diagnosis and treatment, lost production time, wasted energy and materials, manufacturing rejects, and product liability. The U.S. National Institute of Standards and Technology and national standards laboratories around the world distribute Standard Reference Materials, such as metals, chemicals, rubber, plastics, engineering materials, radioactive substances, and environmental and clinical standards that can be used to test the accuracy of analytical procedures.[1]

For example, in treating patients with epilepsy, physicians depend on laboratory tests to measure concentrations of anticonvulsant drugs in blood serum. Drug levels that are too low lead to seizures; high levels are toxic. Because tests of identical serum specimens at different laboratories were giving an unacceptably wide range of results, the National Institute of Standards and Technology developed a Standard Reference Material containing known levels of antiepilepsy drugs in serum. The reference material now enables different laboratories to detect and correct errors in their assay procedures.

Before the introduction of this reference material, five laboratories analyzing identical samples reported a range of results with relative errors of 40% to 110% of the expected value. After distribution of the reference material, the error was reduced to 20% to 40%.

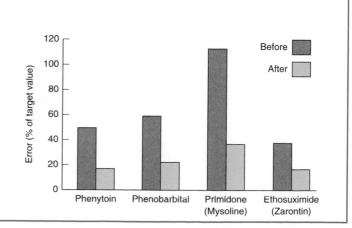

A measurement might be reproducible, but wrong. If you made a mistake preparing a solution for a titration, you might do a series of reproducible titrations but report an incorrect result because the concentration of the titrating solution was not what you intended. In this case, the precision is good but the accuracy is poor. Conversely, it is possible to make poorly reproducible measurements clustered around the correct value. In this case, the precision is poor but the accuracy is good. An ideal procedure is both precise and accurate.

Accuracy is defined as nearness to the "true" value. The word *true* is in quotes because somebody must *measure* the "true" value, and there is error associated with *every* measurement. The "true" value is best obtained by an experienced person using a well-tested procedure. It is desirable to test the result by using different procedures, because, even though each method might be precise, systematic error could lead to poor agreement between methods. Good agreement among several methods affords us confidence, but never proof, that results are accurate.

## Absolute and Relative Uncertainty

An uncertainty of $\pm 0.02$ means that, when the reading is 13.33, the true value could be anywhere in the range 13.31 to 13.35.

**Absolute uncertainty** expresses the margin of uncertainty associated with a measurement. If the estimated uncertainty in reading a calibrated buret is $\pm 0.02$ mL, we say that $\pm 0.02$ mL is the absolute uncertainty associated with the reading.

**Relative uncertainty** compares the size of the absolute uncertainty with the size of its associated measurement. The relative uncertainty of a buret reading of $12.35 \pm 0.02$ mL is a dimensionless quotient:

*Relative uncertainty:*

$$\text{Relative uncertainty} = \frac{\text{absolute uncertainty}}{\text{magnitude of measurement}} \quad (3\text{-}2)$$

$$= \frac{0.02 \text{ mL}}{12.35 \text{ mL}} = 0.002$$

The percent relative uncertainty is simply

If you use a 50-mL buret, design your titration to require 20–40 mL of reagent to produce a small relative uncertainty of 0.1–0.05%.

In a gravimetric analysis, plan to have enough precipitate for a low relative uncertainty. If weighing precision is $\pm 0.3$ mg, a 100-mg precipitate has a relative weighing error of 0.3% and a 300-mg precipitate has an uncertainty of 0.1%.

*Percent relative uncertainty:*

$$\text{Percent relative uncertainty} = 100 \times \text{relative uncertainty} \quad (3\text{-}3)$$

$$= 100 \times 0.002 = 0.2\%$$

If the absolute uncertainty in reading a buret is constant at $\pm 0.02$ mL, the percent relative uncertainty is 0.2% for a volume of 10 mL and 0.1% for a volume of 20 mL.

## ■ ■ ■ 3-4 Propagation of Uncertainty from Random Error

We can usually estimate or measure the random error associated with a measurement, such as the length of an object or the temperature of a solution. The uncertainty might be based on how well we can read an instrument or on our experience with a particular method. If possible, uncertainty is expressed as the *standard deviation* or as a *confidence interval,* which are discussed in Chapter 4. This section applies only to random error. We assume that systematic error has been detected and corrected.

By far, most propagation of uncertainty computations that you will encounter deal with random error, not systematic error. Our goal is always to eliminate systematic error.

For most experiments, we need to perform arithmetic operations on several numbers, each of which has a random error. The most likely uncertainty in the result is not simply the sum of the individual errors, because some of them are likely to be positive and some negative. We expect some cancellation of errors.

### Addition and Subtraction

Suppose you wish to perform the following arithmetic, in which the experimental uncertainties, designated $e_1$, $e_2$, and $e_3$, are given in parentheses.

$$
\begin{array}{l}
\phantom{+}1.76 \ (\pm 0.03) \leftarrow e_1 \\
+\ 1.89 \ (\pm 0.02) \leftarrow e_2 \\
-\ 0.59 \ (\pm 0.02) \leftarrow e_3 \\
\hline
\phantom{+}3.06 \ (\pm e_4)
\end{array}
\quad (3\text{-}4)
$$

The arithmetic answer is 3.06. But what is the uncertainty associated with this result?

For addition and subtraction, the uncertainty in the answer is obtained from the *absolute uncertainties* of the individual terms as follows:

*Uncertainty in addition and subtraction:*

$$e_4 = \sqrt{e_1^2 + e_2^2 + e_3^2} \qquad (3\text{-}5)$$

For addition and subtraction, use *absolute* uncertainty.

For the sum in Equation 3-4, we can write

$$e_4 = \sqrt{(0.03)^2 + (0.02)^2 + (0.02)^2} = 0.04_1$$

The absolute uncertainty $e_4$ is $\pm 0.04$, and we express the answer as $3.06 \pm 0.04$. Although there is only one significant figure in the uncertainty, we wrote it initially as $0.04_1$, with the first insignificant figure subscripted. We retain one or more insignificant figures to avoid introducing round-off errors into later calculations through the number $0.04_1$. The insignificant figure was subscripted to remind us where the last significant figure should be at the conclusion of the calculations.

To find the percent relative uncertainty in the sum of Equation 3-4, we write

$$\text{Percent relative uncertainty} = \frac{0.04_1}{3.06} \times 100 = 1._3\%$$

The uncertainty, $0.04_1$, is $1._3\%$ of the result, 3.06. The subscript 3 in $1._3\%$ is not significant. It is sensible to drop the insignificant figures now and express the final result as

$$3.06\,(\pm 0.04) \qquad \text{(absolute uncertainty)}$$
$$3.06\,(\pm 1\%) \qquad \text{(relative uncertainty)}$$

For addition and subtraction, use absolute uncertainty. Relative uncertainty can be found at the end of the calculation.

### Example  Uncertainty in a Buret Reading

The volume delivered by a buret is the difference between final and initial readings. If the uncertainty in each reading is $\pm 0.02$ mL, what is the uncertainty in the volume delivered?

**Solution**   Suppose that the initial reading is $0.05\,(\pm 0.02)$ mL and the final reading is $17.88\,(\pm 0.02)$ mL. The volume delivered is the difference:

$$
\begin{array}{l}
17.88\,(\pm 0.02) \\
\underline{-\ 0.05\,(\pm 0.02)} \\
17.83\,(\pm e) \qquad e = \sqrt{0.02^2 + 0.02^2} = 0.02_8 \approx 0.03
\end{array}
$$

Regardless of the initial and final readings, if the uncertainty in each one is $\pm 0.02$ mL, the uncertainty in volume delivered is $\pm 0.03$ mL.

## Multiplication and Division

For multiplication and division, first convert all uncertainties into percent relative uncertainties. Then calculate the error of the product or quotient as follows:

*Uncertainty in multiplication and division:*

$$\%e_4 = \sqrt{(\%e_1)^2 + (\%e_2)^2 + (\%e_3)^2} \qquad (3\text{-}6)$$

For multiplication and division, use percent relative uncertainty.

For example, consider the following operations:

$$\frac{1.76\,(\pm 0.03) \times 1.89\,(\pm 0.02)}{0.59\,(\pm 0.02)} = 5.64 \pm e_4$$

First convert absolute uncertainties into percent relative uncertainties.

$$\frac{1.76\,(\pm 1._7\%) \times 1.89\,(\pm 1._1\%)}{0.59\,(\pm 3._4\%)} = 5.64 \pm e_4$$

Then find the percent relative uncertainty of the answer by using Equation 3-6.

$$\%e_4 = \sqrt{(1._7)^2 + (1._1)^2 + (3._4)^2} = 4._0\%$$

The answer is $5.6_4\,(\pm 4._0\%)$.

To convert relative uncertainty into absolute uncertainty, find $4._0\%$ of the answer.

$$4._0\% \times 5.6_4 = 0.04_0 \times 5.6_4 = 0.2_3$$

*Advice*   Retain one or more extra insignificant figures until you have finished your entire calculation. Then round to the correct number of digits. When storing intermediate results in a calculator, keep all digits without rounding.

For multiplication and division, use percent relative uncertainty. Absolute uncertainty can be found at the end of the calculation.

The answer is $5.6_4$ ($\pm 0.2_3$). Finally, drop the insignificant digits.

$$5.6 \ (\pm 0.2) \qquad \text{(absolute uncertainty)}$$
$$5.6 \ (\pm 4\%) \qquad \text{(relative uncertainty)}$$

The denominator of the original problem, 0.59, limits the answer to two digits.

## Mixed Operations

Now consider a computation containing subtraction and division:

$$\frac{[1.76 \ (\pm 0.03) - 0.59 \ (\pm 0.02)]}{1.89 \ (\pm 0.02)} = 0.619_0 \pm \ ?$$

First work out the difference in the numerator, using absolute uncertainties. Thus,

$$1.76 \ (\pm 0.03) - 0.59 \ (\pm 0.02) = 1.17 \ (\pm 0.03_6)$$

because $\sqrt{(0.03)^2 + (0.02)^2} = 0.03_6$.

Then convert into percent relative uncertainties. Thus,

$$\frac{1.17 \ (\pm 0.03_6)}{1.89 \ (\pm 0.02)} = \frac{1.17 \ (\pm 3._1\%)}{1.89 \ (\pm 1._1\%)} = 0.619_0 \ (\pm 3._3\%)$$

because $\sqrt{(3._1\%)^2 + (1._1\%)^2} = 3._3\%$.

The percent relative uncertainty is $3._3\%$, so the absolute uncertainty is $0.03_3 \times 0.619_0 = 0.02_0$. The final answer can be written as

$$0.619 \ (\pm 0.02_0) \qquad \text{(absolute uncertainty)}$$
$$0.619 \ (\pm 3._3\%) \qquad \text{(relative uncertainty)}$$

The result of a calculation ought to be written in a manner consistent with its uncertainty.

Because the uncertainty begins in the 0.01 decimal place, it is reasonable to round the result to the 0.01 decimal place:

$$0.62 \ (\pm 0.02) \qquad \text{(absolute uncertainty)}$$
$$0.62 \ (\pm 3\%) \qquad \text{(relative uncertainty)}$$

## The *Real* Rule for Significant Figures

The real rule: The first uncertain figure is the last significant figure.

*The first digit of the absolute uncertainty is the last significant digit in the answer.* For example, in the quotient

$$\frac{0.002 \ 364 \ (\pm 0.000 \ 003)}{0.025 \ 00 (\pm 0.000 \ 05)} = 0.094 \ 6 \ (\pm 0.000 \ 2)$$

the uncertainty ($\pm 0.000 \ 2$) occurs in the fourth decimal place. Therefore, the answer 0.094 6 is properly expressed with *three* significant figures, even though the original data have four figures. The first uncertain figure of the answer is the last significant figure. The quotient

$$\frac{0.002 \ 664 \ (\pm 0.000 \ 003)}{0.025 \ 00 \ (\pm 0.000 \ 05)} = 0.106 \ 6 \ (\pm 0.000 \ 2)$$

is expressed with *four* significant figures because the uncertainty occurs in the fourth place. The quotient

$$\frac{0.821 \ (\pm 0.002)}{0.803 \ (\pm 0.002)} = 1.022 \ (\pm 0.004)$$

is expressed with *four* figures even though the dividend and divisor each have *three* figures.

It is all right to keep one extra digit when an answer lies between 1 and 2.

Now you can appreciate why *it is all right to keep one extra digit when an answer lies between 1 and 2*. The quotient 82/80 is better written as 1.02 than 1.0. If I write 1.0, you can surmise that the uncertainty is at least $1.0 \pm 0.1 = \pm 10\%$. The actual uncertainty lies in the second decimal place, not the first decimal place.

### Example Significant Figures in Laboratory Work

You prepared a 0.250 M $NH_3$ solution by diluting 8.45 ($\pm 0.04$) mL of 28.0 ($\pm 0.5$) wt% $NH_3$ [density = 0.899 ($\pm 0.003$) g/mL] up to 500.0 ($\pm 0.2$) mL. Find the uncertainty in 0.250 M. The molecular mass of $NH_3$, 17.030 5 g/mol, has negligible uncertainty relative to other uncertainties in this problem.

**Solution** To find the uncertainty in molarity, we need to find the uncertainty in moles delivered to the 500-mL flask. The concentrated reagent contains 0.899 ($\pm$0.003) g of solution per milliliter. Weight percent tells us that the reagent contains 0.280 ($\pm$0.005) g of $NH_3$ per gram of solution. In our calculations, we retain extra insignificant digits and round off only at the end.

$$\begin{array}{l} \text{Grams of } NH_3 \text{ per} \\ \text{mL in concentrated} \\ \text{reagent} \end{array} = 0.899\,(\pm0.003)\,\frac{\text{g solution}}{\text{mL}} \times 0.280\,(\pm0.005)\,\frac{\text{g } NH_3}{\text{g solution}}$$

$$= 0.899\,(\pm0.334\%)\,\frac{\text{g solution}}{\text{mL}} \times 0.280\,(\pm1.79\%)\,\frac{\text{g } NH_3}{\text{g solution}}$$

$$= 0.251\,7\,(\pm1.82\%)\,\frac{\text{g } NH_3}{\text{mL}}$$

because $\sqrt{(0.334\%)^2 + (1.79\%)^2} = 1.82\%$.

Next, we find the moles of ammonia contained in 8.45 ($\pm$0.04) mL of concentrated reagent. The relative uncertainty in volume is $0.04/8.45 = 0.473\%$.

$$\text{mol } NH_3 = \frac{0.251\,7\,(\pm1.82\%)\,\dfrac{\text{g } NH_3}{\text{mL}} \times 8.45\,(\pm0.473\%)\,\text{mL}}{17.030\,5\,(\pm0\%)\,\dfrac{\text{g } NH_3}{\text{mol}}}$$

$$= 0.124\,9\,(\pm1.88\%)\,\text{mol}$$

because $\sqrt{(1.82\%)^2 + (0.473\%)^2 + (0\%)^2} = 1.88\%$.

This much ammonia was diluted to 0.500 0 ($\pm$0.000 2) L. The relative uncertainty in the final volume is $0.000\,2/0.500\,0 = 0.04\%$. The molarity is

$$\frac{\text{mol } NH_3}{\text{L}} = \frac{0.124\,9\,(\pm1.88\%)\,\text{mol}}{0.500\,0\,(\pm0.04\%)\,\text{L}}$$

$$= 0.249\,8\,(\pm1.88\%)\,\text{M}$$

because $\sqrt{(1.88\%)^2 + (0.04\%)^2} = 1.88\%$. The absolute uncertainty is 1.88% of 0.249 8 M = 0.004 7 M. The uncertainty in molarity is in the third decimal place, so our final, rounded answer is

$$[NH_3] = 0.250\,(\pm0.005)\,\text{M}$$

## Exponents and Logarithms

For the function $y = x^a$, the percent relative uncertainty in $y$ ($\%e_y$) is equal to $a$ times the percent relative uncertainty in $x$ ($\%e_x$):

*Uncertainty for*
*powers and roots:*
$$y = x^a \;\Rightarrow\; \%e_y = a(\%e_x) \tag{3-7}$$

For example, if $y = \sqrt{x} = x^{1/2}$, a 2% uncertainty in $x$ will yield a $(\frac{1}{2})(2\%) = 1\%$ uncertainty in $y$. If $y = x^2$, a 3% uncertainty in $x$ leads to a $(2)(3\%) = 6\%$ uncertainty in $y$ (Box 3-2).

If $y$ is the base 10 logarithm of $x$, then the absolute uncertainty in $y$ ($e_y$) is proportional to the relative uncertainty in $x$, which is $e_x/x$:

*Uncertainty for*
*logarithm:*
$$y = \log x \;\Rightarrow\; e_y = \frac{1}{\ln 10}\frac{e_x}{x} \approx 0.434\,29\frac{e_x}{x} \tag{3-8}$$

You should not work with percent relative uncertainty $[100 \times (e_x/x)]$ in calculations with logs and antilogs, because one side of Equation 3-8 has relative uncertainty and the other has absolute uncertainty.

The **natural logarithm** (ln) of $x$ is the number $y$, whose value is such that $x = e^y$, where e ($= 2.718\,28\ldots$) is called the base of the natural logarithm. The absolute uncertainty in $y$ is equal to the relative uncertainty in $x$.

*Uncertainty for*
*natural logarithm:*
$$y = \ln x \;\Rightarrow\; e_y = \frac{e_x}{x} \tag{3-9}$$

Now consider $y = $ antilog $x$, which is the same as saying $y = 10^x$. In this case, the relative uncertainty in $y$ is proportional to the absolute uncertainty in $x$.

---

The rationale for finding the uncertainty in the molecular mass of $NH_3$ is explained in Section 3-5:

N:      14.006 7 $\pm$ 0.000 2
+3H:  +3(1.007 94 $\pm$ 0.000 07)

| | |
|---|---|
| N: | 14.006 7 $\pm$ 0.000 2 |
| +3H: | +3.023 82 $\pm$ 0.000 21 |
| $NH_3$: | 17.030 52 $\pm$ $\sqrt{0.000\,2^2 + 0.000\,21^2}$ |
| = | 17.030 $5_2$ $\pm$ 0.000 $2_9$ |
| = | 17.030 5 $\pm$ 0.000 3 |

Convert absolute uncertainty into percent relative uncertainty for multiplication.

To calculate a power or root on your calculator, use the $y^x$ button. For example, to find a cube root ($y^{1/3}$), raise $y$ to the 0.333 333 333 . . . power with the $y^x$ button.

Use relative uncertainty ($e_x/x$), not percent relative uncertainty [100 $\times$ ($e_x/x$)], in calculations involving log $x$, ln $x$, $10^x$, and $e^x$.

## Box 3-2 Propagation of Uncertainty in the Product $x \cdot x$

Table 3-1 says that the uncertainty in the function $y = x^a$ is $\%e_y = a(\%e_x)$. If $y = x^2$, then $a = 2$ and $\%e_y = 2(\%e_x)$. A 3% uncertainty in $x$ leads to a $(2)(3\%) = 6\%$ uncertainty in $y$.

But what if we just apply the multiplication formula 3-6 to the product $x \cdot x$?

$$x(\pm e_1) \cdot x(\pm e_2) = x^2(\pm e_3)$$

$$\%e_3 = \sqrt{(\%e_1)^2 + (\%e_2)^2} = \sqrt{(3\%)^2 + (3\%)^2} = 4._2\%$$

Which uncertainty is correct, 6% from Table 3-1 or $4._2\%$ from Equation 3-6?

Table 3-1 (6%) is correct. In the formula $y = x^2$, the error in a measured value of $x$ is always positive or always negative. If the true value of $x$ is 1.00 and the measured value is 1.01, the computed value of $x^2$ is $(1.01)^2 = 1.02$. That is, if the measured $x$ is high by 1%, the computed value of $x^2$ is high by 2% because we are multiplying the high value by the high value.

Equation 3-6 presumes that the uncertainty in each factor of the product $x \cdot z$ is random and independent of the other. In the product $x \cdot z$, the measured value of $x$ could be high sometimes and the measured value of $z$ could be low sometimes. In the

majority of cases, the uncertainty in the product $x \cdot z$ is not as great as the uncertainty in $x^2$.

*Example.* The distance traveled by a falling object in time $t$ is $\frac{1}{2}gt^2$, where $g$ is the acceleration of gravity. If $t$ has an uncertainty of 1%, the uncertainty in $t^2$ is $2(\%e_t) = 2(1\%) = 2\%$. The uncertainty in distance computed from $\frac{1}{2}gt^2$ will also be 2%. If you (incorrectly) used Equation 3-6, you would compute an uncertainty in distance of $\sqrt{1\%^2 + 1\%^2} = 1._4\%$.

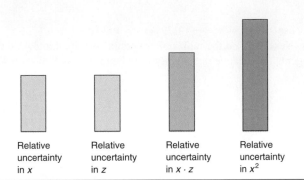

| Relative uncertainty in $x$ | Relative uncertainty in $z$ | Relative uncertainty in $x \cdot z$ | Relative uncertainty in $x^2$ |

Appendix C gives a general rule for propagation of random uncertainty for any function.

| Uncertainty for $10^x$: | $$y = 10^x \quad \Rightarrow \quad \frac{e_y}{y} = (\ln 10)e_x \approx 2.302\ 6\ e_x$$ | (3-10) |

If $y = e^x$, the relative uncertainty in $y$ equals the absolute uncertainty in $x$.

| Uncertainty for $e^x$: | $$y = e^x \quad \Rightarrow \quad \frac{e_y}{y} = e_x$$ | (3-11) |

Table 3-1 summarizes rules for propagation of uncertainty. You need not memorize the rules for exponents, logs, and antilogs, but you should be able to use them.

### Example Uncertainty in H⁺ Concentration

Consider the function $pH = -\log[H^+]$, where $[H^+]$ is the molarity of $H^+$. For $pH = 5.21 \pm 0.03$, find $[H^+]$ and its uncertainty.

**Solution** First solve the equation $pH = -\log[H^+]$ for $[H^+]$: Whenever $a = b$, then $10^a = 10^b$. If $pH = -\log[H^+]$, then $\log[H^+] = -pH$ and $10^{\log[H^+]} = 10^{-pH}$. But $10^{\log[H^+]} = [H^+]$. We therefore need to find the uncertainty in the equation

$$[H^+] = 10^{-pH} = 10^{-(5.21 \pm 0.03)}$$

In Table 3-1, the relevant function is $y = 10^x$, in which $y = [H^+]$ and $x = -(5.21 \pm 0.03)$. For $y = 10^x$, the table tells us that $e_y/y = 2.302\ 6\ e_x$.

$$\frac{e_y}{y} = 2.302\ 6\ e_x = (2.302\ 6)(0.03) = 0.069\ 1 \tag{3-12}$$

The relative uncertainty in $y$ ($= e_y/y$) is 0.069 1. Inserting the value $y = 10^{-5.21} = 6.17 \times 10^{-6}$ into Equation 3-12 gives the answer:

$$\frac{e_y}{y} = \frac{e_y}{6.17 \times 10^{-6}} = 0.069\ 1 \quad \Rightarrow \quad e_y = 4.26 \times 10^{-7}$$

The concentration of $H^+$ is $6.17\ (\pm 0.426) \times 10^{-6} = 6.2\ (\pm 0.4) \times 10^{-6}$ M. An uncertainty of 0.03 in pH gives an uncertainty of 7% in $[H^+]$. Notice that extra digits were retained in the intermediate results and were not rounded off until the final answer.

**Table 3-1** Summary of rules for propagation of uncertainty

| Function | Uncertainty | Function[a] | Uncertainty[b] |
|---|---|---|---|
| $y = x_1 + x_2$ | $e_y = \sqrt{e_{x_1}^2 + e_{x_2}^2}$ | $y = x^a$ | $\%e_y = a\%e_x$ |
| $y = x_1 - x_2$ | $e_y = \sqrt{e_{x_1}^2 + e_{x_2}^2}$ | $y = \log x$ | $e_y = \dfrac{1}{\ln 10}\dfrac{e_x}{x} \approx 0.434\,29\dfrac{e_x}{x}$ |
| $y = x_1 \cdot x_2$ | $\%e_y = \sqrt{\%e_{x_1}^2 + \%e_{x_2}^2}$ | $y = \ln x$ | $e_y = \dfrac{e_x}{x}$ |
| $y = \dfrac{x_1}{x_2}$ | $\%e_y = \sqrt{\%e_{x_1}^2 + \%e_{x_2}^2}$ | $y = 10^x$ | $\dfrac{e_y}{y} = (\ln 10)e_x \approx 2.302\,6\,e_x$ |
| | | $y = e^x$ | $\dfrac{e_y}{y} = e_x$ |

a. x represents a variable and a represents a constant that has no uncertainty,

b. $e_x/x$ is the relative error in x and $\%e_x$ is $100 \times e_x/x$.

## 3-5 Propagation of Uncertainty: Systematic Error

Systematic error occurs in some common situations and is treated differently from random error in arithmetic operations.

### Uncertainty in Molecular Mass

What is the uncertainty in the molecular mass of $O_2$? On the inside cover of this book, we find that the atomic mass of oxygen is $15.999\,4 \pm 0.000\,3$ g/mol. The uncertainty is *not* mainly from random error in measuring the atomic mass. The uncertainty is predominantly from isotopic variation in samples of oxygen from different sources. That is, oxygen from one source could have a mean atomic mass of 15.999 1 and oxygen from another source could have an atomic mass of 15.999 7. The atomic mass of oxygen in a particular lot of reagent has a *systematic* uncertainty. It could be relatively constant at 15.999 7 or 15.999 1, or any value in between, with only a small random variation around the mean value.

If the true mass were 15.999 7, then the mass of $O_2$ is $2 \times 15.999\,7 = 31.999\,4$ g/mol. If the true mass is 15.999 1, then the mass of $O_2$ is $2 \times 15.999\,1 = 31.998\,2$ g/mol. The mass of $O_2$ is somewhere in the range $31.998\,8 \pm 0.000\,6$. The uncertainty of the mass of $n$ atoms is $n \times$ (uncertainty of one atom) $= 2 \times (\pm 0.000\,3) = \pm 0.000\,6$. The uncertainty is not $\pm\sqrt{0.000\,3^2 + 0.000\,3^2} = \pm 0.000\,4_2$. For systematic uncertainty, we add the uncertainties of each term in a sum or difference.

Now let's find the molecular mass of $C_2H_4$:

$$2C:\quad 2(12.010\,7 \pm 0.000\,8) = 24.021\,4 \pm 0.001\,6 \leftarrow 2 \times 0.000\,8$$
$$4H:\quad 4(1.007\,94 \pm 0.000\,07) = \underline{4.031\,76 \pm 0.000\,28} \leftarrow 4 \times 0.000\,07$$
$$28.053\,16 \pm \,? \tag{3-13}$$

Propagation of systematic uncertainty: Uncertainty in mass of $n$ identical atoms = $n \times$ (uncertainty in atomic mass).

For the uncertainty in the sum of the masses of 2C + 4H, we use Equation 3-5, which applies to random error, because the uncertainties in the masses of C and H are independent of each other. One might be positive and one might be negative. So the molecular mass of $C_2H_4$ is

$$28.053\,16 \pm \sqrt{0.001\,6^2 + 0.000\,28^2}$$
$$28.053\,16 \pm 0.001\,6$$
$$28.053 \pm 0.002 \text{ g/mol}$$

We choose to use the rule for propagation of random uncertainty for the sum of atomic masses of different elements.

### Multiple Deliveries from One Pipet

A 25-mL Class A volumetric pipet is certified by the manufacturer to deliver $25.00 \pm 0.03$ mL. The volume delivered by a given pipet is reproducible, but can be anywhere in the range 24.97 to 25.03 mL. The difference between 25.00 mL and the actual volume delivered by a particular pipet is a *systematic* error. It is always the same, within a small random error. You could calibrate a pipet by weighing the water it delivers, as in Section 2-9.

Calibration eliminates systematic error, because we would know that the pipet always delivers, say, $25.991 \pm 0.006$ mL. The remaining uncertainty ($\pm 0.006$ mL) is random error.

If you use an uncalibrated 25-mL Class A volumetric pipet four times to deliver a total of 100 mL, what is the uncertainty in 100 mL? Because the uncertainty is a systematic error, the uncertainty in four pipet volumes is like the uncertainty in the mass of 4 moles of oxygen: The uncertainty is $\pm 4 \times 0.03 = \pm 0.12$ mL, not $\pm \sqrt{0.03^2 + 0.03^2 + 0.03^2 + 0.03^2} = \pm 0.06$ mL.

Calibration improves accuracy. Suppose that a calibrated pipet delivers a mean volume of 24.991 mL with a standard deviation (a random variation) of $\pm 0.006$ mL. If you deliver four aliquots from this pipet, the volume delivered is $4 \times 24.991 = 99.964$ mL and the uncertainty is $\pm \sqrt{0.006^2 + 0.006^2 + 0.006^2 + 0.006^2} = \pm 0.012$ mL.

In this example, calibration reduces the uncertainty from ±0.12 mL to ±0.012 mL.

---

## Terms to Understand

| | | | |
|---|---|---|---|
| absolute uncertainty | determinate error | natural logarithm | significant figure |
| accuracy | indeterminate error | precision | systematic error |
| antilogarithm | logarithm | random error | |
| characteristic | mantissa | relative uncertainty | |

---

## Summary

The number of significant digits in a number is the minimum required to write the number in scientific notation. The first uncertain digit is the last significant figure. In addition and subtraction, the last significant figure is determined by the number with the fewest decimal places (when all exponents are equal). In multiplication and division, the number of figures is usually limited by the factor with the smallest number of digits. The number of figures in the mantissa of the logarithm of a quantity should equal the number of significant figures in the quantity. Random (indeterminate) error affects the precision (reproducibility) of a result, whereas systematic (determinate) error affects the accuracy (nearness to the "true" value). Systematic error can be discovered and eliminated by a clever person, but some random error is always present. For random errors, propagation of uncertainty in addition and subtraction requires absolute uncertainties ($e_3 = \sqrt{e_1^2 + e_2^2}$), whereas multiplication and division utilize relative uncertainties ($\%e_3 = \sqrt{\%e_1^2 + \%e_2^2}$). Other rules for propagation of random error are found in Table 3-1. Always retain more digits than necessary during a calculation and round off to the appropriate number of digits at the end. Systematic error in atomic mass or the volume of a pipet leads to larger uncertainty than we get from random error. We always strive to eliminate systematic errors.

---

## Exercises

**3-A.** Write each answer with a reasonable number of figures. Find the absolute and percent relative uncertainty for each answer.
(a) $[12.41 \,(\pm 0.09) \div 4.16 \,(\pm 0.01)] \times 7.068\,2 \,(\pm 0.000\,4) = ?$
(b) $[3.26 \,(\pm 0.10) \times 8.47 \,(\pm 0.05)] - 0.18 \,(\pm 0.06) = ?$
(c) $6.843 \,(\pm 0.008) \times 10^4 \div [2.09 \,(\pm 0.04) - 1.63 \,(\pm 0.01)] = ?$
(d) $\sqrt{3.24 \pm 0.08} = ?$
(e) $(3.24 \pm 0.08)^4 = ?$
(f) $\log(3.24 \pm 0.08) = ?$
(g) $10^{3.24 \pm 0.08} = ?$

**3-B.** (a) You have a bottle labeled "53.4 ($\pm 0.4$) wt% NaOH—density $= 1.52 \,(\pm 0.01)$ g/mL." How many milliliters of 53.4 wt% NaOH will you need to prepare 2.000 L of 0.169 M NaOH?
(b) If the uncertainty in delivering NaOH is $\pm 0.01$ mL, calculate the absolute uncertainty in the molarity (0.169 M). Assume there is negligible uncertainty in the formula mass of NaOH and in the final volume (2.000 L).

**3-C.** We have a 37.0 ($\pm 0.5$) wt% HCl solution with a density of 1.18 ($\pm 0.01$) g/mL. To deliver 0.050 0 mol of HCl requires 4.18 mL of solution. If the uncertainty that can be tolerated in 0.050 0 mol is $\pm 2\%$, how big can the absolute uncertainty in 4.18 mL be? (*Caution:* In this problem, you have to work backward. You would normally compute the uncertainty in mol HCl from the uncertainty in volume:

$$\text{mol HCl} = \frac{\text{mL solution} \times \dfrac{\text{g solution}}{\text{mL solution}} \times \dfrac{\text{g HCl}}{\text{g solution}}}{\dfrac{\text{g HCl}}{\text{mol HCl}}}$$

But, in this case, we know the uncertainty in mol HCl (2%) and we need to find what uncertainty in mL solution leads to that 2% uncertainty. The arithmetic has the form $a = b \times c \times d$, for which $\%e_a^2 = \%e_b^2 + \%e_c^2 + \%e_d^2$. If we know $\%e_a$, $\%e_c$, and $\%e_d$, we can find $\%e_b$ by subtraction: $\%e_b^2 = \%e_a^2 - \%e_c^2 - \%e_d^2$.)

---

## Problems

### Significant Figures

**3-1.** How many significant figures are there in the following numbers?
(a) 1.903 0    (b) 0.039 10    (c) $1.40 \times 10^4$

**3-2.** Round each number as indicated:
(a) 1.236 7 to 4 significant figures
(b) 1.238 4 to 4 significant figures

(c) 0.135 2 to 3 significant figures
(d) 2.051 to 2 significant figures
(e) 2.005 0 to 3 significant figures

**3-3.** Round each number to three significant figures:
(a) 0.216 74
(b) 0.216 5
(c) 0.216 500 3

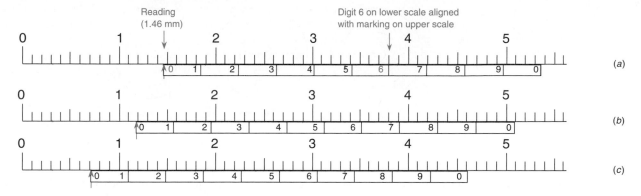

Reading
(1.46 mm)

Digit 6 on lower scale aligned
with marking on upper scale

(a)

(b)

(c)

Figure for Problem 3-4.

**3-4.** *Vernier scale.* The figure above shows a scale found on instruments such as a micrometer caliper used for accurately measuring dimensions of objects. The lower scale slides along the upper scale and is used to interpolate between the markings on the upper scale. In (*a*), the reading (at the left-hand 0 of the lower scale) is between 1.4 and 1.5 on the upper scale. To find the exact reading, observe which mark on the lower scale is aligned with a mark on the upper scale. Because the 6 on the lower scale is aligned with the upper scale, the correct reading is 1.46. Write the correct readings in (*b*) and (*c*) and indicate how many significant figures are in each reading.

**3-5.** Write each answer with the correct number of digits.
(a) $1.021 + 2.69 = 3.711$
(b) $12.3 - 1.63 = 10.67$
(c) $4.34 \times 9.2 = 39.928$
(d) $0.060\ 2 \div (2.113 \times 10^4) = 2.849\ 03 \times 10^{-6}$
(e) $\log(4.218 \times 10^{12}) = ?$
(f) $\text{antilog}(-3.22) = ?$
(g) $10^{2.384} = ?$

**3-6.** Write the formula mass of (a) $BaF_2$ and (b) $C_6H_4O_4$ with a reasonable number of digits. Use the periodic table inside the cover of this book to find atomic masses.

**3-7.** Write each answer with the correct number of significant figures.
(a) $1.0 + 2.1 + 3.4 + 5.8 = 12.300\ 0$
(b) $106.9 - 31.4 = 75.500\ 0$
(c) $107.868 - (2.113 \times 10^2) + (5.623 \times 10^3) = 5\ 519.568$
(d) $(26.14/37.62) \times 4.38 = 3.043\ 413$
(e) $(26.14/(37.62 \times 10^8)) \times (4.38 \times 10^{-2}) = 3.043\ 413 \times 10^{-10}$
(f) $(26.14/3.38) + 4.2 = 11.933\ 7$
(g) $\log(3.98 \times 10^4) = 4.599\ 9$
(h) $10^{-6.31} = 4.897\ 79 \times 10^{-7}$

**3-8.** 🖳 *Controlling the appearance of a graph in Excel.* Figure 3-3 requires gridlines to read the graph for buret corrections. The purpose of this exercise is to format a graph so it looks like Figure 3-3. Follow the procedure in Section 2-11 to make a graph of the data in the following table. The Chart Type is xy Scatter showing data points connected by straight lines. Double click on the x-axis and select the Scale tab. Set Minimum = 0, Maximum = 50, Major unit = 10, and Minor unit = 1. Select the Number tab and highlight Number. Set Decimal places = 0. In a similar manner, set the ordinate to run from $-0.04$ to $+0.05$ with a major unit of 0.02 and a minor unit of 0.01, as in Figure 3-3. The spreadsheet may

overrule you several times. Continue to reset the limits as you want them and click OK each time until the graph looks the way you intend. To add gridlines, click in the graph, go to the CHART menu and select CHART OPTIONS. Select the Gridlines tab and check both sets of Major gridlines and Minor gridlines and click OK. In the CHART OPTIONS menu, select the Legend tab and deselect Show legend. Move the x-axis numbers from the middle of the chart to the bottom as follows: Double click the y-axis (not the x-axis) and select the Scale tab. Set "Value (x) axis crosses at" to $-0.04$. Click OK and the volume labels move beneath the graph. Your graph should look the same as Figure 3-3.

| Volume (mL) | Correction (mL) |
|---|---|
| 0.03 | 0.00 |
| 10.04 | 0.04 |
| 20.03 | 0.02 |
| 29.98 | −0.03 |
| 40.00 | 0.00 |
| 49.97 | 0.03 |

## Types of Error

**3-9.** Why do we use quotation marks around the word *true* in the statement that accuracy refers to how close a measured value is to the "true" value?

**3-10.** Explain the difference between systematic and random errors.

**3-11.** Suppose that in a gravimetric analysis, you forget to dry the filter crucibles before collecting precipitate. After filtering the product, you dry the product and crucible thoroughly before weighing them. Is the mass of product always high or always low? Is the error in mass systematic or random?

**3-12.** State whether the errors in parts (**a**)–(**d**) are random or systematic:
(a) A 25-mL transfer pipet consistently delivers $25.031 \pm 0.009$ mL.
(b) A 10-mL buret consistently delivers $1.98 \pm 0.01$ mL when drained from exactly 0 to exactly 2 mL and consistently delivers 2.03 mL $\pm$ 0.02 mL when drained from 2 to 4 mL.
(c) A 10-mL buret delivered 1.983 9 g of water when drained from exactly 0.00 to 2.00 mL. The next time I delivered water from the 0.00- to the 2.00-mL mark, the delivered mass was 1.990 0 g.
(d) Four consecutive 20.0-$\mu$L injections of a solution into a chromatograph were made and the area of a particular peak was 4 383, 4 410, 4 401, and 4 390 units.

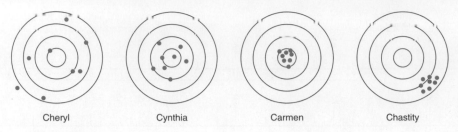

| Cheryl | Cynthia | Carmen | Chastity |

Figure for Problem 3-13.

**3-13.** Cheryl, Cynthia, Carmen, and Chastity shot the targets above at Girl Scout camp. Match each target with the proper description.
(a) accurate and precise
(b) accurate but not precise
(c) precise but not accurate
(d) neither precise nor accurate

**3-14.** Rewrite the number 3.123 56 ($\pm$0.167 89%) in the forms (a) number ($\pm$absolute uncertainty) and (b) number ($\pm$percent relative uncertainty) with an appropriate number of digits.

## Propagation of Uncertainty

**3-15.** Find the absolute and percent relative uncertainty and express each answer with a reasonable number of significant figures.
(a) 6.2 ($\pm$0.2) $-$ 4.1 ($\pm$0.1) = ?
(b) 9.43 ($\pm$0.05) $\times$ 0.016 ($\pm$0.001) = ?
(c) [6.2 ($\pm$0.2) $-$ 4.1 ($\pm$0.1)] $\div$ 9.43 ($\pm$0.05) = ?
(d) 9.43 ($\pm$0.05) $\times$ {[6.2 ($\pm$0.2) $\times$ $10^{-3}$] + [4.1 ($\pm$0.1) $\times$ $10^{-3}$]} = ?

**3-16.** Find the absolute and percent relative uncertainty and express each answer with a reasonable number of significant figures.
(a) 9.23 ($\pm$0.03) + 4.21 ($\pm$0.02) $-$ 3.26 ($\pm$0.06) = ?
(b) 91.3 ($\pm$1.0) $\times$ 40.3 ($\pm$0.2)/21.1 ($\pm$0.2) = ?
(c) [4.97 ($\pm$0.05) $-$ 1.86 ($\pm$0.01)]/21.1 ($\pm$0.2) = ?
(d) 2.016 4 ($\pm$0.000 8) + 1.233 ($\pm$0.002) + 4.61 ($\pm$0.01) = ?
(e) 2.016 4 ($\pm$0.000 8) $\times$ $10^3$ + 1.233 ($\pm$0.002) $\times$ $10^2$ + 4.61 ($\pm$0.01) $\times$ $10^1$ = ?
(f) [3.14 ($\pm$0.05)]$^{1/3}$ = ?
(g) log[3.14 ($\pm$0.05)] = ?

**3-17.** Verify the following calculations:
(a) $\sqrt{3.141\ 5\ (\pm0.001\ 1)}$ = 1.772 $4_3$ ($\pm$0.000 $3_1$)
(b) log[3.141 5 ($\pm$0.001 1)] = 0.497 $1_4$ ($\pm$0.000 $1_5$)
(c) antilog[3.141 5 ($\pm$0.001 1)] = 1.385$_2$($\pm$0.003$_5$) $\times$ $10^3$
(d) ln[3.141 5 ($\pm$0.001 1)] = 1.144 $7_0$($\pm$0.000 $3_5$)
(e) $\log\left(\dfrac{\sqrt{0.104\ (\pm0.006)}}{0.051\ 1\ (\pm0.000\ 9)}\right)$ = 0.80$_0$ ($\pm$0.01$_5$)

**3-18.** Express the molecular mass ($\pm$uncertainty) of $C_9H_9O_6N_3$ with the correct number of significant figures.

**3-19.** (a) Show that the formula mass of NaCl is 58.443 ($\pm$0.002) g/mol.

(b) To prepare a solution of NaCl, you weigh out 2.634 ($\pm$0.002) g and dissolve it in a volumetric flask whose volume is 100.00 ($\pm$0.08) mL. Express the molarity of the solution, along with its uncertainty, with an appropriate number of digits.

**3-20.** What is the true mass of water weighed at 24°C in the air if the apparent mass is 1.034 6 $\pm$ 0.000 2 g? The density of air is 0.001 2 $\pm$ 0.000 1 g/mL and the density of balance weights is 8.0 $\pm$ 0.5 g/mL. The uncertainty in the density of water in Table 2-7 is negligible in comparison to the uncertainty in the density of air.

**3-21.** Twelve dietary iron tablets were analyzed by the gravimetric procedure in Section 1-4 and the final mass of $Fe_2O_3$ (FM 159.688) was 0.277$_4$ $\pm$ 0.001$_8$ g. Find the average mass of Fe per tablet. (Relative uncertainties in atomic masses are small compared with relative uncertainty in the mass of $Fe_2O_3$. Neglect uncertainties in atomic masses in this problem.)

**3-22.** We can measure the concentration of HCl solution (a procedure called *standardizing* the solution) by reaction with pure sodium carbonate: $2H^+ + Na_2CO_3 \rightarrow 2Na^+ + H_2O + CO_2$. A volume of 27.35 $\pm$ 0.04 mL of HCl solution was required for complete reaction with 0.967 4 $\pm$ 0.000 9 g of $Na_2CO_3$ (FM 105.988 $\pm$ 0.001). Find the molarity of the HCl and its absolute uncertainty.

**3-23.** Avogadro's number can be computed from the following measured properties of pure crystalline silicon:[2] (1) atomic mass (obtained from the mass and abundance of each isotope), (2) density of the crystal, (3) size of the unit cell (the smallest repeating unit in the crystal), and (4) number of atoms in the unit cell. For silicon, the mass is $m_{Si}$ = 28.085 384 2 (35) g/mol, where 35 is the standard deviation in the last two digits. The density is $\rho$ = 2.329 031 9 (18) g/cm$^3$, the size of the cubic unit cell is $c_0$ = 5.431 020 36 (33) $\times$ $10^{-8}$ cm, and there are 8 atoms per unit cell. Avogadro's number is computed from the equation

$$N_A = \frac{m_{Si}}{(\rho c_0^3)/8}$$

From the measured properties and their uncertainties (standard deviations), compute Avogadro's number and its uncertainty. To find the uncertainty of $c_0^3$, use the function $y = x^a$ in Table 3-1.

# 4 | Statistics

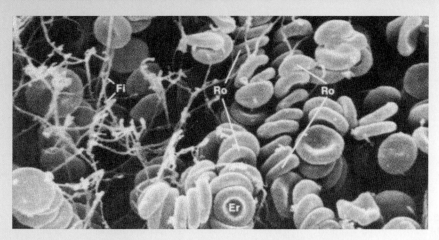

Red blood cells (erythrocytes, Er) tangled in fibrin threads (Fi) in a blood clot. Stacks of erythrocytes in a clot are called a rouleaux formation (Ro). *[From R. H. Kardon, Tissues and Organs (San Francisco: W. H. Freeman, 1978), p. 39.]*

All measurements contain experimental error, so it is never possible to be completely certain of a result. Nevertheless, we often seek the answers to questions such as "Is my red blood cell count today higher than usual?" If today's count is twice as high as usual, it is probably truly higher than normal. But what if the "high" count is not excessively above "normal" counts?

| Count on "normal" days | Today's count |
|---|---|
| 5.1 ⎫ |  |
| 5.3 ⎪ |  |
| 4.8 ⎬ × $10^6$ cells/μL | 5.6 × $10^6$ cells/μL |
| 5.4 ⎪ |  |
| 5.2 ⎭ |  |

The number 5.6 is higher than the five normal values, but the random variation in normal values might lead us to expect that 5.6 will be observed on some "normal" days.

The study of statistics allows us to say that today's value is expected to be observed on 1 out of 20 normal days. It is still up to you to decide what to do with this information.

Experimental measurements always contain some variability, so no conclusion can be drawn with certainty. Statistics gives us tools to accept conclusions that have a high probability of being correct and to reject conclusions that do not.[1,2]

## ■ ■ ■  4-1   Gaussian Distribution

If an experiment is repeated a great many times and if the errors are purely random, then the results tend to cluster symmetrically about the average value (Figure 4-1). The more times the experiment is repeated, the more closely the results approach an ideal smooth curve called the **Gaussian distribution.** In general, we cannot make so many measurements in a lab experiment. We are more likely to repeat an experiment 3 to 5 times than 2 000 times. However, from the small set of results, we can estimate the statistical parameters that describe the large set. We can then make estimates of statistical behavior from the small number of measurements.

We say that the variation in experimental data is *normally distributed* when replicate measurements exhibit the bell-shaped distribution in Figure 4-1. It is equally probable that a measurement will be higher or lower than the mean. The probability of observing any value decreases as its distance from the mean increases.

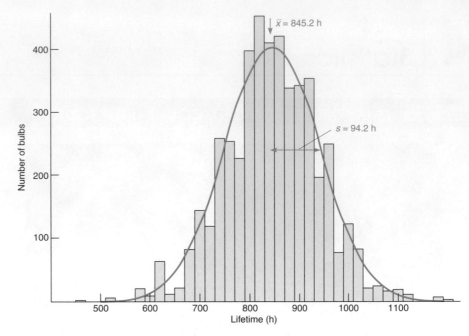

**Figure 4-1** Bar graph and Gaussian curve describing the lifetimes of a hypothetical set of electric light bulbs. The smooth curve has the same mean, standard deviation, and area as the bar graph. Any finite set of data, however, will differ from the bell-shaped curve. The more measurements an investigator makes, the closer they will come to the smooth curve.

## Mean Value and Standard Deviation

In the hypothetical case in Figure 4-1, a manufacturer tested the lifetimes of 4 768 electric light bulbs. The bar graph shows the number of bulbs with a lifetime in each 20-h interval. Lifetimes approximate a Gaussian distribution because variations in the construction of light bulbs, such as filament thickness and quality of attachments, are random. The smooth curve is the Gaussian distribution that best fits the data. Any finite set of data will vary somewhat from the Gaussian curve.

Light bulb lifetimes, and the corresponding Gaussian curve, are characterized by two parameters. The arithmetic **mean,** $\bar{x}$—also called the **average**—is the sum of the measured values divided by $n$, the number of measurements:

*Mean:*
$$\bar{x} = \frac{\sum_i x_i}{n} \qquad (4\text{-}1)$$

where $x_i$ is the lifetime of an individual bulb. The Greek capital sigma, $\Sigma$, means summation: $\Sigma_i x_i = x_1 + x_2 + x_3 + \cdots + x_n$. In Figure 4-1, the mean value is 845.2 h.

The **standard deviation,** $s$, measures how closely the data are clustered about the mean. *The smaller the standard deviation, the more closely the data are clustered about the mean* (Figure 4-2).

*Standard deviation:*
$$s = \sqrt{\frac{\sum_i (x_i - \bar{x})^2}{n - 1}} \qquad (4\text{-}2)$$

In Figure 4-1, $s = 94.2$ h. A set of light bulbs having a small standard deviation in lifetime is more uniformly manufactured than a set with a large standard deviation.

For an *infinite* set of data, the mean is designated by the lowercase Greek letter mu, $\mu$ (the population mean), and the standard deviation is written as a lowercase Greek sigma, $\sigma$ (the population standard deviation). We can never measure $\mu$ and $\sigma$, but the values of $\bar{x}$ and $s$ approach $\mu$ and $\sigma$ as the number of measurements increases.

The quantity $n - 1$ in Equation 4-2 is called the **degrees of freedom.** The square of the standard deviation is called the **variance.** The standard deviation expressed as a percentage of the mean value ($= 100 \times s/\bar{x}$) is called the *relative standard deviation* or the *coefficient of variation.*

---

The mean gives the center of the distribution. The standard deviation measures the width of the distribution.

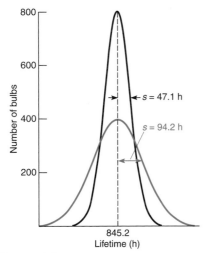

**Figure 4-2** Gaussian curves for two sets of light bulbs, one having a standard deviation half as great as the other. The number of bulbs described by each curve is the same.

An experiment that produces a small standard deviation is more *precise* than one that produces a large standard deviation. Greater precision does not necessarily imply greater *accuracy*, which means nearness to the "truth."

As the number of measurements increases, $\bar{x}$ approaches $\mu$, *if there is no systematic error.*

## Example  Mean and Standard Deviation

Find the average and the standard deviation for 821, 783, 834, and 855.

**Solution**   The average is

$$\bar{x} = \frac{821 + 783 + 834 + 855}{4} = 823._2$$

To avoid accumulating round-off errors, retain one more digit for the average and the standard deviation than was present in the original data. The standard deviation is

$$s = \sqrt{\frac{(821 - 823.2)^2 + (783 - 823.2)^2 + (834 - 823.2)^2 + (855 - 823.2)^2}{(4 - 1)}} = 30._3$$

The average and the standard deviation should both end at the *same decimal place*. For $\bar{x} = 823._2$, we will write $s = 30._3$.

[spreadsheet icon] Spreadsheets have built-in functions for the average and standard deviation. In the spreadsheet in the margin, data points are entered in cells B1 through B4. The average in cell B5 is computed with the statement "= AVERAGE(B1:B4)". B1:B4 means cells B1, B2, B3, and B4. The standard deviation in cell B6 is computed with "= STDEV(B1:B4)".

For ease of reading, cells B5 and B6 were set to display 3 decimal places by use of the CELLS command in the FORMAT menu. A heavy line was placed beneath cell B4 by setting border with the CELLS command of the FORMAT menu.

Learn to use the standard deviation function on your calculator and see that you get $s = 30.269\ 6\ldots$. Do not round off during a calculation. Retain all the extra digits in your calculator.

|   | A | B |
|---|---|---|
| 1 |   | 821 |
| 2 |   | 783 |
| 3 |   | 834 |
| 4 |   | 855 |
| 5 | Average = | 823.250 |
| 6 | Std dev = | 30.270 |

## Significant Figures in Mean and Standard Deviation

We commonly express experimental results in the form: mean $\pm$ standard deviation $= \bar{x} \pm s$. It is sensible to write the results of the preceding example as $823 \pm 30$ or even $8.2\ (\pm 0.3) \times 10^2$ to indicate that the mean has just two significant figures. The expressions $823 \pm 30$ and $8.2\ (\pm 0.3) \times 10^2$ are not suitable for continued calculations in which $\bar{x}$ and $s$ are intermediate results. We will retain one or more insignificant digits to avoid introducing round-off errors into subsequent work. Try not to go into cardiac arrest over significant figures when you see $823._2 \pm 30._3$ as the answer to a problem in this book.

## Standard Deviation and Probability

The formula for a Gaussian curve is

*Gaussian curve:*
$$y = \frac{1}{\sigma\sqrt{2\pi}}\, e^{-(x-\mu)^2/2\sigma^2} \qquad (4\text{-}3)$$

where e ($= 2.718\ 28\ldots$) is the base of the natural logarithm. For a finite set of data, we approximate $\mu$ by $\bar{x}$ and $\sigma$ by $s$. A graph of Equation 4-3 is shown in Figure 4-3, in which the values $\sigma = 1$ and $\mu = 0$ are used for simplicity. The maximum value of $y$ is at $x = \mu$, and the curve is symmetric about $x = \mu$.

It is useful to express deviations from the mean value in multiples, $z$, of the standard deviation. That is, we transform $x$ into $z$, given by

$$z = \frac{x - \mu}{\sigma} \approx \frac{x - \bar{x}}{s} \qquad (4\text{-}4)$$

The probability of measuring $z$ in a certain *range* is equal to the *area* of that range. For example, the probability of observing $z$ between $-2$ and $-1$ is 0.136. This probability corresponds to the shaded area in Figure 4-3. The area under each portion of the Gaussian curve is given in Table 4-1. Because the sum of the probabilities of all the measurements must be unity, the area under the whole curve from $z = -\infty$ to $z = +\infty$ must be unity. The number $1/(\sigma\sqrt{2\pi})$ in Equation 4-3 is called the *normalization factor*. It guarantees that the area under the entire curve is unity. A Gaussian curve with unit area is called a *normal error curve*.

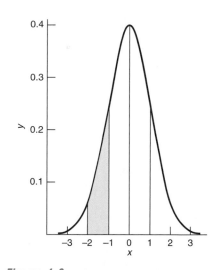

**Figure 4-3**  A Gaussian curve in which $\mu = 0$ and $\sigma = 1$. A Gaussian curve whose area is unity is called a normal error curve. In this case, the abscissa, $x$, is equal to $z$, defined as $z = (x - \mu)/\sigma$.

When $z = +1$, $x$ is one standard deviation above the mean. When $z = -2$, $x$ is two standard deviations below the mean.

## Example  Area Under a Gaussian Curve

Suppose the manufacturer of the bulbs used for Figure 4-1 offers to replace free of charge any bulb that burns out in less than 600 hours. If she plans to sell a million bulbs, how many extra bulbs should she keep available as replacements?

## Table 4-1 Ordinate and area for the normal (Gaussian) error curve,

$$y = \frac{1}{\sqrt{2\pi}} e^{-z^2/2}$$

| $|z|$[a] | $y$ | Area[b] | $|z|$ | $y$ | Area | $|z|$ | $y$ | Area |
|---|---|---|---|---|---|---|---|---|
| 0.0 | 0.398 9 | 0.000 0 | 1.4 | 0.149 7 | 0.419 2 | 2.8 | 0.007 9 | 0.497 4 |
| 0.1 | 0.397 0 | 0.039 8 | 1.5 | 0.129 5 | 0.433 2 | 2.9 | 0.006 0 | 0.498 1 |
| 0.2 | 0.391 0 | 0.079 3 | 1.6 | 0.110 9 | 0.445 2 | 3.0 | 0.004 4 | 0.498 650 |
| 0.3 | 0.381 4 | 0.117 9 | 1.7 | 0.094 1 | 0.455 4 | 3.1 | 0.003 3 | 0.499 032 |
| 0.4 | 0.368 3 | 0.155 4 | 1.8 | 0.079 0 | 0.464 1 | 3.2 | 0.002 4 | 0.499 313 |
| 0.5 | 0.352 1 | 0.191 5 | 1.9 | 0.065 6 | 0.471 3 | 3.3 | 0.001 7 | 0.499 517 |
| 0.6 | 0.333 2 | 0.225 8 | 2.0 | 0.054 0 | 0.477 3 | 3.4 | 0.001 2 | 0.499 663 |
| 0.7 | 0.312 3 | 0.258 0 | 2.1 | 0.044 0 | 0.482 1 | 3.5 | 0.000 9 | 0.499 767 |
| 0.8 | 0.289 7 | 0.288 1 | 2.2 | 0.035 5 | 0.486 1 | 3.6 | 0.000 6 | 0.499 841 |
| 0.9 | 0.266 1 | 0.315 9 | 2.3 | 0.028 3 | 0.489 3 | 3.7 | 0.000 4 | 0.499 904 |
| 1.0 | 0.242 0 | 0.341 3 | 2.4 | 0.022 4 | 0.491 8 | 3.8 | 0.000 3 | 0.499 928 |
| 1.1 | 0.217 9 | 0.364 3 | 2.5 | 0.017 5 | 0.493 8 | 3.9 | 0.000 2 | 0.499 952 |
| 1.2 | 0.194 2 | 0.384 9 | 2.6 | 0.013 6 | 0.495 3 | 4.0 | 0.000 1 | 0.499 968 |
| 1.3 | 0.171 4 | 0.403 2 | 2.7 | 0.010 4 | 0.496 5 | $\infty$ | 0 | 0.5 |

*a.* $z = (x - \mu)/\sigma$.

*b. The area refers to the area between $z = 0$ and $z =$ the value in the table. Thus the area from $z = 0$ to $z = 1.4$ is $0.419\ 2$. The area from $z = -0.7$ to $z = 0$ is the same as from $z = 0$ to $z = 0.7$. The area from $z = -0.5$ to $z = +0.3$ is $(0.191\ 5 + 0.117\ 9) = 0.309\ 4$. The total area between $z = -\infty$ and $z = +\infty$ is unity.*

**Solution** We need to express the desired interval in multiples of the standard deviation and then find the area of the interval in Table 4-1. Because $\bar{x} = 845.2$ and $s = 94.2$, $z = (600 - 845.2)/94.2 = -2.60$. The area under the curve between the mean value and $z = -2.60$ is 0.495 3 in Table 4-1. The entire area from $-\infty$ to the mean value is 0.500 0, so the area from $-\infty$ to $-2.60$ must be $0.500\ 0 - 0.495\ 3 = 0.004\ 7$. The area to the left of 600 hours in Figure 4-1 is only 0.47% of the entire area under the curve. Only 0.47% of the bulbs are expected to fail in fewer than 600 h. If the manufacturer sells 1 million bulbs a year, she should make 4 700 extra bulbs to meet the replacement demand.

**Example**  Using a Spreadsheet to Find Area Beneath a Gaussian Curve

What fraction of bulbs is expected to have a lifetime between 900 and 1 000 h?

**Solution** *We need to find the fraction of the area of the Gaussian curve between $x = 900$ and $x = 1\ 000$ h.* The function NORMDIST in Excel gives the area of the curve from $-\infty$ up to a specified point, $x$. Here is the strategy: We will find the area from $-\infty$ to 900 h, which is the shaded area to the left of 900 h in Figure 4-4. Then we will find the area from $-\infty$ to 1 000 h, which is all the shaded area to the left of 1 000 h in Figure 4-4. The difference between the two is the area from 900 to 1 000 h:

Area from 900 to 1 000 = (area from $-\infty$ to 1 000) $-$ (area from $-\infty$ to 900)  (4-5)

In a spreadsheet, enter the mean in cell A2 and the standard deviation in cell B2. To find the area under the Gaussian curve from $-\infty$ to 900 h in cell C4, we select cell C4 and go to the INSERT menu and choose FUNCTION. In the window that appears, select the Statistical functions and find NORMDIST from the list of possibilities. Double click on NORMDIST and another window appears asking for four values that will be used by NORMDIST. (If you click on help, you will find a cryptic explanation of how to use NORMDIST.)

Values provided to the function NORMDIST($x$,mean,standard_dev,cumulative) are called *arguments* of the function. The first argument is $x$, which is 900. The second argument is the mean, which is 845.2. You can either enter 845.2 for the mean or you can type "A2", which is the cell containing 845.2. The third argument is the standard deviation, for which we enter the value 94.2 or the cell B2. The last argument is called "cumulative." When it has the value TRUE, NORMDIST gives the area under the Gaussian curve. When cumulative is FALSE, NORMDIST gives the ordinate (the $y$-value) of the Gaussian curve. We want area, so enter TRUE. The formula "= NORMDIST(900,$A$2,$B$2,TRUE)" in cell C4

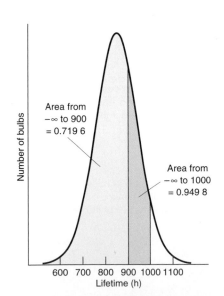

**Figure 4-4** Use of the Gaussian curve to find the fraction of bulbs with a lifetime between 900 and 1 000 h. We find the area between $-\infty$ and 1 000 h and subtract the area between $-\infty$ and 900 h.

| | A | B | C |
|---|---|---|---|
| 1 | Mean = | Std dev = | |
| 2 | 845.2 | 94.2 | |
| 3 | | | |
| 4 | Area from $-\infty$ to 900 = | | 0.7196 |
| 5 | Area from $-\infty$ to 1000 = | | 0.9498 |
| 6 | Area from 900 to 1000 | | 0.2302 |
| 7 | | | |
| 8 | C4 = NORMDIST(900,A2,B2,TRUE) | | |
| 9 | C5 = NORMDIST(1000,A2,B2,TRUE) | | |
| 10 | C6 = C5-C4 | | |

returns the value 0.719 6. This is the area under the Gaussian curve from $-\infty$ to 900 h. To get the area from $-\infty$ to 1 000 h, write "= NORMDIST(1000,$A$2,$B$2,TRUE)" in cell C5. The value returned is 0.949 8. Following Equation 4-5, subtract the areas (C5 − C4) to obtain 0.230 2, which is the area from 900 to 1 000. That is, 23.02% of the area lies in the range 900 to 1 000 h. We expect 23% of the bulbs to have a lifetime of 900 to 1 000 h.

*The standard deviation measures the width of the Gaussian curve.* The larger the value of $\sigma$, the broader the curve. In any Gaussian curve, 68.3% of the area is in the range from $\mu - 1\sigma$ to $\mu + 1\sigma$. That is, more than two-thirds of the measurements are expected to lie within one standard deviation of the mean. Also, 95.5% of the area lies within $\mu \pm 2\sigma$, and 99.7% of the area lies within $\mu \pm 3\sigma$. Suppose that you use two different techniques to measure sulfur in coal: Method A has a standard deviation of 0.4%, and method B has a standard deviation of 1.1%. You can expect that approximately two-thirds of measurements from method A will lie within 0.4% of the mean. For method B, two-thirds will lie within 1.1% of the mean.

The more times you measure a quantity, the more confident you can be that the average value of your measurements is close to the true population mean, $\mu$. Uncertainty decreases in proportion to $1/\sqrt{n}$, where $n$ is the number of measurements. You can decrease the uncertainty of the mean by a factor of 2 ($= \sqrt{4}$) by making 4 times as many measurements and by a factor of 3.16 ($= \sqrt{10}$) by making 10 times as many measurements.

| Range | Percentage of measurements |
|---|---|
| $\mu \pm 1\sigma$ | 68.3 |
| $\mu \pm 2\sigma$ | 95.5 |
| $\mu \pm 3\sigma$ | 99.7 |

To decrease uncertainty by $1/\sqrt{10}$ requires 10 measurements. Instruments with rapid data acquisition allow us to average many experiments in a short time to increase the accuracy of a result.

## 4-2 Confidence Intervals

**Student's *t*** is a statistical tool used most frequently to express confidence intervals and to compare results from different experiments. It is the tool you could use to evaluate the probability that your red blood cell count will be found in a certain range on "normal" days.

"Student" was the pseudonym of W. S. Gosset, whose employer, the Guinness breweries of Ireland, restricted publications for proprietary reasons. Because of the importance of Gosset's work, he was allowed to publish it (*Biometrika* 1908, 6, 1), but under an assumed name.

### Calculating Confidence Intervals

From a limited number of measurements, we cannot find the true population mean, $\mu$, or the true standard deviation, $\sigma$. What we can determine are $\bar{x}$ and $s$, the sample mean and the sample standard deviation. The **confidence interval** is an expression stating that the true mean, $\mu$, is likely to lie within a certain distance from the measured mean, $\bar{x}$. The confidence interval of $\mu$ is given by

*Confidence interval:*
$$\mu = \bar{x} \pm \frac{ts}{\sqrt{n}}$$
(4-6)

where $s$ is the measured standard deviation, $n$ is the number of observations, and $t$ is Student's $t$, taken from Table 4-2.

### Example  Calculating Confidence Intervals

The carbohydrate content of a glycoprotein (a protein with sugars attached to it) is determined to be 12.6, 11.9, 13.0, 12.7, and 12.5 g of carbohydrate per 100 g of protein in replicate analyses. Find the 50% and 90% confidence intervals for the carbohydrate content.

**Table 4-2** Values of Student's *t*

| Degrees of freedom | Confidence level (%) | | | | | | |
|---|---|---|---|---|---|---|---|
| | 50 | 90 | 95 | 98 | 99 | 99.5 | 99.9 |
| 1 | 1.000 | 6.314 | 12.706 | 31.821 | 63.656 | 127.321 | 636.578 |
| 2 | 0.816 | 2.920 | 4.303 | 6.965 | 9.925 | 14.089 | 31.598 |
| 3 | 0.765 | 2.353 | 3.182 | 4.541 | 5.841 | 7.453 | 12.924 |
| 4 | 0.741 | 2.132 | 2.776 | 3.747 | 4.604 | 5.598 | 8.610 |
| 5 | 0.727 | 2.015 | 2.571 | 3.365 | 4.032 | 4.773 | 6.869 |
| 6 | 0.718 | 1.943 | 2.447 | 3.143 | 3.707 | 4.317 | 5.959 |
| 7 | 0.711 | 1.895 | 2.365 | 2.998 | 3.500 | 4.029 | 5.408 |
| 8 | 0.706 | 1.860 | 2.306 | 2.896 | 3.355 | 3.832 | 5.041 |
| 9 | 0.703 | 1.833 | 2.262 | 2.821 | 3.250 | 3.690 | 4.781 |
| 10 | 0.700 | 1.812 | 2.228 | 2.764 | 3.169 | 3.581 | 4.587 |
| 15 | 0.691 | 1.753 | 2.131 | 2.602 | 2.947 | 3.252 | 4.073 |
| 20 | 0.687 | 1.725 | 2.086 | 2.528 | 2.845 | 3.153 | 3.850 |
| 25 | 0.684 | 1.708 | 2.060 | 2.485 | 2.787 | 3.078 | 3.725 |
| 30 | 0.683 | 1.697 | 2.042 | 2.457 | 2.750 | 3.030 | 3.646 |
| 40 | 0.681 | 1.684 | 2.021 | 2.423 | 2.704 | 2.971 | 3.551 |
| 60 | 0.679 | 1.671 | 2.000 | 2.390 | 2.660 | 2.915 | 3.460 |
| 120 | 0.677 | 1.658 | 1.980 | 2.358 | 2.617 | 2.860 | 3.373 |
| ∞ | 0.674 | 1.645 | 1.960 | 2.326 | 2.576 | 2.807 | 3.291 |

*NOTE: In calculating confidence intervals, σ may be substituted for s in Equation 4-6 if you have a great deal of experience with a particular method and have therefore determined its "true" population standard deviation. If σ is used instead of s, the value of t to use in Equation 4-6 comes from the bottom row of Table 4-2.*

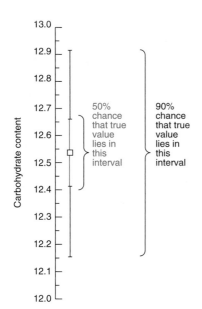

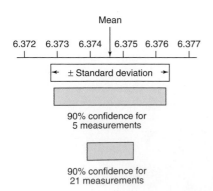

90% confidence for
5 measurements

90% confidence for
21 measurements

**Solution** First calculate $\bar{x}$ (= $12.5_4$) and $s$ (= $0.4_0$) for the five measurements. For the 50% confidence interval, look up *t* in Table 4-2 under 50 and across from *four* degrees of freedom (degrees of freedom = $n - 1$.) The value of *t* is 0.741, so the 50% confidence interval is

$$\mu = \bar{x} \pm \frac{ts}{\sqrt{n}} = 12.5_4 \pm \frac{(0.741)(0.4_0)}{\sqrt{5}} = 12.5_4 \pm 0.1_3$$

The 90% confidence interval is

$$\mu = \bar{x} \pm \frac{ts}{\sqrt{n}} = 12.5_4 \pm \frac{(2.132)(0.4_0)}{\sqrt{5}} = 12.5_4 \pm 0.3_8$$

There is a 50% chance that the true mean, $\mu$, lies within the range $12.5_4 \pm 0.1_3$ ($12.4_1$ to $12.6_7$). There is a 90% chance that $\mu$ lies within the range $12.5_4 \pm 0.3_8$ ($12.1_6$ to $12.9_2$).

## Confidence Intervals as Estimates of Experimental Uncertainty

Chapter 3 gave rules for propagation of uncertainty in calculations. For example, if we were dividing a mass by a volume to find density, the uncertainty in density is derived from the uncertainties in mass and volume. The most common estimates of uncertainty are the standard deviation and the confidence interval.

Suppose you measure the volume of a vessel five times and observe values of 6.375, 6.372, 6.374, 6.377, and 6.375 mL. The average is $\bar{x} = 6.374_6$ mL and the standard deviation is $s = 0.001_8$ mL. You could choose a confidence interval (such as 90%) for the estimate of uncertainty. Using Equation 4-6 with four degrees of freedom, you find that the 90% confidence interval is $\pm ts/\sqrt{n} = \pm(2.132)(0.001_8)/\sqrt{5} = \pm 0.001_7$. By this criterion, the uncertainty in volume is $\pm 0.001_7$ mL.

We can reduce uncertainty by making more measurements. If we make 21 measurements and have the same mean and standard deviation, the 90% confidence interval is reduced from $\pm 0.001_7$ to $\pm ts/\sqrt{n} = \pm(1.725)(0.001_8)/\sqrt{21} = \pm 0.000\ 7$ mL.

Frequently, we use the standard deviation as the estimated uncertainty. For five measurements, we would report a volume of $6.374_6 \pm 0.001_8$ mL. It is good practice to report the number of measurements, so that confidence intervals can be calculated.

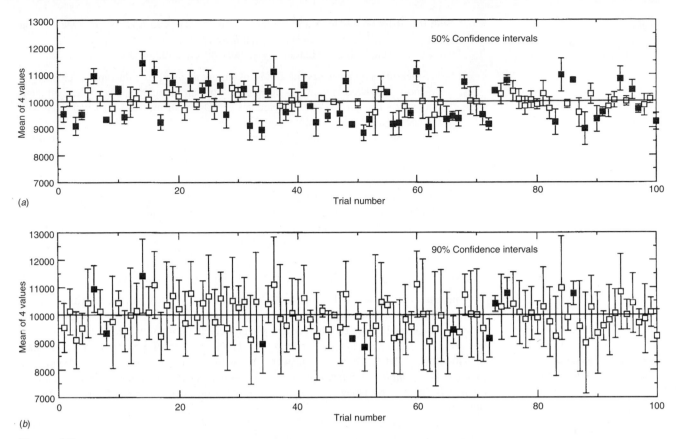

**Figure 4-5** 50% and 90% confidence intervals for the same set of random data. Filled squares are the data points whose confidence interval does not include the true population mean of 10 000.

### The Meaning of a Confidence Interval

Figure 4-5 illustrates the meaning of confidence intervals. A computer chose numbers at random from a Gaussian population with a population mean ($\mu$) of 10 000 and a population standard deviation ($\sigma$) of 1 000 in Equation 4-3. In trial 1, four numbers were chosen, and their mean and standard deviation were calculated with Equations 4-1 and 4-2. The 50% confidence interval was then calculated with Equation 4-6, using $t = 0.765$ from Table 4-2 (50% confidence, 3 degrees of freedom). This trial is plotted as the first point at the left in Figure 4-5a; the square is centered at the mean value of 9 526, and the error bar extends from the lower limit to the upper limit of the 50% confidence interval ($\pm 290$). The experiment was repeated 100 times to produce the points in Figure 4-5a.

The 50% confidence interval is defined such that, if we repeated this experiment an infinite number of times, 50% of the error bars in Figure 4-5a would include the true population mean of 10 000. In fact, I did the experiment 100 times, and 45 of the error bars in Figure 4-5a pass through the horizontal line at 10 000.

Figure 4-5b shows the same experiment with the same set of random numbers, but this time the 90% confidence interval was calculated. For an infinite number of experiments, we would expect 90% of the confidence intervals to include the population mean of 10 000. In Figure 4-5b, 89 of the 100 error bars cross the horizontal line at 10 000.

## ■ ■ ■ ■  4-3  Comparison of Means with Student's *t*

We use a *t* **test** to compare one set of measurements with another to decide whether or not they are "the same." Statisticians say we are testing the *null hypothesis,* which states that the mean values from two sets of measurements are *not* different. Because of inevitable random errors, we do not expect the mean values to be exactly the same, even if we are measuring the same physical quantity. Statistics gives us a probability that the observed difference between two means can arise from purely random measurement error. We customarily reject the null hypothesis if there is less than a 5% chance that the observed difference arises from random

Confidence limits and the *t* test (and, later in this chapter, the *Q* test) assume that data follow a Gaussian distribution. If they do not, different formulas would be required.

error. With this criterion, we have a 95% chance that our conclusion is correct. One time out of 20 when we conclude that two means are not different we will be wrong.

Here are three cases that are handled in slightly different manners:

**Case 1** We measure a quantity several times, obtaining an average value and a standard deviation. We need to compare our answer with an accepted answer. The average is not exactly the same as the accepted answer. Does our measured answer agree with the accepted answer "within experimental error"?

**Case 2** We measure a quantity multiple times by two different methods that give two different answers, each with its own standard deviation. Do the two results agree with each other "within experimental error"?

**Case 3** Sample 1 is measured once by Method 1 and once by Method 2, which do not give exactly the same result. Then a different sample, designated 2, is measured once by Method 1 and once by Method 2; and, again, the results are not exactly equal to each other. The procedure is repeated for $n$ different samples. Do the two methods agree with each other "within experimental error"?

## Case 1. Comparing a Measured Result with a "Known" Value

You purchased a Standard Reference Material (Box 3-1) coal sample certified by the National Institute of Standards and Technology to contain 3.19 wt% sulfur. You are testing a new analytical method to see whether it can reproduce the known value. The measured values are 3.29, 3.22, 3.30, and 3.23 wt% sulfur, giving a mean of $\bar{x} = 3.26_0$ and a standard deviation of $s = 0.04_1$. Does your answer agree with the known answer? To find out, *compute the 95% confidence interval for your answer and see if that range includes the known answer.* If the known answer is not within your 95% confidence interval, then the results do not agree.

So here we go. For 4 measurements, there are 3 degrees of freedom and $t_{95\%} = 3.182$ in Table 4-2. The 95% confidence interval is

> If the "known" answer does not lie within the 95% confidence interval, then the two methods give "different" results.

> Retain many digits in this calculation.

$$\mu = \bar{x} \pm \frac{ts}{\sqrt{n}} = 3.26_0 \pm \frac{(3.182)(0.04_1)}{\sqrt{4}} = 3.26_0 \pm 0.06_5 \qquad (4\text{-}7)$$

$$95\% \text{ confidence interval} = 3.19_5 \text{ to } 3.32_5 \text{ wt\%}$$

The known answer (3.19 wt%) is just outside the 95% confidence interval. Therefore we conclude that there is less than a 5% chance that our method agrees with the known answer.

We conclude that our method gives a "different" result from the known result. However, in this case, the 95% confidence interval is so close to including the known result that it would be prudent to make more measurements before concluding that our new method is not accurate.

## Case 2. Comparing Replicate Measurements

We can use a $t$ test to decide whether two sets of replicate measurements give "the same" or "different" results, within a stated confidence level. An example comes from the work of Lord Rayleigh (John W. Strutt), who is remembered today for landmark investigations of light scattering, blackbody radiation, and elastic waves in solids. His Nobel Prize in 1904 was received for discovering the inert gas argon. This discovery occurred when he noticed a small discrepancy between two sets of measurements of the density of nitrogen gas.

In Rayleigh's time, it was known that dry air was composed of about one-fifth oxygen and four-fifths nitrogen. Rayleigh removed all $O_2$ from air by mixing the air sample with red-hot copper (to make solid CuO). He then measured the density of the remaining gas by collecting it in a fixed volume at constant temperature and pressure. He also prepared the same volume of pure $N_2$ by chemical decomposition of nitrous oxide ($N_2O$), nitric oxide (NO), or ammonium nitrite ($NH_4^+NO_2^-$). Table 4-3 and Figure 4-6 show the mass of gas collected in each experiment. The average mass collected from air (2.310 11 g) is 0.46% greater than the average mass of the same volume of gas from chemical sources (2.299 47 g).

If Rayleigh's measurements had not been performed with care, this difference might have been attributed to experimental error. Instead, Rayleigh understood that the discrepancy was outside his margin of error, and he postulated that gas collected from the air was a mixture of nitrogen with a small amount of a heavier gas, which turned out to be argon.

Let's see how to use a $t$ test to decide whether gas isolated from air is "significantly" heavier than nitrogen isolated from chemical sources. In this case, we have two sets of mea-

**Table 4-3** Masses of gas isolated by Lord Rayleigh

| From air (g) | From chemical decomposition (g) |
|---|---|
| 2.310 17 | 2.301 43 |
| 2.309 86 | 2.298 90 |
| 2.310 10 | 2.298 16 |
| 2.310 01 | 2.301 82 |
| 2.310 24 | 2.298 69 |
| 2.310 10 | 2.299 40 |
| 2.310 28 | 2.298 49 |
| — | 2.298 89 |
| Average | |
| 2.310 11 | 2.299 47 |
| Standard deviation | |
| 0.000 14_3 | 0.001 38 |

SOURCE: R. D. Larsen, *J. Chem. Ed.* **1990**, *67*, 925; see also C. J. Giunta, *J. Chem. Ed.* **1998**, *75*, 1322.

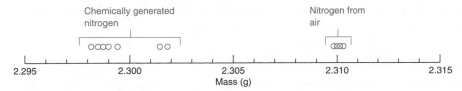

**Figure 4-6** Lord Rayleigh's measurements of the mass of constant volumes of gas (at constant temperature and pressure) isolated by removing oxygen from air or generated by decomposition of nitrogen compounds. Rayleigh recognized that the difference between the two clusters was outside of his experimental error and deduced that a heavier component, which turned out to be argon, was present in gas isolated from air.

surements, each with its own uncertainty and no "known" value. We assume that the population standard deviation ($\sigma$) for each method is essentially the same.

For two sets of data consisting of $n_1$ and $n_2$ measurements (with averages $\bar{x}_1$ and $\bar{x}_2$), we calculate a value of $t$ with the formula

$$t_{\text{calculated}} = \frac{|\bar{x}_1 - \bar{x}_2|}{s_{\text{pooled}}} \sqrt{\frac{n_1 n_2}{n_1 + n_2}} \qquad (4\text{-}8)$$

where $|\bar{x}_1 - \bar{x}_2|$ is the absolute value of the difference (a positive number) and

$$s_{\text{pooled}} = \sqrt{\frac{\sum_{\text{set 1}}(x_i - \bar{x}_1)^2 + \sum_{\text{set 2}}(x_j - \bar{x}_2)^2}{n_1 + n_2 - 2}} = \sqrt{\frac{s_1^2(n_1 - 1) + s_2^2(n_2 - 1)}{n_1 + n_2 - 2}} \qquad (4\text{-}9)$$

Here $s_{\text{pooled}}$ is a *pooled* standard deviation making use of both sets of data. $t_{\text{calculated}}$ from Equation 4-8 is compared with $t$ in Table 4-2 for $n_1 + n_2 - 2$ degrees of freedom. *If $t_{\text{calculated}}$ is greater than $t_{\text{table}}$ at the 95% confidence level, the two results are considered to be different.*

> If $t_{\text{calculated}} > t_{\text{table}}$ (95%), the difference is significant.

---

**Example** Is Lord Rayleigh's Gas from Air Denser than N$_2$ from Chemicals?

The average mass of gas from air in Table 4-3 is $\bar{x}_1 = 2.310\ 11$ g, with a standard deviation of $s_1 = 0.000\ 14_3$ (for $n_1 = 7$ measurements). The mass of gas from chemical sources is $\bar{x}_2 = 2.299\ 47$ g, with $s_2 = 0.001\ 38$ ($n_2 = 8$ measurements).

**Solution** To answer the question, we calculate $s_{\text{pooled}}$ with Equation 4-9,

$$s_{\text{pooled}} = \sqrt{\frac{(0.000\ 14_3)^2(7 - 1) + (0.001\ 38)^2(8 - 1)}{7 + 8 - 2}} = 0.001\ 02$$

and $t_{\text{calculated}}$ with Equation 4-8:

$$t_{\text{calculated}} = \frac{|2.310\ 11 - 2.299\ 47|}{0.001\ 02} \sqrt{\frac{7 \cdot 8}{7 + 8}} = 20.2$$

For $7 + 8 - 2 = 13$ degrees of freedom in Table 4-2, $t_{\text{table}}$ lies between 2.228 and 2.131 for 95% confidence. Because $t_{\text{calculated}} > t_{\text{table}}$, the difference is significant. In fact, $t_{\text{table}}$ for 99.9% confidence is ~4.3. The difference is significant beyond the 99.9% confidence level. Our eyes were not lying to us in Figure 4-6: Gas from the air is undoubtedly denser than N$_2$ from chemical sources. This observation led Rayleigh to discover a heavy constituent of air.

---

Equations 4-8 and 4-9 assume that the population standard deviation is the same for both sets of measurements. If this is not true, then we use the equations

$$t_{\text{calculated}} = \frac{|\bar{x}_1 - \bar{x}_2|}{\sqrt{s_1^2/n_1 + s_2^2/n_2}} \qquad (4\text{-}8a)$$

$$\text{Degrees of freedom} = \left\{ \frac{(s_1^2/n_1 + s_2^2/n_2)^2}{\left(\dfrac{(s_1^2/n_1)^2}{n_1 + 1} + \dfrac{(s_2^2/n_2)^2}{n_2 + 1}\right)} \right\} - 2 \qquad (4\text{-}9a)$$

> Section 4-4 explains how to use the *F* test to see if two standard deviations are "the same" or "different."

For Rayleigh's data in Figure 4-6, we suspect that the population standard deviation from air is smaller than that from chemical sources. Using Equations 4-8a and 4-9a, we find that $t_{calculated} = 21.7$ and degrees of freedom $= 7.22 \approx 7$. This value of $t_{calculated}$ still far exceeds values in Table 4-2 for 7 degrees of freedom at 95% or 99.9% confidence.

## Case 3. Paired $t$ test for Comparing Individual Differences

In this case, we use two methods to make single measurements on several different samples. No measurement has been duplicated. Do the two methods give the same answer "within experimental error"? Figure 4-7 shows measurements of aluminum in 11 samples of drinking water. Results for Method 1 are in column B and results for Method 2 are in column C. For each sample, the two results are similar, but not identical.

To see if there is a significant difference between the methods, we use the paired $t$ test. First, column D computes the difference ($d_i$) between the two results for each sample. The mean of the 11 differences ($\bar{d} = -2.4_{91}$) is computed in cell D16 and the standard deviation of the 11 differences ($s_d$) is computed in cell D17.

$$s_d = \sqrt{\frac{\sum (d_i - \bar{d})^2}{n - 1}} \tag{4-10}$$

$$s_d = \sqrt{\frac{(-3.0 - \bar{d})^2 + (4.8 - \bar{d})^2 + \cdots + (0.2 - \bar{d})^2 + (-11.6 - \bar{d})^2}{11 - 1}} = 6.7_{48}$$

Once you have the mean and standard deviation, compute $t_{calculated}$ with the formula

$$t_{calculated} = \frac{|\bar{d}|}{s_d}\sqrt{n} \tag{4-11}$$

where $|\bar{d}|$ is the absolute value of the mean difference, so that $t_{calculated}$ is always positive. Inserting the mean and standard deviation into Equation 4-11 gives

$$t_{calculated} = \frac{2.4_{91}}{6.7_{48}}\sqrt{11} = 1.2_{24}$$

You may appreciate Box 4-1 at this time.

We find that $t_{calculated}$ ($1.2_{24}$) is less than $t_{table}$ (2.228) listed in Table 4-2 for 95% confidence and 10 degrees of freedom. *There is less than a 95% chance that the two results are different.*

|  | A | B | C | D |
|---|---|---|---|---|
| 1 | Comparison of two methods for measuring Al | | | |
| 2 | | | | |
| 3 | Sample | Method 1 | Method 2 | Difference |
| 4 | number | (μg/L) | (μg/L) | (d$_i$) |
| 5 | 1 | 17.2 | 14.2 | −3.0 |
| 6 | 2 | 23.1 | 27.9 | 4.8 |
| 7 | 3 | 28.5 | 21.2 | −7.3 |
| 8 | 4 | 15.3 | 15.9 | 0.6 |
| 9 | 5 | 23.1 | 32.1 | 9.0 |
| 10 | 6 | 32.5 | 22.0 | −10.5 |
| 11 | 7 | 39.5 | 37.0 | −2.5 |
| 12 | 8 | 38.7 | 41.5 | 2.8 |
| 13 | 9 | 52.5 | 42.6 | −9.9 |
| 14 | 10 | 42.6 | 42.8 | 0.2 |
| 15 | 11 | 52.7 | 41.1 | −11.6 |
| 16 | | | mean = | −2.491 |
| 17 | | | std dev = | 6.748 |
| 18 | | | t$_{calculated}$ = | 1.224 |
| 19 | D5 = C5-B5 | | | |
| 20 | D16 = AVERAGE(D5:D15) | | | |
| 21 | D17 = STDEV(D5:D15) | | | |
| 22 | D18 = ABS(D16)*SQRT(A15)/D17 | | | |
| 23 | ABS = absolute value | | | |

*Figure 4-7* Measurement of Al by 2 methods. [Data from P. T. Srinivasan, T. Viraraghavan, and K. S. Subramanian, "Method Development for Drinking Water Aluminum Measurement Using a Graphite Furnace Atomic Absorption Spectrophotometer," Am. Lab., February 2000, p. 76.]

As a person who will either derive or use analytical results, you should be aware of this warning:[3]

Analytical chemists must always emphasize to the public that *the single most important characteristic of any result obtained from one or more analytical measurements is an adequate statement of its uncertainty interval.* Lawyers usually attempt to dispense with uncertainty and try to obtain unequivocal statements; therefore, an uncertainty interval must be clearly defined in cases involving litigation and/or enforcement proceedings. Otherwise, a value of 1.001 without a specified uncertainty, for example, may be viewed as legally exceeding a permissible level of 1.

Some legal limits make no scientific sense. The Delaney Amendment to the U.S. Federal Food, Drug, and Cosmetic Act of 1958 stated that "no additive [in processed food] shall be deemed to be safe if it is found to induce cancer when ingested by man or animal." This statement meant that no detectable level of any carcinogenic (cancer-causing) pesticide may remain in processed foods, even if the level is far below that which can be shown to cause cancer. The law was passed at a time when the sensitivity of analytical procedures was relatively poor, so the detection limit was relatively high. As sensitivity improved, concentrations of detectable chemical residues decreased by $10^3$ to $10^6$. A concentration that was acceptable in 1965 was $10^6$ times above the legal limit in 1995, regardless of whether there was any evidence that such a low level is harmful. In 1996, Congress changed the law so that the allowed level probably will be set at a concentration that produces less than one excess cancer per million persons exposed. Unfortunately, the scientific basis for predicting effects of low-level exposure on human health is slim.[4]

## 4-4  Comparison of Standard Deviations with the *F* Test

To decide whether Rayleigh's two sets of nitrogen masses in Figure 4-6 are "significantly" different from each other, we used the *t* test. If the standard deviations of two data sets are not significantly different from each other, then we use Equation 4-8 for the *t* test. If the standard deviations are significantly different, then we use Equation 4-8a instead.

The **F test** tells us whether two standard deviations are "significantly" different from each other. *F* is the quotient of the squares of the standard deviations:

$$F_{calculated} = \frac{s_1^2}{s_2^2} \tag{4-12}$$

We always put the larger standard deviation in the numerator so that $F \geq 1$. If $F_{calculated} > F_{table}$ in Table 4-4, then the difference is significant.

Use the *F* test for Case 2 in comparison of means in Section 4-3.

If $F_{calculated} > F_{table}$ (95%), the standard deviations are significantly different from each other.
If $F_{calculated} < F_{table}$, use Equation 4-8.
If $F_{calculated} > F_{table}$, use Equation 4-8a.

The square of the standard deviation is called the *variance*.

**Table 4-4**  Critical values of $F = s_1^2/s_2^2$ at 95% confidence level

| Degrees of freedom for $s_2$ | Degrees of freedom for $s_1$ | | | | | | | | | | | | | |
|---|---|---|---|---|---|---|---|---|---|---|---|---|---|---|
| | 2 | 3 | 4 | 5 | 6 | 7 | 8 | 9 | 10 | 12 | 15 | 20 | 30 | ∞ |
| 2 | 19.0 | 19.2 | 19.2 | 19.3 | 19.3 | 19.4 | 19.4 | 19.4 | 19.4 | 19.4 | 19.4 | 19.4 | 19.5 | 19.5 |
| 3 | 9.55 | 9.28 | 9.12 | 9.01 | 8.94 | 8.89 | 8.84 | 8.81 | 8.79 | 8.74 | 8.70 | 8.66 | 8.62 | 8.53 |
| 4 | 6.94 | 6.59 | 6.39 | 6.26 | 6.16 | 6.09 | 6.04 | 6.00 | 5.96 | 5.91 | 5.86 | 5.80 | 5.75 | 5.63 |
| 5 | 5.79 | 5.41 | 5.19 | 5.05 | 4.95 | 4.88 | 4.82 | 4.77 | 4.74 | 4.68 | 4.62 | 4.56 | 4.50 | 4.36 |
| 6 | 5.14 | 4.76 | 4.53 | 4.39 | 4.28 | 4.21 | 4.15 | 4.10 | 4.06 | 4.00 | 3.94 | 3.87 | 3.81 | 3.67 |
| 7 | 4.74 | 4.35 | 4.12 | 3.97 | 3.87 | 3.79 | 3.73 | 3.68 | 3.64 | 3.58 | 3.51 | 3.44 | 3.38 | 3.23 |
| 8 | 4.46 | 4.07 | 3.84 | 3.69 | 3.58 | 3.50 | 3.44 | 3.39 | 3.35 | 3.28 | 3.22 | 3.15 | 3.08 | 2.93 |
| 9 | 4.26 | 3.86 | 3.63 | 3.48 | 3.37 | 3.29 | 3.23 | 3.18 | 3.14 | 3.07 | 3.01 | 2.94 | 2.86 | 2.71 |
| 10 | 4.10 | 3.71 | 3.48 | 3.33 | 3.22 | 3.14 | 3.07 | 3.02 | 2.98 | 2.91 | 2.84 | 2.77 | 2.70 | 2.54 |
| 11 | 3.98 | 3.59 | 3.36 | 3.20 | 3.10 | 3.01 | 2.95 | 2.90 | 2.85 | 2.79 | 2.72 | 2.65 | 2.57 | 2.40 |
| 12 | 3.88 | 3.49 | 3.26 | 3.11 | 3.00 | 2.91 | 2.85 | 2.80 | 2.75 | 2.69 | 2.62 | 2.54 | 2.47 | 2.30 |
| 13 | 3.81 | 3.41 | 3.18 | 3.02 | 2.92 | 2.83 | 2.77 | 2.71 | 2.67 | 2.60 | 2.53 | 2.46 | 2.38 | 2.21 |
| 14 | 3.74 | 3.34 | 3.11 | 2.96 | 2.85 | 2.76 | 2.70 | 2.65 | 2.60 | 2.53 | 2.46 | 2.39 | 2.31 | 2.13 |
| 15 | 3.68 | 3.29 | 3.06 | 2.90 | 2.79 | 2.71 | 2.64 | 2.59 | 2.54 | 2.48 | 2.40 | 2.33 | 2.25 | 2.07 |
| 16 | 3.63 | 3.24 | 3.01 | 2.85 | 2.74 | 2.66 | 2.59 | 2.54 | 2.49 | 2.42 | 2.35 | 2.28 | 2.19 | 2.01 |
| 17 | 3.59 | 3.20 | 2.96 | 2.81 | 2.70 | 2.61 | 2.55 | 2.49 | 2.45 | 2.38 | 2.31 | 2.23 | 2.15 | 1.96 |
| 18 | 3.56 | 3.16 | 2.93 | 2.77 | 2.66 | 2.58 | 2.51 | 2.46 | 2.41 | 2.34 | 2.27 | 2.19 | 2.11 | 1.92 |
| 19 | 3.52 | 3.13 | 2.90 | 2.74 | 2.63 | 2.54 | 2.48 | 2.42 | 2.38 | 2.31 | 2.23 | 2.16 | 2.07 | 1.88 |
| 20 | 3.49 | 3.10 | 2.87 | 2.71 | 2.60 | 2.51 | 2.45 | 2.39 | 2.35 | 2.28 | 2.20 | 2.12 | 2.04 | 1.84 |
| 30 | 3.32 | 2.92 | 2.69 | 2.53 | 2.42 | 2.33 | 2.27 | 2.21 | 2.16 | 2.09 | 2.01 | 1.93 | 1.84 | 1.62 |
| ∞ | 3.00 | 2.60 | 2.37 | 2.21 | 2.10 | 2.01 | 1.94 | 1.88 | 1.83 | 1.75 | 1.67 | 1.57 | 1.46 | 1.00 |

**Example** Are the Two Standard Deviations of Rayleigh's Data Significantly Different from Each Other?

In Table 4-3, the larger standard deviation is $s_1 = 0.001\ 38$ ($n_1 = 8$ measurements) and the smaller standard deviation is $s_2 = 0.000\ 14_3$ ($n_2 = 7$ measurements).

**Solution** To answer the question, find $F$ with Equation 4-12:

$$F_{calculated} = \frac{s_1^2}{s_2^2} = \frac{(0.001\ 38)^2}{(0.000\ 14_3)^2} = 93._1$$

In Table 4-4, look for $F_{table}$ in the column with 7 degrees of freedom for $s_1$ (because degrees of freedom $= n - 1$) and the row with 6 degrees of freedom for $s_2$. *Because $F_{calculated}$ ($= 93.1$) $> F_{table}$ ($= 4.21$), the standard deviations are different from each other above the 95% confidence level.* The obvious difference in scatter of the two data sets in Figure 4-6 is highly significant.

## 4-5 *t* Tests with a Spreadsheet

Excel has built-in procedures for conducting tests with Student's *t*. To compare Rayleigh's two sets of results in Table 4-3, enter his data in columns B and C of a spreadsheet (Figure 4-8). In rows 13 and 14, we computed the averages and standard deviations, but we did not need to do this.

In the TOOLS menu, you might find DATA ANALYSIS. If not, select ADD-INS in the TOOLS menu and find ANALYSIS TOOLPACK. Put an x beside ANALYSIS TOOLPACK and click OK. DATA ANALYSIS will then be available in the TOOLS menu.

Returning to Figure 4-8, we want to know whether the mean values of the two sets of data are statistically the same or not. In the TOOLS menu, select DATA ANALYSIS. In the window that appears, select t-Test: Two-Sample Assuming Equal Variances. Click OK. The next window asks you in which cells the data are located. Write B5:B12 for Variable 1 and

| | A | B | C | D | E | F | G |
|---|---|---|---|---|---|---|---|
| 1 | Analysis of Rayleigh's Data | | | | t-Test: Two-Sample Assuming Equal Variances | | |
| 2 | | | | | | *Variable 1* | *Variable 2* |
| 3 | | Mass of gas (g) collected from | | | Mean | 2.310109 | 2.299473 |
| 4 | | air | chemical | | Variance | 2.03E-08 | 1.9E-06 |
| 5 | | 2.31017 | 2.30143 | | Observations | 7 | 8 |
| 6 | | 2.30986 | 2.29890 | | Pooled Variance | 1.03E-06 | |
| 7 | | 2.31010 | 2.29816 | | Hypothesized Mean Diff | 0 | |
| 8 | | 2.31001 | 2.30182 | | df | 13 | |
| 9 | | 2.31024 | 2.29869 | | t Stat | 20.21372 | |
| 10 | | 2.31010 | 2.29940 | | P(T<=t) one-tail | 1.66E-11 | |
| 11 | | 2.31028 | 2.29849 | | t Critical one-tail | 1.770932 | |
| 12 | | | 2.29889 | | P(T<=t) two-tail | 3.32E-11 | |
| 13 | Average | 2.31011 | 2.29947 | | t Critical two-tail | 2.160368 | |
| 14 | Std Dev | 0.00014 | 0.00138 | | | | |
| 15 | | | | | t-Test: Two-Sample Assuming Unequal Variances | | |
| 16 | B13 = AVERAGE(B5:B12) | | | | | *Variable 1* | *Variable 2* |
| 17 | B14 = STDEV(B5:B12) | | | | Mean | 2.310109 | 2.299473 |
| 18 | | | | | Variance | 2.03E-08 | 1.9E-06 |
| 19 | | | | | Observations | 7 | 8 |
| 20 | | | | | Hypothesized Mean Diff | 0 | |
| 21 | | | | | df | 7 | |
| 22 | | | | | t Stat | 21.68022 | |
| 23 | | | | | P(T<=t) one-tail | 5.6E-08 | |
| 24 | | | | | t Critical one-tail | 1.894578 | |
| 25 | | | | | P(T<=t) two-tail | 1.12E-07 | |
| 26 | | | | | t Critical two-tail | 2.364623 | |

*Figure 4-8* Spreadsheet for comparing mean values of Rayleigh's measurements in Table 4-3.

C5:C12 for Variable 2. The routine ignores the blank space in cell B12. For the Hypothesized Mean Difference enter 0 and for Alpha enter 0.05. Alpha is the level of probability to which we are testing the difference in the means. With Alpha = 0.05, we are at the 95% confidence level. For Output Range, select cell E1 and click OK.

Excel now goes to work and prints results in cells E1 to G13 of Figure 4-8. Mean values are in cells F3 and G3. Cells F4 and G4 give *variance,* which is the square of the standard deviation. Cell F6 gives *pooled variance* computed with Equation 4-9. That equation was painful to use by hand. Cell F8 shows degrees of freedom ($df = 13$) and $t_{calculated} = 20.2$ from Equation 4-8 appears in cell F9.

At this point in Section 4-3, we consulted Table 4-2 to find that $t_{table}$ lies between 2.228 and 2.131 for 95% confidence and 13 degrees of freedom. Excel gives us the critical value of $t$ (2.160) in cell F13 of Figure 4-8. Because $t_{calculated}$ (= 20.2) > $t_{table}$ (= 2.160), we conclude that the two means are not the same. The difference is significant. Cell F12 states that the probability of observing these two mean values and standard deviations by random chance if the mean values were really the same is $3.32 \times 10^{-11}$. The difference is *highly* significant. For any value of $P \leq 0.05$ in cell F12, we would reject the null hypothesis and conclude that the means are not the same.

Note 5 for this chapter (in the Notes and References section of the book) explains what is meant by 1-tail and 2-tail in the Excel output in Figure 4-8. We use the 2-tailed test in this book.

The *F* test in Equation 4-12 told us that the standard deviations of Rayleigh's two experiments are different. Therefore, we can select the other *t* test found in the TOOLS menu in the DATA ANALYSIS choices. Select t-Test: Two-Sample Assuming Unequal Variances and fill in the blanks exactly as before. Results based on Equations 4-8a and 4-9a are displayed in cells E15 to G26 of Figure 4-8. Just as we found in Section 4-3, the degrees of freedom are $df = 7$ (cell F21) and $t_{calculated} = 21.7$ (cell F22). Because $t_{calculated}$ is greater than the critical value of $t$ (2.36 in cell F26), we reject the null hypothesis and conclude that the two means *are* significantly different.

## ▪▪▪▪ 4-6  *Q* Test for Bad Data

Sometimes one datum is inconsistent with the remaining data. You can use the ***Q* test** to help decide whether to retain or discard a questionable datum. Consider the five results 12.53, 12.56, 12.47, 12.67, and 12.48. Is 12.67 a "bad point"? To apply the *Q* test, arrange the data in order of increasing value and calculate *Q,* defined as

$$Q_{calculated} = \frac{gap}{range} \qquad (4\text{-}13)$$

The *range* is the total spread of the data. The *gap* is the difference between the questionable point and the nearest value.

If $Q_{calculated} > Q_{table}$, the questionable point should be discarded. In the preceding example, $Q_{calculated} = 0.11/0.20 = 0.55$. In Table 4-5, we find $Q_{table} = 0.64$. *Because $Q_{calculated} < Q_{table}$, the questionable point should be retained.* There is more than a 10% chance that the value 12.67 is a member of the same population as the other four numbers.

Some people would argue that you should never discard a datum unless you know that there was an error in the procedure that led to that particular measurement. Others would repeat the questionable measurement several more times to gain higher confidence that one measurement is really out of line (or not). The decision is yours and it is subjective.

## ▪▪▪▪ 4-7  The Method of Least Squares

For most chemical analyses, the response of the procedure must be evaluated for known quantities of analyte (called *standards*) so that the response to an unknown quantity can be interpreted. For this purpose, we commonly prepare a **calibration curve,** such as the one for caffeine in Figure 0-7. Most often, we work in a region where the calibration curve is a straight line.

We use the **method of least squares** to draw the "best" straight line through experimental data points that have some scatter and do not lie perfectly on a straight line.[6] The best line will be such that some of the points lie above and some lie below the line. We will learn to

Table 4-5 is based on 90% confidence. If $Q_{calculated} > Q_{table}$, discard the questionable point.

**Table 4-5**  Values of *Q* for rejection of data

| *Q* (90% confidence)[a] | Number of observations |
|---|---|
| 0.76 | 4 |
| 0.64 | 5 |
| 0.56 | 6 |
| 0.51 | 7 |
| 0.47 | 8 |
| 0.44 | 9 |
| 0.41 | 10 |

*a. Q = gap/range. If $Q_{calculated} > Q_{table}$, the value in question can be rejected with 90% confidence.*

SOURCE: *R. B. Dean and W. J. Dixon, Anal. Chem.* **1951,** *23, 636; see also D. R. Rorabacher, Anal. Chem.* **1991,** *63, 139.*

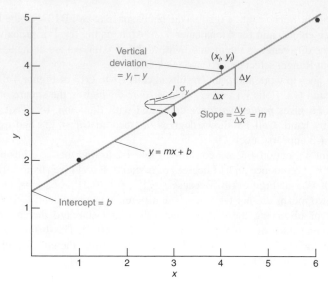

**Figure 4-9** Least-squares curve fitting. The points (1,2) and (6,5) do not fall exactly on the solid line, but they are too close to the line to show their deviations. The Gaussian curve drawn over the point (3,3) is a schematic indication of the fact that each value of $y_i$ is normally distributed about the straight line. That is, the most probable value of $y$ will fall on the line, but there is a finite probability of measuring $y$ some distance from the line.

estimate the uncertainty in a chemical analysis from the uncertainties in the calibration curve and in the measured response to replicate samples of unknown.

## Finding the Equation of the Line

The procedure we use assumes that the errors in the $y$ values are substantially greater than the errors in the $x$ values.[7] This condition is usually true in a calibration curve in which the experimental response ($y$ values) is less certain than the quantity of analyte ($x$ values). A second assumption is that uncertainties (standard deviations) in all the $y$ values are similar.

Suppose we seek to draw the best straight line through the points in Figure 4-9 by minimizing the vertical deviations between the points and the line. We minimize only the vertical deviations because we assume that uncertainties in $y$ values are much greater than uncertainties in $x$ values.

Let the equation of the line be

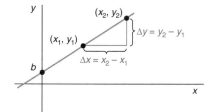

*Equation of straight line:* $$y = mx + b \qquad (4\text{-}14)$$

**Equation for a straight line:** $y = mx + b$

**Slope** ($m$) $= \dfrac{\Delta y}{\Delta x} = \dfrac{y_2 - y_1}{x_2 - x_1}$

*y*-Intercept ($b$) = crossing point on *y*-axis

in which $m$ is the **slope** and $b$ is the **y-intercept.** The vertical deviation for the point ($x_i$, $y_i$) in Figure 4-9 is $y_i - y$, where $y$ is the ordinate of the straight line when $x = x_i$.

$$\text{Vertical deviation} = d_i = y_i - y = y_i - (mx_i + b) \qquad (4\text{-}15)$$

Some of the deviations are positive and some are negative. Because we wish to minimize the magnitude of the deviations irrespective of their signs, we square all the deviations so that we are dealing only with positive numbers:

$$d_i^2 = (y_i - y)^2 = (y_i - mx_i - b)^2$$

Because we minimize the squares of the deviations, this is called the *method of least squares.* It can be shown that minimizing the squares of the deviations (rather than simply their magnitudes) corresponds to assuming that the set of $y$ values is the most probable set.

Finding values of $m$ and $b$ that minimize the sum of the squares of the vertical deviations involves some calculus, which we omit. We will express the final solution for slope and intercept in terms of *determinants,* which summarize certain arithmetic operations. The

To evaluate the determinant, multiply the diagonal elements $e \times h$ and then subtract the product of the other diagonal elements $f \times g$.

**determinant** $\begin{vmatrix} e & f \\ g & h \end{vmatrix}$ represents the value $eh - fg$. So, for example,

$$\begin{vmatrix} 6 & 5 \\ 4 & 3 \end{vmatrix} = (6 \times 3) - (5 \times 4) = -2$$

**Table 4-6** Calculations for least-squares analysis

| $x_i$ | $y_i$ | $x_i y_i$ | $x_i^2$ | $d_i (= y_i - mx_i - b)$ | $d_i^2$ |
|---|---|---|---|---|---|
| 1 | 2 | 2 | 1 | 0.038 46 | 0.001 479 3 |
| 3 | 3 | 9 | 9 | −0.192 31 | 0.036 982 |
| 4 | 4 | 16 | 16 | 0.192 31 | 0.036 982 |
| 6 | 5 | 30 | 36 | −0.038 46 | 0.001 479 3 |
| $\Sigma x_i = 14$ | $\Sigma y_i = 14$ | $\Sigma(x_i y_i) = 57$ | $\Sigma(x_i^2) = 62$ | | $\Sigma(d_i^2) = 0.076\ 923$ |

The slope and the intercept of the "best" straight line are found to be

$$\text{Least-squares} \left\{ \begin{array}{ll} \text{Slope:} & m = \begin{vmatrix} \Sigma(x_i y_i) & \Sigma x_i \\ \Sigma y_i & n \end{vmatrix} \div D \qquad (4\text{-}16) \\[2em] \text{Intercept:} & b = \begin{vmatrix} \Sigma(x_i^2) & \Sigma(x_i y_i) \\ \Sigma x_i & \Sigma y_i \end{vmatrix} \div D \qquad (4\text{-}17) \end{array} \right.$$

Translation of least-squares equations:

$$m = \frac{n\Sigma(x_i y_i) - \Sigma x_i \Sigma y_i}{n\Sigma(x_i^2) - (\Sigma x_i)^2}$$

$$b = \frac{\Sigma(x_i^2)\Sigma y_i - (\Sigma x_i y_i)\Sigma x_i}{n\Sigma(x_i^2) - (\Sigma x_i)^2}$$

where $D$ is

$$D = \begin{vmatrix} \Sigma(x_i^2) & \Sigma x_i \\ \Sigma x_i & n \end{vmatrix} \qquad (4\text{-}18)$$

and $n$ is the number of points.

Let's use these equations to find the slope and intercept of the best straight line through the four points in Figure 4-9. The work is set out in Table 4-6. Noting that $n = 4$ and putting the various sums into the determinants in Equations 4-16, 4-17, and 4-18 gives

$$m = \begin{vmatrix} 57 & 14 \\ 14 & 4 \end{vmatrix} \div \begin{vmatrix} 62 & 14 \\ 14 & 4 \end{vmatrix} = \frac{(57 \times 4) - (14 \times 14)}{(62 \times 4) - (14 \times 14)} = \frac{32}{52} = 0.615\ 38$$

$$b = \begin{vmatrix} 62 & 57 \\ 14 & 14 \end{vmatrix} \div \begin{vmatrix} 62 & 14 \\ 14 & 4 \end{vmatrix} = \frac{(62 \times 14) - (57 \times 14)}{(62 \times 4) - (14 \times 14)} = \frac{70}{52} = 1.346\ 15$$

The equation of the best straight line through the points in Figure 4-9 is therefore

$$y = 0.615\ 38x + 1.346\ 15$$

We tackle the question of significant figures for $m$ and $b$ in the next section.

### Example 🖩 Finding Slope and Intercept with a Spreadsheet

Your scientific calculator has a procedure for computing the slope and intercept of a set of $(x,y)$ data, and you should learn how to use that procedure. Alternatively, Excel has functions called SLOPE and INTERCEPT whose use is illustrated here:

| | A | B | C | D | E | F |
|---|---|---|---|---|---|---|
| 1 | x | y | | | Formulas: | |
| 2 | 1 | 2 | | slope = | | |
| 3 | 3 | 3 | | 0.61538 | D3 = SLOPE(B2:B5,A2:A5) | |
| 4 | 4 | 4 | | intercept = | | |
| 5 | 6 | 5 | | 1.34615 | D5 = INTERCEPT(B2:B5,A2:A5) | |

The slope in cell D3 is computed with the formula "= SLOPE(B2:B5,A2:A5)", where B2:B5 is the range containing the $y$ values and A2:A5 is the range containing $x$ values.

### How Reliable Are Least-Squares Parameters?

To estimate the uncertainties (expressed as standard deviations) in the slope and intercept, an uncertainty analysis must be performed on Equations 4-16 and 4-17. Because the uncertainties in $m$ and $b$ are related to the uncertainty in measuring each value of $y$, we first estimate the standard deviation that describes the population of $y$ values. This standard deviation, $\sigma_y$, characterizes the little Gaussian curve inscribed in Figure 4-9.

We estimate $\sigma_y$, the population standard deviation of all $y$ values, by calculating $s_y$, the standard deviation, for the four measured values of $y$. The deviation of each value of $y_i$ from the center of its Gaussian curve is $d_i = y_i - y = y_i - (mx_i + b)$. The standard deviation of these vertical deviations is

Equation 4-19 is analogous to Equation 4-2.

$$\sigma_y \approx s_y = \sqrt{\frac{\sum (d_i - \bar{d})^2}{(\text{degrees of freedom})}} \qquad (4\text{-}19)$$

But the average deviation, $\bar{d}$, is 0 for the best straight line, so the numerator of Equation 4-19 reduces to $\sum(d_i^2)$.

The *degrees of freedom* is the number of independent pieces of information available. For $n$ data points, there are $n$ degrees of freedom. If you were calculating the standard deviation of $n$ points, you would first find the average to use in Equation 4-2. This leaves $n - 1$ degrees of freedom in Equation 4-2 because only $n - 1$ pieces of information are available in addition to the average. If you know $n - 1$ values and you also know their average, then the $n$th value is fixed and you can calculate it.

For Equation 4-19, we began with $n$ points. Two degrees of freedom were lost in determining the slope and the intercept. Therefore, $n - 2$ degrees of freedom remain. Equation 4-19 becomes

If you know $\bar{x}$ and $n - 1$ of the individual values, you can calculate the $n$th value. Therefore, the problem has just $n - 1$ degrees of freedom once $\bar{x}$ is known.

$$s_y = \sqrt{\frac{\sum(d_i^2)}{n - 2}} \qquad (4\text{-}20)$$

where $d_i$ is given by Equation 4-15.

Uncertainty analysis for Equations 4-16 and 4-17 leads to the following results:

$$\text{Standard deviation of slope and intercept} \begin{cases} s_m^2 = \dfrac{s_y^2 n}{D} & (4\text{-}21) \\[2mm] s_b^2 = \dfrac{s_y^2 \sum(x_i^2)}{D} & (4\text{-}22) \end{cases}$$

where $s_m$ is an estimate of the standard deviation of the slope, $s_b$ is an estimate of the standard deviation of the intercept, $s_y$ is given by Equation 4-20, and $D$ is given by Equation 4-18.

At last, we can assign significant figures to the slope and the intercept in Figure 4-9. In Table 4-6, we see that $\sum(d_i^2) = 0.076\ 923$. Putting this number into Equation 4-20 gives

$$s_y^2 = \frac{0.076\ 923}{4 - 2} = 0.038\ 462$$

Now, we can plug numbers into Equation 4-21 and 4-22 to find

$$s_m^2 = \frac{s_y^2 n}{D} = \frac{(0.038\ 462)(4)}{52} = 0.002\ 958\ 6 \Rightarrow s_m = 0.054\ 39$$

$$s_b^2 = \frac{s_y^2 \sum(x_i^2)}{D} = \frac{(0.038\ 462)(62)}{52} = 0.045\ 859 \Rightarrow s_b = 0.214\ 15$$

Combining the results for $m$, $s_m$, $b$, and $s_b$, we write

The first digit of the uncertainty is the last significant figure. We often retain extra, insignificant digits to prevent round-off errors in further calculations.

$$\text{Slope:} \quad \frac{0.615\ 38}{\pm 0.054\ 39} = 0.62 \pm 0.05 \quad \text{or} \quad 0.61_5 \pm 0.05_4 \qquad (4\text{-}23)$$

$$\text{Intercept:} \quad \frac{1.346\ 15}{\pm 0.214\ 15} = 1.3 \pm 0.2 \quad \text{or} \quad 1.3_5 \pm 0.2_1 \qquad (4\text{-}24)$$

where the uncertainties represent one standard deviation. *The first decimal place of the standard deviation is the last significant figure of the slope or intercept.* Many scientists write results such as $1.35 \pm 0.21$ to avoid excessive round-off.

The 95% confidence interval for the slope is
$$\pm t s_m = \pm(4.303)(0.054) = \pm 0.23$$
based on degrees of freedom $= n - 2 = 2$. The confidence interval is $\pm t s_m$, not $\pm t s_m / \sqrt{n}$, because $\sqrt{n}$ is already implicit in $s_m$.

If you want to express the uncertainty as a confidence interval instead of one standard deviation, multiply the uncertainties in Equations 4-23 and 4-24 by the appropriate value of Student's $t$ from Table 4-2 for $n - 2$ degrees of freedom.

### Example    Finding $s_y$, $s_m$, and $s_b$ with a Spreadsheet

The Excel function LINEST returns the slope and intercept and their uncertainties in a table (a *matrix*). As an example, enter $x$ and $y$ values in columns A and B. Then highlight the 3-row × 2-column region E3:F5 with your mouse. This block of cells is selected to contain the output of the LINEST function. Under the INSERT menu, select FUNCTION. In the window that appears, go to Statistical and double click on LINEST. The new window asks for four inputs to the function. For $y$ values, enter B2:B5. Then enter A2:A5 for $x$ values. The next two entries are both "TRUE". The first TRUE tells Excel that we want to compute the $y$-intercept of the least-squares line and not force the intercept to be 0. The second TRUE tells Excel to print out the standard deviations as well as the slope and intercept. The formula you have just entered is "= LINEST(B2:B5,A2:A5,TRUE,TRUE)". Click OK and the slope appears in cell E3.

|    | A | B | C | D | E | F | G |
|----|---|---|---|---|---|---|---|
| 1  | x | y |   |   | Output from LINEST | | |
| 2  | 1 | 2 |   |   | Slope | Intercept | |
| 3  | 3 | 3 |   | Parameter | 0.61538 | 1.34615 | |
| 4  | 4 | 4 |   | Std Dev | 0.05439 | 0.21414 | |
| 5  | 6 | 5 |   | R^2 | 0.98462 | 0.19612 | Std Dev (y) |
| 6  |   |   |   |   |   |   |   |
| 7  | Highlight cells E3:F5 | | | | | | |
| 8  | Type "= LINEST(B2:B5,A2:A5,TRUE,TRUE)" | | | | | | |
| 9  | Press CTRL+SHIFT+ENTER (on PC) | | | | | | |
| 10 | Press COMMAND+RETURN (on Mac) | | | | | | |

The output of LINEST should be a matrix, not a single number. What went wrong? To tell the computer that you want a matrix, go back and highlight cells E3:F5. "= LINEST(B2:B5,A2:A5,TRUE,TRUE)" appears once again in the formula line. Now press CONTROL+SHIFT+ENTER on a PC or COMMAND(⌘)+RETURN on a Mac. Excel dutifully prints out a matrix in cells E3:F5. Write labels around the block to indicate what is in each cell. The slope and intercept are on the top line. The second line contains $s_m$ and $s_b$. Cell F5 contains $s_y$ and cell E5 contains a quantity called $R^2$, which is defined in Equation 5-2 and is a measure of the goodness of fit of the data to the line. The closer $R^2$ is to unity, the better the fit.

## ▮▮▮▮  4-8  Calibration Curves

A *calibration curve* shows the response of an analytical method to known quantities of analyte.[8] Table 4-7 gives real data from a protein analysis that produces a colored product. A *spectrophotometer* measures the absorbance of light, which is proportional to the quantity of protein analyzed. Solutions containing known concentrations of analyte are called **standard solutions.** Solutions containing all the reagents and solvents used in the analysis, but no deliberately added analyte, are called **blank solutions.** Blanks measure the response of the analytical procedure to impurities or interfering species in the reagents.

When we scan across the three absorbance values in each row of Table 4-7, the number 0.392 seems out of line: It is inconsistent with the other values for 15.0 μg, and the range of values for the 15.0-μg samples is much bigger than the range for the other samples. The linear relation between the average values of absorbance up to the 20.0-μg sample also indicates that the value 0.392 is in error (Figure 4-10). We choose to omit 0.392 from subsequent calculations.

It is reasonable to ask whether all three absorbances for the 25.0-μg samples are low for some unknown reason, because this point falls below the straight line in Figure 4-10. Repetition of this analysis shows that the 25.0-μg point is consistently below the straight line and there is nothing "wrong" with the data in Table 4-7.

Sections 18-1 and 18-2 discuss absorption of light and define the term *absorbance*. We will use concepts from these two sections throughout this book. You may want to read these sections for background.

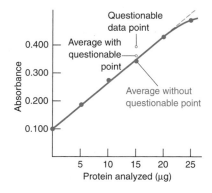

**Figure 4-10** Average absorbance values in Table 4-7 versus micrograms of protein analyzed. Averages for 0 to 20 μg of protein lie on a straight line if the questionable datum 0.392 at 15 μm is omitted.

**Table 4-7**  Spectrophotometer data used to construct calibration curve

| Amount of protein (μg) | Absorbance of independent samples | | | Range | Corrected absorbance | | |
|---|---|---|---|---|---|---|---|
| 0    | 0.099 | 0.099 | 0.100   | 0.001 | $-0.003_3$ | $-0.003_3$ | $0.000_7$ |
| 5.0  | 0.185 | 0.187 | 0.188   | 0.003 | $0.085_7$  | $0.087_7$  | $0.088_7$ |
| 10.0 | 0.282 | 0.272 | 0.272   | 0.010 | $0.182_7$  | $0.172_7$  | $0.172_7$ |
| 15.0 | 0.345 | 0.347 | (0.392) | 0.047 | $0.245_7$  | $0.247_7$  | —         |
| 20.0 | 0.425 | 0.425 | 0.430   | 0.005 | $0.325_7$  | $0.325_7$  | $0.330_7$ |
| 25.0 | 0.483 | 0.488 | 0.496   | 0.013 | $0.383_7$  | $0.388_7$  | $0.396_7$ |

## Constructing a Calibration Curve

We adopt the following procedure for constructing a calibration curve:

**Step 1** Prepare known samples of analyte covering a range of concentrations expected for unknowns. Measure the response of the analytical procedure to these standards to generate data like the left half of Table 4-7.

Absorbance of the blank can arise from the color of starting reagents, reactions of impurities, and reactions of interfering species. Blank values can vary from one set of reagents to another, but corrected absorbance should not.

**Step 2** Subtract the average absorbance ($0.099_3$) of the *blank* samples from each measured absorbance to obtain *corrected absorbance*. The blank measures the response of the procedure when no protein is present.

**Step 3** Make a graph of corrected absorbance versus quantity of protein analyzed (Figure 4-11). Use the least-squares procedure to find the best straight line through the linear portion of the data, up to and including 20.0 μg of protein (14 points, including the 3 corrected blanks, in the shaded portion of Table 4-7). Find the slope and intercept and uncertainties with Equations 4-16, 4-17, 4-20, 4-21, and 4-22. The results are

$$m = 0.016\ 3_0 \qquad s_m = 0.000\ 2_2 \qquad s_y = 0.005_9$$

$$b = 0.004_7 \qquad s_b = 0.002_6$$

Equation of calibration line:

$$y(\pm s_y) = [m(\pm s_m)]x + [b(\pm s_b)]$$

The equation of the linear calibration line is

$$\underbrace{\text{absorbance}}_{y} = m \times \underbrace{(\text{μg of protein})}_{x} + b$$

$$= (0.016\ 3_0)(\text{μg of protein}) + 0.004_7 \qquad (4\text{-}25)$$

where $y$ is the corrected absorbance (= observed absorbance − blank absorbance).

**Step 4** If you analyze an unknown solution at a future time, run a blank at the same time. Subtract the new blank absorbance from the unknown absorbance to obtain the corrected absorbance.

> **Example** Using a Linear Calibration Curve
>
> An unknown protein sample gave an absorbance of 0.406, and a blank had an absorbance of 0.104. How many micrograms of protein are in the unknown?
>
> **Solution** The corrected absorbance is $0.406 - 0.104 = 0.302$, which lies on the linear portion of the calibration curve in Figure 4-11. Equation 4-25 therefore becomes
>
> $$\text{μg of protein} = \frac{\text{absorbance} - 0.004_7}{0.016\ 3_0} = \frac{0.302 - 0.004_7}{0.016\ 3_0} = 18.2_4\ \text{μg} \qquad (4\text{-}26)$$

We prefer calibration procedures with a **linear response,** in which the corrected analytical signal (= signal from sample − signal from blank) is proportional to the quantity of analyte. Although we try to work in the linear range, you can obtain valid results beyond the

*Figure 4-11* Calibration curve for protein analysis in Table 4-7. The equation of the solid straight line fitting the 14 data points (open circles) from 0 to 20 μg, derived by the method of least squares, is $y = 0.016\ 3_0\ (\pm 0.000\ 2_2)x + 0.004_7\ (\pm 0.002_6)$. The standard deviation of $y$ is $s_y = 0.005_9$. The equation of the dashed quadratic curve that fits all 17 data points from 0 to 25 μg, determined by a nonlinear least-squares procedure[6] is $y = -1.1_7\ (\pm 0.2_1) \times 10^{-4}x^2 + 0.018\ 5_8\ (\pm 0.000\ 4_6)x - 0.000\ 7\ (\pm 0.001\ 0)$, with $s_y = 0.004_6$.

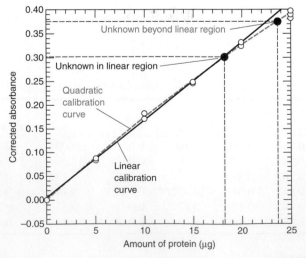

Consider an unknown whose corrected absorbance of 0.375 lies beyond the linear region in Figure 4-11. We can fit all the data points with the quadratic equation[6]

$$y = -1.17 \times 10^{-4}x^2 + 0.018\,58x - 0.000\,7 \qquad \text{(A)}$$

To find the quantity of protein, substitute the measured absorbance into Equation A:

$$0.375 = -1.17 \times 10^{-4}x^2 + 0.018\,58x - 0.000\,7$$

This equation can be rearranged to

$$1.17 \times 10^{-4}x^2 - 0.018\,58x + 0.375\,7 = 0$$

which is a quadratic equation of the form

$$ax^2 + bx + c = 0$$

whose two possible solutions are

$$x = \frac{-b + \sqrt{b^2 - 4ac}}{2a} \qquad x = \frac{-b - \sqrt{b^2 - 4ac}}{2a}$$

Substituting $a = 1.17 \times 10^{-4}$, $b = -0.018\,58$, and $c = 0.375\,7$ into these equations gives

$$x = 135\ \mu g \qquad x = 23.8\ \mu g$$

Figure 4-11 tells us that the correct choice is 23.8 μg, not 135 μg.

---

linear region (>20 μg) in Figure 4-11. The dashed curve that goes up to 25 μg of protein comes from a least-squares fit of the data to the equation $y = ax^2 + bx + c$ (Box 4-2).

The **linear range** of an analytical method is the analyte concentration range over which response is proportional to concentration. A related quantity defined in Figure 4-12 is **dynamic range**—the concentration range over which there is a measurable response to analyte, even if the response is not linear.

Before using your calculator or computer to find the least-squares straight line, make a graph of your data. The graph gives you an opportunity to reject bad data or stimulus to repeat a measurement or to decide that a straight line is not an appropriate function. *Examine your data for sensibility.*

It is not reliable to extrapolate any calibration curve, linear or nonlinear, beyond the measured range of standards. Measure standards in the entire concentration range of interest.

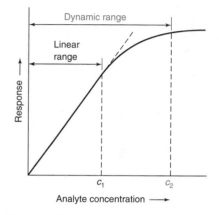

**Figure 4-12** Calibration curve illustrating linear and dynamic ranges.

### Propagation of Uncertainty with a Calibration Curve

In the preceding example, an unknown with a corrected absorbance of $y = 0.302$ had a protein content of $x = 18.2_4\ \mu g$. What is the uncertainty in the number $18.2_4$? A full treatment of the propagation of uncertainty gives the following results:[1,9]

$$\text{Uncertainty in } x\ (= s_x) = \frac{s_y}{|m|} \sqrt{\frac{1}{k} + \frac{1}{n} + \frac{(y - \bar{y})^2}{m^2 \Sigma (x_i - \bar{x})^2}} \qquad \text{(4-27)}$$

$y$ = absorbance of unknown = 0.302
$x_i$ = μg of protein in standards in Table 4-7
  = (0, 0, 0, 5.0, 5.0, 5.0, 10.0, 10.0, 10.0, 15.0, 15.0, 20.0, 20.0, 20.0)
$\bar{y}$ = average of 14 $y$ values = $0.161_8$
$\bar{x}$ = average of 14 $x$ values = $9.64_3\ \mu g$

where $s_y$ is the standard deviation of $y$ (Equation 4-20), $|m|$ is the absolute value of the slope, $k$ is the number of replicate measurements of the unknown, $n$ is the number of data points for the calibration line (14 in Table 4-7), $\bar{y}$ is the mean value of $y$ for the points on the calibration line, $x_i$ are the individual values of $x$ for the points on the calibration line, and $\bar{x}$ is the mean value of $x$ for the points on the calibration line. For a single measurement of the unknown, $k = 1$ and Equation 4-27 gives $s_x = \pm 0.3_9\ \mu g$. If you measure four replicate unknowns ($k = 4$) and the average corrected absorbance is 0.302, the uncertainty is reduced from $\pm 0.3_9$ to $\pm 0.2_3\ \mu g$.

The confidence interval for $x$ is $\pm t s_x$, where $t$ is Student's $t$ (Table 4-2) for $n - 2$ degrees of freedom. If $s_x = 0.2_3\ \mu g$ and $n = 14$ points (12 degrees of freedom), the 95% confidence interval for $x$ is $\pm t s_x = \pm (2.179)(0.2_3) = \pm 0.5_0\ \mu g$.

To find values of $t$ that are not in Table 4-2, use the Excel function TINV. For 12 degrees of freedom and 95% confidence, the function TINV(0.05,12) returns $t = 2.179$.

## 4-9   A Spreadsheet for Least Squares

Figure 4-13 implements least-squares analysis, including propagation of error with Equation 4-27. Enter values of $x$ and $y$ in columns B and C. Then select cells B10:C12. Enter the formula = LINEST(C4:C7,B4:B7,TRUE,TRUE) and press CONTROL+SHIFT+ENTER on a PC or COMMAND⌘+RETURN on a Mac. LINEST returns $m$, $b$, $s_m$, $s_b$, $R^2$, and $s_y$ in cells B10:C12. Write labels in cells A10:A12 and D10:D12 so you know what the numbers in cells B10:C12 mean.

Cell B14 gives the number of data points with the formula = COUNT(B4:B7). Cell B15 computes the mean value of $y$. Cell B16 computes the sum $\Sigma (x_i - \bar{x})^2$ that we need for Equation 4-27. This sum is common enough that Excel has a built in function called DEVSQ that you can find in the Statistics menu of the INSERT FUNCTION menu.

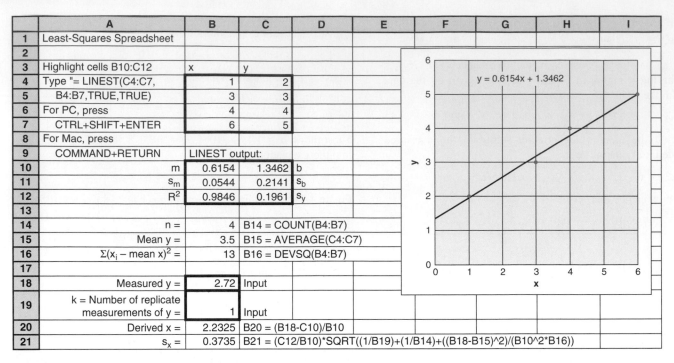

| | A | B | C | D | E | F | G | H | I |
|---|---|---|---|---|---|---|---|---|---|
| 1 | Least-Squares Spreadsheet | | | | | | | | |
| 2 | | | | | | | | | |
| 3 | Highlight cells B10:C12 | x | y | | | | | | |
| 4 | Type "= LINEST(C4:C7, | 1 | 2 | | | | | | |
| 5 | B4:B7,TRUE,TRUE) | 3 | 3 | | | | | | |
| 6 | For PC, press | 4 | 4 | | | | | | |
| 7 | CTRL+SHIFT+ENTER | 6 | 5 | | | | | | |
| 8 | For Mac, press | | | | | | | | |
| 9 | COMMAND+RETURN | LINEST output: | | | | | | | |
| 10 | m | 0.6154 | 1.3462 | b | | | | | |
| 11 | $s_m$ | 0.0544 | 0.2141 | $s_b$ | | | | | |
| 12 | $R^2$ | 0.9846 | 0.1961 | $s_y$ | | | | | |
| 13 | | | | | | | | | |
| 14 | n = | 4 | B14 = COUNT(B4:B7) | | | | | | |
| 15 | Mean y = | 3.5 | B15 = AVERAGE(C4:C7) | | | | | | |
| 16 | $\Sigma(x_i - \text{mean } x)^2 =$ | 13 | B16 = DEVSQ(B4:B7) | | | | | | |
| 17 | | | | | | | | | |
| 18 | Measured y = | 2.72 | Input | | | | | | |
| 19 | k = Number of replicate measurements of y = | 1 | Input | | | | | | |
| 20 | Derived x = | 2.2325 | B20 = (B18-C10)/B10 | | | | | | |
| 21 | $s_x =$ | 0.3735 | B21 = (C12/B10)*SQRT((1/B19)+(1/B14)+((B18-B15)^2)/(B10^2*B16)) | | | | | | |

**Figure 4-13** Spreadsheet for linear least-squares analysis.

Enter the measured mean value of $y$ for replicate measurements of the unknown in cell B18. In cell B19, enter the number of replicate measurements of the unknown. Cell B20 computes the value of $x$ corresponding to the measured mean value of $y$. Cell B21 uses Equation 4-27 to find the uncertainty (the standard deviation) in the value of $x$ for the unknown. If you want a confidence interval for $x$, multiply $s_x$ times Student's $t$ from Table 4-2 for $n - 2$ degrees of freedom and the desired confidence level.

We always want a graph to see if the calibration points lie on a straight line. Follow the instructions in Section 2-11 to plot the calibration data. To add a straight line, click on one data point and they will all be highlighted. Go to the CHART menu and select ADD TRENDLINE. In some versions of Excel there is no CHART menu. In this case, go to the INSERT menu and select TRENDLINE. In the window that appears, select Linear. Go to Options in the TRENDLINE box and select Display Equation on Chart. When you click OK, the least-squares straight line and its equation appear on the graph. Double click on the line and you can adjust its thickness and appearance. Double clicking on the equation allows you to modify its format. Double click on the straight line and select Options. In the Forecast box, you can extend the trendline Forward and Backward as far as you like.

95% confidence interval for $x$ in Figure 4-13:

$x \pm ts_x = 2.232\ 5 \pm (4.303)(0.373\ 5)$
$= 2.2 \pm 1.6$

(degrees of freedom $= n - 2 = 2$)

## Terms to Understand

| | | | |
|---|---|---|---|
| average | dynamic range | mean | Student's $t$ |
| blank solution | $F$ test | method of least squares | $t$ test |
| calibration curve | Gaussian distribution | $Q$ test | variance |
| confidence interval | intercept | slope | |
| degrees of freedom | linear range | standard deviation | |
| determinant | linear response | standard solution | |

## Summary

The results of many measurements of an experimental quantity follow a Gaussian distribution. The measured mean, $\bar{x}$, approaches the true mean, $\mu$, as the number of measurements becomes very large. The broader the distribution, the greater is $\sigma$, the standard deviation. For $n$ measurements, an estimate of the standard deviation is given by $s = \sqrt{[\Sigma(x_i - \bar{x})^2]/(n - 1)}$. About two-thirds of all measurements lie within $\pm 1\sigma$, and 95% lie within $2\sigma$. The probability of observing a value within a certain interval is proportional to the area of that interval, given in Table 4-1.

After you select a confidence level, Student's $t$ is used to find confidence intervals ($\mu = \bar{x} \pm ts/\sqrt{n}$) and to compare mean values measured by different methods. The $F$ test is used to decide whether two standard deviations are significantly different from each other. The $Q$ test helps you to decide whether or not a questionable datum should be discarded. It is best to repeat the measurement several times to increase the probability that your decision is correct.

A calibration curve shows the response of a chemical analysis to known quantities (standard solutions) of analyte. When there is a linear response, the corrected analytical signal (= signal from sample − signal from blank) is proportional to the quantity of analyte. Blank solutions are prepared from the same reagents and solvents used to prepare standards and unknowns, but blanks have no intentionally added analyte. The blank tells us the response of the procedure to impurities or interfering species in the reagents. The blank value is subtracted from measured values of standards prior to constructing the calibration curve. The blank value is subtracted from the response of an unknown prior to computing the quantity of analyte in the unknown.

The method of least squares is used to determine the equation of the "best" straight line through experimental data points. Equations 4-16 to 4-18 and 4-20 to 4-22 provide the least-squares slope and intercept and their standard deviations. Equation 4-27 estimates the uncertainty in $x$ from a measured value of $y$ with a calibration curve. A spreadsheet greatly simplifies least-squares calculations.

## Exercises

**4-A.** For the numbers 116.0, 97.9, 114.2, 106.8, and 108.3, find the mean, standard deviation, range, and 90% confidence interval for the mean. Using the $Q$ test, decide whether the number 97.9 should be discarded.

**4-B.** 🖳 *Spreadsheet for standard deviation.* Let's create a spreadsheet to compute the mean and standard deviation of a column of numbers in two different ways. The spreadsheet below is a template for this exercise.

(a) Reproduce the template on your spreadsheet. Cells B4 to B8 contain the data ($x$ values) whose mean and standard deviation we will compute.

(b) Write a formula in cell B9 to compute the sum of numbers in B4 to B8.

(c) Write a formula in cell B10 to compute the mean value.

(d) Write a formula in cell C4 to compute ($x$ − mean), where $x$ is in cell B4 and the mean is in cell B10. Use the FILL DOWN command to compute values in cells C5 to C8.

(e) Write a formula in cell D4 to compute the square of the value in cell C4. Use the FILL DOWN command to compute values in cells D5 to D8.

(f) Write a formula in cell D9 to compute the sum of the numbers in cells D4 to D8.

(g) Write a formula in cell B11 to compute the standard deviation.

(h) Use cells B13 to B18 to document your formulas.

(i) Now we are going to simplify life by using formulas built into the spreadsheet. In cell B21 type "= SUM(B4:B8)", which means find the sum of numbers in cells B4 to B8. Cell B21 should display the same number as cell B9. In general, you will not know what functions are available and how to write them. Find the FUNCTION menu (under the INSERT menu in Excel) and find SUM in this menu.

(j) Select cell B22. Go to the FUNCTION menu and find AVERAGE. When you type "= AVERAGE(B4:B8)" in cell B22, its value should be the same as B10.

(k) For cell B23, find the standard deviation function ("=STDEV(B4:B8)") and see that the value agrees with cell B11.

**4-C.** Use Table 4-1 for this exercise. Suppose that the mileage at which 10 000 sets of automobile brakes had been 80% worn through was recorded. The average was 62 700, and the standard deviation was 10 400 miles.

(a) What fraction of brakes is expected to be 80% worn in less than 40 860 miles?

(b) What fraction is expected to be 80% worn at a mileage between 57 500 and 71 020 miles?

**4-D.** 🖳 Use the NORMDIST spreadsheet function to answer these questions about the brakes described in Exercise 4-C:

(a) What fraction of brakes is expected to be 80% worn in less than 45 800 miles?

(b) What fraction is expected to be 80% worn at a mileage between 60 000 and 70 000 miles?

**4-E.** A reliable assay shows that the ATP (adenosine triphosphate) content of a certain cell type is 111 μmol/100 mL. You developed a new assay, which gave the following values for replicate analyses: 117, 119, 111, 115, 120 μmol/100 mL (average $= 116._4$). Can you be 95% confident that your result differs from the "known" value?

**4-F.** Traces of toxic, man-made hexachlorohexanes in North Sea sediments were extracted by a known process and by two new procedures, and measured by chromatography.[10]

| Method | Concentration found (pg/g) | Standard deviation (pg/g) | Number of replications |
|---|---|---|---|
| Conventional | 34.4 | 3.6 | 6 |
| Procedure A | 42.9 | 1.2 | 6 |
| Procedure B | 51.1 | 4.6 | 6 |

(a) Are the concentrations parts per million, parts per billion, or something else?

(b) Is the standard deviation for procedure B significantly different from that of the conventional procedure?

| | A | B | C | D |
|---|---|---|---|---|
| 1 | Computing standard deviation | | | |
| 2 | | | | |
| 3 | | Data = x | x-mean | (x-mean)^2 |
| 4 | | 17.4 | | |
| 5 | | 18.1 | | |
| 6 | | 18.2 | | |
| 7 | | 17.9 | | |
| 8 | | 17.6 | | |
| 9 | sum = | | | |
| 10 | mean = | | | |
| 11 | std dev = | | | |
| 12 | | | | |
| 13 | Formulas: | B9 = | | |
| 14 | | B10 = | | |
| 15 | | B11 = | | |
| 16 | | C4 = | | |
| 17 | | D4 = | | |
| 18 | | D9 = | | |
| 19 | | | | |
| 20 | Calculations using built-in functions: | | | |
| 21 | sum = | | | |
| 22 | mean = | | | |
| 23 | std dev = | | | |

Spreadsheet for Exercise 4-B.

(c) Is the mean concentration found by procedure B significantly different from that of the conventional procedure?

(d) Answer the same two questions as parts (b) and (c) to compare procedure A with the conventional procedure.

4-G. 📊 Calibration curve. (You can do this exercise with your calculator, but it is more easily done by the spreadsheet in Figure 4-13). In the Bradford protein determination, the color of a dye changes from brown to blue when it binds to protein. Absorbance of light is measured.

| Protein (µg): | 0.00 | 9.36 | 18.72 | 28.08 | 37.44 |
|---|---|---|---|---|---|
| Absorbance at 595 nm: | 0.466 | 0.676 | 0.883 | 1.086 | 1.280 |

(a) Find the equation of the least-squares straight line through these points in the form $y = [m(\pm s_m)]x + [b(+s_b)]$ with a reasonable number of significant figures.

(b) Make a graph showing the experimental data and the calculated straight line.

(c) An unknown protein sample gave an absorbance of 0.973. Calculate the number of micrograms of protein in the unknown and estimate its uncertainty.

---

## Problems

### Gaussian Distribution

**4-1.** What is the relation between the standard deviation and the precision of a procedure? What is the relation between standard deviation and accuracy?

**4-2.** Use Table 4-1 to state what fraction of a Gaussian population lies within the following intervals:

(a) $\mu \pm \sigma$   (b) $\mu \pm 2\sigma$   (c) $\mu$ to $+\sigma$   (d) $\mu$ to $+0.5\sigma$
(e) $-\sigma$ to $-0.5\sigma$

**4-3.** The ratio of the number of atoms of the isotopes $^{69}$Ga and $^{71}$Ga in eight samples from different sources was measured in an effort to understand differences in reported values of the atomic mass of gallium:[11]

| Sample | $^{69}$Ga/$^{71}$Ga | Sample | $^{69}$Ga/$^{71}$Ga |
|---|---|---|---|
| 1 | 1.526 60 | 5 | 1.528 94 |
| 2 | 1.529 74 | 6 | 1.528 04 |
| 3 | 1.525 92 | 7 | 1.526 85 |
| 4 | 1.527 31 | 8 | 1.527 93 |

Find the (a) mean, (b) standard deviation, and (c) variance.

**4-4.** (a) Calculate the fraction of bulbs in Figure 4-1 expected to have a lifetime greater than 1005.3 h.

(b) What fraction of bulbs is expected to have a lifetime between 798.1 and 901.7 h?

(c) 📊 Use the Excel NORMDIST function to find the fraction of bulbs expected to have a lifetime between 800 and 900 h.

**4-5.** 📊 Blood plasma proteins from patients with malignant breast tumors differ from proteins from healthy people in their solubility in the presence of various polymers. When the polymers dextran and poly(ethylene glycol) are dissolved in water, a two-phase mixture is formed. When plasma proteins are added, they distribute themselves differently between the two phases. The distribution coefficient $(K)$ for any substance is defined as $K =$ [concentration of the substance in phase A]/[concentration of the substance in phase B]. Proteins from healthy people have a mean distribution coefficient of 0.75 with a standard deviation of 0.07. For cancer victims the mean is 0.92 with a standard deviation of 0.11.

(a) Suppose that partition coefficient were used as a diagnostic tool and a positive indication of cancer is taken as $K \geq 0.92$. What fraction of people with tumors would have a false negative indication of cancer because $K < 0.92$?

(b) What fraction of healthy people would have a false positive indication of cancer? This is the fraction of healthy people with $K \geq 0.92$, shown by the shaded area in the graph below. Estimate an answer with Table 4-1 and obtain a more exact result with the NORMDIST function in Excel.

(c) *Extra credit!* Vary the first argument of the NORMDIST function to select a distribution coefficient that would identify 75% of people with tumors. That is, 75% of patients with tumors would have $K$ above the selected distribution coefficient. With this value of $K$, what fraction of healthy people would have a false positive result indicating they have a tumor?

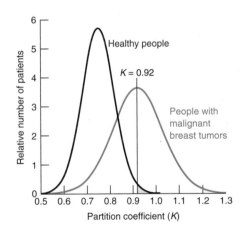

Figure for Problem 4-5. Distribution coefficients of plasma proteins from healthy people and from people with malignant breast tumors. *[Data from B. Y. Zaslavsky, "Bionalytical Applications of Partitioning in Aqueous Polymer Two-Phase Systems," Anal. Chem. **1992**, 64, 765A.]*

**4-6.** 📊 The equation for the Gaussian curve in Figure 4-1 is

$$y = \frac{(\text{total bulbs})(\text{hours per bar})}{s\sqrt{2\pi}} e^{-(x-\bar{x})^2/2s^2}$$

where $\bar{x}$ is the mean value (845.2 h), $s$ is the standard deviation (94.2 h), total bulbs = 4 768, and hours per bar (= 20) is the width of each bar in the bar chart in Figure 4-1. Set up a spreadsheet like the one on the next page to calculate the coordinates of the Gaussian curve in Figure 4-1 from 500 to 1 200 h in 25-h intervals. Note the heavy use of parentheses in the formula at the bottom of the spreadsheet to force the computer to do the arithmetic as intended. Use a computer to graph your results.

| | A | B | C |
|---|---|---|---|
| 1 | Gaussian curve for light bulbs (Fig 4-1) | | |
| 2 | | | |
| 3 | mean = | x (hours) | y (bulbs) |
| 4 | 845.2 | 500 | 0.49 |
| 5 | std dev = | 525 | 1.25 |
| 6 | 94.2 | 550 | 2.98 |
| 7 | total bulbs = | 600 | 13.64 |
| 8 | 4768 | 700 | 123.11 |
| 9 | hours per bar = | 800 | 359.94 |
| 10 | 20 | 845.2 | 403.85 |
| 11 | sqrt(2 pi) = | 900 | 340.99 |
| 12 | 2.506628 | 1000 | 104.67 |
| 13 | | 1100 | 10.41 |
| 14 | | 1200 | 0.34 |
| 15 | Formula for cell C4 = | | |
| 16 | ($A$8*$A$10/($A$6*$A$12))* | | |
| 17 | EXP(-((B4-$A$4)^2)/(2*$A$6^2)) | | |

Spreadsheet for Problem 4-6.

**4-7.** [spreadsheet icon] Repeat Problem 4-6 but use the values 50, 100, and 150 for the standard deviation. Superimpose all three curves on a single graph.

## Confidence Intervals, $t$ Test, $F$ test, and $Q$ Test

**4-8.** What is the meaning of a confidence interval?

**4-9.** What fraction of the vertical bars in Figure 4-5a is expected to include the population mean (10 000) if many experiments are carried out? Why are the 90% confidence interval bars longer than the 50% bars in Figure 4-5?

**4-10.** List the three different cases that we studied for comparison of means, and write the equations used in each case.

**4-11.** The percentage of an additive in gasoline was measured six times with the following results: 0.13, 0.12, 0.16, 0.17, 0.20, 0.11%. Find the 90% and 99% confidence intervals for the percentage of the additive.

**4-12.** Sample 8 of Problem 4-3 was analyzed seven times, with $\bar{x} = 1.527\ 93$ and $s = 0.000\ 07$. Find the 99% confidence interval for sample 8.

**4-13.** A trainee in a medical lab will be released to work on her own when her results agree with those of an experienced worker at the 95% confidence level. Results for a blood urea nitrogen analysis are shown below.

Trainee: $\bar{x} = 14.5_7$ mg/dL $\quad s = 0.5_3$ mg/dL $\quad n = 6$ samples

Experienced worker: $\bar{x} = 13.9_5$ mg/dL $\quad s = 0.4_2$ mg/dL $\quad n = 5$ samples

(a) What does the abbreviation dL stand for?
(b) Should the trainee be released to work alone?

**4-14.** The Ti content (wt%) of five different ore samples (each with a different Ti content) was measured by each of two methods. Do the two analytical techniques give results that are significantly different at the 95% confidence level?

| Sample | Method 1 | Method 2 |
|---|---|---|
| A | 0.013 4 | 0.013 5 |
| B | 0.014 4 | 0.015 6 |
| C | 0.012 6 | 0.013 7 |
| D | 0.012 5 | 0.013 7 |
| E | 0.013 7 | 0.013 6 |

**4-15.** [spreadsheet icon] Now we use a built-in routine in Excel for the paired $t$ test to see if the two methods in Problem 4-14 produce significantly different results. Enter the data for Methods 1 and 2 into two columns of a spreadsheet. Under the TOOLS menu, select DATA ANALYSIS. If DATA ANALYSIS does not appear, select ADD-INS. Select ANALYSIS TOOLPACK, click OK, and DATA ANALYSIS will be loaded into the TOOLS menu. In the DATA ANALYSIS window, select t-Test: Paired Two Sample for Means. Follow the instructions of Section 4-5 and the routine will print out a variety of information including $t_{calculated}$ (which is labeled t Stat) and $t_{table}$ (which is labeled t Critical two-tail). You should reproduce the results of Problem 4-14.

**4-16.** [spreadsheet icon] Lithium isotope ratios are important to medicine, geology, astrophysics, and nuclear chemistry. Measurements of the $^6Li/^7Li$ ratio in a Standard Reference Material are given here.[12] Do the two methods give statistically equivalent results?

| Method 1 | Method 2 |
|---|---|
| 0.082 601 | 0.081 83 |
| 0.082 621 | 0.081 86 |
| 0.082 589 | 0.082 05 |
| 0.082 617 | 0.082 06 |
| 0.082 598 | 0.082 15 |
| | 0.082 08 |

**4-17.** If you measure a quantity four times and the standard deviation is 1.0% of the average, can you be 90% confident that the true value is within 1.2% of the measured average?

**4-18.** Students measured the concentration of HCl in a solution by titrating with different indicators to find the end point.[13]

| Indicator | Mean HCl concentration (M) ($\pm$ standard deviation) | Number of measurements |
|---|---|---|
| 1. Bromothymol blue | 0.095 65 $\pm$ 0.002 25 | 28 |
| 2. Methyl red | 0.086 86 $\pm$ 0.000 98 | 18 |
| 3. Bromocresol green | 0.086 41 $\pm$ 0.001 13 | 29 |

Is the difference between indicators 1 and 2 significant at the 95% confidence level? Answer the same question for indicators 2 and 3.

**4-19.** Hydrocarbons in the cab of an automobile were measured during trips on the New Jersey Turnpike and trips through the Lincoln Tunnel connecting New York and New Jersey.[14] The total concentrations ($\pm$standard deviations) of $m$- and $p$-xylene were

Turnpike: 31.4 $\pm$ 30.0 $\mu$g/m$^3$ (32 measurements)

Tunnel: 52.9 $\pm$ 29.8 $\mu$g/m$^3$ (32 measurements)

Do these results differ at the 95% confidence level? At the 99% confidence level?

**4-20.** A Standard Reference Material is certified to contain 94.6 ppm of an organic contaminant in soil. Your analysis gives values of 98.6, 98.4, 97.2, 94.6, and 96.2 ppm. Do your results differ from the expected result at the 95% confidence level? If you made one more measurement and found 94.5, would your conclusion change?

**4-21.** Nitrite ($NO_2^-$) was measured by two methods in rainwater and unchlorinated drinking water.[15] The results $\pm$ standard deviation (number of samples) are given in the following table.

|                  | Gas chromatography              | Spectrophotometry               |
|------------------|---------------------------------|---------------------------------|
| Rainwater:       | $0.069 \pm 0.005$ mg/L ($n = 7$) | $0.063 \pm 0.008$ mg/L ($n = 5$) |
| Drinking water:  | $0.078 \pm 0.007$ mg/L ($n = 5$) | $0.087 \pm 0.008$ mg/L ($n = 5$) |

(a) Do the two methods agree with each other at the 95% confidence level for both rainwater and drinking water?

(b) For each method, does the drinking water contain significantly more nitrite than the rainwater (at the 95% confidence level)?

**4-22.** Using the $Q$ test, decide whether the value 216 should be rejected from the set of results 192, 216, 202, 195, and 204.

### Linear Least Squares

**4-23.** A straight line is drawn through the points $(3.0, -3.87 \times 10^4)$, $(10.0, -12.99 \times 10^4)$, $(20.0, -25.93 \times 10^4)$, $(30.0, -38.89 \times 10^4)$, and $(40.0, -51.96 \times 10^4)$ to give $m = -1.298\ 72 \times 10^4$, $b = 256.695$, $s_m = 13.190$, $s_b = 323.57$, and $s_y = 392.9$. Express the slope and intercept and their uncertainties with reasonable significant figures.

**4-24.** Here is a least-squares problem that you can do by hand with a calculator. Find the slope and intercept and their standard deviations for the straight line drawn through the points $(x,y) = (0,1)$, $(2,2)$, and $(3,3)$. Make a graph showing the three points and the line. Place error bars $(\pm s_y)$ on the points.

**4-25.** Set up a spreadsheet to reproduce Figure 4-13. *Add error bars:* Double click on a data point on the graph and select Y Error Bars. Check Custom and enter the value of $s_y$ in each box for the $+$ and $-$ error. Better yet, enter the cell containing $s_y$ in both boxes.

**4-26.** *Excel LINEST function.* Enter the data from Problem 4-23 in a spreadsheet and use the LINEST function to find the slope and intercept and standard deviations. Use Excel to draw a graph of the data and add a TRENDLINE.

### Calibration Curves

**4-27.** Explain the following statement: "The validity of a chemical analysis ultimately depends on measuring the response of the analytical procedure to known standards."

**4-28.** Suppose that you carry out an analytical procedure to generate a calibration curve like that shown in Figure 4-11. Then you analyze an unknown and find an absorbance that gives a negative concentration for the analyte. What does this mean?

**4-29.** Using the calibration curve in Figure 4-11, find the quantity of unknown protein that gives a measured absorbance of 0.264 when a blank has an absorbance of 0.095.

**4-30.** Consider the least-squares problem in Figure 4-9.

(a) Suppose that a single new measurement produces a $y$ value of 2.58. Find the corresponding $x$ value and its uncertainty.

(b) Suppose you measure $y$ four times and the average is 2.58. Calculate the uncertainty based on four measurements, not one.

**4-31.** Consider the linear calibration curve in Figure 4-11, which is derived from the 14 corrected absorbances in the shaded region at the right side of Table 4-7. Create a least-squares spreadsheet like Figure 4-13 to compute the equation of the line and the standard deviations of the parameters. Suppose that you find absorbance values of 0.265, 0.269, 0.272, and 0.258 for four identical samples of unknown and absorbances of 0.099, 0.091, 0.101, and 0.097 for four blanks. Find the corrected absorbance by subtracting the average blank from the average absorbance of the unknown. Calculate the amount of protein and its uncertainty in the unknown.

**4-32.** Here are mass spectrometric signals for methane in $H_2$:

| $CH_4$ (vol%): | 0 | 0.062 | 0.122 | 0.245 | 0.486 | 0.971 | 1.921 |
|---|---|---|---|---|---|---|---|
| Signal (mV): | 9.1 | 47.5 | 95.6 | 193.8 | 387.5 | 812.5 | 1 671.9 |

(a) Subtract the blank value (9.1) from all other values. Then use the method of least squares to find the slope and intercept and their uncertainties. Construct a calibration curve.

(b) Replicate measurements of an unknown gave 152.1, 154.9, 153.9, and 155.1 mV, and a blank gave 8.2, 9.4, 10.6, and 7.8 mV. Subtract the average blank from the average unknown to find the average corrected signal for the unknown.

(c) Find the concentration of the unknown and its uncertainty.

**4-33.** The figure gives replicate measurements of As(III) concentration by an electrochemical method.

(a) Using a millimeter ruler, measure each peak height to the near-

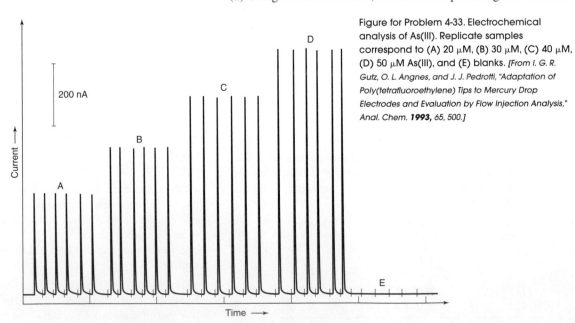

Figure for Problem 4-33. Electrochemical analysis of As(III). Replicate samples correspond to (A) 20 $\mu$M, (B) 30 $\mu$M, (C) 40 $\mu$M, (D) 50 $\mu$M As(III), and (E) blanks. *[From I. G. R. Gutz, O. L. Angnes, and J. J. Pedrotti, "Adaptation of Poly(tetrafluoroethylene) Tips to Mercury Drop Electrodes and Evaluation by Flow Injection Analysis," Anal. Chem. **1993**, 65, 500.]*

est 0.1 mm. Noting the length that corresponds to 200 nA in the figure, make a table showing the observed current (nA) for each concentration (μM) of As(III). The blanks appear to be near 0, so we will disregard them in this problem.

(b) Construct a calibration curve with 24 points (A–D) and find the slope and intercept and their uncertainties, using the method of least squares.

(c) Calculate the concentration (and its uncertainty) of As(III) in an unknown that gave a mean current of 501 nA from six measurements.

**4-34. Nonlinear calibration curve.** Following the procedure in Box 4-2, find how many micrograms (μg) of protein are contained in a sample with a corrected absorbance of 0.350 in Figure 4-11.

**4-35. Logarithmic calibration curve.** Calibration data spanning five orders of magnitude for an electrochemical determination of $p$-nitrophenol are given in the table below. (The blank has already been subtracted from the measured current.) If you try to plot these data on a linear graph extending from 0 to 310 μg/mL and from 0 to 5 260 nA, most of the points will be bunched up near the origin. To handle data with such a large range, a logarithmic plot is helpful.

| $p$-Nitrophenol (μg/mL) | Current (nA) | $p$-Nitrophenol (μg/mL) | Current (nA) |
|---|---|---|---|
| 0.010 0 | 0.215 | 3.00 | 66.7 |
| 0.029 9 | 0.846 | 10.4 | 224 |
| 0.117 | 2.65 | 31.2 | 621 |
| 0.311 | 7.41 | 107 | 2 020 |
| 1.02 | 20.8 | 310 | 5 260 |

*Data from Figure 4 of L. R. Taylor, Am. Lab., February 1993, p. 44.*

(a) Make a graph of log(current) versus log(concentration). Over what range is the log-log calibration linear?

(b) Find the equation of the line in the form $\log(\text{current}) = m \times \log(\text{concentration}) + b$.

(c) Find the concentration of $p$-nitrophenol corresponding to a signal of 99.9 nA.

**4-36. Confidence interval for calibration curve.** To use a calibration curve based on $n$ points, we measure a new value of $y$ and calculate the corresponding value of $x$. The one-standard-deviation uncertainty in $x$, $s_x$, is given by Equation 4-27. We express a *confidence interval* for $x$, using Student's $t$:

$$\text{Confidence interval} = x \pm ts_x$$

where $t$ is taken from Table 4-2 for $n - 2$ degrees of freedom.

A calibration curve based on $n = 10$ known points was used to measure the protein in an unknown. The results were protein = $15.2_2(\pm 0.4_6)$ μg, where $s_x = 0.4_6$ μg. Find the 90% and 99% confidence intervals for protein in the unknown.

# 5 Quality Assurance and Calibration Methods

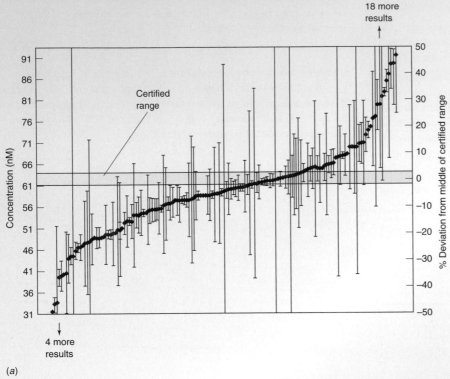
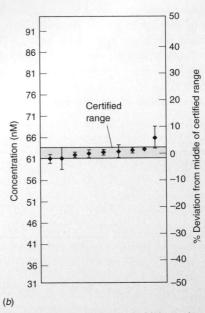

(a)

(b)

(a) Scattered measurements of Pb in river water by different laboratories, each of which employed a recognized quality management system. (b) Reproducible results from national measurement institutes. [From P. De Bièvre and P. D. P. Taylor, "'Demonstration' vs. 'Designation' of Measurement Competence: The Need to Link Accreditation to Metrology," Fresenius J. Anal. Chem. **2000**, 368, 567.]

The Institute for Reference Materials and Measurements in Belgium conducts an International Measurement Evaluation Program to allow laboratories to assess the reliability of their analyses. Panel *a* shows results for lead in river water. Of 181 labs, 18 reported results more than 50% above and 4 reported results more than 50% below the certified level of $62.3 \pm 1.3$ nM. Though most labs in the study employed recognized quality management procedures, a large fraction of results did not include the certified range. Panel *b* shows that when this same river water was analyzed by nine different national measurement institutes, where the most care is taken, all results were close to the certified range.

This example illustrates that there is no guarantee that results are reliable, even if they are obtained by "accredited" laboratories using accepted procedures. A good way to assess the reliability of a lab working for you is to provide the lab with "blind" samples—similar to your unknowns—for which you know the "right" answer, but the analyst does not. If the lab does not find the known result, there is a problem. Periodic blind check samples are required to demonstrate continuing reliability.

Data quality standards:
• Get the right data
• Get the data right
• Keep the data right

[Nancy W. Wentworth,
U.S. Environmental Protection Agency[1]]

**Q**uality assurance is what we do to get the right answer for our purpose. The answer should have sufficient accuracy and precision to support subsequent decisions. There is no point in spending extra money to obtain a more accurate or more precise answer if it is not necessary. This chapter describes basic issues and procedures in quality assurance[2] and introduces two more calibration methods. In Chapter 4, we discussed how to make a calibration curve. In this chapter, we describe *standard addition* and *internal standards*.

# ■ ■ ■ 5-1 Basics of Quality Assurance

"Suppose you are cooking for some friends. While making spaghetti sauce, you taste it, season it, taste it some more. Each tasting is a sampling event with a quality control test. You can taste the whole batch because there is only one batch. Now suppose you run a spaghetti sauce plant that makes 1 000 jars a day. You can't taste each one, so you decide to taste three a day, one each at 11 A.M., 2 P.M., and 5 P.M. If the three jars all taste OK, you conclude all 1 000 are OK. Unfortunately, that may not be true, but the relative risk—that a jar has too much or too little seasoning—is not very important because you agree to refund the money of any customer who is dissatisfied. If the number of refunds is small, say, 100 a year, there is no apparent benefit in tasting 4 jars a day." There would be 365 additional tests to avoid refunds on 100 jars, giving a net loss of 265 jars worth of profit.

In analytical chemistry, the product is not spaghetti sauce, but, rather, raw data, treated data, and results. *Raw data* are individual values of a measured quantity, such as peak areas from a chromatogram or volumes from a buret. *Treated data* are concentrations or amounts found by applying a calibration procedure to the raw data. *Results* are what we ultimately report, such as the mean, standard deviation, and confidence interval, after applying statistics to treated data.

Quotation from Ed Urbansky. Section 5-1 is adapted from a description written by Ed Urbansky.

| Current: | Taste 3 jars each day = 1 095 jars/year Return 100 jars/year to unhappy customers |
|---|---|
| Proposed: | Taste 4 jars each day = 1 460 jars/year Return 0 jars/year because of better quality control |
| Difference: | 1 460 − 1 195 = 265 jars/year |

*Raw data:* individual measurements
*Treated data:* concentrations derived from raw data by use of calibration method
*Results:* quantities reported after statistical analysis of treated data

## Use Objectives

If you manufacture a drug whose therapeutic dose is just a little less than the lethal dose, you should be more careful than if you make spaghetti sauce. The kind of data that you collect and the way in which you collect them depend on how you plan to use those data. An important goal of quality assurance is making sure that results meet the customer's needs. A bathroom scale does not have to measure mass to the nearest milligram, but a drug tablet required to contain 2 mg of active ingredient probably cannot contain $2 \pm 1$ mg. Writing clear, concise **use objectives** for data and results is a critical step in quality assurance and helps prevent misuse of data and results.

Here is an example of a use objective. Drinking water is usually disinfected with chlorine, which kills microorganisms. Unfortunately, chlorine also reacts with organic matter in water to produce "disinfection by-products"—compounds that might harm humans. A disinfection facility was planning to introduce a new chlorination process and wrote the following analytical use objective:

> Analytical data and results shall be used to determine whether the modified chlorination process results in at least a 10% reduction of formation of selected disinfection by-products.

The new process was expected to decrease the disinfection by-products. The use objective says that uncertainty in the analysis must be small enough so that a 10% decrease in selected by-products is clearly distinguishable from experimental error. In other words, is an observed decrease of 10% real?

*Use objective:* states purpose for which results will be used

## Specifications

Once you have use objectives, you are ready to write **specifications** stating how good the numbers need to be and what precautions are required in the analytical procedure. How shall samples be taken and how many are needed? Are special precautions required to protect samples and ensure that they are not degraded? Within practical restraints, such as cost, time, and limited amounts of material available for analysis, what level of accuracy and precision will satisfy the use objectives? What rate of false positives or false negatives is acceptable? These questions need to be answered in detailed specifications.

*Quality assurance begins with sampling.* We must collect representative samples and analyte must be preserved after sample is collected. If our sample is not representative or if analyte is lost after collection, then even the most accurate analysis is meaningless.

What do we mean by *false positives* and *false negatives*? Suppose you must certify that a contaminant in drinking water is below a legal limit. A **false positive** says that the concentration exceeds the legal limit when, in fact, the concentration is below the limit. A **false negative** says that the concentration is below the limit when it is actually above the limit. Even well-executed procedures produce some false conclusions because of the statistical nature of sampling and measurement. More-stringent procedures are required to obtain lower rates of false conclusions. For drinking water, it is likely to be more important to have a low rate

*Specifications might include:*
* sampling requirements
* accuracy and precision
* rate of false results
* selectivity
* sensitivity
* acceptable blank values
* recovery of fortification
* calibration checks
* quality control samples

of false negatives than a low rate of false positives. It would be worse to certify that contaminated water is safe than to certify that safe water is contaminated.

In choosing a method, we also consider selectivity and sensitivity. **Selectivity** (also called *specificity*) means being able to distinguish analyte from other species in the sample (avoiding interference). **Sensitivity** is the capability of responding reliably and measurably to changes in analyte concentration. A method must have a *detection limit* (discussed in Section 5-2) lower than the concentrations to be measured.

Specifications could include required accuracy and precision, reagent purity, tolerances for apparatus, the use of Standard Reference Materials, and acceptable values for blanks. *Standard Reference Materials* (Box 3-1) contain certified levels of analyte in realistic materials that you might be analyzing, such as blood or coal or metal alloys. Your analytical method should produce an answer acceptably close to the certified level or there is something wrong with the accuracy of your method.

Blanks account for interference by other species in the sample and for traces of analyte found in reagents used for sample preservation, preparation, and analysis. Frequent measurements of blanks detect whether analyte from previous samples is carried into subsequent analyses by adhering to vessels or instruments.

A **method blank** is a sample containing all components except analyte, and it is taken through all steps of the analytical procedure. We subtract the response of the method blank from the response of a real sample prior to calculating the quantity of analyte in the sample. A **reagent blank** is similar to a method blank, but it has not been subjected to all sample preparation procedures. The method blank is a more complete estimate of the blank contribution to the analytical response.

A **field blank** is similar to a method blank, but it has been exposed to the site of sampling. For example, to analyze particulates in air, a certain volume of air could be sucked through a filter, which is then dissolved and analyzed. A field blank would be a filter carried to the collection site in the same package with the collection filters. The filter for the blank would be taken out of its package in the field and placed in the same kind of sealed container used for collection filters. The difference between the blank and the collection filters is that air was not sucked through the blank filter. Volatile organic compounds encountered during transportation or in the field are conceivable contaminants of a field blank.

Another performance requirement often specified is *spike recovery*. Sometimes, response to analyte is affected by something else in the sample. We use the word **matrix** to refer to everything else in the sample other than analyte. A **spike,** also called a *fortification,* is a known quantity of analyte added to a sample to test whether the response to a sample is the same as that expected from a calibration curve. Spiked samples are analyzed in the same manner as unknowns. For example, if drinking water is found to contain 10.0 μg/L of nitrate, a spike of 5.0 μg/L could be added. Ideally, the concentration in the spiked portion found by analysis will be 15.0 μg/L. If a number other than 15.0 μg/L is found, then the matrix could be interfering with the analysis.

### Example  Spike Recovery

Let $C$ stand for concentration. One definition of spike recovery is

$$\% \text{ recovery} = \frac{C_{\text{spiked sample}} - C_{\text{unspiked sample}}}{C_{\text{added}}} \tag{5-1}$$

An unknown was found to contain 10.0 μg of analyte per liter. A spike of 5.0 μg/L was added to a replicate portion of unknown. Analysis of the spiked sample gave a concentration of 14.6 μg/L. Find the percent recovery of the spike.

**Solution**  The percent of the spike found by analysis is

$$\% \text{ recovery} = \frac{14.6 \ \mu g/L - 10.0 \ \mu g/L}{5.0 \ \mu g/L} \times 100 = 92\%$$

If the acceptable recovery is specified to be in the range from 96% to 104%, then 92% is unacceptable. Something in your method or techniques needs improvement.

When dealing with large numbers of samples and replicates, we perform periodic calibration checks to make sure that our instrument continues to work properly and the calibration remains valid. In a **calibration check,** we analyze solutions formulated to contain known concentrations of analyte. A specification might, for example, call for one calibration check for every 10 samples. Solutions for calibration checks should be different from the ones used to prepare the original calibration curve. This practice helps to verify that the initial calibration standards were made properly.

**Performance test samples** (also called *quality control samples* or *blind samples*) are a quality control measure to help eliminate bias introduced by the analyst knowing the concentration of the calibration check sample. These samples of known composition are provided to the analyst as unknowns. Results are then compared with the known values, usually by a quality assurance manager.

Together, raw data and results from calibration checks, spike recoveries, quality control samples, and blanks are used to gauge accuracy. Analytical performance on replicate samples and replicate portions of the same sample measures precision. Fortification also helps ensure that qualitative identification of analyte is correct. If you spike the unknown in Figure 0-5 with extra caffeine and the area of a chromatographic peak not thought to be caffeine increases, then you have misidentified the caffeine peak.

**Standard operating procedures** stating what steps will be taken and how they will be carried out are the bulwark of quality assurance. For example, if a reagent has "gone bad" for some reason, control experiments built into your normal procedures should detect that something is wrong and your results should not be reported. It is implicit that everyone follows the standard operating procedures. Adhering to these procedures guards against the normal human desire to take shortcuts based on assumptions that could be false.

A meaningful analysis requires a meaningful sample that represents what is to be analyzed. It must be stored in containers and under conditions so that relevant chemical characteristics do not change. Protection might be needed to prevent oxidation, photodecomposition, or growth of organisms. The *chain of custody* is the trail followed by a sample from the time it is collected to the time it is analyzed and, possibly, archived. Documents are signed each time the material changes hands to indicate who is responsible for the sample. Each person in the chain of custody follows a written procedure telling how the sample is to be handled and stored. Each person receiving a sample should inspect it to see that it is in the expected condition in an appropriate container. If the original sample was a homogeneous

*To gauge accuracy:*
• calibration checks
• fortification recoveries
• quality control samples
• blanks
*To gauge precision:*
• replicate samples
• replicate portions of same sample

---

## Box 5-1  Control Charts

A **control chart** is a visual representation of confidence intervals for a Gaussian distribution. The chart warns us when a monitored property strays dangerously far from an intended *target value.*

Consider a manufacturer making vitamin C tablets intended to have a target value of μ milligrams of vitamin C per tablet. Many analyses over a long time tell us the population standard deviation, σ, associated with the manufacturing process.

For quality control, 25 tablets are removed at random from the manufacturing line each hour and analyzed. The mean value of vitamin C in the 25 tablets is shown by a data point on the control chart.

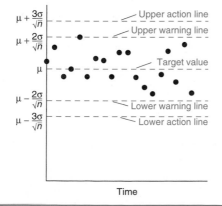

For a Gaussian distribution, 95.5% of all observations are within $\pm 2\sigma/\sqrt{n}$ from the mean and 99.7% are within $\pm 3\sigma/\sqrt{n}$, where $n$ is the number of tablets (= 25) that are averaged each hour. The $\pm 2\sigma/\sqrt{n}$ limits are *warning lines* and the $\pm 3\sigma/\sqrt{n}$ limits are *action lines.* We expect ~4.5% of measurements to be outside the warning lines and ~0.3% to be outside the action lines. It is unlikely that we would observe two consecutive measurements at the warning line (probability = $0.045 \times 0.045 = 0.002\,0$).

The following conditions are considered to be so unlikely that if they occur the process should be shut down for troubleshooting:

• 1 observation outside the action lines
• 2 out of 3 consecutive measurements between the warning and action lines
• 7 consecutive measurements all above or all below the center line
• 6 consecutive measurements all increasing or all decreasing, wherever they are located
• 14 consecutive points alternating up and down, regardless of where they are located
• an obvious nonrandom pattern

For quality assessment of an analytical process, a control chart could show the relative deviation of measured values of calibration check samples or quality control samples from their known values. Another control chart could display the precision of replicate analyses of unknowns or standards as a function of time.

**Table 5-1** Quality assurance process

| Question | Actions |
|---|---|
| *Use Objectives*<br>Why do you want the data and results and how will you use the results? | • Write use objectives |
| *Specifications*<br>How good do the numbers have to be? | • Write specifications<br>• Pick methods to meet specifications<br>• Consider sampling, precision, accuracy, selectivity, sensitivity, detection limit, robustness, rate of false results<br>• Employ blanks, fortification, calibration checks, quality control samples, and control charts to monitor performance<br>• Write and follow standard operating procedures |
| *Assessment*<br>Were the specifications achieved? | • Compare data and results with specifications<br>• Document procedures and keep records suitable to meet use objectives<br>• Verify that use objectives were met |

liquid, but it contains a precipitate when you receive it, the standard operating procedure may dictate that you reject that sample.

Standard operating procedures specify how instruments are to be maintained and calibrated to ensure their reliability. Many labs have their own standard practices, such as recording temperatures of refrigerators, calibrating balances, conducting routine instrument maintenance, or replacing reagents. These practices are part of the overall quality management plan. The rationale behind standard practices is that some equipment is used by many people for different analyses. We save money by having one program to ensure that the most rigorous needs are met.

## Assessment

**Assessment** is the process of (1) collecting data to show that analytical procedures are operating within specified limits and (2) verifying that final results meet use objectives.

Documentation is critical for assessment. Standard *protocols* provide directions for what must be documented and how the documentation is to be done, including how to record information in notebooks. For labs that rely on manuals of standard practices, it is imperative that tasks done to comply with the manuals be monitored and recorded. *Control charts* (Box 5-1) can be used to monitor performance on blanks, calibration checks, and spiked samples to see if results are stable over time or to compare the work of different employees. Control charts can also monitor sensitivity or selectivity, especially if a laboratory encounters a wide variety of matrixes.

Government agencies such as the U.S. Environmental Protection Agency set requirements for quality assurance for their own labs and for certification of other labs. Published standard methods specify precision, accuracy, numbers of blanks, replicates, and calibration checks. To monitor drinking water, regulations state how often and how many samples are to be taken. Documentation is necessary to demonstrate that all requirements have been met. Table 5-1 summarizes the quality assurance process.

## ■ ■ ■ 5-2 Method Validation

**Method validation** is the process of proving that an analytical method is acceptable for its intended purpose.[3] In pharmaceutical chemistry, method validation requirements for regulatory submission include studies of *method specificity, linearity, accuracy, precision, range, limit of detection, limit of quantitation,* and *robustness.*

## Specificity

**Specificity** is the ability of an analytical method to distinguish the analyte from everything else that might be in the sample. *Electrophoresis* is an analytical method in which substances

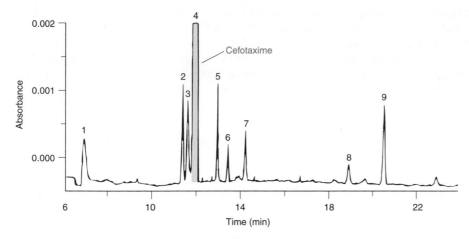

*Figure 5-1* Electropherogram of the drug cefotaxime (peak 4) spiked with known impurities (peaks 2, 3, 5–9) from synthesis of the drug. Peak 1 is a marker for electroosmotic flow. Smaller peaks from unknown impurities also are observed. Separation was performed by micellar electrophoretic capillary chromatography (Section 26-6). *[From H. Fabre and K. D. Altria, "Key Points for Validating CE Methods, Particularly in Pharmaceutical Analysis," LCGC 2001, 19, 498.]*

are separated from one another by their differing rates of migration in a strong electric field. An *electropherogram* is a graph of detector response versus time in an electrophoretic separation. Figure 5-1 shows an electropherogram of the drug cefotaxime (peak 4) spiked with 0.2 wt% of known impurities normally present from the synthesis. A reasonable requirement for specificity might be that there is baseline separation of analyte (cefotaxime) from all impurities that might be present. *Baseline separation* means that the detector signal returns to its baseline before the next compound reaches the detector.

In Figure 5-1, impurity peak 3 is not completely resolved from the cefotaxime. In this case, another reasonable criterion for specificity might be that unresolved impurities at their maximum expected concentration will not affect the assay of cefotaxime by more than 0.5%. If we were trying to measure impurities, as opposed to assaying cefotaxime, a reasonable criterion for specificity is that all impurity peaks having >0.1% of the area in the electropherogram are baseline separated from cefotaxime.

When developing an assay, we need to decide what impurities to deliberately add to test for specificity. For analysis of a drug formulation, we would want to compare the pure drug with one containing additions of all possible synthetic by-products and intermediates, degradation products, and excipients. (*Excipients* are substances added to give desirable form or consistency.) Degradation products might be introduced by exposing the pure material to heat, light, humidity, acid, base, and oxidants to decompose ~20% of the original material.

## Linearity

**Linearity** measures how well a calibration curve follows a straight line. If you know the target concentration of analyte in a drug preparation, for example, test the calibration curve for linearity with five standard solutions spanning the range from 0.5 to 1.5 times the expected analyte concentration. Each standard should be prepared and analyzed three times. (This procedure requires $3 \times 5 = 15$ standards plus three blanks.) To prepare a calibration curve for an impurity that might be present at, say 0.1 to 1 wt%, you might prepare a calibration curve with five standards spanning the range 0.05 to 2 wt%.

A superficial, but common, measure of linearity is the *square of the correlation coefficient, $R^2$*:

*Square of correlation coefficient:*
$$R^2 = \frac{[\Sigma(x_i - \bar{x})(y_i - \bar{y})]^2}{\Sigma(x_i - \bar{x})^2 \Sigma(y_i - \bar{y})^2} \qquad (5\text{-}2)$$

where $\bar{x}$ is the mean of all the $x$ values and $\bar{y}$ is the mean of all the $y$ values. An easy way to find $R^2$ is with the LINEST function in Excel. In the example on page 69, values of $x$ and $y$ are entered in columns A and B. LINEST produces a table in cells E3:F5 that contains $R^2$ in cell E5. $R^2$ must be very close to 1 to represent a linear fit. For a major component of an unknown, a value of $R^2$ above 0.995 or, perhaps, 0.999, is deemed a good fit for many purposes.[4] For the data in Figure 4-9, which do not lie very close to the straight line, $R^2 = 0.985$.

$R^2$ can be used as a diagnostic. If $R^2$ decreases after a method is established, something has gone wrong with the procedure.

Another criterion for linearity is that the *y*-intercept of the calibration curve (after the response of the blank has been subtracted from each standard) should be close to 0. An acceptable degree of "closeness to 0" might be 2% of the response for the target value of analyte. For the assay of impurities, which are present at concentrations lower than that of the major component, an acceptable value of $R^2$ might be $\geq 0.98$ for the range 0.1 to 2 wt% and the *y*-intercept should be $\leq 10\%$ of the response for the 2 wt% standard.

## Accuracy

*Accuracy* is "nearness to the truth." Ways to demonstrate accuracy include

1. Analyze a *Standard Reference Material* (Box 3-1) in a matrix similar to that of your unknown. Your method should find the certified value for analyte in the reference material, within the precision (random uncertainty) of your method.
2. Compare results from two or more different analytical methods. They should agree within their expected precision.
3. Analyze a blank sample spiked with a known addition of analyte. The matrix must be the same as your unknown. When assaying a major component, three replicate samples at each of three levels ranging from 0.5 to 1.5 times the expected sample concentration are customary. For impurities, spikes could cover three levels spanning an expected range of concentrations, such as 0.1 to 2 wt%.
4. If you cannot prepare a blank with the same matrix as the unknown, then it is appropriate to make *standard additions* (Section 5-3) of analyte to the unknown. An accurate assay will find the known amount of analyte that was added.

Spiking is the most common method to evaluate accuracy because reference materials are not usually available and a second analytical method may not be readily available. Spiking ensures that the matrix remains nearly constant.

An example of a specification for accuracy is that the analysis will recover $100 \pm 2\%$ of the spike of a major constituent. For an impurity, the specification might be that recovery is within 0.1 wt% absolute or $\pm 10\%$ relative.

## Precision

*Precision* is the reproducibility of a result. *Instrument precision,* also called *injection precision,* is the reproducibility observed when the same quantity of one sample is repeatedly introduced ($\geq 10$ times) into an instrument. Variability could arise from variation in the injected quantity and variation of instrument response.

*Intra-assay precision* is evaluated by analyzing aliquots of a homogeneous material several times by one person on one day with the same equipment. Each analysis is independent, so the intra-assay precision is telling us how reproducible the analytical method can be. Intra-assay variability is greater than instrument variability, because more steps are involved. Examples of specifications might be that instrument precision is $\leq 1\%$ and intra-assay precision is $\leq 2\%$.

*Ruggedness,* also called *intermediate precision,* is the variation observed when an assay is performed by different people on different instruments on different days in the same lab. Each analysis might incorporate fresh reagents and different chromatography columns.

*Interlaboratory precision* is the most general measure of reproducibility observed when aliquots of the same sample are analyzed by different people in different laboratories. Interlaboratory precision becomes poorer as the level of analyte decreases (Box 5-2).

## Range

**Range** is the concentration interval over which linearity, accuracy, and precision are all acceptable. An example of a specification for range for a major component of a mixture is the concentration interval providing a correlation coefficient of $R^2 \geq 0.995$ (a measure of linearity), spike recovery of $100 \pm 2\%$ (a measure of accuracy), and interlaboratory precision of $\pm 3\%$. For an impurity, an acceptable range might provide a correlation coefficient of $R^2 \geq 0.98$, spike recovery of $100 \pm 10\%$, and interlaboratory precision of $\pm 15\%$.

## Limits of Detection and Quantitation

The **detection limit** (also called the *lower limit of detection*) is the smallest quantity of analyte that is "significantly different" from the blank.[6] Here is a procedure that produces a

Autosamplers in chromatography and graphite furnace atomic spectroscopy, for example, have improved injection precision by a factor of 3–10 compared with that attained by humans.

Confusing terms:

*Linear range:* concentration range over which calibration curve is linear (Figure 4-12)

*Dynamic range:* concentration range over which there is a measurable response (Figure 4-12)

*Range:* concentration range over which linearity, accuracy, and precision meet specifications for analytical method

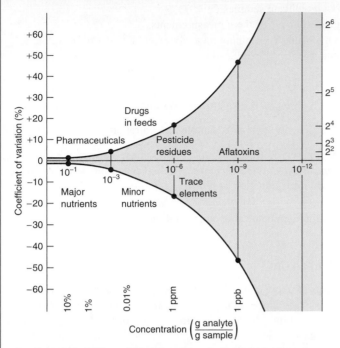

Coefficient of variation of interlaboratory results as a function of sample concentration (expressed as g analyte/g sample). The shaded region has been referred to as the "Horwitz trumpet" because of the way it flares open. [From W. Horwitz, "Evaluation of Analytical Methods Used for Regulation of Foods and Drugs," Anal. Chem. **1982**, 54, 67A.]

*Interlaboratory tests* are routinely used to validate new analytical procedures—especially those intended for regulatory use. Typically, 5 to 10 laboratories are given identical samples and the same written procedure. If all results are "similar," and there is no serious systematic error, then the method is considered "reliable."

The **coefficient of variation** (CV) is the standard deviation divided by the mean: $CV = s/\bar{x}$. Usually the coefficient of variation is expressed as a percentage of the mean: $CV(\%) = 100 \times s/\bar{x}$. The smaller the coefficient of variation, the more precise is a set of measurements.

In reviewing more than 150 interlaboratory studies with different analytes measured by different techniques, it was observed that the coefficient of variation of mean values reported by different laboratories increased as analyte concentration decreased. At best, the coefficient of variation never seemed to be better than[5]

*Horwitz curve:*     $CV(\%) \approx 2^{(1-0.5 \log C)}$

where $C$ is the fraction of analyte in the sample ($C$ = g analyte/g sample). The coefficient of variation within a laboratory is about one-half to two-thirds of the between-laboratory variation. Experimental results varied from the idealized curve by about a factor of 2 in the vertical direction and a factor of 10 in the horizontal direction. About 5–15% of all interlaboratory results were "outliers"—clearly outside the cluster of other results. This incidence of outliers is above the statistical expectation.

When the concentration of analyte is 1 ppm, the coefficient of variation between laboratories is 16%. When the concentration is 1 ppb, the coefficient of variation is 45%. If, perchance, you become a regulation writer one day, acceptable analyte levels should allow for variation among laboratories. The Gaussian distribution tells us that approximately 5% of measurements lie above $\bar{x} + 1.65s$ (Section 4-1). If the target allowable level of analyte is 1.0 ppb, the allowed observed amount might be set at $1 + 1.65 \times 0.45$ ppb, or about 1.7 ppb. This level gives a 5% rate of false positives that exceed the allowed value even though the true value is below 1.0 ppb.

detection limit with ~99% chance of being greater than the blank. That is, only ~1% of samples containing no analyte will give a signal greater than the detection limit (Figure 5-2). We assume that the standard deviation of the signal from samples near the detection limit is similar to the standard deviation from blanks.

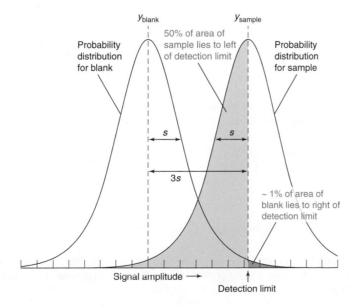

*Figure 5-2* Detection limit. Curves show distribution of measurements expected for a blank and a sample whose concentration is at the detection limit. The area of any region is proportional to the number of measurements in that region. Only ~1% of measurements for a blank are expected to exceed the detection limit. However, 50% of measurements for a sample containing analyte at the detection limit will be below the detection limit. There is a 1% chance of concluding that a blank has analyte above the detection limit. If a sample contains analyte at the detection limit, there is a 50% chance of concluding that analyte is *absent* because its signal is below the detection limit. Curves in this figure are Student's *t* distributions, which are broader than the Gaussian distribution.

1. After estimating the detection limit from previous experience with the method, prepare a sample whose concentration is ~1 to 5 times the detection limit.
2. Measure the signal from $n$ replicate samples ($n \geq 7$).
3. Compute the standard deviation ($s$) of the $n$ measurements.
4. Measure the signal from $n$ blanks (containing no analyte) and find the mean value, $y_{blank}$.
5. The minimum detectable signal, $y_{dl}$, is defined as

*Signal detection limit:* $$y_{dl} = y_{blank} + 3s \qquad (5\text{-}3)$$

6. The corrected signal, $y_{sample} - y_{blank}$, is proportional to sample concentration:

*Calibration line:* $$y_{sample} - y_{blank} = m \times \text{sample concentration} \qquad (5\text{-}4)$$

where $y_{sample}$ is the signal observed for the sample and $m$ is the slope of the linear calibration curve. The *minimum detectable concentration,* also called the detection limit, is obtained by substituting $y_{dl}$ from Equation 5-3 for $y_{sample}$ in Equation 5-4:

*Detection limit:* $$\text{Minimum detectable concentration} = \frac{3s}{m} \qquad (5\text{-}5)$$

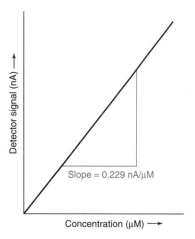

Slope = 0.229 nA/μM

Concentration (μM) ⟶

**Example** Detection Limit

From previous measurements of a low concentration of analyte, the signal detection limit was estimated to be in the low nanoampere range. Signals from seven replicate samples with a concentration about three times the detection limit were 5.0, 5.0, 5.2, 4.2, 4.6, 6.0, and 4.9 nA. Reagent blanks gave values of 1.4, 2.2, 1.7, 0.9, 0.4, 1.5, and 0.7 nA. The slope of the calibration curve for higher concentrations is $m = 0.229$ nA/μM. **(a)** Find the signal detection limit and the minimum detectable concentration. **(b)** What is the concentration of analyte in a sample that gave a signal of 7.0 nA?

**Solution** **(a)** First compute the mean for the blanks and the standard deviation of the samples. Retain extra, insignificant digits to reduce round-off errors.

$$\text{Blank:} \quad \text{Average} = y_{blank} = 1.2_6 \text{ nA}$$

$$\text{Sample:} \quad \text{Standard deviation} = s = 0.5_6 \text{ nA}$$

The signal detection limit from Equation 5-3 is

$$y_{dl} = y_{blank} + 3s = 1.2_6 \text{ nA} + (3)(0.5_6 \text{ nA}) = 2.9_4 \text{ nA}$$

The minimum detectable concentration is obtained from Equation 5-5:

$$\text{Detection limit} = \frac{3s}{m} = \frac{(3)(0.5_6 \text{ nA})}{0.229 \text{ nA/μM}} = 7._3 \text{ μM}$$

**(b)** To find the concentration of a sample whose signal is 7.0 nA, use Equation 5-4:

$$y_{sample} - y_{blank} = m \times \text{concentration}$$

$$\Rightarrow \text{concentration} = \frac{y_{sample} - y_{blank}}{m} = \frac{7.0 \text{ nA} - 1.2_6 \text{ nA}}{0.229 \text{ nA/μM}} = 25._1 \text{ μM}$$

The lower limit of detection given in Equation 5-5 is $3s/m$, where $s$ is the standard deviation of a low-concentration sample and $m$ is the slope of the calibration curve. The standard deviation is a measure of the *noise* (random variation) in a blank or a small signal. When the signal is 3 times as great as the noise, it is readily detectable, but still too small for accurate measurement. A signal that is 10 times as great as the noise is defined as the **lower limit of quantitation,** or the smallest amount that can be measured with reasonable accuracy.

$$\text{Detection limit} \equiv \frac{3s}{m}$$

$$\text{Quantitation limit} \equiv \frac{10s}{m}$$

The symbol $\equiv$ means "is defined as."

$$\text{Lower limit of quantitation} \equiv \frac{10s}{m} \qquad (5\text{-}6)$$

Another common way to define detection limit is from the least-squares equation of a calibration curve: signal detection limit $= b + 3s_y$, where $b$ is the $y$-intercept and $s_y$ is given by Equation 4-20.

The *instrument detection limit* is obtained by replicate measurements ($n \geq 7$) of aliquots from one sample. The *method detection limit,* which is greater than the instrument detection limit, is obtained by preparing $n \geq 7$ individual samples and analyzing each one once.

The **reporting limit** is the concentration below which regulatory rules say that a given analyte is reported as "not detected." "Not detected" does not mean that analyte was not observed, but that it is below a prescribed level. Reporting limits are at least 5 to 10 times higher than the detection limit, so detecting analyte at the reporting limit is not ambiguous. Reporting limits are dictated by regulations and by data quality objectives. For example, it is harder to analyze ions in hazardous waste than in drinking water, because the matrix in waste is so much more complicated and concentrated. Therefore reporting limits for ions in hazardous waste are set higher than reporting limits for ions in drinking water. To validate an analytical procedure, samples are spiked with analytes at their reporting limits, and the assay must return accurate measurements at these levels. The method should be validated for each different matrix that is encountered.

### Robustness

**Robustness** is the ability of an analytical method to be unaffected by small, deliberate changes in operating parameters. For example, a chromatographic method is robust if it gives acceptable results when small changes are made in solvent composition, pH, buffer concentration, temperature, injection volume, and detector wavelength. In tests for robustness, the organic solvent content in the mobile phase could be varied by $\pm 2\%$, the eluent pH varied by $\pm 0.1$, and column temperature varied by $\pm 5°C$. If acceptable results are obtained, the written procedure should state that these variations are tolerable. Capillary electrophoresis requires such small volumes that a given solution could conceivably be used for months before it is used up. Therefore solution stability (shelf life) should be evaluated for robustness.

### Optimization of Analytical Methods

Operating parameters usually need to be optimized when we develop an analytical method. The least efficient way to do this is to vary one parameter at a time while keeping everything else constant. More-efficient procedures are called *fractional factorial experimental design*[7] and *simplex optimization.*[8] Section 7-8 provides an example of the efficient design of a titration experiment.

## ▰ ▰ ▰ ▰  5-3  Standard Addition[9]

In **standard addition,** known quantities of analyte are added to the unknown. From the increase in signal, we deduce how much analyte was in the original unknown. This method requires a linear response to analyte.

Standard addition is especially appropriate when the sample composition is unknown or complex and affects the analytical signal. The *matrix* is everything in the unknown, other than analyte. A **matrix effect** is a change in the analytical signal caused by anything in the sample other than analyte.

Figure 5-3 shows a strong matrix effect in the analysis of perchlorate ($ClO_4^-$) by mass spectrometry. Perchlorate at a level above 18 $\mu g/L$ in drinking water is of concern because it can reduce thyroid hormone production. Standard solutions of $ClO_4^-$ in pure water gave the upper calibration curve in Figure 5-3. The response to standard solutions with the same concentrations of $ClO_4^-$ in groundwater was 15 times less, as shown in the lower curve. Reduction of the $ClO_4^-$ signal is a *matrix effect* attributed to other anions present in the groundwater.

Different groundwaters have different concentrations of many anions, so there is no way to construct a calibration curve for this analysis that would apply to more than one specific groundwater. Hence, the method of standard addition is required. When we add a small volume of concentrated standard to an existing unknown, we do not change the concentration of the matrix very much.

Consider a standard addition in which a sample with unknown initial concentration of analyte $[X]_i$ gives a signal intensity $I_X$. Then a known concentration of standard, S, is added to an aliquot of the sample and a signal $I_{S+X}$ is observed for this second solution. Addition of standard to the unknown changes the concentration of the original analyte because of dilution. Let's call the diluted concentration of analyte $[X]_f$, where "f" stands for "final." We designate the concentration of standard in the final solution as $[S]_f$. (Bear in mind that the chemical species X and S are the same.)

The matrix affects the magnitude of the analytical signal. In standard addition, all samples are in the same matrix.

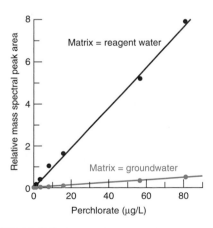

*Figure 5-3* Calibration curves for perchlorate in pure water and in groundwater. [Data from C. J. Koester, H. R. Beller, and R. U. Halden, "Analysis of Perchlorate in Groundwater by Electrospray Ionization Mass Spectrometry/Mass Spectrometry," Environ. Sci. Technol. **2000,** 34, 1862.]

Signal is directly proportional to analyte concentration, so

$$\frac{\text{Concentration of analyte in initial solution}}{\text{Concentration of analyte plus standard in final solution}} = \frac{\text{signal from initial solution}}{\text{signal from final solution}}$$

**Derivation of Equation 5-7:**

$I_X = k[X]_i$, where $k$ is a constant of proportionality

$I_{S+X} = k([S]_f + [X]_f)$, where $k$ is the same constant

Dividing one equation by the other gives

$$\frac{I_X}{I_{S+X}} = \frac{k[X]_i}{k([S]_f + [X]_f)} = \frac{[X]_i}{[S]_f + [X]_f}$$

*Standard addition equation:*  $\dfrac{[X]_i}{[S]_f + [X]_f} = \dfrac{I_X}{I_{S+X}}$    (5-7)

For an initial volume $V_0$ of unknown and added volume $V_s$ of standard with concentration $[S]_i$, the total volume is $V = V_0 + V_s$ and the concentrations in Equation 5-7 are

$$[X]_f = [X]_i\left(\frac{V_0}{V}\right) \qquad\qquad [S]_f = [S]_i\left(\frac{V_s}{V}\right) \qquad (5\text{-}8)$$
           ↑                                          ↑

The quotient (initial volume/final volume), which relates final concentration to initial concentration, is called the **dilution factor.** It comes directly from Equation 1-3.

By expressing the diluted concentration of analyte, $[X]_f$, in terms of the initial concentration of analyte, $[X]_i$, we can solve for $[X]_i$, because everything else in Equation 5-7 is known.

> **Example** Standard Addition
>
> Serum containing $Na^+$ gave a signal of 4.27 mV in an atomic emission analysis. Then 5.00 mL of 2.08 M NaCl were added to 95.0 mL of serum. This spiked serum gave a signal of 7.98 mV. Find the original concentration of $Na^+$ in the serum.
>
> **Solution**   From Equation 5-8, the final concentration of $Na^+$ after dilution with the standard is $[X]_f = [X]_i(V_0/V) = [X]_i(95.0\text{ mL}/100.0\text{ mL})$. The final concentration of added standard is $[S]_f = [S]_i(V_s/V) = (2.08\text{ M})(5.00\text{ mL}/100.0\text{ mL}) = 0.104\text{ M}$. Equation 5-7 becomes
>
> $$\frac{[Na^+]_i}{[0.104\text{ M}] + 0.950[Na^+]_i} = \frac{4.27\text{ mV}}{7.98\text{ mV}} \Rightarrow [Na^+]_i = 0.113\text{ M}$$

### A Graphic Procedure for Standard Addition

There are two common methods to perform standard addition. In the method illustrated in Figure 5-4, equal volumes of unknown are pipetted into several volumetric flasks. Then

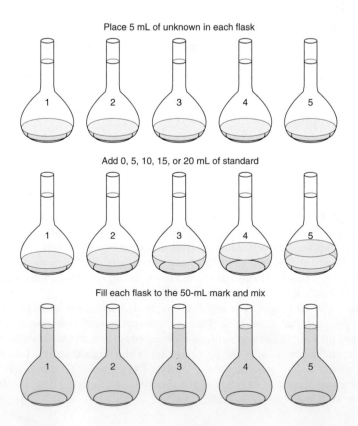

*Figure 5-4* Standard addition experiment with constant total volume.

| | A | B | C | D | E |
|---|---|---|---|---|---|
| 1 | Vitamin C standard addition experiment | | | | |
| 2 | Add 0.279 M ascorbic acid to 50.0 mL orange juice | | | | |
| 3 | | | | | |
| 4 | | Vs = | | | |
| 5 | Vo (mL) = | mL ascorbic | I(s+x) = | x-axis function | y-axis function |
| 6 | 50 | acid added | signal ($\mu$A) | Si*Vs/Vo | I(s+x)*V/Vo |
| 7 | [S]i (mM) = | 0.000 | 1.78 | 0.000 | 1.780 |
| 8 | 279 | 0.050 | 2.00 | 0.279 | 2.002 |
| 9 | | 0.250 | 2.81 | 1.395 | 2.824 |
| 10 | | 0.400 | 3.35 | 2.232 | 3.377 |
| 11 | | 0.550 | 3.88 | 3.069 | 3.923 |
| 12 | | 0.700 | 4.37 | 3.906 | 4.431 |
| 13 | | 0.850 | 4.86 | 4.743 | 4.943 |
| 14 | | 1.000 | 5.33 | 5.580 | 5.437 |
| 15 | | 1.150 | 5.82 | 6.417 | 5.954 |
| 16 | | | | | |
| 17 | D7 = $A$8*B7/$A$6 | | E7 = C7*($A$6+B7)/$A$6 | | |

**Figure 5-5** Data for standard addition experiment with variable total volume.

increasing volumes of standard are added to each flask and each is diluted to the same final volume. Every flask now contains the same concentration of unknown and differing concentrations of standard. (Remember that the standard is the same substance as the unknown.) For each flask, a measurement of analytical signal, $I_{S+X}$, is then made. The added standard should increase the analytical signal by a factor of 1.5 to 3. The method in Figure 5-4 is necessary when the analysis consumes some of the solution.

If the analysis does not consume solution, then we do not have to prepare individual samples for every measurement. In this case, we begin with an unknown solution and measure the analytical signal. Then we add a small volume of concentrated standard and measure the signal again. We add several more small volumes of standard and measure the signal after each addition. The standard should be concentrated so that only small volumes are added and the sample matrix is not appreciably altered.

Figure 5-5 shows data for an experiment in which ascorbic acid (vitamin C) was measured in orange juice by an electrochemical method. The current between a pair of electrodes immersed in the juice is proportional to the concentration of ascorbic acid. Eight standard additions increased the current from 1.78 to 5.82 $\mu$A (column C), which is at the upper end of the desired range of 1.5- to 3-fold increase in analytical signal.

Both methods of standard addition can be analyzed with the graph in Figure 5-6. The theoretical response to the additions is derived by substituting expressions for $[X]_f$ and $[S]_f$

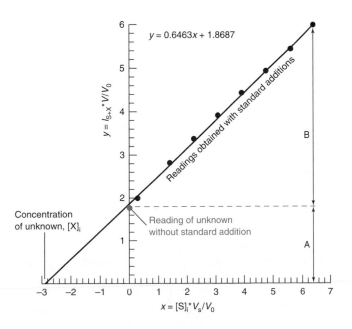

$y = 0.6463x + 1.8687$

$y = I_{S+x} * V/V_0$

Readings obtained with standard additions

Concentration of unknown, $[X]_i$

Reading of unknown without standard addition

$x = [S]_i * V_s/V_0$

**Figure 5-6** Graphical treatment of standard addition data from Figure 5-5. Standard additions should increase the analytical signal to between 1.5 and 3 times its original value (that is, B = 0.5 A to 2A).

Plot $I_{S+X}\left(\dfrac{V}{V_0}\right)$ versus $[S]_i\left(\dfrac{V_s}{V_0}\right)$

x-intercept is $[X]_i$

The equation of the line in Figure 5-6 is $y = mx + b$. The x-intercept is obtained by setting $y = 0$:

$$0 = mx + b$$
$$x = -b/m$$

Standard addition:
For constant total volume, you can plot $I_{S+X}$ versus $[S]_f$ and the x-intercept is $[X]_f$. If you plot $I_{S+X}\left(\dfrac{V}{V_0}\right)$ versus $[S]_i\left(\dfrac{V_s}{V_0}\right)$, the x-intercept is $[X]_i$.

In *standard addition*, the standard is the same substance as the analyte.
An *internal standard* is a different substance from the analyte.

When the relative response of an instrument to analyte and standard remains constant over a range of concentrations, we say there is a linear response. For critical work, this assumption should be verified because it is not always true.[11]

If the detector responds equally to standard and analyte, $F = 1$. If the detector responds twice as much to analyte as to standard, $F = 2$. If the detector responds half as much to analyte as to standard, $F = 0.5$.

---

from Equations 5-8 into Equation 5-7. Following a little rearrangement, we find

*For successive standard additions to one solution:*
$$I_{S+X}\left(\frac{V}{V_0}\right) = I_X + \underbrace{\frac{I_X}{[X]_i}[S]_i\left(\frac{V_s}{V_0}\right)}_{} \tag{5-9}$$

$\underbrace{\phantom{xxxxxxxx}}_{\text{Function to plot on y-axis}}$  $\underbrace{\phantom{xxxxxxxx}}_{\text{Function to plot on x-axis}}$

A graph of $I_{S+X}(V/V_0)$ (the *corrected response*) on the y-axis versus $[S]_i(V_s/V_0)$ on the x-axis should be a straight line. The data plotted in Figure 5-6 are computed in columns D and E of Figure 5-5. The right side of Equation 5-9 is 0 when $[S]_i(V_s/V_0) = -[X]_i$. The magnitude of the intercept on the x-axis is the *original* concentration of unknown, $[X]_i = 2.89$ mM in Figure 5-6.

If all samples for the standard addition experiment are made up to a *constant final volume*, as in Figure 5-4, an alternate way to handle the data is to graph the signal $I_{S+X}$ versus the concentration of diluted standard, $[S]_f$. In this case, the x-intercept of the graph is the *final* concentration of unknown, $[X]_f$, after dilution to the final sample volume.

The uncertainty in the x-intercept is[10]

$$\begin{array}{c}\text{Standard deviation}\\\text{of x-intercept}\end{array} = \frac{s_y}{|m|}\sqrt{\frac{1}{n} + \frac{\bar{y}^2}{m^2\Sigma(x_i - \bar{x})^2}} \tag{5-10}$$

where $s_y$ is the standard deviation of $y$ (Equation 4-20), $|m|$ is the absolute value of the slope of the least-squares line (Equation 4-16), $n$ is the number of data points (nine in Figure 5-6), $\bar{y}$ is the mean value of $y$ for the nine points, $x_i$ are the individual values of $x$ for the nine points, and $\bar{x}$ is the mean value of $x$ for the nine points. For the points in Figure 5-6, the uncertainty in the x-intercept is $0.09_8$ mM.

The confidence interval is $\pm t \times$ (standard deviation of x-intercept), where $t$ is Student's $t$ (Table 4-2) for $n - 2$ degrees of freedom. The 95% confidence interval for the intercept in Figure 5-6 is $\pm(2.365)(0.09_8$ mM$) = \pm0.23$ mM. The value $t = 2.365$ was taken from Table 4-2 for $9 - 2 = 7$ degrees of freedom.

## ■■■■ 5-4 Internal Standards

An **internal standard** is a known amount of a compound, different from analyte, that is added to the unknown. Signal from analyte is compared with signal from the internal standard to find out how much analyte is present.

Internal standards are especially useful for analyses in which the quantity of sample analyzed or the instrument response varies slightly from run to run for reasons that are difficult to control. For example, gas or liquid flow rates that vary by a few percent in a chromatography experiment (Figure 0-4) could change the detector response. A calibration curve is accurate only for the one set of conditions under which it was obtained. However, the *relative* response of the detector to the analyte and standard is usually constant over a wide range of conditions. If signal from the standard increases by 8.4% because of a change in flow rate, signal from the analyte usually increases by 8.4% also. As long as the concentration of standard is known, the correct concentration of analyte can be derived. Internal standards are widely used in chromatography because the small quantity of sample solution injected into the chromatograph is not reproducible.

Internal standards are also desirable when sample loss can occur during sample preparation steps prior to analysis. If a known quantity of standard is added to the unknown prior to any manipulations, the ratio of standard to analyte remains constant because the same fraction of each is lost in any operation.

To use an internal standard, we prepare a known mixture of standard and analyte to measure the relative response of the detector to the two species. In the chromatogram in Figure 5-7, the area under each peak is proportional to the concentration of the species injected into the column. However, the detector generally has a different response to each component. For example, if both analyte (X) and internal standard (S) have concentrations of 10.0 mM, the area under the analyte peak might be 2.30 times greater than the area under the standard peak. We say that the **response factor**, $F$, is 2.30 times greater for X than for S.

*Response factor:*
$$\frac{\text{Area of analyte signal}}{\text{Concentration of analyte}} = F\left(\frac{\text{area of standard signal}}{\text{concentration of standard}}\right) \tag{5-11}$$

$$\frac{A_X}{[X]} = F\left(\frac{A_S}{[S]}\right)$$

[X] and [S] are the concentrations of analyte and standard *after they have been mixed together*. Equation 5-11 is predicated on linear response to both the analyte and the standard.

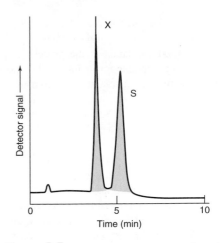

**Example** Using an Internal Standard

In a preliminary experiment, a solution containing 0.083 7 M X and 0.066 6 M S gave peak areas of $A_X$ = 423 and $A_S$ = 347. (Areas are measured in arbitrary units by the instrument's computer.) To analyze the unknown, 10.0 mL of 0.146 M S were added to 10.0 mL of unknown, and the mixture was diluted to 25.0 mL in a volumetric flask. This mixture gave the chromatogram in Figure 5-7, for which $A_X$ = 553 and $A_S$ = 582. Find the concentration of X in the unknown.

**Solution** First use the standard mixture to find the response factor in Equation 5-11:

*Standard mixture:*
$$\frac{A_X}{[X]} = F\left(\frac{A_S}{[S]}\right)$$

$$\frac{423}{0.083\ 7} = F\left(\frac{347}{0.066\ 6}\right) \Rightarrow F = 0.970_0$$

In the mixture of unknown plus standard, the concentration of S is

$$[S] = \underbrace{(0.146\ M)}_{\substack{\text{Initial} \\ \text{concentration}}} \underbrace{\left(\frac{10.0}{25.0}\right)}_{\substack{\text{Dilution} \\ \text{factor}}} = 0.058\ 4\ M$$

Using the known response factor, we substitute back into Equation 5-11 to find the concentration of unknown in the mixture:

*Unknown mixture:*
$$\frac{A_X}{[X]} = F\left(\frac{A_S}{[S]}\right)$$

$$\frac{553}{[X]} = 0.970_0\left(\frac{582}{0.058\ 4}\right) \Rightarrow [X] = 0.057\ 2_1\ M$$

Because X was diluted from 10.0 to 25.0 mL when the mixture with S was prepared, the original concentration of X in the unknown was (25.0 mL/10.0 mL)(0.057 2$_1$ M) = 0.143 M.

**Figure 5-7** Chromatographic separation of unknown (X) and internal standard (S). A known amount of S was added to the unknown. The relative areas of the signals from X and S allow us to find out how much X is in the mixture. It is necessary first to measure the relative response of the detector to each compound.

The dilution factor $\dfrac{\text{initial volume}}{\text{final volume}}$ converts initial concentration into final concentration.

---

## Terms to Understand

| | | | |
|---|---|---|---|
| assessment | field blank | performance test sample | sensitivity |
| calibration check | internal standard | quality assurance | specification |
| coefficient of variation | linearity | range | specificity |
| control chart | lower limit of quantitation | reagent blank | spike |
| detection limit | matrix | reporting limit | standard addition |
| dilution factor | matrix effect | response factor | standard operating procedure |
| false negative | method blank | robustness | use objective |
| false positive | method validation | selectivity | |

---

## Summary

Quality assurance is what we do to get the right answer for our purpose. We begin by writing use objectives, from which specifications for data quality can be derived. Specifications could include requirements for sampling, accuracy, precision, specificity, detection limit, standards, and blank values. For any meaningful analysis, we must first collect a representative sample. In an analysis of standards, the method must produce a result that is acceptably close to the known value. A method blank contains all components except analyte, and it is taken through all steps of the analytical procedure. We subtract the response of the method blank from the response of a real sample prior to calculating the quantity of analyte in the sample. A field blank tells us if analyte is inadvertently picked up by exposure to field conditions. Accuracy can be assessed by analyzing certified standards, by calibration checks performed by the analyst, with spikes made by the analyst, and by blind quality control samples. Written standard operating procedures must be followed rigorously to avoid inadvertent changes in procedure that could affect the outcome. Assessment is the process of (1) collecting data to show that analytical procedures are operating within specified limits and (2) verifying that final results meet use objectives. Control charts can be used to monitor accuracy, precision, or instrument performance as a function of time.

Method validation is the process of proving that an analytical method is acceptable for its intended purpose. In validating a method, we typically demonstrate that requirements are met for specificity, linearity, accuracy, precision, range, limit of detection, limit of quantitation, and robustness. Specificity is the ability to distinguish analyte from anything else. Linearity is usually measured by the square of the correlation coefficient for the calibration curve. Types of precision include instrument precision, intra-assay precision, ruggedness (intermediate precision), and, most generally, interlaboratory precision. The "Horwitz trumpet" is an empirical statement that precision becomes poorer as analyte concentration decreases. Range is the concentration interval over which linearity, accuracy, and precision are acceptable. The detection limit is usually taken as 3 times the standard deviation of the blank. The lower limit of quantitation is 10 times the standard deviation of the blank. The reporting limit is the concentration below which regulations say that analyte is reported as "not detected," even when it is observed. Robustness is the ability of an analytical method to be unaffected by small changes in operating parameters.

A standard addition is a known quantity of analyte added to an unknown to increase the concentration of analyte. Standard additions are especially useful when matrix effects are important. A matrix effect is a change in the analytical signal caused by anything in the sample other than analyte. You should be able to use Equation 5-7 to compute the quantity of analyte in a standard addition experiment. Equation 5-9 is used with multiple standard additions to construct the graph in Figure 5-6, in which the x-intercept gives us the concentration of analyte.

An internal standard is a known amount of a compound, different from analyte, that is added to the unknown. Signal from analyte is compared with signal from the internal standard to find out how much analyte is present. Internal standards are useful when the quantity of sample analyzed is not reproducible, when instrument response varies from run to run, or when sample losses occur in sample preparation. The response factor in Equation 5-11 is the relative response to analyte and standard.

## Exercises

**5-A.** *Detection limit.* In spectrophotometry, we measure the concentration of analyte by its absorbance of light. A low-concentration sample was prepared and nine replicate measurements gave absorbances of 0.004 7, 0.005 4, 0.006 2, 0.006 0, 0.004 6, 0.005 6, 0.005 2, 0.004 4, and 0.005 8. Nine reagent blanks gave values of 0.000 6, 0.001 2, 0.002 2, 0.000 5, 0.001 6, 0.000 8, 0.001 7, 0.001 0, and 0.001 1.

(a) Find the absorbance detection limit with Equation 5-3.

(b) The calibration curve is a graph of absorbance versus concentration. Absorbance is a dimensionless quantity. The slope of the calibration curve is $m = 2.24 \times 10^4 \, M^{-1}$. Find the concentration detection limit with Equation 5-5.

(c) Find the lower limit of quantitation with Equation 5-6.

**5-B.** *Standard addition.* An unknown sample of $Ni^{2+}$ gave a current of 2.36 μA in an electrochemical analysis. When 0.500 mL of solution containing 0.028 7 M $Ni^{2+}$ was added to 25.0 mL of unknown, the current increased to 3.79 μA.

(a) Denoting the initial, unknown concentration as $[Ni^{2+}]_i$, write an expression for the final concentration, $[Ni^{2+}]_f$, after 25.0 mL of unknown was mixed with 0.500 mL of standard. Use the dilution factor for this calculation.

(b) In a similar manner, write the final concentration of added standard $Ni^{2+}$, designated as $[S]_f$.

(c) Find $[Ni^{2+}]_i$ in the unknown.

**5-C.** *Internal standard.* A solution was prepared by mixing 5.00 mL of unknown (element X) with 2.00 mL of solution containing 4.13 μg of standard (element S) per milliliter, and diluting to 10.0 mL. The measured signal ratio in an atomic absorption experiment was (signal due to X)/(signal due to S) = 0.808. In a separate experiment, for equal concentrations of X and S, the signal due to X was found to be 1.31 times more intense than the signal due to S. Find the concentration of X in the unknown.

**5-D.** In Figure 5-6, the x-intercept is $-2.89$ mM and its standard deviation is $0.09_8$ mM. Find the 90% and 99% confidence intervals for the intercept.

**5-E.** *Control chart.* Volatile compounds in human blood serum were measured by purge and trap gas chromatography/mass spectrometry. For quality control, serum was periodically spiked with a constant amount of 1,2-dichlorobenzene and the concentration (ng/g = ppb) was measured. Find the mean and standard deviation for the following spike data and prepare a control chart. State whether or not the observations (Obs.) meet each of the criteria for stability of a control chart.

| Day | Obs. (ppb) | Day | Obs. (ppb) | Day | Obs. (ppb) | Day | Obs. (ppb) | Day | Obs. (ppb) |
|---|---|---|---|---|---|---|---|---|---|
| 0 | 1.05 | 91 | 1.13 | 147 | 0.83 | 212 | 1.03 | 290 | 1.04 |
| 1 | 0.70 | 101 | 1.64 | 149 | 0.88 | 218 | 0.90 | 294 | 0.85 |
| 3 | 0.42 | 104 | 0.79 | 154 | 0.89 | 220 | 0.86 | 296 | 0.59 |
| 6 | 0.95 | 106 | 0.66 | 156 | 0.72 | 237 | 1.05 | 300 | 0.83 |
| 7 | 0.55 | 112 | 0.88 | 161 | 1.18 | 251 | 0.79 | 302 | 0.67 |
| 30 | 0.68 | 113 | 0.79 | 167 | 0.75 | 259 | 0.94 | 304 | 0.66 |
| 70 | 0.83 | 115 | 1.07 | 175 | 0.76 | 262 | 0.77 | 308 | 1.04 |
| 72 | 0.97 | 119 | 0.60 | 182 | 0.93 | 277 | 0.85 | 311 | 0.86 |
| 76 | 0.60 | 125 | 0.80 | 185 | 0.72 | 282 | 0.72 | 317 | 0.88 |
| 80 | 0.87 | 128 | 0.81 | 189 | 0.87 | 286 | 0.68 | 321 | 0.67 |
| 84 | 1.03 | 134 | 0.84 | 199 | 0.85 | 288 | 0.86 | 323 | 0.68 |

SOURCE: D. L. Ashley, M. A. Bonin, F. L. Cardinali, J. M. McCraw, J. S. Holler, L. L. Needham, and D. G. Patterson, Jr., "Determining Volatile Organic Compounds in Blood by Using Purge and Trap Gas Chromatography/Mass Spectrometry," *Anal. Chem.* **1992**, *64*, 1021.

## Problems

### Quality Assurance and Method Validation

**5-1.** Explain the meaning of the quotation at the beginning of this chapter: "Get the right data. Get the data right. Keep the data right."

**5-2.** What are the three parts of quality assurance? What questions are asked in each part and what actions are taken in each part?

**5-3.** How can you validate precision and accuracy?

**5-4.** Distinguish *raw data, treated data,* and *results.*

**5-5.** What is the difference between a *calibration check* and a *performance test sample?*

**5-6.** What is the purpose of a blank? Distinguish *method blank, reagent blank,* and *field blank.*

**5-7.** Distinguish *linear range, dynamic range,* and *range.*

**5-8.** What is the difference between a *false positive* and a *false negative?*

**5-9.** Consider a sample that contains analyte at the detection limit defined in Figure 5-2. Explain the following statements: There is approximately a 1% chance of falsely concluding that a sample containing no analyte contains analyte above the detection limit. There is a 50% chance of concluding that a sample that really contains analyte at the detection limit does not contain analyte above the detection limit.

**5-10.** How is a control chart used? State six indications that a process is going out of control.

**5-11.** Here is a use objective for a chemical analysis to be performed at a drinking water purification plant: "Data and results collected quarterly shall be used to determine whether the concentrations of haloacetates in the treated water demonstrate compliance with the levels set by the Stage 1 Disinfection By-products Rule using Method 552.2" (a specification that sets precision, accuracy, and other requirements). Which of the following questions best summarizes the meaning of the use objective?

(a) Are haloacetate concentrations known within specified precision and accuracy?

(b) Are any haloacetates detectable in the water?

(c) Do any haloacetate concentrations exceed the regulatory limit?

**5-12.** What is the difference between an instrument detection limit and a method detection limit? What is the difference between robustness and ruggedness?

**5-13.** Define the following terms: instrument precision, injection precision, intra-assay precision, ruggedness, intermediate precision, and interlaboratory precision.

**5-14.** 🖩 *Correlation coefficient.* Syntetic data are given in the table at the top of the next column for calibration curves in which 1% or 10% random Gaussian noise was superimposed on linear data that follow the equation $y = 26.4x + 1.37$. Prepare a graph of signal versus concentration for each data set and find the least-squares straight line and correlation coefficient, $R^2$. An easy way to do this is to enter the data into an Excel spreadsheet in columns A, B, and C. In the INSERT menu, select CHART and proceed to plot the data in columns A and B in an XY Scatter chart without a line. Then click on a data point to highlight all the data points. In the CHART menu, select ADD TRENDLINE. (If your version of Excel does not have a CHART menu, look for TRENDLINE in the INSERT menu.) In the window that appears, select Linear. Go to the Options tab and select Display Equation and Display R-Squared. Click OK and you will see the least-squares line, its equation, and the value of $R^2$ on the graph. Repeat the same process for data in columns A and C.

| Concentration | Signal (1% noise) | Signal (10% noise) |
|---|---|---|
| 0 | 1 | 1 |
| 10 | 263 | 284 |
| 20 | 531 | 615 |
| 30 | 801 | 900 |
| 40 | 1 053 | 1 190 |
| 50 | 1 333 | 1 513 |
| 60 | 1 587 | 1 574 |
| 70 | 1 842 | 1 846 |
| 80 | 2 114 | 1 988 |
| 90 | 2 391 | 1 974 |
| 100 | 2 562 | 2 504 |

**5-15.** In a murder trial in the 1990s, the defendant's blood was found at the crime scene. The prosecutor argued that blood was left by the defendant during the crime. The defense argued that police "planted" the defendant's blood from a sample collected later. Blood is normally collected in a vial containing the metal-binding compound EDTA as an anticoagulant with a concentration of ~4.5 mM after the vial has been filled with blood. At the time of the trial, procedures to measure EDTA in blood were not well established. Even though the amount of EDTA found in the crime-scene blood was orders of magnitude below 4.5 mM, the jury acquitted the defendant. This trial motivated the development of a new method to measure EDTA in blood.

(a) *Precision and accuracy.* To measure accuracy and precision of the method, blood was fortified with EDTA to known levels.

$$\text{Accuracy} = 100 \times \frac{\text{mean value found} - \text{known value}}{\text{known value}}$$

$$\text{Precision} = 100 \times \frac{\text{standard deviation}}{\text{mean}} \equiv \textit{coefficient of variation}$$

For each of the three spike levels in the table below, find the precision and accuracy of the quality control samples.

EDTA measurements (ng/mL) at three fortification levels

| Spike: 22.2 ng/mL | 88.2 ng/mL | 314 ng/mL |
|---|---|---|
| Found: 33.3 | 83.6 | 322 |
| 19.5 | 69.0 | 305 |
| 23.9 | 83.4 | 282 |
| 20.8 | 100.0 | 329 |
| 20.8 | 76.4 | 276 |

SOURCE: R. L. Sheppard and J. Henion, *Anal. Chem.* **1997,** *69,* 477A, 2901.

(b) *Detection and quantitation limits.* Low concentrations of EDTA near the detection limit gave the following dimensionless instrument readings: 175, 104, 164, 193, 131, 189, 155, 133, 151, and 176. Ten blanks had a mean reading of $45._0$. The slope of the calibration curve is $1.75 \times 10^9 \, M^{-1}$. Estimate the signal and concentration detection limits and the lower limit of quantitation for EDTA.

**5-16.** (a) From Box 5-2, estimate the minimum expected coefficient of variation, CV(%), for interlaboratory results when the analyte concentration is (i) 1 wt% or (ii) 1 part per trillion.

(b) The coefficient of variation within a laboratory is typically ~0.5–0.7 of the between-laboratory variation. If your class analyzes an unknown containing 10 wt% $NH_3$, what is the minimum expected coefficient of variation for the class?

**5-17.** The experimental data below show how the coefficient of variation increases as analyte concentration decreases in studies conducted in single laboratories. Plot these two sets of data on one graph and plot the Horwitz equation from Box 5-2 on the same graph. These data give an idea of the variability in experimental precision.

| HPLC determination of spiramycin antibiotic | | Gas chromatographic determination of ethanol | |
|---|---|---|---|
| $C\left(\dfrac{\text{g analyte}}{\text{g sample}}\right)$ | CV(%) | $C\left(\dfrac{\text{g analyte}}{\text{g sample}}\right)$ | CV(%) |
| 0.000 1 | 39.7 | 0.000 01 | 3.58 |
| 0.000 2 | 23.4 | 0.000 10 | 1.30 |
| 0.000 5 | 20.7 | 0.000 25 | 0.65 |
| 0.001 | 12.3 | 0.001 00 | 0.55 |
| 0.002 5 | 5.3 | | |
| 0.005 | 2.6 | | |
| 0.01 | 1.6 | | |

SOURCE: J. Vial and A. Jardy, "Experimental Comparison of Different Approaches to Estimate LOD and LOQ of an HPLC Method," Anal. Chem. **1999**, 71, 2672; and J. M. Green, "A Practical Guide to Analytical Method Validation," Anal. Chem. **1996**, 68, 305A.

**5-18.** *Spike recovery and detection limit.* Species of arsenic found in drinking water include $AsO_3^{3-}$ (arsenite), $AsO_4^{3-}$ (arsenate), $(CH_3)_2AsO_2^-$ (dimethylarsinate), and $(CH_3)AsO_3^{2-}$ (methylarsonate). Pure water containing no arsenic was spiked with 0.40 μg arsenate/L. Seven replicate determinations gave 0.39, 0.40, 0.38, 0.41, 0.36, 0.35, and 0.39 μg/L.[12] Find the mean percent recovery of the spike and the concentration detection limit (μg/L).

**5-19.** *Detection limit.* Low concentrations of Ni-EDTA near the detection limit gave the following counts in a mass spectral measurement: 175, 104, 164, 193, 131, 189, 155, 133, 151, 176. Ten measurements of a blank had a mean of 45 counts. A sample containing 1.00 μM Ni-EDTA gave 1 797 counts. Estimate the detection limit for Ni-EDTA.

**5-20.** *Detection limit.* A sensitive chromatographic method was developed to measure sub-part-per-billion levels of the disinfectant by-products iodate ($IO_3^-$), chlorite ($ClO_2^-$), and bromate ($BrO_3^-$) in drinking water. As the oxyhalides emerge from the column, they react with $Br^-$ to make $Br_3^-$, which is measured by its strong absorption at 267 nm. For example, each mole of bromate makes 3 mol of $Br_3^-$ by the reaction $BrO_3^- + 8Br^- + 6H^+ \rightarrow 3Br_3^- + 3H_2O$.

Bromate near its detection limit gave the following chromatographic peak heights and standard deviations. The blank is 0 because chromatographic peak height is measured from the baseline adjacent to the peak. For each concentration, estimate the detection limit. Find the mean detection limit.

| Concentration (μg/L) | Peak height (arbitrary units) | Relative standard deviation (%) | Number of measurements |
|---|---|---|---|
| 0.2 | 17 | 14.4 | 8 |
| 0.5 | 31 | 6.8 | 7 |
| 1.0 | 56 | 3.2 | 7 |
| 2.0 | 111 | 1.9 | 7 |

SOURCE: H. S. Weinberg and H. Yamada, "Post-Ion-Chromatography Derivatization for the Determination of Oxyhalides at Sub-PPB Levels in Drinking Water," Anal. Chem. **1998**, 70, 1.

**5-21.** Olympic athletes are tested to see that they are not using illegal performance-enhancing drugs. Suppose that urine samples are taken and analyzed and the rate of false positive results is 1%. Sup-

pose also that it is too expensive to refine the method to reduce the rate of false positive results. We certainly do not want to accuse innocent people of using illegal drugs. What can you do to reduce the rate of false accusations even though the test always has a false positive rate of 1%?

**Standard Addition**

**5-22.** Why is it desirable in the method of standard addition to add a small volume of concentrated standard rather than a large volume of dilute standard?

**5-23.** An unknown sample of $Cu^{2+}$ gave an absorbance of 0.262 in an atomic absorption analysis. Then 1.00 mL of solution containing 100.0 ppm (= μg/mL) $Cu^{2+}$ was mixed with 95.0 mL of unknown, and the mixture was diluted to 100.0 mL in a volumetric flask. The absorbance of the new solution was 0.500.

(a) Denoting the initial, unknown concentration as $[Cu^{2+}]_i$, write an expression for the final concentration, $[Cu^{2+}]_f$, after dilution. Units of concentration are ppm.

(b) In a similar manner, write the final concentration of added standard $Cu^{2+}$, designated as $[S]_f$.

(c) Find $[Cu^{2+}]_i$ in the unknown.

**5-24.** ▦ *Standard addition graph.* Tooth enamel consists mainly of the mineral calcium hydroxyapatite, $Ca_{10}(PO_4)_6(OH)_2$. Trace elements in teeth of archaeological specimens provide anthropologists with clues about diet and diseases of ancient people. Students at Hamline University measured strontium in enamel from extracted wisdom teeth by atomic absorption spectroscopy. Solutions with a constant total volume of 10.0 mL contained 0.750 mg of dissolved tooth enamel plus variable concentrations of added Sr.

| Added Sr (ng/mL = ppb) | Signal (arbitrary units) |
|---|---|
| 0 | 28.0 |
| 2.50 | 34.3 |
| 5.00 | 42.8 |
| 7.50 | 51.5 |
| 10.00 | 58.6 |

SOURCE: V. J. Porter, P. M. Sanft, J. C. Dempich, D.D. Dettmer, A. E. Erickson, N. A. Dubauskie, S. T. Myster, E. H. Matts, and E. T. Smith, "Elemental Analysis of Wisdom Teeth by Atomic Spectroscopy Using Standard Addition," J. Chem. Ed. **2002**, 79, 1114.

(a) Find the concentration of Sr and its uncertainty in the 10-mL sample solution in parts per billion = ng/mL.

(b) Find the concentration of Sr in tooth enamel in parts per million = μg/g.

(c) If the standard addition intercept is the major source of uncertainty, find the uncertainty in the concentration of Sr in tooth enamel in parts per million.

(d) Find the 95% confidence interval for Sr in tooth enamel.

**5-25.** Europium is a lanthanide element found in parts per billion levels in natural waters. It can be measured from the intensity of orange light emitted when a solution is illuminated with ultraviolet radiation. Certain organic compounds that bind Eu(III) are required to enhance the emission. The figure shows standard addition experiments in which 10.00 mL of sample and 20.00 mL containing a large excess of organic additive were placed in 50-mL volumetric flasks. Then Eu(III) standards (0, 5.00, 10.00, or 15.00 mL) were added and the flasks were diluted to 50.0 mL with $H_2O$. Standards added to tap water contained 0.152 ng/mL (ppb) of Eu(III), but those added to pond water were 100 times more concentrated (15.2 ng/mL).

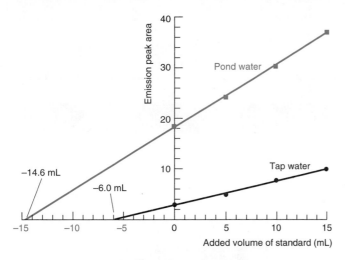

Standard addition of Eu(III) to pond water or tap water. [Data from A. L. Jenkins and G. M. Murray, "Enhanced Luminescence of Lanthanides," J. Chem. Ed. **1998**, 75, 227.]

(a) From the x-intercepts in the graph, calculate the concentration of Eu(III) (ng/mL) in pond water and tap water.

(b) For tap water, the emission peak area increases by 4.61 units when 10.00 mL of 0.152 ng/mL standard are added. This response is 4.61 units/1.52 ng = 3.03 units per ng of Eu(III). For pond water, the response is 12.5 units when 10.00 mL of 15.2 ng/mL standard are added, or 0.082 2 units per ng. How would you explain these observations? Why was standard addition necessary for this analysis?

5-26. 🖾 *Standard addition graph.* Students performed an experiment like that in Figure 5-4 in which each flask contained 25.00 mL of serum, varying additions of 2.640 M NaCl standard, and a total volume of 50.00 mL.

| Flask | Volume of standard (mL) | Na$^+$ atomic emission signal (mV) |
|---|---|---|
| 1 | 0 | 3.13 |
| 2 | 1.000 | 5.40 |
| 3 | 2.000 | 7.89 |
| 4 | 3.000 | 10.30 |
| 5 | 4.000 | 12.48 |

(a) Prepare a graph of Equation 5-9 and find [Na$^+$] in the serum. When you graph $I_{S+X}(V/V_0)$ versus $[S]_i(V_s/V_0)$, the x-intercept is $[X]_i$, not $[X]_f$.

(b) Find the standard deviation and 95% confidence interval for [Na$^+$].

5-27. 🖾 *Standard addition graph.* An assay for substance X is based on its ability to catalyze a reaction that produces radioactive Y. The quantity of Y produced in a fixed time is proportional to the concentration of X in the solution. An unknown containing X in a complex, unknown matrix with an initial volume of 50.0 mL was treated with increments of standard 0.531 M X and the following results were obtained. Prepare a graph of Equation 5-9 and find [X]

and its standard deviation in the original unknown. Also find the 95% confidence interval for [X].

| Volume of added X (μL): | 0 | 100.0 | 200.0 | 300.0 | 400.0 |
|---|---|---|---|---|---|
| Counts/min of radioactive Y: | 1 084 | 1 844 | 2 473 | 3 266 | 4 010 |

**Internal Standards**

5-28. State when standard additions and internal standards, instead of a calibration curve, are desirable, and why.

5-29. A solution containing 3.47 mM X (analyte) and 1.72 mM S (standard) gave peak areas of 3 473 and 10 222, respectively, in a chromatographic analysis. Then 1.00 mL of 8.47 mM S was added to 5.00 mL of unknown X, and the mixture was diluted to 10.0 mL. This solution gave peak areas of 5 428 and 4 431 for X and S, respectively.

(a) Calculate the response factor for the analyte.

(b) Find the concentration of S (mM) in the 10.0 mL of mixed solution.

(c) Find the concentration of X (mM) in the 10.0 mL of mixed solution.

(d) Find the concentration of X in the original unknown.

5-30. Chloroform is an internal standard in the determination of the pesticide DDT in a polarographic analysis in which each compound is reduced at an electrode surface. A mixture containing 0.500 mM chloroform and 0.800 mM DDT gave signals of 15.3 μA for chloroform and 10.1 μA for DDT. An unknown solution (10.0 mL) containing DDT was placed in a 100-mL volumetric flask and 10.2 μL of chloroform (FM 119.39, density = 1.484 g/mL) were added. After dilution to the mark with solvent, polarographic signals of 29.4 and 8.7 μA were observed for the chloroform and DDT, respectively. Find the concentration of DDT in the unknown.

5-31. *Verifying constant response for an internal standard.* When we develop a method using an internal standard, it is important to verify that the response factor is constant over the calibration range. Data are shown below for a chromatographic analysis of naphthalene ($C_{10}H_8$), using deuterated naphthalene ($C_{10}D_8$ in which D is the isotope $^2H$) as an internal standard. The two compounds emerge from the column at almost identical times and are measured by a mass spectrometer, which distinguishes them by molecular mass. From the definition of response factor in Equation 5-11, we can write

$$\frac{\text{Area of analyte signal}}{\text{Area of standard signal}} = F\left(\frac{\text{concentration of analyte}}{\text{concentration of standard}}\right)$$

Prepare a graph of peak area ratio ($C_{10}H_8/C_{10}D_8$) versus concentration ratio $[C_{10}H_8]/[C_{10}D_8]$ and find the slope, which is the response factor. Evaluate $F$ for each of the three samples and find the standard deviation of $F$ to see how "constant" it is.

| Sample | $C_{10}H_8$ (ppm) | $C_{10}D_8$ (ppm) | $C_{10}H_8$ peak area | $C_{10}D_8$ peak area |
|---|---|---|---|---|
| 1 | 1.0 | 10.0 | 303 | 2 992 |
| 2 | 5.0 | 10.0 | 3 519 | 6 141 |
| 3 | 10.0 | 10.0 | 3 023 | 2 819 |

*The volume of solution injected into the column was different in all three runs.*

## CHEMICAL EQUILIBRIUM IN THE ENVIRONMENT

Great Barrier Reef and other coral reefs are threatened with extinction by rising atmospheric $CO_2$. *[Copyright Jon Arnold Images/Alamy.]*

Paper mill on the Potomac River near Westernport, Maryland, neutralizes acid mine drainage in the water. Upstream of the mill, the river is acidic and lifeless; below the mill, the river teems with life. *[Photo courtesy C. Dalpra, Potomac River Basin Commission.]*

Part of the North Branch of the Potomac River runs crystal clear through the scenic Appalachian Mountains, but it is lifeless—a victim of acid drainage from abandoned coal mines. As the river passes a paper mill and a wastewater treatment plant near Westernport, Maryland, the pH rises from an acidic, lethal value of 4.5 to a neutral value of 7.2, at which fish and plants thrive. This happy "accident" comes about because calcium carbonate exiting the paper mill equilibrates with massive quantities of carbon dioxide from bacterial respiration at the sewage treatment plant. The resulting soluble bicarbonate neutralizes the acidic river and restores life downstream of the plant.[1]

$$CaCO_3(s) + CO_2(aq) + H_2O(l) \rightleftharpoons Ca^{2+}(aq) + 2HCO_3^-(aq)$$
Calcium carbonate                                   Dissolved calcium bicarbonate

$$HCO_3^-(aq) + H^+(aq) \xrightarrow{neutralization} CO_2(g) + H_2O(l)$$
Bicarbonate      Acid in river

The chemistry that helps the Potomac River endangers coral reefs, which arc largely $CaCO_3$. Burning of fossil fuel has increased $CO_2$ in the atmosphere from 280 ppm when Captain Cook first sighted the Great Barrier Reef in 1770 to 380 ppm today (Box 20-1). Increased $CO_2$ in the atmosphere adds $CO_2$ to the ocean, dissolving $CaCO_3$ from coral. Rising $CO_2$ and, perhaps, rising atmospheric temperature from the greenhouse effect threaten coral reefs with extinction.[2] $CO_2$ has lowered the average pH of the ocean from its preindustrial value of 8.16 to 8.04 today.[3] Without changes in our activities, the pH could be 7.7 by 2100.

Equilibria govern diverse phenomena from the folding of proteins to the action of acid rain on minerals to the aqueous reactions used in analytical chemistry. This chapter introduces equilibria for the solubility of ionic compounds, complex formation, and acid-base reactions. Chemical equilibrium provides a foundation not only for chemical analysis, but also for other subjects such as biochemistry, geology, and oceanography.

# 6-1 The Equilibrium Constant

For the reaction

$$aA + bB \rightleftharpoons cC + dD \qquad (6\text{-}1)$$

we write the **equilibrium constant**, $K$, in the form

*Equilibrium constant:*
$$K = \frac{[C]^c[D]^d}{[A]^a[B]^b} \qquad (6\text{-}2)$$

Equation 6-2, also called the *law of mass action*, was formulated by the Norwegians C. M. Guldenberg and P. Waage and first published in 1864. Their derivation was based on the idea that the forward and reverse rates of a reaction at equilibrium must be equal.[4]

where the lowercase superscript letters denote stoichiometry coefficients and each capital letter stands for a chemical species. The symbol [A] stands for the concentration of A relative to its standard state (defined next). By definition, *a reaction is favored whenever $K > 1$*.

In the thermodynamic derivation of the equilibrium constant, each quantity in Equation 6-2 is expressed as the *ratio* of the concentration of a species to its concentration in its **standard state.** For solutes, the standard state is 1 M. For gases, the standard state is 1 bar ($\equiv 10^5$ Pa; 1 atm $\equiv 1.013\,25$ bar), and for solids and liquids, the standard states are the pure solid or liquid. It is understood (but rarely written) that [A] in Equation 6-2 really means [A]/(1 M) if A is a solute. If D is a gas, [D] really means (pressure of D in bars)/(1 bar). To emphasize that [D] means pressure of D, we usually write $P_D$ in place of [D]. The terms of Equation 6-2 are actually dimensionless; therefore, all equilibrium constants are dimensionless.

The equilibrium constant is more correctly expressed as a ratio of *activities* rather than of concentrations. We reserve the discussion of activity for Chapter 8.

For the ratios [A]/(1 M) and [D]/(1 bar) to be dimensionless, [A] *must* be expressed in moles per liter (M), and [D] *must* be expressed in bars. If C were a pure liquid or solid, the ratio [C]/(concentration of C in its standard state) would be unity (1) because the standard state is the pure liquid or solid. If [C] is a solvent, the concentration is so close to that of pure liquid C that the value of [C] is still essentially 1.

Equilibrium constants are dimensionless.

*The take-home lesson is this:* When you evaluate an equilibrium constant,

1. Concentrations of solutes should be expressed as moles per liter.
2. Concentrations of gases should be expressed in bars.
3. Concentrations of pure solids, pure liquids, and solvents are omitted because they are unity.

These conventions are arbitrary, but you must use them if you wish to use tabulated values of equilibrium constants, standard reduction potentials, and free energies.

Equilibrium constants are dimensionless but, when specifying concentrations, you must use units of molarity (M) for solutes and bars for gases.

## Manipulating Equilibrium Constants

Consider the reaction

$$HA \rightleftharpoons H^+ + A^- \qquad K_1 = \frac{[H^+][A^-]}{[HA]}$$

Throughout this book, assume that all species in chemical equations are in aqueous solution, unless otherwise specified.

*If the direction of a reaction is reversed, the new value of K is simply the reciprocal of the original value of K.*

*Equilibrium constant for reverse reaction:*
$$H^+ + A^- \rightleftharpoons HA \qquad K_1' = \frac{[HA]}{[H^+][A^-]} = 1/K_1$$

*If two reactions are added, the new K is the product of the two individual values:*

$$
\begin{array}{lll}
HA & \rightleftharpoons \cancel{[H^+]} + A^- & K_1 \\
\cancel{[H^+]} + C & \rightleftharpoons CH^+ & K_2 \\
\hline
HA + C & \rightleftharpoons A^- + CH^+ & K_3
\end{array}
$$

*Equilibrium constant for sum of reactions:*
$$K_3 = K_1 K_2 = \frac{\cancel{[H^+]}[A^-]}{[HA]} \cdot \frac{[CH^+]}{\cancel{[H^+]}[C]} = \frac{[A^-][CH^+]}{[HA][C]}$$

If a reaction is reversed, then $K' = 1/K$. If two reactions are added, then $K_3 = K_1 K_2$.

If *n* reactions are added, the overall equilibrium constant is the product of *n* individual equilibrium constants.

### Example  Combining Equilibrium Constants

The equilibrium constant for the reaction $H_2O \rightleftharpoons H^+ + OH^-$ is called $K_w (= [H^+][OH^-])$ and has the value $1.0 \times 10^{-14}$ at 25°C. Given that $K_{NH_3} = 1.8 \times 10^{-5}$ for the reaction $NH_3(aq) + H_2O \rightleftharpoons NH_4^+ + OH^-$, find $K$ for the reaction $NH_4^+ \rightleftharpoons NH_3(aq) + H^+$.

**Solution**  The third reaction can be obtained by reversing the second reaction and adding it to the first reaction:

$$H_2O \rightleftharpoons H^+ + OH^- \qquad\qquad K = K_w$$

$$\underline{NH_4^+ + OH^- \rightleftharpoons NH_3(aq) + H_2O \qquad\qquad K = 1/K_{NH_3}}$$

$$NH_4^+ \rightleftharpoons H^+ + NH_3(aq) \qquad\qquad K = K_w \cdot \frac{1}{K_{NH_3}} = 5.6 \times 10^{-10}$$

## ■■■■  6-2  Equilibrium and Thermodynamics

Equilibrium is controlled by the thermodynamics of a chemical reaction. The heat absorbed or released (*enthalpy*) and the degree of disorder of reactants and products (*entropy*) independently contribute to the degree to which the reaction is favored or disfavored.

### Enthalpy

$\Delta H = (+)$

**Heat is absorbed**
**Endothermic**

$\Delta H = (-)$

**Heat is liberated**
**Exothermic**

The **enthalpy change,** $\Delta H$, for a reaction is the heat absorbed or released when the reaction takes place under constant applied pressure.[5] The *standard enthalpy change*, $\Delta H°$, refers to the heat absorbed when all reactants and products are in their standard states:[†]

$$HCl(g) \rightleftharpoons H^+(aq) + Cl^-(aq) \qquad \Delta H° = -74.85 \text{ kJ/mol at } 25°C \qquad (6\text{-}3)$$

The negative sign of $\Delta H°$ indicates that heat is released by Reaction 6-3—the solution becomes warmer. For other reactions, $\Delta H$ is positive, which means that heat is absorbed. Consequently, the solution gets colder during the reaction. A reaction for which $\Delta H$ is positive is said to be **endothermic.** Whenever $\Delta H$ is negative, the reaction is **exothermic.**

### Entropy

$\Delta S = (+)$

**Products more disordered**
**than reactants**

$\Delta S = (-)$

**Products less disordered**
**than reactants**

The **entropy,** $S$, of a substance is a measure of its "disorder," which we will not attempt to define in a quantitative way. The greater the disorder, the greater the entropy. In general, a gas is more disordered (has higher entropy) than a liquid, which is more disordered than a solid. Ions in aqueous solution are normally more disordered than in their solid salt:

$$KCl(s) \rightleftharpoons K^+(aq) + Cl^-(aq) \qquad \Delta S° = +76.4 \text{ J/(K·mol) at } 25°C \qquad (6\text{-}4)$$

$\Delta S°$ is the change in entropy (entropy of products minus entropy of reactants) when all species are in their standard states. The positive value of $\Delta S°$ indicates that a mole of $K^+(aq)$ plus a mole of $Cl^-(aq)$ is more disordered than a mole of $KCl(s)$. For Reaction 6-3, $\Delta S° = -130.4 \text{ J/(K·mol)}$ at 25°C. The aqueous ions are less disordered than gaseous HCl.

### Free Energy

Systems at constant temperature and pressure, which are common laboratory conditions, have a tendency toward lower enthalpy and higher entropy. A chemical reaction is driven toward the formation of products by a *negative* value of $\Delta H$ (heat given off) or a *positive* value of $\Delta S$ (more disorder) or both. When $\Delta H$ is negative and $\Delta S$ is positive, the reaction is clearly favored. When $\Delta H$ is positive and $\Delta S$ is negative, the reaction is clearly disfavored.

When $\Delta H$ and $\Delta S$ are both positive or both negative, what decides whether a reaction will be favored? The change in **Gibbs free energy,** $\Delta G$, is the arbiter between opposing tendencies of $\Delta H$ and $\Delta S$. At constant temperature, $T$,

*Free energy:* $$\boxed{\Delta G = \Delta H - T\Delta S} \qquad (6\text{-}5)$$

*A reaction is favored if $\Delta G$ is negative.*

For the dissociation of HCl (Reaction 6-3) when all species are in their standard states, $\Delta H°$ favors the reaction and $\Delta S°$ disfavors it. To find the net result, we evaluate $\Delta G°$:

$$\Delta G° = \Delta H° - T\Delta S°$$
$$= (-74.85 \times 10^3 \text{ J/mol}) - (298.15 \text{ K})(-130.4 \text{ J/K·mol})$$
$$= -35.97 \text{ kJ/mol}$$

Note that 25.00°C = 298.15 K.

---

[†]The definition of the standard state contains subtleties beyond the scope of this book. For Reaction 6-3, the standard state of $H^+$ or $Cl^-$ is the hypothetical state in which each ion is present at a concentration of 1 M but behaves as if it were in an infinitely dilute solution. That is, the standard concentration is 1 M, but the standard behavior is what would be observed in a very dilute solution in which each ion is unaffected by surrounding ions.

$\Delta G°$ is negative, so the reaction is favored when all species are in their standard states. The favorable influence of $\Delta H°$ is greater than the unfavorable influence of $\Delta S°$ in this case.

The point of discussing free energy is to relate the equilibrium constant to the energetics ($\Delta H°$ and $\Delta S°$) of a reaction. The equilibrium constant depends on $\Delta G°$ in the following manner:

*Free energy and equilibrium:*
$$K = e^{-\Delta G°/RT}$$
(6-6)

**Challenge** Satisfy yourself that $K > 1$ if $\Delta G°$ is negative.

where $R$ is the gas constant [$= 8.314\,472$ J/(K·mol)] and $T$ is temperature (kelvin). The more negative the value of $\Delta G°$, the larger is the equilibrium constant. For Reaction 6-3,

$$K = e^{-(-35.97 \times 10^3 \text{ J/mol})/[8.314\,472 \text{ J/(K·mol)}](298.15 \text{ K})} = 2.00 \times 10^6$$

Because the equilibrium constant is large, $HCl(g)$ is very soluble in water and is nearly completely ionized to $H^+$ and $Cl^-$ when it dissolves.

To summarize, a chemical reaction is favored by the liberation of heat ($\Delta H$ negative) and by an increase in disorder ($\Delta S$ positive). $\Delta G$ takes both effects into account to determine whether or not a reaction is favorable. We say that a reaction is *spontaneous* under standard conditions if $\Delta G°$ is negative or, equivalently, if $K > 1$. The reaction is not spontaneous if $\Delta G°$ is positive ($K < 1$). You should be able to calculate $K$ from $\Delta G°$ and vice versa.

$\Delta G = (+)$
Reaction is disfavored
$\Delta G = (-)$
Reaction is favored

## Le Châtelier's Principle

Suppose that a system at equilibrium is subjected to a change that disturbs the system. **Le Châtelier's principle** states that the direction in which the system proceeds back to equilibrium is such that the change is partially offset.

To see what this statement means, let's see what happens when we attempt to change the concentration of one species in the reaction

$$\underset{\text{Bromate}}{BrO_3^-} + \underset{\text{Chromium(III)}}{2Cr^{3+}} + 4H_2O \rightleftharpoons Br^- + \underset{\text{Dichromate}}{Cr_2O_7^{2-}} + 8H^+$$
(6-7)

$$K = \frac{[Br^-][Cr_2O_7^{2-}][H^+]^8}{[BrO_3^-][Cr^{3+}]^2} = 1 \times 10^{11} \text{ at } 25°C$$

Notice that $H_2O$ is omitted from $K$ because it is the solvent.

In one particular equilibrium state of this system, the following concentrations exist: $[H^+] = 5.0$ M, $[Cr_2O_7^{2-}] = 0.10$ M, $[Cr^{3+}] = 0.003\,0$ M, $[Br^-] = 1.0$ M, and $[BrO_3^-] = 0.043$ M. Suppose that the equilibrium is disturbed by adding dichromate to the solution to increase the concentration of $[Cr_2O_7^{2-}]$ from 0.10 to 0.20 M. In what direction will the reaction proceed to reach equilibrium?

According to the principle of Le Châtelier, the reaction should go back to the left to partially offset the increase in dichromate, which appears on the right side of Reaction 6-7. We can verify this algebraically by setting up the **reaction quotient,** $Q$, which has the same form as the equilibrium constant. The only difference is that $Q$ is evaluated with whatever concentrations happen to exist, even though the solution is not at equilibrium. When the system reaches equilibrium, $Q = K$. For Reaction 6-7,

The reaction quotient has the same form as the equilibrium constant, but the concentrations are generally not the equilibrium concentrations.

$$Q = \frac{(1.0)(0.20)(5.0)^8}{(0.043)(0.003\,0)^2} = 2 \times 10^{11} > K$$

Because $Q > K$, the reaction must go to the left to decrease the numerator and increase the denominator, until $Q = K$.

If $Q < K$, then the reaction must proceed to the right to reach equilibrium. If $Q > K$, then the reaction must proceed to the left to reach equilibrium.

1. If a reaction is at equilibrium and products are added (or reactants are removed), the reaction goes to the left,
2. If a reaction is at equilibrium and reactants are added (or products are removed), the reaction goes to the right.

When the temperature of a system is changed, so is the equilibrium constant. Equations 6-5 and 6-6 can be combined to predict the effect of temperature on $K$:

$$K = e^{-\Delta G°/RT} = e^{-(\Delta H° - T\Delta S°)/RT} = e^{(-\Delta H°/RT + \Delta S°/R)}$$
$$= e^{-\Delta H°/RT} \cdot e^{\Delta S°/R}$$
(6-8)

The term $e^{\Delta S°/R}$ is independent of $T$ (at least over a limited temperature range in which $\Delta S°$ is constant). The term $e^{-\Delta H°/RT}$ increases with increasing temperature if $\Delta H°$ is positive and decreases if $\Delta H°$ is negative. Therefore,

1. The equilibrium constant of an endothermic reaction ($\Delta H° = +$) increases if the temperature is raised.
2. The equilibrium constant of an exothermic reaction ($\Delta H° = -$) decreases if the temperature is raised.

These statements can be understood in terms of Le Châtelier's principle as follows. Consider an endothermic reaction:

$$\text{Heat} + \text{reactants} \rightleftharpoons \text{products}$$

If the temperature is raised, then heat is added to the system. The reaction proceeds to the right to partially offset this change.[6]

In dealing with equilibrium problems, we are making *thermodynamic* predictions, not *kinetic* predictions. We are calculating what must happen for a system to reach equilibrium, but not how long it will take. Some reactions are over in an instant; others will not reach equilibrium in a million years. For example, dynamite remains unchanged indefinitely, until a spark sets off the spontaneous, explosive decomposition. The size of an equilibrium constant tells us nothing about the rate (the kinetics) of the reaction. A large equilibrium constant does not imply that a reaction is fast.

Heat can be treated as if it were a reactant in an endothermic reaction and a product in an exothermic reaction.

# ■ ■ ■ ■ 6-3 Solubility Product

In chemical analysis, we encounter solubility in precipitation titrations, electrochemical reference cells, and gravimetric analysis. The effect of acid on the solubility of minerals and the effect of atmospheric $CO_2$ on the solubility (and death) of coral reefs are important in environmental science.

The **solubility product** is the equilibrium constant for the reaction in which a solid salt dissolves to give its constituent ions in solution. Solid is omitted from the equilibrium constant because it is in its standard state. Appendix F lists solubility products.

As an example, consider the dissolution of mercury(I) chloride ($Hg_2Cl_2$, also called mercurous chloride) in water. The reaction is

$$Hg_2Cl_2(s) \rightleftharpoons Hg_2^{2+} + 2Cl^- \tag{6-9}$$

for which the solubility product, $K_{sp}$, is

$$K_{sp} = [Hg_2^{2+}][Cl^-]^2 = 1.2 \times 10^{-18} \tag{6-10}$$

A solution containing excess, undissolved solid is said to be **saturated** with that solid. The solution contains all the solid capable of being dissolved under the prevailing conditions.

The physical meaning of the solubility product is this: If an aqueous solution is left in contact with excess solid $Hg_2Cl_2$, the solid will dissolve until the condition $[Hg_2^{2+}][Cl^-]^2 = K_{sp}$ is satisfied. Thereafter, the amount of undissolved solid remains constant. Unless excess solid remains, there is no guarantee that $[Hg_2^{2+}][Cl^-]^2 = K_{sp}$. If $Hg_2^{2+}$ and $Cl^-$ are mixed together (with appropriate counterions) such that the product $[Hg_2^{2+}][Cl^-]^2$ exceeds $K_{sp}$, then $Hg_2Cl_2$ will precipitate.

We most commonly use the solubility product to find the concentration of one ion when the concentration of the other is known or fixed by some means. For example, what is the concentration of $Hg_2^{2+}$ in equilibrium with 0.10 M $Cl^-$ in a solution of KCl containing excess, undissolved $Hg_2Cl_2(s)$? To answer this question, we rearrange Equation 6-10 to find

$$[Hg_2^{2+}] = \frac{K_{sp}}{[Cl^-]^2} = \frac{1.2 \times 10^{-18}}{0.10^2} = 1.2 \times 10^{-16} \text{ M}$$

Because $Hg_2Cl_2$ is so slightly soluble, additional $Cl^-$ obtained from $Hg_2Cl_2$ is negligible compared with 0.10 M $Cl^-$.

The solubility product does not tell the entire story of solubility. In addition to complications described in Box 6-1, most salts form soluble *ion pairs* to some extent. That is, $MX(s)$ can give $MX(aq)$ as well as $M^+(aq)$ and $X^-(aq)$. In a saturated solution of $CaSO_4$, for example, two-thirds of the dissolved calcium is $Ca^{2+}$ and one-third is $CaSO_4(aq)$.[7] The

*Mercurous ion*, $Hg_2^{2+}$, is a *dimer* (pronounced DIE mer), which means that it consists of two identical units bound together:

250 pm

$[Hg-Hg]^{2+}$

+1 oxidation state of mercury

$OH^-$, $S^{2-}$, and $CN^-$ stabilize Hg(II), thereby converting Hg(I) into Hg(0) and Hg(II):

$$Hg_2^{2+} + 2CN^- \rightarrow Hg(CN)_2(aq) + Hg(l)$$
$$\text{Hg(I)} \qquad\qquad \text{Hg(II)} \qquad \text{Hg(0)}$$

**Disproportionation** is the process in which an element in an intermediate oxidation state gives products in both higher and lower oxidation states.

A *salt* is any ionic solid, such as $Hg_2Cl_2$ or $CaSO_4$.

$CaSO_4(aq)$ **ion pair** is a closely associated pair of ions that behaves as one species in solution. Appendix J and Box 8-1 provide information on ion pairs.[8]

## Common Ion Effect

For the ionic solubility reaction

$$CaSO_4(s) \rightleftharpoons Ca^{2+} + SO_4^{2-} \qquad K_{sp} = 2.4 \times 10^{-5}$$

the product $[Ca^{2+}][SO_4^{2-}]$ is constant at equilibrium in the presence of excess solid $CaSO_4$. If the concentration of $Ca^{2+}$ were increased by adding another source of $Ca^{2+}$, such as $CaCl_2$, then the concentration of $SO_4^{2-}$ must decrease so that the product $[Ca^{2+}][SO_4^{2-}]$ remains constant. In other words, less $CaSO_4(s)$ will dissolve if $Ca^{2+}$ or $SO_4^{2-}$ is already present from some other source. Figure 6-1 shows how the solubility of $CaSO_4$ decreases in the presence of dissolved $CaCl_2$.

This application of Le Châtelier's principle is called the **common ion effect.** *A salt will be less soluble if one of its constituent ions is already present in the solution.*

> *Common ion effect:* **A salt is less soluble if one of its ions is already present in the solution. Demonstration 6-1 illustrates the common ion effect.**

## Separation by Precipitation

Precipitations can sometimes be used to separate ions from each other.[11] For example, consider a solution containing lead(II) ($Pb^{2+}$) and mercury(I) ($Hg_2^{2+}$) ions, each at a concentration of 0.010 M. Each forms an insoluble iodide (Figure 6-2), but the mercury(I) iodide is considerably less soluble, as indicated by the smaller value of $K_{sp}$:

$$PbI_2(s) \rightleftharpoons Pb^{2+} + 2I^- \qquad K_{sp} = 7.9 \times 10^{-9}$$
$$Hg_2I_2(s) \rightleftharpoons Hg_2^{2+} + 2I^- \qquad K_{sp} = 4.6 \times 10^{-29}$$

> The smaller $K_{sp}$ implies a lower solubility for $Hg_2I_2$ because the stoichiometries of the two reactions are the same. If the stoichiometries were different, it does not follow that the smaller $K_{sp}$ would imply lower solubility.

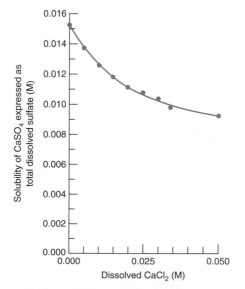

*Figure 6-1* Solubility of $CaSO_4$ in solutions containing dissolved $CaCl_2$. Solubility is expressed as total dissolved sulfate, which is present as free $SO_4^{2-}$ and as the ion pair, $CaSO_4(aq)$. Data from reference 7.

*Figure 6-2* A yellow solid, lead(II) iodide ($PbI_2$), precipitates when a colorless solution of lead nitrate ($Pb(NO_3)_2$) is added to a colorless solution of potassium iodide (KI). *[Photo by Chip Clark.]*

Fill two large test tubes about one-third full with saturated aqueous KCl containing no excess solid. The solubility of KCl is approximately 3.7 M, so the solubility product (ignoring activity effects introduced later) is

$$K_{sp} \approx [K^+][Cl^-] = (3.7)(3.7) = 13.7$$

Now add 1/3 of a test tube of 6 M HCl to one test tube and an equal volume of 12 M HCl to the other. Even though a common ion, $Cl^-$, is added in each case, KCl precipitates only in one tube.

To understand your observations, calculate the concentrations of $K^+$ and $Cl^-$ in each tube after HCl addition. Then evaluate the reaction quotient, $Q = [K^+][Cl^-]$, for each tube. Explain your observations.

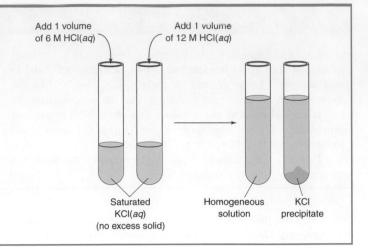

Add 1 volume of 6 M HCl(aq)  Add 1 volume of 12 M HCl(aq)

Saturated KCl(aq) (no excess solid)    Homogeneous solution    KCl precipitate

Is it possible to lower the concentration of $Hg_2^{2+}$ by 99.990% by selective precipitation with $I^-$, without precipitating $Pb^{2+}$?

We are asking whether we can lower $[Hg_2^{2+}]$ to 0.010% of 0.010 M = $1.0 \times 10^{-6}$ M without precipitating $Pb^{2+}$. Here is the experiment: We add enough $I^-$ to precipitate 99.990% of $Hg_2^{2+}$ if all the $I^-$ reacts with $Hg_2^{2+}$ and none reacts with $Pb^{2+}$. To see if any $Pb^{2+}$ should precipitate, we need to know the concentration of $I^-$ in equilibrium with precipitated $Hg_2I_2(s)$ plus the remaining $1.0 \times 10^{-6}$ M $Hg_2^{2+}$.

$$Hg_2I_2(s) \underset{K_{sp}}{\overset{K_{sp}}{\rightleftharpoons}} Hg_2^{2+} + 2I^-$$

$$[Hg_2^{2+}][I^-]^2 = K_{sp}$$

$$(1.0 \times 10^{-6})[I^-]^2 = 4.6 \times 10^{-29}$$

$$[I^-] = \sqrt{\frac{4.6 \times 10^{-29}}{1.0 \times 10^{-6}}} = 6.8 \times 10^{-12}\ M$$

Will this concentration of $I^-$ cause 0.010 M $Pb^{2+}$ to precipitate? We answer this question by seeing if the solubility product of $PbI_2$ is exceeded.

$$Q = [Pb^{2+}][I^-]^2 = (0.010)(6.8 \times 10^{-12})^2$$
$$= 4.6 \times 10^{-25} < K_{sp}\,(\text{for } PbI_2)$$

The reaction quotient, $Q$, is $4.6 \times 10^{-25}$, which is less than $K_{sp}(= 7.9 \times 10^{-9})$ for $PbI_2$. Therefore, $Pb^{2+}$ will not precipitate and separation of $Pb^{2+}$ and $Hg_2^{2+}$ is feasible. We predict that adding $I^-$ to a solution of $Pb^{2+}$ and $Hg_2^{2+}$ will precipitate virtually all the mercury(I) before any lead(II) precipitates.

Life should be so easy! We have just made a thermodynamic prediction. If the system comes to equilibrium, we can achieve the desired separation. However, occasionally one substance *coprecipitates* with the other. In **coprecipitation,** a substance whose solubility is not exceeded precipitates along with another substance whose solubility is exceeded. For example, some $Pb^{2+}$ might become adsorbed on the surface of the $Hg_2I_2$ crystal or might even occupy sites within the crystal. Our calculation says that the separation is worth trying. However, *only an experiment can show whether or not the separation actually works.*

*Question* If you want to know whether a small amount of $Pb^{2+}$ coprecipitates with $Hg_2I_2$, should you measure the $Pb^{2+}$ concentration in the mother liquor (the solution) or the $Pb^{2+}$ concentration in the precipitate? Which measurement is more sensitive? By "sensitive," we mean responsive to a small amount of coprecipitation.
(Answer: Measure $Pb^{2+}$ in precipitate.)

## ▪ ▪ ▪ ▪ 6-4 Complex Formation

If anion $X^-$ precipitates metal $M^+$, it is often observed that a high concentration of $X^-$ causes solid MX to redissolve. The increased solubility arises from the formation of **complex ions,** such as $MX_2^-$, which consist of two or more simple ions bonded to each other.

### Lewis Acids and Bases

In complex ions such as $PbI^+$, $PbI_3^-$, and $PbI_4^{2-}$, iodide is said to be the *ligand* of $Pb^{2+}$. A **ligand** is any atom or group of atoms attached to the species of interest. We say that $Pb^{2+}$

acts as a *Lewis acid* and $I^-$ acts as a *Lewis base* in these complexes. A **Lewis acid** accepts a pair of electrons from a **Lewis base** when the two form a bond:

$$^{++}Pb\ \square + \square\ \ddot{\underset{..}{I}}:^- \quad \rightarrow \quad [Pb-\ddot{\underset{..}{I}}:]^+$$

Room to accept electrons

A pair of electrons to be donated

Lewis acid + Lewis base ⇌ adduct
Electron pair   Electron pair
acceptor    donor

The product of the reaction between a Lewis acid and a Lewis base is called an *adduct*. The bond between a Lewis acid and a Lewis base is called a *dative* or *coordinate covalent* bond.

## Effect of Complex Ion Formation on Solubility

If $Pb^{2+}$ and $I^-$ only reacted to form solid $PbI_2$, then the solubility of $Pb^{2+}$ would always be very low in the presence of excess $I^-$.

$$PbI_2(s) \xrightleftharpoons{K_{sp}} Pb^{2+} + 2I^- \qquad K_{sp} = [Pb^{2+}][I^-]^2 = 7.9 \times 10^{-9} \qquad (6\text{-}11)$$

The observation, however, is that high concentrations of $I^-$ cause solid $PbI_2$ to dissolve. We explain this by the formation of a series of complex ions:

$$Pb^{2+} + I^- \xrightleftharpoons{K_1} PbI^+ \qquad K_1 = [PbI^+]/[Pb^{2+}][I^-] = 1.0 \times 10^2 \qquad (6\text{-}12)$$

$$Pb^{2+} + 2I^- \xrightleftharpoons{\beta_2} PbI_2(aq) \qquad \beta_2 = [PbI_2(aq)]/[Pb^{2+}][I^-]^2 = 1.4 \times 10^3 \qquad (6\text{-}13)$$

$$Pb^{2+} + 3I^- \xrightleftharpoons{\beta_3} PbI_3^- \qquad \beta_3 = [PbI_3^-]/[Pb^{2+}][I^-]^3 = 8.3 \times 10^3 \qquad (6\text{-}14)$$

$$Pb^{2+} + 4I^- \xrightleftharpoons{\beta_4} PbI_4^{2-} \qquad \beta_4 = [PbI_4^{2-}]/[Pb^{2+}][I^-]^4 = 3.0 \times 10^4 \qquad (6\text{-}15)$$

The species $PbI_2(aq)$ in Reaction 6-13 is *dissolved* $PbI_2$, containing two iodine atoms bound to a lead atom. Reaction 6-13 is *not* the reverse of Reaction 6-11, in which the species is solid $PbI_2$.

At low $I^-$ concentrations, the solubility of lead is governed by precipitation of $PbI_2(s)$. However, at high $I^-$ concentrations, Reactions 6-12 through 6-15 are driven to the right (Le Châtelier's principle), and the total concentration of dissolved lead is considerably greater than that of $Pb^{2+}$ alone (Figure 6-3).

Classroom demonstration of complex equilibria: A. R. Johnson, T. M. McQueen, and K. T. Rodolfa, "Species Distribution Diagrams in the Copper-Ammonia System," *J. Chem. Ed.* **2005,** *82,* 408.

Notation for these equilibrium constants is discussed in Box 6-2.

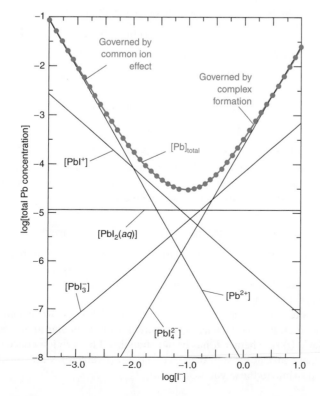

*Figure 6-3* Total solubility of lead(II) (curve with circles) and solubilities of individual species (straight lines) as a function of the concentration of free iodide. To the left of the minimum, $[Pb]_{total}$ is governed by the solubility product for $PbI_2(s)$. As $[I^-]$ is increased, $[Pb]_{total}$ decreases because of the common ion effect. At high values of $[I^-]$, $PbI_2(s)$ redissolves because it reacts with $I^-$ to form soluble complex ions, such as $PbI_4^{2-}$. Note logarithmic scales. The solution is made slightly acidic so that $[PbOH^+]$ is negligible.

## Box 6-2 Notation for Formation Constants

*Formation constants* are the equilibrium constants for complex ion formation. The **stepwise formation constants,** designated $K_i$, are defined as follows:

$$M + X \xrightleftharpoons{K_1} MX \qquad K_1 = [MX]/[M][X]$$

$$MX + X \xrightleftharpoons{K_2} MX_2 \qquad K_2 = [MX_2]/[MX][X]$$

$$MX_{n-1} + X \xrightleftharpoons{K_n} MX_n \qquad K_n = [MX_n]/[MX_{n-1}][X]$$

The **overall,** or **cumulative, formation constants** are denoted $\beta_i$:

$$M + 2X \xrightleftharpoons{\beta_2} MX_2 \qquad \beta_2 = [MX_2]/[M][X]^2$$

$$M + nX \xrightleftharpoons{\beta_n} MX_n \qquad \beta_n = [MX_n]/[M][X]^n$$

A useful relation is that $\beta_n = K_1 K_2 \cdots K_n$.

---

A most useful characteristic of chemical equilibrium is that *all equilibria are satisfied simultaneously.* If we know the concentration of $I^-$, we can calculate the concentration of $Pb^{2+}$ by substituting this value into the equilibrium constant expression for Reaction 6-11, regardless of whether there are other reactions involving $Pb^{2+}$. *The concentration of $Pb^{2+}$ that satisfies any one equilibrium must satisfy all equilibria. There can be only one concentration of $Pb^{2+}$ in the solution.*

### Example Effect of $I^-$ on the Solubility of $Pb^{2+}$

Find the concentrations of $PbI^+$, $PbI_2(aq)$, $PbI_3^-$, and $PbI_4^{2-}$ in a solution saturated with $PbI_2(s)$ and containing dissolved $I^-$ with a concentration of **(a)** 0.001 0 M and **(b)** 1.0 M.

**Solution** **(a)** From $K_{sp}$ for Reaction 6-11, we calculate

$$[Pb^{2+}] = K_{sp}/[I^-]^2 = (7.9 \times 10^{-9})/(0.001\ 0)^2 = 7.9 \times 10^{-3}\ M$$

From Reactions 6-12 through 6-15, we then calculate the concentrations of the other lead-containing species:

$$[PbI^+] = K_1[Pb^{2+}][I^-] = (1.0 \times 10^2)(7.9 \times 10^{-3})(1.0 \times 10^{-3})$$
$$= 7.9 \times 10^{-4}\ M$$

$$[PbI_2(aq)] = \beta_2[Pb^{2+}][I^-]^2 = 1.1 \times 10^{-5}\ M$$

$$[PbI_3^-] = \beta_3[Pb^{2+}][I^-]^3 = 6.6 \times 10^{-8}\ M$$

$$[PbI_4^{2-}] = \beta_4[Pb^{2+}][I^-]^4 = 2.4 \times 10^{-10}\ M$$

**(b)** If, instead, we take $[I^-] = 1.0\ M$, then analogous computations show that

$$[Pb^{2+}] = 7.9 \times 10^{-9}\ M \qquad [PbI_3^-] = 6.6 \times 10^{-5}\ M$$

$$[PbI^+] = 7.9 \times 10^{-7}\ M \qquad [PbI_4^{2-}] = 2.4 \times 10^{-4}\ M$$

$$[PbI_2(aq)] = 1.1 \times 10^{-5}\ M$$

The total concentration of dissolved lead in the preceding example is

$$[Pb]_{total} = [Pb^{2+}] + [PbI^+] + [PbI_2(aq)] + [PbI_3^-] + [PbI_4^{2-}]$$

When $[I^-] = 10^{-3}\ M$, $[Pb]_{total} = 8.7 \times 10^{-3}\ M$, of which 91% is $Pb^{2+}$. As $[I^-]$ increases, $[Pb]_{total}$ decreases by the common ion effect operating in Reaction 6-11. However, at sufficiently high $[I^-]$, complex formation takes over and $[Pb]_{total}$ increases (Figure 6-3). When $[I^-] = 1.0\ M$, $[Pb]_{total} = 3.2 \times 10^{-4}\ M$, of which 76% is $PbI_4^{2-}$.

### Tabulated Equilibrium Constants Are Usually Not "Constant"

If you look up the equilibrium constant of a chemical reaction in two different books, there is an excellent chance that the values will be different (sometimes by a factor of 10 or more).[12] This discrepancy occurs because the constant may have been determined under different conditions and, perhaps, by using different techniques.

A common source of variation in the reported value of $K$ is the ionic composition of the solution. It is important to note whether $K$ is reported for a particular ionic composition (for example, 1 M $NaClO_4$) or whether $K$ has been extrapolated to zero ion concentration. If you need an equilibrium constant for your own work, choose a value of $K$ measured under conditions as close as possible to those you will employ.

The effect of dissolved ions on chemical equilibria is the subject of Chapter 8.

## ■ ■ ■ 6-5 Protic Acids and Bases

Understanding the behavior of acids and bases is essential to every branch of science having anything to do with chemistry. In analytical chemistry, we almost always need to account for the effect of pH on analytical reactions involving complex formation or oxidation-reduction. pH can affect molecular charge and shape—factors that help determine which molecules can be separated from others in chromatography and electrophoresis and which molecules will be detected in some types of mass spectrometry.

In aqueous chemistry, an **acid** is a substance that increases the concentration of $H_3O^+$ **(hydronium ion)** when added to water. Conversely, a **base** decreases the concentration of $H_3O^+$. We will see that a decrease in $H_3O^+$ concentration necessarily requires an increase in $OH^-$ concentration. Therefore, a base increases the concentration of $OH^-$ in aqueous solution.

The word *protic* refers to chemistry involving transfer of $H^+$ from one molecule to another. The species $H^+$ is also called a *proton* because it is what remains when a hydrogen atom loses its electron. Hydronium ion, $H_3O^+$, is a combination of $H^+$ with $H_2O$. Although $H_3O^+$ is a more accurate representation than $H^+$ for the hydrogen ion in aqueous solution, we will use $H_3O^+$ and $H^+$ interchangeably in this book.

### Brønsted-Lowry Acids and Bases

Brønsted and Lowry classified *acids* as proton donors and *bases* as *proton acceptors*. HCl is an acid (a proton donor) and it increases the concentration of $H_3O^+$ in water:

$$HCl + H_2O \rightleftharpoons H_3O^+ + Cl^-$$

The Brønsted-Lowry definition does not require that $H_3O^+$ be formed. This definition can therefore be extended to nonaqueous solvents and even to the gas phase:

$$\underset{\substack{\text{Hydrochloric acid}\\\text{(acid)}}}{HCl(g)} + \underset{\substack{\text{Ammonia}\\\text{(base)}}}{NH_3(g)} \rightleftharpoons \underset{\substack{\text{Ammonium chloride}\\\text{(salt)}}}{NH_4^+Cl^-(s)}$$

> Brønsted-Lowry acid: proton donor
> Brønsted-Lowry base: proton acceptor
> J. N. Brønsted of the University of Copenhagen published his definition of acids and bases in 1923.

For the remainder of this book, when we speak of acids and bases, we are speaking of Brønsted-Lowry acids and bases.

### Salts

Any ionic solid, such as ammonium chloride, is called a **salt.** In a formal sense, a salt can be thought of as the product of an acid-base reaction. When an acid and base react, they are said to **neutralize** each other. Most salts containing cations and anions with a single positive and negative charge are strong electrolytes—they dissociate nearly completely into ions in dilute aqueous solution. Thus, ammonium chloride gives $NH_4^+$ and $Cl^-$ in water:

$$NH_4^+Cl^-(s) \rightarrow NH_4^+(aq) + Cl^-(aq)$$

### Conjugate Acids and Bases

The products of a reaction between an acid and a base are also classified as acids and bases:

> Conjugate acids and bases are related by the gain or loss of one proton. In these structures, a solid wedge is a bond coming out of the plane of the page and a dashed wedge is a bond to an atom behind the page.

Acetate is a base because it can accept a proton to make acetic acid. Methylammonium ion is an acid because it can donate a proton and become methylamine. Acetic acid and the acetate ion are said to be a **conjugate acid-base pair.** Methylamine and methylammonium ion are likewise conjugate. *Conjugate acids and bases are related to each other by the gain or loss of one $H^+$.*

We will write H⁺ when we really mean $H_3O^+$.

It is certain that the proton does not exist by itself in water. The simplest formula found in some crystalline salts is $H_3O^+$. For example, crystals of perchloric acid monohydrate contain pyramidal hydronium (also called *hydroxonium*) ions:

$$HClO_4 \cdot H_2O \quad \text{is really}$$

Hydronium    Perchlorate

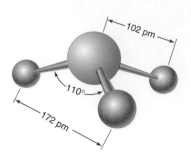

*Figure 6-4* Typical structure of hydronium ion, $H_3O^+$, found in many crystals.[13] The bond enthalpy (heat needed to break a bond) of the OH bond of $H_3O^+$ is 544 kJ/mol, about 84 kJ/mol greater than the OH bond enthalpy in $H_2O$.

The formula $HClO_4 \cdot H_2O$ is a way of specifying the composition of the substance when we are ignorant of its structure. A more accurate formula would be $H_3O^+ClO_4^-$.

Average dimensions of the $H_3O^+$ cation in many crystals are shown in Figure 6-4. In aqueous solution, $H_3O^+$ is tightly associated with three molecules of water through exceptionally strong hydrogen bonds (Figure 6-5). The $H_5O_2^+$ cation is another simple species in which a hydrogen ion is shared by two water molecules.[15]

$$H \quad \quad \quad \quad H$$
$$\diagdown \quad \quad \quad \diagup$$
$$O \cdots H \cdots O$$
$$\diagup \quad \quad \quad \diagdown$$
$$H \quad \leftarrow 243 \text{ pm} \rightarrow \quad H$$

In the gas phase, $H_3O^+$ can be found inside a dodecahedral shell of 20 water molecules (Figure 6-6).

The ion $H_3O_2^-$ $(OH^- \cdot H_2O)$ has been observed by X-ray crystallography.[16] The central $O \cdots H \cdots O$ linkage contains the shortest hydrogen bond involving $H_2O$ that has ever been observed.

$$\quad \quad \quad \quad \quad H$$
$$\quad \quad \quad \quad \quad \diagup$$
$$O \cdots II \cdots O$$
$$\diagup$$
$$H \quad \leftarrow 229 \text{ pm} \rightarrow$$

We will ordinarily write H⁺ in most chemical equations, although we really mean $H_3O^+$. To emphasize the chemistry of water, we will write $H_3O^+$. For example, water can be either an acid or a base. Water is an acid with respect to methoxide:

$$H{-}\overset{..}{\underset{..}{O}}{-}H \ + \ CH_3{-}\overset{..}{\underset{..}{O}}{}^- \ \rightleftharpoons \ H{-}\overset{..}{\underset{..}{O}}{}^- \ + \ CH_3{-}\overset{..}{\underset{..}{O}}{-}H$$

Water          Methoxide          Hydroxide          Methanol

But with respect to hydrogen bromide, water is a base:

$$H_2O \ + \ HBr \ \rightleftharpoons \ H_3O^+ \ + \ Br^-$$

Water     Hydrogen     Hydronium     Bromide
          bromide        ion

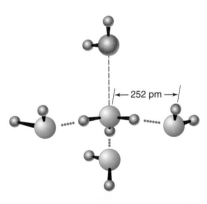

*Figure 6-5* Environment of aqueous $H_3O^+$.[13] Three $H_2O$ molecules are bound to $H_3O^+$ by strong hydrogen bonds (dotted lines), and one $H_2O$ (at the top) is held by weaker ion-dipole attraction (dashed line). The O—H···O hydrogen-bonded distance of 252 pm (picometers, $10^{-12}$ m) compares with an O—H···O distance of 283 pm between hydrogen-bonded water molecules. The discrete cation $(H_2O)_3H_3O^+$ found in some crystals is similar in structure to $(H_2O)_4H_3O^+$ with the weakly bonded $H_2O$ at the top removed.[14]

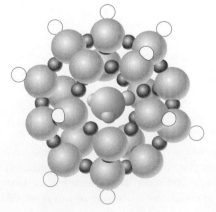

*Figure 6-6* In the gas phase, $H_3O^+$ can be tightly surrounded by 20 molecules of $H_2O$ in a regular dodecahedron held together by 30 hydrogen bonds. Small dark and white atoms are H. Dark H atoms are hydrogen bonded. *[From S. Wei, Z. Shi, and A. W. Castleman, Jr., J. Chem. Phys.* **1991**, *94, 3268, and Chem. Eng. News, 8 April 1991.]*

CHAPTER 6 Chemical Equilibrium

## Autoprotolysis

Water undergoes self-ionization, called **autoprotolysis,** in which it acts as both an acid and a base:

$$H_2O + H_2O \rightleftharpoons H_3O^+ + OH^- \qquad (6\text{-}16)$$

or

$$H_2O \rightleftharpoons H^+ + OH^- \qquad (6\text{-}17)$$

Reactions 6-16 and 6-17 mean the same thing.

   **Protic solvents** have a reactive $H^+$, and all protic solvents undergo autoprotolysis. An example is acetic acid:

$$2CH_3\overset{O}{\overset{\|}{C}}OH \rightleftharpoons CH_3\overset{OH}{\underset{OH}{\overset{+}{C}}} + CH_3\overset{O}{\overset{\|}{C}}-O^- \quad \text{(in acetic acid)} \qquad (6\text{-}18)$$

The extent of these reactions is very small. The *autoprotolysis constants* (equilibrium constants) for Reactions 6-17 and 6-18 are $1.0 \times 10^{-14}$ and $3.5 \times 10^{-15}$, respectively, at 25°C.

Examples of protic solvents (acidic proton bold):

$$\mathbf{H_2O} \qquad CH_3CH_2O\mathbf{H}$$
Water        Ethanol

Examples of aprotic solvents (no acidic protons):

$$CH_3CH_2OCH_2CH_3 \qquad CH_3CN$$
Diethyl ether        Acetonitrile

## ■■■ 6-6  pH

The autoprotolysis constant for $H_2O$ has the special symbol $K_w$, where "w" stands for water:

*Autoprotolysis of water:*
$$H_2O \xrightarrow{K_w} H^+ + OH^- \qquad K_w = [H^+][OH^-] \qquad (6\text{-}19)$$

Table 6-1 shows how $K_w$ varies with temperature. Its value at 25.0°C is $1.01 \times 10^{-14}$.

Recall that $H_2O$ (the solvent) is omitted from the equilibrium constant. The value $K_w = 1.0 \times 10^{-14}$ at 25° C is accurate enough for our purposes in this book.

---

**Example**  Concentration of $H^+$ and $OH^-$ in Pure Water at 25°C

Calculate the concentrations of $H^+$ and $OH^-$ in pure water at 25°C.

**Solution**   The stoichiometry of Reaction 6-19 tells us that $H^+$ and $OH^-$ are produced in a 1:1 molar ratio. Their concentrations must be equal. Calling each concentration $x$, we can write

$$K_w = 1.0 \times 10^{-14} = [H^+][OH^-] = [x][x] \Rightarrow x = 1.0 \times 10^{-7}\ M$$

The concentrations of $H^+$ and $OH^-$ are both $1.0 \times 10^{-7}$ M in pure water.

---

**Example**  Concentration of $OH^-$ When $[H^+]$ Is Known

What is the concentration of $OH^-$ if $[H^+] = 1.0 \times 10^{-3}$ M? (From now on, assume that the temperature is 25°C unless otherwise stated.)

**Solution**   Putting $[H^+] = 1.0 \times 10^{-3}$ M into the $K_w$ expression gives

$$K_w = 1.0 \times 10^{-14} = (1.0 \times 10^{-3})[OH^-] \Rightarrow [OH^-] = 1.0 \times 10^{-11}\ M$$

A concentration of $[H^+] = 1.0 \times 10^{-3}$ M gives $[OH^-] = 1.0 \times 10^{-11}$ M. *As the concentration of $H^+$ increases, the concentration of $OH^-$ necessarily decreases, and vice versa.* A concentration of $[OH^-] = 1.0 \times 10^{-3}$ M gives $[H^+] = 1.0 \times 10^{-11}$ M.

---

**Table 6-1**  Temperature dependence of $K_w{}^a$

| Temperature (°C) | $K_w$ | $pK_w = -\log K_w$ | Temperature (°C) | $K_w$ | $pK_w = -\log K_w$ |
|---|---|---|---|---|---|
| 0 | $1.15 \times 10^{-15}$ | 14.938 | 40 | $2.88 \times 10^{-14}$ | 13.541 |
| 5 | $1.88 \times 10^{-15}$ | 14.726 | 45 | $3.94 \times 10^{-14}$ | 13.405 |
| 10 | $2.97 \times 10^{-15}$ | 14.527 | 50 | $5.31 \times 10^{-14}$ | 13.275 |
| 15 | $4.57 \times 10^{-15}$ | 14.340 | 100 | $5.43 \times 10^{-13}$ | 12.265 |
| 20 | $6.88 \times 10^{-15}$ | 14.163 | 150 | $2.30 \times 10^{-12}$ | 11.638 |
| 25 | $1.01 \times 10^{-14}$ | 13.995 | 200 | $5.14 \times 10^{-12}$ | 11.289 |
| 30 | $1.46 \times 10^{-14}$ | 13.836 | 250 | $6.44 \times 10^{-12}$ | 11.191 |
| 35 | $2.07 \times 10^{-14}$ | 13.685 | 300 | $3.93 \times 10^{-12}$ | 11.406 |

a. Concentrations in the product $[H^+][OH^-]$ in this table are expressed in molality rather than in molarity. Accuracy of log $K_w$ is ±0.01. To convert molality (mol/kg) into molarity (mol/L), multiply by the density of $H_2O$ at each temperature. At 25°C, $K_w = 10^{-13.995}$ $(mol/kg)^2(0.997\ 05\ kg/L)^2 = 10^{-13.998}\ (mol/L)$.

SOURCE: W. L. Marshall and E. U. Franck, "Ion Product of Water Substance, 0–1 000°C, 1–10,000 Bars," J. Phys. Chem. Ref. Data **1981,** 10, 295.

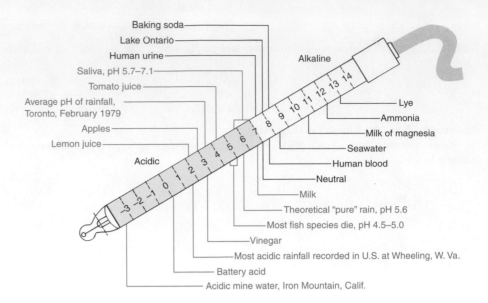

**Figure 6-7** pH of various substances. [From *Chem. Eng. News*, 14 September 1981.] The most acidic rainfall (Box 15-1) is a stronger acid than lemon juice. The most acidic natural waters known are mine waters, with total dissolved metal concentrations of 200 g/L and sulfate concentrations of 760 g/L. The pH of this water, −3.6, does not mean that [H⁺] = 10³·⁶ M = 4 000 M! It means that the *activity* of H⁺ (discussed in Chapter 8) is 10³·⁶. [D. K. Nordstrom, C. N. Alpers, C. J. Placek, and D. W. Blowes, "Negative pH and Extremely Acidic Mine Waters from Iron Mountain, California," *Environ. Sci. Technol.* **2000**, *34*, 254.]

An approximate definition of **pH** is the negative logarithm of the H⁺ concentration.

*Approximate definition of pH:*

$$\text{pH} \approx -\log[\text{H}^+] \qquad (6\text{-}20)$$

Chapter 8 defines pH more accurately in terms of *activities*, but, for most purposes, Equation 6-20 is a good working definition. In pure water at 25°C with [H⁺] = $1.0 \times 10^{-7}$ M, the pH is $-\log(1.0 \times 10^{-7}) = 7.00$. If the concentration of OH⁻ is $1.0 \times 10^{-3}$ M, then [H⁺] = $1.0 \times 10^{-11}$ M and the pH is 11.00.

A useful relation between the concentrations of H⁺ and OH⁻ is

$$\text{pH} + \text{pOH} = -\log(K_w) = 14.00 \text{ at } 25°\text{C} \qquad (6\text{-}21)$$

where pOH = $-\log[\text{OH}^-]$, just as pH = $-\log[\text{H}^+]$. This is a fancy way of saying that, if pH = 3.58, then pOH = 14.00 − 3.58 = 10.42, or [OH⁻] = $10^{-10.42} = 3.8 \times 10^{-11}$ M.

A solution is **acidic** if [H⁺] > [OH⁻]. A solution is **basic** if [H⁺] < [OH⁻]. At 25°C, an acidic solution has a pH below 7, and a basic solution has a pH above 7.

pH values for various common substances are shown in Figure 6-7.

Although pH generally falls in the range 0 to 14, these are not the limits of pH. A pH of −1.00, for example, means $-\log[\text{H}^+] = -1.00$; or [H⁺] = 10 M. This concentration is easily attained in a concentrated solution of a strong acid such as HCl.

### Is There Such a Thing as Pure Water?

In most labs, the answer is "No." Pure water at 25°C should have a pH of 7.00. Distilled water from the tap in most labs is acidic because it contains $CO_2$ from the atmosphere. $CO_2$ is an acid by virtue of the reaction

$$\text{CO}_2 + \text{H}_2\text{O} \rightleftharpoons \underset{\text{Bicarbonate}}{\text{HCO}_3^-} + \text{H}^+ \qquad (6\text{-}22)$$

$CO_2$ can be largely removed by boiling water and then protecting it from the atmosphere.

A century ago, careful measurements of the conductivity of water were made by Friedrich Kohlrausch and his students. To remove impurities, they found it necessary to distill the water *42 consecutive times* under vacuum to reduce conductivity to a limiting value.

## 6-7 Strengths of Acids and Bases

Acids and bases are commonly classified as strong or weak, depending on whether they react nearly "completely" or only "partially" to produce H⁺ or OH⁻. Because there is a continuous range of possibilities for "partial" reaction, there is no sharp distinction between weak

---

pH ≈ −log[H⁺]. The term "pH" was introduced in 1909 by the Danish biochemist S. P. L. Sørensen, who called it the "hydrogen ion exponent."[17]

Take the log of both sides of the $K_w$ expression to derive Equation 6-21:

$$K_w = [\text{H}^+][\text{OH}^-]$$
$$\log K_w = \log[\text{H}^+] + \log[\text{OH}^-]$$
$$-\log K_w = -\log[\text{H}^+] - \log[\text{OH}^-]$$
$$14.00 = \text{pH} + \text{pOH} \text{ (at 25°C)}$$

pH is usually measured with a *glass electrode*, whose operation is described in Chapter 15.

**Table 6-2** Common strong acids and bases

| Formula | Name |
|---|---|
| *Acids* | |
| HCl | Hydrochloric acid (hydrogen chloride) |
| HBr | Hydrogen bromide |
| HI | Hydrogen iodide |
| $H_2SO_4$[a] | Sulfuric acid |
| $HNO_3$ | Nitric acid |
| $HClO_4$ | Perchloric acid |
| *Bases* | |
| LiOH | Lithium hydroxide |
| NaOH | Sodium hydroxide |
| KOH | Potassium hydroxide |
| RbOH | Rubidium hydroxide |
| CsOH | Cesium hydroxide |
| $R_4NOH$[b] | Quaternary ammonium hydroxide |

a. For $H_2SO_4$, only the first proton ionization is complete. Dissociation of the second proton has an equilibrium constant of $1.0 \times 10^{-2}$.

b. This is a general formula for any hydroxide salt of an ammonium cation containing four organic groups. An example is tetrabutylammonium hydroxide: $(CH_3CH_2CH_2CH_2)_4N^+OH^-$.

and strong. However, some compounds react so completely that they are easily classified as strong acids or bases—and, by convention, everything else is termed weak.

## Strong Acids and Bases

Common strong acids and bases are listed in Table 6-2, which you need to memorize. By definition, a strong acid or base is completely dissociated in aqueous solution. That is, the equilibrium constants for the following reactions are large.

$$HCl(aq) \rightleftharpoons H^+ + Cl^-$$

$$KOH(aq) \rightleftharpoons K^+ + OH^-$$

Virtually no undissociated HCl or KOH exists in aqueous solution. Demonstration 6-2 shows one consequence of the strong-acid behavior of HCl.

Even though the hydrogen halides HCl, HBr, and HI are strong acids, HF is *not* a strong acid, as explained in Box 6-3. For most purposes, the hydroxides of the alkaline earth metals ($Mg^{2+}$, $Ca^{2+}$, $Sr^{2+}$, and $Ba^{2+}$) can be considered to be strong bases, although they are far less soluble than alkali metal hydroxides and have some tendency to form $MOH^+$ complexes (Table 6-3).

**Table 6-3** Equilibria of alkaline earth metal hydroxides

$$M(OH)_2(s) \rightleftharpoons M^{2+} + 2OH^-$$
$$K_{sp} = [M^{2+}][OH^-]^2$$
$$M^{2+} + OH^- \rightleftharpoons MOH^+$$
$$K_1 = [MOH^+]/[M^{2-}][OH^-]$$

| Metal | log $K_{sp}$ | log $K_1$ |
|---|---|---|
| $Mg^{2+}$ | −11.15 | 2.58 |
| $Ca^{2+}$ | −5.19 | 1.30 |
| $Sr^{2+}$ | — | 0.82 |
| $Ba^{2+}$ | — | 0.64 |

NOTE: 25°C and ionic strength = 0.

---

### Demonstration 6-2   The HCl Fountain

The complete dissociation of HCl into $H^+$ and $Cl^-$ makes $HCl(g)$ extremely soluble in water.

| | | |
|---|---|---|
| $HCl(g) \rightleftharpoons HCl(aq)$ | (A) |
| $HCl(aq) \rightleftharpoons H^+(aq) + Cl^-(aq)$ | (B) |
| Net reaction:   $HCl(g) \rightleftharpoons H^+(aq) + Cl^-(aq)$ | (C) |

Because the equilibrium of Reaction B lies far to the right, it pulls Reaction A to the right as well.

*Challenge*   The standard free energy change ($\Delta G°$) for Reaction C is −36.0 kJ/mol. Show that the equilibrium constant is $2.0 \times 10^6$.

The extreme solubility of $HCl(g)$ in water is the basis for the HCl fountain,[18] assembled as shown below. In Figure *a*, an inverted 250-mL round-bottom flask containing air is set up with its inlet tube leading to a source of $HCl(g)$ and its outlet tube directed into an inverted bottle of water. As HCl is admitted to the flask, air is displaced. When the bottle is filled with air, the flask is filled mostly with $HCl(g)$.

The hoses are disconnected and replaced with a beaker of indicator and a rubber bulb (Figure *b*). For an indicator, we use slightly alkaline methyl purple, which is green above pH 5.4 and purple below pH 4.8. When ~1 mL of water is squirted from the rubber bulb into the flask, a vacuum is created and indicator solution is drawn up into the flask, making a fascinating fountain (Color Plate 1).

*Question*   Why is a vacuum created when water is squirted into the flask and why does the indicator change color when it enters the flask?

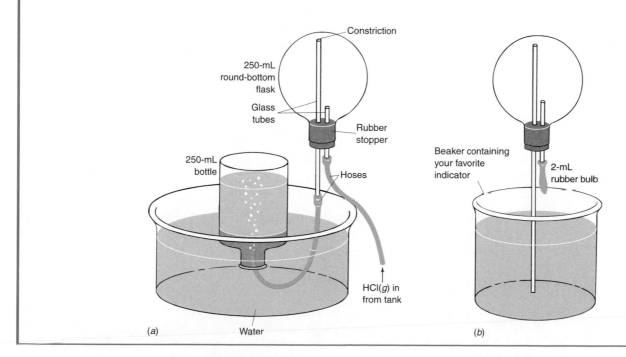

(a)   Water

(b)

**Box 6-3** The Strange Behavior of Hydrofluoric Acid[13]

The hydrogen halides HCl, HBr, and HI are all strong acids, which means that the reactions

$$HX(aq) + H_2O \rightarrow H_3O^+ + X^-$$

(X = Cl, Br, I) all go to completion. Why, then, does HF behave as a weak acid?

The answer is odd. First, HF does completely give up its proton to $H_2O$:

$$HF(aq) \rightarrow H_3O^+ + F^-$$
$$\text{Hydronium} \quad \text{Fluoride}$$
$$\text{ion} \quad \text{ion}$$

But fluoride forms the strongest hydrogen bond of any ion. The hydronium ion remains tightly associated with $F^-$ through a hydrogen bond. We call such an association an **ion pair.**

$$H_3O^+ + F^- \rightleftharpoons F^- \cdots H_3O^+$$
$$\text{An ion pair}$$

Ion pairs are common in aqueous solutions of any ion with a charge greater than 1. Ion pairs are the rule in nonaqueous solvents, which cannot promote ion dissociation as well as water.

Thus, HF does not behave as a strong acid, because $F^-$ and $H_3O^+$ remain associated with each other. Dissolving one mole of the strong acid HCl in water creates one mole of free $H_3O^+$. Dissolving one mole of the "weak" acid HF in water creates little free $H_3O^+$.

HF is not unique in its propensity to form ion pairs. Many moderately strong acids, such as those below, are thought to exist predominantly as ion pairs in aqueous solution (HA + $H_2O$ $\rightleftharpoons$ $A^- \cdots H_3O^+$).[19]

$$CF_3CO_2H$$
Trifluoroacetic acid
$$K_a = 0.31$$

Squaric acid
$$K_a = 0.29$$

## Weak Acids and Bases

All weak acids, denoted HA, react with water by donating a proton to $H_2O$:

*Dissociation of weak acid:* $\quad HA + H_2O \overset{K_a}{\rightleftharpoons} H_3O^+ + A^-$ (6-23)

which means exactly the same as

*Acid dissociation constant:* $K_a = \dfrac{[H^+][A^-]}{[HA]}$

*Dissociation of weak acid:* $\quad HA \overset{K_a}{\rightleftharpoons} H^+ + A^- \qquad K_a = \dfrac{[H^+][A^-]}{[HA]}$ (6-24)

The equilibrium constant is called $K_a$, the **acid dissociation constant**. By definition, a weak acid is one that is only partially dissociated in water. This definition means that $K_a$ is "small" for a weak acid.

Weak bases, B, react with water by abstracting a proton from $H_2O$:

*Base hydrolysis constant:* $K_b = \dfrac{[BH^+][OH^-]}{[B]}$

*Hydrolysis* refers to any reaction with water.

*Base hydrolysis:* $\quad B + H_2O \overset{K_b}{\rightleftharpoons} BH^+ + OH^- \qquad K_b = \dfrac{[BH^+][OH^-]}{[B]}$ (6-25)

The equilibrium constant $K_b$ is called the **base hydrolysis constant.** By definition, a weak base is one for which $K_b$ is "small."

## Common Classes of Weak Acids and Bases

Acetic acid is a typical weak acid.

$$CH_3-C \overset{O}{\underset{O-H}{}} \rightleftharpoons CH_3-C \overset{O}{\underset{O^-}{}} + H^+ \qquad K_a = 1.75 \times 10^{-5}$$ (6-26)

Acetic acid
(HA)

Acetate
($A^-$)

Acetic acid is a representative carboxylic acid, which has the general formula $RCO_2H$, where R is an organic substituent. *Most* **carboxylic acids** *are weak acids, and most* **carboxylate anions** *are weak bases.*

$$R-C \overset{O}{\underset{O-H}{}} \qquad R-C \overset{O}{\underset{O^-}{}}$$

A carboxylic acid
(weak base, HA)

A carboxylate anion
(weak base, $A^-$)

Methylamine is a typical weak base.

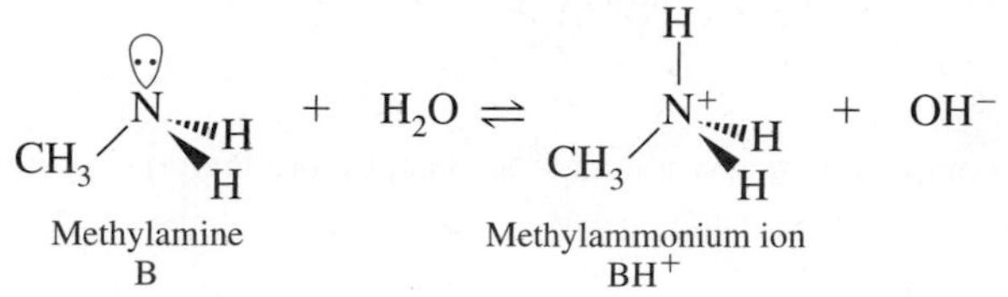

$$\underset{\substack{\text{Methylamine} \\ \text{B}}}{CH_3\overset{\displaystyle\cdot\cdot}{N}H_2} + H_2O \rightleftharpoons \underset{\substack{\text{Methylammonium ion} \\ BH^+}}{CH_3\overset{+}{N}H_3} + OH^- \qquad K_b = 4.47 \times 10^{-4} \quad (6\text{-}27)$$

Amines are nitrogen-containing compounds:

$$\overset{\displaystyle\cdot\cdot}{R N H_2} \quad \text{a primary amine} \qquad\qquad RNH_3^+$$

$$R_2\overset{\displaystyle\cdot\cdot}{N}H \quad \text{a secondary amine} \qquad\quad R_2NH_2^+ \Bigg\}\ \text{ammonium ions}$$

$$R_3\overset{\displaystyle\cdot\cdot}{N} \quad \text{a tertiary amine} \qquad\qquad R_3NH^+$$

**Amines** *are weak bases, and* **ammonium ions** *are weak acids.* The "parent" of all amines is ammonia, $NH_3$. When a base such as methylamine reacts with water, the product is the conjugate acid. That is, methylammonium ion produced in Reaction 6-27 is a weak acid:

$$\underset{BH^+}{CH_3\overset{+}{N}H_3} \overset{K_a}{\rightleftharpoons} \underset{B}{CH_3\overset{\displaystyle\cdot}{N}H_2} + H^+ \qquad K_a = 2.26 \times 10^{-11} \qquad (6\text{-}28)$$

The methylammonium ion is the conjugate acid of methylamine.

You should learn to recognize whether a compound is acidic or basic. The salt methylammonium chloride, for example, dissociates in aqueous solution to give methylammonium cation and chloride:

$$\underset{\substack{\text{Methylammonium} \\ \text{chloride}}}{CH_3\overset{+}{N}H_3Cl^-(s)} \rightarrow CH_3\overset{+}{N}H_3(aq) + Cl^-(aq) \qquad (6\text{-}29)$$

Methylammonium ion, being the conjugate acid of methylamine, is a weak acid (Reaction 6-28). Chloride is the conjugate base of HCl, a strong acid. In other words, $Cl^-$ *has virtually no tendency to associate with* $H^+$, or else HCl would not be a strong acid. Methylammonium chloride is acidic because methylammonium ion is an acid and $Cl^-$ is not a base.

Metal cations, $M^{n+}$, constitute another common class of weak acids.[20] Figure 6-8 shows acid dissociation constants for the reaction

$$M^{n+} + H_2O \overset{K_a}{\rightleftharpoons} MOH^{(n-1)+} + H^+$$

Carboxylic acids ($RCO_2H$) and ammonium ions ($R_3NH^+$) are weak acids. Carboxylate anions ($RCO_2^-$) and amines ($R_3N$) are weak bases.

Although we will usually write a **base** as **B** and an **acid** as **HA**, it is important to realize that **BH⁺** is also an **acid** and **A⁻** is also a **base.**

Methylammonium chloride is a weak acid because
1. It dissociates into $CH_3NH_3^+$ and $Cl^-$.
2. $CH_3NH_3^+$ is a weak acid, being conjugate to $CH_3NH_2$, a weak base.
3. $Cl^-$ has no basic properties. It is conjugate to HCl, a strong acid. That is, HCl dissociates completely.

*Challenge* Phenol ($C_6H_5OH$) is a weak acid. Explain why a solution of the ionic compound potassium phenolate ($C_6H_5O^-K^+$) is basic.

Aqueous metal ions are associated with (*hydrated* by) several $H_2O$ molecules, so a more accurate way to write the acid dissociation reaction is

$$M(H_2O)_x^{n+} \rightleftharpoons M(H_2O)_{x-1}(OH)^{(n-1)+} + H^+$$

| G1 | G2 | G3 | G4 | G5 | G6 | G7 | G8 | G9 | G10 | G11 | G12 | G13 | G14 | G15 |
|---|---|---|---|---|---|---|---|---|---|---|---|---|---|---|
| Li⁺ 13.64 | Be | | | | | | | | | | | | | |
| Na⁺ 13.9 | Mg²⁺ 11.4 | | | | | | | | | | | Al³⁺ 5.00 | | |
| K | Ca²⁺ 12.70 | Sc³⁺ 4.3 | Ti³⁺ 1.3 | VO²⁺ 5.7 | Cr²⁺ 5.5ᵃ; Cr³⁺ 3.66 | Mn²⁺ 10.6 | Fe²⁺ 9.4; Fe³⁺ 2.19 | Co²⁺ 9.7; Co³⁺ 0.5ᵇ | Ni²⁺ 9.9 | Cu²⁺ 7.5 | Zn²⁺ 9.0 | Ga³⁺ 2.6 | Ge | |
| Rb | Sr 13.18 | Y³⁺ 7.7 | Zr⁴⁺ −0.3 | Nb | Mo | Tc | Ru | Rh³⁺ 3.33ᶜ | Pd²⁺ 1.0 | Ag⁺ 12.0 | Cd²⁺ 10.1 | In³⁺ 3.9 | Sn²⁺ 3.4 | Sb |
| Cs | Ba²⁺ 13.36 | La³⁺ 8.5 | Hf | Ta | W | Re | Os | Ir | Pt | Au | Hg₂²⁺ 5.3ᵈ; Hg²⁺ 3.40 | Tl⁺ 13.21 | Pb²⁺ 7.6 | Bi³⁺ 1.1 |

(Arrow labeled "Stronger acid" pointing right, above the table.)

| Ce³⁺ 9.1ᵇ | Pr³⁺ 9.4ᵇ | Nd³⁺ 8.7ᵇ | Pm | Sm³⁺ 8.6ᵇ | Eu³⁺ 8.6ᵈ | Gd³⁺ 9.1ᵇ | Tb³⁺ 8.4ᵈ | Dy³⁺ 8.4ᵈ | Ho³⁺ 8.3 | Er³⁺ 9.1ᵇ | Tm³⁺ 8.2ᵈ | Yb³⁺ 8.4ᵇ | Lu³⁺ 8.2ᵈ |
|---|---|---|---|---|---|---|---|---|---|---|---|---|---|

Ionic strength = 0 unless noted by superscript
a. Ionic strength = 1 M  b. Ionic strength = 3 M  c. Ionic strength = 2.5 M  d. Ionic strength = 0.5 M

*Figure 6-8* Acid dissociation constants ($-\log K_a$) for aqueous metal ions: $M^{n+} + H_2O \overset{K_a}{\rightleftharpoons} MOH^{(n-1)+} + H^+$. For example, for Li⁺, $K_a = 10^{-13.64}$. In Chapter 9, we will learn that the numbers in this table are called $pK_a$. Darkest shades are strongest acids. [Data from R. M. Smith, A. E. Martell, and R. J. Motekaitis, NIST Critical Stability Constants of Metal Complexes Database 46 (Gaithersburg, MD: National Institute of Standards and Technology, 2001).]

Monovalent metal ions are very weak acids ($Na^+$, $K_a = 10^{-13.9}$). Divalent ions tend to be stronger ($Fe^{2+}$, $K_a = 10^{-9.4}$) and trivalent ions are stronger yet ($Fe^{3+}$, $K_a = 10^{-2.19}$).

## Polyprotic Acids and Bases

**Polyprotic acids and bases** are compounds that can donate or accept more than one proton. For example, oxalic acid is diprotic and phosphate is tribasic:

*Notation for acid and base equilibrium constants:* $K_{a1}$ refers to the acidic species with the most protons and $K_{b1}$ refers to the basic species with the fewest protons. The subscript "a" in acid dissociation constants will usually be omitted.

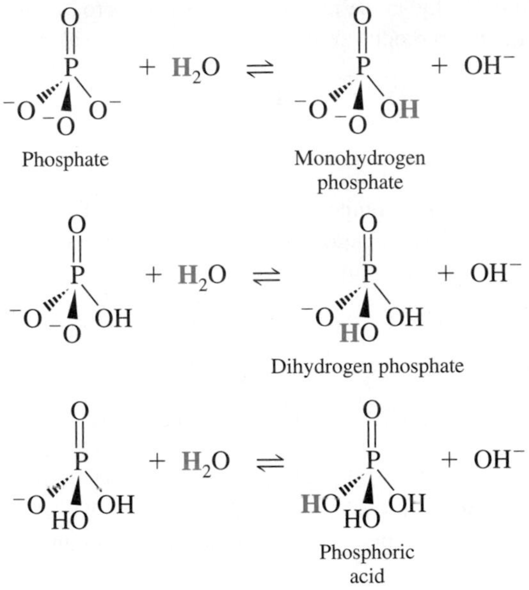

$$HOCCOH \rightleftharpoons H^+ + {}^-OCCOH \qquad K_{a1} = 5.37 \times 10^{-2} \quad (6\text{-}30)$$

Oxalic acid / Monohydrogen oxalate

$$^-OCCOH \rightleftharpoons H^+ + {}^-OCCO^- \qquad K_{a2} = 5.42 \times 10^{-5} \quad (6\text{-}31)$$

Oxalate

$$\text{Phosphate} + H_2O \rightleftharpoons \text{Monohydrogen phosphate} + OH^- \qquad K_{b1} = 2.3 \times 10^{-2} \quad (6\text{-}32)$$

$$+ H_2O \rightleftharpoons + OH^- \qquad K_{b2} = 1.60 \times 10^{-7} \quad (6\text{-}33)$$

Dihydrogen phosphate

$$+ H_2O \rightleftharpoons + OH^- \qquad K_{b3} = 1.42 \times 10^{-12} \quad (6\text{-}34)$$

Phosphoric acid

The standard notation for successive acid dissociation constants of a polyprotic acid is $K_1$, $K_2$, $K_3$, and so on, with the subscript "a" usually omitted. We retain or omit the subscript as dictated by clarity. For successive base hydrolysis constants, we retain the subscript "b." The preceding examples illustrate that $K_{a1}$ (or $K_1$) *refers to the acidic species with the most protons, and $K_{b1}$ refers to the basic species with the least number of protons.* Carbonic acid, a very important diprotic carboxylic acid derived from $CO_2$, is described in Box 6-4.

## Relation Between $K_a$ and $K_b$

A most important relation exists between $K_a$ and $K_b$ of a conjugate acid-base pair in aqueous solution. We can derive this result with the acid HA and its conjugate base $A^-$.

$$HA \rightleftharpoons H^+ + A^- \qquad K_a = \frac{[H^+][A^-]}{[HA]}$$

$$A^- + H_2O \rightleftharpoons HA + OH^- \qquad K_b = \frac{[HA][OH^-]}{[A^-]}$$

$$\overline{H_2O \rightleftharpoons H^+ + OH^-} \qquad K_w = K_a \cdot K_b$$

$$= \frac{[H^+][A^-]}{[HA]} \cdot \frac{[HA][OH^-]}{[A^-]}$$

When the reactions are added, their equilibrium constants are multiplied to give

$K_a \cdot K_b = K_w$ for a conjugate acid-base pair in aqueous solution.

*Relation between $K_a$ and $K_b$ for a conjugate pair:* $\boxed{K_a \cdot K_b = K_w}$ $\qquad (6\text{-}35)$

Equation 6-35 applies to any acid and its conjugate base in aqueous solution.

CHAPTER 6 Chemical Equilibrium

Carbonic acid is formed by the reaction of carbon dioxide with water:

$$CO_2(g) \rightleftharpoons CO_2(aq) \qquad K = \frac{[CO_2(aq)]}{P_{CO_2}} = 0.034\ 4$$

$$CO_2(aq) + H_2O \rightleftharpoons \underset{\text{Carbonic acid}}{\overset{\displaystyle O \atop \displaystyle \| \atop \displaystyle C}{HO \qquad OH}}$$

$$K = \frac{[H_2CO_3]}{[CO_2(aq)]} \approx 0.002$$

$$H_2CO_3 \rightleftharpoons \underset{\text{Bicarbonate}}{HCO_3^-} + H^+ \qquad K_{a1} = 4.46 \times 10^{-7}$$

$$HCO_3^- \rightleftharpoons \underset{\text{Carbonate}}{CO_3^{2-}} + H^+ \qquad K_{a2} = 4.69 \times 10^{-11}$$

Its behavior as a diprotic acid appears anomalous at first, because the value of $K_{a1}$ is about $10^2$ to $10^4$ times smaller than $K_a$ for other carboxylic acids.

| $CH_3CO_2H$ | $HCO_2H$ |
|---|---|
| Acetic acid | Formic acid |
| $K_a = 1.75 \times 10^{-5}$ | $K_a = 1.80 \times 10^{-4}$ |
| $N\equiv CCH_2CO_2H$ | $HOCH_2CO_2H$ |
| Cyanoacetic acid | Glycolic acid |
| $K_a = 3.37 \times 10^{-3}$ | $K_a = 1.48 \times 10^{-4}$ |

The reason for this anomaly is not that $H_2CO_3$ is unusual but, rather, that the value commonly given for $K_{a1}$ applies to the equation

$$\text{All dissolved } CO_2 \rightleftharpoons HCO_3^- + H^+$$
$$(= CO_2(aq) + H_2CO_3)$$

$$K_{a1} = \frac{[HCO_3^-][H^+]}{[CO_2(aq) + H_2CO_3]} = 4.46 \times 10^{-7}$$

Only about 0.2% of dissolved $CO_2$ is in the form $H_2CO_3$. When the true value of $[H_2CO_3]$ is used instead of the value $[H_2CO_3 + CO_2(aq)]$, the value of the equilibrium constant becomes

$$K_{a1} = \frac{[HCO_3^-][H^+]}{H_2CO_3} = 2 \times 10^{-4}$$

The hydration of $CO_2$ (reaction of $CO_2$ with $H_2O$) and dehydration of $H_2CO_3$ are slow reactions, which can be demonstrated in a classroom.[21] Living cells utilize the enzyme *carbonic anhydrase* to speed the rate at which $H_2CO_3$ and $CO_2$ equilibrate in order to process this key metabolite. The enzyme provides an environment just right for the reaction of $CO_2$ with $OH^-$, lowering the *activation energy* (the energy barrier for the reaction) from 50 down to 26 kJ/mol and increasing the rate of reaction by more than a factor of $10^6$.

---

**Example** Finding $K_b$ for the Conjugate Base

$K_a$ for acetic acid is $1.75 \times 10^{-5}$ (Reaction 6-26). Find $K_b$ for acetate ion.

**Solution** This one is trivial:[†]

$$K_b = \frac{K_w}{K_a} = \frac{1.0 \times 10^{-14}}{1.75 \times 10^{-5}} = 5.7 \times 10^{-10}$$

**Example** Finding $K_a$ for the Conjugate Acid

$K_b$ for methylamine is $4.47 \times 10^{-4}$ (Reaction 6-27). Find $K_a$ for methylammonium ion.

**Solution** Once again,

$$K_a = \frac{K_w}{K_b} = \frac{1.0 \times 10^{-14}}{4.47 \times 10^{-4}} = 2.2 \times 10^{-11}$$

For a diprotic acid, we can derive relations between each of two acids and their conjugate bases:

$$\begin{array}{ll} H_2A \rightleftharpoons H^+ + HA^- & K_{a1} \\ \underline{HA^- + H_2O \rightleftharpoons H_2A + OH^-} & K_{b2} \\ H_2O \rightleftharpoons H^+ + OH^- & K_w \end{array} \qquad \begin{array}{ll} HA^- \rightleftharpoons H^+ + A^{2-} & K_{a2} \\ \underline{A^{2-} + H_2O \rightleftharpoons HA^- + OH^-} & K_{b1} \\ H_2O \rightleftharpoons H^+ + OH^- & K_w \end{array}$$

---

[†]In this text, we use $K_w = 10^{-14.00} = 1.0 \times 10^{-14}$. The more accurate value from Table 6-1 is $K_w = 10^{-13.995}$. For acetic acid with $K_a = 10^{-4.756}$, the accurate value of $K_b$ is $10^{-(13.995-4.756)} = 10^{-9.239} = 5.77 \times 10^{-10}$.

The final results are

*General relation between $K_a$ and $K_b$:*

$$K_{a1} \cdot K_{b2} = K_w \tag{6-36}$$

$$K_{a2} \cdot K_{b1} = K_w \tag{6-37}$$

*Challenge* Derive the following results for a triprotic acid:

$$K_{a1} \cdot K_{b3} = K_w \tag{6-38}$$

$$K_{a2} \cdot K_{b2} = K_w \tag{6-39}$$

$$K_{a3} \cdot K_{b1} = K_w \tag{6-40}$$

## Shorthand for Organic Structures

We are beginning to encounter organic (carbon-containing) compounds in this book. Chemists and biochemists use simple conventions for drawing molecules to avoid writing every atom. Each vertex of a structure is understood to be a carbon atom, unless otherwise labeled. In the shorthand, we usually omit bonds from carbon to hydrogen. Carbon forms four chemical bonds. If you see carbon forming fewer than four bonds, the remaining bonds are assumed to go to hydrogen atoms that are not written. Here is an example:

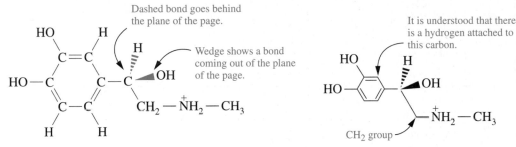

Epinephrine (also called adrenaline)          Shorthand drawing of epinephrine

Benzene, $C_6H_6$, has two equivalent resonance structures, so all C—C bonds are equivalent. We often draw benzene rings with a circle in place of three double bonds.

Benzene
$C_6H_6$

The shorthand shows that the carbon atom at the top right of the six-membered benzene ring forms three bonds to other carbon atoms (one single bond and one double bond), so there must be a hydrogen atom attached to this carbon atom. The carbon atom at the left side of the benzene ring forms three bonds to other carbon atoms and one bond to an oxygen atom. There is no hidden hydrogen atom attached to this carbon. In the $CH_2$ group adjacent to nitrogen, both hydrogen atoms are omitted in the shorthand structure.

## 6-8 Solving Equilibrium Problems with a Concentration Table and a Spreadsheet

Suppose that we prepare 1.00 L of solution containing 0.001 0 mol $Br_2(aq)$, 0.005 0 mol $IO_3^-$, 0.020 mol $Br^-$, 1.00 mol $H^+$, and excess solid $I_2$ and that the reaction is

$$I_2(s) + 5Br_2(aq) + 6H_2O \rightleftharpoons 2IO_3^- + 10Br^- + 12H^+ \qquad K = 1 \times 10^{-19} \tag{6-41}$$

What are the concentrations when the solution comes to equilibrium?

We answer this complicated question by considering stoichiometry and then by using a spreadsheet to find a numerical solution. First, let's evaluate the reaction quotient to find out in which direction the reaction must proceed to reach equilibrium:

Omit solid ($I_2$) and solvent ($H_2O$) from the reaction quotient.

$$Q = \frac{[IO_3^-]^2[Br^-]^{10}[H^+]^{12}}{[Br_2(aq)]^5} = \frac{[0.005\ 0]^2[0.020]^{10}[1.00]^{12}}{[0.001\ 0]^5} = 2.5_6 \times 10^{-7} > K$$

Because $Q > K$, the reaction must go back to the left to reach equililbrium.

The key step in solving the problem is to prepare a table showing initial concentrations and final concentrations after equilibrium is attained. Let's say that $x$ mol of $IO_3^-$ are consumed in the reaction. From the stoichiometry of Reaction 6-41, we know that 10 mol of $Br^-$ are consumed for every 2 mol of $IO_3^-$. If $x$ mol of $IO_3^-$ are consumed, then $\frac{10}{2}x = 5x$ mol of $Br^-$ are consumed. Also, $\frac{12}{2}x = 6x$ mol of $H^+$ must be consumed and $\frac{5}{2}x$ mol of $Br_2(aq)$ must be made. Our table looks like this:

$$\left(\frac{10 \text{ mol Br}^-}{2 \text{ mol IO}_3^-}\right)(x \text{ mol IO}_3^-) = 5x \text{ mol Br}^-$$

|  | $I_2(s)$ | $+$ | $5Br_2(aq)$ | $+$ | $6H_2O$ | $\rightleftharpoons$ | $2IO_3^-$ | $+$ | $10Br^-$ | $+$ | $12H^+$ |
|---|---|---|---|---|---|---|---|---|---|---|---|
| Initial concentration | excess | | 0.001 0 | | | | 0.005 0 | | 0.020 | | 1.00 |
| Final concentration | excess | | 0.001 0 + 2.5x | | | | 0.005 0 − x | | 0.020 − 5x | | 1.00 − 6x |

Be sure you can derive each concentration. The volume is 1 L, so moles and moles/L are the same in this example.

The equilibrium expression is

$$K = \frac{[IO_3^-]^2[Br^-]^{10}[H^+]^{12}}{[Br_2(aq)]^5} = \frac{[0.005\ 0 - x]^2[0.020 - 5x]^{10}[1.00 - 6x]^{12}}{[0.001\ 0 + 2.5x]^5} = 1 \times 10^{-19}$$

$$(6\text{-}42)$$

We can find a value of $x$ that satisfies this equation by trial-and-error guessing with the spreadsheet in Figure 6-9. In column A, enter the chemical species and, in column B, enter the initial concentrations. Cell B8, which we will not use, contains the value of the equilibrium constant just as a reminder. In cell B11 we *guess* a value for $x$. We know that $x$ cannot exceed the initial concentration of $IO_3^-$, so we guess $x = 0.001$. Cells C4:C7 give the final concentrations computed from initial concentrations and the guessed value of $x$. Cell C11 computes the reaction quotient, $Q$, from the final concentrations in cells C4:C7.

Our strategy is to vary $x$ in cell B11 until $Q = K = 1 \times 10^{-19}$ in cell C11. Columns F, G, and H are not part of the spreadsheet, but show the progress of guessing. When we guess $x = 0.001$, then $Q = 1.63 \times 10^{-11}$, which is greater than $K$. Guessing $x = 0.003$ gives $Q = 7.08 \times 10^{-19}$, which is still greater than $K$, but getting closer. Guessing $x = 0.003\ 15$ gives $Q = 9.50 \times 10^{-20}$, which is less than $K$. Therefore $x$ must be less than 0.003 15. Columns F and G show that we can find $x$ with several significant digits after a few guesses.

## Using Excel GOAL SEEK

The Excel routine GOAL SEEK solves Equation 6-42 automatically by varying $x$ systematically until the equation is satisfied. With the spreadsheet in Figure 6-9, select OPTIONS from the TOOLS menu. Select the Calculations tab and set Maximum change to 1e-24. This setting makes the computation fine enough to compute a reaction quotient of $1 \times 10^{-19}$. Highlight cell C11 and select GOAL SEEK from the TOOLS menu. In the GOAL SEEK window, fill in the underlined values: Set cell C11 To value 1e-19 By changing cell B11. When you click OK, GOAL SEEK finds $x = 0.003\ 146\ 45$ in cell B11, which gives $Q = 1.000\ 0 \times 10^{-19}$ in cell C11.

|  | A | B | C | D | E | F | G | H |
|---|---|---|---|---|---|---|---|---|
| 1 | Solving an Equation by Trial and Error | | | | | | | |
| 2 | | | | | | Guessed | Computed | Q compared |
| 3 | | Initial | Final | | | value of x | Q | to K |
| 4 | $[IO_3^-] =$ | 0.005 | 0.0040000 | =B4−B11 | | 0.001 | 1.6343E-11 | too big |
| 5 | $[Br^-] =$ | 0.02 | 0.0150000 | =B5−5*B11 | | 0.003 | 7.0795E-19 | too big |
| 6 | $[H^+] =$ | 1 | 0.9940000 | =B6−6*B11 | | 0.0031 | 1.9131E-19 | too big |
| 7 | $[Br_2(aq)] =$ | 0.001 | 0.0035000 | =B7+2.5*B11 | | 0.00315 | 9.505E-20 | too small |
| 8 | $K_{eq} =$ | 1.00E-19 | | | | 0.00314 | 1.0962E-19 | too big |
| 9 | | | | | | 0.003145 | 1.0209E-19 | too big |
| 10 | | x | Q | | | 0.003146 | 1.0064E-19 | too big |
| 11 | | 0.001 | 1.6343E-11 | | | 0.0031464 | 1.0007E-19 | very close |
| 12 | C11 = $C$4^2*$C$5^10*$C$6^12/$C$7^5 | | | | | | | |

*Figure 6-9* Spreadsheet for solving Equation 6-42.

## Terms to Understand

| | | | |
|---|---|---|---|
| acid | Brønsted-Lowry base | entropy | overall formation constant |
| acid dissociation constant ($K_a$) | carboxylate anion | equilibrium constant | pH |
| acidic solution | carboxylic acid | exothermic | polyprotic acid |
| amine | common ion effect | Gibbs free energy | polyprotic base |
| ammonium ion | complex ion | hydronium ion | protic solvent |
| aprotic solvent | conjugate acid-base pair | ion pair | reaction quotient |
| autoprotolysis | coprecipitation | Le Châtelier's principle | salt |
| base | cumulative formation constant | Lewis acid | saturated solution |
| base hydrolysis constant ($K_b$) | disproportionation | Lewis base | solubility product |
| basic solution | endothermic | ligand | standard state |
| Brønsted-Lowry acid | enthalpy change | neutralization | stepwise formation constant |

## Summary

For the reaction $aA + bB \rightleftharpoons cC + dD$, the equilibrium constant is $K = [C]^c[D]^d/[A]^a[B]^b$. Solute concentrations should be expressed in moles per liter; gas concentrations should be in bars; and the concentrations of pure solids, liquids, and solvents are omitted. If the direction of a reaction is changed, $K' = 1/K$. If two reactions are added, $K_3 = K_1 K_2$. The equilibrium constant can be calculated from the free-energy change for a chemical reaction: $K = e^{-\Delta G°/RT}$. The equation $\Delta G = \Delta H - T\Delta S$ summarizes the observations that a reaction is favored if it liberates heat (exothermic, negative $\Delta H$) or increases disorder (positive $\Delta S$). Le Châtelier's principle predicts the effect on a chemical reaction when reactants or products are added or temperature is changed. The reaction quotient, $Q$, tells how a system must change to reach equilibrium.

The solubility product is the equilibrium constant for the dissolution of a solid salt into its constituent ions in aqueous solution. The common ion effect is the observation that, if one of the ions of that salt is already present in the solution, the solubility of a salt is decreased. Sometimes, we can selectively precipitate one ion from a solution containing other ions by adding a suitable counterion. At high concentration of ligand, a precipitated metal ion may redissolve by forming soluble complex ions. In a metal-ion complex, the metal is a Lewis acid (electron pair acceptor) and the ligand is a Lewis base (electron pair donor).

Brønsted-Lowry acids are proton donors, and Brønsted-Lowry bases are proton acceptors. An acid increases the concentration of $H_3O^+$ in aqueous solution, and a base increases the concentration of $OH^-$. An acid-base pair related through the gain or loss of a single proton is described as conjugate. When a proton is transferred from one molecule to another molecule of a protic solvent, the reaction is called autoprotolysis.

The definition of pH is $pH = -\log[H^+]$ (which will be modified to include activity later). $K_a$ is the equilibrium constant for the dissociation of an acid: $HA + H_2O \rightleftharpoons H_3O^+ + A^-$. $K_b$ is the base hydrolysis constant for the reaction $B + H_2O \rightleftharpoons BH^+ + OH^-$. When either $K_a$ or $K_b$ is large, the acid or base is said to be strong; otherwise, the acid or base is weak. Common strong acids and bases are listed in Table 6-2, which you should memorize. The most common weak acids are carboxylic acids ($RCO_2H$), and the most common weak bases are amines ($R_3N$:). Carboxylate anions ($RCO_2^-$) are weak bases, and ammonium ions ($R_3NH^+$) are weak acids. Metal cations also are weak acids. For a conjugate acid-base pair in water, $K_a \cdot K_b = K_w$. For polyprotic acids, we denote the successive acid dissociation constants as $K_{a1}, K_{a2}, K_{a3}, \cdots$, or just $K_1, K_2, K_3, \cdots$. For polybasic species, we denote successive hydrolysis constants $K_{b1}, K_{b2}, K_{b3}, \cdots$. For a diprotic system, the relations between successive acid and base equilibrium constants are $K_{a1} \cdot K_{b2} = K_w$ and $K_{a2} \cdot K_{b1} = K_w$. For a triprotic system the relations are $K_{a1} \cdot K_{b3} = K_w$, $K_{a2} \cdot K_{b2} = K_w$, and $K_{a3} \cdot K_{b1} = K_w$.

In the chemists' shorthand for organic structures, each vertex is a carbon atom. If fewer than four bonds to that carbon are shown, it is understood that H atoms are attached to the carbon so that it makes four bonds.

You can solve for equilibrium concentrations in a chemical reaction by first creating a table of initial and final concentrations of each species. Then find the final concentrations by systemic guessing with a spreadsheet or by using the Excel GOAL SEEK routine.

## Exercises

**6-A.** Consider the following equilibria, in which all ions are aqueous:

(1) $Ag^+ + Cl^- \rightleftharpoons AgCl(aq)$        $K = 2.0 \times 10^3$

(2) $AgCl(aq) + Cl^- \rightleftharpoons AgCl_2^-$      $K = 9.3 \times 10^1$

(3) $AgCl(s) \rightleftharpoons Ag^+ + Cl^-$        $K = 1.8 \times 10^{-10}$

(a) Calculate the numerical value of the equilibrium constant for the reaction $AgCl(s) \rightleftharpoons AgCl(aq)$.

(b) Calculate the concentration of $AgCl(aq)$ in equilibrium with excess undissolved solid AgCl.

(c) Find the numerical value of $K$ for the reaction $AgCl_2^- \rightleftharpoons AgCl(s) + Cl^-$.

**6-B.** Reaction 6-7 is allowed to come to equilibrium in a solution initially containing 0.010 0 M $BrO_3^-$, 0.010 0 M $Cr^{3+}$, and 1.00 M $H^+$. To find the concentrations at equilibrium, we can construct a table showing initial and final concentrations. We use the stoichiometry coefficients of the reaction to say that, if $x$ mol of $Br^-$ are created, then we must also make $x$ mol of $Cr_2O_7^{2-}$ and $8x$ mol of $H^+$. To produce $x$ mol of $Br^-$, we must have consumed $x$ mol of $BrO_3^-$ and $2x$ mol of $Cr^{3+}$.

(a) Write the equilibrium constant expression that you would use to solve for $x$ to find the concentrations at equilibrium. Do not try to solve the equation.

| | $BrO_3^-$ | + | $2Cr^{3+}$ | + | $4H_2O$ | $\rightleftharpoons$ | $Br^-$ | + | $Cr_2O_7^{2-}$ | + | $8H^+$ |
|---|---|---|---|---|---|---|---|---|---|---|---|
| Initial concentration | 0.010 0 | | 0.010 0 | | | | | | | | 1.00 |
| Final concentration | 0.010 0 − $x$ | | 0.010 0 − 2$x$ | | | | $x$ | | $x$ | | 1.00 + 8$x$ |

**(b)** Since $K = 1 \times 10^{11}$ for Reaction 6-7, it is reasonable to guess that the reaction will go nearly "to completion." That is, we expect both the concentration of $Br^-$ and $Cr_2O_7^{2-}$ to be close to 0.005 00 M at equilibrium. (Why?) That is, $x \approx 0.005\ 00$ M. With this value of $x$, $[H^+] = 1.00 + 8x = 1.04$ M and $[BrO_3^-] = 0.010\ 0 - x = 0.005\ 0$ M. However, we cannot say $[Cr^{3+}] = 0.010\ 0 - 2x = 0$, because there must be some small concentration of $Cr^{3+}$ at equilibrium. Write $[Cr^{3+}]$ for the concentration of $Cr^{3+}$ and solve for $[Cr^{3+}]$. $Cr^{3+}$ is the *limiting reagent* in this example. The reaction uses up $Cr^{3+}$ before consuming $BrO_3^-$.

**6-C.** Find $[La^{3+}]$ in the solution when excess solid lanthanum iodate, $La(IO_3)_3$, is stirred with 0.050 M $LiIO_3$ until the system reaches equilibrium. Assume that $IO_3^-$ from $La(IO_3)_3$ is negligible compared with $IO_3^-$ from $LiIO_3$.

**6-D.** Which will be more soluble (moles of metal dissolved per liter of solution), $Ba(IO_3)_2$ ($K_{sp} = 1.5 \times 10^{-9}$) or $Ca(IO_3)_2$ ($K_{sp} = 7.1 \times 10^{-7}$)? Give an example of a chemical reaction that might occur that would reverse the predicted solubilities.

**6-E.** Fe(III) precipitates from acidic solution by addition of $OH^-$ to form $Fe(OH)_3(s)$. At what concentration of $OH^-$ will $[Fe(III)]$ be reduced to $1.0 \times 10^{-10}$ M? If Fe(II) is used instead, what concentration of $OH^-$ will reduce $[Fe(II)]$ to $1.0 \times 10^{-10}$ M?

**6-F.** Is it possible to precipitate 99.0% of 0.010 M $Ce^{3+}$ by adding oxalate ($C_2O_4^{2-}$) without precipitating 0.010 M $Ca^{2+}$?

$$CaC_2O_4 \qquad K_{sp} = 1.3 \times 10^{-8}$$
$$Ce_2(C_2O_4)_3 \qquad K_{sp} = 3 \times 10^{-29}$$

**6-G.** For a solution of $Ni^{2+}$ and ethylenediamine, the following equilibrium constants apply at 20°C:

$$Ni^{2+} + H_2NCH_2CH_2NH_2 \rightleftharpoons Ni(en)^{2+} \quad \log K_1 = 7.52$$

Ethylenediamine
(abbreviated en)

$$Ni(en)^{2+} + en \rightleftharpoons Ni(en)_2^{2+} \qquad \log K_2 = 6.32$$
$$Ni(en)_2^{2+} + en \rightleftharpoons Ni(en)_3^{2+} \qquad \log K_3 = 4.49$$

Calculate the concentration of free $Ni^{2+}$ in a solution prepared by mixing 0.100 mol of en plus 1.00 mL of 0.010 0 M $Ni^{2+}$ and diluting to 1.00 L with dilute base (which keeps all the en in its unprotonated form). Assume that nearly all the nickel is in the form $Ni(en)_3^{2+}$ so $[Ni(en)_3^{2+}] = 1.00 \times 10^{-5}$ M. Calculate the concentrations of $Ni(en)^{2+}$ and $Ni(en)_2^{2+}$ to verify that they are negligible in comparison with $Ni(en)_3^{2+}$.

**6-H.** If each of the following compounds is dissolved in water, will the solution be acidic, basic, or neutral?

(a) $Na^+Br^-$    (d) $K_3PO_4$    (f) $(CH_3)_4N^+\!\!\bigcirc\!\!-CO_2^-$

(b) $Na^+CH_3CO_2^-$    (e) $(CH_3)_4N^+Cl^-$

(c) $NH_4^+Cl^-$    (g) $Fe(NO_3)_3$

**6-I.** Succinic acid dissociates as follows:

$$K_1 = 6.2 \times 10^{-5}$$
$$K_2 = 2.3 \times 10^{-6}$$

Calculate $K_{b1}$ and $K_{b2}$ for the following reactions:

**6-J.** Histidine is a triprotic amino acid:

$$K_1 = 3 \times 10^{-2}$$
$$K_2 = 8.5 \times 10^{-7}$$
$$K_3 = 4.6 \times 10^{-10}$$

What is the value of the equilibrium constant for the reaction

**6-K.** **(a)** From $K_w$ in Table 6-1, calculate the pH of pure water at 0°, 20°, and 40°C.
**(b)** For the reaction $D_2O \rightleftharpoons D^+ + OD^-$, $K = [D^+][OD^-] = 1.35 \times 10^{-15}$ at 25°C. In this equation, D stands for deuterium, which is the isotope $^2H$. What is the pD ($= -\log[D^+]$) for neutral $D_2O$?

---

## Problems

### Equilibrium and Thermodynamics

**6-1.** To evaluate the equilibrium constant in Equation 6-2, we must express concentrations of solutes in mol/L, gases in bars, and omit solids, liquids, and solvents. Explain why.

**6-2.** Why do we say that the equilibrium constant for the reaction $H_2O \rightleftharpoons H^+ + OH^-$ (or any other reaction) is dimensionless?

**6-3.** Predictions about the direction of a reaction based on Gibbs free energy or Le Châtelier's principle are said to be *thermodynamic*, not *kinetic*. Explain what this means.

**6-4.** Write the expression for the equilibrium constant for each of the following reactions. Write the pressure of a gaseous molecule, X, as $P_X$.
(a) $3Ag^+(aq) + PO_4^{3-}(aq) \rightleftharpoons Ag_3PO_4(s)$
(b) $C_6H_6(l) + \frac{15}{2}O_2(g) \rightleftharpoons 3H_2O(l) + 6CO_2(g)$

**6-5.** For the reaction $2A(g) + B(aq) + 3C(l) \rightleftharpoons D(s) + 3E(g)$, the concentrations at equilibrium are found to be

A: $2.8 \times 10^3$ Pa    C: 12.8 M    E: $3.6 \times 10^4$ Torr
B: $1.2 \times 10^{-2}$ M    D: 16.5 M

Find the numerical value of the equilibrium constant that would appear in a conventional table of equilibrium constants.

**6-6.** From the equations

$$HOCl \rightleftharpoons H^+ + OCl^- \qquad\qquad K = 3.0 \times 10^{-8}$$
$$HOCl + OBr^- \rightleftharpoons HOBr + OCl^- \qquad K = 15$$

find the value of $K$ for the reaction $HOBr \rightleftharpoons H^+ + OBr^-$. All species are aqueous.

**6-7.** (a) A favorable entropy change occurs when $\Delta S$ is positive. Does the order of the system increase or decrease when $\Delta S$ is positive?
(b) A favorable enthalpy change occurs when $\Delta H$ is negative. Does the system absorb heat or give off heat when $\Delta H$ is negative?
(c) Write the relation between $\Delta G$, $\Delta H$, and $\Delta S$. Use the results of parts (a) and (b) to state whether $\Delta G$ must be positive or negative for a spontaneous change.

**6-8.** For the reaction $HCO_3^- \rightleftharpoons H^+ + CO_3^{2-}$, $\Delta G^\circ = +59.0$ kJ/mol at 298.15 K. Find the value of $K$ for the reaction.

**6-9.** The formation of tetrafluoroethylene from its elements is highly exothermic:

$$2F_2(g) \ + \ 2C(s) \ \rightleftharpoons \ F_2C{=}CF_2(g)$$
$$\quad\text{Fluorine}\qquad\quad\text{Graphite}\qquad\text{Tetrafluoroethylene}$$

(a) If a mixture of $F_2$, graphite, and $C_2F_4$ is at equilibrium in a closed container, will the reaction go to the right or to the left if $F_2$ is added?
(b) Rare bacteria from the planet Teflon eat $C_2F_4$ and make Teflon for their cell walls. Will the reaction go to the right or to the left if these bacteria are added?

Teflon

(c) Will the reaction go right or left if solid graphite is added? (Neglect any effect of increased pressure due to the decreased volume in the vessel when solid is added.)
(d) Will the reaction go right or left if the container is crushed to one-eighth of its original volume?
(e) Does the equilibrium constant become larger or smaller if the container is heated?

**6-10.** When $BaCl_2 \cdot H_2O(s)$ is dried in an oven, it loses gaseous water:

$$BaCl_2 \cdot H_2O(s) \rightleftharpoons BaCl_2(s) + H_2O(g)$$
$$\Delta H^\circ = 63.11 \text{ kJ/mol at } 25°C$$
$$\Delta S^\circ = +148 \text{ J/(K·mol) at } 25°C$$

(a) Write the equilibrium constant for this reaction. Calculate the vapor pressure of gaseous $H_2O$ above $BaCl_2 \cdot H_2O$ at 298 K.
(b) Assuming that $\Delta H^\circ$ and $\Delta S^\circ$ are not temperature dependent (a poor assumption), estimate the temperature at which the vapor pressure of $H_2O(g)$ above $BaCl_2 \cdot H_2O(s)$ will be 1 bar.

**6-11.** The equilibrium constant for the reaction of ammonia with water has the following values in the range 5°–10°C:

$$NH_3(aq) + H_2O \rightleftharpoons NH_4^+ + OH^-$$
$$K_b = 1.479 \times 10^{-5} \text{ at } 5°C$$
$$K_b = 1.570 \times 10^{-5} \text{ at } 10°C$$

(a) Assuming that $\Delta H^\circ$ and $\Delta S^\circ$ are constant in the interval 5°–10°C, use Equation 6-8 to find $\Delta H^\circ$ for the reaction in this temperature range.
(b) Describe how Equation 6-8 could be used to make a linear graph to determine $\Delta H^\circ$, if $\Delta H^\circ$ and $\Delta S^\circ$ were constant over some range of temperature.

**6-12.** For the reaction $H_2(g) + Br_2(g) \rightleftharpoons 2HBr(g)$, $K = 7.2 \times 10^{-4}$ at 1 362 K and $\Delta H^\circ$ is positive. A vessel is charged with 48.0 Pa HBr, 1 370 Pa $H_2$, and 3 310 Pa $Br_2$ at 1 362 K.
(a) Will the reaction proceed to the left or to the right to reach equilibrium?
(b) Calculate the pressure (in pascals) of each species in the vessel at equilibrium.
(c) If the mixture at equilibrium is compressed to half its original volume, will the reaction proceed to the left or the right to reestablish equilibrium?
(d) If the mixture at equilibrium is heated from 1 362 to 1 407 K, will HBr be formed or consumed in order to reestablish equilibrium?

**6-13.** *Henry's law* states that the concentration of a gas dissolved in a liquid is proportional to the pressure of the gas. This law is a consequence of the equilibrium

$$X(g) \overset{K_h}{\rightleftharpoons} X(aq) \qquad K_h = \frac{[X]}{P_X}$$

where $K_h$ is called the Henry's law constant. (The same law applies to solvents other than water, but the value of $K_h$ is different for each solvent.) For the gasoline additive MTBE, $K_h = 1.71$ M/bar. Suppose we have a closed container with aqueous solution and air in equilibrium. If the concentration of MTBE in the liquid is found to be $1.00 \times 10^2$ ppm (= 100 µg MTBE/g solution ≈ 100 µg/mL), what is the pressure of MTBE in the air?

$$CH_3{-}O{-}C(CH_3)_3 \quad \text{Methyl-}t\text{-butylether (MTBE, FM 88.15)}$$

## Solubility Product

**6-14.** Find the concentration of $Cu^{2+}$ in equilibrium with $CuBr(s)$ and 0.10 M $Br^-$.

**6-15.** What concentration of $Fe(CN)_6^{4-}$ (ferrocyanide) is in equilibrium with 1.0 µM $Ag^+$ and $Ag_4Fe(CN)_6(s)$. Express your answer with a prefix from Table 1-3.

**6-16.** Find $[Cu^{2+}]$ in a solution saturated with $Cu_4(OH)_6(SO_4)$ if $[OH^-]$ is *fixed* at $1.0 \times 10^{-6}$ M. Note that $Cu_4(OH)_6(SO_4)$ gives 1 mol of $SO_4^{2-}$ for 4 mol of $Cu^{2+}$.

$$Cu_4(OH)_6(SO_4)(s) \rightleftharpoons 4Cu^{2+} + 6OH^- + SO_4^{2-}$$
$$K_{sp} = 2.3 \times 10^{-69}$$

**6-17.** (a) From the solubility product of zinc ferrocyanide, $Zn_2Fe(CN)_6$, calculate the concentration of $Fe(CN)_6^{4-}$ in 0.10 mM $ZnSO_4$ saturated with $Zn_2Fe(CN)_6$. Assume that $Zn_2Fe(CN)_6$ is a negligible source of $Zn^{2+}$.
(b) What concentration of $K_4Fe(CN)_6$ should be in a suspension of solid $Zn_2Fe(CN)_6$ in water to give $[Zn^{2+}] = 5.0 \times 10^{-7}$ M?

**6-18.** Solubility products predict that cation $A^{3+}$ can be 99.999% separated from cation $B^{2+}$ by precipitation with anion $X^-$. When the separation is tried, we find 0.2% contamination of $AX_3(s)$ with $B^{2+}$. Explain what might be happening.

**6-19.** A solution contains 0.050 0 M $Ca^{2+}$ and 0.030 0 M $Ag^+$. Can 99% of $Ca^{2+}$ be precipitated by sulfate without precipitating $Ag^+$? What will be the concentration of $Ca^{2+}$ when $Ag_2SO_4$ begins to precipitate?

**6-20.** A solution contains 0.010 M $Ba^{2+}$ and 0.010 M $Ag^+$. Can 99.90% of either ion be precipitated by chromate ($CrO_4^{2-}$) without precipitating the other metal ion?

**6-21.** If a solution containing 0.10 M $Cl^-$, $Br^-$, $I^-$, and $CrO_4^{2-}$ is treated with $Ag^+$, in what order will the anions precipitate?

### Complex Formation

**6-22.** Explain why the total solubility of lead in Figure 6-3 first decreases and then increases as $[I^-]$ increases. Give an example of the chemistry in each of the two domains.

**6-23.** Identify the Lewis acids in the following reactions:
(a) $BF_3 + NH_3 \rightleftharpoons F_3\bar{B}-\overset{+}{N}H_3$
(b) $F^- + AsF_5 \rightleftharpoons AsF_6^-$

**6-24.** The cumulative formation constant for $SnCl_2(aq)$ in 1.0 M $NaNO_3$ is $\beta_2 = 12$. Find the concentration of $SnCl_2(aq)$ for a solution in which the concentrations of $Sn^{2+}$ and $Cl^-$ are both somehow fixed at 0.20 M.

**6-25.** Given the following equilibria, calculate the concentrations of each zinc-containing species in a solution saturated with $Zn(OH)_2(s)$ and containing $[OH^-]$ at a fixed concentration of $3.2 \times 10^{-7}$ M.

| | |
|---|---|
| $Zn(OH)_2(s)$ | $K_{sp} = 3.0 \times 10^{-16}$ |
| $Zn(OH)^+$ | $\beta_1 = 2.5 \times 10^4$ |
| $Zn(OH)_3^-$ | $\beta_3 = 7.2 \times 10^{15}$ |
| $Zn(OH)_4^{2-}$ | $\beta_4 = 2.8 \times 10^{15}$ |

**6-26.** Although KOH, RbOH, and CsOH have little association between metal and hydroxide in aqueous solution, $Li^+$ and $Na^+$ do form complexes with $OH^-$.

$$Li^+ + OH^- \rightleftharpoons LiOH(aq) \qquad K_1 = \frac{[LiOH(aq)]}{[Li^+][OH^-]} = 0.83$$

$$Na^+ + OH^- \rightleftharpoons NaOH(aq) \qquad K_1 = 0.20$$

Prepare a table like the one in Exercise 6-B showing initial and final concentrations of $Na^+$, $OH^-$, and $NaOH(aq)$ in 1 F NaOH solution. Calculate the fraction of sodium in the form $NaOH(aq)$ at equilibrium.

**6-27.** In Figure 6-3, the concentration of $PbI_2(aq)$ is independent of $[I^-]$. Use any of the equilibrium constants for Reactions 6-11 through 6-15 to find the equilibrium constant for the reaction $PbI_2(s) \rightleftharpoons PbI_2(aq)$, which is equal to the concentration of $PbI_2(aq)$.

**6-28.** 🖩 Consider the following equilibria:[22]

| | |
|---|---|
| $AgI(s) \rightleftharpoons Ag^+ + I^-$ | $K_{sp} = 4.5 \times 10^{-17}$ |
| $Ag^+ + I^- \rightleftharpoons AgI(aq)$ | $\beta_1 = 1.3 \times 10^8$ |
| $Ag^+ + 2I^- \rightleftharpoons AgI_2^-$ | $\beta_2 = 9.0 \times 10^{10}$ |
| $Ag^+ + 3I^- \rightleftharpoons AgI_3^{2-}$ | $\beta_3 = 5.6 \times 10^{13}$ |
| $Ag^+ + 4I^- \rightleftharpoons AgI_4^{3-}$ | $\beta_4 = 2.5 \times 10^{14}$ |
| $2Ag^+ + 6I^- \rightleftharpoons Ag_2I_6^{4-}$ | $K_{26} = 7.6 \times 10^{29}$ |
| $3Ag^+ + 8I^- \rightleftharpoons Ag_3I_8^{5-}$ | $K_{38} = 2.3 \times 10^{46}$ |

Prepare a spreadsheet in which $\log[I^-]$ varies from $-8$ to 0 in increments of 0.5. Calculate the concentrations of all species and the total dissolved silver ($[Ag]_{total}$), and prepare a graph analogous to Figure 6-3. In calculating $[Ag]_{total}$, note that 1 mol of $Ag_2I_6^{4-}$ has 2 mol of Ag and that 1 mol of $Ag_3I_8^{5-}$ has 3 mol of Ag. Suggested

headings for spreadsheet columns: $\log[I^-]$, $[I^-]$, $[Ag^+]$, $[AgI(aq)]$, $[AgI_2^-]$, $[AgI_3^{2-}]$, $[AgI_4^{3-}]$, $[Ag_2I_6^{4-}]$, $[Ag_3I_8^{5-}]$, $[Ag]_{total}$, $\log[Ag^+]$, $\log[AgI(aq)]$, $\log[AgI_2^-]$, $\log[AgI_3^{2-}]$, $\log[AgI_4^{3-}]$, $\log[Ag_2I_6^{4-}]$, $\log[Ag_3I_8^{5-}]$, $\log[Ag]_{total}$. Note that Excel's base 10 logarithm is Log10($x$) and the antilogarithm is $10^x$.

### Acids and Bases

**6-29.** Distinguish Lewis acids and bases from Brønsted-Lowry acids and bases. Give an example of each.

**6-30.** Fill in the blanks:
(a) The product of a reaction between a Lewis acid and a Lewis base is called _____.
(b) The bond between a Lewis acid and a Lewis base is called _____ or _____.
(c) Brønsted-Lowry acids and bases related by gain or loss of one proton are said to be _____.
(d) A solution is *acidic* if _____. A solution is *basic* if _____.

**6-31.** Why is the pH of distilled water usually <7? How can you prevent this from happening?

**6-32.** Gaseous $SO_2$ is created by combustion of sulfur-containing fuels, especially coal. Explain how $SO_2$ in the atmosphere makes acidic rain.

**6-33.** Use electron dot structures to show why tetramethylammonium hydroxide, $(CH_3)_4N^+OH^-$, is an ionic compound. That is, show why hydroxide is not covalently bound to the rest of the molecule.

**6-34.** Identify the Brønsted-Lowry acids among the reactants in the following reactions:
(a) $KCN + HI \rightleftharpoons HCN + KI$
(b) $PO_4^{3-} + H_2O \rightleftharpoons HPO_4^{2-} + OH^-$

**6-35.** Write the autoprotolysis reaction of $H_2SO_4$.

**6-36.** Identify the conjugate acid-base pairs in the following reactions:
(a) $H_3\overset{+}{N}CH_2CH_2\overset{+}{N}H_3 + H_2O \rightleftharpoons H_3\overset{+}{N}CH_2CH_2NH_2 + H_3O^+$

(b) ⬡—$CO_2H$ + ⬡N $\rightleftharpoons$ ⬡—$CO_2^-$ + ⬡$NH^+$

    Benzoic acid        Pyridine        Benzoate      Pyridinium

### pH

**6-37.** Calculate $[H^+]$ and the pH of the following solutions
(a) 0.010 M $HNO_3$      (d) 3.0 M HCl
(b) 0.035 M KOH      (e) 0.010 M $[(CH_3)_4N^+]OH^-$
(c) 0.030 M HCl            Tetramethylammonium hydroxide

**6-38.** Use Table 6-1 to calculate the pH of pure water at (a) 25°C and (b) 100°C.

**6-39.** The equilibrium constant for the reaction $H_2O \rightleftharpoons H^+ + OH^-$ is $1.0 \times 10^{-14}$ at 25°C. What is the value of $K$ for the reaction $4H_2O \rightleftharpoons 4H^+ + 4OH^-$?

**6-40.** An acidic solution containing 0.010 M $La^{3+}$ is treated with NaOH until $La(OH)_3$ precipitates. At what pH does this occur?

**6-41.** Use Le Châtelier's principle and $K_w$ in Table 6-1 to decide whether the autoprotolysis of water is endothermic or exothermic at (a) 25°C; (b) 100°C; (c) 300°C.

### Strengths of Acids and Bases

**6-42.** Make a list of the common strong acids and strong bases. Memorize this list.

**6-43.** Write the formulas and names for three classes of weak acids and two classes of weak bases.

**6-44.** Write the $K_a$ reaction for trichloroacetic acid, $Cl_3CCO_2H$, for anilinium ion, ⟨benzene⟩$-\overset{+}{N}H_3$, and for lanthanum ion, $La^{3+}$.

**6-45.** Write the $K_b$ reactions for pyridine and for sodium 2-mercapto-ethanol.

⟨pyridine structure⟩N:          $HOCH_2CH_2\overset{..}{\underset{..}{S}}:^- Na^+$

    Pyridine       Sodium 2-mercaptoethanol

**6-46.** Write the $K_a$ and $K_b$ reactions of $NaHCO_3$.

**6-47.** Write the stepwise acid-base reactions for the following ions in water. Write the correct symbol (for example, $K_{b1}$) for the equilibrium constant for each reaction.

(a)  $H_3\overset{+}{N}CH_2CH_2\overset{+}{N}H_3$   (b)

    Ethylenediammonium ion

(b)
$$^-OCCH_2CO^-$$
with two C=O (O double bonds above each C)

Malonate ion

**6-48.** Which is a stronger acid, **(a)** or **(b)**?

(a)      O
       ‖
   $Cl_2HCCOH$
Dichloroacetic acid
$K_a = 8 \times 10^{-2}$

(b)      O
       ‖
   $ClH_2CCOH$
Chloroacetic acid
$K_a = 1.36 \times 10^{-3}$

Which is a stronger base, **(c)** or **(d)**?

(c)  $H_2NNH_2$
Hydrazine
$K_b = 1.1 \times 10^{-6}$

(d)      O
       ‖
   $H_2NCNH_2$
Urea
$K_b = 1.5 \times 10^{-14}$

**6-49.** Write the $K_b$ reaction of $CN^-$. Given that the $K_a$ value for HCN is $6.2 \times 10^{-10}$, calculate $K_b$ for $CN^-$.

**6-50.** Write the $K_{a2}$ reaction of phosphoric acid ($H_3PO_4$) and the $K_{b2}$ reaction of disodium oxalate ($Na_2C_2O_4$).

**6-51.** From the $K_b$ values for phosphate in Equations 6-32 through 6-34, calculate the three $K_a$ values of phosphoric acid.

**6-52.** From the following equilibrium constants, calculate the equilibrium constant for the reaction $HO_2CCO_2H \rightleftharpoons 2H^+ + C_2O_4^{2-}$.

$$\underset{\text{Oxalic acid}}{HOCCOH} \rightleftharpoons H^+ + HOCCO^- \qquad K_1 = 5.4 \times 10^{-2}$$
(each C bearing two O, double-bonded)

$$HOCCO^- \rightleftharpoons H^+ + {}^-OCCO^- \qquad K_2 = 5.4 \times 10^{-5}$$
Oxalate
(each C bearing two O, double-bonded)

**6-53.** **(a)** Using only $K_{sp}$ from Table 6-3, calculate how many moles of $Ca(OH)_2$ will dissolve in 1.00 L of water.
**(b)** How will the solubility calculated in part **(a)** be affected by the $K_1$ reaction in Table 6-3?

**6-54.** The planet Aragonose (which is made mostly of the mineral aragonite, whose composition is $CaCO_3$) has an atmosphere containing methane and carbon dioxide, each at a pressure of 0.10 bar. The oceans are saturated with aragonite and have a concentration of $H^+$ equal to $1.8 \times 10^{-7}$ M. Given the following equilibria, calculate how many grams of calcium are contained in 2.00 L of Aragonose seawater.

$$CaCO_3(s, \text{aragonite}) \rightleftharpoons Ca^{2+}(aq) + CO_3^{2-}(aq)$$
$$K_{sp} = 6.0 \times 10^{-9}$$

$$CO_2(g) \rightleftharpoons CO_2(aq)$$
$$K_{CO_2} = 3.4 \times 10^{-2}$$

$$CO_2(aq) + H_2O(l) \rightleftharpoons HCO_3^-(aq) + H^+(aq)$$
$$K_1 = 4.5 \times 10^{-7}$$

$$HCO_3^-(aq) \rightleftharpoons H^+(aq) + CO_3^{2-}(aq)$$
$$K_2 = 4.7 \times 10^{-11}$$

*Don't panic!* Reverse the first reaction, add all the reactions together, and see what cancels.

**Solving Numerical Equilibrium Problems**

**6-55.**   Reaction 6-41 came to equilibrium in a flask containing $Br_2(aq)$, $IO_3^-$, $Br^-$, and $H^+$, each at a concentration of 5.00 mM, plus excess solid $I_2$. In which direction must the reaction proceed to reach equilibrium? Find the concentrations at equilibrium.

**6-56.**   Consider the reaction

$$IO_3^- + 5I^- + 6H^+ \rightleftharpoons 3I_2(aq) + 3H_2O \qquad K = 3 \times 10^{48}$$

in a solution initially containing 1.00 mM $IO_3^-$, 1.00 mM $I^-$, and 1.00 mM $H^+$. Find the concentrations of reactants and products when the solution reaches equilibrium.

# 7 | Let the Titrations Begin

| Descroizilles (1806) Pour out liquid | Gay-Lussac (1824) Blow out liquid | Henry (1846) Copper stopcock | Mohr (1855) Compression clip | Mohr (1855) Glass stopcock |

Burets have terrorized students of analytical chemistry for two centuries. The original buret of Descroizilles was used in nearly the same manner as a graduated cylinder. Hole b at the top was covered with a finger to admit air for fine control of liquid poured from spout a. In Gay-Lussac's buret, liquid was poured from the bent glass side tube. A cork fitted with a short glass tube and rubber hose was inserted into the top of the buret. Liquid flow from the side tube was controlled by blowing into the rubber tube. Henry described the first glass buret with a copper stopcock, but this device was never widely accepted. Mohr's buret with a brass pinchclamp on an India rubber tube (called caoutchouc, pronounced KOO-chook) dominated volumetric analysis for 100 years. The only common analytical reagent that was incompatible with the rubber was potassium permanganate. The glass stopcock similar to today's Teflon stopcock was a contemporary of the pinchclamp, but the stopcock was not perfected and not widely used until the mid-twentieth century.

Procedures in which we measure the volume of reagent needed to react with analyte are called **volumetric analysis.** In this chapter, we discuss principles that apply to all volumetric procedures and then focus on precipitation titrations. We also introduce spectrophotometric titrations, which are especially useful in biochemistry.

## ■■■ 7-1 Titrations

In a **titration,** increments of reagent solution—the **titrant**—are added to analyte until their reaction is complete. From the quantity of titrant required, we can calculate the quantity of analyte that must have been present. Titrant is usually delivered from a buret (Figure 7-1).

The principal requirements for a titration reaction are that it have a large equilibrium constant and proceed rapidly. That is, each increment of titrant should be completely and quickly consumed by analyte until the analyte is used up. The most common titrations are based on acid-base, oxidation-reduction, complex formation, or precipitation reactions.

The **equivalence point** occurs when the quantity of added titrant is the exact amount necessary for stoichiometric reaction with the analyte. For example, 5 mol of oxalic acid react with 2 mol of permanganate in hot acidic solution:

$$5HO-\underset{\substack{\|\\O}}{C}-\underset{\substack{\|\\O}}{C}-OH + 2MnO_4^- + 6H^+ \longrightarrow 10CO_2 + 2Mn^{2+} + 8H_2O \quad (7\text{-}1)$$

**Analyte**      **Titrant**

Oxalic acid     Permanganate

(colorless)     (purple)          (colorless)  (colorless)

If the unknown contains 5.000 mmol of oxalic acid, the equivalence point is reached when 2.000 mmol of $MnO_4^-$ have been added.

The equivalence point is the ideal (theoretical) result we seek in a titration. What we actually measure is the **end point**, which is marked by a sudden change in a physical property of the solution. In Reaction 7-1, a convenient end point is the abrupt appearance of the purple color of permanganate in the flask. Prior to the equivalence point, all permanganate is consumed by oxalic acid, and the titration solution remains colorless. After the equivalence point, unreacted $MnO_4^-$ accumulates until there is enough to see. The *first trace* of purple color is the end point. The better your eyes, the closer will be your measured end point to the true equivalence point. Here, the end point cannot exactly equal the equivalence point, because extra $MnO_4^-$, beyond that needed to react with oxalic acid, is required to exhibit purple color.

Methods for determining when the analyte has been consumed include (1) detecting a sudden change in the voltage or current between a pair of electrodes (Figure 7-9), (2) observing an indicator color change (Color Plate 2), and (3) monitoring absorption of light (Figure 7-5). An **indicator** is a compound with a physical property (usually color) that changes abruptly near the equivalence point. The change is caused by the disappearance of analyte or the appearance of excess titrant.

The difference between the end point and the equivalence point is an inescapable **titration error.** By choosing a physical property whose change is easily observed (such as pH or the color of an indicator), we find that the end point can be very close to the equivalence point. We estimate the titration error with a **blank titration,** in which we carry out the same procedure *without* analyte. For example, we can titrate a solution containing no oxalic acid to see how much $MnO_4^-$ is needed to produce observable purple color. We then subtract this volume of $MnO_4^-$ from the volume observed in the analytical titration.

The validity of an analytical result depends on knowing the amount of one of the reactants used. If a titrant is prepared by dissolving a weighed amount of pure reagent in a known volume of solution, its concentration can be calculated. We call such a reagent a **primary standard** because it is pure enough to be weighed and used directly. A primary standard should be 99.9% pure, or better. It should not decompose under ordinary storage, and it should be stable when dried by heat or vacuum, because drying is required to remove traces of water adsorbed from the atmosphere. Primary standards for many elements are given in Appendix K. Box 7-1 discusses reagent purity. Box 3-1 described Standard Reference Materials that allow laboratories to test the accuracy of their procedures.

Many reagents used as titrants, such as HCl, are not available as primary standards. Instead, we prepare titrant with approximately the desired concentration and use it to titrate a primary standard. By this procedure, called **standardization,** we determine the concentration of titrant. We then say that the titrant is a **standard solution.** The validity of the analytical result ultimately depends on knowing the composition of a primary standard.

In a **direct titration,** we add titrant to analyte until the reaction is complete. Occasionally, we perform a **back titration,** in which we add a known *excess* of one standard reagent to the analyte. Then we titrate the excess reagent with a second standard reagent. A back titration is useful when its end point is clearer than the end point of the direct titration or when an excess of the first reagent is required for complete reaction with analyte. To grasp the difference between direct and back titrations, consider first the addition of permanganate titrant to oxalic acid analyte in Reaction 7-1; this reaction is a direct titration. Alternatively, to perform a back titration, we could add a known *excess* of permanganate to consume oxalic acid. Then we could back-titrate the excess permanganate with standard $Fe^{2+}$ to measure how much permanganate was left after reaction with the oxalic acid.

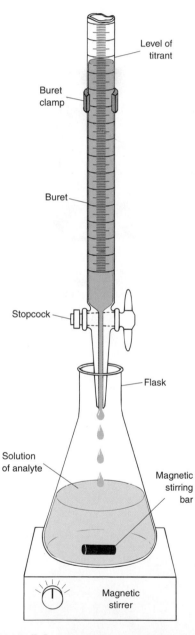

*Figure 7-1* Typical setup for a titration. The analyte is contained in the flask, and the titrant is in the buret. The stirring bar is a magnet coated with Teflon, which is inert to most solutions. The bar is spun by a rotating magnet inside the stirring motor.

(Labels: Level of titrant; Buret clamp; Buret; Stopcock; Flask; Solution of analyte; Magnetic stirring bar; Magnetic stirrer)

Box 7-1    Reagent Chemicals and Primary Standards

Chemicals are sold in many grades of purity. For analytical chemistry, we usually use **reagent-grade** chemicals meeting purity requirements set by the American Chemical Society (ACS) Committee on Analytical Reagents.[2] Sometimes "reagent grade" simply meets purity standards set by the manufacturer. An actual lot analysis for specified impurities should appear on the reagent bottle. For example, here is a lot analysis of zinc sulfate:

| $ZnSO_4$ | ACS Reagent | Lot Analysis: |
|---|---|---|
| Assay: 100.6% | Fe: 0.000 5% | Ca: 0.001% |
| Insoluble matter: 0.002% | Pb: 0.002 8% | Mg: 0.000 3% |
| pH of 5% solution at 25°C: 5.6 | Mn: 0.6 ppm | K: 0.002% |
| Ammonium: 0.000 8% | Nitrate: 0.000 4% | Na: 0.003% |
| Chloride: 1.5 ppm | | |

The assay value of 100.6% means that a specified analysis for one of the major components produced 100.6% of the theoretical value. For example, if $ZnSO_4$ is contaminated with the lower molecular mass $Zn(OH)_2$, the assay for $Zn^{2+}$ will be higher than the value for pure $ZnSO_4$. Less pure chemicals, generally unsuitable for analytical chemistry, carry designations such as "chemically pure" (CP), "practical," "purified," or "technical."

A few chemicals are sold in high enough purity to be *primary standard grade*. Whereas reagent-grade potassium dichromate has a lot assay of $\geq 99.0\%$, primary standard grade $K_2Cr_2O_7$ must be in the range 99.95–100.05%. Besides high purity, a key quality of primary standards is that they are indefinitely stable.

For **trace analysis** (analysis of species at ppm and lower levels), impurities in reagent chemicals must be extremely low. For this purpose, we use very-high-purity, expensive grades of acids such as "trace metal grade" $HNO_3$ or HCl to dissolve samples. We must pay careful attention to reagents and vessels whose impurity levels could be greater than the quantity of analyte we seek to measure.

To protect the purity of chemical reagents, you should

* Avoid putting a spatula into a bottle. Instead, pour chemical out of the bottle into a clean container (or onto weighing paper) and dispense the chemical from the clean container.
* Never pour unused chemical back into the reagent bottle.
* Replace the cap on the bottle immediately to keep dust out.
* Never put a glass stopper from a liquid-reagent container down on the lab bench. Either hold the stopper or place it in a clean place (such as a clean beaker) while you dispense reagent.
* Store chemicals in a cool, dark place. Do not expose them unnecessarily to sunlight.

In a **gravimetric titration,** titrant is measured by mass, not volume. Titrant concentration is expressed as moles of reagent per kilogram of solution. Precision is improved from 0.3% attainable with a buret to 0.1% with a balance. Experiments by Guenther and by Butler and Swift on the Web site for this book (www.whfreeman.com/qca) provide examples. In a gravimetric titration, there is no need for a buret. Titrant can be delivered from a squirt bottle or a pipet. "Gravimetric titrations should become the gold standard, and volumetric glassware should be seen in museums only."[3]

## ▪▪▪▪ 7-2 Titration Calculations

Here are some examples to illustrate stoichiometry calculations in volumetric analysis. The key step is to *relate moles of titrant to moles of analyte*. We also introduce the Kjeldahl titration as a representative volumetric procedure.

**Example** Standardization of Titrant Followed by Analysis of Unknown

The calcium content of urine can be determined by the following procedure:

**Step 1**    Precipitate $Ca^{2+}$ with oxalate in basic solution:

$$Ca^{2+} + C_2O_4^{2-} \rightarrow Ca(C_2O_4) \cdot H_2O(s)$$

$$\text{Oxalate} \qquad\qquad \text{Calcium oxalate}$$

**Step 2**    Wash the precipitate with ice-cold water to remove free oxalate, and dissolve the solid in acid to obtain $Ca^{2+}$ and $H_2C_2O_4$ in solution.

**Step 3**    Heat the solution to 60°C and titrate the oxalate with standardized potassium permanganate until the purple end point of Reaction 7-1 is observed.

Standardization    Suppose that 0.356 2 g of $Na_2C_2O_4$ is dissolved in a 250.0-mL volumetric flask. If 10.00 mL of this solution require 48.36 mL of $KMnO_4$ solution for titration, what is the molarity of the permanganate solution?

**Solution**  The concentration of the oxalate solution is

$$\frac{0.356\ 2\ \text{g Na}_2\text{C}_2\text{O}_4\ /\ (134.00\ \text{g Na}_2\text{C}_2\text{O}_4/\text{mol})}{0.250\ 0\ \text{L}} = 0.010\ 63_3\ \text{M}$$

The moles of $\text{C}_2\text{O}_4^{2-}$ in 10.00 mL are $(0.010\ 63_3\ \text{mol/L})(0.010\ 00\ \text{L}) = 1.063_3 \times 10^{-4}\ \text{mol} = 0.106\ 3_3\ \text{mmol}$. Reaction 7-1 requires 2 mol of permanganate for 5 mol of oxalate, so the $\text{MnO}_4^-$ delivered must have been

Reaction 7-1 requires 2 mol $\text{MnO}_4^-$ for 5 mol $\text{C}_2\text{O}_4^{2-}$.

$$\text{Moles of MnO}_4^- = \left(\frac{2\ \text{mol MnO}_4^-}{5\ \text{mol C}_2\text{O}_4^{2-}}\right)(\text{mol C}_2\text{O}_4^{2-}) = 0.042\ 53_1\ \text{mmol}$$

The concentration of $\text{MnO}_4^-$ in the titrant is therefore

Note that $\dfrac{\text{mmol}}{\text{mL}}$ is the same as $\dfrac{\text{mol}}{\text{L}}$.

$$\text{Molarity of MnO}_4^- = \frac{0.042\ 53_1\ \text{mmol}}{48.36\ \text{mL}} = 8.794_7 \times 10^{-4}\ \text{M}$$

**Analysis of Unknown**  Calcium in a 5.00-mL urine sample was precipitated with $\text{C}_2\text{O}_4^{2-}$, redissolved, and then required 16.17 mL of standard $\text{MnO}_4^-$ solution. Find the concentration of $\text{Ca}^{2+}$ in the urine.

**Solution**  In 16.17 mL of $\text{MnO}_4^-$, there are $(0.016\ 17\ \text{L})(8.794_7 \times 10^{-4}\ \text{mol/L}) = 1.422_1 \times 10^{-5}\ \text{mol MnO}_4^-$. This quantity will react with

Reaction 7-1 requires 5 mol $\text{C}_2\text{O}_4^{2-}$ for 2 mol $\text{MnO}_4^-$.

$$\text{Moles of C}_2\text{O}_4^{2-} = \left(\frac{5\ \text{mol C}_2\text{O}_4^{2-}}{2\ \text{mol MnO}_4^-}\right)(\text{mol MnO}_4^-) = 0.035\ 55_3\ \text{mmol}$$

Because there is one oxalate ion for each calcium ion in $\text{Ca}(\text{C}_2\text{O}_4) \cdot \text{H}_2\text{O}$, there must have been 0.035 55$_3$ mmol of $\text{Ca}^{2+}$ in 5.00 mL of urine:

$$[\text{Ca}^{2+}] = \frac{0.035\ 55_3\ \text{mmol}}{5.00\ \text{mL}} = 0.007\ 11_1\ \text{M}$$

**Example**  Titration of a Mixture

A solid mixture weighing 1.372 g containing only sodium carbonate and sodium bicarbonate required 29.11 mL of 0.734 4 M HCl for complete titration:

$$\text{Na}_2\text{CO}_3 + 2\text{HCl} \rightarrow 2\text{NaCl}(aq) + \text{H}_2\text{O} + \text{CO}_2$$
FM 105.99

$$\text{NaHCO}_3 + \text{HCl} \rightarrow \text{NaCl}(aq) + \text{H}_2\text{O} + \text{CO}_2$$
FM 84.01

Find the mass of each component of the mixture.

Solving for two unknowns requires two independent pieces of information. Here we have the mass of the mixture and the volume of titrant.

**Solution**  Let's denote the grams of $\text{Na}_2\text{CO}_3$ by $x$ and grams of $\text{NaHCO}_3$ by $1.372 - x$. The moles of each component must be

$$\text{Moles of Na}_2\text{CO}_3 = \frac{x\ \text{g}}{105.99\ \text{g/mol}} \qquad \text{Moles of NaHCO}_3 = \frac{(1.372 - x)\ \text{g}}{84.01\ \text{g/mol}}$$

We know that the total moles of HCl used were $(0.029\ 11\ \text{L})(0.734\ 4\ \text{M}) = 0.021\ 38\ \text{mol}$. From the stoichiometry of the two reactions, we can say that

$$2\ (\text{mol Na}_2\text{CO}_3) + \text{mol NaHCO}_3 = 0.021\ 38$$

$$2\left(\frac{x}{105.99}\right) + \frac{1.372 - x}{84.01} = 0.021\ 38 \ \Rightarrow\ x = 0.724\ \text{g}$$

The mixture contains 0.724 g of $\text{Na}_2\text{CO}_3$ and $1.372 - 0.724 = 0.648$ g of $\text{NaHCO}_3$.

## Kjeldahl Nitrogen Analysis

Developed in 1883, the **Kjeldahl nitrogen analysis** remains one of the most accurate and widely used methods for determining nitrogen in substances such as protein, milk, cereal,

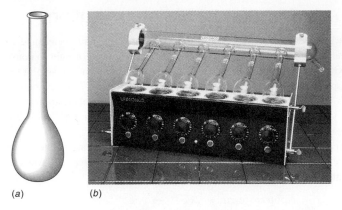

(a)                    (b)

*Figure 7-2* (a) Kjeldahl digestion flask with long neck to minimize loss from spattering.
(b) Six-port manifold for multiple samples provides for exhaust of fumes. *[Courtesy Fisher Scientific, Pittsburgh, PA.]*

and flour.[4] The solid is first *digested* (decomposed and dissolved) in boiling sulfuric acid, which converts nitrogen into ammonium ion, $NH_4^+$, and oxidizes other elements present:

$$\text{Kjeldahl digestion:} \qquad \text{Organic C, H, N} \xrightarrow[H_2SO_4]{\text{boiling}} NH_4^+ + CO_2 + H_2O \qquad (7\text{-}2)$$

Each atom of nitrogen in starting material is converted into one $NH_4^+$ ion.

Mercury, copper, and selenium compounds catalyze the digestion. To speed the reaction, the boiling point of concentrated (98 wt%) sulfuric acid (338°C) is raised by adding $K_2SO_4$. Digestion is carried out in a long-neck *Kjeldahl flask* (Figure 7-2) that prevents loss of sample from spattering. Alternative digestion procedures employ $H_2SO_4$ plus $H_2O_2$ or $K_2S_2O_8$ plus NaOH[5] in a microwave bomb (a pressurized vessel shown in Figure 28-8).

After digestion is complete, the solution containing $NH_4^+$ is made basic, and the liberated $NH_3$ is distilled (with a large excess of steam) into a receiver containing a known amount of HCl (Figure 7-3). Excess, unreacted HCl is then titrated with standard NaOH to determine how much HCl was consumed by $NH_3$.

Neutralization of $NH_4^+$: $\qquad\qquad NH_4^+ + OH^- \rightarrow NH_3(g) + H_2O \qquad$ (7-3)

Distillation of $NH_3$ into standard HCl: $\qquad NH_3 + H^+ \rightarrow NH_4^+ \qquad\qquad\qquad$ (7-4)

Titration of unreacted HCl with NaOH: $\qquad H^+ + OH^- \rightarrow H_2O \qquad\qquad\qquad$ (7-5)

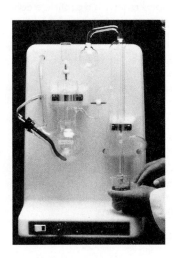

*Figure 7-3* Kjeldahl distillation unit employs electric immersion heater in flask at left to carry out distillation in 5 min. Beaker at right collects liberated $NH_3$ in standard HCl. *[Courtesy Fisher Scientific, Pittsburgh, PA.]*

### Example  Kjeldahl Analysis

A typical protein contains 16.2 wt% nitrogen. A 0.500-mL aliquot of protein solution was digested, and the liberated $NH_3$ was distilled into 10.00 mL of 0.021 40 M HCl. The unreacted HCl required 3.26 mL of 0.019 8 M NaOH for complete titration. Find the concentration of protein (mg protein/mL) in the original sample.

**Solution**   The initial amount of HCl in the receiver was $(10.00 \text{ mL})(0.021\,40 \text{ mmol/mL}) = 0.214\,0$ mmol. The NaOH required for titration of unreacted HCl in Reaction 7-5 was $(3.26 \text{ mL})(0.019\,8 \text{ mmol/mL}) = 0.064\,5$ mmol. The difference, $0.214\,0 - 0.064\,5 = 0.149\,5$ mmol, must be the quantity of $NH_3$ produced in Reaction 7-3 and distilled into the HCl.

Because 1 mol of N in the protein produces 1 mol of $NH_3$, there must have been 0.149 5 mmol of N in the protein, corresponding to

$$(0.149\,5 \text{ mmol})\left(14.006\,74 \frac{\text{mg N}}{\text{mmol}}\right) = 2.093 \text{ mg N}$$

If the protein contains 16.2 wt% N, there must be

$$\frac{2.093 \text{ mg N}}{0.162 \text{ mg N/mg protein}} = 12.9 \text{ mg protein} \Rightarrow \frac{12.9 \text{ mg protein}}{0.500 \text{ mL}} = 25.8 \frac{\text{mg protein}}{\text{mL}}$$

| Protein source | Weight % nitrogen |
|---|---|
| Meat | 16.0 |
| Blood plasma | 15.3 |
| Milk | 15.6 |
| Flour | 17.5 |
| Egg | 14.9 |

*SOURCE: D. J. Holme and H. Peck, Analytical Biochemistry, 3rd ed. (New York: Addison Wesley Longman, 1998), p. 388.*

Absorption of light is discussed in Sections 18-1 and 18-2.

We can follow the course of a titration by any means that allows us to determine when the equivalence point has been reached. In Chapter 4, we used absorption of light to construct a calibration curve. Now we will use absorption of light to monitor the progress of a titration.

A solution of the iron-transport protein, transferrin (Figure 7-4), can be titrated with iron to measure the transferrin content. Transferrin without iron, called apotransferrin, is colorless. Each molecule, with a molecular mass of 81 000, binds two $Fe^{3+}$ ions. When iron binds to the protein, a red color with an absorbance maximum at a wavelength of 465 nm develops. The absorbance is proportional to the concentration of iron bound to the protein. Therefore, the absorbance may be used to follow the course of a titration of an unknown amount of apotransferrin with a standard solution of $Fe^{3+}$.

$$\text{Apotransferrin} + 2Fe^{3+} \rightarrow (Fe^{3+})_2\text{transferrin} \qquad (7\text{-}6)$$
$$\text{(colorless)} \qquad\qquad\qquad\qquad \text{(red)}$$

**Figure 7-4** Each of the two Fe-binding sites of transferrin is located in a cleft in the protein. $Fe^{3+}$ binds to one N atom from the amino acid histidine and three O atoms from tyrosine and aspartic acid. The fifth and sixth ligand sites of the metal are occupied by O atoms from a carbonate anion ($CO_3^{2-}$) anchored in place by electrostatic interaction with the positively charged amino acid arginine and by hydrogen bonding to the protein helix. When transferrin is taken up by a cell, it is brought to a compartment whose pH is lowered to 5.5. $H^+$ then reacts with the carbonate ligand to make $HCO_3^-$ and $H_2CO_3$, thereby releasing $Fe^{3+}$ from the protein. [Adapted from E. N. Baker, B. F. Anderson, H. M. Baker, M. Haridas, G. E. Norris, S. V. Rumball, and C. A. Smith, "Metal and Anion Binding Sites in Lactoferrin and Related Proteins," Pure Appl. Chem. **1990**, 62, 1067.]

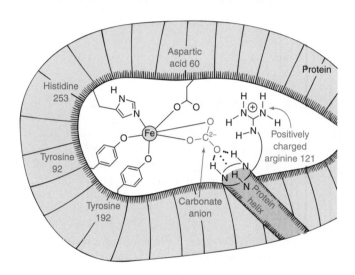

Ferric nitrilotriacetate is soluble at neutral pH. In the absence of nitrilotriacetate, $Fe^{3+}$ precipitates as $Fe(OH)_3$ in neutral solution. Nitrilotriacetate binds $Fe^{3+}$ through four atoms, shown in **bold** type:

Nitrilotriacetate anion

Figure 7-5 shows the titration of 2.000 mL of apotransferrin with $1.79 \times 10^{-3}$ M ferric nitrilotriacetate solution. As iron is added to the protein, red color develops and absorbance increases. When the protein is saturated with iron, no further color can form, and the curve levels off. The extrapolated intersection of the two straight portions of the titration curve at 203 μL in Figure 7-5 is taken as the end point. The absorbance continues to rise slowly after the equivalence point because ferric nitrilotriacetate has some absorbance at 465 nm.

To construct the graph in Figure 7-5, dilution must be considered because the volume is different at each point. Each point on the graph represents the absorbance that would be observed *if the solution had not been diluted from its original volume of 2.000 mL.*

$$\text{Corrected absorbance} = \left(\frac{\text{total volume}}{\text{initial volume}}\right)(\text{observed absorbance}) \qquad (7\text{-}7)$$

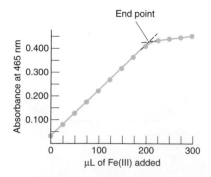

**Figure 7-5** Spectrophotometric titration of transferrin with ferric nitrilotriacetate. Absorbance is corrected as if no dilution had taken place. The initial absorbance of the solution, before iron is added, is due to a colored impurity.

**Example** Correcting Absorbance for the Effect of Dilution

The absorbance measured after adding 125 μL (= 0.125 mL) of ferric nitrilotriacetate to 2.000 mL of apotransferrin was 0.260. Calculate the corrected absorbance that should be plotted in Figure 7-5.

**Solution** The total volume was $2.000 + 0.125 = 2.125$ mL. If the volume had been 2.000 mL, the absorbance would have been greater than 0.260 by a factor of 2.125/2.000.

$$\text{Corrected absorbance} = \left(\frac{2.125 \text{ mL}}{2.000 \text{ mL}}\right)(0.260) = 0.276$$

The absorbance plotted in Figure 7-5 is 0.276.

## ■ ■ ■ 7-4 The Precipitation Titration Curve

We now turn our attention to details of precipitation titrations as an illustration of principles that underlie all titrations. We first study how concentrations of analyte and titrant vary during a titration and then derive equations that can be used to predict titration curves. One reason to calculate titration curves is to understand the chemistry that occurs during titrations. A second reason is to learn how experimental control can be exerted to influence the quality of an analytical titration. For example, certain titrations conducted at the wrong pH could give no discernible end point. In precipitation titrations, the concentrations of analyte and titrant and the size of $K_{sp}$ influence the sharpness of the end point. For acid-base titrations (Chapter 11) and oxidation-reduction titrations (Chapter 16), the theoretical titration curve enables us to choose an appropriate indicator.

The *titration curve* is a graph showing how the concentration of one of the reactants varies as titrant is added. Because concentration varies over many orders of magnitude, it is most useful to plot the p function:

*p function:*
$$pX = -\log_{10}[X] \qquad (7\text{-}8)$$

In Chapter 8, we define the p function correctly in terms of activities instead of concentrations. For now, we use pX = −log[X].

where [X] is the concentration of X.

Consider the titration of 25.00 mL of 0.100 0 M $I^-$ with 0.050 00 M $Ag^+$,

$$I^- + Ag^+ \rightarrow AgI(s) \qquad (7\text{-}9)$$

and suppose that we are monitoring the $Ag^+$ concentration with an electrode. Reaction 7-9 is the reverse of the dissolution of $AgI(s)$, whose solubility product is rather small:

$$AgI(s) \rightleftharpoons Ag^+ + I^- \qquad K_{sp} = [Ag^+][I^-] = 8.3 \times 10^{-17} \qquad (7\text{-}10)$$

Because the equilibrium constant for the titration reaction 7-9 is large ($K = 1/K_{sp} = 1.2 \times 10^{16}$), the equilibrium lies far to the right. It is reasonable to say that each aliquot of $Ag^+$ reacts completely with $I^-$, leaving only a tiny amount of $Ag^+$ in solution. At the equivalence point, there will be a sudden increase in the $Ag^+$ concentration because all the $I^-$ has been consumed and we are now adding $Ag^+$ directly to the solution.

What volume of $Ag^+$ titrant is needed to reach the equivalence point? We calculate this volume, designated $V_e$, with the fact that 1 mol of $Ag^+$ reacts with 1 mol of $I^-$.

$V_e$ = volume of titrant at equivalence point

$$\underbrace{(0.025\ 00\ L)(0.100\ 0\ mol\ I^-/L)}_{mol\ I^-} = \underbrace{(V_e)(0.050\ 00\ mol\ Ag^+/L)}_{mol\ Ag^+}$$

$$\Rightarrow V_e = 0.050\ 00\ L = 50.00\ mL$$

The titration curve has three distinct regions, depending on whether we are before, at, or after the equivalence point. Let's consider each region separately.

Eventually, we will derive a single, unified equation for a spreadsheet that treats all regions of the titration curve. To understand the chemistry of the titration, it is sensible to break the curve into three regions described by approximate equations that are easy to use with a calculator.

### Before the Equivalence Point

Suppose that 10.00 mL of $Ag^+$ have been added. There are more moles of $I^-$ than $Ag^+$ at this point, so virtually all $Ag^+$ is "used up" to make $AgI(s)$. We want to find the small concentration of $Ag^+$ remaining in solution after reaction with $I^-$. Imagine that Reaction 7-9 has gone to completion and that some AgI redissolves (Reaction 7-10). The solubility of $Ag^+$ is determined by the concentration of free $I^-$ remaining in the solution:

$$[Ag^+] = \frac{K_{sp}}{[I^-]} \qquad (7\text{-}11)$$

When $V < V_e$, the concentration of unreacted $I^-$ regulates the solubility of AgI.

Free $I^-$ is overwhelmingly from the $I^-$ that has not been precipitated by 10.00 mL of $Ag^+$. By comparison, $I^-$ from dissolution of $AgI(s)$ is negligible.

So let's find the concentration of unprecipitated $I^-$:

Moles of $I^-$ = original moles of $I^-$ − moles of $Ag^+$ added

$$= (0.025\ 00\ L)(0.100\ mol/L) - (0.010\ 00\ L)(0.050\ 00\ mol/L)$$

$$= 0.002\ 000\ mol\ I^-$$

The volume is 0.035 00 L (25.00 mL + 10.00 mL), so the concentration is

$$[I^-] = \frac{0.002\ 000\ mol\ I^-}{0.035\ 00\ L} = 0.057\ 14\ M \qquad (7\text{-}12)$$

The concentration of $Ag^+$ in equilibrium with this much $I^-$ is

$$[Ag^+] = \frac{K_{sp}}{[I^-]} = \frac{8.3 \times 10^{-17}}{0.057\ 14} = 1.4_5 \times 10^{-15}\ M \tag{7-13}$$

Finally, the p function we seek is

$$pAg^+ = -log[Ag^+] = -log(1.4_5 \times 10^{-15}) = 14.84 \tag{7-14}$$

There are two significant figures in the concentration of $Ag^+$ because there are two significant figures in $K_{sp}$. The two figures in $[Ag^+]$ translate into two figures in the *mantissa* of the p function, which is correctly written as 14.84.

The preceding step-by-step calculation is a tedious way to find the concentration of $I^-$. Here is a streamlined procedure that is well worth learning. Bear in mind that $V_e = 50.00$ mL. When 10.00 mL of $Ag^+$ have been added, the reaction is one-fifth complete because 10.00 mL out of the 50.00 mL of $Ag^+$ needed for complete reaction have been added. Therefore, four-fifths of the $I^-$ is unreacted. If there were no dilution, $[I^-]$ would be four-fifths of its original value. However, the original volume of 25.00 mL has been increased to 35.00 mL. If no $I^-$ had been consumed, the concentration would be the original value of $[I^-]$ times (25.00/35.00). Accounting for both the reaction and the dilution, we can write

$$[I^-] = \underbrace{\left(\frac{4.000}{5.000}\right)}_{\substack{\text{Fraction} \\ \text{remaining}}} \underbrace{(0.100\ 0\ M)}_{\substack{\text{Original} \\ \text{concentration}}} \underbrace{\left(\frac{25.00}{35.00}\right)}_{\substack{\text{Dilution} \\ \text{factor}}} = 0.057\ 14\ M$$

Original volume of $I^-$

Total volume of solution

This is the same result found in Equation 7-12.

> ### Example  Using the Streamlined Calculation
>
> Let's calculate $pAg^+$ when $V_{Ag^+}$ (the volume added from the buret) is 49.00 mL.
>
> **Solution**  Because $V_e = 50.00$ mL, the fraction of $I^-$ reacted is 49.00/50.00, and the fraction remaining is 1.00/50.00. The total volume is $25.00 + 49.00 = 74.00$ mL.
>
> $$[I^-] = \underbrace{\left(\frac{1.00}{50.00}\right)}_{\substack{\text{Fraction} \\ \text{remaining}}} \underbrace{(0.100\ 0\ M)}_{\substack{\text{Original} \\ \text{concentration}}} \underbrace{\left(\frac{25.00}{74.00}\right)}_{\substack{\text{Dilution} \\ \text{factor}}} = 6.76 \times 10^{-4}\ M$$
>
> $$[Ag^+] = K_{sp}/[I^-] = (8.3 \times 10^{-17})/(6.76 \times 10^{-4}) = 1.2_3 \times 10^{-13}\ M$$
> $$pAg^+ = -log[Ag^+] = 12.91$$
>
> The concentration of $Ag^+$ is negligible compared with the concentration of unreacted $I^-$, even though the titration is 98% complete.

## At the Equivalence Point

Now we have added exactly enough $Ag^+$ to react with all the $I^-$. We can imagine that all the AgI precipitates and some redissolves to give equal concentrations of $Ag^+$ and $I^-$. The value of $pAg^+$ is found by setting $[Ag^+] = [I^-] = x$ in the solubility product:

$$[Ag^+][I^-] = K_{sp}$$
$$(x)(x) = 8.3 \times 10^{-17} \Rightarrow x = 9.1 \times 10^{-9} \Rightarrow pAg^+ = -log\ x = 8.04$$

This value of $pAg^+$ is independent of the original concentrations or volumes.

## After the Equivalence Point

Virtually all $Ag^+$ added *before* the equivalence point has precipitated, so $[Ag^+]$ is determined by $Ag^+$ added *after* the equivalence point. Suppose that $V_{Ag^+} = 52.00$ mL. The volume past the equivalence point is 2.00 mL. The calculation proceeds as follows:

$$\text{Moles of } Ag^+ = (0.002\ 00\ L)(0.050\ 00\ mol\ Ag^+/L) = 0.000\ 100\ mol$$
$$[Ag^+] = (0.000\ 100\ mol)/(0.077\ 00\ L) = 1.30 \times 10^{-3}\ M \Rightarrow pAg^+ = 2.89$$

Total volume = 77.00 mL

---

**Left margin notes:**

$log(1.4_5 \times 10^{-15}) = 14.84$

Two significant    Two digits in
figures            mantissa

Significant figures in logarithms were discussed in Section 3-2.

Streamlined calculation *well worth using.*

When $V = V_e$, $[Ag^+]$ is determined by the solubility of pure AgI. *This problem is the same as if we had just added AgI(s) to water.*

When $V > V_e$, $[Ag^+]$ is determined by the excess $Ag^+$ added from the buret.

We could justify three significant figures for the mantissa of pAg$^+$, because there are now three significant figures in [Ag$^+$]. For consistency with earlier results, we retain only two figures. We will generally express p functions in this book with two decimal places.

For a streamlined calculation, the concentration of Ag$^+$ in the buret is 0.050 00 M, and 2.00 mL of titrant are being diluted to (25.00 + 52.00) = 77.00 mL. Hence, [Ag$^+$] is

$$[Ag^+] = \underbrace{(0.050\ 00\ M)}_{\substack{\text{Original}\\\text{concentration}\\\text{of Ag}^+}} \underbrace{\left(\frac{2.00}{77.00}\right)}_{\substack{\text{Dilution}\\\text{factor}}} = 1.30 \times 10^{-3}\ M$$

Volume of excess Ag$^+$

Total volume of solution

The streamlined calculation.

## The Shape of the Titration Curve

Titration curves in Figure 7-6 illustrate the effect of reactant concentration. The equivalence point is the steepest point of the curve. It is the point of maximum slope (a negative slope in this case) and is therefore an inflection point (at which the second derivative is 0):

Steepest slope: $\dfrac{dy}{dx}$ reaches its greatest value

Inflection point: $\dfrac{d^2y}{dx^2} = 0$

In titrations involving 1:1 stoichiometry of reactants, the equivalence point is the steepest point of the titration curve. This is true of acid-base, complexometric, and redox titrations as well. For stoichiometries other than 1:1, such as $2Ag^+ + CrO_4^{2-} \rightarrow Ag_2CrO_4(s)$, the curve is not symmetric near the equivalence point. The equivalence point is not at the center of the steepest section of the curve, and it is not an inflection point. In practice, conditions are chosen such that titration curves are steep enough for the steepest point to be a good estimate of the equivalence point, regardless of the stoichiometry.

A *complexometric* titration involves complex ion formation between titrant and analyte.

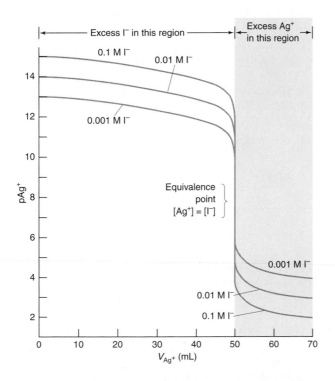

*Figure 7-6* Titration curves showing the effect of diluting the reactants.

Outer curve: 25.00 mL of 0.100 0 M I$^-$ titrated with 0.050 00 M Ag$^+$

Middle curve: 25.00 mL of 0.010 00 M I$^-$ titrated with 0.005 000 M Ag$^+$

Inner curve: 25.00 mL of 0.001 000 M I$^-$ titrated with 0.000 500 0 M Ag$^+$

Figure 7-7 illustrates how $K_{sp}$ affects the titration of halide ions. The least soluble product, AgI, gives the sharpest change at the equivalence point. However, even for AgCl, the curve is steep enough to locate the equivalence point accurately. The larger the equilibrium constant for a titration reaction, the more pronounced will be the change in concentration near the equivalence point.

At the equivalence point, the titration curve is steepest for the least soluble precipitate.

**Figure 7-7** Titration curves showing the effect of $K_{sp}$. Each curve is calculated for 25.00 mL of 0.100 0 M halide titrated with 0.050 00 M Ag+. Equivalence points are marked by arrows.

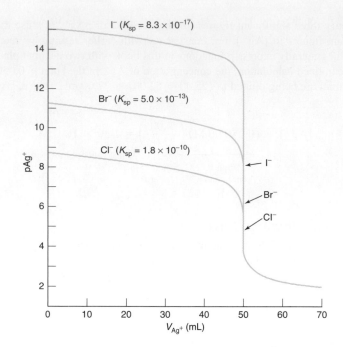

**Example** Calculating Concentrations During a Precipitation Titration

25.00 mL of 0.041 32 M $Hg_2(NO_3)_2$ were titrated with 0.057 89 M $KIO_3$.

$$Hg_2^{2+} + 2IO_3^- \rightarrow Hg_2(IO_3)_2(s)$$
$$\text{Iodate}$$

For $Hg_2(IO_3)_2$, $K_{sp} = 1.3 \times 10^{-18}$. Find $[Hg_2^{2+}]$ in the solution after addition of **(a)** 34.00 mL of $KIO_3$; **(b)** 36.00 mL of $KIO_3$; and **(c)** at the equivalence point.

**Solution** The volume of iodate needed to reach the equivalence point is found as follows:

$$\text{Moles of } IO_3^- = \left(\frac{2 \text{ mol } IO_3^-}{1 \text{ mol } Hg_2^{2+}}\right)(\text{moles of } Hg_2^{2+})$$

$$\underbrace{(V_e)(0.057\ 89 \text{ M})}_{\text{Moles of } IO_3^-} = \underbrace{2(25.00 \text{ mL})(0.041\ 32 \text{ M})}_{\text{Moles of } Hg_2^{2+}} \Rightarrow V_e = 35.69 \text{ mL}$$

**(a)** When $V = 34.00$ mL, the precipitation of $Hg_2^{2+}$ is not yet complete.

$$[Hg_2^{2+}] = \underbrace{\left(\frac{35.69 - 34.00}{35.69}\right)}_{\substack{\text{Fraction} \\ \text{remaining}}} \underbrace{(0.041\ 32 \text{ M})}_{\substack{\text{Original} \\ \text{concentration} \\ \text{of } Hg_2^{2+}}} \underbrace{\left(\frac{25.00}{25.00 + 34.00}\right)}_{\substack{\text{Dilution} \\ \text{factor}}} = 8.29 \times 10^{-4} \text{ M}$$

Original volume of $Hg_2^{2+}$

Total volume of solution

**(b)** When $V = 36.00$ mL, the precipitation is complete. We have gone $(36.00 - 35.69) = 0.31$ mL *past* the equivalence point. The concentration of excess $IO_3^-$ is

$$[IO_3^-] = \underbrace{(0.057\ 89 \text{ M})}_{\substack{\text{Original} \\ \text{concentration} \\ \text{of } IO_3^-}} \underbrace{\left(\frac{0.31}{25.00 + 36.00}\right)}_{\substack{\text{Dilution} \\ \text{factor}}} = 2.9 \times 10^{-4} \text{ M}$$

Volume of excess $IO_3^-$

Total volume of solution

The concentration of $Hg_2^{2+}$ in equilibrium with solid $Hg_2(IO_3)_2$ plus this much $IO_3^-$ is

$$[Hg_2^{2+}] = \frac{K_{sp}}{[IO_3^-]^2} = \frac{1.3 \times 10^{-18}}{(2.9 \times 10^{-4})^2} = 1.5 \times 10^{-11} \text{ M}$$

**(c)** At the equivalence point, there is exactly enough $IO_3^-$ to react with all $Hg_2^{2+}$. We can imagine that all of the ions precipitate and then some $Hg_2(IO_3)_2(s)$ redissolves, giving two moles of iodate for each mole of mercurous ion:

$$Hg_2(IO_3)_2(s) \rightleftharpoons \underset{x}{Hg_2^{2+}} + \underset{2x}{2IO_3^-}$$

$$(x)(2x)^2 = K_{sp} \Rightarrow x = [Hg_2^{2+}] = 6.9 \times 10^{-7} \text{ M}$$

The preceding calculations assume that the only chemistry that occurs is the reaction of anion with cation to precipitate solid salt. If other reactions occur, such as complex formation or ion pair formation, we would have to modify the calculations.

## ▮▮▮ 7-5 Titration of a Mixture

If a mixture of two ions is titrated, the less soluble precipitate forms first. If the solubilities are sufficiently different, the first precipitation is nearly complete before the second commences.

Consider the addition of $AgNO_3$ to a solution containing KI and KCl. Because $K_{sp}(AgI) \ll K_{sp}(AgCl)$, AgI precipitates first. When precipitation of $I^-$ is almost complete, the concentration of $Ag^+$ abruptly increases and AgCl begins to precipitate. When $Cl^-$ is consumed, another abrupt increase in $[Ag^+]$ occurs. We expect to see two breaks in the titration curve. The first corresponds to the AgI equivalence point, and the second to the AgCl equivalence point.

Figure 7-8 shows an experimental curve for this titration. The apparatus used to measure the curve is shown in Figure 7-9, and the theory of how this system measures $Ag^+$ concentration is discussed in Chapter 15.

The $I^-$ end point is taken as the intersection of the steep and nearly horizontal curves shown in the inset of Figure 7-8. Precipitation of $I^-$ is not quite complete when $Cl^-$ begins to precipitate. (The way we know that $I^-$ precipitation is not complete is by a calculation. That's what these obnoxious calculations are for!) Therefore, the end of the steep portion (the intersection) is a better approximation of the equivalence point than is the middle of the steep section. The $Cl^-$ end point is taken as the midpoint of the second steep section, at 47.41 mL.

A liquid containing suspended particles is said to be *turbid* because the particles scatter light.

The product with the smaller $K_{sp}$ precipitates first, if the stoichiometry of the precipitates is the same. Precipitation of $I^-$ and $Cl^-$ with $Ag^+$ produces two breaks in the titration curve. The first corresponds to the reaction of $I^-$ and the second to the reaction of $Cl^-$.

Before $Cl^-$ precipitates, the calculations for AgI precipitation are just like those in Section 7-4.

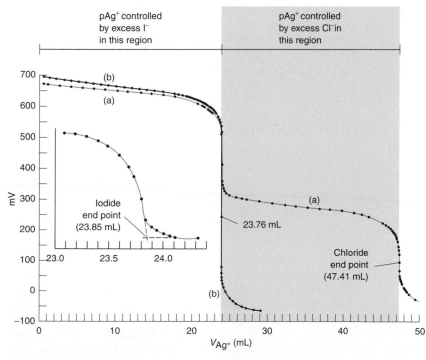

**Figure 7-8** Experimental titration curves. (a) Titration curve for 40.00 mL of 0.050 2 M KI plus 0.050 0 M KCl titrated with 0.084 5 M AgNO₃. The inset is an expanded view of the region near the first equivalence point. (b) Titration curve for 20.00 mL of 0.100 4 M I⁻ titrated with 0.084 5 M Ag⁺.

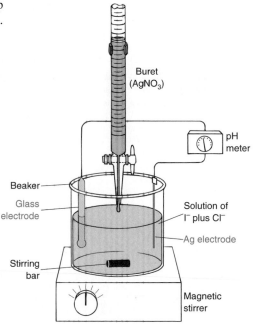

**Figure 7-9** Apparatus for measuring the titration curves in Figure 7-8. The silver electrode responds to changes in $Ag^+$ concentration, and the glass electrode provides a constant reference potential in this experiment. The measured voltage changes by approximately 59 mV for each factor-of-10 change in $[Ag^+]$. All solutions, including AgNO₃, were maintained at pH 2.0 by using 0.010 M sulfate buffer prepared from H₂SO₄ and KOH.

The moles of $Cl^-$ in the sample equal the moles of $Ag^+$ delivered between the first and second end points. That is, it requires 23.85 mL of $Ag^+$ to precipitate $I^-$, and $(47.41 - 23.85) = 23.56$ mL of $Ag^+$ to precipitate $Cl^-$.

Comparing the $I^-/Cl^-$ and pure $I^-$ titration curves in Figure 7-8 shows that the $I^-$ end point is 0.38% too high in the $I^-/Cl^-$ titration. We expect the first end point at 23.76 mL, but it is observed at 23.85 mL. Two factors contribute to this high value. One is random experimental error, which is always present. This discrepancy is as likely to be positive as negative. However, the end point in some titrations, especially $Br^-/Cl^-$ titrations, is systematically 0 to 3% high, depending on conditions. This error is attributed to *coprecipitation* of AgCl with AgBr. **Coprecipitation** means that, even though the solubility of AgCl has not been exceeded, some $Cl^-$ becomes attached to AgBr as it precipitates and carries down an equivalent amount of $Ag^+$. A high concentration of nitrate reduces coprecipitation, perhaps because $NO_3^-$ competes with $Cl^-$ for binding sites.

The second end point in Figure 7-8 corresponds to complete precipitation of both halides. It is observed at the expected value of $V_{Ag^+}$. The concentration of $Cl^-$, found from the *difference* between the two end points, will be slightly low in Figure 7-8, because the first end point is slightly high.

## ■ ■ ■ 🖩 7-6 Calculating Titration Curves with a Spreadsheet

By now you should understand the chemistry that occurs at different stages of a precipitation titration, and you should know how to calculate the shape of a titration curve. We now introduce spreadsheet calculations that are more powerful than hand calculations and less prone to error. If a spreadsheet is not available, you can skip this section with no loss in continuity.

Consider the addition of $V_M$ liters of cation $M^+$ (whose initial concentration is $C_M^0$) to $V_X^0$ liters of solution containing anion $X^-$ with a concentration $C_X^0$.

$$M^+ \quad + \quad X^- \quad \overset{K_{sp}}{\rightleftharpoons} \quad MX(s) \qquad (7\text{-}15)$$

$$\underset{\substack{\text{Titrant} \\ C_M^0, V_M}}{} \qquad \underset{\substack{\text{Analyte} \\ C_X^0, V_X^0}}{}$$

The total moles of added M ($= C_M^0 \cdot V_M$) must equal the moles of $M^+$ in solution ($= [M^+](V_M + V_X^0)$ plus the moles of precipitated $MX(s)$. (This equality is called a *mass balance*, even though it is really a *mole balance*). In a similar manner, we can write a mass balance for X.

The *mass balance* states that the moles of an element in all species in a mixture equal the total moles of that element delivered to the solution.

*Mass balance for M:* $\qquad C_M^0 \cdot V_M = \underbrace{[M^+](V_M + V_X^0)}_{} + \text{mol } MX(s) \qquad (7\text{-}16)$

$$\underset{\substack{\text{Total moles} \\ \text{of added M}}}{} \qquad \underset{\substack{\text{Moles of M} \\ \text{in solution}}}{} \qquad \underset{\substack{\text{Moles of M in} \\ \text{precipitate}}}{}$$

*Mass balance for X:* $\qquad C_X^0 \cdot V_X^0 = \underbrace{[X^-](V_M + V_X^0)}_{} + \text{mol } MX(s) \qquad (7\text{-}17)$

$$\underset{\substack{\text{Total moles} \\ \text{of added X}}}{} \qquad \underset{\substack{\text{Moles of X} \\ \text{in solution}}}{} \qquad \underset{\substack{\text{Moles of X in} \\ \text{precipitate}}}{}$$

Now equate mol $MX(s)$ from Equation 7-16 with mol $MX(s)$ from Equation 7-17:

$$C_M^0 \cdot V_M - [M^+](V_M + V_X^0) = C_X^0 \cdot V_X^0 - [X^-](V_M + V_X^0)$$

which can be rearranged to

*Precipitation of $X^-$ with $M^+$:*
$$V_M = V_X^0 \left( \frac{C_X^0 + [M^+] - [X^-]}{C_M^0 - [M^+] + [X^-]} \right) \qquad (7\text{-}18)$$

A supplementary section at www.whfreeman.com/qca derives a spreadsheet equation for the titration of a mixture, such as that in Figure 7-8.

Equation 7-18 relates the volume of added $M^+$ to $[M^+]$, $[X^-]$, and the constants $V_X^0$, $C_X^0$, and $C_M^0$. To use Equation 7-18 in a spreadsheet, *enter values of pM and compute corresponding values of $V_M$*, as shown in Figure 7-10 for the iodide titration of Figure 7-7. This is backward from the way you normally calculate a titration curve in which $V_M$ would be input and pM would be output. Column C of Figure 7-10 is calculated with the formula $[M^+] = 10^{-pM}$, and column D is given by $[X^-] = K_{sp}/[M^+]$. Column E is calculated from Equation 7-18. The first input value of pM (15.08) was selected by trial and error to produce a small $V_M$. You can start wherever you like. If your initial value of pM is before the true starting point, then $V_M$ in column E will be negative. In practice, you will want more points than we have shown so that you can plot an accurate titration curve.

| | A | B | C | D | E |
|---|---|---|---|---|---|
| 1 | Titration of I- with Ag+ | | | | |
| 2 | | | | | |
| 3 | Ksp(AgI) = | pAg | [Ag+] | [I-] | Vm |
| 4 | 8.30E-17 | 15.08 | 8.32E-16 | 9.98E-02 | 0.035 |
| 5 | Vo = | 15 | 1.00E-15 | 8.30E-02 | 3.195 |
| 6 | 25 | 14 | 1.00E-14 | 8.30E-03 | 39.322 |
| 7 | Co(I) = | 12 | 1.00E-12 | 8.30E-05 | 49.876 |
| 8 | 0.1 | 10 | 1.00E-10 | 8.30E-07 | 49.999 |
| 9 | Co(Ag) = | 8 | 1.00E-08 | 8.30E-09 | 50.000 |
| 10 | 0.05 | 6 | 1.00E-06 | 8.30E-11 | 50.001 |
| 11 | | 4 | 1.00E-04 | 8.30E-13 | 50.150 |
| 12 | | 3 | 1.00E-03 | 8.30E-14 | 51.531 |
| 13 | | 2 | 1.00E-02 | 8.30E-15 | 68.750 |
| 14 | C4 = 10^-B4 | | | | |
| 15 | D4 = $A$4/C4 | | | | |
| 16 | E4 = $A$6*($A$8+C4-D4)/($A$10-C4+D4) | | | | |

# ▮▮▮ 7-7 End-Point Detection

End-point detection for precipitation titrations commonly relies on electrodes (as in Figure 7-9) or indicators. This section discusses two indicator methods applied to the titration of $Cl^-$ with $Ag^+$. Titrations with $Ag^+$ are called **argentometric titrations.**

1. **Volhard titration:** formation of a soluble, colored complex at the end point
2. **Fajans titration:** adsorption of a colored indicator on the precipitate at the end point

## Volhard Titration

The Volhard titration is a titration of $Ag^+$ in $HNO_3$ solution. For $Cl^-$, a back titration is necessary. First, $Cl^-$ is precipitated by a known, excess quantity of standard $AgNO_3$.

$$Ag^+ + Cl^- \rightarrow AgCl(s)$$

The AgCl is filtered and washed, and excess $Ag^+$ in the combined filtrate is titrated with standard KSCN (potassium thiocyanate) in the presence of $Fe^{3+}$.

$$Ag^+ + SCN^- \rightarrow AgSCN(s)$$

When all $Ag^+$ has been consumed, $SCN^-$ reacts with $Fe^{3+}$ to form a red complex.

$$Fe^{3+} + SCN^- \rightarrow FeSCN^{2+}$$
$$\text{Red}$$

The appearance of red color is the end point. Knowing how much $SCN^-$ was required for the back titration tells us how much $Ag^+$ was left over from the reaction with $Cl^-$. The total amount of $Ag^+$ is known, so the amount consumed by $Cl^-$ can be calculated.

In the analysis of $Cl^-$ by the Volhard method, the end point would slowly fade if the AgCl were not filtered off, because AgCl is more soluble than AgSCN. The AgCl slowly dissolves and is replaced by AgSCN. To eliminate this secondary reaction, we filter the AgCl and titrate only the $Ag^+$ in the filtrate. $Br^-$ and $I^-$, whose silver salts are *less* soluble than AgSCN, can be titrated by the Volhard method without isolating the silver halide precipitate.

## Fajans Titration

The Fajans titration uses an **adsorption indicator.** To see how this works, consider the electric charge at the surface of a precipitate. When $Ag^+$ is added to $Cl^-$, there is excess $Cl^-$ in solution prior to the equivalence point. Some $Cl^-$ is adsorbed on the AgCl surface, imparting a negative charge to the crystal (Figure 7-11a). After the equivalence point, there is excess $Ag^+$ in solution. Adsorption of $Ag^+$ onto the AgCl surface places positive charge on the precipitate (Figure 7-11b). The abrupt change from negative to positive occurs at the equivalence point.

Common adsorption indicators are anionic dyes, which are attracted to the positively charged particles produced immediately after the equivalence point. Adsorption of the negatively charged dye onto the positively charged surface changes the color of the dye. The color change is the end point in the titration. Because the indicator reacts with the precipitate

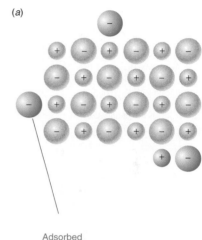

(a)

Adsorbed ions

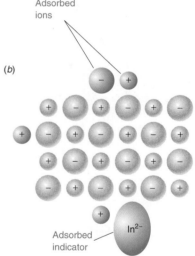

(b)

Adsorbed indicator

In²⁻

**Figure 7-11** Ions from a solution are adsorbed on the surface of a growing crystallite. (*a*) A crystal growing in the presence of excess lattice anions (anions that belong in the crystal) will have a slight negative charge because anions are predominantly adsorbed. (*b*) A crystal growing in the presence of excess lattice cations will have a slight positive charge and can therefore adsorb a negative indicator ion. Anions and cations in the solution that do not belong in the crystal lattice are less likely to be adsorbed than are ions belonging to the lattice. These diagrams omit other ions in solution. Overall, each solution plus its growing crystallites must have zero total charge.

The Fajans titration of $Cl^-$ with $Ag^+$ convincingly demonstrates indicator end points in precipitation titrations. Dissolve 0.5 g of NaCl plus 0.15 g of dextrin in 400 mL of water. The purpose of the dextrin is to retard coagulation of the AgCl precipitate. Add 1 mL of dichlorofluorescein indicator solution containing either 1 mg/mL of dichlorofluorescein in 95% aqueous ethanol or 1 mg/mL of the sodium salt in water. Titrate the NaCl solution with a solution containing 2 g of $AgNO_3$ in 30 mL $H_2O$. About 20 mL are required to reach the end point.

Color Plate 2a shows the yellow color of the indicator in the NaCl solution prior to the titration. Color Plate 2b shows the milky white appearance of the AgCl suspension during titration, before the end point is reached. The pink suspension in Color Plate 2c appears at the end point, when the anionic indicator becomes adsorbed on the cationic particles of precipitate.

Dichlorofluorescein

Tetrabromofluorescein (eosin)

surface, we want as much surface area as possible. To attain maximum surface area, we use conditions that keep the particles as small as possible, because small particles have more surface area than an equal volume of large particles. Low electrolyte concentration helps prevent coagulation of the precipitate and maintains small particle size.

The most common indicator for AgCl is dichlorofluorescein. This dye is greenish yellow in solution, but turns pink when adsorbed onto AgCl (Demonstration 7-1). Because the indicator is a weak acid and must be present in its anionic form, the pH of the reaction must be controlled. The dye eosin is useful in the titration of $Br^-$, $I^-$, and $SCN^-$. It gives a sharper end point than dichlorofluorescein and is more sensitive (that is, less halide can be titrated). It cannot be used for AgCl, because eosin is more strongly bound than $Cl^-$ to AgCl. Eosin binds to AgCl crystallites even before the particles become positively charged.

In all argentometric titrations, but especially with adsorption indicators, strong light (such as daylight through a window) should be avoided. Light decomposes silver salts, and adsorbed indicators are especially light sensitive.

Applications of precipitation titrations are listed in Table 7-1. Whereas the Volhard method is an argentometric titration, the Fajans method has wider applications. Because the Volhard titration is carried out in acidic solution (typically 0.2 M $HNO_3$), it avoids certain interferences that affect other titrations. Silver salts of $CO_3^{2-}$, $C_2O_4^{2-}$, and $AsO_4^{3-}$ are soluble in acidic solution, so these anions do not interfere.

**Table 7-1** Applications of precipitation titrations

| Species analyzed | Notes |
| --- | --- |
| | *Volhard Method* |
| $Br^-$, $I^-$, $SCN^-$, $CNO^-$, $AsO_4^{3-}$ | Precipitate removal is unnecessary. |
| $Cl^-$, $PO_4^{3-}$, $CN^-$, $C_2O_4^{2-}$, $CO_3^{2-}$, $S^{2-}$, $CrO_4^{2-}$ | Precipitate removal required. |
| $BH_4^-$ | Back titration of $Ag^+$ left after reaction with $BH_4^-$: $BH_4^- + 8Ag^+ + 8OH^- \rightarrow 8Ag(s) + H_2BO_3^- + 5H_2O$ |
| $K^+$ | $K^+$ is first precipitated with a known excess of $(C_6H_5)_4B^-$. Remaining $(C_6H_5)_4B^-$ is precipitated with a known excess of $Ag^+$. Unreacted $Ag^+$ is then titrated with $SCN^-$. |
| | *Fajans Method* |
| $Cl^-$, $Br^-$, $I^-$, $SCN^-$, $Fe(CN)_6^{4-}$ | Titration with $Ag^+$. Detection with dyes such as fluoresccin, dichlorofluorescein, eosin, bromophenol blue. |
| $F^-$ | Titration with $Th(NO_3)_4$ to produce $ThF_4$. End point detected with alizarin red S. |
| $Zn^{2+}$ | Titration with $K_4Fe(CN)_6$ to produce $K_2Zn_3[Fe(CN)_6]_2$. End-point detection with diphenylamine. |
| $SO_4^{2-}$ | Titration with $Ba(OH)_2$ in 50 vol% aqueous methanol using alizarin red S as indicator. |
| $Hg_2^{2+}$ | Titration with NaCl to produce $Hg_2Cl_2$. End point detected with bromophenol blue. |
| $PO_4^{3-}$, $C_2O_4^{2-}$ | Titration with $Pb(CH_3CO_2)_2$ to give $Pb_3(PO_4)_2$ or $PbC_2O_4$. End point detected with dibromofluorescein ($PO_4^{3-}$) or fluorescein ($C_2O_4^{2-}$). |

## 7-8 Efficiency in Experimental Design[6]

Efficient experimental design is intended to provide the maximum amount of information in the fewest number of trials. Suppose we have three different unknown solutions of acid, designated A, B, and C. If we titrate each one once with base, we find its concentration, but have no estimate of uncertainty. If we titrate each solution three times, for a total of nine measurements, we would find each concentration and its standard deviation.

| | A | B | C | D | E |
|---|---|---|---|---|---|
| 1 | Experimental Design | | | | |
| 2 | | | | | |
| 3 | Volumes of unknown acids (mL) | | | mL NaOH | mmol |
| 4 | A | B | C | (0.1204 M) | NaOH |
| 5 | 2 | 2 | 2 | 23.29 | 2.804 |
| 6 | 2 | 3 | 1 | 20.01 | 2.409 |
| 7 | 3 | 1 | 2 | 21.72 | 2.615 |
| 8 | 1 | 2 | 3 | 28.51 | 3.433 |
| 9 | 2 | 2 | 2 | 23.26 | 2.801 |
| 10 | | | | | |
| 11 | | | [C] | [B] | [A] |
| 12 | | Molarity | 0.8099 | 0.4001 | 0.1962 |
| 13 | | Std. dev. | 0.0062 | 0.0062 | 0.0062 |
| 14 | | | 0.9994 | 0.0130 | #N/A |
| 15 | | | $R^2$ | $S_y$ | |
| 16 | Highlight cells C12:E14 | | | | |
| 17 | Type "= LINEST(E5:E9,A5:C9,FALSE,TRUE)" | | | | |
| 18 | Press CTRL +SHIFT+ENTER (on PC) | | | | |
| 19 | Press COMMAND+RETURN (on Mac) | | | | |

**Figure 7-12** Spreadsheet for efficient experimental design uses Excel LINEST routine to fit the function $y = m_A x_A + m_B x_B + m_C x_C$ to experimental data by a least-squares procedure.

A more efficient experimental design provides concentrations and standard deviations in fewer than nine experiments. One of many efficient designs is shown in Figure 7-12. Instead of titrating each acid by itself, we titrate mixtures of the acids. For example, in row 5 of the spreadsheet, a mixture containing 2 mL A, 2 mL B, and 2 mL C required 23.29 mL of 0.120 4 M NaOH, which amounts to 2.804 mmol of $OH^-$. In row 6, the acid mixture contained 2 mL A, 3 mL B, and 1 mL C. Other permutations are titrated in rows 7 and 8. Then row 5 is repeated independently in row 9. Column E gives mmol of base for each run.

For each experiment, mmol of base consumed equals mmol of acid in the mixture:

$$\underbrace{mmol\ OH^-}_{y} = \underbrace{[A]V_A}_{m_A x_A} + \underbrace{[B]V_B}_{m_B x_B} + \underbrace{[C]V_C}_{m_C x_C} \qquad (7\text{-}19)$$

where [A] is the concentration of acid A (mol/L) and $V_A$ is the volume of A in mL. Rows 5 through 9 of the spreadsheet are equivalent to the following equalities:

$$\left.\begin{array}{l} 2.804 = [A] \cdot 2 + [B] \cdot 2 + [C] \cdot 2 \\ 2.409 = [A] \cdot 2 + [B] \cdot 3 + [C] \cdot 1 \\ 2.615 = [A] \cdot 3 + [B] \cdot 1 + [C] \cdot 2 \\ 3.433 = [A] \cdot 1 + [B] \cdot 2 + [C] \cdot 3 \\ 2.801 = [A] \cdot 2 + [B] \cdot 2 + [C] \cdot 2 \end{array}\right\} \qquad (7\text{-}20)$$

Our problem is to find the best values for the molarities [A], [B], and [C].

As luck would have it, Excel will find these values for us by the least-squares procedure LINEST. On page 69, we used LINEST to find the slope and intercept in the equation $y = mx + b$. In Figure 7-12, we use LINEST to find the slopes for $y = m_A x_A + m_B x_B + m_C x_C + b$ (where the intercept, $b$, is 0). To execute LINEST, highlight cells C12:E14 and type "= LINEST(E5:E9,A5:C9,FALSE,TRUE)". Then press CTRL + SHIFT +ENTER on a PC or COMMAND(⌘)+RETURN on a Mac. The first argument of LINEST, E5:E9, contains values of $y$ (= mmol $OH^-$). The second argument, A5:C9, contains values of $x$ (= volumes of acid). The third argument (FALSE) tells the computer to set the intercept ($b$) to zero and the fourth argument (TRUE) says that we want some statistics to be computed.

Excel finds the least-squares slopes in row 12 and their uncertainties in row 13. These slopes are the molarites [C], [B], and [A], in reverse order. So, for example, cells C12 and C13 tell us that the molarity of acid C is $0.809_9 \pm 0.006_2$ M.

We require at least $n$ equations to solve for $n$ unknowns. In this example, we have five equations (7-20), but just three unknowns ([A], [B], [C]). The two extra equations allow us to estimate uncertainty in the unknowns. If you do more experiments, you will generally decrease the uncertainty in concentration. An efficient experimental design with five experiments leaves us with more uncertainty than if we had done nine experiments. However, we have cut our effort nearly in half with the efficient design.

Acids could be delivered by transfer pipets whose tolerances are given in Table 2-4. So 2 mL means 2.000 ± 0.006 mL.

$$\left(\frac{mol}{L}\right) \cdot mL = mmol$$

For five equations and three unknowns, there are 5 − 3 = 2 *degrees of freedom*. With no degrees of freedom, there is no information with which to estimate uncertainty.

## Summary

The volume of reagent (titrant) required for stoichiometric reaction of analyte is measured in volumetric analysis. The stoichiometric point of the reaction is called the equivalence point. What we measure by an abrupt change in a physical property (such as the color of an indicator or the potential of an electrode) is the end point. The difference between the end point and the equivalence point is a titration error. This error can be reduced by subtracting results of a blank titration, in which the same procedure is carried out in the absence of analyte, or by standardizing the titrant, using the same reaction and a similar volume as that used for analyte.

The validity of an analytical result depends on knowing the amount of a primary standard. A solution with an approximately desired concentration can be standardized by titrating a primary standard. In a direct titration, titrant is added to analyte until the reaction is complete. In a back titration, a known excess of reagent is added to analyte, and the excess is titrated with a second standard reagent. Calculations of volumetric analysis relate the known moles of titrant to the unknown moles of analyte.

In Kjeldahl analysis, a nitrogen-containing organic compound is digested with a catalyst in boiling $H_2SO_4$. Nitrogen is converted into ammonium ion, which is converted into ammonia with base and distilled into standard HCl. The excess, unreacted HCl tells us how much nitrogen was present in the original analyte.

In a spectrophotometric titration, absorbance of light is monitored as titrant is added. For many reactions, there is an abrupt change in absorbance when the equivalence point is reached. The Fajans titration is based on the adsorption of a charged indicator onto the charged surface of the precipitate after the equivalence point. The Volhard titration, used to measure $Ag^+$, is based on the reaction of $Fe^{3+}$ with $SCN^-$ after the precipitation of AgSCN is complete.

Concentrations of reactants and products during a precipitation titration are calculated in three regions. Before the equivalence point, there is excess analyte. Its concentration is the product (fraction remaining) × (original concentration) × (dilution factor). The concentration of titrant can be found from the solubility product of the precipitate and the known concentration of excess analyte. At the equivalence point, concentrations of both reactants are governed by the solubility product. After the equivalence point, the concentration of analyte can be determined from the solubility product of precipitate and the known concentration of excess titrant.

Efficient experimental design decreases the number of experiments needed to obtain required information and an estimate of uncertainty in that information. A trade-off is that the fewer experiments we do, the greater the uncertainty in the results.

## Exercises

**7-A.** Ascorbic acid (vitamin C) reacts with $I_3^-$ according to the equation

Ascorbic acid

Dehydroascorbic acid

Starch is used as an indicator in the reaction. The end point is marked by the appearance of a deep blue starch-iodine complex when the first fraction of a drop of unreacted $I_3^-$ remains in the solution.

(a) If 29.41 mL of $I_3^-$ solution is required to react with 0.197 0 g of pure ascorbic acid, what is the molarity of the $I_3^-$ solution?

(b) A vitamin C tablet containing ascorbic acid plus an inert binder was ground to a powder, and 0.424 2 g was titrated by 31.63 mL of $I_3^-$. Find the weight percent of ascorbic acid in the tablet.

**7-B.** A solution of NaOH was standardized by titration of a known quantity of the primary standard, potassium hydrogen phthalate:

Potassium hydrogen phthalate
FM 204.221

The NaOH was then used to find the concentration of an unknown solution of $H_2SO_4$:

$$H_2SO_4 + 2NaOH \rightarrow Na_2SO_4 + 2H_2O$$

(a) Titration of 0.824 g of potassium hydrogen phthalate required 38.314 g of NaOH solution to reach the end point detected by phenolphthalein indicator. Find the concentration of NaOH (mol NaOH / kg solution).

(b) A 10.00-mL aliquot of $H_2SO_4$ solution required 57.911 g of NaOH solution to reach the phenolphthalein end point. Find the molarity of $H_2SO_4$.

**7-C.** A solid sample weighing 0.237 6 g contained only malonic acid and aniline hydrochloride. It required 34.02 mL of 0.087 71 M NaOH to neutralize the sample. Find the weight percent of each component in the solid mixture. The reactions are

$$CH_2(CO_2H)_2 + 2OH^- \rightarrow CH_2(CO_2^-)_2 + 2H_2O$$

Malonic acid    Malonate
FM 104.06

$$\text{\Large\textcircled{}}-NH_3^+Cl^- + OH^- \rightarrow \text{\Large\textcircled{}}-NH_2 + H_2O + Cl^-$$

Aniline hydrochloride            Aniline
FM 129.59

**7-D.** Semi-xylenol orange is a yellow compound at pH 5.9 but turns red when it reacts with $Pb^{2+}$. A 2.025-mL sample of semi-xylenol orange at pH 5.9 was titrated with $7.515 \times 10^{-4}$ M $Pb(NO_3)_2$, with the following results:

| Total μL Pb²⁺ added | Absorbance at 490 nm wavelength | Total μL Pb²⁺ added | Absorbance at 490 nm wavelength |
|---|---|---|---|
| 0.0 | 0.227 | 42.0 | 0.425 |
| 6.0 | 0.256 | 48.0 | 0.445 |
| 12.0 | 0.286 | 54.0 | 0.448 |
| 18.0 | 0.316 | 60.0 | 0.449 |
| 24.0 | 0.345 | 70.0 | 0.450 |
| 30.0 | 0.370 | 80.0 | 0.447 |
| 36.0 | 0.399 | | |

Make a graph of absorbance versus microliters of $Pb^{2+}$ added. Be sure to correct the absorbances for dilution. Corrected absorbance is what would be observed if the volume were not changed from its initial value of 2.025 mL. Assuming that the reaction of semi-xylenol orange with $Pb^{2+}$ has a 1:1 stoichiometry, find the molarity of semi-xylenol orange in the original solution.

**7-E.** A 50.0-mL sample of 0.080 0 M KSCN is titrated with 0.040 0 M $Cu^+$. The solubility product of CuSCN is $4.8 \times 10^{-15}$. At each of the following volumes of titrant, calculate $pCu^+$, and construct a graph of $pCu^+$ versus milliliters of $Cu^+$ added: 0.10, 10.0, 25.0, 50.0, 75.0, 95.0, 99.0, 100.0, 100.1, 101.0, 110.0 mL.

**7-F.** Construct a graph of $pAg^+$ versus milliliters of $Ag^+$ for the titration of 40.00 mL of solution containing 0.050 00 M $Br^-$ and 0.050 00 M $Cl^-$. The titrant is 0.084 54 M $AgNO_3$. Calculate $pAg^+$ at the following volumes: 2.00, 10.00, 22.00, 23.00, 24.00, 30.00, 40.00 mL, second equivalence point, 50.00 mL.

**7-G.** Consider the titration of 50.00 ($\pm0.05$) mL of a mixture of $I^-$ and $SCN^-$ with 0.068 3 ($\pm0.000$ 1) M $Ag^+$. The first equivalence point is observed at 12.6 ($\pm0.4$) mL, and the second occurs at 27.7 ($\pm0.3$) mL.
(a) Find the molarity and the uncertainty in molarity of thiocyanate in the original mixture.
(b) Suppose that the uncertainties are all the same, except that the uncertainty of the first equivalence point (12.6 ± ? mL) is variable. What is the maximum uncertainty (milliliters) of the first equivalence point if the uncertainty in $SCN^-$ molarity is to be $\leq4.0\%$?

## Problems

### Volumetric Procedures and Calculations

**7-1.** Explain the following statement: "The validity of an analytical result ultimately depends on knowing the composition of some primary standard."

**7-2.** Distinguish between the terms *end point* and *equivalence point*.

**7-3.** How does a blank titration reduce titration error?

**7-4.** What is the difference between a direct titration and a back titration?

**7-5.** What is the difference between a reagent-grade chemical and a primary standard?

**7-6.** Why are ultrapure acid solvents required to dissolve samples for trace analysis?

**7-7.** How many milliliters of 0.100 M KI are needed to react with 40.0 mL of 0.040 0 M $Hg_2(NO_3)_2$ if the reaction is $Hg_2^{2+} + 2I^- \rightarrow Hg_2I_2(s)$?

**7-8.** For Reaction 7-1, how many milliliters of 0.165 0 M $KMnO_4$ are needed to react with 108.0 mL of 0.165 0 M oxalic acid? How many milliliters of 0.165 0 M oxalic acid are required to react with 108.0 mL of 0.165 0 M $KMnO_4$?

**7-9.** Ammonia reacts with hypobromite, $OBr^-$, by the reaction $2NH_3 + 3OBr^- \rightarrow N_2 + 3Br^- + 3H_2O$. What is the molarity of a hypobromite solution if 1.00 mL of the $OBr^-$ solution reacts with 1.69 mg of $NH_3$?

**7-10.** Sulfamic acid is a primary standard that can be used to standardize NaOH.

$$^+H_3NSO_3^- + OH^- \rightarrow H_2NSO_3^- + H_2O$$

Sulfamic acid
FM 97.094

What is the molarity of a sodium hydroxide solution if 34.26 mL reacts with 0.333 7 g of sulfamic acid?

**7-11.** Limestone consists mainly of the mineral calcite, $CaCO_3$. The carbonate content of 0.541 3 g of powdered limestone was measured by suspending the powder in water, adding 10.00 mL of 1.396 M HCl, and heating to dissolve the solid and expel $CO_2$:

$$CaCO_3(s) + 2H^+ \rightarrow Ca^{2+} + CO_2\uparrow + H_2O$$

Calcium carbonate
FM 100.087

The excess acid required 39.96 mL of 0.100 4 M NaOH for complete titration to a phenolphthalein end point. Find the weight percent of calcite in the limestone.

**7-12.** The Kjeldahl procedure was used to analyze 256 μL of a solution containing 37.9 mg protein/mL. The liberated $NH_3$ was collected in 5.00 mL of 0.033 6 M HCl, and the remaining acid required 6.34 mL of 0.010 M NaOH for complete titration. What is the weight percent of nitrogen in the protein?

**7-13.** Arsenic(III) oxide ($As_2O_3$) is available in pure form and is a useful (but carcinogenic) primary standard for oxidizing agents such as $MnO_4^-$. The $As_2O_3$ is dissolved in base and then titrated with $MnO_4^-$ in acidic solution. A small amount of iodide ($I^-$) or iodate ($IO_3^-$) is used to catalyze the reaction between $H_3AsO_3$ and $MnO_4^-$.

$$As_2O_3 + 4OH^- \rightleftharpoons 2HAsO_3^{2-} + H_2O$$
$$HAsO_3^{2-} + 2H^+ \rightleftharpoons H_3AsO_3$$
$$5H_3AsO_3 + 2MnO_4^- + 6H^+ \rightarrow 5H_3AsO_4 + 2Mn^{2+} + 3H_2O$$

(a) A 3.214-g aliquot of $KMnO_4$ (FM 158.034) was dissolved in 1.000 L of water, heated to cause any reactions with impurities to occur, cooled, and filtered. What is the theoretical molarity of this solution if no $MnO_4$ was consumed by impurities?
(b) What mass of $As_2O_3$ (FM 197.84) would be just sufficient to react with 25.00 mL of the $KMnO_4$ solution in part (a)?

(c) It was found that 0.146 8 g of $As_2O_3$ required 29.98 mL of $KMnO_4$ solution for the faint color of unreacted $MnO_4^-$ to appear. In a blank titration, 0.03 mL of $MnO_4^-$ was required to produce enough color to be seen. Calculate the molarity of the permanganate solution.

**7-14.** A 0.238 6-g sample contained only NaCl and KBr. It was dissolved in water and required 48.40 mL of 0.048 37 M $AgNO_3$ for complete titration of both halides [giving $AgCl(s)$ and $AgBr(s)$]. Calculate the weight percent of Br in the solid sample.

**7-15.** A solid mixture weighing 0.054 85 g contained only ferrous ammonium sulfate and ferrous chloride. The sample was dissolved in 1 M $H_2SO_4$, and the $Fe^{2+}$ required 13.39 mL of 0.012 34 M $Ce^{4+}$ for complete oxidation to $Fe^{3+}$ ($Ce^{4+} + Fe^{2+} \rightarrow Ce^{3+} + Fe^{3+}$). Calculate the weight percent of Cl in the original sample.

$$FeSO_4 \cdot (NH_4)_2SO_4 \cdot 6H_2O \qquad FeCl_2 \cdot 6H_2O$$
Ferrous ammonium sulfate  Ferrous chloride
FM 392.13  FM 234.84

**7-16.** A cyanide solution with a volume of 12.73 mL was treated with 25.00 mL of $Ni^{2+}$ solution (containing excess $Ni^{2+}$) to convert the cyanide into tetracyanonickelate(II):

$$4CN^- + Ni^{2+} \rightarrow Ni(CN)_4^{2-}$$

The excess $Ni^{2+}$ was then titrated with 10.15 mL of 0.013 07 M ethylenediaminetetraacetic acid (EDTA):

$$Ni^{2+} + EDTA^{4-} \rightarrow Ni(EDTA)^{2-}$$

$Ni(CN)_4^{2-}$ does not react with EDTA. If 39.35 mL of EDTA were required to react with 30.10 mL of the original $Ni^{2+}$ solution, calculate the molarity of $CN^-$ in the 12.73-mL cyanide sample.

**7-17.** *Managing a salt-water aquarium.* A tank at the New Jersey State Aquarium has a volume of 2.9 million liters.[7] Bacteria are used to remove nitrate that would otherwise build up to toxic levels. Aquarium water is first pumped into a 2 700-L deaeration tank containing bacteria that consume $O_2$ in the presence of added methanol:

$$2CH_3OH + 3O_2 \xrightarrow{\text{bacteria}} 2CO_2 + 4H_2O \qquad (1)$$
Methanol

Anoxic (deoxygenated) water from the deaeration tank flows into a 1 500-L denitrification reactor containing colonies of *Pseudomonas* bacteria in a porous medium. Methanol is injected continuously and nitrate is converted into nitrite and then into nitrogen:

$$3NO_3^- + CH_3OH \xrightarrow{\text{bacteria}} 3NO_2^- + CO_2 + 2H_2O \qquad (2)$$
Nitrate  Nitrite

$$2NO_2^- + CH_3OH \xrightarrow{\text{bacteria}} N_2 + CO_2 + H_2O + 2OH^- \qquad (3)$$

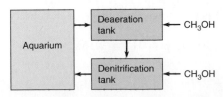

(a) Deaeration can be thought of as a slow, bacteria-mediated titration of $O_2$ by $CH_3OH$. The concentration of $O_2$ in sea water at 24°C is 220 μM. How many liters of $CH_3OH$ (FM 32.04, density = 0.791 g/mL) are required by Reaction 1 for 2.9 million liters of aquarium water?

(b) Write the net reaction showing nitrate plus methanol going to nitrogen. How many liters of $CH_3OH$ are required by the net reaction for 2.9 million liters of aquarium water with a nitrate concentration of 8 100 μM?

(c) In addition to consuming methanol for Reactions 1 through 3, the bacteria require 30% more methanol for their own growth. What is the total volume of methanol required to denitrify 2.9 million liters of aquarium water?

### Spectrophotometric Titrations

**7-18.** Why does the slope of the absorbance-versus-volume graph in Figure 7-5 change abruptly at the equivalence point?

**7-19.** A 2.00-mL solution of apotransferrin was titrated as illustrated in Figure 7-5. It required 163 μL of 1.43 mM ferric nitrilotriacetate to reach the end point.

(a) How many moles of Fe(III) (= ferric nitrilotriacetate) were required to reach the end point?

(b) Each apotransferrin molecule binds two ferric ions. Find the concentration of apotransferrin in the 2.00-mL solution.

**7-20.** The iron-binding site of transferrin in Figure 7-4 can accommodate certain other metal ions besides $Fe^{3+}$ and certain other anions besides $CO_3^{2-}$. Data are given in the table for the titration of transferrin (3.57 mg in 2.00 mL) with 6.64 mM $Ga^{3+}$ solution in the presence of the anion oxalate, $C_2O_4^{2-}$, and in the absence of a suitable anion. Prepare a graph similar to Figure 7-5, showing both sets of data. Indicate the theoretical equivalence point for the binding of one and two $Ga^{3+}$ ions per molecule of protein and the observed end point. How many $Ga^{3+}$ ions are bound to transferrin in the presence and in the absence of oxalate?

| Titration in presence of $C_2O_4^{2-}$ | | Titration in absence of anion | |
|---|---|---|---|
| Total μL $Ga^{3+}$ added | Absorbance at 241 nm | Total μL $Ga^{3+}$ added | Absorbance at 241 nm |
| 0.0 | 0.044 | 0.0 | 0.000 |
| 2.0 | 0.143 | 2.0 | 0.007 |
| 4.0 | 0.222 | 6.0 | 0.012 |
| 6.0 | 0.306 | 10.0 | 0.019 |
| 8.0 | 0.381 | 14.0 | 0.024 |
| 10.0 | 0.452 | 18.0 | 0.030 |
| 12.0 | 0.508 | 22.0 | 0.035 |
| 14.0 | 0.541 | 26.0 | 0.037 |
| 16.0 | 0.558 | | |
| 18.0 | 0.562 | | |
| 21.0 | 0.569 | | |
| 24.0 | 0.576 | | |

### Shape of a Precipitation Curve

**7-21.** Describe the chemistry that occurs in each of the following regions in curve (*a*) in Figure 7-8: (i) before the first equivalence point; (ii) at the first equivalence point; (iii) between the first and second equivalence points; (iv) at the second equivalence point; and (v) past the second equivalence point. For each region except (ii), write the equation that you would use to calculate $[Ag^+]$.

**7-22.** Consider the titration of 25.00 mL of 0.082 30 M KI with 0.051 10 M $AgNO_3$. Calculate $pAg^+$ at the following volumes of added $AgNO_3$: (a) 39.00 mL; (b) $V_e$; (c) 44.30 mL.

**7-23.** The text claims that precipitation of $I^-$ is not complete before $Cl^-$ begins to precipitate in the titration in Figure 7-8. Calculate the concentration of $Ag^+$ at the equivalence point in the titration of $I^-$ alone. Show that this concentration of $Ag^+$ will precipitate $Cl^-$.

**7-24.** A 25.00-mL solution containing 0.031 10 M $Na_2C_2O_4$ was titrated with 0.025 70 M $Ca(NO_3)_2$ to precipitate calcium oxalate: $Ca^{2+} + C_2O_4^{2-} \rightarrow CaC_2O_4(s)$. Find $pCa^{2+}$ at the following volumes of $Ca(NO_3)_2$: **(a)** 10.00; **(b)** $V_e$; **(c)** 35.00 mL.

**7-25.** In precipitation titrations of halides by $Ag^+$, the ion pair $AgX(aq)(X = Cl, Br, I)$ is in equilibrium with the precipitate. Use Appendix J to find the concentrations of $AgCl(aq)$, $AgBr(aq)$, and $AgI(aq)$ during the precipitations.

### Titration of a Mixture

**7-26.** A procedure[8] for determining halogens in organic compounds uses an argentometric titration. To 50 mL of anhydrous ether is added a carefully weighed sample (10–100 mg) of unknown, plus 2 mL of sodium dispersion and 1 mL of methanol. (Sodium dispersion is finely divided solid sodium suspended in oil. With methanol, it makes sodium methoxide, $CH_3O^-Na^+$, which attacks the organic compound, liberating halides.) Excess sodium is destroyed by slow addition of 2-propanol, after which 100 mL of water are added. (Sodium should not be treated directly with water, because the $H_2$ produced can explode in the presence of $O_2$: $2Na + 2H_2O \rightarrow 2NaOH + H_2$.) This procedure gives a two-phase mixture, with an ether layer floating on top of the aqueous layer that contains the halide salts. The aqueous layer is adjusted to pH 4 and titrated with $Ag^+$, using the electrodes in Figure 7-9. How much 0.025 70 M $AgNO_3$ solution will be required to reach each equivalence point when 82.67 mg of 1-bromo-4-chlorobutane ($BrCH_2CH_2CH_2CH_2Cl$; FM 171.46) are analyzed?

**7-27.** Calculate $pAg^+$ at the following points in titration (a) in Figure 7-8: **(a)** 10.00 mL; **(b)** 20.00 mL; **(c)** 30.00 mL; **(d)** second equivalence point; **(e)** 50.00 mL.

**7-28.** A mixture having a volume of 10.00 mL and containing 0.100 0 M $Ag^+$ and 0.100 0 M $Hg_2^{2+}$ was titrated with 0.100 0 M KCN to precipitate $Hg_2(CN)_2$ and AgCN.
**(a)** Calculate $pCN^-$ at each of the following volumes of added KCN: 5.00, 10.00, 15.00, 19.90, 20.10, 25.00, 30.00, 35.00 mL.
**(b)** Should any AgCN be precipitated at 19.90 mL?

### Using Spreadsheets

**7-29.** Derive an expression analogous to Equation 7-18 for the titration of $M^+$(concentration = $C_M^0$, volume = $V_M^0$) with $X^-$

(titrant concentration = $C_X^0$). Your equation should allow you to compute the volume of titrant ($V_X$) as a function of $[X^-]$.

**7-30.** [▦] Use Equation 7-18 to reproduce the curves in Figure 7-7. Plot your results on a single graph.

**7-31.** Consider precipitation of $X^{x-}$ with $M^{m+}$:

$$xM^{m+} + mX^{x-} \rightleftharpoons M_xX_m(s) \qquad K_{sp} = [M^{m+}]^x[X^{x-}]^m$$

Write mass balance equations for M and X and derive the equation

$$V_M = V_X^0 \left( \frac{xC_X^0 + m[M^{m+}] - x[X^{x-}]}{mC_M^0 - m[M^{m+}] + x[X^{x-}]} \right)$$

where $[X^{x-}] = (K_{sp}/[M^{m+}]^x)^{1/m}$.

**7-32.** [▦] Use the equation in Problem 7-31 to calculate the titration curve for 10.0 mL of 0.100 M $CrO_4^{2-}$ titrated with 0.100 M $Ag^+$ to produce $Ag_2CrO_4(s)$.

### End-Point Detection

**7-33.** Why does the surface charge of a precipitate change sign at the equivalence point?

**7-34.** Examine the procedure in Table 7-1 for the Fajans titration of $Zn^{2+}$. Do you expect the charge on the precipitate to be positive or negative after the equivalence point?

**7-35.** Describe how to analyze a solution of NaI by using the Volhard titration.

**7-36.** A 30.00-mL solution of $I^-$ was treated with 50.00 mL of 0.365 0 M $AgNO_3$. $AgI(s)$ was filtered off, and the filtrate (plus $Fe^{3+}$) was titrated with 0.287 0 M KSCN. When 37.60 mL had been added, the solution turned red. How many milligrams of $I^-$ were in the original solution?

### Experiment Design

**7-37.** [▦] Acid-base titrations were carried out in a similar manner to those in Section 7-8. Volumes and results are shown in the following table.[6] Use the Excel LINEST procedure to find the concentrations of acids A, B, and C and estimate their uncertainty.

| Acid volume (mL) | | | mmol of $OH^-$ |
| A | B | C | required |
|---|---|---|---|
| 2 | 2 | 2 | 3.015 |
| 0 | 2 | 2 | 1.385 |
| 2 | 0 | 2 | 2.180 |
| 2 | 2 | 0 | 2.548 |
| 2 | 2 | 2 | 3.140 |

# 8 Activity and the Systematic Treatment of Equilibrium

### Estimated Number of Waters of Hydration*

| Molecule | Tightly bound $H_2O$ |
|---|---|
| $CH_3CH_2CH_3$ | 0 |
| $C_6H_6$ | 0 |
| $CH_3CH_2Cl$ | 0 |
| $CH_3CH_2SH$ | 0 |
| $CH_3-O-CH_3$ | 1 |
| $CH_3CH_2OH$ | 1 |
| $(CH_3)_2C=O$ | 1.5 |
| $CH_3CH=O$ | 1.5 |
| $CH_3CO_2H$ | 2 |
| $CH_3C\equiv N$ | 3 |
| $CH_3\overset{\overset{O}{\|\|}}{C}NHCH_3$ | 4 |
| $CH_3NO_2$ | 5 |
| $CH_3CO_2^-$ | 5 |
| $CH_3NH_2$ | 6 |
| $CH_3SO_3H$ | 7 |
| $NH_3$ | 9 |
| $CH_3SO_3^-$ | 10 |
| $NH_4^+$ | 12 |

*From S. Fu and C. A. Lucy, "Prediction of Electrophoretic Mobilities," Anal. Chem. **1998**, 70, 173.

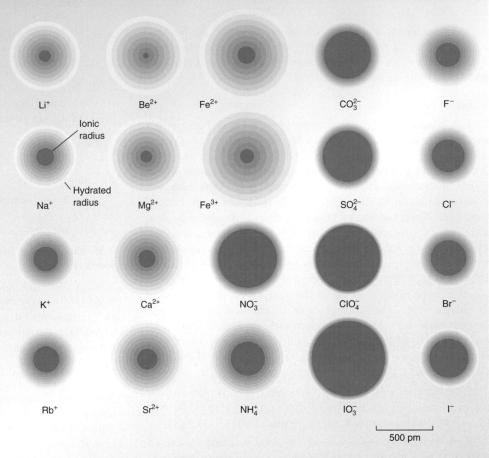

Ionic and hydrated radii of several ions. Smaller, more highly charged ions bind water molecules more tightly and behave as larger hydrated species.[3]

Ions and molecules in solution are surrounded by an organized sheath of solvent molecules. The oxygen atom of $H_2O$ has a partial negative charge and each hydrogen atom has half as much positive charge.

Water binds to cations through the oxygen atom. The first coordination sphere of $Li^+$, for example, is composed of ~4 $H_2O$ molecules.[1] $Cl^-$ binds ~6 $H_2O$ molecules through hydrogen atoms.[1,2] $H_2O$ exchanges rapidly between bulk solvent and ion-coordination sites.

Ionic radii in the figure are measured by X-ray diffraction of ions in crystals. Hydrated radii are estimated from diffusion coefficients of ions in solution and from the mobilities of aqueous ions in an electric field.[3,4] Smaller, more highly charged ions bind more water molecules and behave as larger species in solution. The *activity* of aqueous ions, which we study in this chapter, is related to the size of the hydrated species.

In Chapter 6, we wrote the equilibrium constant for a reaction in the form

$$Fe^{3+} + SCN^- \rightleftharpoons Fe(SCN)^{2+} \qquad K = \frac{[Fe(SCN)^{2+}]}{[Fe^{3+}][SCN^-]} \qquad (8\text{-}1)$$

Pale yellow · Colorless · Red

Figure 8-1, Demonstration 8-1, and Color Plate 3 show that the concentration quotient in Equation 8-1 decreases if you add the "inert" salt $KNO_3$ to the solution. That is, the equilibrium "constant" is not really constant. This chapter explains why concentrations are replaced by *activities* in the equilibrium constant and how activities are used.

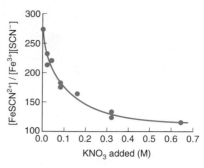

**Figure 8-1** Student data showing that the equilibrium quotient of concentrations for the reaction $Fe^{3+} + SCN^- \rightleftharpoons Fe(SCN)^{2+}$ decreases as potassium nitrate is added to the solution. Color Plate 3 shows the fading of the red color of $Fe(SCN)^{2+}$ after $KNO_3$ has been added. Problem 13-11 gives more information on this chemical system. [*From R. J. Stolzberg, "Discovering a Change in Equilibrium Constant with Change in Ionic Strength," J. Chem. Ed.* **1999,** *76, 640.*]

# ■■■ 8-1 The Effect of Ionic Strength on Solubility of Salts

Consider a saturated solution of $CaSO_4$ in distilled water.

$$CaSO_4(s) \rightleftharpoons Ca^{2+} + SO_4^{2-} \qquad K_{sp} = 2.4 \times 10^{-5} \qquad (8\text{-}2)$$

Figure 6-1 showed that the solubility is 0.015 M. The dissolved species are mainly 0.010 M $Ca^{2+}$, 0.010 M $SO_4^{2-}$, and 0.005 M $CaSO_4(aq)$ (an ion pair).

Now an interesting effect is observed when a salt such as $KNO_3$ is added to the solution. Neither $K^+$ nor $NO_3^-$ reacts with either $Ca^{2+}$ or $SO_4^{2-}$. Yet, when 0.050 M $KNO_3$ is added to the saturated solution of $CaSO_4$, more solid dissolves until the concentrations of $Ca^{2+}$ and $SO_4^{2-}$ have each increased by about 30%.

In general, adding an "inert" salt ($KNO_3$) to a sparingly soluble salt ($CaSO_4$) increases the solubility of the sparingly soluble salt. By "inert," we mean a salt whose ions do not react with the other ions. When we add salt to a solution, we say that the *ionic strength* of the solution increases. The definition of ionic strength will be given shortly.

Addition of an "inert" salt increases the solubility of an ionic compound.

## The Explanation

Why does the solubility increase when salts are added to the solution? Consider one particular $Ca^{2+}$ ion and one particular $SO_4^{2-}$ ion in the solution. The $SO_4^{2-}$ ion is surrounded by cations ($K^+$, $Ca^{2+}$) and anions ($NO_3^-$, $SO_4^{2-}$) in the solution. However, for the average anion, there will be more cations than anions near it because cations are attracted to the anion, but anions are repelled. These interactions create a region of net positive charge around any particular anion. We call this region the **ionic atmosphere** (Figure 8-2). Ions continually diffuse into and out of the ionic atmosphere. The net charge in the atmosphere, averaged over time, is less than the charge of the anion at the center. Similarly, an atmosphere of negative charge surrounds any cation in solution.

The ionic atmosphere attenuates (decreases) the attraction between ions. The cation plus its negative atmosphere has less positive charge than the cation alone. The anion plus its ionic atmosphere has less negative charge than the anion alone. The net attraction between the cation with its ionic atmosphere and the anion with its ionic atmosphere is smaller than it would be between pure cation and anion in the absence of ionic atmospheres. *The greater the ionic strength of a solution, the higher the charge in the ionic atmosphere. Each ion-plus-atmosphere contains less net charge and there is less attraction between any particular cation and anion.*

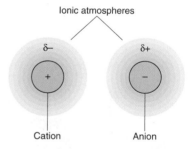

**Figure 8-2** An ionic atmosphere, shown as a spherical cloud of charge $\delta+$ or $\delta-$, surrounds ions in solution. The charge of the atmosphere is less than the charge of the central ion. The greater the ionic strength of the solution, the greater the charge in each ionic atmosphere.

---

### Demonstration 8-1 Effect of Ionic Strength on Ion Dissociation[5]

This experiment demonstrates the effect of ionic strength on the dissociation of the red iron(III) thiocyanate complex:

$$Fe(SCN)^{2+} \rightleftharpoons Fe^{3+} + SCN^-$$

Red · Pale yellow · Colorless

Prepare a solution of 1 mM $FeCl_3$ by dissolving 0.27 g of $FeCl_3 \cdot 6H_2O$ in 1 L of water containing 3 drops of 15 M (concentrated) $HNO_3$. The acid slows the precipitation of $Fe(OH)_3$, which occurs in a few days and necessitates the preparation of fresh solution for this demonstration.

To demonstrate the effect of ionic strength on the dissociation reaction, mix 300 mL of 1 mM $FeCl_3$ with 300 mL of 1.5 mM $NH_4SCN$ or $KSCN$. Divide the pale red solution into two equal portions and add 12 g of $KNO_3$ to one of them to increase the ionic strength to 0.4 M. As $KNO_3$ dissolves, the red $Fe(SCN)^{2+}$ complex dissociates and the color fades (Color Plate 3).

Addition of a few crystals of $NH_4SCN$ or $KSCN$ to either solution drives the reaction toward formation of $Fe(SCN)^{2+}$, thereby intensifying the red color. This reaction demonstrates Le Châtelier's principle—adding a product creates more reactant.

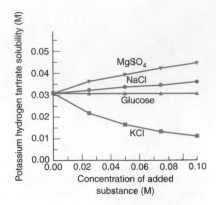

**Figure 8-3** Solubility of potassium hydrogen tartrate increases when the salts MgSO₄ or NaCl are added. There is no effect when the neutral compound glucose is added. Addition of KCl decreases the solubility. *(Why?)* [From C. J. Marzzacco, "Effect of Salts and Nonelectrolytes on the Solubility of Potassium Bitartrate," J. Chem. Ed. **1998**, 75, 1628.]

Increasing ionic strength therefore reduces the attraction between any particular $Ca^{2+}$ ion and any $SO_4^{2-}$ ion, relative to their attraction for each other in distilled water. The effect is to reduce their tendency to come together, thereby *increasing* the solubility of $CaSO_4$.

Increasing ionic strength promotes dissociation into ions. Thus, each of the following reactions is driven to the right if the ionic strength is raised from, say, 0.01 to 0.1 M:

$$Fe(SCN)^{2+} \rightleftharpoons Fe^{3+} + SCN^-$$
Thiocyanate

$$\bigcirc\!\!-OH \rightleftharpoons \bigcirc\!\!-O^- + H^+$$
Phenol          Phenolate

$$\underset{\text{Potassium hydrogen tartrate}}{HO_2CCHCHCO_2K(s)} \rightleftharpoons HO_2CCHCHCO_2^- + K^+$$

Figure 8-3 shows the effect of added salt on the solubility of potassium hydrogen tartrate.

## What Do We Mean by "Ionic Strength"?

**Ionic strength,** $\mu$, is a measure of the total concentration of ions in solution. The more highly charged an ion, the more it is counted.

*Ionic strength:*
$$\mu = \frac{1}{2}(c_1 z_1^2 + c_2 z_2^2 + \cdots) = \frac{1}{2}\sum_i c_i z_i^2 \qquad (8\text{-}3)$$

where $c_i$ is the concentration of the $i$th species and $z_i$ is its charge. The sum extends over *all* ions in solution.

---

**Example** Calculation of Ionic Strength

Find the ionic strength of **(a)** 0.10 M $NaNO_3$; **(b)** 0.010 M $Na_2SO_4$; and **(c)** 0.020 M KBr plus 0.010 M $Na_2SO_4$.

**Solution**

**(a)** $\mu = \frac{1}{2}\{[Na^+]\cdot(+1)^2 + [NO_3^-]\cdot(-1)^2\}$

$= \frac{1}{2}\{0.10\cdot 1 + 0.10\cdot 1\} = 0.10\ M$

**(b)** $\mu = \frac{1}{2}\{[Na^+]\cdot(+1)^2 + [SO_4^{2-}]\cdot(-2)^2\}$

$= \frac{1}{2}\{(0.020\cdot 1) + (0.010\cdot 4)\} = 0.030\ M$

Note that $[Na^+] = 0.020\ M$ because there are two moles of $Na^+$ per mole of $Na_2SO_4$.

**(c)** $\mu = \frac{1}{2}\{[K^+]\cdot(+1)^2 + [Br^-]\cdot(-1)^2 + [Na^+]\cdot(+1)^2 + [SO_4^{2-}]\cdot(-2)^2\}$

$= \frac{1}{2}\{(0.020\cdot 1) + (0.020\cdot 1) + (0.020\cdot 1) + (0.010\cdot 4)\} = 0.050\ M$

---

| Electrolyte | Molarity | Ionic strength |
|---|---|---|
| 1:1 | M | M |
| 2:1 | M | 3 M |
| 3:1 | M | 6 M |
| 2:2 | M | 4 M |

$NaNO_3$ is called a 1:1 electrolyte because the cation and the anion both have a charge of 1. For 1:1 electrolytes, the ionic strength equals the molarity. For other stoichiometries (such as the 2:1 electrolyte $Na_2SO_4$), the ionic strength is greater than the molarity.

Computing the ionic strength of any but the most dilute solutions is complicated because salts with ions of charge $\geq 2$ are not fully dissociated. In Box 8-1 we find that, at a formal concentration of 0.025 M $MgSO_4$, 35% of $Mg^{2+}$ is bound in the ion pair, $MgSO_4(aq)$. The higher the concentration and the higher the ionic charge, the more the ion pairing. There is no simple way to find the ionic strength of 0.025 M $MgSO_4$.

## Box 8-1   Salts with Ions of Charge $\geq |2|$ Do Not Fully Dissociate[6]

Salts composed of cations with a charge of $+1$ and anions with a charge of $-1$ dissociate almost completely at concentrations $<0.1$ M. Salts containing ions with a charge $\geq 2$ are less dissociated, even in dilute solution. Appendix J gives formation constants for *ion pairing*:

*Ion pair formation constant:*

$$M^{n+}(aq) + L^{m-}(aq) \rightleftharpoons \underset{\text{Ion pair}}{M^{n+}L^{m-}(aq)}$$

$$K = \frac{[ML]\gamma_{ML}}{[M]\gamma_M[L]\gamma_L}$$

where the $\gamma_i$ are activity coefficients. With constants from Appendix J, activity coefficients from Equation 8-6, craft, and persistence, you might calculate the following percentages of ion pairing in 0.025 F solutions:

Percentage of metal ion bound as ion pair in 0.025 F $M_xL_y$ solution[a]

| M \\ L | $Cl^-$ | $SO_4^{2-}$ |
|---|---|---|
| $Na^+$ | 0.6% | 4% |
| $Mg^{2+}$ | 8% | 35% |

a. The size of ML was taken as 500 pm to compute its activity coefficient.

The table tells us that 0.025 F NaCl is only 0.6% associated as $Na^+Cl^-(aq)$ and $Na_2SO_4$ is 4% associated as $NaSO_4^-(aq)$. For $MgSO_4$, 35% is ion paired. A solution of 0.025 F $MgSO_4$ contains 0.016 M $Mg^{2+}$, 0.016 M $SO_4^{2-}$, and 0.009 M $MgSO_4(aq)$. The ionic strength of 0.025 F $MgSO_4$ is not 0.10 M, but just 0.065 M. Problem 8-26 provides an example of the type of calculation in this box.

## ■ ■ ■ ■   8-2   Activity Coefficients

The equilibrium constant expression in Equation 8-1 does not predict any effect of ionic strength on a chemical reaction. To account for the effect of ionic strength, concentrations are replaced by **activities:**

*Activity of C:*

$$\mathcal{A}_C = [C]\gamma_C \qquad\qquad (8\text{-}4)$$

Activity of C     Concentration of C     Activity coefficient of C

The activity of species C is its concentration multiplied by its **activity coefficient.** The activity coefficient measures the deviation of behavior from ideality. If the activity coefficient were 1, then the behavior would be ideal and the form of the equilibrium constant in Equation 8-1 would be correct.

> Do not confuse the terms *activity* and *activity coefficient.*

The correct form of the equilibrium constant is

*General form of equilibrium constant:*

$$K = \frac{\mathcal{A}_C^c\mathcal{A}_D^d}{\mathcal{A}_A^a\mathcal{A}_B^b} = \frac{[C]^c\gamma_C^c[D]^d\gamma_D^d}{[A]^a\gamma_A^a[B]^b\gamma_B^b} \qquad\qquad (8\text{-}5)$$

> Equation 8-5 is the "real" equilibrium constant. Equation 6-2, the concentration quotient, $K_c$, did not include activity coefficients:
>
> $$K_c = \frac{[C]^c[D]^d}{[A]^a[B]^b} \qquad (6\text{-}2)$$

Equation 8-5 allows for the effect of ionic strength on a chemical equilibrium because the activity coefficients depend on ionic strength.

For Reaction 8-2, the equilibrium constant is

$$K_{sp} = \mathcal{A}_{Ca^{2+}}\mathcal{A}_{SO_4^{2-}} = [Ca^{2+}]\gamma_{Ca^{2+}}[SO_4^{2-}]\gamma_{SO_4^{2-}}$$

If the concentrations of $Ca^{2+}$ and $SO_4^{2-}$ are to *increase* when a second salt is added to increase ionic strength, the activity coefficients must *decrease* with increasing ionic strength.

At low ionic strength, activity coefficients approach unity, and the thermodynamic equilibrium constant (8-5) approaches the "concentration" equilibrium constant (6-2). One way to measure a thermodynamic equilibrium constant is to measure the concentration ratio (6-2) at successively lower ionic strengths and extrapolate to zero ionic strength. Commonly, tabulated equilibrium constants are not thermodynamic constants but just the concentration ratio (6-2) measured under a particular set of conditions.

### Example   Exponents of Activity Coefficients

Write the solubility product expression for $La_2(SO_4)_3$ with activity coefficients.

**Solution**   Exponents of activity coefficients are the same as exponents of concentrations:

$$K_{sp} = \mathcal{A}_{La^{3+}}^2\mathcal{A}_{SO_4^{2-}}^3 = [La^{3+}]^2\gamma_{La^{3+}}^2[SO_4^{2-}]^3\gamma_{SO_4^{2-}}^3$$

## Activity Coefficients of Ions

The ionic atmosphere model leads to the **extended Debye-Hückel equation**, relating activity coefficients to ionic strength:

Extended Debye-Hückel equation:

$$\log \gamma = \frac{-0.51z^2\sqrt{\mu}}{1 + (\alpha\sqrt{\mu}/305)} \quad \text{(at 25°C)} \qquad (8\text{-}6)$$

1 pm (picometer) = $10^{-12}$ m

In Equation 8-6, $\gamma$ is the activity coefficient of an ion of charge $\pm z$ and size $\alpha$ (picometers, pm) in an aqueous solution of ionic strength $\mu$. The equation works fairly well for $\mu \leq 0.1$ M. To find activity coefficients for ionic strengths above 0.1 M (up to molalities of 2–6 mol/kg for many salts), more complicated *Pitzer equations* are usually used.[7]

Table 8-1 lists sizes ($\alpha$) and activity coefficients of many ions. All ions of the same size and charge appear in the same group and have the same activity coefficients. For example, $Ba^{2+}$ and succinate ion [$^-O_2CCH_2CH_2CO_2^-$, listed as $(CH_2CO_2^-)_2$] each have a size of 500 pm and are listed among the charge $= \pm 2$ ions. At an ionic strength of 0.001 M, both of these ions have an activity coefficient of 0.868.

**Table 8-1** Activity coefficients for aqueous solutions at 25°C

| Ion | Ion size ($\alpha$, pm) | Ionic strength ($\mu$, M) 0.001 | 0.005 | 0.01 | 0.05 | 0.1 |
|---|---|---|---|---|---|---|
| *Charge = ±1* | | | | | | |
| $H^+$ | 900 | 0.967 | 0.933 | 0.914 | 0.86 | 0.83 |
| $(C_6H_5)_2CHCO_2^-$, $(C_3H_7)_4N^+$ | 800 | 0.966 | 0.931 | 0.912 | 0.85 | 0.82 |
| $(O_2N)_3C_6H_2O^-$, $(C_3H_7)_3NH^+$, $CH_3OC_6H_4CO_2^-$ | 700 | 0.965 | 0.930 | 0.909 | 0.845 | 0.81 |
| $Li^+$, $C_6H_5CO_2^-$, $HOC_6H_4CO_2^-$, $ClC_6H_4CO_2^-$, $C_6H_5CH_2CO_2^-$, $CH_2=CHCH_2CO_2^-$, $(CH_3)_2CHCH_2CO_2^-$, $(CH_3CH_2)_4N^+$, $(C_3H_7)_2NH_2^+$ | 600 | 0.965 | 0.929 | 0.907 | 0.835 | 0.80 |
| $Cl_2CHCO_2^-$, $Cl_3CCO_2^-$, $(CH_3CH_2)_3NH^+$, $(C_3H_7)NH_3^+$ | 500 | 0.964 | 0.928 | 0.904 | 0.83 | 0.79 |
| $Na^+$, $CdCl^+$, $ClO_2^-$, $IO_3^-$, $HCO_3^-$, $H_2PO_4^-$, $HSO_3^-$, $H_2AsO_4^-$, $Co(NH_3)_4(NO_2)_2^+$, $CH_3CO_2^-$, $ClCH_2CO_2^-$, $(CH_3)_4N^+$, $(CH_3CH_2)_2NH_2^+$, $H_2NCH_2CO_2^-$ | 450 | 0.964 | 0.928 | 0.902 | 0.82 | 0.775 |
| $^+H_3NCH_2CO_2H$, $(CH_3)_3NH^+$, $CH_3CH_2NH_3^+$ | 400 | 0.964 | 0.927 | 0.901 | 0.815 | 0.77 |
| $OH^-$, $F^-$, $SCN^-$, $OCN^-$, $HS^-$, $ClO_3^-$, $ClO_4^-$, $BrO_3^-$, $IO_4^-$, $MnO_4^-$, $HCO_2^-$, $H_2citrate^-$, $CH_3NH_3^+$, $(CH_3)_2NH_2^+$ | 350 | 0.964 | 0.926 | 0.900 | 0.81 | 0.76 |
| $K^+$, $Cl^-$, $Br^-$, $I^-$, $CN^-$, $NO_2^-$, $NO_3^-$ | 300 | 0.964 | 0.925 | 0.899 | 0.805 | 0.755 |
| $Rb^+$, $Cs^+$, $NH_4^+$, $Tl^+$, $Ag^+$ | 250 | 0.964 | 0.924 | 0.898 | 0.80 | 0.75 |
| *Charge = ±2* | | | | | | |
| $Mg^{2+}$, $Be^{2+}$ | 800 | 0.872 | 0.755 | 0.69 | 0.52 | 0.45 |
| $CH_2(CH_2CH_2CO_2^-)_2$, $(CH_2CH_2CH_2CO_2^-)_2$ | 700 | 0.872 | 0.755 | 0.685 | 0.50 | 0.425 |
| $Ca^{2+}$, $Cu^{2+}$, $Zn^{2+}$, $Sn^{2+}$, $Mn^{2+}$, $Fe^{2+}$, $Ni^{2+}$, $Co^{2+}$, $C_6H_4(CO_2^-)_2$, $H_2C(CH_2CO_2^-)_2$, $(CH_2CH_2CO_2^-)_2$ | 600 | 0.870 | 0.749 | 0.675 | 0.485 | 0.405 |
| $Sr^{2+}$, $Ba^{2+}$, $Cd^{2+}$, $Hg^{2+}$, $S^{2-}$, $S_2O_4^{2-}$, $WO_4^{2-}$, $H_2C(CO_2^-)_2$, $(CH_2CO_2^-)_2$, $(CHOHCO_2^-)_2$ | 500 | 0.868 | 0.744 | 0.67 | 0.465 | 0.38 |
| $Pb^{2+}$, $CO_3^{2-}$, $SO_3^{2-}$, $MoO_4^{2-}$, $Co(NH_3)_5Cl^{2+}$, $Fe(CN)_5NO^{2-}$, $C_2O_4^{2-}$, $Hcitrate^{2-}$ | 450 | 0.867 | 0.742 | 0.665 | 0.455 | 0.37 |
| $Hg_2^{2+}$, $SO_4^{2-}$, $S_2O_3^{2-}$, $S_2O_6^{2-}$, $S_2O_8^{2-}$, $SeO_4^{2-}$, $CrO_4^{2-}$, $HPO_4^{2-}$ | 400 | 0.867 | 0.740 | 0.660 | 0.445 | 0.355 |
| *Charge = ±3* | | | | | | |
| $Al^{3+}$, $Fe^{3+}$, $Cr^{3+}$, $Sc^{3+}$, $Y^{3+}$, $In^{3+}$, lanthanides[a] | 900 | 0.738 | 0.54 | 0.445 | 0.245 | 0.18 |
| $citrate^{3-}$ | 500 | 0.728 | 0.51 | 0.405 | 0.18 | 0.115 |
| $PO_4^{3-}$, $Fe(CN)_6^{3-}$, $Cr(NH_3)_6^{3+}$, $Co(NH_3)_6^{3+}$, $Co(NH_3)_5H_2O^{3+}$ | 400 | 0.725 | 0.505 | 0.395 | 0.16 | 0.095 |
| *Charge = ±4* | | | | | | |
| $Th^{4+}$, $Zr^{4+}$, $Ce^{4+}$, $Sn^{4+}$ | 1 100 | 0.588 | 0.35 | 0.255 | 0.10 | 0.065 |
| $Fe(CN)_6^{4-}$ | 500 | 0.57 | 0.31 | 0.20 | 0.048 | 0.021 |

a. Lanthanides are elements 57–71 in the periodic table.

SOURCE: J. Kielland, J. Am. Chem. Soc. **1937**, 59, 1675.

The ion size $\alpha$ in Equation 8-6 is an empirical parameter that provides agreement between measured activity coefficients and ionic strength up to $\mu \approx 0.1$ M. In theory, $\alpha$ is the diameter of the hydrated ion.[8] However, sizes in Table 8-1 cannot be taken literally. For example, the diameter of $Cs^+$ ion in crystals is 340 pm. The hydrated $Cs^+$ ion must be larger than the ion in the crystal, but the size of $Cs^+$ in Table 8-1 is only 250 pm.

Even though ion sizes in Table 8-1 are empirical parameters, trends among sizes are sensible. Small, highly charged ions bind solvent more tightly and have larger effective sizes than do larger or less highly charged ions. For example, the order of sizes in Table 8-1 is $Li^+ > Na^+ > K^+ > Rb^+$, even though crystallographic radii are $Li^+ < Na^+ < K^+ < Rb^+$.

Ionic and hydrated ion sizes are shown at the opening of this chapter.

### Effect of Ionic Strength, Ion Charge, and Ion Size on the Activity Coefficient

Over the range of ionic strengths from 0 to 0.1 M, the effect of each variable on activity coefficients is as follows:

1. As ionic strength increases, the activity coefficient decreases (Figure 8-4). The activity coefficient ($\gamma$) approaches unity as the ionic strength ($\mu$) approaches 0.
2. As the magnitude of the charge of the ion increases, the departure of its activity coefficient from unity increases. Activity corrections are more important for ions with a charge of $\pm 3$ than for ions with a charge of $\pm 1$ (Figure 8-4).
3. The smaller the ion size ($\alpha$), the more important activity effects become.

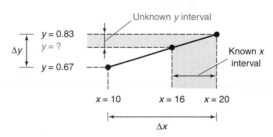

**Figure 8-4** Activity coefficients for differently charged ions with a constant ionic size ($\alpha$) of 500 pm. At zero ionic strength, $\gamma = 1$. The greater the charge of the ion, the more rapidly $\gamma$ decreases as ionic strength increases. Note that the abscissa is logarithmic.

> **Example** Using Table 8-1
>
> Find the activity coefficient of $Ca^{2+}$ in a solution of 3.3 mM $CaCl_2$.
>
> **Solution**  The ionic strength is
>
> $$\mu = \frac{1}{2}\{[Ca^{2+}] \cdot 2^2 + [Cl^-] \cdot (-1)^2\}$$
>
> $$= \frac{1}{2}\{(0.003\ 3) \cdot 4 + (0.006\ 6) \cdot 1\} = 0.010\ M$$
>
> In Table 8-1, $Ca^{2+}$ is listed under the charge $\pm 2$ and has a size of 600 pm. Thus $\gamma = 0.675$ when $\mu = 0.010$ M.

### How to Interpolate

If you need to find an activity coefficient for an ionic strength that is between values in Table 8-1, you can use Equation 8-6. Alternatively, in the absence of a spreadsheet, it is usually easier to interpolate than to use Equation 8-6. In *linear interpolation,* we assume that values between two entries of a table lie on a straight line. For example, consider a table in which $y = 0.67$ when $x = 10$ and $y = 0.83$ when $x = 20$. What is the value of $y$ when $x = 16$?

*Interpolation* is the estimation of a number that lies *between* two values in a table. Estimating a number that lies *beyond* values in a table is called *extrapolation.*

| x value | y value |
|---------|---------|
| 10 | 0.67 |
| 16 | ? |
| 20 | 0.83 |

To interpolate a value of $y$, we can set up a proportion:

*Interpolation:*

$$\frac{\text{Unknown } y \text{ interval}}{\Delta y} = \frac{\text{known } x \text{ interval}}{\Delta x}$$ (8-7)

This calculation is equivalent to saying:

"16 is 60% of the way from 10 to 20. Therefore the y value will be 60% of the way from 0.67 to 0.83."

$$\frac{0.83 - y}{0.83 - 0.67} = \frac{20 - 16}{20 - 10} \Rightarrow y = 0.76_6$$

For $x = 16$, our estimate of $y$ is $0.76_6$.

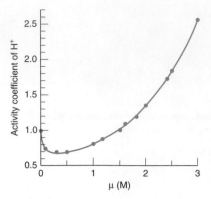

**Figure 8-5** Activity coefficient of $H^+$ in solutions containing 0.010 0 M $HClO_4$ and varying amounts of $NaClO_4$. *[Data derived from L. Pezza, M. Molina, M. de Moraes, C. B. Melios, and J. O. Tognolli, Talanta* **1996,** *43, 1689.]* The authoritative source on electrolyte solutions is H. S. Harned and B. B. Owen, *The Physical Chemistry of Electrolyte Solutions* (New York: Reinhold, 1958).

For neutral species, $\mathcal{A}_C \approx [C]$. A more accurate relation is $\log \gamma = k\mu$, where $k \approx 0$ for ion pairs, $k \approx 0.11$ for $NH_3$ and $CO_2$, and $k \approx 0.2$ for organic molecules. For an ionic strength of $\mu = 0.1$ M, $\gamma \approx 1.00$ for ion pairs, $\gamma \approx 1.03$ for $NH_3$, and $\gamma \approx 1.05$ for organic molecules.

For gases, $\mathcal{A} \approx P$ (bar).

At high ionic strength, $\gamma$ increases with increasing $\mu$.

Note that the activity coefficient of $F^-$ is squared.

---

**Example** Interpolating Activity Coefficients

Calculate the activity coefficient of $H^+$ when $\mu = 0.025$ M.

**Solution** $H^+$ is the first entry in Table 8-1.

| $\mu$ | $\gamma$ for $H^+$ |
|-------|--------------------|
| 0.01  | 0.914              |
| 0.025 | ?                  |
| 0.05  | 0.86               |

The linear interpolation is set up as follows:

$$\frac{\text{Unknown } \gamma \text{ interval}}{\Delta \gamma} = \frac{\text{known } \mu \text{ interval}}{\Delta \mu}$$

$$\frac{0.86 - \gamma}{0.86 - 0.914} = \frac{0.05 - 0.025}{0.05 - 0.01}$$

$$\approx \gamma = 0.89_4$$

Another Solution  A better and slightly more tedious calculation uses Equation 8-6, with the ion size $\alpha = 900$ pm listed for $H^+$ in Table 8-1:

$$\log \gamma_{H^+} = \frac{(-0.51)(1^2)\sqrt{0.025}}{1 + (900\sqrt{0.025}\,/\,305)} = -0.054_{98}$$

$$\gamma_{H^+} = 10^{-0.054_{98}} = 0.88_1$$

## Activity Coefficients of Nonionic Compounds

Neutral molecules, such as benzene and acetic acid, have no ionic atmosphere because they have no charge. To a good approximation, their activity coefficients are unity when the ionic strength is less than 0.1 M. In this book, we set $\gamma = 1$ for neutral molecules. That is, *the activity of a neutral molecule will be assumed to be equal to its concentration.*

For gases such as $H_2$, the activity is written

$$\mathcal{A}_{H_2} = P_{H_2}\gamma_{H_2}$$

where $P_{H_2}$ is pressure in bars. The activity of a gas is called its *fugacity*, and the activity coefficient is called the *fugacity coefficient*. Deviation of gas behavior from the ideal gas law results in deviation of the fugacity coefficient from unity. For gases at or below 1 bar, $\gamma \approx 1$. Therefore, for all gases, *we will set $\mathcal{A} = P(bar)$.*

## High Ionic Strengths

Above an ionic strength of approximately 1 M, activity coefficients of most ions increase, as shown for $H^+$ in $NaClO_4$ solutions in Figure 8-5. We should not be too surprised that activity coefficients in concentrated salt solutions are not the same as those in dilute aqueous solution. The "solvent" is no longer $H_2O$ but, rather, a mixture of $H_2O$ and $NaClO_4$. Hereafter, we limit our attention to dilute aqueous solutions.

---

**Example** Using Activity Coefficients

Find the concentration of $Ca^{2+}$ in equilibrium with 0.050 M NaF saturated with $CaF_2$. The solubility of $CaF_2$ is small, so the concentration of $F^-$ is 0.050 M from NaF.

**Solution** We find $[Ca^{2+}]$ from the solubility product expression, including activity coefficients. The ionic strength of 0.050 M NaF is 0.050 M. At $\mu = 0.050\ 0$ M in Table 8-1, we find $\gamma_{Ca^{2+}} = 0.485$ and $\gamma_{F^-} = 0.81$.

$$K_{sp} = [Ca^{2+}]\gamma_{Ca^{2+}}[F^-]^2\gamma_{F^-}^2$$

$$3.9 \times 10^{-11} = [Ca^{2+}](0.485)(0.050)^2(0.81)^2$$

$$[Ca^{2+}] = 4.9 \times 10^{-8}\ M$$

# ■ ■ ■ 8-3 pH Revisited

The definition $pH \approx -\log[H^+]$ in Chapter 6 is not exact. The real definition is

$$pH = -\log \mathcal{A}_{H^+} = -\log[H^+]\gamma_{H^+} \qquad (8\text{-}8)$$

The real definition of pH!

When we measure pH with a pH meter, we are measuring the negative logarithm of the hydrogen ion *activity,* not its concentration.

### Example pH of Pure Water at 25°C

Let's calculate the pH of pure water by using activity coefficients.

**Solution**  The relevant equilibrium is

$$H_2O \overset{K_w}{\rightleftharpoons} H^+ + OH^- \qquad (8\text{-}9)$$

$$K_w = \mathcal{A}_{H^+}\mathcal{A}_{OH^-} = [H^+]\gamma_{H^+}[OH^-]\gamma_{OH^-} \qquad (8\text{-}10)$$

$H^+$ and $OH^-$ are produced in a 1:1 mole ratio, so their concentrations must be equal. Calling each concentration $x$, we write

$$K_w = 1.0 \times 10^{-14} = (x)\gamma_{H^+}(x)\gamma_{OH^-}$$

But the ionic strength of pure water is so small that it is reasonable to guess that $\gamma_{H^+} = \gamma_{OH^-} = 1$. Using these values in the preceding equation gives

$$1.0 \times 10^{-14} = x^2 \Rightarrow x = 1.0 \times 10^{-7} \text{ M}$$

The concentrations of $H^+$ and $OH^-$ are both $1.0 \times 10^{-7}$ M. The ionic strength is $1.0 \times 10^{-7}$ M, so each activity coefficient is very close to 1.00. The pH is

$$pH = -\log[H^+]\gamma_{H^+} = -\log(1.0 \times 10^{-7})(1.00) = 7.00$$

### Example pH of Water Containing a Salt

Now let's calculate the pH of water containing 0.10 M KCl at 25°C.

**Solution**  Reaction 8-9 tells us that $[H^+] = [OH^-]$. However, the ionic strength of 0.10 M KCl is 0.10 M. The activity coefficients of $H^+$ and $OH^-$ in Table 8-1 are 0.83 and 0.76, respectively, when $\mu = 0.10$ M. Putting these values into Equation 8-10 gives

$$K_w = [H^+]\gamma_{H^+}[OH^-]\gamma_{OH^-}$$

$$1.0 \times 10^{-14} = (x)(0.83)(x)(0.76)$$

$$x = 1.26 \times 10^{-7} \text{ M}$$

The concentrations of $H^+$ and $OH^-$ are equal and are both greater than $1.0 \times 10^{-7}$ M. The activities of $H^+$ and $OH^-$ are not equal in this solution:

$$\mathcal{A}_{H^+} = [H^+]\gamma_{H^+} = (1.26 \times 10^{-7})(0.83) = 1.05 \times 10^{-7}$$

$$\mathcal{A}_{OH^-} = [OH^-]\gamma_{OH^-} = (1.26 \times 10^{-7})(0.76) = 0.96 \times 10^{-7}$$

Finally, we calculate $pH = -\log \mathcal{A}_{H^+} = -\log(1.05 \times 10^{-7}) = 6.98$.

The pH of water changes from 7.00 to 6.98 when we add 0.10 M KCl. KCl is not an acid or a base. The pH changes because KCl affects the activities of $H^+$ and $OH^-$. The pH change of 0.02 unit lies at the limit of accuracy of pH measurements and is hardly important. However, the *concentration* of $H^+$ in 0.10 M KCl ($1.26 \times 10^{-7}$ M) is 26% greater than the concentration of $H^+$ in pure water ($1.00 \times 10^{-7}$ M).

# ■ ■ ■ 8-4 Systematic Treatment of Equilibrium

The *systematic treatment of equilibrium* is a way to deal with all types of chemical equilibria, regardless of their complexity. After setting up general equations, we often introduce specific conditions or judicious approximations that allow simplification. Even simplified calculations are usually very tedious, so we make liberal use of spreadsheets for numerical solutions. When you have mastered the systematic treatment of equilibrium, you should be able to explore the behavior of complex systems.

The systematic procedure is to write as many independent algebraic equations as there are unknowns (species) in the problem. The equations are generated by writing all the chemical equilibrium conditions plus two more: the balances of charge and of mass. There is only one charge balance in a given system, but there could be several mass balances.

## Charge Balance

The **charge balance** is an algebraic statement of electroneutrality: *The sum of the positive charges in solution equals the sum of the negative charges in solution.*

Suppose that a solution contains the following ionic species: $H^+$, $OH^-$, $K^+$, $H_2PO_4^-$, $HPO_4^{2-}$, and $PO_4^{3-}$. The charge balance is

$$[H^+] + [K^+] = [OH^-] + [H_2PO_4^-] + 2[HPO_4^{2-}] + 3[PO_4^{3-}] \qquad \text{(8-11)}$$

This statement says that the total charge contributed by $H^+$ and $K^+$ equals the magnitude of the charge contributed by all of the anions on the right side of the equation. *The coefficient in front of each species always equals the magnitude of the charge on the ion.* This statement is true because a mole of, say, $PO_4^{3-}$ contributes three moles of negative charge. If $[PO_4^{3-}] = 0.01$ M, the negative charge is $3[PO_4^{3-}] = 3(0.01) = 0.03$ M.

Equation 8-11 appears unbalanced to many people. "The right side of the equation has much more charge than the left side!" you might think. But you would be wrong.

For example, consider a solution prepared by weighing out 0.025 0 mol of $KH_2PO_4$ plus 0.030 0 mol of KOH and diluting to 1.00 L. The concentrations of the species at equilibrium are calculated to be

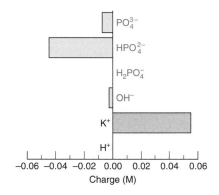

$$[H^+] = 5.1 \times 10^{-12} \text{ M} \qquad [H_2PO_4^-] = 1.3 \times 10^{-6} \text{ M}$$
$$[K^+] = 0.055\,0 \text{ M} \qquad [HPO_4^{2-}] = 0.022\,0 \text{ M}$$
$$[OH^-] = 0.002\,0 \text{ M} \qquad [PO_4^{3-}] = 0.003\,0 \text{ M}$$

This calculation, which you should be able to do when you have finished studying acids and bases, takes into account the reaction of $OH^-$ with $H_2PO_4^-$ to produce $HPO_4^{2-}$ and $PO_4^{3-}$.

Are the charges balanced? Yes, indeed. Plugging into Equation 8-11, we find

$$[H^+] + [K^+] = [OH^-] + [H_2PO_4^-] + 2[HPO_4^{2-}] + 3[PO_4^{3-}]$$
$$5.1 \times 10^{-12} + 0.055\,0 = 0.002\,0 + 1.3 \times 10^{-6} + 2(0.022\,0) + 3(0.003\,0)$$
$$0.055\,0 \text{ M} = 0.055\,0 \text{ M}$$

*Figure 8-6* Charge contributed by each ion in 1.00 L of solution containing 0.025 0 mol $KH_2PO_4$ plus 0.030 0 mol KOH. The total positive charge equals the total negative charge.

The total positive charge is 0.055 0 M, and the total negative charge also is 0.055 0 M (Figure 8-6). Charges must balance in every solution. Otherwise, a beaker with excess positive charge would glide across the lab bench and smash into another beaker with excess negative charge.

The general form of the charge balance for any solution is

*Charge balance:* $\qquad n_1[C_1] + n_2[C_2] + \cdots = m_1[A_1] + m_2[A_2] + \cdots \qquad$ (8-12)

where [C] is the concentration of a cation, $n$ is the charge of the cation, [A] is the concentration of an anion, and $m$ is the magnitude of the charge of the anion.

### Example Writing a Charge Balance

Write the charge balance for a solution containing $H_2O$, $H^+$, $OH^-$, $ClO_4^-$, $Fe(CN)_6^{3-}$, $CN^-$, $Fe^{3+}$, $Mg^{2+}$, $CH_3OH$, HCN, $NH_3$, and $NH_4^+$.

**Solution** Neutral species ($H_2O$, $CH_3OH$, HCN, and $NH_3$) contribute no charge, so the charge balance is

$$[H^+] + 3[Fe^{3+}] + 2[Mg^{2+}] + [NH_4^+] = [OH^-] + [ClO_4^-] + 3[Fe(CN)_6^{3-}] + [CN^-]$$

## Mass Balance

The **mass balance,** also called the *material balance,* is a statement of the conservation of matter. The mass balance states that *the quantity of all species in a solution containing a* *particular atom (or group of atoms) must equal the amount of that atom (or group) delivered to the solution.* It is easier to see this relation through examples than by a general statement.

Suppose that a solution is prepared by dissolving 0.050 mol of acetic acid in water to give a total volume of 1.00 L. The acetic acid partially dissociates into acetate:

$$CH_3CO_2H \rightleftharpoons CH_3CO_2^- + H^+$$
$$\text{Acetic acid} \qquad \text{Acetate}$$

The mass balance states that the quantity of dissociated and undissociated acetic acid in the solution must equal the amount of acetic acid put into the solution.

*Mass balance for acetic acid in water:*
$$0.050 \text{ M} = [CH_3CO_2H] + [CH_3CO_2^-]$$
$$\underset{\substack{\text{What we put into} \\ \text{the solution}}}{} \quad \underset{\substack{\text{Undissociated} \\ \text{product}}}{} \quad \underset{\substack{\text{Dissociated} \\ \text{product}}}{}$$

When a compound dissociates in several ways, the mass balance must include all the products. Phosphoric acid ($H_3PO_4$), for example, dissociates to $H_2PO_4^-$, $HPO_4^{2-}$, and $PO_4^{3-}$. The mass balance for a solution prepared by dissolving 0.025 0 mol of $H_3PO_4$ in 1.00 L is

$$0.025 \text{ 0 M} = [H_3PO_4] + [H_2PO_4^-] + [HPO_4^{2-}] + [PO_4^{3-}]$$

*Activity coefficients do not appear in the mass balance.* The concentration of each species counts exactly the number of atoms of that species.

---

**Example** Mass Balance When the Total Concentration Is Known

Write the mass balances for $K^+$ and for phosphate in a solution prepared by mixing 0.025 0 mol $KH_2PO_4$ plus 0.030 0 mol $KOH$ and diluting to 1.00 L.

**Solution** The total $K^+$ is 0.025 0 M + 0.030 0 M, so one mass balance is

$$[K^+] = 0.055 \text{ 0 M}$$

The total of *all forms* of phosphate is 0.025 0 M, so the mass balance for phosphate is

$$[H_3PO_4] + [H_2PO_4^-] + [HPO_4^{2-}] + [PO_4^{3-}] = 0.025 \text{ 0 M}$$

---

Now consider a solution prepared by dissolving $La(IO_3)_3$ in water.

$$La(IO_3)_3(s) \xrightleftharpoons{K_{sp}} La^{3+} + 3IO_3^-$$
$$\text{Iodate}$$

We do not know how much $La^{3+}$ or $IO_3^-$ is dissolved, but we do know that there must be three iodate ions for each lanthanum ion dissolved. That is, the iodate concentration must be three times the lanthanum concentration. If $La^{3+}$ and $IO_3^-$ are the only species derived from $La(IO_3)_3$, then the mass balance is

$$[IO_3^-] = 3[La^{3+}]$$

If the solution also contains the ion pair $LaIO_3^{2+}$ and the hydrolysis product $LaOH^{2+}$, the mass balance would be

$$[\text{Total iodate}] = 3[\text{total lanthanum}]$$
$$[IO_3^-] + [LaIO_3^{2+}] = 3\{[La^{3+}] + [LaIO_3^{2+}] + [LaOH^{2+}]\}$$

---

**Example** Mass Balance When the Total Concentration Is Unknown

Write the mass balance for a saturated solution of the slightly soluble salt $Ag_3PO_4$, which produces $PO_4^{3-}$ and $3Ag^+$ when it dissolves.

**Solution** If the phosphate in solution remained as $PO_4^{3-}$, we could write

$$[Ag^+] = 3[PO_4^{3-}]$$

because three silver ions are produced for each phosphate ion. However, phosphate reacts with water to give $HPO_4^{2-}$, $H_2PO_4^-$, and $H_3PO_4$, so the mass balance is

$$[Ag^+] = 3\{[PO_4^{3-}] + [HPO_4^{2-}] + [H_2PO_4^-] + [H_3PO_4]\}$$

That is, the number of atoms of $Ag^+$ must equal three times the total number of atoms of phosphorus, regardless of how many species contain phosphorus.

Atoms of Ag = 3(atoms of P)

Box 8-2 illustrates the operation of a mass balance in natural waters.

---

Box 8-2   Calcium Carbonate Mass Balance in Rivers

$Ca^{2+}$ is the most common cation in rivers and lakes. It comes from dissolution of the mineral calcite by the action of $CO_2$ to produce 2 moles of $HCO_3^-$ for each mole of $Ca^{2+}$:

$$CaCO_3(s) + CO_2(aq) + H_2O \rightleftharpoons Ca^{2+} + 2HCO_3^- \qquad (A)$$
Calcite                                                        Bicarbonate

Near neutral pH, most of the product is bicarbonate, not $CO_3^{2-}$ or $H_2CO_3$. The mass balance for the dissolution of calcite is therefore $[HCO_3^-] \approx 2[Ca^{2+}]$. Indeed, measurements of $Ca^{2+}$ and $HCO_3^-$ in many rivers conform to this mass balance, shown by the straight line on the graph. Rivers such as the Danube, the Mississippi, and the Congo, which lie on the line $[HCO_3^-] = 2[Ca^{2+}]$, appear to be saturated with calcium carbonate. If the river water were in equilibrium with atmospheric $CO_2$ ($P_{CO_2} = 10^{-3.4}$ bar), the concentration of $Ca^{2+}$ would be 20 mg/L (see Problem 8-28). Rivers with more than 20 mg of $Ca^{2+}$ per liter have a higher concentration of dissolved $CO_2$ produced by respiration or from inflow of groundwaters with a high $CO_2$ content. Rivers such as the Nile, the Niger, and the Amazon, for which $2[Ca^{2+}] < [HCO_3^-]$, are not saturated with $CaCO_3$.

Just between 1960 and 2000, $CO_2$ in the atmosphere increased by 17% (Box 20-1)—mostly from our burning of fossil fuel. This increase drives Reaction A to the right and threatens the existence of coral reefs,[9] which are huge, living structures consisting largely of $CaCO_3$. Coral reefs are a unique habitat for many aquatic species.

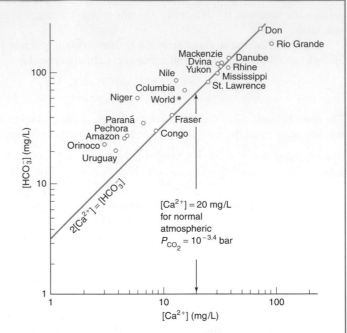

Concentrations of bicarbonate and calcium in many rivers conform to the mass balance for Reaction A: $[HCO_3^-] \approx 2[Ca^{2+}]$. *[Data from W. Stumm and J. J. Morgan, Aquatic Chemistry, 3rd ed. (New York: Wiley-Interscience, 1996), p. 189; and H. D. Holland, The Chemistry of the Atmosphere and Oceans (New York: Wiley-Interscience, 1978).]*

## Systematic Treatment of Equilibrium

Now that we have considered the charge and mass balances, we are ready for the systematic treatment of equilibrium.[10] Here is the general prescription:

**Step 1**  Write the *pertinent reactions.*
**Step 2**  Write the *charge balance* equation.
**Step 3**  Write *mass balance* equations. There may be more than one.
**Step 4**  Write the *equilibrium constant expression* for each chemical reaction. This step is the only one in which activity coefficients appear.
**Step 5**  *Count the equations and unknowns.* There should be as many equations as unknowns (chemical species). If not, you must either find more equilibria or fix some concentrations at known values.
**Step 6**  By hook or by crook, *solve* for all the unknowns.

Steps 1 and 6 are the heart of the problem. Guessing what chemical equilibria exist in a given solution requires a fair degree of chemical intuition. In this book, you will usually be given help with step 1. Unless we know all the relevant equilibria, it is not possible to calculate the composition of a solution correctly. Because we do not know all the chemical reactions, we undoubtedly oversimplify many equilibrium problems.

Step 6 is likely to be your biggest challenge. With $n$ equations involving $n$ unknowns, the problem can always be solved, at least in principle. In the simplest cases, you can do this by hand; but for most problems, approximations are made or a spreadsheet is employed.

## 8-5  Applying the Systematic Treatment of Equilibrium

The systematic treatment of equilibrium is best understood by studying some examples.

### A Simple Example: Ionization of Water

The dissociation of water into $H^+$ and $OH^-$ occurs in every aqueous solution:

$$H_2O \xrightleftharpoons{K_w} H^+ + OH^- \qquad K_w = 1.0 \times 10^{-14} \text{ at } 25°C \qquad (8\text{-}13)$$

Let's apply the systematic treatment to find the concentrations of $H^+$ and $OH^-$ in pure water.

Step 1   Pertinent reactions. The only one is Reaction 8-13.
Step 2   Charge balance. The only ions are $H^+$ and $OH^-$, so the charge balance is

$$[H^+] = [OH^-] \qquad (8\text{-}14)$$

Step 3   Mass balance. Reaction 8-13 creates one $H^+$ for each $OH^-$. The mass balance is simply $[H^+] = [OH^-]$, which is the same as the charge balance for this system.
Step 4   Equilibrium constant expression.

$$K_w = [H^+]\gamma_{H^+}[OH^-]\gamma_{OH^-} = 1.0 \times 10^{-14} \qquad (8\text{-}15)$$

*This is the only step in which activity coefficients enter the problem.*

Step 5   Count equations and unknowns. We have two equations, 8-14 and 8-15, and two unknowns, $[H^+]$ and $[OH^-]$.

We need *n* equations to solve for *n* unknowns.

Step 6   Solve.

Now we must decide what to do about the activity coefficients. We expect that the ionic strength ($\mu$) of pure water will be very low. Therefore, it is reasonable to suppose that $\gamma_{H^+}$ and $\gamma_{OH^-}$ are both unity.

Recall that $\gamma$ approaches 1 as $\mu$ approaches 0.

Putting $[H^+] = [OH^-]$ into Equation 8-15 gives

$$[H^+]\gamma_{H^+}[OH^-]\gamma_{OH^-} = 1.0 \times 10^{-14}$$
$$[H^+] \cdot 1 \cdot [H^+] \cdot 1 = 1.0 \times 10^{-14}$$
$$[H^+] = 1.0 \times 10^{-7}\,M$$

The ionic strength is $10^{-7}$ M. The assumption that $\gamma_{H^+} = \gamma_{OH^-} = 1$ is good.

Also, $[OH^-] = [H^+] = 1.0 \times 10^{-7}$ M. As a reminder, the pH is given by

$$pH = -\log \mathcal{A}_{H^+} = -\log[H^+]\gamma_{H^+} = -\log(1.0 \times 10^{-7})(1) = 7.00$$

This is the correct definition of pH. When we neglect activity coefficients, we will write $pH = -\log[H^+]$.

## Solubility of Calcium Sulfate

Our goal is to find the concentrations of the major species in a saturated solution of $CaSO_4$.

Step 1   Pertinent reactions. Even in such a simple system, there are quite a few reactions:

$$CaSO_4(s) \underset{}{\overset{K_{sp}}{\rightleftharpoons}} Ca^{2+} + SO_4^{2-} \qquad K_{sp} = 2.4 \times 10^{-5} \qquad (8\text{-}16)$$

$$CaSO_4(s) \underset{}{\overset{K_{ion\ pair}}{\rightleftharpoons}} CaSO_4(aq) \qquad K_{ion\ pair} = 5.0 \times 10^{-3} \qquad (8\text{-}17)$$

$$Ca^{2+} + H_2O \underset{}{\overset{K_{acid}}{\rightleftharpoons}} CaOH^+ + H^+ \qquad K_{acid} = 2.0 \times 10^{-13} \qquad (8\text{-}18)$$

$$SO_4^{2-} + H_2O \underset{}{\overset{K_{base}}{\rightleftharpoons}} HSO_4^- + OH^- \qquad K_{base} = 9.8 \times 10^{-13} \qquad (8\text{-}19)$$

$$H_2O \underset{}{\overset{K_w}{\rightleftharpoons}} H^+ + OH^- \qquad K_w = 1.0 \times 10^{-14} \qquad (8\text{-}20)$$

*Caveat Emptor!* In all equilibrium problems, we are limited by how much of the system's chemistry is understood. Unless we know all the relevant equilibria, it is impossible to calculate the composition.

There is no way you can be expected to come up with all of these reactions, so you will be given help with this step.
Step 2   Charge balance. Equating positive and negative charges gives

$$2[Ca^{2+}] + [CaOH^+] + [H^+] = 2[SO_4^{2-}] + [HSO_4^-] + [OH^-] \qquad (8\text{-}21)$$

Multiply $[Ca^{2+}]$ and $[SO_4^{2-}]$ by 2 because 1 mol of each ion has 2 mol of charge.

Step 3   Mass balance. Reaction 8-16 produces 1 mole of sulfate for each mole of calcium. No matter what happens to these ions, the total concentration of all species with sulfate must equal the total concentration of all species with calcium:

$$[\text{Total calcium}] = [\text{total sulfate}]$$
$$[Ca^{2+}] + [CaSO_4(aq)] + [CaOH^+] = [SO_4^{2-}] + [HSO_4^-] + [CaSO_4(aq)] \qquad (8\text{-}22)$$

Step 4   Equilibrium constant expressions. There is one for each chemical reaction.

$$K_{sp} = [Ca^{2+}]\gamma_{Ca^{2+}}[SO_4^{2-}]\gamma_{SO_4^{2-}} = 2.4 \times 10^{-5} \qquad (8\text{-}23)$$

$$K_{ion\ pair} = [CaSO_4(aq)] = 5.0 \times 10^{-3} \qquad (8\text{-}24)$$

$$K_{acid} = \frac{[CaOH^+]\gamma_{CaOH^+}[H^+]\gamma_{H^+}}{[Ca^{2+}]\gamma_{Ca^{2+}}} = 2.0 \times 10^{-13} \qquad (8\text{-}25)$$

$$K_{base} = \frac{[HSO_4^-]\gamma_{HSO_4^-}[OH^-]\gamma_{OH^-}}{[SO_4^{2-}]\gamma_{SO_4^{2-}}} = 9.8 \times 10^{-13} \qquad (8\text{-}26)$$

$$K_w = [H^+]\gamma_{H^+}[OH^-]\gamma_{OH^-} = 1.0 \times 10^{-14} \qquad (8\text{-}27)$$

The activity coefficient of the neutral species $CaSO_4(aq)$ is 1.

Step 4 is the only one where activity coefficients come in.

**Step 5** Count equations and unknowns. There are seven equations (8-21 through 8-27) and seven unknowns: $[Ca^{2+}]$, $[SO_4^{2-}]$, $[CaSO_4(aq)]$, $[CaOH^+]$, $[HSO_4^-]$, $[H^+]$, and $[OH^-]$. In principle, we have all the information necessary to solve the problem.

**Step 6** Solve. Well, this is not easy! We don't know the ionic strength, so we cannot evaluate activity coefficients. Also, where do we start when there are seven unknowns?

Just when the world looks grim, a Good Chemist comes galloping down the hill on her white stallion to rescue us with some hints. "First of all," she says, "Reactions 8-18 and 8-19 have small equilibrium constants, so they are not very important compared with Reactions 8-16 and 8-17. As for activity coefficients, you can begin by setting them all equal to 1 and solving for all concentrations. Then you can calculate the ionic strength and find values for the activity coefficients. With these coefficients, you can solve the problem a second time and find a new set of concentrations. You can repeat this process a few times until reaching a constant answer."

On the advice of the Good Chemist, we neglect the acid-base reactions 8-18 and 8-19. The remaining reactions of calcium sulfate, 8-16 and 8-17, are not reactions with $H_2O$. Therefore, in our approximation, Reaction 8-20 is independent of the calcium sulfate chemistry. Reaction 8-20 produces $[H^+]$ and $[OH^-]$, which we know are $1 \times 10^{-7}$ M. These concentrations of $H^+$ and $OH^-$ are not exactly correct, because we have neglected Reactions 8-18 and 8-19, which do play a role in determining $[H^+]$ and $[OH^-]$.

These approximations leave just Reactions 8-16 and 8-17. In the charge balance 8-21, $CaOH^+$ and $HSO_4^-$ are neglected. Also, $[H^+] = [OH^-]$, so they cancel. The charge balance becomes $2[Ca^{2+}] = 2[SO_4^{2-}]$. In the mass balance 8-22, we discard $CaOH^+$ and $HSO_4^-$ because we neglect Reactions 8-18 and 8-19. The term $[CaSO_4(aq)]$ cancels, leaving $[Ca^{2+}] = [SO_4^{2-}]$, which is the same as the charge balance. We have three unknowns ($[Ca^{2+}]$, $[SO_4^{2-}]$, $[CaSO_4(aq)]$) and three equations:

$$K_{sp} = [Ca^{2+}]\gamma_{Ca^{2+}}[SO_4^{2-}]\gamma_{SO_4^{2-}} = 2.4 \times 10^{-5} \tag{8-23}$$

$$K_{ion\ pair} = [CaSO_4(aq)] = 5.0 \times 10^{-3} \tag{8-24}$$

$$[Ca^{2+}] = [SO_4^{2-}] \tag{8-28}$$

Equation 8-24 says $[CaSO_4(aq)] = 5.0 \times 10^{-3}$ M, so $[CaSO_4(aq)]$ is known.

The simplified problem is reduced to Equations 8-23 and 8-28. We find a first approximate solution by setting activity coefficients to 1:

$$[Ca^{2+}]_1\gamma_{Ca^{2+}}[SO_4^{2-}]_1\gamma_{SO_4^{2-}} = 2.4 \times 10^{-5} \tag{8-29}$$

$$[Ca^{2+}]_1(1)[SO_4^{2-}]_1(1) = 2.4 \times 10^{-5}$$

$[SO_4^{2-}] = [Ca^{2+}]$

$$[Ca^{2+}]_1(1)[Ca^{2+}]_1(1) = 2.4 \times 10^{-5} \Rightarrow [Ca^{2+}]_1 = 4.9 \times 10^{-3}\ M$$

where the subscript 1 means it is our first approximation. If $[Ca^{2+}] = [SO_4^{2-}] = 4.9 \times 10^{-3}$ M, then the ionic strength is $\mu = 4(4.9 \times 10^{-3}\ M) = 0.020$ M. Interpolating in Table 8-1, we find the activity coefficients $\gamma_{Ca^{2+}} = 0.628$ and $\gamma_{SO_4^{2-}} = 0.606$. Putting these coefficients back into Equation 8-29 provides a second approximation:

We are carrying out a method of *successive approximations*. Each cycle is called an *iteration*.

$$[Ca^{2+}]_2(0.628)[Ca^{2+}]_2(0.606) = 2.4 \times 10^{-5} \Rightarrow [Ca^{2+}]_2 = 7.9 \times 10^{-3}\ M$$

$$\Rightarrow \mu = 0.032\ M$$

Repeating this process gives the following results:

| Iteration | $\gamma_{Ca^{2+}}$ | $\gamma_{SO_4^{2-}}$ | $[Ca^{2+}]$ (M) | $\mu$ (M) |
|---|---|---|---|---|
| 1 | 1 | 1 | 0.004 9 | 0.020 |
| 2 | 0.628 | 0.606 | 0.007 9 | 0.032 |
| 3 | 0.570 | 0.542 | 0.008 8 | 0.035 |
| 4 | 0.556 | 0.526 | 0.009 1 | 0.036 |
| 5 | 0.551 | 0.520 | 0.009 2 | 0.037 |
| 6 | 0.547 | 0.515 | 0.009 2 | 0.037 |

Total concentration of dissolved sulfate

$= [SO_4^{2-}] + [CaSO_4(aq)]$

$= 0.009\ 2 + 0.005\ 0 = 0.014\ 2$ M

which is not too far from the measured value in Figure 6-1. With the extended Debye-Hückel equation, instead of interpolation, total dissolved sulfate $= 0.014\ 7$ M.

The 6th iteration gives the same concentration as the 5th, so we have reached a constant answer.

With the advice of the Good Chemist, we simplified the problem tremendously. Now we need to know if her advice was good. From $[Ca^{2+}] = [SO_4^{2-}] = 0.009\ 2$ M and $[H^+] = [OH^-] = 1.0 \times 10^{-7}$ M, we can estimate $[CaOH^+]$ and $[HSO_4^-]$ to see if they are negligible. We are just looking for orders of magnitude, so we ignore activity coefficients. From Equation 8-25, we have

$$\frac{[CaOH^+][H^+]}{[Ca^{2+}]} = 2.0 \times 10^{-13} \qquad (8\text{-}25)$$

$$[CaOH^+] \approx \frac{(2.0 \times 10^{-13})[Ca^{2+}]}{[H^+]} = \frac{(2.0 \times 10^{-13})[0.009\ 2]}{[1.0 \times 10^{-7}]} = 2 \times 10^{-8}\ M$$

From Equation 8-26, we estimate

$$[HSO_4^-] \approx \frac{(9.8 \times 10^{-13})[SO_4^{2-}]}{[OH^-]} = \frac{(9.8 \times 10^{-13})[0.009\ 2]}{[1.0 \times 10^{-7}]} = 9 \times 10^{-8}\ M$$

Both $[CaOH^+]$ and $[HSO_4^-]$ are $\sim 10^5$ times less than $[Ca^{2+}]$ and $[SO_4^{2-}]$, so it was reasonable to neglect $[CaOH^+]$ and $[HSO_4^-]$ in the charge and mass balances.

## We Will Usually Omit Activity Coefficients

Although it is proper to write equilibrium constants in terms of activities, the complexity of manipulating activity coefficients is a nuisance. Most of the time, we will omit activity coefficients unless there is a particular point to be made. Occasional problems will remind you how to use activities.

## Solubility of Magnesium Hydroxide

Let's find the concentrations of species in a saturated solution of $Mg(OH)_2$, given the following chemistry. For simplicity, we ignore activity coefficients.

$$Mg(OH)_2(s) \xrightleftharpoons{K_{sp}} Mg^{2+} + 2OH^- \qquad K_{sp} = [Mg^{2+}][OH^-]^2 = 7.1 \times 10^{-12} \quad (8\text{-}30)$$

$$Mg^{2+} + OH^- \xrightleftharpoons{K_1} MgOH^+ \qquad K_1 = \frac{[MgOH^+]}{[Mg^{2+}][OH^-]} = 3.8 \times 10^2 \quad (8\text{-}31)$$

$$H_2O \xrightleftharpoons{K_w} H^+ + OH^- \qquad K_w = [H^+][OH^-] = 1.0 \times 10^{-14} \quad (8\text{-}32)$$

**Step 1** Pertinent reactions are listed above.
**Step 2** Charge balance: $2[Mg^{2+}] + [MgOH^+] + [H^+] = [OH^-]$  (8-33)
**Step 3** Mass balance. This is a little tricky. From Reaction 8-30, we could say that the concentrations of all species containing $OH^-$ equal two times the concentrations of all magnesium species. However, Reaction 8-32 also creates 1 $OH^-$ for each $H^+$. The mass balance accounts for both sources of $OH^-$:

$$\underbrace{[OH^-] + [MgOH^+]}_{\text{Species containing }OH^-} = 2\underbrace{\{[Mg^{2+}] + [MgOH^+]\}}_{\text{Species containing }Mg^{2+}} + [H^+] \qquad (8\text{-}34)$$

After all this work, Equation 8-34 is equivalent to Equation 8-33.
**Step 4** Equilibrium constant expressions are in Equations 8-30 through 8-32.
**Step 5** Count equations and unknowns. We have four equations (8-30 to 8-33) and four unknowns: $[Mg^{2+}]$, $[MgOH^+]$, $[H^+]$, and $[OH^-]$.
**Step 6** Solve.

Before hitting the algebra, we can make one simplification. The solution must be very basic because we made it from $Mg(OH)_2$. In basic solution, $[OH^-] \gg [H^+]$, so we can neglect $[H^+]$ on the left side of Equation 8-33 in comparison with $[OH^-]$ on the right side. The charge balance simplifies to

$$2[Mg^{2+}] + [MgOH^+] = [OH^-] \qquad (8\text{-}35)$$

From the $K_1$ expression 8-31, we write $[MgOH^+] = K_1[Mg^{2+}][OH^-]$. Substituting this expression for $[MgOH^+]$ into Equation 8-35 gives

$$2[Mg^{2+}] + K_1[Mg^{2+}][OH^-] = [OH^-]$$

which we solve for $[Mg^{2+}]$:

$$[Mg^{2+}] = \frac{[OH^-]}{2 + K_1[OH^-]}$$

Figure 8-7 Spreadsheet for solving Equation 8-36.

| | A | B | C | D |
|---|---|---|---|---|
| 1 | Mg(OH)₂ Solubility | | | |
| 2 | | | | |
| 3 | $K_{sp} =$ | | $[OH^-]_{guess} =$ | $[OH^-]^3/(2 + K_1[OH]) =$ |
| 4 | 7.1E-12 | | 0.0002459 | 7.1000E-12 |
| 5 | $K_1 =$ | | | |
| 6 | 3.8E+02 | | $[Mg^{2+}] =$ | $[MgOH^+] =$ |
| 7 | | | 0.0001174 | 0.0000110 |
| 8 | | | | |
| 9 | D4 = C4^3/(2+A6*C4) | | | |
| 10 | C7 = A4/C4^2 | | | |
| 11 | D8 = A6*C7*C4 | | | |

Substituting this expression for $[Mg^{2+}]$ into the solubility product reduces the equation to a single variable:

$$K_{sp} = [Mg^{2+}][OH^-]^2 = \left(\frac{[OH^-]}{2 + K_1[OH^-]}\right)[OH^-]^2 = \frac{[OH^-]^3}{2 + K_1[OH^-]} \quad (8\text{-}36)$$

We are down to solving the ugly Equation 8-36 for $[OH^-]$. Just when the world looks grim again, the Good Chemist says, "Use a spreadsheet to vary $[OH^-]$ until Equation 8-36 is satisfied." We do this in Figure 8-7, where we *guess* a value of $[OH^-]$ in cell C4 and evaluate the right side of Equation 8-36 in cell D4. When we have guessed the correct value of $[OH^-]$, cell D4 is equal to $K_{sp}$. A better procedure is to use Excel GOAL SEEK, described at the end of Section 6-8, to vary cell C4 until cell D4 is equal to $K_{sp}$. (Before using GOAL SEEK, select OPTIONS from the TOOLS menu. Select the Calculations tab and set Maximum change to 1e-24.) Final results in the spreadsheet are $[OH^-] = 2.5 \times 10^{-4}$ M, $[Mg^{2+}] = 1.2 \times 10^{-4}$ M, and $[MgOH^+] = 1.1 \times 10^{-5}$ M. We also find pH $= -\log[H^+] = -\log(K_w/[OH^-]) = 10.39$.

The final result confirms the approximation we made.

$[H^+] = K_w/[OH^-] = 4.1 \times 10^{-11}$ M $\ll [OH^-]$

## Terms to Understand

| | | |
|---|---|---|
| activity | extended Debye-Hückel | ionic strength |
| activity coefficient | equation | mass balance |
| charge balance | ionic atmosphere | pH |

## Summary

The thermodynamic equilibrium constant for the reaction $aA + bB \rightleftharpoons cC + dD$ is $K = \mathcal{A}_C^c \mathcal{A}_D^d/(\mathcal{A}_A^a \mathcal{A}_B^b)$, where $\mathcal{A}_i$ is the activity of the $i$th species. The activity is the product of the concentration ($c$) and the activity coefficient ($\gamma$): $\mathcal{A}_i = c_i\gamma_i$. For nonionic compounds and gases, $\gamma_i \approx 1$. For ionic species, the activity coefficient depends on the ionic strength, defined as $\mu = \frac{1}{2}\Sigma c_i z_i^2$, where $z_i$ is the charge of an ion. The activity coefficient decreases as ionic strength increases, at least for low ionic strengths ($\leq 0.1$ M). Dissociation of ionic compounds increases with ionic strength because the ionic atmosphere of each ion diminishes the attraction of ions for one another. You should be able to estimate activity coefficients by interpolation in Table 8-1. pH is defined in terms of the activity of $H^+$: pH $= -\log \mathcal{A}_{H^+} = -\log[H^+]\gamma_{H^+}$.

In the systematic treatment of equilibrium, we write pertinent equilibrium expressions, as well as the charge and mass balances. The charge balance states that the sum of all positive charges in solution equals the sum of all negative charges. The mass balance states that the moles of all forms of an element in solution must equal the moles of that element delivered to the solution. We make certain that we have as many equations as unknowns and then solve for the concentrations by using algebra, approximations, spreadsheets, magic, or anything else.

## Exercises

**8-A.** Assuming complete dissociation of the salts, calculate the ionic strength of (a) 0.2 mM $KNO_3$; (b) 0.2 mM $Cs_2CrO_4$; (c) 0.2 mM $MgCl_2$ plus 0.3 mM $AlCl_3$.

**8-B.** Find the activity (not the activity coefficient) of the $(C_3H_7)_4N^+$ (tetrapropylammonium) ion in a solution containing 0.005 0 M $(C_3H_7)_4N^+Br^-$ plus 0.005 0 M $(CH_3)_4N^+Cl^-$.

**8-C.** Using activities, find $[Ag^+]$ in 0.060 M KSCN saturated with $AgSCN(s)$.

**8-D.** Using activities, calculate the pH and concentration of $H^+$ in 0.050 M LiBr at 25°C.

**8-E.** A 40.0-mL solution of 0.040 0 M $Hg_2(NO_3)_2$ was titrated with 60.0 mL of 0.100 M KI to precipitate $Hg_2I_2$ ($K_{sp} = 4.6 \times 10^{-29}$).
(a) What volume of KI is needed to reach the equivalence point?
(b) Calculate the ionic strength of the solution when 60.0 mL of KI have been added.
(c) Using activities, calculate $pHg_2^{2+}$ ($= -\log \mathcal{A}_{Hg_2^{2+}}$) for part (b).

**8-F.** (a) Write the mass balance for $CaCl_2$ in water if the species are $Ca^{2+}$ and $Cl^-$.
(b) Write the mass balance if the species are $Ca^{2+}$, $Cl^-$, $CaCl^+$, and $CaOH^+$.
(c) Write the charge balance for part (b).

**8-G.** Write the charge and mass balances for dissolving $CaF_2$ in water if the reactions are

$$CaF_2(s) \rightleftharpoons Ca^{2+} + 2F^-$$
$$Ca^{2+} + H_2O \rightleftharpoons CaOH^+ + H^+$$
$$Ca^{2+} + F^- \rightleftharpoons CaF^+$$
$$CaF_2(s) \rightleftharpoons CaF_2(aq)$$
$$F^- + H^+ \rightleftharpoons HF(aq)$$
$$HF(aq) + F^- \rightleftharpoons HF_2^-$$

**8-H.** Write charge and mass balances for aqueous $Ca_3(PO_4)_2$ if the species are $Ca^{2+}$, $CaOH^+$, $CaPO_4^-$, $PO_4^{3-}$, $HPO_4^{2-}$, $H_2PO_4^-$, and $H_3PO_4$.

---

## Problems

### Activity Coefficients

**8-1.** Explain why the solubility of an ionic compound increases as the ionic strength of the solution increases (at least up to $\sim 0.5$ M).

**8-2.** Which statements are true: In the ionic strength range 0–0.1 M, activity coefficients decrease with **(a)** increasing ionic strength; **(b)** increasing ionic charge; **(c)** decreasing hydrated radius?

**8-3.** Calculate the ionic strength of **(a)** 0.008 7 M KOH and **(b)** 0.000 2 M $La(IO_3)_3$ (assuming complete dissociation at this low concentration and no hydrolysis reaction to make $LaOH^{2+}$).

**8-4.** Find the activity coefficient of each ion at the indicated ionic strength:

(a) $SO_4^{2-}$          ($\mu = 0.01$ M)
(b) $Sc^{3+}$            ($\mu = 0.005$ M)
(c) $Eu^{3+}$            ($\mu = 0.1$ M)
(d) $(CH_3CH_2)_3NH^+$ ($\mu = 0.05$ M)

**8-5.** Interpolate in Table 8-1 to find the activity coefficient of $H^+$ when $\mu = 0.030$ M.

**8-6.** Calculate the activity coefficient of $Zn^{2+}$ when $\mu = 0.083$ M by using **(a)** Equation 8-6; **(b)** linear interpolation in Table 8-1.

**8-7.** Calculate the activity coefficient of $Al^{3+}$ when $\mu = 0.083$ M by linear interpolation in Table 8-1.

**8-8.** The equilibrium constant for dissolution in water of a nonionic compound, such as diethyl ether $(CH_3CH_2OCH_2CH_3)$, can be written

$$\text{ether}(l) \rightleftharpoons \text{ether}(aq) \qquad K = [\text{ether}(aq)]\gamma_{\text{ether}}$$

At low ionic strength, $\gamma \approx 1$ for neutral compounds. At high ionic strength, most neutral molecules can be *salted out* of aqueous solution. That is, when a high concentration (typically $> 1$ M) of a salt such as NaCl is added to an aqueous solution, neutral molecules usually become *less* soluble. Does the activity coefficient, $\gamma_{\text{ether}}$, increase or decrease at high ionic strength?

**8-9.** Find $[Hg_2^{2+}]$ in saturated $Hg_2Br_2$ in 0.001 00 M KBr.

**8-10.** Find the concentration of $Ba^{2+}$ in a 0.100 M $(CH_3)_4NIO_3$ solution saturated with $Ba(IO_3)_2$.

**8-11.** Find the activity coefficient of $H^+$ in a solution containing 0.010 M HCl plus 0.040 M $KClO_4$. What is the pH of the solution?

**8-12.** Using activities, calculate the pH of a solution containing 0.010 M NaOH plus 0.012 0 M $LiNO_3$. What would be the pH if you neglected activities?

**8-13.** The temperature-dependent form of the extended Debye-Hückel equation 8-6 is

$$\log \gamma = \frac{(-1.825 \times 10^6)(\varepsilon T)^{-3/2} z^2 \sqrt{\mu}}{1 + \alpha \sqrt{\mu}/(2.00\sqrt{\varepsilon T})}$$

where $\varepsilon$ is the (dimensionless) dielectric constant* of water, $T$ is the temperature (K), $z$ is the charge of the ion, $\mu$ is the ionic strength (mol/L), and $\alpha$ is the ion size parameter (pm). The dependence of $\varepsilon$ on temperature is

$$\varepsilon = 79.755e^{(-4.6 \times 10^{-3})(T-293.15)}$$

Calculate the activity coefficient of $SO_4^{2-}$ at 50.00°C when $\mu = 0.100$ M. Compare your value with the one in Table 8-1.

**8-14.** *Extended Debye-Hückel equation.* Use Equation 8-6 to calculate the activity coefficient $(\gamma)$ as a function of ionic strength $(\mu)$ for $\mu = 0.000\ 1, 0.000\ 3, 0.001, 0.003, 0.01, 0.03,$ and 0.1 M.
**(a)** For an ionic charge of $\pm 1$ and a size $\alpha = 400$ pm, make a table of $\gamma$ $(= 10^{\wedge}(\log \gamma))$ for each value of $\mu$.
**(b)** Do the same for ionic charges of $\pm 2, \pm 3,$ and $\pm 4$.
**(c)** Plot $\gamma$ versus $\log \mu$ to obtain a graph similar to Figure 8-4.

### Systematic Treatment of Equilibrium

**8-15.** State the meaning of the charge and mass balance equations.

**8-16.** Why do activity coefficients not appear in the charge and mass balance equations?

**8-17.** Write a charge balance for a solution containing $H^+$, $OH^-$, $Ca^{2+}$, $HCO_3^-$, $CO_3^{2-}$, $Ca(HCO_3)^+$, $Ca(OH)^+$, $K^+$, and $ClO_4^-$.

**8-18.** Write a charge balance for a solution of $H_2SO_4$ in water if the $H_2SO_4$ ionizes to $HSO_4^-$ and $SO_4^{2-}$.

**8-19.** Write the charge balance for an aqueous solution of arsenic acid, $H_3AsO_4$, in which the acid can dissociate to $H_2AsO_4^-$, $HAsO_4^{2-}$, and $AsO_4^{3-}$. Look up the structure of arsenic acid in Appendix G and write the structure of $HAsO_4^{2-}$.

---

**8-I.**    *(Warning: Long problem!)* Using activities, find the concentrations of the major species in 0.10 M $NaClO_4$ saturated with $Mn(OH)_2$. Take the ionic strength to be 0.10 M and suppose that the ion size of $MnOH^+$ is the same as $Mn^{2+}$. Consider just the following chemistry:

$$Mn(OH)_2(s) \overset{K_{sp}}{\rightleftharpoons} Mn^{2+} + 2OH^- \qquad K_{sp} = 1.6 \times 10^{-13}$$
$$Mn^{2+} + OH^- \overset{K_1}{\rightleftharpoons} MnOH^+ \qquad K_1 = 2.5 \times 10^3$$

---

*The dimensionless dielectric constant, $\varepsilon$, measures how well a solvent can separate oppositely charged ions. The force of attraction (newtons) between ions of charge $q_1$ and $q_2$ (coulombs) separated by distance $r$ (meters) is

$$\text{Force} = -(8.988 \times 10^9)\frac{q_1 q_2}{\varepsilon r^2}$$

The larger the value of $\varepsilon$, the smaller the attraction between ions. Water, with $\varepsilon \approx 80$, separates ions very well. Here are some values of $\varepsilon$: methanol, 33; ethanol, 24; benzene, 2; vacuum and air, 1. Ionic compounds dissolved in less polar solvents than water may exist predominantly as ion pairs, not separate ions.

**8-20.** (a) Write the charge and mass balances for a solution made by dissolving $MgBr_2$ to give $Mg^{2+}$, $Br^-$, $MgBr^+$, and $MgOH^+$.
(b) Modify the mass balance if the solution was made by dissolving 0.2 mol $MgBr_2$ in 1 L.

**8-21.** This problem demonstrates what would happen if charge balance did not exist in a solution. The force between two charges was given in the footnote to Problem 8-13. What is the force between two beakers separated by 1.5 m if one contains 250 mL with $1.0 \times 10^{-6}$ M excess negative charge and the other has 250 mL with $1.0 \times 10^{-6}$ M excess positive charge? There are $9.648 \times 10^4$ coulombs per mole of charge. Convert the force from N into pounds with the factor 0.224 8 pounds/N.

**8-22.** For a 0.1 M aqueous solution of sodium acetate, $Na^+CH_3CO_2^-$, one mass balance is simply $[Na^+] = 0.1$ M. Write a mass balance involving acetate.

**8-23.** Consider the dissolution of the compound $X_2Y_3$, which gives $X_2Y_2^{2+}$, $X_2Y^{4+}$, $X_2Y_3(aq)$, and $Y^{2-}$. Use the mass balance to find an expression for $[Y^{2-}]$ in terms of the other concentrations. Simplify your answer as much as possible.

**8-24.** Write a mass balance for a solution of $Fe_2(SO_4)_3$ if the species are $Fe^{3+}$, $Fe(OH)^{2+}$, $Fe(OH)_2^+$, $Fe_2(OH)_2^{4+}$, $FeSO_4^+$, $SO_4^{2-}$, and $HSO_4^-$.

**8-25.** (a) Following the example of $Mg(OH)_2$ in Section 8-5, write the equations needed to find the solubility of $Ca(OH)_2$. *Include activity coefficients* where appropriate. Equilibrium constants are in Appendixes F and I.
(b) Neglecting activity coefficients, compute the concentrations of all species and compute the solubility of $Ca(OH)_2$ in g/L.

**8-26.** Look up the equilibrium constant for the *ion-pairing* reaction $Zn^{2+} + SO_4^{2-} \rightleftharpoons ZnSO_4(aq)$ in Appendix J.
(a) Use the systematic treatment of equilibrium to find $[Zn^{2+}]$ in 0.010 F $ZnSO_4$. Neglect activity coefficients and any other reactions.
(b) Use the answer from part (a) to compute the ionic strength and activity coefficients of $Zn^{2+}$ and $SO_4^{2-}$. Then repeat the calculation using activity coefficients. Repeat the procedure two more times to find a good estimate of $[Zn^{2+}]$. What percentage is ion paired? What is the ionic strength of the solution?

**8-27.** 🖳 *Finding solubility by iteration.* Use the systematic treatment of equilibrium to find the concentrations of the major species in a saturated aqueous solution of LiF. Consider the following reactions

$$LiF(s) \rightleftharpoons Li^+ + F^- \qquad K_{sp} = [Li^+]\gamma_{Li^+}[F^-]\gamma_{F^-} = 0.001\ 7$$
$$LiF(s) \rightleftharpoons LiF(aq) \qquad K_{ion\ pair} = LiF(aq)\gamma_{LiF(aq)} = 0.002\ 9$$

(a) Initially, set the ionic strength to 0 and solve for all the concentrations. Then compute the ionic strength and activity coefficients and find new concentrations. Use several iterations to home in on the correct solution.

(b) In the systematic treatment, you will find that the calculation simplifies to $[Li^+] = [F^-] = \sqrt{K_{sp}/(\gamma_{Li^+}\gamma_{F^-})}$. Set up the following spreadsheet, in which ionic strength in cell B4 is initially given the value 0. Activity coefficients in cells B6 and B8 are from the extended Debye-Hückel equation. Cell B10 computes $[Li^+] = [F^-] = \sqrt{K_{sp}/(\gamma_{Li^+}\gamma_{F^-})}$. With 0 in cell B4, your spreadsheet should compute 1 in cells B6 and B8 and $[Li^+] = [F^-] = 0.041\ 23$ M in cell B10.

| | A | B | C |
|---|---|---|---|
| 1 | Spreadsheet for iterative LiF solubility computation | | |
| 2 | | | |
| 3 | Size (pm) of $Li^+$ = | Ionic strength = | |
| 4 | 600 | 0.00000 | |
| 5 | Size (pm) of $F^-$ = | Activity coeff $Li^+$ = | |
| 6 | 350 | 1 | |
| 7 | $K_{sp}$ = | Activity coeff $F^-$ = | |
| 8 | 0.0017 | 1 | |
| 9 | | $[Li^+]$ = $[F]$ = | |
| 10 | | 0.04123 | |
| 11 | | | |
| 12 | B6 = 10^((−0.51)*SQRT(B4)/(1+A4*SQRT(B4)/305)) | | |
| 13 | B8 = 10^((−0.51)*SQRT(B4)/(1+A6*SQRT(B4)/305)) | | |
| 14 | B10 = SQRT(A8/(B6*B8)) | | |

Because the ionic strength of a 1:1 electrolyte is equal to the concentration, copy the value 0.041 23 from cell B10 into cell B4. (To transfer a numerical value, rather than a formula, COPY cell B10 and then highlight cell B4. In the EDIT menu, select PASTE SPECIAL and then choose Value. The numerical value from B10 will be pasted into B4.) This procedure gives new activity coefficients in cells B6 and B8 and a new concentration in cell B10. Copy the new concentration from cell B10 into cell B4 and repeat this procedure several times until you have a constant answer.

(c) *Using circular references in Excel.* Take your spreadsheet and enter 0 in cell B4. The value 0.041 23 is computed in cell B10. Ideally, you would like to write "=B10" in cell B4 so that the value from B10 would be copied to B4. Excel gives a "circular reference" error message because cell B10 depends on B4 and cell B4 depends on B10. To get around this problem, go to the TOOLS menu and choose OPTIONS. Select Calculation and choose Iteration. Set the maximum change to 0.000 01. Click OK and Excel merrily iterates between cells B10 and B4 until the two values agree within 0.000 01.

**8-28.** *Heterogeneous equilibria and calcite solubility.* If river water in Box 8-2 is saturated with calcite ($CaCO_3$), $[Ca^{2+}]$ is governed by the following equilibria:

$$CaCO_3(s) \rightleftharpoons Ca^{2+} + CO_3^{2-} \qquad K_{sp} = 4.5 \times 10^{-9}$$
$$CO_2(g) \rightleftharpoons CO_2(aq) \qquad K_{CO_2} = 0.032$$
$$CO_2(aq) + H_2O \rightleftharpoons HCO_3^- + H^+ \qquad K_1 = 4.46 \times 10^{-7}$$
$$HCO_3^- \rightleftharpoons CO_3^{2-} + H^+ \qquad K_2 = 4.69 \times 10^{-11}$$

(a) From these reactions, find the equilibrium constant for the reaction

$$CaCO_3(s) + CO_2(aq) + H_2O \rightleftharpoons Ca^{2+} + 2HCO_3^- \qquad K = ? \quad (A)$$

(b) The mass balance for Reaction A is $[HCO_3^-] = 2[Ca^{2+}]$. Find $[Ca^{2+}]$ (in mol/L and in mg/L) in equilibrium with atmospheric $CO_2$ if $P_{CO_2} = 3.8 \times 10^{-4}$ bar. Locate this point on the line in Box 8-2. [Atmospheric $CO_2$ is increasing so rapidly (Box 20-1) that $P_{CO_2}$ needs to be revised between editions of this book. We are changing our environment—with consequences left to the next generation.]

(c) The concentration of $Ca^{2+}$ in the Don River is 80 mg/L. What effective $P_{CO_2}$ is in equilibrium with this much $Ca^{2+}$? How can the river have this much $CO_2$?

**8-29.** (*Warning:* This is a long problem, best done by five people, each completing one line of the table.) A fluoride ion-selective electrode responds to fluoride activity above $\mathcal{A}_{F^-} \approx 10^{-6}$ in the following manner at 25°C:

$$\text{Potential (mV)} = \text{constant} - 59.16 \log \mathcal{A}_{F^-}$$

The constant depends on the reference electrode used with the fluoride electrode. A 20.00-mL solution containing 0.060 0 M NaF was titrated with 0.020 0 M $La(NO_3)_3$ to precipitate $LaF_3$. Just considering the precipitation reaction, fill in the following table and plot the expected titration curve, assuming that the constant is $-20.0$ mV.

| $La(NO_3)_3$ (mL) | Ionic strength (M) | $[F^-]$ (M) | $\gamma_{F^-}$ | Potential (mV) |
|---|---|---|---|---|
| 0 | 0.060 0 | 0.060 0 | 0.80 | 58.0 |
| 5.00 | | | | |
| 10.00 | | | | |
| 19.00 | | | | |
| 20.00 | | | | |
| 22.00 | | | | |

If you use a spreadsheet for this exercise, compute activity coefficients with the extended Debye-Hückel equation and compute many more points. (You can look up the results of a similar titration to compare with your calculations.[11])

# 9 Monoprotic Acid-Base Equilibria

(*a*) Mouse macrophage engulfs two foreign red blood cells as phagocytosis begins. *[From J. P. Revel, in B. Alberts, D. Bray, J. Lewis, M. Raff, K. Roberts, and J. D. Watson, Molecular Biology of the Cell, 2nd ed. (New York: Garland Publishing, 1989.]* (*b*) Macrophages with ingested 1.6-μm-diameter fluorescent beads. (*c*) Fluorescence image of panel *b*. *[From K. P. McNamara, T. Nguyen, G. Dumitrascu, J. Ji, N. Rosenzweig, and Z. Rosenzweig, "Synthesis, Characterization, and Application of Fluorescence Sensing Lipobeads for Intracellular pH Measurements," Anal. Chem. **2001**, 73, 3240.]*

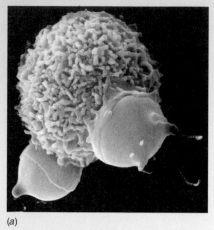
(a)

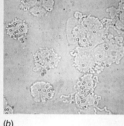
(b)

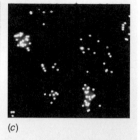
(c)

*Macrophages* are white blood cells that fight infection by ingesting and dissolving foreign cells—a process called *phagocytosis*. The compartment containing the ingested foreign cell merges with compartments called *lysosomes*, which contain digestive enzymes that are most active in acid. Low enzyme activity above pH 7 protects the cell from enzymes that leak into the cell.

One way to measure pH inside the compartment containing the ingested particle and digestive enzymes is to present macrophages with polystyrene beads coated with a lipid membrane to which fluorescent (light-emitting) dyes are covalently bound. Panel *d* shows that fluorescence intensity from the dye fluorescein depends on pH, but fluorescence from tetramethylrhodamine does not. The ratio of emission from the dyes is a measure of pH. Panel *e* shows the fluorescence intensity ratio changing in 3 s as the bead is ingested and the pH around the bead drops from 7.3 to 5.7 to allow digestion to commence.

(*d*) Fluorescence spectra of lipobeads in solutions at pH 5–8. (*e*) pH change during phagocytosis of a single bead by a macrophage. *[From McNamara et al., ibid.]*

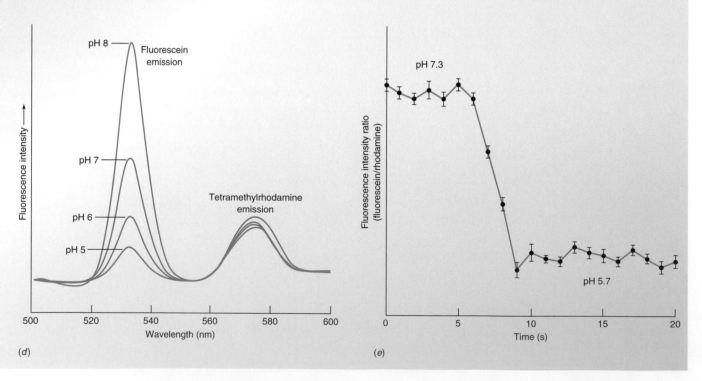

(d)

(e)

cids and bases are essential to virtually every application of chemistry and for the intelligent use of analytical procedures such as chromatography and electrophoresis. It would be difficult to have a meaningful discussion of, say, protein purification or the weathering of rocks without understanding acids and bases. This chapter covers acid-base equilibria and buffers. Chapter 10 treats polyprotic systems involving two or more acidic protons. Nearly every biological macromolecule is polyprotic. Chapter 11 describes acid-base titrations. Now is the time to review fundamentals of acids and bases in Sections 6-5 through 6-7.

# ■■■ 9-1 Strong Acids and Bases

What could be easier than calculating the pH of 0.10 M HBr? HBr is a **strong acid,** so the reaction

$$HBr + H_2O \rightarrow H_3O^+ + Br^-$$

goes to completion, and the concentration of $H_3O^+$ is 0.10 M. It will be our custom to write $H^+$ instead of $H_3O^+$, and we can say

$$pH = -\log[H^+] = -\log(0.10) = 1.00$$

Table 6-2 gave a list of strong acids and bases that you must memorize.

Equilibrium constants for the reaction[1]

$$HX + H_2O \rightleftharpoons H_3O^+ + X^-$$

| | |
|---|---|
| HCl | $K_a = 10^{3.9}$ |
| HBr | $K_a = 10^{5.8}$ |
| HI | $K_a = 10^{10.4}$ |
| $HNO_3$ | $K_a = 10^{1.4}$ |

$HNO_3$ is discussed in Box 9-1.

### Example Activity Coefficient in a Strong-Acid Calculation

Calculate the pH of 0.10 M HBr, using activity coefficients.

**Solution** The ionic strength of 0.10 M HBr is 0.10 M, at which the activity coefficient of $H^+$ is 0.83 (Table 8-1). Remember that pH is $-\log \mathcal{A}_{H^+}$, not $-\log[H^+]$:

$$pH = -\log[H^+]\gamma_{H^+} = -\log(0.10)(0.83) = 1.08$$

---

## Box 9-1 Concentrated $HNO_3$ Is Only Slightly Dissociated[2]

Strong acids in dilute solution are essentially completely dissociated. As concentration increases, the degree of dissociation decreases. The figure shows a *Raman spectrum* of solutions of nitric acid of increasing concentration. The spectrum measures scattering of light whose energy corresponds to vibrational energies of molecules. The sharp signal at 1 049 cm$^{-1}$ in the spectrum of 5.1 M $NaNO_3$ is characteristic of free $NO_3^-$ anion.

A 10.0 M $HNO_3$ solution has a strong signal at 1 049 cm$^{-1}$, arising from $NO_3^-$ from dissociated acid. Bands denoted by asterisks arise from *undissociated* $HNO_3$. As concentration increases, the 1 049 cm$^{-1}$ signal disappears and signals attributed to undissociated $HNO_3$ increase. The graph shows the fraction of dissociation deduced from spectroscopic measurements. It is instructive to realize that, in 20 M $HNO_3$, there are fewer $H_2O$ molecules than there are molecules of $HNO_3$. Dissociation decreases because there is not enough solvent to stabilize the free ions.

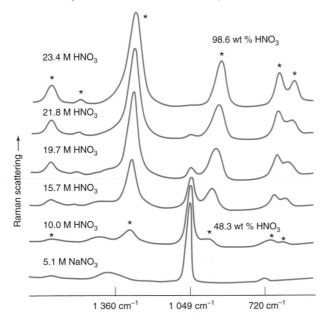

Raman spectrum of aqueous $HNO_3$ at 25°C. Signals at 1 360, 1 049, and 720 cm$^{-1}$ arise from $NO_3^-$ anion. Signals denoted by asterisks are from undissociated $HNO_3$. The wavenumber unit, cm$^{-1}$, is 1/wavelength.

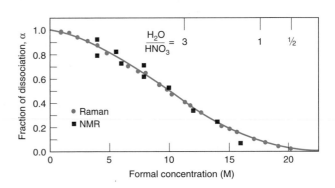

| Temperature (°C) | Acid dissociation constant ($K_a$) |
|---|---|
| 0 | 46.8 |
| 25 | 26.8 |
| 50 | 14.9 |

Now that we have reminded you of activity coefficients, you can breathe a sigh of relief because we will neglect activity coefficients unless there is a specific point to be made.

How do we calculate the pH of 0.10 M KOH? KOH is a **strong base** (completely dissociated), so $[OH^-] = 0.10$ M. Using $K_w = [H^+][OH^-]$, we write

From [OH⁻], you can always find [H⁺]:

$$[H^+] = \frac{K_w}{[OH^-]}$$

$$[H^+] = \frac{K_w}{[OH^-]} = \frac{1.0 \times 10^{-14}}{0.10} = 1.0 \times 10^{-13} \text{ M}$$

$$pH = -\log[H^+] = 13.00$$

Finding the pH of other concentrations of KOH is pretty trivial:

| $[OH^-]$ (M) | $[H^+]$ (M) | pH |
|---|---|---|
| $10^{-3.00}$ | $10^{-11.00}$ | 11.00 |
| $10^{-4.00}$ | $10^{-10.00}$ | 10.00 |
| $10^{-5.00}$ | $10^{-9.00}$ | 9.00 |

A generally useful relation is

pH + pOH = p$K_w$ = 14.00 at 25°C

The temperature dependence of $K_w$ was given in Table 6-1.

*Relation between pH and pOH:*

$$pH + pOH = -\log K_w = 14.00 \text{ at } 25°C \quad (9\text{-}1)$$

## The Dilemma

Well, life seems simple enough so far. Now we ask, "What is the pH of $1.0 \times 10^{-8}$ M KOH?" Applying our usual reasoning, we calculate

Adding base to water cannot *lower* the pH. (Lower pH is more *acidic*.) There must be something wrong.

$$[H^+] = K_w/(1.0 \times 10^{-8}) = 1.0 \times 10^{-6} \text{ M} \Rightarrow pH = 6.00$$

But how can the base KOH produce an acidic solution (pH < 7) when dissolved in pure water? It's impossible.

## The Cure

Clearly, there is something wrong with our calculation. In particular, we have not considered the contribution of $OH^-$ from the ionization of water. In pure water, $[OH^-] = 1.0 \times 10^{-7}$ M, which is greater than the amount of KOH added to the solution. To handle this problem, we resort to the systematic treatment of equilibrium.

**Step 1** *Pertinent reactions.* The only one is $H_2O \overset{K_w}{\rightleftharpoons} H^+ + OH^-$.

**Step 2** *Charge balance.* The species in solution are $K^+$, $OH^-$, and $H^+$. So,

$$[K^+] + [H^+] = [OH^-] \quad (9\text{-}2)$$

**Step 3** *Mass balance.* All $K^+$ comes from the KOH, so $[K^+] = 1.0 \times 10^{-8}$ M.

If we had been using activities, step 4 is the only point at which activity coefficients would have entered.

**Step 4** *Equilibrium constant expression.* $K_w = [H^+][OH^-] = 1.0 \times 10^{-14}$.

**Step 5** *Count.* There are three equations and three unknowns ($[H^+]$, $[OH^-]$, $[K^+]$), so we have enough information to solve the problem.

**Step 6** *Solve.* Because we are seeking the pH, let's set $[H^+] = x$. Writing $[K^+] = 1.0 \times 10^{-8}$ M in Equation 9-2, we get

$$[OH^-] = [K^+] + [H^+] = 1.0 \times 10^{-8} + x$$

Using this expression for $[OH^-]$ in the $K_w$ equilibrium enables us to solve the problem:

$$[H^+][OH^-] = K_w$$

$$(x)(1.0 \times 10^{-8} + x) = 1.0 \times 10^{-14}$$

$$x^2 + (1.0 \times 10^{-8})x - (1.0 \times 10^{-14}) = 0$$

*Solution of a quadratic equation:*

$$ax^2 + bx + c = 0$$

$$x = \frac{-b \pm \sqrt{b^2 - 4ac}}{2a}$$

Retain all digits in your calculator because $b^2$ is sometimes nearly equal to $4ac$. If you round off before computing $b^2 - 4ac$, your answer may be garbage.

$$x = \frac{-1.0 \times 10^{-8} \pm \sqrt{(1.0 \times 10^{-8})^2 - 4(1)(-1.0 \times 10^{-14})}}{2(1)}$$

$$= 9.6 \times 10^{-8} \text{ M, or } -1.1 \times 10^{-7} \text{ M}$$

Rejecting the negative concentration, we conclude that

$$[H^+] = 9.6 \times 10^{-8} \text{ M} \Rightarrow pH = -\log[H^+] = 7.02$$

This pH is eminently reasonable, because $10^{-8}$ M KOH should be very slightly basic.

Figure 9-1 shows the pH calculated for different concentrations of strong base or strong acid in water. There are three regions:

1. When the concentration is "high" ($\geq 10^{-6}$ M), pH is calculated by just considering the added $H^+$ or $OH^-$. That is, the pH of $10^{-5.00}$ M KOH *is* 9.00.
2. When the concentration is "low" ($\leq 10^{-8}$ M), the pH is 7.00. We have not added enough acid or base to change the pH of the water itself.
3. At intermediate concentrations of $10^{-6}$ to $10^{-8}$ M, the effects of water ionization and the added acid or base are comparable. Only in this region is a systematic equilibrium calculation necessary.

Region 1 is the only practical case. Unless you were to protect $10^{-7}$ M KOH from the air, the pH would be overwhelmingly governed by dissolved $CO_2$, not KOH. To obtain a pH near 7, we use a buffer, not a strong acid or base.

## Water Almost Never Produces $10^{-7}$ M $H^+$ and $10^{-7}$ M $OH^-$

The misconception that dissociation of water always produces $10^{-7}$ M $H^+$ and $10^{-7}$ M $OH^-$ is true *only* in pure water with no added acid or base. Any acid or base suppresses water ionization, as predicted by Le Châtelier's principle. In $10^{-4}$ M HBr, for example, the pH is 4. The concentration of $OH^-$ is $[OH^-] = K_w/[H^+] = 10^{-10}$ M. But the only source of $[OH^-]$ is the dissociation of water. If water produces only $10^{-10}$ M $OH^-$, it must also produce only $10^{-10}$ M $H^+$ because it makes one $H^+$ for every $OH^-$. In $10^{-4}$ M HBr solution, water dissociation produces only $10^{-10}$ M $OH^-$ and $10^{-10}$ M $H^+$.

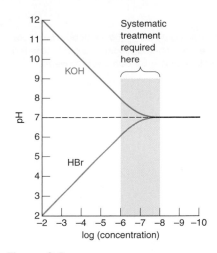

**Figure 9-1** Calculated pH as a function of concentration of strong acid or strong base in water.

*Question* What concentrations of $H^+$ and $OH^-$ are produced by $H_2O$ dissociation in 0.01 M NaOH?

## ■ ■ ■ 9-2 Weak Acids and Bases

Let's review the meaning of the **acid dissociation constant,** $K_a$, for the acid HA:

*Weak-acid equilibrium:*
$$HA \xrightleftharpoons{K_a} H^+ + A^- \qquad K_a = \frac{[H^+][A^-]}{[HA]} \qquad (9\text{-}3)$$

Of course, you know that $K_a$ should really be expressed in terms of activities, not concentrations: $K_a = \mathcal{A}_{H^+}\mathcal{A}_{A^-}/\mathcal{A}_{HA}$.

A **weak acid** is one that is not completely dissociated. That is, Reaction 9-3 does not go to completion. For a base, B, the **base hydrolysis constant,** $K_b$, is defined by the reaction

*Hydrolysis* refers to a reaction with water.

*Weak-base equilibrium:*
$$B + H_2O \xrightleftharpoons{K_b} BH^+ + OH^- \qquad K_b = \frac{[BH^+][OH^-]}{[B]} \qquad (9\text{-}4)$$

A **weak base** is one for which Reaction 9-4 does not go to completion.

**p$K$** is the negative logarithm of an equilibrium constant:

$$pK_w = -\log K_w$$
$$pK_a = -\log K_a$$
$$pK_b = -\log K_b$$

As $K$ increases, p$K$ decreases, and vice versa. Comparing formic and benzoic acids, we see that formic acid is stronger, with a larger $K_a$ and smaller p$K_a$, than benzoic acid.

$$\underset{\text{Formic acid}}{\overset{\displaystyle O \atop \displaystyle \|}{HCOH}} \rightleftharpoons H^+ + \underset{\text{Formate}}{\overset{\displaystyle O \atop \displaystyle \|}{HCO^-}}$$

$$K_a = 1.80 \times 10^{-4}$$
$$\mathbf{p}K_\mathbf{a} = 3.744$$

$$\underset{\text{Benzoic acid}}{\bigcirc\!\!-\!\!\overset{\displaystyle O \atop \displaystyle \|}{COH}} \rightleftharpoons \underset{\text{Benzoate}}{\bigcirc\!\!-\!\!\overset{\displaystyle O \atop \displaystyle \|}{CO^-}} + H^+$$

$$K_a = 6.28 \times 10^{-5}$$
$$\mathbf{p}K_\mathbf{a} = 4.202$$

As $K_a$ increases, p$K_a$ decreases. The smaller p$K_a$ is, the stronger the acid is.

The acid HA and its corresponding base, $A^-$, are said to be a **conjugate acid-base pair,** because they are related by the gain or loss of a proton. Similarly, B and $BH^+$ are a conjugate pair. The important relation between $K_a$ and $K_b$ for a conjugate acid-base pair, derived in Equation 6-35, is

HA and $A^-$ are a *conjugate acid-base pair.* B and $BH^+$ also are conjugate.

*Relation between $K_a$ and $K_b$ for conjugate pair:*
$$K_a \cdot K_b = K_w \qquad (9\text{-}5)$$

The conjugate base of a weak acid is a weak base. The conjugate acid of a weak base is a weak acid. *Weak is conjugate to weak.*

*The conjugate base of a weak acid is a weak base. The conjugate acid of a weak base is a weak acid.* Consider a weak acid, HA, with $K_a = 10^{-4}$. The conjugate base, $A^-$, has $K_b = K_w/K_a = 10^{-10}$. That is, if HA is a weak acid, $A^-$ is a weak base. If $K_a$ were $10^{-5}$, then $K_b$ would be $10^{-9}$. As HA becomes a weaker acid, $A^-$ becomes a stronger base (but never a strong base). Conversely, the greater the acid strength of HA, the less the base strength of $A^-$. However, if either $A^-$ or HA is weak, so is its conjugate. If HA is strong (such as HCl), its conjugate base ($Cl^-$) is *so* weak that it is not a base at all in water.

## Using Appendix G

Appendix G lists acid dissociation constants. Each compound is shown in its *fully protonated form.* Diethylamine, for example, is shown as $(CH_3CH_2)_2NH_2^+$, which is really the diethylammonium ion. The value of $K_a$ ($1.0 \times 10^{-11}$) given for diethylamine is actually $K_a$ for the diethylammonium ion. To find $K_b$ for diethylamine, we write $K_b = K_w/K_a = 1.0 \times 10^{-14}/ 1.0 \times 10^{-11} = 1.0 \times 10^{-3}$.

For polyprotic acids and bases, several $K_a$ values are given. Pyridoxal phosphate is given in its fully protonated form as follows:[3]

Pyridoxyl phosphate is a derivative of vitamin $B_6$.

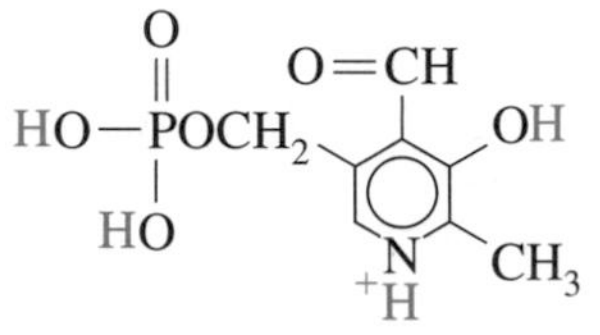

| | $pK_a$ | $K_a$ |
|---|---|---|
| | 1.4 (POH) | 0.04 |
| | 3.44 (OH) | $3.6 \times 10^{-4}$ |
| | 6.01 (POH) | $9.8 \times 10^{-7}$ |
| | 8.45 (NH) | $3.5 \times 10^{-9}$ |

$pK_1$ (1.4) is for dissociation of one of the phosphate protons, and $pK_2$ (3.44) is for the hydroxyl proton. The third most acidic proton is the other phosphate proton, for which $pK_3 = 6.01$, and the $NH^+$ group is the least acidic ($pK_4 = 8.45$).

Species drawn in Appendix G are fully protonated. If a structure in Appendix G has a charge other than 0, it is not the structure that belongs with the name in the appendix. *Names refer to neutral molecules.* The neutral molecule pyridoxal phosphate is not the species drawn above, which has a $+1$ charge. The neutral molecule pyridoxal phosphate is

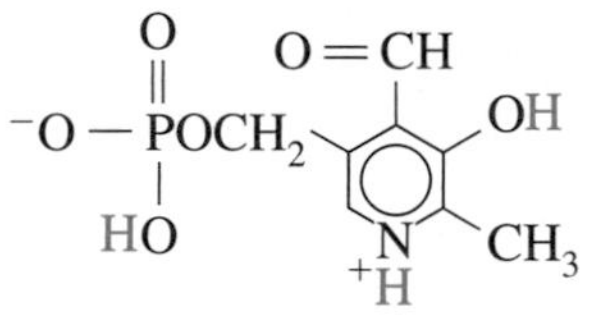

We took away a POH proton, not the $NH^+$ proton, because POH is the most acidic group in the molecule ($pK_a = 1.4$).

As another example, consider the molecule piperazine:

Structure shown for piperazine in Appendix G

Actual structure of piperazine, which *must be neutral*

## ▪▪▪▪ 9-3 Weak-Acid Equilibria

The acetyl $\left( CH_3\overset{O}{\overset{\|}{C}} - \right)$ derivative of *o*-hydroxybenzoic acid is the active ingredient in aspirin.

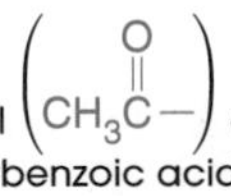

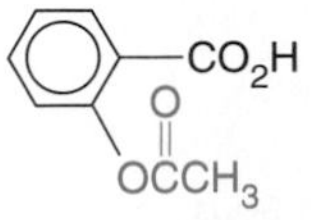

Acetylsalicylic acid

Let's compare the ionization of *ortho*- and *para*-hydroxybenzoic acids:

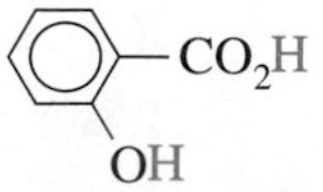

*o*-Hydroxybenzoic acid
(salicylic acid)
$pK_a = 2.97$

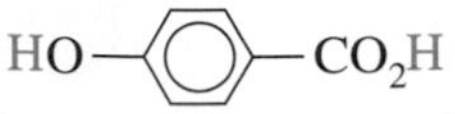

*p*-Hydroxybenzoic acid
$pK_a = 4.54$

Why is the *ortho* isomer 30 times more acidic than the *para* isomer? Any effect that increases the stability of the product of a reaction drives the reaction forward. In the *ortho* isomer, the product of the acid dissociation reaction can form a strong, internal hydrogen bond.

The *para* isomer cannot form such a bond because the —OH and —CO$_2^-$ groups are too far apart. By stabilizing the product, the internal hydrogen bond is thought to make *o*-hydroxy-benzoic acid more acidic than *p*-hydroxybenzoic acid.

## A Typical Weak-Acid Problem

The problem is to find the pH of a solution of the weak acid HA, given the formal concentration of HA and the value of $K_a$.[4] Let's call the formal concentration F and use the systematic treatment of equilibrium:

Reactions:  $\qquad$ $HA \overset{K_a}{\rightleftharpoons} H^+ + A^- \qquad H_2O \overset{K_w}{\rightleftharpoons} H^+ + OH^-$

Charge balance: $\qquad$ $[H^+] = [A^-] + [OH^-]$ $\qquad$ (9-6)

Mass balance: $\qquad$ $F = [A^-] + [HA]$ $\qquad$ (9-7)

Equilibrium expressions: $\qquad$ $K_a = \dfrac{[H^+][A^-]}{[HA]}$ $\qquad$ (9-8)

$\qquad\qquad\qquad\qquad$ $K_w = [H^+][OH^-]$

> *Formal concentration* is the total number of moles of a compound dissolved in a liter. The formal concentration of a weak acid is the total amount of HA placed in the solution, regardless of the fact that some has changed into A$^-$.

There are four equations and four unknowns ($[A^-]$, $[HA]$, $[H^+]$, $[OH^-]$), so the problem is solved if we can just do the algebra.

But it's not so easy to solve these simultaneous equations. If you combine them, you will discover that a cubic equation results. At this point, the Good Chemist rides down again from the mountain on her white stallion to rescue us and cries, "Wait! There is no reason to solve a cubic equation. We can make an excellent, simplifying approximation. (Besides, I have trouble solving cubic equations.)"

For any respectable weak acid, $[H^+]$ from HA will be much greater than $[H^+]$ from H$_2$O. When HA dissociates, it produces A$^-$. When H$_2$O dissociates, it produces OH$^-$. If dissociation of HA is much greater than H$_2$O dissociation, then $[A^-] \gg [OH^-]$, and Equation 9-6 reduces to

$$[H^+] \approx [A^-] \qquad (9\text{-}9)$$

To solve the problem, first set $[H^+] = x$. Equation 9-9 says that $[A^-]$ also is equal to $x$. Equation 9-7 says that $[HA] = F - [A^-] = F - x$. Putting these expressions into Equation 9-8 gives

$$K_a = \frac{[H^+][A^-]}{[HA]} = \frac{(x)(x)}{F - x}$$

> $x = [H^+]$ in weak-acid problems.

Setting $F = 0.050\ 0$ M and $K_a = 1.0_7 \times 10^{-3}$ for *o*-hydroxybenzoic acid, we can solve the equation, because it is just a quadratic equation.

$$\frac{x^2}{0.050\ 0 - x} = 1.0_7 \times 10^{-3}$$

$$x^2 + (1.07 \times 10^{-3})x - 5.35 \times 10^{-5} = 0$$

$$x = 6.8_0 \times 10^{-3} \text{ (negative root rejected)}$$

$$[H^+] = [A^-] = x = 6.8_0 \times 10^{-3} \text{ M}$$

$$[HA] = F - x = 0.043_2 \text{ M}$$

$$pH = -\log x = 2.17$$

Was the approximation $[H^+] \approx [A^-]$ justified? The calculated pH is 2.17, which means that $[OH^-] = K_w/[H^+] = 1.5 \times 10^{-12}$ M.

$$[A^-] \text{ (from HA dissociation)} = 6.8 \times 10^{-3} \text{ M}$$

$$\Rightarrow [H^+] \text{ from HA dissociation} = 6.8 \times 10^{-3} \text{ M}$$

$$[OH^-] \text{ (from H}_2\text{O dissociation)} = 1.5 \times 10^{-12} \text{ M}$$

$$\Rightarrow [H^+] \text{ from H}_2\text{O dissociation} = 1.5 \times 10^{-12} \text{ M}$$

The assumption that H$^+$ is derived mainly from HA is excellent.

> For uniformity, we will usually express pH to the 0.01 decimal place, regardless of the number of places justified by significant figures.

> In a solution of a weak acid, H$^+$ is derived almost entirely from HA, not from H$_2$O.

## Fraction of Dissociation

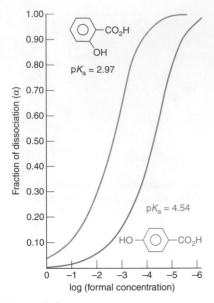

**Figure 9-2** Fraction of dissociation of a weak electrolyte increases as electrolyte is diluted. The stronger acid is more dissociated than the weaker acid at all concentrations.

The **fraction of dissociation,** $\alpha$, is defined as the fraction of the acid in the form $A^-$:

*Fraction of dissociation of an acid:*

$$\alpha = \frac{[A^-]}{[A^-] + [HA]} = \frac{x}{x + (F - x)} = \frac{x}{F} \qquad (9\text{-}10)$$

For 0.050 0 M *o*-hydroxybenzoic acid, we find

$$\alpha = \frac{6.8 \times 10^{-3}\ \text{M}}{0.050\ 0\ \text{M}} = 0.14$$

That is, the acid is 14% dissociated at a formal concentration of 0.050 0 M.

The variation of $\alpha$ with formal concentration is shown in Figure 9-2. **Weak electrolytes** (compounds that are only partially dissociated) dissociate more as they are diluted. *o*-Hydroxybenzoic acid is more dissociated than *p*-hydroxybenzoic acid at the same formal concentration because the *ortho* isomer is a stronger acid. Box 9-2 and Demonstration 9-1 illustrate weak-acid properties.

## The Essence of a Weak-Acid Problem

When faced with finding the pH of a weak acid, you should immediately realize that $[H^+] = [A^-] = x$ and proceed to set up and solve the equation

*Equation for weak acids:*

$$\frac{[H^+][A^-]}{[HA]} = \frac{x^2}{F - x} = K_a \qquad (9\text{-}11)$$

where F is the formal concentration of HA. The approximation $[H^+] = [A^-]$ would be poor only if the acid were too dilute or too weak, neither of which constitutes a practical problem.

---

## Box 9-2  Dyeing Fabrics and the Fraction of Dissociation[5]

Cotton fabrics are largely cellulose, a polymer with repeating units of the sugar glucose:

Structure of cellulose. Hydrogen bonding between glucose units helps make the structure rigid.

Dyes are colored molecules that can form covalent bonds to fabric. For example, Procion Brilliant Blue M-R is a dye with a blue *chromophore* (the colored part) attached to a reactive dichlorotriazine ring:

Blue chromophore       Procion Brilliant Blue M-R fabric dye

Oxygen atoms of the $-CH_2OH$ groups on cellulose can replace Cl atoms of the dye to form covalent bonds that fix the dye permanently to the fabric:

Chemically reactive form of cellulose is deprotonated anion

After the fabric has been dyed in cold water, excess dye is removed with a hot wash. During the hot wash, the second Cl group of the dye is replaced by a second cellulose or by water (giving dye$-$OH).

The chemically reactive form of cellulose is the conjugate base:

$$\text{Cellulose}-CH_2OH \xrightleftharpoons{K_a \approx 10^{-15}} \text{cellulose}-CH_2O^- + H^+$$
$$\qquad\qquad ROH \qquad\qquad\qquad\qquad RO^-$$

To promote dissociation of the cellulose$-CH_2OH$ proton, dyeing is carried out in sodium carbonate solution with a pH around 10.6. The fraction of reactive cellulose is given by the fraction of dissociation of the weak acid at pH 10.6:

$$\text{Fraction of dissociation} = \frac{[RO^-]}{[ROH] + [RO^-]} \approx \frac{[RO^-]}{[ROH]}$$

Because the fraction of dissociation of the very weak acid is so small, $[ROH] \gg [RO^-]$ in the denominator, which is therefore approximately just $[ROH]$. The quotient $[RO^-]/[ROH]$ can be calculated from $K_a$ and the pH:

$$K_a = \frac{[RO^-][H^+]}{[ROH]} \Rightarrow \frac{[RO^-]}{[ROH]} = \frac{K_a}{[H^+]} \approx \frac{10^{-15}}{10^{-10.6}}$$
$$= 10^{-4.4} \approx \text{fraction of dissociation}$$

Only about one cellulose$-CH_2OH$ group in $10^4$ is in the reactive form at pH 10.6.

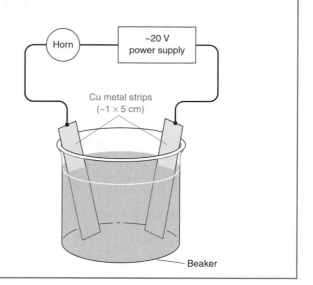
### Example   A Weak-Acid Problem

Find the pH of 0.050 M trimethylammonium chloride.

$$
\left[
\begin{array}{c}
H \\
| \\
N \\
H_3C \cdots\! \overset{\displaystyle |}{} \!\! CH_3 \\
H_3C
\end{array}
\right]^{+} \quad Cl^-
$$

Trimethylammonium chloride

**Solution**   We assume that ammonium halide salts are completely dissociated to give $(CH_3)_3NH^+$ and $Cl^-$.* We then recognize that trimethylammonium ion is a weak acid, being the conjugate acid of trimethylamine, $(CH_3)_3N$, a weak base. $Cl^-$ has no basic or acidic properties and should be ignored. In Appendix G, we find trimethylammonium ion listed as trimethylamine, but drawn as trimethylammonium ion, with $pK_a = 9.799$. So,

Cl⁻ has no acidic or basic properties because it is the conjugate base of the strong acid HCl. If Cl⁻ had appreciable basicity, HCl would not be completely dissociated.

$$
K_a = 10^{-pK_a} = 10^{-9.799} = 1.59 \times 10^{-10}
$$

From here, everything is downhill.

$$
(CH_3)_3NH^+ \xrightarrow{K_a} (CH_3)_3N + H^+
$$
$$
\quad F - x \qquad\qquad x \qquad x
$$

$$
\frac{x^2}{0.050 - x} = 1.59 \times 10^{-10} \tag{9-12}
$$

$$
x = 2.8 \times 10^{-6}\,M \Rightarrow pH = 5.55
$$

---

*$R_4N^+X^-$ salts are not completely dissociated, because there are some *ion pairs*, $R_4N^+X^-(aq)$ (Box 8-1). Equilibrium constants for $R_4N^+ + X^- \rightleftharpoons R_4N^+X^-(aq)$ are given below. For 0.050 F solutions, the fraction of ion pairing is 4% in $(CH_3)_4N^+Br^-$, 7% in $(CH_3CH_2)_4N^+Br^-$, and 9% in $(CH_3CH_2CH_2)_4N^+Br^-$.

| $R_4N^+$ | $X^-$ | $K_{ion\ pair}$ | $R_4N^+$ | $X^-$ | $K_{ion\ pair}$ |
|---|---|---|---|---|---|
| $Me_4N^+$ | $Cl^-$ | 1.1 | $Me_4N^+$ | $I^-$ | 2.0 |
| $Bu_4N^+$ | $Cl^-$ | 2.5 | $Et_4N^+$ | $I^-$ | 2.9 |
| $Me_4N^+$ | $Br^-$ | 1.4 | $Pr_4N^+$ | $I^-$ | 4.6 |
| $Et_4N^+$ | $Br^-$ | 2.4 | $Bu_4N^+$ | $I^-$ | 6.0 |
| $Pr_4N^+$ | $Br^-$ | 3.1 | | | |

$Me = CH_3-$, $Et = CH_3CH_2-$, $Pr = CH_3CH_2CH_2-$, $Bu = CH_3CH_2CH_2CH_2-$

*A handy tip:* Equation 9-11 can always be solved with the quadratic formula. However, an easier method worth trying first is to neglect $x$ in the denominator. If $x$ comes out much smaller than F, then your approximation was good and you need not use the quadratic formula. For Equation 9-12, the approximation works like this:

$$\frac{x^2}{0.050 - x} \approx \frac{x^2}{0.050} = 1.59 \times 10^{-10} \Rightarrow x = \sqrt{(0.050)(1.59 \times 10^{-10})} = 2.8 \times 10^{-6}\,M$$

The approximate solution ($x \approx 2.8 \times 10^{-6}$) is much smaller than the term 0.050 in the denominator of Equation 9-12. Therefore, the approximate solution is fine. A reasonable rule of thumb is to accept the approximation if $x$ comes out to be less than 1% of F.

## ■ ■ ■ 9-4   Weak-Base Equilibria

The treatment of weak bases is almost the same as that of weak acids.

$$B + H_2O \overset{K_b}{\rightleftharpoons} BH^+ + OH^- \qquad K_b = \frac{[BH^+][OH^-]}{[B]}$$

As $K_b$ increases, $pK_b$ decreases and the base becomes stronger.

We suppose that nearly all $OH^-$ comes from the reaction of $B + H_2O$, and little comes from dissociation of $H_2O$. Setting $[OH^-] = x$, we must also set $[BH^+] = x$, because one $BH^+$ is produced for each $OH^-$. Calling the formal concentration of base F ($= [B] + [BH^+]$), we write

$$[B] = F - [BH^+] = F - x$$

Plugging these values into the $K_b$ equilibrium expression, we get

A weak-base problem has the same algebra as a weak-acid problem, except $K = K_b$ and $x = [OH^-]$.

*Equation for weak base:*   $$\frac{[BH^+][OH^-]}{[B]} = \frac{x^2}{F - x} = K_b \qquad (9\text{-}13)$$

which looks a lot like a weak-acid problem, except that now $x = [OH^-]$.

### A Typical Weak-Base Problem

Consider the commonly occurring weak base cocaine.

Cocaine

$$K_b = 2.6 \times 10^{-6}$$

If the formal concentration is 0.037 2 M, the problem is formulated as follows:

$$
\begin{array}{ccccccc}
B & + & H_2O & \rightleftharpoons & BH^+ & + & OH^- \\
0.037\,2 - x & & & & x & & x
\end{array}
$$

$$\frac{x^2}{0.037\,2 - x} = 2.6 \times 10^{-6} \Rightarrow x = 3.1 \times 10^{-4}$$

Because $x = [OH^-]$, we can write

$$[H^+] = K_w/[OH^-] = 1.0 \times 10^{-14}/3.1 \times 10^{-4} = 3.2 \times 10^{-11}$$
$$pH = -\log[H^+] = 10.49$$

This is a reasonable pH for a weak base.

What fraction of cocaine has reacted with water? We can formulate $\alpha$ for a base, called the **fraction of association:**

*Question* What concentration of $OH^-$ is produced by $H_2O$ dissociation in this solution? Was it justified to neglect water dissociation as a source of $OH^-$?

For a base, $\alpha$ is the fraction that has reacted with water.

*Fraction of association of a base:*   $$\alpha = \frac{[BH^+]}{[BH^+] + [B]} = \frac{x}{F} = 0.008\,3 \qquad (9\text{-}14)$$

Only 0.83% of the base has reacted.

166

CHAPTER 9 Monoprotic Acid-Base Equilibria

## Conjugate Acids and Bases—Revisited

Earlier, we noted that **the conjugate base of a weak acid is a weak base,** and **the conjugate acid of a weak base is a weak acid.** We also derived an exceedingly important relation between the equilibrium constants for a conjugate acid-base pair: $K_a \cdot K_b = K_w$.

In Section 9-3, we considered $o$- and $p$-hydroxybenzoic acids, designated HA. Now consider their conjugate bases. For example, the salt sodium $o$-hydroxybenzoate dissolves to give $Na^+$ (which has no acid-base chemistry) and $o$-hydroxybenzoate, which is a weak base.

The acid-base chemistry is the reaction of $o$-hydroxybenzoate with water:

$$
\underset{\substack{A^- (o\text{-hydroxybenzoate}) \\ F - x}}{\text{[benzene ring]}-CO_2^-} + H_2O \rightleftharpoons \underset{\substack{HA \\ x}}{\text{[benzene ring]}-CO_2H} + \underset{x}{OH^-} \tag{9-15}
$$

$$
\frac{x^2}{F - x} = K_b
$$

From the value of $K_a$ for each isomer, we can calculate $K_b$ for the conjugate base.

| Isomer of hydroxybenzoic acid | $K_a$ | $K_b (= K_w/K_a)$ |
|---|---|---|
| *ortho* | $1.0_7 \times 10^{-3}$ | $9.3 \times 10^{-12}$ |
| *para* | $2.9 \times 10^{-5}$ | $3.5 \times 10^{-10}$ |

Using each value of $K_b$ and letting F = 0.050 0 M, we find

$$\text{pH of 0.050 0 M } o\text{-hydroxybenzoate} = 7.83$$

$$\text{pH of 0.050 0 M } p\text{-hydroxybenzoate} = 8.62$$

These are reasonable pH values for solutions of weak bases. Furthermore, as expected, the conjugate base of the stronger acid is the weaker base.

---

### Example A Weak-Base Problem

Find the pH of 0.10 M ammonia.

**Solution** When ammonia is dissolved in water, its reaction is

$$
\underset{\substack{\text{Ammonia} \\ F - x}}{NH_3} + H_2O \xrightarrow{K_b} \underset{\substack{\text{Ammonium} \\ \text{ion} \\ x}}{NH_4^+} + \underset{x}{OH^-}
$$

In Appendix G, we find ammonium ion, $NH_4^+$, listed next to ammonia. $pK_a$ for ammonium ion is 9.245. Therefore, $K_b$ for $NH_3$ is

$$
K_b = \frac{K_w}{K_a} = \frac{10^{-14.00}}{10^{-9.245}} = 1.76 \times 10^{-5}
$$

To find the pH of 0.10 M $NH_3$, we set up and solve the equation

$$
\frac{[NH_4^+][OH^-]}{[NH_3]} = \frac{x^2}{0.10 - x} = K_b = 1.76 \times 10^{-5}
$$

$$
x = [OH^-] = 1.3_2 \times 10^{-3} \text{ M}
$$

$$
[H^+] = \frac{K_w}{[OH^-]} = 7.6 \times 10^{-12} \text{ M} \Rightarrow pH = -\log [H^+] = 11.12
$$

---

### 9-5 Buffers

*A buffered solution resists changes in pH when acids or bases are added or when dilution occurs.* The **buffer** is a mixture of an acid and its conjugate base. There must be comparable amounts of the conjugate acid and base (say, within a factor of 10) to exert significant buffering.

The importance of buffers in all areas of science is immense. At the outset of this chapter, we saw that digestive enzymes in lysosomes operate best in acid, which allows a cell to protect itself from its own enzymes. If enzymes leak into the buffered, neutral cytoplasm, they have low reactivity and do less damage to the cell than they would at their optimum pH. Figure 9-3 shows the pH dependence of an enzyme-catalyzed reaction that is fastest near

---

HA and $A^-$ are a conjugate acid-base pair. So are $BH^+$ and B.

In aqueous solution, [structure]—$CO_2Na$ with OH

gives [structure]—$CO_2^-$ + $Na^+$ with OH
*o*-Hydroxybenzoate

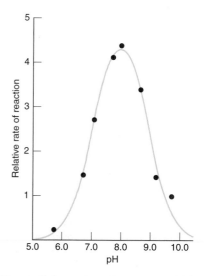

**Figure 9-3** pH dependence of the rate of cleavage of an amide bond by the enzyme chymotrypsin, which helps digest proteins in your intestine. [M. L. Bender, G. E. Clement, F. J. Kézdy, and H. A. Heck, "The Correlation of the pH (pD) Dependence and the Stepwise Mechanism of α-Chymotrypsin-Catalyzed-Reactions," *J. Am. Chem. Soc.* **1964,** 86, 3680.]

$$\underset{\text{Amide bond}}{RC\underset{|}{\overset{\overset{\textstyle O}{\|}}{-}}NHR'}$$

pH 8.0. For an organism to survive, it must control the pH of each subcellular compartment so that each reaction proceeds at the proper rate.

## Mixing a Weak Acid and Its Conjugate Base

When you mix a weak acid with its conjugate base, you get what you mix!

*If you mix A moles of a weak acid with B moles of its conjugate base, the moles of acid remain close to A and the moles of base remain close to B.* Very little reaction occurs to change either concentration.

To understand why this should be so, look at the $K_a$ and $K_b$ reactions in terms of Le Châtelier's principle. Consider an acid with $pK_a = 4.00$ and its conjugate base with $pK_b = 10.00$. Let's calculate the fraction of acid that dissociates in a 0.10 M solution of HA.

$$\underset{0.10-x}{HA} \rightleftharpoons \underset{x}{H^+} + \underset{x}{A^-} \qquad pK_a = 4.00$$

$$\frac{x^2}{F-x} = K_a \Rightarrow x = 3.1 \times 10^{-3}$$

$$\text{Fraction of dissociation} = \alpha = \frac{x}{F} = 0.031$$

The acid is only 3.1% dissociated under these conditions.

In a solution containing 0.10 mol of $A^-$ dissolved in 1.00 L, the extent of reaction of $A^-$ with water is even smaller.

$$\underset{0.10-x}{A^-} + H_2O \rightleftharpoons \underset{x}{HA} + \underset{x}{OH^-} \qquad pK_b = 10.00$$

$$\frac{x^2}{F-x} = K_b \Rightarrow x = 3.2 \times 10^{-6}$$

$$\text{Fraction of association} = \alpha = \frac{x}{F} = 3.2 \times 10^{-5}$$

The approximation that the concentrations of HA and $A^-$ remain unchanged breaks down for dilute solutions or at extremes of pH. We will test the validity of the approximation at the end of this chapter.

*HA dissociates very little, and adding extra $A^-$ to the solution will make the HA dissociate even less. Similarly, $A^-$ does not react very much with water, and adding extra HA makes $A^-$ react even less.* If 0.050 mol of $A^-$ plus 0.036 mol of HA are added to water, there will be close to 0.050 mol of $A^-$ and close to 0.036 mol of HA in the solution at equilibrium.

## Henderson-Hasselbalch Equation

The central equation for buffers is the **Henderson-Hasselbalch equation,** which is merely a rearranged form of the $K_a$ equilibrium expression.

$$K_a = \frac{[H^+][A^-]}{[HA]}$$

$\log xy = \log x + \log y$

$$\log K_a = \log \frac{[H^+][A^-]}{[HA]} = \log[H^+] + \log \frac{[A^-]}{[HA]}$$

$$\underbrace{-\log[H^+]}_{pH} = \underbrace{-\log K_a}_{pK_a} + \log \frac{[A^-]}{[HA]}$$

L. J. Henderson was a physician who wrote $[H^+] = K_a[\text{acid}]/[\text{salt}]$ in a physiology article in 1908, a year before the word "buffer" and the concept of pH were invented by the biochemist S. P. L. Sørensen. Henderson's contribution was the approximation of setting [acid] equal to the concentration of HA placed in solution and [salt] equal to the concentration of $A^-$ placed in solution. In 1916, K. A. Hasselbalch wrote what we call the Henderson-Hasselbalch equation in a biochemical journal.[8]

*Henderson-Hasselbalch equation for an acid:*

$$pH = pK_a + \log\left(\frac{[A^-]}{[HA]}\right) \tag{9-16}$$

The Henderson-Hasselbalch equation tells us the pH of a solution, provided we know the ratio of the concentrations of conjugate acid and base, as well as $pK_a$ for the acid. If a solution is prepared from the weak base B and its conjugate acid, the analogous equation is

*Henderson-Hasselbalch equation for a base:*

$$pH = pK_a + \log\left(\frac{[B]}{[BH^+]}\right) \qquad \begin{array}{l}\checkmark \; pK_a \text{ applies to} \\ \textit{this} \text{ acid}\end{array} \tag{9-17}$$

Equations 9-16 and 9-17 are sensible only when the base ($A^-$ or B) is in the *numerator.* When the concentration of base increases, the log term increases and the pH increases.

where $pK_a$ is the acid dissociation constant of the weak acid $BH^+$. The important features of Equations 9-16 and 9-17 are that the base ($A^-$ or B) appears in the numerator of both equations, and the equilibrium constant is $K_a$ of the acid in the denominator.

*Challenge* Show that, when activities are included, the Henderson-Hasselbalch equation is

$$pH = pK_a + \log \frac{[A^-]\gamma_{A^-}}{[HA]\gamma_{HA}} \qquad (9\text{-}18)$$

## Properties of the Henderson-Hasselbalch Equation

In Equation 9-16, we see that, if $[A^-] = [HA]$, then $pH = pK_a$.

$$pH = pK_a + \log \frac{[A^-]}{[HA]} = pK_a + \log 1 = pK_a$$

When $[A^-] = [HA]$, $pH = pK_a$.

Regardless of how complex a solution may be, whenever $pH = pK_a$, $[A^-]$ must equal $[HA]$. This relation is true because *all equilibria must be satisfied simultaneously in any solution at equilibrium.* If there are 10 different acids and bases in the solution, the 10 forms of Equation 9-16 must all give the same pH, because **there can be only one concentration of H$^+$ in a solution.**

Another feature of the Henderson-Hasselbalch equation is that, for every power-of-10 change in the ratio $[A^-]/[HA]$, the pH changes by one unit (Table 9-1). As the base ($A^-$) increases, the pH goes up. As the acid (HA) increases, the pH goes down. For any conjugate acid-base pair, you can say, for example, that, if $pH = pK_a - 1$, there must be 10 times as much HA as $A^-$. Ten-elevenths is in the form HA and one-eleventh is in the form $A^-$.

**Table 9-1** Effect of $[A^-]/[HA]$ on pH

| $[A^-]/[HA]$ | pH |
|---|---|
| 100:1 | $pK_a + 2$ |
| 10:1 | $pK_a + 1$ |
| 1:1 | $pK_a$ |
| 1:10 | $pK_a - 1$ |
| 1:100 | $pK_a - 2$ |

**Example** Using the Henderson-Hasselbalch Equation

Sodium hypochlorite (NaOCl, the active ingredient of almost all bleaches) was dissolved in a solution buffered to pH 6.20. Find the ratio $[OCl^-]/[HOCl]$ in this solution.

**Solution** In Appendix G, we find that $pK_a = 7.53$ for hypochlorous acid, HOCl. The pH is known, so the ratio $[OCl^-]/[HOCl]$ can be calculated from the Henderson-Hasselbalch equation.

$$HOCl \rightleftharpoons H^+ + OCl^-$$

$$pH = pK_a + \log \frac{[OCl^-]}{[HOCl]}$$

$$6.20 = 7.53 + \log \frac{[OCl^-]}{[HOCl]}$$

$$-1.33 = \log \frac{[OCl^-]}{[HOCl]}$$

$$10^{-1.33} = 10^{\log([OCl^-]/[HOCl])} = \frac{[OCl^-]}{[HOCl]}$$

$10^{\log z} = z$

$$0.047 = \frac{[OCl^-]}{[HOCl]}$$

Finding the ratio $[OCl^-]/[HOCl]$ requires knowing only the pH and the $pK_a$. We do not need to know how much NaOCl was added, or the volume.

## A Buffer in Action

For illustration, we choose a widely used buffer called "tris," which is short for tris(hydroxy-methyl)aminomethane.

$$BH^+ \qquad\rightleftharpoons\qquad B \qquad + H^+$$
$$pK_a = 8.072 \qquad\qquad \text{This form is "tris"}$$

In Appendix G, we find $pK_a$ for the conjugate acid of tris listed as 8.072. An example of a salt containing the $BH^+$ cation is tris hydrochloride, which is $BH^+Cl^-$. When $BH^+Cl^-$ is dissolved in water, it dissociates to $BH^+$ and $Cl^-$.

## Example A Buffer Solution

Find the pH of a solution prepared by dissolving 12.43 g of tris (FM 121.135) plus 4.67 g of tris hydrochloride (FM 157.596) in 1.00 L of water.

**Solution**   The concentrations of B and $BH^+$ added to the solution are

$$[B] = \frac{12.43 \text{ g/L}}{121.135 \text{ g/mol}} = 0.102\ 6 \text{ M} \qquad [BH^+] = \frac{4.67 \text{ g/L}}{157.596 \text{ g/mol}} = 0.029\ 6 \text{ M}$$

Assuming that what we mixed stays in the same form, we plug these concentrations into the Henderson-Hasselbalch equation to find the pH:

$$pH = pK_a + \log \frac{[B]}{[BH^+]} = 8.072 + \log \frac{0.102\ 6}{0.029\ 6} = 8.61$$

The pH of a buffer is nearly independent of volume.

Notice that *the volume of solution is irrelevant,* because volume cancels in the numerator and denominator of the log term:

$$pH = pK_a + \log \frac{\text{moles of B/\cancel{L \text{ of solution}}}}{\text{moles of } BH^+/\cancel{L \text{ of solution}}}$$

$$= pK_a + \log \frac{\text{moles of B}}{\text{moles of } BH^+}$$

## Example Effect of Adding Acid to a Buffer

If we add 12.0 mL of 1.00 M HCl to the solution used in the previous example, what will be the new pH?

**Solution**   The key to this problem is to realize that, *when a strong acid is added to a weak base, they react completely to give* $BH^+$ (Box 9-3). We are adding 12.0 mL of 1.00 M HCl, which contains $(0.012\ 0 \text{ L})(1.00 \text{ mol/L}) = 0.012\ 0$ mol of $H^+$. This much $H^+$ will consume 0.012 0 mol of B to create 0.012 0 mol of $BH^+$, which is shown conveniently in a little table:

| | B | + | $H^+$ | $\rightarrow$ | $BH^+$ |
|---|---|---|---|---|---|
| | Tris | | From HCl | | |
| Initial moles | 0.102 6 | | 0.012 0 | | 0.029 6 |
| Final moles | 0.090 6 | | — | | 0.041 6 |
| | $(0.102\ 6 - 0.012\ 0)$ | | | | $(0.029\ 6 + 0.012\ 0)$ |

---

## Box 9-3 Strong Plus Weak Reacts Completely

A strong acid reacts with a weak base essentially "completely" because the equilibrium constant is large.

$$B + H^+ \rightleftharpoons BH^+ \qquad K = \frac{1}{K_a(\text{for } BH^+)}$$

Weak   Strong
base    acid

If B is tris(hydroxymethyl)aminomethane, the equilibrium constant for reaction with HCl is

$$K = \frac{1}{K_a} = \frac{1}{10^{-8.072}} = 1.2 \times 10^8$$

A strong base reacts "completely" with a weak acid because the equilibrium constant is, again, very large.

$$OH^- + HA \rightleftharpoons A^- + H_2O \qquad K = \frac{1}{K_b(\text{for } A^-)}$$

Strong   Weak
base     acid

If HA is acetic acid, then the equilibrium constant for reaction with NaOH is

$$K = \frac{1}{K_b} = \frac{K_a(\text{for HA})}{K_w} = 1.7 \times 10^9$$

The reaction of a strong acid with a strong base is even more complete than a strong plus weak reaction:

$$H^+ + OH^- \rightleftharpoons H_2O \qquad K = \frac{1}{K_w} = 10^{14}$$

Strong   Strong
acid     base

If you mix a strong acid, a strong base, a weak acid, and a weak base, the strong acid and base will neutralize each other until one is used up. The remaining strong acid or base will then react with the weak base or weak acid.

A buffer resists changes in pH because the added acid or base is consumed by the buffer. As the buffer is used up, it becomes less resistant to changes in pH.

In this demonstration,[9] a mixture containing approximately a 10:1 mole ratio of $HSO_3^-$:$SO_3^{2-}$ is prepared. Because $pK_a$ for $HSO_3^-$ is 7.2, the pH should be approximately

$$pH = pK_a + \log\frac{[SO_3^{2-}]}{[HSO_3^-]} = 7.2 + \log\frac{1}{10} = 6.2$$

When formaldehyde is added, the net reaction is the consumption of $HSO_3^-$, but not of $SO_3^{2-}$:

$$H_2C{=}O + HSO_3^- \rightarrow H_2C\!\!\begin{array}{c}O^-\\ \diagdown\\ SO_3H\end{array} \rightarrow H_2C\!\!\begin{array}{c}OH\\ \diagdown\\ SO_3^-\end{array} \quad (A)$$

Formaldehyde    Bisulfite

$$H_2C{=}O + SO_3^{2-} \rightarrow H_2C\!\!\begin{array}{c}O^-\\ \diagdown\\ SO_3^-\end{array}$$

Sulfite

$$(B)$$

$$H_2C\!\!\begin{array}{c}O^-\\ \diagdown\\ SO_3^-\end{array} + HSO_3^- \rightarrow H_2C\!\!\begin{array}{c}OH\\ \diagdown\\ SO_3^-\end{array} + SO_3^{2-}$$

(In sequence A, bisulfite is consumed directly. In sequence B, the net reaction is destruction of $HSO_3^-$, with no change in the $SO_3^{2-}$ concentration.)

We can prepare a table showing how the pH should change as the $HSO_3^-$ reacts.

| Percentage of reaction completed | $[SO_3^{2-}]$:$[HSO_3^-]$ | Calculated pH |
|---|---|---|
| 0 | 1:10 | 6.2 |
| 90 | 1:1 | 7.2 |
| 99 | 1:0.1 | 8.2 |
| 99.9 | 1:0.01 | 9.2 |
| 99.99 | 1:0.001 | 10.2 |

Through 90% completion, the pH should rise by just 1 unit. In the next 9% of the reaction, the pH will rise by another unit. At the end of the reaction, the change in pH is very abrupt.

In the formaldehyde clock reaction, formaldehyde is added to a solution containing $HSO_3^-$, $SO_3^{2-}$, and phenolphthalein indicator. Phenolphthalein is colorless below a pH of 8 and red above this pH. The solution remains colorless for more than a minute. Suddenly the pH shoots up and the liquid turns pink. Monitoring the pH with a glass electrode gave the results in the graph.

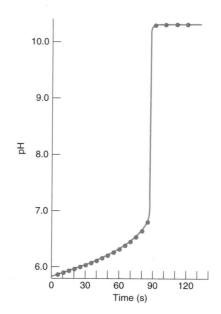

Graph of pH versus time in the formaldehyde clock reaction.

*Procedure:* All solutions should be fresh. Prepare a solution of formaldehyde by diluting 9 mL of 37 wt% formaldehyde to 100 mL. Dissolve 1.5 g of $NaHSO_3$[10] and 0.18 g of $Na_2SO_3$ in 400 mL of water, and add ~1 mL of phenolphthalein solution (Table 11-4). Add 23 mL of formaldehyde solution to the well-stirred buffer solution to initiate the clock reaction. The time of reaction can be adjusted by changing the temperature, concentrations, or volume.

The information in the table allows us to calculate the pH:

$$pH = pK_a + \log\frac{\text{moles of B}}{\text{moles of BH}^+}$$

$$= 8.072 + \log\frac{0.090\,6}{0.041\,6} = 8.41$$

The volume of the solution is irrelevant.

We see that *the pH of a buffer does not change very much when a limited amount of strong acid or base is added.* Addition of 12.0 mL of 1.00 M HCl changed the pH from 8.61 to 8.41. Addition of 12.0 mL of 1.00 M HCl to 1.00 L of unbuffered solution would have lowered the pH to 1.93.

But *why* does a buffer resist changes in pH? It does so because the strong acid or base is consumed by B or $BH^+$. If you add HCl to tris, B is converted into $BH^+$. If you add NaOH, $BH^+$ is converted into B. As long as you don't use up the B or $BH^+$ by adding too much HCl or NaOH, the log term of the Henderson-Hasselbalch equation does not change very much and the pH does not change very much. Demonstration 9-2 illustrates what happens when the buffer does get used up. The buffer has its maximum capacity to resist changes of pH when $pH = pK_a$. We will return to this point later.

*Question*   Does the pH change in the right direction when HCl is added?

A buffer resists changes in pH . . .

. . . because the buffer consumes the added acid or base.

**Example** Calculating How to Prepare a Buffer Solution

How many milliliters of 0.500 M NaOH should be added to 10.0 g of tris hydrochloride to give a pH of 7.60 in a final volume of 250 mL?

**Solution** The moles of tris hydrochloride in 10.0 g are $(10.0 \text{ g})/(157.596 \text{ g/mol}) = 0.063\ 5$. We can make a table to help solve the problem:

| Reaction with OH⁻: | BH⁺ | + | OH⁻ | → | B |
|---|---|---|---|---|---|
| Initial moles | 0.063 5 | | $x$ | | — |
| Final moles | 0.063 5 − $x$ | | — | | $x$ |

The Henderson-Hasselbalch equation allows us to find $x$, because we know pH and $pK_a$.

$$\text{pH} = pK_a + \log \frac{\text{mol B}}{\text{mol BH}^+}$$

$$7.60 = 8.072 + \log \frac{x}{0.063\ 5 - x}$$

$$-0.472 = \log \frac{x}{0.063\ 5 - x}$$

$$10^{-0.472} = \frac{x}{0.063\ 5 - x} \Rightarrow x = 0.016\ 0 \text{ mol}$$

This many moles of NaOH is contained in

$$\frac{0.016\ 0 \text{ mol}}{0.500 \text{ mol/L}} = 0.032\ 0 \text{ L} = 32.0 \text{ mL}$$

## Preparing a Buffer in Real Life!

If you really wanted to prepare a tris buffer of pH 7.60, you would *not* do it by calculating what to mix. Suppose that you wish to prepare 1.00 L of buffer containing 0.100 M tris at a pH of 7.60. You have available solid tris hydrochloride and approximately 1 M NaOH. Here's how I would do it:

1. Weigh out 0.100 mol of tris hydrochloride and dissolve it in a beaker containing about 800 mL of water.
2. Place a calibrated pH electrode in the solution and monitor the pH.
3. Add NaOH solution until the pH is exactly 7.60.
4. Transfer the solution to a volumetric flask and wash the beaker a few times. Add the washings to the volumetric flask.
5. Dilute to the mark and mix.

You do not simply add the calculated quantity of NaOH, because it would not give exactly the desired pH. The reason for using 800 mL of water in the first step is so that the volume will be reasonably close to the final volume during pH adjustment. Otherwise, the pH will change slightly when the sample is diluted to its final volume and the ionic strength changes.

## Buffer Capacity[11]

**Buffer capacity,** β, is a measure of how well a solution resists changes in pH when strong acid or base is added. Buffer capacity is defined as

*Buffer capacity:* $$\beta = \frac{dC_b}{d\text{pH}} = -\frac{dC_a}{d\text{pH}} \qquad (9\text{-}19)$$

where $C_a$ and $C_b$ are the number of moles of strong acid and strong base per liter needed to produce a unit change in pH. The greater the buffer capacity, the more resistant the solution is to pH change.

Figure 9-4a shows $C_b$ versus pH for a solution containing 0.100 F HA with $pK_a = 5.00$. The ordinate ($C_b$) is the formal concentration of strong base needed to be mixed with 0.100 F HA to give the indicated pH. For example, a solution containing 0.050 F OH⁻ plus 0.100 F HA would have a pH of 5.00 (neglecting activities).

Figure 9-4b, which is the derivative of the upper curve, shows buffer capacity for the same system. Buffer capacity reaches a maximum when pH = $pK_a$. That is, *a buffer is most effective in resisting changes in pH when pH = $pK_a$ (that is, when [HA] = [A⁻]).*

---

Reasons why a calculation would be wrong:

1. You might have ignored activity coefficients.
2. The temperature might not be just right.
3. The approximations that [HA] = $F_{HA}$ and [A⁻] = $F_{A^-}$ could be in error.
4. The $pK_a$ reported for tris in your favorite table is probably not what you would measure in your lab.
5. You will probably make an arithmetic error anyway.

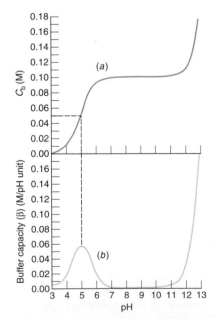

*Figure 9-4* (a) $C_b$ versus pH for a solution containing 0.100 F HA with $pK_a = 5.00$. (b) Buffer capacity versus pH for the same system reaches a maximum when pH = $pK_a$. The lower curve is the derivative of the upper curve.

In choosing a buffer, *seek one whose* $pK_a$ *is as close as possible to the desired pH. The useful pH range of a buffer is usually considered to be* $pK_a \pm 1$ *pH unit.* Outside this range, there is not enough of either the weak acid or the weak base to react with added base or acid. Buffer capacity can be increased by increasing the concentration of the buffer.

Buffer capacity in Figure 9-4b continues upward at high pH (and at low pH, which is not shown) simply because there is a high concentration of $OH^-$ at high pH (or $H^+$ at low pH). Addition of a small amount of acid or base to a large amount of $OH^-$ (or $H^+$) will not have a large effect on pH. A solution of high pH is buffered by the $H_2O/OH^-$ conjugate acid–conjugate base pair. A solution of low pH is buffered by the $H_3O^+/H_2O$ conjugate acid–conjugate base pair.

Table 9-2 lists $pK_a$ values for common buffers that are widely used in biochemistry. The measurement of pH with glass electrodes, and the buffers used by the U.S. National Institute of Standards and Technology to define the pH scale, are described in Chapter 15.

## Buffer pH Depends on Ionic Strength and Temperature

The *correct* Henderson-Hasselbalch equation, 9-18, includes activity coefficients. Failure to include activity coefficients is the principal reason why calculated pH values do not agree with measured values. The $H_2PO_4^-/HPO_4^{2-}$ buffer has a $pK_a$ of 7.20 at 0 ionic strength. At 0.1 M ionic strength, the pH of a 1:1 mole mixture of $H_2PO_4^-$ and $HPO_4^{2-}$ is 6.86 (see Problem 9-43). Molecular biology lab manuals list $pK_a$ for phosphoric acid as 6.86, which is representative for ionic strengths employed in the lab. As another example of ionic strength effects, when a 0.5 M stock solution of phosphate buffer at pH 6.6 is diluted to 0.05 M, the pH rises to 6.9—a rather significant effect.

Buffer $pK_a$ depends on temperature. Tris has an exceptionally large dependence, $-0.028$ $pK_a$ units per degree, near room temperature. A solution of tris with pH 8.07 at 25°C will have pH $\approx$ 8.7 at 4°C and pH $\approx$ 7.7 at 37°C.

Changing ionic strength changes pH.

Changing temperature changes pH.

## When What You Mix Is Not What You Get

In dilute solution, or at extremes of pH, the concentrations of HA and $A^-$ are not equal to their formal concentrations. Suppose we mix $F_{HA}$ moles of HA and $F_{A^-}$ moles of the salt $Na^+A^-$. The mass and charge balances are

What you mix is not what you get in dilute solutions or at extremes of pH.

$$\text{Mass balance:} \quad F_{HA} + F_{A^-} = [HA] + [A^-]$$
$$\text{Charge balance:} \quad [Na^+] + [H^+] = [OH^-] + [A^-]$$

The substitution $[Na^+] = F_{A^-}$ and a little algebra lead to the equations

$$[HA] = F_{HA} - [H^+] + [OH^-] \tag{9-20}$$
$$[A^-] = F_{A^-} + [H^+] - [OH^-] \tag{9-21}$$

So far we have assumed that $[HA] \approx F_{HA}$ and $[A^-] \approx F_{A^-}$, and we used these values in the Henderson-Hasselbalch equation. A more rigorous procedure is to use Equations 9-20 and 9-21. If $F_{HA}$ or $F_{A^-}$ is small, or if $[H^+]$ or $[OH^-]$ is large, then the approximations $[HA] \approx F_{HA}$ and $[A^-] \approx F_{A^-}$ are not good. In acidic solutions, $[H^+] \gg [OH^-]$, so $[OH^-]$ can be ignored in Equations 9-20 and 9-21. In basic solutions, $[H^+]$ can be neglected.

**Example** A Dilute Buffer Prepared from a Moderately Strong Acid

What will be the pH if 0.010 0 mol of HA (with $pK_a = 2.00$) and 0.010 0 mol of $A^-$ are dissolved in water to make 1.00 L of solution?

**Solution** Because the solution is acidic (pH $\approx pK_a = 2.00$), we neglect $[OH^-]$ in Equations 9-20 and 9-21. Setting $[H^+] = x$ in Equations 9-20 and 9-21, we use the $K_a$ equation to find $[H^+]$:

$$\begin{array}{ccccc} HA & \rightleftharpoons & H^+ & + & A^- \\ 0.010\ 0 - x & & x & & 0.010\ 0 + x \end{array}$$

$$K_a = \frac{[H^+][A^-]}{[HA]} = \frac{(x)(0.010\ 0 + x)}{(0.010 - x)} = 10^{-2.00} \tag{9-22}$$

$$\Rightarrow x = 0.004\ 14\ M \Rightarrow pH = -\log[H^+] = 2.38$$

The concentrations of HA and $A^-$ are not what we mixed:

$$[HA] = F_{HA} - [H^+] = 0.005\ 86\ M$$
$$[A^-] = F_{A^-} + [H^+] = 0.014\ 1\ M$$

HA in this solution is more than 40% dissociated. The acid is too strong for the approximation $[HA] \approx F_{HA}$.

In this example, HA is too strong and the concentrations are too low for HA and $A^-$ to be equal to their formal concentrations.

**Table 9-2**  Structures and $pK_a$ values for common buffers[a,b,c,d]

| Name | Structure | $pK_a^e$ | Formula mass | $\Delta(pK_a)/\Delta T$ $(K^{-1})$ |
|---|---|---|---|---|
| N-2-Acetamidoiminodiacetic acid (ADA) | $\overset{\displaystyle O}{\overset{\|}{H_2NCCH_2}}\overset{+}{N}H \begin{smallmatrix} CH_2CO_2H \\ \\ CH_2CO_2H \end{smallmatrix}$ | 1.59 $(CO_2H)$ | 190.15 | — |
| N-Tris(hydroxymethyl)methylglycine (TRICINE) | $(HOCH_2)_3C\overset{+}{N}H_2CH_2CO_2H$ | 2.02 $(CO_2H)$ | 179.17 | −0.003 |
| Phosphoric acid | $H_3PO_4$ | 2.15 $(pK_1)$ | 98.00 | 0.005 |
| ADA | (see above) | 2.48 $(CO_2H)$ | 190.15 | — |
| Piperazine-N,N'-bis(2-ethanesulfonic acid) (PIPES) | $^-O_3SCH_2CH_2\overset{+}{N}H\ \overset{+}{H}NCH_2CH_2SO_3^-$ | 2.67 $(pK_1)$ | 302.37 | — |
| Citric acid | $HO_2CCH_2\overset{\overset{\displaystyle OH}{\|}}{\underset{\underset{\displaystyle CO_2H}{\|}}{C}}CH_2CO_2H$ | 3.13 $(pK_1)$ | 192.12 | −0.002 |
| Glycylglycine | $H_3\overset{+}{N}CH_2\overset{\overset{\displaystyle O}{\|}}{C}NHCH_2CO_2H$ | 3.14 $(CO_2H)$ | 132.12 | 0.000 |
| Piperazine-N,N'-bis(3-propanesulfonic acid) (PIPPS) | $^-O_3SCH_2CH_2CH_2\overset{+}{N}H\ \overset{+}{H}NCH_2CH_2CH_2SO_3^-$ | 3.79 $(pK_1)$ | 330.42 | — |
| Piperazine-N,N'-bis(4-butanesulfonic acid) (PIPBS) | $^-O_3SCH_2CH_2CH_2CH_2\overset{+}{N}H\ \overset{+}{H}NCH_2CH_2CH_2CH_2SO_3^-$ | 4.29 $(pK_1)$ | 358.47 | — |
| N,N'-Diethylpiperazine dihydrochloride (DEPP·2HCl) | $CH_3CH_2\overset{+}{N}H\ \overset{+}{H}NCH_2CH_3\cdot2Cl^-$ | 4.48 $(pK_1)$ | 215.16 | — |
| Citric acid | (see above) | 4.76 $(pK_2)$ | 192.12 | −0.001 |
| Acetic acid | $CH_3CO_2H$ | 4.76 | 60.05 | 0.000 |
| N,N'-Diethylethylenediamine-N,N'-bis(3-propanesulfonic (DESPEN) | $^-O_3SCH_2CH_2CH_2\overset{\overset{+}{N}}{\underset{\underset{\displaystyle CH_3CH_2}{\|}}{}}HCH_2CH_2\overset{+}{H}\overset{\overset{}{N}}{\underset{\underset{\displaystyle CH_2CH_3}{\|}}{}}CH_2CH_2CH_2SO_3^-$ | 5.62 $(pK_1)$ | 360.49 | — |
| 2-(N-Morpholino)ethanesulfonic acid (MES) | $O\underset{}{\bigcirc}\overset{+}{N}HCH_2CH_2SO_3^-$ | 6.27 | 195.24 | −0.009 |
| Citric acid | (see above) | 6.40 $(pK_3)$ | 192.12 | 0.002 |
| N,N,N',N'-Tetraethylethylenediamine dihydrochloride (TEEN·2HCl) | $(CH_3CH_2)_2\overset{+}{N}HCH_2CH_2\overset{+}{H}N(CH_2CH_3)_2\cdot2Cl^-$ | 6.58 $(pK_1)$ | 245.23 | — |
| 1,3-Bis[tris(hydroxymethyl)methylamino]propane hydrochloride (BIS-TRIS propane·2HCl) | $(HOCH_2)_3C\overset{+}{N}H_2(CH_2)_3\overset{+}{N}H_2C(CH_2OH)_3\cdot2Cl^-$ | 6.65 $(pK_1)$ | 355.26 | — |
| ADA | (see above) | 6.84 (NH) | 190.15 | −0.007 |

a. The protonated form of each molecule is shown. Acidic hydrogen atoms are shown in *bold* type.

b. Many buffers in this table are widely used in biomedical research because of their weak metal binding and physiologic inertness (C. L. Bering, J. Chem. Ed. **1987**, 64, 803). In one study, where MES and MOPS had no discernible affinity for $Cu^{2+}$, a minor impurity in HEPES and HEPPS had a strong affinity for $Cu^{2+}$ and MOPSO bound $Cu^{2+}$ stoichiometrically (H. E. Marsh, Y.-P. Chin, L. Sigg, R. Hari, and H. Xu, Anal. Chem. **2003**, 75, 671). ADA, BICINE, ACES, and TES have some metal-binding ability (R. Nakon and C. R. Krishnamoorthy, Science **1983**, 221, 749). Lutidine buffers for the pH range 3 to 8 with limited metal-binding power have been described by U. Bips, H. Elias, M. Hauröder, G. Kleinhans, S. Pfeifer, and K. J. Wannowius, Inorg. Chem. **1983**, 22, 3862.

c. Some data from R. N. Goldberg, N. Kishore, and R. M. Lennen, J. Phys. Chem. Ref. Data **2002**, 31, 231. This paper gives the temperature dependence of $pK_a$.

d. Temperature and ionic strength dependence of $pK_a$ for buffers: HEPES—D. Feng, W. F. Koch, and Y. C. Wu, Anal. Chem. **1989**, 61, 1400; MOPSO—Y. C. Wu, P. A. Berezansky, D. Feng, and W. F. Koch, Anal. Chem. **1993**, 65, 1084; ACES and CHES—R. N. Roy, J. Bice, J. Greer, J. A. Carlsten, J. Smithson, W. S. Good, C. P. Moore, L. N. Roy, and K. M. Kuhler, J. Chem. Eng. Data **1997**, 42, 41; TEMN, TEEN, DEPP, DESPEN, PIPES, PIPPS, PIPBS, MES, MOPS, and MOBS—A. Kandegedara and D. B. Rorabacher, Anal. Chem. **1999**, 71, 3140. This last set of buffers was specifically developed for low metal-binding ability (Q. Yu, A. Kandegedara, Y. Xu, and D. B. Rorabacher, Anal. Biochem. **1997**, 253, 50).

e. $pK_a$ is generally for 25°C and zero ionic strength.

| Name | Structure | p$K_a^e$ | Formula mass | $\Delta(pK_a)/\Delta T$ $(K^{-1})$ |
|---|---|---|---|---|
| $N$-2-Acetamido-2-aminoethanesulfonic acid (ACES) | $\overset{\displaystyle O}{\overset{\displaystyle \|}{H_2NCCH_2}}\overset{+}{N}H_2CH_2CH_2SO_3^-$ | 6.85 | 182.20 | $-0.018$ |
| 3-($N$-Morpholino)-2-hydroxypropanesulfonic acid (MOPSO) | $O\bigcirc\overset{+}{N}HCH_2\overset{\overset{\displaystyle OH}{\displaystyle \|}}{C}HCH_2SO_3^-$ | 6.90 | 225.26 | $-0.015$ |
| Imidazole hydrochloride | $HN^+ \cdots \cdot Cl^-$ | 6.99 | 104.54 | $-0.022$ |
| PIPES | (see above) | 7.14 (p$K_2$) | 302.37 | $-0.007$ |
| 3-($N$-Morpholino)propanesulfonic acid (MOPS) | $O\bigcirc\overset{+}{N}HCH_2CH_2CH_2SO_3^-$ | 7.18 | 209.26 | $-0.012$ |
| Phosphoric acid | $H_3PO_4$ | 7.20 (p$K_2$) | 98.00 | $-0.002$ |
| 4-($N$-Morpholino)butanesulfonic acid (MOBS) | $O\bigcirc\overset{+}{N}HCH_2CH_2CH_2CH_2SO_3^-$ | 7.48 | 223.29 | — |
| $N$-Tris(hydroxymethyl)methyl-2-amino-ethanesulfonic acid (TES) | $(HOCH_2)_3C\overset{+}{N}H_2CH_2CH_2SO_3^-$ | 7.55 | 229.25 | $-0.019$ |
| $N$-2-Hydroxyethylpiperazine-$N'$-2-ethanesulfonic acid (HEPES) | $HOCH_2CH_2N\bigcirc\overset{+}{N}HCH_2CH_2SO_3^-$ | 7.56 | 238.30 | $-0.012$ |
| PIPPS | (see above) | 7.97 (p$K_2$) | 330.42 | — |
| $N$-2-Hydroxyethylpiperazine-$N'$-3-propanesulfonic acid (HEPPS) | $HOCH_2CH_2N\bigcirc\overset{+}{N}HCH_2CH_2CH_2SO_3^-$ | 7.96 | 252.33 | $-0.013$ |
| Glycine amide hydrochloride | $H_3\overset{+}{N}CH_2\overset{\overset{\displaystyle O}{\displaystyle \|}}{C}NH_2 \cdot Cl^-$ | 8.04 | 110.54 | — |
| Tris(hydroxymethyl)aminomethane hydrochloride (TRIS hydrochloride) | $(HOCH_2)_3C\overset{+}{N}H_3 \cdot Cl^-$ | 8.07 | 157.60 | $-0.028$ |
| TRICINE | (see above) | 8.14 (NH) | 179.17 | $-0.018$ |
| Glycylglycine | (see above) | 8.26 (NH) | 132.12 | $-0.026$ |
| $N,N$-Bis(2-hydroxyethyl)glycine (BICINE) | $(HOCH_2CH_2)_2\overset{+}{N}HCH_2CO_2^-$ | 8.33 | 163.17 | $-0.015$ |
| PIPBS | (see above) | 8.55 (p$K_2$) | 358.47 | — |
| DEPP·2HCl | (see above) | 8.58 (p$K_2$) | 207.10 | — |
| DESPEN | (see above) | 9.06 (p$K_2$) | 360.49 | — |
| BIS-TRIS propane·2HCl | (see above) | 9.10 (p$K_2$) | 355.26 | — |
| Ammonia | $NH_4^+$ | 9.24 | 17.03 | $-0.031$ |
| Boric acid | $B(OH)_3$ | 9.24 (p$K_1$) | 61.83 | $-0.008$ |
| Cyclohexylaminoethanesulfonic acid (CHES) | $\bigcirc\!\!-\overset{+}{N}H_2CH_2CH_2SO_3^-$ | 9.39 | 207.29 | $-0.023$ |
| TEEN·2HCl | (see above) | 9.88 (p$K_2$) | 245.23 | — |
| 3-(Cyclohexylamino) propanesulfonic acid (CAPS) | $\bigcirc\!\!-\overset{+}{N}H_2CH_2CH_2CH_2SO_3^-$ | 10.50 | 221.32 | $-0.028$ |
| $N,N,N',N'$-Tetraethylmethylenediamine dihydrochloride (TEMN·2HCl) | $(CH_3CH_2)_2\overset{+}{N}HCH_2CH_2\overset{+}{H}N(CH_2CH_3)_2 \cdot 2Cl^-$ | 11.01 (p$K_2$) | 231.21 | — |
| Phosphoric acid | $H_3PO_4$ | 12.35 (p$K_3$) | 98.00 | $-0.009$ |
| Boric acid | $B(OH)_3$ | 12.74 (p$K_2$) | 61.83 | — |

The Henderson-Hasselbalch equation (with activity coefficients) is *always* true, because it is just a rearrangement of the $K_a$ equilibrium expression. Approximations that are not always true are the statements $[HA] \approx F_{HA}$ and $[A^-] \approx F_{A^-}$.

In summary, a buffer consists of a mixture of a weak acid and its conjugate base. The buffer is most useful when $pH \approx pK_a$. Over a reasonable range of concentration, the pH of a buffer is nearly independent of concentration. A buffer resists changes in pH because it reacts with added acids or bases. If too much acid or base is added, the buffer will be consumed and will no longer resist changes in pH.

---

**Example** 🔢 Excel's Goal Seek Tool and Naming of Cells

We saw in Section 6-8 that GOAL SEEK finds solutions to numerical equations. In setting up Equation 9-22, we made the (superb) approximation $[H^+] \gg [OH^-]$ and neglected $[OH^-]$. With GOAL SEEK, it is easy to use Equations 9-20 and 9-21 without approximations:

$$K_a = \frac{[H^+][A^-]}{[HA]} = \frac{[H^+](F_{A^-} + [H^+] - [OH^-])}{F_{HA} - [H^+] + [OH^-]} \qquad (9\text{-}23)$$

The spreadsheet illustrates GOAL SEEK and the naming of cells to make formulas more meaningful. In column A, enter labels for $K_a$, $K_w$, $F_{HA}$, $F_{A^-}$, $[H^+]$, and $[OH^-]$. Write numerical values for $K_a$, $K_w$, $F_{HA}$, and $F_{A^-}$ in B1:B4. In cell B5, enter a *guess* for $[H^+]$.

| | A | B | C | D | E |
|---|---|---|---|---|---|
| 1 | Ka = | 0.01 | | Reaction quotient | |
| 2 | Kw = | 1.00E-14 | | for Ka = | |
| 3 | FHA = | 0.01 | | [H+][A-]/[HA]= | |
| 4 | FA = | 0.01 | | 0.001222222 | |
| 5 | [H+] = | 1.000E-03 | <-Initial guess | | |
| 6 | [OH-] = | 1E-11 | | D4 = H*(FA+H-OH)/(FHA-H+OH) | |
| 7 | | B6 = Kw/H | | | |

Now we want to name cells Bl through B6. Select cell Bl, go to the INSERT menu, select NAME and then DEFINE. The window will ask if you want to use the name "Ka" that appears in cell Al. If you like this name, click OK. By this procedure, name the other cells in column B "Kw", "FHA", "FA", "H", and "OH". Now when you write a formula referring to cell B2, you can write Kw instead of B2. Kw is an *absolute reference* to cell $B$2.

In cell B6 enter the formula "=Kw/H" and Excel returns the value 1E-11 for $[OH^-]$. The beauty of naming cells is that "=Kw/H" is easier to understand than "=$B$2/$B$5".

In cell D4, write "=H*(FA+H-OH)/(FHA-H+OH)", which is the quotient in Equation 9-23. Excel returns the value 0.001 222 based on the guess $[H^+] = 0.001$ in cell B5.

Now use GOAL SEEK to vary $[H^+]$ in cell B5 until the reaction quotient in cell D4 equals 0.01, which is the value of $K_a$. In the TOOLS menu, select OPTIONS and go to Calculation. Set Maximum Change to 1e-7 to find an answer within 0.000 000 1 of the desired answer. In the TOOLS menu, select GOAL SEEK. For Set cell, enter D4. For To value, enter 0.01. For By changing cell, enter H. We just told GOAL SEEK to vary H (cell B5) until cell D4 equals $0.01 \pm 0.000\,000\,1$. Click OK and the answer $H = 4.142 \times 10^{-3}$ appears in cell B5. Different starting guesses for H might give negative solutions or might not reach a solution. Only one positive value of H satisfies Equation 9-23.

---

## Terms to Understand

| | | | |
|---|---|---|---|
| acid dissociation constant, $K_a$ | fraction of association, $\alpha$ | $pK$ | weak base |
| base hydrolysis constant, $K_b$ | (of a base) | strong acid | weak electrolyte |
| buffer | fraction of dissociation, $\alpha$ | strong base | |
| buffer capacity | (of an acid) | weak acid | |
| conjugate acid-base pair | Henderson-Hasselbalch equation | | |

---

## Summary

*Strong acids or bases.* For practical concentrations ($\geq 10^{-6}$ M), pH or pOH can be found by inspection. When the concentration is near $10^{-7}$ M, we use the systematic treatment of equilibrium to calculate pH. At still lower concentrations, the pH is 7.00, set by autoprotolysis of the solvent.

*Weak acids.* For the reaction $HA \rightleftharpoons H^+ + A^-$, we set up and solve the equation $K_a = x^2/(F - x)$, where $[H^+] = [A^-] = x$, and $[HA] = F - x$. The fraction of dissociation is given by $\alpha = [A^-]/([HA] + [A^-]) = x/F$. The term $pK_a$ is defined as $pK_a = -\log K_a$.

*Weak bases.* For the reaction $B + H_2O \rightleftharpoons BH^+ + OH^-$, we set up and solve the equation $K_b = x^2/(F - x)$, where $[OH^-] = [BH^+] = x$, and $[B] = F - x$. The conjugate acid of a weak base is a weak acid, and the conjugate base of a weak acid is a weak base. For a conjugate acid-base pair, $K_a \cdot K_b = K_w$.

*Buffers.* A buffer is a mixture of a weak acid and its conjugate base. It resists changes in pH because it reacts with added acid or base. The pH is given by the Henderson-Hasselbalch equation:

$$pH = pK_a + \log \frac{[A^-]}{[HA]}$$

where $pK_a$ applies to the species in the denominator. The concentrations of HA and $A^-$ are essentially unchanged from those used to prepare the solution. The pH of a buffer is nearly independent of dilution, but buffer capacity increases as the concentration of buffer increases. The maximum buffer capacity is at $pH = pK_a$, and the useful range is $pH = pK_a \pm 1$.

The conjugate base of a weak acid is a weak base. The weaker the acid, the stronger the base. However, if one member of a conjugate pair is weak, so is its conjugate. The relation between $K_a$ for an acid and $K_b$ for its conjugate base in aqueous solution is $K_a \cdot K_b = K_w$. When a strong acid (or base) is added to a weak base (or acid), they react nearly completely.

## Exercises

**9-A.** Using activity coefficients correctly, find the pH of $1.0 \times 10^{-2}$ M NaOH.

**9-B.** Calculate the pH of
(a) $1.0 \times 10^{-8}$ M HBr
(b) $1.0 \times 10^{-8}$ M $H_2SO_4$ ($H_2SO_4$ dissociates completely to $2H^+$ plus $SO_4^{2-}$ at this low concentration).

**9-C.** What is the pH of a solution prepared by dissolving 1.23 g of 2-nitrophenol (FM 139.11) in 0.250 L?

**9-D.** The pH of 0.010 M $o$-cresol is 6.16. Find $pK_a$ for this weak acid.

$o$-Cresol

**9-E.** Calculate the limiting value of the fraction of dissociation $(\alpha)$ of a weak acid ($pK_a = 5.00$) as the concentration of HA approaches 0. Repeat the same calculation for $pK_a = 9.00$.

**9-F.** Find the pH of 0.050 M sodium butanoate (the sodium salt of butanoic acid, also called butyric acid).

**9-G.** The pH of 0.10 M ethylamine is 11.82.
(a) Without referring to Appendix G, find $K_b$ for ethylamine.
(b) Using results from part **(a)**, calculate the pH of 0.10 M ethylammonium chloride.

**9-H.** Which of the following bases would be most suitable for preparing a buffer of pH 9.00: (i) $NH_3$ (ammonia, $K_b = 1.76 \times 10^{-5}$); (ii) $C_6H_5NH_2$ (aniline, $K_b = 3.99 \times 10^{-10}$); (iii) $H_2NNH_2$ (hydrazine, $K_b = 1.05 \times 10^{-6}$); (iv) $C_5H_5N$ (pyridine, $K_b = 1.58 \times 10^{-9}$)?

**9-I.** A solution contains 63 different conjugate acid-base pairs. Among them is acrylic acid and acrylate ion, with the equilibrium ratio [acrylate]/[acrylic acid] $= 0.75$. What is the pH of the solution?

$$H_2C{=}CHCO_2H \qquad pK_a = 4.25$$
Acrylic acid

**9-J.** (a) Find the pH of a solution prepared by dissolving 1.00 g of glycine amide hydrochloride (Table 9-2) plus 1.00 g of glycine amide in 0.100 L.

Glycine amide
$C_2H_6N_2O$
FM 74.08

(b) How many grams of glycine amide should be added to 1.00 g of glycine amide hydrochloride to give 100 mL of solution with pH 8.00?
(c) What would be the pH if the solution in part **(a)** were mixed with 5.00 mL of 0.100 M HCl?
(d) What would be the pH if the solution in part **(c)** were mixed with 10.00 mL of 0.100 M NaOH?
(e) What would be the pH if the solution in part **(a)** were mixed with 90.46 mL of 0.100 M NaOH? (This is exactly the quantity of NaOH required to neutralize the glycine amide hydrochloride.)

**9-K.** A solution with an ionic strength of 0.10 M containing 0.010 0 M phenylhydrazine has a pH of 8.13. Using activity coefficients correctly, find $pK_a$ for the phenylhydrazinium ion found in phenylhydrazine hydrochloride. Assume that $\gamma_{BH^+} = 0.80$.

Phenylhydrazine
B

Phenylhydrazine hydrochloride
$BH^+Cl^-$

**9-L.** Use the GOAL SEEK spreadsheet at the end of the chapter to find the pH of 1.00 L of solution containing 0.030 mol HA ($pK_a = 2.50$) and 0.015 mol NaA. What would the pH be with the approximations [HA] = 0.030 and $[A^-]$ = 0.015?

## Problems

### Strong Acids and Bases

**9-1.** Why doesn't water dissociate to produce $10^{-7}$ M $H^+$ and $10^{-7}$ M $OH^-$ when some HBr is added?

**9-2.** Calculate the pH of (a) $1.0 \times 10^{-3}$ M HBr; (b) $1.0 \times 10^{-2}$ M KOH.

**9-3.** Calculate the pH of $5.0 \times 10^{-8}$ M $HClO_4$. What fraction of the total $H^+$ in this solution is derived from dissociation of water?

**9-4.** (a) The measured pH of 0.100 M HCl at 25°C is 1.092. From this information, calculate the activity coefficient of $H^+$ and compare your answer with that in Table 8-1.
(b) The measured pH of 0.010 0 M HCl + 0.090 0 M KCl at 25°C is 2.102. From this information, calculate the activity coefficient of $H^+$ in this solution.
(c) The ionic strengths of the solutions in parts **(a)** and **(b)** are the same. What can you conclude about the dependence of activity coefficients on the particular ions in a solution?

## Weak-Acid Equilibria

**9-5.** Write the chemical reaction whose equilibrium constant is
(a) $K_a$ for benzoic acid, $C_6H_5CO_2H$
(b) $K_b$ for benzoate ion, $C_6H_5CO_2^-$
(c) $K_b$ for aniline, $C_6H_5NH_2$
(d) $K_a$ for anilinium ion, $C_6H_5NH_3^+$

**9-6.** Find the pH and fraction of dissociation ($\alpha$) of a 0.100 M solution of the weak acid HA with $K_a = 1.00 \times 10^{-5}$.

**9-7.** $BH^+ClO_4^-$ is a salt formed from the base B ($K_b = 1.00 \times 10^{-4}$) and perchloric acid. It dissociates into $BH^+$, a weak acid, and $ClO_4^-$, which is neither an acid nor a base. Find the pH of 0.100 M $BH^+ClO_4^-$.

**9-8.** Find the pH and concentrations of $(CH_3)_3N$ and $(CH_3)_3NH^+$ in a 0.060 M solution of trimethylammonium chloride.

**9-9.** Use the reaction quotient, $Q$, to explain why the fraction of dissociation of weak acid, HA, increases when the solution is diluted by a factor of 2.

**9-10.** *When is a weak acid weak and when is a weak acid strong?* Show that the weak acid HA will be 92% dissociated when dissolved in water if the formal concentration is one-tenth of $K_a$ (F = $K_a/10$). Show that the fraction of dissociation is 27% when F = $10K_a$. At what formal concentration will the acid be 99% dissociated? Compare your answer with the left-hand curve in Figure 9-2.

**9-11.** A 0.045 0 M solution of benzoic acid has a pH of 2.78. Calculate $pK_a$ for this acid.

**9-12.** A 0.045 0 M solution of HA is 0.60% dissociated. Calculate $pK_a$ for this acid.

**9-13.** Barbituric acid dissociates as follows:

Barbituric acid
HA

$K_a = 9.8 \times 10^{-5}$

$A^-$

(a) Calculate the pH and fraction of dissociation of $10^{-2.00}$ M barbituric acid.
(b) Calculate the pH and fraction of dissociation of $10^{-10.00}$ M barbituric acid.

**9-14.** Using activity coefficients, find the pH and fraction of dissociation of 50.0 mM hydroxybenzene (phenol) in 0.050 M LiBr. Take the size of $C_6H_5O^-$ to be 600 pm.

**9-15.** $Cr^{3+}$ is acidic by virtue of the hydrolysis reaction

$$Cr^{3+} + H_2O \xrightleftharpoons{K_{a1}} Cr(OH)^{2+} + H^+$$

[Further reactions produce $Cr(OH)_2^+$, $Cr(OH)_3$, and $Cr(OH)_4^-$.] Find the value of $K_{a1}$ in Figure 6-8. Considering only the $K_{a1}$ reaction, find the pH of 0.010 M $Cr(ClO_4)_3$. What fraction of chromium is in the form $Cr(OH)^{2+}$?

**9-16.** From the dissociation constant of $HNO_3$ at 25°C in Box 9-1, find the percent dissociated in 0.100 M $HNO_3$ and in 1.00 M $HNO_3$.

**9-17.** 🖩 *Excel* GOAL SEEK. Solve the equation $x^2/(F - x) = K$ by using GOAL SEEK (Section 6-8). Guess a value of $x$ in cell A4 and evaluate $x^2/(F - x)$ in cell B4. Use GOAL SEEK to vary the value of $x$ until $x^2/(F - x)$ is equal to $K$. Use your spreadsheet to check your answer to Problem 9-6.

| | A | B |
|---|---|---|
| 1 | Using Excel GOAL SEEK | |
| 2 | | |
| 3 | x = | $x^2$/(F-x) = |
| 4 | 0.01 | 1.1111E-03 |
| 5 | F = | |
| 6 | 0.1 | |

## Weak-Base Equilibria

**9-18.** Covalent compounds generally have higher vapor pressure than ionic compounds. The "fishy" smell of fish arises from amines in the fish. Explain why squeezing lemon (which is acidic) onto fish reduces the fishy smell (and taste).

**9-19.** Find the pH and fraction of association ($\alpha$) of a 0.100 M solution of the weak base B with $K_b = 1.00 \times 10^{-5}$.

**9-20.** Find the pH and concentrations of $(CH_3)_3N$ and $(CH_3)_3NH^+$ in a 0.060 M solution of trimethylamine.

**9-21.** Find the pH of 0.050 M NaCN.

**9-22.** Calculate the fraction of association ($\alpha$) for $1.00 \times 10^{-1}$, $1.00 \times 10^{-2}$, and $1.00 \times 10^{-12}$ M sodium acetate. Does $\alpha$ increase or decrease with dilution?

**9-23.** A 0.10 M solution of a base has pH = 9.28. Find $K_b$.

**9-24.** A 0.10 M solution of a base is 2.0% hydrolyzed ($\alpha = 0.020$). Find $K_b$.

**9-25.** Show that the limiting fraction of association of a base in water, as the concentration of base approaches 0, is $\alpha = 10^7 K_b/(1 + 10^7 K_b)$. Find the limiting value of $\alpha$ for $K_b = 10^{-4}$ and for $K_b = 10^{-10}$.

## Buffers

**9-26.** Describe how to prepare 100 mL of 0.200 M acetate buffer, pH 5.00, starting with pure liquid acetic acid and solutions containing ~3 M HCl and ~3 M NaOH.

**9-27.** Why is the pH of a buffer nearly independent of concentration?

**9-28.** Why does buffer capacity increase as the concentration of buffer increases?

**9-29.** Why does buffer capacity increase as a solution becomes very acidic (pH ≈ 1) or very basic (pH ≈ 13)?

**9-30.** Why is the buffer capacity maximum when pH = $pK_a$?

**9-31.** Explain the following statement: The Henderson-Hasselbalch equation (with activity coefficients) is *always* true; what may not be correct are the values of [$A^-$] and [HA] that we choose to use in the equation.

**9-32.** Which of the following acids would be most suitable for preparing a buffer of pH 3.10: (i) hydrogen peroxide; (ii) propanoic acid; (iii) cyanoacetic acid; (iv) 4-aminobenzenesulfonic acid?

**9-33.** A buffer was prepared by dissolving 0.100 mol of the weak acid HA ($K_a = 1.00 \times 10^{-5}$) plus 0.050 mol of its conjugate base $Na^+A^-$ in 1.00 L. Find the pH.

**9-34.** Write the Henderson-Hasselbalch equation for a solution of formic acid. Calculate the quotient [$HCO_2^-$]/[$HCO_2H$] at (a) pH 3.000; (b) pH 3.744; (c) pH 4.000.

**9-35.** Given that $pK_b$ for nitrite ion ($NO_2^-$) is 10.85, find the quotient [$HNO_2$]/[$NO_2^-$] in a solution of sodium nitrite at (a) pH 2.00; (b) pH 10.00.

**9-36.** (a) Would you need NaOH or HCl to bring the pH of 0.050 0 M HEPES (Table 9-2) to 7.45?

(b) Describe how to prepare 0.250 L of 0.050 0 M HEPES, pH 7.45.

**9-37.** How many milliliters of 0.246 M $HNO_3$ should be added to 213 mL of 0.006 66 M 2,2′-bipyridine to give a pH of 4.19?

**9-38.** (a) Write the chemical reactions whose equilibrium constants are $K_b$ and $K_a$ for imidazole and imidazole hydrochloride, respectively.

(b) Calculate the pH of a solution prepared by mixing 1.00 g of imidazole with 1.00 g of imidazole hydrochloride and diluting to 100.0 mL.

(c) Calculate the pH of the solution if 2.30 mL of 1.07 M $HClO_4$ are added.

(d) How many milliliters of 1.07 M $HClO_4$ should be added to 1.00 g of imidazole to give a pH of 6.993?

**9-39.** Calculate the pH of a solution prepared by mixing 0.080 0 mol of chloroacetic acid plus 0.040 0 mol of sodium chloroacetate in 1.00 L of water.

(a) First do the calculation by assuming that the concentrations of HA and $A^-$ equal their formal concentrations.

(b) Then do the calculation, using the real values of [HA] and $[A^-]$ in the solution.

(c) Using first your head, and then the Henderson-Hasselbalch equation, find the pH of a solution prepared by dissolving all the following compounds in one beaker containing a total volume of 1.00 L: 0.180 mol $ClCH_2CO_2H$, 0.020 mol $ClCH_2CO_2Na$, 0.080 mol $HNO_3$, and 0.080 mol $Ca(OH)_2$. Assume that $Ca(OH)_2$ dissociates completely.

**9-40.** Calculate how many milliliters of 0.626 M KOH should be added to 5.00 g of MOBS (Table 9-2) to give a pH of 7.40.

**9-41.** (a) Use Equations 9-20 and 9-21 to find the pH and concentrations of HA and $A^-$ in a solution prepared by mixing 0.002 00 mol of acetic acid plus 0.004 00 mol of sodium acetate in 1.00 L of water.

(b) ▦ After working part (a) by hand, use Excel GOAL SEEK to find the same answers.

**9-42.** (a) Calculate the pH of a solution prepared by mixing 0.010 0 mol of the base B ($K_b = 10^{-2.00}$) with 0.020 0 mol of $BH^+Br^-$ and diluting to 1.00 L. First calculate the pH by assuming [B] = 0.010 0 and $[BH^+]$ = 0.020 0 M. Compare this answer with the pH calculated without making such an assumption.

(b) ▦ After working part (a) by hand, use Excel GOAL SEEK to find the same answers.

**9-43.** *Effect of ionic strength on $pK_a$.* $K_a$ for the $H_2PO_4^-/HPO_4^{2-}$ buffer is

$$K_a = \frac{[HPO_4^{2-}][H^+]\gamma_{HPO_4^{2-}}\gamma_{H^+}}{[H_2PO_4^-]\gamma_{H_2PO_4^-}} = 10^{-7.20}$$

If you mix a 1:1 mole ratio of $H_2PO_4^-$ and $HPO_4^{2-}$ at 0 ionic strength, the pH is 7.20. Find the pH of a 1:1 mixture of $H_2PO_4^-$ and $HPO_4^{2-}$ at an ionic strength of 0.10. Remember that $pH = -\log \mathcal{A}_{H^+} = -\log[H^+]\gamma_{H^+}$.

**9-44.** *Systematic treatment of equilibrium.* The acidity of $Al^{3+}$ is determined by the following reactions. Write the equations needed to find the pH of $Al(ClO_4)_3$ at a formal concentration F.

$$Al^{3+} + H_2O \underset{}{\overset{\beta_1}{\rightleftharpoons}} AlOH^{2+} + H^+$$

$$Al^{3+} + 2H_2O \underset{}{\overset{\beta_2}{\rightleftharpoons}} Al(OH)_2^+ + 2H^+$$

$$2Al^{3+} + 2H_2O \underset{}{\overset{K_{22}}{\rightleftharpoons}} Al_2(OH)_2^{4+} + 2H^+$$

$$Al^{3+} + 3H_2O \underset{}{\overset{\beta_3}{\rightleftharpoons}} Al(OH)_3(aq) + 3H^+$$

$$Al^{3+} + 4H_2O \underset{}{\overset{\beta_4}{\rightleftharpoons}} Al(OH)_4^- + 4H^+$$

$$3Al^{3+} + 4H_2O \underset{}{\overset{K_{43}}{\rightleftharpoons}} Al_3(OH)_4^{5+} + 4H^+$$

# 10 Polyprotic Acid-Base Equilibria

## ▪▪▪ PROTEINS ARE POLYPROTIC ACIDS AND BASES

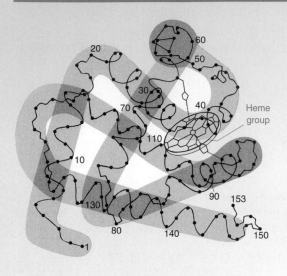

Myoglobin

Heme

Polypeptide backbone of myoglobin. Substituents are deleted for clarity. The heme group contains an iron atom that binds $O_2$, CO, and other small molecules. *[From M. F. Perutz, "The Hemoglobin Molecule." Copyright © 1964 by Scientific American, Inc. All rights reserved.]*

*Proteins* perform biological functions such as structural support, catalysis of chemical reactions, immune response to foreign substances, transport of molecules across membranes, and control of genetic expression. The three-dimensional structure and function of a protein is determined by the sequence of *amino acids* from which the protein is made. The diagram below shows how amino acids are connected to make a *polypeptide*. Of the 20 common amino acids, three have basic substituents and four have acidic substituents. Myoglobin, shown above, folds into several helical (spiral) regions that control access of oxygen and other small molecules to the heme group, whose function is to store $O_2$ in muscle cells. Of the 153 amino acids in sperm-whale myoglobin, 35 have basic side groups and 23 are acidic.

$$\overset{+}{H_3N}-\overset{\overset{\displaystyle H}{|}}{\underset{\underset{\displaystyle R_1}{|}}{C}}-CO_2^- \;+\; \overset{+}{H_3N}-\overset{\overset{\displaystyle H}{|}}{\underset{\underset{\displaystyle R_2}{|}}{C}}-CO_2^- \;+\; \overset{+}{H_3N}-\overset{\overset{\displaystyle H}{|}}{\underset{\underset{\displaystyle R_3}{|}}{C}}-CO_2^- \quad \text{Amino acids}$$

A side group (also called a substituent)

$-2H_2O$

$$\overset{+}{H_3N}-\overset{\overset{\displaystyle H}{|}}{\underset{\underset{\displaystyle R_1}{|}}{C}}-\overset{\overset{\displaystyle O}{\|}}{C}-N-\overset{\overset{\displaystyle H}{|}}{\underset{\underset{\displaystyle R_2}{|}}{C}}-\overset{\overset{\displaystyle O}{\|}}{C}-N-\overset{\overset{\displaystyle H}{|}}{\underset{\underset{\displaystyle R_3}{|}}{C}}-CO_2^-$$

A polypeptide (A long polypeptide is called a protein.)

N-terminal residue

Peptide bond

C-terminal residue

We now introduce **polyprotic** systems—acids or bases that can donate or accept more than one proton. After we have studied **diprotic** systems (with two acidic or basic sites), the extension to three or more acidic sites is straightforward. Then we step back and take a qualitative look at the big picture and think about which species are dominant at any given pH.

The **amino acid** building blocks of proteins have the general structure

Ammonium group ⟶ $H_3\overset{+}{N}$
CH—R ⟵ Substituent
Carboxyl group ⟶ $^-O—C$
‖
O

where R is a different group for each amino acid. The carboxyl group, drawn here in its ionized (basic) form, is a stronger acid than the ammonium group. Therefore, the nonionized form rearranges spontaneously to the **zwitterion:**

$H_2N$
CH—R ⟶
$HO_2C$

$H_3\overset{+}{N}$
CH—R
$^-O_2C$
Zwitterion

A *zwitterion* is a molecule with both positive and negative charges.

At low pH, both the ammonium group and the carboxyl group are protonated. At high pH, neither is protonated. Acid dissociation constants of amino acids are listed in Table 10-1, where each compound is drawn in its fully protonated form.

$pK_a$ values of amino acids in living cells are somewhat different from those in Table 10-1 because physiologic temperature is not 25°C and the ionic strength is not 0.

Zwitterions are stabilized in solution by interactions of $-NH_3^+$ and $-CO_2^-$ with water. The zwitterion is also the stable form of the amino acid in the solid state, where hydrogen bonding from $-NH_3^+$ to $-CO_2^-$ of neighboring molecules occurs. In the gas phase, there are no neighbors to stabilize the charges, so the nonionized structure with intramolecular hydrogen bonding from $-NH_2$ to a carboxyl oxygen predominates:

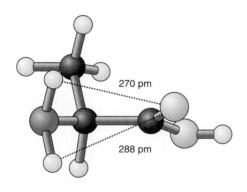

270 pm

288 pm

Gas-phase structure of alanine determined by microwave spectroscopy. [From S. Blanco, A. Lesarri, J. C. López, and J. L. Alonso, "The Gas-Phase Structure of Alanine," J. Am. Chem. Soc. **2004**, 126, 11675.]

Our discussion will focus on the amino acid leucine, designated HL.

$$H_3\overset{+}{N}CHCO_2H \underset{\phantom{}}{\overset{pK_{a1}\,=\,2.328}{\rightleftharpoons}} H_3\overset{+}{N}CHCO_2^- \underset{\phantom{}}{\overset{pK_{a2}\,=\,9.744}{\rightleftharpoons}} H_2NCHCO_2^-$$

$H_2L^+$

Leucine
HL

$L^-$

The substituent R in leucine is an isobutyl group: $(CH_3)_2CHCH_2-$

The equilibrium constants refer to the following reactions:

*Diprotic acid:*     $H_2L^+ \rightleftharpoons HL + H^+$     $K_{a1} \equiv K_1$     (10-1)

$HL \rightleftharpoons L^- + H^+$     $K_{a2} \equiv K_2$     (10-2)

*Diprotic base:*     $L^- + H_2O \rightleftharpoons HL + OH^-$     $K_{b1}$     (10-3)

$HL + H_2O \rightleftharpoons H_2L^+ + OH^-$     $K_{b2}$     (10-4)

We customarily omit the subscript "a" in $K_{a1}$ and $K_{a2}$. We will always write the subscript "b" in $K_{b1}$ and $K_{b2}$.

Recall that the relations between acid and base equilibrium constants are

*Relations between $K_a$ and $K_b$:*

$$K_{a1} \cdot K_{b2} = K_w$$     (10-5)

$$K_{a2} \cdot K_{b1} = K_w$$     (10-6)

**Table 10-1**  Acid dissociation constants of amino acids

| Amino acid[a] | Substituent[a] | Carboxylic acid[b] $pK_a$ | Ammonium[b] $pK_a$ | Substituent[b] $pK_a$ | Formula mass |
|---|---|---|---|---|---|
| Alanine (A) | $-CH_3$ | 2.344 | 9.868 | | 89.09 |
| Arginine (R) | $-CH_2CH_2CH_2NHC \overset{\overset{+}{N}H_2}{\underset{NH_2}{\big\|}}$ | 1.823 | 8.991 | (12.1[c]) | 174.20 |
| Asparagine (N) | $-CH_2\overset{O}{\overset{\|}{C}}NH_2$ | 2.16[c] | 8.73[c] | | 132.12 |
| Aspartic acid (D) | $-CH_2CO_2H$ | 1.990 | 10.002 | 3.900 | 133.10 |
| Cysteine (C) | $-CH_2SH$ | (1.7) | 10.74 | 8.36 | 121.16 |
| Glutamic acid (E) | $-CH_2CH_2CO_2H$ | 2.16 | 9.96 | 4.30 | 147.13 |
| Glutamine (Q) | $-CH_2CH_2\overset{O}{\overset{\|}{C}}NH_2$ | 2.19[c] | 9.00[c] | | 146.15 |
| Glycine (G) | $-H$ | 2.350 | 9.778 | | 75.07 |
| Histidine (H) | $-CH_2-$ imidazole | (1.6) | 9.28 | 5.97 | 155.16 |
| Isoleucine (I) | $-CH(CH_3)(CH_2CH_3)$ | 2.318 | 9.758 | | 131.17 |
| Leucine (L) | $-CH_2CH(CH_3)_2$ | 2.328 | 9.744 | | 131.17 |
| Lysine (K) | $-CH_2CH_2CH_2CH_2NH_3^+$ | (1.77) | 9.07 | 10.82 | 146.19 |
| Methionine (M) | $-CH_2CH_2SCH_3$ | 2.18[c] | 9.08[c] | | 149.21 |
| Phenylalanine (F) | $-CH_2-$ phenyl | 2.20 | 9.31 | | 165.19 |
| Proline (P) | Structure of entire amino acid | 1.952 | 10.640 | | 115.13 |
| Serine (S) | $-CH_2OH$ | 2.187 | 9.209 | | 105.09 |
| Threonine (T) | $-CH(CH_3)(OH)$ | 2.088 | 9.100 | | 119.12 |
| Tryptophan (W) | $-CH_2-$ indole | 2.37[c] | 9.33[c] | | 204.23 |
| Tyrosine (Y) | $-CH_2-$ phenyl$-OH$ | 2.41[c] | 8.67[c] | 11.01[c] | 181.19 |
| Valine (V) | $-CH(CH_3)_2$ | 2.286 | 9.719 | | 117.15 |

a. The acidic protons are shown in bold type. Each amino acid is written in its fully protonated form. Standard abbreviations are shown in parentheses.

b. $pK_a$ values refer to 25°C and zero ionic strength unless marked by c. Values considered to be uncertain are enclosed in parentheses.

c. For these entries, the ionic strength is 0.1 M, and the constant refers to a product of concentrations instead of activities.

SOURCE: A. E. Martell and R. J. Motekaitis, NIST Database 46 (Gaithersburg, MD: National Institute of Standards and Technology, 2001).

We now set out to calculate the pH and composition of individual solutions of 0.050 0 M $H_2L^+$, 0.050 0 M HL, and 0.050 0 M $L^-$. Our methods are general. They do not depend on the charge type of the acids and bases. That is, *we would use the same procedure to find the pH of the diprotic $H_2A$, where A is anything, or $H_2L^+$, where HL is leucine.*

## The Acidic Form, $H_2L^+$

Easy stuff.

Leucine hydrochloride contains the protonated species $H_2L^+$, which can dissociate twice (Reactions 10-1 and 10-2). Because $K_1 = 4.70 \times 10^{-3}$, $H_2L^+$ is a weak acid. HL is an even weaker acid, because $K_2 = 1.80 \times 10^{-10}$. It appears that the $H_2L^+$ will dissociate only partly, and the resulting HL will hardly dissociate at all. For this reason, we make the (superb) approximation that a solution of $H_2L^+$ behaves as a monoprotic acid, with $K_a = K_1$.

With this approximation, the calculation of the pH of 0.050 0 M $H_2L^+$ is a trivial matter.

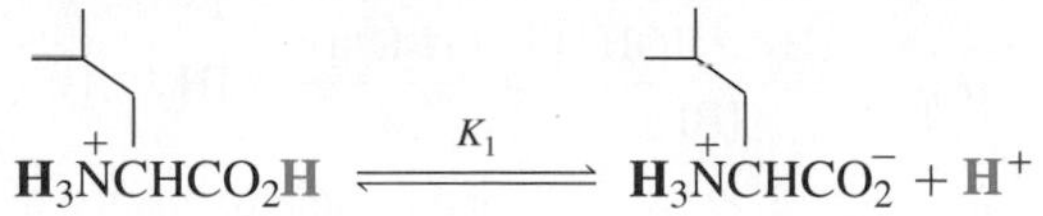

$$\underset{\substack{H_3\overset{+}{N}CHCO_2H \\ H_2L^+ \\ 0.050\ 0 - x}}{} \overset{K_1}{\rightleftharpoons} \underset{\substack{H_3\overset{+}{N}CHCO_2^- \\ HL \\ x}}{} + \underset{\substack{H^+ \\ \\ x}}{H^+}$$

$$K_a - K_1 = 4.70 \times 10^{-3}$$

$$\frac{x^2}{F - x} = K_a \Rightarrow x = 1.32 \times 10^{-2}\ M$$

$$[HL] = x = 1.32 \times 10^{-2}\ M$$

$$[H^+] = x = 1.32 \times 10^{-2}\ M \Rightarrow pH = 1.88$$

$$[H_2L^+] = F - x = 3.68 \times 10^{-2}\ M$$

What is the concentration of $L^-$ in the solution? We have already assumed that it is very small, but it cannot be 0. We can calculate $[L^-]$ from the $K_{a2}$ equation, with the concentrations of HL and $H^+$ that we just computed.

$$K_{a2} = \frac{[H^+][L^-]}{[HL]} \Rightarrow [L^-] = \frac{K_{a2}[HL]}{[H^+]} \qquad (10\text{-}7)$$

$$[L^-] = \frac{(1.80 \times 10^{-10})(1.32 \times 10^{-2})}{(1.32 \times 10^{-2})} = 1.80 \times 10^{-10}\ M \ (= K_{a2})$$

The approximation $[H^+] \approx [HL]$ reduces Equation 10-7 to $[L^-] = K_{a2} = 1.80 \times 10^{-10}\ M$.

Our approximation is confirmed by this last result. The concentration of $L^-$ is about eight orders of magnitude smaller than that of HL. The dissociation of HL is indeed negligible relative to the dissociation of $H_2L^+$. For most diprotic acids, $K_1$ is sufficiently larger than $K_2$ for this approximation to be valid. Even if $K_2$ were just 10 times less than $K_1$, $[H^+]$ calculated by ignoring the second ionization would be in error by only 4%. The error in pH would be only 0.01 pH unit. In summary, *a solution of a diprotic acid behaves like a solution of a monoprotic acid, with $K_a = K_{a1}$.*

## The Basic Form, $L^-$

The species $L^-$, found in a salt such as sodium leucinate, can be prepared by treating leucine (HL) with an equimolar quantity of NaOH. Dissolving sodium leucinate in water gives a solution of $L^-$, the fully basic species. $K_b$ values for this dibasic anion are

$$L^- + H_2O \rightleftharpoons HL + OH^- \qquad K_{b1} = K_w/K_{a2} = 5.55 \times 10^{-5}$$

$$HL + H_2O \rightleftharpoons H_2L^+ + OH^- \qquad K_{b2} = K_w/K_{a1} = 2.13 \times 10^{-12}$$

$K_{b1}$ tells us that $L^-$ will not *hydrolyze* (react with water) very much to give HL. Furthermore, $K_{b2}$ tells us that the resulting HL is such a weak base that hardly any further reaction to make $H_2L^+$ will occur.

We therefore treat $L^-$ as a monobasic species, with $K_b = K_{b1}$. The results of this (fantastic) approximation are outlined as follows.

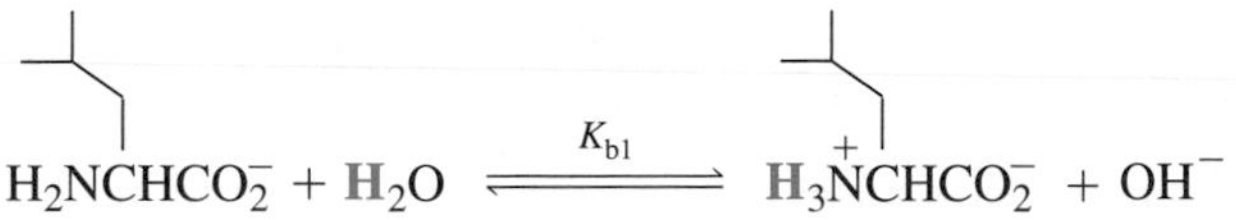

$$\underset{\substack{H_2NCHCO_2^- \\ L^- \\ 0.050\ 0 - x}}{} + H_2O \overset{K_{b1}}{\rightleftharpoons} \underset{\substack{H_3\overset{+}{N}CHCO_2^- \\ HL \\ x}}{} + \underset{\substack{OH^- \\ \\ x}}{OH^-}$$

$$K_b = K_{b1} = \frac{K_w}{K_{a2}} = 5.55 \times 10^{-5}$$

$$\frac{x^2}{F - x} = 5.55 \times 10^{-5} \Rightarrow x = 1.64 \times 10^{-3}\ M$$

$$[HL] = x = 1.64 \times 10^{-3}\ M$$

$$[H^+] = K_w/[OH^-] = K_w/x = 6.11 \times 10^{-12}\ M \Rightarrow pH = 11.21$$

$$[L^-] = F - x = 4.84 \times 10^{-2}\ M$$

**Side notes (right margin):**

$H_2L^+$ can be treated as monoprotic with $K_a = K_{a1}$.

Solve for $x$ with the quadratic equation.

More easy stuff.

**Hydrolysis** is the reaction of anything with water. Specifically, the reaction
$L^- + H_2O \rightleftharpoons HL + OH^-$ is called hydrolysis.

$L^-$ can be treated as monobasic with $K_b = K_{b1}$.

The concentration of $H_2L^+$ can be found from the $K_{b2}$ (or $K_{a1}$) equilibrium.

$$K_{b2} = \frac{[H_2L^+][OH^-]}{[HL]} = \frac{[H_2L^+]x}{x} = [H_2L^+]$$

We find that $[H_2L^+] = K_{b2} = 2.13 \times 10^{-12}$ M, and the approximation that $[H_2L^+]$ is insignificant relative to $[HL]$ is well justified. In summary, if there is any reasonable separation between $K_{a1}$ and $K_{a2}$ (and, therefore, between $K_{b1}$ and $K_{b2}$), *the fully basic form of a diprotic acid can be treated as monobasic, with $K_b = K_{b1}$.*

## The Intermediate Form, HL

A solution prepared from leucine, HL, is more complicated than one prepared from either $H_2L^+$ or $L^-$, because HL is both an acid and a base.

$$HL \rightleftharpoons H^+ + L^- \qquad\qquad K_a = K_{a2} = 1.80 \times 10^{-10} \qquad (10\text{-}8)$$

$$HL + H_2O \rightleftharpoons H_2L^+ + OH^- \qquad K_b = K_{b2} = 2.13 \times 10^{-12} \qquad (10\text{-}9)$$

A molecule that can both donate and accept a proton is said to be **amphiprotic.** The acid dissociation reaction (10-8) has a larger equilibrium constant than the base hydrolysis reaction (10-9), so we expect that a solution of leucine will be acidic.

However, we cannot simply ignore Reaction 10-9, even if $K_a$ and $K_b$ differ by several orders of magnitude. Both reactions proceed to nearly equal extent, because $H^+$ produced in Reaction 10-8 reacts with $OH^-$ from Reaction 10-9, thereby driving Reaction 10-9 to the right.

To treat this case, we resort to the systematic treatment of equilibrium. The procedure is applied to leucine, whose intermediate form (HL) has no net charge. However, the results apply to the intermediate form of *any* diprotic acid, regardless of its charge.

For Reactions 10-8 and 10-9, the charge balance is

$$[H^+] + [H_2L^+] = [L^-] + [OH^-] \quad \text{or} \quad [H_2L^+] - [L^-] + [H^+] - [OH^-] = 0$$

From the acid dissociation equilibria, we replace $[H_2L^+]$ with $[HL][H^+]/K_1$, and $[L^-]$ with $[HL]K_2/[H^+]$. Also, we can always write $[OH^-] = K_w/[H^+]$. Putting these expressions into the charge balance gives

$$\frac{[HL][H^+]}{K_1} - \frac{[HL]K_2}{[H^+]} + [H^+] - \frac{K_w}{[H^+]} = 0$$

which can be solved for $[H^+]$. First, multiply all terms by $[H^+]$:

$$\frac{[HL][H^+]^2}{K_1} - [HL]K_2 + [H^+]^2 - K_w = 0$$

Then rearrange and factor out $[H^+]^2$:

$$[H^+]^2 \left( \frac{[HL]}{K_1} + 1 \right) = K_2[HL] + K_w$$

$$[H^+]^2 = \frac{K_2[HL] + K_w}{\dfrac{[HL]}{K_1} + 1}$$

Now multiply the numerator and denominator by $K_1$ and take the square root:

$$[H^+] = \sqrt{\frac{K_1K_2[HL] + K_1K_w}{K_1 + [HL]}} \qquad (10\text{-}10)$$

Up to this point, we have made no approximations, except to neglect activity coefficients. We solved for $[H^+]$ in terms of known constants plus the single unknown, $[HL]$. Where do we proceed from here?

Just as we are feeling desperate, the Good Chemist gallops down from the mountains on her white stallion to provide the missing insight: "The major species is HL, because it is both a weak acid and a weak base. Neither Reaction 10-8 nor Reaction 10-9 goes very far. For $[HL]$ in Equation 10-10, you can simply substitute the formal concentration, 0.050 0 M."

Taking the Good Chemist's advice, we write Equation 10-10 in its most useful form.

*Intermediate form of diprotic acid:* $\qquad [H^+] \approx \sqrt{\dfrac{K_1K_2F + K_1K_w}{K_1 + F}} \qquad (10\text{-}11)$

where F is the formal concentration of HL ($= 0.050\ 0$ M in the present case).

A tougher problem.

HL is both an acid and a base.

The missing insight!

$K_1$ and $K_2$ in this equation are both *acid* dissociation constants ($K_{a1}$ and $K_{a2}$.)

CHAPTER 10 Polyprotic Acid-Base Equilibria

At long last, we can calculate the pH of 0.050 0 M leucine:

$$[H^+] = \sqrt{\frac{(4.70 \times 10^{-3})(1.80 \times 10^{-10})(0.050\ 0) + (4.70 \times 10^{-3})(1.0 \times 10^{-14})}{4.70 \times 10^{-3} + 0.050\ 0}}$$

$$= 8.80 \times 10^{-7}\ M \Rightarrow pH = 6.06$$

The concentrations of $H_2L^+$ and $L^-$ can be found from the $K_1$ and $K_2$ equilibria, using $[H^+] = 8.80 \times 10^{-7}$ M and $[HL] = 0.050\ 0$ M.

$$[H_2L^+] = \frac{[H^+][HL]}{K_1} = \frac{(8.80 \times 10^{-7})(0.050\ 0)}{4.70 \times 10^{-3}} = 9.36 \times 10^{-6}\ M$$

$$[L^-] = \frac{K_2[HL]}{[H^+]} = \frac{(1.80 \times 10^{-10})(0.050\ 0)}{8.80 \times 10^{-7}} = 1.02 \times 10^{-5}\ M$$

Was the approximation $[HL] \approx 0.050\ 0$ M a good one? It certainly was, because $[H_2L^+] (= 9.36 \times 10^{-6}\ M)$ and $[L^-] (= 1.02 \times 10^{-5}\ M)$ are small in comparison with $[HL]$ ($\approx 0.050\ 0$ M). Nearly all the leucine remained in the form HL. Note also that $[H_2L^+]$ is nearly equal to $[L^-]$. This result confirms that Reactions 10-8 and 10-9 proceed equally, even though $K_a$ is 84 times bigger than $K_b$ for leucine.

> If $[H_2L^+] + [L^-]$ is not much less than $[HL]$ and if you wish to refine your values of $[H_2L^+]$ and $[L^-]$, try the method in Box 10-1.

A summary of results for leucine is given here. Notice the relative concentrations of $H_2L^+$, HL, and $L^-$ in each solution and notice the pH of each solution.

| Solution | pH | $[H^+]$ (M) | $[H_2L^+]$ (M) | $[HL]$ (M) | $[L^-]$ (M) |
|---|---|---|---|---|---|
| 0.050 0 M $H_2A$ | 1.88 | $1.32 \times 10^{-2}$ | $3.68 \times 10^{-2}$ | $1.32 \times 10^{-2}$ | $1.80 \times 10^{-10}$ |
| 0.050 0 M $HA^-$ | 6.06 | $8.80 \times 10^{-7}$ | $9.36 \times 10^{-6}$ | $5.00 \times 10^{-2}$ | $1.02 \times 10^{-5}$ |
| 0.050 0 M $HA^{2-}$ | 11.21 | $6.08 \times 10^{-12}$ | $2.13 \times 10^{-12}$ | $1.64 \times 10^{-3}$ | $4.84 \times 10^{-2}$ |

## Simplified Calculation for the Intermediate Form

Usually Equation 10-11 is a fair-to-excellent approximation. It applies to the intermediate form of any diprotic acid, regardless of its charge type.

An even simpler form of Equation 10-11 results from two conditions that usually exist. First, if $K_2F \gg K_w$, the second term in the numerator of Equation 10-11 can be dropped.

$$[H^+] \approx \sqrt{\frac{K_1 K_2 F + K_1 K_w}{K_1 + F}}$$

Then, if $K_1 \ll F$, the first term in the denominator also can be neglected.

$$[H^+] \approx \sqrt{\frac{K_1 K_2 F}{K_1 + F}}$$

Canceling F in the numerator and denominator gives

$$[H^+] \approx \sqrt{K_1 K_2}$$

or

$$\log[H^+] \approx \tfrac{1}{2}(\log K_1 + \log K_2)$$

$$-\log[H^+] \approx -\tfrac{1}{2}(\log K_1 + \log K_2)$$

> Recall that $\log(x^{1/2}) = \tfrac{1}{2}\log x$ and $\log xy = \log x + \log y$.

*Intermediate form of diprotic acid:* $\qquad pH = \tfrac{1}{2}(pK_1 + pK_2)$ $\qquad\qquad$ (10-12)

> The pH of the intermediate form of a diprotic acid is close to midway between the two $pK_a$ values and is almost independent of concentration.

Equation 10-12 is a good one to keep in your head. It gives a pH of 6.04 for leucine, compared with pH = 6.06 from Equation 10-11. Equation 10-12 says that *the pH of the intermediate form of a diprotic acid is close to midway between $pK_1$ and $pK_2$, regardless of the formal concentration.*

### Example pH of the Intermediate Form of a Diprotic Acid

Potassium hydrogen phthalate, KHP, is a salt of the intermediate form of phthalic acid. Calculate the pH of both 0.10 M and 0.010 M KHP.

Phthalic acid — $H_2P$ $\qquad$ Monohydrogen phthalate — $HP^-$ $\qquad$ Phthalate — $P^{2-}$

(Potassium hydrogen phthalate = $K^+HP^-$)

## Box 10-1 Successive Approximations

The method of *successive approximations* is a good way to deal with difficult equations that do not have simple solutions. For example, Equation 10-11 is not a good approximation when the concentration of the intermediate species of a diprotic acid is not close to F, the formal concentration. This situation arises when $K_1$ and $K_2$ are nearly equal and F is small. Consider a solution of $1.00 \times 10^{-3}$ M HM$^-$, the intermediate form of malic acid.

Malic acid
$H_2M$

HM$^-$

M$^{2-}$

For a first approximation, assume that $[HM^-] \approx 1.00 \times 10^{-3}$ M. Plugging this value into Equation 10-10, we calculate first approximations for $[H^+]$, $[H_2M]$, and $[M^{2-}]$.

$$[H^+]_1 = \sqrt{\frac{K_1 K_2 (0.001\ 00) + K_1 K_w}{K_1 + (0.001\ 00)}} = 4.53 \times 10^{-5}\ M$$

$$\Rightarrow [H_2M]_1 = \frac{[H^+][HM^-]}{K_1}$$

$$= \frac{(4.53 \times 10^{-5})(1.00 \times 10^{-3})}{3.5 \times 10^{-4}} = 1.29 \times 10^{-4}\ M$$

$$[M^{2-}]_1 = \frac{K_2[HM^-]}{[H^+]} = \frac{(7.9 \times 10^{-6})(1.00 \times 10^{-3})}{4.53 \times 10^{-5}}$$

$$= 1.75 \times 10^{-4}\ M$$

Clearly, $[H_2M]$ and $[M^{2-}]$ are not negligible relative to $F = 1.00 \times 10^{-3}$ M, so we need to revise our estimate of $[HM^-]$. The mass balance gives us a second approximation:

$$[HM^-]_2 = F - [H_2M]_1 - [M^{2-}]_1$$
$$= 0.001\ 00 - 0.000\ 129 - 0.000\ 175 = 0.000\ 696\ M$$

Inserting $[HM^-]_2 = 0.000\ 696$ into Equation 10-10 gives

$$[H^+]_2 = \sqrt{\frac{K_1 K_2 (0.000\ 696) + K_1 K_w}{K_1 + (0.000\ 696)}} = 4.29 \times 10^{-5}\ M$$

$$\Rightarrow [H_2M]_2 = 8.53 \times 10^{-5}\ M$$

$$[M^{2-}]_2 = 1.28 \times 10^{-4}\ M$$

$[H_2M]_2$ and $[M^{2-}]_2$ can be used to calculate a third approximation for $[HM^-]$:

$$[HM^-]_3 = F - [H_2M]_2 - [M^{2-}]_2 = 0.000\ 786\ M$$

Plugging $[HM^-]_3$ into Equation 10-10 gives

$$[H^+]_3 = 4.37 \times 10^{-5}$$

and the procedure can be repeated to get

$$[H^+]_4 = 4.35 \times 10^{-5}$$

We are homing in on an estimate of $[H^+]$ in which the precision is already less than 1%. The fourth approximation gives pH = 4.36, compared with pH = 4.34 from the first approximation and pH = 4.28 from the formula pH $\approx \frac{1}{2}(pK_1 + pK_2)$. Considering the uncertainty in pH measurements, all this calculation was hardly worth the effort. However, the concentration $[HM^-]_5$ is 0.000 768 M, which is 23% less than the original estimate of 0.001 00 M. Successive approximations can be carried out by hand, but the process is more easily and reliably performed with a spreadsheet.

---

**Solution** With Equation 10-12, the pH of potassium hydrogen phthalate is estimated as $\frac{1}{2}(pK_1 + pK_2) = 4.18$, regardless of concentration. With Equation 10-11, we calculate pH = 4.18 for 0.10 M K$^+$HP$^-$ and pH = 4.20 for 0.010 M K$^+$HP$^-$.

**Advice** When faced with the intermediate form of a diprotic acid, use Equation 10-11 to calculate the pH. The answer should be close to $\frac{1}{2}(pK_1 + pK_2)$.

### Summary of Diprotic Acid Calculations

Here is how we calculate the pH and composition of solutions prepared from different forms of a diprotic acid ($H_2A$, HA$^-$, or A$^{2-}$).

**Solution of $H_2A$**

1. Treat $H_2A$ as a monoprotic acid with $K_a = K_1$ to find $[H^+]$, $[HA^-]$, and $[H_2A]$.

$$H_2A \underset{}{\overset{K_1}{\rightleftharpoons}} H^+ + HA^- \qquad \frac{x^2}{F - x} = K_1$$
$$F - x \qquad\quad x \qquad x$$

2. Use the $K_2$ equilibrium to solve for $[A^{2-}]$.

$$[A^{2-}] = \frac{K_2[\cancel{HA^-}]}{\cancel{[H^+]}} = K_2$$

**Solution of HA⁻**

1. Use the approximation $[HA^-] \approx F$ and find the pH with Equation 10-11.

$$[H^+] = \sqrt{\frac{K_1 K_2 F + K_1 K_w}{K_1 + F}}$$

The pH should be close to $\frac{1}{2}(pK_1 + pK_2)$.

2. With $[H^+]$ from step 1 and $[HA^-] \approx F$, solve for $[H_2A]$ and $[A^{2-}]$, using the $K_1$ and $K_2$ equilibria.

$$[H_2A] = \frac{[HA^-][H^+]}{K_1} \qquad [A^{2-}] = \frac{K_2[HA^-]}{[H^+]}$$

**Solution of A²⁻**

1. Treat $A^{2-}$ as monobasic, with $K_b = K_{b1} = K_w/K_{a2}$ to find $[A^{2-}]$, $[HA^-]$, and $[H^+]$.

$$A^{2-} + H_2O \overset{K_{b1}}{\rightleftharpoons} HA^- + OH^- \qquad \frac{x^2}{F - x} = K_{b1} = \frac{K_w}{K_{a2}}$$
$$F - x \qquad\qquad x \qquad x$$

$$[H^+] = \frac{K_w}{[OH^-]} = \frac{K_w}{x}$$

2. Use the $K_1$ equilibrium to solve for $[H_2A]$.

$$[H_2A] = \frac{[HA^-][H^+]}{K_{a1}} = \frac{\cancel{[HA^-]}(K_w/\cancel{[OH^-]})}{K_{a1}} = K_{b2}$$

The calculations we have been doing are really important to understand and use. However, we should not get too cocky about our great powers, because there could be equilibria we have not considered. For example, $Na^+$ or $K^+$ in solutions of $HA^-$ or $A^{2-}$ form weak ion pairs that we have neglected:[1]

$$K^+ + A^{2-} \rightleftharpoons \{K^+A^{2-}\}$$
$$K^+ + HA^- \rightleftharpoons \{K^+HA^-\}$$

## ▮▮▮ 10-2 Diprotic Buffers

A buffer made from a diprotic (or polyprotic) acid is treated in the same way as a buffer made from a monoprotic acid. For the acid $H_2A$, we can write *two* Henderson-Hasselbalch equations, both of which are *always* true. If we happen to know $[H_2A]$ and $[HA^-]$, then we will use the $pK_1$ equation. If we know $[HA^-]$ and $[A^{2-}]$, we will use the $pK_2$ equation.

$$pH = pK_1 + \log\frac{[HA^-]}{[H_2A]} \qquad pH = pK_2 + \log\frac{[A^{2-}]}{[HA^-]}$$

All Henderson-Hasselbalch equations (with activity coefficients) are always true for a solution at equilibrium.

---

**Example** A Diprotic Buffer System

Find the pH of a solution prepared by dissolving 1.00 g of potassium hydrogen phthalate and 1.20 g of disodium phthalate in 50.0 mL of water.

**Solution** Monohydrogen phthalate and phthalate were shown in the preceding example. The formula masses are $KHP = C_8H_5O_4K = 204.221$ and $Na_2P = C_8H_4O_4Na_2 = 210.094$. We know $[HP^-]$ and $[P^{2-}]$, so we use the $pK_2$ Henderson-Hasselbalch equation to find the pH:

$$pH = pK_2 + \log\frac{[P^{2-}]}{[HP^-]} = 5.408 + \log\frac{(1.20\ g)/(210.094\ g/mol)}{(1.00\ g)/(204.221\ g/mol)} = 5.47$$

$K_2$ is the acid dissociation constant of $HP^-$, which appears in the denominator of the log term. Notice that the volume of solution was not used to answer the question.

---

**Example** Preparing a Buffer in a Diprotic System

How many milliliters of 0.800 M KOH should be added to 3.38 g of oxalic acid to give a pH of 4.40 when diluted to 500 mL?

$$\begin{array}{c} O\ O \\ \parallel\ \parallel \\ HOCCOH \end{array} \qquad \begin{array}{l} pK_1 = 1.27 \\ pK_2 = 4.266 \end{array}$$
Oxalic acid
$(H_2Ox)$
Formula mass = 90.035

**Solution** The desired pH is above $pK_2$. We know that a 1:1 mole ratio of $HOx^- : Ox^{2-}$ would have pH = $pK_2 = 4.266$. If the pH is to be 4.40, there must be more $Ox^{2-}$ than

$HOx^-$ present. We must add enough base to convert all of the $H_2Ox$ into $HOx^-$, plus enough additional base to convert the right amount of $HOx^-$ into $Ox^{2-}$.

$$H_2Ox + OH^- \rightarrow HOx^- + H_2O$$

$$\uparrow$$

$$pH \approx \tfrac{1}{2}(pK_1 + pK_2) = 2.77$$

$$HOx^- + OH^- \rightarrow Ox^{2-} + H_2O$$

A 1:1 mixture would have
pH = $pK_2$ = 4.266

In 3.38 g of $H_2Ox$, there are $0.037\,5_4$ mol. The volume of 0.800 M KOH needed to react with this much $H_2Ox$ to make $HOx^-$ is $(0.037\,5_4 \text{ mol})/(0.800 \text{ M}) = 46.9_2$ mL.
To produce a pH of 4.40 requires more $OH^-$:

|              | $HOx^-$        | + | $OH^-$ | $\rightarrow$ | $Ox^{2-}$ |
|--------------|----------------|---|--------|---------------|-----------|
| Initial moles | $0.037\,5_4$  |   | $x$    |               | —         |
| Final moles   | $0.037\,5_4 - x$ |  | —      |               | $x$       |

$$pH = pK_2 + \log \frac{[Ox^{2-}]}{[HOx^-]}$$

$$4.40 = 4.266 + \log \frac{x}{0.037\,5_4 - x} \Rightarrow x = 0.021\,6_4 \text{ mol}$$

The volume of KOH needed to deliver $0.021\,6_4$ mole is $(0.021\,6_4 \text{ mol})/(0.800 \text{ M}) = 27.0_5$ mL. The total volume of KOH needed to bring the pH to 4.40 is $46.9_2 + 27.0_5 = 73.9_7$ mL.

## ▪ ▪ ▪ 10-3  Polyprotic Acids and Bases

The treatment of diprotic acids and bases can be extended to polyprotic systems. By way of review, let's write the pertinent equilibria for a triprotic system.

$$H_3A \rightleftharpoons H_2A^- + H^+ \qquad K_{a1} = K_1$$
$$H_2A^- \rightleftharpoons HA^{2-} + H^+ \qquad K_{a2} = K_2$$
$$HA^{2-} \rightleftharpoons A^{3-} + H^+ \qquad K_{a3} = K_3$$
$$A^{3-} + H_2O \rightleftharpoons HA^{2-} + OH^- \qquad K_{b1} = \frac{K_w}{K_{a3}}$$
$$HA^{2-} + H_2O \rightleftharpoons H_2A^- + OH^- \qquad K_{b2} = \frac{K_w}{K_{a2}}$$
$$H_2A^- + H_2O \rightleftharpoons H_3A + OH^- \qquad K_{b3} = \frac{K_w}{K_{a1}}$$

We deal with triprotic systems as follows:

1. $H_3A$ is treated as a monoprotic weak acid, with $K_a = K_1$.
2. $H_2A^-$ is treated as the intermediate form of a diprotic acid.

The K values in Equations 10-13 and 10-14 are $K_a$ values for the triprotic acid.

$$[H^+] \approx \sqrt{\frac{K_1 K_2 F + K_1 K_w}{K_1 + F}} \qquad (10\text{-}13)$$

3. $HA^{2-}$ is also treated as the intermediate form of a diprotic acid. However, $HA^{2-}$ is "surrounded" by $H_2A^-$ and $A^{3-}$, so the equilibrium constants to use are $K_2$ and $K_3$, instead of $K_1$ and $K_2$.

$$[H^+] \approx \sqrt{\frac{K_2 K_3 F + K_2 K_w}{K_2 + F}} \qquad (10\text{-}14)$$

4. $A^{3-}$ is treated as monobasic, with $K_b = K_{b1} = K_w/K_{a3}$.

**Example** A Triprotic System

Find the pH of 0.10 M $H_3His^{2+}$, 0.10 M $H_2His^+$, 0.10 M HHis, and 0.10 M His$^-$, where His stands for the amino acid histidine.

H$_3$His$^{2+}$    $pK_1 = 1.6$    H$_2$His$^+$

$pK_2 = 5.97$

His$^-$    $pK_3 = 9.28$    HHis Histidine

**Solution** *0.10 M $H_3His^{2+}$:* Treating $H_3His^{2+}$ as a monoprotic acid, we write

$$H_3His^{2+} \rightleftharpoons H_2His^+ + H^+$$
$$F - x \qquad x \qquad x$$

$$\frac{x^2}{F - x} = K_1 = 10^{-1.6} \Rightarrow x = 3._9 \times 10^{-2} \text{ M} \Rightarrow \text{pH} = 1.41$$

*0.10 M $H_2His^+$:* Using Equation 10-13, we find

$$[H^+] = \sqrt{\frac{(10^{-1.6})(10^{-5.97})(0.10) + (10^{-1.6})(1.0 \times 10^{-14})}{10^{-1.6} + 0.10}}$$
$$= 1.4_7 \times 10^{-4} \text{ M} \Rightarrow \text{pH} = 3.83$$

which is close to $\frac{1}{2}(pK_1 + pK_2) = 3.78$.

*0.10 M HHis:* Equation 10-14 gives

$$[H^+] = \sqrt{\frac{(10^{-5.97})(10^{-9.28})(0.10) + (10^{-5.97})(1.0 \times 10^{-14})}{10^{-5.97} + 0.10}}$$
$$= 2.3_7 \times 10^{-8} \text{ M} \Rightarrow \text{pH} = 7.62$$

which is the same as $\frac{1}{2}(pK_2 + pK_3) = 7.62$.

*0.10 M His$^-$:* Treating His$^-$ as monobasic, we can write

$$His^- + H_2O \rightleftharpoons HHis + OH^-$$
$$F - x \qquad\qquad x \qquad x$$

$$\frac{x^2}{F - x} = K_{b1} = \frac{K_w}{K_{a3}} = 1.9 \times 10^{-5} \Rightarrow x = 1.3_7 \times 10^{-3} \text{ M}$$
$$\text{pH} = -\log\left(\frac{K_w}{x}\right) = 11.14$$

We have reduced acid-base problems to just three types. When you encounter an acid or base, decide whether you are dealing with an *acidic, basic,* or *intermediate* form. Then do the appropriate arithmetic to answer the question at hand.

Three forms of acids and bases:
• acidic
• basic
• intermediate (amphiprotic)

## 10-4  Which Is the Principal Species?

An example of when you need to know principal species is when you design a chromatographic or electrophoretic separation. You would use different strategies for separating cations, anions, and neutral compounds.

At $pH = pK_a$, $[A^-] = [HA]$ because
$$pH = pK_a + \log \frac{[A^-]}{[HA]}.$$

| pH | Major species |
|----|----|
| $< pK_a$ | HA |
| $> pK_a$ | $A^-$ |

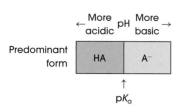

| pH | Major species |
|----|----|
| $pH < pK_1$ | $H_2A$ |
| $pK_1 < pH < pK_2$ | $HA^-$ |
| $pH > pK_2$ | $A^{2-}$ |

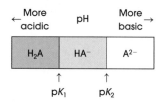

We are often faced with the problem of identifying which species of acid, base, or intermediate is predominant under given conditions. A simple example is, "What is the principal form of benzoic acid at pH 8?"

$$\bigcirc\!\!-CO_2H \quad pK_a = 4.20$$
Benzoic acid

$pK_a$ for benzoic acid is 4.20. This means that, at pH 4.20, there is a 1:1 mixture of benzoic acid (HA) and benzoate ion ($A^-$). At $pH = pK_a + 1$ (= 5.20), the quotient $[A^-]/[HA]$ is 10:1. At $pH = pK_a + 2$ (= 6.20), the quotient $[A^-]/[HA]$ is 100:1. As pH increases, the quotient $[A^-]/[HA]$ increases still further.

For a monoprotic system, the basic species, $A^-$, is the predominant form when $pH > pK_a$. The acidic species, HA, is the predominant form when $pH < pK_a$. The predominant form of benzoic acid at pH 8 is the benzoate anion, $C_6H_5CO_2^-$.

---

**Example** Principal Species—Which One and How Much?

What is the predominant form of ammonia in a solution at pH 7.0? Approximately what fraction is in this form?

**Solution**   In Appendix G, we find $pK_a = 9.24$ for the ammonium ion ($NH_4^+$, the conjugate acid of ammonia, $NH_3$). At $pH = 9.24$, $[NH_4^+] = [NH_3]$. Below pH 9.24, $NH_4^+$ will be the predominant form. Because $pH = 7.0$ is about 2 pH units below $pK_a$, the quotient $[NH_4^+]/[NH_3]$ will be about 100:1. More than 99% is in the form $NH_4^+$.

For polyprotic systems, our reasoning is similar, but there are several values of $pK_a$. Consider oxalic acid, $H_2Ox$, with $pK_1 = 1.27$ and $pK_2 = 4.27$. At $pH = pK_1$, $[H_2Ox] = [HOx^-]$. At $pH = pK_2$, $[HOx^-] = [Ox^{2-}]$. The chart in the margin shows the major species in each pH region.

---

**Example** Principal Species in a Polyprotic System

The amino acid arginine has the following forms:

$$\underset{H_3Arg^{2+}}{\underset{CO_2H}{\overset{NH_3^+\xleftarrow{\;\alpha\;}\;\text{Substituent}}{|}}\,CHCH_2CH_2CH_2NHC\underset{NH_2}{\overset{NH_2^+}{\diagup\!\!\diagdown}}} \quad \xrightleftharpoons{pK_1 = 1.82} \quad \underset{H_2Arg^+}{\underset{CO_2^-}{\overset{NH_3^+}{|}}\,CHCH_2CH_2CH_2NHC\underset{NH_2}{\overset{NH_2^+}{\diagup\!\!\diagdown}}} \quad \xrightleftharpoons{pK_2 = 8.99}$$

$$\underset{\underset{Arginine}{HArg}}{\underset{CO_2^-}{\overset{NH_2}{|}}\,CHCH_2CH_2CH_2NHC\underset{NH_2}{\overset{NH_2^+}{\diagup\!\!\diagdown}}} \quad \xrightleftharpoons{pK_3 = 12.1} \quad \underset{Arg^-}{\underset{CO_2^-}{\overset{NH_2}{|}}\,CHCH_2CH_2CH_2NHC\underset{NH_2}{\overset{NH}{\diagup\!\!\diagdown}}}$$

Appendix G tells us that the $\alpha$-ammonium group (at the left) is more acidic than the substituent (at the right). What is the principal form of arginine at pH 10.0? Approximately what fraction is in this form? What is the second most abundant form at this pH?

**Solution**   We know that, at $pH = pK_2 = 8.99$, $[H_2Arg^+] = [HArg]$. At $pH = pK_3 = 12.1$, $[HArg] = [Arg^-]$. At $pH = 10.0$, the major species is HArg. Because pH 10.0 is about one pH unit higher than $pK_2$, we can say that $[HArg]/[H_2Arg^+] \approx 10:1$. About 90% of arginine is in the form HArg. The second most important species is $H_2Arg^+$, which makes up about 10% of the arginine.

---

**Example** More on Polyprotic Systems

In the pH range 1.82 to 8.99, $H_2Arg^+$ is the principal form of arginine. Which is the second most prominent species at pH 6.0? at pH 5.0?

**Solution**   We know that the pH of the pure intermediate (amphiprotic) species, $H_2Arg^+$, is

$$\text{pH of } H_2Arg^+ \approx \tfrac{1}{2}(pK_1 + pK_2) = 5.40$$

Above pH 5.40 (and below pH $= pK_2$), HArg, the conjugate base of $H_2Arg^+$, will be the second most important species. Below pH 5.40 (and above pH $= pK_1$), $H_3Arg^{2+}$ will be the second most important species.

Figure 10-1 summarizes how we think of a triprotic system. We determine the principal species by comparing the pH of the solution with the $pK_a$ values.

Go back and read the Example "Preparing a Buffer in a Diprotic System" on page 187. See if it makes more sense now.

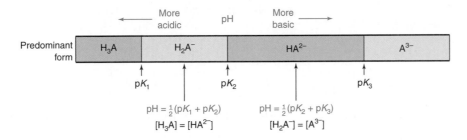

**Figure 10-1** The predominant molecular form of a triprotic system ($H_3A$) in the various pH intervals.

**Speciation** describes the distribution of analyte among possible species. For an acid or base, speciation describes how much of each protonated form is present. When you ingest inorganic arsenic ($AsO(OH)_3$ and $As(OH)_3$) from drinking water, it is methylated to species such as $(CH_3)AsO(OH)_2$, $(CH_3)As(OH)_2$, $(CH_3)_2AsO(OH)$, $(CH_3)_2As(OH)$, $(CH_3)_3AsO$, and $(CH_3)_3As$. Speciation describes the forms and quantities that are present.

## ■■■ 10-5  Fractional Composition Equations

We now derive equations that give the fraction of each species of acid or base at a given pH. These equations will be useful for acid-base and EDTA titrations, as well as for electrochemical equilibria. They will be of key value in Chapter 13.

### Monoprotic Systems

Our goal is to find an expression for the fraction of an acid in each form (HA and $A^-$) as a function of pH. We can do this by combining the equilibrium constant with the mass balance. Consider an acid with formal concentration F:

$$HA \overset{K_a}{\rightleftharpoons} H^+ + A^- \qquad K_a = \frac{[H^+][A^-]}{[HA]}$$

$$\text{Mass balance:} \quad F = [HA] + [A^-]$$

Rearranging the mass balance gives $[A^-] = F - [HA]$, which can be plugged into the $K_a$ expression to give

$$K_a = \frac{[H^+](F - [HA])}{[HA]}$$

or, with a little algebra,

$$[HA] = \frac{[H^+]F}{[H^+] + K_a} \tag{10-15}$$

The *fraction* of molecules in the form HA is called $\alpha_{HA}$.

$$\alpha_{HA} = \frac{[HA]}{[HA] + [A^-]} = \frac{[HA]}{F} \tag{10-16}$$

Dividing Equation 10-15 by F gives

*Fraction in the form HA:* $\qquad \alpha_{HA} = \frac{[HA]}{F} = \frac{[H^+]}{[H^+] + K_a} \tag{10-17}$

In a similar manner, the fraction in the form $A^-$, designated $\alpha_{A^-}$, can be obtained:

*Fraction in the form $A^-$:* $\qquad \alpha_{A^-} = \frac{[A^-]}{F} = \frac{K_a}{[H^+] + K_a} \tag{10-18}$

$\alpha_{HA}$ = fraction of species in the form HA
$\alpha_{A^-}$ = fraction of species in the form $A^-$

$$\alpha_{HA} + \alpha_{A^-} = 1$$

The fraction denoted here as $\alpha_{A^-}$ is the same thing we called the *fraction of dissociation* ($\alpha$) previously.

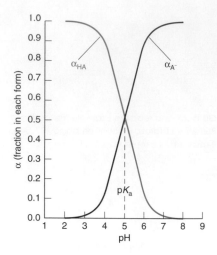

**Figure 10-2** Fractional composition diagram of a monoprotic system with $pK_a = 5.00$. Below pH 5, HA is the dominant form, whereas above pH 5 $A^-$ dominates.

$\alpha_{H_2A}$ = fraction of species in the form $H_2A$
$\alpha_{HA^-}$ = fraction of species in the form $HA^-$
$\alpha_{A^{2-}}$ = fraction of species in the form $A^{2-}$
$\alpha_{H_2A} + \alpha_{HA^-} + \alpha_{A^{2-}} = 1$

The general form of $\alpha$ for the polyprotic acid $H_nA$ is

$$\alpha_{H_nA} = \frac{[H^+]^n}{D}$$

$$\alpha_{H_{n-1}A} = \frac{K_1[H^+]^{n-1}}{D}$$

$$\alpha_{H_{n-j}A} = \frac{K_1K_2\cdots K_j[H^+]^{n-j}}{D}$$

where $D = [H^+]^n + K_1[H^+]^{n-1} + K_1K_2[H^+]^{n-2} + \cdots + K_1K_2K_3\cdots K_n$.

How to apply fractional composition equations to bases

Figure 10-2 shows $\alpha_{HA}$ and $\alpha_{A^-}$ for a system with $pK_a = 5.00$. At low pH, almost all of the acid is in the form HA. At high pH, almost everything is in the form $A^-$.

## Diprotic Systems

The derivation of fractional composition equations for a diprotic system follows the same pattern used for the monoprotic system.

$$H_2A \overset{K_1}{\rightleftharpoons} H^+ + HA^-$$

$$HA^- \overset{K_2}{\rightleftharpoons} H^+ + A^{2-}$$

$$K_1 = \frac{[H^+][HA^-]}{[H_2A]} \Rightarrow [HA^-] = [H_2A]\frac{K_1}{[H^+]}$$

$$K_2 = \frac{[H^+][A^{2-}]}{[HA^-]} \Rightarrow [A^{2-}] = [HA^-]\frac{K_2}{[H^+]} = [H_2A]\frac{K_1K_2}{[H^+]^2}$$

Mass balance: $\quad F = [H_2A] + [HA^-] + [A^{2-}]$

$$F = [H_2A] + \frac{K_1}{[H^+]}[H_2A] + \frac{K_1K_2}{[H^+]^2}[H_2A]$$

$$F = [H_2A]\left(1 + \frac{K_1}{[H^+]} + \frac{K_1K_2}{[H^+]^2}\right)$$

For a diprotic system, we designate the fraction in the form $H_2A$ as $\boldsymbol{\alpha_{H_2A}}$, the fraction in the form $HA^-$ as $\boldsymbol{\alpha_{HA^-}}$, and the fraction in the form $A^{2-}$ as $\boldsymbol{\alpha_{A^{2-}}}$. From the definition of $\alpha_{H_2A}$, we can write

*Fraction in the form $H_2A$:* $\quad \alpha_{H_2A} = \frac{[H_2A]}{F} = \frac{[H^+]^2}{[H^+]^2 + [H^+]K_1 + K_1K_2}$ $\quad$ (10-19)

In a similar manner, we can derive the following equations:

*Fraction in the form $HA^-$:* $\quad \alpha_{HA^-} = \frac{[HA^-]}{F} = \frac{K_1[H^+]}{[H^+]^2 + [H^+]K_1 + K_1K_2}$ $\quad$ (10-20)

*Fraction in the form $A^{2-}$:* $\quad \alpha_{A^{2-}} = \frac{[A^{2-}]}{F} = \frac{K_1K_2}{[H^+]^2 + [H^+]K_1 + K_1K_2}$ $\quad$ (10-21)

Figure 10-3 shows the fractions $\alpha_{H_2A}$, $\alpha_{HA^-}$, and $\alpha_{A^{2-}}$ for fumaric acid, whose two $pK_a$ values are only 1.46 units apart. $\alpha_{HA^-}$ rises only to 0.73 because the two $pK$ values are so close together. There are substantial quantities of $H_2A$ and $A^{2-}$ in the region $pK_1 < pH < pK_2$.

Equations 10-19 through 10-21 apply equally well to B, $BH^+$, and $BH_2^{2+}$ obtained by dissolving the base B in water. The fraction $\alpha_{H_2A}$ applies to the acidic form $BH_2^{2+}$. $\alpha_{HA^-}$ applies to $BH^+$, and $\alpha_{A^{2-}}$ applies to B. The constants $K_1$ and $K_2$ are the *acid* dissociation constants of $BH_2^{2+}$ ($K_1 = K_w/K_{b2}$ and $K_2 = K_w/K_{b1}$).

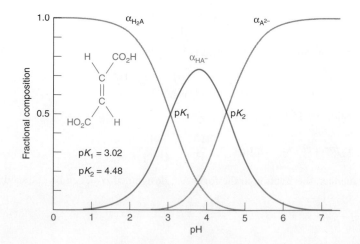

**Figure 10-3** Fractional composition diagram for fumaric acid (*trans*-butenedioic acid). At low pH, $H_2A$ is the dominant form. At intermediate pH, $HA^-$ is dominant; and, at high pH, $A^{2-}$ dominates. Because $pK_1$ and $pK_2$ are not separated very much, the fraction of $HA^-$ never gets very close to unity.

Biochemists speak of the isoelectric or isoionic pH of polyprotic molecules, such as proteins. These terms can be understood in terms of a diprotic system, such as the amino acid alanine.

$$\underset{\substack{\text{Alanine cation}\\ H_2A^+}}{H_3\overset{+}{N}CHCO_2H} \;\rightleftharpoons\; \underset{\substack{\text{Neutral zwitterion}\\ HA}}{H_3\overset{+}{N}CHCO_2^-} + H^+ \qquad pK_1 = 2.34$$

(with CH₃ groups attached)

$$\underset{}{H_3\overset{+}{N}CHCO_2^-} \;\rightleftharpoons\; \underset{\substack{\text{Alanine anion}\\ A^-}}{H_2NCHCO_2^-} + H^+ \qquad pK_2 = 9.87$$

The **isoionic point** (or isoionic pH) is the pH obtained when the pure, neutral polyprotic acid HA (the neutral zwitterion) is dissolved in water. The only ions are $H_2A^+$, $A^-$, $H^+$, and $OH^-$. Most alanine is in the form HA, and the concentrations of $H_2A^+$ and $A^-$ are *not* equal to each other.

The **isoelectric point** (or isoelectric pH) is the pH at which the *average* charge of the polyprotic acid is 0. Most of the molecules are in the uncharged form HA, and the concentrations of $H_2A^+$ and $A^-$ *are* equal to each other. There is always some $H_2A^+$ and some $A^-$ in equilibrium with HA.

When alanine is dissolved in water, the pH of the solution, by definition, is the *isoionic* pH. Because alanine (HA) is the intermediate form of the diprotic acid, $H_2A^+$, $[H^+]$ is given by

*Isoionic point:* $$[H^+] = \sqrt{\frac{K_1K_2F + K_1K_w}{K_1 + F}} \qquad (10\text{-}22)$$

> *Isoionic pH* is the pH of the pure, neutral, polyprotic acid.

> *Isoelectric pH* is the pH at which average charge of the polyprotic acid is 0.

> Alanine is the intermediate form of a diprotic acid, so we use Equation 10-11 to find the pH.

where F is the formal concentration of alanine. For 0.10 M alanine, the isoionic pH is found from

$$[H^+] = \sqrt{\frac{K_1K_2(0.10) + K_1K_w}{K_1 + (0.10)}} = 7.7 \times 10^{-7}\,M \;\Rightarrow\; pH = 6.11$$

From $[H^+]$, $K_1$, and $K_2$, you could calculate $[H_2A^+] = 1.68 \times 10^{-5}\,M$ and $[A^-] = 1.76 \times 10^{-5}\,M$ for pure alanine in water (the *isoionic* solution). There is a slight excess of $A^-$ because HA is a slightly stronger acid than it is a base. It dissociates to make $A^-$ a little more than it reacts with water to make $H_2A^+$.

The *isoelectric* point is the pH at which $[H_2A^+] = [A^-]$, and, therefore, the average charge of alanine is 0. To go from the *isoionic* solution (pure HA in water) to the isoelectric solution, we could add just enough strong acid to decrease $[A^-]$ and increase $[H_2A^+]$ until they are equal. Adding acid necessarily lowers the pH. For alanine, the isoelectric pH must be lower than the isoionic pH.

We calculate the isoelectric pH by first writing expressions for $[H_2A^+]$ and $[A^-]$:

$$[H_2A^+] = \frac{[HA][H^+]}{K_1} \qquad [A^-] = \frac{K_2[HA]}{[H^+]}$$

Setting $[H_2A^+] = [A^-]$, we find

$$\frac{[HA][H^+]}{K_1} = \frac{K_2[HA]}{[H^+]} \;\Rightarrow\; [H^+] = \sqrt{K_1K_2}$$

which gives

*Isoelectric point:* $$pH = \tfrac{1}{2}(pK_1 + pK_2) \qquad (10\text{-}23)$$

> The isoelectric point is midway between the two $pK_a$ values "surrounding" the neutral, intermediate species.

For a diprotic amino acid, the isoelectric pH is halfway between the two $pK_a$ values. The isoelectric pH of alanine is $\tfrac{1}{2}(2.34 + 9.87) = 6.10$.

The isoelectric and isoionic points for a polyprotic acid are almost the same. At the isoelectric pH, the average charge of the molecule is 0; thus $[H_2A^+] = [A^-]$ and $pH = \tfrac{1}{2}(pK_1 + pK_2)$. At the isoionic point, the pH is given by Equation 10-22, and $[H_2A^+]$ is not exactly equal to $[A^-]$.

## Box 10-2 Isoelectric Focusing

At its *isoelectric point,* the average charge of all forms of a protein is 0. It will therefore not migrate in an electric field at its isoelectric pH. This effect is the basis of a sensitive technique of protein separation called **isoelectric focusing.** A protein mixture is subjected to a strong electric field in a medium specifically designed to have a pH gradient. Positively charged molecules move toward the negative pole and negatively charged molecules move toward the positive pole. Each protein migrates until it reaches the point where the pH is the same as its isoelectric pH. At this point, the protein has no net charge and no longer moves. Each protein in the mixture is focused in one region at its isoelectric pH.

An example of isoelectric focusing is shown at the left below. A mixture of seven proteins (and some impurities) was applied to a polyacrylamide gel containing a mixture of polyprotic compounds called *ampholytes.* Each end of the gel was in contact with a conducting solution and several hundred volts were applied across the length of the gel. The ampholytes migrated until they formed a stable gradient ranging from ~pH 3 at one end of the gel to ~pH 10 at the other.[2] Proteins migrated until they reached regions with their isoelectric pH (called pI), at which point they had no net charge and ceased migrating. If a molecule diffuses out of its isoelectric region, it becomes charged and immediately migrates back to its isoelectric zone. When migration was finished, proteins were precipitated in place, and they were stained with a dye to make their positions visible.

The stained gel is shown below the graph. A spectrophotometer scan of the dye peaks is shown on the graph, and a profile of measured pH is plotted. Each dark band of stained protein gives an absorbance peak. The instrument that measures absorbance as a function of position along the gel is called a *densitometer.*

The figure at the lower right shows separation of whole yeast cells at three different stages of growth (called early log, mid log, and stationary) by isoelectric focusing in a capillary tube. Acid-base properties (and, therefore, pI) of cell surfaces change during growth of the colony. Capillary isoelectric focusing is a form of capillary electrophoresis discussed in Chapter 26.

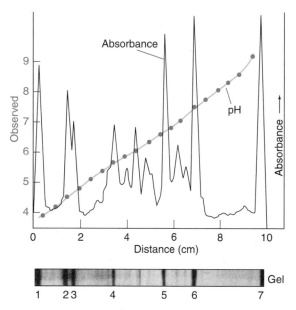

Isoelectric focusing of a mixture of proteins: (1) soybean trypsin inhibitor; (2) β-lactoglobulin A; (3) β-lactoglobulin B; (4) ovotransferrin; (5) horse myoglobin; (6) whale myoglobin; and (7) cytochrome c. [Courtesy BioRad Laboratories, Richmond, CA.]

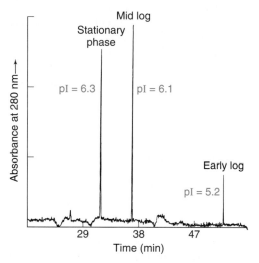

Capillary isoelectric focusing of whole yeast cells taken from three growth stages. After cells have been focused at their isoelectric pH, the inlet end of the capillary was elevated and liquid drained out of the capillary past an ultraviolet detector, creating three peaks observed here. The abscissa is the time required for the bands to reach the detector. [From R. Shen, S. J. Berger, and R. D. Smith, "Capillary Isoelectric Focusing of Yeast Cells," Anal. Chem. **2000,** 72, 4603.]

For a protein, the *isoionic* pH is the pH of a solution of pure protein with no other ions except $H^+$ and $OH^-$. Proteins are usually isolated in a charged form together with counterions such as $Na^+$, $NH_4^+$, or $Cl^-$. When the protein is subjected to intensive *dialysis* (Demonstration 27-1) against pure water, the pH in the protein compartment approaches the isoionic point if the counterions are free to pass through the semipermeable dialysis membrane that retains the protein. The *isoelectric* point is the pH at which the protein has no net charge. Box 10-2 tells how proteins can be separated on the basis of their different isoelectric points.

Related properties—of interest to geology, environmental science, and ceramics—are the surface acidity of a solid[3] and the pH of zero charge. Mineral, clay, and even organic surfaces behave as acids and bases. The silica ($SiO_2$) surface of sand or glass can be simplistically thought of as a diprotic acid:

Problem 10-40 gives experimental data from which $K_{a1}$ and $K_{a2}$ can be calculated for river sediment.

$$\equiv Si-OH_2^+ \xrightleftharpoons{K_{a1}} \equiv Si-OH + H^+ \qquad K_{a1} = \frac{\{SiOH\}[H^+]}{\{SiOH_2^+\}} \qquad (10\text{-}24)$$

$$\equiv Si-OH \xrightleftharpoons{K_{a2}} \equiv Si-O^- + H^+ \qquad K_{a2} = \frac{\{SiO^-\}[H^+]}{\{SiOH\}} \qquad (10\text{-}25)$$

The notation $\equiv$Si represents a surface silicon atom. Silanol groups ($\equiv$Si—OH) can donate or accept a proton to give the surface a negative or positive charge. In the equilibrium constants, the concentrations of the surface species $\{SiOH_2^+\}$, $\{SiOH\}$, and $\{SiO^-\}$ are measured in moles per gram of solid.

The *pH of zero charge* is the pH at which $\{SiOH_2^+\} = \{SiO^-\}$ and, therefore, the surface has no net charge. Like the isoelectric point of a diprotic acid, the pH of zero charge is $\frac{1}{2}(pK_{a1} + pK_{a2})$. *Colloidal particles* (those with diameters in the range 1–100 nm) tend to remain dispersed when they are charged, but they *flocculate* (come together and precipitate) near the pH of zero charge. In capillary electrophoresis (Chapter 26), the surface charge of the silica capillary governs the rate at which solvent moves through the capillary.

## Terms to Understand

| | | | |
|---|---|---|---|
| amino acid | hydrolysis | isoionic point | speciation |
| amphiprotic | isoelectric focusing | polyprotic acids and bases | zwitterion |
| diprotic acids and bases | isoelectric point | | |

## Summary

Diprotic acids and bases fall into three categories:

1. The fully acidic form, $H_2A$, behaves as a monoprotic acid, $H_2A \rightleftharpoons H^+ + HA^-$, for which we solve the equation $K_{a1} = x^2/(F - x)$, where $[H^+] = [HA^-] = x$, and $[H_2A] = F - x$. From $[HA^-]$ and $[H^+]$, $[A^{2-}]$ can be found from the $K_{a2}$ equilibrium.

2. The fully basic form, $A^{2-}$, behaves as a base, $A^{2-} + H_2O \rightleftharpoons HA^- + OH^-$, for which we solve the equation $K_{b1} = x^2/(F - x)$, where $[OH^-] = [HA^-] = x$, and $[A^{2-}] = F - x$. From these concentrations, $[H_2A]$ can be found from the $K_{a1}$ or $K_{b2}$ equilibria.

3. The intermediate (amphiprotic) form, $HA^-$, is both an acid and a base. Its pH is given by

$$[H^+] = \sqrt{\frac{K_1 K_2 F + K_1 K_w}{K_1 + F}}$$

where $K_1$ and $K_2$ are acid dissociation constants for $H_2A$, and F is the formal concentration of the intermediate. In most cases, this equation reduces to the form pH $\approx \frac{1}{2}(pK_1 + pK_2)$, in which pH is independent of concentration.

In triprotic systems, there are two intermediate forms. The pH of each is found with an equation analogous to that for the intermediate form of a diprotic system. Triprotic systems also have one fully acidic

and one fully basic form; these species can be treated as monoprotic for the purpose of calculating pH. For polyprotic buffers, we write the appropriate Henderson-Hasselbalch equation connecting the two principal species in the system. The pK in this equation is the one that applies to the acid in the denominator of the log term.

The principal species of a monoprotic or polyprotic system is found by comparing the pH with the various $pK_a$ values. For pH $< pK_1$, the fully protonated species, $H_nA$, is the predominant form. For $pK_1 <$ pH $< pK_2$, the form $H_{n-1}A^-$ is favored; and, at each successive pK value, the next deprotonated species becomes principal. Finally, at pH values higher than the highest pK, the fully basic form ($A^{n-}$) is dominant. The fractional composition of a solution is expressed by $\alpha$, given in Equations 10-17 and 10-18 for a monoprotic system and Equations 10-19 through 10-21 for a diprotic system.

The isoelectric pH of a polyprotic compound is that pH at which the average charge of all species is 0. For a diprotic amino acid whose amphiprotic form is neutral, the isoelectric pH is given by pH $= \frac{1}{2}(pK_1 + pK_2)$. The isoionic pH of a polyprotic species is the pH that would exist in a solution containing only the ions derived from the neutral polyprotic species and from $H_2O$. For a diprotic amino acid whose amphiprotic form is neutral, the isoionic pH is found from $[H^+] = \sqrt{(K_1 K_2 F + K_1 K_w)/(K_1 + F)}$, where F is the formal concentration of the amino acid.

## Exercises

**10-A.** Find the pH and the concentrations of $H_2SO_3$, $HSO_3^-$, and $SO_3^{2-}$ in each solution: **(a)** 0.050 M $H_2SO_3$; **(b)** 0.050 M $NaHSO_3$; **(c)** 0.050 M $Na_2SO_3$.

**10-B.** **(a)** How many grams of $NaHCO_3$ (FM 84.007) must be added to 4.00 g of $K_2CO_3$ (FM 138.206) to give a pH of 10.80 in 500 mL of water?
**(b)** What will be the pH if 100 mL of 0.100 M HCl are added to the solution in part **(a)**?
**(c)** How many milliliters of 0.320 M $HNO_3$ should be added to 4.00 g of $K_2CO_3$ to give a pH of 10.00 in 250 mL?

**10-C.** How many milliliters of 0.800 M KOH should be added to 5.02 g of 1,5-pentanedioic acid ($C_5H_8O_4$, FM 132.11) to give a pH of 4.40 when diluted to 250 mL?

**10-D.** Calculate the pH of a 0.010 M solution of each amino acid in the form drawn here.

(a) Glutamine
$$H_3\overset{+}{N}CHCO_2^-$$
with side chain: $CH_2$—$CH_2$—$C(=O)$—$NH_2$

(b) Cysteine
$$H_3\overset{+}{N}CHCO_2^-$$
with side chain: $CH_2$—$S^-$

(c) Arginine
$$H_2NCHCO_2^-$$
with side chain: $CH_2$—$CH_2$—$CH_2$—$NH$—$C(NH_2)(=\overset{+}{N}H_2)$

**10-E. (a)** Draw the structure of the predominant form (principal species) of 1,3-dihydroxybenzene at pH 9.00 and at pH 11.00.
**(b)** What is the second most prominent species at each pH?
**(c)** Calculate the percentage in the major form at each pH.

**10-F.** Draw the structures of the predominant forms of glutamic acid and tyrosine at pH 9.0 and pH 10.0. What is the second most abundant species at each pH?

**10-G.** Calculate the isoionic pH of 0.010 M lysine.

**10-H.** Neutral lysine can be written HL. The other forms of lysine are $H_3L^{2+}$, $H_2L^+$, and $L^-$. The isoelectric point is the pH at which the *average* charge of lysine is 0. Therefore, at the isoelectric point, $2[H_3L^{2+}] + [H_2L^+] = [L^-]$. Use this condition to calculate the isoelectric pH of lysine.

---

## Problems

### Diprotic Acids and Bases

**10-1.** Consider $HA^-$, the intermediate form of a diprotic acid. $K_a$ for this species is $10^{-4}$ and $K_b$ is $10^{-8}$. Nonetheless, the $K_a$ and $K_b$ reactions proceed to nearly the same extent when NaHA is dissolved in water. Explain.

**10-2.** Write the general structure of an amino acid. Why do some amino acids in Table 10-1 have two p$K$ values and others three?

**10-3.** Write the chemical reactions whose equilibrium constants are $K_{b1}$ and $K_{b2}$ for the amino acid proline. Find the values of $K_{b1}$ and $K_{b2}$.

**10-4.** Consider the diprotic acid $H_2A$ with $K_1 = 1.00 \times 10^{-4}$ and $K_2 = 1.00 \times 10^{-8}$. Find the pH and concentrations of $H_2A$, $HA^-$, and $A^{2-}$ in **(a)** 0.100 M $H_2A$; **(b)** 0.100 M NaHA; **(c)** 0.100 M $Na_2A$.

**10-5.** We will abbreviate malonic acid, $CH_2(CO_2H)_2$, as $H_2M$. Find the pH and concentrations of $H_2M$, $HM^-$, and $M^{2-}$ in **(a)** 0.100 M $H_2M$; **(b)** 0.100 M NaHM; **(c)** 0.100 M $Na_2M$.

**10-6.** Calculate the pH of 0.300 M piperazine. Calculate the concentration of each form of piperazine in this solution.

**10-7.** Use the method of Box 10-1 to calculate the concentrations of $H^+$, $H_2A$, $HA^-$, and $A^{2-}$ in 0.001 00 M monosodium oxalate, NaHA.

**10-8.** *Activity.* In this problem, we calculate the pH of the intermediate form of a diprotic acid, taking activities into account.
**(a)** Including activity coefficients, derive Equation 10-11 for potassium hydrogen phthalate ($K^+HP^-$ in the example following Equation 10-12).
**(b)** Calculate the pH of 0.050 M KHP, using the results in part **(a)**. Assume that the sizes of both $HP^-$ and $P^{2-}$ are 600 pm.

**10-9.** ▦ *Intermediate form of diprotic acid.* Use the method in Box 10-1 to find the pH and concentration of $HA^-$ in a 0.01 F solution of the amphiprotic salt $Na^+HA^-$ derived from the diprotic acid $H_2A$ with p$K_1 = 4$ and **(a)** p$K_2 = 8$ or **(b)** p$K_2 = 5$.

**10-10.** *Heterogeneous equilibrium.* $CO_2$ dissolves in water to give "carbonic acid" (which is mostly dissolved $CO_2$, as described in Box 6-4).

$$CO_2(g) \rightleftharpoons CO_2(aq) \qquad K = 10^{-1.5}$$

(The equilibrium constant is called the *Henry's law constant* for carbon dioxide, because Henry's law states that the solubility of a gas in a liquid is proportional to the pressure of the gas.) The acid dissociation constants listed for "carbonic acid" in Appendix G apply to $CO_2(aq)$. Given that $P_{CO_2}$ in the atmosphere is $10^{-3.4}$ atm, find the pH of water in equilibrium with the atmosphere.

### Diprotic Buffers

**10-11.** How many grams of $Na_2CO_3$ (FM 105.99) should be mixed with 5.00 g of $NaHCO_3$ (FM 84.01) to produce 100 mL of buffer with pH 10.00?

**10-12.** How many milliliters of 0.202 M NaOH should be added to 25.0 mL of 0.023 3 M salicylic acid (2-hydroxybenzoic acid) to adjust the pH to 3.50?

**10-13.** Describe how you would prepare exactly 100 mL of 0.100 M picolinate buffer, pH 5.50. Possible starting materials are pure picolinic acid (pyridine-2-carboxylic acid, FM 123.11), 1.0 M HCl, and 1.0 M NaOH. Approximately how many milliliters of the HCl or NaOH will be required?

**10-14.** How many grams of $Na_2SO_4$ (FM 142.04) should be added to how many grams of sulfuric acid (FM 98.08) to give 1.00 L of buffer with pH 2.80 and a total sulfur ($= SO_4^{2-} + HSO_4^- + H_2SO_4$) concentration of 0.200 M?

### Polyprotic Acids and Bases

**10-15.** Phosphate, present to an extent of 0.01 M, is one of the main buffers in blood plasma, whose pH is 7.45. Would phosphate be as useful if the plasma pH were 8.5?

**10-16.** Starting with the fully protonated species, write the stepwise acid dissociation reactions of the amino acids glutamic acid and tyrosine. Be sure to remove the protons in the correct order. Which species are the neutral molecules that we call glutamic acid and tyrosine?

**10-17. (a)** Calculate the quotient $[H_3PO_4]/[H_2PO_4^-]$ in 0.050 0 M $KH_2PO_4$.
**(b)** Find the same quotient for 0.050 0 M $K_2HPO_4$.

**10-18. (a)** Which two of the following compounds would you mix to make a buffer of pH 7.45: $H_3PO_4$ (FM 98.00), $NaH_2PO_4$ (FM 119.98), $Na_2HPO_4$ (FM 141.96), and $Na_3PO_4$ (FM 163.94)?
**(b)** If you wanted to prepare 1.00 L of buffer with a total phosphate concentration of 0.050 0 M, how many grams of each of the two selected compounds would you mix together?
**(c)** If you did what you calculated in part **(b),** you would not get a pH of exactly 7.45. Explain how you would really prepare this buffer in the lab.

**10-19.** Find the pH and the concentration of each species of lysine in a solution of 0.010 0 M lysine · HCl, lysine monohydrochloride.

**10-20.** How many milliliters of 1.00 M KOH should be added to 100 mL of solution containing 10.0 g of histidine hydrochloride (His · HCl, FM 191.62) to get a pH of 9.30?

**10-21. (a)** Using activity coefficients, calculate the pH of a solution containing a 2.00:1.00 mole ratio of $HC^{2-}:C^{3-}$, where $H_3C$ is citric acid. Assume that ionic strength $= 0.010$ M.
**(b)** What will be the pH if the ionic strength is raised to 0.10 M and the mole ratio $HC^{2-}:C^{3-}$ is kept constant?

### Which Is the Principal Species?

**10-22.** The acid HA has p$K_a = 7.00$.
**(a)** Which is the principal species, HA or $A^-$, at pH 6.00?
**(b)** Which is the principal species at pH 8.00?
**(c)** What is the quotient $[A^-]/[HA]$ at pH 7.00? at pH 6.00?

**10-23.** The diprotic acid $H_2A$ has $pK_1 = 4.00$ and $pK_2 = 8.00$.
(a) At what pH is $[H_2A] = [HA^-]$?
(b) At what pH is $[HA^-] = [A^{2-}]$?
(c) Which is the principal species at pH 2.00: $H_2A$, $HA^-$, or $A^{2-}$?
(d) Which is the principal species at pH 6.00?
(e) Which is the principal species at pH 10.00?

**10-24.** The base B has $pK_b = 5.00$.
(a) What is the value of $pK_a$ for the acid $BH^+$?
(b) At what pH is $[BH^+] = [B]$?
(c) Which is the principal species at pH 7.00: B or $BH^+$?
(d) What is the quotient $[B]/[BH^+]$ at pH 12.00?

**10-25.** Draw the structure of the predominant form of pyridoxal-5-phosphate at pH 7.00.

## Fractional Composition Equations

**10-26.** The acid HA has $pK_a = 4.00$. Use Equations 10-17 and 10-18 to find the fraction in the form HA and the fraction in the form $A^-$ at pH = 5.00. Does your answer agree with what you expect for the quotient $[A^-]/[HA]$ at pH 5.00?

**10-27.** A dibasic compound, B, has $pK_{b1} = 4.00$ and $pK_{b2} = 6.00$. Find the fraction in the form $BH_2^{2+}$ at pH 7.00, using Equation 10-19. Note that $K_1$ and $K_2$ in Equation 10-19 are acid dissociation constants for $BH_2^{2+}$ ($K_1 = K_w/K_{b2}$ and $K_2 = K_w/K_{b1}$).

**10-28.** What fraction of ethane-1,2-dithiol is in each form ($H_2A$, $HA^-$, $A^{2-}$) at pH 8.00? at pH 10.00?

**10-29.** Calculate $\alpha_{H_2A}$, $\alpha_{HA^-}$, and $\alpha_{A^{2-}}$ for *cis*-butenedioic acid at pH 1.00, 1.92, 6.00, 6.27, and 10.00.

**10-30.** (a) Derive equations for $\alpha_{H_3A}$, $\alpha_{H_2A^-}$, $\alpha_{HA^{2-}}$, and $\alpha_{A^{3-}}$ for a triprotic system.
(b) Calculate the values of these fractions for phosphoric acid at pH 7.00.

**10-31.** A solution containing acetic acid, oxalic acid, ammonia, and pyridine has a pH of 9.00. What fraction of ammonia is not protonated?

**10-32.** A solution was prepared from 10.0 mL of 0.100 M cacodylic acid and 10.0 mL of 0.080 0 M NaOH. To this mixture was added 1.00 mL of $1.27 \times 10^{-6}$ M morphine. Calling morphine B, calculate the fraction of morphine present in the form $BH^+$.

$$O$$
$$\|$$
$$(CH_3)_2AsOH$$

Cacodylic acid
$K_a = 6.4 \times 10^{-7}$

Morphine
$K_b = 1.6 \times 10^{-6}$

**10-33.** *Fractional composition in a diprotic system.* Create a spreadsheet that uses Equations 10-19 through 10-21 to compute the three curves in Figure 10-3. Use graphics software to plot these three curves in a beautifully labeled figure.

**10-34.** *Fractional composition in a triprotic system.* For a triprotic system, the fractional composition equations are

$$\alpha_{H_3A} = \frac{[H^+]^3}{D} \qquad \alpha_{HA^{2-}} = \frac{K_1K_2[H^+]}{D}$$

$$\alpha_{H_2A^-} = \frac{K_1[H^+]^2}{D} \qquad \alpha_{A^{3-}} = \frac{K_1K_2K_3}{D}$$

where $D = [H^+]^3 + K_1[H^+]^2 + K_1K_2[H^+] + K_1K_2K_3$. Use these equations to create a fractional composition diagram analogous to Figure 10-3 for the amino acid tyrosine. What is the fraction of each species at pH 10.00?

**10-35.** *Fractional composition in a tetraprotic system.* Prepare a fractional composition diagram analogous to Figure 10-3 for the tetraprotic system derived from hydrolysis of $Cr^{3+}$:

$$Cr^{3+} + H_2O \rightleftharpoons Cr(OH)^{2+} + H^+ \qquad K_{a1} = 10^{-3.80}$$

$$Cr(OH)^{2+} + H_2O \rightleftharpoons Cr(OH)_2^+ + H^+ \qquad K_{a2} = 10^{-6.40}$$

$$Cr(OH)_2^+ + H_2O \rightleftharpoons Cr(OH)_3(aq) + H^+ \quad K_{a3} = 10^{-6.40}$$

$$Cr(OH)_3(aq) + H_2O \rightleftharpoons Cr(OH)_4^- + H^+ \quad K_{a4} = 10^{-11.40}$$

(Yes, the values of $K_{a2}$ and $K_{a3}$ are equal.)
(a) Use these equilibrium constants to prepare a fractional composition diagram for this tetraprotic system.
(b) You should do this part with your head and your calculator, not your spreadsheet. The solubility of $Cr(OH)_3$ is given by

$$Cr(OH)_3(s) \rightleftharpoons Cr(OH)_3(aq) \qquad K = 10^{-6.84}$$

What concentration of $Cr(OH)_3(aq)$ is in equilibrium with solid $Cr(OH)_3(s)$?
(c) What concentration of $Cr(OH)^{2+}$ is in equilibrium with $Cr(OH)_3(s)$ if the solution pH is adjusted to 4.00?

## Isoelectric and Isoionic pH

**10-36.** What is the difference between the isoelectric pH and the isoionic pH of a protein with many different acidic and basic substituents?

**10-37.** What is wrong with the following statement: At its isoelectric point, the charge on all molecules of a particular protein is 0.

**10-38.** Calculate the isoelectric and isoionic pH of 0.010 M threonine.

**10-39.** Explain how isoelectric focusing works.

**10-40.** *pH of zero charge.* Data in the spreadsheets on the next page were obtained by adding 0.100 M $HNO_3$ or 0.100 M NaOH to a stirred mixture of 10.0 g of dry, sieved river sediment in 100 mL of 0.10 M $NaNO_3$ to maintain constant ionic strength. pH was recorded for each addition after 10 min for equilibration.
We analyze the data in terms of Reactions 10-24 and 10-25. Let the total concentration of surface species (mol/g) be $\{S\}_{tot} = \{SiOH_2^+\} + \{SiOH\} + \{SiO^-\}$. $\{S\}_{tot}$ was taken as three times the moles of $Co(NH_3)_6^{3+}$ that exchanged with sediment in 3 h. The surface charge of the solid (mol/g) is $Q = \{SiOH_2^+\} - \{SiO^-\}$. When a concentration $C_a$ (mol/L) of $HNO_3$ or a concentration $C_b$ of NaOH is added to sediment, the positive surface charge on the solid is

$$Q = \frac{C_a - C_b + [OH^-] - [H^+]}{m} \tag{A}$$

where $m$ is the solid concentration (g/L). For addition of $HNO_3$, $C_b = 0$. For addition of NaOH, $C_a = 0$. For pH well below the pH of zero charge, the charged surface species is $\equiv Si-OH_2^+$, so $Q = \{\equiv Si-OH_2^+\}$. Therefore,

$$K_{a1} = \frac{\{SiOH\}[H^+]}{\{SiOH_2^+\}} = \frac{(\{S\}_{tot} - Q)[H^+]}{Q} \tag{B}$$

For pH well above the pH of zero charge, the charged surface species is $\equiv$Si—O$^-$, so $Q = \{\equiv$Si—O$^-\}$. Therefore,

$$K_{a2} = \frac{\{SiO^-\}[H^+]}{\{SiOH\}} = \frac{Q[H^+]}{\{S\}_{tot} - Q} \qquad (C)$$

In the HNO$_3$ spreadsheet, $C_a = 0.100*V/(100 + V)$, where $V$ is mL of added HNO$_3$. $[H^+] = (10^{-pH})/\gamma_{H^+}$, where $\gamma_{H^+}$ is the activity coefficient of H$^+$ at an ionic strength of 0.1 M. $[OH^-] = (10^{pH-14})/\gamma_{OH^-}$, where $\gamma_{OH^-}$ is the activity coefficient of OH$^-$. $Q$ is given by Equation A and $K_{a1}$ is given by Equation B with $\{S\}_{tot} = 1.30 \times 10^{-4}$ mol/g. In the NaOH spreadsheet, $C_b = 0.100*V/(100 + V)$, where $V$ is mL of added NaOH. $[H^+]$, $[OH^-]$, and $Q$ have the same formulas as in the HNO$_3$ spreadsheet, and $K_{a2}$ is given by Equation C.

Complete the spreadsheets and prepare a graph of p$K_a$ versus $Q$ for each titration. Fit the points with a straight line and find the intercept at $Q = 0$. The intercept of the HNO$_3$ graph is taken as $K_{a1}$ and the intercept of the NaOH graph is taken as $K_{a2}$. Find the pH of zero charge, given by $\frac{1}{2}(pK_{a1} + pK_{a2})$.

| | A | B | C | D | E | F | G | H | I |
|---|---|---|---|---|---|---|---|---|---|
| 1 | Titration of sediment with HNO$_3$ | | | | | | | | |
| 2 | $S_{tot}$ = | HNO$_3$ (mL) | $C_a$ (M) | pH | [H$^+$] | [OH$^-$] | Q (mol/g) | $K_{a1}$ | p$K_{a1}$ |
| 3 | 1.30E-04 | 0.1 | 0.00010 | 5.33 | 5.64E-06 | 2.81E-09 | 9.43E-06 | 7.21E-05 | 4.14 |
| 4 | mol/g | 0.2 | | 4.86 | | | | | |
| 5 | $\gamma(H^+)$ = | 0.5 | | 4.47 | | | | | |
| 6 | 0.83 | 0.6 | | 4.32 | | | | | |
| 7 | $\gamma(OH^-)$ = | 0.8 | | 4.02 | | | | | |
| 8 | 0.76 | 1.0 | | 3.72 | | | | | |
| 9 | grams of | 1.2 | | 3.54 | | | | | |
| 10 | soil = | 1.4 | | 3.41 | | | | | |
| 11 | 10 | 1.6 | | 3.35 | | | | | |
| 12 | | 2.0 | | 3.15 | | | | | |

| | A | B | C | D | E | F | G | H | I |
|---|---|---|---|---|---|---|---|---|---|
| 1 | Titration of sediment with NaOH | | | | | | | | |
| 2 | $S_{tot}$ = | NaOH (mL) | $C_b$ (M) | pH | [H$^+$] | [OH$^-$] | Q (mol/g) | $K_{a2}$ | p$K_{a2}$ |
| 3 | 1.30E-04 | 0.1 | 0.00010 | 6.92 | 1.45E-07 | 1.09E-07 | −9.99E-06 | 1.03E-08 | 7.99 |
| 4 | mol/g | 0.2 | | 7.07 | | | | | |
| 5 | $\gamma(H^+)$ = | 0.3 | | 7.20 | | | | | |
| 6 | 0.83 | 0.4 | | 7.31 | | | | | |
| 7 | $\gamma(OH^-)$ = | 0.5 | | 7.56 | | | | | |
| 8 | 0.76 | 0.6 | | 7.66 | | | | | |
| 9 | grams of | 0.7 | | 7.80 | | | | | |
| 10 | soil = | 0.8 | | 7.93 | | | | | |
| 11 | 10 | 0.9 | | 8.06 | | | | | |
| 12 | | 1.0 | | 8.13 | | | | | |

Spreadsheets for titration of river sediment with HNO$_3$ or NaOH. [Data from M. Davranche, S. Lacour, F. Bordas, and J.-C. Bollinger, "Determination of the Surface Chemical Properties of Natural Solids," J. Chem. Ed. **2003**, 80, 76.]

# 11 | Acid-Base Titrations

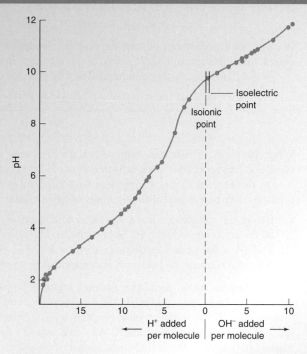

Acid-base titration of the enzyme ribonuclease. The isoionic point is the pH of the pure protein with no ions present except $H^+$ and $OH^-$. The isoelectric point is the pH at which the average charge on the protein is 0. [C. T. Tanford and J. D. Hauenstein, "Hydrogen Ion Equilibria of Ribonuclease," J. Am. Chem. Soc. **1956**, 78, 5287.]

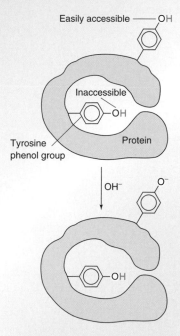

If tyrosine is buried deep inside the protein, it is not readily accessible and a high concentration of $OH^-$ is required to remove the proton from the phenol group.

Ribonuclease is an enzyme with 124 amino acids. Its function is to cleave ribonucleic acid (RNA) into small fragments. A solution containing pure protein, with no other ions present except $H^+$ and $OH^-$ derived from the protein and water, is said to be *isoionic*. From this point near pH 9.6 in the graph, the protein can be titrated with acid or base. Of the 124 amino acids, 16 can be protonated by acid and 20 can lose protons to added base. From the shape of the titration curve, it is possible to deduce the approximate $pK_a$ for each titratable group.[1,2] This information provides insight into the environment of that amino acid in the protein. In ribonuclease, three tyrosine residues have "normal" values of $pK_a$ ($\approx 10$) (Table 10-1) and three others have $pK_a > 12$. The interpretation is that three tyrosine groups are accessible to $OH^-$, and three are buried inside the protein where they cannot be easily titrated. The solid line in the illustration is calculated from $pK_a$ values for all titratable groups.

Theoretical titration curves for enzymes can be calculated from known crystal structures and first principles of electrostatics. Key amino acids at the active site have significantly perturbed $pK_a$ values and unusual regions in which they are partially protonated over a wide pH region.[3] In principle, such titration calculations can identify the active site of a protein whose structure is known, but whose function is not.

**F**rom an acid-base titration curve, we can deduce the quantities and $pK_a$ values of acidic and basic substances in a mixture. In medicinal chemistry, the $pK_a$ and *lipophilicity* of a candidate drug predict how easily it will cross cell membranes. We saw in Chapter 10 that from $pK_a$ and pH, we can compute the charge of a polyprotic acid. Usually, the more highly charged a drug, the harder it is to cross a cell membrane. In this chapter, we learn how to predict the shapes of titration curves and how to find end points with electrodes or indicators.

*Lipophilicity* is a measure of solubility in nonpolar solvents. It is determined from the equilibrium distribution of a drug between water and octanol.

Drug(*aq*) $\rightleftharpoons$ drug (*in octanol*)

$$\text{Lipophilicity} = \log \left( \frac{[\text{drug } (in\ octanol)]}{[\text{drug}(aq)]} \right)$$

For each type of titration in this chapter, *our goal is to construct a graph showing how the pH changes as titrant is added.* If you can do this, then you understand what is happening during the titration, and you will be able to interpret an experimental titration curve.

The first step in each case is to write the chemical reaction between titrant and analyte. Then use that reaction to calculate the composition and pH after each addition of titrant. As a simple example, let's focus on the titration of 50.00 mL of 0.020 00 M KOH with 0.100 0 M HBr. The chemical reaction between titrant and analyte is merely

$$H^+ + OH^- \rightarrow H_2O \qquad K = 1/K_w = 10^{14}$$

Because the equilibrium constant for this reaction is $10^{14}$, it is fair to say that it "goes to completion." *Any amount of $H^+$ added will consume a stoichiometric amount of $OH^-$.*

It is useful to know the volume of HBr ($V_e$) needed to reach the equivalence point:

$$\underbrace{(V_e \text{ (mL)})(0.100 \text{ 0 M})}_{\substack{\text{mmol of HBr} \\ \text{at equivalence point}}} = \underbrace{(50.00 \text{ mL})(0.020 \text{ 00 M})}_{\substack{\text{mmol of } OH^- \\ \text{being titrated}}} \Rightarrow V_e = 10.00 \text{ mL}$$

When 10.00 mL of HBr have been added, the titration is complete. Prior to this point, there is excess, unreacted $OH^-$ present. After $V_e$, there is excess $H^+$ in the solution.

In the titration of any strong base with any strong acid, there are three regions of the titration curve that require different kinds of calculations:

1. Before the equivalence point, the pH is determined by excess $OH^-$ in the solution.
2. At the equivalence point, $H^+$ is just sufficient to react with all $OH^-$ to make $H_2O$. The pH is determined by dissociation of water.
3. After the equivalence point, pH is determined by excess $H^+$ in the solution.

We will do one sample calculation for each region. Complete results are shown in Table 11-1 and Figure 11-1. As a reminder, the *equivalence point* occurs when the added titrant is exactly enough for stoichiometric reaction with the analyte. The equivalence point is the ideal result we seek in a titration. What we actually measure is the *end point,* which is marked by a sudden physical change, such as indicator color or an electrode potential.

**First, write the reaction between *titrant* and *analyte*.**

**The titration reaction.**

**Table 11-1** Calculation of the titration curve for 50.00 mL of 0.020 00 M KOH treated with 0.100 0 M HBr

| | mL HBr added ($V_a$) | Concentration of unreacted $OH^-$ (M) | Concentration of excess $H^+$ (M) | pH |
|---|---|---|---|---|
| | 0.00 | 0.020 0 | | 12.30 |
| | 1.00 | 0.017 6 | | 12.24 |
| | 2.00 | 0.015 4 | | 12.18 |
| | 3.00 | 0.013 2 | | 12.12 |
| | 4.00 | 0.011 1 | | 12.04 |
| Region 1 | 5.00 | 0.009 09 | | 11.95 |
| (excess $OH^-$) | 6.00 | 0.007 14 | | 11.85 |
| | 7.00 | 0.005 26 | | 11.72 |
| | 8.00 | 0.003 45 | | 11.53 |
| | 9.00 | 0.001 69 | | 11.22 |
| | 9.50 | 0.000 840 | | 10.92 |
| | 9.90 | 0.000 167 | | 10.22 |
| | 9.99 | 0.000 016 6 | | 9.22 |
| Region 2 | 10.00 | — | — | 7.00 |
| | 10.01 | | 0.000 016 7 | 4.78 |
| | 10.10 | | 0.000 166 | 3.78 |
| | 10.50 | | 0.000 826 | 3.08 |
| | 11.00 | | 0.001 64 | 2.79 |
| Region 3 | 12.00 | | 0.003 23 | 2.49 |
| (excess $H^+$) | 13.00 | | 0.004 76 | 2.32 |
| | 14.00 | | 0.006 25 | 2.20 |
| | 15.00 | | 0.007 69 | 2.11 |
| | 16.00 | | 0.009 09 | 2.04 |

*Figure 11-1* Calculated titration curve, showing how pH changes as 0.100 0 M HBr is added to 50.00 mL of 0.020 00 M KOH. The equivalence point is also an inflection point.

## Region 1: Before the Equivalence Point

When 3.00 mL of HBr have been added, the reaction is three-tenths complete because $V_e = 10.00$ mL. The fraction of $OH^-$ left unreacted is seven-tenths. The concentration of $OH^-$ remaining in the flask is

Before the equivalence point, there is excess $OH^-$.

$$[OH^-] = \underbrace{\left(\frac{10.00 - 3.00}{10.00}\right)}_{\substack{\text{Fraction} \\ \text{of } OH^- \\ \text{remaining}}} \underbrace{(0.020\ 00\ M)}_{\substack{\text{Initial} \\ \text{concentration} \\ \text{of } OH^-}} \underbrace{\left(\frac{\overset{\substack{\text{Initial volume} \\ \text{of } OH^-}}{\downarrow}}{50.00}{50.00 + 3.00}\right)}_{\substack{\text{Dilution} \\ \text{factor}}} = 0.013\ 2\ M \quad (11\text{-}1)$$

Total volume of solution

This "streamlined" calculation was first described on page 128.

$$[H^+] = \frac{K_w}{[OH^-]} = \frac{1.0 \times 10^{-14}}{0.013\ 2} = 7.5_8 \times 10^{-13}\ M \Rightarrow pH = 12.12$$

Equation 11-1 is an example of the streamlined calculation introduced in Section 7-4 in connection with precipitation titrations. This equation tells us that the concentration of $OH^-$ is equal to a certain fraction of the initial concentration, with a correction for dilution. The dilution factor equals the initial volume of analyte divided by the total volume of solution.

In Table 11-1, the volume of acid added is designated $V_a$. pH is expressed to the 0.01 decimal place, regardless of what is justified by significant figures. We do this for the sake of consistency and also because 0.01 is near the limit of accuracy in pH measurements.

Challenge  Using a setup similar to Equation 11-1, calculate $[OH^-]$ when 6.00 mL of HBr have been added. Check your pH against the value in Table 11-1.

## Region 2: At the Equivalence Point

Region 2 is the equivalence point, where just enough $H^+$ has been added to consume $OH^-$. We could prepare the same solution by dissolving KBr in water. The pH is determined by the dissociation of water:

$$H_2O \rightleftharpoons \underset{x}{H^+} + \underset{x}{OH^-}$$

$$K_w = x^2 \Rightarrow x = 1.00 \times 10^{-7}\ M \Rightarrow pH = 7.00$$

At the equivalence point, pH = 7.00, but *only* in a strong acid–strong base reaction.

The pH at the equivalence point in the titration of any strong base (or acid) with strong acid (or base) will be 7.00 at 25°C.

As we will soon discover, *the pH is **not** 7.00 at the equivalence point in the titration of weak acids or bases.* The pH is 7.00 only if the titrant and analyte are both strong.

## Region 3: After the Equivalence Point

Beyond the equivalence point, we are adding excess HBr to the solution. The concentration of excess $H^+$ at, say, 10.50 mL is given by

$$[H^+] = \underbrace{(0.100\ 0\ M)}_{\substack{\text{Initial} \\ \text{concentration} \\ \text{of } H^+}} \underbrace{\left(\frac{\overset{\substack{\text{Volume of} \\ \text{excess } H^+}}{\downarrow}}{0.50}{50.00 + 10.50}\right)}_{\substack{\text{Dilution} \\ \text{factor}}} = 8.26 \times 10^{-4}\ M$$

Total volume of solution

After the equivalence point, there is excess $H^+$.

$$pH = -\log[H^+] = 3.08$$

At $V_a = 10.50$ mL, there is an excess of just $V_a - V_e = 10.50 - 10.00 = 0.50$ mL of HBr. That is the reason why 0.50 appears in the dilution factor.

## The Titration Curve

The complete titration curve in Figure 11-1 exhibits a rapid change in pH near the equivalence point. The equivalence point is where the slope $(dpH/dV_a)$ is greatest (and the second derivative is 0, which makes it an *inflection point*). To repeat an important statement, the pH at the equivalence point is 7.00 *only* in a strong-acid–strong-base titration. If one or both of the reactants are weak, the equivalence point pH is *not* 7.00.

## ◼◼◼◼ 11-2 Titration of Weak Acid with Strong Base

The titration of a weak acid with a strong base allows us to put all our knowledge of acid-base chemistry to work. The example we examine is the titration of 50.00 mL of 0.020 00 M MES with 0.100 0 M NaOH. MES is an abbreviation for 2-(N-morpholino)ethanesulfonic acid, which is a weak acid with $pK_a = 6.27$.

The *titration reaction* is

$$O\!\!\nearrow\!\!\searrow\!\!\overset{+}{N}HCH_2CH_2SO_3^- \ + \ OH^- \ \rightarrow \ O\!\!\nearrow\!\!\searrow\!\!NCH_2CH_2SO_3^- \ + \ H_2O \qquad (11\text{-}2)$$

$$\underset{\text{MES, } pK_a = 6.27}{\underset{\text{HA}}{}} \qquad\qquad\qquad\qquad\qquad \text{A}^-$$

> **Always start by writing the titration reaction.**

Reaction 11-2 is the reverse of the $K_b$ reaction for the base $A^-$. Therefore, the equilibrium constant for Reaction 11-2 is $K = 1/K_b = 1/(K_w/K_a \text{ (for HA)}) = 5.4 \times 10^7$. The equilibrium constant is so large that we can say that the reaction goes "to completion" after each addition of $OH^-$. As we saw in Box 9-3, *strong plus weak react completely.*

> **Strong + weak → complete reaction**

Let's first calculate the volume of base, $V_b$, needed to reach the equivalence point:

$$\underbrace{(V_b \text{ (mL)})(0.100\ 0 \text{ M})}_{\text{mmol of base}} = \underbrace{(50.00 \text{ mL})(0.020\ 00 \text{ M})}_{\text{mmol of HA}} \Rightarrow V_b = 10.00 \text{ mL}$$

The titration calculations for this problem are of four types:

1. Before any base is added, the solution contains just HA in water. This is a weak acid whose pH is determined by the equilibrium

$$HA \underset{}{\overset{K_a}{\rightleftharpoons}} H^+ + A^-$$

2. From the first addition of NaOH until immediately before the equivalence point, there is a mixture of unreacted HA plus the $A^-$ produced by Reaction 11-2. *Aha! A buffer!* We can use the Henderson-Hasselbalch equation to find the pH.

3. At the equivalence point, "all" HA has been converted into $A^-$. The same solution could have been made by dissolving $A^-$ in water. We have a weak base whose pH is determined by the reaction

$$A^- + H_2O \overset{K_b}{\rightleftharpoons} HA + OH^-$$

4. Beyond the equivalence point, excess NaOH is being added to a solution of $A^-$. To a good approximation, pH is determined by the strong base. We calculate the pH as if we had simply added excess NaOH to water. We neglect the tiny effect of $A^-$.

### Region 1: Before Base Is Added

Before adding any base, we have a solution of 0.020 00 M HA with $pK_a = 6.27$. This is simply a weak-acid problem.

> **The initial solution contains just the *weak acid* HA.**

$$HA \rightleftharpoons H^+ + A^- \qquad K_a = 10^{-6.27}$$
$$\phantom{HA \rightleftharpoons}\ F-x \quad\ x \quad\ x$$

$$\frac{x^2}{0.020\ 00 - x} = K_a \Rightarrow x = 1.03 \times 10^{-4} \Rightarrow pH = 3.99$$

### Region 2: Before the Equivalence Point

Adding $OH^-$ creates a mixture of HA and $A^-$. This mixture is a buffer whose pH can be calculated with the Henderson-Hasselbalch equation (9-16) from the quotient $[A^-]/[HA]$.

> **Before the equivalence point, there is a mixture of HA and $A^-$, which is a *buffer*.**

Suppose we wish to calculate $[A^-]/[HA]$ when 3.00 mL of $OH^-$ have been added. Because $V_e = 10.00$ mL, we have added enough base to react with three-tenths of the HA. We can make a table showing the relative concentrations before and after the reaction:

> **We only need *relative* concentrations because pH depends on the quotient $[A^-]/[HA]$.**

| Titration reaction: | HA | + OH$^-$ | → A$^-$ | + H$_2$O |
|---|---|---|---|---|
| Relative initial quantities (HA ≡ 1) | 1 | $\frac{3}{10}$ | — | — |
| Relative final quantities | $\frac{7}{10}$ | — | $\frac{3}{10}$ | — |

Once we know the *quotient* $[A^-]/[HA]$ in any solution, we know its pH:

$$pH = pK_a + \log\left(\frac{[A^-]}{[HA]}\right) = 6.27 + \log\left(\frac{3/10}{7/10}\right) = 5.90$$

The point at which the volume of titrant is $\frac{1}{2}V_e$ is a special one in any titration.

| Titration reaction: | | HA | + | OH⁻ | → | A⁻ | + | H₂O |
|---|---|---|---|---|---|---|---|---|
| Relative initial quantities | | 1 | | $\frac{1}{2}$ | | — | | — |
| Relative final quantities | | $\frac{1}{2}$ | | — | | $\frac{1}{2}$ | | — |

$$pH = pK_a + \log\left(\frac{1/2}{1/2}\right) = pK_a$$

When $V_b = \frac{1}{2}V_e$, pH = $pK_a$.
This is a landmark in any titration.

When the volume of titrant is $\frac{1}{2}V_e$, pH = $pK_a$ for the acid HA (neglecting activity coefficients). From an experimental titration curve, you can find the approximate value of $pK_a$ by reading the pH when $V_b = \frac{1}{2}V_e$, where $V_b$ is the volume of added base. (To find the true value of $pK_a$ requires activity coefficients.)

**Advice** As soon as you recognize a mixture of HA and A⁻ in any solution, *you have a buffer!* You can calculate the pH from the quotient $[A^-]/[HA]$.

$$pH = pK_a + \log\left(\frac{[A^-]}{[HA]}\right)$$

*Learn to recognize buffers!* They lurk in every corner of acid-base chemistry.


## Region 3: At the Equivalence Point

At the equivalence point, there is exactly enough NaOH to consume HA.

At the equivalence point, HA has been converted into A⁻, a *weak base*.

| Titration reaction: | | HA | + | OH⁻ | → | A⁻ | + | H₂O |
|---|---|---|---|---|---|---|---|---|
| Relative initial quantities | | 1 | | 1 | | — | | — |
| Relative initial quantities | | — | | — | | 1 | | — |

The solution contains "just" A⁻. We could have prepared the same solution by dissolving the salt Na⁺A⁻ in distilled water. *A solution of Na⁺A⁻ is merely a solution of a weak base.*

To compute the pH of a weak base, we write the reaction of the weak base with water:

$$A^- + H_2O \rightleftharpoons HA + OH^- \qquad K_b = \frac{K_w}{K_a}$$
$$\phantom{A^- +} F - x \phantom{H_2O \rightleftharpoons} x \phantom{HA +} x$$

The only tricky point is that the formal concentration of A⁻ is no longer 0.020 00 M, which was the initial concentration of HA. The A⁻ has been diluted by NaOH from the buret:

$$F' = \underbrace{(0.020\ 00\ \text{M})}_{\substack{\text{Initial} \\ \text{concentration} \\ \text{of HA}}} \underbrace{\left(\frac{50.00}{50.00\ +\ 10.00}\right)}_{\substack{\text{Dilution} \\ \text{factor}}} = 0.016\ 7\ \text{M}$$

(Initial volume of HA / Total volume of solution)

With this value of F′, we can solve the problem:

$$\frac{x^2}{F' - x} = K_b = \frac{K_w}{K_a} = 1.86 \times 10^{-8} \Rightarrow x = 1.76 \times 10^{-5}\ \text{M}$$

$$pH = -\log[H^+] = -\log\left(\frac{K_w}{x}\right) = 9.25$$

The pH at the equivalence point in this titration is 9.25. **It is not 7.00.** The equivalence-point pH will *always* be above 7 for the titration of a weak acid, because the acid is converted into its conjugate base at the equivalence point.

The pH is always higher than 7 at the equivalence point in the titration of a weak acid with a strong base.

## Region 4: After the Equivalence Point

Here we assume that the pH is governed by the excess OH⁻.

Now we are adding NaOH to a solution of $A^-$. NaOH is so much stronger a base than $A^-$ that it is fair to say that the pH is determined by the excess $OH^-$.

Let's calculate the pH when $V_b = 10.10$ mL, which is just 0.10 mL past $V_e$. The concentration of excess $OH^-$ is

$$[OH^-] = \overset{\text{Initial concentration of } OH^-}{(0.100\ 0\ M)}\overset{\text{Dilution factor}}{\left(\frac{\overset{\text{Volume of excess } OH^-}{0.10}}{\underset{\text{Total volume of solution}}{50.00 + 10.00}}\right)} = 1.66 \times 10^{-4}\ M$$

**Challenge** Compare the concentration of $OH^-$ from excess titrant at $V_b = 10.10$ mL to the concentration of $OH^-$ from hydrolysis of $A^-$. Satisfy yourself that it is fair to neglect the contribution of $A^-$ to the pH after the equivalence point.

$$pH = -\log\left(\frac{K_w}{[OH^-]}\right) = 10.22$$

### The Titration Curve

Landmarks in a titration:
At $V_b = V_e$, the curve is steepest.
At $V_b = \frac{1}{2}V_e$, pH = p$K_a$ and the slope is minimal.

Calculations for the titration of MES with NaOH are shown in Table 11-2. The calculated titration curve in Figure 11-2 has two easily identified points. One is the equivalence point, which is the steepest part of the curve. The other landmark is the point where $V_b = \frac{1}{2}V_e$ and pH = p$K_a$. This latter point is also an inflection point, having the minimum slope.

If you look back at Figure 9-4b, you will note that the maximum *buffer capacity* occurs when pH = p$K_a$. That is, the solution is most resistant to pH changes when pH = p$K_a$ (and $V_b = \frac{1}{2}V_e$); the slope ($d\text{pH}/dV_b$) is therefore at its minimum.

The *buffer capacity* measures the ability of the solution to resist changes in pH.

Figure 11-3 shows how the titration curve depends on the acid dissociation constant of HA and on the concentrations of reactants. As HA becomes a weaker acid, or as the concentrations of analyte and titrant decrease, the inflection near the equivalence point decreases, until the equivalence point becomes too shallow to detect. *It is not practical to titrate an acid or base when its strength is too weak or its concentration too dilute.*

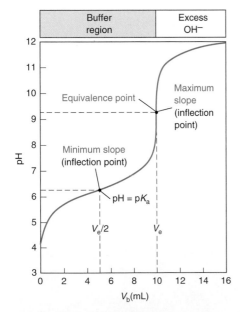

**Figure 11-2** Calculated titration curve for the reaction of 50.00 mL of 0.020 00 M MES with 0.100 0 M NaOH. Landmarks occur at half of the equivalence volume (pH = p$K_a$) and at the equivalence point, which is the steepest part of the curve.

**Table 11-2** Calculation of the titration curve for 50.00 mL of 0.020 00 M MES treated with 0.100 0 M NaOH

|  | mL base added ($V_b$) | pH |
|---|---|---|
| Region 1 (weak acid) | 0.00 | 3.99 |
| | 0.50 | 4.99 |
| | 1.00 | 5.32 |
| | 2.00 | 5.67 |
| | 3.00 | 5.90 |
| | 4.00 | 6.09 |
| | 5.00 | 6.27 |
| Region 2 (buffer) | 6.00 | 6.45 |
| | 7.00 | 6.64 |
| | 8.00 | 6.87 |
| | 9.00 | 7.22 |
| | 9.50 | 7.55 |
| | 9.90 | 8.27 |
| Region 3 (weak base) | 10.00 | 9.25 |
| | 10.10 | 10.22 |
| | 10.50 | 10.91 |
| | 11.00 | 11.21 |
| Region 4 (excess OH⁻) | 12.00 | 11.50 |
| | 13.00 | 11.67 |
| | 14.00 | 11.79 |
| | 15.00 | 11.88 |
| | 16.00 | 11.95 |

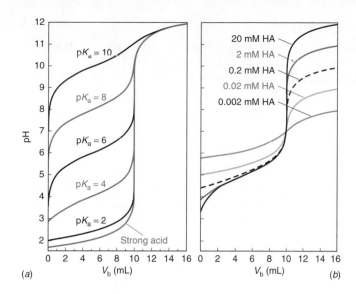

**Figure 11-3** (a) Calculated curves showing the titration of 50.0 mL of 0.020 0 M HA with 0.100 M NaOH. (b) Calculated curves showing the titration of 50.0 mL of HA ($pK_a = 5$) with NaOH whose concentration is five times greater than that of HA. As the acid becomes weaker or more dilute, the end point becomes less distinct.

## ■ ■ ■ 11-3 Titration of Weak Base with Strong Acid

The titration of a weak base with a strong acid is just the reverse of the titration of a weak acid with a strong base. The *titration reaction* is

$$B + H^+ \rightarrow BH^+$$

Because the reactants are a weak base and a strong acid, the reaction goes essentially to completion after each addition of acid. There are four distinct regions of the titration curve:

1. Before acid is added, the solution contains just the weak base, B, in water. The pH is determined by the $K_b$ reaction.

$$\underset{F-x}{B} + H_2O \underset{}{\overset{K_b}{\rightleftharpoons}} \underset{x}{BH^+} + \underset{x}{OH^-}$$

When $V_a$ (= volume of added acid) = 0, we have a *weak-base* problem.

2. Between the initial point and the equivalence point, there is a mixture of B and BH⁺—*Aha! A buffer!* The pH is computed by using

$$pH = pK_a \text{ (for } BH^+\text{)} + \log\left(\frac{[B]}{[BH^+]}\right)$$

When $0 < V_a < V_e$, we have a *buffer*.

In adding acid (increasing $V_a$), we reach the special point where $V_a = \frac{1}{2}V_e$ and $pH = pK_a$ (for $BH^+$). As before, $pK_a$ (and therefore $pK_b$) can be determined easily from the titration curve.

3. At the equivalence point, B has been converted into $BH^+$, a weak acid. The pH is calculated by considering the acid dissociation reaction of $BH^+$.

$$\underset{F'-x}{BH^+} \rightleftharpoons \underset{x}{B} + \underset{x}{H^+} \qquad K_a = \frac{K_w}{K_b}$$

When $V_a = V_e$, the solution contains the *weak acid* BH⁺.

The formal concentration of $BH^+$, $F'$, is not the original formal concentration of B, because some dilution has occurred. The solution contains $BH^+$ at the equivalence point, so it is acidic. *The pH at the equivalence point must be below 7.*

4. After the equivalence point, the excess strong acid determines the pH. We neglect the contribution of weak acid, $BH^+$.

For $V_a > V_e$, there is excess *strong acid*.

---

**Example** Titration of Pyridine with HCl

Consider the titration of 25.00 mL of 0.083 64 M pyridine with 0.106 7 M HCl.

$$\text{(pyridine structure) N:} \qquad K_b = 1.59 \times 10^{-9} \quad \Rightarrow \quad K_a = \frac{K_w}{K_b} = 6.31 \times 10^{-6} \qquad pK_a = 5.20$$

Pyridine

The titration reaction is

$$\text{C}_6\text{H}_5\text{N:} + \text{H}^+ \rightarrow \text{C}_6\text{H}_5\text{NH}^+$$

and the equivalence point occurs at 19.60 mL:

$$\underbrace{(V_e(\text{mL}))(0.106\ 7\ \text{M})}_{\text{mmol of HCl}} = \underbrace{(25.00\ \text{mL})(0.083\ 64\ \text{M})}_{\text{mmol of pyridine}} \Rightarrow V_e = 19.60\ \text{mL}$$

Find the pH when $V_a = 4.63$ mL.

**Solution**  Part of the pyridine has been neutralized, so there is a mixture of pyridine and pyridinium ion—*Aha! A buffer!* The fraction of pyridine that has been titrated is $4.63/19.60 = 0.236$, because it takes 19.60 mL to titrate the whole sample. The fraction of pyridine remaining is $1 - 0.236 = 0.764$. The pH is

$$\text{pH} = \text{p}K_a + \log\left(\frac{[\text{B}]}{[\text{BH}^+]}\right)$$

$$= 5.20 + \log\frac{0.764}{0.236} = 5.71$$

## ■ ■ ■ 11-4  Titrations in Diprotic Systems

The principles developed for titrations of monoprotic acids and bases are readily extended to titrations of polyprotic acids and bases. We will examine two cases.

### A Typical Case

The upper curve in Figure 11-4 is calculated for the titration of 10.0 mL of 0.100 M base (B) with 0.100 M HCl. The base is dibasic, with $\text{p}K_{b1} = 4.00$ and $\text{p}K_{b2} = 9.00$. The titration curve has reasonably sharp breaks at both equivalence points, corresponding to the reactions

$$\text{B} + \text{H}^+ \rightarrow \text{BH}^+$$
$$\text{BH}^+ + \text{H}^+ \rightarrow \text{BH}_2^{2+}$$

The volume at the first equivalence point is 10.00 mL because

$$\underbrace{(V_e\ (\text{mL}))(0.100\ \text{M})}_{\text{mmol of HCl}} = \underbrace{(10.00\ \text{mL})(0.100\ 0\ \text{M})}_{\text{mmol of B}} \Rightarrow V_e = 10.00\ \text{mL}$$

$V_{e2} = 2V_{e1}$, always.

The volume at the second equivalence point must be $2V_e$, because the second reaction requires the same number of moles of HCl as the first reaction.

The pH calculations are similar to those for corresponding points in the titration of a monobasic compound. Let's examine points A through E in Figure 11-4.

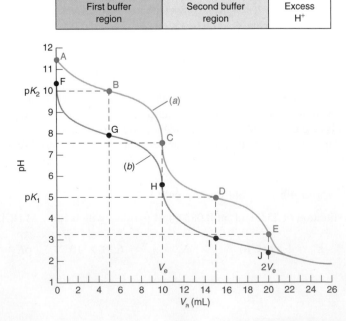

*Figure 11-4*  (*a*) Titration of 10.0 mL of 0.100 M base (p$K_{b1}$ = 4.00, p$K_{b2}$ = 9.00) with 0.100 M HCl. The two equivalence points are C and E. Points B and D are the half-neutralization points, whose pH values equal p$K_{a2}$ and p$K_{a1}$, respectively. (*b*) Titration of 10.0 mL of 0.100 M nicotine (p$K_{b1}$ = 6.15, p$K_{b2}$ = 10.85) with 0.100 M HCl. There is no sharp break at the second equivalence point, J, because the pH is too low.

## Point A

Before acid is added, the solution contains just weak base, B, whose pH is governed by the reaction

$$B + H_2O \xrightleftharpoons{K_{b1}} BH^+ + OH^-$$
$$\phantom{B +} 0.100 - x \phantom{H_2O \xrightleftharpoons{K_{b1}} BH^+ +} x \phantom{+} x$$

$$\frac{x^2}{0.100 - x} = 1.00 \times 10^{-4} \Rightarrow x = 3.11 \times 10^{-3}$$

$$[H^+] = \frac{K_w}{x} \Rightarrow pH = 11.49$$

Recall that the fully basic form of a dibasic compound can be treated as if it were monobasic. (The $K_{b2}$ reaction can be neglected.)

## Point B

At any point between A (the initial point) and C (the first equivalence point), we have a buffer containing B and $BH^+$. Point B is halfway to the equivalence point, so $[B] = [BH^+]$. The pH is calculated from the Henderson-Hasselbalch equation for *the weak acid*, $BH^+$, whose acid dissociation constant is $K_{a2}$ (for $BH_2^{2+}$) $= K_w/K_{b1} = 10^{-10.00}$.

Of course, you remember that
$$K_{a2} = \frac{K_w}{K_{b1}} \qquad K_{a1} = \frac{K_w}{K_{b2}}$$

$$pH = pK_{a2} + \log\frac{[B]}{[BH^+]} = 10.00 + \log 1 = 10.00$$

So the pH at point B is just $pK_{a2}$.

To calculate the quotient $[B]/[BH^+]$ at any point in the buffer region, just find what fraction of the way from point A to point C the titration has progressed. For example, if $V_a = 1.5$ mL, then

$$\frac{[B]}{[BH^+]} = \frac{8.5}{1.5}$$

because 10.0 mL are required to reach the equivalence point and we have added just 1.5 mL. The pH at $V_a = 1.5$ mL is

$$pH = 10.00 + \log\frac{8.5}{1.5} = 10.75$$

## Point C

At the first equivalence point, *B has been converted into $BH^+$, the intermediate form of the diprotic acid, $BH_2^{2+}$. $BH^+$ is both an acid and a base.* From Section 10-1, we know that

$$[H^+] \approx \sqrt{\frac{K_1 K_2 F + K_1 K_w}{K_1 + F}} \qquad \text{(11-3)}$$

$BH^+$ is the *intermediate form* of a diprotic acid.

$pH \approx \frac{1}{2}(pK_1 + pK_2)$

where $K_1$ and $K_2$ are the acid dissociation constants of $BH_2^{2+}$.

The formal concentration of $BH^+$ is calculated by considering dilution of the original solution of B.

$$F = (0.100 \text{ M})\left(\frac{10.0}{20.0}\right) = 0.050\ 0 \text{ M}$$

Initial volume of B ↙ · Original concentration of B · Dilution factor · Total volume of solution ↖

Plugging all the numbers into Equation 11-3 gives

$$[H^+] = \sqrt{\frac{(10^{-5})(10^{-10})(0.050\ 0) + (10^{-5})(10^{-14})}{10^{-5} + 0.050\ 0}} = 3.16 \times 10^{-8}$$

$$pH = 7.50$$

Note that, in this example, $pH = \frac{1}{2}(pK_{a1} + pK_{a2})$.

Point C in Figure 11-4 shows where the intermediate form of a diprotic acid lies on a titration curve. This is the *least-buffered* point on the whole curve, because the pH changes most rapidly if small amounts of acid or base are added. There is a misconception that the intermediate form of a diprotic acid behaves as a buffer when, in fact, it is the *worst choice* for a buffer.

The intermediate form of a polyprotic acid is the worst possible choice for a buffer.

**Challenge** Show that if $V_a$ were 17.2 mL, the ratio in the log term would be

$$\frac{[BH^+]}{[BH_2^{2+}]} = \frac{20.0 - 17.2}{17.2 - 10.0} = \frac{2.8}{7.2}$$

### Point D

At any point between C and E, there is a buffer containing $BH^+$ (the base) and $BH_2^{2+}$ (the acid). When $V_a = 15.0$ mL, $[BH^+] = [BH_2^{2+}]$ and

$$pH = pK_{a1} + \log\frac{[BH^+]}{[BH_2^{2+}]} = 5.00 + \log 1 = 5.00$$

### Point E

Point E is the second equivalence point, at which the solution is formally the same as one prepared by dissolving $BH_2Cl_2$ in water. The formal concentration of $BH_2^{2+}$ is

$$F = (0.100 \text{ M})\underset{\underset{\text{Total volume of solution}}{\nearrow}}{\overset{\overset{\text{Original volume of B}}{\searrow}}{\left(\frac{10.0}{30.0}\right)}} = 0.033\ 3 \text{ M}$$

The pH is determined by the acid dissociation reaction of $BH_2^{2+}$.

$$BH_2^{2+} \rightleftharpoons BH^+ + H^+ \qquad K_{a1} = \frac{K_w}{K_{b2}}$$
$$\phantom{BH_2^{2+}}\ \ \, F - x \quad\ \ x \quad\ \ x$$

At the second equivalence point, we have made $BH_2^{2+}$, which can be treated as a monoprotic weak acid.

$$\frac{x^2}{0.033\ 3 - x} = 1.0 \times 10^{-5} \Rightarrow x = 5.72 \times 10^{-4} \Rightarrow pH = 3.24$$

Beyond the second equivalence point ($V_a > 20.0$ mL), the pH of the solution can be calculated from the volume of strong acid added to the solution. For example, at $V_a = 25.00$ mL, there is an excess of 5.00 mL of 0.100 M HCl in a total volume of $10.00 + 25.00 = 35.00$ mL. The pH is found by writing

$$[H^+] = (0.100 \text{ M})\left(\frac{5.00}{35.00}\right) = 1.43 \times 10^{-2} \text{ M} \Rightarrow pH = 1.85$$

### Blurred End Points

When the pH is too low or too high or when $pK_a$ values are too close together, end points are obscured.

Titrations of many diprotic acids or bases show two clear end points, as in curve *a* in Figure 11-4. Some titrations do not show both end points, as illustrated by curve *b*, which is calculated for the titration of 10.0 mL of 0.100 M nicotine ($pK_{b1} = 6.15$, $pK_{b2} = 10.85$) with 0.100 M HCl. The two reactions are

Nicotine (B) $\longrightarrow$ $BH^+$ $\longrightarrow$ $BH_2^{2+}$

There is no perceptible break at the second equivalence point (J), because $BH_2^{2+}$ is too strong an acid (or, equivalently, $BH^+$ is too weak a base). As the titration approaches low pH ($\lesssim 3$), the approximation that HCl reacts completely with $BH^+$ to give $BH_2^{2+}$ breaks down. To calculate pH between points I and J requires the systematic treatment of equilibrium. Later in this chapter, we describe how to calculate the whole curve with a spreadsheet.

In the titration of ribonuclease at the beginning of this chapter, there is a continuous change in pH, with no clear breaks. The reason is that 29 groups are titrated in the pH interval shown. The 29 end points are so close together that a nearly uniform rise results. The curve can be analyzed to find the many $pK_a$ values, but this analysis requires a computer; and, even then, the individual $pK_a$ values will not be determined very precisely.

## ■■■ 11-5 Finding the End Point with a pH Electrode

Box 11-1 illustrates an important application of acid-base titrations in environmental analysis.

Titrations are most commonly performed either to find out how much analyte is present or to measure equilibrium constants of the analyte. We can obtain the information necessary for both purposes by monitoring the pH of the solution as the titration is performed.

Box 11-1   Alkalinity and Acidity

*Alkalinity* is defined as the capacity of natural water to react with $H^+$ to reach pH 4.5, which is the second equivalence point in the titration of carbonate ($CO_3^{2-}$) with $H^+$. To a good approximation, alkalinity is determined by $OH^-$, $CO_3^{2-}$, and $HCO_3^-$:

$$\text{Alkalinity} \approx [OH^-] + 2[CO_3^{2-}] + [HCO_3^-]$$

When water whose pH is greater than 4.5 is titrated with acid to pH 4.5 (measured with a pH meter), all $OH^-$, $CO_3^{2-}$, and $HCO_3^-$ will have reacted. Other basic species also react, but $OH^-$, $CO_3^{2-}$, and $HCO_3^-$ account for most of the alkalinity in most water samples. Alkalinity is normally expressed as millimoles of $H^+$ needed to bring 1 L of water to pH 4.5.

Alkalinity and *hardness* (dissolved $Ca^{2+}$ and $Mg^{2+}$, Box 12-3) are important characteristics of irrigation water. Alkalinity in excess of the $Ca^{2+} + Mg^{2+}$ content is called "residual sodium carbonate." Water with a residual sodium carbonate content equivalent to $\geq 2.5$ mmol $H^+$/L is not suitable for irrigation. Residual sodium carbonate between 1.25 and 2.5 mmol $H^+$/L is marginal, whereas $\leq 1.25$ mmol $H^+$/L is suitable for irrigation.

*Acidity* of natural waters refers to the total acid content that can be titrated to pH 8.3 with NaOH. This pH is the second equivalence point for titration of carbonic acid ($H_2CO_3$) with $OH^-$. Almost all weak acids in the water also will be titrated in this procedure. Acidity is expressed as millimoles of $OH^-$ needed to bring 1 L of water to pH 8.3.

Figure 11-5 shows an *autotitrator*, which performs the entire operation automatically.[4] Titrant from the plastic bottle at the rear is dispensed in small increments by a syringe pump while pH is measured by electrodes in the beaker of analyte on the stirrer. (We will learn how these electrodes work in Chapter 15.) The instrument waits for pH to stabilize after each addition, before adding the next increment. The end point is computed automatically by finding the maximum slope in the titration curve.

Figure 11-6a shows experimental results for the manual titration of a hexaprotic weak acid, $H_6A$, with NaOH. Because the compound is difficult to purify, only a tiny amount was available for titration. Just 1.430 mg was dissolved in 1.00 mL of water and titrated with microliter quantities of 0.065 92 M NaOH, delivered with a Hamilton syringe.

Figure 11-6a shows two clear breaks, near 90 and 120 μL, which correspond to titration of the *third* and *fourth* protons of $H_6A$.

$$H_4A^{2-} + OH^- \rightarrow H_3A^{3-} + H_2O \quad (\sim 90 \text{ μL equivalence point})$$
$$H_3A^{3-} + OH^- \rightarrow H_2A^{4-} + H_2O \quad (\sim 120 \text{ μL equivalence point})$$

The first two and last two equivalence points are unrecognizable, because they occur at too low or too high a pH.

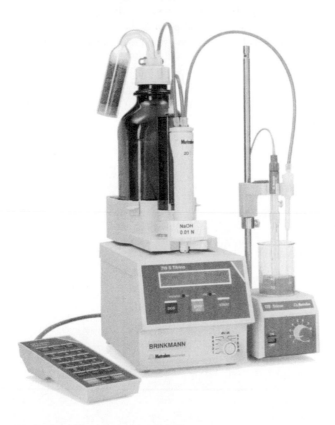

*Figure 11-5* Autotitrator delivers titrant from the bottle at the back to the beaker of analyte on the stirring motor at the right. Electrodes in the beaker monitor pH or concentrations of specific ions. Volume and pH readings can go directly to the spreadsheet program in a computer. [Brinkman Instruments, Westbury, NY.]

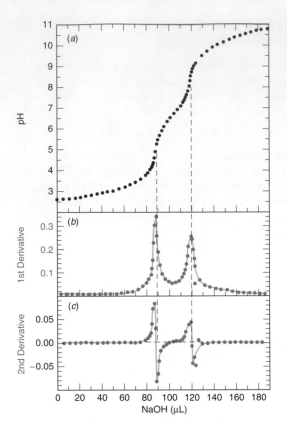

**Figure 11-6** (*a*) Experimental points in the titration of 1.430 mg of xylenol orange, a hexaprotic acid, dissolved in 1.000 mL of aqueous 0.10 M NaNO₃. The titrant was 0.065 92 M NaOH. (*b*) The first derivative, ΔpH/ΔV, of the titration curve. (*c*) The second derivative, Δ(ΔpH/ΔV)/ΔV, which is the derivative of the curve in panel *b*. Derivatives for the first end point are calculated in Table 11-3. End points are taken as maxima in the derivative curve and zero crossings of the second derivative.

## Using Derivatives to Find the End Point

The end point has maximum slope and the second derivative is 0.

The end point is taken as the volume where the slope ($d$pH/$dV$) of the titration curve is greatest. The slope (first derivative) in Figure 11-6b is calculated in Table 11-3. The first two columns contain experimental volume and pH. (The pH meter was precise to three digits, even though accuracy ends in the second decimal place.) To compute the first derivative, each pair of volumes is averaged and the quantity ΔpH/ΔV is calculated. ΔpH is the change in pH between consecutive readings and ΔV is the change in volume between consecutive additions. Figure 11-6c and the last two columns of Table 11-3 give the second derivative, computed in an analogous manner. The end point is the volume at which the second derivative is 0. Figure 11-7 allows us to make good estimates of the end points.

**Table 11-3**  Computation of first and second derivatives for a titration curve

| μL NaOH | pH | First derivative | | Second derivative | |
| --- | --- | --- | --- | --- | --- |
| | | μL | $\dfrac{\Delta \text{pH}}{\Delta \mu \text{L}}$ | μL | $\dfrac{\Delta(\Delta \text{pH}/\Delta \mu \text{L})}{\Delta \mu \text{L}}$ |
| 85.0 | 4.245 | | | | |
| | | 85.5 | 0.155 | | |
| 86.0 | 4.400 | | | 86.0 | 0.071 0 |
| | | 86.5 | 0.226 | | |
| 87.0 | 4.626 | | | 87.0 | 0.081 0 |
| | | 87.5 | 0.307 | | |
| 88.0 | 4.933 | | | 88.0 | 0.033 0 |
| | | 88.5 | 0.340 | | |
| 89.0 | 5.273 | | | 89.0 | −0.083 0 |
| | | 89.5 | 0.257 | | |
| 90.0 | 5.530 | | | 90.0 | −0.068 0 |
| | | 90.5 | 0.189 | | |
| 91.0 | 5.719 | | | 91.25 | −0.039 0 |
| | | 92.0 | 0.130 | | |
| 93.0 | 5.980 | | | | |

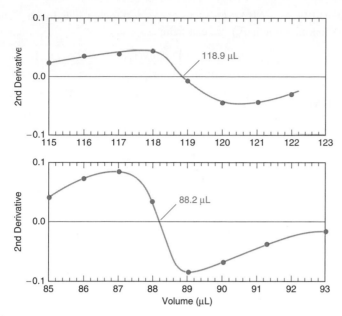

**Figure 11-7** Enlargement of the end-point regions in the second derivative curve shown in Figure 11-6c.

---

**Example** Computing Derivatives of a Titration Curve

Let's see how the first and second derivatives in Table 11-3 are calculated.

**Solution** The first number in the third column, 85.5, is the average of the first two volumes (85.0 and 86.0) in the first column. The derivative $\Delta pH/\Delta V$ is calculated from the first two pH values and the first two volumes:

$$\frac{\Delta pH}{\Delta V} = \frac{4.400 - 4.245}{86.0 - 85.0} = 0.155$$

The coordinates ($x = 85.5$, $y = 0.155$) are one point in the graph of the first derivative in Figure 11-6.

The second derivative is computed from the first derivative. The first entry in the fifth column of Table 11-3 is 86.0, which is the average of 85.5 and 86.5. The second derivative is

$$\frac{\Delta(\Delta pH/\Delta V)}{\Delta V} = \frac{0.226 - 0.155}{86.5 - 85.5} = 0.071$$

The coordinates ($x = 86.0$, $y = 0.071$) are plotted in the second derivative graph at the bottom of Figure 11-6. These calculations are tedious by hand, but trivial with a spreadsheet.

## Using a Gran Plot to Find the End Point[5,6]

A problem with using derivatives to find the end point is that titration data are least accurate right near the end point, because buffering is minimal and electrode response is sluggish. A **Gran plot** uses data from before the end point (typically from 0.8 $V_e$ or 0.9 $V_e$ up to $V_e$) to locate the end point.

Consider the titration of a weak acid, HA:

$$HA \rightleftharpoons H^+ + A^- \qquad K_a = \frac{[H^+]\gamma_{H^+}[A^-]\gamma_{A^-}}{[HA]\gamma_{HA}}$$

It will be necessary to include activity coefficients in this discussion because a pH electrode responds to hydrogen ion *activity*, not concentration.

At any point between the initial point and the end point of the titration, it is usually a good approximation to say that each mole of NaOH converts 1 mol of HA into 1 mol of $A^-$.

A related method uses data from the middle of the titration (not near the equivalence point) to deduce $V_e$ and $K_a$.[7]

Strong plus weak react completely.

If we have titrated $V_a$ mL of HA (whose formal concentration is $F_a$) with $V_b$ mL of NaOH (whose formal concentration is $F_b$), we can write

$$[A^-] = \frac{\text{moles of OH}^- \text{ delivered}}{\text{total volume}} = \frac{V_b F_b}{V_b + V_a}$$

$$[HA] = \frac{\text{original moles of HA} - \text{moles of OH}^-}{\text{total volume}} = \frac{V_a F_a - V_b F_b}{V_a + V_b}$$

Substitution of these values of $[A^-]$ and $[HA]$ into the equilibrium expression gives

$$K_a = \frac{[H^+]\gamma_{H^+} V_b F_b \gamma_{A^-}}{(V_a F_a - V_b F_b)\gamma_{HA}}$$

which can be rearranged to

$$\underbrace{V_b[H^+]\gamma_{H^+}}_{10^{-pH}} = \frac{\gamma_{HA}}{\gamma_{A^-}} K_a \left( \frac{V_a F_a - V_b F_b}{F_b} \right) \qquad (11\text{-}4)$$

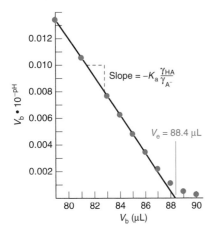

$$\mathcal{A}_{H^+} = [H^+]\gamma_{H^+} = 10^{-pH}$$

The term on the left is $V_b 10^{-pH}$, because $[H^+]\gamma_{H^+} = 10^{-pH}$. The term in parentheses on the right is

$$\frac{V_a F_a - V_b F_b}{F_b} = \frac{V_a F_a}{F_b} - V_b = V_e - V_b$$

$$V_a F_a = V_e F_b \Rightarrow V_e = \frac{V_a F_a}{F_b}$$

Equation 11-4 can, therefore, be written in the form

*Gran plot equation:* $\qquad V_b 10^{-pH} = \frac{\gamma_{HA}}{\gamma_{A^-}} K_a (V_e - V_b) \qquad (11\text{-}5)$

A graph of $V_b 10^{-pH}$ versus $V_b$ is called a *Gran plot*. If $\gamma_{HA}/\gamma_{A^-}$ is constant, the graph is a straight line with a slope of $-K_a \gamma_{HA}/\gamma_{A^-}$ and an $x$-intercept of $V_e$. Figure 11-8 shows a Gran plot for the titration in Figure 11-6. Any units can be used for $V_b$, but the same units must be used on both axes. In Figure 11-8, $V_b$ is expressed in microliters on both axes.

*Gran plot:*
Plot $V_b 10^{-pH}$ versus $V_b$
$x$-intercept $= V_e$
Slope $= -K_a \gamma_{HA}/\gamma_{A^-}$

The beauty of a Gran plot is that it enables us to use data taken *before* the end point to find the end point. The slope of the Gran plot enables us to find $K_a$. Although we derived the Gran function for a monoprotic acid, the same plot ($V_b 10^{-pH}$ versus $V_b$) applies to polyprotic acids (such as $H_6A$ in Figure 11-6).

The Gran function, $V_b 10^{-pH}$, does not actually go to 0, because $10^{-pH}$ is never 0. The curve must be extrapolated to find $V_e$. The reason the function does not reach 0 is that we have used the approximation that every mole of $OH^-$ generates 1 mol of $A^-$, which is not true as $V_b$ approaches $V_e$. Only the linear portion of the Gran plot is used.

Another source of curvature in the Gran plot is changing ionic strength, which causes $\gamma_{HA}/\gamma_{A^-}$ to vary. In Figure 11-6, this variation was avoided by maintaining nearly constant ionic strength with $NaNO_3$. Even without added salt, the last 10–20% of data before $V_e$ gives a fairly straight line because the quotient $\gamma_{HA}/\gamma_{A^-}$ does not change very much. The Gran plot in the acidic region gives accurate results even if $CO_2$ is dissolved in the strong-base titrant. The Gran plot in the basic region can be used to measure $CO_2$ in the strong base.[5]

*Figure 11-8* Gran plot for the first equivalence point of Figure 11-6. This plot gives an estimate of $V_e$ that differs from that in Figure 11-7 by 0.2 μL (88.4 versus 88.2 μL). The last 10–20% of volume prior to $V_e$ is normally used for a Gran plot.

**Challenge** Show that when weak base, B, is titrated with a strong acid, the Gran function is

$$V_a 10^{+pH} = \left( \frac{1}{K_a} \cdot \frac{\gamma_B}{\gamma_{BH^+}} \right)(V_e - V_a) \qquad (11\text{-}6)$$

where $V_a$ is the volume of strong acid and $K_a$ is the acid dissociation constant of $BH^+$. A graph of $V_a 10^{+pH}$ versus $V_a$ should be a straight line with a slope of $-\gamma_B/(\gamma_{BH^+} K_a)$ and an $x$-intercept of $V_e$.

## ■■■■ 11-6 Finding the End Point with Indicators

An indicator is an acid or a base whose various protonated forms have different colors.

An acid-base **indicator** is itself an acid or base whose various protonated species have different colors. An example is thymol blue.

The structures at the top of the page (Thymol blue equilibria) are shown with:

HO ... OH  Red (R) Thymol blue  $\xrightarrow{pK_1 = 1.7}$  Yellow ($Y^-$)  $\xrightarrow{pK_2 = 8.9}$  Blue ($B^{2-}$)

Below pH 1.7, the predominant species is red; between pH 1.7 and pH 8.9, the predominant species is yellow; and above pH 8.9, the predominant species is blue (Color Plate 4). For simplicity, we designate the three species R, $Y^-$, and $B^{2-}$.

The equilibrium between R and $Y^-$ can be written

$$R \xrightleftharpoons{K_1} Y^- + H^+ \qquad pH = pK_1 + \log\frac{[Y^-]}{[R]} \qquad (11\text{-}7)$$

| pH | $[Y^-]:[R]$ | Color |
|----|-------------|-------|
| 0.7 | 1:10 | red |
| 1.7 | 1:1 | orange |
| 2.7 | 10:1 | yellow |

At pH 1.7 ($= pK_1$), there will be a 1:1 mixture of the yellow and red species, which appears orange. As a crude rule of thumb, we can say that the solution will appear red when $[Y^-]/[R] \lesssim 1/10$ and yellow when $[Y^-]/[R] \gtrsim 10/1$. From Equation 11-7, we see that the solution will be red when $pH \approx pK_1 - 1$ and yellow when $pH \approx pK_1 + 1$. In tables of indicator colors, thymol blue is listed as red below pH 1.2 and yellow above pH 2.8. The pH values predicted by our rule of thumb are 0.7 and 2.7. Between pH 1.2 and 2.8, the indicator exhibits various shades of orange. The pH range (1.2 to 2.8) over which the color changes is called the **transition range.** Whereas most indicators have a single color change, thymol blue has another transition, from yellow to blue, between pH 8.0 and pH 9.6. In this range, various shades of green are seen.

Acid-base indicator color changes are featured in Demonstration 11-1. Box 11-2 shows how optical absorption by indicators allows us to measure pH.

## Choosing an Indicator

A titration curve for which pH = 5.54 at the equivalence point is shown in Figure 11-9. An indicator with a color change near this pH would be useful in determining the end point of the titration. In Figure 11-9, the pH drops steeply (from 7 to 4) over a small volume interval. Therefore, any indicator with a color change in this pH interval would provide a fair approximation to the equivalence point. The closer the point of color change is to pH 5.54, the more accurate will be the end point. The difference between the observed end point (color change) and the true equivalence point is called the **indicator error.**

One of the most common indicators is phenolphthalein, usually used for its colorless-to-pink transition at pH 8.0–9.6.

Colorless phenolphthalein pH < 8.0

$2H^+ \parallel 2OH^-$

Pink phenolphthalein pH > 9.6

In strong acid, the colorless form of phenolphthalein turns orange-red. In strong base, the red species loses its color.[8]

Orange-red (in 65–98% $H_2SO_4$)

Colorless (pH > 11)

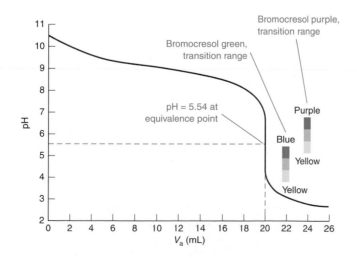

**Figure 11-9** Calculated titration curve for the reaction of 100 mL of 0.010 0 M base ($pK_b = 5.00$) with 0.050 0 M HCl.

This one is just plain fun.[9] Place 900 mL of water and a magnetic stirring bar in each of two 1-L graduated cylinders. Add 10 mL of 1 M $NH_3$ to each. Then put 2 mL of phenolphthalein indicator solution in one and 2 mL of bromothymol blue indicator solution in the other. Both indicators will have the color of their basic species.

Drop a few chunks of Dry Ice (solid $CO_2$) into each cylinder. As the $CO_2$ bubbles through each cylinder, the solutions become more acidic. First, the pink phenolphthalein color disappears. After some time, the pH drops low enough for bromothymol blue to change from blue to green, but not all the way to yellow.

Add 20 mL of 6 M HCl to *the bottom* of each cylinder, using a length of Tygon tubing attached to a funnel. Then stir each solution for a few seconds on a magnetic stirrer. Explain what happens. The sequence of events is shown in Color Plate 5.

---

## Box 11-2   What Does a Negative pH Mean?

In the 1930s, Louis Hammett and his students measured the strengths of very weak acids and bases, using a weak reference base (B), such as *p*-nitroaniline ($pK_a = 0.99$), whose base strength could be measured in aqueous solution.

$$\text{\textit{p}-Nitroanilinium ion} \quad \underset{BH^+}{} \xrightleftharpoons[]{pK_a = 0.99} \quad \text{\textit{p}-Nitroaniline} + H^+ \quad \underset{B}{}$$

Suppose that some *p*-nitroaniline and a second base, C, are dissolved in a strong acid, such as 2 M HCl. The $pK_a$ of $CH^+$ can be measured relative to that of $BH^+$ by first writing a Henderson-Hasselbalch equation for each acid:

$$pH = pK_a \,(\text{for } BH^+) + \log \frac{[B]\gamma_B}{[BH^+]\gamma_{BH^+}}$$

$$pH = pK_a \,(\text{for } CH^+) + \log \frac{[C]\gamma_C}{[CH^+]\gamma_{CH^+}}$$

Setting the two equations equal (because there is only one pH) gives

$$\underbrace{pK_a(\text{for } CH^+) - pK_a(\text{for } BH^+)}_{\Delta pK_a} = \log \frac{[B][CH^+]}{[C][BH^+]} + \log \frac{\gamma_B \gamma_{CH^+}}{\gamma_C \gamma_{BH^+}}$$

The ratio of activity coefficients is close to unity, so the second term on the right is close to 0. Neglecting this last term gives an operationally useful result:

$$\Delta pK_a \approx \log \frac{[B][CH^+]}{[C][BH^+]}$$

That is, if you have a way to find the concentrations of B, $BH^+$, C, and $CH^+$ and if you know $pK_a$ for $BH^+$, then you can find $pK_a$ for $CH^+$.

Concentrations can be measured with a spectrophotometer[10] or by nuclear magnetic resonance,[11] so $pK_a$ for $CH^+$ can be determined. Then, with $CH^+$ as the reference, the $pK_a$ for another compound, $DH^+$, can be measured. This procedure can be extended to measure the strengths of successively weaker bases (such as nitrobenzene, $pK_a = -11.38$), far too weak to be protonated in water.

The acidity of a solvent that protonates the weak base, B, is defined as the **Hammett acidity function:**

*Hammett acidity function:*
$$H_0 = pK_a \,(\text{for } BH^+) + \log \frac{[B]}{[BH^+]}$$

For dilute aqueous solutions, $H_0$ approaches pH. For concentrated acids, $H_0$ is a measure of the acid strength. The weaker a base, B, the stronger the acidity of the solvent must be to protonate the base. Acidity of strongly acidic solvents is now measured more conveniently by electrochemical methods.[12]

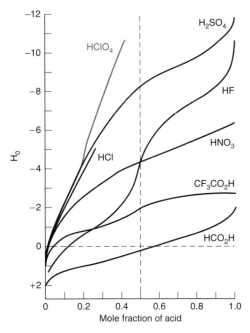

Hammett acidity function, $H_0$, for aqueous solutions of acids. [*Data from R. A. Cox and K. Yates, "Acidity Functions," Can. J. Chem.* **1983**, *61, 2225.*]

When we refer to *negative* pH, we usually mean $H_0$ values. For example, as measured by its ability to protonate very weak bases, 8 M $HClO_4$ has a "pH" close to $-4$. The figure shows that $HClO_4$ is a stronger acid than other mineral acids. Values of $H_0$ for several powerfully acidic solvents are tabulated here.

| Acid | Name | $H_0$ |
|---|---|---|
| $H_2SO_4$ (100%) | sulfuric acid | $-11.93$ |
| $H_2SO_4 \cdot SO_3$ | fuming sulfuric acid (oleum) | $-14.14$ |
| $HSO_3F$ | fluorosulfuric acid | $-15.07$ |
| $HSO_3F$ + 10% $SbF_5$ | "super acid" | $-18.94$ |
| $HSO_3F$ + 7% $SbF_5 \cdot 3SO_3$ | — | $-19.35$ |

If you dump half a bottle of indicator into your reaction, you will introduce a different indicator error. Because indicators are acids or bases, they react with analyte or titrant. The moles of indicator must be negligible relative to the moles of analyte. Never use more than a few drops of dilute indicator solution.

Several of the indicators in Table 11-4 would be useful for the titration in Figure 11-9. If bromocresol purple were used, we would use the purple-to-yellow color change as the end point. The last trace of purple should disappear near pH 5.2, which is quite close to the true equivalence point in Figure 11-9. If bromocresol green were used as the indicator, a color change from blue to green (= yellow + blue) would mark the end point (Box 11-3).

In general, *we seek an indicator whose transition range overlaps the steepest part of the titration curve as closely as possible.* The steepness of the titration curve near the equivalence point in Figure 11-9 ensures that the indicator error caused by the noncoincidence of the end point and equivalence point will not be large. If the indicator end point

Choose an indicator whose transition range overlaps the steepest part of the titration curve.

**Table 11-4** Common Indicators

| Indicator | Transition range (pH) | Acid color | Base color | Preparation |
|---|---|---|---|---|
| Methyl violet | 0.0–1.6 | Yellow | Violet | 0.05 wt% in $H_2O$ |
| Cresol red | 0.2–1.8 | Red | Yellow | 0.1 g in 26.2 mL 0.01 M NaOH. Then add ~225 mL $H_2O$. |
| Thymol blue | 1.2–2.8 | Red | Yellow | 0.1 g in 21.5 mL 0.01 M NaOH. Then add ~225 mL $H_2O$. |
| Cresol purple | 1.2–2.8 | Red | Yellow | 0.1 g in 26.2 mL 0.01 M NaOH. Then add ~225 mL $H_2O$. |
| Erythrosine, disodium | 2.2–3.6 | Orange | Red | 0.1 wt% in $H_2O$ |
| Methyl orange | 3.1–4.4 | Red | Yellow | 0.01 wt% in $H_2O$ |
| Congo red | 3.0–5.0 | Violet | Red | 0.1 wt% in $H_2O$ |
| Ethyl orange | 3.4–4.8 | Red | Yellow | 0.1 wt% in $H_2O$ |
| Bromocresol green | 3.8–5.4 | Yellow | Blue | 0.1 g in 14.3 mL 0.01 M NaOH. Then add ~225 mL $H_2O$. |
| Methyl red | 4.8–6.0 | Red | Yellow | 0.02 g in 60 mL ethanol. Then add 40 mL $H_2O$. |
| Chlorophenol red | 4.8–6.4 | Yellow | Red | 0.1 g in 23.6 mL 0.01 M NaOH. Then add ~225 mL $H_2O$. |
| Bromocresol purple | 5.2–6.8 | Yellow | Purple | 0.1 g in 18.5 mL 0.01 M NaOH. Then add ~225 mL $H_2O$. |
| *p*-Nitrophenol | 5.6–7.6 | Colorless | Yellow | 0.1 wt% in $H_2O$ |
| Litmus | 5.0–8.0 | Red | Blue | 0.1 wt% in $H_2O$ |
| Bromothymol blue | 6.0–7.6 | Yellow | Blue | 0.1 g in 16.0 mL 0.01 M NaOH. Then add ~225 mL $H_2O$. |
| Phenol red | 6.4–8.0 | Yellow | Red | 0.1 g in 28.2 mL 0.01 M NaOH. Then add ~225 mL $H_2O$. |
| Neutral red | 6.8–8.0 | Red | Yellow | 0.01 g in 50 mL ethanol. Then add 50 mL $H_2O$. |
| Cresol red | 7.2–8.8 | Yellow | Red | See above. |
| α-Naphtholphthalein | 7.3–8.7 | Pink | Green | 0.1 g in 50 mL ethanol. Then add 50 mL $H_2O$. |
| Cresol purple | 7.6–9.2 | Yellow | Purple | See above. |
| Thymol blue | 8.0–9.6 | Yellow | Blue | See above. |
| Phenolphthalein | 8.0–9.6 | Colorless | Red | 0.05 g in 50 mL ethanol. Then add 50 mL $H_2O$. |
| Thymolphthalein | 8.3–10.5 | Colorless | Blue | 0.04 g in 50 mL ethanol. Then add 50 mL $H_2O$. |
| Alizarin yellow | 10.1–12.0 | Yellow | Orange-red | 0.01 wt% in $H_2O$ |
| Nitramine | 10.8–13.0 | Colorless | Orange-brown | 0.1 g in 70 mL ethanol. Then add 30 mL $H_2O$. |
| Tropaeolin O | 11.1–12.7 | Yellow | Orange | 0.1 wt% in $H_2O$ |

Box 11-3   World Record Small Titration

The titration of 29 fmol (f = femto = $10^{-15}$) of $HNO_3$ in a 1.9-pL (p = pico = $10^{-12}$) water drop under a layer of hexane in a Petri dish was carried out with 2% precision, using KOH delivered by diffusion from the 1-$\mu$m-diameter tip of a glass capillary tube. The agar plug in the tip of the pipet regulates diffusion of titrant, resulting in a constant delivery rate on the order of fmol/s.[13] A mixture of the indicators bromothymol blue and bromocresol purple, with a distinct yellow-to-purple transition at pH 6.7, allowed the end point to be observed through a microscope by a video camera. Metal ions in picoliter volumes could be titrated with the reagent EDTA, using indicators or electrodes to locate the end point.[14]

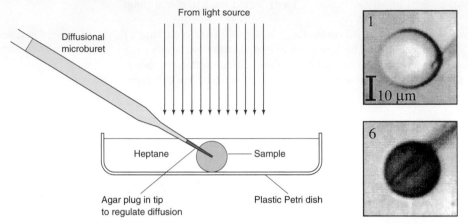

Titration of femtoliter samples. The frames to the right show a larger 8.7-pL droplet before and after the end point, with the microburet making contact at the right side. [M. Gratzl and C. Yi, "Diffusional Microtitration: Acid/Base Titrations in Pico- and Femtoliter Samples," Anal. Chem. **1993**, 65, 2085. Photo courtesy M. Gratzl, Case Western Reserve University.]

were at pH 6.4 (instead of 5.54), the error in $V_e$ would be only 0.25% in this particular case. You can estimate the indicator error by calculating the volume of titrant required to attain pH 6.4 instead of pH 5.54.

## ■■■■ 11-7   Practical Notes

Procedures for preparing standard acid and base are given at the end of this chapter.

Acids and bases in Table 11-5 can be obtained pure enough to be *primary standards.*[17] NaOH and KOH are not primary standards, because they contain carbonate (from reaction with atmospheric $CO_2$) and adsorbed water. Solutions of NaOH and KOH must be standardized against a primary standard such as potassium hydrogen phthalate. Solutions of NaOH for titrations are prepared by diluting a stock solution of 50 wt% aqueous NaOH. Sodium carbonate is insoluble in this stock solution and settles to the bottom.

Alkaline solutions (for example, 0.1 M NaOH) must be protected from the atmosphere; otherwise they absorb $CO_2$:

$$OH^- + CO_2 \rightarrow HCO_3^-$$

$CO_2$ changes the concentration of strong base over a period of time and decreases the extent of reaction near the end point in the titration of weak acids. If solutions are kept in tightly capped polyethylene bottles, they can be used for about a week with little change.

Strongly basic solutions attack glass and are best stored in plastic containers. Such solutions should not be kept in a buret longer than necessary. Boiling 0.01 M NaOH in a flask for 1 h decreases the molarity by 10%, owing to the reaction of $OH^-$ with glass.[18]

## ■■■■ 11-8   The Leveling Effect

*The strongest acid that can exist in water is $H_3O^+$ and the strongest base is $OH^-$.* If an acid stronger than $H_3O^+$ is dissolved in water, it protonates $H_2O$ to make $H_3O^+$. If a base stronger than $OH^-$ is dissolved in water, it deprotonates $H_2O$ to make $OH^-$. Because of this **leveling effect,** $HClO_4$ and HCl behave as if they had the same acid strength; both are *leveled* to $H_3O^+$:

$$HClO_4 + H_2O \rightarrow H_3O^+ + ClO_4^-$$
$$HCl + H_2O \rightarrow H_3O^+ + Cl^-$$

**Table 11-5** Primary standards

| Compound | Density (g/mL) for buoyancy corrections | Notes |
|---|---|---|
| ACIDS | | |
| Potassium hydrogen phthalate FM 204.221 | 1.64 | The pure commercial material is dried at 105°C and used to standardize base. A phenolphthalein end point is satisfactory. |
| HCl Hydrochloric acid FM 36.461 | — | HCl and water distill as an *azeotrope* (a mixture) whose composition ($\sim$6 M) depends on pressure. The composition is tabulated as a function of the pressure during distillation. See Problem 11-55 for more information. |
| $KH(IO_3)_2$ Potassium hydrogen iodate FM 389.912 | — | This is a strong acid, so any indicator with an end point between $\sim$5 and $\sim$9 is adequate. |
| Sulfosalicylic acid double salt FM 550.639 | — | 1 mol of commercial grade sulfosalicylic acid is combined with 0.75 mol of reagent-grade $KHCO_3$, recrystallized several times from water, and dried at 110°C to produce the double salt with 3 $K^+$ ions and one titratable $H^+$.[15] Phenolphthalein is used as the indicator for titration with NaOH. |
| $H_3\overset{+}{N}SO_3^-$ Sulfamic acid FM 97.094 | 2.15 | Sulfamic acid is a strong acid with one acidic proton, so any indicator with an end point between $\sim$5 and $\sim$9 is suitable. |
| BASES | | |
| $H_2NC(CH_2OH)_3$ Tris(hydroxymethyl)aminomethane (also called tris or tham) FM 121.135 | 1.33 | The pure commercial material is dried at 100°–103°C and titrated with strong acid. The end point is in the range pH 4.5–5. $$H_2NC(CH_2OH)_3 + H^+ \rightarrow H_3\overset{+}{N}C(CH_2OH)_3$$ |
| HgO Mercuric oxide FM 216.59 | 11.1 | Pure HgO is dissolved in a large excess of $I^-$ or $Br^-$, whereupon 2 $OH^-$ are liberated: $$HgO + 4I^- + H_2O \rightarrow HgI_4^{2-} + 2OH^-$$ The base is titrated, using an indicator end point. |
| $Na_2CO_3$ Sodium carbonate FM 105.988 | 2.53 | Primary standard grade $Na_2CO_3$ is commercially available. Alternatively, recrystallized $NaHCO_3$ can be heated for 1 h at 260°–270°C to produce pure $Na_2CO_3$. Sodium carbonate is titrated with acid to an end point of pH 4–5. Just before the end point, the solution is boiled to expel $CO_2$. |
| $Na_2B_4O_7 \cdot 10H_2O$ Borax FM 381.372 | 1.73 | The recrystallized material is dried in a chamber containing an aqueous solution saturated with NaCl and sucrose. This procedure gives the decahydrate in pure form.[16] The standard is titrated with acid to a methyl red end point. $$\text{“}B_4O_7 \cdot 10H_2O^{2-}\text{”} + 2H^+ \rightarrow 4B(OH)_3 + 5H_2O$$ |

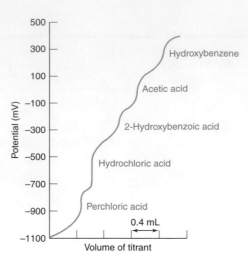

**Question** Where do you think the end point for the acid $H_3O^+ClO_4^-$ would come in Figure 11-10?

**Figure 11-10** Titration of a mixture of acids with tetrabutylammonium hydroxide in methyl isobutyl ketone solvent shows that the order of acid strength is $HClO_4 > HCl > $ 2-hydroxybenzoic acid > acetic acid > hydroxybenzene. Measurements were made with a glass electrode and a platinum reference electrode. The ordinate is proportional to pH, with increasing pH as the potential becomes more positive. *[D. B. Bruss and G. E. A. Wyld, "Methyl Isobutyl Ketone as a Wide-Range Solvent for Titration of Acid Mixtures and Nitrogen Bases," Anal. Chem.* **1957,** *29, 232.]*

In acetic acid solution, $HClO_4$ is a stronger acid than HCl; but, in aqueous solution, both are leveled to the strength of $H_3O^+$.

In acetic acid solvent, which is less basic than $H_2O$, $HClO_4$ and HCl are not leveled to the same strength:

$$HClO_4 \ + \ CH_3CO_2H \ \rightleftharpoons \ CH_3CO_2H_2^+ \ + \ ClO_4^- \qquad K = 1.3 \times 10^{-5}$$
$$\text{Acetic acid solvent}$$

$$HCl + CH_3CO_2H \rightleftharpoons CH_3CO_2H_2^+ + Cl^- \qquad\qquad K = 2.8 \times 10^{-9}$$

The equilibrium constants show that $HClO_4$ is a stronger acid than HCl in acetic acid solvent.

Figure 11-10 shows a titration curve for a mixture of five acids titrated with 0.2 M tetrabutylammonium hydroxide in methyl isobutyl ketone solvent. This solvent is not protonated to a great extent by any of the acids. We see that perchloric acid is a stronger acid than HCl in this solvent as well.

Now consider a base such as urea, $(H_2N)_2C=O$ ($K_b = 1.3 \times 10^{-14}$), that is too weak to give a distinct end point when titrated with a strong acid in water.

$$\text{Titration with } HClO_4 \text{ in } H_2O: \qquad B + H_3O^+ \rightleftharpoons BH^+ + H_2O$$

*The end point cannot be recognized, because the equilibrium constant for the titration reaction is not large enough. If an acid stronger than $H_3O^+$ were available, the titration reaction might have an equilibrium constant large enough to give a distinct end point.* If the same base were dissolved in acetic acid and titrated with $HClO_4$ in acetic acid, a clear end point might be observed. The reaction

A base too weak to be titrated by $H_3O^+$ in water might be titrated by $HClO_4$ in acetic acid solvent.

$$\text{Titration with } HClO_4 \text{ in } CH_3CO_2H: \qquad B + HClO_4 \rightleftharpoons \underbrace{BH^+ClO_4^-}_{\text{An ion pair}}$$

Dielectric constant is discussed in Problem 8-13.

might have a large equilibrium constant, because $HClO_4$ is a much stronger acid than $H_3O^+$. (The product in this reaction is written as an ion pair because the *dielectric constant* of acetic acid is too low to allow ions to separate extensively.) Titrations that are not feasible in water might be feasible in other solvents.[19]

## ▪ ▪ ▪ ▪ ▦ 11-9 Calculating Titration Curves with Spreadsheets

This chapter has been critical for developing your understanding of the chemistry that occurs during titrations. However, the approximations we used are of limited value when concentrations are too dilute or equilibrium constants are not of the right magnitude or $K_a$ values are too closely spaced, like those in a protein. This section develops equations to deal with titrations in a general manner, using spreadsheets.[20] You should understand the principles well enough to derive equations to treat future problems that you encounter.

### Titrating a Weak Acid with a Strong Base

Consider the titration of a volume $V_a$ of acid HA (initial concentration $C_a$) with a volume $V_b$ of NaOH of concentration $C_b$. The charge balance for this solution is

$$\text{Charge balance:} \qquad [H^+] + [Na^+] = [A^-] + [OH^-]$$

and the concentration of $Na^+$ is just

$$[Na^+] = \frac{C_b V_b}{V_a + V_b}$$

because we have diluted $C_b V_b$ moles of NaOH to a total volume of $V_a + V_b$. Similarly, the formal concentration of the weak acid is

$$F_{HA} = [HA] + [A^-] = \frac{C_a V_a}{V_a + V_b}$$

because we have diluted $C_a V_a$ moles of HA to a total volume of $V_a + V_b$.

Now we use the fractional composition equations from Chapter 10. Equation 10-18 told us that

$$[A^-] = \alpha_{A^-} \cdot F_{HA} = \frac{\alpha_{A^-} \cdot C_a V_a}{V_a + V_b} \tag{11-8}$$

where $\alpha_{A^-} = K_a/([H^+] + K_a)$ and $K_a$ is the acid dissociation constant of HA. Substituting for $[Na^+]$ and $[A^-]$ in the charge balance gives

$$[H^+] + \frac{C_b V_b}{V_a + V_b} = \frac{\alpha_{A^-} \cdot C_a V_a}{V_a + V_b} + [OH^-]$$

which you can rearrange to the form

*Fraction of titration for*
*weak acid by strong base:*

$$\phi \equiv \frac{C_b V_b}{C_a V_a} = \frac{\alpha_{A^-} - \dfrac{[H^+] - [OH^-]}{C_a}}{1 + \dfrac{[H^+] - [OH^-]}{C_b}} \tag{11-9}$$

$\alpha_{A^-}$ = fraction of acid in the form $A^-$:

$$\alpha_{A^-} = \frac{[A^-]}{F_{HA}}$$

$\phi = C_b V_b / C_a V_a$ is the fraction of the way to the equivalence point:

| $\phi$ | Volume of base |
| --- | --- |
| 0.5 | $V_b = \frac{1}{2} V_e$ |
| 1 | $V_b = V_e$ |
| 2 | $V_b = 2 V_e$ |

At last! Equation 11-9 is really useful. It relates the volume of titrant $(V_b)$ to the pH and a bunch of constants. The quantity $\phi$, which is the quotient $C_b V_b / C_a V_a$, is the fraction of the way to the equivalence point, $V_e$. When $\phi = 1$, the volume of base added, $V_b$, is equal to $V_e$. Equation 11-9 works backward from the way you are accustomed to thinking, because you need to put in pH (on the right) to get out volume (on the left). Let me say that again: We *put in a concentration of $H^+$* and *get out the volume of titrant* that produces that concentration.

Let's set up a spreadsheet to use Equation 11-9 to calculate the titration curve for 50.00 mL of the weak acid 0.020 00 M MES with 0.100 0 M NaOH, which was shown in Figure 11-2 and Table 11-2. The equivalence volume is $V_e = 10.00$ mL. We use the quantities in Equation 11-9 as follows:

$C_b = 0.1$      $[H^+] = 10^{-pH}$

$C_a = 0.02$      $[OH^-] = K_w/[H^+]$

$V_a = 50$

$K_a = 5.3_7 \times 10^{-7}$    $\alpha_{A^-} = \dfrac{K_a}{[H^+] + K_a}$

$K_w = 10^{-14}$

pH is the input      $V_b = \dfrac{\phi C_a V_a}{C_b}$ is the output

2-(N-Morpholino)ethanesulfonic acid
MES, $pK_a = 6.27$

The input to the spreadsheet in Figure 11-11 is pH in column B and the output is $V_b$ in column G. From the pH, the values of $[H^+]$, $[OH^-]$, and $\alpha_{A^-}$ are computed in columns C, D, and E. Equation 11-9 is used in column F to find the fraction of titration, $\phi$. From this value, we calculate the volume of titrant, $V_b$, in column G.

How do we know what pH values to put in? Trial-and-error allows us to find the starting pH, by putting in a pH and seeing whether $V_b$ is positive or negative. In a few tries, it is easy to home in on the pH at which $V_b = 0$. In Figure 11-11, a pH of 3.90 is too low, because $\phi$ and $V$ are both negative. Input values of pH are spaced as closely as you like, so that you can generate a smooth titration curve. To save space, we show only a few points in Figure 11-11, including the midpoint (pH 6.27 $\Rightarrow V_b = 5.00$ mL) and the end point (pH 9.25 $\Rightarrow V_b = 10.00$ mL). This spreadsheet reproduces Table 11-2 without approximations other than neglecting activity coefficients. It gives correct results even when the approximations used in Table 11-2 fail.

In Figure 11-11, you could use Excel GOAL SEEK described on page 115 to vary the pH in cell B5 until $V_b$ in cell G5 is 0.

Figure 11-11 Spreadsheet that uses Equation 11-9 to calculate the titration curve for 50 mL of the weak acid 0.02 M MES ($pK_a = 6.27$) treated with 0.1 M NaOH. We provide pH as input in column B and the spreadsheet tells us what volume of base is required to generate that pH.

| | A | B | C | D | E | F | G |
|---|---|---|---|---|---|---|---|
| 1 | Titration of weak acid with strong base | | | | | | |
| 2 | | | | | | | |
| 3 | $C_b =$ | pH | [H$^+$] | [OH$^-$] | $\alpha(A^-)$ | $\phi$ | $V_b$ (mL) |
| 4 | 0.1 | 3.90 | 1.26E-04 | 7.94E-11 | 0.004 | -0.002 | -0.020 |
| 5 | $C_a =$ | 3.99 | 1.02E-04 | 9.77E-11 | 0.005 | 0.000 | 0.001 |
| 6 | 0.02 | 5.00 | 1.00E-05 | 1.00E-09 | 0.051 | 0.050 | 0.505 |
| 7 | $V_a =$ | 6.00 | 1.00E-06 | 1.00E-08 | 0.349 | 0.349 | 3.493 |
| 8 | 50 | 6.27 | 5.37E-07 | 1.86E-08 | 0.500 | 0.500 | 5.000 |
| 9 | $K_a =$ | 7.00 | 1.00E-07 | 1.00E-07 | 0.843 | 0.843 | 8.430 |
| 10 | 5.37E-07 | 8.00 | 1.00E-08 | 1.00E-06 | 0.982 | 0.982 | 9.818 |
| 11 | $K_w =$ | 9.00 | 1.00E-09 | 1.00E-05 | 0.998 | 0.999 | 9.987 |
| 12 | 1.E-14 | 9.25 | 5.62E-10 | 1.78E-05 | 0.999 | 1.000 | 10.000 |
| 13 | | 10.00 | 1.00E-10 | 1.00E-04 | 1.000 | 1.006 | 10.058 |
| 14 | | 11.00 | 1.00E-11 | 1.00E-03 | 1.000 | 1.061 | 10.606 |
| 15 | | 12.00 | 1.00E-12 | 1.00E-02 | 1.000 | 1.667 | 16.667 |
| 16 | | | | | | | |
| 17 | C4 = 10^-B4 | | | F4 = (E4-(C4-D4)/\$A\$6)/(1+(C4-D4)/\$A\$4) | | | |
| 18 | D4 = \$A\$12/C4 | | | G4 = F4*\$A\$6*\$A\$8/\$A\$4 | | | |
| 19 | E4 = \$A\$10/(C4+\$A\$10) | | | | | | |

## Titrating a Weak Acid with a Weak Base

Now consider the titration of $V_a$ mL of acid HA (initial concentration $C_a$) with $V_b$ mL of base B whose concentration is $C_b$. Let the acid dissociation constant of HA be $K_a$ and the acid dissociation constant of BH$^+$ be $K_{BH^+}$. The charge balance is

Charge balance:    $[H^+] + [BH^+] = [A^-] + [OH^-]$

As before, we can say that $[A^-] = \alpha_{A^-} \cdot F_{HA}$, where $\alpha_{A^-} = K_a/([H^+] + K_a)$ and $F_{HA} = C_a V_a/(V_a + V_b)$.

We can write an analogous expression for [BH$^+$], which is a monoprotic weak acid. If the acid were HA, we would use Equation 10-17 to say

$\alpha_{HA}$ = fraction of acid in the form HA

$\alpha_{HA} = \dfrac{[HA]}{F_{HA}}$

$[HA] = \alpha_{HA}F_{HA}$    $\alpha_{HA} = \dfrac{[H^+]}{[H^+] + K_a}$

where $K_a$ applies to the acid HA. For the weak acid BH$^+$, we write

$\alpha_{BH^+}$ = fraction of base in the form BH$^+$

$\alpha_{BH^+} = \dfrac{[BH^+]}{F_B}$

$[BH^+] = \alpha_{BH^+} \cdot F_B$    $\alpha_{BH^+} = \dfrac{[H^+]}{[H^+] + K_{BH^+}}$

where the formal concentration of base is $F_B = C_b V_b/(V_a + V_b)$.

Substituting for [BH$^+$] and [A$^-$] in the charge balance gives

$$[H^+] + \frac{\alpha_{BH^+} \cdot C_b V_b}{V_a + V_b} = \frac{\alpha_{A^-} \cdot C_a V_a}{V_a + V_b} + [OH^-]$$

which can be rearranged to the useful result

Fraction of titration for weak acid by weak base:

$$\phi = \frac{C_b V_b}{C_a V_a} = \frac{\alpha_{A^-} - \dfrac{[H^+] - [OH^-]}{C_a}}{\alpha_{BH^+} + \dfrac{[H^+] - [OH^-]}{C_b}} \qquad (11\text{-}10)$$

Equation 11-10 for a weak base looks just like Equation 11-9 for a strong base, except that $\alpha_{BH^+}$ replaces 1 in the denominator.

Table 11-6 gives useful equations derived by writing a charge balance and substituting fractional compositions for various concentrations. For titration of the diprotic acid, H$_2$A, $\phi$ is the fraction of the way to the first equivalence point. When $\phi = 2$, we are at the second equivalence point. It should not surprise you that, when $\phi = 0.5$, pH $\approx pK_1$ and, when $\phi = 1.5$, pH $\approx pK_2$. When $\phi = 1$, we have the intermediate HA$^-$ and pH $\approx \frac{1}{2}(pK_1 + pK_2)$.

220

**Table 11-6** Titration equations for spreadsheets

Titrating strong acid with strong base

$$\phi = \frac{C_b V_b}{C_a V_a} = \frac{1 - \dfrac{[H^+] - [OH^-]}{C_a}}{1 + \dfrac{[H^+] - [OH^-]}{C_b}}$$

Titrating strong base with strong acid

$$\phi = \frac{C_a V_a}{C_b V_b} = \frac{1 + \dfrac{[H^+] - [OH^-]}{C_b}}{1 - \dfrac{[H^+] - [OH^-]}{C_a}}$$

Titrating weak acid (HA) with weak base (B)

$$\phi = \frac{C_b V_b}{C_a V_a} = \frac{\alpha_{A^-} - \dfrac{[H^+] - [OH^-]}{C_a}}{\alpha_{BH^+} + \dfrac{[H^+] - [OH^-]}{C_b}}$$

Titrating $H_2A$ with strong base ($\rightarrow \rightarrow A^{2-}$)

$$\phi = \frac{C_b V_b}{C_a V_a} = \frac{\alpha_{HA^-} + 2\alpha_{A^{2-}} - \dfrac{[H^+] - [OH^-]}{C_a}}{1 + \dfrac{[H^+] - [OH^-]}{C_b}}$$

Titrating dibasic B with strong acid ($\rightarrow \rightarrow BH_2^{2+}$)

$$\phi = \frac{C_a V_a}{C_b V_b} = \frac{\alpha_{BH^+} + 2\alpha_{BH_2^{2+}} + \dfrac{[H^+] - [OH^-]}{C_b}}{1 - \dfrac{[H^+] - [OH^-]}{C_a}}$$

Titrating weak acid (HA) with strong base

$$\phi = \frac{C_b V_b}{C_a V_a} = \frac{\alpha_{A^-} - \dfrac{[H^+] - [OH^-]}{C_a}}{1 + \dfrac{[H^+] - [OH^-]}{C_b}}$$

Titrating weak base (B) with strong acid

$$\phi = \frac{C_a V_a}{C_b V_b} = \frac{\alpha_{BH^+} + \dfrac{[H^+] - [OH^-]}{C_b}}{1 - \dfrac{[H^+] - [OH^-]}{C_a}}$$

Titrating weak base (B) with weak acid (HA)

$$\phi = \frac{C_a V_a}{C_b V_b} = \frac{\alpha_{BH^+} + \dfrac{[H^+] - [OH^-]}{C_b}}{\alpha_{A^-} - \dfrac{[H^+] - [OH^-]}{C_a}}$$

Titrating $H_3A$ with strong base ($\rightarrow \rightarrow \rightarrow A^{3-}$)

$$\phi = \frac{C_b V_b}{C_a V_a} = \frac{\alpha_{H_2A^-} + 2\alpha_{HA^{2-}} + 3\alpha_{A^{3-}} - \dfrac{[H^+] - [OH^-]}{C_a}}{1 + \dfrac{[H^+] - [OH^-]}{C_b}}$$

Titrating tribasic B with strong acid ($\rightarrow \rightarrow \rightarrow BH_3^{3+}$)

$$\phi = \frac{C_a V_a}{C_b V_b} = \frac{\alpha_{BH^+} + 2\alpha_{BH_2^{2+}} + 3\alpha_{BH_3^{3+}} + \dfrac{[H^+] - [OH^-]}{C_b}}{1 - \dfrac{[H^+] - [OH^-]}{C_a}}$$

SYMBOLS

$\phi$ = fraction of the way to the first equivalence point
$C_a$ = initial concentration of acid
$C_b$ = initial concentration of base

$\alpha$ = fraction of dissociation of acid or fraction of association of base
$V_a$ = volume of acid
$V_b$ = volume of base

CALCULATION OF $\alpha$

*Monoprotic systems*

$$\alpha_{HA} = \frac{[H^+]}{[H^+] + K_a}$$

$$\alpha_{BH^+} = \frac{[H^+]}{[H^+] + K_{BH^+}}$$

$$\alpha_{A^-} = \frac{K_a}{[H^+] + K_a}$$

$$\alpha_B = \frac{K_{BH^+}}{[H^+] + K_{BH^+}}$$

SYMBOLS
$K_a$ = acid dissociation constant of HA
$K_{BH^+}$ = acid dissociation constant of $BH^+$ ($= K_w/K_b$)

*Diprotic systems*

$$\alpha_{H_2A} = \frac{[H^+]^2}{[H^+]^2 + [H^+]K_1 + K_1 K_2}$$

$$\alpha_{BH_2^{2+}} = \frac{[H^+]^2}{[H^+]^2 + [H^+]K_1 + K_1 K_2}$$

$$\alpha_{HA^-} = \frac{[H^+]K_1}{[H^+]^2 + [H^+]K_1 + K_1 K_2}$$

$$\alpha_{BH^+} = \frac{[H^+]K_1}{[H^+]^2 + [H^+]K_1 + K_1 K_2}$$

$$\alpha_{A^{2-}} = \frac{K_1 K_2}{[H^+]^2 + [H^+]K_1 + K_1 K_2}$$

$$\alpha_B = \frac{K_1 K_2}{[H^+]^2 + [H^+]K_1 + K_1 K_2}$$

SYMBOLS
$K_1$ and $K_2$ for the acid are the acid dissociation constants of $H_2A$ and $HA^-$, respectively.
$K_1$ and $K_2$ for the base refer to the acid dissociation constants of $BH_2^{2+}$ and $BH^+$, respectively: $K_1 = K_w/K_{b2}$; $K_2 = K_w/K_{b1}$

*Triprotic systems*

$$\alpha_{H_3A} = \frac{[H^+]^3}{[H^+]^3 + [H^+]^2 K_1 + [H^+]K_1 K_2 + K_1 K_2 K_3}$$

$$\alpha_{HA^{2-}} = \frac{[H^+]K_1 K_2}{[H^+]^3 + [H^+]^2 K_1 + [H^+]K_1 K_2 + K_1 K_2 K_3}$$

$$\alpha_{H_2A^-} = \frac{[H^+]^2 K_1}{[H^+]^3 + [H^+]^2 K_1 + [H^+]K_1 K_2 + K_1 K_2 K_3}$$

$$\alpha_{A^{3-}} = \frac{K_1 K_2 K_3}{[H^+]^3 + [H^+]^2 K_1 + [H^+]K_1 K_2 + K_1 K_2 K_3}$$

## Terms to Understand

Gran plot

Hammett acidity function

indicator

indicator error

leveling effect

transition range

## Summary

Key equations used to calculate titration curves:

*Strong acid/strong base titration*

$$H^+ + OH^- \rightarrow H_2O$$

pH is determined by the concentration of excess unreacted $H^+$ or $OH^-$

*Weak acid titrated with $OH^-$*

$$HA + OH^- \rightarrow A^- + H_2O \quad (V_e = \text{equivalence volume})$$
$$(V_b = \text{volume of added base})$$

$V_b = 0$: pH determined by $K_a$ ($HA \xrightleftharpoons{K_a} H^+ + A^-$)

$0 < V_b < V_e$: pH $= pK_a + \log([A^-]/[HA])$

pH $= pK_a$ when $V_b = \frac{1}{2}V_e$ (neglecting activity)

At $V_e$: pH governed by $K_b$ ($A^- + H_2O \xrightleftharpoons{K_b} HA + OH^-$)

After $V_e$: pH is determined by excess $OH^-$

*Weak base titrated with $H^+$*

$$B + H^+ \rightarrow BH^+ \quad (V_e = \text{equivalence volume})$$
$$(V_a = \text{volume of added acid})$$

$V_a = 0$: pH determined by $K_b$ ($B + H_2O \xrightleftharpoons{K_b} BH^+ + OH^-$)

$0 < V_a < V_e$: pH $= pK_{BH^+} + \log([B]/[BH^+])$

pH $= pK_{BH^+}$ when $V_a = \frac{1}{2}V_e$

At $V_e$: pH governed by $K_{BH^+}$ ($BH^+ \xrightleftharpoons{K_{BH^+}} B + H^+$)

After $V_e$: pH is determined by excess $H^+$

*$H_2A$ titrated with $OH^-$*

$$H_2A \xrightarrow{OH^-} HA^- \xrightarrow{OH^-} A^{2-}$$

Equivalence volumes: $V_{e2} = 2V_{e1}$

$V_b = 0$: pH determined by $K_1$ ($H_2A \xrightleftharpoons{K_1} H^+ + HA^-$)

$0 < V_b < V_{e1}$: pH $= pK_1 + \log([HA^-]/[H_2A])$

pH $= pK_1$ when $V_b = \frac{1}{2}V_{e1}$

At $V_{e1}$: $[H^+] = \sqrt{\dfrac{K_1K_2F' + K_1K_w}{K_1 + F'}}$

$\Rightarrow$ pH $\approx \frac{1}{2}(pK_1 + pK_2)$

$F' = $ formal concentration of $HA^-$

$V_{e1} < V_b < V_{e2}$: pH $= pK_2 + \log([A^{2-}]/[HA^-])$

pH $= pK_2$ when $V_b = \frac{3}{2}V_{e1}$

At $V_{e2}$: pH governed by $K_{b1}$

$$(A^{2-} + H_2O \xrightleftharpoons{K_{b1}} HA^- + OH^-)$$

After $V_{e2}$: pH is determined by excess $OH^-$

*Behavior of derivatives at the equivalent point*

First derivative: $\Delta pH/\Delta V$ has greatest magnitude

Second derivative: $\Delta(\Delta pH/\Delta V)/\Delta V = 0$

*Gran plot*

Plot $V_b \cdot 10^{-pH}$ versus $V_b$

x-intercept $= V_e$; slope $= -K_a\gamma_{HA}/\gamma_{A^-}$

$K_a = $ acid dissociation constant

$\gamma = $ activity coefficient

*Choosing an indicator:* Color transition range should match pH at $V_e$. Preferably the color change should occur entirely within the steep portion of the titration curve.

## Exercises

**11-A.** Calculate the pH at each of the following points in the titration of 50.00 mL of 0.010 0 M NaOH with 0.100 M HCl. Volume of acid added: 0.00, 1.00, 2.00, 3.00, 4.00, 4.50, 4.90, 4.99, 5.00, 5.01, 5.10, 5.50, 6.00, 8.00, and 10.00 mL. Make a graph of pH versus volume of HCl added.

**11-B.** Calculate the pH at each point listed for the titration of 50.0 mL of 0.050 0 M formic acid with 0.050 0 M KOH. The points to calculate are $V_b$ = 0.0, 10.0, 20.0, 25.0, 30.0, 40.0, 45.0, 48.0, 49.0, 49.5, 50.0, 50.5, 51.0, 52.0, 55.0, and 60.0 mL. Draw a graph of pH versus $V_b$.

**11-C.** Calculate the pH at each point listed for the titration of 100.0 mL of 0.100 M cocaine (Section 9-4, $K_b = 2.6 \times 10^{-6}$) with 0.200 M $HNO_3$. The points to calculate are $V_a$ = 0.0, 10.0, 20.0, 25.0, 30.0, 40.0, 49.0, 49.9, 50.0, 50.1, 51.0, and 60.0 mL. Draw a graph of pH versus $V_a$.

**11-D.** Consider the titration of 50.0 mL of 0.050 0 M malonic acid with 0.100 M NaOH. Calculate the pH at each point listed and sketch the titration curve: $V_b$ = 0.0, 8.0, 12.5, 19.3, 25.0, 37.5, 50.0, and 56.3 mL.

**11-E.** Write the chemical reactions (including structures of reactants and products) that occur when the amino acid histidine is titrated with perchloric acid. (Histidine is a molecule with no net charge.) A solution containing 25.0 mL of 0.050 0 M histidine was titrated with 0.050 0 M $HClO_4$. Calculate the pH at the following values of $V_a$: 0, 4.0, 12.5, 25.0, 26.0, and 50.0 mL.

**11-F.** Select indicators from Table 11-4 that would be useful for the titrations in Figures 11-1 and 11-2 and the $pK_a = 8$ curve in Figure 11-3. Select a different indicator for each titration and state what color change you would use as the end point.

**11-G.** When 100.0 mL of a weak acid was titrated with 0.093 81 M NaOH, 27.63 mL were required to reach the equivalence point. The pH at the equivalence point was 10.99. What was the pH when only 19.47 mL of NaOH had been added?

**11-H.** A 0.100 M solution of the weak acid HA was titrated with 0.100 M NaOH. The pH measured when $V_b = \frac{1}{2}V_e$ was 4.62. Using activity coefficients, calculate $pK_a$. The size of the $A^-$ anion is 450 pm.

**11-I.** *Finding the end point from pH measurements.* Here are data points around the second apparent end point in Figure 11-6:

| $V_b$ ($\mu$L) | pH | $V_b$ ($\mu$L) | pH |
|---|---|---|---|
| 107 | 6.921 | 117 | 7.878 |
| 110 | 7.117 | 118 | 8.090 |
| 113 | 7.359 | 119 | 8.343 |
| 114 | 7.457 | 120 | 8.591 |
| 115 | 7.569 | 121 | 8.794 |
| 116 | 7.705 | 122 | 8.952 |

(a) Prepare a table analogous to Table 11-3, showing the first and second derivatives. Plot both derivatives versus $V_b$ and locate the end point in each plot.

(b) Prepare a Gran plot analogous to Figure 11-8. Use the least-squares procedure to find the best straight line and find the end point. You will have to use your judgment as to which points lie on the "straight" line.

**11-J.** *Indicator error.* Consider the titration in Figure 11-2 in which the equivalence point pH in Table 11-2 is 9.25 at a volume of 10.00 mL.

(a) Suppose you used the yellow-to-blue transition of thymol blue indicator to find the end point. According to Table 11-4, the last trace of green disappears near pH 9.6. What volume of base is required to reach pH 9.6? The difference between this volume and 10 mL is the indicator error.

(b) If you used cresol red with a color change at pH 8.8, what would be the indicator error?

**11-K.** *Spectrophotometry with indicators.** Acid-base indicators are themselves acids or bases. Consider an indicator, HIn, which dissociates according to the equation

$$HIn \overset{K_a}{\rightleftharpoons} H^+ + In^-$$

The molar absorptivity, $\varepsilon$, is $2\,080\ M^{-1}\,cm^{-1}$ for HIn and $14\,200\ M^{-1}\,cm^{-1}$ for $In^-$, at a wavelength of 440 nm.

(a) Write an expression for the absorbance of a solution containing HIn at a concentration [HIn] and $In^-$ at a concentration $[In^-]$ in a cell of pathlength 1.00 cm. The total absorbance is the sum of absorbances of each component.

(b) A solution containing indicator at a formal concentration of $1.84 \times 10^{-4}$ M is adjusted to pH 6.23 and found to exhibit an absorbance of 0.868 at 440 nm. Calculate $pK_a$ for this indicator.

---

*This problem is based on Beer's law in Section 18-2.

---

## Problems

### Titration of Strong Acid with Strong Base

**11-1.** Distinguish the terms *end point* and *equivalence point*.

**11-2.** Consider the titration of 100.0 mL of 0.100 M NaOH with 1.00 M HBr. Find the pH at the following volumes of acid added and make a graph of pH versus $V_a$: $V_a$ = 0, 1, 5, 9, 9.9, 10, 10.1, and 12 mL.

**11-3.** Why does an acid-base titration curve (pH versus volume of titrant) have an abrupt change at the equivalence point?

### Titration of Weak Acid with Strong Base

**11-4.** Sketch the general appearance of the curve for the titration of a weak acid with a strong base. Explain (in words) what chemistry governs the pH in each of the four distinct regions of the curve.

**11-5.** Why is it not practical to titrate an acid or base that is too weak or too dilute?

**11-6.** A weak acid HA ($pK_a$ = 5.00) was titrated with 1.00 M KOH. The acid solution had a volume of 100.0 mL and a molarity of 0.100 M. Find the pH at the following volumes of base added and make a graph of pH versus $V_b$: $V_b$ = 0, 1, 5, 9, 9.9, 10, 10.1, and 12 mL.

**11-7.** Consider the titration of the weak acid HA with NaOH. At what fraction of $V_e$ does pH = $pK_a$ − 1? At what fraction of $V_e$ does pH = $pK_a$ + 1? Use these two points, plus $V_b$ = 0, $\frac{1}{2}V_e$, $V_e$, and $1.2V_e$ to sketch the titration curve for the reaction of 100 mL of 0.100 M anilinium bromide ("aminobenzene · HBr") with 0.100 M NaOH.

**11-8.** What is the pH at the equivalence point when 0.100 M hydroxyacetic acid is titrated with 0.050 0 M KOH?

**11-9.** Find the equilibrium constant for the reaction of MES (Table 9-2) with NaOH.

**11-10.** When 22.63 mL of aqueous NaOH was added to 1.214 g of cyclohexylaminoethanesulfonic acid (FM 207.29, structure in Table 9-2) dissolved in 41.37 mL of water, the pH was 9.24. Calculate the molarity of the NaOH.

**11-11.** *Use activity coefficients* to calculate the pH after 10.0 mL of 0.100 M trimethylammonium bromide was titrated with 4.0 mL of 0.100 M NaOH.

### Titration of Weak Base with Strong Acid

**11-12.** Sketch the general appearance of the curve for the titration of a weak base with a strong acid. Explain (in words) what chemistry governs the pH in each of the four distinct regions of the curve.

**11-13.** Why is the equivalence point pH necessarily below 7 when a weak base is titrated with strong acid?

**11-14.** A 100.0-mL aliquot of 0.100 M weak base B ($pK_b$ = 5.00) was titrated with 1.00 M HClO$_4$. Find the pH at the following volumes of acid added and make a graph of pH versus $V_a$: $V_a$ = 0, 1, 5, 9, 9.9, 10, 10.1, and 12 mL.

**11-15.** At what point in the titration of a weak base with a strong acid is the maximum buffer capacity reached? This is the point at which a given small addition of acid causes the least pH change.

**11-16.** What is the equilibrium constant for the reaction between benzylamine and HCl?

**11-17.** A 50.0-mL solution of 0.031 9 M benzylamine was titrated with 0.050 0 M HCl. Calculate the pH at the following volumes of added acid: $V_a$ = 0, 12.0, $\frac{1}{2}V_e$, 30.0, $V_e$, and 35.0 mL.

**11-18.** Calculate the pH of a solution made by mixing 50.00 mL of 0.100 M NaCN with

(a) 4.20 mL of 0.438 M HClO$_4$

(b) 11.82 mL of 0.438 M HClO$_4$

(c) What is the pH at the equivalence point with 0.438 M HClO$_4$?

### Titrations in Diprotic Systems

**11-19.** Sketch the general appearance of the curve for the titration of a weak diprotic acid with NaOH. Explain (in words) what chemistry governs the pH in each distinct region of the curve.

**11-20.** The opening page of this chapter shows the titration curve for an enzyme. Is the average charge of the protein positive, negative, or neutral at its isoionic point? How do you know?

**11-21.** The base $Na^+A^-$, whose anion is dibasic, was titrated with HCl to give curve *b* in Figure 11-4. Is point H, the first equivalence point, the isoelectric point or the isoionic point?

**11-22.** The figure compares the titration of a monoprotic weak acid with a monoprotic weak base and the titration of a diprotic acid with strong base.

**(a)** Write the reaction between the weak acid and the weak base and show that the equilibrium constant is $10^{7.78}$. This large value means that the reaction goes "to completion" after each addition of reagent.

**(b)** Why does $pK_2$ intersect the upper curve at $\frac{3}{2}V_e$ and the lower curve at $2V_e$? On the lower curve, "$pK_2$" is $pK_a$ for the acid, $BH^+$.

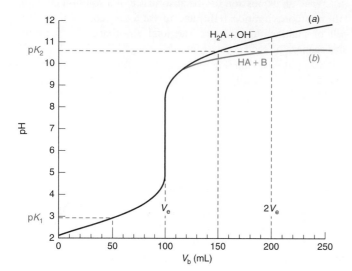

*(a)* Titration of 100 mL of 0.050 M $H_2A$ ($pK_1 = 2.86$, $pK_2 = 10.64$) with 0.050 M NaOH. *(b)* Titration of 100 mL of the weak acid HA (0.050 M, $pK_a = 2.86$) with the weak base B (0.050 M, $pK_b = 3.36$).

**11-23.** The dibasic compound B ($pK_{b1} = 4.00$, $pK_{b2} = 8.00$) was titrated with 1.00 M HCl. The initial solution of B was 0.100 M and had a volume of 100.0 mL. Find the pH at the following volumes of acid added and make a graph of pH versus $V_a$: $V_a = 0, 1, 5, 9, 10, 11, 15, 19, 20,$ and 22 mL.

**11-24.** A 100.0-mL aliquot of 0.100 M diprotic acid $H_2A$ ($pK_1 = 4.00$, $pK_2 = 8.00$) was titrated with 1.00 M NaOH. Find the pH at the following volumes of base added and make a graph of pH versus $V_b$: $V_b = 0, 1, 5, 9, 10, 11, 15, 19, 20,$ and 22 mL.

**11-25.** Calculate the pH at 10.0-mL intervals (from 0 to 100 mL) in the titration of 40.0 mL of 0.100 M piperazine with 0.100 M HCl. Make a graph of pH versus $V_a$.

**11-26.** Calculate the pH when 25.0 mL of 0.020 0 M 2-aminophenol has been titrated with 10.9 mL of 0.015 0 M $HClO_4$.

**11-27.** Consider the titration of 50.0 mL of 0.100 M sodium glycinate, $H_2NCH_2CO_2Na$, with 0.100 M HCl.

**(a)** Calculate the pH at the second equivalence point.

**(b)** Show that our approximate method of calculations gives incorrect (physically unreasonable) values of pH at $V_a = 90.0$ and $V_a = 101.0$ mL.

**11-28.** A solution containing 0.100 M glutamic acid (the molecule with no net charge) was titrated to its first equivalence point with 0.025 0 M RbOH.

**(a)** Draw the structures of reactants and products.

**(b)** Calculate the pH at the first equivalence point.

**11-29.** Find the pH of the solution when 0.010 0 M tyrosine is titrated to the equivalence point with 0.004 00 M $HClO_4$.

**11-30.** This problem deals with the amino acid cysteine, which we will abbreviate $H_2C$.

**(a)** A 0.030 0 M solution was prepared by dissolving dipotassium cysteine, $K_2C$, in water. Then 40.0 mL of this solution were titrated with 0.060 0 M $HClO_4$. Calculate the pH at the first equivalence point.

**(b)** Calculate the quotient $[C^{2-}]/[HC^-]$ in a solution of 0.050 0 M cysteinium bromide (the salt $H_3C^+Br^-$).

**11-31.** How many grams of dipotassium oxalate (FM 166.22) should be added to 20.0 mL of 0.800 M $HClO_4$ to give a pH of 4.40 when the solution is diluted to 500 mL?

**11-32.** When 5.00 mL of 0.103 2 M NaOH were added to 0.112 3 g of alanine (FM 89.093) in 100.0 mL of 0.10 M $KNO_3$, the measured pH was 9.57. *Use activity coefficients* to find $pK_2$ for alanine. Consider the ionic strength of the solution to be 0.10 M and consider each ionic form of alanine to have an activity coefficient of 0.77.

### Finding the End Point with a pH Electrode

**11-33.** What is a Gran plot used for?

**11-34.** Data for the titration of 100.00 mL of a weak acid by NaOH are given below. Find the end point by preparing a Gran plot, using the last 10% of the volume prior to $V_e$.

| mL NaOH | pH | mL NaOH | pH | mL NaOH | pH |
|---|---|---|---|---|---|
| 0.00 | 4.14 | 20.75 | 6.09 | 22.70 | 6.70 |
| 1.31 | 4.30 | 21.01 | 6.14 | 22.76 | 6.74 |
| 2.34 | 4.44 | 21.10 | 6.15 | 22.80 | 6.78 |
| 3.91 | 4.61 | 21.13 | 6.16 | 22.85 | 6.82 |
| 5.93 | 4.79 | 21.20 | 6.17 | 22.91 | 6.86 |
| 7.90 | 4.95 | 21.30 | 6.19 | 22.97 | 6.92 |
| 11.35 | 5.19 | 21.41 | 6.22 | 23.01 | 6.98 |
| 13.46 | 5.35 | 21.51 | 6.25 | 23.11 | 7.11 |
| 15.50 | 5.50 | 21.61 | 6.27 | 23.17 | 7.20 |
| 16.92 | 5.63 | 21.77 | 6.32 | 23.21 | 7.30 |
| 18.00 | 5.71 | 21.93 | 6.37 | 23.30 | 7.49 |
| 18.35 | 5.77 | 22.10 | 6.42 | 23.32 | 7.74 |
| 18.95 | 5.82 | 22.27 | 6.48 | 23.40 | 8.30 |
| 19.43 | 5.89 | 22.37 | 6.53 | 23.46 | 9.21 |
| 19.93 | 5.95 | 22.48 | 6.58 | 23.55 | 9.86 |
| 20.48 | 6.04 | 22.57 | 6.63 | | |

**11-35.** Prepare a second derivative graph to find the end point from the following titration data.

| mL NaOH | pH | mL NaOH | pH | mL NaOH | pH |
|---|---|---|---|---|---|
| 10.679 | 7.643 | 10.725 | 6.222 | 10.750 | 4.444 |
| 10.696 | 7.447 | 10.729 | 5.402 | 10.765 | 4.227 |
| 10.713 | 7.091 | 10.733 | 4.993 | | |
| 10.721 | 6.700 | 10.738 | 4.761 | | |

### Finding the End Point with Indicators

**11-36.** Explain the origin of the rule of thumb that indicator color changes occur at $pK_{HIn} \pm 1$.

**11-37.** Why does a properly chosen indicator change color near the equivalence point in a titration?

**11-38.** The pH of microscopic vesicles (compartments) in living cells can be estimated by infusing an indicator (HIn) into the compartment and measuring the quotient $[In^-]/[HIn]$ from the spectrum of the indicator inside the vesicle. Explain how this tells us the pH.

**11-39.** Write the formula of a compound with a negative $pK_a$.

**11-40.** Consider the titration in Figure 11-2, for which the pH at the equivalence point is calculated to be 9.25. If thymol blue is used as an indicator, what color will be observed through most of the titration prior to the equivalence point? At the equivalence point? After the equivalence point?

**11-41.** What color do you expect to observe for cresol purple indicator (Table 11-4) at the following pH values? (a) 1.0; (b) 2.0; (c) 3.0.

**11-42.** Cresol red has *two* transition ranges listed in Table 11-4. What color would you expect it to be at the following pH values? (a) 0; (b) 1; (c) 6; (d) 9.

**11-43.** Would the indicator bromocresol green, with a transition range of pH 3.8–5.4, ever be useful in the titration of a weak acid with a strong base?

**11-44.** (a) What is the pH at the equivalence point when 0.030 0 M NaF is titrated with 0.060 0 M $HClO_4$?
(b) Why would an indicator end point probably not be useful in this titration?

**11-45.** A titration curve for $NaCO_3$ titrated with HCl is shown here. Suppose that *both* phenolphthalein and bromocresol green are present in the titration solution. State what colors you expect to observe at the following volumes of added HCl: (a) 2mL; (b) 10 mL; (c) 19 mL.

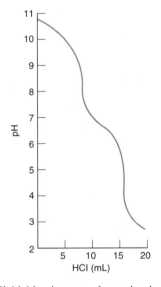

**11-46.** In the Kjeldahl nitrogen determination (Reactions 7-3 through 7-5), the final product is a solution of $NH_4^+$ ions in HCl solution. It is necessary to titrate the HCl without titrating the $NH_4^+$ ions.
(a) Calculate the pH of pure 0.010 M $NH_4Cl$.
(b) Select an indicator that would allow you to titrate HCl but not $NH_4^+$.

**11-47.** A 10.231-g sample of window cleaner containing ammonia was diluted with 39.466 g of water. Then 4.373 g of solution was titrated with 14.22 mL of 0.106 3 M HCl to reach a bromocresol green end point. Find the weight percent of $NH_3$ (FM 17.031) in the cleaner.

### Practical Notes and Leveling Effect

**11-48.** Give the name and formula of a primary standard used to standardize (a) HCl and (b) NaOH.

**11-49.** Why is it more accurate to use a primary standard with a high equivalent mass (the mass required to provide or consume 1 mol of $H^+$) than one with a low equivalent mass?

**11-50.** Explain how to use potassium hydrogen phthalate to standardize a solution of NaOH.

**11-51.** A solution was prepared from 1.023 g of the primary standard tris (Table 11-5) plus 99.367 g of water; 4.963 g of the solution was titrated with 5.262 g of aqueous $HNO_3$ to reach the methyl red end point. Calculate the concentration of the $HNO_3$ (expressed as mol $HNO_3$/kg solution).

**11-52.** The balance says that you have weighed out 1.023 g of tris to standardize a solution of HCl. Use the buoyancy correction in Section 2-3 and the density in Table 11-5 to determine how many grams you have really weighed out. The volume of HCl required to react with the tris was 28.37 mL. Does the buoyancy correction introduce a random or a systematic error into the calculated molarity of HCl? What is the magnitude of the error expressed as a percentage? Is the calculated molarity of HCl higher or lower than the true molarity?

**11-53.** A solution was prepared by dissolving 0.194 7 g of HgO (Table 11-5) in 20 mL of water containing 4 g of KBr. Titration with HCl required 17.98 mL to reach a phenolphthalein end point. Calculate the molarity of the HCl.

**11-54.** How many grams of potassium hydrogen phthalate should be weighed into a flask to standardize ~0.05 M NaOH if you wish to use ~30 mL of base for the titration?

**11-55.** Constant-boiling aqueous HCl can be used as a primary standard for acid-base titrations. When ~20 wt% HCl (FM 36.461) is distilled, the composition of the distillate varies in a regular manner with the barometric pressure:

| P (Torr) | $HCl^a$ (g/100 g solution) |
| --- | --- |
| 770 | 20.196 |
| 760 | 20.220 |
| 750 | 20.244 |
| 740 | 20.268 |
| 730 | 20.292 |

a. The composition of distillate is from C. W. Foulk and M. Hollingsworth, J. Am. Chem. Soc. **1923**, 45, 1223, with numbers corrected for the current values of atomic masses.

Suppose that constant-boiling HCl was collected at a pressure of 746 Torr.
(a) Make a graph of the data in the table to find the weight percent of HCl collected at 746 Torr.
(b) What mass of distillate (weighed in air, using weights whose density is 8.0 g/mL) should be dissolved in 1.000 0 L to give 0.100 00 M HCl? The density of distillate over the whole range in the table is close to 1.096 g/mL. You will need this density to change the mass measured in vacuum to mass measured in air. See Section 2-3 for buoyancy corrections.

**11-56.** What is meant by the leveling effect?

**11-57.** Considering the following $pK_a$ values,[21] explain why dilute sodium methoxide ($NaOCH_3$) and sodium ethoxide ($NaOCH_2CH_3$) are leveled to the same base strength in aqueous solution. Write the chemical reactions that occur when these bases are added to water.

| | |
| --- | --- |
| $CH_3OH$ | $pK_a = 15.54$ |
| $CH_3CH_2OH$ | $pK_a = 16.0$ |
| HOH $pK_a = 15.74$ | (for $K_a = [H^+][OH^-]/[H_2O]$) |

**11-58.** The base B is too weak to titrate in aqueous solution.
(a) Which solvent, pyridine or acetic acid, would be more suitable for the titration of B with $HClO_4$? Why?

(b) Which solvent would be more suitable for the titration of a very weak acid with tetrabutylammonium hydroxide? Why?

**11-59.** Explain why sodium amide ($NaNH_2$) and phenyl lithium ($C_6H_5Li$) are leveled to the same base strength in aqueous solution. Write the chemical reactions that occur when these are added to water.

**11-60.** Pyridine is half protonated in aqueous phosphate buffer at pH 5.2. If you mix 45 mL of phosphate buffer with 55 mL of methanol, the buffer must have a pH of 3.2 to half protonate pyridine. Suggest a reason why.

### Calculating Titration Curves with Spreadsheets

**11-61.** Derive the following equation analogous to those in Table 11-6 for the titration of potassium hydrogen phthalate ($K^+HP^-$) with NaOH:

$$\phi = \frac{C_bV_b}{C_aV_a} = \frac{\alpha_{HP^-} + 2\alpha_{P^{2-}} - 1 - \dfrac{[H^+] - [OH^-]}{C_a}}{1 + \dfrac{[H^+] - [OH^-]}{C_b}}$$

**11-62.** ▦ *Effect of $pK_a$ in the titration of weak acid with strong base.* Use Equation 11-9 with a spreadsheet such as the one shown in Figure 11-11 to compute and plot the family of curves at the left side of Figure 11-3. For a strong acid, choose a large $K_a$, such as $K_a = 10^2$ or $pK_a = -2$.

**11-63.** ▦ *Effect of concentration in the titration of weak acid with strong base.* Use your spreadsheet from Problem 11-62 to prepare a family of titration curves for $pK_a = 6$, with the following combinations of concentrations: (a) $C_a = 20$ mM, $C_b = 100$ mM; (b) $C_a = 2$ mM, $C_b = 10$ mM; (c) $C_a = 0.2$ mM, $C_b = 1$ mM.

**11-64.** ▦ *Effect of $pK_b$ in the titration of weak base with strong acid.* Using the appropriate equation in Table 11-6, prepare a spreadsheet to compute and plot a family of curves analogous to the left part of Figure 11-3 for the titration of 50.0 mL of 0.020 0 M B ($pK_b = -2.00$, 2.00, 4.00, 6.00, 8.00, and 10.00) with 0.100 M HCl. (The value $pK_b = -2.00$ represents a strong base.) In the expression for $\alpha_{BH^+}$, $K_{BH^+} = K_w/K_b$.

**11-65.** ▦ *Titrating weak acid with weak base.*
(a) Use a spreadsheet to prepare a family of graphs for the titration of 50.0 mL of 0.020 0 M HA ($pK_a = 4.00$) with 0.100 M B ($pK_b = 3.00$, 6.00, and 9.00).
(b) Write the acid-base reaction that occurs when acetic acid and sodium benzoate (the salt of benzoic acid) are mixed, and find the equilibrium constant for the reaction. Find the pH of a solution prepared by mixing 212 mL of 0.200 M acetic acid with 325 mL of 0.050 0 M sodium benzoate.

**11-66.** ▦ *Titrating diprotic acid with strong base.* Use a spreadsheet to prepare a family of graphs for the titration of 50.0 mL of 0.020 0 M $H_2A$ with 0.100 M NaOH. Consider the following cases: (a) $pK_1 = 4.00$, $pK_2 = 8.00$; (b) $pK_1 = 4.00$, $pK_2 = 6.00$; (c) $pK_1 = 4.00$, $pK_2 = 5.00$.

**11-67.** ▦ *Titrating nicotine with strong acid.* Prepare a spreadsheet to reproduce the lower curve in Figure 11-4.

**11-68.** ▦ *Titrating triprotic acid with strong base.* Prepare a spreadsheet to graph the titration of 50.0 mL of 0.020 0 M histidine · 2HCl with 0.100 M NaOH. Treat histidine · 2HCl with the triprotic acid equation in Table 11-6.

**11-69.** ▦ *A tetraprotic system.* Write an equation for the titration of tetrabasic base with strong acid ($B + H^+ \rightarrow \rightarrow \rightarrow \rightarrow BH_4^{4+}$). You can do this by inspection of Table 11-6 or you can derive it from the charge balance for the titration reaction. Use a spreadsheet to graph the titration of 50.0 mL of 0.020 0 M sodium pyrophosphate ($Na_4P_2O_7$) with 0.100 M $HClO_4$. Pyrophosphate is the anion of pyrophosphoric acid.

### Using Beer's Law with Indicators

**11-70.** Spectrophotometric properties of a particular indicator are given below:[†]

$$HIn \underset{}{\overset{pK_a = 7.95}{\rightleftharpoons}} In^- + H^+$$

$\lambda_{max} = 395$ nm  $\qquad\qquad$ $\lambda_{max} = 604$ nm
$\varepsilon_{395} = 1.80 \times 10^4$ M$^{-1}$ cm$^{-1}$ $\qquad$ $\varepsilon_{604} = 4.97 \times 10^4$ M$^{-1}$ cm$^{-1}$
$\varepsilon_{604} = 0$

A solution with a volume of 20.0 mL containing $1.40 \times 10^{-5}$ M indicator plus 0.050 0 M benzene-1,2,3-tricarboxylic acid was treated with 20.0 mL of aqueous KOH. The resulting solution had an absorbance at 604 nm of 0.118 in a 1.00-cm cell. Calculate the molarity of the KOH solution.

**11-71.** A certain acid-base indicator exists in three colored forms:[†]

$$H_2In \overset{pK_1 = 1.00}{\rightleftharpoons} HIn^- \overset{pK_2 = 7.95}{\rightleftharpoons} In^{2-}$$

| | | |
|---|---|---|
| $\lambda_{max} = 520$ nm | $\lambda_{max} = 435$ nm | $\lambda_{max} = 572$ nm |
| $\varepsilon_{520} = 5.00 \times 10^4$ | $\varepsilon_{435} = 1.80 \times 10^4$ | $\varepsilon_{572} = 4.97 \times 10^4$ |
| Red | Yellow | Red |
| $\varepsilon_{435} = 1.67 \times 10^4$ | $\varepsilon_{520} = 2.13 \times 10^3$ | $\varepsilon_{520} = 2.50 \times 10^4$ |
| $\varepsilon_{572} = 2.03 \times 10^4$ | $\varepsilon_{572} = 2.00 \times 10^2$ | $\varepsilon_{435} = 1.15 \times 10^4$ |

The units of molar absorptivity, $\varepsilon$, are M$^{-1}$ cm$^{-1}$. A solution containing 10.0 mL of $5.00 \times 10^{-4}$ M indicator was mixed with 90.0 mL of 0.1 M phosphate buffer (pH 7.50). Calculate the absorbance of this solution at 435 nm in a 1.00-cm cell.

---

[†]This problem is based on Beer's law in Section 18-2.

---

## Reference Procedure: Preparing Standard Acid and Base

### Standard 0.1 M NaOH

1. Prepare 50 wt% aqueous NaOH solution in advance and allow the $Na_2CO_3$ precipitate to settle overnight. ($Na_2CO_3$ is insoluble in this solution.) Store the solution in a tightly sealed polyethylene bottle and avoid disturbing the precipitate when supernate is taken. The density is close to 1.50 g of solution per milliliter.

2. Dry primary-standard-grade potassium hydrogen phthalate for 1 h at 110°C and store it in a desiccator.

$$\text{(structure) } CO_2^-K^+ , CO_2H + NaOH \longrightarrow CO_2^-K^+ , CO_2^-Na^+ + H_2O$$

Potassium hydrogen phthalate
FM 204.221

3. Boil 1 L of water for 5 min to expel $CO_2$. Pour the water into a polyethylene bottle, which should be tightly capped whenever possible. Calculate the volume of 50 wt% NaOH needed (~5.3 mL) to produce 1 L of ~0.1 M NaOH. Use a graduated cylinder to transfer this much NaOH to the bottle of water. Mix well and allow the solution to cool to room temperature (preferably overnight).

4. Weigh out four ~0.51-g portions of potassium hydrogen phthalate and dissolve each in ~25 mL of distilled water in a 125-mL flask. Each sample should require ~25 mL of 0.1 M NaOH. Add 3 drops of phenolphthalein indicator (Table 11-4) to each, and titrate one of them rapidly to find the approximate end point. The buret should have a loosely fitted cap to minimize entry of $CO_2$.

5. Calculate the volume of NaOH required for each of the other three samples and titrate them carefully. During each titration, periodically tilt and rotate the flask to wash liquid from the walls into the solution. When very near the end, deliver less than 1 drop of titrant at a time. To do this, carefully suspend a fraction of a drop from the buret tip, touch it to the inside wall of the flask, wash it into the bulk solution by careful tilting, and swirl the solution. The end point is the first appearance of pink color that persists for 15 s. The color will slowly fade as $CO_2$ from air dissolves in the solution.

6. Calculate the average molarity ($\bar{x}$), the standard deviation ($s$), and the relative standard deviation ($s/\bar{x}$). If you have used some care, the relative standard deviation should be <0.2%.

## Standard 0.1 M HCl

1. The inside cover of this book tells us that 8.2 mL of ~37 wt% HCl should be added to 1 L of water to produce ~0.1 M HCl. Prepare this solution in a capped polyethylene bottle, using a graduated cylinder to deliver the HCl.

2. Dry primary-standard-grade $Na_2CO_3$ for 1 h at 110°C and cool it in a desiccator.

3. Weigh four samples containing enough $Na_2CO_3$ to react with ~25 mL of 0.1 M HCl from a buret and place each in a 125-mL flask. When ready to titrate each one, dissolve it in ~25 mL of distilled water.

$$2HCl \quad + \quad Na_2CO_3 \quad \rightarrow \quad CO_2 \quad + \quad 2NaCl \quad + \quad H_2O$$
$$FM\ 105.988$$

Add 3 drops of bromocresol green indicator (Table 11-4) and titrate one sample rapidly to a green color to find the approximate end point.

4. Carefully titrate each of the other samples until it just turns from blue to green. Then boil the solution to expel $CO_2$. The solution should return to a blue color. Carefully add HCl from the buret until the solution turns green again.

5. Titrate one blank prepared from 3 drops of indicator plus 50 mL of 0.05 M NaCl. Subtract the blank volume of HCl from that required to titrate $Na_2CO_3$.

6. Calculate the mean HCl molarity, standard deviation, and relative standard deviation.

# 12 | EDTA Titrations

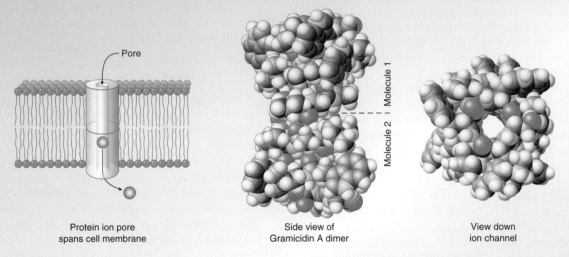

Pore

Protein ion pore
spans cell membrane

Molecule 1
Molecule 2

Side view of
Gramicidin A dimer

View down
ion channel

*Left and center:* Two gramicidin A molecules associate to span a cell membrane. *Right:* Axial view showing ion channel. [*Structure from B. Roux, "Computational Studies of the Gramicidin Channel," Acc. Chem. Res.* **2002,** *35, 366, based on solid-state nuclear magnetic resonance. Schematic at left from L. Stryer, Biochemistry, 4th ed. (New York: W. H. Freeman, 1995).]*

O
‖
C — N
|
H

Amide
group

Gramicidin A is an antibiotic that kills cells by making their membranes permeable to $Na^+$ and $K^+$. Gramicidin A is made of 15 amino acids wound into a helix with a 0.4-nm-diameter channel through the middle. The channel is lined by polar amide groups and the outside of gramicidin is covered by nonpolar hydrocarbons (page 182). *Polar* groups have positive and negative regions that attract neighboring molecules by electrostatic forces. *Nonpolar* groups have little charge separation and are soluble inside the nonpolar cell membrane.

Metal cations dissolve in water and are said to be *hydrophilic* ("water loving"). Cell membranes exclude water and are described as *hydrophobic* ("water hating"). Gramicidin A lodges in the cell membrane because the outside of the molecule is hydrophobic. $Na^+$ and $K^+$ pass through each hydrophilic pore at a rate of $10^7$ ions/s. The pore is selective for monovalent cations; it excludes anions and more highly charged cations.

Part of the Nobel Prize in Chemistry in 2003 was awarded to Roderick MacKinnon for elucidating the structure of potassium channels that selectively permit $K^+$ to pass through membranes of cells such as nerves.[1] Amide oxygen atoms of the protein backbone in the channel are spaced just right to replace waters of hydration from $K(H_2O)_6^+$. There is little change in energy when hydrated $K^+$ sheds $H_2O$ and binds inside the channel. The spacing of amide oxygens is too great by 0.04 nm to displace $H_2O$ from $Na(H_2O)_6^+$. Hydrated $Na^+$ remains outside the channel while hydrated $K^+$ sheds $H_2O$ and binds inside the channel. $K^+$ passes at a rate of $10^8$ ions/s per channel—100 times faster than $Na^+$.

Dissolved nickel in the south San Francisco Bay reaches 110 nM in the summer.

**E**DTA is a merciful abbreviation for *ethylenediaminetetraacetic acid,* a compound that forms strong 1:1 complexes with most metal ions (Figure 12-1) and finds wide use in quantitative analysis. EDTA plays a larger role as a strong metal-binding agent in industrial processes and in products such as detergents, cleaning agents, and food additives that prevent metal-catalyzed oxidation of food. EDTA is an emerging player in environmental chemistry.[2] For example, the majority of nickel discharged into San Francisco Bay and a significant fraction of the iron, lead, copper, and zinc are EDTA complexes that pass unscathed through wastewater treatment plants.

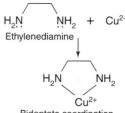

## Figure 12-1

**Figure 12-1** EDTA forms strong 1:1 complexes with most metal ions, binding through four oxygen and two nitrogen atoms. The six-coordinate geometry of $Mn^{2+}$-EDTA found in the compound $KMnEDTA \cdot 2H_2O$ was deduced from X-ray crystallography. [*J. Stein, J. P. Fackler, Jr., G. J. McClune, J. A. Fee, and L. T. Chan, "Reactions of Mn-EDTA and MnCyDTA Complexes with $O_2^-$. X-Ray Structure of KMnEDTA · 2H₂O," Inorg. Chem.* **1979**, *18*, 3511.]

Mn ◯    O ◯    C ●    N ◯

EDTA⁴⁻    + Mn²⁺ ⟶

# ■ ■ ■ ■ 12-1 Metal-Chelate Complexes

Metal ions are **Lewis acids,** accepting electron pairs from electron-donating ligands that are **Lewis bases.** Cyanide ($CN^-$) is called a **monodentate** ligand because it binds to a metal ion through only one atom (the carbon atom). Most transition metal ions bind six ligand atoms. A ligand that attaches to a metal ion through more than one ligand atom is said to be **multidentate** ("many toothed"), or a **chelating ligand** (pronounced KEE-late-ing).

A simple chelating ligand is 1,2-diaminoethane ($H_2NCH_2CH_2NH_2$ also called ethylenediamine), whose binding to a metal ion is shown in the margin. We say that ethylenediamine is *bidentate* because it binds to the metal through two ligand atoms.

The **chelate effect** is the ability of multidentate ligands to form more stable metal complexes than those formed by similar monodentate ligands.[4] For example, the reaction of $Cd(H_2O)_6^{2+}$ with two molecules of ethylenediamine is more favorable than its reaction with four molecules of methylamine:

$$Cd(H_2O)_6^{2+} + 2H_2N{-}{-}NH_2 \rightleftharpoons \left[ \begin{array}{c} \text{Cd(en)}_2(\text{OH}_2)_2 \end{array} \right]^{2+} + 4H_2O$$

Ethylenediamine

$$K \equiv \beta_2 = 8 \times 10^9 \qquad (12\text{-}1)$$

$$Cd(H_2O)_6^{2+} + 4CH_3\ddot{N}H_2 \rightleftharpoons \left[ \begin{array}{c} \text{Cd(CH}_3\text{NH}_2)_4(\text{OH}_2)_2 \end{array} \right]^{2+} + 4H_2O$$

Methylamine

$$K \equiv \beta_4 = 4 \times 10^6 \qquad (12\text{-}2)$$

At pH 12 in the presence of 2 M ethylenediamine and 4 M methylamine, the quotient $[Cd(\text{ethylenediamine})_2^{2+}]/[Cd(\text{methylamine})_4^{2+}]$ is 30.

An important *tetradentate* ligand is adenosine triphosphate (ATP), which binds to divalent metal ions (such as $Mg^{2+}$, $Mn^{2+}$, $Co^{2+}$, and $Ni^{2+}$) through four of their six coordination positions (Figure 12-2). The fifth and sixth positions are occupied by water molecules. The biologically active form of ATP is generally the $Mg^{2+}$ complex.

The *octadentate* ligand in Figure 12-3 is being evaluated as an anticancer agent.[5] This chelate binds a metal through four N atoms and four O atoms. The chelate is covalently attached to a *monoclonal antibody,* which is a protein produced by one specific type of cell in response to one specific foreign substance called an *antigen.* In this case, the antibody binds to a specific feature of a tumor cell. The chelate attached to the antibody carries a tightly bound,

---

$$Ag^+ + 2:\bar{C}{\equiv}N: \rightleftharpoons$$

Lewis acid    Lewis base
(electron pair   (electron pair
acceptor)      donor)

$$(:N{\equiv}C{-}Ag{-}C{\equiv}N:)^-$$

Complex ion

$$H_2N{-}{-}NH_2 + Cu^{2+}$$

Ethylenediamine

$$H_2N{-}{-}NH_2$$
$$Cu^{2+}$$

Bidentate coordination

The term "chelate" is derived from the great claw, or *chela* (from the Greek *chē lē*), of the lobster.[3]

*Chelate effect:* A bidentate ligand forms a more stable complex than do two monodentate ligands.

Nomenclature for formation constants ($K$ and $\beta$) was discussed in Box 6-2.

We have drawn *trans* isomers of the octahedral complexes (with $H_2O$ ligands at opposite poles), but *cis* isomers also are possible (with $H_2O$ ligands adjacent to each other).

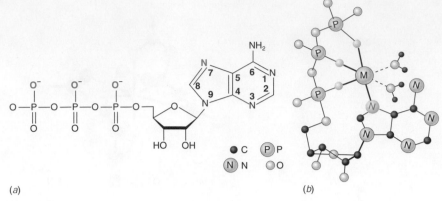

*Figure 12-2* (*a*) Structure of adenosine triphosphate (ATP), with ligand atoms shown in color. (*b*) Possible structure of a metal-ATP complex; the metal, M, has four bonds to ATP and two bonds to $H_2O$ ligands.

Antigen-binding sites

Antibody

Chelate

*Figure 12-3* Synthetic chelate covalently attached to an antibody carries a metal isotope (M) to deliver lethal doses of radiation to tumor cells.

short-lived radioisotope such as $^{90}Y^{3+}$ or $^{177}Lu^{3+}$, which delivers lethal doses of radiation to the tumor. Another important medical application of chelates is described in Box 12-1 (page 232).

Aminocarboxylic acids in Figure 12-4 are synthetic chelating agents. Amine N atoms and carboxylate O atoms are the potential ligand atoms in these molecules (Figures 12-5 and 12-6). When these molecules bind to a metal ion, the ligand atoms lose their protons.

A titration based on complex formation is called a **complexometric titration.** Ligands other than NTA in Figure 12-4 form strong 1:1 complexes with all metal ions except univalent ions such as $Li^+$, $Na^+$, and $K^+$. *The stoichiometry is 1:1 regardless of the charge on the ion.*

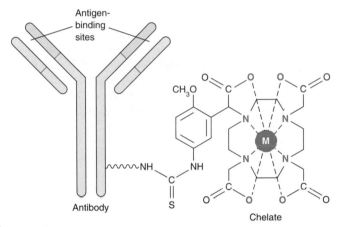

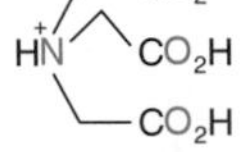

**NTA**
Nitrilotriacetic acid

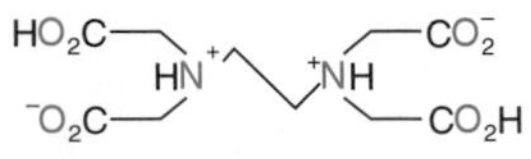

**EDTA**
Ethylenediaminetetraacetic acid
(also called ethylenedinitrilotetraacetic acid)

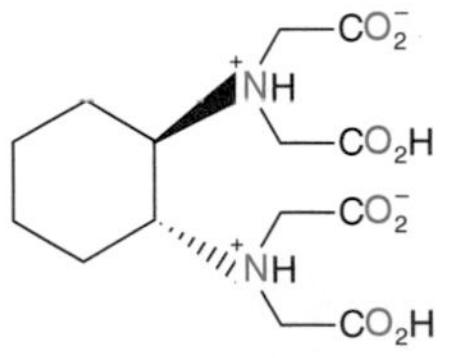

**DCTA**
*trans*-1,2-Diaminocyclohexanetetraacetic acid

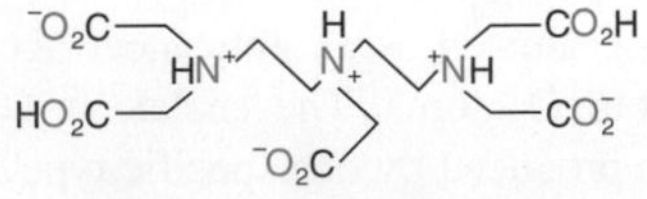

**DTPA**
Diethylenetriaminepentaacetic acid

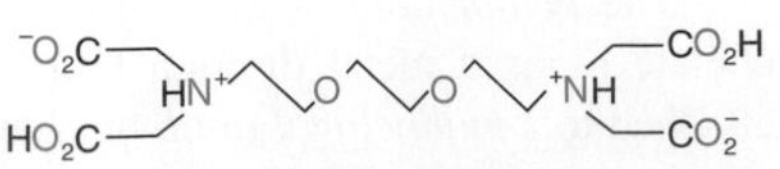

**EGTA**
bis(Aminoethyl)glycolether-*N,N,N',N'*-tetraacetic acid

*Figure 12-4* Structures of analytically useful chelating agents. Nitrilotriacetic acid (NTA) tends to form 2:1 (ligand:metal) complexes with metal ions, whereas the others form 1:1 complexes.

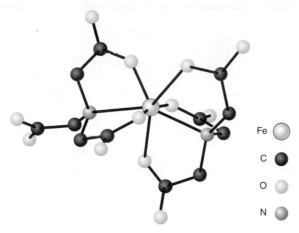

**Figure 12-5** Structure of Fe(NTA)$_2^{3-}$ in the salt Na$_3$[Fe(NTA)$_2$] · 5H$_2$O. The ligand at the right binds to Fe through three O atoms and one N atom. The other ligand uses two O atoms and one N atom. Its third carboxylate group is uncoordinated. The Fe atom is seven coordinate. *[W. Clegg, A. K. Powell, and M. J. Ware, "Structure of Na$_3$ [Fe(NTA)$_2$] · 5H$_2$O," Acta Crystallogr. **1984**, C40, 1822.]*

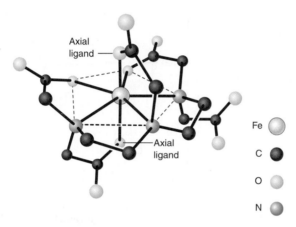

**Figure 12-6** Structure of Fe(DTPA)$^{2-}$ found in the salt Na$_2$[Fe(DTPA)] · 2H$_2$O. The seven-coordinate pentagonal bipyramidal coordination environment of the iron atom features three N and two O ligands in the equatorial plane (dashed lines) and two axial O ligands. The axial Fe–O bond lengths are 11 to 19 pm shorter than those of the more crowded equatorial Fe–O bonds. One carboxyl group of the ligand is uncoordinated. *[D. C. Finnen, A. A. Pinkerton, W. R. Dunham, R. H. Sands, and M. O. Funk, Jr., "Structures and Spectroscopic Characterization of Fe(III)-DTPA Complexes," Inorg. Chem. **1991**, 30, 3960.]*

## ■ ■ ■ ■ 12-2 EDTA

EDTA is, by far, the most widely used chelator in analytical chemistry. By direct titration or through an indirect sequence of reactions, virtually every element of the periodic table can be measured with EDTA.

*One mole of EDTA reacts with one mole of metal ion.*

### Acid-Base Properties

EDTA is a hexaprotic system, designated H$_6$Y$^{2+}$. The highlighted, acidic hydrogen atoms are the ones that are lost upon metal-complex formation.

HO$_2$CCH$_2$⟍    ⟋CH$_2$CO$_2$H
       $\overset{+}{H}$NCH$_2$CH$_2$$\overset{+}{N}$H
HO$_2$CCH$_2$⟋    ⟍CH$_2$CO$_2$H
       H$_6$Y$^{2+}$

| | |
|---|---|
| p$K_1$ = 0.0 (CO$_2$H) | p$K_4$ = 2.69 (CO$_2$H) |
| p$K_2$ = 1.5 (CO$_2$H) | p$K_5$ = 6.13 (NH$^+$) |
| p$K_3$ = 2.00 (CO$_2$H) | p$K_6$ = 10.37 (NH$^+$) |

p$K$ applies at 25°C and $\mu$ = 0.1 M, except p$K_1$ applies at $\mu$ = 1 M

The first four p$K$ values apply to carboxyl protons, and the last two are for the ammonium protons.[11] The neutral acid is tetraprotic, with the formula H$_4$Y. A commonly used reagent is the disodium salt, Na$_2$H$_2$Y · 2H$_2$O.[12]

## Box 12-1    Chelation Therapy and Thalassemia

Oxygen is carried in blood by the iron-containing protein hemoglobin, which consists of two pairs of subunits, designated $\alpha$ and $\beta$. $\beta$-Thalassemia major is a genetic disease in which the $\beta$ subunits of hemoglobin are not synthesized in adequate quantities. Children afflicted with this disease survive only with frequent transfusions of normal red blood cells. The problem with this treatment is that patients accumulate 4–8 g of iron per year from the hemoglobin in the transfused cells. The body has no mechanism for excreting large quantities of iron, and most patients die by age 20 from the toxic effects of iron overload.

To enhance iron excretion, intensive chelation therapy is used. The most successful drug is *desferrioxamine B*, a powerful $Fe^{3+}$-chelator produced by the microbe *Streptomyces pilosus*.[6] The formation constant for the Fe(III) complex, called ferrioxamine B, is $10^{30.6}$. Used in conjunction with ascorbic acid—vitamin C, a reducing agent that reduces $Fe^{3+}$ to the more soluble $Fe^{2+}$—desferrioxamine clears several grams of iron per year from an overloaded patient. The ferrioxamine complex is excreted in the urine.

Desferrioxamine reduces the incidence of heart and liver disease in thalassemia patients. In patients for whom desferrioxamine effectively controls iron overload, there is a 91% rate of cardiac disease–free survival after 15 years of therapy.[7] There are negative effects of desferrioxamine treatment; for example, too high a dose stunts a child's growth.

Desferrioxamine is expensive and must be taken by continuous injection. It is not absorbed through the intestine. Many potent iron chelators have been tested to find an effective one that can be taken orally, but only the drug deferiprone is currently used orally.[8,9] In the long term, bone marrow transplants or gene therapy[10] might cure the disease.

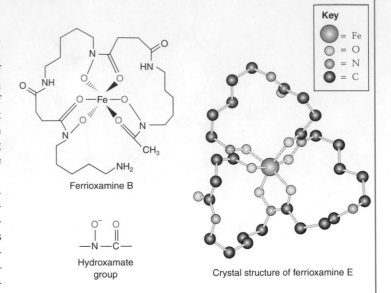

Ferrioxamine B

Hydroxamate group

Crystal structure of ferrioxamine E

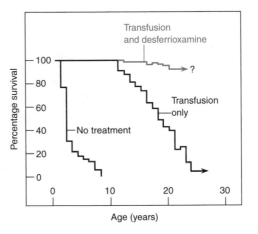

Deferiprone

Structures above show the iron complex ferrioxamine B and the related compound, ferrioxamine E, in which the chelate has a cyclic structure. Graph shows success of transfusions and transfusions plus chelation therapy. *[Crystal structure kindly provided by M. Neu, Los Alamos National Laboratory, based on D. Van der Helm and M. Poling, J. Am. Chem. Soc.* **1976**, *98, 82. Graph from P. S. Dobbin and R. C. Hider, "Iron Chelation Therapy," Chem. Br.* **1990**, *26, 565.]*

The fraction of EDTA in each of its protonated forms is plotted in Figure 12-7. As in Section 10-5, we can define $\alpha$ for each species as the fraction of EDTA in that form. For example, $\alpha_{Y^{4-}}$ is defined as

*Fraction of EDTA in the form $Y^{4-}$:*

$$\alpha_{Y^{4-}} = \frac{[Y^{4-}]}{[H_6Y^{2+}] + [H_5Y^+] + [H_4Y] + [H_3Y^-] + [H_2Y^{2-}] + [HY^{3-}] + [Y^{4-}]}$$

$$\alpha_{Y^{4-}} = \frac{[Y^{4-}]}{[\text{EDTA}]} \tag{12-3}$$

where [EDTA] is the total concentration of all *free* EDTA species in the solution. By "free," we mean EDTA not complexed to metal ions. Following the derivation in Section 10-5, we can show that $\alpha_{Y^{4-}}$ is given by

$$\alpha_{Y^{4-}} = K_1K_2K_3K_4K_5K_6/\{[H^+]^6 + [H^+]^5K_1 + [H^+]^4K_1K_2 + [H^+]^3K_1K_2K_3$$
$$+ [H^+]^2K_1K_2K_3K_4 + [H^+]K_1K_2K_3K_4K_5 + K_1K_2K_3K_4K_5K_6\} \tag{12-4}$$

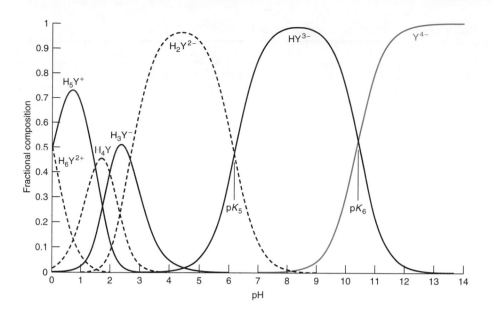

Figure 12-7 Fractional composition diagram for EDTA.

Table 12-1 gives values for $\alpha_{Y^{4-}}$ as a function of pH.

---

**Example** What Does $\alpha_{Y^{4-}}$ Mean?

The fraction of all free EDTA in the form $Y^{4-}$ is called $\alpha_{Y^{4-}}$. At pH 6.00 at a formal concentration of 0.10 M, the composition of an EDTA solution is

$[H_6Y^{2+}] = 8.9 \times 10^{-20}$ M     $[H_5Y^+] = 8.9 \times 10^{-14}$ M     $[H_4Y] = 2.8 \times 10^{-7}$ M

$[H_3Y^-] = 2.8 \times 10^{-5}$ M     $[H_2Y^{2-}] = 0.057$ M     $[HY^{3-}] = 0.043$ M

$[Y^{4-}] = 1.8 \times 10^{-6}$ M

Find $\alpha_{Y^{4-}}$.

**Solution** $\alpha_{Y^{4-}}$ is the fraction in the form $Y^{4-}$:

$$\alpha_{Y^{4-}} = \frac{[Y^{4-}]}{[H_6Y^{2+}] + [H_5Y^+] + [H_4Y] + [H_3Y^-] + [H_2Y^{2-}] + [HY^{3-}] + [Y^{4-}]}$$

$$= \frac{1.8 \times 10^{-6}}{(8.9 \times 10^{-20}) + (8.9 \times 10^{-14}) + (2.8 \times 10^{-7}) + (2.8 \times 10^{-5}) + (0.057) + (0.043) + (1.8 \times 10^{-6})}$$

$$= 1.8 \times 10^{-5}$$

**Question** From Figure 12-7, which species has the greatest concentration at pH 6? At pH 7? At pH 11?

## EDTA Complexes

The equilibrium constant for the reaction of a metal with a ligand is called the **formation constant,** $K_f$, or the **stability constant:**

*Formation constant:*     $M^{n+} + Y^{4-} \rightleftharpoons MY^{n-4}$     $K_f = \dfrac{[MY^{n-4}]}{[M^{n+}][Y^{4-}]}$     (12-5)

Note that $K_f$ for EDTA is defined in terms of the species $Y^{4-}$ reacting with the metal ion. The equilibrium constant could have been defined for any of the other six forms of EDTA in the solution. Equation 12-5 should not be interpreted to mean that only $Y^{4-}$ reacts with metal ions. Table 12-2 shows that formation constants for most EDTA complexes are quite large and tend to be larger for more positively charged cations.

In many complexes, EDTA engulfs the metal ion, forming the six-coordinate species in Figure 12-1. If you try to build a space-filling model of a six-coordinate metal-EDTA complex, you will find that there is considerable strain in the chelate rings. This strain is relieved when the O ligands are drawn back toward the N atoms. Such distortion opens up a seventh

**Table 12-1** Values of $\alpha_{Y^{4-}}$ for EDTA at 20°C and $\mu = 0.10$ M

| pH | $\alpha_{Y^{4-}}$ |
|---|---|
| 0 | $1.3 \times 10^{-23}$ |
| 1 | $1.4 \times 10^{-18}$ |
| 2 | $2.6 \times 10^{-14}$ |
| 3 | $2.1 \times 10^{-11}$ |
| 4 | $3.0 \times 10^{-9}$ |
| 5 | $2.9 \times 10^{-7}$ |
| 6 | $1.8 \times 10^{-5}$ |
| 7 | $3.8 \times 10^{-4}$ |
| 8 | $4.2 \times 10^{-3}$ |
| 9 | $0.041$ |
| 10 | $0.30$ |
| 11 | $0.81$ |
| 12 | $0.98$ |
| 13 | $1.00$ |
| 14 | $1.00$ |

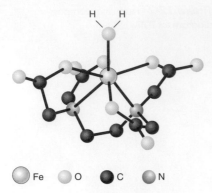

*Figure 12-8* Seven-coordinate geometry of Fe(EDTA)(H₂O)⁻. Other metal ions that form seven-coordinate EDTA complexes include $Fe^{2+}$, $Mg^{2+}$, $Cd^{2+}$, $Co^{2+}$, $Mn^{2+}$, $Ru^{3+}$, $Cr^{3+}$, $Co^{3+}$, $V^{3+}$, $Ti^{3+}$, $In^{3+}$, $Sn^{4+}$, $Os^{4+}$, and $Ti^{4+}$. Some of these same ions also form six-coordinate EDTA complexes. Eight-coordinate complexes are formed by $Ca^{2+}$, $Er^{3+}$, $Yb^{3+}$, and $Zr^{4+}$. [T. Mizuta, J. Wang, and K. Miyoshi, " A 7-Coordinate Structure of Fe(III)-EDTA," Bull. Chem. Soc. Japan **1993**, 66, 2547.]

Only some of the free EDTA is in the form $Y^{4-}$.

**Table 12-2**  Formation constants for metal-EDTA complexes

| Ion | log $K_f$ | Ion | log $K_f$ | Ion | log $K_f$ |
|-----|-----------|-----|-----------|-----|-----------|
| $Li^+$ | 2.95 | $V^{3+}$ | 25.9[a] | $Tl^{3+}$ | 35.3 |
| $Na^+$ | 1.86 | $Cr^{3+}$ | 23.4[a] | $Bi^{3+}$ | 27.8[a] |
| $K^+$ | 0.8 | $Mn^{3+}$ | 25.2 | $Ce^{3+}$ | 15.93 |
| $Be^{2+}$ | 9.7 | $Fe^{3+}$ | 25.1 | $Pr^{3+}$ | 16.30 |
| $Mg^{2+}$ | 8.79 | $Co^{3+}$ | 41.4 | $Nd^{3+}$ | 16.51 |
| $Ca^{2+}$ | 10.65 | $Zr^{4+}$ | 29.3 | $Pm^{3+}$ | 16.9 |
| $Sr^{2+}$ | 8.72 | $Hf^{4+}$ | 29.5 | $Sm^{3+}$ | 17.06 |
| $Ba^{2+}$ | 7.88 | $VO^{2+}$ | 18.7 | $Eu^{3+}$ | 17.25 |
| $Ra^{2+}$ | 7.4 | $VO_2^+$ | 15.5 | $Gd^{3+}$ | 17.35 |
| $Sc^{3+}$ | 23.1[a] | $Ag^+$ | 7.20 | $Tb^{3+}$ | 17.87 |
| $Y^{3+}$ | 18.08 | $Tl^+$ | 6.41 | $Dy^{3+}$ | 18.30 |
| $La^{3+}$ | 15.36 | $Pd^{2+}$ | 25.6[a] | $Ho^{3+}$ | 18.56 |
| $V^{2+}$ | 12.7[a] | $Zn^{2+}$ | 16.5 | $Er^{3+}$ | 18.89 |
| $Cr^{2+}$ | 13.6[a] | $Cd^{2+}$ | 16.5 | $Tm^{3+}$ | 19.32 |
| $Mn^{2+}$ | 13.89 | $Hg^{2+}$ | 21.5 | $Yb^{3+}$ | 19.49 |
| $Fe^{2+}$ | 14.30 | $Sn^{2+}$ | 18.3[b] | $Lu^{3+}$ | 19.74 |
| $Co^{2+}$ | 16.45 | $Pb^{2+}$ | 18.0 | $Th^{4+}$ | 23.2 |
| $Ni^{2+}$ | 18.4 | $Al^{3+}$ | 16.4 | $U^{4+}$ | 25.7 |
| $Cu^{2+}$ | 18.78 | $Ga^{3+}$ | 21.7 | | |
| $Ti^{3+}$ | 21.3 | $In^{3+}$ | 24.9 | | |

NOTE: The stability constant is the equilibrium constant for the reaction $M^{n+} + Y^{4-} \rightleftharpoons MY^{n-4}$. Values in table apply at 25°C and ionic strength 0.1 M unless otherwise indicated.

a. 20°C, ionic strength = 0.1 M.        b. 20°C, ionic strength = 1 M.

SOURCE: A. E. Martell, R. M. Smith, and R. J. Motekaitis, NIST Critically Selected Stability Constants of Metal Complexes, NIST Standard Reference Database 46, Gaithersburg, MD 2001.

coordination position, which can be occupied by $H_2O$, as in Figure 12-8. In some complexes, such as $Ca(EDTA)(H_2O)_2^{2-}$, the metal ion is so large that it accommodates eight ligand atoms.[13] Larger metal ions require more ligand atoms. In the Fe(III) complex ferrioxamine E, whose structure is shown in Box 12-1, the metal coordination sphere is completed by six ligands. When Fe(III) is replace by the larger Pu(IV), three $H_2O$ molecules are required to complete the coordination sphere for a total of nine ligands bound to Pu(IV).[14]

Even if $H_2O$ is attached to the metal ion, the formation constant is still given by Equation 12-5. The relation remains true because solvent ($H_2O$) is omitted from the reaction quotient.

## Conditional Formation Constant

The formation constant $K_f = [MY^{n-4}]/[M^{n+}][Y^{4-}]$ describes the reaction between $Y^{4-}$ and a metal ion. As you see in Figure 12-7, most EDTA is not $Y^{4-}$ below pH 10.37. The species $HY^{3-}$, $H_2Y^{2-}$, and so on, predominate at lower pH. From the definition $\alpha_{Y^{4-}} = [Y^{4-}]/[EDTA]$, we can express the concentration of $Y^{4-}$ as

$$[Y^{4-}] = \alpha_{Y^{4-}}[EDTA]$$

where [EDTA] is the total concentration of all EDTA species not bound to metal ion.

The formation constant can now be rewritten as

$$K_f = \frac{[MY^{n-4}]}{[M^{n+}][Y^{4-}]} = \frac{[MY^{n-4}]}{[M^{n+}]\alpha_{Y^{4-}}[EDTA]}$$

If the pH is fixed by a buffer, then $\alpha_{Y^{4-}}$ is a constant that can be combined with $K_f$:

*Conditional formation constant:*    $$K_f' = \alpha_{Y^{4-}}K_f = \frac{[MY^{n-4}]}{[M^{n+}][EDTA]}$$    (12-6)

The number $K_f' = \alpha_{Y^{4-}}K_f$ is called the **conditional formation constant,** or the *effective formation constant.* It describes the formation of $MY^{n-4}$ at any particular pH. Later, after we learn to use Equation 12-6, we will modify it to allow for the possibility that not all the metal ion is in the form $M^{n+}$.

The conditional formation constant allows us to look at EDTA complex formation as if the uncomplexed EDTA were all in one form:

$$M^{n+} + EDTA \rightleftharpoons MY^{n-4} \qquad K'_f = \alpha_{Y^{4-}} K_f$$

At any given pH, we can find $\alpha_{Y^{4-}}$ and evaluate $K'_f$.

With the conditional formation constant, we can treat EDTA complex formation as if all the free EDTA were in one form.

### Example  Using the Conditional Formation Constant

The formation constant in Table 12-2 for $CaY^{2-}$ is $10^{10.65}$. Calculate the concentrations of free $Ca^{2+}$ in a solution of 0.10 M $CaY^{2-}$ at pH 10.00 and at pH 6.00.

**Solution**  The complex formation reactions is

$$Ca^{2+} + EDTA \rightleftharpoons CaY^{2-} \qquad K'_f = \alpha_{Y^{4-}} K_f$$

where EDTA on the left side of the equation refers to all forms of unbound EDTA ($= Y^{4-}, HY^{3-}, H_2Y^{2-}, H_3Y^-$, and so on). Using $\alpha_{Y^{4-}}$ from Table 12-1, we find

$$\text{At pH 10.00:} \quad K'_f = (0.30)(10^{10.65}) = 1.3_4 \times 10^{10}$$
$$\text{At pH 6.00:} \quad K'_f = (1.8 \times 10^{-5})(10^{10.65}) = 8.0 \times 10^5$$

Dissociation of $CaY^{2-}$ must produce equal quantities of $Ca^{2+}$ and EDTA, so we can write

|  | $Ca^{2+}$ | + | EDTA | $\rightleftharpoons$ | $CaY^{2-}$ |
|---|---|---|---|---|---|
| Initial concentration (M) | 0 | | 0 | | 0.10 |
| Final concentration (M) | $x$ | | $x$ | | $0.10 - x$ |

$$\frac{[CaY^{2-}]}{[Ca^{2+}][EDTA]} = \frac{0.10 - x}{x^2} = K'_f = 1.3_4 \times 10^{10} \quad \text{at pH 10.00}$$

$$= 8.0 \times 10^5 \quad \text{at pH 6.00}$$

Solving for $x$ ($= [Ca^{2+}]$), we find $[Ca^{2+}] = 2.7 \times 10^{-6}$ M at pH 10.00 and $3.5 \times 10^{-4}$ M at pH 6.00. *Using the conditional formation constant at a fixed pH, we treat the dissociated EDTA as if it were a single species.*

You can see from the example that a metal-EDTA complex becomes less stable at lower pH. For a titration reaction to be effective, it must go "to completion" (say, 99.9%), which means that the equilibrium constant is large—the analyte and titrant are essentially completely reacted at the equivalence point. Figure 12-9 shows how pH affects the titration of $Ca^{2+}$ with EDTA. Below pH $\approx 8$, the end point is not sharp enough to allow accurate determination. The conditional formation constant for $CaY^{2-}$ is just too small for "complete" reaction at low pH.

pH can be used to select which metals will be titrated by EDTA and which will not. Metals with higher formation constants can be titrated at lower pH. If a solution containing both $Fe^{3+}$ and $Ca^{2+}$ is titrated at pH 4, $Fe^{3+}$ is titrated without interference from $Ca^{2+}$.

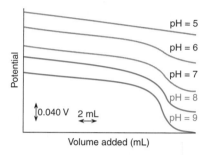

**Figure 12-9**  Titration of $Ca^{2+}$ with EDTA as a function of pH. As the pH is lowered, the end point becomes less distinct. The potential was measured with mercury and calomel electrodes, as described in Exercise 15-B in Chapter 15. [C. N. Reilley and R. W. Schmid, "Chelometric Titration with Potentiometric End Point Detection: Mercury as a pM Indicator Electrode," Anal. Chem. **1958**, 30, 947.]

## ■■■■ 12-3  EDTA Titration Curves[15]

In this section, we calculate the concentration of free $M^{n+}$ during its titration with EDTA. The titration reaction is

$$M^{n+} + EDTA \rightleftharpoons MY^{n-4} \qquad K'_f = \alpha_{Y^{4-}} K_f \qquad (12\text{-}7)$$

If $K'_f$ is large, we can consider the reaction to be complete at each point in the titration.

The titration curve is a graph of pM ($= -\log[M^{n+}]$) versus the volume of added EDTA. The curve is analogous to plotting pH versus volume of titrant in an acid-base titration. There are three natural regions of the titration curve in Figure 12-10.

### Region 1: Before the Equivalence Point
In this region, there is excess $M^{n+}$ left in solution after the EDTA has been consumed. The concentration of free metal ion is equal to the concentration of excess, unreacted $M^{n+}$. The dissociation of $MY^{n-4}$ is negligible.

### Region 2: At the Equivalence Point
There is exactly as much EDTA as metal in the solution. We can treat the solution as if it had been made by dissolving pure $MY^{n-4}$. Some free $M^{n+}$ is generated by the slight dissociation of $MY^{n-4}$:

$$MY^{n-4} \rightleftharpoons M^{n+} + EDTA$$

In this reaction, EDTA refers to the total concentration of free EDTA in all its forms. At the equivalence point, $[M^{n+}] = [EDTA]$.

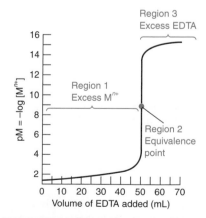

**Figure 12-10**  Three regions in an EDTA titration illustrated for reaction of 50.0 mL of 0.050 0 M $M^{n+}$ with 0.050 0 M EDTA, assuming $K'_f = 1.15 \times 10^{16}$. The concentration of free $M^{n+}$ decreases as the titration proceeds.

### Region 3: After the Equivalence Point

Now there is excess EDTA, and virtually all the metal ion is in the form $MY^{n-4}$. The concentration of free EDTA can be equated to the concentration of excess EDTA added after the equivalence point.

### Titration Calculations

Let's calculate the shape of the titration curve for the reaction of 50.0 mL of 0.040 0 M $Ca^{2+}$ (buffered to pH 10.00) with 0.080 0 M EDTA:

$$Ca^{2+} + EDTA \rightarrow CaY^{2-}$$

The value of $\alpha_{Y^{4-}}$ comes from Table 12-1.

$$K'_f = \alpha_{Y^{4-}} K_f = (0.30)(10^{10.65}) = 1.3_4 \times 10^{10}$$

Because $K'_f$ is large, it is reasonable to say that the reaction goes to completion with each addition of titrant. We want to make a graph in which $pCa^{2+} (= -\log[Ca^{2+}])$ is plotted versus milliliters of added EDTA. The equivalence volume is 25.0 mL.

### Region 1: Before the Equivalence Point

Before the equivalence point, there is excess unreacted $Ca^{2+}$.

Consider the addition of 5.0 mL of EDTA. Because the equivalence point requires 25.0 mL of EDTA, one-fifth of the $Ca^{2+}$ will be consumed and four-fifths remains.

$$[Ca^{2+}] = \underbrace{\left(\frac{25.0 - 5.0}{25.0}\right)}_{\substack{\text{Fraction} \\ \text{remaining} \\ (= 4/5)}} \underbrace{(0.040\ 0)}_{\substack{\text{Original} \\ \text{concentration} \\ \text{of } Ca^{2+}}} \underbrace{\left(\frac{50.0}{55.0}\right)}_{\substack{\text{Dilution} \\ \text{factor}}}$$

(Initial volume of $Ca^{2+}$ / Total volume of solution)

$$= 0.029\ 1\ M \Rightarrow pCa^{2+} = -\log[Ca^{2+}] = 1.54$$

In a similar manner, we could calculate $pCa^{2+}$ for any volume of EDTA less than 25.0 mL.

### Region 2: At the Equivalence Point

At the equivalence point, the major species is $CaY^{2-}$, in equilibrium with small, equal amounts of free $Ca^{2+}$ and EDTA.

Virtually all the metal is in the form $CaY^{2-}$. Assuming negligible dissociation, we find the concentration of $CaY^{2-}$ to be equal to the original concentration of $Ca^{2+}$, with a correction for dilution.

$$[CaY^{2-}] = \underbrace{(0.040\ 0\ M)}_{\substack{\text{Original} \\ \text{concentration} \\ \text{of } Ca^{2+}}} \underbrace{\left(\frac{50.0}{75.0}\right)}_{\substack{\text{Dilution} \\ \text{factor}}} = 0.026\ 7\ M$$

(Initial volume of $Ca^{2+}$ / Total volume of solution)

The concentration of free $Ca^{2+}$ is small and unknown. We can write

|  | $Ca^{2+}$ | + | EDTA | $\rightleftharpoons$ | $CaY^{2-}$ |
|---|---|---|---|---|---|
| Initial concentration (M) | — | | — | | 0.026 7 |
| Final concentration (M) | $x$ | | $x$ | | $0.026\ 7 - x$ |

[EDTA] refers to the total concentration of all forms of EDTA not bound to metal.

$$\frac{[CaY^{2-}]}{[Ca^{2+}][EDTA]} = K'_f = 1.3_4 \times 10^{10}$$

$$\frac{0.026\ 7 - x}{x^2} = 1.3_4 \times 10^{10} \Rightarrow x = 1.4 \times 10^{-6}\ M$$

$$pCa^{2+} = -\log x = 5.85$$

### Region 3: After the Equivalence Point

After the equivalence point, virtually all the metal is present as $CaY^{2-}$. There is a known excess of EDTA present. A small amount of free $Ca^{2+}$ exists in equilibrium with the $CaY^{2-}$ and EDTA.

In this region, virtually all the metal is in the form $CaY^{2-}$, and there is excess, unreacted EDTA. The concentrations of $CaY^{2-}$ and excess EDTA are easily calculated. For example, at 26.0 mL, there is 1.0 mL of excess EDTA.

$$[\text{EDTA}] = (0.080\ 0)\left(\frac{1.0}{76.0}\right) = 1.05 \times 10^{-3}\ \text{M}$$

Volume of excess EDTA

Original concentration of EDTA — Dilution factor — Total volume of solution

$$[\text{CaY}^{2-}] = (0.040\ 0)\left(\frac{50.0}{76.0}\right) = 2.63 \times 10^{-2}\ \text{M}$$

Original volume of $Ca^{2+}$

Original concentration of $Ca^{2+}$ — Dilution factor — Total volume of solution

The concentration of $Ca^{2+}$ is governed by

$$\frac{[\text{CaY}^{2-}]}{[\text{Ca}^{2+}][\text{EDTA}]} = K_f' = 1.3_4 \times 10^{10}$$

$$\frac{[2.63 \times 10^{-2}]}{[\text{Ca}^{2+}](1.05 \times 10^{-3})} = 1.3_4 \times 10^{10}$$

$$[\text{Ca}^{2+}] = 1.9 \times 10^{-9}\ \text{M} \Rightarrow p\text{Ca}^{2+} = 8.73$$

The same sort of calculation can be used for any volume past the equivalence point.

### The Titration Curve

Calculated titration curves for $Ca^{2+}$ and $Sr^{2+}$ in Figure 12-11 show a distinct break at the equivalence point, where the slope is greatest. The $Ca^{2+}$ end point is more distinct than the $Sr^{2+}$ end point because the conditional formation constant, $\alpha_{Y^{4-}} K_f$, for $CaY^{2-}$ is greater than that of $SrY^{2-}$. If the pH is lowered, the conditional formation constant decreases (because $\alpha_{Y^{4-}}$ decreases), and the end point becomes less distinct, as we saw in Figure 12-9. The pH cannot be raised arbitrarily high because metal hydroxide might precipitate.

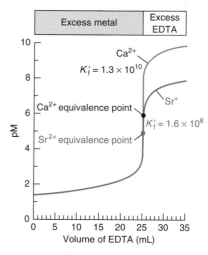

**Figure 12-11** Theoretical titration curves for the reaction of 50.0 mL of 0.040 0 M metal ion with 0.080 0 M EDTA at pH 10.00.

### ▪▪▪▪ 📑 12-4  Do It with a Spreadsheet

Let's see how to reproduce the EDTA titration curves in Figure 12-11 by using one equation that applies to the entire titration. Because the reactions are carried out at fixed pH, the equilibria and mass balances are sufficient to solve for all unknowns.

Consider the titration of metal ion M (concentration = $C_m$, initial volume = $V_m$) with a solution of ligand L (concentration = $C_l$, volume added = $V_l$) to form a 1:1 complex:

$$\text{M} + \text{L} \rightleftharpoons \text{ML} \qquad K_f = \frac{[\text{ML}]}{[\text{M}][\text{L}]} \Rightarrow [\text{ML}] = K_f[\text{M}][\text{L}] \qquad (12\text{-}8)$$

The mass balances for metal and ligand are

Mass balance for M:  $$[\text{M}] + [\text{ML}] = \frac{C_m V_m}{V_m + V_l}$$

Mass balance for L:  $$[\text{L}] + [\text{ML}] = \frac{C_l V_l}{V_m + V_l}$$

Substituting $K_f[\text{M}][\text{L}]$ (from Equation 12-8) for [ML] in the mass balances gives

$$[\text{M}](1 + K_f[\text{L}]) = \frac{C_m V_m}{V_m + V_l} \qquad (12\text{-}9)$$

$$[\text{L}](1 + K_f[\text{M}]) = \frac{C_l V_l}{V_m + V_l} \Rightarrow [\text{L}] = \frac{\dfrac{C_l V_l}{V_m + V_l}}{1 + K_f[\text{M}]} \qquad (12\text{-}10)$$

Now substitute the expression for [L] from Equation 12-10 back into Equation 12-9

$$[\text{M}]\left(1 + K_f \frac{\dfrac{C_l V_l}{V_m + V_l}}{1 + K_f[\text{M}]}\right) = \frac{C_m V_m}{V_m + V_l}$$

Total metal concentration

$$= \frac{\text{initial moles of metal}}{\text{total volume}}$$

$$= \frac{C_m V_m}{V_m + V_l}$$

Total ligand concentration

$$= \frac{\text{moles of ligand added}}{\text{total volume}}$$

$$= \frac{C_l V_l}{V_m + V_l}$$

**Figure 12-12** Spreadsheet for the titration of 50.0 mL of 0.040 0 M $Mg^{2+}$ with 0.080 0 M EDTA at pH 10.00. This spreadsheet reproduces calculations of Section 12-3. pM was varied by trial-and-error to find volumes of 5.00, 25.00, and 26.00 mL used in the preceding section. Better yet, use GOAL SEEK (page 115) to vary pM in cell B9 until the volume in cell E9 is 25.000 mL.

|    | A | B | C | D | E |
|----|---|---|---|---|---|
| 1  | Titration of 50 mL of 0.04 M $Ca^{2+}$ with 0.08 M EDTA | | | | |
| 2  | | | | | |
| 3  | $C_m$ = | pM | M | Phi | V(ligand) |
| 4  | 0.04 | 1.398 | 4.00E-02 | 0.000 | 0.002 |
| 5  | $V_m$ = | 1.537 | 2.90E-02 | 0.201 | 5.026 |
| 6  | 50 | 2.00 | 1.00E-02 | 0.667 | 16.667 |
| 7  | C(ligand) = | 3.00 | 1.00E-03 | 0.963 | 24.074 |
| 8  | 0.08 | 4.00 | 1.00E-04 | 0.996 | 24.906 |
| 9  | $K_f'$ = | 5.85 | 1.41E-06 | 1.000 | 25.0000 |
| 10 | 1.34E+10 | 7.00 | 1.00E-07 | 1.001 | 25.019 |
| 11 | | 8.00 | 1.00E-08 | 1.007 | 25.187 |
| 12 | | 8.73 | 1.86E-09 | 1.040 | 26.002 |
| 13 | C4 = 10^-B4 | | | | |
| 14 | Equation 12-11: | | | | |
| 15 | D4 = (1+$A$10*C4-(C4+C4*C4*$A$10)/$A$4)/ | | | | |
| 16 | (C4*$A$10+(C4+C4*C4*$A$10)/$A$8) | | | | |
| 17 | E4 = D4*$A$4*$A$6/$A$8 | | | | |

and do about five lines of algebra to solve for the fraction of titration $\phi$:

*Spreadsheet equation for titration of M with L:*

$$\phi = \frac{C_1 V_1}{C_m V_m} = \frac{1 + K_f[M] - \dfrac{[M] + K_f[M]^2}{C_m}}{K_f[M] + \dfrac{[M] + K_f[M]^2}{C_1}} \quad (12\text{-}11)$$

Replace $K_f$ by $K_f'$ if L = EDTA.

As in acid-base titrations in Table 11-6, $\phi$ is the fraction of the way to the equivalence point. When $\phi = 1$, $V_1 = V_e$. When $\phi = \frac{1}{2}$, $V_1 = \frac{1}{2}V_e$. And so on.

For a titration with EDTA, you can follow the derivation through and find that the formation constant, $K_f$, should be replaced in Equation 12-11 by the conditional formation constant, $K_f'$, which applies at the fixed pH of the titration. Figure 12-12 shows a spreadsheet in which Equation 12-11 is used to calculate the $Ca^{2+}$ titration curve in Figure 12-11. As in acid-base titrations, your input in column B is pM and the output in column E is volume of titrant. To find the initial point, vary pM until $V_1$ is close to 0.

If you reverse the process and titrate ligand with metal ion, the fraction of the way to the equivalence point is the inverse of the fraction in Equation 12-11:

*Spreadsheet equation for titration of L with M:*

$$\phi = \frac{C_m V_m}{C_1 V_1} = \frac{K_f[M] + \dfrac{[M] + K_f[M]^2}{C_1}}{1 + K_f[M] - \dfrac{[M] + K_f[M]^2}{C_m}} \quad (12\text{-}12)$$

Replace $K_f$ by $K_f'$ if L = EDTA.

## ■ ■ ■ ■ 12-5 Auxiliary Complexing Agents

EDTA titrations in this chapter were selected such that the metal hydroxide would not precipitate at the chosen pH. To permit many metals to be titrated in alkaline solutions with EDTA, we use an **auxiliary complexing agent.** This is a ligand, such as ammonia, tartrate, citrate, or triethanolamine, that binds the metal strongly enough to prevent metal hydroxide from precipitating but weakly enough to give up the metal when EDTA is added. $Zn^{2+}$ is usually titrated in ammonia buffer, which fixes the pH and complexes the metal ion to keep it in solution. Let's see how this works.

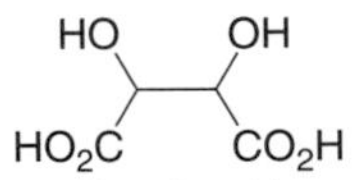

### Metal-Ligand Equilibria[16]

Consider a metal ion that forms two complexes with the auxiliary complexing ligand L:

$$M + L \rightleftharpoons ML \qquad \beta_1 = \frac{[ML]}{[M][L]} \quad (12\text{-}13)$$

$$M + 2L \rightleftharpoons ML_2 \qquad \beta_2 = \frac{[ML_2]}{[M][L]^2} \quad (12\text{-}14)$$

Tartaric acid

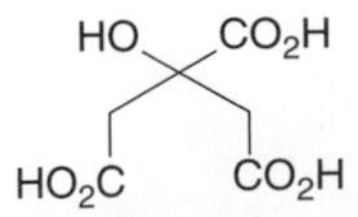

Citric acid

$N(CH_2CH_2OH)_3$

Triethanolamine

The equilibrium constants, $\beta_i$, are called *overall* or **cumulative formation constants.** The fraction of metal ion in the uncomplexed state, M, can be expressed as

$$\alpha_M = \frac{[M]}{C_M} \qquad (12\text{-}15)$$

where $C_M$ refers to the total concentration of all forms of M (= M, ML, and $ML_2$ in this case).

Now let's find a useful expression for $\alpha_M$. The mass balance for metal is

$$C_M = [M] + [ML] + [ML_2]$$

Equations 12-13 and 12-14 allow us to say $[ML] = \beta_1[M][L]$ and $[ML_2] = \beta_2[M][L]^2$. Therefore,

$$C_M = [M] + \beta_1[M][L] + \beta_2[M][L]^2$$
$$= [M]\{1 + \beta_1[L] + \beta_2[L]^2\}$$

Substituting this last result into Equation 12-15 gives the desired result:

*Fraction of free*  
*metal ion:*  $\alpha_M = \dfrac{[M]}{[M]\{1 + \beta_1[L] + \beta_2[L]^2\}} = \dfrac{1}{1 + \beta_1[L] + \beta_2[L]^2} \qquad (12\text{-}16)$

> Overall ($\beta$) and stepwise ($K$) formation constants were distinguished in Box 6-2. The relation between $\beta_n$ and the stepwise formation constants is
>
> $\beta_n = K_1 K_2 K_3 \cdots K_n$

> If the metal forms more than two complexes, Equation 12-16 takes the form
>
> $$\alpha_M = \frac{1}{1 + \beta_1[L] + \beta_2[L]^2 + \cdots + \beta_n[L]^n}$$

### Example  Ammonia Complexes of Zinc

$Zn^{2+}$ and $NH_3$ form the complexes $Zn(NH_3)^{2+}$, $Zn(NH_3)_2^{2+}$, $Zn(NH_3)_3^{2+}$, and $Zn(NH_3)_4^{2+}$. If the concentration of free, *unprotonated* $NH_3$ is 0.10 M, find the fraction of zinc in the form $Zn^{2+}$. (At any pH, there will also be some $NH_4^+$ in equilibrium with $NH_3$.)

**Solution**  Appendix I gives formation constants for the complexes $Zn(NH_3)^{2+}$ ($\beta_1 = 10^{2.18}$), $Zn(NH_3)_2^{2+}$ ($\beta_2 = 10^{4.43}$), $Zn(NH_3)_3^{2+}$ ($\beta_3 = 10^{6.74}$), and $Zn(NH_3)_4^{2+}$ ($\beta_4 = 10^{8.70}$). The appropriate form of Equation 12-16 is

$$\alpha_{Zn^{2+}} = \frac{1}{1 + \beta_1[L] + \beta_2[L]^2 + \beta_3[L]^3 + \beta_4[L]^4} \qquad (12\text{-}17)$$

Equation 12-17 gives the fraction of zinc in the form $Zn^{2+}$. Putting in [L] = 0.10 M and the four values of $\beta_i$ gives $\alpha_{Zn^{2+}} = 1.8 \times 10^{-5}$, which means there is very little free $Zn^{2+}$ in the presence of 0.10 M $NH_3$.

### EDTA Titration with an Auxiliary Complexing Agent

Now consider a titration of $Zn^{2+}$ by EDTA in the presence of $NH_3$. The extension of Equation 12-7 requires a new conditional formation constant to account for the fact that only some of the EDTA is in the form $Y^{4-}$ and only some of the zinc not bound to EDTA is in the form $Zn^{2+}$:

$$K_f'' = \alpha_{Zn^{2+}} \alpha_{Y^{4-}} K_f \qquad (12\text{-}18)$$

In this expression, $\alpha_{Zn^{2+}}$ is given by Equation 12-17 and $\alpha_{Y^{4-}}$ is given by Equation 12-4. For particular values of pH and $[NH_3]$, we can compute $K_f''$ and proceed with titration calculations as in Section 12-3, substituting $K_f''$ for $K_f'$. An assumption in this process is that EDTA is a much stronger complexing agent than ammonia, so essentially all the EDTA binds $Zn^{2+}$ until the metal ion is consumed.

> $K_f''$ is the effective formation constant at a fixed pH and fixed concentration of auxiliary complexing agent. Box 12-2 describes the influence of metal ion hydrolysis on the effective formation constant.

### Example  EDTA Titration in the Presence of Ammonia

Consider the titration of 50.0 mL of $1.00 \times 10^{-3}$ M $Zn^{2+}$ with $1.00 \times 10^{-3}$ M EDTA at pH 10.00 in the presence of 0.10 M $NH_3$. (This is the concentration of $NH_3$. There is also $NH_4^+$ in the solution.) The equivalence point is at 50.0 mL. Find $pZn^{2+}$ after addition of 20.0, 50.0, and 60.0 mL of EDTA.

**Solution**  In Equation 12-17, we found that $\alpha_{Zn^{2+}} = 1.8 \times 10^{-5}$. Table 12-1 tells us that $\alpha_{Y^{4-}} = 0.30$. Therefore, the conditional formation constant is

$$K_f'' = \alpha_{Zn^{2+}} \alpha_{Y^{4-}} K_f = (1.8 \times 10^{-5})(0.30)(10^{16.5}) = 1.7 \times 10^{11}$$

**(a)** *Before the equivalence point—20.0 mL:* Because the equivalence point is 50.0 mL, the fraction of $Zn^{2+}$ remaining is 30.0/50.0. The dilution factor is 50.0/70.0. Therefore, the concentration of zinc not bound to EDTA is

$$C_{Zn^{2+}} = \left(\frac{30.0}{50.0}\right)(1.00 \times 10^{-3}\,\text{M})\left(\frac{50.0}{70.0}\right) = 4.3 \times 10^{-4}\,\text{M}$$

Equation 12-18 states that the effective (conditional) formation constant for an EDTA complex is the product of the formation constant, $K_f$, times the fraction of metal in the form $M^{m+}$ times the fraction of EDTA in the form $Y^{4-}$: $K_f'' = \alpha_{M^{m+}} \alpha_{Y^{4-}} K_f$. Table 12-1 told us that $\alpha_{Y^{4-}}$ increases with pH until it levels off at 1 near pH 11.

In Section 12-3 we had no auxiliary complexing ligand and we implicitly assumed that $\alpha_{M^{m+}} = 1$. In fact, metal ions react with water to form $M(OH)_n$ species. Combinations of pH and metal ion in Section 12-3 were selected so that hydrolysis to $M(OH)_n$ is negligible. We can find such conditions for most $M^{2+}$ ions, but not for $M^{3+}$ or $M^{4+}$. Even in acidic solution, $Fe^{3+}$ hydrolyzes to $Fe(OH)^{2+}$ and $Fe(OH)_2^+$.[17] (Appendix I gives formation constants for hydroxide complexes.) The graph shows that $\alpha_{Fe^{3+}}$ is close to 1 between pH 1 and 2 (log $\alpha_{Fe^{3+}} \approx 0$), but then drops as hydrolysis occurs. At pH 5, the fraction of Fe(III) in the form $Fe^{3+}$ is $\sim 10^{-5}$.

The effective formation constant for $FeY^-$ in the graph has three contributions:

$$K_f''' = \frac{\alpha_{Fe^{3+}} \alpha_{Y^{4-}}}{\alpha_{FeY^-}} K_f$$

As pH increases, $\alpha_{Y^{4-}}$ increases, which increases $K_f'''$. As pH increases, metal hydrolysis occurs, so $\alpha_{Fe^{3+}}$ decreases. The increase in $\alpha_{Y^{4-}}$ is canceled by the decrease in $\alpha_{Fe^{3+}}$, so $K_f'''$ is nearly constant above pH 3. The third contribution to $K_f'''$ is $\alpha_{FeY^-}$, which is the fraction of the EDTA complex in the form $FeY^-$. At low pH, some of the complex gains a proton to form FeHY, which decreases $\alpha_{FeY^-}$ near pH 1. In the pH range 2 to 5, $\alpha_{FeY^-}$ is nearly constant at 1. In neutral and basic solution, complexes such as $Fe(OH)Y^{2-}$ and $[Fe(OH)Y]_2^{4-}$ are formed and $\alpha_{FeY^-}$ decreases.

*Take-home message:* In this book, we restrict ourselves to cases in which there is no hydrolysis and $\alpha_{M^{m+}}$ is controlled by a deliberately added auxiliary ligand. In reality, hydrolysis of $M^{m+}$ and MY influences most EDTA titrations and makes the theoretical analysis more complicated than we pretend in this chapter.

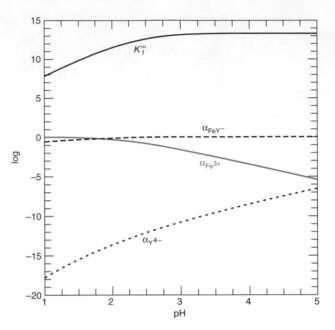

Contributions of $\alpha_{Y^{4-}}$, $\alpha_{Fe^{3+}}$, and $\alpha_{FeY^-}$ to the effective formation constant, $K_f'''$, for $FeY^-$. Curves were calculated by considering the species $H_6Y^{2+}$, $H_5Y^+$, $H_4Y$, $H_3Y^-$, $H_2Y^{2-}$, $HY^{3-}$, $Y^{4-}$, $Fe^{3+}$, $Fe(OH)^{2+}$, $Fe(OH)_2^+$, $FeY^-$, and FeHY.

---

The relation $[Zn^{2+}] = \alpha_{Zn^{2+}} C_{Zn^{2+}}$ follows from Equation 12-15.

However, nearly all zinc not bound to EDTA is bound to $NH_3$. The concentration of free $Zn^{2+}$ is

$$[Zn^{2+}] = \alpha_{Zn^{2+}} C_{Zn^{2+}} = (1.8 \times 10^{-5})(4.3 \times 10^{-4}) = 7.7 \times 10^{-9} \text{ M}$$

$$\Rightarrow pZn^{2+} = -\log[Zn^{2+}] = 8.11$$

Let's try a reality check. The product $[Zn^{2+}][OH^-]^2$ is $[10^{-8.11}][10^{-4.00}]^2 = 10^{-16.11}$, which does not exceed the solubility product of $Zn(OH)_2$ ($K_{sp} = 10^{-15.52}$).

**(b)** *At the equivalence point—50.0 mL:* At the equivalence point, the dilution factor is 50.0/100.0, so $[ZnY^{2-}] = (50.0/100.0)(1.00 \times 10^{-3} \text{ M}) = 5.00 \times 10^{-4}$ M. As in Section 12-3, we write

| | $C_{Zn^{2+}}$ | + | EDTA | $\rightleftharpoons$ | $ZnY^{2-}$ |
|---|---|---|---|---|---|
| Initial concentration (M) | 0 | | 0 | | $5.00 \times 10^{-4}$ |
| Final concentration (M) | $x$ | | $x$ | | $5.00 \times 10^{-4} - x$ |

$$K_f'' = 1.7 \times 10^{11} = \frac{[ZnY^{2-}]}{[C_{Zn^{2+}}][\text{EDTA}]} = \frac{5.00 \times 10^{-4} - x}{x^2}$$

$$\Rightarrow x = C_{Zn^{2+}} = 5.4 \times 10^{-8} \text{ M}$$

$$[Zn^{2+}] = \alpha_{Zn^{2+}} C_{Zn^{2+}} = (1.8 \times 10^{-5})(5.4 \times 10^{-8}) = 9.7 \times 10^{-13} \text{ M}$$

$$\Rightarrow pZn^{2+} = -\log[Zn^{2+}] = 12.01$$

**(c)** *After the equivalence point—60.0 mL:* Almost all zinc is in the form $ZnY^{2-}$. With a dilution factor of 50.0/110.0 for zinc, we find

$$[ZnY^{2-}] = \left(\frac{50.0}{110.0}\right)(1.00 \times 10^{-3} \text{ M}) = 4.5 \times 10^{-4} \text{ M}$$

We also know the concentration of excess EDTA, whose dilution factor is 10.0/110.0:

$$[\text{EDTA}] = \left(\frac{10.0}{110.0}\right)(1.00 \times 10^{-3}\,\text{M}) = 9.1 \times 10^{-5}\,\text{M}$$

Once we know $[\text{ZnY}^{2-}]$ and $[\text{EDTA}]$, we can use the equilibrium constant to find $[\text{Zn}^{2+}]$:

$$\frac{[\text{ZnY}^{2-}]}{[\text{Zn}^{2+}][\text{EDTA}]} = \alpha_{\text{Y}^{4-}}\,K_f = K_f' = (0.30)(10^{16.5}) = 9.5 \times 10^{15}$$

$$\frac{[4.5 \times 10^{-4}]}{[\text{Zn}^{2+}][9.1 \times 10^{-5}]} = 9.5 \times 10^{15} \Rightarrow [\text{Zn}^{2+}] = 5.2 \times 10^{-16}\,\text{M}$$

$$\Rightarrow p\text{Zn}^{2+} = 15.28$$

Note that past the equivalence point the problem did not depend on the presence of $NH_3$, because we knew the concentrations of both $[\text{ZnY}^{2-}]$ and $[\text{EDTA}]$.

Figure 12-13 compares the calculated titration curves for $Zn^{2+}$ in the presence of different concentrations of auxiliary complexing agent. The greater the concentration of $NH_3$, the smaller the change of $p\text{Zn}^{2+}$ near the equivalence point. The auxiliary ligand must be kept below the level that would obliterate the end point of the titration. Color Plate 6 shows the appearance of a $Cu^{2+}$-ammonia solution during an EDTA titration.

### ■ ■ ■  12-6  Metal Ion Indicators

The most common technique to detect the end point in EDTA titrations is to use a metal ion indicator. Alternatives include a mercury electrode (Figure 12-9 and Exercise 15-B) and an ion-selective electrode (Section 15-6). A pH electrode will follow the course of the titration in unbuffered solution, because $H_2Y^{2-}$ releases $2H^+$ when it forms a metal complex.

**Metal ion indicators** (Table 12-3) are compounds whose color changes when they bind to a metal ion. *Useful indicators must bind metal less strongly than EDTA does.*

A typical titration is illustrated by the reaction of $Mg^{2+}$ with EDTA at pH 10, using Calmagite as the indicator.

$$\underset{\text{Red}}{\text{MgIn}} + \underset{\text{Colorless}}{\text{EDTA}} \rightarrow \underset{\text{Colorless}}{\text{MgEDTA}} + \underset{\text{Blue}}{\text{In}} \qquad (12\text{-}19)$$

At the start of the experiment, a small amount of indicator (In) is added to the colorless solution of $Mg^{2+}$ to form a red complex. As EDTA is added, it reacts first with free, colorless $Mg^{2+}$. When free $Mg^{2+}$ is used up, the last EDTA added before the equivalence point displaces indicator from the red MgIn complex. The change from the red MgIn to blue unbound In signals the end point of the titration (Demonstration 12-1).

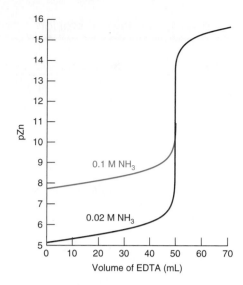

**Figure 12-13** Titration curves for the reaction of 50.0 mL of $1.00 \times 10^{-3}$ M $Zn^{2+}$ with $1.00 \times 10^{-3}$ M EDTA at pH 10.00 in the presence of two different concentrations of $NH_3$.

End-point detection methods:
1. Metal ion indicators
2. Mercury electrode
3. Ion-selective electrode
4. Glass (pH) electrode

The indicator must release its metal to EDTA.

---

### Demonstration 12-1  Metal Ion Indicator Color Changes

This demonstration illustrates the color change associated with Reaction 12-19 and shows how a second dye can be added to produce a more easily detected color change.

STOCK SOLUTIONS

Buffer (pH 10.0): Add 142 mL of concentrated (14.5 M) aqueous ammonia to 17.5 g of ammonium chloride and dilute to 250 mL with water.
$MgCl_2$:  0.05 M
EDTA:   0.05 M $Na_2H_2EDTA \cdot 2H_2O$

Prepare a solution containing 25 mL of $MgCl_2$, 5 mL of buffer, and 300 mL of water. Add six drops of Eriochrome black T or Calmagite indicator (Table 12-3) and titrate with EDTA. Note the color change from wine red to pale blue at the end point (Color Plate 7a). The spectroscopic change accompanying the color change is shown in the figure at the right.

For some observers, the change of indicator color is not as sharp as desired. The colors can be affected by adding an "inert"

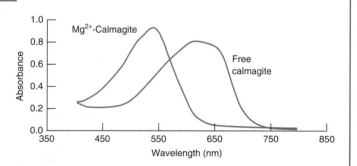

Visible spectra of $Mg^{2+}$-Calmagite and free Calmagite at pH 10 in ammonia buffer. *[From C. E. Dahm, J. W. Hall, and B. E. Mattioni, "A Laser Pointer-Based Spectrometer for Endpoint Detection of EDTA Titrations," J. Chem. Ed.* **2004**, *81, 1787.]*

dye whose color alters the appearance of the solution before and after the titration. Adding 3 mL of methyl red (Table 11-4) (or many other yellow dyes) produces an orange color prior to the end point and a green color after it. This sequence of colors is shown in Color Plate 7b.

**Table 12-3** Common metal ion indicators

| Name | Structure | $pK_a$ | Color of free indicator | Color of metal ion complex |
|---|---|---|---|---|
| Calmagite | (H$_2$In$^-$) | $pK_2 = 8.1$<br>$pK_3 = 12.4$ | H$_2$In$^-$ red<br>HIn$^{2-}$ blue<br>In$^{3-}$ orange | Wine red |
| Eriochrome black T | (H$_2$In$^-$) | $pK_2 = 6.3$<br>$pK_3 = 11.6$ | H$_2$In$^-$ red<br>HIn$^{2-}$ blue<br>In$^{3-}$ orange | Wine red |
| Murexide | (H$_4$In$^-$) | $pK_2 = 9.2$<br>$pK_3 = 10.9$ | H$_4$In$^-$ red-violet<br>H$_3$In$^{2-}$ violet<br>H$_2$In$^{3-}$ blue | Yellow (with Co$^{2+}$, Ni$^{2+}$, Cu$^{2+}$); red with Ca$^{2+}$ |
| Xylenol orange | (H$_3$In$^{3-}$) | $pK_2 = 2.32$<br>$pK_3 = 2.85$<br>$pK_4 = 6.70$<br>$pK_5 = 10.47$<br>$pK_6 = 12.23$ | H$_5$In$^-$ yellow<br>H$_4$In$^{2-}$ yellow<br>H$_3$In$^{3-}$ yellow<br>H$_2$In$^{4-}$ violet<br>HIn$^{5-}$ violet<br>In$^{6-}$ violet | Red |
| Pyrocatechol violet | (H$_3$In$^-$) | $pK_1 = 0.2$<br>$pK_2 = 7.8$<br>$pK_3 = 9.8$<br>$pK_4 = 11.7$ | H$_4$In red<br>H$_3$In$^-$ yellow<br>H$_2$In$^{2-}$ violet<br>HIn$^{3-}$ red-purple | Blue |

*PREPARATION AND STABILITY:*

*Calmagite: 0.05 g/100 mL H$_2$O; solution is stable for a year in the dark.*

*Eriochrome black T: Dissolve 0.1 g of the solid in 7.5 mL of triethanolamine plus 2.5 mL of absolute ethanol; solution is stable for months; best used for titrations above pH 6.5.*

*Murexide: Grind 10 mg of murexide with 5 g of reagent NaCl in a clean mortar; use 0.2–0.4 g of the mixture for each titration.*

*Xylenol orange: 0.5 g/100 mL H$_2$O; solution is stable indefinitely.*

*Pyrocatechol violet: 0.1 g/100 mL; solution is stable for several weeks.*

Most metal ion indicators are also acid-base indicators, with $pK_a$ values listed in Table 12-3. Because the color of free indicator is pH dependent, most indicators can be used only in certain pH ranges. For example, xylenol orange (pronounced ZY-leen-ol) changes from yellow to red when it binds to a metal ion at pH 5.5. This is an easy color change to observe. At pH 7.5, the change is from violet to red and rather difficult to see. A spectrophotometer can measure the color change, but it is more convenient if we can see it. Figure 12-14 shows pH ranges in which many metals can be titrated and indicators that are useful in different ranges.

The indicator must give up its metal ion to the EDTA. If a metal does not freely dissociate from an indicator, the metal is said to **block** the indicator. Eriochrome black T is blocked by Cu$^{2+}$, Ni$^{2+}$, Co$^{2+}$, Cr$^{3+}$, Fe$^{3+}$, and Al$^{3+}$. It cannot be used for the direct titration of any of these metals. It can be used for a back titration, however. For example, excess standard EDTA can be added to Cu$^{2+}$. Then indicator is added and the excess EDTA is back-titrated with Mg$^{2+}$.

**Question** What will the color change be when the back titration is performed?

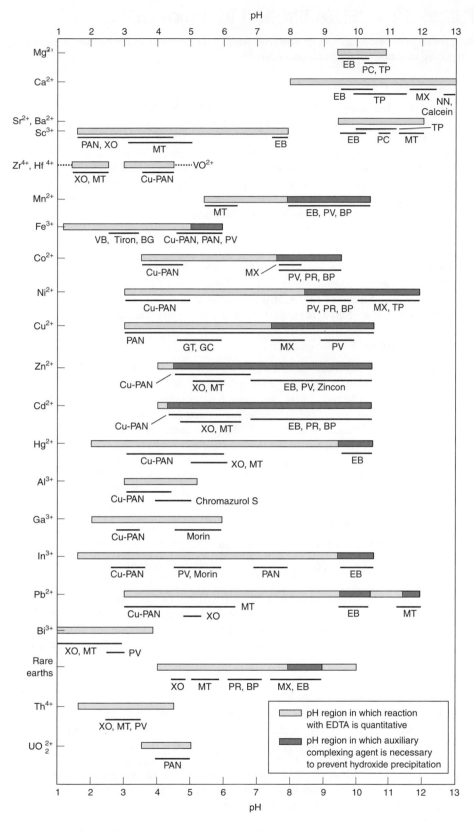

**Figure 12-14** Guide to EDTA titrations of common metals. Light color shows pH region in which reaction with EDTA is quantitative. Dark color shows pH region in which auxiliary complexing agent is required to prevent metal from precipitating. Calmagite is more stable than Eriochrome black T (EB) and can be substituted for EB. *[Adapted from K. Ueno, "Guide for Selecting Conditions of EDTA Titrations," J. Chem. Ed.* **1965**, *42, 432.]*

**Abbreviations for indicators:**

BG, Bindschedler's green leuco base

BP, Bromopyrogallol red

EB, Eriochrome black T

GC, Glycinecresol red

GT, Glycinethymol blue

MT, Methylthymol blue

MX, Murexide

NN, Patton & Reeder's dye

PAN, Pyridylazonaphthol

Cu-PAN, PAN plus Cu-EDTA

PC, o-Cresolphthalein complexone

PR, Pyrogallol red

PV, Pyrocatechol violet

TP, Thymolphthalein complexone

VB, Variamine blue B base

XO, Xylenol orange

## 12-7  EDTA Titration Techniques

Because so many elements can be analyzed with EDTA, there is extensive literature dealing with many variations of the basic procedure.[15,19]

### Direct Titration

In a **direct titration,** analyte is titrated with standard EDTA. The analyte is buffered to a pH at which the conditional formation constant for the metal-EDTA complex is large and the color of the free indicator is distinctly different from that of the metal-indicator complex.

*Auxiliary complexing agents* such as $NH_3$, tartrate, citrate, or triethanolamine may be employed to prevent metal ion from precipitating in the absence of EDTA. For example, $Pb^{2+}$ is titrated in $NH_3$ buffer at pH 10 in the presence of tartrate, which complexes $Pb^{2+}$ and does not allow $Pb(OH)_2$ to precipitate. The lead-tartrate complex must be less stable than the lead-EDTA complex, or the titration would not be feasible.

### Back Titration

In a **back titration,** a known excess of EDTA is added to the analyte. Excess EDTA is then titrated with a standard solution of a second metal ion. A back titration is necessary if analyte precipitates in the absence of EDTA, if it reacts too slowly with EDTA, or if it blocks the indicator. The metal ion for the back titration must not displace analyte from EDTA.

> ### Example  A Back Titration
>
> $Ni^{2+}$ can be analyzed by a back titration by using standard $Zn^{2+}$ at pH 5.5 with xylenol orange indicator. A solution containing 25.00 mL of $Ni^{2+}$ in dilute HCl is treated with 25.00 mL of 0.052 83 M $Na_2EDTA$. The solution is neutralized with NaOH, and the pH is adjusted to 5.5 with acetate buffer. The solution turns yellow when a few drops of indicator are added. Titration with 0.022 99 M $Zn^{2+}$ requires 17.61 mL to reach the red end point. What is the molarity of $Ni^{2+}$ in the unknown?
>
> **Solution**   The unknown was treated with 25.00 mL of 0.052 83 M EDTA, which contains (25.00 mL)(0.052 83 M) = 1.320 8 mmol of EDTA. Back titration required (17.61 mL)(0.022 99 M) = 0.404 9 mmol of $Zn^{2+}$. Because 1 mol of EDTA reacts with 1 mol of any metal ion, there must have been
>
> $$1.320\ 8\ \text{mmol EDTA} - 0.404\ 9\ \text{mmol } Zn^{2+} = 0.915\ 9\ \text{mmol } Ni^{2+}$$
>
> The concentration of $Ni^{2+}$ is 0.915 9 mmol/25.00 mL = 0.036 64 M.

Back titration prevents precipitation of analyte. For example, $Al(OH)_3$ precipitates at pH 7 in the absence of EDTA. An acidic solution of $Al^{3+}$ can be treated with excess EDTA, adjusted to pH 7–8 with sodium acetate, and boiled to ensure complete formation of stable, soluble $Al(EDTA)^-$. The solution is then cooled, Eriochrome black T indicator is added, and back titration with standard $Zn^{2+}$ is performed.

### Displacement Titration

$Hg^{2+}$ does not have a satisfactory indicator, but a **displacement titration** is feasible. $Hg^{2+}$ is treated with excess $Mg(EDTA)^{2-}$ to displace $Mg^{2+}$, which is titrated with standard EDTA.

$$M^{n+} + MgY^{2-} \rightarrow MY^{n-4} + Mg^{2+} \qquad (12\text{-}20)$$

The conditional formation constant for $Hg(EDTA)^{2-}$ must be greater than $K_f'$ for $Mg(EDTA)^{2-}$, or else $Mg^{2+}$ is not displaced from $Mg(EDTA)^{2-}$.

There is no suitable indicator for $Ag^+$. However, $Ag^{2+}$ will displace $Ni^{2+}$ from tetracyanonickelate(II) ion:

$$2Ag^+ + Ni(CN)_4^{2-} \rightarrow 2Ag(CN)_2^- + Ni^{2+}$$

The liberated $Ni^{2+}$ can then be titrated with EDTA to find out how much $Ag^+$ was added.

### Indirect Titration

Anions that precipitate with certain metal ions can be analyzed with EDTA by **indirect titration.** For example, sulfate can be analyzed by precipitation with excess $Ba^{2+}$ at pH 1.

*Phytoremediation.*[18] An approach to removing toxic metals from contaminated soil is to grow plants that accumulate 1–15 g metal/g dry mass of plant. The plant is harvested to recover metals such as Pb, Cd, and Ni. Phytoremediation is greatly enhanced by adding EDTA to mobilize insoluble metals. Unfortunately, rain spreads soluble metal-EDTA complexes through the soil. The natural chelate EDDS mobilizes metal and biodegrades before it can spread very far.

*S,S*-Ethylenediaminedisuccinic acid (EDDS)

*Hardness* is the total concentration of alkaline earth (Group 2) ions, which are mainly $Ca^{2+}$ and $Mg^{2+}$, in water. Hardness is commonly expressed as the equivalent number of milligrams of $CaCO_3$ per liter. Thus, if $[Ca^{2+}] + [Mg^{2+}] = 1$ mM, we would say that the hardness is 100 mg $CaCO_3$ per liter because 100 mg $CaCO_3 = 1$ mmol $CaCO_3$. Water whose hardness is less than 60 mg $CaCO_3$ per liter is considered to be "soft." If the hardness is above 270 mg/L, the water is considered to be "hard."

Hard water reacts with soap to form insoluble curds:

$$Ca^{2+} + 2RCO_2^- \rightarrow Ca(RCO_2)_2(s) \qquad \text{(A)}$$

$$\underset{\text{Soap}}{\phantom{Ca^{2+}}} \qquad \underset{\text{Precipitate}}{\phantom{+}}$$

R is a long-chain hydrocarbon such as $C_{17}H_{35}-$

Enough soap to consume $Ca^{2+}$ and $Mg^{2+}$ must be used before soap is useful for cleaning. Hard water leaves solid deposits called *scale* on pipes when it evaporates. It is not believed that hard water is unhealthy. Hardness is beneficial in irrigation water because alkaline earth ions tend to *flocculate* (cause to aggregate) colloidal particles in soil and thereby increase the permeability of the soil to water. Soft water etches concrete, plaster, and grout.

To measure hardness, the sample is treated with ascorbic acid (or hydroxylamine) to reduce $Fe^{3+}$ to $Fe^{2+}$ and with cyanide to mask $Fe^{2+}$, $Cu^+$, and several other minor metal ions. Titration with EDTA at pH 10 in $NH_3$ buffer then gives the total concentrations of $Ca^{2+}$ and $Mg^{2+}$. $Ca^{2+}$ can be determined separately if the titration is carried out at pH 13 without ammonia. At this pH, $Mg(OH)_2$ precipitates and is inaccessible to EDTA. Interference by many metal ions can be reduced by the right choice of indicators.[21]

Insoluble carbonates are converted into soluble bicarbonates by excess carbon dioxide:

$$CaCO_3(s) + CO_2 + H_2O \rightarrow Ca(HCO_3)_2(aq) \qquad \text{(B)}$$

Heat converts bicarbonate into carbonate (driving off $CO_2$) and precipitates $CaCO_3$ scale that clogs boiler pipes. The fraction of hardness due to $Ca(HCO_3)_2(aq)$ is called *temporary hardness* because this calcium is lost (by precipitation of $CaCO_3$) upon heating. Hardness arising from other salts (mainly dissolved $CaSO_4$) is called *permanent hardness* because it is not removed by heating.

---

The $BaSO_4(s)$ is washed and then boiled with excess EDTA at pH 10 to bring $Ba^{2+}$ back into solution as $Ba(EDTA)^{2-}$. Excess EDTA is back-titrated with $Mg^{2+}$.

Alternatively, an anion can be precipitated with excess metal ion. The precipitate is filtered and washed, and excess metal ion in the filtrate is titrated with EDTA. Anions such as $CO_3^{2-}$, $CrO_4^{2-}$, $S^{2-}$, and $SO_4^{2-}$ can be determined by indirect titration with EDTA.[20]

## Masking

A **masking agent** is a reagent that protects some component of the analyte from reaction with EDTA. For example, $Al^{3+}$ in a mixture of $Mg^{2+}$ and $Al^{3+}$ can be measured by first masking the $Al^{3+}$ with $F^-$, thereby leaving only the $Mg^{2+}$ to react with EDTA.

Cyanide masks $Cd^{2+}$, $Zn^{2+}$, $Hg^{2+}$, $Co^{2+}$, $Cu^+$, $Ag^+$, $Ni^{2+}$, $Pd^{2+}$, $Pt^{2+}$, $Fe^{2+}$, and $Fe^{3+}$, but not $Mg^{2+}$, $Ca^{2+}$, $Mn^{2+}$, or $Pb^{2+}$. When cyanide is added to a solution containing $Cd^{2+}$ and $Pb^{2+}$, only $Pb^{2+}$ reacts with EDTA. (*Caution:* Cyanide forms toxic gaseous HCN below pH 11. Cyanide solutions should be strongly basic and only handled in a hood.) Fluoride masks $Al^{3+}$, $Fe^{3+}$, $Ti^{4+}$, and $Be^{2+}$. (*Caution:* HF formed by $F^-$ in acidic solution is extremely hazardous and should not contact skin and eyes. It may not be immediately painful, but the affected area should be flooded with water and then treated with calcium gluconate gel that you have on hand *before* the accident. First aid providers must wear rubber gloves to protect themselves.) Triethanolamine masks $Al^{3+}$, $Fe^{3+}$, and $Mn^{2+}$; and 2,3-dimercapto-1-propanol masks $Bi^{3+}$, $Cd^{2+}$, $Cu^{2+}$, $Hg^{2+}$, and $Pb^{2+}$.

**Demasking** releases metal ion from a masking agent. Cyanide complexes can be demasked with formaldehyde:

Masking prevents one species from interfering in the analysis of another. Masking is not restricted to EDTA titrations. Box 12-3 gives an important application of masking.

SH
|
$HOCH_2CHCH_2SH$
2,3-Dimercapto-1-propanol

$$M(CN)_m^{n-m} + mH_2CO + mH^+ \rightarrow mH_2C\overset{OH}{\underset{CN}{\diagup\diagdown}} + M^{n+}$$

Formaldehyde

Thiourea masks $Cu^{2+}$ by reducing it to $Cu^+$ and complexing the $Cu^+$. Copper can be liberated from thiourea by oxidation with $H_2O_2$. Selectivity afforded by masking, demasking, and pH control allows individual components of complex mixtures of metal ions to be analyzed by EDTA titration.

---

## Terms to Understand

| | | | |
|---|---|---|---|
| auxiliary complexing agent | complexometric titration | displacement titration | masking agent |
| back titration | conditional formation constant | formation constant | metal ion indicator |
| blocking | cumulative formation constant | indirect titration | monodentate |
| chelate effect | demasking | Lewis acid | multidentate |
| chelating ligand | direct titration | Lewis base | stability constant |

## Summary

In a complexometric titration, analyte and titrant form a complex ion, and the equilibrium constant is called the formation constant, $K_f$. Chelating (multidentate) ligands form more stable complexes than do monodentate ligands, because the entropy of complex formation favors the binding of one large ligand rather than many small ligands. Synthetic aminocarboxylic acids such as EDTA have large metal-binding constants.

Formation constants for EDTA are expressed in terms of $[Y^{4-}]$, even though there are six protonated forms of EDTA. Because the fraction ($\alpha_{Y^{4-}}$) of free EDTA in the form $Y^{4-}$ depends on pH, we define a conditional (or effective) formation constant as $K_f' = \alpha_{Y^{4-}} K_f = [MY^{n-4}]/[M^{n+}][EDTA]$. This constant describes the hypothetical reaction $M^{n+} + EDTA \rightleftharpoons MY^{n-4}$, where EDTA refers to all forms of EDTA not bound to metal ion. Titration calculations fall into three categories. When excess unreacted $M^{n+}$ is present, pM is calculated directly from $pM = -\log[M^{n+}]$. When excess EDTA is present, we know both $[MY^{n-4}]$ and $[EDTA]$, so $[M^{n+}]$ can be calculated from the conditional formation constant. At the equivalence point, the condition $[M^{n+}] = [EDTA]$ allows us to solve for $[M^{n+}]$. A single spreadsheet equation applies in all three regions of the titration curve.

The greater the effective formation constant, the sharper is the EDTA titration curve. Addition of auxiliary complexing agents, which compete with EDTA for the metal ion and thereby limit the sharpness of the titration curve, is often necessary to keep the metal in solution. Calculations for a solution containing EDTA and an auxiliary complexing agent utilize the conditional formation constant $K_f'' = \alpha_M \alpha_{Y^{4-}} K_f$, where $\alpha_M$ is the fraction of free metal ion not complexed by the auxiliary ligand.

For end-point detection, we commonly use metal ion indicators, a glass electrode, an ion-selective electrode, or a mercury electrode. When a direct titration is not suitable, because the analyte is unstable, reacts slowly with EDTA, or has no suitable indicator, a back titration of excess EDTA or a displacement titration of $Mg(EDTA)^{2-}$ may be feasible. Masking prevents interference by unwanted species. Indirect EDTA titrations are available for many anions and other species that do not react directly with the reagent.

## Exercises

**12-A.** Potassium ion in a 250.0 ($\pm 0.1$)-mL water sample was precipitated with sodium tetraphenylborate:

$$K^+ + (C_6H_5)_4B^- \rightarrow KB(C_6H_5)_4(s)$$

The precipitate was filtered, washed, dissolved in an organic solvent, and treated with excess $Hg(EDTA)^{2-}$:

$$4HgY^{2-} + (C_6H_5)_4B^- + 4H_2O \rightarrow$$
$$H_3BO_3 + 4C_6H_5Hg^+ + 4HY^{3-} + OH^-$$

The liberated EDTA was titrated with 28.73 ($\pm 0.03$) mL of 0.043 7 ($\pm 0.000\ 1$) M $Zn^{2+}$. Find $[K^+]$ (and its absolute uncertainty) in the original sample.

**12-B.** A 25.00-mL sample containing $Fe^{3+}$ and $Cu^{2+}$ required 16.06 mL of 0.050 83 M EDTA for complete titration. A 50.00-mL sample of the unknown was treated with $NH_4F$ to protect the $Fe^{3+}$. Then $Cu^{2+}$ was reduced and masked by thiourea. Addition of 25.00 mL of 0.050 83 M EDTA liberated $Fe^{3+}$ from its fluoride complex to form an EDTA complex. The excess EDTA required 19.77 mL of 0.018 83 M $Pb^{2+}$ to reach a xylenol orange end point. Find $[Cu^{2+}]$ in the unknown.

**12-C.** Calculate $pCu^{2+}$ (to the 0.01 decimal place) at each of the following points in the titration of 50.0 mL of 0.040 0 M EDTA with 0.080 0 M $Cu(NO_3)_2$ at pH 5.00: 0.1, 5.0, 10.0, 15.0, 20.0, 24.0, 25.0, 26.0, and 30.0 mL. Make a graph of $pCu^{2+}$ versus volume of titrant.

**12-D.** Calculate the concentration of $H_2Y^{2-}$ at the equivalence point in Exercise 12-C.

**12-E.** Suppose that 0.010 0 M $Mn^{2+}$ is titrated with 0.005 00 M EDTA at pH 7.00.
(a) What is the concentration of free $Mn^{2+}$ at the equivalence point?
(b) What is the quotient $[H_3Y^-]/[H_2Y^{2-}]$ in the solution when the titration is just 63.7% of the way to the equivalence point?

**12-F.** A solution containing 20.0 mL of $1.00 \times 10^{-3}$ M $Co^{2+}$ in the presence of 0.10 M $C_2O_4^{2-}$ at pH 9.00 was titrated with $1.00 \times 10^{-2}$ M EDTA. Using formation constants from Appendix I, calculate $pCo^{2+}$ for the following volumes of added EDTA: 0, 1.00, 2.00, and 3.00 mL. Consider the concentration of $C_2O_4^{2-}$ to be fixed at 0.10 M. Sketch a graph of $pCo^{2+}$ versus milliliters of added EDTA.

**12-G.** Iminodiacetic acid forms 2:1 complexes with many metal ions:

$$\underset{\text{CH}_2\text{CO}_2\text{H}}{\overset{\text{CH}_2\text{CO}_2\text{H}}{\text{H}_2\overset{+}{\text{N}}}} \equiv H_3X^+$$

$$\alpha_{X^{2-}} = \frac{[X^{2-}]}{[H_3X^+] + [H_2X] + [HX^-] + [X^{2-}]}$$
$$Cu^{2+} + 2X^{2-} \rightleftharpoons CuX_2^{2-} \qquad K = 3.5 \times 10^{16}$$

A solution of volume 25.0 mL containing 0.120 M iminodiacetic acid buffered to pH 7.00 was titrated with 25.0 mL of 0.050 0 M $Cu^{2+}$. Given that $\alpha_{X^{2-}} = 4.6 \times 10^{-3}$ at pH 7.00, calculate the concentration of $Cu^{2+}$ in the resulting solution.

## Problems

### EDTA

**12-1.** What is the chelate effect?

**12-2.** State (in words) what $\alpha_{Y^{4-}}$ means. Calculate $\alpha_{Y^{4-}}$ for EDTA at (a) pH 3.50 and (b) pH 10.50.

**12-3.** (a) Find the conditional formation constant for $Mg(EDTA)^{2-}$ at pH 9.00.
(b) Find the concentration of free $Mg^{2+}$ in 0.050 M $Na_2[Mg(EDTA)]$ at pH 9.00.

**12-4.** *Metal ion buffers.* By analogy to a hydrogen ion buffer, a metal ion buffer tends to maintain a particular metal ion concentration in solution. A mixture of the acid HA and its conjugate base $A^-$ maintains $[H^+]$ defined by the equation $K_a = [A^-][H^+]/[HA]$. A mixture of $CaY^{2-}$ and $Y^{4-}$ serves as a $Ca^{2+}$ buffer governed by the equation $1/K_f' = [EDTA][Ca^{2+}]/[CaY^{2-}]$. How many grams of $Na_2EDTA \cdot 2H_2O$ (FM 372.23) should be mixed with 1.95 g of $Ca(NO_3)_2 \cdot 2H_2O$ (FM 200.12) in a 500-mL volumetric flask to give a buffer with $pCa^{2+} = 9.00$ at pH 9.00?

## EDTA Titration Curves

**12-5.** The ion $M^{n+}$ (100.0 mL of 0.050 0 M metal ion buffered to pH 9.00) was titrated with 0.050 0 M EDTA.
(a) What is the equivalence volume, $V_e$, in milliliters?
(b) Calculate the concentration of $M^{n+}$ at $V = \frac{1}{2}V_e$.
(c) What fraction $(\alpha_{Y^{4-}})$ of free EDTA is in the form $Y^{4-}$ at pH 9.00?
(d) The formation constant $(K_f)$ is $10^{12.00}$. Calculate the value of the conditional formation constant $K_f'\,(= \alpha_{Y^{4-}} K_f)$.
(e) Calculate the concentration of $M^{n+}$ at $V = V_e$.
(f) What is the concentration of $M^{n+}$ at $V = 1.100\,V_e$?

**12-6.** Calculate $pCo^{2+}$ at each of the following points in the titration of 25.00 mL of 0.020 26 M $Co^{2+}$ by 0.038 55 M EDTA at pH 6.00:
(a) 12.00 mL; (b) $V_e$; (c) 14.00 mL.

**12-7.** Consider the titration of 25.0 mL of 0.020 0 M $MnSO_4$ with 0.010 0 M EDTA in a solution buffered to pH 8.00. Calculate $pMn^{2+}$ at the following volumes of added EDTA and sketch the titration curve:
(a) 0 mL        (d) 49.0 mL        (g) 50.1 mL
(b) 20.0 mL     (e) 49.9 mL        (h) 55.0 mL
(c) 40.0 mL     (f) 50.0 mL        (i) 60.0 mL

**12-8.** Using the same volumes as in Problem 12-7, calculate $pCa^{2+}$ for the titration of 25.00 mL of 0.020 00 M EDTA with 0.010 00 M $CaSO_4$ at pH 10.00.

**12-9.** Calculate the molarity of $HY^{3-}$ in a solution prepared by mixing 10.00 mL of 0.010 0 M $VOSO_4$, 9.90 mL of 0.010 0 M EDTA, and 10.0 mL of buffer with a pH of 4.00.

**12-10.** 🖳 *Titration of metal ion with EDTA.* Use Equation 12-11 to compute curves (pM versus mL of EDTA added) for the titration of 10.00 mL of 1.00 mM $M^{2+}$ ($= Cd^{2+}$ or $Cu^{2+}$) with 10.0 mM EDTA at pH 5.00. Plot both curves on one graph.

**12-11.** 🖳 *Effect of pH on the EDTA titration.* Use Equation 12-11 to compute curves ($pCa^{2+}$ versus mL of EDTA added) for the titration of 10.00 mL of 1.00 mM $Ca^{2+}$ with 1.00 mM EDTA at pH 5.00, 6.00, 7.00, 8.00, and 9.00. Plot all curves on one graph and compare your results with Figure 12-9.

**12-12.** 🖳 *Titration of EDTA with metal ion.* Use Equation 12-12 to reproduce the results of Exercise 12-C.

### Auxiliary Complexing Agents

**12-13.** State the purpose of an auxiliary complexing agent and give an example of its use.

**12-14.** According to Appendix I, $Cu^{2+}$ forms two complexes with acetate:
$$Cu^{2+} + CH_3CO_2^- \rightleftharpoons Cu(CH_3CO_2)^+ \qquad \beta_1\,(= K_1)$$
$$Cu^{2+} + 2CH_3CO_2^- \rightleftharpoons Cu(CH_3CO_2)_2(aq) \qquad \beta_2$$
(a) Referring to Box 6-2, find $K_2$ for the reaction
$$Cu(CH_3CO_2)^+ + CH_3CO_2^- \rightleftharpoons Cu(CH_3CO_2)_2(aq) \quad K_2$$
(b) Consider 1.00 L of solution prepared by mixing $1.00 \times 10^{-4}$ mol $Cu(ClO_4)_2$ and 0.100 mol $CH_3CO_2Na$. Use Equation 12-16 to find the fraction of copper in the form $Cu^{2+}$.

**12-15.** Calculate $pCu^{2+}$ at each of the following points in the titration of 50.00 mL of 0.001 00 M $Cu^{2+}$ with 0.001 00 M EDTA at pH 11.00 in a solution whose $NH_3$ concentration is somehow *fixed* at 0.100 M:
(a) 0 mL        (c) 45.00 mL        (e) 55.00 mL
(b) 1.00 mL     (d) 50.00 mL

**12-16.** Consider the derivation of the fraction $\alpha_M$ in Equation 12-16.
(a) Derive the following expressions for the fractions $\alpha_{ML}$ and $\alpha_{ML_2}$:
$$\alpha_{ML} = \frac{\beta_1[L]}{1 + \beta_1[L] + \beta_2[L]^2} \qquad \alpha_{ML_2} = \frac{\beta_2[L]^2}{1 + \beta_1[L] + \beta_2[L]^2}$$
(b) Calculate the values of $\alpha_{ML}$ and $\alpha_{ML_2}$ for the conditions in Problem 12-14.

**12-17.** *Microequilibrium constants for binding of metal to a protein.* The iron-transport protein, transferrin, has two distinguishable metal-binding sites, designated a and b. The *microequilibrium* formation constants for each site are defined as follows:

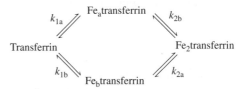

For example, the formation constant $k_{1a}$ refers to the reaction $Fe^{3+}$ + transferrin $\rightleftharpoons$ $Fe_a$transferrin, in which the metal ion binds to site a:
$$k_{1a} = \frac{[Fe_a\text{transferrin}]}{[Fe^{3+}][\text{transferrin}]}$$

(a) Write the chemical reactions corresponding to the conventional macroscopic formation constants, $K_1$ and $K_2$.
(b) Show that $K_1 = k_{1a} + k_{1b}$ and $K_2^{-1} = k_{2a}^{-1} + k_{2b}^{-1}$.
(c) Show that $k_{1a}k_{2b} = k_{1b}k_{2a}$. This expression means that, if you know any three of the microequilibrium constants, you automatically know the fourth one.
(d) *A challenge to your sanity:* From the equilibrium constants below, *find the equilibrium fraction of each of the four species (shown in the diagram) in circulating blood that is 40% saturated with iron* (that is, Fe/transferrin = 0.80, because each protein can bind 2Fe).

| Effective formation constants for blood plasma at pH 7.4 | |
|---|---|
| $k_{1a} = 6.0 \times 10^{22}$ | $k_{2a} = 2.4 \times 10^{22}$ |
| $k_{1b} = 1.0 \times 10^{22}$ | $k_{2b} = 4.2 \times 10^{21}$ |
| $K_1 = 7.0 \times 10^{22}$ | $K_2 = 3.6 \times 10^{21}$ |

The binding constants are so large that you may assume that there is negligible free $Fe^{3+}$. To get started, let's use the abbreviations [T] =[transferrin], [FeT] = $[Fe_aT] + [Fe_bT]$, and $[Fe_2T]$ = $[Fe_2$transferrin]. Now we can write

Mass balance for protein:  [T] + [FeT] + $[Fe_2T]$ = 1    (A)

Mass balance for iron:
$$\frac{[FeT] + 2[Fe_2T]}{[T] + [FeT] + [Fe_2T]} = [FeT] + 2[Fe_2T] = 0.8 \qquad (B)$$

Combined equilibria:  $\dfrac{K_1}{K_2} = 19._{44} = \dfrac{[FeT]^2}{[T][Fe_2T]}$    (C)

Now you have three equations with three unknowns and should be able to tackle this problem.

**12-18.** *Spreadsheet equation for auxiliary complexing agent.* Consider the titration of metal M (concentration = $C_m$, initial volume = $V_m$) with a solution of EDTA (concentration = $C_{EDTA}$, volume added = $V_{EDTA}$) in the presence of an auxiliary complexing ligand (such as ammonia). Follow the general procedure of the derivation in Section 12-4 to show that the master equation for the titration is

$$\phi = \frac{C_{EDTA}V_{EDTA}}{C_m V_m} = \frac{1 + K_f''[M]_{tot} - \dfrac{[M]_{tot} + K_f''[M]_{tot}^2}{C_m}}{K_f''[M]_{tot} + \dfrac{[M]_{tot} + K_f''[M]_{tot}^2}{C_{EDTA}}}$$

where $K_f''$ is the conditional formation constant in the presence of auxiliary complexing agent at the fixed pH of the titration (Equation 12-18) and $[M]_{tot}$ is the total concentration of metal not bound to EDTA. $[M]_{tot}$ is the same as $C_M$ in Equation 12-15. (We changed the symbol to avoid confusing $C_M$ and $C_m$.) The equation above is the same as Equation 12-11, with $[M]$ replaced by $[M]_{tot}$ and $K_f$ replaced by $K_f''$.

**12-19.** ▦ *Auxiliary complexing agent.* Use the equation derived in Problem 12-18 for this exercise.
(a) Prepare a spreadsheet to reproduce the 20-, 50-, and 60-mL points in the EDTA titration of $Zn^{2+}$ in the presence of $NH_3$ in the example in Section 12-5.
(b) Use your spreadsheet to plot the curve for the titration of 50.00 mL of 0.050 0 M $Ni^{2+}$ by 0.100 M EDTA at pH 11.00 in the presence of 0.100 M oxalate.

**12-20.** *Spreadsheet equation for formation of the complexes ML and $ML_2$.* Consider the titration of metal M (concentration $= C_m$, initial volume $= V_m$) with ligand L (concentration $= C_l$, volume added $= V_l$), which can form 1:1 and 2:1 complexes:

$$M + L \rightleftharpoons ML \qquad \beta_1 = \frac{[ML]}{[M][L]}$$

$$M + 2L \rightleftharpoons ML_2 \qquad \beta_2 = \frac{[ML_2]}{[M][L]^2}$$

Let $\alpha_M$ be the fraction of metal in the form M, $\alpha_{ML}$ be the fraction in the form ML, and $\alpha_{ML_2}$ be the fraction in the form $ML_2$. Following the derivation in Section 12-5, you could show that these fractions are given by

$$\alpha_M = \frac{1}{1 + \beta_1[L] + \beta_2[L]^2} \qquad \alpha_{ML} = \frac{\beta_1[L]}{1 + \beta_1[L] + \beta_2[L]^2}$$

$$\alpha_{ML_2} = \frac{\beta_2[L]^2}{1 + \beta_1[L] + \beta_2[L]^2}$$

The concentrations of ML and $ML_2$ are

$$[ML] = \alpha_{ML}\frac{C_m V_m}{V_m + V_l} \qquad [ML_2] = \alpha_{ML_2}\frac{C_m V_m}{V_m + V_l}$$

because $C_m V_m/(V_m + V_l)$ is the total concentration of all metal in the solution. The mass balance for ligand is

$$[L] + [ML] + 2[ML_2] = \frac{C_l V_l}{V_m + V_l}$$

By substituting expressions for [ML] and $[ML_2]$ into the mass balance, show that the master equation for a titration of metal by ligand is

$$\phi = \frac{C_l V_l}{C_m V_m} = \frac{\alpha_{ML} + 2\alpha_{ML_2} + ([L]/C_m)}{1 - ([L]/C_l)}$$

**12-21.** ▦ *Titration of M with L to form ML and $ML_2$.* Use the equation derived in Problem 12-20, where M is $Cu^{2+}$ and L is acetate. Consider the addition of 0.500 M acetate to 10.00 mL of

0.050 0 M $Cu^{2+}$ at pH 7.00 (so that all ligand is present as $CH_3CO_2^-$, not $CH_3CO_2H$). Formation constants for $Cu(CH_3CO_2)^+$ and $Cu(CH_3CO_2)_2$ are given in Appendix I. Construct a spreadsheet in which the input is pL and the output is [L], $V_l$, [M], [ML], and $[ML_2]$. Prepare a graph showing the concentrations of L, M, ML, and $ML_2$ as $V_l$ ranges from 0 to 3 mL.

**Metal Ion Indicators**

**12-22.** Explain why the change from red to blue in Reaction 12-19 occurs suddenly at the equivalence point instead of gradually throughout the entire titration.

**12-23.** List four methods for detecting the end point of an EDTA titration.

**12-24.** Calcium ion was titrated with EDTA at pH 11, using Calmagite as indicator (Table 12-3). Which is the principal species of Calmagite at pH 11? What color was observed before the equivalence point? After the equivalence point?

**12-25.** Pyrocatechol violet (Table 12-3) is to be used as a metal ion indicator in an EDTA titration. The procedure is as follows:

1. Add a known excess of EDTA to the unknown metal ion.
2. Adjust the pH with a suitable buffer.
3. Back-titrate the excess chelate with standard $Al^{3+}$.

From the following available buffers, select the best buffer, and then state what color change will be observed at the end point. Explain your answer.
(a) pH 6–7     (b) pH 7–8     (c) pH 8–9     (d) pH 9–10

**EDTA Titration Techniques**

**12-26.** Give three circumstances in which an EDTA back titration might be necessary.

**12-27.** Describe what is done in a displacement titration and give an example.

**12-28.** Give an example of the use of a masking agent.

**12-29.** What is meant by water hardness? Explain the difference between temporary and permanent hardness.

**12-30.** How many milliliters of 0.050 0 M EDTA are required to react with 50.0 mL of 0.010 0 M $Ca^{2+}$? With 50.0 mL of 0.010 0 M $Al^{3+}$?

**12-31.** A 50.0-mL sample containing $Ni^{2+}$ was treated with 25.0 mL of 0.050 0 M EDTA to complex all the $Ni^{2+}$ and leave excess EDTA in solution. The excess EDTA was then back-titrated, requiring 5.00 mL of 0.050 0 M $Zn^{2+}$. What was the concentration of $Ni^{2+}$ in the original solution?

**12-32.** A 50.0-mL aliquot of solution containing 0.450 g of $MgSO_4$ (FM 120.37) in 0.500 L required 37.6 mL of EDTA solution for titration. How many milligrams of $CaCO_3$ (FM 100.09) will react with 1.00 mL of this EDTA solution?

**12-33.** A 1.000-mL sample of unknown containing $Co^{2+}$ and $Ni^{2+}$ was treated with 25.00 mL of 0.038 72 M EDTA. Back titration with 0.021 27 M $Zn^{2+}$ at pH 5 required 23.54 mL to reach the xylenol orange end point. A 2.000-mL sample of unknown was passed through an ion-exchange column that retards $Co^{2+}$ more than $Ni^{2+}$. The $Ni^{2+}$ that passed through the column was treated with 25.00 mL of 0.038 72 M EDTA and required 25.63 mL of 0.021 27 M $Zn^{2+}$ for back titration. The $Co^{2+}$ emerged from the column later. It, too, was treated with 25.00 mL of 0.038 72 M EDTA. How many milliliters of 0.021 27 M $Zn^{2+}$ will be required for back titration?

**12-34.** A 50.0-mL solution containing $Ni^{2+}$ and $Zn^{2+}$ was treated with 25.0 mL of 0.045 2 M EDTA to bind all the metal. The excess unreacted EDTA required 12.4 mL of 0.012 3 M $Mg^{2+}$ for complete reaction. An excess of the reagent 2,3-dimercapto-1-propanol was then added to displace the EDTA from zinc. Another 29.2 mL of $Mg^{2+}$ were required for reaction with the liberated EDTA. Calculate the molarity of $Ni^{2+}$ and $Zn^{2+}$ in the original solution.

**12-35.** Sulfide ion was determined by indirect titration with EDTA. To a solution containing 25.00 mL of 0.043 32 M $Cu(ClO_4)_2$ plus 15 mL of 1 M acetate buffer (pH 4.5) were added 25.00 mL of unknown sulfide solution with vigorous stirring. The CuS precipitate was filtered and washed with hot water. Then ammonia was added to the filtrate (which contained excess $Cu^{2+}$) until the blue color of $Cu(NH_3)_4^{2+}$ was observed. Titration with 0.039 27 M EDTA required 12.11 mL to reach the murexide end point. Calculate the molarity of sulfide in the unknown.

**12-36.** *Indirect EDTA determination of cesium.* Cesium ion does not form a strong EDTA complex, but it can be analyzed by adding a known excess of $NaBiI_4$ in cold concentrated acetic acid containing excess NaI. Solid $Cs_3Bi_2I_9$ is precipitated, filtered, and removed. The excess yellow $BiI_4^-$ is then titrated with EDTA. The end point occurs when the yellow color disappears. (Sodium thiosulfate is used in the reaction to prevent the liberated $I^-$ from being oxidized to yellow aqueous $I_2$ by $O_2$ from the air.) The precipitation is fairly selective for $Cs^+$. The ions $Li^+$, $Na^+$, $K^+$, and low concentrations of $Rb^+$ do not interfere, although $Tl^+$ does. Suppose that 25.00 mL of unknown containing $Cs^+$ were treated with 25.00 mL of 0.086 40 M $NaBiI_4$ and the unreacted $BiI_4^-$ required 14.24 mL of 0.043 7 M EDTA for complete titration. Find the concentration of $Cs^+$ in the unknown.

**12-37.** The sulfur content of insoluble sulfides that do not readily dissolve in acid can be measured by oxidation with $Br_2$ to $SO_4^{2-}$.[22] Metal ions are then replaced with $H^+$ by an ion-exchange column, and sulfate is precipitated as $BaSO_4$ with a known excess of $BaCl_2$. The excess $Ba^{2+}$ is then titrated with EDTA to determine how much was present. (To make the indicator end point clearer, a small, known quantity of $Zn^{2+}$ also is added. The EDTA titrates both the $Ba^{2+}$ and the $Zn^{2+}$.) Knowing the excess $Ba^{2+}$, we can calculate how much sulfur was in the original material. To analyze the mineral sphalerite (ZnS, FM 97.46), 5.89 mg of powdered solid were suspended in a mixture of $CCl_4$ and $H_2O$ containing 1.5 mmol $Br_2$. After 1 h at 20°C and 2 h at 50°C, the powder dissolved and the solvent and excess $Br_2$ were removed by heating. The residue was dissolved in 3 mL of water and passed through an ion-exchange column to replace $Zn^{2+}$ with $H^+$. Then 5.000 mL of 0.014 63 M

$BaCl_2$ were added to precipitate all sulfate as $BaSO_4$. After the addition of 1.000 mL of 0.010 00 M $ZnCl_2$ and 3 mL of ammonia buffer, pH 10, the excess $Ba^{2+}$ and $Zn^{2+}$ required 2.39 mL of 0.009 63 M EDTA to reach the Calmagite end point. Find the weight percent of sulfur in the sphalerite. What is the theoretical value?

**12-38.** *pH-stat titration.* The graph shows titrations of ~50 μmol of $Cu^{2+}$ or $Mg^{2+}$ with 0.100 0 M $Na_2H_2EDTA$. The $Cu^{2+}$ solution contained 2 mM acetate buffer at an initial pH of 4.00. As EDTA was added, an instrument called a pH-stat automatically added 0.100 0 M NaOH to maintain the pH at a constant value of 4.00. The $Mg^{2+}$ solution contained 5 mM triethanolamine buffer at an initial pH of 8.00. As EDTA was added, the pH-stat added NaOH to maintain pH 8.00. In both titrations, the ionic strength was nearly constant near 0.10 M.

(a) The end point of the $Cu^{2+}$ titration is the intersection of the lines at 486 μL of EDTA. Explain why the volume of NaOH injected by the pH-stat increases up to $V_e$ with a slope of 1.88 μL NaOH/μL EDTA and then the slope becomes 0.00 after $V_e$.

(b) The end point of the $Mg^{2+}$ titration is the intersection of the two dashed lines at 504 μL of EDTA. Explain why the volume of NaOH injected by the pH-stat continues to increase past $V_e$, but the slope changes from 1.97 before $V_e$ to 0.99 after $V_e$.

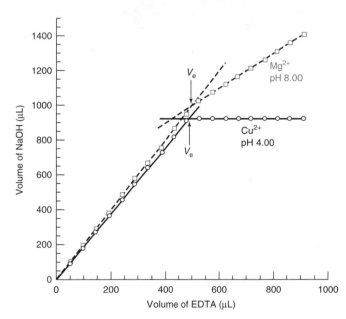

pH-stat titration. [Data from C. Maccà, L. Soldà, and M. Zancato, "pH-Stat Techniques in Titrimetric Analysis," Anal. Chim. Acta **2002**, 470, 277.]

St. Paul's Cathedral, London. *[Pictor International/ Picture Quest.]*

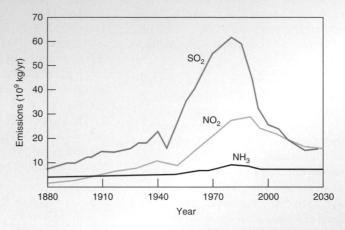

Emissions over Europe estimated by Schöpp and colleagues. *[From R. F. Wright, T. Larssen, L. Camarero, B. J. Crosby, R. C. Ferrier, R. Helliwell, M. Forsius, A. Jenkins, J. Kopáček, V. Majer, F. Moldan, M. Posch, M. Rogora, and W. Schöpp, "Recovery of Acidified European Surface Waters," Environ. Sci. Technol.* **2005,** *39, 64A.]*

Limestone and marble are building materials whose main constituent is calcite, the common crystalline form of calcium carbonate. This mineral is not very soluble in neutral or basic solution ($K_{sp} = 4.5 \times 10^{-9}$), but it dissolves in acid by virtue of two **coupled equilibria,** in which the reactions have a species in common—carbonate in this case:

$$CaCO_3(s) \rightleftharpoons Ca^{2+} + CO_3^{2-}$$
Calcite                        Carbonate

$$CO_3^{2-} + H^+ \rightleftharpoons HCO_3^-$$
Bicarbonate

Carbonate produced in the first reaction is protonated to form bicarbonate in the second reaction. Le Châtelier's principle tells us that, if we remove a product of the first reaction, we will draw the reaction to the right, making calcite more soluble. This chapter deals with coupled equilibria in chemical systems.

Between 1980 and 1990, $\frac{1}{2}$ mm of the thickness of the external stone walls of St. Paul's Cathedral in London was dissolved by acidic rainfall. A corner of the building facing a power station dissolved at 10 times the rate of the rest of the building until the station was closed. The power station and other industries that burn coal emit $SO_2$, which is a major source of acid rain (described in Box 15-1). Loss of heavy industry and laws limiting emissions decreased atmospheric $SO_2$ from as high as 100 ppb in the 1970s to 10 ppb in 2000. Correspondingly, only $\frac{1}{4}$ mm of St. Paul's external stone disappeared between 1990 and 2000.[1,2]

**T**his optional chapter provides tools to compute the concentrations of species in systems with many simultaneous equilibria.[3] The most important tool is the systematic treatment of equilibrium from Chapter 8. The other tool is a spreadsheet for numerical solution of the equilibrium equations. We will also see how to incorporate activity coefficients into equilibrium calculations. Later chapters in this book do not depend on this chapter.

# 13-1 General Approach to Acid-Base Systems

We first illustrate a general approach to find the concentrations of species in mixtures of acids and bases. Consider a solution made by dissolving 20.0 mmol sodium tartrate ($Na^+HT^-$), 15.0 mmol pyridinium chloride ($PyH^+Cl^-$), and 10.0 mmol KOH in a volume of 1.00 L. The problem is to find the pH and concentrations of all species in the solution.

The approach to equilibrium problems in this chapter is adapted from Julian Roberts, University of Redlands.

|  |  |
|---|---|
| D-Tartaric acid | Pyridinium chloride |
| $H_2T$ | $PyH^+Cl^-$ |
| $pK_1 = 3.036$, $pK_2 = 4.366$ | $pK_a = 5.20$ |

For this example, we designate the two acid dissociation constants of $H_2T$ as $K_1$ and $K_2$. We designate the acid dissociation constant of $PyH^+$ as $K_a$.

The chemical reactions and equilibrium constants at 0 ionic strength are

$$H_2T \rightleftharpoons HT^- + H^+ \qquad K_1 = 10^{-3.036} \qquad \text{(13-1)}$$

$$HT^- \rightleftharpoons T^{2-} + H^+ \qquad K_2 = 10^{-4.366} \qquad \text{(13-2)}$$

$$PyH^+ \rightleftharpoons Py + H^+ \qquad K_a = 10^{-5.20} \qquad \text{(13-3)}$$

$$H_2O \rightleftharpoons H^+ + OH^- \qquad K_w = 10^{-14.00} \qquad \text{(13-4)}$$

The charge balance is

$$[H^+] + [PyH^+] + [Na^+] + [K^+] = [OH^-] + [HT^-] + 2[T^{2-}] + [Cl^-] \qquad \text{(13-5)}$$

and there are several mass balances:

There is a factor of 2 in front of $[T^{2-}]$ because the ion has a charge $-2$.
1 M $T^{2-}$ contributes a charge of 2 M.

$$[Na^+] = 0.020\,0\ M \qquad [K^+] = 0.010\,0\ M \qquad [Cl^-] = 0.015\,0\ M$$

$$[H_2T] + [HT^-] + [T^{2-}] = 0.020\,0\ M \qquad [PyH^+] + [Py] = 0.015\,0\ M$$

There are 10 independent equations and 10 species, so we have enough information to solve for all the concentrations.

There is a systematic way to handle this problem without algebraic gymnastics.

**Step 1** Write a *fractional composition equation* from Section 10-5 for each acid or base that appears in the charge balance.

**Step 2** Substitute the fractional composition expressions into the charge balance and enter known values for $[Na^+]$, $[K^+]$, and $[Cl^-]$. Also, write $[OH^-] = K_w/[H^+]$. At this point, you will have a complicated equation in which the only variable is $[H^+]$.

**Step 3** Use your trusty spreadsheet to solve for $[H^+]$.

"Independent" equations cannot be derived from one another. As a trivial example, the equations $a = b + c$ and $2a = 2b + 2c$ are not independent. The three equilibrium expressions for $K_a$, $K_b$, and $K_w$ for a weak acid and its conjugate base provide only two independent equations because we can derive $K_b$ from $K_a$ and $K_w$: $K_b = K_w/K_a$.

Here is a recap of the fractional composition equations from Section 10-5 for *any* monoprotic acid HA and *any* diprotic acid $H_2A$.

*Monoprotic system:*
$$[HA] = \alpha_{HA}F_{HA} = \frac{[H^+]F_{HA}}{[H^+] + K_a} \qquad \text{(13-6a)}$$

$F_{HA} = [HA] + [A^-]$

$$[A^-] = \alpha_{A^-}F_{HA} = \frac{K_a F_{HA}}{[H^+] + K_a} \qquad \text{(13-6b)}$$

*Diprotic system:*
$$[H_2A] = \alpha_{H_2A}F_{H_2A} = \frac{[H^+]^2 F_{H_2A}}{[H^+]^2 + [H^+]K_1 + K_1K_2} \qquad \text{(13-7a)}$$

$F_{H_2A} = [H_2A] + [HA^-] + [A^{2-}]$

$$[HA^-] = \alpha_{HA^-}F_{H_2A} = \frac{K_1[H^+]F_{H_2A}}{[H^+]^2 + [H^+]K_1 + K_1K_2} \qquad \text{(13-7b)}$$

Table 11-6 gave fractional composition equations for $H_3A$.

$$[A^{2-}] = \alpha_{A^{2-}}F_{H_2A} = \frac{K_1K_2F_{H_2A}}{[H^+]^2 + [H^+]K_1 + K_1K_2} \qquad \text{(13-7c)}$$

In each equation, $\alpha_i$ is the fraction in each form. For example, $\alpha_{A^{2-}}$ is the fraction of diprotic acid in the form $A^{2-}$. When we multiply $\alpha_{A^{2-}}$ times $F_{H_2A}$ (the total or formal concentration of $H_2A$), the product is the concentration of $A^{2-}$.

## Applying the General Procedure

Now let's apply the general procedure to the mixture of $0.020\,0\,\text{M}$ sodium tartrate ($\text{Na}^+\text{HT}^-$), $0.015\,0\,\text{M}$ pyridinium chloride ($\text{PyH}^+\text{Cl}^-$), and $0.010\,0\,\text{M}$ KOH. We designate the formal concentrations as $F_{\text{H}_2\text{T}} = 0.020\,0\,\text{M}$ and $F_{\text{PH}^+} = 0.015\,0\,\text{M}$.

**Step 1** Write a *fractional composition equation* for each acid or base that appears in the charge balance.

$$[\text{PyH}^+] = \alpha_{\text{PyH}^+}F_{\text{PyH}^+} = \frac{[\text{H}^+]F_{\text{PyH}^+}}{[\text{H}^+] + K_a} \tag{13-8}$$

$$[\text{HT}^-] = \alpha_{\text{HT}^-}F_{\text{H}_2\text{T}} = \frac{K_1[\text{H}^+]F_{\text{H}_2\text{T}}}{[\text{H}^+]^2 + [\text{H}^+]K_1 + K_1K_2} \tag{13-9}$$

$$[\text{T}^{2-}] = \alpha_{\text{T}^{2-}}F_{\text{H}_2\text{T}} = \frac{K_1K_2F_{\text{H}_2\text{T}}}{[\text{H}^+]^2 + [\text{H}^+]K_1 + K_1K_2} \tag{13-10}$$

All quantities on the right side of these expressions are known, except for $[\text{H}^+]$.

**Step 2** Substitute the fractional composition expressions into the charge balance 13-5. Enter values for $[\text{Na}^+]$, $[\text{K}^+]$, and $[\text{Cl}^-]$, and write $[\text{OH}^-] = K_w/[\text{H}^+]$.

$$[\text{H}^+] + [\text{PyH}^+] + [\text{Na}^+] + [\text{K}^+] = [\text{OH}^-] + [\text{HT}^-] + 2[\text{T}^{2-}] + [\text{Cl}^-] \tag{13-5}$$

$$[\text{H}^+] + \alpha_{\text{PyH}^+}F_{\text{PyH}^+} + [0.020\,0] + [0.010\,0]$$
$$= \frac{K_w}{[\text{H}^+]} + \alpha_{\text{HT}^-}F_{\text{H}_2\text{T}} + 2\alpha_{\text{T}^{2-}}F_{\text{H}_2\text{T}} + [0.015\,0] \tag{13-11}$$

$K_a$, $K_1$, $K_2$, and $[\text{H}^+]$ are contained in the $\alpha$ expressions. The only variable in Equation 13-11 is $[\text{H}^+]$.

**Step 3** The spreadsheet in Figure 13-1 solves Equation 13-11 for $[\text{H}^+]$.

Key step: *Guess* a value for $[\text{H}^+]$ and use Excel SOLVER to vary $[\text{H}^+]$ until it satisfies the charge balance.

In Figure 13-1, shaded boxes contain input. Everything else is computed by the spreadsheet. Values for $F_{\text{H}_2\text{T}}$, $pK_1$, $pK_2$, $F_{\text{PyH}^+}$, $pK_a$, and $[\text{K}^+]$ were given in the problem. The initial value of pH in cell H13 is a *guess*. We will use Excel SOLVER to vary pH until the sum of

| | A | B | C | D | E | F | G | H | I |
|---|---|---|---|---|---|---|---|---|---|
| 1 | Mixture of 0.020 M Na$^+$HT$^-$, 0.015 M PyH$^+$Cl$^-$, and 0.010 M KOH | | | | | | | | |
| 2 | | | | | | | | | |
| 3 | F$_{H2T}$ = | 0.020 | | F$_{PyH+}$ = | 0.015 | | [K$^+$] = | 0.010 | |
| 4 | pK$_1$ = | 3.036 | | pK$_a$ = | 5.20 | | K$_w$ = | 1.00E-14 | |
| 5 | pK$_2$ = | 4.366 | | K$_a$ = | 6.31E-06 | | | | |
| 6 | K$_1$ = | 9.20E-04 | | | | | | | |
| 7 | K$_2$ = | 4.31E-05 | | | | | | | |
| 8 | | | | | | | | | |
| 9 | Species in charge balance: | | | | | | Other concentrations: | | |
| 10 | [H$^+$] = | 1.00E-06 | | [OH$^-$] = | 1.00E-08 | | [H$_2$T] = | 4.93E-07 | |
| 11 | [PyH$^+$] = | 2.05E-03 | | [HT$^-$] = | 4.54E-04 | | [Py] = | 1.29E-02 | |
| 12 | [Na$^+$] = | 0.020 | | [T$^{2-}$] = | 1.95E-02 | | | | |
| 13 | [K$^+$] = | 0.010 | | [Cl$^-$] = | 0.015 | | pH = | 6.000 | ← initial value |
| 14 | | | | | | | | | is a guess |
| 15 | Positive charge minus negative charge = | | | | -2.25E-02 | ← vary pH in H13 with Solver to make this 0 | | | |
| 16 | | | | | E15 = B10+B11+B12+B13-E10-E11-2*E12-E13 | | | | |
| 17 | Check: [PyH$^+$] + [Py] = | | | 0.01500 | (= B11+H11) | | | | |
| 18 | Check: [H$_2$T] + [HT$^-$] + [T$^{2-}$] = | | | 0.02000 | (= H10+E11+E12) | | | | |
| 19 | | | | | | | | | |
| 20 | Formulas: | | | | | | | | |
| 21 | B6 = 10^-B4 | | B7 = 10^-B5 | | E5 = 10^-E4 | | E10 = H4/B10 | | |
| 22 | B10 = 10^-H13 | | B12 = B3 | | B13 = H3 | | E13 = E3 | | |
| 23 | E11 = B6*B10*B3/(B10^2+B10*B6+B6*B7) | | | | | | B11 = B10*E3/(B10+E5) | | |
| 24 | E12 = B6*B7*B3/(B10^2+B10*B6+B6*B7) | | | | | | H11 = E5*E3/(B10+E5) | | |
| 25 | H10 = B10^2*B3/(B10^2+B10*B6+B6*B7) | | | | | | | | |

*Figure 13-1* Spreadsheet for mixture of acids and bases uses SOLVER to find the value of pH in cell H13 that satisfies the charge balance in cell E15. The sums [PyH$^+$] + [Py] in cell D17 and [H$_2$T] + [HT$^-$] + [T$^{2-}$] in cell D18 are computed to verify that the formulas for each of the species do not have mistakes. These sums are independent of pH.

charges in cell E15 is 0. Species in the charge balance are in cells B10:E13. $[H^+]$ in cell B10 is computed from the pH we guessed in cell H13. $[PyH^+]$ in cell B11 is computed with Equation 13-8. Known values are entered for $[Na^+]$, $[K^+]$, and $[Cl^-]$. $[OH^-]$ is computed from $K_W/[H^+]$. $[HT^-]$ and $[T^{2-}]$ in cells E11 and E12 are computed with Equations 13-9 and 13-10.

The sum of charges, $[H^+] + [PyH^+] + [Na^+] + [K^+] - [OH^-] - [HT^-] - 2[T^{2-}] - [Cl^-]$, is computed in cell E15. If we had guessed the correct pH in cell H13, the sum of the charges would be 0. Instead, the sum is $-2.25 \times 10^{-2}$ M. We use Excel SOLVER to vary pH in cell H13 until the sum of charges in cell E15 is 0.

## Using Excel SOLVER

Select SOLVER on the TOOLS menu and a window similar to the one in Figure 19-4 appears. In this window, Set Target Cell E15 Equal To Value of 0 By Changing Cells H13. Then click Solve. SOLVER will vary the pH in cell H13 to make the net charge in cell E15 equal to 0. Starting with a pH of 6 in cell H13, SOLVER returns a net charge of $\sim 10^{-6}$ in cell E15 by adjusting the pH in cell H13 to 4.298.

The reason the charge is $10^{-6}$ instead of 0 is that the default precision of SOLVER is $10^{-6}$. To get a charge closer to 0, select SOLVER and choose Options in the SOLVER window. Precision will probably have the default setting of 0.000 001. Enter 1e-16 for Precision and click OK. Run SOLVER again. This time, the charge in cell E15 is reduced to $\sim 10^{-16}$. The pH in cell H13 is still 4.298 (to three decimal places). The difference in pH required to reduce the net charge from $10^{-6}$ to $10^{-16}$ is not perceptible in the third decimal place of pH. Many chemistry problems involve very large or very small numbers for which you might need to adjust the precision of SOLVER. The concentrations after executing SOLVER are

| | A | B | C | D | E | F | G | H |
|---|---|---|---|---|---|---|---|---|
| 9 | Species in charge balance: | | | | | | Other concentrations: | |
| 10 | $[H^+] =$ | 5.04E-05 | | $[OH^-] =$ | 1.99E-10 | | $[H_2T] =$ | 5.73E-04 |
| 11 | $[PyH^+] =$ | 1.33E-02 | | $[HT^-] =$ | 1.05E-02 | | $[Py] =$ | 1.67E-03 |
| 12 | $[Na^+] =$ | 0.020 | | $[T^{2-}] =$ | 8.95E-03 | | | |
| 13 | $[K^+] =$ | 0.010 | | $[Cl^-] =$ | 0.015 | | pH = | 4.298 |
| 14 | | | | | | | | |
| 15 | Positive charge minus negative charge = | | | 9.71E-17 | | | | |

## Ignorance Is Bliss: A Complication of Ion Pairing

We should not get too cocky with our newfound power to handle complex problems, because we have oversimplified the actual situation. For one thing, we have not included activity coefficients, which ordinarily affect the answer by a few tenths of a pH unit. In Section 13-2, we will show how to incorporate activity coefficients.

Even with activity coefficients, we are always limited by chemistry that we do not know. In the mixture of sodium hydrogen tartrate ($Na^+HT^-$), pyridinium chloride ($PyH^+Cl^-$), and KOH, several possible ion-pair equilibria are

$$Na^+ + T^{2-} \rightleftharpoons NaT^- \qquad K_{NaT^-} = \frac{[NaT^-]}{[Na^+][T^{2-}]} = 8 \qquad (13\text{-}12)$$

$$Na^+ + HT^- \rightleftharpoons NaHT \qquad K_{NaHT} = \frac{[NaHT]}{[Na^+][HT^-]} = 1.6 \qquad (13\text{-}13)$$

$$Na^+ + Py \rightleftharpoons PyNa^+ \qquad K_{PyNa^+} = 1.0 \qquad (13\text{-}14)$$

$$K^+ + T^2 \rightleftharpoons KT^- \qquad K_{KT} = 3$$

$$K^+ + HT^- \rightleftharpoons KHT \qquad K_{KHT} = ?$$

$$PyH^+ + Cl^- \rightleftharpoons PyH^+Cl^- \qquad K_{PyHCl} = ?$$

$$PyH^+ + T^{2-} \rightleftharpoons PyHT^- \qquad K_{PyHT^-} = ?$$

Some equilibrium constants at 0 ionic strength are listed above. Values for the other reactions are not available, but there is no reason to believe that the reactions do not occur.

Equilibrium constants from A. E. Martell, R. M. Smith, and R. J. Motekaitis, NIST Standard Reference Database 46, Version 6.0, 2001.

How could we add ion pairing to our spreadsheet? For simplicity, we outline how to add just Reactions 13-12 and 13-13. With these reactions, the mass balance for sodium is

$$[Na^+] + [NaT^-] + [NaHT] = F_{Na} = F_{H_2T} = 0.020\ 0\ M \qquad (13\text{-}15)$$

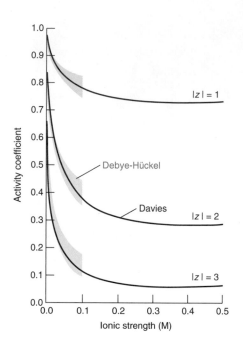

**Figure 13-2** Activity coefficients from extended Debye-Hückel and Davies equations. Shaded areas give Debye-Hückel activity coefficients for the range of ion sizes in Table 8-1.

A 0.5 wt% solution of $KH_2PO_4$ has a *molarity* of 0.028 13 mol/L and a *molality* of 0.028 20 mol/kg. The difference is 0.25%.

$K_2'$ gives the *concentration quotient*

$$\frac{[HPO_4^{2-}][H^+]}{[H_2PO_4^-]}$$

at a specified ionic strength.

From the ion-pair equilibria, we write $[NaT^-] = K_{NaT^-}[Na^+][T^{2-}]$ and $[NaHT] = K_{NaHT}[Na^+][HT^-]$. With these substitutions for $[NaT^-]$ and $[NaHT]$ in the mass balance for sodium, we find an expression for $[Na^+]$:

$$[Na^+] = \frac{F_{H_2T}}{1 + K_{NaT^-}[T^{2-}] + K_{NaHT}[HT^-]} \tag{13-16}$$

A bigger nuisance when considering ion pairs is that the fractional composition equations for $[H_2T]$, $[HT^-]$, and $[T^{2-}]$ also are changed because the mass balance for $H_2T$ now has five species in it instead of three:

$$F_{H_2T} = [H_2T] + [HT^-] + [T^{2-}] + [NaT^-] + [NaHT] \tag{13-17}$$

We must derive new equations analogous to 13-9 and 13-10 from the mass balance 13-17.

The new fractional composition equations are somewhat messy, so we reserve this case for Problem 13-18. The end result is that adding the ion-pair equilibria 13-12 and 13-13 changes the calculated pH from 4.30 to 4.26. This change is not large, so neglecting ion pairs with small equilibrium constants does not lead to serious error. We find that 7% of sodium is tied up in ion pairs. *Our ability to compute the distribution of species in a solution is limited by our knowledge of relevant equilibria.*

## ◼◼◼◼ 13-2 Activity Coefficients

Even if we know all reactions and equilibrium constants for a given system, we cannot compute concentrations accurately without activity coefficients. Chapter 8 gave the extended Debye-Hückel equation 8-6 for activity coefficients with size parameters in Table 8-1. Many ions of interest are not in Table 8-1 and we do not know their size parameter. Therefore we introduce the *Davies equation,* which has no size parameter:

*Davies equation:* $\qquad \log \gamma = -0.51z^2 \left( \frac{\sqrt{\mu}}{1 + \sqrt{\mu}} - 0.3\mu \right) \qquad$ (at 25°C) $\qquad$ (13-18)

where $\gamma$ is the activity coefficient for an ion of charge $z$ at ionic strength $\mu$. Equation 13-18 is used up to $\mu \approx 0.5$ M (Figure 13-2) but is more accurate at lower ionic strength. For best accuracy, Pitzer's equations are used (Chapter 8, reference 7).

Now consider a primary standard buffer containing 0.025 0 $m$ $KH_2PO_4$ and 0.025 0 $m$ $Na_2HPO_4$. Its pH at 25°C is 6.865 ± 0.006.[4] The concentration unit, $m$, is *molality,* which means moles of solute per kilogram of solvent. For precise chemical measurements, concentrations are often expressed in molality, rather than molarity, because molality is independent of temperature. Tabulated equilibrium constants usually apply to molality, not molarity. Uncertainties in equilibrium constants are usually sufficiently great so that the ~0.3% difference between molality and molarity of dilute solutions is unimportant.

Acid-base equilibrium constants for $H_3PO_4$ at $\mu = 0$ at 25°C are

$$H_3PO_4 \rightleftharpoons H_2PO_4^- + H^+ \qquad K_1 = \frac{[H_2PO_4^-]\gamma_{H_2PO_4^-}[H^+]\gamma_{H^+}}{[H_3PO_4]\gamma_{H_3PO_4}} = 10^{-2.148} \tag{13-19}$$

$$H_2PO_4^- \rightleftharpoons HPO_4^{2-} + H^+ \qquad K_2 = \frac{[HPO_4^{2-}]\gamma_{HPO_4^{2-}}[H^+]\gamma_{H^+}}{[H_2PO_4^-]\gamma_{H_2PO_4^-}} = 10^{-7.198} \tag{13-20}$$

$$HPO_4^{2-} \rightleftharpoons PO_4^{3-} + H^+ \qquad K_3 = \frac{[PO_4^{3-}]\gamma_{PO_4^{3-}}[H^+]\gamma_{H^+}}{[HPO_4^{2-}]\gamma_{HPO_4^{2-}}} = 10^{-12.375} \tag{13-21}$$

Equilibrium constants can be determined by measuring concentration quotients at several low ionic strengths and extrapolating to 0 ionic strength.

For $\mu \neq 0$, we rearrange the equilibrium constant expression to incorporate activity coefficients into an effective equilibrium constant, $K'$, at the given ionic strength.

$$K_1' = K_1 \left( \frac{\gamma_{H_3PO_4}}{\gamma_{H_2PO_4^-}\gamma_{H^+}} \right) = \frac{[H_2PO_4^-][H^+]}{[H_3PO_4]} \tag{13-22}$$

$$K_2' = K_2 \left( \frac{\gamma_{H_2PO_4^-}}{\gamma_{HPO_4^{2-}}\gamma_{H^+}} \right) = \frac{[HPO_4^{2-}][H^+]}{[H_2PO_4^-]} \tag{13-23}$$

$$K_3' = K_3 \left( \frac{\gamma_{HPO_4^{2-}}}{\gamma_{PO_4^{3-}}\gamma_{H^+}} \right) = \frac{[PO_4^{3-}][H^+]}{[HPO_4^{2-}]} \quad \text{(13-24)}$$

For ionic species, we compute activity coefficients with the Davies equation 13-18. For the neutral species $H_3PO_4$, we assume that $\gamma \approx 1.00$.

Now we remind ourselves of the equations for water ionization:

$$H_2O \rightleftharpoons H^+ + OH^- \qquad K_w = [H^+]\gamma_{H^+}[OH^-]\gamma_{OH^-} = 10^{-13.995}$$

Values of $K_w$ are found in Table 6-1.

$$K_w' = \frac{K_w}{\gamma_{H^+}\gamma_{OH^-}} = [H^+][OH^-] \Rightarrow [OH^-] = K_w'/[H^+] \quad \text{(13-25)}$$

$$pH = -\log([H^+]\gamma_{H^+}) \quad \text{(13-26)}$$

Here is how we use effective equilibrium constants:

**Step 1** Solve the acid-base problem with the constants $K_1$, $K_2$, and $K_3$, which apply at $\mu = 0$. Activity coefficients are considered to be unity in this first step.

**Step 2** From the results of step 1, compute the ionic strength. Then use the Davies equation to compute activity coefficients. With these activity coefficients, compute the effective equilibrium constants, $K_1'$, $K_2'$, $K_3'$, and $K_w'$.

**Step 3** Solve the acid-base problem again, this time with $K_1'$, $K_2'$, $K_3'$, and $K_w'$.

**Step 4** From the results of step 3, compute a new ionic strength and a new set of $K'$ values. Repeat the process several times until the ionic strength is constant.

Let's find the pH of $0.025\,0\ m$ $KH_2PO_4$ plus $0.025\,0\ m$ $Na_2HPO_4$. The chemical reactions are 13-19 through 13-21 plus water ionization. The mass balances are $[K^+] = 0.025\,0\ m$, $[Na^+] = 0.050\,0\ m$, and total phosphate $\equiv F_{H_3P} = 0.050\,0\ m$. The charge balance is

$$[Na^+] + [K^+] + [H^+] = [H_2PO_4^-] + 2[HPO_4^{2-}] + 3[PO_4^{3-}] + [OH^-] \quad \text{(13-27)}$$

Our strategy is to substitute expressions into the charge balance to obtain an equation in which the only variable is $[H^+]$. For this purpose, we use the fractional composition equations for the triprotic acid, $H_3PO_4$, which we abbreviate as $H_3P$:

$$[P^{3-}] = \alpha_{P^{3-}}F_{H_3P} = \frac{K_1'K_2'K_3'F_{H_3P}}{[H^+]^3 + [H^+]^2K_1' + [H^+]K_1'K_2' + K_1'K_2'K_3'} \quad \text{(13-28)}$$

$$[HP^{2-}] = \alpha_{HP^{2-}}F_{H_3P} = \frac{[H^+]K_1'K_2'F_{H_3P}}{[H^+]^3 + [H^+]^2K_1' + [H^+]K_1'K_2' + K_1'K_2'K_3'} \quad \text{(13-29)}$$

$$[H_2P^-] = \alpha_{H_2P^-}F_{H_3P} = \frac{[H^+]^2K_1'F_{H_3P}}{[H^+]^3 + [H^+]^2K_1' + [H^+]K_1'K_2' + K_1'K_2'K_3'} \quad \text{(13-30)}$$

$$[H_3P] = \alpha_{H_3P}F_{H_3P} = \frac{[H^+]^3F_{H_3P}}{[H^+]^3 + [H^+]^2K_1' + [H^+]K_1'K_2' + K_1'K_2'K_3'} \quad \text{(13-31)}$$

Figure 13-3 puts everything together in a spreadsheet. Input values for $F_{KH_2PO_4}$, $F_{Na_2HPO_4}$, $pK_1$, $pK_2$, $pK_3$, and $pK_w$ are in the shaded cells. We *guess* a value for pH in cell H15 and write the initial ionic strength of 0 in cell C19. Cells A9:H10 compute activities with the Davies equation. With $\mu = 0$, all activity coefficients are 1. Cells A13:H16 compute concentrations. $[H^+]$ in cell B13 is $(10^{-pH})/\gamma_{H^+} = (10^{\wedge}\text{-H15})/B9$. Cell E18 computes the sum of charges.

Not shown in Figure 13-3 is that an initial guess of pH $= 7$ in cell H15 gives a net charge of $0.005\,6\ m$ in cell E18. Excel SOLVER was then used to vary pH in cell H15 to produce a net charge near 0 in cell E18. For this purpose, the precision was set to 1e-16 in SOLVER Options. Figure 13-3 shows that executing SOLVER gives pH $= 7.198$ in cell H15 and a net charge of $-2 \times 10^{-17}\ m$ in cell E18. The calculated ionic strength in cell C20 is $0.100\ m$.

For the second iteration, we write the ionic strength $0.100\ m$ in cell C19 of Figure 13-4. This ionic strength automatically gives new activity coefficients in cells A9:H10 and new effective equilibrium constants in cells H3:H6. The sum of charges in cell E18 is no longer near zero. Application of SOLVER to vary pH in cell H15 to try to get 0 net charge produces a net charge of $9 \times 10^{-17}\ m$ in cell E18. The new ionic strength in cell C20 remains $0.100\ m$, so we are done. When the new ionic strength in cell C20 equals the old ionic strength in cell C19 (say, to three decimal places), there is no need for further iteration.

The final pH in cell H15 of Figure 13-4 is 6.876, which differs from the certified pH of 6.865 by 0.011. This difference is about as close an agreement between a measured and

**13-2  Activity Coefficients**

Figure 13-3 Spreadsheet solved for the system 0.025 0 $m$ $KH_2PO_4$ plus 0.025 0 $m$ $Na_2HPO_4$ with initial ionic strength = 0 and activity coefficients = 1.

| | A | B | C | D | E | F | G | H |
|---|---|---|---|---|---|---|---|---|
| 1 | Mixture of $KH_2PO_4$ and $Na_2HPO_4$ including activity coefficients from Davies equation | | | | | | | |
| 2 | | | | | | | | |
| 3 | $F_{KH2PO4}$ = | 0.0250 | | $pK_1$ = | 2.148 | | $K_1'$ = | 7.11E-03 |
| 4 | $F_{Na2HPO4}$ = | 0.0250 | | $pK_2$ = | 7.198 | | $K_2'$ = | 6.34E-08 |
| 5 | $F_{H3P}$ = | 0.0500 | (=B3+B4) | $pK_3$ = | 12.375 | | $K_3'$ = | 4.22E-13 |
| 6 | | | | $pK_w$ = | 13.995 | | $K_w'$ = | 1.01E-14 |
| 7 | | | | | | | | |
| 8 | Activity coefficients: | | | | | | | |
| 9 | $H^+$ = | 1.00 | | $H_3P$ = | 1.00 | (fixed at 1) | $HP^{2-}$ = | 1.00 |
| 10 | $OH^-$ = | 1.00 | | $H_2P^-$ = | 1.00 | | $P^{3-}$ = | 1.00 |
| 11 | | | | | | | | |
| 12 | Species in charge balance: | | | | | | Other concentrations: | |
| 13 | $[H^+]$ = | 6.34E-08 | | $[OH^-]$ = | 1.60E-07 | | $[H_3P]$ = | 2.23E-07 |
| 14 | $[Na^+]$ = | 0.050000 | | $[H_2P^-]$ = | 2.50E-02 | | | |
| 15 | $[K^+]$ = | 0.025000 | | $[HP^{2-}]$ = | 2.50E-02 | | pH = | 7.198 |
| 16 | | | | $[P^{3-}]$ = | 1.66E-07 | | ↑ initial value is a guess | |
| 17 | | | | | | | | |
| 18 | Positive charge minus negative charge = | | | | -2.27E-17 | | | |
| 19 | Ionic strength = | | 0.0000 | ←initial value is 0 | | | | |
| 20 | New ionic strength = | | 0.1000 | ←substitute this value into cell C19 for next iteration | | | | |
| 21 | | | | | | | | |
| 22 | Formulas: | | | | | | | |
| 23 | H3 = 10^-E3*E9/(E10*B9) | | | | H4 = 10^-E4*E10/(H9*B9) | | | |
| 24 | H5 = 10^-E5*H9/(H10*B9) | | | | H6 = 10^-E6/(B9*B10) | | | |
| 25 | B9 = B10 = E10 = 10^(-0.51*1^2*(SQRT($C$19)/(1+SQRT($C$19))-0.3*$C$19)) | | | | | | | |
| 26 | H9 = 10^(-0.51*2^2*(SQRT($C$19)/(1+SQRT($C$19))-0.3*$C$19)) | | | | | | | |
| 27 | H10 = 10^(-0.51*3^2*(SQRT($C$19)/(1+SQRT($C$19))-0.3*$C$19)) | | | | | | | |
| 28 | B13 = (10^-H15)/B9 | | B14 = 2*B4 | | B15 = B3 | | E13 = H6/(B13) | |
| 29 | E14 = B13^2*H3*B5/(B13^3+B13^2*H3+B13*H3*H4+H3*H4*H5) | | | | | | | |
| 30 | E15 = B13*H3*H4*B5/(B13^3+B13^2*H3+B13*H3*H4+H3*H4*H5) | | | | | | | |
| 31 | E16 = H3*H4*H5*B5/(B13^3+B13^2*H3+B13*H3*H4+H3*H4*H5) | | | | | | | |
| 32 | H13 = B13^3*B5/(B13^3+B13^2*H3+B13*H3*H4+H3*H4*H5) | | | | | | | |
| 33 | E18 = B13+B14+B15-E13-E14-2*E15-3*E16 | | | | | | | |
| 34 | C20 = 0.5*(B13+B14+B15+E13+E14+4*E15+9*E16) | | | | | | | |

| | A | B | C | D | E | F | G | H |
|---|---|---|---|---|---|---|---|---|
| 1 | Mixture of $KH_2PO_4$ and $Na_2HPO_4$ including activity coefficients from Davies equation | | | | | | | |
| 2 | | | | | | | | |
| 3 | $F_{KH2PO4}$ = | 0.0250 | | $pK_1$ = | 2.148 | | $K_1'$ = | 1.17E-02 |
| 4 | $F_{Na2HPO4}$ = | 0.0250 | | $pK_2$ = | 7.198 | | $K_2'$ = | 1.70E-07 |
| 5 | $F_{H3P}$ = | 0.0500 | (=B3+B4) | $pK_3$ = | 12.375 | | $K_3'$ = | 1.86E-12 |
| 6 | | | | $pK_w$ = | 13.995 | | $K_w'$ = | 1.66E-14 |
| 7 | | | | | | | | |
| 8 | Activity coefficients: | | | | | | | |
| 9 | $H^+$ = | 0.78 | | $H_3P$ = | 1.00 | (fixed at 1) | $HP^{2-}$ = | 0.37 |
| 10 | $OH^-$ = | 0.78 | | $H_2P^-$ = | 0.78 | | $P^{3-}$ = | 0.11 |
| 11 | | | | | | | | |
| 12 | Species in charge balance: | | | | | | Other concentrations: | |
| 13 | $[H^+]$ = | 1.70E-07 | | $[OH^-]$ = | 9.74E-08 | | $[H_3P]$ = | 3.65E-07 |
| 14 | $[Na^+]$ = | 0.050000 | | $[H_2P^-]$ = | 2.50E-02 | | | |
| 15 | $[K^+]$ = | 0.025000 | | $[HP^{2-}]$ = | 2.50E-02 | | pH = | 6.876 |
| 16 | | | | $[P^{3-}]$ = | 2.73E-07 | | ↑ initial value is a guess | |
| 17 | | | | | | | | |
| 18 | Positive charge minus negative charge = | | | | 8.56E-17 | | | |
| 19 | Ionic strength = | | 0.1000 | ← initial value is 0 | | | | |
| 20 | New ionic strength = | | 0.1000 | ← substitute this value into cell C19 for next iteration | | | | |

Figure 13-4 Second iteration of spreadsheet solved for the system 0.025 0 $m$ $KH_2PO_4$ plus 0.025 0 $m$ $Na_2HPO_4$ with ionic strength = 0.100 from the first iteration.

computed pH as you will encounter. If we had used extended Debye-Hückel activity coefficients for $\mu = 0.1\ m$ in Table 8-1, the computed pH would be 6.859, differing from the stated pH by just 0.006.

Sometimes SOLVER cannot find a solution if Precision is set too small in the Options window. You can make the Precision larger and see if SOLVER can find a solution. You can also try a different initial guess for pH.

### Back to Basics

A spreadsheet operating on the charge balance to reduce the net charge to zero is an excellent, general method for solving complex equilibrium problems. However, we did learn how to find the pH of a mixture of $KH_2PO_4$ plus $Na_2HPO_4$ back in Chapter 9 by a simple, less rigorous method. Recall that, when we mix a weak acid ($H_2PO_4^-$) and its conjugate base ($HPO_4^{2-}$), *what we mix is what we get*. The pH is estimated from the Henderson-Hasselbalch equation 9-18 with activity coefficients:

$$\text{pH} = pK_a + \log\frac{[\text{A}^-]\gamma_{\text{A}^-}}{[\text{HA}]\gamma_{\text{HA}}} = pK_2 + \log\frac{[\text{HPO}_4^{2-}]\gamma_{\text{HPO}_4^{2-}}}{[\text{H}_2\text{PO}_4^-]\gamma_{\text{H}_2\text{PO}_4^-}} \tag{9-18}$$

For $0.025\ m$ $KH_2PO_4$ plus $0.025\ m$ $Na_2HPO_4$, the ionic strength is

$$\mu = \frac{1}{2}\sum_i c_i z_i^2 = \frac{1}{2}([\text{K}^+]\cdot(+1)^2 + [\text{H}_2\text{PO}_4^-]\cdot(-1)^2 + [\text{Na}^+]\cdot(+1)^2 + [\text{HPO}_4^{2-}]\cdot(-2)^2)$$

$$= \frac{1}{2}([0.025]\cdot 1 + [0.025]\cdot 1 + [0.050]\cdot 1 + [0.025]\cdot 4) = 0.100\ m$$

In Table 8-1, the activity coefficients at $\mu = 0.1\ m$ are 0.775 for $H_2PO_4^-$ and 0.355 for $HPO_4^{2-}$. Inserting these values into Equation 9-18 gives

$$\text{pH} = 7.198 + \log\frac{[0.025]0.355}{[0.025]0.775} = 6.859$$

$pK_a = pK_2$ in Equation 9-18 applies at $\mu = 0$.

The answer is the same that we generate with the spreadsheet because the approximation that what we mix is what we get is excellent in this case.

So you already knew how to find the pH of this buffer by a simple calculation. The value of the general method with the charge balance in the spreadsheet is that it applies even when what you mix is not what you get, because the concentrations are very low or $K_2$ is not so small or there are additional equilibria.

### Ignorance Is Still Bliss

Even in such a simple solution as $KH_2PO_4$ plus $Na_2HPO_4$, for which we were justifiably proud of computing an accurate pH, we have overlooked numerous ion-pair equilibria:

$$\text{PO}_4^{3-} + \text{Na}^+ \rightleftharpoons \text{NaPO}_4^{2-} \qquad K = 27 \qquad \text{HPO}_4^{2-} + \text{Na}^+ \rightleftharpoons \text{NaHPO}_4^- \qquad K = 12$$

$$\text{H}_2\text{PO}_4^- + \text{Na}^+ \rightleftharpoons \text{NaHPO}_4 \qquad K = 2 \qquad \text{NaPO}_4^{2-} + \text{Na}^+ \rightleftharpoons \text{Na}_2\text{PO}_4^- \qquad K = 14$$

$$\text{Na}_2\text{PO}_4^- + \text{H}^+ \rightleftharpoons \text{Na}_2\text{HPO}_4 \qquad K = 5.4 \times 10^{10}$$

There is an analogous set of reactions for $K^+$, whose equilibrium constants are similar to those for $Na^+$.

The reliability of the calculated concentrations depends on knowing all relevant equilibria and having the fortitude to include them all in the computation, which is by no means trivial.

## ■■■ 13-3 Dependence of Solubility on pH

An important example of the effect of pH on solubility is tooth decay. Tooth enamel contains the mineral hydroxyapatite, which is insoluble near neutral pH, but dissolves in acid because both phosphate and hydroxide in the hydroxyapatite react with $H^+$:

$$\text{Ca}_{10}(\text{PO}_4)_6(\text{OH})_2(s) + 14\text{H}^+ \rightleftharpoons 10\text{Ca}^{2+} + 6\text{H}_2\text{PO}_4^- + 2\text{H}_2\text{O}$$
Calcium hydroxyapatite

Bacteria on the surface of our teeth metabolize sugars to produce lactic acid, which lowers the pH enough to slowly dissolve tooth enamel. Fluoride inhibits tooth decay because it forms fluorapatite, $Ca_{10}(PO_4)_6F_2$, which is more acid resistant than hydroxyapatite.

L-Lactic acid

## Solubility of CaF₂

The mineral *fluorite*, $CaF_2$, in Figure 13-5 has a cubic crystal structure and often cleaves to form nearly perfect octahedra (eight-sided solids with equilateral triangular faces). Depending on impurities, the mineral takes on a variety of colors and may fluoresce when irradiated with an ultraviolet lamp.

The solubility of $CaF_2$ is governed by $K_{sp}$ for the salt, hydrolysis of $F^-$ and of $Ca^{2+}$, and by ion pairing between $Ca^{2+}$ and $F^-$:

Solubility products are in Appendix F. The acid dissociation constant of HF is from Appendix G. The hydrolysis constant for Ca is the inverse of the formation constant of CaOH⁺ in Appendix I. The ion pair formation constant for CaF⁺ is listed in Appendix J.

$$CaF_2(s) \rightleftharpoons Ca^{2+} + 2F^- \qquad K_{sp} = [Ca^{2+}][F^-]^2 = 3.2 \times 10^{-11} \quad (13\text{-}32)$$

$$HF \rightleftharpoons H^+ + F^- \qquad K_{HF} = \frac{[H^+][F^-]}{[HF]} = 6.8 \times 10^{-4} \quad (13\text{-}33)$$

$$Ca^{2+} + H_2O \rightleftharpoons CaOH^+ + H^+ \qquad K_a = \frac{[CaOH^+][H^+]}{[Ca^{2+}]} = 2 \times 10^{-13} \quad (13\text{-}34)$$

$$Ca^{2+} + F^- \rightleftharpoons CaF^+ \qquad K_{ip} = \frac{[CaF^+]}{[Ca^{2+}][F^-]} = 4.3 \quad (13\text{-}35)$$

$$H_2O \rightleftharpoons H^+ + OH^- \qquad K_w = [H^+][OH^-] = 1.0 \times 10^{-14} \quad (13\text{-}36)$$

The charge balance is

*Charge balance:* $\quad [H^+] + 2[Ca^{2+}] + [CaOH^+] + [CaF^+] = [OH^-] + [F^-] \qquad (13\text{-}37)$

To find the mass balance, we need to realize that all of the calcium and fluoride species come from $CaF_2$. Therefore, the total fluoride equals two times the total calcium:

$$2[\text{total calcium species}] = [\text{total fluoride species}]$$

$$2\{[Ca^{2+}] + [CaOH^+] + [CaF^+]\} = [F^-] + [HF] + [CaF^+]$$

*Mass balance:* $\quad 2[Ca^{2+}] + 2[CaOH^+] + [CaF^+] = [F^-] + [HF] \qquad (13\text{-}38)$

There are seven independent equations and seven unknowns, so we have enough information.

How to use activity coefficients

For simplicity, we omit activity coefficients, but you do know how to use them. You would solve the problem with all activity coefficients equal to 1, find the ionic strength, and then compute activity coefficients with the Davies equation. Then you would compute effective equilibrium constants incorporating activity coefficients and solve the problem again. After each iteration, you would find a new ionic strength and a new set of activity coefficients. Repeat the process until ionic strength is constant. Wow! You are smart!

We want to reduce seven equations with seven unknowns to one equation with one unknown—but this is not easy. However, we can express all concentrations in terms of $[H^+]$

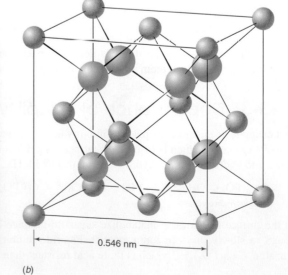

*Figure 13-5* (*a*) Crystals of the mineral fluorite, $CaF_2$. (*b*) Each $Ca^{2+}$ ion is surrounded by eight $F^-$ ions at the corners of a cube. Each $F^-$ ion is surrounded by four $Ca^{2+}$ ions at the corners of a tetrahedron. If you imagine the next unit cell above this one, you should see that the $Ca^{2+}$ ion at the center of the top face of the cell shown here is adjacent to four $F^-$ ions in this cell and four $F^-$ ions in the next cell above it.

F

Ca

|⟵ 0.546 nm ⟶|

(a)          (b)

and $[F^-]$, so we can reduce the mass and charge balance to two equations with two unknowns. Here are the substitutions from the equilibrium expressions:

$$[Ca^{2+}] = K_{sp}/[F^-]^2 \qquad\qquad [HF] = [H^+][F^-]/K_{HF}$$

$$[CaOH^+] = \frac{K_a[Ca^{2+}]}{[H^+]} = \frac{K_aK_{sp}}{[H^+][F^-]^2} \qquad [CaF^+] = K_{ip}[Ca^{2+}][F^-] = \frac{K_{ip}K_{sp}}{[F^-]}$$

The first expression for $[CaOH^+]$ is from Equation 13-34. The second expression is obtained by substituting $[Ca^{2+}] = K_{sp}/[F^-]^2$ into the first equation.

Putting these expressions into the charge balance gives

$$[H^+] + 2[Ca^{2+}] + [CaOH^+] + [CaF^+] - [OH^-] - [F^-] = 0 \qquad \textbf{(13-39a)}$$

$$[H^+] + \frac{2K_{sp}}{[F^-]^2} + \frac{K_aK_{sp}}{[H^+][F^-]^2} + \frac{K_{ip}K_{sp}}{[F^-]} - \frac{K_w}{[H^+]} - [F^-] = 0 \qquad \textbf{(13-39b)}$$

Substitution in the mass balance gives

$$2[Ca^{2+}] + 2[CaOH^+] + [CaF^+] - [F^-] - [HF] = 0$$

$$\frac{2K_{sp}}{[F^-]^2} + \frac{2K_aK_{sp}}{[H^+][F^-]^2} + \frac{K_{ip}K_{sp}}{[F^-]} - [F^-] - \frac{[H^+][F^-]}{K_{HF}} = 0 \qquad \textbf{(13-40)}$$

We could do some messy algebra to solve Equation 13-39b for $[H^+]$ and substitute this expression for $[H^+]$ into Equation 13-40 to get an equation with $[F^-]$ as the only unknown.

Instead, we will use a spreadsheet for a numerical solution. *Suppose that the pH is somehow fixed by a buffer.* The mass balance relating calcium to fluoride is still correct. However, the charge balance is no longer valid, because the buffer introduces additional ions into the solution. We will find a way around the loss of the charge balance shortly.

When buffer is added, the mass balance still applies but the original charge balance is no longer correct.

In the spreadsheet in Figure 13-6, enter pH in column A. Compute $[H^+] = 10^{-pH}$ in column B and *guess* a value for $[F^-]$ in column C. With these values of $[H^+]$ and $[F^-]$, compute

| | A | B | C | D | E | F | G | H | I | J |
|---|---|---|---|---|---|---|---|---|---|---|
| 1 | Finding the concentrations of species in saturated calcium fluoride solution | | | | | | | | | |
| 2 | | | | | | | | | | |
| 3 | $K_{sp} =$ | 3.2E-11 | | Mass balance: | | | | | | |
| 4 | $K_{HF} =$ | 6.8E-04 | | $\dfrac{2K_{sp}}{[F^-]^2} + \dfrac{2K_aK_{sp}}{[H^+][F^-]^2} + \dfrac{K_{ip}K_{sp}}{[F^-]} - [F^-] - \dfrac{[H^+][F^-]}{K_{HF}} = 0$ | | | | | | |
| 5 | $K_a =$ | 2.E-13 | | | | | | | | |
| 6 | $K_{ip} =$ | 4.3 | | | | | | | | |
| 7 | $K_w =$ | 1.0E-14 | | Mass | | | | | | |
| 8 | Input | | $[F^-]$ found by | balance | | | | | | Sum of |
| 9 | pH | $[H^+]$ | SOLVER | sum | $[Ca^{2+}]$ | $[CaOH^+]$ | $[CaF^+]$ | $[HF]$ | $[OH^-]$ | charges |
| 10 | 0 | 1.E+00 | 3.517E-05 | 5.6E-17 | 2.6E-02 | 5.17E-15 | 3.9E-06 | 5.17E-02 | 1.0E-14 | 1.1E+00 |
| 11 | 1 | 1.E-01 | 7.561E-05 | 0.0E+00 | 5.6E-03 | 1.12E-14 | 1.8E-06 | 1.11E-02 | 1.0E-13 | 1.1E-01 |
| 12 | 2 | 1.E-02 | 1.597E-04 | -7.6E-17 | 1.3E-03 | 2.51E-14 | 8.6E-07 | 2.35E-03 | 1.0E-12 | 1.2E-02 |
| 13 | 3 | 1.E-03 | 2.960E-04 | -2.8E-17 | 3.7E-04 | 7.31E-14 | 4.6E-07 | 4.35E-04 | 1.0E-11 | 1.4E-03 |
| 14 | 4 | 1.E-04 | 3.822E-04 | -1.2E-17 | 2.2E-04 | 4.38E-13 | 3.6E-07 | 5.62E-05 | 1.0E-10 | 1.6E-04 |
| 15 | 5 | 1.E-05 | 3.982E-04 | -1.5E-17 | 2.0E-04 | 4.04E-12 | 3.5E-07 | 5.86E-06 | 1.0E-09 | 1.6E-05 |
| 16 | 6 | 1.E-06 | 3.999E-04 | -7.8E-18 | 2.0E-04 | 4.00E-11 | 3.4E-07 | 5.88E-07 | 1.0E-08 | 1.6E-06 |
| 17 | 7 | 1.E-07 | 4.001E-04 | -1.0E-17 | 2.0E-04 | 4.00E-10 | 3.4E-07 | 5.88E-08 | 1.0E-07 | 5.8E-08 |
| 18 | 8 | 1.E-08 | 4.001E-04 | -1.0E-17 | 2.0E-04 | 4.00E-09 | 3.4E-07 | 5.88E-09 | 1.0E-06 | -9.9E-07 |
| 19 | 9 | 1.E-09 | 4.001E-04 | -1.1E-17 | 2.0E-04 | 4.00E-08 | 3.4E-07 | 5.88E-10 | 1.0E-05 | -1.0E-05 |
| 20 | 10 | 1.E-10 | 4.004E-04 | -2.5E-17 | 2.0E-04 | 3.99E-07 | 3.4E-07 | 5.89E-11 | 1.0E-04 | -1.0E-04 |
| 21 | 11 | 1.E-11 | 4.028E-04 | -8.7E-18 | 2.0E-04 | 3.95E-06 | 3.4E-07 | 5.92E-12 | 1.0E-03 | -1.0E-03 |
| 22 | 12 | 1.E-12 | 4.252E-04 | -2.5E-17 | 1.8E-04 | 3.54E-05 | 3.2E-07 | 6.25E-13 | 1.0E-02 | -1.0E-02 |
| 23 | 13 | 1.E-13 | 5.770E-04 | -5.5E-17 | 9.6E-05 | 1.92E-04 | 2.4E-07 | 8.48E-14 | 1.0E-01 | -1.0E-01 |
| 24 | 14 | 1.E-14 | 1.104E-03 | -2.0E-17 | 2.6E-05 | 5.25E-04 | 1.2E-07 | 1.62E-14 | 1.0E+00 | -1.0E+00 |
| 25 | | | | | | | | | | |
| 26 | B10 = 10^-A10 | | | E10 = $B$3/C10^2 | | | F10 = $B$5*$B$3/(B10*C10^2) | | | |
| 27 | G10 = $B$6*$B$3/C10 | | | H10 = B10*C10/$B$4 | | | I10 = $B$7/B10 | | | |
| 28 | D10 = 2*$B$3/C10^2+2*$B$5*$B$3/(B10*C10^2)+$B$6*$B$3/C10-C10-B10*C10/$B$4 | | | | | | | | | |
| 29 | J10 = B10+2*E10+F10+G10-I10-C10 | | | | | | | | | |

*Figure 13-6* Spreadsheet using SOLVER for saturated $CaF_2$ at fixed pH values.

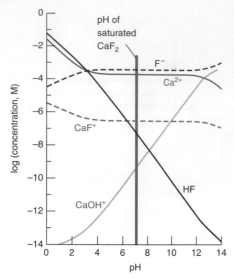

**Figure 13-7**  pH dependence of species in a saturated solution of $CaF_2$. As pH is lowered, $H^+$ reacts with $F^-$ to make HF, and $[Ca^{2+}]$ increases. Note the logarithmic ordinate.

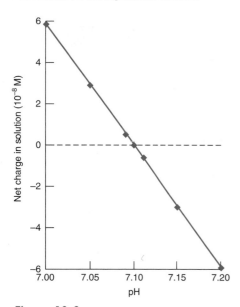

**Figure 13-8**  Net charge in a saturated solution of $CaF_2$ due to $H^+$, $Ca^{2+}$, $CaOH^+$, $CaF^+$, $OH^-$, and $[F^-]$ as a function of pH. The net charge is 0 in unbuffered solution at pH 7.10.

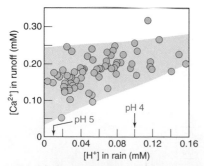

**Figure 13-9**  Measured calcium in acid rain runoff from marble stone (which is largely $CaCO_3$) roughly increases as $[H^+]$ in the rain increases. [Data from P. A. Baedecker and M. M. Reddy, "The Erosion of Carbonate Stone by Acid Rain," J. Chem. Ed. **1993**, 70, 104.]

the mass balance 13-40 in column D. Then use SOLVER to vary $[F^-]$ in column C to make the mass balance 0 in column D.

Each row of the spreadsheet must be dealt with separately. For example, in row 10, the pH was set to 0 in cell A10 and the initial *guessed* value of $[F^-]$ in cell C10 was 0.000 1 M. Before executing SOLVER, Precision was set to 1e-16 in SOLVER Options. In SOLVER, Set Target Cell <u>D10</u> Equal to Value of <u>0</u> By Changing Cells <u>C10</u>. SOLVER changes the value of $[F^-]$ in cell C10 to 3.517E-5 to satisfy the mass balance in cell D10. With the correct value of $[F^-]$ in cell C10, the concentrations of $[Ca^{2+}]$, $[CaOH^+]$, $[CaF^+]$, $[HF]$, and $[OH^-]$ in columns E through I must be correct.

Figure 13-7 shows how the concentrations vary with pH. At low pH, $H^+$ reacts with $F^-$ to produce HF and increases the solubility of $CaF_2$. The species $CaF^+$ and $CaOH^+$ are minor at most pH values, but $CaOH^+$ becomes the major form of calcium above pH 12.7, which is $pK_a$ for Reaction 13-34. A reaction that we did not consider was precipitation of $Ca(OH)_2(s)$. Comparison of the product $[Ca^{2+}][OH^-]^2$ with $K_{sp}$ for $Ca(OH)_2$ indicates that $Ca(OH)_2$ should precipitate at a pH between 13 and 14.

What would be the pH if we had not added buffer? Column J in Figure 13-6 gives the net charge from Equation 13-39a, which includes all ions other than buffer. The pH of unbuffered solution is the pH at which column J is 0. The net charge goes through 0 near pH 7. Finer computations in Figure 13-8 show that the net charge is 0 at pH 7.10. A saturated solution of $CaF_2$ is predicted to have a pH of 7.10 (ignoring activity coefficients).

## Acid Rain Dissolves Minerals and Creates Environmental Hazards

In general, salts of basic ions such as $F^-$, $OH^-$, $S^{2-}$, $CO_3^{2-}$, $C_2O_4^{2-}$, and $PO_4^{3-}$ have increased solubility at low pH, because the anions react with $H^+$. Figure 13-9 shows that marble, which is largely $CaCO_3$, dissolves more readily as the acidity of rain increases. Much of the acid in rain comes from $SO_2$ emissions from combustion of fuels containing sulfur, and from nitrogen oxides produced by all types of combustion. $SO_2$, for example, reacts in the air to make sulfuric acid ($SO_2 + H_2O \rightarrow H_2SO_3 \xrightarrow{oxidation} H_2SO_4$), which finds its way back to the ground in rainfall.

Aluminum is the third most abundant element on earth (after oxygen and silicon), but it is tightly locked into insoluble minerals such as kaolinite ($Al_2(OH)_4Si_2O_5$) and bauxite (AlOOH). Acid rain from human activities is a recent change for our planet that is introducing soluble forms of aluminum (and lead and mercury) into the environment.[5] Figure 13-10 shows that, below pH 5, aluminum is mobilized from minerals and its concentration in lake water rises rapidly. At a concentration of 130 μg/L, aluminum kills fish. In humans, high concentrations of aluminum cause dementia, softening of bones, and anemia. Aluminum is suspected as a possible cause of Alzheimer's disease. Although metallic elements from minerals are liberated by acid, the concentration and availability of metal ions in the environment tend to be regulated by organic matter that binds metal ions.[6]

## Solubility of Barium Oxalate

Now consider the dissolution of $Ba(C_2O_4)$, whose anion is *dibasic* and whose cation is a weak acid.[7] The chemistry in this system is

$$Ba(C_2O_4)(s) \rightleftharpoons Ba^{2+} + \underset{\text{Oxalate}}{C_2O_4^{2-}} \qquad K_{sp} = [Ba^{2+}][C_2O_4^{2-}] = 1.0 \times 10^{-6} \quad \textbf{(13-41)}$$

$$H_2C_2O_4 \rightleftharpoons HC_2O_4^- + H^+ \qquad K_1 = \frac{[H^+][HC_2O_4^-]}{[H_2C_2O_4]} = 5.4 \times 10^{-2} \quad \textbf{(13-42)}$$

$$HC_2O_4^- \rightleftharpoons C_2O_4^{2-} + H^+ \qquad K_2 = \frac{[H^+][C_2O_4^{2-}]}{[HC_2O_4^-]} = 5.42 \times 10^{-5} \quad \textbf{(13-43)}$$

$$Ba^{2+} + H_2O \rightleftharpoons BaOH^+ + H^+ \qquad K_a = \frac{[H^+][BaOH^+]}{[Ba^{2+}]} = 4.4 \times 10^{-14} \quad \textbf{(13-44)}$$

$$Ba^{2+} + C_2O_4^{2-} \rightleftharpoons Ba(C_2O_4)(aq) \qquad K_{ip} = \frac{[Ba(C_2O_4)(aq)]}{[Ba^{2+}][C_2O_4^{2-}]} = 2.1 \times 10^2 \quad \textbf{(13-45)}$$

The value of $K_{sp}$ applies at $\mu = 0.1$ M and 20°C. $K_{ip}$ is for $\mu = 0$ and 18°C. $K_1$, $K_2$, and $K_a$ apply at $\mu = 0$ and 25°C. For lack of better information, we will use this mixed set of equilibrium constants.

The charge balance is

*Charge balance:*  $[H^+] + 2[Ba^{2+}] + [BaOH^+] = [OH^-] + [HC_2O_4^-] + 2[C_2O_4^{2-}]$  **(13-46)**

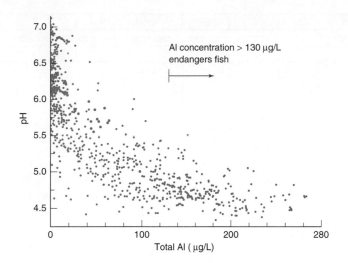

Figure 13-10 Relation of total aluminum (including dissolved and suspended species) in 1 000 Norwegian lakes as a function of the pH of the lake water. The more acidic the water, the greater the aluminum concentration. *[From G. Howells, Acid Rain and Acid Waters, 2nd ed. (Hertfordshire: Ellis Horwood, 1995).]*

The mass balance states that the total moles of barium equal the total moles of oxalate:

$$[\text{Total barium}] = [\text{total oxalate}]$$

$$[\text{Ba}^{2+}] + [\text{BaOH}^+] + \cancel{[\text{Ba(C}_2\text{O}_4)(aq)]} = [\text{H}_2\text{C}_2\text{O}_4] + [\text{HC}_2\text{O}_4^-]$$
$$+ [\text{C}_2\text{O}_4^{2-}] + \cancel{[\text{Ba(C}_2\text{O}_4)(aq)]}$$

*Mass balance:* $\underbrace{[\text{Ba}^{2+}] + [\text{BaOH}^+]}_{F_{\text{Ba}}} = \underbrace{[\text{H}_2\text{C}_2\text{O}_4] + [\text{HC}_2\text{O}_4^-] + [\text{C}_2\text{O}_4^{2-}]}_{F_{\text{H}_2\text{Ox}}}$ (13-47)

We are defining $F_{\text{Ba}}$ and $F_{\text{H}_2\text{Ox}}$ to *exclude* the ion pair $\text{Ba(C}_2\text{O}_4)(aq)$.

There are eight unknowns and eight independent equations (including $[\text{OH}^-] = K_w/[\text{H}^+]$), so we have enough information to find the concentrations of all species.

We deal with ion pairing by adding Reactions 13-41 and 13-45 to find

$$\text{Ba(C}_2\text{O}_4)(s) \rightleftharpoons \text{Ba(C}_2\text{O}_4)(aq) \qquad K = [\text{Ba(C}_2\text{O}_4)(aq)] = K_{sp}K_{ip} = 2.1 \times 10^{-4}$$
$$(13\text{-}48)$$

Therefore, $[\text{Ba(C}_2\text{O}_4)(aq)] = 2.1 \times 10^{-4}$ M as long as undissolved $\text{Ba(C}_2\text{O}_4)(s)$ is present.

Now for our old friends, the fractional composition equations. Abbreviating oxalic acid as $\text{H}_2\text{Ox}$, we can write

The ion pair $\text{Ba(C}_2\text{O}_4)(aq)$ has a constant concentration in this system.

$$[\text{H}_2\text{Ox}] = \alpha_{\text{H}_2\text{Ox}}F_{\text{H}_2\text{Ox}} = \frac{[\text{H}^+]^2 F_{\text{H}_2\text{Ox}}}{[\text{H}^+]^2 + [\text{H}^+]K_1 + K_1K_2} \qquad (13\text{-}49)$$

$F_{\text{H}_2\text{Ox}} = [\text{H}_2\text{C}_2\text{O}_4] + [\text{HC}_2\text{O}_4^-] + [\text{C}_2\text{O}_4^{2-}]$

$$[\text{HOx}^-] = \alpha_{\text{HOx}^-}F_{\text{H}_2\text{Ox}} = \frac{K_1[\text{H}^+]F_{\text{H}_2\text{Ox}}}{[\text{H}^+]^2 + [\text{H}^+]K_1 + K_1K_2} \qquad (13\text{-}50)$$

$$[\text{Ox}^{2-}] = \alpha_{\text{Ox}^{2-}}F_{\text{H}_2\text{Ox}} = \frac{K_1K_2F_{\text{H}_2\text{Ox}}}{[\text{H}^+]^2 + [\text{H}^+]K_1 + K_1K_2} \qquad (13\text{-}51)$$

Also, $\text{Ba}^{2+}$ and $\text{BaOH}^+$ are a conjugate acid-base pair. $\text{Ba}^{2+}$ behaves as a monoprotic acid, HA, and $\text{BaOH}^+$ is the conjugate base $\text{A}^-$.

$$[\text{Ba}^{2+}] = \alpha_{\text{Ba}^{2+}}F_{\text{Ba}} = \frac{[\text{H}^+]F_{\text{Ba}}}{[\text{H}^+] + K_a} \qquad (13\text{-}52)$$

$F_{\text{Ba}} = [\text{Ba}^{2+}] + [\text{BaOH}^+]$

$$[\text{BaOH}^+] = \alpha_{\text{BaOH}^+}F_{\text{Ba}} = \frac{K_aF_{\text{Ba}}}{[\text{H}^+] + K_a} \qquad (13\text{-}53)$$

Suppose that the pH is fixed by adding a buffer (and therefore the charge balance 13-46 is no longer valid). From $K_{sp}$, we can write

$$K_{sp} = [\text{Ba}^{2+}][\text{C}_2\text{O}_4^{2-}] = \alpha_{\text{Ba}^{2+}}F_{\text{Ba}}\alpha_{\text{Ox}^{2-}}F_{\text{H}_2\text{Ox}}$$

But the mass balance 13-47 told us that $F_{\text{Ba}} = F_{\text{H}_2\text{Ox}}$. Therefore,

$$K_{sp} = \alpha_{\text{Ba}^{2+}}F_{\text{Ba}}\alpha_{\text{Ox}^{2-}}F_{\text{H}_2\text{Ox}} = \alpha_{\text{Ba}^{2+}}F_{\text{Ba}}\alpha_{\text{Ox}^{2-}}F_{\text{Ba}}$$

$$\Rightarrow F_{\text{Ba}} = \sqrt{\frac{K_{sp}}{\alpha_{\text{Ba}^{2+}}\alpha_{\text{Ox}^{2-}}}} \qquad (13\text{-}54)$$

13-3 Dependence of Solubility on pH

| | A | B | C | D | E | F | G | H |
|---|---|---|---|---|---|---|---|---|
| 1 | Finding the concentrations of species in saturated barium oxalate solution | | | | | | | |
| 2 | | | | | | | | |
| 3 | $K_{sp}$ = | 1.0E-06 | | $K_1$ = | 5.4E-02 | | $K_{ip}$ = | 2.1E+02 |
| 4 | $K_a$ = | 4.4E-14 | | $K_2$ = | 5.42E-05 | | $K_w$ = | 1.0E-14 |
| 5 | | | | | | | | $F_{Ba}$ |
| 6 | pH | [H$^+$] | $\alpha$(H$_2$Ox) | $\alpha$(HOx$^-$) | $\alpha$(Ox$^{2-}$) | $\alpha$(Ba$^{2+}$) | $\alpha$(BaOH$^+$) | = $F_{H_2Ox}$ |
| 7 | 0 | 1.E+00 | 9.5E-01 | 5.1E-02 | 2.8E-06 | 1.0E+00 | 4.4E-14 | 6.0E-01 |
| 8 | 2 | 1.E-02 | 1.6E-01 | 8.4E-01 | 4.6E-03 | 1.0E+00 | 4.4E-12 | 1.5E-02 |
| 9 | 4 | 1.E-04 | 1.2E-03 | 6.5E-01 | 3.5E-01 | 1.0E+00 | 4.4E-10 | 1.7E-03 |
| 10 | 6 | 1.E-06 | 3.4E-07 | 1.8E-02 | 9.8E-01 | 1.0E+00 | 4.4E-08 | 1.0E-03 |
| 11 | 7.643 | 2.E-08 | 1.8E-10 | 4.2E-04 | 1.0E+00 | 1.0E+00 | 1.9E-06 | 1.0E-03 |
| 12 | 8 | 1.E-08 | 3.4E-11 | 1.8E-04 | 1.0E+00 | 1.0E+00 | 4.4E-06 | 1.0E-03 |
| 13 | 10 | 1.E-10 | 3.4E-15 | 1.8E-06 | 1.0E+00 | 1.0E+00 | 4.4E-04 | 1.0E-03 |
| 14 | 12 | 1.E-12 | 3.4E-19 | 1.8E-08 | 1.0E+00 | 9.6E-01 | 4.2E-02 | 1.0E-03 |
| 15 | 14 | 1.E-14 | 3.4E-23 | 1.8E-10 | 1.0E+00 | 1.9E-01 | 8.1E-01 | 2.3E-03 |
| 16 | | | | | | | | |
| 17 | | | | | | | | net |
| 18 | pH | [Ba$^{2+}$] | [BaOH$^+$] | [H$_2$Ox] | [HOx$^-$] | [Ox$^{2-}$] | [OH$^-$] | charge |
| 19 | 0 | 6.0E-01 | 2.6E-14 | 5.7E-01 | 3.1E-02 | 1.7E-06 | 1.0E-14 | 2.2E+00 |
| 20 | 2 | 1.5E-02 | 6.5E-14 | 2.3E-03 | 1.2E-02 | 6.7E-05 | 1.0E-12 | 2.7E-02 |
| 21 | 4 | 1.7E-03 | 7.4E-13 | 2.0E-06 | 1.1E-03 | 5.9E-04 | 1.0E-10 | 1.2E-03 |
| 22 | 6 | 1.0E-03 | 4.4E-11 | 3.4E-10 | 1.8E-05 | 9.9E-04 | 1.0E-08 | 1.9E-05 |
| 23 | 7.643 | 1.0E-03 | 1.9E-09 | 1.8E-13 | 4.2E-07 | 1.0E-03 | 4.4E-07 | 6.9E-18 |
| 24 | 8 | 1.0E-03 | 4.4E-09 | 3.4E-14 | 1.8E-07 | 1.0E-03 | 1.0E-06 | -8.1E-07 |
| 25 | 10 | 1.0E-03 | 4.4E-07 | 3.4E-18 | 1.8E-09 | 1.0E-03 | 1.0E-04 | -1.0E-04 |
| 26 | 12 | 9.8E-04 | 4.3E-05 | 3.5E-22 | 1.9E-11 | 1.0E-03 | 1.0E-02 | -1.0E-02 |
| 27 | 14 | 4.3E-04 | 1.9E-03 | 7.9E-26 | 4.3E-13 | 2.3E-03 | 1.0E+00 | -1.0E+00 |
| 28 | | | | | | | | |
| 29 | B7 = 10^-A7 | | | | | | B19 = F7*H7 | |
| 30 | C7 = $B7^2/($B7^2+$B7*$E$3+$E$3*$E$4) | | | | | | C19 = G7*H7 | |
| 31 | D7 = $B7*$E$3/($B7^2+$B7*$E$3+$E$3*$E$4) | | | | | | D19 = C7*H7 | |
| 32 | E7  $E$3*$E$4/($B7^2+$B7*$E$3+$E$3*$E$4) | | | | | | E19 = D7*H7 | |
| 33 | F7 = B7/(B7+$B$4) | | | | | | F19 = E7*H7 | |
| 34 | G7 = $B$4/(B7+$B$4) | | | | | | G19 = $H$4/B7 | |
| 35 | H7 = SQRT($B$3/(E7*F7)) | | H19 = B7+2*B19+C19-G19-E19-2*F19 | | | | | |

*Figure 13-11*  Spreadsheet for saturated BaC$_2$O$_4$. SOLVER was used to find the pH in cell A11 necessary to make the net charge 0 in cell H23.

In the spreadsheet in Figure 13-11, pH is specified in column A. From this pH, plus $K_1$ and $K_2$, the fractions $\alpha_{H_2Ox}$, $\alpha_{HOx^-}$, and $\alpha_{Ox^{2-}}$ are computed with Equations 13-49 through 13-51 in columns C, D, and E. From the pH and $K_a$, the fractions $\alpha_{Ba^{2+}}$ and $\alpha_{BaOH^+}$ are computed with Equations 13-52 and 13-53 in columns F and G. The total concentrations of barium and oxalate, $F_{Ba}$ and $F_{H_2Ox}$, are equal and calculated in column H from Equation 13-54. In a real spreadsheet, we would have continued to the right with column I. To fit on this page, the spreadsheet was continued in row 18. In this lower section, the concentrations of [Ba$^{2+}$] and [BaOH$^+$] are computed with Equations 13-52 and 13-53. [H$_2$C$_2$O$_4$], [HC$_2$O$_4^-$], and [C$_2$O$_4^{2-}$] are found with Equations 13-49 through 13-51.

The net charge (= [H$^+$] + 2[Ba$^{2+}$] + [BaOH$^+$] − [OH$^-$] − [HC$_2$O$_4^-$] − 2[C$_2$O$_4^{2-}$]) is computed beginning in cell H19. If we had not added buffer to fix the pH, the net charge would be 0. Net charge goes from positive to negative between pH 6 and 8. By using SOLVER, we find the pH in cell A11 that makes the net charge 0 in cell H23 (with SOLVER Precision = 1e-16). This pH, 7.64, is the pH of unbuffered solution.

Results in Figure 13-12 show that the solubility of barium oxalate is steady at $10^{-3}$ M in the middle pH range. Solubility increases below pH 5 because C$_2$O$_4^{2-}$ reacts with H$^+$ to make HC$_2$O$_4^-$. Solubility increases above pH 13 because Ba$^{2+}$ reacts with OH$^-$ to make BaOH$^+$.

A final point is to see that the solubility of Ba(OH)$_2(s)$ is not exceeded. Evaluation of the product [Ba$^{2+}$][OH$^-$]$^2$ shows that $K_{sp} = 3 \times 10^{-4}$ is not exceeded below pH 13.9. We predict that Ba(OH)$_2(s)$ will begin to precipitate at pH 13.9.

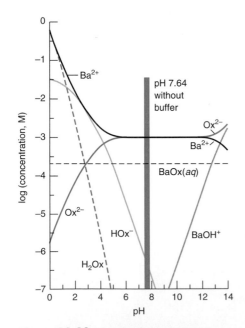

*Figure 13-12*  pH dependence of the concentrations of species in saturated BaC$_2$O$_4$. As pH is lowered, H$^+$ reacts with C$_2$O$_4^{2-}$ to make HC$_2$O$_4^-$ and H$_2$C$_2$O$_4$, and the concentration of Ba$^{2+}$ increases.

## 13-4 Analyzing Acid-Base Titrations with Difference Plots[8]

A *difference plot,* also called a *Bjerrum plot,* is an excellent means to extract metal-ligand formation constants or acid dissociation constants from titration data obtained with electrodes. We will apply the difference plot to an acid-base titration curve.

We derive the key equation for a diprotic acid, $H_2A$, and extend it to a general acid, $H_nA$. The mean fraction of protons bound to $H_2A$ ranges from 0 to 2 and is defined as

$$\bar{n}_H = \frac{\text{moles of bound } H^+}{\text{total moles of weak acid}} = \frac{2[H_2A] + [HA^-]}{[H_2A] + [HA^-] + [A^{2-}]} \qquad (13\text{-}55)$$

We can measure $\bar{n}_H$ by a titration beginning with a mixture of $A$ mmol of $H_2A$ and $C$ mmol of HCl in $V_0$ mL. The reason why we might add HCl is to increase the degree of protonation of $H_2A$, which is partially dissociated in the absence of HCl. We titrate the solution with standard NaOH whose concentration is $C_b$ mol/L. After adding $v$ mL of NaOH, the number of mmol of $Na^+$ in the solution is $C_b v$.

To maintain a nearly constant ionic strength, the solution of $H_2A$ plus HCl contains 0.10 M KCl and the concentrations of $H_2A$ and HCl are much less than 0.10 M. NaOH is sufficiently concentrated so that the added volume is small compared to $V_0$.

The charge balance for the titration solution is

$$[H^+] + [Na^+] + [K^+] = [OH^-] + [Cl^-]_{HCl} + [Cl^-]_{KCl} + [HA^-] + 2[A^{2-}]$$

where $[Cl^-]_{HCl}$ is from HCl and $[Cl^-]_{KCl}$ is from KCl. But $[K^+] = [Cl^-]_{KCl}$, so we cancel these terms. The net charge balance is

$$[H^+] + [Na^+] = [OH^-] + [Cl^-]_{HCl} + [HA^-] + 2[A^{2-}] \qquad (13\text{-}56)$$

The denominator of Equation 13-55 is $F_{H_2A} = [H_2A] + [HA^-] + [A^{2-}]$. The numerator can be written as $2F_{H_2A} - [HA^-] - 2[A^{2-}]$. Therefore

$$\bar{n}_H = \frac{2F_{H_2A} - [HA^-] - 2[A^{2-}]}{F_{H_2A}} \qquad (13\text{-}57)$$

From Equation 13-56, we can write: $-[HA^-] - 2[A^{2-}] = [OH^-] + [Cl^-]_{HCl} - [H^+] - [Na^+]$. Substituting this expression into the numerator of Equation 13-57 gives

$$\bar{n}_H = \frac{2F_{H_2A} + [OH^-] + [Cl^-]_{HCl} - [H^+] - [Na^+]}{F_{H_2A}}$$

$$= 2 + \frac{[OH^-] + [Cl^-]_{HCl} - [H^+] - [Na^+]}{F_{H_2A}}$$

For the general polyprotic acid $H_nA$, the mean fraction of bound protons turns out to be

$$\bar{n}_H = n + \frac{[OH^-] + [Cl^-]_{HCl} - [H^+] - [Na^+]}{F_{H_nA}} \qquad (13\text{-}58)$$

Each term on the right side of Equation 13-58 is known during the titration. From the reagents that were mixed, we can say

$$F_{H_2A} = \frac{\text{mmol } H_2A}{\text{total volume}} = \frac{A}{V_0 + v} \qquad [Cl^-]_{HCl} = \frac{\text{mmol HCl}}{\text{total volume}} = \frac{C}{V_0 + v}$$

$$[Na^+] = \frac{\text{mmol NaOH}}{\text{total volume}} = \frac{C_b v}{V_0 + v}$$

$[H^+]$ and $[OH^-]$ are measured with a pH electrode and calculated as follows: Let the effective value of $K_w$ that applies at $\mu = 0.10$ M be $K'_w = K_w/(\gamma_{H^+}\gamma_{OH^-}) = [H^+][OH^-]$ (Equation 13-25). Remembering that $pH = -\log([H^+]\gamma_{H^+})$, we write

$$[H^+] = \frac{10^{-pH}}{\gamma_{H^+}} \qquad [OH^-] = \frac{K'_w}{[H^+]} = 10^{(pH - pK'_w)} \cdot \gamma_{H^+}$$

Substituting into Equation 13-58 gives the measured fraction of bound protons:

$$\bar{n}_H(\text{measured}) = n + \frac{10^{(pH - pK'_w)} \cdot \gamma_{H^+} + C/(V_0 + v) - (10^{-pH})/\gamma_{H^+} - C_b v/(V_0 + v)}{A/(V_0 + v)}$$

$$(13\text{-}59)$$

Niels Bjerrum (1879-1958) was a Danish physical chemist who made fundamental contributions to inorganic coordination chemistry and is responsible for much of our understanding of acids and bases and titration curves.[9]

Experimental fraction of protons bound to polyprotic acid

In acid-base titrations, a **difference plot,** or *Bjerrum plot,* is a graph of the mean fraction of protons bound to an acid versus pH. The mean fraction is $\bar{n}_H$ calculated with Equation 13-59. For complex formation, the difference plot gives the mean number of ligands bound to a metal versus pL$(= -\log[\text{ligand}])$.

Equation 13-59 gives the measured value of $\bar{n}_H$. What is the theoretical value? For a diprotic acid, the theoretical mean fraction of bound protons is

$$\bar{n}_H(\text{theoretical}) = 2\alpha_{H_2A} + \alpha_{HA^-} \qquad (13\text{-}60)$$

If Equation 13-60 is not obvious, you can derive it from Equation 13-55 by writing the right-hand side as

$$\frac{2[H_2A] + [HA^-]}{[H_2A] + [HA^-] + [A^{2-}]}$$
$$= \frac{2[H_2A] + [HA^-]}{F_{H_2A}}$$
$$= \frac{2[H_2A]}{F_{H_2A}} + \frac{[HA^-]}{F_{H_2A}}$$
$$= 2\alpha_{H_2A} + \alpha_{HA^-}$$

where $\alpha_{H_2A}$ is the fraction of acid in the form $H_2A$ and $\alpha_{HA^-}$ is the fraction in the form $HA^-$. You should be able to write expressions for $\alpha_{H_2A}$ and $\alpha_{HA^-}$ in your sleep by now.

$$\alpha_{H_2A} = \frac{[H^+]^2}{[H^+]^2 + [H^+]K_1 + K_1K_2} \qquad \alpha_{HA^-} = \frac{[H^+]K_1}{[H^+]^2 + [H^+]K_1 + K_1K_2} \qquad (13\text{-}61)$$

We extract $K_1$ and $K_2$ from an experimental titration by constructing a difference plot with Equation 13-59. This plot is a graph of $\bar{n}_H(\text{measured})$ versus pH. We then fit the theoretical curve (Equation 13-60) to the experimental curve by the method of least squares to find the values of $K_1$ and $K_2$ that minimize the sum of the squares of the residuals:

Best values of $K_1$ and $K_2$ minimize the sum of the squares of the residuals.

$$\Sigma(\text{residuals})^2 = \Sigma[\bar{n}_H(\text{measured}) - \bar{n}_H(\text{theoretical})]^2 \qquad (13\text{-}62)$$

$H_3NCH_2CO_2H$

Glycine
$pK_1 = 2.35$ at $\mu = 0$
$pK_2 = 9.78$ at $\mu = 0$

Experimental data for a titration of the amino acid glycine are given in Figure 13-13. The initial 40.0-mL solution contained 0.190 mmol of glycine plus 0.232 mmol of HCl to increase the fraction of fully protonated $^+H_3NCH_2CO_2H$. Aliquots of 0.490 5 M NaOH were added and the pH was measured after each addition. Volumes and pH are listed in columns A and B beginning in row 16. pH was precise to the 0.001 decimal place, but the accuracy of pH measurement is, at best, ±0.02.

Input values of concentration, volume, and moles are in cells B3:B6 in Figure 13-13. Cell B7 has the value 2 to indicate that glycine is a diprotic acid. Cell B8 has the activity coefficient of $H^+$ computed with the Davies equation, 13-18. Cell B9 begins with the effec-

| | A | B | C | D | E | F | G | H | I |
|---|---|---|---|---|---|---|---|---|---|
| 1 | Difference plot for glycine | | | | | | | | |
| 2 | | | | C16 = 10^-B16/$B$8 | | | | | |
| 3 | Titrant NaOH = | 0.4905 | $C_b$ (M) | D16 = 10^-$B$9/C16 | | | | | |
| 4 | Initial volume = | 40 | $V_0$ (mL) | E16 = $B$7+($B$6-$B$3*A16-(C16-D16)*($B$4+A16))/$B$5 | | | | | |
| 5 | Glycine = | 0.190 | L (mmol) | F16 + $C16^2/($C16^2+$C16*$E$10+$E$10*$E$11) | | | | | |
| 6 | HCl added = | 0.232 | A (mmol) | G16 = $C16*$E$10/($C16^2+$C16*$E$10+$E$10*$E$11) | | | | | |
| 7 | Number of H$^+$ = | 2 | n | H16 = 2*F16+G16 | | | | | |
| 8 | Activity coeff = | 0.78 | $\gamma_H$ | I16 = (E16-H16)^2 | | | | | |
| 9 | $pK_w'$ = | 13.807 | | | | | | | |
| 10 | $pK_1$ = | 2.312 | | $K_1$ = | 0.0048713 | = 10^-B10 | | | |
| 11 | $pK_2$ = | 9.625 | | $K_2$ = | 2.371E-10 | = 10^-B11 | | | |
| 12 | $\Sigma(\text{resid})^2$ = | 0.0048 | = sum of column I | | | | | | |
| 13 | | | | | | | | | |
| 14 | v | pH | [H$^+$] = | [OH$^-$] = | Measured | | | Theoretical | (residuals)$^2$ = |
| 15 | mL NaOH | | (10$^{-pH}$)/$\gamma_H$ | (10$^{-pK_w}$)/[H$^+$] | $n_H$ | $\alpha_{H2A}$ | $\alpha_{HA^-}$ | $n_H$ | $(n_{meas} - n_{thoer})^2$ |
| 16 | 0.00 | 2.234 | 7.48E-03 | 2.08E-12 | 1.646 | 0.606 | 0.394 | 1.606 | 0.001656 |
| 17 | 0.02 | 2.244 | 7.31E-03 | 2.13E-12 | 1.630 | 0.600 | 0.400 | 1.600 | 0.000879 |
| 18 | 0.04 | 2.254 | 7.14E-03 | 2.18E-12 | 1.612 | 0.595 | 0.405 | 1.595 | 0.000319 |
| 19 | 0.06 | 2.266 | 6.95E-03 | 2.24E-12 | 1.601 | 0.588 | 0.412 | 1.588 | 0.000174 |
| 20 | 0.08 | 2.278 | 6.76E-03 | 2.30E-12 | 1.589 | 0.581 | 0.419 | 1.581 | 0.000056 |
| 21 | 0.10 | 2.291 | 6.56E-03 | 2.38E-12 | 1.578 | 0.574 | 0.426 | 1.574 | 0.000020 |
| 22 | : | | | | | | | | |
| 23 | 0.50 | 2.675 | 2.71E-03 | 5.75E-12 | 1.353 | 0.357 | 0.643 | 1.357 | 0.000022 |
| 24 | : | | | | | | | | |
| 25 | 1.56 | 11.492 | 4.13E-12 | 3.77E-03 | 0.016 | 0.000 | 0.017 | 0.017 | 0.000000 |
| 26 | 1.58 | 11.519 | 3.88E-12 | 4.01E-03 | 0.018 | 0.000 | 0.016 | 0.016 | 0.000004 |
| 27 | 1.60 | 11.541 | 3.69E-12 | 4.22E-03 | 0.015 | 0.000 | 0.015 | 0.015 | 0.000000 |

*Figure 13-13* Spreadsheet for difference plot of the titration of 0.190 mmol glycine plus 0.232 mmol HCl in 40.0 mL with 0.490 5 M NaOH. Cells A16:B27 give only a fraction of the experimental data. Complete data listed in Problem 13-15 provided by A. Kraft, Heriot-Watt University.

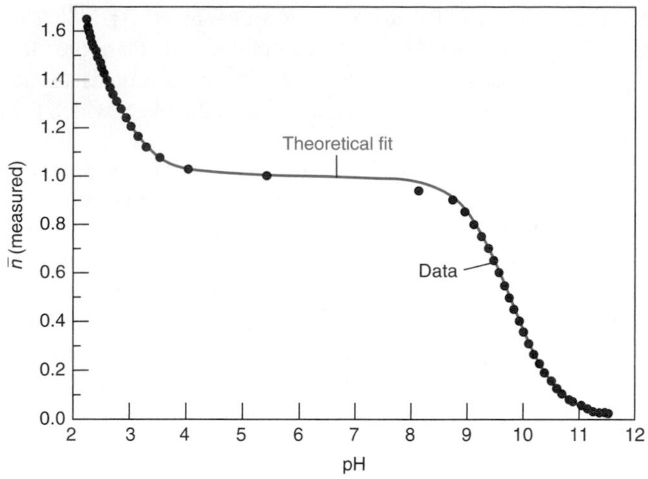

*Figure 13-14* Bjerrum difference plot for the titration of glycine. Many experimental points are omitted from the figure for clarity.

tive value of $pK'_w = 13.797$ in 0.1 M KCl.[10] We allowed the spreadsheet to vary $pK'_w$ for best fit of the experimental data, giving 13.807 in cell B9. Cells B10 and B11 began with estimates of $pK_1$ and $pK_2$ for glycine. We used the values 2.35 and 9.78 from Table 10-1, which apply at $\mu = 0$. As explained in the next subsection, we use SOLVER to vary $pK_1$, $pK_2$, and $pK'_w$ for best fit of the experimental data, giving 2.312 and 9.625 in cells B10 and B11.

The spreadsheet in Figure 13-13 computes [H$^+$] and [OH$^-$] in columns C and D, beginning in row 16. The mean fraction of protonation, $\bar{n}_H$(measured) from Equation 13-59, is in column E. The Bjerrum difference plot in Figure 13-14 shows $\bar{n}_H$(measured) versus pH. Values of $\alpha_{H_2A}$ and $\alpha_{HA^-}$ from Equations 13-61 are computed in columns F and G, and $\bar{n}_H$(theoretical) was computed with Equation 13-60 in column H. Column I contains the squares of the residuals, $[\bar{n}_H(\text{measured}) - \bar{n}_H(\text{theoretical})]^2$. The sum of the squares of the residuals is in cell B12.

## Using Excel SOLVER to Optimize More Than One Parameter

We want values of $pK'_w$, $pK_1$, and $pK_2$ that minimize the sum of squares of residuals in cell B12. Select SOLVER from the TOOLS menu. In the SOLVER window, Set Target Cell <u>B12</u> Equal to <u>Min</u> By Changing Cells <u>B9, B10, B11</u>. Then click Solve and SOLVER finds the best values in cells B9, B10, and B11 to minimize the sum of squares of residuals in cell B12. Starting with 13.797, 2.35, and 9.78 in cells B9, B10, and B11 gives a sum of squares of residuals equal to 0.110 in cell B12. After SOLVER is executed, cells B9, B10, and B11 become 13.807, 2.312, and 9.625. The sum in cell B12 is reduced to 0.0048. When you use SOLVER to optimize several parameters at once, it is a good idea to try different starting values to see if the same solution is reached. Sometimes a local minimum can be reached that is not as low as might be reached elsewhere in parameter space.

The theoretical curve $\bar{n}_H(\text{theoretical}) = 2\alpha_{H_2A} + \alpha_{HA^-}$ using the results of SOLVER are contained in columns F, G, and H in Figure 13-13 and shown by the solid curve in Figure 13-14. The curve fits the experimental data very well, suggesting that we have found reliable values of $pK_1$ and $pK_2$.

It may seem inappropriate to allow $pK'_w$ to vary, because we claimed to know $pK'_w$ at the outset. The change of $pK'_w$ from 13.797 to 13.807 significantly improved the fit. The value 13.797 gave values of $\bar{n}_H$(measured) that became level near 0.04 at the end of the titration in Figure 13-14. This behavior is qualitatively incorrect, because $\bar{n}_H$ must approach 0 at high pH. A small change in $pK'_w$ markedly improves the fit as $\bar{n}_H$ approaches 0.

Strictly speaking, $pK_1$ and $pK_2$ should have primes to indicate that they apply in 0.10 M KCl. We left off the primes to avoid complicating the symbols. We did distinguish $K_w$, which applies at $\mu = 0$, from $K'_w$, which applies at $\mu = 0.10$ M.

## Terms to Understand

coupled equilibria                    difference plot

## Summary

Coupled equilibria are reversible reactions that have a species in common. Therefore each reaction has an effect on the other.

The general treatment of acid-base systems begins with charge and mass balances and equilibrium expressions. There should be as many independent equations as chemical species. Substitute a frac-tional composition equation for each acid or base into the charge balance. After you enter known concentrations of species such as Na$^+$ and Cl$^-$ and substitute $K_w/[\text{H}^+]$ for [OH$^-$], the remaining variable should be [H$^+$]. Use Excel SOLVER to find [H$^+$] and then solve for all other concentrations from [H$^+$]. If there are equilibria in addition to acid-base reactions, such as ion pairing, then you need the full systematic treatment of equilibrium. Make maximum use of fractional composition equations to simplify the problem.

To use activity coefficients, first solve the equilibrium problem with all activity coefficients equal to unity. From the resulting concentrations, compute the ionic strength and use the Davies equation to find activity coefficients. With activity coefficients, calculate the effective equilibrium constant $K'$ for each chemical reaction. $K'$ is the equilibrium quotient of concentrations at a particular ionic strength. Solve the problem again with $K'$ values and find a new ionic strength. Repeat the cycle until the concentrations reach constant values.

We considered solubility problems in which the cation and anion could each undergo one or more acid-base reactions and in which ion pairing could occur. Substitute fractional composition expressions for all acid-base species into the mass balance. In some systems, such as barium oxalate, the resulting equation contains the formal concentrations of anion and cation and [H⁺]. The solubility product provides a relation between the formal concentrations of anion and cation, so you can eliminate one of them from the mass balance. By assuming a value for $[H^+]$, you can solve for the remaining formal concentration and, therefore, for all concentrations. By this means, find the composition as a function of pH. The pH of the unbuffered solution is the pH at which the charge balance is satisfied.

In the more difficult calcium fluoride solubility problem, ion pairing prevented us from reducing the equations to one unknown. Instead, $[H^+]$ and $[F^-]$ were both unknown. However, for a fixed value of $[H^+]$, we could use SOLVER to solve for $[F^-]$. The correct pH is the one at which the charge balance is satisfied.

To extract acid dissociation constants from an acid-base titration curve, we can construct a difference plot, or Bjerrum plot, which is a graph of the mean fraction of bound protons, $\bar{n}_H$, versus pH. This mean fraction can be measured from the quantities of reagents that were mixed and the measured pH. The theoretical shape of the difference plot is an expression in terms of fractional compositions. Use Excel SOLVER to vary equilibrium constants to obtain the best fit of the theoretical curve to the measured points. This process minimizes the sum of squares $[\bar{n}_H(\text{measured}) - \bar{n}_H(\text{theoretical})]^2$.

## Exercises

*Instructors:* Most of these exercises are quite long. Please be kind when you assign them.

**13-A.** Neglecting activity coefficients and ion pairing, find the pH and concentrations of species in 1.00 L of solution containing 0.010 mol hydroxybenzene (HA), 0.030 mol dimethylamine (B), and 0.015 mol HCl.

**13-B.** Repeat Exercise 13-A with activity coefficients from the Davies equation.

**13-C.** (a) Neglecting activity coefficients and ion pairing, find the pH and concentrations of species in 1.00 L of solution containing 0.040 mol 2-aminobenzoic acid (a neutral molecule, HA), 0.020 mol dimethylamine (B), and 0.015 mol HCl.
(b) What fraction of HA is in each of its three forms? What fraction of B is in each of its two forms? Compare your answers with what you would find if HCl reacted with B and then excess B reacted with HA. What pH do you predict from this simple estimate?

**13-D.** Include activity coefficients from the Davies equation to find the pH and concentrations of species in the mixture of sodium tartrate, pyridinium chloride, and KOH in Section 13-1. Consider only Reactions 13-1 through 13-4.

**13-E.** (a) Using the ion-pair equilibrium constant in Appendix J, with activity coefficients = 1, find the concentrations of species in 0.025 M MgSO₄. Hydrolysis of the cation and anion near neutral pH is negligible. Only consider ion-pair formation. You can solve this problem exactly with a quadratic equation. Alternatively, if you use SOLVER, set Precision to 1e-6 (not 1e-16) in the SOLVER Options. If Precision is much smaller, SOLVER does not find a satisfactory solution. The success of SOLVER in this problem depends on how close your initial guess is to the correct answer.
(b) Compute a new ionic strength and repeat part (a) with new activity coefficients from the Davies equation. Perform several iterations until the ionic strength is constant. The fraction of ion pairing that you find should be close to that in Box 8-1, which was derived with Debye-Hückel activity coefficients.
(c) We naively assign the ionic strength of 0.025 M MgSO₄ to be 0.10 M. What is the actual ionic strength of this solution?

**13-F.** (a) Find the concentrations of species in saturated CaF₂ as a function of pH by using Reactions 13-32 through 13-36 and adding the following reaction:

$$HF(aq) + F^- \rightleftharpoons HF_2^- \qquad K_{HF_2} = 10^{0.58}$$

Do not include activity coefficients. Produce a graph similar to Figure 13-7.
(b) Find the pH and concentrations of species in saturated CaF₂.

**13-G.** Make a graph of $[Ag^+]$, $[AgOH(aq)]$, $[CN^-]$, and $[HCN]$ as a function of pH in a saturated solution of AgCN. Consider the following equilibria and do not consider activity coefficients. Find the pH if no buffer were added.

| | |
|---|---|
| AgCN(s) | $pK_{sp} = 15.66$ |
| $HCN(aq) \rightleftharpoons CN^- + H^+$ | $pK_{HCN} = 9.21$ |
| $Ag^+ + H_2O \rightleftharpoons AgOH(aq) + H^+$ | $pK_{Ag} = 12.0$ |

**13-H.** *Difference plot.* A solution containing 3.96 mmol acetic acid plus 0.484 mmol HCl in 200 mL of 0.10 M KCl was titrated with 0.490 5 M NaOH to measure $K_a$ for acetic acid.
(a) Write expressions for the experimental mean fraction of protonation, $\bar{n}_H(\text{measured})$, and the theoretical mean fraction of protonation, $\bar{n}_H(\text{theoretical})$.
(b) From the following data, prepare a graph of $\bar{n}_H(\text{measured})$ versus pH. Find the best values of $pK_a$ and $pK'_w$ by minimizing the sum of the squares of the residuals, $\Sigma[\bar{n}_H(\text{measured}) - \bar{n}_H(\text{theoretical})]^2$.

| V (mL) | pH | V (mL) | pH | V (mL) | pH | V (mL) | pH |
|---|---|---|---|---|---|---|---|
| 0.00 | 2.79 | 2.70 | 4.25 | 5.40 | 4.92 | 8.10 | 5.76 |
| 0.30 | 2.89 | 3.00 | 4.35 | 5.70 | 4.98 | 8.40 | 5.97 |
| 0.60 | 3.06 | 3.30 | 4.42 | 6.00 | 5.05 | 8.70 | 6.28 |
| 0.90 | 3.26 | 3.60 | 4.50 | 6.30 | 5.12 | 9.00 | 7.23 |
| 1.20 | 3.48 | 3.90 | 4.58 | 6.60 | 5.21 | 9.30 | 10.14 |
| 1.50 | 3.72 | 4.20 | 4.67 | 6.90 | 5.29 | 9.60 | 10.85 |
| 1.80 | 3.87 | 4.50 | 4.72 | 7.20 | 5.38 | 9.90 | 11.20 |
| 2.10 | 4.01 | 4.80 | 4.78 | 7.50 | 5.49 | 10.20 | 11.39 |
| 2.40 | 4.15 | 5.10 | 4.85 | 7.80 | 5.61 | 10.50 | 11.54 |

*Data from A. Kraft, J. Chem. Ed. 2003, 80, 554.*

## Problems

*Instructors:* Most of these problems are quite long. Please be kind when you assign them.

**13-1.** Why does the solubility of a salt of a basic anion increase with decreasing pH? Write chemical reactions for the minerals galena (PbS) and cerussite ($PbCO_3$) to explain how acid rain mobilizes traces of metal from relatively inert forms into the environment, where the metals can be taken up by plants and animals.

**13-2.** ▦ (a) Considering just acid-base chemistry, not ion pairing and not activity coefficients, use the systematic treatment of equilibrium to find the pH of 1.00 L of solution containing 0.010 0 mol hydroxybenzene (HA) and 0.005 0 mol KOH.

(b) What pH would you have predicted from your knowledge of Chapter 9?

(c) Find the pH if [HA] and [KOH] were both reduced by a factor of 100.

**13-3.** ▦ Repeat part (a) of Problem 13-2 with Davies activity coefficients. Remember that $pH = -\log([H^+]\gamma_{H^+})$.

**13-4.** From $pK_1$ and $pK_2$ for glycine at $\mu = 0$ in Table 10-1, compute $pK_1'$ and $pK_2'$, which apply at $\mu = 0.1$ M. Use the Davies equation for activity coefficients. Compare your answer with experimental values in cells B10 and B11 of Figure 13-13.

**13-5.** ▦ Considering just acid-base chemistry, not ion pairing and not activity coefficients, use the systematic treatment of equilibrium to find the pH and concentrations of species in 1.00 L of solution containing 0.100 mol ethylenediamine and 0.035 mol HBr. Compare the pH with that found by the methods of Chapter 11.

**13-6.** ▦ Considering just acid-base chemistry, not ion pairing and not activity coefficients, find the pH and concentrations of species in 1.00 L of solution containing 0.040 mol benzene-1,2,3-tricarboxylic acid ($H_3A$), 0.030 mol imidazole (a neutral molecule, HB), and 0.035 mol NaOH.

**13-7.** ▦ Considering just acid-base chemistry, not ion pairing and not activity coefficients, find the pH and concentrations of species in 1.00 L of solution containing 0.020 mol arginine, 0.030 mol glutamic acid, and 0.005 mol KOH.

**13-8.** ▦ Solve Problem 13-7 by using Davies activity coefficients.

**13-9.** ▦ A solution containing 0.008 695 $m$ $KH_2PO_4$ and 0.030 43 $m$ $Na_2HPO_4$ is a primary standard buffer with a stated pH of 7.413 at 25°C. Calculate the pH of this solution by using the systematic treatment of equilibrium with activity coefficients from (a) the Davies equation and (b) the extended Debye-Hückel equation.

**13-10.** ▦ Considering just acid-base chemistry, not ion pairing and not activity coefficients, find the pH and composition of 1.00 L of solution containing 0.040 mol $H_4$EDTA (EDTA = ethylenedinitrilotetraacetic acid ≡ $H_4A$), 0.030 mol lysine (neutral molecule ≡ HL), and 0.050 mol NaOH.

**13-11.** ▦ The solution containing no added $KNO_3$ for Figure 8-1 contains 5.0 mM $Fe(NO_3)_3$, 5.0 μM NaSCN, and 15 mM $HNO_3$. We will use Davies activity coefficients to find the concentrations of all species by using the following reactions:

$$Fe^{3+} + SCN^- \rightleftharpoons Fe(SCN)^{2+} \qquad \log\beta_1 = 3.03 \ (\mu = 0)$$
$$Fe^{3+} + 2SCN^- \rightleftharpoons Fe(SCN)_2^+ \qquad \log\beta_2 = 4.6 \ (\mu = 0)$$
$$Fe^{3+} + H_2O \rightleftharpoons FeOH^{2+} + H^+ \qquad pK_a = 2.195 \ (\mu = 0)$$

This problem is harder than you might think, so we will guide you through the steps,

(a) Write the four equilibrium expressions (including $K_w$). Express the effective equilibrium constants in terms of equilibrium constants and activity coefficients. For example, $K_w' = K_w/\gamma_{H^+}\gamma_{OH^-}$. Write expressions for [Fe(SCN)$^{2+}$], [Fe(SCN)$_2^+$], [FeOH$^{2+}$], and [OH$^-$] in terms of [Fe$^{3+}$], [SCN$^-$], and [H$^+$].

(b) Write the charge balance.

(c) Write mass balances for iron, thiocyanate, Na$^+$, and NO$_3^-$.

(d) Substitute expressions from part (a) into the mass balance for thiocyanate to find an expression for [Fe$^{3+}$] in terms of [SCN$^-$].

(e) Substitute expressions from part (a) into the mass balance for iron to find an expression for [H$^+$] in terms of [Fe$^{3+}$] and [SCN$^-$].

(f) In a spreadsheet, guess a value for [SCN$^-$]. From this value, compute [Fe$^{3+}$] from part (d). From [SCN$^-$] and [Fe$^{3+}$], compute [H$^+$] from part (e).

(g) Compute the concentrations of all other species from [SCN$^-$], [Fe$^{3+}$], and [H$^+$].

(h) Compute the net charge. We want to vary [SCN$^-$] to reduce the net charge to 0. When the net charge is 0, you have the correct concentrations. SOLVER did not work well for these equations. You have several recourses. One is to use GOAL SEEK, which is similar to SOLVER but only handles 1 unknown. Start with $\mu = 0$ and guess [SCN$^-$] = 1e-6. In the TOOLS menu, select OPTIONS and Calculation. Set Maximum change to 1e-16. In the TOOLS menu, select GOAL SEEK. Set cell: {net charge} To value: 0 By changing cell: {[SCN$^-$]}. Click OK and GOAL SEEK finds a solution. Calculate the new ionic strength and activity coefficients from the Davies equation. Calculate effective equilibrium constants. Repeat the process until $\mu$ is constant to three significant digits. If GOAL SEEK fails to work at some stage, vary [SCN$^-$] by hand to reduce the net charge to near 0. When you are close to a correct answer, GOAL SEEK can take you the rest of the way. When you have a solution, verify that the mass balances are satisfied and compute $pH = -\log([H^+]\gamma_{H^+})$.

(i) Find the quotient [Fe(SCN)$^{2+}$]/({[Fe$^{3+}$] + [FeOH$^{2+}$]}[SCN$^-$]). This is the point for [KNO$_3$] = 0 in Figure 8-1. Compare your answer with Figure 8-1. The ordinate of Figure 8-1 is labeled [Fe(SCN)$^{2+}$]/([Fe$^{3+}$][SCN$^-$]), but [Fe$^{3+}$] really refers to the total concentration of iron not bound to thiocyanate.

(j) Find the quotient in part (i) when the solution also contains 0.20 M $KNO_3$. Compare your answer with Figure 8-1.

**13-12.** ▦ (a) Follow the steps of Problem 13-11 to solve this one. From the following equilibria, find the concentrations of species and the pH of 1.0 mM $La_2(SO_4)_3$. Use Davies activity coefficients. You may wish to use GOAL SEEK in place of SOLVER as described in Problem 13-11.

$$La^{3+} + SO_4^{2-} \rightleftharpoons La(SO_4)^+ \qquad \beta_1 = 10^{3.64} \ (\mu = 0)$$
$$La^{3+} + 2SO_4^{2-} \rightleftharpoons La(SO_4)_2^- \qquad \beta_2 = 10^{5.3} \ (\mu = 0)$$
$$La^{3+} + H_2O \rightleftharpoons LaOH^{2+} + H^+ \qquad K_a = 10^{-8.5} \ (\mu = 0)$$

(b) If $La_2(SO_4)_3$ were a strong electrolyte, what would be the ionic strength of 1.0 mM $La_2(SO_4)_3$? What is the actual ionic strength of this solution?

(c) What fraction of lanthanum is La$^{3+}$?

(d) Why did we not consider hydrolysis of SO$_4^{2-}$ to give HSO$_4^-$?

(e) Will La(OH)$_3$(s) precipitate in this solution?

**13-13.** 🖩 Find the composition of a saturated solution of AgCN containing 0.10 M KCN adjusted to pH 12.00 with NaOH. Consider the following equilibria and use Davies activity coefficients.

AgCN(s)                                       $pK_{sp}$ = 15.66

$HCN(aq) \rightleftharpoons CN^- + H^+$            $pK_{HCN}$ = 9.21

$Ag^+ + H_2O \rightleftharpoons AgOH(aq) + H^+$    $pK_{Ag}$ = 12.0

$Ag^+ + CN^- + OH^- \rightleftharpoons Ag(OH)(CN)^-$  $\log K_{AgOHCN}$ = 13.22

$Ag^+ + 2CN^- \rightleftharpoons Ag(CN)_2^-$      $\log \beta_2$ = 20.48

$Ag^+ + 3CN^- \rightleftharpoons Ag(CN)_3^{2-}$   $\log \beta_3$ = 21.7

*Suggested procedure:* Let $[CN^-]$ be the master variable. We know $[H^+]$ from the pH and we can find $[Ag^+] = K'_{sp}/[CN^-]$. (i) Use equilibrium expressions to find $[OH^-]$, $[HCN]$, $[AgOH]$, $[Ag(OH)(CN)]$, $[Ag(CN)_2^-]$, and $[Ag(CN)_3^{2-}]$ in terms of $[CN^-]$, $[Ag^+]$, and $[H^+]$. (ii) Key mass balance: {total silver} + $[K^+]$ = {total cyanide}. (iii) Guess that ionic strength = 0.10 M and compute activity coefficients. (iv) Compute $[H^+] = 10^{-pH}/\gamma_{H^+}$. (v) Guess a value of $[CN^-]$ and compute the concentrations of all species in the mass balance. (vi) Use SOLVER to vary $[CN^-]$ to satisfy the mass balance. (vii) Compute $[Na^+]$ from the charge balance. (viii) Compute the ionic strength and perform several iterations of the entire process until the ionic strength no longer changes.

**13-14.** 🖩 Consider the reactions of $Fe^{2+}$ with the amino acid glycine:

$Fe^{2+} + G^- \rightleftharpoons FeG^+$         $\log \beta_1$ = 4.31

$Fe^{2+} + 2G^- \rightleftharpoons FeG_2(aq)$    $\log \beta_2$ = 7.65

$Fe^{2+} + 3G^- \rightleftharpoons FeG_3^-$      $\log \beta_3$ = 8.87

$Fe^{2+} + H_2O \rightleftharpoons FeOH^+ + H^+$  $pK_a$ = 9.4

$^+H_3NCH_2CO_2H$ glycine, $H_2G^+$  $pK_1$ = 2.350, $pK_2$ = 9.778

Suppose that 0.050 mol of $FeG_2$ is dissolved in 1.00 L and enough HCl is added to adjust the pH to 8.50. Use Davies activity coefficients to find the composition of the solution. What fraction of iron is in each of its forms and what fraction of glycine is in each of its forms? From the distribution of species, explain the principal chemistry that requires addition of HCl to obtain a pH of 8.50.

*Suggested procedure:* (i) Use equilibrium expressions to write all concentrations in terms of $[G^-]$, $[Fe^{2+}]$, and $[H^+]$. (ii) Write one mass balance for iron and one for glycine. (iii) Substitute expressions from step (i) into the mass balance for iron to find $[Fe^{2+}]$ in terms of $[G^-]$ and $[H^+]$. (iv) Guess an ionic strength of 0.01 M and compute activity coefficients. (v) Compute $[H^+] = 10^{-pH}/\gamma_{H^+}$. (vi) Guess a value of $[G^-]$ and compute all concentrations in the mass balance for glycine. (vii) Use SOLVER to vary $[G^-]$ to satisfy the mass balance for glycine. (viii) Compute $[Cl^-]$ from the charge balance. (ix) Compute the ionic strength and perform several iterations of the entire process until the ionic strength no longer changes.

**13-15.** 🖩 Data for the glycine difference plot in Figure 13-14 are given in the table.
**(a)** Reproduce the spreadsheet in Figure 13-13 and show that you get the same values of $pK_1$ and $pK_2$ in cells B10 and B11 after executing SOLVER. Start with different values of $pK_1$ and $pK_2$ and see if SOLVER finds the same solutions.
**(b)** Use SOLVER to find the best values of $pK_1$ and $pK_2$ while fixing $pK'_w$ at its expected value of 13.797. Describe how $\bar{n}_H$(measured) behaves when $pK'_w$ is fixed.

| V (mL) | pH | V (mL) | pH | V (mL) | pH | V (mL) | pH |
|---|---|---|---|---|---|---|---|
| 0.00 | 2.234 | 0.40 | 2.550 | 0.80 | 3.528 | 1.20 | 10.383 |
| 0.02 | 2.244 | 0.42 | 2.572 | 0.82 | 3.713 | 1.22 | 10.488 |
| 0.04 | 2.254 | 0.44 | 2.596 | 0.84 | 4.026 | 1.24 | 10.595 |
| 0.06 | 2.266 | 0.46 | 2.620 | 0.86 | 5.408 | 1.26 | 10.697 |
| 0.08 | 2.278 | 0.48 | 2.646 | 0.88 | 8.419 | 1.28 | 10.795 |
| 0.10 | 2.291 | 0.50 | 2.675 | 0.90 | 8.727 | 1.30 | 10.884 |
| 0.12 | 2.304 | 0.52 | 2.702 | 0.92 | 8.955 | 1.32 | 10.966 |
| 0.14 | 2.318 | 0.54 | 2.736 | 0.94 | 9.117 | 1.34 | 11.037 |
| 0.16 | 2.333 | 0.56 | 2.768 | 0.96 | 9.250 | 1.36 | 11.101 |
| 0.18 | 2.348 | 0.58 | 2.802 | 0.98 | 9.365 | 1.38 | 11.158 |
| 0.20 | 2.363 | 0.60 | 2.838 | 1.00 | 9.467 | 1.40 | 11.209 |
| 0.22 | 2.380 | 0.62 | 2.877 | 1.02 | 9.565 | 1.42 | 11.255 |
| 0.24 | 2.397 | 0.64 | 2.920 | 1.04 | 9.660 | 1.44 | 11.296 |
| 0.26 | 2.413 | 0.66 | 2.966 | 1.06 | 9.745 | 1.46 | 11.335 |
| 0.28 | 2.429 | 0.68 | 3.017 | 1.08 | 9.830 | 1.48 | 11.371 |
| 0.30 | 2.448 | 0.70 | 3.073 | 1.10 | 9.913 | 1.50 | 11.405 |
| 0.32 | 2.467 | 0.72 | 3.136 | 1.12 | 10.000 | 1.52 | 11.436 |
| 0.34 | 2.487 | 0.74 | 3.207 | 1.14 | 10.090 | 1.54 | 11.466 |
| 0.36 | 2.506 | 0.76 | 3.291 | 1.16 | 10.183 | 1.56 | 11.492 |
| 0.38 | 2.528 | 0.78 | 3.396 | 1.18 | 10.280 | 1.58 | 11.519 |
|  |  |  |  |  |  | 1.60 | 11.541 |

*Data from A. Kraft, J. Chem. Ed.* **2003**, *80, 554.*

**13-16.** 🖩 *Difference plot.* A solution containing 0.139 mmol of the triprotic acid tris(2-aminoethyl)amine·3HCl plus 0.115 mmol HCl in 40 mL of 0.10 M KCl was titrated with 0.490 5 M NaOH to measure acid dissociation constants.

$N(CH_2CH_2NH_3^+)_3 \cdot 3Cl^-$     Tris(2-aminoethyl)amine·3HCl
$H_3A^{3+} \cdot 3Cl^-$

**(a)** Write expressions for the experimental mean fraction of protonation, $\bar{n}_H$(measured), and the theoretical mean fraction of protonation, $\bar{n}_H$(theoretical).
**(b)** From the following data, prepare a graph of $\bar{n}_H$(measured) versus pH. Find the best values of $pK_1$, $pK_2$, $pK_3$, and $pK'_w$ by minimizing the sum of the squares of the residuals, $\Sigma[\bar{n}_H$(measured) $- \bar{n}_H$(theoretical)$]^2$.

| V (mL) | pH | V (mL) | pH | V (mL) | pH | V (mL) | pH |
|---|---|---|---|---|---|---|---|
| 0.00 | 2.709 | 0.36 | 8.283 | 0.72 | 9.687 | 1.08 | 10.826 |
| 0.02 | 2.743 | 0.38 | 8.393 | 0.74 | 9.748 | 1.10 | 10.892 |
| 0.04 | 2.781 | 0.40 | 8.497 | 0.76 | 9.806 | 1.12 | 10.955 |
| 0.06 | 2.826 | 0.42 | 8.592 | 0.78 | 9.864 | 1.14 | 11.019 |
| 0.08 | 2.877 | 0.44 | 8.681 | 0.80 | 9.926 | 1.16 | 11.075 |
| 0.10 | 2.937 | 0.46 | 8.768 | 0.82 | 9.984 | 1.18 | 11.128 |
| 0.12 | 3.007 | 0.48 | 8.851 | 0.84 | 10.042 | 1.20 | 11.179 |
| 0.14 | 3.097 | 0.50 | 8.932 | 0.86 | 10.106 | 1.22 | 11.224 |
| 0.16 | 3.211 | 0.52 | 9.011 | 0.88 | 10.167 | 1.24 | 11.268 |
| 0.18 | 3.366 | 0.54 | 9.087 | 0.90 | 10.230 | 1.26 | 11.306 |
| 0.20 | 3.608 | 0.56 | 9.158 | 0.92 | 10.293 | 1.28 | 11.344 |
| 0.22 | 4.146 | 0.58 | 9.231 | 0.94 | 10.358 | 1.30 | 11.378 |
| 0.24 | 5.807 | 0.60 | 9.299 | 0.96 | 10.414 | 1.32 | 11.410 |
| 0.26 | 6.953 | 0.62 | 9.367 | 0.98 | 10.476 | 1.34 | 11.439 |
| 0.28 | 7.523 | 0.64 | 9.436 | 1.00 | 10.545 | 1.36 | 11.468 |
| 0.30 | 7.809 | 0.66 | 9.502 | 1.02 | 10.615 | 1.38 | 11.496 |
| 0.32 | 8.003 | 0.68 | 9.564 | 1.04 | 10.686 | 1.40 | 11.521 |
| 0.34 | 8.158 | 0.70 | 9.626 | 1.06 | 10.756 |  |  |

*Data from A. Kraft, J. Chem. Ed.* **2003**, *80, 554.*

**(c)** Create a fractional composition graph showing the fractions of $H_3A^{3+}$, $H_2A^{2+}$, $HA^+$, and A as a function of pH.

**13-17.** ▦ **(a)** For the following reactions, prepare a diagram showing log(concentration) versus pH for all species in the pH range 2 to 12 for a solution made by dissolving 0.025 mol $CuSO_4$ in 1.00 L. Equilibrium constants apply at $\mu = 0.1$ M, which you should assume is constant. Do not use activity coefficients, because the equilibrium constants already apply at $\mu = 0.1$ M.

$$Cu^{2+} + SO_4^{2-} \rightleftharpoons CuSO_4(aq) \qquad \log K'_{ip} = 1.26$$
$$HSO_4^- \rightleftharpoons SO_4^{2-} + H^+ \qquad \log K'_a = -1.54$$
$$Cu^{2+} + OH^- \rightleftharpoons CuOH^+ \qquad \log \beta'_1 = 6.1$$
$$Cu^{2+} + 2OH^- \rightleftharpoons Cu(OH)_2(aq) \qquad \log \beta'_2 = 11.2*$$
$$Cu^{2+} + 3OH^- \rightleftharpoons Cu(OH)_3^- \qquad \log \beta'_3 = 14.5*$$
$$Cu^{2+} + 4OH^- \rightleftharpoons Cu(OH)_4^{2-} \qquad \log \beta'_4 = 15.6*$$
$$2Cu^{2+} + OH^- \rightleftharpoons Cu_2(OH)^{3+} \qquad \log \beta'_{12} = 8.2*$$
$$2Cu^{2+} + 2OH^- \rightleftharpoons Cu_2(OH)_2^{2+} \qquad \log \beta'_{22} = 16.8$$
$$3Cu^{2+} + 4OH^- \rightleftharpoons Cu_3(OH)_4^{2+} \qquad \log \beta'_{43} = 33.5$$
$$H_2O \rightleftharpoons H^+ + OH^- \qquad \log K'_w = -13.79$$

*Asterisk indicates that equilibrium constant was estimated for $\mu = 0.1$ M from data reported for a different ionic strength.*

Remember that pH $= -\log([H^+]\gamma_{H^+})$ and $[OH^-] = K'_w/[H^+]$. Use $\gamma_{H^+} = 0.78$ for this problem.

*Recommended procedure:* From the mass balance for sulfate, find an expression for $[SO_4^{2-}]$ in terms of $[Cu^{2+}]$ and $[H^+]$. Set up a spreadsheet with pH values between 2 and 12 in column A. From pH, compute $[H^+]$ and $[OH^-]$ in columns B and C. *Guess* a value for $[Cu^{2+}]$ in column D. From $[H^+]$ and $[Cu^{2+}]$, calculate $[SO_4^{2-}]$ from the equation derived from the mass balance for sulfate. From $[H^+]$, $[OH^-]$, $[Cu^{2+}]$, and $[SO_4^{2-}]$, calculate all other concentrations by using equilibrium expressions. Find the total concentration of

Cu by adding the concentrations of all the species. Use SOLVER to vary $[Cu^{2+}]$ in column D so that the total concentration of copper is 0.025 M. You need to use SOLVER for each line in the spreadsheet with a different pH.

**(b)** Find the pH of 0.025 M $CuSO_4$ if nothing is added to adjust the pH. This is the pH at which the net charge is 0.

**(c)** At what pH does each of the following salts precipitate from 0.025 M $CuSO_4$?

$$Cu(OH)_{1.5}(SO_4)_{0.25}(s) \rightleftharpoons Cu^{2+} + 1.5OH^- + 0.25SO_4^{2-}$$
$$\log K'_{sp}(OH\text{-}SO_4) = -16.68$$
$$Cu(OH)_2(s) \rightleftharpoons Cu^{2+} + 2OH^- \qquad \log K'_{sp}(OH) = -18.7*$$
$$CuO(s) + H_2O \rightleftharpoons Cu^{2+} + 2OH^- \qquad \log K'_{sp}(O) = -19.7*$$

**13-18.** ▦ *Ion pairing in acid-base systems.* This problem incorporates ion-pair equilibria 13-12 and 13-13 into the acid-base chemistry of Section 13-1.

**(a)** From the mass balance 13-15, derive Equation 13-16.

**(b)** Substitute equilibrium expressions into mass balance 13-17 to derive an expression for $[T^{2-}]$ in terms of $[H^+]$, $[Na^+]$, and various equilibrium constants.

**(c)** With the same approach as in part **(b)**, derive expressions for $[HT^-]$ and $[H_2T]$.

**(d)** Add the species $[NaT^-]$ and $[NaHT]$ to the spreadsheet in Figure 13-1 and compute the composition and pH of the solution. Compute $[Na^+]$ with Equation 13-16. Compute $[H_2T]$, $[HT^-]$, and $[T^{2-}]$ from the expressions derived in parts **(b)** and **(c)**. Excel will indicate a *circular reference* problem because, for example, the formula for $[Na^+]$ depends on $[T^{2-}]$ and the formula for $[T^{2-}]$ depends on $[Na^+]$. Select the TOOLS menu and go to OPTIONS. Select the Calculations tab and choose Iteration. Click OK and then use SOLVER to find the pH in cell H13 that reduces the net charge in cell E15 to (near) zero.

# 14 Fundamentals of Electrochemistry

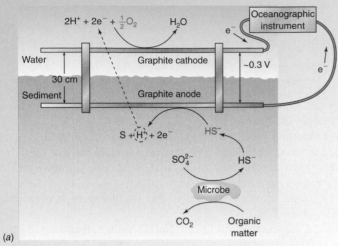

(a)

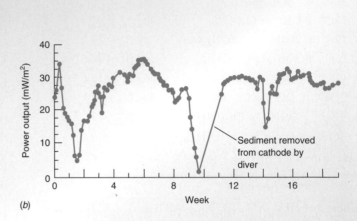

(b)

(a) Fuel cell formed by electrodes above and below the sediment–water interface on the ocean floor. *[Adapted from C. E. Reimers, L. M. Tender, S. Fertig, and W. Wang, "Harvesting Energy from the Marine Sediment–Water Interface, Environ. Sci. Technol. **2001**, 35, 192.]* (b) Power output from prototype fuel cell. (c) Sulfide in sediment pore water is depleted adjacent to anode. *[Charts from L. M. Tender, C. E. Reimers, H. A. Stecher III, D. W. Holmes, D. R. Bond, D. A. Lowry, K. Pilobello, S. J. Fertig, and D. R. Lovley, "Harnessing Microbially Generated Power on the Seafloor," Nat. Biotechnol. **2002**, 20, 821.]*

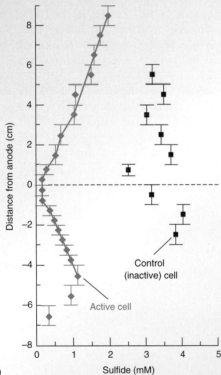

(c)

Microbes residing in sediment beneath oceans and lakes derive energy by oxidizing organic matter. $O_2$ is available as the oxidant at the sediment–water interface, but it is depleted within millimeters below the interface. Nitrate and Fe(III) oxidants are available in the first few centimeters of sediment. When they are exhausted, sulfate becomes the predominant oxidant for a distance of ~1 m. The sulfate reduction product, $HS^-$, is released in millimolar concentrations into solution in the sediment pores.

Two large, perforated graphite electrodes positioned above and below the sediment–water interface act as a fuel cell—a device that derives electricity from a continuous flow of fuel and oxidizer. The electrode in the water is in an oxidizing environment, and the electrode in the sediment is in a reducing environment. A prototype cell in a coastal salt marsh generated $25~mW/m^2$ at 0.27 V over a 4-month period (panel b). Power losses occurred three times when the cathode became covered with sediment. Panel c shows depletion of sulfide in the sediment near the anode, confirming that sulfide is a fuel consumed at this electrode. Oxidation of sulfide accounts for half of the power output. The other half is likely derived from oxidation of acetate (from organic matter) by microorganisms colonizing the anode, with direct transfer of electrons to the anode. Suitably large electrodes could one day provide power to operate remote oceanographic instruments.

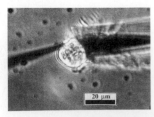

Electrodes next to an adrenal cell measure release of the hormone epinephrine. *[Photo courtesy R. M. Wightman, University of North Carolina.]*

**A** major branch of analytical chemistry uses electrical measurements of chemical processes at the surface of an electrode for analytical purposes. For example, hormones released from a single cell can be measured in this manner. Principles developed in this chapter provide a foundation for potentiometry, redox titrations, electrogravimetric and coulometric analysis, voltammetry, and amperometry in the following chapters.[1,2]

## ■■■ 14-1 Basic Concepts

A **redox reaction** involves transfer of electrons from one species to another. A species is said to be **oxidized** when it *loses electrons*. It is **reduced** when it *gains electrons*. An **oxidizing**

**agent,** also called an **oxidant,** takes electrons from another substance and becomes reduced. A **reducing agent,** also called a **reductant,** gives electrons to another substance and is oxidized in the process. In the reaction

$$Fe^{3+} \ + \ V^{2+} \ \rightarrow \ Fe^{2+} \ + \ V^{3+} \qquad (14\text{-}1)$$

Oxidizing   Reducing
agent        agent

$Fe^{3+}$ is the oxidizing agent because it takes an electron from $V^{2+}$. $V^{2+}$ is the reducing agent because it gives an electron to $Fe^{3+}$. $Fe^{3+}$ is reduced, and $V^{2+}$ is oxidized as the reaction proceeds from left to right. Appendix D reviews oxidation numbers and balancing of redox equations.

*Oxidation:* loss of electrons
*Reduction:* gain of electrons
*Oxidizing agent:* takes electrons
*Reducing agent:* gives electrons

$Fe^{3+} + e^- \rightarrow Fe^{2+}$
$V^{2+} \rightarrow V^{3+} + e^-$

## Chemistry and Electricity

When electrons from a redox reaction flow through an electric circuit, we can learn something about the reaction by measuring current and voltage. Electric current is proportional to the rate of reaction, and the cell voltage is proportional to the free-energy change for the electrochemical reaction. In techniques such as voltammetry, the voltage can be used to identify reactants.

## Electric Charge

Electric charge, $q$, is measured in **coulombs** (C). The magnitude of the charge of a single electron is $1.602 \times 10^{-19}$ C, so a mole of electrons has a charge of $(1.602 \times 10^{-19} \text{ C})(6.022 \times 10^{23} \text{ mol}^{-1}) = 9.649 \times 10^4$ C, which is called the **Faraday constant,** $F$.

$$\text{Faraday constant } (F) = \frac{\text{coulombs}}{\text{mole of electrons}}$$

*Relation between charge and moles:*

$$q \ = \ n \ \cdot \ F \qquad (14\text{-}2)$$

Coulombs   mol $e^-$   $\dfrac{\text{Coulombs}}{\text{mol } e^-}$

where $n$ is the number of moles of electrons transferred.

> **Example** Relating Coulombs to Quantity of Reaction
>
> If 5.585 g of $Fe^{3+}$ were reduced in Reaction 14-1, how many coulombs of charge must have been transferred from $V^{2+}$ to $Fe^{3+}$?
>
> **Solution** First, we find that 5.585 g of $Fe^{3+}$ equal 0.100 0 mol of $Fe^{3+}$. Because each $Fe^{3+}$ ion requires one electron in Reaction 14-1, 0.100 0 mol of electrons must have been transferred. Using the Faraday constant, we find that 0.100 0 mol of electrons corresponds to
>
> $$q = nF = (0.100\ 0 \text{ mol } e^-)\left(9.649 \times 10^4 \frac{C}{\text{mol } e^-}\right) = 9.649 \times 10^3 \text{ C}$$

## Electric Current

The quantity of charge flowing each second through a circuit is called the **current.** The unit of current is the **ampere,** abbreviated A. A current of 1 ampere represents a charge of 1 coulomb per second flowing past a point in a circuit.

> **Example** Relating Current to Rate of Reaction
>
> Suppose that electrons are forced into a platinum wire immersed in a solution containing $Sn^{4+}$ (Figure 14-1), which is reduced to $Sn^{2+}$ at a constant rate of 4.24 mmol/h. How much current passes through the solution?
>
> **Solution** *Two* electrons are required to reduce *one* $Sn^{4+}$ ion:
>
> $$Sn^{4+} + 2e^- \rightarrow Sn^{2+}$$
>
> If $Sn^{4+}$ reacts at a rate of 4.24 mmol/h, electrons flow at a rate of 2(4.24 mmol/h) = 8.48 mmol/h, which corresponds to
>
> $$\frac{8.48 \text{ mmol } e^-/h}{3\ 600 \text{ s/h}} = 2.356 \times 10^{-3} \frac{\text{mmol } e^-}{s} = 2.356 \times 10^{-6} \frac{\text{mol } e^-}{s}$$

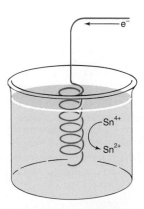

**Figure 14-1** Electrons flowing into a coil of Pt wire at which $Sn^{4+}$ ions in solution are reduced to $Sn^{2+}$. This process could not happen by itself, because there is no complete circuit. If $Sn^{4+}$ is to be reduced at this Pt electrode, some other species must be oxidized at some other place.

To find the current, we convert moles of electrons per second into coulombs per second:

$$\text{Current} = \frac{\text{charge}}{\text{time}} = \frac{\text{coulombs}}{\text{second}} = \frac{\text{moles e}^-}{\text{second}} \cdot \frac{\text{coulombs}}{\text{mole}}$$

$$= \left(2.356 \times 10^{-6} \frac{\text{mol}}{\text{s}}\right)\left(9.649 \times 10^4 \frac{\text{C}}{\text{mol}}\right)$$

$$= 0.227 \text{ C/s} = 0.227 \text{ A}$$

A current of 0.227 A is 227 mA.

In Figure 14-1, we encountered a Pt **electrode,** which conducts electrons into or out of a chemical species in the redox reaction. Platinum is a common *inert* electrode. It does not participate in the redox chemistry except as a conductor of electrons.

## Voltage, Work, and Free Energy

It costs energy to move like charges toward one another. Energy is released when opposite charges move toward each other.

The difference in **electric potential,** $E$, between two points is the work needed (or that can be done) when moving an electric charge from one point to the other. *Potential difference* is measured in **volts** (V). The greater the potential difference between two points, the stronger will be the "push" on a charged particle traveling between those points.

A good analogy for understanding current and potential is to think of water flowing through a garden hose (Figure 14-2). Current is the electric charge flowing past a point in a wire each second. Current is analogous to the volume of water flowing past a point in the hose each second. Potential is a measure of the force pushing on the electrons. The greater the force, the more current flows. Potential is analogous to the pressure on the water in the hose. The greater the pressure, the faster the water flows.

When a charge, $q$, moves through a potential difference, $E$, the work done is

*Relation between work and voltage:*

$$\text{Work} = E \cdot q \qquad (14\text{-}3)$$

Joules     Volts    Coulombs

$1 \text{ V} = 1 \text{ J/C}$

Work has the dimensions of energy, whose units are **joules** (J). One *joule* of energy is gained or lost when 1 *coulomb* of charge moves between points whose potentials differ by 1 *volt*. Equation 14-3 tells us that the dimensions of volts are joules per coulomb.

> ### Example   Electrical Work
>
> How much work can be done if 2.4 mmol of electrons fall through a potential difference of 0.27 V in the ocean-floor fuel cell at the opening of this chapter?
>
> **Solution**   To use Equation 14-3, we must convert moles of electrons into coulombs of charge. The relation is simply
>
> $$q = nF = (2.4 \times 10^{-3} \text{ mol})(9.649 \times 10^4 \text{ C/mol}) = 2.3 \times 10^2 \text{ C}$$
>
> The work that could be done is
>
> $$\text{Work} = E \cdot q = (0.27 \text{ V})(2.3 \times 10^2 \text{ C}) = 62 \text{ J}$$

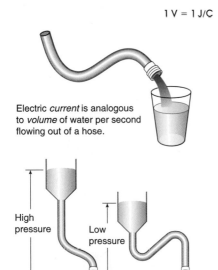

Electric *current* is analogous to *volume* of water per second flowing out of a hose.

High pressure    Low pressure

Electric *potential* is analogous to the hydrostatic *pressure* pushing water through a hose. High pressure gives high flow.

**Figure 14-2**   Analogy between the flow of water through a hose and the flow of electricity through a wire.

In the garden hose analogy, suppose that one end of a hose is raised 1 m above the other end and a volume of 1 L of water flows through the hose. The flowing water goes through a mechanical device to do a certain amount of work. If one end of the hose is raised 2 m above the other, the amount of work that can be done by the falling water is twice as great. The elevation difference between the ends of the hose is analogous to electric potential difference and the volume of water is analogous to electric charge. The greater the electric potential difference between two points in a circuit, the more work can be done by the charge flowing between those two points.

The free-energy change, $\Delta G$, for a chemical reaction conducted reversibly at constant temperature and pressure equals the maximum possible electrical work that can be done by the reaction on its surroundings:

Section 6-2 gave a brief discussion of $\Delta G$.

$$\text{Work done on surroundings} = -\Delta G \qquad (14\text{-}4)$$

The negative sign in Equation 14-4 indicates that the free energy of a system decreases when the work is done on the surroundings.

Combining Equations 14-2, 14-3, and 14-4 produces a relation of utmost importance to chemistry:

$$\Delta G = -\text{work} = -E \cdot q$$

*Relation between free-energy difference and electric potential difference:*

$$\Delta G = -nFE \qquad \text{(14-5)}$$

$q = nF$

Equation 14-5 relates the free-energy change of a chemical reaction to the electrical potential difference (that is, the voltage) that can be generated by the reaction.

## Ohm's Law

**Ohm's law** states that current, $I$, is directly proportional to the potential difference (voltage) across a circuit and inversely proportional to the **resistance, $R$,** of the circuit.

*Ohm's law:*

$$I = \frac{E}{R} \qquad \text{(14-6)}$$

The greater the voltage, the more current will flow. The greater the resistance, the less current will flow.

Units of resistance are **ohms,** assigned the Greek symbol $\Omega$ (omega). A current of 1 ampere flows through a circuit with a potential difference of 1 volt if the resistance of the circuit is 1 ohm. From Equation 14-6, the unit A is equivalent to V/$\Omega$.

Box 14-1 shows measurements of the resistance of single molecules by measuring current and voltage and applying Ohm's law.

## Power

**Power, $P$,** is the work done per unit time. The SI unit of power is J/s, better known as the **watt** (W).

$$P = \frac{\text{work}}{\text{s}} = \frac{E \cdot q}{\text{s}} = E \cdot \frac{q}{\text{s}} \qquad \text{(14-7)}$$

---

### Box 14-1 Molecular Wire[3]

The electrical conductance of a single molecule suspended between two gold electrodes is known from measurement of voltage and current by applying Ohm's law. Conductance is 1/resistance, so it has the units 1/ohm $\equiv$ siemens (S).

To make molecular junctions, a gold scanning tunneling microscope tip was moved in and out of contact with a gold substrate in the presence of a solution containing a test molecule terminated by thiol (—SH) groups. Thiols spontaneously bind to gold, forming bridges such as that shown here. Nanoampere currents were observed with a potential difference of 0.1 V between the gold surfaces.

The graph shows four observations of conductance as the scanning tunneling microscope tip was pulled away from the Au substrate. Conductance plateaus are observed at multiples of 19 nS. An interpretation is that a single molecule connecting two Au surfaces has a conductance of 19 nS (or a resistance of 50 MΩ). If two molecules form parallel bridges, conductance increases to 38 nS. Three molecules give a conductance of 57 nS. If there are three bridges and the electrodes are pulled apart, one of the bridges breaks and conductance drops to 38 nS. When the second bridge breaks, conductance drops to 19 nS. The exact conductance varies because the environment of each molecule on the Au surface is not identical. A histogram of >500 observations in the inset shows peaks at 19, 38, and

57 nS. By contrast, the conductance of a chain of Au atoms is 77 μS, or 4 000 times greater than 19 nS.

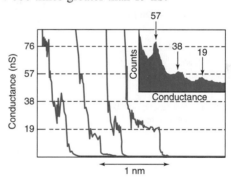

Change in conductance as Au scanning tunneling microscope tip immersed in dithiol solution is withdrawn from Au substrate. *[From X. Xiao, B. Xu, and N. Tao, "Conductance Titration of Single-Peptide Molecules," J. Am. Chem. Soc.* **2004,** *126, 5370.]*

Data in the chart were obtained at pH 2, at which the amino group is protonated to $-NH_3^+$. When the pH was raised to 10, the amino group was neutral. Conductance plateaus became multiples of 9 nS. The conductivity of the molecule changes from 9 nS when it is neutral to 19 nS when there is a positive change on the molecule. The positive charge facilitates electron transfer across the molecule.

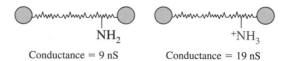

Positive charge in the bridging molecule increases conductivity by a factor of 2.

---

Power (watts) = work per second
$P = E \cdot I = (IR) \cdot I = I^2R$

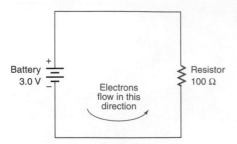

**Figure 14-3** A circuit with a battery and a resistor. Benjamin Franklin investigated static electricity in the 1740s.[4] He thought electricity was a fluid that flows from a silk cloth to a glass rod when the rod is rubbed with the cloth. We now know that electrons flow from glass to silk. However, Franklin's convention for the direction of electric current has been retained, so we say that current flows from positive to negative—in the opposite direction of electron flow.

*F* is the Faraday constant (96 485 C/mol).

*E* is *electric potential difference* measured in volts (V). *E* is the work (J) needed to move a coulomb of positive charge from one point to the other.

*n* is moles of charge moved through the potential difference, *E*.

*I* is *electric current*, measured in amperes (A). It is coulombs per second moving past a point in the circuit. *R* is resistance in ohms ($\Omega$). Units: A = V/$\Omega$.

Power is work per unit time (J/s) done by electricity moving through a circuit.

Because *q*/s is the current, *I*, we can write

$$P = E \cdot I \qquad (14\text{-}8)$$

A cell capable of delivering 1 ampere at a potential difference of 1 volt has a power output of 1 watt.

**Example** Using Ohm's Law

In the circuit in Figure 14-3, the battery generates a potential difference of 3.0 V, and the resistor has a resistance of 100 $\Omega$. We assume that the resistance of the wire connecting the battery and the resistor is negligible. How much current and how much power are delivered by the battery in this circuit?

**Solution** The current in this circuit is

$$I = \frac{E}{R} = \frac{3.0\ \text{V}}{100\ \Omega} = 0.030\ \text{A} = 30\ \text{mA}$$

The power produced by the battery must be

$$P = E \cdot I = (3.0\ \text{V})(0.030\ \text{A}) = 90\ \text{mW}$$

What happens to the power generated by the circuit? *The energy appears as heat in the resistor.* The power (90 mW) equals the rate at which heat is produced in the resistor.

Here is a summary of symbols, units, and relations from the last few pages:

*Relation between charge and moles:*

$$q = n \cdot F$$

Charge    Moles    C/mole
(coulombs, C)

*Relation between work and voltage:*

$$\text{Work} = E \cdot q$$

Joules, J    Volts, V   Coulombs

*Relation between free-energy difference and electric potential difference:*

$$\Delta G = -nFE$$

Joules

*Ohm's law:*

$$I = E\ /\ R$$

Current    Volts   Resistance
(A)        (V)     (ohms, $\Omega$)

*Electric power:*

$$P = \frac{\text{work}}{\text{s}} = E \cdot I$$

Power      J/s        Volts    Amperes
(watts, W)

# 14-2  Galvanic Cells

A galvanic cell uses a spontaneous chemical reaction to generate electricity.

A **galvanic cell** (also called a *voltaic cell*) uses a *spontaneous* chemical reaction to generate electricity. To accomplish this, one reagent must be oxidized and another must be reduced. The two cannot be in contact, or electrons would flow directly from the reducing agent to the oxidizing agent. Instead, the oxidizing and reducing agents are physically separated, and electrons are forced to flow through an external circuit to go from one reactant to the other. Batteries[5] and fuel cells[6] are galvanic cells that consume their reactants to generate electricity. A battery has a static compartment filled with reactants. In a fuel cell, fresh reactants flow past the electrodes and products are continuously flushed from the cell.

## A Cell in Action

Figure 14-4 shows a galvanic cell with two electrodes suspended in a solution of $CdCl_2$. One electrode is cadmium; the other is metallic silver coated with solid AgCl. The reactions are

Reduction:     $2AgCl(s) + 2e^- \rightleftharpoons 2Ag(s) + 2Cl^-(aq)$

Oxidation:     $Cd(s) \rightleftharpoons Cd^{2+}(aq) + 2e^-$

Net reaction:  $Cd(s) + 2AgCl(s) \rightleftharpoons Cd^{2+}(aq) + 2Ag(s) + 2Cl^-(aq)$

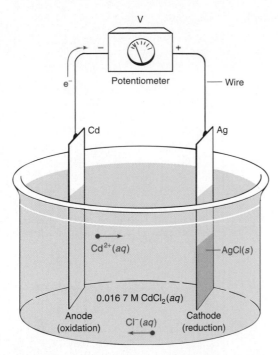

**Figure 14-4** A simple galvanic cell. The **potentiometer** measures voltage. It has positive and negative terminals. When electrons flow into the negative terminal, as in this illustration, the voltage is positive.

The net reaction is composed of a reduction and an oxidation, each of which is called a **half-reaction.** The two half-reactions are written with equal numbers of electrons so that their sum includes no free electrons.

Oxidation of Cd metal to $Cd^{2+}(aq)$ provides electrons that flow through the circuit to the Ag electrode in Figure 14-4. At the Ag surface, $Ag^+$ (from AgCl) is reduced to $Ag(s)$. Chloride from AgCl goes into solution. The free-energy change for the net reaction, $-150$ kJ per mole of Cd, provides the driving force that pushes electrons through the circuit.

Recall that $\Delta G$ is *negative* for a spontaneous reaction.

---

**Example** Voltage Produced by a Chemical Reaction

Calculate the voltage that would be measured by the potentiometer in Figure 14-4.

**Solution** Because $\Delta G = -150$ kJ/mol of Cd, we can use Equation 14-5 (where $n$ is the number of moles of electrons transferred in the balanced net reaction) to write

$$E = -\frac{\Delta G}{nF} = -\frac{-150 \times 10^3 \text{ J}}{(2 \text{ mol})\left(9.649 \times 10^4 \dfrac{\text{C}}{\text{mol}}\right)}$$

Reminder: 1 J/C = 1 volt

$$= +0.777 \text{ J/C} = +0.777 \text{ V}$$

A spontaneous chemical reaction (negative $\Delta G$) produces a *positive voltage.*

---

Chemists define the electrode at which *reduction* occurs as the **cathode.** The **anode** is the electrode at which *oxidation* occurs. In Figure 14-4, Ag is the cathode because reduction takes place at its surface ($2AgCl + 2e^- \rightarrow 2Ag + 2Cl^-$), and Cd is the anode because it is oxidized ($Cd \rightarrow Cd^{2+} + 2e^-$).

*Cathode:* where reduction occurs
*Anode:* where oxidation occurs

## Salt Bridge

Consider the cell in Figure 14-5, in which the reactions are intended to be

| | |
|---|---|
| Cathode: | $2Ag^+(aq) + 2e^- \rightleftharpoons 2Ag(s)$ |
| Anode: | $Cd(s) \rightleftharpoons Cd^{2+}(aq) + 2e^-$ |
| Net reaction: | $Cd(s) + 2Ag^+(aq) \rightleftharpoons Cd^{2+}(aq) + 2Ag(s)$     (14-9) |

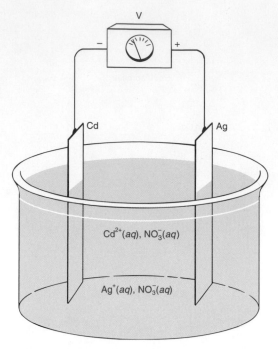

**Figure 14-5** A cell that will not work. The solution contains $Cd(NO_3)_2$ and $AgNO_3$.

The cell in Figure 14-5 is *short-circuited*.

A salt bridge maintains electroneutrality (no charge buildup) throughout the cell. See Demonstration 14-1.

The net reaction is spontaneous, but little current flows through the circuit because $Ag^+$ is not forced to be reduced at the Ag electrode. Aqueous $Ag^+$ can react directly at the $Cd(s)$ surface, giving the same net reaction with no flow of electrons through the external circuit.

We can separate the reactants into two *half-cells*[9] if we connect the two halves with a **salt bridge,** as shown in Figure 14-6. The salt bridge is a U-shaped tube filled with a gel containing a high concentration of $KNO_3$ (or other electrolyte that does not affect the cell reaction). The ends of the bridge are porous glass disks that allow ions to diffuse but minimize mixing of the solutions inside and outside the bridge. When the galvanic cell is operating, $K^+$ from the bridge migrates into the cathode compartment and a small amount of $NO_3^-$ migrates from the cathode into the bridge. Ion migration offsets the charge buildup that would otherwise occur as electrons flow into the silver electrode. The migration of ions out of the bridge is greater than the migration of ions into the bridge because the salt concentration in the

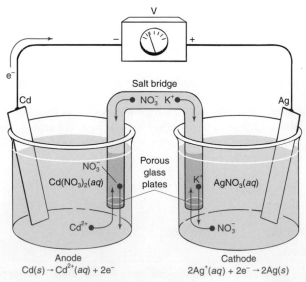

**Figure 14-6** A cell that works—thanks to the salt bridge!

A salt bridge is an ionic medium with a *semipermeable* barrier on each end. Small molecules and ions can cross a semipermeable barrier, but large molecules cannot. Demonstrate a "proper" salt bridge by filling a U-tube with agar and KCl as described in the text and construct the cell shown here.

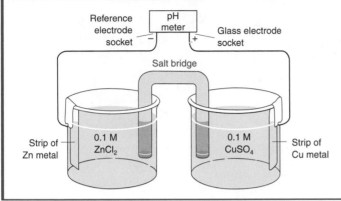

The pH meter is a potentiometer whose negative terminal is the reference electrode socket.

Write the half-reactions for this cell and use the Nernst equation to calculate the theoretical voltage. Measure the voltage with a conventional salt bridge. Then replace the salt bridge with filter paper soaked in NaCl solution and measure the voltage again. Finally, replace the filter-paper salt bridge with two fingers and measure the voltage again. A human is just a bag of salt housed in a semipermeable membrane. Small differences in voltage observed when the salt bridge is replaced can be attributed to the junction potential discussed in Section 15-3. To prove that it is hard to distinguish a chemistry instructor from a hot dog, use a hot dog as a salt bridge[7] and measure the voltage again.

**Challenge** One hundred eighty students at Virginia Tech made a salt bridge by holding hands.[8] Their resistance was lowered from $10^6$ $\Omega$ per student to $10^4$ $\Omega$ per student by wetting everyone's hands. Can your class beat this record?

---

bridge is much higher than the concentration in the half-cells. At the left side of the salt bridge, $NO_3^-$ migrates into the anode compartment and a little $Cd^{2+}$ migrates into the bridge to prevent buildup of positive charge.

For reactions that do not involve $Ag^+$ or other species that react with $Cl^-$, the salt bridge usually contains KCl electrolyte. A typical salt bridge is prepared by heating 3 g of agar with 30 g of KCl in 100 mL of water until a clear solution is obtained. The solution is poured into the U-tube and allowed to gel. The bridge is stored in saturated aqueous KCl.

### Line Notation

Electrochemical cells are described by a notation employing just two symbols:

| phase boundary       ‖ salt bridge

The cell in Figure 14-4 is represented by the *line diagram*

$$Cd(s) \mid CdCl_2(aq) \mid AgCl(s) \mid Ag(s)$$

Each phase boundary is indicated by a vertical line. The electrodes are shown at the extreme left- and right-hand sides of the line diagram. The cell in Figure 14-6 is

$$Cd(s) \mid Cd(NO_3)_2(aq) \parallel AgNO_3(aq) \mid Ag(s)$$

The salt bridge symbol ‖ represents the two phase boundaries on either side of the bridge.

## ▪▪▪ 14-3 Standard Potentials

The voltage measured in Figure 14-6 is the difference in electric potential between the Ag electrode on the right and the Cd electrode on the left. Voltage tells us how much work can be done by electrons flowing from one side to the other (Equation 14-3). The potentiometer (voltmeter) indicates a positive voltage when electrons flow into the negative terminal, as in Figure 14-6. If electrons flow the other way, the voltage is negative.

Sometimes the negative terminal of a voltmeter is labeled "common." It may be colored black and the positive terminal red. When a pH meter with a BNC socket is used as a potentiometer, the center wire is the positive input and the outer connection is the negative input. In older pH meters, the negative terminal is the narrow receptacle to which the reference electrode is connected.

To predict the voltage that will be observed when different half-cells are connected to each other, the **standard reduction potential**, $E°$, for each half-cell is measured by an experiment shown in an idealized form in Figure 14-7. The half-reaction of interest in this diagram is

$$Ag^+ + e^- \rightleftharpoons Ag(s) \qquad \textbf{(14-10)}$$

which occurs in the half-cell at the right connected to the *positive* terminal of the potentiometer. *Standard* means that the activities of all species are unity. For Reaction 14-10, this means that $\mathcal{A}_{Ag^+} = 1$ and, by definition, the activity of $Ag(s)$ also is unity.

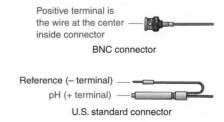

Positive terminal is the wire at the center inside connector

**BNC connector**

Reference (– terminal)
pH (+ terminal)

**U.S. standard connector**

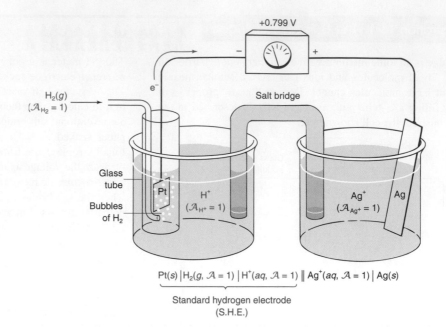

**Figure 14-7** Cell used to measure the standard potential of the reaction $Ag^+ + e^- \rightleftharpoons Ag(s)$. This cell is hypothetical because it is usually not possible to adjust the activity of a species to 1.

+0.799 V

$H_2(g)$
$(\mathcal{A}_{H_2} = 1)$

$e^-$

Salt bridge

Glass tube

Bubbles of $H_2$

Pt

$H^+$
$(\mathcal{A}_{H^+} = 1)$

$Ag^+$
$(\mathcal{A}_{Ag^+} = 1)$

Ag

$Pt(s) \,|\, H_2(g, \mathcal{A} = 1) \,|\, H^+(aq, \mathcal{A} = 1) \,\|\, Ag^+(aq, \mathcal{A} = 1) \,|\, Ag(s)$

Standard hydrogen electrode
(S.H.E.)

**Question** What is the pH of the standard hydrogen electrode?

We will write all half-reactions as *reductions*. By convention, $E° = 0$ for S.H.E.

Walther Nernst appears to have been the first to assign the potential of the hydrogen electrode as 0 in 1897.[10]

We always attach the left-hand electrode to the negative terminal of the potentiometer and the right-hand electrode to the positive terminal. The voltage on the meter is the difference:

Voltage = right-hand electrode potential − left-hand electrode potential

**Challenge** Draw a picture of the cell S.H.E. ‖ $Cd^{2+}$ (aq, $\mathcal{A} = 1$) | $Cd(s)$ and show the direction of electron flow.

The left half-cell, connected to the *negative* terminal of the potentiometer, is called the **standard hydrogen electrode** (S.H.E.). It consists of a catalytic Pt surface in contact with an acidic solution in which $\mathcal{A}_{H^+} = 1$. A stream of $H_2(g)$ bubbled through the electrode saturates the solution with $H_2(aq)$. The activity of $H_2(g)$ is unity if the pressure of $H_2(g)$ is 1 bar. The reaction that comes to equilibrium at the surface of the Pt electrode is

$$H^+(aq, \mathcal{A} = 1) + e^- \rightleftharpoons \tfrac{1}{2}H_2(g, \mathcal{A} = 1) \qquad (14\text{-}11)$$

We *arbitrarily* assign a potential of 0 to the standard hydrogen electrode at 25°C. The voltage measured by the meter in Figure 14-7 can therefore be *assigned* to Reaction 14-10, which occurs in the right half-cell. The measured value $E° = +0.799$ V is the standard reduction potential for Reaction 14-10. The positive sign tells us that electrons flow from left to right through the meter.

We can arbitrarily *assign* a potential to Reaction 14-11 because it serves as a reference point from which we can measure other half-cell potentials. An analogy is the arbitrary assignment of 0°C to the freezing point of water. Relative to this freezing point, hexane boils at 69° and benzene boils at 80°. The difference between boiling points is $80° − 69° = 11°$. If we had assigned the freezing point of water to be 200°C instead of 0°C, hexane boils at 269° and benzene boils at 280°. The difference between boiling points is still 11°. Regardless of where we set zero on the scale, differences between points remain constant.

The line notation for the cell in Figure 14-7 is

$$Pt(s) \,|\, H_2(g, \mathcal{A} = 1) \,|\, H^+(aq, \mathcal{A} = 1) \,\|\, Ag^+(aq, \mathcal{A} = 1) \,|\, Ag(s)$$

or

$$\text{S.H.E.} \,\|\, Ag^+(aq, \mathcal{A} = 1) \,|\, Ag(s)$$

*By convention, the left-hand electrode (Pt) is attached to the negative (reference) terminal of the potentiometer and the right-hand electrode is attached to the positive terminal.* A standard reduction potential is really a potential *difference* between the potential of the reaction of interest and the potential of S.H.E, which we have arbitrarily set to 0.

To measure the standard potential of the half-reaction

$$Cd^{2+} + 2e^- \rightleftharpoons Cd(s) \qquad (14\text{-}12)$$

we construct the cell

$$\text{S.H.E.} \,\|\, Cd^+(aq, \mathcal{A} = 1) \,|\, Cd(s)$$

with the cadmium half-cell connected to the positive terminal of the potentiometer. In this case, we observe a *negative* voltage of −0.402 V. The negative sign means that electrons flow from Cd to Pt, a direction opposite that of the cell in Figure 14-7.

Appendix H contains standard reduction potentials arranged alphabetically by element. If the half-reactions were arranged according to descending value of $E°$ (as in Table 14-1),

**Table 14-1** Ordered redox potentials

| | Oxidizing agent | Reducing agent | $E°(V)$ |
|---|---|---|---|
| | $F_2(g) + 2e^- \rightleftharpoons 2F^-$ | | 2.890 |
| | $O_3(g) + 2H^+ + 2e^- \rightleftharpoons O_2(g) + H_2O$ | | 2.075 |
| | $MnO_4^- + 8H^+ + 5e^- \rightleftharpoons Mn^{2+} + 4H_2O$ | | 1.507 |
| | $Ag^+ + e^- \rightleftharpoons Ag(s)$ | | 0.799 |
| | $Cu^{2+} + 2e^- \rightleftharpoons Cu(s)$ | | 0.339 |
| | $2H^+ + 2e^- \rightleftharpoons H_2(g)$ | | 0.000 |
| | $Cd^{2+} + 2e^- \rightleftharpoons Cd(s)$ | | -0.402 |
| | $K^+ + e^- \rightleftharpoons K(s)$ | | -2.936 |
| | $Li^+ + e^- \rightleftharpoons Li(s)$ | | -3.040 |

(left axis: Oxidizing power increases ↑; right axis: Reducing power increases ↓)

**Question** The potential for the reaction $K^+ + e^- \rightleftharpoons K(s)$ is $-2.936$ V. This means that $K^+$ is a very poor oxidizing agent. (It does not readily accept electrons.) Does this imply that $K^+$ is therefore a good reducing agent?

*Answer:* No! To be a good reducing agent, $K^+$ would have to give up electrons easily (forming $K^{2+}$), which it cannot do. (*But* the large negative reduction potential does imply that $K(s)$ is a good reducing agent.)

we would find the strongest oxidizing agents at the upper left and the strongest reducing agents at the lower right. If we connected the two half-cells represented by Reactions 14-10 and 14-12, $Ag^+$ would be reduced to $Ag(s)$ as $Cd(s)$ is oxidized to $Cd^{2+}$.

## ■ ■ ■ 14-4 Nernst Equation

Le Châtelier's principle tells us that increasing reactant concentrations drives a reaction to the right and increasing the product concentrations drives a reaction to the left. The net driving force for a reaction is expressed by the **Nernst equation,** whose two terms include the driving force under standard conditions ($E°$, which applies when all activities are unity) and a term showing the dependence on reagent concentrations. The Nernst equation tells us the potential of a cell whose reagents are not all at unit activity.

A reaction is spontaneous if $\Delta G$ is negative and $E$ is positive. $\Delta G°$ and $E°$ refer to the free-energy change and potential when the activities of reactants and products are unity. $\Delta G° = -nFE°$.

### Nernst Equation for a Half-Reaction

For the half-reaction

$$aA + ne^- \rightleftharpoons bB$$

the Nernst equation giving the half-cell potential, $E$, is

*Nernst equation:*
$$E = E° - \frac{RT}{nF}\ln\frac{\mathcal{A}_B^b}{\mathcal{A}_A^a}$$
(14-13)

**Challenge** Show that Le Châtelier's principle requires a negative sign in front of the reaction quotient term in the Nernst equation.
*Hint:* The more favorable a reaction, the more positive is $E$.

where $E°$ = standard reduction potential ($\mathcal{A}_A = \mathcal{A}_B = 1$)

$R$ = gas constant (8.314 J/(K · mol) = 8.314 (V · C)/(K · mol))

$T$ = temperature (K)

$n$ = number of electrons in the half-reaction

$F$ = Faraday constant ($9.649 \times 10^4$ C/mol)

$\mathcal{A}_i$ = activity of species $i$

The logarithmic term in the Nernst equation is the **reaction quotient, $Q$.**

$$Q = \mathcal{A}_B^b/\mathcal{A}_A^a$$
(14-14)

$Q$ has the same form as the equilibrium constant, but the activities need not have their equilibrium values. Pure solids, pure liquids, and solvents are omitted from $Q$ because their activities are unity (or close to unity); concentrations of solutes are expressed as moles per liter, and concentrations of gases are expressed as pressures in bars. When all activities are unity, $Q = 1$ and $\ln Q = 0$, thus giving $E = E°$.

Converting the natural logarithm in Equation 14-13 into the base 10 logarithm and inserting $T = 298.15$ K ($25.00°C$) gives the most useful form of the Nernst equation:

*Nernst equation at 25°C:*
$$E = E° - \frac{0.059\ 16\ \text{V}}{n} \log \frac{\mathcal{A}_B^b}{\mathcal{A}_A^a}$$
(14-15)

The potential changes by $59.16/n$ mV for each factor-of-10 change in $Q$.

### Example  Writing the Nernst Equation for a Half-Reaction

Let's write the Nernst equation for the reduction of white phosphorus to phosphine gas:

$$\tfrac{1}{4}P_4(s, \text{white}) + 3H^+ + 3e^- \rightleftharpoons PH_3(g) \qquad E° = -0.046 \text{ V}$$

**Solution**   We omit solids from the reaction quotient, and the concentration of a gas is expressed as the pressure of the gas. Therefore, the Nernst equation is

$$E = -0.046 - \frac{0.059\ 16}{3} \log \frac{P_{PH_3}}{[H^+]^3}$$

### Example  Multiplying a Half-Reaction Does Not Change $E°$

If we multiply a half-reaction by any factor, $E°$ does not change. However, the factor $n$ before the log term and the form of the reaction quotient, $Q$, do change. Let's write the Nernst equation for the reaction in the preceding example, multiplied by 2:

$$\tfrac{1}{2}P_4(s, \text{white}) + 6H^+ + 6e^- \rightleftharpoons 2PH_3(g) \qquad E° = -0.046 \text{ V}$$

**Solution**

$$E = -0.046 - \frac{0.059\ 16}{6} \log \frac{P_{PH_3}^2}{[H^+]^6}$$

Even though this Nernst equation does not look like the one in the preceding example, Box 14-2 shows that the numerical value of $E$ is unchanged. The squared term in the reaction quotient cancels the doubled value of $n$ in front of the log term.

## Nernst Equation for a Complete Reaction

In Figure 14-6, the negative terminal of the potentiometer is connected to the Cd electrode and the positive terminal is connected to the Ag electrode. The voltage, $E$, is the difference between the potentials of the two electrodes:

*Nernst equation for a complete cell:*   $\boxed{E = E_+ - E_-}$
(14-16)

where $E_+$ is the potential of the electrode attached to the positive terminal of the potentiometer and $E_-$ is the potential of the electrode attached to the negative terminal. The potential

---

### Box 14-2  $E°$ and the Cell Voltage Do Not Depend on How You Write the Cell Reaction

*Multiplying a half-reaction by any number does not change the standard reduction potential, $E°$.* The potential difference between two points is the work done *per coulomb of charge* carried through that potential difference ($E = \text{work}/q$). Work per coulomb is the same whether 0.1, 2.3, or $10^4$ coulombs have been transferred. The total work is different in each case, but work per coulomb is constant. Therefore, we do not double $E°$ if we multiply a half-reaction by 2.

*Multiplying a half-reaction by any number does not change the half-cell potential, $E$.* Consider a half-cell reaction written with either one or two electrons:

$$Ag^+ + e^- \rightleftharpoons Ag(s) \qquad E = E° - 0.059\ 16 \log\left(\frac{1}{[Ag^+]}\right)$$

$$2Ag^+ + 2e^- \rightleftharpoons 2Ag(s) \qquad E = E° - \frac{0.059\ 16}{2} \log\left(\frac{1}{[Ag^+]^2}\right)$$

The two expressions are equal because $\log a^b = b \log a$:

$$\frac{0.059\ 16}{2} \log\left(\frac{1}{[Ag^+]^2}\right) = \frac{2 \times 0.059\ 16}{2} \log\left(\frac{1}{[Ag^+]}\right)$$

$$= 0.059\ 16 \log\left(\frac{1}{[Ag^+]}\right)$$

The exponent in the log term is canceled by the factor $1/n$ that precedes the log term. The cell potential cannot depend on how you write the reaction.

CHAPTER 14 Fundamentals of Electrochemistry

of each half-reaction (*written as a reduction*) is governed by a Nernst equation like Equation 14-13, and the voltage for the complete reaction is the difference between the two half-cell potentials.

Here is a procedure for writing a net cell reaction and finding its voltage:

**Step 1** Write *reduction* half-reactions for both half-cells and find $E°$ for each in Appendix H. Multiply the half-reactions as necessary so that they both contain the same number of electrons. When you multiply a reaction, you *do not* multiply $E°$.

**Step 2** Write a Nernst equation for the right half-cell, which is attached to the positive terminal of the potentiometer. This is $E_+$.

**Step 3** Write a Nernst equation for the left half-cell, which is attached to the negative terminal of the potentiometer. This is $E_-$.

**Step 4** Find the net cell voltage by subtraction: $E = E_+ - E_-$.

**Step 5** To write a balanced net cell reaction, subtract the left half-reaction from the right half-reaction. (Subtraction is equivalent to reversing the left-half reaction and adding.)

If the net cell voltage, $E(= E_+ - E_-)$, is positive, then the net cell reaction is spontaneous in the forward direction. If the net cell voltage is negative, then the reaction is spontaneous in the reverse direction.

$E > 0$: net cell reaction goes →
$E < 0$: net cell reaction goes ←

---

## Example Nernst Equation for a Complete Reaction

Find the voltage of the cell in Figure 14-6 if the right half-cell contains 0.50 M $AgNO_3(aq)$ and the left half-cell contains 0.010 M $Cd(NO_3)_2(aq)$. Write the net cell reaction and state whether it is spontaneous in the forward or reverse direction.

### Solution

**Step 1** Right electrode:     $2Ag^+ + 2e^- \rightleftharpoons 2Ag(s)$     $E_+° = 0.799$ V

        Left electrode:     $Cd^{2+} + 2e^- \rightleftharpoons Cd(s)$     $E_-° = -0.402$ V

**Step 2** Nernst equation for right electrode:

$$E_+ = E_+° - \frac{0.059\ 16}{2} \log \frac{1}{[Ag^+]^2} = 0.799 - \frac{0.059\ 16}{2} \log \frac{1}{[0.50]^2} = 0.781\ V$$

Pure solids, pure liquids, and solvents are omitted from Q.

**Step 3** Nernst equation for left electrode:

$$E_- = E_-° - \frac{0.059\ 16}{2} \log \frac{1}{[Cd^{2+}]} = -0.402 - \frac{0.059\ 16}{2} \log \frac{1}{[0.010]} = -0.461\ V$$

**Step 4** Cell voltage:     $E = E_+ - E_- = 0.781 - (-0.461) = +1.242$ V

**Step 5** Net cell reaction:

$$\begin{array}{r} 2Ag^+ + 2e^- \rightleftharpoons 2Ag(s) \\ - \quad \underline{Cd^{2+} + 2e^- \rightleftharpoons Cd(s)} \\ \hline Cd(s) + 2Ag^+ \rightleftharpoons Cd^{2+} + 2Ag(s) \end{array}$$

Subtracting a reaction is the same as reversing the reaction and adding.

Because the voltage is positive, the net reaction is spontaneous in the forward direction. $Cd(s)$ is oxidized and $Ag^+$ is reduced. Electrons flow from the left-hand electrode to the right-hand electrode.

What if you had written the Nernst equation for the right half-cell with just one electron instead of two?

$$Ag^+ + e^- \rightleftharpoons Ag(s)$$

Would the net cell voltage be different from what we calculated? It better not be, because the chemistry is still the same. Box 14-2 shows that *neither $E°$ nor $E$ depends on how we write the reaction*. Box 14-3 shows how to derive standard reduction potentials for half-reactions that are the sum of other half-reactions.

## An Intuitive Way to Think About Cell Potentials[2]

In the preceding example, we found that $E$ for the silver half-cell was 0.781 V and $E$ for the cadmium half-cell was $-0.461$ V. Place these values on the number line in

## Box 14-3  Latimer Diagrams: How to Find $E°$ for a New Half-Reaction

A **Latimer diagram** displays standard reduction potentials $E°$, connecting various oxidation states of an element.[11] For example, in acid solution, the following standard reduction potentials are observed:

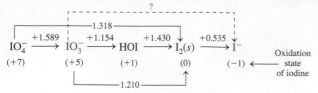

As an example to understand what the Latimer diagram means, the notation $IO_3^- \xrightarrow{1.154} HOI$ stands for the balanced equation

$$IO_3^- + 5H^+ + 4e^- \rightleftharpoons HOI + 2H_2O \qquad E° = +1.154 \text{ V}$$

We can derive reduction potentials for arrows that are not shown in the diagram by using $\Delta G°$. For example, the reaction shown by the dashed line in the Latimer diagram is

$$IO_3^- + 6H^+ + 6e^- \rightleftharpoons I^- + 3H_2O$$

To find $E°$ for this reaction, express the reaction as a sum of reactions whose potentials are known.

The standard free-energy change, $\Delta G°$, for a reaction is given by

$$\Delta G° = -nFE°$$

*When two reactions are added to give a third reaction, the sum of the individual $\Delta G°$ values must equal the overall value of $\Delta G°$.*

To apply free energy to our problem, we write two reactions whose sum is the desired reaction:

$$IO_3^- + 6H^+ + 5e^- \xrightarrow{E_1° = 1.210} \tfrac{1}{2}I_2(s) + 3H_2O$$
$$\Delta G_1° = -5F(1.210)$$

$$\tfrac{1}{2}I_2(s) + e^- \xrightarrow{E_2° = 0.535} I^- \qquad \Delta G_2° = -1F(0.535)$$

$$IO_3^- + 6H^+ + 6e^- \xrightarrow{E_3° = ?} I^- + 3H_2O \qquad \Delta G_3° = -6FE_3°$$

But, because $\Delta G_1° + \Delta G_2° = \Delta G_3°$, we can solve for $E_3°$:

$$\Delta G_3° = \Delta G_1° + \Delta G_2°$$

$$-6FE_3° = -5F(1.210) - 1F(0.535)$$

$$E_3° = \frac{5(1.210) + 1(0.535)}{6} = 1.098 \text{ V}$$

---

Figure 14-8 and note that *electrons always flow toward a more positive potential.* Therefore, electrons in the circuit flow from cadmium to silver. The separation of the two half-cells is 1.242 V. This diagram works the same way even if both half-cell potentials are positive or both are negative. Electrons always flow toward more positive potential.

Electrons flow toward more positive potential.

### Different Descriptions of the Same Reaction

In Figure 14-4, the right half-reaction can be written

$$AgCl(s) + e^- \rightleftharpoons Ag(s) + Cl^- \qquad E_+° = 0.222 \text{ V} \qquad \textbf{(14-17)}$$

$$E_+ = E_+° - 0.059\,16 \log[Cl^-] = 0.222 - 0.059\,16 \log(0.033\,4) = 0.309_3 \text{ V} \qquad \textbf{(14-18)}$$

The $Cl^-$ in the silver half-reaction was derived from $0.016\,7$ M $CdCl_2(aq)$.

Suppose that a different, less handsome, author had written this book and had chosen to describe the half-reaction differently:

$$Ag^+ + e^- \rightleftharpoons Ag(s) \qquad E_+° = 0.799 \text{ V} \qquad \textbf{(14-19)}$$

*This description is just as valid as the previous one.* In both cases, Ag(I) is reduced to Ag(0).

If the two descriptions are equally valid, then they should predict the same voltage. The Nernst equation for Reaction 14-19 is

$$E_+ = 0.799 - 0.059\,16 \log \frac{1}{[Ag^+]}$$

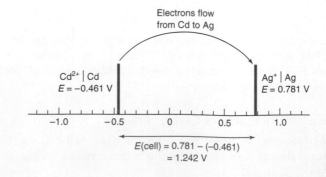

*Figure 14-8*  An intuitive view of cell potentials.[2] Electrons always flow to the right in this diagram.

To find the concentration of $Ag^+$, we use the solubility product for AgCl. Because the cell contains 0.033 4 M $Cl^-$ and solid AgCl, we can say

$$[Ag^+] = \frac{K_{sp}(\text{for AgCl})}{[Cl^-]} = \frac{1.8 \times 10^{-10}}{0.033\ 4} = 5.4 \times 10^{-9}\ M$$

$K_{sp} = [Ag^+][Cl^-]$

Putting this value into the Nernst equation gives

$$E_+ = 0.799 - 0.059\ 16\ \log\frac{1}{5.4 \times 10^{-9}} = 0.309_9\ V$$

which differs from the value calculated in Equation 14-18 because of the accuracy of $K_{sp}$ and the neglect of activity coefficients. Equations 14-17 and 14-19 give the same voltage because they describe the same cell.

The cell voltage cannot depend on how we write the reaction!

## Advice for Finding Relevant Half-Reactions

When faced with a cell drawing or a line diagram, first write reduction reactions for each half-cell. To do this, *look in the cell for an element in two oxidation states*. For the cell

How to figure out the half-cell reactions

$$Pb(s)\ |\ PbF_2(s)\ |\ F^-(aq)\ \|\ Cu^{2+}(aq)\ |\ Cu(s)$$

we see Pb in two oxidation states, as $Pb(s)$ and $PbF_2(s)$, and Cu in two oxidation states, as $Cu^{2+}$ and $Cu(s)$. Thus, the half-reactions are

| Right half-cell: | $Cu^{2+} + 2e^- \rightleftharpoons Cu(s)$ | |
| --- | --- | --- |
| Left half-cell: | $PbF_2(s) + 2e^- \rightleftharpoons Pb(s) + 2F^-$ | **(14-20)** |

You might have chosen to write the Pb half-reaction as

| Left half-cell: | $Pb^{2+} + 2e^- \rightleftharpoons Pb(s)$ | **(14-21)** |
| --- | --- | --- |

because you know that, if $PbF_2(s)$ is present, there must be some $Pb^{2+}$ in the solution. Reactions 14-20 and 14-21 are equally valid and should predict the same cell voltage. Your choice of reactions depends on whether the $F^-$ or $Pb^{2+}$ concentration is easier to figure out.

We described the left half-cell in terms of a redox reaction involving Pb because Pb is the element that appears in two oxidation states. We would not write a reaction such as $F_2(g) + 2e^- \rightleftharpoons 2F^-$, because $F_2(g)$ is not shown in the line diagram of the cell.

Don't invent species not shown in the cell. Use what is shown in the line diagram to select the half-reactions.

## The Nernst Equation Is Used in Measuring Standard Reduction Potentials

The standard reduction potential would be observed if the half-cell of interest (with unit activities) were connected to a standard hydrogen electrode, as it is in Figure 14-7. It is nearly impossible to construct such a cell, because we have no way to adjust concentrations and ionic strength to give unit activities. In reality, activities less than unity are used in each half-cell, and the Nernst equation is employed to extract the value of $E°$ from the cell voltage.[12] In the hydrogen electrode, standard buffers with known pH (Table 15-3) are used to obtain known activities of $H^+$.

Problem 14-20 gives an example of the use of the Nernst equation to find $E°$.

## ■ ■ ■ 14-5 $E°$ and the Equilibrium Constant

*A galvanic cell produces electricity because the cell reaction is not at equilibrium.* The potentiometer allows negligible current (Box 14-4), so concentrations in the cell remain unchanged. If we replaced the potentiometer with a wire, there would be much more current and concentrations would change until the cell reached equilibrium. At that point, nothing would drive the reaction, and $E$ would be 0. When a battery (which is a galvanic cell) runs down to 0 V, the chemicals inside have reached equilibrium and the battery is "dead."

At equilibrium, $E$ (not $E°$) = 0.

Now let's relate $E$ for a whole cell to the reaction quotient, $Q$, for the net cell reaction. For the two half-reactions

| Right electrode: | $aA + ne^- \rightleftharpoons cC$ | $E°_+$ |
| --- | --- | --- |
| Left electrode: | $dD + ne^- \rightleftharpoons bB$ | $E°_-$ |

Box 14-4  Concentrations in the Operating Cell

Why doesn't operation of a cell change the concentrations in the cell? Cell voltage is measured under conditions of *negligible current flow*. The resistance of a high-quality pH meter is $10^{13}\ \Omega$. If you use this meter to measure a potential of 1 V, the current is

$$I = \frac{E}{R} = \frac{1\ \text{V}}{10^{13}\ \Omega} = 10^{-13}\ \text{A}$$

If the cell in Figure 14-6 produces 50 mV, the current through the circuit is $0.050\ \text{V}/10^{13}\ \Omega = 5 \times 10^{-15}\ \text{A}$. This value corresponds to a flow of

$$\frac{5 \times 10^{-15}\ \text{C/s}}{9.649 \times 10^{4}\ \text{C/mol}} = 5 \times 10^{-20}\ \text{mol e}^{-}/\text{s}$$

The rate at which $Cd^{2+}$ is produced is $2.5 \times 10^{-20}$ mol/s, which has a negligible effect on the cadmium concentration in the cell. *The meter measures the voltage of the cell without affecting concentrations in the cell.*

If the salt bridge were left in a real cell for very long, concentrations and ionic strength would change because of diffusion between each compartment and the salt bridge. We assume that cells are set up for such a short time that mixing does not happen.

---

the Nernst equation looks like this:

$$E = E_{+} - E_{-} = E_{+}^{\circ} - \frac{0.059\ 16}{n} \log \frac{\mathcal{A}_{\text{C}}^{c}}{\mathcal{A}_{\text{A}}^{a}} - \left( E_{-}^{\circ} - \frac{0.059\ 16}{n} \log \frac{\mathcal{A}_{\text{B}}^{b}}{\mathcal{A}_{\text{D}}^{d}} \right)$$

$$E = \underbrace{(E_{+}^{\circ} - E_{-}^{\circ})}_{E^{\circ}} - \frac{0.059\ 16}{n} \log \underbrace{\frac{\mathcal{A}_{\text{C}}^{c}\mathcal{A}_{\text{D}}^{d}}{\mathcal{A}_{\text{A}}^{a}\mathcal{A}_{\text{B}}^{b}}}_{Q} = E^{\circ} - \frac{0.059\ 16}{n} \log Q \qquad \textbf{(14-22)}$$

$\log a + \log b = \log ab$

Equation 14-22 is true at any time. In the special case when the cell is at equilibrium, $E = 0$ and $Q = K$, the equilibrium constant. Therefore, Equation 14-22 is transformed into these most important forms at equilibrium:

To go from Equation 14-23 to 14-24:

$$\frac{0.059\ 16}{n} \log K = E^{\circ}$$

$$\log K = \frac{nE^{\circ}}{0.059\ 16}$$

$$10^{\log K} = 10^{nE^{\circ}/0.059\ 16}$$

$$K = 10^{nE^{\circ}/0.059\ 16}$$

*Finding $E^{\circ}$ from $K$:*   $\boxed{E^{\circ} = \dfrac{0.059\ 16}{n} \log K}$   (at 25°C)   **(14-23)**

*Finding $K$ from $E^{\circ}$:*   $\boxed{K = 10^{nE^{\circ}/0.059\ 16}}$   (at 25°C)   **(14-24)**

Equation 14-24 allows us to deduce the equilibrium constant from $E^{\circ}$. Alternatively, we can find $E^{\circ}$ from $K$ with Equation 14-23.

---

We associate $E_{-}^{\circ}$ with the half-reaction that must be *reversed* to get the desired net reaction.

### Example  Using $E^{\circ}$ to Find the Equilibrium Constant

Find the equilibrium constant for the reaction

$$Cu(s) + 2Fe^{3+} \rightleftharpoons 2Fe^{2+} + Cu^{2+}$$

**Solution**   The reaction is divided into two half-reactions found in Appendix H:

$$\begin{array}{lr} 2Fe^{3+} + 2e^{-} \rightleftharpoons 2Fe^{2+} & E_{+}^{\circ} = 0.771\ \text{V} \\ - \quad Cu^{2+} + 2e^{-} \rightleftharpoons Cu(s) & E_{-}^{\circ} = 0.339\ \text{V} \\ \hline Cu(s) + 2Fe^{3+} \rightleftharpoons 2Fe^{2+} + Cu^{2+} & \end{array}$$

Then we find $E^{\circ}$ for the net reaction

$$E^{\circ} = E_{+}^{\circ} - E_{-}^{\circ} = 0.771 - 0.339 = 0.432\ \text{V}$$

and compute the equilibrium constant with Equation 14-24:

$$K = 10^{(2)(0.432)/(0.059\ 16)} = 4 \times 10^{14}$$

Significant figures for logs and exponents were discussed in Section 3-2.

A modest value of $E^{\circ}$ produces a large equilibrium constant. The value of $K$ is correctly expressed with one significant figure, because $E^{\circ}$ has three digits. Two are used for the exponent (14), and only one is left for the multiplier (4).

## Finding $K$ for Net Reactions That Are Not Redox Reactions

Consider the following half-reactions whose difference is the solubility reaction for iron(II) carbonate (which is not a redox reaction):

$$
\begin{array}{lll}
& \text{FeCO}_3(s) + 2e^- \rightleftharpoons \text{Fe}(s) + \text{CO}_3^{2-} & E_+^\circ = -0.756 \text{ V} \\
- & \text{Fe}^{2+} + 2e^- \rightleftharpoons \text{Fe}(s) & E_-^\circ = -0.44 \text{ V} \\
\hline
& \text{FeCO}_3(s) \rightleftharpoons \text{Fe}^{2+} + \text{CO}_3^{2-} & E^\circ = -0.756 - (-0.44) = -0.31_6 \text{ V} \\
& \quad \text{Iron(II)} & \\
& \quad \text{carbonate} &
\end{array}
$$

$$
K = K_{sp} = 10^{(2)(-0.31_6)/(0.059\ 16)} = 10^{-11}
$$

By finding $E^\circ$ for the net reaction, we can compute $K_{sp}$ for iron(II) carbonate. Potentiometric measurements allow us to find equilibrium constants that are too small or too large to measure by determining concentrations of reactants and products directly.

"Wait!" you protest. "How can there be a redox potential for a reaction that is not a redox reaction?" Box 14-3 shows that the redox potential is just another way of expressing the free-energy change of the reaction. The more energetically favorable the reaction (the more negative $\Delta G^\circ$), the more positive is $E^\circ$.

The general form of a problem involving the relation between $E^\circ$ values for half-reactions and $K$ for a net reaction is

$$
\begin{array}{lll}
& \text{Half-reaction:} & E_+^\circ \\
- & \text{Half-reaction:} & E_-^\circ \\
\hline
& \text{Net reaction:} & E^\circ = E_+^\circ - E_-^\circ \quad K = 10^{nE^\circ/0.059\ 16}
\end{array}
$$

If you know $E_-^\circ$ and $E_+^\circ$, you can find $E^\circ$ and $K$ for the net cell reaction. Alternatively, if you know $E^\circ$ and either $E_-^\circ$ or $E_+^\circ$, you can find the missing standard potential. If you know $K$, you can calculate $E^\circ$ and use it to find either $E_-^\circ$ or $E_+^\circ$, provided you know one of them.

---

> **Example** Relating $E^\circ$ and $K$
>
> From the formation constant of Ni(glycine)$_2$ plus $E^\circ$ for the Ni$^{2+}$ | Ni($s$) couple,
>
> $$
> \begin{array}{ll}
> \text{Ni}^{2+} + 2 \text{ glycine}^- \rightleftharpoons \text{Ni(glycine)}_2 & K \equiv \beta_2 = 1.2 \times 10^{11} \\
> \text{Ni}^{2+} + 2e^- \rightleftharpoons \text{Ni}(s) & E^\circ = -0.236 \text{ V}
> \end{array}
> $$
>
> deduce the value of $E^\circ$ for the reaction
>
> $$
> \text{Ni(glycine)}_2 + 2e^- \rightleftharpoons \text{Ni}(s) + 2 \text{ glycine}^- \qquad \textbf{(14-25)}
> $$
>
> **Solution**  We need to see the relations among the three reactions:
>
> $$
> \begin{array}{lll}
> & \text{Ni}^{2+} + 2e^- \rightleftharpoons \text{Ni}(s) & E_+^\circ = -0.236 \text{ V} \\
> - & \text{Ni(glycine)}_2 + 2e^- \rightleftharpoons \text{Ni}(s) + 2 \text{ glycine}^- & E_-^\circ = ? \\
> \hline
> & \text{Ni}^{2+} + 2 \text{ glycine}^- \rightleftharpoons \text{Ni(glycine)}_2 & E^\circ = ? \quad K = 1.2 \times 10^{11}
> \end{array}
> $$
>
> We know that $E_+^\circ - E_-^\circ$ must equal $E^\circ$, so we can deduce the value of $E_-^\circ$ if we can find $E^\circ$. But $E^\circ$ can be determined from the equilibrium constant for the net reaction:
>
> $$
> E^\circ = \frac{0.059\ 16}{n} \log K = \frac{0.059\ 16}{2} \log(1.2 \times 10^{11}) = 0.328 \text{ V}
> $$
>
> Hence, the standard reduction potential for half-reaction 14-25 is
>
> $$
> E_-^\circ = E_+^\circ - E^\circ = -0.236 - 0.328 = -0.564 \text{ V}
> $$

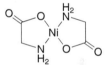

Possible structure
of Ni(glycine)$_2$

---

## ▪▪▪ 14-6  Cells as Chemical Probes[13]

It is essential to distinguish two classes of equilibria associated with galvanic cells:

1. Equilibrium *between* the two half-cells
2. Equilibrium *within* each half-cell

If a galvanic cell has a nonzero voltage, then the net cell reaction is not at equilibrium. We say that equilibrium *between* the two half-cells has not been established.

$E^\circ$ for dissolution of iron(II) carbonate is negative, which means that the reaction is "not spontaneous." "Not spontaneous" simply means $K < 1$. The reaction proceeds until the concentrations of reactants and products satisfy the equilibrium condition.

Any two of the three $E^\circ$ values allow us to calculate the third value.

The more negative potential of $-0.564$ V for reducing Ni(glycine)$_2$ as compared to $-0.236$ V for reducing Ni$^{2+}$ tells us that it is harder to reduce Ni(glycine)$_2$ than Ni$^{2+}$. Ni$^{2+}$ is stabilized with respect to reduction by complexation with glycine.

**14-6  Cells as Chemical Probes**

285

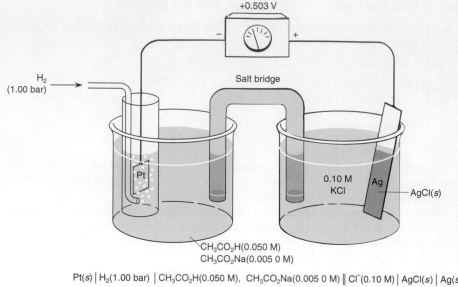

**Figure 14-9** This galvanic cell can be used to measure the pH of the left half-cell.

$$Pt(s) \mid H_2(1.00 \text{ bar}) \mid \underbrace{CH_3CO_2H(0.050 \text{ M}), \ CH_3CO_2Na(0.005\ 0 \text{ M})}_{\text{Aha! A buffer!}} \parallel Cl^-(0.10 \text{ M}) \mid AgCl(s) \mid Ag(s)$$

A chemical reaction that can occur *within one half-cell* will reach equilibrium and is assumed to remain at equilibrium. Such a reaction is not the net cell reaction.

*We allow half-cells to stand long enough to come to chemical equilibrium **within** each half-cell.* For example, in the right-hand half-cell in Figure 14-9, the reaction

$$AgCl(s) \rightleftharpoons Ag^+(aq) + Cl^-(aq)$$

is at equilibrium. It is not part of the net cell reaction. It is simply a reaction that occurs when $AgCl(s)$ is in contact with water. In the left half-cell, the reaction

$$CH_3CO_2H \rightleftharpoons CH_3CO_2^- + H^+$$

has also come to equilibrium. Neither of them is part of the net cell redox reaction.

The reaction for the right half-cell of Figure 14-9 is

$$AgCl(s) + e^- \rightleftharpoons Ag(s) + Cl^-(aq, 0.10 \text{ M}) \qquad E_+^\circ = 0.222 \text{ V}$$

But what is the reaction in the left half-cell? The only element we find in two oxidation states is hydrogen. We see that $H_2(g)$ bubbles into the cell, and we also realize that every aqueous solution contains $H^+$. Therefore, hydrogen is present in two oxidation states, and the half-reaction can be written as

$$2H^+(aq, ? \text{ M}) + 2e^- \rightleftharpoons H_2(g, 1.00 \text{ bar}) \qquad E_-^\circ = 0$$

The net cell reaction is not at equilibrium, because the measured voltage is 0.503 V, not 0 V.

The Nernst equation for the cell reaction is

$$E = E_+ - E_- = (0.222 - 0.059\ 16 \log[Cl^-]) - \left(0 - \frac{0.059\ 16}{2} \log \frac{P_{H_2}}{[H^+]^2}\right)$$

When we put in all the known quantities, we discover that the only unknown is $[H^+]$. *The measured voltage therefore allows us to find $[H^+]$ in the left half-cell:*

$$0.503 = (0.222 - 0.059\ 16 \log[0.10]) - \left(0 - \frac{0.059\ 16}{2} \log \frac{1.00}{[H^+]^2}\right)$$

$$\Rightarrow [H^+] = 1.8 \times 10^{-4} \text{ M}$$

This, in turn, allows us to evaluate the equilibrium constant for the acid-base reaction that has come to equilibrium in the left half-cell:

**Question** Why can we assume that the concentrations of acetic acid and acetate ion are equal to their initial (formal) concentrations?

$$K_a = \frac{[CH_3CO_2^-][H^+]}{[CH_3CO_2H]} = \frac{(0.005\ 0)(1.8 \times 10^{-4})}{0.050} = 1.8 \times 10^{-5}$$

The cell in Figure 14-9 acts as a *probe* to measure $[H^+]$ in the left half-cell. Using this type of cell, we could determine the equilibrium constant for acid dissociation or base hydrolysis in the left half-cell.

## Survival Tips

Problems in this chapter include some brainbusters designed to bring together your knowledge of electrochemistry, chemical equilibrium, solubility, complex formation, and acid-base chemistry. They require you to find the equilibrium constant for a reaction that occurs in only one half-cell. The reaction of interest is not the net cell reaction and is not a redox reaction. Here is a good approach:

**Step 1**  Write the two half-reactions and their standard potentials. If you choose a half-reaction for which you cannot find $E°$, then find another way to write the reaction.

**Step 2**  Write a Nernst equation for the net reaction and put in all the known quantities. If all is well, there will be only one unknown in the equation.

**Step 3**  Solve for the unknown concentration and use that concentration to solve the chemical equilibrium problem that was originally posed.

The half-reactions that you write *must* involve species that appear in two oxidation states in the cell.

---

**Example**  Analyzing a Very Complicated Cell

The cell in Figure 14-10 measures the formation constant ($K_f$) of $Hg(EDTA)^{2-}$. The right-hand compartment contains 0.500 mmol of $Hg^{2+}$ and 2.00 mmol of EDTA in 0.100 L buffered to pH 6.00. The voltage is +0.342 V. Find the value of $K_f$ for $Hg(EDTA)^{2-}$.

**Solution**

**Step 1**  The left half-cell is a standard hydrogen electrode for which we can say $E_- = 0$. In the right half-cell, mercury is in two oxidation states. So let's write the half-reaction

$$Hg^{2+} + 2e^- \rightleftharpoons Hg(l) \qquad E_+° = 0.852 \text{ V}$$

$$E_+ = 0.852 - \frac{0.059\ 16}{2} \log\left(\frac{1}{[Hg^{2+}]}\right)$$

In the right half-cell, the reaction between $Hg^{2+}$ and EDTA is

$$Hg^{2+} + Y^{4-} \overset{K_f}{\rightleftharpoons} HgY^{2-}$$

Because we expect $K_f$ to be large, we assume that virtually all the $Hg^{2+}$ has reacted to make $HgY^{2-}$. Therefore, the concentration of $HgY^{2-}$ is 0.500 mmol/100 mL = 0.005 00 M. The remaining EDTA has a total concentration of $(2.00 - 0.50)$ mmol/100 mL = 0.015 0 M. The right-hand compartment therefore contains 0.005 00 M $HgY^{2-}$, 0.015 0 M EDTA, and a small, unknown concentration of $Hg^{2+}$.

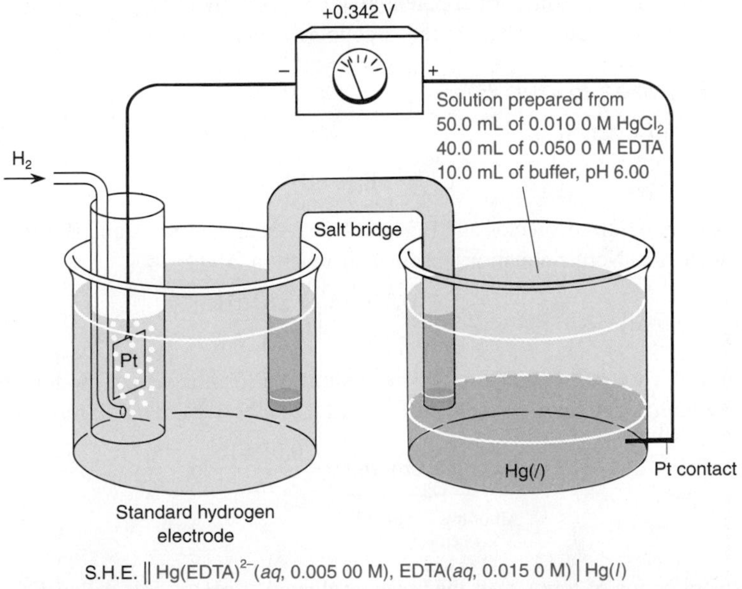

*Figure 14-10*  A galvanic cell that can be used to measure the formation constant for $Hg(EDTA)^{2-}$.

The formation constant for $HgY^{2-}$ can be written

$$K_f = \frac{[HgY^{2-}]}{[Hg^{2+}][Y^{4-}]} = \frac{[HgY^{2-}]}{[Hg^{2+}]\alpha_{Y^{4-}}[EDTA]}$$

where [EDTA] is the formal concentration of EDTA not bound to metal. In this cell, $[EDTA] = 0.015\ 0$ M. The fraction of EDTA in the form $Y^{4-}$ is $\alpha_{Y^{4-}}$ (Section 12-2). Because we know that $[HgY^{2-}] = 0.005\ 00$ M, all we need to find is $[Hg^{2+}]$ in order to evaluate $K_f$.

**Step 2** The Nernst equation for the net cell reaction is

$$E = 0.342 = E_+ - E_- = \left(0.852 - \frac{0.059\ 16}{2}\log\left(\frac{1}{[Hg^{2+}]}\right)\right) - (0)$$

in which the only unknown is $[Hg^{2+}]$.

**Step 3** Now we solve the Nernst equation to find $[Hg^{2+}] = 5._7 \times 10^{-18}$ M, and this value of $[Hg^{2+}]$ allows us to evaluate the formation constant for $HgY^{2-}$:

$$K_f = \frac{[HgY^{2-}]}{[Hg^{2+}]\alpha_{Y^{4-}}[EDTA]} = \frac{(0.005\ 00)}{(5._7 \times 10^{-18})(1.8 \times 10^{-5})(0.015\ 0)}$$

$$= 3 \times 10^{21}$$

The mixture of EDTA plus $Hg(EDTA)^{2-}$ in the cathode serves as a mercuric ion "buffer" that fixes the concentration of $Hg^{2+}$. This concentration, in turn, determines the cell voltage.

# ■ ■ ■ 14-7 Biochemists Use $E^{\circ\prime}$

Perhaps the most important redox reactions in living organisms are in respiration, in which molecules of food are oxidized by $O_2$ to yield energy or metabolic intermediates. The standard reduction potentials that we have been using so far apply to systems in which all activities of reactants and products are unity. If $H^+$ is involved in the reaction, $E^\circ$ applies when pH = 0 ($\mathcal{A}_{H^+} = 1$). *Whenever $H^+$ appears in a redox reaction, or whenever reactants or products are acids or bases, reduction potentials are pH dependent.*

Because the pH inside a plant or animal cell is about 7, reduction potentials that apply at pH 0 are not particularly appropriate. For example, at pH 0, ascorbic acid (vitamin C) is a more powerful reducing agent than succinic acid. However, at pH 7, this order is reversed. It is the reducing strength at pH 7, not at pH 0, that is relevant to a living cell.

The *standard potential* for a redox reaction is defined for a galvanic cell in which all activities are unity. The **formal potential** is the reduction potential that applies under a *specified* set of conditions (including pH, ionic strength, and concentration of complexing agents). Biochemists call the formal potential at pH 7 $E^{\circ\prime}$ (read "$E$ zero prime"). Table 14-2 lists $E^{\circ\prime}$ values for various biological redox couples.

## Relation Between $E^\circ$ and $E^{\circ\prime}$

Consider the half-reaction

$$a\text{A} + ne^- \rightleftharpoons b\text{B} + m\text{H}^+ \qquad E^\circ$$

in which A is an oxidized species and B is a reduced species. Both A and B might be acids or bases, as well. The Nernst equation for this half-reaction is

$$E = E^\circ - \frac{0.059\ 16}{n}\log\frac{[B]^b[H^+]^m}{[A]^a}$$

To find $E^{\circ\prime}$, we rearrange the Nernst equation to a form in which the log term contains only the *formal concentrations* of A and B raised to the powers $a$ and $b$, respectively.

*Recipe for $E^{\circ\prime}$:* $\qquad E = \underbrace{E^\circ + \text{other terms}}_{\substack{\text{All of this is called } E^{\circ\prime} \\ \text{when pH} = 7}} - \frac{0.059\ 16}{n}\log\frac{F_B^b}{F_A^a}$ **(14-26)**

The entire collection of terms over the brace, evaluated at pH = 7, is called $E^{\circ\prime}$.

To convert [A] or [B] into $F_A$ or $F_B$, we use fractional composition equations (Section 10-5), which relate the formal (that is, total) concentration of *all* forms of an acid or a

---

Margin notes:

Recall that $[Y^{4-}] = \alpha_{Y^{4-}}[EDTA]$.

The value of $\alpha_{Y^{4-}}$ comes from Table 12-1.

The formal potential at pH = 7 is called $E^{\circ\prime}$.

If we had included activity coefficients, they would appear in $E^{\circ\prime}$ also.

**Table 14-2** Reduction potentials of biological interest

| Reaction | $E°$ (V) | $E°'$ (V) |
|---|---|---|
| $O_2 + 4H^+ + 4e^- \rightleftharpoons 2H_2O$ | +1.229 | +0.816 |
| $Fe^{3+} + e^- \rightleftharpoons Fe^{2+}$ | +0.771 | +0.771 |
| $I_2 + 2e^- = 2I^-$ | +0.535 | +0.535 |
| Cytochrome $a$ $(Fe^{3+})$ + $e^- \rightleftharpoons$ cytochrome $a$ $(Fe^{2+})$ | +0.290 | +0.290 |
| $O_2(g) + 2H^+ + 2e^- \rightleftharpoons H_2O_2$ | +0.695 | +0.281 |
| Cytochrome $c$ $(Fe^{3+})$ + $e^- \rightleftharpoons$ cytochrome $c$ $(Fe^{2+})$ | — | +0.254 |
| 2,6-Dichlorophenolindophenol + $2H^+$ + $2e^- \rightleftharpoons$ reduced 2,6-dichlorophenolindophenol | — | +0.22 |
| Dehydroascorbate + $2H^+$ + $2e^- \rightleftharpoons$ ascorbate + $H_2O$ | +0.390 | +0.058 |
| Fumarate + $2H^+$ + $2e^- \rightleftharpoons$ succinate | +0.433 | +0.031 |
| Methylene blue + $2H^+$ + $2e^- \rightleftharpoons$ reduced product | +0.532 | +0.011 |
| Glyoxylate + $2H^+$ + $2e^- \rightleftharpoons$ glycolate | — | -0.090 |
| Oxaloacetate + $2H^+$ + $2e^- \rightleftharpoons$ malate | +0.330 | -0.102 |
| Pyruvate + $2H^+$ + $2e^- \rightleftharpoons$ lactate | +0.224 | -0.190 |
| Riboflavin + $2H^+$ + $2e^- \rightleftharpoons$ reduced riboflavin | — | -0.208 |
| FAD + $2H^+$ + $2e^- \rightleftharpoons FADH_2$ | — | -0.219 |
| (Glutathione-S)$_2$ + $2H^+$ + $2e^- \rightleftharpoons$ 2 glutathione-SH | — | -0.23 |
| Safranine T + $2e^- \rightleftharpoons$ leucosafranine T | -0.235 | -0.289 |
| $(C_6H_5S)_2 + 2H^+ + 2e^- \rightleftharpoons 2C_6H_5SH$ | — | -0.30 |
| $NAD^+ + H^+ + 2e^- \rightleftharpoons$ NADH | -0.105 | -0.320 |
| $NADP^+ + H^+ + 2e^- \rightleftharpoons$ NADPH | — | -0.324 |
| Cystine + $2H^+$ + $2e^- \rightleftharpoons$ 2 cysteine | — | -0.340 |
| Acetoacetate + $2H^+$ + $2e^- \rightleftharpoons$ L-β-hydroxybutyrate | — | -0.346 |
| Xanthine + $2H^+$ + $2e^- \rightleftharpoons$ hypoxanthine + $H_2O$ | — | -0.371 |
| $2H^+ + 2e^- \rightleftharpoons H_2$ | 0.000 | -0.414 |
| Gluconate + $2H^+$ + $2e^- \rightleftharpoons$ glucose + $H_2O$ | — | -0.44 |
| $SO_4^{2-} + 2e^- + 2H^+ \rightleftharpoons SO_3^{2-} + H_2O$ | — | -0.454 |
| $2SO_3^{2-} + 2e^- + 4H^+ \rightleftharpoons S_2O_4^{2-} + 2H_2O$ | — | -0.527 |

base to its concentration in a *particular* form:

*Monoprotic system:*

$$[HA] = \alpha_{HA}F = \frac{[H^+]F}{[H^+] + K_a} \qquad (14\text{-}27)$$

$$[A^-] = \alpha_{A^-}F = \frac{K_aF}{[H^+] + K_a} \qquad (14\text{-}28)$$

*Diprotic system:*

$$[H_2A] = \alpha_{H_2A}F = \frac{[H^+]^2F}{[H^+]^2 + [H^+]K_1 + K_1K_2} \qquad (14\text{-}29)$$

$$[HA^-] = \alpha_{HA^-}F = \frac{K_1[H^+]F}{[H^+]^2 + [H^+]K_1 + K_1K_2} \qquad (14\text{-}30)$$

$$[A^{2-}] = \alpha_{A^{2-}}F = \frac{K_1K_2F}{[H^+]^2 + [H^+]K_1 + K_1K_2} \qquad (14\text{-}31)$$

where F is the formal concentration of HA or $H_2A$, $K_a$ is the acid dissociation constant for HA, and $K_1$ and $K_2$ are the acid dissociation constants for $H_2A$.

For a monoprotic acid:

$$F = [HA] + [A^-]$$

For a diprotic acid:

$$F = [H_2A] + [HA^-] + [A^{2-}]$$

> **Example** Finding the Formal Potential
>
> Find $E°'$ for the reaction 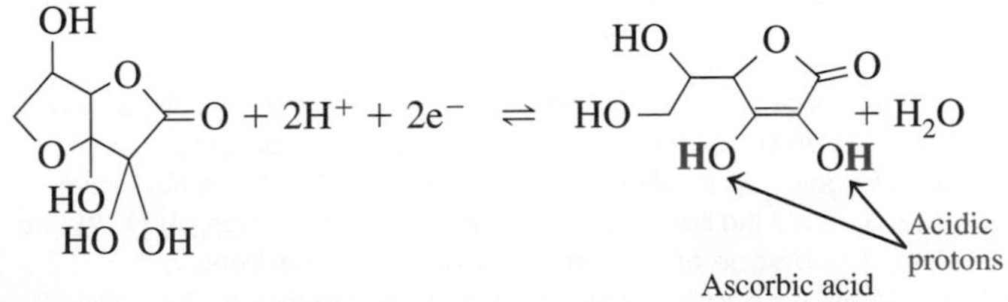
>
>
> $$E° = 0.390 \text{ V} \quad (14\text{-}32)$$
>
> Acidic protons
>
> Dehydroascorbic acid (oxidized)
>
> Ascorbic acid (vitamin C) (reduced)
>
> $pK_1 = 4.10 \quad pK_2 = 11.79$

**Solution**  Abbreviating dehydroascorbic acid as D, and ascorbic acid as $H_2A$, we rewrite the reduction as

$$D + 2H^+ + 2e^- \rightleftharpoons H_2A + H_2O$$

for which the Nernst equation is

$$E = E° - \frac{0.059\ 16}{2} \log \frac{[H_2A]}{[D][H^+]^2} \tag{14-33}$$

D is not an acid or a base, so its formal concentration equals its molar concentration: $F_D = [D]$. For the diprotic acid $H_2A$, we use Equation 14-29 to express $[H_2A]$ in terms of $F_{H_2A}$:

$$[H_2A] = \frac{[H^+]^2\ F_{H_2A}}{[H^+]^2 + [H^+]K_1 + K_1K_2}$$

Putting these values into Equation 14-33 gives

$$E = E° - \frac{0.059\ 16}{2} \log \left( \frac{\dfrac{[H^+]^2\ F_{H_2A}}{[H^+]^2 + [H^+]K_1 + K_1K_2}}{F_D[H^+]^2} \right)$$

which can be rearranged to the form

$$E = \underbrace{E° - \frac{0.059\ 16}{2} \log \frac{1}{[H^+]^2 + [H^+]K_1 + K_1K_2}}_{\substack{\text{Formal potential } (= E°' \text{ if pH} = 7) \\ = +0.062\ \text{V}}} - \frac{0.059\ 16}{2} \log \frac{F_{H_2A}}{F_D} \tag{14-34}$$

Putting the values of $E°$, $K_1$, and $K_2$ into Equation 14-34 and setting $[H^+] = 10^{-7.00}$, we find $E°' = +0.062$ V.

Curve *a* in Figure 14-11 shows how the calculated formal potential for Reaction 14-32 depends on pH. The potential decreases as the pH increases, until pH $\approx$ pK$_2$. Above pK$_2$,

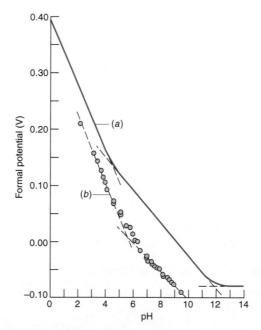

**Figure 14-11**  Reduction potential of ascorbic acid, showing its dependence on pH. (*a*) Graph of the function labeled formal potential in Equation 14-34. (*b*) Experimental polarographic half-wave reduction potential of ascorbic acid in a medium of ionic strength = 0.2 M. The half-wave potential, discussed in Chapter 17, is nearly the same as the formal potential. At high pH (>12), the half-wave potential does not level off to a slope of 0, as Equation 14-34 predicts. Instead, a hydrolysis reaction of ascorbic acid occurs and the chemistry is more complex than Reaction 14-32. [J. J. Ruiz, A. Aldaz, and M. Dominguez, *Can. J. Chem.* **1977**, *55*, 2799; ibid. **1978**, *56*, 1533.]

CHAPTER 14 Fundamentals of Electrochemistry

$A^{2-}$ is the dominant form of ascorbic acid, and no protons are involved in the net redox reaction. Therefore, the potential becomes independent of pH.

A biological example of $E^{\circ\prime}$ is the reduction of Fe(III) in the protein transferrin, which was introduced in Figure 7-4. This protein has two Fe(III)-binding sites, one in each half of the molecule designated C and N for the carboxyl and amino terminals of the peptide chain. Transferrin carries Fe(III) through the blood to cells that require iron. Membranes of these cells have a receptor that binds Fe(III)-transferrin and takes it into a compartment called an endosome into which $H^+$ is pumped to lower the pH to ~5.8. Iron is released from transferrin in the endosome and continues into the cell as Fe(II) attached to an intracellular metal-transport protein. The entire cycle of transferrin uptake, metal removal, and transferrin release back to the bloodstream takes 1–2 min. The time required for Fe(III) to dissociate from transferrin at pH 5.8 is ~6 min, which is too long to account for release in the endosome. The reduction potential of Fe(III)-transferrin at pH 5.8 is $E^{\circ\prime} = -0.52$ V, which is too low for physiologic reductants to reach.

The mystery of how Fe(III) is released from transferrin in the endosome was solved by measuring $E^{\circ\prime}$ for the Fe(III)-transferrin-receptor complex at pH 5.8. To simplify the chemistry, transferrin was cleaved and only the C-terminal half of the protein (designated $\mathrm{Trf_C}$) was used. Figure 14-12 shows measurements of $\log\{[\mathrm{Fe(III)Trf_C}]/[\mathrm{Fe(II)Trf_C}]\}$ for free protein and for the protein-receptor complex. In Equation 14-26, $E = E^{\circ\prime}$ when the log term is zero (that is, when $[\mathrm{Fe(III)Trf_C}] = [\mathrm{Fe(II)Trf_C}]$). Figure 14-12 shows that $E^{\circ\prime}$ for $\mathrm{Fe(III)Trf_C}$ is near $-0.50$ V, but $E^{\circ\prime}$ for the $\mathrm{Fe(III)Trf_C}$-receptor complex is $-0.29$ V. The reducing agents NADH and NADPH in Table 14-2 are strong enough to reduce $\mathrm{Fe(III)Trf_C}$ bound to its receptor at pH 5.8, but not strong enough to reduce free Fe(III)-transferrin.

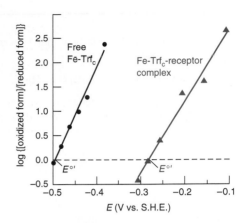

**Figure 14-12** Spectroscopic measurement of $\log\{[\mathrm{Fe(III)Trf_C}]/[\mathrm{Fe(II)Trf_C}]\}$ versus potential at pH 5.8. [S. Dhungana, C. H. Taboy, O. Zak, M. Larvie, A. L. Crumbliss, and P. Aisen, "Redox Properties of Human Transferrin Bound to Its Receptor," Biochemistry, **2004**, 43, 205.]

## Terms to Understand

| | | | |
|---|---|---|---|
| ampere | formal potential | oxidation | resistance |
| anode | galvanic cell | oxidizing agent | salt bridge |
| cathode | half-reaction | potentiometer | standard hydrogen electrode |
| coulomb | joule | power | standard reduction potential |
| current | Latimer diagram | reaction quotient | volt |
| $E^{\circ\prime}$ | Nernst equation | redox reaction | watt |
| electric potential | ohm | reducing agent | |
| electrode | Ohm's law | reductant | |
| Faraday constant | oxidant | reduction | |

## Summary

Work done when a charge of $q$ coulombs passes through a potential difference of $E$ volts is work $= E \cdot q$. The maximum work that can be done on the surroundings by a spontaneous chemical reaction is related to the free-energy change for the reaction: work $= -\Delta G$. If the chemical change produces a potential difference, $E$, the relation between free energy and the potential difference is $\Delta G = -nFE$. Ohm's law ($I = E/R$) describes the relation between current, voltage, and resistance in an electric circuit. It can be combined with the definitions of work and power ($P = $ work per second) to give $P = E \cdot I = I^2R$.

A galvanic cell uses a spontaneous redox reaction to produce electricity. The electrode at which oxidation occurs is the anode, and the electrode at which reduction occurs is the cathode. Two half-cells are usually separated by a salt bridge that allows ions to migrate from one side to the other to maintain charge neutrality but prevents reactants in the two half-cells from mixing. The standard reduction potential of a half-reaction is measured by pairing that half-reaction with a standard hydrogen electrode. The term "standard" means that activities of reactants and products are unity. If several half-reactions are added to give another half-reaction, the standard potential of the net half-reaction can be found by equating the free energy of the net half-reaction to the sum of free energies of the component half-reactions.

The voltage for a complete reaction is the difference between the potentials of the two half-reactions: $E = E_+ - E_-$, where $E_+$ is the potential of the half-cell connected to the positive terminal of the potentiometer and $E_-$ is the potential of the half-cell connected to the negative terminal. The potential of each half-reaction is given by the Nernst equation: $E = E^\circ - (0.059\,16/n) \log Q$ (at 25°C), where each reaction is written as a reduction and $Q$ is the reaction quotient. The reaction quotient has the same form as the equilibrium constant, but it is evaluated with concentrations existing at the time of interest. Electrons flow through the circuit from the electrode with the more negative potential to the electrode with the more positive potential.

Complex equilibria can be studied by making them part of an electrochemical cell. If we measure the voltage and know the concentrations (activities) of all but one of the reactants and products, the Nernst equation allows us to compute the concentration of the unknown species. The electrochemical cell serves as a probe for that species.

Biochemists use the formal potential of a half-reaction at pH 7 ($E^{\circ\prime}$) instead of the standard potential ($E^\circ$), which applies at pH 0. $E^{\circ\prime}$ is found by writing the Nernst equation for the half-reaction and grouping together all terms except the logarithm containing the formal concentrations of reactant and product. The combination of terms, evaluated at pH 7, is $E^{\circ\prime}$.

## Exercises

**14-A.** A mercury cell formerly used to power heart pacemakers has the following reaction:

$$Zn(s) + HgO(s) \rightarrow ZnO(s) + Hg(l) \qquad E° = 1.35 \text{ V}$$

If the power required to operate the pacemaker is 0.010 0 W, how many kilograms of HgO (FM 216.59) will be consumed in 365 days? How many pounds of HgO is this? (1 pound = 453.6 g)

**14-B.** Calculate $E°$ and $K$ for each of the following reactions.
(a) $I_2(s) + 5Br_2(aq) + 6H_2O \rightleftharpoons 2IO_3^- + 10Br^- + 12H^+$
(b) $Cr^{2+} + Fe(s) \rightleftharpoons Fe^{2+} + Cr(s)$
(c) $Mg(s) + Cl_2(g) \rightleftharpoons Mg^{2+} + 2Cl^-$
(d) $5MnO_2(s) + 4H^+ \rightleftharpoons 3Mn^{2+} + 2MnO_4^- + 2H_2O$
(e) $Ag^+ + 2S_2O_3^{2-} \rightleftharpoons Ag(S_2O_3)_2^{3-}$
(f) $CuI(s) \rightleftharpoons Cu^+ + I^-$

**14-C.** Calculate the voltage of each of the following cells. With the reasoning in Figure 14-8, state the direction of electron flow.
(a) $Fe(s) \mid FeBr_2(0.010 \text{ M}) \parallel NaBr(0.050 \text{ M}) \mid Br_2(l) \mid Pt(s)$
(b) $Cu(s) \mid Cu(NO_3)_2(0.020 \text{ M}) \parallel Fe(NO_3)_2(0.050 \text{ M}) \mid Fe(s)$
(c) $Hg(l) \mid Hg_2Cl_2(s) \mid KCl(0.060 \text{ M}) \parallel KCl(0.040 \text{ M}) \mid$ $Cl_2(g, 0.50 \text{ bar}) \mid Pt(s)$

**14-D.** The left half-reaction of the cell drawn here can be written in *either* of two ways:

$$AgI(s) + e^- \rightleftharpoons Ag(s) + I^- \qquad (1)$$

or

$$Ag^+ + e^- \rightleftharpoons Ag(s) \qquad (2)$$

The right half-cell reaction is

$$H^+ + e^- \rightleftharpoons \tfrac{1}{2}H_2(g) \qquad (3)$$

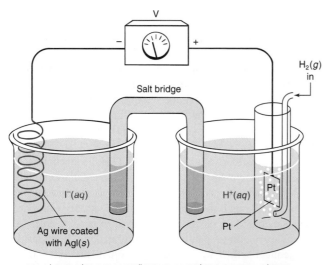

Ag(s) | AgI(s) | NaI (0.10 M) ‖ HCl (0.10 M) | H₂(g, 0.20 bar) | Pt(s)

(a) Using Reactions 2 and 3, calculate $E°$ and write the Nernst equation for the cell.
(b) Use the value of $K_{sp}$ for AgI to compute $[Ag^+]$ and find the cell voltage. By the reasoning in Figure 14-8, in which direction do electrons flow?
(c) Suppose, instead, that you wish to describe the cell with Reactions 1 and 3. We know that the cell voltage ($E$, not $E°$) must be the same, no matter which description we use. Write the Nernst equation for Reactions 1 and 3 and use it to find $E°$ in Reaction 1. Compare your answer with the value in Appendix H.

**14-E.** Calculate the voltage of the cell

$$Cu(s) \mid Cu^{2+}(0.030 \text{ M}) \parallel K^+Ag(CN)_2^-(0.010 \text{ M}),$$
$$HCN(0.10 \text{ F}), \text{ buffer to pH } 8.21 \mid Ag(s)$$

by considering the following reactions:

$$Ag(CN)_2^- + e^- \rightleftharpoons Ag(s) + 2CN^- \qquad E° = -0.310 \text{ V}$$
$$HCN \rightleftharpoons H^+ + CN^- \qquad pK_a = 9.21$$

By the reasoning in Figure 14-8, in which direction do electrons flow?

**14-F.** (a) Write a balanced equation for the reaction $PuO_2^+ \rightarrow Pu^{4+}$ and calculate $E°$ for the reaction.

$$PuO_2^{2+} \xrightarrow{+0.966} PuO_2^+ \xrightarrow{\text{?}} Pu^{4+} \xrightarrow{+1.006} Pu^{3+}$$
$$\underset{1.021}{\underline{\qquad\qquad}}$$

(b) Predict whether an equimolar mixture of $PuO_2^{2+}$ and $PuO_2^+$ will oxidize $H_2O$ to $O_2$ at a pH of 2.00 and $P_{O_2} = 0.20$ bar. Will $O_2$ be liberated at pH 7.00?

**14-G.** Calculate the voltage of the following cell, in which KHP is potassium hydrogen phthalate, the monopotassium salt of phthalic acid.

$$Hg(l) \mid Hg_2Cl_2(s) \mid KCl(0.10 \text{ M}) \parallel KHP(0.050 \text{ M}) \mid$$
$$H_2(g, 1.00 \text{ bar}) \mid Pt(s)$$

**14-H.** The following cell has a voltage of 0.083 V:

$$Hg(l) \mid Hg(NO_3)_2(0.001 \ 0 \text{ M}), KI(0.500 \text{ M}) \parallel \text{S.H.E.}$$

From this voltage, calculate the equilibrium constant for the reaction

$$Hg^{2+} + 4I^- \rightleftharpoons HgI_4^{2-}$$

In 0.5 M KI, virtually all the mercury is present as $HgI_4^{2-}$.

**14-I.** The formation constant for $Cu(EDTA)^{2-}$ is $6.3 \times 10^{18}$, and $E°$ is $+0.339$ V for the reaction $Cu^{2+} + 2e^- \rightleftharpoons Cu(s)$. From this information, find $E°$ for the reaction

$$CuY^{2-} + 2e^- \rightleftharpoons Cu(s) + Y^{4-}$$

**14-J.** On the basis of the following reaction, state which compound, $H_2(g)$ or glucose, is the more powerful reducing agent at pH = 0.00.

$$\begin{array}{c}
CO_2H \\
| \\
HCOH \\
| \\
HOCH \\
| \\
HCOH \\
| \\
HCOH \\
| \\
CH_2OH \\
\text{Gluconic acid} \\
pK_a = 3.56
\end{array}
\quad + 2H^+ + 2e^- \xrightleftharpoons{E°' = -0.45 \text{ V}} \quad
\begin{array}{c}
CHO \\
| \\
HCOH \\
| \\
HOCH \\
| \\
HCOH \\
| \\
HCOH \\
| \\
CH_2OH \\
\text{Glucose} \\
\text{(no acidic protons)}
\end{array}$$

**14-K.** Living cells convert energy derived from sunlight or combustion of food into energy-rich ATP (adenosine triphosphate) molecules. For ATP synthesis, $\Delta G = +34.5$ kJ/mol. This energy is then made available to the cell when ATP is hydrolyzed to ADP (adenosine diphosphate). In animals, ATP is synthesized when protons pass through a complex enzyme in the mitochondrial membrane.[14] Two factors account for the movement of protons through this enzyme into the mitochondrion (see the figure): (1) $[H^+]$ is higher outside the mitochondrion than inside because protons are *pumped* out of the mitochondrion by enzymes that catalyze the oxidation of food. (2) The inside of the mitochondrion is negatively charged with respect to the outside.

(a) The synthesis of one ATP molecule requires $2H^+$ to pass through the phosphorylation enzyme. The difference in free energy when a molecule travels from a region of high activity to a region of low activity is

$$\Delta G = -RT \ln \frac{\mathscr{A}_{high}}{\mathscr{A}_{low}}$$

How big must the pH difference be (at 298 K) if the passage of two protons is to provide enough energy to synthesize one ATP molecule?

(b) pH differences this large have not been observed in mitochondria. How great an electric potential difference between inside and outside is necessary for the movement of two protons to provide energy to synthesize ATP? In answering this question, neglect any contribution from the pH difference.

(c) The energy for ATP synthesis is thought to be provided by *both* the pH difference and the electric potential. If the pH difference is 1.00 pH unit, what is the magnitude of the potential difference?

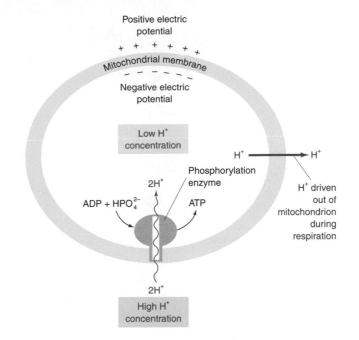

---

## Problems

### Basic Concepts

**14-1.** Explain the difference between electric charge ($q$, coulombs), electric current ($I$, amperes), and electric potential ($E$, volts).

**14-2.** (a) How many electrons are in one coulomb?
(b) How many coulombs are in one mole of charge?

**14-3.** The basal rate of consumption of $O_2$ by a 70-kg human is about 16 mol of $O_2$ per day. This $O_2$ oxidizes food and is reduced to $H_2O$, providing energy for the organism:

$$O_2 + 4H^+ + 4e^- \rightleftharpoons 2H_2O$$

(a) To what current (in amperes = C/s) does this respiration rate correspond? (Current is defined by the flow of electrons from food to $O_2$.)
(b) Compare your answer in part (a) with the current drawn by a refrigerator using $5.00 \times 10^2$ W at 115 V. Remember that power (in watts) = work/s = $E \cdot I$.
(c) If the electrons flow from nicotinamide adenine dinucleotide (NADH) to $O_2$, they experience a potential drop of 1.1 V. What is the power output (in watts) of our human friend?

**14-4.** A 6.00-V battery is connected across a 2.00-k$\Omega$ resistor.
(a) How many electrons per second flow through the circuit?
(b) How many joules of heat are produced for each electron?
(c) If the circuit operates for 30.0 min, how many moles of electrons will have flowed through the resistor?
(d) What voltage would the battery need to deliver for the power to be $1.00 \times 10^2$ W?

**14-5.** Consider the redox reaction

$$I_2 + 2S_2O_3^{2-} \rightleftharpoons 2I^- + S_4O_6^{2-}$$
$$\text{Thiosulfate} \qquad \text{Tetrathionate}$$

(a) Identify the oxidizing agent on the left side of the reaction and write a balanced oxidation half-reaction.
(b) Identify the reducing agent of the left side of the reaction and write a balanced reduction half-reaction.
(c) How many coulombs of charge are passed from reductant to oxidant when 1.00 g of thiosulfate reacts?
(d) If the rate of reaction is 1.00 g of thiosulfate consumed per minute, what current (in amperes) flows from reductant to oxidant?

**14-6.** The space shuttle's expendable booster engines derive their power from solid reactants:

$$6NH_4^+ClO_4^-(s) + 10Al(s) \rightarrow 3N_2(g) + 9H_2O(g)$$
$$\text{FM 117.49} \qquad\qquad\qquad + 5Al_2O_3(s) + 6HCl(g)$$

(a) Find the oxidation numbers of the elements N, Cl, and Al in reactants and products. Which reactants act as reducing agents and which act as oxidants?
(b) The heat of reaction is $-9\,334$ kJ for every 10 mol of Al consumed. Express this as heat released per gram of total reactants.

### Galvanic Cells

**14-7.** Explain how a galvanic cell uses a spontaneous chemical reaction to generate electricity.

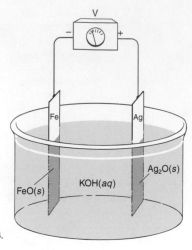
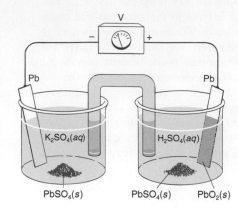

Galvanic cells for Problem 14-8.

**14-8.** Write a line notation and two reduction half-reactions for each cell pictured above.

**14-9.** Draw a picture of the following cell and write reduction half-reactions for each electrode:

$$Pt(s) \mid Fe^{3+}(aq), Fe^{2+}(aq) \parallel$$
$$Cr_2O_7^{2-}(aq), Cr^{3+}(aq), HA(aq) \mid Pt(s)$$

**14-10.** Consider the rechargeable battery

$$Zn(s) \mid ZnCl_2(aq) \parallel Cl^-(aq) \mid Cl_2(l) \mid C(s)$$

**(a)** Write reduction half-reactions for each electrode. From which electrode will electrons flow from the battery into a circuit if the electrode potentials are not too different from $E^\circ$ values?

**(b)** If the battery delivers a constant current of $1.00 \times 10^3$ A for 1.00 h, how many kilograms of $Cl_2$ will be consumed?

**14-11. (a)** Organic matter whose composition is approximately $CH_2O$ falls to the ocean floor near continental areas at a rate of 2 to 10 mol of carbon per $m^2$ per year. The net cell reaction at the opening of this chapter is $CH_2O + O_2 \rightarrow CO_2 + H_2O$. Write balanced half-reactions for this net reaction.

**(b)** If organic matter is completely consumed, how many coulombs of charge would flow in 1 year through a cell whose electrodes occupy 1 $m^2$? How much steady electric current (C/s) does this represent?

**(c)** If the current flows through a potential difference of 0.3 V, how much power would be generated?

**(d)** Write the two electrode reactions and explain why one is different from the answer to part **(a)**.

## Standard Potentials

**14-12.** Which will be the strongest oxidizing agent under standard conditions (that is, all activities = 1): $HNO_2$, Se, $UO_2^{2+}$, $Cl_2$, $H_2SO_3$, or $MnO_2$?

**14-13. (a)** Cyanide ion causes $E^\circ$ for Fe(III) to decrease:

$$Fe^{3+} + e^- \rightleftharpoons Fe^{2+} \qquad E^\circ = 0.771 \text{ V}$$
Ferric              Ferrous

$$Fe(CN)_6^{3-} + e^- \rightleftharpoons Fe(CN)_6^{4-} \qquad E^\circ = 0.356 \text{ V}$$
Ferricyanide              Ferrocyanide

Which ion, Fe(III) or Fe(II), is stabilized more by complexing with $CN^-$?

**(b)** Using Appendix H, answer the same question when the ligand is phenanthroline instead of cyanide.

Phenanthroline

## Nernst Equation

**14-14.** What is the difference between $E$ and $E^\circ$ for a redox reaction? Which one runs down to 0 when the complete cell comes to equilibrium?

**14-15. (a)** Use the Nernst equation to write the spontaneous chemical reaction that occurs in the cell in Demonstration 14-1.

**(b)** If you use your fingers as a salt bridge in Demonstration 14-1, will your body take in $Cu^{2+}$ or $Zn^{2+}$?

**14-16.** Write the Nernst equation for the following half-reaction and find $E$ when pH = 3.00 and $P_{AsH_3}$ = 1.0 mbar.

$$As(s) + 3H^+ + 3e^- \rightleftharpoons AsH_3(g) \qquad E^\circ = -0.238 \text{ V}$$
Arsine

**14-17. (a)** Write the line notation for the following cell.

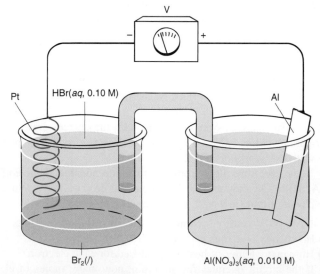

**(b)** Calculate the potential of each half-cell and the cell voltage, $E$. In which direction will electrons flow through the circuit? Write the spontaneous net cell reaction.

**(c)** The left half-cell was loaded with 14.3 mL of $Br_2(l)$ (density = 3.12 g/mL). The aluminum electrode contains 12.0 g of Al. Which element, $Br_2$ or Al, is the limiting reagent? (That is, which reagent will be used up first?)

**(d)** If the cell is somehow operated under conditions in which it produces a constant voltage of 1.50 V, how much electrical work will have been done when 0.231 mL of $Br_2(l)$ has been consumed?

**(e)** If the potentiometer is replaced by a 1.20-k$\Omega$ resistor and if the heat dissipated by the resistor is $1.00 \times 10^{-4}$ J/s, at what rate (grams per second) is Al(s) dissolving? (In this question, the voltage is not 1.50 V.)

**14-18.** A nickel-metal hydride rechargeable battery for laptop computers is based on the following chemistry:

Cathode:

$$NiOOH(s) + H_2O + e^- \underset{\text{charge}}{\overset{\text{discharge}}{\rightleftharpoons}} Ni(OH)_2(s) + OH^-$$

Anode:

$$MH(s) + OH^- \underset{\text{charge}}{\overset{\text{discharge}}{\rightleftharpoons}} M(s) + H_2O + e^-$$

The anode material, MH, is a transition metal hydride or rare earth alloy hydride. Explain why the voltage remains nearly constant during the entire discharge cycle.

**14-19.** Suppose that the concentrations of NaF and KCl were each 0.10 M in the cell

$$Pb(s) \mid PbF_2(s) \mid F^-(aq) \parallel Cl^-(aq) \mid AgCl(s) \mid Ag(s)$$

**(a)** Using the half-reactions $2AgCl(s) + 2e^- \rightleftharpoons 2Ag(s) + 2Cl^-$ and $PbF_2(s) + 2e^- \rightleftharpoons Pb(s) + 2F^-$, calculate the cell voltage.

**(b)** Now calculate the cell voltage by using the reactions $2Ag^+ + 2e^- \rightleftharpoons 2Ag(s)$ and $Pb^{2+} + 2e^- \rightleftharpoons Pb(s)$. For this part, you will need the solubility products for $PbF_2$ and AgCl.

**14-20.** The following cell was set up to measure the standard reduction potential of the $Ag^+ \mid Ag$ couple:

$$Pt(s) \mid HCl(0.010\ 00\ M),\ H_2(g) \parallel AgNO_3(0.010\ 00\ M) \mid Ag(s)$$

The temperature was 25°C (the standard condition) and atmospheric pressure was 751.0 Torr. Because the vapor pressure of water is 23.8 Torr at 25°C, $P_{H_2}$ in the cell was $751.0 - 23.8 = 727.2$ Torr. The Nernst equation for the cell, including activity coefficients, is derived as follows:

Right electrode: $\quad Ag^+ + e^- \rightleftharpoons Ag(s) \qquad E_+^\circ = E_{Ag^+|Ag}^\circ$

Left electrode: $\quad H^+ + e^- \rightleftharpoons \frac{1}{2}H_2(g) \qquad E_-^\circ = 0\ V$

$$E_+ = E_{Ag^+|Ag}^\circ - 0.059\ 16 \log\left(\frac{1}{[Ag^+]\gamma_{Ag^+}}\right)$$

$$E_- = 0 - 0.059\ 16 \log\left(\frac{P_{H_2}^{1/2}}{[H^+]\gamma_{H^+}}\right)$$

$$E = E_+ - E_- = E_{Ag^+|Ag}^\circ - 0.059\ 16 \log\left(\frac{[H^+]\gamma_{H^+}}{P_{H_2}^{1/2}[Ag^+]\gamma_{Ag^+}}\right)$$

Given a measured cell voltage of $+0.798\ 3$ V and using activity coefficients from Table 8-1, find $E_{Ag^+|Ag}^\circ$. Be sure to express $P_{H_2}$ in bar in the reaction quotient.

**14-21.** Write a balanced chemical equation (in acidic solution) for the reaction represented by the question mark on the lower arrow.[15] Calculate $E^\circ$ for the reaction.

$$
\begin{array}{c}
\xrightarrow{\hspace{3cm}} \ \ 1.441 \ \ \downarrow \\
BrO_3^- \xrightarrow{1.491} HOBr \xrightarrow{1.584} Br_2(aq) \xrightarrow{1.098} Br^- \\
\underset{?}{\underleftarrow{\hspace{4cm}}}\uparrow
\end{array}
$$

**14-22.** What must be the relation between $E_1^\circ$ and $E_2^\circ$ if the species $X^+$ is to disproportionate spontaneously under standard conditions to $X^{3+}$ and X(s)? Write a balanced equation for the disproportionation.

$$X^{3+} \xrightarrow{E_1^\circ} X^+ \xrightarrow{E_2^\circ} X(s)$$

**14-23.** *Including activities,* calculate the voltage of the cell Ni(s) $\mid$ $NiSO_4(0.002\ 0\ M) \parallel CuCl_2(0.003\ 0\ M) \mid Cu(s)$. Assume that the salts are completely dissociated (that is, neglect ion-pair formation).

### Relation of $E^\circ$ and the Equilibrium Constant

**14-24.** For the reaction $CO + \frac{1}{2}O_2 \rightleftharpoons CO_2$, $\Delta G^\circ = -257$ kJ per mole of CO at 298 K. Find $E^\circ$ and the equilibrium constant for the reaction.

**14-25.** Calculate $E^\circ$, $\Delta G^\circ$, and K for the following reactions.
**(a)** $4Co^{3+} + 2H_2O \rightleftharpoons 4Co^{2+} + O_2(g) + 4H^+$
**(b)** $Ag(S_2O_3)_2^{3-} + Fe(CN)_6^{4-} \rightleftharpoons Ag(s) + 2S_2O_3^{2-} + Fe(CN)_6^{3-}$

**14-26.** A solution contains 0.100 M $Ce^{3+}$, $1.00 \times 10^{-4}$ M $Ce^{4+}$, $1.00 \times 10^{-4}$ M $Mn^{2+}$, 0.100 M $MnO_4^-$, and 1.00 M $HClO_4$.
**(a)** Write a balanced net reaction that can occur between species in this solution.
**(b)** Calculate $\Delta G^\circ$ and K for the reaction.
**(c)** Calculate E for the conditions given.
**(d)** Calculate $\Delta G$ for the conditions given.
**(e)** At what pH would the given concentrations of $Ce^{4+}$, $Ce^{3+}$, $Mn^{2+}$, and $MnO_4^-$ be in equilibrium at 298 K?

**14-27.** For the cell $Pt(s) \mid VO^{2+}(0.116\ M),\ V^{3+}(0.116\ M)$, $H^+(1.57\ M) \parallel Sn^{2+}(0.031\ 8\ M),\ Sn^{4+}(0.031\ 8\ M) \mid Pt(s)$, $E$(not $E^\circ$) $= -0.289$ V. Write the net cell reaction and calculate its equilibrium constant. Do not use $E^\circ$ values from Appendix H to answer this question.

**14-28.** Calculate $E^\circ$ for the half-reaction $Pd(OH)_2(s) + 2e^- \rightleftharpoons Pd(s) + 2OH^-$ given that $K_{sp}$ for $Pd(OH)_2$ is $3 \times 10^{-28}$ and $E^\circ = 0.915$ V for the reaction $Pd^{2+} + 2e^- \rightleftharpoons Pd(s)$.

**14-29.** From the standard potentials for reduction of $Br_2(aq)$ and $Br_2(l)$ in Appendix H, calculate the solubility of $Br_2$ in water at 25°C. Express your answer as g/L.

**14-30.** Given the following information, calculate the standard potential for the reaction $FeY^- + e^- \rightleftharpoons FeY^{2-}$, where Y is EDTA.

$$FeY^- + e^- \rightleftharpoons Fe^{2+} + Y^{4-} \qquad E^\circ = -0.730\ V$$

$$FeY^{2-}: \qquad K_f = 2.0 \times 10^{14}$$

$$FeY^-: \qquad K_f = 1.3 \times 10^{25}$$

**14-31.** For modest temperature excursions away from 25°C, the change in $E^\circ$ for a half-reaction can be written in the form

$$E^\circ(T) = E^\circ + \left(\frac{dE^\circ}{dT}\right)\Delta T$$

where $E°(T)$ is the standard reduction potential at temperature $T(°C)$, and $\Delta T$ is $(T - 25)$. For the reaction $Al^{3+} + 3e^- \rightleftharpoons Al(s)$, $dE°/dT = 0.533$ mV/K near 25°C. Find $E°$ for this half-reaction at 50°C.

**14-32.** This problem is slightly tricky. Calculate $E°$, $\Delta G°$, and $K$ for the reaction

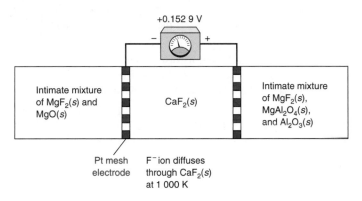

2Cu²⁺ + 2I⁻ + HO—⟨⟩—OH ⇌

Hydroquinone

2CuI(s) + O=⟨⟩=O + 2H⁺

Quinone

which is the sum of *three* half-reactions listed in Appendix H. Use $\Delta G°(= -nFE°)$ for each of the half-reactions to find $\Delta G°$ for the net reaction. Note that, if you reverse the direction of a reaction, you reverse the sign of $\Delta G°$.

**14-33.** *Thermodynamics of a solid-state reaction.* The following electrochemical cell is reversible at 1 000 K in an atmosphere of flowing $O_2(g)$:[16]

Left half-cell:   $MgF_2(s) + \frac{1}{2}O_2(g) + 2e^- \rightleftharpoons MgO(s) + 2F^-$

Right half-cell:   $MgF_2(s) + Al_2O_3(s) + \frac{1}{2}O_2(g) + 2e^- \rightleftharpoons$

$MgAl_2O_4(s) + 2F^-$

**(a)** Write a Nernst equation for each half-cell. Write the net reaction and its Nernst equation. The activity of $O_2(g)$ is the same on both sides and the activity of $F^-$ is the same on both sides, governed by $F^-$ ions diffusing through $CaF_2(s)$. Show that the observed voltage is $E°$ for the net reaction.
**(b)** From the relation $\Delta G° = -nFE°$, find $\Delta G°$ for the net reaction. Note that 1 V = 1 J/C.
**(c)** The cell voltage in the temperature range $T = 900$ to $1\,250$ K is $E(V) = 0.122\,3 + 3.06 \times 10^{-5}\,T$. Assuming that $\Delta H°$ and $\Delta S°$ are constant, find $\Delta H°$ and $\Delta S°$ from the relation $\Delta G° = \Delta H° - T\Delta S°$.

**Using Cells as Chemical Probes**

**14-34.** Using the cell in Figure 14-10 as an example, explain what we mean when we say that there is equilibrium *within* each half-cell but not necessarily *between* the two half-cells.

**14-35.** The cell $Pt(s) \,|\, H_2(g, 1.00\text{ bar}) \,|\, H^+(aq, pH = 3.60) \,\|\, Cl^-(aq, x\text{ M}) \,|\, AgCl(s) \,|\, Ag(s)$ can be used as a probe to find the concentration of $Cl^-$ in the right compartment.
**(a)** Write reactions for each half-cell, a balanced net cell reaction, and the Nernst equation for the net cell reaction.
**(b)** Given a measured cell voltage of 0.485 V, find $[Cl^-]$ in the right compartment.

**14-36.** The quinhydrone electrode was introduced in 1921 as a means of measuring pH.[17]

$Pt(s) \,|\, $ 1:1 mole ratio of quinone(aq) and hydroquinone(aq),

unknown pH $\|\, Cl^-(aq, 0.50\text{ M}) \,|\, Hg_2Cl_2(s) \,|\, Hg(l) \,|\, Pt(s)$

The solution whose pH is to be measured is placed in the left half-cell, which also contains a 1:1 mole ratio of quinone and hydroquinone. The half-cell reaction is

O=⟨⟩=O + 2H⁺ + 2e⁻ ⇌ HO—⟨⟩—OH

Quinone                                    Hydroquinone

**(a)** Write reactions for each half-cell and for the whole cell. Write the Nernst equation for the whole cell.
**(b)** Ignoring activities, rearrange the Nernst equation to the form $E(\text{cell}) = A + (B \cdot pH)$, where A and B are constants. Calculate A and B at 25°C.
**(c)** If the pH were 4.50, in which direction would electrons flow through the potentiometer?

**14-37.** The voltage for the following cell is 0.490 V. Find $K_b$ for the organic base $RNH_2$.

$Pt(s) \,|\, H_2(1.00\text{ bar}) \,|\,$

$RNH_2(aq, 0.10\text{ M}), RNH_3^+Cl^-(aq, 0.050\text{ M}) \,\|\, \text{S.H.E.}$

**14-38.** The voltage of the cell shown here is −0.246 V. The right half-cell contains the metal ion, $M^{2+}$, whose standard reduction potential is −0.266 V.

$$M^{2+} + 2e^- \rightleftharpoons M(s) \qquad E° = -0.266\text{ V}$$

Calculate $K_f$ for the metal-EDTA complex.

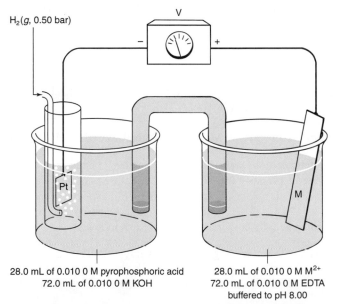

28.0 mL of 0.010 0 M pyrophosphoric acid      28.0 mL of 0.010 0 M M²⁺
72.0 mL of 0.010 0 M KOH                       72.0 mL of 0.010 0 M EDTA
                                               buffered to pH 8.00

**14-39.** The following cell was constructed to find the difference in $K_{sp}$ between two naturally occurring forms of $CaCO_3(s)$, called *calcite* and *aragonite*.[18]

$Pb(s) \,|\, PbCO_3(s) \,|\, CaCO_3(s, \text{calcite}) \,|\, \text{buffer(pH 7.00)} \,\|$

$\text{buffer(pH 7.00)} \,|\, CaCO_3(s, \text{aragonite}) \,|\, PbCO_3(s) \,|\, Pb(s)$

Each compartment of the cell contains a mixture of solid $PbCO_3$ ($K_{sp} = 7.4 \times 10^{-14}$) and either calcite or aragonite, both of which have $K_{sp} \approx 5 \times 10^{-9}$. Each solution was buffered to pH 7.00 with an inert buffer, and the cell was completely isolated from atmospheric $CO_2$. The measured cell voltage was −1.8 mV. Find the ratio of solubility products, $K_{sp}$ (for calcite)/$K_{sp}$ (for aragonite)

**14-40.** *Do not ignore activity coefficients in this problem.* If the voltage for the following cell is 0.512 V, find $K_{sp}$ for $Cu(IO_3)_2$. Neglect any ion pairing.

$$Ni(s) \mid NiSO_4(0.002\ 5\ M) \parallel KIO_3(0.10\ M) \mid Cu(IO_3)_2(s) \mid Cu(s)$$

### Biochemists Use $E^{\circ\prime}$

**14-41.** Explain what $E^{\circ\prime}$ is and why it is preferred over $E^\circ$ in biochemistry.

**14-42.** We are going to find $E^{\circ\prime}$ for the reaction $C_2H_2(g) + 2H^+ + 2e^- \rightleftharpoons C_2H_4(g)$.
(a) Write the Nernst equation for the half-reaction, using $E^\circ$ from Appendix H.
(b) Rearrange the Nernst equation to the form

$$E = E^\circ + \text{other terms} - \frac{0.059\ 16}{2} \log\left(\frac{P_{C_2H_4}}{P_{C_2H_2}}\right)$$

(c) The quantity $(E^\circ + \text{other terms})$ is $E^{\circ\prime}$. Evaluate $E^{\circ\prime}$ for pH = 7.00.

**14-43.** Evaluate $E^{\circ\prime}$ for the half-reaction $(CN)_2(g) + 2H^+ + 2e^- \rightleftharpoons 2HCN(aq)$.

**14-44.** Calculate $E^{\circ\prime}$ for the reaction

$$\underset{\text{Oxalic acid}}{H_2C_2O_4} + 2H^+ + 2e^- \rightleftharpoons \underset{\text{Formic acid}}{2HCO_2H} \qquad E^\circ = 0.204\ V$$

**14-45.** HOx is a monoprotic acid with $K_a = 1.4 \times 10^{-5}$ and $H_2Red^-$ is a diprotic acid with $K_1 = 3.6 \times 10^{-4}$ and $K_2 = 8.1 \times 10^{-8}$. Find $E^\circ$ for the reaction

$$HOx + e^- \rightleftharpoons H_2Red^- \qquad E^{\circ\prime} = 0.062\ V$$

**14-46.** Given the following information, find $K_a$ for nitrous acid, $HNO_2$.

$$NO_3^- + 3H^+ + 2e^- \rightleftharpoons HNO_2 + H_2O \qquad E^\circ = 0.940\ V$$
$$E^{\circ\prime} = 0.433\ V$$

**14-47.** Using the reaction

$$HPO_4^{2-} + 2H^+ + 2e^- \rightleftharpoons HPO_3^{2-} + H_2O \qquad E^\circ = -0.234\ V$$

and acid dissociation constants from Appendix G, calculate $E^\circ$ for the reaction

$$H_2PO_4^- + H^+ + 2e^- \rightleftharpoons HPO_3^{2-} + H_2O$$

**14-48.** This problem requires knowledge of Beer's law from Chapter 18. The oxidized form (Ox) of a flavoprotein that functions as a one-electron reducing agent has a molar absorptivity ($\varepsilon$) of $1.12 \times 10^4\ M^{-1} \cdot cm^{-1}$ at 457 nm at pH 7.00. For the reduced form (Red), $\varepsilon = 3.82 \times 10^3\ M^{-1} \cdot cm^{-1}$ at 457 nm at pH 7.00.

$$Ox + e^- \rightleftharpoons Red \qquad E^{\circ\prime} = -0.128\ V$$

The substrate (S) is the molecule reduced by the protein.

$$Red + S \rightleftharpoons Ox + S^-$$

Both S and $S^-$ are colorless. A solution at pH 7.00 was prepared by mixing enough protein plus substrate (Red + S) to produce initial concentrations [Red] = [S] = $5.70 \times 10^{-5}\ M$. The absorbance at 457 nm was 0.500 in a 1.00-cm cell.
(a) Calculate the concentrations of Ox and Red from the absorbance data.
(b) Calculate the concentrations of S and $S^-$.
(c) Calculate the value of $E^{\circ\prime}$ for the reaction $S + e^- \rightleftharpoons S^-$.

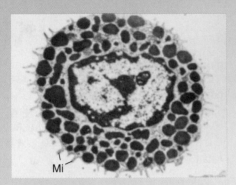

Mast cell (noncirculating white blood cell) with dark granules containing regulatory molecules such as heparin. (Mi indicates details of a cell's surface called microvilli and microfolds.) *[From R. G. Kessel and R. H. Kardon, Tissues and Organs (San Francisco: W. H. Freeman, 1978), p. 14; D. Lagunoff, J. Invest. Dermatol.* **1972**, *58, 296.]*

Partial structure of heparin, showing negatively charged substituents. Heparin is a *proteoglycan*, containing 95% polysaccharide (sugar) and 5% protein. *[Structure from R. Sasisekharan, Massachusetts Institute of Technology.]*

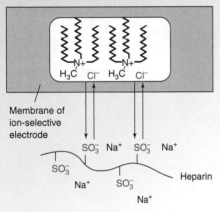

Ion exchange between heparin and Cl⁻ associated with tetraalkylammonium ions in the membrane of the ion-selective electrode. Ion-selective electrodes described in this chapter reach an equilibrium electric potential at the membrane-solution interface. Unlike those electrodes, the heparin sensor has a nonequilibrium response that depends on the steady-state flux of heparin to the membrane surface a few minutes after the membrane is exposed to heparin.[1]

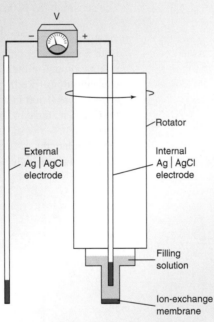

Rotating ion-selective heparin electrode. Figure 17-12 shows that rotation brings analyte to the electrode surface by convection.

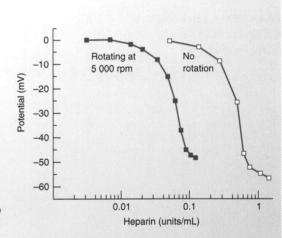

Rotation of heparin electrode improves sensitivity by an order of magnitude. *[From Q. Ye and M. E. Meyerhoff, "Rotating Electrode Potentiometry: Lowering the Detection Limit of Nonequilibrium Polyion-Sensitive Membrane Electrodes," Anal. Chem.* **2001**, *73, 332.]*

Blood clotting must be precisely regulated in the human body to minimize *hemorrhage* (bleeding) in an injury and *thrombosis* (uncontrolled clotting) in healthy tissue. Clotting is regulated by numerous substances, including *heparin,* a highly negatively charged molecule secreted by mast cells near the walls of blood vessels.[2]

Heparin administered during surgery to inhibit clotting must be monitored to prevent uncontrolled bleeding. Until recently, only clotting time could be measured; there was no direct, rapid determination of heparin at physiologic levels. Now, a rotating ion-selective electrode is available that can respond to clinically relevant concentrations of heparin in real time during surgery.[3] In this chapter, we study the principles governing electrodes that respond selectively to particular analytes.

Clever chemists have designed electrodes that respond selectively to specific analytes in solution or in the gas phase. Typical ion-selective electrodes are about the size of your pen. Really clever chemists created ion-sensing field effect transistors that are just hundreds of micrometers in size and can be inserted into a blood vessel. The use of electrodes to measure voltages that provide chemical information is called **potentiometry.**

In the simplest case, analyte is an *electroactive species* that is part of a galvanic cell. An **electroactive species** is one that can donate or accept electrons at an electrode. We turn the unknown solution into a half-cell by inserting an electrode, such as a Pt wire, that can transfer electrons to or from the analyte. Because this electrode responds to analyte, it is called the **indicator electrode.** We then connect this half-cell to a second half-cell by a salt bridge. The second half-cell has a fixed composition, so it has a constant potential. Because of its constant potential, the second half-cell is called a **reference electrode.** The cell voltage is the difference between the variable potential of the analyte half-cell and the constant potential of the reference electrode.

Dan received intravenous heparin in December 2004, but never stopped typing!
*[Photo from Sally Harris.]*

## ■ ■ ■ 15-1  Reference Electrodes

Suppose you want to measure the relative amounts of $Fe^{2+}$ and $Fe^{3+}$ in a solution. You can make this solution part of a galvanic cell by inserting a Pt wire and connecting the cell to a constant-potential half-cell by a salt bridge, as shown in Figure 15-1.

The two half-reactions (written as *reductions*) are

| | | |
|---|---|---|
| Right electrode: | $Fe^{3+} + e^- \rightleftharpoons Fe^{2+}$ | $E_+^\circ = 0.771$ V |
| Left electrode: | $AgCl(s) + e^- \rightleftharpoons Ag(s) + Cl^-$ | $E_-^\circ = 0.222$ V |

*Indicator electrode:* responds to analyte activity
*Reference electrode:* maintains a fixed (reference) potential

The electrode potentials are

$$E_+ = 0.771 - 0.059\ 16 \log\left(\frac{[Fe^{2+}]}{[Fe^{3+}]}\right)$$

$$E_- = 0.222 - 0.059\ 16 \log[Cl^-]$$

and the cell voltage is the difference $E_+ - E_-$:

$$E = \left\{0.771 - 0.059\ 16 \log\left(\frac{[Fe^{2+}]}{[Fe^{3+}]}\right)\right\} - \{0.222 - 0.059\ 16 \log[Cl^-]\}$$

$E_+$ is the potential of the electrode attached to the positive input of the potentiometer. $E_-$ is the potential of the electrode attached to the negative input of the potentiometer.

But $[Cl^-]$ in the left half-cell is constant, fixed by the solubility of KCl, with which the solution is saturated. Therefore, the cell voltage changes only when the quotient $[Fe^{2+}]/[Fe^{3+}]$ changes.

The half-cell on the left in Figure 15-1 can be thought of as a *reference electrode*. We can picture the cell and salt bridge enclosed by the dashed line as a single unit dipped into the

The voltage really tells us the quotient of *activities*, $\mathcal{A}_{Fe^{2+}}/\mathcal{A}_{Fe^{3+}}$. However, we will neglect activity coefficients and write the Nernst equation with concentrations instead of activities.

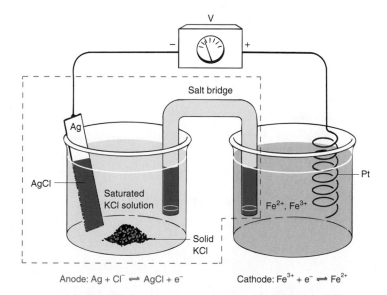

Anode: $Ag + Cl^- \rightleftharpoons AgCl + e^-$      Cathode: $Fe^{3+} + e^- \rightleftharpoons Fe^{2+}$

$Ag(s) \mid AgCl(s) \mid Cl^-(aq) \parallel Fe^{2+}(aq), Fe^{3+}(aq) \mid Pt(s)$

*Figure 15-1* A galvanic cell that can be used to measure the quotient $[Fe^{2+}]/[Fe^{3+}]$ in the right half-cell. The Pt wire is the *indicator electrode,* and the entire left half-cell plus salt bridge (enclosed by the dashed line) can be considered a *reference electrode.*

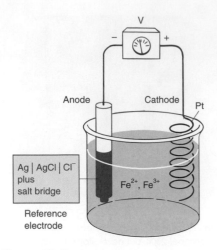

**Figure 15-2** Another view of Figure 15-1. The contents of the dashed box in Figure 15-1 are now considered to be a reference electrode dipped into the analyte solution.

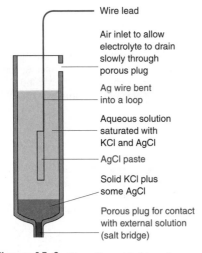

**Figure 15-3** Silver-silver chloride reference electrode.

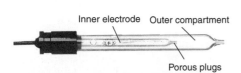

**Figure 15-4** Double-junction reference electrode. The inner electrode is the same as the one in Figure 15-3. The solution in the outer compartment is compatible with analyte solution. For example, if you do not want $Cl^-$ to contact the analyte, the outer electrode can be filled with $KNO_3$ solution. The inner and outer solutions slowly mix, so the outer compartment must be refilled periodically with fresh $KNO_3$ solution. *[Courtesy Fisher Scientific, Pittsburgh, PA.]*

analyte solution, as shown in Figure 15-2. The Pt wire is the indicator electrode, whose potential responds to the quotient $[Fe^{2+}]/[Fe^{3+}]$. The reference electrode completes the redox reaction and provides a *constant potential* to the left side of the potentiometer. Changes in the cell voltage result from changes in the quotient $[Fe^{2+}]/[Fe^{3+}]$.

## Silver-Silver Chloride Reference Electrode[4]

The half-cell enclosed by the dashed line in Figure 15-1 is called a **silver-silver chloride electrode.** Figure 15-3 shows how the electrode is reconstructed as a thin tube that can be dipped into an analyte solution. Figure 15-4 shows a *double-junction electrode* that minimizes contact between analyte solution and KCl from the electrode. The silver-silver chloride and calomel reference electrodes (described soon) are used because they are convenient. A standard hydrogen electrode (S.H.E.) is difficult to use because it requires $H_2$ gas and a freshly prepared catalytic Pt surface that is easily poisoned in many solutions.

The standard reduction potential for the $AgCl|Ag$ couple is $+0.222$ V at 25°C. This would be the potential of a silver-silver chloride electrode if $\mathcal{A}_{Cl^-}$ were unity. But the activity of $Cl^-$ in a saturated solution of KCl at 25°C is not unity, and the potential of the electrode in Figure 15-3 is $+0.197$ V with respect to S.H.E. at 25°C.

*Ag|AgCl electrode:* $\quad\quad AgCl(s) + e^- \rightleftharpoons Ag(s) + Cl^- \quad\quad E° = +0.222$ V

$$E(\text{saturated KCl}) = +0.197 \text{ V}$$

A problem with reference electrodes is that porous plugs become clogged, thus causing sluggish, unstable electrical response. Some designs incorporate a free-flowing capillary in place of the porous plug. Other designs allow you to force fresh solution from the electrode through the electrode-analyte junction prior to a measurement.

## Calomel Electrode

The **calomel electrode** in Figure 15-5 is based on the reaction

*Calomel electrode:* $\quad\quad \frac{1}{2}Hg_2Cl_2(s) + e^- \rightleftharpoons Hg(l) + Cl^- \quad\quad E° = +0.268$ V

$$\underset{\substack{\text{Mercury(I) chloride} \\ \text{(calomel)}}}{} \quad\quad E(\text{saturated KCl}) = +0.241 \text{ V}$$

The standard potential ($E°$) for this reaction is $+0.268$ V. If the cell is saturated with KCl at 25°C, the potential is $+0.241$ V. A calomel electrode saturated with KCl is called a **saturated calomel electrode,** abbreviated **S.C.E.** The advantage in using saturated KCl is that $[Cl^-]$ does not change if some liquid evaporates.

## Voltage Conversions Between Different Reference Scales

If an electrode has a potential of $-0.461$ V with respect to a calomel electrode, what is the potential with respect to a silver-silver chloride electrode? What would be the potential with respect to the standard hydrogen electrode?

To answer these questions, consider this diagram, which shows the positions of the calomel and silver-silver chloride electrodes with respect to the standard hydrogen electrode:

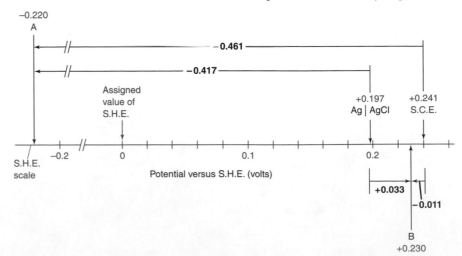

You can see that point A, which is $-0.461$ V from S.C.E., is $-0.417$ V from the silver-silver chloride electrode and $-0.220$ V with respect to S.H.E. Point B, whose potential is

+0.033 V from silver-silver chloride, is −0.011 V from S.C.E. and +0.230 V from S.H.E. By keeping this diagram in mind, you can convert potentials from one scale to another.

## ■ ■ ■ 15-2 Indicator Electrodes

We will study two broad classes of indicator electrodes. *Metal electrodes* described in this section develop an electric potential in response to a redox reaction at the metal surface. *Ion-selective electrodes,* described later, are not based on redox processes. Instead, selective binding of one type of ion to a membrane generates an electric potential.

The most common metal indicator electrode is platinum, which is relatively *inert*—it does not participate in many chemical reactions. Its purpose is simply to transmit electrons to or from species in solution. Gold electrodes are even more inert than Pt. Various types of carbon are used as indicator electrodes because the rates of many redox reactions on the carbon surface are fast. A metal electrode works best when its surface is large and clean. To clean the electrode, dip it briefly in hot 8 M $HNO_3$ and rinse with distilled water.

Figure 15-6 shows how a silver electrode can be used with a reference electrode to measure $Ag^+$ concentration.[5] The reaction at the Ag indicator electrode is

$$Ag^+ + e^- \rightleftharpoons Ag(s) \qquad E_+^\circ = 0.799 \text{ V}$$

The calomel reference half-cell reaction is

$$Hg_2Cl_2(s) + 2e^- \rightleftharpoons 2Hg(l) + 2Cl^- \qquad E_- = 0.241 \text{ V}$$

and the reference potential ($E_-$, not $E_-^\circ$) is fixed at 0.241 V because the reference cell is saturated with KCl. The Nernst equation for the entire cell is therefore

$$E = E_+ - E_- = \left\{0.799 - 0.059\,16 \log\left(\frac{1}{[Ag^+]}\right)\right\} - \left\{0.241\right\}$$

$$\underbrace{\phantom{0.799 - 0.059\,16 \log\left(\frac{1}{[Ag^+]}\right)}}_{\substack{\text{Potential of } Ag \mid Ag^+ \\ \text{indicator electrode}}} \qquad \underbrace{\phantom{0.241}}_{\substack{\text{Potential of} \\ \text{S.C.E. reference electrode}}}$$

$$E = 0.558 + 0.059\,16 \log[Ag^+] \qquad \qquad (15\text{-}1)$$

That is, the voltage of the cell in Figure 15-6 provides a measure of $[Ag^+]$. Ideally, the voltage changes by 59.16 mV (at 25°C) for each factor-of-10 change in $[Ag^+]$.

In Figure 7-9, we used a silver indicator electrode and a *glass* reference electrode. The glass electrode responds to pH and the cell in Figure 7-9 contains a buffer to maintain constant pH. Therefore, the glass electrode remains at a constant potential; it is being used in an unconventional way as a reference electrode.

---

### Example Potentiometric Precipitation Titration

A 100.0-mL solution containing 0.100 0 M NaCl was titrated with 0.100 0 M $AgNO_3$, and the voltage of the cell shown in Figure 15-6 was monitored. Calculate the voltage after the addition of 65.0 mL of $AgNO_3$.

**Solution** The titration reaction is

$$Ag^+ + Cl^- \rightarrow AgCl(s)$$

for which the equivalence point is 100.0 mL. At 65.0 mL, 65.0% of $Cl^-$ has precipitated and 35.0% remains in solution:

$$[Cl^-] = \underbrace{(0.350)}_{\substack{\text{Fraction} \\ \text{remaining}}} \underbrace{(0.100\,0 \text{ M})}_{\substack{\text{Original} \\ \text{concentration} \\ \text{of } Cl^-}} \overbrace{\left(\frac{100.0}{165.0}\right)}^{\substack{\text{Initial volume} \\ \text{of } Cl^-}} = 0.021\,2 \text{ M}$$

$$\underbrace{\phantom{\left(\frac{100.0}{165.0}\right)}}_{\substack{\text{Dilution} \\ \text{factor}}} \underbrace{\phantom{xxxx}}_{\substack{\text{Total volume} \\ \text{of solution}}}$$

To find the cell voltage in Equation 15-1, we need to know $[Ag^+]$:

$$[Ag^+][Cl^-] = K_{sp} \Rightarrow [Ag^+] = \frac{K_{sp}}{[Cl^-]} = \frac{1.8 \times 10^{-10}}{0.021\,2} = 8.5 \times 10^{-9} \text{ M}$$

The cell voltage is therefore

$$E = 0.558 + 0.059\,16 \log(8.5 \times 10^{-9}) = 0.081 \text{ V}$$

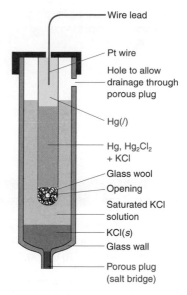

**Figure 15-5** A saturated calomel electrode (S.C.E.).

Wire lead
Pt wire
Hole to allow drainage through porous plug
Hg($l$)
Hg, $Hg_2Cl_2$ + KCl
Glass wool
Opening
Saturated KCl solution
KCl($s$)
Glass wall
Porous plug (salt bridge)

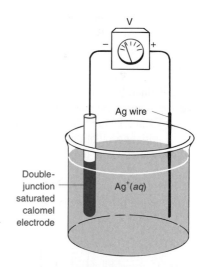

**Figure 15-6** Use of Ag and calomel electrodes to measure $[Ag^+]$. The calomel electrode has a double junction, like that in Figure 15-4. The outer compartment of the electrode is filled with $KNO_3$, so there is no direct contact between $Cl^-$ in the inner compartment and $Ag^+$ in the beaker.

Ag wire
Double-junction saturated calomel electrode
$Ag^+(aq)$

The cell responds to a change in [Cl⁻], which necessarily changes [Ag⁺] because [Ag⁺][Cl⁻] = $K_{sp}$.

We see from the example that *a silver electrode is also a halide electrode, if solid silver halide is present.*[6] If the solution contains AgCl(s), we substitute $[Ag^+] = K_{sp}/[Cl^-]$ into Equation 15-1 to find an expression relating the cell voltage to [Cl⁻]:

$$E = 0.558 + 0.059\ 16 \log\left(\frac{K_{sp}}{[Cl^-]}\right)$$

Demonstration 15-1 is a great example of indicator and reference electrodes.

Metals, including Ag, Cu, Zn, Cd, and Hg, can be used as indicator electrodes for their aqueous ions. Most metals, however, are unsuitable for this purpose, because the equilibrium $M^{n+} + ne^- \rightleftharpoons M$ is not readily established at the metal surface.

## Demonstration 15-1 Potentiometry with an Oscillating Reaction[7]

The Belousov-Zhabotinskii reaction is a cerium-catalyzed oxidation of malonic acid by bromate, in which the quotient $[Ce^{3+}]/[Ce^{4+}]$ oscillates by a factor of 10 to 100.[8]

$$3CH_2(CO_2H)_2 + 2BrO_3^- + 2H^+ \rightarrow$$
Malonic acid     Bromate

$$2BrCH(CO_2H)_2 + 3CO_2 + 4H_2O$$
Bromomalonic acid

When the $Ce^{4+}$ concentration is high, the solution is yellow. When $Ce^{3+}$ predominates, the solution is colorless. With redox indicators (Section 16-2), this reaction oscillates through a sequence of colors.[9]

Oscillation between yellow and colorless is set up in a 300-mL beaker with the following solutions:

160 mL of 1.5 M $H_2SO_4$
 40 mL of 2 M malonic acid
 30 mL of 0.5 M $NaBrO_3$ (or saturated $KBrO_3$)
  4 mL of saturated ceric ammonium sulfate,
   ($Ce(SO_4)_2 \cdot 2(NH_4)_2SO_4 \cdot 2H_2O$)

After an induction period of 5 to 10 min with magnetic stirring, oscillations can be initiated by adding 1 mL of ceric ammonium sulfate solution. The reaction may need more $Ce^{4+}$ over a 5-min period to initiate oscillations.

A galvanic cell is built around the reaction as shown in the figure. The quotient $[Ce^{3+}]/[Ce^{4+}]$ is monitored by Pt and calomel electrodes. You should be able to write the cell reactions and a Nernst equation for this experiment.

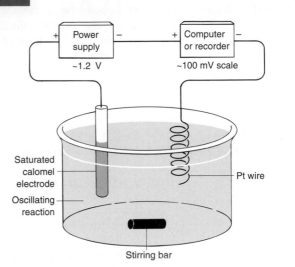

Apparatus used to monitor the quotient $[Ce^{3+}]/[Ce^{4+}]$ for an oscillating reaction. [*The idea for this demonstration came from George Rossman, California Institute of Technology.*]

In place of a potentiometer (a pH meter), use a computer or recorder to show the oscillations. Because the potential oscillates over a range of ~100 mV but is centered near ~1.2 V, the cell voltage is offset by ~1.2 V with any available power supply.[10] Trace *a* shows what is usually observed. The potential changes rapidly during the abrupt colorless-to-yellow change and gradually during the gentle yellow-to-colorless change. Trace *b* shows two different cycles superimposed in the same solution. This unusual event occurred in a reaction that had been oscillating normally for about 30 min.[11]

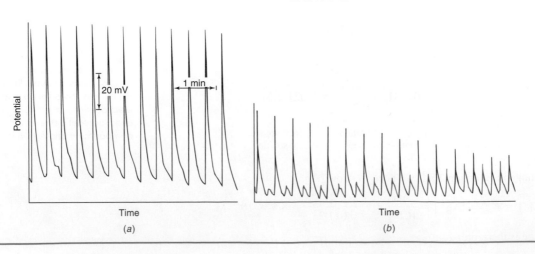

## 15-3 What Is a Junction Potential?

Whenever dissimilar electrolyte solutions are in contact, a voltage difference called the **junction potential** develops at their interface. This small voltage (usually a few millivolts) is found at each end of a salt bridge connecting two half-cells. *The junction potential puts a fundamental limitation on the accuracy of direct potentiometric measurements,* because we usually do not know the contribution of the junction to the measured voltage.

To see why the junction potential occurs, consider a solution of NaCl in contact with distilled water (Figure 15-7). $Na^+$ and $Cl^-$ ions begin to diffuse from the NaCl solution into the water. However, $Cl^-$ ion has a greater **mobility** than $Na^+$. That is, $Cl^-$ diffuses faster than $Na^+$. As a result, a region rich in $Cl^-$, with excess negative charge, develops at the front. Behind it is a positively charged region depleted of $Cl^-$. The result is an electric potential difference at the junction of the NaCl and $H_2O$ phases. The junction potential opposes the movement of $Cl^-$ and accelerates the movement of $Na^+$. The steady-state junction potential represents a balance between the unequal mobilities that create a charge imbalance and the tendency of the resulting charge imbalance to retard the movement of $Cl^-$.

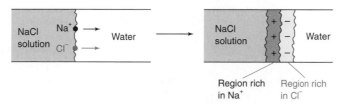

**Figure 15-7** Development of the junction potential caused by unequal mobilities of $Na^+$ and $Cl^-$.

Table 15-1 shows mobilities of several ions and Table 15-2 lists some liquid junction potentials. Saturated KCl is used in a salt bridge because $K^+$ and $Cl^-$ have similar mobilities. Junction potentials at the two interfaces of a KCl salt bridge are slight.

Nonetheless, the junction potential of 0.1 M HCl | 3.5 M KCl is 3.1 mV. A pH electrode has a response of 59 mV per pH unit. A pH electrode dipped into 0.1 M HCl will have a junction potential of ~3 mV, or an error of 0.05 pH units (12% error in $[H^+]$).

---

### Example Junction Potential

A 0.1 M NaCl solution was placed in contact with a 0.1 M $NaNO_3$ solution. Which side of the junction is positive?

**Solution** Because $[Na^+]$ is equal on both sides, there is no net diffusion of $Na^+$ across the junction. However, $Cl^-$ diffuses into $NaNO_3$, and $NO_3^-$ diffuses into NaCl. The mobility of $Cl^-$ is greater than that of $NO_3^-$, so the NaCl region will be depleted of $Cl^-$ faster than the $NaNO_3$ region will be depleted of $NO_3^-$. The $NaNO_3$ side will become negative, and the NaCl side will become positive.

---

## 15-4 How Ion-Selective Electrodes Work[12]

**Ion-selective electrodes,** discussed in the remainder of this chapter, respond selectively to one ion. These electrodes are fundamentally different from metal electrodes in that ion-selective electrodes do not involve redox processes. The key feature of an ideal ion-selective electrode is a thin membrane capable of binding only the intended ion.

Consider the *liquid-based ion-selective electrode* in Figure 15-8a. The electrode is said to be "liquid based" because the ion-selective membrane is a *hydrophobic* organic polymer impregnated with a viscous organic solution containing an ion exchanger and, sometimes, a ligand that selectively binds the analyte cation, $C^+$. The inside of the electrode contains filling solution with the ions $C^+(aq)$ and $B^-(aq)$. The outside of the electrode is immersed in analyte solution containing $C^+(aq)$, $A^-(aq)$, and, perhaps, other ions. Ideally, it does not matter what $A^-$ and $B^-$ and the other ions are. The electric potential difference (the voltage) across the ion-selective membrane is measured by two reference electrodes, which might be Ag | AgCl. If the concentration (really, the activity) of $C^+$ in the analyte solution changes, the

$E_{observed} = E_{cell} + E_{junction}$

Because the junction potential is usually unknown, $E_{cell}$ is uncertain.

**Table 15-1** Mobilities of ions in water at 25°C

| Ion | Mobility $[m^2/(s \cdot V)]^a$ |
| --- | --- |
| $H^+$ | $36.30 \times 10^{-8}$ |
| $Rb^+$ | $7.92 \times 10^{-8}$ |
| $K^+$ | $7.62 \times 10^{-8}$ |
| $NH_4^+$ | $7.61 \times 10^{-8}$ |
| $La^{3+}$ | $7.21 \times 10^{-8}$ |
| $Ba^{2+}$ | $6.59 \times 10^{-8}$ |
| $Ag^+$ | $6.42 \times 10^{-8}$ |
| $Ca^{2+}$ | $6.12 \times 10^{-8}$ |
| $Cu^{2+}$ | $5.56 \times 10^{-8}$ |
| $Na^+$ | $5.19 \times 10^{-8}$ |
| $Li^+$ | $4.01 \times 10^{-8}$ |
| $OH^-$ | $20.50 \times 10^{-8}$ |
| $Fe(CN)_6^{4-}$ | $11.45 \times 10^{-8}$ |
| $Fe(CN)_6^{3-}$ | $10.47 \times 10^{-8}$ |
| $SO_4^{2-}$ | $8.27 \times 10^{-8}$ |
| $Br^-$ | $8.13 \times 10^{-8}$ |
| $I^-$ | $7.96 \times 10^{-8}$ |
| $Cl^-$ | $7.91 \times 10^{-8}$ |
| $NO_3^-$ | $7.40 \times 10^{-8}$ |
| $ClO_4^-$ | $7.05 \times 10^{-8}$ |
| $F^-$ | $5.70 \times 10^{-8}$ |
| $HCO_3^-$ | $4.61 \times 10^{-8}$ |
| $CH_3CO_2^-$ | $4.24 \times 10^{-8}$ |

*a. The mobility of an ion is the terminal velocity that the particle achieves in an electric field of 1 V/m. Mobility = velocity/field. The units of mobility are therefore $(m/s)/(V/m) = m^2/(s \cdot V)$.*

**Table 15-2** Liquid junction potentials at 25°C

| Junction | Potential (mV) |
| --- | --- |
| 0.1 M NaCl \| 0.1 M KCl | $-6.4$ |
| 0.1 M NaCl \| 3.5 M KCl | $-0.2$ |
| 1 M NaCl \| 3.5 M KCl | $-1.9$ |
| 0.1 M HCl \| 0.1 M KCl | $+27$ |
| 0.1 M HCl \| 3.5 M KCl | $+3.1$ |

*NOTE: A positive sign means that the right side of the junction becomes positive with respect to the left side.*

Hydrophobic: "water hating" (does not mix with water)

**Figure 15-8** (*a*) Ion-selective electrode immersed in aqueous solution containing analyte cation, $C^+$. Typically, the membrane is made of poly(vinyl chloride) impregnated with the *plasticizer* dioctyl sebacate, a nonpolar liquid that softens the membrane and dissolves the ion-selective ionophore (L), the complex ($LC^+$), and a hydrophobic anion ($R^-$). (*b*) Close-up of membrane. Ellipses encircling pairs of ions are a guide for the eye to count the charge in each phase. Bold colored ions represent excess charge in each phase. The electric potential difference across each surface of the membrane depends on the activity of analyte ion in the aqueous solution contacting the membrane.

L has some ability to bind other ions besides $C^+$, so those other ions interfere to some extent with the measurement of $C^+$. An ion-selective electrode uses a ligand with a strong preference to bind the desired ion.

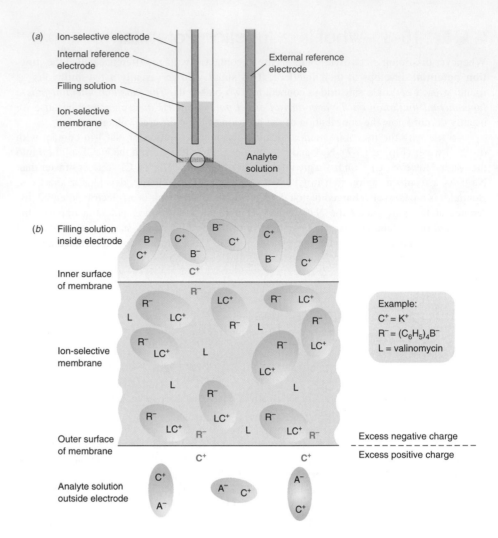

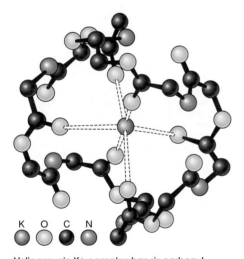

K   O   C   N

Valinomycin-$K^+$ complex has six carbonyl oxygen atoms in octahedral coordination around $K^+$. *[From L. Stryer, Biochemistry, 4th ed (New York: W. H. Freeman, 1995), p. 273.]*

The region of charge imbalance extends just a few nanometers into the membrane and into the neighboring solution.

voltage measured between the two reference electrodes also changes. By using a calibration curve, the voltage tells us the activity of $C^+$ in the analyte solution.

Figure 15-8b shows how the electrode works. The key in this example is the ligand, L (called an *ionophore*), that is soluble inside the membrane and selectively binds analyte ion. In a potassium ion-selective electrode, for example, L could be valinomycin, a natural antibiotic secreted by certain microorganisms to carry $K^+$ ion across cell membranes. *The ligand, L, is chosen to have a high affinity for analyte cation, $C^+$, and low affinity for other ions.* In an ideal electrode, L binds only $C^+$. Real electrodes always have some affinity for other cations, so these cations interfere to some degree with the measurement of $C^+$. For charge neutrality, the membrane also contains a hydrophobic anion, $R^-$, such as tetraphenylborate, $(C_6H_5)_4B^-$, that is soluble in the membrane and poorly soluble in water.

Almost all analyte ion inside the membrane in Figure 15-8b is bound in the complex $LC^+$, which is in equilibrium with a small amount of free $C^+$ in the membrane. The membrane also contains excess free L. $C^+$ can diffuse across the interface. In an ideal electrode, $R^-$ cannot leave the membrane, because it is not soluble in water, and the aqueous anion $A^-$ cannot enter the membrane, because it is not soluble in the organic phase. As soon as a few $C^+$ ions diffuse from the membrane into the aqueous phase, there is excess positive charge in the aqueous phase. This imbalance creates an electric potential difference that opposes diffusion of more $C^+$ into the aqueous phase.

When $C^+$ diffuses from a region of activity $\mathcal{A}_m$ in the membrane to a region of activity $\mathcal{A}_o$ in the outer solution, the free-energy change is

$$\Delta G = \underbrace{\Delta G_{\text{solvation}}}_{\substack{\Delta G \text{ due to change} \\ \text{in solvent}}} - \underbrace{RT \ln\left(\frac{\mathcal{A}_m}{\mathcal{A}_o}\right)}_{\substack{\Delta G \text{ due to} \\ \text{change in activity} \\ \text{(concentration)}}}$$

where $R$ is the gas constant and $T$ is temperature (K). $\Delta G_{\text{solvation}}$ is the change in solvation energy when the environment around $C^+$ changes from the organic liquid in the membrane to the aqueous solution outside the membrane. The term $-RT \ln(\mathcal{A}_m/\mathcal{A}_o)$ gives the free energy change when a species diffuses between regions of different activities (concentrations). In the absence of a phase boundary, $\Delta G$ would always be negative when a species diffuses from a region of high activity to one of lower activity.

The driving force for diffusion of $C^+$ from the membrane to the aqueous solution is the favorable solvation of the ion by water. As $C^+$ diffuses from the membrane into the water, there is a buildup of positive charge in the water immediately adjacent to the membrane. The charge separation creates an electric potential difference ($E_{\text{outer}}$) across the membrane. The free-energy difference for $C^+$ in the two phases is $\Delta G = -nFE_{\text{outer}}$, where $F$ is the Faraday constant and $n$ is the charge of the ion. At equilibrium, the net change in free energy for diffusion of $C^+$ across the membrane boundary must be 0:

$$\underbrace{\Delta G_{\text{solvation}} - RT \ln\left(\frac{\mathcal{A}_m}{\mathcal{A}_o}\right)}_{\substack{\Delta G \text{ due to transfer between phases} \\ \text{and activity difference}}} + \underbrace{(-nFE_{\text{outer}})}_{\substack{\Delta G \text{ due to} \\ \text{charge imbalance}}} = 0$$

Solving for $E_{\text{outer}}$, we find that the electric potential difference across the boundary between the membrane and the outer aqueous solution in Figure 15-8b is

*Electric potential difference across phase boundary between membrane and analyte:*

$$E_{\text{outer}} = \frac{\Delta G_{\text{solvation}}}{nF} - \left(\frac{RT}{nF}\right)\ln\left(\frac{\mathcal{A}_m}{\mathcal{A}_o}\right) \qquad (15\text{-}2)$$

There is also a potential difference $E_{\text{inner}}$ at the boundary between the inner filling solution and the membrane, with terms analogous to those in Equation 15-2.

The potential difference between the outer analyte solution and the inner filling solution is the difference $E = E_{\text{outer}} - E_{\text{inner}}$. In Equation 15-2, $E_{\text{outer}}$ depends on the activities of $C^+$ in the analyte solution and in the membrane near its outer surface. $E_{\text{inner}}$ is constant because the activity of $C^+$ in the filling solution is constant.

But the activity of $C^+$ in the membrane ($\mathcal{A}_m$) is very nearly constant for the following reason: The high concentration of $LC^+$ in the membrane is in equilibrium with free L and a small concentration of free $C^+$ in the membrane. The hydrophobic anion $R^-$ is poorly soluble in water and therefore cannot leave the membrane. *Very little* $C^+$ can diffuse out of the membrane because each $C^+$ that enters the aqueous phase leaves behind one $R^-$ in the membrane. (This separation of charge is the source of the potential difference at the phase boundary.) As soon as a tiny fraction of $C^+$ diffuses from the membrane into solution, further diffusion is prevented by excess positive charge in the solution near the membrane.

So the potential difference between the outer and the inner solutions is

$$E = E_{\text{outer}} - E_{\text{inner}} = \frac{\Delta G_{\text{solvation}}}{nF} - \left(\frac{RT}{nF}\right)\ln\left(\frac{\mathcal{A}_m}{\mathcal{A}_o}\right) - E_{\text{inner}}$$

$$E = \underbrace{\frac{\Delta G_{\text{solvation}}}{nF}}_{\text{Constant}} + \left(\frac{RT}{nF}\right)\ln \mathcal{A}_o - \underbrace{\left(\frac{RT}{nF}\right)\ln \mathcal{A}_m}_{\text{Constant}} - \underbrace{E_{\text{inner}}}_{\text{Constant}}$$

$$\ln\frac{x}{y} = \ln x - \ln y$$

Combining the constant terms, we find that the potential difference across the membrane depends only on the activity of analyte in the outer solution:

$$E = \text{constant} + \left(\frac{RT}{nF}\right)\ln \mathcal{A}_o$$

Converting ln into log and inserting values of $R$, $T$, and $F$ gives a useful expression for the potential difference across the membrane:

From Appendix A,

$$\log x = \left(\frac{1}{\ln 10}\right)\ln x = 0.434\ 3 \ln x.$$

*Electric potential difference for ion-selective electrode:*

$$E = \text{constant} + \frac{0.059\ 16}{n}\log \mathcal{A}_o \text{ (volts at 25°C)} \qquad (15\text{-}3)$$

where $n$ is the charge of the analyte ion and $\mathcal{A}_o$ is its activity in the outer (unknown) solution. Equation 15-3 applies to any ion-selective electrode, including a glass pH electrode. If the analyte is an anion, the sign of $n$ is negative. Later, we will modify the equation to account for interfering ions.

A difference of 59.16 mV (at 25°C) builds up across a glass pH electrode for every factor-of-10 change in activity of $H^+$ in the analyte solution. Because a factor-of-10 difference in activity of $H^+$ is 1 pH unit, a difference of, say, 4.00 pH units would lead to a potential difference of $4.00 \times 59.16 = 237$ mV. The charge of a calcium ion is $n = 2$, so a potential difference of $59.16/2 = 29.58$ mV is expected for every factor-of-10 change in activity of $Ca^{2+}$ in the analyte measured with a calcium ion-selective electrode.

# ■ ■ ■ ■ 15-5 pH Measurement with a Glass Electrode

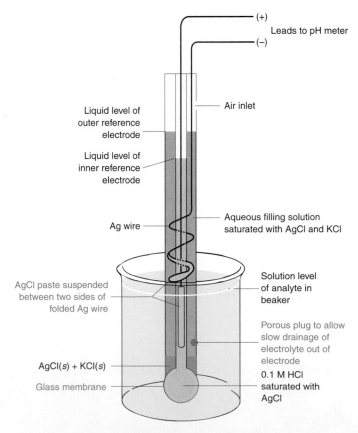

The **glass electrode** used to measure pH is the most common *ion-selective electrode*. A typical pH **combination electrode,** incorporating both glass and reference electrodes in one body, is shown in Figure 15-9. A line diagram of this cell can be written as follows:

Glass membrane<br>selectively binds $H^+$

$$\text{Ag}(s) \mid \text{AgCl}(s) \mid \text{Cl}^-(aq) \parallel \text{H}^+(aq, \text{outside}) \vdots \text{H}^+(aq, \text{inside}), \text{Cl}^-(aq) \mid \text{AgCl}(s) \mid \text{Ag}(s)$$

| Outer reference electrode | $H^+$ outside glass electrode (analyte solution) | $H^+$ inside glass electrode | Inner reference electrode |

The pH-sensitive part of the electrode is the thin glass bulb or cone at the bottom of the electrodes in Figures 15-9 and 15-10. The reference electrode at the left of the preceding line diagram is the coiled Ag | AgCl electrode in the combination electrode in Figure 15-9. The reference electrode at the right side of the line diagram is the straight Ag | AgCl electrode at the center of the electrode in Figure 15-9. The two reference electrodes measure the electric

M. Cremer at the Institute of Physiology at Munich discovered in 1906 that a potential difference of 0.2 V developed across a glass membrane with acid on one side and neutral saline solution on the other. The student Klemensiewicz, working with F. Haber in Karlsruhe in 1908, improved the glass electrode and carried out the first acid-base titration to be monitored with a glass electrode.[13]

pH electrodes other than glass can be used when glass is unsuitable.[14]

(+)
Leads to pH meter
(−)

Liquid level of outer reference electrode

Liquid level of inner reference electrode

Air inlet

Ag wire

Aqueous filling solution saturated with AgCl and KCl

AgCl paste suspended between two sides of folded Ag wire

Solution level of analyte in beaker

Porous plug to allow slow drainage of electrolyte out of electrode

AgCl(s) + KCl(s)

0.1 M HCl saturated with AgCl

Glass membrane

**Figure 15-9** Diagram of a glass combination electrode with a silver-silver chloride reference electrode. The glass electrode is immersed in a solution of unknown pH so that the porous plug on the lower right is below the surface of the liquid. The two silver electrodes measure the voltage across the glass membrane.

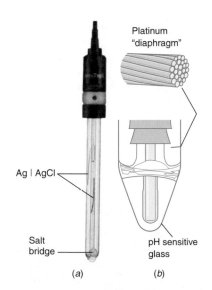

Platinum "diaphragm"

Ag | AgCl

Salt bridge

pH sensitive glass

(a)          (b)

**Figure 15-10** (a) Glass-body combination electrode with pH-sensitive glass bulb at the bottom. The porous ceramic plug (the salt bridge) connects analyte solution to the reference electrode. Two silver wires coated with AgCl are visible inside the electrode. [Courtesy Fisher Scientific, Pittsburgh PA.] (b) A pH electrode with a platinum diaphragm (a bundle of Pt wires), which is said to be less prone to clogging than a ceramic plug. [W. Knappek, Am. Lab. News Ed. July **2003**, p. 14.]

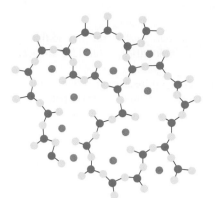

Figure 15-11 Schematic structure of glass, which consists of an irregular network of $SiO_4$ tetrahedra connected through oxygen atoms. ● = O, ● = Si, ● = cation. Cations such as $Li^+$, $Na^+$, $K^+$, and $Ca^{2+}$ are coordinated to the oxygen atoms. The silicate network is not planar. This diagram is a projection of each tetrahedron onto the plane of the page.
[Adapted from G. A. Perley, "Glasses for Measurement of pH," Anal. Chem. **1949**, 21, 394.]

So little current flows across a glass electrode that it was not practical when discovered in 1906. One of the first people to use a vacuum tube amplifier to measure pH with a glass electrode was an undergraduate, W. H. Wright at the University of Illinois in 1928, who knew about electronics from amateur radio. Arnold Beckman at Caltech invented a portable, rugged, vacuum tube pH meter in 1935, which revolutionized chemical instrumentation.[15]

potential difference across the glass membrane. The salt bridge in the line diagram is the porous plug at the bottom-right side of the combination electrode in Figure 15-9.

Figure 15-11 shows the irregular structure of the silicate lattice in glass. Negatively charged oxygen atoms in glass can bind to cations of suitable size. Monovalent cations, particularly $Na^+$, can move sluggishly through the silicate lattice. A schematic cross section of the glass membrane of a pH electrode is shown in Figure 15-12. The two surfaces swell as they absorb water. Metal ions in these *hydrated gel* regions of the membrane diffuse out of the glass and into solution. $H^+$ can diffuse into the membrane to replace the metal ions. The reaction in which $H^+$ replaces cations in the glass is an **ion-exchange equilibrium** (Figure 15-13). A pH electrode responds selectively to $H^+$ because $H^+$ is the only ion that binds significantly to the hydrated gel layer.

To perform an electrical measurement, at least some tiny current must flow through the entire circuit—even across the glass pH electrode membrane. Studies with tritium (radioactive $^3H$) show that $H^+$ does not cross the glass membrane. However, $Na^+$ sluggishly crosses the membrane. The $H^+$-sensitive membrane may be thought of as two surfaces electrically connected by $Na^+$ transport. The membrane's resistance is typically $10^8$ $\Omega$, so little current actually flows across it.

The potential difference between inner and outer silver-silver chloride electrodes in Figure 15-9 depends on the chloride concentration in each electrode compartment and on the potential difference across the glass membrane. Because $[Cl^-]$ is fixed in each compartment and because $[H^+]$ is fixed on the inside of the glass membrane, the only variable is the pH of analyte solution outside the glass membrane. Equation 15-3 states that *the voltage of the ideal pH electrode changes by 59.16 mV for every pH-unit change of analyte activity at 25°C.*

The response of real glass electrodes is described by the Nernst-like equation

*Response of*     $E = \text{constant} + \beta(0.059\ 16) \log \mathcal{A}_{H^+}(\text{outside})$
*glass electrode:*   $E = \text{constant} - \beta(0.059\ 16)\, \text{pH}(\text{outside})$     (at 25°C)     (15-4)

The value of $\beta$, the *electromotive efficiency*, is close to 1.00 (typically >0.98). We measure the constant and $\beta$ when we calibrate the electrode in solutions of known pH.

Electrodes

Beckman's pH meter

The pH electrode measures $H^+$ activity, not $H^+$ concentration.

The number 0.059 16 V is $\dfrac{RT \ln 10}{F}$, where $R$ is the gas constant, $T$ is temperature, and $F$ is the Faraday constant.

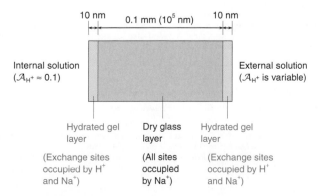

Figure 15-12 Schematic cross section of the glass membrane of a pH electrode.

| 10 nm | 0.1 mm ($10^5$ nm) | 10 nm |

Internal solution ($\mathcal{A}_{H^+} \approx 0.1$)          External solution ($\mathcal{A}_{H^+}$ is variable)

| Hydrated gel layer | Dry glass layer | Hydrated gel layer |
|---|---|---|
| (Exchange sites occupied by $H^+$ and $Na^+$) | (All sites occupied by $Na^+$) | (Exchange sites occupied by $H^+$ and $Na^+$) |

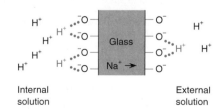

Internal solution          External solution

Figure 15-13 Ion-exchange equilibria on surfaces of a glass membrane: $H^+$ replaces metal ions bound to the negatively charged oxygen atoms. The pH of the internal solution is fixed. As the pH of the external solution (the sample) changes, the electric potential difference across the glass membrane changes.

A pH electrode **must** be calibrated before it can be used. It should be calibrated about every 2 h during sustained use. The pH of the calibration standards should bracket the pH of the unknown.

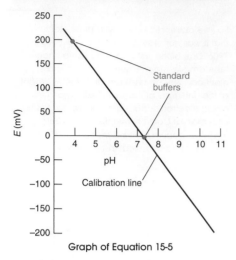

Graph of Equation 15-5

## Calibrating a Glass Electrode

A pH electrode should be calibrated with two (or more) standard buffers selected so that the pH of the unknown lies within the range of the standards. Standards in Table 15-3 are accurate to ±0.01 pH unit.[16]

When you calibrate an electrode with standard buffers, you measure a voltage with the electrode in each buffer. The pH of buffer S1 is $pH_{S1}$ and the measured electrode potential in this buffer is $E_{S1}$. The pH of buffer S2 is $pH_{S2}$ and the measured electrode potential is $E_{S2}$. The equation of the line through the two standard points is

$$\frac{E - E_{S1}}{pH - pH_{S1}} = \frac{E_{S2} - E_{S1}}{pH_{S2} - pH_{S1}} \tag{15-5}$$

The slope of the line is $\Delta E/\Delta pH = (E_{S2} - E_{S1})/(pH_{S2} - pH_{S1})$, which is 59.16 mV/pH unit at 25°C for an ideal electrode and $\beta(59.16)$ mV/pH unit for a real electrode, where $\beta$ is the correction factor in Equation 15-4.

To measure the pH of an unknown, measure the potential of the unknown with the calibrated electrode and find pH by substitution in Equation 15-5.

$$\frac{E_{unknown} - E_{S1}}{pH_{unknown} - pH_{S1}} = \frac{E_{S2} - E_{S1}}{pH_{S2} - pH_{S1}} \tag{15-6}$$

**Table 15-3** pH values of National Institute of Standards and Technology buffers

| Temperature (°C) | Saturated (25°C) potassium hydrogen tartrate (1) | 0.05 m potassium dihydrogen citrate (2) | 0.05 m potassium hydrogen phthalate (3) | 0.08 m MOPSO 0.08 m NaMOPSO 0.08 m NaCl (4) |
|---|---|---|---|---|
| 0 | — | 3.863 | 4.003 | 7.268 |
| 5 | — | 3.840 | 3.999 | 7.182 |
| 10 | — | 3.820 | 3.998 | 7.098 |
| 15 | — | 3.802 | 3.999 | 7.018 |
| 20 | — | 3.788 | 4.002 | 6.940 |
| 25 | 3.557 | 3.776 | 4.008 | 6.865 |
| 30 | 3.552 | 3.766 | 4.015 | 6.792 |
| 35 | 3.549 | 3.759 | 4.024 | 6.722 |
| 37 | 3.548 | 3.756 | 4.028 | 6.695 |
| 40 | 3.547 | 3.753 | 4.035 | 6.654 |
| 45 | 3.547 | 3.750 | 4.047 | 6.588 |
| 50 | 3.549 | 3.749 | 4.060 | 6.524 |
| 55 | 3.554 | — | 4.075 | — |
| 60 | 3.560 | — | 4.091 | — |
| 70 | 3.580 | — | 4.126 | — |
| 80 | 3.609 | — | 4.164 | — |
| 90 | 3.650 | — | 4.205 | — |
| 95 | 3.674 | — | 4.227 | — |

NOTE: The designation m stands for molality. Masses in the buffer recipes below are apparent masses measured in air.

In the buffer solution preparations, it is essential to use high-purity materials and freshly distilled or deionized water of resistivity greater than 2 000 ohm · m. Solutions having pH 6 or above should be stored in plastic containers, preferably ones with an NaOH trap to prevent ingress of atmospheric carbon dioxide. They can normally be kept for 2–3 weeks, or slightly longer in a refrigerator. Buffer materials in this table are available as Standard Reference Materials from the National Institute of Standards and Technology. http://ts.nist.gov/srm. pH standards for $D_2O$ and aqueous-organic solutions can be found in P. R. Mussini, T. Mussini, and S. Rondinini, Pure Appl. Chem. **1997**, 69, 1007.

1. Saturated (25°C) potassium hydrogen tartrate, $KHC_4H_4O_6$. An excess of the salt is shaken with water, and it can be stored in this way. Before use, it should be filtered or decanted at a temperature between 22° and 28°C.

2. 0.05 m potassium dihydrogen citrate, $KH_2C_6H_5O_7$. Dissolve 11.41 g of the salt in 1 L of solution at 25°C.

3. 0.05 m potassium hydrogen phthalate. Although this is not usually essential, the crystals may be dried at 100°C for 1 h, then cooled in a desiccator. At 25°C, 10.12 g $C_6H_4(CO_2H)(CO_2K)$ are dissolved in water, and the solution made up to 1 L.

4. 0.08 m MOPSO ((3-N-morpholino)-2-hydroxypropanesulfonic acid, Table 9-2), 0.08 m sodium salt of MOPSO, 0.08 m NaCl. Buffers 4 and 6 are recommended for 2-point standardization of electrodes for pH measurements of physiologic fluids. MOPSO is crystallized twice from 70 wt% ethanol and dried in vacuum at 50°C for 24 h. NaCl is dried at 110°C for 4 h. $Na^+MOPSO^-$ may be prepared by neutralization of MOPSO with standard NaOH. The sodium salt is also available as a Standard Reference Material. Dissolve 18.00 g MOPSO, 19.76 g $Na^+MOPSO^-$, and 4.674 g NaCl in 1.000 kg $H_2O$.

Alas, modern pH meters are "black boxes" that do these calculations for us by applying Equations 15-5 and 15-6 and automatically displaying pH.

Before using a pH electrode, be sure that the air inlet near the upper end of the electrode in Figure 15-9 is not capped. (This hole is capped during storage to prevent evaporation of the reference electrode filling solution.) Wash the electrode with distilled water and gently *blot* it dry with a tissue. Do not *wipe* it, because this action might produce a static charge on the glass.

To calibrate the electrode, dip it in a standard buffer whose pH is near 7 and allow the electrode to equilibrate with stirring for at least a minute. Following the manufacturer's instructions, press a key that might say "calibrate" or "read" on a microprocessor-controlled meter or adjust the reading of an analog meter to indicate the pH of the standard buffer. Then wash the electrode with water, blot it dry, and immerse it in a second standard whose pH is farther from 7 than the pH of the first standard. Enter the second buffer on the meter. Finally, dip the electrode in the unknown, stir the liquid, allow the reading to stabilize, and read the pH.

Store a glass electrode in aqueous solution to prevent dehydration of the glass. Ideally, the solution should be similar to that inside the reference compartment of the electrode. If the electrode has dried, recondition it in dilute acid for several hours. If the electrode is to be used above pH 9, soak it in a high-pH buffer. (The field effect transistor pH electrode in Section 15-8 is stored dry. Prior to use, scrub it gently with a soft brush and soak it in pH 7 buffer for 10 min.)

> Do not leave a glass electrode out of water (or in a nonaqueous solvent) any longer than necessary.

**Table 15-3 (Continued)**   pH values of National Institute of Standards and Technology buffers

| 0.025 $m$ potassium dihydrogen phosphate 0.025 $m$ disodium hydrogen phosphate (5) | 0.08 $m$ HEPES 0.08 $m$ NaHEPES 0.08 $m$ NaCl (6) | 0.008 695 $m$ potassium dihydrogen phosphate 0.030 43 $m$ disodium hydrogen phosphate (7) | 0.01 $m$ borax (8) | 0.025 $m$ sodium bicarbonate 0.025 $m$ sodium carbonate (9) |
|---|---|---|---|---|
| 6.984 | 7.853 | 7.534 | 9.464 | 10.317 |
| 6.951 | 7.782 | 7.500 | 9.395 | 10.245 |
| 6.923 | 7.713 | 7.472 | 9.332 | 10.179 |
| 6.900 | 7.646 | 7.448 | 9.276 | 10.118 |
| 6.881 | 7.580 | 7.429 | 9.225 | 10.062 |
| 6.865 | 7.516 | 7.413 | 9.180 | 10.012 |
| 6.853 | 7.454 | 7.400 | 9.139 | 9.966 |
| 6.844 | 7.393 | 7.389 | 9.102 | 9.925 |
| 6.840 | 7.370 | 7.384 | 9.081 | 9.910 |
| 6.838 | 7.335 | 7.380 | 9.068 | 9.889 |
| 6.834 | 7.278 | 7.373 | 9.038 | 9.856 |
| 6.833 | 7.223 | 7.367 | 9.011 | 9.828 |
| 6.834 | — | — | 8.985 | — |
| 6.836 | — | — | 8.962 | — |
| 6.845 | — | — | 8.921 | — |
| 6.859 | — | — | 8.885 | — |
| 6.877 | — | — | 8.850 | — |
| 6.886 | — | — | 8.833 | — |

5. *0.025 m disodium hydrogen phosphate, 0.025 m potassium dihydrogen phosphate. The anhydrous salts are best; each should be dried for 2 h at 120°C and cooled in a desiccator, because they are slightly hygroscopic. Higher drying temperatures should be avoided to prevent formation of condensed phosphates. Dissolve 3.53 g Na₂HPO₄ and 3.39 g KH₂PO₄ in water to give 1 L of solution at 25°C.*

6. *0.08 m HEPES (N-2-hydroxyethylpiperazine-N'-2-ethanesulfonic acid, Table 9-2), 0.08 m sodium salt of HEPES, 0.08 m NaCl. Buffers 4 and 6 are recommended for 2-point standardization of electrodes for pH measurements of physiologic fluids. HEPES is crystallized twice from 80 wt% ethanol and dried in vacuum at 50°C for 24 h. NaCl is dried at 110°C for 4 h. Na⁺HEPES⁻ may be prepared by neutralization of HEPES with standard NaOH. The sodium salt is also available as a Standard Reference Material. Dissolve 19.04 g HEPES, 20.80 g Na⁺HEPES⁻, and 4.674 g NaCl in 1.000 kg H₂O.*

7. *0.008 695 m potassium dihydrogen phosphate, 0.030 43 m disodium hydrogen phosphate. Prepare like Buffer 5; dissolve 1.179 g KH₂PO₄ and 4.30 g Na₂HPO₄ in water to give 1 L of solution at 25°C.*

8. *0.01 m sodium tetraborate decahydrate. Dissolve 3.80 g Na₂B₄O₇ ·10H₂O in water to give 1 L of solution. This borax solution is particularly susceptible to pH change from carbon dioxide absorption, and it should be correspondingly protected.*

9. *0.025 m sodium bicarbonate, 0.025 m sodium carbonate. Primary standard grade Na₂CO₃ is dried at 250°C for 90 min and stored over CaCl₂ and Drierite. Reagent-grade NaHCO₃ is dried over molecular sieves and Drierite for 2 days at room temperature. Do not heat NaHCO₃, or it may decompose to Na₂CO₃. Dissolve 2.092 g of NaHCO₃ and 2.640 g of Na₂CO₃ in 1 L of solution at 25°C.*

SOURCES: *R. G. Bates, J. Res. Natl. Bureau Stds.* **1962,** *66A, 179; B. R. Staples and R. G. Bates, J. Res. Natl. Bureau Stds.* **1969,** *73A, 37. Data on HEPES and MOPSO are from Y. C. Wu, P. A. Berezansky, D. Feng, and W. F. Koch, Anal. Chem.* **1993,** *65, 1084, and D. Feng, W. F. Koch, and Y. C. Wu, Anal. Chem.* **1989,** *61, 1400. Instructions for preparing some of these solutions are from G. Mattock in C. N. Reilley, ed., Advances in Analytical Chemistry and Instrumentation (New York: Wiley, 1963), Vol. 2, p. 45. See also R. G. Bates, Determination of pH: Theory and Practice, 2nd ed. (New York: Wiley, 1973), Chap. 4.*

Combustion products from automobiles and factories include nitrogen oxides and sulfur dioxide, which react with water in the atmosphere to produce acids.[17]

$$SO_2 + H_2O \rightarrow H_2SO_3 \xrightarrow{\text{oxidation}} H_2SO_4$$
Sulfurous acid    Sulfuric acid

*Acid rain* in North America is most severe in the east, downwind of many coal-fired power plants. In the 3-year period 1995–1997, after $SO_2$ emissions were limited by a new law, concentrations of $SO_4^{2-}$ and $H^+$ in precipitation decreased by 10–25% in the eastern United States.[18]

Acid rain threatens lakes and forests throughout the world. Monitoring the pH of rainwater is a critical component of programs to measure and reduce the production of acid rain.

To identify systematic errors in the measurement of pH of rainwater, a careful study was conducted with 17 laboratories.[19] Eight samples were provided to each laboratory, along with instructions on how to conduct the measurements. Each laboratory used two buffers to standardize pH meters. Sixteen laboratories successfully measured the pH of Unknown A (within $\pm 0.02$ pH unit), which was 4.008 at 25°C. One lab whose measurement was 0.04 pH unit low had a faulty commercial standard buffer.

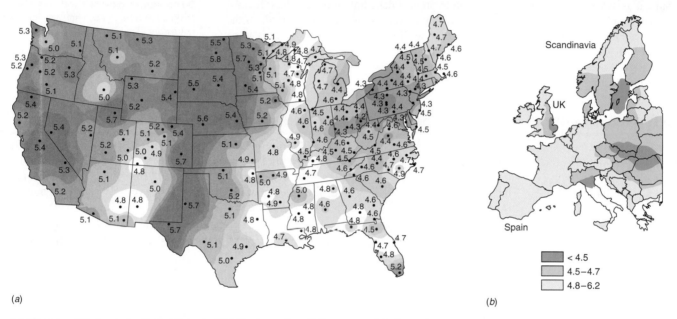

(a)

(b)

(a) pH of precipitation in the United States in 2001. The lower the pH, the more acidic the water. *[From National Atmospheric Deposition Program (NRSP-3)/National Trends Network (2002). Illinois State Water Survey, 2204 Griffith Dr. Champaign, IL 61820. See also http://nadp.sws.uiuc.edu and www.epa.gov/acidrain.]*
(b) pH of rain in Europe. No values are reported for Italy and Greece. *[From H. Rodhe, F. Dentener, and M. Schulz, Environ. Sci. Technol.* **2002,** *36, 4382.]*

Panel (*c*) shows typical results for the pH of rainwater. The average of the 17 measurements is given by the horizontal line at pH 4.14 and the letters s, t, u, v, w, x, y, and z identify types of pH electrodes. Types s and w had relatively large systematic errors. The type s electrode was a combination electrode (Figure 15-9) containing a reference electrode liquid junction with an exceptionally large area. Electrode type w had a reference electrode filled with a gel.

A hypothesis was that variations in the liquid junction potential (Section 15-3) led to variations among the pH measurements.

Standard buffers have ionic strengths of 0.05 to 0.1 M, whereas rainwater samples have ionic strengths two or more orders of magnitude lower. To see whether junction potential caused systematic errors, $2 \times 10^{-4}$ M HCl was used as a pH standard in place of high ionic strength buffers. Panel (*d*) shows good results from all but the first lab. The standard deviation of 17 measurements was reduced from 0.077 pH unit (with standard buffer) to 0.029 pH unit (with HCl standard). It was concluded that junction potential caused most of the variability between labs and that a low ionic strength standard is appropriate for rainwater pH measurements.[20,21]

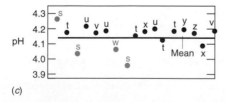

(c)

(c) pH of rainwater from identical samples measured at 17 different labs using standard buffers for calibration. Letters designate different types of pH electrodes.

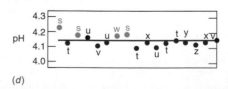

(d)

(d) Rainwater pH measured by using low ionic strength HCl for calibration.

If electrode response becomes sluggish or if an electrode cannot be calibrated properly, try soaking it in 6 M HCl, followed by water. As a last resort, soak the electrode in 20 wt% aqueous ammonium bifluoride, $NH_4HF_2$, for 1 min in a plastic beaker. This reagent dissolves glass and exposes fresh surface. Wash the electrode with water and try calibrating it again. Avoid contact with ammonium bifluoride, which produces a painful HF burn.

### Errors in pH Measurement

1. *Standards.* A pH measurement cannot be more accurate than our standards, which are typically ±0.01 pH unit.
2. *Junction potential.* A *junction potential* exists at the porous plug near the bottom of the electrode in Figure 15-9. If the ionic composition of the analyte solution is different from that of the standard buffer, the junction potential will change *even if the pH of the two solutions is the same* (Box 15-1). This effect gives an uncertainty of at least ~0.01 pH unit.
3. *Junction potential drift.* Most combination electrodes have a Ag | AgCl reference electrode containing saturated KCl solution. More than 350 mg Ag/L dissolve in the KCl, mainly as $AgCl_4^{3-}$ and $AgCl_3^{2-}$. In the porous plug, KCl is diluted and AgCl can precipitate. If analyte solution contains a reducing agent, $Ag(s)$ also can precipitate in the plug. Both effects change the junction potential, causing a slow drift of the pH reading (solid colored circles in Figure 15-14). You can compensate for this error by recalibrating the electrode every 2 h.

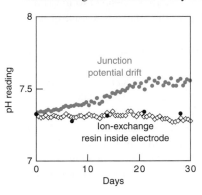

**Figure 15-14** Solid colored circles show the drift in apparent pH of a low-conductivity industrial water supply measured continuously by a single electrode. Individual measurements with a freshly calibrated electrode (black circles) demonstrate that the pH is not drifting. Drift is attributed to slow clogging of the electrode's porous plug with AgCl(s). When a cation-exchange resin was placed inside the reference electrode near the porous plug, Ag(I) was bound by the resin and did not precipitate. This electrode gave the drift-free, continuous reading shown by open diamonds. *[From S. Ito, H. Hachiya, K. Baba, Y. Asano, and H. Wada, "Improvement of the Ag | AgCl Reference Electrode and Its Application to pH Measurement," Talanta* **1995,** *42, 1685.]*

4. *Sodium error.* When [$H^+$] is very low and [$Na^+$] is high, the electrode responds to $Na^+$ and the apparent pH is lower than the true pH. This is called the **sodium error** or *alkaline error* (Figure 15-15).
5. *Acid error.* In strong acid, the measured pH is higher than the actual pH, perhaps because the glass is saturated with $H^+$ and cannot be further protonated (Figure 15-15).
6. *Equilibration time.* It takes time for an electrode to equilibrate with a solution. A well-buffered solution requires ~30 s with adequate stirring. A poorly buffered solution (such as one near the equivalence point of a titration) needs many minutes.
7. *Hydration of glass.* A dry electrode requires several hours of soaking before it responds to $H^+$ correctly.
8. *Temperature.* A pH meter should be calibrated at the same temperature at which the measurement will be made.
9. *Cleaning.* If an electrode has been exposed to a hydrophobic liquid, such as oil, it should be cleaned with a solvent that will dissolve the liquid and then conditioned well in aqueous solution. The reading of an improperly cleaned electrode can drift for hours while the electrode re-equilibrates with aqueous solution.

Errors 1 and 2 limit the accuracy of pH measurement with the glass electrode to ±0.02 pH unit, at best. Measurement of pH *differences* between solutions can be accurate to about ±0.002 pH unit, but knowledge of the true pH will still be at least an order of magnitude more uncertain. An uncertainty of ±0.02 pH unit corresponds to an uncertainty of ±5% in $\mathcal{A}_{H^+}$.

### ■ ■ ■ ■ 15-6 Ion-Selective Electrodes[23]

A critically ill patient is wheeled into the emergency room, and the doctor needs blood chemistry information quickly to help her make a diagnosis. Analytes in Table 15-4 are part of the critical care profile of blood chemistry. Every analyte in the table can be measured by

---

The apparent pH will charge if the ionic composition of the analyte changes, even when the actual pH is constant.

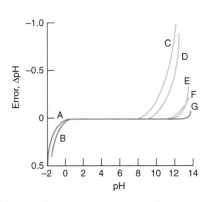

**Figure 15-15** Acid and alkaline errors of some glass electrodes. A: Corning 015, $H_2SO_4$. B: Corning 015, HCl. C: Corning 015, 1 M $Na^+$. D: Beckman-GP, 1 M $Na^+$. E: L & N Black Dot, 1 M $Na^+$. F: Beckman Type E, 1 M $Na^+$. G: Ross electrode.[22] *[From R. G. Bates, Determination of pH: Theory and Practice, 2nd ed. (New York: Wiley, 1973). Ross electrode data are from Orion, Ross pH Electrode Instruction Manual.]*

**Challenge** Use Equation 15-4 to show that the potential of the glass electrode changes by 1.3 mV when $\mathcal{A}_{H^+}$ changes by 5.0%. Show that 1.3 mV = 0.02 pH unit.

*Moral:* A small uncertainty in voltage (1.3 mV) or pH (0.02 unit) corresponds to a large uncertainty (5%) in analyte concentration. Similar uncertainties arise in other potentiometric measurements.

electrochemical means. Ion-selective electrodes are the method of choice for $Na^+$, $K^+$, $Cl^-$, pH, and $P_{CO_2}$. The "Chem 7" test constitutes up to 70% of tests performed in the hospital lab. It measures $Na^+$, $K^+$, $Cl^-$, total $CO_2$, glucose, urea, and creatinine, four of which are analyzed with ion-selective electrodes.

Most ion-selective electrodes fall into one of the following classes:

1.  *Glass membranes* for $H^+$ and certain monovalent cations
2.  *Solid-state electrodes* based on inorganic salt crystals
3.  *Liquid-based electrodes* using a hydrophobic polymer membrane saturated with a hydrophobic liquid ion exchanger
4.  *Compound electrodes* with a species-selective electrode enclosed by a membrane that is able to separate that species from others or that generates the species in a chemical reaction.

We will discuss the last three types of electrodes as soon as we introduce the language for characterizing the selectivity of an electrode.

## Selectivity Coefficient

No electrode responds exclusively to one kind of ion, but the glass pH electrode is among the most selective. Sodium ion is the principal interfering species, and its effect on the pH reading is only significant when $[H^+] \lesssim \sim 10^{-12}$ M and $[Na^+] \gtrsim \sim 10^{-2}$ M (Figure 15-15).

An electrode intended to measure ion A also responds to ion X. The **selectivity coefficient** gives the relative response of the electrode to different species with the same charge:

*Selectivity coefficient:*
$$k_{A,X} = \frac{\text{response to X}}{\text{response to A}} \qquad (15\text{-}7)$$

The smaller the selectivity coefficient, the less the interference by X. A $K^+$ ion-selective electrode that uses the chelator valinomycin as the liquid ion exchanger has selectivity coefficients $k_{K^+,Na^+} = 1 \times 10^{-5}$, $k_{K^+,Cs^+} = 0.44$, and $k_{K^+,Rb^+} = 2.8$. These coefficients tell us that $Na^+$ hardly interferes with the measurement of $K^+$, but $Cs^+$ and $Rb^+$ interfere strongly. In fact, the electrode responds more to $Rb^+$ than to $K^+$.

For interfering ions with the same charge as the primary ion, the response of ion-selective electrodes is described by the equation[12,24]

*Response of ion-selective electrode:*
$$E = \text{constant} \pm \beta \frac{0.059\,16}{n} \log\left[ \mathcal{A}_A + \sum_X (k_{A,X}\mathcal{A}_X) \right] \qquad (15\text{-}8)$$

where $\mathcal{A}_A$ is the activity of the primary ion (A), $\mathcal{A}_X$ is the activity of an interfering species (X), and $k_{A,X}$ is the selectivity coefficient. The magnitude of the charge of A is $n$. If the ion-selective electrode is connected to the positive terminal of the potentiometer, the sign of the log term is positive if A is a cation and negative if A is an anion. $\beta$ is near 1 for most electrodes.

---

**Example** Using the Selectivity Coefficient

A fluoride ion-selective electrode has a selectivity coefficient $k_{F^-,OH^-} = 0.1$. What will be the change in electrode potential when $1.0 \times 10^{-4}$ M $F^-$ at pH 5.5 is raised to pH 10.5?

**Solution** Using Equation 15-8 with $\beta = 1$, we find that the potential with negligible $OH^-$ at pH 5.5 is

$$E = \text{constant} - 0.059\,16 \log[1.0 \times 10^{-4}] = \text{constant} + 236.6 \text{ mV}$$

At pH 10.50, $[OH^-] = 3.2 \times 10^{-4}$ M, so the electrode potential is

$$E = \text{constant} - 0.059\,16 \log[1.0 \times 10^{-4} + (0.1)(3.2 \times 10^{-4})]$$
$$= \text{constant} + 229.5 \text{ mV}$$

The change is $229.5 - 236.6 = -7.1$ mV, which is quite significant. If you didn't know about the pH change, you would think that the concentration of $F^-$ had increased by 32%.

---

**Table 15-4**  Critical care profile

| Function | Analyte |
|---|---|
| Conduction | $K^+$, $Ca^{2+}$ |
| Contraction | $Ca^{2+}$, $Mg^{2+}$ |
| Energy level | Glucose, $P_{O_2}$, lactate, hematocrit |
| Ventilation | $P_{O_2}$, $P_{CO_2}$ |
| Perfusion | Lactate, $SO_2\%$, hematocrit |
| Acid-base | pH, $P_{CO_2}$, $HCO_3^-$ |
| Osmolality | $Na^+$, glucose |
| Electrolyte balance | $Na^+$, $K^+$, $Ca^{2+}$, $Mg^{2+}$ |
| Renal function | Blood urea nitrogen, creatinine |

SOURCE: C. C. Young, "Evolution of Blood Chemistry Analyzers Based on Ion Selective Electrodes," *J. Chem. Ed.* **1997**, *74*, 177.

### Reminder: How Ion-Selective Electrodes Work

In Figure 15-8, analyte ions equilibrate with ion-exchange sites at the outer surface of the ion-selective membrane. Diffusion of analyte ions out of the membrane creates a slight charge imbalance (an electric potential difference) across the interface between the membrane and the analyte solution. Changes in analyte ion concentration in the solution change the potential difference across the outer boundary of the ion-selective membrane. By using a calibration curve, we can relate the potential difference to analyte concentration.

An ion-selective electrode responds to the activity of *free analyte*, not complexed analyte. For example, when the $Pb^{2+}$ in tap water at pH 8 was measured with a sensitive ion-selective electrode, the result was $[Pb^{2+}] = 2 \times 10^{-10}$ M.[25] When lead in the same tap water was measured by inductively coupled plasma–mass spectrometry (Section 21-6), the result was more than 10 times greater: $3 \times 10^{-9}$ M. The discrepancy arose because the inductively coupled plasma measures *all* lead and the ion-selective electrode measures *free* $Pb^{2+}$. In tap water at pH 8, much of the lead is complexed by $CO_3^{2-}$, $OH^-$, and other anions. When the pH of tap water was adjusted to 4, $Pb^{2+}$ dissociated from its complexes and the concentration indicated by the ion-selective electrode was $3 \times 10^{-9}$ M—equal to that measured by inductively coupled plasma.

Analyte ions establish an ion-exchange equilibrium at the surface of the ion-selective membrane. Other ions that bind to the same sites interfere with the measurement.

The ion-selective electrode responds to $Pb^{2+}$ with little response to $Pb(OH)^+$ or $Pb(CO_3)(aq)$.

### Solid-State Electrodes

A **solid-state ion-selective electrode** based on an inorganic crystal is shown in Figure 15-16. A common electrode of this type is the fluoride electrode, employing a crystal of $LaF_3$ doped with $Eu^{2+}$. *Doping* means adding a small amount of $Eu^{2+}$ in place of $La^{3+}$. The filling solution contains 0.1 M NaF and 0.1 M NaCl. Fluoride electrodes are used to monitor and control the fluoridation of municipal water supplies.

To conduct a tiny electric current, $F^-$ migrates across the $LaF_3$ crystal, as shown in Figure 15-17. Anion vacancies are created within the crystal when we dope $LaF_3$ with $EuF_2$. An adjacent fluoride ion can jump into the vacancy, leaving a new vacancy behind. In this manner, $F^-$ diffuses from one side to the other.

By analogy with the pH electrode, the response of the $F^-$ electrode is

*Response of $F^-$ electrode:*  $\quad E = \text{constant} - \beta(0.059\ 16) \log \mathcal{A}_{F^-}(\text{outside})$  (15-9)

where $\beta$ is close to 1.00. Equations 15-9 and 15-4 have opposite signs before the log term because one equation applies to anions and one to cations. The $F^-$ electrode has a nearly Nernstian response over a $F^-$ concentration range from about $10^{-6}$ M to 1 M (Figure 15-18). The electrode is more responsive to $F^-$ than to most other ions by $>1\ 000$. The only interfering species is $OH^-$, for which the selectivity coefficient is $k_{F^-,OH^-} = 0.1$. At low pH, $F^-$ is converted into HF ($pK_a = 3.17$), to which the electrode is insensitive.

A routine procedure for measuring $F^-$ is to dilute the unknown in a high ionic strength buffer containing acetic acid, sodium citrate, NaCl, and NaOH to adjust the pH to 5.5. The buffer keeps all standards and unknowns at a constant ionic strength, so the fluoride activity coefficient is constant in all solutions (and can therefore be ignored).

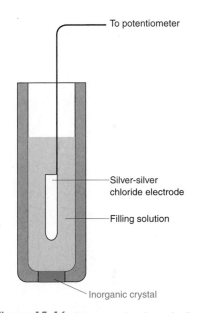

**Figure 15-16** Diagram of an ion-selective electrode employing an inorganic salt crystal as the ion-selective membrane.

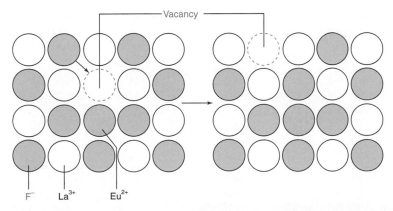

**Figure 15-17** Migration of $F^-$ through $LaF_3$ doped with $EuF_2$. Because $Eu^{2+}$ has less charge than $La^{3+}$, an anion vacancy occurs for every $Eu^{2+}$. A neighboring $F^-$ can jump into the vacancy, thereby moving the vacancy to another site. Repetition of this process moves $F^-$ through the lattice.

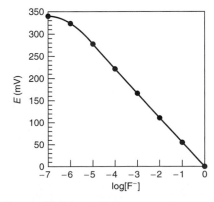

**Figure 15-18** Calibration curve for fluoride ion-selective electrode. [Data from M. S. Frant and J. W. Ross, Jr., "Electrode for Sensing Fluoride Ion Activity in Solution," Science **1966**, 154, 1553.]

$$E = \text{constant} - \beta(0.059\ 16)\log[F^-]\gamma_{F^-}$$

$$= \underbrace{\text{constant} - \beta(0.059\ 16)\log\gamma_{F^-}}_{\substack{\text{This expression is constant because} \\ \gamma_{F^-} \text{ is constant at constant ionic strength}}} - \beta(0.059\ 16)\log[F^-]$$

At pH 5.5, there is no interference by $OH^-$ and little conversion of $F^-$ into HF. Citrate complexes $Fe^{3+}$ and $Al^{3+}$, which would otherwise bind $F^-$ and interfere with the analysis.

---

**Example** Response of an Ion-Selective Electrode

When a fluoride electrode was immersed in standard solutions (maintained at a constant ionic strength of 0.1 M with $NaNO_3$), the following potentials (versus S.C.E.) were observed:

| $[F^-]$ (M) | $E$ (mV) |
|---|---|
| $1.00 \times 10^{-5}$ | 100.0 |
| $1.00 \times 10^{-4}$ | 41.5 |
| $1.00 \times 10^{-3}$ | −17.0 |

Because the ionic strength is constant, the response should depend on the logarithm of the $F^-$ *concentration*. Find $[F^-]$ in an unknown that gave a potential of 0.0 mV.

**Solution** We fit the calibration data with Equation 15-9:

$$E = m \underbrace{\log[F^-]}_{x} + b$$

where $E$ is $y$.

Plotting $E$ versus $\log[F^-]$ gives a straight line with a slope $m = -58.5$ mV and a $y$-intercept $b = -192.5$ mV. Setting $E = 0.0$ mV, we solve for $[F^-]$:

$$0.0 = (-58.5)\log[F^-] - 192.5 \Rightarrow [F^-] = 5.1 \times 10^{-4}\ M$$

---

Another common inorganic crystal electrode uses $Ag_2S$ for the membrane. This electrode responds to $Ag^+$ and to $S^{2-}$. If we dope the electrode with CuS, CdS, or PbS, we can prepare electrodes sensitive to $Cu^{2+}$, $Cd^{2+}$, or $Pb^{2+}$, respectively (Table 15-5).

Figure 15-19 illustrates the mechanism by which a CdS crystal responds selectively to certain ions. The crystal can be cleaved to expose planes of Cd atoms or S atoms. The Cd plane in Figure 15-19a selectively adsorbs $HS^-$ ions, whereas an S plane does not interact strongly with $HS^-$. Figure 15-19b shows strong response of the exposed Cd face to $HS^-$, but only weak response when the S face is exposed. The opposite behavior is observed in response to $Cd^{2+}$ ions. The partial response of the S face to $HS^-$ in the upper curve is attributed to about 10% of the exposed atoms actually being Cd instead of S.

**Table 15-5** Properties of solid-state ion-selective electrodes

| Ion | Concentration range (M) | Membrane material | pH range | Interfering species |
|---|---|---|---|---|
| $F^-$ | $10^{-6}$–1 | $LaF_3$ | 5–8 | $OH^-$ (0.1 M) |
| $Cl^-$ | $10^{-4}$–1 | AgCl | 2–11 | $CN^-$, $S^{2-}$, $I^-$, $S_2O_3^{2-}$, $Br^-$ |
| $Br^-$ | $10^{-5}$–1 | AgBr | 2–12 | $CN^-$, $S^{2-}$, $I^-$ |
| $I^-$ | $10^{-6}$–1 | AgI | 3–12 | $S^{2-}$ |
| $SCN^-$ | $10^{-5}$–1 | AgSCN | 2–12 | $S^{2-}$, $I^-$, $CN^-$, $Br^-$, $S_2O_3^{2-}$ |
| $CN^-$ | $10^{-6}$–$10^{-2}$ | AgI | 11–13 | $S^{2-}$, $I^-$ |
| $S^{2-}$ | $10^{-5}$–1 | $Ag_2S$ | 13–14 | |

## Liquid-Based Ion-Selective Electrodes

A **liquid-based ion-selective electrode** is similar to the solid-state electrode in Figure 15-16, except that the liquid-based electrode has a hydrophobic membrane impregnated with

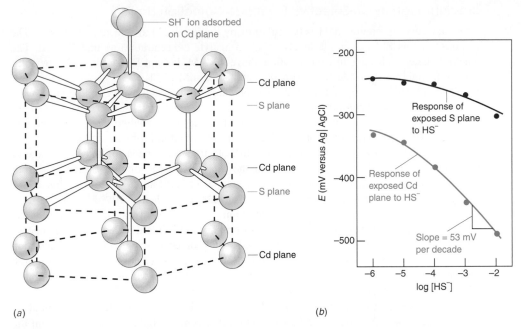

(a)                                                                 (b)

*Figure 15-19* (a) Crystal structure of hexagonal CdS, showing alternating planes of Cd and S along the vertical axis (the crystal c-axis). HS⁻ is shown adsorbed to the uppermost Cd plane. (b) Potentiometric response of exposed crystal faces to HS⁻. *[From K. Uosaki, Y. Shigematsu, H. Kita, Y. Umezawa, and R. Souda, "Crystal-Face-Specific Response of a Single-Crystal Cadmium Sulfide Based Ion-Selective Electrode," Anal. Chem.* **1989**, *61, 1980.]*

a hydrophobic ion exchanger (called an *ionophore*) that is selective for analyte ion (Figure 15-20). The response of a $Ca^{2+}$ ion-selective electrode is given by

*Response of $Ca^{2+}$*
*electrode:* $$E = \text{constant} + \beta\left(\frac{0.059\ 16}{2}\right)\log \mathcal{A}_{Ca^{2+}}(\text{outside}) \qquad (15\text{-}10)$$

where $\beta$ is close to 1.00. Equations 15-10 and 15-9 have different signs before the log term, because one involves an anion and the other a cation. Note also that the charge of $Ca^{2+}$ requires a factor of 2 in the denominator before the logarithm.

The membrane at the bottom of the electrode in Figure 15-20 is made from poly(vinyl chloride) impregnated with ion exchanger. One particular $Ca^{2+}$ liquid ion exchanger consists of the neutral hydrophobic ligand (L) and a salt of the hydrophobic anion ($Na^+R^-$) dissolved in a hydrophobic liquid:

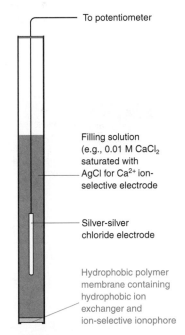

*Figure 15-20* Calcium ion-selective electrode with a liquid ion exchanger.

Hydrophobic anion (R⁻)
Tetrakis[3,5-bis(trifluoromethyl)phenyl]borate

Hydrophobic liquid solvent
2-Nitrophenyl octyl ether

Hydrophobic $Ca^{2+}$-binding ligand (L)
*N, N*-Dicyclohexyl-*N'*,*N'*-dioctadecyl-3-oxapentanediamide

The most serious interference for this $Ca^{2+}$ electrode comes from $Sr^{2+}$. The selectivity coefficient in Equation 15-7 is $k_{Ca^{2+},Sr^{2+}} = 0.13$, which means that the response to $Sr^{2+}$ is 13% as great as the response to an equal concentration of $Ca^{2+}$. For most cations, $k < 10^{-3}$.

The heparin-sensitive electrode at the beginning of this chapter is similar to the $Ca^{2+}$ electrode, but the ion exchanger in the membrane is tridodecylmethylammonium chloride, $(C_{12}H_{25})_3NCH_3^+Cl^-$. $Cl^-$ in the membrane exchanges with the negatively charged heparin.

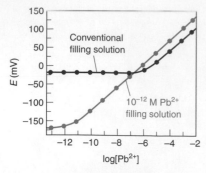

**Figure 15-21** Response of $Pb^{2+}$ liquid-based ion-selective electrode with (*black curve*) conventional filling solution containing 0.5 mM $Pb^{2+}$ or (*colored curve*) metal ion buffer filling solution in which $[Pb^{2+}] = 10^{-12}$ M. [From T. Sokalski, A. Ceresa, T. Zwickl, and E. Pretsch, "Large Improvement of the Lower Detection Limit of Ion-Selective Polymer Membrane Electrodes," J. Am. Chem. Soc. **1997**, 119, 11347.]

Lower detection limits for $Ca^{2+}$, $Cd^{2+}$, $Ag^+$, $K^+$, $Na^+$, $I^-$, and $ClO_4^-$ ion-selective electrodes were demonstrated when concentrations were reduced in the internal filling solutions.[26] A future improvement will come when the ionophore is dissolved in a conductive polymer in direct electrical contact with a metal conductor.[27] This electrode entirely omits the inner filling solution.

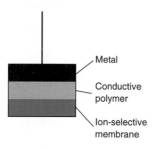

Metal

Conductive polymer

Ion-selective membrane

## Breakthrough in Ion-Selective Electrode Detection Limits[25]

The black curve in Figure 15-21 is typical of many liquid-based ion-selective electrodes. The response of this $Pb^{2+}$ electrode levels off at an analyte concentration around $10^{-6}$ M. The electrode detects changes in concentration above $10^{-6}$ M, but not below $10^{-6}$ M. The solution in the internal electrode compartment contains 0.5 mM $PbCl_2$.

The colored curve in Figure 15-21 was obtained with the same electrode components, but the internal filling solution was replaced by a *metal ion buffer* (Section 15-7) that fixes $[Pb^{2+}]$ at $10^{-12}$ M. Now the electrode responds to changes in analyte $Pb^{2+}$ concentration down to $\sim 10^{-11}$ M. The buffer for this electrode is described in Problem 15-43.

The sensitivity of liquid-based ion-selective electrodes has been limited by leakage of the primary ion ($Pb^{2+}$, in this case) from the internal filling solution through the ion-exchange membrane. Leakage provides a substantial concentration of primary ion at the external surface of the membrane. If analyte concentration is below $10^{-6}$ M, leakage from the electrode maintains an effective concentration near $10^{-6}$ M at the outer surface of the electrode. If we can lower the concentration of primary ion inside the electrode, the concentration of leaking ion outside the membrane is reduced by orders of magnitude and the detection limit is correspondingly reduced. The sensitivity of a solid-state electrode cannot be lowered by changing the filling solution, because analyte concentration is governed by the solubility of the inorganic salt crystal forming the ion-sensitive membrane.

The response of the electrode with $10^{-12}$ M $Pb^{2+}$ in the filling solution is limited by interference from $Na^+$ from the internal solution that contains 0.05 M $Na_2EDTA$ from the metal ion buffer. When the filling solution is buffered to $10^{-12}$ M $Pb^{2+}$, the apparent selectivity coefficient (Equation 15-7) is decreased by one to five orders of magnitude for most interfering cations. Not only is the detection limit for $Pb^{2+}$ improved by $10^5$, but the observed selectivity for $Pb^{2+}$ over other cations increases by several orders of magnitude.

## Compound Electrodes

**Compound electrodes** contain a conventional electrode surrounded by a membrane that isolates (or generates) the analyte to which the electrode responds. The $CO_2$ gas-sensing electrode in Figure 15-22 consists of an ordinary glass pH electrode surrounded by a thin layer of electrolyte solution enclosed in a semipermeable membrane made of rubber, Teflon, or polyethylene. A $Ag \mid AgCl$ reference electrode is immersed in the electrolyte solution. When $CO_2$ diffuses through the semipermeable membrane, it lowers the pH in the electrolyte. The response of the glass electrode to the change in pH is a measure of the $CO_2$ concentration outside the electrode.[28] Other acidic or basic gases, including $NH_3$, $SO_2$, $H_2S$, $NO_x$ (nitrogen

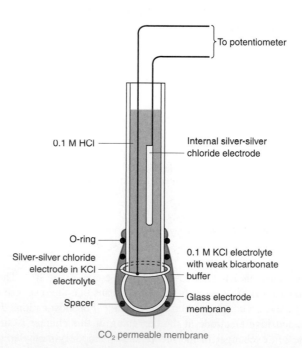

To potentiometer

0.1 M HCl

Internal silver-silver chloride electrode

O-ring

Silver-silver chloride electrode in KCl electrolyte

Spacer

0.1 M KCl electrolyte with weak bicarbonate buffer

Glass electrode membrane

$CO_2$ permeable membrane

**Figure 15-22** $CO_2$ gas-sensing electrode. The membrane is stretched taut and there is a thin layer of electrolyte between the membrane and the glass bulb.

oxides), and $HN_3$ (hydrazoic acid) can be detected in the same manner. These electrodes can be used to measure gases in solution or *in the gas phase.*

Numerous ingenious compound electrodes using *enzymes* have been built.[29] These devices contain a conventional electrode coated with an enzyme that catalyzes a reaction of the analyte. The product of the reaction is detected by the electrode.

Acidic or basic gases are detected by a pH electrode surrounded by an electrolyte and enclosed in a gas-permeable membrane.

# ■ ■ ■ 15-7   Using Ion-Selective Electrodes

Ion-selective electrodes respond linearly to the logarithm of analyte activity over four to six orders of magnitude. Electrodes do not consume unknowns, and they introduce negligible contamination. Response time is seconds or minutes, so electrodes are used to monitor flow streams in industrial applications. Color and turbidity do not hinder electrodes. Microelectrodes can be used inside living cells.

Precision is rarely better than 1%, and usually it is worse. Electrodes can be fouled by proteins or other organic solutes, which lead to sluggish, drifting response. Certain ions interfere with or poison particular electrodes. Some electrodes are fragile and have limited shelf life.

Electrodes respond to the *activity* of *uncomplexed* analyte ion. Therefore, ligands must be absent or masked. Because we usually wish to know concentrations, not activities, an inert salt is often used to bring all standards and samples to a high, constant ionic strength. If activity coefficients are constant, the electrode potential gives concentrations directly.

Advantages of ion-selective electrodes:
1. Linear response to log $\mathcal{A}$ over a wide range
2. Nondestructive
3. Noncontaminating
4. Short response time
5. Unaffected by color or turbidity

A 1-mV error in potential corresponds to a 4% error in monovalent ion activity. A 5-mV error corresponds to a 22% error. The relative error *doubles* for divalent ions and *triples* for trivalent ions.

Electrodes respond to the *activity* of *uncomplexed* ion. If ionic strength is constant, concentration is proportional to activity and the electrode measures concentration.

## Standard Addition with Ion-Selective Electrodes

When ion-selective electrodes are used, it is important that the composition of the standard solution closely approximates the composition of the unknown. The medium in which the analyte exists is called the **matrix.** For complex or unknown matrixes, the **standard addition** method (Section 5-3) can be used. In this technique, the electrode is immersed in unknown and the potential is recorded. Then a small volume of standard solution is added, so as not to perturb the ionic strength of the unknown. The change in potential tells how the electrode responds to analyte and, therefore, how much analyte was in the unknown. It is best to add several successive aliquots and use a graphical procedure to extrapolate back to the concentration of unknown. Standard addition is best if the additions increase analyte to 1.5 to 3 times its original concentration.

The graphical procedure is based on the equation for the response of the ion-selective electrode, which we will write in the form

$$E = k + \beta \left( \frac{RT \ln 10}{nF} \right) \log[X] \qquad (15\text{-}11)$$

where $E$ is the meter reading in volts and $[X]$ is the concentration of analyte. This reading is the difference in potential of the ion-selective electrode and the reference electrode. The constants $k$ and $\beta$ depend on the particular ion-selective electrode. The factor $(RT/F) \ln 10$ is 0.059 16 V at 298.15 K. If $\beta = 1$, then the response is Nernstian. We abbreviate the term $(\beta RT/nF) \ln 10$ as $S$.

Let the initial volume of unknown be $V_0$ and the initial concentration of analyte be $c_X$. Let the volume of added standard be $V_S$ and the concentration of standard be $c_S$. Then the total concentration of analyte after standard is added is $(V_0 c_X + V_S c_S)/(V_0 + V_S)$. Substituting this expression for $[X]$ in Equation 15-11 and doing some rearrangement give

*Standard addition plot for ion-selective electrode:*

$$\underbrace{(V_0 + V_S)10^{E/S}}_{y} = \underbrace{10^{k/S}V_0 c_X}_{b} + \underbrace{10^{k/S}c_S V_S}_{m} \overset{\uparrow}{\underset{x}{}} \qquad (15\text{-}12)$$

A graph of $(V_0 + V_S)10^{E/S}$ on the $y$-axis versus $V_S$ on the $x$-axis has a slope $m = 10^{k/S}c_S$ and a $y$-intercept of $10^{k/S}V_0 c_X$. The $x$-intercept is found by setting $y = 0$:

$$x\text{-intercept} = -\frac{b}{m} = -\frac{10^{k/S}V_0 c_X}{10^{k/S}c_S} = -\frac{V_0 c_X}{c_S} \qquad (15\text{-}13)$$

Equation 15-13 gives us the concentration of unknown, $c_X$, from $V_0$, $c_S$, and the $x$-intercept. Figure 15-23 is a graph of Equation 15-12 used in Exercise 15-F.

A weakness of standard addition with ion-selective electrodes is that we cannot measure $\beta$ in Equation 15-11 in the unknown matrix. We could measure $\beta$ in a series of standard solutions (not containing unknown) and use this value to compute S in the function

$R$ = gas constant
$T$ = temperature (K)
$n$ = charge of the ion being detected
$F$ = Faraday constant

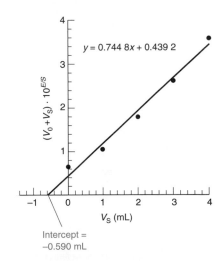

$y = 0.744\ 8x + 0.439\ 2$

Intercept = −0.590 mL

**Figure 15-23** Standard addition graph for ion-selective electrode based on Equation 15-12. [Data from G. Li, B. J. Polk, L. A. Meazell, and D. W. Hatchett, "ISE Analysis of Hydrogen Sulfide in Cigarette Smoke," J. Chem. Ed. **2000**, 77, 1049.]

$(V_0 + V_S)10^{E/S}$ in Equation 15-12. Another procedure is to add a concentrated, known matrix to the unknown and to all standards so that the matrix is essentially the same in all solutions.

### Metal Ion Buffers

It is pointless to dilute $CaCl_2$ to $10^{-6}$ M for standardizing an ion-selective electrode. At this low concentration, $Ca^{2+}$ will be lost by adsorption on glass or reaction with impurities.

An alternative is to prepare a **metal ion buffer** from the metal and a suitable ligand. For example, consider the reaction of $Ca^{2+}$ with nitrilotriacetic acid (NTA) at a pH high enough for NTA to be in its fully basic form ($NTA^{3-}$):

$$Ca^{2+} + NTA^{3-} \rightleftharpoons CaNTA^-$$

$$K_f = \frac{[CaNTA^-]}{[Ca^{2+}][NTA^{3-}]} = 10^{6.46} \text{ in 0.1 M } KNO_3 \qquad (15\text{-}14)$$

If equal concentrations of $NTA^{3-}$ and $CaNTA^-$ are present in a solution,

$$[Ca^{2+}] = \frac{[\cancel{CaNTA^-}]}{K_f[\cancel{NTA^{3-}}]} = 10^{-6.46} \text{ M}$$

> ### Example  Preparing a Metal Ion Buffer
>
> What concentration of $NTA^{3-}$ should be added to $1.0_0 \times 10^{-2}$ M $CaNTA^-$ in 0.1 M $KNO_3$ to give $[Ca^{2+}] = 1.0_0 \times 10^{-6}$ M?
>
> **Solution**  From Equation 15-14, we write
>
> $$[NTA^{3-}] = \frac{[CaNTA^-]}{K_f[Ca^{2+}]} = \frac{1.00 \times 10^{-2}}{(10^{6.46})(1.00 \times 10^{-6})} = 3.4_7 \times 10^{-3} \text{ M}$$
>
> These are practical concentrations of $CaNTA^-$ and of $NTA^{3-}$.

A metal ion buffer is the only way to get the concentration $[Pb^{2+}] = 10^{-12}$ M for the filling solution in the electrode in Figure 15-21.

## ■ ■ ■ ■  15-8  Solid-State Chemical Sensors

Solid-state chemical sensors are fabricated by the same technology used for microelectronic chips. The field effect transistor (FET) is the heart of commercially available sensors such as the pH electrode in Figure 15-24.

### Semiconductors and Diodes

**Semiconductors** such as Si (Figure 15-25), Ge, and GaAs are materials whose electrical *resistivity*[30] lies between those of conductors and insulators. The four valence electrons of the pure materials are all involved in bonds between atoms (Figure 15-26a). A phosphorus impurity, with five valence electrons, provides one extra **conduction electron** that is free to move through the crystal (Figure 15-26b). An aluminum impurity has one less bonding electron than necessary, creating a vacancy called a **hole**, which behaves as a positive charge carrier. When a neighboring electron fills the hole, a new hole appears at an adjacent position (Figure 15-26c). A semiconductor with excess conduction electrons is called *n-type,* and one with excess holes is called *p-type.*

### Sidebar (left margin)

Glass vessels are not used for very dilute solutions, because ions are adsorbed on the glass. Plastic bottles are better than glass for dilute solutions. Adding strong acid (0.1–1 M) to any solution helps minimize adsorption of cations on the walls of the container, because $H^+$ competes with other cations for ion-exchange sites.

$$HN(CH_2CO_2H)_3^+$$

$$H_4NTA^+$$

$pK_1 = 1.1 \qquad pK_3 = 2.940$
$pK_2 = 1.650 \qquad pK_4 = 10.334$

If the pH were not so high that all ligand is in the form $NTA^{3-}$, you would need to calculate the fraction in the form $NTA^{3-}$ for use in Equation 15-14. This approach is used with $\alpha_{Y^{4-}}$ for EDTA in Section 12-2.

(a)

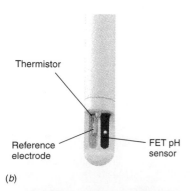

Thermistor

Reference electrode

FET pH sensor

(b)

**Figure 15-24**  (a) pH-sensitive field effect transistor. (b) Combination pH electrode based on field effect transistor. The thermistor senses temperature and is used for automatic temperature compensation. [*Courtesy Sentron, Federal Way, WA.*]

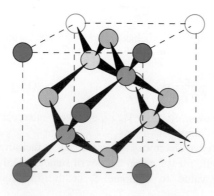

**Figure 15-25**  Diamondlike structure of silicon. Each atom is tetrahedrally bonded to four neighbors, with a Si–Si distance of 235 pm.

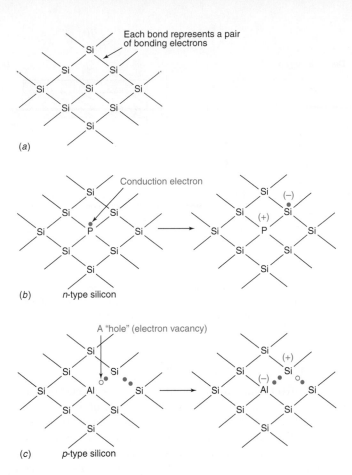

Each bond represents a pair
of bonding electrons

(a)

Conduction electron

(b)    *n*-type silicon

A "hole" (electron vacancy)

(c)    *p*-type silicon

**Figure 15-26** (*a*) The electrons of pure
silicon are all involved in the sigma-bonding
framework. (*b*) An impurity atom such as
phosphorus adds one extra electron (●),
which is relatively free to move through the
crystal. (*c*) An aluminum impurity atom lacks
one electron needed for the sigma-bonding
framework. The hole (○) introduced by the
Al atom can be occupied by an electron from
a neighboring bond, effectively moving
the hole to the neighboring bond.

A **diode** is a *pn* junction (Figure 15-27a). If *n*-Si is made negative with respect to *p*-Si,
electrons flow from the external circuit into the *n*-Si. At the *pn* junction, electrons and holes
combine. As electrons move from the *p*-Si into the circuit, a fresh supply of holes is created
in the *p*-Si. The net result is that current flows when *n*-Si is negative with respect to *p*-Si. The
diode is said to be *forward biased*.

If the polarity is reversed (Figure 15-27b), electrons are drawn out of *n*-Si and holes are
drawn out of *p*-Si, leaving a thin *depletion region* devoid of charge carriers near the *pn* junction. The diode is *reverse biased* and does not conduct current in the reverse direction.

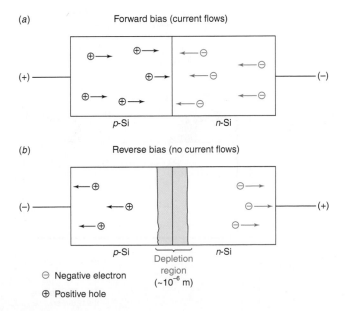

(a)    Forward bias (current flows)

(+)                                    (−)

*p*-Si          *n*-Si

(b)    Reverse bias (no current flows)

(−)                                    (+)

*p*-Si          *n*-Si

Depletion
region
($\sim 10^{-6}$ m)

⊖ Negative electron

⊕ Positive hole

An *activation energy* is needed to get the
charge carriers to move across the diode.
For Si, ~0.6 V of forward bias is required
before current will flow. For Ge, the
requirement is ~0.2 V.

For moderate reverse bias voltages, no current
flows. If the voltage is sufficiently negative,
*breakdown* occurs and current flows in the
reverse direction.

**Figure 15-27** Behavior of a *pn* junction, showing that current (*a*) can flow under forward bias
conditions, but (*b*) is prevented from flowing under reverse bias.

**15-8  Solid-State Chemical Sensors**

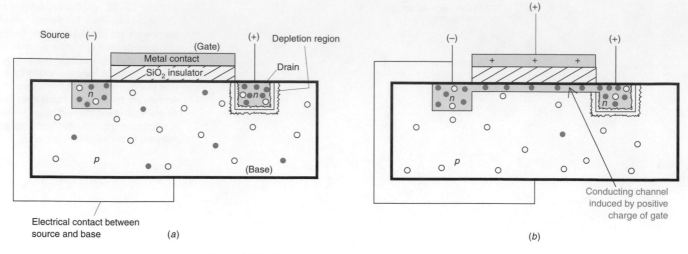

Source (−)    (Gate)    (+)   Depletion region

Metal contact

SiO₂ insulator

Drain

n

n

p

(Base)

Electrical contact between source and base

(a)

(+)

(−)        (+)

+   +   +

n

n

p

Conducting channel induced by positive charge of gate

(b)

**Figure 15-28**   Operation of a field effect transistor. (*a*) Nearly random distribution of holes and electrons in the base in the absence of gate potential. (*b*) Positive gate potential attracts electrons that form a conductive channel beneath the gate. Current can flow through this channel between source and drain.

## Chemical-Sensing Field Effect Transistor

The *base* of the **field effect transistor** in Figure 15-28 is constructed of *p*-Si with two *n*-type regions called *source* and *drain*. An insulating surface layer of SiO₂ is overcoated by a conductive metal *gate* between source and drain. The source and the base are held at the same electric potential. When a voltage is applied between source and drain (Figure 15-28a), little current flows because the drain-base interface is a *pn* junction in reverse bias.

If the gate is made positive, electrons from the base are attracted toward the gate and form a conductive channel between source and drain (Figure 15-28b). The current increases as the gate is made more positive. *The potential of the gate therefore regulates the current flow between source and drain.*

The *essential feature of the chemical-sensing field effect transistor in Figure 15-29 is the chemically sensitive layer over the gate.* An example is a layer of AgBr. When exposed to silver nitrate solution, Ag⁺ is adsorbed on the AgBr (Figure 7-11), giving it a positive charge and increasing current between source and drain. *The voltage that must be applied by an*

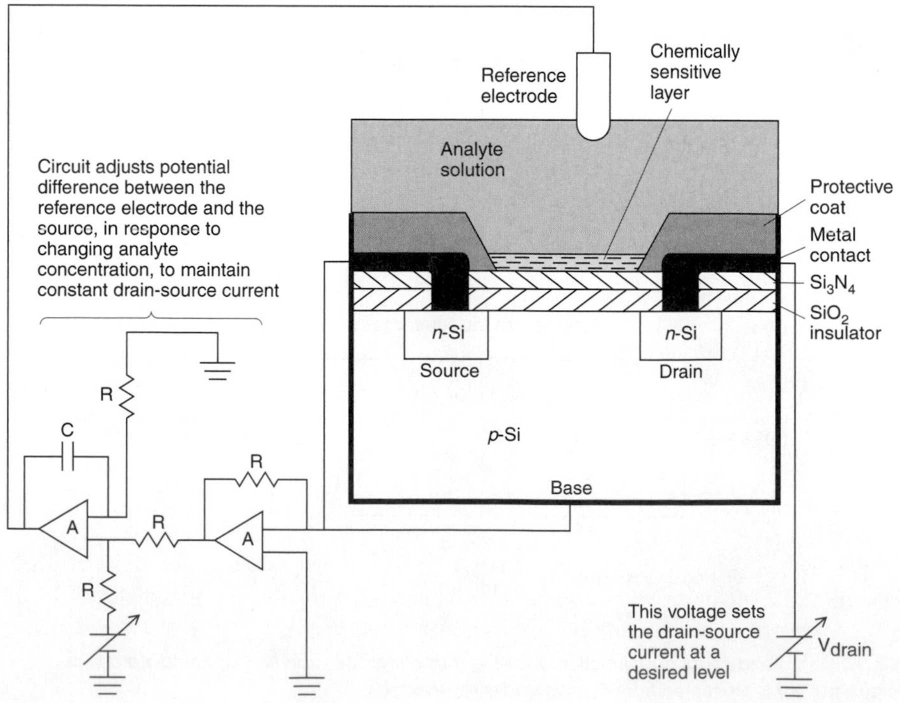

> The more positive the gate, the more current can flow between source and drain.

**Figure 15-29** Operation of a chemical-sensing field effect transistor. The transistor is coated with an insulating SiO₂ layer and a second layer of Si₃N₄ (silicon nitride), which is impervious to ions and improves electrical stability. The circuit at the lower left adjusts the potential difference between the reference electrode and the source in response to changes in the analyte solution such that a constant drain-source current is maintained.

Reference electrode

Chemically sensitive layer

Analyte solution

Protective coat

Metal contact

Si₃N₄

SiO₂ insulator

Circuit adjusts potential difference between the reference electrode and the source, in response to changing analyte concentration, to maintain constant drain-source current

*n*-Si

*n*-Si

Source

Drain

R

C

R

R

A

R

A

*p*-Si

Base

R

This voltage sets the drain-source current at a desired level

V$_{drain}$

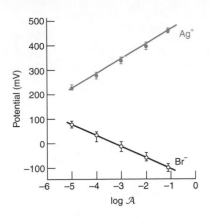

*Figure 15-30* Response of a silver bromide-coated field effect transistor. Error bars are 95% confidence intervals for data obtained from 195 sensors prepared from different chips. *[From R. P. Buck and D. E. Hackleman, "Field Effect Potentiometric Sensors," Anal. Chem. **1977**, 49, 2315.]*

*external circuit to bring the current back to its initial value is the response to Ag⁺.* Figure 15-30 shows that $Ag^+$ makes the gate more positive and $Br^-$ makes the gate more negative. The response is close to 59 mV for a 10-fold concentration change. The transistor is smaller (Figure 15-24) and more rugged than ion-selective electrodes. The sensing surface is typically only 1 mm². Chemical-sensing field effect transistors have been designed for species such as $H^+$,[31] $NH_4^+$, $Ca^{2+}$, $NO_3^-$,[32] $CO_2$,[33] pesticides (detection limit $= 10^{-12}$ M),[34,35] adrenaline,[36] and nicotinamide adenine dinucleotides [NAD(P)⁺ and NAD(P)H].[37]

## Terms to Understand

| | | | |
|---|---|---|---|
| calomel electrode | hole | metal ion buffer | silver-silver chloride electrode |
| combination electrode | indicator electrode | mobility | sodium error |
| compound electrode | ion-exchange equilibrium | potentiometry | solid-state ion-selective |
| conduction electron | ion-selective electrode | reference electrode | electrode |
| diode | junction potential | saturated calomel electrode | standard addition |
| electroactive species | liquid-based ion-selective | (S.C.E.) | |
| field effect transistor | electrode | selectivity coefficient | |
| glass electrode | matrix | semiconductor | |

## Summary

In potentiometric measurements, the indicator electrode responds to changes in the activity of analyte, and the reference electrode is a self-contained half-cell with a constant potential. The most common reference electrodes are calomel and silver-silver chloride. Common indicator electrodes include (1) the inert Pt electrode, (2) a silver electrode responsive to $Ag^+$, halides, and other ions that react with $Ag^+$, and (3) ion-selective electrodes. Unknown junction potentials at liquid-liquid interfaces limit the accuracy of most potentiometric measurements.

Ion-selective electrodes, including the glass pH electrode, respond mainly to one ion that is selectively bound to the ion-exchange membrane of the electrode. The potential difference across the membrane, $E$, depends on the activity, $\mathcal{A}_o$, of the target ion in the external analyte solution. At 25°C, the relation is $E(V) =$ constant $+$ $(0.059\ 16/n) \ln \mathcal{A}_o$, where $n$ is the charge of the target ion. For interfering ions (X) with the same charge as the primary ion (A), the response of ion-selective electrodes is $E =$ constant $\pm \beta(0.059\ 16/n)$ $\log[\mathcal{A}_A + \Sigma k_{A,X}\mathcal{A}_X]$, where $k_{A,X}$ is the selectivity coefficient for each species and $\beta$ is close to 1. Common ion-selective electrodes can be classified as solid state, liquid based, and compound. Quantitation with ion-selective electrodes is usually done with a calibration curve or by the method of standard addition. Metal ion buffers are appropriate for establishing and maintaining low concentrations of ions. A chemical-sensing field effect transistor is a solid-state device that uses a chemically sensitive coating to alter the electrical properties of a semiconductor in response to changes in the chemical environment.

## Exercises

**15-A.** The apparatus in Figure 7-9 was used to monitor the titration of 100.0 mL of solution containing 50.0 mL of 0.100 M $AgNO_3$ and 50.0 mL of 0.100 M $TlNO_3$. The titrant was 0.200 M NaBr. Suppose that the glass electrode (used as a *reference electrode* in this experiment) gives a constant potential of +0.200 V.

The glass electrode is attached to the *positive* terminal of the pH meter, and the silver wire to the negative terminal. Calculate the cell voltage at each volume of NaBr, and sketch the titration curve: 1.0, 15.0, 24.0, 24.9, 25.2, 35.0, 50.0, and 60.0 mL.

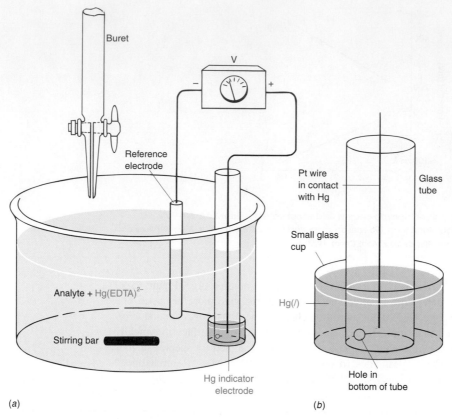

(a) Apparatus for Exercise 15-B. (b) Enlarged view of mercury electrode.

**15-B.** The apparatus in the figure above can follow the course of an EDTA titration and was used to generate the curves in Figure 12-9. The heart of the cell is a pool of liquid Hg in contact with the solution and with a Pt wire. A small amount of $HgY^{2-}$ added to the analyte equilibrates with a very tiny amount of $Hg^{2+}$:

$$Hg^{2+} + Y^{4-} \rightleftharpoons HgY^{-2}$$

$$K_f = \frac{[HgY^{2-}]}{[Hg^{2+}][Y^{4-}]} = 10^{21.5} \qquad (A)$$

The redox equilibrium $Hg^{2+} + 2e^- \rightleftharpoons Hg(l)$ is established rapidly at the surface of the Hg electrode, so the Nernst equation for the cell can be written in the form

$$E = E_+ - E_- = \left(0.852 - \frac{0.059\ 16}{2}\log\left(\frac{1}{[Hg^{2+}]}\right)\right) - E_- \qquad (B)$$

where $E_-$ is the constant potential of the reference electrode. From Equation A, $[Hg^{2+}] = [HgY^{2-}]/K_f[Y^{4-}]$, and this can be substituted into Equation B to give

$$E = 0.852 - \frac{0.059\ 16}{2}\log\left(\frac{[Y^{4-}]K_f}{[HgY^{2-}]}\right) - E_-$$

$$= 0.852 - E_- - \frac{0.059\ 16}{2}\log\left(\frac{K_f}{[HgY^{2-}]}\right) - \frac{0.059\ 16}{2}\log[Y^{4-}] \qquad (C)$$

where $K_f$ is the formation constant for $HgY^{2-}$. This apparatus thus responds to the changing EDTA concentration during an EDTA titration.

Suppose that 50.0 mL of 0.010 0 M $MgSO_4$ are titrated with 0.020 0 M EDTA at pH 10.0, using the apparatus in the figure with an S.C.E. reference electrode. Analyte contains $1.0 \times 10^{-4}$ M $Hg(EDTA)^{2-}$ added at the beginning of the titration. Calculate the cell voltage at the following volumes of added EDTA, and draw a graph of millivolts versus milliliters: 0, 10.0, 20.0, 24.9, 25.0, and 26.0 mL.

**15-C.** A solid-state fluoride ion-selective electrode responds to $F^-$, but not to HF. It also responds to hydroxide ion at high concentration when $[OH^-] \approx [F^-]/10$. Suppose that such an electrode gave a potential of +100 mV (versus S.C.E.) in $10^{-5}$ M NaF and +41 mV in $10^{-4}$ M NaF. Sketch qualitatively how the potential would vary if the electrode were immersed in $10^{-5}$ M NaF and the pH ranged from 1 to 13.

**15-D.** One commercial glass-membrane sodium ion-selective electrode has a selectivity coefficient $k_{Na^+,H^+} = 36$. When this electrode was immersed in 1.00 mM NaCl at pH 8.00, a potential of −38 mV (versus S.C.E.) was recorded.

(a) Neglecting activity coefficients and assuming $\beta = 1$ in Equation 15-8, calculate the potential if the electrode were immersed in 5.00 mM NaCl at pH 8.00.

(b) What would the potential be for 1.00 mM NaCl at pH 3.87? You can see that pH is a critical variable for the sodium electrode.

**15-E.** An ammonia gas-sensing electrode gave the following calibration points when all solutions contained 1 M NaOH.

| $NH_3$ (M) | $E$ (mV) | $NH_3$ (M) | $E$ (mV) |
|---|---|---|---|
| $1.00 \times 10^{-5}$ | 268.0 | $5.00 \times 10^{-4}$ | 368.0 |
| $5.00 \times 10^{-5}$ | 310.0 | $1.00 \times 10^{-3}$ | 386.4 |
| $1.00 \times 10^{-4}$ | 326.8 | $5.00 \times 10^{-3}$ | 427.6 |

A dry food sample weighing 312.4 mg was digested by the Kjeldahl procedure (Section 7-2) to convert all nitrogen into $NH_4^+$. The digestion solution was diluted to 1.00 L, and 20.0 mL were transferred to a 100-mL volumetric flask. The 20.0-mL aliquot was treated with 10.0 mL of 10.0 M NaOH plus enough NaI to complex the Hg catalyst from the digestion and diluted to 100.0 mL. When measured with the ammonia electrode, this solution gave a reading of 339.3 mV. Calculate the wt% of nitrogen in the food sample.

**15-F.** $H_2S$ from cigarette smoke was collected by bubbling smoke through aqueous NaOH and measured with a sulfide ion-selective electrode. Standard additions of volume $V_S$ containing $Na_2S$ at concentration $c_S = 1.78$ mM were then made to $V_0 = 25.0$ mL of unknown and the electrode response, $E$, was measured.

| $V_S$ (mL) | $E$ (V) | $V_S$ (mL) | $E$ (V) |
|---|---|---|---|
| 0 | 0.046 5 | 3.00 | 0.030 0 |
| 1.00 | 0.040 7 | 4.00 | 0.026 5 |
| 2.00 | 0.034 4 | | |

From a separate calibration curve, it was found that $\beta = 0.985$ in Equation 15-11. Using $T = 298.15$ K and $n = -2$ (the charge of $S^{2-}$), prepare a standard addition graph with Equation 15-12 and find the concentration of sulfide in the unknown.

## Problems

### Reference Electrodes

**15-1. (a)** Write the half-reactions for the silver-silver chloride and calomel reference electrodes.
**(b)** Predict the voltage for the following cell.

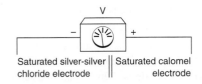

Saturated silver-silver chloride electrode ‖ Saturated calomel electrode

**15-2.** Convert the following potentials. The Ag | AgCl and calomel reference electrodes are saturated with KCl.
**(a)** 0.523 V versus S.H.E. = ? versus Ag | AgCl
**(b)** −0.111 V versus Ag | AgCl = ? versus S.H.E.
**(c)** −0.222 V versus S.C.E. = ? versus S.H.E.
**(d)** 0.023 V versus Ag | AgCl = ? versus S.C.E.
**(e)** −0.023 V versus S.C.E. = ? versus Ag | AgCl

**15-3.** Suppose that the silver-silver chloride electrode in Figure 15-2 is replaced by a saturated calomel electrode. Calculate the cell voltage if $[Fe^{2+}]/[Fe^{3+}] = 2.5 \times 10^{-3}$.

**15-4.** For a silver-silver chloride electrode, the following potentials are observed:

$$E° = 0.222 \text{ V} \qquad E(\text{saturated KCl}) = 0.197 \text{ V}$$

*Predict* the value of $E$ for a calomel electrode saturated with KCl, given that $E°$ for the calomel electrode is 0.268 V. (Your answer will not be exactly the value 0.241 used in this book.)

**15-5.** Using the following potentials, calculate the *activity* of $Cl^-$ in 1 M KCl.

$$E°(\text{calomel electrode}) = 0.268 \text{ V}$$

$$E(\text{calomel electrode, 1 M KCl}) = 0.280 \text{ V}$$

### Indicator Electrodes

**15-6.** A cell was prepared by dipping a Cu wire and a saturated calomel electrode into 0.10 M $CuSO_4$ solution. The Cu wire was attached to the positive terminal of a potentiometer and the calomel electrode was attached to the negative terminal.
**(a)** Write a half-reaction for the Cu electrode.
**(b)** Write the Nernst equation for the Cu electrode.
**(c)** Calculate the cell voltage.

**15-7.** Explain why a silver electrode can be an indicator electrode for $Ag^+$ and for halides.

**15-8.** A 10.0-mL solution of 0.050 0 M $AgNO_3$ was titrated with 0.025 0 M NaBr in the cell

$$\text{S.C.E.} \parallel \text{titration solution} \mid Ag(s)$$

Find the cell voltage for 0.1, 10.0, 20.0, and 30.0 mL of titrant.

**15-9.** A solution prepared by mixing 25.0 mL of 0.200 M KI with 25.0 mL of 0.200 M NaCl was titrated with 0.100 M $AgNO_3$ in the following cell:

$$\text{S.C.E.} \parallel \text{titration solution} \mid Ag(s)$$

Call the solubility products of AgI and AgCl $K_I$ and $K_{Cl}$, respectively. The answers to parts **(a)** and **(b)** should be expressions containing these constants.
**(a)** Calculate $[Ag^+]$ when 25.0 mL of titrant have been added.
**(b)** Calculate $[Ag^+]$ when 75.0 mL of titrant have been added.
**(c)** Write an expression showing how the cell voltage depends on $[Ag^+]$.
**(d)** The titration curve is shown in the figure below. Calculate the numerical value of the quotient $K_{Cl}/K_I$.

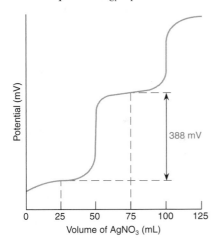

**15-10.** A solution containing 50.0 mL of 0.100 M EDTA buffered to pH 10.00 was titrated with 50.0 mL of 0.020 0 M $Hg(ClO_4)_2$ in the cell shown in Exercise 15-B:

$$\text{S.C.E.} \parallel \text{titration solution} \mid Hg(l)$$

From the cell voltage $E = -0.027$ V, find the formation constant of $Hg(EDTA)^{2-}$.

**15-11.** Consider the cell S.C.E. ‖ cell solution | $Pt(s)$, whose voltage is −0.126 V. The cell solution contains 2.00 mmol of $Fe(NH_4)_2(SO_4)_2$, 1.00 mmol of $FeCl_3$, 4.00 mmol of $Na_2EDTA$, and lots of buffer, pH 6.78, in a volume of 1.00 L.
**(a)** Write a reaction for the right half-cell.
**(b)** Find the value of $[Fe^{2+}]/[Fe^{3+}]$ in the cell solution. (This expression gives the ratio of *uncomplexed* ions.)
**(c)** Find the quotient of formation constants, $(K_f$ for $FeEDTA^-)/(K_f$ for $FeEDTA^{2-})$.

**15-12.** Here's a cell you'll really like:

$$Ag(s) \mid AgCl(s) \mid KCl(aq, \text{saturated}) \parallel \text{cell solution} \mid Cu(s)$$

The cell solution was made by mixing

25.0 mL of 4.00 mM KCN

25.0 mL of 4.00 mM $KCu(CN)_2$

25.0 mL of 0.400 M acid, HA, with $pK_a = 9.50$

25.0 mL of KOH solution

The measured voltage was $-0.440$ V. *Calculate the molarity of the KOH solution.* Assume that essentially all copper(I) is $Cu(CN)_2^-$. A little HCN comes from the reaction of KCN with HA. Neglect the small amount of HA consumed by reaction with HCN. For the right half-cell, the reaction is $Cu(CN)_2^- + e^- \rightleftharpoons Cu(s) + 2CN^-$.

## Junction Potential

**15-13.** What causes a junction potential? How does this potential limit the accuracy of potentiometric analyses? Identify a cell in the illustrations in Section 14-2 that has no junction potential.

**15-14.** Why is the 0.1 M HCl | 0.1 M KCl junction potential of opposite sign and greater magnitude than the 0.1 M NaCl | 0.1 M KCl potential in Table 15-2?

**15-15.** Which side of the liquid junction 0.1 M $KNO_3$ | 0.1 M NaCl will be negative?

**15-16.** How many seconds will it take for (a) $H^+$ and (b) $NO_3^-$ to migrate a distance of 12.0 cm in a field of $7.80 \times 10^3$ V/m? (Refer to the footnote in Table 15-1.)

**15-17.** Suppose that an ideal hypothetical cell such as that in Figure 14-7 were set up to measure $E°$ for the half-reaction $Ag^+ + e^- \rightleftharpoons Ag(s)$.
(a) Calculate the equilibrium constant for the net cell reaction.
(b) If there were a junction potential of $+2$ mV (increasing $E$ from 0.799 to 0.801 V), by what percentage would the calculated equilibrium constant increase?
(c) Answer parts (a) and (b), using 0.100 V instead of 0.799 V for the value of $E°$ for the silver reaction.

**15-18.** Explain how the cell $Ag(s) | AgCl(s) | 0.1$ M HCl | 0.1 M KCl | $AgCl(s) | Ag(s)$ can be used to measure the 0.1 M HCl | 0.1 M KCl junction potential.

**15-19.** ▦ The junction potential, $E_j$, between solutions $\alpha$ and $\beta$ can be estimated with the Henderson equation:

$$E_j \approx \frac{\sum_i \dfrac{|z_i|u_i}{z_i}[C_i(\beta) - C_i(\alpha)]}{\sum_i |z_i|u_i[C_i(\beta) - C_i(\alpha)]} \frac{RT}{F} \ln \frac{\sum_i |z_i|u_i C_i(\alpha)}{\sum_i |z_i|u_i C_i(\beta)}$$

where $z_i$ is the charge of species $i$, $u_i$ is the *mobility* of species $i$ (Table 15-1), $C_i(\alpha)$ is the concentration of species $i$ in phase $\alpha$, and $C_i(\beta)$ is the concentration in phase $\beta$. (Activity coefficients are neglected in this equation.)
(a) Using your calculator, show that the junction potential of 0.1 M HCl | 0.1 M KCl is 26.9 mV at 25°C. (Note that $(RT/F) \ln x = 0.059\ 16 \log x$.)
(b) Set up a spreadsheet to reproduce the result in part (a). Then use your spreadsheet to compute and plot the junction potential for 0.1 M HCl | $x$ M KCl, where $x$ varies from 1 mM to 4 M.

(c) Use your spreadsheet to explore the behavior of the junction potential for $y$ M HCl | $x$ M KCl, where $y = 10^{-4}$, $10^{-3}$, $10^{-2}$, and $10^{-1}$ M and $x = 1$ mM or 4 M.

## pH Measurement with a Glass Electrode

**15-20.** Describe how you would calibrate a pH electrode and measure the pH of blood (which is ~7.5) *at 37°C.* Use the standard buffers in Table 15-3.

**15-21.** List the sources of error associated with pH measurement with the glass electrode.

**15-22.** If electrode C in Figure 15-15 were placed in a solution of pH 11.0, what would the pH reading be?

**15-23.** Which National Institute of Standards and Technology buffer(s) would you use to calibrate an electrode for pH measurements in the range 3–4?

**15-24.** Why do glass pH electrodes tend to indicate a pH lower than the actual pH in strongly basic solution?

**15-25.** Suppose that the Ag | AgCl outer electrode in Figure 15-9 is filled with 0.1 M NaCl instead of saturated KCl. Suppose that the electrode is calibrated in a dilute buffer containing 0.1 M KCl at pH 6.54 at 25°C. The electrode is then dipped in a second buffer *at the same pH* and same temperature, but containing 3.5 M KCl. Use Table 15-2 to estimate how much the indicated pH will change.

**15-26.** (a) When the difference in pH across the membrane of a glass electrode at 25°C is 4.63 pH units, how much voltage is generated by the pH gradient?
(b) What would the voltage be for the same pH difference at 37°C?

**15-27.** When calibrating a glass electrode, 0.025 $m$ potassium dihydrogen phosphate/0.025 $m$ disodium hydrogen phosphate buffer (Table 15-3) gave a reading of $-18.3$ mV at 20°C and 0.05 $m$ potassium hydrogen phthalate buffer gave a reading of $+146.3$ mV. What is the pH of an unknown giving a reading of $+50.0$ mV? What is the slope of the calibration curve (mV/pH unit) and what is the theoretical slope at 20°C? Find the value of $\beta$ in Equation 15-4.

**15-28.** *Activity problem.* The 0.025 0 $m$ $KH_2PO_4$/0.025 0 $m$ $Na_2HPO_4$ buffer (5) in Table 15-3 has a pH of 6.865 at 25°C.
(a) Show that the ionic strength of the buffer is $\mu = 0.100$ $m$.
(b) From the pH and $K_2$ for phosphoric acid, find the quotient of activity coefficients, $\gamma_{HPO_4^{2-}}/\gamma_{H_2PO_4^-}$, at $\mu = 0.100$ $m$.
(c) You have the urgent need to prepare a pH 7.000 buffer to be used as a calibration standard.[38] You can use the activity coefficient ratio from part (b) to accurately prepare such a buffer if the ionic strength is kept at 0.100 $m$. What molalities of $KH_2PO_4$ and $Na_2HPO_4$ should be mixed to obtain a pH of 7.000 and $\mu = 0.100$ $m$?

## Ion-Selective Electrodes

**15-29.** Explain the principle of operation of ion-selective electrodes. How does a compound electrode differ from a simple ion-selective electrode?

**15-30.** What does the selectivity coefficient tell us? Is it better to have a large or a small selectivity coefficient?

**15-31.** What makes a liquid-based ion-selective electrode specific for one analyte?

**15-32.** Why is it preferable to use a metal ion buffer to achieve $pM = 8$, rather than just dissolving enough M to give a $10^{-8}$ M solution?

**15-33.** To determine the *concentration* of a dilute analyte with an ion-selective electrode, why do we use standards with a constant, high concentration of an inert salt?

**15-34.** A cyanide ion-selective electrode obeys the equation

$$E = \text{constant} - 0.059\,16\log[CN^-]$$

The potential was $-0.230$ V when the electrode was immersed in 1.00 mM NaCN.

(a) Evaluate the constant in the preceding equation.
(b) Using the result from part (a), find $[CN^-]$ if $E = -0.300$ V.
(c) Without using the constant from part (a), find $[CN^-]$ if $E = -0.300$ V.

**15-35.** By how many volts will the potential of a $Mg^{2+}$ ion-selective electrode change if the electrode is removed from $1.00 \times 10^{-4}$ M $MgCl_2$ and placed in $1.00 \times 10^{-3}$ M $MgCl_2$?

**15-36.** When measured with a $F^-$ ion-selective electrode with a Nernstian response at 25°C, the potential due to $F^-$ in unfluoridated groundwater in Foxboro, Massachusetts, was 40.0 mV more positive than the potential of tap water in Providence, Rhode Island. Providence maintains its fluoridated water at the recommended level of $1.00 \pm 0.05$ mg $F^-$/L. What is the concentration of $F^-$ in mg/L in groundwater in Foxboro? (Disregard the uncertainty.)

**15-37.** The selectivities of a $Li^+$ ion-selective electrode are indicated on the following diagram. Which alkali metal (Group 1) ion causes the most interference? Which alkaline earth (Group 2) ion causes the most interference? How much greater must be $[K^+]$ than $[Li^+]$ for the two ions to give equal response?

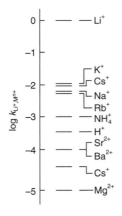

**15-38.** A metal ion buffer was prepared from 0.030 M ML and 0.020 M L, where ML is a metal-ligand complex and L is free ligand.

$$M + L \rightleftharpoons ML \qquad K_f = 4.0 \times 10^8$$

Calculate the concentration of free metal ion, M, in this buffer.

**15-39.** The following data were obtained when a $Ca^{2+}_2$ ion-selective electrode was immersed in standard solutions whose ionic strength was constant at 2.0 M.

| $Ca^{2+}$ (M) | $E$ (mV) |
|---|---|
| $3.38 \times 10^{-5}$ | $-74.8$ |
| $3.38 \times 10^{-4}$ | $-46.4$ |
| $3.38 \times 10^{-3}$ | $-18.7$ |
| $3.38 \times 10^{-2}$ | $+10.0$ |
| $3.38 \times 10^{-1}$ | $+37.7$ |

(a) Prepare a calibration curve and find the least-squares slope and intercept and their standard deviations.
(b) Calculate the value of $\beta$ in Equation 15-10.
(c) Calculate $[Ca^{2+}]$ (and its associated uncertainty) of a sample that gave a reading of $-22.5\ (\pm 0.3)$ mV in four replicate measurements.

**15-40.** The selectivity coefficient, $k_{Li^+,H^+}$, for a $Li^+$ ion-selective electrode is $4 \times 10^{-4}$. When this electrode is placed in $3.44 \times 10^{-4}$ M $Li^+$ solution at pH 7.2, the potential is $-0.333$ V versus S.C.E. What would be the potential if the pH were lowered to 1.1 and the ionic strength were kept constant?

**15-41.** *Standard addition.* A particular $CO_2$ compound electrode like the one in Figure 15-22 obeys the equation $E = \text{constant} - [\beta RT\,(\ln 10)/2F]\log[CO_2]$, where $R$ is the gas constant, $T$ is temperature (303.15 K), $F$ is the Faraday constant, and $\beta = 0.933$ (measured from a calibration curve). $[CO_2]$ represents all forms of dissolved carbon dioxide at the pH of the experiment, which was 5.0. Standard additions of volume $V_S$ containing a standard concentration $c_S = 0.020\,0$ M $NaHCO_3$ were made to an unknown solution whose initial volume was $V_0 = 55.0$ mL.

| $V_S$ (mL) | $E$ (V) | $V_S$ (mL) | $E$ (V) |
|---|---|---|---|
| 0 | 0.079 0 | 0.300 | 0.058 8 |
| 0.100 | 0.072 4 | 0.800 | 0.050 9 |
| 0.200 | 0.065 3 | | |

Prepare a standard addition graph with Equation 15-12 and find $[CO_2]$ in the unknown.

**15-42.** A $Ca^{2+}$ ion-selective electrode was calibrated in metal ion buffers whose ionic strength was fixed at 0.50 M. Using the following electrode readings, write an equation for the response of the electrode to $Ca^{2+}$ and $Mg^{2+}$.

| $[Ca^{2+}]$ (M) | $[Mg^{2+}]$ (M) | mV |
|---|---|---|
| $1.00 \times 10^{-6}$ | 0 | $-52.6$ |
| $2.43 \times 10^{-4}$ | 0 | $+16.1$ |
| $1.00 \times 10^{-6}$ | $3.68 \times 10^{-3}$ | $-38.0$ |

**15-43.** The $Pb^{2+}$ ion buffer used inside the electrode for the colored curve in Figure 15-21 was prepared by mixing 1.0 mL of 0.10 M $Pb(NO_3)_2$ with 100.0 mL of 0.050 M $Na_2EDTA$. At the measured pH of 4.34, $\alpha_{Y^{4-}} = 1.46 \times 10^{-8}$ (Equation 12-3). Show that $[Pb^{2+}] = 1.4 \times 10^{-12}$ M.

**15-44.** A lead ion buffer was prepared by mixing 0.100 mmol of $Pb(NO_3)_2$ with 2.00 mmol of $Na_2C_2O_4$ in a volume of 10.0 mL.

(a) Given the following equilibrium, find the concentration of free $Pb^{2+}$ in this solution.

$$Pb^{2+} + 2C_2O_4^{2-} \rightleftharpoons Pb(C_2O_4)_2^{2-} \qquad K = \beta_2 = 10^{6.54}$$

(b) How many mmol of $Na_2C_2O_4$ should be used to give $[Pb^{2+}] = 1.00 \times 10^{-7}$ M?

**15-45.** Solutions with a wide range of $Hg^{2+}$ concentrations were prepared to calibrate an experimental $Hg^{2+}$ ion-selective electrode. For the range $10^{-5} < [Hg^{2+}] < 10^{-1}$ M, $Hg(NO_3)_2$ was used directly. The range $10^{-11} < [Hg^{2+}] < 10^{-6}$ M could be covered by the buffer system $HgCl_2(s) + KCl(aq)$ (based on $pK_{sp}$ for $HgCl_2 = 13.16$). The range $10^{-15} < [Hg^{2+}] < 10^{-11}$ M was obtained with $HgBr_2(s) + KBr(aq)$ (based on $pK_{sp}$ for $HgBr_2 = 17.43$). The resulting calibration curve is shown in the figure. Calibration points for the $HgCl_2/KCl$ buffer are not in line with the other data. Suggest a possible explanation.

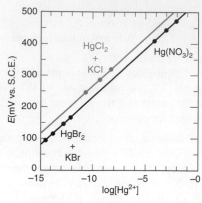

$Hg^{2+}$ ion-selective electrode calibration curve from J. A. Shatkin, H. S. Brown, and S. Licht, "Composite Graphite Ion Selective Electrode Array Potentiometry for the Detection of Mercury and Other Relevant Ions in Aquatic Systems," *Anal. Chem.* 1995, 67, 1147. It was not stated in the paper, but we presume that all solutions had the same ionic strength.

**15-46.** *Activity problem.* Citric acid is a triprotic acid ($H_3A$) whose anion ($A^{3-}$) forms stable complexes with many metal ions. $Ca^{2+}$ forms a 1:1 complex with citrate under the conditions of this problem.

$$Ca^{2+} + A^{3-} \xrightleftharpoons{K_f} CaA^-$$

A calcium ion-selective electrode gave a calibration curve similar to Figure B-2 in Appendix B, with a slope of 29.58 mV. When the electrode was immersed in a solution having $\mathcal{A}_{Ca^{2+}} = 1.00 \times 10^{-3}$, the reading was $+2.06$ mV. Calcium citrate solution was prepared by mixing equal volumes of solutions 1 and 2.

Solution 1:
$$[Ca^{2+}] = 1.00 \times 10^{-3} \text{ M, pH} = 8.00, \mu = 0.10 \text{ M}$$

Solution 2:
$$[Citrate]_{total} = 1.00 \times 10^{-3} \text{ M, pH} = 8.00, \mu = 0.10 \text{ M}$$

When the electrode was immersed in the calcium citrate solution, the reading was $-25.90$ mV.

(a) Calculate the activity of $Ca^{2+}$ in the calcium citrate solution.

(b) Calculate the formation constant, $K_f$, for $CaA^-$. Assume that the size of $CaA^-$ is 500 pm. At pH 8.00 and $\mu = 0.10$ M, the fraction of free citrate in the form $A^{3-}$ is 0.998.

**Solid-State Chemical Sensors**

**15-47.** What does analyte do to a chemical-sensing field effect transistor to produce a signal related to the activity of analyte?

# 16 Redox Titrations

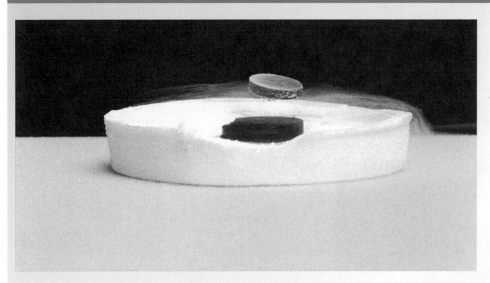

Permanent magnet levitates above superconducting disk cooled in a pool of liquid nitrogen. Redox titrations are crucial in measuring the chemical composition of a superconductor. *[Photo courtesy D. Cornelius, Michelson Laboratory, with materials from T. Vanderah.]*

*Superconductors* are materials that lose all electric resistance when cooled below a critical temperature. Prior to 1987, all known superconductors required cooling to temperatures near that of liquid helium (4 K), a process that is costly and impractical for all but a few applications. In 1987, a giant step was taken when "high-temperature" superconductors that retain their superconductivity above the boiling point of liquid nitrogen (77 K) were discovered. The most startling characteristic of a superconductor is magnetic levitation, shown above. When a magnetic field is applied to a superconductor, current flows in the outer skin of the material such that the applied magnetic field is exactly canceled by the induced magnetic field, and the net field inside the specimen is zero. Expulsion of a magnetic field from a superconductor is called the *Meissner effect*.

A prototypical high-temperature superconductor is yttrium barium copper oxide, $YBa_2Cu_3O_7$, in which two-thirds of the copper is in the $+2$ oxidation state and one-third is in the unusual $+3$ state. Another example is $Bi_2Sr_2(Ca_{0.8}Y_{0.2})Cu_2O_{8.295}$, in which the average oxidation state of copper is $+2.105$ and the average oxidation state of bismuth is $+3.090$ (which is formally a mixture of $Bi^{3+}$ and $Bi^{5+}$). The most reliable means to unravel these complex formulas is through "wet" oxidation-reduction titrations, described in this chapter.

A **redox titration** is based on an oxidation-reduction reaction between analyte and titrant. In addition to the many common analytes in chemistry, biology, and environmental and materials science that can be measured by redox titrations, exotic oxidation states of elements in uncommon materials such as superconductors and laser materials are measured by redox titrations. For example, chromium added to laser crystals to increase their efficiency is found in the common oxidation states $+3$ and $+6$, and the unusual $+4$ state. A redox titration is a good way to unravel the nature of this complex mixture of chromium ions.

This chapter introduces the theory of redox titrations and discusses some common reagents. A few of the oxidants and reductants in Table 16-1 can be used as titrants.[2] Most reductants react with $O_2$ and require protection from air to be used as titrants.

Iron and its compounds are environmentally acceptable redox agents that are finding increased use in remediating toxic waste in groundwaters:[1]

$$CrO_4^{2-} + Fe(0) + 4H_2O \rightarrow$$

Dissolved chromate (carcinogen)    Iron particles (reductant)

$$Cr(OH)_3(s) + Fe(OH)_3(s) + 2OH^-$$

*Mixed hydroxide precipitate (relatively safe)*

$$3H_2S + 8HFeO_4^- + 6H_2O \rightarrow$$

Pollutant    Ferrate(VI) (oxidant)

$$8Fe(OH)_3(s) + 3SO_4^{2-} + 2OH^-$$

*(safe products)*

## Table 16-1  Oxidizing and reducing agents

| Oxidants | | Reductants | |
|---|---|---|---|
| $BiO_3^-$ | Bismuthate | 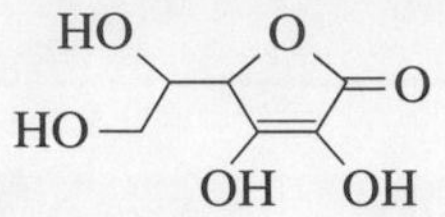 | Ascorbic acid (vitamin C) |
| $BrO_3^-$ | Bromate | | |
| $Br_2$ | Bromine | | |
| $Ce^{4+}$ | Ceric | | |
| $CH_3\text{—}\langle\ \rangle\text{—}SO_2NCl^-Na^+$ | Chloramine T | $BH_4^-$ | Borohydride |
| | | $Cr^{2+}$ | Chromous |
| $Cl_2$ | Chlorine | $S_2O_4^{2-}$ | Dithionite |
| $ClO_2$ | Chlorine dioxide | $Fe^{2+}$ | Ferrous |
| $Cr_2O_7^{2-}$ | Dichromate | $N_2H_4$ | Hydrazine |
| $FeO_4^{2-}$ | Ferrate(VI) | | |
| $H_2O_2$ | Hydrogen peroxide | | Hydroquinone |
| $Fe^{2+} + H_2O_2$ | Fenton reagent[3] | $NH_2OH$ | Hydroxylamine |
| $OCl^-$ | Hypochlorite | $H_3PO_2$ | Hypophosphorous acid |
| $IO_3^-$ | Iodate | | |
| $I_2$ | Iodine | | |
| $Pb(acetate)_4$ | Lead(IV) acetate | 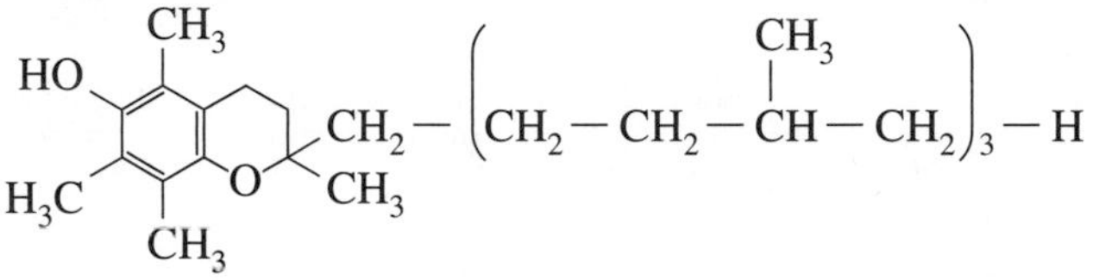 | Retinol (vitamin A) |
| $HNO_3$ | Nitric acid | | |
| O | Atomic oxygen | $Sn^{2+}$ | Stannous |
| $O_2$ | Dioxygen (oxygen) | $SO_3^{2-}$ | Sulfite |
| $O_3$ | Ozone | $SO_2$ | Sulfur dioxide |
| $HClO_4$ | Perchloric acid | $S_2O_3^{2-}$ | Thiosulfate |
| $IO_4^-$ | Periodate | | |
| $MnO_4^-$ | Permanganate | | |
| $S_2O_8^{2-}$ | Peroxydisulfate | | |
| | | | $\alpha$-Tocopherol (vitamin E) |

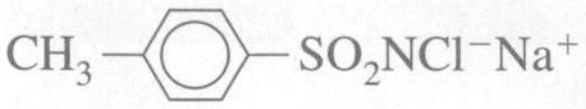

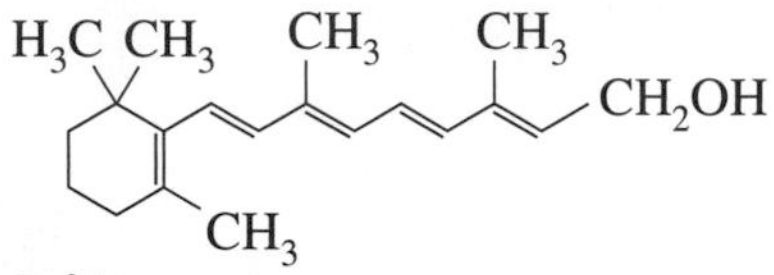

## ■ ■ ■ ■  16-1  The Shape of a Redox Titration Curve

Consider the titration of iron(II) with standard cerium(IV), monitored potentiometrically with Pt and calomel electrodes as shown in Figure 16-1. The titration reaction is

> The titration reaction goes to completion after each addition of titrant. The equilibrium constant is $K = 10^{nE°/0.059\ 16}$ at 25°C.

$$\text{Titration reaction:}\qquad Ce^{4+} + Fe^{2+} \rightarrow Ce^{3+} + Fe^{3+} \qquad (16\text{-}1)$$

Ceric    Ferrous    Cerous    Ferric
titrant    analyte

The way in which we write redox reactions is misleading. Reaction 16-1 appears as if an electron moves from $Fe^{2+}$ to $Ce^{4+}$ to give $Fe^{3+}$ and $Ce^{2+}$. In fact, this reaction and many others are thought to proceed through atom transfer, not electron transfer.* In this case, a hydrogen atom (a proton plus an electron) could be transferred from aqueous $Fe^{2+}$ to aqueous $Ce^{4+}$ species:

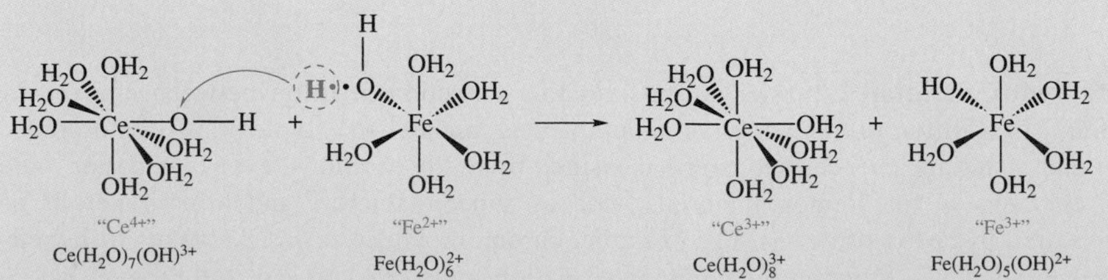

"Ce⁴⁺"  $Ce(H_2O)_7(OH)^{3+}$     "Fe²⁺"  $Fe(H_2O)_6^{2+}$     "Ce³⁺"  $Ce(H_2O)_8^{3+}$     "Fe³⁺"  $Fe(H_2O)_5(OH)^{2+}$

Other common redox reactions between metallic species could proceed through transfer of oxygen atoms or halogen atoms to effect net electron transfer from one metal to another.

*D. T. Sawyer, "Conceptual Considerations in Molecular Science," *J. Chem. Ed.*. **2005**, *82*, 985.

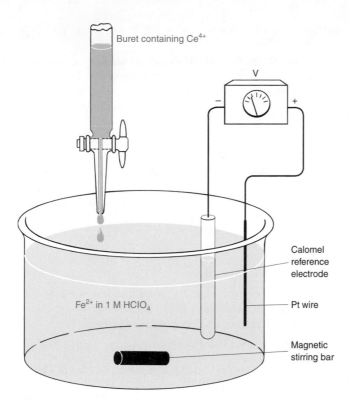

Buret containing $Ce^{4+}$

V

Calomel reference electrode

Pt wire

$Fe^{2+}$ in 1 M $HClO_4$

Magnetic stirring bar

*Figure 16-1* Apparatus for potentiometric titration of $Fe^{2+}$ with $Ce^{4+}$.

for which $K \approx 10^{16}$ in 1 M $HClO_4$. Each mole of ceric ion oxidizes 1 mol of ferrous ion rapidly and quantitatively. The titration reaction creates a mixture of $Ce^{4+}$, $Ce^{3+}$, $Fe^{2+}$, and $Fe^{3+}$ in the beaker in Figure 16-1.

At the Pt indicator electrode, *two* reactions come to equilibrium:

| Indicator half-reaction: | $Fe^{3+} + e^- \rightleftharpoons Fe^{2+}$ | $E° = 0.767$ V | **(16-2)** |
| Indicator half-reaction: | $Ce^{4+} + e^- \rightleftharpoons Ce^{3+}$ | $E° = 1.70$ V | **(16-3)** |

Equilibria 16-2 and 16-3 are both established at the Pt electrode.

The potentials cited here are the formal potentials that apply in 1 M $HClO_4$. The Pt indicator electrode responds to the relative concentrations (really, activities) of $Ce^{4+}$ and $Ce^{3+}$ or $Fe^{3+}$ and $Fe^{2+}$.

We now set out to calculate how the cell voltage changes as $Fe^{2+}$ is titrated with $Ce^{4+}$. The titration curve has three regions.

## Region 1: Before the Equivalence Point

As each aliquot of $Ce^{4+}$ is added, titration reaction 16-1 consumes the $Ce^{4+}$ and creates an equal number of moles of $Ce^{3+}$ and $Fe^{3+}$. Prior to the equivalence point, excess unreacted $Fe^{2+}$ remains in the solution. Therefore, we can find the concentrations of $Fe^{2+}$ and $Fe^{3+}$ without difficulty. On the other hand, we cannot find the concentration of $Ce^{4+}$ without solving a fancy little equilibrium problem. Because the amounts of $Fe^{2+}$ and $Fe^{3+}$ are both known, it is *convenient* to calculate the cell voltage by using Reaction 16-2 instead of Reaction 16-3.

We can use either Reaction 16-2 or Reaction 16-3 to describe the cell voltage at any time. However, because we know the concentrations of $Fe^{2+}$ and $Fe^{3+}$, it is more convenient for now to use Reaction 16-2.

$$E = E_+ - E_- \quad \text{(16-4)}$$

$$E = \left[ 0.767 - 0.059\,16 \log\left( \frac{[Fe^{2+}]}{[Fe^{3+}]} \right) \right] - 0.241 \quad \text{(16-5)}$$

↑
Formal potential for $Fe^{3+}$ reduction in 1 M $HClO_4$

↑
Potential of saturated calomel electrode

$E_+$ is the potential of the Pt electrode connected to the positive terminal of the potentiometer in Figure 16-1. $E_-$ is the potential of the calomel electrode connected to the negative terminal.

$$E = 0.526 - 0.059\,16 \log\left( \frac{[Fe^{2+}]}{[Fe^{3+}]} \right) \quad \text{(16-6)}$$

One special point is reached before the equivalence point. When the volume of titrant is one-half of the amount required to reach the equivalence point ($V = \frac{1}{2}V_e$), $[Fe^{3+}] = [Fe^{2+}]$. In this case, the log term is 0, and $E_+ = E°$ for the $Fe^{3+} | Fe^{2+}$ couple. *The point at which*

For Reaction 16-2, $E_+ = E°(Fe^{3+} | Fe^{2+})$ when $V = \frac{1}{2}V_e$.

$V = \frac{1}{2}V_e$ is analogous to the point at which $pH = pK_a$ when $V = \frac{1}{2}V_e$ in an acid-base titration.

The voltage at zero titrant volume cannot be calculated, because we do not know how much $Fe^{3+}$ is present. If $[Fe^{3+}] = 0$, the voltage calculated with Equation 16-6 would be $-\infty$. In fact, there must be some $Fe^{3+}$ in each reagent, either as an impurity or from oxidation of $Fe^{2+}$ by atmospheric oxygen. In any case, the voltage could never be lower than that needed to reduce the solvent ($H_2O + e^- \rightarrow \frac{1}{2}H_2 + OH^-$).

## Region 2: At the Equivalence Point

Exactly enough $Ce^{4+}$ has been added to react with all the $Fe^{2+}$. Virtually all cerium is in the form $Ce^{3+}$, and virtually all iron is in the form $Fe^{3+}$. Tiny amounts of $Ce^{4+}$ and $Fe^{2+}$ are present at equilibrium. From the stoichiometry of Reaction 16-1, we can say that

$$[Ce^{3+}] = [Fe^{3+}] \tag{16-7}$$

and

$$[Ce^{4+}] = [Fe^{2+}] \tag{16-8}$$

To understand why Equations 16-7 and 16-8 are true, imagine that *all* the cerium and the iron have been converted into $Ce^{3+}$ and $Fe^{3+}$. Because we are at the equivalence point, $[Ce^{3+}] = [Fe^{3+}]$. Now let Reaction 16-1 come to equilibrium:

$$Fe^{3+} + Ce^{3+} \rightleftharpoons Fe^{2+} + Ce^{4+} \qquad \text{(reverse of Reaction 16-1)}$$

If a little bit of $Fe^{3+}$ goes back to $Fe^{2+}$, an equal number of moles of $Ce^{4+}$ must be made. So $[Ce^{4+}] = [Fe^{2+}]$.

At the equivalence point, we use both Reactions 16-2 and 16-3 to calculate the cell voltage. This is strictly a matter of algebraic convenience.

At any time, Reactions 16-2 and 16-3 are *both* in equilibrium at the Pt electrode. At the equivalence point, it is *convenient* to use both reactions to describe the cell voltage. The Nernst equations for these reactions are

$$E_+ = 0.767 - 0.059\ 16 \log\left(\frac{[Fe^{2+}]}{[Fe^{3+}]}\right) \tag{16-9}$$

$$E_+ = 1.70 - 0.059\ 16 \log\left(\frac{[Ce^{3+}]}{[Ce^{4+}]}\right) \tag{16-10}$$

Here is where we stand: Equations 16-9 and 16-10 are both statements of algebraic truth. But neither one alone allows us to find $E_+$, because we do not know exactly what tiny concentrations of $Fe^{2+}$ and $Ce^{4+}$ are present. It is possible to solve the four simultaneous equations 16-7 through 16-10 by first *adding* Equations 16-9 and 16-10:

log *a* + log *b* = log *ab*

$$2E_+ = 0.767 + 1.70 - 0.059\ 16 \log\left(\frac{[Fe^{2+}]}{[Fe^{3+}]}\right) - 0.059\ 16 \log\left(\frac{[Ce^{3+}]}{[Ce^{4+}]}\right)$$

$$2E_+ = 2.46_7 - 0.059\ 16 \log\left(\frac{[Fe^{2+}][Ce^{3+}]}{[Fe^{3+}][Ce^{4+}]}\right)$$

But, because $[Ce^{3+}] = [Fe^{3+}]$ and $[Ce^{4+}] = [Fe^{2+}]$ at the equivalence point, the ratio of concentrations in the log term is unity. Therefore, the logarithm is 0 and

$$2E_+ = 2.46_7 \text{ V} \Rightarrow E_+ = 1.23 \text{ V}$$

The cell voltage is

$$E = E_+ - E(\text{calomel}) = 1.23 - 0.241 = 0.99 \text{ V} \tag{16-11}$$

In this particular titration, the equivalence-point voltage is independent of the concentrations and volumes of the reactants.

## Region 3: After the Equivalence Point

After the equivalence point, we use Reaction 16-3 because we can easily calculate the concentrations of $Ce^{3+}$ and $Ce^{4+}$. It is not convenient to use Reaction 16-2, because we do not know the concentration of $Fe^{2+}$, which has been "used up."

Now virtually all iron atoms are $Fe^{3+}$. The moles of $Ce^{3+}$ equal the moles of $Fe^{3+}$, and there is a known excess of unreacted $Ce^{4+}$. Because we know both $[Ce^{3+}]$ and $[Ce^{4+}]$, it is *convenient* to use Reaction 16-3 to describe the chemistry at the Pt electrode:

$$E = E_+ - E(\text{calomel}) = \left[1.70 - 0.059\ 16 \log\left(\frac{[Ce^{3+}]}{[Ce^{4+}]}\right)\right] - 0.241 \tag{16-12}$$

At the special point when $V = 2V_e$, $[Ce^{3+}] = [Ce^{4+}]$ and $E_+ = E°(Ce^{4+} \mid Ce^{3+}) = 1.70$ V.

Before the equivalence point, the indicator electrode potential is fairly steady near $E°(Fe^{3+} | Fe^{2+}) = 0.77$ V.[4] After the equivalence point, the indicator electrode potential levels off near $E°(Ce^{4+} | Ce^{3+}) = 1.70$ V. At the equivalence point, there is a rapid rise in voltage.

### Example  Potentiometric Redox Titration

Suppose that we titrate 100.0 mL of 0.050 0 M $Fe^{2+}$ with 0.100 M $Ce^{4+}$, using the cell in Figure 16-1. The equivalence point occurs when $V_{Ce^{4+}} = 50.0$ mL. Calculate the cell voltage at 36.0, 50.0, and 63.0 mL.

**Solution**  *At 36.0 mL:*  This is 36.0/50.0 of the way to the equivalence point. Therefore, 36.0/50.0 of the iron is in the form $Fe^{3+}$ and 14.0/50.0 is in the form $Fe^{2+}$. Putting $[Fe^{2+}]/[Fe^{3+}] = 14.0/36.0$ into Equation 16-6 gives $E = 0.550$ V.

*At 50.0 mL:*  Equation 16-11 tells us that the cell voltage at the equivalence point is 0.99 V, regardless of the concentrations of reagents for this particular titration.

*At 63.0 mL:*  The first 50.0 mL of cerium have been converted into $Ce^{3+}$. Because 13.0 mL of excess $Ce^{4+}$ has been added, $[Ce^{3+}]/[Ce^{4+}] = 50.0/13.0$ in Equation 16-12, and $E = 1.424$ V.

## Shapes of Redox Titration Curves

The calculations described above allow us to plot the titration curve for Reaction 16-1 in Figure 16-2, which shows potential as a function of the volume of added titrant. The equivalence point is marked by a steep rise in voltage. The value of $E_+$ at $\frac{1}{2}V_e$ is the formal potential of the $Fe^{3+} | Fe^{2+}$ couple, because the quotient $[Fe^{2+}]/[Fe^{3+}]$ is unity at this point. The voltage at any point in this titration depends only on the *ratio* of reactants; their *concentrations* do not figure in any calculations in this example. We expect, therefore, that the curve in Figure 16-2 will be independent of dilution. We should observe the same curve if both reactants were diluted by a factor of 10.

For Reaction 16-1, the titration curve in Figure 16-2 is symmetric near the equivalence point because the reaction stoichiometry is 1:1. Figure 16-3 shows the curve calculated for the titration of $Tl^+$ by $IO_3^-$ in 1.00 M HCl.

$$IO_3^- + 2Tl^+ + 2Cl^- + 6H^+ \rightarrow ICl_2^- + 2Tl^{3+} + 3H_2O \qquad (16\text{-}13)$$

The curve is *not symmetric about the equivalence point,* because the stoichiometry of reactants is 2:1, not 1:1. Still, the curve is so steep near the equivalence point that negligible error is introduced if the center of the steep part is taken as the end point. Demonstration 16-1 provides an example of an asymmetric titration curve whose shape also depends on the pH of the reaction medium.

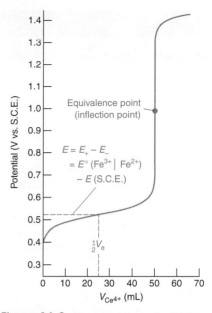

**Figure 16-2** Theoretical curve for titration of 100.0 mL of 0.050 0 M $Fe^{2+}$ with 0.100 M $Ce^{4+}$ in 1 M $HClO_4$. You cannot calculate the potential for zero titrant, but you can start at a small volume such as $V_{Ce^{4+}} = 0.1$ mL.

Anyone with a serious need to calculate redox titration curves should use a spreadsheet with a more general set of equations than we use in this section.[5] The supplement at www.freeman.com/qca explains how to use spreadsheets to compute redox titration curves.

The curve in Figure 16-2 is essentially independent of the concentrations of analyte and titrant. The curve is symmetric near $V_e$ because the stoichiometry is 1:1.

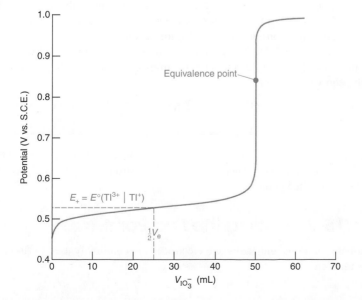

**Figure 16-3** Theoretical curve for titration of 100.0 mL of 0.010 0 M $Tl^+$ with 0.010 0 M $IO_3^-$ in 1.00 M HCl. The equivalence point at 0.842 V is not at the center of the steep part of the curve. When the stoichiometry of the reaction is not 1:1, the curve is not symmetric.

This reaction illustrates many principles of potentiometric titrations:

$$MnO_4^- + 5Fe^{2+} + 8H^+ \rightarrow Mn^{2+} + 5Fe^{3+} + 4H_2O \qquad (A)$$

$$\underset{\text{Titrant}}{\phantom{MnO_4^-}} \quad \underset{\text{Analyte}}{\phantom{5Fe^{2+}}}$$

Dissolve 0.60 g of $Fe(NH_4)_2(SO_4)_2 \cdot 6H_2O$ (FM 392.13; 1.5 mmol) in 400 mL of 1 M $H_2SO_4$. Titrate the well-stirred solution with 0.02 M $KMnO_4$ ($V_e \approx 15$ mL), using Pt and calomel electrodes with a pH meter as a potentiometer. Before use, zero the meter by connecting the two inputs directly to each other and setting the millivolt scale to 0.

Calculate some points on the theoretical titration curve before performing the experiment. Then compare the theoretical and experimental results. Also note the coincidence of the potentiometric and visual end points.

**Question**  Potassium permanganate is purple, and all the other species in this titration are colorless (or very faintly colored). What color change is expected at the equivalence point?

To calculate the theoretical curve, we use the following half-reactions:

$$Fe^{3+} + e^- \rightleftharpoons Fe^{2+} \qquad E° = 0.68 \text{ V in 1 M } H_2SO_4 \quad (B)$$

$$MnO_4^- + 8H^+ + 5e^- \rightarrow Mn^{2+} + 4H_2O \qquad E° = 1.507 \text{ V} \quad (C)$$

Prior to the equivalence point, calculations are similar to those in Section 16-1 for the titration of $Fe^{2+}$ by $Ce^{4+}$, but with $E° = 0.68$ V. After the equivalence point, you can find the potential by using Reaction C. For example, suppose that you titrate 0.400 L of 3.75 mM $Fe^{2+}$ with 0.020 0 M $KMnO_4$. From the stoichiometry of Reaction A, the equivalence point is $V_e = 15.0$ mL. When you have added 17.0 mL of $KMnO_4$, the concentrations of species in Reaction C are $[Mn^{2+}] = 0.719$ mM, $[MnO_4^-] = 0.095\ 9$ mM, and $[H^+] = 0.959$ M (neglecting the small quantity of $H^+$ consumed in the titration). The voltage should be

$$E = E_+ - E(\text{calomel})$$

$$= \left[ 1.507 - \frac{0.059\ 16}{5} \log\left( \frac{[Mn^{2+}]}{[MnO_4^-][H^+]^8} \right) \right] - 0.241$$

$$= \left[ 1.507 - \frac{0.059\ 16}{5} \log\left( \frac{7.19 \times 10^{-4}}{(9.59 \times 10^{-5})(0.959)^8} \right) \right] - 0.241$$

$$= 1.254 \text{ V}$$

To calculate the voltage at the equivalence point, we add the Nernst equations for Reactions B and C, as we did for the cerium and iron reactions in Section 16-1. Before doing so, however, multiply the permanganate equation by 5 so that we can add the log terms:

$$E_+ = 0.68 - 0.059\ 16 \log\left( \frac{[Fe^{2+}]}{[Fe^{3+}]} \right)$$

$$5E_+ = 5\left[ 1.507 - \frac{0.059\ 16}{5} \log\left( \frac{[Mn^{2+}]}{[MnO_4^-][H^+]^8} \right) \right]$$

Now add the two equations to get

$$6E_+ = 8.215 - 0.059\ 16 \log\left( \frac{[Mn^{2+}][Fe^{2+}]}{[MnO_4^-][Fe^{3+}][H^+]^8} \right) \quad (D)$$

But the stoichiometry of the titration reaction A tells us that at the equivalence point $[Fe^{3+}] = 5[Mn^{2+}]$ and $[Fe^{2+}] = 5[MnO_4^-]$. Substituting these values into Equation D gives

$$6E_+ = 8.215 - 0.059\ 16 \log\left( \frac{[Mn^{2+}](5[MnO_4^-])}{[MnO_4^-](5[Mn^{2+}])[H^+]^8} \right)$$

$$= 8.215 - 0.059\ 16 \log\left( \frac{1}{[H^+]^8} \right) \quad (E)$$

Inserting the concentration of $[H^+]$, which is $(400/415)(1.00 \text{ M}) = 0.964$ M, we find

$$6E_+ = 8.215 - 0.059\ 16 \log\left( \frac{1}{(0.964)^8} \right) \Rightarrow E_+ = 1.368 \text{ V}$$

The predicted cell voltage at $V_e$ is $E = E_+ - E(\text{calomel}) = 1.368 - 0.241 = 1.127$ V.

---

The voltage change near the equivalence point increases as the difference between $E°$ of the two redox couples in the titration increases. The larger the difference in $E°$, the greater the equilibrium constant for the titration reaction. For Figure 16-2, half-reactions 16-2 and 16-3 differ by 0.93 V, and there is a large break at the equivalence point in the titration curve. For Figure 16-3, the half-reactions differ by 0.47 V, so there is a smaller break at the equivalence point.

$$IO_3^- + 2Cl^- + 6H^+ + 4e^- \rightleftharpoons ICl_2^- + 3H_2O \qquad E° = 1.24 \text{ V}$$

$$Tl^{3+} + 2e^- \rightleftharpoons Tl^+ \qquad E° = 0.77 \text{ V}$$

You would not choose a weak acid to titrate a weak base, because the break at $V_e$ would not be very large.

Clearest results are achieved with the strongest oxidizing and reducing agents. The same rule applies to acid-base titrations where strong-acid or strong-base titrants give the sharpest break at the equivalence point.

## ■ ■ ■  16-2  Finding the End Point

As in acid-base titrations, indicators and electrodes are commonly used to find the end point of a redox titration.

## Redox Indicators

A **redox indicator** is a compound that changes color when it goes from its oxidized to its reduced state. The indicator ferroin changes from pale blue (almost colorless) to red.

$$\begin{bmatrix} \text{(ferroin ligand)}\text{Fe(III)} \end{bmatrix}^{3+}_{3} + e^- \rightleftharpoons \begin{bmatrix} \text{(ferroin ligand)}\text{Fe(II)} \end{bmatrix}^{2+}_{3}$$

Oxidized ferroin  (pale blue)  In(oxidized)       Reduced ferroin  (red)  In(reduced)

To predict the potential range over which the indicator color will change, we first write a Nernst equation for the indicator.

$$\text{In(oxidized)} + n e^- \rightleftharpoons \text{In(reduced)}$$

$$E = E^\circ - \frac{0.059\ 16}{n} \log\left(\frac{[\text{In(reduced)}]}{[\text{In(oxidized)}]}\right) \qquad (16\text{-}14)$$

As with acid-base indicators, the color of In(reduced) will be observed when

$$\frac{[\text{In(reduced)}]}{[\text{In(oxidized)}]} \gtrsim \frac{10}{1}$$

and the color of In(oxidized) will be observed when

$$\frac{[\text{In(reduced)}]}{[\text{In(oxidized)}]} \lesssim \frac{1}{10}$$

Putting these quotients into Equation 16-14 tells us that the color change will occur over the range

$$E = \left(E^\circ \pm \frac{0.059\ 16}{n}\right) \text{volts}$$

A redox indicator changes color over a range of $\pm(59/n)$ mV, centered at $E^\circ$ for the indicator. $n$ is the number of electrons in the indicator half-reaction.

For ferroin, with $E^\circ = 1.147$ V (Table 16-2), we expect the color change to occur in the approximate range 1.088 V to 1.206 V with respect to the standard hydrogen electrode. With a saturated calomel reference electrode, the indicator transition range will be

$$\begin{pmatrix} \text{Indicator transition} \\ \text{range versus calomel} \\ \text{electrode (S.C.E.)} \end{pmatrix} = \begin{pmatrix} \text{transition range} \\ \text{versus standard hydrogen} \\ \text{electrode (S.H.E.)} \end{pmatrix} - E(\text{calomel}) \quad (16\text{-}15)$$

$$= (1.088 \text{ to } 1.206) - (0.241)$$

$$= 0.847 \text{ to } 0.965 \text{ V (versus S.C.E.)}$$

See the diagram on page 300 for a better understanding of Equation 16-15.

Ferroin would therefore be a useful indicator for the titrations in Figures 16-2 and 16-3.

The indicator transition range should overlap the steep part of the titration curve.

**Table 16-2**  Redox indicators

| Indicator | Color | | $E^\circ$ |
| | Oxidized | Reduced | |
|---|---|---|---|
| Phenosafranine | Red | Colorless | 0.28 |
| Indigo tetrasulfonate | Blue | Colorless | 0.36 |
| Methylene blue | Blue | Colorless | 0.53 |
| Diphenylamine | Violet | Colorless | 0.75 |
| 4′-Ethoxy-2,4-diaminoazobenzene | Yellow | Red | 0.76 |
| Diphenylamine sulfonic acid | Red-violet | Colorless | 0.85 |
| Diphenylbenzidine sulfonic acid | Violet | Colorless | 0.87 |
| Tris(2,2′-bipyridine)iron | Pale blue | Red | 1.120 |
| Tris(1,10-phenanthroline)iron (ferroin) | Pale blue | Red | 1.147 |
| Tris(5-nitro-1,10-phenanthroline)iron | Pale blue | Red-violet | 1.25 |
| Tris(2,2′-bipyridine)ruthenium | Pale blue | Yellow | 1.29 |

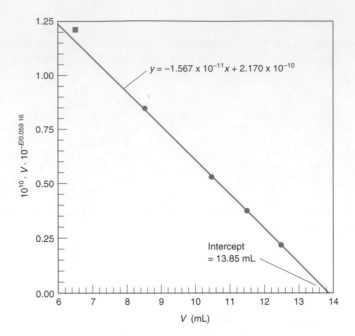

**Figure 16-4** Gran plot for titration of $Fe^{2+}$ by $Ce^{4+}$ in Exercise 16-E.[6] The line was fit to the four points shown by circles. In the function on the ordinate, the value of $n$ is 1. Numerical values were multiplied by $10^{10}$ for ease of display. This multiplication does not change the intercept.

$y = -1.567 \times 10^{-11}x + 2.170 \times 10^{-10}$

Intercept = 13.85 mL

The larger the difference in standard potential between titrant and analyte, the greater the break in the titration curve at the equivalence point. A redox titration is usually feasible if the difference between analyte and titrant is $\gtrsim 0.2$ V. However, the end point of such a titration is not very sharp and is best detected potentiometrically. If the difference in formal potentials is $\gtrsim 0.4$ V, then a redox indicator usually gives a satisfactory end point.

## Gran Plot

With the apparatus in Figure 16-1, we measure electrode potential, $E$, versus volume of titrant, $V$, during a redox titration. The end point is the maximum of the first derivative, $\Delta E / \Delta V$, or the zero crossing of the second derivative, $\Delta(\Delta E / \Delta V) / \Delta V$ (Figure 11-6).

A more accurate way to use potentiometric data is to prepare a Gran plot[6,7] as we did for acid-base titrations in Section 11-5. The Gran plot uses data from well before the equivalence point ($V_e$) to locate $V_e$. Potentiometric data taken close to $V_e$ are the least accurate because electrodes are slow to equilibrate with species in solution when one member of a redox couple is nearly used up.

For the oxidation of $Fe^{2+}$ to $Fe^{3+}$, the potential prior to $V_e$ is

The constant 0.059 16 V is $(RT \ln 10)/nF$, where $R$ is the gas constant, $T$ is 298.15 K, $F$ is the Faraday constant, and $n$ is the number of electrons in the $Fe^{3+} \mid Fe^{2+}$ redox half-reaction ($n = 1$). For $T \neq 298.15$ K or $n \neq 1$, the number 0.059 16 will change.

$$E = \left[ E^{\circ\prime} - 0.059\ 16 \log\left(\frac{[Fe^{2+}]}{[Fe^{3+}]}\right) \right] - E_{\text{ref}} \qquad (16\text{-}16)$$

where $E^{\circ\prime}$ is the formal potential for $Fe^{3+} \mid Fe^{2+}$ and $E_{\text{ref}}$ is the potential of the reference electrode (which we have also been calling $E_-$). If the volume of analyte is $V_0$ and the volume of titrant is $V$ and if the reaction goes "to completion" with each addition of titrant, we can show that $[Fe^{2+}]/[Fe^{3+}] = (V_e - V)/V$. Substituting this expression into Equation 16-16 and rearranging, we eventually stumble on a linear equation:

$$\underbrace{V\ 10^{-nE/0.059\ 16}}_{y} = \underbrace{V_e\ 10^{-n(E_{\text{ref}} - E^{\circ\prime})/0.059\ 16}}_{b} - \underbrace{V}_{x}\ \underbrace{10^{-n(E_{\text{ref}} - E^{\circ\prime})/0.059\ 16}}_{m} \qquad (16\text{-}17)$$

A graph of $V\ 10^{-nE/0.059\ 16}$ versus $V$ should be a straight line whose $x$-intercept is $V_e$ (Figure 16-4). If the ionic strength of the reaction is constant, the activity coefficients are constant, and Equation 16-17 gives a straight line over a wide volume range. If ionic strength varies as titrant is added, we just use the last 10–20% of the data prior to $V_e$.

## The Starch-Iodine Complex

Numerous analytical procedures are based on redox titrations involving iodine. Starch[8] is the indicator of choice for these procedures because it forms an intense blue complex with iodine. Starch is not a redox indicator; it responds specifically to the presence of $I_2$, not to a change in redox potential.

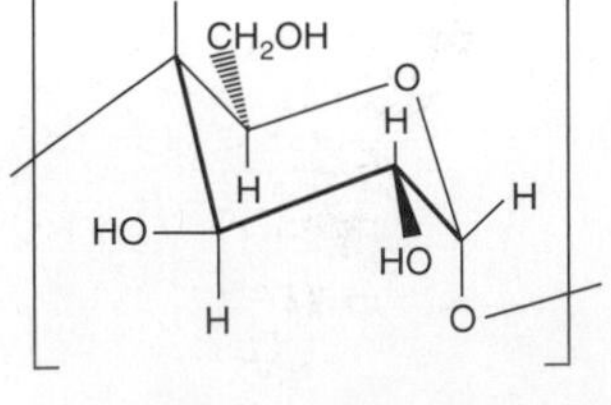

The active fraction of starch is amylose, a polymer of the sugar $\alpha$-D-glucose, with the repeating unit shown in Figure 16-5. Small molecules can fit into the center of the coiled,

**Figure 16-5** Structure of the repeating unit of amylose.

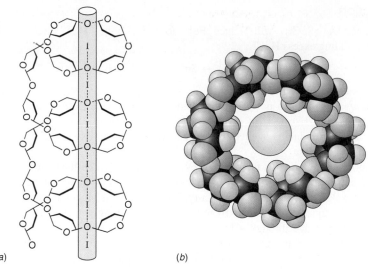

(a)                                    (b)

**Figure 16-6** (a) Schematic structure of the starch-iodine complex. The amylose chain forms a helix around $I_6$ units. *[Adapted from A. T. Calabrese and A. Khan, "Amylose-Iodine Complex Formation with KI: Evidence for Absence of Iodide Ions Within the Complex," J. Polymer Sci.* **1999**, *A37, 2711.]* (b) View down the starch helix, showing iodine inside the helix.[8] *[Figure kindly provided by R. D. Hancock, Power Engineering, Salt Lake City.]*

helical polymer in Figure 16-6. In the presence of starch, iodine forms $I_6$ chains inside the amylose helix and the color turns dark blue.

$$I\cdot I\cdot I\cdot I\cdot I\cdot I\cdot I$$

Starch is readily biodegraded, so it should be freshly dissolved or the solution should contain a preservative, such as $HgI_2$ (~1 mg/100 mL) or thymol. A hydrolysis product of starch is glucose, which is a reducing agent. Therefore, partially hydrolyzed starch solution can be a source of error in a redox titration.

## ■■■■ 16-3 Adjustment of Analyte Oxidation State

Sometimes we need to adjust the oxidation state of analyte before it can be titrated. For example, $Mn^{2+}$ can be **preoxidized** to $MnO_4^-$ and then titrated with standard $Fe^{2+}$. Preadjustment must be quantitative, and you must eliminate excess preadjustment reagent.

Excess preadjustment reagent must be destroyed so that it will not interfere in the subsequent titration.

### Preoxidation

Several powerful oxidants can be easily removed after preoxidation. *Peroxydisulfate* ($S_2O_8^{2-}$, also called *persulfate*) is a strong oxidant that requires $Ag^+$ as a catalyst.

$$S_2O_8^{2-} + Ag^+ \rightarrow SO_4^{2-} + \underbrace{SO_4^- + Ag^{2+}}_{\substack{\text{Two powerful} \\ \text{oxidants}}}$$

Excess reagent is destroyed by boiling the solution after oxidation of analyte is complete.

$$2S_2O_8^{2-} + 2H_2O \xrightarrow{\text{boiling}} 4SO_4^{2-} + O_2 + 4H^+$$

The $S_2O_8^{2-}$ and $Ag^+$ mixture oxidizes $Mn^{2+}$ to $MnO_4^-$, $Ce^{3+}$ to $Ce^{4+}$, $Cr^{3+}$ to $Cr_2O_7^{2-}$, and $VO^{2+}$ to $VO_2^+$.

*Silver(II) oxide* (AgO) dissolves in concentrated mineral acids to give $Ag^{2+}$, with oxidizing power similar to $S_2O_8^{2-}$ plus $Ag^+$. Excess $Ag^{2+}$ can be removed by boiling:

$$4Ag^{2+} + 2H_2O \xrightarrow{\text{boiling}} 4Ag^+ + O_2 + 4H^+$$

Solid *sodium bismuthate* ($NaBiO_3$) has an oxidizing strength similar to that of $Ag^{2+}$ and $S_2O_8^{2-}$. Excess solid oxidant is removed by filtration.

*Hydrogen peroxide* is a good oxidant in basic solution. It can transform $Co^{2+}$ into $Co^{3+}$, $Fe^{2+}$ into $Fe^{3+}$, and $Mn^{2+}$ into $MnO_2$. In acidic solution it can *reduce* $Cr_2O_7^{2-}$ to $Cr^{3+}$ and $MnO_4^-$ to $Mn^{2+}$. Excess $H_2O_2$ spontaneously **disproportionates** in boiling water.

$$2H_2O_2 \xrightarrow{\text{boiling}} O_2 + 2H_2O$$

In *disproportionation*, a reactant is converted into products in higher and lower oxidation states. The compound oxidizes and reduces *itself*.

**Challenge** Write a half-reaction in which $H_2O_2$ behaves as an oxidant and a half-reaction in which it behaves as a reductant.

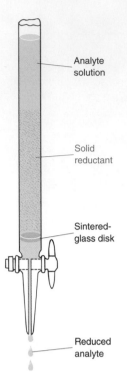

Analyte solution

Solid reductant

Sintered-glass disk

Reduced analyte

*Figure 16-7* A column filled with a solid reagent used for prereduction of analyte is called a *reductor*. Often the analyte is drawn through by suction.

## Prereduction

*Stannous chloride* ($SnCl_2$) will reduce $Fe^{3+}$ to $Fe^{2+}$ in hot HCl. Excess reductant is destroyed by adding excess $HgCl_2$:

$$Sn^{2+} + 2HgCl_2 \rightarrow Sn^{4+} + Hg_2Cl_2 + 2Cl^-$$

The $Fe^{2+}$ is then titrated with an oxidant.

*Chromous chloride* is a powerful reductant sometimes used to **prereduce** analyte to a lower oxidation state. Excess $Cr^{2+}$ is oxidized by atmospheric $O_2$. *Sulfur dioxide* and *hydrogen sulfide* are mild reducing agents that can be expelled by boiling an acidic solution after the reduction is complete.

An important prereduction technique uses a column packed with a solid reducing agent. Figure 16-7 shows the *Jones reductor,* which contains zinc coated with zinc *amalgam*. An **amalgam** is a solution of anything in mercury. The amalgam is prepared by mixing granular zinc with 2 wt% aqueous $HgCl_2$ for 10 min and then washing with water. You can reduce $Fe^{3+}$ to $Fe^{2+}$ by passage through a Jones reductor, using 1 M $H_2SO_4$ as solvent. Wash the column well with water and titrate the combined washings with standard $MnO_4^-$, $Ce^{4+}$, or $Cr_2O_7^{2-}$. Perform a blank determination with a solution passed through the reductor in the same manner as the unknown.

Most reduced analytes are reoxidized by atmospheric oxygen. To prevent oxidation, the reduced analyte can be collected in an acidic solution of $Fe^{3+}$. Ferric ion is reduced to $Fe^{2+}$, which is stable in acid. The $Fe^{2+}$ is then titrated with an oxidant. By this means, elements such as Cr, Ti, V, and Mo can be indirectly analyzed.

Zinc is a powerful reducing agent, with $E° = -0.764$ for the reaction $Zn^{2+} + 2e^- \rightleftharpoons Zn(s)$, so the Jones reductor is not very selective. The *Walden reductor,* filled with solid Ag and 1 M HCl, is more selective. The reduction potential for Ag | AgCl (0.222 V) is high enough that species such as $Cr^{3+}$ and $TiO^{2+}$ are not reduced and therefore do not interfere in the analysis of a metal such as $Fe^{3+}$. Another selective reductor uses granular Cd metal. In determining levels of nitrogen oxides for air-pollution monitoring,[9] the gases are first converted into $NO_3^-$, which is not easy to analyze. Passing nitrate through a Cd-filled column reduces $NO_3^-$ to $NO_2^-$, for which a convenient spectrophotometric analysis is available.

## ◼◼◼◻ 16-4 Oxidation with Potassium Permanganate

Potassium permanganate ($KMnO_4$) is a strong oxidant with an intense violet color. In strongly acidic solutions (pH $\lesssim$ 1), it is reduced to colorless $Mn^{2+}$.

$$\underset{\text{Permanganate}}{MnO_4^-} + 8H^+ + 5e^- \rightleftharpoons \underset{\text{Manganous}}{Mn^{2+}} + 4H_2O \qquad E° = 1.507 \text{ V}$$

In neutral or alkaline solution, the product is the brown solid, $MnO_2$.

$$MnO_4^- + 4H^+ + 3e^- \rightleftharpoons \underset{\substack{\text{Manganese} \\ \text{dioxide}}}{MnO_2(s)} + 2H_2O \qquad E° = 1.692 \text{ V}$$

In strongly alkaline solution (2 M NaOH), green manganate ion is produced.

$$MnO_4^- + e^- \rightleftharpoons \underset{\text{Manganate}}{MnO_4^{2-}} \qquad E° = 0.56 \text{ V}$$

> $KMnO_4$ serves as its own indicator in acidic solution.

Representative permanganate titrations are listed in Table 16-3. For titrations in strongly acidic solution, $KMnO_4$ serves as its own indicator because the product, $Mn^{2+}$, is colorless (Color Plate 8). The end point is taken as the first persistent appearance of pale pink $MnO_4^-$. If the titrant is too dilute to be seen, an indicator such as ferroin can be used.

### Preparation and Standardization

> $KMnO_4$ is not a primary standard.

Potassium permanganate is not a primary standard because traces of $MnO_2$ are invariably present. In addition, distilled water usually contains enough organic impurities to reduce some freshly dissolved $MnO_4^-$ to $MnO_2$. To prepare a 0.02 M stock solution, dissolve $KMnO_4$ in distilled water, boil it for an hour to hasten the reaction between $MnO_4^-$ and organic impurities, and filter the resulting mixture through a clean, sintered-glass filter to remove precipitated $MnO_2$. Do not use filter paper (organic matter!) for the filtration. Store the reagent in a dark glass bottle. Aqueous $KMnO_4$ is unstable by virtue of the reaction

$$4MnO_4^- + 2H_2O \rightarrow 4MnO_2(s) + 3O_2 + 4OH^-$$

**Table 16-3** Analytical applications of permanganate titrations

| Species analyzed | Oxidation reaction | Notes |
|---|---|---|
| $Fe^{2+}$ | $Fe^{2+} \rightleftharpoons Fe^{3+} + e^-$ | $Fe^{3+}$ is reduced to $Fe^{2+}$ with $Sn^{2+}$ or a Jones reductor. Titration is carried out in 1 M $H_2SO_4$ or 1 M HCl containing $Mn^{2+}$, $H_3PO_4$, and $H_2SO_4$. $Mn^{2+}$ inhibits oxidation of $Cl^-$ by $MnO_4$. $H_3PO_4$ complexes $Fe^{3+}$ to prevent formation of yellow $Fe^{3+}$-chloride complexes. |
| $H_2C_2O_4$ | $H_2C_2O_4 \rightleftharpoons 2CO_2 + 2H^+ + 2e^-$ | Add 95% of titrant at 25°C, then complete titration at 55°–60°C. |
| $Br^-$ | $Br^- \rightleftharpoons \frac{1}{2}Br_2(g) + e^-$ | Titrate in boiling 2 M $H_2SO_4$ to remove $Br_2(g)$. |
| $H_2O_2$ | $H_2O_2 \rightleftharpoons O_2(g) + 2H^+ + 2e^-$ | Titrate in 1 M $H_2SO_4$. |
| $HNO_2$ | $HNO_2 + H_2O \rightleftharpoons NO_3^- + 3H^+ + 2e^-$ | Add excess standard $KMnO_4$ and back-titrate after 15 min at 40°C with $Fe^{2+}$. |
| $As^{3+}$ | $H_3AsO_3 + H_2O \rightleftharpoons H_3AsO_4 + 2H^+ + 2e^-$ | Titrate in 1 M HCl with KI or ICl catalyst. |
| $Sb^{3+}$ | $H_3SbO_3 + H_2O \rightleftharpoons H_3SbO_4 + 2H^+ + 2e^-$ | Titrate in 2 M HCl. |
| $Mo^{3+}$ | $Mo^{3+} + 2H_2O \rightleftharpoons MoO_2^{2+} + 4H^+ + 3e^-$ | Reduce Mo in a Jones reductor, and run the $Mo^{3+}$ into excess $Fe^{3+}$ in 1 M $H_2SO_4$. Titrate the $Fe^{2+}$ formed. |
| $W^{3+}$ | $W^{3+} + 2H_2O \rightleftharpoons WO_2^{2+} + 4H^+ + 3e^-$ | Reduce W with Pb(Hg) at 50°C and titrate in 1 M HCl. |
| $U^{4+}$ | $U^{4+} + 2H_2O \rightleftharpoons UO_2^{2+} + 4H^+ + 2e^-$ | Reduce U to $U^{3+}$ with a Jones reductor. Expose to air to produce $U^{4+}$, which is titrated in 1 M $H_2SO_4$. |
| $Ti^{3+}$ | $Ti^{3+} + H_2O \rightleftharpoons TiO^{2+} + 2H^+ + e^-$ | Reduce Ti to $Ti^{3+}$ with a Jones reductor, and run the $Ti^{3+}$ into excess $Fe^{3+}$ in 1 M $H_2SO_4$. Titrate the $Fe^{2+}$ that is formed. |
| $Mg^{2+}$, $Ca^{2+}$, $Sr^{2+}$, $Ba^{2+}$, $Zn^{2+}$, $Co^{2+}$, $La^{3+}$, $Th^{4+}$, $Pb^{2+}$, $Ce^{3+}$, $BiO^+$, $Ag^+$ | $H_2C_2O_4 \rightleftharpoons 2CO_2 + 2H^+ + 2e^-$ | Precipitate the metal oxalate. Dissolve in acid and titrate the $H_2C_2O_4$. |
| $S_2O_8^{2-}$ | $S_2O_8^{2-} + 2Fe^{2+} + 2H^+ \rightleftharpoons 2Fe^{3+} + 2HSO_4^-$ | Peroxydisulfate is added to excess standard $Fe^{2+}$ containing $H_3PO_4$. Unreacted $Fe^{2+}$ is titrated with $MnO_4^-$. |
| $PO_4^{3-}$ | $Mo^{3+} + 2H_2O \rightleftharpoons MoO_2^{2+} + 4H^+ + 3e^-$ | $(NH_4)_3PO_4 \cdot 12MoO_3$ is precipitated and dissolved in $H_2SO_4$. The Mo(VI) is reduced (as above) and titrated. |

which is slow in the absence of $MnO_2$, $Mn^{2+}$, heat, light, acids, and bases. Permanganate should be standardized often for the most accurate work. Prepare and standardize fresh dilute solutions from 0.02 M stock solution, using water distilled from alkaline $KMnO_4$.

Potassium permanganate can be standardized by titration of sodium oxalate ($Na_2C_2O_4$) by Reaction 7-1 or pure electrolytic iron wire. Dissolve dry (105°C, 2 h) sodium oxalate (available in a 99.9–99.95% pure form) in 1 M $H_2SO_4$ and treat it with 90–95% of the required $KMnO_4$ solution at room temperature. Then warm the solution to 55–60°C and complete the titration by slow addition of $KMnO_4$. Subtract a blank value to account for the quantity of titrant (usually one drop) needed to impart a pink color to the solution.

If pure Fe wire is used as a standard, dissolve it in warm 1.5 M $H_2SO_4$ under $N_2$. The product is $Fe^{2+}$, and the cooled solution can be used to standardize $KMnO_4$ (or other oxidants) with no special precautions. Adding 5 mL of 86 wt% $H_3PO_4$ per 100 mL of solution masks the yellow color of $Fe^{3+}$ and makes the end point easier to see. Ferrous ammonium sulfate, $Fe(NH_4)_2(SO_4)_2 \cdot 6H_2O$, and ferrous ethylenediammonium sulfate, $Fe(H_3NCH_2CH_2NH_3)(SO_4)_2 \cdot 2H_2O$, are sufficiently pure to be standards for most purposes.

## ■ ■ ■ 16-5  Oxidation with $Ce^{4+}$

Reduction of $Ce^{4+}$ to $Ce^{3+}$ proceeds cleanly in acidic solutions. The aquo ion, $Ce(H_2O)_n^{4+}$, probably does not exist in any of these solutions, because Ce(IV) binds anions ($ClO_4^-$, $SO_4^{2-}$, $NO_3^-$, $Cl^-$) very strongly. Variation of the $Ce^{4+} | Ce^{3+}$ formal potential with the medium is indicative of these interactions:

$$Ce^{4+} + e^- \rightleftharpoons Ce^{3+} \qquad \text{Formal potential} \begin{cases} 1.70 \text{ V in 1 F } HClO_4 \\ 1.61 \text{ V in 1 F } HNO_3 \\ 1.47 \text{ V in 1 F } HCl \\ 1.44 \text{ V in 1 F } H_2SO_4 \end{cases}$$

The varying formal potential implies that different cerium species are present in each solution.

$Ce^{4+}$ is yellow and $Ce^{3+}$ is colorless, but the color change is not distinct enough for cerium to be its own indicator. Ferroin and other substituted phenanthroline redox indicators (Table 16-2) are well suited to titrations with $Ce^{4+}$.

$Ce^{4+}$ can be used in place of $KMnO_4$ in most procedures. In the oscillating reaction in Demonstration 15-1, $Ce^{4+}$ oxidizes malonic acid to $CO_2$ and formic acid:

$$CH_2(CO_2H)_2 + 2H_2O + 6Ce^{4+} \rightarrow 2CO_2 + HCO_2H + 6Ce^{3+} + 6H^+$$

Malonic acid                                                   Formic acid

This reaction can be used for quantitative analysis of malonic acid by heating a sample in 4 M $HClO_4$ with excess standard $Ce^{4+}$ and back-titrating unreacted $Ce^{4+}$ with $Fe^{2+}$. Analogous procedures are available for many alcohols, aldehydes, ketones, and carboxylic acids.

## Preparation and Standardization

Primary-standard-grade ammonium hexanitratocerate(IV), $(NH_4)_2Ce(NO_3)_6$, can be dissolved in 1 M $H_2SO_4$ and used directly. Although the oxidizing strength of $Ce^{4+}$ is greater in $HClO_4$ or $HNO_3$, these solutions undergo slow photochemical decomposition with concomitant oxidation of water. $Ce^{4+}$ in $H_2SO_4$ is stable indefinitely, despite the fact that the reduction potential of 1.44 V is great enough to oxidize $H_2O$ to $O_2$. The reaction with water is very slow, even though it is thermodynamically favorable. Solutions in HCl are unstable because $Cl^-$ is oxidized to $Cl_2$—rapidly when the solution is hot. Sulfuric acid solutions of $Ce^{4+}$ can be used for titrations of unknowns in HCl because the reaction with analyte is fast, whereas reaction with $Cl^-$ is slow. Less expensive salts, including $Ce(HSO_4)_4$, $(NH_4)_4Ce(SO_4)_4 \cdot 2H_2O$, and

## Box 16-1    Environmental Carbon Analysis and Oxygen Demand

Drinking water and industrial waste streams are partially characterized and regulated on the basis of their carbon content and oxygen demand.[10] *Inorganic carbon* (IC) is the $CO_2(g)$ liberated when water is acidified to pH < 2 with $H_3PO_4$ and purged with Ar or $N_2$. IC corresponds to $CO_3^{2-}$ and $HCO_3^-$ in the sample. After inorganic carbon is removed, *total organic carbon* (TOC) is the $CO_2$ produced by oxidizing the remaining organic matter:

TOC analysis:     Organic carbon $\xrightarrow[\text{metal catalyst}]{O_2/\sim700°C} CO_2$

*Total carbon* (TC) is defined as the sum TC = TOC + IC.

Instruments using different oxidation techniques produce different values for TOC, because not all organic matter is oxidized by each technique. The current state of the art is such that TOC is really defined by the result obtained with a particular instrument.

Commercial instruments that measure TOC by thermal oxidation have detection limits of 4–50 ppb (4–50 μg C/L). A typical 20-μL water sample is analyzed in 3 min, using infrared absorption to measure $CO_2$. Other instruments oxidize organic matter by irradiating a suspension of solid $TiO_2$ catalyst (0.2 g/L) in water at pH 3.5 with ultraviolet light.[11] Light creates electron-hole pairs (Section 15-8) in the $TiO_2$. Holes oxidize $H_2O$ to hydroxyl radical (HO·), a powerful oxidant that converts organic carbon into $CO_2$, which is measured by the electrical conductivity of carbonic acid.[12] Color Plate 9 shows an instrument in which $K_2S_2O_8$ in acid is exposed to ultraviolet radiation to generate sulfate radical ($SO_4^-$), which oxidizes organic matter to $CO_2$. (Pure $TiO_2$ hardly absorbs visible light, so it cannot use sunlight efficiently. By doping $TiO_2$ with ~1 wt% carbon, the efficiency of using visible light is markedly increased.[13])

TOC is widely used to determine compliance with discharge laws. Municipal and industrial wastewater typically has TOC > 1 mg C/mL. Tap water TOC is 50–500 ng C/mL. High-purity water for the electronics industry has TOC < 1 ng C/mL.

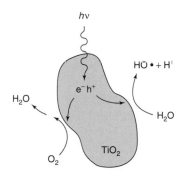

Ultraviolet radiation absorbed by $TiO_2$ creates an electron-hole pair. The hole oxidizes $H_2O$ to the powerful oxidant HO·. The electron reduces $O_2$ to $H_2O$ by a chain of reactions. $TiO_2$ is a catalyst and $O_2$ is consumed in the net reaction: organic C + $O_2 \rightarrow CO_2$.

*Total oxygen demand* (TOD) tells us how much $O_2$ is required for complete combustion of pollutants in a waste stream. A volume of $N_2$ containing a known quantity of $O_2$ is mixed with the sample and complete combustion is carried out. The remaining $O_2$ is measured by a potentiometric sensor (Box 17-1). Different species in the waste stream consume different amounts of $O_2$. For example, urea consumes five times as much $O_2$ as formic acid does. Species such as $NH_3$ and $H_2S$ also contribute to TOD.

Pollutants can be oxidized by refluxing with dichromate ($Cr_2O_7^{2-}$). *Chemical oxygen demand* (COD) is defined as the $O_2$ that is chemically equivalent to the $Cr_2O_7^{2-}$ consumed in this process. Each $Cr_2O_7^{2-}$ consumes $6e^-$ (to make $2Cr^{3+}$) and each $O_2$ consumes $4e^-$ (to make $H_2O$). Therefore, 1 mol of $Cr_2O_7^{2-}$ is chemically equivalent to 1.5 mol of $O_2$ for this computation. COD analysis is carried out by refluxing polluted water for 2 h with excess standard $Cr_2O_7^{2-}$ in $H_2SO_4$ solution containing $Ag^+$ catalyst. Unreacted $Cr_2O_7^{2-}$ is measured by titration with standard $Fe^{2+}$ or by spectrophotometry. Permits for industry may include COD limits for waste streams. "Oxidizability," which is used in

$CeO_2 \cdot xH_2O$ (also called $Ce(OH)_4$), are adequate for preparing titrants that are subsequently standardized with $Na_2C_2O_4$ or Fe as described for $MnO_4^-$.

## 16-6 Oxidation with Potassium Dichromate

In acidic solution, orange dichromate ion is a powerful oxidant that is reduced to chromic ion:

$$Cr_2O_7^{2-} + 14H^+ + 6e^- \rightleftharpoons 2Cr^{3+} + 7H_2O \qquad E° = 1.36 \text{ V}$$

Dichromate            Chromic

In 1 M HCl, the formal potential is just 1.00 V and, in 2 M $H_2SO_4$, it is 1.11 V; so dichromate is a less powerful oxidizing agent than $MnO_4^-$ or $Ce^{4+}$. In basic solution, $Cr_2O_7^{2-}$ is converted into yellow chromate ion ($CrO_4^{2-}$), whose oxidizing power is nil:

$$CrO_4^{2-} + 4H_2O + 3e^- \rightleftharpoons Cr(OH)_3(s, \text{hydrated}) + 5OH^- \qquad E° = -0.12 \text{ V}$$

Potassium dichromate, $K_2Cr_2O_7$, is a primary standard. Its solutions are stable, and it is cheap. Because $Cr_2O_7^{2-}$ is orange and complexes of $Cr^{3+}$ range from green to violet, indicators with distinctive color changes, such as diphenylamine sulfonic acid or diphenylbenzidine sulfonic acid, are used to find a dichromate end point. Alternatively, reactions can be monitored with Pt and calomel electrodes.

$K_2Cr_2O_7$ is not as strong an oxidant as $KMnO_4$ or $Ce^{4+}$. It is employed chiefly for the determination of $Fe^{2+}$ and, indirectly, for species that will oxidize $Fe^{2+}$ to $Fe^{3+}$. For indirect analyses, the unknown is treated with a measured excess of $Fe^{2+}$. Then unreacted $Fe^{2+}$ is titrated with $K_2Cr_2O_7$. For example, $ClO_3^-$, $NO_3^-$, $MnO_4^-$, and organic peroxides can be analyzed this way. Box 16-1 describes the use of dichromate in water pollution analysis.

Cr(VI) waste is toxic and should not be poured down the drain. See Box 2-1.

$$^-O_3S-\bigcirc-NH-\bigcirc-\bigcirc-NH-\bigcirc-SO_3^-$$

Diphenylbenzidine sulfonate (reduced, colorless)

$\downarrow$

$$^-O_3S-\bigcirc-N=\bigcirc=\bigcirc=N-\bigcirc-SO_3^-$$

Diphenylbenzidine sulfonate (oxidized, violet)

$$+ \ 2H^+ + 2e^-$$

---

Europe, is analogous to COD. Oxidizability is measured by refluxing with permanganate in acid solution at 100°C for 10 min. Each $MnO_4^-$ consumes five electrons and is chemically equivalent to 1.25 mol of $O_2$. Electrochemical methods based on photooxidation with $TiO_2$ could replace cumbersome refluxing with $Cr_2O_7^{2-}$ or $MnO_4^-$. Problem 17-20 describes one proven method.

*Biochemical oxygen demand* (BOD) is the $O_2$ required for biochemical degradation of organic matter by microorganisms. The procedure calls for incubating a sealed container of wastewater with no extra air space for 5 days at 20°C in the dark while microbes metabolize organic compounds in the waste. The $O_2$ dissolved in the solution is measured before and after the incubation. The difference is BOD.[15] BOD also measures species such as $HS^-$ and $Fe^{2+}$ that may be in the water. Inhibitors are added to prevent oxidation of nitrogen species such as $NH_3$. There is great interest in developing a rapid analysis to provide information equivalent to BOD. This goal could be achieved by substituting ferricyanide ($Fe(CN)_6^{3-}$) for $O_2$ as the electron sink for bacterial degradation of organic matter. Ferricyanide requires just 3 h and results are similar to that of the 5-day standard procedure.[16]

*Bound nitrogen* includes all nitrogen-containing compounds, except $N_2$, dissolved in water. Kjeldahl nitrogen analysis, described in Section 7-2, is excellent for amines and amides but fails to respond to many other forms of nitrogen. An automated combustion analyzer converts almost all forms of nitrogen in aqueous samples into NO, which is measured by chemiluminescence after reaction with ozone:[17]

*Bound nitrogen analysis:*

$$\text{Nitrogen compounds} \xrightarrow[\text{catalyst}]{O_2/\sim 1\,000°C} NO$$

$$\underset{\text{Nitrous oxide}}{NO} + \underset{\text{Ozone}}{O_3} \rightarrow \underset{\substack{\text{Nitric oxide} \\ \text{Excited electronic state}}}{NO_2^*}$$

$$NO_2^* \rightarrow NO_2 + \underset{\substack{\text{Characteristic} \\ \text{light emission}}}{h\nu}$$

Among common forms of nitrogen, only azide ($N_3^-$) and hydrazines ($RNHNH_2$) are not quantitatively converted into NO by this combustion method. Bound nitrogen measurements are required for compliance with wastewater discharge regulations.

Here is a "green" idea: $TiO_2$ can be blended into polyvinyl chloride (PVC) plastic so that the plastic degrades in sunlight.[14] Ordinary PVC lasts many years in municipal landfills after it is discarded. $TiO_2$-blended PVC would decompose in a short time. [*Courtesy H. Hidaka and S. Horikoshi, Meisei University, Tokyo.*]

$TiO_2$-blended PVC before irradiation

After irradiation for 20 days

When a reducing analyte is titrated directly with iodine (to produce $I^-$), the method is called *iodimetry*. In *iodometry*, an oxidizing analyte is added to excess $I^-$ to produce iodine, which is then titrated with standard thiosulfate solution.

Molecular iodine is only slightly soluble in water ($1.3 \times 10^{-3}$ M at 20°C), but its solubility is enhanced by complexation with iodide.

$$I_2(aq) + I^- \rightleftharpoons I_3^- \qquad K = 7 \times 10^2$$

Iodine     Iodide     Triiodide

A typical 0.05 M solution of $I_3^-$ for titrations is prepared by dissolving 0.12 mol of KI plus 0.05 mol of $I_2$ in 1 L of water. When we speak of using iodine as a titrant, we almost always mean that we are using a solution of $I_2$ plus excess $I^-$.

## Use of Starch Indicator

As described in Section 16-2, starch is used as an indicator for iodine. In a solution with no other colored species, it is possible to see the color of ~5 μM $I_3^-$. With starch, the limit of detection is extended by about a factor of 10.

In iodimetry (titration *with* $I_3^-$), starch can be added at the beginning of the titration. The first drop of excess $I_3^-$ after the equivalence point causes the solution to turn dark blue. In iodometry (titration *of* $I_3^-$), $I_3^-$ is present throughout the reaction up to the equivalence point. *Starch should not be added to such a reaction until immediately before the equivalence point* (as detected visually, by fading of the $I_3^-$; Color Plate 10). Otherwise some iodine tends to remain bound to starch particles after the equivalence point is reached.

Starch-iodine complexation is temperature dependent. At 50°C, the color is only one-tenth as intense as at 25°C. If maximum sensitivity is required, cooling in ice water is recommended.[20] Organic solvents decrease the affinity of iodine for starch and markedly reduce the utility of the indicator.

## Preparation and Standardization of $I_3^-$ Solutions

Triiodide ($I_3^-$) is prepared by dissolving solid $I_2$ in excess KI. Sublimed $I_2$ is pure enough to be a primary standard, but it is seldom used as a standard because it evaporates while it is being weighed. Instead, the approximate amount is rapidly weighed, and the solution of $I_3^-$ is standardized with a pure sample of analyte or $Na_2S_2O_3$.

Acidic solutions of $I_3^-$ are unstable because the excess $I^-$ is slowly oxidized by air:

$$6I^- + O_2 + 4H^+ \rightarrow 2I_3^- + 2H_2O$$

In neutral solutions, oxidation is insignificant in the absence of heat, light, and metal ions. At pH $\gtrsim$ 11, triiodide disproportionates to hypoiodous acid, iodate, and iodide.

An excellent way to prepare standard $I_3^-$ is to add a weighed quantity of potassium iodate to a small excess of KI.[21] Then add excess strong acid (giving pH $\approx$ 1) to produce $I_3^-$ by quantitative reverse disproportionation:

$$IO_3^- + 8I^- + 6H^+ \rightleftharpoons 3I_3^- + 3H_2O \qquad \text{(16-18)}$$

($KIO_3$ primary standard)

Freshly acidified iodate plus iodide can be used to standardize thiosulfate. The $I_3^-$ must be used immediately or else it is oxidized by air. The disadvantage of $KIO_3$ is its low molecular mass relative to the number of electrons it accepts. This property leads to a larger-than-desirable relative weighing error in preparing solutions.

## Use of Sodium Thiosulfate

Sodium thiosulfate is the almost universal titrant for triiodide. In neutral or acidic solution, triiodide oxidizes thiosulfate to tetrathionate:

$$I_3^- + 2S_2O_3^{2-} \rightleftharpoons 3I^- + O = \overset{\displaystyle O}{\underset{\displaystyle O^-}{\overset{\|}{\underset{|}{S}}}} - S - S - \overset{\displaystyle O}{\underset{\displaystyle O^-}{\overset{\|}{\underset{|}{S}}}} = O \qquad \text{(16-19)}$$

Thiosulfate                Tetrathionate

In basic solution, $I_3^-$ disproportionates to $I^-$ and HOI, which can oxidize $S_2O_3^{2-}$ to $SO_4^{2-}$. Reaction 16-19 needs to be carried out below pH 9. The common form of thiosulfate, $Na_2S_2O_3 \cdot 5H_2O$, is not pure enough to be a primary standard. Instead, thiosulfate is usually standardized by reaction with a fresh solution of $I_3^-$ prepared from $KIO_3$ plus KI or a solution of $I_3^-$ standardized with $As_4O_6$.

Anhydrous, primary standard $Na_2S_2O_3$ can be prepared from the pentahydrate.[22]

A stable solution of $Na_2S_2O_3$ can be prepared by dissolving the reagent in high-quality, freshly boiled distilled water. Dissolved $CO_2$ makes the solution acidic and promotes disproportionation of $S_2O_3^{2-}$:

$$S_2O_3^{2-} + H^+ \rightleftharpoons \underset{\text{Bisulfite}}{HSO_3^-} + \underset{\text{Sulfur}}{S(s)} \qquad (16\text{-}20)$$

and metal ions catalyze atmospheric oxidation of thiosulfate:

$$2Cu^{2+} + 2S_2O_3^{2-} \rightarrow 2Cu^+ + S_4O_6^{2-}$$

$$2Cu^+ + \tfrac{1}{2}O_2 + 2H^+ \rightarrow 2Cu^{2+} + H_2O$$

Thiosulfate solutions should be stored in the dark. Addition of 0.1 g of sodium carbonate per liter maintains the pH in an optimum range for stability of the solution. Three drops of chloroform should also be added to each bottle of thiosulfate solution to help prevent bacterial growth. An acidic solution of thiosulfate is unstable, but the reagent can be used to titrate $I_3^-$ in acidic solution because the reaction with triiodide is faster than Reaction 16-20.

## Analytical Applications of Iodine

Reducing agents can be titrated directly with standard $I_3^-$ in the presence of starch, until reaching the intense blue starch-iodine end point (Table 16-4). An example is the iodimetric

Reducing agent $+ I_3^- \rightarrow 3I^-$

**Table 16-4** Titrations with standard triiodide (iodimetric titrations)

| Species analyzed | Oxidation reaction | Notes |
|---|---|---|
| $As^{3+}$ | $H_3AsO_3 + H_2O \rightleftharpoons H_3AsO_4 + 2H^+ + 2e^-$ | Titrate directly in $NaHCO_3$ solution with $I_3^-$. |
| $Sn^{2+}$ | $SnCl_4^{2-} + 2Cl^- \rightleftharpoons SnCl_6^{2-} + 2e^-$ | Sn(IV) is reduced to Sn(II) with granular Pb or Ni in 1 M HCl and titrated in the absence of oxygen. |
| $N_2H_4$ | $N_2H_4 \rightleftharpoons N_2 + 4H^+ + 4e^-$ | Titrate in $NaHCO_3$ solution. |
| $SO_2$ | $SO_2 + H_2O \rightleftharpoons H_2SO_3$ <br> $H_2SO_3 + H_2O \rightleftharpoons SO_4^{2-} + 4H^+ + 2e^-$ | Add $SO_2$ (or $H_2SO_3$ or $HSO_3^-$ or $SO_3^{2-}$) to excess standard $I_3^-$ in dilute acid and back-titrate unreacted $I_3^-$ with standard thiosulfate. |
| $H_2S$ | $H_2S \rightleftharpoons S(s) + 2H^+ + 2e^-$ | Add $H_2S$ to excess $I_3^-$ in 1 M HCl and back-titrate with thiosulfate. |
| $Zn^{2+}, Cd^{2+}, Hg^{2+}, Pb^{2+}$ | $M^{2+} + H_2S \rightarrow MS(s) + 2H^+$ <br> $MS(s) \rightleftharpoons M^{2+} + S + 2e^-$ | Precipitate and wash metal sulfide. Dissolve in 3 M HCl with excess standard $I_3^-$ and back-titrate with thiosulfate. |
| Cysteine, glutathione, thioglycolic acid, mercaptoethanol | $2RSH \rightleftharpoons RSSR + 2H^+ + 2e^-$ | Titrate the sulfhydryl compound at pH 4–5 with $I_3^-$. |
| HCN | $I_2 + HCN \rightleftharpoons ICN + I^- + H^+$ | Titrate in carbonate-bicarbonate buffer, using $p$-xylene as an extraction indicator. |
| $H_2C{=}O$ | $H_2CO + 3OH^- \rightleftharpoons HCO_2^- + 2H_2O + 2e^-$ | Add excess $I_3^-$ plus NaOH to the unknown. After 5 min, add HCl and back-titrate with thiosulfate. |
| Glucose (and other reducing sugars) | $\overset{\displaystyle O}{\overset{\displaystyle \|}{RCH}} + 3OH^- \rightleftharpoons RCO_2^- + 2H_2O + 2e^-$ | Add excess $I_3^-$ plus NaOH to the sample. After 5 min, add HCl and back-titrate with thiosulfate. |
| Ascorbic acid (vitamin C) | Ascorbate $+ H_2O \rightleftharpoons$ dehydroascorbate $+ 2H^+ + 2e^-$ | Titrate directly with $I_3^-$. |
| $H_3PO_3$ | $H_3PO_3 + H_2O \rightleftharpoons H_3PO_4 + 2H^+ + 2e^-$ | Titrate in $NaHCO_3$ solution. |

**Box 16-2** Iodometric Analysis of High-Temperature Superconductors

An important application of superconductors (see beginning of this chapter) is in powerful electromagnets needed for medical magnetic resonance imaging. Ordinary conductors in such magnets require a huge amount of electric power. Because electricity moves through a superconductor with no resistance, the voltage can be removed from the electromagnetic coil once the current has started. Current continues to flow, and the power consumption is *0* because the resistance is 0.

A breakthrough in superconductor technology came with the discovery[24] of yttrium barium copper oxide, $YBa_2Cu_3O_7$, whose crystal structure is shown here. When heated, the material readily loses oxygen atoms from the Cu–O chains, and any composition between $YBa_2Cu_3O_7$ and $YBa_2Cu_3O_6$ is observable.

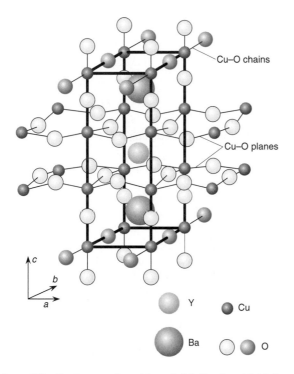

Structure of $YBa_2Cu_3O_7$, reproduced from G. F. Holland and A. M. Stacy, "Physical Properties of the Quaternary Oxide Superconductor $YBa_2Cu_3O_x$," *Acc. Chem. Res.* **1988,** *21,* 8. One-dimensional Cu–O chains (shown in color) run along the crystallographic *b*-axis, and two-dimensional Cu–O sheets lie in the *a–b* plane. Loss of colored oxygen atoms from the chains at elevated temperature results in $YBa_2Cu_3O_6$.

When high-temperature superconductors were discovered, the oxygen content in the formula $YBa_2Cu_3O_x$ was unknown. $YBa_2Cu_3O_7$ represents an unusual set of oxidation states, because the common states of yttrium and barium are $Y^{3+}$ and $Ba^{2+}$, and

the common states of copper are $Cu^{2+}$ and $Cu^+$. If all the copper were $Cu^{2+}$, the formula of the superconductor would be $(Y^{3+})(Ba^{2+})_2(Cu^{2+})_3(O^{2-})_{6.5}$, with a cation charge of $+13$ and an anion charge of $-13$. If $Cu^+$ is present, the oxygen content would be less than 6.5 per formula unit. The composition $YBa_2Cu_3O_7$ requires $Cu^{3+}$, which is rather rare. Formally, $YBa_2Cu_3O_7$ can be thought of as $(Y^{3+})(Ba^{2+})_2(Cu^{2+})_2(Cu^{3+})(O^{2-})_7$ with a cation charge of $+14$ and an anion charge of $-14$.

Redox titrations proved to be the most reliable way to measure the oxidation state of copper and thereby deduce the oxygen content of $YBa_2Cu_3O_x$.[25] An iodometric method includes two experiments. In Experiment A, $YBa_2Cu_3O_x$ is dissolved in dilute acid, in which $Cu^{3+}$ is converted into $Cu^{2+}$. For simplicity, we write the equations for the formula $YBa_2Cu_3O_7$, but you could balance these equations for $x \neq 7$.[26]

$$YBa_2Cu_3O_7 + 13H^+ \rightarrow$$
$$Y^{3+} + 2Ba^{2+} + 3Cu^{2+} + \tfrac{13}{2}H_2O + \tfrac{1}{4}O_2 \qquad (1)$$

The total copper content is measured by treatment with iodide

$$3Cu^{2+} + \tfrac{15}{2}I^- \rightarrow 3CuI(s) + \tfrac{3}{2}I_3^- \qquad (2)$$

and titration of the liberated triiodide with standard thiosulfate by Reaction 16-19. Each mole of Cu in $YBa_2Cu_3O_7$ is equivalent to 1 mol of $S_2O_3^{2-}$ in Experiment A.

In Experiment B, $YBa_2Cu_3O_x$ is dissolved in dilute acid containing $I^-$. Each mole of $Cu^{3+}$ produces 1 mol of $I_3^-$, and each mole of $Cu^{2+}$ produces 0.5 mol of $I_3^-$.

$$Cu^{3+} + 4I^- \rightarrow CuI(s) + I_3^- \qquad (3)$$
$$Cu^{2+} + \tfrac{5}{2}I^- \rightarrow CuI(s) + \tfrac{1}{2}I_3^- \qquad (4)$$

The moles of $S_2O_3^{2-}$ required in Experiment A equal the total moles of Cu in the superconductor. The difference in $S_2O_3^{2-}$ required between Experiments B and A gives the $Cu^{3+}$ content. From this difference, you can find $x$ in the formula $YBa_2Cu_3O_x$.[27]

Although we can balance cation and anion charges in the formula $YBa_2Cu_3O_7$ by including $Cu^{3+}$ in the formula, there is no evidence for discrete $Cu^{3+}$ ions in the crystal. There is also no evidence that some of the oxygen is in the form of peroxide, $O_2^{2-}$, which also would balance the cation and anion charges. The best description of the valence state in the solid crystal involves electrons and holes delocalized in the Cu–O planes and chains. Nonetheless, the formal designation of $Cu^{3+}$ and Equations 1 through 4 accurately describe the redox chemistry of $YBa_2Cu_3O_7$. Problem 16-34 describes titrations that separately measure the oxidation numbers of Cu and Bi in superconductors such as $Bi_2Sr_2(Ca_{0.8}Y_{0.2})Cu_2O_{8.295}$.

determination of vitamin C:

$$\text{Ascorbic acid (vitamin C)} + I_3^- + H_2O \rightarrow \text{Dehydroascorbic acid}^{23} + 3I^- + 2H^+$$

Ascorbic acid
(vitamin C)

Dehydroascorbic
acid[23]

Oxidizing agents can be treated with excess $I^-$ to produce $I_3^-$ (Table 16-5, Box 16-2). The iodometric analysis is completed by titrating the liberated $I_3^-$ with standard thiosulfate. Starch is not added until just before the end point.

Oxidizing agent + $3I^- \rightarrow I_3^-$

**Table 16-5** Titration of $I_3^-$ produced by analyte (iodometric titrations)

| Species analyzed | Reaction | Notes |
|---|---|---|
| $Cl_2$ | $Cl_2 + 3I^- \rightleftharpoons 2Cl^- + I_3^-$ | Reaction in dilute acid. |
| $HOCl$ | $HOCl + H^+ + 3I^- \rightleftharpoons Cl^- + I_3^- + H_2O$ | Reaction in 0.5 M $H_2SO_4$. |
| $Br_2$ | $Br_2 + 3I^- \rightleftharpoons 2Br^- + I_3^-$ | Reaction in dilute acid. |
| $BrO_3^-$ | $BrO_3^- + 6H^+ + 9I^- \rightleftharpoons Br^- + 3I_3^- + 3H_2O$ | Reaction in 0.5 M $H_2SO_4$. |
| $IO_3^-$ | $2IO_3^- + 16I^- + 12H^+ \rightleftharpoons 6I_3^- + 6H_2O$ | Reaction in 0.5 M HCl. |
| $IO_4^-$ | $2IO_4^- + 22I^- + 16H^+ \rightleftharpoons 8I_3^- + 8H_2O$ | Reaction in 0.5 M HCl. |
| $O_2$ | $O_2 + 4Mn(OH)_2 + 2H_2O \rightleftharpoons 4Mn(OH)_3$ <br> $2Mn(OH)_3 + 6H^+ + 3I^- \rightleftharpoons 2Mn^{2+} + I_3^- + 6H_2O$ | The sample is treated with $Mn^{2+}$, NaOH, and KI. After 1 min, it is acidified with $H_2SO_4$, and the $I_3^-$ is titrated. |
| $H_2O_2$ | $H_2O_2 + 3I^- + 2H^+ \rightleftharpoons I_3^- + 2H_2O$ | Reaction in 1 M $H_2SO_4$ with $NH_4MoO_3$ catalyst. |
| $O_3{}^a$ | $O_3 + 3I^- + 2H^+ \rightleftharpoons O_2 + I_3^- + H_2O$ | $O_3$ is passed through neutral 2 wt% KI solution. Add $H_2SO_4$ and titrate. |
| $NO_2^-$ | $2HNO_2 + 2H^+ + 3I^- \rightleftharpoons 2NO + I_3^- + 2H_2O$ | The nitric oxide is removed (by bubbling $CO_2$ generated in situ) prior to titration of $I_3^-$. |
| $As^{5+}$ | $H_3AsO_4 + 2H^+ + 3I^- \rightleftharpoons H_3AsO_3 + I_3^- + H_2O$ | Reaction in 5 M HCl. |
| $S_2O_8^{2-}$ | $S_2O_8^{2-} + 3I^- \rightleftharpoons 2SO_4^{2-} + I_3^-$ | Reaction in neutral solution. Then acidify and titrate. |
| $Cu^{2+}$ | $2Cu^{2+} + 5I^- \rightleftharpoons 2CuI(s) + I_3^-$ | $NH_4HF_2$ is used as a buffer. |
| $Fe(CN)_6^{3-}$ | $2Fe(CN)_6^{3-} + 3I^- \rightleftharpoons 2Fe(CN)_6^{4-} + I_3^-$ | Reaction in 1 M HCl. |
| $MnO_4^-$ | $2MnO_4^- + 16H^+ + 15I^- \rightleftharpoons 2Mn^{2+} + 5I_3^- + 8H_2O$ | Reaction in 0.1 M HCl. |
| $MnO_2$ | $MnO_2(s) + 4H^+ + 3I^- \rightleftharpoons Mn^{2+} + I_3^- + 2H_2O$ | Reaction in 0.5 M $H_3PO_4$ or HCl. |
| $Cr_2O_7^{2-}$ | $Cr_2O_7^{2-} + 14H^+ + 9I^- \rightleftharpoons 2Cr^{3+} + 3I_3^- + 7H_2O$ | Reaction in 0.4 M HCl requires 5 min for completion and is particularly sensitive to air oxidation. |
| $Ce^{4+}$ | $2Ce^{4+} + 3I^- \rightleftharpoons 2Ce^{3+} + I_3^-$ | Reaction in 1 M $H_2SO_4$. |

a. The pH must be $\geq 7$ when $O_3$ is added to $I^-$. In acidic solution, each $O_3$ produces 1.25 $I_3^-$, not 1 $I_3^-$. [N. V. Klassen, D. Marchington, and H. C. E. McGowan, Anal. Chem. **1994**, 66, 2921.]

## Terms to Understand

| | | |
|---|---|---|
| amalgam | preoxidation | redox indicator |
| disproportionation | prereduction | redox titration |

## Summary

Redox titrations are based on an oxidation-reduction reaction between analyte and titrant. Sometimes a quantitative chemical preoxidation (with reagents such as $S_2O_8^{2-}$, AgO, $NaBiO_3$, or $H_2O_2$) or prereduction (with reagents such as $SnCl_2$, $CrCl_2$, $SO_2$, $H_2S$, or a metallic reductor column) is necessary to adjust the oxidation state of the analyte prior to analysis. The end point of a redox titration is commonly detected by potentiometry or with a redox indicator. A useful indicator must have a transition range ($= E°$(indicator) $\pm 0.059\ 16/n$ V) that overlaps the abrupt change in potential of the titration curve.

The greater the difference in reduction potential between analyte and titrant, the sharper will be the end point. Plateaus before and after the equivalence point are centered near $E°$(analyte) and $E°$(titrant). Prior to the equivalence point, the half-reaction involving analyte is used to find the voltage because the concentrations of both the oxidized and the reduced forms of analyte are known. After the equivalence point, the half-reaction involving titrant is employed. At the equivalence point, both half-reactions are used simultaneously to find the voltage.

Common oxidizing titrants include $KMnO_4$, $Ce^{4+}$, and $K_2Cr_2O_7$. Many procedures are based on oxidation with $I_3^-$ or titration of $I_3^-$ liberated in a chemical reaction.

## Exercises

**16-A.** A 20.0-mL solution of 0.005 00 M $Sn^{2+}$ in 1 M HCl was titrated with 0.020 0 M $Ce^{4+}$ to give $Sn^{4+}$ and $Ce^{3+}$. Calculate the potential (versus S.C.E.) at the following volumes of $Ce^{4+}$: 0.100, 1.00, 5.00, 9.50, 10.00, 10.10, and 12.00 mL. Sketch the titration curve.

**16-B.** Would indigo tetrasulfonate be a suitable redox indicator for the titration of $Fe(CN)_6^{4-}$ with $Tl^{3+}$ in 1 M HCl? (*Hint:* The potential at the equivalence point must be between the potentials for each redox couple.)

**16-C.** Compute the titration curve for Demonstration 16-1, in which 400.0 mL of 3.75 mM $Fe^{2+}$ are titrated with 20.0 mM $MnO_4^-$ at a *fixed pH* of 0.00 in 1 M $H_2SO_4$. Calculate the potential versus S.C.E. at titrant volumes of 1.0, 7.5, 14.0, 15.0, 16.0, and 30.0 mL and sketch the titration curve.

**16-D.** A 128.6-mg sample of a protein (FM 58 600) was treated with 2.000 mL of 0.048 7 M $NaIO_4$ to react with all the serine and threonine residues.

$$
\begin{array}{cc}
\underset{\substack{\text{HOCH} \\ | \\ \overset{+}{H_3}\text{NCHCO}_2^-}}{\overset{\text{R}}{|}} + IO_4^- \rightarrow & \underset{\substack{\text{O=CH} \\ + \\ \text{O=CH} \\ | \\ \text{CO}_2^-}}{\overset{\text{R}}{|}} + NH_4^+ + IO_3^-
\end{array}
$$

Serine (R = H)
Threonine (R = $CH_3$)

The solution was then treated with excess iodide to convert the unreacted periodate into iodine:

$$IO_4^- + 3I^- + H_2O \rightleftharpoons IO_3^- + I_3^- + 2OH^-$$

Titration of the iodine required 823 μL of 0.098 8 M thiosulfate. Calculate the number of serine plus threonine residues per molecule of protein. Answer to the nearest integer.

**16-E.** A titration of 50.0 mL of unknown $Fe^{2+}$ with 0.100 M $Ce^{4+}$ at 25°C monitored with Pt and calomel electrodes gave data in the following table.[6] Prepare a Gran plot and decide which data lie on a straight line. Find the x-intercept of this line, which is the equivalence volume. Calculate the molarity of $Fe^{2+}$ in the unknown.

| Titrant volume, V (mL) | E (volts) |
|---|---|
| 6.50 | 0.635 |
| 8.50 | 0.651 |
| 10.50 | 0.669 |
| 11.50 | 0.680 |
| 12.50 | 0.696 |

## Problems

### Shape of a Redox Titration Curve

**16-1.** Consider the titration in Figure 16-2.
(a) Write a balanced titration reaction.
(b) Write two different half-reactions for the indicator electrode.
(c) Write two different Nernst equations for the cell voltage.
(d) Calculate E at the following volumes of $Ce^{4+}$: 10.0, 25.0, 49.0, 50.0, 51.0, 60.0, and 100.0 mL. Compare your results with Figure 16-2.

**16-2.** Consider the titration of 100.0 mL of 0.010 0 M $Ce^{4+}$ in 1 M $HClO_4$ by 0.040 0 M $Cu^+$ to give $Ce^{3+}$ and $Cu^{2+}$, using Pt and saturated Ag | AgCl electrodes to find the end point.
(a) Write a balanced titration reaction.
(b) Write two different half-reactions for the indicator electrode.
(c) Write two different Nernst equations for the cell voltage.
(d) Calculate E at the following volumes of $Cu^+$: 1.00, 12.5, 24.5, 25.0, 25.5, 30.0, and 50.0 mL. Sketch the titration curve.

**16-3.** Consider the titration of 25.0 mL of 0.010 0 M $Sn^{2+}$ by 0.050 0 M $Tl^{3+}$ in 1 M HCl, using Pt and saturated calomel electrodes to find the end point.
(a) Write a balanced titration reaction.
(b) Write two different half-reactions for the indicator electrode.
(c) Write two different Nernst equations for the cell voltage.
(d) Calculate E at the following volumes of $Tl^{3+}$: 1.00, 2.50, 4.90, 5.00, 5.10, and 10.0 mL. Sketch the titration curve.

**16-4.** Ascorbic acid (0.010 0 M) was added to 10.0 mL of 0.020 0 M $Fe^{3+}$ in a solution buffered to pH 0.30, and the potential was monitored with Pt and saturated Ag | AgCl electrodes.

Dehydroascorbic acid $+ 2H^+ + 2e^- \rightleftharpoons$ ascorbic acid $+ H_2O$
$$E° = 0.390 \text{ V}$$

(a) Write a balanced equation for the titration reaction.
(b) Using $E° = 0.767$ V for the $Fe^{3+} | Fe^{2+}$ couple, calculate the cell voltage when 5.0, 10.0, and 15.0 mL of ascorbic acid have been added. (*Hint:* Refer to the calculations in Demonstration 16-1.)

**16-5.** Consider the titration of 25.0 mL of 0.050 0 M $Sn^{2+}$ with 0.100 M $Fe^{3+}$ in 1 M HCl to give $Fe^{2+}$ and $Sn^{4+}$, using Pt and calomel electrodes.
(a) Write a balanced titration reaction.
(b) Write two half-reactions for the indicator electrode.
(c) Write two Nernst equations for the cell voltage.
(d) Calculate E at the following volumes of $Fe^{3+}$: 1.0, 12.5, 24.0, 25.0, 26.0, and 30.0 mL. Sketch the titration curve.

### Finding the End Point

**16-6.** Select indicators from Table 16-2 that would be suitable for finding the end point in Figure 16-3. What color changes would be observed?

**16-7.** Would tris(2,2'-bipyridine)iron be a useful indicator for the titration of $Sn^{2+}$ with $Mn(EDTA)^-$? (*Hint:* The potential at the equivalence point must be between the potentials for each redox couple.)

### Adjustment of Analyte Oxidation State

**16-8.** Explain what we mean by *preoxidation* and *prereduction*. Why is it important to be able to destroy the reagents used for these purposes?

**16-9.** Write balanced reactions for the destruction of $S_2O_8^{2-}$, $Ag^{2+}$, and $H_2O_2$ by boiling.

**16-10.** What is a Jones reductor and what is it used for?

**16-11.** Why don't $Cr^{3+}$ and $TiO^{2+}$ interfere in the analysis of $Fe^{3+}$ when a Walden reductor, instead of a Jones reductor, is used for prereduction?

## Redox Reactions of $KMnO_4$, Ce(IV), and $K_2Cr_2O_7$

**16-12.** From information in Table 16-3, explain how you would use $KMnO_4$ to find the content of $(NH_4)_2S_2O_8$ in a solid mixture with $(NH_4)_2SO_4$. What is the purpose of phosphoric acid in the procedure?

**16-13.** Write balanced half-reactions in which $MnO_4^-$ acts as an oxidant at (a) pH = 0; (b) pH = 10; (c) pH = 15.

**16-14.** When 25.00 mL of unknown were passed through a Jones reductor, molybdate ion $(MoO_4^{2-})$ was converted into $Mo^{3+}$. The filtrate required 16.43 mL of 0.010 33 M $KMnO_4$ to reach the purple end point.

$$MnO_4^- + Mo^{3+} \rightarrow Mn^{2+} + MoO_2^{2+}$$

A blank required 0.04 mL. Balance the reaction and find the molarity of molybdate in the unknown.

**16-15.** A 25.00-mL volume of commercial hydrogen peroxide solution was diluted to 250.0 mL in a volumetric flask. Then 25.00 mL of the diluted solution was mixed with 200 mL of water and 20 mL of 3 M $H_2SO_4$ and titrated with 0.021 23 M $KMnO_4$. The first pink color was observed with 27.66 mL of titrant. A blank prepared from water in place of $H_2O_2$ required 0.04 mL to give visible pink color. Using the $H_2O_2$ reaction in Table 16-3, find the molarity of the commercial $H_2O_2$.

**16-16.** Two possible reactions of $MnO_4^-$ with $H_2O_2$ to produce $O_2$ and $Mn^{2+}$ are:

Scheme 1:      $MnO_4^- \rightarrow Mn^{2+}$

                $H_2O_2 \rightarrow O_2$

Scheme 2:      $MnO_4^- \rightarrow O_2 + Mn^{2+}$

                $H_2O_2 \rightarrow H_2O$

(a) Complete the half-reactions for both schemes by adding $e^-$, $H_2O$, and $H^+$ and write a balanced net equation for each scheme.
(b) Sodium peroxyborate tetrahydrate, $NaBO_3 \cdot 4H_2O$ (FM 153.86), produces $H_2O_2$ when dissolved in acid: $BO_3^- + 2H_2O \rightarrow H_2O_2 + H_2BO_3^-$. To decide whether Scheme 1 or Scheme 2 occurs, students at the U.S. Naval Academy[28] weighed 1.023 g $NaBO_3 \cdot 4H_2O$ into a 100-mL volumetric flask, added 20 mL of 1 M $H_2SO_4$, and diluted to the mark with $H_2O$. Then they titrated 10.00 mL of this solution with 0.010 46 M $KMnO_4$ until the first pale pink color persisted. How many mL of $KMnO_4$ are required in Scheme 1 and in Scheme 2? (The Scheme 1 stoichiometry was observed.)

**16-17.** A 50.00-mL sample containing $La^{3+}$ was treated with sodium oxalate to precipitate $La_2(C_2O_4)_3$, which was washed, dissolved in acid, and titrated with 18.04 mL of 0.006 363 M $KMnO_4$. Calculate the molarity of $La^{3+}$ in the unknown.

**16-18.** An aqueous glycerol solution weighing 100.0 mg was treated with 50.0 mL of 0.083 7 M $Ce^{4+}$ in 4 M $HClO_4$ at 60°C for 15 min to oxidize the glycerol to formic acid:

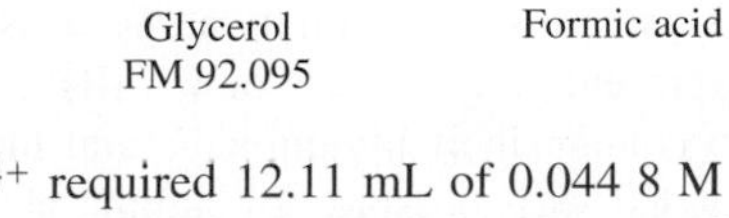

$$
\begin{array}{ccc}
CH_2 & -CH- & CH_2 \\
| & | & | \\
OH & OH & OH
\end{array}
\qquad HCO_2H
$$

         Glycerol           Formic acid
        FM 92.095

The excess $Ce^{4+}$ required 12.11 mL of 0.044 8 M $Fe^{2+}$ to reach a ferroin end point. What is the weight percent of glycerol in the unknown?

**16-19.** Nitrite $(NO_2^-)$ can be determined by oxidation with excess $Ce^{4+}$, followed by back titration of the unreacted $Ce^{4+}$. A 4.030-g sample of solid containing only $NaNO_2$ (FM 68.995) and $NaNO_3$ was dissolved in 500.0 mL. A 25.00-mL sample of this solution was treated with 50.00 mL of 0.118 6 M $Ce^{4+}$ in strong acid for 5 min, and the excess $Ce^{4+}$ was back-titrated with 31.13 mL of 0.042 89 M ferrous ammonium sulfate.

$$2Ce^{4+} + NO_2^- + H_2O \rightarrow 2Ce^{3+} + NO_3^- + 2H^+$$
$$Ce^{4+} + Fe^{2+} \rightarrow Ce^{3+} + Fe^{3+}$$

Calculate the weight percent of $NaNO_2$ in the solid.

**16-20.** Calcium fluorapatite $(Ca_{10}(PO_4)_6F_2$, FM 1 008.6) laser crystals were doped with chromium to improve their efficiency. It was suspected that the chromium could be in the +4 oxidation state.

1. To measure the total oxidizing power of chromium in the material, a crystal was dissolved in 2.9 M $HClO_4$ at 100°C, cooled to 20°C, and titrated with standard $Fe^{2+}$, using Pt and Ag | AgCl electrodes to find the end point. Chromium oxidized above the +3 state should oxidize an equivalent amount of $Fe^{2+}$ in this step. That is, $Cr^{4+}$ would consume one $Fe^{2+}$, and each atom of $Cr^{6+}$ in $Cr_2O_7^{2-}$ would consume three $Fe^{2+}$:

$$Cr^{4+} + Fe^{2+} \rightarrow Cr^{3+} + Fe^{3+}$$
$$\tfrac{1}{2}Cr_2O_7^{2-} + 3Fe^{2+} \rightarrow Cr^{3+} + 3Fe^{3+}$$

2. In a second step, the total chromium content was measured by dissolving a crystal in 2.9 M $HClO_4$ at 100°C and cooling to 20°C. Excess $S_2O_8^{2-}$ and $Ag^+$ were then added to oxidize all chromium to $Cr_2O_7^{2-}$. Unreacted $S_2O_8^{2-}$ was destroyed by boiling, and the remaining solution was titrated with standard $Fe^{2+}$. In this step, each Cr in the original unknown reacts with three $Fe^{2+}$.

$$Cr^{x+} \xrightarrow{S_2O_8^{2-}} Cr_2O_7^{2-}$$
$$\tfrac{1}{2}Cr_2O_7^{2-} + 3Fe^{2+} \rightarrow Cr^{3+} + 3Fe^{3+}$$

In step 1, 0.437 5 g of laser crystal required 0.498 mL of 2.786 mM $Fe^{2+}$ (prepared by dissolving $Fe(NH_4)_2(SO_4)_2 \cdot 6H_2O$ in 2 M $HClO_4$). In step 2, 0.156 6 g of crystal required 0.703 mL of the same $Fe^{2+}$ solution. Find the average oxidation number of Cr in the crystal and find the total micrograms of Cr per gram of crystal.

### Methods Involving Iodine

**16-21.** Why is iodine almost always used in a solution containing excess $I^-$?

**16-22.** State two ways to make standard triiodide solution.

**16-23.** In which technique, iodimetry or iodometry, is starch indicator not added until just before the end point? Why?

**16-24.** (a) Potassium iodate solution was prepared by dissolving 1.022 g of $KIO_3$ (FM 214.00) in a 500-mL volumetric flask. Then 50.00 mL of the solution was pipetted into a flask and treated with excess KI (2 g) and acid (10 mL of 0.5 M $H_2SO_4$). How many moles of $I_3^-$ are created by the reaction?
(b) The triiodide from part (a) reacted with 37.66 mL of $Na_2S_2O_3$ solution. What is the concentration of the $Na_2S_2O_3$ solution?
(c) A 1.223-g sample of solid containing ascorbic acid and inert ingredients was dissolved in dilute $H_2SO_4$ and treated with 2 g of KI and 50.00 mL of $KIO_3$ solution from part (a). Excess triiodide

required 14.22 mL of $Na_2S_2O_3$ solution from part **(b).** Find the weight percent of ascorbic acid (FM 176.13) in the unknown.

**(d)** Does it matter whether starch indicator is added at the beginning or near the end point in the titration in part **(c)**?

**16-25.** A 3.026-g portion of a copper(II) salt was dissolved in a 250-mL volumetric flask. A 50.0-mL aliquot was analyzed by adding 1 g of KI and titrating the liberated iodine with 23.33 mL of 0.046 68 M $Na_2S_2O_3$. Find the weight percent of Cu in the salt. Should starch indicator be added to this titration at the beginning or just before the end point?

**16-26.** $H_2S$ was measured by slowly adding 25.00 mL of aqueous $H_2S$ to 25.00 mL of acidified standard 0.010 44 M $I_3^-$ to precipitate elemental sulfur. (If $[H_2S] > 0.01$ M, then precipitated sulfur traps some $I_3^-$ solution, which is not subsequently titrated.) The remaining $I_3^-$ was titrated with 14.44 mL of 0.009 336 M $Na_2S_2O_3$. Find the molarity of the $H_2S$ solution. Should starch indicator be added to this titration at the beginning or just before the end point?

**16-27.** From the following reduction potentials,

$$I_2(s) + 2e^- \rightleftharpoons 2I^- \qquad E° = 0.535 \text{ V}$$
$$I_2(aq) + 2e^- \rightleftharpoons 2I^- \qquad E° = 0.620 \text{ V}$$
$$I_3^- + 2e^- \rightleftharpoons 3I^- \qquad E° = 0.535 \text{ V}$$

**(a)** Calculate the equilibrium constant for $I_2(aq) + I^- \rightleftharpoons I_3^-$.
**(b)** Calculate the equilibrium constant for $I_2(s) + I^- \rightleftharpoons I_3^-$.
**(c)** Calculate the solubility (g/L) of $I_2(s)$ in water.

**16-28.** The Kjeldahl analysis in Section 7-2 is used to measure the nitrogen content of organic compounds, which are digested in boiling sulfuric acid to decompose to ammonia, which, in turn, is distilled into standard acid. The remaining acid is then back-titrated with base. Kjeldahl himself had difficulty in 1880 discerning by lamplight the methyl red indicator end point in the back titration. He could have refrained from working at night, but instead he chose to complete the analysis differently. After distilling the ammonia into standard sulfuric acid, he added a mixture of $KIO_3$ and KI to the acid. The liberated iodine was then titrated with thiosulfate, using starch for easy end-point detection—even by lamplight.[29] Explain how the thiosulfate titration is related to the nitrogen content of the unknown. Derive a relation between moles of $NH_3$ liberated in the digestion and moles of thiosulfate required for titration of iodine.

**16-29.** Some people have an allergic reaction to the food preservative sulfite ($SO_3^{2-}$). Sulfite in wine was measured by the following procedure: To 50.0 mL of wine were added 5.00 mL of solution containing $(0.804\ 3\ \text{g KIO}_3 + 5\ \text{g KI})/100\ \text{mL}$. Acidification with 1.0 mL of 6.0 M $H_2SO_4$ quantitatively converted $IO_3^-$ into $I_3^-$. The $I_3^-$ reacted with $SO_3^{2-}$ to generate $SO_4^{2-}$, leaving excess $I_3^-$ in solution. The excess $I_3^-$ required 12.86 mL of 0.048 18 M $Na_2S_2O_3$ to reach a starch end point.

**(a)** Write the reaction that occurs when $H_2SO_4$ is added to $KIO_3 + KI$ and explain why 5 g KI were added to the stock solution. Is it necessary to measure out 5 g very accurately? Is it necessary to measure 1.0 mL of $H_2SO_4$ very accurately?

**(b)** Write a balanced reaction between $I_3^-$ and sulfite.

**(c)** Find the concentration of sulfite in the wine. Express your answer in mol/L and in mg $SO_3^{2-}$ per liter.

**(d)** *t test.* Another wine was found to contain 277.7 mg $SO_3^{2-}$/L with a standard deviation of $\pm 2.2$ mg/L for three determinations

by the iodimetric method. A spectrophotometric method gave $273.2 \pm 2.1$ mg/L in three determinations. Are these results significantly different at the 95% confidence level?

**16-30.** Potassium bromate, $KBrO_3$, is a primary standard for the generation of $Br_2$ in acidic solution:

$$BrO_3^- + 5Br^- + 6H^+ \rightleftharpoons 3Br_2(aq) + 3H_2O$$

The $Br_2$ can be used to analyze many unsaturated organic compounds. $Al^{3+}$ was analyzed as follows: An unknown was treated with 8-hydroxyquinoline (oxine) at pH 5 to precipitate aluminum oxinate, $Al(C_9H_6ON)_3$. The precipitate was washed, dissolved in warm HCl containing excess KBr, and treated with 25.00 mL of 0.020 00 M $KBrO_3$.

The excess $Br_2$ was reduced with KI, which was converted into $I_3^-$. The $I_3^-$ required 8.83 mL of 0.051 13 M $Na_2S_2O_3$ to reach a starch end point. How many milligrams of Al were in the unknown?

**16-31.** *Iodometric analysis of high-temperature superconductor.* The procedure in Box 16-2 was carried out to find the effective copper oxidation state, and therefore the number of oxygen atoms, in the formula $YBa_2Cu_3O_{7-z}$, where $z$ ranges from 0 to 0.5.

**(a)** In Experiment A of Box 16-2, 1.00 g of superconductor required 4.55 mmol of $S_2O_3^{2-}$. In Experiment B, 1.00 g of superconductor required 5.68 mmol of $S_2O_3^{2-}$. Calculate the value of $z$ in the formula $YBa_2Cu_3O_{7-z}$ (FM $666.246 - 15.999\ 4z$).

**(b)** *Propagation of uncertainty.* In several replications of Experiment A, the thiosulfate required was 4.55 ($\pm 0.10$) mmol of $S_2O_3^{2-}$ per gram of $YBa_2Cu_3O_{7-z}$. In Experiment B, the thiosulfate required was 5.68 ($\pm 0.05$) mmol of $S_2O_3^{2-}$ per gram. Calculate the uncertainty of $x$ in the formula $YBa_2Cu_3O_x$.

**16-32.** Here is a description of an analytical procedure for superconductors containing unknown quantities of Cu(I), Cu(II), Cu(III), and peroxide ($O_2^{2-}$):[30] "The possible trivalent copper and/or peroxide-type oxygen are reduced by Cu(I) when dissolving the sample (*ca.* 50 mg) in deoxygenated HCl solution (1 M) containing a known excess of monovalent copper ions (*ca.* 25 mg CuCl). On the other hand, if the sample itself contained monovalent copper, the amount of Cu(I) in the solution would increase upon dissolving the sample. The excess Cu(I) was then determined by coulometric back-titration . . . in an argon atmosphere." The abbreviation "*ca.*" means "approximately." Coulometry is an electrochemical method in which the electrons liberated in the reaction $Cu^+ \rightarrow Cu^{2+} + e^-$ are measured from the charge flowing through an electrode. Explain with your own words and equations how this analysis works.

**16-33.** $Li_{1+y}CoO_2$ is an anode for high-energy-density lithium batteries. Cobalt is present as a mixture of Co(III) and Co(II). Most preparations also contain inert lithium salts and moisture. To find the stoichiometry, Co was measured by atomic absorption and its average oxidation state was measured by a potentiometric titration.[31] For the titration, 25.00 mg of solid were dissolved under $N_2$ in 5.000 mL containing 0.100 0 M $Fe^{2+}$ in 6 M $H_2SO_4$ plus 6 M $H_3PO_4$ to give a clear pink solution:

$$Co^{3+} + Fe^{2+} \rightarrow Co^{2+} + Fe^{3+}$$

Unreacted $Fe^{2+}$ required 3.228 mL of 0.015 93 M $K_2Cr_2O_7$ for complete titration.

(a) How many mmol of $Co^{3+}$ are contained in 25.00 mg of the material?

(b) Atomic absorption found 56.4 wt% Co in the solid. What is the average oxidation state of Co?

(c) Find $y$ in the formula $Li_{1+y}CoO_2$.

(d) What is the theoretical quotient wt% Li/wt% Co in the solid? The observed quotient, after washing away inert lithium salts, was $0.138\ 8 \pm 0.000\ 6$. Is the observed quotient consistent with the average cobalt oxidation state?

**16-34.** *Warning! The Surgeon General has determined that this problem is hazardous to your health.* The oxidation numbers of Cu and Bi in high-temperature superconductors of the type $Bi_2Sr_2(Ca_{0.8}Y_{0.2})Cu_2O_x$ (which could contain $Cu^{2+}$, $Cu^{3+}$, $Bi^{3+}$, and $Bi^{5+}$) can be measured by the following procedure.[32] In Experiment A, the superconductor is dissolved in 1 M HCl containing excess 2 mM CuCl. $Bi^{5+}$ (written as $BiO_3^-$) and $Cu^{3+}$ consume $Cu^+$ to make $Cu^{2+}$:

$$BiO_3^- + 2Cu^+ + 4H^+ \rightarrow BiO^+ + 2Cu^{2+} + 2H_2O \quad (1)$$

$$Cu^{3+} + Cu^+ \rightarrow 2Cu^{2+} \quad (2)$$

The excess, unreacted $Cu^+$ is then titrated by *coulometry,* which is described in Chapter 17. In Experiment B, the superconductor is dissolved in 1 M HCl containing excess 1 mM $FeCl_2 \cdot 4H_2O$. $Bi^{5+}$ reacts with the $Fe^{2+}$ but not with $Cu^{3+}$.[33]

$$BiO_3^- + 2Fe^{2+} + 4H^+ \rightarrow BiO^+ + 2Fe^{3+} + 2H_2O \quad (3)$$

$$Cu^{3+} + \tfrac{1}{2}H_2O \rightarrow Cu^{2+} + \tfrac{1}{4}O_2 + H^+ \quad (4)$$

The excess, unreacted $Fe^{2+}$ is then titrated by coulometry. The total oxidation number of Cu + Bi is measured in Experiment A, and the oxidation number of Bi is determined in Experiment B. The difference gives the oxidation number of Cu.

(a) In Experiment A, a sample of $Bi_2Sr_2CaCu_2O_x$ (FM 760.37 + 15.999 4$x$) (containing no yttrium) weighing 102.3 mg was dissolved in 100.0 mL of 1 M HCl containing 2.000 mM CuCl. After reaction with the superconductor, coulometry detected 0.108 5 mmol of unreacted $Cu^+$ in the solution. In Experiment B, 94.6 mg of superconductor were dissolved in 100.0 mL of 1 M HCl containing 1.000 mM $FeCl_2 \cdot 4H_2O$. After reaction with the superconductor, coulometry detected 0.057 7 mmol of unreacted $Fe^{2+}$. Find the average oxidation numbers of Bi and Cu in the superconductor and the oxygen stoichiometry coefficient, $x$.

(b) Find the uncertainties in the oxidation numbers and $x$ if the quantities in Experiment A are 102.3 ($\pm$0.2) mg and 0.108 5 ($\pm$0.000 7) mmol and the quantities in Experiment B are 94.6 ($\pm$0.2) mg and 0.057 7 ($\pm$0.000 7) mmol. Assume negligible uncertainty in other quantities.

# 17 Electroanalytical Techniques

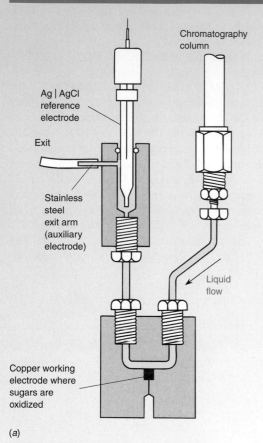

(a)

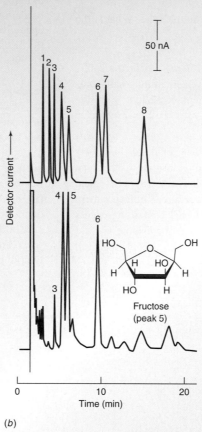

(b)

(a) Electrochemical detector measures sugars emerging from a chromatography column. Sugars are oxidized at the Cu electrode, whose potential is regulated with respect to the Ag | AgCl reference electrode. Reduction ($H_2O + e^- \rightarrow \frac{1}{2} H_2 + OH^-$) occurs at the stainless steel exit arm, and electric current is measured between Cu and steel. [Adapted from Bioanalytical Systems, West Lafayette, IN.]

(b) Anion-exchange separation of sugars in 0.1 M NaOH with CarboPac PA1 column. Upper chromatogram shows a standard mixture of (1) fucose, (2) methylglucose, (3) arabinose, (4) glucose, (5) fructose, (6) lactose, (7) sucrose, and (8) cellobiose. Lower chromatogram was obtained with Bud Dry beer diluted by a factor of 100 with water and filtered through a 0.45-μm membrane to remove particles. [P. Luo, M. Z. Luo, and R. P. Baldwin, "Determination of Sugars in Food Products," J. Chem. Ed. 1993, 70, 679.]

You can measure sugars in your favorite beverage by separating them by anion-exchange chromatography (Chapter 26) in strongly basic solution and detecting them with an electrode as they emerge from the column.[1] The —OH groups of sugars such as fructose, whose structure is shown in the chromatogram, partially dissociate to —O⁻ anions in 0.1 M NaOH. Anions are separated from one another when they pass through a column that has fixed positive charges. As sugars emerge from the column, they are detected by oxidation at a Cu electrode poised at a potential of +0.55 V versus Ag | AgCl. The chromatogram is a graph of detector current versus time. Each sugar gives a peak whose area is proportional to the moles exiting the column.

| Brand | Sugar concentration (g/L) | | | |
|---|---|---|---|---|
| | Glucose | Fructose | Lactose | Maltose |
| Budweiser | 0.54 | 0.26 | 0.84 | 2.05 |
| Bud Dry | 0.14 | 0.29 | 0.46 | — |
| Coca-Cola | 45.1 | 68.4 | — | 1.04 |
| Pepsi | 44.0 | 42.9 | — | 1.06 |
| Diet Pepsi | 0.03 | 0.01 | — | — |

Previous chapters dealt with *potentiometry*—in which voltage was measured in the absence of significant electric current. Now we consider electroanalytical methods in which current is essential to the measurement.[3] Techniques in this chapter are all examples of **electrolysis**—the process in which a chemical reaction is forced to occur at an electrode by an imposed voltage (Demonstration 17-1).

## ■ ■ ■ 17-1 Fundamentals of Electrolysis

Suppose we dip Cu and Pt electrodes into a solution of $Cu^{2+}$ and force electric current through to deposit copper metal at the cathode and to liberate $O_2$ at the anode.

Cathode: $\qquad Cu^{2+} + 2e^- \rightleftharpoons Cu(s)$

Anode: $\qquad H_2O \rightleftharpoons \frac{1}{2}O_2(g) + 2H^+ + 2e^-$ $\qquad$ (17-1)

Net reaction: $\qquad H_2O + Cu^{2+} \rightleftharpoons Cu(s) + \frac{1}{2}O_2(g) + 2H^+$

Figure 17-1 shows how we might conduct the experiment. The potentiometer measures the voltage applied by the power source between the two electrodes. The ammeter measures the current flowing through the circuit.

The electrode at which the reaction of interest occurs is called the **working electrode.** In Figure 17-1, we happen to be interested in reduction of $Cu^{2+}$, so Cu is the working electrode. The other electrode is called the *counter electrode*. We adopt the convention that *current is positive if reduction occurs at the working electrode.*

### Current Measures the Rate of Reaction

If a current $I$ flows for a time $t$, the charge $q$ passing any point in the circuit is

*Relation of charge to current and time:*
$$\boxed{q = I \cdot t}$$ $\qquad$ (17-2)

Coulombs $\quad$ Amperes · Seconds

The number of moles of electrons is

$$\text{Moles of } e^- = \frac{\text{coulombs}}{\text{coulombs/mole}} = \frac{I \cdot t}{F}$$

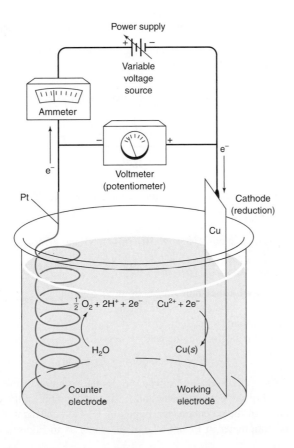

Power supply

Variable voltage source

Ammeter

$e^-$

Voltmeter (potentiometer)

Pt

$e^-$

Cathode (reduction)

Cu

$\frac{1}{2}O_2 + 2H^+ + 2e^-$ $\qquad$ $Cu^{2+} + 2e^-$

$H_2O$ $\qquad$ $Cu(s)$

Counter electrode $\qquad$ Working electrode

Electrolytic production of aluminum by the Hall-Heroult process consumes ~5% of the electrical output of the United States! $Al^{3+}$ in a molten solution of $Al_2O_3$ and cryolite ($Na_3AlF_6$) is reduced to Al at the cathode of a cell that typically draws 250 kA. This process was invented by Charles Hall in 1886 when he was 22 years old, just after graduating form Oberlin College.[2]

Charles Martin Hall. *[Photo courtesy of Alcoa.]*

*Convention: cathodic current* is considered *positive.*

An **ampere** is an electric current of 1 coulomb per second

A **coulomb** contains $6.241\ 5 \times 10^{18}$ electrons

Faraday constant:

$\qquad F = 9.648\ 5 \times 10^4$ C/mol

$$\text{Moles of electrons} = \frac{I \cdot t}{F}$$

*Figure 17-1* Electrolysis experiment. The power supply ⊣⊢ is a variable voltage source. The potentiometer measures voltage and the ammeter measures current.

Approximately 7% of electric power in the United States goes into electrolytic chemical production. The electrolysis apparatus pictured here consists of a sheet of Al foil taped or cemented to a wood surface. Any size will work, but an area about 15 cm on a side is convenient for a classroom demonstration. Tape to the metal foil (at one edge only) a sandwich consisting of filter paper,

printer paper, and another sheet of filter paper. Make a stylus from Cu wire (18 gauge or thicker) looped at one end and passed through a length of glass tubing.

Prepare a fresh solution from 1.6 g of KI, 20 mL of water, 5 mL of 1 wt% starch solution, and 5 mL of phenolphthalein indicator solution. (If the solution darkens after several days, decolorize it with a few drops of dilute $Na_2S_2O_3$.) Soak the three layers of paper with the KI-starch-phenolphthalein solution. Connect the stylus and foil to a 12-V DC power source, and write on the paper with the stylus.

When the stylus is the cathode, water is reduced to $H_2$ plus $OH^-$ and pink color appears from the reaction of $OH^-$ with phenolphthalein.

$$\text{Cathode:} \qquad H_2O + e^- \rightarrow \tfrac{1}{2}H_2(g) + OH^-$$

When the polarity is reversed and the stylus is the anode, $I^-$ is oxidized to $I_2$ and a black (very dark blue) color appears from the reaction of $I_2$ with starch.

$$\text{Anode:} \qquad I^- \rightarrow \tfrac{1}{2}I_2 + e^-$$

Pick up the top sheet of filter paper and the printer paper, and you will discover that the writing appears in the opposite color on the bottom sheet of filter paper (Color Plate 11).

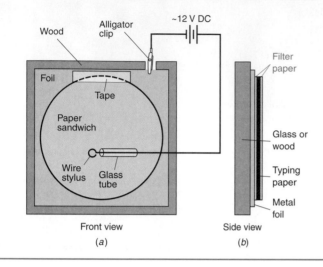

Front view
(a)

Side view
(b)

---

If a reaction requires $n$ electrons per molecule, the quantity reacting in time $t$ is

*Relation of moles to current and time:*

$$\text{Moles reacted} = \frac{I \cdot t}{nF} \qquad (17\text{-}3)$$

### Example  Relating Current, Time, and Amount of Reaction

If a current of 0.17 A flows for 16 min through the cell in Figure 17-1, how many grams of Cu(s) will be deposited?

**Solution**  We first calculate the moles of $e^-$ flowing through the cell:

$$\text{Moles of } e^- = \frac{I \cdot t}{F} = \frac{\left(0.17\dfrac{C}{s}\right)(16 \text{ min})\left(60\dfrac{s}{min}\right)}{96\,485\left(\dfrac{C}{mol}\right)} = 1.6_9 \times 10^{-3} \text{ mol}$$

The cathode half-reaction requires $2e^-$ for each Cu deposited. Therefore,

$$\text{Moles of Cu}(s) = \tfrac{1}{2}(\text{moles of } e^-) = 8.4_5 \times 10^{-4} \text{ mol}$$

The mass of Cu(s) deposited is $(8.4_5 \times 10^{-4} \text{ mol})(63.546 \text{ g/mol}) = 0.054 \text{ g}$.

### Voltage Changes When Current Flows

Figure 17-1 is drawn with the same conventions as Figures 14-4 and 14-6. The cathode—where reduction occurs—is at the right side of the figure. The positive terminal of the potentiometer is on the right-hand side.

If electric current is negligible, the cell voltage is

$$E = E(\text{cathode}) - E(\text{anode}) \qquad (17\text{-}4)$$

In Chapter 14, we wrote $E = E_+ - E_-$, where $E_+$ is the potential of the electrode attached to the positive terminal of the potentiometer and $E_-$ is the potential of the electrode attached to the negative terminal of the potentiometer. Equation 17-4 is equivalent to $E = E_+ - E_-$. The polarity of the potentiometer in Figure 17-1 is the same as in Figures 14-4 and 14-6. In an

To use $E = E(\text{cathode}) - E(\text{anode})$, you must write both reactions as *reductions*. $E(\text{cathode}) - E(\text{anode})$ is called the *open-circuit* voltage. It is the (imaginary) voltage that would be measured if there were no electrical connection between cathode and anode, so no current could flow.

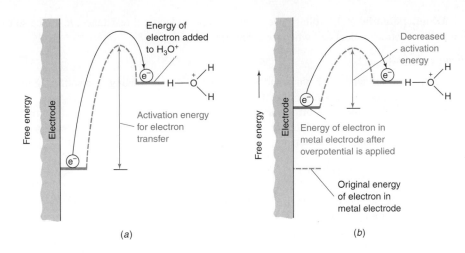

**Figure 17-2** (a) Schematic energy profile for electron transfer from a metal to $H_3O^+$, leading to liberation of $H_2$. (b) Applying a potential to the metal raises the energy of the electron in the metal and decreases the activation energy for electron transfer.

electrolysis, electrons come from the negative terminal of the power supply into the cathode of the electrolysis cell. $E$(cathode) is the potential of the electrode connected to the negative terminal of the power supply and $E$(anode) is the potential of the electrode connected to the positive terminal.

The *cathode* is the electrode connected to the *negative* terminal of the power supply.

If Reaction 17-1 contains 0.2 M $Cu^{2+}$ and 1.0 M $H^+$ and liberates $O_2$ at a pressure of 1.0 bar, we find

$$E = \underbrace{\left\{0.339 - \frac{0.059\,16}{2}\log\left(\frac{1}{[Cu^{2+}]}\right)\right\}}_{E(\text{cathode})} - \underbrace{\left\{1.229 - \frac{0.059\,16}{2}\log\left(\frac{1}{P_{O_2}^{1/2}[H^+]^2}\right)\right\}}_{E(\text{anode})}$$

$$= \left\{0.339 - \frac{0.059\,16}{2}\log\left(\frac{1}{[0.20]}\right)\right\} - \left\{1.229 - \frac{0.059\,16}{2}\log\left(\frac{1}{(1.0)^{1/2}[1.0]^2}\right)\right\}$$

$$= 0.318 - 1.229 = -0.911 \text{ V}$$

This is the voltage that would be read on the potentiometer in Figure 17-1 if there were negligible current. The voltage is negative because the positive terminal of the potentiometer is connected to the negative side of the power supply. The free-energy change computed in the margin is positive because the reaction is not spontaneous. We need the power supply to force the reaction to occur. If current is not negligible, *overpotential, ohmic potential,* and *concentration polarization* can change the voltage required to drive the reaction.

Free-energy change for Reaction 17-1:

$$\Delta G = -nFE = -nF(-0.911 \text{ V})$$

$$= -(2)\left(96\,485\,\frac{C}{mol}\right)(-0.911 \text{ V})$$

$$= +1.76 \times 10^5 \text{ C} \cdot \text{V/mol}$$

$$= +1.76 \times 10^5 \text{ J/mol} = 176 \text{ kJ/mol}$$

Note that $C \times V = J$.

**Overpotential** is the voltage required to overcome the *activation energy* for a reaction at an electrode (Figure 17-2).[5] The faster you wish to drive the reaction, the greater the overpotential that must be applied. Electric current is a measure of the rate of electron transfer. Applying a greater overpotential will sustain a higher *current density* (current per unit area of electrode surface, $A/m^2$). Table 17-1 shows that the overpotential for liberation of $H_2$ at a Cu surface must be increased from 0.479 to 1.254 V to increase the current density from 10 $A/m^2$ to 10 000 $A/m^2$. Activation energy depends on the nature of the surface. $H_2$ is evolved at a Pt surface with little overpotential, whereas a Hg surface requires ~1 V to drive the reaction.

**Table 17-1** Overpotential (V) for gas evolution at various current densities at 25°C

| Electrode | 10 A/m² | | 100 A/m² | | 1 000 A/m² | | 10 000 A/m² | |
|---|---|---|---|---|---|---|---|---|
| | H₂ | O₂ | H₂ | O₂ | H₂ | O₂ | H₂ | O₂ |
| Platinized Pt | 0.015 4 | 0.398 | 0.030 0 | 0.521 | 0.040 5 | 0.638 | 0.048 3 | 0.766 |
| Smooth Pt | 0.024 | 0.721 | 0.068 | 0.85 | 0.288 | 1.28 | 0.676 | 1.49 |
| Cu | 0.479 | 0.422 | 0.584 | 0.580 | 0.801 | 0.660 | 1.254 | 0.793 |
| Ag | 0.475 1 | 0.580 | 0.761 8 | 0.729 | 0.874 9 | 0.984 | 1.089 0 | 1.131 |
| Au | 0.241 | 0.673 | 0.390 | 0.963 | 0.588 | 1.244 | 0.798 | 1.63 |
| Graphite | 0.599 5 | | 0.778 8 | | 0.977 4 | | 1.220 0 | |
| Pb | 0.52 | | 1.090 | | 1.179 | | 1.262 | |
| Zn | 0.716 | | 0.746 | | 1.064 | | 1.229 | |
| Hg | 0.9 | | 1.0 | | 1.1 | | 1.1 | |

SOURCE: *International Critical Tables* **1929**, 6, 339. *This reference also gives overpotentials for $Cl_2$, $Br_2$, and $I_2$.*

**Ohmic potential** is the voltage needed to overcome electrical resistance, $R$, of the solution in the electrochemical cell when a current, $I$, is flowing:

*Ohmic potential:* $$E_{\text{ohmic}} = IR \qquad (17\text{-}5)$$

Resistance is measured in ohms, whose symbol is capital Greek omega, $\Omega$.

If the cell has a resistance of 2 ohms and a current of 20 mA is flowing, the voltage required to overcome the resistance is $E = (2\ \Omega)(20\ \text{mA}) = 0.040\ \text{V}$.

**Concentration polarization** occurs when the concentrations of reactants or products are not the same at the surface of the electrode as they are in bulk solution. For Reaction 17-1, the Nernst equation should be written

Electrodes respond to concentrations of reactants and products adjacent to the electrode, not to concentrations in the bulk solution.

$$E(\text{cathode}) = 0.339 - \frac{0.059\ 16}{2} \log\left(\frac{1}{[\text{Cu}^{2+}]_s}\right)$$

where $[\text{Cu}^{2+}]_s$ is the concentration in the solution *at the surface of the electrode*. If reduction of $\text{Cu}^{2+}$ occurs rapidly, $[\text{Cu}^{2+}]_s$ could be very small because $\text{Cu}^{2+}$ cannot diffuse to the electrode as fast as it is consumed. As $[\text{Cu}^{2+}]_s$ decreases, $E(\text{cathode})$ becomes more negative.

If $[\text{Cu}^{2+}]_s$ were reduced from 0.2 M to 2 $\mu$M, $E$(cathode) would change from 0.318 to 0.170 V.

Overpotential, ohmic potential, and concentration polarization make electrolysis more difficult. They drive the cell voltage more negative, requiring more voltage from the power supply in Figure 17-1 to drive the reaction forward.

$$E = \underbrace{E(\text{cathode}) - E(\text{anode})}_{} - IR - \text{overpotentials} \qquad (17\text{-}6)$$

These terms include the effects of concentration polarization

There can be concentration polarization and overpotential at both the cathode and the anode.

---

**Example** Effects of Ohmic Potential, Overpotential, and Concentration Polarization

Suppose we wish to electrolyze $\text{I}^-$ to $\text{I}_3^-$ in a 0.10 M KI solution containing $3.0 \times 10^{-5}$ M $\text{I}_3^-$ at pH 10.00 with $P_{\text{H}_2}$ fixed at 1.00 bar.

$$3\text{I}^- + 2\text{H}_2\text{O} \rightarrow \text{I}_3^- + \text{H}_2(g) + 2\text{OH}^-$$

**(a)** Find the cell voltage if no current is flowing. **(b)** Then suppose that electrolysis increases $[\text{I}_3^-]_s$ to $3.0 \times 10^{-4}$ M, but other concentrations are unaffected. Suppose that the cell resistance is 2.0 $\Omega$, the current is 63 mA, the cathode overpotential is 0.382 V, and the anode overpotential is 0.025 V. What voltage is needed to drive the reaction?

**Solution** **(a)** The open-circuit voltage is $E(\text{cathode}) - E(\text{anode})$:

Cathode: $\quad 2\text{H}_2\text{O} + 2e^- \rightarrow \text{H}_2(g) + 2\text{OH}^- \qquad E° = -0.828\ \text{V}$

Anode: $\qquad\qquad \text{I}_3^- + 2e^- \rightarrow 3\text{I}^- \qquad\qquad E° = 0.535\ \text{V}$

$$E(\text{cathode}) = -0.828 - \frac{0.059\ 16}{2} \log(P_{\text{H}_2}[\text{OH}^-]^2)$$

$$= -0.828 - \frac{0.059\ 16}{2} \log[(1.00)(1.0 \times 10^{-4})^2] = -0.591\ \text{V}$$

$$E(\text{anode}) = 0.535 - \frac{0.059\ 16}{2} \log\left(\frac{[\text{I}^-]^3}{[\text{I}_3^-]}\right)$$

$$= 0.535 - \frac{0.059\ 16}{2} \log\left(\frac{[0.10]^3}{[3.0 \times 10^{-5}]}\right) = 0.490\ \text{V}$$

$$E = E(\text{cathode}) - E(\text{anode}) = -1.081\ \text{V}$$

We would have to apply $-1.081$ V to force the reaction to occur.
**(b)** Now $E(\text{cathode})$ is unchanged but $E(\text{anode})$ changes because $[\text{I}_3^-]_s$ is different from $[\text{I}_3^-]$ in bulk solution.

$$E(\text{anode}) = 0.535 - \frac{0.059\ 16}{2} \log\left(\frac{[0.10]^3}{[3.0 \times 10^{-4}]}\right) = 0.520\ \text{V}$$

$$E = E(\text{cathode}) - E(\text{anode}) - IR - \text{overpotentials}$$

$$= -0.591\ \text{V} - 0.520\ \text{V} - (2.0\ \Omega)(0.063\ \text{A}) - 0.382\ \text{V} - 0.025\ \text{V}$$

$$= -1.644\ \text{V}$$

Instead of $-1.081$ V, we need to apply $-1.644$ V to drive the reaction.

## Controlled-Potential Electrolysis with a Three-Electrode Cell

An **electroactive species** is one that can be oxidized or reduced at an electrode. We regulate the potential of the working electrode to control which electroactive species react and which do not. Metal electrodes are said to be **polarizable,** which means that their potentials are easily changed when small currents flow. A reference electrode such as calomel or Ag | AgCl is said to be **nonpolarizable,** because its potential does not vary much unless a significant current is flowing. Ideally, we want to measure the potential of a polarizable working electrode with respect to a nonpolarizable reference electrode. How can we do this if there is to be significant current at the working electrode and negligible current at the reference electrode?

The answer is to introduce a third electrode (Figure 17-3). The **working electrode** is the one at which the reaction of interest occurs. A calomel or other **reference electrode** is used to measure the potential of the working electrode. The **auxiliary electrode** (the *counterelectrode*) is the current-supporting partner of the working electrode. Current flows between the working and the auxiliary electrodes. Negligible current flows through the reference electrode, so its potential is unaffected by ohmic potential, concentration polarization, and overpotential. It truly maintains a constant reference potential. In **controlled-potential electrolysis,** the voltage difference between working and reference electrodes in a three-electrode cell is regulated by an electronic device called a **potentiostat.**

Concentration polarization and overpotential can both occur at the working and auxiliary electrodes. There is an ohmic potential drop between working and auxiliary electrodes. To obtain the best measurement of the working electrode potential, the reference electrode should be placed as close as possible to the working electrode (Figure 17-4).

*Working electrode:* where the analytical reaction occurs
*Auxiliary electrode:* the other electrode needed for current flow
*Reference electrode:* used to measure the potential of the working electrode

The chromatographic detector at the opening of this chapter has a Cu working electrode, a stainless steel auxiliary electrode, and a Ag | AgCl reference electrode.

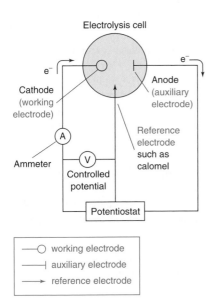

**Figure 17-3** Circuit used for controlled-potential electrolysis with a three-electrode cell.

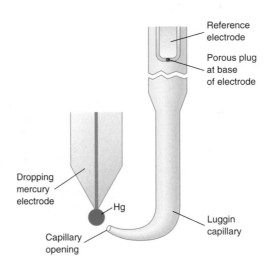

**Figure 17-4** Use of a Luggin capillary to position a reference electrode as close as possible to the working electrode (shown as a dropping mercury electrode in this illustration). The capillary, with an opening of ~0.2 mm, is filled with the same electrolyte that is in the analyte solution. The reference electrode is in contact with the capillary solution. There is negligible current flow in the capillary, so there is negligible ohmic loss between the tip of the capillary and the reference electrode.

## ▮▮▮▮ 17-2 Electrogravimetric Analysis

In **electrogravimetric analysis,** analyte is quantitatively deposited on an electrode by electrolysis. The electrode is weighed before and after deposition. The increase in mass tells us how much analyte was deposited. We can measure $Cu^{2+}$ in a solution by reducing it to $Cu(s)$ on a clean Pt gauze cathode with a large surface area (Figure 17-5). $O_2$ is liberated at the counter electrode.

How do you know when electrolysis is complete? One way is to observe the disappearance of color in a solution from which a colored species such as $Cu^{2+}$ or $Co^{2+}$ is removed. Another way is to expose most, but not all, of the surface of the cathode to the solution during electrolysis. To test whether or not the reaction is complete, raise the beaker or add water so that fresh surface of the cathode is exposed to the solution. After an additional period of electrolysis (15 min, say), see if the newly exposed electrode surface has a deposit. If it does,

Test for completion of the deposition:
• disappearance of color
• no deposition on freshly exposed electrode surface
• qualitative test for analyte in solution

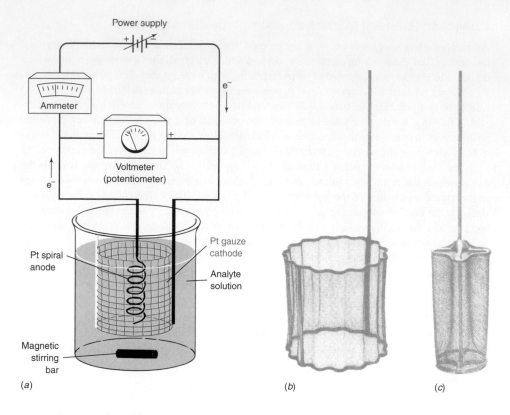

Figure 17-5 (a) Electrogravimetric analysis. Analyte is deposited on the large Pt gauze electrode. If analyte is to be oxidized, rather than reduced, the polarity of the power supply is reversed so that deposition still occurs on the large electrode. (b) Outer Pt gauze electrode. (c) Optional inner Pt gauze electrode designed to be spun by a motor in place of magnetic stirring.

repeat the procedure. If not, the electrolysis is done. A third method is to remove a drop of solution and perform a qualitative test for analyte.

In the preceding section, we calculated that $-0.911$ V needs to be applied between the electrodes to deposit Cu(s) on the cathode. The actual behavior of the electrolysis in Figure 17-6 shows that nothing special happens at $-0.911$ V. Near $-2$ V, the reaction begins in earnest. At low voltage, a small *residual current* is observed from reduction at the cathode and an equal amount of oxidation at the anode. Reduction might involve traces of dissolved $O_2$, impurities such as $Fe^{3+}$, or surface oxide on the electrode.

Table 17-1 shows that an overpotential of $\sim1$ V is required for $O_2$ formation at the Pt anode. Overpotential is the main reason why not much happens in Figure 17-6 until $-2$ V is applied. Beyond $-2$ V, the rate of reaction (the current) increases steadily. Around $-4.6$ V, the current increases more rapidly with the onset of reduction of $H_3O^+$ to $H_2$. Gas bubbles at the electrode interfere with deposition of solids.

The voltage between the two electrodes is

$$E = E(\text{cathode}) - E(\text{anode}) - IR - \text{overpotentials} \qquad (17\text{-}6)$$

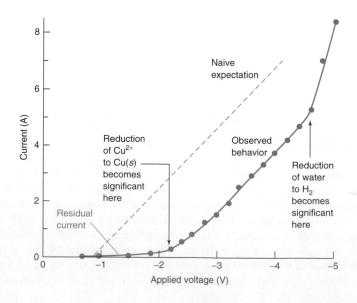

Figure 17-6 Observed current-voltage relation for electrolysis of 0.2 M $CuSO_4$ in 1 M $HClO_4$ under $N_2$, using the apparatus in Figure 17-5.

Suppose we hold the applied potential at $E = -2.0$ V. As $Cu^{2+}$ is used up, the current decreases and both ohmic and overpotentials decrease in magnitude. $E$(anode) is fairly constant because of the high concentration of solvent being oxidized at the anode ($H_2O \rightarrow \frac{1}{2}O_2 + 2H^+ + 2e^-$). If $E$ and $E$(anode) are constant and if $IR$ and overpotentials decrease in magnitude, then *E(cathode) must become more negative* to maintain the algebraic equality in Equation 17-6. $E$(cathode) drops in Figure 17-7 to $-0.4$ V, at which $H_3O^+$ is reduced to $H_2$. As $E$(cathode) falls from $+0.3$ V to $-0.4$ V, other ions such as $Co^{2+}$, $Sn^{2+}$, and $Ni^{2+}$ can be reduced. In general, then, *when the applied voltage is constant, the cathode potential drifts to more negative values and solutes more easily reduced than $H^+$ will be electrolyzed.*

To prevent the cathode potential from becoming so negative that unintended ions are reduced, a cathodic **depolarizer** such as $NO_3^-$ can be added to the solution. The cathodic depolarizer is more easily reduced than $H_3O^+$:

$$NO_3^- + 10H^+ + 8e^- \rightarrow NH_4^+ + 3H_2O$$

Alternatively, we can use a three-electrode cell (Figure 17-3) with a potentiostat to control the cathode potential and prevent unwanted side reactions.

> A *cathodic depolarizer* is reduced in preference to solvent. For *oxidation* reactions, *anodic depolarizers* include $N_2H_4$ (hydrazine) and $NH_2OH$ (hydroxylamine).

---

**Example** Controlled-Potential Electrolysis

What cathode potential is required to reduce 99.99% of 0.10 M $Cu^{2+}$ to $Cu(s)$? Is it possible to remove this $Cu^{2+}$ without reducing 0.10 M $Sn^{2+}$ in the same solution?

$$Cu^{2+} + 2e^- \rightleftharpoons Cu(s) \qquad E° = 0.339 \text{ V}$$
$$Sn^{2+} + 2e^- \rightleftharpoons Sn(s) \qquad E° = -0.141 \text{ V}$$

**Solution** If 99.99% of $Cu^{2+}$ were reduced, the concentration of remaining $Cu^{2+}$ would be $1.0 \times 10^{-5}$ M, and the required cathode potential would be

$$E(\text{cathode}) = 0.339 - \frac{0.059\ 16}{2} \log\left(\underbrace{\frac{1}{1.0 \times 10^{-5}}}_{[Cu^{2+}]}\right) = 0.19 \text{ V}$$

The cathode potential required to reduce $Sn^{2+}$ is

$$E(\text{cathode, for reduction of } Sn^{2+}) = -0.141 - \frac{0.059\ 16}{2} \log\left(\underbrace{\frac{1}{0.10}}_{[Sn^{2+}]}\right) = -0.17 \text{ V}$$

We do not expect reduction of $Sn^{2+}$ at a cathode potential more positive than $-0.17$ V. The reduction of 99.99% of $Cu^{2+}$ without reducing $Sn^{2+}$ appears feasible.

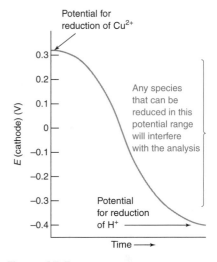

**Figure 17-7** $E$(cathode) becomes more negative with time when electrolysis is conducted in a two-electrode cell with a constant voltage between the electrodes.

## 17-3 Coulometry

**Coulometry** is a chemical analysis based on counting the electrons used in a reaction. For example, cyclohexene can be titrated with $Br_2$ generated by electrolytic oxidation of $Br^-$:

$$2Br^- \rightarrow Br_2 + 2e^- \qquad (17-7)$$

$$Br_2 + \phantom{xx} \longrightarrow \phantom{xxxx} \qquad (17-8)$$

Cyclohexene     *trans*-1,2-Dibromocyclohexane

> *Coulometric methods* are based on measuring the number of electrons that participate in a chemical reaction.

The initial solution contains an unknown quantity of cyclohexene and a large amount of $Br^-$. When Reaction 17-7 has generated just enough $Br_2$ to react with all the cyclohexene, the moles of electrons liberated in Reaction 17-7 are equal to twice the moles of $Br_2$ and therefore twice the moles of cyclohexene.

The reaction is carried out *at a constant current* with the apparatus in Figure 17-8. $Br_2$ generated at the Pt anode at the left reacts with cyclohexene. When cyclohexene is consumed, the concentration of $Br_2$ suddenly rises, signaling the end of the reaction.

The rise in $Br_2$ concentration is detected by measuring the current between the two detector electrodes at the right in Figure 17-8. A voltage of 0.25 V applied between these two electrodes is not enough to electrolyze any solute, so only a tiny current of $<1$ μA flows

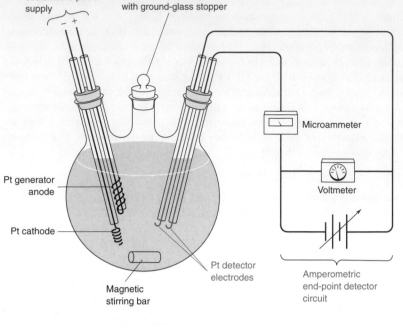

**Figure 17-8** Apparatus for coulometric titration of cyclohexene with $Br_2$. The solution contains cyclohexene, 0.15 M KBr, and 3 mM mercuric acetate in a mixed solvent of acetic acid, methanol, and water. Mercuric acetate catalyzes the addition of $Br_2$ to the olefin. [Adapted from D. H. Evans, "Coulometric Titration of Cyclohexene with Bromine," J. Chem. Ed. **1968,** 45, 88.]

To constant-current coulometric power supply

Port for adding reagents, with ground-glass stopper

Microammeter

Voltmeter

Pt generator anode

Pt cathode

Pt detector electrodes

Amperometric end-point detector circuit

Magnetic stirring bar

Both $Br_2$ and $Br^-$ must be present for the detector half-reactions to occur. Prior to the equivalence point, there is $Br^-$, but virtually no $Br_2$.

through the microammeter. At the equivalence point, cyclohexene is consumed, $[Br_2]$ suddenly increases, and detector current flows by virtue of the reactions:

Detector anode: $\quad\quad 2Br^- \rightarrow Br_2 + 2e^-$

Detector cathode: $\quad\quad Br_2 + 2e^- \rightarrow 2Br^-$

In practice, enough $Br_2$ is first generated in the absence of cyclohexene to give a detector current of 20.0 µA. When cyclohexene is added, the current decreases to a tiny value because $Br_2$ is consumed. $Br_2$ is then generated by the coulometric circuit, and the end point is taken when the detector again reaches 20.0 µA. Because the reaction is begun with $Br_2$ present, impurities that can react with $Br_2$ before analyte is added are eliminated.

The electrolysis current (not to be confused with the detector current) for the $Br_2$-generating electrodes can be controlled by a hand-operated switch. As the detector current approaches 20.0 µA, you close the switch for shorter and shorter intervals. This practice is analogous to adding titrant dropwise from a buret near the end of a titration. The switch in the coulometer circuit serves as a "stopcock" for addition of $Br_2$ to the reaction.

**Example** Coulometric Titration

A 2.000-mL volume containing 0.611 3 mg of cyclohexene/mL is to be titrated in Figure 17-8. With a constant current of 4.825 mA, how much time is required for complete titration?

**Solution** The quantity of cyclohexene is

$$\frac{(2.000 \text{ mL})(0.611\ 3 \text{ mg/mL})}{(82.146 \text{ mg/mmol})} = 0.014\ 88 \text{ mmol}$$

In Reactions 17-7 and 17-8, each mole of cyclohexene requires 1 mol of $Br_2$, which requires 2 mol of electrons. For 0.014 88 mmol of cyclohexene to react, 0.029 76 mmol of electrons must flow. From Equation 17-3,

$$\text{Moles of } e^- = \frac{I \cdot t}{F} \Rightarrow t = \frac{(\text{moles of } e^-)F}{I}$$

$$t = \frac{(0.029\ 76 \times 10^{-3} \text{ mol})(96\ 485 \text{ C/mol})}{(4.825 \times 10^{-3} \text{ C/s})} = 595.1 \text{ s}$$

It will require just under 10 min to complete the reaction.

Advantages of coulometry:
• precision
• sensitivity
• generation of unstable reagents in situ (in place)

Commercial coulometers deliver electrons with an accuracy of $\sim 0.1\%$. With extreme care, the Faraday constant has been measured to within several parts per million by coulom-

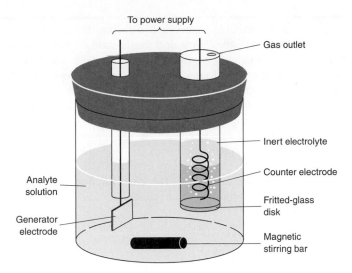

To power supply

Gas outlet

Inert electrolyte

Counter electrode

Fritted-glass disk

Magnetic stirring bar

Analyte solution

Generator electrode

**Figure 17-9** Isolating the counter electrode from analyte. Ions can flow through the porous fritted-glass disk. The liquid level in the counter electrode compartment should be higher than the liquid in the reactor so that analyte solution does not flow into the compartment.

etry.[6] Automated coulometers commonly generate $H^+$, $OH^-$, $Ag^+$, and $I_2$ to titrate analytes including $CO_2$, sulfites in food, and sulfide in wastewater.[7] Unstable reagents such as $Ag^{2+}$, $Cu^+$, $Mn^{3+}$, and $Ti^{3+}$ can be generated and used in situ.

In Figure 17-8, the reactive species ($Br_2$) is generated at the anode. The cathode products ($H_2$ from solvent and Hg from the catalyst) do not interfere with the reaction of $Br_2$ and cyclohexene. In some cases, however, $H_2$ or Hg could react with analyte. Then it is desirable to separate the counter electrode from the analyte, using the cell in Figure 17-9. $H_2$ bubbles innocuously out of the cathode chamber without mixing with the bulk solution.

The Latin *in situ* means "in place." Reagent is used right where it is generated.

### Types of Coulometry

Coulometry employs either a *constant current* or a *controlled potential*. Constant-current methods, like the preceding $Br_2$/cyclohexene example, are called **coulometric titrations.** If we know the current and the time of reaction, we know how many coulombs have been delivered from Equation 17-2: $q = I \cdot t$.

Controlled-potential coulometry in a three-electrode cell is more selective than constant-current coulometry. Because the working electrode potential is constant, current decreases exponentially as analyte concentration decreases. Charge is measured by integrating current over the time of the reaction:

$$q = \int_0^t I \, dt \qquad (17\text{-}9)$$

The number of coulombs is equal to the area under a curve of current versus time. Problem 17-20 provides an example.

In controlled-potential coulometry, current decays exponentially. You can approach the equivalence point by letting the current decay to an arbitrarily set value. For example, the current (*above* the residual current) will ideally be 0.1% of its initial value when 99.9% of the analyte has been consumed.

## 17-4 Amperometry

In **amperometry,** we measure the electric current between a pair of electrodes that are driving an electrolysis reaction. One reactant is the intended analyte and the measured current is proportional to the concentration of analyte. The measurement of dissolved $O_2$ with the **Clark electrode** in Box 17-1 is based on amperometry. Numerous *biosensors* also employ amperometry. **Biosensors**[8,9] use biological components such as *enzymes, antibodies,* or DNA for highly selective response to one analyte. Biosensors can be based on any kind of analytical signal, but electrical and optical signals are most common. A different kind of sensor based on conductivity—the "electronic nose"—is described in Box 17-2 (page 360).

*Amperometry:* Electric current is proportional to the concentration of analyte.
*Coulometry:* Total number of electrons (= current × time) tells us how much analyte is present.

*Enzyme:* A protein that catalyzes a biochemical reaction. The enzyme increases the rate of reaction by many orders of magnitude.
*Antibody:* A protein that binds to a specific target molecule called an *antigen.* Antibodies bind to foreign cells that infect your body to initiate their destruction or identify them for attack by immune system cells.

### Blood Glucose Monitor

People with diabetes may need to monitor their blood sugar (glucose) level several times a day to control their disease through diet and insulin injections. Figure 17-10 shows a home glucose monitor featuring a disposable test strip with two carbon working electrodes and a Ag | AgCl reference electrode. As little as 4 μL of blood applied in the circular opening

Box 17-1 Oxygen Sensors

$O_2$ in solution is measured amperometrically with a **Clark electrode,** in which a Pt cathode is held at $-0.6$ V with respect to a Ag | AgCl anode. The cell is covered by a semipermeable membrane, across which $O_2$ can diffuse in a few seconds. The current is proportional to the dissolved $O_2$ concentration.

$$\text{Cathode reaction:} \quad O_2 + 4H^+ + 4e^- \rightleftharpoons 2H_2O$$

The electrode must be calibrated in solutions of known $O_2$ concentration.

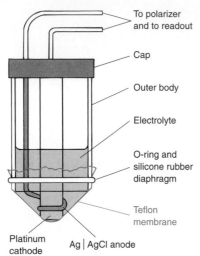

A Clark electrode can fit into the tip of a surgical catheter that is stored in a dry, sterile state. When inserted through the umbilical artery of a newborn infant, water diffuses in and activates the electrode. By this means, blood $O_2$ is monitored to detect respiratory distress. For longer-term monitoring ($\sim 1$ day), $O_2$-sensing catheters can be coated with a nitric oxide (NO)–releasing polymer that inhibits blood clotting on the sensor.[10] Micron-size amperometric $O_2$ sensors have been designed for insertion into single cells.[11] The graph shows the gradient of dissolved $O_2$ measured by a micro-Clark electrode next to a cluster of pancreatic cells. The concentration next to the cells is low because the cells consume $O_2$.

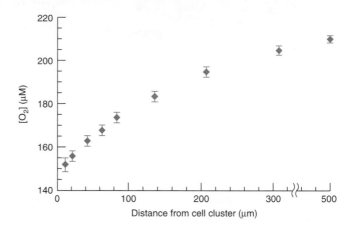

Clark oxygen electrode. [D. T. Sawyer, A. Sobkowiak, and J. L. Roberts, Jr., Electrochemistry for Chemists, 2nd ed. (New York: Wiley, 1995).] A modern, commercial oxygen electrode is a three-electrode design with a Au cathode, a Ag anode, a Ag | AgBr reference electrode, and a 50-µm-thick fluorinated ethylene-propylene polymer membrane. Leland Clark, who invented the Clark oxygen electrode, also invented the glucose monitor and the heart-lung machine.

Gradient of dissolved $O_2$ near a cluster of cells measured by micro oxygen electrode. [S.-K. Jung, J. R. Trimarchi, R. H. Sanger, and P. J. S. Smith, "Development and Application of a Self-Referencing Glucose Microsensor for the Measurement of Glucose Consumption by Pancreatic β-Cells," Anal. Chem. **2001,** 73, 3759.]

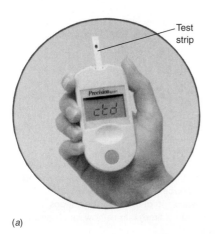

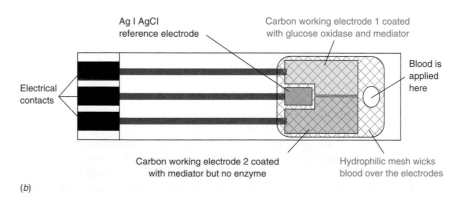

(a)  (b)

*Figure 17-10* (a) Personal glucose monitor used by diabetics to measure blood sugar level. (b) Details of disposable test strip to which a drop of blood is applied. [Courtesy Abbott Laboratories MediSense Products, Bedford, MA.]

The catalytic converter of an automobile decreases emissions of CO, nitrogen oxides ($NO_x$), and hydrocarbons. When engine exhaust enters the converter, CO and hydrocarbons are *oxidized* to $CO_2$ and nitrogen oxides are *reduced* to $N_2$ by Pt and Rh catalysts. The balance between oxidation and reduction requires an optimized air-to-fuel ratio. Too much air prohibits the reduction of $NO_x$ and too little air prevents complete combustion of CO and hydrocarbons. A potentiometric $O_2$ sensor is critical for maintaining the correct air-to-fuel ratio. This sensor contains a layer of $Y_2O_3$-doped $ZrO_2$ coated with porous Pt electrodes. Just as $LaF_3$ doped with $EuF_2$ contains mobile $F^-$ ions (Figure 15-17), $ZrO_2$ doped with $Y_2O_3$ has oxide vacancies that allow $O^{2-}$ ions to diffuse through the solid at ~600°C. One side of the sensor is exposed to exhaust gas and the other side is exposed to air. If $P_{O_2}$ is higher on the left side than on the right, $O_2$ diffuses through the porous Pt electrode at the left and is reduced to $O^{2-}$ at the surface of the $Y_2O_3$-doped $ZrO_2$. Oxide ions diffuse through the solid and are oxidized at the right-hand electrode. This galvanic cell produces a voltage difference between the two electrodes that is measured by a potentiometer.[12]

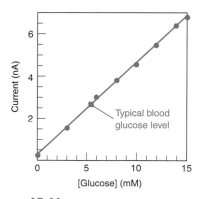

Construction of zirconia $O_2$ sensor for automobiles. [From J. T. Woestman and E. M. Logothetis, The Industrial Physicist (supplement to Physics Today), December 1995, p. 20.]

Redox chemistry at the two surfaces of the $O_2$ sensor. The voltage difference between the two electrodes is governed by the Nernst equation: $\Delta V = (RT/2F) \ln\{P_{O_2}(\text{left})/P_{O_2}(\text{right})\}$, where $R$ is the gas constant, $T$ is the sensor temperature, and $F$ is the Faraday constant.

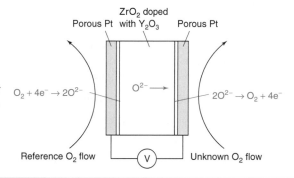

at the right of the figure is wicked over all three electrodes by a thin *hydrophilic* ("water-loving") mesh. A 20-s measurement begins when the liquid reaches the reference electrode.

Working electrode 1 is coated with the enzyme *glucose oxidase* and a *mediator*, which we describe soon. The enzyme catalyzes the reaction of glucose with $O_2$:

*Reaction in coating above working electrode 1:*

Glucose + $O_2$ $\xrightarrow{\text{glucose oxidase}}$ Gluconolactone + $H_2O_2$    (17-10)

In the absence of enzyme, the oxidation of glucose is negligible.

Early glucose monitors measured $H_2O_2$ from Reaction 17-10 by oxidation at a single working electrode, which was held at +0.6 V versus Ag | AgCl:

*Reaction at working electrode 1:*
$$H_2O_2 \rightarrow O_2 + 2H^+ + 2e^- \qquad (17\text{-}11)$$

The current is proportional to the concentration of $H_2O_2$, which, in turn, is proportional to the glucose concentration in blood (Figure 17-11).

One problem with early glucose monitors is that their response depended on the concentration of $O_2$ in the enzyme layer, because $O_2$ participates in Reaction 17-10. If the $O_2$ concentration is low, the monitor responds as though the glucose concentration were low.

**Figure 17-11** Response of an amperometric glucose electrode with dissolved $O_2$ concentration corresponding to an oxygen pressure of 0.027 bar, which is 20% lower than the typical concentration in subcutaneous tissue. [Data from S.-K. Jung and G. W. Wilson, "Polymeric Mercaptosilane-Modified Platinum Electrodes for Elimination of Interferants in Glucose Biosensors," Anal. Chem. **1996**, 68, 591.]

In the "old days," chemists prided themselves on their ability to identify compounds by odor. Smelling unknown chemicals is a bad idea because some vapors are toxic. Chemists are developing "electronic noses" to recognize odors to assess the freshness of meat, to find out if fruit is internally bruised, and to detect adulteration of food products.[13]

One approach to recognize vapors is to coat interdigitated electrodes with an electronically conductive polymer, such as a derivative of polypyrrole.

Polypyrrole

When gaseous odor molecules are absorbed by the polymer, the electrical conductivity of the polymer changes. Different gases affect conductivity in different ways. Other sensor coatings are polymers containing conductive particles of silver or graphite.

When the polymer absorbs small molecules, it swells and the conductivity decreases.

Pt/Au interdigitated electrodes

Interdigitated electrodes coated with conductive polymer to create an electronic nose. The conductivity of the polymer changes when it absorbs odor molecules. The spacing between "fingers" is ~0.25 mm.

One commercial "nose" has 32 sets of electrodes, each coated with a different polymer. The sensor yields 32 different responses when exposed to a vapor. The 32 changes are a "fingerprint" of the vapor. The electronic nose must be "trained" to recognize an odor by its characteristic fingerprint. Pattern recognition algorithms are employed for this purpose. Other electronic noses are based on changes in optical absorption of polymers at the tips of optical fibers and on changes at the gates of field effect transistors (Section 15-8).

---

> A *mediator* transports electrons between the analyte and the working electrode. The mediator undergoes no net reaction itself.

A good way to reduce $O_2$ dependence is to incorporate into the enzyme layer a species that substitutes for $O_2$ in Reaction 17-10. A substance that transports electrons between the analyte (glucose, in this case) and the electrode is called a **mediator.**

*Reaction in coating above working electrode 1:*

$$Glucose + 2 \left[ Fe(CH_3) \right]^+ \xrightarrow[\text{oxidase}]{\text{glucose}} gluconolactone + 2\ Fe(CH_3) + 2H^+ \qquad (17\text{-}12)$$

1,1'-Dimethylferricinium cation
Mediator

1,1'-Dimethylferrocene

> Ferrocene contains flat, five-membered rings, similar to benzene. Each ring formally carries one negative charge, so the oxidation state of Fe, which sits between the rings, is +2. This molecule is called a *sandwich complex.*

The mediator consumed in Reaction 17-12 is then regenerated at the working electrode:

*Reaction at working electrode 1:*

$$Fe(CH_3) \xrightarrow[-e^-]{\text{working electrode}} \left[ Fe(CH_3) \right]^+ \qquad (17\text{-}13)$$

> The mediator lowers the required working electrode potential from 0.6 V to 0.2 V versus Ag | AgCl, thereby improving the stability of the sensor and eliminating some interference by other species in the blood.

The current at the working electrode is proportional to the concentration of ferrocene, which, in turn, is proportional to the concentration of glucose in the blood.

One problem with glucose monitors is that species such as ascorbic acid (vitamin C), uric acid, and acetaminophen (Tylenol) found in blood can be oxidized at the same potential required to oxidize the mediator in Reaction 17-13. To correct for this interference, the test strip in Figure 17-10 has a second indicator electrode coated with mediator, *but not with glucose oxidase.* Interfering species that are reduced at electrode 1 are also reduced at electrode 2. The current due to glucose is the current at electrode 1 *minus* the current at electrode 2 (both measured with respect to the reference electrode). Now you see why the test strip has two working electrodes.

> A cleverly modified amperometric sensor measures glucose at a concentration of 2 fM in a 30-μL volume containing just 36 000 molecules of glucose.[14]

A challenge is to manufacture glucose monitors in such a reproducible manner that they do not require calibration. A user expects to add a drop of blood to the test strip and get a reliable reading without first constructing a calibration curve from known concentrations of glucose in blood. Each lot of test strips is highly reproducible and calibrated at the factory.

CHAPTER 17 Electroanalytical Techniques

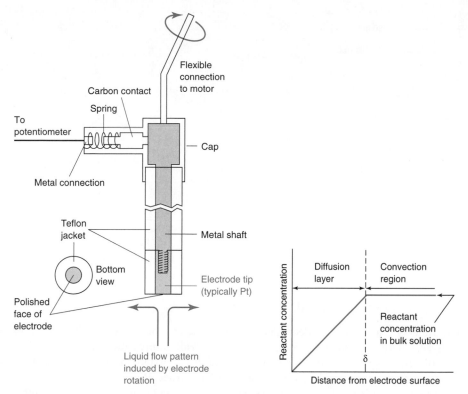

**Figure 17-12** (*a*) Rotating disk electrode. Only the polished bottom surface of the electrode, which is typically 5 mm in diameter, contacts the solution. (*b*) Schematic concentration profile of analyte near the surface of the rotating disk electrode when the potential is great enough to reduce the concentration of analyte to 0 at the electrode surface.

## Rotating Disk Electrode

A molecule has three ways to reach the surface of an electrode: (1) *diffusion* through a concentration gradient; (2) *convection,* which is the movement of bulk fluid by physical means such a stirring or boiling; and (3) *migration,* which is the attraction or repulsion of an ion by a charged surface. A common working electrode for amperometry is the **rotating disk electrode,** for which convection and diffusion control the flux of analyte to the electrode.[15]

Three ways for analyte to reach an electrode:
• diffusion
• convection
• migration

When the electrode in Figure 17-12a is spun at ~1 000 revolutions per minute, a vortex is established that brings analyte near the electrode very rapidly by convection. If the potential is great enough, analyte reacts very rapidly at the electrode, reducing the concentration right at the surface to near 0. The resulting concentration gradient is shown schematically in Figure 17-12b. Analyte must traverse the final, short distance (~10–100 μm) by diffusion alone.

The rate at which analyte diffuses from bulk solution to the surface of the electrode is proportional to the concentration difference between the two regions:

$$\text{Current} \propto \text{rate of diffusion} \propto [C]_0 - [C]_s \tag{17-14}$$

The symbol ∝ means "is proportional to."

where $[C]_0$ is the concentration in bulk solution and $[C]_s$ is the concentration at the surface of the electrode. At sufficiently great potential, the rate of reaction at the electrode is so fast that $[C]_s \ll [C]_0$ and Equation 17-14 reduces to

$$\text{Limiting current} = \text{diffusion current} \propto [C]_0 \tag{17-15}$$

The limiting current is called the **diffusion current** because it is governed by the rate at which analyte can diffuse to the electrode. The proportionality of diffusion current to bulk-solute concentration is the basis for quantitative analysis by amperometry and, in the next section, voltammetry.

The faster a rotating disk electrode spins, the thinner is the diffusion layer in Figure 17-12b and the greater is the diffusion current. A rapidly rotating Pt electrode can measure 20 nM $H_2O_2$ in rainwater.[16] $H_2O_2$ is oxidized to $O_2$ at +0.4 V (versus S.C.E.) at the Pt surface and the current is proportional to $[H_2O_2]$ in the rainwater.

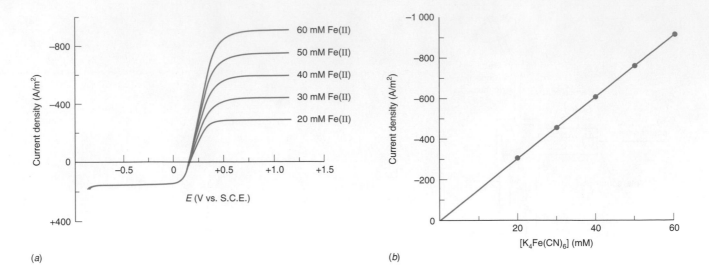

(a)                                                                (b)

**Figure 17-13** (*a*) Voltammograms for a mixture of 10 mM $K_3Fe(CN)_6$ and 20–60 mM $K_4Fe(CN)_6$ in 0.1 M $Na_2SO_4$ at a glassy carbon rotating electrode. Rotation speed = 2 000 revolutions/min and voltage sweep rate = 5 mV/s. (*b*) Dependence of limiting current on $K_4Fe(CN)_6$ concentration. *[From J. Nikolic, E. Expósito, J. Iniesta, J. González-Garcia, and V. Montiel, "Theoretical Concepts and Applications of a Rotating Disk Electrode," J. Chem. Ed.* **2000,** *77, 1191.]*

### ■ ■ ■ 17-5 Voltammetry

**Voltammetry** is a collection of techniques in which the relation between current and voltage is observed during electrochemical processes.[17] The **voltammogram** in Figure 17-13a is a graph of current versus working electrode potential for a mixture of ferricyanide and ferrocyanide being oxidized or reduced at a rotating disk electrode. By convention, current is positive when analyte is reduced at the working electrode. The limiting (diffusion) current for oxidation of $Fe(CN)_6^{4-}$ is observed at potentials above $+0.5$ V (versus S.C.E.).

$$Fe(CN)_6^{4-} \rightarrow Fe(CN)_6^{3-} + e^-$$

Ferrocyanide      Ferricyanide
Fe(II)             Fe(III)

In this region, current is governed by the rate at which $Fe(CN)_6^{4-}$ diffuses to the electrode. Figure 17-13b shows that this current is proportional to $[Fe(CN)_6^{4-}]$ in bulk solution. Below 0 V, there is another plateau corresponding to the diffusion current for reduction of $Fe(CN)_6^{3-}$, whose concentration is constant in all the solutions.

#### Polarography

Polarography has been largely replaced by voltammetry with electrode materials that do not present the toxicity hazard of mercury. Principles described for the mercury electrode apply to other electrodes. Mercury is still the electrode of choice for stripping analysis, which is the most sensitive voltammetric technique. For cleaning up mercury spills, see note 18.

Voltammetry conducted with a **dropping-mercury electrode,** is called **polarography** (Figure 17-14). The dispenser suspends one drop of mercury from the bottom of the capillary. After current and voltage are measured, the drop is mechanically dislodged. Then a fresh drop is suspended and the next measurement is made. Freshly exposed Hg yields repro-

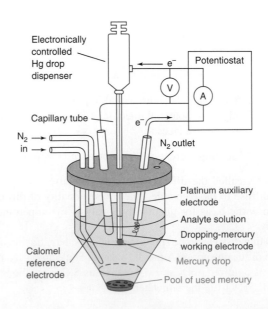

**Figure 17-14** Polarography apparatus featuring a dropping-mercury working electrode. Polarography was invented by Jaroslav Heyrovský in 1922, for which he received a Nobel Prize in 1959.

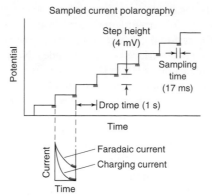

Sampled current polarography

Step height
(4 mV)

Sampling
time
(17 ms)

Drop time (1 s)

Time

Faradaic current

Charging current

Time

**Figure 17-15** *Staircase voltage profile used in* sampled current polarography. *Current is measured only during the intervals shown by heavy, colored lines. Potential is scanned toward more negative values as the experiment progresses. Lower graph shows that charging current decays more rapidly than faradaic current after each voltage step.*

ducible current-potential behavior. The current for other electrodes, such as Pt, depends on surface condition.

The vast majority of reactions studied with the Hg electrode are reductions. At a Pt surface, reduction of $H^+$ competes with reduction of many analytes:

$$2H^+ + 2e^- \rightarrow H_2(g) \qquad E° = 0$$

Table 17-1 showed that there is a large *overpotential* for reduction of $H^+$ at the Hg surface. Reactions that are thermodynamically less favorable than reduction of $H^+$ can be carried out without competitive reduction of $H^+$. In neutral or basic solutions, even alkali metal (Group 1) cations are reduced more easily than $H^+$. Furthermore, reduction of a metal into a mercury *amalgam* is more favorable than reduction to the solid state:

$$K^+ + e^- \rightarrow K(s) \qquad E° = -2.936 \text{ V}$$
$$K^+ + e^- + Hg \rightarrow K(in\ Hg) \qquad E° = -1.975 \text{ V}$$

An **amalgam** is anything dissolved in Hg.

Mercury is not useful for studying oxidations because Hg is oxidized in noncomplexing media near $+0.25$ V (versus S.C.E.). If the concentration of $Cl^-$ is 1 M, Hg is oxidized near 0 V because Hg(II) is stabilized by $Cl^-$:

$$Hg(l) + 4Cl^- \rightleftharpoons HgCl_4^{2-} + 2e^-$$

To study oxidation by voltammetry, Pt, Au, C, or diamond working electrodes in appropriate solvents provide a wide range of accessible redox potentials.

One way to conduct a measurement is by **sampled current polarography** with the *staircase voltage ramp* in Figure 17-15. After each drop of Hg is dispensed, the potential is made more negative by 4 mV. After a wait of almost 1 s, current is measured during the last 17 ms of the life of each Hg drop. The **polarographic wave** in Figure 17-16a results from reduction of $Cd^{2+}$ analyte to form an amalgam:

$$Cd^{2+} + 2e^- \rightarrow Cd(in\ Hg)$$

The potential at which half the maximum current is reached in Figure 17-16a, called the **half-wave potential** ($E_{1/2}$), is characteristic of a given analyte in a given medium and can be used for qualitative analysis. For electrode reactions in which reactants and products are both in solution, such as $Fe^{3+} + e^- \rightleftharpoons Fe^{2+}$, $E_{1/2}$ is nearly equal to $E°$ for the half-reaction.

Approximate working electrode potential range (vs. S.C.E.) in 1 M $H_2SO_4$:

| | |
|---|---|
| Pt | $-0.2$ to $+0.9$ V |
| Au | $-0.3$ to $+1.4$ V |
| Hg | $-1.3$ to $+0.1$ V |
| Glassy carbon | $-0.8$ to $+1.1$ V |
| B-doped diamond[19] | $-1.5$ to $+1.7$ V |
| Fluorinated B-doped diamond[20] | $-2.5$ to $+2.5$ V |

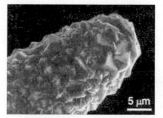

5 µm

Boron-doped diamond coating on Pt electrode. [From J. Cvačka et al., Anal. Chem. **2003**, 75, 2678. Courtesy G. M. Swain, Michigan State University.]

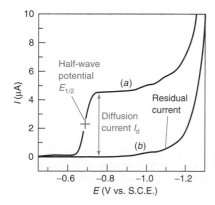

10

8

6

4

2

0

$I$ (µA)

Half-wave potential $E_{1/2}$

(a)

Residual current

Diffusion current $I_d$

(b)

$-0.6$ $-0.8$ $-1.0$ $-1.2$
$E$ (V vs. S.C.E.)

**Figure 17-16** *Sampled current polarograms of* (a) 5 mM $Cd^{2+}$ in 1 M HCl and (b) 1 M HCl alone.

Polarograms in the older literature have large oscillations superimposed on the wave in Figure 17-16a. For the first 50 years of polarography, current was measured continuously as Hg flowed from a capillary tube. Each drop grew until it fell off and was replaced by a new drop. The current oscillated from a low value when the drop was small to a high value when the drop was big.

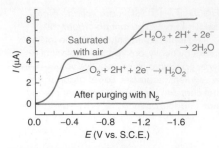

**Figure 17-17** Sampled current polarogram of 0.1 M KCl saturated with air and after bubbling $N_2$ through to remove $O_2$.

*Faradaic current:* due to redox reaction at the electrode

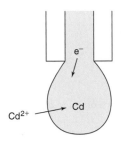

*Charging current:* due to electrostatic attraction or repulsion of ions in solution and electrons in the electrode

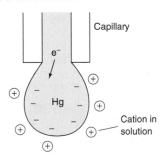

Faradaic current is the signal of interest. Charging current obscures the signal of interest, so we seek to minimize charging current.

For quantitative analysis, the *diffusion current* in the plateau region is proportional to the concentration of analyte. Diffusion current is measured from the baseline recorded without analyte in Figure 17-16b. The **residual current** in the absence of analyte is due to reduction of impurities in solution and on the surface of the electrodes. Near $-1.2$ V in Figure 17-16, current increases rapidly as reduction of $H^+$ to $H_2$ commences.

For quantitative analysis, the limiting current should be controlled by the rate at which analyte can diffuse to the electrode. We minimize convection by using an unstirred solution. We minimize migration (electrostatic attraction of analyte) by using a high concentration of *supporting electrolyte,* such as 1 M HCl in Figure 17-16.

Oxygen must be absent because $O_2$ gives two polarographic waves when it is reduced to $H_2O_2$ and then to $H_2O$ (Figure 17-17). Typically, $N_2$ is bubbled through analyte solution for 10 min to remove $O_2$.[21] Then $N_2$ flow in the gas phase is continuted to keep $O_2$ out. The liquid should not be purged with $N_2$ during a measurement, because we do not want convection of analyte to the electrode.

## Faradaic and Charging Currents

The current we seek to measure in voltammetry is **faradaic current** due to reduction or oxidation of analyte at the working electrode. In Figure 17-16a, faradaic current comes from reduction of $Cd^{2+}$ at the Hg electrode. Another current, called **charging current** (or *capacitor current* or *condenser current*) interferes with every measurement. We step the working electrode to a more negative potential by forcing electrons into the electrode from the potentiostat. In response, cations in solution flow toward the electrode, and anions flow away from the electrode (Box 17-3). This flow of ions and electrons, called the *charging current,* is not from redox reactions. We try to minimize charging current, which obscures faradaic current. The charging current usually controls the detection limit in voltammetry.

The little graph at the bottom of Figure 17-15 shows the behavior of faradaic and charging currents after each potential step. Faradaic current decays because analyte cannot diffuse to the electrode fast enough to sustain the high reaction rate. Charging current decays even faster because ions near the electrode redistribute themselves rapidly. A second after each potential step, the faradaic current is still significant and the charging current is very small.

## Square Wave Voltammetry

The most efficient voltage profile for voltammetry, called **square wave voltammetry,** uses the waveform in Figure 17-18, which consists of a square wave superimposed on a staircase.[22] During each cathodic pulse, there is a rush of analyte to be reduced at the electrode

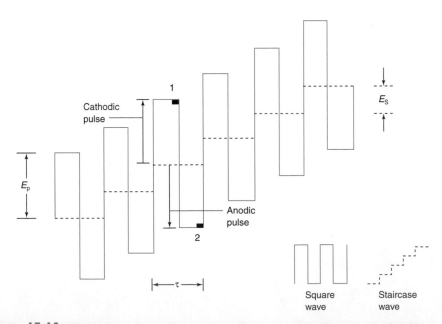

**Figure 17-18** Waveform for square wave voltammetry. Typical parameters are pulse height $(E_p) = 25$ mV, step height $(E_s) = 10$ mV, and pulse period $(\tau) = 5$ ms. Current is measured in regions 1 and 2. Optimum values are $E_p = 50/n$ mV and $E_s = 10/n$ mV, where $n$ is the number of electrons in the half-reaction.

When a power supply pumps electrons into or out of an electrode, the charged surface of the electrode attracts ions of opposite charge. The charged electrode and the oppositely charged ions next to it constitute the **electric double layer.**

A given solution has a *potential of zero charge* at which there is no excess charge on the electrode. This potential is −0.58 V (versus a calomel electrode containing 1 M KCl) for a mercury electrode immersed in 0.1 M KBr. It shifts to −0.72 V for the same electrode in 0.1 M KI.

The first layer of molecules at the surface of the electrode is *specifically adsorbed* by van der Waals and electrostatic forces. The adsorbed solute could be neutral molecules, anions, or cations. Iodide is more strongly adsorbed than bromide, so the potential of zero charge for KI is more negative than for KBr: A

more negative potential is required to expel adsorbed iodide from the electrode surface.

The next layer beyond the specifically adsorbed layer is rich in cations attracted by the negative electrode. The excess of cations decreases with increasing distance from the electrode. This region, whose composition is different from that of bulk solution, is called the *diffuse part of the double layer* and is typically 0.3–10 nm thick. The thickness is controlled by the balance between attraction toward the electrode and randomization by thermal motion.

When a species is created or destroyed by an electrochemical reaction, its concentration near the electrode is different from its concentration in bulk solution (Figure 17-12 and Color Plate 12). The region containing excess product or decreased reactant is called the *diffusion layer* (not to be confused with the diffuse part of the double layer).

Electrode-solution interface. The tightly adsorbed inner layer (also called the *compact, Helmholtz,* or *Stern* layer) may include solvent and any solute molecules. Cations in the inner layer do not completely balance the charge of the electrode. Therefore, excess cations are required in the *diffuse part of the double layer* for charge balance.

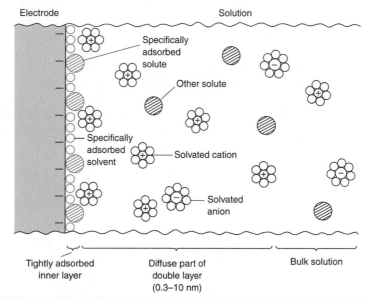

surface. During the anodic pulse, analyte that was just reduced is reoxidized. The square wave polarogram in Figure 17-19 is the *difference* in current between intervals 1 and 2 in Figure 17-18. Electrons flow from the electrode to analyte at point 1 and in the reverse direction at point 2. Because the two currents have opposite signs, their difference is larger than either current alone. When the difference is plotted, the shape of the square wave polarogram in Figure 17-19 is essentially the derivative of the sampled current polarogram.

The signal in square wave voltammetry is increased relative to a sampled current voltammogram and the wave becomes peak shaped. The signal is increased because reduced product from each cathodic pulse is right at the surface of the electrode waiting to be oxidized

**Advantages of square wave voltammetry:**
- increased signal
- derivative (peak) shape permits better resolution of neighboring signals
- faster measurement

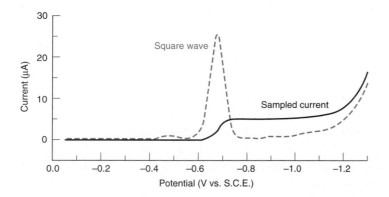

*Figure 17-19* Comparison of polarograms of 5 mM $Cd^{2+}$ in 1 M HCl. Waveforms are shown in Figures 17-15 and 17-18. Sampled current: drop time = 1 s, step height = 4 mV, sampling time = 17 ms. Square wave: drop time = 1 s, step height = 4 mV, pulse period = 67 ms, pulse height = 25 mV, sampling time = 17 ms.

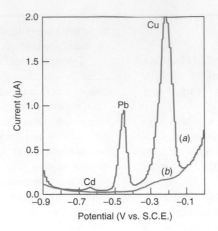

**Figure 17-20** (a) Anodic stripping voltammogram of honey dissolved in water and acidified to pH 1.2 with HCl. Cd, Pb, and Cu were reduced from solution into a thin film of Hg for 5 min at −1.4 V (versus S.C.E.) prior to recording the voltammogram. (b) Voltammogram obtained without 5-min reduction step. The concentrations of Cd and Pb in the honey were 7 and 27 ng/g (ppb), respectively. Precision was 2–4%. *[From Y. Li, F. Wahdat, and R. Neeb, "Digestion-Free Determination of Heavy Metals in Honey," Fresenius J. Anal. Chem. **1995**, 351, 678.]*

by the next anodic pulse. Each anodic pulse provides a high concentration of reactant at the surface of the electrode for the next cathodic pulse. The detection limit is reduced from $\sim10^{-5}$ M for sampled current voltammetry to $\sim10^{-7}$ M in square wave voltammetry. Because it is easier to resolve neighboring peaks than neighboring waves, square wave voltammetry can resolve species whose half-wave potentials differ by $\sim0.05$ V, whereas the potentials must differ by $\sim0.2$ V to be resolved in sampled current voltammetery. Square wave voltammetry is much faster than other voltammetric techniques. The square wave polarogram in Figure 17-19 was recorded in one-fifteenth of the time required for the sampled current polarogram. In principle, the shorter the pulse period, $\tau$, in Figure 17-18, the greater the current that will be observed. With $\tau = 5$ ms (a practical lower limit) and $E_s = 10$ mV, an entire square wave polarogram with a 1-V width is obtained with *one drop of Hg in 0.5 s.* Such rapid sweeps allow voltammograms to be recorded on individual components as they emerge from a chromatography column.

## Stripping Analysis[23]

In **stripping analysis,** analyte from a dilute solution is concentrated into a thin film of Hg or other electrode material, usually by electroreduction. The electroactive species is then *stripped* from the electrode by reversing the direction of the voltage sweep. The potential becomes more *positive, oxidizing* the species back into solution. Current measured during oxidation is proportional to the quantity of analyte that was deposited. Figure 17-20 shows an anodic stripping voltammogram of Cd, Pb, and Cu from honey.

Stripping is the most sensitive voltammetric technique (Table 17-2), because analyte is concentrated from a dilute solution. The longer the period of concentration, the more sensitive is the analysis. Only a fraction of analyte from the solution is deposited, so deposition must be done for a reproducible time (such as 5 min) with reproducible stirring.

Sensitivity can be extended by several means. *Ultrasonic vibration* during electrodeposition lowers the detection limit and enables the analysis of some analytes in difficult matrices.[24] In another approach, the detection limit for Fe(III) in seawater is $10^{-11}$ M with a *catalytic stripping*

*Stripping analysis:*

1. Concentrate analyte into a thin film of Hg by reduction.
2. Reoxidize analyte by making the potential more positive.
3. Measure polarographic signal during oxidation.

**Table 17-2** Detection limits for stripping analysis

| Analyte | Stripping mode | Detection limit |
|---|---|---|
| Ag$^+$ | Anodic | $2 \times 10^{-12}$ M[a] |
| Testosterone | Anodic | $2 \times 10^{-10}$ M[b] |
| I$^-$ | Cathodic | $1 \times 10^{-10}$ M[c] |
| DNA or RNA | Cathodic | 2–5 pg/mL[d] |
| Fe$^{3+}$ | Cathodic | $1 \times 10^{-11}$ M[e] |

*a. S. Dong and Y. Wang, Anal. Chim. Acta **1988**, 212, 341.*

*b. J. Wang, "Adsorptive Stripping Voltammetry." EG&G Princeton Applied Research Application Note A-7 (1985).*

*c. G. W. Luther III, C. Branson Swartz, and W. J. Ullman, Anal. Chem. **1988**, 60, 1721. I$^-$ is deposited onto the mercury drop by anodic oxidation:*

$$Hg(l) + I^- = \tfrac{1}{2}Hg_2I_2(adsorbed\ on\ Hg) + e^-$$

*d. S. Reher, Y. Lepka, and G. Schwedt, Fresenius J. Anal. Chem. **2000**, 368, 720; J. Wang, Anal. Chim. Acta **2003**, 500, 247.*

*e. H. Obata and C. M. G. van den Berg, Anal. Chem. **2001**, 73, 2522.*

 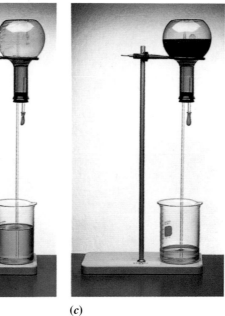

**Color Plate 1**   **HCl Fountain (Demonstration 6-2)** (*a*) Basic indicator solution in beaker. (*b*) Indicator is drawn into flask and changes to acidic color. (*c*) Solution levels at end of experiment.

(*a*)　　　　　(*b*)　　　　　(*c*)

(*a*)　　　　　(*b*)　　　　　(*c*)

**Color Plate 2**   **Fajans Titration of Cl⁻ with AgNO₃ Using Dichlorofluorescein (Demonstration 7-1)**
(*a*) Indicator before beginning titration. (*b*) AgCl precipitate before end point. (*c*) Indicator adsorbed on precipitate after end point.

(*a*)　　　　　(*b*)

**Color Plate 3**   **Effect of Ionic Strength on Ionic Dissociation (Demonstration 8-1)**
(*a*) Two beakers containing identical solutions with $FeSCN^{2+}$, $Fe^{3+}$, and $SCN^-$. (*b*) The red color of $Fe(SCN)^{2+}$ fades when $KNO_3$ is added to the right-hand beaker because the equilibrium $Fe^{3+} + SCN^- \rightleftharpoons Fe(SCN)^{2+}$ shifts to the left.

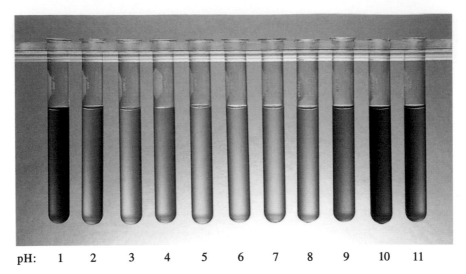

pH:  1    2    3    4    5    6    7    8    9    10    11

**Color Plate 4  Thymol Blue (Section 11-6)**  Acid-base indicator thymol blue between pH 1 and 11. The p$K$ values are 1.7 and 8.9.

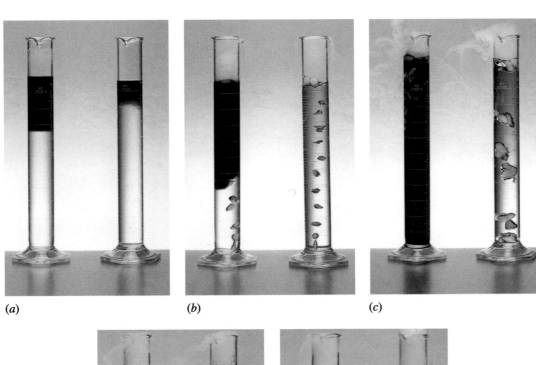

(*a*)                    (*b*)                    (*c*)

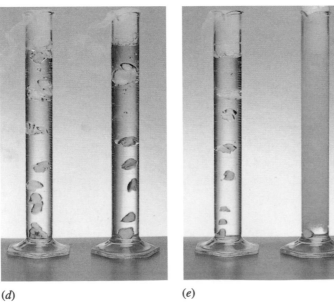

(*d*)                    (*e*)

**Color Plate 5  Indicators and Acidity of CO$_2$ (Demonstration 11-1)**  (*a*) Cylinders before adding Dry Ice. Ethanol indicator solutions of phenolphthalein (left) and bromothymol blue (right) have not yet mixed with entire cylinder. (*b*) Adding Dry Ice causes bubbling and mixing. (*c*) Further mixing. (*d*) Phenolphthalein changes to its colorless acidic form. Color of bromothymol blue is due to mixture of acidic and basic forms. (*e*) After addition of HCl and stirring of right-hand cylinder, bubbles of CO$_2$ can be seen leaving solution, and indicator changes completely to its acidic color.

**Color Plate 6** Titration of Cu(II) with EDTA, Using Auxiliary Complexing Agent (Section 12-5)
0.02 M CuSO$_4$ before titration (left). Color of Cu(II)-ammonia complex after adding ammonia buffer, pH 10 (center). End-point color when all ammonia ligands have been displaced by EDTA (right).

*(a)*

*(b)*

**Color Plate 7** Titration of Mg$^{2+}$ by EDTA, Using Eriochrome Black T Indicator (Demonstration 12-1)
(*a*) Before (left), near (center), and after (right) equivalence point. (*b*) Same titration with methyl red added as inert dye to alter colors.

**Color Plate 8** Titration of VO$^{2+}$ with Potassium Permanganate (Section 16-4) Blue VO$^{2+}$ solution prior to titration (left). Mixture of blue VO$^{2+}$ and yellow VO$_2^+$ observed during titration (center). Dark color of MnO$_4^-$ at end point (right).

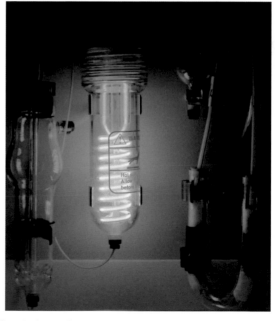

**Color Plate 9** Photolytic Environmental Carbon Analyzer (Box 16-1) A measured water sample is injected into the chamber at the left, where it is acidified with H$_3$PO$_4$ and sparged (bubbled with Ar or N$_2$) to remove CO$_2$ derived from HCO$_3^-$ and CO$_3^{2-}$. The CO$_2$ is measured by its infrared absorbance. The sample is then forced into the digestion chamber, where S$_2$O$_8^{2-}$ is added and the sample is exposed to ultraviolet radiation from an immersion lamp (the coil at the center of the photo). Sulfate radicals (SO$_4^-$) formed by irradiation oxidize most organic compounds to CO$_2$, which is measured by infrared absorbance. The U-tube at the right contains Sn and Cu granules to scavenge volatile acids, such as HCl and HBr, liberated in the digestion. [Photo courtesy Ed Urbansky, U.S. Environmental Protection Agency, Cincinnati, OH.]

**Color Plate 10** **Iodometric Titration (Section 16-7)**
$I_3^-$ solution (left). $I_3^-$ solution before end point in titration
with $S_2O_3^{2-}$ (left center). $I_3^-$ solution immediately before
end point with starch indicator present (right center). At
the end point (right).

(a)

(b)

(c)

**Color Plate 11** **Electrochemical Writing (Demonstration 17-1)** (a) Stylus used
as cathode. (b) Stylus used as anode. (c) The polarity of the foil backing is opposite that
of the stylus and produces reverse color on the bottom sheet of paper.

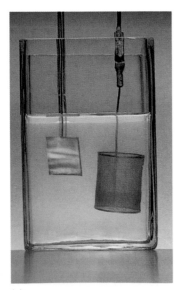

(a)

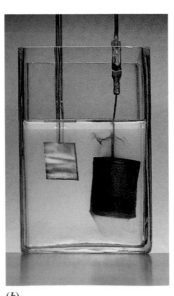

(b)

**Color Plate 12** **Formation of Diffusion Layer
During Electrolysis (Box 17-3)** (a) Cu electrode (flat
plate, left) and Pt electrode (mesh basket, right)
immersed in solution containing KI and starch, with
no electric current. (b) Starch-iodine complex forms at
surface of Pt anode when current flows.

**Color Plate 13** **Grating Dispersion**
**(Section 18-2)** Visible spectrum produced
by a grating inside a spectrophotometer.

**Color Plate 14** Beer's Law (Section 18-2)
$Fe(phenanthroline)_3^{2+}$ standards for spectrophotometric analysis. Volumetric flasks contain
$Fe(phenanthroline)_3^{2+}$ with Fe concentrations
ranging from 1 mg/L (left) to 10 mg/L (right). The
absorbance, as evidenced by the intensity of the
color, is proportional to the iron concentration.

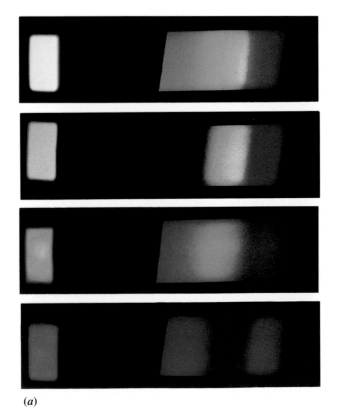

(a)

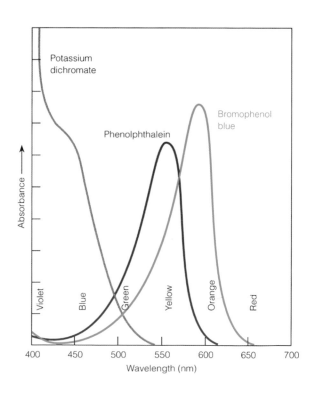

(b)

**Color Plate 15** **Absorption Spectra (Demonstration 18-1)** (a) Projected visible spectra of
(from top to bottom) white light, potassium dichromate, bromophenol blue, and phenolphthalein.
(b) Visible absorption spectra of the same compounds recorded with a spectrophotometer.

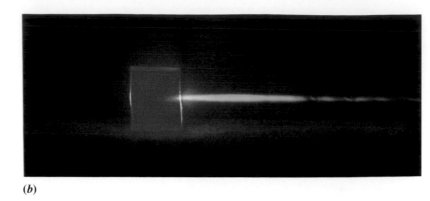

(a)          (b)

**Color Plate 16** **Luminescence (Section 18-6)** (a) Green crystal of yttrium aluminum garnet containing a small amount of $Cr^{3+}$. (b) When irradiated with high-intensity blue light from a laser at the right side, $Cr^{3+}$ absorbs blue light and emits lower energy red light. When the laser is removed, the crystal appears green again. [Courtesy M. Seltzer, Michelson Laboratory.]

**Color Plate 17** **Quenching of Ru(II) Luminescence by $O_2$ (Section 19-6)** Left: Orange-red luminescence from ~5 μM (bipyridyl)$_3$RuCl$_2$ in methanol after most of the air was removed by bubbling with Dry Ice. Right: Luminescence is quenched (decreased) after bubbling $O_2$ through the solution for 30 s.

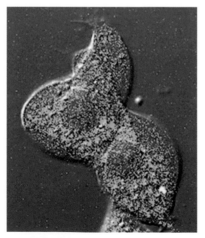 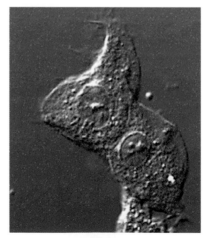

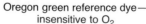
Oregon green reference dye—        Ru(II) dye—
insensitive to $O_2$        sensitive to $O_2$

**Color Plate 18** **Fluorescence from $O_2$-indicator beads inside living cells. (Section 19-6)** Green light emitted from the dye Oregon Green is independent of local $O_2$ concentration. Red light emitted from tris(4,7-diphenyl-1,10-phenanthroline)ruthenium(II) chloride decreases in the presence of $O_2$. The ratio of intensities at the red and green wavelengths tells us the concentration of $O_2$ inside the cell. [Courtesy R. Kopelman and E. Monson, University of Michigan. From H. Xu, J. W. Aylott, R. Kopelman, T. J. Miller, and M. A. Philbert, "Real-Time Method for Determination of $O_2$ Inside Living Cells Using Optical Nanosensors," *Anal. Chem.* **2001**, *73*, 4124.]

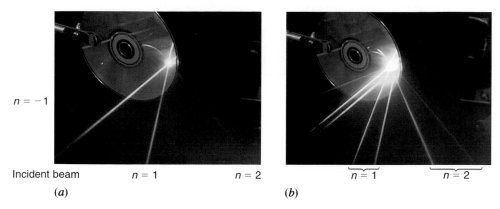

$n = -1$

Incident beam      $n = 1$      $n = 2$           $n = 1$      $n = 2$

(*a*)                       (*b*)

**Color Plate 19**   **Laser Diffraction from a Compact Disk (Section 20-2)**   The grooves in an audio compact disk or a computer compact disk have a spacing of 1.6 μm. (*a*) When a red laser strikes disk at normal incidence ($\theta = 0$ in Figure 20-6 and Equation 20-2) three diffracted beams with orders $n = +1$, $+2$, and $-1$ are observed. (*b*) Red and green lasers strike the disk at normal incidence. Green light has a shorter wavelength than red light, and so, according to Equation 20-2, green light is diffracted at smaller angle ($\phi$). Beams have been made visible by "fog" from liquid nitrogen. [Courtesy J. Tellinghuisen, Vanderbilt University. See J. Tellinghuisen, "Exploring the Diffraction Grating Using a He-Ne Laser and a CD-ROM," *J. Chem. Ed.* **2002**, *79*, 703.]

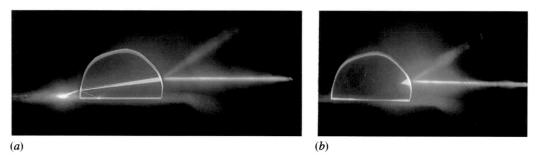

(*a*)                       (*b*)

**Color Plate 20**   **Transmission, Reflection, Refraction, Absorption, and Luminescence (Section 20-4)**   (*a*) Blue-green laser is directed into a crystal of yttrium aluminum garnet doped with $Er^{3+}$, which emits yellow light. Light entering the crystal from the right is refracted (bent) and partially reflected at the right-hand surface of the crystal. The laser beam appears yellow inside the crystal because of luminescence from $Er^{3+}$. As it exits the crystal at the left side, the laser beam is refracted again, and partially reflected back into the crystal. (*b*) Same experiment, but with blue light instead of blue-green light. Blue light is absorbed by $Er^{3+}$ and does not penetrate very far into the crystal. [Courtesy M. Seltzer, Michelson Laboratory.]

**Color Plate 21**   **Multiple Internal Reflections in a Bulk Crystal (Section 20-4)**   Multiple internal reflections observed when blue laser light enters a crystal of yttrium aluminum garnet doped with $Ho^{3+}$. The beam entering from the right is mostly reflected back into the crystal at each face, creating a zigzag pattern inside the crystal. Part of the light is transmitted out of the crystal at each face. In an optical fiber, the angle of incidence is such that the beam is totally reflected within the fiber. [Courtesy M. Seltzer, Michelson Laboratory.]

25 μm

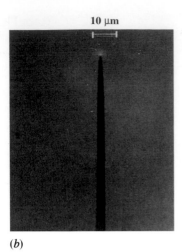

10 μm

(a)                    (b)

**Color Plate 22** **Oxygen Optode (Section 20-4)**
(*a*) Sensor prepared from a 100-μm-diameter
optical fiber. The active layer at the end contains
tris(1,10-phenanthroline)Ru(II) chloride dissolved
in polyacrylamide that is covalently bound to the
fiber. Light coming down the fiber excites the Ru
compound, which emits characteristic orange-red light
that is collected with a microscope. When immersed in a
sample containing $O_2$, emission is decreased. The
decrease is a measure of $O_2$ concentration. (*b*) An optode
with a submicrometer tip pulled from a larger fiber. This
fiber can detect 10 amol of $O_2$. [From Z. Rosenzweig and
R. Kopelman, "Development of a Submicrometer Optical
Fiber Oxygen Sensor," *Anal. Chem.* **1995**, *67*, 2650.]

**Color Plate 23** **Polychromator for Inductively Coupled Plasma Atomic Emission
Spectrometer with One Detector for Each Element (Section 21-4)** Light emitted by a
sample in the plasma enters the polychromator at the right and is dispersed into its
component wavelengths by grating at the bottom of the diagram. Each different emission
wavelength (shown schematically by colored lines) is diffracted at a different angle and
directed to a different photomultiplier detector on the focal curve. Each detector sees
only one preselected element, and all elements are measured simultaneously. [Courtesy
TJA Solutions, Franklin, MA.]

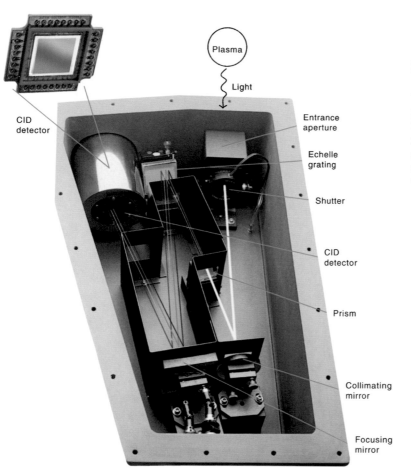

**Color Plate 24** **Polychromator for Inductively Coupled Plasma Atomic Emission Spectrometer with One Detector for All Elements (Section 21-4)** Light emitted by a sample in the plasma enters the polychromator at the upper right and is dispersed vertically by a prism and then horizontally by a grating. The resulting two-dimensional pattern of wavelengths from 165 to 1 000 nm is detected by a charge injection device detector with 262 000 pixels. All elements are detected simultaneously. [Courtesy TJA Solutions, Franklin, MA.]

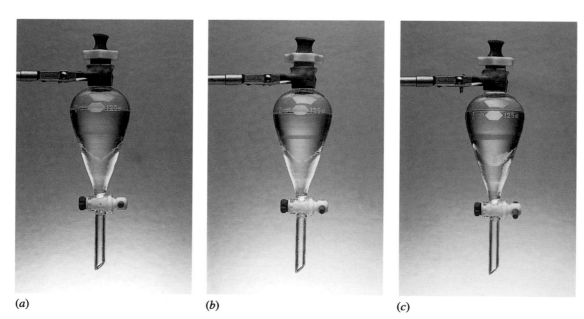

(*a*)              (*b*)              (*c*)

**Color Plate 25** **Extraction of Uranyl Nitrate from Water into Ether (Section 23-1)**
(*a*) Separatory funnel with lower aqueous layer containing yellow 1 M $UO_2(NO_3)_2$ (plus 3 M $HNO_3$ and 4 M $Ca(NO_3)_2$) beneath colorless diethyl ether layer prior to mixing.
(*b*) Yellow uranyl nitrate is distributed in both layers after shaking. (*c*) After eight extractions with ether, almost all yellow uranyl nitrate has been removed from the aqueous layer.

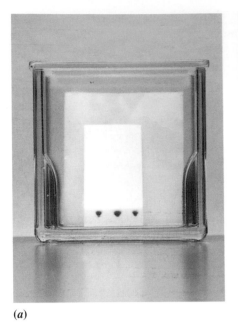
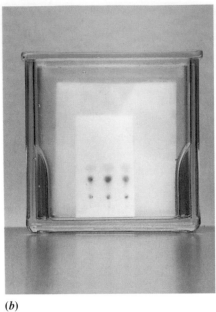

(a)  (b)

**Color Plate 26** **Thin-Layer Chromatography (Section 25-1)** The mixture to be separated is placed in tiny spots near the base of a plastic or glass plate coated with an adsorptive stationary phase. When the plate is placed in a shallow pool of solvent in a closed chamber, liquid migrates *up* the plate by capillary action. Different components of the mixture are carried along by the solvent to different extents, depending on how strongly they are adsorbed on the stationary phase. The stronger the adsorption, the slower a component travels. (a) Solvent ascends past a mixture of dyes near the bottom of the plate. (b) Separation achieved after solvent has ascended most of the way up the plate.

(a)  (b)

(c)

**Color Plate 27** **Supercritical Carbon Dioxide (Box 25-2)** (a) Liquid carbon dioxide in a 60-mL steel chamber at 30°C and 6.9 MPa. The red color is from a little $I_2$ added to the liquid to make it visible. (b) Beginning of the supercritical phase transition as the temperature is raised. (c) Single-phase supercritical carbon dioxide. [From H. Black, "Supercritical $CO_2$: 'The Greener Solvent'," *Environ. Sci. Technol.* **1996**, *30*, 124A. Photos courtesy D. Pesiri and W. Tumas, Los Alamos National Laboratory.] For a classroom demonstration of the critical transition of $SF_6$, see R. Chang and J. F. Skinner, "A Lecture Demonstration of the Critical Phenomenon," *J. Chem. Ed.* **1992**, *69*, 158.

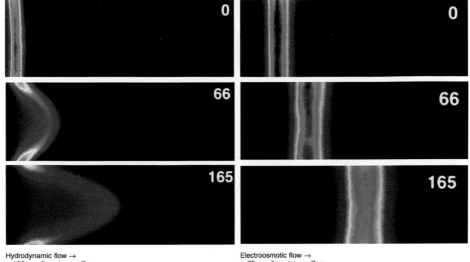

Hydrodynamic flow →
100-μm-diameter capillary

Electroosmotic flow →
75-μm-diameter capillary

**Color Plate 28** **Velocity Profiles for Hydrodynamic and Electroosmotic Flow (Section 26-5)** A fluorescent dye was imaged inside a capillary tube at times 0, 66, and 165 ms after initiating flow. The highest concentration of dye is represented by blue and the lowest concentration is red in these images in which different colors are assigned to different fluorescence intensities. [From P. H. Paul, M. G. Garguilo, and D. J. Rakestraw, "Imaging of Pressure- and Electrokinetically Driven Flows Through Open Capillaries," *Anal. Chem.* **1998,** *70,* 2459. See also D. Ross, T. J. Johnson, and L. E. Locascio, "Imaging Electroosmotic Flow in Plastic Microchannels," *Anal. Chem.* **2001,** *73,* 2509.]

Base number

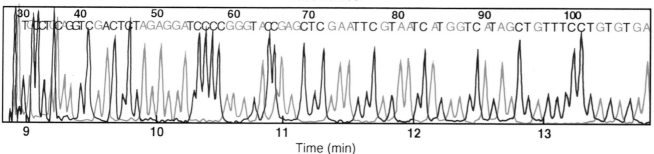

Time (min)

Base number

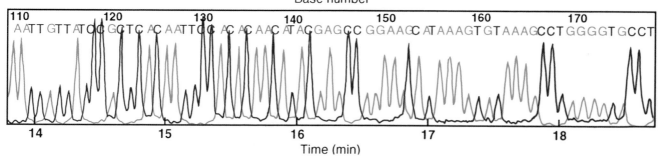

Time (min)

**Color Plate 29** **DNA Sequencing by Capillary Gel Electrophoresis with Fluorescent Labels (Section 26-6)** Tall red peaks correspond to chains terminating in cytosine and short red peaks correspond to thymine. Tall blue peaks arise from adenine and short blue peaks indicate guanine. Two different fluorescent labels and two fluorescence wavelengths were required to generate this information. [From M. C. Ruiz-Martinez, J. Berka, A. Belenkii, F. Foret, A. W. Miller, and B. L. Karger, "DNA Sequencing by Capillary Electrophoresis," *Anal. Chem.* **1993,** *65,* 2851.]

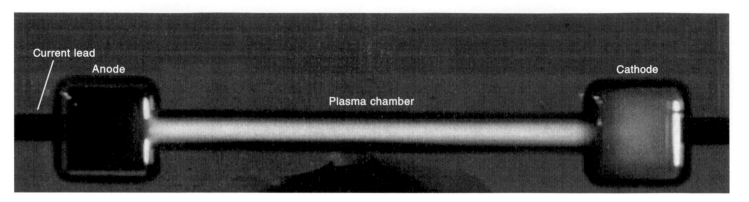

Current lead
Anode

Plasma chamber

Cathode

**Color Plate 30** **Plasma on a Chip (Section 26-6)** One of the most exotic devices incorporated into a chip to date is a direct current helium plasma with a 50-nL volume operating at 0.1 bar at 850 V and 60 μA. Hydrocarbons in a gas stream passing through the plasma decompose into fragments that emit characteristic frequencies of light. The lifetime of this experimental device was limited to 2 h by sputtering of the cathode (loss of electrode material) caused by He$^+$ ions bombarding the cathode. [From J. C. T. Eijkel, H. Stoeri, and A. Manz, "A Molecular Emission Detector on a Chip Employing a Direct Current Microplasma," *Anal. Chem.* **1999,** *71,* 2600. Photo courtesy A. Manz, Imperial College, London.]

(a)  (b)  (c)

**Color Plate 31** **Colloids and Dialysis (Demonstration 27-1)** (*a*) Colloidal Fe(III) (left) and ordinary aqueous Fe(III) (right). (*b*) Dialysis bags containing colloidal Fe(III) (left) and a solution of Cu(II) (right) immediately after placement in flasks of water. (*c*) After 24 h of dialysis, the Cu(II) has diffused out and is dispersed uniformly between the bag and the flask, but the colloidal Fe(III) remains inside the bag.

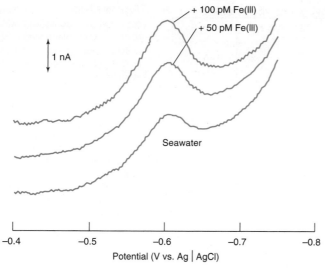

+ 100 pM Fe(III)

+ 50 pM Fe(III)

1 nA

Seawater

| | | | | |
|-0.4|-0.5|-0.6|-0.7|-0.8|

Potential (V vs. Ag | AgCl)

**Figure 17-21** Sampled current cathodic stripping voltammogram of Fe(III) in seawater plus two standard additions of 50 pM Fe(III). *[From H. Obata and C. M G. van den Berg, "Determination of Picomolar Levels of Iron in Seawater Using Catalytic Cathodic Stripping Voltammetry," Anal. Chem. 2001, 73, 2522.]*

process. First, 20 μM 2,3-dihydroxynaphthalene (L), 20 mM bromate ($BrO_3^-$), and pH 8.0 buffer are added to seawater that has been purged with $N_2$ to remove $O_2$. Dihydroxynaphthalene forms a complex, $L_nFe(III)$, which adsorbs onto a mercury drop electrode poised at −0.1 V versus Ag | AgCl during 60 s of vigorous stirring. After stirring is stopped and the solution becomes stationary, the potential is scanned from −0.1 to −0.8 V to give the lower trace in Figure 17-21. At −0.6 V, Fe(III) is reduced to Fe(II), which begins to diffuse away from the electrode. Before Fe(II) goes very far, $BrO_3^-$ oxidizes Fe(II) back to Fe(III), which becomes readsorbed and available to be reduced again. The cathodic stripping current is 290 times greater in the presence of 20 mM $BrO_3^-$ than without $BrO_3^-$. Fe(II) is a catalyst for the net reduction of $BrO_3^-$.

$$L_nFe(III)_{adsorbed} + e^- \xrightarrow[\text{stripping}]{\text{cathodic}} L_nFe(II)_{\text{in diffusion layer}}$$

$$\swarrow BrO_3^-$$

$$L_nFe(III)_{\text{in diffusion layer}}$$

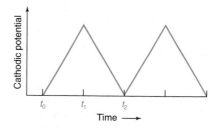

OH

OH

2,3-Dihydroxynaphthalene
(Presumably, the polarizable pi electron cloud of naphthalene is bound to the polarizable Hg by van der Waals forces)

## Cyclic Voltammetry

In **cyclic voltammetry,** we apply the triangular waveform in Figure 17-22 to the working electrode. After the application of a linear voltage ramp between times $t_0$ and $t_1$ (typically a few seconds), the ramp is reversed to bring the potential back to its initial value at time $t_2$. The cycle may be repeated many times.

The initial portion of the cyclic voltammogram in Figure 17-23, beginning at $t_0$, exhibits a *cathodic wave*. Instead of leveling off at the top of the wave, current decreases at more negative potential because analyte becomes depleted near the electrode. Diffusion is too slow to replenish analyte near the electrode. At the time of peak voltage ($t_1$) in Figure 17-23, the cathodic current has decayed to a small value. After $t_1$, the potential is reversed and, eventually, reduced product near the electrode is oxidized, thereby giving rise to an *anodic wave*. Finally, as the reduced product is depleted, the anodic current decays back toward its initial value at $t_2$.

Figure 17-23a illustrates a *reversible* reaction that is fast enough to maintain equilibrium concentrations of reactant and product *at the electrode surface*. The peak anodic and peak cathodic currents have equal magnitudes in a reversible process, and

$$E_{pa} - E_{pc} = \frac{2.22RT}{nF} = \frac{57.0}{n} \text{ (mV)} \quad \text{(at 25°C)} \quad (17\text{-}16)$$

where $E_{pa}$ and $E_{pc}$ are the potentials at which the *peak* anodic and *peak* cathodic currents are observed and $n$ is the number of electrons in the half-reaction. The half-wave potential, $E_{1/2}$, lies midway between the two peak potentials. For an irreversible reaction, the cathodic and anodic peaks are drawn out and more separated (Figure 17-23b). At the limit of irreversibility, where the oxidation is very slow, no anodic peak is seen.

Current decreases after the cathodic peak because of *concentration polarization*.

Cathodic potential

Time →
$t_0$  $t_1$  $t_2$

**Figure 17-22** Waveform for cyclic voltammetry. Corresponding times are indicated in Figure 17-23.

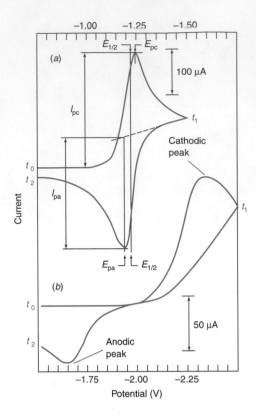

**Figure 17-23** Cyclic voltammograms of (*a*) 1 mM $O_2$ in acetonitrile with 0.10 M $(C_2H_5)_4N^+ClO_4^-$ electrolyte and (*b*) 0.060 mM 2-nitropropane in acetonitrile with 0.10 M $(n\text{-}C_7H_{15})_4N^+ClO_4^-$ electrolyte. The reaction in curve *a* is

$$O_2 + e^- \rightleftharpoons O_2^-$$
$$\text{Superoxide}$$

Working electrode, Hg; reference electrode, $Ag \mid 0.001\,M\,AgNO_3(aq) \mid 0.10\,M\,(C_2H_5)_4N^+ClO_4^-$ in acetonitrile; scan rate = 100 V/s. $I_{pa}$ is the peak anodic current and $I_{pc}$ is the peak cathodic current. $E_{pa}$ and $E_{pc}$ are the potentials at which these currents are observed. [*From D. H. Evans, K. M. O'Connell, R. A. Petersen, and M. J. Kelly, "Cyclic Voltammetry," J. Chem. Ed.* **1983**, *60, 290.*]

For a reversible reaction, the peak current ($I_{pc}$, amperes) for the forward sweep of the first cycle is proportional to the concentration of analyte and the square root of sweep rate:

$$I_{pc} = (2.69 \times 10^8) n^{3/2} ACD^{1/2} v^{1/2} \qquad \text{(at 25°C)} \qquad \textbf{(17-17)}$$

where $n$ is the number of electrons in the half-reaction, $A$ is the area of the electrode ($m^2$), $C$ is the concentration (mol/L), $D$ is the diffusion coefficient of the electroactive species ($m^2$/s), and $v$ is sweep rate (V/s). The faster the sweep rate, the greater the peak current, as long as the reaction remains reversible. If the electroactive species is adsorbed on the electrode, the peak current is proportional to $v$ rather than $\sqrt{v}$.

For catalytic stripping in Figure 17-21, current was proportional to $\sqrt{v}$, consistent with the rate-limiting step being diffusion of $BrO_3^-$ to the electrode. If the rate-limiting step were reduction of Fe(III) adsorbed on the electrode, the peak current would have been proportional to $v$.

Cyclic voltammetry is used to characterize the redox behavior of compounds such as $C_{60}$ in Figure 17-24 and to elucidate the kinetics of electrode reactions.[25]

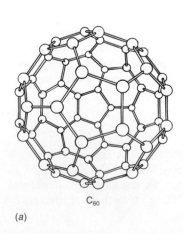

$C_{60}$

(*a*)

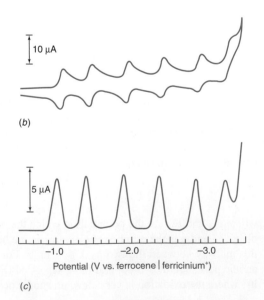

**Figure 17-24** (*a*) Structure of $C_{60}$ (buckminsterfullerene). (*b*) Cyclic voltammetry and (*c*) polarogram of 0.8 mM $C_{60}$, showing six waves for reduction to $C_{60}^-, C_{60}^{2-}, \ldots, C_{60}^{6-}$. The acetonitrile/toluene solution was at $-10°C$ with $(n\text{-}C_4H_9)_4N^+PF_6^-$ supporting electrolyte. The reference electrode contains the ferrocene | ferricinium$^+$ redox couple. Ferrocene is $(C_5H_5)_2Fe$ and ferricinium cation is $(C_5H_5)_2Fe^+$. The structure of ferrocene was shown in Reaction 17-12. [*From Q. Xie, E. Pérez-Cordero, and L. Echegoyen, "Electrochemical Detection of $C_{60}^{6-}$ and $C_{70}^{6-}$," J. Am. Chem. Soc.* **1992**, *114, 3978.*]

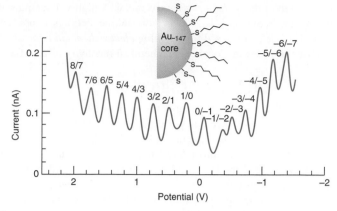

**Figure 17-25** Voltammogram of gold nanoparticles (Au$_{\sim 147}$ capped with ~50 hexanethiol molecules) in 1,2-dichloroethane solution recorded with 25-μm-diameter Pt working electrode. The nanoparticle exhibits oxidation states from −7 to +8 over the potential range of this scan. Supporting electrolyte: 10 mM [$(C_6H_5)_3P{=}N{=}P(C_6H_5)_3$]$^+$[$(C_6F_5)_4B$]$^-$. Potential measured versus "quasi-reference" electrode—a silver wire whose potential is ~0.1 V versus Ag | AgCl. *[From B. M. Quinn, P. Liljeroth, V. Ruiz, T. Laaksonen, and K. Kontturi, "Electrochemical Resolution of 15 Oxidation States for Monolayer Protected Gold Nanoparticles," J. Am. Chem. Soc. 2003, 125, 6644.]*

4 μm

**Figure 17-26** Electron micrograph of the tip of a Nafion-coated carbon fiber electrode. The carbon inside the electrode has a diameter of 10 μm. Nafion permits cations to pass but excludes anions. *[Photo courtesy R. M. Wightman. From R. M. Wightman, L. J. May, and A. C. Michael, "Detection of Dopamine Dynamics in the Brain," Anal. Chem. 1988, 60, 769A.]*

## Microelectrodes

*Microelectrodes* have working dimensions from a few tens of micrometers down to nanometers.[26] The surface area of the electrode is small, so the current is tiny. With a small current, the ohmic drop ($= IR$) in a highly resistive medium is small, allowing microelectrodes to be used in poorly conducting nonaqueous media (Figure 17-25). The electrical capacitance of the double layer (Box 17-3) of a microelectrode is very small. Low capacitance gives low background charging current relative to the faradaic current of a redox reaction, lowering the detection limit by as much as three orders of magnitude over conventional electrodes. Low capacitance also enables the potential to be varied at rates up to 500 kV/s, thus allowing short-lived species with lifetimes of less than 1 μs to be studied.

Sufficiently small electrodes fit inside a living cell. A carbon fiber coated with insulating polymer provides a 1-μm-diameter working electrode at the exposed tip. Carbon is fouled by adsorption of organic molecules inside living cells, so a thin layer of Pt or Au is electrolytically plated onto the exposed carbon. The metallized electrode is cleaned in situ (inside the cell during an experiment) by an anodic voltage pulse that desorbs surface-bound species and produces an oxide layer on the metal, followed by a cathodic pulse to reduce the oxide.[27]

Figure 17-26 shows a carbon fiber coated with a cation-exchange membrane called Nafion. This membrane, whose structure is shown in Problem 17-7, has fixed negative charges. Cations diffuse through the membrane, but anions are excluded. The electrode can measure the cationic neurotransmitter dopamine in a rat brain.[28] Negatively charged ascorbate, which ordinarily interferes with dopamine analysis, is excluded by Nafion. The response to dopamine is 1 000 times higher than the response to ascorbate.

The gold electrode in Figure 17-27 has a spherical tip with a diameter of 300 nm. The structure is coated with insulating paint deposited by electrophoresis, in which positively

**Advantages of ultramicroelectrodes:**
- Fit into small places
- Useful in resistive, nonaqueous media (because of small ohmic losses)
- Rapid voltage scans (possible because of small double-layer capacitance) allow short-lived species to be studied
- Sensitivity increased by orders of magnitude because of low charging current

Dopamine

$$+ 2H^+ + 2e^-$$

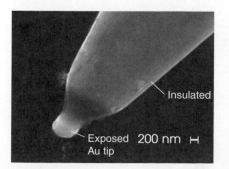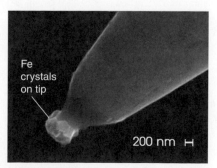

**Figure 17-27** Gold microelectrode with spherical tip. *[From J. Abbou, C. Demaille, M. Druet, and J. Moiroux, "Fabrication of Submicrometer-Sized Gold Electrodes of Controlled Geometry for Scanning Electrochemical-Atomic Force Microscopy," Anal. Chem. 2002, 74, 6355.]*

charged paint is attracted to the negatively charged metal. A high voltage pulse then exposes the spherical tip. When $FeSO_4(aq)$ is reduced, crystalline $Fe(s)$ is deposited only at the spherical tip. This electrode can be used for *scanning electrochemical microscopy,* in which a microelectrode is used to probe the local electrochemical reactivity of a surface.[29]

# ■ ■ ■ ■ 17-6 Karl Fischer Titration of H₂O

The **Karl Fischer titration,**[30] which measures traces of water in transformer oil, solvents, foods, polymers, and other substances, is performed half a million times each day.[31] The titration is usually performed by delivering titrant from an automated buret or by coulometric generation of titrant. The volumetric procedure tends to be appropriate for larger amounts of water (but can go as low as ~1 mg $H_2O$) and the coulometric procedure tends to be appropriate for smaller amounts of water.

We illustrate the coulometric procedure in Figure 17-28, in which the main compartment contains anode solution plus unknown. The smaller compartment at the left has an internal Pt electrode immersed in cathode solution and an external Pt electrode immersed in the anode solution of the main compartment. The two compartments are separated by an ion-permeable membrane. Two Pt electrodes are used for end-point detection.

Anode solution contains an alcohol, a base, $SO_2$, $I^-$, and possibly another organic solvent. Methanol and diethylene glycol monomethyl ether ($CH_3OCH_2CH_2OCH_2CH_2OH$) are typical alcohols. Typical bases are imidazole and diethanolamine. The organic solvent may contain chloroform, formamide, or other solvents. The trend is to avoid chlorinated solvents because of their environmental hazards. When analyzing nonpolar substances such as transformer oil, sufficient solvent, such as chloroform, should be used to make the reaction homogeneous. Otherwise, moisture trapped in oily emulsions is inaccessible. (An *emulsion* is a fine suspension of liquid-phase droplets in another liquid.)

The anode at the lower left in Figure 17-28 generates $I_2$ by oxidation of $I^-$. In the presence of $H_2O$, reactions occur between the alcohol (ROH), base (B), $SO_2$, and $I_2$.

$$ROH + SO_2 + B \rightarrow BH^+ + ROSO_2^- \qquad (17\text{-}18)$$

$$H_2O + I_2 + ROSO_2^- + 2B \rightarrow ROSO_3^- + 2BH^+I^- \qquad (17\text{-}19)$$

The net reaction is oxidation of $SO_2$ by $I_2$, with formation of $ROSO_3^-$. One mole of $I_2$ is consumed for each mole of $H_2O$ when the solvent is methanol. In other solvents, the stoichiometry can be more complex.[31]

In a typical procedure, the main compartment in Figure 17-28 is filled with anode solution and the coulometric generator is filled with cathode solution that may contain reagents designed to be reduced at the cathode. Current is run until moisture in the main compartment is consumed, as indicated by the end-point detection system described after the Example. An unknown is injected through the septum and the coulometer is run again until moisture has been consumed. Two moles of electrons correspond to 1 mol of $H_2O$ if the $I_2$:$H_2O$ stoichiometry is 1:1.

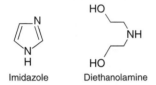

Imidazole          Diethanolamine

pH is maintained in the range 4 to 7. Above pH 8, nonstoichiometric side reactions occur. Below pH 3, the reaction is very slow.

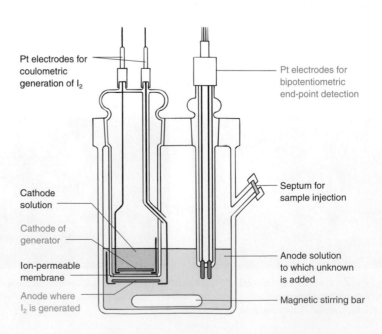

*Figure 17-28* Apparatus for coulometric Karl Fischer titration.

It is routine to standardize Karl Fischer reagents, or even a coulometer, with a standard such as lincomycin hydrochloride monohydrate, which contains 3.91 wt% $H_2O$. The coulometer is run until the end point is reached, indicating that the Karl Fischer reagent is dry. A port is opened briefly to add solid lincomycin, which is then titrated to the same end point. Then an unknown is added and titrated in the same manner. Find the wt% $H_2O$ in the unknown.

| Milligrams lincomycin | $\mu$g $H_2O$ observed | $\mu$g $H_2O$ theoretical | Difference ($\mu$g) = blank correction |
|---|---|---|---|
| 3.89 | 172.4 | 152.1 | $172.4 - 152.1 = 20.3$ |
| 13.64 | 556.3 | 533.3 | $556.3 - 533.3 = 23.0$ |
| 19.25 | 771.4 | 752.7 | $771.4 - 752.7 = 18.7$ |
| | | | Average = 20.7 |

| Milligrams unknown | $\mu$g $H_2O$ observed | $\mu$g $H_2O$ corrected (= observed − 20.7) | wt% $H_2O$ in unknown |
|---|---|---|---|
| 24.17 | 540.8 | 520.1 | 520.1 $\mu$g/24.17 mg = 2.15% |
| 17.08 | 387.6 | 366.9 | 366.9 $\mu$g/17.08 mg = 2.15% |

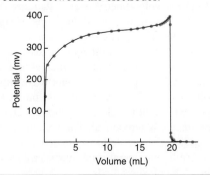

Lincomycin hydrochloride monohydrate
$C_{18}H_{37}N_2O_7SCl$, FM 461.01
3.91 wt% $H_2O$

*SOURCE: Data from W. C. Schinzer, Pfizer Co., Michigan Pharmaceutical Sciences, Portage, MI.*

**Solution** For lincomycin, we observe ~20.7 $\mu$g more $H_2O$ than expected, *independent of the sample size.* Excess $H_2O$ comes from the atmosphere when the port is opened to add solid. To determine moisture in unknowns, subtract this blank from the total moisture titrated. This procedure can generate very reproducible data.

A **bipotentiometric** measurement is the most common way to detect the end point of a Karl Fischer titration. The detector circuit maintains a *constant current* (usually 5 or 10 $\mu$A) between the two detector electrodes at the right in Figure 17-28 while measuring the voltage needed to sustain the current. Prior to the equivalence point, the solution contains $I^-$, but little $I_2$ (which is consumed in Reaction 17-19 as fast as it is generated). To maintain a current of 10 $\mu$A, the cathode potential must be negative enough to reduce some component of the solvent system. At the equivalence point, excess $I_2$ suddenly appears and current can be carried at very low voltage by Reactions A and B in Demonstration 17-2. The abrupt voltage drop marks the end point.

---

## Demonstration 17-2 The Karl Fischer Jacks of a pH Meter

Some pH meters contain a pair of sockets at the back labeled "K-F" or "Karl Fischer." When the manufacturer's instructions are followed, a constant current (usually about 10 $\mu$A) is applied across these terminals. To perform a bipotentiometric titration, Pt electrodes are connected to the K-F sockets. The meter is set to the millivolt scale, which displays the voltage needed to maintain the constant current between the electrodes.

The figure shows the results of a bipotentiometric titration of ascorbic acid with $I_3^-$. Ascorbic acid (146 mg) was dissolved in 200 mL of water in a 400-mL beaker. Two Pt electrodes were attached to the K-F outlets of the pH meter and spaced about 4 cm apart in the magnetically stirred solution. The solution was titrated with 0.04 M $I_3^-$ (prepared by dissolving 2.4 g of KI plus 1.2 g of $I_2$ in 100 mL of water), and the voltage was recorded after each addition. Prior to the equivalence point, all the $I_3^-$ is reduced to $I^-$ by the excess ascorbic acid. Reaction B can occur, but Reaction A cannot. A voltage of about 300 mV is required to support a constant current of 10 $\mu$A. (The ascorbate | dehydroascorbate couple does not react at a Pt electrode and cannot carry current.) After the equivalence point, excess $I_3^-$ is present, so Reactions A and B both occur, and the voltage drops precipitously.

Cathode: $I_3^- + 2e^- \rightarrow 3I^-$ (A)

Anode: $3I^- \rightarrow I_3^- + 2e^-$ (B)

A trend in Karl Fischer coulometric instrumentation is to eliminate the separate cathode compartment in Figure 17-28 to reduce conditioning time required before samples can be analyzed and to eliminate clogging of the membrane.[32] The challenge is to minimize interference by products of the cathodic reaction.

End points in Karl Fischer titrations tend to drift because of slow reactions and leakage of water into the cell from the air. Some instruments measure the rate at which $I_2$ must be generated to maintain the end point and compare this rate with that measured before sample was added. Other instruments allow you to set a "persistence of end point" time, typically 5 to 60 s, during which the detector voltage must be stable to define the end point.

A round robin study of accuracy and precision of the coulometric procedure identified sources of systematic error.[33] In some labs, either the instruments were inaccurate or workers did not measure the quantity of standards accurately. In other cases, the solvent was not appropriate. Commercial reagents are designed for Karl Fischer analysis. Reagents recommended by the instrument manufacturer should be used with each instrument.

## Terms to Understand

| | | | |
|---|---|---|---|
| amalgam | coulometric titration | half-wave potential | residual current |
| ampere | coulometry | Karl Fischer titration | rotating disk electrode |
| amperometry | cyclic voltammetry | mediator | sampled current polarography |
| auxiliary electrode | depolarizer | nonpolarizable electrode | square wave voltammetry |
| biosensor | diffusion current | ohmic potential | stripping analysis |
| bipotentiometric titration | dropping-mercury electrode | overpotential | voltammetry |
| charging current | electric double layer | polarizable electrode | voltammogram |
| Clark electrode | electroactive species | polarographic wave | working electrode |
| concentration polarization | electrogravimetric analysis | polarography | |
| controlled-potential electrolysis | electrolysis | potentiostat | |
| coulomb | faradaic current | reference electrode | |

## Summary

In electrolysis, a chemical reaction is forced to occur by the flow of electricity through a cell. The moles of electrons flowing through the cell are $It/F$, where $I$ is current, $t$ is time, and $F$ is the Faraday constant. The magnitude of the voltage that must be applied to an electrolysis cell is $E = E(\text{cathode}) - E(\text{anode}) - IR - \text{overpotentials}$.

1. Overpotential is the voltage required to overcome the activation energy of an electrode reaction. A greater overpotential is required to drive a reaction at a faster rate.
2. Ohmic potential ($= IR$) is that voltage needed to overcome internal resistance of the cell.
3. Concentration polarization occurs when the concentration of electroactive species near an electrode is not the same as its concentration in bulk solution. Concentration polarization is embedded in the terms $E(\text{cathode})$ and $E(\text{anode})$.

Overpotential, ohmic potential, and concentration polarization always oppose the desired reaction and require a greater voltage to be applied for electrolysis.

Controlled-potential electrolysis is conducted in a three-electrode cell in which the potential of the working electrode is measured with respect to a reference electrode to which negligible current flows. Current flows between the working and auxiliary electrodes.

In electrogravimetric analysis, analyte is deposited on an electrode, whose increase in mass is then measured. With a constant voltage in a two-electrode cell, electrolysis is not very selective, because the working electrode potential changes as the reaction proceeds.

In coulometry, the moles of electrons needed for a chemical reaction are measured. In a coulometric (constant current) titration, the time needed for complete reaction measures the number of electrons consumed. Controlled-potential coulometry is more selective than constant-current coulometry, but slower. Electrons consumed in the reaction are measured by integration of the current-versus-time curve.

In amperometry, the current at the working electrode is proportional to analyte concentration. The amperometric glucose monitor generates $H_2O_2$ by enzymatic oxidation of glucose and the $H_2O_2$ is measured by oxidation at an electrode. A mediator is employed to rapidly shuttle electrons between electrode and analyte.

Voltammetry is a collection of methods in which the dependence of current on the applied potential of the working electrode is observed. Polarography is voltammetry with a dropping-mercury working electrode. This electrode gives reproducible results because fresh surface is always exposed. Hg is useful for reductions because the high overpotential for $H^+$ reduction on Hg prevents interference by $H^+$ reduction. Oxidations are usually studied with other electrodes because Hg is readily oxidized. For quantitative analysis, the diffusion current is proportional to analyte concentration if there is a sufficient concentration of supporting electrolyte. The half-wave potential is characteristic of a particular analyte in a particular medium.

Sampled current voltammetry uses a staircase voltage profile for measurements with successive, static drops of Hg. One second after each voltage step, charging current is nearly 0, but there is still substantial faradaic current from the redox reaction.

Square wave voltammetry achieves increased sensitivity and a derivative peak shape by applying a square wave superimposed on a staircase voltage ramp. With each cathodic pulse, there is a rush of analyte to be reduced at the electrode surface. During the anodic pulse, reduced analyte is reoxidized. The voltammogram is the difference between the cathodic and the anodic currents. Square wave voltammetry permits fast, real-time measurements not possible with other electrochemical methods.

Stripping is the most sensitive form of voltammetry. In anodic stripping polarography, analyte is concentrated into a single drop of mercury by reduction at a fixed voltage for a fixed time. The potential is then made more positive, and current is measured as analyte is reoxidized. In cyclic voltammetry, a triangular waveform is applied, and cathodic and anodic processes are observed in succession. Microelectrodes fit into small places and their low current allows them to be used in resistive, nonaqueous media. Their low capacitance increases sensitivity by reducing charging current and permits rapid voltage scanning, which allows very short-lived species to be studied.

The Karl Fischer titration of water uses a buret to deliver reagent or coulometry to generate reagent. In bipotentiometric end-point detection, the voltage needed to maintain a constant current between two Pt electrodes is measured. The voltage changes abruptly at the equivalence point, when one member of a redox couple is either created or destroyed.

## Exercises

**17-A.** A dilute $Na_2SO_4$ solution is to be electrolyzed with a pair of smooth Pt electrodes at a current density of $100 \ A/m^2$ and a current of $0.100 \ A$. The products are $H_2(g)$ and $O_2(g)$ at $1.00$ bar. Calculate the required voltage if the cell resistance is $2.00 \ \Omega$ and there is no concentration polarization. What would your answer be if the Pt electrodes were replaced by Au electrodes?

**17-B.** **(a)** At what cathode potential will $Sb(s)$ deposition commence from $0.010 \ M \ SbO^+$ solution at pH $0.00$? Express this potential versus S.H.E. and versus $Ag \mid AgCl$.

$$SbO^+ + 2H^+ + 3e^- \rightleftharpoons Sb(s) + H_2O \qquad E° = 0.208 \ V$$

**(b)** What percentage of $0.10 \ M \ Cu^{2+}$ could be reduced electrolytically to $Cu(s)$ before $0.010 \ M \ SbO^+$ in the same solution begins to be reduced at pH $0.00$?

**17-C.** Calculate the cathode potential (versus S.C.E.) needed to reduce cobalt(II) to $1.0 \ \mu M$ in each of the following solutions. In each case, $Co(s)$ is the product of the reaction.
**(a)** $0.10 \ M \ HClO_4$
**(b)** $0.10 \ M \ C_2O_4^{2-}$

$$Co(C_2O_4)_2^{2-} + 2e^- \rightleftharpoons Co(s) + 2C_2O_4^{2-} \qquad E° = -0.474 \ V$$

This question is asking you to find the potential at which $[Co(C_2O_4)_2^{2-}]$ will be $1.0 \ \mu M$.
**(c)** $0.10 \ M$ EDTA at pH $7.00$

**17-D.** Ions that react with $Ag^+$ can be determined electrogravimetrically by deposition on a silver working anode:

$$Ag(s) + X^- \rightarrow AgX(s) + e^-$$

**(a)** What will be the final mass of a silver anode used to electrolyze $75.00 \ mL$ of $0.023 \ 80 \ M$ KSCN if the initial mass of the anode is $12.463 \ 8 \ g$?
**(b)** At what electrolysis voltage (versus S.C.E.) will $AgBr(s)$ be deposited from $0.10 \ M \ Br^-$? (Consider negligible current flow, so that there is no ohmic potential, concentration polarization, or overpotential.)
**(c)** Is it theoretically possible to separate $99.99\%$ of $0.10 \ M$ KI from $0.10 \ M$ KBr by controlled-potential electrolysis?

**17-E.** Chlorine has been used for decades to disinfect drinking water. An undesirable side effect of this treatment is reaction with organic impurities to create organochlorine compounds, some of which could be toxic. Monitoring total organic halide (designated TOX) is required for many water providers. A standard procedure for TOX is to pass water through activated charcoal that adsorbs organic compounds. Then the charcoal is combusted to liberate hydrogen halides:

$$\text{Organic halide (RX)} \xrightarrow{O_2/800°C} CO_2 + H_2O + HX$$

HX is absorbed into aqueous solution and measured by coulometric titration with a silver anode:

$$X^-(aq) + Ag(s) \rightarrow AgX(s) + e^-$$

When $1.00 \ L$ of drinking water was analyzed, a current of $4.23 \ mA$ was required for $387 \ s$. A blank prepared by oxidizing charcoal required $6 \ s$ at $4.23 \ mA$. Express the TOX of the drinking water as $\mu mol$ halogen/L. If all halogen is chlorine, express the TOX as $\mu g \ Cl/L$.

**17-F.** $Cd^{2+}$ was used as an internal standard in the analysis of $Pb^{2+}$ by square wave polarography. $Cd^{2+}$ gives a reduction wave at $-0.60 \ (\pm 0.02) \ V$ and $Pb^{2+}$ gives a reduction wave at $-0.40 \ (\pm 0.02) \ V$. It was first verified that the ratio of peak heights is proportional to the ratio of concentrations over the whole range employed in the experiment. Here are results for known and unknown mixtures:

| Analyte | Concentration (M) | Current ($\mu A$) |
|---|---|---|
| *Known* | | |
| $Cd^{2+}$ | $3.23 \ (\pm 0.01) \times 10^{-5}$ | $1.64 \ (\pm 0.03)$ |
| $Pb^{2+}$ | $4.18 \ (\pm 0.01) \times 10^{-5}$ | $1.58 \ (\pm 0.03)$ |
| *Unknown + Internal Standard* | | |
| $Cd^{2+}$ | ? | $2.00 \ (\pm 0.03)$ |
| $Pb^{2+}$ | ? | $3.00 \ (\pm 0.03)$ |

The unknown mixture was prepared by mixing $25.00 \ (\pm 0.05) \ mL$ of unknown (containing only $Pb^{2+}$) plus $10.00 \ (\pm 0.05) \ mL$ of $3.23 \ (\pm 0.01) \times 10^{-4} \ M \ Cd^{2+}$ and diluting to $50.00 \ (\pm 0.05) \ mL$.
**(a)** Disregarding uncertainties, find $[Pb^{2+}]$ in the undiluted unknown.
**(b)** Find the absolute uncertainty for the answer to part **(a)**.

**17-G.** Consider the cyclic voltammogram of the $Co^{3+}$ compound $Co(B_9C_2H_{11})_2^-$. Suggest a chemical reaction to account for each wave. Are the reactions reversible? How many electrons are involved in each step? Sketch the sampled current and square wave polarograms expected for this compound.

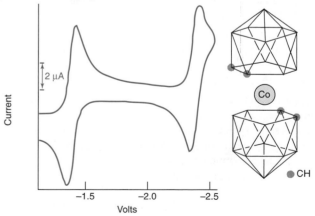

Cyclic voltammogram of $Co(B_9C_2H_{11})_2^-$. [W. E. Geiger, Jr., W. L. Bowden, and N. El Murr, "An Electrochemical Study of the Protonation Site of the Cobaltocene Anion and of Cyclopentadienylcobalt(I) Dicarbollides," Inorg. Chem. **1979**, 18, 2358.]

| $E_{1/2}$ (V vs. S.C.E.) | $I_{pa}/I_{pc}$ | $E_{pa} - E_{pc}$ (mV) |
|---|---|---|
| $-1.38$ | $1.01$ | $60$ |
| $-2.38$ | $1.00$ | $60$ |

**17-H.** In a coulometric Karl Fischer water analysis, 25.00 mL of pure "dry" methanol required 4.23 C to generate enough $I_2$ to react with the residual $H_2O$ in the methanol. A suspension of 0.847 6 g of finely ground polymeric material in 25.00 mL of the same "dry" methanol required 63.16 C. Find the weight percent of $H_2O$ in the polymer.

## Problems

### Fundamentals of Electrolysis

**17-1.** The figure below shows the behavior of Pt and Ag cathodes at which reduction of $H_3O^+$ to $H_2(g)$ occurs. Explain why the two curves are not superimposed.

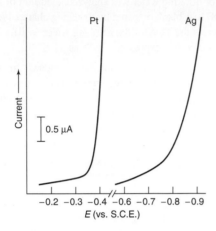

Current versus voltage for Pt and Ag electrodes in $O_2$-free, aqueous $H_2SO_4$ adjusted to pH 3.2. *[From D. Marín, F. Mendicuti, and C. Teijeiro, "An Electrochemistry Experiment: Hydrogen Evolution Reaction on Different Electrodes," J. Chem. Ed.* **1994**, *71, A277.]*

**17-2.** How many hours are required for 0.100 mol of electrons to flow through a circuit if the current is 1.00 A?

**17-3.** The standard free-energy change for the formation of $H_2(g)$ + $\frac{1}{2}O_2(g)$ from $H_2O(l)$ is $\Delta G° = +237.13$ kJ. The reactions are

Cathode: $2H_2O + 2e^- \rightleftharpoons H_2(g) + 2OH^-$

Anode: $H_2O \rightleftharpoons \frac{1}{2}O_2(g) + 2H^+ + 2e^-$

Calculate the standard voltage ($E°$) needed to decompose water into its elements by electrolysis. What does the word *standard* mean in this question?

**17-4.** Consider the following electrolysis reactions.

Cathode: $H_2O(l) + e^- \rightleftharpoons \frac{1}{2}H_2(g, 1.0 \text{ bar}) + OH^-(aq, 0.10 \text{ M})$

Anode: $Br^-(aq, 0.10 \text{ M}) \rightleftharpoons \frac{1}{2}Br_2(l) + e^-$

**(a)** Calculate the voltage needed to drive the net reaction if current is negligible.
**(b)** Suppose that the cell has a resistance of 2.0 Ω and a current of 100 mA. How much voltage is needed to overcome the cell resistance? This is the ohmic potential.
**(c)** Suppose that the anode reaction has an overpotential of 0.20 V and that the cathode overpotential is 0.40 V. What voltage is needed to overcome these effects combined with those of parts **(a)** and **(b)**?
**(d)** Suppose that concentration polarization occurs. The concentration of $OH^-$ at the cathode surface increases to 1.0 M and the concentration of $Br^-$ at the anode surface decreases to 0.010 M. What voltage is needed to overcome these effects combined with those of parts **(b)** and **(c)**?

**17-5.** Which voltage, $V_1$ or $V_2$, in the diagram, is constant in controlled-potential electrolysis?

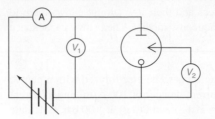

Diagram for Problem 17-5.

**17-6.** The Weston cell shown here is a stable voltage standard formerly used in potentiometers. (The potentiometer compares an unknown voltage with that of the standard. In contrast with the conditions of this problem, very little current may be drawn from the cell if it is to be an accurate voltage standard.)

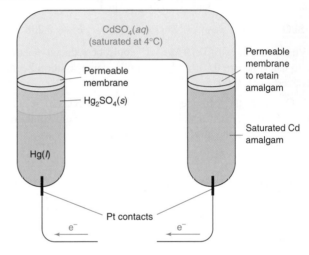

$Hg_2SO_4 + Cd(\text{in Hg}) \rightleftharpoons 2Hg + CdSO_4$

A saturated Weston cell containing excess solid $CdSO_4$ is a more precise voltage standard than the unsaturated cell but is more sensitive to temperature and mechanical shock and cannot be easily incorporated into portable equipment.

**(a)** How much work (J) can be done by the Weston cell if the voltage is 1.02 V and 1.00 mL of Hg (density = 13.53 g/mL) is deposited?
**(b)** If the cell passes current through a 100-Ω resistor that dissipates heat at a rate of 0.209 J/min, how many grams of Cd are oxidized each hour? (This question is not meant to be consistent with part **(a).** The voltage is no longer 1.02 volts.)

**17-7.** The chlor-alkali process,[34] in which seawater is electrolyzed to produce $Cl_2$ and NaOH, is the second most important commercial electrolysis, behind production of aluminum.

Anode: $Cl^- \rightarrow \frac{1}{2}Cl_2 + e^-$

Hg Cathode: $Na^+ + H_2O + e^- \rightarrow NaOH + \frac{1}{2}H_2$

The semipermeable Nafion membrane used to separate the anode and cathode compartments resists chemical attack. Its anionic side chains permit conduction of $Na^+$, but not anions. The cathode compartment is loaded with pure water, and the anode compartment contains seawater from which $Ca^{2+}$ and $Mg^{2+}$ have been removed. Explain how the membrane allows NaOH to be formed free of NaCl.

$$-[(CF_2CF_2)_n-CFCF_2]_x-$$
$$|$$
$$O$$
$$|$$
$$CF_2$$
$$|$$
$$CFCF_3 \qquad \text{Nafion}$$
$$|$$
$$O$$
$$|$$
$$CF_2CF_2SO_3^-Na^+$$

**17-8.** The lead-acid battery in a car has six cells in series, each delivering close to 2.0 V for a total of 12 V when the battery is discharging. To recharge the battery requires ~2.4 V per cell, or ~14 V for the entire battery.[35] Explain these observations in terms of Equation 17-6.

### Electrogravimetric Analysis

**17-9.** A 0.326 8-g unknown containing $Pb(CH_3CHOHCO_2)_2$ (lead lactate, FM 385.3) plus inert material was electrolyzed to produce 0.111 1 g of $PbO_2$ (FM 239.2). Was the $PbO_2$ deposited at the anode or at the cathode? Find the weight percent of lead lactate in the unknown.

**17-10.** A solution of $Sn^{2+}$ is to be electrolyzed to reduce the $Sn^{2+}$ to $Sn(s)$. Calculate the cathode potential (versus S.H.E.) needed to reduce the $Sn^{2+}$ concentration to $1.0 \times 10^{-8}$ M if no concentration polarization occurs. What would be the potential versus S.C.E. instead of S.H.E? Would the potential be more positive or more negative if concentration polarization occurred?

**17-11.** What cathode potential (versus S.H.E.) is required to reduce 99.99% of Cd(II) from a solution containing 0.10 M Cd(II) in 1.0 M ammonia if there is negligible current? Consider the following reactions and assume that nearly all Cd(II) is in the form $Cd(NH_3)_4^{2+}$.

$$Cd^{2+} + 4NH_3 \rightleftharpoons Cd(NH_3)_4^{2+} \qquad \beta_4 = 3.6 \times 10^6$$
$$Cd^{2+} + 2e^- \rightleftharpoons Cd(s) \qquad E° = -0.402 \text{ V}$$

### Coulometry

**17-12.** Explain how the amperometric end-point detector in Figure 17-8 operates.

**17-13.** What does a mediator do?

**17-14.** The sensitivity of a coulometer is governed by the delivery of its minimum current for its minimum time. Suppose that 5 mA can be delivered for 0.1 s.
(a) How many moles of electrons are delivered by 5 mA for 0.1 s?
(b) How many milliliters of a 0.01 M solution of a two-electron reducing agent are required to deliver the same number of electrons?

**17-15.** The experiment in Figure 17-8 required 5.32 mA for 964 s for complete reaction of a 5.00-mL aliquot of unknown cyclohexene solution.
(a) How many moles of electrons passed through the cell?
(b) How many moles of cyclohexene reacted?
(c) What was the molarity of cyclohexene in the unknown?

**17-16.** $H_2S(aq)$ can be analyzed by titration with coulometrically generated $I_2$.

$$H_2S + I_2 \rightarrow S(s) + 2H^+ + 2I^-$$

To 50.00 mL of sample were added 4 g of KI. Electrolysis required 812 s at 52.6 mA. Calculate the concentration of $H_2S(\mu g/mL)$ in the sample.

**17-17.** $Ti^{3+}$ is to be generated in 0.10 M $HClO_4$ solution for coulometric reduction of azobenzene.

$$TiO^{2+} + 2H^+ + e^- \rightleftharpoons Ti^{3+} + H_2O \qquad E° = 0.100 \text{ V}$$
$$4Ti^{3+} + \underset{\text{Azobenzene}}{C_6H_5N{=}NC_6H_5} + 4H_2O \rightarrow$$
$$\underset{\text{Aniline}}{2C_6H_5NH_2} + 4TiO^{2+} + 4H^+$$

At the counter electrode, water is oxidized, and $O_2$ is liberated at a pressure of 0.20 bar. Both electrodes are made of smooth Pt, and each has a total surface area of 1.00 cm². The rate of reduction of the azobenzene is 25.9 nmol/s, and the resistance of the solution between the generator electrodes is 52.4 Ω.
(a) Calculate the current density (A/m²) at the electrode surface. Use Table 17-1 to estimate the overpotential for $O_2$ liberation.
(b) Calculate the cathode potential (versus S.H.E.) assuming that $[TiO^{2+}]_{surface} = [TiO^{2+}]_{bulk} = 0.050$ M and $[Ti^{3+}]_{surface} = 0.10$ M.
(c) Calculate the anode potential (versus S.H.E.).
(d) What should the applied voltage be?

**17-18.** In an extremely accurate measurement of the Faraday constant, a pure silver anode was oxidized to $Ag^+$ with a constant current of 0.203 639 0 ($\pm$0.000 000 4) A for 18 000.075 ($\pm$0.010) s to give a mass loss of 4.097 900 ($\pm$0.000 003) g from the anode. Given that the atomic mass of Ag is 107.868 2 ($\pm$0.000 2), find the value of the Faraday constant and its uncertainty.

**17-19.** *Coulometric titration of sulfite in wine.*[36] Sulfur dioxide is added to many foods as a preservative. In aqueous solution, the following species are in equilibrium:

$$\underset{\text{Sulfur dioxide}}{SO_2} \rightleftharpoons \underset{\text{Sulfurous acid}}{H_2SO_3} \rightleftharpoons \underset{\text{Bisulfite}}{HSO_3^-} \rightleftharpoons \underset{\text{Sulfite}}{SO_3^{2-}} \quad (A)$$

Bisulfite reacts with aldehydes in food near neutral pH:

Sulfite is released from the adduct in 2 M NaOH and can be analyzed by its reaction with $I_3^-$ to give $I^-$ and sulfate. Excess $I_3^-$ must be present for quantitative reaction.

Here is a coulometric procedure for analysis of total sulfite in white wine. Total sulfite means all species in Reaction (A) and the adduct in Reaction (B). We use white wine so that we can see the color of a starch-iodine end point.

1. Mix 9.00 mL of wine plus 0.8 g NaOH and dilute to 10.00 mL. The NaOH releases sulfite from its organic adducts.
2. Generate $I_3^-$ at the working electrode (the anode) by passing a known current for a known time through the cell in Figure 17-9. The cell contains 30 mL of 1 M acetate buffer (pH 3.7) plus 0.1 M KI. The reaction in the cathode compartment is reduction of $H_2O$ to $H_2 + OH^-$. The frit retards diffusion of $OH^-$ into the main compartment, where it would react with $I_3^-$ to give $IO^-$.
3. Generate $I_3^-$ at the anode with a current of 10.00 mA for 4.00 min.
4. Inject 2.000 mL of the wine/NaOH solution into the cell, where the sulfite reacts with $I_3^-$, leaving excess $I_3^-$.
5. Add 0.500 mL of 0.050 7 M thiosulfate to consume $I_3^-$ by Reaction 16-19 and leave excess thiosulfate.

6. Add starch indicator to the cell and generate fresh $I_3^-$ with a constant current of 10.0 mA. A time of 131 s was required to consume excess thiosulfate and reach the starch end point.

(a) In what pH range is each form of sulfurous acid predominant?
(b) Write balanced half-reactions for the anode and cathode.
(c) At pH 3.7, the dominant form of sulfurous acid is $HSO_3^-$ and the dominant form of sulfuric acid is $SO_4^{2-}$. Write balanced reactions between $I_3^-$ and $HSO_3^-$ and between $I_3^-$ and thiosulfate.
(d) Find the concentration of total sulfite in undiluted wine.

**17-20.** *Chemical oxygen demand by coulometry.* An electrochemical device incorporating photooxidation on a $TiO_2$ surface could replace refluxing with $Cr_2O_7^{2-}$ to measure chemical oxygen demand (Box 16-1). The diagram shows a working electrode held at $+0.3$ V versus Ag | AgCl and coated with nanoparticles of $TiO_2$. Upon ultraviolet irradiation, electrons and holes are generated in the $TiO_2$. Holes oxidize organic matter at the surface. Electrons reduce $H_2O$ at the auxiliary electrode in a compartment connected to the working compartment by a salt bridge. The sample compartment is only 0.18 mm thick with a volume of 13.5 μL. It requires ~1 min for all organic matter to diffuse to the $TiO_2$ surface and be exhaustively oxidized.

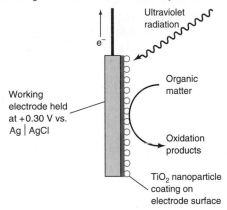

The blank curve in the graph below shows the response when the sample compartment contains just electrolyte. Before irradiation, no current is observed. Ultraviolet radiation causes a spike in the current with a rapid decrease to a steady level near 40 μA. This current arises from oxidation of water at the $TiO_2$ surface under ultraviolet exposure. The upper curve shows the same experiment, but with

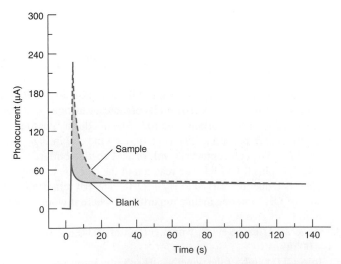

Photocurrent response for sample and blank. Both solutions contain 2 M $NaNO_3$. [From H. Zhao, D. Jiang, S. Zhang, K. Catterall, and R. John, "Development of a Direct Photoelectrochemical Method for Determination of Chemical Oxygen Demand," *Anal. Chem.* **2004**, *76*, 155.]

wastewater in the sample compartment. The increased current arises from oxidation of organic matter. When the organic matter is consumed, the current decreases to the blank level. The area between the two curves tells us how many electrons flow from oxidation of organic matter in the sample.

(a) Balance the oxidation half-reaction that occurs in this cell:

$$C_cH_hO_oN_nX_x + AH_2O \rightarrow BCO_2 + CX^- + DNH_3 + EH^+ + Fe^-$$

where X is any halogen. Express the stoichiometry coefficients A, B, C, D, E, and F in terms of c, h, o, n, and x.
(b) How many molecules of $O_2$ are required to balance the half-reaction in part (a) by reduction of oxygen ($O_2 + 4H^+ + 4e^- \rightarrow 2H_2O$)?
(c) The area between the two curves in the graph is $\int_0^\infty (I_{sample} - I_{blank})dt = 9.43$ mC. This is the number of electrons liberated by complete oxidation of the sample. How many moles of $O_2$ would be required for the same oxidation?
(d) Chemical oxygen demand (COD) is expressed as mg of $O_2$ required to oxidize 1 L of sample. Find the COD for this sample.
(e) If the only oxidizable substance in the sample were $C_9H_6NO_2ClBr_2$, what is its concentration in mol/L?

**Amperometry**

**17-21.** What is a Clark electrode, and how does it work?

**17-22.** (a) How does the glucose monitor work?
(b) Why is a mediator advantageous in the glucose monitor?

**17-23.** For a rotating disk electrode operating at sufficiently great potential, the redox reaction rate is governed by the rate at which analyte diffuses through the diffusion layer to the electrode (Figure 17-12b). The thickness of the diffusion layer is

$$\delta = 1.61D^{1/3}\nu^{1/6}\omega^{-1/2}$$

where D is the diffusion coefficient of reactant ($m^2/s$), $\nu$ is the kinematic viscosity of the liquid (= viscosity/density = $m^2/s$), and ω is the rotation rate (radians/s) of the electrode. There are $2\pi$ radians in a circle. The current density ($A/m^2$) is

$$\text{Current density} = 0.62nFD^{2/3}\nu^{-1/6}\omega^{1/2}C_0$$

where n is the number of electrons in the half-reaction, F is the Faraday constant, and $C_0$ is the concentration of the electroactive species in bulk solution ($mol/m^3$, not mol/L). Consider the oxidation of $Fe(CN)_6^{4-}$ in a solution of 10.0 mM $K_3Fe(CN)_6$ + 50.0 mM $K_4Fe(CN)_6$ at $+0.90$ V (versus S.C.E.) at a rotation speed of $2.00 \times 10^3$ revolutions per minute.[15] The diffusion coefficient of $Fe(CN)_6^{4-}$ is $2.5 \times 10^{-9}$ $m^2/s$, and the kinematic viscosity is $1.1 \times 10^{-6}$ $m^2/s$. Calculate the thickness of the diffusion layer and the current density. If you are careful, the current density should look like the value in Figure 17-13b.

**Voltammetry**

**17-24.** In 1 M $NH_3$/1 M $NH_4Cl$ solution, $Cu^{2+}$ is reduced to $Cu^+$ near $-0.3$ (versus S.C.E.), and $Cu^+$ is reduced to Cu(*in Hg*) near $-0.6$ V.

(a) Sketch a qualitative sampled current polarogram for a solution of $Cu^+$.
(b) Sketch a polarogram for a solution of $Cu^{2+}$.
(c) Suppose that Pt, instead of Hg, were used as the working electrode. Which, if any, reduction potential would you expect to change?

**17-25. (a)** Explain the difference between charging current and faradaic current.
**(b)** What is the purpose of waiting 1 s after a voltage pulse before measuring current in sampled current voltammetry?
**(c)** Why is square wave voltammetry more sensitive than sampled current voltammetry?

**17-26.** Suppose that the diffusion current in a polarogram for reduction of $Cd^{2+}$ at a mercury electrode is 14 μA. If the solution contains 25 mL of 0.50 mM $Cd^{2+}$, what percentage of $Cd^{2+}$ is reduced in the 3.4 min required to scan from $-0.6$ to $-1.2$ V?

**17-27.** The drug Librium gives a polarographic wave with $E_{1/2} = -0.265$ V (versus S.C.E.) in 0.05 M $H_2SO_4$. A 50.0-mL sample containing Librium gave a wave height of 0.37 μA. When 2.00 mL of 3.00 mM Librium in 0.05 M $H_2SO_4$ were added to the sample, the wave height increased to 0.80 μA. Find the molarity of Librium in the unknown.

**17-28.** Explain what is done in anodic stripping voltammetry. Why is stripping the most sensitive polarographic technique?

**17-29.** The figure below shows a series of standard additions of $Cu^{2+}$ to acidified tap water measured by anodic stripping voltammetry at an iridium electrode. The unknown and all standard additions were made up to the same final volume.
**(a)** What chemical reaction occurs during the concentration stage of the analysis?
**(b)** What chemical reaction occurs during the stripping stage of the analysis?
**(c)** Find the concentration of $Cu^{2+}$ in the tap water.

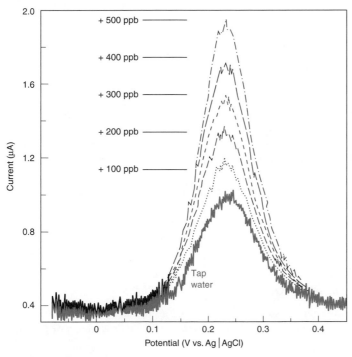

Anodic stripping voltammograms of tap water and five standard additions of 100 ppb $Cu^{2+}$. [From M. A. Nolan and S. P. Kounaves, "Microfabricated Array of Ir Microdisks for Determination of $Cu^{2+}$ or $Hg^{2+}$ Using Square Wave Stripping Voltammetry," Anal. Chem. **1999**, 71, 3567.]

**17-30.** From the two standard additions of 50 pm Fe(III) in Figure 17-21, find the concentration of Fe(III) in the seawater. Estimate where the baseline should be drawn for each trace and measure the peak height from the baseline. Consider the volume to be constant for all three solutions.

**17-31.** The cyclic voltammogram of the antibiotic chloramphenicol (abbreviated $RNO_2$) is shown here. The scan was started at 0 V, and potential was swept toward negative voltage. The first cathodic wave, A, is from the reaction $RNO_2 + 4e^- + 4H^+ \rightarrow RNHOH + H_2O$. Explain what happens at peaks B and C by using the reaction $RNO + 2e^- + 2H^+ \rightleftharpoons RNHOH$. Why was peak C not seen in the initial scan?

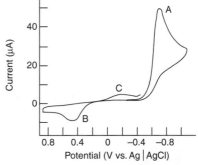

Cyclic voltammogram of $3.7 \times 10^{-4}$ chloramphenicol in 0.1 M acetate buffer, pH 4.62. The voltage of the carbon paste working electrode was scanned at a rate of 350 mV/s. [P. T. Kissinger and W. R. Heineman, "Cyclic Voltammetry," J. Chem. Ed. **1983**, 60, 702.]

**17-32.** Peak current ($I_p$) and scan rate ($v$) are listed for cyclic voltammetry (Fe(II) → Fe(III)) of a water-soluble ferrocene derivative in 0.1 M NaCl.[37]

| Scan rate (V/s) | Peak anodic current (μA) |
|---|---|
| 0.019 2 | 2.18 |
| 0.048 9 | 3.46 |
| 0.075 1 | 4.17 |
| 0.125 | 5.66 |
| 0.175 | 6.54 |
| 0.251 | 7.55 |

If a graph of $I_p$ versus $\sqrt{v}$ gives a straight line, then the reaction is diffusion controlled. Prepare such a graph and use it to find the diffusion coefficient of the reactant for this one-electron oxidation. The area of the working electrode is 0.020 1 $cm^2$ and the concentration of reactant is 1.00 mM.

**17-33.** What are the advantages of using a microelectrode for voltammetric measurements?

**17-34.** What is the purpose of the Nafion membrane in Figure 17-26?

### Karl Fischer Titration

**17-35.** Write the chemical reactions that show that 1 mol of $I_2$ is required for 1 mol of $H_2O$ in a Karl Fischer titration.

**17-36.** Explain how the end point is detected in a Karl Fischer titration in Figure 17-28.

# 18 Fundamentals of Spectrophotometry

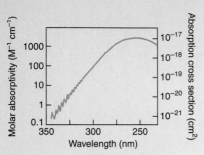

Spectrum of ozone, showing maximum absorption of ultraviolet radiation at a wavelength near 260 nm. [Adapted from R. P. Wayne, Chemistry of Atmospheres (Oxford: Clarendon Press, 1991).]

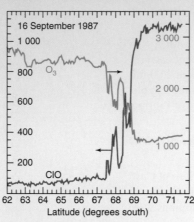

Spectroscopically measured concentrations of $O_3$ and $ClO$ (measured in ppb = nL/L) in the stratosphere near the South Pole in 1987. The loss of $O_3$ at latitudes where $ClO$ has a high concentration is consistent with the known chemistry of catalytic destruction of $O_3$ by halogen radicals. [From J. G. Anderson, W. H. Brune, and M. H. Proffitt, J. Geophys. Res. **1989**, 94D, 11465.]

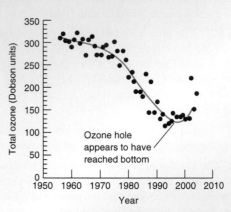

Mean atmospheric $O_3$ at Halley in Antarctica in October. Dobson units are defined in Problem 18-13. [From J. D. Shanklin, British Antarctic Survey, http://www.antarctica.ac.uk/met/jds/ozone/]

Ozone, formed at altitudes of 20 to 40 km by the action of solar ultraviolet radiation ($h\nu$) on $O_2$, absorbs ultraviolet radiation that causes sunburns and skin cancer.

$$O_2 \xrightarrow{h\nu} 2O \qquad O + O_2 \rightarrow O_3$$
$$\text{Ozone}$$

In 1985, the British Antarctic Survey reported that the total ozone in the Antarctic stratosphere had decreased by 50% in early spring, relative to levels observed in the preceding 20 years. Ground, airborne, and satellite observations have since shown that this "ozone hole" occurs only in early spring (Figure 1-1) and continued to deepen until the year 2000.

An explanation begins with chlorofluorocarbons such as Freon-12 ($CCl_2F_2$), formerly used in refrigerators and air conditioners. These long-lived compounds, which are not found in nature,[2] diffuse to the stratosphere, where they catalyze ozone decomposition.

(1) $\quad CCl_2F_2 \xrightarrow{h\nu} CClF_2 + Cl \qquad$ Photochemical $Cl$ formation

(2) $\quad Cl + O_3 \rightarrow ClO + O_2$

(3) $\quad O_3 \xrightarrow{h\nu} O + O_2 \qquad\qquad$ Net reaction of (2)–(4): Catalytic $O_3$ destruction

(4) $\quad O + ClO \rightarrow Cl + O_2 \qquad\qquad 2O_3 \rightarrow 3O_2$

$Cl$ produced in step 4 reacts again in step 2, so a single $Cl$ atom can destroy $>10^5$ molecules of $O_3$. The chain is terminated when $Cl$ or $ClO$ reacts with hydrocarbons or $NO_2$ to form $HCl$ or $ClONO_2$.

Stratospheric clouds[3] formed during the antarctic winter catalyze the reaction of $HCl$ with $ClONO_2$ to form $Cl_2$, which is split by sunlight into $Cl$ atoms to initiate $O_3$ destruction:

$$HCl + ClONO_2 \xrightarrow[\text{polar clouds}]{\text{surface of}} Cl_2 + HNO_3 \qquad Cl_2 \xrightarrow{h\nu} 2Cl$$

Polar stratospheric clouds require winter cold to form. Only when the sun is rising in September and October, and clouds are still present, are conditions right for $O_3$ destruction.

To protect life from ultraviolet radiation, international treaties now ban or phase out chlorofluorocarbons, and there is an effort to find safe substitutes.

$S$pectrophotometry is any technique that uses light to measure chemical concentrations. A procedure based on absorption of visible light is called *colorimetry*. The most-cited article in the journal *Analytical Chemistry* from 1945 to 1999 describes a colorimetric method by which biochemists measure sugars.[4]

After the discovery of the Antarctic ozone "hole" in 1985, atmospheric chemist Susan Solomon led the first expedition in 1986 specifically intended to make chemical measurements of the Antarctic atmosphere by using balloons and ground-based spectroscopy. The expedition discovered that ozone depletion occurred after polar sunrise and that the concentration of chemically active chlorine in the stratosphere was ~100 times greater than had been predicted from gas-phase chemistry. Solomon's group identified chlorine as the culprit in ozone destruction and polar stratospheric clouds as the catalytic surface for the release of so much chlorine.

## ■ ■ ■ ■ 18-1 Properties of Light

It is convenient to describe light in terms of both particles and waves. Light waves consist of perpendicular, oscillating electric and magnetic fields. For simplicity, a *plane-polarized* wave is shown in Figure 18-1. In this figure, the electric field is in the $xy$ plane, and the magnetic field is in the $xz$ plane. **Wavelength,** $\lambda$, is the crest-to-crest distance between waves. **Frequency,** $\nu$, is the number of complete oscillations that the wave makes each second. The unit of frequency is second$^{-1}$. One oscillation per second is also called one **hertz** (Hz). A frequency of $10^6$ s$^{-1}$ is therefore said to be $10^6$ Hz, or 1 *megahertz* (MHz).

The relation between frequency and wavelength is

*Relation between frequency and wavelength:*

$$\nu\lambda = c \tag{18-1}$$

where $c$ is the speed of light ($2.998 \times 10^8$ m/s in vacuum). In a medium other than vacuum, the speed of light is $c/n$, where $n$ is the **refractive index** of that medium. For visible wavelengths in most substances, $n > 1$, so visible light travels more slowly through matter than through vacuum. When light moves between media with different refractive indexes, the frequency remains constant but the wavelength changes.

With regard to energy, it is more convenient to think of light as particles called **photons.** Each photon carries the energy, $E$, which is given by

*Relation between energy and frequency:*

$$E = h\nu \tag{18-2}$$

where $h$ is *Planck's constant* ($= 6.626 \times 10^{-34}$ J · s).

Equation 18-2 states that energy is proportional to frequency. Combining Equations 18-1 and 18-2, we can write

$$E = \frac{hc}{\lambda} = hc\tilde{\nu} \tag{18-3}$$

where $\tilde{\nu}$ ($= 1/\lambda$) is called **wavenumber.** Energy is inversely proportional to wavelength and directly proportional to wavenumber. Red light, with a longer wavelength than blue light, is less energetic than blue light. The most common unit for wavenumber in the literature is cm$^{-1}$, read "reciprocal centimeters" or "wavenumbers."

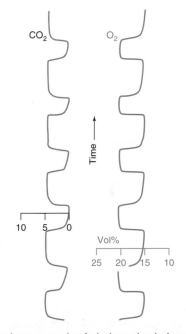

Here is an example of what spectrophotometry can do. These traces show the continuous measurement of $O_2$ and $CO_2$ in the airway of a healthy person. $O_2$ is detected by ultraviolet absorption at a wavelength of 147 nm. $CO_2$ is detected by infrared absorption at a wavenumber of $2.3 \times 10^3$ cm$^{-1}$. [P. B. Arnoudse, H. L. Pardue, J. D. Bourland, R. Miller, and L. A. Geddes, "Breath-by-Breath Determination of $O_2$ and $CO_2$ Based on Nondispersive Absorption Measurements," *Anal. Chem.* **1992**, *64*, 200.]

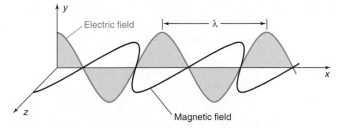

*Figure 18-1* Plane-polarized electromagnetic radiation of wavelength $\lambda$, propagating along the $x$-axis. The electric field of plane-polarized light is confined to a single plane. Ordinary, unpolarized light has electric field components in all planes.

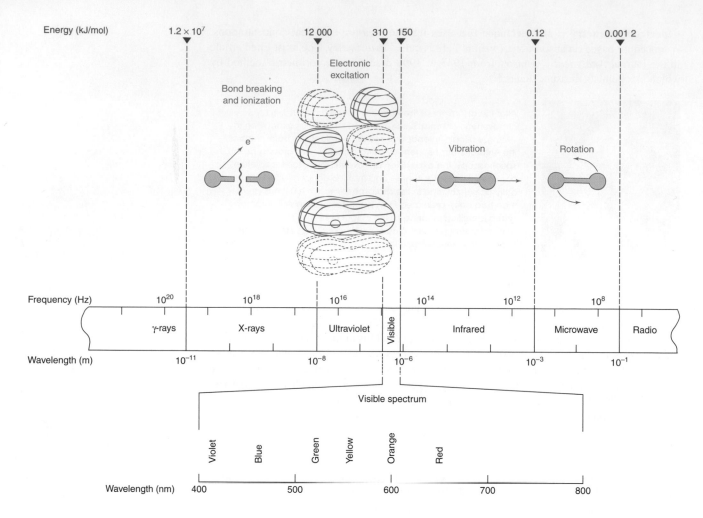

Energy (kJ/mol)

| 1.2 × 10⁷ | 12 000 | 310 | 150 | 0.12 | 0.001 2 |

Bond breaking and ionization

Electronic excitation

Vibration

Rotation

Frequency (Hz)

$10^{20}$  $10^{18}$  $10^{16}$  $10^{14}$  $10^{12}$  $10^8$

| γ-rays | X-rays | Ultraviolet | Visible | Infrared | Microwave | Radio |

Wavelength (m)

$10^{-11}$   $10^{-8}$   $10^{-6}$   $10^{-3}$   $10^{-1}$

Visible spectrum

Violet | Blue | Green | Yellow | Orange | Red

Wavelength (nm)  400  500  600  700  800

**Figure 18-2** Electromagnetic spectrum, showing representative molecular processes that occur when light in each region is absorbed. The visible spectrum spans the wavelength range 380–780 nanometers (1 nm = $10^{-9}$ m).

Regions of the **electromagnetic spectrum** are labeled in Figure 18-2. Names are historical. There are no abrupt changes in characteristics as we go from one region to the next, such as visible to infrared. Visible light—the kind of electromagnetic radiation we see—represents only a small fraction of the electromagnetic spectrum.

## ■ ■ ■ 18-2 Absorption of Light

When a molecule absorbs a photon, the energy of the molecule increases. We say that the molecule is promoted to an **excited state** (Figure 18-3). If a molecule emits a photon, the energy of the molecule is lowered. The lowest energy state of a molecule is called the **ground state.** Figure 18-2 indicates that microwave radiation stimulates rotation of molecules when it is absorbed. Infrared radiation stimulates vibrations. Visible and ultraviolet radiation promote electrons to higher-energy orbitals. X-rays and short-wavelength ultraviolet radiation break chemical bonds and ionize molecules. Medical X-rays damage the human body, so your exposure should be minimized.

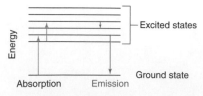

**Figure 18-3** Absorption of light increases the energy of a molecule. Emission of light decreases its energy.

**Example** Photon Energies

By how many kilojoules per mole is the energy of $O_2$ increased when it absorbs ultraviolet radiation with a wavelength of 147 nm? How much is the energy of $CO_2$ increased when it absorbs infrared radiation with a wavenumber of 2 300 cm⁻¹?

**Solution**   For the ultraviolet radiation, the energy increase is

$$\Delta E = h\nu = h\frac{c}{\lambda}$$

$$= (6.626 \times 10^{-34} \text{ J} \cdot \text{s})\left[\frac{(2.998 \times 10^8 \text{ m/s})}{(147 \text{ nm})(10^{-9} \text{ m/nm})}\right] = 1.35 \times 10^{-18} \text{ J/molecule}$$

$$(1.35 \times 10^{-18} \text{ J/molecule})(6.022 \times 10^{23} \text{ molecules/mol}) = 814 \text{ kJ/mol}$$

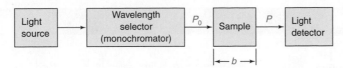

**Figure 18-4** Schematic diagram of a single-beam spectrophotometric experiment. $P_0$, irradiance of beam entering sample; $P$, irradiance of beam emerging from sample; $b$, length of path through sample.

This is enough energy to break the O=O bond in oxygen. For $CO_2$, the energy increase is

$$\Delta E = h\nu = h\frac{c}{\lambda} = hc\tilde{\nu} \qquad \left(\text{recall that } \tilde{\nu} = \frac{1}{\lambda}\right)$$

$$= (6.626 \times 10^{-34}\,\text{J}\cdot\text{s})(2.998 \times 10^8\,\text{m/s})(2\,300\,\text{cm}^{-1})(100\,\text{cm/m})$$

$$= 4.6 \times 10^{-20}\,\text{J/molecule} = 28\,\text{kJ/mol}$$

Infrared absorption increases the amplitude of the vibrations of the $CO_2$ bonds.

When light is absorbed by a sample, the *irradiance* of the beam of light is decreased. **Irradiance,** $P$, is the energy per second per unit area of the light beam. A rudimentary spectrophotometric experiment is illustrated in Figure 18-4. Light is passed through a **monochromator** (a prism, a grating, or even a filter) to select one wavelength (Color Plate 13). Light of a single wavelength is said to be **monochromatic,** which means "one color." The monochromatic light, with irradiance $P_0$, strikes a sample of length $b$. The irradiance of the beam emerging from the other side of the sample is $P$. Some of the light may be absorbed by the sample, so $P \le P_0$.

**Transmittance,** $T$, is defined as the fraction of the original light that passes through the sample.

*Transmittance:*
$$T = \frac{P}{P_0} \tag{18-4}$$

Therefore, $T$ has the range 0 to 1. The *percent transmittance* is simply $100T$ and ranges between 0 and 100%. **Absorbance** is defined as

*Absorbance:*
$$A = \log\left(\frac{P_0}{P}\right) = -\log T \tag{18-5}$$

When no light is absorbed, $P = P_0$ and $A = 0$. If 90% of the light is absorbed, 10% is transmitted and $P = P_0/10$. This ratio gives $A = 1$. If only 1% of the light is transmitted, $A = 2$. Absorbance is sometimes called *optical density.*

Absorbance is so important because it is directly proportional to the concentration, $c$, of the light-absorbing species in the sample (Color Plate 14).

*Beer's law:*
$$A = \varepsilon bc \tag{18-6}$$

Equation 18-6, which is the heart of spectrophotometry as applied to analytical chemistry, is called the *Beer-Lambert law*,[6] or simply **Beer's law.** Absorbance is dimensionless, but some people write "absorbance units" after absorbance. The concentration of the sample, $c$, is usually given in units of moles per liter (M). The pathlength, $b$, is commonly expressed in centimeters. The quantity $\varepsilon$ (epsilon) is called the **molar absorptivity** (or *extinction coefficient* in the older literature) and has the units $M^{-1}\,cm^{-1}$ to make the product $\varepsilon bc$ dimensionless. Molar absorptivity is the characteristic of a substance that tells how much light is absorbed at a particular wavelength.

> **Irradiance** is the energy per unit time per unit area in the light beam (watts per square meter, W/m²). The terms *intensity* or *radiant power* have been used for this same quantity.
>
> **Monochromatic light** consists of "one color" (one wavelength). The better the monochromator, the narrower is the range of wavelengths in the emerging beam.

Relation between transmittance and absorbance:

| $P/P_0$ | % $T$ | $A$ |
|---------|-------|-----|
| 1 | 100 | 0 |
| 0.1 | 10 | 1 |
| 0.01 | 1 | 2 |

Box 18-1 explains why absorbance, not transmittance, is directly proportional to concentration.

---

**Example** Absorbance, Transmittance, and Beer's Law

Find the absorbance and transmittance of a 0.002 40 M solution of a substance with a molar absorptivity of 313 $M^{-1}\,cm^{-1}$ in a cell with a 2.00-cm pathlength.

**Solution** Equation 18-6 gives us the absorbance.

$$A = \varepsilon bc = (313\,M^{-1}\,cm^{-1})(2.00\,\text{cm})(0.002\,40\,M) = 1.50$$

Beer's law, Equation 18-6, states that *absorbance* is directly proportional to the concentration of the absorbing species. The fraction of light passing through a sample (the *transmittance*) is related logarithmically, not linearly, to the sample concentration. Why should this be?

Imagine light of irradiance $P$ passing through an *infinitesimally thin* layer of solution whose thickness is $dx$. A physical model of the absorption process suggests that, within the infinitesimally thin layer, decrease in power ($dP$) ought to be proportional to the incident power ($P$), to the concentration of absorbing species ($c$), and to the thickness of the section ($dx$):

$$dP = -\beta P c \, dx \qquad\qquad (A)$$

where $\beta$ is a constant of proportionality and the minus sign indicates a decrease in $P$ as $x$ increases. The rationale for saying that the decrease in power is proportional to the incident power may be understood from a numerical example. If 1 photon out of 1 000 incident photons is absorbed in a thin layer, we would expect that 2 out of 2 000 incident photons would be absorbed. The decrease in photons (power) is proportional to the incident flux of photons (power).

Equation A can be rearranged and integrated to find an expression for $P$:

$$-\frac{dP}{P} = \beta c \, dx \Rightarrow -\int_{P_0}^{P} \frac{dP}{P} = \beta c \int_{0}^{b} dx$$

The limits of integration are $P = P_0$ at $x = 0$ and $P = P$ at $x = b$.

$$-\ln P - (-\ln P_0) = \beta cb \Rightarrow \ln\left(\frac{P_0}{P}\right) = \beta cb$$

Convert ln into log, using the relation $\ln z = (\ln 10)(\log z)$, to obtain Beer's law:

$$\underbrace{\log\left(\frac{P_0}{P}\right)}_{\text{Absorbance}} = \underbrace{\left(\frac{\beta}{\ln 10}\right)}_{\text{Constant} \equiv \varepsilon} cb \Rightarrow A = \varepsilon cb$$

The logarithmic relation of $P_0/P$ to concentration arises because, in each infinitesimal portion of the total volume, *the decrease in power is proportional to the power incident upon that section.* As light travels through the sample, the power loss in each succeeding layer decreases, because the magnitude of the incident power that reaches each layer is decreasing. Molar absorptivity ranges from 0 (if the probability for photon absorption is 0) to approximately $10^5 \, M^{-1} \, cm^{-1}$ (when the probability for photon absorption approaches unity).

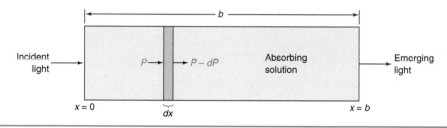

Transmittance is obtained from Equation 18-5 by raising 10 to the power equal to the expression on each side of the equation:

$$\log T = -A$$

$$T = 10^{\log T} = 10^{-A} = 10^{-1.50} = 0.031\ 6$$

Just 3.16% of the incident light emerges from this solution.

*If $x = y$, $10^x = 10^y$.*

Equation 18-6 could be written

$$A_\lambda = \varepsilon_\lambda bc$$

because $A$ and $\varepsilon$ depend on the wavelength of light. The quantity $\varepsilon$ is simply a coefficient of proportionality between absorbance and the product $bc$. The greater the molar absorptivity, the greater the absorbance. An **absorption spectrum** is a graph showing how $A$ (or $\varepsilon$) varies with wavelength. Demonstration 18-1 illustrates the meaning of an absorption spectrum.

*The plural of "spectrum" is "spectra."*

The part of a molecule responsible for light absorption is called a **chromophore.** Any substance that absorbs visible light appears colored when white light is transmitted through it or reflected from it. (White light contains all the colors in the visible spectrum.) The substance absorbs certain wavelengths of the white light, and our eyes detect the wavelengths that are not absorbed. Table 18-1 gives a rough guide to colors. The observed color is called the *complement* of the absorbed color. For example, bromophenol blue has maximum absorbance at 591 nm and its observed color is blue.

*The color of a solution is the complement of the color of the light that it absorbs.*
*The color we perceive depends not only on the wavelength of light, but on its intensity.*

## When Beer's Law Fails

Beer's law states that absorbance is proportional to the concentration of the absorbing species. It applies to *monochromatic* radiation[8] and it works very well for dilute solutions ($\leqslant 0.01$ M) of most substances.

**Table 18-1**  Colors of visible light

| Wavelength of maximum absorption (nm) | Color absorbed | Color observed |
|---|---|---|
| 380—420 | Violet | Green-yellow |
| 420—440 | Violet-blue | Yellow |
| 440—470 | Blue | Orange |
| 470—500 | Blue-green | Red |
| 500—520 | Green | Purple |
| 520—550 | Yellow-green | Violet |
| 550—580 | Yellow | Violet-blue |
| 580—620 | Orange | Blue |
| 620—680 | Red | Blue-green |
| 680—780 | Red | Green |

In concentrated solutions, solute molecules influence one another as a result of their proximity. When solute molecules get close to one another, their properties (including molar absorptivity) change somewhat. At very high concentration, the solute *becomes* the solvent. Properties of a molecule are not exactly the same in different solvents. Nonabsorbing solutes in a solution can also interact with the absorbing species and alter the absorptivity.

If the absorbing molecule participates in a concentration-dependent chemical equilibrium, the absorptivity changes with concentration. For example, in concentrated solution, a weak acid, HA, may be mostly undissociated. As the solution is diluted, dissociation increases. If the absorptivity of $A^-$ is not the same as that of HA, the solution will appear not to obey Beer's law as it is diluted.

> Beer's law works for monochromatic radiation passing through a dilute solution in which the absorbing species is not participating in a concentration-dependent equilibrium.

## 18-3  Measuring Absorbance

The minimum requirements for a spectrophotometer (a device to measure absorbance of light) were shown in Figure 18-4. Light from a continuous source is passed through a monochromator, which selects a narrow band of wavelengths from the incident beam. This

---

**Demonstration 18-1  Absorption Spectra**

The spectrum of visible light can be projected on a screen in a darkened room in the following manner:[7] Mount four layers of plastic diffraction grating* on a cardboard frame having a square hole large enough to cover the lens of an overhead projector. Tape the assembly over the projector lens facing the screen. Place an opaque cardboard surface with two 1 × 3 cm slits on the working surface of the projector.

When the lamp is turned on, the white image of each slit is projected on the center of the screen. A visible spectrum appears on

*Edmund Scientific Co., www.edmundoptics.com, catalog no. NT40-267.

either side of each image. By placing a beaker of colored solution over one slit, you can see its color projected on the screen where the white image previously appeared. The spectrum beside the colored image loses intensity at wavelengths absorbed by the colored species.

Color Plate 15a shows the spectrum of white light and the spectra of three different colored solutions. You can see that potassium dichromate, which appears orange or yellow, absorbs blue wavelengths. Bromophenol blue absorbs orange wavelengths and appears blue to our eyes. Phenolphthalein absorbs the center of the visible spectrum. For comparison, spectra of these three solutions recorded with a spectrophotometer are shown in Color Plate 15b.

This same setup can be used to demonstrate fluorescence and the properties of colors.[7]

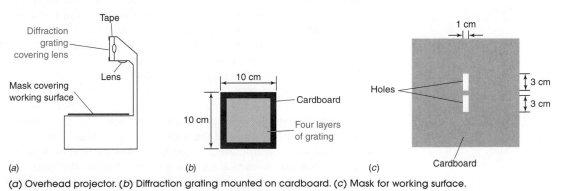

(a) Overhead projector. (b) Diffraction grating mounted on cardboard. (c) Mask for working surface.

---

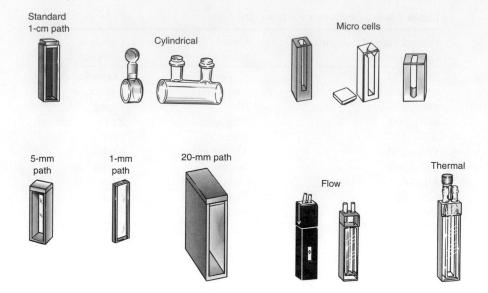

**Figure 18-5** Common cuvets for visible and ultraviolet spectroscopy. Flow cells permit continuous flow of solution through the cell. In the thermal cell, liquid from a constant-temperature bath flows through the cell jacket to maintain a desired temperature.
*[Courtesy A. H. Thomas Co., Philadelphia, PA.]*

Approximate low-energy cutoff for common infrared windows:

| | |
|---|---|
| sapphire ($Al_2O_3$) | 1 500 cm$^{-1}$ |
| NaCl | 650 cm$^{-1}$ |
| KBr | 350 cm$^{-1}$ |
| AgCl | 350 cm$^{-1}$ |
| CsBr | 250 cm$^{-1}$ |
| CsI | 200 cm$^{-1}$ |

"monochromatic" light travels through a sample of pathlength $b$, and the irradiance of the emergent light is measured.

For visible and ultraviolet spectroscopy, a liquid sample is usually contained in a cell called a **cuvet** that has flat, fused-silica ($SiO_2$) faces (Figure 18-5). Glass is suitable for visible, but not for ultraviolet spectroscopy, because it absorbs ultraviolet radiation. The most common cuvets have a 1.000-cm pathlength and are sold in matched sets for sample and reference.

For infrared measurements, cells are commonly constructed of NaCl or KBr. For the 400 to 50 cm$^{-1}$ far-infrared region, polyethylene is a transparent window. Solid samples are commonly ground to a fine powder, which can be added to mineral oil (a viscous hydrocarbon also called Nujol) to give a dispersion that is called a *mull* and is pressed between two KBr plates. The analyte spectrum is obscured in a few regions in which the mineral oil absorbs infrared radiation. Alternatively, a 1 wt% mixture of solid sample with KBr can be ground to a fine powder and pressed into a translucent pellet at a pressure of ~60 MPa (600 bar). Solids and powders can also be examined by *diffuse reflectance,* in which reflected infrared radiation, instead of transmitted infrared radiation, is observed. Wavelengths absorbed by the sample are not reflected as well as other wavelengths. This technique is sensitive only to the surface of the sample.

Figure 18-4 outlines a *single-beam* instrument, which has only one beam of light. We do not measure the incident irradiance, $P_0$, directly. Rather, the irradiance of light passing through a reference cuvet containing pure solvent (or a reagent blank) is *defined* as $P_0$. This cuvet is then removed and replaced by an identical one containing sample. The irradiance of light striking the detector after passing through the sample is the quantity $P$. Knowing both $P$ and $P_0$ allows $T$ or $A$ to be determined. The reference cuvet compensates for reflection, scattering, and absorption by the cuvet and solvent. A *double-beam* instrument, which splits the light to pass alternately between sample and reference cuvets, is described in Chapter 20.

In recording an absorbance spectrum, first record a *baseline spectrum* with reference solutions (pure solvent or a reagent blank) in both cuvets. If the instrument were perfect, the baseline would be 0 everywhere. In our imperfect world, the baseline usually exhibits small positive and negative absorbance. We subtract the baseline absorbance from the sample absorbance to obtain the true absorbance at each wavelength.

For spectrophotometric analysis, we normally choose the wavelength of maximum absorbance for two reasons: (1) The sensitivity of the analysis is greatest at maximum absorbance (that is, we get the maximum response for a given concentration of analyte). (2) The curve is relatively flat at the maximum, so there is little variation in absorbance if the monochromator drifts a little or if the width of the transmitted band changes slightly. Beer's law is obeyed when the absorbance is constant across the selected wave band.

Spectrophotometers are most accurate at intermediate levels of absorbance ($A \approx 0.4-0.9$). If too little light gets through the sample (high absorbance), the intensity is hard to measure. If too much light gets through (low absorbance), it is hard to distinguish the difference between the sample and the reference. It is desirable to adjust the sample concentration so that absorbance falls in an intermediate range. Compartments must be tightly closed to avoid stray light, which leads to false readings.

Figure 18-6 shows measured spectrophotometer errors. Electrical noise was only modestly dependent on sample absorbance. The largest source of imprecision for $A < 0.6$ was irreproducible positioning of the cuvet in the sample holder, despite care in placing the cuvet. The resulting error curve reaches a minimum near $A = 0.6$. In Section 5-2, we learned that *the detection limit for an analytical procedure is determined by the reproducibility of the measurement.* The less noise, the lower the concentration of analyte that can be detected.

Keep containers covered to avoid dust, which scatters light and appears to the spectrophotometer to be absorbance. Filtering the final solution through a very fine filter may be necessary in critical work. Handle cuvets with a tissue to avoid putting fingerprints on the faces—fingerprints scatter and absorb light—and keep your cuvets scrupulously clean.

Slight mismatch between sample and reference cuvets, over which you have little control, leads to systematic errors in spectrophotometry. For best precision, you should place cuvets in the spectrophotometer as reproducibly as possible. Random variation in absorbance arises from slight misplacement of the cuvet in its holder, or turning a flat cuvet around by 180°, or rotation of a circular cuvet.

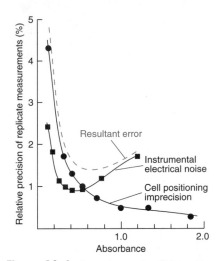

Figure 18-6  Errors in spectrophotometric measurements due to dark current noise and cell positioning imprecision in a research-quality instrument. [Data from L. D. Rothman, S. R. Crouch, and J. D. Ingle, Jr., "Theoretical and Experimental Investigation of Factors Affecting Precision in Molecular Absorption Spectrophotometry," Anal. Chem. **1975**, 47, 1226.]

## ■ ■ ■ 18-4  Beer's Law in Chemical Analysis

For a compound to be analyzed by spectrophotometry, it must absorb light, and this absorption should be distinguishable from that due to other substances in the sample. Because most compounds absorb ultraviolet radiation, measurements in this region of the spectrum tend to be inconclusive, and analysis is usually restricted to the visible spectrum. If there are no interfering species, however, ultraviolet absorbance is satisfactory. Proteins are normally assayed in the ultraviolet region at 280 nm because the aromatic groups present in virtually every protein have an absorbance maximum at 280 nm.

### Example  Measuring Benzene in Hexane

**(a)** Pure hexane has negligible ultraviolet absorbance above a wavelength of 200 nm. A solution prepared by dissolving 25.8 mg of benzene ($C_6H_6$, FM 78.11) in hexane and diluting to 250.0 mL had an absorption peak at 256 nm and an absorbance of 0.266 in a 1.000-cm cell. Find the molar absorptivity of benzene at this wavelength.

This example illustrates the measurement of molar absorptivity from a single solution. It is better to measure several concentrations to obtain a more reliable absorptivity and to demonstrate that Beer's law is obeyed.

**Solution**  The concentration of benzene is

$$[C_6H_6] = \frac{(0.025\ 8\ \text{g})/(78.11\ \text{g/mol})}{0.250\ 0\ \text{L}} = 1.32_1 \times 10^{-3}\ \text{M}$$

We find the molar absorptivity from Beer's law:

$$\text{Molar absorptivity} = \varepsilon = \frac{A}{bc} = \frac{(0.266)}{(1.000\ \text{cm})(1.32_1 \times 10^{-3}\ \text{M})} = 201._3\ \text{M}^{-1}\ \text{cm}^{-1}$$

**(b)** A sample of hexane contaminated with benzene had an absorbance of 0.070 at 256 nm in a cuvet with a 5.000-cm pathlength. Find the concentration of benzene in mg/L.

**Solution**  Using Beer's law with the molar absorptivity from part **(a)**, we find:

$$[C_6H_6] = \frac{A}{\varepsilon b} = \frac{0.070}{(201._3\ \text{M}^{-1}\ \text{cm}^{-1})(5.000\ \text{cm})} = 6.9_5 \times 10^{-5}\ \text{M}$$

$$[C_6H_6] = \left(6.9_5 \times 10^{-5}\ \frac{\text{mol}}{\text{L}}\right)\left(78.11 \times 10^3\ \frac{\text{mg}}{\text{mol}}\right) = 5.4_2\ \frac{\text{mg}}{\text{L}}$$

### Serum Iron Determination

Iron for biosynthesis is transported through the bloodstream by the protein transferrin. The following procedure measures the Fe content of transferrin.[10] This analysis requires only about 1 µg for an accuracy of 2–5%. Human blood usually contains about 45 vol% cells and 55 vol% plasma (liquid). If blood is collected without an anticoagulant, the blood clots, and the liquid that remains is called *serum*. Serum normally contains about 1 µg of Fe/mL attached to transferrin.

For a laboratory experiment or research project, you can measure total iron and $Fe^{3+}$ at ppb levels in rainwater.[9] $Fe^{3+}$ decreases during daylight hours when sunlight plus reducing agents (organic matter?) convert $Fe^{3+}$ into $Fe^{2+}$. At night, oxidation by $O_2$ restores the $Fe^{3+}$.

HO
OH
O
O
HO
OH
**Ascorbic acid**
**(vitamin C)**

*Supernate:* the liquid layer above the solid that collects at the bottom of a tube during centrifugation.

The blank described here contains all sources of absorbance other than analyte. An alternative blank for some analyses contains analyte but no color-forming reagent. The choice of blank depends on which species interfere at the analytical wavelength.

The serum iron determination has three steps:

**Step 1** Reduce $Fe^{3+}$ in transferrin to $Fe^{2+}$, which is released from the protein. Commonly employed reducing agents are hydroxylamine hydrochloride ($NH_3OH^+Cl^-$), thioglycolic acid, or ascorbic acid.

$$2Fe^{3+} + 2HSCH_2CO_2H \rightarrow 2Fe^{2+} + HO_2CCH_2S-SCH_2CO_2H + 2H^+$$
$$\text{Thioglycolic acid}$$

**Step 2** Add trichloroacetic acid ($Cl_3CCO_2H$) to precipitate proteins, leaving $Fe^{2+}$ in solution. Centrifuge the mixture to remove the precipitate. If protein were left in the solution, it would partially precipitate in the final solution. Light scattered by the precipitate would be mistaken for absorbance.

$$\text{Protein}(aq) \xrightarrow{CCl_3CO_2H} \text{protein}(s)$$

**Step 3** Transfer a measured volume of supernatant liquid from step 2 to a fresh vessel and add buffer plus excess ferrozine to form a purple complex. Measure the absorbance at the 562-nm peak (Figure 18-7). The buffer provides a pH at which complex formation is complete.

$$Fe^{2+} + 3\text{ferrozine}^{2-} \rightarrow (\text{ferrozine})_3Fe^{4-}$$
$$\text{Purple complex}$$
$$\lambda_{max} = 562 \text{ nm}$$

In most spectrophotometric analyses, it is important to prepare a **reagent blank** containing all reagents, but with analyte replaced by distilled water. Any absorbance of the blank is due to the color of uncomplexed ferrozine plus the color caused by the iron impurities in the reagents and glassware. *Subtract the blank absorbance from the absorbance of samples and standards before doing any calculations.*

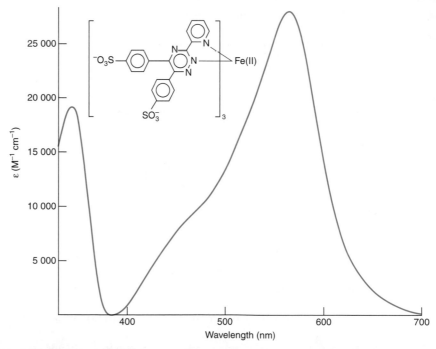

*Figure 18-7* Visible absorption spectrum of the complex (ferrozine)$_3$Fe(II) used in the colorimetric analysis of iron.

Use a series of iron standards for a *calibration curve* (Figure 18-8) to show that Beer's law is obeyed. Standards are prepared by the same procedure as unknowns. The absorbance of the unknown should fall within the region covered by the standards. Pure iron wire (with a shiny, rust-free surface) is dissolved in acid to prepare accurate iron standards. Ferrous ammonium sulfate ($Fe(NH_4)_2(SO_4)_2 \cdot 6H_2O$) and ferrous ethylenediammonium sulfate ($Fe(H_3NCH_2CH_2NH_3)(SO_4)_2 \cdot 4H_2O$) are suitable standards for less accurate work.

If unknowns and standards are prepared with identical volumes, then the quantity of iron in the unknown can be calculated from the least-squares equation for the calibration line. For example, in Figure 18-8, if the unknown has an absorbance of 0.357 (after subtracting the absorbance of the blank), the sample contains 3.59 μg of iron.

In the preceding iron determination, the results would be about 10% high because serum copper also forms a colored complex with ferrozine. Interference is eliminated if neocuproine or thiourea is added. These reagents **mask** $Cu^+$ by forming strong complexes that prevent $Cu^+$ from reacting with ferrozine.

Neocuproine   $2$  $+ Cu^+ \longrightarrow$   Tightly bound complex with little absorbance at 562 nm

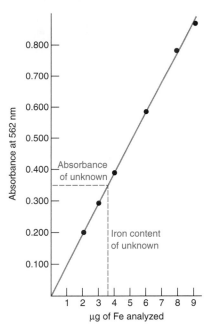

Figure 18-8   Calibration curve showing the validity of Beer's law for the (ferrozine)$_3$Fe(II) complex used in serum iron determination. Each sample was diluted to a final volume of 5.00 mL. Therefore, 1.00 μg of iron is equivalent to a concentration of $3.58 \times 10^{-6}$ M.

## Example Serum Iron Analysis

Serum iron and standard iron solutions were analyzed as follows:

**Step 1**  To 1.00 mL of sample, add 2.00 mL of reducing agent and 2.00 mL of acid to reduce and release Fe from transferrin.

**Step 2**  Precipitate proteins with 1.00 mL of 30 wt% trichloroacetic acid. For calculations, assume that the volume is $1.00 + 2.00 + 2.00 + 1.00 = 6.00$ mL (no change in volume due to mixing). Centrifuge the mixture to remove protein.

**Step 3**  Transfer 4.00 mL of supernatant liquid to a fresh test tube and add 1.00 mL of solution containing ferrozine and buffer. Measure the absorbance after 10 min.

**Step 4**  To establish the calibration curve in Figure 18-8, dilute 1.00 mL of standard iron (2–9 μg) to 6.00 mL with other reagents. Transfer 4.00 mL (containing 4.00/6.00 of the Fe) to a new vessel and dilute with 1.00 mL of ferrozine plus buffer.

The blank absorbance was 0.038 at 562 nm in a 1.000-cm cell. A serum sample had an absorbance of 0.129. After the blank was subtracted from each standard absorbance, the points in Figure 18-8 were obtained. The least-squares line through the standard points is

$$\text{Absorbance} = 0.067_0 \times (\text{μg Fe in initial sample}) + 0.001_5$$

According to Beer's law, the intercept should be 0, not $0.001_5$. However, we will use the observed intercept for our analysis. Find the concentration of iron in the serum.

**Solution**   Rearranging the least-squares equation of the calibration line and inserting the corrected absorbance (observed − blank = $0.129 - 0.038 = 0.091$) of unknown, we find

$$\text{μg Fe in unknown} = \frac{\text{absorbance} - 0.001_5}{0.067_0} = \frac{0.091 - 0.001_5}{0.067_0} = 1.33_6 \text{ μg}$$

The concentration of Fe in the serum is

$$[\text{Fe}] = \text{moles of Fe/liters of serum}$$

$$= \left(\frac{1.33_6 \times 10^{-6} \text{ g Fe}}{55.845 \text{ g Fe/mol Fe}}\right) \Big/ (1.00 \times 10^{-3} \text{ L}) = 2.39 \times 10^{-5} \text{ M}$$

To find the uncertainty in μg Fe, use Equation 4-27.

## ▮ ▮ ▮ 18-5   What Happens When a Molecule Absorbs Light?

When a molecule absorbs a photon, the molecule is promoted to a more energetic *excited state* (Figure 18-3). Conversely, when a molecule emits a photon, the energy of the molecule falls by an amount equal to the energy of the photon.

For example, consider formaldehyde in Figure 18-9a. In its ground state, the molecule is planar, with a double bond between carbon and oxygen. From the electron dot description of formaldehyde, we expect two pairs of nonbonding electrons to be localized on the oxygen atom. The double bond consists of a sigma bond between carbon and oxygen and a pi bond made from the $2p_y$ (out-of-plane) atomic orbitals of carbon and oxygen.

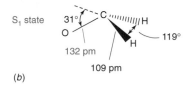

Figure 18-9   Geometry of formaldehyde. (a) Ground state. (b) Lowest excited singlet state.

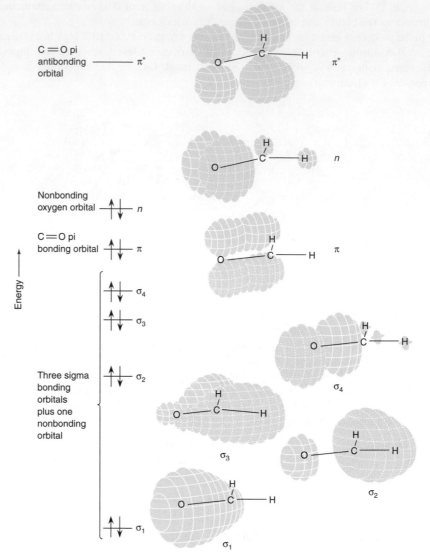

C═O pi antibonding orbital —— $\pi^*$

Nonbonding oxygen orbital $n$

C═O pi bonding orbital $\pi$

Energy

$\sigma_4$

$\sigma_3$

Three sigma bonding orbitals plus one nonbonding orbital

$\sigma_2$

$\sigma_1$

$\pi^*$

$n$

$\pi$

$\sigma_4$

$\sigma_3$

$\sigma_2$

$\sigma_1$

**Figure 18-10** Molecular orbital diagram of formaldehyde, showing energy levels and orbital shapes. The coordinate system of the molecule was shown in Figure 18-9. *[From W. L. Jorgensen and L. Salem, The Organic Chemist's Book of Orbitals (New York: Academic Press, 1973).]*

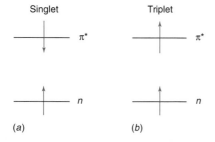

Singlet    Triplet

$\pi^*$    $\pi^*$

$n$    $n$

(a)    (b)

**Figure 18-11** Two possible electronic states arise from an $n \rightarrow \pi^*$ transition. (*a*) Excited singlet state, $S_1$. (*b*) Excited triplet state, $T_1$.

The terms "singlet" and "triplet" are used because a triplet state splits into three slightly different energy levels in a magnetic field, but a singlet state is not split.

## Electronic States of Formaldehyde

**Molecular orbitals** describe the distribution of electrons in a molecule, just as *atomic orbitals* describe the distribution of electrons in an atom. In the molecular orbital diagram for formaldehyde in Figure 18-10, one of the nonbonding orbitals of oxygen is mixed with the three sigma bonding orbitals. These four orbitals, labeled $\sigma_1$ through $\sigma_4$, are each occupied by a pair of electrons with opposite spin (spin quantum numbers $= +\frac{1}{2}$ and $-\frac{1}{2}$). At higher energy is an occupied pi bonding orbital ($\pi$), made of the $p_y$ atomic orbitals of carbon and oxygen. The highest-energy occupied orbital is the nonbonding orbital ($n$), composed principally of the oxygen $2p_x$ atomic orbital. The lowest-energy unoccupied orbital is the pi antibonding orbital ($\pi^*$). Electrons in this orbital produce repulsion, rather than attraction, between carbon and oxygen atoms.

In an **electronic transition,** an electron from one molecular orbital moves to another orbital, with a concomitant increase or decrease in the energy of the molecule. The lowest-energy electronic transition of formaldehyde promotes a nonbonding ($n$) electron to the antibonding pi orbital ($\pi^*$).[11] There are two possible transitions, depending on the spin quantum numbers in the excited state (Figure 18-11). The state in which the spins are opposed is called a **singlet state.** If the spins are parallel, we have a **triplet state.**

The lowest-energy excited singlet and triplet states are called $S_1$ and $T_1$, respectively. In general, $T_1$ has lower energy than $S_1$. In formaldehyde, the transition $n \rightarrow \pi^*(T_1)$ requires the absorption of visible light with a wavelength of 397 nm. The $n \rightarrow \pi^*(S_1)$ transition occurs when ultraviolet radiation with a wavelength of 355 nm is absorbed.

With an electronic transition near 397 nm, you might expect from Table 18-1 that formaldehyde appears green-yellow. In fact, formaldehyde is colorless, because the probability of undergoing any transition between singlet and triplet states (such as $n(S_0) \rightarrow \pi^*(T_1)$) is exceedingly small. Formaldehyde absorbs so little light at 397 nm that our eyes do not detect any absorbance at all. Singlet-to-singlet transitions such as $n(S_0) \rightarrow \pi^*(S_1)$ are much more probable, and the ultraviolet absorption is more intense.

Although formaldehyde is planar in its ground state, it has a pyramidal structure in both the $S_1$ (Figure 18-9b) and $T_1$ excited states. Promotion of a nonbonding electron to an antibonding C—O orbital lengthens the C—O bond and changes the molecular geometry.

Remember, the shorter the wavelength of electromagnetic radiation, the greater the energy.

## Vibrational and Rotational States of Formaldehyde

Absorption of visible or ultraviolet radiation promotes electrons to higher-energy orbitals in formaldehyde. Infrared and microwave radiation are not energetic enough to induce electronic transitions, but they can change the vibrational or rotational motion of the molecule.

Each of the four atoms of formaldehyde can move along three axes in space, so the entire molecule can move in $4 \times 3 = 12$ different ways. Three of these motions correspond to translation of the entire molecule in the $x$, $y$, and $z$ directions. Another three motions correspond to rotation about the $x$-, $y$-, and $z$-axes of the molecule. The remaining six motions are vibrations shown in Figure 18-12.

A nonlinear molecule with $n$ atoms has $3n - 6$ vibrational modes and three rotations. A linear molecule can rotate about only two axes; it therefore has $3n - 5$ vibrational modes and two rotations.

When formaldehyde absorbs an infrared photon with a wavenumber of 1 251 cm$^{-1}$ (= 14.97 kJ/mol), the asymmetric bending vibration in Figure 18-12 is stimulated: Oscillations of the atoms are increased in amplitude, and the energy of the molecule increases.

The C—O stretching vibration of formaldehyde is reduced from 1 746 cm$^{-1}$ in the $S_0$ state to 1 183 cm$^{-1}$ in the $S_1$ state because the strength of the C—O bond decreases when the antibonding $\pi^*$ orbital is populated.

Spacings between rotational energy levels of a molecule are even smaller than vibrational energy spacings. A molecule in the rotational ground state could absorb microwave photons with energies of 0.029 07 or 0.087 16 kJ/mol (wavelengths of 4.115 or 1.372 mm) to be promoted to the two lowest excited states. Absorption of microwave radiation causes the molecule to rotate faster than it does in its ground state.

Your microwave oven heats food by transferring rotational energy to water molecules in the food.

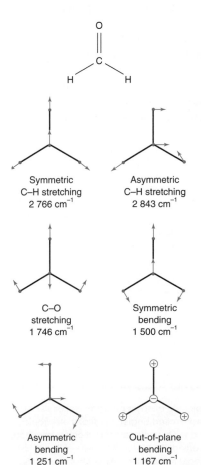

Symmetric
C–H stretching
2 766 cm$^{-1}$

Asymmetric
C–H stretching
2 843 cm$^{-1}$

C–O
stretching
1 746 cm$^{-1}$

Symmetric
bending
1 500 cm$^{-1}$

Asymmetric
bending
1 251 cm$^{-1}$

Out-of-plane
bending
1 167 cm$^{-1}$

*Figure 18-12* The six kinds of vibrations of formaldehyde. The wavenumber of infrared radiation needed to stimulate each kind of motion is given in units of reciprocal centimeters, cm$^{-1}$.

## Combined Electronic, Vibrational, and Rotational Transitions

In general, when a molecule absorbs light having sufficient energy to cause an electronic transition, **vibrational** and **rotational transitions**—that is, changes in the vibrational and rotational states—occur as well. Formaldehyde can absorb one photon with just the right energy to cause the following simultaneous changes: (1) a transition from the $S_0$ to the $S_1$ electronic state, (2) a change in vibrational energy from the ground vibrational state of $S_0$ to an excited vibrational state of $S_1$, and (3) a transition from one rotational state of $S_0$ to a different rotational state of $S_1$.

Electronic absorption bands are usually broad (Figure 18-7) because many different vibrational and rotational levels are available at slightly different energies. A molecule can absorb photons with a wide range of energies and still be promoted from the ground electronic state to one particular excited electronic state.

## What Happens to Absorbed Energy?

Suppose that absorption promotes a molecule from the ground electronic state, $S_0$, to a vibrationally and rotationally excited level of the excited electronic state $S_1$ (Figure 18-13). Usually, the first process after absorption is *vibrational relaxation* to the lowest vibrational level of $S_1$. In this *radiationless* transition, labeled $R_1$ in Figure 18-13, vibrational energy is transferred to other molecules (solvent, for example) through collisions, not by emission of a photon. The net effect is to convert part of the energy of the absorbed photon into heat spread through the entire medium.

At the $S_1$ level, several events can happen. The molecule could enter a highly excited vibrational level of $S_0$ having the same energy as $S_1$. This is called *internal conversion* (IC). From this excited state, the molecule can relax back to the ground vibrational state and transfer its energy to neighboring molecules through collisions. This radiationless process is labeled $R_2$. If a molecule follows the path $A–R_1–IC–R_2$ in Figure 18-13, the entire energy of the photon will have been converted into heat.

Alternatively, the molecule could cross from $S_1$ into an excited vibrational level of $T_1$. Such an event is known as *intersystem crossing* (ISC). After the radiationless vibrational relaxation $R_3$, the molecule finds itself at the lowest vibrational level of $T_1$. From here, the molecule might undergo a second intersystem crossing to $S_0$, followed by the radiationless relaxation $R_4$. All processes mentioned so far simply convert light into heat.

A molecule could also relax from $S_1$ or $T_1$ to $S_0$ by emitting a photon. The radiational transition $S_1 \rightarrow S_0$ is called **fluorescence** (Box 18-2), and the radiational transition $T_1 \rightarrow S_0$ is called **phosphorescence.** The relative rates of internal conversion, intersystem crossing, fluorescence, and phosphorescence depend on the molecule, the solvent, and conditions such as temperature and pressure. The energy of phosphorescence is less than the energy of fluorescence, so phosphorescence comes at longer wavelengths than fluorescence (Figure 18-14).

Vibrational transitions usually involve simultaneous rotational transitions. Electronic transitions usually involve simultaneous vibrational and rotational transitions.

*Internal conversion:* a radiationless transition between states with the same spin quantum numbers (e.g., $S_1 \rightarrow S_0$).

*Intersystem crossing:* a radiationless transition between states with different spin quantum numbers (e.g., $T_1 \rightarrow S_0$).

*Fluorescence:* the emission of a photon during a transition between states with the same spin quantum numbers (e.g., $S_1 \rightarrow S_0$).
*Phosphorescence:* the emission of a photon during a transition between states with different spin quantum numbers (e.g., $T_1 \rightarrow S_0$).

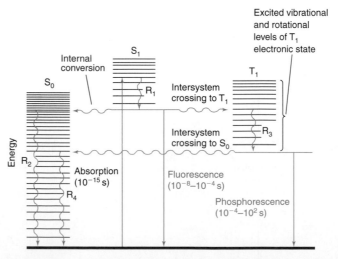

**Figure 18-13** Physical processes that can occur after a molecule absorbs an ultraviolet or visible photon. $S_0$ is the ground electronic state. $S_1$ and $T_1$ are the lowest excited singlet and triplet electronic states. Straight arrows represent processes involving photons, and wavy arrows are radiationless transitions. R denotes vibrational relaxation. Absorption could terminate in any of the vibrational levels of $S_1$, not just the one shown. Fluorescence and phosphorescence can terminate in any of the vibrational levels of $S_0$.

A fluorescent lamp is a glass tube filled with Hg vapor. The inner walls are coated with a *phosphor* (luminescent substance) consisting of calcium halophosphate ($Ca_5(PO_4)_3F_{1-x}Cl_x$) doped with $Mn^{2+}$ and $Sb^{3+}$. Hg atoms, promoted to an excited state by electric current passing through the lamp, emit ultraviolet radiation at 185 and 254 nm, and a series of visible lines in panel *a*. This radiation is absorbed by $Sb^{3+}$, and some of the energy is passed on to $Mn^{2+}$. $Sb^{3+}$ emits blue light, and $Mn^{2+}$ emits yellow light (panel *b*), with the combined emission appearing white.

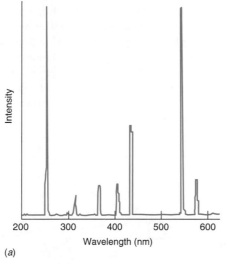

(a)

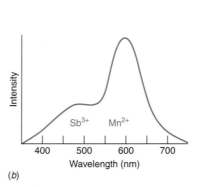

(b)

Emission spectrum of Hg vapor (*a*) and phosphor (*b*) in fluorescent lamps. [From S. R. Goode and L. A. Metz, "Emission Spectroscopy in the Undergraduate Laboratory," J. Chem. Ed. **2003,** 80, 1455; J. A. DeLuca, "An Introduction to Luminescence in Inorganic Solids," J. Chem. Ed. **1980,** 57, 541.]

A 13-W fluorescent light that fits in a standard screw-in socket provides the same light as the 60-W bulb it replaces. The expected lifetime of the fluorescent lamp is 10 000 h, whereas that of the incandescent bulb is 750 h. The fluorescent light is more expensive than the incandescent bulb but saves a great deal of electricity and money over its lifetime.

Most white fabrics also fluoresce. Just for fun, turn on an ultraviolet lamp in a darkened room containing several people (*but do not look directly at the lamp*). You will discover emission from white fabrics (shirts, pants, shoelaces, unmentionables) containing fluorescent compounds to enhance whiteness. You might see fluorescence from teeth and from recently bruised areas of skin that show no surface damage. Many demonstrations of fluorescence and phosphorescence have been described.[12]

A fluorescent whitener from laundry detergent

Fluorescence and phosphorescence are relatively rare. Molecules generally decay from the excited state by radiationless transitions. The *lifetime* of fluorescence is always very short ($10^{-8}$ to $10^{-4}$ s). The lifetime of phosphorescence is much longer ($10^{-4}$ to $10^2$ s). Therefore, phosphorescence is even rarer than fluorescence, because a molecule in the $T_1$ state has a good chance of undergoing intersystem crossing to $S_0$ before phosphorescence can occur.

The *lifetime* of a state is the time needed for the population of that state to decay to 1/e of its initial value, where e is the base of natural logarithms.

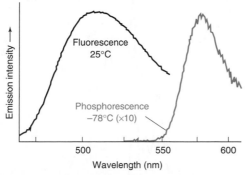

Figure 18-14  Example showing that phosphorescence comes at lower energy than fluorescence from the same molecule. The phosphorescence is ~10 times weaker than the fluorescence and is observed only when the sample is cooled. [Data from J. C. Fister III, J. M. Harris, D. Rank, and W. Wacholtz, "Molecular Photophysics of Acridine Yellow Studied by Phosphorescence and Delayed Fluorescence," J. Chem. Ed. **1997,** 74, 1208.]

## 18-6 Luminescence

Fluorescence and phosphorescence are examples of **luminescence,** which is emission of light from an excited state of a molecule. Luminescence is inherently more sensitive than absorption. Imagine yourself in a stadium at night; the lights are off, but each of the 50 000 raving fans is holding a lighted candle. If 500 people blow out their candles, you will hardly notice the difference. Now imagine that the stadium is completely dark; then 500 people suddenly light their candles. In this case, the effect is dramatic. The first example is analogous to changing transmittance from 100% to 99%, which is equivalent to an absorbance of $-\log 0.99 = 0.004\ 4$. It is hard to measure such a small absorbance because the background is so bright. The second example is analogous to observing fluorescence from 1% of the molecules in a sample. Against the black background, the fluorescence is striking.

Luminescence is sensitive enough to observe *single molecules*.[13] Figure 18-15 shows *observed* tracks of two molecules of the highly fluorescent Rhodamine 6G at 0.78-s intervals in a thin layer of silica gel. These direct observations confirm the "random walk" of diffusing molecules postulated by Albert Einstein in 1905.

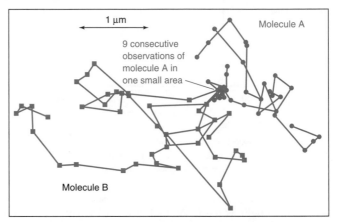

Rhodamine 6G

**Figure 18-15** Tracks of two molecules of 20 pM rhodamine 6G in silica gel observed by fluorescence integrated over 0.20-s periods at 0.78-s intervals. Some points are not connected, because the molecule disappeared above or below the focal plane in the 0.45-μm-thick film and was not observed in a particular observation interval. In the nine periods when molecule A was in one location, it might have been adsorbed to a particle of silica. An individual molecule emits thousands of photons in 0.2 s as the molecule cycles between ground and excited states. Only a fraction of these photons reaches the detector, which generates a burst of ~10–50 electrons. *[From K. S. McCain, D. C. Hanley, and J. M. Harris, "Single-Molecule Fluorescence Trajectories for Investigating Molecular Transport in Thin Silica Sol-Gel Films," Anal. Chem. **2003**, 75, 4351.]*

### Relation Between Absorption and Emission Spectra

Figure 18-13 shows that fluorescence and phosphorescence come at lower energy than absorption (the *excitation energy*). That is, molecules emit radiation at longer wavelengths than the radiation they absorb. Examples are shown in Figure 18-16 and Color Plate 16.

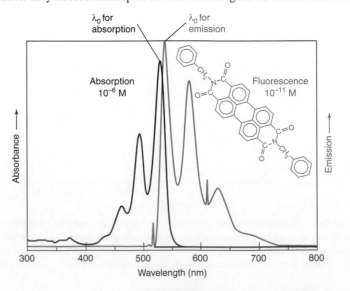

**Figure 18-16** Absorption (black line) and emission (colored line) of bis(benzylimido) perylene in dichloromethane solution, illustrating the approximate mirror image relation between absorption and emission. The $10^{-11}$ M solution used for emission had an average of just 10 analyte molecules in the volume probed by the 514-nm excitation laser. *[From P. J. G. Goulet, N. P. W. Pieczonka, and R. F. Aroca, "Overtones and Combinations in Single-Molecule Surface-Enhanced Resonance Raman Scattering Spectra," Anal. Chem. **2003**, 75, 1918.]*

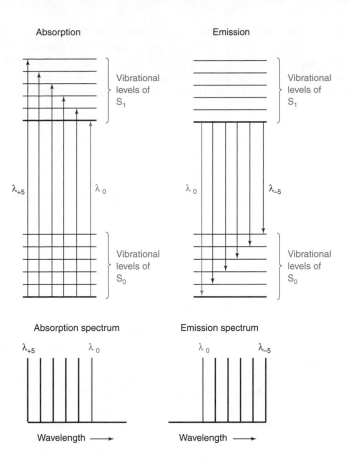

**Figure 18-17** Energy-level diagram showing why structure is seen in the absorption and emission spectra and why the spectra are roughly mirror images of each other. In absorption, wavelength $\lambda_0$ comes at lowest energy, and $\lambda_{+5}$ is at highest energy. In emission, wavelength $\lambda_0$ comes at highest energy, and $\lambda_{-5}$ is at lowest energy.

Figure 18-17 explains why emission comes at lower energy than absorption and why the emission spectrum is roughly the mirror image of the absorption spectrum. In absorption, wavelength $\lambda_0$ corresponds to a transition from the ground vibrational level of $S_0$ to the lowest vibrational level of $S_1$. Absorption maxima at higher energy (shorter wavelength) correspond to the $S_0$-to-$S_1$ transition accompanied by absorption of one or more quanta of vibrational energy. In polar solvents, vibrational structure is usually broadened beyond recognition, and only a broad envelope of absorption is observed. In less polar or nonpolar solvents, vibrational structure is observed.

After absorption, the vibrationally excited $S_1$ molecule relaxes back to the lowest vibrational level of $S_1$ prior to emitting any radiation. Emission from $S_1$ in Figure 18-17 can go to any vibrational level of $S_0$. The highest-energy transition comes at wavelength $\lambda_0$, with a series of peaks following at longer wavelength. The absorption and emission spectra will have an approximate mirror image relation if the spacings between vibrational levels are roughly equal and if the transition probabilities are similar.

The $\lambda_0$ transitions in Figure 18-16 (and later in Figure 18-21) do not exactly overlap. In the emission spectrum, $\lambda_0$ comes at slightly lower energy than in the absorption spectrum. The reason is seen in Figure 18-18. A molecule absorbing radiation is initially in its electronic ground state, $S_0$. This molecule possesses a certain geometry and solvation. Suppose that the excited state is $S_1$. The electronic transition is faster than the vibrational motion of atoms or the translational motion of solvent molecules. When radiation is first absorbed, the excited $S_1$ molecule still possesses its $S_0$ geometry and solvation. Shortly

Electronic transitions are so fast, relative to nuclear motion, that each atom has nearly the same position and momentum before and after a transition. This is called the *Franck-Condon principle*.

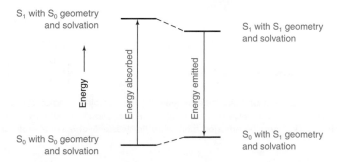

**Figure 18-18** Diagram showing why the $\lambda_0$ transitions do not exactly overlap in Figures 18-16 and 18-21.

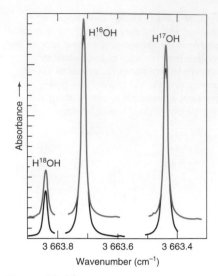

**Figure 18-19** Laser spectroscopy of water vapor showing individual rotational transitions of $H_2^{16}O$, $H_2^{17}O$, and $H_2^{18}O$. The upper trace is from a standard water sample and the lower trace is from an unknown. Relative peak areas in the two spectra provide isotope ratios to an accuracy of 0.1%. [E. R. Th. Kerstel, R. van Trigt, N. Dam, J. Reuss, and H. A. J. Meijer, "Simultaneous Determination of $^2H/^1H$, $^{17}O/^{16}O$, and $^{18}O/^{16}O$ Isotope Abundance Ratios in Water by Means of Laser Spectrometry," Anal. Chem. **1999**, 71, 5297.]

*Emission spectrum:* constant $\lambda_{ex}$ and variable $\lambda_{em}$

*Excitation spectrum:* variable $\lambda_{ex}$ and constant $\lambda_{em}$

after excitation, the geometry and solvation change to their most favorable values for the $S_1$ state. This rearrangement lowers the energy of the excited molecule. When an $S_1$ molecule fluoresces, it returns to the $S_0$ state with $S_1$ geometry and solvation. This unstable configuration must have a higher energy than that of an $S_0$ molecule with $S_0$ geometry and solvation. The net effect in Figure 18-18 is that the $\lambda_0$ emission energy is less than the $\lambda_0$ excitation energy.

Solution-phase spectra are broadened because absorbing molecules are surrounded by solvent molecules with a variety of orientations that create slightly different energy levels for different absorbing molecules. Simple gas-phase molecules, which are not in close contact with neighbors and which have a limited number of energy levels, have extremely sharp absorptions. In Figure 18-19, individual rotational transitions of gaseous $H_2^{16}O$, $H_2^{17}O$, and $H_2^{18}O$ are readily distinguished even though they are just 0.2 cm$^{-1}$ apart. A tunable infrared laser was required as a light source with a sufficiently narrow linewidth to obtain this spectrum. From the absorbance of each peak, we can measure the relative amounts of each isotope. Isotope ratios in cores drilled from polar ice caps provide a record of the earth's temperature for 250 000 years (Box 18-3). Mass spectrometry (Chapter 22) is normally used to measure isotope ratios, but laser spectroscopy is an alternative.

## Excitation and Emission Spectra

An emission experiment is shown in Figure 18-20. An excitation wavelength ($\lambda_{ex}$) is selected by one monochromator, and luminescence is observed through a second monochromator, usually positioned at 90° to the incident light to minimize the intensity of scattered light reaching the detector. If we hold the excitation wavelength fixed and scan through the emitted radiation, an **emission spectrum** such as Figure 18-14 is produced. An emission spectrum is a graph of emission intensity versus emission wavelength.

An **excitation spectrum** is measured by varying the excitation wavelength and measuring emitted light at one particular wavelength ($\lambda_{em}$). An excitation spectrum is a graph of emission intensity versus excitation wavelength (Figure 18-21). *An excitation spectrum looks very much like an absorption spectrum* because, the greater the absorbance at the excitation wavelength, the more molecules are promoted to the excited state and the more emission will be observed.

In emission spectroscopy, we measure emitted irradiance rather than the fraction of incident irradiance striking the detector. Detector response varies with wavelength, so the recorded emission spectrum is not a true profile of emitted irradiance versus emission wavelength. For analytical measurements employing a single emission wavelength, this effect is inconsequential. If a true profile is required, it is necessary to calibrate the detector for the wavelength dependence of its response.

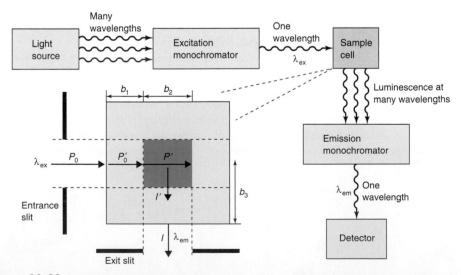

**Figure 18-20** Essentials of a luminescence experiment. The sample is irradiated at one wavelength and emission is observed over a range of wavelengths. The excitation monochromator selects the excitation wavelength ($\lambda_{ex}$) and the emission monochromator selects one wavelength at a time ($\lambda_{em}$) to observe.

Natural oxygen is 99.8% $^{16}O$, 0.2% $^{18}O$, and 0.04% $^{17}O$. Small variations in isotope ratios in $H_2O$ in the ocean, the atmosphere, and rain and snow depend on local temperature. The warmer the air, the more enriched it is in $H_2{}^{18}O$. The $^{18}O$ content of snow preserved in glaciers provides a record of air temperatures for ~250 000 years in 2-km-long cores drilled from ice sheets in Greenland and Antarctica. Depth in the ice pack has been correlated with age in years by a variety of measurements.

The figure shows variations in the $^{18}O$ content of ice deposited in Greenland over the past 160 000 years. The measured quantities are depth in the ice and $\delta^{18}O$, defined as $1\,000 \times [R_{ice} - R_{ref}]/R_{ref}$, where $R_{ice}$ is the ratio $^{18}O/^{16}O$ in ice and $R_{ref}$ is the $^{18}O/^{16}O$ ratio in a reference material, which is modern seawater. $\delta^{18}O$ is the change in $^{18}O/^{16}O$ ratio between ice and the reference material, measured in parts per thousand (‰). The derived quantities are age of the ice and the air temperature at which the snow fell.

The past 10 000 years have been an exceptionally stable, high-temperature period. Over most of the ice-core record, temperature switched between two "quasi-stationary climate stages."[14] From 20 000 to 80 000 years ago, the warm-climate air temperature was about $-35°C$ and the cold temperature was near $-42°C$. Transitions between the two climates required as little as 10–20 years, and each climate persisted between 70 and 5 000 years.

Similar changes are seen in other terrestrial and marine geological records from the North Atlantic region, but not from the Antarctic. Circulation changes in the North Atlantic Ocean are conjectured to be the driving force for the climate changes.

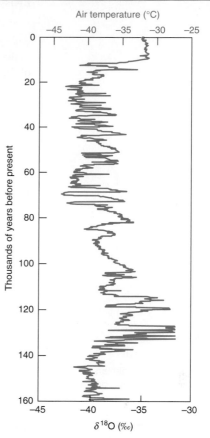

$^{18}O$ content of ice core drilled from Summit in central Greenland.
*[M. Anklin et al., "Climate Instability During the Last Interglacial Period Recorded in the GRIP Ice Core," Nature 1993, 364, 203.]*

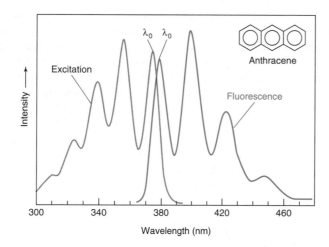

*Figure 18-21* Excitation and emission spectra of anthracene have the same mirror image relation as the absorption and emission spectra in Figure 18-16. An excitation spectrum is nearly the same as an absorption spectrum. *[C. M. Byron and T. C. Werner, "Experiments in Synchronous Fluorescence Spectroscopy for the Undergraduate Instrumental Chemistry Course," J. Chem. Ed. 1991, 68, 433.]*

## Luminescence Intensity

A simplified view of processes occurring during a luminescence measurement is shown in the enlarged sample cell at the lower left of Figure 18-20. We expect emission to be proportional to the irradiance absorbed by the sample. In Figure 18-20, the irradiance (W/m²) incident on the sample cell is $P_0$. Some is absorbed over the pathlength $b_1$, so the irradiance striking the central region of the cell is

$$\text{Irradiance striking central region} = P_0' = P_0 10^{-\varepsilon_{ex}b_1c} \qquad (18\text{-}7)$$

where $\varepsilon_{ex}$ is the molar absorptivity at the wavelength $\lambda_{ex}$ and $c$ is the concentration of analyte. The irradiance of the beam when it has traveled the additional distance $b_2$ is

$$P' = P_0' 10^{-\varepsilon_{ex}b_2c} \qquad (18\text{-}8)$$

Equation 18-7 comes from Equations 18-5 and 18-6. If species other than analyte absorbed at the wavelengths of interest, we would have to include them.

Emission intensity $I$ is proportional to the irradiance absorbed in the central region of the cell:

$$\text{Emission intensity} = I' = k'(P_0' - P') \quad (18\text{-}9)$$

where $k'$ is a constant of proportionality. Not all radiation emitted from the center of the cell in the direction of the exit slit is observed. Some is absorbed between the center and the edge of the cell. Emission intensity $I$ emerging from the cell is given by Beer's law:

$$I = I'\,10^{-\varepsilon_{em}b_3 c} \quad (18\text{-}10)$$

where $\varepsilon_{em}$ is the molar absorptivity at the emission wavelength and $b_3$ is the distance from the center to the edge of the cell.

Combining Equations 18-9 and 18-10 gives an expression for emission intensity:

$$I = k'(P_0' - P')10^{-\varepsilon_{em}b_3 c} \quad (18\text{-}11)$$

Substituting expressions for $P_0'$ and $P'$ from Equations 18-7 and 18-8, we obtain a relation between incident irradiance and emission intensity:

$$
\begin{aligned}
I &= k'(P_0 10^{-\varepsilon_{ex}b_1 c} - P_0 10^{-\varepsilon_{ex}b_1 c}\,10^{-\varepsilon_{ex}b_2 c})10^{-\varepsilon_{em}b_3 c} \\
&= k'P_0 \underbrace{10^{-\varepsilon_{ex}b_1 c}}_{\substack{\text{Loss of intensity} \\ \text{in region 1}}} \underbrace{(1 - 10^{-\varepsilon_{ex}b_2 c})}_{\substack{\text{Absorption} \\ \text{in region 2}}} \underbrace{10^{-\varepsilon_{em}b_3 c}}_{\substack{\text{Loss of intensity} \\ \text{in region 3}}}
\end{aligned} \quad (18\text{-}12)
$$

Consider the limit of low concentration, which means that the exponents $\varepsilon_{ex}b_1 c$, $\varepsilon_{ex}b_2 c$, and $\varepsilon_{em}b_3 c$ are all very small. The terms $10^{-\varepsilon_{ex}b_1 c}$, $10^{-\varepsilon_{ex}b_2 c}$, and $10^{-\varepsilon_{em}b_3 c}$ are all close to unity. We can replace $10^{-\varepsilon_{ex}b_1 c}$ and $10^{-\varepsilon_{em}b_3 c}$ with 1 in Equation 18-12. We cannot replace $10^{-\varepsilon_{ex}b_2 c}$ with 1, because we are subtracting this term from 1 and would be left with 0. Instead, we expand $10^{-\varepsilon_{ex}b_2 c}$ in a power series:

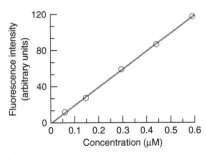

The series 18-13 follows from the relation $10^{-A} = (e^{\ln 10})^{-A} = e^{-A\ln 10}$ and the expansion of $e^x$:

$$e^x = 1 + \frac{x^1}{1!} + \frac{x^2}{2!} + \frac{x^3}{3!} + \cdots$$

$$10^{-\varepsilon_{ex}b_2 c} = 1 - \varepsilon_{ex}b_2 c \ln 10 + \frac{(\varepsilon_{ex}b_2 c \ln 10)^2}{2!} - \frac{(\varepsilon_{ex}b_2 c \ln 10)^3}{3!} + \cdots \quad (18\text{-}13)$$

Each term of Equation 18-13 becomes smaller and smaller, so we just keep the first two terms. The central factor in Equation 18-12 becomes $(1 - 10^{-\varepsilon_{ex}b_2 c}) = (1 - [1 - \varepsilon_{ex}b_2 c \ln 10]) = \varepsilon_{ex}b_2 c \ln 10$ and the entire equation can be written

*Emission intensity at low concentration:*
$$k'P_0(\varepsilon_{ex}b_2 c \ln 10) = \boxed{I = kP_0 c} \quad (18\text{-}14)$$

where $k = k'\varepsilon_{ex}b_2 \ln 10$ is a constant.

Equation 18-14 says that, *at low concentration, emission intensity is proportional to analyte concentration.* Data for anthracene in Figure 18-22 are linear below $10^{-6}$ M. Blank samples invariably scatter light and must be run in every analysis. Equation 18-14 tells us that doubling the incident irradiance ($P_0$) will double the emission intensity (up to a point). In contrast, doubling $P_0$ has no effect on absorbance, which is a *ratio* of two intensities. The sensitivity of a luminescence measurement can be increased by more than a factor of 3 by the simple expedient of using a mirror coating on the two walls of the sample cell opposite the slits in Figure 18-20.[15]

For higher concentrations, we need all the terms in Equation 18-12. As concentration increases, a peak emission is reached. Then emission decreases because absorption increases more rapidly than the emission. We say the emission is *quenched* (decreased) by **self-absorption,** which is the absorption of excitation or emission energy by analyte molecules in the solution. At high concentration, even the *shape* of the emission spectrum can change, because absorption and emission both depend on wavelength.

## Example: Fluorimetric Assay of Selenium in Brazil Nuts

Selenium is a trace element essential to life. For example, the selenium-containing enzyme glutathione peroxidase catalyzes the destruction of peroxides (ROOH) that are harmful to cells. Conversely, at high concentration, selenium can be toxic.

To measure selenium in Brazil nuts, 0.1 g of nut is digested with 2.5 mL of 70 wt% $HNO_3$ in a Teflon *bomb* (Figure 28-8) in a microwave oven. Hydrogen selenate ($H_2SeO_4$) in the digest is reduced to hydrogen selenite ($H_2SeO_3$) with hydroxylamine ($NH_2OH$). Selenite is then **derivatized** to form a fluorescent product that is extracted into cyclohexane.

*Derivatization:* the chemical alteration of analyte so that it can be detected easily or separated easily from other species.

$$\text{2,3-Diaminonaphthalene} + H_2SeO_3 \xrightarrow[50°C]{pH = 2} \text{Fluorescent product} + 3H_2O \quad (18\text{-}15)$$

2,3-Diaminonaphthalene        Fluorescent product

**Figure 18-22** Linear calibration curve for fluorescence of anthracene measured at the wavelength of maximum fluorescence in Figure 18-21. *[C. M. Byron and T. C. Werner, "Experiments in Synchronous Fluorescence Spectroscopy for the Undergraduate Instrumental Chemistry Course," J. Chem. Ed.* **1991,** *68, 433.]*

Maximum response of the fluorescent product was observed with an excitation wavelength of 378 nm and an emission wavelength of 518 nm in Figure 18-23. Emission is proportional to concentration only up to ~0.1 μg Se/mL. Beyond 0.1 μg Se/mL, the response becomes curved, eventually reaches a maximum, and finally *decreases* with increasing concentration as self-absorption dominates. This behavior is predicted by Equation 18-12.

## Luminescence in Analytical Chemistry

Some analytes, such as riboflavin (vitamin $B_2$)[16] and polycyclic aromatic compounds (an important class of carcinogens), are naturally fluorescent and can be analyzed directly. Most compounds are not luminescent. However, coupling to a fluorescent moiety provides a route to sensitive analyses. *Fluorescein* is a strongly fluorescent compound that can be coupled to many molecules for analytical purposes. Fluorescent labeling of fingerprints is a powerful tool in forensic analysis.[17] Sensor molecules whose luminescence responds selectively to a variety of simple cations and anions are available.[18] $Ca^{2+}$ can be measured from the fluorescence of a complex it forms with a derivative of fluorescein called calcein.

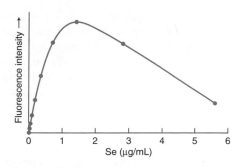

**Figure 18-23** Fluorescence calibration curve for the selenium-containing product in Reaction 18-15. The curvature and maximum are due to self-absorption. [*From M.-C. Sheffield and T. M. Nahir, "Analysis of Selenium in Brazil Nuts by Microwave Digestion and Fluorescence Detection," J. Chem. Ed.* **2002**, *79, 1345.*]

Fluorescein

Chelating aminodiacetate group

Calcein

Luminol
(5-amino-2,3-dihydro-1,4-phthalazinedione)

oxidant (such as NO or $H_2O_2$)

$OH^-$, metal catalyst

$+ N_2 +$ blue light

Molecular biologists use *DNA microarrays* ("gene chips") to monitor gene expression and mutations and to detect and identify pathological microorganisms.[19] A single chip can contain thousands of known single-strand DNA sequences in known locations. The chip is incubated with unknown single-strand DNA that has been tagged with fluorescent labels. After the unknown DNA has bound to its complementary strands on the chip, the amount bound to each spot on the chip is measured by fluorescence intensity.

**Chemiluminescence**—the emission of light arising from a chemical reaction—is valuable in chemical analysis.[20] Light from a firefly or light stick is chemiluminescence.[21] $Ca^{2+}$ within cellular organelles such as mitochondria can be monitored by observing light emitted when $Ca^{2+}$ binds to the protein aequorin obtained from jellyfish.[22] Nitric oxide (NO), which transmits signals between cells, can be measured at the parts per billion level by its chemiluminescent reaction with the compound luminol.[23] The hormone thyroxine can be measured over the concentration range $10^{-14}$ to $10^{-7}$ M by a chemiluminescent method.[24]

## Terms to Understand

| | | | |
|---|---|---|---|
| absorbance | emission spectrum | masking | rotational transition |
| absorption spectrum | excitation spectrum | molar absorptivity | self-absorption |
| Beer's law | excited state | molecular orbital | singlet state |
| chemiluminescence | fluorescence | monochromatic light | spectrophotometry |
| chromophore | frequency | monochromator | transmittance |
| cuvet | ground state | phosphorescence | triplet state |
| derivatization | hertz | photon | vibrational transition |
| electromagnetic spectrum | irradiance | reagent blank | wavelength |
| electronic transition | luminescence | refractive index | wavenumber |

## Summary

Light can be thought of as waves whose wavelength ($\lambda$) and frequency ($\nu$) have the important relation $\lambda\nu = c$, where $c$ is the speed of light. Alternatively, light may be viewed as consisting of photons whose energy ($E$) is given by $E = h\nu = hc/\lambda = hc\tilde{\nu}$, where $h$ is Planck's constant and $\tilde{\nu}$ ($= 1/\lambda$) is the wavenumber. Absorption of light is commonly measured by absorbance ($A$) or transmittance ($T$), defined as $A = \log(P_0/P)$ and $T = P/P_0$, where $P_0$ is the incident irradiance and $P$ is the exiting irradiance. Absorption spectroscopy is useful in quantitative analysis because absorbance is proportional to the concentration of the absorbing species in dilute solution (Beer's law): $A = \varepsilon bc$. In this equation, $b$ is pathlength, $c$ is concentration, and $\varepsilon$ is the molar absorptivity (a constant of proportionality).

Basic components of a spectrophotometer include a radiation source, a monochromator, a sample cell, and a detector. To minimize errors in spectrophotometry, samples should be free of particles, cuvets must be clean, and they must be positioned reproducibly in the sample holder. Measurements should be made at a wavelength of maximum absorbance. Instrument errors tend to be minimized if the absorbance falls in the range $A \approx 0.4-0.9$.

When a molecule absorbs light, it is promoted to an excited state from which it returns to the ground state by radiationless processes or by fluorescence (singlet → singlet emission) or phosphorescence (triplet → singlet emission). Emission intensity is proportional to concentration at low concentration. At sufficiently high concentration, emission decreases because of self-absorption by the analyte. An excitation spectrum (a graph of emission intensity versus excitation wavelength) is similar to an absorption spectrum (a graph of absorbance versus wavelength). An emission spectrum (a graph of emission intensity versus emission wavelength) is observed at lower energy than the absorption spectrum and tends to be the mirror image of the absorption spectrum. A molecule that is not fluorescent can be analyzed by attaching a fluorescent group to it. Light emitted by a chemical reaction—chemiluminescence—also is used for quantitative analysis.

## Exercises

**18-A.** **(a)** What value of absorbance corresponds to 45.0% $T$?
**(b)** If a 0.010 0 M solution exhibits 45.0% $T$ at some wavelength, what will be the percent transmittance for a 0.020 0 M solution of the same substance?

**18-B.** **(a)** A $3.96 \times 10^{-4}$ M solution of compound A exhibited an absorbance of 0.624 at 238 nm in a 1.000-cm cuvet; a blank solution containing only solvent had an absorbance of 0.029 at the same wavelength. Find the molar absorptivity of compound A.
**(b)** The absorbance of an unknown solution of compound A in the same solvent and cuvet was 0.375 at 238 nm. Find the concentration of A in the unknown.
**(c)** A concentrated solution of compound A in the same solvent was diluted from an initial volume of 2.00 mL to a final volume of 25.00 mL and then had an absorbance of 0.733. What is the concentration of A in the concentrated solution?

**18-C.** Ammonia can be determined spectrophotometrically by reaction with phenol in the presence of hypochlorite ($OCl^-$):

$$2 \; \text{Phenol (colorless)} -OH + NH_3 \xrightarrow{OCl^-} \text{Blue product, } \lambda_{max} = 625 \text{ nm}$$

Phenol (colorless)    Ammonia (colorless)    Blue product, $\lambda_{max} = 625$ nm

A 4.37-mg sample of protein was chemically digested to convert its nitrogen into ammonia and then diluted to 100.0 mL. Then 10.0 mL of the solution were placed in a 50-mL volumetric flask and treated with 5 mL of phenol solution plus 2 mL of sodium hypochlorite solution. The sample was diluted to 50.0 mL, and the absorbance at 625 nm was measured in a 1.00-cm cuvet after 30 min. For reference, a standard solution was prepared from 0.010 0 g of $NH_4Cl$ (FM 53.49) dissolved in 1.00 L of water. Then 10.0 mL of this standard were placed in a 50-mL volumetric flask and analyzed in the same manner as the unknown. A reagent blank was prepared by using distilled water in place of unknown.

| Sample | Absorbance at 625 nm |
|---|---|
| Blank | 0.140 |
| Reference | 0.308 |
| Unknown | 0.592 |

**(a)** Calculate the molar absorptivity of the blue product.
**(b)** Calculate the weight percent of nitrogen in the protein.

**18-D.** $Cu^+$ reacts with neocuproine to form the colored complex (neocuproine)$_2Cu^+$, with an absorption maximum at 454 nm. Neocuproine is particularly useful because it reacts with few other metals. The copper complex is soluble in 3-methyl-1-butanol (isoamyl alcohol), an organic solvent that does not dissolve appreciably in water. In other words, when isoamyl alcohol is added to water, a two-layered mixture results, with the denser water layer at the bottom. If (neocuproine)$_2Cu^+$ is present, virtually all of it goes into the organic phase. For the purpose of this problem, assume that the isoamyl alcohol does not dissolve in the water at all and that all the colored complex will be in the organic phase. Suppose that the following procedure is carried out:

1. A rock containing copper is pulverized, and all metals are extracted from it with strong acid. The acidic solution is neutralized with base and made up to 250.0 mL in flask A.
2. Next, 10.00 mL of the solution are transferred to flask B and treated with 10.00 mL of a reducing agent to reduce all $Cu^{2+}$ to $Cu^+$. Then 10.00 mL of buffer is added to bring the pH to a value suitable for complex formation with neocuproine.
3. After that, 15.00 mL of this solution are withdrawn and placed in flask C. To the flask are added 10.00 mL of an aqueous solution containing neocuproine and 20.00 mL of isoamyl alcohol. After the mixture has been shaken well and the phases allowed to separate, all (neocuproine)$_2Cu^+$ is in the organic phase.
4. A few milliliters of the upper layer are withdrawn, and the absorbance at 454 nm is measured in a 1.00-cm cell. A blank carried through the same procedure gives an absorbance of 0.056.

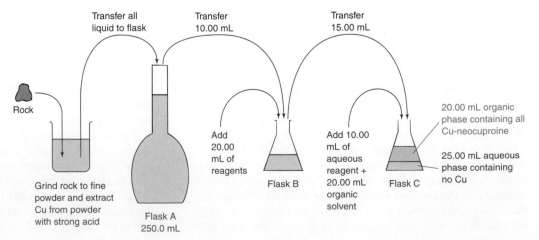

Figure for Exercise 18-D.

(a) Suppose that the rock contained 1.00 mg of Cu. What will be the concentration of Cu (moles per liter) in the isoamyl alcohol phase?
(b) If the molar absorptivity of (neocuproine)$_2$Cu$^+$ is $7.90 \times 10^3$ M$^{-1}$ cm$^{-1}$, what will be the observed absorbance? Remember that a blank carried through the same procedure gave an absorbance of 0.056.
(c) A rock is analyzed and found to give a final absorbance of 0.874 (uncorrected for the blank). How many milligrams of Cu are in the rock?

18-E. ▦  The figure below shows an infrared absorption peak for solutions containing 10 to 50 vol% acetone ([CH$_3$]$_2$C=O) in water. The shape of the spectrum is somewhat different for each mixture, which shifts the *baseline* (dotted line) and position of the absorption peak. The baseline is the blank absorbance that must be subtracted from each peak absorbance to obtain corrected absorbance.

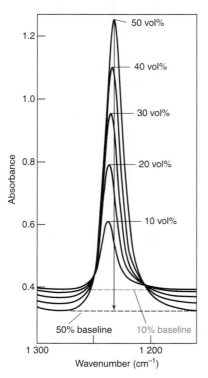

(a) Using a millimeter ruler, measure to the nearest 0.1 mm how many millimeters correspond to 0.8 absorbance units (from 0.4 to 1.2) on the ordinate (*y*-axis) of the spectrum.
(b) Measure the corrected height of the peak corresponding to 50 vol% acetone, using the distance from the baseline marked "50%." By comparing your measurement in part (b) with that in part (a), find the corrected absorbance of the 50 vol% acetone peak. For example, if 0.8 absorbance unit = 58.4 mm and the peak height is 63.7 mm, then the corrected absorbance is found from the proportion

$$\frac{\text{Corrected peak height (mm)}}{\text{Length for 0.8 absorbance unit (mm)}} = \frac{\text{corrected peak absorbance}}{0.8 \text{ absorbance unit}}$$

$$\frac{63.7 \text{ mm}}{58.4 \text{ mm}} = \frac{\text{corrected peak absorbance}}{0.8 \text{ absorbance unit}} \Rightarrow$$

$$\text{corrected peak absorbance} = 0.873$$

(c) The baselines for 10 and 50 vol% acetone are shown in the figure. Draw baselines for the 20, 30, and 40 vol% acetone spectra and find the corrected absorbance for the 10–40 vol% solutions.
(d) Make a calibration curve showing corrected absorbance versus volume percent of acetone. (This curve does not go through the origin.)
(e) Find the least-squares equation of the calibration line, including uncertainties in the slope and intercept.
(f) Find vol% acetone in a solution for which three replicate measurements gave an average absorbance of 0.611. Use Equation 4-27 to find the uncertainty.

Figure for Exercise 18-E. Infrared absorption spectra of 10–50 vol% acetone in water. Vertical arrow shows corrected absorbance for 50 vol% acetone, which is obtained by subtracting baseline absorbance from peak absorbance. [Spectra from A. Afran, "FTIR Absorbance Linearity of Square Column Attenuated Total Reflectance," Am. Lab. February 1993, p. 40MMM.]

---

## Problems

### Properties of Light

**18-1.** Fill in the blanks.
(a) If you double the frequency of electromagnetic radiation, you _____ the energy.
(b) If you double the wavelength, you _____ the energy.
(c) If you double the wavenumber, you _____ the energy.

**18-2.** (a) How much energy (in kilojoules) is carried by one mole of photons of red light with $\lambda = 650$ nm?
(b) How many kilojoules are carried by one mole of photons of violet light with $\lambda = 400$ nm?

**18-3.** Calculate the frequency (Hz), wavenumber (cm$^{-1}$), and energy (J/photon and J/[mol of photons]) of visible light with a wavelength of 562 nm.

**18-4.** Which molecular processes correspond to the energies of microwave, infrared, visible, and ultraviolet photons?

**18-5.** The characteristic orange light produced by sodium in a flame is due to an intense emission called the sodium D line. This "line" is actually a doublet, with wavelengths (measured in vacuum) of 589.157 88 and 589.755 37 nm. The index of refraction of air at a wavelength near 589 nm is 1.000 292 6. Calculate the frequency, wavelength, and wavenumber of each component of the D line, measured in air.

### Absorption of Light and Measuring Absorbance

**18-6.** Explain the difference between transmittance, absorbance, and molar absorptivity. Which one is proportional to concentration?

**18-7.** What is an absorption spectrum?

**18-8.** Why does a compound whose visible absorption maximum is at 480 nm (blue-green) appear to be red?

**18-9.** Why is it most accurate to measure absorbances in the range $A = 0.4–0.9$?

**18-10.** The absorbance of a $2.31 \times 10^{-5}$ M solution of a compound is 0.822 at a wavelength of 266 nm in a 1.00-cm cell. Calculate the molar absorptivity at 266 nm.

**18-11.** What color would you expect to observe for a solution of the ion Fe(ferrozine)$_3^{4-}$, which has a visible absorbance maximum at 562 nm?

**18-12.** When I was a boy, Uncle Wilbur let me watch as he analyzed the iron content of runoff from his banana ranch. A 25.0-mL sample was acidified with nitric acid and treated with excess KSCN to form a red complex. (KSCN itself is colorless.) The solution was then diluted to 100.0 mL and put in a variable-pathlength cell. For comparison, a 10.0-mL reference sample of $6.80 \times 10^{-4}$ M $Fe^{3+}$ was treated with $HNO_3$ and KSCN and diluted to 50.0 mL. The reference was placed in a cell with a 1.00-cm light path. The runoff sample exhibited the same absorbance as the reference when the pathlength of the runoff cell was 2.48 cm. What was the concentration of iron in Uncle Wilbur's runoff?

**18-13.** The *absorption cross section* on the ordinate of the ozone absorption spectrum at the beginning of this chapter is defined by the relation

$$\text{Transmittance } (T) = e^{-n\sigma b}$$

where $n$ is the number of absorbing molecules per cubic centimeter, $\sigma$ is the absorption cross section ($cm^2$), and $b$ is the pathlength (cm). The total ozone in the atmosphere is approximately $8 \times 10^{18}$ molecules above each square centimeter of the earth's surface (from the surface up to the top of the atmosphere). If this were compressed into a 1-cm-thick layer, the concentration would be $8 \times 10^{18}$ molecules/$cm^3$.

(a) Using the ozone spectrum at the beginning of the chapter, estimate the transmittance and absorbance of this 1-$cm^3$ sample at 325 and 300 nm.

(b) Sunburns are caused by radiation in the 295- to 310-nm region. At the center of this region, the transmittance of atmospheric ozone is 0.14. Calculate the absorption cross section for $T = 0.14$, $n = 8 \times 10^{18}$ molecules/$cm^3$, and $b = 1$ cm. By what percentage does the transmittance increase if the ozone concentration decreases by 1% to $7.92 \times 10^{18}$ molecules/$cm^3$?

(c) Atmospheric $O_3$ is measured in *Dobson units* (1 unit = $2.69 \times 10^{16}$ molecules $O_3$ above each $cm^2$ of Earth's surface). (Dobson unit $\equiv$ thickness [in hundredths of a millimeter] that $O_3$ column would occupy if it were compressed to 1 atm at 0°C.) The graph shows variations in $O_3$ concentration as a function of latitude and season. Using an absorption cross section of $2.5 \times 10^{-19}$ $cm^2$, calculate the transmittance in the winter and in the summer at 30°–50° N latitude, at which $O_3$ varies from 290 to 350 Dobson units. By what percentage is the ultraviolet transmittance greater in winter than in summer?

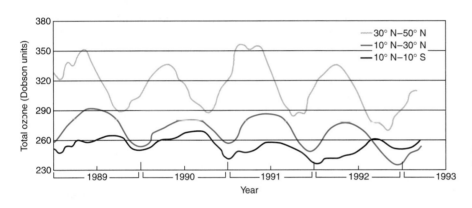

Figure for Problem 18-13. Variation in atmospheric ozone at different latitudes. *[From P. S. Zurer, Chem. Eng. News, 24 May 1993, p. 8.]*

**18-14.** 📠 *Measuring heat of vaporization by spectrophotometry.* The vapor pressure of a liquid increases with temperature according to the approximate equation:

$$\ln P_{vapor} = -\frac{\Delta H_{vap}}{RT} + \text{constant}$$

where $\Delta H_{vap}$ is the heat of vaporization (required to vaporize 1 mol of liquid) and $T$ is kelvins. Over a limited range of temperature, $\Delta H_{vap}$ is constant. The absorbance of a gas is proportional to its pressure: $A = kP_{vapor}$, where $k$ is a constant that is nearly independent of temperature over a limited temperature range. The absorbance of toluene vapor as a function of temperature is in the table below.

| Temperature (K) | Absorbance at 259.4 nm |
|---|---|
| 307.7 | 0.473 |
| 306.7 | 0.460 |
| 303.0 | 0.366 |
| 299.3 | 0.313 |
| 298.9 | 0.298 |
| 293.8 | 0.227 |

*Data from G. Marin-Puga, M. Guzman L., and F. Hevia, "Determination of Enthalpy of Vaporization of Pure Liquids by UV Spectrometry," J. Chem. Ed. **1995**, 72, 91.*

(a) From a graph of $\ln A$ versus $1/T$, find the heat of vaporization of toluene. What are the units of $\Delta H_{vap}$?

(b) The vapor pressure of toluene is 40.0 Torr at 31.8°C. Find the molar absorptivity if the measurements reported in the table were made in a 1.00-cm cell.

**Beer's Law in Chemical Analysis**

**18-15.** What is the purpose of neocuproine in the serum iron analysis?

**18-16.** A compound with a molecular mass of 292.16 was dissolved in a 5-mL volumetric flask. A 1.00-mL aliquot was withdrawn, placed in a 10-mL volumetric flask, and diluted to the mark. The absorbance at 340 nm was 0.427 in a 1.000-cm cuvet. The molar absorptivity for this compound at 340 nm is $\varepsilon_{340} = 6\,130$ $M^{-1}$ $cm^{-1}$.
(a) Calculate the concentration of compound in the cuvet.
(b) What was the concentration of compound in the 5-mL flask?
(c) How many milligrams of compound were used to make the 5-mL solution?

**18-17.** If a sample for spectrophotometric analysis is placed in a 10-cm cell, the absorbance will be 10 times greater than the absorbance in a 1-cm cell. Will the absorbance of the reagent-blank solution also be increased by a factor of 10?

**18-18.** You have been sent to India to investigate the occurrence of goiter disease attributed to iodine deficiency. As part of your investigation, you must make field measurements of traces of iodide ($I^-$) in groundwater. The procedure is to oxidize ($I^-$) to $I_2$ and convert the $I_2$ into an intensely colored complex with the dye brilliant green in the organic solvent toluene.

(a) A $3.15 \times 10^{-6}$ M solution of the colored complex exhibited an absorbance of 0.267 at 635 nm in a 1.000-cm cuvet. A blank solution made from distilled water in place of groundwater had an absorbance of 0.019. Find the molar absorptivity of the colored complex.

(b) The absorbance of an unknown solution prepared from groundwater was 0.175. Find the concentration of the unknown.

**18-19.** Nitrite ion, $NO_2^-$, is a preservative for bacon and other foods, but it is potentially carcinogenic. A spectrophotometric determination of $NO_2^-$ makes use of the following reactions:

$$HO_3S-\bigcirc-NH_2 + NO_2^- + 2H^+ \longrightarrow$$

Sulfanilic acid

$$HO_3S-\bigcirc-\overset{+}{N}\equiv N + 2H_2O$$

$$HO_3S-\bigcirc-\overset{+}{N}\equiv N + \bigcirc-NH_2 \longrightarrow$$

1-Aminonaphthalene

$$HO_3S-\bigcirc-N=N-\bigcirc-NH_2 + H^+$$

Colored product, $\lambda_{max} = 520$ nm

Here is an abbreviated procedure for the determination:

1. To 50.0 mL of unknown solution containing nitrite is added 1.00 mL of sulfanilic acid solution.
2. After 10 min, 2.00 mL of 1-aminonaphthalene solution and 1.00 mL of buffer are added.
3. After 15 min, the absorbance is read at 520 nm in a 5.00-cm cell.

The following solutions were analyzed:

A. 50.0 mL of food extract known to contain no nitrite (that is, a negligible amount); final absorbance = 0.153.
B. 50.0 mL of food extract suspected of containing nitrite; final absorbance = 0.622.
C. Same as B, but with 10.0 μL of $7.50 \times 10^{-3}$ M NaNO$_2$ added to the 50.0-mL sample; final absorbance = 0.967.

(a) Calculate the molar absorptivity, $\varepsilon$, of the colored product. Remember that a 5.00-cm cell was used.

(b) How many micrograms of $NO_2^-$ were present in 50.0 mL of food extract?

## Luminescence

**18-20.** In formaldehyde, the transition $n \rightarrow \pi^*(T_1)$ occurs at 397 nm, and the $n \rightarrow \pi^*(S_1)$ transition comes at 355 nm. What is the difference in energy (kJ/mol) between the $S_1$ and $T_1$ states? This difference is due to the different electron spins in the two states.

**18-21.** What is the difference between fluorescence and phosphorescence?

**18-22.** What is the difference between luminescence and chemiluminescence?

**18-23.** Consider a molecule that can fluoresce from the $S_1$ state and phosphoresce from the $T_1$ state. Which is emitted at longer wavelength, fluorescence or phosphorescence? Make a sketch showing absorption, fluorescence, and phosphorescence on a single spectrum.

**18-24.** What is the difference between a fluorescence excitation spectrum and a fluorescence emission spectrum? Which one resembles an absorption spectrum?

**18-25.** Excitation and fluorescence spectra of anthracene are shown in Figure 18-21. The wavelengths of maximum absorption and emission are approximately 357 and 402 mm. The molar absorptivities at these wavelengths are approximately $\varepsilon_{ex} = 9.0 \times 10^3$ M$^{-1}$ cm$^{-1}$ and $\varepsilon_{em} = 5 \times 10^1$ M$^{-1}$ cm$^{-1}$. Consider a fluorescence experiment in Figure 18-20 with cell dimensions $b_1 = 0.30$ cm, $b_2 = 0.40$ cm, and $b_3 = 0.5$ cm. Calculate the relative fluorescence intensity with Equation 18-12 as a function of concentration over the range $10^{-8}$ to $10^{-3}$ M. Explain the shape of the curve. Up to approximately what concentration is fluorescence proportional to concentration (within 5%)? Is the calibration range in Figure 18-22 sensible?

**18-26.** *Standard addition.* Selenium from 0.108 g of Brazil nuts was converted into the fluorescent products in Reaction 18-15, which was extracted into 10.0 mL of cyclohexane. Then 2.00 mL of the cyclohexane solution was placed in a cuvet for fluorescence measurement. Standard additions of fluorescent product containing 1.40 μg Se/mL are given in the table below. Construct a standard addition graph like Figure 5-6 to find the concentration of Se in the 2.00-mL unknown solution. Find the wt% of Se in the nuts and its uncertainty and 95% confidence interval.

| Volume of standard added (μL) | Fluorescence intensity (arbitrary units) |
|---|---|
| 0 | 41.4 |
| 10.0 | 49.2 |
| 20.0 | 56.4 |
| 30.0 | 63.8 |
| 40.0 | 70.3 |

# 19 | Applications of Spectrophotometry

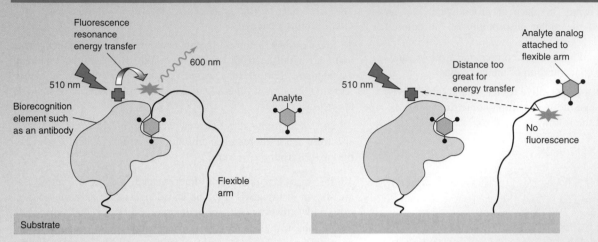

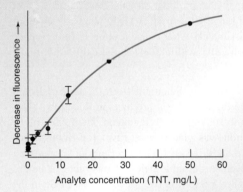

🔲 Radiant energy absorber (donor)

✴ Radiant energy emitter (acceptor)

Response of biosensor to TNT is a *decrease* in fluorescence with increasing concentration of analyte. *[From I. L. Medintz, E. R. Goldman, M. E. Lassman, A. Hayhurst, A. W. Kusterbeck, and J. R. Deschamps, "Self-Assembled TNT Biosensor Based on Modular Multifunctional Surface-Tethered Components," Anal. Chem.* **2005**, *77, 365.]*

This biosensor consists of two surface-tethered components. The biorecognition element can be an antibody (Section 19-5), DNA, RNA, or carbohydrate with specific affinity for analyte. A structural analog of the analyte is bound to a flexible arm adjacent to the recognition element. In the absence of analyte, the tethered analog binds to the recognition element.

A chromophore 🔲 that efficiently absorbs radiant energy is attached to the recognition element adjacent to the recognition site. The fluorescent chromophore ✴ is attached to the flexible arm adjacent to the analyte structural analog. At the left side of the figure, absorbing and emitting chromophores are close to each other. Radiant energy with a wavelength of 510 nm is absorbed by 🔲 and efficiently transferred through space to ✴, which fluoresces strongly at 600 nm. This *fluorescence resonance energy transfer* decreases with the 6th power of the distance between the donor and acceptor.[1]

When analyte is added, it displaces the tethered analog from this biorecognition element. The higher the concentration of analyte, the more displacement occurs. When the analog is displaced from the binding site, ✴ and 🔲 are no longer close enough for energy transfer, and fluorescence is decreased. In the graph, analyte is trinitrotoluene (the explosive, TNT) and the detection limit is 0.1 mg/L (0.1 ppm). After a measurement, the sensor is washed to remove analyte and then reused for more analyses.

---

**T**his chapter describes several applications of absorption and emission of electromagnetic radiation in chemical analysis. Another application—spectrophotometric titrations—was already covered in Section 7-3. We also use Excel SOLVER and spreadsheet matrix manipulations as powerful tools for numerical analysis.

## ■■■■ 19-1 Analysis of a Mixture

*The absorbance of a solution at any wavelength is the sum of absorbances of all the species in the solution.*

Absorbance is additive.

*Absorbance of a mixture:*
$$A = \varepsilon_X b[X] + \varepsilon_Y b[Y] + \varepsilon_Z b[Z] + \cdots \tag{19-1}$$

where $\varepsilon$ is the molar absorptivity of each species at the wavelength in question and $b$ is the cell pathlength (Figure 18-4). If we know the spectra of the pure components, we can mathematically disassemble the spectrum of a mixture into those of its components. In acid-base chemistry, such a procedure allows us to measure the concentrations of the acidic and basic forms of an indicator. This information, combined with the Henderson-Hasselbalch equation (9-16), provides a precise measurement of pH by spectrophotometry.[2]

For a mixture of compounds X and Y, two cases are distinguished. In Figure 19-1a, the absorption bands of pure X and pure Y overlap significantly everywhere. This case is best treated by a spreadsheet procedure that uses measurements at many wavelengths. In Figure 19-1b, X and Y have relatively little overlap in some regions. We can analyze this case with a pair of measurements at wavelength $\lambda'$, where X makes the major contribution, and wavelength $\lambda''$, where Y makes the major contribution.

### What to Do When the Individual Spectra Overlap

We will apply a really simple least-squares analysis of Equation 19-1 to the spectrum of an unknown mixture of $H_2O_2$ complexes of Ti(IV) and V(V) in Figure 19-2. The figure also shows spectra of standard 1.32 mM Ti(IV) and 1.89 mM V(V).

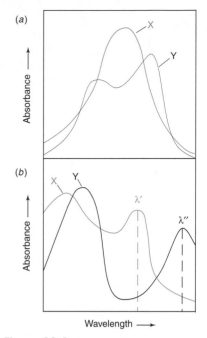

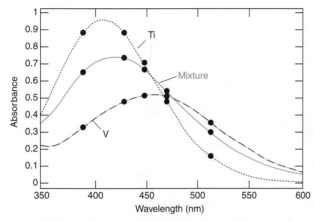

**Figure 19-1** Two cases for analysis of a mixture. (*a*) Spectra of the pure components have substantial overlap. (*b*) Regions exist in which each component makes the major contribution.

**Figure 19-2** Visible spectra of 1.32 mM Ti(IV), 1.89 mM V(V), and an unknown mixture containing both ions. All solutions contain 0.5 wt% $H_2O_2$ and 0.01 M $H_2SO_4$. Absorbance at points shown by dots is listed in Figure 19-3. [*From M. Blanco, H. Iturriaga, S. Maspoch, and P. Tarín, "A Simple Method for Spectrophotometric Determination of Two Components with Overlapped Spectra," J. Chem. Ed.* **1989,** *66, 178.*]

First, we measure the absorbance of each standard at several wavelengths shown by dots in Figure 19-2. Results are listed in columns A–C of Figure 19-3. Concentrations of the standards are entered in cells A14 and A16 and the pathlength (1.000 cm) is indicated in cell A19.

Measure absorbance at more wavelengths than there are components in the mixture. For a two-component mixture, use at least three wavelengths. More wavelengths increase the accuracy.

|   | A | B | C | D | E | F | G | H |
|---|---|---|---|---|---|---|---|---|
| 1 | Analysis of a Mixture When You Have More Data Points Than Components of the Mixture | | | | | | | |
| 2 | | | | Measured | | | | |
| 3 | | | | Absorbance | | | Calculated | |
| 4 | Wave- | Absorbance of Standard: | | of Mixture | Molar Absorptivity | | Absorbance | |
| 5 | length | Titanium | Vanadium | Am | Titanium | Vanadium | Acalc | [Acalc-Am]^2 |
| 6 | 390 | 0.895 | 0.326 | 0.651 | 678.0 | 172.5 | 0.8505 | 3.981E-02 |
| 7 | 430 | 0.884 | 0.497 | 0.743 | 669.7 | 263.0 | 0.9327 | 3.597E-02 |
| 8 | 450 | 0.694 | 0.528 | 0.665 | 525.8 | 279.4 | 0.8051 | 1.963E-02 |
| 9 | 470 | 0.481 | 0.512 | 0.547 | 364.4 | 270.9 | 0.6353 | 7.796E-03 |
| 10 | 510 | 0.173 | 0.374 | 0.314 | 131.1 | 197.9 | 0.3289 | 2.233E-04 |
| 11 | | | | | | | sum = | 1.034E-01 |
| 12 | Standards | | Concentrations in the mixture | | | | | |
| 13 | [Ti](M) = | | (to be found by Solver) | | | | | |
| 14 | 0.00132 | | [Ti] = | 0.001000 | | | | |
| 15 | [V](M) = | | [V] = | 0.001000 | | | | |
| 16 | 0.00189 | | | | | | | |
| 17 | Pathlength | | E6 = B6/($A$19*$A$14) | | | | | |
| 18 | (cm) = | | F6 = C6/($A$19*$A$16) | | | | | |
| 19 | 1.000 | | G6 = E6*$A$19*$D$14+F6*$A$19*$D$15 | | | | | |
| 20 | | | H6 = (G6-D6)^2 | | | | | |

**Figure 19-3** Spreadsheet using SOLVER to analyze the mixture in Figure 19-2.

Calling the two components X (= Ti) and Y (= V), we find the molar absorptivity of each component at each wavelength from Beer's law:

$$\varepsilon_X = \frac{A_{X_s}}{b[X]_s} \qquad \varepsilon_Y = \frac{A_{Y_s}}{b[Y]_s} \qquad (19\text{-}2)$$

where $A_{X_s}$ is the absorbance of the standard and $[X]_s$ is the concentration of the standard. This computation is shown in columns E and F of Figure 19-3. The measured absorbance of the unknown mixture at each wavelength, $A_m$, is listed in column D. At each wavelength, this absorbance is the sum of absorbances of the components:

$$A_m = \varepsilon_X b[X] + \varepsilon_Y b[Y] \qquad (19\text{-}3)$$

However, we do not know the concentrations [X] and [Y] in the mixture.

To find [X] and [Y], we begin by *guessing* concentrations and inserting them in cells D14 and D15. The guesses do not have to be close to correct values. We arbitrarily chose 0.001 M for both guesses. The calculated absorbance of the mixture is then computed in column G from the equation

$$A_{calc} = \varepsilon_X b[X]_{guess} + \varepsilon_Y b[Y]_{guess} \qquad (19\text{-}4)$$

For example, $A_{calc}$ in cell G6 = (678.0)(1.000)[0.001] + (172.5)(1.000)[0.001]. Column H gives the square of the difference between calculated and measured absorbance.

Column H contains $(A_{calc} - A_m)^2$

The least-squares condition is to minimize the sum of squares $(A_{calc} - A_m)^2$ by varying the concentrations $[X]_{guess}$ and $[Y]_{guess}$. The "best" values of $[X]_{guess}$ and $[Y]_{guess}$ in cells D14 and D15 are those that minimize the sum of squares in cell H11.

Excel has a powerful tool called SOLVER that carries out the minimization for us. You will probably see SOLVER in the TOOLS menu. If not, in the TOOLS menu, go to ADD-INS and select Solver Add-In. Click OK. SOLVER will be loaded and will then appear in the TOOLS menu.

Highlight cell H11 in Figure 19-3, go to the TOOLS menu, and select SOLVER. The window in Figure 19-4 will appear. Enter "H11" in Set Target Cell. Then select the button that says Min. Enter "D14,D15" in By Changing Cells. We just told SOLVER to minimize cell H11 by changing cells D14 and D15. Click Solve. After a little work, SOLVER finds the values 0.000 670 in cell D14 and 0.001 123 in cell D15. The sum of squares in cell H11 is reduced from 0.103 to 0.000 028. Cells D14 and D15 now tell us that [Ti(IV)] = 0.670 mM and [V(V)] = 1.123 mM in the mixture.

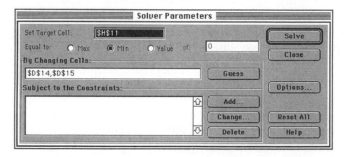

*Figure 19-4*  SOLVER window in Excel.

This procedure is readily extended to mixtures containing more than two components. The more points you measure, the more accurate the result is likely to be.

## What to Do When the Individual Spectra Are Well Resolved

If spectra of the individual components of a mixture are moderately resolved from one another, as at wavelengths $\lambda'$ and $\lambda''$ in Figure 19-1b, we can solve two simultaneous equations to find the concentrations in the mixture. The absorbance of the mixture at any wavelength is the sum of absorbances of each component at that wavelength. For wavelengths $\lambda'$ and $\lambda''$,

$$A' = \varepsilon_X' b[X] + \varepsilon_Y' b[Y]$$
$$A'' = \varepsilon_X'' b[X] + \varepsilon_Y'' b[Y] \qquad (19\text{-}5)$$

where the $\varepsilon$ values apply to each species at each wavelength. The absorptivities of X and Y at wavelengths $\lambda'$ and $\lambda''$ must be measured in separate experiments.

We can solve the Equations 19-5 for the two unknowns [X] and [Y]. The result is

*Analysis of a mixture when spectra are resolved:*

$$[X] = \frac{\begin{vmatrix} A' & \varepsilon_Y'b \\ A'' & \varepsilon_Y''b \end{vmatrix}}{\begin{vmatrix} \varepsilon_X'b & \varepsilon_Y'b \\ \varepsilon_X''b & \varepsilon_Y''b \end{vmatrix}} \qquad [Y] = \frac{\begin{vmatrix} \varepsilon_X'b & A' \\ \varepsilon_X''b & A'' \end{vmatrix}}{\begin{vmatrix} \varepsilon_X'b & \varepsilon_Y'b \\ \varepsilon_X''b & \varepsilon_Y''b \end{vmatrix}} \qquad (19\text{-}6)$$

In Equations 19-6, each symbol $\begin{vmatrix} a & b \\ c & d \end{vmatrix}$ is a *determinant*. It is a shorthand way of writing the product $a \times d - b \times c$. Thus, $\begin{vmatrix} 1 & 2 \\ 3 & 4 \end{vmatrix}$ means $1 \times 4 - 2 \times 3 = -2$.

$$\begin{vmatrix} a & b \\ c & d \end{vmatrix} = a \times d - b \times c$$

### Example  Analysis of a Mixture, Using Equations 19-6

The molar absorptivities of compounds X and Y were measured with pure samples of each:

|  | $\varepsilon$ (M$^{-1}$ cm$^{-1}$) | |
| --- | --- | --- |
| $\lambda$ (nm) | X | Y |
| 272 | 16 400 | 3 870 |
| 327 | 3 990 | 6 420 |

A mixture of compounds X and Y in a 1.000-cm cell had an absorbance of 0.957 at 272 nm and 0.559 at 327 nm. Find the concentrations of X and Y in the mixture.

**Solution**  Using Equations 19-6 and setting $b = 1.000$, we find

$$[X] = \frac{\begin{vmatrix} 0.957 & 3\,870 \\ 0.559 & 6\,420 \end{vmatrix}}{\begin{vmatrix} 16\,400 & 3\,870 \\ 3\,990 & 6\,420 \end{vmatrix}} = \frac{(0.957)(6\,420) - (3\,870)(0.559)}{(16\,400)(6\,420) - (3\,870)(3\,990)} = 4.43 \times 10^{-5}\,\text{M}$$

$$[Y] = \frac{\begin{vmatrix} 16\,400 & 0.957 \\ 3\,990 & 0.559 \end{vmatrix}}{\begin{vmatrix} 16\,400 & 3\,870 \\ 3\,990 & 6\,420 \end{vmatrix}} = 5.95 \times 10^{-5}\,\text{M}$$

To analyze a mixture of two compounds, it is necessary to measure absorbance at two wavelengths and to know $\varepsilon$ at each wavelength for each compound. Similarly, a mixture of $n$ components can be analyzed by making $n$ absorbance measurements at $n$ wavelengths.

### Solving Simultaneous Linear Equations with Excel

Excel solves systems of linear equations with a single statement. If the following matrix mathematics is unfamiliar to you, disregard it. The important result is the template in Figure 19-5 for solving simultaneous equations. You can use this template by following the instructions in the last paragraph of this section, even if the math is not familiar.

|  | A | B | C | D | E | F | G |
| --- | --- | --- | --- | --- | --- | --- | --- |
| 1 | Solving Simultaneous Linear Equations with Excel Matrix Operations | | | | | | |
| 2 | | | | | | | |
| 3 | Wavelength | Coefficient Matrix | | Absorbance | | Concentrations | |
| 4 | | | | of unknown | | in mixture | |
| 5 | 272 | 16440 | 3870 | 0.957 | | 4.4178E-05 | ← [X] |
| 6 | 327 | 3990 | 6420 | 0.559 | | 5.9615E-05 | ← [Y] |
| 7 | | K | | A | | C | |
| 8 | | | | | | | |
| 9 | 1. Enter matrix of coefficients εb in cells B5:C6 | | | | | | |
| 10 | 2. Enter absorbance of unknown at each wavelength (cells D5:D6) | | | | | | |
| 11 | 3. Highlight block of blank cells required for solution (F5 and F6) | | | | | | |
| 12 | 4. Type the formula "= MMULT(MINVERSE(B5:C6),D5:D6)" | | | | | | |
| 13 | 5. Press CONTROL+SHIFT+ENTER on a PC or COMMAND+RETURN on a Mac | | | | | | |
| 14 | 6. Behold! The answer appears in cells F5 and F6 | | | | | | |

*Figure 19-5*  Solving simultaneous linear equations with Excel.

The simultaneous equations of the preceding example are

$$A' = \varepsilon'_X b[X] + \varepsilon'_Y b[Y] \qquad 0.957 = 16\,440[X] + 3\,870[Y]$$
$$A'' = \varepsilon''_X b[X] + \varepsilon''_Y b[Y] \quad \Rightarrow \quad 0.559 = \phantom{0}3\,990[X] + 6\,420[Y]$$

which can be written in matrix notation in the form

$$\underbrace{\begin{bmatrix} 0.957 \\ 0.559 \end{bmatrix}}_{\mathbf{A}} = \underbrace{\begin{bmatrix} 16\,400 & 3\,870 \\ 3\,990 & 6\,420 \end{bmatrix}}_{\mathbf{K}} \underbrace{\begin{bmatrix} [X] \\ [Y] \end{bmatrix}}_{\mathbf{C}} \tag{19-7}$$

**K** is the *matrix* of molar absorptivity times pathlength, $\varepsilon b$. **A** is the matrix of absorbance of the unknown. A matrix, such as **A,** with only one column or one row is called a *vector*. **C** is the vector of unknown concentrations.

A matrix $\mathbf{K}^{-1}$, called the *inverse* of **K,** is such that the products $\mathbf{KK}^{-1}$ or $\mathbf{K}^{-1}\mathbf{K}$ are equal to a unit matrix with 1's on the diagonal and 0 elsewhere.[3] We can solve Equation 19-7 for the concentration vector, **C,** by multiplying both sides of the equation by $\mathbf{K}^{-1}$:

$$\mathbf{KC} = \mathbf{A}$$
$$\underbrace{\mathbf{K}^{-1}\mathbf{KC}}_{= \mathbf{C}} = \mathbf{K}^{-1}\mathbf{A}$$

*The algorithm for solving simultaneous equations is really simple: Find the inverse matrix $\mathbf{K}^{-1}$ and multiply it times A. The product is C, the concentrations in the unknown mixture.*

In Figure 19-5, we enter the wavelengths in column A just to keep track of information. We will not use these wavelengths for computation. Enter the products $\varepsilon b$ for pure X in column B and $\varepsilon b$ for pure Y in column C. The array in cells B5:C6 is the matrix **K.** The Excel function MINVERSE(B5:C6) gives the inverse matrix, $\mathbf{K}^{-1}$. The function MMULT(matrix 1, matrix 2) gives the product of two matrices (or a matrix and a vector). The concentration vector, **C,** is equal to $\mathbf{K}^{-1}\mathbf{A}$, which we get with the single statement

$$= \text{MMULT}(\underbrace{\text{MINVERSE(B5:C6)}}_{\mathbf{K}^{-1}}, \underbrace{\text{D5:D6}}_{\mathbf{A}})$$

To use the template in Figure 19-5, enter the coefficients, $\varepsilon b$, measured from the pure compounds, in cells B5:C6. Enter the absorbance of the unknown in cells D5:D6. Highlight cells F5:F6 and type the formula "=MMULT(MINVERSE(B5:C6), D5:D6)". Press CONTROL+SHIFT+ENTER on a PC or COMMAND(⌘)+RETURN on a Mac. The concentrations [X] and [Y] in the mixture now appear in cells F5:F6.

## Isosbestic Points

Often one absorbing species, X, is converted into another absorbing species, Y, in the course of a chemical reaction. This transformation leads to an obvious behavior shown in Figure 19-6. If the spectra of pure X and pure Y cross each other at any wavelength, then every spectrum recorded during this chemical reaction will cross at that same point, called an **isosbestic point.** *An isosbestic point is good evidence that only two principal species are present.*[4]

---

*The product $\mathbf{K}^{-1}\mathbf{KC}$ is just $\mathbf{C}$:*

$$\underbrace{\begin{bmatrix} 1 & 0 \\ 0 & 1 \end{bmatrix} \begin{bmatrix} [X] \\ [Y] \end{bmatrix}}_{\mathbf{K}^{-1}\mathbf{K} \qquad \mathbf{C}} = \underbrace{\begin{bmatrix} [X] \\ [Y] \end{bmatrix}}_{\mathbf{C}}$$

Procedure for solving simultaneous equations with Excel

---

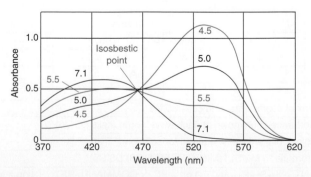

**Figure 19-6** Absorption spectrum of $3.7 \times 10^{-4}$ M methyl red between pH 4.5 and 7.1. *[From E. J. King, Acid-Base Equilibria (Oxford: Pergamon Press, 1965).]*

Consider methyl red, an indicator that changes between red (HIn) and yellow (In⁻) near pH 5.1:

$$(CH_3)_2N-\bigcirc-N \quad \overset{\displaystyle CO_2^-}{\underset{\displaystyle \overset{+}{N}H}{\bigcirc}} \quad \overset{pK_2 \approx 5.1}{\rightleftharpoons} \quad (CH_3)_2N-\bigcirc-N \quad \overset{\displaystyle CO_2^-}{\underset{\displaystyle N}{\bigcirc}}$$

HIn                In⁻
(red)              (yellow)

Because the spectra of HIn and In⁻ (at the same concentration) happen to cross at 465 nm, all spectra in Figure 19-6 cross at this point. (If the spectra of HIn and In⁻ crossed at several points, there would be several isosbestic points.)

To see why there is an isosbestic point, we write an equation for the absorbance of the solution at 465 nm:

$$A^{465} = \varepsilon_{HIn}^{465}b[HIn] + \varepsilon_{In^-}^{465}b[In^-] \qquad (19\text{-}8)$$

But, because the spectra of pure HIn and pure In⁻ (at the same concentration) cross at 465 nm, $\varepsilon_{HIn}^{465}$ must be equal to $\varepsilon_{In^-}^{465}$. Setting $\varepsilon_{HIn}^{465} = \varepsilon_{In^-}^{465} = \varepsilon^{465}$, we can factor Equation 19-8:

$$A^{465} = \varepsilon^{465}b([HIn] + [In^-]) \qquad (19\text{-}9)$$

In Figure 19-6, all solutions contain the same total concentration of methyl red ($= [HIn] + [In^-]$). Only the pH varies. Therefore, the sum of concentrations in Equation 19-9 is constant, and $A^{465}$ is constant.

> An isosbestic point occurs when $\varepsilon_X = \varepsilon_Y$ and $[X] + [Y]$ is constant.

# ■ ■ ■ 19-2  Measuring an Equilibrium Constant: The Scatchard Plot

To measure an equilibrium constant, we measure concentrations (actually activities) of species at equilibrium. This section shows how spectrophotometry can be used for this purpose.

Let's examine the equilibrium in which the species P and X react to form PX.

$$P + X \rightleftharpoons PX \qquad (19\text{-}10)$$

> Absorbance is proportional to *concentration* (not activity), so concentrations must be converted into activities to get true equilibrium constants.

Neglecting activity coefficients, we can write

$$K = \frac{[PX]}{[P][X]} \qquad (19\text{-}11)$$

Consider a series of solutions in which increments of X are added to a constant amount of P. Letting $P_0$ be the total concentration of P (in the forms P and PX), we can write

$$[P] = P_0 - [PX] \qquad (19\text{-}12)$$

> Clearing the cobwebs from your brain, you realize that Equation 19-12 is a mass balance.

Now the equilibrium expression, Equation 19-11, can be rearranged as follows:

$$\frac{[PX]}{[X]} = K[P] = K(P_0 - [PX]) \qquad (19\text{-}13)$$

A graph of $[PX]/[X]$ versus $[PX]$ has a slope of $-K$ and is called a **Scatchard plot.**[5] It is widely used in biochemistry to measure equilibrium constants (Figure 19-7).

> A *Scatchard plot* is a graph of $[PX]/[X]$ versus $[PX]$. The slope is $-K$.

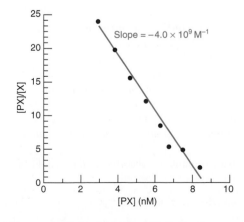

*Figure 19-7*  Scatchard plot for binding of antigen (X) to antibody (P). The antibody binds the explosive trinitrotoluene (TNT). The antigen is a fluorescent analog of TNT. From the slope, the binding constant for the reaction $P + X \rightleftharpoons PX$ is $K = 4.0 \times 10^9 \text{ M}^{-1}$. [Derived from Figure 4 of A. Bromberg and R. A. Mathies, "Homogeneous Immunoassay for Detection of TNT on a Capillary Electrophoresis Chip," Anal. Chem. **2003**, 75, 1188.]

If we know [PX], we can find [X] with the mass balance

$$X_0 = [\text{total X}] = [PX] + [X]$$

To measure [PX], we might use spectrophotometric absorbance. Suppose that P and PX each have some absorbance at wavelength $\lambda$, but X has no absorbance at this wavelength. For simplicity, let all measurements be made in a cell of pathlength 1.000 cm so that we can omit $b(= 1.000 \text{ cm})$ when writing Beer's law.

The absorbance at some wavelength is the sum of absorbances of PX and P:

$$A = \varepsilon_{PX}[PX] + \varepsilon_P[P]$$

Substituting $[P] = P_0 - [PX]$, we can write

$$A = \varepsilon_{PX}[PX] + \underbrace{\varepsilon_P P_0}_{A_0} - \varepsilon_P[PX] \qquad (19\text{-}14)$$

But $\varepsilon_P P_0$ is $A_0$, the initial absorbance before any X is added. Therefore,

$$A = [PX](\varepsilon_{PX} - \varepsilon_P) + A_0 \Rightarrow [PX] = \frac{\Delta A}{\Delta \varepsilon} \qquad (19\text{-}15)$$

where $\Delta\varepsilon = \varepsilon_{PX} - \varepsilon_P$ and $\Delta A(= A - A_0)$ is the observed absorbance after each addition of X minus the initial absorbance.

Substituting [PX] from Equation 19-15 into Equation 19-13 gives

*Scatchard equation:*
$$\frac{\Delta A}{[X]} = K\Delta\varepsilon P_0 - K\Delta A \qquad (19\text{-}16)$$

Problem 19-13 gives an alternate way to find $K$ by using Excel SOLVER.

A graph of $\Delta A/[X]$ versus $\Delta A$ should be a straight line with a slope of $-K$. Absorbance measured while P is titrated with X can be used to find $K$ for the reaction of X with P.

Two cases commonly arise in the application of Equation 19-16. If $K$ is small, then large concentrations of X are needed to produce PX. Therefore, $X_0 \gg P_0$, and $[X] \approx X_0$. Alternatively, if $K$ is not small, then $[X] \neq X_0$, and [X] must be measured, either at another wavelength or by measuring a different physical property.

Errors in a Scatchard plot could be substantial. We define the fraction of saturation as

Equation 19-13 can be recast as $S/[X] = K(1 - S)$. Plot $S/[X]$ versus $S$.

$$\text{Fraction of saturation} = S = \frac{[PX]}{P_0} \qquad (19\text{-}17)$$

The most accurate data are obtained for $0.2 \leq S \leq 0.8$.[6] A range representing ~75% of the total saturation curve should be measured before concluding that equilibrium (19-10) is obeyed. People have made mistakes by exploring too little of the binding curve.

## ■ ■ ■ 19-3 The Method of Continuous Variation

Suppose that several complexes can form between species P and X:

$$P + X \rightleftharpoons PX \qquad (19\text{-}18)$$

$$P + 2X \rightleftharpoons PX_2 \qquad (19\text{-}19)$$

$$P + 3X \rightleftharpoons PX_3 \qquad (19\text{-}20)$$

If one complex (say, $PX_2$) predominates, the **method of continuous variation** (also called *Job's method*) allows us to identify the stoichiometry of the predominant complex.

The classical procedure is to mix P and X and dilute to constant volume so that the total concentration $[P] + [X]$ is constant. For example, 2.50 mM solutions of P and X could be mixed as shown in Table 19-1 to give various X:P ratios, but constant total concentration. The absorbance of each solution is measured, typically at $\lambda_{max}$ for the complex, and a graph is made showing *corrected* absorbance (defined in Equation 19-21) versus mole fraction of X. *Maximum absorbance is reached at the composition corresponding to the stoichiometry of the predominant complex.*

For the reaction $P + nX \rightleftharpoons PX_n$, you could show that $[PX_n]$ reaches a maximum when the initial concentrations have the ratio $[X]_0 = n[P]_0$. To do this, write $K = [PX_n]/\{([P]_0 - [PX_n])([X]_0 - n[PX_n])\}$ and set the partial derivatives $\partial[PX_n]/\partial[P]_0$ and $\partial[PX_n]/\partial[P]_0$ equal to 0.

Corrected absorbance is the measured absorbance minus the absorbance that would be produced by free P and free X alone:

$$\text{Corrected absorbance} = \text{measured absorbance} - \varepsilon_P b P_T - \varepsilon_X b X_T \qquad (19\text{-}21)$$

**Table 19-1** Solutions for the method of continuous variation

| mL of<br>2.50 mM P | mL of<br>2.50 mM X | Mole ratio<br>(X:P) | Mole fraction of X<br>$\left(\dfrac{\text{mol X}}{\text{mol X + mol P}}\right)$ |
|---|---|---|---|
| 1.00 | 9.00 | 9.00:1 | 0.900 |
| 2.00 | 8.00 | 4.00:1 | 0.800 |
| 2.50 | 7.50 | 3.00:1 | 0.750 |
| 3.33 | 6.67 | 2.00:1 | 0.667 |
| 4.00 | 6.00 | 1.50:1 | 0.600 |
| 5.00 | 5.00 | 1.00:1 | 0.500 |
| 6.00 | 4.00 | 1:1.50 | 0.400 |
| 6.67 | 3.33 | 1:2.00 | 0.333 |
| 7.50 | 2.50 | 1:3.00 | 0.250 |
| 8.00 | 2.00 | 1:4.00 | 0.200 |
| 9.00 | 1.00 | 1:9.00 | 0.100 |

*NOTE: All solutions are diluted to a total volume of 25.0 mL with a buffer.*

where $\varepsilon_P$ and $\varepsilon_X$ are the molar absorptivities of pure P and pure X, $b$ is the pathlength, and $P_T$ and $X_T$ are the total concentrations of P and X in the solution. For the first solution in Table 19-1, $P_T = (1.00/25.0)(2.50\ \text{mM}) = 0.100\ \text{mM}$ and $X_T = (9.00/25.0)(2.50\ \text{mM}) = 0.900\ \text{mM}$. If P and X do not absorb at the wavelength of interest, no absorbance correction is needed.

Maximum absorbance occurs at the mole fraction of X corresponding to the stoichiometry of the complex (Figure 19-8). If the predominant complex is $PX_2$, the maximum occurs at mole fraction of $X = 2/(2 + 1) = 0.667$.

$$\text{Mole fraction of X in } P_aX_b = \frac{b}{b + a} \quad (= 0.667 \text{ when } b = 2 \text{ and } a = 1)$$

If $P_3X$ were predominant, the maximum would occur at (mole fraction of X) $= 1/(1 + 3) = 0.250$.

Here are some precautions for the method of continuous variation:

1.  Verify that the complex follows Beer's law.
2.  Use constant ionic strength and pH, if applicable.
3.  Take readings at more than one wavelength; the maximum should occur at the same mole fraction for each wavelength.
4.  Do experiments at different total concentrations of P + X. If a second set of solutions were prepared in the proportions given in Table 19-1, but from stock concentrations of 5.00 mM, the maximum should still occur at the same mole fraction.

*Method of continuous variation:*

$$P + nX \rightleftharpoons PX_n$$

*Maximum absorbance occurs when (mole fraction of X) = $n/(n + 1)$.*

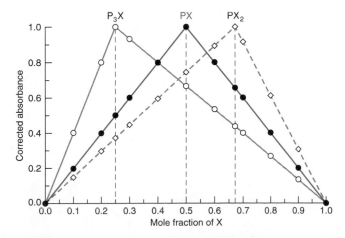

**Figure 19-8** Ideal behavior of Job plots for formation of the complexes $P_3X$, PX, and $PX_2$.

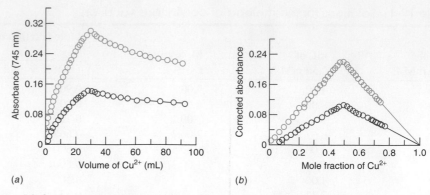

(a)                                    (b)

**Figure 19-9** (*a*) Spectrophotometric titration of 30.0 mL of EDTA in acetate buffer with $CuSO_4$ in the same buffer. Upper curve: [EDTA] = $[Cu^{2+}]$ = 5.00 mM. Lower curve: [EDTA] = $[Cu^{2+}]$ = 2.50 mM. The absorbance has not been "corrected" in any way. (*b*) Transformation of data into mole fraction format. The absorbance of free $CuSO_4$ at the same formal concentration has been subtracted from each point in panel *a*. EDTA is transparent at this wavelength. *[From Z. D. Hill and P. MacCarthy, "Novel Approach to Job's Method," J. Chem. Ed. **1986**, 63, 162.]*

The method of continuous variation can be carried out with many separate solutions, as in Table 19-1. However, a titration is more sensible. Figure 19-9a shows a titration of EDTA with $Cu^{2+}$. In Figure 19-9b, the abscissa has been transformed into mole fraction of $Cu^{2+}$ (= [moles of $Cu^{2+}$]/[moles of $Cu^{2+}$ + moles of EDTA]) instead of volume of $Cu^{2+}$. The sharp maximum at a mole fraction of 0.5 indicates formation of a 1:1 complex. If the equilibrium constant is not large, the maximum is more curved than in Figure 19-9b. The curvature can be used to estimate the equilibrium constant.[7]

# 19-4  Flow Injection Analysis[8]

In **flow injection analysis,** a sample is injected into a moving liquid stream to which various reagents can be added. After a suitable time, the reacted sample reaches a detector, which is usually a spectrophotometric cell. Flow injection is widely used in medical and pharmaceutical analysis, water analysis, and industrial process control.

In the Figure 19-10a, a solvent carrier stream is pumped continuously through the sample injector, in which 40 to 200 μL of sample is added to the stream. A reagent stream is then combined with the carrier stream, and the resulting solution passes through a coil to give time for reaction. The concentration of analyte in the sample is determined by measuring absorbance of the stream in a cell like that shown in Figure 25-19. Typical flow rates are 0.5–2.5 mL/min, and the diameter of the inert Teflon tubing from which the system is constructed is about 0.5 mm. Coils are 10–200 cm in length to allow suitable reaction times. Replicate analyses can be completed in 20–60 s.

Figure 19-10b gives an idea of possible variations. Reagents 1 and 2 are mixed and added to the sample in the carrier before reagent 3 is added. Commercial apparatus (Figure 19-11) allows different flow paths to be readily assembled. Other schemes are based on catalysis by analyte in an unknown.[9] The more analyte, the further the analytical reaction proceeds in a fixed time and the greater the concentration of product that is measured. Reagent streams can be passed through reactive columns, ion exchangers, dialysis tubes, gas diffusers, photochemical reactors, and solvent extractors. Detectors might employ absorbance,[10]

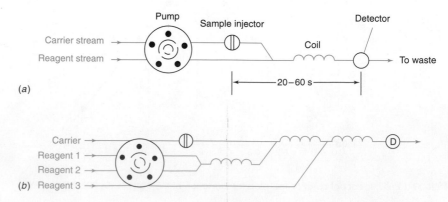

**Figure 19-10** Schematic diagrams of flow injection analysis, showing two different reagent addition schemes.

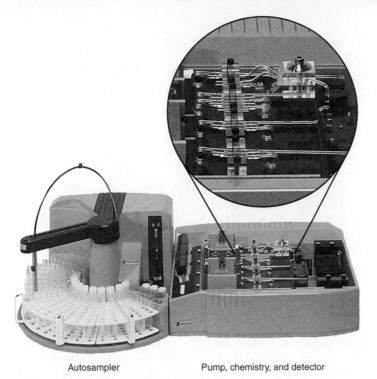

Autosampler      Pump, chemistry, and detector

**Figure 19-11** Flow injection analysis system with enlarged view of chemistry section. Different modular chemistry units are installed for different analyses. *[Courtesy Skalar, Inc., Norcross, GA.]*

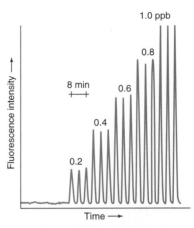

**Figure 19-12** Flow injection analysis of ppb levels of $H_2O_2$ in air based on formation of a fluorescent product. *[From J. Li and P. K. Dasgupta, "Measurement of Atmospheric $H_2O_2$ and Hydroxymethyl Hydroperoxide with a Diffusion Scrubber and Light Emitting Diode-Liquid Core Waveguide-Based Fluorometry," Anal. Chem.* **2000,** *72, 5338.]*

luminescence,[11] chemiluminescence,[12] or electrochemistry. A key feature of flow injection is rapid, repetitive analysis (Figure 19-12).

## 19-5 Immunoassays and Aptamers

An important application of fluorescence is in **immunoassays,** which employ antibodies to detect analyte. An *antibody* is a protein produced by the immune system of an animal in response to a foreign molecule called an *antigen*. The antibody recognizes the antigen that stimulated synthesis of the antibody. The formation constant for the antibody-antigen complex is very large, whereas the binding of the antibody to other molecules is weak.

Figure 19-13 illustrates the principle of an *enzyme-linked immunosorbent assay,* abbreviated ELISA in biochemical literature. Antibody 1, which is specific for the analyte of interest (the antigen), is bound to a polymeric support. In steps 1 and 2, analyte is incubated with the polymer-bound antibody to form a complex. The fraction of antibody sites that bind analyte is proportional to the concentration of analyte in the unknown. The surface is then

Rosalyn Yalow received the Nobel Prize in medicine in 1977 for developing immunoassay techniques in the 1950s, using proteins labeled with radioactive [131]I to enable their detection.[13] Yalow, a physicist, worked with Solomon Berson, a medical doctor, in this pioneering effort.

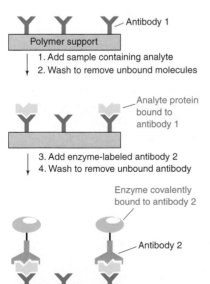

**Figure 19-13** Enzyme-linked immunosorbent assay. Antibody 1, which is specific for the analyte of interest, is bound to a polymer support and treated with unknown. After excess, unbound molecules have been washed away, the analyte remains bound to antibody 1. The bound analyte is then treated with antibody 2, which recognizes a different site on the analyte and to which an enzyme is covalently attached. After unbound material has been washed away, each molecule of analyte is coupled to an enzyme that will be used in Figure 19-14.

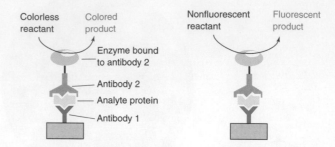

*Figure 19-14* Enzyme bound to antibody 2 can catalyze reactions that produce colored or fluorescent products. Each molecule of analyte bound in the immunoassay leads to many molecules of colored or fluorescent product that are easily measured.

washed to remove unbound substances. In steps 3 and 4, the antibody-antigen complex is treated with antibody 2, which recognizes a different region of the analyte. Antibody 2 was specially prepared for the immunoassay by covalent attachment of an enzyme that will be used later in the process. Again, excess unbound substances are washed away.

The enzyme attached to antibody 2 is critical for quantitative analysis. Figure 19-14 shows two ways in which the enzyme can be used. The enzyme can transform a colorless reactant into a colored product. Because one enzyme molecule catalyzes the same reaction many times, many molecules of colored product are created for each analyte molecule. The enzyme thereby *amplifies* the signal in the chemical analysis. The higher the concentration of analyte in the original unknown, the more enzyme is bound and the greater the extent of the enzyme-catalyzed reaction. Alternatively, the enzyme can convert a nonfluorescent reactant into a fluorescent product. Colorimetric and fluorometric enzyme-linked immunosorbent assays are sensitive to less than a nanogram of analyte. Pregnancy tests are based on the immunoassay of a placental protein in urine.

## Immunoassays in Environmental Analysis

Commercial immunoassay kits are available for screening and analysis of pesticides, industrial chemicals, explosives, and microbial toxins at the parts per trillion to parts per million levels in groundwater, soil, and food.[14] An advantage of screening in the field is that uncontaminated regions that require no further attention are readily identified. An immunoassay can be 20–40 times less expensive than a chromatographic analysis and can be completed in 0.3–3 h in the field, using 1-mL samples. Chromatography generally must be done in a lab and might require several days, because analyte must first be extracted or concentrated from liter-quantity samples to obtain a sufficient concentration.

## Time-Resolved Fluorescence Immunoassays[15]

The sensitivity of fluorescence immunoassays can be enhanced by a factor of 100 (to detect $10^{-13}$ M analyte) with time-resolved measurements of luminescence from the lanthanide ion $Eu^{3+}$. Organic chromophores such as fluorescein are plagued by background fluorescence at 350–600 nm from solvent, solutes, and particles. This background decays to a negligible level within 100 μs after excitation. Sharp luminescence at 615 nm from $Eu^{3+}$, however, has a much longer lifetime, falling to 1/e ($= 37\%$) of its initial intensity in approximately 700 μs. In a time-resolved fluorescence measurement (Figure 19-15), luminescence is measured between 200 and 600 μs after a brief laser pulse at 340 nm. The next pulse is flashed at 1 000 μs, and the cycle is repeated approximately 1 000 times per second. Rejecting emission within 200 μs of the excitation eliminates most of the background fluorescence.

Figure 19-16 shows how $Eu^{3+}$ can be incorporated into an immunoassay. A chelating group that binds lanthanide ions is attached to antibody 2 in Figure 19-13. While bound to the antibody, $Eu^{3+}$ has weak luminescence. After all steps in Figure 19-13 have been completed, the pH of the solution is lowered in the presence of a soluble chelator that extracts $Eu^{3+}$ into solution. Strong luminescence from the soluble metal ion is then easily detected by a time-resolved measurement.

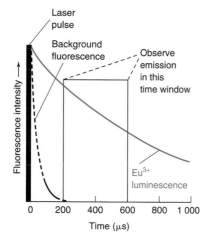

*Figure 19-15* Emission intensity in a time-resolved fluorescence experiment.

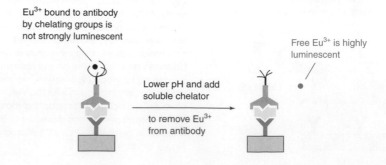

*Figure 19-16* Antibody 2 in the immunosorbent assay of Figure 19-13 can be labeled with $Eu^{3+}$, which is not strongly luminescent when bound to the antibody. To complete the analysis, the pH of the solution is lowered to liberate strongly luminescent $Eu^{3+}$ ion.

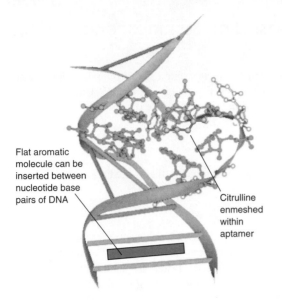

Flat aromatic molecule can be inserted between nucleotide base pairs of DNA

Citrulline enmeshed within aptamer

**Figure 19-17** Aptamer that specifically binds citrulline within a pocket in a short stretch of RNA. Long straight lines represent hydrogen-bonded nucleotide bases. The three-dimensional structure was deduced from nuclear magnetic resonance.
[M. Famulok, G. Mayer, and M. Blind, "Nucleic Acid Aptamers—From Selection in Vitro to Applications in Vivo," Acc. Chem. Res. **2000**, 33, 591.]

Citrulline

## Aptamers: Synthetic Nucleic Acid "Antibodies"

**Aptamers** are ~15–40-base-long pieces of DNA (deoxyribonucleic acid) or RNA (ribonucleic acid) that strongly and selectively bind to a specific molecule. An aptamer for a desired target molecule is chosen from a pool of random DNA or RNA sequences by successive cycles of binding to the target, removing unbound material, and replicating the bound nucleic acid.[16] Once the sequence of nucleic acids in an aptamer for a specific target is known, that aptamer can be synthesized in large quantities. The aptamer behaves as a custom-made, synthetic "antibody." Aptamers can bind to small sections of macromolecules, such as proteins, or they can engulf an entire small molecule, as in Figure 19-17.

Figure 19-17 also indicates that flat, aromatic molecules can be inserted between the flat, hydrogen-bonded nucleotide base pairs of DNA. We say that the flat molecule is *intercalated* between the base pairs. One of many analytical applications of aptamers utilizes a "molecular light switch" intercalated in the aptamer.

The ruthenium complex in the margin is slightly luminescent in aqueous solution but is strongly luminescent when intercalated in DNA. To measure the protein immunoglobulin E (IgE), an aptamer is selected that binds IgE. Excess ruthenium complex is added to the aptamer, so that the ruthenium is highly luminescent. When IgE is added to the intercalated aptamer, the aptamer binds to IgE and the ruthenium complex is expelled. Each "light switch" molecule that is expelled loses most of its luminescence. Figure 19-18 shows the decrease in luminescence with increasing concentration of IgE. At low concentrations, the effect is linear and can be used to measure the concentration of added IgE.

Unlike antibodies, which are fragile proteins that must be refrigerated for storage, aptamers are stable organic molecules with a long shelf life at room temperature. Aptamers have tremendous potential for use in highly specific chemical sensors.

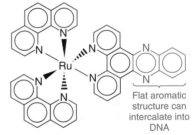

Flat aromatic structure can intercalate into DNA

"Molecular light switch"

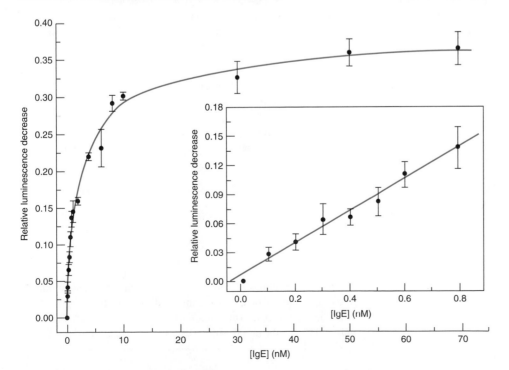

**Figure 19-18** Response of "molecular light switch" to immunoglobulin E (IgE). Aptamer concentration is 5 nM and ruthenium complex concentration is 40 nM. Addition of IgE displaces the ruthenium complex from the aptamer and decreases luminescence at 610 nm. Excitation wavelength = 450 nm.
[From Y. Jiang, X. Fang, and C. Bai, "Signaling Aptamer/Protein Binding by a Molecular Light Switch Complex," Anal. Chem. **2004**, 76, 5230.]

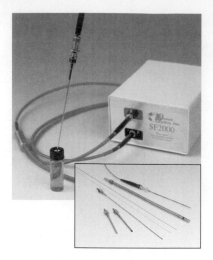

Figure 19-19 Fiber-optic sensor measures $O_2$ by its ability to quench the luminescence of Ru(II) at the tip of the fiber. A blue-light-emitting diode provides excitation energy. *[Courtesy Ocean Optics, Dunedin, FL.]*

# ■■■ 19-6 Sensors Based on Luminescence Quenching

When a molecule absorbs a photon, it is promoted to an excited state from which it can lose energy as heat or it can emit a lower-energy photon (Figure 18-13). Box 19-1 describes how absorbed light can be converted into electricity. In this section, we discuss how excited-state molecules can serve as chemical sensors (Figure 19-19).

## Luminescence Quenching

Suppose that the molecule M absorbs light and is promoted to the excited state M*:

Absorption: $\qquad M + h\nu \rightarrow M^*$ $\qquad$ Rate $= \dfrac{d[M^*]}{dt} = k_a[M]$

The rate at which M* is created, $d[M^*]/dt$, is proportional to the concentration of M. The rate constant, $k_a$, depends on the intensity of illumination and the absorptivity of M. The more intense the light and the more efficiently it is absorbed, the faster M* will be created.

Following absorption, M* can emit a photon and return to the ground state:

Emission: $\qquad M^* \rightarrow M + h\nu$ $\qquad$ Rate $= -\dfrac{d[M^*]}{dt} = k_e[M^*]$

The rate at which M* is lost is proportional to the concentration of M*. Alternatively, the excited molecule can lose energy in the form of heat:

Deactivation: $\qquad M^* \rightarrow M + heat$ $\qquad$ Rate $= -\dfrac{d[M^*]}{dt} = k_d[M^*]$

---

## Box 19-1 Converting Light into Electricity

Earth's deserts receive 250–300 $W/m^2$ of solar irradiance. If solar energy could be used with 10% efficiency, 4% of the sunlight falling on the deserts would provide all the energy used in the world in 2000. The solar cell[17] described here has a conversion efficiency close to 7%.

Sunlight enters the cell from the left through a transparent electrode made of fluorine-doped tin oxide. The electrode is coated with a 10-μm-thick layer of nanometer-size $TiO_2$ particles that are each coated with a *sensitizer*. The sensitizer is a Ru(II) complex that absorbs a large fraction of visible light. A layer of sensitizer coated on a planar electrode is so thin that it would absorb just 1% of the light. The nano-size $TiO_2$ particles have so much surface area that there is enough sensitizer to absorb 99% of the light entering the cell.

When Ru(II) sensitizer absorbs light, it is promoted to an excited state from which an electron is injected into the *conduction band* of the semiconductor $TiO_2$ within ~50 fs. This is one of the

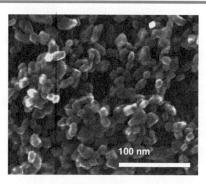

Scanning electron micrograph of sintered, nanocrystalline $TiO_2$. *Sintering is a heat treatment at 450°C that causes small particles to grow together by forming bridges ("necks") between particles. [From A. Hagfeldt and M. Grätzel, "Molecular Photovoltaics," Acc. Chem. Res. 2000, 33, 269.]*

fastest chemical processes ever studied. Instead of flowing back from $TiO_2$ to Ru(III), most electrons flow through an external circuit (where they can do useful work) to the Pt electrode at the right side of the cell. At the Pt surface, $I_3^-$ is reduced to $I^-$. The cycle is completed when $I^-$ reduces Ru(III) back to Ru(II):

$$Ru(II) + h\nu \rightarrow Ru(II)^* \qquad (\text{* denotes excited state})$$
$$Ru(II)^* \rightarrow Ru(III) + e^- \quad (\text{injected into } TiO_2)$$

$e^-$ flows through circuit from tin oxide electrode to Pt

at Pt electrode: $I_3^- + 2e^- \rightarrow 3I^-$

$$3I^- + 2Ru(III) \rightarrow I_3^- + 2Ru(II)$$

The graph shows the efficiency with which photons incident on the left side of the cell are converted into electrons in the circuit. The sensitizer absorbs much of the solar spectrum. The cell retains 92% of its initial efficiency after 1 000 h at 80°C. Dye-

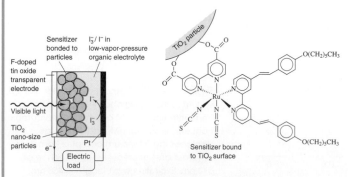

Essential features of a photocell based on sensitizer-coated nanometer-size $TiO_2$ particles.

Still another possibility is that the excited molecule can transfer energy to a different molecule, called a *quencher* (Q), to promote the quencher to an excited state (Q*):

Quenching:     $M^* + Q \rightarrow M + Q^*$          Rate $= -\dfrac{d[M^*]}{dt} = k_q[M^*][Q]$

> **Quenching** is the process whereby emission from an excited molecule is decreased by energy transfer to another molecule (the quencher).

The excited quencher can then lose its energy by a variety of processes.

Under constant illumination, the system soon reaches a steady state in which the concentrations of M* and M remain constant. In the steady state, the rate of appearance of M* must equal the rate of destruction of M*. The rate of appearance is

$$\text{Rate of appearance of } M^* = \dfrac{d[M^*]}{dt} = k_a[M]$$

The rate of disappearance is the sum of the rates of emission, deactivation, and quenching

$$\text{Rate of disappearance of } M^* = k_e[M^*] + k_d[M^*] + k_q[M^*][Q]$$

Setting the rates of appearance and disappearance equal to each other gives

$$k_a[M] = k_e[M^*] + k_d[M^*] + k_q[M^*][Q] \qquad (19\text{-}22)$$

The **quantum yield** for a photochemical process is the fraction of absorbed photons that produce a desired result. If the result occurs every time a photon is absorbed, then the quantum yield is unity. The quantum yield is a number between 0 and 1.

The quantum yield for emission from M* is the rate of emission divided by the rate of absorption. In the absence of quencher, we designate this quantum yield $\Phi_0$:

$$\Phi_0 = \dfrac{\text{photons emitted per second}}{\text{photons absorbed per second}} = \dfrac{\text{emission rate}}{\text{absorption rate}} = \dfrac{k_e[M^*]}{k_a[M]}$$

---

sensitized photocells suitable for external use on buildings are now commercially available.[18] The competing technology, which is more efficient at converting sunlight into electricity, is photovoltaic cells made of silicon.

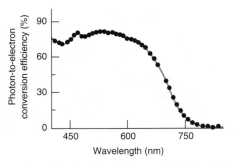

*Photoaction spectrum* showing the efficiency with which photons incident on the solar cell are converted into electrons in the circuit. *[From P. Wang, C. Klein, R. Humphry-Baker, S. M. Zakeeruddin, and M. Grätzel, "A High Molecular Extinction Coefficient Sensitizer for Stable Dye-Sensitized Solar Cells," J. Am. Chem. Soc.* **2005,** *127, 808.]*

Converting sunlight directly into electricity is good, but converting sunlight into a fuel such as $H_2$ or $CH_3OH$ would be even better. The fuel could be used when it is needed, not just when the sun is shining. Green leaves of plants use sunlight to reduce $CO_2$ to carbohydrates, a fuel that is oxidized back to $CO_2$ by animals and plants to provide energy.

We are rapidly exhausting the earth's supply of fossil fuels (coal, oil, and gas), which are also raw materials for plastics, fabrics, and many essential items. We face a major problem when the supply of raw materials is gone. Moreover, burning fossil fuels increases $CO_2$ in the atmosphere, which threatens to alter the climate.

Why do we burn irreplaceable raw materials? The shortsighted answer is that electricity from fossil-fuel-burning generators is cheaper than energy from renewable sources, such as solar, wind, and geothermal. The chart shows the estimated costs of generating electricity in the year 2000. Only hydroelectric power is less expensive than fossil-fuel generators. The chart also shows estimated $CO_2$ emissions from each source of power. Fossil fuels are, by far, the worst. We will probably consume fossil fuel until this increasingly scarce resource becomes more expensive than the cost of renewable energy—which is likely to happen in your lifetime.

Nuclear energy is cost competitive with fossil fuel, has very low greenhouse gas emission, and creates far less air pollution. However, fear of potential accidents and intractable issues of waste containment have prevented construction of nuclear power plants in the United States for three decades.

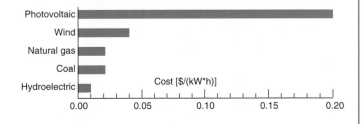

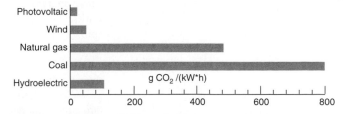

Estimated cost and $CO_2$ emissions for building 1.3-GW (gigawatt) power plants in 2000 and operating them for 20 years. *[Data from S. Pacca and A. Horvath, "Greenhouse Gas Emissions from Building and Operating Electric Power Plants in the Upper Colorado River Basin," Environ. Sci. Technol.* **2002,** *36, 3194.]*

Substituting in the value of $k_a[M]$ from Equation 19-22 and setting $[Q] = 0$ gives an expression for the quantum yield of emission in the steady state:

$$\Phi_0 = \frac{k_e[M^*]}{k_e[M^*] + k_d[M^*] + k_q[M^*][0]} = \frac{k_e}{k_e + k_d} \qquad (19\text{-}23)$$

If $[Q] \neq 0$, then the quantum yield for emission ($\Phi_Q$) is

$$\Phi_Q = \frac{k_e[M^*]}{k_e[M^*] + k_d[M^*] + k_q[M^*][Q]} = \frac{k_e}{k_e + k_d + k_q[Q]} \qquad (19\text{-}24)$$

In luminescence-quenching experiments, we measure emission in the absence and presence of a quencher. Equations 19-23 and 19-24 tell us that the relative yields are

*Stern-Volmer equation:* $\quad \dfrac{\Phi_0}{\Phi_Q} = \dfrac{k_e + k_d + k_q[Q]}{k_e + k_d} = 1 + \left(\dfrac{k_q}{k_e + k_d}\right)[Q] \qquad (19\text{-}25)$

The *Stern-Volmer equation* says that, if we measure relative emission ($\Phi_0/\Phi_Q$) as a function of quencher concentration and plot this quantity versus $[Q]$, we should observe a straight line. The quantity $\Phi_0/\Phi_Q$ in Equation 19-25 is equivalent to $I_0/I_Q$, where $I_0$ is the emission intensity in the absence of quencher and $I_Q$ is the intensity in the presence of quencher.

### A Luminescent Intracellular $O_2$ Sensor

We illustrate our discussion with Ru(II) complexes, which strongly absorb visible light, efficiently emit light at significantly longer wavelengths than they absorb, are stable for long periods, and have a long-lived excited state whose emission is quenched by $O_2$ (Color Plate 17).[19] A widely used luminescent Ru(II) complex is $\text{Ru(dpp)}_3^{2+} \cdot 2Cl^-$.

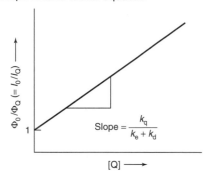

$$\equiv \quad (\text{dpp})_3\text{Ru(II)}$$

dpp = 4,7-diphenyl-1,10-phenanthroline

Oxygen is a good quencher because its ground state has two unpaired electrons—it is a *triplet* state designated $^3O_2$. $O_2$ has a low-lying *singlet* state with no unpaired electrons. Figure 18-13 showed that the lowest-lying excited state of many molecules is a triplet. This triplet excited state $^3M^*$ can transfer energy to $^3O_2$ to produce a ground-state singlet molecule and excited $^1O_2^*$.

$$^3M^* \quad + \quad ^3O_2 \quad \rightarrow \quad ^1M \quad + \quad ^1O_2^*$$

| Excited state | Ground state | Ground state | Excited state |

There are two electron spins up and two down in both the reactants and products. This energy transfer therefore conserves the net spin and is more rapid than processes that change the spin.

Color Plate 18 shows tiny dots of light from fluorescent silica beads shot into living cells by a "gene-gun" that is usually used to infect cells with DNA-coated particles. Two dyes are bound inside each porous bead whose size is 100 to 600 nm.

When illuminated with blue light, one dye emits green light near 525 nm and the other emits red light near 610 nm (Figure 19-20). The green dye is not affected by $O_2$, but the red ruthenium dye is affected. From the ratio of red-to-green emission intensity, we can calculate the concentration of $O_2$ in the vicinity of the beads. Results show that more than three-fourths of available intracellular $O_2$ is consumed by the cells within 2 min at 21°C after removing $O_2$

---

Quantum yield for emission in the absence of quenching ($\Phi_0$).

Quenching reduces the quantum yield for emission ($\Phi_Q < \Phi_0$).

Graph of Stern-Volmer equation:

Slope $= \dfrac{k_q}{k_e + k_d}$

The ground state of Ru(II) is a singlet, and the lowest excited state is a triplet. When Ru(II) absorbs visible light, it is promoted to the luminescent triplet state. $O_2$ quenches the luminescence by providing a radiationless pathway by which the triplet is converted into the ground-state singlet.

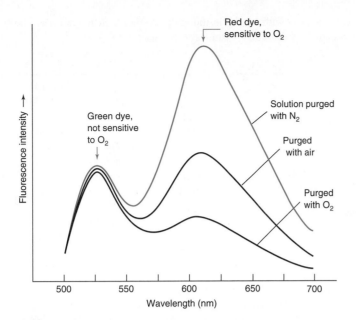

**Figure 19-20** Fluorescence from $O_2$-indicator beads showing constant intensity near 525 nm and variable intensity near 610 nm. The ratio of emission intensity at these two wavelengths is related to $O_2$ concentration. *[From H. Xu, J. W. Aylott, R. Kopelman, T. J. Miller, and M. A. Philbert, "Real-Time Method for Determination of $O_2$ Inside Living Cells Using Optical Nanosensors," Anal. Chem.* **2001**, *73, 4124.]*

from the external liquid medium. The sensitivity of intracellular $O_2$ measurement has been increased further by using luminescent Pt-based dyes.[20]

| Conditions | Intracellular $O_2$ (ppm) |
|---|---|
| Air-saturated buffer solution | $8.8 \pm 0.8$ |
| Cells in air-saturated buffer | $7.9 \pm 2.1$ |
| Cells in buffer 25 s after removing $O_2$ | $6.5 \pm 1.7$ |
| Cells in buffer 120 s after removing $O_2$ | $\leq 1.5$ |

## Terms to Understand

| | | | |
|---|---|---|---|
| aptamer | immunoassay | method of continuous variation | quenching |
| flow injection analysis | isosbestic point | quantum yield | Scatchard plot |

## Summary

The absorbance of a mixture is the sum of absorbances of the individual components. At a minimum, you should be able to find the concentrations of two species in a mixture by writing and solving two simultaneous equations for absorbance at two wavelengths. This procedure is most accurate if the two absorption spectra have regions where they do not overlap very much. With a spreadsheet, you should be able to use matrix operations to solve *n* simultaneous Beer's law equations for *n* components in a solution, with measurements at *n* wavelengths. You should be able to use Excel SOLVER to decompose a spectrum into a sum of spectra of the components by minimizing the function $(A_{calc} - A_m)^2$.

Isosbestic (crossing) points are observed when a solution contains variable proportions of two components with a constant total concentration. A Scatchard plot is used to measure an equilibrium constant, and the method of continuous variation allows us to determine the stoichiometry of a complex. In flow injection analysis, sample injected into a carrier stream is mixed with a color-forming reagent and passed into a flow-through detector.

Immunoassays use antibodies to detect the analyte of interest. In an enzyme-linked immunosorbent assay, signal is amplified by coupling analyte to an enzyme that catalyzes many cycles of a reaction that produces a colored or fluorescent product. Time-resolved fluorescence measurements provide sensitivity by separating analyte fluorescence in time and wavelength from background fluorescence. Aptamers are short pieces of DNA or RNA that are selected to bind tightly to a target molecule that can be small or large. Once an aptamer is identified for a particular target, it can be synthesized and used in place of antibodies in chemical analyses.

Luminescence intensity is proportional to the concentration of the emitting species if the concentration is low enough. We can measure some analytes, such as $O_2$, by their ability to quench (decrease) the luminescence of another compound.

## Exercises

**19-A.** 🖩 This problem can be worked with Equations 19-6 on a calculator or with the spreadsheet in Figure 19-5. Transferrin is the iron-transport protein found in blood. It has a molecular mass of 81 000 and carries two $Fe^{3+}$ ions. Desferrioxamine B is a chelator used to treat patients with iron overload (Box 12-1). It has a molecular mass of about 650 and can bind one $Fe^{3+}$. Desferrioxamine can take iron from many sites within the body and is excreted (with its iron) through the kidneys. The molar absorptivities of these compounds (saturated with iron) at two wavelengths are given in the table below. Both compounds are colorless (no visible absorption) in the absence of iron.

|  | $\varepsilon(M^{-1}\,cm^{-1})$ | |
|---|---|---|
| $\lambda$ (nm) | Transferrin | Desferrioxamine |
| 428 | 3 540 | 2 730 |
| 470 | 4 170 | 2 290 |

(a) A solution of transferrin exhibits an absorbance of 0.463 at 470 nm in a 1.000-cm cell. Calculate the concentration of transferrin in milligrams per milliliter and the concentration of bound iron in micrograms per milliliter.

(b) After adding desferrioxamine (which dilutes the sample), the absorbance at 470 nm was 0.424, and the absorbance at 428 nm was 0.401. Calculate the fraction of iron in transferrin and the fraction in desferrioxamine. Remember that transferrin binds two iron atoms and desferrioxamine binds only one.

**19-B.** 🖿 The spreadsheet lists molar absorptivities of three dyes and the absorbance of a mixture of the dyes at visible wavelengths. Use the least-squares procedure in Figure 19-3 to find the concentration of each dye in the mixture.

|  | A | B | C | D | E |
|---|---|---|---|---|---|
| 1 | Mixture of Dyes | | | | |
| 2 | | | | | Absorbance |
| 3 | Wavelength | ——— | Molar Absorptivity | ——— | of mixture |
| 4 | (nm) | Tartrazine | Sunset Yellow | Ponceau 4R | Am |
| 5 | 350 | 6.229E+03 | 2.019E+03 | 4.172E+03 | 0.557 |
| 6 | 375 | 1.324E+04 | 4.474E+03 | 2.313E+03 | 0.853 |
| 7 | 400 | 2.144E+04 | 7.403E+03 | 3.310E+03 | 1.332 |
| 8 | 425 | 2.514E+04 | 8.551E+03 | 4.534E+03 | 1.603 |
| 9 | 450 | 2.200E+04 | 1.275E+04 | 6.575E+03 | 1.792 |
| 10 | 475 | 1.055E+04 | 1.940E+04 | 1.229E+04 | 2.006 |
| 11 | 500 | 1.403E+03 | 1.869E+04 | 1.673E+04 | 1.821 |
| 12 | 525 | 0.000E+00 | 7.641E+03 | 1.528E+04 | 1.155 |
| 13 | 550 | 0.000E+00 | 3.959E+02 | 9.522E+03 | 0.445 |
| 14 | 575 | 0.000E+00 | 0.000E+00 | 1.814E+03 | 0.084 |

*Data from J. J. B. Nevado, J. R. Flores, and M. J. V. Llerena, "Simultaneous Spectrophotometric Determination of Tartrazine, Sunset Yellow, and Ponceau 4R in Commercial Products," Fresenius J. Anal. Chem.* **1998**, *361, 465.*

**19-C.** Compound P, which absorbs light at 305 nm, was titrated with X, which does not absorb at this wavelength. The product, PX, also absorbs at 305 nm. Absorbance was measured in a 1.000-cm cell, and the concentration of free X was determined independently, with results in the table. Prepare a Scatchard plot and find the equilibrium constant for the reaction X + P $\rightleftharpoons$ PX.

| Experiment | $P_0$ (M) | $X_0$ (M) | A | [X] (M) |
|---|---|---|---|---|
| 0 | 0.010 0 | 0 | 0.213 | 0 |
| 1 | 0.010 0 | 0.001 00 | 0.303 | $4.42 \times 10^{-6}$ |
| 2 | 0.010 0 | 0.002 00 | 0.394 | $9.10 \times 10^{-6}$ |
| 3 | 0.010 0 | 0.003 00 | 0.484 | $1.60 \times 10^{-5}$ |
| 4 | 0.010 0 | 0.004 00 | 0.575 | $2.47 \times 10^{-5}$ |
| 5 | 0.010 0 | 0.005 00 | 0.663 | $3.57 \times 10^{-5}$ |
| 6 | 0.010 0 | 0.006 00 | 0.752 | $5.52 \times 10^{-5}$ |
| 7 | 0.010 0 | 0.007 00 | 0.840 | $8.20 \times 10^{-5}$ |
| 8 | 0.010 0 | 0.008 00 | 0.926 | $1.42 \times 10^{-4}$ |
| 9 | 0.010 0 | 0.009 00 | 1.006 | $2.69 \times 10^{-4}$ |
| 10 | 0.010 0 | 0.010 00 | 1.066 | $5.87 \times 10^{-4}$ |
| 11 | 0.010 0 | 0.020 00 | 1.117 | $9.66 \times 10^{-3}$ |

**19-D.** Complex formation by 3-aminopyridine and picric acid in chloroform solution gives a yellow product with an absorbance maximum at 400 nm. Neither starting material absorbs significantly at this wavelength. Stock solutions containing $1.00 \times 10^{-4}$ M of each compound were mixed as indicated, and the absorbances of the mixtures were recorded. Prepare a graph of absorbance versus mole fraction of 3-aminopyridine and find the stoichiometry of the complex.

| Picric acid (mL) | 3-Aminopyridine (mL) | Absorbance at 400 nm |
|---|---|---|
| 2.70 | 0.30 | 0.106 |
| 2.40 | 0.60 | 0.214 |
| 2.10 | 0.90 | 0.311 |
| 1.80 | 1.20 | 0.402 |
| 1.50 | 1.50 | 0.442 |
| 1.20 | 1.80 | 0.404 |
| 0.90 | 2.10 | 0.318 |
| 0.60 | 2.40 | 0.222 |
| 0.30 | 2.70 | 0.110 |

*SOURCE: Data from E. Bruneau, D. Lavabre, G. Levy, and J. C. Micheau, "Quantitative Analysis of Continuous-Variation Plots with a Comparison of Several Methods," J. Chem. Ed.* **1992**, *69, 833.*

## Problems

### Analysis of a Mixture

**19-1.** 🖿 This problem can be worked by calculator or with the spreadsheet in Figure 19-5. Consider compounds X and Y in the example labeled "Analysis of a Mixture, Using Equations 19-6" in Section 19-1. Find [X] and [Y] in a solution whose absorbance is 0.233 at 272 nm and 0.200 at 327 nm in a 0.100-cm cell.

**19-2.** 🖿 The figure (see next page) shows spectra of $1.00 \times 10^{-4}$ M $MnO_4^-$, $1.00 \times 10^{-4}$ M $Cr_2O_7^{2-}$, and an unknown mixture of both, all in 1.000-cm-pathlength cells. Absorbances are given in the table. Use the least-squares procedure in Figure 19-3 to find the concentration of each species in the mixture.

| Wavelength (nm) | $MnO_4^-$ standard | $Cr_2O_7^{2-}$ standard | Mixture |
|---|---|---|---|
| 266 | 0.042 | 0.410 | 0.766 |
| 288 | 0.082 | 0.283 | 0.571 |
| 320 | 0.168 | 0.158 | 0.422 |
| 350 | 0.125 | 0.318 | 0.672 |
| 360 | 0.056 | 0.181 | 0.366 |

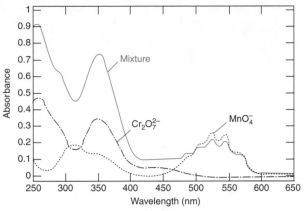

Visible spectrum of $MnO_4^-$, $Cr_2O_7^{2-}$, and an unknown mixture containing both ions. [From M. Blanco, H. Iturriaga, S. Maspoch, and P. Tarín, "A Simple Method for Spectrophotometric Determination of Two-Components with Overlapped Spectra," J. Chem. Ed. **1989**, 66, 178.]

**19-3.** When are isosbestic points observed and why?

**19-4.** The metal ion indicator xylenol orange (Table 12-3) is yellow at pH 6 ($\lambda_{max} = 439$ nm). The spectral changes that occur as $VO^{2+}$ is added to the indicator at pH 6 are shown below. The mole ratio $VO^{2+}$/xylenol orange at each point is

| Trace | Mole ratio | Trace | Mole ratio | Trace | Mole ratio |
|---|---|---|---|---|---|
| 0 | 0 | 6 | 0.60 | 12 | 1.3 |
| 1 | 0.10 | 7 | 0.70 | 13 | 1.5 |
| 2 | 0.20 | 8 | 0.80 | 14 | 2.0 |
| 3 | 0.30 | 9 | 0.90 | 15 | 3.1 |
| 4 | 0.40 | 10 | 1.0 | 16 | 4.1 |
| 5 | 0.50 | 11 | 1.1 | | |

Suggest a sequence of chemical reactions to explain the spectral changes, especially the isosbestic points at 457 and 528 nm.

**19-5.** Infrared spectra are customarily recorded on a transmittance scale so that weak and strong bands can be displayed on the same scale. The region near $2\,000$ cm$^{-1}$ in the infrared spectra of compounds A and B is shown in the figure. Note that absorption corresponds to a downward peak on this scale. The spectra were recorded from a 0.010 0 M solution of each, in cells with 0.005 00-cm pathlengths. A mixture of A and B in a 0.005 00-cm cell gave a transmittance of 34.0% at $2\,022$ cm$^{-1}$ and 38.3% at $1\,993$ cm$^{-1}$. Find the concentrations of A and B.

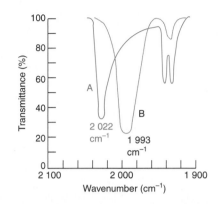

| Wavenumber | Pure A | Pure B |
|---|---|---|
| $2\,022$ cm$^{-1}$ | 31.0% $T$ | 97.4% $T$ |
| $1\,993$ cm$^{-1}$ | 79.7% $T$ | 20.0% $T$ |

**19-6.** 🖿 Spectroscopic data for the indicators thymol blue (TB), semithymol blue (STB), and methylthymol blue (MTB) are shown in the table. A solution of TB, STB, and MTB in a 1.000-cm cuvet had absorbances of 0.412 at 455 nm, 0.350 at 485 nm, and 0.632 at 545 nm. Modify the spreadsheet in Figure 19-5 to handle three

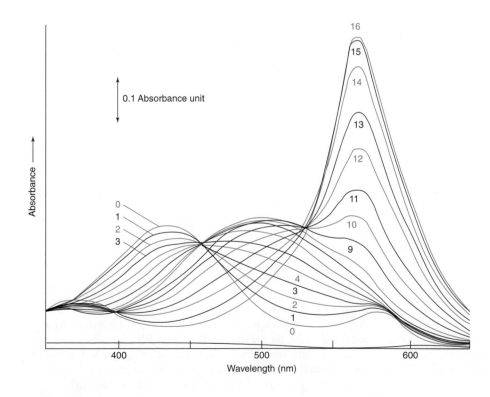

Figure for Problem 19-4. Absorption spectra for the reaction of xylenol orange with $VO^{2+}$ at pH 6.0. [From D. C. Harris and M. H. Gelb, "Binding of Xylenol Orange to Transferrin: Demonstration of Metal-Anion Linkage," Biochim. Biophys. Acta **1980**, 623, 1.]

simultaneous equations and find [TB], [STB], and [MTB] in the mixture.

| | $\varepsilon(M^{-1}\,cm^{-1})$ | | |
|---|---|---|---|
| $\lambda$ (nm) | TB | STB | MTB |
| 455 | 4 800 | 11 100 | 18 900 |
| 485 | 7 350 | 11 200 | 11 800 |
| 545 | 36 400 | 13 900 | 4 450 |

SOURCE: Data from S. Kiciak, H. Gontarz, and E. Krzyżanowska, "Monitoring the Synthesis of Semimethylthymol Blue and Methylthymol Blue," Talanta **1995**, 42, 1245.

**19-7.** 🖳 The table below gives the product $\varepsilon b$ for four pure compounds and a mixture at infrared wavelengths. Modify Figure 19-5 to solve four equations and find the concentration of each compound. You can treat the coefficient matrix as if it were molar absorptivity because the pathlength was constant (but unknown) for all measurements.

| Wavelength | Coefficient matrix ($\varepsilon b$) | | | | Absorbance |
|---|---|---|---|---|---|
| ($\mu$m) | p-xylene | m-xylene | o-xylene | ethylbenzene | of unknown |
| 12.5 | 1.5020 | 0.0514 | 0 | 0.0408 | 0.1013 |
| 13.0 | 0.0261 | 1.1516 | 0 | 0.0820 | 0.09943 |
| 13.4 | 0.0342 | 0.0355 | 2.532 | 0.2933 | 0.2194 |
| 14.3 | 0.0340 | 0.0684 | 0 | 0.3470 | 0.03396 |

Data from Z. Zdravkovski, "Mathcad in Chemistry Calculations. II. Arrays," J. Chem. Ed. **1992**, 69, 242A.

**19-8.** A solution was prepared by mixing 25.00 mL of 0.080 0 M aniline, 25.00 mL of 0.060 0 M sulfanilic acid, and 1.00 mL of $1.23 \times 10^{-4}$ M HIn and then diluting to 100.0 mL. (HIn stands for protonated indicator.)

Anilinium ion
$pK_a = 4.601$

Sulfanilic acid
$pK_a = 3.232$

$$HIn \rightleftharpoons H^+ + In^-$$

$\varepsilon_{325} = 2.45 \times 10^4\,M^{-1}\,cm^{-1}$  $\varepsilon_{325} = 4.39 \times 10^3\,M^{-1}\,cm^{-1}$
$\varepsilon_{550} = 2.26 \times 10^4\,M^{-1}\,cm^{-1}$  $\varepsilon_{550} = 1.53 \times 10^4\,M^{-1}\,cm^{-1}$

The absorbance measured at 550 nm in a *5.00-cm* cell was 0.110. Find the concentrations of HIn and ln$^-$ and $pK_a$ for HIn.

**19-9.** *Chemical equilibrium and analysis of a mixture.* A remote optical sensor for $CO_2$ in the ocean was designed to operate without the need for calibration.[21] The sensor compartment is separated from seawater by a silicone membrane through which $CO_2$, but not dissolved ions, can diffuse. Inside the sensor, $CO_2$ equilibrates with $HCO_3^-$ and $CO_3^{2-}$. For each measurement, the sensor is flushed with fresh solution containing 50.0 $\mu$M bromothymol blue indicator (NaHIn) and 42.0 $\mu$M NaOH. All indicator is in the form HIn$^-$ or In$^{2-}$ near neutral pH, so we can write two mass balances: (1) [HIn$^-$] + [In$^{2-}$] = $F_{In}$ = 50.0 $\mu$M and (2) [Na$^+$] = $F_{Na}$ = 50.0 $\mu$M + 42.0 $\mu$M = 92.0 $\mu$M. HIn$^-$ has an absorbance maximum at 434 nm and In$^{2-}$ has a maximum at 620 nm. The sensor measures the absorbance *ratio* $R_A = A_{620}/A_{434}$ reproducibly without need for calibration. From this ratio, we can find [$CO_2(aq)$] in the seawater as outlined here:

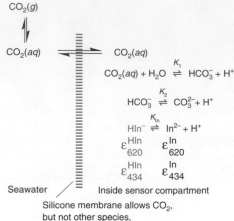

Seawater / Inside sensor compartment
Silicone membrane allows $CO_2$, but not other species, to diffuse across

**(a)** From Beer's law for the mixture, write equations for [HIn$^-$] and [In$^{2-}$] in terms of the absorbances at 620 and 434 nm ($A_{620}$ and $A_{434}$). Then show that

$$\frac{[In^{2-}]}{[HIn^-]} = \frac{R_A\varepsilon_{434}^{HIn^-} - \varepsilon_{620}^{HIn^-}}{\varepsilon_{620}^{In^{2-}} - R_A\varepsilon_{434}^{In^{2-}}} \equiv R_{In} \qquad (A)$$

**(b)** From the mass balance (1) and the acid dissociation constant $K_{In}$, show that

$$[HIn^-] = \frac{F_{In}}{R_{In} + 1} \qquad (B)$$

$$[In^{2-}] = \frac{K_{In}F_{In}}{[H^+](R_{In} + 1)} \qquad (C)$$

**(c)** Show that [H$^+$] = $K_{In}/R_{In}$. $\qquad (D)$
**(d)** From the carbonic acid dissociation equilibria, show that

$$[HCO_3^-] = \frac{K_1[CO_2(aq)]}{[H^+]} \qquad (E)$$

$$[CO_3^{2-}] = \frac{K_1K_2[CO_2(aq)]}{[H^+]^2} \qquad (F)$$

**(e)** Write the charge balance for the solution in the sensor compartment. Substitute in expressions B, C, E, and F for [HIn$^-$], [In$^{2-}$], [HCO$_3^-$], and [CO$_3^{2-}$].
**(f)** Suppose that the various constants have the following values:

$\varepsilon_{434}^{HIn^-} = 8.00 \times 10^3\,M^{-1}\,cm^{-1}$  $K_1 = 3.0 \times 10^{-7}$
$\varepsilon_{620}^{HIn^-} = 0$  $K_2 = 3.3 \times 10^{-11}$
$\varepsilon_{434}^{In^{2-}} = 1.90 \times 10^3\,M^{-1}\,cm^{-1}$  $K_{In} = 2.0 \times 10^{-7}$
$\varepsilon_{620}^{In^{2-}} = 1.70 \times 10^4\,M^{-1}\,cm^{-1}$  $K_w = 6.7 \times 10^{-15}$

From the measured absorbance ratio $R_A = A_{620}/A_{434} = 2.84$, find [$CO_2(aq)$] in the seawater.
**(g)** Approximately what is the ionic strength inside the sensor compartment? Were we justified in neglecting activity coefficients in working this problem?

**Measuring an Equilibrium Constant**

**19-10.** Figure 19-7 is a Scatchard plot for the addition of 0–20 nM antigen X to a fixed concentration of antibody P ($P_0 = 10$ nM). Prepare a Scatchard plot from the data in the table (on the next page) and find $K$ for the reaction P + X $\rightleftharpoons$ PX. The table gives measured concentrations of unbound X and the complex PX. It is recommended that the fraction of saturation should span the range ~0.2–0.8. What is the range of the fraction of saturation for the data?

| [X] (nM) | [PX] (nM) | [X] (nM) | [PX] (nM) |
|---|---|---|---|
| $0.12_0$ | $2.8_7$ | $0.73_1$ | $6.2_9$ |
| $0.19_2$ | $3.8_0$ | $1.2_2$ | $6.7_7$ |
| $0.29_6$ | $4.6_6$ | $1.5_0$ | $7.5_2$ |
| $0.45_0$ | $5.5_4$ | $3.6_1$ | $8.4_5$ |

**19-11.** Compound P was titrated with X to form the complex PX. A series of solutions was prepared with the total concentration of P remaining fixed at $1.00 \times 10^{-5}$ M. Both P and X have no visible absorbance, but PX has an absorption maximum at 437 nm. The table below shows how the absorbance at 437 nm in a 5.00-cm cell depends on the total concentration of added X ($X_T = [X] + [PX]$).

| $X_T$ (M) | $A$ | $X_T$ (M) | $A$ |
|---|---|---|---|
| 0 | 0.000 | 0.020 0 | 0.535 |
| 0.002 00 | 0.125 | 0.040 0 | 0.631 |
| 0.004 00 | 0.213 | 0.060 0 | 0.700 |
| 0.006 00 | 0.286 | 0.080 0 | 0.708 |
| 0.008 00 | 0.342 | 0.100 | 0.765 |
| 0.010 00 | 0.406 | | |

(a) Make a Scatchard plot of $\Delta A/[X]$ versus $\Delta A$. In this plot, [X] refers to the species X, not to $X_T$. However, because $X_T \gg [P]$, we can safely say that $[X] \approx X_T$ in this experiment.
(b) From the slope of the graph, find the equilibrium constant, $K$.

**19-12.** Iodine reacts with mesitylene to form a complex with an absorption maximum at 332 nm in $CCl_4$ solution:

$$I_2 + \text{(Mesitylene)} = \text{(Complex)} \qquad K = \frac{[\text{complex}]}{[I_2][\text{mesitylene}]}$$

Iodine    Mesitylene    Complex
$\varepsilon_{332} \approx 0$    $\varepsilon_{332} \approx 0$    $\lambda_{max} = 332$ nm

(a) Given that the product absorbs at 332 nm, but neither reactant has significant absorbance at this wavelength, use the equilibrium constant, $K$, and Beer's law to show that

$$\frac{A}{[\text{Mesitylene}][I_2]_{tot}} = K\varepsilon - \frac{KA}{[I_2]_{tot}}$$

where $A$ is the absorbance at 332 nm, $\varepsilon$ is the molar absorptivity of the complex at 332 nm, [mesitylene] is the concentration of free mesitylene, and $[I_2]_{tot}$ is the total concentration of iodine in the solution ($= [I_2] + [\text{complex}]$). The cell pathlength is 1.000 cm.
(b) Spectrophotometric data for this reaction are shown in the table below. Because $[\text{mesitylene}]_{tot} \gg [I_2]$, we can say that $[\text{mesitylene}] \approx [\text{mesitylene}]_{tot}$. Prepare a graph of $A/([\text{mesitylene}][I_2]_{tot})$ versus $A/[I_2]_{tot}$ and find the equilibrium constant and molar absorptivity of the complex.

| $[\text{Mesitylene}]_{tot}$ (M) | $[I_2]_{tot}$ (M) | Absorbance at 332 nm |
|---|---|---|
| 1.690 | $7.817 \times 10^{-5}$ | 0.369 |
| 0.921 8 | $2.558 \times 10^{-4}$ | 0.822 |
| 0.633 8 | $3.224 \times 10^{-4}$ | 0.787 |
| 0.482 9 | $3.573 \times 10^{-4}$ | 0.703 |
| 0.390 0 | $3.788 \times 10^{-4}$ | 0.624 |
| 0.327 1 | $3.934 \times 10^{-4}$ | 0.556 |

SOURCE: Data from P. J. Ogren and J. R. Norton, "Applying a Simple Linear Least-Squares Algorithm to Data with Uncertainties in Both Variables," *J. Chem. Ed.* **1992**, *69*, A130.

**19-13.**    Now we use SOLVER to find $K$ for Problem 19-12. The only absorbing species at 332 nm is the complex, so, from Beer's law, $[\text{complex}] = A/\varepsilon$ (because pathlength = 1.000 cm). $I_2$ is either free or bound in the complex, so $[I_2] = [I_2]_{tot} - [\text{complex}]$. There is a huge excess of mesitylene, so $[\text{mesitylene}] \approx [\text{mesitylene}]_{tot}$.

$$K = \frac{[\text{complex}]}{[I_2][\text{mesitylene}]} = \frac{A/\varepsilon}{([I_2]_{tot} - A/\varepsilon)[\text{mesitylene}]_{tot}}$$

The spreadsheet below shows some of the data. You will need to use all the data. Column A contains [mesitylene] and column B contains $[I_2]_{tot}$. Column C lists the measured absorbance. *Guess* a value of the molar absorptivity of the complex, $\varepsilon$, in cell A7. Then compute the concentration of the complex ($= A/\varepsilon$) in column D. The equilibrium constant in column E is given by E2 = [complex]/([I_2][mesitylene]) = (D2)/((B2-D2)*A2).

| | A | B | C | D | E |
|---|---|---|---|---|---|
| 1 | [Mesitylene] | [I2]tot | A | [Complex] = A/$\varepsilon$ | Keq |
| 2 | 1.6900 | 7.817E-05 | 0.369 | 7.380E-05 | 9.99282 |
| 3 | 0.9218 | 2.558E-04 | 0.822 | 1.644E-04 | 1.95128 |
| 4 | 0.6338 | 3.224E-04 | 0.787 | 1.574E-04 | 1.50511 |
| 5 | | | | Average = | 3.54144 |
| 6 | Guess for ε: | | | Standard Dev = | 3.32038 |
| 7 | 5.000E+03 | | | Stdev/Average = | 0.93758 |

What should we minimize with SOLVER? We want to vary $\varepsilon$ in cell A7 until the values of $K$ in column E are as constant as possible. We would like to minimize a function like $\sum(K_i - K_{average})^2$, where $K_i$ is the value in each line of the table and $K_{average}$ is the average of all computed values. The problem with $\sum(K_i - K_{average})^2$ is that we can minimize this function simply by making $K_i$ very small, but not necessarily constant. What we really want is for all the $K_i$ to be clustered around the mean value. A good way to do this is to minimize the *relative standard deviation* of the $K_i$, which is (standard deviation)/average. In cell E5, we compute the average value of $K$ and in cell E6, the standard deviation. Cell E7 contains the relative standard deviation. Use SOLVER to minimize cell E7 by varying cell A7. Compare your answer with that of Problem 19-12.

**Method of Continuous Variation**

**19-14.** *Method of continuous variation.* Make a graph of absorbance versus mole fraction of thiocyanate for the data in the table.

| mL $Fe^{3+}$ solution[a] | mL $SCN^-$ solution[b] | Absorbance at 455 nm |
|---|---|---|
| 30.00 | 0 | 0.001 |
| 27.00 | 3.00 | 0.122 |
| 24.00 | 6.00 | 0.226 |
| 21.00 | 9.00 | 0.293 |
| 18.00 | 12.00 | 0.331 |
| 15.00 | 15.00 | 0.346 |
| 12.00 | 18.00 | 0.327 |
| 9.00 | 21.00 | 0.286 |
| 6.00 | 24.00 | 0.214 |
| 3.00 | 27.00 | 0.109 |
| 0 | 30.00 | 0.002 |

a. $Fe^{3+}$ solution: 1.00 mM $Fe(NO_3)_3$ + 10.0 mM $HNO_3$.

b. $SCN^-$ solution: 1.00 mM KSCN + 15.0 mM HCl.

SOURCE: Data from Z. D. Hill and P. MacCarthy, "Novel Approach to Job's Method," *J. Chem. Ed.* **1986**, *63*, 162.

(a) What is the stoichiometry of the predominant $Fe(SCN)_n^{3-n}$ species?

(b) Why is the peak not as sharp as those in Figure 19-8?

(c) Why does one solution contain 10.0 mM acid and the other 15.0 mM acid?

**19-15.** Simulating a Job's plot. Consider the reaction $A + 2B \rightleftharpoons AB_2$, for which $K = [AB_2]/[A][B]^2$. Suppose that the following mixtures of A and B at a fixed total concentration of $10^{-4}$ M are prepared:

| $[A]_{total}$ (M) | $[B]_{total}$ (M) | $[A]_{total}$ (M) | $[B]_{total}$ (M) |
|---|---|---|---|
| $1.00 \times 10^{-5}$ | $9.00 \times 10^{-5}$ | $5.00 \times 10^{-5}$ | $5.00 \times 10^{-5}$ |
| $2.00 \times 10^{-5}$ | $8.00 \times 10^{-5}$ | $6.00 \times 10^{-5}$ | $4.00 \times 10^{-5}$ |
| $2.50 \times 10^{-5}$ | $7.50 \times 10^{-5}$ | $7.00 \times 10^{-5}$ | $3.00 \times 10^{-5}$ |
| $3.00 \times 10^{-5}$ | $7.00 \times 10^{-5}$ | $8.00 \times 10^{-5}$ | $2.00 \times 10^{-5}$ |
| $3.33 \times 10^{-5}$ | $6.67 \times 10^{-5}$ | $9.00 \times 10^{-5}$ | $1.00 \times 10^{-5}$ |
| $4.00 \times 10^{-5}$ | $6.00 \times 10^{-5}$ | | |

(a) Prepare a spreadsheet to find the concentration of $AB_2$ for each mixture, for equilibrium constants of $K = 10^6$, $10^7$, and $10^8$. One way to do this is to enter the values of $[A]_{total}$ and $[B]_{total}$ in columns A and B, respectively. Then put a trial (guessed) value of $[AB_2]$ in column C. From the mass balances $[A]_{total} = [A] + [AB_2]$ and $[B]_{total} = [B] + 2[AB_2]$, we can write $K = [AB_2]/[A][B]^2 = [AB_2]/\{([A]_{total} - [AB_2])([B]_{total} - 2[AB_2])^2\}$. In column D, enter the reaction quotient $[AB_2]/[A][B]^2$. For example, cell D2 has the formula "=C2/((A2-C2)(B2-2*C2)^2)". Then vary the value of $[AB_2]$ in cell C2 with SOLVER until the reaction quotient in cell D2 is equal to the desired equilibrium constant (such as $10^8$).

(b) Prepare a graph by the method of continuous variation in which you plot $[AB_2]$ versus mole fraction of A for each equilibrium constant. Explain the shapes of the curves.

**19-16.** A study was conducted with derivatives of the DNA nucleotide bases adenine and thymine bound inside micelles (Box 26-1) in aqueous solution.

Adenine derivative     Thymine derivative     Sodium dodecyl sulfate

Sodium dodecyl sulfate forms micelles with the hydrocarbon tails pointed inward and ionic headgroups exposed to water. It was hypothesized that the bases would form a 1:1 hydrogen-bonded complex inside the micelle as they do in DNA.

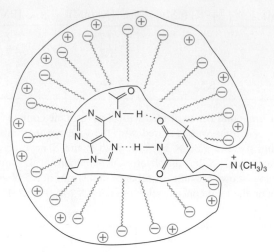

Hydrogen-bonded base pair inside micelle with hydrocarbon tails anchoring the bases to the micelle.

To test the hypothesis, aliquots of 5.0 mM adenine derivative were mixed with aliquots of 5.0 mM thymine derivative in proportions shown in the table. Each solution also contained 20 mM sodium dodecyl sulfate. The concentration of product measured by nuclear magnetic resonance also is shown in the table. Are the results consistent with formation of a 1:1 complex? Explain your answer.

| Adenine volume (mL) | Thymine volume (mL) | Concentration of product (mM) |
|---|---|---|
| 0.450 | 0.050 | $0.118 \pm 0.009$ |
| 0.400 | 0.100 | $0.202 \pm 0.038$ |
| 0.350 | 0.150 | $0.265 \pm 0.021$ |
| 0.300 | 0.200 | $0.307 \pm 0.032$ |
| 0.250 | 0.250 | $0.312 \pm 0.060$ |
| 0.200 | 0.300 | $0.296 \pm 0.073$ |
| 0.150 | 0.350 | $0.260 \pm 0.122$ |
| 0.100 | 0.400 | $0.187 \pm 0.110$ |
| 0.050 | 0.450 | $0.103 \pm 0.104$ |

### Luminescence and Immunoassay

**19-17.** Explain how signal amplification is achieved in enzyme-linked immunosorbent assays.

**19-18.** What is the advantage of a time-resolved emission measurement with $Eu^{3+}$ versus measurement of fluorescence from organic chromophores?

**19-19.** Here is an immunoassay for explosives such as trinitrotoluene (TNT).[22]

Trinitrotoluene (TNT)

Fluorescently labeled TNT

1. A column is prepared with covalently bound antibodies to TNT. Fluorescence-labeled TNT is passed through the column to saturate the antibodies with labeled TNT. The column is washed with excess solvent until no fluorescence is detected at the outlet.

2. Air containing TNT vapors is drawn through a sampling device that concentrates traces of the gaseous vapors into distilled water.

3. Aliquots of water from the sampler are injected into the antibody column. The fluorescence intensity of the liquid exiting the column is proportional to the concentration of TNT in the water from the sampler over the range 20 to 1 200 ng/mL.

Draw pictures showing the state of the column in steps 1 and 3 and explain how this detector works.

**19-20.** The graph below shows the effect of pH on quenching of luminescence of tris(2,2'-bipyridine)Ru(II) by 2,6-dimethylphenol. The ordinate, $K_{SV}$, is the collection of constants, $k_q/(k_e + k_d)$, in the Stern-Volmer equation 19-25. The greater $K_{SV}$, the greater the quenching. Suggest an explanation for the shape of the graph and estimate $pK_a$ for 2,6-dimethylphenol.

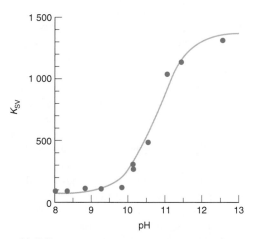

Tris(2,2'-bipyridine)ruthenium(II)

2,6-Dimethylphenol

Quenching of tris(2,2'-bipyridine)Ru(II) by 2,6-dimethylphenol as a function of pH. [Data from H. Gsponer, G. A. Argüello, and G. A. Argüello, "Determinations of $pK_a$ from Luminescence Quenching Data," J. Chem. Ed. **1997**, 74, 968.]

**19-21.** *Fluorescence quenching in micelles.* Consider an aqueous solution with a high concentration of micelles (Box 26-1) and relatively low concentrations of the fluorescent molecule pyrene and a quencher (cetylpyridinium chloride, designated Q), both of which dissolve in the micelles.

$CH_2(CH_2)_{14}CH_3$

$Cl^-$

Pyrene
(fluorescent species)

Cetylpyridinium chloride
(quencher, Q)

Quenching occurs if pyrene and Q are in the same micelle. Let the total concentration of quencher be [Q] and the concentration of micelles be [M]. The average number of quenchers per micelle is $\bar{Q} = [Q]/[M]$. If Q is randomly distributed among the micelles, then the probability that a particular micelle has $n$ molecules of Q is given by the *Poisson distribution:*[23]

$$\text{Probability of } n \text{ molecules of Q in micelle} \equiv P_n = \frac{\bar{Q}^n}{n!} e^{-\bar{Q}} \quad (1)$$

where $n!$ is $n$ factorial ($= n[n-1][n-2]\dots[1]$). The probability that there are no molecules of Q in a micelle is

$$\text{Probability of } 0 \text{ molecules of Q in micelle} = \frac{\bar{Q}^0}{0!} e^{-\bar{Q}} = e^{-\bar{Q}} \quad (2)$$

because $0! \equiv 1$.

Let $I_0$ be the fluorescence intensity of pyrene in the absence of Q and let $I_Q$ be the intensity in the presence of Q (both measured at the same concentration of micelles). The quotient $I_Q/I_0$ must be $e^{-\bar{Q}}$, which is the probability that a micelle does not possess a quencher molecule. Substituting $\bar{Q} = [Q]/[M]$ gives

$$I_Q/I_0 = e^{-\bar{Q}} = e^{-[Q]/[M]} \quad (3)$$

Micelles are made of the surfactant molecule, sodium dodecyl sulfate (shown in Problem 19-16). When surfactant is added to a solution, no micelles form until a minimum concentration called the *critical micelle concentration* (CMC) is attained. When the total concentration of surfactant, [S], exceeds the critical concentration, then the surfactant found in micelles is [S] − [CMC]. The molar concentration of micelles is

$$[M] = \frac{[S] - [CMC]}{N_{av}} \quad (4)$$

where $N_{av}$ is the average number of molecules of surfactant in each micelle.

Combining Equations 3 and 4 gives an expression for fluorescence as a function of total quencher concentration, [Q]:

$$\ln \frac{I_0}{I_Q} = \frac{[Q]N_{av}}{[S] - [CMC]} \quad (5)$$

By measuring fluorescence intensity as a function of [Q] at fixed [S], we can find the average number of molecules of S per micelle if we know the critical micelle concentration (which is independently measured in solutions of S). The table below gives data for 3.8 μM pyrene in a micellar solution with a total concentration of sodium dodecyl sulfate [S] = 20.8 mM.

| Q (μM) | $I_0/I_Q$ | Q (μM) | $I_0/I_Q$ | Q (μM) | $I_0/I_Q$ |
|---|---|---|---|---|---|
| 0 | 1 | 158 | 2.03 | 316 | 4.04 |
| 53 | 1.28 | 210 | 2.60 | 366 | 5.02 |
| 105 | 1.61 | 262 | 3.30 | 418 | 6.32 |

SOURCE: *Data from M. F. R. Prieto, M. C. R. Rodríguez, M. M. González, A. M. R. Rodríguez, and J. C. M. Fernández, "Fluorescene Quenching in Microheterogeneous Media," J. Chem. Ed. 1995, 72, 662.*

(a) If micelles were not present, quenching would be expected to follow the Stern-Volmer equation 19-25. Show that the graph of $I_0/I_Q$ versus [Q] is not linear.

(b) The critical micelle concentration is 8.1 mM. Prepare a graph of $\ln (I_0/I_Q)$ versus [Q]. Use Equation 5 to find $N_{av}$, the average number of sodium dodecyl sulfate molecules per micelle.

(c) Find the concentration of micelles, [M], and the average number of molecules of Q per micelle, $\bar{Q}$, when [Q] = 0.200 mM.

(d) Compute the fractions of micelles containing 0, 1, and 2 molecules of Q when [Q] = 0.200 mM.

# 20 Spectrophotometers

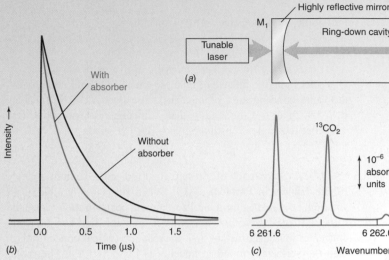

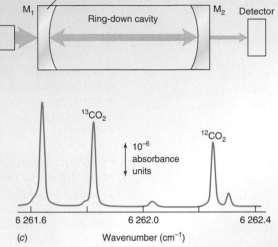

Cavity ring-down spectrum of ~3 mbar of $CO_2$, which is similar to the concentration in human breath. [From E. R. Crosson, K. N. Ricci, B. A. Richman, F. C. Chilese, T. G. Owano, R. A. Provencal, M. W. Todd, J. Glasser, A. A. Kachanov, B. A. Paldus, T. G. Spence, and R. N. Zare, "Stable Isotope Ratios Using Cavity Ring-Down Spectroscopy: Determination of $^{13}C/^{12}C$ for Carbon Dioxide in Human Breath," Anal. Chem. **2002**, 74, 2003.]

Cavity ring-down can measure absorbance as low as $\sim 10^{-6}$ and has the potential to provide sensitive detectors for chromatography.[1] In part (a), a laser pulse is directed into a cavity with mirrors on both ends. If the mirror reflectivity is 99.98%, then 0.02% of the power penetrates mirror $M_1$ and enters the cavity. The laser is shut off and light inside the cavity bounces back and forth, losing 0.02% of its intensity each time it strikes a mirror. A detector outside mirror $M_2$ measures light leaking through $M_2$. Graph (b) shows decay of the detector signal from a cavity containing a nonabsorbing liquid. If an absorbing species is present, decay is faster because signal is lost by absorption during each pass between the mirrors. The difference in signal decay time with and without absorber provides a measure of absorbance. Because decay time for an individual laser pulse is measured, fluctuation in light intensity from pulse to pulse has little influence on the measured absorbance. The effective pathlength is $\sim 10^3$ times the length of the cavity because light makes $\sim 10^3$ passes between mirrors during the measurement.

Spectrum (c) shows absorbance measured for $CO_2(g)$ with the natural mixture of 98.9% $^{12}C$ and 1.1% $^{13}C$. Peaks arise from transitions between rotational levels of two vibrational states. The spectral region was chosen to include a strong absorption of $^{13}CO_2$ and a weak absorption of $^{12}CO_2$, so that the intensities of the isotopic peaks are similar. Each point in the spectrum was obtained by varying the laser frequency.

The areas of the $^{13}CO_2$ and $^{12}CO_2$ peaks from human breath were used to determine whether a patient was infected with *Helicobacter pylori,* a bacterium that causes ulcers. After ingesting $^{13}C$-urea, *H. pylori* converts $^{13}C$-urea into $^{13}CO_2$, which appears in the patient's breath. The ratio $^{13}C/^{12}C$ in the breath of an infected person increases by 1–5%, whereas the ratio $^{13}C/^{12}C$ from an uninfected person is constant to within 0.1%.

$T = P/P_0$
$A = -\log T = \varepsilon bc$
$\varepsilon =$ molar absorptivity
$b =$ pathlength
$c =$ concentration

**F**igure 18-4 showed the essential features of a *single-beam spectrophotometer*. Light from a *source* is separated into narrow bands of wavelength by a *monochromator*, passed through a sample, and measured by a *detector*. We measure the *irradiance* ($P_0$, watts/m²) striking the detector with a *reference* cell (a solvent blank or reagent blank) in the sample compartment. When the reference is replaced by the sample of interest, some radiation is usually absorbed and the irradiance striking the detector, $P$, is smaller than $P_0$. The quotient $P/P_0$, which is a number between 0 and 1, is the *transmittance, T*. *Absorbance*, which is proportional to concentration, is $A = \log P_0/P = -\log T$.

A single-beam spectrophotometer is inconvenient because the sample and reference must be placed alternately in the beam. For measurements at multiple wavelengths, the reference must be run at each wavelength. A single-beam instrument is poorly suited to measuring

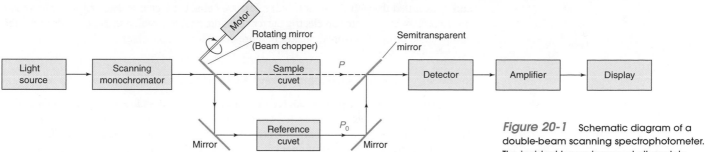

absorbance as a function of time, as in a kinetics experiment, because both the source intensity and the detector response slowly drift.

Figure 20-1 shows a *double-beam spectrophotometer,* in which light alternately passes through the sample and reference (blank), directed by a rotating mirror (the *chopper*) into and out of the light path. When light passes through the sample, the detector measures irradiance $P$. When the chopper diverts the beam through the reference cuvet, the detector measures $P_0$. The beam is chopped several times per second, and the circuitry automatically compares $P$ and $P_0$ to obtain transmittance and absorbance. This procedure provides automatic correction for changes of the source intensity and detector response with time and wavelength, because the power emerging from the two samples is compared so frequently. Most research-quality spectrophotometers provide automatic wavelength scanning and continuous recording of absorbance versus wavelength.

It is routine to first record a *baseline* spectrum with reference solution in both cuvets. The baseline absorbance at each wavelength is then subtracted from the measured absorbance of the sample to obtain the true absorbance of the sample at each wavelength.

A double-beam ultraviolet-visible spectrophotometer is shown in Figure 20-2. Visible light comes from a quartz-halogen lamp (like that in an automobile headlight), and the ultraviolet source is a deuterium arc lamp that emits in the range 200 to 400 nm. Only one lamp is used at a time. Grating 1 selects a narrow band of wavelengths to enter the monochromator, which selects an even narrower band to pass through the sample. After being chopped

**Figure 20-2** (*a*) Varian Cary 3E Ultraviolet-Visible Spectrophotometer. (*b*) Optical train. *[Courtesy Varian Australia Pty, Ltd., Victoria, Australia.]*

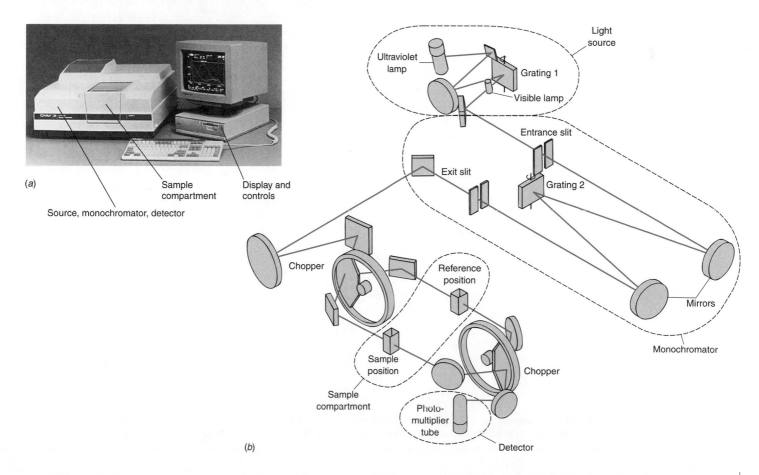

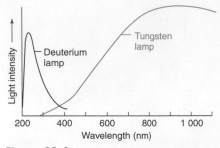

Figure 20-3 Intensity of a tungsten filament at 3 200 K and a deuterium arc lamp.

and transmitted through the sample and reference cells, the signal is detected by a *photomultiplier tube*, which creates an electric current proportional to irradiance. Now we describe the components of the spectrophotometer in more detail.

## ■■■■ 20-1 Lamps and Lasers: Sources of Light

A *tungsten lamp* is an excellent source of continuous visible and near-infrared radiation. A typical tungsten filament operates at a temperature near 3 000 K and produces useful radiation in the range 320 to 2 500 nm (Figure 20-3). This range covers the entire visible region and parts of the infrared and ultraviolet regions as well. Ultraviolet spectroscopy normally employs a *deuterium arc lamp* in which an electric discharge (a spark) causes $D_2$ to dissoci-

---

### Box 20-1 Blackbody Radiation and the Greenhouse Effect

When an object is heated, it emits radiation—it glows. Even at room temperature, objects radiate at infrared frequencies. Imagine a hollow sphere whose inside surface is perfectly black. That is, the surface absorbs all radiation striking it. If the sphere is at constant temperature, it must emit as much radiation as it absorbs. If a small hole were made in the wall, we would observe that the escaping radiation has a continuous spectral distribution. The object is called a *blackbody,* and the radiation is called **blackbody radiation.** Emission from real objects such as the tungsten filament of a light bulb resembles that from an ideal blackbody.

The power per unit area radiating from the surface of an object is called the *exitance* (or *emittance*), $M$, and is given by

*Exitance from blackbody:* $\qquad M = \sigma T^4$

where $\sigma$ is the Stefan-Boltzmann constant ($5.669\ 8 \times 10^{-8}$ W/($m^2 \cdot K^4$)). A blackbody whose temperature is 1 000 K radiates $5.67 \times 10^4$ watts per square meter of surface area. If the temperature is doubled, the exitance increases by a factor of $2^4 = 16$.

The blackbody emission spectrum at right changes with temperature, as shown in the graph. Near 300 K, maximum emission occurs at infrared wavelengths. The outer region of the sun behaves like a blackbody with a temperature near 5 800 K, emitting mainly *visible* light.

Of the solar flux of 1 368 W/m² reaching the upper atmosphere, 23% is absorbed by the atmosphere and 25% is reflected back into space. Earth absorbs 48% of the solar flux and reflects 4%. Radiation reaching Earth should be just enough to keep the surface temperature at 254 K, which would not support life as we know it. Why does the average temperature of Earth's surface stay at a comfortable 287 K?

The blackbody curves tell us that Earth radiates mainly *infrared* radiation, rather than visible light. Although the atmosphere is transparent to incoming visible light, it strongly absorbs outgoing infrared radiation. The main absorbers, called *greenhouse gases,* are water[3] and $CO_2$ and, to lesser extents, $O_3$, $CH_4$, chlorofluorocarbons, and $N_2O$. Radiation emitted from Earth is absorbed by the atmosphere and part of it is reradiated back to Earth. The atmosphere behaves like an insulating blanket, main-

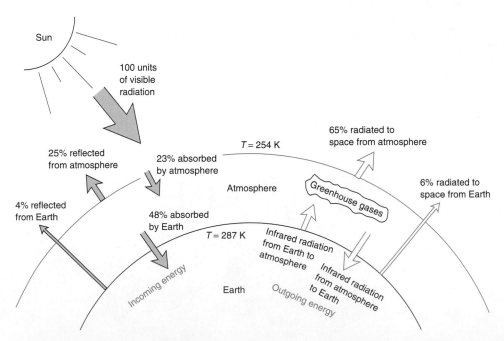

Balance between energy reaching Earth from the sun and energy reradiated to space. Exchange of infrared radiation between Earth and its atmosphere keeps Earth's surface 33 K warmer than the upper atmosphere.

ate and emit ultraviolet radiation from 200 to 400 nm (Figure 20-3). In a typical ultraviolet-visible spectrophotometer, a change is made between deuterium and tungsten lamps when passing through 360 nm, so that the source with the highest intensity is always employed. For selected visible and ultraviolet frequencies, electric discharge (spark) lamps filled with mercury vapor (Box 18-2) or xenon gas are widely used. Light-emitting diodes provide narrow bands of visible and near-infrared (close to visible) radiation.[2]

Infrared radiation in the range 4 000 to 200 cm$^{-1}$ is commonly obtained from a silicon carbide *globar,* heated to near 1 500 K by an electric current. The globar emits radiation with approximately the same spectrum as a *blackbody* at 1 000 K (Box 20-1).

**Lasers** provide isolated lines of a single wavelength for many applications. A laser with a wavelength of 3 μm might have a **bandwidth** (range of wavelengths) of $3 \times 10^{-14}$ to

Ultraviolet radiation is harmful to the naked eye. Do not view an ultraviolet source without protection.

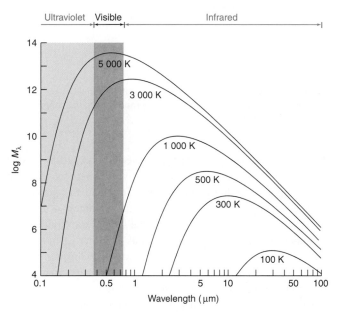

Spectral distribution of blackbody radiation. The family of curves is called the *Planck distribution* after Max Planck, who derived the law governing blackbody radiation. Note that both axes are logarithmic.

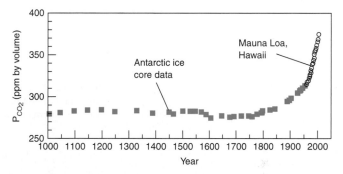

Atmospheric $CO_2$ measured from $CO_2$ in antarctic ice cores (squares) and from atmospheric measurements (circles) at Mauna Loa in Hawaii. *[Antarctic data from D. M. Etheridge, L. P. Steele, R. L. Langenfelds, R. J. Francey, J.-M. Barnola, and V. I. Morgan, in Trends: A Compendium of Data on Global Change, Carbon Dioxide Information Analysis Center, Oak Ridge National Laboratory, Oak Ridge, TN, 1998. Hawaii data from C. D. Keeling and T. P. Whorf, Scripps Institution of Oceanography, http://cdiac.esd.ornl.gov/ndps/ndp001.html]*

taining Earth's surface temperature 33 K warmer than the temperature of the upper atmosphere.[4]

Man's activities since the dawn of the Industrial Revolution have increased atmospheric carbon dioxide levels through the

burning of fossil fuel. Our ability to predict how Earth will respond to the rise in greenhouse gas concentration is primitive, but there is evidence that the temperature of the earth's surface has risen in the past century.[5] Will there be disastrous climatic changes? Will there be compensating responses that lead to little temperature change? We cannot accurately answer these questions, but prudence suggests that we should avoid making such large relative changes in our atmosphere.

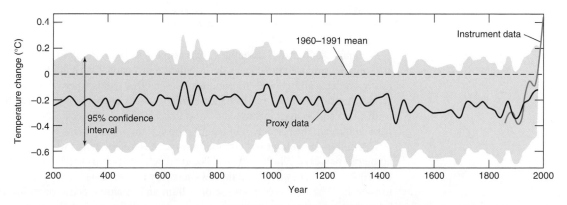

Estimate of global temperature based on proxy data such as tree rings and isotope ratios in sediments and ice cores. The 1990s was the warmest decade in 2000 years. *[From M. E. Mann and P. D. Jones, "Global Surface Temperatures Over the Past Two Millennia," Geophys. Res. Lett. **2003**, 30, 1820.]*

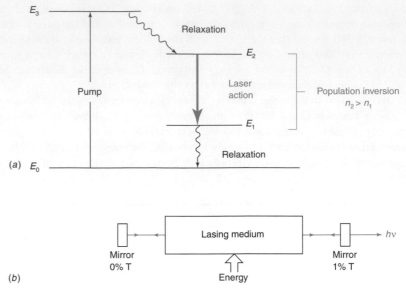

**Figure 20-4** (*a*) Energy-level diagram illustrating the principle of operation of a laser. (*b*) Basic components of a laser. The population inversion is created in the lasing medium. Pump energy might be derived from intense lamps or an electric discharge.

*Properties of laser light:*

| | |
|---|---|
| monochromatic: | one wavelength |
| extremely bright: | high power at one wavelength |
| collimated: | parallel rays |
| polarized: | electric field oscillates in one plane |
| coherent: | all waves in phase |

$3 \times 10^{-8}$ μm. The bandwidth is measured where the radiant power falls to half of its maximum value. The brightness of a low-power laser at its output wavelength is $10^{13}$ times greater than that of the sun at its brightest (yellow) wavelength. (Of course, the sun emits all wavelengths, whereas the laser emits only a narrow band. The total brightness of the sun is much greater than that of the laser.) The angular divergence of the laser beam from its direction of travel is typically less than 0.05°, a property that allows us to illuminate a small target. Laser light is typically *plane polarized,* with the electric field oscillating in one plane perpendicular to the direction of travel (Figure 18-1). Laser light is *coherent,* which means that all waves emerging from the laser oscillate in phase with one another.

A necessary condition for lasing is *population inversion,* in which a higher energy state has a greater population, $n$, than a lower energy state in the lasing medium. In Figure 20-4a, this condition occurs when the population of state $E_2$ exceeds that of $E_1$. Molecules in ground state $E_0$ of the lasing medium are *pumped* to excited state $E_3$ by broadband radiation from a powerful lamp or by an electric discharge. Molecules in state $E_3$ rapidly relax to $E_2$, which has a relatively long lifetime. After a molecule in $E_2$ decays to $E_1$, it rapidly relaxes to the ground state, $E_0$ (thereby keeping the population of $E_2$ greater than the population of $E_1$).

A photon with an energy that exactly spans two states can be absorbed to raise a molecule to an excited state. Alternatively, that same photon can stimulate the excited molecule to emit a photon and return to the lower state. This is called *stimulated emission.* When a photon emitted by a molecule falling from $E_2$ to $E_1$ strikes another molecule in $E_2$, a second photon can be emitted with the same phase and polarization as the incident photon. If there is a population inversion ($n_2 > n_1$), one photon stimulates the emission of many photons as it travels through the laser.

Figure 20-4b shows essential components of a laser. Pump energy directed through the side of the lasing medium creates the population inversion. One end of the laser cavity is a mirror that reflects all light (0% transmittance). The other end is a partially transparent mirror that reflects most light (1% transmittance). Photons with energy $E_2 - E_1$ that bounce back and forth between the mirrors stimulate an avalanche of new photons. The small fraction of light passing through the partially transparent mirror at the right is the useful output of the laser.

A helium-neon laser is a common source of red light with a wavelength of 632.8 nm and an output power of 0.1–25 mW. An electric discharge pumps helium atoms to state $E_3$ in Figure 20-4. The excited helium transfers energy by colliding with a neon atom, raising the neon to state $E_2$. The high concentration of helium and intense electric pumping create a population inversion among neon atoms.

In a *laser diode,* population inversion of charge carriers in a semiconductor is achieved by a very high electric field across a *pn* junction in gallium arsenide.[6] Most laser diodes operate at red and near-infrared wavelengths (680–1 550 nm).

# ▪▪▪ 20-2 Monochromators

A **monochromator** disperses light into its component wavelengths and selects a narrow band of wavelengths to pass on to the sample or detector. The monochromator in Figure 20-2 consists of entrance and exit slits, mirrors, and a *grating* to disperse the light. *Prisms* were used instead of gratings in older instruments.

## Gratings[7]

A **grating** is a reflective or transmissive optical component with a series of closely ruled lines. When light is reflected from or transmitted through the grating, each line behaves as a separate source of radiation. Different wavelengths of light are reflected or transmitted at different angles from the grating (Color Plate 13). The bending of light rays by a grating is called **diffraction.** (The bending of light rays by a prism or lens, which is called *refraction,* is discussed in Section 20-4.)

In the grating monochromator in Figure 20-5, *polychromatic* radiation from the entrance slit is *collimated* (made into a beam of parallel rays) by a concave mirror. These rays fall on a reflection grating, whereupon different wavelengths are diffracted at different angles. The light strikes a second concave mirror, which focuses each wavelength at a different point on the focal plane. The orientation of the reflection grating directs only one narrow band of wavelengths to the exit slit of the monochromator. Rotation of the grating allows different wavelengths to pass through the exit slit.

*Grating:* optical element with closely spaced lines
*Diffraction:* bending of light by a grating
*Refraction:* bending of light by a lens or prism

*Polychromatic:* many wavelengths
*Monochromatic:* one wavelength

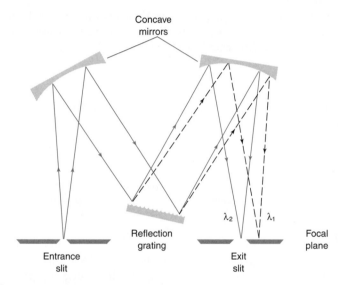

**Figure 20-5** Czerny-Turner grating monochromator.

The reflection grating in Figure 20-6 is ruled with a series of closely spaced, parallel grooves with a repeat distance $d$. The grating is coated with aluminum to make it reflective. A thin protective layer of silica ($SiO_2$) on top of the aluminum protects the metal surface from oxidizing, which would reduce its reflectivity. When light is reflected from the grating, each groove behaves as a source of radiation. When adjacent light rays are in phase, they

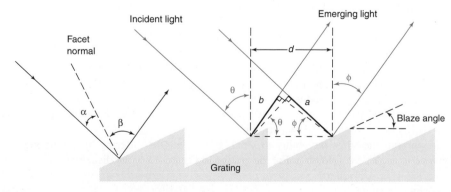

**Figure 20-6** Principle of a reflection grating.

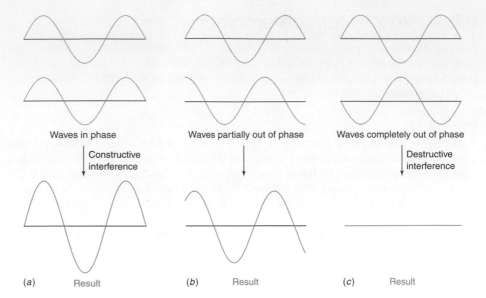

Waves in phase   Waves partially out of phase   Waves completely out of phase

Constructive interference   Destructive interference

(a) Result   (b) Result   (c) Result

reinforce one another. When they are not in phase, they partially or completely cancel one another (Figure 20-7).

Consider the incident and emerging light rays shown in Figure 20-6. Fully constructive interference occurs when the difference in length of the two paths is an integral multiple of the wavelength of light. The difference in pathlength is the distance $a - b$ in Figure 20-6. Constructive interference occurs if

$$n\lambda = a - b \qquad (20\text{-}1)$$

where the diffraction order $n = \pm1, \pm2, \pm3, \pm4, \ldots$ The interference maximum for which $n = \pm1$ is called *first-order diffraction*. When $n = \pm2$, we have *second-order diffraction*, and so on.

In Figure 20-6, the incident angle $\theta$ is defined to be positive. The diffraction angle $\phi$ goes in the opposite direction from $\theta$, so, by convention, $\phi$ is negative. It is possible for $\phi$ to be on the same side of the normal as $\theta$, in which case $\phi$ would be positive. In Figure 20-6, $a = d \sin \theta$ and $b = -d \sin \phi$ (because $\phi$ is negative and $\sin \phi$ is negative). Substituting into Equation 20-1 gives the condition for constructive interference:

*Grating equation:* $\qquad n\lambda = d(\sin \theta + \sin \phi) \qquad (20\text{-}2)$

where $d$ is the distance between adjacent grooves. For each incident angle, $\theta$, there is a series of reflection angles, $\phi$, at which a given wavelength will produce maximum constructive interference (Color Plate 19).

### Resolution, Dispersion, and Efficiency of a Grating

**Resolution** measures the ability to separate two closely spaced peaks. The greater the resolution, the smaller is the difference ($\Delta\lambda$) between two wavelengths that can be distinguished from each other. The precise definition (which is beyond the scope of this discussion) means that the valley between the two peaks is about three-fourths of the height of the peaks when they are just barely resolved. The resolution of a grating is given by

*Resolution of grating:* $\qquad \dfrac{\lambda}{\Delta\lambda} = nN \qquad (20\text{-}3)$

where $\lambda$ is wavelength, $n$ is the diffraction order in Equation 20-2, and $N$ is the number of grooves of the grating that are illuminated. The more grooves in a grating, the better the resolution between closely spaced wavelengths. Equation 20-3 tells us that, if we desire a first-order resolution of $10^4$, there must be $10^4$ grooves in the grating. If the grating has a ruled length of 10 cm, we require $10^3$ grooves/cm.

**Dispersion** measures the ability to separate wavelengths differing by $\Delta\lambda$ through the difference in angle, $\Delta\phi$ (radians). For the grating in Figure 20-6, the dispersion is

*Dispersion of grating:* $\qquad \dfrac{\Delta\phi}{\Delta\lambda} = \dfrac{n}{d \cos \phi} \qquad (20\text{-}4)$

*Resolution:* ability to distinguish two closely spaced peaks

*Dispersion:* ability to produce angular separation of adjacent wavelengths

where $n$ is the diffraction order. Dispersion and resolution both increase with decreasing groove spacing. Equation 20-4 tells us that a grating with $10^3$ grooves/cm provides a resolution of 0.102 radians (5.8°) per micrometer of wavelength if $n = 1$ and $\phi = 10°$. Wavelengths differing by 1 μm would be separated by an angle of 5.8°.

Decreasing the monochromator exit slit width decreases the bandwidth of radiation and decreases the energy reaching the detector. Thus, *resolution of closely spaced absorption bands is achieved at the expense of decreased signal-to-noise ratio*. For quantitative analysis, a monochromator bandwidth that is ≲1/5 of the width of the absorption band (measured at half the peak height) is reasonable.

Trade-off between resolution and signal: The narrower the exit slit, the greater the resolution and the noisier the signal.

The relative *efficiency* of a grating (which is typically 45–80%) is defined as

$$\text{Relative efficiency} = \frac{E_\lambda^n(\text{grating})}{E_\lambda(\text{mirror})} \qquad (20\text{-}5)$$

where $E_\lambda^n(\text{grating})$ is the irradiance at a particular wavelength diffracted in the order of interest, $n$, and $E_\lambda(\text{mirror})$ is the irradiance at the same wavelength that would be reflected by a mirror with the same coating as the grating. Efficiency is partially controlled by the *blaze angle* at which the grooves are cut in Figure 20-6. To direct a certain wavelength into the diffraction order of interest, the blaze angle is chosen such that $\alpha = \beta$ in Figure 20-6, because this condition gives maximum reflection. Each grating is optimized for a limited range of wavelengths, so a spectrophotometer may require several different gratings to scan through its entire spectral range.

For a *flat* reflective surface, the angle of incidence ($\alpha$ in Figure 20-6) is equal to the angle of reflection ($\beta$).

## Choosing the Monochromator Bandwidth

The wider the exit slit in Figure 20-5, the wider the band of wavelengths selected by the monochromator. We usually measure slit width in terms of the bandwidth of radiation selected by the slit. Instead of saying that a slit is 0.3 mm wide, we might say that the *bandwidth* getting through the slit is 1.0 nm.

A wide slit increases the energy reaching the detector and gives a high signal-to-noise ratio, leading to good precision in measuring absorbance. However, Figure 20-8 shows that, if the bandwidth is large relative to the width of the peak being measured, peak shape is distorted. We choose a bandwidth as wide as the spectrum permits to allow the most possible light to reach the detector. A monochromator bandwidth that is 1/5 as wide as the absorption peak generally gives acceptably small distortion of the peak shape.[8]

Monochromator bandwidth should be as large as possible, but small compared with the width of the peak being measured.

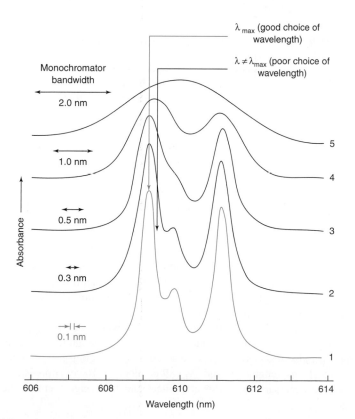

**Figure 20-8** Increasing monochromator bandwidth broadens the bands and decreases the apparent absorbance of $Pr^{3+}$ in a crystal of yttrium aluminum garnet (a laser material). *[Courtesy M. D. Seltzer, Michelson Laboratory, China Lake, CA.]*

High-quality spectrometers could have two monochromators in series (called a *double monochromator*) to reduce stray light. Unwanted radiation that passes through the first monochromator is rejected by the second monochromator.

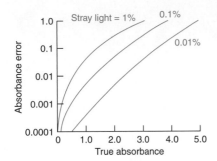

**Figure 20-9** Absorbance error introduced by different levels of stray light. Stray light is expressed as a percentage of the irradiance incident on the sample. [*M. R. Sharp, "Stray Light in UV-VIS Spectrophotometers," Anal. Chem.* **1984**, *53*, 339A.]

## Stray Light

In every instrument, some **stray light** (wavelengths outside the bandwidth expected from the monochromator) reaches the detector. Stray light coming through the monochromator from the light source arises from diffraction into unwanted orders and angles and unintended scattering from optical components and walls. Stray light can also come from outside the instrument if the sample compartment is not perfectly sealed. Entry holes for tubing or electrical wires required at the sample for some experiments should be sealed to reduce stray light. Error from stray light is most serious when the sample absorbance is high (Figure 20-9) because the stray light constitutes a large fraction of the light reaching the detector.

### Example Stray Light

If the true absorbance of a sample is 2.00 and there is 1.0% stray light, find the apparent absorbance.

**Solution** A true absorbance of 2.00 means that the true transmittance is $T = 10^{-A} = 10^{-2.00} = 0.010 = 1.0\%$. Transmittance is the irradiance passing through the sample, $P$, divided by the irradiance passing through the reference, $P_0$: $T = P/P_0$. If stray light with irradiance $S$ passes through both the sample and the reference, the apparent transmittance is

$$\text{Apparent transmittance} = \frac{P + S}{P_0 + S} \tag{20-6}$$

If $P/P_0 = 0.010$ and there is 1.0% stray light, then $S = 0.010$ and the apparent transmittance is

$$\text{Apparent transmittance} = \frac{P + S}{P_0 + S} = \frac{0.010 + 0.010}{1 + 0.010} = 0.019_8$$

The apparent absorbance is $-\log T = -\log(0.019_8) = 1.70$, instead of 2.00.

Research-quality instruments provide $<0.1\%$ stray light, and sometimes much less.

Table 20-1 gives the absorbance of a solution that you can prepare to test the accuracy of absorbance measurements on your spectrophotometer. Absorbance accuracy is affected by all components of the spectrophotometer, as well as stray light. Standards are also available to measure wavelength accuracy.[9]

**Table 20-1** Calibration standard for ultraviolet absorbance

| Wavelength (nm) | Absorbance of $K_2Cr_2O_7$ (60.06 mg/L) in 5.0 mM $H_2SO_4$ in 1-cm cell |
|---|---|
| 235 | $0.748 \pm 0.010$ |
| 257 | $0.865 \pm 0.010$ |
| 313 | $0.292 \pm 0.010$ |
| 350 | $0.640 \pm 0.010$ |

SOURCE: *S. Ebel, "Validation of Analysis Methods," Fresenius J. Anal. Chem.* **1992**, *342, 769.*

## Filters

Filters permit certain bands of wavelength to pass through.

It is frequently necessary to *filter* (remove) wide bands of radiation from a signal. For example, the grating monochromator in Figure 20-5 directs first-order diffraction of a small wavelength band to the exit slit. (By "first order," we mean diffraction for which $n = 1$ in Equation 20-2.) Let $\lambda_1$ be the wavelength whose first-order diffraction reaches the exit slit. Inspection of Equation 20-2 shows that, if $n = 2$, the wavelength $\frac{1}{2}\lambda_1$ also reaches the same exit slit because $\frac{1}{2}\lambda_1$ gives constructive interference at the same angle as $\lambda_1$ does. For $n = 3$, the wavelength $\frac{1}{3}\lambda_1$ also reaches the slit. One solution for selecting just $\lambda_1$ is to place a filter in the beam, so that wavelengths $\frac{1}{2}\lambda_1$ and $\frac{1}{3}\lambda_1$ will be blocked. To cover a wide range of wavelengths, it may be necessary to use several filters and to change them as the wavelength region changes.

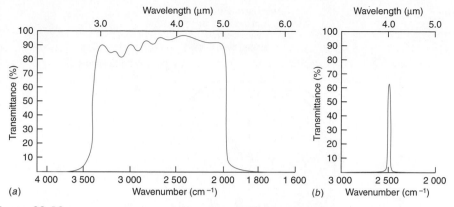

**Figure 20-10** Transmission spectra of interference filters. (*a*) Wide band-pass filter has ~90% transmission in the 3- to 5-μm wavelength range but <0.01% transmittance outside this range. (*b*) Narrow band-pass filter has a transmission width of 0.1 μm centered at 4 μm. *[Courtesy Barr Associates, Westford, MA.]*

The simplest filter is colored glass, which absorbs a broad portion of the spectrum and transmits other portions. Table 18-1 showed the relation between absorbed and transmitted colors. For finer control, *interference filters* are constructed to pass radiation in the region of interest and reflect other wavelengths (Figure 20-10). These devices derive their performance from constructive or destructive interference of light waves within the filter.

## ■ ■ ■ ■ 20-3 Detectors

A detector produces an electric signal when it is struck by photons. For example, a **phototube** emits electrons from a photosensitive, negatively charged surface (the cathode) when struck by visible light or ultraviolet radiation. The electrons flow through a vacuum to a positively charged collector whose current is proportional to the radiation intensity.

Figure 20-11 shows that detector response depends on the wavelength of the incident photons. For example, for a given radiant power (W/m²) of 420-nm light, the S-20 photomultiplier produces a current about four times greater than the current produced for the same radiant power of 300-nm radiation. The response below 280 nm and above 800 nm is essentially 0. In a single-beam spectrophotometer, the 100% transmittance control must be readjusted each time the wavelength is changed. This calibration adjusts the spectrophotometer to the maximum detector output that can be obtained at each wavelength. Subsequent readings are scaled to the 100% reading.

> Detector response is a function of wavelength of incident light.

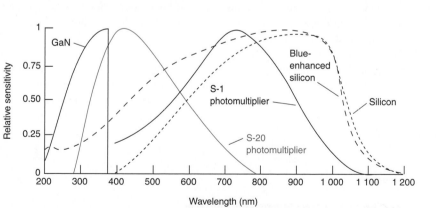

**Figure 20-11** Detector response. The greater the sensitivity, the greater the current or voltage produced by the detector for a given incident irradiance (W/m²) of photons. Each curve is normalized to a maximum value of 1. *[Courtesy Barr Associates, Westford, MA. GaN data from APA Optics, Blaine, MN.]*

### Photomultiplier Tube

A **photomultiplier tube** (Figure 20-12) is a very sensitive device in which electrons emitted from the photosensitive surface strike a second surface, called a *dynode,* which is positive with respect to the photosensitive emitter. Electrons are accelerated and strike the dynode

**Figure 20-12** Diagram of a photomultiplier tube with nine dynodes. Amplification of the signal occurs at each dynode, which is approximately 90 volts more positive than the previous dynode.

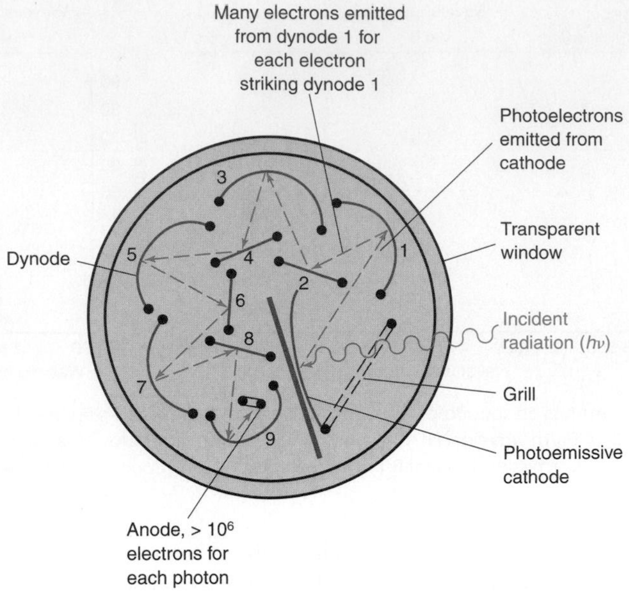

Many electrons emitted from dynode 1 for each electron striking dynode 1

Photoelectrons emitted from cathode

Transparent window

Dynode

Incident radiation ($h\nu$)

Grill

Photoemissive cathode

Anode, $> 10^6$ electrons for each photon

with more than their original kinetic energy. Each energetic electron knocks more than one electron from the dynode. These new electrons are accelerated toward a second dynode, which is more positive than the first dynode. Upon striking the second dynode, even more electrons are knocked off and accelerated toward a third dynode. This process is repeated several times, so more than $10^6$ electrons are finally collected for each photon striking the first surface. Extremely low light intensities are translated into measurable electric signals. As sensitive as a photomultiplier is, your eye is even more sensitive (Box 20-2).

## Photodiode Array

Conventional spectrophotometers scan through a spectrum one wavelength at a time. Newer instruments record the entire spectrum at once in a fraction of a second. One application of rapid scanning is chromatography, in which the full spectrum of a compound is recorded in seconds as it emerges from the chromatography column.

At the heart of rapid spectroscopy is the **photodiode array** shown in Figure 20-13 (or the charge coupled device described later). Rows of *p*-type silicon on a substrate (the underlying body) of *n*-type silicon create a series of *pn* junction diodes. A reverse bias is applied to each diode, drawing electrons and holes away from the junction. There is a depletion region at each junction, in which there are few electrons and holes. The junction acts as a capacitor, with charge stored on either side of the depletion region. At the beginning of the measurement cycle, each diode is fully charged.

When radiation strikes the semiconductor, free electrons and holes are created and migrate to regions of opposite charge, partially discharging the capacitor. The more radiation that strikes each diode, the less charge remains at the end of the measurement. The longer the array is irradiated between readings, the more each capacitor is discharged. The state of each capacitor is determined at the end of the cycle by measuring the current needed to recharge the capacitor.

For a refresher on semiconductors, see Section 15-8.

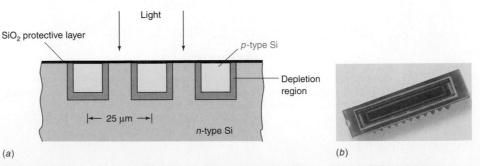

Light

SiO$_2$ protective layer

*p*-type Si

Depletion region

|← 25 μm →|

*n*-type Si

(a)

(b)

**Figure 20-13** (*a*) Schematic cross-sectional view of photodiode array. (*b*) Photograph of array with 1 024 elements, each 25 μm wide and 2.5 mm high. The central black rectangle is the photosensitive area. The entire chip is 5 cm in length. *[Courtesy Oriel Corporation, Stratford, CT.]*

The retina at the back of your eye contains photosensitive cells, called *rods* and *cones,* that are sensitive to levels of light varying over seven orders of magnitude. Light impinging on these cells is translated into nerve impulses that are transmitted by the optic nerve to the brain. Rod cells detect the dimmest light but cannot distinguish colors. Cone cells operate in bright light and give us color vision.

A stack of about 1 000 *disks* in each rod cell contains the light-sensing protein *rhodopsin,*[10] in which the chromophore 11-*cis*-retinal (from vitamin A) is attached to the protein *opsin.* When light is absorbed by rhodopsin, a series of rapid transformations releases all-*trans*-retinal. At this stage, the pigment is *bleached* (loses all color) and cannot respond to more light until retinal isomerizes back to the 11-*cis* form and recombines with the protein.

In the dark, there is a continuous flow of $10^9$ Na$^+$ ions per second out of the rod cell's inner segment, through the adjoining medium, and into the cell's outer segment. An energy-dependent process using adenosine triphosphate (ATP) and oxygen pumps Na$^+$ out of the cell. Another process involving a molecule called cyclic GMP keeps the gates of the outer segment open for ions to flow back into the cell. When light is absorbed and rhodopsin is bleached, a series of reactions leads to destruction of cyclic GMP and shutdown of the channels through which Na$^+$ flows into the cell. A single photon reduces the ion current by 3%—corresponding to a decreased current of $3 \times 10^7$ ions per second. This *amplification* is greater than that of a photomultiplier tube, which is one of the most sensitive man-made photodetectors. The ion current returns to its dark value as the protein and retinal recombine and cyclic GMP is restored to its initial concentration.

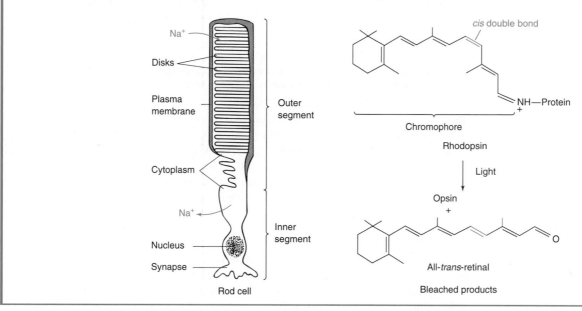

In a *dispersive spectrometer* (Figure 20-1), only one narrow band of wavelengths reaches the detector at any time. In a *photodiode array spectrophotometer* (Figure 20-14), all wavelengths are recorded simultaneously, allowing more rapid acquisition of the spectrum or higher signal-to-noise ratio, or some combination of both. In the photodiode array spectrophotometer in Figure 20-14, *white light* (with all wavelengths) passes through the sample. The light then enters a **polychromator,** which disperses the light into its component wavelengths and directs the light at the diode array. Each diode receives a *different wavelength,* and all wavelengths are measured simultaneously. Resolution depends on how closely spaced the diodes are and how much dispersion is produced by the polychromator. The spectrum in Figure 21-26 (in the next chapter) was produced with a photodiode array detector.

Photodiode arrays allow faster spectral acquisition (~1 s) than dispersive instruments (which require several minutes). Photodiode array instruments have almost no moving parts, so they are more rugged than dispersive instruments that must rotate the grating and change

> A photodiode array spectrophotometer measures all wavelengths at once, giving faster acquisition and higher signal-to-noise ratio.

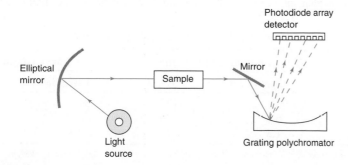

*Figure 20-14* Schematic design of photodiode array spectrophotometer.

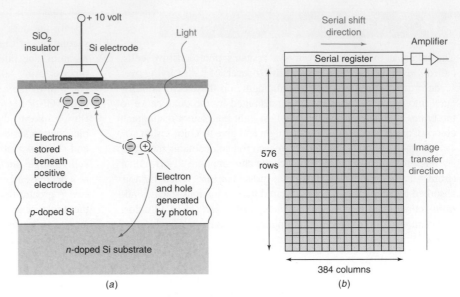

*Figure 20-15* Schematic representation of a charge coupled device. (*a*) Cross-sectional view, indicating charge generation and storage in each pixel. (*b*) Top view, showing two-dimensional nature of an array. An actual array is about the size of a postage stamp.

filters to scan through the spectrum. The resolution of ~0.1 nm attainable with a dispersive instrument and the wavelength accuracy are better than those of a photodiode array (~0.5–1.5 nm resolution). Stray light is less in a dispersive instrument than in a photodiode instrument, giving the dispersive instrument a greater dynamic range for measuring high absorbance. Stray light in a photodiode array instrument is not substantially increased when the sample compartment is open. In a dispersive instrument, the compartment must be tightly closed.

## Charge Coupled Device[11]

A **charge coupled device** is an extremely sensitive detector that stores photo-generated charge in a two-dimensional array. The device in Figure 20-15a is constructed of *p*-doped Si on an *n*-doped substrate. The structure is capped with an insulating layer of $SiO_2$, on top of which is placed a pattern of conducting Si electrodes. When light is absorbed in the *p*-doped region, an electron is introduced into the conduction band and a hole is left in the valence band. The electron is attracted to the region beneath the positive electrode, where it is stored. The hole migrates to the *n*-doped substrate, where it combines with an electron. Each electrode can store $10^5$ electrons before electrons spill out into adjacent elements.

The charge coupled device is a two-dimensional array, as shown in Figure 20-15b. After the desired observation time, electrons stored in each *pixel* (picture element) of the top row are moved into the serial register at the top and then moved, one pixel at a time, to the top right position, where the charge is read out. Then the next row is moved up and read out, and the sequence is repeated until the entire array has been read. The transfer of stored charges is carried out by an array of electrodes considerably more complex than we have indicated in Figure 20-15a. Charge transfer from one pixel to the next is extremely efficient, with a loss of approximately five of every million electrons.

The minimum detectable signal for visible light in Table 20-2 is 17 photons/s. The sensitivity of the charge coupled device is derived from its high *quantum efficiency* (electrons generated per incident photon), low background electrical noise (thermally generated free

Digital cameras use charge coupled devices to record the image.

Electrons from adjacent pixels can be combined to create a single, larger picture element. This process, called *binning*, increases the sensitivity of the charge coupled device at the expense of resolution.

**Table 20-2** Minimum detectable signal (photons/s/detector element) of ultraviolet/visible detectors

| Signal acquisition time (s) | Photodiode array | | Photomultiplier tube | | Charge coupled device | |
|---|---|---|---|---|---|---|
| | **Ultraviolet** | **Visible** | **Ultraviolet** | **Visible** | **Ultraviolet** | **Visible** |
| 1 | 6 000 | 3 300 | 30 | 122 | 31 | 17 |
| 10 | 671 | 363 | 6.3 | 26 | 3.1 | 1.7 |
| 100 | 112 | 62 | 1.8 | 7.3 | 0.3 | 0.2 |

SOURCE: *R. B. Bilhorn, J. V. Sweedler, P. M. Epperson, and M. B. Denton, "Charge Transfer Device Detectors for Analytical Optical Spectroscopy," Appl. Spectros.* **1987**, *41, 1114.*

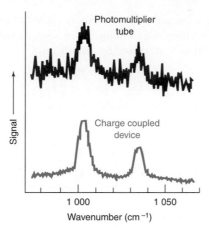

Signal

Photomultiplier tube

Charge coupled device

1 000    1 050

Wavenumber (cm$^{-1}$)

*Figure 20-16* Comparison of spectra recorded in 5 min by a photomultiplier tube and a charge coupled device. *[From P. M. Epperson, J. V. Sweedler, R. B. Bilhorn, G. R. Sims, and M. B. Denton, "Applications of Charge Transfer Devices in Spectroscopy," Anal. Chem. **1988**, 60, 327A.]*

electrons), and low noise associated with readout. Figure 20-16 compares spectra recorded under the same conditions by a photomultiplier tube and by a charge coupled device. A photomultiplier is very sensitive, but a charge coupled device is even better.

### Infrared Detectors

Detectors for visible and ultraviolet radiation rely on incoming photons to eject electrons from a photosensitive surface or to promote electrons from the valence band of silicon to the conduction band. Infrared photons do not have sufficient energy to generate a signal in either kind of detector. Therefore, other kinds of devices are used for infrared detection.

A **thermocouple** is a junction between two different electrical conductors. Electrons have lower free energy in one conductor than in the other, so they flow from one to the other until the resulting voltage difference prevents further flow. The junction potential is temperature dependent because electrons flow back to the high-energy conductor at higher temperature. If a thermocouple is blackened to absorb radiation, its temperature (and hence voltage) becomes sensitive to radiation. A typical sensitivity is 6 V per watt of radiation absorbed.

A **ferroelectric material,** such as deuterated triglycine sulfate, has a permanent electric polarization because of alignment of the molecules in the crystal. One face of the crystal is positively charged and the opposite face is negative. The polarization is temperature dependent, and its variation with temperature is called the *pyroelectric effect.* When the crystal absorbs infrared radiation, its temperature and polarization change. The voltage change is the signal in a pyroelectric detector. Deuterated triglycine sulfate is a common detector in Fourier transform spectrometers, described later in this chapter.

A **photoconductive detector** is a semiconductor whose conductivity increases when infrared radiation excites electrons from the valence band to the conduction band. **Photovoltaic detectors** contain *pn* junctions, across which an electric field exists. Absorption of infrared radiation creates electrons and holes, which are attracted to opposite sides of the junction and which change the voltage across the junction. Mercury cadmium telluride ($Hg_{1-x}Cd_xTe$, $0 < x < 1$) is a detector material whose sensitivity to different wavelengths is affected by the stoichiometry coefficient, $x$. Photoconductive and photovoltaic devices can be cooled to 77 K (liquid nitrogen temperature) to reduce thermal electric noise by more than an order of magnitude.

### ■ ■ ■ 20-4   Optical Sensors

An *optode* is a chemical sensor based on an *optical fiber*. To understand how optodes work, we first need to know a little about refraction of light.

### Refraction

The speed of light in a medium of **refractive index** $n$ is $c/n$, where $c$ is the speed of light in vacuum. That is, for vacuum, $n = 1$. Refractive index is commonly measured at 20°C at the wavelength of the sodium D line ($\lambda = 589.3$ nm). The frequency of light, $v$, inside a medium does not change from the frequency in vacuum. Because the speed of light, $c/n$, inside a medium decreases from that of vacuum, the wavelength decreases so that $\lambda v = c/n$.

In a *ferroelectric material,* the dipole moments of molecules remain aligned in the absence of an external field. This alignment gives the material a permanent electric polarization.

Infrared radiation promotes electrons from the valence band of silicon to the conduction band. Semiconductors that are used as infrared detectors have smaller band gaps than silicon.

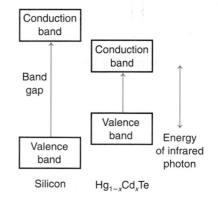

Refractive index at sodium D line:

| | |
|---|---|
| vacuum | 1 |
| air (0°, 1 bar) | 1.000 29 |
| water | 1.33 |
| fused silica | 1.46 |
| benzene | 1.50 |
| bromine | 1.66 |
| iodine | 3.34 |

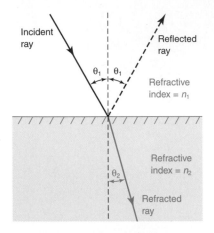

Figure 20-17 Illustration of Snell's law: $n_1 \sin \theta_1 = n_2 \sin \theta_2$. When light passes from air into any medium, the greater the refractive index of the medium, the smaller is $\theta_2$.

Light traveling from a region of high refractive index ($n_1$) to a region of low refractive index ($n_2$) is totally reflected if the angle of incidence exceeds the critical angle given by $\sin \theta_{critical} = n_2/n_1$.

When light is reflected, the angle of reflection is equal to the angle of incidence (Figure 20-17). When light passes from one medium into another, its path is bent (Color Plate 20). This bending, called **refraction,** is described by **Snell's law:**

Snell's law: $$n_1 \sin \theta_1 = n_2 \sin \theta_2 \qquad (20\text{-}7)$$

where $n_1$ and $n_2$ are the refractive indexes of the two media and $\theta_1$ and $\theta_2$ are angles defined in Figure 20-17.

### Example Refraction of Light by Water

Visible light travels from air (medium 1) into water (medium 2) at a 45° angle ($\theta_1$ in Figure 20-17). At what angle, $\theta_2$, does the light ray pass through the water?

**Solution** The refractive index for air is close to 1, and for water, 1.33. From Snell's law,

$$(1.00)(\sin 45°) = (1.33)(\sin \theta_2) \Rightarrow \theta_2 = 32°$$

What is $\theta_2$ if the incident ray is perpendicular to the surface (that is, $\theta_1 = 0°$)?

$$(1.00)(\sin 0°) = (1.33)(\sin \theta_2) \Rightarrow \theta_2 = 0°$$

A perpendicular ray is not refracted.

### Optical Fibers

**Optical fibers** carry light by *total internal reflection.* Optical fibers are replacing electric wires for communication because fibers are immune to electrical noise, transmit data at a higher rate, and can handle more signals. Optical fibers can bring an optical signal from inside a chemical reactor out to a spectrophotometer for process monitoring.

A flexible optical fiber has a high-refractive-index, transparent core enclosed in a lower-refractive-index, transparent cladding (Figure 20-18a). The cladding is enclosed in a protective plastic jacket. The core and the coating can be made from glass or polymer.

Consider the light ray striking the wall of the core in Figure 20-18b at the angle of incidence $\theta_i$. Part of the ray is reflected inside the core, and part might be transmitted into the cladding at the angle of refraction, $\theta_r$ (Color Plate 21). If the index of refraction of the core is $n_1$ and the index of the cladding is $n_2$, Snell's law (Equation 20-7) tells us that

$$n_1 \sin \theta_i = n_2 \sin \theta_r \Rightarrow \sin \theta_r = \frac{n_1}{n_2} \sin \theta_i \qquad (20\text{-}8)$$

If $(n_1/n_2) \sin \theta_i$ is greater than 1, no light is transmitted into the cladding because $\sin \theta_r$ cannot be greater than 1. In such a case, $\theta_i$ exceeds the *critical angle* for total internal reflection. *If $n_1/n_2 > 1$, there is a range of angles $\theta_i$ in which essentially all light is reflected at the walls of the core, and a negligible amount enters the cladding.* All rays entering one end of the fiber within a certain cone of acceptance emerge from the other end of the fiber with little loss.

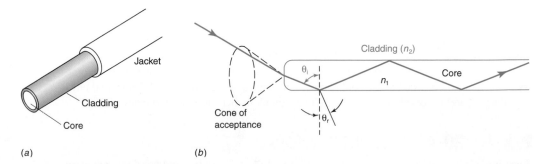

Figure 20-18 (a) Optical fiber construction and (b) principle of operation. Any light ray entering within the cone of acceptance will be totally reflected at the wall of the fiber.

(a)    (b)

### Optodes

We can create optical sensors for specific analytes by placing a chemically sensitive layer at the end of the fiber. An optical fiber sensor is called an **optode** (or *optrode*), derived from the words "optical" and "electrode." Optodes have been designed to respond to diverse analytes such as sulfites in food, nitric oxide in cells, and explosives in groundwater.[12]

The end of the $O_2$ optode in Color Plate 22 is coated with a Ru(II) complex in a layer of polymer. Luminescence from Ru(II) is *quenched* (decreased) by $O_2$, as discussed in Section 19-6. The optode is inserted into a sample as small as 100 fL ($100 \times 10^{-15}$ L) on the

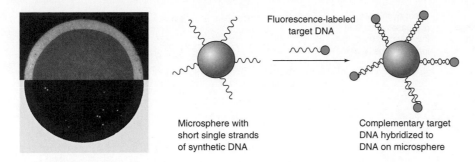

**Figure 20-19** Fiber-optic sensor for detection of specific DNA sequences. Upper half shows circular wells in etched tip of bundle. Fluorescence image in lower half identifies wells to which fluorescent target DNA has bound. *[From J. R. Epstein, M. Lee, and D. R. Walt, "High-Density Fiber-Optic Genosensor Microsphere Array Capable of Zeptomole Detection Limits," Anal. Chem.* **2002,** *74, 1836. Photo courtesy D. R. Walt, Tufts University.]*

Fluorescence-labeled target DNA

Microsphere with short single strands of synthetic DNA

Complementary target DNA hybridized to DNA on microsphere

stage of a microscope. The degree of quenching tells us the concentration of $O_2$ in the sample. The detection limit is 10 amol of $O_2$. By incorporating the enzyme glucose oxidase (Equation 17-10), the optode becomes a glucose sensor with a detection limit of 1 fmol of glucose.[13] An optode for measuring biochemical oxygen demand (Box 16-1) uses yeast cells immobilized in a membrane to consume $O_2$ and Ru(II) luminescence for measuring $O_2$.[14]

The upper part of the photo in Figure 20-19 shows the end of a 0.5-mm-diameter bundle of ~6 000 optical fibers etched briefly with HF. The glass core dissolves more rapidly than the cladding, leaving a small well at the end of each fiber. A droplet of liquid containing 3-μm microspheres functionalized with synthetic single strands of DNA is evaporated on the tip of the fiber, leaving microspheres in many of the wells. Excess microspheres not contained in wells are washed away. The synthetic DNA is complementary to desired analytical target DNA, such as part of the gene associated with the disease cystic fibrosis.

The fiber with its microspheres is incubated with known fluorescence-labeled target strands of DNA. Figure 20-19 shows that 10 of the 6 000 wells become fluorescent, indicating that they contain microspheres that bind the known target. The image of the fluorescent wells is made with a camera and optics at the far end of the fiber bundle. The target is removed by washing under conditions that dissociate double strands of DNA. The fiber bundle is then incubated with unknown fluorescence-labeled DNA. A new image is then recorded. If the unknown contained the target sequence, the same wells will light up. Luminescence intensity is a measure of the concentration of target DNA. Different wells can contain different synthetic DNA strands to detect different targets simultaneously. The detection limit is 600 *molecules* of DNA. A variant of this approach can study the biochemistry of individual live cells in each well.[15]

### Attenuated Total Reflectance

Figure 20-18 shows total internal reflection of a light ray bouncing back and forth as it travels along an optical fiber. The same behavior is observed in a flat layer of material whose refractive index, $n_1$, is greater than the refractive index of the surroundings, $n_2$. A planar layer in which light is totally reflected is called a **waveguide.** A chemical sensor can be fabricated by placing a chemically sensitive layer on a waveguide.[16]

When the light wave in Figure 20-18 strikes the wall, the ray is totally reflected if $\theta_i$ exceeds the critical angle given by $\sin \theta_{critical} = n_2/n_1$. Even though light is totally reflected, the electric field of the light penetrates the cladding to some extent. Figure 20-20 shows that, when the oscillating electric field encounters the reflective interface, the field dies out exponentially inside the cladding. The part of the light that penetrates the wall of an optical fiber or waveguide is called an *evanescent wave.*

**Question** How many molecules are in 10 amol?

*Evanescent* means "vanishing" or "fleeting." Light "escapes" from the waveguide, but it vanishes over a short distance.

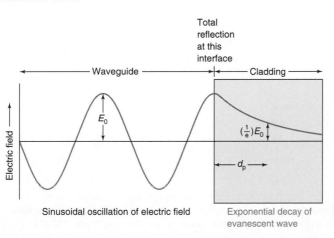

Total reflection at this interface

Waveguide — Cladding

$E_0$

$(\frac{1}{e})E_0$

$d_p$

Electric field

Sinusoidal oscillation of electric field

Exponential decay of evanescent wave

**Figure 20-20** Behavior of an electromagnetic wave when it strikes a surface from which it is totally reflected. The field penetrates the reflective barrier and dies out exponentially.

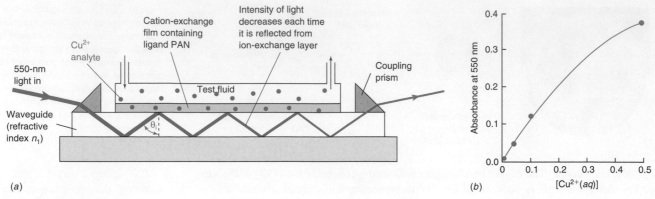

(a)

(b)

**Figure 20-21** Attenuated total reflectance sensor for Cu²⁺. Light traveling through the waveguide is attenuated (decreased) in the cation-exchange film in the presence of Cu²⁺. *[Data from T. Shtoyko, I. Zudans, C. J. Seliskar, W. R. Heineman, and J. N. Richardson, "An Attenuated Total Reflectance Sensor for Copper," J. Chem. Ed.* **2004**, *81, 1617.]*

1-(2-Pyridylazo)-2-naphthol
(PAN)

The electric field, $E$, of the evanescent wave in Figure 20-20 is given by

$$E = E_0 e^{-x/d_p} \qquad \left( d_p = \frac{\lambda/n_1}{2\pi\sqrt{\sin^2\theta_i - (n_2/n_1)^2}} \right) \qquad (20\text{-}9)$$

where $E_0$ is the magnitude of the field at the reflective interface, $x$ is the distance into the cladding, and $\lambda$ is the wavelength of light in a vacuum. The *penetration depth*, $d_p$, is the distance at which the evanescent field dies down to $1/e$ of its value at the interface. Consider a waveguide with $n_1 = 1.70$ and $n_2 = 1.45$, giving a critical angle of 58.5°. If light with a wavelength of 590 nm has a 70° angle of incidence, then the penetration depth is 140 nm. This depth is great enough to allow the light to interact with many layers of large molecules, such as proteins, whose dimensions are on the order of 10 nm.

Figure 20-21a shows a sensor based on **attenuated total reflectance** of light passing through a coated waveguide. "Attenuated" means "decreased." When light passes through the waveguide, it is totally reflected at both surfaces. On the upper surface is a thin layer of a low-refractive-index cation-exchange membrane impregnated with the metal-binding ligand PAN. When test fluid containing Cu²⁺ is circulated over the membrane, Cu(PAN)⁺ is formed in the membrane. Figure 20-21b shows that the absorbance of Cu(PAN)⁺ increases until the membrane is saturated with Cu²⁺ above 0.5 mM.

### Surface Plasmon Resonance[17]

Conduction electrons in a metal are nearly free to move within the metal in response to an applied electric field. A *surface plasma wave,* also called a *surface plasmon,* is an electromagnetic wave that propagates along the boundary between a metal and a *dielectric* (an electrical insulator). The electromagnetic field decreases exponentially into both layers but is concentrated in the dielectric layer.

Figure 20-22a shows essentials of one common **surface plasmon resonance** measurement. Monochromatic light whose electric field oscillates in the plane of the page is directed into a prism whose bottom face is coated with a thin layer (~50 nm) of gold. The bottom surface of the gold is coated with a chemical layer (~2–20 nm) that selectively binds an analyte of inter-

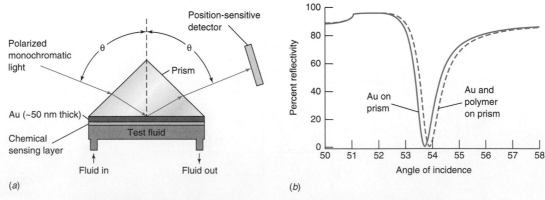

(a)

(b)

**Figure 20-22** (a) Essentials of a surface plasmon resonance measurement. (b) Reflectivity versus angle, θ. *[Adapted from J. M. Brockman, B. P. Nelson, and R. M. Corn, "Surface Plasmon Resonance Imaging Measurements of Ultrathin Organic Films," Ann. Rev. Phys. Chem.* **2000**, *51, 41.]*

est. At low angles of incidence, $\theta$, much but not all of the light is reflected by the gold. When $\theta$ increases to the critical angle for total internal reflection, the reflectivity is ideally 100%. As $\theta$ increases further, a surface plasmon (oscillating electron cloud) is set up, absorbing energy from the incident light. Because some of the energy is absorbed in the gold layer, the reflectivity decreases from 100%. There is a small range of angles at which the plasmon is in resonance with the incident light, creating the sharp dip in the curve in Figure 20-22b. As $\theta$ increases beyond the resonance condition, less energy is absorbed and the reflectivity increases.

The angle at which reflectivity is minimum depends on the refractive indexes of all the layers in Figure 20-22a. When analyte binds to the chemically sensitive layer, the refractive index of that layer changes slightly and the angle of minimum reflectivity changes slightly. Commercial instruments can measure changes in the surface plasmon resonance angle with a precision of $\sim 10^{-4}$ to $10^{-5}$ degrees.

For biosensors, the chemically sensitive layer might contain an antibody or antigen, DNA or RNA, a protein, or a carbohydrate that has a selective interaction with some analyte. An example of a synthetic sensing layer is a *molecularly imprinted polymer* (Box 26-2) with a polyacrylamide backbone (Figure 26-12) and phenylboronic acid substituents. When the polymer is synthesized in the presence of the intended analyte nicotinamide adenine dinucleotide ($NAD^+$), pockets are formed in which amide and boronic acid groups are properly disposed for hydrogen bonding to $NAD^+$. The $NAD^+$ can be removed from the polymer by treatment with $NH_3$, leaving behind empty pockets that selectively bind $NAD^+$, as shown below.

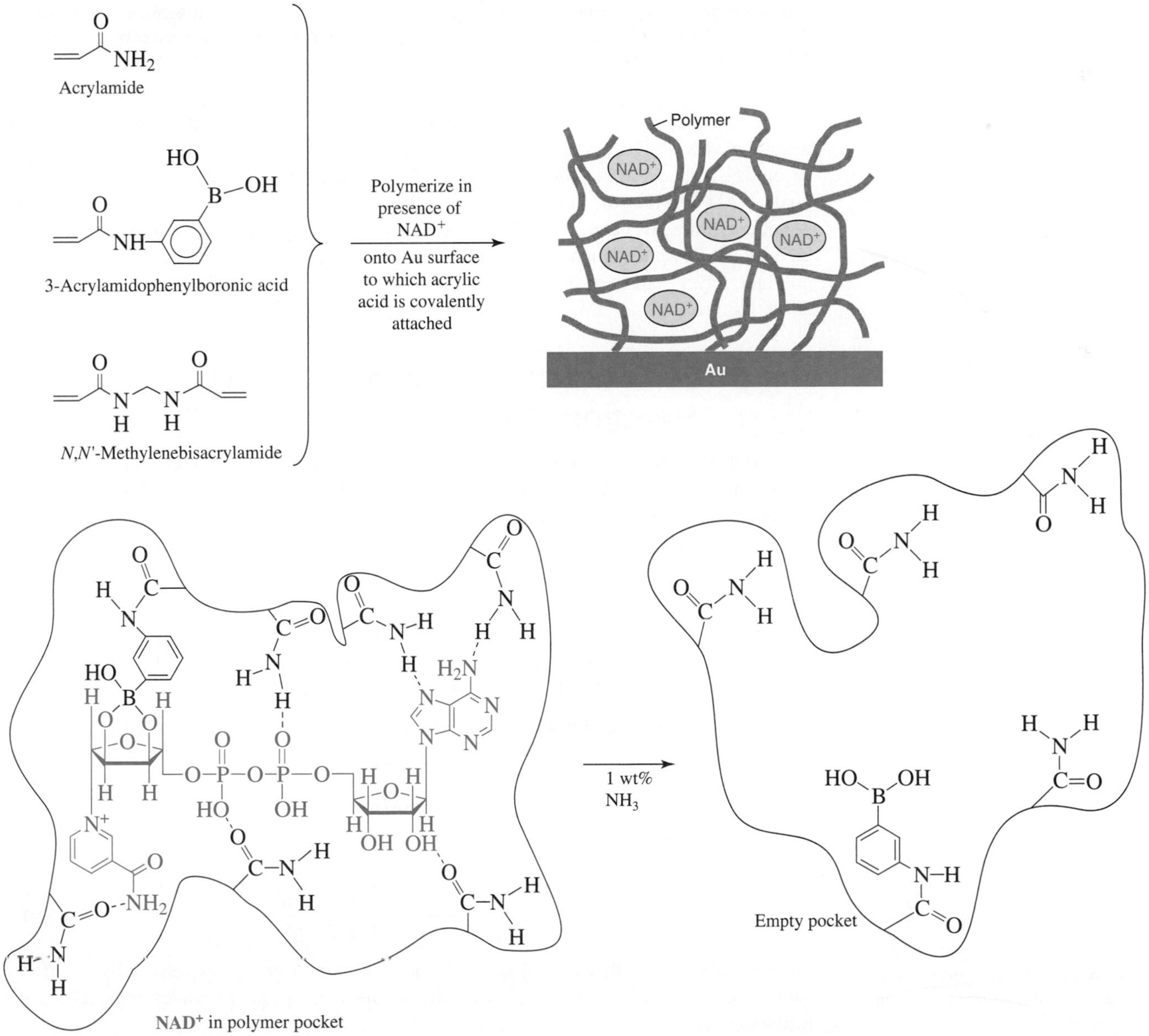

Acrylamide

3-Acrylamidophenylboronic acid

N,N'-Methylenebisacrylamide

Polymerize in presence of $NAD^+$

onto Au surface to which acrylic acid is covalently attached

Polymer

$NAD^+$

Au

$NAD^+$ in polymer pocket

1 wt% $NH_3$

Empty pocket

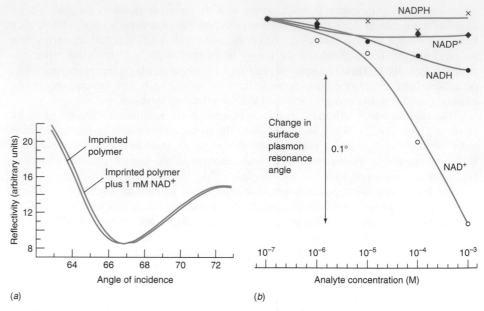

(a)                                                    (b)

**Figure 20-23** (*a*) Surface plasmon resonance spectrum of sensor coated with molecularly imprinted polymer that selectively binds NAD$^+$. (*b*) Response of sensor to four similar molecules shows largest response to NAD$^+$, which was the template for polymerization. *[From O. A. Raitman, V. I. Chegel, A. B. Kharitonov, M. Zayats, E. Katz, and I. Willner, "Analysis of NAD(P)$^+$ and NAD(P)H Cofactors by Means of Imprinted Polymers Associated with Au Surfaces: A Surface Plasmon Resonance Study," Anal. Chim. Acta* **2004,** *504, 101.]*

The surface plasmon resonance minimum reflectivity in Figure 20-23 shifts by ∼0.15° when 1 mM NAD$^+$ binds to the imprinted polymer. The shift is not as great for the related species NADH, NADP$^+$, and NADPH, confirming that the imprinted polymer selectively binds NAD$^+$. When the observed reflectivity was fitted to the theoretical response, the polymer film was calculated to be $22 \pm 3$ nm thick and had a binding capacity of 2.26 µg NAD$^+$/cm$^2$. When 1 mM NAD$^+$ binds to the polymer, the refractive index of the polymer layer changes from 1.45 to 1.40 and the layer thickness increases by $3.0 \pm 0.2$ nm.

## ■ ■ ■  20-5  Fourier Transform Infrared Spectroscopy[18]

A photodiode array or charge coupled device can measure an entire spectrum at once. The spectrum is spread into its component wavelengths, and each wavelength is directed onto one detector element. For the infrared region, the most important method for observing the entire spectrum at once is *Fourier transform spectroscopy.*

### Fourier Analysis

**Fourier analysis** is a procedure in which a curve is decomposed into a sum of sine and cosine terms, called a *Fourier series.* To analyze the curve in Figure 20-24, which spans the interval $x_1 = 0$ to $x_2 = 10$, the Fourier series has the form.

*Fourier series:*  $y = a_0 \sin(0\omega x) + b_0 \cos(0\omega x) + a_1 \sin(1\omega x) + b_1 \cos(1\omega x)$

$$+ a_2 \sin(2\omega x) + b_2 \cos(2\omega x) + \cdots$$

$$= \sum_{n=0}^{\infty} [a_n \sin(n\omega x) + b_n \cos(n\omega x)] \tag{20-10}$$

where

$$\omega = \frac{2\pi}{x_2 - x_1} = \frac{2\pi}{10 - 0} = \frac{\pi}{5} \tag{20-11}$$

Equation 20-10 says that the value of $y$ for any value of $x$ can be expressed by an infinite sum of sine and cosine waves. Successive terms correspond to waves with increasing frequency.

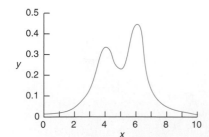

**Figure 20-24** A curve to be decomposed into a sum of sine and cosine terms by Fourier analysis.

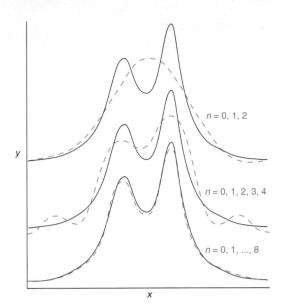

**Figure 20-25** Fourier series reconstruction of the curve in Figure 20-24. Solid line is the original curve and dashed lines are made from a series of $n = 0$ to $n = 2, 4,$ or $8$ in Equation 20-10. Coefficients $a_n$ and $b_n$ are given in Table 20-3.

**Table 20-3** Fourier coefficients for Figure 20-25

| $n$ | $a_n$ | $b_n$ |
|---|---|---|
| 0 | 0 | 0.136 912 |
| 1 | −0.006 906 | −0.160 994 |
| 2 | 0.015 185 | 0.037 705 |
| 3 | −0.014 397 | 0.024 718 |
| 4 | 0.007 860 | −0.043 718 |
| 5 | 0.000 089 | 0.034 864 |
| 6 | −0.004 813 | −0.018 858 |
| 7 | 0.006 059 | 0.004 580 |
| 8 | −0.004 399 | 0.003 019 |

Figure 20-25 shows how sequences of three, five, or nine sine and cosine waves give better and better approximations to the curve in Figure 20-24. The coefficients $a_n$ and $b_n$ required to construct the curves in Figure 20-25 are given in Table 20-3.

## Interferometry

The heart of a Fourier transform infrared spectrophotometer is the **interferometer** in Figure 20-26. Radiation from the source at the left strikes a *beamsplitter,* which transmits some light and reflects some light. For the sake of this discussion, consider a beam of monochromatic radiation. (In fact, the Fourier transform spectrophotometer uses a continuum source of infrared radiation, not a monochromatic source.) For simplicity, suppose that the beamsplitter reflects half of the light and transmits half. When light strikes the beamsplitter at point O, some is reflected to a stationary mirror at a distance OS and some is transmitted to a movable mirror at a distance OM. The rays reflected by the mirrors travel back to the beamsplitter, where half of each ray is transmitted and half is reflected. One recombined ray travels in the direction of the detector, and another heads back to the source.

In general, the paths OM and OS are not equal, so the two waves reaching the detector are not in phase. If the two waves are in phase, they interfere constructively to give a wave

*Albert Michelson developed the interferometer about 1880 and conducted the Michelson-Morley experiment in 1887, in which it was found that the speed of light is independent of the motion of the source and the observer. This crucial experiment led Einstein to the theory of relativity. Michelson also used the interferometer to create the predecessor of today's length standard based on the wavelength of light. He received the Nobel Prize in 1907 "for precision optical instruments and the spectroscopic and metrological investigations carried out with their aid."*

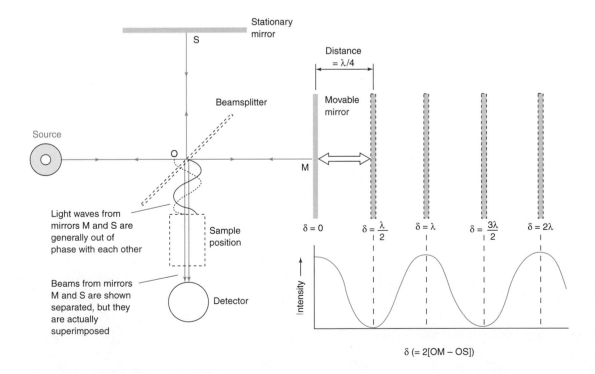

**Figure 20-26** Schematic diagram of Michelson interferometer. Detector response as a function of retardation ($= 2$[OM − OS]) is shown for monochromatic incident radiation of wavelength $\lambda$.

Spectrum

Interferogram

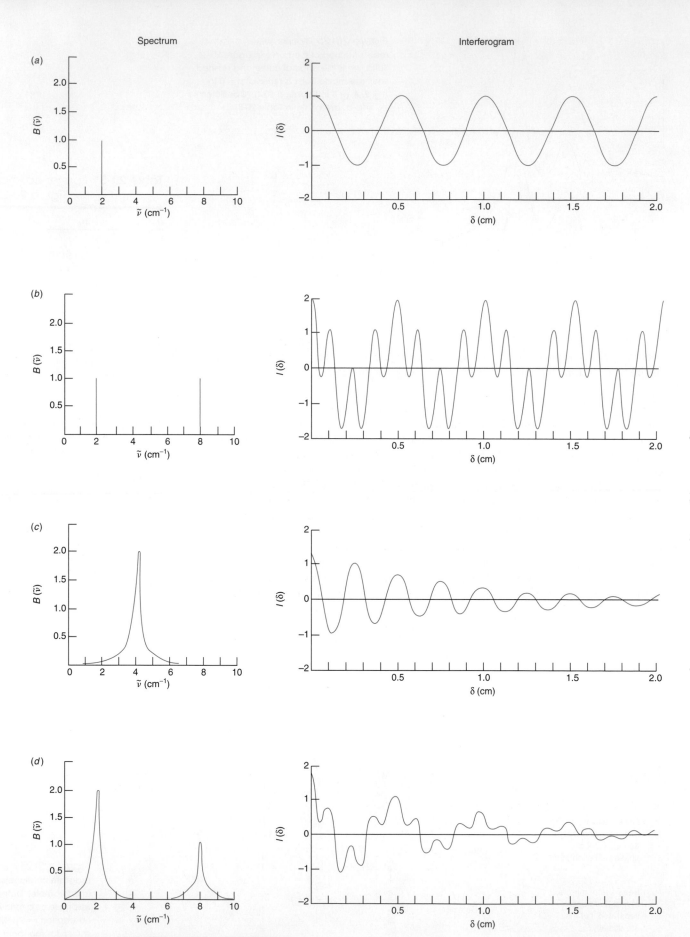

*Figure 20-27*  Interferograms produced by different spectra.

with twice the amplitude, as shown in Figure 20-7. If the waves are one-half wavelength (180°) out of phase, they interfere destructively and cancel. For any intermediate phase difference, there is partial cancellation.

The difference in pathlength followed by the two waves in Figure 20-26 is $2(OM - OS)$. This difference is called the *retardation, $\delta$*. Constructive interference occurs whenever $\delta$ is an integral multiple of the wavelength, $\lambda$. A minimum appears when $\delta$ is a half-integral multiple of $\lambda$. If mirror M moves away from the beamsplitter at a constant speed, light reaching the detector goes through a sequence of maxima and minima as the interference alternates between constructive and destructive phases.

A graph of output light intensity versus retardation, $\delta$, is called an **interferogram.** If the light from the source is monochromatic, the interferogram is a simple cosine wave:

$$I(\delta) = B(\tilde{\nu}) \cos\left(\frac{2\pi\delta}{\lambda}\right) = B(\tilde{\nu}) \cos(2\pi\tilde{\nu}\delta) \qquad (20\text{-}12)$$

where $I(\delta)$ is the intensity of light reaching the detector and $\tilde{\nu}$ is the wavenumber ($= 1/\lambda$) of the light. Clearly, $I$ is a function of the retardation, $\delta$. $B(\tilde{\nu})$ is a constant that accounts for the intensity of the light source, efficiency of the beamsplitter (which never gives exactly 50% reflection and 50% transmission), and response of the detector. All these factors depend on $\tilde{\nu}$. For monochromatic light, there is only one value of $\tilde{\nu}$.

Figure 20-27a shows the interferogram produced by monochromatic radiation of wavenumber $\tilde{\nu}_0 = 2 \text{ cm}^{-1}$. The wavelength (repeat distance) of the interferogram can be seen in the figure to be $\lambda = 0.5 \text{ cm}$, which is equal to $1/\tilde{\nu}_0 = 1/(2 \text{ cm}^{-1})$. Figure 20-27b shows the interferogram that results from a source with two monochromatic waves $\tilde{\nu}_0 = 2$ and $\tilde{\nu}_0 = 8 \text{ cm}^{-1}$) with relative intensities 1:1. A short wave oscillation ($\lambda = \frac{1}{8}$cm) is superimposed on a long wave oscillation ($\lambda = \frac{1}{2}$ cm). The interferogram is a sum of two terms:

$$I(\delta) = B_1 \cos(2\pi\tilde{\nu}_1\delta) + B_2 \cos(2\pi\tilde{\nu}_2\delta) \qquad (20\text{-}13)$$

where $B_1 = 1$, $\tilde{\nu}_1 = 2 \text{ cm}^{-1}$, $B_2 = 1$, and $\tilde{\nu}_2 = 8 \text{ cm}^{-1}$.

Fourier analysis decomposes a curve into its component wavelengths. Fourier analysis of the interferogram in Figure 20-27a gives the (trivial) result that the interferogram is made from a single wavelength function, with $\lambda = \frac{1}{2}$ cm. Fourier analysis of the interferogram in Figure 20-27b gives the slightly more interesting result that the interferogram is composed of two wavelengths ($\lambda = \frac{1}{2}$ and $\lambda = \frac{1}{8}$ cm) with relative contributions 1:1. We say that the spectrum is the *Fourier transform* of the interferogram.

The interferogram in Figure 20-27c is derived from a spectrum with an absorption band centered at $\tilde{\nu}_0 = 4 \text{ cm}^{-1}$. The interferogram is the sum of contributions from all source wavelengths. The Fourier transform of the interferogram in Figure 20-27c is indeed the third spectrum in Figure 20-27c. That is, decomposition of the interferogram into its component wavelengths gives back the band centered at $\tilde{\nu}_0 = 4 \text{ cm}^{-1}$. *Fourier analysis of the interferogram gives back the intensities of its component wavelengths.*

The interferogram in Figure 20-27d comes from two absorption bands in the spectrum at the left. The Fourier transform of this interferogram gives back the spectrum to its left.

Fourier analysis of the interferogram gives back the spectrum from which the interferogram is made. *The spectrum is the Fourier transform of the interferogram.*

## Fourier Transform Spectroscopy

In a Fourier transform spectrometer, the sample is usually placed between the interferometer and the detector, as in Figures 20-26 and 20-28. Because the sample absorbs certain wavelengths, *the interferogram contains the spectrum of the source minus the spectrum of the sample.* An interferogram of a reference sample containing the cell and solvent is first recorded and transformed into a spectrum. Then the interferogram of a sample in the same solvent is recorded and transformed into a spectrum. The quotient of the sample spectrum divided by the reference spectrum is the transmission spectrum of the sample (Figure 20-29, page 447). The quotient is the same as computing $P/P_0$ to find transmittance. $P_0$ is the irradiance received at the detector through the reference, and $P$ is the irradiance received after passage through the sample.

The interferogram loses intensity at those wavelengths absorbed by the sample.

The interferogram is recorded at discrete intervals. The *resolution* of the spectrum (ability to discern closely spaced peaks) is approximately equal to $(1/\Delta) \text{ cm}^{-1}$, where $\Delta$ is the maximum retardation. If the mirror travel is $\pm 2$ cm, the retardation is $\pm 4$ cm and the resolution is $1/(4 \text{ cm}) = 0.25 \text{ cm}^{-1}$.

Resolution $\approx 1/\Delta \text{ cm}^{-1}$
$\Delta$ = maximum retardation

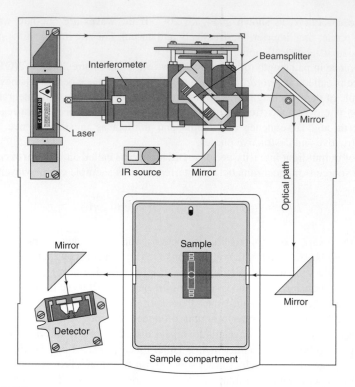

*Figure 20-28* Layout of Fourier transform infrared spectrometer. *[Courtesy Nicolet, Madison, WI.]*

For a spectral range of $\Delta\tilde{\nu}$ cm$^{-1}$, points must be taken at retardation intervals of $1/(2\Delta\tilde{\nu})$.

The wavelength range of the spectrum is determined by how the interferogram is sampled. The closer the spacing between data points, the greater the wavelength range of the spectrum. To cover a range of $\Delta\tilde{\nu}$ wavenumbers requires sampling the interferogram at retardation intervals of $\delta = 1/(2\Delta\tilde{\nu})$. If $\Delta\tilde{\nu}$ is 4 000 cm$^{-1}$, sampling must occur at intervals of $\delta = 1/(2 \cdot 4\,000\ \text{cm}^{-1}) = 1.25 \times 10^{-4}\ \text{cm} = 1.25\ \mu\text{m}$. This sampling interval corresponds to a mirror motion of 0.625 $\mu$m. For every centimeter of mirror travel, $1.6 \times 10^4$ data points must be collected. If the mirror moves at a rate of 2 mm per second, the data collection rate would be 3 200 points per second.

The source, beamsplitter, and detector each limit the usable wavelength range. Clearly, the instrument cannot respond to a wavelength that is absorbed by the beamsplitter or a wavelength to which the detector does not respond. The beamsplitter for the mid-infrared region ($\sim$4 000 to 400 cm$^{-1}$) is typically a layer of germanium evaporated onto a KBr plate. For longer wavelengths ($\tilde{\nu} < 400$ cm$^{-1}$), a film of the organic polymer Mylar is a suitable beamsplitter.

To control the sampling interval for the interferogram, a monochromatic visible laser beam is passed through the interferometer along with the polychromatic infrared light (Figure 20-28). The laser gives destructive interference whenever the retardation is a half-integral multiple of the laser wavelength. These zeros in the laser signal, observed with a visible detector, are used to control sampling of the infrared interferogram. For example, an infrared data point might be taken at every second zero point of the visible-light interferogram. The precision with which the laser frequency is known gives an accuracy of 0.01 cm$^{-1}$ in the infrared spectrum, which is a 100-fold improvement over the accuracy of dispersive (grating) instruments.

## Advantages of Fourier Transform Spectroscopy

Compared with dispersive instruments, the Fourier transform spectrometer offers an improved signal-to-noise ratio at a given resolution, much better frequency accuracy, speed, and built-in data-handling capabilities. The signal-to-noise improvement comes mainly because the Fourier transform spectrometer uses energy from the entire spectrum, instead of analyzing a sequence of small wavebands available from a monochromator. Precise reproduction of wavenumber position from one spectrum to the next allows Fourier transform instruments to average signals from multiple scans to further improve the signal-to-noise ratio.

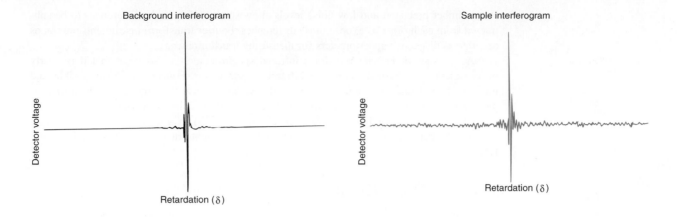

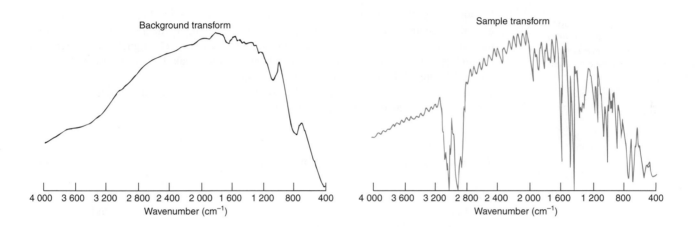

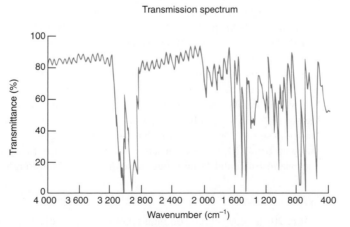

**Figure 20-29** Fourier transform infrared spectrum of polystyrene film. The Fourier transform of the background interferogram gives a spectrum determined by the source intensity, beamsplitter efficiency, detector response, and absorption by traces of $H_2O$ and $CO_2$ in the atmosphere. The sample compartment is purged with dry $N_2$ to reduce the levels of $H_2O$ and $CO_2$. The transform of the sample interferogram is a measure of all the instrumental factors, plus absorption by the sample. The transmission spectrum is obtained by dividing the sample transform by the background transform. Each interferogram is an average of 32 scans and contains 4 096 data points, giving a resolution of 4 cm$^{-1}$. The mirror velocity was 0.693 cm/s. *[Courtesy M. P. Nadler, Michelson Laboratory, China Lake, CA.]*

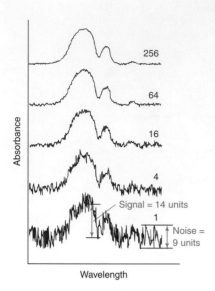

**Figure 20-30** Effect of signal averaging on a simulated noisy spectrum. Labels refer to number of scans averaged. [From R. Q. Thompson, "Experiments in Software Data Handling," J. Chem. Ed. **1985**, 62, 866.]

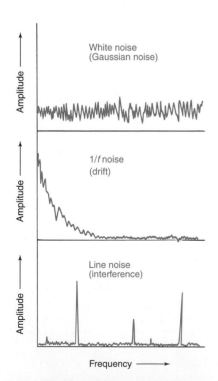

**Figure 20-31** Three types of noise in electrical instruments. White noise is always present. A beam chopping frequency can be selected to reduce 1/f noise and line noise to insignificant values.

Wavenumber precision and low noise levels allow spectra with slight differences to be subtracted from each other to expose those differences. Fourier transform instruments are not as accurate as dispersive spectrometers for measuring transmittance.

Advantages of Fourier transform infrared spectrometers are so great that it is nearly impossible to purchase a dispersive infrared spectrometer. Fourier transform visible and ultraviolet spectrometers are not commercially available, because of the requirement to sample the interferometer at intervals of $\delta = 1/(2\Delta\tilde{\nu})$. For visible spectroscopy, $\Delta\tilde{\nu}$ could be 25 000 cm$^{-1}$ (corresponding to 400 nm), giving $\delta = 0.2$ μm and a mirror movement of 0.1 μm between data points. Such fine control over significant ranges of mirror motion is not feasible.

# ■ ■ ■ ■ 20-6 Dealing with Noise[19]

An advantage of Fourier transform spectroscopy is that the entire interferogram is recorded in a few seconds and stored in a computer. The signal-to-noise ratio can be improved by collecting tens or hundreds of interferograms and averaging them.

## Signal Averaging

**Signal averaging** can improve the quality of data, as illustrated in Figure 20-30.[20] The lowest trace contains a great deal of noise. A simple way to estimate the noise level is to measure the maximum amplitude of the noise in a region free of signal. The signal is measured from the middle of the baseline noise to the middle of the noisy peak. By this criterion, the lowest trace of Figure 20-30 has a signal-to-noise ratio of 14/9 = 1.6.

A more common measurement of noise, which requires a digitized signal and a computer, is the **root-mean-square** (rms) **noise,** defined as

$$\textit{Root-mean-square noise:} \qquad \text{rms noise} = \sqrt{\frac{\sum_i (A_i - \overline{A})^2}{n}} \qquad (20\text{-}14)$$

where $A_i$ is the measured signal for the $i$th data point, $\overline{A}$ is the mean signal, and $n$ is the number of data points. For a large number of data points, the rms noise is the standard deviation of the noise. It is best to apply Equation 20-14 where the signal is flat, as it is at the left and right sides of Figure 20-30. If you must use data at the center of Figure 20-30, the average $\overline{A}$ must be fitted to the continuously rising or falling signal. The rms noise is ~5 times less than the peak-to-peak noise. If we used rms noise instead of peak-to-peak noise, we would say that the noise in the bottom spectrum of Figure 20-30 is 9/5 = 1.8 and the signal-to-noise ratio is 14/1.8 = 7.8. Clearly, the signal-to-noise ratio depends on how you define noise.

Consider what happens if you record the spectrum twice and add the results. The signal is the same in both spectra and adds to give twice the value of each spectrum. If $n$ spectra are added, the signal will be $n$ times as large as in the first spectrum. Noise is random, so it may be positive or negative at any point. It turns out that, if $n$ spectra are added, the noise increases in proportion to $\sqrt{n}$. Because the signal increases in proportion to $n$, the signal-to-noise ratio increases in proportion to $n/\sqrt{n} = \sqrt{n}$.

When we average $n$ spectra, the signal-to-noise ratio is improved by $\sqrt{n}$. To improve the signal-to-noise ratio by a factor of 2 requires averaging four spectra. To improve the signal-to-noise ratio by a factor of 10 requires averaging 100 spectra. Spectroscopists record as many as $10^4$ to $10^5$ scans to observe weak signals. It is rarely possible to do better than this because instrumental instabilities cause drift in addition to random noise.

## Types of Noise

Figure 20-31 shows three common types of noise in electrical instruments.[21] The upper trace is ordinary, random *white noise* (also called *Gaussian noise*) arising from causes such as the random motion of electrons in a circuit. The second trace shows *1/f noise,* also called *drift,* which is greatest at zero frequency and decreases in proportion to 1/frequency. An example of low-frequency noise in laboratory instruments is flickering or drifting of a light source in a spectrophotometer or a flame in atomic spectroscopy. Drift arises from causes such as slow changes in instrument components with temperature and age and variation of power-line voltage to an instrument. The classical way to detect and account for drift is to periodically measure standards and correct the instrument reading for any observed change.[22] The bottom trace contains *line noise* (also called *interference* or *whistle noise*) at discrete frequencies such as the 60-Hz transmission-line frequency or the 0.2-Hz vibrational frequency when elephants walk through the basement of your building.

**Question** By what factor should the signal-to-noise ratio be improved when 16 spectra are averaged? Measure the noise level in Figure 20-30 to test your prediction.

## Beam Chopping

The spectrophotometer in Figure 20-1 has a rotating mirror called a *chopper* that alternately sends light through the sample and reference cells. Chopping allows both cells to be sampled almost continuously, and it provides a means of noise reduction. **Beam chopping** moves the analytical signal from zero frequency to the frequency of the chopper. The chopping frequency can be selected so that $1/f$ noise and line noise are minimal. High-frequency detector circuits are required to take advantage of beam chopping.

---

## Terms to Understand

| | | | |
|---|---|---|---|
| attenuated total reflectance | Fourier analysis | photoconductive detector | resolution |
| bandwidth | grating | photodiode array | root-mean-square (rms) noise |
| beam chopping | interferogram | photomultiplier tube | signal averaging |
| blackbody radiation | interferometer | phototube | Snell's law |
| charge coupled device | laser | photovoltaic detector | stray light |
| diffraction | monochromator | polychromator | surface plasmon resonance |
| dispersion | optical fiber | refraction | thermocouple |
| ferroelectric material | optode | refractive index | waveguide |

---

## Summary

Components of spectrophotometers include the source, sample cell, monochromator, and detector. Tungsten and deuterium lamps provide visible and ultraviolet radiation; a silicon carbide globar is a good infrared source. Tungsten and silicon carbide lamps behave approximately as blackbodies, which are objects that absorb all light striking them. Emission of radiant energy from the surface of the blackbody is proportional to the fourth power of temperature and shifts to shorter wavelengths as temperature increases. Lasers provide high-intensity, coherent, monochromatic radiation by stimulated emission from a medium in which an excited state has been pumped to a higher population than that of a lower state. Sample cells must be transparent to the radiation of interest. A reference sample compensates for reflection and scattering by the cell and solvent. A grating monochromator disperses light into its component wavelengths. The finer a grating is ruled, the higher the resolution and the greater the dispersion of wavelengths over angles. Narrow slits improve resolution but increase noise, because less light reaches the detector. A bandwidth that is 1/5 of the width of the spectral peak is a good compromise between maximizing signal-to-noise ratio and minimizing peak distortion. Stray light introduces absorbance errors that are most serious when the transmittance of a sample is very low. Filters pass wide bands of wavelength and reject other bands.

A photomultiplier tube is a sensitive detector of visible and ultraviolet radiation; photons cause electrons to be ejected from a metallic cathode. The signal is amplified at each successive dynode on which the photoelectrons impinge. Photodiode arrays and charge coupled devices are solid-state detectors in which photons create electrons and holes in semiconductor materials. Coupled to a polychromator, these devices can record all wavelengths of a spectrum simultaneously, with resolution limited by the number and spacing of detector elements. Common infrared detectors include thermocouples, ferroelectric materials, and photoconductive and photovoltaic devices.

When light passes from a region of refractive index $n_1$ to a region of refractive index $n_2$, the angle of refraction ($\theta_2$) is related to the angle of incidence ($\theta_1$) by Snell's law: $n_1 \sin \theta_1 = n_2 \sin \theta_2$. Optical fibers and flat waveguides transmit light by a series of total internal reflections. Optodes are sensors based on optical fibers. Some optodes have a layer of material whose absorbance or fluorescence changes in the presence of analyte. Light can be carried to and from the tip by the optical fiber. When light is transmitted through a fiber optic or a waveguide by total internal reflectance, some light, called the evanescent wave, penetrates through the reflective interface during each reflection. In attenuated total reflectance devices, the waveguide is coated with a substance that absorbs light in the presence of analyte. In a surface plasmon resonance sensor, we measure the change in the angle of minimum reflectance from a gold film coated with a chemically sensitive layer on the back face of a prism.

Fourier analysis decomposes a signal into its component wavelengths. An interferometer contains a beamsplitter, a stationary mirror, and a movable mirror. Reflection of light from the two mirrors creates an interferogram. Fourier analysis of the interferogram tells us what frequencies went into the interferogram. In a Fourier transform spectrophotometer, the interferogram of the source is first measured without sample. Then sample is placed in the beam and a second interferogram is recorded. The transforms of the interferograms tell what amount of light at each frequency reaches the detector with and without the sample. The quotient of the two transforms is the transmission spectrum. The resolution of a Fourier transform spectrum is approximately $1/\Delta$, where $\Delta$ is the maximum retardation. To cover a wavenumber range $\Delta\tilde{\nu}$ requires sampling the interferogram at intervals of $\delta = 1/(2\Delta\tilde{\nu})$. If you average $n$ scans, the signal-to-noise ratio should increase by $\sqrt{n}$. Beam chopping in a dual-beam spectrophotometer reduces $1/f$ and line noise.

## Exercises

**20-A.** **(a)** If a diffraction grating has a resolution of $10^4$, is it possible to distinguish two spectral lines with wavelengths of 10.00 and 10.01 μm?

**(b)** With a resolution of $10^4$, how close in wavenumbers ($cm^{-1}$) is the closest line to 1 000 $cm^{-1}$ that can barely be resolved?

**(c)** Calculate the resolution of a 5.0-cm-long grating ruled at 250 lines/mm for first-order ($n = 1$) diffraction and tenth-order ($n = 10$) diffraction.

**(d)** Find the angular dispersion ($\Delta\phi$, in radians and degrees) between light rays with wavenumbers of 1 000 and 1 001 $cm^{-1}$ for second-order diffraction ($n = 2$) from a grating with 250 lines/mm and $\phi = 30°$.

**20-B.** The true absorbance of a sample is 1.000, but the monochromator passes 1.0% stray light. Add the light coming through the sample to the stray light to find the apparent transmittance of the sample. Convert this back into absorbance and find the relative error in the calculated concentration of the sample.

**20-C.** Refer to the Fourier transform infrared spectrum in Figure 20-29.

**(a)** The interferogram was sampled at retardation intervals of $1.266\ 0 \times 10^{-4}$ cm. What is the theoretical wavenumber range (0 to ?) of the spectrum?

**(b)** A total of 4 096 data points were collected from $\delta = -\Delta$ to $\delta = +\Delta$. Compute the value of $\Delta$, the maximum retardation.

**(c)** Calculate the approximate resolution of the spectrum.

**(d)** How many microseconds elapse between each datum?

**(e)** How many seconds were required to record each interferogram once?

**(f)** What kind of beamsplitter is typically used for the region 400 to 4 000 $cm^{-1}$? Why is the region below 400 $cm^{-1}$ not observed?

**20-D.** The table below shows signal-to-noise ratios recorded in a nuclear magnetic resonance experiment. Construct graphs of **(a)** signal-to-noise ratio versus $n$ and **(b)** signal-to-noise ratio versus $\sqrt{n}$, where $n$ is the number of scans. Draw error bars corresponding to the standard deviation at each point. Is the signal-to-noise ratio proportional to $\sqrt{n}$? Find the 95% confidence interval for each row of the table.

Signal-to-noise ratio at the aromatic protons of 1% ethylbenzene in $CCl_4$

| Number of experiments | Number of accumulations | Signal-to-noise ratio | Standard deviation |
|---|---|---|---|
| 8 | 1 | 18.9 | 1.9 |
| 6 | 4 | 36.4 | 3.7 |
| 6 | 9 | 47.3 | 4.9 |
| 8 | 16 | 66.7 | 7.0 |
| 6 | 25 | 84.6 | 8.6 |
| 6 | 36 | 107.2 | 10.7 |
| 6 | 49 | 130.3 | 13.3 |
| 4 | 64 | 143.2 | 15.1 |
| 4 | 81 | 146.2 | 15.0 |
| 4 | 100 | 159.4 | 17.1 |

*SOURCE: Data from M. Henner, P. Levior, and B. Ancian, "An NMR Spectrometer-Computer Interface Experiment," J. Chem. Ed. **1979**, 56, 685.*

## Problems

### The Spectrophotometer

**20-1.** Describe the role of each component of the spectrophotometer in Figure 20-1.

**20-2.** Explain how a laser generates light. List important properties of laser light.

**20-3.** Would you use a tungsten or a deuterium lamp as a source of 300-nm radiation?

**20-4.** Which variables increase the resolution of a grating?

**20-5.** What is the role of a filter in a grating monochromator?

**20-6.** What are the advantages and disadvantages of decreasing monochromator slit width?

**20-7.** Deuterated triglycine sulfate (abbreviated DTGS) is a common ferroelectric infrared detector material. Explain how it works.

**20-8.** Consider a reflection grating operating with an incident angle of 40° in Figure 20-6.

**(a)** How many lines per centimeter should be etched in the grating if the first-order diffraction angle for 600 nm (visible) light is to be $-30°$?

**(b)** Answer the same question for 1 000 $cm^{-1}$ (infrared) light.

**20-9.** Show that a grating with $10^3$ grooves/cm provides a dispersion of 5.8° per μm of wavelength if $n = 1$ and $\phi = 10°$ in Equation 20-4.

**20-10.** **(a)** What resolution is required for a diffraction grating to resolve wavelengths of 512.23 and 512.26 nm?

**(b)** With a resolution of $10^4$, how close in nm is the closest line to 512.23 nm that can barely be resolved?

**(c)** Calculate the fourth-order resolution of a grating that is 8.00 cm long and is ruled at 185 lines/mm.

**(d)** Find the angular dispersion ($\Delta\phi$) between light rays with wavelengths of 512.23 and 512.26 nm for first-order diffraction ($n = 1$) and thirtieth-order diffraction from a grating with 250 lines/mm and $\phi = 3.0°$.

**20-11.** The true absorbance of a sample is 1.500, but 0.50% stray light reaches the detector. Find the apparent transmittance and apparent absorbance of the sample.

**20-12.** The pathlength of a cell for infrared spectroscopy can be measured by counting *interference fringes* (ripples in the transmission spectrum). The spectrum below shows 30 interference maxima between 1 906 and 698 $cm^{-1}$ obtained by placing an empty KBr cell in a spectrophotometer.

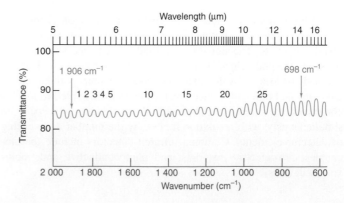

The fringes arise because light reflected from the cell compartment interferes constructively or destructively with the unreflected beam.

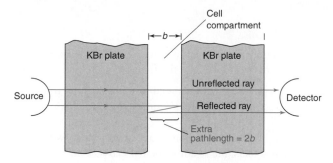

If the reflected beam travels an extra distance $\lambda$, it will interfere constructively with the unreflected beam. If the reflection pathlength is $\lambda/2$, destructive interference occurs. Peaks therefore arise when $m\lambda = 2b$ and troughs occur when $m\lambda/2 = 2b$, where $m$ is an integer. If the medium between the KBr plates has refractive index $n$, the wavelength in the medium is $\lambda/n$, so the equations become $m\lambda/n = 2b$ and $m\lambda/2n = 2b$. The cell pathlength can be shown to be given by

$$b = \frac{N}{2n} \cdot \frac{\lambda_1\lambda_2}{\lambda_2 - \lambda_1} = \frac{N}{2n} \cdot \frac{1}{\tilde{\nu}_1 - \tilde{\nu}_2}$$

where $N$ maxima occur between wavelengths $\lambda_1$ and $\lambda_2$. Calculate the pathlength of the cell that gave the interference fringes shown earlier.

**20-13.** Calculate the power per unit area (the exitance, W/m²) radiating from a blackbody at 77 K (liquid nitrogen temperature) and at 298 K (room temperature).

**20-14.** The exitance (power per unit area per unit wavelength) from a blackbody (Box 20-1) is given by the *Planck distribution:*

$$M_\lambda = \frac{2\pi hc^2}{\lambda^5}\left(\frac{1}{e^{hc/\lambda kT} - 1}\right)$$

where $\lambda$ is wavelength, $T$ is temperature (K), $h$ is Planck's constant, $c$ is the speed of light, and $k$ is Boltzmann's constant. The area under each curve between two wavelengths in the blackbody graph in Box 20-1 is equal to the power per unit area (W/m²) emitted between those two wavelengths. We find the area by integrating the Planck function between wavelengths $\lambda_1$ and $\lambda_2$:

$$\text{Power emitted} = \int_{\lambda_1}^{\lambda_2} M_\lambda \, d\lambda$$

For a narrow wavelength range, $\Delta\lambda$, the value of $M_\lambda$ is nearly constant and the power emitted is simply the product $M_\lambda\Delta\lambda$.
(a) Evaluate $M_\lambda$ at $\lambda = 2.00$ μm and at $\lambda = 10.00$ μm at $T = 1\,000$ K.
(b) Calculate the power emitted per square meter at $1\,000$ K in the interval $\lambda = 1.99$ μm to $\lambda = 2.01$ μm by evaluating the product $M_\lambda\Delta\lambda$, where $\Delta\lambda = 0.02$ μm.
(c) Repeat part (b) for the interval 9.99 to 10.01 μm.
(d) The quantity $[M_\lambda\,(\lambda = 2\,\mu m)]/[M_\lambda\,(\lambda = 10\,\mu m)]$ is the relative exitance at the two wavelengths. Compare the relative exitance at these two wavelengths at $1\,000$ K with the relative exitance at 100 K. What does your answer mean?

**20-15.** In the cavity ring-down measurement at the opening of this chapter, absorbance is given by

$$A = \frac{L}{c \ln 10}\left(\frac{1}{\tau} - \frac{1}{\tau_0}\right)$$

where $L$ is the length of the cavity between mirrors, $c$ is the speed of light, $\tau$ is the ring-down lifetime with sample in the cavity, and $\tau_0$ is the ring-down lifetime with no sample in the cavity. Ring-down lifetime is obtained by fitting the observed ring-down signal intensity $I$ to an exponential decay of the form $I = I_0 e^{-t/\tau}$, where $I_0$ is the initial intensity and $t$ is time. A measurement of $CO_2$ is made at a wavelength absorbed by the molecule. The ring-down lifetime for a 21.0-cm-long empty cavity is 18.52 μs and 16.06 μs for a cavity containing $CO_2$. Find the absorbance of $CO_2$ at this wavelength.

### Optical Sensors

**20-16.** Light passes from benzene (medium 1) to water (medium 2) in Figure 20-17 at (a) $\theta_1 = 30°$ or (b) $\theta_1 = 0°$. Find the angle $\theta_2$ in each case.

**20-17.** Explain how an optical fiber works. Why does the fiber still work when it is bent?

**20-18.** (a) Explain how the attenuated total reflection sensor in Figure 20-21 works.
(b) For a given angle of incidence, the sensitivity of the attenuated total reflectance sensor increases as the thickness of the waveguide decreases. Explain why. (The waveguide can be less than 1 μm thick.)

**20-19.** (a) Find the critical value of $\theta_i$ in Figure 20-18 beyond which there is total internal reflection in a $ZrF_4$-based infrared optical fiber whose core refractive index is 1.52 and whose cladding refractive index is 1.50.
(b) The loss of radiant power (due to absorption and scatter) in an optical fiber of length $\ell$ is expressed in decibels per meter (dB/m), defined as

$$\frac{\text{Power out}}{\text{Power in}} = 10^{-\ell(\text{dB/m})/10}$$

Calculate the quotient power out/power in for a 20.0-m-long fiber with a loss of 0.010 0 dB/m.

**20-20.** Find the minimum angle $\theta_i$ for total reflection in the optical fiber in Figure 20-18 if the index of refraction of the cladding is 1.400 and the index of refraction of the core is (a) 1.600 or (b) 1.800.

**20-21.** The prism shown below is used to totally reflect light at a 90° angle. No surface of this prism is silvered. Use Snell's law to explain why total reflection occurs. What is the minimum refractive index of the prism for total reflection?

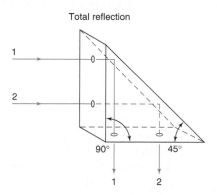

Total reflection

**20-22.** Here is an extremely sensitive method for measuring nitrite ($NO_2^-$) down to 1 nM in natural waters. The water sample is treated with sulfanilamide and *N*-(1-naphthylethylenediamine) in acid solution to produce a colored product with a molar absorptivity of $4.5 \times 10^4 \, M^{-1} \, cm^{-1}$ at 540 nm. The colored solution is pumped into a 4.5-meter-long, coiled Teflon tube whose fluorocarbon wall has a refractive index of 1.29. The aqueous solution inside the tube has a refractive index near 1.33. The colored solution is pumped through the coiled tube. An optical fiber delivers white light into one end of the tube, and an optical fiber at the other end leads to a polychromator and detector.

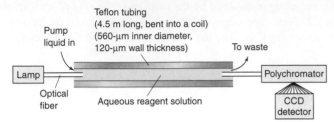

Long-pathlength spectrometer. *[Adapted from W. Yao, R. H. Byrne, and R. D. Waterbury, "Determination of Nanomolar Concentrations of Nitrite and Nitrate Using Long Path Length Absorbance Spectroscopy," Environ. Sci. Technol.* **1998**, *32, 2646.]*

(a) Explain the purpose of the coiled Teflon tube and explain how it functions.
(b) What is the critical angle of incidence for total internal reflection at the Teflon | water interface?
(c) What is the predicted absorbance of a 1.0 nM solution of colored reagent?

**20-23.** (a) A particular silica glass waveguide is reported to have a loss coefficient of 0.050 dB/cm (power out/power in, defined in Problem 20-19) for 514-nm-wavelength light. The thickness of the waveguide is 0.60 μm and the length is 3.0 cm. The angle of incidence ($\theta_i$ in Figure 20-21) is 70°. What fraction of the incident radiant intensity is transmitted through the waveguide?
(b) If the index of refraction of the waveguide material is 1.5, what is the wavelength of light inside the waveguide? What is the frequency?

**20-24.** ▦  The variation of refractive index *n*, with wavelength for fused silica is given by

$$n^2 - 1 = \frac{(0.696 \, 166 \, 3)\lambda^2}{\lambda^2 - (0.068 \, 404 \, 3)^2} + \frac{(0.407 \, 942 \, 6)\lambda^2}{\lambda^2 - (0.116 \, 241 \, 4)^2}$$
$$+ \frac{(0.897 \, 479 \, 4)\lambda^2}{\lambda^2 - (9.896 \, 161)^2}$$

where λ is expressed in μm.
(a) Make a graph of *n* versus λ with points at the following wavelengths: 0.2, 0.4, 0.6, 0.8, 1, 2, 3, 4, 5, and 6 μm.
(b) The ability of a prism to spread apart (disperse) neighboring wavelengths increases as the slope $dn/d\lambda$ increases. Is the dispersion of fused silica greater for blue light or red light?

### Fourier Transform Spectroscopy

**20-25.** The interferometer mirror of a Fourier transform infrared spectrophotometer travels $\pm 1$ cm.
(a) How many centimeters is the maximum retardation, $\Delta$?
(b) State what is meant by resolution.
(c) What is the approximate resolution ($cm^{-1}$) of the instrument?
(d) At what retardation interval, $\delta$, must the interferogram be sampled (converted into digital form) to cover a spectral range of 0 to $2 \, 000 \, cm^{-1}$?

**20-26.** Explain why the transmission spectrum in Figure 20-29 is calculated from the quotient (sample transform)/(background transform) instead of the difference (sample transform) − (background transform).

### Dealing with Noise

**20-27.** A spectrum has a signal-to-noise ratio of 8/1. How many spectra must be averaged to increase the signal-to-noise ratio to 20/1?

**20-28.** A measurement with a signal-to-noise ratio of 100/1 can be thought of as a signal, *S*, with 1% uncertainty, *e*. That is, the measurement is $S \pm e = 100 \pm 1$.
(a) Use the rules for propagation of uncertainty to show that, if you add two such signals, the result is total signal = $200 \pm \sqrt{2}$, giving a signal-to-noise ratio of $200/\sqrt{2} = 141/1$.
(b) Show that, if you add four such measurements, the signal-to-noise ratio increases to 200/1.
(c) Show that averaging *n* measurements increases the signal-to-noise ratio by a factor of $\sqrt{n}$ compared with the value for one measurement.

**20-29.** Results of an electrochemical experiment are shown in the figure below. In each case, a voltage is applied between two electrodes at time = 20 s and the absorbance of a solution decreases until the voltage is stepped back to its initial value at time = 60 s. The upper traces show the average results for 100, 300, and 1 000 repetitions of the experiment. The measured signal-to-rms noise ratio in the upper trace is 60.0. Predict the signal-to-noise ratios expected for 300, 100, and 1 cycle and compare your answers with the observed values in the figure.

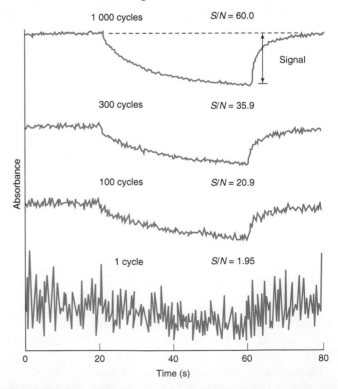

Signal averaging in an experiment in which absorbance is measured after the electric potential is changed at 20 s. *[From A. F. Slaterbeck, T. H. Ridgeway, C. J. Seliskar, and W. R. Heineman, "Spectroelectrochemical Sensing Based on Multimode Selectivity Simultaneously Achievable in a Single Device," Anal. Chem.* **1999**, *71, 1196.]*

# 21 | Atomic Spectroscopy

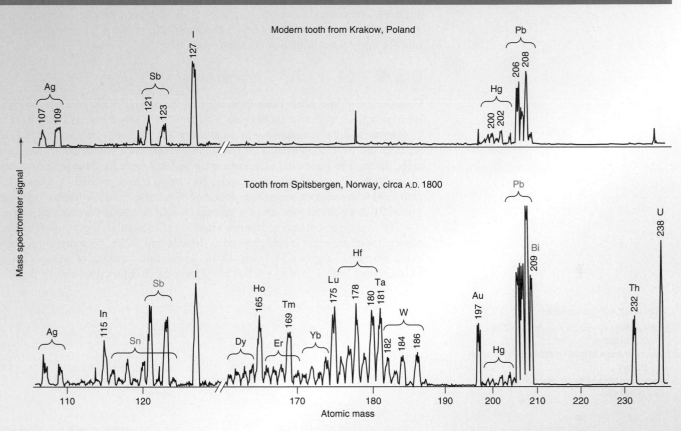

In *atomic spectroscopy*, a substance is decomposed into atoms in a flame, furnace, or *plasma*. (A **plasma** is a gas that is hot enough to contain ions and free electrons.) Each element is measured by absorption or emission of ultraviolet or visible radiation by the gaseous atoms. To measure trace elements in a tooth, tiny portions of the tooth are vaporized (*ablated*) by a laser pulse[1] and swept into a plasma. The plasma ionizes some of the atoms, which pass into a mass spectrometer that separates ions by their mass and measures their quantity.

Elements are incorporated into teeth from the diet or by inhalation. The figure shows trace element profiles measured by laser ablation–plasma ionization–mass spectrometry of the dentine of teeth from a modern person and one who lived in Scandinavia about A.D. 1800. The contrast is striking. The old tooth contains significant amounts of tin and bismuth, which are nearly absent in the modern tooth. The old tooth contains more lead and antimony than the modern tooth. Tin and lead are constituents of pewter, which was used for cooking vessels and utensils. Bismuth and antimony also might come from pewter.

Even more striking in the old tooth is the abundance of rare earths (dysprosium, holmium, erbium, thulium, ytterbium, and lutetium) and the elements tantalum, tungsten, gold, thorium, and uranium. Rare earth minerals are found in Scandinavia (in fact, many rare earth elements were discovered there), but what were they used for? Did people prepare food with them? Did they somehow get into the food chain?

Trace element profile of a tooth from a modern man and from a person who lived in Scandinavia 200 years ago. [From A. Cox, F. Keenan, M. Cooke, and J. Appleton, "Trace Element Profiling of Dental Tissues Using Laser Ablation Inductively Coupled Plasma–Mass Spectrometry," *Fresenius J. Anal. Chem.* **1996**, *354, 254.*]

Inductively coupled argon plasma atomizes substances at 6 000 K.

In *atomic spectroscopy,* samples are vaporized at 2 000–8 000 K and decompose into atoms. Concentrations of atoms in the vapor are measured by emission or absorption of characteristic wavelengths of radiation. Because of its high sensitivity, its ability to distinguish one element from another in a complex sample, its ability to perform simultaneous multielement analyses, and the ease with which many samples can be automatically analyzed, atomic spectroscopy is a principal tool of analytical chemistry.[2]

Analyte is measured at parts per million ($\mu$g/g) to parts per trillion (pg/g) levels. To analyze major constituents, the sample must be diluted to reduce concentrations to the parts per million level. As we saw in the analysis of teeth, trace constituents can be measured directly without *preconcentration*. The precision of atomic spectroscopy, typically 1–2%, is not as good as that of some wet chemical methods. The equipment is expensive, but widely available. Unknowns, standards, and blanks can be loaded into an *autosampler,* which is a turntable that automatically rotates each sample into position for analysis. The instrument runs for many hours without human intervention.

*Preconcentration:* concentrating a dilute analyte to a level high enough to be analyzed

# ■ ■ ■ 21-1  An Overview

Consider the **atomic absorption** experiment in the middle of Figure 21-1.[3] In Figure 21-2, a liquid sample is aspirated (sucked) into a flame whose temperature is 2 000–3 000 K. Liquid evaporates and the remaining solid is **atomized** (broken into atoms) in the flame, which replaces the cuvet in conventional spectrophotometry. The pathlength of the flame is typically 10 cm. The *hollow-cathode lamp* at the left in Figure 21-2 has an iron cathode. When the cathode is bombarded with energetic $Ne^+$ or $Ar^+$ ions, excited Fe atoms vaporize and emit light with the same frequencies absorbed by analyte Fe in the flame. At the right side of Figure 21-2, a detector measures the amount of light that passes through the flame.

An important difference between atomic and molecular spectroscopy is the width of absorption or emission bands. Spectra of liquids and solids typically have bandwidths of ~100 nm, as in Figures 18-7 and 18-14. In contrast, spectra of gaseous atoms consist of sharp lines with widths of ~0.001 nm (Figure 21-3). Lines are so sharp that there is usu-

Types of atomic spectroscopy:

1. *absorption* of sharp lines from hollow-cathode lamp
2. *fluorescence* following absorption of laser radiation
3. *emission* from a thermally populated excited state

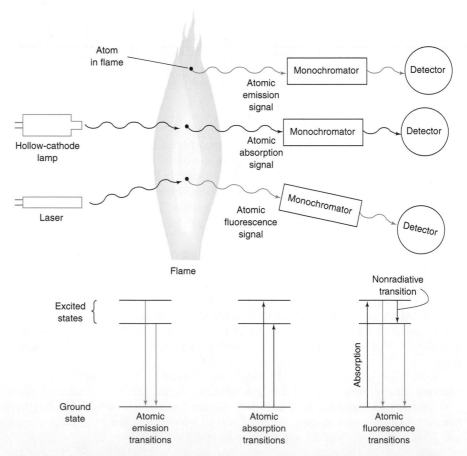

**Figure 21-1**  Absorption, emission, and fluorescence by atoms in a flame. In atomic absorption, atoms absorb part of the light from the source and the remainder of the light reaches the detector. Atomic emission comes from atoms that are in an excited state because of the high thermal energy of the flame. To observe atomic fluorescence, atoms are excited by an external lamp or laser. An excited atom can fall to a lower state and emit radiation.

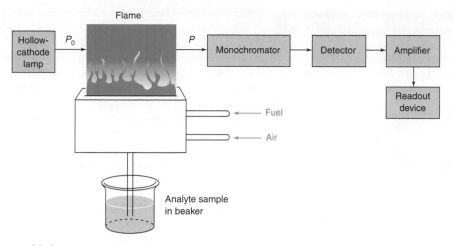

*Figure 21-2* Atomic absorption experiment. As in Figure 18-4, transmittance = $T = P/P_0$ and absorbance = $A = -\log T$. In practice, $P_0$ is the irradiance reaching the detector when no sample is going into the flame, and $P$ is measured while sample is present.

ally little overlap between the spectra of different elements in the same sample. Therefore, some instruments can measure more than 70 elements simultaneously. We will see later that sharp analyte absorption lines require that the light source also have sharp lines.

Figure 21-1 also illustrates an **atomic fluorescence** experiment. Atoms in the flame are irradiated by a laser to promote them to an excited electronic state from which they can fluoresce to return to the ground state. Figure 21-4 shows atomic fluorescence from 2 ppb of lead in tap water. Atomic fluorescence is potentially a thousand times more sensitive than atomic absorption, but equipment for atomic fluorescence is not common. An important example of atomic fluorescence is in the analysis of mercury (Box 21-1).

By contrast, **atomic emission** (Figure 21-1) is widely used.[5] Collisions in the very hot *plasma* promote some atoms to excited electronic states from which they can emit photons to

Fluorescence is more sensitive than absorption because we can observe a weak fluorescence signal above a dark background. In absorption, we are looking for small differences between large amounts of light reaching the detector.

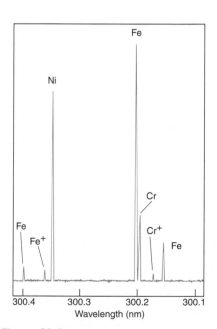

*Figure 21-3* A portion of the emission spectrum of a steel hollow-cathode lamp, showing lines from gaseous Fe, Ni, and Cr atoms and weak lines from $Cr^+$ and $Fe^+$ ions. The monochromator resolution is 0.001 nm, which is comparable to the true linewidths.

[From A. P. Thorne, "Fourier Transform Spectrometry in the Ultraviolet," Anal. Chem. **1991**, 63, 57A.]

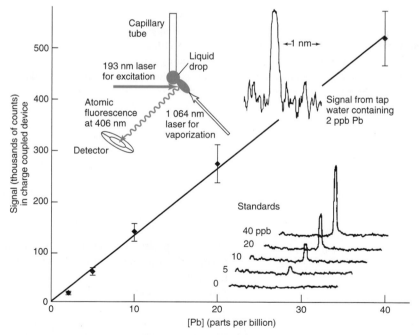

*Figure 21-4* Atomic fluorescence from Pb at 405.8 nm. Water containing parts per billion (ppb) of colloidal $PbCO_3$ was ejected from a capillary tube and exposed to a 6-ns pulse of 1 064-nm laser radiation focused on the drop. This pulse created a plume of vapor moving toward the laser. After 2.5 μs, the plume was exposed to a 193-nm laser pulse, creating excited Pb atoms whose fluorescence was measured for 0.1 μs with an optical system whose resolution was 0.2 nm. The figure shows a calibration curve constructed from colloidal $PbCO_3$ standards and the signal from tap water containing 2 ppb Pb. [From S. K. Ho and N. H. Cheung, "Sub-Part-per-Billion Analysis of Aqueous Lead Colloids by ArF Laser Induced Atomic Fluorescence," Anal. Chem. **2005**, 77, 193.]

**Box 21-1** Mercury Analysis by Cold Vapor Atomic Fluorescence

Mercury is a volatile toxic pollutant. The map shows Hg(0) concentrations in the air near Earth's surface. Mercury is also found as Hg(II)(aq) in clouds and on particles in the atmosphere.

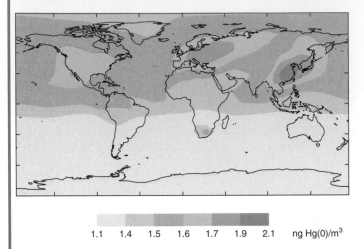

Global annual average surface concentration of Hg(0). [From C. Seigneur, K. Vijayaraghavan, P. Karamchandani, and C. Scott, "Global Source Attribution of Mercury Deposition in the United States," Environ. Sci. Technol. **2004**, 38, 555.]

Approximately two-thirds of atmospheric mercury comes from human activities, including coal burning, waste incineration, and $Cl_2$ production by the chlor-alkali process (Problem 17-7).

A sensitive method to measure mercury in matrices such as water, soil, and fish generates Hg(g), which is measured by atomic absorption or fluorescence. For most environmental samples, automated digestion/analysis equipment is available.[4] For water analysis by one standard method, all mercury is first oxidized to Hg(II) with BrCl in the purge flask in the drawing. Halogens are reduced with hydroxylamine ($NH_2OH$), and Hg(II) is reduced to Hg(0) with $SnCl_2$. Hg(0) is then purged from solution by bubbling purified Ar or $N_2$. Hg(0) is collected at room temperature in

the sample trap, which contains gold-coated silica sand. Hg binds to Au while other gases in the purge stream pass through. The sample trap is then heated to 450°C to release Hg(g), which is caught in the analytical trap at room temperature. Two traps are used so that all other gaseous impurities are removed prior to analysis. Hg(g) is then released from the analytical trap by heating and flows into the fluorescence cell. Fluorescence intensity strongly depends on gaseous impurities that can quench the emission from Hg.

The lower limit of quantitation is ~0.5 ng/L (parts per trillion). To measure such small quantities requires extraordinary care at every stage of analysis to prevent contamination. Mercury amalgam fillings in a worker's teeth can contaminate samples exposed to exhaled breath.

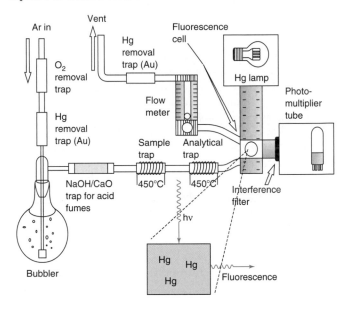

Mercury analysis by U.S. Environmental Protection Agency Method 1631.

---

return to lower-energy states. No lamp is required. Emission intensity is proportional to the concentration of the element in the sample. Emission from atoms in a plasma is now the dominant form of atomic spectroscopy.

## 21-2 Atomization: Flames, Furnaces, and Plasmas

In atomic spectroscopy, analyte is *atomized* in a flame, an electrically heated furnace, or a plasma. Flames were used for decades, but they have been replaced by the inductively coupled plasma and the graphite furnace. We begin our discussion with flames because they are still common in teaching labs.

### Flames

Most flame spectrometers use a **premix burner,** such as that in Figure 21-5, in which fuel, oxidant, and sample are mixed before introduction into the flame. Sample solution is drawn into the *pneumatic nebulizer* by the rapid flow of oxidant (usually air) past the tip of the sample capillary. Liquid breaks into a fine mist as it leaves the capillary. The spray is directed against a glass bead, upon which the droplets break into smaller particles. The formation of small droplets is termed **nebulization.** A fine suspension of liquid (or solid) particles in a gas is called an **aerosol.** The nebulizer creates an aerosol from the liquid sample. The mist, oxi-

Organic solvents with surface tensions lower than that of water are excellent for atomic spectroscopy because they form smaller droplets, thus leading to more efficient atomization.

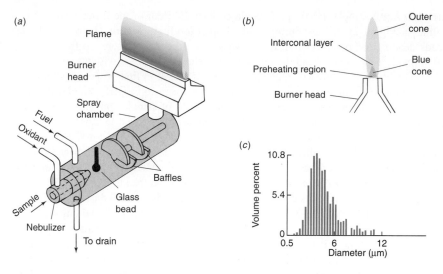

*(a)* Flame

Burner head

Spray chamber

Fuel

Oxidant

Sample

Nebulizer

Baffles

Glass bead

To drain

*(b)*

Outer cone

Interconal layer

Preheating region

Blue cone

Burner head

*(c)*

Volume percent

10.8

5.4

0

0.5    6    12
Diameter (μm)

**Figure 21-5** (*a*) Premix burner. (*b*) End view of flame. The slot in the burner head is about 0.5 mm wide. (*c*) Distribution of droplet sizes produced by a particular nebulizer. *[From R. H. Clifford, I. Ishii, A. Montaser, and G. A. Meyer, "Droplet-Size and Velocity Distributions of Aerosols from Commonly Used Nebulizers," Anal. Chem. **1990**, 62, 390.]*

dant, and fuel flow past baffles that promote further mixing and block large droplets of liquid. Excess liquid collects at the bottom of the spray chamber and flows out to a drain. Aerosol reaching the flame contains only about 5% of the initial sample.

The most common fuel-oxidizer combination is acetylene and air, which produces a flame temperature of 2 400–2 700 K (Table 21-1). If a hotter flame is required to atomize high-boiling elements (called *refractory* elements), acetylene and nitrous oxide are usually used. In the flame profile in Figure 21-5b, gas entering the preheating region is heated by conduction and radiation from the primary reaction zone (the blue cone in the flame). Combustion is completed in the outer cone, where surrounding air is drawn into the flame. Flames emit light that must be subtracted from the total signal to obtain the analyte signal.

**Table 21-1**  Maximum flame temperatures

| Fuel | Oxidant | Temperature (K) |
| --- | --- | --- |
| Acetylene, HC≡CH | Air | 2 400–2 700 |
| Acetylene | Nitrous oxide, $N_2O$ | 2 900–3 100 |
| Acetylene | Oxygen | 3 300–3 400 |
| Hydrogen | Air | 2 300–2 400 |
| Hydrogen | Oxygen | 2 800–3 000 |
| Cyanogen, N≡C—C≡N | Oxygen | 4 800 |

Droplets entering the flame evaporate; then the remaining solid vaporizes and decomposes into atoms. Many elements form oxides and hydroxides in the outer cone. Molecules do not have the same spectra as atoms, so the atomic signal is lowered. Molecules also emit broad radiation that must be subtracted from the sharp atomic signals. If the flame is relatively rich in fuel (a "rich" flame), excess carbon tends to reduce metal oxides and hydroxides and thereby increases sensitivity. A "lean" flame, with excess oxidant, is hotter. Different elements require either rich or lean flames for best analysis. The height in the flame at which maximum atomic absorption or emission is observed depends on the element being measured and the flow rates of sample, fuel, and oxidant.[6]

## Furnaces[7]

An electrically heated **graphite furnace** is more sensitive than a flame and requires less sample. From 1 to 100 μL of sample are injected into the furnace through the hole at the center of Figure 21-6. Light from a hollow-cathode lamp travels through windows at each end of the graphite tube. To prevent oxidation of the graphite, Ar gas is passed over the furnace, and the maximum recommended temperature is 2 550°C for not more than 7 s.

In flame spectroscopy, the *residence time* of analyte in the optical path is <1 s as it rises through the flame. A graphite furnace confines the atomized sample in the optical path for several seconds, thereby affording higher sensitivity. Whereas 1–2 mL is the minimum volume of solution necessary for flame analysis, as little as 1 μL is adequate for a furnace. Precision is rarely better than 5–10% with manual sample injection, but automated injection improves reproducibility to ~1%.

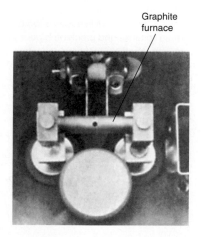

Graphite furnace

**Figure 21-6**  An electrically heated graphite furnace for atomic spectroscopy (~38 mm long, in this case). *[Courtesy Instrumentation Laboratory, Wilmington, MA.]*

Furnaces offer increased sensitivity and require less sample than a flame.

When you inject sample, the droplet should contact the floor of the furnace and remain in a small area (Figure 21-7a). If you inject the droplet too high (Figure 21-7b), it splashes and spreads, leading to poor precision. In the worst case, the drop adheres to the tip of the pipet and is finally deposited around the injector hole when the pipet is withdrawn.

Compared with flames, furnaces require more operator skill to find proper conditions for each type of sample. The furnace is heated in three or more steps to properly atomize the sample. To measure Fe in the iron-storage protein ferritin, 10 μL of sample containing ~0.1 ppm Fe are injected into the furnace at ~90°C. The furnace is programmed to *dry* the sample at 125°C for 20 s to remove solvent. Drying is followed by 60 s of *charring* at 1 400°C to destroy organic matter. Charring is also called *pyrolysis,* which means decomposing with heat. Charring creates smoke that would interfere with the Fe determination. After charring, the sample is atomized at 2 100°C for 10 s. Absorbance reaches a maximum and then decreases as Fe evaporates from the oven. The analytical signal is the time-integrated absorbance during atomization. After atomization, the furnace is heated to 2 500°C for 3 s to clean out any remaining residue.

The furnace is purged with Ar or $N_2$ during each step except atomization to remove volatile material. Gas flow is halted during atomization to avoid blowing analyte out of the furnace. When developing a method for a new kind of sample, it is important to record the signal as a function of time, because signals are also observed from smoke during charring and from the glow of the red-hot oven in the last part of atomization. A skilled operator must interpret which signal is due to analyte so that the right peak is integrated.

The furnace in Figure 21-8a performs better than a simple graphite tube. Sample is injected onto a platform that is heated by radiation from the furnace wall, so its temperature lags behind that of the wall. Analyte does not vaporize until the wall reaches constant temperature (Figure 21-8b). At constant furnace temperature, the area under the absorbance peak in Figure 21-8b is a reliable measure of the analyte. A heating rate of 2 000 K/s rapidly dissociates molecules and increases the concentration of free atoms in the furnace.

The furnace in Figure 21-8a is heated *transversely* (from side to side) to provide nearly uniform temperature over the whole furnace. In furnaces with *longitudinal* (end-to-end) heating, the center of the furnace is hotter than the ends. Atoms from the central region condense at the ends, where they can vaporize during the next sample run. Interference from previous runs, called a *memory effect,* is reduced in a transversely heated furnace. To further reduce memory effects, ordinary graphite is coated with a dense layer of *pyrolytic graphite* formed by thermal decomposition of an organic vapor. The coating seals the relatively porous graphite, so it cannot absorb foreign atoms.

A sample can be *preconcentrated* by injecting and evaporating multiple aliquots in the graphite furnace prior to analysis.[8] To measure traces of As in drinking water, a 30-μL

---

You must determine reasonable time and temperature for each stage of the analysis. Once a program is established, it can be applied to a large number of similar samples.

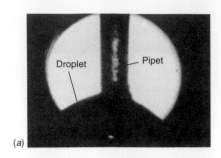

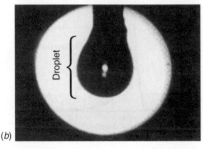

**Figure 21-7** (*a*) The correct position for injecting sample into a graphite furnace deposits the droplet in a small volume on the floor of the furnace. (*b*) If injection is too high, the sample splatters and precision is poor. *[From P. K. Booth, "Improvements in Method Development for Graphite Furnace Atomic Absorption Spectrometry," Am. Lab. February **1995**, p. 48X.]*

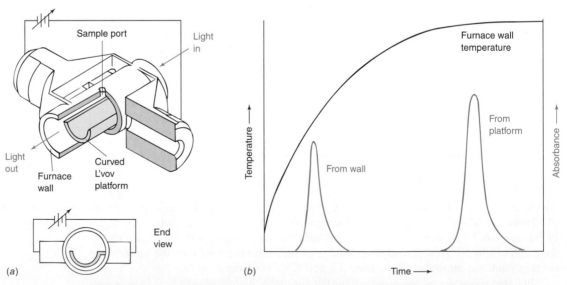

**Figure 21-8** (*a*) Transversely heated graphite furnace maintains nearly constant temperature over its whole length, thereby reducing memory effect from previous runs. The *L'vov platform* is uniformly heated by radiation from the outer wall, not by conduction. The platform is attached to the wall by one small connection that is hidden from view. *[Courtesy Perkin-Elmer Corp., Norwalk, CT.]* (*b*) Heating profiles comparing analyte evaporation from wall and from platform. *[From W. Slavin, "Atomic Absorption Spectroscopy," Anal. Chem. **1982**, 54, 685A.]*

458    **CHAPTER 21** Atomic Spectroscopy

aliquot of water plus matrix modifier was injected and evaporated. The procedure was repeated five more times so that the total sample volume was 180 μL. The detection limit for As was 0.3 μg/L (parts per billion). Without preconcentration, the detection limit would have been 1.8 μg/L. This increased capability is critical because As is considered to be a health hazard at concentrations of just a few parts per billion.

Liquid samples are ordinarily used in furnaces. However, in *direct solid sampling,* a solid is analyzed without sample preparation (Figure 21-9). For example, trace impurities in tungsten powder used to make components for industry can be analyzed by weighing 0.1 to 100 mg of powder onto a graphite platform.[9] The platform is transferred into the furnace and heated to 2 600°C to atomize impurities in the tungsten, but not the tungsten itself, which melts at 3 410°C. After several runs, residual tungsten is scraped off the platform, which could be reused 400 times. Because so much more sample is analyzed when solid is injected than when liquid is injected, detection limits for trace impurities can be as much as 100 times lower than those obtained for liquid injection. For example, Zn could be detected at a level of 10 pg/g (10 parts per trillion) when 100 mg of tungsten were analyzed. Calibration curves are obtained by injecting standard solutions of the trace elements and analyzing them as conventional liquids. Results obtained from direct solid sampling are in good agreement with results obtained by laboriously dissolving the solid. Other solids that have been analyzed by direct solid sampling include graphite, silicon carbide, cement, and river sediments.[10]

## Matrix Modifiers for Furnaces

Everything in a sample other than analyte is called the **matrix.** Ideally, the matrix decomposes and vaporizes during the charring step. A **matrix modifier** is a substance added to the sample to reduce the loss of analyte during the charring by making the matrix more volatile or the analyte less volatile.

The matrix modifier ammonium nitrate can be added to seawater to increase the volatility of the matrix NaCl. Figure 21-10a shows a graphite furnace heating profile used to analyze Mn in seawater. When 0.5 M NaCl solution is subjected to this profile, signals are observed at the analytical wavelength of Mn, (Figure 21-10b). Much of the apparent absorbance is probably due to optical scatter by smoke created by heating NaCl. Absorption at the start of the atomization step interferes with the measurement of Mn. Adding $NH_4NO_3$ to the sample in Figure 21-10c reduces matrix absorption peaks. $NH_4NO_3$ plus NaCl give $NH_4Cl$ and $NaNO_3$, which cleanly evaporate instead of making smoke.

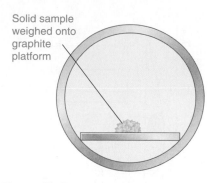

**Figure 21-9** Direct solid sampling showing end view of furnace.

Solid sample weighed onto graphite platform

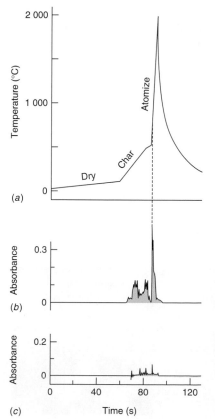

**Figure 21-10** Reduction of interference by using a matrix modifier. (*a*) Graphite furnace temperature profile for analysis of Mn in seawater. (*b*) Absorbance profile when 10 μL of 0.5 M reagent-grade NaCl is subjected to the temperature profile in panel *a.* Absorbance is monitored at the Mn wavelength of 279.5 nm with a bandwidth of 0.5 nm. (*c*) Reduced absorbance from 10 μL of 0.5 M NaCl plus 10 μL of 50 wt% $NH_4NO_3$ matrix modifier. *[From M. N. Quigley and F. Vernon, "Matrix Modification Experiment for Electrothermal Atomic Absorption Spectrophotometry," J. Chem. Ed.* **1996**, *73, 980.]*

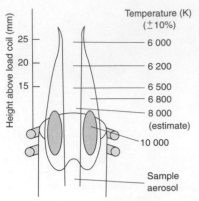

**Figure 21-11** Temperature profile of inductively coupled plasma. *[From V. A. Fassel,"Simultaneous or Sequential Determination of the Elements at All Concentration Levels," Anal. Chem. **1979**, 51, 1290A.]*

A **piezoelectric crystal** is one whose dimensions change in an applied electric field. A sinusoidal voltage applied between two faces of the crystal causes it to oscillate. Quartz is the most common piezoelectric material.

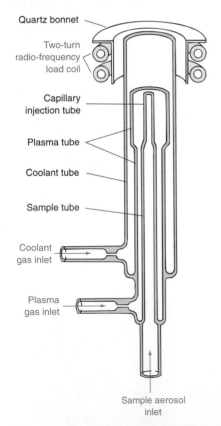

**Figure 21-12** Inductively coupled plasma burner. *[From R. N. Savage and G. M. Hieftje, "Miniature Inductively Coupled Plasma Source for Atomic Emission Spectrometry," Anal. Chem. **1979**, 51, 408.]*

The matrix modifier $Pd(NO_3)_2$ is added to seawater to decrease the volatility of the analyte Sb. In the absence of modifier, 90% of Sb is lost during charring at 1 250°C. With the modifier, the seawater matrix can be largely evaporated at 1 400°C without loss of Sb.[11]

When a graphite furnace is used, it is important to monitor the absorption signal as a function of time, as in Figure 21-10b. Peak shapes help you decide how to adjust time and temperature for each step to obtain a clean signal from analyte. Also, a graphite furnace has a finite lifetime. Degradation of peak shape or a large change in the slope of the calibration curve tells you that it is time to change the furnace.

The matrix modifier $Mg(NO_3)_2$ raises the temperature for atomization of Al analyte.[12] At high temperature, $Mg(NO_3)_2$ decomposes to $MgO(g)$ and Al is converted into $Al_2O_3$. At sufficiently high temperature, $Al_2O_3$ decomposes to Al and O, and Al evaporates. Evaporation of Al is retarded by $MgO(g)$ by virtue of the reaction

$$3MgO(g) + 2Al(s) \rightleftharpoons 3Mg(g) + Al_2O_3(s) \qquad (21-1)$$

When MgO has evaporated, Reaction 21-1 no longer occurs and $Al_2O_3$ decomposes and evaporates. A matrix modifier that raises the boiling temperature of analyte allows a higher charring temperature to be used to remove matrix without losing analyte.

## Inductively Coupled Plasmas

The **inductively coupled plasma**[13] shown at the beginning of the chapter is twice as hot as a combustion flame (Figure 21-11). The high temperature, stability, and relatively inert Ar environment in the plasma eliminate much of the interference encountered with flames. Simultaneous multielement analysis, described in Section 21-4, is routine for inductively coupled plasma atomic emission spectroscopy, which has replaced flame atomic absorption. The plasma instrument costs more to purchase and operate than a flame instrument.

The cross-sectional view of an inductively coupled plasma burner in Figure 21-12 shows two turns of a 27- or 41-MHz induction coil wrapped around the upper opening of the quartz apparatus. High-purity Ar gas is fed through the plasma gas inlet. After a spark from a Tesla coil ionizes Ar, free electrons are accelerated by the radio-frequency field. Electrons collide with atoms and transfer their energy to the entire gas, maintaining a temperature of 6 000 to 10 000 K. The quartz torch is protected from overheating by Ar coolant gas.

The concentration of analyte needed for adequate signal is reduced by an order of magnitude with an *ultrasonic nebulizer* (Figure 21-13), in which sample solution is directed onto a *piezoelectric* crystal oscillating at 1 MHz. The vibrating crystal creates a fine aerosol that is carried by a stream of Ar through a heated tube, where solvent evaporates. The stream then passes through a refrigerated zone in which solvent condenses and is removed. Analyte reaches the plasma flame as an aerosol of dry, solid particles. Plasma energy is not needed to evaporate solvent, so more energy is available for atomization. Also, a larger fraction of the sample reaches the plasma than with a conventional nebulizer.

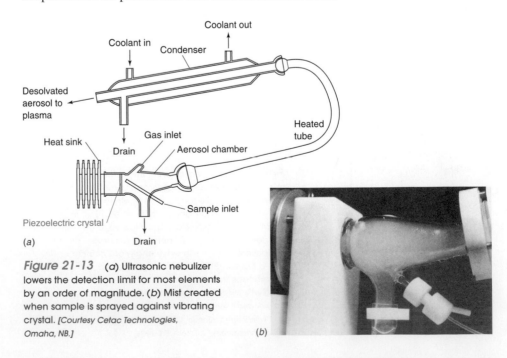

**Figure 21-13** (*a*) Ultrasonic nebulizer lowers the detection limit for most elements by an order of magnitude. (*b*) Mist created when sample is sprayed against vibrating crystal. *[Courtesy Cetac Technologies, Omaha, NB.]*

**Table 21-2** Comparison of detection limits for Ni+ ion at 231 nm

| Technique[a] | Detection limits for different instruments (ng/mL) |
|---|---|
| ICP/atomic emission (pneumatic nebulizer) | 3–50 |
| ICP/atomic emission (ultrasonic nebulizer) | 0.3–4 |
| Graphite furnace/ atomic absorption | 0.02–0.06 |
| ICP/mass spectrometry | 0.001–0.2 |

a. ICP = inductively coupled plasma.

SOURCE: J. M. Mermet and E. Poussel, "ICP Emission Spectrometers: Analytical Figures of Merit," Appl. Spectros. **1995**, 49, 12A.

Sensitivity with an inductively coupled plasma is further enhanced by a factor of 3–10 by observing emission along the length of the plasma instead of across the diameter of the plasma. Additional sensitivity is obtained by detecting ions with a mass spectrometer instead of by optical emission (Table 21-2), as described in Section 21-6.

## ■ ■ ■ 21-3 How Temperature Affects Atomic Spectroscopy

Temperature determines the degree to which a sample breaks down into atoms and the extent to which a given atom is found in its ground, excited, or ionized states. Each of these effects influences the strength of the signal we observe.

### The Boltzmann Distribution

Consider an atom with energy levels $E_0$ and $E^*$ separated by $\Delta E$ (Figure 21-14). An atom (or molecule) may have more than one state at a given energy. Figure 21-14 shows three states at $E^*$ and two at $E_0$. The number of states at each energy is called the *degeneracy*. We will call the degeneracies $g_0$ and $g^*$.

The **Boltzmann distribution** describes the relative populations of different states at thermal equilibrium. If equilibrium exists (which is not true in the blue cone of a flame but is probably true above the blue cone), the relative population ($N^*/N_0$) of any two states is

*Boltzmann distribution:*
$$\frac{N^*}{N_0} = \frac{g^*}{g_0} e^{-\Delta E/kT} \tag{21-2}$$

where $T$ is temperature (K) and $k$ is Boltzmann's constant ($1.381 \times 10^{-23}$ J/K).

### Effect of Temperature on Excited-State Population

The lowest excited state of a sodium atom lies $3.371 \times 10^{-19}$ J/atom above the ground state. The degeneracy of the excited state is 2, whereas that of the ground state is 1. The fraction of Na in the excited state in an acetylene-air flame at 2 600 K is, from Equation 21-2,

$$\frac{N^*}{N_0} = \left(\frac{2}{1}\right) e^{-(3.371 \times 10^{-19} \text{ J})/[(1.381 \times 10^{-23} \text{ J/K}) \times (2\,600 \text{ K})]} = 1.67 \times 10^{-4}$$

That is, less than 0.02% of the atoms are in the excited state.

If the temperature were 2 610 K, the fraction of atoms in the excited state would be

$$\frac{N^*}{N_0} = \left(\frac{2}{1}\right) e^{-(3.371 \times 10^{-19} \text{ J})/[(1.381 \times 10^{-23} \text{ J/K}) \times (2\,610 \text{ K})]} = 1.74 \times 10^{-4}$$

The fraction of atoms in the excited state is still less than 0.02%, but that fraction has increased by $100(1.74 - 1.67)/1.67 = 4\%$.

### The Effect of Temperature on Absorption and Emission

We see that more than 99.98% of the sodium atoms are in their ground state at 2 600 K. *Varying the temperature by 10 K hardly affects the ground-state population and would not noticeably affect the signal in atomic absorption.*

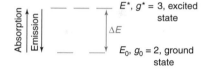

*Figure 21-14* Two energy levels with different degeneracies. Ground-state atoms can absorb light to be promoted to the excited state. Excited-state atoms can emit light to return to the ground state.

The Boltzmann distribution applies to a system at thermal equilibrium.

A 10-K temperature rise changes the excited-state population by 4% in this example.

**Table 21-3** Effect of energy difference and temperature on population of excited states

| Wavelength difference of states (nm) | Energy difference of states (J/atom) | Excited-state fraction ($N^*/N_0$)[a] | |
|---|---|---|---|
| | | 2 500 K | 6 000 K |
| 250 | $7.95 \times 10^{-19}$ | $1.0 \times 10^{-10}$ | $6.8 \times 10^{-5}$ |
| 500 | $3.97 \times 10^{-19}$ | $1.0 \times 10^{-5}$ | $8.3 \times 10^{-3}$ |
| 750 | $2.65 \times 10^{-19}$ | $4.6 \times 10^{-4}$ | $4.1 \times 10^{-2}$ |

a. Based on the equation $N^*/N_0 = (g^*/g_0)e^{-\Delta E/kT}$ in which $g^* = g_0 = 1$.

How would emission intensity be affected by a 10 K rise in temperature? In Figure 21-14, absorption arises from ground-state atoms, but emission arises from excited-state atoms. Emission intensity is proportional to the population of the excited state. *Because the excited-state population changes by 4% when the temperature rises 10 K, emission intensity rises by 4%.* It is critical in atomic *emission* spectroscopy that the flame be very stable or emission intensity will vary significantly. In atomic *absorption* spectroscopy, temperature variation is important but not as critical.

Almost all atomic emission is carried out with an inductively coupled plasma, whose temperature is more stable than that of a flame. The plasma is normally used for emission, not absorption, because it is so hot that there is a substantial population of excited-state atoms and ions. Table 21-3 compares excited-state populations for a flame at 2 500 K and a plasma at 6 000 K. Although the fraction of excited atoms is small, each atom emits many photons per second because it is rapidly promoted back to the excited state by collisions.

Energy levels of halogen atoms (F, Cl, Br, I) are so high that they emit ultraviolet radiation below 200 nm. This spectral region is called *vacuum ultraviolet* because radiation below 200 nm is absorbed by $O_2$, so spectrometers for the far-ultraviolet were customarily evacuated. Some plasma emission spectrometers are now purged with $N_2$ to exclude air so that the region 130 to 200 nm is accessible and Cl, Br, I, P, and S can be analyzed.[14]

## ■ ■ ■ 21-4 Instrumentation

Fundamental requirements for an atomic absorption experiment are shown in Figure 21-2. Principal differences between atomic and ordinary molecular spectroscopy lie in the light source (or lack of a light source in atomic emission), the sample container (the flame, furnace, or plasma), and the need to subtract background emission.

### Atomic Linewidths[15]

Beer's law requires that the linewidth of the radiation source should be substantially narrower than the linewidth of the absorbing sample. Otherwise, the measured absorbance will not be proportional to the sample concentration. Atomic absorption lines are very sharp, with an intrinsic width of only $\sim 10^{-4}$ nm.

Linewidth is governed by the **Heisenberg uncertainty principle,** which says that the shorter the lifetime of the excited state, the more uncertain is its energy:

*Heisenberg uncertainty principle:* $\qquad \delta E \delta t \gtrsim \dfrac{h}{4\pi}$ $\qquad\qquad$ (21-3)

where $\delta E$ is the uncertainty in the energy difference between ground and excited states, $\delta t$ is the lifetime of the excited state before it decays to the ground state, and $h$ is Planck's constant. Equation 21-3 says that the uncertainty in the energy difference between two states multiplied by the lifetime of the excited state is at least as big as $h/4\pi$. If $\delta t$ decreases, then $\delta E$ increases. The lifetime of an excited state of an isolated gaseous atom is $\sim 10^{-9}$ s. Therefore, the uncertainty in its energy is

$$\delta E \gtrsim \frac{h}{4\pi\delta t} = \frac{6.6 \times 10^{-34}\,\text{J} \cdot \text{s}}{4\pi(10^{-9}\,\text{s})} \approx 10^{-25}\,\text{J}$$

Suppose that the energy difference ($\Delta E$) between the ground and the excited states of an atom corresponds to visible light with a wavelength of $\lambda = 500$ nm. This energy difference is $\Delta E = hc/\lambda = 4.0 \times 10^{-19}$ J (Equation 18-3, $c$ is the speed of light). The relative uncertainty in the energy difference is $\delta E/\Delta E \gtrsim (10^{-25}\,\text{J})/(4.0 \times 10^{-19}\,\text{J}) \approx 2 \times 10^{-7}$.

The relative uncertainty in wavelength ($\delta\lambda/\lambda$) is the same as the relative uncertainty in energy:

$$\frac{\delta\lambda}{\lambda} = \frac{\delta E}{\Delta E} \gtrsim 2 \times 10^{-7} \Rightarrow \delta\lambda \gtrsim 10^{-4} \text{ nm} \qquad (21\text{-}4)$$

The inherent linewidth of an atomic absorption or emission signal is $\sim 10^{-4}$ nm because of the short lifetime of the excited state.

Two mechanisms broaden the lines to $10^{-3}$ to $10^{-2}$ nm in atomic spectroscopy. One is the **Doppler effect.** An atom moving toward the radiation source samples the oscillating electromagnetic wave more frequently than one moving away from the source (Figure 21-15). That is, an atom moving toward the source "sees" higher-frequency light than that encountered by one moving away. In the laboratory frame of reference, the atom moving toward the source absorbs lower-frequency light than that absorbed by the one moving away. The linewidth, $\delta\lambda$, due to the Doppler effect, is

*Doppler linewidth:* $\qquad\qquad \delta\lambda \approx \lambda(7 \times 10^{-7})\sqrt{\dfrac{T}{M}} \qquad\qquad (21\text{-}5)$

where $T$ is temperature (K) and $M$ is the mass of the atom in atomic mass units. For an emission line near $\lambda = 300$ nm from Fe ($M = 56$ atomic mass units) at 2 500 K in Figure 21-3, the Doppler linewidth is $(300 \text{ nm})(7 \times 10^{-7})\sqrt{2\,500/56} = 0.001\,4$ nm, which is an order of magnitude greater than the natural linewidth.

Linewidth is also affected by **pressure broadening** from collisions between atoms. Collisions shorten the lifetime of the excited state. The uncertainty in the frequency of atomic absorption and emission lines is roughly numerically equal to the collision frequency between atoms and is proportional to pressure. The Doppler effect and pressure broadening are similar in magnitude and yield linewidths of $10^{-3}$ to $10^{-2}$ nm in atomic spectroscopy.

## Hollow-Cathode Lamp

Monochromators generally cannot isolate lines narrower than $10^{-3}$ to $10^{-2}$ nm. To produce narrow lines of the correct frequency, we use a **hollow-cathode lamp** containing a vapor of the same element as that being analyzed.

A hollow-cathode lamp, in Figure 21-16, is filled with Ne or Ar at a pressure of $\sim 130$–700 Pa (1–5 Torr). The cathode is made of the element whose emission lines we want. When $\sim 500$ V are applied between the anode and the cathode, gas is ionized and positive ions are accelerated toward the cathode. After ionization occurs, the lamp is maintained at a constant current of 2–30 mA by a lower voltage. Cations strike the cathode with enough energy to "sputter" metal atoms from the cathode into the gas phase. Gaseous atoms excited by collisions with high-energy electrons emit photons. This atomic radiation has the same frequency absorbed by analyte in the flame or furnace. Atoms in the lamp are cooler than atoms in a flame, so lamp emission is sufficiently narrower than the absorption bandwidth of atoms in the flame to be nearly "monochromatic" (Figure 21-17). The purpose of a monochromator in atomic spectroscopy is to select one line from the hollow-cathode lamp and to reject as much emission from the flame or furnace as possible. A different lamp is usually required for each element, although some lamps are made with more than one element in the cathode.

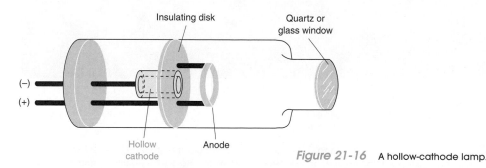

Figure 21-16 A hollow-cathode lamp.

Labels: Insulating disk, Quartz or glass window, (−), (+), Hollow cathode, Anode

## Multielement Detection with the Inductively Coupled Plasma

An inductively coupled plasma emission spectrometer does not require any lamps and can measure as many as $\sim 70$ elements simultaneously. Color Plates 23 and 24 illustrate two designs for multielement analysis. In Plate 23, atomic emission enters the polychromator and

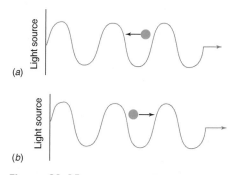

$\Delta\lambda$ is the width of an absorption or emission line measured at half the height of the peak.

Figure 21-15 The Doppler effect. A molecule moving (a) toward the radiation source "feels" the electromagnetic field oscillate more often than one moving (b) away from the source.

Doppler and pressure effects broaden the atomic lines by one to two orders of magnitude relative to their inherent linewidths.

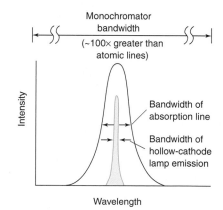

Figure 21-17 Relative bandwidths of hollow-cathode emission, atomic absorption, and a monochromator. Linewidths are measured at half the signal height. The linewidth from the hollow cathode is relatively narrow because the gas temperature in the lamp is lower than a flame temperature (so there is less Doppler broadening) and the pressure in the lamp is lower than the pressure in a flame (so there is less pressure broadening).

Labels: Monochromator bandwidth (~100× greater than atomic lines), Intensity, Bandwidth of absorption line, Bandwidth of hollow-cathode lamp emission, Wavelength

is dispersed into its component wavelengths by the grating at the bottom. One photomultiplier detector (Figure 20-12) is required at the correct position for each element to be analyzed.

In Plate 24, atomic emission entering from the top right is reflected by a collimating mirror (which makes the light rays parallel), dispersed in the vertical plane by a prism, and then dispersed in the horizontal plane by a grating. Dispersed radiation lands on a *charge injection device* (CID) detector, which is related to the charge coupled device (CCD) in Figure 20-15. Different wavelengths are spread over the 262 000 pixels of the CID shown at the upper left of Plate 24. In a CCD detector, each pixel must be read one at a time in row-by-row order. Each pixel of a CID detector can be read individually at any time. Another advantage of the CID detector over the CCD detector is that strong signals in one pixel are less prone to spill over into neighboring pixels (a process called *blooming* in CCD detectors). CID detectors can therefore measure weak emission signals adjacent to strong signals. Figure 21-18 shows an actual spectrum as seen by a CID detector.

The spectrometer in Color Plate 24 is purged with $N_2$ or Ar to exclude $O_2$, thereby allowing ultraviolet wavelengths in the 100–200 nm range to be observed. This spectral region permits more sensitive detection of some elements that are normally detected at longer wavelengths and allows halogens, P, S, and N to be measured (with poor detection limits of tens of parts per million). These nonmetallic elements cannot be observed at wavelengths above 200 nm. The photomultiplier spectrometer in Color Plate 23 is more expensive and complicated than the CID spectrometer in Color Plate 24 but provides lower detection limits because a photomultiplier tube is more sensitive than a CID detector.

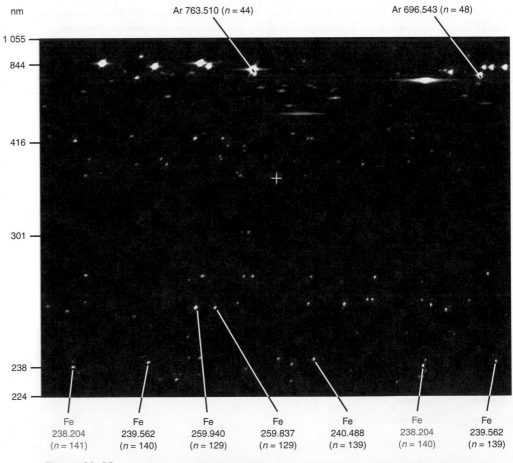

**Figure 21-18** "Constellation image" of inductively coupled plasma emission from 200 μg Fe/mL seen by charge injection detector. Almost all peaks are from iron. Horizontally blurred "galaxies" near the top are Ar plasma emission. A prism spreads wavelengths of 200–400 nm over most of the detector. Wavelengths > 400 nm are bunched together at the top. A grating provides high resolution in the horizontal direction. Selected peaks are labeled with wavelength (in nanometers) and diffraction order (*n* in Equation 20-1) in parentheses. Two Fe peaks labeled in color at the lower left and lower right are both the same wavelength (238.204 nm) diffracted into different orders by the grating. *[Courtesy M. D. Seltzer, Michelson Laboratory, China Lake, CA.]*

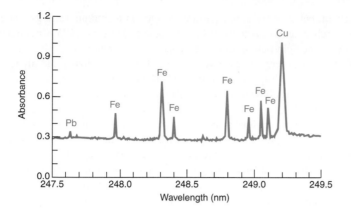

Figure 21-19 Graphite furnace absorption spectrum of bronze dissolved in $HNO_3$. [From B. T. Jones, B. W. Smith, and J. D. Winefordner, "Continuum Source Atomic Absorption Spectrometry in a Graphite Furnace with Photodiode Array Detection," Anal. Chem. **1989,** 61, 1670.]

## Background Correction

Atomic spectroscopy must provide **background correction** to distinguish analyte signal from absorption, emission, and optical scattering of the sample matrix, the flame, plasma, or white-hot graphite furnace. Figure 21-19 shows the spectrum of a sample analyzed in a graphite furnace. Sharp atomic signals with a maximum absorbance near 1.0 are superimposed on a broad background with an absorbance of 0.3. If we did not subtract the background absorbance, significant errors would result. Background correction is critical for graphite furnaces, which tend to contain residual smoke from charring. Optical scatter from smoke must be distinguished from optical absorption by analyte.

Figure 21-20 shows how background is subtracted in an emission spectrum collected with a charge injection device detector. The figure shows 15 pixels from one row of the detector centered on an analytical peak. (The spectrum is manipulated by a computer algorithm to make it look smooth, even though the original data consist of a single reading for each pixel and would look like a bar graph.) Pixels 7 and 8 were selected to represent the peak. Pixels 1 and 2 represent the baseline at the left and pixels 14 and 15 represent the baseline at the right. The mean baseline is the average of pixels 1, 2, 14, and 15. The mean peak amplitude is the average of pixels 7 and 8. The corrected peak height is the mean peak amplitude minus the mean baseline amplitude.

For atomic absorption, *beam chopping* or electrical *modulation* of the hollow-cathode lamp (pulsing it on and off) can distinguish the signal of the flame from the atomic line at the same wavelength. Figure 21-21 shows light from the lamp being periodically blocked by a rotating chopper. Signal reaching the detector while the beam is blocked must be from flame emission. Signal reaching the detector when the beam is not blocked is from the lamp and the flame. The difference between the two signals is the desired analytical signal.

Beam chopping corrects for flame emission but not for scattering. Most spectrometers provide an additional means to correct for scattering and broad background absorption. Deuterium lamps and Zeeman correction systems are most common.

For *deuterium lamp background correction,* broad emission from a $D_2$ lamp (Figure 20-3) is passed through the flame in alternation with that from the hollow cathode. The monochromator bandwidth is so wide that a negligible fraction of $D_2$ radiation is absorbed by the analyte atomic absorption line. Light from the hollow-cathode lamp is absorbed by analyte and absorbed and scattered by background. Light from the $D_2$ lamp is absorbed and scattered only by background. The difference between absorbance measured with the hollow-cathode and absorbance measured with the $D_2$ lamp is the absorbance due to analyte.

Background signal arises from absorption, emission, or scatter by everything in the sample besides analyte (the *matrix*), as well as from absorption, emission, or scatter by the flame, the plasma, or the furnace.

Background correction methods:
• adjacent pixels of CID display
• beam chopping
• $D_2$ lamp
• Zeeman

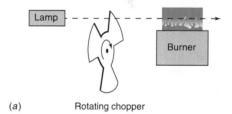

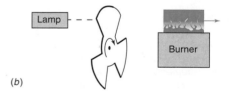

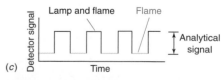

Figure 21-21 Operation of a beam chopper for subtracting the signal due to flame background emission. (*a*) Lamp and flame emission reach detector. (*b*) Only flame emission reaches detector. (*c*) Resulting square wave signal.

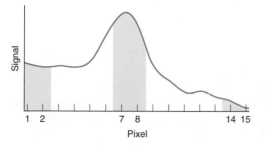

Figure 21-20 Data from charge injection detector illustrating baseline correction in plasma emission spectrometry. The mean value of pixels on either side of a peak is subtracted from the mean value of pixels under the peak. [*Courtesy M. D. Seltzer, Michelson Laboratory, China Lake, CA.*]

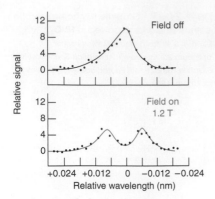

**Figure 21-22** Zeeman effect on Co fluorescence in a graphite furnace with excitation at 301 nm and detection at 341 nm. The magnetic field strength for the lower spectrum is 1.2 tesla. *[From J. P. Dougherty, F. R. Preli, Jr., J. T. McCaffrey, M. D. Seltzer, and R. G. Michel, "Instrumentation for Zeeman Electrothermal Atomizer Laser Excited Atomic Fluorescence Spectrometry," Anal. Chem.* **1987,** *59, 1112.]*

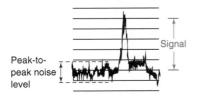

**Figure 21-23** Measurement of peak-to-peak noise level and signal level. The signal is measured from its base at the midpoint of the noise component along the slightly slanted baseline. This sample exhibits a signal-to-noise ratio of 2.4.

**Figure 21-24** Flame, furnace, and inductively coupled plasma emission and inductively coupled plasma—mass spectrometry detection limits (ng/g = ppb) with instruments from GBC Scientific Equipment, Australia. *[Flame, furnace, ICP from R. J. Gill, Am. Lab. November 1993, 24F. ICP–MS from T. T. Nham, Am. Lab. August 1998, 17A. Data for Cl, Br, and I are from reference 14.]* Accurate quantitative analysis requires concentrations 10–100 times greater than the detection limit.

An excellent, but expensive, background correction technique for a graphite furnace for many elements relies on the *Zeeman effect* (pronounced ZAY-mon). When a magnetic field is applied parallel to the light path through a furnace, the absorption (or emission) line of analyte atoms is split into three components. Two are shifted to slightly lower and higher wavelengths (Figure 21-22), and one component is unshifted. The unshifted component does not have the correct electromagnetic polarization to absorb light traveling parallel to the magnetic field and is therefore "invisible."

To use the Zeeman effect for background correction, a strong magnetic field is pulsed on and off. Sample and background are observed when the field is off, and background alone is observed when the field is on. The difference is the corrected signal.

The advantage of the Zeeman background correction is that it operates at the analytical wavelength. In contrast, $D_2$ background correction is made over a broad band. A structured or sloping background is averaged by this process, potentially misrepresenting the true background signal at the analytical wavelength. The background correction algorithm illustrated in Figure 21-20 is similar to $D_2$ background correction, but the wavelength range in Figure 21-20 is restricted to the immediate vicinity of the analytical peak.

## Detection Limits

One definition of **detection limit** is the concentration of an element that gives a signal equal to two times the peak-to-peak noise level of the baseline (Figure 21-23). The baseline noise level should be measured for a blank sample.[16]

Figure 21-24 compares detection limits for flame, furnace, and inductively coupled plasma analyses on instruments from one manufacturer. The detection limit for furnaces is typically two orders of magnitude lower than that observed with a flame because the sample is confined in the small volume of the furnace for a relatively long time. For instruments in Figure 21-24, detection limits for the inductively coupled plasma are intermediate between the flame and the furnace. With ultrasonic nebulization and axial plasma viewing, the sensitivity of the inductively coupled plasma is close to that of the graphite furnace.

Commercial standard solutions for flame atomic absorption are not necessarily suitable for plasma and furnace analyses. The latter methods require purer grades of water and acids for standard solutions and, especially, for dilutions. For the most sensitive analyses, solutions are prepared in a dust-free environment (a clean room with a filtered air supply) to reduce background contamination that *will be* detected by your instruments.

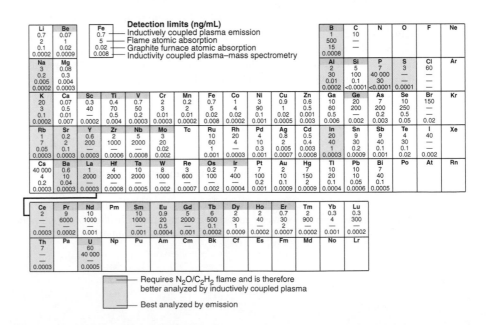

Detection limits (ng/mL)

Legend (order of values in each cell):
- Inductively coupled plasma emission
- Flame atomic absorption
- Graphite furnace atomic absorption
- Inductivity coupled plasma–mass spectrometry

| Element | ICP emission | Flame AA | Graphite furnace AA | ICP–MS |
|---|---|---|---|---|
| Li | 0.7 | 2 | 0.1 | 0.0002 |
| Be | 0.07 | 1 | 0.02 | 0.0009 |
| B | 1 | 500 | 15 | 0.0008 |
| C | 10 | — | | |
| N | | — | | |
| O | | — | | |
| F | | — | | |
| Ne | | | | |
| Na | 3 | 0.2 | 0.005 | 0.0002 |
| Mg | 0.08 | 0.3 | 0.004 | 0.0003 |
| Al | 2 | 30 | 0.01 | 0.0002 |
| Si | 5 | 100 | 0.1 | <0.0001 |
| P | 7 | 40 000 | 30 | <0.0001 |
| S | 3 | — | — | 0.0001 |
| Cl | 60 | — | — | — |
| Ar | | | | |
| K | 20 | 3 | 0.1 | 0.0002 |
| Ca | 0.07 | 0.5 | 0.01 | 0.007 |
| Sc | 0.3 | 40 | — | 0.0002 |
| Ti | 0.4 | 70 | 0.5 | 0.004 |
| V | 0.7 | 50 | 0.2 | 0.0003 |
| Cr | 2 | 3 | 0.01 | 0.0003 |
| Mn | 0.2 | 2 | 0.01 | 0.0002 |
| Fe | 0.7 | 5 | 0.02 | 0.008 |
| Co | 1 | 4 | 0.02 | 0.0002 |
| Ni | 3 | 90 | 0.1 | 0.008 |
| Cu | 0.9 | 1 | 0.02 | 0.0002 |
| Zn | 0.6 | 0.5 | 0.001 | 0.001 |
| Ga | 10 | 60 | 0.5 | 0.006 |
| Ge | 20 | 200 | — | 0.002 |
| As | 7 | 200 | 0.2 | 0.003 |
| Se | 10 | 250 | 0.5 | 0.05 |
| Br | 150 | — | — | 0.02 |
| Kr | | | | |
| Rb | 1 | 7 | 0.05 | 0.0003 |
| Sr | 0.2 | 2 | 0.1 | 0.0003 |
| Y | 0.6 | 200 | — | 0.0003 |
| Zr | 2 | 1000 | — | 0.0006 |
| Nb | 5 | 2000 | — | 0.0008 |
| Mo | 3 | 20 | 0.02 | 0.002 |
| Tc | | | | |
| Ru | 10 | 60 | 1 | 0.001 |
| Rh | 20 | 4 | — | 0.0003 |
| Pd | 4 | 10 | 0.3 | 0.001 |
| Ag | 0.8 | 2 | 0.005 | 0.0007 |
| Cd | 0.5 | 0.4 | 0.003 | 0.0008 |
| In | 20 | 40 | 1 | 0.0003 |
| Sn | 9 | 30 | 0.2 | 0.0009 |
| Sb | 9 | 40 | 0.1 | 0.001 |
| Te | 4 | 30 | 0.1 | 0.02 |
| I | 40 | — | — | 0.002 |
| Xe | | | | |
| Cs | 40 000 | 4 | 0.2 | 0.0003 |
| Ba | 0.6 | 10 | 0.04 | 0.0003 |
| La | — | 2000 | — | 0.0003 |
| Hf | 4 | 2000 | — | 0.0008 |
| Ta | 10 | 2000 | — | 0.0005 |
| W | 8 | 1000 | — | 0.0005 |
| Re | 3 | 600 | — | 0.002 |
| Os | 0.2 | 100 | — | 0.0007 |
| Ir | 7 | 400 | — | 0.002 |
| Pt | 7 | 100 | 2 | 0.0004 |
| Au | 2 | 10 | 0.1 | 0.001 |
| Hg | 7 | 150 | 2 | 0.0009 |
| Tl | 10 | 20 | 0.1 | 0.0009 |
| Pb | 10 | 10 | 0.05 | 0.0004 |
| Bi | 7 | 40 | 0.1 | 0.0006 |
| Po | | | | 0.0005 |
| At | | | | |
| Rn | | | | |
| Ce | 2 | — | — | 0.0003 |
| Pr | 9 | 6000 | — | 0.0002 |
| Nd | 10 | 1000 | — | 0.001 |
| Pm | | | | |
| Sm | 10 | 1000 | — | 0.001 |
| Eu | 0.9 | 20 | — | 0.0004 |
| Gd | 5 | 2000 | — | 0.001 |
| Tb | 6 | 500 | — | 0.0002 |
| Dy | 2 | 30 | — | 0.0009 |
| Ho | 2 | 40 | — | 0.0002 |
| Er | 0.7 | 30 | — | 0.0007 |
| Tm | 2 | 900 | — | 0.0002 |
| Yb | 0.3 | 4 | — | 0.001 |
| Lu | 0.3 | 300 | — | 0.0002 |
| Th | 7 | — | — | 0.0003 |
| Pa | | | | |
| U | 60 | 40 000 | — | 0.0005 |

- Requires $N_2O/C_2H_2$ flame and is therefore better analyzed by inductively coupled plasma
- Best analyzed by emission

## ■ ■ ■ ■ 21-5 Interference

*Interference* is any effect that changes the signal while analyte concentration remains unchanged. Interference can be corrected by removing the source of interference or by preparing standards that exhibit the same interference.

## Types of Interference

**Spectral interference** refers to the overlap of analyte signal with signals due to other elements or molecules in the sample or with signals due to the flame or furnace. Interference from the flame can be subtracted by using $D_2$ or Zeeman background correction. The best means of dealing with overlap between lines of different elements in the sample is to choose another wavelength for analysis. High-resolution spectrometers eliminate interference from other elements by resolving closely spaced lines (Figure 21-25).

Elements that form very stable diatomic oxides are incompletely atomized at the temperature of the flame or furnace. The spectrum of a molecule is much broader and more complex than that of an atom, because vibrational and rotational transitions are combined with electronic transitions (Section 18-5). The broad spectrum leads to spectral interference at many wavelengths. Figure 21-26 shows a plasma containing Y and Ba atoms as well as YO molecules. Note how broad the molecular emission is relative to the atomic emission.

When trace impurities in tungsten powder are analyzed by graphite furnace atomic absorption using direct solid sampling (page 459), $WO_3$ from the surface of the powder sublimes and fills the furnace with vapor, creating spectral interference throughout the visible and ultraviolet regions. By heating the powder under $H_2$ at $1\ 000-1\ 200°C$ in the furnace prior to atomization, $WO_3$ is reduced to metallic tungsten and the interference is eliminated.[9]

**Chemical interference** is caused by any component of the sample that decreases the extent of atomization of analyte. For example, $SO_4^{2-}$ and $PO_4^{3-}$ hinder the atomization of $Ca^{2+}$, perhaps by forming nonvolatile salts. **Releasing agents** are chemicals that are added to a sample to decrease chemical interference. EDTA and 8-hydroxyquinoline protect $Ca^{2+}$ from the interfering effects of $SO_4^{2-}$ and $PO_4^{3-}$. $La^{3+}$ also can be used as a releasing agent, apparently because it preferentially reacts with $PO_4^{3-}$ and frees the $Ca^{2+}$. A fuel-rich flame reduces certain oxidized analyte species that would otherwise hinder atomization. Higher flame temperatures eliminate many kinds of chemical interference.

**Ionization interference** can be a problem in the analysis of alkali metals at relatively low temperature and in the analyses of other elements at higher temperature. For any element, we can write a gas-phase ionization reaction:

$$M(g) \rightleftharpoons M^+(g) + e^-(g) \qquad K = \frac{[M^+][e^-]}{[M]} \qquad (21\text{-}6)$$

Because alkali metals have low ionization potentials, they are most extensively ionized. At 2 450 K and a pressure of 0.1 Pa, sodium is 5% ionized. With its lower ionization potential, potassium is 33% ionized. Ionized atoms have different energy levels from those of neutral atoms, so the desired signal is decreased. If there is a strong signal from the ion, you could use the ion signal rather than the atomic signal.

An **ionization suppressor** decreases the extent of ionization of analyte. For example, in the analysis of potassium, it is recommended that solutions contain 1 000 ppm of CsCl, because cesium is more easily ionized than potassium. By producing a high concentration of electrons in the flame, ionization of Cs suppresses ionization of K. Ionization suppression is desirable in a low-temperature flame in which we want to observe neutral atoms.

The *method of standard addition* (Section 5-3) compensates for many types of interference by adding known quantities of analyte to the unknown in its complex matrix. For example, Figure 21-27 shows the analysis of strontium in aquarium water by standard addition.

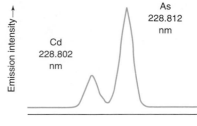

**Figure 21-25** A Cd line at 228.802 nm causes spectral interference with the As line at 228.812 nm in most spectrometers. With sufficiently high resolution, the two peaks are separated and there is no interference. The instrument used for this spectrum has a 1-m Czerny-Turner monochromator (Figure 20-5) with a resolution of 0.005 nm from 160 to 320 nm and 0.010 nm from 320 to 800 nm. *[Courtesy Jobin Yvon Horiba Group, Longjumeau Cedex, France.]*

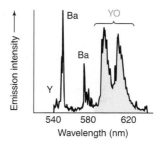

**Figure 21-26** Emission from a plasma produced by laser irradiation of the high-temperature superconductor $YBa_2Cu_3O_7$. Solid is vaporized by the laser and excited atoms and molecules such as YO emit light at characteristic wavelengths. *[From W. A. Weimer, "Plasma Emission from Laser Ablation of $YBa_2Cu_3O_7$," Appl. Phys. Lett. **1988**, 52, 2171.]*

Le Châtelier's principle tells us that adding electrons to the right side of Reaction 21-6 drives the reaction back to the left.

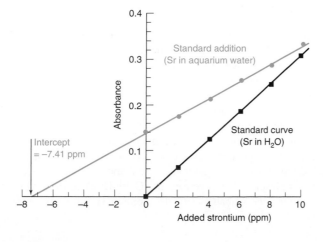

**Figure 21-27** Atomic absorption calibration curve for Sr added to distilled water and standard addition of Sr to aquarium water. All solutions are made up to a constant volume, so the ordinate is the concentration of added Sr. *[Data from L. D. Gilles de Pelichy, C. Adams, and E. T. Smith, "Analysis of Sr in Marine Aquariums by Atomic Absorption Spectroscopy," J. Chem. Ed. **1997**, 74, 1192.]*

The slope of the standard addition curve is 0.018 8 absorbance units/ppm. If, instead, Sr is added to distilled water, the slope is 0.030 8 absorbance units/ppm. That is, in distilled water, the absorbance increases 0.030 8/0.018 8 = 1.64 times more than it does in aquarium water for each addition of standard Sr. We attribute the lower response in aquarium water to interference by other species that are present. The absolute value of the $x$-intercept of the standard addition curve, 7.41 ppm, is a reliable measure of Sr in the aquarium.

## Virtues of the Inductively Coupled Plasma

An inductively coupled argon plasma eliminates many common interferences. The plasma is twice as hot as a conventional flame, and the residence time of analyte in the flame is about twice as long. Therefore, atomization is more complete and signal is enhanced. Formation of analyte oxides and hydroxides is negligible. The plasma is remarkably free of background radiation 15–35 mm above the load coil where sample emission is observed.

In flame emission spectroscopy, the concentration of electronically excited atoms in the cooler, outer part of the flame is lower than in the warmer, central part of the flame. Emission from the central region is absorbed in the outer region. This **self-absorption** increases with increasing concentration of analyte and gives nonlinear calibration curves. In a plasma, the temperature is more uniform, and self-absorption is not nearly so important. Plasma emission calibration curves are linear over five orders of magnitude. In flames and furnaces, the linear range is about two orders of magnitude. For inductively coupled plasma–mass spectrometry, the linear range is eight orders of magnitude (Table 21-4).

**Table 21-4** Comparison of atomic analysis methods

|  | Flame absorption | Furnace absorption | Plasma emission | Plasma–mass spectrometry |
|---|---|---|---|---|
| Detection limits (ng/g) | 10–1 000 | 0.01–1 | 0.1–10 | 0.000 01–0.000 1 |
| Linear range | $10^2$ | $10^2$ | $10^5$ | $10^8$ |
| Precision |  |  |  |  |
| short term (5–10 min) | 0.1–1% | 0.5–5% | 0.1–2% | 0.5–2% |
| long term (hours) | 1–10% | 1–10% | 1–5% | <5% |
| Interferences |  |  |  |  |
| spectral | very few | very few | many | few |
| chemical | many | very many | very few | some |
| mass | — | — | — | many |
| Sample throughput | 10–15 s/element | 3–4 min/element | 6–60 elements/min | all elements in 2–5 min |
| Dissolved solid | 0.5–5% | >20% slurries and solids | 1–20% | 0.1–0.4% |
| Sample volume | large | very small | medium | medium |
| Purchase cost | 1 | 2 | 4–9 | 10–15 |

SOURCE: Adapted from TJA Solutions, Franklin, MA.

## ■ ■ ■ ■ 21-6 Inductively Coupled Plasma–Mass Spectrometry

The ionization energy of Ar is 15.8 electron volts (eV), which is higher than those of all elements except He, Ne, and F. In an Ar plasma, analyte elements can be ionized by collisions with $Ar^+$, excited Ar atoms, or energetic electrons. In atomic emission spectroscopy, we usually observe the more abundant neutral atoms, M. However, the plasma can be directed into a mass spectrometer (Chapter 22), which separates and measures ions according to their mass-to-charge ratio.[17] For the most accurate measurements of isotope ratios, the mass spectrometer has one detector for each desired isotope.[18]

The trace element profile of teeth at the opening of this chapter was obtained by inductively coupled plasma–mass spectrometry. Figure 21-28 shows an example in which coffee beans were extracted with trace-metal-grade nitric acid and the aqueous extract was analyzed by inductively coupled plasma–mass spectrometry. Coffee brewed from either bean contains ~15 ng Pb/mL. However, the Cuban beans also contain Hg at a concentration similar to that of Pb.

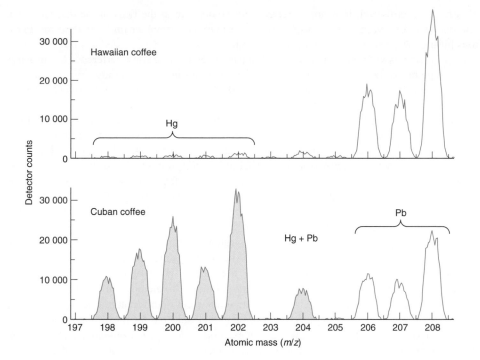

Figure 21-28 Partial elemental profile of coffee beans by inductively coupled plasma–mass spectrometry. Both beans have similar Pb content, but the Cuban beans have a much higher Hg content than the Hawaiian beans. A blank has not been subtracted from either spectrum, so the small amount of Hg in the upper spectrum could be in the blank. [Courtesy G. S. Ostrom and M. D. Seltzer, Michelson Laboratory, China Lake, CA.]

The difficulty in sampling anything with a mass spectrometer is that the spectrometer requires high vacuum to avoid collisions between ions and background gas molecules that divert the ions from their trajectory in a magnetic field. Figure 21-29 shows an example of an interface between a horizontal Ar plasma and a mass spectrometer. The plasma at the left is directed onto a water-cooled Ni sampling cone with a 1-mm-diameter orifice through which a fraction of the plasma can pass. Behind the sampling cone is a water-cooled skimmer cone with an even smaller orifice. The extraction lens behind the skimmer cone has a high negative potential to attract positive ions from the plasma. The pressure is reduced in each successive section of the instrument.

From the skimmer cone, ions enter a collision cell that might contain $H_2$ or He or both. The collision cell guides ions to the entrance of the mass separator and reduces the spread of ion kinetic energies by a factor of 10. More important, molecular ions such as $ArO^+$, $Ar_2^+$, and $ArCl^+$ dissociate when they strike the gas in the collision cell. If these species did not dissociate, they would interfere with the analysis of elements of the same mass. For example, $^{40}Ar^{16}O^+$ has nearly the same mass as $^{56}Fe^+$, and $^{40}Ar_2^+$ has nearly the same mass as $^{80}Se^+$. Interference by ions of similar mass is called **isobaric interference.** Doubly ionized $^{138}Ba^{2+}$ interferes with $^{69}Ga^+$ because each has approximately the same mass-to-charge ratio ($138/2 = 69/1$). High-resolution mass spectrometers[19] eliminate isobaric interference by separating species such as $^{40}Ar^{16}O^+$ and $^{56}Fe^+$, but most systems do not have sufficient resolution to separate these species. After dissociation in the collision cell, ions are separated

Ar is an "inert" gas with virtually no chemistry. However, $Ar^+$ has the same electronic configuration as Cl, and its chemistry is similar to that of halogens.

$^{40}Ar^{16}O^+$ and $^{56}Fe^+$ differ by 0.02 atomic mass units.

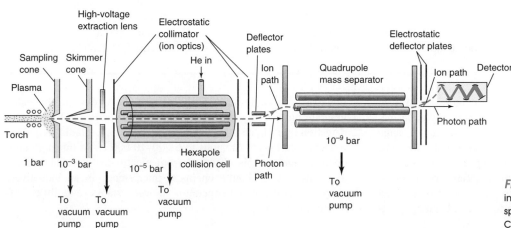

Figure 21-29 Interface between inductively coupled plasma and mass spectrometer. [Courtesy TJA Solutions, Franklin, MA.] Chapter 22 discusses mass spectrometry.

by a mass spectrometer. Ions are deflected into the detector at the right of the diagram, and photons from the plasma that transit through the mass spectrometer miss the detector. Photons hitting the detector would generate a signal.

For elements with multiple isotopes, you can check for isobaric interference by measuring isotope ratios. For example, if the ratio of Se isotopes agrees with those found in nature ($^{74}Se : ^{76}Se : ^{77}Se : ^{78}Se : ^{80}Se : ^{82}Se = 0.008\ 7 : 0.090 : 0.078 : 0.235 : 0.498 : 0.092$), then it is unlikely that there is interference at any of these masses.

Detection limits for inductively coupled plasma–mass spectrometry are so low (Figure 21-24 and Table 21-2) as to tax the cleanliness of reagents, glassware, and procedures. Solutions must be made from extremely pure water and trace-metal-grade $HNO_3$ in Teflon or polyethylene vessels protected from dust. HCl and $H_2SO_4$ are avoided because they create isobaric interferences. The plasma–mass spectrometer interface cannot tolerate high concentrations of dissolved solids that clog the orifice of the sampling cone. The plasma reduces organic matter to carbon that can clog the orifice. Organic material can be analyzed if some $O_2$ is fed into the plasma to oxidize the carbon.

Matrix effects on the yield of ions in the plasma are important, so calibration standards should be in the same matrix as the unknown. Alternatively, internal standards can be used if they have nearly the same ionization energy as the analyte. For example, Tm can be used as an internal standard for U. The ionization energies of these two elements are 5.81 and 6.08 eV, respectively, so they should ionize to nearly the same extent in different matrices. If possible, internal standards with just one major isotope should be selected for maximum response.

At the opening of this chapter, we saw an example of *laser ablation–inductively coupled plasma–mass spectrometry*[20] for the analysis of teeth. In laser **ablation,** a pulsed, high-energy laser beam is focused onto a microscopic spot on a solid sample, creating an explosion of particles, atoms, electrons, and ions into the gas phase. Material is typically removed to a depth of 0.02 to 5 $\mu m$ by each pulse (Figure 21-30). Ablation product generated in a sealed chamber is swept by Ar or He through a Teflon-coated tube into the plasma for analysis by mass spectrometry or atomic emission.

A major problem for quantitative analysis by laser ablation is that elements can be selectively ablated, selectively transported to the plasma, or selectively atomized in the plasma. Therefore, the relative numbers of ions detected are not necessarily equal to relative amounts of the elements in the solid sample. The most reliable—but usually unattainable—calibration method is to make a standard sample containing elements of interest in the same matrix as the unknown. Mussel shell in Figure 21-30 is composed principally of $CaCO_3$. A calibration standard was made by dissolving known quantities of metals with a large excess of $Ca^{2+}$ in acid solution and precipitating everything with $CO_3^{2-}$. The carbonate precipitate was washed, dried, and pressed into a dense pellet whose ablation behavior is similar to that of mussel shell. In the absence of *matrix-matched standards,* results from laser ablation can be compared with results obtained by completely digesting a material and analyzing the homogeneous solution.

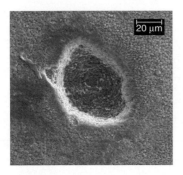

20 μm

**Figure 21-30** Microscopic crater ablated into a mussel shell by 10 pulses from a 266-nm laser with a beam energy of 4.5 mJ per 10-ns pulse and a repetition rate of 10 Hz. *[From V. R. Bellotto and N. Miekely, "Improvements in Calibration Procedures for the Quantitative Determination of Trace Elements in Carbonate Material (Mussel Shells) by Laser Ablation ICP-MS," Fresenius J. Anal. Chem.* **2000,** *367, 635.]*

---

## Terms to Understand

| | | | |
|---|---|---|---|
| ablation | Boltzmann distribution | inductively coupled plasma | piezoelectric crystal |
| aerosol | chemical interference | ionization interference | plasma |
| atomic absorption | detection limit | ionization suppressor | premix burner |
| atomic emission | Doppler effect | isobaric interference | pressure broadening |
| atomic fluorescence | graphite furnace | matrix | releasing agent |
| atomization | Heisenberg uncertainty principle | matrix modifier | self-absorption |
| background correction | hollow-cathode lamp | nebulization | spectral interference |

---

## Summary

In atomic spectroscopy, absorption, emission, or fluorescence from gaseous atoms is measured. Liquids may be atomized by a plasma, a furnace, or a flame. Flame temperatures are usually in the range 2 300–3 400 K. The choice of fuel and oxidant determines the temperature of the flame and affects the extent of spectral, chemical, or ionization interference that will be encountered. Temperature instability affects atomization in atomic absorption and has an even larger effect on atomic emission, because the excited-state population is exponentially sensitive to temperature. An electrically heated graphite furnace requires less sample than a flame and has a lower detection limit. In an inductively coupled plasma, a radio-frequency induction coil heats $Ar^+$ ions to 6 000–10 000 K. At this high temperature, emission is observed from electronically excited atoms and ions. There is little chemical interference in an inductively coupled plasma, the temperature is very stable, and little self-absorption is observed.

Plasma emission spectroscopy does not require a light source and is capable of measuring ~70 elements simultaneously with a charge injection device detector. Background correction for a given emission peak is based on subtracting the intensity of neighboring pixels in the detector. The lowest detection limits are obtained by directing the plasma into a mass spectrometer that separates and measures ions from the plasma. In flame and furnace atomic absorption spectroscopy, a hollow-cathode lamp made of the analyte element provides spectral lines sharper than those of the atomic vapor. The inherent linewidth of atomic lines is limited by the Heisenberg uncertainty principle. Lines in a flame, furnace, or plasma are broadened by a factor of 10–100 by the Doppler effect and by atomic collisions. Correction for background emission from the flame is possible by electrically pulsing the lamp on and off or mechanically chopping the beam. Light scattering and spectral background can be subtracted by measuring absorption of a deuterium lamp or by Zeeman background correction, in which the atomic energy levels are alternately shifted in and out of resonance with the lamp frequency by a magnetic field. Chemical interference can be reduced by addition of releasing agents, which prevent the analyte from reacting with interfering species. Ionization interference in flames is suppressed by adding easily ionized elements such as Cs.

In inductively coupled plasma–mass spectrometry, isobaric interference occurs between species with the same mass and charge. Interference can be eliminated if the mass spectral resolution is sufficiently great or by dissociating an interfering polyatomic species with a collision cell. When laser ablation is used to sample a solid, matrix-matched standards are often necessary for quantitative analysis.

## Exercises

**21-A.** Li was determined by atomic emission with the method of standard addition. Use a graph similar to Figure 5-6 to find the concentration of Li and its uncertainty in pure unknown. The Li standard contained 1.62 $\mu$g Li/mL.

| Unknown (mL) | Standard (mL) | Final volume (mL) | Emission intensity (arbitrary units) |
|---|---|---|---|
| 10.00 | 0.00 | 100.0 | 309 |
| 10.00 | 5.00 | 100.0 | 452 |
| 10.00 | 10.00 | 100.0 | 600 |
| 10.00 | 15.00 | 100.0 | 765 |
| 10.00 | 20.00 | 100.0 | 906 |

**21-B.** Mn was used as an internal standard for measuring Fe by atomic absorption. A standard mixture containing 2.00 $\mu$g Mn/mL and 2.50 $\mu$g Fe/mL gave a quotient (Fe signal/Mn signal) = 1.05/1.00. A mixture with a volume of 6.00 mL was prepared by mixing 5.00 mL of unknown Fe solution with 1.00 mL containing 13.5 $\mu$g Mn/mL. The absorbance of this mixture at the Mn wavelength was 0.128, and the absorbance at the Fe wavelength was 0.185. Find the molarity of the unknown Fe solution.

**21-C.** The atomic absorption signal shown below was obtained with 0.048 5 $\mu$g Fe/mL in a graphite furnace. Estimate the detection limit for Fe, defined for this problem as the concentration of Fe that gives a signal-to-noise ratio of 2.

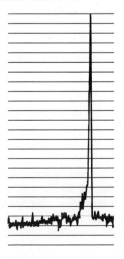

**21-D.** The measurement of Li in brine (salt water) is used by geochemists to help determine the origin of this fluid in oil fields. Flame atomic emission and absorption of Li are subject to interference by scattering, ionization, and overlapping spectral emission from other elements. Atomic absorption analysis of replicate samples of a marine sediment gave the results in the table below.

| Sample and treatment | Li found ($\mu$g/g) | Analytical method | Flame type |
|---|---|---|---|
| 1. None | 25.1 | standard curve | air/$C_2H_2$ |
| 2. Dilute to 1/10 with $H_2O$ | 64.8 | standard curve | air/$C_2H_2$ |
| 3. Dilute to 1/10 with $H_2O$ | 82.5 | standard addition | air/$C_2H_2$ |
| 4. None | 77.3 | standard curve | $N_2O$/$C_2H_2$ |
| 5. Dilute to 1/10 with $H_2O$ | 79.6 | standard curve | $N_2O$/$C_2H_2$ |
| 6. Dilute to 1/10 with $H_2O$ | 80.4 | standard addition | $N_2O$/$C_2H_2$ |

SOURCE: B. Baraj, L. F. H. Niencheski, R. D. Trapaga, R. G. França, V. Cocoli, and D. Robinson, "Interference in the Flame Atomic Absorption Determination of Li," *Fresenius J. Anal. Chem.* **1999**, *364*, 678.

(a) Suggest a reason for the increasing apparent concentration of Li in samples 1 through 3.

(b) Why do samples 4 through 6 give an almost constant result?

(c) What value would you recommend for the real concentration of Li in the sample?

## Problems

### Techniques of Atomic Spectroscopy

**21-1.** In which technique, atomic absorption or atomic emission, is flame temperature stability more critical? Why?

**21-2.** State the advantages and disadvantages of a furnace compared with a flame in atomic absorption spectroscopy.

**21-3.** Figure 21-10 shows a temperature profile for a furnace atomic absorption experiment. Explain the purpose of each different part of the heating profile.

**21-4.** State the advantages and disadvantages of the inductively coupled plasma compared with conventional flames in atomic spectroscopy.

**21-5.** Explain what is meant by the Doppler effect. Rationalize why Doppler broadening increases with increasing temperature and decreasing mass in Equation 21-5.

**21-6.** Explain how the following background correction techniques work: (a) beam chopping; (b) deuterium lamp; (c) Zeeman.

**21-7.** Explain what is meant by spectral, chemical, ionization, and isobaric interference.

**21-8.** (a) Explain the purpose of the collision cell in Figure 21-29.
(b) An unusual use of a collision cell is to intentionally *create* molecules. In geologic strontium isotopic analysis, there is isobaric interference between $^{87}Rb^+$ and $^{87}Sr^+$. Collision with $CH_3F$ converts $Sr^+$ into $SrF^+$. How does this reaction eliminate the isobaric interference?

**21-9.** The laser atomic fluorescence excitation and emission spectra of sodium in an air-acetylene flame are shown below. In the *excitation* spectrum, the laser (bandwidth = 0.03 nm) was scanned through various wavelengths while the detector monochromator (bandwidth = 1.6 nm) was held fixed near 589 nm. In the *emission* spectrum, the laser was fixed at 589.0 nm, and the detector monochromator wavelength was varied. Explain why the emission spectrum gives one broad band, whereas the excitation spectrum gives two sharp lines. How can the excitation linewidths be much narrower than the detector monochromator bandwidth?

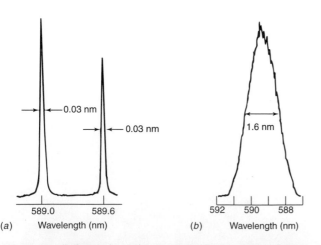

(a) Wavelength (nm)

(b) Wavelength (nm)

Fluorescence excitation and emission spectra of the two sodium D lines in an air-acetylene flame. (a) In the excitation spectrum, the laser was scanned. (b) In the emission spectrum, the monochromator was scanned. The monochromator slit width was the same for both spectra. [From S. J. Weeks, H. Haraguchi, and J. D. Winefordner, "Improvement of Detection Limits in Laser-Excited Atomic Fluorescence Flame Spectrometry," Anal. Chem. **1978**, 50, 360.]

**21-10.** Concentrations (pg per g of snow) of metals by atomic fluorescence in the Agassiz Ice Cap in Greenland for the period 1988–1992 are:[21] Pb, $1.0_4 (\pm 0.1_7) \times 10^2$; Tl, $0.43 \pm 0.08_7$; Cd, $3.5 \pm 0.8_7$; Zn, $1.7_4 (\pm 0.2_6) \times 10^2$; and Al, $6._1 (\pm 1._7) \times 10^3$. The mean annual snowfall was 11.5 g/cm². Calculate the mean annual flux of each metal in units of ng/cm². Flux means how much metal falls on each cm².

**21-11.** Calculate the emission wavelength (nm) of excited atoms that lie $3.371 \times 10^{-19}$ J per molecule above the ground state.

**21-12.** Derive the entries for 500 nm in Table 21-3. Find $N^*/N_0$ at 6 000 K if $g^* = 3$ and $g_0 = 1$.

**21-13.** Calculate the Doppler linewidth for the 589-nm line of Na and for the 254-nm line of Hg, both at 2 000 K.

**21-14.** The first excited state of Ca is reached by absorption of 422.7-nm light.
(a) Find the energy difference (kJ/mol) between ground and excited states.
(b) The degeneracies are $g^*/g_0 = 3$ for Ca. Find $N^*/N_0$ at 2 500 K.
(c) By what percentage will $N^*/N_0$ change with a 15-K rise in temperature?
(d) Find $N^*/N_0$ at 6 000 K.

**21-15.** An *electron volt* (eV) is the energy change of an electron moved through a potential difference of 1 volt: eV = $(1.602 \times 10^{-19}$ C$)(1$ V$) = 1.602 \times 10^{-19}$ J per electron = 96.49 KJ per mole of electrons. Use the Boltzmann distribution to fill in the table and explain why Br is not readily observed in atomic absorption or atomic emission.

| Element: | Na | Cu | Br |
|---|---|---|---|
| Excited state energy (eV): | 2.10 | 3.78 | 8.04 |
| Wavelength (nm): | | | |
| Degeneracy ratio $(g^*/g_0)$: | 3 | 3 | 2/3 |
| $N^*/N_0$ at 2 600 K in flame: | | | |
| $N^*/N_0$ at 6 000 K in plasma: | | | |

**21-16.** MgO prevents premature evaporation of Al in a furnace by maintaining the aluminum as $Al_2O_3$. Another type of matrix modifier prevents loss of signal from the atom X that readily forms the molecular carbide XC in a graphite furnace (a source of carbon). For example, adding yttrium to a sample containing barium increases the Ba signal by 30%. The bond dissociation energy of YC is greater than that of BaC. Explain what is happening to increase the Ba signal.

### Quantitative Analysis by Atomic Spectroscopy

**21-17.** Why is an internal standard most appropriate for quantitative analysis when unavoidable sample losses are expected during sample preparation?

**21-18.** *Standard addition.* To measure Ca in breakfast cereal, 0.521 6 g of crushed Cheerios was ashed in a crucible at 600°C in air for 2 h.[22] The residue was dissolved in 6 M HCl, quantitatively transferred to a volumetric flask, and diluted to 100.0 mL. Then 5.00-mL aliquots were transferred to 50-mL volumetric flasks. Each was treated with standard $Ca^{2+}$ (containing 20.0 μg/mL), diluted to volume with $H_2O$, and analyzed by flame atomic absorption. Construct a standard addition graph and use the method of least squares to find the x-intercept and its uncertainty. Find wt% Ca in Cheerios and its uncertainty.

| Ca²⁺ standard (mL) | Absorbance | Ca²⁺ standard (mL) | Absorbance |
|---|---|---|---|
| 0 | 0.151 | 8.00 | 0.388 |
| 1.00 | 0.185 | 10.00 | 0.445 |
| 3.00 | 0.247 | 15.00 | 0.572 |
| 5.00 | 0.300 | 20.00 | 0.723 |

**21-19.** *Internal standard.* A solution was prepared by mixing 10.00 mL of unknown (X) with 5.00 mL of standard (S) containing 8.24 μg S/mL and diluting the mixture to 50.0 mL. The measured signal quotient was (signal due to X/signal due to S) = 1.690/1.000.

(a) In a separate experiment in which the concentrations of X and S were equal, the quotient was (signal due to X/signal due to S) = 0.930/1.000. What is the concentration of X in the unknown?

(b) Answer the same question if, in a separate experiment in which the concentration of X was 3.42 times the concentration of S, the quotient was (signal due to X/signal due to S) = 0.930/1.000.

**21-20.** 🖳 Potassium standards gave the following emission intensities at 404.3 nm. Emission from the unknown was 417. Find [K⁺] and its uncertainty in the unknown.

| Sample (μg K/mL): | 0 | 5.00 | 10.00 | 20.00 | 30.00 |
|---|---|---|---|---|---|
| Relative emission: | 0 | 124 | 243 | 486 | 712 |

**21-21.** Free cyanide in aqueous solution can be determined indirectly by atomic absorption on the basis of its ability to dissolve silver as it passes through a porous silver membrane filter at pH 12.[23]

$$4Ag(s) + 8CN^- + 2H_2O + O_2 \rightarrow 4Ag(CN)_2^- + 4OH^-$$

A series of silver standards gave a linear calibration curve in flame atomic absorption with a slope of 807 meter units per ppm Ag in the standard. (The meter units are arbitrary numbers proportional to absorbance, and ppm refers to μg Ag/mL.) An unknown cyanide solution passed through the silver membrane gave a meter reading of 198 units. Find the molarity of CN⁻ in the unknown.

**21-22.** *Quality assurance.* Tin is leached (dissolved) into canned foods from the tin-plated steel can.[24] For analysis by inductively coupled plasma atomic emission, food is digested by microwave heating in a Teflon bomb (Figure 28-8) in three steps with HNO₃, H₂O₂, and HCl.

(a) CsCl is added to the final solution at a concentration of 1 g/L. What is the purpose of the CsCl?

(b) 🖳 Calibration data are shown in the table. Find the slope and intercept and their standard deviations and $R^2$, which is a measure of the goodness of fit of the data to a line. Draw the calibration curve.

| Sn (μg/L) | Emission at 189.927 nm |
|---|---|
| 0 | 4.0 |
| 10.0 | 8.5 |
| 20.0 | 19.6 |
| 30.0 | 23.6 |
| 40.0 | 31.1 |
| 60.0 | 41.7 |
| 100.0 | 78.8 |
| 200.0 | 159.1 |

(c) Interference by high concentrations of other elements was assessed at different emission lines of Sn. Foods containing little tin were digested and spiked with Sn at 100.0 μg/L. Then other elements were deliberately added. The following table shows selected results. Which elements interfere at each of the two wavelengths? Which wavelength is preferred for the analysis?

| Element added at 50 mg/L | Sn found (μg/L) with 189.927-nm emission line | Sn found (μg/L) with 235.485-nm emission line |
|---|---|---|
| None | 100.0 | 100.0 |
| Ca | 96.4 | 104.2 |
| Mg | 98.9 | 92.6 |
| P | 106.7 | 104.6 |
| Si | 105.7 | 102.9 |
| Cu | 100.9 | 116.2 |
| Fe | 103.3 | intense emission |
| Mn | 99.5 | 126.3 |
| Zn | 105.3 | 112.8 |
| Cr | 102.8 | 76.4 |

(d) *Limits of detection and quantitation.* The slope of the calibration curve in part (a) is 0.782 units per (μg/L) of Sn. Food containing little Sn gave a mean signal of 5.1 units for seven replicates. Food spiked with 30.0 μg Sn/L gave a mean signal of 29.3 units with a standard deviation of 2.4 units for seven replicates. Use Equations 5-5 and 5-6 to estimate the limits of detection and quantitation.

(e) A 2.0-g food sample was digested and eventually diluted to 50 mL for analysis. Express the limit of quantitation from part (d) in terms of mg Sn/kg food.

# 22 Mass Spectrometry

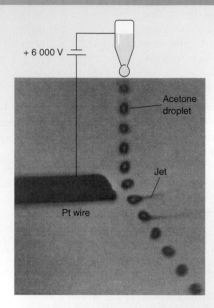

+ 6 000 V

Acetone droplet

Jet

Pt wire

Deflection and disintegration of liquid droplets falling past a wire held at +6 000 V. *[From D. B. Hager and N. J. Dovichi, "Behavior of Microscopic Liquid Droplets Near a Strong Electrostatic Field: Droplet Electrospray," Anal. Chem. **1994,** 66, 1593. See also D. B. Hager, N. J. Dovichi, J. Klassen, and P. Kebarle, "Droplet Electrospray Mass Spectrometry," Anal. Chem. **1994,** 66, 3944.]*

One method for expelling charged protein molecules into the gas phase for mass spectrometry is called *electrospray.* In the experiment shown here, droplets of acetone with a diameter of 16 μm fall past a Pt wire held at +6 000 V with respect to the nozzle from which the droplets came. High voltage creates a glowing electric corona discharge (a plasma containing electrons and positive ions) around the wire, but the discharge is not visible in this photograph. Droplets falling through the discharge become positively charged and are repelled by the wire, deflecting their path to the right. When positively charged droplets come close to the wire, we see a fine stream of liquid jetting out away from the positively charged wire. Microscopic droplets in the fine spray rapidly evaporate. If the liquid had been an aqueous protein solution, the water would evaporate, leaving charged protein molecules in the gas phase.

Francis W. Aston (1877–1945) developed a "mass spectrograph" in 1919 that could separate ions differing in mass by as little as 1% and focus them onto a photographic plate. Aston immediately found that neon consists of two isotopes ($^{20}$Ne and $^{22}$Ne) and went on to discover 212 of the 281 naturally occurring isotopes. He received the Nobel Prize for chemistry in 1922.

*Mass spectrometry* has long been used to measure isotopes and decipher organic structures. Oxygen isotopes in ice cores record the history of Earth's climate (Box 18-3). The constancy of the $^{18}O/^{16}O$ ratio in certain dinosaur bones strongly suggests that these species were warm blooded.[1] Mass spectrometry can elucidate the amino acid sequence in a protein,[2] the sequence of nucleic acids in DNA, the structure of a complex carbohydrate, and the types of lipids in a single organism. Mass spectrometry can measure masses of individual cells[3] and classify cells.[4] Mass spectrometry is the most powerful detector for chromatography, offering both qualitative and quantitative information, providing high sensitivity, and distinguishing different substances with the same retention time.

## 22-1 What Is Mass Spectrometry?

**Mass spectrometry** is a technique for studying the masses of atoms or molecules or fragments of molecules.[5] To obtain a mass spectrum, gaseous species desorbed from condensed phases are ionized; the ions are accelerated by an electric field and then separated according to their mass-to-charge ratio, $m/z$. If all charges are +1, then $m/z$ is numerically equal to the mass. If an ion has a charge of +2, for example, then $m/z$ is 1/2 of the mass. The **mass spectrum** in Figure 22-1 displays detector response versus $m/z$, showing four natural isotopes of $Pb^+$ ions. The area of each peak is proportional to the abundance of each isotope. Box 22-1 defines *nominal mass,* which is the mass we usually speak of in this chapter.

Figure 22-2 shows a **magnetic sector mass spectrometer,** which uses a magnetic field to allow ions of a selected $m/z$ to pass from the ion source to the detector.[6] Gaseous molecules entering at the upper left are converted into ions (usually positive ions), accelerated by an electric field, and expelled into the analyzer tube, where they encounter a magnetic field perpendicular to their direction of travel. The tube is maintained under high vacuum ($\sim 10^{-5}$ Pa) so that ions are not deflected by collision with background gas molecules. The magnet deflects ions toward the detector at the far end of the tube (see Box 22-2). Heavy ions are not deflected enough and light ones are deflected too much to reach the detector. The spectrum of masses is obtained by varying the magnetic field strength.

At the *electron multiplier* detector, each arriving ion starts a cascade of electrons, just as a photon starts a cascade of electrons in a photomultiplier tube (Figure 20-12). A series of dynodes multiplies the number of electrons by $\sim 10^5$ before they reach the anode where current is measured. The mass spectrum shows detector current as a function of $m/z$ selected by the magnetic field.

Mass spectrometers work equally well for negative and positive ions by reversing voltages where the ions are formed and detected. To detect negative ions, a *conversion dynode* with a positive potential is placed before the conventional detector. When bombarded by negative ions, this dynode liberates positive ions that are accelerated into the electron multiplier, which amplifies the signal.

## Electron Ionization

Molecules entering the ion source in Figure 22-2 are converted into ions by **electron ionization.** Electrons emitted from a hot filament (like the one in a light bulb) are accelerated through 70 V before interacting with incoming molecules. Some ($\sim 0.01\%$) molecules (M) absorb as much as 12–15 electron volts (1 eV = 96.5 kJ/mol), which is enough for ionization:

$$M \ + \ e^- \ \rightarrow \ M^{+\cdot} \ + \ e^- \ + \ e^-$$

70 eV     Molecular ion  ~55 eV     0.1 eV

Figure 22-1 Mass spectrum showing natural isotopes of Pb observed as an impurity in brass. *[From Y. Su, Y. Duan, and Z. Jin, "Development and Evaluation of a Glow Discharge Microwave-Induced Plasma Tandem Source for Time-of-Flight Mass Spectrometry," Anal. Chem.* **2000,** *72, 5600.]* The variability of isotopic abundances in Pb from natural sources creates a large uncertainty in the atomic mass (207.2 ± 0.1) in the periodic table.

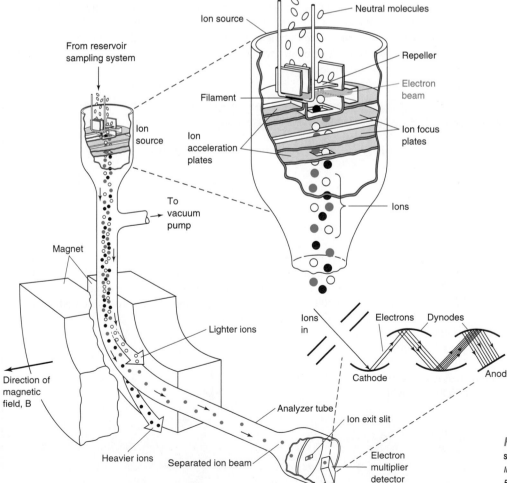

Figure 22-2 Magnetic sector mass spectrometer. *[Adapted from F. W. McLafferty, Interpretation of Mass Spectra (New York: Benjamin, 1966).]*

A reasonable match of the experimental spectrum to one in the computer library is **not** proof of molecular structure—it is just a clue.[7] You must be able to explain all major peaks (and even minor peaks at high *m/z*) in the spectrum in terms of the proposed structure, and you should obtain a matching spectrum from an authentic sample before reaching a conclusion. The authentic sample must have the same chromatographic retention time as the proposed unknown. Many isomers produce nearly identical mass spectra.

Almost all stable molecules have an even number of electrons. When one electron is lost, the resulting cation with one unpaired electron is designated $M^{+\cdot}$, the **molecular ion.** After ionization, $M^{+\cdot}$ usually has enough residual internal energy (~1 eV) to break into fragments.

A small positive potential on the repeller plate of the ion source pushes ions toward the analyzer tube, and a small potential on the ion focus plates creates a focused beam. High voltage (~1 000–10 000 V) between the ion acceleration plates imparts a high velocity to ions as they are expelled from the bottom of the ion gun.

The electron kinetic energy of 70 eV is much greater than the ionization energy of molecules. Consider formaldehyde in Figure 22-3, whose molecular orbitals were shown in Figure 18-10. The most easily lost electron comes from a nonbonding ("lone pair") orbital centered on oxygen, with an ionization energy of 11.0 eV. To remove a pi bonding electron from neutral formaldehyde requires 14.1 eV, and to remove the highest-energy sigma bonding electron from the neutral molecule requires 15.9 eV.

Interaction with a 70-eV electron will most likely remove the electron with lowest ionization energy. The resulting molecular ion, $M^{+\cdot}$, can have so much extra energy that it breaks into fragments. There might be so little $M^{+\cdot}$ that its peak is small or absent in the mass spectrum. The electron ionization mass spectrum at the left side of Figure 22-4 does not exhibit an $M^{+\cdot}$ peak that would be at *m/z* 226. Instead, there are peaks at *m/z* 197, 156, 141, 112, 98, 69, and 55, arising from fragmentation of $M^{+\cdot}$. These peaks provide clues about the structure of the molecule. A computer search is commonly used to match the spectrum of an unknown to similar spectra in a library.[8]

## Box 22-2   How Ions of Different Masses Are Separated by a Magnetic Field

Electron ionization in the ion source of the mass spectrometer in Figure 22-2 creates positive ions, $M^{z+}$, with different masses. Let the mass of an ion be $m$ and let its charge be $+ze$ (where $e$ is the magnitude of the charge of an electron). When the ion is accelerated through a potential difference $V$ by the ion acceleration plates, it acquires a kinetic energy equal to the electric potential difference:

$$\underset{\substack{\text{Kinetic energy} \\ (\text{v} = \text{velocity})}}{\tfrac{1}{2}m\text{v}^2} = \underset{\substack{\text{Potential energy}}}{zeV} \quad \Rightarrow \quad \text{v} = \sqrt{\frac{2zeV}{m}} \qquad \text{(A)}$$

An ion with charge $ze$ and velocity v traveling perpendicular to a magnetic field $B$ experiences a force $zevB$ that is perpendicular to both the velocity vector and the magnetic field vector. This force deflects the ion through a circular path of radius $r$. The centripetal force ($mv^2/r$) required to deflect the particle is provided by the magnetic field.

$$\underset{\substack{\text{Centripetal} \\ \text{force}}}{\frac{m\text{v}^2}{r}} = \underset{\substack{\text{Magnetic} \\ \text{force}}}{zevB} \quad \Rightarrow \quad \text{v} = \frac{zeBr}{m} \qquad \text{(B)}$$

Equating velocities from Equations A and B gives

$$\frac{zeBr}{m} = \sqrt{\frac{2zeV}{m}} \quad \Rightarrow \quad \boxed{\frac{m}{z} = \frac{eB^2r^2}{2V}} \qquad \text{(C)}$$

Equation C tells us the radius of curvature of the path traveled by an ion with mass $m$ and charge $z$. The radius of curvature is fixed by the geometry of the hardware. Ions can be selected to reach the detector by adjusting the magnetic field, $B$, or the accelerating voltage, $V$. Normally, $B$ is varied to select ions and $V$ is fixed near 3 000 V. Transmission of ions and detector response both decrease when $V$ is decreased.

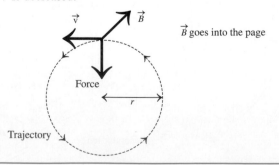

$\vec{B}$ goes into the page

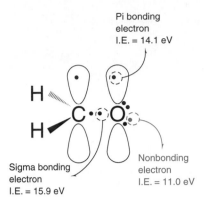

Pi bonding
electron
I.E. = 14.1 eV

Sigma bonding
electron
I.E. = 15.9 eV

Nonbonding
electron
I.E. = 11.0 eV

**Figure 22-3** Ionization energies (I.E.) of valence electrons in formaldehyde. *[Data from C. R. Brundle, M. B. Robin, N. A. Kuebler, and H. Basch, "Perfluoro Effects in Photoelectron Spectroscopy," J. Am. Chem. Soc. 1972, 94, 1451.]* Here are other first ionization energies:

| | |
|---|---|
| $CH_3CH_2CH_2CH_3$ | 10.6 eV (sigma) |
| $CH_2{=}CHCH_2CH_3$ | 9.6 eV (pi) |
| $(CH_3CH_2)_2\ddot{O}$ | 9.6 eV (nonbonding) |
| ⬡NH | 8.6 eV (nonbonding) |
| ⬡ | 9.2 eV (pi) |

If you lower the kinetic energy of electrons in the ionization source to, say, 20 eV, there will be a lower yield of ions and less fragmentation. You would likely observe a greater abundance of molecular ions. We customarily used 70 eV because it gives reproducible fragmentation patterns that can be compared with library spectra.

The most intense peak in a mass spectrum is called the **base peak.** Intensities of other peaks are expressed as a percentage of the base peak intensity. In the electron ionization spectrum in Figure 22-4, the base peak is at *m/z* 141.

**Chemical ionization** produces less fragmentation than electron ionization. For chemical ionization, the ionization source is filled with a *reagent gas* such as methane, isobutane, or ammonia, at a pressure of ~1 mbar. Energetic electrons (100–200 eV) convert $CH_4$ into a variety of reactive products:

$$CH_4 + e^- \rightarrow CH_4^{+\bullet} + 2e^-$$
$$CH_4^{+\bullet} + CH_4 \rightarrow CH_5^+ + {}^\bullet CH_3$$
$$CH_4^{+\bullet} \rightarrow CH_3^+ + H^\bullet$$
$$CH_3^+ + CH_4 \rightarrow C_2H_5^+ + H_2$$

$CH_5^+$ is a potent proton donor that reacts with analyte to give the *protonated molecule,* $MH^+$, which is usually the most abundant ion in the chemical ionization mass spectrum.

$$CH_5^+ + M \rightarrow CH_4 + MH^+$$

In the chemical ionization spectrum in Figure 22-4, $MH^+$ at *m/z* 227 is the second strongest peak and there are fewer fragments than in the electron ionization spectrum.

Ammonia or isobutane are used in place of $CH_4$ to reduce the fragmentation of $MH^+$. These reagents bind $H^+$ more strongly than $CH_4$ does and impart less energy to $MH^+$ when the proton is transferred to M. $NO^+$ is a mild, versatile ionization reagent generated from NO by radioactive ${}^{210}Po$.[10] Negative chemical ionization reagents include $O_2^-$, $F^-$, and $SF_6^-$.[10]

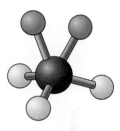

$CH_5^+$ is described as a $CH_3$ tripod with an added $\mathbf{H_2}$ unit. [**H**—C—**H**] is held together by two electrons distributed over three atoms. Atoms of the $\mathbf{H_2}$ unit rapidly exchange with atoms of the $CH_3$ unit.[9]

The *molecular ion,* $M^{+\bullet}$, can be formed by reactions such as

$$CH_4^{+\bullet} + M \rightarrow CH_4 + M^{+\bullet}$$

$MH^+$ is the *protonated molecule,* not the molecular ion.

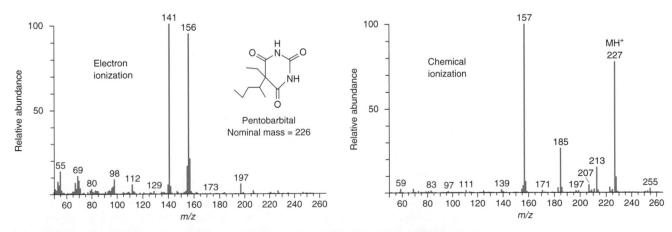

**Figure 22-4** Mass spectra of the sedative pentobarbital, by using electron ionization (left) or chemical ionization (right). The molecular ion ($M^{+\bullet}$, *m/z* 226) is not evident with electron ionization. The dominant ion from chemical ionization is $MH^+$. The peak at *m/z* 255 in the chemical ionization spectrum is from $M(C_2H_5)^+$. *[Courtesy Varian Associates, Sunnyvale, CA.]*

Pentobarbital
Nominal mass = 226

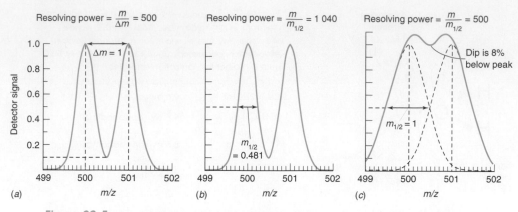

$$\text{Resolving power} = \frac{m}{\Delta m} = 500 \qquad \text{Resolving power} = \frac{m}{m_{1/2}} = 1\,040 \qquad \text{Resolving power} = \frac{m}{m_{1/2}} = 500$$

**Figure 22-5** Resolving power. (*a*) By one definition, the resolving power is $m/\Delta m = 500/1 = 500$. (*b*) By a second definition, the resolving power for the same pair of peaks is $m/m_{1/2} = 500/0.481 = 1\,040$. (*c*) With the second definition, two peaks at $m/z$ 500 and 501 are just barely discernible if the resolving power is 500.

## Resolving Power

Mass spectra in Figure 22-4 are computer-generated bar graphs. In contrast, Figure 22-1 shows the actual detector signal. Each mass spectral peak has a width that limits how closely two peaks could be spaced and still be resolved. If peaks are too close, they appear to be a single peak.

The higher the **resolving power** of a mass spectrometer, the better it is able to separate two peaks with similar mass.

$$\text{Resolving power} = \underbrace{\frac{m}{\Delta m}}_{\text{Figure 22-5a}} \quad \text{or} \quad \underbrace{\frac{m}{m_{1/2}}}_{\text{Figure 22-5b}} \qquad (22\text{-}1)$$

> The expression $m/m_{1/2}$ gives a value twice as great as $m/\Delta m$ for resolving power.
>
> *Resolution*, which is the inverse of resolving power ($\Delta m/m$), is the smallest mass difference that can be detected as separate peaks. Resolving power is a big number, and resolution is a small number.

where $m$ is the smaller value of $m/z$. The denominator is defined in two different ways. In Figure 22-5a, the denominator is the separation of the two peaks ($\Delta m$) when the overlap at their base is 10% as high as the peak. The resolving power in Figure 22-5a is $m/\Delta m = 500/1.00 = 5.00 \times 10^2$. In Figure 22-5b, the denominator is taken as $m_{1/2}$, the width of the peak at half the maximum height, which is 0.481 Da (dalton, see Box 22-1). By this definition, the resolving power is $m/m_{1/2} = 500/0.481 = 1.04 \times 10^3$ for the same two peaks. Figure 22-5c shows that, when the expression $m/m_{1/2}$ gives a resolving power of $5.00 \times 10^2$, two peaks at $m/z$ 500 and 501 are barely discernible. Specify which definition you use when expressing resolving power.

### Example Resolving Power

Using the $^{208}$Pb peak in Figure 22-1, find the resolving power from the expression $m/m_{1/2}$.

**Solution** The width at half height is 0.146 $m/z$ units. Therefore,

$$\text{Resolving power} = \frac{m}{m_{1/2}} = \frac{208}{0.146} = 1.42 \times 10^3$$

An instrument with a resolving power of $1.42 \times 10^3$ separates peaks well near $m/z$ 200 but gives a barely discernible separation at $m/z$ 1 420.

## ▪▪▪ 22-2 Oh, Mass Spectrum, Speak to Me!*

*An expression borrowed from O. David Sparkman, a master teacher.

Each mass spectrum has a story to tell. The molecular ion, M$^{+\cdot}$, tells us the molecular mass of an unknown. Unfortunately, with electron ionization, some compounds do not exhibit a molecular ion, because M$^{+\cdot}$ breaks apart so efficiently. However, the fragments provide the most valuable clues to the structure of an unknown. To find the molecular mass, we can obtain a chemical ionization mass spectrum, which usually has a strong peak for MH$^+$.

The **nitrogen rule** helps us propose compositions for molecular ions: If a compound has an odd number of nitrogen atoms—in addition to any number of C, H, halogens, O, S, Si, and P—then M$^{+\cdot}$ has an odd nominal mass. For a compound with an even number of nitrogen atoms (0, 2, 4, and so on), M$^{+\cdot}$ has an even nominal mass. A molecular ion at $m/z$ 128 can have 0 or 2 N atoms, but it cannot have 1 N atom.

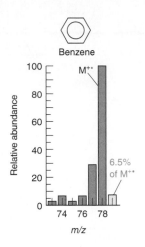
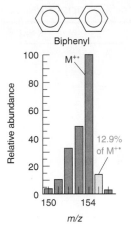

**Figure 22-6** Electron ionization (70 eV) mass spectra of molecular ion region of benzene ($C_6H_6$) and biphenyl ($C_{12}H_{10}$). *[From NIST/EPA/NIH Mass Spectral Database.[8]]*

## Molecular Ion and Isotope Patterns

Electron ionization of aromatic compounds (those containing benzene rings) usually gives significant intensity for $M^{+\cdot}$. $M^{+\cdot}$ is the base peak (most intense) in the spectra of benzene and biphenyl in Figure 22-6.

The next higher mass peak, $M+1$, provides information on elemental composition. Table 22-1 lists the natural abundance of several isotopes. For carbon, 98.93% of atoms are $^{12}C$ and 1.07% are $^{13}C$. Nearly all hydrogen is $^1H$, with 0.012% $^2H$. Applying the factors in Table 22-2 to $C_nH_m$, the intensity of the $M+1$ peak should be

*Intensity of M+1 relative to molecular ion for $C_nH_m$*

$$\text{Intensity} = \underbrace{n \times 1.08\%}_{\text{From }^{13}C} + \underbrace{m \times 0.012\%}_{\text{From }^2H} \qquad (22\text{-}2)$$

Although we write the molecular ion as $M^{+\cdot}$, we use the notation $M+1$ and $M-29$ for other peaks, without indicating positive charge. $M+1$ refers to an ion with a mass one unit greater than that of $M^{+\cdot}$.

## Table 22-1 Isotopes of selected elements

| Element | Mass number | Mass (Da)[a] | Abundance (atom %)[b] | Element | Mass number | Mass (Da)[a] | Abundance (atom %)[b] |
|---|---|---|---|---|---|---|---|
| Proton | 1 | 1.007 276 467 | — | Cl | 35 | 34.968 85 | 75.78 |
| Neutron | 1 | 1.008 664 916 | — | | 37 | 36.965 90 | 24.22 |
| Electron | — | 0.000 548 580 | — | Ar | 36 | 35.967 55 | 0.336 |
| H | 1 | 1.007 825 | 99.988 | | 38 | 37.962 73 | 0.063 |
| | 2 | 2.014 10 | 0.012 | | 40 | 39.962 38 | 99.600 |
| B | 10 | 10.012 94 | 19.9 | Fe | 54 | 53.939 61 | 5.845 |
| | 11 | 11.009 31 | 80.1 | | 56 | 55.934 94 | 91.754 |
| C | 12 | 12(exact) | 98.93 | | 57 | 56.935 40 | 2.119 |
| | 13 | 13.003 35 | 1.07 | | 58 | 57.933 28 | 0.282 |
| N | 14 | 14.003 07 | 99.632 | Br | 79 | 78.918 34 | 50.69 |
| | 15 | 15.000 11 | 0.368 | | 81 | 80.916 29 | 49.31 |
| O | 16 | 15.994 91 | 99.757 | I | 127 | 126.904 47 | 100 |
| | 17 | 16.999 13 | 0.038 | Hg | 196 | 195.965 81 | 0.15 |
| | 18 | 17.999 16 | 0.205 | | 198 | 197.966 75 | 9.97 |
| F | 19 | 18.998 40 | 100 | | 199 | 198.968 26 | 16.87 |
| Si | 28 | 27.976 93 | 92.230 | | 200 | 199.968 31 | 23.10 |
| | 29 | 28.976 49 | 4.683 | | 201 | 200.970 29 | 13.18 |
| | 30 | 29.973 77 | 3.087 | | 202 | 201.970 63 | 29.86 |
| P | 31 | 30.973 76 | 100 | | 204 | 203.973 48 | 6.87 |
| S | 32 | 31.972 07 | 94.93 | Pb | 204 | 203.973 03 | 1.4 |
| | 33 | 32.971 46 | 0.76 | | 206 | 205.974 45 | 24.1 |
| | 34 | 33.967 87 | 4.29 | | 207 | 206.975 88 | 22.1 |
| | 36 | 35.967 08 | 0.02 | | 208 | 207.976 64 | 52.4 |

*a. 1 dalton (Da) ≡ 1/12 of the mass of $^{12}C$ = 1.660 538 86 (28) × $10^{-27}$ kg (from http://physics.nist.gov/constants, June 2004). Nuclide masses from G. Audi, A. H. Wapsta, and C. Thibault, Nucl. Phys. **2003**, A729, 337, found at www.nndc.bnl.gov/masses/ June 2004. This source provides more significant figures for atomic mass than are cited in this table.*

*b. Abundance is representative of what is found in nature. Significant variations are observed. For example, $^{18}O$ in natural substances has been found in the range 0.188 to 0.222 atom%. The latest list of isotope abundances, which is slightly different from this table, is found in J. K. Böhlke et al., "Isotopic Compositions of the Elements, 2001," J. Phys. Chem. Ref. Data **2005**, 34, 57.*

**Table 22-2** Isotope abundance factors (%) for interpreting mass spectra

| Element | X+1 | X+2 | X+3 | X+4 | X+5 | X+6 |
|---|---|---|---|---|---|---|
| H | $0.012n$ | | | | | |
| C | $1.08n$ | $0.0058n(n-1)$ | | | | |
| N | $0.369n$ | | | | | |
| O | $0.038n$ | $0.205n$ | | | | |
| F | $0$ | | | | | |
| Si | $5.08n$ | $3.35n$ | $0.170n(n-1)$ | $0.056n(n-1)$ | | |
| P | $0$ | | | | | |
| S | $0.801n$ | $4.52n$ | $0.036n(n-1)$ | $0.102n(n-1)$ | | |
| Cl | — | $32.0n$ | — | $5.11n(n-1)$ | — | $0.544n(n-1)(n-2)$ |
| Br | — | $97.3n$ | — | $47.3n(n-1)$ | — | $15.3n(n-1)(n-2)$ |
| I | $0$ | | | | | |

*EXAMPLE: For a peak at m/z = X containing n carbon atoms, the intensity from carbon at X+1 is n × 1.08% of the intensity at X. The intensity at X+2 is n(n − 1) × 0.005 8% of the intensity at X. Contributions from isotopes of other atoms in the ion are additive.*

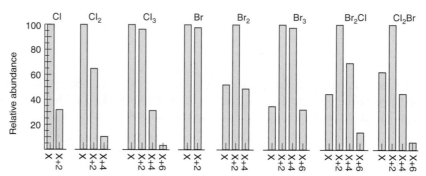

**Figure 22-7** Calculated isotopic patterns for species containing Cl and Br.

For benzene, $C_6H_6$, $M^{+\cdot}$ is observed at $m/z$ 78. The predicted intensity at $m/z$ 79 is $6 \times 1.08 + 6 \times 0.012 = 6.55\%$ of the abundance of $M^{+\cdot}$. The observed intensity in Figure 22-6 is 6.5%. An intensity within ±10% of the expected value (5.9 to 7.2, in this case) is within the precision of ordinary mass spectrometers. For biphenyl ($C_{12}H_{10}$), we predict that M+1 should have $12 \times 1.08 + 10 \times 0.012 = 13.1\%$ of the intensity of $M^{+\cdot}$. The observed value is 12.9%.

### Example Listening to a Mass Spectrum

In the chemical ionization mass spectrum of pentobarbital in Figure 22-4, the peak with the most significant intensity at the high end of the mass spectrum at $m/z$ 227 is suspected to be $MH^+$. If this is so, then the nominal mass of M is 226. The nitrogen rule tells us that a molecule with an even mass must have an even number of nitrogen atoms. If you know from elemental analysis that the compound contains only C, H, N, and O, how many atoms of carbon would you suspect are in the molecule?

**Solution** The $m/z$ 228 peak has 12.0% of the height of the $m/z$ 227 peak. Table 22-2 tells us that $n$ carbon atoms will contribute $n \times 1.08\%$ intensity at $m/z$ 228 from $^{13}C$. Contributions from $^2H$ and $^{17}O$ are small. $^{15}N$ makes a larger contribution, but there are probably few atoms of nitrogen in the compound. Our first guess is

$$\text{Number of C atoms} = \frac{\text{observed peak intensity for M+1}}{\text{contribution per carbon atom}} = \frac{12.0\%}{1.08\%} = 11.1 \approx 11$$

The actual composition of $MH^+$ at $m/z$ 227 is $C_{11}H_{19}O_3N_2$. From the factors in Table 22-2, the theoretical intensity at $m/z$ 228 is

$$\text{Intensity} = \underbrace{11 \times 1.08\%}_{^{13}C} + \underbrace{19 \times 0.012\%}_{^2H} + \underbrace{3 \times 0.038\%}_{^{17}O} + \underbrace{2 \times 0.369\%}_{^{15}N} = 13.0\%$$

The theoretical intensity is within the 10% uncertainty of the observed value of 12.0%.

Ions containing Cl or Br have distinctive isotopic peaks shown in Figure 22-7.[11] In the mass spectrum of 1-bromobutane in Figure 22-8, two nearly equal peaks at $m/z$ 136 and 138

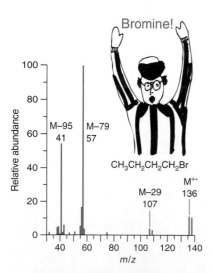

**Figure 22-8** Electron ionization mass spectrum (70 eV) of 1-bromobutane. *[From A. Illies, P. B. Shevlin, G. Childers, M. Peschke, and J. Tsai, "Mass Spectrometry for Large Undergraduate Laboratory Sections," J. Chem. Ed. 1995, 72, 717. Referee from Maddy Harris.]*

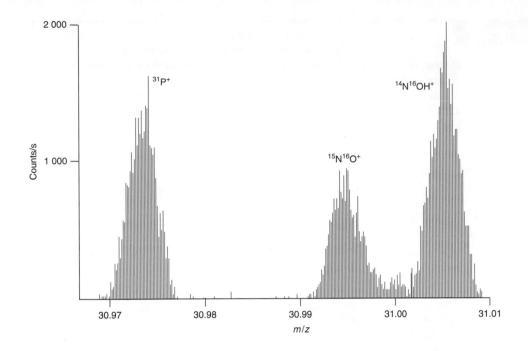

**Figure 22-9** High-resolution double-focusing sector-field mass spectrum distinguishing $^{31}P^+$, $^{15}N^{16}O^+$, and $^{14}N^{16}OH^+$ with a resolution of $m/\Delta m = 4\,000$. The spectrum is a bar graph of detector counts/s at each increment of $m/z$. [From J. S. Becker, S. F. Boulyga, C. Pickhardt, J. Becker, S. Buddrus, and M. Przybylski, "Determination of Phosphorus in Small Amounts of Protein Samples by ICP-MS," Anal. Bioanal. Chem. **2003**, 375, 561.]

are a strong indication that the molecular ion contains one Br atom. The fragment at $m/z$ 107 has a nearly equal partner at $m/z$ 109, strongly suggesting that this fragment ion contains Br. Box 22-3 describes other information available from isotope ratios.

## High-Resolution Mass Spectrometry

An ion at $m/z$ 84 could have a variety of elemental compositions, such as $C_5H_8O^+ = 84.056\,96$ Da or $C_6H_{12}^+ = 84.093\,35$ Da. We can determine which composition is correct if the spectrometer can distinguish sufficiently small differences in mass. With a *double-focusing mass spectrometer* or an accurate *time-of-flight mass spectrometer* (Section 22-3), we can resolve differences of 0.001 at $m/z$ 100. Figure 22-9 shows a high-resolution spectrum that distinguishes $^{31}P^+$, $^{15}N^{16}O^+$, and $^{14}N^{16}OH^+$—all with a nominal mass of 31.

To ensure accurate mass measurement, spectrometers are calibrated with compounds such as perfluorokerosene ($CF_3(CF_2)_nCF_3$) or perfluorotributylamine (($CF_3CF_2CF_2CF_2)_3N$). In high-resolution spectra, the exact masses of fluorocarbon fragments are slightly lower than those of ions containing C, H, O, N, and S. For high-resolution work, standards should be run with the unknown.

## Rings + Double Bonds

If we know the composition of a molecular ion, and we want to propose its structure, a handy equation that gives the number of **rings + double bonds** (R + DB) is

*Rings + double bonds formula:*

$$R + DB = c - h/2 + n/2 + 1 \qquad (22\text{-}3)$$

where $c$ is the number of Group 14 atoms (C, Si, and so on, which all make four bonds), $h$ is the number of (H + halogen) atoms (which make one bond), and $n$ is the number of Group 15 atoms (N, P, As, and so on, which make three bonds). Group 16 atoms (O, S, and so on, which make two bonds) do not enter the formula. Here is an example:

$R + DB = c - h/2 + n/2 + 1$

$R + DB = (14 + 1) - \dfrac{22 + 1 + 1}{2} + \dfrac{1 + 1}{2} + 1 = 5$

The molecule has one ring + four double bonds.
Note that $c$ includes C + Si, $h$ includes H + Cl + Br, and $n$ includes N + As.

$$\underbrace{C_{14}Si}_{c}\ \underbrace{H_{22}Cl\,Br}_{h}\ \underbrace{NAsO_3S}_{n}$$

Predicted masses from Table 22-1 and masses observed by high-resolution mass spectrometry:

$C_5H_8O^{+\bullet}$:

| | |
|---|---|
| 5C | $5 \times 12.000\,00$ |
| 8H | $+8 \times 1.007\,825$ |
| 1O | $+1 \times 15.994\,91$ |
| $-e^-$ | $-1 \times 0.000\,55$ |
| | 84.056 96 |
| observed:[15] | 84.059 1 |

$C_6H_{12}^{+\bullet}$:

| | |
|---|---|
| 6C | $6 \times 12.000\,00$ |
| 12H | $12 \times 1.007\,825$ |
| $-e^-$ | $-1 \times 0.000\,55$ |
| | 84.093 35 |
| observed:[15] | 84.093 9 |

If P makes more than three bonds or if S makes more than two bonds, then Equation 22-3 does not include these extra bonds. Examples that violate Equation 22-3:

Box 22-3   Isotope Ratio Mass Spectrometry

The opening of Chapter 24 describes the analysis of cholesterol from ancient human bones by *isotope ratio mass spectrometry*.[12] A compound eluted from a gas chromatography column is passed through a combustion furnace loaded with a metal catalyst (such as CuO/Pt at 820°C) to oxidize organic compounds to $CO_2$. The $H_2O$ combustion by-product is removed by passage through a Nafion fluorocarbon tube (page 375). Water can diffuse through the membrane, but other combustion products are retained. $CO_2$ then enters a mass spectrometer that monitors $m/z$ 44 ($^{12}CO_2$) and 45 ($^{13}CO_2$). One detector is dedicated to each ion.

Natural carbon is composed of 98.9% $^{12}C$ and 1.1% $^{13}C$. The chart shows consistent, small variations in $^{13}C$ from natural

sources. The standard used to measure carbon isotope ratios is calcium carbonate from the Pee Dee belemnite (fossil shell) formation in South Carolina (designated PDB) with a ratio $R_{PDB} = {}^{13}C/{}^{12}C = 0.011\ 237_2$. (The composition is accurate to four significant figures, but small differences are precise to six significant figures.) The $\delta^{13}C$ scale expresses small variations in isotopic compositions:

$$\delta^{13}C\ (\text{parts per thousand, ‰}) = 1\ 000\left(\frac{R_{sample} - R_{PDB}}{R_{PDB}}\right)$$

$\delta^{13}C$ of natural materials provides information about their biological and geographic origins.[13,14]

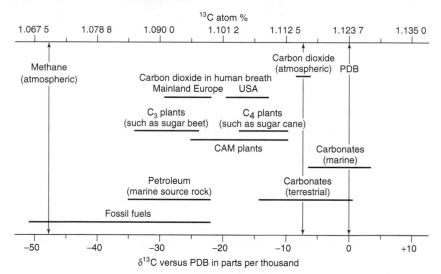

Variation in $^{13}C$ from natural sources. $C_3$, $C_4$, and CAM are types of plants with distinct metabolic pathways leading to different $^{13}C$ incorporation. *[Adapted from W. Meier-Augenstein, LCGC **1997**, 15, 244.]*

## Identifying the Molecular Ion Peak

Figure 22-10 shows electron ionization mass spectra of isomers with the elemental composition $C_6H_{12}O$. $M^{+\cdot}$ is marked by the solid triangle at $m/z$ 100.

If these spectra had been obtained from unknowns, how would we know that the peak at $m/z$ 100 represents the molecular ion? Here are some guidelines:

**Identifying the molecular ion**

1. $M^{+\cdot}$ will be at the highest $m/z$ value of any of the "significant" peaks in the spectrum that cannot be attributed to isotopes or background. "Background" arises from sources such as pump oil in the spectrometer and stationary phase from a gas chromatography column. Experience is required to recognize these signals. With electron ionization, the molecular ion peak intensity is often no greater than 5–20% of the base peak and might not represent more than 1% of all ions.
2. Intensities of isotopic peaks at $M+1$, $M+2$, and so on, must be consistent with the proposed formula.
3. The peak for the heaviest fragment ion should not correspond to an improbable mass loss from $M^{+\cdot}$. It is rare to find a mass loss in the range 3–14 or 21–25 Da. Common losses include 15 ($CH_3$), 17 (OH or $NH_3$), 18 ($H_2O$), 29 ($C_2H_5$), 31 ($OCH_3$), 43 Da ($CH_3CO$ or $C_3H_7$), and many others. If you think that $m/z$ 150 represents $M^{+\cdot}$, but there is a significant peak at $m/z$ 145, then $M^{+\cdot}$ is not assigned correctly, because a mass loss of 5 Da is improbable. Alternatively, the two peaks represent ions from different compounds, or both peaks represent fragments from a compound whose mass is greater than 150 Da.
4. If a fragment ion is known to contain, say, three atoms of element X, then there must be at least three atoms of X in the molecular ion.

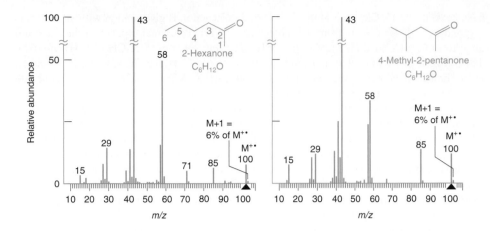

**Figure 22-10** Electron ionization (70 eV) mass spectra of isomeric ketones with the composition $C_6H_{12}O$. *[From NIST/EPA/NIH Mass Spectral Database.[8]]*

In both spectra in Figure 22-10, the peak at highest $m/z$ with "significant" intensity is $m/z$ 100. The next-highest significant peak is $m/z$ 85, representing a loss of 15 Da (probably $CH_3$). Peaks at $m/z$ 85 and 100 are consistent with $m/z$ 100 representing $M^{+\cdot}$. If $M^{+\cdot}$ has an even mass, the nitrogen rule says there cannot be an odd number of N atoms in the molecule.

In both spectra, the M+1 peak has 6% of the intensity of the $M^{+\cdot}$ peak, with just one significant digit in the measurement. From the intensity of the peak for M+1, we estimate the number of C atoms:

$$\text{Number of C atoms} = \frac{\text{observed (M+1)/}M^{+\cdot}\text{ intensity}}{\text{contribution per carbon atom}} = \frac{6\%}{1.08\%} = 5._6 \approx 6$$

If there are six C atoms and no N atoms, a possible composition is $C_6H_{12}O$. The expected intensity of M+1 is

$$\text{Intensity} = \underbrace{6 \times 1.08\%}_{^{13}C} + \underbrace{12 \times 0.012\%}_{^{2}H} + \underbrace{1 \times 0.038\%}_{^{17}O} = 6.7\% \text{ of } M^{+\cdot}$$

The observed intensity of 6% for M+1 is within experimental error of 6.7%, so $C_6H_{12}O$ is consistent with the data so far. $C_6H_{12}O$ requires one ring or double bond.

### Interpreting Fragmentation Patterns

Consider how the 2-hexanone molecular ion can break apart to give the many peaks in Figure 22-10. Reactions A and B in Figure 22-11 show $M^{+\cdot}$ derived from loss of a nonbonding electron from oxygen, which has the lowest ionization energy. In Reaction A, the C—C bond adjacent to C=O splits so that one electron goes to each C atom. The products are a neutral butyl radical ($^{\cdot}C_4H_9$) and $CH_3CO^{+\cdot}$. Only the ion is detected by the mass spectrometer, giving the

The small peak at $m/z$ 86 is not a loss of 14 Da from $M^{+\cdot}$. It is the isotopic partner of the $m/z$ 85 peak containing $^{13}C$.

We see no evidence at M+1, M+2, and M+3 for the elements Cl, Br, Si, or S.

**Challenge** What intensities would you see at M+2 and M+3 if there were one Cl, one Br, one Si, or one S in the molecule?

$$R + DB = c - h/2 + n/2 + 1$$
$$= 6 - 12/2 + 0 + 1 = 1$$

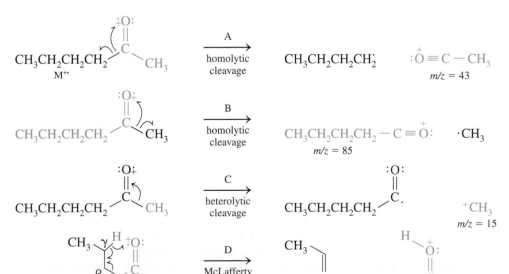

$\curvearrowright$ is transfer of one electron

$\curvearrowright$ is transfer of two electrons

Types of bond breaking:
- *Homolytic cleavage:* 1 electron remains with each fragment
- *Heterolytic cleavage:* Both electrons stay with one fragment

In general, the most intense peaks correspond to the most stable fragments.

**Figure 22-11** Four possible fragmentation pathways for the molecular ion of 2-hexanone.

base peak at $m/z$ 43. Cleavage of the C—C bond in Reaction B gives an ion with $m/z$ 85, corresponding to loss of $\cdot CH_3$ from the molecular ion. Two other major peaks in the spectrum arise from $C_4$—$C_5$ bond cleavage to give $CH_3CH_2^+$ ($m/z$ 29) and $^+CH_2CH_2COCH_3$ ($m/z$ 71). The nitrogen rule told us that molecules containing only C, H, halogens, O, S, Si, P, and an even number of N atoms (such as 0) have an even mass. A fragment of a neutral molecule that is missing a H atom must have an odd mass.

We have yet to account for the second-tallest peak at $m/z$ 58, which is special because it has an *even* mass. The molecular ion has an even mass (100 Da). Radical fragments, such as $CH_3CH_2^+$, have an *odd* mass. All of the fragments discussed so far have an odd mass. The peak at $m/z$ 58 results from the loss of a *neutral* molecule with an *even* mass of 42 Da.

Reaction D in Figure 22-11 shows a common rearrangement that leads to loss of a neutral molecule with even mass. In ketones with a H atom on the $\gamma$ carbon atom, the H atom can be transferred to $O^+$. Concomitantly, the $C_\alpha$—$C_\beta$ bond cleaves and a neutral molecule of propene ($CH_3CH=CH_2$, 42 Da) is lost. The resulting ion has a mass of 58 Da.

The spectra in Figure 22-10 allow us to distinguish one isomer of $C_6H_{12}O$ from the other. The main difference between the spectra is a peak at $m/z$ 71 for 2-hexanone that is absent in 4-methyl-2-pentanone. The peak at $m/z$ 71 results from loss of ethyl radical, $CH_3CH_2^+$, from $M^{+\cdot}$. Ethyl radical is derived from carbon atoms 5 and 6 of 2-hexanone. There is no simple way for an ethyl radical to cleave from 4-methyl-2-pentanone. The diagram in the margin shows how peaks at $m/z$ 15, 85, 43, and 57 can arise from breaking bonds in 4-methyl-2-pentanone. A rearrangement like that at the bottom of Figure 22-11 accounts for $m/z$ 58.

Other major peaks in Figure 22-10 might be $CH_2=C=O^{+\cdot}$ ($m/z$ 42), $C_3H_5^+$ ($m/z$ 41), $C_3H_3^+$ ($m/z$ 39), $C_2H_5^+$ or $HC\equiv O^+$ ($m/z$ 29), and $C_2H_3^+$ ($m/z$ 27). Small fragments are common in many spectra and not very useful for structure determination.

Interpretation of mass spectra to elucidate molecular structures is an important field.[16] Fragmentation patterns can even unravel the structures of large biological molecules.

## ◼◼▮▮ 22-3 Types of Mass Spectrometers

Figure 22-2 shows a *magnetic sector mass spectrometer* in which ions with different mass, but constant kinetic energy, are separated by their trajectories in a magnetic field. The kinetic energy is imparted to the ions by the voltage between the acceleration plates in the ion source. Ions are created from the neutral molecule with a small spread of kinetic energies and are accelerated to different extents depending on where in the ion source they were formed.

The resolving power of a mass spectrometer is limited by the variation in kinetic energy of ions emerging from the source, which is typically $\sim 0.1\%$. This variation limits the resolving power to $\sim 1\,000$, corresponding to a resolution of 0.1 at $m/z$ 100. In a **double-focusing mass spectrometer,** ions ejected from the source pass through an electric sector as well as a magnetic sector (Figure 22-12). With both sectors in series, it is possible to achieve a resolving power of $\sim 10^5$, corresponding to a resolution of 0.001 at $m/z$ 100.

### Transmission Quadrupole Mass Spectrometer

Magnetic sector and double-focusing mass spectrometers are not detectors of choice for chromatography. Figure 22-13 shows a **transmission quadrupole mass spectrometer**[17] connected to an open tubular gas chromatography column to record multiple spectra from each component as it is eluted. Species exiting the chromatography column pass through a heated connector into the electron ionization source, which is pumped to maintain a pressure of $\sim 10^{-9}$ bar by using a high-speed turbomolecular or oil diffusion pump. Ions are accelerated through a potential of 5–15 V before entering the quadrupole filter.

The quadrupole is one of the most common mass separators because of its low cost. It consists of four parallel metal rods to which are applied both a constant voltage and a radio-frequency oscillating voltage. The electric field deflects ions in complex trajectories as they migrate from the ionization chamber toward the detector, allowing only ions with one particular mass-to-charge ratio to reach the detector. Other ions (nonresonant ions) collide with the rods and are lost before they reach the detector. Rapidly varying voltages select ions of different masses to reach the detector. Transmission quadrupoles can record 2–8 spectra per second, covering a range as high as 4 000 $m/z$ units. They typically can resolve peaks separated by $m/z$ 0.3. Unlike magnetic sector and double-focusing instruments, which operate at *constant resolving power,* transmission quadrupole mass spectrometers operate at *constant resolution.* That is, ions of $m/z$ 100 and 101 are separated to the same degree as ions of $m/z$ 500 and 501.

Bond cleavages of 4-methyl-2-pentanone leading to $m/z$ 15, 85, 43, and 57:

**Challenge** Draw a rearrangement like Reaction D in Figure 22-11 to show how $m/z$ 58 arises.

***Figure 22-12*** Electrostatic sector of a double-focusing mass spectrometer. Positive ions are attracted toward the negative plate. Trajectories of high-energy ions are changed less than trajectories of low-energy ions. Ions reaching the exit slit have a narrow range of kinetic energies.

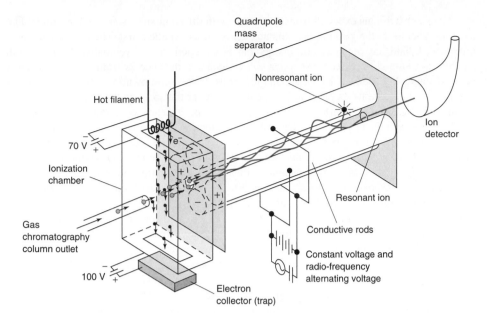

Hot filament

70 V

Ionization
chamber

Gas
chromatography
column outlet

100 V

Electron
collector (trap)

Quadrupole
mass
separator

Nonresonant ion

Ion
detector

Resonant ion

Conductive rods

Constant voltage and
radio-frequency
alternating voltage

*Figure 22-13* Quadrupole mass spectrometer. Ideally, the rods should have a hyperbolic cross section on the surfaces that face one another.

## Time-of-Flight Mass Spectrometer

The principle of the **time-of-flight mass spectrometer** is shown in Figure 22-14.[18] The ion source is at the upper left. About 3 000 to 20 000 times per second, a voltage of 5 000 V is applied to the backplate to accelerate ions toward the right and expel them from the ion source into the drift region, where there is no electric or magnetic field and no further acceleration. Ideally, all have the same kinetic energy, which is $\frac{1}{2}mv^2$, where $m$ is the mass of the ion and v is its velocity. If ions have the same kinetic energy, but different masses, the lighter ones travel faster than the heavier ones. In its simplest incarnation, the time-of-flight mass spectrometer is just a long, straight, evacuated tube with the source at one end and the detector at the other end. Ions expelled from the source drift to the detector in order of increasing mass, because the lighter ones travel faster.

The time-of-flight mass spectrometer, like magnetic sector and double-focusing instruments, operates at constant resolving power. Figure 22-14 shows an instrument designed for improved resolving power. The main limitation on resolving power is that all ions do not emerge from the source with the same kinetic energy. An ion formed close to the backplate is accelerated through a higher voltage difference than one formed near the grid, so the ion near the backplate gains more kinetic energy. Also, there is some distribution of kinetic energies among the ions, even in the absence of the accelerating voltage. Heavier ions with more than average kinetic energy reach the detector at the same time as lighter ions with less than average energy.

Ideally, the kinetic energy of an ion expelled from the source is $zeV$, where $ze$ is the charge of the ion and $V$ is the voltage on the backplate.

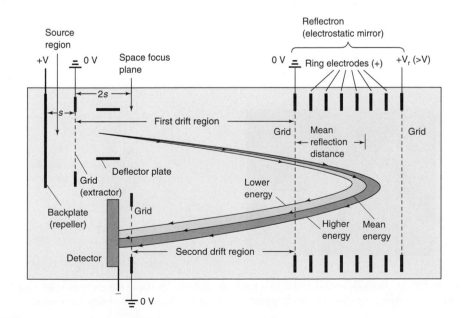

*Figure 22-14* Time-of-flight mass spectrometer. Positive ions are accelerated out of the source by voltage $+V$ periodically applied to the backplate. Light ions travel faster and reach the detector sooner than heavier ions.

Consider two ions expelled from the source with different speeds at different times. The ion formed close to the grid has less kinetic energy but is expelled first. The ion formed close to the backplate has more kinetic energy but is expelled later. Eventually, the faster ion catches up to the slower ion at the *space focus plane* at a distance $2s$ from the grid (where $s$ is the distance between the backplate and the grid). It can be shown that all ions of a given mass reach the space focus plane at the same time. After the space focus plane, ions diverge again, with faster ions overtaking slower ones. In the absence of countermeasures, ions would be badly spread by the time they reach the detector and resolving power would be low.

To improve resolving power, ions are turned around by the "reflectron" at the right side of Figure 22-14. The reflectron is a series of hollow rings held at increasingly positive potential, terminated by a grid whose potential is more positive than the accelerating potential on the backplate of the source. Ions entering the reflectron are slowed down, stopped, and reflected back to the left. The more kinetic energy an ion had when it entered the reflectron, the farther it penetrates before it is turned around. Reflected ions reach a new space focus plane at the grid in front of the detector. *All ions of the same mass reach this grid at the same time, regardless of their initial kinetic energies.*

Resolving power can be as high as 1 000 to 25 000, and $m/z$ accuracy is $\sim$0.001. Other advantages of the time-of-flight spectrometer are its high acquisition rate ($10^2$ to $10^4$ spectra/s) and capability for measuring very high masses ($m/z \approx 10^6$). However, the time-of-flight mass spectrometer requires a lower operating pressure than do the transmission quadrupole and magnetic sector instruments ($10^{-12}$ bar versus $10^{-9}$ bar).

## Quadrupole Ion-Trap Mass Spectrometer

The **quadrupole ion-trap mass spectrometer,**[19] which is one of the most widely used mass separators, is a compact device that is well suited as a chromatography detector. In the internal ionization quadrupole ion trap in Figure 22-15, substances emerging from the chromatography column enter the cavity of the mass analyzer from the lower left through a heated transfer line. The gate electrode periodically admits electrons from the filament at the top into the cavity through holes in the end cap. Molecules undergo electron ionization in the cavity formed by the two end caps and a ring electrode, all of which are electrically isolated from one another. Alternatively, chemical ionization is achieved by adding a reagent gas such as methane to the cavity.

A constant-frequency radio-frequency voltage applied to the central ring electrode causes ions to circulate in stable trajectories around the cavity. Increasing the amplitude of the radio-frequency voltage expels ions of a particular $m/z$ value by sending them into unstable trajectories that pass through the exit holes in the end caps. Ions expelled through the lower end cap in Figure 22-15 are captured by the electron multiplier and detected with high sensitivity ($\sim$1–10 pg). Scans from $m/z$ 10 to 650 can be conducted eight times per second. Resolving power is 1 000–4 000, $m/z$ accuracy is 0.1, and the maximum $m/z$ is approximately 20 000. In other mass analyzers, only a small fraction of ions reach the detector. With an ion trap, half of the ions of all $m/z$ values reach the detector, giving this spectrometer 10–100 times more sensitivity than the transmission quadrupole.

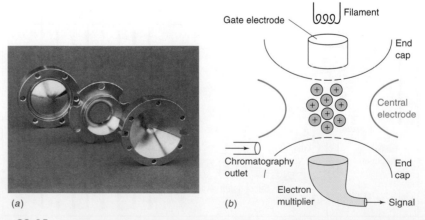

*Figure 22-15* Ion-trap mass spectrometer. (*a*) Mass analyzer consists of two end caps (left and right) and a central ring electrode. (*b*) Schematic diagram. *[Courtesy Varian Associates, Sunnyvale, CA.]*

## Ion Mobility Spectrometer[20]

More than $10^4$ **ion mobility spectrometers** are deployed at airport security checkpoints to detect explosives, and perhaps $10^5$ hand-portable devices are used by military and civil defense personnel. Although functionally similar to mass spectrometers, mobility spectrometers are operated in air at ambient pressure and ion mobility spectrometry is *not* a form of mass spectrometry. Ion mobility spectrometry does not measure molecular mass and provides no structural information. However, it is so widely used that we introduce it here.

*Electrophoresis,* which is discussed in Chapter 26, is the migration of ions in solution under the influence of an electric field. Ion mobility spectrometry is *gas-phase electrophoresis,* which separates ions according to their size-to-charge ratio. Unlike mass spectrometry, ion mobility spectrometry is capable of separating isomers.

At first sight, the ion mobility spectrometer in Figure 22-16a reminds us of a time-of-flight spectrometer. In portable units, the drift tube is 5 to 10 cm long. Typically, sample adsorbed on a cotton swab is placed in the heated anvil at the left to desorb analyte vapor. Dry air doped with a chemical ionization reagent (such as $Cl_2$ for anions and acetone or $NH_3$ for cations) sweeps the vapor through a tube containing 10 millicuries of $^{63}Ni$. Reagent gas ionized by β-emission from $^{63}Ni$ reacts with analyte to generate analyte ions.

A spectral scan is initiated when a ~250-μs voltage pulse on a gating grid admits a packet of ions into the drift tube. An electric field of 200–300 V/cm in the drift region is established by potential differences on the drift rings. The field causes either cations or anions to drift to the right at ~1 to 2 m/s. Ions are retarded by collisions with gas molecules at atmospheric pressure. Each ion travels at its own speed equal to $KE$, where $E$ is the electric field and $K$ is the *mobility*. Small ions have greater mobility than large ions of the same charge because large ions experience more drag.

The ion mobility "spectrum" in Figure 22-16b is a graph of detector response versus drift time for several explosives. Peak area is proportional to the number of ions. Peaks are

Small ions drift faster than large ions. For a description of ion mobility spectrometry in electrophoretic terms, see problem 26–46.

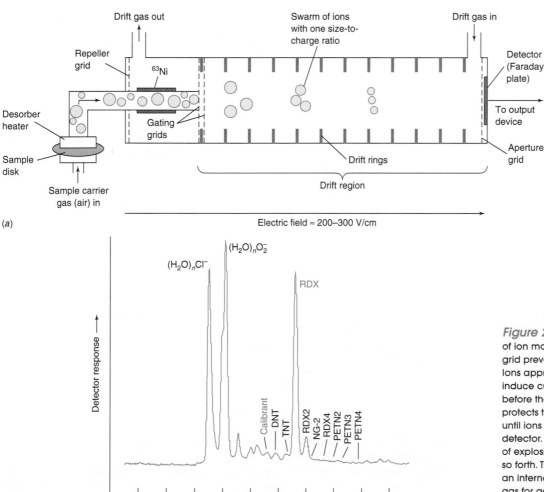

(a)

(b)

**Figure 22-16** (*a*) Schematic illustration of ion mobility spectrometer. The aperture grid prevents serious line broadening. Ions approaching a bare detector plate induce current, which appears as signal before the arrival of the ions. The aperture grid protects the detector from induced current until ions are between the grid and the detector. (*b*) Negative ion mobility spectrum of explosives designated RDX, TNT, PETN, and so forth. The calibrant 4-nitrobenzonitrile is an internal mobility standard. $Cl_2$ is the reagent gas for generating anions by chemical ionization. *[Courtesy W. R. Stott, Smiths Detection, Toronto.]*

identified by their mobility, which is reproducibly measured relative to that of an internal standard. Runs are repeated ~20 times per second. The displayed spectrum averages many runs and requires 2–5 s to obtain.

Detection limits are 0.1 to 1 pg for compounds with favorable ionization chemistry. Mobility spectrometers have limited resolution, but false positives are minimized by combining mobility determinations with selective ionization.

## ■ ■ ■ 22-4   Chromatography–Mass Spectrometry

Mass spectrometry is widely used as the detector in chromatography to provide both qualitative and quantitative information. The spectrometer can be highly selective for the analyte of interest. This selectivity eases the requirements for sample preparation or complete chromatographic separation of components in a mixture, and it increases signal-to-noise ratio.

Mass spectrometry requires high vacuum to prevent molecular collisions during ion separation. Chromatography is inherently a high-pressure technique. The problem in marrying the two techniques is to remove the huge excess of matter between the chromatograph and the spectrometer. Gas chromatography felicitously evolved to employ narrow capillary columns whose eluate does not overwhelm the pumping capabilities of the mass spectrometer vacuum system. The capillary column is connected directly to the inlet of the mass spectrometer through a heated transfer line, as in Figures 22-13 and 22-15.[21]

Liquid chromatography creates a huge volume of gas when solvent vaporizes at the interface between the column and the mass spectrometer.[22] Most of this gas must be removed prior to ion separation. Nonvolatile mobile-phase additives (such as phosphate buffer), which are commonly used in chromatography, need to be avoided when using mass spectrometry. *Pneumatically assisted electrospray* and *atmospheric pressure chemical ionization* are dominant methods for introducing eluate from liquid chromatography into a mass spectrometer.

### Electrospray

Pneumatically assisted **electrospray**,[24] also called *ion spray,* is illustrated in Figure 22-17a. Liquid from the chromatography column enters the steel nebulizer capillary at the upper left, along with a coaxial flow of $N_2$ gas. For positive ion mass spectrometry, the nebulizer is held at 0 V and the spray chamber is held at $-3\,500$ V. For negative ion mass spectrometry, all voltages would be reversed. The strong electric field at the nebulizer outlet, combined with the coaxial flow of $N_2$ gas, creates a fine aerosol of charged particles.

*Ions that vaporize from aerosol droplets were already in solution in the chromatography column.* For example, protonated bases $(BH^+)$ and ionized acids $(A^-)$ can be observed. Other gas-phase ions arise from complexation between analyte, M (which could be neutral or charged), and stable ions from the solution. Examples include

$$MH^+(mass = M + 1) \qquad M(K^+)(mass = M + 39)$$
$$M(NH_4^+)(mass = M + 18) \qquad M(HCO_2^-)(mass = M + 45)$$
$$M(Na^+)(mass = M + 23) \qquad M(CH_3CO_2^-)(mass = M + 59)$$

Positive ions from the aerosol are attracted toward the glass capillary leading into the mass spectrometer by an even more negative potential of $-4\,500$ V. Gas flowing from atmospheric pressure in the spray chamber transports ions to the right through the capillary to its exit, where the pressure is reduced to ~3 mbar by a vacuum pump.

Figure 22-17b provides more detail on ionization. Voltage imposed between the steel nebulizer capillary and the inlet to the mass spectrometer creates excess charge in the liquid by redox reactions. If the nebulizer is positively biased, oxidation enriches the liquid in positive ions by reactions such as

$$Fe(s) \rightarrow Fe^{2+} + 2e^-$$
$$H_2O(l) \rightarrow \tfrac{1}{2}O_2 + 2H^+ + 2e^-$$

Electrons from the oxidation flow through the external circuit and eventually neutralize gaseous positive ions at the inlet to the mass spectrometer. It is possible for analyte to be chemically altered by species such as $HO^\bullet$ generated during electrospray.[25]

Charged liquid exiting the capillary forms a cone and then a fine filament and finally breaks into a spray of fine droplets (see Figure 22-17c and the opening of this chapter). A droplet shrinks to ~1 μm by solvent evaporation until the repulsive force of the excess charge equals the cohesive force of surface tension. At that point, the droplet breaks up by

Volatile buffer components and additives for liquid chromatography that are compatible with mass spectrometry include $NH_3$, $HCO_2H$, $CH_3CO_2H$, $CCl_3CO_2H$, $(CH_3)_3N$, and $(C_2H_5)_3N$. Avoid additive concentrations >20 mM.

J. B. Fenn received part of the Nobel Prize in chemistry in 2002[23] for electrospray ionization. K. Tanaka received part of the same prize for matrix-assisted laser desorption/ionization, described in Box 22-4.

You, the chemist, need to adjust the pH of the chromatography solvent to favor $BH^+$ or $A^-$ for mass spectrometric detection.

Electrospray requires low-ionic-strength solvent so that buffer ions do not overwhelm analyte ions in the mass spectrum. Low-surface-tension organic solvent is better than water. In reversed-phase chromatography (Section 25-1), it is good to use a stationary phase that strongly retains analyte so that a high fraction of organic solvent can be used. A flow rate of 0.05 to 0.4 mL/min is best for electrospray.

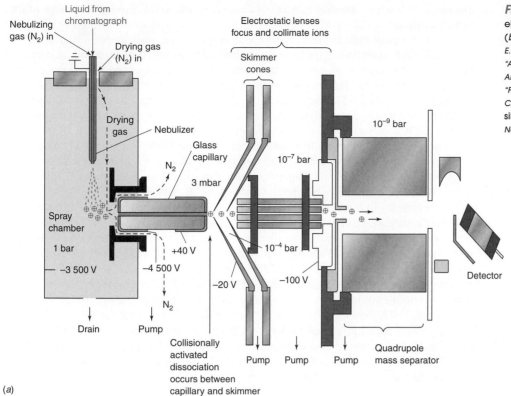

(a)

Figure 22-17 (a) Pneumatically assisted electrospray interface for mass spectrometry. (b) Gas-phase ion formation. [Adapted from E. C. Huang, T. Wachs, J. J. Conboy, and J. D. Henion, "Atmospheric Pressure Ionization Mass Spectrometry," Anal. Chem. **1990,** 62, 713A, and P. Kebarle and L. Tang, "From Ions in Solution to Ions in the Gas Phase," Anal. Chem. **1993,** 65, 972A.] (c) Electrospray from a silica capillary. [Courtesy R. D. Smith, Pacific Northwest Laboratory, Richland, WA.]

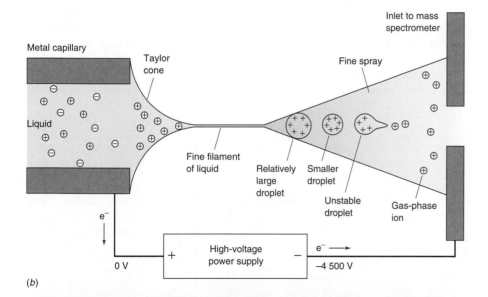

(b)

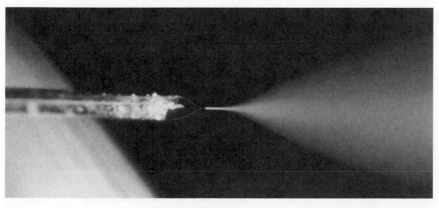

(c)

Collisionally activated dissociation of acetaminophen:

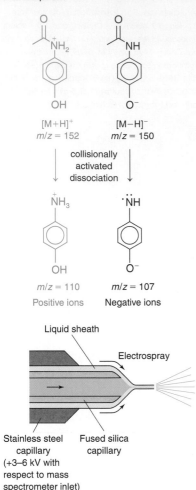

[M+H]⁺
$m/z = 152$

[M−H]⁻
$m/z = 150$

collisionally
activated
dissociation

$m/z = 110$
Positive ions

$m/z = 107$
Negative ions

Liquid sheath

Electrospray

Stainless steel
capillary
(+3–6 kV with
respect to mass
spectrometer inlet)

Fused silica
capillary

*Figure 22-18* Electrospray interface for capillary electrophoresis/mass spectrometry.

Unlike electrospray, atmospheric pressure chemical ionization *does create gaseous ions from neutral analyte molecules.* Analyte must have some volatility. For nonvolatile molecules such as sugars and proteins, electrospray can be used.

ejecting tiny droplets with diameters of ~10 nm. They evaporate, leaving their cargo of ions in the gas phase.

In electrospray, little fragmentation of analyte occurs and mass spectra are simple. Fragmentation can be intentionally increased by **collisionally activated dissociation** (also referred to as *collision-induced dissociation*) in the region between the glass capillary and the skimmer cone in Figure 22-17a. The pressure in this region is ~3 mbar and the background gas is mainly $N_2$. In Figure 22-17a, the outlet of the glass capillary is coated with a metal layer that is held at +40 V. The potential difference between the metal skimmer cone and the capillary is $-20 - (40) = -60$ V. Positive ions accelerated through 60 V collide with $N_2$ molecules with enough energy to break into a few fragments. Adjusting the skimmer cone voltage controls the degree of fragmentation. A small voltage difference favors molecular ions, whereas a large voltage difference creates fragments that aid in identification of analyte. Collisionally activated dissociation also tends to break apart complexes such as $M(Na^+)$.

For example, with a cone voltage difference of −20 V, the positive ion spectrum of the drug acetaminophen exhibits a base peak at $m/z$ 152 for the protonated molecule, $[M+H]^+$ (colored species in the margin). A smaller peak at $m/z$ 110 probably corresponds to the fragment shown in the margin. When the cone voltage difference is −50 V, collisionally activated dissociation decreases the peak at $m/z$ 152 and increases the fragment peak at $m/z$ 110. The negative ion spectrum has a large peak at $m/z$ 150 for $[M−H]^-$. As the cone voltage difference is raised from +20 V to +50 V, this peak decreases and a fragment at $m/z$ 107 increases.

Figure 22-18 shows an electrospray interface for capillary electrophoresis. The silica capillary is contained in a stainless steel capillary held at the required outlet potential for electrophoresis. The steel makes electrical contact with the liquid inside the silica by a liquid sheath flowing between the capillaries. The sheath liquid, which is typically a mixed organic/aqueous solvent, constitutes ~90% of the aerosol.

## Atmospheric Pressure Chemical Ionization

In **atmospheric pressure chemical ionization,** heat and a coaxial flow of $N_2$ convert eluate into a fine mist from which solvent and analyte evaporate (Figure 22-19). Like chemical ionization in the ion source of a mass spectrometer, atmospheric pressure chemical ionization creates new ions from gas-phase reactions between ions and molecules. The distinguishing feature of this technique is that a high voltage is applied to a metal needle in the path of the aerosol. An electric *corona* (a plasma containing charged particles) forms around the needle,

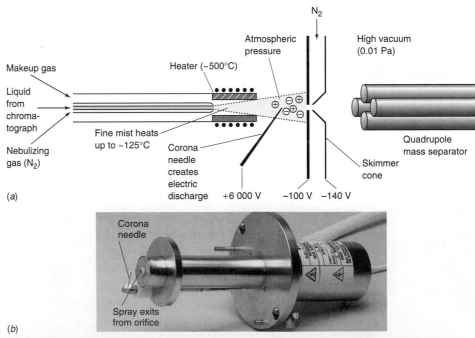

*Figure 22-19* (a) Atmospheric pressure chemical ionization interface between a liquid chromatography column and a mass spectrometer. A fine aerosol is produced by the nebulizing gas flow and the heater. The electric discharge from the corona needle creates gaseous ions from the analyte. [Adapted from E. C. Huang, T. Wachs, J. J. Conboy, and J. D. Henion, "Atmospheric Pressure Ionization Mass Spectrometry," Anal. Chem. **1990,** 62, 713A.] (b) Atmospheric pressure chemical ionization probe. [Courtesy Shimadzu Scientific Instruments, Columbia, MD.]

injecting electrons into the aerosol and creating ions. For example, protonated analyte, MH$^+$, can be formed in the following manner:

$$N_2 + e^- \rightarrow N_2^{+\cdot} + 2e^-$$
$$N_2^{+\cdot} + 2N_2 \rightarrow N_4^{+\cdot} + N_2$$
$$N_4^{+\cdot} + H_2O \rightarrow H_2O^{+\cdot} + 2N_2$$
$$H_2O^{+\cdot} + H_2O \rightarrow H_3O^+ + {}^\cdot OH$$
$$H_3O^+ + nH_2O \rightarrow H_3O^+(H_2O)_n$$
$$H_3O^+(H_2O)_n + M \rightarrow MH^+ + (n+1)H_2O$$

Analyte M might also form a negative ion by electron capture:

$$M + e^- \rightarrow M^{-\cdot}$$

A molecule, X—Y, in the eluate might create a negative ion by the reaction

$$X-Y + e^- \rightarrow X^\cdot + Y^-$$

The species Y$^-$ could abstract a proton from a weakly acidic analyte, AH:

$$AH + Y^- \rightarrow A^- + HY$$

Atmospheric pressure chemical ionization handles a variety of analytes and accepts chromatography flow rates up to 2 mL/min. Generally, to be observed, analyte M must be capable of forming the protonated ion, MH$^+$. Atmospheric pressure chemical ionization tends to produce single-charge ions and is unsuitable for macromolecules such as proteins. There is little fragmentation, but the skimmer cone voltage difference can be adjusted to favor *collisionally activated dissociation* into a small number of fragments.

The cluster ion, $H_3O^+(H_2O)_{20}(g)$, was shown in Figure 6-6.

At least one commercial detector responds to an increased range of analytes by simultaneous electrospray and atmospheric pressure chemical ionization. Ions produced by electrospray are routed directly to the mass analyzer. Neutral molecules left after electrospray then undergo atmospheric pressure chemical ionization.[26]

### Selected Ion Monitoring and Selected Reaction Monitoring

*Selected ion monitoring* and *selected reaction monitoring* increase the selectivity of mass spectrometry for individual analytes and improve the sensitivity by decreasing the response to everything else (that is, decreasing background noise). Figure 22-20a shows a liquid

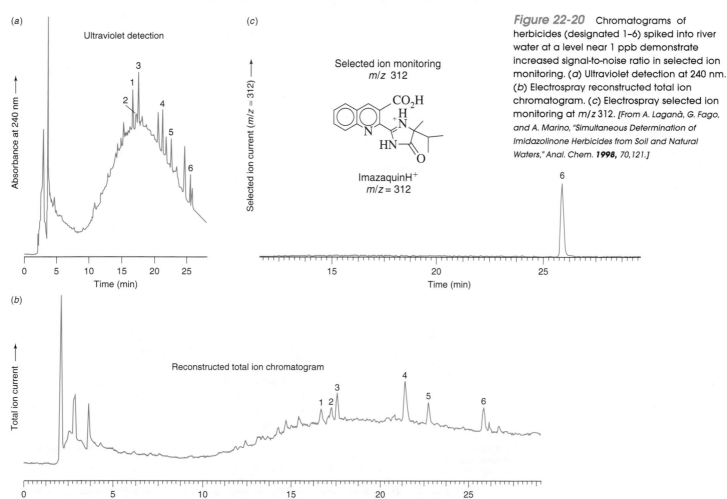

**Figure 22-20** Chromatograms of herbicides (designated 1–6) spiked into river water at a level near 1 ppb demonstrate increased signal-to-noise ratio in selected ion monitoring. (*a*) Ultraviolet detection at 240 nm. (*b*) Electrospray reconstructed total ion chromatogram. (*c*) Electrospray selected ion monitoring at *m/z* 312. [From A. Laganà, G. Fago, and A. Marino, "Simultaneous Determination of Imidazolinone Herbicides from Soil and Natural Waters," Anal. Chem. **1998**, 70,121.]

chromatogram using ultraviolet absorbance to detect a mixture of herbicides (designated 1–6) spiked at a level near 1 ppb into river water. (*Spiked* means deliberately added.) The broad hump underlying the analytes represents many natural substances in the river water. The simplest way to use a mass spectrometer as a chromatographic detector is to substitute it for the ultraviolet monitor and add up the total current of all ions of all masses detected above a selected value. This **reconstructed total ion chromatogram** is shown in Figure 22-20b, which is just as congested as the ultraviolet chromatogram because all substances emerging at any moment contribute to the signal. This chromatogram is "reconstructed" by a computer from individual mass spectra recorded during chromatography.

To be more selective, we use **selected ion monitoring** in which the mass spectrometer is set to monitor just a few values of *m/z* (never more than four or five in any time interval). Figure 22-20c shows the **selected ion chromatogram** in which just *m/z* 312 is monitored. The signal corresponds to MH$^+$ from herbicide 6, which is imazaquin. The signal-to-noise ratio in selected ion monitoring is greater than the signal-to-noise ratio in chromatogram *a* or *b* because (1) most of the spectral acquisition time is spent collecting data in a small mass range and (2) little but the intended analyte gives a signal at *m/z* 312.

Selectivity and signal-to-noise ratio can be increased markedly by **selected reaction monitoring,** illustrated in Figure 22-21 with a *triple quadrupole mass spectrometer.* A mixture of ions enters quadrupole Q1, which passes just one selected **precursor ion** to the second stage, Q2. The second stage passes all ions of all masses straight on to the third stage, Q3. However, while inside Q2, which is called a *collision cell,* the precursor ion collides with $N_2$ or Ar at a pressure of ~$10^{-8}$ to $10^{-6}$ bar and breaks into fragments called **product ions.** Quadrupole Q3 allows only specific product ions to reach the detector.

Selected reaction monitoring is extremely selective for the analyte of interest. An example is provided by the analysis of human estrogens in sewage at parts per trillion levels (ng/L). Estrogens are hormones in the ovarian cycle. The synthetic estrogen 17α-ethinylestradiol (designated EE$_2$) is a contraceptive. Even at parts per trillion levels, some estrogens can provoke reproductive disturbances in fish.

A project in Italy measured the estrogens entering the aquatic environment from human waste. Now think about this problem. Sewage contains thousands of organic compounds—many at high concentrations. To measure nanograms of one analyte was beyond the capabilities of analytical chemistry until recently. Some *sample preparation* was necessary to remove polar compounds from the less polar analyte and to *preconcentrate* analyte. Raw sewage (150 mL) was filtered to remove particles >1.5 μm and then passed through a *solid-phase extraction* cartridge (Section 28-3) containing carbon adsorbent that retained analyte. The cartridge was washed with polar solvents to remove polar materials. A fraction containing estrogens was taken from the cartridge by a mixture of dichloromethane and methanol. This fraction was evaporated and dissolved in 200 μL of aqueous solution containing another estrogen as internal standard. A volume of 50 μL was injected for chromatography.

Selected reaction monitoring is one of several techniques carried out by consecutive mass filters. These techniques are collectively called *tandem mass spectrometry* or *mass spectrometry/mass spectrometry* or simply *MS/MS.* The process in Q2 is *collisionally activated dissociation.*

In *preconcentration,* analyte collected from 150 mL was dissolved in 200 μL. The concentration increased by a factor of

$$\frac{150 \times 10^{-3}\,L}{200 \times 10^{-6}\,L} = 750$$

3.6 ng/L in sewage became 750 × 3.6 ng/L = 2.7 μg/L for chromatography.

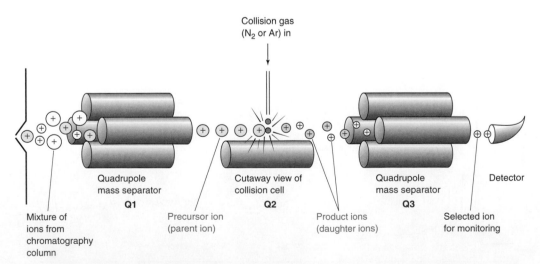

Collision gas (N$_2$ or Ar) in

Quadrupole mass separator **Q1**

Cutaway view of collision cell **Q2**

Quadrupole mass separator **Q3**

Detector

Mixture of ions from chromatography column

Precursor ion (parent ion)

Product ions (daughter ions)

Selected ion for monitoring

*Figure 22-21* Principle of selected reaction monitoring, also called tandem mass spectrometry, mass spectrometry/mass spectrometry, or MS/MS.

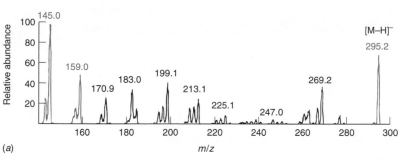

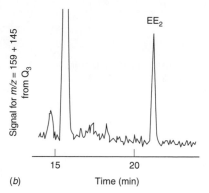

(a)  (b)

**Figure 22-22** (a) Electrospray tandem mass spectrum of pure estrogen $EE_2$. The ion $[M-H]^-$ ($m/z$ 295) was selected by quadrupole Q1 in Figure 22-21 and dissociated in Q2; the full spectrum of fragments was measured by Q3. (b) Selected reaction monitoring chromatogram showing the elution of 3.6 ng/L of estrogen $EE_2$ extracted from sewage. The signal is the sum of $m/z$ 159 + 145 from Q3 when $m/z$ 295 was selected by Q1. *[From C. Baronti, R. Curini, G. D'Ascenzo, A. di Corcia, A. Gentili, and R. Samperi, "Monitoring Estrogens at Activated Sludge Sewage Treatment Plants and in a Receiving River Water," Environ. Sci. Technol. **2000**, 34, 5059.]*

Figure 22-22a shows the collisionally activated dissociation mass spectrum of the deprotonated molecule of estrogen $EE_2$. The *precursor ion* $[M-H]^-$ ($m/z$ 295) obtained by electrospray was isolated by mass separator Q1 in Figure 22-21 and sent into Q2 for *collisionally activated dissociation.* Then all fragments $>m/z$ 140 were analyzed by Q3 to give the mass spectrum in Figure 22-22a. For subsequent *selected reaction monitoring,* only the *product ions* at $m/z$ 159 and 145 would be selected by Q3. The chromatogram in Figure 22-22b shows the signal from these product ions when $m/z$ 295 was selected by Q1. From the area of the peak for $EE_2$, its concentration in the sewage is calculated to be 3.6 ng/L. Amazingly, there are other compounds that contribute mass spectral signals for this same set of masses (295 → 159 + 145) at elution times of 15–18 min. $EE_2$ was identified by its retention time and its complete mass spectrum.

## Electrospray of Proteins

Electrospray is well suited for the study of charged macromolecules such as proteins.[29,30] It has been used to study intact viruses with masses up to 40 MDa.[31] A typical protein has carboxylic acid and amine side chains (Table 10-1) that give it a net positive or negative charge, depending on pH. Electrospray (Figure 22-17) ejects preexisting ions from solution into the gas phase.

Each peak in the mass spectrum of the protein cytochrome $c$ in Figure 22-23 arises from molecules with different numbers of protons, $MH_n^{n+}$.[32] Although we have labeled charges on four of the peaks, we do not know what these charges are without analyzing the spectrum. If we can find the charge for each species, we can find the molecular mass, $M$, of the neutral protein.

Box 22-4 describes MALDI, the other most useful way to introduce proteins into a mass spectrometer.

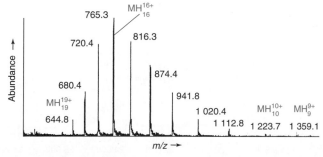

**Figure 22-23** Electrospray quadrupole mass spectrum of acidified bovine cytochrome $c$ showing peaks from species with different numbers of protons, $MH_n^{n+}$. *[From T. Wachs and J. Henion, "Electrospray Device for Coupling Microscale Separations and Other Miniaturized Devices with Electrospray Mass Spectrometry," Anal. Chem. **2001**, 73, 632.]*

To find the charge, consider a peak with $m/z = m_n$ derived from the neutral molecule plus $n$ protons:

The mass is the sum of the masses of protein, $M$, plus $n$ atoms of hydrogen ($n \times 1.008$).

$$m_n = \frac{mass}{charge} = \frac{M + n(1.008)}{n} = \frac{M}{n} + 1.008 \quad \Rightarrow \quad \boxed{m_n - 1.008 = \frac{M}{n}} \quad (22\text{-}4)$$

The next peak at lower $m/z$ should have $n + 1$ protons and a charge of $n + 1$. For this peak,

$$m_{n+1} = \frac{M + (n + 1)(1.008)}{n + 1} = \frac{M}{n + 1} + 1.008 \quad \Rightarrow \quad \boxed{m_{n+1} - 1.008 = \frac{M}{n + 1}} \quad (22\text{-}5)$$

Forming the quotient of the expressions in the boxes, we get

$$\frac{m_n - 1.008}{m_{n+1} - 1.008} = \frac{M/n}{M/(n + 1)} = \frac{n + 1}{n} \quad (22\text{-}6)$$

## Box 22-4   Matrix-Assisted Laser Desorption/Ionization

Major methods for introducing proteins and other macromolecules into mass spectrometers are electrospray and **matrix-assisted laser desorption/ionization** (MALDI).[18,27] Most often, MALDI is used with a time-of-flight mass spectrometer, which can measure $m/z$ up to $10^6$. Typically, 1 μL of a 10 μM solution of analyte is mixed with 1 μL of a 1–100 mM solution of an ultraviolet-absorbing compound such as 2,5-dihydroxybenzoic acid (the *matrix*) directly on a probe that fits into the source of the spectrometer. Evaporation of the liquid leaves an intimate mixture of fine crystals of matrix plus analyte.

To introduce ions into the gas phase for mass spectrometry, a brief infrared or ultraviolet pulse (600 ps) from a laser is directed onto the sample. The matrix vaporizes and expands into the gas phase, carrying analyte along with it. The high matrix/sample ratio inhibits association between analyte molecules and provides protonated or ionic species that transfer charge to analyte, much of which carries a single charge. Shortly after ions expand into the source, a voltage pulse applied to the backplate expels ions into the spectrometer. The resolving power is $10^3$–$10^4$ and mass accuracy can be 0.005–0.05%. The spectrum below shows proteins from milk that has not undergone any sample preparation except for mixing with the matrix. It is possible to map differences in chemical composition in different regions by directing a laser at different parts of fixed cells, such as neurons.[28]

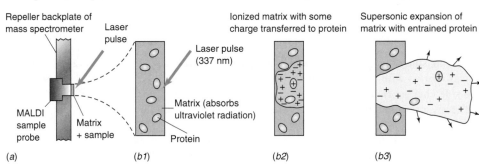

Sequence of events in matrix-assisted laser desorption/ionization. (*a*) Dried mixture of analyte and matrix on sample probe inserted into backplate of ion source. (*b1*) Enlarged view of laser pulse striking sample. (*b2*) Matrix is ionized and vaporized by laser and transfers some charge to analyte. (*b3*) Vapor expands in a supersonic plume.

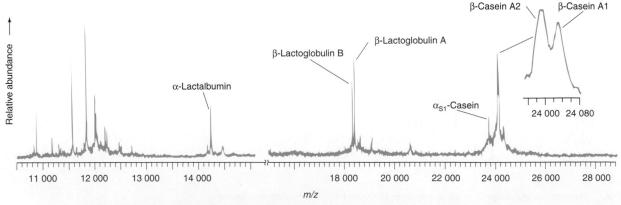

Partial mass spectrum of cow's milk (containing 2% milk fat) observed by MALDI/time-of-flight mass spectrometry.
[From R. M. Whittal and L. Li, "Time-Lag Focusing MALDI-TOF Mass Spectrometry," Am. Lab. December 1997, p. 30.]

**Table 22-3** Analysis of electrospray mass spectrum of cytochrome $c$ in Figure 22-23

| Observed $m/z \equiv m_n$ | $m_{n+1} - 1.008$ | $m_n - m_{n+1}$ | Charge $= n = \dfrac{m_{n+1} - 1.008}{m_n - m_{n+1}}$ | Molecular mass $= n \times (m_n - 1.008)$ |
|---|---|---|---|---|
| 1 359.1 | 1 222.7 | 135.4 | $9.03 \approx 9$ | $1.222_3 \times 10^4$ |
| 1 223.7 | 1 111.8 | 110.9 | $10.03 \approx 10$ | $1.222_7 \times 10^4$ |
| 1 112.8 | 1 019.4 | 92.4 | $11.03 \approx 11$ | $1.223_0 \times 10^4$ |
| 1 020.4 | 940.8 | 78.6 | $11.97 \approx 12$ | $1.223_3 \times 10^4$ |
| 941.8 | 873.4 | 67.4 | $12.96 \approx 13$ | $1.223_0 \times 10^4$ |
| 874.4 | 815.3 | 58.1 | $14.03 \approx 14$ | $1.222_7 \times 10^4$ |
| 816.3 | 764.3 | 51.0 | $14.99 \approx 15$ | $1.222_9 \times 10^4$ |
| 765.3 | 719.4 | 44.9 | $16.02 \approx 16$ | $1.222_9 \times 10^4$ |
| 720.4 | 679.4 | 40.0 | $16.98 \approx 17$ | $1.223_0 \times 10^4$ |
| 680.4 | 643.8 | 35.6 | $18.08 \approx 18$ | $1.222_9 \times 10^4$ |
| 644.8 | | | 19 | $1.223_2 \times 10^4$ |
| | | | mean $= 1.222_9 \, (\pm 0.000\,3) \times 10^4$ | |

SOURCE: *Data from T. Wachs and J. Henion, "Electrospray Device for Coupling Microscale Separations and Other Miniaturized Devices with Electrospray Mass Spectrometry," Anal. Chem.* **2001**, *73, 632. See also M. Mann, C. K. Meng, and J. B. Fenn, "Interpreting Mass Spectra of Multiply Charged Ions," Anal. Chem.* **1989**, *61, 1702.*

Solving Equation 22-6 for $n$ gives the charge on peak $m_n$:

$$n = \frac{m_{n+1} - 1.008}{m_n - m_{n+1}} \qquad (22\text{-}7)$$

Table 22-3 finds the charge of each peak, $n$, with Equation 22-7. The value of $n$ is computed in the fourth column. The charge of the first peak at $m/z$ 1 359.1 is $n = 9$. We assign this peak as $\text{MH}_9^{9+}$. The next peak at $m/z$ 1 223.7 is $\text{MH}_{10}^{10+}$, and so on.

From any peak, we can find the mass of the neutral molecule by rearranging the right side of Equation 22-4:

$$M = n \times (m_n - 1.008) \qquad (22\text{-}8)$$

Masses computed with Equation 22-8 appear in the last column of Table 22-3. A limitation on the accuracy of molecular mass determination is the accuracy of the $m/z$ scale of the mass spectrum. Proteins of known mass can be used to calibrate the instrument.

## Terms to Understand

| | | | |
|---|---|---|---|
| atmospheric pressure chemical ionization | electrospray | molecular mass | rings + double bonds formula |
| atomic mass | ion mobility spectrometer | nitrogen rule | selected ion chromatogram |
| base peak | magnetic sector mass spectrometer | nominal mass | selected ion monitoring |
| chemical ionization | mass spectrometry | precursor ion | selected reaction monitoring |
| collisionally activated dissociation | mass spectrum | product ion | time-of-flight mass spectrometer |
| double-focusing mass spectrometer | matrix-assisted laser desorption/ionization (MALDI) | quadrupole ion-trap mass spectrometer | transmission quadrupole mass spectrometer |
| electron ionization | molecular ion | reconstructed total ion chromatogram | |
| | | resolving power | |

## Summary

Ions are created or desorbed in the ion source of a mass spectrometer. Neutral molecules are converted into ions by electron ionization (which produces a molecular ion, $M^{+\cdot}$, and many fragments) or by chemical ionization (which tends to create $MH^+$ and few fragments). A magnetic sector mass spectrometer separates gaseous ions by accelerating them in an electric field and deflecting ions of different mass-to-charge ($m/z$) ratio through different arcs. Ions are detected by an electron multiplier, which works like a photomultiplier tube. The mass spectrum is a graph of detector response versus $m/z$ value. A double-focusing mass spectrometer attains high resolution by employing an electric sector with the magnetic sector to select ions with a narrow range of kinetic energy. Other mass separators include the transmission quadrupole mass spectrometer, the time-of-flight mass spectrometer, and the quadrupole ion trap. The time-of-flight instrument is capable of high acquisition rates and has a nearly unlimited upper mass range. Resolving power is defined as $m/\Delta m$ or $m/m_{1/2}$, where $m$ is the mass being measured, $\Delta m$ is the difference in mass between two peaks that are separated with a 10% valley between them, and $m_{1/2}$ is the width of a peak at half-height. An ion mobility spectrometer separates gas-phase ions by their different mobilities when migrating in an electric field through gas at atmospheric pressure.

In a mass spectrum, the molecular ion is at the highest $m/z$ value of any "significant" peak that cannot be attributed to isotopes or background signals. For a given composition, you should be able to predict the relative intensities of the isotopic peaks at M+1, M+2, and so on. Among common elements, Cl and Br have particularly diagnostic isotope patterns. From a molecular composition, the rings + double bonds equation helps us propose structures. An organic compound with an odd number of nitrogen atoms will have an odd mass. Fragment ions arising from bond cleavage and rearrangements provide clues to molecular structure.

Gas emerging from a capillary gas chromatography column can go directly into the ion source of a well-pumped mass spectrometer to provide qualitative and quantitative information about the components of a mixture. For liquid chromatography, atmospheric pressure chemical ionization utilizes a corona discharge needle to create a variety of gaseous ions. Alternatively, electrospray employs high voltage at the exit of the column, combined with coaxial $N_2$ gas flow, to create a fine aerosol containing charged species that were already present in the liquid phase. Analyte is often associated with other ions to give species such as $[MNa]^+$ or $[M(CH_3CO_2)]^-$. Control of pH helps ensure that selected analytes are in anionic or cationic form. Both atmospheric pressure chemical ionization and electrospray tend to produce unfragmented ions. Collisionally activated dissociation to produce fragment ions is controlled by the cone voltage at the mass spectrometer inlet. Electrospray of proteins typically creates an array of highly charged ions such as $MH_n^{n+}$. Matrix-assisted laser desorption/ionization is a gentle way to produce predominantly singly charged, intact protein ions for mass spectrometry.

A reconstructed total ion chromatogram shows the signal from all ions above a chosen $m/z$ emerging from chromatography as a function of time. Selected ion monitoring of one or a few values of $m/z$ is selective for one or a few analytes. In selected reaction monitoring, a precursor ion isolated by one mass filter passes into a collision cell in which it breaks into products. One (or more) product ion is then selected by a second mass filter for passage to the detector. This process is extremely selective for just one analyte and vastly increases the signal-to-noise ratio for this analyte.

## Exercises

**22-A.** Measure the width at half-height of the peak at $m/z$ 53 and calculate the resolving power of the spectrometer from the expression $m/m_{1/2}$. Would you expect to be able to resolve two peaks at 100 and 101 Da?

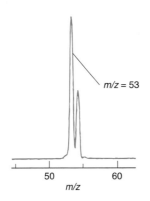

$m/z = 53$

50    60
$m/z$

Mass spectrum from V. J. Angelico, S. A. Mitchell, and V. H. Wysocki, "Low-Energy Ion–Surface Reactions of Pyrazine with Two Classes of Self-Assembled Monolayers," Anal. Chem. **2000**, 72, 2603.

**22-B.** What resolving power is required to distinguish $CH_3CH_2^+$ from $HC{\equiv}O^+$?

**22-C.** *Isotope patterns.* Consider an element with two isotopes whose natural abundances are $a$ and $b$ $(a + b = 1)$. If there are $n$ atoms of the element in a compound, the probability of finding each combination of isotopes is derived from the expansion of the binomial $(a + b)^n$. For carbon, the abundances are $a = 0.989\ 3$ for $^{12}C$ and $b = 0.010\ 7$ for $^{13}C$. The probability of finding 2 $^{12}C$ atoms in acetylene, $HC{\equiv}CH$, is given by the first term of the expansion of $(a + b)^2 = a^2 + 2ab + b^2$. The value of $a^2$ is $(0.989\ 3)^2 = 0.978\ 7$, so the probability of finding 2 $^{12}C$ atoms in acetylene is $0.978\ 7$. The probability of finding $1\ ^{12}C + 1\ ^{13}C$ is $2ab = 2(0.989\ 3)(0.010\ 7) = 0.021\ 2$. The probability of finding 2 $^{13}C$ is $(0.010\ 7)^2 = 0.000\ 114$. The molecular ion, by definition, contains 2 $^{12}C$ atoms. The M+1 peak contains $1\ ^{12}C + 1\ ^{13}C$. The intensity of M+1 relative to $M^{+\cdot}$ will be $(0.021\ 2)/(0.978\ 7) = 0.021\ 7$. (We are ignoring $^2H$ because its natural abundance is small.) Predict the relative amounts of $C_6H_4{}^{35}Cl_2$, $C_6H_4{}^{35}Cl^{37}Cl$, and $C_6H_4{}^{37}Cl_2$ in 1,2-dichlorobenzene. Draw a stick diagram of the distribution, like Figure 22-6.

**22-D.** (a) Find the number of rings plus double bonds in a molecule with the composition $C_{14}H_{12}$ and draw one plausible structure. (b) For an ion or radical, the rings + double bonds formula gives noninteger answers because the formula is based on valences in neutral molecules with all electrons paired. How many rings plus double bonds are predicted for $C_4H_{10}NO^+$? Draw one structure for $C_4H_{10}NO^+$.

**22-E.** (a) Spectra A and B belong to the isomers of $C_6H_{12}O$ below. Explain how you can tell which isomer goes with each spectrum.

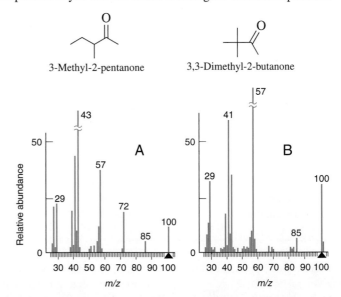

3-Methyl-2-pentanone          3,3-Dimethyl-2-butanone

Mass spectra of isomeric ketones with the composition $C_6H_{12}O$ from NIST/EPA/NIH Mass Spectral Database.[8]

(b) The intensity of the M+1 peak at $m/z$ 101 must be incorrect in both spectra. It is entirely missing in spectrum A and too intense (15.6% of intensity of $M^{+\cdot}$) in spectrum B. What should be the intensity of M+1 relative to $M^{+\cdot}$ for the composition $C_6H_{12}O$?

**22-F.** (This is a long exercise suitable for group work.) Relative intensities for the molecular ion region of several compounds are listed in parts **(a)–(d)** and shown in the figure. Suggest a composition for each molecule and calculate the expected isotopic peak intensities.

**(a)** $m/z$ (intensity): 94 (999), 95 (68), 96 (3)

**(b)** $m/z$ (intensity): 156 (566), 157 (46), 158 (520), 159 (35)

**(c)** $m/z$ (intensity): 224 (791), 225 (63), 226 (754), 227 (60), 228 (264), 229 (19), 230 (29)

**(d)** $m/z$ (intensity): 154 (122), 155 (9), 156 (12) (*Hint:* Contains sulfur.)

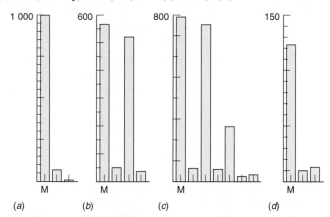

(a)     (b)     (c)     (d)

Spectral data for Exercise 22-F from NIST/EPA/NIH Mass Spectral Database.[8]

**22-G.** *Protein molecular mass from electrospray.* The enzyme lysozyme[33] exhibits $MH_n^{n+}$ peaks at $m/z = 1\ 789.1$, $1\ 590.4$, $1\ 431.5$, $1\ 301.5$, and $1\ 193.1$. Follow the procedure of Table 22-3 to find the mean molecular mass and its standard deviation.

**22-H.** *Quantitative analysis by selected ion monitoring.* Caffeine in beverages and urine can be measured by adding caffeine-$D_3$ as an internal standard and using selected ion monitoring to measure each compound by gas chromatography. The figure below shows mass chromatograms of caffeine ($m/z$ 194) and caffeine-$D_3$ ($m/z$ 197), which have nearly the same retention time.

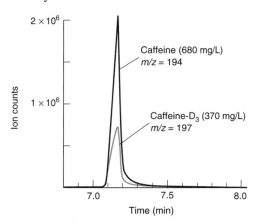

Selected ion monitoring mass chromatogram showing caffeine and caffeine-$D_3$ eluted from a capillary gas chromatography column. *[From D. W. Hill, B. T. McSharry, and L. S. Trzupek, "Quantitative Analysis by Isotopic Dilution Using Mass Spectrometry," J. Chem. Ed. **1988**, 65, 907.]*

Caffeine ($M = 194$ Da)     Caffeine-$D_3$ ($M = 197$ Da)

Suppose that the following data were obtained for standard mixtures:

| Caffeine (mg/L) | Caffeine-$D_3$ (mg/L) | Caffeine peak area | Caffeine-$D_3$ peak area |
|---|---|---|---|
| $13.60 \times 10^2$ | $3.70 \times 10^2$ | 11 438 | 2 992 |
| $6.80 \times 10^2$ | $3.70 \times 10^2$ | 6 068 | 3 237 |
| $3.40 \times 10^2$ | $3.70 \times 10^2$ | 2 755 | 2 819 |

NOTE: *Injected volume was different in all three runs.*

**(a)** Compute the mean response factor in the equation

$$\frac{\text{Area of analyte signal}}{\text{Area of standard signal}} = F\left(\frac{\text{concentration of analyte}}{\text{concentration of standard}}\right)$$

**(b)** For analysis of a cola beverage, 1.000 mL of beverage was treated with 50.0 µL of standard solution containing 1.11 g/L caffeine-$D_3$ in methanol. The combined solution was passed through a solid-phase extraction cartridge that retains caffeine. Polar solutes were washed off with water. Then the caffeine was washed off the cartridge with an organic solvent and the solvent was evaporated to dryness. The residue was dissolved in 50 µL of methanol for gas chromatography. Peak areas were 1 144 for $m/z$ 197 and 1 733 for $m/z$ 194. Find the concentration of caffeine (mg/L) in the beverage.

---

## Problems

**What Is Mass Spectrometry?**

**22-1.** Briefly describe how a magnetic sector mass spectrometer works.

**22-2.** How are ions created for each of the mass spectra in Figure 22-4? Why are the two spectra so different?

**22-3.** Define the unit *dalton.* From this definition, compute the mass of 1 Da in grams. The mean of 60 measurements of the mass of individual *E. coli* cells vaporized by MALDI and measured with a quadrupole ion trap was 5.03 ($\pm$0.14) $\times$ $10^{10}$ Da.[3] Express this mass in femtograms.

**22-4.** Nickel has two major and three minor isotopes. For the purpose of this problem, suppose that the *only* isotopes are $^{58}$Ni and $^{60}$Ni. The atomic mass of $^{58}$Ni is 57.935 3 Da and the mass of $^{60}$Ni is 59.933 2 Da. From the amplitude of the peaks in the spectrum below, calculate the atomic mass of Ni and compare your answer with the value in the periodic table.

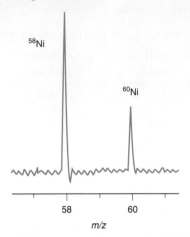

Mass spectrum from Y. Su, Y. Duan, and Z. Jin, "Helium Plasma Source Time-of-Flight Mass Spectrometry: Off-Cone Sampling for Elemental Analysis," *Anal. Chem.* **2000,** *72,* 2455.

**22-5.** Measure the width at half-height of the tallest peak in the following spectrum and calculate the resolving power of the spectrometer from the expression $m/m_{1/2}$. Would you expect to be able to distinguish two peaks at 10 000 and 10 001 Da?

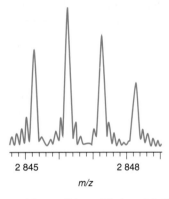

MALDI mass spectrum of the peptide melittin from P. B. O'Connor and C. E. Costello, "Application of Multishot Acquisition in Fourier Transform Mass Spectrometry," *Anal. Chem.* **2000,** *72,* 5125.

**22-6.** The two peaks near $m/z$ 31.00 in Figure 22-9 differ in mass by 0.010 Da. *Estimate* the resolving power of the spectrometer from the expression $m/\Delta m$ without making any measurements in the figure.

**22-7.** The highest-resolution mass spectra are obtained by Fourier transform ion cyclotron resonance mass spectrometry.[34] Molecular ions of two peptides (chains of seven amino acids) differing in mass by 0.000 45 Da were separated with a 10% valley between them. The ions each have a mass of 906.49 Da and a width at half-height of 0.000 27 Da. Compute the resolving power by the 10% valley formula and by the half-width formula. Compare the difference in mass of these two compounds with the mass of an electron.

### The Mass Spectrum

**22-8.** The mass of a fragment ion in a high-resolution spectrum is 83.086 5 Da. Which composition, $C_5H_7O^+$ or $C_6H_{11}^+$, better matches the observed mass?

**22-9.** Calculate the theoretical masses of the species in Figure 22-9 and compare your answers with the values observed in the figure.

**22-10.** *Isotope patterns.* Referring to Exercise 22-C, predict the relative amounts of $C_2H_2{}^{79}Br_2$, $C_2H_2{}^{79}Br^{81}Br$, and $C_2H_2{}^{81}Br_2$ in 1,2-dibromoethylene. Compare your answer with Figure 22-7.

**22-11.** *Isotope patterns.* Referring to Exercise 22-C, predict the relative abundances of $^{10}B_2H_6$, $^{10}B^{11}BH_6$, and $^{11}B_2H_6$ for diborane ($B_2H_6$).

**22-12.** Find the number of rings plus double bonds in molecules with the following compositions and draw one plausible structure for each: **(a)** $C_{11}H_{18}N_2O_3$; **(b)** $C_{12}H_{15}BrNPOS$; **(c)** a fragment in a mass spectrum with the composition $C_3H_5^+$.

**22-13.** (Each part of this problem is quite long and best worked by groups of students.) Peak intensities of the molecular ion region are listed in parts **(a)–(g)** and shown in the figure below. Identify which peak represents the molecular ion, suggest a composition for it, and calculate the expected isotopic peak intensities. Restrict your attention to elements in Table 22-1.

**(a)** $m/z$ (intensity): 112 (999), 113 (69), 114 (329), 115 (21)

**(b)** $m/z$ (intensity): 146 (999), 147 (56), 148 (624), 149 (33), 150 (99), 151 (5)

**(c)** $m/z$ (intensity): 90 (2), 91 (13), 92 (96), 93 (999), 94 (71), 95 (2)

**(d)** $m/z$ (intensity): 226 (4), 227 (6), 228 (130), 229 (215), 230 (291), 231 (168), 232 (366), 233 (2), 234 (83) (Calculate expected intensities from isotopes of the major element present.)

**(e)** $m/z$ (intensity): 172 (531), 173 (12), 174 (999), 175 (10), 176 (497)

**(f)** $m/z$ (intensity): 177 (3), 178 (9), 179 (422), 180 (999), 181 (138), 182 (9)

**(g)** $m/z$ (intensity): 182 (4), 183 (1), 184 (83), 185 (16), 186 (999), 187 (132), 188 (10)

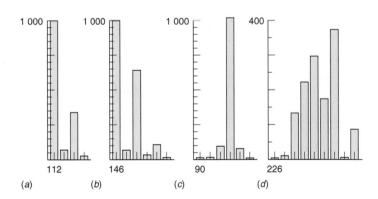

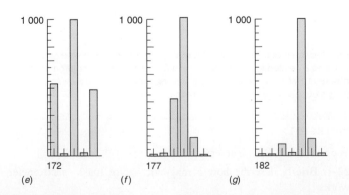

Spectral data for Problem 22-13 from NIST/EPA/NIH Mass Spectral Database.[8]

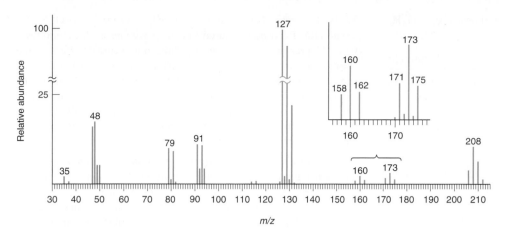

Mass spectrum for Problem 22-14 from NIST/EPA/NIH Mass Spectral Database.[8]

**22-14.** Suggest a composition for the halogen compound whose mass spectrum is shown above. Assign each of the major peaks.

**22-15.** The chart in Box 22-3 shows that $CO_2$ in human breath in the United States has a value of $\delta^{13}C$ different from that of $CO_2$ in human breath in mainland Europe. Suggest an explanation.

**22-16. (a)** The mass of $^1H$ in Table 22-1 is 1.007 825 Da. Compare it with the sum of the masses of a proton and an electron given in the table. **(b)** $^2H$ (deuterium) contains one proton, one neutron, and one electron. Compare the sum of the masses of these three particles with the mass of $^2H$.
**(c)** The discrepancy in part **(b)** comes from the conversion of mass into binding energy that holds the nucleus together. The relation of mass, $m$, to energy, $E$, is $E = mc^2$, where $c$ is the speed of light. From the discrepancy in part **(b)**, calculate the binding energy of $^2H$ in joules and in kJ/mol. (1 Da = $1.660\ 538\ 73 \times 10^{-27}$ kg.)
**(d)** The binding energy (ionization energy) of the electron in a hydrogen or deuterium atom is 13.6 eV. Use Table 1-4 to convert this number into kJ/mol and compare it with the binding energy of the $^2H$ nucleus.
**(e)** A typical bond dissociation energy in a molecule is 400 kJ/mol. How many times larger is the nuclear binding energy of $^2H$ than a bond energy?

**22-17.** *Isotope patterns.* From the natural abundance of $^{79}Br$ and $^{81}Br$, predict the relative amounts of $CH^{79}Br_3$, $CH^{79}Br_2^{81}Br$, $CH^{79}Br^{81}Br_2$, and $CH^{81}Br_3$. As explained in Exercise 22-C, the fraction of each isotopic molecule comes from the expansion of $(a + b)^3$, where $a$ is the abundance of $^{79}Br$ and $b$ is the abundance of $^{81}Br$. Note that

$$(a + b)^n = a^n + na^{n-1}b + \frac{n(n-1)}{2!}a^{n-2}b^2 + \frac{n(n-1)(n-2)}{3!}a^{n-3}b^3 + \cdots.$$ Compare your answer with Figure 22-7.

**22-18.** 🖳 *Isotope patterns.* (Caution: This problem could lead to serious brain injury.) For an element with three isotopes with abundances $a$, $b$, and $c$, the distribution of isotopes in a molecule with $n$ atoms is based on the expansion of $(a + b + c)^n$. Predict what the mass spectrum of $Si_2$ will look like.

## Mass Spectrometers

**22-19.** Explain how a double-focusing mass spectrometer achieves high resolution.

**22-20.** A limitation on how many spectra per second can be recorded by a time-of-flight mass spectrometer is the time it takes the slowest ions to go from the source to the detector. Suppose we want to scan up to $m/z$ 500. Calculate the speed of this heaviest ion if it is accelerated through 5.00 kV in the source. How long would

it take to drift 2.00 m through a spectrometer? At what frequency could you record spectra if a new extraction cycle were begun each time this heaviest ion reached the detector? What would be the frequency if you wanted to scan up to $m/z$ 1 000?

**22-21.** What is the purpose of the reflectron in a time-of-flight mass spectrometer?

**22-22.** The *mean free path* is the average distance a molecule travels before colliding with another molecule. The mean free path, $\lambda$, is given by $\lambda = kT/(\sqrt{2}\ \sigma P)$, where $k$ is Boltzmann's constant, $T$ is the temperature (K), $P$ is the pressure (Pa), and $\sigma$ is the collision cross section. For a molecule with a diameter $d$, the collision cross section is $\pi d^2$. The collision cross section is the area swept out by the molecule within which it will strike any other molecule it encounters. The magnetic sector mass spectrometer is maintained at a pressure of $\sim 10^{-5}$ Pa so that ions do not collide with (and deflect) each other as they travel through the mass analyzer. What is the mean free path of a molecule with a diameter of 1 nm at 300 K in the mass analyzer?

## Chromatography–Mass Spectrometry

**22-23.** Which liquid chromatography–mass spectrometry interface, atmospheric pressure chemical ionization or electrospray, requires analyte ions to be in solution prior to the interface? How does the other interface create gaseous ions from neutral species in solution?

**22-24.** What is collisionally activated dissociation? At what points in a mass spectrometer does it occur?

**22-25.** What is the difference between a total ion chromatogram and a selected ion chromatogram?

**22-26.** What is selected reaction monitoring? Why is it also called MS/MS? Why does it improve the signal-to-noise ratio for a particular analyte?

**22-27. (a)** To detect the drug ibuprofen by liquid chromatography–mass spectrometry, would you choose the positive or negative ion mode for the spectrometer? Would you choose acidic or neutral chromatography solvent? State your reasons.

Ibuprofen (FM 206)

**(b)** If the unfragmented ion has an intensity of 100, what should be the intensity of M+1?

**22-28.** An electrospray/transmission quadrupole mass spectrum of the $\alpha$ chain of hemoglobin from acidic solution exhibits nine peaks corresponding to $MH_n^{n+}$. Find the charge, $n$, for peaks A–I. Calculate the molecular mass of the neutral protein, M, from peaks A, B, G, H, and I, and find the mean value.

| Peak | m/z | Amplitude | Peak | m/z | Amplitude |
|------|------|-----------|------|------------|-----------|
| A | 1 261.5 | 0.024 | F | not stated | 1.000 |
| B | 1 164.6 | 0.209 | G | 834.3 | 0.959 |
| C | not stated | 0.528 | H | 797.1 | 0.546 |
| D | not stated | 0.922 | I | 757.2 | 0.189 |
| E | not stated | 0.959 | | | |

**22-29.** The molecular ion region in the mass spectrum of a large molecule, such as a protein, consists of a cluster of peaks differing by 1 Da. The reason is that a molecule with many atoms has a high probability of containing one or several atoms of $^{13}C$, $^{15}N$, $^{18}O$, $^2H$, and $^{34}S$. In fact, the probability of finding a molecule with only $^{12}C$, $^{14}N$, $^{16}O$, $^1H$, and $^{32}S$ may be so small that the nominal molecular ion is not observed. The electrospray mass spectrum of the rat protein interleukin-8 consists of a series of clusters of peaks arising from intact molecular ions with different charge. One cluster has peaks at m/z 1 961.12, 1 961.35, 1 961.63, 1 961.88, 1 962.12 (tallest peak), 1 962.36, 1 962.60, 1 962.87, 1 963.10, 1 963.34, 1 963.59, 1 963.85, and 1 964.09. These peaks correspond to isotopic ions differing by 1 Da. From the observed peak separation, find the charge of the ions in this cluster. From m/z of the tallest peak, estimate the molecular mass of the protein.

**22-30.** Phytoplankton at the ocean surface maintain the fluidity of their cell membranes by altering their lipid (fat) composition when the temperature changes. When the ocean temperature is high, plankton synthesize relatively more 37:2 than 37:3.[35]

$$37:2 = C_{37}H_{70}O$$

$$37:3 = C_{37}H_{68}O$$

After they die, plankton sink to the ocean floor and end up buried in sediment. The deeper we sample a sediment, the farther back into time we delve. By measuring the relative quantities of cell-membrane compounds at different depths in the sediment, we can infer the temperature of the ocean long ago.

The molecular ion regions of the chemical ionization mass spectra of 37:2 and 37:3 are listed in the table. Predict the expected intensities of M, M+1, and M+2 for each of the four species listed. Include contributions from C, H, O, and N, as appropriate. Compare your predictions with the observed values. Discrepant intensities in these data are typical unless care is taken to obtain high-quality data.

| Compound | Species in mass spectrum | M | M+1 | M+2 |
|----------|--------------------------|------|------|------|
| 37:3 | [MNH$_4$]$^+$ (m/z 546)[a] | 100 | 35.8 | 7.0 |
| 37:3 | [MH]$^+$ (m/z 529)[b] | 100 | 23.0 | 8.0 |
| 37:2 | [MNH$_4$]$^+$ (m/z 548)[a] | 100 | 40.8 | 3.7 |
| 37:2 | [MH]$^+$ (m/z 531)[b] | 100 | 33.4 | 8.4 |

*a. Chemical ionization with ammonia.*
*b. Chemical ionization with isobutane.*

**22-31.** Chlorate ($ClO_3^-$), chlorite ($ClO_2^-$), bromate ($BrO_3^-$), and iodate ($IO_3^-$) can be measured in drinking water at the 1-ppb level with 1% precision by selected reaction monitoring.[36] Chlorate and chlorite arise from $ClO_2$ used as a disinfectant. Bromate and iodate can be formed from $Br^-$ or $I^-$ when water is disinfected with ozone ($O_3$). For the highly selective measurement of chlorate, the negative ion selected by Q1 in Figure 22-21 is m/z 83 and the negative ion selected by Q3 is m/z 67. Explain how this measurement works and how it distinguishes $ClO_3^-$ from $ClO_2^-$, $BrO_3^-$, and $IO_3^-$.

**22-32.** *Quantitative analysis by isotope dilution.* In *isotope dilution,* a known amount of an unusual isotope (called the *spike*) is added to an unknown as an internal standard for quantitative analysis. After the mixture has been homogenized, some of the element of interest must be isolated. The ratio of the isotopes is then measured. From this ratio, the quantity of the element in the original unknown can be calculated.

Natural vanadium has atom fractions $^{51}V = 0.997\ 5$ and $^{50}V = 0.002\ 5$. The atom fraction is defined as

$$\text{Atom fraction of } ^{51}V = \frac{\text{atoms of } ^{51}V}{\text{atoms of } ^{50}V + \text{atoms of } ^{51}V}$$

A spike enriched in $^{50}V$ has atom fractions $^{51}V = 0.639\ 1$ and $^{50}V = 0.360\ 9$.

(a) Let isotope A be $^{51}V$ and isotope B be $^{50}V$. Let $A_x$ be the atom fraction of isotope A (= atoms of A/[atoms of A + atoms of B]) in an unknown. Let $B_x$ be the atom fraction of B in an unknown. Let $A_s$ and $B_s$ be the corresponding atom fractions in a spike. Let $C_x$ be the total concentration of all isotopes of vanadium ($\mu$mol/g) in the unknown and let $C_s$ be the concentration in the spike. Let $m_x$ be the mass of unknown and $m_s$ be the mass of spike. After $m_x$ grams of unknown are mixed with $m_s$ grams of spike, the ratio of isotopes in the mixture is found to be $R$. Show that

$$R = \frac{\text{mol A}}{\text{mol B}} = \frac{A_x C_x m_x + A_s C_s m_s}{B_x C_x m_x + B_s C_s m_s} \quad \text{(A)}$$

(b) Solve Equation A for $C_x$ to show that

$$C_x = \left(\frac{C_s m_s}{m_x}\right)\left(\frac{A_s - RB_s}{RB_x - A_x}\right) \quad \text{(B)}$$

(c) A 0.401 67-g sample of crude oil containing an unknown concentration of natural vanadium was mixed with 0.419 46 g of spike containing 2.243 5 $\mu$mol V/g enriched with $^{50}V$ (atom fractions: $^{51}V = 0.639\ 1$, $^{50}V = 0.360\ 9$).[37] After dissolution and equilibration of the oil and the spike, some of the vanadium was isolated by ion-exchange chromatography. The measured isotope ratio in the isolated vanadium was $R = {}^{51}V/{}^{50}V = 10.545$. Find the concentration of vanadium ($\mu$mol/g) in the crude oil.

(d) Examine the calculation in part (c) and express the answer with the correct number of significant figures.

**22-33.** *A literature project.* Until the 1960s, dinosaurs were thought to be cold-blooded animals, which means they could not regulate their body temperature. Reference 1 describes how the $^{18}O/^{16}O$ ratio in dinosaur bones suggests that some species were warm blooded. Find reference 1, preferably at http://pubs.acs.org/ac if your institution has an electronic subscription to *Analytical Chemistry*. Explain how the $^{18}O/^{16}O$ ratio implies that an animal is warm or cold blooded. Explain the criteria that were used to determine the likelihood that $^{18}O/^{16}O$ in bone phosphate was altered after the dinosaur died. Describe how bone samples were prepared for analysis of oxygen isotopes and state the results of the measurements.

## ■ ■ ■ MEASURING SILICONES LEAKING FROM BREAST IMPLANTS

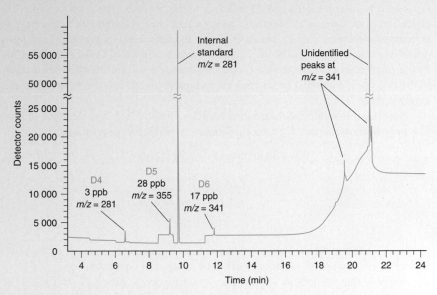

Selected ion monitoring gas chromatogram of plasma extract showing traces of siloxanes 5 years after 5-year-old breast implants were removed. *[From D. Flassbeck, B. Pfleiderer, R. Grumping, and A. V. Hirner, "Determination of Low Molecular Weight Silicones in Women After Exposure to Breast Implants by GC/MS," Anal. Chem.* **2001,** *73, 606.]*

| Name: | D3 | D4 | D5 | D6 |
|---|---|---|---|---|
| Molecular mass: | 222 | 296 | 370 | 444 |
| Main fragments: | **207** | **281** | 73, 267, **355** | 73, **341**, 429 |

High-molecular-mass poly(dimethylsiloxane), $[(CH_3)_2SiO]_n$, is used as a stationary phase in gas chromatography (Table 24-1) and as the gel in breast implants. Approximately 1–2% of silicones in breast implants are low-molecular-mass materials that can leak from intact implants and travel through the circulatory and lymph systems to take up residence in lipid-rich tissues.

Gas chromatography with selected ion mass spectrometric detection (Section 22-4) provides a specific, sensitive means to measure silicones. Analytes were extracted with hexane from 1 mL of blood plasma to which the internal standard, $[(CH_3)_3SiO]_4Si$, was added. Each analyte was monitored at the mass of its most abundant fragment. The chromatogram on this page would be enormously more complex if detection were not by selected ion monitoring. By setting the mass spectrometer to observe only each intended analyte near its known retention time, everything else eluted from the column becomes invisible.

The observation of silicones at part per billion levels in human tissue does not necessarily imply a health risk. Risk must be assessed in medical studies that make use of the analytical data.

In the vast majority of real analytical problems, we must separate, identify, and measure one or more components from a complex mixture. This chapter discusses fundamentals of analytical separations, and the next three chapters describe specific methods.

## ■ ■ ■ 23-1 Solvent Extraction

**Extraction** is the transfer of a solute from one phase to another. Common reasons to carry out an extraction in analytical chemistry are to isolate or concentrate the desired analyte or to separate it from species that would interfere in the analysis. The most common case is the extraction of an aqueous solution with an organic solvent. Diethyl ether, toluene, and hexane are common solvents that are *immiscible* with and less dense than water. They form a separate phase that floats on top of the aqueous phase, as shown in Color Plate 25. Chloroform, dichloromethane, and carbon tetrachloride are common solvents that are denser than water.* In the two-phase mixture, one phase is predominantly water and the other phase is predominantly organic.

Suppose that solute S is partitioned between phases 1 and 2, as depicted in Figure 23-1. The **partition coefficient,** $K$, is the equilibrium constant for the reaction

$$S \text{ (in phase 1)} \rightleftharpoons S \text{ (in phase 2)}$$

*Partition coefficient:*

$$K = \frac{\mathcal{A}_{S_2}}{\mathcal{A}_{S_1}} \approx \frac{[S]_2}{[S]_1} \tag{23-1}$$

where $\mathcal{A}_{S_1}$ refers to the activity of solute in phase 1. Lacking knowledge of the activity coefficients, we will write the partition coefficient in terms of concentrations.

Suppose that solute S in $V_1$ mL of solvent 1 (water) is extracted with $V_2$ mL of solvent 2 (toluene). Let $m$ be the moles of S in the system and let $q$ be the fraction of S remaining in phase 1 at equilibrium. The molarity in phase 1 is therefore $qm/V_1$. The fraction of total solute transferred to phase 2 is $(1 - q)$, and the molarity in phase 2 is $(1 - q)m/V_2$. Therefore,

$$K = \frac{[S]_2}{[S]_1} = \frac{(1 - q)m/V_2}{qm/V_1}$$

from which we can solve for $q$:

$$\begin{matrix} \text{Fraction remaining in phase 1} \\ \text{after 1 extraction} \end{matrix} = q = \frac{V_1}{V_1 + KV_2} \tag{23-2}$$

Equation 23-2 says that the fraction of solute remaining in the water (phase 1) depends on the partition coefficient and the volumes. If the phases are separated and fresh toluene (solvent 2) is added, the fraction of solute remaining in the water at equilibrium will be

$$\begin{matrix} \text{Fraction remaining in phase 1} \\ \text{after 2 extractions} \end{matrix} = q \cdot q = \left( \frac{V_1}{V_1 + KV_2} \right)^2$$

After $n$ extractions with volume $V_2$, the fraction remaining in the water is

$$\begin{matrix} \text{Fraction remaining in phase 1} \\ \text{after } n \text{ extractions} \end{matrix} = q^n = \left( \frac{V_1}{V_1 + KV_2} \right)^n \tag{23-3}$$

### Example Extraction Efficiency

Solute A has a partition coefficient of 3 between toluene and water, with three times as much in the toluene phase. Suppose that 100 mL of a 0.010 M aqueous solution of A is extracted with toluene. What fraction of A remains in the aqueous phase **(a)** if one extraction with 500 mL is performed or **(b)** if five extractions with 100 mL are performed?

---

*Whenever a choice exists between $CHCl_3$ and $CCl_4$, the less toxic $CHCl_3$ should be chosen. Hexane and toluene are greatly preferred over benzene, which is a carcinogen.

---

## Margin notes

Two liquids are **miscible** if they form a single phase when they are mixed in any ratio. *Immiscible* liquids remain in separate phases. Organic solvents with low polarity are generally immiscible with water, which is highly polar.

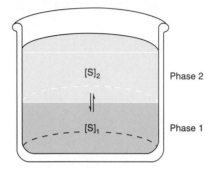

**Figure 23-1** Partitioning of a solute between two liquid phases.

For simplicity, we assume that the two phases are not soluble in each other. A more realistic treatment considers that most liquids are partially soluble in each other.[1]

The larger the partition coefficient, the less solute remains in phase 1.

Extraction example: If $q = \frac{1}{4}$, then $\frac{1}{4}$ of the solute remains in phase 1 after one extraction. A second extraction reduces the concentration to $\frac{1}{4}$ of the value after the first extraction = $(\frac{1}{4})(\frac{1}{4}) = \frac{1}{16}$ of the initial concentration.

**Solution** (a) With water as phase 1 and toluene as phase 2, Equation 23-2 says that, after a 500-mL extraction, the fraction remaining in the aqueous phase is

$$q = \frac{100}{100 + (3)(500)} = 0.062 \approx 6\%$$

(b) With five 100-mL extractions, the fraction remaining is given by Equation 23-3:

$$\text{Fraction remaining} = \left(\frac{100}{100 + (3)(100)}\right)^5 = 0.000\ 98 \approx 0.1\%$$

*It is more efficient to do several small extractions than one big extraction.*

Many small extractions are much more effective than a few large extractions.

## pH Effects

If a solute is an acid or base, its charge changes as the pH is changed. Usually, a neutral species is more soluble in an organic solvent and a charged species is more soluble in aqueous solution. Consider a basic amine whose neutral form, B, has partition coefficient $K$ between aqueous phase 1 and organic phase 2. Suppose that the conjugate acid, $BH^+$, is soluble *only* in aqueous phase 1. Let's denote its acid dissociation constant as $K_a$. The **distribution coefficient, $D$,** is defined as

*Distribution coefficient:* $\qquad D = \dfrac{\text{total concentration in phase 2}}{\text{total concentration in phase 1}}$  (23-4)

which becomes

$$D = \frac{[B]_2}{[B]_1 + [BH^+]_1} \qquad (23\text{-}5)$$

Substituting $K = [B]_2/[B]_1$ and $K_a = [H^+][B]_1/[BH^+]_1$ into Equation 23-5 leads to

*Distribution of base between two phases:* $\qquad D = \dfrac{K \cdot K_a}{K_a + [H^+]} = K \cdot \alpha_B \qquad (23\text{-}6)$

$$\alpha_B = \frac{[B]_{aq}}{[B]_{aq} + [BH^+]_{aq}}$$

$\alpha_B$ is the same as $\alpha_{A^-}$ in Equation 10-18.

where $\alpha_B$ is the fraction of weak base in the neutral form, B, in the aqueous phase. *The distribution coefficient D is used in place of the partition coefficient K in Equation 23-2 when dealing with a species that has more than one chemical form, such as B and $BH^+$.*

Charged species tend to be more soluble in water than in organic solvent. To extract a base into water, use a pH low enough to convert B into $BH^+$ (Figure 23-2). To extract the acid HA into water, use a pH high enough to convert HA into $A^-$.

> **Challenge** Suppose that the acid HA (with dissociation constant $K_a$) is partitioned between aqueous phase 1 and organic phase 2. Calling the partition coefficient $K$ for HA and assuming that $A^-$ is not soluble in the organic phase, show that the distribution coefficient is given by
>
> *Distribution of acid between two phases:* $\qquad D = \dfrac{K \cdot [H^+]}{[H^+] + K_a} = K \cdot \alpha_{HA} \qquad (23\text{-}7)$
>
> where $\alpha_{HA}$ is the fraction of weak acid in the form HA in the aqueous phase.

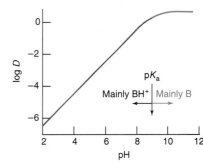

**Figure 23-2** Effect of pH on the distribution coefficient for the extraction of a base into an organic solvent. In this example, $K = 3.0$ and $pK_a$ for $BH^+$ is 9.00.

### Example Effect of pH on Extraction

Suppose that the partition coefficient for an amine, B, is $K = 3.0$ and the acid dissociation constant of $BH^+$ is $K_a = 1.0 \times 10^{-9}$. If 50 mL of 0.010 M aqueous amine is extracted with 100 mL of solvent, what will be the formal concentration remaining in the aqueous phase (a) at pH 10.00? (b) at pH 8.00?

**Solution** (a) At pH 10.00, $D = KK_a/(K_a + [H^+]) = (3.0)(1.0 \times 10^{-9})/(1.0 \times 10^{-9} + 1.0 \times 10^{-10}) = 2.73$. *Using D in place of K*, Equation 23-2 says that the fraction remaining in the aqueous phase is

$$q = \frac{50}{50 + (2.73)(100)} = 0.15 \Rightarrow 15\% \text{ left in water}$$

The concentration of amine in the aqueous phase is 15% of 0.010 M = 0.001 5 M.

**(b)** At pH 8.00, $D = (3.0)(1.0 \times 10^{-9})/(1.0 \times 10^{-9} + 1.0 \times 10^{-8}) = 0.273$. Therefore,

$$q = \frac{50}{50 + (0.273)(100)} = 0.65 \Rightarrow 65\% \text{ left in water}$$

The concentration in the aqueous phase is 0.006 5 M. At pH 10, the base is predominantly in the form B and is extracted into the organic solvent. At pH 8, it is in the form $BH^+$ and remains in the water.

## Extraction with a Metal Chelator

Most complexes that can be extracted into organic solvents are neutral. Charged complexes, such as $Fe(EDTA)^-$ or $Fe(1,10\text{-phenanthroline})_3^{2+}$, are not very soluble in organic solvents. One scheme for separating metal ions from one another is to selectively complex one ion with an organic ligand and extract it into an organic solvent. Ligands such as dithizone (Demonstration 23-1), 8-hydroxyquinoline, and cupferron are commonly employed.

Each ligand can be represented as a weak acid, HL, which loses one proton when it binds to a metal ion through the atoms shown in **bold** type.

$$HL(aq) \rightleftharpoons H^+(aq) + L^-(aq) \qquad K_a = \frac{[H^+]_{aq}[L^-]_{aq}}{[HL]_{aq}} \tag{23-8}$$

$$nL^-(aq) + M^{n+}(aq) \rightleftharpoons ML_n(aq) \qquad \beta = \frac{[ML_n]_{aq}}{[M^{n+}]_{aq}[L^-]_{aq}^n} \tag{23-9}$$

Each of these ligands can react with many different metal ions, but some selectivity is achieved by controlling the pH.

8-Hydroxyquinoline
(oxine)

Cupferron

β is the overall formation constant defined in Box 6-2.

---

## Demonstration 23-1   Extraction with Dithizone

Dithizone (diphenylthiocarbazone) is a green compound that is soluble in nonpolar organic solvents and insoluble in water below pH 7. It forms red, hydrophobic complexes with most di- and trivalent metal ions.

Dithizone
(green)                                (colorless)

Metal complex
(red)

Dithizone is used for analytical extractions, for colorimetric determinations of metal ions, and for removing traces of metals from aqueous buffers.

In the last application, an aqueous buffer is extracted repeatedly with a green solution of dithizone in $CHCl_3$. As long as the

organic phase turns red, metal ions are being extracted from the buffer. When the extracts are green, the last traces of metal ions have been removed. In Figure 23-4, we see that only certain metal ions can be extracted at a given pH.

You can demonstrate the equilibrium between green ligand and red complex by using three large test tubes sealed with tightly fitting rubber stoppers. Place some hexane plus a few milliliters of dithizone solution (prepared by dissolving 1 mg of dithizone in 100 mL of $CHCl_3$) in each test tube. Add distilled water to tube A, tap water to tube B, and 2 mM $Pb(NO_3)_2$ to tube C. After shaking and settling, tubes B and C contain a red upper phase, whereas A remains green.

Proton equilibrium in the dithizone reaction is shown by adding a few drops of 1 M HCl to tube C. After shaking, the dithizone turns green again. Competition with a stronger ligand is shown by adding a few drops of 0.05 M EDTA solution to tube B. Again, shaking causes a reversion to the green color.

### Practicing "Green" Chemistry

Chemical procedures that produce less waste or less hazardous waste are said to be "green" because they reduce harmful environmental effects. In chemical analyses with dithizone, you can substitute aqueous micelles (Box 26-1) for the organic phase (which has traditionally been chloroform, $CHCl_3$) to eliminate chlorinated solvent and the tedious extraction.[2] For example, a solution containing 5.0 wt% of the micelle-forming surfactant Triton X-100 dissolves $8.3 \times 10^{-5}$ M dithizone at 25°C and pH $< 7$. The concentration of dithizone inside the micelles, which constitute a small fraction of the volume of solution, is much greater than $8.3 \times 10^{-5}$ M. Aqueous micellar solutions of dithizone can be used for the spectrophotometric analysis of metals such as Zn(II), Cd(II), Hg(II), Cu(II), and Pb(II) with results comparable to those obtained with an organic solvent.

Let's derive an equation for the distribution coefficient of a metal between two phases when essentially all the metal in the aqueous phase (*aq*) is in the form $M^{n+}$ and all the metal in the organic phase (*org*) is in the form $ML_n$ (Figure 23-3). We define the partition coefficients for ligand and complex as follows:

$$HL(aq) \rightleftharpoons HL(org) \qquad K_L = \frac{[HL]_{org}}{[HL]_{aq}} \qquad (23\text{-}10)$$

$$ML_n(aq) \rightleftharpoons ML_n(org) \qquad K_M = \frac{[ML_n]_{org}}{[ML_n]_{aq}} \qquad (23\text{-}11)$$

The distribution coefficient we seek is

$$D = \frac{[\text{total metal}]_{org}}{[\text{total metal}]_{aq}} \approx \frac{[ML_n]_{org}}{[M^{n+}]_{aq}} \qquad (23\text{-}12)$$

From Equations 23-11 and 23-9, we can write

$$[ML_n]_{org} = K_M[ML_n]_{aq} = K_M\beta[M^{n+}]_{aq}[L^-]_{aq}^n$$

Using $[L^-]_{aq}$ from Equation 23-8 gives

$$[ML_n]_{org} = \frac{K_M\beta[M^{n+}]_{aq}K_a^n[HL]_{aq}^n}{[H^+]_{aq}^n}$$

Putting this value of $[ML_n]_{org}$ into Equation 23-12 gives

$$D \approx \frac{K_M\beta K_a^n[HL]_{aq}^n}{[H^+]_{aq}^n}$$

Because most HL is in the organic phase, we substitute $[HL]_{aq} = [HL]_{org}/K_L$ to produce the most useful expression for the distribution coefficient:

*Distribution of metal-chelate complex between phases:*
$$D \approx \frac{K_M\beta K_a^n}{K_L^n}\frac{[HL]_{org}^n}{[H^+]_{aq}^n} \qquad (23\text{-}13)$$

We see that the distribution coefficient for metal ion extraction depends on pH and ligand concentration. It is often possible to select a pH where $D$ is large for one metal and small for another. For example, Figure 23-4 shows that $Cu^{2+}$ could be separated from $Pb^{2+}$ and $Zn^{2+}$ by extraction with dithizone at pH 5. Demonstration 23-1 illustrates the pH dependence of an extraction with dithizone. Box 23-1 describes *crown ethers* that are used to extract polar reagents into nonpolar solvents for chemical reactions.

$M^{n+}$ is in the aqueous phase and $ML_n$ is in the organic phase.

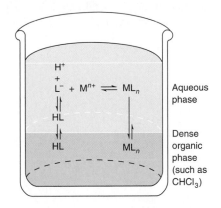

**Figure 23-3** Extraction of a metal ion with a chelator. The predominant form of metal in the aqueous phase is $M^{n+}$, and the predominant form in the organic phase is $ML_n$.

You can select a pH to bring the metal into either phase.

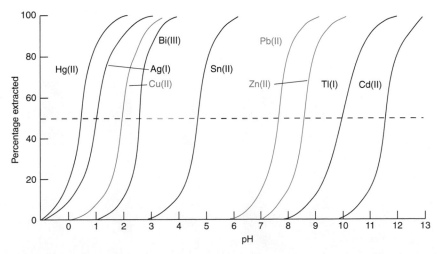

**Figure 23-4** Extraction of metal ions by dithizone into $CCl_4$. At pH 5, $Cu^{2+}$ is completely extracted into $CCl_4$, whereas $Pb^{2+}$ and $Zn^{2+}$ remain in the aqueous phase. [Adapted from G. H. Morrison and H. Freiser in C. L. Wilson and D. Wilson, eds., Comprehensive Analytical Chemistry, Vol. IA (New York: Elsevier, 1959).]

**Box 23-1   Crown Ethers**

*Crown ethers* are synthetic compounds that envelop metal ions (especially alkali metal cations) in a pocket of oxygen ligands. Crown ethers are used as *phase transfer catalysts* because they

can extract water-soluble ionic reagents into nonpolar solvents, where reaction with hydrophobic compounds can occur. In the potassium complex of dibenzo-30-crown-10, $K^+$ is engulfed by 10 oxygen atoms, with K—O distances averaging 288 pm. Only the hydrophobic outside of the complex is exposed to solvent.

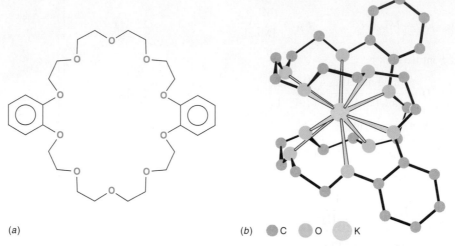

(a)                                                      (b)  ● C  ● O  ● K

(*a*) Molecular structure of dibenzo-30-crown-10. (*b*) Three-dimensional structure of its $K^+$ complex.
*[Adapted from M. A. Bush and M. R. Truter, "Crystal Structures of Alkali-Metal Complexes with Cyclic Polyethers,"*
*J. Chem. Soc. Chem. Commun.* **1970**, *1439.]*

# ■ ■ ■ ■  23-2   What Is Chromatography?

In 1903, M. Tswett first applied adsorption chromatography to the separation of plant pigments, using a hydrocarbon solvent and inulin powder (a carbohydrate) as stationary phase. The separation of colored bands led to the name *chromatography*, from the Greek word *chromatos*, meaning "color." Tswett later found that $CaCO_3$ or sucrose also could be used as stationary phases.[3]

Chromatography operates on the same principle as extraction, but one phase is held in place while the other moves past it.[4,5] Figure 23-5 shows a solution containing solutes A and B placed on top of a column packed with solid particles and filled with solvent. When the outlet is opened, solutes A and B flow down into the column. Fresh solvent is then applied to the top of the column and the mixture is washed down the column by continuous solvent flow. If solute A is more strongly adsorbed than solute B on the solid particles, then solute A spends a smaller fraction of the time free in solution. Solute A moves down the column more slowly than solute B and emerges at the bottom after solute B. We have just separated a mixture into its components by *chromatography*.

The **mobile phase** (the solvent moving through the column) in chromatography is either a liquid or a gas. The **stationary phase** (the one that stays in place inside the column) is most commonly a viscous liquid chemically bonded to the inside of a capillary tube or onto the surface of solid particles packed in the column. Alternatively, as in Figure 23-5, the

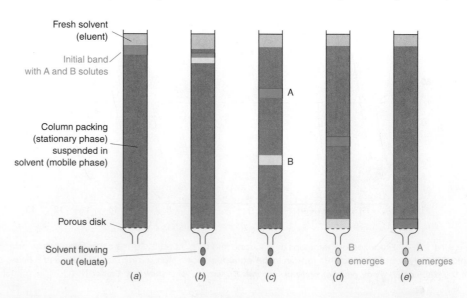

*Figure 23-5*  The idea behind chromatography: Solute A, with a greater affinity than solute B for the stationary phase, remains on the column longer.

solid particles themselves may be the stationary phase. In any case, the partitioning of solutes between mobile and stationary phases gives rise to separation.

Fluid entering the column is called **eluent.** Fluid emerging from the end of the column is called **eluate:**

$$\text{eluent} \atop \text{in} \quad \rightarrow \quad \overline{\text{COLUMN}} \quad \rightarrow \quad \text{eluate} \atop \text{out}$$

*elu**ent**—in*
*elu**ate**—out*

The process of passing liquid or gas through a chromatography column is called **elution.**

Columns are either **packed** or **open tubular.** A packed column is filled with particles of stationary phase, as in Figure 23-5. An open tubular column is a narrow, hollow capillary with stationary phase coated on the inside walls.

## Types of Chromatography

Chromatography is divided into categories on the basis of the mechanism of interaction of the solute with the stationary phase, as shown in Figure 23-6.

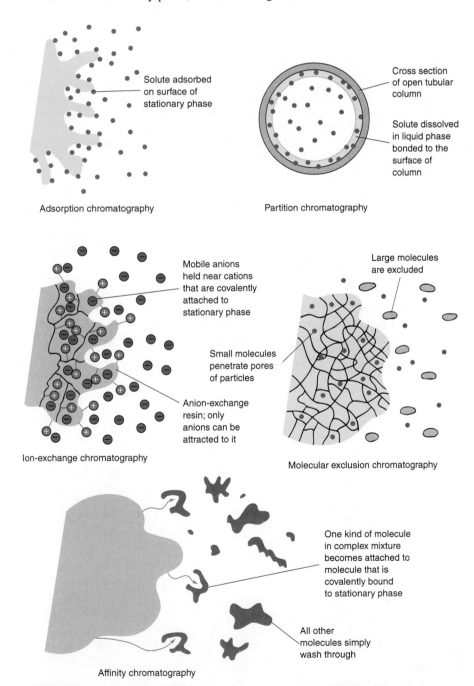

**Figure 23-6** Major types of chromatography.

**Adsorption chromatography.** A solid stationary phase and a liquid or gaseous mobile phase are used. Solute is adsorbed on the surface of the solid particles. The more strongly a solute is adsorbed, the slower it travels through the column.

**Partition chromatography.** A liquid stationary phase is bonded to a solid surface, which is typically the inside of the silica ($SiO_2$) chromatography column in gas chromatography. Solute equilibrates between the stationary liquid and the mobile phase, which is a flowing gas in gas chromatography.

**Ion-exchange chromatography.** Anions such as $-SO_3^-$ or cations such as $-N(CH_3)_3^+$ are covalently attached to the stationary solid phase, usually a *resin,* in this type of chromatography. Solute ions of the opposite charge are attracted to the stationary phase by electrostatic force. The mobile phase is a liquid.

**Molecular exclusion chromatography.** Also called **gel filtration** or **gel permeation** chromatography, this technique separates molecules by size, with the larger solutes passing through most quickly. In the ideal case of molecular exclusion, there is no attractive interaction between the stationary phase and the solute. Rather, the liquid or gaseous mobile phase passes through a porous gel. The pores are small enough to exclude large solute molecules but not small ones. Large molecules stream past without entering the pores. Small molecules take longer to pass through the column because they enter the gel and therefore must flow through a larger volume before leaving the column.

**Affinity chromatography.** This most selective kind of chromatography employs specific interactions between one kind of solute molecule and a second molecule that is covalently attached (immobilized) to the stationary phase. For example, the immobilized molecule might be an antibody to a particular protein. When a mixture containing a thousand proteins is passed through the column, only the one protein that reacts with the antibody binds to the column. After all other solutes have been washed from the column, the desired protein is dislodged by changing the pH or ionic strength.

## ■ ■ ■ ■ 23-3 A Plumber's View of Chromatography

The speed of the mobile phase passing through a chromatography column is expressed either as a volume flow rate or as a linear flow rate. Consider a liquid chromatography experiment in which the column has an inner diameter of 0.60 cm (radius $\equiv r = 0.30$ cm) and the mobile phase occupies 20% of the column volume. Each centimeter of column length has a volume of ($\pi r^2 \times$ length) $= \pi(0.30$ cm$)^2(1$ cm$) = 0.283$ mL, of which 20% ($= 0.056\ 5$ mL) is mobile phase (solvent). The **volume flow rate,** such as 0.30 mL/min, tells how many milliliters of solvent per minute travel through the column. The **linear flow rate** tells how many centimeters are traveled in 1 min by the solvent. Because 1 cm of column length contains 0.056 5 mL of mobile phase, 0.30 mL would occupy (0.30 mL)/(0.056 5 mL/cm) = 5.3 cm of column length. The linear flow rate corresponding to 0.30 mL/min is 5.3 cm/min.

### The Chromatogram

Solutes eluted from a chromatography column are observed with various detectors described in later chapters. A **chromatogram** is a graph showing the detector response as a function of elution time. Figure 23-7 shows what might be observed when a mixture of octane, nonane, and an unknown is separated by gas chromatography, which is described in Chapter 24. The

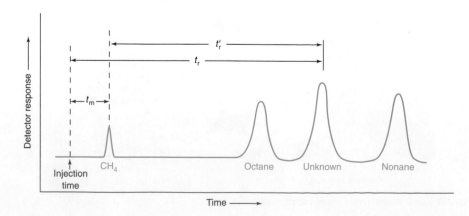

*Figure 23-7* Schematic gas chromatogram showing measurement of retention times.

**retention time,** $t_r$, for each component is the time needed after injection of the mixture onto the column until that component reaches the detector. **Retention volume,** $V_r$, is the volume of mobile phase required to elute a particular solute from the column.

Unretained mobile phase travels through the column in the minimum possible time, designated $t_m$. The **adjusted retention time,** $t_r'$, for a solute is the additional time required for solute to travel the length of the column beyond the time required by unretained solvent:

*Adjusted retention time:* $$t_r' = t_r - t_m \qquad (23\text{-}14)$$

In gas chromatography, $t_m$ is usually taken as the time needed for $CH_4$ to travel through the column (Figure 23-7).

For any two components 1 and 2, the **relative retention,** $\alpha$, is the ratio of their adjusted retention times:

*Relative retention:* $$\alpha = \frac{t_{r2}'}{t_{r1}'} \qquad (23\text{-}15)$$

where $t_{r2}' > t_{r1}'$, so $\alpha > 1$. The greater the relative retention, the greater the separation between two components. Relative retention is fairly independent of flow rate and can therefore be used to help identify peaks when the flow rate changes.

For each peak in the chromatogram, the **capacity factor,** $k'$, is defined as

*Capacity factor:* $$k' = \frac{t_r - t_m}{t_m} \qquad (23\text{-}16)$$

Capacity factor is also called *retention factor, capacity ratio,* or *partition ratio* and is frequently written as $k$ instead of $k'$.

The longer a component is retained by the column, the greater is the capacity factor. To monitor the performance of a particular column, it is good practice to periodically measure the capacity factor of a standard, the number of plates (Equation 23-28), and peak asymmetry (Figure 23-13). Changes in these parameters indicate degradation of the column.

---

**Example** Retention Parameters

A mixture of benzene, toluene, and methane was injected into a gas chromatograph. Methane gave a sharp spike in 42 s, whereas benzene required 251 s and toluene was eluted in 333 s. Find the adjusted retention time and capacity factor for each solute and the relative retention.

**Solution** The adjusted retention times are

Benzene: $t_r' = t_r - t_m = 251 - 42 = 209$ s    Toluene: $t_r' = 333 - 42 = 291$ s

The capacity factors are

Benzene: $k' = \dfrac{t_r - t_m}{t_m} = \dfrac{251 - 42}{42} = 5.0$    Toluene: $k' = \dfrac{333 - 42}{42} = 6.9$

The relative retention is expressed as a number greater than unity:

$$\alpha = \frac{t_r'(\text{toluene})}{t_r'(\text{benzene})} = \frac{333 - 42}{251 - 42} = 1.39$$

---

## Relation Between Retention Time and the Partition Coefficient

The capacity factor in Equation 23-16 is equivalent to

$$k' = \frac{\text{time solute spends in stationary phase}}{\text{time solute spends in mobile phase}} \qquad (23\text{-}17)$$

Let's see why this is true. If the solute spends all its time in the mobile phase and none in the stationary phase, it would be eluted in time $t_m$, by definition. Putting $t_r = t_m$ into Equation 23-16 gives $k' = 0$, because solute spends no time in the stationary phase. Suppose that solute spends equal time in the stationary and mobile phases. The retention time would then be $t_r = 2t_m$ and $k' = (2t_m - t_m)/t_m = 1$. If solute spends three times as much time in the stationary phase as in the mobile phase, $t_r = 4t_m$ and $k' = (4t_m - t_m)/t_m = 3$.

If solute spends three times as much time in the stationary phase as in the mobile phase, there will be three times as many moles of solute in the stationary phase as in the mobile phase at any time. The quotient in Equation 23-17 is equivalent to

$$\frac{\text{Time solute spends in stationary phase}}{\text{Time solute spends in mobile phase}} = \frac{\text{moles of solute in stationary phase}}{\text{moles of solute in mobile phase}}$$

$$k' = \frac{C_s V_s}{C_m V_m} \qquad (23\text{-}18)$$

where $C_s$ is the concentration of solute in the stationary phase, $V_s$ is the volume of the stationary phase, $C_m$ is the concentration of solute in the mobile phase, and $V_m$ is the volume of the mobile phase.

The quotient $C_s/C_m$ is the ratio of concentrations of solute in the stationary and mobile phases. If the column is run slowly enough to be near equilibrium, the quotient $C_s/C_m$ is the *partition coefficient, K,* introduced in connection with solvent extraction. Therefore, we cast Equation 23-18 in the form

*Relation of retention time to partition coefficient:* $\qquad k' = K\dfrac{V_s}{V_m} \overset{\text{Eq. 23-16}}{=} \dfrac{t_r - t_m}{t_m} = \dfrac{t_r'}{t_m} \qquad (23\text{-}19)$

which relates the retention time to the partition coefficient and the volumes of stationary and mobile phases. Because $t_r' \propto k' \propto K$, relative retention can also be expressed as

*Relative retention:* $\qquad \alpha = \dfrac{t_{r2}'}{t_{r1}'} = \dfrac{k_2'}{k_1'} = \dfrac{K_2}{K_1} \qquad (23\text{-}20)$

That is, the relative retention of two solutes is proportional to the ratio of their partition coefficients. This relation is the physical basis of chromatography.

> **Partition coefficient** $= K = \dfrac{C_s}{C_m}$

> *Physical basis of chromatography*
> The greater the ratio of partition coefficients between mobile and stationary phases, the greater the separation between two components of a mixture.

---

### Example  Retention Time and the Partition Coefficient

In the preceding example, methane gave a sharp spike in 42 s, whereas benzene required 251 s. The open tubular chromatography column has an inner diameter of 250 $\mu$m and is coated on the inside with a layer of stationary phase 1.0 $\mu$m thick. Estimate the partition coefficient ($K = C_s/C_m$) for benzene between stationary and mobile phases and state what fraction of the time benzene spends in the mobile phase.

**Solution**  We need to calculate the relative volumes of the stationary and mobile phases. The column is an open tube with a thin coating of stationary phase on the inside wall.

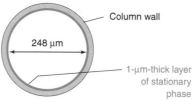

Radius of hollow cavity: $r_1 = 124$ $\mu$m
Radius to middle of stationary phase:
  $r_2 = 124.5$ $\mu$m

Cross-sectional area of column $= \pi r_1^2$
  $= \pi(124\ \mu\text{m})^2 = 4.83 \times 10^4\ \mu\text{m}^2$

Cross-sectional area of coating $\approx 2\pi r_2 \times$ thickness
  $= 2\pi(124.5\ \mu\text{m})(1.0\ \mu\text{m}) = 7.8 \times 10^2\ \mu\text{m}^2$

The relative volumes of the phases are proportional to the relative cross-sectional areas of the phases. Therefore, $V_s/V_m = (7.8 \times 10^2\ \mu\text{m}^2)/(4.83 \times 10^4\ \mu\text{m}^2) = 0.016\ 1$. In the preceding example, we found that the capacity factor for benzene is

$$k' = \frac{t_r - t_m}{t_m} = \frac{251 - 42}{42} = 5.0$$

Substituting this value into Equation 23-19 gives the partition coefficient:

$$k' = K\frac{V_s}{V_m} \Rightarrow 5.0 = K\,(0.016\ 1) \Rightarrow K = 310$$

To find the fraction of time spent in the mobile phase, we use Equations 23-16 and 23-17:

$$k' = \frac{\text{time in stationary phase}}{\text{time in mobile phase}} = \frac{t_r - t_m}{t_m} = \frac{t_s}{t_m} \Rightarrow t_s = k' t_m$$

where $t_s$ is the time in the stationary phase. The fraction of time in the mobile phase is

$$\text{Fraction of time in mobile phase} = \frac{t_m}{t_s + t_m} = \frac{t_m}{k' t_m + t_m} = \frac{1}{k' + 1} = \frac{1}{5.0 + 1} = 0.17$$

*Retention volume, $V_r$,* is the volume of mobile phase required to elute a particular solute from the column:

*Retention volume:*
$$V_r = t_r \cdot u_v \tag{23-21}$$

where $u_v$ is the volume flow rate (volume per unit time) of the mobile phase. The retention volume of a particular solute is constant over a range of flow rates.

### Scaling Up

We normally carry out chromatography for *analytical* purposes (to separate and identify or measure the components of a mixture) or for *preparative* purposes (to purify a significant quantity of a component of a mixture). Analytical chromatography is usually performed with thin columns that provide good separation. For preparative chromatography, we use fatter columns that can handle more load (Figure 23-8).[6] Preparative chromatography is especially important in the pharmaceutical industry, which can afford the high cost of separating compounds such as *optical isomers* of drugs (Boxes 24-1 and 25-2).

If you have developed a chromatographic procedure to separate 2 mg of a mixture on a column with a diameter of 1.0 cm, what size column should you use to separate 20 mg of the mixture? The most straightforward way to scale up is to maintain the same column length and to increase the cross-sectional area to maintain a constant ratio of sample mass to column volume. Because cross-sectional area is $\pi r^2$, where $r$ is the column radius, the desired diameter is given by

*Scaling equation:*
$$\frac{\text{Large mass}}{\text{Small mass}} = \left(\frac{\text{large column radius}}{\text{small column radius}}\right)^2 \tag{23-22}$$

$$\frac{20 \text{ mg}}{2 \text{ mg}} = \left(\frac{\text{large column radius}}{0.50 \text{ cm}}\right)^2$$

$$\text{Large column radius} = 1.58 \text{ cm}$$

A column with a diameter near 3 cm would be appropriate.

To reproduce the conditions of the smaller column in the larger column, the *linear flow rate* (not the volume flow rate) should be kept constant. Because the area (and hence volume) of the large column is 10 times greater than that of the small column in this example, the volume flow rate should be 10 times greater to maintain a constant linear flow rate. If the small column had a volume flow rate of 0.3 mL/min, the large column should be run at 3 mL/min.

### ■ ■ ■ 23-4 Efficiency of Separation

Two factors contribute to how well compounds are separated by chromatography. One is the difference in elution times between peaks: The farther apart, the better their separation. The other factor is how broad the peaks are: The wider the peaks, the poorer their separation. This section discusses how we measure the efficiency of a separation.

### Resolution

Solute moving through a chromatography column tends to spread into a Gaussian shape with standard deviation $\sigma$ (Figure 23-9). The longer a solute spends passing through a column, the broader the band becomes. Common measures of breadth are (1) the width $w_{1/2}$ measured at a height equal to half of the peak height and (2) the width $w$ at the baseline between tangents drawn to the steepest parts of the peak. From Equation 4-3 for a Gaussian peak, it is possible to show that $w_{1/2} = 2.35\sigma$ and $w = 4\sigma$.

In chromatography, the **resolution** of two peaks from each other is defined as

*Resolution:*
$$\text{Resolution} = \frac{\Delta t_r}{w_{av}} = \frac{\Delta V_r}{w_{av}} = \frac{0.589\Delta t_r}{w_{1/2av}} \tag{23-23}$$

where $\Delta t_r$ or $\Delta V_r$ is the separation between peaks (in units of time or volume) and $w_{av}$ is the average width of the two peaks in corresponding units. (Peak width is measured at the base, as shown in Figure 23-9.) Alternatively, the last expression in Equation 23-23 uses

Volume is proportional to time, so any ratio of times can be written as the corresponding ratio of volumes. For example,

$$k' = \frac{t_r - t_m}{t_m} = \frac{V_r - V_m}{V_m}$$

where $V_r$ is the retention volume for solute and $V_m$ is the elution volume for an unretained component. $V_m$ is equal to the volume of mobile phase in the column.

Scaling rules:
- Keep column length constant.
- Cross-sectional area of column $\propto$ mass of analyte:
$$\frac{\text{Mass}_2}{\text{Mass}_1} = \left(\frac{\text{radius}_2}{\text{radius}_1}\right)^2$$
(The symbol $\propto$ means "is proportional to.")
- Maintain constant linear flow rate:
$$\frac{\text{Volume flow}_2}{\text{Volume flow}_1} = \left(\frac{\text{radius}_2}{\text{radius}_1}\right)^2$$
- Sample volume applied to column $\propto$ mass of analyte.
- If you do change the length of the column, then the mass of sample can be increased in proportion to the increase in length.

**Figure 23-8** Industrial-scale preparative chromatography column can purify a kilogram of material. The column volume is 300 L. *[Courtesy Prochrom, Inc., Indianapolis, IN.]*

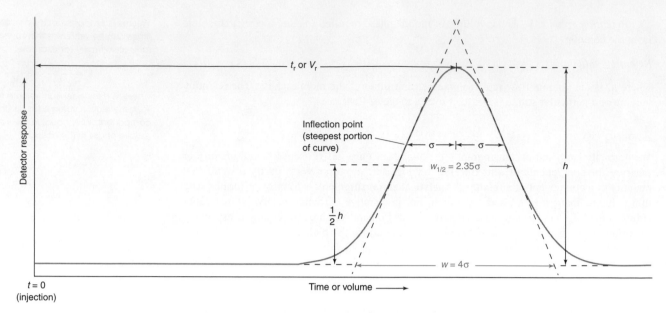

**Figure 23-9** Idealized Gaussian chromatogram showing how $w$ and $w_{1/2}$ are measured. The value of $w$ is obtained by extrapolating the tangents to the inflection points down to the baseline.

$w_{1/2av}$, the width at half-height of Gaussian peaks. Figure 23-10 shows the overlap of two peaks with different degrees of resolution. For quantitative analysis, a resolution $>1.5$ is highly desirable.

**Example** Measuring Resolution

A peak with a retention time of 407 s has a width at the base of 13 s. A neighboring peak is eluted at 424 s with a width of 16 s. Find the resolution for these two components.

**Solution** Resolution $= \dfrac{\Delta t_r}{w_{av}} = \dfrac{424 - 407}{\frac{1}{2}(13 + 16)} = 1.1_7$

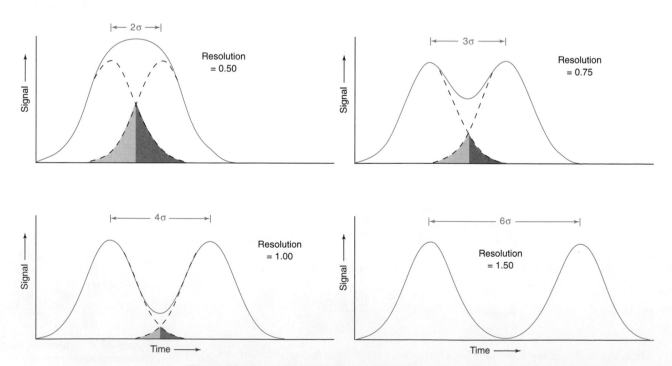

**Figure 23-10** Resolution of Gaussian peaks of equal area and amplitude. Dashed lines show individual peaks, and solid lines are the sum of two peaks. Overlapping area is shaded.

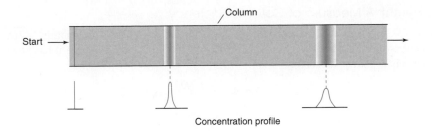

Start →                                            Column

Concentration profile

## Diffusion

A band of solute broadens as it moves through a chromatography column. Ideally, an infinitely narrow band applied to the inlet of the column emerges with a Gaussian shape at the outlet (Figure 23-11). In less ideal circumstances, the band becomes asymmetric.

One main cause of band spreading is *diffusion*. The **diffusion coefficient** measures the rate at which a substance moves randomly from a region of high concentration to a region of lower concentration. Figure 23-12 shows spontaneous diffusion of solute across a plane with a concentration gradient $dc/dx$. The number of moles crossing each square meter per second, called the *flux, J*, is proportional to the concentration gradient:

*Definition of diffusion coefficient:*

$$\text{Flux}\left(\frac{\text{mol}}{\text{m}^2 \cdot \text{s}}\right) \equiv J = -D\frac{dc}{dx} \qquad (23\text{-}24)$$

Equation 23-24 is called *Fick's first law of diffusion*. If concentration is expressed as $\text{mol/m}^3$, the units of $D$ are $\text{m}^2/\text{s}$.

The constant of proportionality, $D$, is the diffusion coefficient, and the negative sign is necessary because the net flux is from the region of high concentration to the region of low concentration. Table 23-1 shows that diffusion in liquids is $10^4$ times slower than diffusion in gases. Macromolecules such as ribonuclease and albumin diffuse 10 to 100 times slower than small molecules.

If solute begins its journey through a column in an infinitely sharp layer with $m$ moles per unit cross-sectional area of the column and spreads by diffusion as it travels, then the Gaussian profile of the band is described by

*Broadening of chromatography band by diffusion:*

$$c = \frac{m}{\sqrt{4\pi Dt}}e^{-x^2/(4Dt)} \qquad (23\text{-}25)$$

where $c$ is concentration $(\text{mol/m}^3)$, $t$ is time, and $x$ is the distance along the column from the current center of the band. (The band center is always $x = 0$ in this equation.) Comparison of Equations 23-25 and 4-3 shows that the standard deviation of the band is

*Standard deviation of band:*

$$\sigma = \sqrt{2Dt} \qquad (23\text{-}26)$$

Bandwidth $\propto \sqrt{t}$. If the elution time increases by a factor of 4, diffusion will broaden the band by a factor of 2.

**Table 23-1** Representative diffusion coefficients at 298 K

| Solute | Solvent | Diffusion coefficient $(\text{m}^2/\text{s})$ |
|---|---|---|
| $H_2O$ | $H_2O$ | $2.3 \times 10^{-9}$ |
| Sucrose | $H_2O$ | $0.52 \times 10^{-9}$ |
| Glycine | $H_2O$ | $1.1 \times 10^{-9}$ |
| $CH_3OH$ | $H_2O$ | $1.6 \times 10^{-9}$ |
| Ribonuclease (FM 13 700) | $H_2O$ (293 K) | $0.12 \times 10^{-9}$ |
| Serum albumin (FM 65 000) | $H_2O$ (293 K) | $0.059 \times 10^{-9}$ |
| $I_2$ | Hexane | $4.0 \times 10^{-9}$ |
| $CCl_4$ | Heptane | $3.2 \times 10^{-9}$ |
| $N_2$ | $CCl_4$ | $3.4 \times 10^{-9}$ |
| $CS_2(g)$ | Air (293 K) | $1.0 \times 10^{-5}$ |
| $O_2(g)$ | Air (273 K) | $1.8 \times 10^{-5}$ |
| $H^+$ | $H_2O$ | $9.3 \times 10^{-9}$ |
| $OH^-$ | $H_2O$ | $5.3 \times 10^{-9}$ |
| $Li^+$ | $H_2O$ | $1.0 \times 10^{-9}$ |
| $Na^+$ | $H_2O$ | $1.3 \times 10^{-9}$ |
| $K^+$ | $H_2O$ | $2.0 \times 10^{-9}$ |
| $Cl^-$ | $H_2O$ | $2.0 \times 10^{-9}$ |
| $I^-$ | $H_2O$ | $2.0 \times 10^{-9}$ |

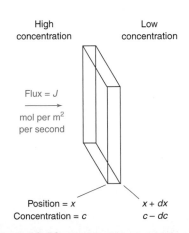

High concentration          Low concentration

Flux = J
mol per m²
per second

Position = x          x + dx
Concentration = c    c − dc

**Figure 23-12** The flux of molecules diffusing across a plane of unit area is proportional to the concentration gradient and to the diffusion coefficient: $J = -D(dc/dx)$.

## Plate Height: A Measure of Column Efficiency

$u_x$ = linear flow rate (distance/time)
$u_v$ = volume flow rate (volume/time)

Equation 23-26 tells us that the standard deviation for diffusive band spreading is $\sqrt{2Dt}$. If solute has traveled a distance $x$ at the linear flow rate $u_x$ (m/s), then the time it has been on the column is $t = x/u_x$. Therefore

$$\sigma^2 = 2Dt = 2D\frac{x}{u_x} = \underbrace{\left(\frac{2D}{u_x}\right)}_{\text{Plate height} \equiv H}x = Hx$$

The term *plate height* comes from distillation theory. Some high-performance distillation columns contain discrete units called plates, in which liquid and vapor equilibrate with each other. As a teenager, A. J. P. Martin, coinventor of partition chromatography, built distillation columns in discrete sections from coffee cans. (We don't know what he was distilling!) When he formulated the theory of partition chromatography, he adopted terms from distillation theory.

*Plate height:* $\qquad\qquad H = \sigma^2/x \qquad\qquad$ (23-27)

**Plate height** is the constant of proportionality between the variance, $\sigma^2$, of the band and the distance it has traveled, $x$. The name came from the theory of distillation in which separation could be performed in discrete stages called plates. Plate height is also called the *height equivalent to a theoretical plate*. Plate height is approximately the length of column required for one equilibration of solute between mobile and stationary phases. We explore this concept further in Box 23-2. *The smaller the plate height, the narrower the bandwidth.*

The ability of a column to separate components of a mixture is improved by decreasing plate height. An efficient column has more theoretical plates than an inefficient column. Different solutes passing through the same column have different plate heights because they have different diffusion coefficients. Plate heights are ~0.1 to 1 mm in gas chromatography, ~10 μm in high-performance liquid chromatography, and <1 μm in capillary electrophoresis.

> Small plate height ⇒
> narrow peaks ⇒
> better separations

Plate height is the length $\sigma^2/x$, where $\sigma$ is the standard deviation of the Gaussian band in Figure 23-9 and $x$ is the distance traveled. For solute emerging from a column of length $L$, the number of plates, $N$, in the entire column is the length $L$ divided by the plate height:

$$N = \frac{L}{H} = \frac{Lx}{\sigma^2} = \frac{L^2}{\sigma^2} = \frac{16L^2}{w^2}$$

because $x = L$ and $\sigma = w/4$. In this expression, $w$ has units of length and the number of plates is dimensionless. If we express $L$ and $w$ (or $\sigma$) in units of time instead of length, $N$ is still dimensionless. We obtain the most useful expression for $N$ by writing

Choose a peak with a capacity factor greater than 5 when you measure plate height for a column.

*Number of plates on column:* $\qquad N = \dfrac{16t_r^2}{w^2} = \dfrac{t_r^2}{\sigma^2} \qquad$ (23-28a)

where $t_r$ is the retention time of the peak and $w$ is the width at the base in Figure 23-9 *in units of time*. If we use the width at half-height instead of the width at the base, we get

**Challenge** If $N$ is constant, show that the width of a chromatographic peak increases with increasing retention time. That is, successive peaks in a chromatogram should be increasingly broad.

*Number of plates on column:* $\qquad N = \dfrac{5.55t_r^2}{w_{1/2}^2} \qquad$ (23-28b)

### Example  Measuring Plates

A solute with a retention time of 407 s has a width at the base of 13 s on a column 12.2 m long. Find the number of plates and plate height.

**Solution**

$$N = \frac{16t_r^2}{w^2} = \frac{16 \cdot 407^2}{13^2} = 1.57 \times 10^4$$

$$H = \frac{L}{N} = \frac{12.2\ \text{m}}{1.57 \times 10^4} = 0.78\ \text{mm}$$

To estimate the number of theoretical plates for the asymmetric peak in Figure 23-13, draw a horizontal line across the band at 1/10th of the maximum height. The quantities A and B can then be measured, and the number of plates is[7]

All quantities must be measured in the same units, such as minutes or centimeters of chart paper.

$$N \approx \frac{41.7(t_r/w_{0.1})^2}{A/B + 1.25} \qquad (23\text{-}29)$$

where $w_{0.1}$ is the width at 1/10th height ($= A + B$).

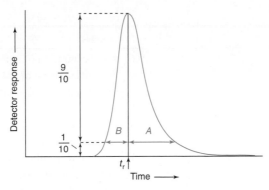

**Figure 23-13** Asymmetric peak showing parameters used to estimate the number of theoretical plates.

## Factors Affecting Resolution

For two closely spaced peaks, the relation between plates and resolution is[8]

$$\text{Resolution} = \frac{\sqrt{N}}{4}(\gamma - 1) \qquad (23\text{-}30)$$

where $N$ is the number of theoretical plates and $\gamma$ is the **separation factor,** defined as the quotient of linear velocities of the two solutes ($\gamma > 1$). Retention time is inversely proportional to velocity. If the retention times for two components are $t_A = 341$ s and $t_B = 348$ s, $\gamma = 348/341 = 1.021$.

One important feature of Equation 23-30 is that resolution is proportional to $\sqrt{N}$. Therefore, *doubling the column length increases resolution by* $\sqrt{2}$. Figure 23-14 shows the effect of column length on the separation of L-phenylalanine (Table 10-1) from L-phenylalanine-D$_5$, in which the phenyl ring bears five deuterium atoms. The mixture was passed through a pair of liquid chromatography columns by an ingenious valving system that recycled the mixture through the same two columns over and over. After 1 pass in Figure 23-14, the relative retention, $\alpha$, is only 1.03. By 15 passes, baseline separation has been achieved. The inset of Figure 23-14 shows that the square of resolution is proportional to the number of passes, as predicted by Equation 23-30.

Separation factor: $\gamma = \dfrac{u_A}{u_B} = \dfrac{t_B}{t_A}$

$u_A, u_B$ = linear velocities of components A and B

$t_A, t_B$ = retention times of components A and B

Resolution $\propto \sqrt{N} \propto \sqrt{L}$

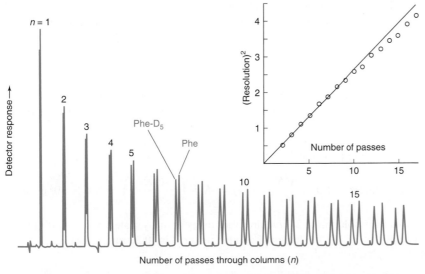

**Figure 23-14** Separation of 0.5 mM L-phenylalanine and 0.5 mM L-phenylalanine-D$_5$ by repeated passes through a pair of Hypersil C8 liquid chromatography columns (25 cm × 4.6 mm) eluted with 10:90 acetonitrile:water containing 25 mM Na$_2$SO$_4$ and 0.1% trifluoroacetic acid in the water. *[From K. Lan and J. W. Jorgensen, "Pressure-Induced Retention Variations in Reversed-Phase Alternate-Pumping Recycle Chromatography," Anal. Chem.* **1998**, *70, 2773.]*

**Table 23-2** Summary of chromatography equations

| Quantity | Equation | Parameters |
|---|---|---|
| Partition coefficient | $K = C_s/C_m$ | $C_s$ = concentration of solute in stationary phase |
| | | $C_m$ = concentration of solute in mobile phase |
| Adjusted retention time | $t'_r = t_r - t_m$ | $t_r$ = retention time of solute of interest |
| | | $t_m$ = retention time of unretained solute |
| Retention volume | $V_r = t_r \cdot u_v$ | $u_v$ = volume flow rate = volume/unit time |
| Capacity factor | $k' = t'_r/t_m = KV_s/V_m$ | $V_s$ = volume of stationary phase |
| | | $V_m$ = volume of mobile phase |
| | $k' = \dfrac{t_s}{t_m}$ | $t_s$ = time solute spends in stationary phase |
| | | $t_m$ = time solute spends in mobile phase |
| Relative retention | $\alpha = \dfrac{t'_{r2}}{t'_{r1}} = \dfrac{k'_2}{k'_1} = \dfrac{K_2}{K_1}$ | Subscripts 1 and 2 refer to two solutes |
| Separation factor | $\gamma = t_2/t_1 \ (\gamma > 1)$ | $t_2$ = retention time of solute 2 |
| | | $t_1$ = retention time of solute 1 |
| Number of plates | $N = \dfrac{16t_r^2}{w^2} = \dfrac{5.55t_r^2}{w_{1/2}^2}$ | $w$ = width at base |
| | | $w_{1/2}$ = width at half-height |
| Plate height | $H = \dfrac{\sigma^2}{x} = \dfrac{L}{N}$ | $\sigma$ = standard deviation of band |
| | | $x$ = distance traveled by center of band |
| | | $L$ = length of column |
| | | $N$ = number of plates on column |
| Resolution | $\text{Resolution} = \dfrac{\Delta t_r}{w_{av}} = \dfrac{\Delta V_r}{w_{av}}$ | $\Delta t_r$ = difference in retention times |
| | | $\Delta V_r$ = difference in retention volumes |
| | | $w_{av}$ = average width measured at baseline in same units as numerator (time or volume) |
| | $\text{Resolution} = \dfrac{\sqrt{N}}{4}(\gamma - 1)$ | $N$ = number of plates |
| | | $\gamma$ = separation factor ($\gamma > 1$) |

Equation 23-30 also tells us that resolution increases as the separation factor $\gamma$ increases. The separation factor is the relative velocity of the two components through the column. The way to change relative velocity is to change the stationary phase in gas chromatography or either the stationary or the mobile phase in liquid chromatography. Important equations from chromatography are summarized in Table 23-2.

> **Example** Plates Needed for Desired Resolution
>
> Two solutes have a separation factor of $\gamma = 1.06$. How many plates are required to give a resolution of 1.0? of 2.0? If the plate height is 0.20 mm, how long must the column be for a resolution of 1.0?
>
> **Solution** We use Equation 23-30:
>
> $$\text{Resolution} = 1.0 = \frac{\sqrt{N}}{4}(\gamma - 1) \Rightarrow N = \left(\frac{4(1.0)}{1.06 - 1}\right)^2 = 4.4 \times 10^3 \text{ plates}$$
>
> To double the resolution to 2.0 requires four times as many plates = $1.8 \times 10^4$ plates. For a resolution of 1.0, the length of column required is (0.20 mm/plate) $\times$ ($4.4 \times 10^3$ plates) = 0.88 m.

## ■■■ 23-5 Why Bands Spread[9]

A band of solute invariably spreads as it travels through a chromatography column (Figure 23-11) and emerges at the detector with a standard deviation $\sigma$. Each individual mechanism contributing to broadening produces a standard deviation $\sigma_i$. The observed variance ($\sigma_{obs}^2$) is the sum of variances from all contributing mechanisms:

*Variance is additive:*
$$\sigma_{obs}^2 = \sigma_1^2 + \sigma_2^2 + \sigma_3^2 + \cdots = \sum \sigma_i^2 \qquad (23\text{-}31)$$

*Variance is additive, but standard deviation is not.*

## Broadening Outside the Column

Some factors outside the column contribute to broadening. For example, solute cannot be applied to the column in an infinitesimally thin zone, so the band has a finite width even before it enters the column. If the band is applied as a plug of width $\Delta t$ (measured in units of time), the contribution to the variance of the final bandwidth is

*Variance due to injection or detection:*
$$\sigma_{injection}^2 = \sigma_{detector}^2 = \frac{(\Delta t)^2}{12} \qquad (23\text{-}32)$$

The same relation holds for broadening in a detector that requires a time $\Delta t$ for the sample to pass through. Sometimes on-column detection is possible, which eliminates band spreading in a detector.

---

**Example** Band Spreading Before and After the Column

A band from a column eluted at a rate of 1.35 mL/min has a width at half-height of 16.3 s. The sample was applied as a sharp plug with a volume of 0.30 mL, and the detector volume is 0.20 mL. Find the variances introduced by injection and detection. What would be the width at half-height if broadening occurred only on the column?

**Solution** Figure 23-9 tells us that the width at half-height is $w_{1/2} = 2.35\sigma$. Therefore, the observed total variance is

$$\sigma_{obs}^2 = \left(\frac{w_{1/2}}{2.35}\right)^2 = \left(\frac{16.3}{2.35}\right)^2 = 48.11 \text{ s}^2$$

The contribution from injection is $\sigma_{injection}^2 = (\Delta t)_{injection}^2/12$. The time of injection is $\Delta t_{injection} = (0.30 \text{ mL})/(1.35 \text{ mL/min}) = 0.222 \text{ min} = 13.3 \text{ s}$. Therefore,

$$\sigma_{injection}^2 = \frac{\Delta t_{injection}^2}{12} = \frac{13.3^2}{12} = 14.81 \text{ s}^2$$

The time spent in the detector is $\Delta t_{detector} = (0.20 \text{ mL})/(1.35 \text{ mL/min}) = 8.89 \text{ s}$, so $\sigma_{detector}^2 = (\Delta t)_{detector}^2/12 = 6.58 \text{ s}^2$. The observed variance is

$$\sigma_{obs}^2 = \sigma_{column}^2 + \sigma_{injection}^2 + \sigma_{detector}^2$$
$$48.11 = \sigma_{column}^2 + 14.81 + 6.58 \Rightarrow \sigma_{column} = 5.17 \text{ s}$$

The width due to column broadening alone is $w_{1/2} = 2.35\sigma_{column} = 12.1 \text{ s}$, which is about three-fourths of the observed width.

---

The worst possible band spreading occurs in the large dead space beneath some rudimentary benchtop chromatography columns where each new drop exiting the column mixes with a significant volume of eluate already in the dead space. To minimize band spreading, dead spaces and tubing lengths should be minimized. Sample should be applied uniformly in a narrow zone and allowed to enter the column before mixing with eluent.

## Plate Height Equation

Plate height, $H$, is proportional to the variance of a chromatographic band (Equation 23-27): The smaller the plate height, the narrower the band. The **van Deemter equation** tells us how the column and flow rate affect the plate height:

*van Deemter equation for plate height:*
$$H \approx A + \frac{B}{u_x} + Cu_x \qquad (23\text{-}33)$$

$$\underset{\substack{\text{Multiple} \\ \text{paths}}}{} \quad \underset{\substack{\text{Longitudinal} \\ \text{diffusion}}}{} \quad \underset{\substack{\text{Equilibration} \\ \text{time}}}{}$$

Packed columns: $A, B, C \neq 0$
Open tubular columns: $A = 0$
Capillary electrophoresis: $A = C = 0$

where $u_x$ is the linear flow rate and $A$, $B$, and $C$ are constants for a given column and stationary phase. Changing the column and stationary phase changes $A$, $B$, and $C$. The van Deemter

**23-5 Why Bands Spread**

517

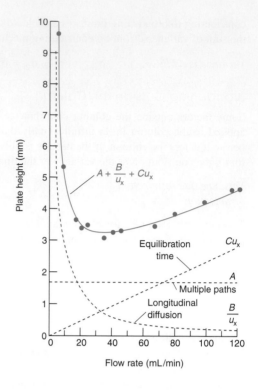

**Figure 23-15** Application of van Deemter equation to gas chromatography: $A = 1.65$ mm, $B = 25.8$ mm $\cdot$ mL/min, and $C = 0.023\ 6$ mm $\cdot$ min/mL. [Experimental points from H. W. Moody, "The Evaluation of the Parameters in the van Deemter Equation," J. Chem. Ed. **1982**, 59, 290.]

equation says there are band-broadening mechanisms that are proportional to flow rate, inversely proportional to flow rate, and independent of flow rate (Figure 23-15).

In packed columns, all three terms contribute to band broadening. For open tubular columns, the multiple path term, $A$, is 0, so bandwidth decreases and resolution increases. In capillary electrophoresis (Chapter 26), both $A$ and $C$ go to 0, thereby reducing plate height to submicron values and providing extraordinary separation powers.

## Longitudinal Diffusion

If you could apply a thin, disk-shaped band of solute to the center of a column, the band would slowly broaden as molecules diffuse from the high concentration within the band to regions of lower concentration on the edges of the band. Diffusional broadening of a band, called **longitudinal diffusion** because it takes place along the axis of the column, occurs while the band is transported along the column by the flow of solvent (Figure 23-16).

The term $B/u_x$ in Equation 23-33 arises from longitudinal diffusion. The faster the linear flow, the less time is spent in the column and the less diffusional broadening occurs. Equation 23-26 told us that the variance resulting from diffusion is

$$\sigma^2 = 2D_m t = \frac{2D_m L}{u_x}$$

*Plate height due to longitudinal diffusion:*

$$H_D = \frac{\sigma^2}{L} = \frac{2D_m}{u_x} \equiv \frac{B}{u_x} \qquad (23\text{-}34)$$

> Because longitudinal diffusion in a gas is much faster than diffusion in a liquid, the optimum linear flow rate in gas chromatography is higher than in liquid chromatography.

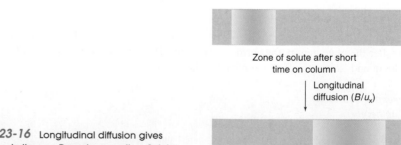

**Figure 23-16** Longitudinal diffusion gives rise to $B/u_x$ in the van Deemter equation. Solute continuously diffuses away from the concentrated center of its zone. The faster the flow, the less time is spent on the column and the less longitudinal diffusion occurs.

CHAPTER 23 Introduction to Analytical Separations

where $D_m$ is the diffusion coefficient of solute in the mobile phase, $t$ is time, and $H_D$ is the plate height due to longitudinal diffusion. The time needed to travel the length of the column is $L/u_x$, where $L$ is the column length and $u_x$ is the linear flow rate.

## Finite Equilibration Time Between Phases

The term $Cu_x$ in Equation 23-33 comes from the finite time required for solute to equilibrate between mobile and stationary phases. Although some solute is stuck in the stationary phase, the remainder in the mobile phase moves forward, thereby resulting in spreading of the overall zone of solute (Figure 23-17).

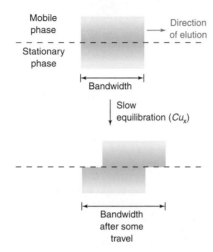

**Figure 23-17** The finite time required for solute to equilibrate between mobile and stationary phases gives rise to $Cu_x$ in the van Deemter equation. The slower the linear flow, the more complete equilibration is and the less zone broadening occurs.

The plate height from finite equilibration time, also called the *mass transfer term,* is

*Plate height due to finite equilibration time:*
$$H_{\text{mass transfer}} = Cu_x = (C_s + C_m)u_x \qquad (23\text{-}35)$$

where $C_s$ describes the rate of mass transfer through the stationary phase and $C_m$ describes mass transfer through the mobile phase. Specific equations for $C_s$ and $C_m$ depend on the type of chromatography.

For gas chromatography in an open tubular column, the terms are

*Mass transfer in stationary phase:*
$$C_s = \frac{2k'}{3(k'+1)^2}\frac{d^2}{D_s} \qquad (23\text{-}35a)$$

*Mass transfer in mobile phase:*
$$C_m = \frac{1+6k'+11k'^2}{24(k'+1)^2}\frac{r^2}{D_m} \qquad (23\text{-}35b)$$

where $k'$ is the capacity factor, $d$ is the thickness of stationary phase, $D_s$ is the diffusion coefficient of solute in the stationary phase, $r$ is the column radius, and $D_m$ is the diffusion coefficient of solute in the mobile phase. Decreasing stationary phase thickness, $d$, reduces plate height and increases efficiency because solute can diffuse faster from the farthest depths of the stationary phase into the mobile phase. Decreasing column radius, $r$, reduces plate height by decreasing the distance through which solute must diffuse to reach the stationary phase.

Mass transfer plate height is also decreased by increasing temperature, which increases the diffusion coefficient of solute in the stationary phase. In Figure 23-18, raising the temperature allows the linear flow rate to be increased by a factor of 5 while maintaining acceptable resolution. Resolution is maintained because of the increased rate of mass transfer between phases at elevated temperature. Many common silica-based stationary phases for liquid chromatography are not stable at elevated temperature. The zirconia ($ZrO_2$)-based material in Figure 23-18 is used because it is stable.

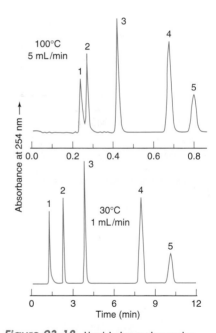

**Figure 23-18** Liquid chromatography showing decreased analysis time when temperature is raised from 30° to 100°C. 1, uracil; 2, *p*-nitroaniline; 3, methyl benzoate; 4, phenetole; 5, toluene. The 4.6-mm-diameter × 100-cm-long column was packed with 4.5-μm-diameter zirconia ($ZrO_2$) coated with 2.1 wt% polybutadiene and eluted with 20 vol% acetonitrile in water. *[J. Li, Y. Hu, and P. W. Carr, "Fast Separations at Elevated Temperatures on Polybutadiene-Coated Zirconia Reversed-Phase Material," Anal. Chem.* **1997,** *69, 3884.]* For a silica-based stationary phase, temperature is usually kept below 60°C to prevent hydrolysis of the silica.

## Multiple Flow Paths

The term *A* was formerly called the *eddy diffusion* term.

The term *A* in the van Deemter equation (23-33) arises from multiple effects for which the theory is murky. Figure 23-19 is a pictorial explanation of one effect. Because some flow paths are longer than others, molecules entering the column at the same time on the left are eluted at different times on the right. For simplicity, we approximate many different effects by the constant *A* in Equation 23-33.

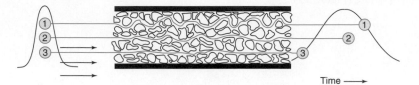

**Figure 23-19** Band spreading from multiple flow paths. The smaller the stationary phase particles, the less serious this problem is. This process is absent in an open tubular column. *[Adapted from H. M. McNair and E. J. Bonelli, Basic Gas Chromatography (Palo Alto, CA: Varian Instrument Division, 1968).]*

## Advantages of Open Tubular Columns

Compared with packed columns, open tubular columns can provide
1. higher resolution
2. shorter analysis time
3. increased sensitivity
4. lower sample capacity

In gas chromatography, we have a choice of using open tubular columns or packed columns. For similar analysis times, open tubular columns provide higher resolution and increased sensitivity to small quantities of analyte. Open tubular columns have small sample capacity, so they are not useful for preparative separations.

For a given pressure, flow rate is proportional to the cross-sectional area of the column and inversely proportional to column length:

$$\text{Flow} \propto \frac{\text{area}}{\text{length}}$$

Particles in a packed column resist flow of the mobile phase, so the linear flow rate cannot be very fast. For the same length of column and applied pressure, the linear flow rate in an open tubular column is much higher than that of a packed column. Therefore, the open tubular column can be made 100 times longer than the packed column, to give a similar pressure drop and linear flow rate. If plate height is the same, the longer column provides 100 times more theoretical plates, yielding $\sqrt{100} = 10$ times more resolution.

Compared with packed columns, open tubular columns allow
1. increased linear flow rate or a longer column or both
2. decreased plate height, which means higher resolution

Plate height is reduced in an open tubular column because band spreading by multiple flow paths (Figure 23-19) cannot occur. In the van Deemter curve for the packed column in Figure 23-15, the *A* term accounts for half of the plate height at the most efficient flow rate (minimum *H*) near 30 mL/min. If *A* were deleted, the number of plates on the column would be doubled. To obtain high performance from an open tubular column, the radius of the column must be small and the stationary phase must be as thin as possible to ensure rapid exchange of solute between mobile and stationary phases.

Table 23-3 compares the performances of packed and open tubular gas chromatography columns with the same stationary phase. For similar analysis times, the open tubular column gives resolution seven times better (10.6 versus 1.5) than that of the packed

**Table 23-3** Comparison of packed and wall-coated open tubular column performance[a]

| Property | Packed | Open tubular |
|---|---|---|
| Column length, *L* | 2.4 m | 100 m |
| Linear gas velocity | 8 cm/s | 16 cm/s |
| Plate height for methyl oleate | 0.73 mm | 0.34 mm |
| Capacity factor, $k'$, for methyl oleate | 58.6 | 2.7 |
| Theoretical plates, *N* | 3 290 | 294 000 |
| Resolution of methyl stearate and methyl oleate | 1.5 | 10.6 |
| Retention time of methyl oleate | 29.8 min | 38.5 min |

a. Methyl stearate ($CH_3(CH_2)_{16}CO_2CH_3$) and methyl oleate (cis-$CH_3(CH_2)_7CH{=}CH(CH_2)_7CO_2CH_3$) were separated on columns with poly(diethylene glycol succinate) stationary phase at 180°C.

SOURCE: L. S. Ettre. Introduction to Open Tubular Columns (Norwalk, CT: Perkin-Elmer Corp., 1979), p. 26.

column. Alternatively, speed could be traded for resolution. If the open tubular column were reduced to 5 m in length, the same solutes could be separated with a resolution of 1.5, but the time would be reduced from 38.5 to 0.83 min.

## A Touch of Reality: Asymmetric Bandshapes

A Gaussian bandshape results when the partition coefficient, $K (= C_s/C_m)$, is independent of the concentration of solute on the column. In real columns, $K$ changes as the concentration of solute increases, and bandshapes are skewed.[10] A graph of $C_s$ versus $C_m$ (at a given temperature) is called an *isotherm*. Three common isotherms and their resulting bandshapes are shown in Figure 23-20. The center isotherm is the ideal one, leading to a symmetric peak.

The upper isotherm in Figure 23-20 arises from an *overloaded* column in which too much solute has been applied to the column. As the concentration of solute increases, the solute becomes more and more soluble in the stationary phase. There is so much solute in the stationary phase that the stationary phase begins to resemble solute. (There is a rule of thumb in chemistry that "like dissolves like.") The front of an overloaded peak has gradually increasing concentration. As the concentration increases, the band becomes overloaded. The solute is so soluble in the overloaded zone that little solute trails behind the peak. The band emerges gradually from the column but ends abruptly.

The lower isotherm in Figure 23-20 arises when small quantities of solute are retained more strongly than large quantities. It leads to a long "tail" of gradually decreasing concentration after the peak.

$C_s$ = concentration of solute in stationary phase

$C_m$ = concentration of solute in mobile phase

Overloading produces a gradual rise and an abrupt fall of the chromatographic peak.

A long tail occurs when some sites retain solute more strongly than other sites.

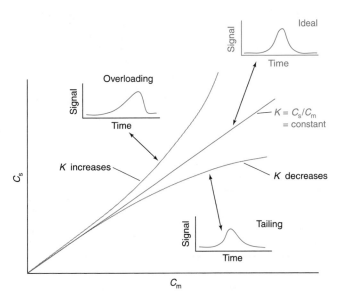

**Figure 23-20** Common isotherms and their resulting chromatographic bandshapes.

Sites that bind solute strongly cause tailing. Silica surfaces of columns and stationary phase particles have hydroxyl groups that form hydrogen bonds with polar solutes, thereby leading to serious tailing. **Silanization** reduces tailing by blocking the hydroxyl groups with nonpolar trimethylsilyl groups:

$$\underset{\substack{\text{Solid phase} \\ \text{with exposed} \\ \text{—OH groups}}}{\overset{\displaystyle OH \qquad OH}{-Si-O-Si-}} + \underset{\text{Hexamethyldisilazane}}{(CH_3)_3SiNHSi(CH_3)_3} \rightarrow \underset{\text{Protected surface}}{\overset{\displaystyle (CH_3)_3SiO \qquad OSi(CH_3)_3}{-Si-O-Si-}} + NH_3 \quad (23\text{-}36)$$

Glass and silica columns used for gas and liquid chromatography can also be silanized to minimize interaction of the solute with active sites on the walls.

Now that you have been exposed to many concepts, you might want to read about a microscopic model of chromatography in Box 23-2.

## Box 23-2   Microscopic Description of Chromatography

A *stochastic* theory provides a simple model for chromatography.[11] The term "stochastic" implies the presence of a random variable. The model supposes that, as a molecule travels through a column, it spends an average time $\tau_m$ in the mobile phase between adsorption events. The time between desorption and the next adsorption is random, but the *average* time is $\tau_m$. The average time spent adsorbed to the stationary phase between one adsorption and one desorption is $\tau_s$. While the molecule is adsorbed on the stationary phase, it does not move. When the molecule is in the mobile phase, it moves with the speed $u_x$ of the mobile phase. The probability that an adsorption or desorption occurs in a given time follows the Poisson distribution, which was described briefly in Problem 19-21.

We assume that all molecules spend total time $t_m$ in the mobile phase. This is the retention time of unretained solute. Important results of the stochastic model are:

- A solute molecule is adsorbed and desorbed an average of $n$ times as it flows through the column, where $n = t_m/\tau_m$.

- The adjusted retention time for a solute is

$$t_r' = n\tau_s \qquad (A)$$

This is the average time that the solute is bound to the stationary phase during its transit through the column.

- The width (standard deviation) of a peak due to effects of the stationary phase is

$$\sigma = \tau_s\sqrt{2n} \qquad (B)$$

Consider the idealized chromatogram in the illustration, with one unretained component and two retained substances, A and B. The chromatographic parameters are representative of a high-performance liquid chromatography separation on a 15-cm-long × 0.39-cm-diameter column packed with 5-μm-diameter spherical particles of $C_{18}$-silica (Section 25-1). With a volume flow rate of 1.0 mL/min, the linear velocity is $u_x = 2.4$ mm/s. From the measured width at half-height ($w_{1/2}$), the standard deviation ($\sigma$) of a Gaussian peaks is computed from $w_{1/2} = 2.35\sigma$ (Figure 23-9). The plate number for components A and B, computed with Equation 23-28, is $N = (t_r/\sigma)^2 = 1.00 \times 10^4$.

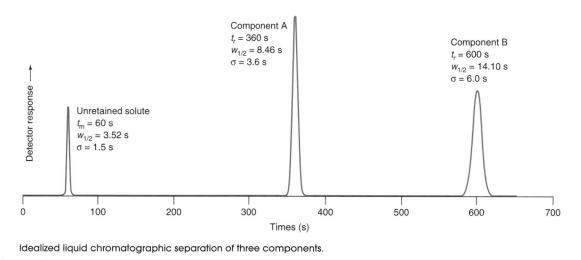

**Component A**
$t_r = 360$ s
$w_{1/2} = 8.46$ s
$\sigma = 3.6$ s

**Component B**
$t_r = 600$ s
$w_{1/2} = 14.10$ s
$\sigma = 6.0$ s

**Unretained solute**
$t_m = 60$ s
$w_{1/2} = 3.52$ s
$\sigma = 1.5$ s

Detector response →

Times (s)

Idealized liquid chromatographic separation of three components.

## Terms to Understand

| | | | |
|---|---|---|---|
| adjusted retention time | elution | molecular exclusion | retention time |
| adsorption chromatography | extraction | chromatography | retention volume |
| affinity chromatography | gel filtration chromatography | open tubular column | separation factor |
| capacity factor | gel permeation chromatography | packed column | silanization |
| chromatogram | ion-exchange chromatography | partition chromatography | stationary phase |
| diffusion coefficient | linear flow rate | partition coefficient | van Deemter equation |
| distribution coefficient | longitudinal diffusion | plate height | volume flow rate |
| eluate | miscible | relative retention | |
| eluent | mobile phase | resolution | |

## Summary

A solute can be extracted from one phase into another in which it is more soluble. The ratio of solute concentrations in each phase at equilibrium is called the partition coefficient. If more than one form of the solute exists, we use a distribution coefficient instead of a partition coefficient. We derived equations relating the fraction of solute extracted to the partition or distribution coefficient, volumes, and pH. Many small extractions are more effective than a few large extractions.

A metal chelator, soluble only in organic solvents, can extract metal ions from aqueous solutions, with selectivity achieved by adjusting pH.

In adsorption and partition chromatography, a continuous equilibration of solute between mobile and stationary phases occurs. Eluent goes into a column and eluate comes out. Columns may be packed with stationary phase or may be open tubular, with stationary phase bonded to the inner wall. In ion-exchange chromatogra-

The stochastic model applies to processes involving the stationary phase. To analyze the chromatogram, we need to subtract contributions to peak broadening from dispersion in the mobile phase and extra-column effects such as finite injection width and finite detector volume. These effects account for the width of the unretained peak. To subtract the unwanted effects, we write

$$\sigma^2_{observed} = \sigma^2_{stationary\ phase} + \sigma^2_{unretained\ peak}$$

$$\sigma^2_{stationary\ phase} = \sigma^2_{observed} - \sigma^2_{unretained\ peak}$$

For component A, $\sigma^2_{stationary\ phase} = \sigma^2_{observed} - \sigma^2_{unretained\ peak} = (3.6\ s)^2 - (1.5\ s)^2 \Rightarrow \sigma_{stationary\ phase} = 3.27\ s$. For component B, we find $\sigma_{stationary\ phase} = 5.81\ s$. The adjusted retention time for component A is $t'_r = t_r - t_m = 360 - 60 = 300\ s$. For component B, $t'_r = 600 - 60 = 540\ s$.

Now we use $t'_r$ and $\sigma(= \sigma_{stationary\ phase})$ for each component to find physically meaningful parameters. Combining Equations A and B, we find

$$n = 2\left(\frac{t'_r}{\sigma}\right)^2 \qquad \tau_s = \left(\frac{\sigma^2}{2t'_r}\right)$$

and we already knew that $\tau_m = t_m/n$. From the parameters in the illustration, we compute the results in the table.

| | Component A | Component B |
|---|---|---|
| $n$ | 16 800 | 17 300 |
| $\tau_s$ | 17.8 ms | 31.2 ms |
| $\tau_m$ | 3.6 ms | 3.5 ms |
| Distance between adsorptions ($= u_x\tau_m$) | 8.6 μm | 8.4 μm |

We see that both components spend nearly the same time (~3.5 ms) in the mobile phase between adsorption events. Component A spends an average of 17.8 ms bound to the stationary phase each time it is adsorbed, and component B spends an average of 31.2 ms. This difference in $\tau_s$ is the reason why A and B are separated from each other.

During its transit through the column, each substance becomes adsorbed $n \approx 17\,000$ times. The distance traveled between adsorptions is ~8.6 μm. The chromatogram was simulated for a column with $N = 10\,000$ theoretical plates. The plate height is 15 cm/(10 000 plates) = 15 μm. In Section 23-4, we stated that plate height is approximately the length of column required for one equilibration of solute between mobile and stationary phases. From the stochastic theory in this example, we find that there are approximately two equilibrations with the stationary phase in each length corresponding to the plate height.

The time required for a solute to flow past a given particle of stationary phase whose diameter is 5 μm is (5 μm)/(2.4 mm/s) = 2.1 ms. The stochastic theory predicts that the fraction of time that a molecule in the mobile phase will travel *less than* distance $d$ is $1 - e^{-d/\tau_m} = 1 - e^{-(2.1\ ms)/(3.5\ ms)} = 0.55$. That is, approximately half of the time, a solute molecule does not travel as far as the next particle of stationary phase before becoming adsorbed again to the same particle from which it just desorbed. If we lined up spherical particles of stationary phase, it would take 30 000 particles to cover the 15-cm length of the column. Each solute molecule binds ~17 000 times as it transits the column, and half of those binding steps are to the same particle from which it just desorbed.

The simple model provides microscopic insight into events that occur during chromatography. The model omits some phenomena that occur in real columns. For example, in a porous stationary phase, the mobile phase could be stagnant inside the pores. When a molecule enters such a pore, it will adsorb and desorb many times from the same particle before escaping from the pore.

phy, the solute is attracted to the stationary phase by coulombic forces. In molecular exclusion chromatography, the fraction of stationary phase volume available to solute decreases as the size of the solute molecules increases. Affinity chromatography relies on specific, noncovalent interactions between the stationary phase and one solute in a complex mixture.

The relative retention of two components is the quotient of their adjusted retention times. The capacity factor for a single component is the adjusted retention time divided by the elution time for solvent. Capacity factor gives the ratio of time spent by solute in the stationary phase to time spent in the mobile phase. When a separation is scaled up from a small load to a large load, the cross-sectional area of the column should be increased in proportion to the loading. Column length and linear flow rate are held constant.

Plate height ($H = \sigma^2/x$) is related to the breadth of a band emerging from the column. The smaller the plate height, the sharper the band. The number of plates for a Gaussian peak is $N = 5.55\ t_r^2/w_{1/2}^2$. Plate height is approximately the length of column required for one equilibrium of solute between mobile and stationary phases. Resolution of neighboring peaks is the difference in retention time divided by the average width (measured at the base-line, $w = 4\sigma$). Resolution is proportional to $\sqrt{N}$ and increases with the separation factor, $\gamma$, which is the quotient of linear velocities of two components. Doubling the length of a column increases resolution by $\sqrt{2}$.

The standard deviation of a diffusing band of solute is $\sigma = \sqrt{2Dt}$, where $D$ is the diffusion coefficient and $t$ is time. The van Deemter equation describes band broadening on the chromatographic column: $H \approx A + B/u_x + Cu_x$, where $H$ is plate height, $u_x$ is linear flow rate, and $A$, $B$, and $C$ are constants. The first term represents irregular flow paths; the second, longitudinal diffusion; and the third, the finite rate of transfer of solute between mobile and stationary phases. The optimum flow rate, which minimizes plate height, is faster for gas chromatography than for liquid chromatography. The number of plates and the optimal flow rate increase as the stationary phase particle size decreases. In gas chromatography, open tubular columns can provide higher resolution or shorter analysis times than packed columns. Bands spread during injection and detection, as well as during passage through the column. The observed variance of the band is the sum of the variances for all mechanisms of spreading. Overloading and tailing can be corrected by using smaller samples and by masking strong adsorption sites on the stationary phase.

## Exercises

**23-A.** Consider a chromatography experiment in which two components with capacity factors $k_1' = 4.00$ and $k_2' = 5.00$ are injected into a column with $N = 1.00 \times 10^3$ theoretical plates. The retention time for the less-retained component is $t_{r1} = 10.0$ min.

(a) Calculate $t_m$ and $t_{r2}$. Find $w_{1/2}$ (width at half-height) and $w$ (width at the base) for each peak.

(b) Using graph paper, sketch the chromatogram analogous to Figure 23-7, supposing that the two peaks have the same amplitude (height). Draw the half-widths accurately.

(c) Calculate the resolution of the two peaks and compare this value with those drawn in Figure 23-10.

**23-B.** A solute with a partition coefficient of 4.0 is extracted from 10 mL of phase 1 into phase 2.

(a) What volume of phase 2 is needed to extract 99% of the solute in one extraction?

(b) What is the total volume of solvent 2 needed to remove 99% of the solute in three equal extractions instead?

**23-C.** (a) Find the capacity factors for octane and nonane in Figure 23-7.

(b) Find the ratio

$$\frac{\text{Time octane spends in stationary phase}}{\text{Total time octane spends on column}}$$

(c) Find the relative retention for octane and nonane.

(d) Find the partition coefficient for octane by assuming that the volume of the stationary phase equals half the volume of the mobile phase.

**23-D.** A gas chromatogram of a mixture of toluene and ethyl acetate is shown here.

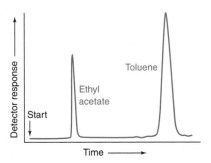

(a) Use the width of each peak (measured at the base) to calculate the number of theoretical plates in the column. Estimate all lengths to the nearest 0.1 mm.

(b) Using the width of the toluene peak at its base, calculate the width expected at half-height. Compare the measured and calculated values. When the thickness of the pen trace is significant relative to the length being measured, it is important to take the pen width into account. You can measure from the edge of one trace to the corresponding edge of the other trace, as shown in the diagram in the next column.

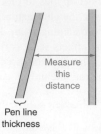

Measure this distance

Pen line thickness

**23-E.** The three chromatograms shown below were obtained with 2.5, 1.0, and 0.4 μL of ethyl acetate injected on the same column under the same conditions. Explain why the peak becomes less symmetric with increasing sample size.

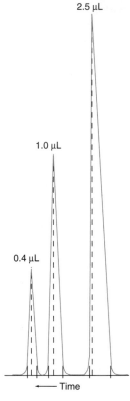

**23-F.** The relative retention for two compounds in gas chromatography is 1.068 on a column with a plate height of 0.520 mm. The capacity factor for compound 1 is 5.16.

(a) Find the separation factor ($\gamma$) for the two compounds.

(b) What length of column will separate the compounds with a resolution of 1.00?

(c) The retention time for air ($t_m$) is 2.00 min. If the number of plates is the same for both compounds, find $t_r$ and $w_{1/2}$ for each peak.

(d) If the ratio of stationary phase to mobile phase is 0.30, find the partition coefficient for compound 1.

## Problems

### Solvent Extraction

**23-1.** If you are extracting a substance from water into ether, is it more effective to do one extraction with 300 mL of ether or three extractions with 100 mL?

**23-2.** If you wish to extract aqueous acetic acid into hexane, is it more effective to adjust the aqueous phase to pH 3 or pH 8?

**23-3.** Why is it difficult to extract the EDTA complex of aluminum into an organic solvent but easy to extract the 8-hydroxyquinoline complex?

**23-4.** Why is the extraction of a metal ion into an organic solvent with 8-hydroxyquinoline more complete at higher pH?

**23-5.** The distribution coefficient for extraction of a metal complex from aqueous to organic solvents is $D = [\text{total metal}]_{\text{org}}/$

[total metal]$_{aq}$. Give physical reasons why $\beta$ and $K_a$ appear in the numerator of Equation 23-13, but $K_L$ and $[H^+]_{aq}$ appear in the denominator.

**23-6.** Give a physical interpretation of Equations 23-6 and 23-7 in terms of the fractional composition equations for a monoprotic acid discussed in Section 10-5.

**23-7.** Solute S has a partition coefficient of 4.0 between water (phase 1) and chloroform (phase 2) in Equation 23-1.
(a) Calculate the concentration of S in chloroform if $[S(aq)]$ is 0.020 M.
(b) If the volume of water is 80.0 mL and the volume of chloroform is 10.0 mL, find the quotient (mol S in chloroform)/(mol S in water).

**23-8.** The solute in Problem 23-7 is initially dissolved in 80.0 mL of water. It is then extracted six times with 10.0-mL portions of chloroform. Find the fraction of solute remaining in the aqueous phase.

**23-9.** The weak base B ($K_b = 1.0 \times 10^{-5}$) equilibrates between water (phase 1) and benzene (phase 2).
(a) Define the distribution coefficient, $D$, for this system.
(b) Explain the difference between $D$ and $K$, the partition coefficient.
(c) Calculate $D$ at pH 8.00 if $K = 50.0$.
(d) Will $D$ be greater or less at pH 10 than at pH 8? Explain why.

**23-10.** Consider the extraction of $M^{n+}$ from aqueous solution into organic solution by reaction with protonated ligand, HL:

$$M^{n+}(aq) + nHL(org) \rightleftharpoons ML_n(org) + nH^+(aq)$$

$$K_{extraction} = \frac{[ML_n]_{org}[H^+]_{aq}^n}{[M^{n+}]_{aq}[HL]_{org}^n}$$

Rewrite Equation 23-13 in terms of $K_{extraction}$ and express $K_{extraction}$ in terms of the constants in Equation 23-13. Give a physical reason why each constant increases or decreases $K_{extraction}$.

**23-11.** Butanoic acid has a partition coefficient of 3.0 (favoring benzene) when distributed between water and benzene. Find the formal concentration of butanoic acid in each phase when 100 mL of 0.10 M aqueous butanoic acid is extracted with 25 mL of benzene (a) at pH 4.00 and (b) at pH 10.00.

**23-12.** For a given value of $[HL]_{org}$ in Equation 23-13, over what pH range (how many pH units) will $D$ change from 0.01 to 100 if $n = 2$?

**23-13.** For the extraction of $Cu^{2+}$ by dithizone in $CCl_4$, $K_L = 1.1 \times 10^4$, $K_M = 7 \times 10^4$, $K_a = 3 \times 10^{-5}$, $\beta = 5 \times 10^{22}$, and $n = 2$.
(a) Calculate the distribution coefficient for extraction of 0.1 $\mu$M $Cu^{2+}$ into $CCl_4$ by 0.1 mM dithizone at pH 1.0 and pH 4.0.
(b) If 100 mL of 0.1 $\mu$M aqueous $Cu^{2+}$ are extracted once with 10 mL of 0.1 mM dithizone at pH 1.0, what fraction of $Cu^{2+}$ remains in the aqueous phase?

**23-14.** Consider the extraction of 100.0 mL of $M^{2+}(aq)$ by 2.0 mL of $1 \times 10^{-5}$ dithizone in $CHCl_3$, for which $K_L = 1.1 \times 10^4$, $K_M = 7 \times 10^4$, $K_a = 3 \times 10^{-5}$, $\beta = 5 \times 10^{18}$, and $n = 2$.
(a) Derive an expression for the fraction of metal ion extracted into the organic phase, in terms of the distribution coefficient and volumes of the two phases.

(b) ▦ Prepare a graph of the percentage of metal ion extracted over the pH range 0 to 5.

**A Plumber's View of Chromatography**

**23-15.** Match the terms in the first list with the characteristics in the second list.

1. adsorption chromatography
2. partition chromatography
3. ion-exchange chromatography
4. molecular exclusion chromatography
5. affinity chromatography

A. Ions in mobile phase are attracted to counterions covalently attached to stationary phase.
B. Solute in mobile phase is attracted to specific groups covalently attached to stationary phase.
C. Solute equilibrates between mobile phase and surface of stationary phase.
D. Solute equilibrates between mobile phase and film of liquid attached to stationary phase.
E. Different-sized solutes penetrate voids in stationary phase to different extents. Largest solutes are eluted first.

**23-16.** The partition coefficient for a solute in chromatography is $K = C_s/C_m$, where $C_s$ is the concentration in the stationary phase and $C_m$ is the concentration in the mobile phase. The larger the partition coefficient, the longer it takes a solute to be eluted. Explain why.

**23-17.** (a) Write the meaning of the capacity factor, $k'$, in terms of time spent by solute in each phase.
(b) Write an expression in terms of $k'$ for the fraction of time spent by a solute molecule in the mobile phase.
(c) The *retention ratio* in chromatography is defined as

$$R = \frac{\text{time for solvent to pass through column}}{\text{time for solute to pass through column}} = \frac{t_m}{t_r}$$

Show that $R$ is related to the capacity factor by the equation $R = 1/(k' + 1)$.

**23-18.** (a) A chromatography column with a length of 10.3 cm and inner diameter of 4.61 mm is packed with a stationary phase that occupies 61.0% of the volume. If the volume flow rate is 1.13 mL/min, find the linear flow rate in cm/min.
(b) How long does it take for solvent (which is the same as unretained solute) to pass through the column?
(c) Find the retention time for a solute with a capacity factor of 10.0.

**23-19.** An open tubular column is 30.1 m long and has an inner diameter of 0.530 mm. It is coated on the inside wall with a layer of stationary phase that is 3.1 $\mu$m thick. Unretained solute passes through in 2.16 min, whereas a particular solute has a retention time of 17.32 min.
(a) Find the linear and volume flow rates.
(b) Find the capacity factor for the solute and the fraction of time spent in the stationary phase.
(c) Find the partition coefficient, $C_s/C_m$, for this solute.

**23-20.** A chromatographic procedure separates 4.0 mg of unknown mixture on a column with a length of 40 cm and a diameter of 0.85 cm.
(a) What size column would you use to separate 100 mg of the same mixture?
(b) If the flow is 0.22 mL/min on the small column, what volume flow rate should be used on the large column?
(c) If mobile phase occupies 35% of the column volume, calculate the linear flow rates for the small column and the large column.

**23-21.** Solvent passes through a column in 3.0 min, but solute requires 9.0 min.
**(a)** Calculate the capacity factor, $k'$.
**(b)** What fraction of time is the solute in the mobile phase in the column?
**(c)** The volume of stationary phase is 1/10th of the volume of the mobile phase in the column ($V_s = 0.10\, V_m$). Find the partition coefficient, $K$, for this system.

**23-22.** Solvent occupies 15% of the volume of a chromatography column whose inner diameter is 3.0 mm. If the volume flow rate is 0.2 mL/min, find the linear flow rate.

**23-23.** Consider a chromatography column in which $V_s = V_m/5$. Find the capacity factor if $K = 3$ and if $K = 30$.

**23-24.** The retention volume of a solute is 76.2 mL for a column with $V_m = 16.6$ mL and $V_s = 12.7$ mL. Calculate the capacity factor and the partition coefficient for this solute.

**23-25.** An open tubular column has a diameter of 207 μm and the thickness of the stationary phase on the inner wall is 0.50 μm. Unretained solute passes through in 63 s and a particular solute emerges in 433 s. Find the partition coefficient for this solute and find the fraction of time spent in the stationary phase.

**23-26.** Isotopic compounds are separated in Figure 23-14 by repeated passage through a pair of columns. Each cycle in the figure represents one pass through total length $L = 50$ cm containing $N$ theoretical plates. The separation factor is $\gamma = 1.018$.
**(a)** The observed resolution after 10 cycles is 1.60. Calculate the number of theoretical plates, $N$, in column length $L$. The mixture has passed through length $10L$ in 10 cycles.
**(b)** Find the plate height in μm.
**(c)** Predict the resolution expected after just two cycles. The observed value is 0.71.

**Efficiency and Band Spreading**

**23-27.** Chromatograms of compounds A and B were obtained at the same flow rate with two columns of equal length.

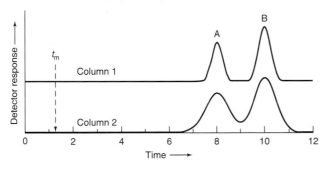

**(a)** Which column has more theoretical plates?
**(b)** Which column has a larger plate height?
**(c)** Which column gives higher resolution?
**(d)** Which column gives a greater relative retention?
**(e)** Which compound has a higher capacity factor?
**(f)** Which compound has a greater partition coefficient?
**(g)** What is the numerical value of the separation factor?

**23-28.** Why does plate height depend on linear flow rate, not volume flow rate?

**23-29.** Which column is more efficient: plate height = **(a)** 0.1 mm or **(b)** 1 mm?

**23-30.** Why is longitudinal diffusion a more serious problem in gas chromatography than in liquid chromatography?

**23-31.** In chromatography, why is the optimal flow rate greater if the stationary phase particle size is smaller?

**23-32.** What is the optimal flow rate in Figure 23-15 for best separation of solutes?

**23-33.** Explain how silanization reduces tailing of chromatographic peaks.

**23-34.** Describe how nonlinear partition isotherms lead to non-Gaussian bandshapes. Draw the bandshape produced by an overloaded column and a column with tailing.

**23-35.** A separation of 2.5 mg of an unknown mixture has been optimized on a column of length $L$ and diameter $d$. Explain why you might not achieve the same resolution for 5.0 mg on a column of length $2L$ and diameter $d$.

**23-36.** An infinitely sharp zone of solute is placed at the center of a column at time $t = 0$. After diffusion for time $t_1$, the standard deviation of the Gaussian band is 1.0 mm. After 20 min more, at time $t_2$, the standard deviation is 2.0 mm. What will be the width after another 20 min, at time $t_3$?

**23-37.** A chromatogram with ideal Gaussian bands has $t_r = 9.0$ min and $w_{1/2} = 2.0$ min.
**(a)** How many theoretical plates are present?
**(b)** Find the plate height if the column is 10 cm long.

**23-38. (a)** The asymmetric chromatogram in Figure 23-13 has a retention time equal to 15 min, and the values of A and B are 33 and 11 s, respectively. Find the number of theoretical plates.
**(b)** The width of the Gaussian peak in Figure 23-9 at a height equal to 1/10th of the peak height is $4.297\sigma$. Suppose that the peak in part **(a)** is symmetric with A = B = 22 s. Use Equations 23-28 and 23-29 to find the plate number.

**23-39.** Two chromatographic peaks with widths, $w$, of 6 min are eluted at 24 and 29 min. Which diagram in Figure 23-10 will most closely resemble the chromatogram?

**23-40.** A chromatographic band has a width, $w$, of 4.0 mL and a retention volume of 49 mL. What width is expected for a band with a retention volume of 127 mL? Assume that the only band spreading occurs on the column itself.

**23-41.** A band from a column eluted at 0.66 mL/min has a width at half-height, $w_{1/2}$, of 39.6 s. The sample was applied as a sharp plug with a volume of 0.40 mL, and the detector volume is 0.25 mL. Find the variances introduced by injection and detection. What would be $w_{1/2}$ if the only broadening occurred on the column?

**23-42.** Two compounds with partition coefficients of 15 and 18 are to be separated on a column with $V_m/V_s = 3.0$ and $t_m = 1.0$ min. Calculate the number of theoretical plates needed to produce a resolution of 1.5.

**23-43. (a)** Calculate the number of theoretical plates needed to achieve a resolution of 2.0 for compounds with relative retention times of $t_2/t_1 = 1.01$, 1.05, or 1.10.
**(b)** How can you increase $N$ and $\gamma = t_2/t_1$ in a chromatography experiment?

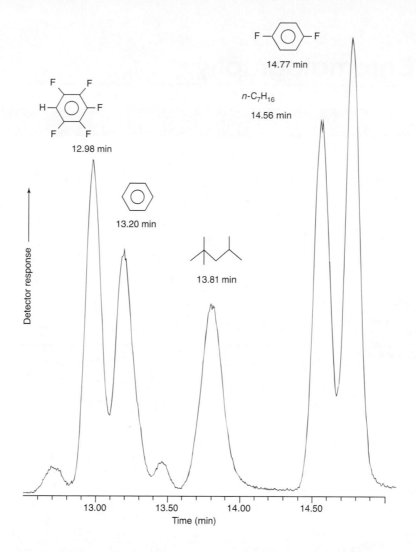

14.77 min

F—⬡—F

n-C₇H₁₆
14.56 min

F   F
H—⬡—F
F   F
12.98 min

⬡
13.20 min

13.81 min

13.00     13.50     14.00     14.50
Time (min)

Detector response

Figure for Problem 23-44

**23-44.** Consider the peaks for pentafluorobenzene and benzene in the chromatogram shown above. The elution time for unretained solute is 1.06 min. The open tubular column is 30.0 m in length and 0.530 mm in diameter, with a layer of stationary phase 3.0 μm thick on the inner wall.

(a) Find the adjusted retention times and capacity factors for both compounds.

(b) Find the relative retention, $\alpha$.

(c) Find the separation factor, $\gamma$.

(d) Measuring $w_{1/2}$ on the chromatogram, find the number of plates, $N_1$ and $N_2$, and the plate height for these two compounds.

(e) Measuring the width, $w$, at the baseline on the chromatogram, find the number of plates for these two compounds.

(f) Use your answer to part (e) to find the resolution between the two peaks.

(g) Using the number of plates [$N = \sqrt{N_1N_2}$, with values from part (e)] and the observed separation factor, calculate what the resolution should be and compare your answer with the measured resolution in part (f).

**23-45.** 🖾 A layer with negligible thickness containing 10.0 nmol of methanol ($D = 1.6 \times 10^{-9}$ m²/s) was placed in a tube of water

5.00 cm in diameter and allowed to spread by diffusion. Using Equation 23-25, prepare a graph showing the Gaussian concentration profile of the methanol zone after 1.00, 10.0, and 100 min. Prepare a second graph showing the same experiment with the enzyme ribonuclease ($D = 0.12 \times 10^{-9}$ m²/s).

**23-46.** 🖾 A 0.25-mm-diameter open tubular gas chromatography column is coated with stationary phase that is 0.25 μm thick. The diffusion coefficient for a compound with a capacity factor $k' = 10$ is $D_m = 1.0 \times 10^{-5}$ m²/s in the gas phase and $D_s = 1.0 \times 10^{-9}$ m²/s in the stationary phase. Consider longitudinal diffusion and finite equilibration time in the mobile and stationary phases as sources of broadening. Prepare a graph showing the plate height from each of these three sources and the total plate height as a function of linear flow rate (from 2 cm/s to 1 m/s). Then change the stationary phase thickness to 2 μm and repeat the calculations. Explain the difference in the two results.

**23-47.** 🖾 Consider two Gaussian peaks with relative areas 4:1. Construct a set of graphs to show the overlapping peaks if the resolution is 0.5, 1, or 2.

# 24 Gas Chromatography

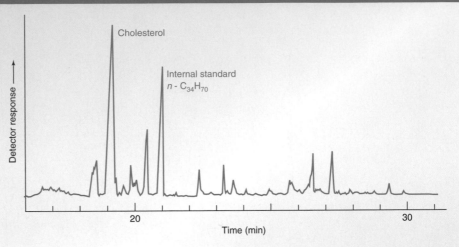

Gas chromatogram of cholesterol and other lipids extracted from bones and derivatized with trimethylsilyl ($(CH_3)_3Si—$) groups to increase volatility for chromatography. Bone contains 2 to 50 μg cholesterol/gram of dry bone. [From A. W. Stott and R. P. Evershed, "δ$^{13}$C Analysis of Cholesterol Preserved in Archaeological Bones and Teeth," Anal. Chem. **1996**, 68, 4402.]

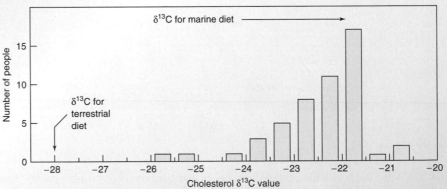

$^{13}$C content of cholesterol from bones of 50 people who lived on British coast in the years A.D. 500–1800. δ$^{13}$C is the deviation of the atomic ratio $^{13}C/^{12}C$ from that of a standard material, measured in parts per thousand. [Data from A. W. Stott and R. P. Evershed, "δ$^{13}$C Analysis of Cholesterol Preserved in Archaeological Bones and Teeth," Anal. Chem. **1996**, 68, 4402.]

The $^{13}$C content of cholesterol preserved in ancient bones provides information on the diets of people who lived long ago. Approximately 1.1% of natural carbon is $^{13}$C and 98.9% is $^{12}$C. Different types of plants and animals have consistent, slightly different ratios of $^{13}C/^{12}C$, which reflect their biosynthetic pathways.

To learn whether people from the ancient British coastal town of Barton-on-Humber ate mainly plants or fish, cholesterol from the bones of 50 people was extracted with organic solvent, isolated by gas chromatography, combusted to convert it into $CO_2$, and its $^{13}C/^{12}C$ ratio measured by mass spectrometry. Observed $^{13}C/^{12}C$ ratios differ from that of a standard material by about −21 to −24 parts per thousand. A diet of local plants yields a δ$^{13}$C value (defined in Box 22-3) in cholesterol of −28 parts per thousand. Values more positive than −28 parts per thousand are indicative of a marine diet. It appears that the population got much of its food from the sea.

Chapter 23 gave a foundation for understanding chromatographic separations. Chapters 24 through 26 discuss specific methods and instrumentation. The goal is for you to understand how chromatographic methods work and what parameters you can control for best results.

## ■■■ 24-1 The Separation Process in Gas Chromatography

Gas chromatography:

*mobile phase:* **gas**

*stationary phase:* usually a nonvolatile liquid, but sometimes a solid

*analyte:* gas or volatile liquid

In **gas chromatography,**[1,2] gaseous analyte is transported through the column by a gaseous mobile phase, called the **carrier gas.** In *gas-liquid partition chromatography,* the stationary phase is a nonvolatile liquid bonded to the inside of the column or to a fine solid support

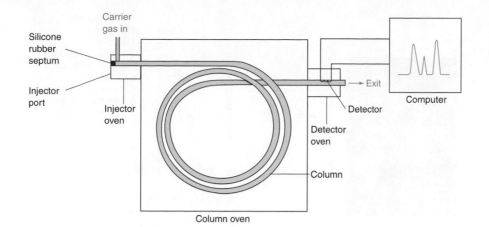

*Figure 24-1* Schematic diagram of a gas chromatograph.

(Figure 23-6, upper right). In *gas-solid adsorption chromatography,* analyte is adsorbed directly on solid particles of stationary phase (Figure 23-6, upper left).

In the schematic *gas chromatograph* in Figure 24-1, volatile liquid or gaseous sample is injected through a **septum** (a rubber disk) into a heated port, in which it rapidly evaporates. Vapor is swept through the column by He, $N_2$, or $H_2$ carrier gas, and separated analytes flow through a detector, whose response is displayed on a computer. The column must be hot enough to provide sufficient vapor pressure for analytes to be eluted in a reasonable time. The detector is maintained at a higher temperature than the column so that all analytes will be gaseous.

The choice of carrier gas depends on the detector and the desired separation efficiency and speed.

## Open Tubular Columns

The vast majority of analyses use long, narrow **open tubular columns** (Figure 24-2) made of fused silica ($SiO_2$) and coated with polyimide (a plastic capable of withstanding 350°C) for support and protection from atmospheric moisture.[3] As discussed in Section 23-5, open tubular columns offer higher resolution, shorter analysis time, and greater sensitivity than packed columns, but have less sample capacity.

The *wall-coated* column in Figure 24-2c features a 0.1- to 5-μm-thick film of stationary liquid phase on the inner wall of the column. A *support-coated* column has solid particles coated with stationary liquid phase and attached to the inner wall. In the *porous-layer* column

Compared with packed columns, open tubular columns offer

1. higher resolution
2. shorter analysis time
3. greater sensitivity
4. lower sample capacity

*Wall-coated open tubular column (WCOT):* liquid stationary phase on inside wall of column
*Support-coated open tubular column (SCOT):* liquid stationary phase coated on solid support attached to inside wall of column
*Porous-layer open tubular column (PLOT):* **solid** stationary phase on inside wall of column

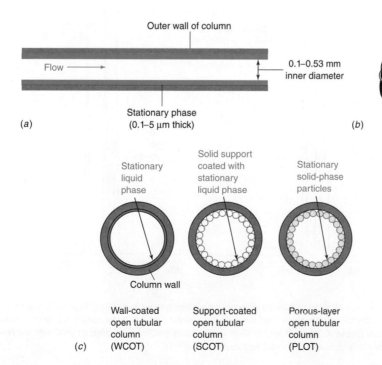

*Figure 24-2* (*a*) Typical dimensions of open tubular gas chromatography column. (*b*) Fused silica column with a cage diameter of 0.2 m and column length of 15–100 m. (*c*) Cross-sectional view of wall-coated, support-coated, and porous-layer columns.

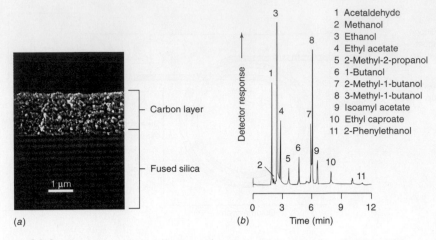

(a)                                    (b)        Time (min)

**Figure 24-3**   (*a*) Porous carbon stationary phase (2 μm thick) on inside wall of fused silica open tubular column. (*b*) Chromatogram of vapors from the headspace of a beer can, obtained with 0.25-mm-diameter × 30-m-long porous carbon column operated at 30°C for 2 min and then ramped up to 160°C at 20°/min. [*Courtesy Alltech Associates, State College, PA.*]

in Figure 24-3, solid particles *are* the active stationary phase. With their high surface area, support-coated columns can handle larger samples than can wall-coated columns. The performance of support-coated columns is intermediate between those of wall-coated columns and packed columns.

Column inner diameters are typically 0.10 to 0.53 mm and lengths are 15 to 100 m, with 30 m being common. Narrow columns provide higher resolution than wider columns (Figure 24-4) but require higher operating pressure and have less sample capacity. Diameters of 0.32 mm or greater tend to overload the gas-handling system of a mass spectrometer, so the gas stream must be split and only a fraction is sent to the mass spectrometer. The number of theoretical plates, $N$, on a column is proportional to the length. In Equation 23-30, resolution is proportional to $\sqrt{N}$ and, therefore, to the square root of column length (Figure 24-5).

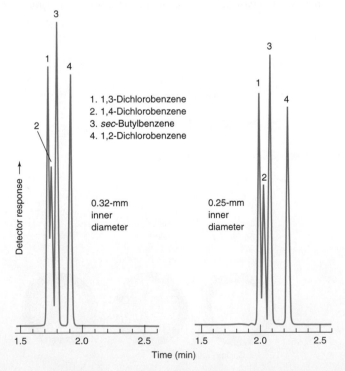

1. 1,3-Dichlorobenzene
2. 1,4-Dichlorobenzene
3. *sec*-Butylbenzene
4. 1,2-Dichlorobenzene

0.32-mm inner diameter

0.25-mm inner diameter

Time (min)

**Figure 24-4**   Effect of open tubular column inner diameter on resolution. Narrower columns provide higher resolution. Notice the increased resolution of peaks 1 and 2 in the narrow column. Conditions: DB-1 stationary phase (0.25 μm thick) in 15-m wall-coated column operated at 95°C with He linear velocity of 34 cm/s. [*Courtesy J&W Scientific, Folsom, CA.*]

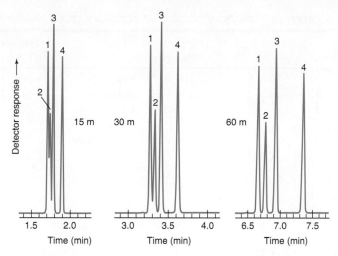

*Figure 24-5* Resolution increases in proportion to the square root of column length. Notice the increased resolution of peaks 1 and 2 as length is increased. Conditions: DB-1 stationary phase (0.25 μm thick) in 0.32-mm-diameter wall-coated column operated at 95°C with He linear velocity of 34 cm/s. Compounds 1–4 are the same as in Figure 24-4. *[Courtesy J&W Scientific, Folsom, CA.]*

At the constant linear velocity in Figure 24-6, increasing the thickness of the stationary phase increases retention time and sample capacity and increases the resolution of early-eluting peaks with a capacity factor of $k' \lesssim 5$. (Capacity factor was defined in Equation 23-16). Thick films of stationary phase can shield analytes from the silica surface and reduce *tailing* (Figure 23-20) but can also increase bleed (decomposition and evaporation) of the stationary phase at elevated temperature. A thickness of 0.25 μm is standard, but thicker films are used for volatile analytes.

The choice of liquid stationary phase (Table 24-1) is based on the rule "like dissolves like." Nonpolar columns are best for nonpolar solutes (Table 24-2). Columns of intermediate polarity are best for intermediate polarity solutes, and strongly polar columns are best for strongly polar solutes. As a column ages, stationary phase bakes off, surface silanol groups (Si—O—H) are exposed, and tailing increases. Exposure to $O_2$ from air at high temperatures also degrades the column. To reduce the tendency of stationary phase to bleed from the column at high temperature, it is usually *bonded* (covalently attached) to the silica surface or covalently *cross-linked* to itself. Box 24-1 describes *chiral* (optically active) bonded phases for separating optical isomers.

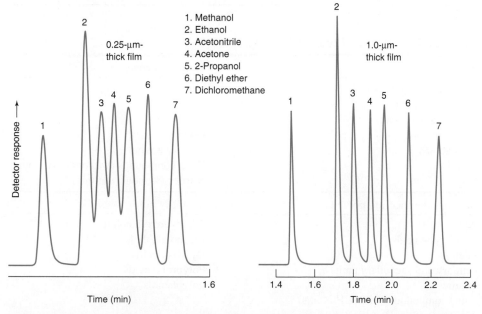

1. Methanol
2. Ethanol
3. Acetonitrile
4. Acetone
5. 2-Propanol
6. Diethyl ether
7. Dichloromethane

*Figure 24-6* Effect of stationary phase thickness on open tubular column performance. Increasing thickness increases retention time and increases resolution of early-eluting peaks. Conditions: DB-1 stationary phase in 15-m-long × 0.32-mm-diameter wall-coated column operated at 40°C with He linear velocity of 38 cm/s. *[Courtesy J&W Scientific, Folsom, CA.]*

**Table 24-1**  Common stationary phases in capillary gas chromatography

| Structure | Polarity | | Temperature range (°C) |
|---|---|---|---|
| (Diphenyl)$_x$(dimethyl)$_{1-x}$ polysiloxane | $x = 0$ | Nonpolar | $-60°$–$320°$ |
| | $x = 0.05$ | Nonpolar | $-60°$–$320°$ |
| | $x = 0.35$ | Intermediate polarity | $0°$–$300°$ |
| | $x = 0.65$ | Intermediate polarity | $50°$–$370°$ |
| (Cyanopropylphenyl)$_{0.14}$ (dimethyl)$_{0.86}$ polysiloxane | Intermediate polarity | | $-20°$–$280°$ |
| Carbowax (poly(ethylene glycol)) | Strongly polar | | $40°$–$250°$ |
| (Biscyanopropyl)$_{0.9}$ (cyanopropylphenyl)$_{0.1}$ polysiloxane | Strongly polar | | $0°$–$275°$ |

The structures in the left column are:

- $\left[\text{O—Si(diphenyl)}\right]_x \left[\text{O—Si(CH}_3\text{)}_2\right]_{1-x}$  (Diphenyl)$_x$(dimethyl)$_{1-x}$ polysiloxane
- $\left[\text{O—Si(CN-propyl)(phenyl)}\right]_{0.14} \left[\text{O—Si(CH}_3\text{)}_2\right]_{0.86}$  (Cyanopropylphenyl)$_{0.14}$ (dimethyl)$_{0.86}$ polysiloxane
- $+(\text{CH}_2\text{CH}_2\text{—O})_n$  Carbowax (poly(ethylene glycol))
- $\left[\text{O—Si(biscyanopropyl)}\right]_{0.9} \left[\text{O—Si(cyanopropyl)(phenyl)}\right]_{0.1}$  (Biscyanopropyl)$_{0.9}$ (cyanopropylphenyl)$_{0.1}$ polysiloxane

**Table 24-2**  Polarity of solutes

| Nonpolar | Weak intermediate polarity |
|---|---|
| Saturated hydrocarbons | Ethers |
| Olefinic hydrocarbons | Ketones |
| Aromatic hydrocarbons | Aldehydes |
| Halocarbons | Esters |
| Mercaptans | Tertiary amines |
| Sulfides | Nitro compounds (without $\alpha$-H atoms) |
| CS$_2$ | Nitriles (without $\alpha$-atoms) |

| Strong intermediate polarity | Strongly polar |
|---|---|
| Alcohols | Polyhydroxyalcohols |
| Carboxylic acids | Amino alcohols |
| Phenols | Hydroxy acids |
| Primary and secondary amines | Polyprotic acids |
| Oximes | Polyphenols |
| Nitro compounds (with $\alpha$-H atoms) | |
| Nitriles (with $\alpha$-H atoms) | |

SOURCE: Adapted from H. M. McNair and E. J. Bonelli, *Basic Gas Chromatography* (Palo Alto, CA: Varian Instrument Division, 1968).

## Box 24-1 Chiral Phases for Separating Optical Isomers

*Optical isomers*—also called *enantiomers*—are mirror image compounds that cannot be superimposed. For example, the natural amino acid building blocks of proteins are L-amino acids.

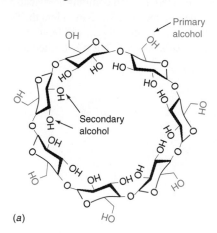

Enantiomers of an amino acid

Volatile derivative for gas chromatography

Chromatography with a *chiral* (optically active) stationary phase is one of the few ways to separate enantiomers. We can estimate ages of fossils up to 500 million years old by measuring the fraction of amino acid that has transformed into the D enantiomer in a fossil.[4,5] Amino acids do not have enough vapor pressure for gas chromatography. A volatile derivative suitable for gas chromatography is shown in the figure above.[6]

Common chiral stationary phases for gas chromatography have *cyclodextrins* bonded to a conventional polysiloxane stationary phase.[7,8] Cyclodextrins are naturally occurring cyclic sugars. β-Cyclodextrin has a 0.78-nm-diameter opening into a chiral, hydrophobic cavity. The hydroxyls are capped with alkyl groups to decrease the polarity of the faces.[9]

Enantiomers have different affinities for the cyclodextrin cavity, so they separate as they travel through the chromatography column. The chromatogram below shows a chiral separation of a by-product found in pesticides.

Programmed temperature (120°–200°C) chiral separation on a 0.25-mm × 25-m open tubular column with a 0.25-μm-thick stationary phase containing 10 wt% fully methylated β-cyclodextrin chemically bonded to dimethyl polysiloxane. [From W. Vetter and W. Jun, "Elucidation of a Polychlorinated Bipyrrole Structure Using Enantioselective GC," Anal. Chem. **2002,** *74, 4287.*]

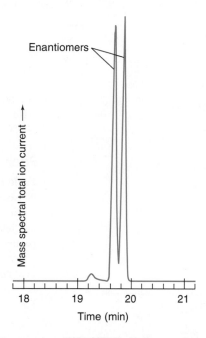

(*a*) Structure of β-cyclodextrin, a cyclic sugar made of seven glucose molecules. (α-Cyclodextrin contains six monomers and γ-cyclodextrin contains eight.) (*b*) Primary hydroxyl groups lie on one face and the secondary hydroxyl groups lie on the other face.

Chlorinated pesticide impurity. The two rings are perpendicular to each other. The mirror images are not superimposable, because there is no free rotation about the C–N bond between the rings.

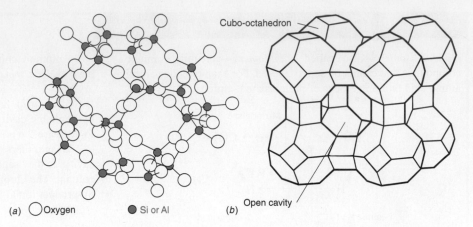

Cubo-octahedron

Open cavity

(*a*) ○ Oxygen    ● Si or Al    (*b*)

Among the solids used for porous-layer open tubular columns, alumina ($Al_2O_3$) can separate hydrocarbons in gas-solid adsorption chromatography. **Molecular sieves** (Figure 24-7) are inorganic or organic materials with cavities into which small molecules enter and are partially retained.[10] Molecules such as $H_2$, $O_2$, $N_2$, $CO_2$, and $CH_4$ can be separated from one another. Gases can be dried by passage through traps containing molecular sieves because water is strongly retained. Inorganic sieves can be regenerated (freed of water) by heating to 300°C in vacuum or under flowing $N_2$.

## Packed Columns

**Packed columns** contain fine particles of solid support coated with nonvolatile liquid stationary phase, or the solid itself may be the stationary phase. Compared with open tubular columns, packed columns provide greater sample capacity but give broader peaks, longer retention times, and less resolution. (Compare Figure 24-8 with Figure 24-3.) Despite their inferior resolution, packed columns are used for preparative separations, which require a great deal of stationary phase, or to separate gases that are poorly retained. Packed columns are usually made of stainless steel or glass and are typically 3–6 mm in diameter and 1–5 m in length. The solid support is often silica that is *silanized* (Reaction 23-36) to reduce hydrogen bonding to polar solutes. For tenaciously binding solutes, Teflon is a useful support, but it is limited to <200°C.

Teflon is a chemically inert polymer with the structure $-CF_2-CF_2-CF_2-CF_2-$.

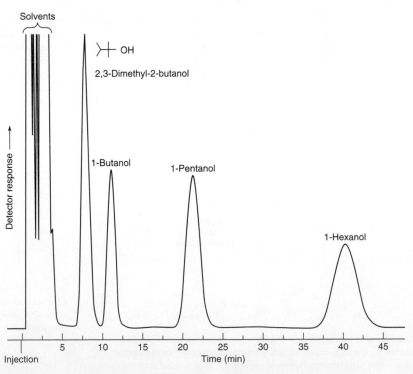

**Figure 24-8** Chromatogram of alcohol mixture at 40°C using packed column (2 mm inner diameter × 76 cm long) containing 20% Carbowax 20M on Gas-Chrom R support and flame ionization detector. *[Courtesy Norman Pearson.]*

In a packed column, uniform particle size decreases the multiple path term in the van Deemter equation (23-33), thereby reducing plate height and increasing resolution. Small particle size decreases the time required for solute equilibration, thereby improving column efficiency. However, the smaller the particle size, the less space between particles and the more pressure required to force mobile phase through the column. Particle size is expressed in micrometers or as a *mesh size,* which refers to the size of screens through which the particles are passed or retained (Table 28-2). A 100/200 mesh particle passes through a 100 mesh screen, but not through a 200 mesh screen. The mesh number equals the number of openings per linear inch of screen.

## Retention Index

Figure 24-9 illustrates how the relative retention times of polar and nonpolar solutes change as the polarity of the stationary phase changes. In Figure 24-9a, 10 compounds are eluted nearly in order of increasing boiling point from a nonpolar stationary phase. The principal determinant of retention on this column is the volatility of the solutes. In Figure 24-9b, the strongly polar stationary phase strongly retains the polar solutes. The three alcohols are the last to be eluted, following the three ketones, which follow four alkanes. Hydrogen bonding to the stationary phase is probably the strongest force leading to retention. Dipole interactions of the ketones are the second strongest force.

The Kovats **retention index,** $I$, for a linear alkane equals 100 times the number of carbon atoms. For octane, $I = 800$; and for nonane, $I = 900$. A compound eluted between octane and nonane (Figure 23-7) has a retention index between 800 and 900 computed by the formula

*Retention index:* $$I = 100\left[ n + (N - n)\frac{\log t_r'(\text{unknown}) - \log t_r'(n)}{\log t_r'(N) - \log t_r'(n)} \right] \qquad (24\text{-}1)$$

where $n$ is the number of carbon atoms in the *smaller* alkane; $N$ is the number of carbon atoms in the *larger* alkane; $t_r'(n)$ is the adjusted retention time of the *smaller* alkane; and $t_r'(N)$ is the adjusted retention time of the *larger* alkane.

Retention index relates the retention time of a solute to the retention times of linear alkanes.

Adjusted retention time $= t_r' = t_r - t_m$

$t_r$ = retention time for solute
$t_m$ = time for unretained solute ($CH_4$) to pass through column

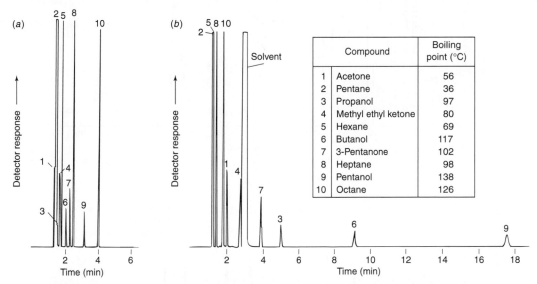

| Compound | | Boiling point (°C) |
|---|---|---|
| 1 | Acetone | 56 |
| 2 | Pentane | 36 |
| 3 | Propanol | 97 |
| 4 | Methyl ethyl ketone | 80 |
| 5 | Hexane | 69 |
| 6 | Butanol | 117 |
| 7 | 3-Pentanone | 102 |
| 8 | Heptane | 98 |
| 9 | Pentanol | 138 |
| 10 | Octane | 126 |

**Figure 24-9** Separation of 10 compounds on (*a*) nonpolar poly(dimethylsiloxane) and (*b*) strongly polar poly(ethylene glycol) 1-μm-thick stationary phases in 0.32-mm-diameter × 30-m-long open tubular columns at 70°C. *[Courtesy Restek Co., Bellefonte, PA.]*

### Example  Retention Index

If the retention times in Figure 23-7 are $t_r(CH_4) = 0.5$ min, $t_r(\text{octane}) = 14.3$ min, $t_r(\text{unknown}) = 15.7$ min, and $t_r(\text{nonane}) = 18.5$ min, find the retention index for the unknown.

**Solution**  The index is computed with Equation 24-1:

$$I = 100\left[ 8 + (9 - 8)\frac{\log 15.2 - \log 13.8}{\log 18.0 - \log 13.8} \right] = 836$$

The retention index of 657 for benzene on poly(dimethylsiloxane) in Table 24-3 means that benzene is eluted between hexane ($I \equiv 600$) and heptane ($I \equiv 700$) from this nonpolar stationary phase. Nitropropane is eluted just after heptane on the same column. As we go down the table, the stationary phases become more polar. For (biscyanopropyl)$_{0.9}$(cyanopropylphenyl)$_{0.1}$-polysiloxane at the bottom of the table, benzene is eluted after decane, and nitropropane is eluted after $n$-$C_{14}H_{30}$.

## Temperature and Pressure Programming

In **temperature programming,** the temperature of a column is raised *during* the separation to increase solute vapor pressure and decrease retention times of late-eluting components. At a constant temperature of 150°C, the more volatile compounds in Figure 24-10 emerge close together, and less volatile compounds may not even be eluted from the column. If the temperature is increased from 50° to 250°C at a rate of 8°/min, all compounds are eluted and the separation of peaks is fairly uniform. Do not raise the temperature so high that analytes and stationary phase decompose.

Many chromatographs are equipped with electronic pressure control of the carrier gas. Increasing the inlet pressure increases the flow of mobile phase and decreases retention time. In some cases, programmed pressure can be used instead of programmed temperature to reduce retention times of late-eluting components. At the end of a run, the pressure can be rapidly reduced to its initial value for the next run. Time is not wasted waiting for a hot column to cool before the next injection. Programmed pressure is useful for analytes that cannot tolerate high temperature.

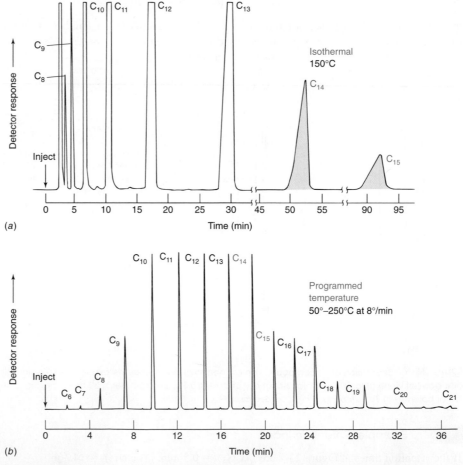

*Figure 24-10* Comparison of (*a*) isothermal (constant temperature) and (*b*) programmed temperature chromatography. Each sample contains linear alkanes run on a 1.6-mm-diameter × 6-m-long packed column containing 3% Apiezon L (liquid phase) on 100/120 mesh VarAport 30 solid support with He flow rate of 10 mL/min. Detector sensitivity is 16 times greater in panel *a* than in panel *b*. [From H. M. McNair and E. J. Bonelli, Basic Gas Chromatography (Palo Alto, CA: Varian Instrument Division, 1968).]

**Table 24-3** Retention indexes for several compounds on common stationary phases

| | Retention index[a] | | | | |
|---|---|---|---|---|---|
| **Phase** | **Benzene**<br>**b.p. 80°C** | **Butanol**<br>**b.p. 117°C** | **2-Pentanone**<br>**b.p. 102°C** | **1-Nitropropane**<br>**b.p. 132°C** | **Pyridine**<br>**b.p. 116°C** |
| Poly(dimethylsiloxane) | 657 | 648 | 670 | 708 | 737 |
| (Diphenyl)$_{0.05}$(dimethyl)$_{0.95}$-<br>polysiloxane | 672 | 664 | 691 | 745 | 761 |
| (Diphenyl)$_{0.35}$(dimethyl)$_{0.65}$-<br>polysiloxane | 754 | 717 | 777 | 871 | 879 |
| (Cyanopropylphenyl)$_{0.14}$-<br>(dimethyl)$_{0.86}$polysiloxane | 726 | 773 | 784 | 880 | 852 |
| (Diphenyl)$_{0.65}$(dimethyl)$_{0.35}$-<br>polysiloxane | 797 | 779 | 824 | 941 | 943 |
| Poly(ethylene glycol) | 956 | 1 142 | 987 | 1 217 | 1 185 |
| (Biscyanopropyl)$_{0.9}$-<br>(cyanopropylphenyl)$_{0.1}$-<br>polysiloxane | 1 061 | 1 232 | 1 174 | 1 409 | 1 331 |

a. *For reference, boiling points (b.p.) for various alkanes are hexane, 69°C; heptane, 98°C; octane, 126°C; nonane, 151°C; decane, 174°C; undecane, 196°C. Retention indexes for the straight-chain alkanes are fixed values and do not vary with the stationary phase: hexane, 600; heptane, 700; octane, 800; nonane, 900; decane, 1 000; undecane, 1 100.*

SOURCE: *Restek Chromatography Products Catalog, 1993–94, Bellefonte, PA.*

## Carrier Gas

Helium is the most common carrier gas and is compatible with most detectors. For a flame ionization detector, $N_2$ gives a lower detection limit than He. Figure 24-11 shows that $H_2$, He, and $N_2$ give essentially the same optimal plate height (0.3 mm) at significantly different flow rates. Optimal flow rate increases in the order $N_2 < He < H_2$. Fastest separations can be achieved with $H_2$ as carrier gas, and $H_2$ can be run much faster than its optimal velocity with little penalty in resolution.[11] Figure 24-12 shows the effect of carrier gas on the separation of two compounds on the same column with the same temperature program.

There are drawbacks to using $H_2$. It can catalytically react with unsaturated compounds on metal surfaces, and it cannot be used with a mass spectrometric detector, because $H_2$ breaks down vacuum pump oil in the detector. The main reason why $H_2$ was not used more often in the

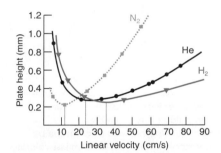

**Figure 24-11** Van Deemter curves for gas chromatography of *n*-$C_{17}H_{36}$ at 175°C using $N_2$, He, or $H_2$ in a 0.25-mm-diameter × 25-m-long wall-coated column with OV-101 stationary phase. *[From R. R. Freeman, ed., High Resolution Gas Chromatography (Palo Alto, CA: Hewlett Packard Co., 1981).]*

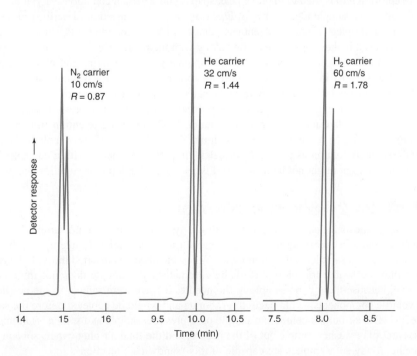

**Figure 24-12** Separation of two polyaromatic hydrocarbons on a wall-coated open tubular column with different carrier gases. Resolution, *R*, increases and analysis time decreases as we change from $N_2$ to He to $H_2$ carrier gas. *[Courtesy J&W Scientific, Folsom, CA.]*

past is that it forms explosive mixtures in air when $H_2$ is greater than 4 vol%. Flow rates in capillary chromatography are unlikely to create a dangerous concentration of $H_2$. Electrolytic generators produce high-purity $H_2$ and eliminate the need for tanks of compressed $H_2$.

$H_2$ and He give better resolution (smaller plate height) than $N_2$ at high flow rate because solutes diffuse more rapidly through $H_2$ and He than through $N_2$. The more rapidly a solute diffuses between phases, the smaller is the mass transfer ($Cu_x$) term in the van Deemter equation (23-33). Equations 23-35a and 23-35b describe the effects of the finite rate of mass transfer in an open tubular column. If the stationary phase is thin enough ($\leq 0.5$ μm), mass transfer is dominated by slow diffusion through the *mobile phase* rather than through the *stationary phase*. That is, $C_s \ll C_m$ in Equations 23-35a and 23-35b. For a column of a given radius, $r$, and a solute of a given capacity factor, $k'$, the only variable affecting the rate of mass transfer in the mobile phase (Equation 23-35b) is the diffusion coefficient of solute through the mobile phase. Diffusion coefficients follow the order $H_2 > He > N_2$.

Most analyses are run at carrier gas velocities that are 1.5 to 2 times as great as the optimum velocity at the minimum of the van Deemter curve. The higher velocity is chosen to give the maximum efficiency (most theoretical plates) per unit time. A small decrease in resolution is tolerated in return for faster analyses.

Gas flow through a narrow column may be too low for best detector performance. Therefore, the optimum gas for separation is used in the column and the best gas for detection (called *makeup gas*) is added between the column and the detector.

Impurities in the carrier gas degrade the stationary phase. High-quality gases should be used, and even they should be passed through purifiers to remove $O_2$, $H_2O$, and traces of organic compounds prior to entering the column. Steel or copper tubing, rather than plastic or rubber tubing, should be used for gas lines because metals are less permeable to air and do not release volatile contaminants into the gas stream.

## Guard Columns and Retention Gaps

In gas chromatography, a *guard column* and a *retention gap* are each typically a 3- to 10-meter length of empty capillary in front of the capillary chromatography column. The capillary is silanized so that solutes are not retained by the bare silica wall. Physically, the guard column and the retention gap are identical, but they are employed for different purposes.

Guard column: accumulates nonvolatile substances that would contaminate chromatography column
Retention gap: improves peak shape by separating volatile solvent from less volatile solutes prior to chromatography

The purpose of a **guard column** is to accumulate nonvolatile substances that would otherwise contaminate the chromatography column and degrade its performance. The guard column is normally the same diameter as the chromatography column. Periodically, the beginning of the guard column should be cut off to eliminate the nonvolatile residues. Trim the guard column when you observe irregular peak shapes from a column that had been producing symmetric peaks.

A **retention gap** is used to improve peak shapes under certain conditions. If you introduce a large volume of sample ($>2$ μL) by *splitless* or *on-column* injection (described in the next section), microdroplets of liquid solvent can persist inside the column for the first few meters. Solutes dissolved in the droplets are carried along with them and give rise to a series of ragged bands. The retention gap allows solvent to evaporate prior to entering the chromatography column. Use at least 1 m of retention gap per microliter of solvent. Even small volumes of solvent that have a very different polarity from the stationary phase can cause irregular solute peak shapes. The retention gap helps separate solvent from solute to improve peak shapes.

We calculate plate number, $N$, with Equation 23-28 by using retention time and peak width. Plate height, $H$, is the length of the column, $L$, divided by $N$. Do not include the retention gap or guard column as part of $L$ for calculating $H$.[12] For peaks with a capacity factor $k' < 5$, plate height might not be meaningful when a guard column or retention gap is used.

## ■ ■ ■ 24-2 Sample Injection[13]

Figure 24-13 shows a good technique for using a syringe to inject liquid sample into a gas chromatograph. After cleaning the syringe several times with solvent, take up air, then solvent, then air, then sample, and then more air. When the needle is inserted through the rubber septum into the heated injection port of the chromatograph, sample does not immediately evaporate, because there is no sample in the needle. If there were sample in the needle, the most volatile components would begin to evaporate and would be depleted before the sample is injected. The air bubble behind the sample plug prevents sample and solvent from mixing. The solvent plug washes sample out of the needle, and the final air plug expels solvent from the needle. Many autosamplers are capable of this "sandwich" injection.

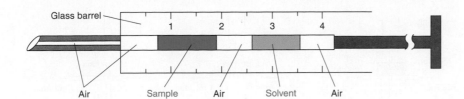

**Figure 24-13** "Sandwich" injection technique. *[Adapted from J. T. Watson, Introduction to Mass Spectrometry, 3rd ed. (Philadelphia: Lippincott-Raven, 1997).]*

An injection port with a silanized glass liner is shown in Figure 24-14. Carrier gas sweeps vaporized sample from the port into the chromatography column. For analytical chromatography, the injected volume is typically 0.1–2 μL of liquid sample. Gases are injected by a gas-tight syringe into the same kind of sample loop used in liquid chromatography (Figure 25-18). Decomposed sample, nonvolatile components, and septum debris accumulate in the glass liner, which is periodically replaced. The liner must seal properly or carrier gas will bypass the liner. The lifetime of the rubber septum could be as little as 20 manual injections or ~100 autosampler injections.

Different injection port liners are designed for split, splitless, and on-column injection and for use with solid-phase microextraction.

## Split Injection

If analytes of interest constitute >0.1% of the sample, **split injection** is usually preferred. For high-resolution work, best results are obtained with the smallest amount of sample (≤1 μL) that can be adequately detected—preferably containing ≤1 ng of each component. A complete injection contains too much material for a 0.32-mm-diameter or smaller column. A split injection delivers only 0.2–2% of the sample to the column. In Figure 24-14, sample is injected rapidly (<1 s) through the septum into the evaporation zone. The injector temperature is kept high (for example, 350°C) to promote fast evaporation. A brisk flow of carrier gas sweeps the sample through the *mixing chamber,* where complete vaporization and good mixing occur. At the split point, a small fraction of vapor enters the chromatography column, but most passes through needle valve 2 to a waste vent. The pressure regulator leading to needle valve 2 controls the fraction of sample discarded. The proportion of sample that does not reach the column is called the *split ratio* and typically ranges from 50:1 to 600:1. After sample has been flushed from the injection port (~30 s), needle valve 2 is closed and carrier gas flow is correspondingly reduced. Quantitative analysis with split injection can be inaccurate because the split ratio is not reproducible from run to run.

Injection into open tubular columns:

*split:* routine for introducing small sample volume into open tubular column

*splitless:* best for trace levels of high-boiling solutes in low-boiling solvents

*on-column:* best for thermally unstable solutes and high-boiling solvents; best for quantitative analysis

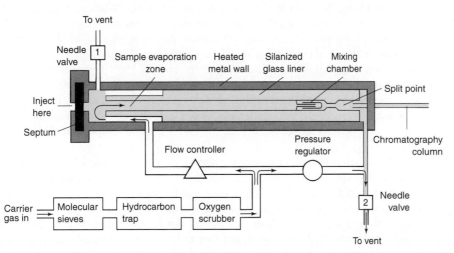

**Figure 24-14** Injection port for split injection into an open tubular column. The glass liner is slowly contaminated by nonvolatile and decomposed samples and must be replaced periodically. For splitless injection, the glass liner is a straight tube with no mixing chamber. For dirty samples, split injection is used and a packing material can be placed inside the liner to adsorb undesirable components of the sample.

A 1-μL liquid injection creates roughly 0.5 mL of gas volume, which can fill the glass liner in Figure 24-14. Some vapor can escape backward toward the septum. Lower-boiling components evaporate first and are more likely to escape than components with higher boiling points. The injection port temperature should be high enough to minimize this fractionation of the sample. However, if the injector temperature is too high, decomposition can occur. During injection and chromatography, *septum purge* gas flow through needle valve 1 in Figure 24-14 is run at ~1 mL/min to remove excess sample vapor and gas that bleeds from the hot rubber septum.

## Splitless Injection

For trace analysis of analytes that are less than 0.01% of the sample, **splitless injection** is appropriate. The same port shown for split injection in Figure 24-14 is used. However, the

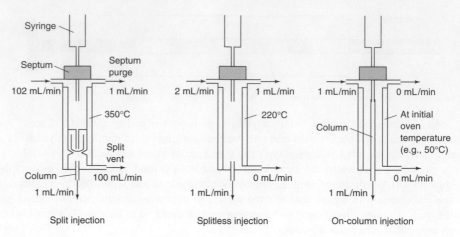

**Figure 24-15** Representative injection conditions for split, splitless, and on-column injection into an open tubular column.

glass liner is a straight, empty tube with no mixing chamber, as shown in Figure 24-15. A large volume (~2 μL) of dilute solution in a low-boiling solvent is injected slowly (~2 s) into the liner, with the split vent closed. Slow flow through the septum purge is maintained during injection and chromatography to remove any vapors that escape from the injection liner. Injector temperature for splitless injection is lower (~220°C) than that for split injection, because the sample spends more time in the port and we do not want it to decompose. The residence time of the sample in the glass liner is ~1 min, because carrier gas flows through the liner at the column flow rate, which is ~1 mL/min. In splitless injection, ~80% of the sample is applied to the column, and little fractionation occurs during injection.

The initial column temperature is set 40°C *below* the boiling point of the solvent, which therefore condenses at the beginning of the column. As solutes catch up with the condensed plug of solvent, they are trapped in the solvent in a narrow band at the beginning of the column. This **solvent trapping** leads to sharp chromatographic peaks. Without solvent trapping, the bands could not be sharper than the 1-min injection time. Chromatography is initiated by raising the column temperature to vaporize the solvent trapped at the head of the column.

An alternative means of condensing solutes in a narrow band at the beginning of the column is called **cold trapping.** The initial column temperature is 150°C lower than the boiling points of the solutes of interest. Solvent and low-boiling components are eluted rapidly, but high-boiling solutes remain in a narrow band at the beginning of the column. The column is then rapidly warmed to initiate chromatography of the high-boiling solutes. For low-boiling solutes, *cryogenic focusing* is required, with an initial column temperature below room temperature.

Figure 24-16 shows effects of operating parameters in split and splitless injections. Experiment A is a standard split injection with brisk flow through the split vent in Figure 24-15. The column was kept at 75°C. The injection liner was purged rapidly by carrier gas, and peaks are quite sharp. Experiment B shows the same sample injected in the same way, except the split vent was closed. Then the injection liner was purged slowly, and sample was applied to the column over a long time. Peaks are broad, and they tail badly because fresh carrier gas continuously mixes with vapor in the injector, making it more and more dilute but never completely flushing the sample from the injector. Peak areas in B are much greater than those in A because the entire sample reaches the column in B, whereas only a small fraction of sample reaches the column in A.

Experiment C is the same as B, but the split vent was opened after 30 s to rapidly purge all vapors from the injection liner. The bands in chromatogram C would be similar to those in B, but the bands are truncated after 30 s. Experiment D was the same as C, except that the column was initially cooled to 25°C to trap solvent and solutes at the beginning of the column. This is the correct condition for splitless injection. Solute peaks are sharp because the solutes were applied to the column in a narrow band of trapped solvent. Detector response in D is different from A–C. Actual peak areas in D are greater than those in A because most of the sample is applied to the column in D, but only a small fraction is applied in A. To make experiment D a proper splitless injection, the sample would need to be much more dilute.

For solvent trapping, sample should contain $10^4$ times as much solvent as analyte and column temperature should be 40°C below the solvent's boiling point.

For cold trapping, stationary phase film thickness must be $\geq 2$ μm.

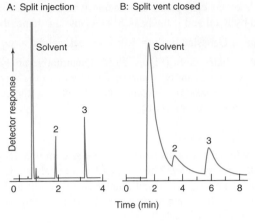

A: Split injection  B: Split vent closed

Solvent

Detector response →

Time (min)

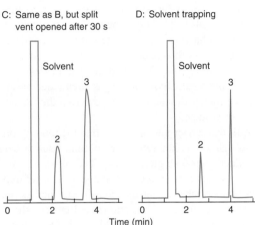

C: Same as B, but split vent opened after 30 s  D: Solvent trapping

Solvent  Solvent

Time (min)

**Figure 24-16** Split and splitless injections of a solution containing 1 vol% methyl isobutyl ketone (b.p. 118°C) and 1 vol% *p*-xylene (b.p. 138°C) in dichloromethane (b.p. 40°C) on a BP-10 moderately polar cyanopropyl phenyl methyl silicone open tubular column (0.22 mm diameter × 10 m long, film thickness = 0.25 μm, column temperature = 75°C). Vertical scale is the same for A–C. In D, signal heights should be multiplied by 2.33 to be on the same scale as A–C. *[From P. J. Marriott and P. D. Carpenter, "Capillary Gas Chromatography Injection," J. Chem. Ed. **1996**, 73, 96.]*

## On-Column Injection

**On-column injection** is used for samples that decompose above their boiling points and is preferred for quantitative analysis. Solution is injected directly into the column, without going through a hot injector (Figure 24-15). The initial column temperature is low enough to condense solutes in a narrow zone. Warming the column initiates chromatography. Samples are subjected to the lowest possible temperature in this procedure, and little loss of any solute occurs. The needle of a standard microliter syringe fits inside a 0.53-mm-diameter column, but this column does not give the best resolution. For 0.20- to 0.32-mm-diameter columns, which give better resolution, special syringes with thin silica needles are required.

## ■ ■ ■ 24-3 Detectors

For *qualitative analysis,* two detectors that can identify compounds are the mass spectrometer (Section 22-4) and the Fourier transform infrared spectrometer (Section 20-5). A peak can be identified by comparing its spectrum with a library of spectra recorded in a computer. For mass spectral identification, sometimes two prominent peaks are selected in the electron ionization spectrum. The *quantitation ion* is used for quantitative analysis. The *confirmation ion* is used for qualitative identification. For example, the confirmation ion might be expected to be 65% as abundant as the quantitation ion. If the observed abundance is not close to 65%, then we suspect that the compound is misidentified.

Another method to identify a peak is to compare its retention time with that of an authentic sample of the suspected compound. The most reliable way to compare retention times is by **co-chromatography,** in which an authentic compound is added to the unknown. If the added compound is identical with a component of the unknown, then the relative area of that one peak will increase. Identification is tentative only when carried out with one column, but it is firmer when carried out on several columns with different stationary phases.

*Quantitative analysis* is based on the area of a chromatographic peak. In the *linear response* concentration range, *the area of a peak is proportional to the quantity of that component.* In most instruments, peak area is measured automatically by a computer. Judgment is

*Linear response* means that peak area is proportional to analyte concentration. For very narrow peaks, peak height is often substituted for peak area.

$$\frac{A_X}{[X]} = F\left(\frac{A_S}{[S]}\right)$$

$A_X$ = area of analyte signal
$A_S$ = area of internal standard
$[X]$ = concentration of analyte
$[S]$ = concentration of standard
$F$ = response factor

**Table 24-4** Thermal conductivity at 273 K and 1 atm

| Gas | Thermal conductivity $J/(K \cdot m \cdot s)$ |
|---|---|
| $H_2$ | 0.170 |
| He | 0.141 |
| $NH_3$ | 0.021 5 |
| $N_2$ | 0.024 3 |
| $C_2H_4$ | 0.017 0 |
| $O_2$ | 0.024 6 |
| Ar | 0.016 2 |
| $C_3H_8$ | 0.015 1 |
| $CO_2$ | 0.014 4 |
| $Cl_2$ | 0.007 6 |

*The energy per unit area per unit time flowing from a hot region to a cold region is given by*

$$\text{Energy flux } (J/m^2 \cdot s) = -\kappa(dT/dx)$$

*where $\kappa$ is the thermal conductivity [units = $J/(K \cdot m \cdot s)$] and $dT/dx$ is the temperature gradient (K/m). Thermal conductivity is to energy flux as the diffusion coefficient is to mass flux.*

Thermal conductivity detector:

- $10^4$ linear response range
- $H_2$ and He give lowest detection limit
- sensitivity increases with
   increasing filament current
   decreasing flow rate
   lower detector block temperature

required to draw baselines beneath peaks and decide where to measure the area.[14] If peak area must be measured by hand and if the peak has a Gaussian shape, then the area is

$$\text{Area of Gaussian peak} = 1.064 \times \text{peak height} \times w_{1/2} \qquad (24\text{-}2)$$

where $w_{1/2}$ is the width at half-height (Figure 23-9). Quantitative analysis is almost always performed by adding a known quantity of *internal standard* to the unknown (Section 5-4). After measurement of the *response factor* with standard mixtures, the equation in the margin is used to measure the quantity of unknown.

## Thermal Conductivity Detector

In the past, **thermal conductivity detectors** were most common in gas chromatography because they are simple and *universal:* They respond to all analytes. Unfortunately, thermal conductivity is not sensitive enough to detect minute quantities of analyte eluted from open tubular columns smaller than 0.53 mm in diameter. Thermal conductivity detectors are still used for 0.53-mm columns and for packed columns.

*Thermal conductivity* measures the ability of a substance to transport heat from a hot region to a cold region (Table 24-4). Helium is the carrier gas commonly used with a thermal conductivity detector. Helium has the second highest thermal conductivity (after $H_2$), so any analyte mixed with helium lowers the conductivity of the gas stream. In Figure 24-17, eluate from the chromatography column flows over a hot tungsten-rhenium filament. When analyte emerges from the column, the conductivity of the gas stream decreases, the filament gets hotter, its electrical resistance increases, and the voltage across the filament changes. The detector measures the change in voltage.

It is common to split the carrier gas into two streams, sending part through the analytical column and part through a matched reference column. Each stream is passed over a different filament or alternated over a single filament. The resistance of the sample filament is measured with respect to that of the reference filament. The reference column minimizes flow differences when temperature is changed. Sensitivity increases with the square of the filament current. However, the maximum recommended current should not be exceeded, to avoid burning out the filament. The filament should be off when carrier gas is not flowing.

The sensitivity of a thermal conductivity detector (but *not* that of the flame ionization detector, described next) is inversely proportional to flow rate: It is more sensitive at a lower flow rate. Sensitivity also increases with increasing temperature differences between the filament and the surrounding block in Figure 24-17. The block should therefore be maintained at the lowest temperature that allows all solutes to remain gaseous.

## Flame Ionization Detector

In the **flame ionization detector** in Figure 24-18, eluate is burned in a mixture of $H_2$ and air. Carbon atoms (except carbonyl and carboxyl carbons) produce CH radicals, which are thought to produce $CHO^+$ ions and electrons in the flame.

$$CH + O \rightarrow CHO^+ + e^-$$

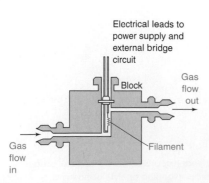

*Figure 24-17* Thermal conductivity detector. *[Courtesy Varian Associates, Palo Alto, CA.]*

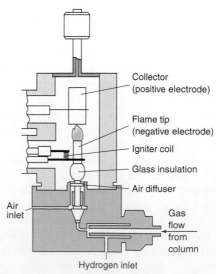

*Figure 24-18* Flame ionization detector. *[Courtesy Varian Associates, Palo Alto, CA.]*

**Table 24-5** Detection limits and linear ranges of gas chromatography detectors

| Detector | Approximate detection limit | Linear range |
|---|---|---|
| Thermal conductivity | 400 pg/mL (propane) | $>10^5$ |
| Flame ionization | 2 pg/s | $>10^7$ |
| Electron capture | As low as 5 fg/s | $10^4$ |
| Flame photometric | <1 pg/s (phosphorus) | $>10^4$ |
| | <10 pg/s (sulfur) | $>10^3$ |
| Nitrogen-phosphorus | 100 fg/s | $10^5$ |
| Sulfur chemiluminescence | 100 fg/s (sulfur) | $10^5$ |
| Photoionization | 25 pg to 50 pg (aromatics) | $>10^5$ |
| Fourier transform infrared | 200 pg to 40 ng | $10^4$ |
| Mass spectrometric | 25 fg to 100 pg | $10^5$ |

SOURCE: *Most data are from D. G. Westmoreland and G. R. Rhodes, "Detectors for Gas Chromatography," Pure Appl. Chem. **1989**, 61, 1147.*

Only about 1 in $10^5$ carbon atoms produces an ion, but ion production is proportional to the number of susceptible carbon atoms entering the flame. In the absence of analyte, $\sim 10^{-14}$ A flows between the flame tip and the collector, which is held at +200 to 300 V with respect to the flame tip. Eluted analytes produce a current of $\sim 10^{-12}$ A, which is converted to voltage, amplified, filtered to remove high-frequency noise, and finally converted to a digital signal.

Response to organic compounds is proportional to solute mass over seven orders of magnitude. The detection limit is $\sim 100$ times smaller than that of the thermal conductivity detector (Table 24-5) and is reduced by 50% when $N_2$ carrier gas is used instead of He. For open tubular columns, $N_2$ makeup gas is added to the $H_2$ or He eluate before it enters the detector. The flame ionization detector is sensitive enough for narrow-bore columns. It responds to most hydrocarbons and is insensitive to nonhydrocarbons such as $H_2$, He, $N_2$, $O_2$, CO, $CO_2$, $H_2O$, $NH_3$, NO, $H_2S$, and $SiF_4$.

Flame ionization detector:

- $N_2$ gives best detection limit
- signal proportional to number of susceptible carbon atoms
- 100-fold better detection than thermal conductivity
- $10^7$ linear response range

## Electron Capture Detector

Most detectors other than flame ionization and thermal conductivity respond to limited classes of analytes. The **electron capture detector** is sensitive to halogen-containing molecules, conjugated carbonyls, nitriles, nitro compounds, and organometallic compounds but relatively insensitive to hydrocarbons, alcohols, and ketones. The carrier or makeup gas must be either $N_2$ or 5% methane in Ar. Moisture decreases the sensitivity. Gas entering the detector is ionized by high-energy electrons ("β-rays") emitted from a foil containing radioactive $^{63}Ni$. Electrons thus formed are attracted to an anode, producing a small, steady current. When analytes with a high electron affinity enter the detector, they capture some electrons. The detector responds by varying the frequency of voltage pulses between anode and cathode to maintain a constant current. The electron capture detector is extremely sensitive (Table 24-5), with a detection limit comparable to that of mass spectrometric selected ion monitoring. Figure 24-19 shows an application in which this detector was used to measure halogenated "greenhouse gases" (Box 20-1) in the lower atmosphere.

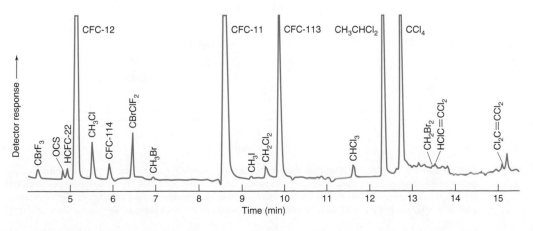

**Figure 24-19** Partial gas chromatogram for which an electron capture detector was used to measure halogenated compounds in air collected by an aircraft at an altitude of 800 m at a location 1 400 km south of New Zealand in 1995. [*From F. S. Rowland, "Stratospheric Ozone Depletion by Chlorofluorocarbons," Angew. Chem. Int. Ed. Engl.* **1996**, *35, 1787.*]

## Other Detectors

Other gas chromatography detectors:

*electron capture:* halogens, conjugated C=O, —C≡N, —NO₂
*nitrogen-phosphorus:* highlights P, N
*flame photometer:* individual selected elements, such as P, S, Sn, Pb
*photoionization:* aromatics, unsaturated compounds
*sulfur chemiluminescence:* S
*nitrogen chemiluminescence:* N
*atomic emission:* most elements (selected individually)
*mass spectrometer:* most analytes
*infrared spectrometer:* most analytes

The *nitrogen-phosphorus detector,* also called an *alkali flame detector,* is a modified flame ionization detector that is especially sensitive to compounds containing N and P.[15] Its response to N and P is $10^4$–$10^6$ times greater than the response to carbon. It is particularly important for drug, pesticide, and herbicide analyses. Ions such as $NO_2^-$, $CN^-$, and $PO_2^-$, produced by these elements when they contact a $Rb_2SO_4$-containing glass bead at the burner tip, create the current that is measured. $N_2$ from air is inert to this detector and does not interfere. The bead must be replaced periodically because $Rb_2SO_4$ is consumed. Figure 24-26 (in the next section) shows a chromatogram from a nitrogen-phosphorus detector.

A *flame photometric detector* measures optical emission from phosphorus, sulfur, lead, tin, or other selected elements. When eluate passes through a $H_2$-air flame, as in the flame ionization detector, excited atoms emit characteristic light. Phosphorus emission at 536 nm or sulfur emission at 394 nm can be isolated by a narrow-band interference filter and detected with a photomultiplier tube.

A *photoionization detector* uses a vacuum ultraviolet source to ionize aromatic and unsaturated compounds, with little response to saturated hydrocarbons or halocarbons. Electrons produced by the ionization are collected and measured.

A *sulfur chemiluminescence detector* takes exhaust from a flame ionization detector, in which sulfur has been oxidized to SO, and mixes it with ozone ($O_3$) to form an excited state of $SO_2$ that emits blue light and ultraviolet radiation. Emission intensity is proportional to the mass of sulfur eluted, regardless of the source, and the response to S is $10^7$ times greater than the response to C (Figure 24-20). A *nitrogen chemiluminescence detector* is analogous. Combustion of eluate at 1 800°C converts nitrogen into NO, which reacts with $O_3$ to create a chemiluminescent product. The response to N is $10^7$ times greater than the response to C.

Reactions thought to give sulfur chemiluminescence:

Sulfur compound $\xrightarrow{H_2 - O_2\ flame}$ SO + products

$SO + O_3 \rightarrow SO_2^* + O_2$
    ($SO_2^*$ = excited state)

$SO_2^* \rightarrow SO_2 + h\nu$

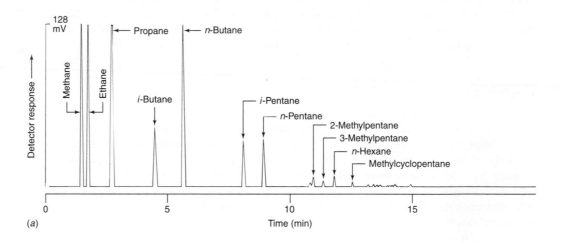

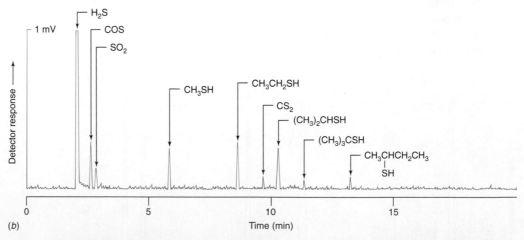

*Figure 24-20* Gas chromatograms showing sulfur compounds in natural gas: (*a*) flame ionization detector response and (*b*) sulfur chemiluminescence detector response. The organosulfur compounds are too dilute to be seen in flame ionization, and the sulfur chemiluminescence is insensitive to hydrocarbons.

[From N. G. Johansen and J. W. Birks, "Determination of Sulfur Compounds in Difficult Matrices," *Am. Lab.* February 1991, 112.]

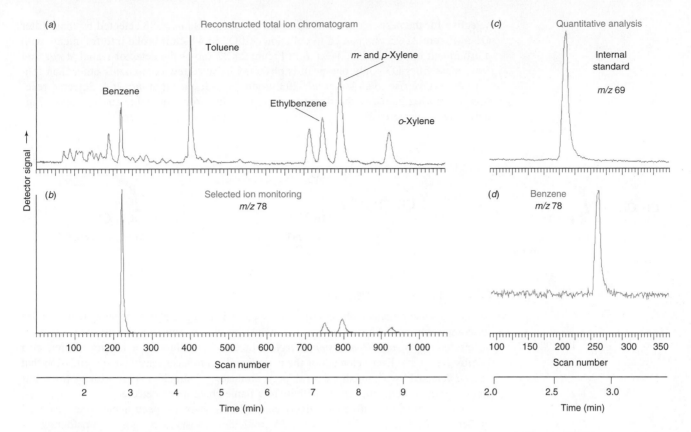

**Figure 24-21** Selected ion monitoring in gas chromatography–mass spectrometry. (*a*) Reconstructed total ion chromatogram of automobile exhaust with electron ionization. (*b*) Selected ion monitoring at *m/z* 78. (*c* and *d*) Quantitative analysis of benzene after adding an internal standard with a prominent ion at *m/z* 69. [*Courtesy Inficon, Syracuse, NY.*]

## Gas Chromatography–Mass Spectrometry

For those who can afford it, mass spectrometry is the detector of choice in chromatography. The mass spectrum is sensitive and provides both qualitative and quantitative information. With *selected ion monitoring* or *selected reaction monitoring* (Section 22-4), we can readily measure one component in a complex chromatogram of poorly separated compounds. Selected ion monitoring lowers the detection limit by a factor of $10^2$–$10^3$ compared with *m/z* scanning, because more time is spent just collecting ions of interest in selected ion monitoring.

Figure 24-21 illustrates **selected ion monitoring.** The *reconstructed total ion chromatogram* in trace *a* was obtained from a portable gas chromatograph–mass spectrometer designed for identification of spills at accident sites. A total of 1 072 spectra of eluate were recorded at equal time intervals between 1 and 10 min. The ordinate in the reconstructed total ion chromatogram is the sum of detector signal for all *m/z* above a selected cutoff. It measures everything eluted from the column. The selected ion chromatogram in trace *b* is obtained by parking the detector at *m/z* 78 and only measuring this one mass. A peak is observed for benzene (nominal mass 78 Da) and minor peaks are observed for benzene derivatives at 7–9 min. For quantitative analysis, an internal standard with a signal at *m/z* 69 is added to the mixture. Even though this internal standard overlaps the congested part of the chromatogram near a retention time of 2 min, the selected ion chromatogram for *m/z* 69 has a single peak in trace *c*. To measure benzene, the area of the *m/z* 78 peak in trace *d* is compared with the area of *m/z* 69 in trace *c*.

**Selected reaction monitoring** is illustrated in Figure 24-22. Trace *a* is the reconstructed total ion chromatogram of extract from an orange peel. To make the analysis

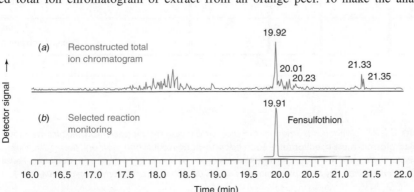

**Figure 24-22** Selected reaction monitoring in gas chromatography–mass spectrometry. (*a*) Reconstructed total ion chromatogram of extract from orange peel with electron ionization. (*b*) Selected reaction monitoring with the precursor ion *m/z* 293 selected by mass filter Q1 in Figure 22-21 and product ion *m/z* 264 selected by mass filter Q3. The chromatogram is a graph of intensity at *m/z* 264 from Q3 versus time. [*Courtesy Thermo Finnigan GC and GC/MS Division, San Jose, CA.*]

selective for the pesticide fensulfothion, the precursor ion *m/z* 293 selected by mass filter Q1 in Figure 22-21 was passed to collision cell Q2 in which it broke into fragments with a prominent ion at *m/z* 264. Trace *b* in Figure 24-22 shows the detector signal at *m/z* 264 from mass filter Q3. Only one peak is observed because few compounds other than fensulfothion give rise to an ion at *m/z* 293 producing a fragment at *m/z* 264. Selected reaction monitoring increases the signal-to-noise ratio in chromatographic analysis and eliminates much interference.

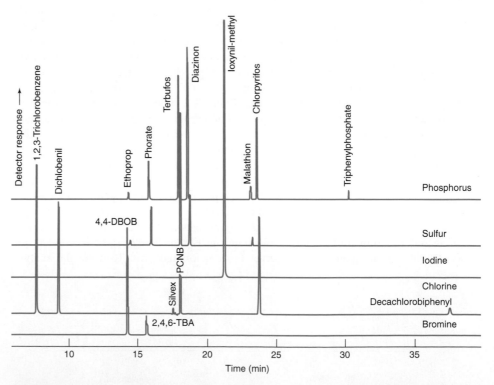

Fensulfothion
(nominal mass 308 Da)

M − 15 (*m/z* 293)
Selected by Q1

M − 15 − 29 (*m/z* 264)
Selected by Q3

## Element-Specific Plasma Detectors

Eluate from a chromatography column can be passed through a plasma to atomize and ionize its components and measure selected elements by atomic emission spectroscopy or mass spectrometry. An *atomic emission detector* directs eluate through a helium plasma in a microwave cavity. Every element of the periodic table produces characteristic emission that can be detected by a photodiode array polychromator (Figure 20-14). Sensitivity for sulfur can be 10 times better than the sensitivity of a flame photometric detector.

The extremely sensitive inductively coupled plasma–mass spectrometer was described in Section 21-6. Figure 24-23 shows 15 pesticides measured by gas chromatography–inductively coupled plasma–mass spectrometry. Eluate was atomized and ionized in the plasma. Ions were measured by a mass spectrometer that could monitor any set of *m/z* values. The figure shows traces for P, S, I, Cl, or Br.

***Figure 24-23*** Extracted element chromatograms produced by gas chromatography–inductively coupled plasma–mass spectrometry. Each trace responds to just one element. *[From D. Profrock, P. Leonhard, S. Wilbur, and A. Prange, "Sensitive, Simultaneous Determination of P, S, Cl, Br, and I Containing Pesticides in Environmental Samples by GC Hyphenated with Collision-Cell ICP-MS," J. Anal. Atom. Spectros. **2004**, 19, 1.]*

## ■ ■ ■ ■ 24-4 Sample Preparation

**Sample preparation** is the process of transforming a sample into a form that is suitable for analysis. This process might entail extracting analyte from a complex matrix, *preconcentrating* very dilute analytes to get a concentration high enough to measure, removing or masking interfering species, or chemically transforming (*derivatizing*) analyte into a more convenient or more easily detected form. Chapter 28 is devoted to sample preparation, so we now describe only techniques that are especially applicable to gas chromatography.

**Solid-phase microextraction** is a method to extract compounds from liquids, air, or even sludge without using any solvent.[16] The key component is a fused-silica fiber coated with a 10- to 100-μm-thick film of stationary phase similar to those used in gas chromatography. Figure 24-24 shows the fiber attached to the base of a syringe with a fixed metal needle. The fiber can be extended from the needle or retracted inside the needle. Figure 24-25 demonstrates the procedure of exposing the fiber to a sample solution (or the gaseous headspace above the liquid) for a fixed time while stirring and, perhaps, heating. It is best to determine by experiment how much time is required for the fiber to become saturated with analyte and to allow this much time for extraction. If you use shorter times, the concentration of analyte in the fiber is likely to vary from sample to sample. Only a fraction of the analyte in the sample is extracted into the fiber. When you extract the headspace, liquid sample should occupy about two-thirds of the vial. Too much headspace volume reduces extraction efficiency.

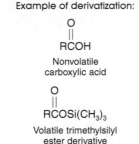

Example of derivatization:

$$\underset{\text{Nonvolatile}}{\overset{\overset{\displaystyle O}{\displaystyle \|}}{RCOH}}$$

Nonvolatile
carboxylic acid

$$\overset{\overset{\displaystyle O}{\displaystyle \|}}{RCOSi(CH_3)_3}$$

Volatile trimethylsilyl
ester derivative

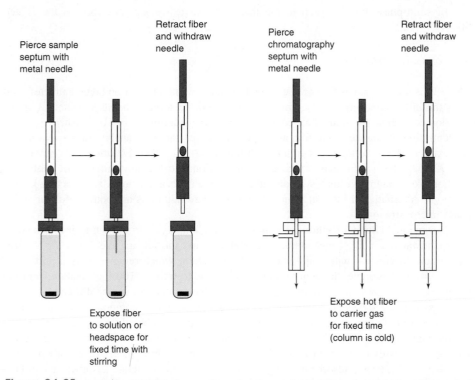

**Figure 24-25**  Sampling by solid-phase microextraction and desorption of analyte from the coated fiber into a gas chromatograph. *[Adapted from Supelco Chromatography Products catalog, Bellefonte, PA.]*

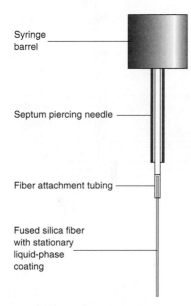

**Figure 24-24**  Syringe for solid-phase microextraction. The fused silica fiber is withdrawn inside the steel needle after sample collection and when the syringe is used to pierce a septum.

After sample collection, retract the fiber and insert the syringe into a gas chromatograph. Extend the fiber into the hot injection liner, where analyte is thermally desorbed from the fiber in splitless mode for a fixed time. Collect desorbed analyte by *cold trapping* (Section 24-2) at the head of the column prior to chromatography. If there will be a long time between sampling and injection, insert the needle into a septum to seal the fiber from the atmosphere. Figure 24-26 shows a chromatogram of chemical warfare nerve agents isolated from seawater by solid-phase microextraction and detected with a nitrogen-phosphorus detector. The chromatogram is deceptively simple because the detector responds only to compounds containing N and P.

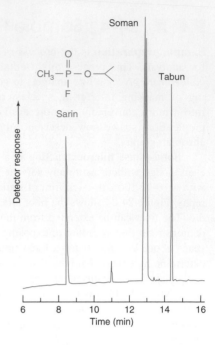

**Figure 24-26** Gas chromatogram of nerve agents sampled by solid-phase microextraction for 30 min from seawater spiked with 60 nL of each agent per liter (60 ppb by volume). The fiber had a 65-μm-thick coating of copoly(dimethylsiloxane/ divinylbenzene). The nitrogen-phosphorus detector had a detection limit of 0.05 ppb. Analytes were desorbed from the fiber for 2 min at 250°C in splitless mode in the injection port. The column temperature was 30°C during desorption and then it was ramped up at 10°C/min during chromatography. The column was 0.32 mm × 30 m with a 1-μm coating of (phenyl)$_{0.05}$(methyl)$_{0.95}$polysiloxane. Soman appears as a split peak because it has two isomers. [From H.-Å. Lakso and W. F. Ng, "Determination of Chemical Warfare Agents in Natural Water Samples by Solid-Phase Microextraction," Anal. Chem. **1997**, 69, 1866.]

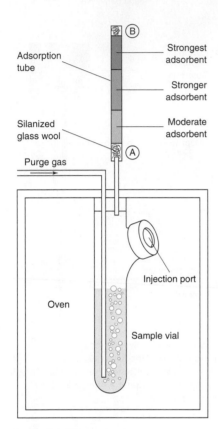

**Figure 24-27** Purge and trap apparatus for extracting volatile substances from a liquid or solid by flowing gas.

Tenax

You need to establish the time and temperature required to purge 100% of the analyte from the sample in separate control experiments.

In solid-phase microextraction, the mass of analyte ($m$, μg) absorbed in the coated fiber is

*Mass of analyte extracted:*
$$m = \frac{KV_f C_0 V_s}{KV_f + V_s} \qquad (24\text{-}3)$$

where $V_f$ is the volume of film on the fiber, $V_s$ is the volume of solution being extracted, and $C_0$ is the initial concentration (μg/mL) of analyte in the solution being extracted. $K$ is the partition coefficient for solute between the film and the solution: $K = C_f/C_s$, where $C_f$ is the concentration of analyte in the film and $C_s$ is the concentration of analyte in the solution. If you extract a large volume of solution such that $V_s \gg KV_f$, then Equation 24-3 reduces to $m = KV_f C_0$. That is, the mass extracted is proportional to the concentration of analyte in solution. For quantitative analysis, you can construct a calibration curve by extracting known solutions. Alternatively, internal standards and standard additions are both useful for solid-phase microextraction.

**Purge and trap** is a method for removing volatile analytes from liquids or solids (such as groundwater or soil), concentrating the analytes, and introducing them into a gas chromatograph. In contrast with solid-phase microextraction, which removes only a portion of analyte from the sample, the goal in purge and trap is to remove 100% of analyte from the sample. Quantitative removal of polar analytes from polar matrices can be difficult.

Figure 24-27 shows apparatus for measuring volatile flavor components in carbonated beverages. Helium purge gas from a stainless steel needle is bubbled through the cola in the sample vial, which is heated to 50°C to aid evaporation of analytes. Purge gas exiting the sample vial passes through an adsorption tube containing three layers of adsorbent compounds with increasing adsorbent strength. For example, the moderate adsorbent could be a nonpolar phenylmethylpolysiloxane, the stronger adsorbent could be the polymer Tenax, and the strongest adsorbent could be carbon molecular sieves.[17]

During the purge and trap process, gas flows through the adsorbent tube from end A to end B in Figure 24-27. After purging all analyte from the sample into the adsorption tube, reverse the gas flow to go from B to A and purge the trap at 25°C to remove as much water or other solvent as possible from the adsorbents. Then connect outlet A of the adsorption tube to the injection port of a gas chromatograph operating in splitless mode and heat the trap to ~200°C. Desorbed analytes flow into the chromatography column, where they are concentrated by cold trapping. After complete desorption from the trap, warm the chromatography column to initiate the separation.

**Thermal desorption** is a method to release volatile compounds from solid samples. A weighed sample is placed in a steel or glass tube and held in place with glass wool. The sample is purged with carrier gas to remove $O_2$, which is vented to the air, not into the chromatography column. The desorption tube is then connected to the chromatography column and heated to release volatile substances that are collected by *cold trapping* at the beginning of the column. The column is then heated rapidly to initiate chromatography.

# ■ ■ ■ 24-5 Method Development in Gas Chromatography

With the bewildering choice of parameters in gas chromatography, is there a rational way to choose a procedure for a particular problem? In general, there are many satisfactory solutions. We now discuss some broad guidelines for selecting a method to use.[18] The order in which decisions should be made is to consider (1) the goal of the analysis, (2) sample preparation, (3) detector, (4) column, and (5) injection.

Order of decisions:
1. goal of analysis
2. sample preparation
3. detector
4. column
5. injection

## Goal of the Analysis

What is required from the analysis? Is it qualitative identification of components in a mixture? Will you require high-resolution separation of everything or do you just need good resolution in a portion of the chromatogram? Can you sacrifice resolution to shorten the analysis time? Do you need quantitative analysis of one or many components? Do you need high precision? Will analytes be present in adequate concentration or do you need preconcentration or a very sensitive detector for ultratrace analysis? How much can the analysis cost? Each of these factors creates trade-offs in selecting techniques.

## Sample Preparation

The key to successful chromatography of a complex sample is to clean it up before it ever sees the column. In Section 24-4, we described solid-phase microextraction, purge and trap, and thermal desorption to isolate volatile components from complex matrices. Other methods include liquid extraction, supercritical fluid extraction, and solid-phase extraction, most of which are described in Chapter 28. These techniques isolate desired analytes from interfering substances, and they can concentrate dilute analytes up to detectable levels. If you do not clean up your samples, chromatograms could contain a broad "forest" of unresolved peaks, and nonvolatile substances will ruin the expensive chromatography column.

Garbage in—garbage out!

## Choosing the Detector

The next step is to choose a detector. Do you need information about everything in the sample or do you want a detector that is specific for a particular element or a particular class of compounds?

The most general purpose detector for open tubular chromatography is a mass spectrometer. Flame ionization is probably the most popular detector, but it mainly responds to hydrocarbons; and Table 24-5 shows that it is not as sensitive as electron capture, nitrogen-phosphorus, or chemiluminescence detectors. The flame ionization detector requires the sample to contain ≥10 ppm of each analyte for split injection. The thermal conductivity detector responds to all classes of compounds, but it is not sensitive enough for high-resolution, narrow-bore, open tubular columns.

Sensitive detectors for ultratrace analysis each respond to a limited class of analytes. The electron capture detector is specific for halogen-containing molecules, nitriles, nitro compounds, and conjugated carbonyls. For split injection, the sample should contain ≥100 ppb of each analyte if an electron capture detector is to be used. A photoionization detector can be specific for aromatic and unsaturated compounds. The nitrogen-phosphorus detector has enhanced response to compounds containing either of these two elements, but it also responds to hydrocarbons. Sulfur and nitrogen chemiluminescence detectors each respond to just one element. Flame photometric detectors are specific for selected elements, such as S, P, Pb, or Sn. A selective detector might be chosen to simplify the chromatogram by not responding to everything that is eluted (as in Figure 24-23). Mass spectrometry with selected reaction monitoring (Figure 24-21) is an excellent way to monitor one analyte of interest in a complex sample.

If qualitative information is required to identify eluates, then mass spectral or infrared detectors are good choices. The infrared detector, like the thermal conductivity detector, is not sensitive enough for high-resolution, narrow-bore, open tubular columns.

24-5 Method Development in Gas Chromatography

549

## Selecting the Column

The basic choices are the stationary phase, column diameter and length, and the thickness of stationary phase. A nonpolar stationary phase in Table 24-1 is most useful. An intermediate polarity stationary phase will handle most separations that the nonpolar column cannot. For highly polar compounds, a strongly polar column might be necessary. Optical isomers and closely related geometric isomers require special stationary phases for separation.

Table 24-6 shows that there are only a few sensible combinations of column diameter and film thickness. Highest resolution is afforded by the narrowest columns. Thin-film, narrow-bore columns are especially good for separating mixtures of high-boiling-point compounds that are retained too strongly on thick-film columns. Short retention times provide high-speed analyses. However, thin-film, narrow-bore columns have very low sample capacity, require high-sensitivity detectors (flame ionization might not be adequate), do not retain low-boiling-point compounds well, and could suffer from exposure of surface active sites on the silica.

Thick-film, narrow-bore columns in Table 24-6 provide a good compromise between resolution and sample capacity. They can be used with most detectors (but usually not thermal conductivity or infrared) and with compounds of high volatility. Retention times are longer than those of thin-film columns. Thick-film, wide-bore columns are required for use with thermal conductivity and infrared detectors. They have high sample capacity and can handle highly volatile compounds, but they give low resolution and have long retention times.

If a particular column meets most of your requirements but does not provide sufficient resolution, then a longer column of the same type could be used. Doubling the length of a column doubles the number of plates and, according to Equation 23-30, increases resolution by $\sqrt{2}$. This is not necessarily the best way to increase resolution, because it doubles the retention time. Using a narrower column increases resolution with no penalty in retention time. Selecting another stationary phase changes the separation factor ($\gamma$ in Equation 23-30) and might resolve components of interest.

If you have measured resolution of a few key components of a mixture under a small number of conditions, commercial software is available to optimize conditions (such as temperature and pressure programming) for the best separation.[19] Just over the horizon, greatly improved separations will be possible by coupling two different columns in series with programmable control of the flow rates (pressure) in each column *during* the separation.[20]

> To improve resolution, use a:
> - longer column
> - narrower column
> - different stationary phase

**Table 24-6** Gas chromatography column comparisons[18]

| Description | Thin-film narrow-bore | Thick-film narrow-bore | Thick-film wide-bore |
|---|---|---|---|
| Inner diameter | 0.10–0.32 mm | 0.25–0.32 mm | 0.53 mm |
| Film thickness | ~0.2 μm | ~1–2 μm | ~2–5 μm |
| Advantages | High resolution<br>Trace analysis<br>Fast separations<br>Low temperatures<br>Elute high-b.p. compounds | Good capacity<br>Good resolution<br>(4 000 plates/m)<br>Easy to use<br>Retains volatile compounds<br>Good for mass spectrometry | High capacity (100 ng/<br>solute)<br>Good for thermal conductivity<br>and infrared detectors<br>Simple injection techniques |
| Disadvantages | Low capacity<br>(≤1 ng/solute)<br>Requires high-sensitivity<br>detector (not mass spectrometry)<br>Surface activity of exposed silica | Moderate resolution<br>Long retention time for<br>high-b.p. compounds | Low resolution<br>(500–2 000 plates/m)<br>Long retention time for<br>high-b.p. compounds |

## Choosing the Injection Method

The last major decision is how to inject the sample. *Split injection* is best for high concentrations of analyte or gas analysis. Quantitative analysis is very poor. Less volatile components can be lost during injection. Split injection offers high resolution and can handle dirty samples if an adsorbent packing is added to the injection liner. Thermally unstable compounds can decompose during the high-temperature injection.

*Splitless injection* is required for very dilute solutions. It offers high resolution but is poor for quantitative analysis because less volatile compounds can be lost during injection. It is better than split injection for compounds of moderate thermal stability because the injection temperature is lower. Splitless injection introduces sample onto the column slowly, so solvent trapping or cold trapping is required. Therefore, splitless injection cannot be used for isothermal chromatography. Samples containing less than 100 ppm of each analyte can be analyzed with a column film thickness <1 μm with splitless injection. Samples containing 100–1 000 ppm of each analyte require a column film thickness ≥1 μm.

*On-column injection* is best for quantitative analysis and for thermally sensitive compounds. It is strictly a low-resolution technique and cannot be used with columns whose inner diameter is less than 0.2 mm. It can handle dilute or concentrated solutions and relatively large or small volumes. Other column requirements are the same as in splitless injection.

Split injection:
- concentrated sample
- high resolution
- dirty samples (use packed liner)
- could cause thermal decomposition

Splitless injection:
- dilute sample
- high resolution
- requires solvent trapping or cold trapping

On-column injection:
- best for quantitative analysis
- thermally sensitive compounds
- low resolution

---

## Terms to Understand

| | | | |
|---|---|---|---|
| carrier gas | molecular sieve | sample preparation | splitless injection |
| co-chromatography | on-column injection | selected ion monitoring | temperature programming |
| cold trapping | open tubular column | selected reaction monitoring | thermal conductivity detector |
| electron capture detector | packed column | septum | thermal desorption |
| flame ionization detector | purge and trap | solid-phase microextraction | |
| gas chromatography | retention gap | solvent trapping | |
| guard column | retention index | split injection | |

---

## Summary

In gas chromatography, a volatile liquid or gaseous solute is carried by a gaseous mobile phase over a stationary phase on the inside of an open tubular column or on a solid support. Long, narrow, fused-silica open tubular columns have low capacity but give excellent separation. They can be wall coated, support coated, or porous layer. Packed columns provide high capacity but poor resolution. Each liquid stationary phase most strongly retains solutes in its own polarity class ("like dissolves like"). Solid stationary phases include porous carbon, alumina, and molecular sieves. The retention index measures elution times in relation to those of linear alkanes. Temperature or pressure programming reduces elution times of strongly retained components. Without compromising separation efficiency, the linear flow rate may be increased when $H_2$ or He, instead of $N_2$, is used as carrier gas. Split injection provides high-resolution separations of relatively concentrated samples. Splitless injection of very dilute samples requires solvent trapping or cold trapping to concentrate solutes at the start of the column (to give sharp bands). On-column injection is best for quantitative analysis and for thermally unstable solutes.

Quantitative analysis is usually done with internal standards in gas chromatography. Co-chromatography of an authentic sample with an unknown on several different columns is useful for qualitative identification of a peak. Mass spectral and infrared detectors provide qualitative information that helps identify peaks from unknown compounds. The mass spectrometer becomes more sensitive and less subject to interference when selected ion monitoring or selected reaction monitoring is employed. Thermal conductivity detection has universal response but is not sensitive enough for narrow-bore, open tubular columns. Flame ionization detection is sensitive enough for most columns and responds to most organic compounds. Electron capture, nitrogen-phosphorus, flame photometry, photoionization, chemiluminescence, and atomic emission detectors are specific for certain classes of compounds or individual elements.

You need to decide the goal of an analysis before developing a chromatographic method. The key to successful chromatography is to have a clean sample. Solid-phase microextraction, purge and trap, and thermal desorption can isolate volatile components from complex matrices. After the sample preparation method has been chosen, the remaining decisions for method development are to select a detector, a column, and the injection method, in that order.

---

## Exercises

**24-A. (a)** When a solution containing 234 mg of butanol (FM 74.12) and 312 mg of hexanol (FM 102.17) in 10.0 mL was separated by gas chromatography, the relative peak areas were butanol:hexanol = 1.00:1.45. Considering butanol to be the internal standard, find the response factor for hexanol.

**(b)** Use Equation 24-2 to estimate the areas of the peaks for butanol and hexanol in Figure 24-8.

**(c)** The solution from which the chromatogram was generated contained 112 mg of butanol. What mass of hexanol was in the solution?

**(d)** What is the largest source in uncertainty in this problem? How great is this uncertainty?

**24-B.** When 1.06 mmol of 1-pentanol and 1.53 mmol of 1-hexanol were separated by gas chromatography, they gave relative peak areas of 922 and 1 570 units, respectively. When 0.57 mmol of pentanol was added to an unknown containing hexanol, the relative chromatographic peak areas were 843:816 (pentanol:hexanol). How much hexanol did the unknown contain?

24-C. (a) In Table 24-3, 2-pentanone has a retention index of 987 on a poly(ethylene glycol) column (also called Carbowax). Between which two straight-chain hydrocarbons is 2-pentanone eluted?
(b) An unretained solute is eluted from a certain column in 1.80 min. Decane ($C_{10}H_{22}$) is eluted in 15.63 min and undecane ($C_{11}H_{24}$) is eluted in 17.22 min. What is the retention time of a compound whose retention index is 1 050?

24-D. For a homologous series of compounds (those with similar structures, but differing by the number of $CH_2$ groups in a chain), log $t_r'$ is usually a linear function of the number of carbon atoms. A compound was known to be a member of the family

$$(CH_3)_2CH(CH_2)_nCH_2OSi(CH_3)_3$$

(a) From the gas chromatographic retention times given here, prepare a graph of log $t_r'$ versus $n$ and estimate the value of $n$ in the chemical formula.

| | | | |
|---|---|---|---|
| $n = 7$ | 4.0 min | $CH_4$ | 1.1 min |
| $n = 8$ | 6.5 min | unknown | 42.5 min |
| $n = 14$ | 86.9 min | | |

(b) Calculate the capacity factor for the unknown.

24-E. Resolution of two peaks depends on the column plate number, $N$, and the separation factor, $\gamma$, as given by Equation 23-30. Suppose you have two peaks with a resolution of 1.0 and you want to increase the resolution to 1.5 for a baseline separation for quantitative analysis (Figure 23-10).
(a) You can increase the resolution to 1.5 by increasing column length. By what factor must the column length be increased? If flow rate is constant, how much longer will the separation require when the length is increased?
(b) You might change the separation factor by choosing a different stationary phase. If $\gamma$ was 1.013, to what value must it be increased to obtain a resolution of 1.5? If you were separating two alcohols with a $(diphenyl)_{0.05}(dimethyl)_{0.95}$ polysiloxane stationary phase (Table 24-1), what stationary phase would you choose to increase $\gamma$? Will this change affect the time required for chromatography?

---

## Problems

24-1. (a) What is the advantage of temperature programming in gas chromatography?
(b) What is the advantage of pressure programming?

24-2. (a) What are the relative advantages and disadvantages of packed and open tubular columns in gas chromatography?
(b) Explain the difference between wall-coated, support-coated, and porous-layer open tubular columns.
(c) What is the advantage of bonding (covalently attaching) the stationary phase to the column wall or cross-linking the stationary phase to itself?

24-3. (a) Why do open tubular columns provide greater resolution than packed columns in gas chromatography?
(b) Why do $H_2$ and He allow more rapid linear flow rates in gas chromatography than $N_2$ does, without loss of column efficiency (Figure 24-12)?

24-4. (a) When would you use split, splitless, or on-column injection in gas chromatography?
(b) Explain how solvent trapping and cold trapping work in splitless injection.

24-5. To which kinds of analytes do the following gas chromatography detectors respond?
(a) thermal conductivity    (f) photoionization
(b) flame ionization    (g) sulfur chemiluminescence
(c) electron capture    (h) atomic emission
(d) flame photometric    (i) mass spectrometer
(e) nitrogen-phosphorus

24-6. Why does a thermal conductivity detector respond to all analytes except the carrier gas? Why isn't the flame ionization detector universal?

24-7. Explain what is displayed in a reconstructed total ion chromatogram, selected ion monitoring, and selected reaction monitoring. Which technique is most selective and which is least selective and why?

24-8. Use Table 24-3 to predict the elution order of the following compounds from columns containing (a) poly(dimethylsiloxane), (b) $(diphenyl)_{0.35}(dimethyl)_{0.65}$polysiloxane, and (c) poly(ethylene glycol): hexane, heptane, octane, benzene, butanol, 2-pentanone.

24-9. Use Table 24-3 to predict the elution order of the following compounds from columns containing (a) poly(dimethylsiloxane), (b) $(diphenyl)_{0.35}(dimethyl)_{0.65}$polysiloxane, and (c) poly(ethylene glycol):

| | | | |
|---|---|---|---|
| 1. | 1-pentanol | 4. | octane |
| | ($n$-$C_5H_{11}OH$, b.p. 138°C) | | ($n$-$C_8H_{18}$, b.p. 126°C) |
| 2. | 2-hexanone | 5. | nonane |
| | ($CH_3C(=O)C_4H_9$, b.p. 128°C) | | ($n$-$C_9H_{20}$, b.p. 151°C) |
| 3. | heptane | 6. | decane |
| | ($n$-$C_7H_{16}$, b.p. 98°C) | | ($n$-$C_{10}H_{22}$, b.p. 174°C) |

24-10. This problem reviews concepts from Chapter 23. An unretained solute passes through a chromatography column in 3.7 min and analyte requires 8.4 min.
(a) Find the adjusted retention time and capacity factor for the analyte.
(b) The volume of the mobile phase is 1.4 times the volume of the stationary phase. Find the partition coefficient for the analyte.

24-11. If retention times in Figure 23-7 are 1.0 min for $CH_4$, 12.0 min for octane, 13.0 min for unknown, and 15.0 min for nonane, find the Kovats retention index for the unknown.

24-12. Retention time depends on temperature, $T$, according to the equation log $t_r' = (a/T) + b$, where $a$ and $b$ are constants for a specific compound on a specific column. A compound is eluted from a gas chromatography column at an adjusted retention time $t_r' = 15.0$ min when the column temperature is 373 K. At 363 K, $t_r' = 20.0$ min. Find the parameters $a$ and $b$ and predict $t_r'$ at 353 K.

24-13. What is the purpose of derivatization in chromatography? Give an example.

**24-14.** Explain how solid-phase microextraction works. Why is cold trapping necessary during injection with this technique? Is all the analyte in an unknown extracted into the fiber in solid-phase microextraction?

**24-15.** Why is splitless injection used with purge and trap sample preparation?

**24-16.** State the order of decisions in method development for gas chromatography.

**24-17. (a)** Why is it illogical to use a thin stationary phase (0.2 μm) in a wide-bore (0.53-mm) open tubular column?
**(b)** Consider a narrow-bore (0.25 mm diameter), thin-film (0.10 μm) column with 5 000 plates per meter. Consider also a wide-bore (0.53-mm diameter), thick-film (5.0 μm) column with 1 500 plates per meter. The density of stationary phase is approximately 1.0 g/mL. What mass of stationary phase is in each column in a length equivalent to one theoretical plate? How many nanograms of analyte can be injected into each column if the mass of analyte is not to exceed 1.0% of the mass of stationary phase in one theoretical plate?

**24-18.** How can you improve the resolution between two closely spaced peaks in gas chromatography?

**24-19. (a)** When a solution containing 234 mg of pentanol (FM 88.15) and 237 mg of 2,3-dimethyl-2-butanol (FM 102.17) in 10.0 mL was separated, the relative peak areas were pentanol:2,3-dimethyl-2-butanol = 0.913:1.00. Considering pentanol to be the internal standard, find the response factor for 2,3-dimethyl-2-butanol.
**(b)** Use Equation 24-2 to find the areas for pentanol and 2,3-dimethyl-2-butanol in Figure 24-8.
**(c)** The concentration of pentanol internal standard in the unknown solution was 93.7 mM. What was the concentration of 2,3-dimethyl-2-butanol?

**24-20.** A standard solution containing $6.3 \times 10^{-8}$ M iodoacetone and $2.0 \times 10^{-7}$ M $p$-dichlorobenzene (an internal standard) gave peak areas of 395 and 787, respectively, in a gas chromatogram. A 3.00-mL unknown solution of iodoacetone was treated with 0.100 mL of $1.6 \times 10^{-5}$ M $p$-dichlorobenzene and the mixture was diluted to 10.00 mL. Gas chromatography gave peak areas of 633 and 520 for iodoacetone and $p$-dichlorobenzene, respectively. Find the concentration of iodoacetone in the 3.00 mL of original unknown.

**24-21.** An unknown compound was co-chromatographed with heptane and decane. The adjusted retention times were heptane, 12.6 min; decane, 22.9 min; unknown, 20.0 min. The retention indexes for heptane and decane are 700 and 1 000, respectively. Find the retention index for the unknown.

**24-22.** The gasoline additive methyl $t$-butyl ether (MTBE) has been leaking into groundwater ever since its introduction in the 1990s. MTBE can be measured at parts per billion levels by solid-phase microextraction from groundwater to which 25% (wt/vol) NaCl has been added (*salting out*, Problem 8-8). After microextraction, analytes are thermally desorbed from the fiber in the port of a gas chromatograph. The figure shows a reconstructed total ion chromatogram and selected ion monitoring of substances desorbed from the extraction fiber.

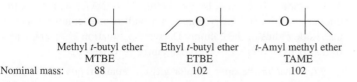

| Methyl $t$-butyl ether MTBE | Ethyl $t$-butyl ether ETBE | $t$-Amyl methyl ether TAME |
|---|---|---|
| Nominal mass: 88 | 102 | 102 |

**(a)** What is the purpose of adding NaCl prior to extraction?
**(b)** What nominal mass is being observed in selected ion monitoring? Why are only three peaks observed?
**(c)** Here is a list of major ions above $m/z$ 50 in the mass spectra. The base (tallest) peak is marked by an asterisk. Given that MTBE and TAME have an intense peak at $m/z$ 73 and there is no significant peak at $m/z$ 73 for ETBE, suggest a structure for $m/z$ 73. Suggest structures for all ions listed in the table.

| MTBE | ETBE | TAME |
|---|---|---|
| 73* | 87 | 87 |
| 57 | 59* | 73* |
| | 57 | 71 |
| | | 55 |

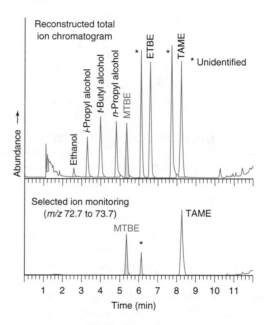

Reconstructed total ion chromatogram and selected ion monitoring of solid-phase microextract of groundwater. Chromatography conditions: 0.32 mm × 30 m column with 5-μm film of poly(dimethylsiloxane). Temperature = 50°C for 4 min, then raised 20°C/min to 90°C, held for 3 min, and then raised 40°C/min to 200°C. [From D. A. Cassada, Y. Zhang, D. D. Snow, and R. F. Spalding, "Trace Analysis of Ethanol, MTBE, and Related Oxygenate Compounds in Water Using Solid-Phase Microextraction and Gas Chromatography/Mass Spectrometry," Anal. Chem. **2000**, 72, 4654.]

**24-23.** Here is a student procedure to measure nicotine in urine. A 1.00-mL sample of biological fluid was placed in a 12-mL vial containing 0.7 g $Na_2CO_3$ powder. After 5.00 μg of the internal standard 5-aminoquinoline were injected, the vial was capped with a Teflon-coated silicone rubber septum. The vial was heated to 80°C for 20 min and then a solid-phase microextraction needle was passed through the septum and left in the headspace for 5.00 min. The fiber was retracted and inserted into a gas chromatograph. Volatile substances were desorbed from the fiber at 250°C for 9.5 min in the injection port while the column was at 60°C. The column temperature was then raised to 260°C at 25°C/min and eluate was monitored by electron ionization mass spectrometry with selected ion monitoring at $m/z$ 84 for nicotine and $m/z$ 144 for internal standard. Calibration data from replicate standard mixtures taken through the same procedure are given in the table.

| Nicotine in urine ($\mu$g/L) | Area ratio $m/z$ 84/144 |
|---|---|
| 12 | $0.05_6$, $0.05_9$ |
| 51 | $0.40_2$, $0.39_1$ |
| 102 | $0.68_4$, $0.66_9$ |
| 157 | $1.01_1$, $1.06_3$ |
| 205 | $1.27_8$, $1.35_5$ |

*Based on A. E. Wittner, D. M. Klinger, X. Fan, M. Lam, D. T. Mathers, and S. A. Mabury, "Quantitative Determination of Nicotine and Cotinine in Urine and Sputum Using a Combined SPME-GC/MS Method," J. Chem. Ed. 2002, 79, 1257.*

(a) Why was the vial heated to 80°C before and during extraction?
(b) Why was the chromatography column kept at 60°C during thermal desorption of the extraction fiber?
(c) Suggest a structure for $m/z$ 84 from nicotine. What is the $m/z$ 144 ion from the internal standard, 5-aminoquinoline?

Nicotine
$C_{10}H_{14}N_2$

5-Aminoquinoline
$C_9H_8N_2$

(d) 🖩 Urine from an adult female nonsmoker had an area ratio $m/z$ 84/144 = 0.51 and 0.53 in replicate determinations. Urine from a nonsmoking girl whose parents are heavy smokers had an area ratio 1.18 and 1.32. Find the nicotine concentration ($\mu$g/L) and its uncertainty in the urine of each person.

**24-24.** Nitric oxide (NO) is a cell-signaling agent in numerous physiologic processes including vasodilation, inhibition of clotting, and inflammation. A sensitive chromatography–mass spectrometry method was developed to measure two of its metabolites, nitrite ($NO_2^-$) and nitrate ($NO_3^-$), in biological fluids. Internal standards, $^{15}NO_2^-$ and $^{15}NO_3^-$, were added to the fluid at concentrations of 80.0 and 800.0 $\mu$M, respectively. The naturally occurring $^{14}NO_2^-$ and $^{14}NO_3^-$ plus the internal standards were then converted into volatile derivatives in aqueous acetone:

Pentafluorobenzyl bromide

$m/z$ 62 for $^{14}N$
$m/z$ 63 for $^{15}N$

$m/z$ 46 for $^{14}N$
$m/z$ 47 for $^{15}N$

Because biological fluids are so complex, the derivatives were first isolated by high-performance liquid chromatography. For quantitative analysis, liquid chromatography peaks corresponding to the two products were injected into a gas chromatograph, ionized by *negative ion* chemical ionization (giving major peaks for $NO_2^-$ and $NO_3^-$), and the products measured by selected ion monitoring. Results are shown in the figure in the next column. If the $^{15}N$ internal standards undergo the same reactions and same separations at the same rate as the $^{14}N$ analytes, then the concentrations of analytes are simply

$$[^{14}NO_x^-] = [^{15}NO_x^-](R - R_{blank})$$

where $R$ is the measured peak area ratio ($m/z$ 46/47 for nitrite and $m/z$ 62/63 for nitrate) and $R_{blank}$ is the measured ratio of peak areas in a blank prepared from the same buffers and reagents with no added nitrate or nitrite. The ratios of peak areas are $m/z$ 46/47 = 0.062 and $m/z$ 62/63 = 0.538. The ratios for the blank were $m/z$ 46/47 = 0.040 and $m/z$ 62/63 = 0.058. Find the concentrations of nitrite and nitrate in the urine.

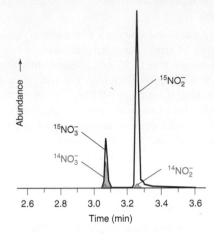

Selected ion chromatogram showing *negative ions* at $m/z$ 46, 47, 62, and 63 obtained by derivatizing nitrite and nitrate plus internal standards ($^{15}NO_2^-$ and $^{15}NO_3^-$) in urine. [From D. Tsikas, "Derivatization and Quantification of Nitrite and Nitrate in Biological Fluids by Gas Chromatography/Mass Spectrometry," Anal. Chem. 2000, 72, 4064.]

**24-25.** *van Deemter equation for open tubular column.* Equation 23-33 contains terms ($A$, $B$, and $C$) describing three band-broadening mechanisms.
(a) Which term is 0 for an open tubular column? Why?
(b) Express the value of $B$ in terms of measurable physical properties.
(c) Express the value of $C$ in terms of measurable physical quantities.
(d) The linear flow rate that produces minimum plate height is found by setting the derivative $dH/du_x = 0$. Find an expression of the minimum plate height in terms of the measurable physical quantities used to answer parts (b) and (c).

**24-26.** *Theoretical performance in gas chromatography.* As the inside radius of an open tubular gas chromatography column is decreased, the maximum possible column efficiency increases and sample capacity decreases. For a thin stationary phase that equilibrates rapidly with analyte, the minimum theoretical plate height is given by

$$\frac{H_{min}}{r} = \sqrt{\frac{1 + 6k' + 11k'^2}{3(1 + k')^2}}$$

where $r$ is the inside radius of the column and $k'$ is the capacity factor.
(a) Find the limit of the square-root term as $k' \to 0$ (unretained solute) and as $k' \to \infty$ (infinitely retained solute).
(b) If the column radius is 0.10 mm, find $H_{min}$ for the two cases in part (a).
(c) What is the maximum number of theoretical plates in a 50-m-long column with a 0.10-mm radius if $k' = 5.0$?
(d) The relation between capacity factor $k'$ and partition coefficient $K$ (Equation 23-19) can also be written $k' = 2tK/r$, where $t$ is the thickness of the stationary phase in a wall-coated column and $r$ is the inside radius of the column. Derive the equation $k' = 2tK/r$ and find $k'$ if $K = 1\,000$, $t = 0.20$ $\mu$m, and $r = 0.10$ mm.

**24-27.** Consider the open tubular gas chromatography of $n\text{-}C_{12}H_{26}$ on a 25-m-long $\times$ 0.53-mm-diameter column of 5% phenyl–95%

methyl polysiloxane with a stationary phase thickness of 3.0 µm with He carrier gas at 125°C. The observed capacity factor for $n$-$C_{12}H_{26}$ is 8.0. Measurements were made of plate height, $H$ (m), at various values of linear velocity, $u_x$ (m/s). A least-squares curve through the data points is given by

$$H = (6.0 \times 10^{-5} \text{ m}^2/\text{s})/u_x + (2.09 \times 10^{-3} \text{ s})u_x$$

From the coefficients of the van Deemter equation, find the diffusion coefficient of $n$-$C_{12}H_{26}$ in the mobile and stationary phases. Why is one of these diffusion coefficients so much greater than the other?

**24-28.** ▦ *Efficiency of solid-phase microextraction.* Equation 24-3 gives the mass of analyte extracted into a solid-phase microextraction fiber as a function of the partition coefficient between the fiber coating and the solution.

(a) A commercial fiber with a 100-µm-thick coating has a film volume of $6.9 \times 10^{-4}$ mL.[21] Suppose that the initial concentration of analyte in solution is $C_0 = 0.10$ µg/mL (100 ppb). Use a spreadsheet to prepare a graph showing the mass of analyte extracted into the fiber as a function of solution volume for partition coefficients of 10 000, 5 000, 1 000, and 100. Let the solution volume vary from 0 to 100 mL.

(b) Evaluate the limit of Equation 24-3 as $V_s$ gets big relative to $KV_f$. Does the extracted mass in your graph approach this limit?

(c) What percentage of the analyte from 10.0 mL of solution is extracted into the fiber when $K = 100$ and when $K = 10\ 000$?

**24-29.** *Mass spectral interpretation.* Box 24-1 shows the separation of enantiomers with the formula $C_9H_4N_2Cl_6$.

(a) Verify that the formula for rings + double bonds (22-3) agrees with the structure.

(b) Find the nominal mass of $C_9H_4N_2Cl_6$ (Box 22-1).

(c) The high-mass region of the electron impact mass spectrum of one enantiomer is shown here. Suggest an assignment for $m/z$ 350, 315, 280, 245, and 210.

(d) The relative abundance of $^{35}Cl$ and $^{37}Cl$ in a molecule containing $n$ Cl atoms is given by the terms of the binomial expansion

$$(a + b)^n = a^n + na^{n-1}b + \frac{n(n-1)}{2!}a^{n-2}b^2 + \frac{n(n-1)(n-2)}{3!}a^{n-3}b^3 + \cdots$$

where $a$ is the natural abundance of $^{35}Cl$ (0.757 7) and

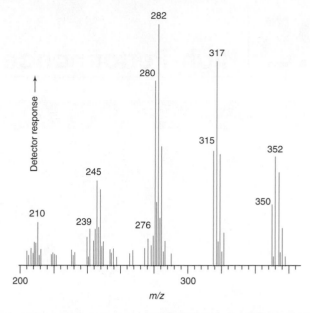

Mass spectrum of one enantiomer of $C_9H_4N_2Cl_6$ from Box 24-1. *[From W. Vetter and W. Jun, "Elucidation of a Polychlorinated Bipyrrole Structure Using Enantioselective GC," Anal. Chem. **2002**, 74, 4287.]*

$b$ is the natural abundance of $^{37}Cl$ (0.242 3). The first term gives the abundance of $^{35}Cl_n{}^{37}Cl_0$, the second term gives $^{35}Cl_{n-1}{}^{37}Cl_1$, and so forth. The spreadsheet shows how to compute the terms of $(a + b)^6$ with Excel in cells D8:D14. In cell D8, the function is "=BINOMDIST (A8,$B$5,$B$3,FALSE)" =BINOMDIST(6,6,0.7577,FALSE), which translates into D8 = $(0.757\ 7)^6$. When you highlight cell D8 and FILL DOWN, the function in cell D9 is "=BINOMDIST(A9,$B$5, $B$3,FALSE)" =BINOMDIST(5,6,0.7577,FALSE), which translates into D9 = $6(0.757\ 7)^5(0.242\ 3)^1$—the second term of the expansion. Column E normalizes the abundances in column D so that the most intense peak is 100. The spreadsheet predicts that a molecule with 6 Cl atoms will have a ratio of intensities M : M+2 : M+4 : M+6 : M+8 : M+10 : M+12 = 52.12 : 100 : 79.95 : 34.09 : 8.18 : 1.05 : 0.06 (if there are no other significant isotopes of other elements in the molecule). Compute the expected abundances of Cl isotopes for species with 5, 4, 3, and 2 Cl atoms and compare your results with the observed clusters in the mass spectrum.

| | A | B | C | D | E |
|---|---|---|---|---|---|
| 1 | Isotopic abundance from binomial distribution | | | | |
| 2 | | | | | |
| 3 | $^{35}Cl$ = | 0.7577 | natural abundance | | |
| 4 | $^{37}Cl$ = | 0.2423 | natural abundance | | |
| 5 | n = | 6 | | | |
| 6 | | | | | Relative |
| 7 | $^{35}Cl$ | Formula | Mass | Abundance | abundance |
| 8 | 6 | $^{35}Cl_6{}^{37}Cl_0$ | M | 0.18923 | 52.12 |
| 9 | 5 | $^{35}Cl_5{}^{37}Cl_1$ | M+2 | 0.36307 | 100.00 |
| 10 | 4 | $^{35}Cl_4{}^{37}Cl_2$ | M+4 | 0.29026 | 79.95 |
| 11 | 3 | $^{35}Cl_3{}^{37}Cl_3$ | M+6 | 0.12376 | 34.09 |
| 12 | 2 | $^{35}Cl_2{}^{37}Cl_4$ | M+8 | 0.02968 | 8.18 |
| 13 | 1 | $^{35}Cl_1{}^{37}Cl_5$ | M+10 | 0.00380 | 1.05 |
| 14 | 0 | $^{35}Cl_0{}^{37}Cl_6$ | M+12 | 0.00020 | 0.06 |
| 15 | | | | | |
| 16 | D8 = BINOMDIST (A8,$B$5,$B$3,FALSE) | | | | |
| 17 | E8 = 100*D8/MAX($D$8:$D$14) | | | | |

Spreadsheet illustrating use of binomial distribution function.

## ■■■ IN VIVO MICRODIALYSIS FOR MEASURING DRUG METABOLISM

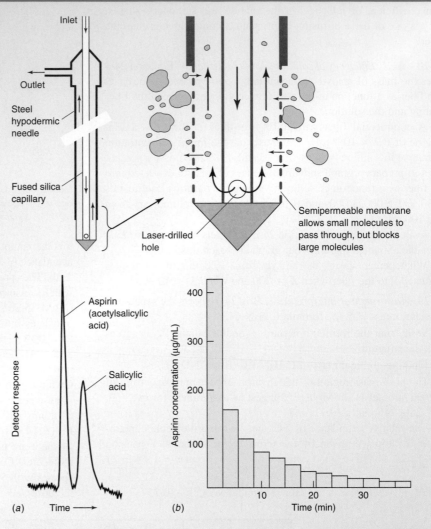

Microdialysis probe. Enlarged view of the lower end shows a semipermeable membrane that allows small molecules to pass in both directions, but excludes large molecules.

(a) Chromatogram of blood microdialysis sample drawn 5 min after intravenous injection of aspirin. (b) Aspirin concentration in blood versus time after injection of aspirin. *[From K. L. Steele, D. O. Scott, and C. E. Lunte, "Pharmacokinetic Studies of Aspirin in Rats Using in Vivo Microdialysis Sampling," Anal. Chim. Acta 1991, 246, 181.]*

**Dialysis** is the process in which small molecules diffuse across a *semipermeable membrane* that has pore sizes large enough to pass small molecules but not large ones. A *microdialysis probe* has a semipermeable membrane attached to the shaft of a hypodermic needle, which can be inserted into an animal. Fluid is pumped through the probe from the inlet to the outlet. Small molecules from the animal diffuse into the probe and are rapidly transported to the outlet. Fluid exiting the probe (*dialysate*) can be analyzed by liquid chromatography.

The chromatogram and bar graph show results of a study of aspirin metabolism in a rat. Aspirin is converted into salicylic acid by enzymes in the bloodstream. To measure the conversion rate, aspirin was injected into a rat and dialysate from a microdialysis probe in a vein of the rat was monitored by liquid chromatography. If you simply withdrew blood for analysis, aspirin would continue to be metabolized by enzymes in the blood. Microdialysis separates the small aspirin molecule from large enzyme molecules.

The first equipment for high-performance liquid chromatography was built by C. Horváth in 1965 at Yale University.

Liquid chromatography is important because most compounds are not sufficiently volatile for gas chromatography. **High-performance liquid chromatography (HPLC)** uses high pressure to force solvent through closed columns containing very fine particles that give high-resolution separations.[1,2] The HPLC system in Figure 25-1 consists of a solvent delivery system, a sample injection valve, a high-pressure column, a detector, and a computer to control the system and display results. Many systems include an oven for temperature control of

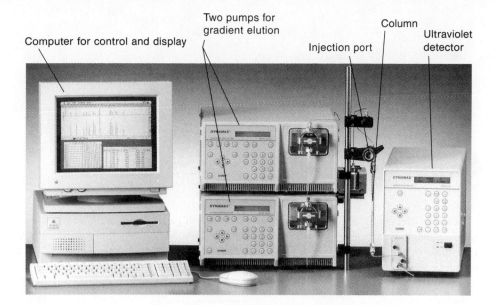

Computer for control and display

Two pumps for gradient elution

Injection port

Column

Ultraviolet detector

**Figure 25-1** Equipment for high-performance liquid chromatography (HPLC). *[Courtesy Rainin Instrument Co., Emeryville, CA.]*

the column. These components and the factors that govern the quality of a chromatographic separation are discussed in this chapter. For now, we restrict ourselves to liquid-liquid partition and liquid-solid adsorption chromatography. Chapter 26 deals with ion-exchange, molecular exclusion, and affinity chromatography.

## 25-1 The Chromatographic Process

Increasing the rate at which solute equilibrates between stationary and mobile phases increases the efficiency of chromatography. For gas chromatography with an open tubular column, rapid equilibration is accomplished by reducing the diameter of the column so that molecules can diffuse quickly between the channel and the stationary phase on the wall. Diffusion in liquids is 100 times slower than diffusion in gases. Therefore, in liquid chromatography, it is not generally feasible to use open tubular columns, because the diameter of the solvent channel is too great to be traversed by a solute molecule in a short time. Liquid chromatography is conducted with packed columns.

Increasing efficiency is equivalent to decreasing plate height, $H$, in the van Deemter equation (23-33):

$$H \approx A + \frac{B}{u_x} + Cu_x$$

$u_x$ = linear flow rate

### Small Particles Give High Efficiency but Require High Pressure

The efficiency of a packed column increases as the size of the stationary phase particles decreases. Typical particle sizes in HPLC are 3–5 μm. Figures 25-2a and b illustrate the increased resolution afforded by decreasing particle size from 4 to 1.7 μm. Plate number increased from 2 000 to 7 500 when the particle size decreased, so the peaks are sharper with the smaller particle size. In Figure 25-2c, a stronger solvent was used to elute the peaks from the column in trace *b* in less time. Decreasing particle size permits us to improve resolution or to maintain the same resolution while decreasing run time.

> **Example** Scaling Relations Between Columns
>
> Commonly, silica particles occupy ~40% of the column volume and solvent occupies ~60% of the column volume, regardless of particle size. The column in Figure 25-2a has an inside diameter of 4.6 mm and was run at a volume flow rate, $u_v$, of 3.0 mL/min with a sample size of 20 μL. The column in Figure 25-2b has a diameter of $d_c$ = 2.1 mm. What flow rate should be used in trace *b* to achieve the same linear velocity, $u_x$, as in trace *a*? What sample volume should be injected?
>
> **Solution** Column volume is proportional to the square of column diameter. Changing the diameter from 4.6 to 2.1 mm reduces volume by a factor of $(2.1/4.6)^2 = 0.208$. Therefore, $u_v$ should be reduced by a factor of 0.208 to maintain the same linear velocity.
>
> $u_v$(small column) = $0.208 \times u_v$(large column) = $(0.208)(3.0 \text{ mL/min})$ = 0.62 mL/min
>
> To maintain the same ratio of injected sample to column volume,
>
> Injection volume in small column = $0.208 \times$ (injection volume in large column)
>
> $= (0.208)(20 \text{ μL}) = 4.2 \text{ μL}$

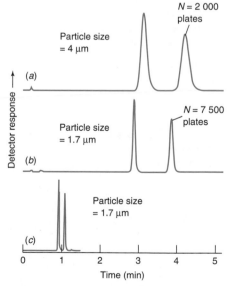

**Figure 25-2** (*a* and *b*) Chromatograms of the same sample run at the same linear velocity on 5.0-cm-long columns packed with $C_{18}$-silica. (*c*) A stronger solvent was used to elute solutes more rapidly from the column in panel *b*. [*Y. Yang and C. C. Hodges, "Assay Transfer from HPLC to UPLC for Higher Analysis Throughput," LCGC Supplement May 2005, p. 31.*]

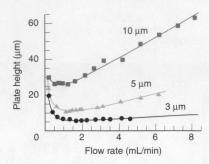

**Figure 25-3** Plate height as a function of flow rate for stationary phase particle diameters of 10, 5, and 3 μm. *[Courtesy Perkin-Elmer Corp., Norwalk, CT.]*

The van Deemter plots in Figure 25-3 show that small particles reduce plate height and that plate height is not very sensitive to increased flow rate when the particles are small. At the optimum flow rate (the minimum in Figure 25-3), the number of theoretical plates in a column of length $L$ (cm) is *approximately*[3]

$$N \approx \frac{3\,000\,L(\text{cm})}{d_{\text{p}}(\mu\text{m})} \tag{25-1}$$

where $d_{\text{p}}$ is the particle diameter in μm. The 5-cm-long column in Figure 25-2a with 4-μm-diameter particles is predicted to provide $\sim(3\,000)(5.0)/4 = 3\,800$ plates. The observed plate number for the second peak is 2 000. It may be that the column was not run at optimum flow rate. When the stationary phase particle diameter is reduced to 1.7 μm, the optimum plate number is expected to be $\sim(3\,000)(5.0)/1.7 = 8\,800$. The observed value is 7 200.

One reason why small particles give better resolution is that they provide more uniform flow through the column, thereby reducing the multiple path term, $A$, in the van Deemter equation (23-33). A second reason is that the distance through which solute must diffuse in the mobile and stationary phases is on the order of the particle size. The smaller the particles, the less distance solute must diffuse. This effect decreases the $C$ term in the van Deemter equation for finite equilibration time. The optimum flow rate for small particles is faster than for large particles because solutes diffuse through smaller distances.

An added benefit of small particle size, coupled with a narrow column and higher flow, is that analyte is not diluted so much as it travels through the column. The limit of quantitation in Figure 25-2c (50 μg/L) is four times lower than the limit of quantitation in Figure 25-2a (200 μg/L).

The penalty for small particle size is resistance to solvent flow. The pressure required to drive solvent through a column is

*Column pressure:*
$$P = f\frac{u_{\text{x}}\eta L}{\pi r^2 d_{\text{p}}^2} \tag{25-2}$$

where $u_{\text{x}}$ is linear flow rate, $\eta$ is the viscosity of the solvent, $L$ is the length of the column, $r$ is column radius, and $d_{\text{p}}$ is the particle diameter. The factor $f$ depends on particle shape and particle packing. The physical significance of Equation 25-2 is that pressure in HPLC is proportional to flow rate and column length and inversely proportional to the square of column radius (or diameter) and the square of particle size. The difference between traces $a$ and $b$ in Figure 25-2 is that particle size was decreased from 4 μm to 1.7 μm and column diameter was decreased from 4.6 to 2.1 mm. Therefore, the required pressure should increase by a factor of $(4.6\text{ mm}/2.1\text{ mm})^2(4\ \mu\text{m}/1.7\ \mu\text{m})^2 = 27$. That is, *27 times more pressure* is required to operate the column in Figure 25-2b.

Until recently, HPLC operated at pressures of $\sim$7–40 MPa (70–400 bar, 1 000–6 000 pounds/inch²) to attain flow rates of $\sim$0.5–5 mL/min. In 2004, commercial equipment became available to employ 1.5- to 2-μm-diameter particles at pressures up to 100 MPa (1 000 bar, 15 000 pounds/inch²). These instruments enable a substantial improvement in resolution or a decrease in run time. Table 25-1 shows theoretical performance for different particle sizes; such performance was realized in research with ultrahigh-pressure equipment.

Smaller particle size leads to

• higher plate number
• higher pressure
• shorter optimum run time
• lower detection limit

*Viscosity* measures resistance of a fluid to flow. The more viscous a liquid, the slower it flows at a given pressure.

**Table 25-1** Performance as a function of particle diameter

| Particle size $d_p$ (μm) | Retention time (min) | Plate number ($N$) | Required pressure (bar) |
|---|---|---|---|
| 5.0 | 30 | 25 000 | 19 |
| 3.0 | 18 | 42 000 | 87 |
| 1.5 | 9 | 83 000 | 700 |
| 1.0 | 6 | 125 000 | 2 300 |

NOTE: *Theoretical performance of 33-μm-diameter × 25-cm-long capillary for minimum plate height for solute with capacity factor k′ = 2 and diffusion coefficient = 6.7 × 10⁻¹⁰ m²/s in water-acetonitrile eluent.*

SOURCE: *J. E. MacNair, K. D. Patel, and J. W. Jorgenson, "Ultrahigh-Pressure Reversed-Phase Capillary Liquid Chromatography with 1.0-μm Particles," Anal. Chem.* **1999**, *71, 700.*

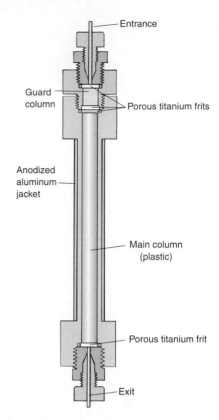

Entrance
Guard column
Porous titanium frits
Anodized aluminum jacket
Main column (plastic)
Porous titanium frit
Exit

**Figure 25-4** HPLC column with replaceable guard column to collect irreversibly adsorbed impurities. Titanium frits distribute the liquid evenly over the diameter of the column. *[Courtesy Upchurch Scientific, Oak Harbor, WA.]*

## The Column

The HPLC equipment in Figure 25-1 uses steel or plastic columns that are 5–30 cm in length, with an inner diameter of 1–5 mm (Figure 25-4). Columns are expensive and easily degraded by dust or particles in the sample or solvent and by irreversible adsorption of impurities from the sample or solvent. Therefore, the entrance to the main column is protected by a short **guard column** containing the same stationary phase as the main column. Fine particles and strongly adsorbed solutes are retained in the guard column, which is periodically replaced.

Heating a chromatography column[4] usually decreases the viscosity of the solvent, thereby reducing the required pressure or permitting faster flow. Increased temperature decreases retention times (Figure 23-18) and improves resolution by hastening diffusion of solutes. However, increased temperature can degrade the stationary phase and decrease column lifetime. When column temperature is not controlled, it fluctuates with the ambient temperature. Using a column heater set 10°C above room temperature improves the reproducibility of retention times and the precision of quantitative analysis. For a heated column, the mobile phase should be passed through a preheating metal coil between the injector and the column so that the solvent and column are at the same temperature. If their temperatures are different, peaks become distorted and retention times change.

The most common HPLC column diameter is 4.6 mm. There is a trend toward narrower columns (2 mm, 1 mm, and capillary columns down to ~25 μm) for several reasons. Narrow columns are more compatible with mass spectrometers, which require low solvent flow. Narrow columns require less sample and produce less waste. Heat generated by friction of solvent flow inside the column is more easily dissipated from a narrow column to maintain isothermal conditions. Instruments must be specially designed to accommodate column diameters <2 mm or else band broadening outside the column becomes significant.

Band broadening outside the column was discussed in Section 23-5.

## The Stationary Phase

The most common support is highly pure, spherical, **microporous particles** of silica that are permeable to solvent and have a surface area of several hundred square meters per gram (Figure 25-5). Most silica should not be used above pH 8, because it dissolves in base. Special grades of silica are stable up to pH 9 or 10. For separation of basic compounds at pH 8–12, polymeric supports such as polystyrene (Figure 26-1) can be used. Stationary phase is covalently attached to the polymer.

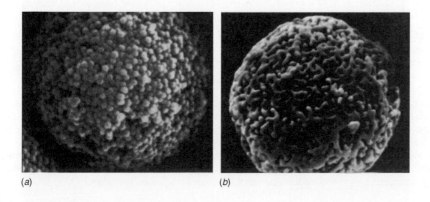

(a)　(b)

**Figure 25-5** Scanning electron micrographs of silica chromatography particles. (*a*) Aggregate of spherical particles with 50% porosity and a surface area of 150 m²/g. (*b*) Spongelike structure with 70% porosity and a surface area of 300 m²/g. Pores are the entryways into the interior of the particles. In both cases, the nominal pore size is 10 nm, but the distribution of pore sizes is greater in the spongelike structure. The spongelike structure also dissolves more readily in base. *[From Hewlett-Packard Co. and R. E. Majors, LCGC May 1997, p. S8.]*

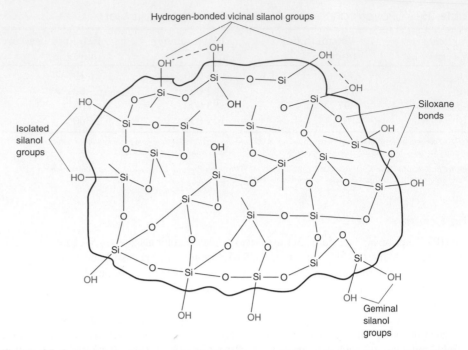

**Figure 25-6** Schematic structure of silica particle. *[From R. E. Majors, LCGC May 1997, p. S8.]* In particles that are <2 μm in diameter, some of the Si—O—Si bridges inside the silica are replaced by Si—CH$_2$—CH$_2$—Si bridges, which provide mechanical rigidity to withstand high pressure.

A silica surface (Figure 25-6) has up to 8 μmol of silanol groups (Si—OH) per square meter. Silanol groups are protonated at pH ~2–3. They dissociate to negative Si—O$^-$ over a broad pH range above 3. Exposed Si—O$^-$ groups strongly retain protonated bases (for example, RNH$_3^+$) and lead to tailing (Figure 25-7). Metallic impurities in silica also cause tailing. Type B silica in Figure 25-7 has fewer exposed silanol groups and less metallic impurities. Type C silica causes even less tailing because 90% of the Si—OH group are replaced by Si—H bonds, which do not retain solutes by hydrogen bonding.

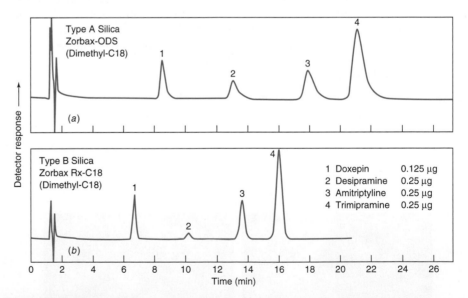

**Figure 25-7** Tailing of amine bases on silica: (*a*) Type A silica support gives distorted peaks. (*b*) Less acidic type B silica with fewer Si—OH groups and less metallic impurity gives symmetric peaks with shorter retention time. In both cases, chromatography was performed with a 0.46 × 15 cm column eluted at 1.0 mL/min at 40°C with 30 vol% acetonitrile/70 vol% sodium phosphate buffer (pH 2.5) containing 0.2 wt% triethylamine and 0.2 wt% trifluoroacetic acid. The detector measured ultraviolet absorbance at 254 nm. Additives such as triethylamine and trifluoroacetic acid are often used to mask strong adsorption sites to reduce tailing. *[From J. J. Kirkland, Am. Lab. June 1994, p. 28K.]*

CHAPTER 25 High-Performance Liquid Chromatography

Bare silica can be used as the stationary phase for adsorption chromatography. Most commonly, liquid-liquid partition chromatography is conducted with a **bonded stationary phase** covalently attached to the silica surface by reactions such as

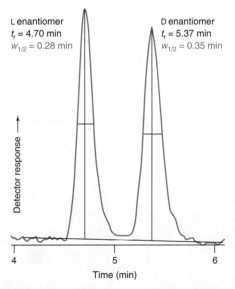

Silica surface

Residual silanol groups on the silica surface are capped with trimethylsilyl groups by reaction with $ClSi(CH_3)_3$ to eliminate polar adsorption sites that cause tailing.

| Common polar phases | | | Common nonpolar phases | | |
|---|---|---|---|---|---|
| $R = (CH_2)_3NH_2$ | | amino | $R = (CH_2)_{17}CH_3$ | | octadecyl |
| $R = (CH_2)_3C \equiv N$ | | cyano | $R = (CH_2)_7CH_3$ | | octyl |
| $R = (CH_2)_2OCH_2CH(OH)CH_2OH$ | | diol | $R = (CH_2)_3C_6H_5$ | | phenyl |

The octadecyl ($C_{18}$) stationary phase is by far the most common in HPLC. It is frequently designated ODS, for octadecylsilane.

There are ~4 μmol of R groups per square meter of support surface area, with little bleeding of the stationary phase from the column during chromatography. For separating optical isomers (Figure 25-8), many optically active R groups, such as the one Exercise 25-B, are commercially available.[5]

The siloxane (Si—O—SiR) bond hydrolyzes below pH 2, so HPLC with a bonded phase on a silica support is generally limited to the pH range 2–8. If bulky isobutyl groups are attached to the silicon atom of the bonded phase (Figure 25-9), the stationary phase is protected from attack by $H_3O^+$ and is stable for long periods at low pH, even at elevated

Bidentate $C_{18}$ stationary phase stable in the pH range 2–11.5:

L enantiomer
$t_r = 4.70$ min
$w_{1/2} = 0.28$ min

D enantiomer
$t_r = 5.37$ min
$w_{1/2} = 0.35$ min

Detector response

Time (min)

**Figure 25-8** Baseline separation of enantiomers of the drug Ritalin by HPLC with a chiral stationary phase. One enantiomer is pharmacologically active for treating attention deficit disorder and narcolepsy. The other enantiomer has little activity but could contribute to undesired side effects. Pharmaceutical companies are moving toward providing enantiomerically pure drugs, which could be safer than mixtures of optical isomers. [From R. Bakhtiar, L. Ramos, and F. L. S. Tse, "Quantification of Methylphenidate in Plasma Using Chiral Liquid-Chromatography/Tandem Mass Spectrometry: Application to Toxicokinetic Studies," Anal. Chim. Acta **2002**, 469, 261.]

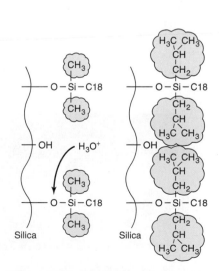

**Figure 25-9** Bulky isobutyl groups protect siloxane bonds from hydrolysis at low pH. [From J. J. Kirkland, Am. Lab. June 1994, p. 28K.]

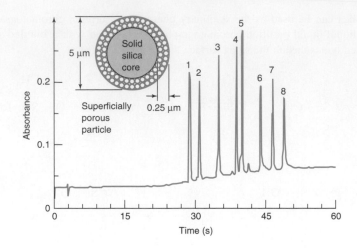

**Figure 25-10** Rapid separation of proteins on superficially porous $C_{18}$-silica in 2.1 × 75 mm column containing Poroshell 300SB-C18. Mobile phase A: 0.1 wt% trifluoroacetic acid in $H_2O$. Mobile phase B: 0.07 wt% trifluoroacetic acid in acetonitrile. Solvent was changed continuously from 95 vol% A/5 vol% B to 100% B over 1 min. Flow = 3 mL/min at 70°C at 26 MPa (260 bar) with ultraviolet detection at 215 nm. Peaks: 1, angiotensin II; 2, neurotensin; 3, ribonuclease; 4, insulin; 5, lysozyme; 6, myoglobin; 7, carbonic anhydrase; 8, ovalbumin. [From R. E. Majors, LCGC Column Technology Supplement June 2004, p. 8K. Courtesy Agilent Technologies.]

Superficially porous particles:
- rapid mass transfer into thin porous layer
- 5-μm diameter does not require high column pressure
- low sample capacity

temperature (for example, pH 0.9 at 90°C).[6] Box 25-1 describes a new kind of silica stationary phase.

Figure 25-10 shows a rapid separation of proteins on **superficially porous particles,** which consist of a 0.25-μm-thick porous silica layer on a 5-μm-diameter nonporous silica core. A stationary phase such as $C_{18}$ is bonded to the thin, porous outer layer. Mass transfer of solute into a 0.25-μm-thick layer is 10 times faster than mass transfer into fully porous particles with a radius of 2.5 μm, enabling high efficiency at high flow rate. Superficially porous particles are especially suitable for separation of macromolecules such as proteins, which diffuse more slowly than small molecules.

Porous graphitic carbon deposited on silica[8] is a stationary phase that exhibits increased retention of nonpolar compounds relative to retention by $C_{18}$. Graphite has high affinity for polar compounds and separates isomers that cannot be separated on $C_{18}$. The stationary phase is stable in 10 M acid and 10 M base.

---

## Box 25-1 Monolithic Silica Columns

Time is money. In commercial laboratories, the faster an analysis can be done, the less it costs. *Monolithic silica columns* enable us to increase the flow rate in liquid chromatography while retaining good separation.[7]

Each column in the left-hand photograph is a single, porous silica rod polymerized from liquid precursors. The adjacent micrographs show the silica skeleton with a network of ~2-μm pores. The inside of the skeleton contains a finer network of ~13-nm pores that are too small to be seen in the micrographs. Approximately 80% of the volume of the rod is empty space. The

surface area is 300 m²/g, which compares favorably with that of excellent stationary phase materials. $C_{18}$ or other bonded phases are attached to the silica for reversed-phase chromatography. After fabrication, the silica rod is tightly encased in a chemically resistant plastic tube made of polyether ether ketone.

Because of its open, rigid structure, solvent flows through the monolithic column with relatively little resistance. The same pressure required to obtain a flow rate of 1 mL/min with 3.5-μm spherical particles provides a flow rate of 9 mL/min in the monolithic column. At 9 mL/min, the plate height in the monolithic column is only 50% greater than the minimum plate height observed at 2 mL/min.

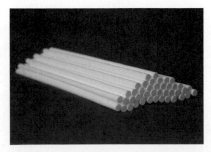

Monolithic silica rods.

[Photos courtesy D. Cunningham, Merck KGaA, Darmstadt, Germany.]

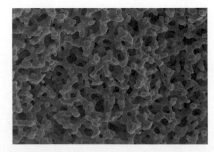

Structure of rod showing ~2-μm pores.

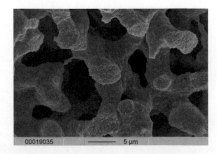

Enlarged view. Invisible 13-nm pores are located within the silica skeleton.

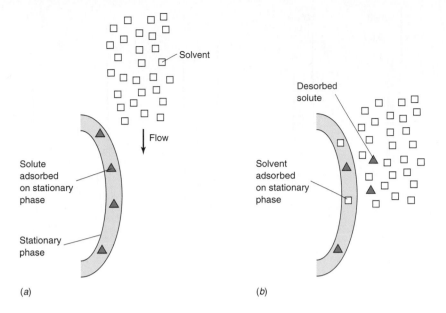

(a)    (b)

## The Elution Process

In *adsorption chromatography,* solvent molecules compete with solute molecules for sites on the stationary phase (Figure 25-11 and Color Plate 26). The relative abilities of different solvents to elute a given solute from the adsorbent are nearly independent of the nature of the solute. Elution occurs when solvent displaces solute from the stationary phase.

An *eluotropic series* ranks solvents by their relative abilities to displace solutes from a given adsorbent. The **eluent strength,** $\varepsilon°$, is a measure of the solvent adsorption energy, with the value for pentane defined as 0 for adsorption on bare silica. Table 25-2 ranks solvents by their eluent strength on bare silica. The more polar the solvent, the greater is its eluent strength for adsorption chromatography with bare silica. The greater the eluent strength, the more rapidly will solutes be eluted from the column.

Adsorption chromatography on bare silica is an example of **normal-phase chromatography,** in which we use a polar stationary phase and a less polar solvent. *A more polar solvent has a higher eluent strength.* **Reversed-phase chromatography** is the more common

*Normal-phase chromatography:*
- polar stationary phase
- more polar solvent has higher eluent strength

*Reversed-phase chromatography:*
- nonpolar stationary phase
- less polar solvent has higher eluent strength

**Table 25-2**  Eluotropic series and ultraviolet cutoff wavelengths of solvents for adsorption chromatography on silica

| Solvent | Eluent strength ($\varepsilon°$) | Ultraviolet cutoff (nm) |
|---|---|---|
| Pentane | 0.00 | 190 |
| Hexane | 0.01 | 195 |
| Heptane | 0.01 | 200 |
| Trichlorotrifluoroethane | 0.02 | 231 |
| Toluene | 0.22 | 284 |
| Chloroform | 0.26 | 245 |
| Dichloromethane | 0.30 | 233 |
| Diethyl ether | 0.43 | 215 |
| Ethyl acetate | 0.48 | 256 |
| Methyl *t*-butyl ether | 0.48 | 210 |
| Dioxane | 0.51 | 215 |
| Acetonitrile | 0.52 | 190 |
| Acetone | 0.53 | 330 |
| Tetrahydrofuran | 0.53 | 212 |
| 2-Propanol | 0.60 | 205 |
| Methanol | 0.70 | 205 |

NOTE: The ultraviolet cutoff for water is 190 nm.

SOURCES: L. R. Snyder, in High-Performance Liquid Chromatography (C. Horváth, ed.), Vol. 3 (New York: Academic Press, 1983); Burdick & Jackson Solvent Guide, 3rd ed. (Muskegon, MI: Burdick & Jackson Laboratories, 1990).

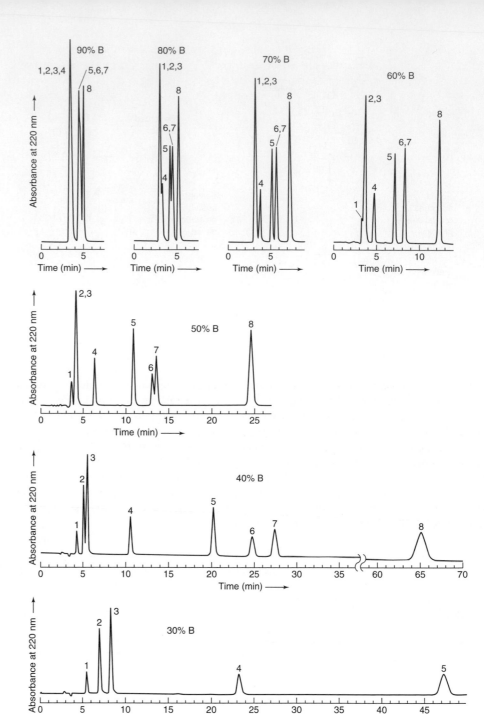

- Aqueous buffers for HPLC are prepared and the pH adjusted *prior* to mixing with organic solvent.[9]
- Ultrapure water for HPLC should be freshly prepared by a purification train or by distillation. Water extracts impurities from polyethylene or glass after storage for a few hours.
- To prepare 70% B, for example, mix 70 mL of B with 30 mL of A. *The result is different* from placing 70 mL of B in a volumetric flask and diluting to 100 mL with A because there is a volume change when A and B are mixed.

*General elution problem:* For a complex mixture, isocratic conditions can often be found to produce adequate separation of early-eluting peaks or late-eluting peaks, but not both. This problem drives us to use gradient elution.

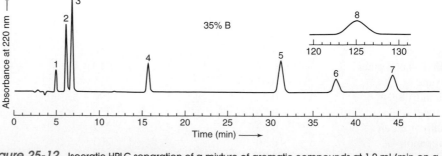

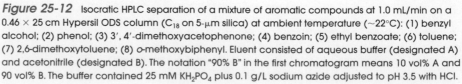

*Figure 25-12* Isocratic HPLC separation of a mixture of aromatic compounds at 1.0 mL/min on a 0.46 × 25 cm Hypersil ODS column ($C_{18}$ on 5-μm silica) at ambient temperature (~22°C): (1) benzyl alcohol; (2) phenol; (3) 3', 4'-dimethoxyacetophenone; (4) benzoin; (5) ethyl benzoate; (6) toluene; (7) 2,6-dimethoxytoluene; (8) *o*-methoxybiphenyl. Eluent consisted of aqueous buffer (designated A) and acetonitrile (designated B). The notation "90% B" in the first chromatogram means 10 vol% A and 90 vol% B. The buffer contained 25 mM $KH_2PO_4$ plus 0.1 g/L sodium azide adjusted to pH 3.5 with HCl.

scheme in which the stationary phase is nonpolar or weakly polar and the solvent is more polar. *A less polar solvent has a higher eluent strength.* Reversed-phase chromatography eliminates peak tailing because the stationary phase has few sites that can strongly adsorb a solute to cause tailing (Figure 23-20). Reversed-phase chromatography is also less sensitive to polar impurities (such as water) in the eluent.

## Isocratic and Gradient Elution

**Isocratic elution** is performed with a single solvent (or constant solvent mixture). If one solvent does not provide sufficiently rapid elution of all components, then **gradient elution** can be used. In this case, increasing amounts of solvent B are added to solvent A to create a continuous gradient.

Figure 25-12 shows the effect of increasing eluent strength in the isocratic elution of eight compounds from a reversed-phase column. In a reversed-phase separation, eluent strength *decreases* as the solvent becomes *more* polar. The first chromatogram (upper left) was obtained with a solvent consisting of 90 vol% acetonitrile and 10 vol% aqueous buffer. Acetonitrile has a high eluent strength, and all compounds are eluted rapidly. Only three peaks are observed because of overlap. It is customary to call the aqueous solvent A and the organic solvent B. The first chromatogram was obtained with 90% B. When eluent strength is reduced by changing the solvent to 80% B, there is slightly more separation and five peaks are observed. At 60% B, we begin to see a sixth peak. At 40% B, there are eight clear peaks, but compounds 2 and 3 are not fully resolved. At 30% B, all peaks would be resolved, but the separation takes too long. Backing up to 35% B (the bottom trace) separates all peaks in a little over 2 h (which is still too long for many purposes).

From the isocratic elutions in Figure 25-12, the gradient in Figure 25-13 was selected to resolve all peaks in 38 min. First, 30% B (B = acetonitrile) was run for 8 min to separate components 1, 2, and 3. The eluent strength was then increased steadily over 5 min to 45% B and held there for 15 min to elute peaks 4 and 5. Finally, the solvent was changed to 80% B over 2 min and held there to elute the last peaks.

*Isocratic elution:* one solvent
*Gradient elution:* continuous change of solvent composition to increase eluent strength. Gradient elution in HPLC is analogous to temperature programming in gas chromatography. Increased eluent strength is required to elute more strongly retained solutes.

Box 25-2 (pages 568–569) describes gradient elution in *supercritical fluid chromatography.*

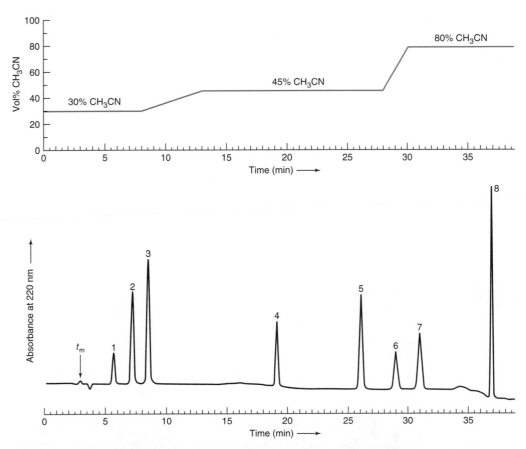

**Figure 25-13** Gradient elution of the same mixture of aromatic compounds in Figure 25-12 with the same column, flow rate, and solvents. The upper trace is the *segmented gradient* profile, so named because it is divided into several different segments.

## Selecting the Separation Mode

There are no firm rules in Figure 25-14. Methods in either part of the diagram may work perfectly well for molecules whose size belongs to the other part.

There can be many ways to separate components of a given mixture. Figure 25-14 is a decision tree for choosing a starting point. If the molecular mass of analyte is below 2 000, use the upper part of the figure; if the molecular mass is greater than 2 000, use the lower part. In either part, the first question is whether the solutes dissolve in water or in organic solvents. Suppose we have a mixture of small molecules (molecular mass <2 000) soluble in dichloromethane. Table 25-2 is essentially a ranking of solvent polarity, with the most-polar solvents at the bottom. The eluent strength of dichloromethane (0.30) is closer to that of chloroform (0.26) than it is to those of alcohols, acetonitrile, or ethyl acetate ($\geq 0.48$). Therefore,

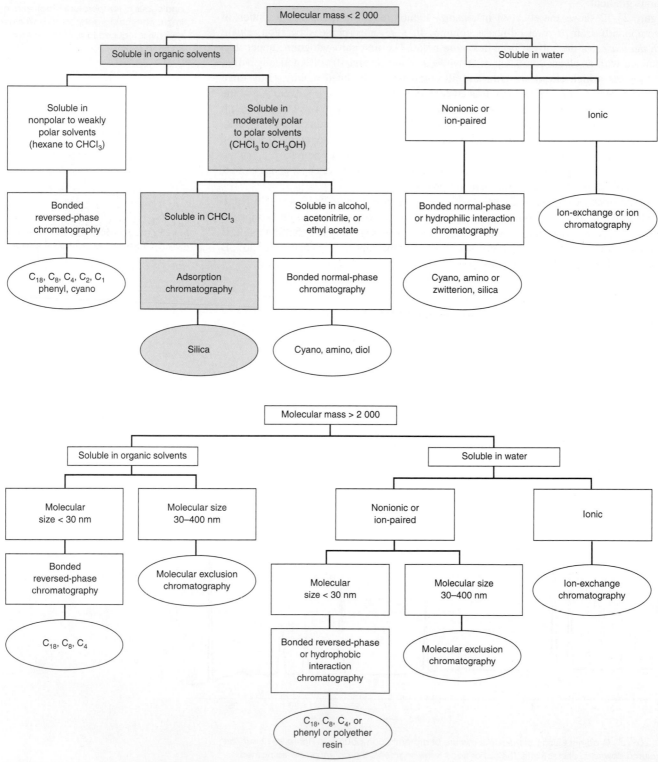

*Figure 25-14* Guide to HPLC mode selection.

Figure 25-14 suggests that we try adsorption chromatography on silica. The decision path is highlighted in color.

If solutes dissolve only in nonpolar or weakly polar solvents, the decision tree suggests that we try reversed-phase chromatography. Our choices include bonded phases containing octadecyl ($C_{18}$), octyl, butyl, ethyl, methyl, phenyl, and cyano groups.

If the molecular masses of solutes are >2 000 and if they are soluble in organic solvents and their molecular diameter is >30 nm, Figure 25-14 tells us to try molecular exclusion chromatography. Stationary phases for this type of separation are described in Chapter 26. If the molecular masses of solutes are >2 000, and they are soluble in water, but not ionic, and have diameters <30 nm, the decision tree says to use reversed-phase chromatography, or *hydrophobic interaction chromatography.*

Hydrophobic interaction chromatography is based on the interaction of a hydrophobic stationary phase with a hydrophobic region of a solute such as a protein. (**Hydrophobic substances** repel water, and their surfaces are not wetted by water. **Hydrophilic substances** are soluble in water or attract water to their surfaces. A protein can have hydrophilic regions that make it soluble in water and hydrophobic regions capable of interacting with a hydrophobic stationary phase.) Two hydrophobic stationary phases are shown in Figure 25-15. The bulk particle can be a polymer with 100-nm pores, through which most molecules can diffuse. The surface is coated with phenyl groups or with poly(ethylene glycol), both of which attract hydrophobic solutes and provide a means of chromatographic separation. Proteins are adsorbed on the hydrophobic surface when the aqueous mobile phase contains high salt concentration. Eluent strength is increased by decreasing the salt concentration.

*Hydrophilic interaction chromatography* is most useful for small molecules that are too polar to be retained by reversed-phase columns. Stationary phases for hydrophilic interaction chromatography, such as those in Figure 25-15, are strongly polar. The mobile phase contains ($\geq$80%) polar organic solvent mixed with water. Solutes equilibrate between the mobile phase and a layer of aqueous phase on the surface of the stationary phase. Eluent strength is increased by *increasing* the fraction of water in the mobile phase. In normal-phase chromatography, the solvent is nonaqueous. To increase eluent strength, we increase the polarity of the nonaqueous solvent. In reversed-phase chromatography, the solvent is usually aqueous, and eluent strength is increased by *decreasing* the fraction of water in the mobile phase to increase the solubility of solutes in the mobile phase.

Water does not *wet* a hydrophobic surface made of carbon nanotubes, so a drop remains almost spherical. The drop would flatten out on a hydrophilic surface, such as glass. [*Courtesy Karen Gleason, Massachusetts Institute of Technology.*[13]]

Stationary phases for hydrophobic interaction chromatography

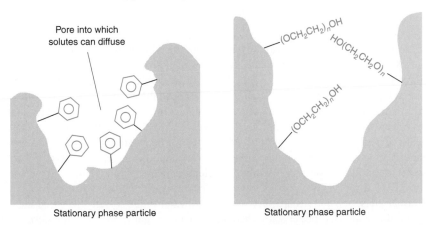

Stationary phases for hydrophilic interaction chromatography

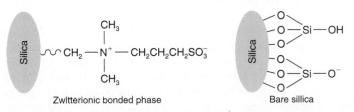

Zwitterionic bonded phase          Bare sillica

*Figure 25-15* Stationary phases for hydrophobic and hydrophilic interaction chromatography.

Box 25-2 "Green" Technology: Supercritical Fluid Chromatography

Preparative chromatography with supercritical $CO_2$ is a "green" technology that reduces the use of organic solvents in the pharmaceutical industry. Though $CO_2$ is not a very good solvent by itself, when mixed with some organic solvent, it is capable of dissolving a variety of compounds. Even without a cosolvent, $CO_2$ can interact with polar solutes as a hydrogen-bond acceptor, a weak Lewis acid, or weak Lewis base.[10]

In the phase diagram, panel (a), solid $CO_2$ (Dry Ice) is in equilibrium with gaseous $CO_2$ at a temperature of $-78.7°C$ and a pressure of 1.00 bar.[11] The solid *sublimes* without turning into liquid. At any temperature above the *triple point* at $-56.6°C$, there is a pressure at which liquid and vapor coexist as separate phases. For example, at 0°C, liquid is in equilibrium with gas at 34.9 bar. Moving up the liquid-gas boundary, we see that two phases always exist until the *critical point* is reached at 31.3°C

and 73.9 bar. *Above this temperature, only one phase exists, no matter what the pressure.* We call this phase a **supercritical fluid** (Color Plate 27). Its density and viscosity are between those of the gas and liquid, as is its ability to act as a solvent.

Critical constants

| Compound | Critical temperature (°C) | Critical pressure (bar) | Critical density (g/mL) |
|---|---|---|---|
| Argon | $-122.5$ | 47 | 0.53 |
| Carbon dioxide | 31.3 | 73.9 | 0.448 |
| Ammonia | 132.3 | 112.8 | 0.24 |
| Water | 374.4 | 229.8 | 0.344 |
| Methanol | 240.5 | 79.9 | 0.272 |
| Diethyl ether | 193.6 | 36.8 | 0.267 |

Supercritical fluid chromatography provides increased speed and resolution, relative to liquid chromatography, because of increased diffusion coefficients of solutes in supercritical fluids. (However, speed and resolution are slower than those of gas chromatography.) Unlike gases, supercritical fluids can dissolve nonvolatile solutes. When the pressure on the supercritical solution is released, the solvent turns to gas, leaving the solute in the gas phase for easy detection. Carbon dioxide is the supercritical fluid of choice for chromatography because it is compatible with flame ionization and ultraviolet detectors, it has a low critical temperature, and it is nontoxic.

Equipment for supercritical fluid chromatography is similar to that for HPLC with packed columns[12] or open tubular columns. Eluent strength is increased in HPLC by gradient elution and in gas chromatography by raising the temperature. In supercritical

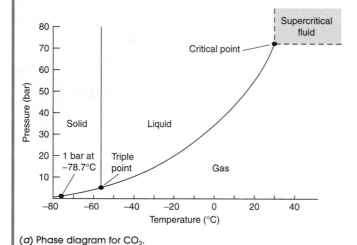

(a) Phase diagram for $CO_2$.

Purging with gas is called *sparging*.

**Question:** Why does water have low eluent strength in reversed-phase separations and high eluent strength in normal-phase separations?

**Reduce Waste Solvent Without Sacrificing Resolution**
- Use shorter columns with smaller-diameter particles.
- Switch from 4.6-mm-diameter column to 2.1 mm.
- For isocratic separations, use an electronic recycler that recycles eluate when no peak is being eluted.

## Solvents

Very pure HPLC-grade (expensive) solvents are required to prevent degradation of costly columns with impurities and to minimize detector background signals from contaminants. A filter is used on the intake tubing in the solvent reservoir to reject micron-size particles. Sample and solvent are passed through a short, expendable *guard column* (Figure 25-4), with the same stationary phase as the analytical column to collect strongly adsorbed species. Even with a guard column, periodic washing of the analytical column is recommended.[14] Before use, solvents are purged with He or evacuated to remove dissolved air. Air bubbles create difficulties with pumps, columns, and detectors. Dissolved $O_2$ absorbs ultraviolet radiation in the 250–200-nm wavelength range,[15] interfering with ultraviolet detection.

Normal-phase separations are very sensitive to water in the solvent. To speed the equilibration of the stationary phase with changing eluents, organic solvents for normal-phase chromatography should be 50% saturated with water. This can be done by adding a few milliliters of water to dry solvent and stirring. Then separate the wet solvent from the excess water and mix the wet solvent with an equal volume of dry solvent.

For gradient elution in reversed-phase separations, 10–20 empty column volumes of initial solvent should be passed through the column after a run to reequilibrate the stationary phase with solvent for the next run. Reequilibration can take as long as the separation. By adding 3 vol% 1-propanol to each solvent (so that there is always 3% 1-propanol at any point in the gradient), you can reduce the reequilibration volume down to 1.5 empty column volumes.[16] Propanol is thought to coat the stationary phase with a monolayer of alcohol that does not change much throughout the gradient.

fluid chromatography, eluent strength is increased by making the solvent *denser* by increasing the pressure. The chromatogram in panel (*b*) illustrates density gradient elution.

Separation of enantiomers is important in the pharmaceutical industry, where one enantiomer could be beneficial and the other could be harmful. Chromatographic separation of enantiomers uses an optically active stationary phase (Box 24-1). Supercritical fluid chromatography with a mixture of $CO_2$ and organic solvent reduces organic solvent use by up to 90%. The low viscosity of the supercritical fluid also permits faster flow rates that increase productivity. The chromatograms in panel (*c*) show a single injection and multiple overlapping injections for the preparative separation of enantiomers of a developmental drug candidate. Automated injection into a conventional laboratory-size column can purify up to ~100 g of enantiomers in several days. Larger industrial columns allow kilogram quantities to be separated, with dramatic savings in organic solvent.

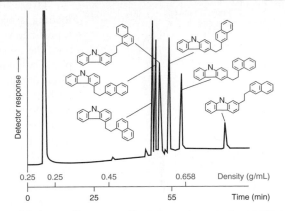

(*b*) Open tubular capillary supercritical chromatogram of aromatic compounds with $CO_2$, using density gradient elution at 140°C. [From R. D. Smith, B. W. Wright, and C. R. Yonker, "Supercritical Fluid Chromatography," *Anal. Chem.* **1988,** *60, 1323A.*]

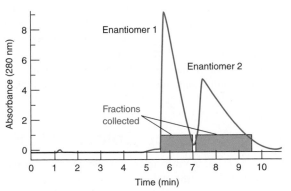

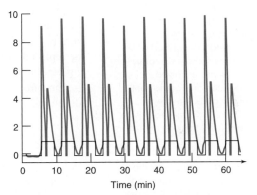

(*c*) Preparative separation of enantiomers by supercritical fluid chromatography. Single injection (left) and multiple overlapping injections (right) on a 2 cm × 25 cm Chiralcel OD column eluted with 68 vol% $CO_2$/32 vol% (25 mM isobutylamine in methanol) at 50 mL/min with outlet pressure = 10 MPa (100 bar). Each 2-mL injection contains 320 mg of solute. [From C. J. Welch, W. R. Leonard, Jr., O. DaSilva, M. Biba, J. Albaneze-Walker, D. W. Henderson, B. Laing, and D. J. Mathre, "Preparative Chiral SFC as a Green Technology for Rapid Access to Enantiopurity in Pharmaceutical Process Research," *LCGC* **2005,** *23, 16A.*]

## Maintaining Symmetric Bandshape

HPLC columns should be capable of providing narrow, symmetric peaks. If a new column does not reproduce the quality of the separation of a standard mixture that the manufacturer states it should and if you have satisfied yourself that the problem is not in the rest of your system, return the column.

> A standard mixture should be injected each day to evaluate the HPLC system. Changes in peak shapes or retention times alert you to a problem.

The asymmetry factor *A/B* in Figure 23-13 should rarely be outside the range 0.9–1.5. Tailing of amines (Figure 25-7) might be eliminated by adding 30 mM triethylamine to the mobile phase. The additive binds to sites on silica that would otherwise strongly bind analyte. Tailing of acidic compounds might be eliminated by adding 30 mM ammonium acetate. For unknown mixtures, 30 mM triethylammonium acetate is useful. If tailing persists, 10 mM dimethyloctylamine or dimethyloctylammonium acetate might be effective. A problem with additives is that they increase the equilibration time required when changing solvents. Improved grades of silica (types B and C) have reduced tailing and, therefore, reduced the need for additives.

Columns should be washed periodically to prevent buildup of strongly adsorbed compounds that can reduce column performance.[14] Tailing or splitting of peaks can occur if the frit at the beginning of the column becomes clogged with particles.[17] You can try to unclog the frit by disconnecting and reversing the column and flushing it with 20–30 mL. The column should not be connected to the detector during reverse flushing. If peaks are still distorted, it is time to discard the column.

Doubled peaks or altered retention times (Figure 25-16) sometimes occur if the solvent in which sample is dissolved has much greater eluent strength than the mobile phase. Try to dissolve the sample in a solvent of lower eluent strength or in the mobile phase itself.

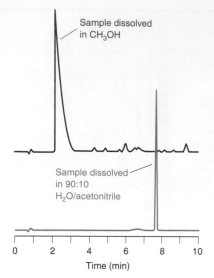

**Figure 25-16** Effect of sample solvent on retention time and peak shape of *n*-butylaniline. Eluent (1 mL/min) is 90:10 (vol/vol) $H_2O$/acetonitrile with 0.1 wt% trifluoroacetic acid. Lower sample was dissolved in eluent. Upper sample was dissolved in methanol, which is a much stronger solvent than eluent. Column: 15 cm × 4.6 mm, 5-$\mu$m-diameter $C_{18}$-silica, 30°C. Injection: 10 $\mu$L containing 0.5 $\mu$g analyte. Ultraviolet detection at 254 nm. *[Courtesy Supelco, Bellefonte, PA.]*

Sample dissolved in $CH_3OH$

Sample dissolved in 90:10 $H_2O$/acetonitrile

The longer the time spent in any part of the chromatographic system, the more a band broadens by diffusion. Section 23-5 discusses band spreading in the injector and detector.

Overloading causes the distorted shape shown in Figure 23-20.[18] To see whether overloading is occurring, reduce the sample mass by a factor of 10 and see whether retention times increase or peaks become narrower. If either change occurs, reduce the mass again until the injection size does not affect retention time and peak shape. In general, reversed-phase columns can handle 1–10 $\mu$g of sample per gram of silica. A 4.6-mm-diameter column contains 1 g of silica in a length of 10 cm. To prevent peak broadening from too large an injection volume, the injected volume should be less than 15% of the peak volume measured at the baseline. For example, if a peak eluted at 1 mL/min has a width of 0.2 min, the peak volume is 0.2 mL. The injection volume should not exceed 15% of 0.2 mL, or 30 $\mu$L.

The volume of a chromatography system outside of the column from the point of injection to the point of detection is called the **dead volume,** or the *extra-column volume.* Excessive dead volume allows bands to broaden by diffusion or mixing. Use short, narrow tubing whenever possible, and be sure that connections are made with matched fittings to minimize dead volume and thereby minimize extra-column band spreading.

HPLC columns have a typical lifetime of 500 to 2 000 injections.[3] You can monitor the health of a column by keeping a record of pressure, resolution, and peak shape. The pressure required to maintain a given flow rate increases as a column ages. System wear becomes serious when pressure exceeds 17 MPa (2 500 pounds/inch²). It is desirable to develop methods in which pressure does not exceed 14 MPa (2 000 pounds/inch²). When the pressure reaches 17 MPa, the in-line 0.5-$\mu$m frit between the autosampler and the column should be replaced. If this does not help, it is probably time for a new column. If you use a column for repetitive analyses, replace the column when the required resolution is lost or when tailing becomes significant. Resolution and tailing criteria should be established during method development.

## ■ ■ ■ 25-2 Injection and Detection in HPLC

We now consider the hardware required to inject sample and solvent onto the column and to detect compounds as they leave the column. Mass spectrometric detection, which is extremely powerful and important, was discussed in Section 22-4.

### Pumps and Injection Valves

The quality of a pump for HPLC is measured by how steady and reproducible a flow it can produce. A fluctuating flow rate can create detector noise that obscures weak signals. Figure 25-17 shows a pump with two sapphire pistons that produce a programmable, constant flow rate up to 10 mL/min at pressures up to 40 MPa (400 bar, 6 000 pounds/inch²). Gradients made from up to four solvents are constructed by proportioning the liquids through a four-way valve at low pressure and then pumping the mixture at high pressure

**Figure 25-17** High-pressure piston pump for HPLC. Solvent at the left passes through an electronic inlet valve synchronized with the large piston and designed to minimize the formation of solvent vapor bubbles during the intake stroke. The spring-loaded outlet valve maintains a constant outlet pressure, and the damper further reduces pressure surges. Pressure surges from the first piston are decreased in the damper that "breathes" against a constant outside pressure. Pressure surges are typically <1% of the operating pressure. As the large piston draws in liquid, the small piston propels liquid to the column. During the return stroke of the small piston, the large piston delivers solvent into the expanding chamber of the small piston. Part of the solvent fills the chamber while the remainder flows to the column. Delivery rate is controlled by the stroke volumes.

*[Courtesy Hewlett-Packard Co., Palo Alto, CA.]*

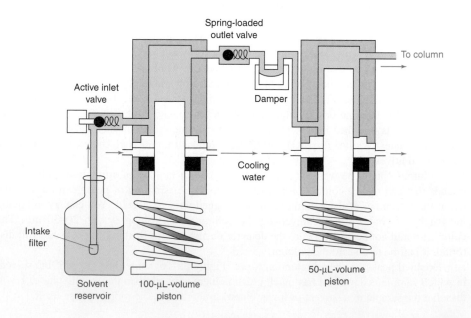

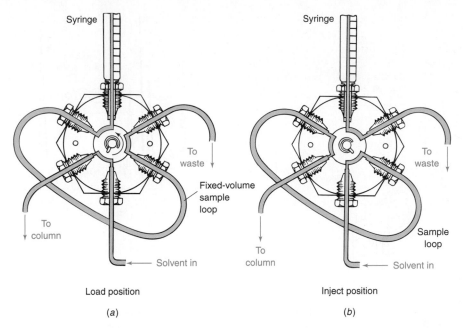

Syringe

To waste

Fixed-volume sample loop

To column

Solvent in

Load position

(a)

Syringe

To waste

Sample loop

To column

Solvent in

Inject position

(b)

**Figure 25-18** Injection valve for HPLC. Replaceable sample loop comes in various fixed-volume sizes.

into the column. The gradient is electronically controlled and programmable in 0.1 vol% increments.

The *injection valve* in Figure 25-18 has interchangeable sample loops, each of which holds a fixed volume. Loops of different sizes hold volumes that range from 2 to 1 000 μL. In the load position, a syringe is used to wash and load the loop with fresh sample at atmospheric pressure. High-pressure flow from the pump to the column passes through the segment of valve at the lower left. When the valve is rotated 60° counterclockwise, the content of the sample loop is injected into the column at high pressure.

- Pass samples through a 0.5-μm filter to remove particles prior to injection.
- The HPLC syringe needle is *blunt*, not pointed, to prevent damage to the injection port.

### Spectrophotometric Detectors

An ideal detector of any type (Table 25-3) is sensitive to low concentrations of every analyte, provides linear response, and does not broaden the eluted peaks. It is also insensitive to changes in temperature and solvent composition. To prevent peak broadening, the detector volume should be less than 20% of the volume of the chromatographic band. Gas bubbles in the detector create noise, so back pressure may be applied to the detector to prevent bubble formation during depressurization of eluate.

An **ultraviolet detector** using a flow cell such as that in Figure 25-19 is the most common HPLC detector, because many solutes absorb ultraviolet light. Simple systems employ the intense 254-nm emission of a mercury lamp. More versatile instruments have deuterium, xenon, or tungsten lamps and a monochromator, so you can choose the optimum ultraviolet

**Table 25-3** Comparison of commercial HPLC detectors

| Detector | Approximate limit of detection[a] (ng) | Useful with gradient? |
|---|---|---|
| Ultraviolet | 0.1–1 | Yes |
| Refractive index | 100–1 000 | No |
| Evaporative light-scattering | 0.1–1 | Yes |
| Electrochemical | 0.01–1 | No |
| Fluorescence | 0.001–0.01 | Yes |
| Nitrogen (N $\xrightarrow{combustion}$ NO $\xrightarrow{O_3}$ NO$_2^*$ → $h\nu$) | 0.3 | Yes |
| Conductivity | 0.5–1 | No |
| Mass spectrometry | 0.1–1 | Yes |
| Fourier transform infrared | 1 000 | Yes |

a. Most detection limits from E. W. Yeung and R. E. Synovec, "Detectors for Liquid Chromatography," *Anal. Chem.* **1986**, 58, 1237A.

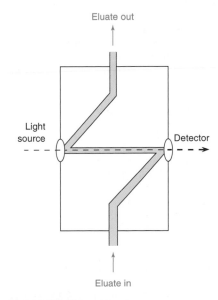

Eluate out

Light source

Detector

Eluate in

**Figure 25-19** Light path in a spectrophotometric micro flow cell. One common cell with a pathlength of 1 cm has a volume of 8 μL. Another cell with a 0.5-cm pathlength contains only 2.5 μL.

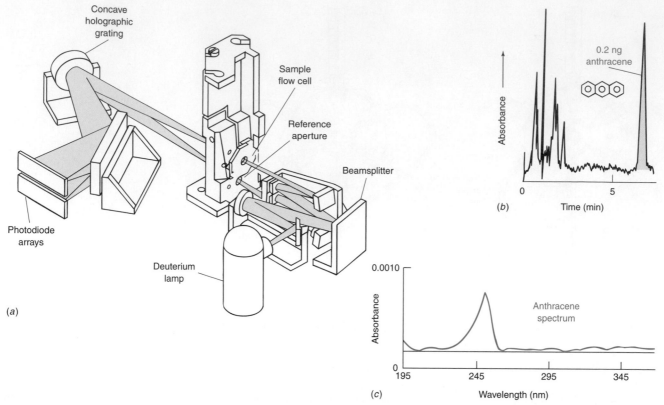

*Figure 25-20* Photodiode array ultraviolet detector for HPLC. (*a*) Dual-beam optical system uses grating polychromator, one diode array for the sample spectrum, and another diode array for the reference spectrum. Photodiode arrays are described in Section 20-3. (*b*) Reversed-phase chromatography (using $C_{18}$-silica) of sample containing 0.2 ng of anthracene, with detection at 250 nm. Full-scale absorbance is 0.001. (*c*) Spectrum of anthracene recorded as it emerged from the column. *[Courtesy Perkin-Elmer Corp., Norwalk, CT.]*

A photodiode array was used to record the ultraviolet spectrum of each peak in Figure 25-12 as it was eluted. By this means, it was possible to determine which compounds were in each peak.

*Linear range:* analyte concentration range over which detector response is proportional to concentration

*Dynamic range:* range over which detector responds in any manner (not necessarily linearly) to changes in analyte concentration (see page 71)

*Detection limit:* concentration of analyte that gives a specified signal-to-noise ratio

or visible wavelength for your analytes. The system in Figure 25-20 uses a *photodiode array* to record the spectrum of each solute as it is eluted. High-quality detectors provide full-scale absorbance ranges from 0.000 5 to 3 absorbance units, with a noise level near 1% of full scale. The linear range extends over five orders of magnitude of solute concentration (which is another way of saying that Beer's law is obeyed over this range). Ultraviolet detectors are good for gradient elution with nonabsorbing solvents. Table 25-2 gave the approximate cut-off wavelengths below which a solvent absorbs too strongly.

*Fluorescence detectors* excite the eluate with a laser and measure fluorescence (Figures 18-13 and 18-20). These detectors are very sensitive but respond only to the few analytes that fluoresce. To increase the utility of fluorescence and electrochemical detectors (described later), fluorescent or electroactive groups can be covalently attached to the analyte.[19] This process of **derivatization** can be performed on the mixture prior to chromatography or by addition of reagents to the eluate between the column and the detector (called *post-column derivatization*). For example, Tb(EDTA)$^-$ can be added to the hormones epinephrine (page 114), norepinephrine, and dopamine emerging from a chromatography column.[20] These compounds form complexes with Tb(III) which can be excited near 300 nm and emit strongly at 500–600 nm. The detection limit is 10 to 100 nM with Tb fluorescence.

## Refractive Index Detector

Refraction is described in Section 20-4.

A **refractive index detector** responds to almost every solute, but its detection limit is about 1 000 times poorer than that of the ultraviolet detector. The deflection-type detector in Figure 25-21 has two triangular 5- to 10-μL compartments through which pure solvent or eluate passes. Collimated (parallel) visible light filtered to remove infrared radiation (which would heat the sample) passes through the cell with pure solvent in both compartments and is directed to a photodiode array by the deflection plate. When solute with a different refractive index enters the cell, the beam is deflected and different pixels of the array are irradiated.

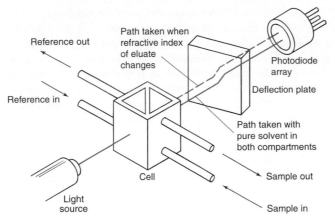

**Figure 25-21** Deflection-type refractive index detector.

Refractive index detectors are useless in gradient elution because it is impossible to match exactly the sample and the reference while the solvent composition is changing. Refractive index detectors are sensitive to changes in pressure and temperature (~0.01°C). Because of their low sensitivity, refractive index detectors are not useful for trace analysis. They also have a small linear range, spanning only a factor of 500 in solute concentration. The primary appeal of this detector is its nearly universal response to all solutes, including those that have little ultraviolet absorption.

## Evaporative Light-Scattering Detector

An **evaporative light-scattering detector** responds to any analyte that is significantly less volatile than the mobile phase.[21] In Figure 25-22, eluate enters the detector at the top. In the nebulizer, eluate is mixed with nitrogen gas and forced through a small-bore needle to form a

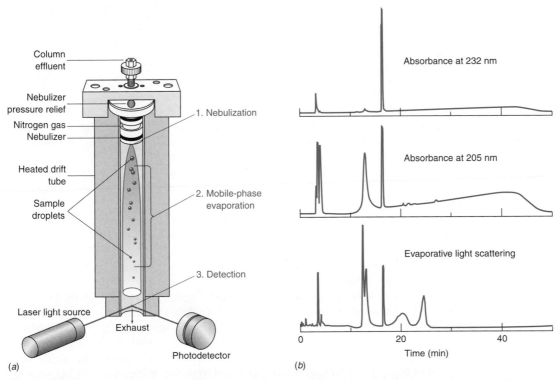

**Figure 25-22** (a) Operation of an evaporative light-scattering detector. *[Courtesy Alltech Associates, Deerfield, IL.]* (b) Comparison of ultraviolet absorbance and evaporative light-scattering detector response. Soluble components were extracted from a drug tablet and separated by reversed-phase liquid chromatography on a 4.6 × 150 mm column containing $C_{18}$-silica with a 30-nm pore size for separating polymers. Solvent A is 0.01 wt% trifluoroacetic acid in $H_2O$. Solvent B is 45 wt% propanol and 0.01 wt% trifluoroacetic acid in $H_2O$. A gradient was run from 10 vol% B up to 90 vol% B at 1 mL/min. Gradient time was not stated. *[From L. A. Doshier, J. Hepp, and K. Benedek, "Method Development Tools for the Analysis of Complex Pharmaceutical Samples," Am. Lab. December 2002, p. 18.]*

uniform dispersion of droplets. Solvent evaporates from the droplets in the heated drift tube, leaving a fine mist of solid particles to enter the detection zone at the bottom. The particles are detected by the light that they scatter from a diode laser to a photodiode.

The evaporative light-scattering detector response is related to the mass of analyte, not to the structure or molecular mass of the analyte. If you see a large peak and a small peak, you can be pretty sure that there is less material in the small peak than in the large peak. With an ultraviolet detector, a small mass of strongly absorbing analyte can give a stronger signal than a large mass of weakly absorbing analyte. The evaporative light-scattering detector response is fit to the equation.

$$\log A = a + b \log m \qquad (25\text{-}3)$$

where $A$ is the area of the signal (in a graph of scattered light intensity versus time), $m$ is the mass of analyte, and $a$ and $b$ are constants to fit the calibration curve.

The evaporative light-scattering detector is compatible with gradient elution. Also, there are no peaks associated with the solvent front, so there is no interference with early-eluting peaks. What do we mean by the solvent front? In Figure 25-13, you can see small positive and negative signals at 3–4 min. These signals arise from changes in the refractive index of the mobile phase due to solvent in which the sample was dissolved. This change displaces the ultraviolet detector signal at the time $t_m$ required for unretained mobile phase to pass through the column. If a peak is eluted close to the time $t_m$, it can be distorted by the solvent front peaks. An evaporative light-scattering detector has no solvent front peaks.

If you use a buffer in the eluent, it must be volatile or else it will evaporate down to solid particles that scatter light and obscure the analyte. Low-concentration buffers made from acetic, formic, or trifluoroacetic acid, ammonium acetate, diammonium phosphate, ammonia, or triethylamine are suitable. Buffers for evaporative light scattering are the same as for mass spectrometric detection.

Figure 25-22b compares evaporative light scattering with ultraviolet absorbance for the detection of soluble components of a drug tablet containing polymers and small-molecule active ingredients. Two components have absorbance at 232 nm. Four or five components are evident with 205-nm detection. The solvent gradient produces a sloping baseline at 205 nm. All components are observed by evaporative light scattering and there is no slope to the baseline. Some broad peaks arise from polymers with a distribution of molecular mass.

## Electrochemical Detector

An **electrochemical detector** responds to analytes that can be oxidized or reduced, such as phenols, aromatic amines, peroxides, mercaptans, ketones, aldehydes, conjugated nitriles, aromatic halogen compounds, and aromatic nitro compounds. The opening of Chapter 17 shows a detector in which eluate is oxidized or reduced at the working electrode. The potential is maintained at a selected value with respect to the Ag | AgCl reference electrode, and current is measured between the working electrode and the stainless steel auxiliary electrode. For oxidizable solutes, copper or glassy carbon working electrodes are common. For reducible solutes, a drop of mercury is a good working electrode. Current is proportional to solute concentration over six orders of magnitude. Aqueous or other polar solvents containing dissolved electrolytes are required, and they must be rigorously free of oxygen. Metal ions extracted from tubing can be masked by adding EDTA to the solvent. The detector is sensitive to flow rate and temperature changes.

*Pulsed electrochemical measurements* at Au or Pt working electrodes expand the classes of detectable compounds to include alcohols, carbohydrates, and sulfur compounds.[22] The electrode is held at +0.8 V (versus a saturated calomel electrode) for 120 ms to oxidatively desorb organic compounds from the electrode surface and to oxidize the metal surface. Then the electrode is brought to −0.6 V for 200 ms to reduce the oxide to pristine metal. The electrode is then brought to a constant working potential (typically in the range +0.4 to −0.4 V), at which analyte is oxidized or reduced. After a delay of 400 ms for the charging current (Figure 17-15) to decay to 0, current is integrated for the next 200 ms to measure analyte. The sequence of pulses just described is then repeated to measure successive data points as eluate emerges from the column. By this means, 10 ppb ethylene glycol ($HOCH_2CH_2OH$) in Figure 25-23 gives a signal-to-noise ratio of 3.

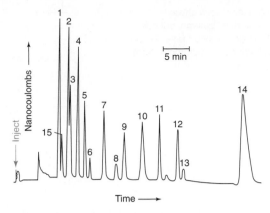

**Figure 25-23** Pulsed electrochemical detection of alcohols separated on Dionex AS-1 anion-exchange column with 0.05 M HClO₄. Peaks: 1, glycerol; 2, ethylene glycol; 3, propylene glycol; 4, methanol; 5, ethanol; 6, 2-propanol; 7, 1-propanol; 8, 2-butanol; 9, 2-methyl-1-propanol; 10, 1-butanol; 11, 3-methyl-1-butanol; 12, 1-pentanol; 13, cyclohexanol; 14, diethylene glycol. *[From D. C. Johnson and W. R. LaCourse, "Liquid Chromatography with Pulsed Electrochemical Detection at Gold and Platinum Electrodes," Anal. Chem. **1990**, 62, 589A.]*

## ■ ■ ■  25-3  Method Development for Reversed-Phase Separations[1]

Many separations encountered in industrial and research laboratories can be handled by reversed-phase chromatography. We now describe a general procedure for developing an isocratic separation of an unknown mixture with a reversed-phase column. The next section deals with gradient separations. In method development, the goals are to obtain adequate separation in a reasonable time. Ideally, the procedure should be *rugged,* which means that the separation will not be seriously degraded by gradual deterioration of the column, *small* variations in solvent composition, pH, and temperature, or use of a different batch of the same stationary phase, perhaps from a different manufacturer. If your column does not have temperature control, you can at least insulate it to reduce temperature fluctuations.

Reversed-phase chromatography is usually adequate to separate mixtures of low-molecular-mass neutral or charged organic compounds. If isomers do not separate well, normal-phase chromatography or porous graphitic carbon is recommended because solutes have stronger, more specific interactions with the stationary phase. For enantiomers, chiral stationary phases (Box 24-1) are required. Separations of inorganic ions, polymers, and biological macromolecules are described in Chapter 26.

As in gas chromatography (Section 24-5), the first steps in method development are to (1) determine the goal of the analysis, (2) select a method of sample preparation to ensure a "clean" sample, and (3) choose a detector that allows you to observe the desired analytes in the mixture. The remainder of method development described in the following sections assumes that steps 1 through 3 have been carried out.

> Desired attributes of a new chromatographic method:
> - adequate resolution of desired analytes
> - short run time
> - rugged (not drastically affected by small variations in conditions)

> Initial steps in method development
> 1. determine goal
> 2. select method of sample preparation
> 3. choose detector

### Criteria for an Adequate Separation

The *capacity factor* (Equation 23-16) is a measure of retention time, $t_r$, in units of the time $t_m$ required for mobile phase or an unretained solute to pass through the column. Reasonable separations demand that the capacity factors for all peaks be in the range 0.5–20. If the capacity factor is too small, the first peak is distorted by the solvent front. If the capacity factor is too great, the run takes too long. In the lowest trace in Figure 25-12, $t_m$ is the time when the first baseline disturbance is observed near 3 min. If you do not observe a baseline disturbance, you can estimate

> Capacity factor: $k' = \dfrac{t_r - t_m}{t_m}$
>
> $t_r$ = retention time of analyte
> $t_m$ = elution time for mobile phase or unretained solute

*Solvent front:*  $\qquad V_m \approx \dfrac{Ld_c^2}{2} \qquad$ or, equivalently,  $\qquad t_m \approx \dfrac{Ld_c^2}{2F} \qquad$ (25-4)

where $V_m$ is the volume at which unretained solute is eluted (= volume at which solvent front appears), $L$ is the length of the column (cm), $d_c$ is the column diameter (cm), and $F$ is

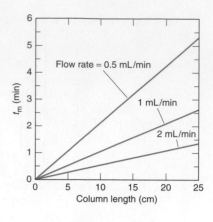

**Figure 25-24** Time ($t_m$) for mobile phase to traverse a 4.6-mm-diameter HPLC column, estimated with the equation $t_m \approx Ld_c^2/(2F)$, where $L$ is the length of the column (cm), $d_c$ is the column diameter (cm), and $F$ is the flow rate (mL/min).

---

the flow rate (mL/min) (Figure 25-24). In reversed-phase chromatography, $t_m$ could be measured by running the unretained solutes uracil (detected at 260 nm) or $NaNO_3$ (detected at 210 nm) through the column.

For quantitation, a minimum resolution (Figure 23-10) of 1.5 between the two closest peaks is desired so that neighboring peaks are separated. For ruggedness, a resolution of 2 is even better. That way, resolution is still adequate if small changes in conditions or slow deterioration of the column degrades the resolution.

Another criterion for a successful chromatographic method is not to exceed the upper operating pressure for your hardware. Keeping pressure below ~15 MPa (150 bar, 2 200 pounds/inch$^2$) prolongs the life of the pump, valves, seals, and autosampler. Pressure can double during the life of a column because of progressive clogging. Establishing an operating pressure of $\leq$15 MPa during method development allows for column degradation.

All peaks (certainly all peaks that need to be measured) should be symmetric, with an asymmetry factor $A/B$ in Figure 23-13 in the range 0.9–1.5. Asymmetric peak shapes should be corrected as described at the end of Section 25-1 before optimizing a separation.

## Optimization with One Organic Solvent

Combinations of acetonitrile, methanol, and tetrahydrofuran with water (or aqueous buffer) provide a sufficient range of dipolar and hydrogen-bonding interactions with solutes to separate a vast number of compounds in reversed-phase chromatography. The first solvent mixture to try is acetonitrile and water. Acetonitrile has low viscosity, which allows relatively low operating pressure, and it permits ultraviolet detection down to 190 nm (Table 25-2). At 190 nm, many analytes have some absorbance. Methanol is the second choice for organic solvent because it has a higher viscosity and longer wavelength ultraviolet cutoff. Tetrahydrofuran is the third choice because it has less usable ultraviolet range, it is slowly oxidized,[24] and it equilibrates more slowly with stationary phase. Starting conditions for reversed-phase HPLC are listed in Table 25-4.

**Attributes of a good separation**
- $0.5 \leq k' \leq 20$
- resolution $\geq 2$
- operating pressure $\leq$ 15 MPa
- $0.9 \leq$ asymmetry factor $\leq 1.5$

**Choice of organic solvent**

1. acetonitrile
2. methanol
3. tetrahydrofuran

To avoid disposing of acetonitrile as hazardous waste, you can hydrolyze it to sodium acetate and flush it down the drain.[23]

---

**Table 25-4** Starting conditions for reversed-phase chromatography[1]

| | |
|---|---|
| Stationary phase: | $C_{18}$ or $C_8$ on 5-$\mu$m-diameter spherical silica particles. Less acidic type B silica is preferred. For operation above 50°C, sterically protected silica (Figure 25-9) is preferred. |
| Column: | 0.46 × 15 cm column for 5-$\mu$m particles[a]<br>0.46 × 7.5 cm column for 3.5-$\mu$m particles (shorter run, same resolution) |
| Flow rate: | 2.0 mL/min |
| Mobile phase: | $CH_3CN/H_2O$ for neutral analytes<br>$CH_3CN$/aqueous buffer[b] for ionic analytes<br>5 vol% $CH_3CN$ in $H_2O$ to 100% $CH_3CN$ for gradient elution |
| Temperature: | 35°–40°C if temperature control is available |
| Sample size: | 25–50 $\mu$L containing ~25–50 $\mu$g of each analyte |

*a. A 0.30 × 15 cm column reduces solvent consumption to $(0.30/0.46)^2 = 43\%$ of the volume required for 0.46-cm diameter, reducing the flow to $(0.43)(2.0$ mL/min$) = 0.86$ mL/min.*

*b. Buffer is 25–50 mM phosphate/pH 2–3 made by treating $H_3PO_4$ with KOH. $K^+$ is more soluble than $Na^+$ in organic solvents and leads to less tailing. Add 0.2 g sodium azide per liter as a preservative if the buffer will not be used quickly.*

CHAPTER 25 High-Performance Liquid Chromatography

Figure 25-12 illustrates a succession of experiments to establish that 35 vol% $CH_3CN$ (designated B) plus 65 vol% buffer is a good solvent to separate the particular mixture of analytes. The initial experiment was done with a high concentration of $CH_3CN$ (90% B) to ensure the elution of all components of the unknown. Then %B was successively lowered to separate all the components. Eluent containing 40% B did not separate peaks 2 and 3 adequately, and 30% B took too long to elute peak 8. Therefore, 35% B was selected.

With 35% B, peak 1 is eluted at 4.9 min and peak 8 is eluted at 125.2 min. The solvent front appears at $t_m = 2.7$ min. Therefore, $k'$ for peak 1 is $(4.9 - 2.7)/2.7 = 0.8$ and $k'$ for peak 8 is $(125.2 - 2.7)/2.7 = 45$. For $k' > 20$, gradient elution (Section 25-4) is indicated. If all peaks in Figure 25-12 could be resolved while maintaining $0.5 \leq k' \leq 20$, then we would have a successful isocratic separation. If we were not concerned about measuring peaks 2 and 3, then 45% B would probably be a good choice.

## Optimization with Two or Three Organic Solvents

Figures 25-25 and 25-26 illustrate a systematic process to develop a separation with combinations of solvents. Method development is finished as soon as the separation meets all your criteria. There is a good chance of attaining adequate separation without going through all the steps.

Step 1   Optimize the separation with acetonitrile/buffer to generate chromatogram A in Figure 25-26.

Step 2   Optimize the separation with methanol/buffer to generate chromatogram B.

Step 3   Optimize the separation with tetrahydrofuran/buffer to generate chromatogram C.

Step 4   Mix the solvents used in A, B, and C, one pair at a time, in 1:1 proportion, to generate chromatograms D, E, and F.

Step 5   Construct a 1:1:1 mixture of the solvents for A, B, and C to generate chromatogram G.

Step 6   If some of the results A through G are almost good enough, select the two best points and mix the solvents to obtain points between those two.

Let's examine Figure 25-26 to see how the systematic procedure works. Step 1 is to generate chromatogram A by varying the proportions of acetonitrile and aqueous buffer (as in Figure 25-12) to obtain the best separation within the constraint that $0.5 \leq k' \leq 20$. At the best composition, 30 vol% acetonitrile/70% buffer, peaks 4 and 5 are not resolved adequately for quantitative analysis.

In HPLC, lowering the flow rate usually improves resolution. For Figure 25-26, we chose a flow rate of 1.0 mL/min, rather than the 2.0 mL/min recommended in Table 25-4. With the lower flow rate, we chose to keep $k' < 10$ (instead of <20) to maintain the run time below ~25 min. In chromatogram A, $k' = 1.1$ for peak 1 and $k' = 8.1$ for peak 7. If the flow rate had been 2.0 mL/min, then $k' < 20$ would give a run time below ~25 min.

Step 2 seeks a methanol/buffer solvent to obtain the best separation at point B. It is not necessary to start all over again with 90% methanol. Figure 25-27 (page 579) allows us to select a methanol/water mixture that has approximately the same eluent strength as a particular acetonitrile/water mixture. A vertical line drawn at 30% acetonitrile (the composition used in chromatogram A) intersects the methanol line near 40%. Therefore, 40% methanol has about the same eluent strength as 30% acetonitrile. The first experiment carried out to establish point B in Figure 25-26 utilized 40% methanol. A little trial-and-error (with 45% methanol and 35% methanol) demonstrated that 40% methanol gave the best separation, but the separation is poor. In chromatogram B in Figure 25-26, the seven components give only five peaks. When we changed from acetonitrile to methanol, the order of elution of some compounds changed.

Step 3 generates chromatogram C in Figure 25-26, using tetrahydrofuran. Figure 25-27 tells us that 22% tetrahydrofuran has the same eluent strength as 30% acetonitrile. When 22% tetrahydrofuran was tried, elution times were too long. Trial-and-error demonstrated that 32% tetrahydrofuran was best. All seven compounds are cleanly separated in chromatogram C in an acceptable time. However, a baseline dip associated with the solvent front between peaks 3 and 1 interferes with quantitative analysis of compound 1. The order of elution with tetrahydrofuran is quite different from the order with acetonitrile. In general, changing solvent is a powerful way to change relative retention of different compounds.

Step 4 generates chromatograms D, E, and F. The composition at D is a 1:1 mixture of the solvents used in A and B. Because we used 30% acetonitrile at A and 40% methanol at B, D was obtained with 15% acetonitrile/20% methanol/65% buffer. Similarly, E was obtained

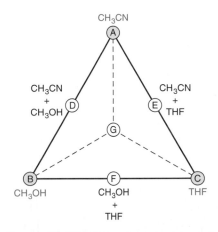

"Rule of Three": Decreasing % B by 10% increases capacity factor, $k'$, by a factor of ~3.

**Figure 25-25** HPLC method development triangle. THF stands for tetrahydrofuran. Figure 25-26 shows how the procedure is applied to a real chromatographic separation.

If the separation does not look promising after step 5, you will need to try a different stationary phase or a different form of chromatography.

The way we know which peak is which is to record the entire ultraviolet spectrum of each peak as it is eluted, using a photodiode array spectrometer.

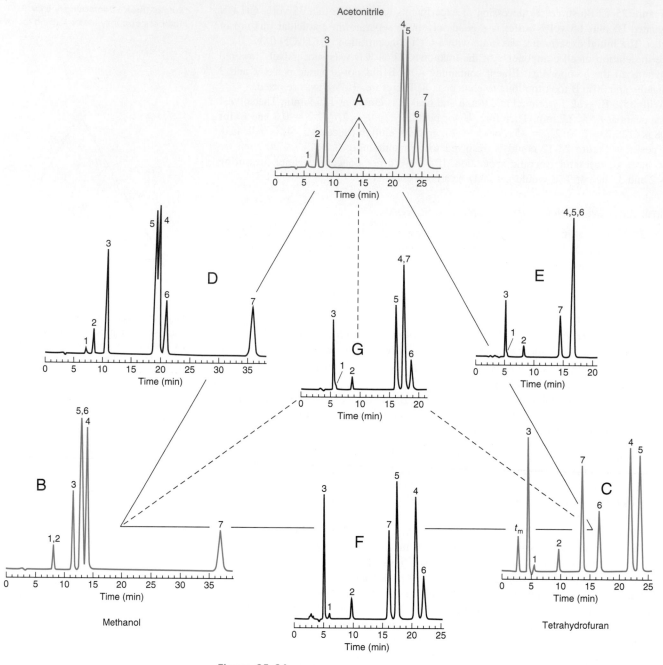

*Figure 25-26* Application of the method development triangle to the separation of seven aromatic compounds by HPLC. Column: $0.46 \times 25$ cm Hypersil ODS ($C_{18}$ on 5-$\mu$m silica) at ambient temperature ($\sim 22°C$). Elution rate was 1.0 mL/min with the following solvents: (A) 30 vol% acetonitrile/70 vol% buffer; (B) 40% methanol/60% buffer; (C) 32% tetrahydrofuran/68% buffer. The aqueous buffer contained 25 mM $KH_2PO_4$ plus 0.1 g/L $NaN_3$ adjusted to pH 3.5 with HCl. Points D, E, and F are midway between the vertices: (D) 15% acetonitrile/20% methanol/65% buffer; (E) 15% acetonitrile/16% tetrahydrofuran/69% buffer; (F) 20% methanol/16% tetrahydrofuran/64% buffer. Point G at the center of the triangle is an equal blend of A, B, and C with the composition 10% acetonitrile/13% methanol/11% tetrahydrofuran/66% buffer. The negative dip in C between peaks 3 and 1 is associated with the solvent front. Peak identities were tracked with a photodiode array ultraviolet spectrophotometer: (1) benzyl alcohol; (2) phenol; (3) 3′,4′-dimethoxyacetophenone; (4) *m*-dinitrobenzene; (5) *p*-dinitrobenzene; (6) *o*-dinitrobenzene; (7) benzoin.

with a 1:1 mixture of the solvents for A and C. F was obtained with a 1:1 mixture of the solvents for B and C.

Chromatogram D is not acceptable, because peaks 4 and 5 overlap. Chromatogram E is terrible because peaks 1 and 3 overlap and peaks 4, 5, and 6 overlap. But (hurrah!) chromatogram F is what we have been looking for: All peaks are separated and the first

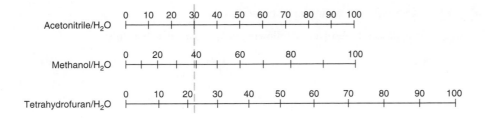

Figure 25-27 Nomograph showing volume percentage of solvents having the same eluent strength. A vertical line intersects each solvent line at the same eluent strength. For example, 30 vol% acetonitrile/70 vol% water has about the same eluent strength as 40 vol% methanol or 22 vol% tetrahydrofuran. [From L. R. Snyder, J. J. Kirkland, and J. L. Glajch, Practical HPLC Method Development (New York: Wiley, 1997).]

peak (3) is adequately removed from the solvent dip. We are finished! The solvent composition at F (20% methanol/16% tetrahydrofuran/64% buffer) does the job. Peaks 4 and 6 have the minimum resolution of 1.8. (We would have preferred resolution $>2.0$.) All peaks are symmetric and within the range $k' = 0.9 - 7.5$. The operating pressure remained reasonable.

If none of the trials had worked, step 5 would generate chromatogram G in Figure 25-26, using a 1:1:1 mixture of solvents from A, B, and C. For completeness, the result at G is shown. Peaks 1 and 3 overlap and peaks 4 and 7 overlap.

If some of the compositions A through G were almost good enough, compositions between the points might have been better. For example, if A and D were almost good enough, perhaps a mixture of the solvents for A and D might have been better.

## Temperature as a Variable

Column temperature affects the relative retention of different compounds and elevated temperature permits high-speed chromatography to be conducted.[25] Figure 25-28 suggests a systematic procedure for method development in which solvent composition and temperature are the two independent variables.[1] For elevated temperature operation, pH should be below 6 to retard dissolution of silica. Alternatively, zirconia-based stationary phases work up to at least 200°C.

## Do It with a Computer!

Method development is vastly simplified by computer simulations using commercial software. With input from a small number of real experiments, a program can predict the effects of solvent composition and temperature in isocratic or gradient separations. You can select optimum conditions in minutes with the computer instead of days in the lab. Of course, you must verify the prediction by a real experiment. Commercial software saves huge expenses in method development in industrial laboratories.

## Choosing a Stationary Phase

$C_{18}$-silica is the most common stationary phase and it separates a wide range of mixtures when the solvent is chosen carefully, as in Figure 25-26. However, it cannot achieve all separations. Table 25-5 is a guide to selecting other bonded phases.

$$\text{Resolution} = \frac{\Delta t_r}{w_{av}} = \frac{\Delta t_r}{1.70 w_{1/2}}$$

$\Delta t_r$ = separation between peaks
$w_{av}$ = average width at baseline
$w_{1/2}$ = average width at half-height

After optimizing the solvent, you might still need to improve resolution. To increase resolution, you can

• decrease flow rate
• increase column length
• decrease particle size

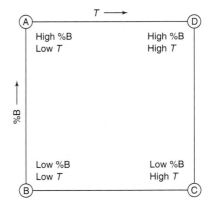

Figure 25-28 Isocratic method development for HPLC, using solvent composition, % B, and temperature, T, as independent variables. % B and T are each varied between selected low and high values. From the appearance of chromatograms resulting from conditions A–D, we can select intermediate conditions to improve the separation.

**Table 25-5** Selection of bonded stationary phases for HPLC

| Bonded group | Polarity | Retention mechanisms | Comments |
|---|---|---|---|
| $C_{18}, C_8, C_4$ | Nonpolar | van der Waals | $C_8$ does not retain hydrophobic compounds as strongly as $C_{18}$ |
| Phenyl | Nonpolar | Hydrophobic and pi-pi | |
| Cyano | Intermediate | Hydrophobic, dipole-dipole, and pi-pi | Resolves polar organic compounds by reversed-phase or normal-phase chromatography |
| Amino | Polar ($-NH_2$) or ionic ($-NH_3^+$) | Dipole-dipole and H-bonding | Normal-phase or ion-exchange separations; separates carbohydrates, polar organic compounds, and inorganic ions; reacts with aldehydes and ketones |
| Bare silica | Very polar | H-bonding | Normal-phase separations |

SOURCE: C. S. Young and R. J. Weigand, "An Efficient Approach to Column Selection in HPLC Method Development," LCGC **2002**, 20, 464.

**Example** Selecting a Bonded Phase

Suggest an approach to resolve the following compounds in a short time.

**1.** Rutin  **2.** Chatechin  **3.** Myricetin

**4.** Quercitin  **5.** Apigenin  **6.** Naringenin

**Solution** Compounds with the most similar structures are likely to be hardest to separate. Of the compounds in the mixture, **5** and **6** are most similar, differing by just one C=C bond. Table 25-5 suggests that a stationary phase with phenyl or cyano groups, which retain analytes by pi-pi interactions, might be good for distinguishing **5** and **6**. For rapid analysis but good resolution, select a short column with small particle size and run it at a rapid flow rate. Figure 25-29 shows that **5** and **6** are resolved by a phenyl column but not by a $C_{18}$ column.

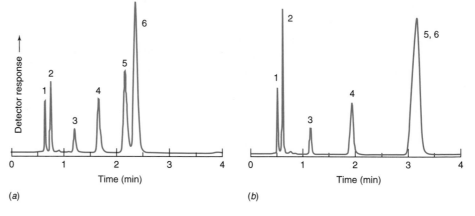

**Figure 25-29** Separation of six compounds on (*a*) phenyl- and (*b*) $C_{18}$-silica columns with 3-μm particle size using 35:65 (vol/vol) acetonitrile/0.2% aqueous trifluoroacetic acid. Column size: 7 × 53 mm; flow rate = 2.5 mL/min. [From C. S. Young and R. J. Weigand, "An Efficient Approach to Column Selection in HPLC Method Development," LCGC **2002**, 20, 464. Courtesy Alltech Associates.]

We have discussed several ways to change the relative retention of two closely spaced peaks. Some methods are harder to implement than others. One suggested order of steps to try—from easiest to hardest—is: (1) change the solvent strength by varying the fraction of each solvent; (2) change the temperature; (3) change the pH (in small steps); (4) use a different solvent; (5) use a different kind of stationary phase.[26]

## ■■■■ 25-4 Gradient Separations

Figure 25-12 shows an isocratic separation of eight compounds that required a run time of more than 2 h. When eluent strength was low enough to resolve early peaks (2 and 3), the elution of later peaks was very slow. To retain the desired resolution but decrease the analysis time, the *segmented gradient* (a gradient with several distinct parts) in Figure 25-13 was selected. Peaks 1–3 were separated with a low eluent strength (30% B). Between 8 and 13 min, B was increased linearly from 30 to 45% to elute the middle peaks. Between 28 and 30 min, B was increased linearly from 45 to 80% to elute the final peaks.

## Dwell Volume and Dwell Time

The volume between the point at which solvents are mixed and the beginning of the column is called the **dwell volume.** The *dwell time,* $t_D$, is the time required for the gradient to reach the column. Dwell volumes range from 0.5 to 10 mL in different systems. For Figure 25-13, the dwell volume is 5 mL, and the flow rate is 1.0 mL/min. Therefore, the dwell time is 5 min. A solvent change initiated at 8 min does not reach the column until 13 min.

Differences in dwell volumes between different systems are an important reason why conditions for gradient separations on one chromatograph do not necessarily transfer to another. It is helpful to state the dwell volume for your system when you report a gradient separation. One way to compensate for dwell volume is to inject sample at the time $t_D$ instead of at $t = 0$.

You can measure dwell volume by first disconnecting the column and connecting the inlet tube directly to the outlet tube. Place water in reservoirs A and B of the solvent delivery system. Add 0.1 vol% acetone to reservoir B. Program the gradient to go from 0 to 100% B in 20 min and begin the gradient at $t = 0$. With the detector set to 260 nm, the response will ideally look like that in Figure 25-30. The delay between the start of the gradient and the first response at the detector is the dwell time, $t_D$.

## Gradient Elution Is a Fine Way to Begin Method Development

The quickest way to survey a new mixture to decide whether to use isocratic or gradient elution is to run a broad gradient, as in Figure 25-31a.[27] This figure shows how the sample mixture in Figure 25-12 is separated by a linear gradient from 10 to 90% acetonitrile in 40 min.

$$t_D = \frac{\text{dwell volume (mL)}}{\text{flow rate (mL/min)}}$$

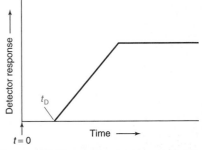

**Figure 25-30** Measurement of dwell time, using nonabsorbing solvent in reservoir A and a weak absorber in reservoir B. The gradient from 0 to 100% B is begun at time $t = 0$ but does not reach the detector until time $t_D$. The column is removed from the system for this measurement. Real response will be rounded instead of having the sharp intersections shown in this illustration.

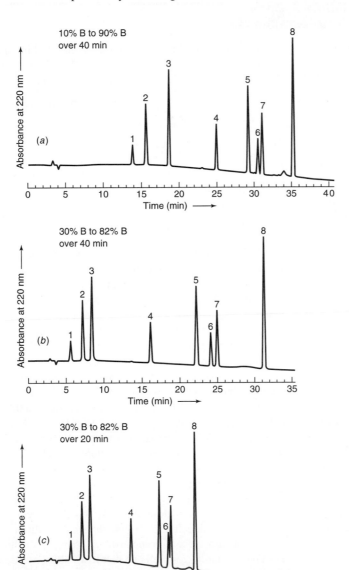

**Figure 25-31** Linear gradient separations of the same mixture used in Figure 25-12 in the same column and solvent system [buffer (solvent A) with acetonitrile (solvent B)] at a flow rate of 1.0 mL/min. The dwell time was 5 min.

- The first run on a new mixture should be a gradient.
- If $\Delta t/t_G > 0.25$, use gradient elution.
- If $\Delta t/t_G < 0.25$, use isocratic elution.
- Isocratic solvent should have composition applied to column halfway through the period $\Delta t$.

The *gradient time*, $t_G$, is the time over which the solvent composition is changed (40 min). Box 25-3 tells you how to select $t_G$. Let $\Delta t$ be the difference in the retention time between the first and the last peak in the chromatogram. In Figure 25-31a, $\Delta t = 35.5 - 14.0 = 21.5$ min. The criterion to choose whether to use a gradient is

$$\text{Use a gradient if } \Delta t/t_G > 0.25 \qquad \text{Use isocratic elution if } \Delta t/t_G < 0.25$$

If all peaks are eluted over a narrow solvent range, then isocratic elution is feasible. If a wide solvent range is required, then gradient elution is more practical. In Figure 25-31a, $\Delta t/t_G = 21.5/40 = 0.54 > 0.25$. Therefore, gradient elution is recommended. Isocratic elution is possible, but the time required in Figure 25-12 is impractically long.

If isocratic elution is indicated because $\Delta t/t_G < 0.25$, then a good starting solvent is the composition at the point halfway through the interval $\Delta t$. That is, if the first peak is eluted at 10 min and the last peak is eluted at 20 min, a reasonable isocratic solvent has the composition at 15 min in the gradient.

## Developing a Gradient Separation

The first run should survey a broad range of eluent strength such as 10 to 90% B in 40 min in Figure 25-31a. Because the dwell time was 5 min and the gradient began at $t = 0$, the gradient did not reach the column until $t = 5$ min. (It would have been better to inject the

---

### Box 25-3 Choosing Gradient Conditions and Scaling Gradients

We now provide equations that allow you to select sensible linear gradient conditions (as in Figure 25-32) and to scale gradients from one column to another.

The retention of a compound in isocratic elution depends on solvent composition:

$$\log k' \approx \log k'_w - S\Phi \qquad (25\text{-}5)$$

where $k'$ is the capacity factor, $k'_w$ would be the capacity factor for 100% aqueous eluent, $\Phi$ is the fraction of organic solvent ($\Phi = 0.4$ for 40 vol% $CH_3CN$/60 vol% $H_2O$), and $S$ is a constant for each compound, with a typical value of ~4 for small molecules. From two experiments with different values of $\Phi$, you can compute $k'_w$ and $S$. These values are input for software that predicts how chromatograms will appear for different conditions.

For gradient elution, an average capacity factor $k^*$ applies to all peaks:

$$k^* = \frac{t_G F}{\Delta\Phi V_m S} \qquad (25\text{-}6)$$

where $t_G$ is gradient time (min), $F$ is flow rate (mL/min), $\Delta\Phi$ is the change in solvent composition during the gradient, $V_m$ is the volume of mobile phase in the column (mL), and $S$ is measured with Equation 25-5. We take $S = 4$ for this discussion.

In isocratic elution, a capacity factor $k' \approx 5$ provides separation from the solvent front and does not require excessive time. For gradient elution, $k^* \approx 5$ is a reasonable starting condition. Let's calculate a sensible gradient time for the experiment in Figure 25-31a, in which we chose a gradient from 10% to 90% B ($\Delta\Phi = 0.8$) in a $0.46 \times 25$ cm column eluted at 1.0 mL/min. From Equation 25-4, $V_m \approx L d_c^2/2 = (25 \text{ cm})(0.46 \text{ cm})^2/2 = 2.6_5$ mL. We calculate the required gradient time by rearranging Equation 25-6:

$$t_G = \frac{k^*\Delta\Phi V_m S}{F} = \frac{(5)(0.8)(2.65 \text{ mL})(4)}{(1.0 \text{ mL/min})} = 42 \text{ min} \qquad (25\text{-}7)$$

A reasonable gradient time would be 42 min. In Figure 25-31a, $t_G$ is 40 min, giving $k^* = 4.7$. In Figure 25-31b, we changed the gradient to $\Delta\Phi = 0.52$, giving better separation:

$$k^* = \frac{t_G F}{\Delta\Phi V_m S} = \frac{(40 \text{ min})(1.0 \text{ mL/min})}{(0.52)(2.65 \text{ mL})(4)} = 7.3$$

The separation is poorer in Figure 25-31c, for which $k^* = 3.6$.

If you have a successful gradient separation and want to transfer it from column 1 to column 2, whose dimensions are different, the scaling relations are:

$$\frac{F_2}{F_1} = \frac{m_2}{m_1} = \frac{d_2}{d_1} = \frac{V_2}{V_1} \qquad (25\text{-}8)$$

where $F$ is volume flow rate (mL/min), $m$ is the mass of sample, $d$ is the delay time before the gradient reaches the column, and $V$ is total column volume. The gradient time, $t_G$, should not be changed. In Figure 25-32, the delay time $d = 5$ min is due to the dwell volume between the mixer and the column. Equation 25-8 tells us to change volume flow rate, sample mass, and dwell time in proportion to column volume. If dwell volume is small in comparison with the volume of solvent on the column, $V_m$, the delay time $d$ can be inconsequential. If dwell volume is large, it becomes an important factor over which you might have little control.

Suppose that you have optimized a gradient on a $0.46 \times 25$ cm column and you want to transfer it to a $0.21 \times 10$ cm column. The quotient $V_2/V_1$ is $(\pi r^2 L)_2/(\pi r^2 L)_1$, where $r$ is column radius and $L$ is column length. For these columns, $V_2/V_1 = 0.083$. Equation 25-8 tells us to decrease the volume flow rate, the sample mass, and the delay time to 0.083 times the values used for the large column. The gradient time should not be changed.

When you make these changes, you will discover that $k^*$ is the same for both columns. If you change a condition that affects $k^*$, you should make a compensating change to restore $k^*$. For example, Equation 25-6 tells us that if we choose to double $t_G$ we should cut the flow rate in half so that the product $t_G F$ is constant and $k^*$ remains constant.

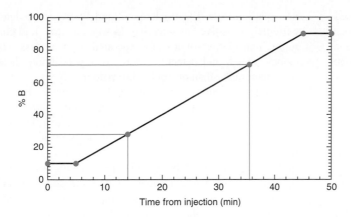

**Figure 25-32** Solvent gradient for Figure 25-31a. The gradient was begun at the time of injection ($t = 0$), but the dwell time was 5 min. Therefore, the solvent was 10% B for the first 5 min. Then the composition increased linearly to 90% B over 40 min. After $t = 45$ min, the composition was constant at 90% B.

sample at $t = 5$ min, but this was not done.) Serendipitously, the first run in Figure 25-31 separated all eight peaks. We could stop at this point if we were willing to settle for a 36-min run time.

The next step in developing a gradient method is to spread the peaks out by choosing a shallower gradient. For a dwell time of 5 min, the gradient profile for Figure 25-31a looks like Figure 25-32. Peak 1 was eluted at 14.0 min when the solvent was 28% B. Peak 8 was eluted near 35.5 min, when the solvent was 71% B. The portions of the gradient from 10 to 28% B and 71 to 90% B were not really needed. Therefore, the second run could be made with a gradient from 28 to 71% B over the same $t_G$ (40 min). Conditions chosen for the run in Figure 25-31b were 30 to 82% B in 40 min. This gradient spread the peaks out and reduced the run time slightly to 32 min.

In Figure 25-31c, we want to see whether a steeper gradient could be used to reduce the run time. The gradient limits were the same as in trace $b$, but $t_G$ was reduced to 20 min. Peaks 6 and 7 are not fully resolved with the shorter gradient time. Trace $b$ represents reasonable conditions for the gradient separation.

If the separation in Figure 25-31b were not acceptable, you could try to improve it by reducing the flow rate or going to a *segmented gradient,* as in Figure 25-13. The rationale for a segmented gradient is to use a good solvent composition for each region of the chromatogram and then to increase % B for the next region. It is easy to experiment with flow rate and gradient profiles. More difficult approaches to improve the separation are to change the solvent, use a longer column, use a smaller particle size, or change the stationary phase.

With all these tools, you can usually find a way to separate the components of a mixture if it does not contain too many compounds. If reversed-phase chromatography fails, normal-phase chromatography or one of the methods in Chapter 26 could be appropriate. Method development is part science, part art, and part luck.

Steps in gradient method development:

1. Run a wide gradient (e.g., 5 to 100% B) over 40–60 min. From this run, decide whether gradient or isocratic elution is best.
2. If gradient elution is chosen, eliminate portions of the gradient prior to the first peak and following the last peak. Use the same gradient time as in step 1.
3. If the separation in step 2 is acceptable, try reducing the gradient time to reduce the run time.

---

## Terms to Understand

| | | | |
|---|---|---|---|
| bonded stationary phase | evaporative light-scattering | hydrophobic substance | supercritical fluid |
| dead volume | detector | isocratic elution | superficially porous particle |
| derivatization | gradient elution | microporous particles | ultraviolet detector |
| dialysis | guard column | normal-phase chromatography | |
| dwell volume | high-performance liquid | refractive index detector | |
| electrochemical detector | chromatography (HPLC) | reversed-phase chromatography | |
| eluent strength | hydrophilic substance | separation factor | |

---

## Summary

In high-performance liquid chromatography (HPLC), solvent is pumped at high pressure through a column containing stationary phase particles with diameters of 1.5–5 $\mu$m. The smaller the particle size, the more efficient the column, but the greater the resistance to flow. Microporous silica particles with a covalently bonded liquid phase such as octadecyl groups ($-C_{18}H_{37}$) are most common. Eluent strength measures the ability of a given solvent to elute solutes from the column. In normal-phase chromatography, the stationary phase is polar and a less polar solvent is used. Eluent strength increases as the polarity of the solvent increases. Reversed-phase chromatography employs a nonpolar stationary phase and polar solvent. Eluent strength increases as the polarity of the solvent decreases. Most separations of organic compounds can be done on reversed-phase columns. Normal-phase chromatography or porous graphitic carbon are good for separating isomers. Chiral phases are used for optical isomers. Techniques for separating inorganic ions, polymers, and biological macromolecules are described in Chapter 26.

If a solution of organic solvent and water is used in reversed-phase chromatography, eluent strength increases as the percentage of organic solvent increases. If the solvent has a fixed composition, the process is called isocratic elution. In gradient elution, eluent strength is increased during chromatography by increasing the percentage of strong solvent.

A short guard column containing the same stationary phase as the analytical column is placed before the analytical column to protect it from contamination with particles or irreversibly adsorbed solutes. A high-quality pump provides smooth solvent flow. The injection valve allows rapid, precise sample introduction. The column is best housed in an oven to maintain a reproducible temperature. Column efficiency increases at elevated temperature because the rate of mass transfer between phases is increased. Mass spectrometric detection provides quantitative and qualitative information for each substance eluted from the column. Ultraviolet detection is most common and it can provide qualitative information if a photodiode array is used to record a full spectrum of each analyte. Refractive index detection has universal response but is not very sensitive. Evaporative light scattering responds to the mass of each

nonvolatile solute. Electrochemical and fluorescence detectors are extremely sensitive but selective. In supercritical fluid chromatography, nonvolatile solutes are separated by a process whose efficiency, speed, and detectors more closely resemble those of gas chromatography than of liquid chromatography.

Steps in method development: (1) determine the goal of the analysis, (2) select a method of sample preparation, (3) choose a detector, and (4) use a systematic procedure to select solvent for isocratic or gradient elution. Aqueous acetonitrile, methanol, and tetrahydrofuran are customary solvents for reversed-phase separations. A separation can be optimized by varying several solvents or by using one solvent and temperature as the principal variables. If further resolution is required, flow rate can be decreased and you can use a longer column with smaller particle size. Criteria for a successful separation are $0.5 \leq k' \leq 20$, resolution $\geq 2.0$, operating pressure $\leq 15$ MPa, and asymmetry factor in the range 0.9–1.5. In gradient elution, solvent composition does not begin to change until the dwell volume has passed from the point of solvent mixing to the start of the column. A wide gradient is a good first choice to determine whether to use isocratic or gradient elution.

## Exercises

**25-A.** A known mixture of compounds A and B gave the following HPLC results:

| Compound | Concentration (mg/mL in mixture) | Peak area (arbitrary units) |
|---|---|---|
| A | 1.03 | 10.86 |
| B | 1.16 | 4.37 |

A solution was prepared by mixing 12.49 mg of B plus 10.00 mL of unknown containing just A and diluting to 25.00 mL. Peak areas of 5.97 and 6.38 were observed for A and B, respectively. Find the concentration of A (mg/mL) in the unknown.

**25-B.** A bonded stationary phase for the separation of optical isomers has the structure

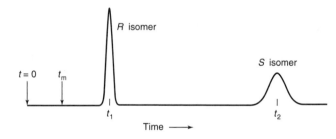

To resolve the enantiomers of amines, alcohols, or thiols, the compounds are first derivatized with a nitroaromatic group that increases their interaction with the bonded phase and makes them observable with a spectrophotometric detector.

When the mixture is eluted with 20 vol% 2-propanol in hexane, the $R$ enantiomer is eluted before the $S$ enantiomer, with the following chromatographic parameters:

$$\text{Resolution} = \frac{\Delta t_r}{w_{av}} = 7.7 \qquad \text{Relative retention } (\alpha) = 4.53$$

$$k' \text{ for } R \text{ isomer} = 1.35 \qquad t_m = 1.00 \text{ min}$$

where $w_{av}$ is the average width of the two Gaussian peaks at their base.

(a) Find $t_1$, $t_2$, and $w_{av}$, with units of minutes.

(b) The width of a peak at half-height is $w_{1/2}$ (Figure 23-9). If the plate number for each peak is the same, find $w_{1/2}$ for each peak.

(c) The area of a Gaussian peak is $1.064 \times$ peak height $\times w_{1/2}$, where $w_{1/2}$ is the width at half-height in Figure 23-9. Given that the areas under the two bands should be equal, find the relative peak heights (height$_R$/height$_S$).

**25-C.** (a) Make a graph showing the retention time of each peak in Figure 25-26 in chromatograms A, D, and B as a function of position along the line AB. Predict the retention times for solvent compositions midway between A and D and midway between D and B. Draw a stick diagram (representing each peak as a vertical line) of the two predicted chromatograms.

(b) What would be the solvent compositions midway between A and D and midway between D and B?

**25-D.** Two peaks emerge from a chromatography column as sketched in the following illustration.

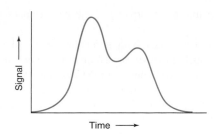

Time →

According to Equation 23-30, the resolution is given by

$$\text{Resolution} = \frac{\sqrt{N}}{4}(\gamma - 1)$$

where $N$ is the number of theoretical plates and $\gamma$ is the **separation factor,** defined as the quotient of linear velocities of the two solutes ($\gamma > 1$).

(a) If you change the solvent or the stationary phase, you will change the separation factor. Sketch the chromatogram if $\gamma$ increases but $N$ is constant.

(b) If you increase the column length or decrease the particle size or flow rate, you can increase the number of plates. Sketch the chromatogram if $N$ increases but $\gamma$ is constant.

## Problems

### High-Performance Liquid Chromatography

**25-1.** (a) Why does eluent strength increase as solvent becomes less polar in reversed-phase chromatography, whereas eluent strength increases as solvent becomes more polar in normal-phase chromatography?
(b) What kind of gradient is used in supercritical fluid chromatography?

**25-2.** Why are the relative eluent strengths of solvents in adsorption chromatography fairly independent of solute?

**25-3.** (a) Why is high pressure needed in HPLC?
(b) What is a bonded phase in liquid chromatography?

**25-4.** (a) Use Equation 25-1 to estimate the length of column required to achieve $1.0 \times 10^4$ plates if the stationary phase particle size is 10.0, 5.0, 3.0, or 1.5 μm.
(b) Why do smaller particles give better resolution?

**25-5.** If a 15-cm-long column has a plate height of 5.0 μm, what will be the half-width (in seconds) of a peak eluted at 10.0 min? If plate height = 25 μm, what will be $w_{1/2}$?

**25-6.** Why are silica stationary phases generally limited to operating in the pH range 2–8? Why does the silica in Figure 25-9 have improved stability at low pH?

**25-7.** How do additives such as triethylamine reduce tailing of certain solutes?

**25-8.** HPLC peaks should generally not have an asymmetry factor $A/B$ in Figure 23-13 outside the range 0.9–1.5.
(a) Sketch the shape of a peak with an asymmetry of 1.8.
(b) What might you do to correct the asymmetry?

**25-9.** (a) Sketch a graph of the van Deemter equation (plate height versus flow rate). What would the curve look like if the multiple path term were 0? If the longitudinal diffusion term were 0? If the finite equilibration time term were 0?
(b) Explain why the van Deemter curve for 3-μm particles in Figure 25-3 is nearly flat at high flow rate. What can you say about each of the terms in the van Deemter equation for 3-μm particles?
(c) Why is superficially porous silica used for separating large molecules, such as proteins?

**25-10.** (a) From $t_r$ and $w_{1/2}$ in Figure 25-8, find $N$ for each peak.
(b) From $t_r$ and $w_{1/2}$, find the resolution.
(c) Use Equation 23-30 with the average $N$ to find the resolution.

**25-11.** (a) According to Equation 25-2, if all conditions are constant, but particle size is reduced from 3 μm to 0.7 μm, by what factor must pressure be increased to maintain constant linear velocity?

(b) If all conditions except pressure are constant, by what factor will linear velocity increase if column pressure is increased by a factor of 10?
(c) With 0.7-μm particles in a 50-μm-diameter × 9-cm-long column, increasing pressure from 70 MPa to 700 MPa decreased analysis time by approximately a factor of 10 while increasing plate count from 12 000 to 45 000.[28] Explain why small particles permit 10-fold faster flow without losing efficiency or, in this case, with improved efficiency.

**25-12.** Using Figure 25-14, suggest which type of liquid chromatography you could use to separate compounds in each of the following categories:
(a) FM <2 000, soluble in octane
(b) FM <2 000, soluble in dioxane
(c) FM <2 000, ionic
(d) FM >2 000, soluble in water, nonionic, size 50 nm
(e) FM >2 000, soluble in water, ionic
(f) FM >2 000, soluble in tetrahydrofuran, size 50 nm

**25-13.** Explain the following statements:
(a) Normal-phase chromatography generally uses nonaqueous solvents, and eluent strength increases as the solvent becomes more polar.
(b) Reversed-phase chromatography generally uses aqueous organic solvents, and eluent strength increases as the fraction of organic solvent increases.
(c) Hydrophilic interaction chromatography uses aqueous organic solvents, and eluent strength increases as the fraction of water in the aqueous solvent increases.

**25-14.** Microporous silica particles with a density of 2.2 g/mL and a diameter of 10 μm have a measured surface area of 300 m²/g. Calculate the surface area of the spherical silica as if it were simply solid particles. What does this calculation tell you about the shape or porosity of the particles?

**25-15.** (a) Nonpolar aromatic compounds were separated by HPLC on an octadecyl ($C_{18}$) bonded phase. The eluent was 65 vol% methanol in water. How would the retention times be affected if 90% methanol were used instead?
(b) Octanoic acid and 1-aminooctane were passed through the same column described in part (a), using an eluent of 20% methanol/80% buffer (pH 3.0). State which compound is expected to be eluted first and why.

$$CH_3CH_2CH_2CH_2CH_2CH_2CH_2CO_2H$$
Octanoic acid

$$CH_3CH_2CH_2CH_2CH_2CH_2CH_2CH_2NH_2$$
1-Aminooctane

**25-16.** Suppose that an HPLC column produces Gaussian peaks. The detector measures absorbance at 254 nm. A sample containing equal moles of compounds A and B was injected into the column. Compound A ($\varepsilon_{254} = 2.26 \times 10^4\,M^{-1}\,cm^{-1}$) has a height $h = 128$ mm and a half-width $w_{1/2} = 10.1$ mm. Compound B ($\varepsilon_{254} = 1.68 \times 10^4\,M^{-1}\,cm^{-1}$) has $w_{1/2} = 7.6$ mm. What is the height of peak B in millimeters?

**25-17.** Capacity factors for three solutes separated on a $C_8$ nonpolar stationary phase are listed in the table. Eluent was a 70:30 (vol/vol) mixture of 50 mM citrate buffer (adjusted to pH with $NH_3$) plus methanol. Draw the dominant species of each compound at each pH in the table and explain the behavior of the capacity factors.

| Analyte | pH 3 | pH 5 | pH 7 |
|---|---|---|---|
| Acetophenone | 4.21 | 4.28 | 4.37 |
| Salicylic acid | 2.97 | 0.65 | 0.62 |
| Nicotine | 0.00 | 0.13 | 3.11 |

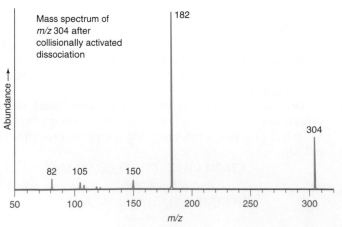

Acetophenone

Salicylic acid
$pK_a = 2.97$

Nicotine
$pK_1 = 3.15$
$pK_2 = 7.85$

**25-18.** Morphine and morphine 3-β-D-glucuronide were separated on two different 4.6-mm-diameter $\times$ 50-mm-long columns with 3-μm particles.[29] Column A was $C_{18}$-silica run at 1.4 mL/min and column B was bare silica run at 2.0 mL/min.

Morphine

Morphine 3-β-D-glucuronide

(a) Estimate the volume, $V_m$, and time, $t_m$, at which unretained solute would emerge from each column. The observed times are 0.65 min for column A and 0.50 min for column B.

(b) Column A was eluted with 2 vol% acetonitrile in water containing 10 mM ammonium formate at pH 3. Morphine 3-β-D-glucuronide

emerged at 1.5 min and morphine at 2.8 min. Explain the order of elution.

(c) Column B was eluted with a 5.0-min gradient beginning at 90 vol% acetonitrile in water and ending at 50 vol% acetonitrile in water. Both solvents contained 10 mM ammonium formate, pH 3. Morphine emerged at 1.3 min and morphine 3-β-D-glucuronide emerged at 2.7 min. Explain the order of elution. Why does the gradient go to decreasing acetonitrile?

(d) Find the capacity factor $k'$ for each solute on column A, using $t_m = 0.65$ min.

(e) From Box 25-3, estimate $k^*$ assuming $S = 4$ and with $t_m = 0.50$ min.

**25-19.** *Chromatography–mass spectrometry.* Cocaine metabolism in rats can be studied by injecting the drug and periodically withdrawing blood to measure levels of metabolites by HPLC–mass spectrometry. For quantitative analysis, isotopically labeled internal standards are mixed with the blood sample. Blood was analyzed by reversed-phase chromatography with an acidic eluent and atmospheric pressure chemical ionization mass spectrometry for detection. The mass spectrum of the collisionally activated dissociation products from the $m/z$ 304 positive ion is shown in the figure. Selected reaction monitoring ($m/z$ 304 from mass filter Q1 and $m/z$ 182 from Q3 in Figure 22-21) gave a single chromatographic peak at 9.22 min for cocaine. The internal standard $^2H_5$-cocaine gave a single peak at 9.19 min for $m/z$ 309 (Q1) $\rightarrow$ 182 (Q3).

(a) Draw the structure of the ion at $m/z$ 304.

(b) Suggest a structure for the ion at $m/z$ 182.

(c) The intense peaks at $m/z$ 182 and 304 do not have $^{13}C$ isotopic partners at $m/z$ 183 and 305. Explain why.

(d) Rat plasma is exceedingly complex. Why does the chromatogram show just one clean peak?

(e) Given that $^2H_5$-cocaine has only two major mass spectral peaks at $m/z$ 309 and 182, which atoms are labeled with deuterium?

(f) Explain how you would use $^2H_5$-cocaine for measuring cocaine in blood.

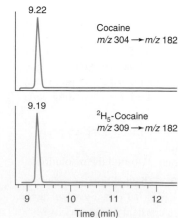

Cocaine

Mass spectrum of $m/z$ 304 after collisionally activated dissociation

Figure for Problem 25-19

*Left:* Mass spectrum of collisionally activated dissociation products from $m/z$ 304 positive ion from atmospheric pressure chemical ionization mass spectrum of cocaine. *Right:* Chromatogram obtained by selected reaction monitoring. [From G. Singh, V. Arora, P. T. Fenn, B. Mets, and I. A. Blair, "Isotope Dilution Liquid Chromatography Tandem Mass Spectrometry Assay for Trace Analysis of Cocaine and Its Metabolites in Plasma," *Anal. Chem.* **1999,** *71,* 2021.]

CHAPTER 25 High-Performance Liquid Chromatography

**25-20.** *Chromatography–mass spectrometry.* Figure 25-8 shows the separation of enantiomers of the drug Ritalin.

Ritalin (methylphenidate)
$C_{14}H_{19}O_2N$
Nominal mass = 233

**(a)** Detection in Figure 25-8 is by atmospheric pressure chemical ionization with selected reaction monitoring of the $m/z$ $234 \rightarrow 84$ transition. Explain how this detection works and propose structures for $m/z$ 234 and $m/z$ 84.

**(b)** For quantitative analysis, the internal standard $^2H_3$-Ritalin with a deuterated methyl group was added. The deuterated enantiomers have the same retention times as the unlabeled enantiomers. Which selected reaction monitoring transition should be monitored to produce a chromatogram of the internal standard in which unlabeled Ritalin will be invisible?

### Method Development

**25-21. (a)** Explain how to measure $k'$ and resolution.
**(b)** State three methods for measuring $t_m$ in reversed-phase chromatography.
**(c)** Estimate $t_m$ for a 0.46 × 15 cm column containing 5-μm particles operating at a flow rate of 1.5 mL/min. Estimate $t_m$ if the particle size were 3.5 μm instead.

**25-22.** What is the difference between dead volume and dwell volume? How do each of these volumes affect a chromatogram?

**25-23.** What does it mean for a separation procedure to be "rugged" and why is it desirable?

**25-24.** What are criteria for an adequate isocratic chromatographic separation?

**25-25.** Explain how to use a gradient for the first run to decide whether isocratic or gradient elution would be more appropriate.

**25-26.** What are the general steps in developing an isocratic separation for reversed-phase chromatography with one organic solvent?

**25-27.** What are the general steps in developing an isocratic separation for reversed-phase chromatography with two organic solvents?

**25-28.** What are the general steps in developing an isocratic separation for reversed-phase chromatography with one organic solvent and temperature as variables?

**25-29.** The "rule of three" states that the capacity factor for a given solute increases *approximately* threefold when the organic phase increases by 10%. In Figure 25-12, $t_m = 2.7$ min. Find $k'$ for peak 5 at 50% B. Predict the retention time for peak 5 at 40% B and compare the observed and predicted times.

**25-30.** Make a graph showing the retention times of peaks 6, 7, and 8 in Figure 25-12 as a function of % acetonitrile (% B). Predict the retention time of peak 8 at 45% B.

**25-31. (a)** Make a graph showing the retention time of each peak in Figure 25-26 in chromatograms B, F, and C as a function of position along the line BC. Predict the retention times for solvent compositions midway between B and F and midway between F and C. Draw a stick diagram (representing each peak as a vertical line) of each of the two predicted chromatograms.
**(b)** What would be the solvent compositions midway between B and F and midway between F and C?

**25-32.** Suppose that in Figure 25-26 the optimum concentrations of solvents at points A, B, and C are 50% acetonitrile, 60% methanol, and 40% tetrahydrofuran, respectively. What will be the solvent compositions at points D, E, F, and G?

**25-33.** A reversed-phase separation of a reaction mixture calls for isocratic elution with 48% methanol/52% water. If you want to change the procedure to use acetonitrile/water, what is a good starting percentage of acetonitrile to try?

**25-34. (a)** When you try separating an unknown mixture by reversed-phase chromatography with 50% acetonitrile/50% water, the peaks are too close together and are eluted in the range $k' = 2–6$. Should you use a higher or lower concentration of acetonitrile in the next run?
**(b)** When you try separating an unknown mixture by normal-phase chromatography with pure 50% hexane/50% methyl *t*-butyl ether, the peaks are too close together and are eluted in the range $k' = 2–6$. Should you use a higher or lower concentration of hexane in the next run?

**25-35.** A mixture of 14 compounds was subjected to a reversed-phase gradient separation going from 5% to 100% acetonitrile with a gradient time of 60 min. The sample was injected at $t =$ dwell time. All peaks were eluted between 22 and 41 min.
**(a)** Is the mixture more suitable for isocratic or gradient elution?
**(b)** If the next run is a gradient, select the starting and ending % acetonitrile and the gradient time.

**25-36. (a)** List ways in which the resolution between two closely spaced peaks might be changed.
**(b)** After optimization of an isocratic elution with several solvents, the resolution of two peaks is 1.2. How might you improve the resolution without changing solvents or the kind of stationary phase?

**25-37. (a)** You wish to use a wide gradient from 5 vol% to 95 vol% B for the first separation of a mixture of small molecules to decide whether to use gradient or isocratic elution. What should be the gradient time, $t_G$, for a 0.46 × 15 cm column containing 3-μm particles with a flow of 1.0 mL/min?
**(b)** You optimized the gradient separation going from 20 vol% to 34 vol% B in 11.5 min at 1.0 mL/min. Find $k^*$ for this optimized separation. To scale up to a 1.0 × 15 cm column, what should be the gradient time and the volume flow rate? If the sample load on the small column was 1 mg, what sample load can be applied to the large column? Verify that $k^*$ is unchanged.

# 26 | Chromatographic Methods and Capillary Electrophoresis

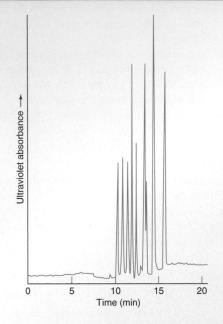

The inside of a monolithic electrochromatography column contains silicate "fingers" coated with stationary phase. Aromatic compounds were separated with a mean plate number of 80 000 in a 50-cm column with 15-kV applied voltage. *[From J. D. Hayes and A. Malik, "Sol-Gel Monolithic Columns with Reversed Electroosmotic Flow for Capillary Electrochromatography," Anal. Chem. **2000**, 72, 4090. Photo courtesy A. Malik, University of South Florida.]*

*Electrochromatography*[1] uses an electric field, rather than pressure, to propel mobile phase through a capillary column with uniform, pluglike flow (Color Plate 28). The capillary in the photograph contains a monolithic silicate structure similar to that in Box 25-1. Polymerization of soluble precursors inside the capillary forms chemical structures such as that shown below. The surface is coated with covalently anchored, positively charged quaternary ammonium groups. Mobile anions in solution provide charge balance. A strong electric field draws anions toward the anode, pulling the entire solution in the capillary along with them.

$C_{18}$ groups attached to the quaternary ammonium cations are the chromatographic stationary phase. Solutes passing through the capillary are separated when they partition between the mobile solvent and the stationary $C_{18}$ phase.

This chapter continues our discussion of chromatographic methods and introduces capillary electrophoresis. In electrophoresis and electrochromatography, an electric field drives liquid through a capillary tube by *electroosmosis*. This process enables us to create miniature analytical chips in which fluids are driven through capillary channels etched into glass or plastic. Chemical reactions and chemical separations are carried out on these chips. In the future, people will carry a "lab on a chip" into the field for investigations that require large instruments today.

## ■ ■ ■ 26-1  Ion-Exchange Chromatography

In **ion-exchange chromatography,** retention is based on the attraction between solute ions and charged sites bound to the stationary phase (Figure 23-6). In **anion exchangers,** positively charged groups on the stationary phase attract solute <u>anions</u>. **Cation exchangers** contain covalently bound, negatively charged sites that attract solute <u>cations</u>.

*Anion* exchangers contain bound *positive* groups.
*Cation* exchangers contain bound *negative* groups.

### Ion Exchangers

**Resins** are amorphous (noncrystalline) particles of organic material. *Polystyrene resins* for ion exchange are made by the copolymerization of styrene and divinylbenzene (Figure 26-1). Divinylbenzene content is varied from 1 to 16% to increase the extent of **cross-linking** of the insoluble hydrocarbon polymer. The benzene rings can be modified to produce a cation-exchange resin, containing sulfonate groups ($-SO_3^-$), or an anion-exchange resin, containing ammonium groups ($-NR_3^+$). If methacrylic acid is used in place of styrene, a polymer with carboxyl groups results.

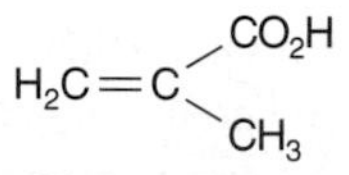

Table 26-1 classifies ion exchangers as strongly or weakly acidic or basic. Sulfonate groups ($-SO_3^-$) of strongly acidic resins remain ionized even in strongly acidic solutions. Carboxyl groups ($-CO_2^-$) of the weakly acidic resins are protonated near pH 4 and lose their cation-exchange capacity. "Strongly basic" quaternary ammonium groups ($-CH_2NR_3^+$) (which are not really basic at all) remain cationic at all values of pH. Weakly basic tertiary

Methacrylic acid

Strongly acidic cation exchangers: $RSO_3^-$
Weakly acidic cation exchangers: $RCO_2^-$
"Strongly basic" anion exchangers: $RNR_3'^+$
Weakly basic anion exchangers: $RNR_2'H^+$

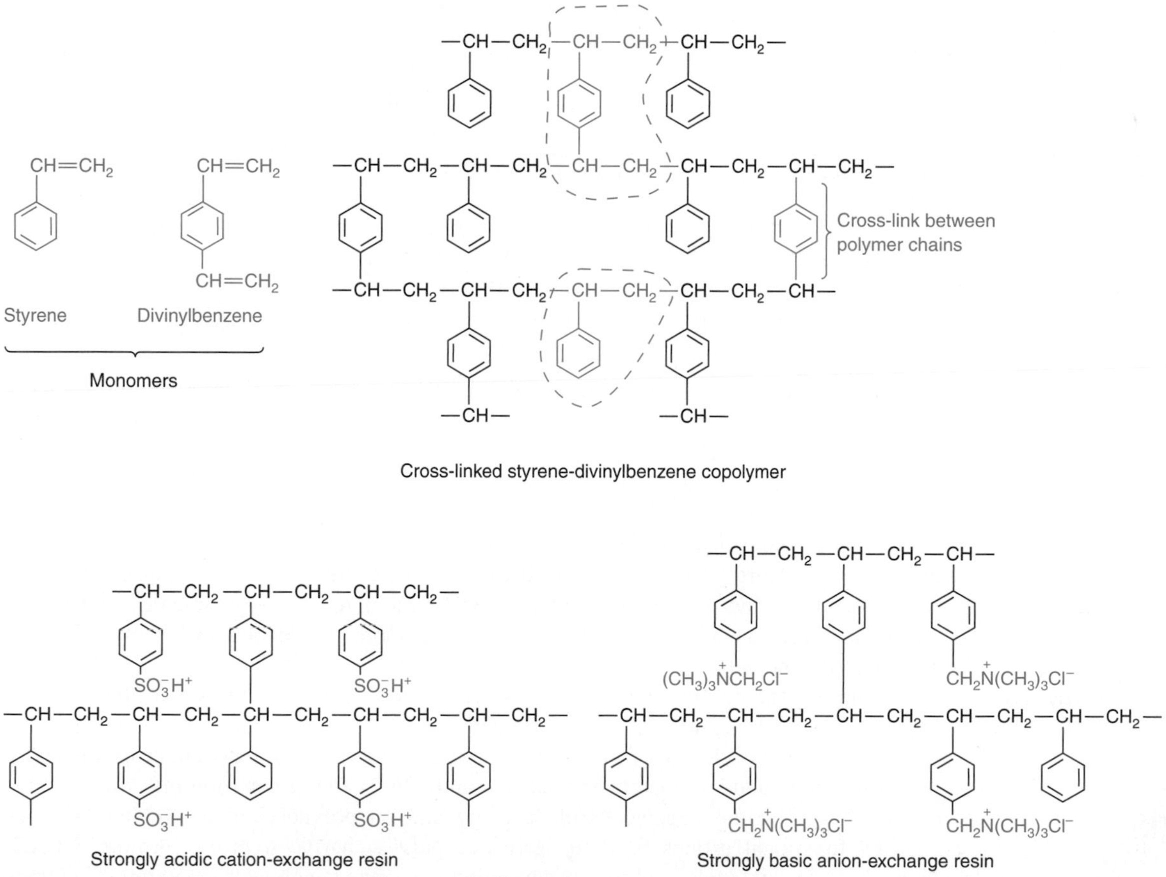

Cross-linked styrene-divinylbenzene copolymer

Styrene    Divinylbenzene
Monomers

Cross-link between polymer chains

Strongly acidic cation-exchange resin

Strongly basic anion-exchange resin

*Figure 26-1*  Structures of styrene-divinylbenzene cross-linked ion-exchange resins.

**Table 26-1**   Ion-exchange resins

| Resin type | Chemical constitution | Usual form as purchased | Common trade names | | Selectivity | Thermal stability |
|---|---|---|---|---|---|---|
| | | | Rohm & Haas | Dow Chemical | | |
| Strongly acidic cation exchanger | Sulfonic acid groups attached to styrene and divinylbenzene copolymer | Aryl—$SO_3^-H^+$ | Amberlite IR-120 | Dowex 50W | $Ag^+ > Rb^+ > Cs^+ >$ $K^+ > NH_4^+ > Na^+ >$ $H^+ > Li^+$ $Zn^{2+} > Cu^{2+} > Ni^{2+}$ $> Co^{2+}$ | Good up to 150°C |
| Weakly acidic cation exchanger | Carboxylic acid groups attached to acrylic and divinylbenzene copolymer | R—$COO^-Na^+$ | Amberlite IRC-50 | — | $H^+ \gg Ag^+ > K^+ >$ $Na^+ > Li^+$ $H^+ \gg Fe^{2+} > Ba^{2+}$ $Sr^{2+} > Ca^{2+} > Mg^{2+}$ | Good up to 100°C |
| Strongly basic anion exchanger | Quaternary ammonium groups attached to styrene and divinylbenzene copolymer | Aryl—$CH_2N(CH_3)_3^+Cl^-$ | Amberlite IRA-400 | Dowex 1 | $I^- > phenolate^- >$ $HSO_4^- > ClO_3^- >$ $NO_3^- > Br^- > CN^-$ $> HSO_3^- > NO_2^- >$ $Cl^- > HCO_3^- > IO_3^-$ $> HCOO^- >$ $acetate^- > OH^- > F^-$ | $OH^-$ form fair up to 50°C $Cl^-$ and other forms good up to 150°C |
| Weakly basic anion exchanger | Polyalkylamine groups attached to styrene and divinylbenzene copolymer | Aryl—$NH(R)_2^+Cl^-$ | Amberlite IR-45 | Dowex 3 | $Aryl—SO_3H > citric >$ $CrO_3 > H_2SO_4 >$ $tartaric > oxalic >$ $H_3PO_4 > H_3AsO_4 >$ $HNO_3 > HI > HBr$ $> HCl > HF >$ $HCO_2H >$ $CH_3CO_2H > H_2CO_3$ | Extensive information not available; tentatively limited to 65°C |

SOURCE: Adapted from J. X. Khym, Analytical Ion-Exchange Procedures in Chemistry and Biology (Englewood Cliffs, NJ: Prentice Hall, 1974).

ammonium ($—CH_2NHR_2^+$) anion exchangers are deprotonated in moderately basic solution and lose their ability to bind anions.

The extent of cross-linking is indicated by the notation "-X$N$" after the name of the resin. For example, Dowex 1-X4 contains 4% divinylbenzene, and Bio-Rad AG 50W-X12 contains 12% divinylbenzene. The resin becomes more rigid and less porous as cross-linking increases. Lightly cross-linked resins permit rapid equilibration of solute between the inside and the outside of the particle. However, resins with little cross-linking swell in water. Hydration decreases both the density of ion-exchange sites and the selectivity of the resin for different ions. More heavily cross-linked resins exhibit less swelling and higher exchange capacity and selectivity, but they have longer equilibration times. The charge density of polystyrene ion exchangers is so great that highly charged macromolecules such as proteins may be irreversibly bound.

Cellulose and dextran ion exchangers, which are polymers of the sugar glucose, possess larger pore sizes and lower charge densities than those of polystyrene resins. They are well suited to ion exchange of macromolecules, such as proteins. Dextran, cross-linked by glycerin, is sold under the name Sephadex (Figure 26-2). Other macroporous ion exchangers are based on the polysaccharide agarose and on polyacrylamide. Table 26-2 lists charged functional groups used to derivatize polysaccharide hydroxyl groups. DEAE-Sephadex, for example, refers to an anion-exchange Sephadex containing diethylaminoethyl groups.

Because they are much softer than polystyrene resins, dextran and its relatives are called **gels**.

Figure 26-12 shows the structure of polyacrylamide.

CHAPTER 26 Chromatographic Methods and Capillary Electrophoresis

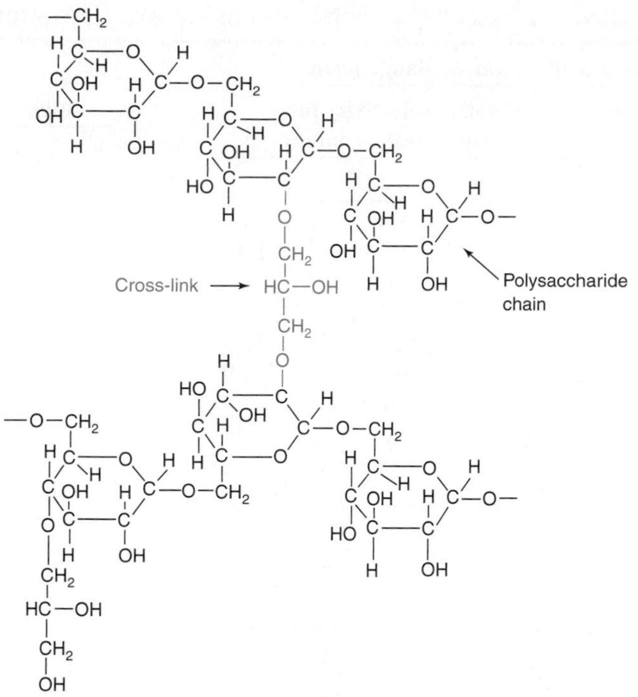

**Figure 26-2** Structure of Sephadex, a cross-linked dextran sold by Pharmacia Fine Chemicals, Piscataway, NJ.

**Table 26-2**  Common active groups of ion-exchange gels

| Type | Abbreviation | Name | Structure |
|---|---|---|---|
| *Cation Exchangers* | | | |
| Strong acid | SP | Sulfopropyl | $-OCH_2CH_2CH_2SO_3H$ |
| | SE | Sulfoethyl | $-OCH_2CH_2SO_3H$ |
| Intermediate acid | P | Phosphate | $-OPO_3H_2$ |
| Weak acid | CM | Carboxymethyl | $-OCH_2CO_2H$ |
| *Anion Exchangers* | | | |
| Strong base | TEAE | Triethylaminoethyl | $-OCH_2CH_2\overset{+}{N}(CH_2CH_3)_3$ |
| | QAE | Diethyl(2-hydroxypropyl) quaternary amino | $-OCH_2CH_2\overset{+}{N}(CH_2CH_3)_2$ $\|$ $CH_2CHOHCH_3$ |
| Intermediate base | DEAE | Diethylaminoethyl | $-OCH_2CH_2N(CH_2CH_3)_2$ |
| | ECTEOLA | Triethanolamine coupled to cellulose through glyceryl chains | |
| | BD | Benzoylated DEAE groups | |
| Weak base | PAB | *p*-Aminobenzyl | $-O-CH_2-\bigcirc-NH_2$ |

## Ion-Exchange Selectivity

Consider the competition of $Na^+$ and $Li^+$ for sites on the cation-exchange resin $R^-$:

*Selectivity coefficient:*

$$R^-Na^+ + Li^+ \rightleftharpoons R^-Li^+ + Na^+ \qquad K = \frac{[R^-Li^+][Na^+]}{[R^-Na^+][Li^+]} \qquad (26\text{-}1)$$

The equilibrium constant is called the **selectivity coefficient,** because it describes the relative selectivity of the resin for $Li^+$ and $Na^+$. Selectivities of polystyrene resins in Table 26-3 tend to increase with the extent of cross-linking, because the pore size of the resin shrinks as cross-linking increases. Ions such as $Li^+$, with a large hydrated radius (Chapter 8 opener), do not have as much access to the resin as smaller ions, such as $Cs^+$, do.

**Table 26-3**  Relative selectivity coefficients of ion-exchange resins

| | Sulfonic acid cation-exchange resin | | | Quarternary ammonium anion-exchange resin | |
| | Relative selectivity for divinylbenzene content | | | | |
| Cation | 4% | 8% | 10% | Anion | Relative selectivity |
|---|---|---|---|---|---|
| $Li^+$ | 1.00 | 1.00 | 1.00 | $F^-$ | 0.09 |
| $H^+$ | 1.30 | 1.26 | 1.45 | $OH^-$ | 0.09 |
| $Na^+$ | 1.49 | 1.88 | 2.23 | $Cl^-$ | 1.0 |
| $NH_4^+$ | 1.75 | 2.22 | 3.07 | $Br^-$ | 2.8 |
| $K^+$ | 2.09 | 2.63 | 4.15 | $NO_3^-$ | 3.8 |
| $Rb^+$ | 2.22 | 2.89 | 4.19 | $I^-$ | 8.7 |
| $Cs^+$ | 2.37 | 2.91 | 4.15 | $ClO_4^-$ | 10.0 |
| $Ag^+$ | 4.00 | 7.36 | 19.4 | | |
| $Tl^+$ | 5.20 | 9.66 | 22.2 | | |

SOURCE: *Amberlite Ion Exchange Resins—Laboratory Guide (Rohm & Haas Co., 1979).*

In general, ion exchangers favor the binding of ions of higher charge, decreased hydrated radius, and increased *polarizability*. A fairly general order of selectivity for cations is

$$Pu^{4+} \gg La^{3+} > Ce^{3+} > Pr^{3+} > Eu^{3+} > Y^{3+} > Sc^{3+} > Al^{3+} \gg$$
$$Ba^{2+} > Pb^{2+} > Sr^{2+} > Ca^{2+} > Ni^{2+} > Cd^{2+} > Cu^{2+} >$$
$$Co^{2+} > Zn^{2+} > Mg^{2+} > UO_2^{2+} \gg Ti^+ > Ag^+ > Rb^+ > K^+ >$$
$$NH_4^+ > Na^+ > H^+ > Li^+$$

Reaction 26-1 can be driven in either direction, even though $Na^+$ is bound more tightly than $Li^+$. Washing a column containing $Na^+$ with a substantial excess of $Li^+$ will replace $Na^+$ with $Li^+$. Washing a column in the $Li^+$ form with $Na^+$ will convert it into the $Na^+$ form.

Ion exchangers loaded with one kind of ion bind small amounts of a different ion nearly quantitatively. $Na^+$-loaded resin will bind small amounts of $Li^+$ nearly quantitatively, even though the selectivity is greater for $Na^+$. The same column binds large quantities of $Ni^{2+}$ or $Fe^{3+}$, because the resin has greater selectivity for these ions than for $Na^+$. Even though $Fe^{3+}$ is bound more tightly than $H^+$, $Fe^{3+}$ can be quantitatively removed from the resin by washing with excess acid.

## Donnan Equilibrium

When an ion exchanger is placed in an electrolyte solution, *the concentration of electrolyte is higher outside the resin than inside it.* The equilibrium between ions in solution and ions inside the resin is called the **Donnan equilibrium.**

Consider a quaternary ammonium anion-exchange resin ($R^+$) in its $Cl^-$ form immersed in a solution of KCl. Let the concentration of an ion inside the resin be $[X]_i$ and the concentration outside the resin be $[X]_o$. It can be shown from thermodynamics that the ion product inside the resin is approximately equal to the product outside the resin:

$$[K^+]_i[Cl^-]_i = [K^+]_o[Cl^-]_o \tag{26-2}$$

From considerations of charge balance, we know that

$$[K^+]_o = [Cl^-]_o \tag{26-3}$$

Inside the resin, there are three charged species, and the charge balance is

$$[R^+]_i + [K^+]_i = [Cl^-]_i \tag{26-4}$$

where $[R^+]$ is the concentration of quaternary ammonium ions attached to the resin. Substituting Equations 26-3 and 26-4 into Equation 26-2 gives

$$[K^+]_i([K^+]_i + [R^+]_i) = [K^+]_o^2 \tag{26-5}$$

which says that $[K^+]_o$ must be greater than $[K^+]_i$.

Suppose that the concentration of cationic sites in the resin is 6.0 M. When the Cl⁻ form of this resin is immersed in 0.050 M KCl, what will be the ratio $[K^+]_o/[K^+]_i$?

**Solution**   Let us assume that $[K^+]_o$ remains 0.050 M. Equation 26-5 gives

$$[K^+]_i([K^+]_i + 6.0) = (0.050)^2 \Rightarrow [K^+]_i = 0.000\ 42\ M$$

The concentration of $K^+$ inside the resin is less than 1% of that outside the resin.

Ions with the *same* charge as the resin are excluded. (The quaternary ammonium resin excludes $K^+$.) The counterion, $Cl^-$, is *not excluded* from the resin. There is no electrostatic barrier to penetration of an anion into the resin. Anion exchange takes place freely in the quaternary ammonium resin even though cations are repelled from the resin.

> The high concentration of positive charges within the resin repels cations from the resin.

The Donnan equilibrium is the basis of *ion-exclusion chromatography*. Because dilute electrolytes are excluded from the resin, they pass through a column faster than nonelectrolytes, such as sugar, which freely penetrates the resin. When a solution of NaCl and sugar is applied to an ion-exchange column, NaCl emerges from the column *before* the sugar.

## Conducting Ion-Exchange Chromatography

Ion-exchange *resins* are used for applications involving small molecules (FM ≲ 500), which can penetrate the small pores of the resin. A mesh size (Table 28-2) of 100/200 is suitable for most work. Higher mesh numbers (smaller particle size) lead to finer separations but slower column operation. For preparative separations, the sample may occupy 10 to 20% of the column volume. Ion-exchange *gels* are used for large molecules (such as proteins and nucleic acids), which cannot penetrate the pores of resins. Separations involving harsh chemical conditions (high temperature, high radiation levels, strongly basic solution, or powerful oxidizing agents) employ *inorganic ion exchangers,* such as hydrous oxides of Zr, Ti, Sn, and W.

> Three classes of ion exchangers:
> 1. resins
> 2. gels
> 3. inorganic exchangers

**Gradient elution** with increasing ionic strength or changing pH is extremely valuable in ion-exchange chromatography. Consider a column to which anion $A^-$ is bound more tightly than anion $B^-$ is. We separate $A^-$ from $B^-$ by elution with $C^-$, which is less tightly bound than either $A^-$ or $B^-$. As the concentration of $C^-$ is increased, $B^-$ is eventually displaced and moves down the column. At a still higher concentration of $C^-$, the anion $A^-$ also is eluted.

> An ionic strength gradient is analogous to a solvent or temperature gradient.

In Figure 26-3, an $[H^+]$ gradient was used for a cation-exchange separation. The column for this separation has nitrilotriacetic acid groups that bind lanthanide cations in the order

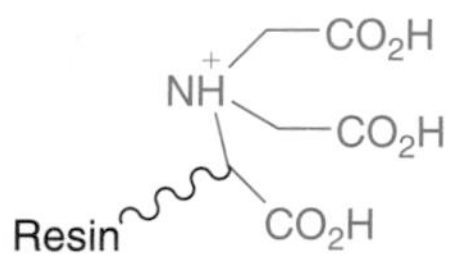

Nitrilotriacetic acid

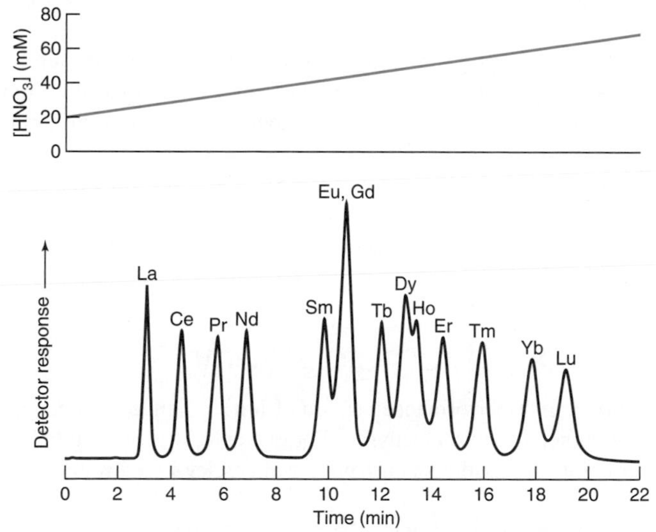

**Figure 26-3** Elution of lanthanide(III) ions from a cation-exchange resin. Eluent was varied from 20 to 80 mM HNO₃ over 25 min. The higher the atomic number of the lanthanide, the smaller is its ionic radius and the more strongly it binds to the resin. Lanthanides were detected by reaction with a color-forming reagent after elution. [From Y. Inoue. H. Kumagai, Y. Shimomura, T. Yokoyama, and T. M. Suzuki, "Ion Chromatographic Separation of Rare-Earth Elements Using a Nitrilotriacetate-Type Chelating Resin as the Stationary Phase," *Anal. Chem.* **1996,** *68,* 1517.]

$Lu^{3+} > Yb^{3+} > \ldots > Ce^{3+} > La^{3+}$. More strongly bound cations require a higher concentration of $H^+$ to be eluted. The order of selectivity coefficients for lanthanides on this column is the reverse of the order listed on page 592.

## Applications of Ion Exchange

Ion exchange can be used to convert one salt into another. For example, we can prepare tetrapropylammonium hydroxide from a tetrapropylammonium salt of some other anion:

$$(CH_3CH_2CH_2)_4N^+I^- \xrightarrow[\text{OH}^- \text{ form}]{\text{anion exchanger}} (CH_3CH_2CH_2)_4N^+OH^-$$

Tetrapropylammonium          Tetrapropylammonium<br>
iodide          hydroxide

Ion exchange is used for **preconcentration** of trace components of a solution to obtain enough for analysis. For example, we can pass a large volume of fresh lake water through a cation-exchange resin in the $H^+$ form to concentrate metal ions onto the resin. Chelex 100, a styrene-divinylbenzene resin containing iminodiacetic acid groups, is noteworthy for its ability to bind transition metal ions. A concentrated solution of metals is obtained by eluting the resin with a small volume of 2 M $HNO_3$, which protonates the iminodiacetate groups.

Ion exchange is used to purify water. **Deionized water** is prepared by passing water through an anion-exchange resin in its $OH^-$ form and a cation-exchange resin in its $H^+$ form. Suppose, for example, that $Cu(NO_3)_2$ is present in the solution. The cation-exchange resin binds $Cu^{2+}$ and replaces it with $2H^+$. The anion-exchange resin binds $NO_3^-$ and replaces it with $OH^-$. The eluate is pure water:

$$\left. \begin{array}{l} Cu^{2+} \xrightarrow{H^+ \text{ ion exchange}} 2H^+ \\[2ex] 2NO_3^- \xrightarrow{OH^- \text{ ion exchange}} 2OH^- \end{array} \right\} \longrightarrow \text{pure } H_2O$$

In many laboratory buildings, tap water is initially purified by passage through activated carbon (which adsorbs organic material) and then by *reverse osmosis*. In this process, water is forced by pressure through a membrane containing pores through which few molecules larger than $H_2O$ can pass. Most ions cannot pass through the pores, because their *hydrated radius* (see opening of Chapter 8) is larger than the pore size. Reverse osmosis removes ~95–99% of ions, organic molecules, bacteria, and particles from water.

"Water polishing" equipment used in many laboratories further purifies the water after reverse osmosis. The water is passed through activated carbon and then through several ion-exchange cartridges that convert ions into $H^+$ and $OH^-$. The resulting high-purity water has a resistivity (Chapter 15, note 30) of 180 000 ohm $\cdot$ m (18 Mohm $\cdot$ cm), with concentrations of individual ions below 1 ng/mL (1 ppb).[2]

Ion-exchange chromatography can be used to separate mirror image ions.[3] The mixture is applied to an ion-exchange column and eluted with one enantiomer of an ion such as tartrate. The two enantiomers of the mixture to be separated are eluted at different rates by the single enantiomer of tartrate.

In the pharmaceutical industry, ion-exchange resins are used for drug stabilization and as aids for tablet disintegration. Ion exchangers are also used for taste masking, for sustained-release products, as topical products for application to skin, and for ophthalmic or nasal delivery.[4]

## 26-2 Ion Chromatography

**Ion chromatography,** a high-performance version of ion-exchange chromatography, is generally the method of choice for anion analysis.[5] For example, it is used in the semiconductor industry to monitor anions and cations at 0.1 ppb levels in deionized water.

### Suppressed-Ion Anion and Cation Chromatography

In **suppressed-ion <u>anion</u> chromatography** (Figure 26-4a), a mixture of <u>anions</u> is separated by ion exchange and detected by electrical conductivity. The key feature of suppressed-ion chromatography is removal of unwanted electrolyte prior to conductivity measurement.

For the sake of illustration, consider a sample containing $NaNO_3$ and $CaSO_4$ injected into the *separator column*—an anion-exchange column in the carbonate form—followed by

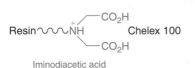

Iminodiacetic acid

Water softeners use ion exchange to remove $Ca^{2+}$ and $Mg^{2+}$ from "hard" water (Box 12-3).

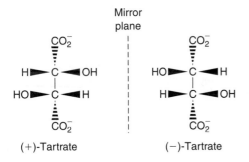

*What are the ions in pristine snow?* Antarctic snow provides a measure of global atmospheric chemistry because there are no local sources of pollution. One study found the following species by ion chromatography:

| Ion | Concentrations observed ($\mu$g/L = ppb) | |
|---|---|---|
| | Minimum | Maximum |
| $F^-$ | 0.10 | 6.20 |
| $Cl^-$ | 25 | 40 100 |
| $Br^-$ | 0.8 | 49.4 |
| $NO_3^-$ | 8.6 | 354 |
| $SO_4^{2-}$ | 10.6 | 4 020 |
| $H_2PO_4^-$ | 1.8 | 49.0 |
| $HCO_2^-$ | 1.1 | 45.7 |
| $CH_3CO_2^-$ | 5.0 | 182 |
| $CH_3SO_3^-$ | 1.1 | 281 |
| $NH_4^+$ | 2.4 | 46.5 |
| $Na^+$ | 15 | 17 050 |
| $K^+$ | 3.1 | 740 |
| $Mg^{2+}$ | 2.7 | 1 450 |
| $Ca^{2+}$ | 12.6 | 1 010 |

SOURCE: R. Udisti, S. Bellandi, and G. Piccardi, "Analysis of Snow from Antarctica," *Fresenius J. Anal. Chem.* **1994,** *349, 289.*

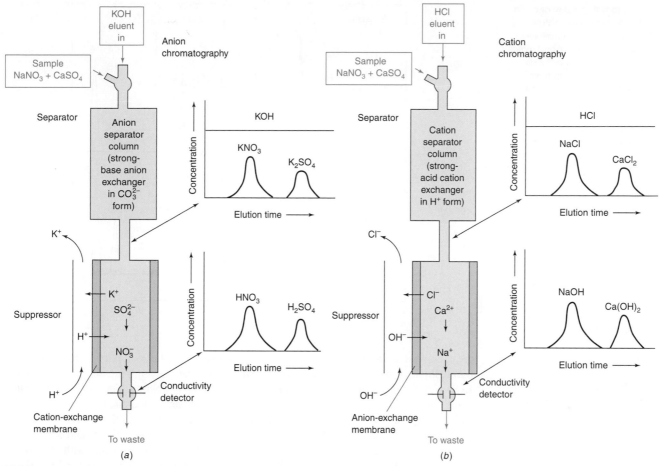

**Figure 26-4** Schematic illustrations of (*a*) suppressed-ion anion chromatography and (*b*) suppressed-ion cation chromatography.

elution with KOH. $NO_3^-$ and $SO_4^{2-}$ equilibrate with the resin and are slowly displaced by the $OH^-$ eluent. $Na^+$ and $Ca^{2+}$ cations are not retained and simply wash through. Eventually, $KNO_3$ and $K_2SO_4$ are eluted from the separator column, as shown in the upper graph of Figure 26-4a. These species cannot be easily detected, however, because the solvent contains a high concentration of KOH, whose high conductivity obscures that of the analyte species.

To remedy this problem, the solution next passes through a *suppressor*, in which cations are replaced by $H^+$. $H^+$ exchanges with $K^+$, in this example, through a cation-exchange membrane in the suppressor. $H^+$ diffuses from high concentration outside the membrane to low concentration inside the membrane. $K^+$ diffuses from high concentration inside to low concentration outside. $K^+$ is carried away outside the membrane, so its concentration is always low on the outside. The net result is that KOH eluent, which has high conductivity, is converted into $H_2O$, which has low conductivity. When analyte is present, $HNO_3$ or $H_2SO_4$ with high conductivity is produced and detected.

*Suppressed-ion <u>cation</u> chromatography* is conducted in a similar manner, but the suppressor replaces $Cl^-$ from eluent with $OH^-$ through an anion-exchange membrane. Figure 26-4b illustrates the separation of $NaNO_3$ and $CaSO_4$. With HCl eluent, NaCl and $CaCl_2$ emerge from the cation-exchange separator column, and NaOH and $Ca(OH)_2$ emerge from the suppressor column. HCl eluate is converted into $H_2O$ in the suppressor.

Figure 26-5 illustrates a student experiment to measure ions in pond water. Eluent for the anion separation was $NaHCO_3/Na_2CO_3$ buffer. The product from eluent after passing through the suppressor is $H_2CO_3$, which has low conductivity.

In automated systems, $H^+$ and $OH^-$ eluents and suppressors are generated by electrolysis of $H_2O$.[6] Figure 26-6 shows a system that generates KOH for >1 000 hours of isocratic or gradient elution before it is necessary to refresh the reagents. Water in the reservoir of aqueous $K_2HPO_4$ decomposes at the metal anode to produce $H^+$ and $O_2(g)$. $H^+$ reacts with $HPO_4^-$ to form $H_2PO_4^{2-}$. For each $H^+$ ion that is generated, one $K^+$ ion migrates through

The separator column separates the analytes, and the suppressor replaces the ionic eluent with a nonionic species.

Benzene-1,4-diammonium cation is a stronger eluent that can be used instead of $H^+$ for suppressed-ion cation chromatography. After suppression, a neutral product is formed:

$$H_3\overset{+}{N}-\!\!\!\bigcirc\!\!\!-\overset{+}{N}H_3 \xrightarrow{OH^-}$$

Benzene-l,4-diammonium ion

$$H_2N-\!\!\!\bigcirc\!\!\!-NH_2$$

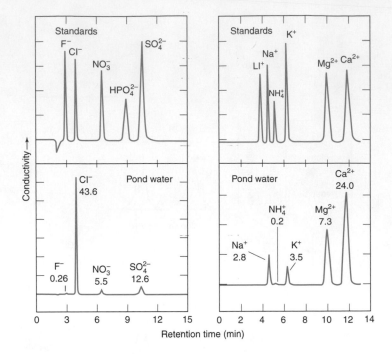

**Figure 26-5** Ion chromatography of pond water. Upper chromatograms were obtained from mixtures of standards. Concentrations of ions in lower chromatograms from pond water are in units of μg/mL (ppm). Anion analysis was done with an IonPac AS14 column with 1.0 mM $NaHCO_3$/3.5 mM $Na_2CO_3$ eluent with ion suppression and conductivity detection. Cation analysis used an IonPac CS12A column with 11 mM $H_2SO_4$ eluent, ion suppression, and conductivity detection. [*From K. Sinniah and K. Piers, "Ion Chromatography: Analysis of Ions in Pond Waters," J Chem. Ed.* **2001,** *78, 358.*]

the cation-exchange barrier membrane, which transports $K^+$, but not anions, and allows negligible liquid to pass. The barrier membrane must withstand the high pressure of liquid in the KOH generation chamber destined for the chromatography column. For each $H^+$ generated at the anode, one $K^+$ flows through the cation-exchange barrier and one $OH^-$ is generated at the cathode. Liquid exiting the KOH generation chamber contains KOH and $H_2$. The stream passes through an anion trap that removes traces of anions such as carbonate and degradation products from the ion-exchange resin. The trap is continuously replenished with electrolytically generated $OH^-$, which is not shown in the diagram. After the anion trap, the liquid flows through a polymeric capillary that is permeable to $H_2$. The $H_2$ diffuses into an external liquid stream and is removed. The concentration of KOH produced by the apparatus in Figure 26-6 is governed by the liquid flow rate and the electric current. With computer control of the power supply, a precise gradient can be generated.

In the past, KOH eluent was usually contaminated with $CO_3^{2-}$. When $CO_3^{2-}$ passes through the suppressor after the ion chromatography column, it is converted into $H_2CO_3$, which has some electrical conductivity that interferes with detection of analytes. In gradient

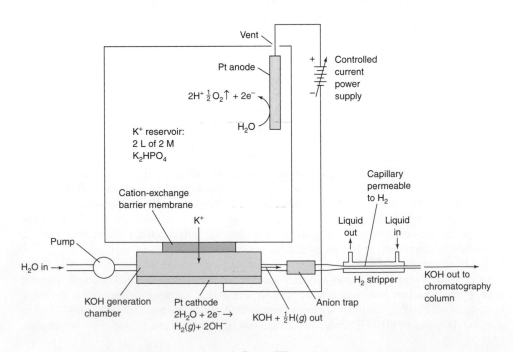

**Figure 26-6** Electrolytic KOH eluent generator for ion chromatography. [*Adapted from Y. Liu, K. Srinivasan, C. Pohl, and N. Avdalovic, "Recent Developments in Electrolytic Devices for Ion Chromatography," J. Biochem. Biophys. Methods* **2004,** *60, 205.*]

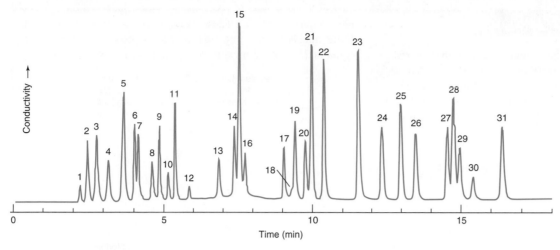

**Figure 26-7** Anion separation by ion chromatography with a gradient of electrolytically generated KOH and conductivity detection after suppression. Column: Dionex IonPac AS11; diameter = 4 mm; flow = 2.0 mL/min. Eluent: 0.5 mM KOH for 2.5 min, 0.5 to 5.0 mM KOH from 2.5 to 6 min; 5.0 to 38.2 mM KOH from 6 to 18 min. Peaks: (1) quinate, (2) $F^-$, (3) acetate, (4) propanoate, (5) formate, (6) methylsulfonate, (7) pyruvate, (8) valerate, (9) chloroacetate, (10) $BrO_3^-$, (11) $Cl^-$ (12) $NO_2^-$, (13) trifluoroacetate, (14) $Br^-$, (15) $NO_3^-$, (16) $ClO_3^-$, (17) selenite, (18) $CO_3^{2-}$, (19) malonate, (20) maleate, (21) $SO_4^{2-}$, (22) $C_2O_4^{2-}$, (23) tungstate, (24) phthalate, (25) phosphate, (26) chromate, (27) citrate, (28) tricarballylate, (29) isocitrate, (30) *cis*-aconitate, (31) *trans*-aconitate.
[Courtesy Dionex Corp., Sunnyvale, CA]

elution with increasing KOH, the concentration of $H_2CO_3$ also increases, and so does the background conductivity. The feedstock for the electrolytic generator is pure $H_2O$ and the product is aqueous KOH containing very little $CO_3^{2-}$ Figure 26-7 shows an impressive separation of 31 anions with a hydroxide gradient.

The suppressors in Figure 26-4 also have been replaced by electrolytic units that generate $H^+$ or $OH^-$ necessary to neutralize the eluate and require only $H_2O$ as feedstock.[6] With electrolytic eluent generation and electrolytic suppression, ion chromatography has been simplified and highly automated. Readily available software can be used to simulate and optimize ion chromatographic separations.[7]

## Ion Chromatography Without Suppression

If the ion-exchange capacity of the separator column is sufficiently low and if dilute eluent is used, ion suppression is unnecessary. Also, anions of weak acids, such as borate, silicate, sulfide, and cyanide, cannot be determined with ion suppression, because these anions are converted into very weakly conductive products (such as $H_2S$).

For *nonsuppressed anion chromatography,* we use a resin with an exchange capacity near 5 μequiv/g, with $10^{-4}$ M $Na^+$ or $K^+$ salts of benzoic, *p*-hydroxybenzoic, or phthalic acid as eluent. These eluents give a low background conductivity, and analyte anions are detected by a small *change* in conductivity as they emerge from the column. By judicious choice of pH, an average eluent charge between 0 and $-2$ can be obtained, which allows control of eluent strength. Even dilute carboxylic acids (which are slightly ionized) are suitable eluents for some separations. *Nonsuppressed cation chromatography* is conducted with dilute $HNO_3$ eluent for monovalent ions and with ethylenediammonium salts ($^+H_2NCH_2CH_2NH_2^+$) for divalent ions.

## Detectors

Conductivity detectors respond to all ions. In suppressed-ion chromatography, it is easy to measure analyte because eluent conductivity is reduced to near 0 by suppression. Suppression also allows us to use eluent concentration gradients.

In nonsuppressed anion chromatography, the conductivity of the analyte anion is higher than that of the eluent, so conductivity increases when analyte emerges from the column. Detection limits are normally in the mid-ppb to low-ppm range but can be lowered by a factor of 10 by using carboxylic acid eluents instead of carboxylate salts.

Benzoate

*p*-Hydroxybenzoate

Phthalate

A **surfactant** is a molecule that accumulates at the interface between two phases and modifies the surface tension. (*Surface tension* is the energy per unit area needed to form a surface or interface.) One common class of surfactants for aqueous solution are molecules with long hydrophobic tails and ionic head groups, such as

Cetyltrimethylammonium bromide
$C_{16}H_{33}N(CH_3)_3{}^+Br^-$

A **micelle** is an aggregate of surfactants. In water, the hydrophobic tails form clusters that are, in effect, little oil drops insulated from the aqueous phase by the ionic head groups. At low concentration, surfactant molecules do not form micelles. When their concentration exceeds the *critical micelle concentration,* spontaneous aggregation into micelles occurs.[8] Isolated surfactant molecules exist in equilibrium with micelles. Nonpolar organic solutes are soluble inside micelles. Cetyltrimethylammonium bromide forms micelles containing ~61 molecules (mass ≈ 22 000 Da) in water at 25°C at a critical micelle concentration of 0.9 mM.

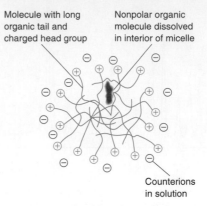

Molecule with long organic tail and charged head group

Nonpolar organic molecule dissolved in interior of micelle

Counterions in solution

Structure of a micelle formed when ionic molecules with long, nonpolar tails aggregate in aqueous solution. The interior of the micelle resembles a nonpolar organic solvent, whereas the exterior charged groups interact strongly with water. [F. M. Menger, R, Zana, and B. Lindman, "Portraying the Structure of Micelles," J. Chem. Ed. **1998,** 75, 115.]

---

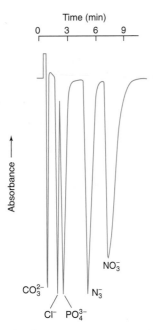

Time (min)

*Figure 26-8* Indirect spectrophotometric detection of transparent ions. The column was eluted with 1 mM sodium phthalate plus 1 mM borate buffer, pH 10. The principle of indirect detection is illustrated in Figure 26-30. [Reproduced from H. Small, "Indirect Photometric Chromotography," Anal. Chem. **1982,** 54, 462.]

Using benzoate or phthalate eluents provides for sensitive ($<1$ ppm) **indirect detection** of anions. In Figure 26-8, eluate has strong, constant ultraviolet absorption. As each analyte emerges, nonabsorbing analyte anion replaces an equivalent amount of absorbing eluent anion. Absorbance therefore *decreases* when analyte appears. For cation chromatography, $CuSO_4$ is a suitable ultraviolet-absorbing eluent.

## Ion-Pair Chromatography

**Ion-pair chromatography** uses a reversed-phase HPLC column instead of an ion-exchange column. To separate a mixture of cations (for example, protonated organic bases), an anionic *surfactant* (Box 26-1) such as $n\text{-}C_8H_{17}SO_3^-$ is added to the mobile phase. The surfactant lodges in the stationary phase, effectively transforming the stationary phase into an ion exchanger (Figure 26-9). When analyte cations pass through the column, they can associate with the sta-

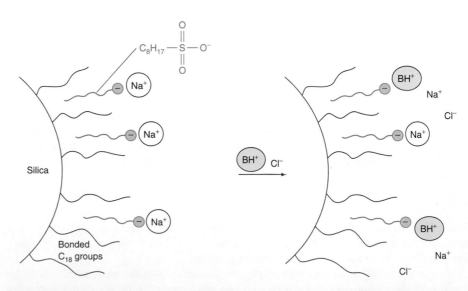

*Figure 26-9* Principle of ion-pair chromatography. The surfactant sodium octanesulfonate added to the mobile phase binds to the nonpolar stationary phase. Negative sulfonate groups protruding from the stationary phase then act as ion-exchange sites for analyte cations such as protonated organic bases, $BH^+$.

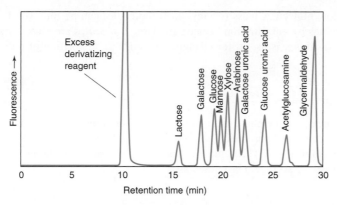

**Figure 26-10** Separation of carbohydrates by ion-pair chromatography. Carbohydrates were *derivatized* by covalently attaching *p*-aminobenzoate ($H_2N-C_6H_4-CO_2^-$), which changes carbohydrates into fluorescent anions. The anions were separated on a $0.30 \times 25$ cm column of AQUA $C_{18}$-silica using tetrabutylammonium cation as the ion-pair reagent. Eluent was a linear 60-min gradient starting with 20 mM aqueous ($n$-$C_4H_9)_4N^+HSO_4^-$, pH 2.0 (solvent A), and ending with 50:50 A:methanol. The method was used to measure carbohydrates at 10–100-ng/mL levels in water leaching from landfills. *[From A. Meyer, C. Raba, and K. Fischer, "Ion-Pair HPLC Determination of Sugars, Amino Sugars, and Uronic Acids," Anal. Chem.* **2001**, *73, 2377.]*

tionary phase by electrostatic attraction to the surfactant anions.[9] The retention mechanism is a mixture of reversed-phase and ion-exchange interactions. To separate analyte anions, tetrabutyl-ammonium salts can be added to the mobile phase as the ion-pair reagent (Figure 26-10).

Ion-pair chromatography is more complex than reversed-phase chromatography because equilibration of the surfactant with the stationary phase is slow, the separation is more sensitive to variations in temperature and pH, and the concentration of surfactant affects the separation. Methanol is the organic solvent of choice because ionic surfactants are more soluble in methanol/water mixtures than in acetonitrile/water mixtures. Strategies for method development analogous to the scheme in Figure 25-28 vary the pH and surfactant concentration with fixed methanol concentration and temperature.[10] Because of the slow equilibration of surfactant with the stationary phase, gradient elution is not recommended in ion-pair chromatography.

## ▇▇▇ 26-3 Molecular Exclusion Chromatography

In **molecular exclusion chromatography** (also called *size exclusion* or **gel filtration** or *gel permeation chromatography*), molecules are separated according to size. Small molecules penetrate the pores in the stationary phase, but large molecules do not (Figure 23-6). Because small molecules must pass through an effectively larger volume, *large molecules are eluted first* (Figure 26-11). This technique is widely used in biochemistry to purify macromolecules.

> Large molecules pass through the column faster than small molecules do.

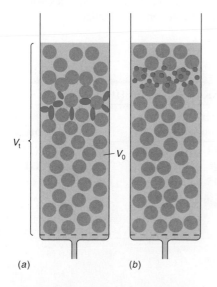

**Figure 26-11** (*a*) Large molecules cannot penetrate the pores of the stationary phase. They are eluted by a volume of solvent equal to the volume of mobile phase. (*b*) Small molecules that can be found inside or outside the gel require a larger volume for elution. $V_t$ is the total column volume occupied by gel plus solvent. $V_0$ is the volume of the mobile phase outside the gel particles.

(*a*)  (*b*)

Salts of low molecular mass (or any small molecule) can be removed from solutions of large molecules by gel filtration because the large molecules are eluted first. This technique, called *desalting,* is useful for changing the buffer composition of a macromolecule solution.

## The Elution Equation

The *total* volume of mobile phase in a chromatography column is $V_m$, which includes solvent inside and outside the gel particles. The volume of mobile phase *outside* the gel particles is called the **void volume,** $V_0$. The volume of solvent *inside* the gel is therefore $V_m - V_0$. The quantity $K_{av}$ (read "$K$ average") is defined as

$$K_{av} = \frac{V_r - V_0}{V_m - V_0} \tag{26-6}$$

In pure molecular exclusion, all molecules are eluted between $K_{av} = 0$ and $K_{av} = 1$.

where $V_r$ is the retention volume for a solute. For a large molecule that does not penetrate the gel, $V_r = V_0$, and $K_{av} = 0$. For a small molecule that freely penetrates the gel, $V_r = V_m$, and $K_{av} = 1$. Molecules of intermediate size penetrate some gel pores, but not others, so $K_{av}$ is between 0 and 1. Ideally, gel penetration is the only mechanism by which molecules are retained in this type of chromatography. In fact, there is always some adsorption, so $K_{av}$ can be greater than 1.

Void volume is measured by passing a large, inert molecule through the column.[11] Its elution volume is defined as $V_0$. Blue Dextran 2000, a blue dye of molecular mass $2 \times 10^6$, is commonly used for this purpose. The volume $V_m$ can be calculated from the measured column bed volume per gram of dry gel. For example, 1 g of dry Sephadex G-100 produces 15 to 20 mL of bed volume when swollen with aqueous solution. The solid phase occupies only ~1 mL of the bed volume, so $V_m$ is 14 to 19 mL, or 93–95% of the total column volume. Different solid phases produce widely varying column bed volumes when swollen with solvent.

## Stationary Phase[12]

Gels for open column, preparative-scale molecular exclusion include Sephadex (Table 26-4), whose structure was given in Figure 26-2, and Bio-Gel P, which is a polyacrylamide cross-linked by $N, N'$-methylenebisacrylamide (Figure 26-12). The smallest pore sizes in highly cross linked gels exclude molecules with a molecular mass $\gtrsim 700$, whereas the largest pore sizes exclude molecules with molecular mass $\gtrsim 10^8$. The finer the particle size of the gel, the greater the resolution and the slower the flow rate of the column.

For HPLC, polystyrene spheres are available with pore sizes ranging from 5 nm up to hundreds of nanometers. Particles with a 5-$\mu$m diameter yield up to 80 000 plates per meter of column length. Silica (Table 26-4) with controlled pore size provides 10 000–16 000 plates per meter. The silica is coated with a hydrophilic phase that minimizes solute adsorption. A hydroxylated polyether resin with a well-defined pore size can be used over the pH range 2–12, whereas silica phases generally cannot be used above pH 8. Particles with different pore sizes can be mixed to give a wider molecular size separation range.

**Table 26-4** Representative molecular exclusion media

| Gel filtration in open columns | | TSK SW silica for HPLC | | |
|---|---|---|---|---|
| Name | Fractionation range for globular proteins (Da) | Name | Pore size (nm) | Fractionation range for globular proteins (Da) |
| Sephadex G-10 | to 700 | G2000SW | 13 | 500–60 000 |
| Sephadex G-25 | 1 000–5 000 | G3000SW | 24 | 1 000–300 000 |
| Sephadex G-50 | 1 500–30 000 | G4000SW | 45 | 5 000–1 000 000 |
| Sephadex G-75 | 3 000–80 000 | G5000SW | 100 | >1 500 000 |
| Sephadex G-100 | 4 000–150 000 | | | |
| Sephadex G-200 | 5 000–600 000 | | | |

NOTE: *Sephadex is manufactured by GE Amersham Biosciences. TSK SW silica is manufactured by Tosoh Corp.*

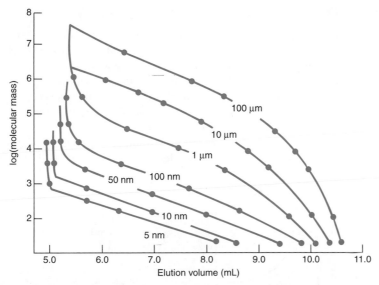

Figure 26-12 Structure of polyacrylamide.

## Molecular Mass Determination

Gel filtration is used mainly to separate molecules of significantly different molecular size (Figure 26-13). For each stationary phase, we construct a calibration curve, which is a graph of log(molecular mass) versus elution volume (Figure 26-14). We estimate the molecular mass of an unknown by comparing its elution volume with those of standards. We must exercise caution in interpreting results, however, because molecules with the same molecular mass but different shapes exhibit different elution characteristics. For proteins, it is important to use an ionic strength high enough ($>0.05$ M) to eliminate electrostatic adsorption of solute by occasional charged sites on the gel.

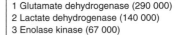

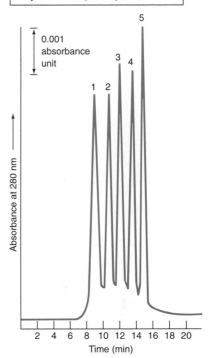

Figure 26-13 Separation of proteins by molecular exclusion chromatography with TSK 3000SW column. *[Courtesy Varian Associates, Palo Alto, CA.]*

Figure 26-14 Molecular mass calibration graph for polystyrene on Beckman μSpherogel molecular exclusion column (0.77 × 30 cm). Resin pore size labeled on the lines ranges from 5 nm to 100 μm. *[Courtesy Anspec Co., Ann Arbor, MI.]*

*Nanoparticles* can be separated by molecular exclusion chromatography just as proteins are separated. Figure 26-15 shows the relation between measured size and retention time of CdSe *quantum dots.* These are particles containing ~2 000 CdSe units in a dense, crystalline core capped by alkyl thiol (RS) groups on Cd and trialkylphosphine ($R_3P$) groups on Se.

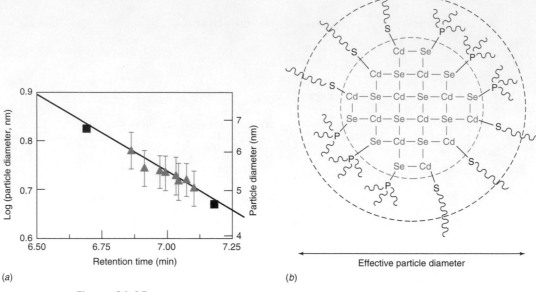

(a)

(b)

*Figure 26-15* Larger CdSe quantum dots are eluted before smaller quantum dots by 0.1 M trioctylphosphine in toluene at 1.0 mL/min in size exclusion chromatography on a 7.5 × 300 mm cross-linked polystyrene column of 100-nm pore size Polymer Labs PLgel 5 μm. Triangles are CdSe and squares are polystyrene calibration standards. The size of the CdSe core was measured with a transmission electron microscope and the length of 1-dodecanethiol endcaps (0.123 nm) was added to the radius. *[Data from K. M. Krueger, A. M. Al-Somali, J. C. Falkner, and V. L. Colvin, "Characterization of Nanocrystalline CdSe by Size Exclusion Chromatography," Anal. Chem.* **2005,** *77, 3511.]*

Quantum dots are useful because the wavelength of their visible emission depends on their size. Size is controlled during synthesis of the quantum dot by reaction time or other conditions. Quantum dots with different sizes can be used as spectroscopic labels in biological experiments.[13]

## ■ ■ ■ 26-4 Affinity Chromatography

**Affinity chromatography** is used to isolate a single compound from a complex mixture. The technique is based on specific binding of that one compound to the stationary phase (Figure 23-6). When sample is passed through the column, only one solute is bound. After everything else has washed through, the one adhering solute is eluted by changing a condition such as pH or ionic strength to weaken its binding. Affinity chromatography is especially applicable in biochemistry and is based on specific interactions between enzymes and substrates, antibodies and antigens, or receptors and hormones.

Figure 26-16 shows the isolation of the protein immunoglobulin G (IgG) by affinity chromatography on a column containing covalently bound *protein A*. Protein A binds to

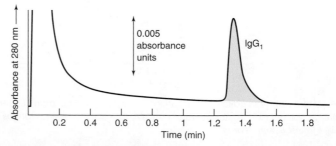

*Figure 26-16* Purification of monoclonal antibody IgG by affinity chromatography on a 0.46 × 5 cm column containing protein A covalently attached to polymer support. Other proteins in the sample are eluted from 0 to 0.3 min at pH 7.6. When eluent pH is lowered to 2.6, IgG is freed from protein A and emerges from the column. *[From B. J. Compton and L. Kreilgaard, "Chromatographic Analysis of Therapeutic Proteins," Anal. Chem.* **1994,** *66, 1175A.]*

Box 26-2  Molecular Imprinting[15]

A **molecularly imprinted polymer** is one that is polymerized in the presence of a template molecule to which components of the polymer have some affinity. When the template is removed, the polymer is "imprinted" with the shape of the template and with complementary functional groups that can bind the template. The imprinted polymer can be a stationary phase in affinity chromatography or a recognition element in a chemical sensor because the polymer preferentially binds the original template molecule. Optical isomers can be separated by passage through a molecularly imprinted polymer for which one of the optical isomers was the template.

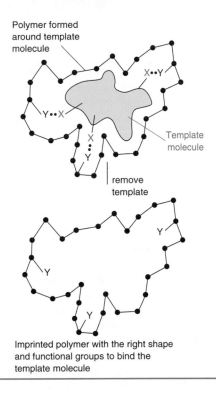

Polymer formed around template molecule

X··Y

Y··X

X
⋮
Y

remove template

Template molecule

Imprinted polymer with the right shape and functional groups to bind the template molecule

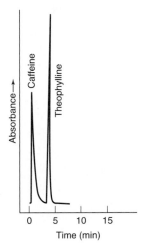

Separation of theophylline and caffeine by column containing polymer imprinted with theophylline. Caffeine washed right through with $CHCl_3$ solvent. Addition of 20 μL of $CH_3OH$ to the solvent breaks hydrogen bonds between theophylline and the polymer and elutes theophylline in a volume of 1 mL. *[From W. M. Mullett and E. P. C. Lai, Anal. Chem.* **1998,** *70, 3636.]*

one specific region of IgG at pH $\gtrsim$ 7.2. When a crude mixture containing IgG and other proteins was passed through the column at pH 7.6, everything except IgG was eluted within 0.3 min. At 1 min, the eluent pH was lowered to 2.6 and IgG was cleanly eluted at 1.3 min.

Optical isomers of a drug can have completely different therapeutic effects. Affinity chromatography can be used to isolate individual optical isomers for drug testing.[14] The drug candidate in the margin has two chiral carbon atoms indicated by colored dots. With two possible geometries at each site, there are four stereoisomers. The mixture of isomers was covalently attached to protein and injected into mice to raise a mixture of antibodies to all stereoisomers. Antibodies are produced by B cells in the spleen. One B cell produces only one kind of antibody. By isolating individual B cells, it is possible to isolate the gene for the antibody to each of the stereoisomers. The gene can be transplanted into *E. coli* cells for mass production of a single kind of antibody, called a *monoclonal antibody.* When the mixture of stereoisomers is passed through a column to which just one kind of antibody is attached, only one of the four stereoisomers is retained. By lowering the pH, the retained isomer is eluted in pure form.

Box 26-2 shows how *molecularly imprinted polymers* can be used as affinity media. *Aptamers* (Section 19-5) are another class of chromatographically useful compounds with high affinity for a selected target.[16]

### 26-5  Principles of Capillary Electrophoresis[17,18]

**Electrophoresis** is the migration of ions in solution under the influence of an electric field. In the **capillary electrophoresis** experiment shown in Figure 26-17, components of a solution are separated by applying a voltage of ~30 kV from end to end of a fused-silica ($SiO_2$) capillary tube that is 50 cm long and has an inner diameter of 25–75 μm. Different solutes

Cations are attracted to the negative terminal (the cathode).

Anions are attracted to the positive terminal (the anode).

**Figure 26-17** Apparatus for capillary electrophoresis. One way to inject sample is to place the capillary in a sample vial and apply pressure to the vial or suction at the outlet of the capillary. The use of an electric field for sample injection is described in the text.

Electrophoresis in glass capillaries was first described by J. W. Jorgenson in 1981.[19]

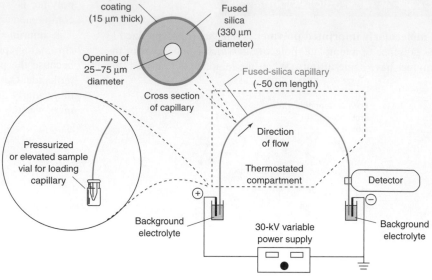

Polyimide coating (15 μm thick)
Fused silica (330 μm diameter)
Opening of 25–75 μm diameter
Cross section of capillary
Fused-silica capillary (~50 cm length)
Direction of flow
Pressurized or elevated sample vial for loading capillary
Thermostated compartment
Detector
Background electrolyte
Background electrolyte
30-kV variable power supply

Electric potential difference = 30 kV

$$\text{Electric field} = \frac{30\,\text{kV}}{0.50\,\text{m}} = 60\,\frac{\text{kV}}{\text{m}}$$

have different *mobilities* and therefore migrate through the capillary at different speeds.[20] Modifications of this experiment described in Section 26-6 allow neutral molecules, as well as ions, to be separated.

Capillaries for electrophoresis can be small enough to insert into single, large, living cells to analyze their contents.[21] Smaller cells can be taken one at a time into the capillary, burst open, and analyzed (Figure 26-18). A single nucleus or a single *vesicle* (a storage compartment found inside a cell) can be taken into a capillary for analysis of its contents.[22] Assays of enzymes from a single cell can be performed with detection limits at the zeptomol ($10^{-21}$ mol) level.[23] Electrophoresis has been used to study the release of neurotransmitters from an isolated retina, which receives and processes visual information.[24] Electrophoresis can be combined with immunoaffinity techniques to isolate a limited class of analytes that are retained by an antibody and then to separate the members of that class.[25]

Capillary electrophoresis provides unprecedented resolution. When we conduct chromatography in a packed column, peaks are broadened by three mechanisms in the van Deemter equation (23-33): multiple flow paths, longitudinal diffusion, and finite rate of mass transfer. An open tubular column eliminates multiple paths and thereby reduces plate height and improves resolution. Capillary electrophoresis reduces plate height further by knocking out the mass transfer term that comes from the finite time needed for solute to equilibrate

(a)

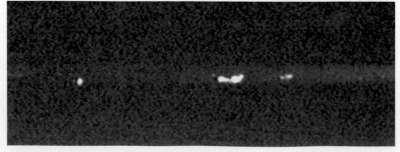

(b)

**Figure 26-18** Micrographs showing (a) fluorescent image of a single, intact cell immediately after injection into the capillary and (b) cellular debris beginning to separate 10 s after application of high voltage. *[From S. N. Krylov, D. A. Starke, E. A. Arriaga, Z. Zhang, N. W. C. Chan, M. M. Palcic, and N. J. Dovichi, "Instrumentation for Chemical Cytometry," Anal. Chem. **2000**, 72, 872.]*

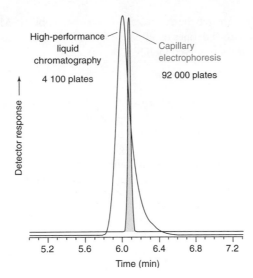

High-performance liquid chromatography
4 100 plates

Capillary electrophoresis
92 000 plates

Detector response →

Time (min)

**Figure 26-19** Comparison of peak widths for benzyl alcohol ($C_6H_5CH_2OH$) in capillary electrophoresis and HPLC. [From S. Fazio, R. Vivilecchia, L. Lesueur, and J, Sheridan, Am. Biotech. Lab. January 1990, p. 10.]

between mobile and stationary phases. In capillary electrophoresis, *there is no stationary phase*. The only fundamental source of broadening under ideal conditions is longitudinal diffusion:

$$H = \cancel{A} + \frac{B}{u_x} + \cancel{Cu_x} \qquad (26\text{-}7)$$

↑
Multiple path term
eliminated by open
tubular column

↑
Mass transfer term
eliminated because there
is no stationary phase

(Other sources of broadening in real systems are mentioned later.) Capillary electrophoresis routinely produces 50 000–500 000 theoretical plates (Figure 26-19), which is an order-of-magnitude better performance than chromatography.

## Electrophoresis

When an ion with charge $q$ (coulombs) is placed in an electric field $E$ (V/m), the force on the ion is $qE$ (newtons). In solution, the retarding frictional force is $fu_{ep}$, where $u_{ep}$ is the velocity of the ion and $f$ is the *friction coefficient*. The subscript "ep" stands for "electrophoresis." The ion quickly reaches a steady speed when the accelerating force equals the frictional force:

$$+ \atop + \atop + \qquad \xleftarrow{fu_{ep}} \; \oplus \; \xrightarrow{qE} \qquad - \atop - \atop - \qquad \begin{array}{ccc} qE & = & fu_{ep} \\ \text{Accelerating} & & \text{Frictional} \\ \text{force} & & \text{force} \end{array}$$

*Electrophoretic mobility:*
$$u_{ep} = \frac{q}{f}E \equiv \mu_{ep}E \qquad (26\text{-}8)$$

↑
Electrophoretic mobility

*Electrophoretic* **mobility** ($\mu_{ep}$) is the constant of proportionality between the speed of the ion and the strength of the electric field. Mobility is proportional to the charge of the ion and inversely proportional to the friction coefficient. For molecules of similar size, the magnitude of the mobility increases with charge:

We encountered *mobility* earlier in connection with junction potentials (Table 15-1).

$$\mu_{ep} = -2.54 \times 10^{-8} \frac{m^2}{V \cdot s}$$

$$\mu_{ep} = -4.69 \times 10^{-8} \frac{m^2}{V \cdot s}$$

$$\mu_{ep} = -5.95 \times 10^{-8} \frac{m^2}{V \cdot s}$$

(solvent is $H_2O$ at 25°C)

For a spherical particle of radius $r$ moving through a fluid of *viscosity* $\eta$, the friction coefficient, $f$, is

*Stokes equation:*
$$f = 6\pi\eta r \qquad (26\text{-}9)$$

*Viscosity* measures resistance to flow in a fluid. The units are kg m⁻¹ s⁻¹. Relative to water, maple syrup is very viscous and hexane has low viscosity.

If you measure the friction coefficient of a molecule with Equation 26-8 and the solution viscosity, then $r$ in Equation 26-9 is the *hydrodynamic radius* of the molecule.

Mobility is $q/f$, so, the greater the radius, the lower the mobility. Most molecules are not spherical, but Equation 26-9 defines an effective *hydrodynamic radius* of a molecule, as if it were a sphere, based on its observed mobility.

## Electroosmosis

The inside wall of a fused-silica capillary is covered with silanol (Si—OH) groups with a negative charge (Si—O⁻) above pH ≈ 2. Figure 26-20a shows the *electric double layer* (Box 17-3) at the wall of the capillary. The double layer consists of fixed negative charge on the wall and excess cations near the wall. A tightly adsorbed, immobile layer of cations partially neutralizes the negative charge. The remaining negative charge is neutralized by mobile cations in the *diffuse part of the double layer* in solution near the wall. The thickness of the diffuse part of the double layer ranges from ~10 nm when the ionic strength is 1 mM to ~0.3 nm when the ionic strength is 1 M.

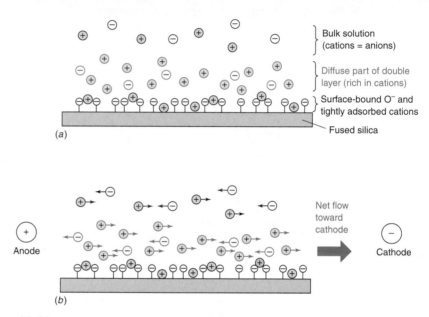

(a)

(b)

**Figure 26-20** (*a*) Electric double layer created by negatively charged silica surface and nearby cations. (*b*) Predominance of cations in diffuse part of the double layer produces net electroosmotic flow toward the cathode when an external field is applied.

Ions in the diffuse part of the double layer adjacent to the capillary wall are the "pump" that drives electroosmotic flow.

In an electric field, cations are attracted to the cathode and anions are attracted to the anode (Figure 26-20b). Excess cations in the diffuse part of the double layer impart net momentum toward the cathode. This pumping action, called **electroosmosis** (or *electroendosmosis*), is driven by cations within ~10 nm of the walls and creates uniform pluglike *electroosmotic flow* of the entire solution toward the cathode (Figure 26-21a). This process is in sharp contrast with *hydrodynamic flow*, which is driven by a pressure difference. In hydro-

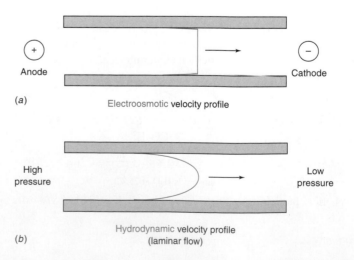

(a) Electroosmotic **velocity** profile

**Figure 26-21** (*a*) Electroosmosis gives uniform flow over more than 99.9% of the cross section of the capillary. The speed decreases immediately adjacent to the capillary wall. (*b*) Parabolic velocity profile of hydrodynamic flow (also called *laminar flow*), with the highest velocity at the center of the tube and zero velocity at the walls. Experimentally observed velocity profiles are shown in Color Plate 28

(b) Hydrodynamic **velocity** profile (laminar flow)

dynamic flow, the velocity profile through a cross section of the fluid is parabolic: It is fastest at the center and slows to 0 at the walls (Figure 26-21b and Color Plate 28).

The constant of proportionality between electroosmotic velocity, $u_{eo}$, and applied field is called *electroosmotic mobility*, $\mu_{eo}$.

*Electroosmotic mobility:*
$$u_{eo} = \mu_{eo}E \qquad (26\text{-}10)$$
$\uparrow$
Electroosmotic mobility
(units = $m^2/[V \cdot s]$)

Electroosmotic mobility is proportional to the surface charge density on the silica and inversely proportional to the square root of ionic strength. Electroosmosis decreases at low pH (Si—O$^-$ → Si—OH decreases surface charge density) and high ionic strength. At pH 9 in 20 mM borate buffer, electroosmotic flow is ~2 mm/s. At pH 3, flow is reduced by an order of magnitude.

Uniform electroosmotic flow contributes to the high resolution of capillary electrophoresis. Any effect that decreases uniformity creates band broadening and decreases resolution. The flow of ions in the capillary generates heat (called *Joule heating*) at a rate of $I^2R$ joules per second, where $I$ is current (A) and $R$ is the resistance of the solution (ohms) (Section 14-1). Typically, the centerline of the capillary channel is 0.02 to 0.3 K hotter than the edge of the channel. Viscosity is lower in the warmer region, disturbing the flat electroosmotic profile of the fluid. Joule heating is not a serious problem in a capillary tube with a diameter of 50 μm, but the temperature gradient would be prohibitive if the diameter were $\geq 1$ mm. Some instruments cool the capillary to reduce the electrical conductivity of solution inside the capillary and prevent runaway Joule heating.

## Mobility

The *apparent* (or observed) *mobility*, $\mu_{app}$, of an ion is the sum of the electrophoretic mobility of the ion plus the electroosmotic mobility of the solution.

*Apparent mobility:*
$$\mu_{app} = \mu_{ep} + \mu_{eo} \qquad (26\text{-}11)$$

For an analyte *cation* moving in the same direction as the electroosmotic flow, $\mu_{ep}$ and $\mu_{eo}$ have the same sign, so $\mu_{app}$ is greater than $\mu_{ep}$. Electrophoresis transports *anions* in the opposite direction from electroosmosis (Figure 26-20b), so for anions the two terms in Equation 26-11 have opposite signs. At neutral or high pH, brisk electroosmosis transports anions to the *cathode* because electroosmosis is usually faster than electrophoresis. At low pH, electroosmosis is weak and anions may never reach the detector. If you want to separate anions at low pH, you can reverse the polarity of the instrument to make the sample side negative and the detector side positive.

The apparent mobility, $\mu_{app}$, of a particular species is the net speed, $u_{net}$, of the species divided by the electric field, $E$:

*Apparent mobility:*
$$\mu_{app} = \frac{u_{net}}{E} = \frac{L_d/t}{V/L_t} \qquad (26\text{-}12)$$

where $L_d$ is the length of column from injection to the detector, $L_t$ is the total length of the column from end to end, $V$ is the voltage applied between the two ends, and $t$ is the time required for solute to migrate from the injection end to the detector. Electroosmotic flow is measured by adding an ultraviolet-absorbing neutral solute to the sample and measuring its *migration time*, $t_{neutral}$, to the detector.

For quantitative analysis by electrophoresis, *normalized peak areas* are required. The normalized peak area is the measured peak area divided by the migration time. In chromatography, each analyte passes through the detector at the same rate, so peak area is proportional to the quantity of analyte. In electrophoresis, analytes with different apparent mobilities pass through the detector at different rates. The higher the apparent mobility, the shorter the migration time and the less time the analyte spends in the detector. To correct for time spent in the detector, divide the peak area for each analyte by its migration time.

Electroosmotic velocity is measured by adding to the sample a *neutral* molecule to which the detector responds.

Electroosmotic velocity = $\dfrac{\text{distance from injector to detector}}{\text{migration time of neutral molecule}}$

The capillary must be thin enough to dissipate heat rapidly. Temperature gradients disturb the flow and reduce resolution.

Speed = $\dfrac{\text{distance to detector}}{\text{migration time}} = \dfrac{L_d}{t}$

Electric field = $\dfrac{\text{applied volts}}{\text{distance between ends of capillary}} = \dfrac{V}{L_t}$

For quantitative analysis, use

$\dfrac{\text{Peak area}}{\text{Migration time}}$

*Electroosmotic mobility* is the speed of the neutral species, $u_{\text{neutral}}$, divided by the electric field:

*Electroosmotic mobility:*
$$\mu_{\text{eo}} = \frac{u_{\text{neutral}}}{E} = \frac{L_d/t_{\text{neutral}}}{V/L_t} \qquad (26\text{-}13)$$

The *electrophoretic mobility* of an analyte is the difference $\mu_{\text{app}} - \mu_{\text{eo}}$.

For maximum precision, mobilities are measured relative to an internal standard. Absolute variation from run to run should not affect relative mobilities, unless there are time-dependent (nonequilibrium) interactions of the solute with the wall.

For molecules of similar size, the magnitude of the electrophoretic mobility increases with charge. A protein "charge ladder" is a synthetic mixture made from a single protein with many different charges. For example, we can obtain such a mixture by acetylating variable numbers of lysine amino acid side chains (Table 10-1) to reduce their charge from $+1$ ($R\!-\!NH_3^+$) to 0 ($R\!-\!NHC(=O)CH_3$).[26]

$$\underset{\text{Lysine}}{\text{Protein}-NH_2} + \underset{\text{Acetic anhydride}}{CH_3\overset{\displaystyle O}{\overset{\|}{C}}-O-\overset{\displaystyle O}{\overset{\|}{C}}CH_3} \longrightarrow \underset{\text{Acetylated product }(-NHAc)}{\text{Protein}-NHCCH_3} + CH_3CO_2H$$

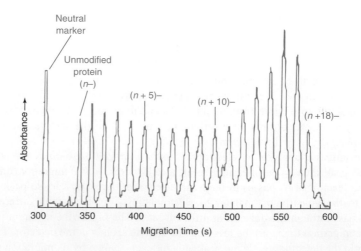

Protein (+4)    acetylation →    +3    +    +2    +    +1    +    0

Lysine $-NH_2$ groups have $pK_a \approx 10.3$. At pH 8.3, 99% of these groups are protonated ($-NH_3^+$). Acetylation gives a mixture with every possible number of modified amino groups from 0 to the total number of lysine residues. This mixture gives the electropherogram in Figure 26-22 with a series of nearly evenly spaced peaks. Each molecule has approximately the same size and shape (and therefore nearly the same friction coefficient) but a different charge.

Problem 26-39 shows how to find the charge of unmodified protein from the charge ladder.

## Example  Mobilities in a Protein Charge Ladder

Bovine carbonic anhydrase is a protein with 18 lysine residues. The 19 peaks in Figure 26-22 arise from unmodified protein ($P^{n-}$) plus protein with every possible degree of acetylation: $P^{(n+1)-}$, $P^{(n+2)-}$, $P^{(n+3)-}$, ... , $P^{(n+18)-}$. The voltage applied to the 0.840-m-long capillary is $2.50 \times 10^4$ V. A neutral marker molecule, carried by electroosmotic flow, requires 308 s to travel 0.640 m from the inlet to the detector. Migration times of $P^{n-}$ and $P^{(n+1)-}$ are 343 s and 355 s, respectively. Find the electroosmotic velocity and electroosmotic mobility. Find the apparent and electrophoretic mobilities of $P^{n-}$ and $P^{(n+1)-}$.

**Figure 26-22** Protein charge ladder. Bovine carbonic anhydrase was acetylated to give species with charges of $n^-$ (unacetylated), $(n+1)^-, (n+2)^-, \ldots, (n+18)^-$ (fully acetylated). Electrophoresis was carried out at pH 8.3 at $2.50 \times 10^4$ V in a capillary with a length of 0.840 m and a distance to the detector of 0.640 m. The neutral, ultraviolet-absorbing marker used to measure electroosmotic flow was mesityl oxide, $(CH_3)_2C=CHC(=O)CH_3$. *[From M. K. Menon and A. L. Zydney, "Determination of Effective Protein Charge by Capillary Electrophoresis," Anal. Chem. **2000,** 72, 5714.]*

**Solution** Electroosmotic velocity is found from the migration time of the neutral marker:

$$\text{Electroosmotic velocity} = \frac{\text{distance to detector } (L_d)}{\text{migration time}} = \frac{0.640 \text{ m}}{308 \text{ s}} = 2.08 \text{ mm/s}$$

Electric field is the voltage divided by the total length, $L_t$, of the column: $E = 25\ 000 \text{ V}/0.840 \text{ m} = 2.98 \times 10^4 \text{ V/m}$. Mobility is the constant of proportionality between velocity and electric field:

$$u_{eo} = \mu_{eo}E \Rightarrow \mu_{eo} = \frac{u_{eo}}{E} = \frac{0.002\ 08 \text{ m/s}}{2.98 \times 10^4 \text{ V/m}} = 6.98 \times 10^{-8} \frac{\text{m}^2}{\text{V} \cdot \text{s}}$$

The mobility of the neutral marker, which we just calculated, is the electroosmotic mobility for the entire solution.

The apparent mobility of $P^{n-}$ is obtained from its migration time:

$$\mu_{app} = \frac{u_{net}}{E} = \frac{0.640 \text{ m}/343 \text{ s}}{2.98 \times 10^4 \text{ V/m}} = 6.26 \times 10^{-8} \frac{\text{m}^2}{\text{V} \cdot \text{s}}$$

Electrophoretic mobility describes the response of the ion to the electric field. We subtract the electroosmotic mobility from the apparent mobility to find electrophoretic mobility:

$$\mu_{app} = \mu_{ep} + \mu_{eo} \Rightarrow \mu_{ep} = \mu_{app} - \mu_{eo}$$

$$= (6.26 - 6.98) \times 10^{-8} \frac{\text{m}^2}{\text{V} \cdot \text{s}} = -0.72 \times 10^{-8} \frac{\text{m}^2}{\text{V} \cdot \text{s}}$$

The electrophoretic mobility is negative because the protein has a negative charge and migrates in a direction opposite that of electroosmotic flow. Electroosmotic flow at pH 8.3 is faster than electromigration, so the protein does get carried to the detector. Similar calculations for the modified protein $P^{(n+1)-}$ give $\mu_{app} = 6.05 \times 10^{-8} \text{ m}^2/(\text{V} \cdot \text{s})$ and $\mu_{ep} = -0.93 \times 10^{-8} \text{ m}^2/(\text{V} \cdot \text{s})$. The electrophoretic mobility of $P^{(n+1)-}$ is more negative than that of $P^{n-}$ because the charge is more negative.

## Theoretical Plates and Resolution

Consider a capillary of length $L_d$ from the inlet to the detector. In Section 23-4, we defined the number of theoretical plates as $N = L_d^2/\sigma^2$, where $\sigma$ is the standard deviation of the band. If the only mechanism of zone broadening is longitudinal diffusion, the standard deviation was given by Equation 23-26: $\sigma = \sqrt{2Dt}$, where $D$ is the diffusion coefficient and $t$ is the migration time ($= L_d/u_{net} = L_d/[\mu_{app}E]$). Combining these equations with the definition of electric field ($E = V/L_t$, where $V$ is the applied voltage) gives an expression for the number of plates:

*Number of plates:*

$$N = \frac{\mu_{app}V}{2D} \frac{L_d}{L_t} \tag{26-14}$$

How many theoretical plates might we hope to attain? Using a typical value of $\mu_{app} = 2 \times 10^{-8} \text{ m}^2/(\text{V} \cdot \text{s})$ (derived for a 10-min migration time in a capillary with $L_t = 60$ cm, $L_d = 50$ cm, and 25 kV) and using diffusion coefficients from Table 23-1, we find

For $K^+$:
$$N = \frac{[2 \times 10^{-8} \text{ m}^2/(\text{V} \cdot \text{s})][25\ 000 \text{ V}]}{2(2 \times 10^{-9} \text{ m}^2/\text{s})} \frac{0.50 \text{ m}}{0.60 \text{ m}} = 1.0 \times 10^5 \text{ plates}$$

For serum albumin:
$$N = \frac{[2 \times 10^{-8} \text{ m}^2/(\text{V} \cdot \text{s})][25\ 000 \text{ V}]}{2(0.059 \times 10^{-9} \text{ m}^2/\text{s})} \frac{0.50 \text{ m}}{0.60 \text{ m}} = 3.5 \times 10^6 \text{ plates}$$

For the small, rapidly diffusing $K^+$ ion, we expect 100 000 plates. For the slowly diffusing protein serum albumin (FM 65 000), we expect more than 3 million plates. High plate count means that bands are very narrow and resolution between adjacent bands is excellent.

In reality, additional sources of zone broadening include the finite width of the injected band (Equation 23-32), a parabolic flow profile from heating inside the capillary, adsorption of solute on the capillary wall (which acts as a stationary phase), the finite length of the detection zone, and mobility mismatch of solute and buffer ions that leads to nonideal elec-

Number of plates: $N = \dfrac{L_d^2}{\sigma^2}$

$L_d$ = distance to detector
$\sigma$ = standard deviation of Gaussian band.

$L_t$ = total length of column

Under special conditions in which a reverse hydrodynamic flow was imposed to slow the passage of analytes through the capillary, up to 17 million plates were observed in the separation of small molecules![27]

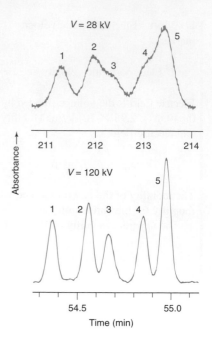

**Figure 26-23** Small region from the electropherogram of a complex mixture shows that increasing voltage increases resolution. All conditions are the same in both runs except voltage, which is ordinarily limited to ~30 kV. Special precautions were required to prevent electric arcing and overheating at 120 kV. *[From K. M. Hutterer and J. W. Jorgenson, "Ultrahigh-voltage Capillary Zone Electrophoresis," Anal. Chem.* **1999,** *77, 1293.]*

trophoretic behavior. If these other factors are properly controlled, ~$10^5$ plates are routinely achieved.

Equation 26-14 says that, for a constant ratio $L_d/L_t$, plate count is independent of capillary length. In contrast with chromatography, longer capillaries in electrophoresis do not give higher resolution.

Equation 26-14 also tells us that, the higher the voltage, the greater the number of plates (Figure 26-23). Voltage is ultimately limited by capillary heating, which produces a parabolic temperature profile that gives band broadening. The optimum voltage is found by making an *Ohm's law plot* of current versus voltage with *background electrolyte* (also called *run buffer*) in the capillary. In the absence of overheating, this curve should be a straight line. The maximum allowable voltage is the value at which the curve deviates from linearity (by, say, 5%). Buffer concentration and composition, thermostat temperature, and active cooling all play roles in how much voltage can be tolerated. Up to a point, higher voltage gives better resolution and faster separations.

As in chromatography, resolution between closely spaced peaks A and B in an electropherogram is related to plate count, $N$, and separation factor, $\gamma$, by Equation 23-30: resolution = $(\sqrt{N}/4)(\gamma - 1)$. The *separation factor* ($\gamma = u_{net,A}/u_{net,B}$) is the quotient of migration times $t_B/t_A$. Increasing $\gamma$ increases separation of peaks, and increasing $N$ decreases their width.

**26-6 Conducting Capillary Electrophoresis**

Clever variations of electrophoresis allow us to separate neutral molecules as well as ions, to separate optical isomers, and to lower detection limits by up to $10^6$. Adaptations of electrophoresis provide a foundation for new technology called "analysis on a chip." In the future, drug discovery and clinical diagnosis will depend on small chips carrying out unprecedented numbers of operations with unprecedented speed.

### Controlling the Environment Inside the Capillary

The inside capillary wall controls the electroosmotic velocity and provides undesired adsorption sites for multiply charged molecules, such as proteins. A fused-silica capillary should be prepared for its first use by washing for 15 min each ($>$20 column volumes) with 1 M NaOH and 0.1 M NaOH, followed by run buffer (~20 mM buffer). For subsequent use at high pH, wash for ~10 s with 0.1 M NaOH, followed by deionized water and then by at least 5 min with run buffer.[28] If the capillary is being run with pH 2.5 phosphate buffer, wash between runs with 1 M phosphoric acid, deionized water, and run buffer.[29] When changing buffers, allow at least 5 min of flow for equilibration. For the pH range 4–6, at which equilibration of the wall with buffer is very slow, the capillary needs frequent regeneration with

*Background electrolyte* (the solution in the capillary and the electrode reservoirs) controls pH and electrolyte composition in the capillary.

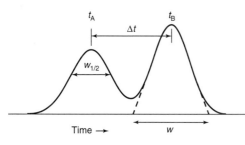

$$\text{Resolution} = \frac{\Delta t}{w_{av}} = \frac{0.589 \Delta t}{w_{1/2av}}$$

$$\text{Resolution} = \frac{\sqrt{N}}{4}(\gamma - 1)$$

$N$ = plate number

$\gamma$ = separation factor

$$= \frac{\text{speed of faster species}}{\text{speed of slower species}}$$

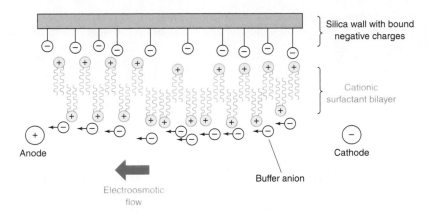

**Figure 26-24** Charge reversal created by a cationic surfactant bilayer coated on the capillary wall. The diffuse part of the double layer contains excess anions, and electroosmotic flow is in the direction opposite that shown in Figure 26-20. The surfactant is the didodecyldimethylammonium ion, $(n\text{-}C_{12}H_{25})N(CH_3)_2^+$ , represented as 〰〰⊕ in the illustration.

0.1 M NaOH if migration times become erratic. Buffer in both reservoirs should be replaced periodically because ions become depleted and electrolysis raises the pH at the cathode and lowers the pH at the anode. The capillary inlet should be ~2 mm away from and below the electrode to minimize entry of electrolytically generated acid or base into the column.[30] Stored capillaries should be filled with distilled water.

Different separations require more or less electroosmotic flow. Small anions with high mobility and highly negatively charged proteins require brisk electroosmotic flow or they will not travel toward the cathode. At pH 2, there is little charge on the silanol groups and little electroosmotic flow. At pH 11, the wall is highly charged and electroosmotic flow is strong. Proteins with many positively charged substituents can bind tightly to negatively charged silica. To control this, 30–60 mM diaminopropane (which gives $^+H_3NCH_2CH_2CH_2NH_3^+$) may be added to the run buffer to neutralize charge on the wall. The wall charge can be reduced to near 0 by covalent attachment of silanes with neutral, hydrophilic substituents. However, many coatings are unstable under alkaline conditions.

You can reverse the direction of electroosmotic flow by adding a cationic surfactant, such as didodecyldimethylammonium bromide, to the run buffer.[31] This molecule has a positive charge at one end and two long hydrocarbon tails. The surfactant coats the negatively charged silica with the tails pointing away from the surface (Figure 26-24). A second layer of surfactant orients itself in the opposite direction so that the tails form a nonpolar hydrocarbon layer. This *bilayer* adheres tightly to the wall of the capillary and effectively reverses the wall charge from negative to positive. Buffer anions create electroosmotic flow from cathode to anode when voltage is applied. Best results are obtained when the capillary is freshly regenerated for each run.

> Example of covalent coating that helps prevent protein from sticking to the capillary and provides reproducible migration times:
>
> O—Si(⋯OCH₃)—O—⌒⌒—O—Polyacrylamide

### Sample Injection and Composition

Injection may be **hydrodynamic** (using a pressure difference between the two ends of the capillary) or **electrokinetic** (using an electric field to drive sample into the capillary). For hydrodynamic injection (Figure 26-17), the capillary is dipped into a sample solution and the injected volume is

> For quantitative analysis, it is critical to use an internal standard because the amount of sample injected into the capillary is not very reproducible.

*Hydrodynamic injection:* $$\text{Volume} = \frac{\Delta P \pi d^4 t}{128 \eta L_t} \qquad (26\text{-}15)$$

where $\Delta P$ is the pressure difference between the ends of the capillary, $d$ is the inner diameter of the capillary, $t$ is the injection time, $\eta$ is the sample viscosity, and $L_t$ is the total length of the capillary.

> **Example** Hydrodynamic Injection Time
>
> How much time is required to inject a sample equal to 1.0% of the length of a 50-cm capillary if the diameter is 50 μm and the pressure difference is $2.0 \times 10^4$ Pa (0.20 bar)? Assume that the viscosity is 0.001 0 kg/(m · s), which is close to the viscosity of water.
>
> **Solution** The injection plug will be 0.50 cm long and occupy a volume of $\pi r^2 \times$ length $= \pi(25 \times 10^{-6} \text{ m})^2 (0.5 \times 10^{-2} \text{ m}) = 9.8 \times 10^{-12} \text{ m}^3$. The required time is
>
> $$t = \frac{128 \eta L_t (\text{volume})}{\Delta P \pi d^4} = \frac{128[0.001\ 0\ \text{kg/(m · s)}](0.50\ \text{m})(9.8 \times 10^{-12}\ \text{m}^3)}{(2.0 \times 10^4\ \text{Pa})\pi(50 \times 10^{-6}\ \text{m})^4} = 5.0\ \text{s}$$
>
> The units work out when we realize that Pa = force/area = $(\text{kg · m/s}^2)/\text{m}^2 = \text{kg/(m · s}^2)$.

For electrokinetic injection, the capillary is dipped in the sample and a voltage is applied between the ends of the capillary. The moles of each ion taken into the capillary in $t$ seconds are

$$\textit{Electrokinetic injection:} \qquad \text{Moles injected} = \mu_{app}\underbrace{\left(E\frac{\kappa_b}{\kappa_s}\right)}_{\text{Effective electric field}}t\pi r^2 C \qquad (26\text{-}16)$$

where $\mu_{app}$ is the apparent mobility of analyte ($= \mu_{ep} + \mu_{eo}$), $E$ is the applied electric field (V/m), $r$ is the capillary radius, $C$ is the sample concentration (mol/m$^3$), and $\kappa_b/\kappa_s$ is the ratio of conductivities of the buffer and sample. One problem with electrokinetic injection is that each analyte has a different mobility. For quantitative analysis, the injected sample does not have the same composition as the original sample. Electrokinetic injection is most useful for capillary gel electrophoresis (described later), in which the liquid in the capillary is too viscous for hydrodynamic injection.

---

**Example** Electrokinetic Injection Time

How much time is required to inject a sample equal to 1.0% of the length of a 50-cm capillary if the diameter is 50 $\mu$m and the injection electric field is 10 kV/m? Assume that the sample has 1/10 of the conductivity of background electrolyte and $\mu_{app} = 2.0 \times 10^{-8}$ m$^2$/(V · s).

**Solution**  Equation 26-16 is simpler than it looks. The factor $\kappa_b/\kappa_s$ is 10 in this case. The length of sample plug injected onto the column is just (sample speed) $\times$ (time) $= \mu_{app}Et$. The desired injection plug will be 0.50 cm long. The required time is

$$t = \frac{\text{plug length}}{\text{speed}} = \frac{\text{plug length}}{\mu_{app}\left(E\dfrac{\kappa_b}{\kappa_s}\right)} = \frac{0.005\,0 \text{ m}}{[2.0 \times 10^{-8} \text{ m}^2/(\text{V} \cdot \text{s})](10\,000 \text{ V/m})(10)} = 2.5 \text{ s}$$

Equation 26-16 multiplies the plug length times its cross-sectional area to find its volume and then multiplies by concentration to find the moles in that volume.

---

## Conductivity Effects: Stacking and Skewed Bands

We choose conditions so that analyte is focused into narrow bands at the start of the capillary by a process called **stacking.** Without stacking, if you inject a zone with a length of 5 mm, no analyte band can be narrower than 5 mm when it reaches the detector.

Stacking depends on the relation between the electric field in the zone of injected sample and that in the background electrolyte on either side of the sample. Optimal buffer concentration in the sample solution is 1/10 of the background electrolyte concentration, and the sample concentration should be 1/500 of the background electrolyte concentration. If the sample has a much lower ionic strength than the run buffer, the sample's conductivity is lower and its resistance is much greater. Electric field is inversely proportional to conductivity: The lower the conductivity, the greater the electric field. The electric field across the sample plug inside the capillary is higher than the electric field in the background electrolyte. Figure 26-25 shows ions in the sample plug migrating very fast, because the electric field is very high. When ions reach the zone boundary, they slow down because the field is lower outside the sample plug. This process of *stacking* continues until most of the analyte cations are concentrated at one end of the sample plug and most of the analyte anions are at the other end. The broad injection becomes concentrated into narrow bands of analyte cations or anions. Figure 26-26 shows signal enhancement by stacking.

If the conductivity of an analyte band is significantly different from the conductivity of the background electrolyte, peak distortion occurs. Figure 26-27 shows a band containing one analyte (as opposed to the sample plug in Figure 26-25, which contains all analytes in the entire injection). Suppose that the background conductivity is greater than the analyte conductivity ($\kappa_b > \kappa_a$). Under these conditions, the electric field is lower outside the analyte band than inside the band. The band migrates to the right in Figure 26-27. An analyte molecule that diffuses past the front on the right suddenly encounters a lower electric field and it slows down. Soon, the analyte zone catches up with the molecule and it is back in the zone. A molecule that diffuses out of the zone on the left encounters a lower electric field and it also slows down. The analyte zone is moving faster than the wayward molecule

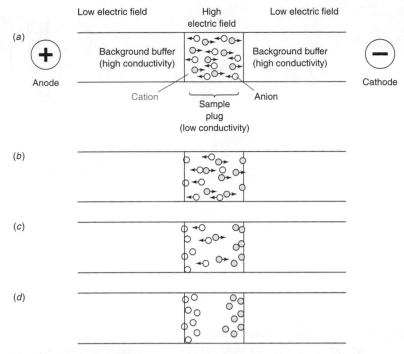

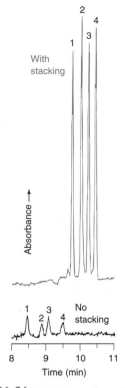

**Figure 26-25** Stacking of anions and cations at opposite ends of the low-conductivity sample plug (zone) occurs because the electric field in the sample plug is much higher than the electric field in the background electrolyte. Time increases from panels *a* to *d*. Electroneutrality is maintained by migration of background electrolyte ions, which are not shown.

and pulls away. This condition leads to a sharp front and a broad tail, as shown in the lower right electropherogram in Figure 26-27. When $\kappa_b < \kappa_a$, we observe the opposite electropherogram.

To minimize band distortion, sample concentration must be much less than the background electrolyte concentration. Otherwise it is necessary to choose a buffer co-ion that has the same mobility as the analyte ion. (The *co-ion* is the buffer ion with the same charge as analyte. The *counterion* has the opposite charge.)

**Figure 26-26** Lower trace: Sample injected electrokinetically for 2 s without stacking is limited in volume to prevent band broadening. Upper trace: With stacking, 15 times more sample could be injected (for 30 s), so the signal is 15 times stronger with no increase in bandwidth. [*From Y. Zhao and C. E. Lunte, "pH-Mediated Field Amplification On-Column Preconcentration of Anions in Physiological Samples for Capillary Electrophoresis," Anal. Chem.* **1999,** *71, 3985.]*

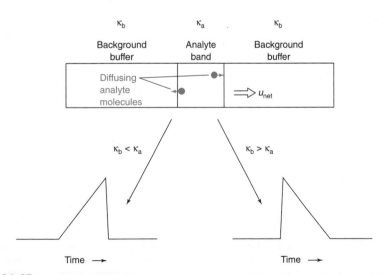

**Figure 26-27** Irregular peak shapes arise when the conductivity of the analyte band, $\kappa_a$, is not the same as the background conductivity, $\kappa_b$.

## Detectors

Water is so transparent that *ultraviolet detectors* can operate at wavelengths as short as 185 nm, where most solutes have strong absorption. To take advantage of short-wavelength ultraviolet detection, background electrolyte must have very low absorption. Borate buffers are com-

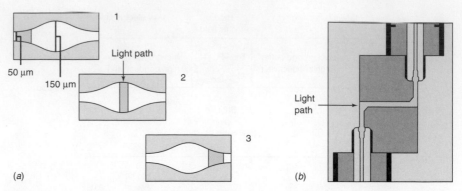

**Figure 26-28** Capillary designs to increase pathlength for measuring ultraviolet absorption, (*a*) Bubble cell. "Plug" of solute zone is maintained as it passes through the bubble, (*b*) Right-angle bend. Light path is made of black fused silica to reduce stray light. Reflective interior serves as a "light pipe" to maximize transmission. Detector response is linear up to 1.4 absorbance units. *[Courtesy Agilent Technologies, Palo Alto, CA.]*

monly used in electrophoresis because of their transparency.[32] Sensitivity is poor because the optical pathlength is only as wide as the capillary, which is 25–75 $\mu$m. Figure 26-28 shows a "bubble cell" that increases the absorbance signal-to-noise ratio by a factor of 3 to 5 and a right-angle bend that increases signal-to-noise by a factor of 10. However, the greater pathlength in the right-angle bend design leads to some band broadening. Successive peaks must be separated by 3 mm or they will overlap in the detector.

*Fluorescence detection* is sensitive to naturally fluorescent analytes or to fluorescent derivatives. *Amperometric detection* is sensitive to analytes that can be oxidized or reduced at an electrode (Figure 26-29). *Conductivity detection* with ion-exchange suppression of the background electrolyte (as in Figure 26-4) can detect small analyte ions at 1–10 ng/mL. *Electrospray mass spectrometry* (Figure 22-18) provides low detection limits and gives qualitative information about analytes.[33]

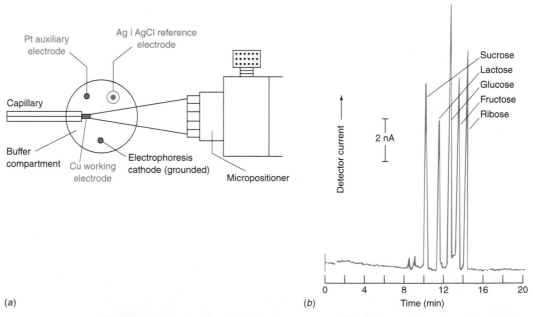

**Figure 26-29** (*a*) Amperometric detection with macroscopic working electrode at the outlet of the capillary. (*b*) Electropherogram of sugars separated in 0.1 M NaOH, in which OH groups are partially ionized, thereby turning the molecules into anions. *[From J. Ye and R. P. Baldwin, "Amperometric Detection in Capillary Electrophoresis with Normal Size Electrodes," Anal. Chem.* **1993**, *65, 3525.]*

Figure 26-30 shows the principle of *indirect detection,*[34] which applies to fluorescence, absorbance, amperometry, conductivity, and other forms of detection. A substance with a steady background signal is added to the background electrolyte. In the analyte band, analyte molecules displace the chromophoric substance, so the detector signal decreases when analyte passes by. Figure 26-31 shows an impressive separation of $Cl^-$ isotopes with indirect

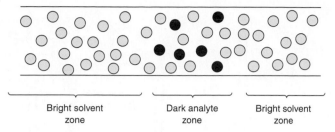

*Figure 26-30* Principle of indirect detection. When analyte emerges from the capillary, the strong background signal decreases.

detection in the presence of the ultraviolet-absorbing anion chromate. Electroneutrality dictates that an analyte band containing $Cl^-$ must have a lower concentration of $CrO_4^{2-}$ than is found in the background electrolyte. With less $CrO_4^{2-}$ to absorb ultraviolet radiation, a negative peak appears when $Cl^-$ reaches the detector. Benzoate and phthalate are other anions useful for this purpose. Detection limits in capillary electrophoresis are generally about an order of magnitude higher than detection limits in ion chromatography but one to two orders of magnitude lower than detection limits for ion-selective electrodes.

Approximate detection limits ($\mu M$) for indirect detection in electrophoresis:

| | |
|---|---|
| Ultraviolet absorption | 1–100 |
| Fluorescence | 0.001–1 |
| Chemiluminescence | 0.001–0.01 |
| Conductivity | 0.01–100 |
| Amperometry | $10^{-5}$–10 |
| Mass spectrometry | 0.001–0.01 |

SOURCE: Mostly from C. Vogt and G. L. Klunder, "Separation of Metal Ions by Capillary Electrophoresis— Diversity, Advantages, and Drawbacks of Detection Methods," Fresenius J. Anal. Chem. **2001**, 370, 316.

*Figure 26-31* Separation of natural isotopes of 0.56 mM $Cl^-$ by capillary electrophoresis with indirect spectrophotometric detection at 254 nm. Background electrolyte contains 5 mM $CrO_4^{2-}$ to provide absorbance at 254 nm and 2 mM borate buffer, pH 9.2. The capillary had a diameter of 75 $\mu m$, a total length of 47 cm (length to detector = 40 cm), and an applied voltage of 20 kV. The difference in electrophoretic mobility of $^{35}Cl^-$ and $^{37}Cl^-$ is just 0.12%. Conditions were adjusted so that electroosmotic flow was nearly equal to and opposite electrophoretic flow. The resulting near-zero net velocity gave the two isotopes maximum time to be separated by their slightly different mobilties. *[From C. A. Lucy and T. L. McDonald, "Separation of Chloride Isotopes by Capillary Electrophoresis Based on the Isotope Effect on Ion Mobility," Anal. Chem. **1995**, 67, 1074.]*

## Micellar Electrokinetic Chromatography[35]

The type of electrophoresis we have been discussing so far is called **capillary zone electrophoresis.** Separation is based on differences in electrophoretic mobility. If the capillary wall is negative, electroosmotic flow is toward the cathode (Figure 26-20) and the order of elution is cations before neutrals before anions. If the capillary wall charge is reversed by coating it with a cationic surfactant (Figure 26-24) and the instrument polarity is reversed, then the order of elution is anions before neutrals before cations. Neither scheme separates neutral molecules from one another.

**Micellar electrokinetic chromatography** separates neutral molecules and ions. We illustrate a case in which the anionic surfactant sodium dodecyl sulfate is present above its *critical micelle concentration* (Box 26-1), so negatively charged micelles are formed.[36] In Figure 26-32, electroosmotic flow is to the right. Electrophoretic migration of the negatively charged micelles is to the left, but net motion is to the right because the electroosmotic flow dominates.

Normal order of elution in capillary zone electrophoresis:

1. cations (highest mobility first)
2. all neutrals (unseparated)
3. anions (highest mobility last)

Sodium dodecyl sulfate ($n\text{-}C_{12}H_{25}OSO_3^- \ Na^+$)

*Micellar electrokinetic chromatography:* The more time the neutral analyte spends inside the micelle, the longer is its migration time. This technique was introduced by S. Terabe in 1984.[37]

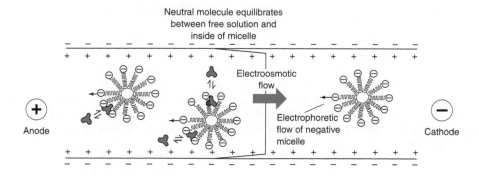

Neutral molecule equilibrates between free solution and inside of micelle

*Figure 26-32* Negatively charged sodium dodecyl sulfate micelles migrate upstream against the electroosmotic flow. Neutral molecules (solid color) are in dynamic equilibrium between free solution and the inside of the micelle. The more time spent in the micelle, the more the neutral molecule lags behind the electroosmotic flow.

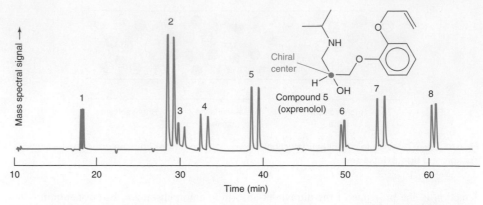

**Figure 26-33** Separation of enantiomers of eight β-blocker drugs by micellar electrokinetic chromatography at pH 8.0 in a 120-cm capillary at 30 kV. Micelles were formed by a polymer surfactant containing L-leucinate substituents for chiral recognition. The structure of one compound is shown. *[From C. Akbay, S, A. A. Rizvi, and S. A. Shamsi, "Simultaneous Enantioseparation and Tandem UV-MS Detection of Eight β-Blockers in Micellar Electrokinetic Chromatography Using a Chiral Molecular Micelle," Anal. Chem. **2005,** 77, 1672.]*

In the absence of micelles, all neutral molecules reach the detector in time $t_0$. Micelles injected with the sample reach the detector in time $t_{mc}$, which is longer than $t_0$ because the micelles migrate upstream. If a neutral molecule equilibrates between free solution and the inside of the micelles, its migration time is increased, because it migrates at the slower rate of the micelle part of the time. The neutral molecule reaches the detector at a time between $t_0$ and $t_{mc}$. *The more time the neutral molecule spends inside the micelle, the longer is its migration time.* Migration times of cations and anions also are affected by micelles, because ions partition between the solution and the micelles and interact electrostatically with the micelles.

Micellar electrokinetic chromatography is a form of chromatography because the micelles behave as a *pseudostationary phase.* Separation of neutral molecules is based on partitioning between the solution and the pseudostationary phase. The mass transfer term $Cu_x$ is no longer 0 in the van Deemter equation 26-7, but mass transfer into the micelles is fairly fast and band broadening is modest.

Imagination runs wild with the variables in micellar electrokinetic chromatography. We can add anionic, cationic, zwitterionic, and neutral surfactants to change the partition coefficients of analytes. (Cationic surfactants also change the charge on the wall and the direction of electroosmotic flow.) We can add solvents such as acetonitrile or *N*-methylformamide to increase the solubility of organic analytes and to change the partition coefficient between the solution and the micelles.[38,39] We can add cyclodextrins (Box 24-1) to separate optical isomers that spend different fractions of the time associated with the cyclodextrins.[40] In Figure 26-33, chiral micelles were used to separate enantiomers of chiral drugs.

*Sweeping* is a method to concentrate analyte by a factor of $10^3$–$10^5$ for trace analysis by micellar electrokinetic chromatography. Recall that *stacking* occurs when analyte in a low-conductivity electrolyte is injected into a capillary containing higher-conductivity electrolyte. The stronger electric field in the low-conductivity region forces ions to migrate rapidly and stack at the boundaries of the two electrolytes (Figure 26-25). In **sweeping,** a migrating ionic reagent such as sodium dodecyl sulfate micelles or a chelator binds analyte and concentrates it into a narrow band.

Figure 26-34 shows how analyte cations can be swept to concentrate them from trace levels to detectable levels. We begin in panel *a* with a capillary filled with 50 mM $H_3PO_4$/ 10% $CH_3OH$. Pressure is applied in panel *b* to inject a plug of 100 mM $H_3PO_4$ to occupy 10% of the column, followed by a short plug of $H_2O$. In panel *c*, analyte cations $C^+$ are injected electrokinetically with positive voltage at the left. The electric field is relatively high in the low-conductivity $H_2O$ and relatively low in the high-conductivity 100 mM $H_3PO_4$. As in Figure 26-25, $C^+$ *stacks* at the interface when the electric field abruptly decreases at the interface between $H_2O$ and $H_3PO_4$. Electrokinetic injection is continued for 200–600 s to fill much of the 100 mM $H_3PO_4$ zone with a high concentration of $C^+$ in panel *d* before disconnecting from the electric field.

*Stacking:* rapid migration of analyte in a low-conductivity region to concentrate at the interface with a high-conductivity region

*Sweeping:* migration of a collector species such as a micelle or chelator to concentrate analyte in a narrow region at the front of the migrating collector

Sweeping was introduced by J. P. Quirino and S. Terabe, "Exceeding 5000-Fold Concentraton of Dilute Analytes in Micellar Electrokinetic Chromatography," *Science* **1998,** *282,* 465.

**CHAPTER 26 Chromatographic Methods and Capillary Electrophoresis**

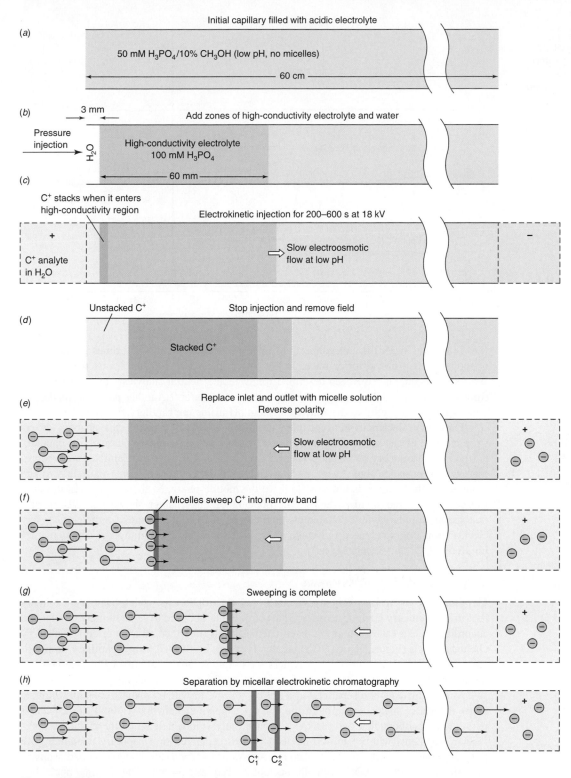

**Figure 26-34** Cation-selective exhaustive injection—sweeping.

In Figure 26-34e, the inlet and outlet solutions are replaced by 50 mM sodium dodecyl sulfate (SDS)/10 mM $H_3PO_4$/10% $CH_3OH$. Reverse voltage is applied with negative at the left. SDS micelles migrate rapidly to the right while electroosmotic flow proceeds slowly to the left. The advancing micelles associate with $C^+$ and *sweep* $C^+$ to the right in a concentrated, narrow band in panel *f*. In panel *g*, all analyte has been concentrated into a narrow band. As micelles continue to migrate to the right in panel *h*, different analyte cations (designated $C_1^+$ and $C_2^+$) separate from each other because $C_2^+$ is more soluble than $C_1^+$ in the micelles.

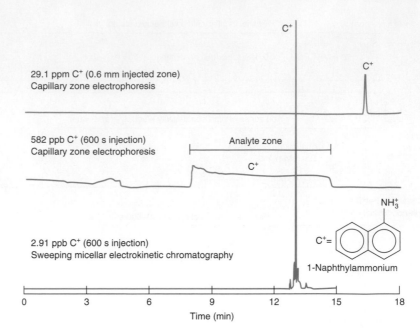

In Figure 26-35, the concentration of injected analyte cation is $10^4$ times greater in the upper trace than in the lower trace. In the lower trace, sweeping increases the signal by $7 \times 10^4$. Procedures have been described for sweeping neutral analytes[41] or anions derived from weak acids[42] and sweeping of metal ions by a chelator.[43] Another powerful method to concentrate analyte utilizes dynamic changes in pH inside the capillary.[44]

**Capillary electrochromatography,** described at the opening of this chapter, differs from micellar electrokinetic chromatography by using a true stationary phase.[45] The capillary is filled with stationary phase and solvent is driven by electroosmosis. Capillary electrochromatography provides about twice as many plates as HPLC for the same particle size and column length. Pressure is not used to drive the mobile phase, so there is no pressure drop associated with the small particles. The opening of this chapter showed an electrochromatography capillary with a continuous stationary phase instead of discrete particles. Capillary electrochromatography has a full range of applications, such as separation of enantiomers,[46] ion exchange,[47] and trace analysis.[48]

## Capillary Gel Electrophoresis

**Capillary gel electrophoresis** is a variant of gel electrophoresis, which has been a primary tool in biochemistry for four decades. Slabs of polymer gel used to separate macromolecules according to size have customarily been chemical gels, in which chains are cross-linked by chemical bonds (Figure 26-36a). Chemical gels cannot be flushed from a capillary, so physical gels (Figure 26-36b) in which polymers are simply entangled are used. Physical gels can be flushed and reloaded to generate a fresh capillary for each separation.

Macromolecules are separated in a gel by *sieving,* in which smaller molecules migrate faster than large molecules through the entangled polymer network. Color Plate 29 shows part of a deoxyribonucleic acid (DNA) sequence analysis in which a mixture of fluorescence-labeled fragments with up to 400 nucleotides was separated in a capillary containing 6% polyacrylamide (Figure 26-12, with no cross-links). DNA with 30 nucleotides had a migration time of 9 min and DNA with 400 nucleotides required 34 min. The terminal nucleotide of each fragment was labeled with one of two fluorescent labels, each of which was detected by fluorescence at two wavelengths. The combination of two labels and two wavelengths allows a unique assignment of each of the four possible nucleotides.[49] Capillary electrophoresis was an enabling technology for determining the sequence of nucleic acids in the human genome.[50]

A beautiful example of sieving in electrophoresis is illustrated in Figure 26-37. Quartz pillars were etched into a 25-μm-wide channel in a quartz plate by lithographic techniques used to make electronic chips. The pillars at the top left of the figure are 200 nm in diameter. A cover was placed over the pillars and the channel was filled with run buffer. Fluorescently labeled DNA fragments applied to the beginning of the channel separated after passing through 8 mm of the channel. The smallest fragments pass through first.

Confusing terms:

*Micellar electrokinetic chromatography:*
electrophoresis with micelles acting as
pseudostationary phase

*Capillary electrochromatography:* similar to
HPLC, except mobile phase is driven by
electroosmosis instead of pressure

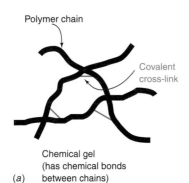

Polymer chain

Covalent
cross-link

Chemical gel
(has chemical bonds
(a)  between chains)

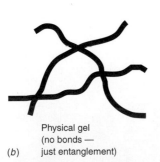

Physical gel
(no bonds —
(b)  just entanglement)

**Figure 26-36** (a) A chemical gel contains
covalent cross-links between different polymer
chains. (b) A physical gel is not cross-linked
but derives its properties from physical
entanglement of the polymers.

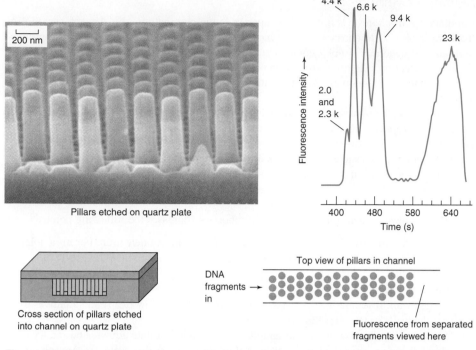

Pillars etched on quartz plate

Cross section of pillars etched
into channel on quartz plate

Top view of pillars in channel

DNA
fragments →
in

Fluorescence from separated
fragments viewed here

**Figure 26-37** Separation of 2 000- to 23 000-Da DNA fragments by sieving through quartz pillars in a capillary electrophoresis channel on a quartz plate. *[From N. Kaji, Y, Tezuka, Y. Takamura, M. Ueda, T. Nishimoto, H. Nakanishi, Y. Horiike, and Y. Baba, "Separation of Long DNA Molecules by Quartz Nanopillar Chips Under a Direct Current Electric Field," Anal. Chem.* **2004,** *76, 15.]*

Biochemists measure molecular mass of proteins by *sodium dodecyl sulfate* (SDS)–*gel electrophoresis.* Proteins are first *denatured* (unfolded to random coils) by reducing their disulfide bonds ($-S-S-$) with excess 2-mercaptoethanol ($HSCH_2CH_2OH$) and adding sodium dodecyl sulfate ($C_{12}H_{25}OSO_3^-Na^+$). Dodecyl sulfate anion coats hydrophobic regions and gives the protein a large negative charge that is approximately proportional to the length of the protein. Denatured proteins are then separated by electrophoresis through a sieving gel. Large molecules are retarded more than small molecules, which is the opposite behavior from size exclusion chromatography. In Figure 26-38, the logarithm of molecular mass of the SDS-coated protein is proportional to 1/(migration time). Absolute migration times vary from run to run, so relative migration times are measured. The relative migration time is the migration time of a protein divided by the migration time of a fast-moving small dye molecule.

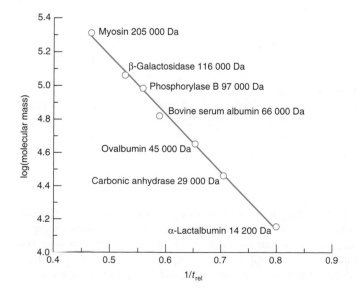

**Figure 26-38** Calibration curve for protein molecular mass in sodium dodecyl sulfate–capillary gel electrophoresis. The abscissa, $t_{rel}$, is the migration time of each protein divided by the migration time of a small dye molecule. *[Data from J. K, Grady, J. Zang, T. M. Laue, P. Arosio, and N. D, Chasteen, "Characterization of the H- and L-Subunit Ratios in Ferritins by Sodium Dodecyl Sulfate–Capillary Gel Electrophoresis," Anal. Biochem.* **2002,** *302, 263.]*

## Method Development

Capillary electrophoresis is not used as much as liquid chromatography. Advantages of electrophoresis relative to chromatography include (1) higher resolution, (2) low waste production, and (3) generally simpler equipment. Drawbacks of electrophoresis include (1) higher limits of detection, (2) run-to-run irreproducibility of migration times, (3) insolubility of some analytes in common electrolyte solutions, and (4) inability to scale up to a preparative separation.

Liquid chromatography is two decades more mature than capillary electrophoresis. As trained "electropherographers" become more common, more separations will be handled by electrophoresis. For example, electrophoresis displaced liquid chromatography as the preferred method for forensic analysis of alkaloids in opium and heroin.[51] The enabling technology for this application was dynamic coating of the capillary between runs to eliminate adsorption of analytes on the silica surface and decrease variations in migration times to less than 0.5%.

Method development for capillary electrophoresis addresses the following points:[52]

1. Select a detection method that provides the required limit of detection. For ultraviolet absorption, select the optimum wavelength. If necessary, use indirect detection or derivatization.
2. If there is a choice, separate analytes as anions, which do not stick to the negatively charged wall. If separating polycations, such as proteins at low pH, select additives to coat the walls or to reverse the wall charge.
3. Dissolve the entire sample. If the sample is not soluble in dilute aqueous buffer, try adding 6 M urea or surfactants. Acetate buffer tends to dissolve more organic solutes than does phosphate buffer. If necessary, nonaqueous solvents can be used.[53]
4. Determine how many peaks are present. Assign each peak by using authentic samples of analytes and diode array ultraviolet detection or mass spectrometry.
5. For complex mixtures, use computerized experimental design to help optimize separation conditions.[54]
6. Use the short end of the capillary in Figure 26-17 for quick scouting runs to determine the direction of migration and whether broad peaks are present. Broad peaks indicate that wall effects are occurring and a coating could be required.
7. See if pH alone provides adequate separation. For acids, begin with 50 mM borate buffer, pH 9.3. For bases, try 50 mM phosphate, pH 2.5. If the separation is not adequate, try adjusting buffer pH close to the average $pK_a$ of the solutes.
8. If pH does not provide adequate separation or if analytes are neutral, use surfactants for micellar electrokinetic capillary chromatography. For chiral solutes, try adding cyclodextrins.
9. Select a capillary wash procedure. If migration times are reproducible from run to run with no wash, then no wash is required. If migration times increase, wash for 5 to 10 s with 0.1 M NaOH, followed by 5 min with buffer. If migration times still drift, try increasing or decreasing the NaOH wash time by a few seconds. If migration times decrease, wash with 0.1 M $H_3PO_4$. If proteins or other cations stick to the walls, try washing with 0.1 M sodium dodecyl sulfate or use a commercial dynamic coating.
10. If necessary, select a sample cleanup method. Cleanup could be required if resolution is poor, if the salt content is high, or if the capillary fouls. Cleanup might involve solid-phase extraction (Section 28-3), protein precipitation, or dialysis (Demonstration 27-1).
11. If the detection limit is not sufficient, select a stacking or sweeping method to concentrate analyte in the capillary.
12. For quantitative analysis, determine the linear range required to measure the least concentrated and most concentrated analytes. If desired, select an internal standard. If the migration time or peak area of the standard changes, then you have an indication that some condition has drifted out of control.

## ■ ■ ■ 26-7 Lab on a Chip

One of the most exciting and rapidly developing areas of analytical chemistry is the "lab on a chip."[55] Liquids can be moved with precise control by electroosmosis or by pressure through micron-size channels etched into glass or plastic chips that are currently about the size of a microscope slide. Chemical reactions can be conducted by moving picoliters of fluid from different reservoirs, mixing them, and subjecting the products to chemical analy-

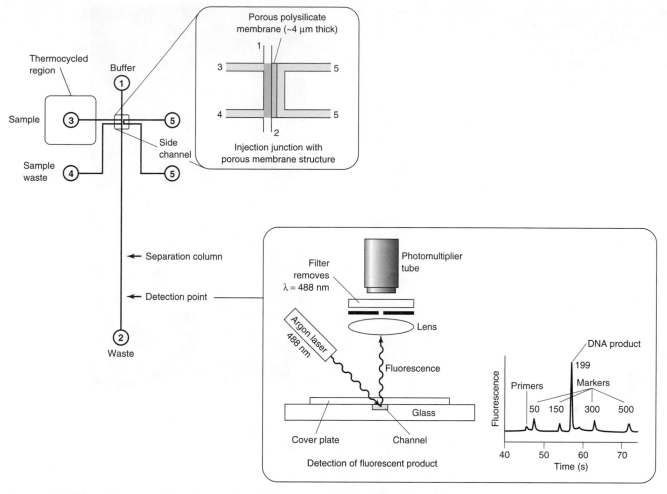

*Figure 26-39* MicroChannel device for replication and analysis of DNA. Channels are 7–10 μm deep and 40–45 μm wide. *[From J. Khandurina, T. E. McKnight, S. C. Jacobson, L. C, Waters, R. S. Foot, and J. M. Ramsey, "Integrated System for Rapid-PCR-Based DNA Analysis in Microfluidic Devices," Anal. Chem.* **2000***, 72, 2995.]*

sis on the chip with a variety of detectors (Color Plate 30).[56] Microreactors consuming micrograms of reactants can be used to optimize conditions for synthetic reactions, exploring dozens of conditions in a short time.[57]

Figure 26-39 shows a chip designed to replicate as few as 15 molecules of DNA by the *polymerase chain reaction* and then separate and characterize the products.[58] A 5-μL DNA sample is placed in reservoir 3, along with a supply of deoxynucleotide triphosphate (DNA building blocks), primers (short pieces of DNA that match sections of the DNA to be replicated), and a heat-resistant polymerase (an enzyme that synthesizes DNA). Reservoir 3 is taken through 25 heating cycles in 20 min. The number of DNA molecules is doubled in each cycle. A voltage applied between points 3 and 4 fills the channel with DNA solution by electroosmosis between these two points. To concentrate DNA in the channel between points 1 and 2, voltage is applied to drive liquid from point 3 to point 5. Small molecules and water can pass through the porous membrane between the 1–2 channel and the 5–5 channel. DNA is too big to pass through the pores and is concentrated in the 1–2 channel.[59] The DNA is then analyzed by gel electrophoresis conducted by applying voltage between points 1 and 2 with laser-induced fluorescence detection near point 2. The electropherogram shows one predominant product with 199 base pairs.

Figure 26-40 shows an integrated plastic biochip capable of conducting genetic analysis of whole blood—including target cell capture, sample preparation, DNA amplification, hybridization with target DNA, and electrochemical detection. This disposable device is inserted into nondisposable hardware that provides electric power, microprocessor control, detector electronics, readout, and a magnet to manipulate beads that capture specific cells in the sample. The ancillary hardware could potentially be shrunk to hand-held size. Applications of this device include pathogenic bacteria detection and genetic screening.

*Figure 26-40* Integrated biochip for DNA analysis of biological samples. Dimensions are 100 × 60 × 2 mm. Two bright circles are piezoelectric vibrators for ultrasonic agitation.
[From R. H. Liu, J. Yang, R. Lenigk, J. Bonanno, and P. Grodzinski, "Self-Contained, Fully Integrated Biochip for Sample Preparation, Polymerase Chain Reaction Amplification, and DNA Microarray Detection," Anal. Chem. **2004**, 76, 1824.]

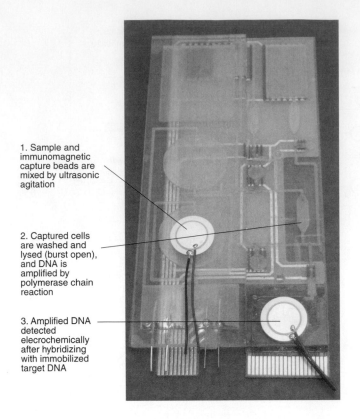

1. Sample and immunomagnetic capture beads are mixed by ultrasonic agitation

2. Captured cells are washed and lysed (burst open), and DNA is amplified by polymerase chain reaction

3. Amplified DNA detected elecrochemically after hybridizing with immobilized target DNA

The biochip in Figure 26-40 is engineered to conduct microfluidic operations at low cost. For example, Figure 26-41 shows how a channel can be closed and then opened. The channel is open to fluid flow at the left. To close the channel, an electric heater melts a small plug of paraffin wax in the side arm. Heat expands air in the side bulb to expel paraffin, which closes the fluid channel. The paraffin cools and the channel remains closed. To open the channel, heat is applied again and the molten paraffin is pushed into the wide region. This valve can close and open the channel only once. Mixing in the biochip in Figure 26-40 is accomplished by two piezoelectric disks that vibrate rapidly when an oscillating electric field is applied. Specific DNA is detected at the end of the process by an electochemical (voltammetric) method.

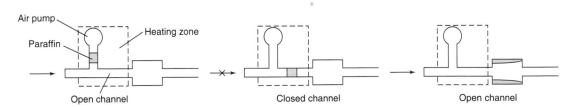

Air pump

Paraffin

Heating zone

Open channel

Closed channel

Open channel

*Figure 26-41* Method for closing and opening a channel in the microfluidic chip in Figure 26-40.

## Terms to Understand

affinity chromatography
anion exchanger
capillary electrochromatography
capillary electrophoresis
capillary gel electrophoresis
capillary zone electrophoresis
cation exchanger
cross-linking
deionized water
Donnan equilibrium

electrokinetic injection
electroosmosis
electrophoresis
gel
gel filtration
gradient elution
hydrodynamic injection
indirect detection
ion chromatography
ion-exchange chromatography

ion-pair chromatography
micellar electrokinetic chromatography
micelle
mobility
molecular exclusion chromatography
molecularly imprinted polymer
preconcentration
resin

selectivity coefficient
stacking
suppressed-ion chromatography
surfactant
sweeping
void volume

## Summary

Ion-exchange chromatography employs resins and gels with covalently bound charged groups that attract solute counterions (and that exclude ions having the same charge as the resin). Polystyrene resins are useful for small ions. Greater cross-linking of the resin increases the capacity, selectivity, and time needed for equilibration. Ion-exchange gels based on cellulose and dextran have large pore sizes and low charge densities, suitable for the separation of macromolecules. Certain inorganic solids have ion-exchange properties and are useful at extremes of temperature or radiation. Ion exchangers operate by the principle of mass action, with a gradient of increasing ionic strength most commonly used to effect a separation.

In suppressed-ion chromatography, a separator column separates ions of interest, and a suppressor membrane converts eluent into a nonionic form so that analytes can be detected by their conductivity. Alternatively, nonsuppressed ion chromatography uses an ion-exchange column and low-concentration eluent. If the eluent absorbs light, indirect spectrophotometric detection is convenient and sensitive. Ion-pair chromatography utilizes an ionic surfactant in the eluent to make a reversed-phase column function as an ion-exchange column.

Molecular exclusion chromatography is based on the inability of large molecules to enter small pores in the stationary phase. Small molecules enter these pores and therefore exhibit longer elution times than large molecules. Molecular exclusion is used for separations based on size and for molecular mass determinations of macromolecules. In affinity chromatography, the stationary phase retains one particular solute in a complex mixture. After all other components have been eluted, the desired species is liberated by a change in conditions.

In capillary zone electrophoresis, ions are separated by differences in mobility in a strong electric field applied between the ends of a silica capillary tube. The greater the charge and the smaller the hydrodynamic radius, the greater the electrophoretic mobility. Normally, the capillary wall is negative, and solution is transported from anode to cathode by electroosmosis of cations in the electric double layer. Solute cations arrive first, followed by neutral species, followed by solute anions (if electroosmosis is stronger than electrophoresis). Apparent mobility is the sum of electrophoretic mobility and electroosmotic mobility (which is the same for all species). Zone dispersion (broadening) arises mainly from longitudinal diffusion and the finite length of the injected sample. Stacking of solute ions in the capillary occurs when the sample has a low conductivity. Electroosmotic flow is reduced at low pH because surface $Si—O^-$ groups are protonated. $Si—O^-$ groups can be masked by polyamine cations, and the wall charge can be reversed by a cationic surfactant that forms a bilayer along the wall. Covalent coatings reduce electroosmosis and wall adsorption. Hydrodynamic sample injection uses pressure or siphoning; electrokinetic injection uses an electric field. Ultraviolet absorbance is commonly used for detection. Micellar electrophoretic chromatography uses micelles as a pseudostationary phase to separate neutral molecules and ions, which can be concentrated by stacking. Micelles can also be used to sweep long plugs of analyte into concentrated bands, thereby lowering detection limits by factors of $10^3$–$10^5$. Capillary electrochromatography is essentially the same as HPLC, but the mobile phase is driven by electroosmosis instead of pressure. Capillary gel electrophoresis separates macromolecules by sieving. In contrast with molecular exclusion chromatography, small molecules move fastest in gel electrophoresis. The "lab on a chip" uses electroosmotic or hydrodynamic liquid flow in lithographically fabricated chips to conduct chemical reactions and chemical analysis.

## Exercises

**26-A.** Vanadyl sulfate (VOSO$_4$, FM 163.00), as supplied commercially, is contaminated with H$_2$SO$_4$ (FM 98.08) and H$_2$O. A solution was prepared by dissolving 0.244 7 g of impure VOSO$_4$ in 50.0 mL of water. Spectrophotometric analysis indicated that the concentration of the blue VO$^{2+}$ ion was 0.024 3 M. A 5.00-mL sample was passed through a cation-exchange column loaded with H$^+$. When VO$^{2+}$ from the 5.00-mL sample became bound to the column, the H$^+$ released required 13.03 mL of 0.022 74 M NaOH for titration. Find the weight percent of each component (VOSO$_4$, H$_2$SO$_4$, and H$_2$O ) in the vanadyl sulfate.

**26-B.** Blue Dextran 2000 was eluted during gel filtration in a volume of 36.4 mL from a 2.0 × 40 cm (diameter × length) column of Sephadex G-50, which fractionates molecules in the molecular mass range 1 500 to 30 000.

(a) At what retention volume would hemoglobin (molecular mass 64 000) be expected?

(b) Suppose that radioactive $^{22}$NaCl, which is not adsorbed on the column, is eluted in a volume of 109.8 mL. What would be the retention volume of a molecule with $K_{av} = 0.65$?

**26-C.** Consider a capillary electrophoresis experiment conducted near pH 9, at which the electroosmotic flow is stronger than the electrophoretic flow.

(a) Draw a picture of the capillary, showing the placement of the anode, cathode, injector, and detector. Show the direction of electroosmotic flow and the direction of electrophoretic flow of a cation and an anion. Show the direction of net flow.

(b) Using Table 15-1, explain why Cl$^-$ has a shorter migration time than I$^-$. Predict whether Br$^-$ will have a shorter migration time than Cl$^-$ or a greater migration time than I$^-$.

(c) Why is the mobility of I$^-$ greater than that of Cl$^-$?

## Problems

### Ion-Exchange and Ion Chromatography

**26-1.** State the purpose of the separator and suppressor in suppressed-ion chromatography. For cation chromatography, why is the suppressor an anion-exchange membrane?

**26-2.** State the effects of increasing cross-linking on an ion-exchange column.

**26-3.** What is deionized water? What kind of impurities are not removed by deionization?

**26-4.** The exchange capacity of an ion-exchange resin can be defined as the number of moles of charged sites per gram of dry resin. Describe how you would measure the exchange capacity of an anion-exchange resin by using standard NaOH, standard HCl, or any other reagent you wish.

**26-5.** Consider a negatively charged protein adsorbed on an anion-exchange gel at pH 8.

(a) How will a gradient of eluent pH (from pH 8 to some lower pH) be useful for eluting the protein? Assume that the ionic strength of the eluent is kept constant.

(b) How would a gradient of ionic strength (at constant pH) be useful for eluting the protein?

**26-6.** What does the designation 200/400 mesh mean on a bottle of chromatography stationary phase? What is the size range of such particles? (See Table 28-2.) Which particles are smaller, 100/200 or 200/400 mesh?

**26-7.** Propose a scheme for separating trimethylamine, dimethylamine, methylamine, and ammonia from one another by ion-exchange chromatography.

**26-8.** Suppose that an ion-exchange resin $(R^-Na^+)$ is immersed in a solution of NaCl. Let the concentration of $R^-$ in the resin be 3.0 M.

(a) What will the ratio $[Cl^-]_o/[Cl^-]_i$ be if $[Cl^-]_o$ is 0.10 M?

(b) What will the ratio $[Cl^-]_o/[Cl^-]_i$ be if $[Cl^-]_o$ is 1.0 M?

(c) Will the fraction of electrolyte inside the resin increase or decrease as the outside concentration of electrolyte increases?

**26-9.** *Material balance.* If you intend to measure all the anions and cations in an unknown, one sanity check on your results is that the total positive charge should equal the total negative charge. Concentrations of anions and cations in pond water in Figure 26-5 are expressed in $\mu g/mL$. Find the total concentration of negative and positive charge (mol/L) to assess the quality of the analysis. What do you conclude about this analysis?

**26-10.** Compounds with $-\overset{\overset{\displaystyle OH}{|}}{C}-\overset{\overset{\displaystyle OH}{|}}{C}-$ or $-\overset{\overset{\displaystyle OH}{|}}{C}-\overset{\overset{\displaystyle NH_2}{|}}{C}-$ linkages can be analyzed by cleavage with periodate. One mole of 1,2-ethanediol consumes one mole of periodate:

$$\begin{matrix} CH_2OH \\ | \\ CH_2OH \end{matrix} + \ IO_4^- \ \rightarrow \ 2CH_2{=}O \ + \ H_2O \ + \ IO_3^-$$

1,2-Ethanediol    Periodate    Formaldehyde              Iodate
FM 62.07

To analyze 1,2-ethanediol, oxidation with excess $IO_4^-$ is followed by passage of the whole reaction solution through an anion-exchange resin that binds both $IO_4^-$ and $IO_3^-$. $IO_3^-$ is then selectively and quantitatively eluted with $NH_4Cl$. Absorbance of the eluate is measured at 232 nm ($\varepsilon = 900 \ M^{-1} \ cm^{-1}$) to find the quantity of $IO_3^-$ produced by the reaction. In one experiment, 0.213 9 g of aqueous 1,2-ethanediol was dissolved in 10.00 mL. Then 1.000 mL of the solution was treated with 3 mL of 0.15 M $KIO_4$ and subjected to ion-exchange separation of $IO_3^-$ from unreacted $IO_4^-$. The eluate (diluted to 250.0 mL) gave $A_{232} = 0.521$ in a 1.000-cm cell, and a blank gave $A_{232} = 0.049$. Find the wt% of 1,2-ethanediol in the original sample.

**26-11.** In ion-exclusion chromatography, ions are separated from nonelectrolytes by an ion-exchange column. Nonelectrolytes penetrate the stationary phase, whereas half of the ions are repelled by the fixed charges. Because electrolytes have access to less of the column volume, they are eluted before nonelectrolytes. The chromatogram here shows the separation of trichloroacetic acid (TCA, $pK_a = 0.5$), dichloroacetic acid (DCA, $pK_a = 1.1$), and

monochloroacetic acid (MCA, $pK_a = 2.86$) by passage through a cation-exchange resin eluted with 0.01 M HCl. Explain why the three acids are separated and why they elute in the order shown.

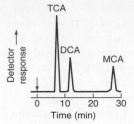

Separation of acids on cation-exchange column. *[From V. T. Turkelson and M. Richards, "Separation of the Citric Acid Cycle Acids by Liquid Chromatography," Anal. Chem.* **1978**, *50, 1420.]*

**26-12.** 🖳 Norepinephrine (NE) in human urine can be assayed by ion-pair chromatography by using an octadecylsilane stationary phase and sodium octyl sulfate as the mobile-phase additive. Electrochemical detection (oxidation at 0.65 V versus Ag|AgCl) is used, with 2,3-dihydroxybenzylamine (DHBA) as internal standard.

Norepinephrine cation (NE)      2,3-Dihydroxybenzylamine cation (DHBA)

$$CH_3CH_2CH_2CH_2CH_2CH_2CH_2CH_2OSO_3^- \ Na^+$$
Sodium octyl sulfate

(a) Explain the physical mechanism by which an ion-pair separation works.

(b) A urine sample containing an unknown amount of NE and a fixed, added concentration of DHBA gave a detector peak height ratio NE/DHBA = 0.298. Then small standard additions of NE were made, with the following results:

| Added concentration of NE (ng/mL) | Peak height ratio NE/DHBA |
|---|---|
| 12 | 0.414 |
| 24 | 0.554 |
| 36 | 0.664 |
| 48 | 0.792 |

Using the graphical treatment shown in Figure 5-6, find the original concentration of NE in the specimen.

**26-13.** Decomposition of dithionite $(S_2O_4^{2-})$ was studied by chromatography on an anion-exchange column eluted with 20 mM trisodium 1,3,6-naphthalenetrisulfonate in 90 vol% $H_2O$/10 vol% $CH_3CN$ with ultraviolet detection at 280 nm. A solution of sodium dithionite stored for 34 days in the absence of air gave five peaks identified as $SO_3^{2-}$, $SO_4^{2-}$, $S_2O_3^{2-}$, $S_2O_4^{2-}$, and $S_2O_5^{2-}$. All the peaks had a *negative* absorbance. Explain why.

**26-14.** (a) Suppose that the reservoir in Figure 26-6 contains 1.5 L of 2.0 M $K_2PO_4$. For how many hours can the reservoir provide 20 mM KOH at a flow of 1.0 mL/min if 75% consumption of $K^+$ in the reservoir is feasible?

(b) What starting and ending current would be required to produce a gradient from 5.0 mM KOH to 0.10 M KOH at 1.0 mL/min flow rate?

**26-15.** The system in Figure 26-6 can be adapted to produce the strong-acid eluent methanesulfonic acid $(CH_3SO_3^-H^+)$. For this purpose, the polarity of the electrodes is reversed and the reservoir can contain $NH_4^+CH_3SO_3^-$. The barrier membrane and the resin bed at the bottom of the figure must both be anion exchangers loaded with $CH_3SO_3^-$. Draw this system and write the chemistry that occurs in each part.

### Molecular Exclusion and Affinity Chromatography

**26-16.** (a) How can molecular exclusion chromatography be used to measure the molecular mass of a protein?
(b) Which pore size in Figure 26-14 is most suitable for chromatography of molecules with molecular mass near 100 000?

**26-17.** A gel filtration column has a radius, $r$, of 0.80 cm and a length, $l$, of 20.0 cm.
(a) Calculate the volume, $V_t$, of the column, which is equal to $\pi r^2 l$.
(b) The void volume, $V_0$, was 18.1 mL and the total volume of mobile phase was 35.8 mL. Find $K_{av}$ for a solute eluted at 27.4 mL.

**26-18.** Ferritin (molecular mass 450 000), transferrin (molecular mass 80 000), and ferric citrate were separated by molecular exclusion chromatography on Bio-Gel P-300. The column had a length of 37 cm and a 1.5-cm diameter. Eluate fractions of 0.65 mL were collected. The maximum of each peak came at the following fractions: ferritin, 22; transferrin, 32; and ferric citrate, 84. (That is, the ferritin peak came at an elution volume of $22 \times 0.65 = 14.3$ mL.) Assuming that ferritin is eluted at the void volume and that ferric citrate is eluted at $V_m$, find $K_{av}$ for transferrin.

**26-19.** (a) The void volume in Figure 26-14 is the volume at which the curves rise vertically at the left. What is the smallest molecular mass of molecules excluded from the 10-nm-pore-size column?
(b) What is the molecular mass of molecules eluted at 6.5 mL from the 10-nm column?

**26-20.** A polystyrene resin molecular exclusion HPLC column has a diameter of 7.8 mm and a length of 30 cm. The solid portions of the gel particles occupy 20% of the volume, the pores occupy 40%, and the volume between particles occupies 40%.
(a) At what volume would totally excluded molecules be expected to emerge?
(b) At what volume would the smallest molecules be expected?
(c) A mixture of polyethylene glycols of various molecular masses is eluted between 23 and 27 mL. What does this imply about the retention mechanism for these solutes on the column?

**26-21.** 🖩 The following substances were separated on a gel filtration column. Estimate the molecular mass of the unknown.

| Compound | $V_r$(mL) | Molecular mass (Da) |
|---|---|---|
| Blue Dextran 2000 | 17.7 | $2 \times 10^6$ |
| Aldolase | 35.6 | 158 000 |
| Catalase | 32.3 | 210 000 |
| Ferritin | 28.6 | 440 000 |
| Thyroglobulin | 25.1 | 669 000 |
| Unknown | 30.3 | ? |

### Capillary Electrophoresis

**26-22.** What is electroosmosis?

**26-23.** Electroosmotic velocities of buffered solutions are shown for a bare silica capillary and one with aminopropyl groups (silica—Si—$CH_2CH_2CH_2NH_2$) covalently attached to the wall. A positive sign means that flow is toward the cathode. Explain the signs and relative magnitudes of the velocities.

| | Electroosmotic velocity (mm/s) for $E = 4.0 \times 10^4$ V/m | |
|---|---|---|
| | pH 10 | pH 2.5 |
| Bare silica | +3.1 | +0.2 |
| Aminopropyl-modified silica | +1.8 | −1.3 |

SOURCE: K. Emoto, J. M. Harris, and M. Van Alstine, "Grafting Poly(ethylene glycol) Epoxide to Amino-Derivatized Quartz: Effect of Temperature and pH on Grafting Density," Anal. Chem. **1996**, 68, 3751.

**26-24.** Fluorescent derivatives of amino acids separated by capillary zone electrophoresis had migration times with the following order: arginine (fastest) < phenylalanine < asparagine < serine < glycine (slowest). Explain why arginine has the shortest migration time.

**26-25.** What is the principal source of zone broadening in ideal capillary electrophoresis?

**26-26.** (a) An electrophoresis channel etched into a glass plate has a rectangular $12 \times 50$ μm cross section. How long (mm) is the sample zone for a 100-pL sample?
(b) If sample travels 24 mm in 8 s to reach the detector, what is the standard deviation in bandwidth contributed by the finite length of the injection zone? (*Hint:* See Equation 23-32.)
(c) If the diffusion coefficient of a solute is $1.0 \times 10^{-8}$ m$^2$/s, what is the diffusional contribution (standard deviation) to band broadening?
(d) What is the expected total bandwidth at the baseline($w = 4\sigma$)?

**26-27.** State three different methods to reduce electroosmotic flow. Why does the direction of electroosmotic flow change when a silica capillary is washed with a cationic surfactant?

**26-28.** Explain how neutral molecules can be separated by micellar electrokinetic chromatography. Why is this a form of chromatography?

**26-29.** (a) What pressure difference is required to inject a sample equal to 1.0% of the length of a 60.0-cm capillary in 4.0 s if the diameter is 50 μm? Assume that the viscosity of the solution is 0.001 0 kg/(m · s).
(b) The pressure exerted by a column of water of height $h$ is $h\rho g$, where $\rho$ is the density of water and $g$ is the acceleration of gravity (9.8 m/s$^2$). To what height would you need to raise the sample vial to create the necessary pressure to load the sample in 4.0 s? Is it possible to raise the inlet of this column to this height? How could you obtain the desired pressure?

**26-30.** (a) How many moles of analyte are present in a 10.0 μM solution that occupies 1.0% of the length of a 25-μm $\times$ 60.0-cm capillary?
(b) What voltage is required to inject this many moles into a capillary in 4.0 s if the sample has 1/10 of the conductivity of background electrolyte, $\mu_{app} = 3.0 \times 10^{-8}$ m$^2$/(V · s), and the sample concentration is 10.0 μM?

**26-31.** Measure the number of plates for the electrophoretic peak in Figure 26-19. Use the formula for asymmetric peaks to find the number of plates for the chromatographic peak.

**26-32. (a)** A long thin molecule has a greater friction coefficient than a short fat molecule. Predict whether fumarate or maleate will have greater electrophoretic mobility.

Fumarate          Maleate

**(b)** Electrophoresis is run with the injection end positive and the detection end negative. At pH 8.5, both anions have a charge of $-2$. The electroosmotic flow from the positive terminal to the negative terminal is greater than the electrophoretic flow, so these two anions have a net migration from the positive to the negative end of the capillary in electrophoresis. From your answer to part **(a)**, predict the order of elution of these two species.

**(c)** At pH 4.0, both anions have a charge close to $-1$, and the electroosmotic flow is weak. Therefore electrophoresis is run with the injection end negative and the detection end positive. The anions migrate from the negative end of the capillary to the positive end. Predict the order of elution.

**26-33. (a)** A particular solution in a particular capillary has an electroosmotic mobility of $1.3 \times 10^{-8} \, m^2/(V \cdot s)$ at pH 2 and $8.1 \times 10^{-8} \, m^2/(V \cdot s)$ at pH 12. How long will it take a neutral solute to travel 52 cm from the injector to the detector if 27 kV is applied across the 62-cm-long capillary tube at pH 2? At pH 12?

**(b)** An analyte anion has an electrophoretic mobility of $-1.6 \times 10^{-8} \, m^2/(V \cdot s)$. How long will it take to reach the detector at pH 2? At pH 12?

**26-34.** Figure 26-23 shows the effect on resolution of increasing voltage from 28 to 120 kV.

**(a)** What is the expected ratio of migration times ($t_{120 \, kV}/t_{28 \, kV}$) in the two experiments? Measure the migration times for peak 1 and find the observed ratio.

**(b)** What is the expected ratio of plates ($N_{120 \, kV}/N_{28 \, kV}$) in the two experiments?

**(c)** What is the expected ratio of bandwidths ($\sigma_{120 \, kV}/\sigma_{28 \, kV}$)?

**(d)** What is the physical reason why increasing voltage decreases bandwidth and increases resolution?

**26-35.** The observed behavior of benzyl alcohol ($C_6H_5CH_2OH$) in capillary electrophoresis is given here. Draw a graph showing the number of plates versus the electric field and explain what happens as the field increases.

| Electric field (V/m) | Number of plates |
|---|---|
| 6 400 | 38 000 |
| 12 700 | 78 000 |
| 19 000 | 96 000 |
| 25 500 | 124 000 |
| 31 700 | 124 000 |
| 38 000 | 96 000 |

**26-36.** Measure the width of the $^{35}Cl^-$ peak at half-height in Figure 26-31 and calculate the plate number. The capillary was 40.0 cm in length. Find the plate height.

**26-37.** The migration time for $Cl^-$ in a capillary zone electrophoresis experiment is 17.12 min and the migration time for $I^-$ is 17.78

min. From mobilities in Table 15-1, predict the migration time of $Br^-$. (The observed value is 19.6 min.)

**26-38.** 🔲 *Molecular mass by sodium dodecylsulfate–gel electrophoresis.* Ferritin is a hollow iron-storage protein[60] consisting of 24 subunits that are a variable mixture of heavy (H) or light (L) chains, arranged in octahedral symmetry. The hollow center, with a diameter of 8 nm, can hold up to 4 500 iron atoms in the approximate form of the mineral ferrihydrite ($5Fe_2O_3 \cdot 9H_2O$). Iron(II) enters the protein through eight pores located on the 3-fold symmetry axes of the octahedron. Oxidation to Fe(III) occurs at catalytic sites on the H chains. Other sites on the inside of the L chains appear to nucleate the crystallization of ferrihydrite.

Migration times for protein standards and the ferritin subunits are given in the table. Prepare a graph of log(molecular mass) versus 1/(relative migration time), where relative migration time = (migration time)/(migration time of marker dye). Compute the molecular mass of the ferritin light and heavy chains. The masses of the chains, computed from amino acid sequences, are 19 766 and 21 099 Da.

| Protein | Molecular mass (Da) | Migration time (min) |
|---|---|---|
| Orange G marker dye | small | 13.17 |
| α-Lactalbumin | 14 200 | 16.46 |
| Carbonic anhydrase | 29 000 | 18.66 |
| Ovalbumin | 45 000 | 20.16 |
| Bovine serum albumin | 66 000 | 22.36 |
| Phosphorylase B | 97 000 | 23.56 |
| β-Galactosidase | 116 000 | 24.97 |
| Myosin | 205 000 | 28.25 |
| Ferritin light chain | | 17.07 |
| Ferritin heavy chain | | 17.97 |

SOURCE: J. K. Grady, J. Zang, T. M. Laue, P. Arosio, and N. D. Chasteen, "Characterization of the H- and L-Subunit Ratios in Ferritins by Sodium Dodecyl Sulfate–Capillary Gel Electrophoresis," Anal. Biochem. **2002**, 302, 263.

**26-39.** 🔲 *Protein charge ladder.* Electrophoretic mobility is proportional to charge. If members of a charge ladder (Figure 26-22) have the same friction coefficient (that is, the same size and shape), then the charge of the unmodified protein divided by its electrophoretic mobility, $z_0/\mu_0$, is equal to the charge of the nth member divided by its electrophoretic mobility ($z_0 + \Delta z_n)/\mu_n$. Setting these two expressions equal to each other and rearranging gives

$$\Delta z_n = z_0 \left( \frac{\mu_n}{\mu_0} - 1 \right)$$

where $z_0$ is the charge of the unmodified protein, $\Delta z_n$ is the charge difference between the nth modified protein and the unmodified protein, $\mu_n$ is the electrophoretic mobility of the nth modified protein, and $\mu_0$ is the electrophoretic mobility of the unmodified protein. The migration time of the neutral marker molecule in Figure 26-22 is 308.5 s. The migration time of the unmodified protein is 343.0 s. Other members of the charge ladder have migration times of 355.4, 368.2, 382.2, 395.5, 409.1, 424.9, 438.5, 453.0, 467.0, 482.0, 496.4, 510.1, 524.1, 536.9, 551.4, 565.1, 577.4, and 588.5 s. Calculate the electrophoretic mobility of each protein and prepare a plot of $\Delta z_n$ versus ($\mu_n/\mu_0$) − 1. If the points lie on a straight line, the slope is the charge of the unmodified protein, $z_0$. Prepare such a plot, suggest an explanation for its shape, and find $z_0$.

**26-40.** *Resolution.* Suppose that the electroosmotic mobility of a solution is $+1.61 \times 10^{-7}$ m$^2$/(V · s). How many plates are required to separate sulfate from bromide with a resolution of 2.0? Use Table 15-1 for mobilities and Equation 23-30 for resolution.

**26-41.** The water-soluble vitamins niacinamide (a neutral compound), riboflavin (a neutral compound), niacin (an anion), and thiamine (a cation) were separated by micellar electrokinetic chromatography in 15 mM borate buffer (pH 8.0) with 50 mM sodium dodecyl sulfate. The migration times were niacinamide (8.1 min), riboflavin (13.0 min), niacin (14.3 min), and thiamine (21.9 min). What would the order have been in the absence of sodium dodecyl sulfate? Which compound is most soluble in the micelles?

**26-42.** When the following three compounds are separated by micellar electrokinetic chromatography at pH 9.6, three peaks are observed. When 10 mM α-cyclodextrin is added to the run buffer, two of the three peaks split into two peaks, giving a total of five peaks. Explain this observation and predict which compound does not split.

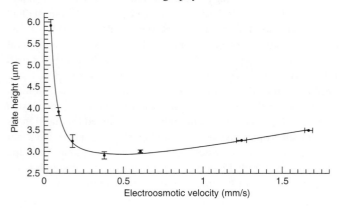

| 1 | 2 | 3 |
| --- | --- | --- |
| Cyclobarbital | Thiopental | Phenobarbital |

**26-43.** A van Deemter plot for the separation of neutral dyes by micellar electrokinetic chromatography follows.[61]

*(plot: Plate height (μm) on y-axis from 2.5 to 6.0 vs Electroosmotic velocity (mm/s) on x-axis from 0 to 1.5)*

**(a)** Explain why plate height increases at low and high velocities.

**(b)** The irregular flow path term, $A$, in the van Deemter equation should really be 0 for the ideal case of micellar electrokinetic chromatography. The observed value of $A$ is 2.32 μm, which accounts for two-thirds of the band broadening at the optimum velocity. Suggest some reasons why $A$ is not 0.

**26-44.** To obtain the best separation of two weak acids in capillary electrophoresis, it makes sense to use the pH at which their charge difference is greatest. Prepare a spreadsheet to examine the charges of malonic and phthalic acid as a function of pH. At what pH is the difference greatest?

**26-45.** *Optimizing a separation of acids.* Benzoic acid containing $^{16}$O can be separated from benzoic acid containing $^{18}$O by electrophoresis at a suitable pH because they have slightly different acid dissociation constants. The difference in mobility is caused by the different fraction of each acid in the anionic form, A$^-$.

Calling this fraction α, we can write

$$H^{16}A \underset{}{\overset{^{16}K}{\rightleftharpoons}} H^+ + {}^{16}A^- \qquad H^{18}A \underset{}{\overset{^{18}K}{\rightleftharpoons}} H^+ + {}^{18}A^-$$

$$^{16}\alpha = \frac{^{16}K}{^{16}K + [H^+]} \qquad ^{18}\alpha = \frac{^{18}K}{^{18}K + [H^+]}$$

where $K$ is the equilibrium constant. The greater the fraction of acid in the form A$^-$, the faster it will migrate in the electric field. It can be shown that, for electrophoresis, the maximum separation will occur when $\Delta\alpha/\sqrt{\overline{\alpha}}$ is a maximum. In this expression, $\Delta\alpha = {}^{16}\alpha - {}^{18}\alpha$, and $\overline{\alpha}$ is the average fraction of dissociation $[= \frac{1}{2}({}^{16}\alpha + {}^{18}\alpha)]$.

**(a)** Let us denote the ratio of acid dissociation constants as $R = {}^{16}K/{}^{18}K$. In general, $R$ will be close to unity. For benzoic acid, $R = 1.020$. Abbreviate $^{16}K$ as $K$ and write $^{18}K = K/R$. Derive an expression for $\Delta\alpha/\sqrt{\overline{\alpha}}$ in terms of $K$, $[H^+]$, and $R$. Because both equilibrium constants are nearly equal ($R$ is close to unity), set $\overline{\alpha}$ equal to $^{16}\alpha$ in your expression.

**(b)** Find the maximum value of $\Delta\alpha/\sqrt{\overline{\alpha}}$ by taking the derivative with respect to $[H^+]$ and setting it equal to 0. Show that the maximum difference in mobility of isotopic benzoic acids occurs when $[H^+] = (K/2R)(1 + \sqrt{1 + 8R})$.

**(c)** Show that, for $R \approx 1$, this expression simplifies to $[H^+] = 2K$, or pH = p$K$ − 0.30. That is, the maximum electrophoretic separation should occur when the column buffer has pH = p$K$ − 0.30, regardless of the exact value of $R$.[62]

**26-46. (a)** *Ion mobility spectrometry* (Section 22-3) is *gas-phase electrophoresis.* Describe how ion mobility spectrometry works and state the analogies between this technique and capillary electrophoresis.

**(b)** As in electrophoresis, the velocity, $u$, of a gas-phase ion is $u = \mu E$, where $\mu$ is the mobility of the ion and $E$ is the electric field ($E = V/L$, where $V$ is the voltage difference across distance $L$). In ion mobility spectrometry, the time to go from the gate to the detector (Figure 22-16a) is called *drift time,* $t_d$. Drift time is related to voltage: $t_d = L/u = L/(\mu E) = L/(\mu(V/L)) = L^2/\mu V$. Plate number is $N = 5.55(t_d/w_{1/2})^2$, where $w_{1/2}$ is the width of the peak at half-height. Ideally, peak width depends only on the width of the gate pulse that admits ions to the drift tube and on diffusive broadening of ions while they migrate:[63]

$$w_{1/2}^2 = \underbrace{t_g^2}_{\substack{\text{Initial width} \\ \text{from gate pulse}}} + \underbrace{\left(\frac{16kT\ln2}{Vez}\right)t_d^2}_{\substack{\text{Diffusive} \\ \text{broadening}}}$$

where $t_g$ is the time that the ion gate is open, $k$ is Boltzmann's constant, $T$ is temperature, $V$ is the potential difference from the gate to the detector, $e$ is the elementary charge, and $z$ is the charge of the ion. Prepare a graph of $N$ versus $V$ ($0 \le V \le 20\,000$) for an ion with $\mu = 8 \times 10^{-5}$ m$^2$/(s · V), and $t_g = 0, 0.05,$ or 0.2 ms at 300 K. Let the length of the drift region be $L = 0.2$ m. Explain the shapes of the curves. What is the disadvantage of using short $t_g$?

**(c)** Why does decreasing $T$ increase $N$?

**(d)** In a well-optimized ion mobility spectrometer, protonated arginine ion ($z = 1$) had a drift time of 24.925 ms and $w_{1/2} = 0.154$ ms at 300 K. Find $N$. For $V = 12\,500$ V and $t_g = 0.05$ ms, what is the theoretical plate number?

**(e)** Resolution is given by $R = (\sqrt{N}/4)(\gamma - 1)$, where $\gamma$ is the ratio of drift times for two components. In a well-optimized ion mobility spectrometer, protonated leucine had $t_d = 22.5$ ms and protonated isoleucine had $t_d = 22.0$ ms. Both had $N \approx 80\,000$. What is the resolution of the two peaks?

# 27 | Gravimetric and Combustion Analysis

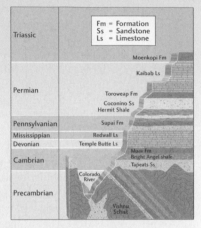

Layers of rock exposed in the Grand Canyon by the erosive action of the Colorado River provide a window on a billion years of Earth's history. *[From J. Grotzinger, T. H. Jordan, F. Press, and R. Siever, Understanding Earth, 5th ed. (New York: W. H. Freeman, 2006), p. 177.]*

In the 1800s, geologists understood that new layers (*strata*) of rock are deposited on top of older layers. Characteristic fossils in each layer helped geologists to identify strata from the same geologic era all around the world. However, the actual age of each layer was unknown.

In 1910, Arthur Holmes, a 20-year-old geology student at Imperial College in London, realized that radioactive decay could be used to measure the age of a rock. Physicists had discovered that U decays with a half life of 4.5 billion years to eight atoms of He and suspected that the final product was Pb. Holmes conjectured that when a U-containing mineral crystallized, it should be relatively free of impurities. Once the mineral solidified, Pb would begin to accumulate. The ratio Pb/U is a "clock" giving the age of the mineral.

Holmes isolated U minerals from a 'Devonian'-age rock. He measured the U content by the rate of production of radioactive Rn gas, which is a decay product along the way to Pb. To measure Pb, he *fused* (Section 28-2) each mineral in borax, dissolved the fused mass in acid, and quantitatively precipitated milligram quantities of $PbSO_4$. The nearly constant ratio Pb/U = 0.046 g/g in 15 minerals was consistent with the hypotheses that Pb is the end product of radioactive decay and that little Pb had been present when the minerals crystallized. More importantly, the average age of the minerals was 370 million years.

**Geologic ages deduced by Holmes in 1911**

| Geologic period | Pb/U (g/g) | Millions of years | Today's accepted value |
| --- | --- | --- | --- |
| Carboniferous | 0.041 | 340 | 362–330 |
| Devonian | 0.045 | 370 | 380–362 |
| Silurian | 0.053 | 430 | 443–418 |
| Precambrian | 0.125–0.20 | 1 025–1 640 | 900–2 500 |

*SOURCE: C. Lewis, The Dating Game (Cambridge: Cambridge University Press, 2000); A. Holmes, "The Association of Lead with Uranium in Rock-Minerals, and Its Application to the Measurement of Geological Time," Proc. Royal Soc. Lond. A 1911, 85, 248.*

In **gravimetric analysis,** the mass of a product is used to calculate the quantity of the original analyte (the species being analyzed). Exceedingly careful gravimetric analysis by T. W. Richards and his colleagues early in the twentieth century determined the atomic masses of

Ag, Cl, and N to six-figure accuracy.[1] This Nobel Prize–winning research allowed the accurate determination of atomic masses of many elements. In **combustion analysis,** a sample is burned in excess oxygen and products are measured. Combustion is typically used to measure C, H, N, S, and halogens in organic compounds. To measure other elements in food, organic matter is burned in a closed system, the products and *ash* (unburned material) are dissolved in acid or base, and measured by inductively coupled plasma with atomic emission or mass spectrometry.

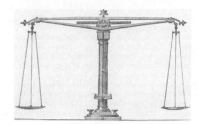

Gravimetry was the main form of chemical analysis in the eighteenth and nineteenth centuries but is too tedious to be a method of choice today. However, gravimetry is still one of the most accurate methods. Standards used to calibrate instruments are frequently derived from gravimetric or titrimetric procedures.

Nineteenth-century balance reproduced from *Fresenius' Quantitative Chemical Analysis,* 2nd American ed., 1881.

## ▪▪▪▪ 27-1  An Example of Gravimetric Analysis

An example of gravimetric analysis is the determination of $Cl^-$ by precipitation with $Ag^+$:

$$Ag^+ + Cl^- \rightarrow AgCl(s) \qquad (27\text{-}1)$$

The mass of AgCl produced tells us how much $Cl^-$ was originally present.

### ▪ Example  A Gravimetric Calculation

A 10.00-mL solution containing $Cl^-$ was treated with excess $AgNO_3$ to precipitate 0.436 8 g of AgCl. What was the molarity of $Cl^-$ in the unknown?

**Solution**   The formula mass of AgCl is 143.321. A precipitate weighing 0.436 8 g contains

$$\frac{0.436\ 8\ \cancel{g\ AgCl}}{143.321\ \cancel{g\ AgCl}/mol\ AgCl} = 3.048 \times 10^{-3}\ mol\ AgCl$$

Because 1 mol of AgCl contains 1 mol of $Cl^-$, there must have been $3.048 \times 10^{-3}$ mol of $Cl^-$ in the unknown.

$$[Cl^-] = \frac{3.048 \times 10^{-3}\ mol}{0.010\ 00\ L} = 0.304\ 8\ M$$

### ▪ Example  Marie Curie's Measurement of the Atomic Mass of Radium

As part of her Ph.D. research (*Radioactive Substances,* 1903), Marie Curie measured the atomic mass of radium, a new, radioactive element that she discovered. She knew that radium is in the same family as barium, so the formula of radium chloride is $RaCl_2$. In one experiment, 0.091 92 g of pure $RaCl_2$ was dissolved and treated with excess $AgNO_3$ to precipitate 0.088 90 g of AgCl. How many moles of $Cl^-$ were in the $RaCl_2$? From this measurement, find the atomic mass of Ra.

Marie and Pierre Curie and Henri Becquerel shared the Nobel Prize in physics in 1903 for pioneering investigations of radioactivity. The Curies needed 4 years to isolate 100 mg of $RaCl_2$ from several tons of ore. Marie received the Nobel Prize in chemistry in 1911 for her isolation of metallic radium. Linus Pauling, John Bardeen, and Frederick Sanger are the only others who received two Nobel Prizes.

**Solution**   AgCl precipitate weighing 0.088 90 g contains

$$\frac{0.088\ 90\ \cancel{g\ AgCl}}{143.321\ \cancel{g\ AgCl}/mol\ AgCl} = 6.202_9 \times 10^{-4}\ mol\ AgCl$$

Because 1 mol of AgCl contains 1 mol of $Cl^-$, there must have been $6.202_9 \times 10^{-4}$ mol of $Cl^-$ in the $RaCl_2$. For 2 mol of Cl, there must be 1 mol of Ra, so

$$mol\ radium = \frac{6.202_9 \times 10^{-4}\ \cancel{mol\ Cl}}{2\ \cancel{mol\ Cl}/mol\ Ra} = 3.101_4 \times 10^{-4}\ mol$$

Let the formula mass of $RaCl_2$ be *x*. We found that 0.091 92 g $RaCl_2$ contains $3.101_4 \times 10^{-4}$ mol $RaCl_2$. Therefore

$$3.101_4 \times 10^{-4}\ mol\ RaCl_2 = \frac{0.091\ 92\ g\ RaCl}{x\ g\ RaCl_2/mol\ RaCl_2}$$

$$x = \frac{0.091\ 92\ g\ RaCl}{3.101_4 \times 10^{-4}\ mol\ RaCl_2} = 296.3_8$$

The atomic mass of Cl is 35.453, so the formula mass of $RaCl_2$ is

Formula mass of $RaCl_2$ = atomic mass of Ra + 2(35.453 g/mol) = $296.3_8$ g/mol

$\Rightarrow$ atomic mass of Ra = 225.5 g/mol

The inside cover of this book lists the atomic number (the integer mass) of the long-lived isotope of Ra, which is 226.

**Table 27-1** Representative gravimetric analyses

| Species analyzed | Precipitated form | Form weighed | Interfering species |
|---|---|---|---|
| $K^+$ | $KB(C_6H_5)_4$ | $KB(C_6H_5)_4$ | $NH_4^+$, $Ag^+$, $Hg^{2+}$, $Tl^+$, $Rb^+$, $Cs^+$ |
| $Mg^{2+}$ | $Mg(NH_4)PO_4 \cdot 6H_2O$ | $Mg_2P_2O_7$ | Many metals except $Na^+$ and $K^+$ |
| $Ca^{2+}$ | $CaC_2O_4 \cdot H_2O$ | $CaCO_3$ or $CaO$ | Many metals except $Mg^{2+}$, $Na^+$, $K^+$ |
| $Ba^{2+}$ | $BaSO_4$ | $BaSO_4$ | $Na^+$, $K^+$, $Li^+$, $Ca^{2+}$, $Al^{3+}$, $Cr^{3+}$, $Fe^{3+}$, $Sr^{2+}$, $Pb^{2+}$, $NO_3^-$ |
| $Ti^{4+}$ | $TiO(5,7\text{-dibromo-8-}$ $hydroxyquinoline)_2$ | Same | $Fe^{3+}$, $Zr^{4+}$, $Cu^{2+}$, $C_2O_4^{2-}$, citrate, HF |
| $VO_4^{3-}$ | $Hg_3VO_4$ | $V_2O_5$ | $Cl^-$, $Br^-$, $I^-$, $SO_4^{2-}$, $CrO_4^{2-}$, $AsO_3^{3-}$, $PO_4^{3-}$ |
| $Cr^{3+}$ | $PbCrO_4$ | $PbCrO_4$ | $Ag^+$, $NH_4^+$ |
| $Mn^{2+}$ | $Mn(NH_4)PO_4 \cdot H_2O$ | $Mn_2P_2O_7$ | Many metals |
| $Fe^{3+}$ | $Fe(HCO_2)_3$ | $Fe_2O_3$ | Many metals |
| $Co^{2+}$ | $Co(1\text{-nitroso-2-naphtholate})_2$ | $CoSO_4$ (by reaction with $H_2SO_4$) | $Fe^{3+}$, $Pd^{2+}$, $Zr^{4+}$ |
| $Ni^{2+}$ | $Ni(dimethylglyoximate)_2$ | Same | $Pd^{2+}$, $Pt^{2+}$, $Bi^{3+}$, $Au^{3+}$ |
| $Cu^{2+}$ | $CuSCN$ | $CuSCN$ | $NH_4^+$, $Pb^{2+}$, $Hg^{2+}$, $Ag^+$ |
| $Zn^{2+}$ | $Zn(NH_4)PO_4 \cdot H_2O$ | $Zn_2P_2O_7$ | Many metals |
| $Ce^{4+}$ | $Ce(IO_3)_4$ | $CeO_2$ | $Th^{4+}$, $Ti^{4+}$, $Zr^{4+}$ |
| $Al^{3+}$ | $Al(8\text{-hydroxyquinolate})_3$ | Same | Many metals |
| $Sn^{4+}$ | $Sn(cupferron)_4$ | $SnO_2$ | $Cu^{2+}$, $Pb^{2+}$, As(III) |
| $Pb^{2+}$ | $PbSO_4$ | $PbSO_4$ | $Ca^{2+}$, $Sr^{2+}$, $Ba^{2+}$, $Hg^{2+}$, $Ag^+$, HCl, $HNO_3$ |
| $NH_4^+$ | $NH_4B(C_6H_5)_4$ | $NH_4B(C_6H_5)_4$ | $K^+$, $Rb^+$, $Cs^+$ |
| $Cl^-$ | $AgCl$ | $AgCl$ | $Br^-$, $I^-$, $SCN^-$, $S^{2-}$, $S_2O_3^{2-}$, $CN^-$ |
| $Br^-$ | $AgBr$ | $AgBr$ | $Cl^-$, $I^-$, $SCN^-$, $S^{2-}$, $S_2O_3^{2-}$, $CN^-$ |
| $I^-$ | $AgI$ | $AgI$ | $Cl^-$, $Br^-$, $SCN^-$, $S^{2-}$, $S_2O_3^{2-}$, $CN^-$ |
| $SCN^-$ | $CuSCN$ | $CuSCN$ | $NH_4^+$, $Pb^{2+}$, $Hg^{2+}$, $Ag^+$ |
| $CN^-$ | $AgCN$ | $AgCN$ | $Cl^-$, $Br^-$, $I^-$, $SCN^-$, $S^{2-}$, $S_2O_3^{2-}$ |
| $F^-$ | $(C_6H_5)_3SnF$ | $(C_6H_5)_3SnF$ | Many metals (except alkali metals), $SiO_4^{4-}$, $CO_3^{2-}$ |
| $ClO_4^-$ | $KClO_4$ | $KClO_4$ | |
| $SO_4^{2-}$ | $BaSO_4$ | $BaSO_4$ | $Na^+$, $K^+$, $Li^+$, $Ca^{2+}$, $Al^{3+}$, $Cr^{3+}$, $Fe^{3+}$, $Sr^{2+}$, $Pb^{2+}$, $NO_3^-$ |
| $PO_4^{3-}$ | $Mg(NH_4)PO_4 \cdot 6H_2O$ | $Mg_2P_2O_7$ | Many metals except $Na^+$, $K^+$ |
| $NO_3^-$ | Nitron nitrate | Nitron nitrate | $ClO_4^-$, $I^-$, $SCN^-$, $CrO_4^{2-}$, $ClO_3^-$, $NO_2^-$, $Br^-$, $C_2O_4^{2-}$ |
| $CO_3^{2-}$ | $CO_2$ (by acidification) | $CO_2$ | (The liberated $CO_2$ is trapped with Ascarite and weighed.) |

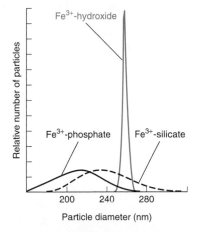

**Figure 27-1** Measured particle-size distribution of colloids formed when $FeSO_4$ was oxidized to $Fe^{3+}$ in $10^{-4}$ M $OH^-$ in the presence of phosphate ($PO_4^{3-}$), silicate ($SiO_4^{4-}$), or no added anions. [From M. L. Magnuson, D. A. Lytle, C. M. Frietch, and C. A. Kelty, "Characterization of Submicron Aqueous Iron(III) Colloids by Sedimentation Field Flow Fractionation," Anal. Chem. **2001**, 73, 4815.]

Supersaturation tends to decrease the particle size of a precipitate.

Representative analytical precipitations are listed in Table 27-1. A few common organic **precipitants** (agents that cause precipitation) are listed in Table 27-2. Conditions must be controlled to selectively precipitate one species. Potentially interfering substances may need to be removed prior to analysis.

## ■ ■ ■ 27-2 Precipitation

The ideal product of a gravimetric analysis should be very pure, insoluble, and easily filterable, and it should possess a known composition. Few substances meet all these requirements, but appropriate techniques can help optimize the properties of gravimetric precipitates. For example, solubility is usually decreased by cooling the solution.

Particles of precipitate should not be so small that they clog or pass through the filter. Large crystals also have less surface area to which impurities can become attached. At the other extreme is a *colloidal suspension* of particles that have diameters in the approximate range 1–500 nm and pass through most filters (Figure 27-1 and Demonstration 27-1). Precipitation conditions have much to do with the resulting particle size.

### Crystal Growth

Crystallization occurs in two phases: nucleation and particle growth. During **nucleation**, molecules in solution come together randomly and form small aggregates. *Particle growth* requires the addition of more molecules to a nucleus to form a crystal. When a solution contains more solute than should be present at equilibrium, the solution is said to be **supersaturated.**

Nucleation proceeds faster than particle growth in a highly supersaturated solution. The result is a suspension of tiny particles or, worse, a colloid. In a less supersaturated solution, nucleation is slower, and the nuclei have a chance to grow into larger, more tractable particles.

**Table 27-2** Common organic precipitating agents

| Name | Structure | Ions precipitated |
|------|-----------|-------------------|
| Dimethylglyoxime | | $Ni^{2+}$, $Pd^{2+}$, $Pt^{2+}$ |
| Cupferron | | $Fe^{3+}$, $VO_2^+$, $Ti^{4+}$, $Zr^{4+}$, $Ce^{4+}$, $Ga^{3+}$, $Sn^{4+}$ |
| 8-Hydroxyquinoline (oxine) | | $Mg^{2+}$, $Zn^{2+}$, $Cu^{2+}$, $Cd^{2+}$, $Pb^{2+}$, $Al^{3+}$, $Fe^{3+}$, $Bi^{3+}$, $Ga^{3+}$, $Th^{4+}$, $Zr^{4+}$, $UO_2^{2+}$, $TiO^{2+}$ |
| Salicylaldoxime | | $Cu^{2+}$, $Pb^{2+}$, $Bi^{3+}$, $Zn^{2+}$, $Ni^{2+}$, $Pd^{2+}$ |
| 1-Nitroso-2-naphthol | | $Co^{2+}$, $Fe^{3+}$, $Pd^{2+}$, $Zr^{4+}$ |
| Nitron | | $NO_3^-$, $ClO_4^-$, $BF_4^-$, $WO_4^{2-}$ |
| Sodium tetraphenylborate | $Na^+B(C_6H_5)_4^-$ | $K^+$, $Rb^+$, $Cs^+$, $NH_4^+$, $Ag^+$, organic ammonium ions |
| Tetraphenylarsonium chloride | $(C_6H_5)_4As^+Cl^-$ | $Cr_2O_7^{2-}$, $MnO_4^-$, $ReO_4^-$, $MoO_4^{2-}$, $WO_4^{2-}$, $ClO_4^-$, $I_3^-$ |

Techniques that promote particle growth include

1. Raising the temperature to increase solubility and thereby decrease supersaturation
2. Adding precipitant slowly with vigorous mixing, to prevent a local, highly supersaturated condition where the stream of precipitant first enters the analyte
3. Using a large volume of solution so the concentrations of analyte and precipitant are low

## Homogeneous Precipitation

In **homogeneous precipitation,** the precipitant is generated slowly by a chemical reaction (Table 27-3). For example, urea decomposes in boiling water to produce $OH^-$:

$$H_2N\text{—}\underset{Urea}{\overset{\overset{\displaystyle O}{\|}}{C}}\text{—}NH_2 + 3H_2O \xrightarrow{heat} CO_2 + 2NH_4^+ + 2OH^- \qquad (27\text{-}2)$$

By this means, the pH of a solution can be raised very gradually. Slow $OH^-$ formation enhances the particle size of Fe(III) formate precipitate:

$$\underset{Formic\ acid}{H\text{—}\overset{\overset{\displaystyle O}{\|}}{C}\text{—}OH} + OH^- \rightarrow \underset{Formate}{HCO_2^-} + H_2O \qquad (27\text{-}3)$$

$$3HCO_2^- + Fe^{3+} \rightarrow \underset{Fe(III)\ formate}{Fe(HCO_2)_3 \cdot nH_2O(s)} \downarrow \qquad (27\text{-}4)$$

**Colloids** are particles with diameters of ~1–500 nm. They are larger than molecules but too small to precipitate. They remain in solution indefinitely, suspended by the Brownian motion (random movement) of solvent molecules.[2]

To prepare colloid iron(III) hydroxide, heat 200 mL of distilled water in a beaker to 70°–90°C and leave an identical beaker of water at room temperature. Add 1 mL of 1 M $FeCl_3$ to each beaker and stir. The warm solution turns brown-red in a few seconds, whereas the cold solution remains yellow (Color Plate 31). The yellow color is characteristic of low-molecular-mass $Fe^{3+}$ compounds. The red color results from colloidal aggregates of $Fe^{3+}$ ions held together by hydroxide, oxide, and some chloride ions. These particles have a molecular mass of $10^5$ and a diameter of 10 nm, and they contain $10^3$ atoms of Fe.

You can demonstrate the size of colloidal particles with a **dialysis** experiment in which two solutions are separated by a *semipermeable membrane* that has pores with diameters of 1–5 nm.[3] Small molecules diffuse through these pores, but large molecules (such as proteins or colloids) cannot. (Collecting biological samples by *microdialysis* was discussed at the opening of Chapter 25.)

Pour some of the brown-red colloidal Fe solution into a dialysis tube knotted at one end; then tie off the other end. Drop the tube into a flask of distilled water to show that the color remains entirely within the bag after several days (Color Plate 31). For comparison, leave an identical bag containing a dark blue solution of 1 M $CuSO_4 \cdot 5H_2O$ in another flask of water. $Cu^{2+}$ diffuses out and the solution in the flask will be a uniform light blue color in 24 h. The yellow food coloring, tartrazine, can be used in place of

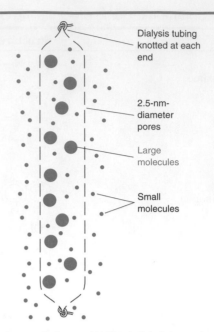

Dialysis tubing knotted at each end

2.5-nm-diameter pores

Large molecules

Small molecules

Large molecules remain trapped inside a dialysis bag, whereas small molecules diffuse through the membrane in both directions.

$Cu^{2+}$. If dialysis is conducted in hot water, it is completed during one class period.[4]

Dialysis is used to treat patients suffering from kidney failure. Blood is passed over a membrane through which metabolic waste products diffuse and are diluted into a large volume of liquid that is discarded. Protein molecules, which are a necessary part of the blood plasma, are too large to cross the membrane and are retained in the blood.

## Precipitation in the Presence of Electrolyte

An *electrolyte* is a compound that dissociates into ions when it dissolves.

Although it is common to find the excess common ion adsorbed on the crystal surface, it is also possible to find other ions selectively adsorbed. In the presence of citrate and sulfate, there is more citrate than sulfate adsorbed on a particle of $BaSO_4$.

Ionic compounds are usually precipitated in the presence of an electrolyte. To understand why, we must discuss how tiny colloidal crystallites *coagulate* (come together) into larger crystals. We illustrate the case of AgCl, which is commonly formed in 0.1 M $HNO_3$.

Figure 27-2 shows a colloidal particle of AgCl growing in a solution containing excess $Ag^+$, $H^+$, and $NO_3^-$. The surface of the particle has excess positive charge due to the **adsorption** of extra silver ions on exposed chloride ions. (To be adsorbed means to be attached to the

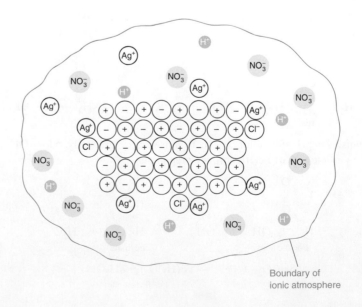

*Figure 27-2*  Colloidal particle of AgCl growing in a solution containing excess $Ag^+$, $H^+$, and $NO_3^-$. The particle has a net positive charge because of adsorbed $Ag^+$ ions. The region of solution surrounding the particle is called the *ionic atmosphere*. It has a net negative charge because the particle attracts anions and repels cations.

Boundary of ionic atmosphere

**Table 27-3** Common reagents used for homogeneous precipitation

| Precipitant | Reagent | Reaction | Some elements precipitated |
|---|---|---|---|
| $OH^-$ | Urea | $(H_2N)_2CO + 3H_2O \rightarrow CO_2 + 2NH_4^+ + 2OH^-$ | Al, Ga, Th, Bi, Fe, Sn |
| $OH^-$ | Potassium cyanate | $HOCN + 2H_2O \rightarrow NH_4^+ + CO_2 + OH^-$<br>Hydrogen cyanate | Cr, Fe |
| $S^{2-}$ | Thioacetamide[a] | $\underset{\substack{\|\| \\ S}}{CH_3CNH_2} + H_2O \rightarrow \underset{\substack{\|\| \\ O}}{CH_3CNH_2} + H_2S$ | Sb, Mo, Cu, Cd |
| $SO_4^{2-}$ | Sulfamic acid | $H_3\overset{+}{N}SO_3^- + H_2O \rightarrow NH_4^+ + SO_4^{2-} + H^+$ | Ba, Ca, Sr, Pb |
| $C_2O_4^{2-}$ | Dimethyl oxalate | $\underset{\substack{\|\|\|\| \\ OO}}{CH_3OCCOCH_3} + 2H_2O \rightarrow 2CH_3OH + C_2O_4^{2-} + 2H^+$ | Ca, Mg, Zn |
| $PO_4^{3-}$ | Trimethyl phosphate | $(CH_3O)_3P{=}O + 3H_2O \rightarrow 3CH_3OH + PO_4^{3-} + 3H^+$ | Zr, Hf |
| $CrO_4^{2-}$ | Chromic ion plus bromate | $2Cr^{3+} + BrO_3^- + 5H_2O \rightarrow 2CrO_4^{2-} + Br^- + 10H^+$ | Pb |
| 8-Hydroxyquinoline | 8-Acetoxyquinoline | | Al, U, Mg, Zn |

a. Hydrogen sulfide is volatile and toxic; it should be handled only in a well-vented hood. Thioacetamide is a carcinogen that should be handled with gloves. If thioacetamide contacts your skin, wash yourself thoroughly immediately. Leftover reagent is destroyed by heating at 50°C with 5 mol of NaOCl per mole of thioacetamide. [H. Elo, J. Chem. Ed. **1987**, 64, A144.]

surface. In contrast, **absorption** means penetration beyond the surface, to the inside.) The positively charged surface attracts anions and repels cations from the *ionic atmosphere* (Figure 8-2) surrounding the particle. The positively charged particle and the negatively charged ionic atmosphere together are called the **electric double layer.**

Colloidal particles must collide with one another to coalesce. However, the negatively charged ionic atmospheres of the particles repel one another. The particles, therefore, must have enough kinetic energy to overcome electrostatic repulsion before they can coalesce.

Heat promotes coalescence by increasing the kinetic energy. Increasing electrolyte concentration ($HNO_3$ for AgCl) decreases the volume of the ionic atmosphere and allows particles to come closer together before electrostatic repulsion becomes significant. For this reason, most gravimetric precipitations are done in the presence of an electrolyte.

## Digestion

After precipitation, most procedures call for a period of standing in the presence of the hot mother liquor. This treatment, called **digestion,** promotes slow recrystallization of the precipitate. Particle size increases and impurities tend to be expelled from the crystal.

Liquid from which a substance precipitates or crystallizes is called the *mother liquor.*

## Purity

*Adsorbed* impurities are bound to the surface of a crystal. *Absorbed* impurities (within the crystal) are classified as *inclusions* or *occlusions*. Inclusions are impurity ions that randomly occupy sites in the crystal lattice normally occupied by ions that belong in the crystal. Inclusions are more likely when the impurity ion has a size and charge similar to those of one of the ions that belongs to the product. Occlusions are pockets of impurity that are literally trapped inside the growing crystal.

Adsorbed, occluded, and included impurities are said to be **coprecipitated.** That is, the impurity is precipitated along with the desired product, even though the solubility of the impurity has not been exceeded. Coprecipitation tends to be worst in colloidal precipitates

One way to remove arsenic from drinking water in Bangladesh is by coprecipitation with $Fe(OH)_3$.[5] Fe(II) or Fe(s) is added to the water and allowed to oxidize in air for several hours to precipitate $Fe(OH)_3$. After filtration through sand to remove solids, the water is drinkable.

*N-p*-Chlorophenylcinnamo-
hydroxamic acid (RH)
(ligand atoms are **bold**)

Ammonium chloride, for example, decomposes as follows when it is heated:

$$NH_4Cl(s) \rightarrow NH_3(g) + HCl(g)$$

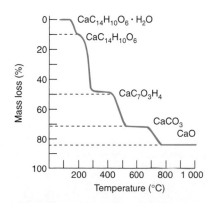

*Figure 27-3* Thermogravimetric curve for calcium salicylate. *[From G. Liptay, ed., Atlas of Thermoanalytical Curves (London: Heyden and Son, 1976).]*

such as $BaSO_4$, $Al(OH)_3$, and $Fe(OH)_3$, which have large surface area. Many procedures call for washing away the mother liquor, redissolving the precipitate, and *reprecipitating* the product. During the second precipitation, the concentration of impurities in solution is lower than during the first precipitation, and the degree of coprecipitation therefore tends to be lower. Occasionally, a trace component is intentionally isolated by coprecipitation with a major component of the solution. The precipitate used to collect the trace component is said to be a *gathering agent,* and the process is called **gathering.**

Some impurities can be treated with a **masking agent** to prevent them from reacting with the precipitant. In the gravimetric analysis of $Be^{2+}$, $Mg^{2+}$, $Ca^{2+}$, or $Ba^{2+}$ with the reagent *N-p*-chlorophenylcinnamohydroxamic acid, impurities such as $Ag^+$, $Mn^{2+}$, $Zn^{2+}$, $Cd^{2+}$, $Hg^{2+}$, $Fe^{2+}$, and $Ga^{3+}$ are kept in solution by excess KCN. $Pb^{2+}$, $Pd^{2+}$, $Sb^{3+}$, $Sn^{2+}$, $Bi^{3+}$, $Zr^{4+}$, $Ti^{4+}$, $V^{5+}$, and $Mo^{6+}$ are masked with a mixture of citrate and oxalate.

$$\underset{\text{Analyte}}{Ca^{2+}} + \underset{\substack{\textit{N-p}\text{-Chlorophenyl-}\\\text{cinnamohydroxamic acid}}}{2RH} \rightarrow \underset{\text{Precipitate}}{CaR_2(s)\downarrow} + 2H^+$$

$$\underset{\text{Impurity}}{Mn^{2+}} + \underset{\text{Masking agent}}{6CN^-} \rightarrow \underset{\text{Stays in solution}}{Mn(CN)_6^{4-}}$$

Even when a precipitate forms in a pure state, impurities might collect on the product while it is standing in the mother liquor. This is called *postprecipitation* and usually involves a supersaturated impurity that does not readily crystallize. An example is the crystallization of $MgC_2O_4$ on $CaC_2O_4$.

Washing a precipitate on a filter helps remove droplets of liquid containing excess solute. Some precipitates can be washed with water, but many require electrolyte to maintain coherence. For these precipitates, the ionic atmosphere is required to neutralize the surface charge of the tiny particles. If electrolyte is washed away with water, the charged solid particles repel one another and the product breaks up. This breaking up, called **peptization,** results in loss of product through the filter. AgCl will peptize if washed with water, so it is washed with dilute $HNO_3$ instead. Electrolyte used for washing must be volatile so that it will be lost during drying. Volatile electrolytes include $HNO_3$, HCl, $NH_4NO_3$, $NH_4Cl$, and $(NH_4)_2CO_3$.

## Product Composition

The final product must have a known, stable composition. A **hygroscopic substance** is one that picks up water from the air and is therefore difficult to weigh accurately. Many precipitates contain a variable quantity of water and must be dried under conditions that give a known (possibly zero) stoichiometry of $H_2O$.

**Ignition** (strong heating) is used to change the chemical form of some precipitates. For example, igniting $Fe(HCO_2)_3 \cdot nH_2O$ at 850°C for 1 h gives $Fe_2O_3$, and igniting $Mg(NH_4)PO_4 \cdot 6H_2O$ at 1 100°C gives $Mg_2P_2O_7$.

In **thermogravimetric analysis,** a substance is heated, and its mass is measured as a function of temperature. Figure 27-3 shows how the composition of calcium salicylate changes in four stages:

(27-5)

The composition of the product depends on the temperature and duration of heating.

## ■■■■ 27-3 Examples of Gravimetric Calculations

We now examine some examples that illustrate how to relate the mass of a gravimetric precipitate to the quantity of the original analyte. The general approach is to relate the moles of product to the moles of reactant.

## Example Relating Mass of Product to Mass of Reactant

The piperazine content of an impure commercial material can be determined by precipitating and weighing the diacetate:[6]

$$:NH\ HN: + 2CH_3CO_2H \rightarrow H_2\overset{+}{N}\ \overset{+}{N}H_2(CH_3CO_2^-)_2 \qquad (27\text{-}6)$$

| Piperazine | Acetic acid | Piperazine diacetate |
|---|---|---|
| FM 86.136 | FM 60.052 | FM 206.240 |

If you were performing this analysis, it would be important to determine that the impurities in the piperazine do not precipitate—otherwise the result will be high.

In one experiment, 0.312 6 g of the sample was dissolved in 25 mL of acetone, and 1 mL of acetic acid was added. After 5 min, the precipitate was filtered, washed with acetone, dried at 110°C, and found to weigh 0.712 1 g. What is the weight percent of piperazine in the commercial material?

**Solution** For each mole of piperazine in the impure material, 1 mol of product is formed.

$$\text{Moles of product} = \frac{0.712\ 1\ \cancel{g}}{206.240\ \cancel{g}/\text{mol}} = 3.453 \times 10^{-3}\ \text{mol}$$

This many moles of piperazine corresponds to

$$\text{Grams of piperazine} = (3.453 \times 10^{-3}\ \cancel{\text{mol}})\left(86.136\ \frac{g}{\cancel{\text{mol}}}\right) = 0.297\ 4\ g$$

which gives

$$\text{Percentage of piperazine in analyte} = \frac{0.297\ 4\ \cancel{g}}{0.312\ 6\ \cancel{g}} \times 100 = 95.14\%$$

An alternative (but equivalent) way to work this problem is to realize that 206.240 g (1 mol) of product will be formed for every 86.136 g (1 mol) of piperazine analyzed. Because 0.712 1 g of product was formed, the amount of reactant is given by

$$\frac{x\ \text{g piperazine}}{0.712\ 1\ \text{g product}} = \frac{86.136\ \text{g piperazine}}{206.243\ \text{g product}}$$

$$\Rightarrow x = \left(\frac{86.136\ \text{g piperazine}}{206.240\ \cancel{\text{g product}}}\right) 0.712\ 1\ \cancel{\text{g product}} = 0.297\ 4\ \text{g piperazine}$$

The quantity 86.136/206.240 is the *gravimetric factor* relating the mass of starting material to the mass of product.

The *gravimetric factor* relates mass of product to mass of analyte.

For a reaction in which the stoichiometric relation between analyte and product is not 1:1, we must use the correct stoichiometry in formulating the gravimetric factor. For example, an unknown containing $Mg^{2+}$ (atomic mass = 24.305 0) can be analyzed gravimetrically to produce magnesium pyrophosphate ($Mg_2P_2O_7$, FM 222.553). The gravimetric factor would be

$$\frac{\text{Grams of Mg in analyte}}{\text{Grams of } Mg_2P_2O_7 \text{ formed}} = \frac{2 \times (24.305\ 0)}{222.553}$$

because it takes 2 mol of $Mg^{2+}$ to make 1 mol of $Mg_2P_2O_7$.

## Example Calculating How Much Precipitant to Use

**(a)** To measure the nickel content in steel, the alloy is dissolved in 12 M HCl and neutralized in the presence of citrate ion, which maintains iron in solution. The slightly basic solution is warmed, and dimethylglyoxime (DMG) is added to precipitate the red DMG-nickel complex quantitatively. The product is filtered, washed with cold water, and dried at 110°C.

$$Ni^{2+} + 2 \quad \longrightarrow \quad + 2H^+ \qquad (27\text{-}7)$$

| DMG | Bis(dimethylglyoximate)nickel(II) |
|---|---|
| FM 58.69 | FM 116.12 |
| | FM 288.91 |

If the nickel content is known to be near 3 wt% and you wish to analyze 1.0 g of steel, what volume of 1.0 wt% alcoholic DMG solution should be used to give a 50% excess of DMG for the analysis? Assume that the density of the alcohol solution is 0.79 g/mL.

**Solution**  Because the Ni content is about 3%, 1.0 g of steel will contain about 0.03 g of Ni, which corresponds to

$$\frac{0.03 \text{ g Ni}}{58.69 \text{ g Ni/mol Ni}} = 5.11 \times 10^{-4} \text{ mol Ni}$$

This amount of metal requires

$$2(5.11 \times 10^{-4} \text{ mol Ni})(116.12 \text{ g DMG/mol Ni}) = 0.119 \text{ g DMG}$$

because 1 mol of $Ni^{2+}$ requires 2 mol of DMG. A 50% excess of DMG would be $(1.5)(0.119 \text{ g}) = 0.178 \text{ g}$. This much DMG is contained in

$$\frac{0.178 \text{ g DMG}}{0.010 \text{ g DMG/g solution}} = 17.8 \text{ g solution}$$

which occupies a volume of

$$\frac{17.8 \text{ g solution}}{0.79 \text{ g solution/mL}} = 23 \text{ mL}$$

**(b)** If 1.163 4 g of steel gives 0.179 5 g of precipitate, what is the percentage of Ni in the steel?

**Solution**  For each mole of Ni in the steel, 1 mol of precipitate will be formed. Therefore, 0.179 5 g of precipitate corresponds to

$$\frac{0.179 \text{ 5 g Ni(DMG)}_2}{288.91 \text{ g Ni(DMG)}_2\text{/mol Ni(DMG)}_2} = 6.213 \times 10^{-4} \text{ mol Ni(DMG)}_2$$

The Ni in the alloy must therefore be

$$(6.213 \times 10^{-4} \text{ mol Ni})\left(58.69 \frac{\text{g}}{\text{mol Ni}}\right) = 0.036 \text{ 46 g}$$

The mass percent of Ni in steel is

$$\frac{0.036 \text{ 46 g Ni}}{1.163 \text{ 4 g steel}} \times 100 = 3.134\%$$

A slightly simpler way to approach this problem comes from realizing that 58.69 g of Ni (1 mol) would give 288.91 g (1 mol) of product. Calling the mass of Ni in the sample $x$, we can write

$$\frac{\text{Grams of Ni analyzed}}{\text{Grams of product formed}} = \frac{x}{0.179 \text{ 5}} = \frac{58.69}{288.91} \Rightarrow \text{Ni} = 0.036 \text{ 46 g}$$

**Example**  A Problem with Two Components

A mixture of the 8-hydroxyquinoline complexes of Al and Mg weighed 1.084 3 g. When ignited in a furnace open to the air, the mixture decomposed, leaving a residue of $Al_2O_3$ and MgO weighing 0.134 4 g. Find the weight percent of $Al(C_9H_6NO)_3$ in the original mixture.

AlQ$_3$  
FM 459.43

MgQ$_2$  
FM 312.61

FM 101.96   FM 40.304

**Solution**  We will abbreviate the 8-hydroxyquinoline anion as Q. Letting the mass of AlQ$_3$ be $x$ and the mass of MgQ$_2$ be $y$, we can write

$$x + y = 1.084 \text{ 3 g}$$

Mass of   Mass of  
AlQ$_3$      MgQ$_2$

The moles of Al are $x/459.43$, and the moles of Mg are $y/312.61$. The moles of $Al_2O_3$ are one-half of the total moles of Al, because it takes 2 mol of Al to make 1 mol of $Al_2O_3$.

$$\text{Moles of } Al_2O_3 = \left(\frac{1}{2}\right)\frac{x}{459.43}$$

The moles of MgO will equal the moles of Mg $= y/312.61$. Now we can write

$$\underbrace{\underbrace{\left(\frac{1}{2}\right)\frac{x}{459.43}}_{\text{Mol } Al_2O_3}\underbrace{(101.96)}_{\frac{g\ Al_2O_3}{mol\ Al_2O_3}}}_{\text{Mass of } Al_2O_3} + \underbrace{\underbrace{\frac{y}{312.61}}_{\text{mol MgO}}\underbrace{(40.304)}_{\frac{g\ MgO}{mol\ MgO}}}_{\text{Mass of MgO}} = 0.134\ 4\ g$$

Substituting $y = 1.084\ 3 - x$ into the preceding equation gives

$$\left(\frac{1}{2}\right)\left(\frac{x}{459.43}\right)(101.96) + \left(\frac{1.084\ 3 - x}{312.61}\right)(40.304) = 0.134\ 4\ g$$

from which we find $x = 0.300\ 3$ g, which is 27.70% of the original mixture.

## 27-4 Combustion Analysis

A historically important form of gravimetric analysis was *combustion analysis,* used to determine the carbon and hydrogen content of organic compounds burned in excess $O_2$ (Figure 27-4). Instead of weighing combustion products, modern instruments use thermal conductivity, infrared absorption, or coulometry (with electrochemically generated reagents) to measure the products.

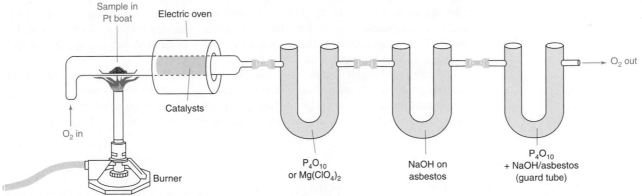

**Figure 27-4** Gravimetric combustion analysis for carbon and hydrogen.

### Gravimetric Combustion Analysis

In gravimetric combustion analysis, partially combusted product is passed through catalysts such as Pt gauze, CuO, $PbO_2$, or $MnO_2$ at elevated temperature to complete the oxidation to $CO_2$ and $H_2O$. The combustion products are flushed through a chamber containing $P_4O_{10}$ ("phosphorus pentoxide"), which absorbs water, and then through a chamber of Ascarite (NaOH on asbestos), which absorbs $CO_2$. The increase in mass of each chamber tells how much hydrogen and carbon, respectively, were initially present. A guard tube prevents atmospheric $H_2O$ or $CO_2$ from entering the chambers.

**Example** Combustion Analysis Calculations

A compound weighing 5.714 mg produced 14.414 mg of $CO_2$ and 2.529 mg of $H_2O$ upon combustion. Find the weight percent of C and H in the sample.

**Solution**  One mole of $CO_2$ contains 1 mol of carbon. Therefore,

$$\text{Moles of C in sample} = \text{moles of } CO_2 \text{ produced}$$

$$= \frac{14.414 \times 10^{-3}\ g\ CO_2}{44.010\ g/mol\ CO_2} = 3.275 \times 10^{-4}\ mol$$

$$\text{Mass of C in sample} = (3.275 \times 10^{-4}\ mol\ C)(12.010\ 7\ g/mol\ C) = 3.934\ mg$$

$$\text{Weight percent of C} = \frac{3.934\ mg\ C}{5.714\ mg\ sample} \times 100 = 68.84\%$$

One mole of $H_2O$ contains 2 mol of H. Therefore,

$$\text{Moles of H in sample} = 2(\text{moles of } H_2O \text{ produced})$$

$$= 2\left(\frac{2.529 \times 10^{-3} \text{ g } H_2O}{18.015 \text{ g/mol } H_2O}\right) = 2.808 \times 10^{-4} \text{ mol}$$

$$\text{Mass of H in sample} = (2.808 \times 10^{-4} \text{ mol H})(1.007\,9 \text{ g/mol H}) = 2.830 \times 10^{-4} \text{ g}$$

$$\text{Weight percent of H} = \frac{0.283\,0 \text{ mg H}}{5.714 \text{ mg sample}} \times 100 = 4.952\%$$

### Combustion Analysis Today[7]

Figure 27-5 shows an instrument that measures C, H, N, and S in a single operation. First, a ~2-mg sample is accurately weighed and sealed in a tin or silver capsule. The analyzer is swept with He gas that has been treated to remove traces of $O_2$, $H_2O$, and $CO_2$. At the start of a run, a measured excess of $O_2$ is added to the He stream. Then the capsule is dropped into a preheated ceramic crucible, where the capsule melts and sample is rapidly oxidized.

$$C, H, N, S \xrightarrow{1\,050°C/O_2} CO_2(g) + H_2O(g) + N_2(g) + \underbrace{SO_2(g) + SO_3(g)}_{95\% \, SO_2}$$

Elemental analyzers use an *oxidation catalyst* to complete the oxidation of sample and a *reduction catalyst* to carry out any required reduction and to remove excess $O_2$.

Products pass through hot $WO_3$ catalyst to complete the combustion of carbon to $CO_2$. In the next zone, metallic Cu at 850°C converts $SO_3$ into $SO_2$ and removes excess $O_2$:

$$Cu + SO_3 \xrightarrow{850°C} SO_2 + CuO(s)$$

$$Cu + \tfrac{1}{2}O_2 \xrightarrow{850°C} CuO(s)$$

The mixture of $CO_2$, $H_2O$, $N_2$, and $SO_2$ is separated by gas chromatography (Figure 27-6), and each component is measured with a thermal conductivity detector (Section 24-3). Figure 27-7 shows a different C,H,N,S analyzer that uses infrared absorbance to measure $CO_2$, $H_2O$, and $SO_2$ and thermal conductivity for $N_2$.

A key to elemental analysis is *dynamic flash combustion,* which creates a short burst of gaseous products, instead of slowly bleeding products out over several minutes. This feature is important because chromatographic analysis requires that the whole sample be injected at once. Otherwise, the injection zone is so broad that the products cannot be separated.

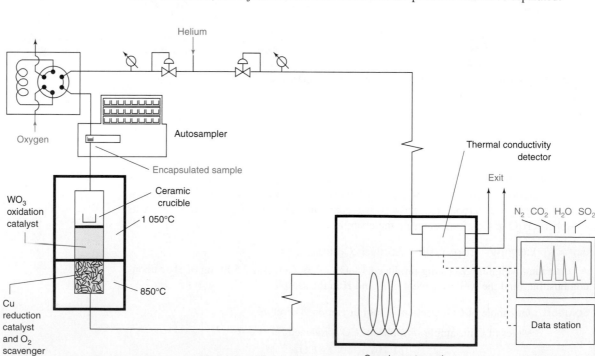

**Figure 27-5** Diagram of C,H,N,S elemental analyzer that uses gas chromatographic separation and thermal conductivity detection. *[Adapted from E. Pella, "Elemental Organic Analysis. 2. State of the Art," Am. Lab, August 1990, p. 28.]*

CHAPTER 27 Gravimetric and Combustion Analysis

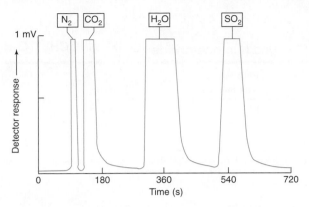

**Figure 27-6** Gas chromatographic trace from elemental analyzer, showing substantially complete separation of combustion products. The area of each peak (when they are not off scale) is proportional to the mass of each product. *[From E. Pella, "Elemental Organic Analysis. 2. State of the Art," Am. Lab. August 1990, p. 28.]*

**Figure 27-7** Combustion analyzer that uses infrared absorbance to measure $CO_2$, $H_2O$, and $SO_2$ and thermal conductivity to measure $N_2$. Three separate infrared cells in series are equipped with filters that isolate wavelengths absorbed by one of the products. Absorbance is integrated over time as the combustion product mixture is swept through each cell. *[Courtesy Leco Corp., St. Joseph, MI.]*

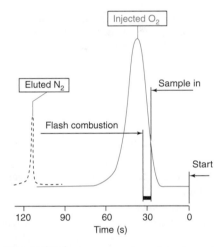

**Figure 27-8** Sequence of events in dynamic flash combustion. *[From E. Pella, "Elemental Organic Analysis. 1. Historical Developments," Am. Lab. February 1990, p. 116.]*

The Sn capsule is oxidized to $SnO_2$, which

1. liberates heat to vaporize and crack (decompose) sample
2. uses available oxygen immediately
3. ensures that sample oxidation occurs in gas phase
4. acts as an oxidation catalyst

In dynamic flash combustion, the tin-encapsulated sample is dropped into the preheated furnace shortly after the flow of a 50 vol% $O_2$/50 vol% He mixture is started (Figure 27-8). The Sn capsule melts at 235°C and is instantly oxidized to $SnO_2$, thereby liberating 594 kJ/mol, and heating the sample to 1 700°–1 800°C. If the sample is dropped in before very much $O_2$ is present, decomposition (cracking) occurs prior to oxidation, which minimizes the formation of nitrogen oxides. (Flammable liquid samples would be admitted prior to any $O_2$ to prevent explosions.)

Analyzers that measure C, H, and N, but not S, use catalysts that are better optimized for this process. The oxidation catalyst is $Cr_2O_3$. The gas then passes through hot $Co_3O_4$ coated with Ag to absorb halogens and sulfur. A hot Cu column then scavenges excess $O_2$.

Oxygen analysis requires a different strategy. The sample is thermally decomposed (a process called **pyrolysis**) in the absence of added $O_2$. Gaseous products are passed through nickelized carbon at 1 075°C to convert oxygen from the compound into CO (not $CO_2$). Other products include $N_2$, $H_2$, $CH_4$, and hydrogen halides. Acidic products are absorbed by NaOH, and the remaining gases are separated and measured by gas chromatography with a thermal conductivity detector.

**Table 27-4** C, H, and N in acetanilide: ![structure] —NHCCH$_3$ (C$_8$H$_9$NO)

| Element | Theoretical value (wt%) | Instrument 1 | Instrument 2 |
|---|---|---|---|
| C | 71.09 | 71.17 ± 0.41 | 71.22 ± 1.1 |
| H | 6.71 | 6.76 ± 0.12 | 6.84 ± 0.10 |
| N | 10.36 | 10.34 ± 0.08 | 10.33 ± 0.13 |

SOURCE: Data from E. M. Hodge, H. P. Patterson, M. C. Williams, and E. S. Gladney, Am. Lab. June 1991. p. 34. Uncertainties are standard deviations from five replicate determinations.

For halogenated compounds, combustion gives $CO_2$, $H_2O$, $N_2$, and HX (X = halogen). The HX is trapped in aqueous solution and titrated with $Ag^+$ ions in a coulometer (Section 17-3). This instrument counts the electrons produced (one electron for each $Ag^+$) during complete reaction with HX.

Table 27-4 shows results of five replicate analyses of pure acetanilide on two different commercial instruments. Chemists consider a result within ±0.3 of the theoretical percentage of an element to be good evidence that the compound has the expected formula. For N in acetanilide, ±0.3 corresponds to a relative error of 0.3/10.36 = 3%, which is not hard to achieve. For C, ±0.3 corresponds to a relative error of 0.3/71.09 = 0.4%, which is not so easy. The standard deviation for C in Instrument 1 is 0.41/71.17 = 0.6%; for Instrument 2, it is 1.1/71.22 = 1.5%.

Silicon compounds, such as SiC, Si$_3$N$_4$, and silicates (rocks), can be analyzed by combustion with elemental fluorine (F$_2$) in a nickel vessel to produce volatile SiF$_4$ and fluorinated products of every element in the periodic table except O, N, He, Ne, Ar, and Kr.[8] All major and minor combustion products can then be measured by mass spectrometry. Nitrogen in Si$_3$N$_4$ and some other metal nitrides can be analyzed by heating to 3 000°C in an inert atmosphere to liberate nitrogen as $N_2$, which is measured by thermal conductivity in apparatus similar to that in Figure 27-7.

F$_2$ is exceedingly reactive and therefore exceedingly dangerous. It must be handled only in systems especially designed for its use.

## Terms to Understand

| | | | |
|---|---|---|---|
| absorption | dialysis | homogeneous precipitation | peptization |
| adsorption | digestion | hygroscopic substance | precipitant |
| colloid | electric double layer | ignition | pyrolysis |
| combustion analysis | gathering | masking agent | supersaturated solution |
| coprecipitation | gravimetric analysis | nucleation | thermogravimetric analysis |

## Summary

Gravimetric analysis is based on the formation of a product whose mass can be related to the mass of analyte. Most commonly, analyte ion is precipitated by a suitable counterion. Measures taken to reduce supersaturation and promote the formation of large, easily filtered particles (as opposed to colloids) include (1) raising the temperature during precipitation, (2) slowly adding and vigorously mixing reagents, (3) maintaining a large sample volume, and (4) using homogeneous precipitation. Precipitates are usually digested in hot mother liquor to promote particle growth and recrystallization. All precipitates are then filtered and washed; some must be washed with a volatile electrolyte to prevent peptization. The product is heated to dryness or ignited to achieve a reproducible, stable composition. Gravimetric calculations relate moles of product to moles of analyte.

In combustion analysis, an organic compound in a tin capsule is rapidly heated with excess oxygen to give predominantly $CO_2$, $H_2O$, $N_2$, $SO_2$, and HX (hydrogen halides). A hot oxidation catalyst completes the process, and hot Cu scavenges excess oxygen. For sulfur analysis, hot copper also converts $SO_3$ into $SO_2$. Products may be separated by gas chromatography and measured by their thermal conductivity. Some instruments use infrared absorption or coulometric reactions (counting electrons with an electronic circuit) to measure the products. Oxygen analysis is done by pyrolysis in the absence of oxygen, a process that ultimately converts oxygen from the compound into CO.

## Exercises

**27-A.** An organic compound with a formula mass of 417 was analyzed for ethoxyl (CH$_3$CH$_2$O—) groups by the reactions

$$ROCH_2CH_3 + HI \rightarrow ROH + CH_3CH_2I$$

$$CH_3CH_2I + Ag^+ + OH^- \rightarrow AgI(s) + CH_3CH_2OH$$

A 25.42-mg sample of compound produced 29.03 mg of AgI. How many ethoxyl groups are there in each molecule?

**27-B.** A 0.649-g sample containing only K$_2$SO$_4$ (FM 174.27) and (NH$_4$)$_2$SO$_4$ (FM 132.14) was dissolved in water and treated with Ba(NO$_3$)$_2$ to precipitate all $SO_4^{2-}$ as BaSO$_4$ (FM 233.39). Find the weight percent of $K_2SO_4$ in the sample if 0.977 g of precipitate was formed.

**27-C.** Consider a mixture of the two solids $BaCl_2 \cdot 2H_2O$ (FM 244.26) and KCl (FM 74.551), in an unknown ratio. (The notation $BaCl_2 \cdot 2H_2O$ means that a crystal is formed with two water molecules for each $BaCl_2$.) When the unknown is heated to 160°C for 1 h, the water of crystallization is driven off:

$$BaCl_2 \cdot 2H_2O(s) \xrightarrow{160°C} BaCl_2(s) + 2H_2O(g)$$

A sample originally weighing 1.783 9 g weighed 1.562 3 g after heating. Calculate the weight percent of Ba, K, and Cl in the original sample.

**27-D.** A mixture containing only aluminum tetrafluoroborate, $Al(BF_4)_3$ (FM 287.39), and magnesium nitrate, $Mg(NO_3)_2$ (FM 148.31), weighed 0.282 8 g. It was dissolved in 1 wt% aqueous HF and treated with nitron solution to precipitate a mixture of nitron tetrafluoroborate and nitron nitrate weighing 1.322 g. Find the weight percent of Mg in the original solid mixture.

Nitron
$C_{20}H_{16}N_4$
FM 312.37

Nitron tetrafluoroborate
$C_{20}H_{17}N_4BF_4$
FM 400.18

Nitron nitrate
$C_{20}H_{17}N_5O_3$
FM 375.39

---

## Problems

### Gravimetric Analysis

**27-1. (a)** What is the difference between absorption and adsorption? **(b)** How is an inclusion different from an occlusion?

**27-2.** State four desirable properties of a gravimetric precipitate.

**27-3.** Why is high relative supersaturation undesirable in a gravimetric precipitation?

**27-4.** What measures can be taken to decrease the relative supersaturation during a precipitation?

**27-5.** Why are many ionic precipitates washed with electrolyte solution instead of pure water?

**27-6.** Why is it less desirable to wash AgCl precipitate with aqueous $NaNO_3$ than with $HNO_3$ solution?

**27-7.** Why would a reprecipitation be employed in a gravimetric analysis?

**27-8.** Explain what is done in thermogravimetric analysis.

**27-9.** Explain how the cantilever at the opening of Chapter 2 measures extremely small mass. The detection limit is near 1 attogram. How many molecules with a molecular mass of 100 are in 1 ag? Express the sensitivity as Hz/ag.

**27-10.** A 50.00-mL solution containing NaBr was treated with excess $AgNO_3$ to precipitate 0.214 6 g of AgBr (FM 187.772). What was the molarity of NaBr in the solution?

**27-11.** To find the $Ce^{4+}$ content of a solid, 4.37 g were dissolved and treated with excess iodate to precipitate $Ce(IO_3)_4$. The precipitate was collected, washed well, dried, and ignited to produce 0.104 g of $CeO_2$ (FM 172.114). What was the weight percent of Ce in the original solid?

**27-12.** Marie Cure dissolved 0.091 92 g of $RaCl_2$ and treated it with excess $AgNO_3$ to precipitate 0.088 90 g of AgCl. In her time (1900), the atomic mass of Ag was known to be 107.8 and that of Cl was 35.4. From these values, find the atomic mass of Ra that Marie Curie would have calculated.

**27-13.** A 0.050 02-g sample of impure piperazine contained 71.29 wt% piperazine (FM 86.136). How many grams of product (FM 206.240) will be formed when this sample is analyzed by Reaction 27-6?

**27-14.** A 1.000-g sample of unknown gave 2.500 g of bis(dimethylglyoximate)nickel(II) (FM 288.91) when analyzed by Reaction 27-7. Find the weight percent of Ni in the unknown.

**27-15.** Name the product obtained in Figure 27-3 when calcium salicylate monohydrate is heated to 550°C and to 1 000°C. Using the formula masses of these products, calculate what mass is expected to remain when 0.635 6 g of calcium salicylate monohydrate is heated to 550°C or 1 000°C.

**27-16.** A method to measure soluble organic carbon in seawater includes oxidation of the organic materials to $CO_2$ with $K_2S_2O_8$, followed by gravimetric determination of the $CO_2$ trapped by a column of NaOH-coated asbestos. A water sample weighing 6.234 g produced 2.378 mg of $CO_2$ (FM 44.010). Calculate the ppm carbon in the seawater.

**27-17.** How many milliliters of 2.15% alcoholic dimethylglyoxime should be used to provide a 50.0% excess for Reaction 27-7 with 0.998 4 g of steel containing 2.07 wt% Ni? Assume that the density of the dimethylglyoxime solution is 0.790 g/mL.

**27-18.** Twenty dietary iron tablets with a total mass of 22.131 g were ground and mixed thoroughly. Then 2.998 g of the powder were dissolved in $HNO_3$ and heated to convert all iron into $Fe^{3+}$. Addition of $NH_3$ precipitated $Fe_2O_3 \cdot xH_2O$, which was ignited to give 0.264 g of $Fe_2O_3$ (FM 159.69). What is the average mass of $Fe_2SO_4 \cdot 7H_2O$ (FM 278.01) in each tablet?

**27-19.** Finely ground mineral (0.632 4 g) was dissolved in 25 mL of boiling 4 M HCl and diluted with 175 mL $H_2O$ containing two drops of methyl red indicator. The solution was heated to 100°C and 50 mL of warm solution containing 2.0 g $(NH_4)_2C_2O_4$ were slowly added to precipitate $CaC_2O_4$. Then 6 M $NH_3$ was added until the indicator changed from red to yellow, showing that the liquid was neutral or slightly basic. After slow cooling for 1 h, the liquid was decanted and the solid transferred to a filter crucible and washed with cold 0.1 wt% $(NH_4)_2C_2O_4$ solution five times until no $Cl^-$ was detected in the filtrate upon addition of $AgNO_3$ solution. The crucible was dried at 105°C for 1 h and then at 500° ± 25°C in a furnace for 2 h.

$$Ca^{2+} + C_2O_4^{2-} \xrightarrow{105°} CaC_2O_4 \cdot H_2O(s) \xrightarrow{500°} CaCO_3(s)$$
FM 40.078                                             100.087

The mass of the empty crucible was 18.231 1 g and the mass of the crucible with $CaCO_3(s)$ was 18.546 7 g.

(a) Find the wt% Ca in the mineral.

(b) Why is the unknown solution heated to boiling and the precipitant solution, $(NH_4)_2C_2O_4$, also heated before slowly mixing the two?

(c) What is the purpose of washing the precipitate with 0.1 wt% $(NH_4)_2C_2O_4$?

(d) What is the purpose of testing the filtrate with $AgNO_3$ solution?

**27-20.** A 1.475-g sample containing $NH_4Cl$ (FM 53.492), $K_2CO_3$ (FM 138.21), and inert ingredients was dissolved to give 0.100 L of solution. A 25.0-mL aliquot was acidified and treated with excess sodium tetraphenylborate, $Na^+B(C_6H_5)_4^-$, to precipitate $K^+$ and $NH_4^+$ ions completely:

$$(C_6H_5)_4B^- + K^+ \rightarrow (C_6H_5)_4BK(s)$$
$$\text{FM 358.33}$$

$$(C_6H_5)_4B^- + NH_4^+ \rightarrow (C_6H_5)_4BNH_4(s)$$
$$\text{FM 337.27}$$

The resulting precipitate amounted to 0.617 g. A fresh 50.0-mL aliquot of the original solution was made alkaline and heated to drive off all the $NH_3$:

$$NH_4^+ + OH^- \rightarrow NH_3(g) + H_2O$$

It was then acidified and treated with sodium tetraphenylborate to give 0.554 g of precipitate. Find the weight percent of $NH_4Cl$ and $K_2CO_3$ in the original solid.

**27-21.** A mixture containing only $Al_2O_3$ (FM 101.96) and $Fe_2O_3$ (FM 159.69) weighs 2.019 g. When heated under a stream of $H_2$, $Al_2O_3$ is unchanged, but $Fe_2O_3$ is converted into metallic Fe plus $H_2O(g)$. If the residue weighs 1.774 g, what is the weight percent of $Fe_2O_3$ in the original mixture?

**27-22.** A solid mixture weighing 0.548 5 g contained only ferrous ammonium sulfate hexahydrate and ferrous chloride hexahydrate. The sample was dissolved in 1 M $H_2SO_4$, oxidized to $Fe^{3+}$ with $H_2O_2$, and precipitated with cupferron. The ferric cupferron complex was ignited to produce 0.167 8 g of ferric oxide, $Fe_2O_3$ (FM 159.69). Calculate the weight percent of Cl in the original sample.

$FeSO_4 \cdot (NH_4)_2SO_4 \cdot 6H_2O$    $FeCl_2 \cdot 6H_2O$

Ferrous ammonium sulfate    Ferrous chloride
hexahydrate    hexahydrate
FM 392.13    FM 234.84

Cupferron
FM 155.16

**27-23.** *Propagation of error.* A mixture containing only silver nitrate and mercurous nitrate was dissolved in water and treated with excess sodium cobalticyanide, $Na_3[Co(CN)_6]$ to precipitate both cobalticyanide salts:

AgNO$_3$          FM 169.873
Ag$_3$[Co(CN)$_6$]     FM 538.643
Hg$_2$(NO$_3$)$_2$      FM 525.19
(Hg$_2$)$_3$[Co(CN)$_6$]$_2$  FM 1 633.62

(a) The unknown weighed 0.432 1 g and the product weighed 0.451 5 g. Find wt% AgNO$_3$ in the unknown. *Caution:* Keep all the

digits in your calculator or else serious rounding errors may occur. Do not round off until the end.

(b) Even a skilled analyst is not likely to have less than a 0.3% error in isolating the precipitate. Suppose that there is negligible error in all quantities, except the mass of product. Suppose that the mass of product has an uncertainty of 0.30%. Calculate the relative uncertainty in the mass of $AgNO_3$ in the unknown.

**27-24.** The thermogravimetric trace below shows mass loss by $Y_2(OH)_5Cl \cdot xH_2O$ upon heating. In the first step, waters of hydration are lost to give ~8.1% mass loss. After a second decomposition step, 19.2% of the original mass is lost. Finally, the composition stabilizes at $Y_2O_3$ above 800°C.

(a) Find $x$ in the formula $Y_2(OH)_5Cl \cdot xH_2O$. Because the 8.1% mass loss is not accurately defined in the experiment, use the 31.8% total mass loss for your calculation.

(b) Suggest a formula for the material remaining at the 19.2% plateau. Be sure that the charges of all ions in your formula sum to 0. The cation is $Y^{3+}$.

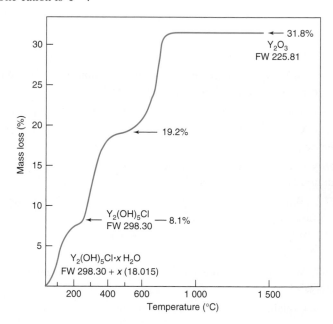

Thermogravimetric analysis of $Y_2(OH)_5Cl \cdot xH_2O$ [From T. Hours, P. Bergez, J. Charpin, A. Larbot, C. Guizard, and L. Cot, "Preparation and Characterization of Yttrium Oxide by a Sol-Gel Process," Ceramic Bull. **1992**, 71, 200.]

**27-25.** When the *high-temperature superconductor* yttrium barium copper oxide (see beginning of Chapter 16 and Box 16-2) is heated under flowing $H_2$, the solid remaining at 1 000°C is a mixture of $Y_2O_3$, BaO, and Cu. The starting material has the formula $YBa_2Cu_3O_{7-x}$, in which the oxygen stoichiometry varies between 7 and 6.5 ($x = 0$ to 0.5).

$$YBa_2Cu_3O_{7-x}(s) + (3.5 - x)H_2(g) \xrightarrow{1\,000°C}$$
$$\underset{\substack{\text{FM} \\ 666.19 - 16.00x}}{}$$

$$\tfrac{1}{2}Y_2O_3(s) + 2BaO(s) + 3Cu(s) + \underbrace{(3.5 - x)H_2O(g)}_{YBa_2Cu_3O_{3.5}}$$

(a) *Thermogravimetric analysis.* When 34.397 mg of $YBa_2Cu_3O_{7-x}$ were subjected to this analysis, 31.661 mg of solid remained after heating to 1 000°C. Find the value of $x$ in $YBa_2Cu_3O_{7-x}$.

(b) *Propagation of error.* Suppose that the uncertainty in each mass in part (a) is $\pm0.002$ mg. Find the uncertainty in the value of $x$.

**27-26.** *Man in the vat problem.*[9] Long ago, a workman at a dye factory fell into a vat containing hot, concentrated sulfuric and nitric acids. He dissolved completely! Because nobody witnessed the accident, it was necessary to prove that he fell in so that the man's wife could collect his insurance money. The man weighed 70 kg, and a human body contains ~6.3 parts per thousand (mg/g) phosphorus. The acid in the vat was analyzed for phosphorus to see whether it contained a dissolved human.

(a) The vat contained $8.00 \times 10^3$ L of liquid, and a 100.0-mL sample was analyzed. If the man did fall into the vat, what is the expected quantity of phosphorus in 100.0 mL?

(b) The 100.0-mL sample was treated with a molybdate reagent that precipitated ammonium phosphomolybdate, $(NH_4)_3 [P(Mo_{12}O_{40})] \cdot 12H_2O$. This substance was dried at 110°C to remove waters of hydration and heated to 400°C until it reached the constant composition $P_2O_5 \cdot 24MoO_3$, which weighed 0.371 8 g. When a fresh mixture of the same acids (not from the vat) was treated in the same manner, 0.033 1 g of $P_2O_5 \cdot 24MoO_3$ (FM 3 596.46) was produced. This *blank determination* gives the amount of phosphorus in the starting reagents. The $P_2O_5 \cdot 24MoO_3$ that could have come from the dissolved man is therefore $0.371\ 8 - 0.033\ 1 = 0.338\ 7$ g. How much phosphorus was present in the 100.0-mL sample? Is this quantity consistent with a dissolved man?

**27-27.** Some analyte ions can be gathered by precipitating $LaPO_4$.[10] To 100.0 mL of sample were added 2 mL containing 5 mg $La^{3+}$/mL in 0.6 M HCl and 0.3 mL of 0.5 M $H_3PO_4$. The pH was adjusted to 3.0 by adding $NH_3$. The precipitate was allowed to settle and collected on a filter with 0.2-μm pore size. The filter, which has negligible volume (<0.01 mL) was placed in a 10-mL volumetric flask and treated with 1 mL of 16 M $HNO_3$ to dissolve the precipitate. The flask was made up to 10 mL with $H_2O$.

(a) A 100.0-mL distilled water sample was spiked with 10.0 μg of each of the following elements: $Fe^{3+}$, $Pb^{2+}$, $Cd^{2+}$, $In^{3+}$, $Cr^{3+}$, $Mn^{2+}$, $Co^{2+}$, $Ni^{2+}$, and $Cu^{2+}$. Atomic emission analysis of the final solution in the 10-mL flask found $[Fe^{3+}] = 17.6$ μM, $[Pb^{2+}] = 5.02$ μM, $[Cd^{2+}] = 8.77$ μM, $[In^{3+}] = 8.50$ μM, $[Cr^{3+}] < 0.05$ μM, $[Mn^{2+}] = 6.64$ μM, $[Co^{2+}] = 1.09$ μM, $[Ni^{2+}] < 0.05$ μM, and $[Cu^{2+}] = 6.96$ μM. Find % recovery of each element ($= 100 \times$ μg found/μg added).

(b) Which of the tested cations are quantitatively gathered?

(c) Gathered ions are preconcentrated from dilute sample into a more concentrated final solution that is analyzed. By what factor are the elements preconcentrated in this procedure?

## Combustion Analysis

**27-28.** What is the difference between combustion and pyrolysis?

**27-29.** What is the purpose of the $WO_3$ and Cu in Figure 27-5?

**27-30.** Why is tin used to encapsulate a sample for combustion analysis?

**27-31.** Why is sample dropped into the preheated furnace before the oxygen concentration reaches its peak in Figure 27-8?

**27-32.** Write a balanced equation for the combustion of benzoic acid, $C_6H_5CO_2H$, to give $CO_2$ and $H_2O$. How many milligrams of $CO_2$ and of $H_2O$ will be produced by the combustion of 4.635 mg of benzoic acid?

**27-33.** Write a balanced equation for the combustion of $C_8H_7NO_2SBrCl$ in a C,H,N,S elemental analyzer.

**27-34.** Combustion analysis of a compound known to contain just C, H, N, and O demonstrated that it is 46.21 wt% C, 9.02 wt% H, 13.74 wt% N, and, by difference, $100 - 46.21 - 9.02 - 13.74 = 31.03\%$ O. This means that 100 g of unknown would contain 46.21 g of C, 9.02 g of H, and so on. Find the atomic ratio C:H:N:O and express it as the lowest reasonable integer ratio.

**27-35.** A mixture weighing 7.290 mg contained only cyclohexane, $C_6H_{12}$ (FM 84.159), and oxirane, $C_2H_4O$ (FM 44.053). When the mixture was analyzed by combustion analysis, 21.999 mg of $CO_2$ (FM 44.010) was produced. Find the weight percent of oxirane in the mixture.

**27-36.** Use the uncertainties from Instrument 1 in Table 27-4 to estimate the uncertainties in the stoichiometry coefficients in the formula $C_8H_{h\pm x}N_{n\pm y}$.

**27-37.** One way to determine sulfur is by combustion analysis, which produces a mixture of $SO_2$ and $SO_3$ that can be passed through $H_2O_2$ to convert both into $H_2SO_4$, which is titrated with standard base. When 6.123 mg of a substance were burned, the $H_2SO_4$ required 3.01 mL of 0.015 76 M NaOH for titration. Find wt% sulfur in the sample.

**27-38.** *Statistics of coprecipitation.*[11] In Experiment 1, 200.0 mL of solution containing 10.0 mg of $SO_4^{2-}$ (from $Na_2SO_4$) were treated with excess $BaCl_2$ solution to precipitate $BaSO_4$ containing some coprecipitated $Cl^-$. To find out how much coprecipitated $Cl^-$ was present, the precipitate was dissolved in 35 mL of 98 wt% $H_2SO_4$ and boiled to liberate HCl, which was removed by bubbling $N_2$ gas through the $H_2SO_4$. The $HCl/N_2$ stream was passed into a reagent solution that reacted with $Cl^-$ to give a color that was measured. Ten replicate trials gave values of 7.8, 9.8, 7.8, 7.8, 7.8, 7.8, 13.7, 12.7, 13.7, and 12.7 μmol $Cl^-$. Experiment 2 was identical to the first one, except that the 200.0-mL solution also contained 6.0 g of $Cl^-$ (from NaCl). Ten replicate trials gave 7.8, 10.8, 8.8, 7.8, 6.9, 8.8, 15.7, 12.7, 13.7, and 14.7 μmol $Cl^-$.

(a) Find the mean, standard deviation, and 95% confidence interval for $Cl^-$ in each experiment.

(b) Is there a significant difference between the two experiments? What does your answer mean?

(c) If there were no coprecipitate, what mass of $BaSO_4$ (FM 233.39) would be expected?

(d) If the coprecipitate is $BaCl_2$ (FM 208.23), what is the average mass of precipitate ($BaSO_4 + BaCl_2$) in Experiment 1? By what percentage is the mass greater than the mass in part (c)?

# 28 Sample Preparation

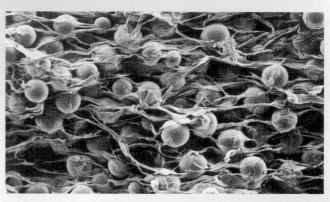

Is it alive? Ion-exchange beads enmeshed in a polytetrafluoroethylene membrane. *[Courtesy Bio-Rad Laboratories, Hercules, CA.]*

Solid-phase extraction membrane. *[Courtesy Alltech Associates, Deerfield, IL.]*

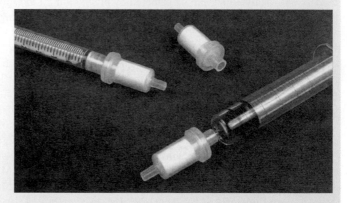

Solid-phase extraction cartridge. *[Courtesy Alltech Associates, Deerfield, IL.]*

Desired analytes or unwanted impurities in liquid samples can be trapped in solid-phase extraction membranes containing any stationary phase used in high-performance liquid chromatography. The micrograph shows 25-µm-diameter ion-exchange beads entangled in a loose membrane of polytetrafluoroethylene (Teflon). The membrane is a circular disk that fits into a plastic holder attached to a syringe. To conduct an extraction, solution from the syringe is simply forced through the disk.

A sandwich made of a cation-exchange membrane loaded with $H^+$ and an anion-exchange membrane loaded with $OH^-$ deionizes a solution, because all entering cations are exchanged for outgoing $H^+$ and all incoming anions are exchanged for outgoing $OH^-$. A cation-exchange membrane loaded with $Ag^+$ selectively removes halides $(X^-)$ from a sample, forming $AgX(s)$ on the membrane. A cation-exchange membrane loaded with $Ba^{2+}$ selectively removes sulfate by forming $BaSO_4(s)$.

*Heterogeneous:* different composition from place to place in a material
*Homogeneous:* same composition everywhere

**A** chemical analysis is meaningless unless you begin with a meaningful sample. To measure cholesterol in a dinosaur skeleton or herbicide in a truckload of oranges, you must have a strategy for selecting a *representative sample* from a *heterogeneous* material. Figure 28-1 shows that the concentration of nitrate in sediment beneath a lake drops by two orders of magnitude in the first 3 mm below the surface. If you want to measure nitrate in sediment, it makes an enormous difference whether you select a core sample that is 1 m deep or skim the top 2 mm of sediment for analysis. **Sampling** is the process of collecting a representative

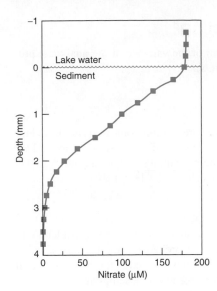

**Figure 28-1** Depth profile of nitrate in sediment from freshwater Lake Søbygård in Denmark. A similar profile was observed in saltwater sediment. Measurements were made with a *biosensor* containing live bacteria that convert $NO_3^-$ into $N_2O$, which was then measured amperometrically by reduction at a silver cathode. *[From L. H. Larsen, T. Kjær, and N. P. Revsbech, "A Microscale $NO_3^-$ Biosensor for Environmental Applications," Anal. Chem. **1997**, 69, 3527.]*

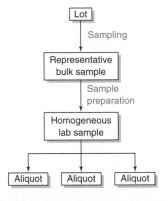

**Figure 28-2** *Sampling* is the process of selecting a representative bulk sample from the lot. *Sample preparation* is the process that converts a bulk sample into a homogeneous laboratory sample. Sample preparation also refers to steps that eliminate interfering species or that concentrate the analyte.

sample for analysis.[1] Real samples also generally require some degree of *sample preparation* to remove substances that interfere in the analysis of the desired analyte and, perhaps, to convert the analyte into a form suitable for analysis.[2]

The terminology of sampling and sample preparation is shown in Figure 28-2. A *lot* is the total material (dinosaur skeleton, truckload of oranges, and so on) from which samples are taken. A *bulk sample* (also called a *gross sample*) is taken from the lot for analysis or *archiving* (storing for future reference). The bulk sample must be representative of the lot, and the choice of bulk sample is critical to producing a valid analysis. Box 0-1 gave a strategy for sampling heterogeneous material.

From the representative bulk sample, a smaller, homogeneous *laboratory sample* is formed that must have the same composition as the bulk sample. For example, we might obtain a laboratory sample by grinding an entire solid bulk sample to a fine powder, mixing thoroughly, and keeping one bottle of powder for testing. Small portions (called *aliquots*) of the laboratory sample are used for individual analyses. *Sample preparation* is the series of steps needed to convert a representative bulk sample into a form suitable for chemical analysis.

Besides choosing a sample judiciously, we must be careful about storing the sample. The composition may change with time after collection because of chemical changes, reaction with air, or interaction of the sample with its container. Glass is a notorious ion exchanger that alters the concentrations of trace ions in solution. Therefore, plastic (especially Teflon) collection bottles are frequently employed. Even these materials can absorb trace levels of analytes. For example, a 0.2 μM $HgCl_2$ solution lost 40–95% of its concentration in 4 h in polyethylene bottles. A 2 μM $Ag^+$ solution in a Teflon bottle lost 2% of its concentration in a day and 28% in a month.[3]

Plastic containers must be washed before use. Table 28-1 shows that manganese in blood serum samples increased by a factor of 7 when stored in unwashed polyethylene containers prior to analysis. In the most demanding *trace analysis* of lead at 1 pg/g in polar ice cores, it was observed that polyethylene containers contributed a measurable flux of 1 fg of lead per $cm^2$ per day even after they had been soaked in acid for 7 months.[4] Steel needles are an avoidable source of metal contamination in biochemical analysis.

A study of mercury in Lake Michigan found levels near 1.6 pM ($1.6 \times 10^{-12}$ M), which is *two orders of magnitude* below concentrations observed in many earlier studies.[5] Previous investigators apparently unknowingly contaminated their samples. A study of handling techniques for the analysis of lead in rivers investigated variations in sample collection, sample containers, protection during transportation from the field to the lab, filtration techniques, chemical preservatives, and preconcentration procedures.[6] Each individual step that deviated from best practice *doubled* the apparent concentration of lead in stream water. Clean rooms with filtered air supplies are essential in trace analysis. Even with the best precautions, the precision of trace analysis becomes poorer as the concentration of analyte decreases (Box 5-2).

"Unless the complete history of any sample is known with certainty, the analyst is well advised not to spend his [or her] time in analyzing it."[7] Your laboratory notebook should describe how a sample was collected and stored and exactly how it was handled, as well as stating how it was analyzed.

**Table 28-1** Manganese concentration of serum stored in washed and unwashed polyethylene containers

| Container[a] | Mn (ng/mL) |
|---|---|
| Unwashed | 0.85 |
| Unwashed | 0.55 |
| Unwashed | 0.20 |
| Unwashed | 0.67 |
| Average | $0.57 \pm 0.27$ |
| | |
| Washed | 0.096 |
| Washed | 0.018 |
| Washed | 0.12 |
| Washed | 0.10 |
| Average | $0.084 \pm 0.045$ |

a. Washed containers were rinsed with water distilled twice from fused-silica vessels, which introduce less contamination into water than does glass.

SOURCE: J. Versieck, Trends Anal. Chem. **1983**, 2, 110.

Variance = (standard deviation)$^2$

For random errors, the overall *variance*, $s_o^2$, is the sum of the variance of the analytical procedure, $s_a^2$, and the variance of the sampling operation, $s_s^2$:

$$\text{Total variance} = \text{analytical variance} + \text{sampling variance}$$

*Additivity of variance:*
$$s_o^2 = s_a^2 + s_s^2 \qquad (28\text{-}1)$$

If either $s_a$ or $s_s$ is sufficiently smaller than the other, there is little point in trying to reduce the smaller one. For example, if $s_s$ is 10% and $s_a$ is 5%, the overall standard deviation is 11% ($\sqrt{0.10^2 + 0.05^2} = 0.11$). A more expensive and time-consuming analytical procedure that reduces $s_a$ to 1% only improves $s_o$ from 11 to 10% ($\sqrt{0.10^2 + 0.01^2} = 0.10$).

## Origin of Sampling Variance

To understand the nature of the uncertainty in selecting a sample for analysis, consider a random mixture of two kinds of solid particles. The theory of probability allows us to state the likelihood that a randomly drawn sample has the same composition as the bulk sample. It may surprise you to learn how large a sample is required for accurate sampling.[9]

Suppose that the mixture contains $n_A$ particles of type A and $n_B$ particles of type B. The probabilities of drawing A or B from the mixture are

$$p = \text{probability of drawing A} = \frac{n_A}{n_A + n_B} \qquad (28\text{-}2)$$

$$q = \text{probability of drawing B} = \frac{n_B}{n_A + n_B} = 1 - p \qquad (28\text{-}3)$$

If $n$ particles are drawn at random, the expected number of particles of type A is $np$ and the standard deviation of many drawings is known from the binomial distribution to be

*Standard deviation in sampling operation:*
$$s_n = \sqrt{npq} \qquad (28\text{-}4)$$

---

**Example** Statistics of Drawing Particles

A mixture contains 1% KCl particles and 99% $KNO_3$ particles. If $10^4$ particles are taken, what is the expected number of KCl particles, and what will be the standard deviation if the experiment is repeated many times?

**Solution** The expected number is just

$$\text{Expected number of KCl particles} = np = (10^4)(0.01) = 100 \text{ particles}$$

and the standard deviation will be

$$\text{Standard deviation} = \sqrt{npq} = \sqrt{(10^4)(0.01)(0.99)} = 9.9$$

*The standard deviation $\sqrt{npq}$ applies to both kinds of particles.* The standard deviation is 9.9% of the expected number of KCl particles, but only 0.1% of the expected number of $KNO_3$ particles ($nq = 9\,900$). If you want to know how much nitrate is in the mixture, this sample is probably sufficient. For chloride, 9.9% uncertainty may not be acceptable.

---

How much sample corresponds to $10^4$ particles? Suppose that the particles are 1-mm-diameter spheres. The volume of a 1-mm-diameter sphere is $\frac{4}{3}\pi(0.5 \text{ mm})^3 = 0.524 \ \mu L$. The density of KCl is 1.984 g/mL and that of $KNO_3$ is 2.109 g/mL, so the average density of the mixture is $(0.01)(1.984) + (0.99)(2.109) = 2.108$ g/mL. The mass of mixture containing $10^4$ particles is $(10^4)(0.524 \times 10^{-3} \text{ mL})(2.108 \text{ g/mL}) = 11.0$ g. *If you take 11.0-g test portions from a larger laboratory sample, the expected sampling standard deviation for chloride is 9.9%. The sampling standard deviation for nitrate will be only 0.1%.*

Wow! I'm surprised that such a large sample has such a large standard deviation.

How can you prepare a mixture of 1-mm-diameter particles? You might make such a mixture by grinding larger particles and passing them through a 16 mesh sieve, whose screen openings are squares with sides of length 1.18 mm (Table 28-2). Particles that pass through the screen are then passed through a 20 mesh sieve, whose openings are 0.85 mm, and

**Table 28-2**  Standard test sieves

| Sieve number | Screen opening (mm) | Sieve number | Screen opening (mm) |
|---|---|---|---|
| 5 | 4.00 | 45 | 0.355 |
| 6 | 3.35 | 50 | 0.300 |
| 7 | 2.80 | 60 | 0.250 |
| 8 | 2.36 | 70 | 0.212 |
| 10 | 2.00 | 80 | 0.180 |
| 12 | 1.70 | 100 | 0.150 |
| 14 | 1.40 | 120 | 0.125 |
| 16 | 1.18 | 140 | 0.106 |
| 18 | 1.00 | 170 | 0.090 |
| 20 | 0.850 | 200 | 0.075 |
| 25 | 0.710 | 230 | 0.063 |
| 30 | 0.600 | 270 | 0.053 |
| 35 | 0.500 | 325 | 0.045 |
| 40 | 0.425 | 400 | 0.038 |

*EXAMPLE: Particles designated 50/100 mesh pass through a 50 mesh sieve but are retained by a 100 mesh sieve. Their size is in the range 0.150–0.300 mm.*

material that does not pass is retained for your sample. This procedure gives particles whose diameters are in the range 0.85–1.18 mm. We refer to the size range as 16/20 *mesh*.

Suppose that much finer particles of 80/120 mesh size (average diameter = 152 μm, average volume = 1.84 nL) were used instead. Now the mass containing $10^4$ particles is reduced from 11.0 to 0.038 8 g. We could analyze a larger sample to reduce the sampling uncertainty for chloride.

---

**Example**  Reducing Sample Uncertainty with a Larger Test Portion

How many grams of 80/120 mesh sample are required to reduce the chloride sampling uncertainty to 1%?

**Solution**  We are looking for a standard deviation of 1% of the number of KCl particles (= 1% of $np$):

$$\sigma_n = \sqrt{npq} = (0.01)np$$

Using $p = 0.01$ and $q = 0.99$, we find $n = 9.9 \times 10^5$ particles. With a particle volume of 1.84 nL and an average density of 2.108 g/mL, the mass required for 1% chloride sampling uncertainty is

$$\text{Mass} = (9.9 \times 10^5 \text{ particles})\left(1.84 \times 10^{-6}\frac{\text{mL}}{\text{particle}}\right)\left(2.108\frac{\text{g}}{\text{mL}}\right) = 3.84 \text{ g}$$

Even with an average particle diameter of 152 μm, we must analyze 3.84 g to reduce the sampling uncertainty to 1%. There is no point using an expensive analytical method with a precision of 0.1%, because the overall uncertainty will still be 1% from sampling uncertainty.

> There is no advantage to reducing the analytical uncertainty if the sampling uncertainty is high, and vice versa.

Sampling uncertainty arises from the random nature of drawing particles from a mixture. If the mixture is a liquid and the particles are molecules, there are about $10^{22}$ particles/mL. It will not require much volume of homogeneous liquid solution to reduce the sampling error to a negligible value. Solids, however, must be ground to very fine dimensions, and large quantities must be used to ensure a small sampling variance. Grinding invariably contaminates the sample with material from the grinding apparatus.

Table 28-3 illustrates another problem with heterogeneous materials. Nickel ore was crushed into small particles that were sieved and analyzed. Parts of the ore that are deficient in nickel are relatively resistant to fracture, so the larger particles do not have the same chemical composition as the smaller particles. It is necessary to grind the entire ore to a fine power to have any hope of obtaining a representative sample.

**Table 28-3**  Nickel content of crushed ore

| Particle mesh size | Ni content (wt%) |
|---|---|
| <230 | 13.52 ± 0.69 |
| 120/230 | 13.20 ± 0.74 |
| 25/120 | 13.22 ± 0.49 |
| 10/25 | 10.54 ± 0.84 |
| >10 | 9.08 ± 0.69 |

*NOTE: Uncertainty is ±1 standard deviation.*

SOURCE: J. G. Dunn, D. N. Phillips, and W. van Bronswijk. "An Exercise to Illustrate the Importance of Sample Preparation in Analytical Chemistry," J. Chem. Ed. **1997**, 74, 1188.

## Choosing a Sample Size

A well-mixed powder containing KCl and $KNO_3$ is an example of a heterogeneous material in which the variation from place to place is random. *How much of a random mixture should be analyzed to reduce the sampling variance for one analysis to a desired level?*

To answer this question, consider Figure 28-3, which shows results for sampling the radioisotope $^{24}Na$ in human liver. The tissue was "homogenized" in a blender but was not truly homogeneous, because it was a suspension of small particles in water. The average number of radioactive counts per second per gram of sample was about 237. When the mass of sample for each analysis was about 0.09 g, the standard deviation (shown by the error bar at the left in the diagram) was ±31 counts per second per gram of homogenate, which is ±13.1% of the mean value (237). When the sample size was increased to about 1.3 g, the standard deviation decreased to ±13 counts/s/g, or ±5.5% of the mean. For a sample size near 5.8 g, the standard deviation was reduced to ±5.7 counts/s/g, or ±2.4% of the mean.

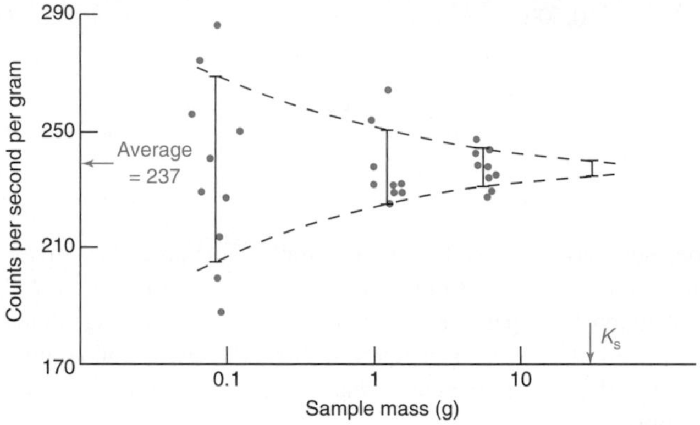

*Figure 28-3* Sampling diagram of experimental results for $^{24}Na$ in liver homogenate. Dots are experimental points, and error bars extend ±1 standard deviation about the mean. Note that the scale on the abscissa is logarithmic. [From B. Kratochvil and J. K. Taylor, "Sampling for Chemical Analysis," *Anal. Chem.* **1981**, 53, 925A; National Bureau of Standards Internal Report 80-2164, 1980, p. 66.]

Equation 28-4 told us that when $n$ particles are drawn from a mixture of two kinds of particles (such as liver tissue particles and droplets of water), the sampling standard deviation will be $\sigma_n = \sqrt{npq}$, where $p$ and $q$ are the fraction of each kind of particle present. The relative standard deviation is $\sigma_n/n = \sqrt{npq}/n = \sqrt{pq/n}$. The relative variance, $(\sigma_n/n)^2$, is therefore

$$\text{Relative variance} \equiv R^2 = \left(\frac{\sigma_n}{n}\right)^2 = \frac{pq}{n} \Rightarrow nR^2 = pq \quad (28\text{-}5)$$

Noting that the mass of sample drawn, $m$, is proportional to the number of particles drawn, we can rewrite Equation 28-5 in the form

*Sampling constant:* $\qquad\qquad mR^2 = K_s \qquad\qquad (28\text{-}6)$

in which $R$ is the relative standard deviation (expressed as a percentage) due to sampling and $K_s$ is called the *sampling constant*. $K_s$ is the mass of sample producing a relative sampling standard deviation of 1%.

Let's see if Equation 28-6 describes Figure 28-3. Table 28-4 shows that $mR^2$ is approximately constant for large samples, but agreement is poor for the smallest sample. Attributing the poor agreement at low mass to random sampling variation, we assign $K_s \approx 36$ g in Equation 28-6. This is the average from the 1.3- and 5.8-g samples in Table 28-4.

**Table 28-4** Calculation of sampling constant for Figure 28-3

| Sample mass, $m$ (g) | Relative standard deviation (%) | $mR^2$ (g) |
|---|---|---|
| 0.09 | 13.1 | 15.4 |
| 1.3 | 5.5 | 39.3 |
| 5.8 | 2.4 | 33.4 |

**Example** Mass of Sample Required to Produce a Given Sampling Variance

What mass in Figure 28-3 will give a sampling standard deviation of ±7%?

**Solution** With the sampling constant $K_s \approx 36$ g, the answer is

$$m = \frac{K_s}{R^2} = \frac{36\text{ g}}{7^2} = 0.73\text{ g}$$

A 0.7-g sample should give ~7% sampling standard deviation. This is strictly a sampling standard deviation. The net variance will be the sum of variances from sampling and from analysis (Equation 28-1).

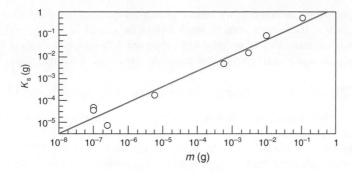

**Figure 28-4** Data are for Mn in powdered algae, showing that the sampling constant, $K_s$, in Equation 28-6 is approximately proportional to sample mass, $m$, over six orders of magnitude of mass. Four different measurement techniques were required to measure Mn in different regions of the wide mass range. Both axes are logarithmic. *[Data from M. Rossbach and E. Zeiller, "Assessment of Element-Specific Homogeneity in Reference Materials Using Microanalytical Techniques," Anal. Bioanal. Chem.* **2003**, *377, 334.]*

Figure 28-4 shows an example in which Equation 28-6 is approximately valid over six orders of magnitude of sample mass.

## Choosing the Number of Replicate Analyses

We just saw that a single 0.7-g sample is expected to give a sampling standard deviation of ±7%. *How many 0.7-g samples must be analyzed to give 95% confidence that the mean is known to within ±4%?* The meaning of 95% confidence is that there is only a 5% chance that the true mean lies more than ±4% away from the measured mean. The question we just asked refers only to sampling uncertainty and assumes that analytical uncertainty is much smaller than sampling uncertainty.

Rearranging Student's *t* Equation 4-6 allows us to answer the question:

*Required number of replicate analyses:*

$$\underbrace{\mu - \bar{x}}_{e} = \frac{ts_s}{\sqrt{n}} \Rightarrow n = \frac{t^2 s_s^2}{e^2} \tag{28-7}$$

The sampling contribution to the overall uncertainty can be reduced by analyzing more samples.

in which $\mu$ is the true population mean, $\bar{x}$ is the measured mean, $n$ is the number of samples needed, $s_s^2$ is the variance of the sampling operation, and $e$ is the sought-for uncertainty. Both the quantities $s_s$ and $e$ must be expressed as absolute uncertainties or both must be expressed as relative uncertainties. Student's $t$ is taken from Table 4-2 for 95% confidence at $n - 1$ degrees of freedom. Because $n$ is not yet known, the value of $t$ for $n = \infty$ can be used to estimate $n$. After a value of $n$ is calculated, the process is repeated a few times until a constant value of $n$ is found.

> ### Example  Sampling a Random Bulk Material
>
> How many 0.7-g samples must be analyzed to give 95% confidence that the mean is known to within ±4%?
>
> **Solution**  A 0.7-g sample gives $s_s = 7\%$, and we are seeking $e = 4\%$. We will express both uncertainties in relative form. Taking $t = 1.960$ (from Table 4-2 for 95% confidence and $\infty$ degrees of freedom) as a starting value, we find
>
> $$n \approx \frac{(1.960)^2 (0.07)^2}{(0.04)^2} = 11.8 \approx 12$$
>
> For $n = 12$, there are 11 degrees of freedom, so a second trial value of Student's $t$ (interpolated from Table 4-2) is 2.209. A second cycle of calculation gives
>
> $$n \approx \frac{(2.209)^2 (0.07)^2}{(0.04)^2} = 14.9 \approx 15$$
>
> For $n = 15$, there are 14 degrees of freedom and $t = 2.150$, which gives
>
> $$n \approx \frac{(2.150)^2 (0.07)^2}{(0.04)^2} = 14.2 \approx 14$$
>
> For $n = 14$, there are 13 degrees of freedom and $t = 2.170$, which gives
>
> $$n \approx \frac{(2.170)^2 (0.07)^2}{(0.04)^2} = 14.4 \approx 14$$
>
> The calculations reach a constant value near $n \approx 14$, so we need about 14 samples of 0.7-g size to determine the mean value to within 4% with 95% confidence.

For the preceding calculations, we needed prior knowledge of the standard deviation. A preliminary study of the sample must be made before the remainder of the analysis can be planned. If there are many similar samples to be analyzed, a thorough analysis of one sample might allow you to plan less thorough—but adequate—analyses of the remainder.

## ▪▪▪ 28-2 Dissolving Samples for Analysis[10]

Once a *bulk sample* is selected, a *laboratory sample* must be prepared for analysis (Figure 28-2). A coarse solid sample should be ground and mixed so that the laboratory sample has the same composition as the bulk sample. Solids are typically dried at 110°C at atmospheric pressure to remove adsorbed water prior to analysis. Temperature-sensitive samples may simply be stored in an environment that brings them to a constant, reproducible moisture level.

The laboratory sample is usually dissolved for analysis. It is important to dissolve the entire sample, or else we cannot be sure that all of the analyte was dissolved. If the sample does not dissolve under mild conditions, *acid digestion* or *fusion* may be used. Organic material may be destroyed by *combustion* (also called *dry ashing*) or *wet ashing* (oxidation with liquid reagents) to place inorganic elements in suitable form for analysis.

### Grinding

Solids can be ground in a **mortar and pestle** like those shown in Figure 28-5. The steel mortar (also called a percussion mortar or "diamond" mortar) is a hardened steel tool into which the sleeve and pestle fit snugly. Materials such as ores and minerals can be crushed by striking the pestle lightly with a hammer. The *agate* mortar (or similar ones made of porcelain, mullite, or alumina) is designed for grinding small particles into a fine powder. Less expensive mortars tend to be more porous and more easily scratched, which leads to contamination of the sample with mortar material or portions of previously ground samples. A ceramic mortar can be cleaned by wiping with a wet tissue and washing with distilled water. Difficult residues can be removed by grinding with 4 M HCl in the mortar or by grinding an abrasive cleaner (such as Ajax), followed by washing with HCl and water. A *boron carbide* mortar and pestle is five times harder than agate and less prone to contaminate the sample.

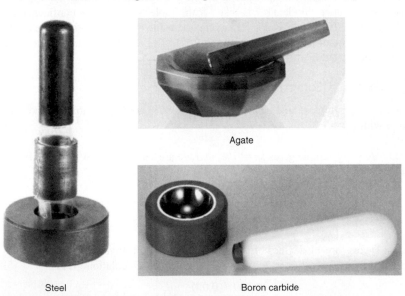

Agate

Steel

Boron carbide

*Figure 28-5* Steel, agate, and boron carbide mortars and pestles. The mortar is the base and the pestle is the grinding tool. In regard to boron carbide, the mortar is a hemispheric shell enclosed in a plastic or aluminum body. The pestle has a boron carbide button at the tip of a plastic handle.
*[Courtesy Thomas Scientific, Swedesboro, NJ, and Spex Industries, Edison, NJ.]*

A **ball mill** is a grinding device in which steel or ceramic balls are rotated inside a drum to crush the sample to a fine powder. Figure 28-6 shows a Wig-L-Bug, which pulverizes a sample by shaking it in a vial with a ball that moves back and forth. For soft materials, plastic vials and balls are appropriate. For harder materials, steel, agate, and tungsten carbide are used. A Shatterbox laboratory mill spins a puck and ring inside a grinding container at 825 revolutions per minute to pulverize up to 100 g of material (Figure 28-7). Tungsten carbide and zirconia containers are used for very hard samples.

*Figure 28-6* Wig-L-Bug sample shaker and polystyrene vial and ball for pulverizing soft materials. The arrow shows the direction of shaking. *[Courtesy Spex Industries, Edison, NJ.]*

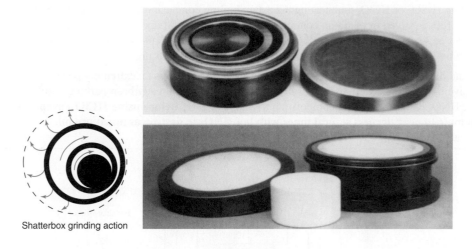

Shatterbox grinding action

*Figure 28-7* Shatterbox laboratory mill spins a puck and ring inside a container at high speed to grind up to 100 mL of sample to a fine powder. *[Courtesy Spex Industries, Edison, NJ.]*

## Dissolving Inorganic Materials with Acids

Table 28-5 lists acids commonly used for dissolving inorganic materials. The nonoxidizing acids HCl, HBr, HF, $H_3PO_4$, dilute $H_2SO_4$, and dilute $HClO_4$ dissolve metals by the redox reaction

$$M + nH^+ \rightarrow M^{n+} + \frac{n}{2}H_2 \qquad (28\text{-}8)$$

Metals with negative reduction potentials should dissolve, although some, such as Al, form a protective oxide coat that inhibits dissolution. Volatile species formed by protonation of anions such as carbonate $(CO_3^{2-} \rightarrow H_2CO_3 \rightarrow CO_2)$, sulfide $(S^{2-} \rightarrow H_2S)$, phosphide $(P^{3-} \rightarrow PH_3)$, fluoride $(F^- \rightarrow HF)$, and borate $(BO_3^{3-} \rightarrow H_3BO_3)$ will be lost from hot acids in open vessels. Volatile metal halides such as $SnCl_4$ and $HgCl_2$ and some molecular oxides such as $OsO_4$ and $RuO_4$ also can be lost. Hot hydrofluoric acid is especially useful for dissolving silicates. Glass or platinum vessels can be used for HCl, HBr, $H_2SO_4$, $H_3PO_4$, and $HClO_4$. HF should be used in Teflon, polyethylene, silver, or platinum vessels. The highest-quality acids must be used to minimize contamination by the concentrated reagent.

Substances that do not dissolve in nonoxidizing acids may dissolve in the oxidizing acids $HNO_3$, hot, concentrated $H_2SO_4$, or hot, concentrated $HClO_4$. Nitric acid attacks most metals, but not Au and Pt, which dissolve in the 3:1 (vol/vol) mixture of HCl:$HNO_3$ called **aqua regia.** Strong oxidants such as $Cl_2$ or $HClO_4$ in HCl dissolve difficult materials such as Ir at elevated temperature. A mixture of $HNO_3$ and HF attacks the refractory carbides, nitrides, and borides of Ti, Zr, Ta, and W. A powerful oxidizing solution known as "piranha solution" is a 1:1 (vol/vol) mixture of 30 wt% $H_2O_2$ plus 98 wt% $H_2SO_4$. Hot, concentrated $HClO_4$ (described later for organic substances) is a **DANGEROUS,** powerful oxidant whose oxidizing power is increased by adding concentrated $H_2SO_4$ and catalysts such as $V_2O_5$ or $CrO_3$.

Digestion is conveniently carried out in a Teflon-lined **bomb** (a sealed vessel) heated in a microwave oven.[11] The vessel in Figure 28-8 has a volume of 23 mL and digests up to 1 g of inorganic material in up to 15 mL of concentrated acid or digests 0.1 g of organic material,

HF causes excruciating burns. Exposure of just 2% of your body to concentrated (48 wt%) HF can kill you. Flood the affected area with water for 5 min and then coat the skin with 2.5% calcium gluconate gel kept in the lab for this purpose, and seek medical help. If the gel is not available, use whatever calcium salt is handy. HF damage can continue to develop days after exposure.

**Table 28-5**  Acids for sample dissolution

| Acid | Typical composition (wt% and density) | Notes |
|---|---|---|
| HCl | 37% 1.19 g/mL | Nonoxidizing acid useful for many metals, oxides, sulfides, carbonates, and phosphates. Constant boiling composition at 109°C is 20% HCl. As, Sb, Ge, and Pb form volatile chlorides that may be lost from an open vessel. |
| HBr | 48–65% | Similar to HCl in solvent properties. Constant boiling composition at 124°C is 48% HBr. |
| $H_2SO_4$ | 95–98% 1.84 g/mL | Good solvent at its boiling point of 338°C. Attacks metals. Dehydrates and oxidizes organic compounds. |
| $H_3PO_4$ | 85% 1.70 g/mL | Hot acid dissolves refractory oxides insoluble in other acids. Becomes anhydrous above 150°C. Dehydrates to pyrophosphoric acid ($H_2PO_3-O-PO_3H_2$) above 200°C and dehydrates further to metaphosphoric acid ($[HPO_3]_n$) above 300°C. |
| HF | 50% 1.16 g/mL | Used primarily to dissolve silicates, making volatile $SiF_4$. This product and excess HF are removed by adding $H_2SO_4$ or $HClO_4$ and heating. Constant boiling composition at 112°C is 38% HF. Used in Teflon, silver, or platinum containers. Extremely harmful upon contact or inhalation. Fluorides of As, B, Ge, Se, Ta, Nb, Ti, and Te are volatile. $LaF_3$, $CaF_2$, and $YF_3$ precipitate. $F^-$ is removed by adding $H_3BO_3$ and taking to dryness with $H_2SO_4$ present. |
| $HClO_4$ | 60–72% 1.54–1.67 g/mL | Cold and dilute acid are not oxidizing, but hot, concentrated acid is an extremely powerful, explosive oxidant, especially useful for organic matter that has already been partially oxidized by hot $HNO_3$. Constant boiling composition at 203°C is 72%. **Before using $HClO_4$, evaporate the sample to near dryness several times with hot $HNO_3$ to destroy as much organic material as possible.** |

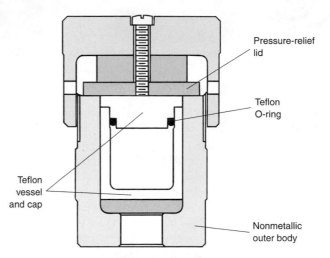

**Figure 28-8**  Microwave digestion bomb lined with Teflon. The outer container retains strength to 150°C but rarely reaches 50°C. [Courtesy Parr Instrument Co., Moline, IL.]

*Figure 28-9*  Automated apparatus that fuses three samples at once over propane burners. It also provides mechanical agitation of the Pt/Au crucibles. Crucibles are shown in the tipped position used to pour contents out after fusion is complete. [Courtesy Spex Industries, Edison, NJ.]

which releases a great deal of $CO_2(g)$. Microwave energy heats the contents to 200°C in a minute. To prevent explosions, the lid releases gas from the vessel if the internal pressure exceeds 8 MPa (80 bar). The bomb cannot be made of metal, which absorbs microwaves. An advantage of a bomb is that it is cooled prior to opening, thus preventing loss of volatile products.

## Dissolving Inorganic Materials by Fusion

Substances that will not dissolve in acid can usually be dissolved by a hot, molten inorganic **flux** (Table 28-6). Finely powdered unknown is mixed with 2 to 20 times its mass of solid flux, and **fusion** (melting) is carried out in a platinum–gold alloy crucible at 300° to 1 200°C in a furnace or over a burner. The apparatus in Figure 28-9 fuses three samples at once over propane burners with mechanical agitation of the crucibles. When the samples are homogeneous, the molten flux is poured into beakers containing 10 wt% aqueous $HNO_3$ to dissolve the product.

Most fusions use lithium tetraborate ($Li_2B_4O_7$, m.p. 930°C), lithium metaborate ($LiBO_2$, m.p. 845°C), or a mixture of the two. A nonwetting agent such as KI can be added to prevent the flux from sticking to the crucible. For example, 0.2 g of cement might be fused with 2 g of $Li_2B_4O_7$ and 30 mg of KI.

A disadvantage of a flux is that impurities are introduced by the large mass of solid reagent. If part of the unknown can be dissolved with acid prior to fusion, it should be dissolved. Then the insoluble component is dissolved with flux and the two portions are combined for analysis.

Basic fluxes in Table 28-6 ($LiBO_2$, $Na_2CO_3$, NaOH, KOH, and $Na_2O_2$) are best used to dissolve acidic oxides of Si and P. Acidic fluxes ($Li_2B_4O_7$, $Na_2B_4O_7$, $K_2S_2O_7$, and $B_2O_3$) are most suitable for basic oxides (including cements and ores) of the alkali metals, alkaline earths, lanthanides, and Al. $KHF_2$ is useful for lanthanide oxides. Sulfides and some oxides, some iron and platinum alloys, and some silicates require an oxidizing flux for dissolution. For this purpose, pure $Na_2O_2$ may be suitable or oxidants such as $KNO_3$, $KClO_3$, or $Na_2O_2$

Fusion is a last resort, because the flux can introduce impurities.

**Table 28-6** Fluxes for sample dissolution

| Flux | Crucible | Uses |
|---|---|---|
| $Na_2CO_3$ | Pt | For dissolving silicates (clays, rocks, minerals, glasses), refractory oxides, insoluble phosphates, and sulfates. |
| $Li_2B_4O_7$ or $LiBO_2$ or $Na_2B_4O_7$ | Pt, graphite Au-Pt alloy, Au-Rh-Pt alloy | Individual or mixed borates are used to dissolve aluminosilicates, carbonates, and samples with high concentrations of basic oxides. $B_4O_7^{2-}$ is called tetraborate and $BO_2^-$ is metaborate. |
| NaOH or KOH | Au, Ag | Dissolves silicates and SiC. Frothing occurs when $H_2O$ is eliminated from flux, so it is best to prefuse flux and then add sample. Analytical capabilities are limited by impurities in NaOH and KOH. |
| $Na_2O_2$ | Zr, Ni | Strong base and powerful oxidant good for silicates not dissolved by $Na_2CO_3$. Useful for iron and chromium alloys. Because it slowly attacks crucibles, a good procedure is to coat the inside of a Ni crucible with molten $Na_2CO_3$, cool, and add $Na_2O_2$. The peroxide melts at lower temperature than the carbonate, which shields the crucible from the melt. |
| $K_2S_2O_7$ | Porcelain, $SiO_2$, Au, Pt | Potassium pyrosulfate ($K_2S_2O_7$) is prepared by heating $KHSO_4$ until all water is lost and foaming ceases. Alternatively, potassium persulfate ($K_2S_2O_8$) decomposes to $K_2S_2O_7$ upon heating. Good for refractory oxides, not silicates. |
| $B_2O_3$ | Pt | Useful for oxides and silicates. Principal advantage is that flux can be completely removed as volatile methyl borate ($[CH_3O]_3B$) by several treatments with HCl in methanol. |
| $Li_2B_4O_7$ + $Li_2SO_4$ (2:1 wt/wt) | Pt | Example of a powerful mixture for dissolving refractory silicates and oxides in 10–20 min at 1 000°C. One gram of flux dissolves 0.1 g of sample. The solidified melt dissolves readily in 20 mL of hot 1.2 M HCl. |

can be added to $Na_2CO_3$. Boric oxide can be converted into $B(OCH_3)_3$ after fusion and completely evaporated. The solidified flux is treated with 100 mL of methanol saturated with HCl gas and heated gently. The procedure is repeated several times, if necessary, to remove all of the boron.

## Decomposition of Organic Substances

Digestion of organic material is classified as either **dry ashing,** when the procedure does not include liquid, or **wet ashing,** when liquid is used. Occasionally, fusion with $Na_2O_2$ (called Parr oxidation) or alkali metals may be carried out in a sealed bomb. Section 27-4 discussed *combustion analysis,* in which C, H, N, S, and halogens are measured.

Convenient *wet-ashing* procedures include microwave digestion with acid in a Teflon bomb (Figure 28-8). For example, 0.25 g of animal tissue can be digested for metal analysis by placing the sample in a 60-mL Teflon vessel containing 1.5 mL of high-purity 70% $HNO_3$ plus 1.5 mL of high-purity 96% $H_2SO_4$ and heating it in a 700-W kitchen microwave oven for 1 min.[12] Teflon bombs fitted with temperature and pressure sensors allow safe, programmable control of digestion conditions. An important wet-ashing process is *Kjeldahl digestion* with $H_2SO_4$ for nitrogen analysis (Section 7-2).

In the *Carius method,* digestion is performed with fuming $HNO_3$ (which contains excess dissolved $NO_2$) in a sealed, heavy-walled glass tube at 200°–300°C. For safety, the glass Carius tube should be contained in a steel vessel pressurized to approximately the same pressure expected inside the glass tube.[13] For trace analysis, sample should be placed inside a fused-silica tube inside the glass tube. Silica provides as little as 1–10% as much extractable metal as glass.[14]

Figure 28-10 shows microwave wet-ashing apparatus. Sulfuric acid or a mixture of $H_2SO_4$ and $HNO_3$ (~15 mL of acid per gram of unknown) is added to an organic substance in a glass digestion tube fitted with the reflux cap. In the first step, the sample is *carbonized* for 10 to 20 min at gentle reflux until all particles have dissolved and the solution has a uniform black appearance. Power is turned off and the sample is allowed to cool for 1–2 min. Next, *oxidation* is carried out by adding $H_2O_2$ or $HNO_3$ through the reflux cap until the color disappears or the solution is just barely tinted. If the solution is not homogeneous, the power is turned up and the sample is heated to bring all solids into solution. Repeated cycles of

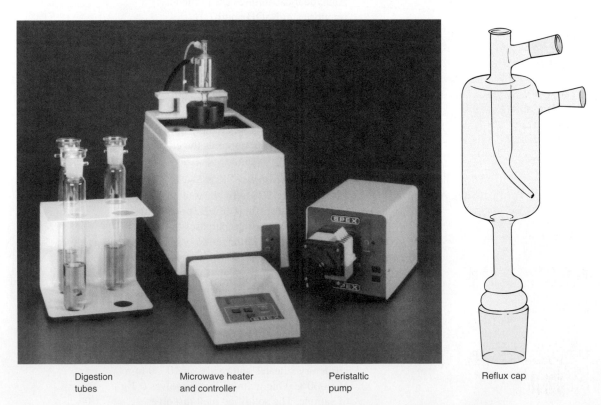

Digestion tubes     Microwave heater and controller     Peristaltic pump     Reflux cap

*Figure 28-10* Microwave apparatus for digesting organic materials by wet ashing.
*[Courtesy Spex Industries, Edison, NJ.]*

CHAPTER 28 Sample Preparation

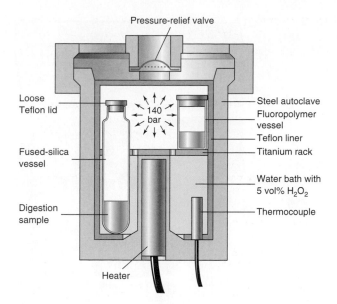

Pressure-relief valve

Loose
Teflon lid

Fused-silica
vessel

Digestion
sample

Heater

Steel autoclave
Fluoropolymer
vessel
Teflon liner
Titanium rack

Water bath with
5 vol% $H_2O_2$

Thermocouple

140
bar

**Figure 28-11** High-pressure autoclave allows digestion up to 270°C without $H_2SO_4$ in open vessels inside autoclave. *[Adaped from B. Maichin, M. Zischka, and G. Knapp, "Pressurized Wet Digestion in Open Vessels," Anal. Bioanal. Chem.* **2003,** *376, 715.]*

oxidation and solubilization may be required. Once conditions for a particular type of material are worked out, the procedure is automated, with power levels and reagent delivery (by the peristaltic pump) programmed into the controller.

The high-pressure asher in Figure 28-11 uses a resistive heating element inside a sealed chamber for digestion at temperatures up to 270°C under a pressure up to 140 bar. High pressure allows acids to be heated to high temperature without boiling. At high temperature, $HNO_3$ oxidizes organic matter without assistance from $H_2SO_4$, which is not as pure as $HNO_3$ and is therefore less suitable for trace analysis. Silica or fluoropolymer vessels inside the sealed chamber are loosely sealed by Teflon caps that permit evolved gases to escape. The bottom of the vessel is filled with 5 vol% $H_2O_2$ in $H_2O$. Hydrogen peroxide reduces nitrogen oxides generated by digestion of organic matter. As an example, a 1-g sample of animal tissue could be digested in a 50-mL fused-silica vessel containing 5 mL of high-purity 70 vol% $HNO_3$ plus 0.2 mL of high-purity 37 vol% HCl. Metallic elements in the digestion solution could be measured at part per billion to part per million levels by inductively coupled plasma–atomic emission.

Wet ashing with refluxing $HNO_3$-$HClO_4$ (Figure 28-12) is a widely applicable, but hazardous, procedure.[15] **Perchloric acid has caused numerous explosions.** Use a good blast shield in a metal-lined fume hood designed for $HClO_4$. First, heat the sample slowly to boiling with $HNO_3$ but *without* $HClO_4$. Boil to near dryness to destroy easily oxidized material that might explode in the presence of $HClO_4$. Add fresh $HNO_3$ and repeat the evaporation several times. After the sample cools to room temperature, add $HClO_4$ and heat again. If possible, $HNO_3$ should be present during the $HClO_4$ treatment. A large excess of $HNO_3$ should be present when oxidizing organic materials.

Bottles of $HClO_4$ should not be stored on wooden shelves, because acid spilled on wood can form explosive cellulose perchlorate esters. Perchloric acid also should not be stored near organic reagents or reducing agents. A reviewer of this book once wrote, "I have seen someone substitute perchloric acid for sulfuric acid in a Jones reductor experiment with spectacular results—no explosion but the tube melted!"

The combination of $Fe^{2+}$ and $H_2O_2$, called *Fenton's reagent,* oxidizes organic material in dilute aqueous solutions. For example, organic components of urine could be destroyed in 30 min at 50°C to release traces of mercury for analysis.[17] To do so, a 50-mL sample was adjusted to pH 3–4 with 0.5 M $H_2SO_4$. Then 50 µL of saturated aqueous ferrous ammonium sulfate, $Fe(NH_4)_2(SO_4)_2$, were added, followed by 100 µL of 30% $H_2O_2$.

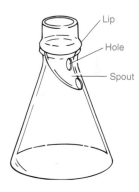

Lip

Hole

Spout

**Figure 28-12** Reflux cap for wet ashing in an Erlenmeyer flask. The hole allows vapor to escape, and the spout is curved to contact the inside of the flask. *[From D. D. Siemer and H. G. Brinkley, "Erlenmeyer Flask-Reflux Cap for Acid Sample Decomposition," Anal. Chem.* **1981,** *53, 750.]*

$HClO_4$ with organic material is an extreme **explosion hazard.** Always oxidize first with $HNO_3$. Always use a blast shield for $HClO_4$.

Fenton's reagent generates OH· radical and, possibly, $Fe^{II}OOH$ as the active species.[16]

## ▪▪▪▪ 28-3 Sample Preparation Techniques

**Sample preparation** is the series of steps required to transform a sample so that it is suitable for analysis. Sample preparation may include dissolving the sample, extracting analyte from a complex matrix, concentrating a dilute analyte to a level that can be measured, chemically converting analyte into a detectable form, and removing or masking interfering species.

Section 24-4 described *solid-phase microextraction, purge and trap,* and *thermal desorption*—sample preparation methods that are especially useful for gas chromatography.

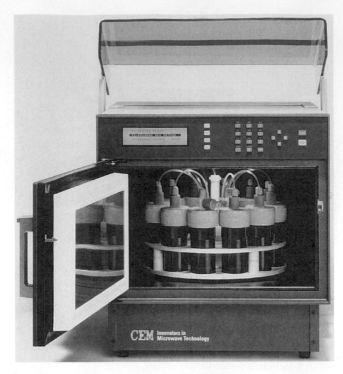

*Figure 28-13* Extraction vessels in a microwave oven that processes up to 12 samples in under 30 min. Each 100-mL vessel has a vent tube that releases vapor if the pressure exceeds 14 bar. Vapors from the chamber are ultimately vented to a fume hood. The temperature inside each vessel can be monitored and used to control the microwave power. *[Courtesy CEM Corp., Matthews, NC.]*

## Liquid Extraction Techniques

In **extraction,** analyte is dissolved in a solvent that does not necessarily dissolve the entire sample and does not decompose the analyte. In a typical *microwave-assisted extraction* of pesticides from soil, a mixture of soil plus acetone and hexane is placed in a Teflon-lined bomb (Figures 28-8 and 28-13) and heated by microwaves to 150°C. This temperature is 50° to 100° higher than the boiling points of solvents at atmospheric pressure. Pesticides dissolve, but the soil remains behind. The liquid is then analyzed by chromatography.

Acetone absorbs microwaves, so it can be heated in a microwave oven. Hexane does not absorb microwaves. To perform an extraction with pure hexane, the liquid is placed in a fluoropolymer insert inside the Teflon vessel in Figure 28-8.[18] The walls of the insert contain carbon black, which absorbs microwaves and heats the solvent.

**Supercritical fluid extraction** uses a *supercritical fluid* (Box 25-2) as the extraction solvent.[20] $CO_2$ is the most common supercritical fluid because it is inexpensive and it eliminates the need for costly disposal of waste organic solvents. Addition of a second solvent such as methanol increases the solubility of polar analytes. Nonpolar substances, such as petroleum hydrocarbons, can be extracted with supercritical argon.[21] The extraction process can be monitored by infrared spectroscopy because Ar has no infrared absorption.

Figure 28-14a shows how a supercritical fluid extraction can be carried out. Pressurized fluid is pumped through a heated extraction vessel. Fluid can be left in contact with the sample for some time or it can be pumped through continuously. At the outlet of the extraction vessel, the fluid flows through a capillary tube to release pressure. Exiting $CO_2$ evaporates, leaving extracted analyte in the collection vessel. Alternatively, the $CO_2$ can be bubbled through a solvent in the collection vessel to leave a solution of analyte.

Figure 28-14b shows the extraction of organic compounds from dust collected with a vacuum cleaner from door mats at the chemistry building of Ohio State University. The chromatogram of the extract in Figure 28-14c exhibits myriad organic compounds that you and I inhale in every breath.

Figure 28-15 shows glassware for continuous *liquid-liquid extraction* of a nonvolatile analyte. In Figure 28-15a, the extracting solvent is denser than the liquid being extracted. Solvent boils from the flask and condenses into the extraction vessel. Dense droplets of

Some chelators can extract metal ions into supercritical $CO_2$ (containing small quantities of methanol or water). The ligand below dissolves lanthanides and actinides:[19]

$CF_2CF_2CF_3$

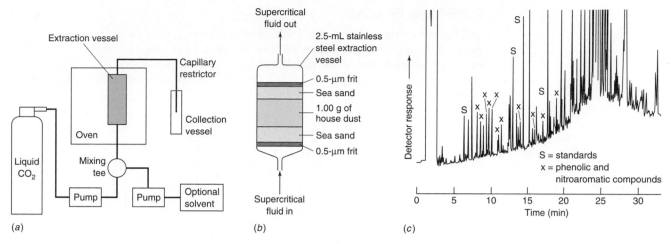

**Figure 28-14** (*a*) Apparatus for supercritical fluid extraction. (*b*) Vessel for extracting house dust at 50°C with 20 mol% methanol/80 mol% $CO_2$ at 24.0 MPa (240 bar). (*c*) Gas chromatogram of $CH_2Cl_2$ solution of extract using a 30 m × 0.25 mm diphenyl$_{0.05}$dimethyl$_{0.95}$siloxane column (1 μm film thickness) with a temperature ranging from 40° to 280°C and flame ionization detection. [*From T. S. Reighard and S. V. Olesik, "Comparison of Supercritical Fluids and Enhanced-Fluidity Liquids for the Extraction of Phenolic Pollutants from House Dust," Anal. Chem.* **1996,** *68, 3612.]*

solvent falling through the liquid column extract the analyte. When the liquid level is high enough, extraction solvent is pushed through the return tube to the solvent reservoir. By this means, analyte is slowly transferred from the light liquid at the left into the dense liquid in the reservoir. Figure 28-15b shows the procedure when the extraction solvent is less dense than the liquid being extracted.

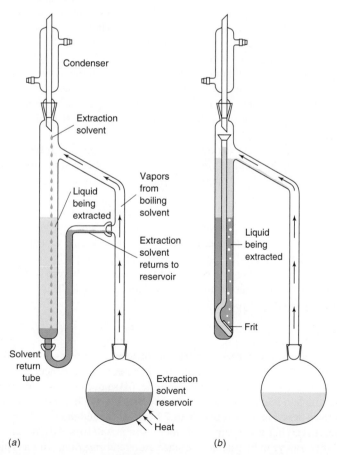

**Figure 28-15** Continuous liquid-liquid extraction apparatus used when extraction solvent is (*a*) denser than the liquid being extracted or (*b*) lighter than the liquid being extracted.

## Solid-Phase Extraction[22]

**Solid-phase extraction** uses a small volume of a chromatographic stationary phase or molecularly imprinted polymer[23] (Box 26-2) to isolate desired analytes from a sample. The extraction removes much of the sample matrix to simplify the analysis. The opening of this chapter shows a solid-phase extraction membrane and extraction cartridges mounted on syringes.

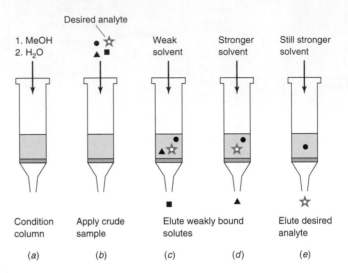

**Figure 28-16** Steps in solid-phase extraction.

Figure 28-16 shows steps in the solid-phase extraction of 10 ng/mL of steroids from urine. First, a syringe containing 1 mL of $C_{18}$-silica is conditioned with 2 mL of methanol to remove adsorbed organic material (Figure 28-16a). Then the column is washed with 2 mL of water. When the 10-mL urine sample is applied, nonpolar components adhere to the $C_{18}$-silica, and polar components pass through (Figure 28-16b). The column is then rinsed with 4 mL of 25 mM borate buffer at pH 8 to remove polar substances (Figure 28-16c). Then rinses with 4 mL of 40 vol% methanol/60% water and 4 mL of 20% acetone/80% water remove less polar substances (Figure 28-16d). Finally, elution with two 0.5-mL aliquots of 73% methanol/27% water washes the steroids from the column (Figure 28-16e).

Figure 28-17 compares chromatograms of the drug naproxen in blood serum with or without sample cleanup by solid-phase extraction. Without cleanup, serum proteins overlap and obscure the signal from naproxen. Solid-phase extraction removes most of the protein.

Solid-phase extractions can reduce solvent consumption in analytical chemistry. For example, a standard procedure approved by the U.S. Environmental Protection Agency for the analysis of pesticides in wastewater requires 200 mL of dichloromethane for the liquid-liquid extraction of 1 L of water. The same analytes can be isolated by solid-phase extraction on $C_{18}$-silica disks. The pesticides are recovered from the disks by supercritical fluid extraction with $CO_2$ that is finally vented into a small volume of hexane. This one kind of analysis can save $10^5$ kg of $CH_2Cl_2$ per year.[24]

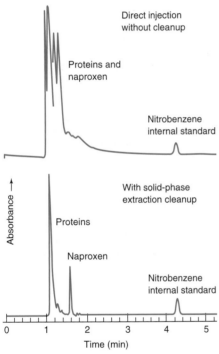

**Figure 28-17** HPLC of naproxen in blood serum with no cleanup (upper trace) or with prior sample cleanup (lower trace) by solid-phase extraction on $C_8$-silica. *[From R. E. Majors and A. D. Broske, "New Directions in Solid-Phase Extraction Particle Design," Am Lab. February 2002, p. 22.]*

Chelex-100 in $Mg^{2+}$ form

## Preconcentration

Trace analysis often requires **preconcentration** of analyte to bring it to a higher concentration prior to analysis. Metal ions in natural waters can be preconcentrated with cation-exchange resin. For example, a 500-mL volume of seawater adjusted to pH 6.5 with ammonium acetate and ammonia was passed through 2 g of Chelex-100 in the $Mg^{2+}$ form to trap all the trace-metal ions. Washing with 2 M $HNO_3$ eluted the metals in a total volume of 10 mL, thereby giving a concentration increase of 500/10 = 50. Metals in the $HNO_3$ solution were then analyzed by graphite furnace atomic absorption, with a typical detection limit for Pb being 15 pg/mL. The detection limit for Pb in the seawater is therefore 50 times lower, or 0.3 pg/mL. Figure 28-18 shows the effect of pH on the recovery of metals

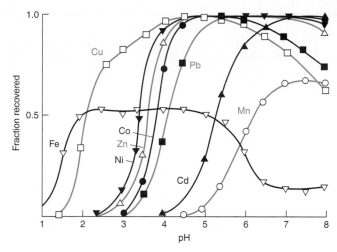

**Figure 28-18** The pH dependence of the recovery of trace metals from seawater by Chelex-100. The graph shows the pH of the seawater when it was passed through the column. [*From S.-C. Pai, Anal. Chim. Acta **1988**, 211, 271.*]

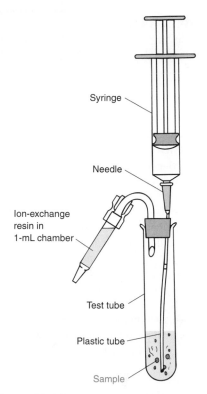

**Figure 28-19** Apparatus for trapping basic or acidic gases by ion exchange. [*From D.D. Siemer, "Ion Exchange Resins for Trapping Gases: Carbonate Determination," Anal. Chem. **1987**, 59, 2439.*]

from seawater. At low pH, $H^+$ competes with the metal ions for ion-exchange sites and prevents complete recovery.

Ion-exchange resins can capture basic or acidic gases. Carbonate liberated as $CO_2$ from $(ZrO)_2CO_3(OH)_2 \cdot xH_2O$ used in nuclear fuel reprocessing can be measured by placing a known amount of powdered solid in the test tube in Figure 28-19 and adding 3 M $HNO_3$. When the solution is purged with $N_2$, $CO_2$ is captured quantitatively by moist anion-exchange resin in the side arm:

$$CO_2 + H_2O \rightarrow H_2CO_3$$
$$2\,\text{resin}^+OH^- + H_2CO_3 \rightarrow (\text{resin}^+)_2CO_3^{2-} + 2H_2O$$

Carbonate is eluted from the resin with 1 M $NaNO_3$ and measured by titration with acid. Table 28-7 gives other applications of this technique.

## Derivatization

**Derivatization** is a procedure in which analyte is chemically modified to make it easier to detect or separate. For example, formaldehyde and other aldehydes and ketones in air, breath, or cigarette smoke[25] can be trapped and derivatized by passing air through a tiny cartridge containing 0.35 g of silica coated with 0.3 wt% 2,4-dinitrophenylhydrazine. Carbonyls are converted into the 2,4-dinitrophenylhydrazone derivative, which is eluted with 5 mL of acetonitrile and analyzed by HPLC. The products are readily detected by their strong ultraviolet absorbance near 360 nm.

**Table 28-7** Use of ion-exchange resin for trapping gases

| Gas | Species trapped | Eluent | Analytical method |
|---|---|---|---|
| $CO_2$ | $CO_3^{2-}$ | 1 M $NaNO_3$ | Titrate with acid |
| $H_2S$ | $S^{2-}$ | 0.5 M $Na_2CO_3$ + $H_2O_2$ | $S^{2-}$ is oxidized to $SO_4^{2-}$ by $H_2O_2$. The sulfate is measured by ion chromatography. |
| $SO_2$ | $SO_3^{2-}$ | 0.5 M $Na_2CO_3$ + $H_2O_2$ | $SO_3^{2-}$ is oxidized to $SO_4^{2-}$ by $H_2O_2$. The sulfate is measured by ion chromatography. |
| HCN | $CN^-$ | 1 M $Na_2SO_4$ | Titration of $CN^-$ with hypobromite: $CN^- + OBr^- \rightarrow CNO^- + Br^-$ |
| $NH_3$ | $NH_4^+$ | 1 M $NaNO_3$ | Colorimetric assay with Nessler's reagent: $2K_2HgI_4 + 2NH_3 \rightarrow NH_2Hg_2I_3 + 4KI + NH_4I$ <br> Nessler's reagent — Absorbs strongly at 400–425 nm |

*SOURCE: D. D. Siemer, "Ion Exchange Resins for Trapping Gases: Carbonate Determination," Anal. Chem. **1987**, 59, 2439.*

## Terms to Understand

aqua regia
ball mill
bomb
derivatization

dry ashing
extraction
flux
fusion

mortar and pestle
preconcentration
sample preparation
sampling

solid-phase extraction
supercritical fluid extraction
wet ashing

## Summary

The variance of an analysis is the sum of the sampling variance and the analytical variance. Sampling variance can be understood in terms of the statistics of selecting particles from a heterogeneous mixture. If the probabilities of selecting two kinds of particles from a two-particle mixture are $p$ and $q$, the standard deviation in selecting $n$ particles is $\sqrt{npq}$. You should be able to use this relation to estimate how large a sample is required to reduce the sampling variance to a desired level. Student's $t$ can be used to estimate how many repetitions of the analysis are required to reach a certain level of confidence in the final result.

Many inorganic materials can be dissolved in strong acids with heating. Glass vessels are often useful, but Teflon, platinum, or silver are required for HF, which dissolves silicates. If a nonoxidizing acid is insufficient, aqua regia or other oxidizing acids may do the job. A Teflon-lined bomb heated in a microwave oven is a convenient means of dissolving difficult samples. If acid digestion fails, fusion in a molten salt will usually work, but the large quantity of flux adds trace impurities. Organic materials are decomposed by wet ashing with hot concentrated acids or by dry ashing with heat.

Analytes can be separated from complex matrices by sample preparation techniques that include liquid extraction, supercritical fluid extraction, and solid-phase extraction. Dilute ionic analytes can be preconcentrated by adsorption onto an ion-exchange resin. Nonionic analytes can be concentrated by solid-phase extraction. Derivatization transforms the analyte into a more easily detected or separated form.

## Exercises

**28-A.** A box contains 120 000 red marbles and 880 000 yellow marbles.

(a) If you draw a random sample of 1 000 marbles from the box, what are the expected numbers of red and yellow marbles?

(b) Now put those marbles back in the box and repeat the experiment. What will be the absolute and relative standard deviations for the numbers in part (a) after many drawings of 1 000 marbles?

(c) What will be the absolute and relative standard deviations after many drawings of 4 000 marbles?

(d) If you quadruple the size of the sample, you decrease the sampling standard deviation by a factor of _____. If you increase the sample size by a factor of $n$, you decrease the sampling standard deviation by a factor of _____.

(e) What sample size is required to reduce the sampling standard deviation of red marbles to $\pm 2\%$?

**28-B.** (a) What mass of sample in Figure 28-3 is expected to give a sampling standard deviation of $\pm 10\%$?

(b) With the mass from part (a), how many samples should be taken to assure 95% confidence that the mean is known to within $\pm 20$ counts per second per gram?

**28-C.** Because they were used in gasoline to boost octane rating, alkyl lead compounds found their way into the environment as toxic pollutants. One way to measure lead compounds in natural waters is by anodic stripping voltammetry (Figure 17-20), in which the lead is first reduced to the element at a mercury electrode and dissolves in the mercury. Reoxidation occurs when the electrode potential is made sufficiently positive, with current proportional to the concentration of dissolved Pb.

Reduction at $-1.2$ V: $(CH_3CH_2)_3PbCl \rightarrow Pb$ (in Hg)

Oxidation at $-0.5$ V: $Pb$ (in Hg) $\rightarrow Pb^{2+}(aq)$

Inorganic (ionic) forms of lead are present in much higher concentrations than alkyl lead in natural samples, so inorganic lead must be removed before analysis of alkyl lead. One way to do this is by *coprecipitation* with $BaSO_4$, as shown in the figure.

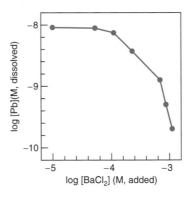

Coprecipitation of $Pb^{2+}$ with $BaSO_4$ from distilled water at pH 2. *[Data from N. Mikac and M. Branica, "Separation of Dissolved Alkyllead and Inorganic Lead Species by Coprecipitation with BaSO₄," Anal. Chim. Acta 1988, 212, 349.]*

In this experiment, $BaCl_2$ added to $Pb^{2+}$ in distilled water at the concentration on the abscissa was precipitated with a 20% excess of $Na_2SO_4$ and the residual concentration of lead was measured by stripping voltammetry. Suggest control experiments that should be performed to show that coprecipitation can be used to remove a large amount of inorganic lead from a small quantity of $(CH_3CH_2)_3PbCl$ in seawater, without removing the alkyl lead compound.

**28-D.** A soil sample contains some acid-soluble inorganic matter, some organic material, and some minerals that do not dissolve in any combination of hot acids that you try. Suggest a procedure for dissolving the entire sample.

## Problems

### Statistics of Sampling

**28-1.** Explain what is meant by the statement "Unless the complete history of any sample is known with certainty, the analyst is well advised not to spend his or her time in analyzing it."

**28-2.** Explain what is meant by "analytical quality" and "data quality" in the following quotation: "We need to update the environmental data quality model to explicitly distinguish *analytical quality* from *data quality*. We need to begin spending as much effort ensuring sample and subsample representativeness for heterogeneous matrices as we spend overseeing the analysis of an extract. We need to stop acting like data variability stemming from laboratory analysis is all-important while variability stemming from the sampling process can be ignored. . . ."[26]

**28-3.** **(a)** In the analysis of a barrel of powder, the standard deviation of the sampling operation is $\pm 4\%$ and the standard deviation of the analytical procedure is $\pm 3\%$. What is the overall standard deviation?

**(b)** To what value must the sampling standard deviation be reduced so that the overall standard deviation is $\pm 4\%$?

**28-4.** What mass of sample in Figure 28-3 is expected to give a sampling standard deviation of $\pm 6\%$?

**28-5.** Explain how to prepare a powder with an average particle diameter near 100 $\mu$m by using sieves from Table 28-2. How would such a particle mesh size be designated?

**28-6.** An example of a mixture of 1-mm-diameter particles of KCl and $KNO_3$ in a number ratio 1:99 follows Equation 28-4. A sample containing $10^4$ particles weighs 11.0 g. What is the expected number and relative standard deviation of KCl particles in a sample weighing $11.0 \times 10^2$ g?

**28-7.** When you flip a coin, the probability of its landing on each side is $p = q = \frac{1}{2}$ in Equations 28-2 and 28-3. If you flip it $n$ times, the expected number of heads equals the expected number of tails $= np = nq = \frac{1}{2}n$. The expected standard deviation for $n$ flips is $\sigma_n = \sqrt{npq}$. From Table 4-1, we expect that 68.3% of the results will lie within $\pm 1\sigma_n$ and 95.5% of the results will lie within $\pm 2\sigma_n$.
**(a)** Find the expected standard deviation for the number of heads in 1 000 coin flips.
**(b)** By interpolation in Table 4-1, find the value of $z$ that includes 90% of the area of the Gaussian curve. We expect that 90% of the results will lie within this number of standard deviations from the mean.
**(c)** If you repeat the 1 000 coin flips many times, what is the expected range for the number of heads that includes 90% of the results? (For example, your answer might be "The range 490 to 510 will be observed 90% of the time.")

**28-8.** In analyzing a lot with random sample variation, you find a sampling standard deviation of $\pm 5\%$. Assuming negligible error in the analytical procedure, how many samples must be analyzed to give 95% confidence that the error in the mean is within $\pm 4\%$ of the true value? Answer the same question for a confidence level of 90%.

**28-9.** In an experiment analogous to that in Figure 28-3, the sampling constant is found to be $K_s = 20$ g.
**(a)** What mass of sample is required for a $\pm 2\%$ sampling standard deviation?

**(b)** How many samples of the size in part **(a)** are required to produce 90% confidence that the mean is known to within 1.5%?

**28-10.** $^{87}Sr/^{86}Sr$ isotopes ratios were measured in polar ice cores of varying sample size. The 95% confidence interval, expressed as a percentage of the mean $^{87}Sr/^{86}Sr$ isotope ratio, decreased with increasing quantity of Sr measured:

| pg Sr | Confidence interval | pg Sr | Confidence interval |
|---|---|---|---|
| 57 | $\pm 0.057\%$ | 506 | $\pm 0.035\%$ |
| 68 | $\pm 0.069\%$ | 515 | $\pm 0.027\%$ |
| 110 | $\pm 0.049\%$ | 916 | $\pm 0.018\%$ |
| 110 | $\pm 0.045\%$ | 955 | $\pm 0.022\%$ |

SOURCE: Data from G. R. Burton, V. I. Morgan, C. F. Boutron, and K. J. R. Rosman, "High-Sensitivity Measurements of Strontium Isotopes in Polar Ice," Anal. Chim. Acta **2002**, 469, 225.

We postulate that the confidence interval is related to the mass of sample by the relation $mR^2 = K_s$, where $m$ is the mass of Sr in picograms, $R$ is the confidence interval expressed as a percentage of the isotope ratio ($R = 0.022$ for the 955-pg sample), and $K_s$ is a constant with units of picograms. Find the average value of $K_s$ and its standard deviation. Justify why we expect $mR^2 = K_s$ to hold if all measurements are made with the same number of replications.

**28-11.** Consider a random mixture containing 4.00 g of $Na_2CO_3$ (density 2.532 g/mL) and 96.00 g of $K_2CO_3$ (density 2.428 g/mL) with a uniform spherical particle radius of 0.075 mm.
**(a)** Calculate the mass of a single particle of $Na_2CO_3$ and the number of particles of $Na_2CO_3$ in the mixture. Do the same for $K_2CO_3$.
**(b)** What is the expected number of particles in 0.100 g of the mixture?
**(c)** Calculate the relative sampling standard deviation in the number of particles of each type in a 0.100-g sample of the mixture.

### Sample Preparation

**28-12.** From their standard reduction potentials, which of the following metals would you expect to dissolve in HCl by the reaction $M + nH^+ \rightarrow M^{n+} + \frac{n}{2}H_2$: Zn, Fe, Co, Al, Hg, Cu, Pt, Au? (When the potential predicts that the element will not dissolve, it probably will not. If it is expected to dissolve, it may dissolve if some other process does not interfere. Predictions based on standard reduction potentials at 25°C are only tentative, because the potentials and activities in hot, concentrated solutions vary widely from those in the table of standard potentials.)

**28-13.** The following wet-ashing procedure was used to measure arsenic in organic soil samples by atomic absorption spectroscopy: A 0.1- to 0.5-g sample was heated in a 150-mL Teflon bomb in a microwave oven for 2.5 min with 3.5 mL of 70% $HNO_3$. After the sample cooled, a mixture containing 3.5 mL of 70% $HNO_3$, 1.5 mL of 70% $HClO_4$, and 1.0 mL of $H_2SO_4$ was added and the sample was reheated for three 2.5-min intervals with 2-min unheated periods in between. The final solution was diluted with 0.2 M HCl for analysis. Explain why $HClO_4$ was not introduced until the second heating.

**28-14.** Barbital can be isolated from urine by solid-phase extraction with $C_{18}$-silica. The barbital is then eluted with 1:1 vol/vol acetone: chloroform. Explain how this procedure works.

Barbital

**28-15.** Referring to Table 28-7, explain how an anion-exchange resin can be used for absorption and analysis of $SO_2$ released by combustion.

**28-16.** In 2002, workers at the Swedish National Food Administration discovered that heated, carbohydrate-rich foods, such as french fries, potato chips, and bread, contain alarming levels (0.1 to 4 $\mu$g/g) of acrylamide, a known carcinogen.[27]

Acrylamide

After the discovery, simplified methods were developed to measure ppm levels of acrylamide in food. In one procedure, 10 g of pulverized, frozen french fries were mixed for 20 min with 50 mL of $H_2O$ to extract acrylamide, which is very soluble in water (216 g/100 mL). The liquid was decanted and centrifuged to remove suspended matter and the internal standard $^2H_3$-acrylamide was added to 1 mL of extract. A solid-phase extraction column containing 100 mg of cation-exchange polymer with protonated sulfonic acid groups ($-SO_3H$) was washed twice with 1-mL portions of methanol and twice with 1-mL portions of water. The aqueous food extract (1 mL) was then passed through the column to bind protonated acrylamide, ($-NH_3^+$) to sulfonate ($-SO_3^-$) on the column. The column was dried for 30 s at 0.3 bar and then acrylamide was eluted with 1 mL of $H_2O$. Eluate was analyzed by liquid chromatography with a polar bonded phase. The chromatograms show the results monitored by ultraviolet absorbance or by mass spectrometry. The retention time of acrylamide is different on the two columns because they have different dimensions and different flow rates.

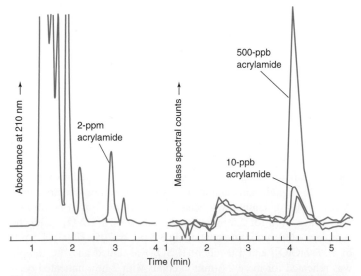

Chromatograms of acrylamide extract after passage through solid-phase extraction column: (left) Phenomenex Synergi Polar-RP 4-$\mu$m column eluted with 96:4 (vol/vol) $H_2O$:$CH_3CN$; (right) Phenomenex Synergi Hydro-RP 4-$\mu$m column eluted with 96:4:0.1 (vol/vol/vol) $H_2O$:$CH_3OH$:$HCO_2H$. *[From L. Peng, T. Farkas, L. Loo, J. Teuscher, and K. Kallury, "Rapid and Reproducible Extraction of Acrylamide in French Fries Using a Single Solid-Phase Sorbent," Am. Lab. New Ed. October 2003, p.10.]*

**(a)** What is the purpose of solid-phase extraction prior to chromatography? How does the ion-exchange sorbent retain acrylamide?

**(b)** Why are there many peaks when chromatography is monitored by ultraviolet absorbance?

**(c)** Mass spectral detection used selected reaction monitoring (Figure 22-21) with the $m/z$ 72 → 55 transition for acrylamide and 75 → 58 for $^2H_3$-acrylamide. Explain how this detection method works and suggest structures for the ions with $m/z$ 72 and 55 from acrylamide.

**(d)** Why does mass spectral detection give just one major peak?

**(e)** How is the internal standard used for quantitation with mass spectral detection?

**(f)** Where does $^2H_3$-acrylamide appear with ultraviolet absorbance? With mass spectral selected reaction monitoring?

**(g)** Why does the mass spectral method give quantitative results even though retention of acrylamide by the ion-exchange sorbent is not quantitative and elution of acrylamide from the sorbent by 1 mL of water might not be quantitative?

**28-17.** Many metals in seawater can be preconcentrated for analysis by coprecipitation with $Ga(OH)_3$. A 200-$\mu$L HCl solution containing 50 $\mu$g of $Ga^{3+}$ is added to 10.00 mL of the seawater. When the pH is brought to 9.1 with NaOH, a jellylike precipitate forms. After centrifugation to pack the precipitate, the water is removed and the gel is washed with water. Then the gel is dissolved in 50 $\mu$L of 1 M $HNO_3$ and aspirated into an inductively coupled plasma for atomic emission analysis. The preconcentration factor is 10 mL/50 $\mu$L = 200. The figure shows elemental concentrations in seawater as a function of depth near hydrothermal vents.

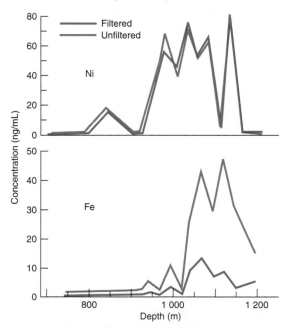

Depth profile of elements in seawater near hydrothermal vents. *[From T. Akagi and H. Haraguchi, "Simultaneous Multielement Determination of Trace Metals Using 10 mL of Seawater by Inductively Coupled Plasma Atomic Emission Spectrometry with Gallium Coprecipitation and Microsampling Technique," Anal. Chem. **1990**, 62, 81.]*

**(a)** What is the atomic ratio (Ga added):(Ni in seawater) for the sample with the highest concentration of Ni?

**(b)** The results given by gray lines were obtained with seawater samples that were not filtered prior to coprecipitation. The colored lines refer to filtered samples. The results for Ni do not vary between the two procedures, but the results for Fe do vary. Explain what this means.

**28-18.** Barium titanate, a ceramic used in electronics, was analyzed by the following procedure: Into a Pt crucible was placed 1.2 g of $Na_2CO_3$ and 0.8 g of $Na_2B_4O_7$ plus 0.314 6 g of unknown. After fusion at 1 000°C in a furnace for 30 min, the cooled solid was extracted with 50 mL of 6 M HCl, transferred to a 100-mL volumetric flask, and diluted to the mark. A 25.00-mL aliquot was treated with 5 mL of 15% tartaric acid (which complexes $Ti^{4+}$ and keeps it in aqueous solution) and 25 mL of ammonia buffer, pH 9.5. The solution was treated with organic reagents that complex $Ba^{2+}$, and the Ba complex was extracted into $CCl_4$. After acidification (to release the $Ba^{2+}$ from its organic complex), the $Ba^{2+}$ was back-extracted into 0.1 M HCl. The final aqueous sample was treated with ammonia buffer and methylthymol blue (a metal ion indicator) and titrated with 32.49 mL of 0.011 44 M EDTA. Find the weight percent of Ba in the ceramic.

**28-19.** Acid-base equilibria of Cr(III) were summarized in Problem 10-35. Cr(VI) in aqueous solution above pH 6 exists as the yellow tetrahedral chromate ion, $CrO_4^{2-}$. Between pH 2 and 6, Cr(VI) exists as an equilibrium mixture of $HCrO_4^-$ and orange-red dichromate, $Cr_2O_7^{2-}$. Cr(VI) is a carcinogen, but Cr(III) is not considered to be as harmful. The following procedure was used to measure Cr(VI) in airborne particulate matter in workplaces.

1. Particles were collected by drawing a known volume of air through a polyvinyl chloride filter with 5-μM pore size.
2. The filter was placed in a centrifuge tube and 10 mL of 0.05 M $(NH_4)_2SO_4$/0.05 M $NH_3$ buffer, pH 8, were added. The immersed filter was agitated by ultrasonic vibration for 30 min at 35°C to extract all Cr(III) and Cr(VI) into solution.
3. A measured volume of extract was passed through a "strongly basic" anion exchanger (Table 26-1) in the $Cl^-$ form. Then the resin was washed with distilled water. Liquid containing Cr(III) from the extract and the wash was discarded.
4. Cr(VI) was then eluted from the column with 0.5 M $(NH_4)_2SO_4$/0.05 M $NH_3$ buffer, pH 8, and collected in a vial.
5. The eluted Cr(VI) solution was acidified with HCl and treated with a solution of 1,5-diphenylcarbazide, a reagent that forms a colored complex with Cr(VI). The concentration of the complex was measured by its visible absorbance.

(a) What are the dominant species of Cr(VI) and Cr(III) at pH 8?
(b) What is the purpose of the anion exchanger in step 3?
(c) Why is a "strongly basic" anion exchanger used instead of a "weakly basic" exchanger?
(d) Why is Cr(VI) eluted in step 4 but not step 3?

**28-20.** The county landfill in the diagram was monitored to verify that toxic compounds were not leaching into the local water supply. Wells drilled at 21 locations were monitored over a year and pollutants were observed only at sites 8, 11, 12, and 13. Monitoring all 21 sites each month is very expensive. Suggest a strategy to use *composite samples* (Box 0-1) made from more than one well at a time to reduce the cost of routine monitoring. How will your scheme affect the minimum detectable level for pollutants at a particular site?

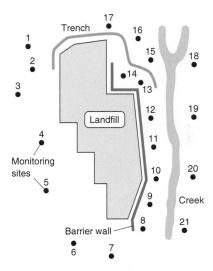

Diagram of county landfill showing the location of wells used to monitor groundwater. [Adapted from P.-C. Li and R. Rajagopal, Am. Environ. Lab. October 1994, p. 37.]

# Notes and References

## Chapter 0

1. J. Stocker, D. Balluch, M. Gsell, H. Harms, J. Feliciano, S. Daunert, K. A. Malik, and J. R. Van der Meer, "Development of a Set of Simple Bacterial Biosensors for Quantitative and Rapid Measurements of Arsenite and Arsenate in Potable Water," *Environ. Sci. Technol.* **2003,** *37,* 4743. [This notation refers to the journal *Environmental Science and Technology,* volume *37,* page 4743, in the year **2003.**]

2. For related biosensors, see A. Roda, M. Guardigli, E. Michelini, M. Mirasoli, and P. Pasini, "Analytical Bioluminescence and Chemiluminescence," *Anal. Chem.* **2003,** *75,* 462A.

3. G. Tannenbaum, "Chocolate: A Marvelous Natural Product of Chemistry," *J. Chem. Ed.* **2004,** *81,* 1131.

4. T. J. Wenzel, "A New Approach to Undergraduate Analytical Chemistry," *Anal. Chem.* **1995,** *67,* 470A. See also T. J. Wenzel, "The Lecture as a Learning Device," *Anal. Chem.* **1999,** *71,* 817A; T. J. Wenzel, "Cooperative Student Activities as Learning Devices," *Anal. Chem.* **2000,** *72,* 293A; T. J. Wenzel, "Practical Tips for Cooperative Learning," *Anal. Chem.* **2000,** *72,* 359A; T. J. Wenzel, "Undergraduate Research as a Capstone Learning Experience," *Anal. Chem.* **2000,** *72,* 547A.

5. W. R. Kreiser and R. A. Martin, Jr., *J. Assoc. Off. Anal. Chem.* **1978,** *61,* 1424; W. R. Kreiser and R. A. Martin, Jr., *J. Assoc. Off. Anal. Chem.* **1980,** *63,* 591. Today you would find many more recent references on caffeine.

6. A good source for many well-tested analytical procedures is W. Horwitz, *Official Methods of Analysis of AOAC International,* 17th ed. (Gaithersburg, MD: AOAC International, 2000).

7. W. Fresenius, "The Position of the Analyst as Expert: Yesterday and Today," *Fresenius J. Anal. Chem.* **2000,** *368,* 548.

## Chapter 1

1. J. R. de Laeter and H. S. Peiser, "A Century of Progress in the Sciences Due to Atomic Weight and Isotopic Composition Measurements," *Anal. Bioanal. Chem.* **2003,** *375,* 62.

2. S. J. Hawkes, "Salts Are Mostly Not Ionized," *J. Chem. Ed.* **1996,** *73,* 421. $MgCl^+$ is called an *ion pair.* See Box 8-1.

3. J. L. Sarmiento and N. Gruber, "Sinks for Anthropogenic Carbon," *Physics Today,* August 2002, p. 30.

4. U. Shahin, S.-M. Yi, R. D. Paode, and T. M. Holsen, "Long-Term Elemental Dry Deposition Fluxes Measured Around Lake Michigan," *Environ. Sci. Technol.* **2000,** *34,* 1887.

## Chapter 2

1. M. Sepaniak, P. Datskos, N. Lavrik, and C. Tipple, "Microcantilever Transducers: A New Approach in Sensor Technology," *Anal. Chem.* **2002,** *74,* 568A; C. Ziegler, "Cantilever-Based Biosensors," *Anal. Bioanal. Chem.* **2004,** *379,* 946.

2. F. Tian, K. M. Hansen, T. L. Ferrell, T. Thundat, and D. C. Hansen, "Dynamic Microcantilever Sensors for Discerning Biomolecular Interactions," *Anal. Chem.* **2005,** *77,* 1601; C. Savran, S. M. Knudsen, A. D. Ellington, and S. R. Manalis, "Micromechanical Detection of Proteins Using Aptamer-Based Receptor Molecules," *Anal. Chem.* **2004,** *76,* 3194; P. Dutta, C. A. Tipple, N. V. Lavrik, P. G. Datskos, H. Hofstetter, O. Hofstetter, and M. J. Sepaniak, "Enantioselective Sensors Based on Antibody-Mediated Nanomechanics," *Anal. Chem.* **2003,** *75,* 2342.

3. A related, but less sensitive technology, uses the *quartz crystal microbalance,* in which the crystal has dimensions of millimeters. When analyte binds to a chemically sensitive layer on a gold coating on the surface of quartz, the oscillating frequency of the crystal is altered. See, for example, B. Zuo, S. Li, Z. Guo, J. Zhang, and C. Chen, "Piezoelectric Immunosensor for SARS-Associated Coronavirus," *Anal. Chem.* **2004,** *76,* 3536; A.-S. Cans, F. Höök, O. Shupliakov, A. G. Ewing, P. S. Eriksson, L. Brodin, and O. Orwar, "Measurement of the Dynamics of Exocytosis and Vesicle Retrieval at Cell Populations Using a Quartz Crystal Microbalance," *Anal. Chem.* **2001,** *73,* 5805; X. Su, R. Robelek, Y. Wu, G. Wang, and W. Knoll, "Detection of Point Mutations and Insertion Mutations in DNA Using a Quartz Crystal Microbalance and MutS, a Mismatch Binding Protein," *Anal. Chem.* **2004,** *76,* 489.

4. Recycling of certain household waste, such as televisions, computers, and rechargeable batteries, is desirable to control toxic pollution. Televisions are a significant source of lead that can leach into groundwater from dump sites. [S. E. Musson, Y.-C. Jang, T. G. Townsend, and I.-H. Chung, "Characterization of Lead Leachability from Cathode Ray Tubes," *Environ. Sci. Technol.* **2000,** *34,* 4376.]

5. Handling and disposing of chemicals is described in *Prudent Practices for Handling Hazardous Chemicals in Laboratories* (1983) and *Prudent Practices for Disposal of Chemicals in Laboratories* (1983), both available from National Academy Press, 2101 Constitution Avenue N.W., Washington, DC 20418. See also P. Patnaik, *A Comprehensive Guide to the Hazardous Properties of Chemical Substances,* 2nd ed. (New York: Wiley, 1999); G. Lunn and E. B. Sansone, *Destruction of Hazardous Chemicals in the Laboratory* (New York: Wiley, 1994); and M. A. Armour, *Hazardous Laboratory Chemical Disposal Guide* (Boca Raton, FL: CRC Press, 1991).

6. To recover gold from electronics, see J. W. Hill and T. A. Lear, "Recovery of Gold from Electronic Scrap," *J. Chem. Ed.* **1988,** *65,* 802. To remove Hg from gold, soak it in a 1:1 mixture of 0.01 M $(NH_4)_2S_2O_8$ and 0.01 M $HNO_3$ [T. Nomura and M. Fujisawa, "Electrolytic Determination of Mercury(II) in Water with a Piezoelectric Quartz Crystal," *Anal. Chim. Acta* **1986,** *182,* 267].

7. Compact disks illustrating basic laboratory techniques are available from http://jchemed.chem.wisc.edu/

8. J. M. Bonicamp, "Weigh This Way," *J. Chem. Ed.* **2002,** *79,* 476.

9. For a demonstration of buoyancy, see K. D. Pinkerton, "Sink or Swim: The Cartesian Diver," *J. Chem. Ed.* **2001,** *78,* 200A (JCE Classroom Activity No. 33).

10. R. Batting and A. G. Williamson, "Single-Pan Balances, Buoyancy, and Gravity or 'A Mass of Confusion,'" *J. Chem. Ed.* **1984,** *61,* 51; J. E. Lewis and L. A. Woolf, "Air Buoyancy Corrections for Single-Pan Balances," *J. Chem. Ed.* **1971,** *48,* 639; F. F. Cantwell, B. Kratochvil, and W. E. Harris, "Air Buoyancy Errors and the Optical Scale of a Constant Load Balance," *Anal. Chem.* **1978,** *50,* 1010; G. D. Chapman, "Weighing with Electronic Balances," National Research Council of Canada, Report NRCC 38659 (1996).

11. Air density (g/L) = $(0.003\ 485\ B - 0.001\ 318\ V)/T$, where $B$ is barometric pressure (Pa), $V$ is the vapor pressure of water in the air (Pa), and $T$ is air temperature (K).

12. U. Henriksson and J. C. Eriksson, "Thermodynamics of Capillary Rise: Why Is the Meniscus Curved?" *J. Chem. Ed.* **2004,** *81,* 150.

13. Cleaning solution is prepared by dissolving 36 g of ammonium peroxydisulfate, $(NH_4)_2S_2O_8$, in a *loosely stoppered* 2.2-L ("one gallon") bottle of 98 wt% sulfuric acid [H. M. Stahr, W. Hyde, and L. Sigler, "Oxidizing Acid Baths—Without Chromate Hazards," *Anal. Chem.* **1982,** *54,* 1456A]. Add $(NH_4)_2S_2O_8$ every few weeks to maintain oxidizing power. Keep it loosely stoppered to prevent gas buildup [P. S. Surdhar, "Laboratory Hazard," *Anal. Chem.* **1992,** *64,* 310A]. The commercial cleaning solution EOSULF (containing the metal binder, EDTA, and a sulfonate detergent) is an alternative for removing baked-on lipid or protein from glassware. [P. L. Manske, T. M. Stimpfel, and E. L. Gershey, "A Less Hazardous Chromic Acid Substitute for Cleaning Glassware," *J. Chem. Ed.* **1990,** *67,* A280.] Another powerful oxidizing cleaning solution, called "piranha solution," is a 1:1 (vol/vol) mixture of 30 wt% $H_2O_2$ plus 98 wt% $H_2SO_4$.

14. W. B. Guenther, "Supertitrations: High-Precision Methods," *J. Chem. Ed.* **1988,** *65,* 1097; D. D. Siemer, S. D. Reeder, and M. A. Wade, "Syringe Buret Adaptor," *J. Chem. Ed.* **1988,** *65,* 467.

15. M. M. Singh, C. McGowan, Z. Szafran, and R. M. Pike, "A Modified Microburet for Microscale Titration," *J. Chem. Ed.* **1998**, *75*, 371; "A Comparative Study of Microscale and Standard Burets," *J. Chem. Ed.* **2000**, *77*, 625.

16. D. R. Burfield and G. Hefter, "Oven Drying of Volumetric Glassware," *J. Chem. Ed.* **1987**, *64*, 1054.

17. M. Connors and R. Curtis, "Pipetting Error," *Am. Lab. News Ed.*, June 1999, p. 20; ibid., December 1999, p. 12; R. H. Curtis and G. Rodrigues, ibid., February 2004, p. 12.

18. Calibrate your micropipet by measuring the mass of water it delivers [B. Kratochvil and N. Motkosky, "Precision and Accuracy of Mechanical-Action Micropipets," *Anal. Chem.* **1987**, *59*, 1064]. A colorimetric calibration kit is available from Artel, Inc., Westbrook, ME; 207-854-0860.

19. E. J. Billo, *Microsoft Excel for Chemists,* 2nd ed. (New York: Wiley, 2001); R. de Levie, *How to Use Excel® in Analytical Chemistry and in General Scientific Data Analysis* (Cambridge: Cambridge University Press, 2001); R. de Levie, *Advanced Excel for Scientific Data Analysis* (Oxford: Oxford University Press, 2004); B. V. Liengme, *A Guide to Microsoft Excel 2002,* 3rd ed. (Amsterdam: Elsevier, 2002).

20. D. Bohrer, P. Cícero do Nascimento, P. Martins, and R. Binotto, "Availability of Aluminum from Glass on an Al Form Ion Exchanger in the Presence of Complexing Agents and Amino Acids," *Anal. Chim. Acta* **2002**, *459*, 267.

## Chapter 3

1. A catalog of Standard Reference Materials is available from SRMINFO@enh.nist.gov. European Reference Materials are available from http://www.erm-crm.org. See H. Emons, J. Marriott, and R. Matchatt, "ERM—A New Landmark for Reference Materials," *Anal. Bioanal. Chem.* **2005**, *381*, 28.

2. P. De Bièvre, S. Valkiers, and P. D. P. Taylor, "The Importance of Avogadro's Constant for Amount-of-Substance Measurements," *Fresenius J. Anal. Chem.* **1998**, *361*, 227.

## Chapter 4

1. Excellent, readable sources on statistics are J. C. Miller and J. N. Miller, *Statistics and Chemometrics for Analytical Chemistry,* 4th ed. (Harlow, UK: Prentice Hall, 2000) and P. C. Meier and R. E. Zünd, *Statistical Methods in Analytical Chemistry,* 2nd ed. (New York: Wiley, 2000).

2. H. Mark and J. Workman, Jr., *Statistics in Spectroscopy,* 2nd ed. (Amsterdam: Elsevier, 2003); J. E. DeMuth, *Basic Statistics and Pharmaceutical Statistical Applications* (New York: Marcel Dekker, 1999); and R. N. Forthofer and E. S. Lee, *Introduction to Biostatistics* (San Diego: Academic Press, 1995).

3. L. H. Keith, W. Crummett, J. Deegan, Jr., R. A. Libby, J. K. Taylor, and G. Wentler, "Principles of Environmental Analysis," *Anal. Chem.* **1983**, *55*, 2210. See also C. A. Kuffner, Jr., E. Marchi, J. M. Morgado, and C. R. Rubio, "Capillary Electrophoresis and Daubert: Time for Admission," *Anal. Chem.* **1996**, *68*, 241A.

4. R. Bate, ed., *What Risk?* (Oxford: Butterworth-Heinemann, 1999).

5. In Figure 4-8, cells F12 and F13 give results of a *2-tailed* significance test, which we used in Section 4-3 without stating so. If we have no reason to presume that one method should give a higher result than the other, we use a 2-tailed test. If we have reason to think that method 1 gives a higher result than method 2, we use the *1-tailed* test whose results are displayed in cells F10 and F11. The 1-tailed test evaluates the null hypothesis that "method 1 does not give a higher result than method 2." Cell F11 says that the critical value of $t$ for the 1-tailed test is 1.77. Because $t_{calculated}$ ($= 20.2$) $> 1.77$, we conclude that the null hypothesis is false. Cell F10 states that the probability of observing these two mean values and standard deviations by random chance if method 1 does not give a higher result than method 2 is $1.66 \times 10^{-11}$.

6. For a comprehensive approach to least-squares fitting of nonlinear curves, including analysis of uncertainty, see J. Tellinghuisen, "Understanding Least Squares Through Monte Carlo Calculations," *J. Chem. Ed.* **2005**, *82*, 157; P. Ogren, B. Davis, and N. Guy, "Curve Fitting, Confidence Intervals and Correlations, and Monte Carlo Visualizations for Multilinear Problems in Chemistry: A General Spreadsheet Approach," *J. Chem. Ed.* **2001**, *78*, 827; see also D. C. Harris, "Nonlinear Least-Squares Curve Fitting with Microsoft Excel Solver," *J. Chem. Ed.* **1998**, *75*, 119; C. Salter and R. de Levie, "Nonlinear Fits of

Standard Curves: A Simple Route to Uncertainties in Unknowns," *J. Chem. Ed.* **2002**, *79*, 268; R. de Levie, "Estimating Parameter Precision in Nonlinear Least Squares with Excel's Solver," *J. Chem. Ed.* **1999**, *76*, 1594; S. E. Feller and C. F. Blaich, "Error Estimates for Fitted Parameters," *J. Chem. Ed.* **2001**, *78*, 409; R. de Levie, "When, Why, and How to Use Weighted Least Squares," *J. Chem. Ed.* **1986**, *63*, 10; and P. J. Ogren and J. R. Norton, "Applying a Simple Linear Least-Squares Algorithm to Data with Uncertainties in Both Variables," *J. Chem. Ed.* **1992**, *69*, A130.

7. In this book, we plot analytical response on the $y$-axis versus concentration on the $x$-axis. The inverse calibration ($y$ = concentration, $x$ = response) is said to provide a less biased estimate of concentration from a measured response. See V. Centner, D. L. Massart, and S. de Jong, "Inverse Calibration Predicts Better Than Classical Calibration," *Fresenius J. Anal. Chem.* **1998**, *361*, 2; D. Grientschnig, "Relation Between Prediction Errors of Inverse and Classical Calibration," *Fresenius J. Anal. Chem.* **2000**, *367*, 497; J. Tellinghuisen, "Inverse vs Classical Calibration for Small Data Sets," *Fresenius J. Anal. Chem.* **2000**, *368*, 585.

8. K. Danzer and L. A. Currie, "Guidelines for Calibration in Analytical Chemistry," *Pure Appl. Chem.* **1998**, *70*, 993.

9. C. Salter, "Error Analysis Using the Variance-Covariance Matrix," *J. Chem. Ed.* **2000**, *77*, 1239. Salter's Equation 8 is equivalent to Equation 4-27, although this is not obvious.

10. D. Sterzenbach, B. W. Wenclawiak, and V. Weigelt, "Determination of Chlorinated Hydrocarbons in Marine Sediments at the Part-per-Trillion Level with Supercritical Fluid Extraction,"*Anal. Chem.* **1997**, *69*, 831.

11. J. W. Gramlich and L. A. Machlan, "Isotopic Variations in Commercial High-Purity Gallium," *Anal. Chem.* **1985**, *57*, 1788.

12. S. Ahmed, N. Jabeen, and E. ur Rehman, "Determination of Lithium Isotopic Composition by Thermal Ionization Mass Spectrometry," *Anal. Chem.* **2002**, *74*, 4133; L. W. Green, J. J. Leppinen, and N. L. Elliot, "Isotopic Analysis of Lithium as Thermal Dilithium Fluoride Ions," *Anal. Chem.* **1988**, *60*, 34.

13. D. T. Harvey, "Statistical Evaluation of Acid/Base Indicators," *J. Chem. Ed.* **1991**, *68*, 329.

14. N. J. Lawryk and C. P. Weisel, "Concentration of Volatile Organic Compounds in the Passenger Compartments of Automobiles," *Environ. Sci. Technol.* **1996**, *30*, 810.

15. I. Sarudi and I. Nagy, "A Gas Chromatographic Method for the Determination of Nitrite in Natural Waters," *Talanta* **1995**, *42*, 1099.

## Chapter 5

1. C. Hogue, "Ferreting Out Erroneous Data," *Chem. Eng. News,* 1 April 2002, p. 49.

2. B. W. Wenclawiak, M. Koch, and E. Hadjicostas, eds., *Quality Assurance in Analytical Chemistry* (Heidelberg: Springer-Verlag, 2004); P. Quevauviller, *Quality Assurance for Water Analysis* (Chichester, UK: Wiley, 2002); M. Valcárcel, *Principles of Analytical Chemistry* (Berlin: Springer-Verlag, 2000); J. Kenkel, *A Primer on Quality in the Analytical Laboratory* (Boca Raton, FL: Lewis Press, 1999); F. E. Prichard, *Quality in the Analytical Chemistry Laboratory* (New York: Wiley, 1995); J. K. Taylor, *Quality Assurance of Chemical Measurements* (Chelsea, MI: Lewis Publishers, 1987).

3. J. M. Green, "A Practical Guide to Analytical Method Validation," *Anal. Chem.* **1996**, *68*, 305A; M. Swartz and I. S. Krull, "Validation of Bioanalytical Methods—Highlights of FDA's Guidance," *LCGC* **2003**, *21*, 136; J. D. Orr, I. S. Krull, and M. E. Swartz, "Validation of Impurity Methods," *LCGC* **2003**, *21*, 626 and 1146; P. Nader, G. Dremmydiotis, J. Tremblay, G. Lévesque, and A. Bartlett, "Validation of Bioanalytical Capillary Gel Electrophoresis Methods for the Determination of Oligonucleotides," *LCGC* **2002**, *20*, 856.

4. R. de Levie, "Two Linear Correlation Coefficients," *J. Chem. Ed.* **2003**, *80*, 1030.

5. W. Horwitz, L. R. Kamps, and K. W. Boyer, *J. Assoc. Off. Anal. Chem.* **1980**, *63*, 1344; W. Horwitz, "Evaluation of Analytical Methods Used for Regulation of Foods and Drugs," *Anal. Chem.* **1982**, *54*, 67A; P. Hall and B. Selinger, "A Statistical Justification Relating Interlaboratory Coefficients of Variation with Concentration Levels," *Anal. Chem.* **1989**, *61*, 1465; and R. Albert and W. Horwitz, "A Heuristic Derivation of the Horwitz Curve," *Anal. Chem.* **1997**, *69*, 789.

6. J. Vial and A. Jardy, "Experimental Comparison of the Different Approaches to Estimate LOD and LOQ of an HPLC Method," *Anal. Chem.* **1999,** *71,* 2672; G. L. Long and J. D. Winefordner, "Limit of Detection," *Anal. Chem.* **1983,** *55,* 713A; W. R. Porter, "Proper Statistical Evaluation of Calibration Data," *Anal. Chem.* **1983,** *55,* 1290A; S. Geiss and J. W. Einmax, "Comparison of Detection Limits in Environmental Analysis," *Fresenius J. Anal. Chem.* **2001,** *370,* 673; M. E. Zorn, R. D. Gibbons, and W. C. Sonzogni, "Evaluation of Approximate Methods for Calculating the Limit of Detection and Limit of Quantitation," *Environ. Sci. Technol.* **1999,** *33,* 2291; J. D. Burdge, D. L. MacTaggart, and S. O. Farwell, "Realistic Detection Limits from Confidence Bands," *J. Chem. Ed.* **1999,** *76,* 434.

7. D. Montgomery, *Design and Analysis of Experiments,* 5th ed. (New York, Wiley, 2001); C. F. Wu and M. Hamada, *Experiments: Planning, Analysis, and Parameter Design Optimization* (New York: Wiley, 2000); M. Anderson and P. Whitcomb, *DoE Simplified: Practical Tools for Effective Experimentation* (Portland, OR: Productivity, Inc., 2000); E. Morgan, *Chemometrics: Experimental Design* (Chichester, UK: Wiley, 1991); D. L. Massart, A. Dijkstra, and L. Kaufman, *Evaluation and Optimization of Laboratory Methods and Analytical Procedures* (Amsterdam: Elsevier, 1978); G. E. P. Box, W. G. Hunter, and J. S. Hunter, *Statistics for Experimenters: An Introduction to Design Data Analysis and Model Building* (New York: Wiley, 1978); R. S. Strange, "Introduction to Experimental Design for Chemists," *J. Chem. Ed.* **1990,** *67,* 113.

8. S. N. Deming and S. L. Morgan, "Simplex Optimization of Variables in Analytical Chemistry," *Anal. Chem.* **1973,** *45,* 278A; D. J. Leggett, "Instrumental Simplex Optimization," *J. Chem. Ed.* **1983,** *60,* 707; S. Srijaranai, R. Burakham, T. Khammeng, and R. L. Deming, "Use of the Simplex Method to Optimize the Mobile Phase for the Micellar Chromatographic Separation of Inorganic Anions," *Anal. Bioanal. Chem.* **2002,** *374,* 145; D. Betteridge, A. P. Wade, and A. G. Howard, "Reflections on the Modified Simplex," *Talanta* **1985,** *32,* 709, 723.

9. M. Bader, "A Systematic Approach to Standard Addition Methods in Instrumental Analysis," *J. Chem. Ed.* **1980,** *57,* 703.

10. G. R. Bruce and P. S. Gill, "Estimates of Precision in a Standard Additions Analysis," *J. Chem. Ed.* **1999,** *76,* 805.

11. J. R. Troost and E. Y. Olavasen ["Gas Chromatographic/Mass Spectrometric Calibration Bias," *Anal. Chem.* **1996,** *68,* 708] discovered that a chromatography procedure from the U.S. Environmental Protection Agency had a nonlinear response on a variety of instruments. The assumption of constant response factor led to errors as great as 40%.

12. J. A. Day, M. Montes-Bayón, A. P. Vonderheide, and J. A. Caruso, "A Study of Method Robustness for Arsenic Speciation in Drinking Water Samples by Anion Exchange HPLC-ICP-MS," *Anal. Bioanal. Chem.* **2002,** *373,* 664.

## Chapter 6

1. D. P. Sheer and D. C. Harris, "Acidity Control in the North Branch Potomac," *J. Water Pollution Control Federation* **1982,** *54,* 1441.

2. R. E. Weston, Jr., "Climate Change and Its Effect on Coral Reefs," *J. Chem. Ed.* **2000,** *77,* 1574.

3. P. D. Thacker, "Global Warming's Other Effects on the Oceans," *Environ. Sci. Technol.* **2005,** *39,* 10A.

4. J. K. Baird, "A Generalized Statement of the Law of Mass Action," *J. Chem. Ed.* **1999,** *76,* 1146; R. de Levie, "What's in a Name?" *J. Chem. Ed.* **2000,** *77,* 610.

5. For thermodynamic data, see N. Jacobson, "Use of Tabulated Thermochemical Data for Pure Compounds," *J. Chem. Ed.* **2001,** *78,* 814; http://webbook.nist.gov/chemistry/ and http://www.crct.polymtl.ca/fact/websites.htm; M. W. Chase, Jr., *NIST-JANAF Thermochemical Tables,* 4th ed; *J. Phys. Chem. Ref. Data,* Monograph 9 (New York: American Chemical Society and American Physical Society, 1998).

6. The solubility of most ionic compounds increases with temperature, despite the fact that the standard heat of solution ($\Delta H°$) is negative for about half of them. Discussions of this seeming contradiction can be found in G. M. Bodner, "On the Misuse of Le Châtelier's Principle for the Prediction of the Temperature Dependence of the Solubility of Salts," *J. Chem. Ed.* **1980,** *57,* 117, and R. S. Treptow, "Le Châtelier's Principle Applied to the Temperature Dependence of Solubility," *J. Chem. Ed.* **1984,** *61,* 499.

7. A. K. Sawyer, "Solubility and $K_{sp}$ of Calcium Sulfate: A General Chemistry Laboratory Experiment," *J. Chem. Ed.* **1983,** *60,* 416.

8. A really good book to read about solubility and all types of equilibrium calculations is W. B. Guenther, *Unified Equilibrium Calculations* (New York: Wiley, 1991).

9. E. Koubek, "Demonstration of the Common Ion Effect," *J. Chem. Ed.* **1993,** *70,* 155.

10. For many *great* chemical demonstrations, see B. Z. Shakhashiri, *Chemical Demonstrations: A Handbook for Teachers of Chemistry* (Madison, WI: University of Wisconsin Press, 1983–1992), 4 volumes. See also L. E. Summerlin and J. L. Ealy, Jr., *Chemical Demonstrations: A Sourcebook for Teachers,* 2nd ed. (Washington, DC: American Chemical Society, 1988).

11. A demonstration of selective precipitation by addition of $Pb^{2+}$ to a solution containing $CO_3^{2-}$ and $I^-$ is described by T. P. Chirpich, "A Simple, Vivid Demonstration of Selective Precipitation," *J. Chem. Ed.* **1988,** *65,* 359.

12. A computer database of critically selected equilibrium constants is found in R. M. Smith, A. E. Martell, and R. J. Motekaitis, *NIST Critical Stability Constants of Metal Complexes Database 46* (Gaithersburg, MD: National Institute of Standards and Technology, 2001). Books dealing with the experimental measurement of equilibrium constants include A. Martell and R. Motekaitis, *Determination and Use of Stability Constants* (New York: VCH, 1992); K. A. Conners, *Binding Constants: The Measurement of Molecular Complex Stability* (New York: Wiley, 1987); and D. J. Leggett, ed., *Computational Methods for the Determination of Formation Constants* (New York: Plenum Press, 1985).

13. P. A. Giguère, "The Great Fallacy of the $H^+$ Ion," *J. Chem. Ed.* **1979,** *56,* 571; P. A. Giguère and S. Turrell, "The Nature of Hydrofluoric Acid: A Spectroscopic Study of the Proton-Transfer Complex, $H_3O^+ \cdot F^-$," *J. Am. Chem. Soc.* **1980,** *102,* 5473.

14. Z. Xie, R. Bau, and C. A. Reed, "A Crystalline $[H_9O_4]^+$ Hydronium Salt with a Weakly Coordinating Anion," *Inorg. Chem.* **1995,** *34,* 5403.

15. F. A. Cotton, C. K. Fair, G. E. Lewis, G. N. Mott, K. K. Ross, A. J. Schultz, and J. M. Williams, "X-Ray and Neutron Diffraction Studies of $[V(H_2O)_6][H_5O_2][CF_3SO_3]_4$," *J. Am. Chem. Soc.* **1984,** *106,* 5319.

16. K. Abu-Dari, K. N. Raymond, and D. P. Freyberg, "The Bihydroxide ($H_3O_2^-$) Anion," *J. Am. Chem. Soc.* **1979,** *101,* 3688.

17. W. B. Jensen, "The Symbol for pH," *J. Chem. Ed.* **2004,** *81,* 21.

18. For a related demonstration with $NH_3$, see M. D. Alexander, "The Ammonia Smoke Fountain," *J. Chem. Ed.* **1999,** *76,* 210; N. C. Thomas, "A Chemiluminescent Ammonia Fountain," *J. Chem. Ed.* **1990,** *67,* 339; and N. Steadman, "Ammonia Fountain Improvements," *J. Chem. Ed.* **1992,** *66,* 764.

19. L. M. Schwartz, "Ion-Pair Complexation in Moderately Strong Aqueous Acids," *J. Chem. Ed.* **1995,** *72,* 823. Even though it is not "free," $H_3O^+$ in ion pairs with anions such as $CF_3CO_2^-$ and $CCl_3CO_2^-$ appears to participate in ionic conductance [R. I. Gelb and J. S. Alper, "Anomalous Conductance in Electrolyte Solutions," *Anal. Chem.* **2000,** *72,* 1322].

20. S. J. Hawkes, "All Positive Ions Give Acid Solutions in Water," *J. Chem. Ed.* **1996,** *73,* 516.

21. M. Kern, "The Hydration of Carbon Dioxide," *J. Chem. Ed.* **1960,** *37,* 14. Great demonstrations with $CO_2$, including one with carbonic anhydrase, are described by J. A. Bell, "Every Year Begins a Millennium," *J. Chem. Ed.* **2000,** *77,* 1098.

22. The equilibrium constants apply in 4 M $NaClO_4$ at 25°C. Structures of some of these complexes are discussed by G. B. Kauffman, M. Karabassi, and G. Bergerhoff, "Metallo Complexes: An Experiment for the Undergraduate Laboratory," *J. Chem. Ed.* **1984,** *61,* 729.

## Chapter 7

1. F. Szabadváry, *History of Analytical Chemistry* (Langhorne, PA: Gordon and Breach, 1992); E. R. Madsen, *The Development of Titrimetric Analysis 'till 1806* (Copenhagen: E. E. C. Gad, 1958); C. Duval, *J. Chem. Ed.* **1951,** *28,* 508; A. Johansson, *Anal. Chim. Acta* **1988,** *206,* 97; and J. T. Stock, "A Backward Look at Scientific Instrumentation." *Anal. Chem.* **1993,** *65,* 344A. Instructions for use of the burets were taken from Szabadváry's *History of Analytical Chemistry* and from O. D. Allen, ed., *Fresenius' Quantitative Chemical Analysis*

(2nd American ed., 1881). Drawings originally appeared in F. A. H. Descroizilles, *Ann. Chim.* **1806,** *60,* 17; J. L. Gay-Lussac, *Ann. Chim. Phys.* **1824,** *26,* 152; É. O. Henry, *J. Chim. Pharm.* **1846,** *6,* 301; F. Mohr, *Lehrbuch der Chemisch-Analytischen Titrirmethode* (Braunschweig, 1855). The picture of the Gay-Lussac buret is from *Fresenius' Quantitative Chemical Analysis.*

2. *Reagent Chemicals,* 9th ed. (Washington, DC: American Chemical Society, 1999).

3. R. W. Ramette, "In Support of Weight Titrations," *J. Chem. Ed.* **2004,** *81,* 1715.

4. The Kjeldahl digestion captures amine ($-NR_2$) or amide ($-C[=O]NR_2$) nitrogens (where R can be H or an organic group), but not oxidized nitrogen such as nitro ($-NO_2$) or azo ($-N=N-$) groups, which must be reduced first to amines or amides.

5. W. Maher, F. Krikowa, D. Wruck, H. Louie, T. Nguyen, and W. Y. Huang, "Determination of Total Phosphorus and Nitrogen in Turbid Waters by Oxidation with Alkaline Potassium Peroxodisulfate," *Anal. Chim. Acta* **2002,** *463,* 283.

6. P. de. B. Harrington, E. Kolbrich, and J. Cline, "Experimental Design and Multiplexed Modeling Using Titrimetry and Spreadsheets," *J. Chem. Ed.* **2002,** *79,* 863.

7. G. Grguric, "Denitrification as a Model Chemical Process," *J. Chem. Ed.* **2002,** *79,* 179.

8. M. L. Ware, M. D. Argentine, and G. W. Rice, "Potentiometric Determination of Halogen Content in Organic Compounds Using Dispersed Sodium Reduction," *Anal. Chem.* **1988,** *60,* 383.

## Chapter 8

1. H. Ohtaki and T. Radnal, "Structure and Dynamics of Hydrated Ions," *Chem. Rev.* **1993,** *93,* 1157.

2. A. G. Sharpe, "The Solvation of Halide Ions and Its Chemical Significance," *J. Chem. Ed.* **1990,** *67,* 309.

3. E. R. Nightingale, Jr., "Phenomenological Theory of Ion Solvation: Effective Radii of Hydrated Ions," *J. Phys. Chem.* **1959,** *63,* 1381.

4. K. H. Stern and E. S. Amis, "Ionic Size," *Chem. Rev.* **1959,** *59,* 1.

5. D. R. Driscol, "'Invitation to Enquiry': The $Fe^{3+}/CNS^-$ Equilibrium," *J. Chem. Ed.* **1979,** *56,* 603.

6. S. J. Hawkes, "Salts Are Mostly NOT Ionized," *J. Chem. Ed.* **1996,** *73,* 421; S. O. Russo and G. I. H. Hanania, "Ion Association, Solubilities, and Reduction Potentials in Aqueous Solution," *J. Chem. Ed.* **1989,** *66,* 148.

7. K. S. Pitzer, *Activity Coefficients in Electrolyte Solutions,* 2nd ed. (Boca Raton, FL: CRC Press, 1991); B. S. Krumgalz, R. Pogorelskii, A. Sokolov, and K. S. Pitzer, "Volumetric Ion Interaction Parameters for Single-Solute Aqueous Electrolyte Solutions at Various Temperatures," *J. Phys. Chem. Ref. Data* **2000,** *29,* 1123.

8. J. Kielland, "Individual Activity Coefficients of Ions in Aqueous Solutions," *J. Am. Chem. Soc.* **1937,** *59,* 1675.

9. R. E. Weston, Jr., "Climate Change and Its Effect on Coral Reefs," *J. Chem. Ed.* **2000,** *77,* 1574; R. A. Feely, C. L. Sabine, K. Lee, W. Berelson, J. Kleypas, Vi. J. Fabry, and F. J. Millero, "Impact of Anthropogenic $CO_2$ on the $CaCO_3$ System in the Oceans," *Science* **2004,** *305,* 362.

10. For more about equilibrium calculations, see W. B. Guenther, *Unified Equilibrium Calculations* (New York: Wiley, 1991); J. N. Butler, *Ionic Equilibrium: Solubility and pH Calculations* (New York: Wiley, 1998); and M. Meloun, *Computation of Solution Equilibria* (New York: Wiley, 1988). For equilibrium calculation software, see http://www.micromath.com/ and http://www.acadsoft.co.uk/

11. M. S. Frant and J. W. Ross, Jr., "Electrode for Sensing Fluoride Ion Activity in Solution," *Science* **1966,** *154,* 1553.

## Chapter 9

1. R. Schmid and A. M. Miah, "The Strength of the Hydrohalic Acids," *J. Chem. Ed.* **2001,** *78,* 116.

2. T. F. Young, L. F. Maranville, and H. M. Smith, "Raman Spectral Investigations of Ionic Equilibria in Solutions of Strong Electrolytes" in W. J. Hamer, ed., *The Structure of Electrolytic Solutions* (New York: Wiley, 1959).

3. Acid dissociation constants do not tell us which protons dissociate in each step. Assignments for pyridoxal phosphate come from nuclear magnetic resonance spectroscopy [B. Szpoganicz and A. E. Martell, "Thermodynamic and Microscopic Equilibrium Constants of Pyridoxal 5′-Phosphate," *J. Am. Chem. Soc.* **1984,** *106,* 5513].

4. For an alternative treatment approach, see H. L. Pardue, I. N. Odeh, and T. M. Tesfai, "Unified Approximations: A New Approach for Monoprotic Weak Acid-Base Equilibria," *J. Chem. Ed.* **2004,** *81,* 1367.

5. M. C. Bonneau, "The Chemistry of Fabric Reactive Dyes," *J. Chem. Ed.* **1995,** *72,* 724.

6. R. T. da Rocha, I. G. R. Gutz, and C. L. do Lago, "A Low-Cost and High-Performance Conductivity Meter," *J. Chem. Ed.* **1997,** *74,* 572; G. Berenato and D. F. Maynard, "A Simple Audio Conductivity Device," *J. Chem. Ed.* **1997,** *74,* 415; S. K. S. Zawacky, "A Cheap, Semiquantitative Hand-Held Conductivity Tester," *J. Chem. Ed.* **1995,** *72,* 728; T. R. Rettich, "An Inexpensive and Easily Constructed Device for Quantitative Conductivity Experiments," *J. Chem. Ed.* **1989,** *66,* 168; and D. A. Katz and C. Willis, "Two Safe Student Conductivity Apparatus," *J. Chem. Ed.* **1994,** *71,* 330.

7. L. R. Kuck, R. D. Godec, P. P. Kosenka, and J. W. Birks, "High-Precision Conductometric Detector for the Measurement of Atmospheric $CO_2$," *Anal. Chem.* **1998,** *70,* 4678.

8. H. N. Po and N. M. Senozan, "The Henderson-Hasselbalch Equation: Its History and Limitations," *J. Chem. Ed.* **2001,** *78,* 1499; R. de Levie, "The Henderson-Hasselbalch Equation: Its History and Limitations," *J. Chem. Ed.* **2003,** *80,* 146.

9. F. B. Dutton and G. Gordon in H. N. Alyea and F. B. Dutton, eds., *Tested Demonstrations in Chemistry,* 6th ed. (Easton, PA: Journal of Chemical Education, 1965), p. 147; R. L. Barrett, "The Formaldehyde Clock Reaction," *J. Chem. Ed.* **1955,** *32,* 78. See also J. J. Fortman and J. A. Schreier, "Some Modified Two-Color Formaldehyde Clock Salutes for Schools with Colors of Gold and Green or Gold and Red," *J. Chem. Ed.* **1991,** *68,* 324; M. G. Burnett, "The Mechanism of the Formaldehyde Clock Reaction," *J. Chem. Ed.* **1982,** *59,* 160; and P. Warneck, "The Formaldehyde-Sulfite Clock Reaction Revisited," *J. Chem. Ed.* **1989,** *66,* 334.

10. The chemical we call sodium bisulfite ($NaHSO_3$) is apparently not the solid in the reagent bottle, which is reported to be sodium metabisulfite ($Na_2S_2O_5$) [D. Tudela, "Solid $NaHSO_3$ Does Not Exist," *J. Chem. Ed.* **2000,** *77,* 830; see also H. D. B. Jenkins and D. Tudela, "New Methods to Estimate Lattice Energies: Application to Bisulfite and Metabisulfite," *J. Chem. Ed.* **2003,** *80,* 1482]. $NaHSO_3$ is produced when $Na_2S_2O_5$ dissolves in water. A reagent bottle I use for the formaldehyde clock reaction is labeled "sodium bisulfite," but no formula is given. The label gives the reagent assay "as $SO_2$: minimum 58.5%." Pure $NaHSO_3$ is equivalent to 61.56 wt% $SO_2$ and pure $Na_2S_2O_5$ is equivalent to 67.40 wt% $SO_2$.

11. E. T. Urbansky and M. R. Schock, "Understanding, Deriving, and Computing Buffer Capacity," *J. Chem. Ed.* **2000,** *77,* 1640.

## Chapter 10

1. P. G. Daniele, "$Na^+$, $K^+$, and $Ca^{2+}$ Complexes of Low Molecular Weight Ligands in Aqueous Solution," *J. Chem. Soc. Dalton Trans.* **1985,** 2353.

2. P. G. Righetti, E. Gianazza, C. Gelfi, M. Chiari, and P. K. Sinha, "Isoelectric Focusing in Immobilized pH Gradients," *Anal. Chem.* **1989,** *61,* 1602.

3. Experiment on surface acidity of a solid: L. Tribe and B. C. Barja, "Adsorption of Phosphate on Goethite," *J. Chem. Ed.* **2004,** *81,* 1624.

## Chapter 11

1. B. Mörnstam, K.-G. Wahlund, and B. Jönsson, "Potentiometric Acid-Base Titration of a Colloidal Solution," *Anal. Chem.* **1997,** *69,* 5037. For titration of entire cell surfaces, see I. Sokolov, D. S. Smith, G. S. Henderson, Y. A. Gorby, and F. G. Ferris, "Cell Surface Electrochemical Heterogeneity of the Fe(III)-Reducing Bacteria *Shewanella putrefaciens,*" *Environ. Sci. Technol.* **2001,** *35,* 341.

2. A method to measure total charge on a protein as it binds to assorted ions is described by M. K. Menon and A. L. Zydney, "Measurement of Protein Charge and Ion Binding Using Capillary Electrophoresis," *Anal. Chem.* **1998,** *70,* 1581.

3. M. J. Ondrechen, J. G. Clifton, and D. Ringe, "THEMATICS: A Simple Computational Predictor of Enzyme Function from Structure," *Proc. Natl. Acad. Sci. USA,* **2001,** *98,* 12473.

4. K. R. Williams, "Automatic Titrators in the Analytical and Physical Chemistry Laboratories," *J. Chem. Ed.* **1998**, *75*, 1133; K. L. Headrick, T. K. Davies, and A. N. Haegele, "A Simple Laboratory-Constructed Automatic Titrator," *J. Chem. Ed.* **2000**, *77*, 389.

5. M. Inoue and Q. Fernando, "Effect of Dissolved $CO_2$ on Gran Plots," *J. Chem. Ed.* **2001**, *78*, 1132; G. Gran, "Equivalence Volumes in Potentiometric Titrations," *Anal. Chim. Acta* **1988**, *206*, 111; F. J. C. Rossotti and H. Rossotti, "Potentiometric Titrations Using Gran Plots," *J. Chem. Ed.* **1965**, *42*, 375; L. M. Schwartz, "Uncertainty of a Titration Equivalence Point," *J. Chem. Ed.* **1992**, *69*, 879. L. M. Schwartz, "Advances in Acid-Base Gran Plot Methodology," *J. Chem. Ed.* **1987**, *64*, 947.

6. M. Rigobello-Masini and J. C. Masini, "Application of Modified Gran Functions and Derivative Methods to Potentiometric Acid Titration Studies of the Distribution of Inorganic Carbon Species in Cultivation Medium of Marine Microalgae," *Anal. Chim. Acta* **2001**, *448*, 239.

7. G. Papanastasiou and I. Ziogas, "Simultaneous Determination of Equivalence Volumes and Acid Dissociation Constants from Potentiometric Titration Data," *Talanta* **1995**, *42*, 827.

8. G. Wittke, "Reactions of Phenolphthalein at Various pH Values," *J. Chem. Ed.* **1983**, *60*, 239.

9. Demonstrations with universal indicator (a mixed indicator with many color changes) are described in J. T. Riley, "Flashy Solutions," *J. Chem. Ed.* **1977**, *54*, 29.

10. T. A. Canada, L. R. Allain, D. B. Beach, and Z. Xue, "High-Acidity Determination in Salt-Containing Acids by Optical Sensors," *Anal. Chem.* **2002**, *74*, 2535.

11. D. Fárcasiu and A. Ghenciu, "Acidity Functions from $^{13}C$-NMR," *J. Am. Chem. Soc.* **1993**, *115*, 10901.

12. B. Hammouti, H. Oudda, A. El Maslout, and A. Benayada, "A Sensor for the In Situ Determination of Acidity Levels in Concentrated Sulfuric Acid," *Fresenius J. Anal. Chem.* **1999**, *365*, 310. For use of glass electrodes to measure pH as low as −4, see D. K. Nordstrom, C. N. Alpers, C. J. Ptacek, and D. W. Blowes, "Negative pH and Extremely Acidic Mine Waters from Iron Mountain, California," *Environ. Sci. Technol.* **2000**, *34*, 254.

13. C. Yi and M. Gratzl, "Diffusional Mictrotitration: Reagent Delivery by a Diffusional Microburet into Microscopic Samples," *Anal. Chem.* **1994**, *66*, 1976; M. Gratzl, H. Lu, T. Matsumoto, C. Yi, and G. R. Bright, "Fine Chemical Manipulations of Microscopic Liquid Samples," *Anal. Chem.* **1999**, *71*, 2751; H. Lu and M. Gratzl, "Optical Detection in Microscopic Domains," *Anal. Chem.* **2000**, *72*, 1569; K. Tohda, H. Lu, Y. Umezawa, and M. Gratzl, "Optical Detection in Microscopic Domains: Monitoring Nonfluorescent Molecules with Fluorescence," *Anal. Chem.* **2001**, *73*, 2070; E. Litborn, M. Stjernström, and J. Roeraade, "Nanoliter Titration Based on Piezoelectric Drop-on-Demand Technology and Laser-Induced Fluorescence Detection," *Anal. Chem.* **1998**, *70*, 4847.

14. C. Yi, D. Huang, and M. Gratzl, "Complexometric Determination of Metal Ions by Microscopic Diffusional Titration," *Anal. Chem.* **1996**, *68*, 1580; H. Xie and M. Gratzl, "Diffusional Titration of Metal Ions in Microliter Samples with Potentiometric Indication," *Anal. Chem.* **1996**, *68*, 3665.

15. R. A. Butler and R. G. Bates, "Double Potassium Salt of Sulfosalicylic Acid in Acidmetry and pH Control," *Anal. Chem.* **1976**, *48*, 1669.

16. Borax goes down to the pentahydrate upon standing: R. Naumann, C. Alexander-Weber, and F. G. K. Baucke, "Limited Stability of the pH Reference Material Sodium Tetraborate Decahydrate ("Borax")," *Fresenius J. Anal. Chem.* **1994**, *350*, 119.

17. Instructions for purifying and using primary standards can be found in the following books: J. A. Dean, *Analytical Chemistry Handbook* (New York: McGraw-Hill, 1995), pp. 3-28 to 3-30; J. Bassett, R. C. Denney, G. H. Jeffery, and J. Mendham, *Vogel's Textbook of Quantitative Inorganic Analysis,* 4th ed. (Essex, UK: Longman, 1978), pp. 296–306; I. M. Kolthoff and V. A. Stenger, *Volumetric Analysis,* Vol. 2 (New York: Wiley-Interscience, 1947).

18. A. A. Smith, "Consumption of Base by Glassware," *J. Chem. Ed.* **1986**, *63*, 85; G. Perera and R. H. Doremus, "Dissolution Rates of Commercial Soda-Lime and Pyrex Borosilicate Glasses," *J. Am. Ceram. Soc.* **1991**, *74*, 1554.

19. J. S. Fritz, *Acid-Base Titrations in Nonaqueous Solvents* (Boston: Allyn & Bacon, 1973); J. Kucharsky and L. Safarik, *Titrations in Non-Aqueous Solvents* (New York: Elsevier, 1963); W. Huber, *Titrations in Nonaqueous Solvents* (New York: Academic Press, 1967); I. Gyenes, *Titration in Non-Aqueous Media* (Princeton, NJ: Van Nostrand, 1967).

20. R. de Levie, "A General Simulator for Acid-Base Titrations," *J. Chem. Ed.* **1999**, *76*, 987; R. de Levie, "Explicit Expressions of the General Form of the Titration Curve in Terms of Concentration," *J. Chem. Ed.* **1993**, *70*, 209; R. de Levie, "General Expressions for Acid-Base Titrations of Arbitrary Mixtures," *Anal. Chem.* **1996**, *68*, 585; R. de Levie, *Principles of Quantitative Chemical Analysis* (New York: McGraw-Hill, 1997).

21. P. Ballinger and F. A. Long, "Acid Ionization Constants of Alcohols," *J. Am. Chem. Soc.* **1960**, *82*, 795.

## Chapter 12

1. R. MacKinnon, "Potassium Channels and the Atomic Basis of Selective Ion Conduction" (Nobel Lecture), *Angew. Chem. Int. Ed.* **2004**, *43*, 4265.

2. W. D. Bedsworth and D. L. Sedlak, "Sources and Environmental Fate of Strongly Complexed Nickel in Estuariane Waters," *Environ. Sci. Technol.* **1999**, *33*, 926; B. Nowack, "Environmental Chemistry of Aminopolycarboxylate Chelating Agents," *Environ. Sci. Technol.* **2002**, *36*, 4009.

3. D. T. Haworth, "Some Linguistic Details on Chelation," *J. Chem. Ed.* **1998**, *75*, 47.

4. The chelate effect is often attributed to a favorable entropy change for multidentate binding. Recent work does not support this explanation: V. Vallet, U. Wahlgren, and I. Grenthe, "Chelate Effect and Thermodynamics of Metal Complex Formation in Solution: A Quantum Chemical Study," *J. Am. Chem. Soc.* **2003**, *125*, 14941.

5. R. Mullin, "Targeted Drugs," *Chem. Eng. News,* 16 May 2005, p. 9; M. R. McDevitt, D. Ma, L. T. Lai, J. Simon, P. Borchardt, R. K. Frank, K. Wu, V. Pellegrini, M. J. Curcio, M. Miederer, N. H. Bander, and D. A. Scheinberg, "Tumor Therapy with Targeted Atomic Nanogenerators," *Science* **2001**, *294*, 1537.

6. Z. Hou, K. N. Raymond, B. O'Sullivan, T. W. Esker, and T. Nishio, "Microbial Macrocyclic Dihydroxamate Chelating Agents," *Inorg. Chem.* **1998**, *37*, 6630.

7. N. F. Olivieri and G. M. Brittenham, "Iron-Chelating Therapy and the Treatment of Thalassemia," *Blood* **1997**, *89*, 739.

8. D. R. Richardson and P. Ponka, "Development of Iron Chelators to Treat Iron Overload Disease," *Am. J. Hematol.* **1998**, *58*, 299.

9. J. Savulescu, "Thalassaemia Major: The Murky Story of Deferiprone," *Br. Med. J.* **2004**, *328*, 358.

10. C. May, S. Rivella, J. Callegari, G. Heller, K. M. L. Gaensler, L. Luzzatto, and M. Sadelain, "Therapeutic Haemoglobin Synthesis in β-Thalassaemic Mice Expressing Lentivirus-Encoded Human β-Globin," *Nature* **2000**, *406*, 82.

11. A. E. Martell, R. M. Smith, and R. J. Motekaitis, *NIST Standard Reference Database 46* (Gaithersburg, MD: National Institute of Standards and Technology, 2001).

12. $H_4Y$ can be dried at 140°C for 2 h and used as a primary standard. It can be dissolved by adding NaOH solution from a plastic container. NaOH solution from a glass bottle should not be used, because it contains alkaline earth metals leached from the glass. Reagent-grade $Na_2H_2Y \cdot 2H_2O$ contains ~0.3% excess water. It may be used in this form with suitable correction for the mass of excess water or dried to the composition $Na_2H_2Y \cdot 2H_2O$ at 80°C. [W. J. Blaedel and H. T. Knight, "Purification and Properties of Disodium Salt of Ethylenediaminetetraacetic Acid as a Primary Standard," *Anal. Chem.* **1954**, *26*, 741.]

13. R. L. Barnett and V. A. Uchtman, "Crystal Structures of Ca(CaEDTA) $\cdot$ 7H_2O and NaCaNTA," *Inorg. Chem.* **1979**, *18*, 2674.

14. S. G. John, C. E. Ruggiero, L. E. Hersman, C.-S. Tung, and M. P. Neu, "Siderophore Mediated Plutonium Accumulation by *Microbacterium flavescens,*" *Environ. Sci. Technol.* **2001**, *35*, 2942.

15. A definitive reference for the theory of EDTA titration curves is A. Ringbom, *Complexation in Analytical Chemistry* (New York: Wiley, 1963).

16. For discussion of metal-ligand equilibria with numerous examples, see P. Letkeman, "Computer-Modeling of Metal Speciation in Human Blood Serum,"

J. Chem. Ed. **1996,** 73, 165; A. Rojas-Hernández, M. T. Ramírez, I. González, and J. G. Ibanez, "Predominance-Zone Diagrams in Solution Chemistry," J. Chem. Ed. **1995,** 72, 1099; and A. Bianchi and E. Garcia-España, "Use of Calculated Species Distribution Diagrams to Analyze Thermodynamic Selectivity," J. Chem. Ed. **1999,** 76, 1727.

17. W. N. Perara and G. Hefter, "Mononuclear Cyano- and Hydroxo-Complexes of Iron(III)," Inorg. Chem. **2003,** 42, 5917.

18. S. Tandy, K. Bossart, R. Mueller, J. Ritschel, L. Hauser, R. Schulin, and B. Nowack, "Extraction of Heavy Metals from Soils Using Biodegradable Chelating Agents," Environ. Sci. Technol. **2004,** 38, 937; B. Kos and D. Leštan, "Induced Phytoextraction/Soil Washing of Lead Using Biodegradable Chelate and Permeable Barriers, Environ. Sci. Technol. **2003,** 37, 624; S. V. Sahi, N. L. Bryant, N. C. Sharma, and S. R. Singh, "Characterization of a Lead Hyperaccumulator Shrub," Environ. Sci. Technol. **2002,** 36, 4676.

19. G. Schwarzenbach and H. Flaschka, Complexometric Titrations, H. M. N. H. Irving, trans. (London: Methuen, 1969); H. A. Flaschka, EDTA Titrations (New York: Pergamon Press, 1959); J. A. Dean, Analytical Chemistry Handbook (New York: McGraw-Hill, 1995); A. E. Martell and R. D. Hancock, Metal Complexes in Aqueous Solution (New York: Plenum Press, 1996).

20. Indirect determinations of monovalent cations are described by I. M. Yurist, M. M. Talmud, and P. M. Zaitsev, "Complexometric Determination of Monovalent Metals," J. Anal. Chem. USSR **1987,** 42, 911.

21. D. P. S. Rathore, P. K. Bhargava, M. Kumar, and R. K. Talra, "Indicators for the Titrimetric Determination of Ca and Total Ca + Mg with EDTA," Anal. Chim. Acta **1993,** 281, 173.

22. T. Darjaa, K. Yamada, N. Sato, T. Fujino, and Y. Waseda, "Determination of Sulfur in Metal Sulfides by Bromine Water-CCl$_4$ Oxidative Dissolution and Modified EDTA Titration," Fresenius J. Anal. Chem. **1998,** 361, 442.

## Chapter 13

1. J. Gorman in Science News, 9 September 2000, p. 165.

2. R. F. Wright et al., "Recovery of Acidified European Surface Waters," Environ. Sci. Technol. **2005,** 39, 64A.

3. Books that will teach you more about equilibrium calculations include W. B. Guenther, Unified Equilibrium Calculations (New York: Wiley, 1991); J. N. Butler, Ionic Equilibrium: Solubility and pH Calculations (New York: Wiley, 1998); and M. Meloun, Computation of Solution Equilibria (New York: Wiley, 1988). For commercial software that performs complex equilibrium calculations, see http://www.micromath.com/ and http://www.acadsoft.co.uk/.

4. R. G. Bates, Determination of pH, 2nd ed. (New York: Wiley, 1973), p. 86, is the authoritative reference on pH. The pH uncertainty of primary standards could be greater than $\pm 0.006$ at temperatures other than 25°C.

5. R. B. Martin, "Aluminum: A Neurotoxic Product of Acid Rain," Acc. Chem. Res. **1994,** 27, 204.

6. R. Jugdaohsingh, M. M. Campbell, R. P. H. Thompson, C. R. McCrohan, K. N. White, and J. J. Powell, "Mucus Secretion by the Freshwater Snail Lymnaea stagnalis Limits Aluminum Concentrations of the Aqueous Environment," Environ. Sci. Technol. **1998,** 32, 2591; M. Ravichandran, G. R. Aiken, M. M. Reddy, and J. N. Ryan, "Enhanced Dissolution of Cinnabar (Mercuric Sulfide) by Dissolved Organic Matter Isolated from the Florida Everglades," Environ. Sci. Technol. **1998,** 32, 3305; S. Sauvé, M. McBride, and W. Hendershot, "Lead Phosphate Solubility in Water and Soil Suspensions," Environ. Sci. Technol. **1998,** 32, 388.

7. Our approach is similar to one by J. L. Guiñón, J. Garcia-Antón, and V. Pérez-Herranz, "Spreadsheet Techniques for Evaluating the Solubility of Sparingly Soluble Salts of Weak Acids," J. Chem. Ed. **1999,** 76, 1157.

8. A. Kraft, "Determination of the p$K_a$ of Multiprotic, Weak Acids by Analyzing Potentiometric Acid-Base Titration Data with Difference Plots," J. Chem. Ed. **2003,** 80, 554.

9. G. B. Kauffman, "Niels Bjerrum: A Centenial Evaluation," J. Chem. Ed. **1980,** 57, 779, 863.

10. Table 6-1 gives p$K_w$ = 13.995 at $\mu = 0$ at 25°C. The expression of $K_w$ to which this value applies is given in terms of molalities, $m$:

$$K_w = \frac{m_{H^+}\gamma_{H^+}m_{OH^-}\gamma_{OH^-}}{\mathcal{A}_{H_2O}} = 10^{-13.995}$$

We want to evaluate $K_w'$ for 0.1 M KCl. The factor for converting molality into molarity in 0.1 M K is 0.994 in Table 12-1-1A of H. S. Harned and B. B. Owen, Physical Chemistry of Electrolyte Solutions, 3rd ed. (New York: Reinhold, 1958), p. 725. The factor $\gamma_{H^+}\gamma_{OH^-}/\mathcal{A}_{H_2O}$ is 0.626 in 0.10 M KCl, interpolated from Table 15-2-1A of Harned and Owen, p. 752. $K_w'$ is the product of concentrations [H$^+$][OH$^-$]:

$$[H^+][OH^-] = \frac{m_{H^+}(0.994)\gamma_{H^+}m_{OH^-}(0.994)\gamma_{OH^-}}{\mathcal{A}_{H_2O}} \frac{\mathcal{A}_{H_2O}}{\gamma_{H^+}\gamma_{OH^-}}$$

$$= 10^{-13.995}(0.994^2)\left(\frac{1}{0.626}\right) = 10^{-13.797}$$

## Chapter 14

1. General treatments of electrochemistry: A. Hamnett, C. H. Hamann, and W. Vielstich, Electrochemistry (New York: Wiley, 1998); Z. Galus, Fundamentals of Electrochemical Analysis (New York: Ellis Horwood, 1994); C. M. A. Brett and A. M. O. Brett, Electrochemistry (Oxford: Oxford University Press, 1993); and H. B. Oldham and J. C. Myland, Fundamentals of Electrochemical Science (San Diego: Academic Press, 1993).

2. K. Rajeshwar and J. G. Ibanez, Environmental Electrochemistry (San Diego: Academic Press, 1997).

3. N. J. Tao, "Measurement and Control of Single-Molecule Conductance," J. Mater. Chem. **2005,** 15, 3260; N. Tao, "Electrochemical Fabrication of Metallic Quantum Wires," J. Chem. Ed. **2005,** 82, 720; S. Lindsay, "Single-Molecule Electronic Measurements with Metal Electrodes," J. Chem. Ed. **2005,** 82, 727; R. A. Wassel and C. B. Gorman, "Establishing the Molecular Basis for Molecular Electronics," Angew. Chem. Int. Ed. **2004,** 43, 5120.

4. S. Weinberg, The Discovery of Subatomic Particles (Cambridge: Cambridge University Press, 2003), pp. 13–16. A wonderful book by a Nobel Prize winner.

5. M. J. Smith and C. A. Vincent, "Structure and Content of Some Primary Batteries," J. Chem. Ed. **2001,** 78, 519; R. S. Treptow, "The Lead-Acid Battery: Its Voltage in Theory and in Practice," J. Chem. Ed. **2002,** 79, 334; M. J. Smith and C. A. Vincent, "Why Do Some Batteries Last Longer Than Others?" J. Chem. Ed. **2002,** 79, 851.

6. O. Zerbinati, "A Direct Methanol Fuel Cell," J. Chem. Ed. **2002,** 79, 829; J. M. Ogden, "Hydrogen: The Fuel of the Future?" Physics Today, April 2002, 69.

7. P. Krause and J. Manion, "A Novel Approach to Teaching Electrochemical Principles," J. Chem. Ed. **1996,** 73, 354.

8. L. P. Silverman and B. B. Bunn, "The World's Longest Human Salt Bridge," J. Chem. Ed. **1992,** 69, 309.

9. Classroom demonstration of half-cells: J. D. Ciparick, "Half Cell Reactions: Do Students Ever See Them?" J. Chem. Ed. **1991,** 68, 247.

10. A. W. von Smolinski, C. E. Moore, and B. Jaselskis, "The Choice of the Hydrogen Electrode as the Base for the Electromotive Series" in Electrochemistry, Past and Present, ACS Symposium Series 390, J. T. Stock and M. V. Orna, eds. (Washington, DC: American Chemical Society, 1989), Chap. 9.

11. H. Frieser, "Enhanced Latimer Potential Diagrams via Spreadsheets," J. Chem. Ed. **1994,** 71, 786.

12. A. Arévalo and G. Pastor, "Verification of the Nernst Equation and Determination of a Standard Electrode Potential," J. Chem. Ed. **1985,** 62, 882.

13. For a classroom demonstration using a cell as a chemical probe, see R. H. Anderson, "An Expanded Silver Ion Equilibria Demonstration: Including Use of the Nernst Equation and Calculation of Nine Equilibrium Constants," J. Chem. Ed. **1993,** 70, 940.

14. J. E. Walker, "ATP Synthesis by Rotary Catalysis (Nobel Lecture)," Angew. Chem. Int. Ed. **1998,** 37, 2309; P. D. Boyer, "Energy, Life, and ATP (Nobel Lecture)," Angew. Chem. Int. Ed. **1998,** 37, 2297; W. S. Allison, "F$_1$-ATPase," Acc. Chem. Res. **1998,** 31, 819.

15. For an advanced equilibrium problem based on the bromine Latimer diagram, see T. Michalowski, "Calculation of pH and Potential E for Bromine Aqueous Solution," J. Chem. Ed. **1994,** 71, 560.

16. K. T. Jacob, K. P. Jayadevan, and Y. Waseda, "Electrochemical Determination of the Gibbs Energy of Formation of MgAl$_2$O$_4$," *J. Am. Ceram. Soc.* **1998,** *81,* 209.

17. J. T. Stock, "Einar Biilmann (1873–1946): pH Determination Made Easy," *J. Chem. Ed.* **1989,** *66,* 910.

18. The cell in this problem would not give an accurate result, because of the *junction potential* at each liquid junction (Section 15-3). A cell without a liquid junction is described by P. A. Rock, "Electrochemical Double Cells," *J. Chem. Ed.* **1975,** *52,* 787.

## Chapter 15

1. B. Fu, E. Bakker, J. H. Yun, V. C. Yang, and M. E. Meyerhoff, "Response Mechanism of Polymer Membrane-Based Potentiometric Polyion Sensors," *Anal. Chem.* **1994,** *66,* 2250.

2. I. Capila and R. J. Linhardt, "Heparin-Protein Interactions," *Angew. Chem. Int. Ed.* **2002,** *41,* 390.

3. W. Qin, W. Zhang, K. P. Xiao, and M. E. Meyerhoff, "Enhanced Sensitivity Electrochemical Assay of Low-Molecular-Weight Heparins Using Rotating Polyion-Sensitive Membrane Electrodes," *Anal. Bioanal. Chem.* **2003,** *377,* 929.

4. Practical aspects of electrode fabrication are discussed in D. T. Sawyer, A. Sobkowiak, and J. L. Roberts, Jr., *Electrochemistry for Chemists,* 2nd ed. (New York: Wiley, 1995); G. A. East and M. A. del Valle, "Easy-to-Make Ag/AgCl Reference Electrode," *J. Chem. Ed.* **2000,** *77,* 97.

5. A demonstration of potentiometry with a silver electrode (or a microscale experiment for general chemistry) is described by D. W. Brooks, D. Epp, and H. B. Brooks, "Small-Scale Potentiometry and Silver One-Pot Reactions," *J. Chem. Ed.* **1995,** *72,* A162.

6. D. Dobčnik, J. Stergulec, and S. Gomišček, "Preparation of an Iodide Ion-Selective Electrode by Chemical Treatment of a Silver Wire," *Fresenius J. Anal. Chem.* **1996,** *354,* 494.

7. I. R. Epstein and J. A. Pojman, *An Introduction to Nonlinear Chemical Dynamics: Oscillations, Waves, Patterns, and Chaos* (New York: Oxford University Press, 1998); I. R. Epstein, K. Kustin, P. De Kepper, and M. Orbán, *Scientific American,* March 1983, p. 112; and H. Degn, "Oscillating Chemical Reactions in Homogeneous Phase," *J. Chem. Ed.* **1972,** *49,* 302.

8. Mechanisms of oscillating reactions are discussed by O. Benini, R. Cervellati, and P. Fetto, "The BZ Reaction: Experimental and Model Studies in the Physical Chemistry Laboratory," *J. Chem. Ed.* **1996,** *73,* 865; R. J. Field and F. W. Schneider, "Oscillating Chemical Reactions and Nonlinear Dynamics," *J. Chem. Ed.* **1989,** *66,* 195; R. M. Noyes, "Some Models of Chemical Oscillators," *J. Chem. Ed.* **1989,** *66,* 190; P. Ruoff, M. Varga, and E. Körös, "How Bromate Oscillators Are Controlled," *Acc. Chem. Res.* **1988,** *21,* 326; M. M. C. Ferriera, W. C. Ferriera, Jr., A. C. S. Lino, and M. E. G. Porto, "Uncovering Oscillations, Complexity, and Chaos in Chemical Kinetics Using *Mathematica,*" *J. Chem. Ed.* **1999,** *76,* 861; G. Schmitz, L. Kolar-Anić, S. Anić, and Z. Čupić, "The Illustration of Multistability," *J. Chem. Ed.* **2000,** *77,* 1502.

9. H. E. Prypsztejn, "Chemiluminescent Oscillating Demonstrations: The Chemical Buoy, the Lighting Wave, and the Ghostly Cylinder," *J. Chem. Ed.* **2005,** *82,* 53; D. Kolb, "Overhead Projector Demonstrations," *J. Chem. Ed.* **1988,** *65,* 1004; R. J. Field, "A Reaction Periodic in Time and Space," *J. Chem. Ed.* **1972,** *49,* 308; J. N. Demas and D. Diemente, "An Oscillating Chemical Reaction with a Luminescent Indicator," *J. Chem. Ed.* **1973,** *50,* 357; J. F. Lefelhocz, "The Color Blind Traffic Light," *J. Chem. Ed.* **1972,** *49,* 312: P. Aroca, Jr., and R. Aroca, "Chemical Oscillations: A Microcomputer-Controlled Experiment," *J. Chem. Ed.* **1987,** *64,* 1017: J. Amrehn, P. Resch, and F. W. Schneider, "Oscillating Chemiluminescence with Luminol in the Continuous Flow Stirred Tank Reactor," *J. Phys. Chem.* **1988,** *92,* 3318; D. Avnir, "Chemically Induced Pulsations of Interfaces: The Mercury Beating Heart," *J. Chem. Ed.* **1989,** *66,* 211; K. Yoshikawa, S. Nakata, M. Yamanaka, and T. Waki, "Amusement with a Salt-Water Oscillator," *J. Chem. Ed.* **1989,** *66,* 205; L. J. Soltzberg, M. M. Boucher, D. M. Crane, and S. S. Pazar, "Far from Equilibrium—The Flashback Oscillator," *J. Chem. Ed.* **1987,** *64,* 1043; S. M. Kaushik, Z. Yuan, and R. M. Noyes, "A Simple Demonstration of a Gas Evolution Oscillator," *J. Chem. Ed.* **1986,** *63,* 76; R. F. Melka, G. Olsen, L. Beavers, and J. A. Draeger, "The Kinetics of Oscillating Reactions," *J. Chem. Ed.* **1992,** *69,* 596; J. M. Merino, "A Simple, Continuous-Flow Stirred-Tank Reactor for the Demonstration and Investigation of Oscillating Reactions," *J. Chem. Ed.* **1992,** *69,* 754.

10. T. Kappes and P. C. Hauser, "A Simple Supplementary Offset Device for Data Acquisition Systems," *J. Chem. Ed.* **1999,** *76,* 1429.

11. [Br$^-$] also oscillates in this experiment. For an [I$^-$] oscillator, see T. S. Briggs and W. C. Rauscher, "An Oscillating Iodine Clock," *J. Chem. Ed.* **1973,** *50,* 496.

12. E. Bakker, P. Bühlmann, and E. Pretsch, "Carrier-Based Ion-Selective Electrodes and Bulk Optodes," *Chem. Rev.* **1997,** *97,* 3083; ibid. **1998,** *98,* 1593.

13. C. E. Moore, B. Jaselskis, and A. von Smolinski, "Development of the Glass Electrode" in *Electrochemistry, Past and Present,* ACS Symposium Series 390, J. T. Stock and M. V. Orna, eds. (Washington, DC: American Chemical Society, 1989), Chap. 19.

14. Ir | IrO$_2$ electrodes (commercially available from Cypress Systems, Lawrence, KS) can measure pH in harsh environments or microscopic spaces [S. A. M. Marzouk, "Improved Electrodeposited Iridium Oxide pH Sensor Fabricated on Etched Titanium Substrates," *Anal. Chem.* **2003,** *75,* 1258; A. N. Bezbaruah and T. C. Zhang, "Fabrication of Anodically Electrodeposited Iridium Oxide Film pH Microelectrodes for Microenvironmental Studies," *Anal. Chem.* **2002,** *74,* 5726; D. O. Wipf, F. Ge, T. W. Spaine, and J. E. Baur, "Microscopic Measurement of pH with IrO$_2$ Microelectrodes," *Anal. Chem.* **2000,** *72,* 4921]. For pH measurement in nanoscopic spaces, see X. Zhang, B. Ogorevc, and J. Wang, "Solid-State pH Nanoelectrode Based on Polyaniline Thin Film Electrodeposited onto Ion-Beam Etched Carbon Fiber," *Anal. Chim. Acta* **2002,** *452,* 1. A ZrO$_2$ electrode can measure pH up to 300°C [L. W. Niedrach, "Electrodes for Potential Measurements in Aqueous Systems at High Temperatures and Pressures," *Angew. Chem.* **1987,** *26,* 161].

15. B. Jaselskis, C. E. Moore, and A. von Smolinski, "Development of the pH Meter" in *Electrochemistry, Past and Present,* ACS Symposium Series 390, J. T. Stock and M. V. Orna, eds. (Washington, DC: American Chemical Society, 1989), Chap. 18.

16. H. B. Kristensen, A. Salomon, and G. Kokholm, "International pH Scales and Certification of pH," *Anal. Chem.* **1991,** *63,* 885A.

17. L. M. Goss, "A Demonstration of Acid Rain and Lake Acidification: Wet Deposition of Sulfur Dioxide," *J. Chem. Ed.* **2003,** *80,* 39.

18. J. A. Lynch, V. C. Bowersox, and J. W. Grimm, "Acid Rain Reduced in Eastern United States," *Environ. Sci. Technol.* **2000,** *34,* 940; R. E. Baumgardner, Jr., T. F. Lavery, C. M. Rogers, and S. S. Isil, "Estimates of the Atmospheric Deposition of Sulfur and Nitrogen Species: Clean Air Status and Trends Network, 1990–2000," *Environ. Sci. Technol.* **2002,** *36,* 2614; www.epa.gov/acidrain

19. W. F. Koch, G. Marinenko, and R. C. Paule, "An Interlaboratory Test of pH Measurements in Rainwater," *J. Res. Natl. Bur. Stds.* **1986,** *91,* 23.

20. *Free diffusion junction* electrodes are designed to minimize junction potentials. The junction is a Teflon capillary tube containing electrolyte that is periodically renewed by syringe.

21. Spectrophotometry with acid-base indictors is another means for measuring the pH of low ionic strength natural waters. C. R. French, J. J. Carr, E. M. Dougherty, L. A. K. Eidson, J. C. Reynolds, and M. D. DeGrandpre, "Spectrophotometric pH Measurements of Freshwater," *Anal. Chim. Acta* **2002,** *453,* 13.

22. The reference electrode in the Ross combination electrode is Pt | I$_2$, I$^-$. This electrode is claimed to give improved precision and accuracy over conventional pH electrodes [R. C. Metcalf, *Analyst* **1987,** *112,* 1573].

23. History of ion-selective electrodes: M. S. Frant, "Where Did Ion Selective Electrodes Come From?" *J. Chem. Ed.* **1997,** *74,* 159; J. Ruzicka, "The Seventies—Golden Age for Ion Selective Electrodes," *J. Chem. Ed.* **1997,** *74,*167; T. S. Light, "Industrial Use and Applications of Ion Selective Electrodes," *J. Chem. Ed.* **1997,** *74,* 171; and C. C. Young, "Evolution of Blood Chemistry Analyzers Based on Ion Selective Electrodes," *J. Chem. Ed.* **1997,** *74,* 177; R. P. Buck and E. Lindner, "Tracing the History of Selective Ion Sensors," *Anal. Chem.* **2001,** *73,* 88A.

24. The Nicolsky-Eisenman equation (15-8) is predicated on equal charges for the primary (A) and interfering (X) ions [E. Bakker, R. Meruva, E. Pretsch, and M. Meyerhoff, "Selectivity of Polymer Membrane-Based Ion-Selective Electrodes," *Anal. Chem.* **1994,** *66,* 3021]. Recipes for measuring selectivity

coefficients for ions of different charge and discussions of subtle pitfalls can be found in E. Bakker, E. Pretsch, and P. Bühlmann, *Anal. Chem.* **2000,** *72,* 1127; and K. Ren, "Selectivity Problems of Membrane Ion-Selective Electrodes," *Fresenius J. Anal. Chem.* **1999,** *365,* 389.

25. E. Bakker and E. Pretsch, "The New Wave of Ion-Selective Electrodes," *Anal. Chem.* **2002,** *74,* 420A.

26. A. Radu, M. Telting-Diaz, and E. Bakker, "Rotating Disk Potentiometry for Inner Solution Optimization of Low-Detection-Limit Ion-Selective Electrodes," *Anal. Chem.* **2003,** *75,* 6922.

27. J. Bobacka, T. Lindfors, A. Lewenstam, and A. Ivaska, "All-Solid-State Ion Sensors Using Conducting Polymers as Ion-to-Electron Transducers," *Am. Lab.,* February 2004, 13; A. Konopka, T. Sokalski, A. Michalska, A. Lewenstam, and M. Maj-Zurawska, "Factors Affecting the Potentiometric Response of All-Solid-State Solvent Polymeric Membrane Calcium-Selective Electrode for Low-Level Measurement," *Anal. Chem.* **2004,** *76,* 6410; M. Fouskaki and N. A. Chaniotakis, "Thick Membrane, Solid Contact Ion Selective Electrode for the Detection of Lead at Picomolar Levels," *Anal. Chem.* **2005,** *77,* 1780.

28. Many other kinds of $CO_2$ electrodes have been described [e.g., J. H. Shin, J. S. Lee, S. H. Choi, D. K. Lee, H. Nam, and G. S. Cha, "A Planar $pCO_2$ Sensor with Enhanced Electrochemical Properties," *Anal. Chem.* **2000,** *72,* 4468]. Carbonate liquid-based ion-selective electrodes also are available: Y. S. Choi, L. Lvova, J. H. Shin, S. H. Oh, C. S. Lee, B. H. Kim, G. S. Cha, and H. Nam, "Determination of Oceanic Carbon Dioxide Using a Carbonate-Selective Electrode," *Anal. Chem.* **2002,** *74,* 2435.

29. J. D. Czaban, "Electrochemical Sensors in Clinical Chemistry," *Anal. Chem.* **1985,** *57,* 345A.

30. *Resistivity,* $\rho$, measures how well a substance retards the flow of electric current when an electric field is applied: $J = E/\rho$, where $J$ is current density (current flowing through a unit cross section of the material, $A/m^2$) and $E$ is electric field (V/m). Units of resistivity are $V \cdot m/A$ or $\Omega \cdot m$, because $\Omega = V/A$, where $\Omega = $ ohm. Conductors have resistivities near $10^{-8} \Omega \cdot m$, semiconductors have resistivities of $10^{-4}$ to $10^7 \Omega \cdot m$, and insulators have resistivities of $10^{12}$ to $10^{20} \Omega \cdot m$. The reciprocal of resistivity is *conductivity.* Resistivity does not depend on the dimensions of the substance. Resistance, $R$, is related to resistivity by the equation $R = \rho l/A$, where $l$ is the length and $A$ is the cross-sectional area of the conducting substance.

31. S.-S. Jan, J.-L. Chiang, Y.-C. Chen, J.-C. Chou, and C.-C Cheng, "Characteristics of the Hydrogen Ion-Sensitive Field Effect Transistors with Sol-Gel-Derived Lead Titanate Gate," *Anal. Chim. Acta* **2002,** *469,* 205.

32. C. Jiménez, I. Marqués, and J. Bartrolí, "Continuous-Flow Systems for On-Line Water Monitoring Using Back-Side Contact ISFET-Based Sensors," *Anal. Chem.* **1996,** *68,* 3801.

33. J. H. Shin, H. J. Lee, C. Y. Kim, B. K. Oh, K. L. Rho, H. Nam, and G. S. Cha, "ISFET-Based Differential $pCO_2$ Sensors Employing a Low-Resistance Gas-Permeable Membrane," *Anal. Chem.* **1996,** *68,* 3166.

34. A. M. Nyamsi Hendji, N. Jaffrezic-Renault, C. Martelet, P. Clechet, A. A. Shul'ga, V. I. Strikha, L. I. Netchiporuk, A. P. Soldatkin, and W. B. Wlodarski, "Sensitive Detection of Pesticides Using Differential ISFET-Based System with Immobilized Cholinesterases," *Anal. Chim. Acta* **1993,** *281,* 3.

35. M. Lahav, A. B. Kharitonov, O. Katz, T. Kunitake, and I. Willner, "Tailored Chemosensors for Chloroaromatic Acids Using Molecular Imprinted $TiO_2$ Thin Films on Ion-Sensitive Field-Effect Transistor," *Anal. Chem.* **2001,** *73,* 720.

36. A. B. Kharitonov, A. N. Shipway, and I. Willner, "An Au Nanoparticle/Bisbipyridinium Cyclophane-Functionalized Ion-Sensitive Field-Effect Transistor for the Sensing of Adrenaline," *Anal. Chem.* **1999,** *71,* 5441.

37. S. P. Pogorelova, M. Zayats, T. Bourenko, A. B. Kharitonov, O. Lioubashevski, E. Katz, and I. Willner, "Analysis of $NAD(P)^+/NAD(P)H$ Cofactors by Imprinted Polymer Membranes Associated with Ion-Sensitive Field-Effect Transistor Devices and Au-Quartz Crystals," *Anal. Chem.* **2003,** *75,* 509.

38. D. C. Jackman, "A Recipe for the Preparation of a pH 7.00 Calibration Buffer," *J. Chem. Ed.* **1993,** *70,* 853.

## Chapter 16

1. T. Astrup, S. L. S. Stipp, and T. H. Christensen, "Immobilization of Chromate from Coal Fly Ash Leachate Using an Attenuating Barrier Containing Zero-Valent Iron," *Environ. Sci. Technol.* **2000,** *34,* 4163; S. H. Joo, A. J. Feitz,

and T. D. Waite, "Oxidative Degradation of the Carbothiolate Herbicide, Molinate, Using Nanoscale Zero-Valent Iron," *Environ. Sci. Technol.* **2004,** *38,* 2242; R. Miehr, P. G. Tratnyek, J. Z. Bandstra, M. M. Scherer, M. J. Alowitz, and E. U. Bylaska, "Diversity of Contaminant Reduction Reactions by Zerovalent Iron: Role of the Reductant," *Environ. Sci. Technol.* **2004,** *38,* 139; V. K. Sharma, C. R. Burnett, D. B. O'Connor, and D. Cabelli, "Iron(VI) and Iron(V) Oxidation of Thiocyanate," *Environ. Sci. Technol.* **2002,** *36,* 4182.

2. Information on redox titrations: J. Bassett, R. C. Denney, G. H. Jeffery, and J. Mendham, *Vogel's Textbook of Inorganic Analysis,* 4th ed. (Essex, UK: Longman, 1978); H. A. Laitinen and W. E. Harris, *Chemical Analysis,* 2nd ed. (New York: McGraw-Hill, 1975); I. M. Kolthoff, R. Belcher, V. A. Stenger, and G. Matsuyama, *Volumetric Analysis,* Vol. 3 (New York: Wiley, 1957); A. Berka, J. Vulterin, and J. Zýka, *Newer Redox Titrants,* H. Weisz, trans. (Oxford: Pergamon, 1965).

3. J. Ermírio, F. Moraes, F. H. Quina, C. A. O. Nascimento, D. N. Silva, and O. Chiavone-Filho, "Treatment of Saline Wastewater Contaminate with Hydrocarbons by the Photo-Fenton Process," *Environ. Sci. Technol.* **2004,** *38,* 1183; B. Gözmen, M. A. Oturan, N. Oturan, and O. Erbatur, "Indirect Electrochemical Treatment of Bisphenol A in Water via Electrochemically Generated Fenton's Reagent," *Environ. Sci. Technol.* **2003,** *37,* 3716; P. S. Zurer, "Ridding the World of Unwanted Chemicals," *Chem. Eng. News,* 2 September 2002, p. 34.

4. Equations 16-9 and 16-10 are analogous to the Henderson-Hasselbalch equation of acid-base buffers. Prior to the equivalence point, the redox titration is *buffered* to a potential near $E_+ = $ formal potential for $Fe^{3+} \mid Fe^{2+}$ by the presence of $Fe^{3+}$ and $Fe^{2+}$. After the equivalence point, the reaction is *buffered* to a potential near $E_+ = $ formal potential for $Ce^{4+} \mid Ce^{3+}$. [R. de Levie "Redox Buffer Strength," *J. Chem. Ed.* **1999,** *76,* 574.]

5. D. W. King, "A General Approach for Calculating Speciation and Poising Capacity of Redox Systems with Multiple Oxidation States: Application to Redox Titrations and the Generation of p$\varepsilon$-pH Diagrams," *J. Chem. Ed.* **2002,** *79,* 1135.

6. T. J. MacDonald, B. J. Barker, and J. A. Caruso, "Computer Evaluation of Titrations by Gran's Method," *J. Chem. Ed.* **1972,** *49,* 200.

7. M. da Conceição Silva Barreto, L. de Lucena Medieros, and P. C. de Holanda Furtado, "Indirect Potentiometric Titration of Fe(III) with Ce(IV) by Gran's Method," *J. Chem. Ed.* **2001,** *78,* 91.

8. R. D. Hancock and B. J. Tarbet, "The Other Double Helix: The Fascinating Chemistry of Starch," *J. Chem. Ed.* **2000,** *77,* 988.

9. J. H. Margeson, J. C. Suggs, and M. R. Midgett, "Reduction of Nitrate to Nitrite with Cadmium," *Anal. Chem.* **1980,** *52,* 1955.

10. E. T. Urbansky, "Total Organic Carbon Analyzers as Tools for Measuring Carbonaceous Matter in Natural Waters," *J. Environ. Monit.* **2001,** *3,* 102. General references on environmental analysis: M. Radojevic and V. N. Bashkin, *Practical Environmental Analysis* (Cambridge: Royal Society of Chemistry, 1999); and D. Perez-Bendito and S. Rubio, *Environmental Analytical Chemistry* (Amsterdam: Elsevier, 1998).

11. L. J. Stolzberg and V. Brown, "Note on Photocatalytic Destruction of Organic Wastes: Methyl Red as a Substrate," *J. Chem. Ed.* **2005,** *82,* 526; J. A. Poce-Fatou, M. L. A. Gil, R. Alcántara, C. Botella, and J. Martin, "Photochemical Reactor for the Study of Kinetics and Adsorption Phenomena," *J. Chem. Ed.* **2004,** *81,* 537; J. C. Yu and L. Y. L. Chan, "Photocatalytic Degradation of a Gaseous Organic Pollutant," *J. Chem. Ed.* **1998,** *75,* 750.

12. R. Dunn, "New Developments in Membrane-Selective Conductometric Instruments for Total Organic Carbon Determination in Water," *Am. Lab.,* September 2004, p. 22.

13. S. Sakthivel and H. Kisch, "Daylight Photocatalysis by Carbon-Modified Titanium Dioxide," *Angew. Chem. Int. Ed.* **2003,** *42,* 4908.

14. S. Horikoshi, N. Serpone, Y. Hisamatsu, and H. Hidaka, "Photocatalyzed Degradation of Polymers in Aqueous Semiconductor Suspensions," *Environ. Sci. Technol.* **1998,** *32,* 4010.

15. BOD and COD procedures are described in *Standard Methods for the Examination of Wastewater,* 20th ed. (Washington, DC: American Public Health Association, 1998), which is the standard reference for water analysis.

16. K. Catterall, H. Zhao, N. Pasco, and R. John, "Development of a Rapid Ferricyanide-Mediated Assay for Biochemical Oxygen Demand Using a Mixed Microbial Consortium," *Anal. Chem.* **2003,** *75,* 2584.

**17.** B. Wallace and M. Purcell, "The Benefits of Nitrogen and Total Organic Carbon Determination by High-Temperature Combustion," *Am. Lab. News Ed.,* February 2003, p. 58.

**18.** W. Gottardi, "Redox-Potentiometric/Titrimetric Analysis of Aqueous Iodine Solutions," *Fresenius J. Anal. Chem.* **1998,** *362,* 263.

**19.** S. C. Petrovic and G. M. Bodner, "An Alternative to Halogenated Solvents for Halogen/Halide Extractions," *J. Chem. Ed.* **1991,** *68,* 509.

**20.** G. L. Hatch, "Effect of Temperature on the Starch-Iodine Spectrophotometric Calibration Line," *Anal. Chem.* **1982,** *54,* 2002.

**21.** Y. Xie, M. R. McDonald, and D. W. Margerum, "Mechanism of the Reaction Between Iodate and Iodide Ions in Acid Solutions," *Inorg. Chem.* **1999,** *38,* 3938.

**22.** Prepare anhydrous $Na_2S_2O_3$ by refluxing 21 g of $Na_2S_2O_3 \cdot 5H_2O$ with 100 mL of methanol for 20 min. Then filter the anhydrous salt, wash with 20 mL of methanol, and dry at 70°C for 30 min. [A. A. Woolf, "Anhydrous Sodium Thiosulfate as a Primary Iodometric Standard," *Anal. Chem.* **1982,** *54,* 2134.]

**23.** J. Hvoslef and B. Pedersen, "The Structure of Dehydroascorbic Acid in Solution," *Acta Chem. Scand.* **1979,** *B33,* 503, and D. T. Sawyer, G. Chiericato, Jr., and T. Tsuchiya, "Oxidation of Ascorbic Acid and Dehydroascorbic Acid by Superoxide in Aprotic Media," *J. Am. Chem. Soc.* **1982,** *104,* 6273.

**24.** R. J. Cava, "Oxide Superconductors," *J. Am. Ceram. Soc.* **2000,** *83,* 5.

**25.** D. C. Harris, M. E. Hills, and T. A. Hewston, "Preparation, Iodometric Analysis, and Classroom Demonstration of Superconductivity in $YBa_2Cu_3O_{8-x}$," *J. Chem. Ed.* **1987,** *64,* 847; D. C. Harris, "Oxidation State Chemical Analysis," in T. A. Vanderah, ed., *Chemistry of Superconductor Materials* (Park Ridge, NJ: Noyes, 1992); B. D. Fahlman, "Superconductor Synthesis—An Improvement," *J. Chem. Ed.* **2001,** *78,* 1182. Superconductor demonstration kits can be purchased from several vendors, including Sargent-Welch, 7400 N. Linder Ave., Skokie, IL 60077-1026.

**26.** Experiments with an $^{18}$O-enriched superconductor show that the $O_2$ evolved in Reaction 1 is all derived from the solid, not from the solvent [M. W. Shafer, R. A. de Groot, M. M. Plechaty, G. J. Scilla, B. L. Olson, and E. I. Cooper, "Evolution and Chemical State of Oxygen upon Acid Dissolution of $YBa_2Cu_3O_{6.98}$," *Mater. Res. Bull.* **1989,** *24,* 687; P. Salvador, E. Fernandez-Sanchez, J. A. Garcia Dominguez, J. Amdor, C. Cascales, and I. Rasines, "Spontaneous $O_2$ Release from $SmBa_2Cu_3O_{7-x}$ High $T_c$ Superconductor in Contact with Water," *Solid State Commun.* **1989,** *70,* 71].

**27.** A more sensitive and elegant iodometric procedure is described by E. H. Appelman, L. R. Morss, A. M. Kini, U. Geiser, A. Umezawa, G. W. Crabtree, and K. D. Carlson, "Oxygen Content of Superconducting Perovskites $La_{2-x}Sr_xCuO_y$ and $YBa_2Cu_3O_y$," *Inorg. Chem.* **1987,** *26,* 3237. This method can be modified by adding standard $Br_2$ to analyze superconductors with oxygen in the range 6.0–6.5, in which there is formally $Cu^+$ and $Cu^{2+}$. The use of electrodes instead of starch to find the end point in iodometric titrations of superconductors is recommended [P. Phinyocheep and I. M. Tang, "Determination of the Hole Concentration (Copper Valency) in the High $T_c$ Superconductors," *J. Chem. Ed.* **1994,** *71,* A115].

**28.** C. L. Copper and E. Koubek, "Analysis of an Oxygen Bleach," *J. Chem. Ed.* **2001,** *78,* 652.

**29.** M. T. Garrett, Jr., and J. F. Stehlik, "Classical Analysis," *Anal. Chem.* **1992,** *64,* 310A.

**30.** K. Peitola, K. Fujinami, M. Karppinen, H. Yamauchi, and L. Niiniströ, "Stoichiometry and Copper Valence in the $Ba_{1-y}CuO_{2+\delta}$ System," *J. Mater. Chem.* **1999,** *9,* 465.

**31.** S. Scaccia and M. Carewska, "Determination of Stoichiometry of $Li_{1+y}CoO_2$ Materials by Flame Atomic Absorption Spectrometry and Automated Potentiometric Titration," *Anal. Chim. Acta* **2002,** *453,* 35.

**32.** M. Karppinen, A. Fukuoka, J. Wang, S. Takano, M. Wakata, T. Ikemachi, and H. Yamauchi, "Valence Studies on Various Superconducting Bismuth and Lead Cuprates and Related Materials," *Physica* **1993,** *C208,* 130.

**33.** The oxygen liberated in Reaction 4 of this problem may be derived from the superconductor, not from solvent water. In any case, $BiO_3^-$ reacts with $Fe^{2+}$, and $Cu^{3+}$ does not, when the sample is dissolved in acid.

## Chapter 17

**1.** T. R. I. Cataldi, C. Campa, and G. E. De Benedetto, "Carbohydrate Analysis by High-Performance Anion-Exchange Chromatography with Pulsed Amperometric Detection," *Fresenius J. Anal. Chem.* **2000,** *368,* 739.

**2.** W. E. Haupin, "Electrochemistry of the Hall-Heroult Process for Aluminum Smelting," *J. Chem. Ed.* **1983,** *60,* 279; N. C. Craig, "Charles Martin Hall—The Young Man, His Mentor, and His Metal," *J. Chem. Ed.* **1986,** *63,* 557.

**3.** A. J. Bard and L. R. Faulkner, *Electrochemical Methods and Applications,* 2nd ed. (New York: Wiley, 2001); J. Wang, *Analytical Electrochemistry,* 2nd ed. (New York: Wiley, 2000); J. O'M. Bockris and A. K. N. Reddy, *Modern Electrochemistry,* 2nd ed. (Dordrecht, Netherlands: Kluwer, 1998–2001), 3 vols.; F. Scholz, ed., *Electroanalytical Methods* (Berlin: Springer-Verlag, 2002); P. Vanysek, *Modern Techniques in Electroanalysis* (New York: Wiley, 1996); A. J. Bard and M. Stratmann, eds., *Encyclopedia of Electrochemistry* (New York: Wiley-VCH, 2002).

**4.** E. C. Gilbert in H. N. Alyea and F. B. Dutton, eds., *Tested Demonstrations in Chemistry* (Easton, PA: Journal of Chemical Education, 1965), p. 145.

**5.** J. O'M. Bockris, "Overpotential: A Lacuna in Scientific Knowledge," *J. Chem. Ed.* **1971,** *48,* 352.

**6.** D. N. Craig, J. I. Hoffman, C. A. Law, and W. J. Hamer, "Determination of the Value of the Faraday with a Silver-Perchloric Acid Coulometer," *J. Res. Natl. Bur. Stds.* **1960,** *64A,* 381; H. Diehl, "High-Precision Coulometry and the Value of the Faraday," *Anal. Chem.* **1979,** *51,* 318A.

**7.** J. Greyson and S. Zeller, "Analytical Coulometry in Monier-Williams Sulfite-in-Food Determinations," *Am. Lab.,* July 1987, p. 44; D. T. Pierce, M. S. Applebee, C. Lacher, and J. Bessie, "Low Parts Per Billion Determination of Sulfide by Coulometric Argentometry," *Environ. Sci. Technol.* **1998,** *32,* 1734.

**8.** A. Mulchandani and O. A. Sadik, eds., *Chemical and Biological Sensors for Environmental Monitoring Biosensors* (Washington, DC: American Chemical Society, 2000); D. Diamond, ed., *Principles of Chemical and Biological Sensors* (New York: Wiley, 1998); A. Cunningham, *Introduction to Bioanalytical Sensors* (New York: Wiley, 1998); G. Ramsay, *Commercial Biosensors: Applications to Clinical, Bioprocess, and Environmental Samples* (New York: Wiley, 1998); K. R. Rogers and A. Mulchandani, eds., *Affinity Biosensors, Techniques and Protocols* (Totowa, NJ: Humana Press, 1998); E. Palaček and M. Fojta, "Detecting DNA Hybridization and Damage," *Anal. Chem.* **2001,** *73,* 75A.

**9.** Examples of biosensors: J. H. Thomas, S. K. Kim, P. J. Hesketh, H. B. Halsall, and W. R. Heineman, "Bead-Based Electrochemical Immunoassay for Bacteriophage MS2," *Anal. Chem.* **2004,** *76,* 2700; Y. Zhang, H.-H. Kim, and A. Heller, "Enzyme-Amplified Amperometric Detection of 3000 Copies of DNA in a 10-µL Droplet at 0.5 fM Concentration," *Anal. Chem.* **2003,** *75,* 3267; A. S. Mittelmann, E. Z. Ron, and J. Rishpon, "Amperometric Quantification of Total Coliforms and Specific Detection of *E. coli*," *Anal. Chem.* **2002,** *74,* 903; A. Avramescu, S. Andreescu, T. Noguer, C. Bala, D. Andreescu, and J.-L. Marty, "Biosensors Designed for Environmental and Food Quality Control," *Anal. Bioanal. Chem.* **2002,** *374,* 25; T. M. O'Regan, L. J. O'Riordan, M. Pravda, C. K. O'Sullivan, and G. G. Guilbault, "Direct Detection of Myoglobin in Whole Blood Using a Disposable Amperometric Immunosensor," *Anal. Chim. Acta* **2002,** *460,* 141.

**10.** M. C. Frost, S. M. Rudich, H. Zhang, M. A. Maraschio, and M. E. Meyerhoff, "In Vivo Biocompatibility and Analytical Performance of Intravascular Amperometric Oxygen Sensors Prepared with Improved Nitric Oxide-Releasing Silicone Rubber Coating," *Anal. Chem.* **2002,** *74,* 5942.

**11.** S.-K. Jung, W. Gorski, C. A. Aspinwall, L. M. Kauri, and R. T. Kennedy, "Oxygen Microsensor and Its Application to Single Cells and Mouse Pancreatic Islets," *Anal. Chem.* **1999,** *71,* 3642.

**12.** When current is *applied* to a $Y_2O_3$-doped $ZrO_2$ oxygen electrode, the electrode becomes a coulometric *generator* of $O_2$ and has been used for oxidative titrations at 600°C. [K. Hirakata, W. E. Rhine, and M. J. Cima, "Carbon Determination in Small Amounts of Ceramic Powder by Coulombic Titration," *J. Am. Ceram. Soc.* **1995,** *78,* 2834.] The solid electrode material can be used for $O_2$ filtration, $O_2$ generation, or $O_2$ removal. [S. P. S. Badwal and F. T. Ciachhi, "Ceramic Membrane Technologies for Oxygen Separation," *Adv. Mater.* **2001,** *13,* 993.]

**13.** J. Yinon, "Detection of Explosives by Electronic Noses," *Anal. Chem.* **2003,** *75,* 99A; M. C. C. Oliveros, J. L. P. Pavón, C. G. Pinto, M. E. F. Laespada, B. M. Cordero, and M. Forina, "Electronic Nose Based on Metal Oxide Semiconductor Sensors as a Fast Alternative for the Detection of Adulteration of Virgin Olive Oils," *Anal. Chim. Acta* **2002,** *459,* 219; C. L. Honeybourne, "Organic Vapor Sensors for Food Quality Assessment," *J. Chem. Ed.* **2000,** *77,* 338; E. Zubritsky, "E-Noses Keep an Eye on the Future," *Anal. Chem.* **2000,** *72,*

421A; D. J. Strike, M. G. H. Meijerink, and M. Koudelka-Hep, "Electronic Noses—A Mini-Review," *Fresenius J. Anal. Chem.* **1999,** *364,* 499.

**14.** N. Mano and A. Heller, "Detection of Glucose at 2 fM Concentration," *Anal. Chem.* **2005,** *77,* 729.

**15.** J. Nikolic, E. Expósito, J. Iniesta, J. González-Garcia, and V. Montiel, "Theoretical Concepts and Applications of a Rotating Disk Electrode," *J. Chem. Ed.* **2000,** *77,* 1191.

**16.** J. Lagrange and P. Lagrange, "Voltammetric Method for the Determination of $H_2O_2$ in Rainwater," *Fresenius J. Anal. Chem.* **1991,** *339,* 452.

**17.** A. J. Bard and C. G. Zoski, "Voltammetry Retrospective," *Anal. Chem.* **2000,** *72,* 346A; A. M. Bond, *Broadening Electrochemical Horizons* (Oxford: Oxford University Press, 2002).

**18.** To clean a mercury spill, consolidate the droplets with a piece of cardboard. Then suck the mercury into a filter flask with an aspirator. A disposable Pasteur pipet attached to a hose makes a good vacuum cleaner. To remove residual mercury, sprinkle elemental zinc powder on the surface and dampen the powder with 5% aqueous $H_2SO_4$ to make a paste. Mercury dissolves in the zinc. After working the paste into contaminated areas with a sponge or brush, allow the paste to dry and sweep it up. Discard the powder as contaminated mercury waste. This procedure is better than sprinkling sulfur on the spill. Sulfur coats the mercury but does not react with the bulk of the droplet [D. N. Easton, "Management and Control of Hg Exposure," *Am. Lab.,* July 1988, p. 66].

**19.** Boron-doped chemical-vapor-deposited diamond is an exceptionally inert carbon electrode with a very wide potential window and very low voltammetric background current. [A. E. Fischer, Y. Show, and G. M. Swain, "Electrochemical Performance of Diamond Thin-Film Electrodes from Different Commercial Sources," *Anal. Chem.* **2004,** *76,* 2553.]

**20.** S. Ferro and A. De Battisti, "The 5-V Window of Polarizability of Fluorinated Diamond Electrodes in Aqueous Solution," *Anal. Chem.* **2003,** *75,* 7040.

**21.** To remove traces of $O_2$ from $N_2$ for voltammetry, the gas is bubbled through two consecutive columns of liquid. The first column removes $O_2$ by reaction with $V^{2+}$, and the second saturates the gas stream with water at the same vapor pressure as that found in the voltammetry cell. The second column contains the same supporting electrolyte solution used for voltammetry. The first column is prepared by boiling 2 g of $NH_4VO_3$ (ammonium metavanadate) with 25 mL of 12 M HCl and reducing to $V^{2+}$ with zinc amalgam. (Amalgam is prepared by covering granulated Zn with 2 wt% $HgCl_2$ solution and stirring for 10 min to reduce $Hg^{2+}$ to Hg, which coats the Zn. The liquid is decanted, and the amalgam is washed three times with water by decantation. Amalgamation increases the overpotential for $H^+$ reduction at the Zn surface, so the Zn is not wasted by reaction with acid.) The blue or green oxidized vanadium solution turns violet upon reduction. When the violet color is exhausted through use, it can be regenerated by adding more zinc amalgam or HCl or both. Two $V^{2+}$ bubble tubes can be used in series (in addition to a third tube with supporting electrolyte). When $V^{2+}$ in the first tube is expended, the second tube is still effective.

**22.** J. G. Osteryoung and R. A. Osteryoung, "Square Wave Voltammetry," *Anal. Chem.* **1985,** *57,* 101A; J. G. Osteryoung, "Voltammetry for the Future," *Acc. Chem. Res.* **1993,** *26,* 77.

**23.** J. Wang, *Stripping Analysis: Principles, Instrumentation and Applications* (Deerfield Beach, FL: VCH, 1984); Kh. Z. Brainina, N. A. Malakhova, and N. Yu. Stojko, "Stripping Voltammetry in Environmental and Food Analysis," *Fresenius J. Anal. Chem.* **2000,** *368,* 307.

**24.** A. J. Saterlay and R. G. Compton, "Sonoelectroanalysis—An Overview," *Fresenius J. Anal. Chem.* **2000,** *367,* 308; A. O. Sim, C. E. Banks, and R. G. Compton, "Sonically Assisted Electroanalytial Detection of Ultratrace Arsenic," *Anal. Chem.* **2004,** *76,* 5051.

**25.** P. Zanello, *Inorganic Electrochemistry: Theory, Practice and Application* (Cambridge: Royal Society of Chemistry, 2003); G. A. Mabbott, "An Introduction to Cyclic Voltammetry," *J. Chem. Ed.* **1983,** *60,* 697; P. T. Kissinger and W. R. Heineman, "Cyclic Voltammetry," *J. Chem. Ed.* **1983,** *60,* 702; D. H. Evans, K. M. O'Connell, R. A. Petersen, and M. J. Kelly, "Cyclic Voltammetry," *J. Chem. Ed.* **1983,** *60,* 290; H. H. Thorp, "Electrochemistry of Proton-Coupled Redox Reactions," *J. Chem. Ed.* **1992,** *69,* 251.

**26.** J. J. Watkins, B. Zhang, and H. S. White, "Electrochemistry at Nanometer-Scaled Electrodes," *J. Chem. Ed.* **2005,** *82,* 713; R. J. Forster, "Microelectrodes:

New Dimensions in Electrochemistry," *Chem. Soc. Rev.* **1994,** 289; S. Ching, R. Dudek, and E. Tabet, "Cyclic Voltammetry with Ultramicroelectrodes," *J. Chem. Ed.* **1994,** *71,* 602; E. Howard and J. Cassidy, "Analysis with Microelectrodes Using Microsoft Excel Solver," *J. Chem. Ed.* **2000,** *77,* 409.

**27.** T. K. Chen, Y. Y. Lau, D. K. Y. Wong, and A. G. Ewing, "Pulse Voltammetry in Single Cells Using Platinum Microelectrodes," *Anal. Chem.* **1992,** *64,* 1264.

**28.** A. J. Cunningham and J. B. Justice, Jr., "Approaches to Voltammetric and Chromatographic Monitoring of Neurochemicals in Vivo," *J. Chem. Ed.* **1987,** *64,* A34. Another Nafion-coated electrode can detect $10^{-20}$ mol of the neurotransmitter nitric oxide within a single cell [T. Malinski and Z. Taha, "Nitric Oxide Release from a Single Cell Measured *in Situ* by a Porphyrinic-Based Microsensor," *Nature* **1992,** *358,* 676].

**29.** A. J. Bard and M. V. Mirkin, eds., *Scanning Electrochemical Microscopy* (New York: Marcel Dekker, 2001).

**30.** S. K. MacLeod, "Moisture Determination Using Karl Fischer Titrations," *Anal. Chem.* **1991,** *63,* 557A.

**31.** S. Grünke and G. Wünsch, "Kinetics and Stoichiometry in the Karl Fischer Solution," *Fresenius J. Anal. Chem.* **2000,** *368,* 139.

**32.** A. Cedergren and S. Jonsson, "Progress in Karl Fischer Coulometry Using Diaphragm-Free Cells," *Anal. Chem.* **2001,** *73,* 5611.

**33.** S. A. Margolis and J. B. Angelo, "Interlaboratory Assessment of Measurement Precision and Bias in the Coulometric Karl Fischer Determination of Water," *Anal. Bioanal. Chem.* **2002,** *374,* 505.

**34.** C. M. Sánchez-Sánchez, E. Expósito, A. Frías-Ferrer, J. González-García, V. Montiel, and A. Aldaz, "Chlor-Alkali Industry: A Laboratory Scale Approach," *J. Chem. Ed.* **2004,** *81,* 698; D. J. Wink, "The Conversion of Chemical Energy," *J. Chem. Ed.* **1992,** *69,* 108; S. Venkatesh and B. V. Tilak, "Chlor-Alkali Technology," *J. Chem. Ed.* **1983,** *60,* 276. Nafion is a trademark of Du Pont Co.

**35.** R. S. Treptow, "The Lead-Acid Battery: Its Voltage in Theory and Practice," *J. Chem. Ed.* **2002,** *79,* 334. Includes *activity coefficients* of electrolyte in the battery.

**36.** D. Lowinsohn and M. Bertotti, "Coulometric Titrations in Wine Samples: Determination of S(IV) and the Formation of Adducts," *J. Chem. Ed.* **2002,** *79,* 103. Some species in wine in addition to sulfite react with $I_3^-$. A blank titration to correct for such reactions is described in this article.

**37.** M. E. Gomez and A. E. Kaifer, "Voltammetric Behavior of a Ferrocene Derivative." *J. Chem. Ed.* **1992,** *69,* 502.

## Chapter 18

**1.** R. S. Stolarski, "The Antarctic Ozone Hole," *Scientific American,* January 1988. The 1995 Nobel Prize in Chemistry was shared by Paul Crutzen, Mario Molina, and F. Sherwood Rowland for "their work in atmospheric chemistry, particularly concerning the formation and decomposition of ozone." Their Nobel lectures can be found in P. J. Crutzen, "My Life with $O_3$, $NO_x$, and Other $YZO_x$ Compounds," *Angew. Chem. Int. Ed. Engl.* **1996,** *35,* 1759; M. J. Molina, "Polar Ozone Depletion," ibid., 1779; F. S. Rowland, "Stratospheric Ozone Depletion by Chlorofluorocarbons," ibid., 1787.

**2.** J. H. Butler, M. Battle, M. L. Bender, S. A. Montzka, A. D. Clarke, E. S. Saltzman, C. M. Sucher, J. P. Severinghaus, and J. W. Elkins, "A Record of Atmospheric Halocarbons During the Twentieth Century from Polar Firn Air," *Nature* **1999,** *399,* 749.

**3.** O. B. Toon and R. P. Turco, "Polar Stratospheric Clouds and Ozone Depletion," *Scientific American,* June 1991; A. J. Prenni and M. A. Tolbert, "Studies of Polar Stratospheric Cloud Formation," *Acc. Chem. Res.* **2001,** *234,* 545.

**4.** The most-cited article is by M. Dubois, K. A. Gilles, J. K. Hamilton, P. A. Rebers, and F. Smith, "Colorimetric Method for Determination of Sugars and Related Substances," *Anal. Chem.* **1956,** *28,* 350. [J. Riordon, E. Zubritsky, and A. Newman, "Top 10 Articles," *Anal. Chem.* **2000,** *72,* 324A.]

**5.** Classroom exercise to "derive" Beer's law: R. W. Ricci, M. A. Ditzler, and L. P. Nestor, "Discovering the Beer-Lambert Law," *J. Chem. Ed.* **1994,** *71,* 983. An alternate derivation: W. D. Bare, "A More Pedagogically Sound Treatment of Beer's Law: A Derivation Based on a Corpuscular-Probability Model," *J. Chem. Ed.* **2000,** *77,* 929.

6. D. R. Malinin and J. H. Yoe, "Development of the Laws of Colorimetry: A Historical Sketch," *J. Chem. Ed.* **1961**, *38,* 129. The equation that we call "Beer's law" embodies contributions by P. Bouguer (1698–1758), J. H. Lambert (1728–1777), and A. Beer (1825–1863). Beer published his work in 1852, and similar conclusions were independently reached and published within a few months by F. Bernard.

7. D. H. Alman and F. W. Billmeyer, Jr., "A Simple System for Demonstrations in Spectroscopy," *J. Chem. Ed.* **1976**, *53,* 166. For another approach, see F. H. Juergens, "Spectroscopy in Large Lecture Halls," *J. Chem. Ed.* **1988**, *65,* 266.

8. For Beer's law, "monochromatic" means that the bandwidth of the light must be substantially smaller than the width of the absorption band in the spectrum of the chromophore [W. E. Wentworth, "Dependence of the Beer-Lambert Absorption Law on Monochromatic Radiation," *J. Chem. Ed.* **1966**, *43,* 262].

9. K. S. Patel, A. Shukla, A. Goswami, S. K. Chandavanshi, and P. Hoffmann, "A New Spectrophotometric Method for the Determination of Total and Ferric Iron in Rainwater at the ppb Level," *Fresenius J. Anal. Chem.* **2001**, *369,* 530.

10. D. C. Harris, "Serum Iron Determination: A Sensitive Colorimetric Experiment," *J. Chem. Ed.* **1978**, *55,* 539.

11. A pictorial description of the dynamics of the $n \rightarrow \pi^*$ and $\pi \rightarrow \pi^*$ transitions is given by G. Henderson, "A New Look at Carbonyl Electronic Transitions," *J. Chem. Ed.* **1990**, *67,* 392.

12. J. W. Bozzelli, "A Fluorescence Lecture Demonstration," *J. Chem. Ed.* **1982**, *59,* 787; G. L. Goe, "A Phosphorescence Demonstration," *J. Chem. Ed.* **1972**, *49,* 412; E. M. Schulman, "Room Temperature Phosphorescence," *J. Chem. Ed.* **1976**, *53,* 522; F. B. Bramwell and M. L. Spinner, "Phosphorescence: A Demonstration," *J. Chem. Ed.* **1977**, *54,* 167; S. Roalstad, C. Rue, C. B. LeMaster, and C. Lasko, "A Room-Temperature Emission Lifetime Experiment for the Physical Chemistry Laboratory," *J. Chem. Ed.* **1997**, *74,* 853.

13. C. Zander, J. Enderlein, and R. Keller, eds., *Single Molecule Detection in Solution* (New York: Wiley, 2002); R. A. Keller, W. P. Ambrose, A. A. Arias, H. Cai, S. R. Emory, P. M. Goodwin, and J. H. Jett, "Analytical Applications of Single-Molecule Detection," *Anal. Chem.* **2002**, *74,* 317A; J. Zimmermann, A. van Dorp, and A. Renn, "Fluorescence Microscopy of Single Molecules," *J. Chem. Ed.* **2004**, *81,* 553; T. A Byassee, W. C. W. Chan, and S. Nie, "Probing Single Molecules in Single Living Cells," *Anal. Chem.* **2000**, *72,* 5606.

14. S. J. Johnsen, H. B. Clausen, W. Dansgaard, K. Fuhrer, N. Gundestrup, C. U. Hammer, P. Iversen, J. Jouzel, B. Stauffer, and J. P. Steffensen, "Irregular Glacial Interstadials Recorded in a New Greenland Ice Core," *Nature* **1992**, *359,* 311; W. Dansgaard, S. J. Johnsen, H. B. Clausen, D. Dahl-Jensen, N. S. Gundestrup, C. U. Hammer, C. S. Hvidberg, J. P. Steffensen, A. E. Sveinbjörnsdottir, J. Jouzel, and G. Bond, "Evidence for General Instability of Past Climate from a 250-kyr Ice-Core Record," *Nature* **1993**, *364,* 218; M. Anklin et al., "Climate Instability During the Last Interglacial Period Recorded in the GRIP Ice Core," *Nature* **1993**, *364,* 203.

15. B. Fanget, O. Devos, and M. Draye, "Correction of Inner Filter Effect in Mirror Coating Cells for Trace Level Fluorescence Measurements," *Anal. Chem.* **2003**, *75,* 2790.

16. D. S. Chatellier and H. B. White III, "What Color Is Egg White? A Biochemical Demonstration of the Formation of a Vitamin-Protein Complex Using Fluorescence Quenching," *J. Chem. Ed.* **1988**, *65,* 814.

17. E. R. Menzel, "Detection of Latent Fingerprints by Laser-Excited Luminescence," *Anal. Chem.* **1989**, *61,* 557A.

18. S. O. Obare and C. J. Murphy, "A Two-Color Fluorescent Lithium Ion Sensor," *Inorg. Chem.* **2001**, *40,* 6080; L. Fabbrizzi, N. Marcotte, F. Stomeo, and A. Taglietti, "Pyrophosphate Detection in Water by Fluorescence Competition Assays," *Angew. Chem. Int. Ed.* **2002**, *41,* 3811.

19. J. B. Rampal, ed., *DNA Arrays: Methods and Protocols* (Totowa, NJ: Humana Press, 2001); D. Gerion, F. Chen, B. Kannan, A. Fu, W. J. Parak, D. J. Chen, A. Majumdar, and A. P. Alivisatos, "Room-Temperature Single-Nucleotide Polymorphism and Multiallele DNA Detection Using Fluorescent Nanocrystals and Microarrays," *Anal. Chem.* **2003**, *75,* 4766.

20. A. M. Garcia-Campana and W. R. G. Baeyens, eds., *Chemiluminescence in Analytical Chemistry* (New York: Marcel Dekker, 2001); L. J. Kricka, "Clinical Applications of Chemiluminescence," *Anal. Chim. Acta* **2003**, *500,* 279.

21. C. Salter, K. Range, and G. Salter, "Laser-Induced Fluorescence of Lightsticks," *J. Chem. Ed.* **1999**, *76,* 84; E. Wilson, "Light Sticks," *Chem. Eng. News,* 18 January 1999, p. 65.

22. R. Rizzuto, A. W. M. Simpson, M. Brini, and T. Pozzan, "Rapid Changes of Mitochondrial $Ca^{2+}$ Revealed by Specifically Targeted Recombinant Aequorin," *Nature* **1992**, *358,* 325; A. Toda, P. Pasini, M. Guardigli, M. Baraldini, M. Musiani, and M. Mirasoli, "Bio- and Chemiluminescence in Bioanalysis," *Fresenius J. Anal. Chem.* **2000**, *366,* 752; M. L. Grayeski, "Chemiluminescence Analysis," *Anal. Chem.* **1987**, *59,* 1243A.

23. J. K. Robinson, M. J. Bollinger, and J. W. Birks, "Luminol/$H_2O_2$ Chemiluminescence Detector for the Analysis of NO in Exhaled Breath," *Anal. Chem.* **1999**, *71,* 5131. Many substances can be analyzed by coupling their chemistry to luminol oxidation. See, for example, O. V. Zui and J. W. Birks, "Trace Analysis of Phosphorus in Water by Sorption Preconcentration and Luminol Chemiluminescence," *Anal. Chem.* **2000**, *72,* 1699.

24. J. C. Lewis and S. Daunert, "Bioluminescence Immunoassay for Thyroxine Employing Genetically Engineered Mutant Aequorins Containing Unique Cysteine Residues," *Anal. Chem.* **2001**, *732,* 3227.

## Chapter 19

1. L. Stryer, "Fluorescence Energy Transfer as a Spectroscopic Ruler," *Annu. Rev. Biochem.* **1978**, *47,* 819; C. Berney and G. Danuser, "FRET or No FRET: A Quantitative Comparison," *Biophys. J.* **2003**, *84,* 3992; http://www.probes.com/handbook/.

2. T. R. Martz, J. J. Carr, C. R. French, and M. D. DeGrandpre, "A Submersible Autonomous Sensor for Spectrophotometric pH Measurements of Natural Waters," *Anal. Chem.* **2003**, *75,* 1844; W. Yao and R. H. Byrne, "Spectrophotometric Determination of Freshwater pH Using Bromocresol Purple and Phenol Red," *Environ. Sci. Technol.* **2001**, *35,* 1197; H. Yamazaki, R. P. Sperline, and H. Freiser, "Spectrophotometric Determination of pH and Its Applications to Determination of Thermodynamic Equilibrium Constants," *Anal. Chem.* **1992**, *64,* 2720.

3. To multiply a matrix times a vector, multiply each row of the matrix times each element of the vector as follows:

$$\begin{bmatrix} a_1 & a_2 \\ b_1 & b_2 \end{bmatrix}\begin{bmatrix} X \\ Y \end{bmatrix} = \begin{bmatrix} a_1X + a_2Y \\ b_1X + b_2Y \end{bmatrix}$$

The product is a vector. The product of a matrix times a matrix is another matrix obtained by multiplying rows times columns:

$$\begin{bmatrix} a_1 & a_2 \\ b_1 & b_2 \end{bmatrix}\begin{bmatrix} c_1 & c_2 \\ d_1 & d_2 \end{bmatrix} = \begin{bmatrix} row\ 1 \times column\ 1 & row\ 1 \times column\ 2 \\ row\ 2 \times column\ 1 & row\ 2 \times column\ 2 \end{bmatrix}$$

$$= \begin{bmatrix} a_1c_1 + a_2d_1 & a_1c_2 + a_2d_2 \\ b_1c_1 + b_2d_1 & b_1c_2 + b_2d_2 \end{bmatrix}$$

The matrix **B** below is the inverse of **A** because their product is the unit matrix:

$$\underset{\mathbf{A}}{\begin{bmatrix} 1 & 2 \\ 3 & 4 \end{bmatrix}}\underset{\mathbf{B}}{\begin{bmatrix} -2 & 1 \\ \frac{3}{2} & -\frac{1}{2} \end{bmatrix}} = \begin{bmatrix} 1 \cdot -2 + 2 \cdot \frac{3}{2} & 1 \cdot 1 + 2 \cdot -\frac{1}{2} \\ 3 \cdot -2 + 4 \cdot \frac{3}{2} & 3 \cdot 1 + 4 \cdot -\frac{1}{2} \end{bmatrix} = \underset{\text{Unit matrix}}{\begin{bmatrix} 1 & 0 \\ 0 & 1 \end{bmatrix}}$$

4. Under certain conditions, it is possible for a solution with more than two principal species to exhibit an isosbestic point. See D. V. Stynes, "Misinterpretation of Isosbestic Points: Ambient Properties of Imidazole," *Inorg. Chem.* **1975**, *14,* 453.

5. G. Scatchard, *Ann. N. Y. Acad. Sci.* **1949**, *51,* 660.

6. D. A. Deranleau, "Theory of the Measurement of Weak Molecular Complexes," *J. Am. Chem. Soc.* **1969**, *91,* 4044.

7. E. Bruneau, D. Lavabre, G. Levy, and J. C. Micheau, "Quantitative Analysis of Continuous-Variation Plots with a Comparison of Several Methods," *J. Chem. Ed.* **1992**, *69,* 833; V. M. S. Gil and N. C. Oliveira, "On the Use of the Method of Continuous Variation," *J. Chem. Ed.* **1990**, *67,* 473; Z. D. Hill and P. MacCarthy, "Novel Approach to Job's Method," *J. Chem. Ed.* **1986**, *63,* 162.

8. M. Trojanowicz, *Flow Injection Analysis* (River Edge, NJ: World Scientific Publishing, 2000); M. Valcárcel and M. D. Luque de Castro, *Flow Injection Analysis* (Chichester, UK: Ellis Horwood, 1987); J. Ruzicka and E. H. Hansen,

"Flow Injection Analysis: From Beaker to Microfluidics," *Anal. Chem.* **2000,** *72,* 212A; J. Ruzicka and L. Scampavia, "From Flow Injection to Bead Injection," *Anal. Chem.* **1999,** *71,* 257A.

9. A. A. Ensafi and G. B. Dehaghi, "Ultra-Trace Analysis of Nitrite in Food Samples by Flow Injection with Spectrophotometric Detection," *Fresenius J. Anal. Chem.* **1999,** *363,* 131.

10. D. Thouron, R. Vuillemin, X. Philippon, A. Lourenço, C. Provost, A. Cruzado, and V. Garçon, "An Autonomous Nutrient Analyzer for Oceanic Long-Term in Situ Biogeochemical Monitoring," *Anal. Chem.* **2003,** *75,* 2601.

11. H. Mana and U. Spohn, "Sensitive and Selective Flow Injection Analysis of Hydrogen Sulfite/Sulfur Dioxide by Fluorescence Detection with and without Membrane Separation by Gas Diffusion, *Anal. Chem.* **2001,** *73,* 3187.

12. J. Zheng, S. R. Springston, and J. Weinstein-Lloyd, "Quantitative Analysis of Hydroperoxyl Radical Using Flow Injection Analysis with Chemiluminescence Detection," *Anal. Chem.* **2003,** *75,* 2601; A. R. Bowie, E. P. Achterberg, P. N. Sedwick, S. Ussher, and P. J. Worsfold, "Real-Time Monitoring of Picomolar Concentrations of Iron(II) in Marine Waters Using Automated Flow Injection-Chemiluminescence Instrumentation, *Environ. Sci. Technol.* **2002,** *36,* 4600.

13. R. S. Yalow, "Development and Proliferation of Radioimmunoassay Technology," *J. Chem. Ed.* **1999,** *76,* 767; R. P. Ekins, "Immunoassay, DNA Analysis, and Other Ligand Binding Assay Techniques" *J. Chem. Ed.* **1999,** *76,* 769; E. F. Ullman, "Homogeneous Immunoassays," *J. Chem. Ed.* **1999,** *76,* 781; E. Straus, "Radioimmunoassay of Gastrointestinal Hormones," *J. Chem. Ed.* **1999,** *76,* 788.

14. S. J. Gee, B. D. Hammock, and J. V. Van Emon, *Environmental Immunochemical Analysis for Detection of Pesticides and Other Chemicals: A User's Guide* (Westwood, NJ: Noyes, 1997); E. M. Brun, M. Garcés-García, E. Escuín, S. Morais, R. T. Puchades, and A. Maquieira, "Assessment of Novel Diazinon Immunoasays for Water Analysis," *Environ. Sci. Technol.* **2004,** *38,* 1115; J. Zeravik, K. Skryjová, Z. Nevoranková, and M. Fránek, "Development of Direct ELISA for the Determination of 4-Nonylphenol and Octylphenol," *Anal. Chem.* **2004,** *76,* 1021.

15. E. P. Diamandis and T. K. Christopoulos, "Europium Chelate Labels in Time-Resolved Fluorescence Immunoassays and DNA Hybridization Assays," *Anal. Chem.* **1990,** *62,* 1149A.

16. R. Mukhopadhyay, "Aptamers Are Ready for the Spotlight," *Anal. Chem.* **2005,** *77,* 114A; S. D. Mendonsa and M. T. Bowser, "In Vitro Selection of High-Affinity DNA Ligands for Human IgE Using Capillary Electrophoresis," *Anal. Chem.* **2004,** *76,* 5387.

17. G. J. Meyer, "Efficient Light-to-Electrical Energy Conversion: Nanocrystalline $TiO_2$ Films Modified with Inorganic Sensitizers," *J. Chem. Ed.* **1997,** *74,* 652. For a student experiment to construct a photocell, see G. P. Smestad and M. Grätzel, "Demonstrating Electron Transfer and Nanotechnology: A Natural Dye-Sensitized Nanocrystalline Energy Converter," *J. Chem. Ed.* **1998,** *75,* 752.

18. http://www.greatcell.com

19. K. A. Kneas, W. Xu, J. N. Demas, and B. A. DeGraff, "Dramatic Demonstration of Oxygen Sensing by Luminescence Quenching," *J. Chem. Ed.* **1997,** *74,* 696.

20. Y.-E. L. Koo, Y. Cao, R. Kopelman, S. M. Koo, M. Brasuel, and M. A. Philbert, "Real-Time Measurements of Dissolved Oxygen Inside Living Cells by Organically Modified Silicate Fluorescent Nanosensors," *Anal. Chem.* **2004,** *76,* 2498.

21. M. D. DeGrandpre, M. M. Baehr, and T. R. Hammar, "Calibration-Free Optical Chemical Sensors," *Anal. Chem.* **1999,** *71,* 1152.

22. J. P. Whelan, A. W. Kusterbeck, G. A. Wemhoff, R. Bredehorst, and F. S. Ligler, "Continuous-Flow Immunosensor for Detection of Explosives," *Anal. Chem.* **1993,** *65,* 3561; U. Narang, P. R. Gauger, and F. S. Ligler, "A Displacement Flow Immunosensor for Explosive Detection Using Microcapillaries," *Anal. Chem.* **1997,** *69,* 2779.

23. A Poisson distribution is valid when (a) all possible outcomes are random and independent of one another, (b) the maximum possible value of $n$ is a large number, and (c) the average value of $n$ is a small fraction of the maximum possible value.

## Chapter 20

1. K. L. Bechtel, R. N. Zare, A. A. Kachanov, S. S. Sanders, and B. A. Paldus, "Cavity Ring-Down Spectroscopy for HPLC," *Anal. Chem.* **2005,** *77,* 1177; B. Bahnev, L. van der Sneppen, A. E. Wiskerke, F. Ariese, C. Gooijer, and W. Ubachs, "Miniaturized Cavity Ring-Down Detection in a Liquid Flow Cell," *Anal. Chem.* **2005,** *77,* 1188.

2. A. Bergh, G. Craford, A. Duggal, and R. Haitz, "The Promise and Challenge of Solid-State Lighting," *Physics Today,* December 2001, p. 42; H. Sevian, S. Müller, H. Rudmann, and M. F. Rubner, "Using Organic Light-Emitting Electrochemical Thin-Film Devices to Teach Materials Science," *J. Chem. Ed.* **2004,** *81,* 1620.

3. J. M. Kauffman, "Water in the Atmosphere," *J. Chem. Ed.* **2004,** *81,* 1229.

4. S. K. Lower, "Thermal Physics (and Some Chemistry) of the Atmosphere," *J. Chem. Ed.* **1998,** *75,* 837; W. C. Trogler, "Environmental Chemistry of Trace Atmospheric Gases," *J. Chem. Ed.* **1995,** *72,* 973.

5. B. Hileman, "Climate Change," *Chem. Eng. News,* 15 December 2003, p. 27; E. Raschke, "Is the Additional Greenhouse Effect Already Evident in the Current Climate?" *Fresenius J. Anal. Chem.* **2001,** *371,* 791.

6. M. G. D. Baumann, J. C. Wright, A. B. Ellis, T. Kuech, and G. C. Lisensky, "Diode Lasers," *J. Chem. Ed.* **1992,** *69,* 89; T. Imasaka and N. Ishibashi, "Diode Lasers and Practical Trace Analysis," *Anal. Chem.* **1990,** *62,* 363A.

7. W. E. L. Grossman, "The Optical Characteristics and Production of Diffraction Gratings," *J. Chem. Ed.* **1993,** *70,* 741.

8. G. C.-Y. Chan and W. T. Chan, "Beer's Law Measurements Using Non-Monochromatic Light Sources—A Computer Simulation," *J. Chem. Ed.* **2001,** *78,* 1285.

9. J. C. Travis et al., "Intrinsic Wavelength Standard Absorption Bands in Holmium Oxide Solution for UV/Visible Molecular Absorption Spectrophotometry," *J. Phys. Chem. Ref. Data* **2005,** *34,* 41.

10. K. Palczewski et al., "Crystal Structure of Rhodopsin," *Science* **2000,** *289,* 739.

11. J. M. Harnly and R. E. Fields, "Solid-State Array Detectors for Analytical Spectrometry," *Appl. Spectros.* **1997,** *51,* 334A; Q. S. Hanley, C. W. Earle, F. M. Pennebaker, S. P. Madden, and M. B. Denton, "Charge-Transfer Devices in Analytical Instrumentation," *Anal. Chem.* **1996,** *68,* 661A; J. V. Sweedler, K. L. Ratzlaff, and M. B. Denton, eds., *Charge Transfer Devices in Spectroscopy* (New York: VCH, 1994).

12. S. K. van Bergen, I. B. Bakaltcheva, J. S. Lundgren, and L. C. Shriver-Lake, "On-Site Detection of Explosives in Groundwater with a Fiber Optic Biosensor," *Environ. Sci. Technol.* **2000,** *34,* 704; P. T. Charles, P. R. Gauger, C. H. Patterson, Jr., and A. W. Kusterbeck, "On-Site Immunoanalysis of Nitrate and Nitroaromatic Compounds in Groundwater," *Environ. Sci. Technol.* **2000,** *34,* 4641; W. Tan, R. Kopenman, S. L. R. Barker, and M. T. Miller, "Ultrasmall Optical Sensors for Cellular Measurements," *Anal. Chem.* **1999,** *71,* 606A; S. L. R. Barker, Y. Zhao, M. A. Marletta, and R. Kopelman, "Cellular Applications of a Fiber-Optic Biosensor Based on a Dye-Labeled Guanylate Cyclase," *Anal. Chem.* **1999,** *71,* 2071; M. Kuratli and E. Pretsch, "$SO_2$-Selective Optodes," *Anal. Chem.* **1994,** *66,* 85.

13. Z. Rosenzweig and R. Kopelman, "Analytical Properties and Sensor Size Effects of a Micrometer-Sized Optical Fiber Glucose Biosensor," *Anal. Chem.* **1996,** *68,* 1408.

14. C. Preininger, I. Klimant, and O. S. Wolfbeis, "Optical Fiber Sensor for Biological Oxygen Demand," *Anal. Chem.* **1994,** *66,* 1841.

15. I. Biran and D. R. Walt, "Optical Imaging Fiber-Based Single Live Cell Arrays," *Anal. Chem.* **2002,** *74,* 3046.

16. J. T. Bradshawl, S. B. Mendes, and S. S. Saavedra, "Planar Integrated Optical Waveguide Spectroscopy," *Anal. Chem.* **2005,** *77,* 28A.

17. J. M. Brockman, B. P. Nelson, and R. M. Corn, "Surface Plasmon Resonance Imaging Measurements of Ultrathin Organic Films," *Annu. Rev. Phys. Chem.* **2000,** *51,* 41; J. Homola, "Present and Future of Surface Plasmon Resonance Biosensors," *Anal. Bioanal. Chem.* **2003,** *377,* 528; H. Q. Zhang, S. Boussaad, and N. J. Tao, "High-Performance Differential Surface Plasmon Resonance Sensor Using Quadrant Cell Photodetector," *Rev. Sci. Instrum.* **2003,** *74,* 150; M. Suzuki, F. Ozawa, W. Sugimoto, and S. Aso, "Miniature Surface-Plasmon Resonance Immunosensors—Rapid and Repetitive Procedure," *Anal. Bioanal. Chem.* **2002,** *372,* 301.

18. P. R. Griffiths and J. A. de Haseth, *Fourier Transform Infrared Spectrometry* (New York: Wiley, 1986), and W. D. Perkins, "Fourier Transform-Infrared Spectroscopy," *J. Chem. Ed.* **1986,** *63,* A5; **1987,** *64,* A269, A296.

19. There are many excellent digital and electronic techniques for improving the quality of a spectrum without having to average numerous scans. See T. C. O'Haver, "An Introduction to Signal Processing in Chemical Measurement," *J. Chem. Ed.* **1991,** *68,* A147; M. G. Prais, "Spreadsheet Exercises for Instrumental Analysis," *J. Chem. Ed.* **1992,** *69,* 488; R. Q. Thompson, "Experiments in Software Data Handling," *J. Chem. Ed.* **1985,** *62,* 866; M. P. Eastman, G. Kostal, and T. Mayhew, "An Introduction to Fast Fourier Transforms Through the Study of Oscillating Reactions," *J. Chem. Ed.* **1986,** *63,* 453; and B. H. Vassos and L. López, "Signal-to-Noise Improvement," *J. Chem. Ed.* **1985,** *62,* 542.

20. G. M. Hieftje, "Signal-to-Noise Enhancement Through Instrumental Techniques," *Anal. Chem.* **1972,** *44,* 81A [No. 6], 69A [No. 7]; D. C. Tardy, "Signal Averaging," *J. Chem. Ed.* **1986,** *63,* 648. See also T. Kaneta, "Hadamard Transform CE," *Anal. Chem.* **2001,** *73,* 540A.

21. N. N. Sesi, M. W. Borer, T. K. Starn, and G. M. Hieftje, "A Standard Approach to Collecting and Calculating Noise Amplitude Spectra," *J. Chem. Ed.* **1998,** *75,* 788.

22. M. L. Salit and G. C. Turk, "A Drift Correction Procedure," *Anal. Chem.* **1998,** *70,* 3184.

## Chapter 21

1. E. R. Denoyer, K. J. Fredeen, and J. W. Hager, "Laser Solid Sampling for Inductively Coupled Plasma Mass Spectrometry," *Anal. Chem.* **1991,** *63,* 445A; K. Niemax, "Laser Ablation—Reflections on a Very Complex Technique for Solid Sampling," *Fresenius J. Anal. Chem.* **2001,** *370,* 332. A major challenge for laser ablation–mass spectrometry is quantitative analysis. One scheme achieves semiquantitative analysis without standards by comparing the signal from each element with the total mass spectrometric signal: A. M. Leach and G. M. Hieftje, "Standardless Semiquantitative Analysis of Metals Using Single-Shot Laser Ablation Inductively Couple Plasma Time-of-Flight Mass Spectrometry," *Anal. Chem.* **2001,** *73,* 2959.

2. L. H. J. Lajunen and P. Perämäki, *Spectrochemical Analysis by Atomic Absorption* (Cambridge: Royal Society of Chemistry, 2004).

3. History: A. Walsh, "The Development of Atomic Absorption Methods of Elemental Analysis 1952–1962," *Anal. Chem.* **1991,** *63,* 933A; B. V. L'vov, "Graphite Furnace Atomic Absorption Spectrometry," *Anal. Chem.* **1991,** *63,* 924A.

4. D. L. Pfeil and A. Reed, "Automating the Digestion and Determination of Mercury in a Variety of Environmental Sample Matrices," *Am. Lab.,* March 2002, p. 26.

5. History: R. F. Jarrell, "A Brief History of Atomic Emission Spectrochemical Analysis, 1666–1950," *J. Chem. Ed.* **2000,** *77,* 573; R. F. Jarrell, F. Brech, and M. J. Gustafson, "A History of Thermo Jarrell Ash Corporation and Spectroscopist Richard F. Jarrell," *J. Chem. Ed.* **2000,** *77,* 592; G. M. Hieftje, "Atomic Emission Spectroscopy—It Lasts and Lasts and Lasts," *J. Chem. Ed.* **2000,** *77,* 577.

6. R. J. Stolzberg, "Optimizing Signal-to-Noise Ratio in Flame Atomic Absorption Using Sequential Simplex Optimization," *J. Chem. Ed.* **1999,** *76,* 834.

7. D. J. Butcher and J. Sneddon, *A Practical Guide to Graphite Furnace Atomic Absorption Spectrometry* (New York: Wiley, 1998).

8. J. B. Voit, "Low-Level Determination of Arsenic in Drinking Water," *Am. Lab. News Ed.,* February 2002, p. 62.

9. M. Hornung and V. Krivan, "Determination of Trace Impurities in Tungsten by Direct Solid Sampling Using a Transversely Heated Graphite Tube," *Anal. Chem.* **1998,** *70,* 3444.

10. U. Schäffer and V. Krivan, "Analysis of High Purity Graphite and Silicon Carbide by Direct Solid Sampling Electrothermal Atomic Absorption Spectrometry," *Fresenius J. Anal. Chem.* **2001,** *371,* 859; R. Nowka and H. Müller, "Direct Analysis of Solid Samples by Graphite Furnace Atomic Absorption Spectrometry," *Fresenius J. Anal. Chem.* **1997,** *359,* 132.

11. J. Y. Cabon, "Influence of Experimental Parameters on the Determination of Antimony in Seawater by Atomic Absorption Spectrometry," *Anal. Bioanal. Chem.* **2003,** *374,* 1282.

12. D. L. Styris and D. A. Redfield, "Mechanisms of Graphite Furnace Atomization of Aluminum by Molecular Beam Sampling Mass Spectrometry," *Anal. Chem.* **1987,** *59,* 2891.

13. S. Greenfield, "Invention of the Annular Inductively Coupled Plasma as a Spectroscopic Source," *J. Chem. Ed.* **2000,** *77,* 584.

14. V. B. E. Thomsen, G. J. Roberts, and D. A. Tsourides, "Vacuumless Spectrochemistry in the Vacuum Ultraviolet," *Am. Lab.,* August 1997, p. 18H.

15. V. B. E. Thomsen, "Why Do Spectral Lines Have a Linewidth?" *J. Chem. Ed.* **1995,** *72,* 616.

16. A better measure of noise is the root-mean-square noise (Equation 20-14), which is ~5 times less than the peak-to-peak noise. Therefore 2 times the peak-to-peak noise level is ~10 times the root-mean-square noise. The detection limit that is 2 times the peak-to-peak noise level is close to the limit of quantitation in Equation 5-6. The lesson is that you should define how you express a detection limit when you report one.

17. R. Thomas, *Practical Guide to ICP-MS* (New York: Marcel Dekker, 2004); H. E. Taylor, *Inductively Coupled Plasma-Mass Spectrometry* (San Diego: Academic Press, 2001); C. M. Barshick, D. C. Duckworth, and D. H. Smith, eds., *Inorganic Mass Spectrometry* (New York: Marcel Dekker, 2000); S. J. Hill, ed., *Inductively Coupled Plasma Spectrometry and Its Applications* (Sheffield, UK: Sheffield Academic Press, 1999); A. Montaser, ed., *Inductively Coupled Plasma Mass Spectrometry* (New York: Wiley, 1998); G. Holland and S. D. Tanner, eds., *Plasma Source Mass Spectrometry* (Cambridge: Royal Society of Chemistry, 1997, 1999).

18. M. J. Felton, "Plasma Opens New Doors in Isotope Ratio MS," *Anal. Chem.* **2003,** *75,* 119A.

19. L. Moens and N. Jakubowski, "Double-Focusing Mass Spectrometers in ICPMS," *Anal. Chem.* **1998,** *70,* 251A; F. A. M. Planchon, C. F. Boutron, C. Barbante, E. W. Wolff, G. Cozzi, V. Gaspari, C. P. Ferrari, and P. Cescon, "Ultrasensitive Determination of Heavy Metals at the Sub-pg/g Level in Ultraclean Antarctic Snow Samples by Inductively Coupled Plasma Sector Field Mass Spectrometry," *Anal. Chem. Acta* **2001,** *450,* 193.

20. B. Hattendorf, C. Latkoczy, and D. Günther, "Laser Ablation–ICPMS," *Anal. Chem.* **2003,** *75,* 341A; R. E. Russo, X. Mao, and S. S. Mao, "The Physics of Laser Ablation in Microchemical Analysis," *Anal. Chem.* **2002,** *74,* 71A.

21. V. Cheam, G. Lawson, I. Lechner, and R. Desrosiers, "Recent Metal Pollution in Agassiz Ice Cap," *Environ. Sci. Technol.* **1998,** *32,* 3974.

22. A. Bazzi, B. Kreuz, and J. Fischer, "Determination of Calcium in Cereal with Flame Atomic Absorption Spectroscopy," *J. Chem. Ed.* **2004,** *81,* 1042.

23. J. J. Rosentreter and R. K. Skogerboe, "Trace Determination and Speciation of Cyanide Ion by Atomic Absorption Spectroscopy," *Anal. Chem.* **1991,** *63,* 682.

24. L. Perring and M. Basic-Dvorzak, "Determination of Total Tin in Canned Food Using Inductively Coupled Plasma Atomic Emission Spectroscopy," *Anal. Bioanal. Chem.* **2002,** *374,* 235.

## Chapter 22

1. W. J. Showers, R. Barrick, and B. Genna, "Isotopic Analysis of Dinosaur Bones," *Anal. Chem.* **2002,** *74,* 142A.

2. F. Klink, *Introduction to Protein and Peptide Analysis with Mass Spectrometry* (Fullerton, CA: Academy Savant, 2004), computer training Program CMSP-10.

3. W.-P. Peng, Y.-C. Yang, M.-W. Kang, Y. T. Lee, and H.-C. Chang, "Measuring Masses of Single Bacterial Whole Cells with a Quadrupole Ion Trap," *J. Am. Chem. Soc.* **2004,** *126,* 11766–11767.

4. D. P. Fergenson et al., "Reagentless Detection and Classification of Individual Bioaerosol Particles in Seconds," *Anal. Chem.* **2004,** *76,* 373; J. J. Jones, M. J. Stump, R. C. Fleming, J. O. Lay, Jr., and C. L. Wilkins, "Investigation of MALDI-TOF and FT-MS Techniques of Analysis of *Escherichia coli* Whole Cells," *Anal. Chem.* **2003,** *75,* 1340.

5. J. H. Gross, *Mass Spectrometry: A Textbook* (Berlin: Springer-Verlag, 2004); K. Doward, *Mass Spectrometry: A Foundation Course* (Cambridge: Royal Society of Chemistry, 2004); C. G. Herbert and R. A. W. Johnstone, *Mass Spectrometry Basics* (Boca Raton, FL: CRC Press, 2002); J. Barker, *Mass Spectrometry,* 2nd ed. (Chichester, UK: Wiley, 1999); J. T. Watson, *Introduction to Mass Spectrometry,* 3rd ed. (Philadelphia: Lippincott-Raven, 1997); R. A. W.

Johnstone and M. E. Rose, *Mass Spectrometry for Chemists and Biochemists* (Cambridge: Cambridge University Press, 1996); C. Dass, *Principles and Practice of Biological Mass Spectrometry* (New York: Wiley, 2001).

6. For a demonstration of mass spectrometry, see N. C. Grim and J. L. Sarquis, "Mass Spectrometry Analogy on the Overhead Projector," *J. Chem. Ed.* **1995**, *72*, 930.

7. O. D. Sparkman, "Evaluating Electron Ionization Mass Spectral Library Search Results," *J. Am. Soc. Mass Spectrom.* **1996**, *7*, 313.

8. The NIST/EPA/NIH Mass Spectral Database (SRData@enh.nist.gov) includes more than 100 000 compounds. The Wiley/NIST Registry of MS data is available from Palisade Corp. at http://www.palisade-ms.com. Electron ionization mass spectra of many compounds can be viewed at http://webbook.nist.gov/chemistry.

9. O. Asvany, P. Kumar P, B. Redlich, I. Hegemann, S. Schlemmer, and D. Marx, "Understanding the Infrared Spectrum of Bare $CH_5^+$," *Science* **2005**, *309*, 1219.

10. J. D. Hearn and G. D. Smith, "A Chemical Ionization Mass Spectrometry Method for the Online Analysis of Organic Aerosols," *Anal. Chem.* **2004**, *76*, 2820.

11. R. A. Gross, Jr., "A Mass Spectral Chlorine Rule for Use in Structure Determinations in Sophomore Organic Chemistry," *J. Chem. Ed.* **2004**, *81*, 1161.

12. I. T. Platzner, *Modern Isotope Ratio Mass Spectrometry* (New York: Wiley, 1997).

13. W. Chen and M. V. Orna, "Recent Advances in Archaeological Chemistry," *J. Chem. Ed.* **1996**, *73*, 485.

14. A. M. Pollard and C. Heron, *Archaeological Chemistry* (Cambridge: Royal Society of Chemistry, 1996).

15. J. T. Watson and K. Biemann, "High-Resolution Mass Spectra of Compounds Emerging from a Gas Chromatograph," *Anal. Chem.* **1964**, *36*, 1135. A classic paper on gas chromatography/mass spectrometry.

16. R. M. Smith, *Understanding Mass Spectra: A Basic Approach,* 2nd ed. (Hoboken, NJ: Wiley, 2004).

17. P. E. Miller and M. B. Denton, "The Quadrupole Mass Filter: Basic Operating Concepts," *J. Chem. Ed.* **1986**, *63*, 617; M. Henchman and C. Steel, "Design and Operation of a Portable Quadrupole Mass Spectrometer for the Undergraduate Curriculum," *J. Chem. Ed.* **1998**, *75*, 1042; C. Steel and M. Henchman, "Understanding the Quadrupole Mass Filter Through Computer Simulation," *J. Chem. Ed.* **1998**, *75*, 1049; J. J. Leary and R. L. Schidt, "Quadrupole Mass Spectrometers: An Intuitive Look at the Math," *J. Chem. Ed.* **1996**, *73*, 1142.

18. R. J. Cotter, *Time-of-Flight Mass Spectrometry* (Washington, DC: American Chemical Society, 1997); R. J. Cotter, "The New Time-of-Flight Mass Spectrometry," *Anal. Chem.* **1999**, *71*, 445A.

19. Z. Ziegler, "Ion Traps Come of Age," *Anal. Chem.* **2002**, *74*, 489A; C. M. Henry, "The Incredible Shrinking Mass Spectrometers," *Anal. Chem.* **1999**, *71*, 264A; R. G. Cooks and R. E. Kaiser, Jr., "Quadrupole Ion Trap Mass Spectrometry," *Acc. Chem. Res.* **1990**, *23*, 213.

20. G. A. Eiceman and J. A. Stone, "Ion Mobility Spectrometers in National Defense," *Anal. Chem.* **2004**, *76*, 392A; H. H. Hill, W. F. Siems, R. H. St. Louis, and D. G. McMinn, "Ion Mobility Spectrometry," *Anal. Chem.* **1990**, *62*, 1201A; G. A. Eiceman and Z. Karpas, *Ion Mobility Spectrometry,* 2nd ed. (Boca Raton, FL: Taylor and Francis CRC Press, 2005); R. A. Miller, G. A. Eiceman, E. G. Nazarov, and A. T. King, "A Micro-Machined High-Field Asymmetric Waveform-Ion Mobility Spectrometer (FA-IMS)," *Sens. Actuators B Chem.* **2000**, *67*, 300.

21. H.-J. Hübschmann, *Handbook of GC/MS: Fundamentals and Applications* (Weinheim: Wiley-VCH, 2001); M. Oehme, *Practical Introduction to GC-MS Analysis with Quadrupoles* (Heidelberg: Hüthig Verlag, 1998); F. G. Kitson, B. S. Larsen, and C. N. McEwen, *Gas Chromatography and Mass Spectrometry: A Practical Guide* (San Diego: Academic Press, 1996).

22. B. Ardrey, *Liquid Chromatography-Mass Spectrometry: An Introduction* (Chichester, UK: Wiley, 2003); W. M. A. Niessen, *Liquid Chromatography-Mass Spectrometry,* 2nd ed (New York: Marcel Dekker, 1999); M. S. Lee, *LC/MS Applications in Drug Development* (New York: Wiley, 2002); J. Abian, "Historical Feature: The Coupling of Gas and Liquid Chromatography with Mass Spectrometry," *J. Mass Spectrom.* **1999**, *34*, 157.

23. M. M. Vestling, "Using Mass Spectrometry for Proteins," *J. Chem. Ed.* **2002**, *80*, 122; C. M. Henry, "Winning Ways," *Chem. Eng. News*, 18 November 2002, p. 62.

24. R. B. Cole, ed., *Electrospray Ionization Mass Spectrometry: Fundamentals, Instrumentation, and Applications* (New York: Wiley, 1997).

25. S. Liu, W. J. Griffiths, and J. Sjövall, "On-Column Electrochemical Reactions Accompanying the Electrospray Process," *Anal. Chem.* **2003**, *75*, 1022.

26. W. P. Duncan and P. D. Perkins, "LC-MS with Simultaneous Electrospray and Atmospheric Pressure Chemical Ionization," *Am. Lab.*, March 2005, p. 28.

27. C. Fenselau, "MALDI MS and Strategies for Protein Analysis," *Anal. Chem.* **1997**, *69*, 661A; R. W. Nelson, D. Nedelkov, and K. A. Tubbs, "Biomolecular Interaction Analysis Mass Spectrometery," *Anal. Chem.* **2000**, *72*, 405A; J. J. Thomas, R. Bakhtiar, and G. Siuzdak, "Mass Spectrometry in Viral Proteomics," *Acc. Chem. Res.* **2000**, *33*, 179; A. P. Snyder, *Interpreting Protein Mass Spectra* (Washington, DC: American Chemical Society, 2000); S. C. Moyer and R. J. Cotter, "Atmospheric Pressure MALDI," *Anal. Chem.* **2002**, *74*, 469A.

28. S. S. Rubakhin, W. T. Greenough, and J. V. Sweedler, "Spatial Profiling with MALDI MS: Distribution of Neuropeptides Within Single Neurons," *Anal. Chem.* **2003**, *75*, 5374.

29. S. A. Hofstadler, R. Bakhtiar, and R. D. Smith, "Electrospray Ionization Mass Spectrometry: Instrumentation and Spectral Interpretation," *J. Chem. Ed.* **1996**, *73*, A82.

30. R. Bakhtiar, R. Hofstadler, and R. D. Smith, "Electrospray Ionization Mass Spectrometry: Characterization of Peptides and Proteins," *J. Chem. Ed.* **1996**, *73*, A118; C. E. C. A. Hop and R. Bakhtiar, "Electrospray Ionization Mass Spectrometry: Applications in Inorganic Chemistry and Synthetic Polymer Chemistry," *J. Chem. Ed.* **1996**, *73*, A162.

31. S. D. Fuerstenau, W. H. Benner, J. J. Thomas, C. Brugidou, B. Bothner, and G. Siuzdak, "Mass Spectrometry of an Intact Virus," *Angew. Chem. Int. Ed.* **2001**, *40*, 542.

32. Multiply charged ions can be converted into singly charged ions by passage through a corona discharge between the electrospray nozzle and the mass analyzer. With a time-of-flight spectrometer to measure high *m/z*, singly charged ions simplify protein analysis. D. D. Ebeling, M. S. Westphall, M. Scalf, and L. M. Smith, "Corona Discharge in Charge Reduction Electrospray Mass Spectrometry," *Anal. Chem.* **2000**, *72*, 5158.

33. S. D. Maleknia and K. M. Downard, "Charge Ratio Analysis Method: Approach for the Deconvolution of Electrospray Mass Spectra," *Anal. Chem.* **2005**, *77*, 111.

34. F. He, C. L. Hendrickson, and A. G. Marshall, "Baseline Mass Resolution of Peptide Isobars: A Record for Molecular Mass Resolution," *Anal. Chem.* **2001**, *73*, 647.

35. R. Chaler, J. O. Grimalt, C. Pelejero, and E. Calvo, "Sensitivity Effects in $U_{37}^{k'}$ Paleotemperature Estimation by Chemical Ionization Mass Spectrometry," *Anal. Chem.* **2000**, *72*, 5892.

36. L. Charles and D. Pépin, "Electrospray Ion Chromatography–Tandem Mass Spectrometry of Oxyhalides at Sub-ppb Levels," *Anal. Chem.* **1998**, *70*, 353.

37. J. D. Fassett and P. J. Paulsen, "Isotope Dilution Mass Spectrometry for Accurate Elemental Analysis," *Anal. Chem.* **1989**, *61*, 643A.

## Chapter 23

1. T. Michalowski, "Effect of Mutual Solubility of Solvents in Multiple Liquid-Liquid Extraction," *J. Chem. Ed.* **2002**, *79*, 1267.

2. R. P. Paradkar and R. R. Williams, "Micellar Colorimetric Determination of Dithizone Metal Chelates," *Anal. Chem.* **1994**, *66*, 2752.

3. H. H. Strain and J. Sherma, "M. Tswett: Adsorption Analysis and Chromatographic Methods," *J. Chem. Ed.* **1967**, *44*, 238 (a translation of Tswett's original article); H. H. Strain and J. Sherma, "Michael Tswett's Contributions to Sixty Years of Chromatography," *J. Chem. Ed.* **1967**, *44*, 235; L. S. Ettre, "M. S. Tswett and the Invention of Chromatography," *LCGC* **2003**, *21*, 458; L. S. Ettre, "The Birth of Partition Chromatography," *LCGC* **2001**, *19*, 506; L. S. Ettre, "The Story of Thin-Layer Chromatography," *LCGC* **2001**, *19*, 712; L. S. Ettre, "A. A. Zhuykhovitskii—A Russian Pioneer of Gas Chromatography," *LCGC* **2000**, *18*, 1148; V. R. Meyer, "Michael Tswett and His Method," *Anal. Chem.* **1997**, *69*, 284A; Ya. I. Yashin, "History of Chromatography

(1903–1993)," *Russ. J. Anal. Chem.* **1994,** *49,* 939; K. I. Sakodynskii, "Discovery of Chromatography by M. S. Tsvet," *Russ. J. Anal. Chem.* **1994,** *48,* 897.

4. C. F. Poole, *The Essence of Chromatography* (Amsterdam: Elsevier, 2003) (graduate-level treatment of theory and specific techniques); J. M. Miller, *Chromatography: Concepts and Contrasts,* 2nd ed. (Hoboken, NJ: Wiley, 2005).

5. C. A. Smith and F. W. Villaescusa, "Simulating Chromatographic Separations in the Classroom," *J. Chem. Ed.* 203, 80, 1023.

6. S. Miller, "Prep LC Systems for Chemical Separations," *Anal. Chem.* **2003,** *75,* 477A; R. E. Majors, "The Role of the Column in Preparative HPLC," *LCGC* **2004,** *22,* 416; A. Brandt and S. Kueppers, "Practical Aspects of Preparative HPLC in Pharmaceutical Development and Production," *LCGC* **2002,** *20,* 14.

7. J. P. Foley and J. G. Dorsey, "Equations for Calculation of Chromatographic Figures of Merit for Ideal and Skewed Peaks," *Anal. Chem.* **1983,** *55,* 730; B. A. Bidlingmeyer and F. V. Warren, Jr., "Column Efficiency Measurement," *Anal. Chem.* **1984,** *56,* 1583A.

8. M. T. Bowser, G. M. Bebault, X. Peng, and D. D. Y. Chen, "Redefining the Separation Factor: Pathway to a Unified Separation Science," *Electrophoresis* **1997,** *18,* 2928. The conventional equation is resolution $= \frac{\sqrt{N}}{4}\left(\frac{\alpha - 1}{\alpha}\right)\left(\frac{k'_2}{1 + k'_{av}}\right)$, where $\alpha$ is relative retention, $k'_2$ is the capacity factor for the more retained component, and $k'_{av}$ is the average capacity factor for the two components. This expression is equivalent to $\frac{\sqrt{N}}{4}(\gamma - 1)$ for closely spaced peaks for which $k'_A \approx k'_B \approx k'_{av}$.

9. J. C. Giddings, *Unified Separation Science* (New York: Wiley, 1991); S. J. Hawkes, "Modernization of the van Deemter Equation for Chromatographic Zone Dispersion," *J. Chem. Ed.* **1983,** *60,* 393.

10. For numerical simulation of skewed bandshapes, see S. Sugata and Y. Abe, "An Analogue Column Model for Nonlinear Isotherms: The Test Tube Model," *J. Chem. Ed.* **1997,** *74,* 406, and B. R. Sundheim, "Column Operations: A Spreadsheet Model," *J. Chem. Ed.* **1992,** *69,* 1003.

11. A. Felinger, "Molecular Movement in an HPLC Column: A Stochastic Analysis," *LCGC* **2004,** *22,* 642; J. C. Giddings, *Dynamics of Chromatography* (New York: Marcel Dekker, 1965).

## Chapter 24

1. J. V. Hinshaw and L. S. Ettre, *Introduction to Open Tubular Gas Chromatography* (Cleveland, OH: Advanstar Communications, 1994); H. M. McNair and J. M. Miller, *Basic Gas Chromatography* (New York: Wiley, 1998); R. L. Grob and E. F. Barry, eds., *Modern Practice of Gas Chromatography* (New York: Wiley, 2004).

2. A. Wollrab, "Lecture Experiments in Gas-Liquid Chromatography," *J. Chem. Ed.* **1982,** *59,* 1042; C. E. Bricker, M. A. Taylor, and K. E. Kolb, "Simple Classroom Demonstration of Gas Chromatography," *J. Chem. Ed.* **1981,** *58,* 41.

3. L. S. Ettre, "Evolution of Capillary Columns for Gas Chromatography," *LCGC* **2001,** *19,* 48.

4. V. R. Meyer, "Amino Acid Racemization: A Tool for Fossil Dating," *Chemtech,* July 1992, p. 412. For another application, see T. F. Bidleman and R. L. Falconer, "Using Enantiomers to Trace Pesticide Emissions," *Environ. Sci. Technol.* **1999,** *33,* 206A.

5. A. M. Pollard and C. Heron, *Archaeological Chemistry* (Cambridge: Royal Society of Chemistry, 1996)—a very good book.

6. I. Molnár-Perl, *Quantitation of Amino Acids and Amines by Chromatography* (New York: Elsevier, 2005).

7. T. J. Ward, "Chiral Separations," *Anal. Chem.* **2002,** *74,* 2863; J. Hernández-Benito, M. P. Garcia-Santos, E. O'Brien, E. Calle, and J. Casado, "A Practical Integrated Approach to Supramolecular Chemistry III. Thermodynamics of Inclusion Phenomena," *J. Chem. Ed.* **2004,** *81,* 540; B. D. Wagner, P. J. MacDonald, and M. Wagner, "Visual Demonstration of Supramolecular Chemistry: Fluorescence Enhancement upon Host-Guest Inclusion," *J. Chem. Ed.* **2000,** *77,*178; D. Díaz, I. Vargas-Baca, and J. Graci-Mora, "β-Cyclodextrin Inclusion Complexes with Iodine," *J. Chem. Ed.* **1994,** *71,* 708.

8. *Ionic liquids* are low-melting salts that have very low volatility. An optically active ionic liquid can be the stationary phase for gas chromatographic separation of enantiomers. J. Ding, T. Welton, and D. W. Armstrong, "Chiral Ionic Liquids as Stationary Phases in Gas Chromatography," *Anal. Chem.* **2004,** *76,* 6819.

9. A. Berthod, W. Li, and D. W. Armstrong, "Multiple Enantioselective Retention Mechanisms on Derivatized Cyclodextrin Gas Chromatographic Chiral Stationary Phases," *Anal. Chem.* **1992,** *64,* 873; K. Bester, "Chiral Analysis for Environmental Applications," *Anal. Bioanal. Chem.* **2003,** *376,* 302.

10. E. N. Coker and P. J. Davis, "Experiments with Zeolites at the Secondary-School Level," *J. Chem. Ed.* **1999,** *76,* 1417.

11. In programmed temperature runs with a constant inlet pressure, the flow rate decreases during the run because the viscosity of the carrier gas increases as temperature increases. The effect can be significant (e.g., 30% decrease in linear velocity for a 200° temperature increase), so it is a good idea to set the initial linear velocity *above* the optimum value so it does not decrease too far below the optimum. Equations for calculating flow rates as a function of temperature and pressure are given by J. V. Hinshaw and L. S. Ettre, *Introduction to Open Tubular Gas Chromatography* (Cleveland, OH: Advanstar Communications, 1994), and L. S. Ettre and J. V. Hinshaw, *Basic Relationships of Gas Chromatography* (Cleveland, OH: Advanstar Communications, 1993).

12. J. V. Hinshaw, "The Retention Gap Effect," *LCGC* **2004,** *22,* 624.

13. K. Grob, *Split and Splitless Injection for Quantitative Gas Chromatography* (New York: Wiley, 2001).

14. A. N. Papas and M. F. Delaney, "Evaluation of Chromatographic Integrators and Data Systems," *Anal. Chem.* **1987,** *59,* 55A.

15. B. Erickson, "Measuring Nitrogen and Phosphorus in the Presence of Hydrocarbons," *Anal. Chem.* **1998,** *70,* 599A.

16. J. Pawliszyn, *Solid Phase Microextraction* (New York: Wiley, 1997); S. A. S. Wercinski, ed., *Solid Phase Microextraction: A Practical Guide* (New York: Marcel Dekker, 1999); Z. Zhang, M. J. Yang, and J. Pawliszyn, "Solid-Phase Microextraction," *Anal. Chem.* **1994,** *66,* 844A; P. Mayer, J. Tolls, J. C. M. Hermens, and D. MacKay, "Equilibrium Sampling Devices," *Environ. Sci. Technol.* **2003,** *37,* 185A.

17. K. Dettmer and W. Engewald, "Absorbent Materials Commonly Used in Air Analysis for Adsorptive Enrichment and Thermal Desorption of Volatile Organic Compounds," *Anal. Bioanal. Chem.* **2002,** *373,* 490.

18. S. Cram, "How to Develop, Validate, and Troubleshoot Capillary GC Methods," American Chemical Society Short Course, 1996.

19. J. V. Hinshaw, "Strategies for GC Optimization: Software," *LCGC* **2000,** *18,* 1040. Provides equations for optimizing isothermal gas chromatography with a spreadsheet.

20. R. Sacks, C. Coutant, and A. Grall, "Advancing the Science of Column Selectivity," *Anal. Chem.* **2000,** *72,* 525A.

21. J. J. Langenfeld, S. B. Hawthorne, and D. J. Miller, "Quantitative Analysis of Fuel-Related Hydrocarbons in Surface Water and Wastewater by Solid-Phase Microextraction," *Anal. Chem.* **1996,** *68,* 144.

## Chapter 25

1. L. R. Snyder, J. J. Kirkland, and J. L. Glajch, *Practical HPLC Method Development* (New York: Wiley, 1997). This is the definitive source on method development.

2. S. Kromidas, *More Practical Problem Solving in HPLC* (Weinheim: Wiley-VCH, 2005); S. Kromidas, *Practical Problem Solving in HPLC* (New York: Wiley, 2000); V. R. Meyer, *Practical High-Performance Liquid Chromatography,* 4th ed. (New York: Wiley-VCH, 2004); T. Hanai and R. M. Smith, *HPLC: A Practical Guide* (New York: Springer-Verlag, 1999); V. R. Meyer, *Pitfalls and Errors of HPLC in Pictures* (New York: Wiley, 1998); R. Eksteen, P. Schoenmakers, and N. Miller, eds., *Handbook of HPLC* (New York: Marcel Dekker, 1998); U. D. Neue, *HPLC Columns: Theory, Technology, and Practice* (New York: Wiley, 1997); A. Weston and P. R. Brown, *HPLC and CE* (San Diego: Academic Press, 1997); L. R. Snyder, "HPLC—Past and Present," *Anal. Chem.* **2000,** *72,* 412A.

3. J. W. Dolan, "LC Troubleshooting," *LCGC* **2003,** *21,* 888.

4. J. W. Dolan, "The Importance of Temperature," *LCGC* **2002,** *20,* 524; Y. Yang and D. R. Lynch, Jr., "Stationary Phases for High-Temperature LC Separations," *LCGC* (June 2004 supplement), p. 34.

5. D. W. Armstrong and B. Zhang, "Chiral Stationary Phases for HPLC," *Anal. Chem.* **2001,** *73,* 557A.

6. For high-pH operation of any silica-based stationary phase, temperature should not exceed 40°C, organic buffers should be used instead of phosphate or carbonate, and methanol instead of acetonitrile should be the organic solvent. [J. J. Kirkland, J. D. Martosella, J. W. Henderson, C. H. Dilks, Jr., and J. B. Adams, Jr., "HPLC of Basic Compounds at High pH with a Silica-Based Bidentate-C18 Bonded-Phase Column," *Am. Lab.,* November 1999, p. 22; E. D. Neue, "HPLC Troubleshooting," *Am. Lab.,* February 2002, p. 72.]

7. N. Tanaka, H. Kobayashi, K. Nakanishi, H. Minakuchi, and N. Ishizuka, "Monolithic LC Columns," *Anal. Chem.* **2001,** *73,* 421A; N. Tanaka and H. Kobayashi, "Monolithic Columns for Liquid Chromatogaphy," *Anal. Bioanal. Chem.* **2003,** *376,* 298; F. Svec, T. B. Tennikova, and Z. Deyl, eds., *Monolithic Materials: Preparation, Properties and Applications* (Amsterdam: Elsevier, 2003).

8. P. Ross, "Porous Graphitic Carbon in HPLC," *LCGC* **2000,** *18,* 18; S. Mazan, G. Crétier, N. Gilon, J.-M. Mermet, and J.-L. Rocca, "Porous Graphitic Carbon as Stationary Phase for LC-ICPMS Separation of Arsenic Compounds," *Anal. Chem.* **2002,** *74,* 1281. For another robust stationary phase, see C. J. Dunlap, C. V. McNeff, D. Stoll, and P. W. Carr, "Zirconia Stationary Phases for Extreme Separations," *Anal. Chem.* **2001,** *73,* 599A.

9. A. P. Schellinger and P. W. Carr, "Solubility of Buffers in Aqueous-Organic Eluents for Reversed-Phase Liquid Chromatography," *LCGC* **2004,** *22,* 544; D. Sykora, E. Tesarova, and D. W. Armstrong, "Practical Considerations of the Influence of Organic Modifiers on the Ionization of Analytes and Buffers in Reversed-Phase LC," *LCGC* **2002,** *20,* 974; G. W. Tindall, "Mobile-Phase Buffers. I. The Interpretation of pH in Partially Aqueous Mobile Phases," *LCGC* **2002,** *20,* 102; S. Espinosa, E. Bosch, and M. Rosés, "Acid-Base Constants of Neutral Bases in Acetonitrile-Water Mixtures," *Anal. Chim. Acta* **2002,** *454,* 157.

10. P. Raveendran, Y. Ikushima, and S. L. Wallen, "Polar Attributes of Supercritical Carbon Dioxide," *Acc. Chem. Res.* **2005,** *38,* 478.

11. V. T. Lieu, "Simple Experiment for Demonstration of Phase Diagram of Carbon Dioxide," *J. Chem. Ed.* **1996,** *73,* 837.

12. K. Anton and C. Berger, eds., *Supercritical Fluid Chromatography with Packed Columns—Techniques and Applications* (New York: Marcel Dekker, 1998).

13. K. K. S. Lau, J. Bico, K. B. K. Teo, M. Chhowalla, G. A. J. Amaratunga, W. I. Milne, G. H. McKinley, and K. K. Gleason, "Superhydrophobic Carbon Nanotube Forests," *Nano Lett.* **2003,** *3,* 1701.

14. Remove the guard column before washing the analytical column, so that impurities from the guard column are not washed into the analytical column. Bare silica and cyano- and diol-bonded phases are washed (in order) with heptane, chloroform, ethyl acetate, acetone, ethanol, and water. Then the order is reversed, using dried solvents, to reactivate the column. Use 10 empty column volumes of each solvent. Amino-bonded phases are washed in the same manner as silica, but a 0.5 M ammonia wash is used after water. $C_{18}$ and other nonpolar phases are washed with water, acetonitrile, and chloroform, and then the order is reversed. If this is insufficient, wash with 0.5 M sulfuric acid, and then water. [F. Rabel and K. Palmer, *Am. Lab.,* August 1992, p. 65.] Between uses, reversed-phase columns can be stored in methanol or in water-organic solvent mixtures that do not contain salts. Normal-phase columns should be stored in 2-propanol or hexane. See also R. E. Majors, "The Cleaning and Regeneration of Reversed-Phase HPLC Columns," *LCGC* **2003,** *21,* 19.

15. Y. Egi and A. Ueyanagi, "Ghost Peaks and Aerated Sample Solvent," *LCGC* **1998,** *16,* 112.

16. D. L. Warner and J. G. Dorsey, "Reduction of Total Analysis Time in Gradient Elution, Reversed-Phase Liquid Chromatography," *LCGC* **1997,** *15,* 254.

17. C. Hawkins and J. W. Dolan, "Understanding Split Peaks," *LCGC* **2003,** *21,* 1134.

18. J. W. Dolan, "How Much Is Too Much?" *LCGC* **1999,** *17,* 508.

19. H. Lingeman and W. J. M. Underberg, eds., *Detection-Oriented Derivatization Techniques in Liquid Chromatography* (New York: Marcel Dekker, 1990).

20. M. A. Fotopoulou and P. C. Ioannou, "Post-Column Terbium Complexation and Sensitized Fluorescence for the Determination of Norepinephrine, Epinephrine and Dopamine Using High-Performance Liquid Chromatography," *Anal. Chim. Acta* **2002,** *462,* 179.

21. J. A. Koropchak, S. Sadain, X. Yang, L.-E. Magnusson, M. Heybroek, M. Anisimov, and S. L. Kaufman, "Nanoparticle Detection Technology," *Anal. Chem.* **1999,** *71,* 386A; C. S. Young and J. W. Dolan, "Success with Evaporative Light-Scattering Detection," *LCGC* **2003,** *21,* 120.

22. W. R. LaCourse, *Pulsed Electrochemical Detection in High-Performance Liquid Chromatography* (New York: Wiley, 1997).

23. Acetonitrile can be hydrolyzed to sodium acetate and poured down the drain for disposal: $CH_3CN + NaOH + H_2O \rightarrow CH_3CO_2Na + NH_3$. Dilute the chromatography eluate to 10 vol% $CH_3CN$ with water. To 1.0 L of 10 vol% $CH_3CN$, add 475 mL of 10 M NaOH. The solution can be left at 20°C for 25 days in a hood or heated to 80°C for 70 min to reduce the $CH_3CN$ concentration to 0.025 vol%. Mix the hydrolysate with waste acid so that it is approximately neutral before disposal down the drain. [K. Gilomen, H. P. Stauffer, and V. R. Meyer, *LCGC* **1996,** *14,* 56.]

24. Tetrahydrofuran can be stored for at least half a year without oxidation by adding 25 vol% $H_2O$. [J. Zhao and P. W. Carr, "The Magic of Water in Tetrahydrofuran—Preventing Peroxide Formation," *LCGC* **1999,** *17,* 346.]

25. J. D. Thompson and P. W. Carr, "High-Speed Liquid Chromatography by Simultaneous Optimization of Temperature and Eluent Composition," *Anal. Chem.* **2002,** *74,* 4150.

26. J. W. Dolan, "Resolving Minor Peaks," *LCGC* **2002,** *20,* 594.

27. For more on developing gradient separations, see reference 1 and J. W. Dolan, "The Scouting Gradient Alternative," *LCGC* **2000,** *18,* 478.

28. J. M. Cintrón and L. A. Colón, "Organo-Silica Nano-Particles Used in Ultrahigh-Pressure Liquid Chromatography," *Analyst* **2002,** *127,* 705.

29. E. S. Grumbach, D. M. Wagrowski-Diehl, J. R. Mazzeo, B. Alden, and P. C. Iraneta, "Hydrophilic Interaction Chromatography Using Silica Columns for the Retention of Polar Analytes and Enhanced ESI-MS Sensitivity," *LCGC* **2004,** *22,* 1010.

## Chapter 26

1. K. K. Unger, M. Huber, K. Walhagen, T. P. Hennessy, and M. T. W. Hearn, "A Critical Appraisal of Capillary Electrochromatography," *Anal. Chem.* **2002,** *74,* 200A.

2. B. A. Kyereboah-Taylor, "Ultrapure Water for Ion Chromatography," *Am. Lab.,* August 1995, p. 24; B. M. Stewart and D. Darbouret, "Ultrapure Water for ICP-MS," *Am. Lab. News Ed.,* April 1998, p. 36; A. de Chatellus, "Purification Media," *Am. Lab. News Ed.,* January 1998, p. 8.

3. A mixture of enantiomers of cationic metal complexes was applied to a cation-exchange column and eluted by one enantiomer of tartrate anion. The tartrate has a different ion-pair formation constant with each enantiomer of the metal complex and therefore removes one metal enantiomer from the column before the other metal enantiomer. M. Cantuel, G. Bernardinelli, G. Muller, J. P. Riehl, and C. Piguet, "The First Enantiomerically Pure Helical Noncovalent Tripod for Assembling Nine-Coordinate Lanthanide(III) Podates," *Inorg. Chem.* **2004,** *43,* 1840; Y. Yoshikawa and K. Yamasaki, "Chromatographic Resolution of Metal Complexes on Sephadex Ion Exchangers," *Coord. Chem. Rev.* **1979,** *28,* 2005.

4. D. P. Elder, "Pharmaceutical Applications of Ion-Exchange Resins," *J. Chem. Ed.* **2005,** *82,* 575.

5. J. S. Fritz and D. T. Gjerde, *Ion Chromatography,* 3rd ed. (New York: Wiley-VCH, 2000); J. Weiss, *Handbook of Ion Chromatography* (Weinheim: Wiley-VCH, 2004); P. R. Haddad and P. E. Jackson, *Ion Chromatography: Principles and Applications* (New York: Elsevier, 1990); H. Small, *Ion Chromatography* (New York: Plenum Press, 1989); P. R. Haddad, "Ion Chromatography Retrospective," *Anal. Chem.* **2001,** *73,* 266A; H. Small, "Ion Chromatography: An Account of Its Conception and Early Development," *J. Chem. Ed.* **2004,** *81,* 1277; B. Evans, "The History of Ion Chromatography: The Engineering Perspective," *J. Chem. Ed.* **2004,** *81,* 1285.

6. Y. Liu, K. Srinivasan, C. Pohl, and N. Avdalovic, "Recent Developments in Electrolytic Devices for Ion Chromatography," *J. Biochem. Biophys. Methods* **2004,** *60,* 205.

7. J. E. Madden, M. J. Shaw, G. W. Dicinoski, and P. R. Haddad, "Simulation and Optimization of Retention in Ion Chromatography Using Virtual Column 2 Software," *Anal. Chem.* **2002,** *74,* 6023; P. R. Haddad, M. J. Shaw, J. E. Madden, and G. W. Dicinoski, "Computer-Based Undergraduate Exercise Using Internet-Accessible Simulation Software for the Study of Retention Behavior and Optimization of Separation Conditions in Ion Chromatography," *J. Chem. Ed.* **2004,** *81,* 1293; http://www.virtualcolumn.com.

8. P. C. Schulz and D. Clausse, "An Undergraduate Physical Chemistry Experiment on Surfactants: Electrochemical Study of Commercial Soap,"

*J. Chem. Ed.* **2003,** *80,* 1053; A. Domínguez, A. Fernández, N. González, E. Iglesias, and L. Montenegro, "Determination of Critical Micelle Concentration of Some Surfactants by Three Techniques," *J. Chem. Ed.* **1997,** *74,*1227; K. R. Williams and L. H. Tennant, "Micelles in the Physical/Analytical Chemistry Laboratory: Acid Dissociation of Neutral Red Indicator," *J. Chem. Ed.* **2001,** *78,* 349; S. A. Tucker, V. L. Amszi, and W. E. Acree, Jr., "Studying Acid-Base Equilibria in Two-Phase Solvent Media," *J. Chem. Ed.* **1993,** *70,* 80.

9. M. A. Hervas and C. E. Fabara, "A Simple Demonstration of the Ion-Pairing Effect on the Solubility of Charged Molecules." *J. Chem. Ed.* **1995,** *72,* 437.

10. L. R. Snyder, J. J. Kirkland, and J. L. Glajch, *Practical HPLC Method Development* (New York: Wiley, 1997); J. W. Dolan, "Improving an Ion-Pairing Method," *LCGC* **1996,** *14,* 768.

11. T. W. Perkins, T. W. Root, and E. N. Lightfoot, "Measuring Column Void Volumes with NMR," *Anal. Chem.* **1997,** *69,* 3293.

12. C.-S. Wu, ed., *Handbook of Size Exclusion Chromatography,* 2nd ed. (New York: Marcel Dekker, 2004); R. K. Scopes, *Protein Purification,* 3rd ed. (New York: Springer-Verlag, 1994).

13. M. A. Hahn, J. S. Tabb, and T. D. Krauss, "Detection of Single Bacterial Pathogens with Semiconductor Quantum Dots," *Anal. Chem.* **2005,** *77,* 4861; E. M. Boatman, G. C. Lisensky, and K. J. Nordell, "A Safer, Easier, Faster Synthesis for CdSe Quantum Dot Nanocrystals," *J. Chem. Ed.* **2005,** *82,* 1697; L. D. Winkler, J. F. Arceo, W. C. Hughes, B. A. DeGraff, and B. H. Augustine, "Quantum Dots: An Experiment for Physical or Materials Chemistry," *J. Chem. Ed.* **2005,** *82,* 1700.

14. T. K. Nevanen, H. Simolin, T. Suortti, A. Koivula, and H. Söderlund, "Development of a High-Throughput Format for Solid-Phase Extraction of Enantiomers Using an Immunosorbent in 384-Well Plates," *Anal. Chem.* **2005,** *77,* 3038.

15. E. Turiel and A. Martin-Esteban, "Molecularly Imprinted Polymers: Toward Highly Selective Stationary Phases in Liquid Chromatography and Capillary Electrophoresis," *Anal. Bioanal. Chem.* **2004,** *378,* 1876; K. Haupt, "Molecularly Imprinted Polymers: The Next Generation," *Anal. Chem.* **2003,** *75,* 376A.

16. Q. Deng, I. German, D. Buchanan, and R. T. Kennedy, "Retention and Separation of Adenosine and Analogues by Affinity Chromatography with an Aptamer Stationary Phase," *Anal. Chem.* **2001,** *73,* 5415.

17. R. Weinberger, *Practical Capillary Electrophoresis,* 2nd ed. (San Diego, CA: Academic Press, 2000); T. Wehr, R. Rodríguez-Diaz, and M. Zhu, *Capillary Electrophoresis of Proteins* (New York: Marcel Dekker, 1999); M. G. Khaledi, ed., *High Performance Capillary Electrophoresis: Theory, Techniques, and Applications* (New York: Wiley, 1998); P. Camilleri, ed., *Capillary Electrophoresis: Theory and Practice* (Boca Raton, FL: CRC Press, 1998); J. P. Landers, *Handbook of Capillary Electrophoresis,* 2nd ed. (Boca Raton, FL: CRC Press, 1997); H. Shintani and J. Polonský, *Handbook of Capillary Electrophoresis Applications* (London: Blackie, 1997); J. R. Petersen and A. A. Mohammad, eds. *Clinical and Forensic Applications of Capillary Electrophoresis* (Totowa, NJ: Humana Press, 2001).

18. C. L. Cooper, "Capillary Electrophoresis: Theoretical and Experimental Background," *J. Chem. Ed.* **1998,** *75,* 343; C. L. Cooper and K. W. Whitaker, "Capillary Electrophoresis: Applications," *J. Chem. Ed.* **1998,** *75,* 347.

19. J. W. Jorgensen and K. D. Lukacs, "Zone Electrophoresis in Open Tubular Glass Capillaries," *Anal. Chem.* **1981,** *53,* 1298.

20. For a demonstration of electrophoresis, see J. G. Ibanez, M. M. Singh, R. M. Pike, and Z. Szafran, "Microscale Electrokinetic Processing of Soils," *J. Chem. Ed.* **1998,** *75,* 634.

21. T. M. Olefirowicz and A. G. Ewing, "Capillary Electrophoresis in 2 and 5 μm Diameter Capillaries: Application to Cytoplasmic Analysis," *Anal. Chem.* **1990,** *62,* 1872; Q. Xue and E. S. Yeung, "Variability of Intracellular Lactate Dehydrogenase Isoenzymes in Single Human Erythrocytes," *Anal. Chem.* **1994,** *66,* 1175; K. Bächmann, H. Lochmann, and A. Bazzanella, "Microscale Processes in Single Plant Cells," *Anal. Chem.* **1998,** *70,* 645A; P. D. Floyd, L. L. Moroz, R. Gillette, and J. V. Sweedler, "Capillary Electrophoresis Analysis of NO Synthase Related Metabolites in Single Neurons," *Anal. Chem.* **1998,** *70,* 2243.

22. G. Xiong, Y. Chen, and E. A. Arriaga, "Measuring the Doxorubicin Content of Single Nuclei by Micellar Electrokinetic Capillary Chromatography with Laser-Induced Fluorescence Detection," *Anal. Chem.* **2005,** *77,* 3488; S. J. Lillard, D. T. Chiu, R. H. Scheller, R. N. Zare, S. E. Rodríguez-Cruz, E. R. Williams, O. Orwar, M. Sandberg, and J. A. Lundqvist, "Separation and

Characterization of Amines from Individual Atrial Gland Vesicles of *Aplysia californica,*" *Anal. Chem.* **1998,** *70,* 3517.

23. G. K. Shoemaker, J. Lorieau, L. H. Lau, C. S. Gillmor, and M. M. Palcic, "Multiple Sampling in Single-Cell Enzyme Assays," *Anal. Chem.* **2005,** *77,* 3132; X. Sun and W. Jin, "Catalysis-Electrochemical Determination of Zeptomol Enzyme and Its Application for Single-Cell Analysis," *Anal. Chem.* **2003,** *75,* 6050.

24. K. B. O'Brien, M. Esguerra, R. F. Miller, and M. T. Bowser, "Monitoring Neurotransmitter Release from Isolated Retinas Using Online Microdialysis-Capillary Electrophoresis," *Anal. Chem.* **2004,** *76,* 5069.

25. N. A. Guzman and T. M. Phillips, "Immunoaffinity CE for Proteomics Studies," *Anal. Chem.* **2005,** *77,* 60A.

26. J. D. Carbeck, I. J. Colton, J. Gao, and G. M. Whitesides, "Protein Charge Ladders, Capillary Electrophoresis, and the Role of Electrostatics in Biomolecular Recognition," *Acc. Chem. Res.* **1998,** *31,* 343.

27. C. T. Culbertson and J. W. Jorgenson, "Flow Counterbalanced Capillary Electrophoresis," *Anal. Chem.* **1994,** *66,* 955.

28. R. Weinberger, "Capillary Electrophoresis," *Am. Lab.,* April 2005, p. 25.

29. H. Watley, "Making CE Work—Points to Consider," *LCGC* **1999,** *17,* 426.

30. M. Macka, P. Andersson, and P. R. Haddad, "Changes in Electrolyte pH Due to Electrolysis During Capillary Zone Electrophoresis," *Anal. Chem.* **1998,** *70,* 743.

31. N. E. Baryla, J. E. Melanson, M. T. McDermott, and C. A. Lucy, "Characterization of Surfactant Coatings in Capillary Electrophoresis by Atomic Force Microscopy," *Anal. Chem.* **2001,** *73,* 4558; M. M. Yassine and C. A. Lucy, "Enhanced Stability Self-Assembled Coatings for Protein Separations by Capillary Zone Electrophoresis Through the Use of Long-Chained Surfactants," *Anal. Chem.* **2005,** *77,* 62.

32. At a wavelength of 190 nm, phosphate at pH 7.2 has about three times the absorbance of borate at pH 9.2. Glycine, citrate, HEPES, and TRIS buffers have significant absorption near 210 nm. Borate buffer should be prepared from sodium tetraborate (borax, $Na_2B_4O_7 \cdot 10H_2O$), not from boric acid ($B(OH)_3$), which has somewhat different acid-base chemistry. An ultraviolet-absorbing impurity in borate buffer can be removed by passage through a $C_{18}$ solid-phase extraction column.

33. M. Serwe and G. A. Ross, "Comparison of CE-MS and LC-MS for Peptide Samples," *LCGC* **2000,** *18,* 46.

34. E. S. Yeung and W. G. Kuhr, "Indirect Detection Methods for Capillary Separations," *Anal. Chem.* **1991,** *63,* 275A; J. Ren and X. Huang, "Indirect Chemiluminescence Detection for Capillary Electrophoresis of Cations Using Co(III) as a Probe Ion," *Anal. Chem.* **2001,** *731,* 2663; M. Macka, C. Johns, P. Doble, and P. R. Haddad, "Indirect Photometric Detection in CE Using Buffered Electrolytes," *LCGC* **2001,** *19,* 38, 178.

35. A. Berthod and C. García-Alvarez-Coque, *Micellar Liquid Chromatography* (New York: Marcel Dekker, 2000); S. Terabe, "Micellar Electrokinetic Chromatography," *Anal. Chem.* **2004,** *76,* 240A; U. Pyell, "Micellar Electrokinetic Chromatography—From Theoretical Concepts to Real Samples (Review)," *Fresenius J. Anal. Chem.* **2001,** *371,* 691.

36. Demonstrations: N. Gani and J. Khanam, "Are Surfactant Molecules Really Oriented in the Interface?" *J. Chem. Ed.* **2002,** *79,* 332; C. J. Marzzacco, "The Effect of SDS Micelle on the Rate of a Reaction," *J. Chem. Ed.* **1992,** *69,* 1024; C. P. Palmer, "Demonstrating Chemical and Analytical Concepts Using Electrophoresis and Micellar Electrokinetic Chromatography," *J. Chem. Ed.* **1999,** *76,* 1542.

37. S. Terabe, K. Otsuka, K. Ichikawa, A. Tsuchiya, and T. Ando, "Electrokinetic Separations with Micellar Solutions and Open Tubular Capillaries," *Anal. Chem.* **1984,** *56,* 111; ibid., **1985,** *57,* 834.

38. Polar organic solvents with electrolytes such as sodium *p*-toluenesulfonate are compatible with capillary electrophoresis. Background electrolyte need not be an aqueous solution. [P. B. Wright, A. S. Lister, and J. G. Dorsey, "Behavior and Use of Nonaqueous Media Without Supporting Electrolyte in Capillary Electrophoresis and Capillary Electrochromatography," *Anal. Chem.* **1997,** *69,* 3251; I. E. Valkó, H. Sirén, and M.-L. Riekkola, "Capillary Electrophoresis in Nonaqueous Media: An Overview," *LCGC* **1997,** *15,* 560.]

39. S. Li and S. G. Wever, "Separation of Neutral Compounds in Nonaqueous Solvents by Capillary Zone Electrophoresis," *J. Am. Chem. Soc.* **2000,** *122,* 3787.

40. S. Conradi, C. Vogt, and E. Rohde, "Separation of Enantiomeric Barbiturates by Capillary Electrophoresis Using a Cyclodextrin-Containing Run Buffer," *J. Chem. Ed.* **1997,** *74,* 1122; F.-T. Chen and R. A. Evangelista, "Highly Sulfated Cyclodextrins: The Solution for Chiral Analysis," *Am. Lab.,* August 2002, p. 30; M. A. Schwarz and P. C. Hauser, "Chiral On-Chip Separations of Neurotransmitters," *Anal. Chem.* **2000,** *752,* 4691; W. Zhu and G. Vigh, "A Family of Single-Isomer, Sulfated γ-Cyclodextrin Chiral Resolving Agents for Capillary Electrophoresis," *Anal. Chem.* **2000,** *72,* 310.

41. J. Palmer, N. J. Munro, and J. P. Landers, "A Universal Concept for Stacking Neutral Analytes in Micellar Capillary Electrophoresis," *Anal. Chem.* **1999,** *71,* 1679; J. Palmer, D. S. Burji, N. J. Munro, and J. P. Landers, "Electrokinetic Injection for Stacking Neutral Analytes in Capillary and Microchip Electrophoresis," *Anal. Chem.* **2001,** *73,* 725; J. P. Quirino, S. Terabe, and P. Bocek, "Sweeping of Neutral Analytes in Electrokinetic Chromatography with High-Salt-Containing Matrixes," *Anal. Chem.* **2000,** *72,* 1934.

42. L. Zhu, C. Tu, and H. K. Lee, "On-Line Concentration of Acidic Compounds by Anion-Selective Exhaustive Injection-Sweeping-Micellar Electrokinetic Chromatography," *Anal. Chem.* **2002,** *74,* 5820.

43. K. Isoo and S. Terabe, "Analysis of Metal Ions by Sweeping via Dynamic Complexation and Cation-Selective Exhaustive Injection in Capillary Electrophoresis," *Anal. Chem.* **2003,** *75,* 6789.

44. W. Wei, G. Zue, and E. S. Yeung, "One-Step Concentration of Analytes Based on Dynamic Change in pH in Capillary Zone Electrophoresis," *Anal. Chem.* **2002,** *74,* 934; P. Britz-McKibbin, K. Otsuka, and S. Terabe, "On-Line Focusing of Flavin Derivatives Using Dynamic pH Junction-Sweeping Capillary Electrophoresis with Laser-Induced Fluorescence Detection," *Anal. Chem.* **2002,** *74,* 3736.

45. L. A. Colón, Y. Guo, and A. Fermier, "Capillary Electrochromatography," *Anal. Chem.* **1997,** *69,* 461A.

46. Y. Gong and H. K. Lee, "Application of Cyclam-Capped β-Cyclodextrin-Bonded Silica Particles as a Chiral Stationary Phase in Capillary Electrochromatography for Enantiomeric Separations," *Anal. Chem.* **2003,** *75,* 1348; M. W. Kamande, X. Zhu, C. Kapnissi-Christodoulou, and I. M. Warner, "Chiral Separations Using a Polypeptide and Polymeric Dipeptide Surfactant Polyelectrolyte Multilayer Coating in Open-Tubular Capillary Electrochromatography," *Anal. Chem.* **2004,** *76,* 6681.

47. J. R. Hutchinson, P. Zakaria, A. R. Bowie, M. Macka, N. Avdalovic, and P. R. Haddad, "Anion-Exchange Capillary Electrochromatography and In-Line Sample Preconcentration in Capillary Electrophoresis," *Anal. Chem.* **2005,** *77,* 407.

48. J. P. Quirino, M. T. Duylay, B. D. Bennett, and R. N. Zare, "Strategy for On-Line Preconcentration in Chromatographic Separations," *Anal. Chem.* **2001,** *73,* 5539; J. P. Quirino, M. T. Duylay, and R. N. Zare, "On-Line Preconcentration in Capillary Electrochromatography Using a Porous Monolith Together with Solvent Gradient and Sample Stacking," *Anal. Chem.* **2001,** *73,* 5557.

49. DNA sequencing systems can read 500 000 bases per day: J. P. Smith and V. Henson-Smith, "DNA Sequencers Rely on CE," *Anal. Chem.* **2001,** *73,* 327A; O. Salas-Solano, E. Carrilho, L. Kotler, A. W. Miller, W. Goetzinger, Z. Sosic, and B. L. Karger, "Routine DNA Sequencing of 1 000 Bases in Less Than One Hour," *Anal. Chem.* **1998,** *70,* 3996.

50. E. Zubritsky, "How Analytical Chemists Saved the Human Genome Project," *Anal. Chem.* **2002,** *74,* 23A.

51. R. Weinberger, "An Interview with Ira Lurie of the DEA," *Am. Lab.,* January 2005, p. 6; I. S. Lurie, P. A. Hays, A. D. Garcia, and S. Panicker, "Use of Dynamically Coated Capillaries for the Determination of Heroin, Basic Impurities and Adulterants with Capillary Electrophoresis," *J. Chromatogr. A* **2004,** *1034,* 227.

52. R. Weinberger, "Method Development for Capillary Electrophoresis," *Am. Lab.,* March 2003, p. 54.

53. M. L. Riekkola, M. Jussila, S. P. Possas, and I. E. Valko, "Non-Aqueous Capillary Electrophoresis," *J. Chromatogr. A* **2000,** *892,* 155.

54. A. M. Siouffi and R. Phan-Tan-Luu, "Optimization Methods in Chromatography and Electrophoresis," *J. Chromatogr. A* **2000,** *892,* 75.

55. D. R. Reyes, D. Iossifidis, P.-A. Auroux, and A. Manz, "Micro Total Analysis Systems. 1. Introduction, Theory, and Technology," *Anal. Chem.* **2002,** *74,* 2623; P.-A. Auroux, D. Iossifidis, D. R. Reyes, and A. Manz, "Micro Total Analysis Systems. 2. Analytical Standard Operations and Applications," *Anal. Chem.* **2002,** *74,* 2637; R. T. Kelly and A. T. Wooley, "Microfluidic Systems for Integrated, High-Throughput DNA Analysis," *Anal. Chem.* **2005,** *77,* 96A; C. A. Emrich, H. Tian, I. L. Medintz, and R. A. Mathies, "Microfabricated 384-Lane Capillary Array Electrophoresis Bioanalyzer for Ultrahigh-Throughput Genetic Analysis," *Anal. Chem.* **2002,** *74,* 5076; T. D. Boone, Z. H. Fan, H. H. Hooper, A. J. Ricco, H. Tan, and S. J. Williams, "Plastic Advances Microfluidic Chips," *Anal. Chem.* **2002,** *74,* 78A; R. F. Renzi, J. Stamps, B. A. Horn, S. Ferko, V. A. VanderNoot, J. A. A. West, R. Crocker, B. Wiedenman, D. Yee, and J. A. Fruetel, "Hand-Held Microanalytical Instrument for Chip-Based Electrophoretic Separations of Proteins," *Anal. Chem.* **2005,** *77,* 435; J. G. E. Gardeniers and A. van den Berg, "Lab-on-a-Chip Systems for Biomedical and Environmental Monitoring," *Anal. Bioanal. Chem.* **2004,** *378,* 1700; J. C. McDonald and G. M. Whitesides, "Poly(dimethylsiloxane) as a Material for Fabricating Microfluidic Devices," *Acc. Chem. Res.* **2002,** *35,* 491; Y. Huang, S. Joo, M. Duhon, M. Heller, B. Wallace, and X. Xu, "Dielectrophoretic Cell Separation and Gene Expression Profiling on Microelectronic Chip Arrays," *Anal. Chem.* **2002,** *74,* 3362; D. Figeys and D. Pinto, "Lab-on-a-Chip: A Revolution in Biological and Medical Sciences," *Anal. Chem.* **2000,** *72,* 330A; C. H. Legge, "Chemistry Under the Microscope—Lab-on-a-Chip Technologies," *J. Chem. Ed.* **2002,** *79,* 173.

56. J. A. C. Broekaert, "The Development of Microplasmas for Spectrochemical Analysis," *Anal. Bioanal. Chem.* **2002,** *374,* 182; T. Kitamori, M. Tokeshi, A. Hibara, and K. Sato, "Thermal Lens Microscopy and Microchip Chemistry," *Anal. Chem.* **2004,** *76,* 52A; R. P. Baldwin, T. J. Roussel, Jr., M. M. Crain, V. Bathlagunda, D. J. Jackson, J. Gullapalli, J. A. Conklin, R. Pai, J. F. Naber, K. M. Walsh, and R. S. Keynton, "Fully Integrated On-Chip Electrochemical Detection for Capillary Electrophoresis in a Microfabricated Device," *Anal. Chem.* **2002,** *74,* 3690; M. L. Chabinyc, D. T. Chiu, J. C. McDonald, A. D. Strook, J. F. Christian, A. M. Karger, and G. M. Whitesides, "An Integrated Fluorescence Detection System in Poly(dimethylsiloxane) for Microfluidic Applications," *Anal. Chem.* **2001,** *73,* 4491.

57. D. M. Ratner, E. R. Murphy, M. Jhunjhunwala, D. A. Snyder, K. F. Jensen, and P. H. Seeberger, "Microreactor-Based Reaction Optimization in Organic Chemistry—Glycosylation as a Challenge," *Chem. Commun.* **2005,** 578.

58. M. G. Roper, C. J. Easley, and J. P. Landers, "Advances in Polymerase Chain Reaction on Microfluidic Chips," *Anal. Chem.* **2005,** *77,* 3887; K. D. Dorfman, M. Chabert, J.-H. Codarbox, G. Rousseau, P. de Cremoux, and J.-L. Viovy, "Contamination-Free Continuous Flow Microfluidic Polymerase Chain Reaction for Quantitative and Clinical Applications," *Anal. Chem.* **2005,** *77,* 3700.

59. A similar scheme can be used to concentrate proteins: R. S. Foot, J. Khandurina, S. C. Jacobson, and J. M. Ramsey, "Preconcentration of Proteins on Microfluidic Devices Using Porous Silica Membranes," *Anal. Chem.* **2005,** *77,* 57.

60. N. D. Chasteen and P. M. Harrison, "Mineralization in Ferritin: An Efficient Means of Iron Storage," *J. Struct. Biol.* **1999,** *126,* 182.

61. Data from A. W. Moore, Jr., S. C. Jacobson, and J. M. Ramsey, "Microchip Separations of Neutral Species via Micellar Electrokinetic Capillary Chromotography," *Anal. Chem.* **1995,** *67,* 4184.

62. The optimum pH for separating cations is p$K$ + 0.30: K. K.-C. Yeung and C. A. Lucy, "Isotopic Separation of [$^{14}$N]- and [$^{15}$N] Aniline by Capillary Electrophoresis Using Surfactant-Controlled Reversed Electroosmotic Flow," *Anal. Chem.* **1998,** *70,* 3286.

63. G. R. Asbury and H. H. Hill, Jr., "Evaluation of Ultrahigh Resolution Ion Mobility Spectrometry as an Analytical Separation Device in Chromatographic Terms," *J. Microcolumn Sep.* **2000,** *12,* 172; H. E. Revercomb and E. A. Mason, "Theory of Plasma Chromatography/Gaseous Electrophoresis," *Anal. Chem.* **1975,** *47,* 970.

## Chapter 27

1. T. W. Richards, *Chem. Rev.* **1925,** *1,* 1; C. M. Beck II, "Classical Analysis: A Look at the Past, Present, and Future," *Anal. Chem.* **1994,** *66,* 225A; I. M. Kolthoff, "Analytical Chemistry in the U.S.A. in the First Quarter of This Century," *Anal. Chem.* **1994,** *66,* 241A; D. T. Burns, "Highlights in the History of Quantitation in Chemistry," *Fresenius J. Anal. Chem.* **1990,** *337,* 205; L. Niiniströ, "Analytical Instrumentation in the 18th Century," *Fresenius J. Anal. Chem.* **1990,** *337,* 213.

**2.** For an experiment with colloids, see C. D. Keating, M. D. Musick, M. H. Keefe, and M. J. Natan, "Kinetics and Thermodynamics of Au Colloid Monolayer Self Assembly," *J. Chem. Ed.* **1999,** *76,* 949.

**3.** Cellulose tubing such as catalog number 3787, sold by Thomas Scientific, www.thomassci.com, is adequate for this demonstration.

**4.** M. Suzuki, "The Movement of Molecules and Heat Energy: Two Demonstrative Experiments," *J. Chem. Ed.* **1993,** *70,* 821.

**5.** L. C. Roberts, S. J. Hug, T. Ruettimann, M. Billah, A. W. Khan, and M. T. Rahman, "Arsenic Removal with Iron(II) and Iron(III) Waters with High Silicate and Phosphate Concentrations," *Environ. Sci. Technol.* **2004,** *38,* 307.

**6.** G. W. Latimer, Jr., "Piperazine as the Diacetate," *J. Chem. Ed.* **1966,** *43,* 148; G. R. Bond, *Anal. Chem.* **1962,** *32,* 1332.

**7.** E. Pella, "Elemental Organic Analysis. 1. Historical Developments," *Am. Lab.,* February 1990, p. 116; "Elemental Organic Analysis. 2. State of the Art," August 1990, p. 28.

**8.** K. Russe, H. Kipphardt, and J. A. C. Broekaert, "Determination of Main and Minor Components of Silicon Based Materials by Combustion with $F_2$," *Anal. Chem.* **2000,** *72,* 3875.

**9.** R. W. Ramette, "Stoichiometry to the Rescue (A Calculation Challenge)," *J. Chem. Ed.* **1988,** *65,* 800.

**10.** S. Kagaya, M. Saiki, Z. A. Malek, Y. Araki, and K. Hasegawa, "Coprecipitation with Lanthanum Phosphate as a Technique for Separation and Preconcentration of Iron(III) and Lead," *Fresenius J. Anal. Chem.* **2001,** *371,* 391.

**11.** F. Torrades and M. Castellvi, "Spectrophotometric Determination of $Cl^-$ in $BaSO_4$ Precipitate," *Fresenius J. Anal. Chem.* **1994,** *349,* 734.

## Chapter 28

**1.** E. P. Popek, *Sampling and Analysis of Environmental Chemical Pollutants* (Amsterdam: Academic Press, 2003); P. Gy, *Sampling for Analytical Purposes* (Chichester, UK: Wiley, 1998); B. B. Kebbekus and S. Mitra, *Environmental Chemical Analysis* (London: Blackie, 1998); N. T. Crosby and I. Patel, *General Principles of Good Sampling Practice* (Cambridge: Royal Society of Chemistry, 1995); K. G. Carr-Brion and J. R. P. Clarke, *Sampling Systems for Process Analysis,* 2nd ed. (Oxford: Butterworth-Heinemann, 1996); L. H. Keith, *Environmental Sampling and Analysis: A Practical Guide* (Chelsea, MI: Lewis, 1991); R. F. Cross, "Reducing Sample Size and Obtaining Representative Samples," *LCGC* **2000,** *18,* 468.

**2.** S. Mitra, ed., *Sample Preparation Techniques in Analytical Chemistry* (Hoboken, NJ: Wiley, 2003).

**3.** D. T. Sawyer, A. Sobkowiak, and J. L. Roberts, Jr., *Electrochemistry for Chemists,* 2nd ed. (New York: Wiley, 1995), p. 262.

**4.** P. Vallelonga, K. Van de Velde, J. P. Candelone, C. Ly, K. J. R. Rosman, C. F. Boutron, V. I. Morgan, and D. J. Mackey, "Recent Advances in Measurement of Pb Isotopes in Polar Ice and Snow at Sub-Picogram Per Gram Concentrations Using Thermal Ionisation Mass Spectrometry," *Anal. Chim. Acta* **2002,** *453,* 1.

**5.** R. P. Mason and K. A. Sullivan, "Mercury in Lake Michigan," *Environ. Sci. Technol.* **1997,** *31,* 942.

**6.** G. Benoit, K. S. Hunter, and T. F. Rozan, "Sources of Trace Metal Contamination Artifacts During Collection, Handling, and Analysis of Freshwaters," *Anal. Chem.* **1997,** *69,* 1006.

**7.** R. E. Thiers, *Methods of Biochemical Analysis* (D. Glick, ed.), Vol. 5 (New York: Interscience, 1957), p. 274.

**8.** B. Kratochvil and J. K. Taylor, "Sampling for Chemical Analysis," *Anal. Chem.* **1981,** *53,* 924A; H. A. Laitinen and W. E. Harris, *Chemical Analysis,* 2nd ed. (New York: McGraw-Hill, 1975), Chap. 27; S. K. Thompson, *Sampling* (New York, Wiley, 1992).

**9.** J. E. Vitt and R. C. Engstrom, "Effect of Sample Size on Sampling Error," *J. Chem. Ed.* **1999,** *76,* 99; R. D. Guy, L. Ramaley, and P. D. Wentzell, "Experiment in the Sampling of Solids for Chemical Analysis," *J. Chem. Ed.* **1998,** *75,* 1028. Demonstrations: M. R. Ross, "A Classroom Exercise in Sampling Technique," *J. Chem. Ed.* **2000,** *77,* 1015; J. R. Hartman, "In-Class Experiment on the Importance of Sampling Techniques and Statistical Analysis of Data," *J. Chem. Ed.* **2000,** *77,* 1017.

**10.** D. C. Bogen, *Treatise on Analytical Chemistry,* 2nd ed. (P. J. Elving, E. Grushka, and I. M. Kolthoff, eds.), Part I, Vol. 5 (New York: Wiley, 1982), Chap. 1.

**11.** H. M. Kingston and S. J. Haswell, *Microwave-Enhanced Chemistry* (Washington, DC: American Chemical Society, 1997).; B. D. Zehr, "Development of Inorganic Microwave Dissolutions," *Am. Lab.,* December 1992, p. 24; B. D. Zehr, J. P. VanKuren, and H. M. McMahon, "Inorganic Microwave Digestions Incorporating Bases," *Anal. Chem.* **1994,** *66,* 2194.

**12.** P. Aysola, P. Anderson, and C. H. Langford, "Wet Ashing in Biological Samples in a Microwave Oven Under Pressure Using Poly(tetrafluoroethylene) Vessels," *Anal. Chem.* **1987,** *59,* 1582.

**13.** S. E. Long and W. R. Kelly, "Determination of Mercury in Coal by Isotope Dilution Cold-Vapor Generation Inductively Coupled Plasma Mass Spectrometry," *Anal. Chem.* **2002,** *74,* 1477.

**14.** M. Rehkämper, A. N. Halliday, and R. F. Wentz, "Low-Blank Digestion of Geological Samples for Pt-Group Analysis Using a Modified Carius Tube Design," *Fresenius J. Anal. Chem.* **1998,** *361,* 217.

**15.** A. A. Schilt, *Perchloric Acid and Perchlorates,* 2nd ed. (Powell, OH: GFS Chemicals, 2003).

**16.** C. Walling, "Intermediates in the Reactions of Fenton Type Reagents," *Acc. Chem. Res.* **1998,** *31,* 155; P. A. MacFaul, D. D. M. Wayner, and K. U. Ingold, "A Radical Account of 'Oxygenated Fenton Chemistry,'" *Acc. Chem. Res.* **1998,** *31,* 159; D. T. Sawyer, A. Sobkowiak, and T. Matsushita, "Oxygenated Fenton Chemistry," *Acc. Chem. Res.* **1996,** *29,* 409.

**17.** L. Ping and P. K. Dasgupta, "Determination of Total Mercury in Water and Urine by a Gold Film Sensor Following Fenton's Reagent Digestion," *Anal. Chem.* **1989,** *61,* 1230.

**18.** G. LeBlanc, "Microwave-Accelerated Techniques for Solid Sample Extraction," *LCGC* **1999,** *17,* S30 (June 1999 supplement).

**19.** Y. Lin and C. M. Wai, "Supercritical Fluid Extraction of Lanthanides with Fluorinated β-Diketones and Tributyl Phosphate," *Anal. Chem.* **1994,** *66,* 1971.

**20.** L. T. Taylor, *Supercritical Fluid Extraction* (New York: Wiley, 1996); C. L. Phelps, N. G. Smart, and C. M. Wai, "Past, Present, and Possible Future Applications of Supercritical Fluid Extraction Technology," *J. Chem. Ed.* **1996,** *73,* 1163; M. E. P. McNally, "Advances in Environmental SFE," *Anal. Chem.* **1995,** *67,* 308A; L. T. Taylor, "Strategies for Analytical SFE," *Anal. Chem.* **1995,** *67,* 364A.

**21.** S. Liang and D. C. Tilotta, "Extraction of Petroleum Hydrocarbons from Soil Using Supercritical Argon," *Anal. Chem.* **1998,** *70,* 616.

**22.** E. M. Thurman and M. S. Mills, *Solid-Phase Extraction: Principles and Practice* (New York: Wiley, 1998); N. J. K. Simpson, *Solid-Phase Extraction* (New York: Marcel Dekker, 2000).

**23.** C. Crescenzi, S. Bayoudh, P. A. G. Cormack, T. Klein, and K. Ensing, "Determination of Clenbuterol in Bovine Liver by Combining Matrix Solid-Phase Dispersion and Molecularly Imprinted Solid-Phase Extraction Followed by Liquid Chromatography/Mass Spectrometry," *Anal. Chem.* **2001,** *73,* 2171; W. M. Mullett, P. Martin, and J. Pawliszyn, "In-Tube Molecularly Imprinted Polymer Solid-Phase Microextraction for the Selective Determination of Propranolol," *Anal. Chem.* **2001,** *73,* 2383; N. Masqué, R. M. Marcé, F. Borrull, P. A. G. Cormack, and D. C. Sherrington, "Synthesis and Evaluation of a Molecularly Imprinted Polymer for Solid-Phase Extraction of 4-Nitrophenol from Environmental Water," *Anal. Chem.* **2000,** *72,* 4122.

**24.** R. Hites and V. S. Ong, "Determination of Pesticides and Polychlorinated Biphenyls in Water: A Low-Solvent Method," *Environ. Sci. Technol.* **1995,** *29,* 1259.

**25.** J. W. Wong, K. K. Ngim, T. Shibamoto, S. A. Mabury, J. P. Eiserich, and H. C. H. Yeo, "Determination of Formaldehyde in Cigarette Smoke," *J. Chem. Ed.* **1997,** *74,* 1100. As another example, an analysis of nitrite ($NO_2^-$) in natural waters is based on its reaction with 2,4-dinitrophenylhydrazine to give an azide ($R-N_3$) that is measured by HPLC with ultraviolet detection at 307 nm. [R. J. Kieber and P. J. Seaton, "Determination of Subnanomolar Concentrations of Nitrite in Natural Waters," *Anal. Chem.* **1995,** *67,* 3261.]

**26.** D. M. Crumbling, Letter to Editor, *C&E News,* 12 August 2002, p. 4.

**27.** E. Tarcke, P. Rydberg, P. Karlsson, S. Eriksson, and M. Tornqvist, "Analysis of Acrylamide, a Carcinogen Formed in Heated Foodstuffs," *J. Agric. Food Chem.* **2002,** 50. 4998; B. E. Erickson, "Finding Acrylamide," *Anal. Chem.* **2004,** *76.* 247A. See collection of papers in *J. AOAC Int.,* January/February **2005,** *88,* 227–330.

# Glossary

**ablation** Vaporization of a small volume of material by a laser pulse.

**abscissa** Horizontal ($x$) axis of a graph.

**absolute uncertainty** An expression of the margin of uncertainty associated with a measurement. Absolute error also could refer to the difference between a measured value and the "true" value.

**absorbance, $A$** Defined as $A = \log(P_0/P)$, where $P_0$ is the radiant power of light (power per unit area) striking the sample on one side and $P$ is the radiant power emerging from the other side. Also called *optical density*.

**absorptance, $a$** Fraction of incident radiant power absorbed by a sample.

**absorption** Occurs when a substance is taken up *inside* another. See also *adsorption*.

**absorption coefficient** Light absorbed by a sample is attenuated at the rate $P_2/P_1 = e^{-\alpha b}$, where $P_1$ is the initial radiant power, $P_2$ is the power after traversing a pathlength $b$, and $\alpha$ is called the absorption coefficient.

**absorption spectrum** A graph of absorbance or transmittance of light versus wavelength, frequency, or wavenumber.

**accuracy** A measure of how close a measured value is to the "true" value.

**acid** A substance that increases the concentration of $H^+$ when added to water.

**acid-base titration** One in which the reaction between analyte and titrant is an acid-base reaction.

**acid dissociation constant, $K_a$** Equilibrium constant for the reaction of an acid, HA, with $H_2O$:

$$HA + H_2O \rightleftharpoons A^- + H_3O^+ \qquad K_a = \frac{\mathcal{A}_{A^-}\mathcal{A}_{H_3O^+}}{\mathcal{A}_{HA}}$$

**acid error** Occurs in strongly acidic solutions, where glass electrodes tend to indicate a value of pH that is too high.

**acidic solution** One in which the activity of $H^+$ is greater than the activity of $OH^-$.

**acidity** In natural waters, the quantity of carbonic acid and other dissolved acids that react with strong base when the pH of the sample is raised to 8.3. Expressed as mmol $OH^-$ needed to raise the pH of 1 L to pH 8.3.

**acid wash** Procedure in which glassware is soaked in 3–6 M HCl for >1 h (followed by rinsing well with distilled water and soaking in distilled water) to remove traces of cations adsorbed on the surface of the glass and to replace them with $H^+$.

**activation energy, $E_a$** Energy needed for a process to overcome a barrier that otherwise prevents the process from occurring.

**activity, $\mathcal{A}$** The value that replaces concentration in a thermodynamically correct equilibrium expression. The activity of X is given by $\mathcal{A}_x = [X]\gamma_x$, where $\gamma_x$ is the activity coefficient and $[X]$ is the concentration.

**activity coefficient, $\gamma$** The number by which the concentration must be multiplied to give activity.

**adduct** Product formed when a Lewis base combines with a Lewis acid.

**adjusted retention time $t_r'$** In chromatography, this parameter is $t_r' = t_r - t_m$, where $t_r$ is the retention time of a solute and $t_m$ is the time needed for mobile phase to travel the length of the column.

**adsorption** Occurs when a substance becomes attached to the *surface* of another substance. See also *absorption*.

**adsorption chromatography** A technique in which solute equilibrates between the mobile phase and adsorption sites on the stationary phase.

**adsorption indicator** Used for precipitation titrations, it becomes attached to a precipitate and changes color when the surface charge of the precipitate changes sign at the equivalence point.

**aerosol** A suspension of very small liquid or solid particles in air or gas. Examples include fog and smoke.

**affinity chromatography** A technique in which a particular solute is retained by a column by virtue of a specific interaction with a molecule covalently bound to the stationary phase.

**aliquot** Portion.

**alkali flame detector** Modified flame ionization detector that responds to N and P, which produce ions when they contact a $Rb_2SO_4$-containing glass bead in the flame. Also called *nitrogen-phosphorus detector*.

**alkalimetric titration** With reference to EDTA, this technique involves titration of the protons liberated from EDTA upon binding to a metal.

**alkaline error** See *sodium error*.

**alkalinity** In natural water, the quantity of base (mainly $HCO_3^-$, $CO_3^{2-}$, and $OH^-$) that reacts with strong acid when the pH of the sample is lowered to 4.5. Expressed as mmol $H^+$ needed to lower the pH of 1 L to pH 4.5.

**allosteric interaction** An effect at one part of a molecule caused by a chemical reaction or conformational change at another part of the molecule.

**amalgam** A solution of anything in mercury.

**amine** A compound with the general formula $RNH_2$, $R_2NH$, or $R_3N$, where R is any group of atoms.

**amino acid** One of 20 building blocks of proteins, having the general structure

$$\overset{\displaystyle R}{\underset{}{\overset{|}{H_3\overset{+}{N}CCO_2^-}}}$$

where R is a different substituent for each acid.

**ammonium ion** *The ammonium ion is $NH_4^+$. An* ammonium ion is any ion of the type $RNH_3^+$, $R_2NH_2^+$, $R_3NH^+$, or $R_4N^+$, where R is an organic substituent.

**ampere, $A$** One ampere is the current that will produce a force of exactly $2 \times 10^{-7}$ N/m when that current flows through two "infinitely" long, parallel conductors of negligible cross section, with a spacing of 1 m, in a vacuum.

**amperometric detector** See *electrochemical detector*.

**amperometric titration** One in which the end point is determined by monitoring the current passing between two electrodes immersed in the sample solution and maintained at a constant potential difference.

**amperometry** Measurement of electric current for analytical purposes.

**amphiprotic molecule** One that can act as both a proton donor and a proton acceptor. The intermediate species of polyprotic acids are amphiprotic.

**analysis of variance** A statistical tool used to dissect the overall random error into contributions from several sources.

**analyte** Substance being analyzed.

**analytical chromatography** Chromatography of small quantities of material conducted for the purpose of qualitative or quantitative analysis or both.

**analytical concentration** See *formal concentration*.

**anhydrous** Adjective describing a substance from which all water has been removed.

**anion** A negatively charged ion.

**anion exchanger** An ion exchanger with positively charged groups covalently attached to the support. It can reversibly bind anions.

**anode** Electrode at which oxidation occurs. In electrophoresis, it is the positively charged electrode.

**anodic depolarizer** A molecule that is easily oxidized, thereby preventing the anode potential of an electrochemical cell from becoming too positive.

**anodic wave** In polarography, a flow of current due to oxidation of analyte.

**anolyte** Solution present in the anode chamber of an electrochemical cell.

**antibody** A protein manufactured by an organism to sequester foreign molecules and mark them for destruction.

**antigen** A molecule that is foreign to an organism and that causes antibodies to be made.

**antilogarithm** The antilogarithm of $a$ is $b$ if $10^a = b$.

**antireflection coating** A coating placed on an optical component to diminish reflection. Ideally, the index of refraction of the coating should be $\sqrt{n_1 n_2}$, where $n_1$ is the refractive index of the surrounding medium and $n_2$ is the refractive index of the optical component. The thickness of the coating should be one-fourth of the wavelength of light inside the coating. Antireflection coatings are also made by layers that produce a gradation of refractive index.

**apparent mobility** Constant of proportionality, $\mu_{app}$, between the net speed, $u_{net}$, of an ion in solution and the applied electric field, $E$: $u_{net} = \mu_{app} E$. Apparent mobility is the sum of the electrophoretic and electroosmotic mobilities.

**aprotic solvent** One that cannot donate protons (hydrogen ions) in an acid-base reaction.

**aptamer** A short (15–40 bases long) length of single- or double-stranded DNA (deoxyribonucleic acid) or RNA (ribonucleic acid) that strongly binds to a selected molecule.

**aqua regia** A 3:1 (vol/vol) mixture of concentrated (37 wt%) HCl and concentrated (70 wt%) $HNO_3$.

**aqueous** In water (as an *aqueous* solution).

**aquo ion** The species $M(H_2O)_n^{m+}$, containing just the cation M and its tightly bound water ligands.

**argentometric titration** One using $Ag^+$ ion.

**Arrhenius equation** Empirical relation between the rate constant, $k$, for a chemical reaction and temperature, $T$, in kelvins: $k = Be^{-\Delta G^{\ddagger}/RT}$, where $R$ is the gas constant, $\Delta G^{\ddagger}$ is the free energy of activation for the chemical reaction, and $B$ is a constant called the preexponential factor. This equation is more commonly written in the form $k = Ae^{-E_a/RT}$, where $E_a$ is called the activation energy.

**ashless filter paper** Specially treated paper that leaves a negligible residue after ignition. It is used for gravimetric analysis.

**assessment** In quality assurance, the process of (1) collecting data to show that analytical procedures are operating within specified limits and (2) verifying that final results meet use objectives.

**asymmetry potential** When the activity of analyte is the same on the inside and outside of an ion-selective electrode, there should be no voltage across the membrane. In fact, the two surfaces are never identical, and some voltage (called the asymmetry potential) is usually observed.

**atmosphere, atm** One atm is defined as a pressure of 101 325 Pa. It is equal to the pressure exerted by a column of Hg 760 mm in height at the earth's surface.

**atmospheric pressure chemical ionization** A method to interface liquid chromatography to mass spectrometry. Liquid is nebulized into a fine aerosol by a coaxial gas flow and the application of heat. Electrons from a high-voltage corona discharge create cations and anions from analyte exiting the chromatography column. The most common species observed with this interface is $MH^+$, the protonated analyte, with little fragmentation.

**atomic absorption spectroscopy** A technique in which the absorption of light by free gaseous atoms in a flame or furnace is used to measure the concentration of atoms.

**atomic emission spectroscopy** A technique in which the emission of light by thermally excited atoms in a flame or furnace is used to measure the concentration of atoms.

**atomic fluorescence spectroscopy** A technique in which electronic transitions of atoms in a flame, furnace, or plasma are excited by light, and the fluorescence is observed at a right angle to the incident beam.

**atomic mass** Number of grams of an element containing Avogadro's number of atoms.

**atomization** Process in which a compound is decomposed into its atoms at high temperature.

**attenuated total reflection** An analytical technique based on passage of light through a waveguide or optical fiber by total internal reflection. The absorption of the cladding is sensitive to the presence of analyte. Some of the evanescent wave is absorbed in the cladding during each reflection in the presence of analyte. The more analyte, the more signal is lost.

**autoprotolysis** Reaction of a neutral solvent, in which two molecules of the same species transfer a proton from one to the other, thereby producing ions; e.g., $CH_3OH + CH_3OH \rightleftharpoons CH_3OH_2^+ + CH_3O^-$.

**auxiliary complexing agent** A species, such as ammonia, that is added to a solution to stabilize another species and keep that other species in solution. It binds loosely enough to be displaced by a titrant.

**auxiliary electrode** Current-carrying partner of the working electrode in an electrolysis.

**average** The sum of several values divided by the number of values. Also called *mean*.

**Avogadro's number** The number of atoms in exactly 0.012 kg of $^{12}$C.

**azeotrope** The constant-boiling distillate produced by two liquids. It is of constant composition, containing both substances.

**background buffer** In capillary electrophoresis, the buffer in which separation is carried out. Also called *run buffer*.

**background correction** In atomic spectroscopy, a means of distinguishing signal due to analyte from signal due to absorption, emission, or scattering by the flame, furnace, plasma, or sample matrix.

**back titration** One in which an excess of standard reagent is added to react with analyte. Then the excess reagent is titrated with a second reagent or with a standard solution of analyte.

**ball mill** A drum in which solid sample is ground to fine powder by tumbling with hard ceramic balls.

**band gap** Energy separating the valence band and conduction band in a semiconductor.

**band-pass filter** A filter that allows a band of wavelengths to pass through it while absorbing or reflecting other wavelengths.

**bandwidth** Usually, the range of wavelengths or frequencies of an absorption or emission band, measured at a height equal to half of the peak height. It also refers to the width of radiation emerging from the exit slit of a monochromator.

**base** A substance that decreases the concentration of $H^+$ when added to water.

**base "dissociation" constant** A misnomer for *base hydrolysis constant, $K_b$*.

**base hydrolysis constant, $K_b$** The equilibrium constant for the reaction of a base, B, with $H_2O$:

$$B + H_2O \rightleftharpoons BH^+ + OH^- \qquad K_b = \frac{\mathcal{A}_{BH^+}\mathcal{A}_{OH^-}}{\mathcal{A}_B}$$

**base peak** Most intense peak in a mass spectrum.

**basic solution** One in which the activity of $OH^-$ is greater than the activity of $H^+$.

**beam chopper** A rotating mirror that directs light alternately through the sample and reference cells of a double-beam spectrophotometer.

**beam chopping** A technique using a rotating *beam chopper* to modulate the signal in a spectrophotometer at a frequency at which noise is reduced. In atomic absorption, periodic blocking of the beam allows a distinction to be made between light from the source and light from the flame.

**beamsplitter** A partially reflective, partially transparent plate that reflects some light and transmits some light.

**Beer's law** Relates the absorbance, $A$, of a sample to its concentration, $c$, pathlength, $b$, and molar absorptivity, $\varepsilon$: $A = \varepsilon bc$.

**biamperometric titration** An amperometric titration conducted with two polarizable electrodes held at a constant potential difference.

**bilayer** Formed by a surfactant, a two-dimensional membrane structure in which polar or ionic headgroups are pointing outward and nonpolar tails are pointing inward.

**biochemical oxygen demand, BOD** In a water sample, the quantity of dissolved oxygen consumed by microorganisms during a 5-day incubation in a sealed vessel at 20°C. Oxygen consumption is limited by organic nutrients, so BOD is a measure of pollutant concentration.

**biosensor** Device that uses biological components such as enzymes, antibodies, or DNA, in combination with electrical, optical, or other signals, to achieve a highly selective response to one analyte.

**bipotentiometric titration** A potentiometric titration in which a constant current is passed between two polarizable electrodes immersed in the sample solution. An abrupt change in potential characterizes the end point.

**Bjerrum plot** See *difference plot.*

**blackbody** An ideal surface that absorbs all photons striking it. If the blackbody is at constant temperature, it must emit as much radiant energy as it absorbs.

**blackbody radiation** Radiation emitted by a blackbody. The energy and spectral distribution of the emission depend only on the temperature of the blackbody.

**blank solution** A solution not intended to contain analyte. It could be made from all reagents—except unknown—that would be used in an analytical procedure. Analyte signal measured with a blank solution could be due to impurities in the reagents or, possibly, interference.

**blank titration** One in which a solution containing all reagents except analyte is titrated. The volume of titrant needed in the blank titration should be subtracted from the volume needed to titrate unknown.

**blind sample** See *performance test sample.*

**blocking** Occurs when a metal ion binds tightly to a metal ion indicator. A blocked indicator is unsuitable for a titration because no color change is observed at the end point.

**bolometer** An infrared detector whose electrical resistance changes when it is heated by infrared radiation.

**Boltzmann distribution** Relative population of two states at thermal equilibrium:

$$\frac{N_2}{N_1} = \frac{g_2}{g_1}e^{-(E_2-E_1)/kT}$$

where $N_i$ is the population of the state, $g_i$ is the degeneracy of the state, $E_i$ is the energy of the state, $k$ is Boltzmann's constant, and $T$ is temperature in kelvins. Degeneracy refers to the number of states with the same energy.

**bomb** Sealed vessel for conducting high-temperature, high-pressure reactions.

**bonded stationary phase** In high-performance liquid chromatography, a stationary liquid phase covalently attached to the solid support.

**Brewster window** A flat optical window tilted at an angle such that light whose electric vector is polarized parallel to the plane of the window is 100% transmitted. Light polarized perpendicular to the window is partially reflected. It is used on the ends of a laser to produce light whose electric field oscillates perpendicularly to the long axis of the laser.

**Brønsted-Lowry acid** A proton (hydrogen ion) donor.

**Brønsted-Lowry base** A proton (hydrogen ion) acceptor.

**buffer** A mixture of an acid and its conjugate base. A buffered solution is one that resists changes in pH when acids or bases are added.

**buffer capacity, β** A measure of the ability of a buffer to resist changes in pH. The larger the buffer capacity, the greater the resistance to pH change. The definition of buffer capacity is $\beta = dC_b/dpH = -dC_a/dpH$, where $C_a$ and $C_b$ are the number of moles of strong acid or base per liter needed to produce a unit change in pH. Also called *buffer intensity.*

**buffer intensity** See *buffer capacity.*

**bulk sample** Material taken from lot being analyzed—usually chosen to be representative of the entire lot. Also called *gross sample.*

**bulk solution** Chemists' jargon referring to the main body of a solution. In electrochemistry, we distinguish properties of the bulk solution from properties that may be different in the immediate vicinity of an electrode.

**buoyancy** Upward force exerted on an object in a liquid or gaseous fluid. An object weighed in air appears lighter than its actual mass by an amount equal to the mass of air that it displaces.

**buret** A calibrated glass tube with a stopcock at the bottom. Used to deliver known volumes of liquid.

**calibration** Process of measuring the actual physical quantity (such as mass, volume, force, or electric current) corresponding to an indicated quantity on the scale of an instrument.

**calibration check** In a series of analytical measurements, a calibration check is an analysis of a solution formulated by the analyst to contain a known concentration of analyte. It is the analyst's own check that procedures and instruments are functioning correctly.

**calibration curve** A graph showing the value of some property versus concentration of analyte. When the corresponding property of an unknown is measured, its concentration can be determined from the graph.

**calomel electrode** A common reference electrode based on the half-reaction

$$Hg_2Cl_2(s) + 2e^- \rightleftharpoons 2Hg(l) + 2Cl^-$$

**candela, cd** Luminous intensity, in a given direction, of a source that emits monochromatic radiation of frequency 540 THz and that has a radiant intensity of 1/683 W/sr in that direction.

**capacitor current** See *charging current.*

**capacity factor, $k'$** In chromatography, the adjusted retention time for a peak divided by the time for the mobile phase to travel through the column. Capacity factor is also equal to the ratio of the time spent by the solute in the stationary phase to the time spent in the mobile phase. Also called *retention factor, capacity ratio,* and *partition ratio.*

**capacity ratio** See *capacity factor.*

**capillary constant** The quantity $m^{2/3}t^{1/6}$ characteristic of each dropping Hg electrode. The rate of flow is $m$ (mg/s) and $t$ is the drop interval (s). The capillary constant is proportional to the square root of the Hg height.

**capillary electrochromatography** A version of high-performance liquid chromatography in which mobile phase is driven by electroosmosis instead of a pressure gradient.

**capillary electrophoresis** Separation of a mixture into its components by using a strong electric field imposed between the two ends of a narrow capillary tube filled with electrolyte solution.

**capillary gel electrophoresis** A form of capillary electrophoresis in which the tube is filled with a polymer gel that serves as a sieve for macromolecules. The largest molecules have the slowest migration through the gel.

**capillary zone electrophoresis** A form of capillary electrophoresis in which ionic solutes are separated because of differences in their electrophoretic mobility.

**carboxylate anion** Conjugate base ($RCO_2^-$) of a carboxylic acid.

**carboxylic acid** A molecule with the general structure $RCO_2H$, where R is any group of atoms.

**carcinogen** A cancer-causing agent.

**carrier gas** Mobile-phase gas in gas chromatography.

**catalytic wave** One that results when the product of a polarographic reaction is rapidly regenerated by reaction with another species and the polarographic wave height increases.

**cathode** Electrode at which reduction occurs. In electrophoresis, it is the negatively charged electrode.

**cathodic depolarizer** A molecule that is easily reduced, thereby preventing the cathode potential of an electrochemical cell from becoming very low.

**catholyte** Solution present in the cathode chamber of an electrochemical cell.

**cation** A positively charged ion.

**cation exchanger** An ion exchanger with negatively charged groups covalently attached to the support. It can reversibly bind cations.

**chain of custody** Trail followed by a sample from the time it is collected to the time it is analyzed and, possibly, archived.

**characteristic** The part of a logarithm at the left of the decimal point.

**charge balance** A statement that the sum of all positive charge in solution equals the magnitude of the sum of all negative charge in solution.

**charge coupled device** An extremely sensitive detector in which light creates electrons and holes in a semiconductor material. The electrons are attracted to regions near positive electrodes, where the electrons are "stored" until they are ready to be counted. The number of electrons in each pixel (picture element) is proportional to the number of photons striking the pixel.

**charging current** Electric current arising from charging or discharging of the electric double layer at the electrode-solution interface. Also called *capacitor current* or *condenser current.*

**charring** In a gravimetric analysis, the precipitate (and filter paper) are first dried gently. Then the filter paper is *charred* at intermediate temperature to destroy the paper without letting it inflame. Finally, the precipitate is ignited at high temperature to convert it into its analytical form.

**chelate effect** The observation that a single multidentate ligand forms metal complexes that are more stable than those formed by several individual ligands with the same ligand atoms.

**chelating ligand** A ligand that binds to a metal through more than one atom.

**chemical coulometer** A device that measures the yield of an electrolysis reaction to determine how much electricity has flowed through a circuit.

**chemical interference** In atomic spectroscopy, any chemical reaction that decreases the efficiency of atomization.

**chemical ionization** A gentle method of producing ions for a mass spectrometer without extensive fragmentation of the analyte molecule, M. A reagent gas such as $CH_4$ is bombarded with electrons to make $CH_5^+$, which transfers $H^+$ to M, giving $MH^+$.

**chemical oxygen demand, COD** In a natural water or industrial effluent sample, the quantity of $O_2$ equivalent to the quantity of $K_2Cr_2O_7$ consumed by refluxing the sample with a standard dichromate–sulfuric acid solution containing $Ag^+$ catalyst. Because 1 mol of $K_2Cr_2O_7$ consumes $6e^- (Cr^{6+} \rightarrow Cr^{3+})$, it is equivalent to 1.5 mol of $O_2(O \rightarrow O^{2-})$.

**chemiluminescence** Emission of light by an excited-state product of a chemical reaction.

**chromatogram** A graph showing chromatography detector response as a function of elution time or volume.

**chromatograph** A machine used to perform chromatography.

**chromatography** A technique in which molecules in a mobile phase are separated because of their different affinities for a stationary phase. The greater the affinity for the stationary phase, the longer a molecule is retained.

**chromophore** The part of a molecule responsible for absorption of light of a particular frequency.

**chronoamperometry** A technique in which the potential of a working electrode in an unstirred solution is varied rapidly while the current between the working and the auxiliary electrodes is measured. Suppose that the analyte is reducible and that the potential of the working electrode is made more negative. Initially, no reduction occurs. At a certain potential, the analyte begins to be reduced and the current increases. As the potential becomes more negative, the current increases further until the concentration of analyte at the surface of the electrode is sufficiently depleted. Then the current decreases, even though the potential becomes more negative. The maximum current is proportional to the concentration of analyte in bulk solution.

**chronopotentiometry** A technique in which a constant current is forced to flow between two electrodes. The voltage remains fairly steady until the concentration of an electroactive species becomes depleted. Then the voltage changes rapidly as a new redox reaction assumes the burden of current flow. The elapsed time when the voltage suddenly changes is proportional to the concentration of the initial electroactive species in bulk solution.

**Clark electrode** One that measures the activity of dissolved oxygen by amperometry.

**coagulation** With respect to gravimetric analysis, small crystallites coming together to form larger crystals.

**co-chromatography** Simultaneous chromatography of known compounds with an unknown. If a known and an unknown have the same retention time on several columns, they are probably identical.

**coefficient of variation** The standard deviation, $s$, expressed as a percentage of the mean value $\bar{x}$: coefficient of variation $= 100 \times s/\bar{x}$. Also called *relative standard deviation*.

**coherence** Degree to which electromagnetic waves are in phase with one another. Laser light is highly coherent.

**co-ion** An ion with the same charge as the ion of interest.

**cold trapping** Splitless gas chromatography injection technique in which solute is condensed far below its boiling point in a narrow band at the start of the column.

**collimated light** Light in which all rays travel in parallel paths.

**collimation** Process of making light rays travel parallelly to one another.

**colloid** A dissolved particle with a diameter in the approximate range 1–100 nm. It is too large to be considered one molecule but too small to simply precipitate.

**collisionally activated dissociation** Fragmentation of an ion in a mass spectrometer by high-energy collisions with gas molecules. In atmospheric pressure chemical ionization or electrospray interfaces, collisionally activated dissociation at the inlet to the mass filter can be promoted by varying the cone voltage. In tandem mass spectrometry, dissociation occurs in a collision cell between the two mass separators.

**collision cell** Middle stage of a tandem mass spectrometer in which the precursor ion selected by the first stage is fragmented by collisions with gas molecules.

**combination electrode** Consists of a glass electrode with a concentric reference electrode built on the same body.

**combustion analysis** A technique in which a sample is heated in an atmosphere of $O_2$ to oxidize it to $CO_2$ and $H_2O$, which are collected and weighed or measured by gas chromatography. Modifications permit the simultaneous analysis of N, S, and halogens.

**common ion effect** Occurs when a salt is dissolved in a solution already containing one of the ions of the salt. The salt is less soluble than it would be in a solution without that ion. An application of Le Châtelier's principle.

**complex ion** Historical name for any ion containing two or more ions or molecules that are each stable by themselves; e.g., $CuCl_3^-$ contains $Cu^+ + 3Cl^-$.

**complexometric titration** One in which the reaction between analyte and titrant involves complex formation.

**composite sample** A representative sample prepared from a heterogeneous material. If the material consists of distinct regions, the composite is made of portions of each region, with relative amounts proportional to the size of each region.

**compound electrode** An ion-selective electrode consisting of a conventional electrode surrounded by a barrier that is selectively permeable to the analyte of interest. Alternatively, the barrier region might convert external analyte into a different species, to which the inner electrode is sensitive.

**concentration** An expression of the quantity per unit volume or unit mass of a substance. Common measures of concentration are molarity (mol/L) and molality (mol/kg of solvent).

**concentration cell** A galvanic cell in which both half-reactions are the same, but the concentrations in each half-cell are not identical. The cell reaction increases the concentration of species in one half-cell and decreases the concentration in the other.

**concentration polarization** Occurs when an electrode reaction occurs so rapidly that the concentration of solute near the surface of the electrode is not the same as the concentration in bulk solution.

**condenser current** See *charging current*.

**conditional formation constant** See *effective formation constant*.

**conduction band** Energy levels containing conduction electrons in a semiconductor.

**conduction electron** An electron free to move about within a solid and carry electric current. In a semiconductor, the energies of the conduction electrons are above those of the valence electrons that are localized in chemical bonds. The energy separating the valence and conduction bands is called the band gap.

**conductivity, $\sigma$** Proportionality constant between electric current density, $J(A/m^2)$, and electric field, $E$ (V/m): $J = \sigma E$. Units are $\Omega^{-1}m^{-1}$. Conductivity is the reciprocal of *resistivity*.

**cone voltage** Voltage applied between the *skimmer cone* and a nearby orifice through which gaseous ions flow into the mass separator of a mass spectrometer. The magnitude of the voltage can be increased to promote collisionally activated dissociation of ions prior to mass separation.

**confidence interval** Range of values within which there is a specified probability that the true value lies.

**conjugate acid-base pair** An acid and a base that differ only through the gain or loss of a single proton.

**constant-current electrolysis** Electrolysis in which a constant current flows between working and auxiliary electrodes. As reactants are consumed, an ever-increasing voltage is required to keep the current flowing, so this is the least selective form of electrolysis.

**constant mass** In gravimetric analysis, the product is heated and cooled to room temperature in a desiccator until successive weighings are "constant." There is no standard definition of "constant mass"; but, for ordinary work, it is usually taken to be about $\pm 0.3$ mg. Constancy is usually limited by the irreproducible regain of moisture during cooling and weighing.

**constant-voltage electrolysis** Electrolysis in which a constant voltage is maintained between working and auxiliary electrodes. This is less selective than controlled-potential electrolysis because the potential of the working electrode becomes more extreme as ohmic potential and overpotential change.

**control chart** A graph in which periodic observations of a process are recorded to determine whether the process is within specified control limits.

**controlled-potential electrolysis** A technique for selective reduction (or oxidation), in which the voltage between working and reference electrodes is held constant.

**convection** Process in which solute is carried from one place to another by bulk motion of the solution.

**cooperativity** Interaction between two parts of a molecule such that an event at one part affects the behavior of the other. Consider a molecule, M, with two sites to which two identical species, S, can bind to form MS and $MS_2$. When binding at one site makes binding at the second site more favorable than in the absence of the first binding, there *is positive cooperativity*. When the first binding makes the second binding less favorable, there is *negative cooperativity*. In *noncooperative* binding, neither site affects the other.

**coprecipitation** Occurs when a substance whose solubility is not exceeded precipitates along with one whose solubility is exceeded.

**correlation coefficient** The square of the correlation coefficient, $R^2$, is a measure of goodness of fit of data points to a straight line. The closer $R^2$ is to 1, the better the fit.

**coulomb, C** Amount of charge per second that flows past any point in a circuit when the current is 1 ampere. There are approximately 96 485 coulombs in a mole of electrons.

**coulometric titration** One conducted with a constant current for a measured time.

**coulometry** A technique in which the quantity of analyte is determined by measuring the number of coulombs needed for complete electrolysis.

**countercurrent distribution** A technique in which a series of solvent extractions is used to separate solutes from one another.

**counterelectrode** Current-carrying partner of the working electrode. Also called *auxiliary electrode.*

**counterion** An ion with a charge opposite that of the ion of interest.

**coupled equilibria** Reversible chemical reactions that have a species in common. For example, the product of one reaction could be a reactant in another reaction.

**critical point** Critical temperature and pressure of a substance.

**critical pressure** Pressure above which a fluid cannot be condensed to two phases (liquid and gas), no matter how low the temperature.

**critical temperature** Temperature above which a fluid cannot be condensed to two phases (liquid and gas), no matter how great a pressure is applied.

**cross-linking** Covalent linkage between different strands of a polymer.

**crystallization** Process in which a substance comes out of solution slowly to form a solid with a regular arrangement of atoms.

**cumulative formation constant, $\beta_n$** Equilibrium constant for a reaction of the type $M + nX \rightleftharpoons MX_n$. Also called *overall formation constant.*

**current, $I$** Amount of charge flowing through a circuit per unit time (A/s).

**current density** Electric current per unit area ($A/m^2$).

**cuvet** A cell with transparent walls used to hold samples for spectrophotometric measurements.

**cyclic voltammetry** A polarographic technique with a triangular waveform. Both cathodic and anodic currents are observed for reversible reactions.

**data quality objective** Accuracy, precision, and sampling requirements for an analytical method.

**dead stop end point** End point of a biamperometric titration.

**dead volume** Volume of a chromatography system (not including the column) between the point of injection and the point of detection. Also called *extra-column volume.*

**Debye-Hückel equation** Gives the activity coefficient, $\gamma$, as a function of ionic strength, $\mu$. The extended Debye-Hückel equation, applicable to ionic strengths up to about 0.1 M, is $\log \gamma = [-0.51z^2 \sqrt{\mu}]/[1 + (\alpha \sqrt{\mu}/305)]$, where $z$ is the ionic charge and $\alpha$ is the effective hydrated radius in picometers.

**decant** To pour liquid off a solid or, perhaps, a denser liquid. The denser phase is left behind.

**decomposition potential** In an electrolysis, that voltage at which rapid reaction first begins.

**degree of freedom** In statistics, the number of independent observations on which a result is based.

**deionized water** Water that has been passed through a cation exchanger (in the $H^+$ form) and an anion exchanger (in the $OH^-$ form) to remove ions from the solution.

**deliquescent substance** Like a hygroscopic substance, one that spontaneously picks up water from the air. It can eventually absorb so much water that the substance completely dissolves.

**demasking** Removal of a masking agent from the species protected by the masking agent.

**density** Mass per unit volume.

**depolarizer** A molecule that is oxidized or reduced at a modest potential. It is added to an electrolytic cell to prevent the cathode or anode potential from becoming too extreme.

**derivatization** Chemical alteration to attach a group to a molecule so that the molecule can be detected conveniently. Alternatively, treatment can alter volatility or solubility.

**desalting** Removal of salts (or any small molecules) from a solution of macromolecules. Gel filtration or dialysis is used for desalting.

**desiccant** A drying agent.

**desiccator** A sealed chamber in which samples can be dried in the presence of a desiccant or by vacuum pumping or both.

**detection limit** The smallest quantity of analyte that is "significantly different" from a blank. The detection limit is often taken as the mean signal for blanks plus 3 times the standard deviation of a low-concentration sample. If you have a recorder trace with a signal plus adjacent baseline noise, the detection limit is sometimes taken as twice the peak-to-peak noise level or 10 times the *root-mean-square noise* level (which is 1/5 of the peak-to-peak noise level). Also called *lower limit of detection.*

**determinant** The value of the two-dimensional determinant $\begin{vmatrix} a & b \\ c & d \end{vmatrix}$ is the difference $ad - bc$.

**determinate error** See *systematic error.*

**deuterium arc lamp** Source of broadband ultraviolet radiation. An electric discharge (a spark) in deuterium gas causes $D_2$ molecules to dissociate and emit many wavelengths of radiation.

**dialysis** A technique in which solutions are placed on either side of a semipermeable membrane that allows small molecules, but not large molecules, to cross. Small molecules in the two solutions diffuse across and equilibrate between the two sides. Large molecules are retained on their original side.

**dielectric constant, $\varepsilon$** The electrostatic force, $F$, between two charged particles is given by $F = kq_1q_2/\varepsilon r^2$, where $k$ is a constant, $q_1$ and $q_2$ are the charges, $r$ is the separation between particles, and $\varepsilon$ is the dielectric constant of the medium. The higher the dielectric constant, the less force is exerted by one charged particle on another.

**difference plot** A graph of the mean fraction of protons bound to an acid versus pH. For complex formation, the difference plot gives the mean number of ligands bound to a metal versus pL ($= -\log[\text{ligand concentration}]$). Also called *Bjerrum plot.*

**differential pulse polarography** A technique in which current is measured before and at the end of potential pulses superimposed on a voltage ramp. It is more sensitive than ordinary polarography, and the signal closely approximates the derivative of a polarographic wave.

**diffraction** Occurs when electromagnetic radiation passes through or is reflected from slits with a spacing comparable to the wavelength. Interference of waves from adjacent slits produces a spectrum of radiation, with each wavelength emerging at a different angle.

**diffuse part of the double layer** Region of solution near a charged surface in which excess counterions are attracted to the charge. The thickness of this layer is 0.3–10 nm.

**diffuse reflection** Occurs when a rough surface reflects light in all directions.

**diffusion** Random motion of molecules in a liquid or gas (or, very slowly, in a solid).

**diffusion coefficient, $D$** Defined by Fick's first law of diffusion: $J = -D(dc/dx)$, where $J$ is the rate at which molecules diffuse across a plane of unit area and $dc/dx$ is the concentration gradient in the direction of diffusion.

**diffusion current** In polarography, the current observed when the rate of reaction is limited by the rate of diffusion of analyte to the electrode. Diffusion current = limiting current − residual current.

**diffusion layer** Region near an electrode containing excess product or decreased reactant involved in the electrode reaction. The thickness of this layer can be hundreds of micrometers.

**digestion** Process in which a precipitate is left (usually warm) in the presence of mother liquor to promote particle recrystallization and growth. Purer, more easily filterable crystals result. Also used to describe any chemical treatment in which a substance is decomposed to transform the analyte into a form suitable for analysis.

**dilution factor** Factor (initial volume of reagent)/(total volume of solution) used to multiply the initial concentration of reagent to find the diluted concentration.

**dimer** A molecule made from two identical units.

**diode** A semiconductor device consisting of a *pn* junction through which current can pass in only one direction. Current flows when the *n*-type material is made negative and the *p*-type material is made positive. A voltage sufficient to overcome the activation energy for carrier movement must be supplied before any current flows. For silicon diodes, this voltage is ~0.6 V. If a sufficiently large voltage, called the breakdown voltage, is applied in the reverse direction, current will flow in the wrong direction through the diode.

**diprotic acid** One that can donate two protons.

**direct current polarography** Classical form of polarography, in which a linear voltage ramp is applied to the working electrode.

**direct titration** One in which the analyte is treated with titrant and the volume of titrant required for complete reaction is measured.

**dispersion** A measure of the ability of a monochromator to separate wavelengths differing by $\Delta\lambda$ through the angle $\Delta\phi$. The greater the dispersion, the greater the angle separating two closely spaced wavelengths. For a prism, dispersion refers to the rate of change of refractive index with wavelength, $dn/d\lambda$.

**displacement titration** An EDTA titration procedure in which analyte is treated with excess $MgEDTA^{2-}$ to displace $Mg^{2+}$: $M^{n+} + MgEDTA^{2-} \rightleftharpoons MEDTA^{n-4} + Mg^{2+}$. The liberated $Mg^{2+}$ is then titrated with EDTA. This procedure is useful if there is not a suitable indicator for direct titration of $M^{n+}$.

**disproportionation** A reaction in which an element in one oxidation state gives products containing that element in both higher and lower oxidation states; e.g., $2Cu^+ \rightleftharpoons Cu^{2+} + Cu(s)$.

**distribution coefficient, $D$** For a solute partitioned between two phases, the distribution coefficient is the total concentration of all forms of solute in phase 2 divided by the total concentration in phase 1.

**Donnan equilibrium** Ions of the same charge as those fixed on an exchange resin are repelled from the resin. Thus, anions do not readily penetrate a cation-exchange resin, and cations are repelled from an anion-exchange resin.

**dopant** When small amounts of substance B are added to substance A, we call B a dopant and say that A is doped with B. Doping is done to alter the properties of A.

**Doppler effect** A molecule moving toward a source of radiation experiences a higher frequency than one moving away from the source.

**double-focusing mass spectrometer** A spectrometer that uses electric and magnetic sectors in series to obtain high resolution.

**double layer** See *electric double layer*.

**drift** Slow change in the response of an instrument due to various causes such as changes in electrical components with temperature, variation in power-line voltage to an instrument, and aging of components within instruments. Same as *1/f noise* or *flicker noise*.

**dropping-mercury electrode** One that delivers fresh drops of Hg to a polarographic cell.

**dry ashing** Oxidation of organic matter with $O_2$ at high temperature to leave behind inorganic components for analysis.

**dwell volume** Volume in chromatography between the point of mixing solvents and the start of the column.

**dynamic range** Range of analyte concentration over which a change in concentration gives a change in detector response.

$E°$ Standard reduction potential.

$E°'$ Effective standard reduction potential at pH 7 (or at some other specified conditions).

**EDTA (ethylenediaminetetraacetic acid)** $(HO_2CCH_2)_2NCH_2CH_2N-(CH_2CO_2H)_2$, the most widely used reagent for complexometric titrations. It forms 1:1 complexes with virtually all cations with a charge of 2 or more.

**effective formation constant** Equilibrium constant for formation of a complex under a particular stated set of conditions, such as pH, ionic strength, and concentration of auxiliary complexing species. Also called *conditional formation constant*.

**effervescence** Rapid release of gas with bubbling and hissing.

**efflorescence** Property by which the outer surface or entire mass of a substance turns into powder from loss of water of crystallization.

**effluent** See *eluate*.

**einstein** A mole of photons. The symbol of this unit is the same as its name, einstein.

**electric discharge emission spectroscopy** A technique in which atomization and excitation are stimulated by an electric arc, a spark, or microwave radiation.

**electric double layer** Region comprising the charged surface of an electrode or a particle plus the oppositely charged region of solution adjacent to the surface. Also called *double layer*.

**electric potential** The electric potential (in volts) at a point is the energy (in joules) needed to bring 1 coulomb of positive charge from infinity to that point. The potential difference between two points is the energy needed to transport 1 coulomb of positive charge from the negative point to the positive point.

**electroactive species** Any species that can be oxidized or reduced at an electrode.

**electrocapillary maximum** Potential at which the net charge on a mercury drop from a dropping-mercury electrode is 0 (and the surface tension of the drop is maximal).

**electrochemical detector** Liquid chromatography detector that measures current when an electroactive solute emerges from the column and passes over a working electrode held at a fixed potential with respect to a reference electrode. Also called *amperometric detector*.

**electrochemistry** Use of electrical measurements on a chemical system for analytical purposes. Also refers to use of electricity to drive a chemical reaction or use of a chemical reaction to produce electricity.

**electrode** A device through which electrons flow into or out of chemical species involved in a redox reaction.

**electroendosmosis** See *electroosmosis*.

**electrogravimetric analysis** A technique in which the mass of an electrolytic deposit is used to quantify the analyte.

**electrokinetic injection** In capillary electrophoresis, the use of an electric field to inject sample into the capillary. Because different species have different mobilities, the injected sample does not have the same composition as the original sample.

**electrolysis** Process in which the passage of electric current causes a chemical reaction to occur.

**electrolyte** A substance that produces ions when dissolved.

**electrolytic cell** A cell in which a chemical reaction that would not otherwise occur is driven by a voltage applied between two electrodes.

**electromagnetic spectrum** The whole range of electromagnetic radiation, including visible light, radio waves, X-rays, etc.

**electron capture detector** Gas chromatography detector that is particularly sensitive to compounds with halogen atoms, nitro groups, and other groups with high electron affinity. Makeup gas ($N_2$ or 5% $CH_4$ in Ar) is ionized by β-rays from $^{63}$Ni to liberate electrons that produce a small, steady current. High-electron-affinity analytes capture some of the electrons and reduce the detector current.

**electronic balance** A weighing device that uses an electromagnetic servomotor to balance the load on the pan. The mass of the load is proportional to the current needed to balance it.

**electronic transition** One in which an electron is promoted from one energy level to another.

**electron ionization** Interaction of analyte molecules (M) with high-energy electrons in the ion source of a mass spectrometer to give the cation radical, $M^{+\cdot}$, and fragments derived from $M^{+\cdot}$.

**electron multiplier** An ion detector that works like a photomultiplier tube. Cations striking a cathode liberate electrons. A series of *dynodes* multiplies the number of electrons by $\sim 10^5$ before they reach the anode.

**electroosmosis** Bulk flow of fluid in a capillary tube induced by an electric field. Mobile ions in the diffuse part of the double layer at the wall of the capillary serve as the "pump." Also called *electroendosmosis.*

**electroosmotic flow** Uniform, pluglike flow of fluid in a capillary tube under the influence of an electric field. The greater the charge on the wall of the capillary, the greater the number of counterions in the double layer and the stronger the electroosmotic flow.

**electroosmotic mobility, $\mu_{eo}$** Constant of proportionality between the electroosmotic speed, $u_{eo}$, of a fluid in a capillary and the applied electric field, $E$: $u_{eo} = \mu_{eo}E$.

**electroosmotic velocity** Speed with which solvent flows through a capillary electrophoresis column. It is measured by adding a detectable neutral molecule to the sample. Electroosmotic velocity is the distance from injector to detector divided by the time required for the neutral molecule to reach the detector.

**electropherogram** A graph of detector response versus time for electrophoresis.

**electrophoresis** Migration of ions in solution in an electric field. Cations move toward the cathode and anions move toward the anode.

**electrophoretic mobility, $\mu_{ep}$** Constant of proportionality between the electrophoretic speed, $u_{ep}$, of an ion in solution and the applied electric field, $E$: $u_{ep} = \mu_{ep}E$.

**electrospray** A method for interfacing liquid chromatography to mass spectrometry. A high potential applied to the liquid at the column exit creates charged droplets in a fine aerosol. Gaseous ions are derived from ions that were already in the mobile phase on the column. It is common to observe protonated bases($BH^+$), ionized acids($A^-$), and complexes formed between analyte, M (which could be neutral or charged), and stable ions such as $NH_4^+$, $Na^+$, $HCO_2^-$, or $CH_3CO_2^-$ that were already in solution.

**eluate** What comes out of a chromatography column. Also called *effluent.*

**eluent** Solvent applied to the beginning of a chromatography column.

**eluent strength, $\varepsilon°$** A measure of the ability of a solvent to elute solutes from a chromatography column. Also called *solvent strength,* eluent strength is a measure of the adsorption energy of a solvent on the stationary phase in chromatography.

**eluotropic series** Ranks solvents according to their ability to displace solutes from the stationary phase in adsorption chromatography.

**elution** Process of passing a liquid or a gas through a chromatography column.

**emission spectrum** A graph of luminescence intensity versus luminescence wavelength (or frequency or wavenumber), obtained with a fixed excitation wavelength.

**emissivity** A quotient given by the radiant emission from a real object divided by the radiant emission of a blackbody at the same temperature.

**emulsion** A fine suspension of immiscible liquid droplets in another liquid phase.

**enantiomers** Mirror image isomers that cannot be superimposed on each other. Also called *optical isomers.*

**endergonic reaction** One for which $\Delta G$ is positive; it is not spontaneous.

**endothermic reaction** One for which $\Delta H$ is positive; heat must be supplied to reactants for them to react.

**end point** Point in a titration at which there is a sudden change in a physical property, such as indicator color, pH, conductivity, or absorbance. Used as a measure of the equivalence point.

**enthalpy change, $\Delta H$** The heat absorbed or released when a reaction occurs at constant pressure.

**enthalpy of hydration** Heat liberated when a gaseous species is transferred to water.

**entropy** A measure of the "disorder" of a substance.

**enzyme** A protein that catalyzes a chemical reaction.

**equilibrium** State in which the forward and reverse rates of all reactions are equal, so the concentrations of all species remain constant.

**equilibrium constant, $K$** For the reaction $a\text{A} + b\text{B} \rightleftharpoons c\text{C} + d\text{D}$, $K = \mathcal{A}_C^c\mathcal{A}_D^d/\mathcal{A}_A^a\mathcal{A}_B^b$, where $\mathcal{A}_i$ is the activity of the $i$th species.

**equimolar mixture of compounds** One that contains an equal number of moles of each compound.

**equivalence point** Point in a titration at which the quantity of titrant is exactly sufficient for stoichiometric reaction with the analyte.

**equivalent** For a redox reaction, the amount of reagent that can donate or accept 1 mole of electrons. For an acid-base reaction, the amount of reagent that can donate or accept 1 mole of protons.

**equivalent weight** The mass of substance containing one equivalent.

**evanescent wave** Light that penetrates the walls of an optical fiber or waveguide in which the light travels by total internal reflection.

**evaporative light-scattering detector** A liquid chromatography detector that makes a fine mist of eluate and evaporates solvent from the mist in a heated zone. The remaining particles of liquid or solid solute flow past a laser beam and are detected by their ability to scatter the light.

**excitation spectrum** A graph of luminescence (measured at a fixed wavelength) versus excitation frequency or wavelength. It closely corresponds to an absorption spectrum because the luminescence is generally proportional to the absorbance.

**excited state** Any state of an atom or a molecule having more than the minimum possible energy.

**exergonic reaction** One for which $\Delta G$ is negative; it is spontaneous.

**exitance, $M$** Power per unit area radiating from the surface of an object.

**exothermic reaction** One for which $\Delta H$ is negative; heat is liberated when products are formed.

**extended Debye-Hückel equation** See *Debye-Hückel equation.*

**extensive property** A property of a system or chemical reaction, such as entropy, that depends on the amount of matter in the system; e.g., $\Delta G$, which is twice as large when 2 moles of product are formed as it is when 1 mole is formed. See also *intensive property.*

**extinction coefficient** See *molar absorptivity.*

**extra-column volume** See *dead volume.*

**extraction** Process in which a solute is transferred from one phase to another. Analyte is sometimes removed from a sample by extraction into a solvent that dissolves the analyte.

**extrapolation** Estimation of a value that lies beyond the range of measured data.

**F test** For two variances, $s_1^2$ and $s_2^2$ (with $s_1$ chosen to be the larger of the two), the statistic $F$ is defined as $F = s_1^2/s_2^2$. To decide whether $s_1$ is significantly greater than $s_2$, we compare $F$ with the critical values in a table based on a certain confidence level. If the calculated value of $F$ is greater than the value in the table, the difference is significant.

**Fajans titration** A precipitation titration in which the end point is signaled by adsorption of a colored indicator on the precipitate.

**false negative** A conclusion that the concentration of analyte is below a certain limit when, in fact, the concentration is above the limit.

**false positive** A conclusion that the concentration of analyte exceeds a certain limit when, in fact, the concentration is below the limit.

**Farad, F** Unit of electrical capacitance; 1 farad of capacitance will store 1 coulomb of charge in a potential difference of 1 volt.

**faradaic current** That component of current in an electrochemical cell due to oxidation and reduction reactions.

**Faraday constant** $9.648\,533\,83 \times 10^4$ C/mol of charge.

**Faraday cup** A mass spectrometric ion detector in which each arriving cation is neutralized by an electron. The current required to neutralize the ions is proportional to the number of cations arriving at the Faraday cup.

**Faraday's laws** Two laws stating that the extent of an electrochemical reaction is directly proportional to the quantity of electricity that has passed through the cell. The mass of substance that reacts is proportional to its formula mass and inversely proportional to the number of electrons required in its half-reaction.

**ferroelectric material** A solid with a permanent electric polarization (dipole) in the absence of an external electric field. The polarization results from alignment of molecules within the solid.

**field blank** A blank sample exposed to the environment at the sample collection site and transported in the same manner as other samples between the lab and the field.

**field effect transistor** A semiconductor device in which the electric field between gate and base governs the flow of current between source and drain.

**filtrate** Liquid that passes through a filter.

**fines** The smallest particles of stationary phase used for chromatography. It is desirable to remove the fines before packing a column because they plug the column and retard solvent flow.

**Fischer titration** See *Karl Fischer titration.*

**flame ionization detector** A gas chromatography detector in which solute is burned in a $H_2$-air flame to produce $CHO^+$ ions. The current carried through the flame by these ions is proportional to the concentration of susceptible species in the eluate.

**flame photometer** A device that uses flame atomic emission and a filter photometer to quantify Li, Na, K, and Ca in liquid samples.

**flame photometric detector** Gas chromatography detector that measures optical emission from S and P in $H_2$-$O_2$ flame.

**flow adaptor** An adjustable plungerlike device that may be used on either side of a chromatographic bed to support the bed and to minimize the dead space through which liquid can flow outside of the column bed.

**flow injection analysis** Analytical technique in which sample is injected into a flowing stream. Other reagents also can be injected into the stream, and some type of measurement, such as absorption of light, is made downstream. Because the sample spreads out as it travels, different concentrations of sample are available in different parts of the band when it reaches the detector.

**fluorescence** Process in which a molecule emits a photon $10^{-8}$ to $10^{-4}$ s after absorbing a photon. It results from a transition between states of the same spin multiplicity (e.g., singlet → singlet).

**flux** In sample preparation, flux is the agent used as the medium for a fusion. In transport phenomena, flux is the quantity of whatever you like crossing each unit area in one unit of time. For example, the flux of diffusing molecules could be mol/($m^2 \cdot$ s). Heat flux would be J/($m^2 \cdot$ s).

**formal concentration** The molarity of a substance if it did not change its chemical form on being dissolved. It represents the total number of moles

of substance dissolved in a liter of solution, regardless of any reactions that take place when the solute is dissolved. Also called *analytical concentration.*

**formality, F** See *formal concentration.*

**formal potential** Potential of a half-reaction (relative to a standard hydrogen electrode) when the formal concentrations of reactants and products are unity. Any other conditions (such as pH, ionic strength, and concentrations of ligands) also must be specified.

**formation constant** Equilibrium constant for the reaction of a metal with its ligands to form a metal-ligand complex. Also called *stability constant.*

**formula mass, FM** The mass containing 1 mole of the indicated chemical formula of a substance. For example, the formula mass of $CuSO_4 \cdot 5H_2O$ is the sum of the masses of copper, sulfate, and five water molecules.

**fortification** Same as a *spike*—a deliberate addition of analyte to a sample.

**Fourier analysis** Process of decomposing a function into an infinite series of sine and cosine terms. Because each term represents a certain frequency or wavelength, Fourier analysis decomposes a function into its component frequencies or wavelengths.

**Fourier series** Infinite sum of sine and cosine terms that add to give a particular function in a particular interval.

**fraction of association, $\alpha$** For the reaction of a base (B) with $H_2O$, the fraction of base in the form $BH^+$.

**fraction of dissociation, $\alpha$** For the dissociation of an acid (HA), the fraction of acid in the form $A^-$.

**frequency** The number of oscillations of a wave per second.

**friction coefficient** A molecule migrating through a solution is retarded by a force that is proportional to its speed. The constant of proportionality is the friction coefficient.

**fugacity** The activity of a gas.

**fugacity coefficient** Activity coefficient for a gas.

**fusion** Process in which an otherwise insoluble substance is dissolved in a molten salt such as $Na_2CO_3$, $Na_2O_2$, or KOH. Once the substance has dissolved, the melt is cooled, dissolved in aqueous solution, and analyzed.

**galvanic cell** One that produces electricity by means of a spontaneous chemical reaction. Also called a *voltaic cell.*

**gas chromatography** A form of chromatography in which the mobile phase is a gas.

**gathering** A process in which a trace constituent of a solution is intentionally coprecipitated with a major constituent.

**Gaussian distribution** Theoretical bell-shaped distribution of measurements when all error is random. The center of the curve is the mean, $\mu$, and the width is characterized by the standard deviation, $\sigma$. A *normalized* Gaussian distribution, also called the *normal error curve,* has an area of unity and is given by

$$y = \frac{1}{\sigma\sqrt{2\pi}}\, e^{-(x-\mu)^2/2\sigma^2}$$

**gel** Chromatographic stationary phase particles, such as Sephadex or polyacrylamide, which are soft and pliable.

**gel filtration chromatography** See *molecular exclusion chromatography.*

**gel permeation chromatography** See *molecular exclusion chromatography.*

**geometric mean** For a series of $n$ measurements with the values $x_i$, geometric mean $= \sqrt[n]{x_1 \cdot x_2 \cdots x_n}$.

**Gibbs free energy, G** The change in Gibbs free energy, $\Delta G$, for any process at constant temperature is related to the change in enthalpy, $\Delta H$, and entropy, $\Delta S$, by the equation $\Delta G = \Delta H - T\Delta S$, where $T$ is temperature in kelvins. A process is spontaneous (thermodynamically favorable) if $\Delta G$ is negative.

**glass electrode** One that has a thin glass membrane across which a pH-dependent voltage develops. The voltage (and hence pH) is measured by a pair of reference electrodes on either side of the membrane.

**glassy carbon electrode** An inert carbon electrode, impermeable to gas, and especially well suited as an anode. The isotropic structure (same in all directions) is thought to consist of tangled ribbons of graphitelike sheets, with some cross-linking.

**globar** An infrared radiation source made of a ceramic such as silicon carbide heated by passage of electricity.

**Golay cell** Infrared detector that uses expansion of a gas in a blackened chamber to deform a flexible mirror. Deflection of a beam of light by the mirror changes the power impinging on a phototube.

**Gooch crucible** A short, cup-shaped container with holes at the bottom, used for filtration and ignition of precipitates. For ignition, the crucible is made of porcelain or platinum and lined with a mat of ceramic fibers to retain the precipitate. For precipitates that do not need ignition, the crucible is made of glass and has a porous glass disk instead of holes at the bottom.

**gradient elution** Chromatography in which the composition of the mobile phase is progressively changed to increase the eluent strength of the solvent.

**graduated cylinder** A tube with volume calibrations along its length. Also called *graduate*.

**gram-atom** The amount of an element containing Avogadro's number of atoms; it is the same as a mole of the element.

**Gran plot** A graph such as the plot of $V_b \cdot 10^{-pH}$ versus $V_b$ used to find the end point of a titration. $V_b$ is the volume of base (titrant) added to an acid being titrated. The slope of the linear portion of the graph is related to the dissociation constant of the acid.

**graphite furnace** A hollow graphite rod that can be heated electrically to about 2 500 K to decompose and atomize a sample for atomic spectroscopy.

**grating** Either a reflective or a transmitting surface etched with closely spaced lines; used to disperse light into its component wavelengths.

**gravimetric analysis** Any analytical method that relies on measuring the mass of a substance (such as a precipitate) to complete the analysis.

**gravimetric titration** A titration in which the mass of titrant is measured, instead of the volume. Titrant concentration is conveniently expressed as mol reagent/kg titrant solution. Gravimetric titrations can be more accurate and precise than volumetric titrations.

**gross sample** See *bulk sample*.

**ground state** State of an atom or a molecule with the minimum possible energy.

**guard column** In high-performance liquid chromatography, a short column packed with the same material as the main column, placed between the injector and the main column. The guard column removes impurities that might irreversibly bind to the main column and degrade it. Also called *precolumn*. In gas chromatography, the guard column is a length of empty, silanized capillary ahead of the chromatography column. Nonvolatile residues accumulate in the guard column.

**half-cell** Part of an electrochemical cell in which half of an electrochemical reaction (either the oxidation or the reduction reaction) occurs.

**half-height** Half of the maximum amplitude of a signal.

**half-reaction** Any redox reaction can be conceptually broken into two half-reactions, one involving only oxidation and one involving only reduction.

**half-wave potential** Potential at the midpoint of the rise in the current of a polarographic wave.

**half-width** Width of a signal at its half-height.

**Hall process** Electrolytic production of aluminum metal from a molten solution of $Al_2O_3$ and cryolite ($Na_3AlF_6$).

**Hammett acidity function** The acidity of a solvent that protonates the weak base, B, is called the Hammett acidity function, $H_0$, and is given by

$$H_0 = pK_a(\text{for BH}^+) + \log \frac{[B]}{[BH^+]}$$

For dilute aqueous solutions, $H_0$ approaches pH.

**hanging-drop electrode** One with a stationary drop of Hg that is used for stripping analysis.

**hardness** Total concentration of alkaline earth ions in natural water expressed as mg $CaCO_3$ per liter of water as if all of the alkaline earths present were $CaCO_3$.

**Heisenberg uncertainty principle** Certain pairs of physical quantities cannot be known simultaneously with arbitrary accuracy. If $\delta E$ is the uncertainty in the energy difference between two atomic states and $\delta t$ is the lifetime of the excited state, their product cannot be known more accurately than $\delta E \delta t \gtrsim h/(4\pi)$, where $h$ is Planck's constant. A similar relation holds between the position and the momentum of a particle. If position is known very accurately, then the uncertainty in momentum is large, and vice versa.

**Henderson-Hasselbalch equation** A logarithmic rearranged form of the acid dissociation equilibrium equation:

$$pH = pK_a + \log \frac{[A^-]}{[HA]}$$

**Henry's law** The partial pressure of a gas in equilibrium with gas dissolved in a solution is proportional to the concentration of dissolved gas: $P = k[\text{dissolved gas}]$. The constant $k$ is called the *Henry's law constant*. It is a function of the gas, the liquid, and the temperature.

**hertz, Hz** Unit of frequency, $s^{-1}$.

**heterogeneous** Not uniform throughout.

**HETP, height equivalent to a theoretical plate** The length of a chromatography column divided by the number of theoretical plates in the column.

**hexadentate ligand** One that binds to a metal atom through six ligand atoms.

**high-performance liquid chromatography, HPLC** A chromatographic technique using very small stationary phase particles and high pressure to force solvent through the column.

**hole** Absence of an electron in a semiconductor. When a neighboring electron moves into the hole, a new hole is created where the electron came from. By this means, a hole can move through a solid just as an electron can move through a solid.

**hollow-cathode lamp** One that emits sharp atomic lines characteristic of the element from which the cathode is made.

**homogeneous** Having the same composition everywhere.

**homogeneous precipitation** A technique in which a precipitating agent is generated slowly by a reaction in homogeneous solution, effecting a slow crystallization instead of a rapid precipitation of product.

**hydrated radius** Effective size of an ion or a molecule plus its associated water molecules in solution.

**hydrodynamic flow** Motion of fluid through a tube, driven by a pressure difference. Hydrodynamic flow is usually laminar, in which there is a parabolic profile of velocity vectors, with the highest velocity at the center of the stream and zero velocity at the walls.

**hydrodynamic injection** In capillary electrophoresis, the use of a pressure difference between the two ends of the capillary to inject sample into the capillary. Injection is achieved by applying pressure on one end, by applying suction on one end, or by siphoning.

**hydrodynamic radius** Effective radius of a molecule migrating through a fluid. It is defined by the Stokes equation, in which the friction coefficient is $6\pi\eta r$, where $\eta$ is the viscosity of the fluid and $r$ is the hydrodynamic radius of the molecule.

**hydrolysis** "Reaction with water." The reaction $B + H_2O \rightleftharpoons BH^+ + OH^-$ is often called hydrolysis of a base.

**hydronium ion, $H_3O^+$** What we really mean when we write $H^+(aq)$.

**hydrophilic substance** One that is soluble in water or attracts water to its surface.

**hydrophobic interaction chromatography** Chromatographic separation based on the interaction of a hydrophobic solute with a hydrophobic stationary phase.

**hydrophobic substance** One that is insoluble in water or repels water from its surface.

**hygroscopic substance** One that readily picks up water from the atmosphere.

**ignition** The heating to high temperature of some gravimetric precipitates to convert them into a known, constant composition that can be weighed.

**immunoassay** An analytical measurement using antibodies.

**inclusion** An impurity that occupies lattice sites in a crystal.

**indeterminate error** See *random error*.

**indicator** A compound having a physical property (usually color) that changes abruptly near the equivalence point of a chemical reaction.

**indicator electrode** One that develops a potential whose magnitude depends on the activity of one or more species in contact with the electrode.

**indicator error** Difference between the indicator end point of a titration and the true equivalence point.

**indirect detection** Chromatographic detection based on the *absence* of signal from a background ionic species. For example, in ion chromatography, a light-absorbing ionic species can be added to the eluent. Nonabsorbing analyte replaces an equivalent amount of light-absorbing eluent when analyte emerges from the column, thereby decreasing the absorbance of eluate.

**indirect titration** One that is used when the analyte cannot be directly titrated. For example, analyte A may be precipitated with excess reagent R. The product is filtered, and the excess R washed away. Then AR is dissolved in a new solution, and R can be titrated.

**inductive effect** Attraction of electrons by an electronegative element through the sigma-bonding framework of a molecule.

**inductively coupled plasma** A high-temperature plasma that derives its energy from an oscillating radio-frequency field. It is used to atomize a sample for atomic emission spectroscopy.

**inflection point** One at which the derivative of the slope is 0: $d^2y/dx^2 = 0$. That is, the slope reaches a maximum or minimum value.

**injection precision** See *instrument precision*.

**inner Helmholtz plane** Imaginary plane going through the centers of ions or molecules specifically adsorbed on an electrode.

**inorganic carbon** In a natural water or industrial effluent sample, the quantity of dissolved carbonate and bicarbonate.

**instrument precision** Reproducibility observed when the same quantity of one sample is repeatedly introduced into an instrument. Also called *injection precision*.

**intensity** See *irradiance*.

**intensive property** A property of a system or chemical reaction that does not depend on the amount of matter in the system; e.g., temperature and electric potential. See also *extensive property*.

**intercalation** Binding of a flat, aromatic molecule between the flat, hydrogen-bonded base pairs in DNA or RNA.

**intercept** For a straight line whose equation is $y = mx + b$, $b$ is the intercept. It is the value of $y$ when $x = 0$.

**interference** A phenomenon in which the presence of one substance changes the signal in the analysis of another substance.

**interference filter** A filter that transmits a particular band of wavelengths and reflects others. Transmitted light interferes constructively within the filter, whereas light that is reflected interferes destructively.

**interferogram** A graph of light intensity versus retardation (or time) for the radiation emerging from an interferometer.

**interferometer** A device with a beamsplitter, fixed mirror, and moving mirror that breaks input light into two beams that interfere with each other. The degree of interference depends on the difference in pathlength of the two beams.

**interlaboratory precision** The reproducibility observed when aliquots of the same sample are analyzed by different people in different laboratories.

**intermediate precision** See *ruggedness*.

**internal conversion** A radiationless, isoenergetic, electronic transition between states of the same electron-spin multiplicity.

**internal standard** A known quantity of a compound other than analyte added to a solution containing an unknown quantity of analyte. The concentration of analyte is then measured relative to that of the internal standard.

**interpolation** Estimation of the value of a quantity that lies between two known values.

**intersystem crossing** A radiationless, isoenergetic, electronic transition between states of different electron-spin multiplicity.

**intra-assay precision** Precision observed when analyzing aliquots of a homogeneous material several times by one person on one day with the same equipment.

**iodimetry** Use of triiodide (or iodine) as a titrant.

**iodometry** A technique in which an oxidant is treated with $I^-$ to produce $I_3^-$, which is then titrated (usually with thiosulfate).

**ion chromatography** High-performance liquid chromatography ion-exchange separation of ions. See also *suppressed-ion chromatography* and *single-column ion chromatography*.

**ion-exchange chromatography** A technique in which solute ions are retained by oppositely charged sites in the stationary phase.

**ion-exchange equilibrium** An equilibrium involving replacement of a cation by a different cation or replacement of an anion by a different anion. Usually the ions in these reactions are bound by electrostatic forces.

**ion-exchange membrane** Membrane containing covalently bound charged groups. Oppositely charged ions in solution penetrate the membrane freely, but similarly charged ions tend to be excluded from the membrane by the bound charges.

**ion-exclusion chromatography** A technique in which electrolytes are separated from nonelectrolytes by an ion-exchange resin.

**ionic atmosphere** The region of solution around an ion or a charged particle. It contains an excess of oppositely charged ions.

**ionic radius** Effective size of an ion in a crystal.

**ionic strength, $\mu$** Given by $\mu = \frac{1}{2}\Sigma_i c_i z_i^2$, where $c_i$ is the concentration of the $i$th ion in solution and $z_i$ is the charge on that ion. The sum extends over all ions in solution, including the ions whose activity coefficients are being calculated.

**ionization interference** In atomic spectroscopy, a lowering of signal intensity as a result of ionization of analyte atoms.

**ionization suppressor** An element used in atomic spectroscopy to decrease the extent of ionization of the analyte.

**ion mobility spectrometer** An instrument that measures the drift time of gaseous ions migrating in an electric field against a flow of gas. The "spectrum" of detector current versus drift time is really an electropherogram of a gas.

**ionophore** A molecule with a hydrophobic outside and a polar inside that can engulf an ion and carry the ion through a hydrophobic phase (such as a cell membrane).

**ion pair** A closely associated anion and cation, held together by electrostatic attraction. In solvents less polar than water, ions are usually found as ion pairs.

**ion-pair chromatography** Separation of ions on reversed-phase high-performance liquid chromatography column by adding to the eluent a hydrophobic counterion that pairs with analyte ion and is attracted to stationary phase.

**ion-selective electrode** One whose potential is selectively dependent on the concentration of one particular ion in solution.

**ion spray** See *electrospray*.

**irradiance** Power per unit area ($W/m^2$) of a beam of electromagnetic radiation. Also called *radiant power* or *intensity*.

**isobaric interference** In mass spectrometry, overlap of two peaks with nearly the same mass. For example, $^{41}K^+$ and $^{40}ArH^+$ differ by 0.01 atomic mass unit and appear as a single peak unless the spectrometer resolution is great enough to separate them.

**isocratic elution** Chromatography using a single solvent for the mobile phase.

**isoelectric buffer** A neutral, polyprotic acid occasionally used as a low-conductivity "buffer" for capillary zone electrophoresis. For example, a solution of pure aspartic acid ($pK_1 = 1.99$, $pK_2 = 3.90$, $pK_3 = 10.00$) has pH $= \frac{1}{2}(pK_1 + pK_2) = 2.94$. Calling pure aspartic acid a "buffer" is an oxymoron, because the buffer capacity is *a minimum* at pH 2.94 and increases to maxima at pH 1.99 and 3.90. However, as the pH drifts away from 2.94, the solution gains significant buffer capacity. When electrophoresis is conducted in a background electrolyte of aspartic acid, the pH stays near 2.94 and conductivity remains very low, permitting a high electric field to be used, thus enabling rapid separations.

**isoelectric focusing** A technique in which a sample containing polyprotic molecules is subjected to a strong electric field in a medium with a pH gradient. Each species migrates until it reaches the region of its isoelectric pH. In that region, the molecule has no net charge, ceases to migrate, and remains focused in a narrow band.

**isoelectric pH** That pH at which the average charge of a polyprotic species is 0. Same as *isoelectric point.*

**isoionic pH** The pH of a pure solution of a neutral, polyprotic molecule. The only ions present are $H^+$, $OH^-$, and those derived from the polyprotic species. Same as *isoionic point.*

**isosbestic point** A wavelength at which the absorbance spectra of two species cross each other. The appearance of isosbestic points in a solution in which a chemical reaction is occurring is evidence that there are only two components present, with a constant total concentration.

**isotope ratio mass spectrometry** A mass spectrometric technique designed to provide accurate measurements of the ratio of different ions of a selected element. The instrument has one detector dedicated to each isotope.

**Job's method** See *method of continuous variation.*

**Jones reductor** A column packed with zinc amalgam. An oxidized analyte is passed through to reduce the analyte, which is then titrated with an oxidizing agent.

**joule, J** SI unit of energy. One joule is expended when a force of 1 N acts over a distance of 1 m. This energy is equivalent to that required to raise 102 g (about $\frac{1}{4}$ pound) by 1 m at sea level.

**Joule heating** Heat produced in an electric circuit by the flow of electricity. Power $(J/s) = I^2R$, where $I$ is the current (A) and $R$ is the resistance (ohms).

**junction potential** An electric potential that exists at the junction between two different electrolyte solutions or substances. It arises in solutions as a result of unequal rates of diffusion of different ions.

**Karl Fischer titration** A sensitive technique for determining water, based on the reaction of $H_2O$ with an amine, $I_2$, $SO_2$, and an alcohol.

**kelvin, K** Absolute unit of temperature defined such that the temperature of water at its triple point (where water, ice, and water vapor are at equilibrium) is 273.16 K and the absolute zero of temperature is 0 K.

**Kieselguhr** German term for diatomaceous earth, which was formerly used as a solid support in gas chromatography.

**kilogram, kg** SI unit of mass equal to the mass of a particular Pt-Ir cylinder kept at the International Bureau of Weights and Measures, Sèvres, France.

**kinetic current** A polarographic wave that is affected by the rate of a chemical reaction involving the analyte and some species in the solution.

**kinetic polarization** Occurs whenever an overpotential is associated with an electrode process.

**Kjeldahl nitrogen analysis** Procedure for the analysis of nitrogen in organic compounds. The compound is digested with boiling $H_2SO_4$ to convert nitrogen into $NH_4^+$, which is treated with base and distilled as $NH_3$ into a standard acid solution. The moles of acid consumed equal the moles of $NH_3$ liberated from the compound.

**Kovats index** See *retention index.*

**laboratory sample** Portion of bulk sample taken to the lab for analysis. Must have the same composition as the bulk sample.

**laminar flow** Motion with a parabolic velocity profile of fluid through a tube. Motion is fastest at the center and zero at the walls.

**laser** Source of intense, coherent monochromatic radiation. Light is produced by stimulated emission of radiation from a medium in which an excited state has been pumped to a high population. Coherence means that all light exiting the laser has the same phase.

**Latimer diagram** One that shows the reduction potentials connecting a series of species containing an element in different oxidation states.

**law of mass action** States that, for the chemical reaction $a\text{A} + b\text{B} \rightleftharpoons c\text{C} + d\text{D}$, the condition at equilibrium is $K = \mathcal{A}_\text{C}^c \mathcal{A}_\text{D}^d / \mathcal{A}_\text{A}^a \mathcal{A}_\text{B}^b$, where $\mathcal{A}_i$ is the activity of the $i$th species. The law is usually used in approximate form, in which the activities are replaced by concentrations.

**least squares** Process of fitting a mathematical function to a set of measured points by minimizing the sum of the squares of the distances from the points to the curve.

**Le Châtelier's principle** If a system at equilibrium is disturbed, the direction in which it proceeds back to equilibrium is such that the disturbance is partly offset.

**leveling effect** The strongest acid that can exist in solution is the protonated form of the solvent. A stronger acid will donate its proton to the solvent and be leveled to the acid strength of the protonated solvent. Similarly, the strongest base that can exist in a solvent is the deprotonated form of the solvent.

**Lewis acid** One that can form a chemical bond by sharing a pair of electrons donated by another species.

**Lewis base** One that can form a chemical bond by sharing a pair of its electrons with another species.

**ligand** An atom or a group attached to a central atom in a molecule. The term is often used to mean any group attached to anything else of interest.

**limiting current** In a polarographic experiment, the current that is reached at the plateau of a polarographic wave. See also *diffusion current.*

**limit of quantitation** The minimum signal that can be measured "accurately," often taken as the mean signal for blanks plus 10 times the standard deviation of a low-concentration sample.

**linear flow rate** In chromatography, the distance per unit time traveled by the mobile phase.

**linear interpolation** A form of interpolation in which the variation in some quantity is assumed to be linear. For example, to find the value of $b$ when $a = 32.4$ in the following table,

| $a$: | 32 | 32.4 | 33 |
| --- | --- | --- | --- |
| $b$: | 12.85 | $x$ | 17.96 |

you can set up the proportion

$$\frac{32.4 - 32}{33 - 32} = \frac{x - 12.85}{17.96 - 12.85}$$

which gives $x = 14.89$.

**linearity** A measure of how well data in a graph follow a straight line.

**linear range** Concentration range over which the change in detector response is proportional to the change in analyte concentration.

**linear response** The case in which the analytical signal is directly proportional to the concentration of analyte.

**linear voltage ramp** The linearly increasing potential that is applied to the working electrode in polarography.

**line noise** Noise concentrated at discrete frequencies that come from sources external to an intended measuring system. Common sources include radiation emanating from the 60-Hz power line, vacuum-pump motors, and radio frequency devices. Same as *interference* or *whistle noise.*

**lipid bilayer** Double layer formed by molecules containing hydrophilic headgroup and hydrophobic tail. The tails of the two layers associate with each other and the headgroups face the aqueous solvent.

**liquid-based ion-selective electrode** One that has a hydrophobic membrane separating an inner reference electrode from the analyte solution. The membrane is saturated with a liquid ion exchanger dissolved in a nonpolar solvent. The ion-exchange equilibrium of analyte between the liquid ion exchanger and the aqueous solution gives rise to the electrode potential.

**liquid chromatography** A form of chromatography in which the mobile phase is a liquid.

**liter, L** Common unit of volume equal to exactly 1 000 $cm^3$.

**Littrow prism** A prism with a reflecting back surface.

**logarithm** The base 10 logarithm of $n$ is $a$ if $10^a = n$ (which means $\log n = a$). The natural logarithm of $n$ is $a$ if $e^a = n$ (which means $\ln n = a$). The number e ($=2.718\ 28\ldots$) is called the base of the natural logarithm.

**longitudinal diffusion** Diffusion of solute molecules parallel to the direction of travel through a chromatography column.

**Lorentzian** A function commonly used to describe the shape of a spectroscopic band: amplitude $= A_{max}\Gamma^2/[\Gamma^2 + (\nu - \nu_0)^2]$, where $\nu$ is the frequency (or wavenumber), $\nu_0$ is the frequency (or wavenumber) at the center of the band, $2\Gamma$ is the width at half-height, and $A_{max}$ is the maximum amplitude.

**lot** Entire material that is to be analyzed. Examples are a bottle of reagent, a lake, or a truckload of gravel.

**lower limit of detection** See *detection limit.*

**lower limit of quantitation** Smallest amount of analyte that can be measured with reasonable accuracy. Usually taken as 10 times the standard deviation of a low-concentration sample. Also called *quantitation limit.*

**luminescence** Any emission of light by a molecule.

**L'vov platform** Platform on which sample is placed in a graphite-rod furnace for atomic spectroscopy to prevent sample vaporization before the walls reach constant temperature.

**magnetic sector mass spectrometer** A device that separates gaseous ions that have the same kinetic energy by passing them through a magnetic field perpendicular to their velocity. Trajectories of ions with a certain mass-to-charge ratio are bent exactly enough to reach the detector. Other ions are deflected too much or too little.

**makeup gas** Gas added to the exit stream from a gas chromatography column for the purpose of changing flow rate or gas composition to optimize detection of analyte.

**MALDI** See *matrix-assisted laser desorption/ionization.*

**mantissa** The part of a logarithm to the right of the decimal point.

**Mariotte flask** A reservoir that maintains a constant hydrostatic pressure for liquid chromatography.

**masking** Process of adding a chemical substance (a *masking agent*) to a sample to prevent one or more components from interfering in a chemical analysis.

**masking agent** A reagent that selectively reacts with one (or more) component(s) of a solution to prevent the component(s) from interfering in a chemical analysis.

**mass balance** A statement that the sum of the moles of any element in all of its forms in a solution must equal the moles of that element delivered to the solution.

**mass chromatogram** See *selected ion chromatogram.*

**mass spectrometer** An instrument that converts gaseous molecules into ions, accelerates them in an electric field, separates them according to their mass-to-charge ratio, and detects the amount of each species.

**mass spectrometry** A technique in which gaseous molecules are ionized, accelerated by an electric field, and then separated according to their mass.

**mass spectrometry–mass spectrometry, MS–MS** See *selected reaction monitoring.*

**mass spectrum** In mass spectrometry, a graph showing the relative abundance of each ion as a function of its mass-to-charge ratio.

**mass titration** One in which the mass of titrant, instead of the volume, is measured.

**matrix** The medium containing analyte. For many analyses, it is important that standards be prepared in the same matrix as the unknown.

**matrix-assisted laser desorption/ionization, MALDI** A gentle technique for introducing predominantly singly charged, intact macromolecular ions into the gas phase. An intimate solid mixture of analyte plus a large excess of a small, ultraviolet-absorbing molecule is irradiated by a pulse from an ultraviolet laser. The small molecule (the matrix) absorbs the radiation, becomes ionized, evaporates, and expands in a supersonic jet that carries analyte into the gas phase. Matrix ions apparently transfer charge to the analyte.

**matrix effect** A change in analytical signal caused by anything in the sample other than analyte.

**matrix modifier** Substance added to sample for atomic spectroscopy to make the matrix more volatile or the analyte less volatile so that the matrix evaporates before analyte does.

**maximum suppressor** A surface-active agent (such as the detergent Triton X-100) used to eliminate current maxima in polarography.

**mean** The average of a set of all results.

**mean activity coefficient** For the salt (cation)$_m$(anion)$_n$, the mean activity coefficient, $\gamma_\pm$, is related to the individual ion activity coefficients ($\gamma_+$ and $\gamma_-$) by the equation $\gamma_\pm = (\gamma_+^m \gamma_-^n)^{1/(m+n)}$.

**mechanical balance** A balance having a beam that pivots on a fulcrum. Standard masses are used to measure the mass of an unknown.

**median** For a set of data, that value above and below which there are equal numbers of data.

**mediator** In electrolysis, a molecule that carries electrons between the electrode and the intended analyte. Used when the analyte cannot react directly at the electrode or when analyte concentration is so low that other reagents react instead. Mediator is recycled indefinitely by oxidation or reduction at the counterelectrode.

**meniscus** Curved surface of a liquid.

**mesh size** The number of spacings per linear inch in a standard screen used to sort particles.

**metal ion buffer** Consists of a metal-ligand complex plus excess free ligand. The two serve to fix the concentration of free metal ion through the reaction $M + nL \rightleftharpoons ML_n$.

**metal ion indicator** A compound whose color changes when it binds to a metal ion.

**meter, m** SI unit of length defined as the distance that light travels in a vacuum during $\frac{1}{299\ 792\ 458}$ of a second.

**method blank** A sample without deliberately added analyte. The method blank is taken through all steps of a chemical analysis, including sample preparation.

**method of continuous variation** Procedure for finding the stoichiometry of a complex by preparing a series of solutions with different metal-to-ligand ratios. The ratio at which the extreme response (such as spectrophotometric absorbance) occurs corresponds to the stoichiometry of the complex. Also called *Job's method.*

**method of least squares** Process of fitting a mathematical function to a set of measured points by minimizing the sum of the squares of the distances from the points to the curve.

**method validation** Process of proving that an analytical method is acceptable for its intended purpose.

**micellar electrokinetic chromatography** A form of capillary electrophoresis in which a micelle-forming surfactant is present. Migration times of solutes depend on the fraction of time spent in the micelles.

**micelle** An aggregate of molecules with ionic headgroups and long, nonpolar tails. The inside of the micelle resembles hydrocarbon solvent, whereas the outside interacts strongly with aqueous solution.

**microelectrode** An electrode with a diameter on the order of 10 μm (or less). Microelectrodes fit into small places, such as living cells. Their small current gives rise to little ohmic loss, so they can be used in resistive, nonaqueous media. Small double-layer capacitance allows their voltage to be changed rapidly, permitting short-lived species to be studied.

**microequilibrium constant** An equilibrium constant that describes the reaction of a chemically distinct site in a molecule. For example, a base may be protonated at two distinct sites, each of which has a different equilibrium constant.

**microporous particles** Chromatographic stationary phase consisting of porous particles 3–10 μm in diameter, with high efficiency and high capacity for solute.

**migration** Electrostatically induced motion of ions in a solution under the influence of an electric field.

**miscible liquids** Two liquids that form a single phase when mixed in any ratio.

**mobile phase** In chromatography, the phase that travels through the column.

**mobility** The terminal velocity that an ion reaches in a field of 1 V/m. Velocity = mobility × field.

**modulation amplitude** In polarography, the magnitude of the voltage pulse applied to the working electrode.

**Mohr titration** Argentometric titration conducted in the presence of chromate. The end point is signaled by the formation of red $Ag_2CrO_4(s)$.

**molality** A measure of concentration equal to the number of moles of solute per kilogram of solvent.

**molar absorptivity, ε** Constant of proportionality in Beer's law: $A = \varepsilon bc$, where $A$ is absorbance, $b$ is pathlength, and $c$ is the molarity of the absorbing species. Also called *extinction coefficient.*

**molarity, M** A measure of concentration equal to the number of moles of solute per liter of solution.

**mole, mol** SI unit for the amount of substance that contains as many molecules as there are atoms in 12 g of $^{12}C$. There are approximately $6.022 \times 10^{23}$ molecules per mole.

**molecular exclusion chromatography** A technique in which the stationary phase has a porous structure into which small molecules can enter but large molecules cannot. Molecules are separated by size, with larger molecules moving faster than smaller ones. Also called *gel filtration* or *gel permeation* or *size exclusion chromatography.*

**molecular ion** In mass spectrometry, an ion that has not lost or gained any atoms during ionization.

**molecularly imprinted polymer** A polymer synthesized in the presence of a template molecule. After the template is removed, the polymer has a void with the right shape to hold the template, and polymer functional groups are positioned correctly to bind to template functional groups.

**molecular mass** Number of grams of a substance that contains Avogadro's number of molecules.

**molecular orbital** Describes the distribution of an electron within a molecule.

**molecular sieve** A solid particle with pores the size of small molecules. Zeolites (sodium aluminosilicates) are a common type.

**mole fraction** Number of moles of a substance in a mixture divided by the total number of moles of all components present.

**monochromatic light** Light of a single wavelength (color).

**monochromator** A device (usually a prism, grating, or filter) that disperses light into its component wavelengths and selects a narrow band of wavelengths to pass through the exit slit.

**monodentate ligand** One that binds to a metal ion through only one atom.

**monolithic column** Chromatographic column in which polymerization is conducted inside the column to fill the column with porous stationary phase. Monolithic columns allow faster flow rates because the pore structure is maintained at high pressure.

**mortar and pestle** A mortar is a hard ceramic or steel vessel in which a solid sample is ground with a hard tool called a pestle.

**mother liquor** Solution from which a substance has crystallized.

**mull** A fine dispersion of a solid in an oil.

**multidentate ligand** One that binds to a metal ion through more than one atom.

**natural logarithm** The natural logarithm (ln) of $a$ is $b$ if $e^b = a$. See also *logarithm.*

**nebulization** Process of breaking the liquid sample into a mist of fine droplets.

**nebulizer** In atomic spectroscopy, this device breaks the liquid sample into a mist of fine droplets.

**needle valve** A valve with a tapered plunger that fits into a small orifice to constrict flow.

**nephelometry** A technique in which the intensity of light scattered at 90° by a suspension is measured to determine the concentration of suspended particles. In a precipitation titration, the scattering increases until the equivalence point is reached, and then remains constant.

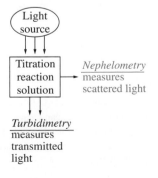

**Nernst equation** Relates the voltage of a cell, $E$, to the activities of reactants and products:

$$E = E° - \frac{RT}{nF} \ln Q$$

where $R$ is the gas constant, $T$ is temperature in kelvins, $F$ is the Faraday constant, $Q$ is the reaction quotient, and $n$ is the number of electrons transferred in the balanced reaction. $E°$ is the cell voltage when all activities are unity.

**neutralization** Process in which a stoichiometric equivalent of acid is added to a base (or vice versa).

**neutron-activation analysis** A technique in which radiation is observed from a sample bombarded by slow neutrons. The radiation gives both qualitative and quantitative information about the sample composition.

**newton, N** SI unit of force. One newton will accelerate a mass of 1 kg by $1 \text{ m/s}^2$.

**nitrogen-phosphorus detector** See *alkali flame detector.*

**nitrogen rule** A compound with an odd number of nitrogen atoms—in addition to C, H, halogens, O, S, Si, and P—will have an odd nominal mass. A compound with an even number of nitrogen atoms (0, 2, 4, etc.) will have an even nominal mass.

**noise** Signals originating from sources other than those intended to be measured. See, for example, *line noise* and *white noise.*

**nominal mass** *Integer* mass of the species with the most abundant isotope of each of the constituent atoms. For C, H, and Br, the most abundant isotopes are $^{12}C$, $^{1}H$, and $^{79}Br$. Therefore, the nominal mass of $C_2H_5Br$ is $(2 \times 12) + (5 \times 1) + (1 \times 79) = 108$.

**nonelectrolyte** A substance that does not dissociate into ions when dissolved.

**nonpolarizable electrode** One whose potential remains nearly constant, even when current flows; e.g., a saturated calomel electrode.

**normal error curve** A Gaussian distribution whose area is unity.

**normal hydrogen electrode, N.H.E.** See *standard hydrogen electrode (S.H.E.).*

**normality** $n$ times the molarity of a redox reagent, where $n$ is the number of electrons donated or accepted by that species in a particular chemical reaction. For acids and bases, it is also $n$ times the molarity, but $n$ is the number of protons donated or accepted by the species.

**normal-phase chromatography** A chromatographic separation utilizing a polar stationary phase and a less polar mobile phase.

**normal pulse polarography** A polarographic technique in which a voltage pulse is applied to each drop of mercury near the end of its lifetime. Current is measured for a short time near the end of each pulse. The voltage is reduced to its baseline value for most of the life of each new drop and the pulse is applied only near the end of the drop life.

**nucleation** Process whereby molecules in solution come together randomly to form small crystalline aggregates that can grow into larger crystals.

**null hypothesis** In statistics, the supposition that two quantities do not differ from each other or that two methods do not give different results.

**occlusion** An impurity that becomes trapped (sometimes with solvent) in a pocket within a growing crystal.

**ohm, Ω** SI unit of electrical resistance. A current of 1 A flows across a potential difference of 1 V if the resistance of the circuit is 1 Ω.

**ohmic potential** Voltage required to overcome the electric resistance of an electrochemical cell.

**Ohm's law** States that the current, $I$, in a circuit is proportional to voltage, $E$, and inversely proportional to resistance, $R$: $I = E/R$.

**Ohm's law plot** In capillary electrophoresis, a graph of current versus applied voltage. The graph deviates from a straight line when Joule heating becomes significant.

**on-column injection** Used in gas chromatography to place a thermally unstable sample directly on the column without excessive heating in an injection port. Solute is condensed at the start of the column by low temperature, and then the temperature is raised to initiate chromatography.

**open tubular column** In chromatography, a capillary column whose walls are coated with stationary phase.

**optical density, OD** See *absorbance.*

**optical fiber** Fiber that carries light by total internal reflection because the transparent core has a higher refractive index than the surrounding cladding.

**optical isomers** See *enantiomers*.

**optode** A sensor based on an optical fiber. Also called *optrode*.

**ordinate** Vertical, *y*, axis of a graph.

**osmolarity** An expression of concentration that gives the total number of particles (ions and molecules) per liter of solution. For nonelectrolytes such as glucose, the osmolarity equals the molarity. For the strong electrolyte $CaCl_2$, the osmolarity is three times the molarity, because each mole of $CaCl_2$ provides 3 mol of ions ($Ca^{2+} + 2Cl^-$).

**outer Helmholtz plane** Imaginary plane passing through centers of hydrated ions just outside the layer of specifically adsorbed molecules on the surface of an electrode.

**overall formation constant, $\beta_n$** See *cumulative formation constant*.

**overpotential** Potential above that expected from the equilibrium potential, concentration polarization, and ohmic potential needed to carry out an electrolytic reaction at a given rate. It is 0 for a reversible reaction.

**oxidant** See *oxidizing agent*.

**oxidation** A loss of electrons or a raising of the oxidation state.

**oxidation number** See *oxidation state*.

**oxidation state** A bookkeeping device used to tell how many electrons have been gained or lost by a neutral atom when it forms a compound. Also called *oxidation number*.

**oxidizability** In a natural water or industrial effluent sample, the quantity of $O_2$ equivalent to the quantity of $KMnO_4$ consumed by refluxing the sample with standard permanganate. Each $KMnO_4$ consumes five electrons and is chemically equivalent to 1.25 mol of $O_2$.

**oxidizing agent** A substance that takes electrons in a chemical reaction. Also called *oxidant*.

**p function** The negative logarithm (base 10) of a quantity: $pX = -\log X$.

**packed column** A chromatography column filled with stationary phase particles.

**parallax** Apparent displacement of an object when the observer changes position. Occurs when the scale of an instrument is viewed from a position that is not perpendicular to the scale. The apparent reading is not the true reading.

**particle growth** Process in which molecules become attached to a crystal to form a larger crystal.

**partition chromatography** A technique in which separation is achieved by equilibration of solute between two phases.

**partition coefficient, $K$** The equilibrium constant for the reaction in which a solute is partitioned between two phases: solute (in phase 1) $\rightleftharpoons$ solute (in phase 2).

**partition ratio** See *capacity factor*.

**pascal, Pa** SI unit of pressure equal to 1 $N/m^2$. There are $10^5$ Pa in 1 bar and 101 325 Pa in 1 atm.

**pellicular particles** A type of stationary phase used in liquid chromatography. Contains a thin layer of liquid coated on a spherical bead. It has high efficiency (low plate height) but low capacity.

**peptization** Occurs when washing some ionic precipitates with distilled water causes the ions that neutralize the charges of individual particles to be washed away. The particles then repel one another, disintegrate, and pass through the filter with the wash liquid.

**performance test sample** In a series of analytical measurements, a performance test sample is inserted to see if the procedure gives correct results when the analyst does not know the right answer. The performance test sample is formulated by someone other than the analyst to contain a known concentration of analyte. Also called a *quality control sample* or *blind sample*.

**permanent hardness** Component of water hardness not due to dissolved alkaline earth bicarbonates. This hardness remains in the water after boiling.

**pH** Defined as $pH = -\log \mathcal{A}_{H^+}$ where $\mathcal{A}_{H^+}$ is the activity of $H^+$. In most approximate applications, the pH is taken as $-\log [H^+]$.

**phase transfer catalysis** A technique in which a compound such as a crown ether is used to extract a reactant from one phase into another in which a chemical reaction can occur.

**pH meter** A potentiometer that can measure voltage when extremely little current is flowing. It is used with a glass electrode to measure pH.

**pH of zero charge** The pH at which the net charge on the surface of a solid is zero.

**phospholipid** A molecule with a phosphate-containing polar headgroup and long hydrocarbon (lipid) tail.

**phosphorescence** Emission of light during a transition between states of different spin multiplicity (e.g., triplet $\rightarrow$ singlet). Phosphorescence is slower than fluorescence, with emission occurring $\sim10^{-4}$ to $10^2$ s after absorption of a photon.

**photochemistry** Chemical reaction initiated by absorption of a photon.

**photoconductive detector** A detector whose conductivity changes when light is absorbed by the detector material.

**photodiode array** An array of semiconductor diodes used to detect light. The array is normally used to detect light that has been spread into its component wavelengths. One small band of wavelengths falls on each detector.

**photomultiplier tube** One in which the cathode emits electrons when struck by light. The electrons then strike a series of dynodes (plates that are positive with respect to the cathode), and more electrons are released each time a dynode is struck. As a result, more than $10^6$ electrons may reach the anode for every photon striking the cathode.

**photon** A "particle" of light with energy $h\nu$, where $h$ is Planck's constant and $\nu$ is the frequency of the light.

**phototube** A vacuum tube with a photoemissive cathode. The electric current flowing between the cathode and the anode is proportional to the intensity of light striking the cathode.

**photovoltaic detector** A photodetector with a junction across which the voltage changes when light is absorbed by the detector material.

**pH-stat** A device that maintains a constant pH in a solution by continually injecting (or electrochemically generating) acid or base to counteract pH changes.

**piezoelectric crystal** A crystal that deforms when an electric field is applied.

**piezoelectric effect** Development of electric charge on the surface of certain crystals when subjected to pressure. Conversely, application of an electric field deforms the crystal.

**pilot ion** In polarography, an internal standard.

**pipet** A glass tube calibrated to deliver a fixed or variable volume of liquid.

**pK** The negative logarithm (base 10) of an equilibrium constant: $pK = -\log K$.

**Planck distribution** Equation giving the spectral distribution of blackbody radiation:

$$M_\lambda = \frac{2\pi hc^2}{\lambda^5}\left(\frac{1}{e^{hc/\lambda kT} - 1}\right)$$

where $h$ is Planck's constant, $c$ is the speed of light, $\lambda$ is the wavelength of light, $k$ is Boltzmann's constant, and $T$ is temperature in kelvins. $M_\lambda$ is the power (watts) per square meter of surface per meter of wavelength radiating from the surface. The integral $\int_{\lambda_1}^{\lambda_2} M_\lambda \, d\lambda$ gives the power emitted per unit area in the wavelength interval from $\lambda_1$ to $\lambda_2$.

**Planck's constant** Fundamental constant of nature equal to the energy of light divided by its frequency: $h = E/\nu \approx 6.626 \times 10^{-34}$ J·s.

**plasma** A gas that is hot enough to contain free ions and electrons, as well as neutral molecules.

**plate height, $H$** Length of a chromatography column divided by the number of theoretical plates in the column.

**polarizability** Proportionality constant relating the induced dipole to the strength of the electric field. When a molecule is placed in an electric field, a dipole is induced in the molecule by attraction of the electrons toward the positive pole and attraction of the nuclei toward the negative pole.

**polarizable electrode** One whose potential can change readily when a small current flows. Examples are Pt or Ag wires used as indicator electrodes.

**polarogram** A graph showing the relation between current and potential during a polarographic experiment.

**polarograph** An instrument used to obtain and record a polarogram.

**polarographic wave** S-shaped increase in current during a redox reaction in polarography.

**polarography** A voltammetry experiment using a dropping-mercury electrode.

**polychromator** A device that spreads light into its component wavelengths and directs each small band of wavelengths to a different region where it is detected by a photodiode array.

**polyprotic acid or base** Compound that can donate or accept more than one proton.

**population inversion** A necessary condition for laser operation in which the population of an excited energy level is greater than that of a lower energy level.

**porous-layer column** Gas chromatography column containing an adsorptive solid phase coated on the inside surface of its wall.

**postprecipitation** Adsorption of otherwise soluble impurities on the surface of a precipitate after the precipitation is over.

**potential** See *electric potential.*

**potentiometer** A device that measures electric potential by balancing it with a known potential of the opposite sign. A potentiometer measures the same quantity as that measured by a voltmeter, but the potentiometer is designed to draw much less current from the circuit being measured.

**potentiometric stripping analysis** Technique in which analyte is electrochemically concentrated onto the working electrode at a controlled potential. The potentiostat is then disconnected and the voltage of the working electrode is measured as a function of time, during which a species in solution oxidizes or reduces the concentrated analyte back into solution. The time required to reach the potentiometric end point is proportional to the quantity of analyte.

**potentiometry** An analytical method in which an electric potential difference (a voltage) of a cell is measured.

**potentiostat** An electronic device that maintains a constant voltage between a pair of electrodes.

**power** Amount of energy per unit time (J/s) being expended.

**ppb, parts per billion** An expression of concentration denoting nanograms ($10^{-9}$ g) of solute per gram of solution.

**ppm, parts per million** An expression of concentration denoting micrograms ($10^{-6}$ g) of solute per gram of solution.

**precipitant** A substance that precipitates a species from solution.

**precipitation** Occurs when a substance leaves solution rapidly (to form either microcrystalline or amorphous solid).

**precipitation titration** One in which the analyte forms a precipitate with the titrant.

**precision** A measure of the reproducibility of a measurement.

**precolumn** See *guard column.*

**preconcentration** Process of concentrating trace components of a mixture prior to their analysis.

**precursor ion** In tandem mass spectrometry (selected reaction monitoring), the ion selected by the first mass separator for fragmentation in the collision cell.

**premix burner** In atomic spectroscopy, one in which the sample is nebulized and simultaneously mixed with fuel and oxidant before being fed into the flame.

**preoxidation** In some redox titrations, adjustment of the analyte oxidation state to a higher value so that it can be titrated with a reducing agent.

**preparative chromatography** Chromatography of large quantities of material conducted for the purpose of isolating pure material.

**prereduction** Process of reducing an analyte to a lower oxidation state prior to performing a titration with an oxidizing agent.

**pressure** Force per unit area, commonly measured in pascals (N/m) or bars.

**pressure broadening** In spectroscopy, line broadening due to collisions between molecules.

**primary standard** A reagent that is pure enough and stable enough to be used directly after weighing. The entire mass is considered to be pure reagent.

**prism** A transparent, triangular solid. Each wavelength of light passing through the prism is bent (refracted) at a different angle as it passes through.

**product** The species created in a chemical reaction. Products appear on the right side of the chemical equation.

**product ion** In tandem mass spectrometry (selected reaction monitoring), the fragment ion from the collision cell selected by the final mass separator for passage through to the detector.

**protic solvent** One with an acidic hydrogen atom.

**protocol** In quality assurance, written directions stating what must be documented and how the documentation is to be done, including how to record information in notebooks.

**proton** The ion $H^+$.

**proton acceptor** A Brønsted-Lowry base: a molecule that combines with $H^+$.

**protonated molecule** In mass spectrometry, the ion $MH^+$ resulting from addition of $H^+$ to the analyte.

**proton donor** A Brønsted-Lowry acid: a molecule that can provide $H^+$ to another molecule.

**purge** To force a fluid (usually gas) to flow through a substance or a chamber, usually to extract something from the substance being purged or to replace the fluid in the chamber with the purging fluid.

**purge and trap** A method for removing volatile analytes from liquids or solids, concentrating the analytes, and introducing them into a gas chromatograph. A carrier gas bubbled through a liquid or solid extracts volatile analytes, which are then trapped in a tube containing adsorbent. After analyte has been collected, the adsorbent tube is heated and purged to desorb the analytes, which are collected by cold trapping at the start of a gas chromatography column.

**pyroelectric effect** Variation with temperature in the electric polarization of a ferroelectric material.

**pyrolysis** Thermal decomposition of a substance.

*Q* **test** Statistical test used to decide whether to discard a datum that appears discrepant.

**quadrupole ion-trap mass spectrometer** An instrument that separates gaseous ions by trapping them in stable trajectories inside a metallic chamber to which a radio-frequency electric field is applied. Application of an oscillating electric field between the ends of the chamber destabilizes the trajectories of ions with a particular mass-to-charge ratio, expelling them from the cavity and into a detector.

**qualitative analysis** Process of determining the identity of the constituents of a substance.

**quality assurance** Quantitative indications that demonstrate whether data requirements have been met. Also refers to the broader process that includes quality control, quality assessment, and documentation of procedures and results designed to ensure adequate data quality.

**quality control** Active measures taken to ensure the required accuracy and precision of a chemical analysis.

**quality control sample** See *performance test sample.*

**quantitation limit** See *lower limit of quantitation.*

**quantitative analysis** Process of measuring how much of a constituent is present in a substance.

**quantitative transfer** To transfer the entire contents from one vessel to another. This process is usually accomplished by rinsing the first vessel several times with fresh liquid and pouring each rinse into the receiving vessel.

**quantum yield** In photochemistry, the fraction of absorbed photons that produce a particular result. For example, if a molecule can isomerize from a cis to a trans isomer when light is absorbed, the quantum yield for isomerization is the number of molecules that isomerize divided by the number that absorb photons. Quantum yield is in the range 0 to 1.

**quaternary ammonium ion** A cation containing four substituents attached to a nitrogen atom; e.g., $(CH_3CH_2)_4N^+$, the tetraethylammonium ion.

**quenching** Process in which emission from an excited molecule is decreased by energy transfer to another molecule called a *quencher.*

**radian, rad** SI unit of plane angle. There are $2\pi$ radians in a complete circle.

**radiant power** See *irradiance.* Also called *intensity.*

**random error** A type of error, which can be either positive or negative and cannot be eliminated, based on the ultimate limitations on a physical measurement. Also called *indeterminate error.*

**random heterogeneous material** A material in which there are differences in composition with no pattern or predictability and on a fine scale. When you collect a portion of the material for analysis, you obtain some of each of the different compositions.

**random sample** Bulk sample constructed by taking portions of the entire lot at random.

**range** Difference between the highest and the lowest values in a set of data. Also called *spread.* With respect to an analytical method, range is the concentration interval over which linearity, accuracy, and precision are all acceptable.

**raw data** Individual values of a measured quantity, such as peak areas from a chromatogram or volumes from a buret.

**reactant** The species consumed in a chemical reaction. It appears on the left side of a chemical equation.

**reaction quotient, $Q$** Expression having the same form as the equilibrium constant for a reaction. However, the reaction quotient is evaluated for a particular set of existing activities (concentrations), which are generally not the equilibrium values. At equilibrium, $Q = K$.

**reagent blank** A solution prepared from all of the reagents, but no analyte. The blank measures the response of the analytical method to impurities in the reagents or any other effects caused by any component other than the analyte.

**reagent gas** In a chemical ionization source for mass spectrometry, reagent gas (normally methane, isobutane, or ammonia at $\sim 1$ mbar) is converted into strongly proton donating species such as $CH_5^+$ by a process beginning with electron ionization. Protonated reagent gas reacts with analyte to create protonated analyte.

**reagent-grade chemical** A high-purity chemical generally suitable for use in quantitative analysis and meeting purity requirements set by organizations such as the American Chemical Society.

**reconstructed total ion chromatogram** In chromatography, a graph of the sum of intensities of all ions detected at all masses (above a selected cutoff) versus time.

**redox couple** A pair of reagents related by electron transfer; e.g., $Fe^{3+} \mid Fe^{2+}$ or $MnO_4^- \mid Mn^{2+}$.

**redox indicator** A compound used to find the end point of a redox titration because its various oxidation states have different colors. The standard potential of the indicator must be such that its color changes near the equivalence point of the titration.

**redox reaction** A chemical reaction in which electrons are transferred from one element to another.

**redox titration** One in which the reaction between analyte and titrant is an oxidation-reduction reaction.

**reduced plate height** In chromatography, the quotient plate height/$d$, where the numerator is the height equivalent to a theoretical plate and the denominator is the diameter of stationary phase particles.

**reducing agent** A substance that donates electrons in a chemical reaction. Also called *reductant.*

**reductant** See *reducing agent.*

**reduction** A gain of electrons or a lowering of the oxidation state.

**reference electrode** One that maintains a constant potential against which the potential of another half-cell may be measured.

**reflectance, $R$** Fraction of incident radiant power reflected by an object.

**refraction** Bending of light when it passes between media with different refractive indexes.

**refractive index, $n$** The speed of light in any medium is $c/n$, where $c$ is the speed of light in vacuum and $n$ is the refractive index of the medium. The refractive index also measures the angle at which a light ray is bent when it passes from one medium into another. Snell's law states that $n_1 \sin \theta_1 = n_2 \sin \theta_2$, where $n_i$ is the refractive index for each medium and $\theta_i$ is the angle of the ray with respect to the normal between the two media.

**refractive index detector** Liquid chromatography detector that measures the change in refractive index of eluate as solutes emerge from the column.

**relative retention** In chromatography, the ratio of adjusted retention times for two components. If component 1 has an adjusted retention time of $t'_{r1}$ and component 2 has an adjusted retention time of $t'_{r2} (> t'_{r1})$, the relative retention is $\alpha = t'_{r2}/t'_{r1}$.

**relative standard deviation** See *coefficient of variation.*

**relative supersaturation** Defined as $(Q - S)/S$, where $S$ is the concentration of solute in a saturated solution and $Q$ is the concentration in a particular supersaturated solution.

**relative uncertainty** Uncertainty of a quantity divided by the value of the quantity. It is usually expressed as a percentage of the measured quantity.

**releasing agent** In atomic spectroscopy, a substance that prevents chemical interference.

**replicate measurements** Repeated measurements of the same quantity.

**reporting limit** Concentration below which regulations dictate that an analyte is reported as "not detected." The reporting limit is typically set 5 to 10 times higher than the detection limit.

**reprecipitation** Sometimes a gravimetric precipitate can be freed of impurities only by redissolving it and reprecipitating it. The impurities are present at lower concentration during the second precipitation and are less likely to coprecipitate.

**residual current** The small current that is observed prior to the decomposition potential in an electrolysis.

**resin** Small, hard particles of an ion exchanger, such as polystyrene with ionic substituents.

**resistance, $R$** A measure of the retarding force opposing the flow of electric current.

**resistivity, $\rho$** A measure of the ability of a material to retard the flow of electric current. $J = E/\rho$, where $J$ is the current density ($A/m^2$) and $E$ is electric field (V/m). Units of resistivity are $V \cdot m/A = ohm \cdot m = \Omega \cdot m$. The resistance, $\Omega$, of a conductor with a given length and cross-sectional area is given by $R = \rho \cdot length/area$.

**resolution** How close two bands in a spectrum or a chromatogram can be to each other and still be seen as two peaks. In chromatography, it is defined as the difference in retention times of adjacent peaks divided by their width.

**resolving power** In mass spectrometry, the value of $m/z$ at which two peaks separated by 1 mass unit are distinguishable. If resolving power is taken as $m/\Delta m$, then the overlap at the base of the peaks is 10% of the peak height. If resolving power is taken as $m/\Delta m_{1/2}$, then the dip between the two peaks is 8% below the peak heights. In these definitions, $\Delta m$ is the separation between the peaks and $\Delta m_{1/2}$ is the width of each peak at half the maximum height.

**response factor, $F$** Relative response of a detector to analyte (X) and internal standard (S): (signal from X)/[X] = $F$(signal from S)/[S]. Once you have measured $F$ with a standard mixture, you can use it to find [X] in an unknown if you know [S] and the quotient (signal from X)/(signal from S).

**results** What we ultimately report after applying statistics to treated data.

**retardation, $\delta$** Difference in pathlength between light striking the stationary and moving mirrors of an interferometer.

**retention factor** See *capacity factor.*

**retention gap** In gas chromatography, a 3- to 10-m length of empty, silanized capillary ahead of the chromatography column. The retention gap improves the peak shape of solutes that elute close to solvent when large volumes of solvent are injected or when the solvent has a very different polarity from that of the stationary phase.

**retention index, $I$** In gas chromatography, the Kovats retention index is a logarithmic scale that relates the retention time of a compound to those of linear alkanes. Pentane would be given an index of 500, hexane 600, heptane 700, etc.

**retention ratio** In chromatography, the time required for solvent to pass through the column divided by the time required for solute to pass through the column.

**retention time** The time, measured from injection, needed for a solute to be eluted from a chromatography column.

**retention volume** The volume of solvent needed to elute a solute from a chromatography column.

**reversed-phase chromatography** A technique in which the stationary phase is less polar than the mobile phase.

**rings + double bonds formula** The number of rings + double bonds in a molecule with the formula $C_cH_hN_nO_x$ is $c - h/2 + n/2 + 1$, where $c$ includes all Group 14 atoms (C, Si, etc., which all make four bonds), $h$ includes H + halogens (which make one bond), and $n$ is the number of Group 15 atoms (N, P, As, etc., which make three bonds). Group 16 atoms (which make two bonds) do not affect the result.

**robustness** Ability of an analytical method to be unaffected by small, deliberate changes in operating parameters.

**root-mean-square (rms) noise** Standard deviation of the noise in a region where the signal is flat:

$$\text{rms noise} = \sqrt{\frac{\sum_i (A_i - \overline{A})^2}{n}}$$

where $A_i$ is the measured signal for the $i$th data point, $\overline{A}$ is the mean signal, and $n$ is the number of data points.

**rotating-disk electrode** A motor-driven electrode with a smooth flat face in contact with the solution. Rapid convection created by rotation brings fresh analyte to the surface of the electrode. A Pt electrode is especially suitable for studying anodic processes, in which a mercury electrode would be too easily oxidized.

**rotational transition** Occurs when a molecule changes its rotation energy.

**rubber policeman** A glass rod with a flattened piece of rubber on the tip. The rubber is used to scrape solid particles from glass surfaces in gravimetric analysis.

**ruggedness** Precision observed when an assay is performed by different people on different instruments on different days in the same lab. Also called *intermediate precision*.

**run buffer** See *background buffer*.

**salt** An ionic solid.

**salt bridge** A conducting ionic medium in contact with two electrolyte solutions. It allows ions to flow without allowing immediate diffusion of one electrolyte solution into the other.

**sample cleanup** Removal of portions of the sample that do not contain analyte and may interfere with analysis.

**sampled current polarography** Polarographic technique in which the voltage is increased for each drop of mercury and current is measured for a short time at the end of each drop life.

**sample preparation** Transforming a sample into a state that is suitable for analysis. This process can include concentrating a dilute analyte and removing or masking interfering species.

**sampling** The process of collecting a representative sample for analysis.

**sampling variance** The square of the standard deviation arising from heterogeneity of the sample, not from the analytical procedure. For inhomogeneous materials, it is necessary to take larger portions or more portions to reduce the uncertainty of composition due to variation from one region to another. Total variance is the sum of variances from sampling and from analysis.

**saturated solution** One that contains the maximum amount of a compound that can dissolve at equilibrium.

**Scatchard plot** A graph used to find the equilibrium constant for a reaction such as $X + P \rightleftharpoons PX$. It is a graph of [PX]/[X] versus [PX] or any functions proportional to these quantities. The magnitude of the slope of the graph is the equilibrium constant.

**S.C.E., saturated calomel electrode** A calomel electrode saturated with KCl. The electrode half-reaction is $Hg_2Cl_2(s) + 2e^- \rightleftharpoons 2Hg(l) + 2Cl^-$.

**second, s** SI unit of time equal to the duration of 9 192 631 770 periods of the radiation corresponding to the transition between two hyperfine levels of the ground state of $^{133}Cs$.

**segregated heterogeneous material** A material in which differences in composition are on a large scale. Different regions have obviously different composition.

**selected ion chromatogram** A graph of detector response versus time when a mass spectrometer monitors just one or a few species of selected mass-to-charge ratio, $m/z$, emerging from a chromatograph.

**selected ion monitoring** Use of a mass spectrometer to monitor species with just one or a few mass-to-charge ratios, $m/z$.

**selected reaction monitoring** A technique in which the precursor ion selected by one mass separator passes through a collision cell in which the precursor breaks into several fragment ions (product ions). A second mass separator then selects one (or a few) of these ions for detection. Selected reaction monitoring improves chromatographic signal-to-noise ratio because it is insensitive to almost everything other than the intended analyte. Also called *mass spectrometry–mass spectrometry (MS–MS)* or *tandem mass spectrometry*.

**selectivity** Capability of an analytical method to distinguish analyte from other species in the sample. Also called *specificity*.

**selectivity coefficient** With respect to an ion-selective electrode, a measure of the relative response of the electrode to two different ions. In ion-exchange chromatography, the selectivity coefficient is the equilibrium constant for displacement of one ion by another from the resin.

**self-absorption** In a luminescence measurement, a high concentration of analyte molecules can absorb excitation energy from excited analyte. If the absorbed energy is dissipated as heat instead of light, fluorescence does not increase in proportion to analyte concentration. Analyte concentration can be so high that fluorescence *decreases* with increasing concentration. In flame emission atomic spectroscopy, there is a lower concentration of excited-state atoms in the cool, outer part of the flame than in the hot, inner flame. The cool atoms can absorb emission from the hot ones and thereby decrease the observed signal.

**semiconductor** A material whose conductivity ($10^{-7}$ to $10^4$ $\Omega^{-1} \cdot m^{-1}$) is intermediate between that of good conductors ($10^8$ $\Omega^{-1} \cdot m^{-1}$) and that of insulators ($10^{-20}$ to $10^{-12}$ $\Omega^{-1} \cdot m^{-1}$).

**semipermeable membrane** A thin layer of material that allows some substances, but not others, to pass across the material. A dialysis membrane allows small molecules to pass, but not large molecules.

**sensitivity** Response of an instrument or method to a given amount of analyte.

**separation factor** For components A and B separated by chromatography or electrophoresis, the separation factor $\gamma$ is the quotient of linear velocities: $\gamma = u_A/u_B = t_B/t_A$, where $u$ is linear velocity and $t$ is retention time.

**separator column** Ion-exchange column used to separate analyte species in ion chromatography.

**septum** A disk, usually made of silicone rubber, covering the injection port of a gas chromatograph. The sample is injected by syringe through the septum.

**SI units** International system of units based on the meter, kilogram, second, ampere, kelvin, candela, mole, radian, and steradian.

**sieving** In electrophoresis, the separation of macromolecules by migration through a polymer gel. Movement of the smallest molecules is fastest and that of the largest is slowest.

**signal averaging** Improvement of a signal by averaging successive scans. The signal increases in proportion to the number of scans accumulated. The noise increases in proportion to the square root of the number of scans. Therefore, the signal-to-noise ratio improves in proportion to the square root of the number of scans collected.

**significant figure** The number of significant digits in a quantity is the minimum number of digits needed to express the quantity in scientific notation. In experimental data, the first uncertain figure is the last significant figure.

**silanization** Treatment of a chromatographic solid support or glass column with hydrophobic silicon compounds that bind to the most reactive Si—OH groups. It reduces irreversible adsorption and tailing of polar solutes.

**silver-silver chloride electrode** A common reference electrode containing a silver wire coated with AgCl paste and dipped in a solution saturated with AgCl and (usually) KCl. The half-reaction is $AgCl(s) + e^- \rightleftharpoons Ag(s) + Cl^-$.

**single-column ion chromatography** Separation of ions on a low-capacity ion-exchange column, using low-ionic-strength eluent.

**single-electrode potential** Voltage measured when the electrode of interest is connected to the positive terminal of a potentiometer and a standard hydrogen electrode is connected to the negative terminal.

**singlet state** One in which all electron spins are paired.

**size exclusion chromatography** See *molecular exclusion chromatography*.

**slope** For a straight line whose equation is $y = mx + b$, the value of $m$ is the slope. It is the ratio $\Delta y/\Delta x$ for any segment of the line.

**slurry** A suspension of a solid in a solvent.

**Smith-Hieftje background correction** In atomic absorption spectroscopy, a method of distinguishing analyte signal from background signal, based on applying a periodic pulse of high current to the hollow-cathode lamp to distort the lamp signal. Signal detected during the current pulse is subtracted from signal detected without the pulse to obtain the corrected response.

**smoothing** Use of a mathematical procedure or electrical filtering to improve the quality of a signal.

**Snell's law** Relates angle of refraction, $\theta_2$, to angle of incidence, $\theta_1$, for light passing from a medium with refractive index $n_1$ to a medium of refractive index $n_2$: $n_1 \sin \theta_1 = n_2 \sin \theta_2$. Angles are measured with respect to the normal to the surface between the two media.

**sodium error** Occurs when a glass pH electrode is placed in a strongly basic solution containing very little $H^+$ and a high concentration of $Na^+$. The electrode begins to respond to $Na^+$ as if it were $H^+$, so the pH reading is lower than the actual pH. Also called *alkaline error*.

**solid-phase extraction** Preconcentration procedure in which a solution is passed through a short column of chromatographic stationary phase, such as $C_{18}$ on silica. Trace solutes adsorbed on the column can be eluted with a small volume of solvent of high eluent strength.

**solid-phase microextraction** Extraction of compounds from liquids or gases into a coated fiber dispensed from a syringe needle. After extraction, the fiber is withdrawn into the needle and the needle is injected through the septum of a chromatograph. The fiber is extended inside the injection port and adsorbed solutes are desorbed by heating (for gas chromatography) or solvent (for liquid chromatography).

**solid-state ion-selective electrode** An ion-selective electrode that has a solid membrane made of an inorganic salt crystal. Ion-exchange equilibria between the solution and the surface of the crystal account for the electrode potential.

**solubility product, $K_{sp}$** Equilibrium constant for the dissociation of a solid salt to give its ions in solution. For the reaction $M_mN_n(s) \rightleftharpoons mM^{n+} + nN^{m-}$, $K_{sp} = \mathcal{A}_{M^{n+}}^m \mathcal{A}_{N^{m-}}^n$ where $\mathcal{A}$ is the activity of each species.

**solute** A minor component of a solution.

**solvation** Interaction of solvent molecules with solute. Solvent molecules orient themselves around solute to minimize the energy through dipole and van der Waals forces.

**solvent** Major constituent of a solution.

**solvent extraction** A method in which a chemical species is transferred from one liquid phase to another. It is used to separate components of a mixture.

**solvent strength** See *eluent strength*.

**solvent trapping** Splitless gas chromatography injection technique in which solvent is condensed near its boiling point at the start of the column. Solutes dissolve in a narrow band in the condensed solvent.

**speciation** Describes the distribution of an element or compound among different chemical forms.

**species** Chemists refer to any element, compound, or ion of interest as a *species*. The word *species* is both singular and plural.

**specifications** In quality assurance, written statements describing how good analytical results need to be and what precautions are required in an analytical method.

**specific adsorption** Process in which molecules are held tightly to a surface by van der Waals or electrostatic forces.

**specific gravity** A dimensionless quantity equal to the mass of a substance divided by the mass of an equal volume of water at 4°C. Specific gravity is virtually identical with density in g/mL.

**specificity** Ability of an analysis to distinguish the intended analyte from anything else that might be in the sample. Also called *selectivity*.

**spectral interference** In atomic spectroscopy, any physical process that affects the light intensity at the analytical wavelength. Created by substances that absorb, scatter, or emit light of the analytical wavelength.

**spectrophotometer** A device used to measure absorption of light. It includes a source of light, a wavelength selector (monochromator), and an electrical means of detecting light.

**spectrophotometric analysis** Any method in which light absorption, emission, reflection, or scattering is used to measure chemical concentrations.

**spectrophotometric titration** One in which absorption of light is used to monitor the progress of a chemical reaction.

**spectrophotometry** In a broad sense, any method using light to measure chemical concentrations.

**specular reflection** Reflection of light at an angle equal to the angle of incidence.

**spike** Addition of a known compound (usually at a known concentration) to an unknown. In isotope dilution mass spectrometry, the spike is the added, unusual isotope. *Spike* is a noun and a verb.

**split injection** Used in capillary gas chromatography to inject a small fraction of sample onto the column while the rest of the sample is discarded.

**splitless injection** Used in capillary gas chromatography for trace analysis and quantitative analysis. The entire sample in a low-boiling solvent is directed to the column, where the sample is concentrated by *solvent trapping* (condensing the solvent below its boiling point) or *cold trapping* (condensing solutes far below their boiling range). The column is then warmed to initiate separation.

**spontaneous process** One that is energetically favorable. It will eventually occur, but thermodynamics makes no prediction about how long it will take.

**spread** See *range*.

**square wave voltammetry** A form of *voltammetry* (measurement of current versus potential in an electrochemical cell) in which the potential waveform consists of a square wave superimposed on a voltage staircase. The technique is faster and more sensitive than voltammetry with other waveforms.

**stability constant** See *formation constant*.

**stacking** In electrophoresis, the process of concentrating ions into a narrow band at the interface of electrolytes of low conductivity and high conductivity. Stacking occurs because the electric field in low-conductivity electrolyte is stronger than the field in high-conductivity electrolyte. Ions in the low-conductivity region migrate rapidly until they reach the interface, where the electric field is much smaller.

**standard addition** A technique in which an analytical signal due to an unknown is first measured. Then a known quantity of analyte is added, and the increase in signal is recorded. From the response, it is possible to calculate what quantity of analyte was in the unknown.

**standard curve** A graph showing the response of an analytical technique to known quantities of analyte.

**standard deviation** A statistic measuring how closely data are clustered about the mean value. For a finite set of data, the standard deviation, $s$, is computed from the formula

$$ s = \sqrt{\frac{\Sigma_i(x_i - \bar{x})^2}{n-1}} = \sqrt{\frac{\Sigma_i(x_i^2)}{n-1} - \frac{(\Sigma_i x_i)^2}{n(n-1)}} $$

where $n$ is the number of results, $x_i$ is an individual result, and $\bar{x}$ is the mean result. For a large number of measurements, $s$ approaches $\sigma$, the true standard deviation of the population, and $\bar{x}$ approaches $\mu$, the true population mean.

**standard hydrogen electrode, S.H.E.** One that contains $H_2(g)$ bubbling over a catalytic Pt surface immersed in aqueous $H^+$. The activities of $H_2$ and $H^+$ are both unity in the hypothetical standard electrode. The reaction is $H^+ + e^- \rightleftharpoons \frac{1}{2}H_2(g)$. Also called *normal hydrogen electrode (N.H.E.)*.

**standardization** Process of determining the concentration of a reagent by reaction with a known quantity of a second reagent.

**standard operating procedure** A written procedure that must be rigorously followed to ensure the quality of a chemical analysis.

**standard reduction potential, $E°$** The voltage that would be measured when a hypothetical cell containing the desired half-reaction (with all species present at unit activity) is connected to a standard hydrogen electrode anode.

**Standard Reference Materials** Certified samples sold by the U.S. National Institute of Standards and Technology (or national measurement institutes of other countries) containing known concentrations or quantities of particular analytes. Used to standardize testing procedures in different laboratories.

**standard solution** A solution whose composition is known by virtue of the way that it was made from a reagent of known purity or by virtue of its reaction with a known quantity of a standard reagent.

**standard state** The standard state of a solute is 1 M and the standard state of a gas is 1 bar. Pure solids and liquids are considered to be in their standard states. In equilibrium constants, dimensionless concentrations are expressed as a ratio of the concentration of each species to its concentration in its standard state.

**stationary phase** In chromatography, the phase that does not move through the column.

**stepwise formation constant, $K_n$** Equilibrium constant for a reaction of the type $ML_{n-1} + L \rightleftharpoons ML_n$.

**steradian, sr** Unit of solid angle. There are $4\pi$ steradians in a complete sphere.

**stimulated emission** Emission of a photon induced by the passage of another photon of the same wavelength.

**stoichiometry** Calculation of quantities of substances participating in a chemical reaction.

**Stokes equation** The friction coefficient for a molecule migrating through solution is $6\pi\eta r$, where $\eta$ is the viscosity of the fluid and $r$ is the hydrodynamic radius (equivalent spherical radius) of the molecule.

**stray light** In spectrophotometry, light reaching the detector that is not part of the narrow set of wavelengths expected from the monochromator.

**stripping analysis** A sensitive polarographic technique in which analyte is concentrated from dilute solution by reduction into a drop (or a film) of Hg. It is then analyzed polarographically during an anodic redissolution process. Some analytes can be oxidatively concentrated onto an electrode other than Hg and stripped in a reductive process.

**strong acids or bases** Those that are completely dissociated (to $H^+$ or $OH^-$) in water.

**strong electrolyte** One that mostly dissociates into ions in solution.

**Student's $t$** A statistical tool used to express confidence intervals and to compare results from different experiments.

**sulfur chemiluminescence detector** Gas chromatography detector for the element sulfur. Exhaust from a flame ionization detector is mixed with $O_3$ to form an excited state of $SO_2$ that emits light, which is detected.

**superconductor** A material that loses all electric resistance when cooled below a critical temperature.

**supercritical fluid** A fluid whose temperature is above its critical temperature and whose pressure is above its critical pressure. It has properties of both a liquid and a gas.

**supercritical fluid chromatography** Chromatography using supercritical fluid as the mobile phase. Capable of highly efficient separations of nonvolatile solutes and able to use detectors suitable for gas or liquid.

**supercritical fluid extraction** Extraction of compounds (usually from solids) with a supercritical fluid solvent.

**superficially porous particle** A stationary phase particle for liquid chromatography containing a thin, porous outer layer and a dense, nonporous core. Mass transfer is faster in the superficially porous particle than in a fully porous particle of the same diameter.

**supernatant liquid** Liquid remaining above the solid after a precipitation. Also called *supernate*.

**supersaturated solution** One that contains more dissolved solute than would be present at equilibrium.

**support-coated column** Open tubular gas chromatography column in which the stationary phase is coated on solid support particles attached to the inside wall of the column.

**supporting electrolyte** An unreactive salt added in high concentration to solutions for voltammetric measurements (such as polarography). The supporting electrolyte carries most of the ion-migration current and therefore decreases the coulombic migration of electroactive species to a negligible level. The electrolyte also decreases the resistance of the solution.

**suppressed-ion chromatography** Separation of ions by using an ion-exchange column followed by a suppressor (membrane or column) to remove ionic eluent.

**suppressor column** Ion-exchange column used in ion chromatography to transform ionic eluent into a nonionic form.

**surface-modified electrode** An electrode whose surface has been changed by a chemical reaction. For example, electroactive materials that react specifically with certain solutes can be attached to the electrode.

**surface plasmon resonance** A sensitive means to measure the binding of molecules to a thin gold layer on the underside of a prism. Light directed through the prism is reflected from the gold surface. There is one narrow range of angles at which reflection is nearly 0 because the gold absorbs the light to set up oscillations (called *plasmons*) of the electron cloud in the metal. When a thin layer of material (such as protein or DNA) binds to the side of the gold away from the prism, the electrical properties of the gold are changed and the reflectivity changes.

**surfactant** A molecule with an ionic or polar headgroup and a long, nonpolar tail. Surfactants aggregate in aqueous solution to form micelles. Surfactants derive their name from the fact that they accumulate at boundaries between polar and nonpolar phases and modify the surface tension, which is the free energy of formation of the surface. Soaps are surfactants.

**sweeping** In capillary electrophoresis, migration of a collector species such as a micelle or chelator to concentrate analyte into a narrow region at the front of the migrating collector species.

**syringe** A device having a calibrated barrel into which liquid is sucked by a plunger. The liquid is expelled through a needle by pushing on the plunger.

**systematic error** Error due to procedural or instrumental factors that cause a measurement to be consistently too large or too small. The error can, in principle, be discovered and corrected. Also called *determinate error*.

**systematic treatment of equilibrium** A method that uses the charge balance, mass balance(s), and equilibria to completely specify a system's composition.

**$t$ test** Statistical test used to decide whether the results of two experiments are within experimental uncertainty of each other. The uncertainty must be specified to within a certain probability.

**tailing** Asymmetric chromatographic band in which the later part elutes very slowly. It often results from adsorption of a solute onto a few active sites on the stationary phase.

**tandem mass spectrometry** See *selected reaction monitoring*.

**tare** As a noun, *tare* is the mass of an empty vessel used to receive a substance to be weighed. As a verb, *tare* means setting the balance reading to 0 when an empty vessel or weighing paper is placed on the pan.

**temperature programming** Raising the temperature of a gas chromatography column during a separation to reduce the retention time of late-eluting components.

**temporary hardness** Component of water hardness due to dissolved alkaline earth bicarbonates. It is temporary because boiling causes precipitation of the carbonates.

**test portion** Part of the laboratory sample used for one analysis. Also called *aliquot*.

**theoretical plate** An imaginary construct in chromatography denoting a segment of a column in which one equilibration of solute occurs between stationary and mobile phases. The number of theoretical plates on a column with Gaussian bandshapes is defined as $N = t_r^2/\sigma^2$, where $t_r$ is the retention time of a peak and $\sigma$ is the standard deviation of the band.

**thermal conductivity, $\kappa$** Rate at which a substance transports heat (energy per unit time per unit area) through a temperature gradient (degrees per unit distance). Energy flow $[J/(s \cdot m^2)] = -\kappa(dT/dx)$, where $\kappa$ is the thermal conductivity $[W/(m \cdot K)]$ and $dT/dx$ is the temperature gradient (K/m).

**thermal conductivity detector** A device that detects substances eluted from a gas chromatography column by measuring changes in the thermal conductivity of the gas stream.

**thermal desorption** A sample preparation technique used in gas chromatography to release volatile substances from a solid sample by heating.

**thermistor** A device whose electrical resistance changes markedly with changes in temperature.

**thermocouple** An electrical junction across which a temperature-dependent voltage exists. Thermocouples are calibrated for measurement of temperature and usually consist of two dissimilar metals in contact with each other.

**thermogravimetric analysis** A technique in which the mass of a substance is measured as the substance is heated. Changes in mass indicate decomposition of the substance, often to well-defined products.

**thermometric titration** One in which the temperature is measured to determine the end point. Most titration reactions are exothermic, so the temperature rises during the reaction and suddenly stops rising when the equivalence point is reached.

**thin-layer chromatography** Liquid chromatography in which the stationary phase is coated on a flat glass or plastic plate. Solute is spotted near the bottom of the plate. The bottom edge of the plate is placed in contact with solvent, which creeps up the plate by capillary action.

**time-of-flight mass spectrometer** Ions of different mass accelerated through the same electric field have different velocities: The lighter ions move faster than the heavier ions. The time-of-flight spectrometer finds the mass-to-charge ratio by measuring the time that each group of ions requires to travel a fixed distance to the detector.

**titer** A measure of concentration, usually defined as how many milligrams of reagent B will react with 1 mL of reagent A. One milliliter of $AgNO_3$ solution with a titer of 1.28 mg NaCl/mL will be consumed by 1.28 mg NaCl in the reaction $Ag^+ + Cl^- \rightarrow AgCl(s)$. The same solution of $AgNO_3$ has a titer of 0.993 mg of $KH_2PO_4$/mL, because 1 mL of AgNO solution will be consumed by 0.993 mg $KH_2PO_4$ to precipitate $Ag_3PO_4$.

**titrant** Substance added to the analyte in a titration.

**titration** A procedure in which one substance (titrant) is carefully added to another (analyte) until complete reaction has occurred. The quantity of titrant required for complete reaction tells how much analyte is present.

**titration curve** A graph showing how the concentration of a reactant or a physical property of the solution varies as one reactant (the titrant) is added to another (the analyte).

**titration error** Difference between the observed end point and the true equivalence point in a titration.

**tolerance** Manufacturer's stated uncertainty in the accuracy of a device such as a buret or volumetric flask. A 100-mL flask with a tolerance of $\pm 0.08$ mL may contain 99.92 to 100.08 mL and be within tolerance.

**total carbon** In a natural water or industrial effluent sample, the quantity of $CO_2$ produced when the sample is completely oxidized by oxygen at 900°C in the presence of a catalyst.

**total ion chromatogram** A graph of detector response versus time when a mass spectrometer monitors all ions above a selected $m/z$ ratio emerging from a chromatograph.

**total organic carbon** In a natural water or industrial effluent sample, the quantity of $CO_2$ produced when the sample is first acidified and purged to remove carbonate and bicarbonate and then completely oxidized by oxygen at 900°C in the presence of a catalyst.

**total oxygen demand** In a natural water or industrial effluent sample, the quantity of $O_2$ required for complete oxidation of species in the water at 900°C in the presence of a catalyst.

**trace analysis** Chemical analysis of very low levels of analyte, typically ppm and lower.

**transition range** For an acid-base indicator, the pH range over which the color change occurs. For a redox indicator, the potential range over which the color change occurs.

**transmission quadrupole mass spectrometer** A mass spectrometer that separates ions by passing them between four metallic cylinders to which are applied direct current and oscillating electric fields. Resonant ions with the right mass-to-charge ratio pass through the chamber to the detector while nonresonant ions are deflected into the cylinders and are lost.

**transmittance,** $T$ Defined as $T = P/P_0$, where $P_0$ is the radiant power of light striking the sample on one side and $P$ is the radiant power of light emerging from the other side of the sample.

**treated data** Concentrations or amounts of analyte found from raw data with a calibration curve or some other calibration method.

**triple point** The one temperature and pressure at which the solid, liquid, and gaseous forms of a substance are in equilibrium with one another.

**triplet state** An electronic state in which there are two unpaired electrons.

**tungsten lamp** An ordinary light bulb in which electricity passing through a tungsten filament heats the wire and causes it to emit visible light.

**turbidimetry** A technique in which the decrease in intensity of light traveling through a turbid solution (a solution containing suspended particles) is measured. The greater the concentration of suspended particles, the less light is transmitted. See diagram under *nephelometry*.

**turbidity** Light-scattering property associated with suspended particles in a liquid. A turbid solution appears cloudy.

**turbidity coefficient** The transmittance of a turbid solution is given by $P/P_0 = e^{-\tau b}$, where $P$ is the transmitted radiant power, $P_0$ is the incident radiant power, $b$ is the pathlength, and $\tau$ is the turbidity coefficient.

**ultraviolet detector** Liquid chromatography detector that measures ultraviolet absorbance of solutes emerging from the column.

**use objectives** In quality assurance, use objectives are a written statement of how results will be used. Use objectives are required before specifications can be written for the method.

**valence band** Energy levels containing valence electrons in a semiconductor. The electrons in these levels are localized in chemical bonds.

**van Deemter equation** Describes the dependence of chromatographic plate height, $H$, on linear flow rate, $u_x$: $H = A + B/u_x + Cu_x$. The constant $A$ depends on band-broadening processes such as multiple flow paths that are independent of flow rate. $B$ depends on the rate of diffusion of solute in the mobile phase. $C$ depends on the rate of mass transfer between the stationary and mobile phases.

**variance,** $\sigma^2$ The square of the standard deviation.

**vibrational transition** Occurs when a molecule changes its vibrational energy.

**viscosity** Resistance to flow in a fluid.

**void volume,** $V_0$ Volume of the mobile phase outside the gel particles in a molecular exclusion chromatography column.

**volatile** Easily vaporized.

**volatilization** Selective removal of a component from a mixture by transforming the component into a volatile (low-boiling) species and removing it by heating, pumping, or bubbling a gas through the mixture.

**Volhard titration** Titration of $Ag^+$ with $SCN^-$ in the presence of $Fe^{3+}$. Formation of red $Fe(SCN)^{2+}$ marks the end point.

**volt, V** Unit of electric potential difference. If the potential difference between two points is 1 volt, then 1 joule of energy is required to move 1 coulomb of charge between the two points.

**voltaic cell** See *galvanic cell*.

**voltammetry** An analytical method in which the relation between current and voltage is observed during an electrochemical reaction.

**voltammogram** A graph of current versus electrode potential in an electrochemical cell.

**volume flow rate** In chromatography, the volume of mobile phase per unit time eluted from the column.

**volume percent** Defined as (volume of solute/volume of solution) $\times$ 100.

**volumetric analysis** A technique in which the volume of material needed to react with the analyte is measured.

**volumetric flask** One having a tall, thin neck with a calibration mark. When the liquid level is at the calibration mark, the flask contains its specified volume of liquid.

**Walden reductor** A column packed with silver and eluted with HCl. Analyte is reduced during passage through the column. The reduced product is titrated with an oxidizing agent.

**wall-coated column** Hollow chromatographic column in which the stationary phase is coated on the inside surface of the wall.

**watt, W** SI unit of power equal to an energy flow of 1 joule per second. When an electric current of 1 ampere flows through a potential difference of 1 volt, the power is 1 watt.

**waveguide** A thin layer or hollow structure in which electromagnetic radiation is totally reflected.

**wavelength, λ** Distance between consecutive crests of a wave.

**wavenumber, $\tilde{\nu}$** Reciprocal of the wavelength, $1/\lambda$.

**weak acids and bases** Those whose dissociation constants are not large.

**weak electrolyte** One that only partly dissociates into ions when it dissolves.

**weighing paper** Paper on which to place a solid reagent on a balance. Weighing paper has a very smooth surface, from which solids fall easily for transfer to a vessel.

**weight percent** (Mass of solute/mass of solution) $\times$ 100.

**weight/volume percent** (Mass of solute/volume of solution) $\times$ 100.

**Weston cell** A stable voltage source based on the reaction $Cd(s) + HgSO_4(aq) \rightleftharpoons CdSO_4(aq) + Hg(l)$. It was formerly used to standardize a potentiometer.

**wet ashing** Destruction of organic matter in a sample by a liquid reagent (such as boiling aqueous $HClO_4$) prior to analysis of an inorganic component.

**white light** Light of all different wavelengths.

**white noise** Random noise, also called *Gaussian noise,* due to random movement of charge carriers in an electric circuit (called *thermal noise, Johnson noise,* or *Nyquist noise*) or from random arrival of photons to a detector (called *shot noise* or *Schottky noise*).

**Wien displacement law** Approximate formula for the wavelength, $\lambda_{max}$, of maximum blackbody emission: $\lambda_{max} \cdot T \approx hc/5k = 2.878 \times 10^{-3}\,\text{m} \cdot \text{K}$, where $T$ is temperature in kelvins, $h$ is Planck's constant, $c$ is the speed of light, and $k$ is Boltzmann's constant. Valid for $T > 100$ K.

**working electrode** One at which the reaction of interest occurs.

**Zeeman background correction** Technique used in atomic spectroscopy in which analyte signals are shifted outside the detector monochromator range by applying a strong magnetic field to the sample. Signal that remains is the background.

**Zeeman effect** Shifting of atomic energy levels in a magnetic field.

**zwitterion** A molecule with a positive charge localized at one position and a negative charge localized at another position.

# APPENDIX A

# Logarithms and Exponents

If $a$ is the base 10 logarithm of $n$ ($a = \log n$), then $n = 10^a$. On a calculator, you find the logarithm of a number by pressing the "log" button. If you know $a = \log n$ and you wish to find $n$, use the "antilog" button or raise 10 to the power $a$:

$$a = \log n$$

$$10^a = 10^{\log n} = n (\Rightarrow n = \text{antilog } a)$$

Natural logarithms (ln) are based on the number e ($= 2.718\,281\ldots$) instead of 10:

$$b = \ln n$$

$$e^b = e^{\ln n} = n$$

On a calculator, you find the ln of $n$ with the "ln" button. To find $n$ when you know $b = \ln n$, use the $e^x$ key.

Here are some useful properties to know:

$$\log (a \cdot b) = \log a + \log b \qquad \log 10^a = a$$

$$\log \left(\frac{a}{b}\right) = \log a - \log b \qquad a^b \cdot a^c = a^{(b+c)}$$

$$\log (a^b) = b \log a \qquad \frac{a^b}{a^c} = a^{(b-c)}$$

## Problems

Test yourself by simplifying each expression as much as possible:

(a) $e^{\ln a}$

(b) $10^{\log a}$

(c) $\log 10^a$

(d) $10^{-\log a}$

(e) $e^{-\ln a^3}$

(f) $e^{\ln a^{-3}}$

(g) $\log (10^{1/a^3})$

(h) $\log (10^{-a^2})$

(i) $\log (10^{a^2 - b})$

(j) $\log (2a^3 \, 10^{b^2})$

(k) $e^{(a + \ln b)}$

(l) $10^{[(\log 3) - (4 \log 2)]}$

*Solving a logarithmic equation:* In working with the Nernst and Henderson-Hasselbalch equations, we will need to solve equations such as

$$a = b - c \log \frac{d}{gx}$$

for the variable, $x$. First isolate the log term:

$$\log \frac{d}{gx} = \frac{(b - a)}{c}$$

Then raise 10 to the value of each side of the equation:

$$10^{\log(d/gx)} = 10^{(b-a)/c}$$

But $10^{\log (d/gx)}$ is just $d/gx$, so

$$\frac{d}{gx} = 10^{(b-a)/c} \quad \Rightarrow \quad x = \frac{d}{g 10^{(b-a)/c}}$$

*Converting between ln x and log x:* The relation between them is derived by writing $x = 10^{\log x}$ and taking ln of both sides:

$$\ln x = \ln (10^{\log x}) = (\log x)(\ln 10)$$

because $\ln a^b = b \ln a$.

## Answers

| | | | |
|---|---|---|---|
| (a) $a$ | (d) $1/a$ | (g) $1/a^3$ | (j) $b^2 + \log(2a^3)$ |
| (b) $a$ | (e) $1/a^3$ | (h) $-a^2$ | (k) $be^a$ |
| (c) $a$ | (f) $1/a^3$ | (i) $a^2 - b$ | (l) $3/16$ |

# APPENDIX B

# Graphs of Straight Lines

The general form of the equation of a straight line is

$$y = mx + b$$

where $m$ = slope = $\dfrac{\Delta y}{\Delta x} = \dfrac{y_2 - y_1}{x_2 - x_1}$

$b$ = intercept on $y$-axis

The meanings of slope and intercept are illustrated in Figure B-1.

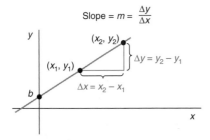

**Figure B-1** Parameters of a straight line.

If you know two points [$(x_1, y_1)$ and $(x_2, y_2)$] that lie on the line, you can generate the equation of the line by noting that the slope is the same for every pair of points on the line. Calling some general point on the line $(x, y)$, we can write

$$\frac{y - y_1}{x - x_1} = \frac{y_2 - y_1}{x_2 - x_1} = m \qquad \text{(B-1)}$$

which can be rearranged to the form

$$y - y_1 = \left(\frac{y_2 - y_1}{x_2 - x_1}\right)(x - x_1)$$

$$y = \underbrace{\left(\frac{y_2 - y_1}{x_2 - x_1}\right)}_{m} x + \underbrace{y_1 - \left(\frac{y_2 - y_1}{x_2 - x_1}\right)x_1}_{b}$$

When you have a series of experimental points that should lie on a line, the best line is generally obtained by the method of least squares, described in Chapter 4. This method gives the slope and the intercept directly. If, instead, you wish to draw the "best" line by eye, you can derive the equation of the line by selecting two points *that lie on the line* and applying Equation B-1.

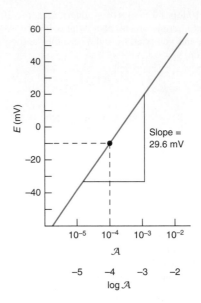

**Figure B-2** A linear graph in which one axis is a logarithmic function.

Sometimes you are presented with a linear plot in which $x$ or $y$ or both are nonlinear functions. An example is shown in Figure B-2, in which the potential of an electrode is expressed as a function of the activity of analyte. Given that the slope is 29.6 mV and the line passes through the point ($\mathcal{A} = 10^{-4}$, $E = -10.2$), find the equation of the line. To do this, first note that the $y$-axis is linear but the $x$-axis is *logarithmic*. That is, the function $E$ versus $\mathcal{A}$ is *not* linear, but $E$ versus log $\mathcal{A}$ is linear. The form of the straight line should therefore be

$$E = \underset{\underset{y}{\uparrow}}{(29.6)} \underset{\underset{m}{\uparrow}}{\log \mathcal{A}} + \underset{\underset{x}{\uparrow}}{b}$$

To find $b$, we can use the coordinates of the one known point in Equation B-1:

$$\frac{y - y_1}{x - x_1} = \frac{E - E_1}{\log \mathcal{A} - \log \mathcal{A}_1} = \frac{E - (-10.2)}{\log \mathcal{A} - \log (10^{-4})} = m = 29.6$$

or

$$E + 10.2 = 29.6 \log \mathcal{A} + (29.6)(4)$$

$$E(\text{mV}) = 29.6\ (\text{mV}) \log \mathcal{A} + 108.2\ (\text{mV})$$

# APPENDIX C

# Propagation of Uncertainty

The rules for propagation of uncertainty in Table 3-1 are special cases of a general formula. Suppose you wish to calculate the function, $F$, of several experimental quantities, $x, y, z, \ldots$ If the errors $(e_x, e_y, e_z, \ldots)$ in $x, y, z, \ldots$ are small, random, and independent of one another, then the uncertainty, $e_F$, in the function $F$ is approximately:*

$$e_F = \sqrt{\left(\frac{\partial F}{\partial x}\right)^2 e_x^2 + \left(\frac{\partial F}{\partial y}\right)^2 e_y^2 + \left(\frac{\partial F}{\partial z}\right)^2 e_z^2 + \ldots} \qquad \text{(C-1)}$$

The quantities in parentheses are partial derivatives, which are calculated in the same manner as ordinary derivatives, except that all but one variable are treated as constants. For example, if $F = 3xy^2$, then $\partial F/\partial x = 3y^2$ and $\partial F/\partial y = (3x)(2y) = 6xy$.

As an example of using Equation C-1, let's find the uncertainty in the function

$$F = x^y = (2.00 \pm 0.02)^{3.00 \pm 0.09}$$

The partial derivatives are

$$\frac{\partial F}{\partial x} = yx^{y-1} \qquad \frac{\partial F}{\partial y} = x^y \ln x$$

Putting these quantities into Equation C-1 gives

$$e_F = \sqrt{(yx^{y-1})^2 e_x^2 + (x^y \ln x)^2 e_y^2}$$

$$= \sqrt{y^2 x^{2y-2}\, e_x^2 + x^{2y}\,(\ln x)^2\, e_y^2}$$

$$= \sqrt{y^2 x^{2y}\left(\frac{e_x}{x}\right)^2 + x^{2y}\,(\ln x)^2\, e_y^2}$$

Multiplying and dividing the second term by $y^2$ allows us to rearrange to a more pleasant form:

$$e_F = \sqrt{y^2 x^{2y}\left(\frac{e_x}{x}\right)^2 + y^2 x^{2y}\,(\ln x)^2\left(\frac{e_y}{y}\right)^2}$$

Removing $\sqrt{y^2 x^{2y}} = yF$ from both terms gives

$$e_F = yF\sqrt{\left(\frac{e_x}{x}\right)^2 + (\ln x)^2\left(\frac{e_y}{y}\right)^2}$$

Now for the number crunching. Disregarding uncertainties for a moment, we know that $F = 2.00^{3.00} = 8.00 \pm ?$ The uncertainty is obtained from the equation above:

$$e_F = 3.00 \cdot 8.00\sqrt{\left(\frac{0.02}{2.00}\right)^2 + (\ln 2.00)^2\left(\frac{0.09}{3.00}\right)^2} = 0.55$$

Reasonable answers are $F = 8.0_0 \pm 0.5_5$ or $8.0 \pm 0.6$.

## Exercises

C-1. Verify the following calculations.
(a) $2.36^{4.39 \pm 0.08} = 43._4 \pm 3._0$
(b) $(2.36 \pm 0.06)^{4.39 \pm 0.08} = 43._4 \pm 5._7$

C-2. For $F = \sin(2\pi xy)$, show that

$$e_F = 2\pi xy \cos(2\pi xy)\sqrt{\left(\frac{e_x}{x}\right)^2 + \left(\frac{e_y}{y}\right)^2}$$

## Covariance in Propagation of Uncertainty

Equation C-1 presumes that errors in $x$, $y$, and $z$ are independent of one another. A common case in which this is not true is when we use the least-squares slope and intercept to compute a new quantity, such as the value of $x$ from an observed value of $y$. In general, uncertainties in the slope and intercept are correlated, so they are not independent errors.

Let's restrict our attention to a function, $F$, of the two experimental parameters, $m$ and $b$, whose uncertainties are $s_m$ and $s_b$. If the uncertainties are correlated, the equation for propagation of uncertainty is[†]

$$e_F = \sqrt{\underbrace{\left(\frac{\partial F}{\partial m}\right)^2 s_m^2 + \left(\frac{\partial F}{\partial b}\right)^2 s_b^2}_{\substack{\text{Variance terms from} \\ \text{Equation C-1}}} + \underbrace{2\left(\frac{\partial F}{\partial m}\right)\left(\frac{\partial F}{\partial b}\right) s_{mb}}_{\substack{\text{Covariance accounts for} \\ \text{correlation of } m \text{ and } b}}} \qquad \text{(C-2)}$$

The last term in Equation C-2 reflects the fact that uncertainties in $m$ and $b$ are not independent of each other. The quantity $s_{mb}$ is called the *covariance* and it can be positive or negative.

In linear least-squares analysis, the variance and covariance are[‡]

*Variance:*

$$s_m^2 = \frac{s_y^2 n}{D} \qquad s_b^2 = \frac{s_y^2 \sum (x_i^2)}{D} \qquad \text{(Equations 4-21 and 4-22)}$$

*Covariance:*

$$s_{mb} = \frac{-s_y^2 \sum (x_i)}{D} \qquad \text{(C-3)}$$

where $s_y^2$ is the square of Equation 4-20, $D$ is given by Equation 4-18, and $n$ is the number of data points.

---

*A numerical recipe for evaluating Equation C-1 with a spreadsheet is given by R. de Levie, *J. Chem. Ed.* **2000,** *77,* 534.

---

[†]E. F. Meyer, *J. Chem. Ed.* **1997,** *74,* 1339.
[‡]C. Salter, *J. Chem. Ed.* **2000,** *77,* 1239.

## Example  Finding the x-Intercept

For the line $y = mx + b$, the x-intercept occurs when $y = 0$, or $x = -b/m$. Let's designate the x-intercept as the function $F = -b/m$. Find the x-intercept and its uncertainty for the least-squares line in Figure 4-9.

**Solution**  The following quantities are computed in Section 4-7:

$$m = 0.615\,38 \quad s_m^2 = 0.002\,958\,6 \quad s_y^2 = 0.038\,462 \quad \sum(x_i) = 14$$
$$b = 1.346\,15 \quad s_b^2 = 0.045\,859 \quad D = 52$$

The covariance in Equation C-3 is therefore

$$s_{mb} = \frac{-s_y^2 \sum(x_i)}{D} = \frac{-(0.038\,462)(14)}{52} = -0.010\,355$$

The x-intercept is just $F = -b/m = -(1.346\,15)/(0.615\,38) = -2.187\,5$.

To find the uncertainty in $F$, we use Equation C-2. The derivatives in C-2 are

$$\frac{\partial F}{\partial m} = \frac{\partial(-b/m)}{\partial m} = \frac{b}{m^2} = \frac{1.346\,15}{0.615\,38^2} = 3.554\,7$$

$$\frac{\partial F}{\partial b} = \frac{\partial(-b/m)}{\partial b} = \frac{-1}{m} = \frac{-1}{0.615\,38} = -1.625\,0$$

Now we can evaluate the uncertainty with Equation C-2:

$$e_F = \sqrt{\left(\frac{\partial F}{\partial m}\right)^2 s_m^2 + \left(\frac{\partial F}{\partial b}\right)^2 s_b^2 + 2\left(\frac{\partial F}{\partial m}\right)\left(\frac{\partial F}{\partial b}\right) s_{mb}}$$

$$= \sqrt{\begin{array}{l} (3.554\,7)^2(0.002\,958\,6) + \\ (-1.625\,0)^2(0.045\,859) + \\ 2(3.554\,7)(-1.625\,0)(-0.010\,355) \end{array}}$$

$$= 0.527\,36$$

The final answer can now be written with a reasonable number of digits:

$$F = -2.187\,5 \pm 0.527\,36 = -2.1_9 \pm 0.5_3$$

If we had used Equation C-1 and ignored the covariance term in Equation C-2, we would have computed an uncertainty of $\pm 0.4_0$.

To learn how to compute variance and covariance and to see how to include weighting factors in least-squares curve fitting, see J. Tellinghuisen, "Understanding Least Squares Through Monte Carlo Calculations," *J. Chem. Ed.* **2005**, *82*, 157.

# APPENDIX D

# Oxidation Numbers and Balancing Redox Equations

The *oxidation number,* or *oxidation state,* is a bookkeeping device used to keep track of the number of electrons formally associated with a particular element. The oxidation number is meant to tell how many electrons have been lost or gained by a neutral atom when it forms a compound. Because oxidation numbers have no real physical meaning, they are somewhat arbitrary, and not all chemists will assign the same oxidation number to a given element in an unusual compound. However, there are some ground rules that provide a useful start.

1. The oxidation number of an element by itself—e.g., $Cu(s)$ or $Cl_2(g)$—is 0.
2. The oxidation number of H is almost always $+1$, except in metal hydrides—e.g., NaH—in which H is $-1$.
3. The oxidation number of oxygen is almost always $-2$. The only common exceptions are peroxides, in which two oxygen atoms are connected and each has an oxidation number of $-1$. Two examples are hydrogen peroxide (H—O—O—H) and its anion (H—O—O$^-$). The oxidation number of oxygen in gaseous $O_2$ is, of course, 0.
4. The alkali metals (Li, Na, K, Rb, Cs, Fr) almost always have an oxidation number of $+1$. The alkaline earth metals (Be, Mg, Ca, Sr, Ba, Ra) are almost always in the $+2$ oxidation state.
5. The halogens (F, Cl, Br, I) are usually in the $-1$ oxidation state. Exceptions occur when two different halogens are bound to each other or when a halogen is bound to more than one atom. When different halogens are bound to each other, we assign the oxidation number $-1$ to the more electronegative halogen.

*The sum of the oxidation numbers of each atom in a molecule must equal the charge of the molecule.* In $H_2O$, for example, we have

$$
\begin{array}{ll}
\text{2 hydrogen} = 2(+1) = & +2 \\
\text{oxygen} = & \underline{-2} \\
\text{net charge} & 0
\end{array}
$$

In $SO_4^{2-}$, sulfur must have an oxidation number of $+6$ so that the sum of the oxidation numbers will be $-2$:

$$
\begin{array}{ll}
\text{oxygen} = 4(-2) = & -8 \\
\text{sulfur} = & \underline{+6} \\
\text{net charge} & -2
\end{array}
$$

In benzene ($C_6H_6$), the oxidation number of each carbon must be $-1$ if hydrogen is assigned the number $+1$. In cyclohexane ($C_6H_{12}$), the oxidation number of each carbon must be $-2$, for the same reason. The carbons in benzene are in a higher oxidation state than those in cyclohexane.

The oxidation number of iodine in $ICl_2^-$ is $+1$. This is unusual, because halogens are usually $-1$. However, because chlorine is more electronegative than iodine, we assign Cl as $-1$, thereby forcing I to be $+1$.

The oxidation number of As in $As_2S_3$ is $+3$, and the value for S is $-2$. This is arbitrary but reasonable. Because S is more electronegative than As, we make S negative and As positive; and, because S is in the same family as oxygen, which is usually $-2$, we assign S as $-2$, thus leaving As as $+3$.

The oxidation number of S in $S_4O_6^{2-}$ (tetrathionate) is $+2.5$. The *fractional oxidation state* comes about because six O atoms contribute $-12$. Because the charge is $-2$, the four S atoms must contribute $+10$. The average oxidation number of S must be $+\frac{10}{4} = 2.5$.

The oxidation number of Fe in $K_3Fe(CN)_6$ is $+3$. To make this assignment, we first recognize cyanide ($CN^-$) as a common ion that carries a charge of $-1$. Six cyanide ions give $-6$, and three potassium ions ($K^+$)

give $+3$. Therefore, Fe should have an oxidation number of $+3$ for the whole formula to be neutral. In this approach, it is not necessary to assign individual oxidation numbers to carbon and nitrogen, as long as we recognize that the charge of CN is $-1$.

## Problems

Answers are given at the end of this appendix.

**D-1.** Write the oxidation state of the boldface atom in each of the following species.

(a) **Ag**Br
(b) **S**$_2$O$_3^{2-}$
(c) **Se**F$_6$
(d) H**S**$_2$O$_3^-$
(e) H**O**$_2$
(f) N**O**
(g) **Cr**$^{3+}$
(h) **Mn**O$_2$
(i) **Pb**(OH)$_3^-$
(j) **Fe**(OH)$_3$
(k) **Cl**O$^-$
(l) K$_4$**Fe**(CN)$_6$
(m) **Cl**O$_2$
(n) **Cl**O$_2^-$
(o) **Mn**(CN)$_6^{4-}$
(p) **N**$_2$
(q) **N**H$_4^+$
(r) **N**$_2$H$_5^+$
(s) H**As**O$_3^{2-}$
(t) **Co**$_2$(CO)$_8$ (CO group is neutral)
(u) (CH$_3$)$_4$**Li**$_4$
(v) **P**$_4$O$_{10}$
(w) **C**$_2$H$_6$O (ethanol, CH$_3$CH$_2$OH)
(x) **V**O(SO$_4$)
(y) **Fe**$_3$O$_4$
(z) **C**$_3$H$_3^+$

Structure:

$$
H - \overset{H}{\underset{H}{\overset{|}{\underset{\phantom{|}}{C}}}} \overset{\displaystyle\diagup C}{\underset{\displaystyle\diagdown C}{\overset{\oplus}{}}}
$$

**D-2.** Identify the oxidizing agent and the reducing agent on the left side of each of the following reactions.

(a) $Cr_2O_7^{2-} + 3Sn^{2+} + 14H^+ \rightarrow 2Cr^{3+} + 3Sn^{4+} + 7H_2O$
(b) $4I^- + O_2 + 4H^+ \rightarrow 2I_2 + 2H_2O$
(c) $5CH_3\overset{O}{\overset{\|}{C}}H + 2MnO_4^- + 6H^+ \rightarrow 5CH_3\overset{O}{\overset{\|}{C}}OH + 2Mn^{2+} + 3H_2O$
(d) $HOCH_2CHOHCH_2OH + 2IO_4^- \rightarrow$
  Glycerol $\quad 2H_2C{=}O + HCO_2H + 2IO_3^- + H_2O$
  Formaldehyde Formic acid
(e) $C_8H_8 + 2Na \rightarrow C_8H_8^{2-} + 2Na^+$
  $C_8H_8$ is cyclooctatetraene with the structure ⬡
(f) $I_2 + OH^- \rightarrow HOI + I^-$
  Hypoiodous
  acid

## Balancing Redox Reactions

To balance a reaction involving oxidation and reduction, we must first identify which element is oxidized and which is reduced. We then break the net reaction into two imaginary *half-reactions,* one of which involves only oxidation and the other only reduction. Although free electrons never appear in a balanced net reaction, they do appear in balanced half-reactions. If we are dealing with aqueous solutions, we proceed to balance each half-reaction, using $H_2O$ and either $H^+$ or $OH^-$, as necessary. *A reaction is balanced when the number of atoms of each element is the same on both sides and the net charge is the same on both sides.*[*]

---

[*]A completely different method for balancing complex redox equations by inspection has been described by D. Kolb, *J. Chem. Ed.* **1981,** *58,* 642. For some challenging problems in balancing redox equations, see R. Stout, *J. Chem. Ed.* **1995,** *72,* 1125.

## Acidic Solutions

Here are the steps we will follow:

1. Assign oxidation numbers to the elements that are oxidized or reduced.
2. Break the reaction into two half-reactions, one involving oxidation and the other reduction.
3. For each half-reaction, balance the number of atoms that are oxidized or reduced.
4. Balance the electrons to account for the change in oxidation number by adding electrons to one side of each half-reaction.
5. Balance oxygen atoms by adding $H_2O$ to one side of each half-reaction.
6. Balance the H atoms by adding $H^+$ to one side of each half-reaction.
7. Multiply each half-reaction by the number of electrons in the other half-reaction so that the number of electrons on each side of the total reaction will cancel. Then add the two half-reactions and simplify to the smallest integral coefficients.

### Example  Balancing a Redox Equation

Balance the following equation using $H^+$, but not $OH^-$:

$$Fe^{2+} + MnO_4^- \rightleftharpoons Fe^{3+} + Mn^{2+}$$
$$\quad +2 \qquad +7 \qquad +3 \qquad +2$$
$$\qquad \text{Permanganate}$$

#### Solution

1. *Assign oxidation numbers.* They are assigned for Fe and Mn in each species in the above reaction.
2. *Break the reaction into two half-reactions.*

   Oxidation half-reaction: $\quad Fe^{2+} \rightleftharpoons Fe^{3+}$
   $$\qquad\qquad\qquad\qquad\qquad +2 \qquad +3$$

   Reduction half-reaction: $\quad MnO_4^- \rightleftharpoons Mn^{2+}$
   $$\qquad\qquad\qquad\qquad\qquad +7 \qquad\quad +2$$

3. *Balance the atoms that are oxidized or reduced.* Because there is only one Fe or Mn in each species on each side of the equation, the atoms of Fe and Mn are already balanced.
4. *Balance electrons.* Electrons are added to account for the change in each oxidation state.

   $$Fe^{2+} \rightleftharpoons Fe^{3+} + e^-$$
   $$MnO_4^- + 5e^- \rightleftharpoons Mn^{2+}$$

   In the second case, we need $5e^-$ on the left side to take Mn from $+7$ to $+2$.
5. *Balance oxygen atoms.* There are no oxygen atoms in the Fe half-reaction. There are four oxygen atoms on the left side of the Mn reaction, so we add four molecules of $H_2O$ to the right side:

   $$MnO_4^- + 5e^- \rightleftharpoons Mn^{2+} + 4H_2O$$

6. *Balance hydrogen atoms.* The Fe equation is already balanced. The Mn equation needs $8H^+$ on the left.

   $$MnO_4^- + 5e^- + 8H^+ \rightarrow Mn^{2+} + 4H_2O$$

   *At this point, each half-reaction must be completely balanced (the same number of atoms and charge on each side) or you have made a mistake.*
7. *Multiply and add the reactions.* We multiply the Fe equation by 5 and the Mn equation by 1 and add:

   $$5Fe^{2+} \rightleftharpoons 5Fe^{3+} + 5e^-$$
   $$\underline{MnO_4^- + 5e^- + 8H^+ \rightleftharpoons Mn^{2+} + 4H_2O}$$
   $$5Fe^{2+} + MnO_4^- + 8H^+ \rightleftharpoons 5Fe^{3+} + Mn^{2+} + 4H_2O$$

The total charge on each side is $+17$, and we find the same number of atoms of each element on each side. The equation is balanced.

### Example  A Reverse Disproportionation

Now try the next reaction, which represents the reverse of a *disproportionation*. (In a disproportionation, an element in one oxidation state reacts to give the same element in higher and lower oxidation states.)

$$I_2 + IO_3^- + Cl^- \rightleftharpoons ICl_2^-$$
$$\quad 0 \qquad +5 \qquad -1 \qquad\quad +1\,-1$$
$$\text{Iodine} \quad \text{Iodate}$$

#### Solution

1. The oxidation numbers are assigned above. Note that chlorine has an oxidation number of $-1$ on both sides of the equation. Only iodine is involved in electron transfer.
2. Oxidation half-reaction: $\quad I_2 \rightleftharpoons ICl_2^-$
   $$\qquad\qquad\qquad\qquad\quad 0 \qquad +1$$

   Reduction half-reaction: $\quad IO_3^- \rightleftharpoons ICl_2^-$
   $$\qquad\qquad\qquad\qquad\qquad +5 \qquad +1$$

3. We need to balance I atoms in the first reaction and add $Cl^-$ to each reaction to balance Cl.

   $$I_2 + 4Cl^- \rightleftharpoons 2ICl_2^-$$
   $$IO_3^- + 2Cl^- \rightleftharpoons ICl_2^-$$

4. Now add electrons to each.

   $$I_2 + 4Cl^- \rightleftharpoons 2ICl_2^- + 2e^-$$
   $$IO_3^- + 2Cl^- + 4e^- \rightleftharpoons ICl_2^-$$

   The first reaction needs $2e^-$ because there are two I atoms, each of which changes from 0 to $+1$.
5. The second reaction needs $3H_2O$ on the right side to balance oxygen atoms.

   $$IO_3^- + 2Cl^- + 4e^- \rightleftharpoons ICl_2^- + 3H_2O$$

6. The first reaction is balanced, but the second needs $6H^+$ on the left.

   $$IO_3^- + 2Cl^- + 4e^- + 6H^+ \rightleftharpoons ICl_2^- + 3H_2O$$

   As a check, the charge on each side of this half-reaction is $-1$, and all atoms are balanced.
7. Multiply and add.

   $$2(I_2 + 4Cl^- \qquad\qquad\qquad \rightleftharpoons 2ICl_2^- + 2e^-)$$
   $$\underline{IO_3^- + 2Cl^- + 4e^- + 6H^+ \rightleftharpoons ICl_2^- + 3H_2O}$$
   $$2I_2 + IO_3^- + 10Cl^- + 6H^+ \rightleftharpoons 5ICl_2^- + 3H_2O \qquad\text{(D-1)}$$

We multiplied the first reaction by 2 so that there would be the same number of electrons in each half-reaction. You could have multiplied the first reaction by 4 and the second by 2, but then all coefficients would simply be doubled. We customarily write the smallest coefficients.

## Basic Solutions

The method many people prefer for basic solutions is to balance the equation first with $H^+$. The answer can then be converted into one in which $OH^-$ is used instead. This is done by adding to each side of the equation a number of hydroxide ions equal to the number of $H^+$ ions appearing in the equation. For example, to balance Equation D-1 with $OH^-$ instead of $H^+$, proceed as follows:

$$2I_2 + IO_3^- + 10Cl^- + 6H^+ \rightleftharpoons 5ICl_2 + 3H_2O$$
$$\qquad\qquad\qquad\qquad +6OH^- \qquad\qquad +6OH^-$$

$$2I_2 + IO_3^- + 10Cl^- + \underbrace{6H^+ + 6OH^-} \rightleftharpoons 5ICl_2 + 3H_2O + 6OH^-$$
$$\qquad\qquad\qquad\qquad\qquad 6H_2O$$
$$\qquad\qquad\qquad\qquad\qquad \Downarrow$$
$$\qquad\qquad\qquad\qquad\qquad 3H_2O$$

Realizing that $6H^+ + 6OH^- = 6H_2O$, and canceling $3H_2O$ on each side, gives the final result:

$$2I_2 + IO_3^- + 10Cl^- + 3H_2O \rightleftharpoons 5ICl_2 + 6OH^-$$

## Problems

**D-3.** Balance the following reactions by using $H^+$ but not $OH^-$.

**(a)** $Fe^{3+} + Hg_2^{2+} \rightleftharpoons Fe^{2+} + Hg^{2+}$
**(b)** $Ag + NO_3^- \rightleftharpoons Ag^+ + NO$
**(c)** $VO^{2+} + Sn^{2+} \rightleftharpoons V^{3+} + Sn^{4+}$
**(d)** $SeO_4^{2-} + Hg + Cl^- \rightleftharpoons SeO_3^{2-} + Hg_2Cl_2$
**(e)** $CuS + NO_3^- \rightleftharpoons Cu^{2+} + SO_4^{2-} + NO$
**(f)** $S_2O_3^{2-} + I_2 \rightleftharpoons I^- + S_4O_6^{2-}$
**(g)** $ClO_3^- + As_2S_3 \rightleftharpoons Cl^- + H_2AsO_4^- + SO_4^{2-}$

**(h)** $Cr_2O_7^{2-} + CH_3CH{\overset{O}{\overset{\|}{}}} \rightleftharpoons CH_3C{\overset{O}{\overset{\|}{}}}OH + Cr^{3+}$

**(i)** $MnO_4^{2-} \rightleftharpoons MnO_2 + MnO_4^-$
**(j)** $Hg_2SO_4 + Ca^{2+} + S_8 \rightleftharpoons Hg_2^{2+} + CaS_2O_3$
**(k)** $ClO_3^- \rightleftharpoons Cl_2 + O_2$

**D-4.** Balance the following equations by using $OH^-$ but not $H^+$.

**(a)** $PbO_2 + Cl^- \rightleftharpoons ClO^- + Pb(OH)_3^-$
**(b)** $HNO_2 + SbO^+ \rightleftharpoons NO + Sb_2O_5$
**(c)** $Ag_2S + CN^- + O_2 \rightleftharpoons S + Ag(CN)_2^- + OH^-$
**(d)** $HO_2^- + Cr(OH)_3^- \rightleftharpoons CrO_4^{2-} + OH^-$
**(e)** $ClO_2 + OH^- \rightleftharpoons ClO_2^- + ClO_3^-$
**(f)** $WO_3^- + O_2 \rightleftharpoons HW_6O_{21}^{5-} + OH^-$
**(g)** $Mn_2O_3 + CN^- \rightleftharpoons Mn(CN)_6^{4-} + (CN)_2$
**(h)** $Cu^{2+} + H_2 \rightleftharpoons Cu + H_2O$
**(i)** $BH_4^- + H_2O \rightleftharpoons H_3BO_3 + H_2$
**(j)** $Mn_2O_3 + Hg + CN^- \rightleftharpoons Mn(CN)_6^{4-} + Hg(CN)_2$

**(k)** $MnO_4^- + HCCH_2CH_2OH{\overset{O}{\overset{\|}{}}} \rightleftharpoons CH_2(CO_2^-)_2 + MnO_2$

**(l)** $K_3V_5O_{14} + HOCH_2CHOHCH_2OH \rightleftharpoons VO(OH)_2 + HCO_2^- + K^+$

## Answers

**D-1.** 

| | | |
|---|---|---|
| **(a)** $+1$ | **(j)** $+3$ | **(s)** $+3$ |
| **(b)** $+2$ | **(k)** $+1$ | **(t)** $0$ |
| **(c)** $+6$ | **(l)** $+2$ | **(u)** $-4$ |
| **(d)** $+2$ | **(m)** $+4$ | **(v)** $+5$ |
| **(e)** $-\frac{1}{2}$ | **(n)** $+3$ | **(w)** $-2$ |
| **(f)** $+2$ | **(o)** $+2$ | **(x)** $+4$ |
| **(g)** $+3$ | **(p)** $0$ | **(y)** $+8/3$ |
| **(h)** $+4$ | **(q)** $-3$ | **(z)** $-2/3$ |
| **(i)** $+2$ | **(r)** $-2$ | |

**D-2.**

| | Oxidizing agent | Reducing agent |
|---|---|---|
| **(a)** | $Cr_2O_7^{2-}$ | $Sn^{2+}$ |
| **(b)** | $O_2$ | $I^-$ |
| **(c)** | $MnO_4^-$ | $CH_3CHO$ |
| **(d)** | $IO_4^-$ | Glycerol |
| **(e)** | $C_8H_8$ | Na |
| **(f)** | $I_2$ | $I_2$ |

Reaction **(f)** is called a *disproportionation,* because an element in one oxidation state is transformed into two different oxidation states—one higher and one lower than the original oxidation state.

**D-3.** **(a)** $2Fe^{3+} + Hg_2^{2+} \rightleftharpoons 2Fe^{2+} + 2Hg^{2+}$
**(b)** $3Ag + NO_3^- + 4H^+ \rightleftharpoons 3Ag^+ + NO + 2H_2O$
**(c)** $4H^+ + 2VO^{2+} + Sn^{2+} \rightleftharpoons 2V^{3+} + Sn^{4+} + 2H_2O$
**(d)** $2Hg + 2Cl^- + SeO_4^{2-} + 2H^+ \rightleftharpoons Hg_2Cl_2 + SeO_3^{2-} + H_2O$
**(e)** $3CuS + 8NO_3^- + 8H^+ \rightleftharpoons 3Cu^{2+} + 3SO_4^{2-} + 8NO + 4H_2O$
**(f)** $2S_2O_3^{2-} + I_2 \rightleftharpoons S_4O_6^{2-} + 2I^-$
**(g)** $14ClO_3^- + 3As_2S_3 + 18H_2O \rightleftharpoons$
$$14Cl^- + 6H_2AsO_4^- + 9SO_4^{2-} + 24H^+$$
**(h)** $Cr_2O_7^{2-} + 3CH_3CHO + 8H^+ \rightleftharpoons 2Cr^{3+} + 3CH_3CO_2H + 4H_2O$
**(i)** $4H^+ + 3MnO_4^{2-} \rightleftharpoons MnO_2 + 2MnO_4^- + 2H_2O$
**(j)** $2Hg_2SO_4 + 3Ca^{2+} + \frac{1}{2}S_8 + H_2O \rightleftharpoons 2Hg_2^{2+} + 3CaS_2O_3 + 2H^+$
**(k)** $2H^+ + 2ClO_3^- \rightleftharpoons Cl_2 + \frac{5}{2}O_2 + H_2O$
The balanced half-reaction for $As_2S_3$ in **(g)** is

$$As_2S_3 + 20H_2O \rightleftharpoons 2H_2AsO_4^- + 3SO_4^{2-} + 28e^- + 36H^+$$
$$\underset{+3\ -2}{} \qquad\qquad \underset{+5}{} \qquad \underset{+6}{}$$

Because $As_2S_3$ is a single compound, we must consider the $As_2S_3 \rightarrow H_2AsO_4^-$ and $As_2S_3 \rightarrow SO_4^{2-}$ reactions together. The net change in oxidation number for the *two* As atoms is $2(5 - 3) = +4$. The net change in oxidation number for the *three* S atoms is $3[6 - (-2)] = +24$. Therefore, $24 + 4 = 28e^-$ are involved in the half-reaction.

**D-4.** **(a)** $H_2O + OH^- + PbO_2 + Cl^- \rightleftharpoons Pb(OH)_3^- + ClO^-$
**(b)** $4HNO_2 + 2SbO^- + 2OH^- \rightleftharpoons 4NO + Sb_2O_5 + 3H_2O$
**(c)** $Ag_2S + 4CN^- + \frac{1}{2}O_2 + H_2O \rightleftharpoons S + 2Ag(CN)_2^- + 2OH^-$
**(d)** $2HO_2^- + Cr(OH)_3^- \rightleftharpoons CrO_4^{2-} + OH^- + 2H_2O$
**(e)** $2ClO_2 + 2OH^- \rightleftharpoons ClO_2^- + ClO_3^- + H_2O$
**(f)** $12WO_3^- + 3O_2 + 2H_2O \rightleftharpoons 2HW_6O_{21}^{5-} + 2OH^-$
**(g)** $Mn_2O_3 + 14CN^- + 3H_2O \rightleftharpoons 2Mn(CN)_6^{4-} + (CN)_2 + 6OH^-$
**(h)** $Cu^{2+} + H_2 + 2OH^- \rightleftharpoons Cu + 2H_2O$
**(i)** $BH_4^- + 4H_2O \rightleftharpoons H_3BO_3 + 4H_2 + OH^-$
**(j)** $3H_2O + Mn_2O_3 + Hg + 14CN^- \rightleftharpoons$
$$2Mn(CN)_6^{4-} + Hg(CN)_2 + 6OH^-$$

**(k)** $2MnO_4^- + HCCH_2CH_2OH{\overset{O}{\overset{\|}{}}} \rightleftharpoons 2MnO_2 + 2H_2O + CH_2(CO_2^-)_2$

For **(k)**, the organic half-reaction is $8OH^- + C_3H_6O_2 \rightleftharpoons C_3H_2O_4^{2-} + 6e^- + 6H_2O$.
**(l)** $32H_2O + 8K_3V_5O_{14} + 5HOCH_2CHOHCH_2OH \rightleftharpoons$
$$40VO(OH)_2 + 15HCO_2^- + 9OH^- + 24K^+$$
For **(l)**, the two half-reactions are $K_3V_5O_{14} + 9H_2O + 5e^- \rightleftharpoons 5VO(OH)_2 + 8OH^- + 3K^+$ and $C_3H_8O_3 + 11OH^- \rightleftharpoons 3HCO_2^- + 8e^- + 8H_2O$.

# APPENDIX E

# Normality

The *normality*, N, of a redox reagent is $n$ times the molarity, where $n$ is the number of electrons donated or accepted by that species in a chemical reaction.

$$N = nM \qquad \text{(E-1)}$$

For example, in the half-reaction

$$MnO_4^- + 8H^+ + 5e^- \rightleftharpoons Mn^{2+} + 4H_2O \qquad \text{(E-2)}$$

the normality of permanganate ion is five times its molarity, because each $MnO_4^-$ accepts $5e^-$. If the molarity of permanganate is 0.1 M, the normality for the reaction

$$MnO_4^- + 5Fe^{2+} + 8H^+ \rightleftharpoons Mn^{2+} + 5Fe^{3+} + 4H_2O \qquad \text{(E-3)}$$

is $5 \times 0.1 = 0.5$ N (read "0.5 normal"). In this reaction, each $Fe^{2+}$ ion donates one electron. The normality of ferrous ion *equals* the molarity of ferrous ion, even though it takes five ferrous ions to balance the reaction.

In the half-reaction

$$MnO_4^- + 4H^+ + 3e^- \rightleftharpoons MnO_2 + 2H_2O \qquad \text{(E-4)}$$

each $MnO_4^-$ ion accepts only *three* electrons. The normality of permanganate for this reaction is equal to three times the molarity of permanganate. A 0.06 N permanganate solution for this reaction contains 0.02 M $MnO_4^-$.

The normality of a solution is a statement of the moles of "reacting units" per liter. One mole of reacting units is called one *equivalent*. Therefore, the units of normality are equivalents per liter (equiv/L). For redox reagents, *one equivalent is the amount of substance that can donate or accept one mole of electrons*. It is possible to speak of equivalents only with respect to a particular half-reaction. For example, in Reaction E-2, there are five equivalents per mole of $MnO_4^-$; but, in Reaction E-4, there are only three equivalents per mole of $MnO_4^-$. The mass of substance containing one equivalent is called the *equivalent mass*. The formula mass of $KMnO_4$ is 158.033 9. The equivalent mass of $KMnO_4$ for Reaction E-2 is 158.033 9/5 = 31.606 8 g/equiv. The equivalent mass of $KMnO_4$ for Reaction E-4 is 158.033 9/3 = 52.678 0 g/equiv.

## Example   Finding Normality

Find the normality of a solution containing 6.34 g of ascorbic acid in 250.0 mL if the relevant half-reaction is

Ascorbic acid          Dehydroascorbic acid
(vitamin C)

**Solution**   The formula mass of ascorbic acid ($C_6H_8O_6$) is 176.124. In 6.34 g, there are $(6.34 \text{ g})/(176.124 \text{ g/mol}) = 3.60 \times 10^{-2}$ mol. Because each mole contains 2 equivalents in this example, 6.34 g = (2 equiv/mol) $(3.60 \times 10^{-2} \text{ mol}) = 7.20 \times 10^{-2}$ equivalent. The normality is $(7.20 \times 10^{-2} \text{ equiv})/(0.250 0 \text{ L}) = 0.288$ N.

## Example   Using Normality

How many grams of potassium oxalate should be dissolved in 500.0 mL to make a 0.100 N solution for titration of $MnO_4^-$?

$$5H_2C_2O_4 + 2MnO_4^- + 6H^+ \rightleftharpoons 2Mn^{2+} + 10CO_2 + 8H_2O \qquad \text{(E-5)}$$

**Solution**   It is first necessary to write the oxalic acid half-reaction:

$$H_2C_2O_4 \rightleftharpoons 2CO_2 + 2H^+ + 2e^-$$

It is apparent that there are two equivalents per mole of oxalic acid. Hence, a 0.100 N solution will be 0.050 0 M:

$$\frac{0.100 \text{ equiv/L}}{2 \text{ equiv/mol}} = 0.050 0 \text{ mol/L} = 0.050 0 \text{ M}$$

Therefore, we must dissolve $(0.050 0 \text{ mol/L})(0.500 0 \text{ L}) = 0.025 0$ mol in 500.0 mL. Because the formula mass of $K_2C_2O_4$ is 166.216, we should use $(0.025 0 \text{ mol}) \times (166.216 \text{ g/mol}) = 4.15$ g of potassium oxalate.

The utility of normality in volumetric analysis lies in the equation

$$N_1V_1 = N_2V_2 \qquad \text{(E-6)}$$

where $N_1$ is the normality of reagent 1, $V_1$ is the volume of reagent 1, $N_2$ is the normality of reagent 2, and $V_2$ is the volume of reagent 2. $V_1$ and $V_2$ may be expressed in any units, as long as the same units are used for both.

## Example   Finding Normality

A solution containing 25.0 mL of oxalic acid required 13.78 mL of 0.041 62 N $KMnO_4$ for titration, according to Reaction E-5. Find the normality and molarity of the oxalic acid.

**Solution**   Setting up Equation E-6, we write

$$N_1(25.0 \text{ mL}) = (0.041 62 \text{ N})(13.78 \text{ mL})$$

$$N_1 = 0.022 94 \text{ equiv/L}$$

Because there are two equivalents per mole of oxalic acid in Reaction E-5,

$$M = \frac{N}{n} = \frac{0.022 94}{2} = 0.011 47 \text{ M}$$

Normality is sometimes used in acid-base or ion-exchange chemistry. With respect to acids and bases, the equivalent mass of a reagent is the amount that can donate or accept 1 mole of $H^+$. With respect to ion exchange, the equivalent mass is the mass of reagent containing 1 mole of charge.

# APPENDIX F

# Solubility Products*

| Formula | $pK_{sp}$ | $K_{sp}$ | Formula | $pK_{sp}$ | $K_{sp}$ |
|---|---|---|---|---|---|
| Azides: L = $N_3^-$ | | | Chromates: L = $CrO_4^{2-}$ | | |
| CuL | 8.31 | $4.9 \times 10^{-9}$ | BaL | 9.67 | $2.1 \times 10^{-10}$ |
| AgL | 8.56 | $2.8 \times 10^{-9}$ | CuL | 5.44 | $3.6 \times 10^{-6}$ |
| $Hg_2L_2$ | 9.15 | $7.1 \times 10^{-10}$ | $Ag_2L$ | 11.92 | $1.2 \times 10^{-12}$ |
| TlL | 3.66 | $2.2 \times 10^{-4}$ | $Hg_2L$ | 8.70 | $2.0 \times 10^{-9}$ |
| $PdL_2(\alpha)$ | 8.57 | $2.7 \times 10^{-9}$ | $Tl_2L$ | 12.01 | $9.8 \times 10^{-13}$ |
| | | | | | |
| Bromates: L = $BrO_3^-$ | | | Cobalticyanides: L = $Co(CN)_6^{3-}$ | | |
| $BaL \cdot H_2O$ (f) | 5.11 | $7.8 \times 10^{-6}$ | $Ag_3L$ | 25.41 | $3.9 \times 10^{-26}$ |
| AgL | 4.26 | $5.5 \times 10^{-5}$ | $(Hg_2)_3L_2$ | 36.72 | $1.9 \times 10^{-37}$ |
| TlL | 3.78 | $1.7 \times 10^{-4}$ | | | |
| $PbL_2$ | 5.10 | $7.9 \times 10^{-6}$ | Cyanides: L = $CN^-$ | | |
| | | | AgL | 15.66 | $2.2 \times 10^{-16}$ |
| Bromides: L = $Br^-$ | | | $Hg_2L_2$ | 39.3 | $5 \times 10^{-40}$ |
| CuL | 8.3 | $5 \times 10^{-9}$ | $ZnL_2$ (h) | 15.5 | $3 \times 10^{-16}$ |
| AgL | 12.30 | $5.0 \times 10^{-13}$ | | | |
| $Hg_2L_2$ | 22.25 | $5.6 \times 10^{-23}$ | Ferrocyanides: L = $Fe(CN)_6^{4-}$ | | |
| TlL | 5.44 | $3.6 \times 10^{-6}$ | $Ag_4L$ | 44.07 | $8.5 \times 10^{-45}$ |
| $HgL_2$ (f) | 18.9 | $1.3 \times 10^{-19}$ | $Zn_2L$ | 15.68 | $2.1 \times 10^{-16}$ |
| $PbL_2$ | 5.68 | $2.1 \times 10^{-6}$ | $Cd_2L$ | 17.38 | $4.2 \times 10^{-18}$ |
| | | | $Pb_2L$ | 18.02 | $9.5 \times 10^{-19}$ |
| Carbonates: L = $CO_3^{2-}$ | | | | | |
| MgL | 7.46 | $3.5 \times 10^{-8}$ | Fluorides: L = $F^-$ | | |
| CaL (calcite) | 8.35 | $4.5 \times 10^{-9}$ | LiL | 2.77 | $1.7 \times 10^{-3}$ |
| CaL (aragonite) | 8.22 | $6.0 \times 10^{-9}$ | $MgL_2$ | 8.13 | $7.4 \times 10^{-9}$ |
| SrL | 9.03 | $9.3 \times 10^{-10}$ | $CaL_2$ | 10.50 | $3.2 \times 10^{-11}$ |
| BaL | 8.30 | $5.0 \times 10^{-9}$ | $SrL_2$ | 8.58 | $2.6 \times 10^{-9}$ |
| $Y_2L_3$ | 30.6 | $2.5 \times 10^{-31}$ | $BaL_2$ | 5.82 | $1.5 \times 10^{-6}$ |
| $La_2L_3$ | 33.4 | $4.0 \times 10^{-34}$ | $LaL_3$ | 18.7 | $2 \times 10^{-19}$ |
| MnL | 9.30 | $5.0 \times 10^{-10}$ | $ThL_4$ | 28.3 | $5 \times 10^{-29}$ |
| FeL | 10.68 | $2.1 \times 10^{-11}$ | $PbL_2$ | 7.44 | $3.6 \times 10^{-8}$ |
| CoL | 9.98 | $1.0 \times 10^{-10}$ | | | |
| NiL | 6.87 | $1.3 \times 10^{-7}$ | Hydroxides: L = $OH^-$ | | |
| CuL | 9.63 | $2.3 \times 10^{-10}$ | $MgL_2$ (amorphous) | 9.2 | $6 \times 10^{-10}$ |
| $Ag_2L$ | 11.09 | $8.1 \times 10^{-12}$ | $MgL_2$ (brucite crystal) | 11.15 | $7.1 \times 10^{-12}$ |
| $Hg_2L$ | 16.05 | $8.9 \times 10^{-17}$ | $CaL_2$ | 5.19 | $6.5 \times 10^{-6}$ |
| ZnL | 10.00 | $1.0 \times 10^{-10}$ | $BaL_2 \cdot 8H_2O$ | 3.6 | $3 \times 10^{-4}$ |
| CdL | 13.74 | $1.8 \times 10^{-14}$ | $YL_3$ | 23.2 | $6 \times 10^{-24}$ |
| PbL | 13.13 | $7.4 \times 10^{-14}$ | $LaL_3$ | 20.7 | $2 \times 10^{-21}$ |
| | | | $CeL_3$ | 21.2 | $6 \times 10^{-22}$ |
| Chlorides: L = $Cl^-$ | | | $UO_2 (\rightleftharpoons U^{4+} + 4OH^-)$ | 56.2 | $6 \times 10^{-57}$ |
| CuL | 6.73 | $1.9 \times 10^{-7}$ | $UO_2L_2 (\rightleftharpoons UO_2^{2+} + 2OH^-)$ | 22.4 | $4 \times 10^{-23}$ |
| AgL | 9.74 | $1.8 \times 10^{-10}$ | $MnL_2$ | 12.8 | $1.6 \times 10^{-13}$ |
| $Hg_2L_2$ | 17.91 | $1.2 \times 10^{-18}$ | $FeL_2$ | 15.1 | $7.9 \times 10^{-16}$ |
| TlL | 3.74 | $1.8 \times 10^{-4}$ | $CoL_2$ | 14.9 | $1.3 \times 10^{-15}$ |
| $PbL_2$ | 4.78 | $1.7 \times 10^{-5}$ | $NiL_2$ | 15.2 | $6 \times 10^{-16}$ |

*The designations $\alpha$, $\beta$, or $\gamma$ after some formulas refer to particular crystalline forms (which are customarily identified by Greek letters). Data for salts except oxalates are taken mainly from A. E. Martell and R. M. Smith, Critical Stability Constants, Vol. 4 (New York: Plenum Press, 1976). Data for oxalates are from L. G. Sillén and A. E. Martell, Stability Constants of Metal-Ion Complexes, Supplement No. 1 (London: The Chemical Society, Special Publication No. 25, 1971). Another source: R. M. H. Verbeeck et al., Inorg. Chem. 1984, 23, 1922.

Conditions are 25°C and zero ionic strength unless otherwise indicated: (a) 19°C; (b) 20°C; (c) 38°C; (d) 0.1 M; (e) 0.2 M; (f) 0.5 M; (g) 1 M; (h) 3 M; (i) 4 M; (j) 5 M.

*(Continued)*

| Formula | $pK_{sp}$ | $K_{sp}$ | Formula | $pK_{sp}$ | $K_{sp}$ |
|---|---|---|---|---|---|
| $CuL_2$ | 19.32 | $4.8 \times 10^{-20}$ | **Phosphates: L = $PO_4^{3-}$** | | |
| $VL_3$ | 34.4 | $4.0 \times 10^{-35}$ | $MgHL \cdot 3H_2O$ ($\rightleftharpoons Mg^{2+} + HL^{2-}$) | 5.78 | $1.7 \times 10^{-6}$ |
| $CrL_3$ (d) | 29.8 | $1.6 \times 10^{-30}$ | $CaHL \cdot 2H_2O$ ($\rightleftharpoons Ca^{2+} + HL^{2-}$) | 6.58 | $2.6 \times 10^{-7}$ |
| $FeL_3$ | 38.8 | $1.6 \times 10^{-39}$ | $SrHL$ ($\rightleftharpoons Sr^{2+} + HL^{2-}$) (b) | 6.92 | $1.2 \times 10^{-7}$ |
| $CoL_3$ (a) | 44.5 | $3 \times 10^{-45}$ | $BaHL$ ($\rightleftharpoons Ba^{2+} + HL^{2-}$) (b) | 7.40 | $4.0 \times 10^{-8}$ |
| $VOL_2$ ($\rightleftharpoons VO^{2+} + 2OH^-$) | 23.5 | $3 \times 10^{-24}$ | $LaL$ (f) | 22.43 | $3.7 \times 10^{-23}$ |
| $PdL_2$ | 28.5 | $3 \times 10^{-29}$ | $Fe_3L_2 \cdot 8H_2O$ | 36.0 | $1 \times 10^{-36}$ |
| $ZnL_2$ (amorphous) | 15.52 | $3.0 \times 10^{-16}$ | $FeL \cdot 2H_2O$ | 26.4 | $4 \times 10^{-27}$ |
| $CdL_2$ ($\beta$) | 14.35 | $4.5 \times 10^{-15}$ | $(VO)_3L_2$ ($\rightleftharpoons 3VO^{2+} + 2L^{3-}$) | 25.1 | $8 \times 10^{-26}$ |
| $HgO$ (red) ($\rightleftharpoons Hg^{2+} + 2OH^-$) | 25.44 | $3.6 \times 10^{-26}$ | $Ag_3L$ | 17.55 | $2.8 \times 10^{-18}$ |
| $Cu_2O$ ($\rightleftharpoons 2Cu^+ + 2OH^-$) | 29.4 | $4 \times 10^{-30}$ | $Hg_2HL$ ($\rightleftharpoons Hg_2^{2+} + HL^{2-}$) | 12.40 | $4.0 \times 10^{-13}$ |
| $Ag_2O$ ($\rightleftharpoons 2Ag^+ + 2OH^-$) | 15.42 | $3.8 \times 10^{-16}$ | $Zn_3L_2 \cdot 4H_2O$ | 35.3 | $5 \times 10^{-36}$ |
| $AuL_3$ | 5.5 | $3 \times 10^{-6}$ | $Pb_3L_2$ (c) | 43.53 | $3.0 \times 10^{-44}$ |
| $AlL_3$ ($\alpha$) | 33.5 | $3 \times 10^{-34}$ | $GaL$ (g) | 21.0 | $1 \times 10^{-21}$ |
| $GaL_3$ (amorphous) | 37 | $10^{-37}$ | $InL$ (g) | 21.63 | $2.3 \times 10^{-22}$ |
| $InL_3$ | 36.9 | $1.3 \times 10^{-37}$ | | | |
| $SnO$ ($\rightleftharpoons Sn^{2+} + 2OH^-$) | 26.2 | $6 \times 10^{-27}$ | **Sulfates: L = $SO_4^{2-}$** | | |
| $PbO$ (yellow) ($\rightleftharpoons Pb^{2+} + 2OH^-$) | 15.1 | $8 \times 10^{-16}$ | $CaL$ | 4.62 | $2.4 \times 10^{-5}$ |
| $PbO$ (red) ($\rightleftharpoons Pb^{2+} + 2OH^-$) | 15.3 | $5 \times 10^{-16}$ | $SrL$ | 6.50 | $3.2 \times 10^{-7}$ |
| | | | $BaL$ | 9.96 | $1.1 \times 10^{-10}$ |
| **Iodates: L = $IO_3^-$** | | | $RaL$ (b) | 10.37 | $4.3 \times 10^{-11}$ |
| $CaL_2$ | 6.15 | $7.1 \times 10^{-7}$ | $Ag_2L$ | 4.83 | $1.5 \times 10^{-5}$ |
| $SrL_2$ | 6.48 | $3.3 \times 10^{-7}$ | $Hg_2L$ | 6.13 | $7.4 \times 10^{-7}$ |
| $BaL_2$ | 8.81 | $1.5 \times 10^{-9}$ | $PbL$ | 6.20 | $6.3 \times 10^{-7}$ |
| $YL_3$ | 10.15 | $7.1 \times 10^{-11}$ | | | |
| $LaL_3$ | 10.99 | $1.0 \times 10^{-11}$ | **Sulfides: L = $S^{2-}$** | | |
| $CeL_3$ | 10.86 | $1.4 \times 10^{-11}$ | $MnL$ (pink) | 10.5 | $3 \times 10^{-11}$ |
| $ThL_4$ (f) | 14.62 | $2.4 \times 10^{-15}$ | $MnL$ (green) | 13.5 | $3 \times 10^{-14}$ |
| $UO_2L_2$ ($\rightleftharpoons UO_2^{2+} + 2IO_3^-$)(e) | 7.01 | $9.8 \times 10^{-8}$ | $FeL$ | 18.1 | $8 \times 10^{-19}$ |
| $CrL_3$ (f) | 5.3 | $5 \times 10^{-6}$ | $CoL$ ($\alpha$) | 21.3 | $5 \times 10^{-22}$ |
| $AgL$ | 7.51 | $3.1 \times 10^{-8}$ | $CoL$ ($\beta$) | 25.6 | $3 \times 10^{-26}$ |
| $Hg_2L_2$ | 17.89 | $1.3 \times 10^{-18}$ | $NiL$ ($\alpha$) | 19.4 | $4 \times 10^{-20}$ |
| $TlL$ | 5.51 | $3.1 \times 10^{-6}$ | $NiL$ ($\beta$) | 24.9 | $1.3 \times 10^{-25}$ |
| $ZnL_2$ | 5.41 | $3.9 \times 10^{-6}$ | $NiL$ ($\gamma$) | 26.6 | $3 \times 10^{-27}$ |
| $CdL_2$ | 7.64 | $2.3 \times 10^{-8}$ | $CuL$ | 36.1 | $8 \times 10^{-37}$ |
| $PbL_2$ | 12.61 | $2.5 \times 10^{-13}$ | $Cu_2L$ | 48.5 | $3 \times 10^{-49}$ |
| | | | $Ag_2L$ | 50.1 | $8 \times 10^{-51}$ |
| **Iodides: L = $I^-$** | | | $Tl_2L$ | 21.2 | $6 \times 10^{-22}$ |
| $CuL$ | 12.0 | $1 \times 10^{-12}$ | $ZnL$ ($\alpha$) | 24.7 | $2 \times 10^{-25}$ |
| $AgL$ | 16.08 | $8.3 \times 10^{-17}$ | $ZnL$ ($\beta$) | 22.5 | $3 \times 10^{-23}$ |
| $CH_3HgL$ ($\rightleftharpoons CH_3Hg^+ + I^-$) (b, g) | 11.46 | $3.5 \times 10^{-12}$ | $CdL$ | 27.0 | $1 \times 10^{-27}$ |
| $CH_3CH_2HgL$ ($\rightleftharpoons CH_3CH_2Hg^+ + I^-$) | 4.11 | $7.8 \times 10^{-5}$ | $HgL$ (black) | 52.7 | $2 \times 10^{-53}$ |
| $TlL$ | 7.23 | $5.9 \times 10^{-8}$ | $HgL$ (red) | 53.3 | $5 \times 10^{-54}$ |
| $Hg_2L_2$ | 28.34 | $4.6 \times 10^{-29}$ | $SnL$ | 25.9 | $1.3 \times 10^{-26}$ |
| $SnL_2$ (i) | 5.08 | $8.3 \times 10^{-6}$ | $PbL$ | 27.5 | $3 \times 10^{-28}$ |
| $PbL_2$ | 8.10 | $7.9 \times 10^{-9}$ | $In_2L_3$ | 69.4 | $4 \times 10^{-70}$ |
| | | | | | |
| **Oxalates: L = $C_2O_4^{2-}$** | | | **Thiocyanates: L = $SCN^-$** | | |
| $CaL$ (b, d) | 7.9 | $1.3 \times 10^{-8}$ | $CuL$ (j) | 13.40 | $4.0 \times 10^{-14}$ |
| $SrL$ (b, d) | 6.4 | $4 \times 10^{-7}$ | $AgL$ | 11.97 | $1.1 \times 10^{-12}$ |
| $BaL$ (b, d) | 6.0 | $1 \times 10^{-6}$ | $Hg_2L_2$ | 19.52 | $3.0 \times 10^{-20}$ |
| $La_2L_3$ (b, d) | 25.0 | $1 \times 10^{-25}$ | $TlL$ | 3.79 | $1.6 \times 10^{-4}$ |
| $ThL_2$ (g) | 21.38 | $4.2 \times 10^{-22}$ | $HgL_2$ | 19.56 | $2.8 \times 10^{-20}$ |
| $UO_2L$ ($\rightleftharpoons UO_2^{2+} + C_2O_4^{2-}$) (b, d) | 8.66 | $2.2 \times 10^{-9}$ | | | |

# APPENDIX G

## Acid Dissociation Constants

| Name | Structure* | $pK_a$† | $K_a$‡ |
|---|---|---|---|
| Acetic acid (ethanoic acid) | $CH_3CO_2H$ | 4.756 | $1.75 \times 10^{-5}$ |
| Alanine | $\begin{array}{c} NH_3^+ \\ | \\ CHCH_3 \\ | \\ CO_2H \end{array}$ | 2.344 ($CO_2H$) <br> 9.868 ($NH_3$) | $4.53 \times 10^{-3}$ <br> $1.36 \times 10^{-10}$ |
| Aminobenzene (aniline) | $C_6H_5{-}NH_3^+$ | 4.601 | $2.51 \times 10^{-5}$ |
| 4-Aminobenzenesulfonic acid (sulfanilic acid) | $^-O_3S{-}C_6H_4{-}NH_3^+$ | 3.232 | $5.86 \times 10^{-4}$ |
| 2-Aminobenzoic acid (anthranilic acid) | (benzene ring with $NH_3^+$ and $CO_2H$) | 2.08 ($CO_2H$) <br> 4.96 ($NH_3$) | $8.3 \times 10^{-3}$ <br> $1.10 \times 10^{-5}$ |
| 2-Aminoethanethiol (2-mercaptoethylamine) | $HSCH_2CH_2NH_3^+$ | 8.21 (SH) ($\mu = 0.1$) <br> 10.73 ($NH_3$) ($\mu = 0.1$) | $6.2 \times 10^{-9}$ <br> $1.86 \times 10^{-11}$ |
| 2-Aminoethanol (ethanolamine) | $HOCH_2CH_2NH_3^+$ | 9.498 | $3.18 \times 10^{-10}$ |
| 2-Aminophenol | (benzene ring with OH and $NH_3^+$) | 4.70 ($NH_3$) (20°) <br> 9.97 (OH) (20°) | $2.0 \times 10^{-5}$ <br> $1.05 \times 10^{-10}$ |
| Ammonia | $NH_4^+$ | 9.245 | $5.69 \times 10^{-10}$ |
| Arginine | $\begin{array}{c} NH_3^+ \\ | \\ CHCH_2CH_2CH_2NHC \\ | \\ CO_2H \end{array} {=}NH_2^+, NH_2$ | 1.823 ($CO_2H$) <br> 8.991 ($NH_3$) <br> (12.1) ($NH_2$) ($\mu = 0.1$) | $1.50 \times 10^{-2}$ <br> $1.02 \times 10^{-9}$ <br> $8 \times 10^{-13}$ |
| Arsenic acid (hydrogen arsenate) | $\begin{array}{c} O \\ \| \\ HO{-}As{-}OH \\ | \\ OH \end{array}$ | 2.31 <br> 7.05 <br> 11.9 | $4.9 \times 10^{-3}$ <br> $8.9 \times 10^{-8}$ <br> $1.3 \times 10^{-12}$ |
| Arsenious acid (hydrogen arsenite) | $As(OH)_3$ | 9.29 | $5.1 \times 10^{-10}$ |
| Asparagine | $\begin{array}{c} NH_3^+ \quad O \\ | \qquad \| \\ CHCH_2CNH_2 \\ | \\ CO_2H \end{array}$ | 2.16 ($CO_2H$) ($\mu = 0.1$) <br> 8.73 ($NH_3$) ($\mu = 0.1$) | $6.9 \times 10^{-3}$ <br> $1.86 \times 10^{-9}$ |

*Each acid is written in its protonated form. The acidic protons are indicated in bold type.

†$pK_a$ values refer to 25°C and zero ionic strength unless otherwise indicated. Values in parentheses are considered to be less reliable. Data are from A. E. Martell, R. M. Smith, and R. J. Motekaitis, NIST Database 46 (Guithersburg, MD: National Institute of Standards and Technology, 2001).

‡The accurate way to calculate $K_b$ for the conjugate base is $pK_b = 13.995 - pK_a$ and $K_b = 10^{-pK_b}$.

(Continued)

| Name | Structure | $pK_a$ | $K_a$ |
|---|---|---|---|
| Aspartic acid | $\begin{array}{c}NH_3^+\\ |\quad\beta\\ CHCH_2CO_2H\\ |\\ \alpha\,CO_2H\end{array}$ | 1.990 ($\alpha$-CO$_2$H)<br>3.900 ($\beta$-CO$_2$H)<br>10.002 (NH$_3$) | $1.02 \times 10^{-2}$<br>$1.26 \times 10^{-4}$<br>$9.95 \times 10^{-11}$ |
| Aziridine<br>(dimethyleneimine) | $\triangleright NH_2^+$ | 8.04 | $9.1 \times 10^{-9}$ |
| Benzene-1,2,3-tricarboxylic acid<br>(hemimellitic acid) | [structure: benzene ring with three $CO_2H$ groups] | 2.86<br>4.30<br>6.28 | $1.38 \times 10^{-3}$<br>$5.0 \times 10^{-5}$<br>$5.2 \times 10^{-7}$ |
| Benzoic acid | [benzene ring]—$CO_2H$ | 4.202 | $6.28 \times 10^{-5}$ |
| Benzylamine | [benzene ring]—$CH_2NH_3^+$ | 9.35 | $4.5 \times 10^{-10}$ |
| 2,2′-Bipyridine | [two pyridine rings, $\overset{+}{N}H$ and N] | 4.34 | $4.6 \times 10^{-5}$ |
| Boric acid<br>(hydrogen borate) | $B(OH)_3$ | 9.237<br>(12.74) (20°)<br>(13.80) (20°) | $5.79 \times 10^{-10}$<br>$1.82 \times 10^{-13}$<br>$1.58 \times 10^{-14}$ |
| Bromoacetic acid | $BrCH_2CO_2H$ | 2.902 | $1.25 \times 10^{-3}$ |
| Butane-2,3-dione dioxime<br>(dimethylglyoxime) | $\begin{array}{c}HON\quad\ NOH\\ \diagdown\quad\diagup\\ CH_3\quad CH_3\end{array}$ | 10.66<br>12.0 | $2.2 \times 10^{-11}$<br>$1 \times 10^{-12}$ |
| Butanoic acid | $CH_3CH_2CH_2CO_2H$ | 4.818 | $1.52 \times 10^{-5}$ |
| cis-Butenedioic acid<br>(maleic acid) | $\begin{array}{c}\diagup CO_2H\\ \diagdown CO_2H\end{array}$ | 1.92<br>6.27 | $1.20 \times 10^{-2}$<br>$5.37 \times 10^{-7}$ |
| trans-Butenedioic acid<br>(fumaric acid) | $\begin{array}{c}\quad\quad CO_2H\\ \diagup\\ HO_2C\end{array}$ | 3.02<br>4.48 | $9.5 \times 10^{-4}$<br>$3.3 \times 10^{-5}$ |
| Butylamine | $CH_3CH_2CH_2CH_2NH_3^+$ | 10.640 | $2.29 \times 10^{-11}$ |
| Carbonic acid*<br>(hydrogen carbonate) | $\begin{array}{c}O\\ \|\|\\ HO-C-OH\end{array}$ | 6.351<br>10.329 | $4.46 \times 10^{-7}$<br>$4.69 \times 10^{-11}$ |
| Chloroacetic acid | $ClCH_2CO_2H$ | 2.865 | $1.36 \times 10^{-3}$ |
| 3-Chloropropanoic acid | $ClCH_2CH_2CO_2H$ | 4.11 | $7.8 \times 10^{-5}$ |
| Chlorous acid<br>(hydrogen chlorite) | $HOCl{=}O$ | 1.96 | $1.10 \times 10^{-2}$ |

*The concentration of "carbonic acid" is considered to be the sum $[H_2CO_3] + [CO_2(aq)]$. See Box 6-4.

**APPENDIX G  Acid Dissociation Constants**

| Name | Structure | $pK_a$ | $K_a$ |
|------|-----------|--------|-------|
| Chromic acid (hydrogen chromate) | $$\begin{array}{c} O \\ \parallel \\ HO-Cr-OH \\ \parallel \\ O \end{array}$$ | $-0.2\ (20°)$<br>$6.51$ | $1.6$<br>$3.1 \times 10^{-7}$ |
| Citric acid (2-hydroxypropane-1,2,3-tricarboxylic acid) | $$\begin{array}{c} CO_2H \\ \mid \\ HO_2CCH_2CCH_2CO_2H \\ \mid \\ OH \end{array}$$ | $3.128$<br>$4.761$<br>$6.396$ | $7.44 \times 10^{-4}$<br>$1.73 \times 10^{-5}$<br>$4.02 \times 10^{-7}$ |
| Cyanoacetic acid | $NCCH_2CO_2H$ | $2.472$ | $3.37 \times 10^{-3}$ |
| Cyclohexylamine | cyclohexyl$-NH_3^+$ | $10.567$ | $2.71 \times 10^{-11}$ |
| Cysteine | $$\begin{array}{c} NH_3^+ \\ \mid \\ CHCH_2SH \\ \mid \\ CO_2H \end{array}$$ | $(1.7)\ (CO_2H)$<br>$8.36\ (SH)$<br>$10.74\ (NH_3)$ | $2 \times 10^{-2}$<br>$4.4 \times 10^{-9}$<br>$1.82 \times 10^{-11}$ |
| Dichloroacetic acid | $Cl_2CHCO_2H$ | $(1.1)$ | $8 \times 10^{-2}$ |
| Diethylamine | $(CH_3CH_2)_2NH_2^+$ | $11.00$ | $1.0 \times 10^{-11}$ |
| 1,2-Dihydroxybenzene (catechol) | benzene with two OH at 1,2 | $9.41$<br>$(13.3)\ (\mu = 0.1)$ | $3.9 \times 10^{-10}$<br>$5.0 \times 10^{-14}$ |
| 1,3-Dihydroxybenzene (resorcinol) | benzene with two OH at 1,3 | $9.30\ (\mu = 0.1)$<br>$11.06\ (\mu = 0.1)$ | $5.0 \times 10^{-10}$<br>$8.7 \times 10^{-12}$ |
| D-2,3-Dihydroxybutanedioic acid (D-tartaric acid) | $$\begin{array}{c} OH \\ \mid \\ HO_2CCHCHCO_2H \\ \mid \\ OH \end{array}$$ | $3.036$<br>$4.366$ | $9.20 \times 10^{-4}$<br>$4.31 \times 10^{-5}$ |
| 2,3-Dimercaptopropanol | $$\begin{array}{c} HOCH_2CHCH_2SH \\ \mid \\ SH \end{array}$$ | $8.63\ (\mu = 0.1)$<br>$10.65\ (\mu = 0.1)$ | $2.3 \times 10^{-9}$<br>$2.2 \times 10^{-11}$ |
| Dimethylamine | $(CH_3)_2NH_2^+$ | $10.774$ | $1.68 \times 10^{-11}$ |
| 2,4-Dinitrophenol | benzene with $NO_2$, $O_2N-$, $-OH$ | $4.114$ | $7.69 \times 10^{-5}$ |
| Ethane-1,2-dithiol | $HSCH_2CH_2SH$ | $8.85\ (30°, \mu = 0.1)$<br>$10.43\ (30°, \mu = 0.1)$ | $1.4 \times 10^{-9}$<br>$3.7 \times 10^{-11}$ |
| Ethylamine | $CH_3CH_2NH_3^+$ | $10.673$ | $2.12 \times 10^{-11}$ |
| Ethylenediamine (1,2-diaminoethane) | $H_3\overset{+}{N}CH_2CH_2\overset{+}{N}H_3$ | $6.848$<br>$9.928$ | $1.42 \times 10^{-7}$<br>$1.18 \times 10^{-10}$ |

(Continued)

| Name | Structure | $pK_a$ | $K_a$ |
|---|---|---|---|
| Ethylenedinitrilotetraacetic acid (EDTA) | $(HO_2CCH_2)_2\overset{+}{N}HCH_2CH_2\overset{+}{N}H(CH_2CO_2H)_2$ | (0.0) $(CO_2H)$ $(\mu = 1.0)$<br>(1.5) $(CO_2H)$ $(\mu = 0.1)$<br>2.00 $(CO_2H)$ $(\mu = 0.1)$<br>2.69 $(CO_2H)$ $(\mu = 0.1)$<br>6.13 (NH) $(\mu = 0.1)$<br>10.37 (NH) $(\mu = 0.1)$ | 1.0<br>0.032<br>0.010<br>0.002 0<br>$7.4 \times 10^{-7}$<br>$4.3 \times 10^{-11}$ |
| Formic acid (methanoic acid) | $HCO_2H$ | 3.744 | $1.80 \times 10^{-4}$ |
| Glutamic acid | $\overset{NH_3^+}{\underset{\alpha\,CO_2H}{\overset{|}{\underset{|}{CHCH_2CH_2CO_2H}}}}$ (γ) | 2.160 $(\alpha\text{-}CO_2H)$<br>4.30 $(\gamma\text{-}CO_2H)$<br>9.96 $(NH_3)$ | $6.92 \times 10^{-3}$<br>$5.0 \times 10^{-5}$<br>$1.10 \times 10^{-10}$ |
| Glutamine | $\overset{NH_3^+}{\underset{CO_2H}{\overset{|}{\underset{|}{CHCH_2CH_2\overset{O}{\overset{||}{C}}NH_2}}}}$ | 2.19 $(CO_2H)$ $(\mu = 0.1)$<br>9.00 $(NH_3)$ $(\mu = 0.1)$ | $6.5 \times 10^{-3}$<br>$1.00 \times 10^{-9}$ |
| Glycine (aminoacetic acid) | $\overset{NH_3^+}{\underset{CO_2H}{\overset{|}{\underset{|}{CH_2}}}}$ | 2.350 $(CO_2H)$<br>9.778 $(NH_3)$ | $4.47 \times 10^{-3}$<br>$1.67 \times 10^{-10}$ |
| Guanidine | $H_2N-\overset{\overset{+NH_2}{||}}{C}-NH_2$ | (13.5) $(\mu = 1.0)$ | $3 \times 10^{-14}$ |
| 1,6-Hexanedioic acid (adipic acid) | $HO_2CCH_2CH_2CH_2CH_2CO_2H$ | 4.424<br>5.420 | $3.77 \times 10^{-5}$<br>$3.80 \times 10^{-6}$ |
| Hexane-2,4-dione | $CH_3\overset{O}{\overset{||}{C}}CH_2\overset{O}{\overset{||}{C}}CH_2CH_3$ | 9.38 | $4.2 \times 10^{-10}$ |
| Histidine | $\overset{NH_3^+}{\underset{CO_2H}{\overset{|}{\underset{|}{CHCH_2}}}}$ (imidazole ring, NH) | 1.6 $(CO_2H)$<br>5.97 (NH)<br>9.28 $(NH_3)$ | $3 \times 10^{-2}$<br>$1.07 \times 10^{-6}$<br>$5.2 \times 10^{-10}$ |
| Hydrazine | $H_3\overset{+}{N}-\overset{+}{N}H_3$ | −0.99<br>8.02 | 9.8<br>$9.5 \times 10^{-9}$ |
| Hydrazoic acid (hydrogen azide) | $HN=\overset{+}{N}=\overset{-}{N}$ | 4.65 | $2.2 \times 10^{-5}$ |
| Hydrogen cyanate | $HOC\equiv N$ | 3.48 | $3.3 \times 10^{-4}$ |
| Hydrogen cyanide | $HC\equiv N$ | 9.21 | $6.2 \times 10^{-10}$ |
| Hydrogen fluoride | $HF$ | 3.17 | $6.8 \times 10^{-4}$ |
| Hydrogen peroxide | $HOOH$ | 11.65 | $2.2 \times 10^{-12}$ |
| Hydrogen sulfide | $H_2S$ | 7.02<br>14.0* | $9.5 \times 10^{-8}$<br>$1.0 \times 10^{-14}$* |
| Hydrogen thiocyanate | $HSC\equiv N$ | (1.1) (20°C) | 0.08 |
| Hydroxyacetic acid (glycolic acid) | $HOCH_2CO_2H$ | 3.832 | $1.48 \times 10^{-4}$ |

*D. J. Phillips and S. L. Phillips. "High Temperature Dissociation Constants of HS⁻ and the Standard Thermodynamic Values for S²⁻,"
J. Chem. Eng. Data **2000**, 45, 981.

| Name | Structure | $pK_a$ | $K_a$ |
|---|---|---|---|
| Hydroxybenzene (phenol) | C6H5—OH | 9.997 | $1.01 \times 10^{-10}$ |
| 2-Hydroxybenzoic acid (salicylic acid) | (ring with CO$_2$H and OH) | 2.972 ($CO_2H$)<br>(13.7) (OH) | $1.07 \times 10^{-3}$<br>$2 \times 10^{-14}$ |
| L-Hydroxybutanedioic acid (malic acid) | $HO_2CCH_2CHCO_2H$ (with OH) | 3.459<br>5.097 | $3.48 \times 10^{-4}$<br>$8.00 \times 10^{-6}$ |
| Hydroxylamine | $HON\overset{+}{H}_3$ | 5.96 (NH)<br>(13.74) (OH) | $1.10 \times 10^{-6}$<br>$1.8 \times 10^{-14}$ |
| 8-Hydroxyquinoline (oxine) | (quinoline structure, HO, $\overset{+}{N}H$) | 4.94 (NH)<br>9.82 (OH) | $1.15 \times 10^{-5}$<br>$1.51 \times 10^{-10}$ |
| Hypobromous acid (hydrogen hypobromite) | HOBr | 8.63 | $2.3 \times 10^{-9}$ |
| Hypochlorous acid (hydrogen hypochlorite) | HOCl | 7.53 | $3.0 \times 10^{-8}$ |
| Hypoiodous acid (hydrogen hypoiodite) | HOI | 10.64 | $2.3 \times 10^{-11}$ |
| Hypophosphorous acid (hydrogen hypophosphite) | $H_2POH$ (with =O) | (1.3) | $5 \times 10^{-2}$ |
| Imidazole (1,3-diazole) | (imidazole ring, $NH^+$, N–H) | 6.993<br>(14.5) | $1.02 \times 10^{-7}$<br>$3 \times 10^{-15}$ |
| Iminodiacetic acid | $H_2\overset{+}{N}(CH_2CO_2H)_2$ | 1.85 ($CO_2H$)<br>2.84 ($CO_2H$)<br>9.79 ($NH_2$) | $1.41 \times 10^{-2}$<br>$1.45 \times 10^{-3}$<br>$1.62 \times 10^{-10}$ |
| Iodic acid (hydrogen iodate) | $HOI=O$ (with =O) | 0.77 | 0.17 |
| Iodoacetic acid | $ICH_2CO_2H$ | 3.175 | $6.68 \times 10^{-4}$ |
| Isoleucine | $CHCH(CH_3)CH_2CH_3$ (with $NH_3^+$ and $CO_2H$) | 2.318 ($CO_2H$)<br>9.758 ($NH_3$) | $4.81 \times 10^{-3}$<br>$1.75 \times 10^{-10}$ |
| Leucine | $CHCH_2CH(CH_3)_2$ (with $NH_3^+$ and $CO_2H$) | 2.328 ($CO_2H$)<br>9.744 ($NH_3$) | $4.70 \times 10^{-3}$<br>$1.80 \times 10^{-10}$ |
| Lysine | $CHCH_2CH_2CH_2CH_2\overset{\epsilon}{N}H_3^+$ (with $\alpha\,NH_3^+$ and $CO_2H$) | (1.77) ($CO_2H$)<br>9.07 ($\alpha$-$NH_3$)<br>10.82 ($\epsilon$-$NH_3$) | $1.7 \times 10^{-2}$<br>$8.5 \times 10^{-10}$<br>$1.51 \times 10^{-11}$ |

*(Continued)*

| Name | Structure | $pK_a$ | $K_a$ |
|---|---|---|---|
| Malonic acid (propanedioic acid) | $HO_2CCH_2CO_2H$ | 2.847<br>5.696 | $1.42 \times 10^{-3}$<br>$2.01 \times 10^{-6}$ |
| Mercaptoacetic acid (thioglycolic acid) | $HSCH_2CO_2H$ | 3.64 ($CO_2H$)<br>10.61 (SH) | $2.3 \times 10^{-4}$<br>$2.5 \times 10^{-11}$ |
| 2-Mercaptoethanol | $HSCH_2CH_2OH$ | $9.7_5$ | $1.8 \times 10^{-10}$ |
| Methionine | $\overset{NH_3^+}{\underset{CO_2H}{\overset{\mid}{C}H}}CH_2CH_2SCH_3$ | 2.18 ($CO_2H$) ($\mu = 0.1$)<br>9.08 ($NH_3$) ($\mu = 0.1$) | $6.6 \times 10^{-3}$<br>$8.3 \times 10^{-10}$ |
| 2-Methoxyaniline (o-anisidine) | | 4.526 | $2.98 \times 10^{-5}$ |
| 4-Methoxyaniline (p-anisidine) | $CH_3O$—⟨⟩—$\overset{+}{N}H_3$ | 5.357 | $4.40 \times 10^{-6}$ |
| Methylamine | $CH_3\overset{+}{N}H_3$ | 10.645 | $2.26 \times 10^{-11}$ |
| 2-Methylaniline (o-toluidine) | | 4.447 | $3.57 \times 10^{-5}$ |
| 4-Methylaniline (p-toluidine) | $CH_3$—⟨⟩—$\overset{+}{N}H_3$ | 5.080 | $8.32 \times 10^{-6}$ |
| 2-Methylphenol (o-cresol) | | 10.31 | $4.9 \times 10^{-11}$ |
| 4-Methylphenol (p-cresol) | $CH_3$—⟨⟩—$OH$ | 10.269 | $5.5 \times 10^{-11}$ |
| Morpholine (perhydro-1,4-oxazine) | | 8.492 | $3.22 \times 10^{-9}$ |
| 1-Naphthoic acid | | 3.67 | $2.1 \times 10^{-4}$ |
| 2-Naphthoic acid | | 4.16 | $6.9 \times 10^{-5}$ |
| 1-Naphthol | | 9.416 | $3.84 \times 10^{-10}$ |
| 2-Naphthol | | 9.573 | $2.67 \times 10^{-10}$ |
| Nitrilotriacetic acid | $H\overset{+}{N}(CH_2CO_2H)_3$ | (1.0) ($CO_2H$) (25°, $\mu = 0.1$)<br>2.0 ($CO_2H$) (25°)<br>2.940 ($CO_2H$) (20°)<br>10.334 (NH) (20°) | 0.10<br>0.010<br>$1.15 \times 10^{-3}$<br>$4.63 \times 10^{-11}$ |

| Name | Structure | $pK_a$ | $K_a$ |
|---|---|---|---|
| 2-Nitrobenzoic acid | NO$_2$ / CO$_2$H (benzene ring) | 2.185 | $6.53 \times 10^{-3}$ |
| 3-Nitrobenzoic acid | NO$_2$ / CO$_2$H (benzene ring) | 3.449 | $3.56 \times 10^{-4}$ |
| 4-Nitrobenzoic acid | $O_2N$—⟨⟩—$CO_2H$ | 3.442 | $3.61 \times 10^{-4}$ |
| Nitroethane | $CH_3CH_2NO_2$ | 8.57 | $2.7 \times 10^{-9}$ |
| 2-Nitrophenol | NO$_2$ / OH (benzene ring) | 7.230 | $5.89 \times 10^{-8}$ |
| 3-Nitrophenol | NO$_2$ / OH (benzene ring) | 8.37 | $4.3 \times 10^{-9}$ |
| 4-Nitrophenol | $O_2N$—⟨⟩—$OH$ | 7.149 | $7.10 \times 10^{-8}$ |
| N-Nitrosophenylhydroxylamine (cupferron) | ⟨⟩—N(NO)(OH) | 4.16 ($\mu = 0.1$) | $6.9 \times 10^{-5}$ |
| Nitrous acid | $HON{=}O$ | 3.15 | $7.1 \times 10^{-4}$ |
| Oxalic acid (ethanedioic acid) | $HO_2CCO_2H$ | 1.27 <br> 4.266 | $5.4 \times 10^{-2}$ <br> $5.42 \times 10^{-5}$ |
| Oxoacetic acid (glyoxylic acid) | $HCCO_2H$ (with C=O) | 3.46 | $3.5 \times 10^{-4}$ |
| Oxobutanedioic acid (oxaloacetic acid) | $HO_2CCH_2CCO_2H$ (with C=O) | 2.56 <br> 4.37 | $2.8 \times 10^{-3}$ <br> $4.3 \times 10^{-5}$ |
| 2-Oxopentanedioic (α-ketoglutaric acid) | $HO_2CCH_2CH_2CCO_2H$ (with C=O) | (1.90) ($\mu = 0.5$) <br> 4.44 ($\mu = 0.5$) | $1.26 \times 10^{-2}$ <br> $3.6 \times 10^{-5}$ |
| 2-Oxopropanoic acid (pyruvic acid) | $CH_3CCO_2H$ (with C=O) | 2.48 | $3.3 \times 10^{-3}$ |
| 1,5-Pentanedioic acid (glutaric acid) | $HO_2CCH_2CH_2CH_2CO_2H$ | 4.345 <br> 5.422 | $4.52 \times 10^{-5}$ <br> $3.78 \times 10^{-6}$ |
| Pentanoic acid (valeric acid) | $CH_3CH_2CH_2CH_2CO_2H$ | 4.843 | $1.44 \times 10^{-5}$ |
| 1,10-Phenanthroline | (phenanthroline, NH⁺ HN⁺) | (1.8) ($\mu = 0.1$) <br> 4.91 | $1.6 \times 10^{-2}$ <br> $1.23 \times 10^{-5}$ |
| Phenylacetic acid | ⟨⟩—$CH_2CO_2H$ | 4.310 | $4.90 \times 10^{-5}$ |

*(Continued)*

| Name | Structure | $pK_a$ | $K_a$ |
|---|---|---|---|
| Phenylalanine | $NH_3^+$ \| $CHCH_2$—(phenyl) \| $CO_2H$ | 2.20 ($CO_2H$) <br> 9.31 ($NH_3$) | $6.3 \times 10^{-3}$ <br> $4.9 \times 10^{-10}$ |
| Phosphoric acid* <br> (hydrogen phosphate) | O ‖ HO—P—OH \| OH | 2.148 <br> 7.198 <br> 12.375 | $7.11 \times 10^{-3}$ <br> $6.34 \times 10^{-8}$ <br> $4.22 \times 10^{-13}$ |
| Phosphorous acid <br> (hydrogen phosphite) | O ‖ HP—OH \| OH | (1.5) <br> 6.78 | $3 \times 10^{-2}$ <br> $1.66 \times 10^{-7}$ |
| Phthalic acid <br> (benzene-1,2-dicarboxylic acid) | (benzene ring with $CO_2H$ and $CO_2H$) | 2.950 <br> 5.408 | $1.12 \times 10^{-3}$ <br> $3.90 \times 10^{-6}$ |
| Piperazine <br> (perhydro-1,4-diazine) | $\overset{+}{H_2}N$—(ring)—$N\overset{+}{H_2}$ | 5.333 <br> 9.731 | $4.65 \times 10^{-6}$ <br> $1.86 \times 10^{-10}$ |
| Piperidine | (ring)$N\overset{+}{H_2}$ | 11.125 | $7.50 \times 10^{-12}$ |
| Proline | (ring with $CO_2H$, $N$, $\overset{+}{H_2}$) | 1.952 ($CO_2H$) <br> 10.640 ($NH_2$) | $1.12 \times 10^{-2}$ <br> $2.29 \times 10^{-11}$ |
| Propanoic acid | $CH_3CH_2CO_2H$ | 4.874 | $1.34 \times 10^{-5}$ |
| Propenoic acid <br> (acrylic acid) | $H_2C{=}CHCO_2H$ | 4.258 | $5.52 \times 10^{-5}$ |
| Propylamine | $CH_3CH_2CH_2NH_3^+$ | 10.566 | $2.72 \times 10^{-11}$ |
| Pyridine <br> (azine) | (ring)$NH^+$ | 5.20 | $6.3 \times 10^{-6}$ |
| Pyridine-2-carboxylic acid <br> (picolinic acid) | (ring)$NH^+$ with $CO_2H$ | (1.01) ($CO_2H$) <br> 5.39 (NH) | $9.8 \times 10^{-2}$ <br> $4.1 \times 10^{-6}$ |
| Pyridine-3-carboxylic acid <br> (nicotinic acid) | $HO_2C$ (ring)$NH^+$ | 2.03 ($CO_2H$) <br> 4.82 (NH) | $9.3 \times 10^{-3}$ <br> $1.51 \times 10^{-5}$ |
| Pyridoxal-5-phosphate | O ‖ HO—POCH$_2$ \| HO (ring with O=CH, OH, $N^+H$, $CH_3$) | 1.4 (POH) ($\mu = 0.1$) <br> 3.44 (OH) ($\mu = 0.1$) <br> 6.01 (POH) ($\mu = 0.1$) <br> 8.45 (NH) ($\mu = 0.1$) | 0.04 <br> $3.6 \times 10^{-4}$ <br> $9.8 \times 10^{-7}$ <br> $3.5 \times 10^{-9}$ |
| Pyrophosphoric acid <br> (hydrogen diphosphate) | O O ‖ ‖ $(HO)_2POP(OH)_2$ | 0.83 <br> 2.26 <br> 6.72 <br> 9.46 | 0.15 <br> $5.5 \times 10^{-3}$ <br> $1.9 \times 10^{-7}$ <br> $3.5 \times 10^{-10}$ |

*$pK_3$ from A. G. Miller and J. W. Macklin, Anal. Chem. 1983, 55, 684.

| Name | Structure | $pK_a$ | $K_a$ |
|------|-----------|--------|-------|
| Pyrrolidine | ⬠NH$_2^+$ | 11.305 | $4.95 \times 10^{-12}$ |
| Serine | NH$_3^+$<br>\|<br>CHCH$_2$OH<br>\|<br>CO$_2$H | 2.187 (CO$_2$H)<br>9.209 (NH$_3$) | $6.50 \times 10^{-3}$<br>$6.18 \times 10^{-10}$ |
| Succinic acid (butanedioic acid) | HO$_2$CCH$_2$CH$_2$CO$_2$H | 4.207<br>5.636 | $6.21 \times 10^{-5}$<br>$2.31 \times 10^{-6}$ |
| Sulfuric acid (hydrogen sulfate) | O<br>\|\|<br>HO—S—OH<br>\|\|<br>O | 1.987 (p$K_2$) | $1.03 \times 10^{-2}$ |
| Sulfurous acid (hydrogen sulfite) | O<br>\|\|<br>HOSOH | 1.857<br>7.172 | $1.39 \times 10^{-2}$<br>$6.73 \times 10^{-8}$ |
| Thiosulfuric acid (hydrogen thiosulfate) | O<br>\|\|<br>HOSSH<br>\|\|<br>O | (0.6)<br>(1.6) | 0.3<br>0.03 |
| Threonine | NH$_3^+$<br>\|<br>CHCHOHCH$_3$<br>\|<br>CO$_2$H | 2.088 (CO$_2$H)<br>9.100 (NH$_3$) | $8.17 \times 10^{-3}$<br>$7.94 \times 10^{-10}$ |
| Trichloroacetic acid | Cl$_3$CCO$_2$H | (0.5) | 0.3 |
| Triethanolamine | (HOCH$_2$CH$_2$)$_3$NH$^+$ | 7.762 | $1.73 \times 10^{-8}$ |
| Triethylamine | (CH$_3$CH$_2$)$_3$NH$^+$ | 10.72 | $1.9 \times 10^{-11}$ |
| 1,2,3-Trihydroxybenzene (pyrogallol) | OH<br>⬡—OH<br>OH | 8.96 (25°, $\mu = 0.1$)<br>11.00 (25°, $\mu = 0.1$)<br>(14.0) (20°, $\mu = 0.1$) | $1.10 \times 10^{-9}$<br>$1.00 \times 10^{-11}$<br>$10^{-14}$ |
| Trimethylamine | (CH$_3$)$_3$NH$^+$ | 9.799 | $1.59 \times 10^{-10}$ |
| Tris(hydroxymethyl)amino-methane (tris or tham) | (HOCH$_2$)$_3$CNH$_3^+$ | 8.072 | $8.47 \times 10^{-9}$ |
| Tryptophan | NH$_3^+$<br>\|<br>CHCH$_2$⬡<br>\|<br>CO$_2$H | 2.37 (CO$_2$H) ($\mu = 0.1$)<br>9.33 (NH$_3$) ($\mu = 0.1$) | $4.3 \times 10^{-3}$<br>$4.7 \times 10^{-10}$ |
| Tyrosine | NH$_3^+$<br>\|<br>CHCH$_2$—⬡—OH<br>\|<br>CO$_2$H | 2.41 (CO$_2$H) ($\mu = 0.1$)<br>8.67 (NH$_3$) ($\mu = 0.1$)<br>11.01 (OH) ($\mu = 0.1$) | $3.9 \times 10^{-3}$<br>$2.1 \times 10^{-9}$<br>$9.8 \times 10^{-12}$ |
| Valine | NH$_3^+$<br>\|<br>CHCH(CH$_3$)$_2$<br>\|<br>CO$_2$H | 2.286 (CO$_2$H)<br>9.719 (NH$_3$) | $5.18 \times 10^{-3}$<br>$1.91 \times 10^{-10}$ |
| Water* | H$_2$O | 13.997 | $1.01 \times 10^{-14}$ |

*The constant given for water is $K_w$.

# Standard Reduction Potentials*

| Reaction | $E°$ (volts) | $dE°/dT$ (mV/K) |
|---|---|---|
| **Aluminum** | | |
| $Al^{3+} + 3e^- \rightleftharpoons Al(s)$ | −1.677 | 0.533 |
| $AlCl^{2+} + 3e^- \rightleftharpoons Al(s) + Cl^-$ | −1.802 | |
| $AlF_6^{3-} + 3e^- \rightleftharpoons Al(s) + 6F^-$ | −2.069 | |
| $Al(OH)_4^- + 3e^- \rightleftharpoons Al(s) + 4OH^-$ | −2.328 | −1.13 |
| **Antimony** | | |
| $SbO^+ + 2H^+ + 3e^- \rightleftharpoons Sb(s) + H_2O$ | 0.208 | |
| $Sb_2O_3(s) + 6H^+ + 6e^- \rightleftharpoons 2Sb(s) + 3H_2O$ | 0.147 | −0.369 |
| $Sb(s) + 3H^+ + 3e^- \rightleftharpoons SbH_3(g)$ | −0.510 | −0.030 |
| **Arsenic** | | |
| $H_3AsO_4 + 2H^+ + 2e^- \rightleftharpoons H_3AsO_3 + H_2O$ | 0.575 | −0.257 |
| $H_3AsO_3 + 3H^+ + 3e^- \rightleftharpoons As(s) + 3H_2O$ | 0.247 5 | −0.505 |
| $As(s) + 3H^+ + 3e^- \rightleftharpoons AsH_3(g)$ | −0.238 | −0.029 |
| **Barium** | | |
| $Ba^{2+} + 2e^- + Hg \rightleftharpoons Ba(in\ Hg)$ | −1.717 | |
| $Ba^{2+} + 2e^- \rightleftharpoons Ba(s)$ | −2.906 | −0.401 |
| **Beryllium** | | |
| $Be^{2+} + 2e^- \rightleftharpoons Be(s)$ | −1.968 | 0.60 |
| **Bismuth** | | |
| $Bi^{3+} + 3e^- \rightleftharpoons Bi(s)$ | 0.308 | 0.18 |
| $BiCl_4^- + 3e^- \rightleftharpoons Bi(s) + 4Cl^-$ | 0.16 | |
| $BiOCl(s) + 2H^+ + 3e^- \rightleftharpoons Bi(s) + H_2O + Cl^-$ | 0.160 | |
| **Boron** | | |
| $2B(s) + 6H^+ + 6e^- \rightleftharpoons B_2H_6(g)$ | −0.150 | −0.296 |
| $B_4O_7^{2-} + 14H^+ + 12e^- \rightleftharpoons 4B(s) + 7H_2O$ | −0.792 | |
| $B(OH)_3 + 3H^+ + 3e^- \rightleftharpoons B(s) + 3H_2O$ | −0.889 | −0.492 |
| **Bromine** | | |
| $BrO_4^- + 2H^+ + 2e^- \rightleftharpoons BrO_3^- + H_2O$ | 1.745 | −0.511 |
| $HOBr + H^+ + e^- \rightleftharpoons \frac{1}{2}Br_2(l) + H_2O$ | 1.584 | −0.75 |
| $BrO_3^- + 6H^+ + 5e^- \rightleftharpoons \frac{1}{2}Br_2(l) + 3H_2O$ | 1.513 | −0.419 |
| $Br_2(aq) + 2e^- \rightleftharpoons 2Br^-$ | 1.098 | −0.499 |
| $Br_2(l) + 2e^- \rightleftharpoons 2Br^-$ | 1.078 | −0.611 |
| $Br_3^- + 2e^- \rightleftharpoons 3Br^-$ | 1.062 | −0.512 |
| $BrO^- + H_2O + 2e^- \rightleftharpoons Br^- + 2OH^-$ | 0.766 | −0.94 |
| $BrO_3^- + 3H_2O + 6e^- \rightleftharpoons Br^- + 6OH^-$ | 0.613 | −1.287 |
| **Cadmium** | | |
| $Cd^{2+} + 2e^- + Hg \rightleftharpoons Cd(in\ Hg)$ | −0.380 | |
| $Cd^{2+} + 2e^- \rightleftharpoons Cd(s)$ | −0.402 | −0.029 |
| $Cd(C_2O_4)(s) + 2e^- \rightleftharpoons Cd(s) + C_2O_4^{2-}$ | −0.522 | |
| $Cd(C_2O_4)_2^{2-} + 2e^- \rightleftharpoons Cd(s) + 2C_2O_4^{2-}$ | −0.572 | |
| $Cd(NH_3)_4^{2+} + 2e^- \rightleftharpoons Cd(s) + 4NH_3$ | −0.613 | |
| $CdS(s) + 2e^- \rightleftharpoons Cd(s) + S^{2-}$ | −1.175 | |
| **Calcium** | | |
| $Ca(s) + 2H^+ + 2e^- \rightleftharpoons CaH_2(s)$ | 0.776 | |
| $Ca^{2+} + 2e^- + Hg \rightleftharpoons Ca(in\ Hg)$ | −2.003 | |
| $Ca^{2+} + 2e^- \rightleftharpoons Ca(s)$ | −2.868 | −0.186 |

*All species are aqueous unless otherwise indicated. The reference state for amalgams is an infinitely dilute solution of the element in Hg. The temperature coefficient, $dE°/dT$, allows us to calculate the standard potential, $E°(T)$, at temperature $T$: $E°(T) = E° + (dE°/dT)\Delta T$, where $\Delta T$ is $T - 298.15$ K. Note the units mV/K for $dE°/dT$. Once you know $E°$ for a net cell reaction at temperature $T$, you can find the equilibrium constant, $K$, for the reaction from the formula $K = 10^{nFE°/RT \ln 10}$, where $n$ is the number of electrons in each half-reaction, $F$ is the Faraday constant, and $R$ is the gas constant.

SOURCES: The most authoritative source is S. G. Bratsch, J. Phys. Chem. Ref. Data **1989**, 18, 1. Additional data come from L. G. Sillen and A. E. Martell, Stability Constants of Metal-Ion Complexes (London: The Chemical Society, Special Publications Nos. 17 and 25. 1964 and 1971); G. Milazzo and S. Caroli, Tables of Standard Electrode Potentials (New York: Wiley, 1978); T. Mussini, P. Longhi, and S. Rondinini, Pure Appl. Chem. **1985**, 57, 169. Another good source is A. J. Bard, R. Parsons, and J. Jordan. Standard Potentials in Aqueous Solution (New York: Marcel Dekker, 1985). Reduction potentials for 1 200 free radical reactions are given by P. Wardman, J. Phys. Chem. Ref. Data **1989**, 18, 1637.

| Reaction | $E°$ (volts) | $dE°/dT$ (mV/K) |
|---|---|---|
| Ca(acetate)$^+$ + 2e$^-$ $\rightleftharpoons$ Ca(s) + acetate$^-$ | −2.891 | |
| CaSO$_4$(s) + 2e$^-$ $\rightleftharpoons$ Ca(s) + SO$_4^{2-}$ | −2.936 | |
| Ca(malonate)(s) + 2e$^-$ $\rightleftharpoons$ Ca(s) + malonate$^{2-}$ | −3.608 | |

**Carbon**

| Reaction | $E°$ (volts) | $dE°/dT$ (mV/K) |
|---|---|---|
| C$_2$H$_2$(g) + 2H$^+$ + 2e$^-$ $\rightleftharpoons$ C$_2$H$_4$(g) | 0.731 | |
| O=⬡=O + 2H$^+$ + 2e$^-$ $\rightleftharpoons$ HO–⬡–OH | 0.700 | |
| CH$_3$OH + 2H$^+$ + 2e$^-$ $\rightleftharpoons$ CH$_4$(g) + H$_2$O | 0.583 | −0.039 |
| Dehydroascorbic acid + 2H$^+$ + 2e$^-$ $\rightleftharpoons$ ascorbic acid + H$_2$O | 0.390 | |
| (CN)$_2$(g) + 2H$^+$ + 2e$^-$ $\rightleftharpoons$ 2HCN(aq) | 0.373 | |
| H$_2$CO + 2H$^+$ + 2e$^-$ $\rightleftharpoons$ CH$_3$OH | 0.237 | −0.51 |
| C(s) + 4H$^+$ + 4e$^-$ $\rightleftharpoons$ CH$_4$(g) | 0.131 5 | −0.209 2 |
| HCO$_2$H + 2H$^+$ + 2e$^-$ $\rightleftharpoons$ H$_2$CO + H$_2$O | −0.029 | −0.63 |
| CO$_2$(g) + 2H$^+$ + 2e$^-$ $\rightleftharpoons$ CO(g) + H$_2$O | −0.103 8 | −0.397 7 |
| CO$_2$(g) + 2H$^+$ + 2e$^-$ $\rightleftharpoons$ HCO$_2$H | −0.114 | −0.94 |
| 2CO$_2$(g) + 2H$^+$ + 2e$^-$ $\rightleftharpoons$ H$_2$C$_2$O$_4$ | −0.432 | −1.76 |

**Cerium**

| Reaction | $E°$ (volts) | | $dE°/dT$ (mV/K) |
|---|---|---|---|
| | 1.72 | | 1.54 |
| | 1.70 | 1 F HClO$_4$ | |
| Ce$^{4+}$ + e$^-$ $\rightleftharpoons$ Ce$^{3+}$ | 1.44 | 1 F H$_2$SO$_4$ | |
| | 1.61 | 1 F HNO$_3$ | |
| | 1.47 | 1 F HCl | |
| Ce$^{3+}$ + 3e$^-$ $\rightleftharpoons$ Ce(s) | −2.336 | | 0.280 |

**Cesium**

| Reaction | $E°$ (volts) | $dE°/dT$ (mV/K) |
|---|---|---|
| Cs$^+$ + e$^-$ + Hg $\rightleftharpoons$ Cs(in Hg) | −1.950 | |
| Cs$^+$ + e$^-$ $\rightleftharpoons$ Cs(s) | −3.026 | −1.172 |

**Chlorine**

| Reaction | $E°$ (volts) | $dE°/dT$ (mV/K) |
|---|---|---|
| HClO$_2$ + 2H$^+$ + 2e$^-$ $\rightleftharpoons$ HOCl + H$_2$O | 1.674 | 0.55 |
| HClO + H$^+$ + e$^-$ $\rightleftharpoons \frac{1}{2}$Cl$_2$(g) + H$_2$O | 1.630 | −0.27 |
| ClO$_3^-$ + 6H$^+$ + 5e$^-$ $\rightleftharpoons \frac{1}{2}$Cl$_2$(g) + 3H$_2$O | 1.458 | −0.347 |
| Cl$_2$(aq) + 2e$^-$ $\rightleftharpoons$ 2Cl$^-$ | 1.396 | −0.72 |
| Cl$_2$(g) + 2e$^-$ $\rightleftharpoons$ 2Cl$^-$ | 1.360 4 | −1.248 |
| ClO$_4^-$ + 2H$^+$ + 2e$^-$ $\rightleftharpoons$ ClO$_3^-$ + H$_2$O | 1.226 | −0.416 |
| ClO$_3^-$ + 3H$^+$ + 2e$^-$ $\rightleftharpoons$ HClO$_2$ + H$_2$O | 1.157 | −0.180 |
| ClO$_3^-$ + 2H$^+$ + e$^-$ $\rightleftharpoons$ ClO$_2$ + H$_2$O | 1.130 | 0.074 |
| ClO$_2$ + e$^-$ $\rightleftharpoons$ ClO$_2^-$ | 1.068 | −1.335 |

**Chromium**

| Reaction | $E°$ (volts) | $dE°/dT$ (mV/K) |
|---|---|---|
| Cr$_2$O$_7^{2-}$ + 14H$^+$ + 6e$^-$ $\rightleftharpoons$ 2Cr$^{3+}$ + 7H$_2$O | 1.36 | −1.32 |
| CrO$_4^{2-}$ + 4H$_2$O + 3e$^-$ $\rightleftharpoons$ Cr(OH)$_3$ (s, hydrated) + 5OH$^-$ | −0.12 | −1.62 |
| Cr$^{3+}$ + e$^-$ $\rightleftharpoons$ Cr$^{2+}$ | −0.42 | 1.4 |
| Cr$^{3+}$ + 3e$^-$ $\rightleftharpoons$ Cr(s) | −0.74 | 0.44 |
| Cr$^{2+}$ + 2e$^-$ $\rightleftharpoons$ Cr(s) | −0.89 | −0.04 |

**Cobalt**

| Reaction | $E°$ (volts) | | $dE°/dT$ (mV/K) |
|---|---|---|---|
| | 1.92 | | 1.23 |
| Co$^{3+}$ + e$^-$ $\rightleftharpoons$ Co$^{2+}$ | 1.817 | 8 F H$_2$SO$_4$ | |
| | 1.850 | 4 F HNO$_3$ | |
| Co(NH$_3$)$_5$(H$_2$O)$^{3+}$ + e$^-$ $\rightleftharpoons$ Co(NH$_3$)$_5$(H$_2$O)$^{2+}$ | 0.37 | 1 F NH$_4$NO$_3$ | |
| Co(NH$_3$)$_6^{3+}$ + e$^-$ $\rightleftharpoons$ Co(NH$_3$)$_6^{2+}$ | 0.1 | | |
| CoOH$^+$ + H$^+$ + 2e$^-$ $\rightleftharpoons$ Co(s) + H$_2$O | 0.003 | | −0.04 |
| Co$^{2+}$ + 2e$^-$ $\rightleftharpoons$ Co(s) | −0.282 | | 0.065 |
| Co(OH)$_2$(s) + 2e$^-$ $\rightleftharpoons$ Co(s) + 2OH$^-$ | −0.746 | | −1.02 |

**Copper**

| Reaction | $E°$ (volts) | $dE°/dT$ (mV/K) |
|---|---|---|
| Cu$^+$ + e$^-$ $\rightleftharpoons$ Cu(s) | 0.518 | −0.754 |
| Cu$^{2+}$ + 2e$^-$ $\rightleftharpoons$ Cu(s) | 0.339 | 0.011 |
| Cu$^{2+}$ + e$^-$ $\rightleftharpoons$ Cu$^+$ | 0.161 | 0.776 |
| CuCl(s) + e$^-$ $\rightleftharpoons$ Cu(s) + Cl$^-$ | 0.137 | |
| Cu(IO$_3$)$_2$(s) + 2e$^-$ $\rightleftharpoons$ Cu(s) + 2IO$_3^-$ | −0.079 | |
| Cu(ethylenediamine)$_2^+$ + e$^-$ $\rightleftharpoons$ Cu(s) + 2 ethylenediamine | −0.119 | |
| CuI(s) + e$^-$ $\rightleftharpoons$ Cu(s) + I$^-$ | −0.185 | |
| Cu(EDTA)$^{2-}$ + 2e$^-$ $\rightleftharpoons$ Cu(s) + EDTA$^{4-}$ | −0.216 | |
| Cu(OH)$_2$(s) + 2e$^-$ $\rightleftharpoons$ Cu(s) + 2OH$^-$ | −0.222 | |
| Cu(CN)$_2^-$ + e$^-$ $\rightleftharpoons$ Cu(s) + 2CN$^-$ | −0.429 | |
| CuCN(s) + e$^-$ $\rightleftharpoons$ Cu(s) + CN$^-$ | −0.639 | |

*(Continued)*

| Reaction | $E°$ (volts) | | $dE°/dT$ (mV/K) |
|---|---|---|---|
| **Dysprosium** | | | |
| $Dy^{3+} + 3e^- \rightleftharpoons Dy(s)$ | $-2.295$ | | $0.373$ |
| **Erbium** | | | |
| $Er^{3+} + 3e^- \rightleftharpoons Er(s)$ | $-2.331$ | | $0.388$ |
| **Europium** | | | |
| $Eu^{3+} + e^- \rightleftharpoons Eu^{2+}$ | $-0.35$ | | $1.53$ |
| $Eu^{3+} + 3e^- \rightleftharpoons Eu(s)$ | $-1.991$ | | $0.338$ |
| $Eu^{2+} + 2e^- \rightleftharpoons Eu(s)$ | $-2.812$ | | $-0.26$ |
| **Fluorine** | | | |
| $F_2(g) + 2e^- \rightleftharpoons 2F^-$ | $2.890$ | | $-1.870$ |
| $F_2O(g) + 2H^+ + 4e^- \rightleftharpoons 2F^- + H_2O$ | $2.168$ | | $-1.208$ |
| **Gadolinium** | | | |
| $Gd^{3+} + 3e^- \rightleftharpoons Gd(s)$ | $-2.279$ | | $0.315$ |
| **Gallium** | | | |
| $Ga^{3+} + 3e^- \rightleftharpoons Ga(s)$ | $-0.549$ | | $0.61$ |
| $GaOOH(s) + H_2O + 3e^- \rightleftharpoons Ga(s) + 3OH^-$ | $-1.320$ | | $-1.08$ |
| **Germanium** | | | |
| $Ge^{2+} + 2e^- \rightleftharpoons Ge(s)$ | $0.1$ | | |
| $H_4GeO_4 + 4H^+ + 4e^- \rightleftharpoons Ge(s) + 4H_2O$ | $-0.039$ | | $-0.429$ |
| **Gold** | | | |
| $Au^+ + e^- \rightleftharpoons Au(s)$ | $1.69$ | | $-1.1$ |
| $Au^{3+} + 2e^- \rightleftharpoons Au^+$ | $1.41$ | | |
| $AuCl_2^- + e^- \rightleftharpoons Au(s) + 2Cl^-$ | $1.154$ | | |
| $AuCl_4^- + 2e^- \rightleftharpoons AuCl_2^- + 2Cl^-$ | $0.926$ | | |
| **Hafnium** | | | |
| $Hf^{4+} + 4e^- \rightleftharpoons Hf(s)$ | $-1.55$ | | $0.68$ |
| $HfO_2(s) + 4H^+ + 4e^- \rightleftharpoons Hf(s) + 2H_2O$ | $-1.591$ | | $-0.355$ |
| **Holmium** | | | |
| $Ho^{3+} + 3e^- \rightleftharpoons Ho(s)$ | $-2.33$ | | $0.371$ |
| **Hydrogen** | | | |
| $2H^+ + 2e^- \rightleftharpoons H_2(g)$ | $0.000\ 0$ | | $0$ |
| $H_2O + e^- \rightleftharpoons \frac{1}{2}H_2(g) + OH^-$ | $-0.828\ 0$ | | $-0.836\ 0$ |
| **Indium** | | | |
| $In^{3+} + 3e^- + Hg \rightleftharpoons In(in\ Hg)$ | $-0.313$ | | |
| $In^{3+} + 3e^- \rightleftharpoons In(s)$ | $-0.338$ | | $0.42$ |
| $In^{3+} + 2e^- \rightleftharpoons In^+$ | $-0.444$ | | |
| $In(OH)_3(s) + 3e^- \rightleftharpoons In(s) + 3OH^-$ | $-0.99$ | | $-0.95$ |
| **Iodine** | | | |
| $IO_4^- + 2H^+ + 2e^- \rightleftharpoons IO_3^- + H_2O$ | $1.589$ | | $-0.85$ |
| $H_5IO_6 + 2H^+ + 2e^- \rightleftharpoons HIO_3 + 3H_2O$ | $1.567$ | | $-0.12$ |
| $HOI + H^+ + e^- \rightleftharpoons \frac{1}{2}I_2(s) + H_2O$ | $1.430$ | | $-0.339$ |
| $ICl_3(s) + 3e^- \rightleftharpoons \frac{1}{2}I_2(s) + 3Cl^-$ | $1.28$ | | |
| $ICl(s) + e^- \rightleftharpoons \frac{1}{2}I_2(s) + Cl^-$ | $1.22$ | | |
| $IO_3^- + 6H^+ + 5e^- \rightleftharpoons \frac{1}{2}I_2(s) + 3H_2O$ | $1.210$ | | $-0.367$ |
| $IO_3^- + 5H^+ + 4e^- \rightleftharpoons HOI + 2H_2O$ | $1.154$ | | $-0.374$ |
| $I_2(aq) + 2e^- \rightleftharpoons 2I^-$ | $0.620$ | | $-0.234$ |
| $I_2(s) + 2e^- \rightleftharpoons 2I^-$ | $0.535$ | | $-0.125$ |
| $I_3^- + 2e^- \rightleftharpoons 3I^-$ | $0.535$ | | $-0.186$ |
| $IO_3^- + 3H_2O + 6e^- \rightleftharpoons I^- + 6OH^-$ | $0.269$ | | $-1.163$ |
| **Iridium** | | | |
| $IrCl_6^{2-} + e^- \rightleftharpoons IrCl_6^{3-}$ | $1.026$ | 1 F HCl | |
| $IrBr_6^{2-} + e^- \rightleftharpoons IrBr_6^{3-}$ | $0.947$ | 2 F NaBr | |
| $IrCl_6^{2-} + 4e^- \rightleftharpoons Ir(s) + 6Cl^-$ | $0.835$ | | |
| $IrO_2(s) + 4H^+ + 4e^- \rightleftharpoons Ir(s) + 2H_2O$ | $0.73$ | | $-0.36$ |
| $IrI_6^{2-} + e^- \rightleftharpoons IrI_6^{3-}$ | $0.485$ | 1 F KI | |
| **Iron** | | | |
| $Fe(phenanthroline)_3^{3+} + e^- \rightleftharpoons Fe(phenanthroline)_3^{2+}$ | $1.147$ | | |
| $Fe(bipyridyl)_3^{3+} + e^- \rightleftharpoons Fe(bipyridyl)_3^{2+}$ | $1.120$ | | |
| $FeOH^{2+} + H^+ + e^- \rightleftharpoons Fe^{2+} + H_2O$ | $0.900$ | | $0.096$ |
| $FeO_4^{2-} + 3H_2O + 3e^- \rightleftharpoons FeOOH(s) + 5OH^-$ | $0.80$ | | $-1.59$ |
| $Fe^{3+} + e^- \rightleftharpoons Fe^{2+}$ | $\begin{cases} 0.771 \\ 0.732 \\ 0.767 \\ 0.746 \end{cases}$ | $\begin{matrix} \\ \text{1 F HCl} \\ \text{1 F HClO}_4 \\ \text{1 F HNO}_3 \end{matrix}$ | $1.175$ |

| Reaction | $E°$ (volts) | $dE°/dT$ (mV/K) |
|---|---|---|
| $FeOOH(s) + 3H^+ + e^- \rightleftharpoons Fe^{2+} + 2H_2O$ | 0.74 | −1.05 |
| $Ferricinium^+ + e^- \rightleftharpoons ferrocene$ | 0.400 | |
| $Fe(CN)_6^{3-} + e^- \rightleftharpoons Fe(CN)_6^{4-}$ | 0.356 | |
| $Fe(glutamate)^{3+} + e^- \rightleftharpoons Fe(glutamate)^{2+}$ | 0.240 | |
| $FeOH^+ + H^+ + 2e^- \rightleftharpoons Fe(s) + H_2O$ | −0.16 | 0.07 |
| $Fe^{2+} + 2e^- \rightleftharpoons Fe(s)$ | −0.44 | 0.07 |
| $FeCO_3(s) + 2e^- \rightleftharpoons Fe(s) + CO_3^{2-}$ | −0.756 | −1.293 |

Lanthanum

| Reaction | $E°$ (volts) | $dE°/dT$ (mV/K) |
|---|---|---|
| $La^{3+} + 3e^- \rightleftharpoons La(s)$ | −2.379 | 0.242 |
| $La(succinate)^+ + 3e^- \rightleftharpoons La(s) + succinate^{2-}$ | −2.601 | |

Lead

| Reaction | $E°$ (volts) | $dE°/dT$ (mV/K) |
|---|---|---|
| $Pb^{4+} + 2e^- \rightleftharpoons Pb^{2+}$ | 1.69   1 F HNO$_3$ | |
| $PbO_2(s) + 4H^+ + SO_4^{2-} + 2e^- \rightleftharpoons PbSO_4(s) + 2H_2O$ | 1.685 | |
| $PbO_2(s) + 4H^+ + 2e^- \rightleftharpoons Pb^{2+} + 2H_2O$ | 1.458 | −0.253 |
| $3PbO_2(s) + 2H_2O + 4e^- \rightleftharpoons Pb_3O_4(s) + 4OH^-$ | 0.269 | −1.136 |
| $Pb_3O_4(s) + H_2O + 2e^- \rightleftharpoons 3PbO(s, red) + 2OH^-$ | 0.224 | −1.211 |
| $Pb_3O_4(s) + H_2O + 2e^- \rightleftharpoons 3PbO(s, yellow) + 2OH^-$ | 0.207 | −1.177 |
| $Pb^{2+} + 2e^- \rightleftharpoons Pb(s)$ | −0.126 | −0.395 |
| $PbF_2(s) + 2e^- \rightleftharpoons Pb(s) + 2F^-$ | −0.350 | |
| $PbSO_4(s) + 2e^- \rightleftharpoons Pb(s) + SO_4^{2-}$ | −0.355 | |

Lithium

| Reaction | $E°$ (volts) | $dE°/dT$ (mV/K) |
|---|---|---|
| $Li^+ + e^- + Hg \rightleftharpoons Li(in\ Hg)$ | −2.195 | |
| $Li^+ + e^- \rightleftharpoons Li(s)$ | −3.040 | −0.514 |

Lutetium

| Reaction | $E°$ (volts) | $dE°/dT$ (mV/K) |
|---|---|---|
| $Lu^{3+} + 3e^- \rightleftharpoons Lu(s)$ | −2.28 | 0.412 |

Magnesium

| Reaction | $E°$ (volts) | $dE°/dT$ (mV/K) |
|---|---|---|
| $Mg^{2+} + 2e^- + Hg \rightleftharpoons Mg(in\ Hg)$ | −1.980 | |
| $Mg(OH)^+ + H^+ + 2e^- \rightleftharpoons Mg(s) + H_2O$ | −2.022 | 0.25 |
| $Mg^{2+} + 2e^- \rightleftharpoons Mg(s)$ | −2.360 | 0.199 |
| $Mg(C_2O_4)(s) + 2e^- \rightleftharpoons Mg(s) + C_2O_4^{2-}$ | −2.493 | |
| $Mg(OH)_2(s) + 2e^- \rightleftharpoons Mg(s) + 2OH^-$ | −2.690 | −0.946 |

Manganese

| Reaction | $E°$ (volts) | $dE°/dT$ (mV/K) |
|---|---|---|
| $MnO_4^- + 4H^+ + 3e^- \rightleftharpoons MnO_2(s) + 2H_2O$ | 1.692 | −0.671 |
| $Mn^{3+} + e^- \rightleftharpoons Mn^{2+}$ | 1.56 | 1.8 |
| $MnO_4^- + 8H^+ + 5e^- \rightleftharpoons Mn^{2+} + 4H_2O$ | 1.507 | −0.646 |
| $Mn_2O_3(s) + 6H^+ + 2e^- \rightleftharpoons 2Mn^{2+} + 3H_2O$ | 1.485 | −0.926 |
| $MnO_2(s) + 4H^+ + 2e^- \rightleftharpoons Mn^{2+} + 2H_2O$ | 1.230 | −0.609 |
| $Mn(EDTA)^- + e^- \rightleftharpoons Mn(EDTA)^{2-}$ | 0.825 | −1.10 |
| $MnO_4^- + e^- \rightleftharpoons MnO_4^{2-}$ | 0.56 | −2.05 |
| $3Mn_2O_3(s) + H_2O + 2e^- \rightleftharpoons 2Mn_3O_4(s) + 2OH^-$ | 0.002 | −1.256 |
| $Mn_3O_4(s) + 4H_2O + 2e^- \rightleftharpoons 3Mn(OH)_2(s) + 2OH^-$ | −0.352 | −1.61 |
| $Mn^{2+} + 2e^- \rightleftharpoons Mn(s)$ | −1.182 | −1.129 |
| $Mn(OH)_2(s) + 2e^- \rightleftharpoons Mn(s) + 2OH^-$ | −1.565 | −1.10 |

Mercury

| Reaction | $E°$ (volts) | $dE°/dT$ (mV/K) |
|---|---|---|
| $2Hg^{2+} + 2e^- \rightleftharpoons Hg_2^{2+}$ | 0.908 | 0.095 |
| $Hg^{2+} + 2e^- \rightleftharpoons Hg(l)$ | 0.852 | −0.116 |
| $Hg_2^{2+} + 2e^- \rightleftharpoons 2Hg(l)$ | 0.796 | −0.327 |
| $Hg_2SO_4(s) + 2e^- \rightleftharpoons 2Hg(l) + SO_4^{2-}$ | 0.614 | |
| $Hg_2Cl_2(s) + 2e^- \rightleftharpoons 2Hg(l) + 2Cl^-$ | $\begin{cases} 0.268 \\ 0.241\ \text{(saturated calomel electrode)} \end{cases}$ | |
| $Hg(OH)_3^- + 2e^- \rightleftharpoons Hg(l) + 3OH^-$ | 0.231 | |
| $Hg(OH)_2 + 2e^- \rightleftharpoons Hg(l) + 2OH^-$ | 0.206 | −1.24 |
| $Hg_2Br_2(s) + 2e^- \rightleftharpoons 2Hg(l) + 2Br^-$ | 0.140 | |
| $HgO(s, yellow) + H_2O + 2e^- \rightleftharpoons Hg(l) + 2OH^-$ | 0.098 3 | −1.125 |
| $HgO(s, red) + H_2O + 2e^- \rightleftharpoons Hg(l) + 2OH^-$ | 0.097 7 | −1.120 6 |

Molybdenum

| Reaction | $E°$ (volts) | $dE°/dT$ (mV/K) |
|---|---|---|
| $MoO_4^{2-} + 2H_2O + 2e^- \rightleftharpoons MoO_2(s) + 4OH^-$ | −0.818 | −1.69 |
| $MoO_4^{2-} + 4H_2O + 6e^- \rightleftharpoons Mo(s) + 8OH^-$ | −0.926 | −1.36 |
| $MoO_2(s) + 2H_2O + 4e^- \rightleftharpoons Mo(s) + 4OH^-$ | −0.980 | −1.196 |

Neodymium

| Reaction | $E°$ (volts) | $dE°/dT$ (mV/K) |
|---|---|---|
| $Nd^{3+} + 3e^- \rightleftharpoons Nd(s)$ | −2.323 | 0.282 |

Neptunium

| Reaction | $E°$ (volts) | $dE°/dT$ (mV/K) |
|---|---|---|
| $NpO_3^+ + 2H^+ + e^- \rightleftharpoons NpO_2^{2+} + H_2O$ | 2.04 | |
| $NpO_2^{2+} + e^- \rightleftharpoons NpO_2^+$ | 1.236 | 0.058 |

*(Continued)*

| Reaction | $E°$ (volts) | $dE°/dT$ (mV/K) |
|---|---|---|
| $NpO_2^+ + 4H^+ + e^- \rightleftharpoons Np^{4+} + 2H_2O$ | 0.567 | −3.30 |
| $Np^{4+} + e^- \rightleftharpoons Np^{3+}$ | 0.157 | 1.53 |
| $Np^{3+} + 3e^- \rightleftharpoons Np(s)$ | −1.768 | 0.18 |
| **Nickel** | | |
| $NiOOH(s) + 3H^+ + e^- \rightleftharpoons Ni^{2+} + 2H_2O$ | 2.05 | −1.17 |
| $Ni^{2+} + 2e^- \rightleftharpoons Ni(s)$ | −0.236 | 0.146 |
| $Ni(CN)_4^{2-} + e^- \rightleftharpoons Ni(CN)_3^{2-} + CN^-$ | −0.401 | |
| $Ni(OH)_2(s) + 2e^- \rightleftharpoons Ni(s) + 2OH^-$ | −0.714 | −1.02 |
| **Niobium** | | |
| $\frac{1}{2}Nb_2O_5(s) + H^+ + e^- \rightleftharpoons NbO_2(s) + \frac{1}{2}H_2O$ | −0.248 | −0.460 |
| $\frac{1}{2}Nb_2O_5(s) + 5H^+ + 5e^- \rightleftharpoons Nb(s) + \frac{5}{2}H_2O$ | −0.601 | −0.381 |
| $NbO_2(s) + 2H^+ + 2e^- \rightleftharpoons NbO(s) + H_2O$ | −0.646 | −0.347 |
| $NbO_2(s) + 4H^+ + 4e^- \rightleftharpoons Nb(s) + 2H_2O$ | −0.690 | −0.361 |
| **Nitrogen** | | |
| $HN_3 + 3H^+ + 2e^- \rightleftharpoons N_2(g) + NH_4^+$ | 2.079 | 0.147 |
| $N_2O(g) + 2H^+ + 2e^- \rightleftharpoons N_2(g) + H_2O$ | 1.769 | −0.461 |
| $2NO(g) + 2H^+ + 2e^- \rightleftharpoons N_2O(g) + H_2O$ | 1.587 | −1.359 |
| $NO^+ + e^- \rightleftharpoons NO(g)$ | 1.46 | |
| $2NH_3OH^+ + H^+ + 2e^- \rightleftharpoons N_2H_5^+ + 2H_2O$ | 1.40 | −0.60 |
| $NH_3OH^+ + 2H^+ + 2e^- \rightleftharpoons NH_4^+ + H_2O$ | 1.33 | −0.44 |
| $N_2H_5^+ + 3H^+ + 2e^- \rightleftharpoons 2NH_4^+$ | 1.250 | −0.28 |
| $HNO_2 + H^+ + e^- \rightleftharpoons NO(g) + H_2O$ | 0.984 | 0.649 |
| $NO_3^- + 4H^+ + 3e^- \rightleftharpoons NO(g) + 2H_2O$ | 0.955 | 0.028 |
| $NO_3^- + 3H^+ + 2e^- \rightleftharpoons HNO_2 + H_2O$ | 0.940 | −0.282 |
| $NO_3^- + 2H^+ + e^- \rightleftharpoons \frac{1}{2}N_2O_4(g) + H_2O$ | 0.798 | 0.107 |
| $N_2(g) + 8H^+ + 6e^- \rightleftharpoons 2NH_4^+$ | 0.274 | −0.616 |
| $N_2(g) + 5H^+ + 4e^- \rightleftharpoons N_2H_5^+$ | −0.214 | −0.78 |
| $N_2(g) + 2H_2O + 4H^+ + 2e^- \rightleftharpoons 2NH_3OH^+$ | −1.83 | −0.96 |
| $\frac{3}{2}N_2(g) + H^+ + e^- \rightleftharpoons HN_3$ | −3.334 | −2.141 |
| **Osmium** | | |
| $OsO_4(s) + 8H^+ + 8e^- \rightleftharpoons Os(s) + 4H_2O$ | 0.834 | −0.458 |
| $OsCl_6^{2-} + e^- \rightleftharpoons OsCl_6^{3-}$ | 0.85　　1 F HCl | |
| **Oxygen** | | |
| $OH + H^+ + e^- \rightleftharpoons H_2O$ | 2.56 | −1.0 |
| $O(g) + 2H^+ + 2e^- \rightleftharpoons H_2O$ | 2.430 1 | −1.148 4 |
| $O_3(g) + 2H^+ + 2e^- \rightleftharpoons O_2(g) + H_2O$ | 2.075 | −0.489 |
| $H_2O_2 + 2H^+ + 2e^- \rightleftharpoons 2H_2O$ | 1.763 | −0.698 |
| $HO_2 + H^+ + e^- \rightleftharpoons H_2O_2$ | 1.44 | −0.7 |
| $\frac{1}{2}O_2(g) + 2H^+ + 2e^- \rightleftharpoons H_2O$ | 1.229 1 | −0.845 6 |
| $O_2(g) + 2H^+ + 2e^- \rightleftharpoons H_2O_2$ | 0.695 | −0.993 |
| $O_2(g) + H^+ + e^- \rightleftharpoons HO_2$ | −0.05 | −1.3 |
| **Palladium** | | |
| $Pd^{2+} + 2e^- \rightleftharpoons Pd(s)$ | 0.915 | 0.12 |
| $PdO(s) + 2H^+ + 2e^- \rightleftharpoons Pd(s) + H_2O$ | 0.79 | −0.33 |
| $PdCl_6^{4-} + 2e^- \rightleftharpoons Pd(s) + 6Cl^-$ | 0.615 | |
| $PdO_2(s) + H_2O + 2e^- \rightleftharpoons PdO(s) + 2OH^-$ | 0.64 | −1.2 |
| **Phosphorus** | | |
| $\frac{1}{4}P_4(s,\ white) + 3H^+ + 3e^- \rightleftharpoons PH_3(g)$ | −0.046 | −0.093 |
| $\frac{1}{4}P_4(s,\ red) + 3H^+ + 3e^- \rightleftharpoons PH_3(g)$ | −0.088 | −0.030 |
| $H_3PO_4 + 2H^+ + 2e^- \rightleftharpoons H_3PO_3 + H_2O$ | −0.30 | −0.36 |
| $H_3PO_4 + 5H^+ + 5e^- \rightleftharpoons \frac{1}{4}P_4(s,\ white) + 4H_2O$ | −0.402 | −0.340 |
| $H_3PO_3 + 2H^+ + 2e^- \rightleftharpoons H_3PO_2 + H_2O$ | −0.48 | −0.37 |
| $H_3PO_2 + H^+ + e^- \rightleftharpoons \frac{1}{4}P_4(s) + 2H_2O$ | −0.51 | |
| **Platinum** | | |
| $Pt^{2+} + 2e^- \rightleftharpoons Pt(s)$ | 1.18 | −0.05 |
| $PtO_2(s) + 4H^+ + 4e^- \rightleftharpoons Pt(s) + 2H_2O$ | 0.92 | −0.36 |
| $PtCl_4^{2-} + 2e^- \rightleftharpoons Pt(s) + 4Cl^-$ | 0.755 | |
| $PtCl_6^{2-} + 2e^- \rightleftharpoons PtCl_4^{2-} + 2Cl^-$ | 0.68 | |
| **Plutonium** | | |
| $PuO_2^+ + e^- \rightleftharpoons PuO_2(s)$ | 1.585 | 0.39 |
| $PuO_2^{2+} + 4H^+ + 2e^- \rightleftharpoons Pu^{4+} + 2H_2O$ | 1.000 | −1.615 1 |
| $Pu^{4+} + e^- \rightleftharpoons Pu^{3+}$ | 1.006 | 1.441 |
| $PuO_2^{2+} + e^- \rightleftharpoons PuO_2^+$ | 0.966 | 0.03 |
| $PuO_2(s) + 4H^+ + 4e^- \rightleftharpoons Pu(s) + 2H_2O$ | −1.369 | −0.38 |
| $Pu^{3+} + 3e^- \rightleftharpoons Pu(s)$ | −1.978 | 0.23 |

| Reaction | $E°$ (volts) | | $dE°/dT$ (mV/K) |
|---|---|---|---|
| **Potassium** | | | |
| $K^+ + e^- + Hg \rightleftharpoons K(in\ Hg)$ | -1.975 | | |
| $K^+ + e^- \rightleftharpoons K(s)$ | -2.936 | | -1.074 |
| **Praseodymium** | | | |
| $Pr^{4+} + e^- \rightleftharpoons Pr^{3+}$ | 3.2 | | 1.4 |
| $Pr^{3+} + 3e^- \rightleftharpoons Pr(s)$ | -2.353 | | 0.291 |
| **Promethium** | | | |
| $Pm^{3+} + 3e^- \rightleftharpoons Pm(s)$ | -2.30 | | 0.29 |
| **Radium** | | | |
| $Ra^{2+} + 2e^- \rightleftharpoons Ra(s)$ | -2.80 | | -0.44 |
| **Rhenium** | | | |
| $ReO_4^- + 2H^+ + e^- \rightleftharpoons ReO_3(s) + H_2O$ | 0.72 | | -1.17 |
| $ReO_4^- + 4H^+ + 3e^- \rightleftharpoons ReO_2(s) + 2H_2O$ | 0.510 | | -0.70 |
| **Rhodium** | | | |
| $Rh^{6+} + 3e^- \rightleftharpoons Rh^{3+}$ | 1.48 | 1 F $HClO_4$ | |
| $Rh^{4+} + e^- \rightleftharpoons Rh^{3+}$ | 1.44 | 3 F $H_2SO_4$ | |
| $RhCl_6^{2-} + e^- \rightleftharpoons RhCl_6^{3-}$ | 1.2 | | |
| $Rh^{3+} + 3e^- \rightleftharpoons Rh(s)$ | 0.76 | | 0.4 |
| $2Rh^{3+} + 2e^- \rightleftharpoons Rh_2^{4+}$ | 0.7 | | |
| $RhCl_6^{3-} + 3e^- \rightleftharpoons Rh(s) + 6Cl^-$ | 0.44 | | |
| **Rubidium** | | | |
| $Rb^+ + e^- + Hg \rightleftharpoons Rb(in\ Hg)$ | -1.970 | | |
| $Rb^+ + e^- \rightleftharpoons Rb(s)$ | -2.943 | | -1.140 |
| **Ruthenium** | | | |
| $RuO_4^- + 6H^+ + 3e^- \rightleftharpoons Ru(OH)_2^{2+} + 2H_2O$ | 1.53 | | |
| $Ru(dipyridyl)_3^{3+} + e^- \rightleftharpoons Ru(dipyridyl)_3^{2+}$ | 1.29 | | |
| $RuO_4(s) + 8H^+ + 8e^- \rightleftharpoons Ru(s) + 4H_2O$ | 1.032 | | -0.467 |
| $Ru^{2+} + 2e^- \rightleftharpoons Ru(s)$ | 0.8 | | |
| $Ru^{3+} + 3e^- \rightleftharpoons Ru(s)$ | 0.60 | | |
| $Ru^{3+} + e^- \rightleftharpoons Ru^{2+}$ | 0.24 | | |
| $Ru(NH_3)_6^{3+} + e^- \rightleftharpoons Ru(NH_3)_6^{2+}$ | 0.214 | | |
| **Samarium** | | | |
| $Sm^{3+} + 3e^- \rightleftharpoons Sm(s)$ | -2.304 | | 0.279 |
| $Sm^{2+} + 2e^- \rightleftharpoons Sm(s)$ | -2.68 | | -0.28 |
| **Scandium** | | | |
| $Sc^{3+} + 3e^- \rightleftharpoons Sc(s)$ | -2.09 | | 0.41 |
| **Selenium** | | | |
| $SeO_4^{2-} + 4H^+ + 2e^- \rightleftharpoons H_2SeO_3 + H_2O$ | 1.150 | | 0.483 |
| $H_2SeO_3 + 4H^+ + 4e^- \rightleftharpoons Se(s) + 3H_2O$ | 0.739 | | -0.562 |
| $Se(s) + 2H^+ + 2e^- \rightleftharpoons H_2Se(g)$ | -0.082 | | 0.238 |
| $Se(s) + 2e^- \rightleftharpoons Se^{2-}$ | -0.67 | | -1.2 |
| **Silicon** | | | |
| $Si(s) + 4H^+ + 4e^- \rightleftharpoons SiH_4(g)$ | -0.147 | | -0.196 |
| $SiO_2(s, quartz) + 4H^+ + 4e^- \rightleftharpoons Si(s) + 2H_2O$ | -0.990 | | -0.374 |
| $SiF_6^{2-} + 4e^- \rightleftharpoons Si(s) + 6F^-$ | -1.24 | | |
| **Silver** | | | |
| $Ag^{2+} + e^- \rightleftharpoons Ag^+$ | 2.000 | 4 F $HClO_4$ | |
| | 1.989 | | 0.99 |
| | 1.929 | 4 F $HNO_3$ | |
| $Ag^{3+} + 2e^- \rightleftharpoons Ag^+$ | 1.9 | | |
| $AgO(s) + H^+ + e^- \rightleftharpoons \frac{1}{2}Ag_2O(s) + \frac{1}{2}H_2O$ | 1.40 | | |
| $Ag^+ + e^- \rightleftharpoons Ag(s)$ | 0.799\ 3 | | -0.989 |
| $Ag_2C_2O_4(s) + 2e^- \rightleftharpoons 2Ag(s) + C_2O_4^{2-}$ | 0.465 | | |
| $AgN_3(s) + e^- \rightleftharpoons Ag(s) + N_3^-$ | 0.293 | | |
| $AgCl(s) + e^- \rightleftharpoons Ag(s) + Cl^-$ | 0.222 | | |
| | 0.197 | saturated KCl | |
| $AgBr(s) + e^- \rightleftharpoons Ag(s) + Br^-$ | 0.071 | | |
| $Ag(S_2O_3)_2^{3-} + e^- \rightleftharpoons Ag(s) + 2S_2O_3^{2-}$ | 0.017 | | |
| $AgI(s) + e^- \rightleftharpoons Ag(s) + I^-$ | -0.152 | | |
| $Ag_2S(s) + H^+ + 2e^- \rightleftharpoons 2Ag(s) + SH^-$ | -0.272 | | |
| **Sodium** | | | |
| $Na^+ + e^- + Hg \rightleftharpoons Na(in\ Hg)$ | -1.959 | | |
| $Na^+ + \frac{1}{2}H_2(g) + e^- \rightleftharpoons NaH(s)$ | -2.367 | | -1.550 |
| $Na^+ + e^- \rightleftharpoons Na(s)$ | -2.714\ 3 | | -0.757 |

(*Continued*)

| Reaction | $E°$ (volts) | | $dE°/dT$ (mV/K) |
|---|---|---|---|
| **Strontium** | | | |
| $Sr^{2+} + 2e^- \rightleftharpoons Sr(s)$ | −2.889 | | −0.237 |
| **Sulfur** | | | |
| $S_2O_8^{2-} + 2e^- \rightleftharpoons 2SO_4^{2-}$ | 2.01 | | |
| $S_2O_6^{2-} + 4H^+ + 2e^- \rightleftharpoons 2H_2SO_3$ | 0.57 | | |
| $4SO_2 + 4H^+ + 6e^- \rightleftharpoons S_4O_6^{2-} + 2H_2O$ | 0.539 | | −1.11 |
| $SO_2 + 4H^+ + 4e^- \rightleftharpoons S(s) + 2H_2O$ | 0.450 | | −0.652 |
| $2H_2SO_3 + 2H^+ + 4e^- \rightleftharpoons S_2O_3^{2-} + 3H_2O$ | 0.40 | | |
| $S(s) + 2H^+ + 2e^- \rightleftharpoons H_2S(g)$ | 0.174 | | 0.224 |
| $S(s) + 2H^+ + 2e^- \rightleftharpoons H_2S(aq)$ | 0.144 | | −0.21 |
| $S_4O_6^{2-} + 2H^+ + 2e^- \rightleftharpoons 2HS_2O_3^-$ | 0.10 | | −0.23 |
| $5S(s) + 2e^- \rightleftharpoons S_5^{2-}$ | −0.340 | | |
| $S(s) + 2e^- \rightleftharpoons S^{2-}$ | −0.476 | | −0.925 |
| $2S(s) + 2e^- \rightleftharpoons S_2^{2-}$ | −0.50 | | −1.16 |
| $2SO_3^{2-} + 3H_2O + 4e^- \rightleftharpoons S_2O_3^{2-} + 6OH^-$ | −0.566 | | −1.06 |
| $SO_3^{2-} + 3H_2O + 4e^- \rightleftharpoons S(s) + 6OH^-$ | −0.659 | | −1.23 |
| $SO_4^{2-} + 4H_2O + 6e^- \rightleftharpoons S(s) + 8OH^-$ | −0.751 | | −1.288 |
| $SO_4^{2-} + H_2O + 2e^- \rightleftharpoons SO_3^{2-} + 2OH^-$ | −0.936 | | −1.41 |
| $2SO_3^{2-} + 2H_2O + 2e^- \rightleftharpoons S_2O_4^{2-} + 4OH^-$ | −1.130 | | −0.85 |
| $2SO_4^{2-} + 2H_2O + 2e^- \rightleftharpoons S_2O_6^{2-} + 4OH^-$ | −1.71 | | −1.00 |
| **Tantalum** | | | |
| $Ta_2O_5(s) + 10H^+ + 10e^- \rightleftharpoons 2Ta(s) + 5H_2O$ | −0.752 | | −0.377 |
| **Technetium** | | | |
| $TcO_4^- + 2H_2O + 3e^- \rightleftharpoons TcO_2(s) + 4OH^-$ | −0.366 | | −1.82 |
| $TcO_4^- + 4H_2O + 7e^- \rightleftharpoons Tc(s) + 8OH^-$ | −0.474 | | −1.46 |
| **Tellurium** | | | |
| $TeO_3^{2-} + 3H_2O + 4e^- \rightleftharpoons Te(s) + 6OH^-$ | −0.47 | | −1.39 |
| $2Te(s) + 2e^- \rightleftharpoons Te_2^{2-}$ | −0.84 | | |
| $Te(s) + 2e^- \rightleftharpoons Te^{2-}$ | −0.90 | | −1.0 |
| **Terbium** | | | |
| $Tb^{4+} + e^- \rightleftharpoons Tb^{3+}$ | 3.1 | | 1.5 |
| $Tb^{3+} + 3e^- \rightleftharpoons Tb(s)$ | −2.28 | | 0.350 |
| **Thallium** | | | |
| | 1.280 | | 0.97 |
| | 0.77 | 1 F HCl | |
| $Tl^{3+} + 2e^- \rightleftharpoons Tl^+$ | 1.22 | 1 F H$_2$SO$_4$ | |
| | 1.23 | 1 F HNO$_3$ | |
| | 1.26 | 1 F HClO$_4$ | |
| $Tl^+ + e^- + Hg \rightleftharpoons Tl(in\ Hg)$ | −0.294 | | |
| $Tl^+ + e^- \rightleftharpoons Tl(s)$ | −0.336 | | −1.312 |
| $TlCl(s) + e^- \rightleftharpoons Tl(s) + Cl^-$ | −0.557 | | |
| **Thorium** | | | |
| $Th^{4+} + 4e^- \rightleftharpoons Th(s)$ | −1.826 | | 0.557 |
| **Thulium** | | | |
| $Tm^{3+} + 3e^- \rightleftharpoons Tm(s)$ | −2.319 | | 0.394 |
| **Tin** | | | |
| $Sn(OH)_3^+ + 3H^+ + 2e^- \rightleftharpoons Sn^{2+} + 3H_2O$ | 0.142 | | |
| $Sn^{4+} + 2e^- \rightleftharpoons Sn^{2+}$ | 0.139 | 1 F HCl | |
| $SnO_2(s) + 4H^+ + 2e^- \rightleftharpoons Sn^{2+} + 2H_2O$ | −0.094 | | −0.31 |
| $Sn^{2+} + 2e^- \rightleftharpoons Sn(s)$ | −0.141 | | −0.32 |
| $SnF_6^{2-} + 4e^- \rightleftharpoons Sn(s) + 6F^-$ | −0.25 | | |
| $Sn(OH)_6^{2-} + 2e^- \rightleftharpoons Sn(OH)_3^- + 3OH^-$ | −0.93 | | |
| $Sn(s) + 4H_2O + 4e^- \rightleftharpoons SnH_4(g) + 4OH^-$ | −1.316 | | −1.057 |
| $SnO_2(s) + H_2O + 2e^- \rightleftharpoons SnO(s) + 2OH^-$ | −0.961 | | −1.129 |
| **Titanium** | | | |
| $TiO^{2+} + 2H^+ + e^- \rightleftharpoons Ti^{3+} + H_2O$ | 0.1 | | −0.6 |
| $Ti^{3+} + e^- \rightleftharpoons Ti^{2+}$ | −0.9 | | 1.5 |
| $TiO_2(s) + 4H^+ + 4e^- \rightleftharpoons Ti(s) + 2H_2O$ | −1.076 | | 0.365 |
| $TiF_6^{2-} + 4e^- \rightleftharpoons Ti(s) + 6F^-$ | −1.191 | | |
| $Ti^{2+} + 2e^- \rightleftharpoons Ti(s)$ | −1.60 | | −0.16 |
| **Tungsten** | | | |
| $W(CN)_8^{3-} + e^- \rightleftharpoons W(CN)_8^{4-}$ | 0.457 | | |
| $W^{6+} + e^- \rightleftharpoons W^{5+}$ | 0.26 | 12 F HCl | |
| $WO_3(s) + 6H^+ + 6e^- \rightleftharpoons W(s) + 3H_2O$ | −0.091 | | −0.389 |

| Reaction | $E°$ (volts) | | $dE°/dT$ (mV/K) |
|---|---|---|---|
| $W^{5+} + e^- \rightleftharpoons W^{4+}$ | $-0.3$ | 12 F HCl | |
| $WO_2(s) + 2H_2O + 4e^- \rightleftharpoons W(s) + 4OH^-$ | $-0.982$ | | $-1.197$ |
| $WO_4^{2-} + 4H_2O + 6e^- \rightleftharpoons W(s) + 8OH^-$ | $-1.060$ | | $-1.36$ |
| **Uranium** | | | |
| $UO_2^+ + 4H^+ + e^- \rightleftharpoons U^{4+} + 2H_2O$ | $0.39$ | | $-3.4$ |
| $UO_2^{2+} + 4H^+ + 2e^- \rightleftharpoons U^{4+} + 2H_2O$ | $0.273$ | | $-1.582$ |
| $UO_2^{2+} + e^- \rightleftharpoons UO_2^+$ | $0.16$ | | $0.2$ |
| $U^{4+} + e^- \rightleftharpoons U^{3+}$ | $-0.577$ | | $1.61$ |
| $U^{3+} + 3e^- \rightleftharpoons U(s)$ | $-1.642$ | | $0.16$ |
| **Vanadium** | | | |
| $VO_2^+ + 2H^+ + e^- \rightleftharpoons VO^{2+} + H_2O$ | $1.001$ | | $-0.901$ |
| $VO^{2+} + 2H^+ + e^- \rightleftharpoons V^{3+} + H_2O$ | $0.337$ | | $-1.6$ |
| $V^{3+} + e^- \rightleftharpoons V^{2+}$ | $-0.255$ | | $1.5$ |
| $V^{2+} + 2e^- \rightleftharpoons V(s)$ | $-1.125$ | | $-0.11$ |
| **Xenon** | | | |
| $H_4XeO_6 + 2H^+ + 2e^- \rightleftharpoons XeO_3 + 3H_2O$ | $2.38$ | | $0.0$ |
| $XeF_2 + 2H^+ + 2e^- \rightleftharpoons Xe(g) + 2HF$ | $2.2$ | | |
| $XeO_3 + 6H^+ + 6e^- \rightleftharpoons Xe(g) + 3H_2O$ | $2.1$ | | $-0.34$ |
| **Ytterbium** | | | |
| $Yb^{3+} + 3e^- \rightleftharpoons Yb(s)$ | $-2.19$ | | $0.363$ |
| $Yb^{2+} + 2e^- \rightleftharpoons Yb(s)$ | $-2.76$ | | $-0.16$ |
| **Yttrium** | | | |
| $Y^{3+} + 3e^- \rightleftharpoons Y(s)$ | $-2.38$ | | $0.034$ |
| **Zinc** | | | |
| $ZnOH^+ + H^+ + 2e^- \rightleftharpoons Zn(s) + H_2O$ | $-0.497$ | | $0.03$ |
| $Zn^{2+} + 2e^- \rightleftharpoons Zn(s)$ | $-0.762$ | | $0.119$ |
| $Zn^{2+} + 2e^- + Hg \rightleftharpoons Zn(in\ Hg)$ | $-0.801$ | | |
| $Zn(NH_3)_4^{2+} + 2e^- \rightleftharpoons Zn(s) + 4NH_3$ | $-1.04$ | | |
| $ZnCO_3(s) + 2e^- \rightleftharpoons Zn(s) + CO_3^{2-}$ | $-1.06$ | | |
| $Zn(OH)_3^- + 2e^- \rightleftharpoons Zn(s) + 3OH^-$ | $-1.183$ | | |
| $Zn(OH)_4^{2-} + 2e^- \rightleftharpoons Zn(s) + 4OH^-$ | $-1.199$ | | |
| $Zn(OH)_2(s) + 2e^- \rightleftharpoons Zn(s) + 2OH^-$ | $-1.249$ | | $-0.999$ |
| $ZnO(s) + H_2O + 2e^- \rightleftharpoons Zn(s) + 2OH^-$ | $-1.260$ | | $-1.160$ |
| $ZnS(s) + 2e^- \rightleftharpoons Zn(s) + S^{2-}$ | $-1.405$ | | |
| **Zirconium** | | | |
| $Zr^{4+} + 4e^- \rightleftharpoons Zr(s)$ | $-1.45$ | | $0.67$ |
| $ZrO_2(s) + 4H^+ + 4e^- \rightleftharpoons Zr(s) + 2H_2O$ | $-1.473$ | | $-0.344$ |

# Formation Constants*

| Reacting ions | $\log \beta_1$ | $\log \beta_2$ | $\log \beta_3$ | $\log \beta_4$ | $\log \beta_5$ | $\log \beta_6$ | Temperature (°C) | Ionic strength ($\mu$, M) |
|---|---|---|---|---|---|---|---|---|
| **Acetate, $CH_3CO_2^-$** | | | | | | | | |
| $Ag^+$ | 0.73 | 0.64 | | | | | 25 | 0 |
| $Ca^{2+}$ | 1.24 | | | | | | 25 | 0 |
| $Cd^{2+}$ | 1.93 | 3.15 | | | | | 25 | 0 |
| $Cu^{2+}$ | 2.23 | 3.63 | | | | | 25 | 0 |
| $Fe^{2+}$ | 1.82 | | | | | | 25 | 0.5 |
| $Fe^{3+}$ | 3.38 | 7.1 | 9.7 | | | | 20 | 0.1 |
| $Mg^{2+}$ | 1.25 | | | | | | 25 | 0 |
| $Mn^{2+}$ | 1.40 | | | | | | 25 | 0 |
| $Na^+$ | −0.18 | | | | | | 25 | 0 |
| $Ni^{2+}$ | 1.43 | | | | | | 25 | 0 |
| $Zn^{2+}$ | 1.28 | 2.09 | | | | | 20 | 0.1 |
| **Ammonia, $NH_3$** | | | | | | | | |
| $Ag^+$ | 3.31 | 7.23 | | | | | 25 | 0 |
| $Cd^{2+}$ | 2.51 | 4.47 | 5.77 | 6.56 | | | 30 | 0 |
| $Co^{2+}$ | 1.99 | 3.50 | 4.43 | 5.07 | 5.13 | 4.39 | 30 | 0 |
| $Cu^{2+}$ | 3.99 | 7.33 | 10.06 | 12.03 | | | 30 | 0 |
| $Hg^{2+}$ | 8.8 | 17.5 | 18.50 | 19.28 | | | 22 | 2 |
| $Ni^{2+}$ | 2.67 | 4.79 | 6.40 | 7.47 | 8.10 | 8.01 | 30 | 0 |
| $Zn^{2+}$ | 2.18 | 4.43 | 6.74 | 8.70 | | | 30 | 0 |
| **Cyanide, $CN^-$** | | | | | | | | |
| $Ag^+$ | | 20 | 21 | | | | 20 | 0 |
| $Cd^{2+}$ | 5.18 | 9.60 | 13.92 | 17.11 | | | 25 | ? |
| $Cu^+$ | | 24 | 28.6 | 30.3 | | | 25 | 0 |
| $Ni^{2+}$ | | | | 30 | | | 25 | 0 |
| $Tl^{3+}$ | 13.21 | 26.50 | 35.17 | 42.61 | | | 25 | 4 |
| $Zn^{2+}$ | | 11.07 | 16.05 | 19.62 | | | 25 | 0 |
| **Ethylenediamine (1,2-diaminoethane), $H_2NCH_2CH_2NH_2$** | | | | | | | | |
| $Ag^+$ | 4.70 | 7.70 | 9.7 | | | | 20 | 0.1 |
| $Cd^{2+}$ | 5.69 | 10.36 | 12.80 | | | | 25 | 0.5 |
| $Cu^{2+}$ | 10.66 | 19.99 | | | | | 20 | 0 |
| $Hg^{2+}$ | 14.3 | 23.3 | 23.2 | | | | 25 | 0.1 |
| $Ni^{2+}$ | 7.52 | 13.84 | 18.33 | | | | 20 | 0 |
| $Zn^{2+}$ | 5.77 | 10.83 | 14.11 | | | | 20 | 0 |
| **Hydroxide, $OH^-$** | | | | | | | | |
| $Ag^+$ | 2.0 | 3.99 | | | | | 25 | 0 |
| $Al^{3+}$ | 9.00 | 17.9 | 25.2 | 33.3 | | | 25 | 0 |
| | $\log \beta_{22} = 20.3$ | $\log \beta_{43} = 42.1$ | | | | | | |
| $Ba^{2+}$ | 0.64 | | | | | | 25 | 0 |
| $Bi^{3+}$ | 12.9 | 23.5 | 33.0 | 34.8 | | | 25 | 0 |
| | $\log \beta_{12\,6} = 165.3\ (\mu = 1)$ | | | | | | | |
| $Be^{2+}$ | 8.6 | 14.4 | 18.8 | 18.6 | | | 25 | 0 |
| | $\log \beta_{12} = 10.82\ (\mu = 0.1)$ | | $\log \beta_{33} = 32.54\ (\mu = 0.1)$ | | $\log \beta_{65} = 66.24\ (\mu = 3)$ | $\log \beta_{86} = 85\ (\mu = 0)$ | | |
| $Ca^{2+}$ | 1.30 | | | | | | 25 | 0 |
| $Cd^{2+}$ | 3.9 | 7.7 | 10.3 ($\mu = 3$) | 12.0 ($\mu = 3$) | | | 25 | 0 |
| | $\log \beta_{12} = 4.6$ | $\log \beta_{44} = 23.2$ | | | | | | |
| $Ce^{3+}$ | 4.9 | | | | | | 25 | 3 |
| | $\log \beta_{22} = 12.4$ | $\log \beta_{53} = 35.1$ | | | | | | |
| $Co^{2+}$ | 4.3 | 9.2 | 10.5 | 9.7 | | | 25 | 0 |
| | $\log \beta_{12} = 3$ | $\log \beta_{44} = 25.5$ | | | | | | |
| $Co^{3+}$ | 13.52 | | | | | | 25 | 3 |

*The overall (cumulative) formation constant, $\beta_n$, is the equilibrium constant for the reaction $M + nL \rightleftharpoons ML_n$: $\beta_n = [ML_n]/([M][L]^n)$. $\beta_n$ is related to stepwise formation constants ($K_i$) by $\beta_n = K_1 K_2 \ldots K_n$ (Box 6-2). $\beta_{nm}$ is the cumulative formation constant for the reaction $mM + nL \rightleftharpoons M_m L_n$: $\beta_{nm} = [M_m L_n]/([M]^m [L]^n)$. The subscript $n$ refers to the ligand and $m$ refers to the metal. Data from L. G. Sillén and A. E. Martell, *Stability Constants of Metal-Ion Complexes* (London: The Chemical Society. Special Publications No. 17 and 25, 1964 and 1971); and A. E. Martell, R. M. Smith, and R. J. Motekaitis, *NIST Critical Stability Constants of Metal Complexes Database 46* (Gaithersburg, MD: National Institute of Standards and Technology, 2001).

| Reacting ions | $\log \beta_1$ | $\log \beta_2$ | $\log \beta_3$ | $\log \beta_4$ | $\log \beta_5$ | $\log \beta_6$ | Temperature (°C) | Ionic strength ($\mu$, M) |
|---|---|---|---|---|---|---|---|---|
| $Cr^{2+}$ | 8.5 | | | | | | 25 | 1 |
| $Cr^{3+}$ | 10.34 | 17.3 | | | | | 25 | 0 |
| | | ($\mu = 0.1$) | | | | | | |
| | $\log \beta_{22} = 24.0$ ($\mu = 1$) | | $\log \beta_{43} = 37.0$ ($\mu = 1$) | $\log \beta_{44} = 50.7$ ($\mu = 2$) | | | | |
| $Cu^{2+}$ | 6.5 | 11.8 | 14.5 | 15.6 | | | 25 | 0 |
| | | | ($\mu = 1$) | ($\mu = 1$) | | | | |
| | $\log \beta_{12} = 8.2$ ($\mu = 3$) | | $\log \beta_{22} = 17.4$ | $\log \beta_{43} = 35.2$ | | | | |
| $Fe^{2+}$ | 4.6 | 7.5 | 13 | 10 | | | 25 | 0 |
| $Fe^{3+}$ | 11.81 | 23.4 | | 34.4 | | | 25 | 0 |
| | $\log \beta_{22} = 25.14$ | | $\log \beta_{43} = 49.7$ | | | | | |
| $Ga^{3+}$ | 11.4 | 22.1 | 31.7 | 39.4 | | | 25 | 0 |
| $Gd^{3+}$ | 4.9 | | | | | | 25 | 3 |
| | $\log \beta_{22} = 14.14$ | | | | | | | |
| $Hf^{4+}$ | 13.7 | | | | 52.8 | | 25 | 0 |
| $Hg_2^{2+}$ | 8.7 | | | | | | 25 | 0.5 |
| | $\log \beta_{12} = 11.5$ ($\mu = 3$) | | $\log \beta_{45} = 48.24$ ($\mu = 3$) | | | | | |
| $Hg^{2+}$ | 10.60 | 21.8 | 20.9 | | | | 25 | 0 |
| | $\log \beta_{12} = 10.7$ | | $\log \beta_{33} = 35.6$ | | | | | |
| $In^{3+}$ | 10.1 | 20.2 | 29.5 | 33.8 | | | 25 | 0 |
| | $\log \beta_{22} = 23.2$ ($\mu = 3$) | | $\log \beta_{44} = 47.8$ ($\mu = 0.1$) | $\log \beta_{64} = 43.1$ ($\mu = 0.1$) | | | | |
| $La^{3+}$ | 5.5 | | | | | | 25 | 0 |
| | $\log \beta_{22} = 10.7$ ($\mu = 3$) | | $\log \beta_{95} = 38.4$ | | | | | |
| $Li^+$ | 0.36 | | | | | | 25 | 0 |
| $Mg^{2+}$ | 2.6 | $-0.3$ | | | | | 25 | 0 |
| | | ($\mu = 3$) | | | | | | |
| | $\log \beta_{44} = 18.1$ ($\mu = 3$) | | | | | | | |
| $Mn^{2+}$ | 3.4 | | | 7.7 | | | 25 | 0 |
| | $\log \beta_{12} = 6.8$ | | $\log \beta_{32} = 18.1$ | | | | | |
| $Na^+$ | 0.1 | | | | | | 25 | 0 |
| $Ni^{2+}$ | 4.1 | 9 | 12 | | | | 25 | 0 |
| | $\log \beta_{12} = 4.7$ ($\mu = 1$) | | $\log \beta_{44} = 28.3$ | | | | | |
| $Pb^{2+}$ | 6.4 | 10.9 | 13.9 | | | | 25 | 0 |
| | $\log \beta_{12} = 7.6$ | $\log \beta_{43} = 32.1$ | $\log \beta_{44} = 36.0$ | $\log \beta_{86} = 68.4$ | | | | |
| $Pd^{2+}$ | 13.0 | 25.8 | | | | | 25 | 0 |
| $Rh^{3+}$ | 10.67 | | | | | | 25 | 2.5 |
| $Sc^{3+}$ | 9.7 | 18.3 | 25.9 | 30 | | | 25 | 0 |
| | $\log \beta_{22} = 22.0$ | | $\log \beta_{53} = 53.8$ | | | | | |
| $Sn^{2+}$ | 10.6 | 20.9 | 25.4 | | | | 25 | 0 |
| | $\log \beta_{22} = 23.2$ | | $\log \beta_{43} = 49.1$ | | | | | |
| $Sr^{2+}$ | 0.82 | | | | | | 25 | 0 |
| $Th^{4+}$ | 10.8 | 21.1 | | 41.1 | | | 25 | 0 |
| | | | | ($\mu = 3$) | | | | |
| | $\log \beta_{22} = 23.6$ ($\mu = 3$) | | $\log \beta_{32} = 33.8$ ($\mu = 3$) | $\log \beta_{53} = 53.7$ ($\mu = 3$) | | | | |
| $Ti^{3+}$ | 12.7 | | | | | | 25 | 0 |
| | $\log \beta_{22} = 24.6$ ($\mu = 1$) | | | | | | | |
| $Tl^+$ | 0.79 | $-0.8$ | | | | | 25 | 0 |
| | | ($\mu = 3$) | | | | | | |
| $Tl^{3+}$ | 13.4 | 26.6 | 38.7 | 41.0 | | | 25 | 0 |
| $U^{4+}$ | 13.4 | | | | | | 25 | 0 |
| $VO^{2+}$ | 8.3 | | | | | | 25 | 0 |
| | $\log \beta_{22} = 21.3$ | | | | | | | |
| $Y^{3+}$ | 6.3 | | | | | | 25 | 0 |
| | $\log \beta_{22} = 13.8$ | $\log \beta_{53} = 38.4$ | | | | | | |
| $Zn^{2+}$ | 5.0 | 10.2 | 13.9 | 15.5 | | | 25 | 0 |
| | $\log \beta_{12} = 5.5$ ($\mu = 3$) | | $\log \beta_{44} = 27.9$ ($\mu = 3$) | | | | | |
| $Zr^{4+}$ | 14.3 | | | | 54.0 | | 25 | 0 |
| | $\log \beta_{43} = 55.4$ | $\log \beta_{84} = 106.0$ | | | | | | |
| **Nitrilotriacetate, $N(CH_2CO_2^-)_3$** | | | | | | | | |
| $Ag^+$ | 5.16 | | | | | | 20 | 0.1 |
| $Al^{3+}$ | 9.5 | | | | | | 20 | 0.1 |
| $Ba^{2+}$ | 4.83 | | | | | | 20 | 0.1 |
| $Ca^{2+}$ | 6.46 | | | | | | 20 | 0.1 |
| $Cd^{2+}$ | 10.0 | 14.6 | | | | | 20 | 0.1 |

*(Continued)*

| Reacting ions | $\log \beta_1$ | $\log \beta_2$ | $\log \beta_3$ | $\log \beta_4$ | $\log \beta_5$ | $\log \beta_6$ | Temperature (°C) | Ionic strength ($\mu$, M) |
|---|---|---|---|---|---|---|---|---|
| $Co^{2+}$ | 10.0 | 13.9 | | | | | 20 | 0.1 |
| $Cu^{2+}$ | 11.5 | 14.8 | | | | | 20 | 0.1 |
| $Fe^{3+}$ | 15.91 | 24.61 | | | | | 20 | 0.1 |
| $Ga^{3+}$ | 13.6 | 21.8 | | | | | 20 | 0.1 |
| $In^{3+}$ | 16.9 | | | | | | 20 | 0.1 |
| $Mg^{2+}$ | 5.46 | | | | | | 20 | 0.1 |
| $Mn^{2+}$ | 7.4 | | | | | | 20 | 0.1 |
| $Ni^{2+}$ | 11.54 | | | | | | 20 | 0.1 |
| $Pb^{2+}$ | 11.47 | | | | | | 20 | 0.1 |
| $Tl^+$ | 4.75 | | | | | | 20 | 0.1 |
| $Zn^{2+}$ | 10.44 | | | | | | 20 | 0.1 |
| **Oxalate, $^-O_2CCO_2^-$** | | | | | | | | |
| $Al^{3+}$ | | | 15.60 | | | | 20 | 0.1 |
| $Ba^{2+}$ | 2.31 | | | | | | 18 | 0 |
| $Ca^{2+}$ | 1.66 | 2.69 | | | | | 25 | 1 |
| $Cd^{2+}$ | 3.71 | | | | | | 20 | 0.1 |
| $Co^{2+}$ | 4.69 | 7.15 | | | | | 25 | 0 |
| $Cu^{2+}$ | 6.23 | 10.27 | | | | | 25 | 0 |
| $Fe^{3+}$ | 7.54 | 14.59 | 20.00 | | | | ? | 0.5 |
| $Ni^{2+}$ | 5.16 | 6.5 | | | | | 25 | 0 |
| $Zn^{2+}$ | 4.85 | 7.6 | | | | | 25 | 0 |

1,10-Phenanthroline,

| Reacting ions | $\log \beta_1$ | $\log \beta_2$ | $\log \beta_3$ | $\log \beta_4$ | $\log \beta_5$ | $\log \beta_6$ | Temperature (°C) | Ionic strength ($\mu$, M) |
|---|---|---|---|---|---|---|---|---|
| $Ag^+$ | 5.02 | 12.07 | | | | | 25 | 0.1 |
| $Ca^{2+}$ | 0.7 | | | | | | 20 | 0.1 |
| $Cd^{2+}$ | 5.17 | 10.00 | 14.25 | | | | 25 | 0.1 |
| $Co^{2+}$ | 7.02 | 13.72 | 20.10 | | | | 25 | 0.1 |
| $Cu^{2+}$ | 8.82 | 15.39 | 20.41 | | | | 25 | 0.1 |
| $Fe^{2+}$ | 5.86 | 11.11 | 21.14 | | | | 25 | 0.1 |
| $Fe^{3+}$ | | | 14.10 | | | | 25 | 0.1 |
| $Hg^{2+}$ | | 19.65 | 23.4 | | | | 20 | 0.1 |
| $Mn^{2+}$ | 4.50 | 8.65 | 12.70 | | | | 25 | 0.1 |
| $Ni^{2+}$ | 8.0 | 16.0 | 23.9 | | | | 25 | 0.1 |
| $Zn^{2+}$ | 6.30 | 11.95 | 17.05 | | | | 25 | 0.1 |

# Logarithm of the Formation Constant for the Reaction $M(aq) + L(aq) \rightleftharpoons ML(aq)$*

| M | $F^-$ | $Cl^-$ | $Br^-$ | $I^-$ | $NO_3^-$ | $ClO_4^-$ | $IO_3^-$ | $SCN^-$ | $SO_4^{2-}$ | $CO_3^{2-}$ |
|---|---|---|---|---|---|---|---|---|---|---|
| $Li^+$ | 0.23 | — | — | — | — | — | — | — | 0.64 | — |
| $Na^+$ | −0.2 | −0.5 | — | — | −0.55 | −0.7 | −0.4 | — | 0.72 | 1.27 |
| $K^+$ | −1.2[a] | −0.5 | — | −0.4 | −0.19 | −0.03 | −0.27 | — | 0.85 | — |
| $Rb^+$ | — | −0.4 | — | 0.04 | −0.08 | 0.15 | −0.19 | — | 0.60 | — |
| $Cs^+$ | — | −0.2 | 0.03 | −0.03 | −0.02 | 0.23 | −0.11 | — | 0.3 | — |
| $Ag^+$ | 0.4 | 3.31 | 4.6 | 6.6 | −0.1 | −0.1 | 0.63 | 4.8 | 1.3 | — |
| $(CH_3)_4N^+$ | — | 0.04 | 0.16 | 0.31 | — | 0.27 | — | — | — | — |
| $Mg^{2+}$ | 2.05 | 0.6 | −1.4[d] | — | — | — | 0.72 | −0.9[d] | 2.23 | 2.92 |
| $Ca^{2+}$ | 0.63 | 0.2[b] | — | — | 0.5 | — | 0.89 | — | 2.36 | 3.20 |
| $Sr^{2+}$ | 0.14 | −0.22[a] | — | — | 0.6 | — | 1.00 | — | 2.2 | 2.81 |
| $Ba^{2+}$ | −0.20 | −0.44[a] | — | — | 0.7 | — | 1.10 | — | 2.2 | 2.71 |
| $Zn^{2+}$ | 1.3 | 0.4 | −0.07 | −1.5[d] | 0.4 | — | — | 1.33 | 2.34 | 4.76 |
| $Cd^{2+}$ | 1.2 | 1.98 | 2.15 | 2.28 | 0.5 | — | 0.51[a] | 1.98 | 2.46 | 3.49[b] |
| $Hg_2^{2+}$ | — | — | — | — | 0.08[f] | — | — | — | 1.30[f] | — |
| $Sn^{2+}$ | — | 1.64 | 1.16 | 0.70[e] | 0.44[a] | — | — | 0.83[a] | — | — |
| $Y^{3+}$ | 4.81 | −0.1[a] | −0.15[a] | — | — | — | — | −0.07[f] | 3.47 | 8.2 |
| $La^{3+}$ | 3.60 | −0.1[a] | — | — | 0.1[a] | — | — | 0.12[a] | 3.64 | 5.6[d] |
| $In^{3+}$ | 4.65 | 2.32[c] | 2.01[c] | 1.64[c] | 0.18 | — | — | — | 3.15 | 1.85[a] |

*Unless otherwise indicated, conditions are 25°C and $\mu = 0$. From A. E. Martell, R. M. Smith, and R. J. Motekaitis, NIST Critical Stability Constants of Metal Complexes Database 46 (Gaithersburg, MD: National Institute of Standards and Technology, 2001).

a. $\mu = 1\ M$; b. $\mu = 0.1\ M$; c. $\mu = 0.7\ M$; d. $\mu = 3\ M$; e. $\mu = 4\ M$; f. $\mu = 0.5\ M$.

# APPENDIX K

# Analytical Standards

The table in this appendix recommends primary standards for many elements. An *elemental assay standard* must contain a known amount of the desired element. A *matrix matching standard* must contain extremely low concentrations of undesired impurities, such as the analyte. If you want to prepare 10 ppm Fe in 10% aqueous NaCl, the NaCl must not contain significant Fe impurity, because the impurity would then have a higher concentration than the deliberately added Fe.

Rather than using compounds in the table, many people purchase certified solutions whose concentrations are traceable to standards from the National Institute of Standards and Technology (NIST or other national institutes of standards). By *NIST traceable,* we mean that the solution has been prepared from a standard material certified by NIST or that it has been compared with an NIST standard by a reliable analytical procedure.

Manufacturers frequently indicate elemental purity by some number of 9s. This deceptive nomenclature is based on the measurement of certain impurities. For example, 99.999% (five 9s) pure Al is certified to contain $\leq 0.001\%$ *metallic* impurities, based on the analysis of other metals present. However, C, H, N, and O are not measured. The Al might contain 0.1% $Al_2O_3$ and still be "five 9s pure." For the most accurate work, the dissolved gas content in solid elements may also be a source of error.

Carbonates, oxides, and other compounds may not possess the expected stoichiometry. For example, $TbO_2$ will have a high Tb content if some $Tb_4O_7$ is present. Ignition in an $O_2$ atmosphere may be helpful, but the final stoichiometry is never guaranteed. Carbonates may contain traces of bicarbonate, oxide, and hydroxide. Firing in a $CO_2$ atmosphere may improve the stoichiometry. Sulfates may contain some $HSO_4^-$. Some chemical analysis may be required to ensure that you know what you are really working with.

Most metal standards dissolve in 6 M HCl or $HNO_3$ or a mixture of the two, possibly with heating. Frothing accompanies dissolution of metals or carbonates in acid, so vessels should be loosely covered by a watchglass or Teflon lid to prevent loss of material. Concentrated $HNO_3$ (16 M) may *passivate* some metals, forming an insoluble oxide coat that prevents dissolution. If you have a choice between using a bulk element or a powder as standards, the bulk form is preferred because it has a smaller surface area on which oxides can form and impurities can be adsorbed. After a pure metal to be used as a standard is cut, it should be etched ("pickled") in a dilute solution of the acid in which it will be dissolved to remove surface oxides and contamination from the cutter. The metal is then washed well with water and dried in a vacuum desiccator.

Dilute solutions of metals are best prepared in Teflon or plastic vessels, because glass is an ion exchanger that can replace analyte species. Specially cleaned glass vials are commercially available for trace organic analysis. Volumetric dilutions are rarely more accurate than 0.1%, so gravimetric dilutions are required for greater accuracy. Of course, weights should be corrected for buoyancy with Equation 2-1. Evaporation of standard solutions is a source of error that is prevented if the mass of the reagent bottle is recorded after each use. If the mass changes between uses, the contents are evaporating.

## Calibration standards

| Element | Source[a] | Purity | Comments[b] |
|---|---|---|---|
| Li | SRM 924 ($Li_2CO_3$) | $100.05 \pm 0.02\%$ | E; dry at 200°C for 4 h. |
| | $Li_2CO_3$ | five–six 9s | M; purity calculated from impurities. Stoichiometry unknown. |
| Na | SRM 919 or 2201 (NaCl) | 99.9% | E; dry for 24 h over $Mg(ClO_4)_2$. |
| | $Na_2CO_3$ | three 9s | M; purity based on metallic impurities. |
| K | SRM 918 (KCl) | 99.9% | E; dry for 24 h over $Mg(ClO_4)_2$. |
| | SRM 999 (KCl) | $52.435 \pm 0.004\%$ K | E; ignite at 500°C for 4 h. |
| | $K_2CO_3$ | five–six 9s | M; purity based on metallic impurities. |
| Rb | SRM 984 (RbCl) | $99.90 \pm 0.02\%$ | E; hygroscopic. Dry for 24 h over $Mg(ClO_4)_2$. |
| | $Rb_2CO_3$ | | M |
| Cs | $Cs_2CO_3$ | | M |
| Be | metal | three 9s | E, M; purity based on metallic impurities. |
| Mg | SRM 929 | $100.1 \pm 0.4\%$ | E; magnesium gluconate clinical standard. |
| | | $5.403 \pm 0.022\%$ Mg | Dry for 24 h over $Mg(ClO_4)_2$. |
| | metal | five 9s | E; purity based on metallic impurities. |
| Ca | SRM 915 ($CaCO_3$) | three 9s | E; use without drying. |
| | $CaCO_3$ | five 9s | E, M; dry at 200°C for 4 h in $CO_2$. User must determine stoichiometry. |
| Sr | SRM 987 ($SrCO_3$) | 99.8% | E; ignite to establish stoichiometry. Dry at 110°C for 1 h. |
| | $SrCO_3$ | five 9s | M; up to 1% off stoichiometry. Ignite to establish stoichiometry. Dry at 200°C for 4 h. |
| Ba | $BaCO_3$ | four–five 9s | M; dry at 200°C for 4 h. |

Transition metals: Use pure metals (usually $\geq$four 9s) for elemental and matrix standards. Assays are based on impurities and do not include dissolved gases.

Lanthanides: Use pure metals (usually $\geq$four 9s) for elemental standards and oxides as matrix standards. Oxides may be difficult to dry and stoichiometry is not certain.

a. SRM is the National Institute of Standards and Technology designation for a Standard Reference Material.

b. E means elemental assay standard; M means matrix matching standard.

SOURCES: J. R. Moody, R. R. Greenberg, K. W. Pratt, and T. C. Rains, "Recommended Inorganic Chemicals for Calibration," Anal. Chem. **1988**, 60, 1203A.

| Element | Source[a] | Purity | Comments[b] |
|---------|-----------|--------|-------------|
| B | SRM 951 ($H_3BO_3$) | $100.00 \pm 0.01$ | E; expose to room humidity ($\sim$35%) for 30 min before use. |
| Al | metal | five 9s | E, M; SRM 1257 Al metal available. |
| Ga | metal | five 9s | E, M; SRM 994 Ga metal available. |
| In | metal | five 9s | E, M |
| Tl | metal | five 9s | E, M; SRM 997 Tl metal available. |
| C | | | No recommendation. |
| Si | metal | six 9s | E, M; SRM 990 $SiO_2$ available. |
| Ge | metal | five 9s | E, M |
| Sn | metal | six 9s | E, M; SRM 741 Sn metal available. |
| Pb | metal | five 9s | E, M; several SRMs available. |
| N | $NH_4Cl$ | six 9s | E; can be prepared from HCl + $NH_3$. |
| | $N_2$ | >three 9s | E |
| | $HNO_3$ | six 9s | M; contaminated with $NO_x$. Purity based on impurities. |
| P | SRM 194 ($NH_4H_2PO_4$) | three 9s | E |
| | $P_2O_5$ | five 9s | E, M; difficult to keep dry. |
| | $H_3PO_4$ | four 9s | E; must titrate 2 hydrogens to be certain of stoichiometry. |
| As | metal | five 9s | E, M |
| | SRM 83d ($As_2O_3$) | $99.992\ 6 \pm 0.003\ 0\%$ | Redox standard. As assay not ensured. |
| Sb | metal | four 9s | E, M |
| Bi | metal | five 9s | E, M |
| O | $H_2O$ | eight 9s | E, M; contains dissolved gases. |
| | $O_2$ | >four 9s | E |
| S | element | six 9s | E, M; difficult to dry. Other sources are $H_2SO_4$, $Na_2SO_4$, and $K_2SO_4$. Stoichiometry must be proved (e.g., no $SO_3^{2-}$ present). |
| Se | metal | five 9s | E, M; SRM 726 Se metal available. |
| Te | metal | five 9s | E, M |
| F | NaF | four 9s | E, M; no good directions for drying. |
| Cl | NaCl | four 9s | E, M; dry for 24 h over $Mg(ClO_4)_2$. Several SRMs (NaCl and KCl) available. |
| Br | KBr | four 9s | E, M; need to dry and demonstrate stoichiometry. |
| | $Br_2$ | four 9s | E |
| I | sublimed $I_2$ | six 9s | E |
| | KI | three 9s | E, M |
| | $KIO_3$ | three 9s | Stoichiometry not ensured. |

# Solutions to Exercises

## Chapter 1

**1-A. (a)** $\dfrac{(25.00\ \text{mL})(0.791\ 4\ \text{g/mL})/(32.042\ \text{g/mol})}{0.500\ 0\ \text{L}} = 1.235\ \text{M}$

**(b)** 500.0 mL of solution weighs $(1.454\ \text{g/mL}) \times (500.0\ \text{mL}) = 727.0\ \text{g}$ and contains 25.00 mL ($= 19.78\ \text{g}$) of methanol. The mass of chloroform in 500.0 mL must be $727.0 - 19.78 = 707.2\ \text{g}$. The molality of methanol is

$$\text{Molality} = \frac{\text{mol methanol}}{\text{kg chloroform}}$$
$$= \frac{(19.78\ \text{g})/(32.042\ \text{g/mol})}{0.707\ 2\ \text{kg}}$$
$$= 0.872\ 9\ m$$

**1-B. (a)** $\left(\dfrac{48.0\ \text{g HBr}}{100.0\ \text{g solution}}\right)\left(1.50\ \dfrac{\text{g solution}}{\text{mL solution}}\right)$

$$= \frac{0.720\ \text{g HBr}}{\text{mL solution}} = \frac{720\ \text{g HBr}}{\text{L solution}}$$
$$= 8.90\ \text{M}$$

**(b)** $\dfrac{36.0\ \text{g HBr}}{0.480\ \text{g HBr/g solution}} = 75.0\ \text{g solution}$

**(c)** $233\ \text{mmol} = 0.233\ \text{mol}$

$\dfrac{0.233\ \text{mol}}{8.90\ \text{mol/L}} = 0.026\ 2\ \text{L} = 26.2\ \text{mL}$

**(d)** $\text{M}_{\text{conc}} \cdot V_{\text{conc}} = \text{M}_{\text{dil}} \cdot V_{\text{dil}}$
$(8.90\ \text{M}) \cdot (x\ \text{mL}) = (0.160\ \text{M}) \cdot (250\ \text{mL}) \Rightarrow x = 4.49\ \text{mL}$

**1-C.** Each mol of $Ca(NO_3)_2$ (FM 164.088) contains 2 mol $NO_3^-$ (FM 62.005), so the fraction of mass that is nitrate is

$$\left(\frac{2\ \text{mol NO}_3^-}{\text{mol Ca(NO}_3)_2}\right)\left(\frac{62.005\ \text{g NO}_3^-/\text{mol NO}_3^-}{164.088\ \text{g Ca(NO}_3)_2/\text{mol Ca(NO}_3)_2}\right)$$
$$= 0.755\ 7\ \frac{\text{g NO}_3^-}{\text{g Ca(NO}_3)_2}$$

If the dissolved $Ca(NO_3)_2$ has a concentration of 12.6 ppm, the concentration of dissolved $NO_3^-$ is $(0.755\ 7)(12.6\ \text{ppm}) = 9.52\ \text{ppm}$.

## Chapter 2

**2-A. (a)** At 15°C, water density $= 0.999\ 102\ 6\ \text{g/mL}$.

$$m = \frac{(5.397\ 4\ \text{g})\left(1 - \dfrac{0.001\ 2\ \text{g/mL}}{8.0\ \text{g/mL}}\right)}{\left(1 - \dfrac{0.001\ 2\ \text{g/mL}}{0.999\ 102\ 6\ \text{g/mL}}\right)} = 5.403\ 1\ \text{g}$$

**(b)** At 25°C, water density $= 0.997\ 047\ 9\ \text{g/mL}$ and $m = 5.403\ 1\ \text{g}$.

**2-B.** Use Equation 2-1 with $m' = 0.296\ 1\ \text{g}$, $d_a = 0.001\ 2\ \text{g/mL}$, $d_w = 8.0\ \text{g/mL}$, and $d = 5.24\ \text{g/mL} \Rightarrow m = 0.296\ 1\ \text{g}$.

**2-C.** $\dfrac{c'}{d'} = \dfrac{c}{d}$

Let the primes stand for 16°C:

$$\Rightarrow \frac{c'\ \text{at}\ 16°\text{C}}{0.998\ 946\ 0\ \text{g/mL}} = \frac{0.051\ 38\ \text{M}}{0.997\ 299\ 5\ \text{g/mL}}$$
$$\Rightarrow c'\ \text{at}\ 16° = 0.051\ 46\ \text{M}$$

**2-D.** Column 3 of Table 2-7 tells us that water occupies 1.003 3 mL/g at 22°C. Therefore, $(15.569\ \text{g}) \times (1.003\ 3\ \text{mL/g}) = 15.620\ \text{mL}$.

## Chapter 3

**3-A. (a)** $[12.41\ (\pm 0.09) \div 4.16\ (\pm 0.01)] \times 7.068\ 2\ (\pm 0.000\ 4)$

$$= \frac{12.41\ (\pm 0.725\%) \times 7.068\ 2\ (\pm 0.005\ 7\%)}{4.16\ (\pm 0.240\%)}$$

$= 21.086\ (\pm 0.764\%)$ (because $\sqrt{0.725^2 + 0.005\ 7^2 + 0.240^2} = 0.764$)
$= 21.0_9\ (\pm 0.1_6)$ or $21.1\ (\pm 0.2)$;

relative uncertainty $= \dfrac{0.1_6}{21.0_9} \times 100 = 0.8\%$

**(b)** $[3.26\ (\pm 0.10) \times 8.47\ (\pm 0.05)] - 0.18\ (\pm 0.06)$
$= [3.26\ (\pm 3.07\%) \times 8.47\ (\pm 0.59\%)] - 0.18\ (\pm 0.06)$
$= [27.612\ (\pm 3.13\%)] - 0.18\ (\pm 0.06)$
$= [27.612\ (\pm 0.864)] - 0.18\ (\pm 0.06)$
$= 27.4_3\ (\pm 0.8_7)$ or $27.4\ (\pm 0.9)$; relative uncertainty $= 3._2\%$

**(c)** $6.843\ (\pm 0.008) \times 10^4 \div \underbrace{[2.09\ (\pm 0.04) - 1.63\ (\pm 0.01)]}_{\text{Combine absolute uncertainties}}$

$= 6.843\ (\pm 0.008) \times 10^4 \div \underbrace{[0.46\ (\pm 0.041\ 2)]}_{\text{Combine relative uncertainties}}$

$= 6.843\ (\pm 0.117\%) \times 10^4 \div [0.46\ (\pm 8.96\%)] = 1.49\ (\pm 8.96\%) \times 10^5$
$= 1.4_9\ (\pm 0.1_3) \times 10^5$; relative uncertainty $= 9._0\%$

**(d)** $\%e_y = \tfrac{1}{2}\%e_x = \tfrac{1}{2}\left(\dfrac{0.08}{3.24} \times 100\right) = 1.235\%$

$(3.24 \pm 0.08)^{1/2} = 1.80 \pm 1.235\%$
$\qquad\qquad = 1.80 \pm 0.02_2\ (\pm 1._2\%)$

**(e)** $\%e_y = 4\%e_x = 4\left(\dfrac{0.08}{3.24} \times 100\right) = 9.877\%$

$(3.24 \pm 0.08)^4 = 110.20 \pm 9.877\%$
$\qquad\qquad = 1.1_0\ (\pm 0.1_1) \times 10^2\ (\pm 9._9\%)$

**(f)** $e_y = 0.434\ 29\dfrac{e_x}{x} = 0.434\ 29\left(\dfrac{0.08}{3.24}\right) = 0.010\ 7$

$\log(3.24 \pm 0.08) = 0.510\ 5 \pm 0.010\ 7$
$\qquad\qquad = 0.51 \pm 0.01\ (\pm 2._1\%)$

**(g)** $\dfrac{e_y}{y} = 2.302\ 6\ e_x = 2.302\ 6\ (0.08) = 0.184$

$10^{3.24 \pm 0.08} = 1.74 \times 10^3 \pm 18.4\%$
$\qquad\qquad = 1.7_4\ (\pm 0.3_2) \times 10^3\ (\pm 18\%)$

**3-B. (a)** 2.000 L of 0.169 M NaOH (FM $= 39.997$) requires $0.338\ \text{mol} = 13.52\ \text{g NaOH}$.

$$\frac{13.52\ \text{g NaOH}}{0.534\ \text{g NaOH/g solution}} = 25.32\ \text{g solution}$$
$$\frac{25.32\ \text{g solution}}{1.52\ \text{g solution/mL solution}} = 16.6_6\ \text{mL}$$

**(b)** Molarity $=$

$$\frac{[16.66\ (\pm 0.10)\ \text{mL}]\left[1.52\ (\pm 0.01)\ \dfrac{\text{g solution}}{\text{mL}}\right] \times \left[0.534\ (\pm 0.004)\ \dfrac{\text{g NaOH}}{\text{g solution}}\right]}{\left(39.997\ 1\ \dfrac{\text{g NaOH}}{\text{mol}}\right)(2.000\ \text{L})}$$

Because the relative errors in formula mass and final volume are negligible ($\approx 0$), we can write

$$\text{Relative error in molarity} = \sqrt{\left(\frac{0.10}{16.66}\right)^2 + \left(\frac{0.01}{1.52}\right)^2 + \left(\frac{0.004}{0.534}\right)^2}$$
$$= 1.16\%$$
$$\text{Molarity} = 0.169\ (\pm 0.002)$$

**3-C.** $0.0500 (\pm 2\%) \text{ mol} =$

$$[4.18 (\pm x) \text{ mL}]\left[1.18 (\pm 0.01) \frac{\text{g solution}}{\text{mL}}\right]$$

$$\frac{\times \left[0.370 (\pm 0.005) \frac{\text{g HCl}}{\text{g solution}}\right]}{36.461 \frac{\text{g HCl}}{\text{mol}}}$$

Error analysis:

$$(0.02)^2 = \left(\frac{x}{4.18}\right)^2 + \left(\frac{0.01}{1.18}\right)^2 + \left(\frac{0.005}{0.370}\right)^2$$

$$x = 0.05 \text{ mL}$$

## Chapter 4

**4-A.** Mean $= \frac{1}{5}(116.0 + 97.9 + 114.2 + 106.8 + 108.3)$

$$= 108.6_4$$

Standard deviation $= \sqrt{\dfrac{(116.0 - 108.6_4)^2 + \cdots + (108.3 - 108.6_4)^2}{5 - 1}}$

$$= 7.1_4$$

90% confidence interval $= 108.6_4 \pm \dfrac{(2.132)(7.1_4)}{\sqrt{5}} = 108.6_4 \pm 6.8_1$

$$Q_{\text{calculated}} = \frac{106.8 - 97.9}{116.0 - 97.9} = 0.49 < Q_{\text{table}} (= 0.64)$$

Therefore, 97.9 should be retained.

**4-B.**

| | A | B | C | D |
|---|---|---|---|---|
| 1 | Computing standard deviation | | | |
| 2 | | | | |
| 3 | | Data = x | x − mean | (x-mean)^2 |
| 4 | | 17.4 | −0.44 | 0.1936 |
| 5 | | 18.1 | 0.26 | 0.0676 |
| 6 | | 18.2 | 0.36 | 0.1296 |
| 7 | | 17.9 | 0.06 | 0.0036 |
| 8 | | 17.6 | −0.24 | 0.0576 |
| 9 | sum = | 89.2 | | 0.452 |
| 10 | mean = | 17.84 | | |
| 11 | std dev = | 0.3362 | | |
| 12 | | | | |
| 13 | Formulas: | B9 = B4+B5+B6+B7+B8 | | |
| 14 | | B10 = B9/5 | | |
| 15 | | B11 = SQRT(D9/(5−1)) | | |
| 16 | | C4 = B4−$B$10 | | |
| 17 | | D4 = C4^2 | | |
| 18 | | D9 = D4+D5+D6+D7+D8 | | |
| 19 | | | | |
| 20 | Calculations using built-in functions: | | | |
| 21 | sum = | 89.2 | | |
| 22 | mean = | 17.84 | | |
| 23 | std dev = | 0.3362 | | |
| 24 | | | | |
| 25 | Formulas: | B21 = SUM(B4:B8) | | |
| 26 | | B22 = AVERAGE(B4:B8) | | |
| 27 | | B23 = STDEV(B4:B8) | | |

**4-C.** **(a)** We need to find the fraction of the area of the Gaussian curve between $x = -\infty$ and $x = 40\,860$ h. When $x = 40\,860$, $z = (40\,860 - 62\,700)/10\,400 = -2.100\,0$. The Gaussian curve is symmetric, so the area from $-\infty$ to $-2.100\,0$ is the same as the area from $2.100\,0$ to $+\infty$. Table 4-1 tells us that the area between $z = 0$ and $x = 2.1$ is $0.482\,1$. Because the area from $z = 0$ to $z = \infty$ is $0.500\,0$, the area from $z = 2.100\,0$ to $z = \infty$ is $0.500\,0 - 0.482\,1 = 0.017\,9$. The fraction of brakes expected to be 80% worn in less than 40 860 miles is $0.017\,9$, or 1.79%.
**(b)** At 57 500 miles, $z = (57\,500 - 62\,700)/10\,400 = -0.500\,0$. At 71 020 miles, $z = (71\,020 - 62\,700)/10\,400 = +0.800\,0$. The area under the Gaussian curve from $z = -0.500\,0$ to $z = 0$ is the same as the area from $z = 0$ to $z = +0.500\,0$, which is $0.191\,5$ in Table 4-1. The area from $z = 0$ to $z = +0.800\,0$ is $0.288\,1$. The total area from $z = -0.500\,0$ to $z = +0.800\,0$ is $0.191\,5 + 0.288\,1 = 0.479\,6$. The fraction expected to be 80% worn between 57 500 and 71 020 miles is $0.479\,6$, or 47.96%.

**4-D.** The answers in cells C4 and C9 of the following spreadsheet are **(a)** 0.052 and **(b)** 0.361.

| | A | B | C |
|---|---|---|---|
| 1 | Mean = | Std dev = | |
| 2 | 62700 | 10400 | |
| 3 | | | |
| 4 | Area from $-\infty$ to 45800 = | | 0.052081 |
| 5 | Area from $-\infty$ to 60000 = | | 0.397580 |
| 6 | Area from $-\infty$ to 70000 = | | 0.758637 |
| 7 | | | |
| 8 | Area from 60000 to 70000 | | |
| 9 | | = C6−C5 = | 0.361056 |
| 10 | | | |
| 11 | Formula: | | |
| 12 | C4 = NORMDIST(45800,A2,B2,TRUE) | | |

**4-E.** For 117, 119, 111, 115, 120 $\mu$mol/100 mL, $\bar{x} = 116._4$ and $s = 3._{58}$. The 95% confidence interval for 4 degrees of freedom is

$$\mu = \bar{x} \pm \frac{ts}{\sqrt{n}} = 116._4 \pm \frac{(2.776)(3._{58})}{\sqrt{5}} = 116._4 \pm 4._4$$

$$= 112._0 \text{ to } 120._8 \ \mu\text{mol}/100 \text{ mL}$$

The 95% confidence interval does not include the accepted value of 111 $\mu$mol/100 mL, so the *difference is significant.*

**4-F.** **(a)** pg/g corresponds to $10^{-12}$ g/g, which is parts per trillion.
**(b)** $F_{\text{calculated}} = 4.6^2/3.6^2 = 1.6_3 < F_{\text{table}} = 5.05$ (for 5 degrees of freedom in both numerator and denominator). Standard deviations are not significantly different at 95% confidence level.
**(c)** Because $F_{\text{calculated}} < F_{\text{table}}$, we can use Equations 4-8 and 4-9.

$$s_{\text{pooled}} = \sqrt{\frac{s_1^2(n_1 - 1) + s_2^2(n_2 - 1)}{n_1 + n_2 - 2}}$$

$$= \sqrt{\frac{4.6^2(6 - 1) + 3.6^2(6 - 1)}{6 + 6 - 2}} = 4.1_3$$

$$t_{\text{calculated}} = \frac{|\bar{x}_1 - \bar{x}_2|}{s_{\text{pooled}}} \sqrt{\frac{n_1 n_2}{n_1 + n_2}} = \frac{|51.1 - 34.4|}{4.1_3} \sqrt{\frac{6 \cdot 6}{6 + 6}} = 7.0_0$$

Because $t_{\text{calculated}} (= 7.0_0) > t_{\text{table}} (= 2.228$ for 10 degrees of freedom), the difference is significant at the 95% confidence level.
**(d)** $F_{\text{calculated}} = 3.6^2/1.2^2 = 9.0_0 > F_{\text{table}} = 5.05$. The standard deviations are significantly different at the 95% confidence level. Therefore, we use Equations 4-8a and 4-9a to compare the means:

$$\text{Degrees of freedom} = \left\{ \frac{(s_1^2/n_1 + s_2^2/n_2)^2}{\dfrac{(s_1^2/n_1)^2}{n_1 + 1} + \dfrac{(s_2^2/n_2)^2}{n_2 + 1}} \right\} - 2$$

$$= \left\{ \frac{(3.6^2/6 + 1.2^2/6)^2}{\left(\frac{(3.6^2/6)^2}{6+1} + \frac{(1.2^2/6)^2}{6+1}\right)} \right\} - 2 = 6.5_4 \approx 7$$

$$t_{\text{calculated}} = \frac{|\bar{x}_1 - \bar{x}_2|}{\sqrt{s_1^2/n_1 + s_2^2/n_2}} = \frac{|34.4 - 42.9|}{\sqrt{3.6^2/6 + 1.2^2/6}} = 5.4_9$$

Because $t_{\text{calculated}}$ (= 5.4_9) > $t_{\text{table}}$ (= 2.365 for 7 degrees of freedom), the difference is significant at the 95% confidence level.

**4-G. (a)**

| $x_i$ | $y_i$ | $x_i y_i$ | $x_i^2$ | $d_i$ | $d_i^2$ |
|---|---|---|---|---|---|
| 0.00 | 0.466 | 0 | 0 | −0.004 6 | $2.12 \times 10^{-5}$ |
| 9.36 | 0.676 | 6.327 | 87.61 | +0.001 6 | $2.58 \times 10^{-6}$ |
| 18.72 | 0.883 | 16.530 | 350.44 | +0.004 8 | $2.31 \times 10^{-5}$ |
| 28.08 | 1.086 | 30.495 | 788.49 | +0.004 0 | $1.61 \times 10^{-5}$ |
| 37.44 | 1.280 | 47.923 | 1 401.75 | −0.005 8 | $3.34 \times 10^{-5}$ |
| Sum: 93.60 | 4.391 | 101.275 | 2 628.29 | | $9.64 \times 10^{-5}$ |

$$D = \begin{vmatrix} \Sigma(x_i^2) & \Sigma x_i \\ \Sigma x_i & n \end{vmatrix}$$
$$= (2\,628.29)(5) - (93.60)(93.60) = 4\,380.5$$

$$m = \begin{vmatrix} \Sigma x_i y_i & \Sigma x_i \\ \Sigma y_i & n \end{vmatrix} \div D$$
$$= \frac{(101.275)(5) - (93.60)(4.391)}{D}$$
$$= 95.377 \div 4\,380.5 = 0.021\,773$$

$$b = \begin{vmatrix} \Sigma(x_i^2) & \Sigma x_i y_i \\ \Sigma x_i & \Sigma y_i \end{vmatrix} \div D$$
$$= \frac{(2\,628.29)(4.391) - (101.275)(93.60)}{D}$$
$$= 2\,061.48 \div 4\,380.5 = 0.470\,60$$

$$s_y^2 = \frac{\Sigma(d_i^2)}{n - 2} = \frac{9.64 \times 10^{-5}}{3}$$
$$= 3.21 \times 10^{-5}; \; s_y = 0.005\,67$$

$$s_m = \sqrt{\frac{s_y^2 n}{D}} = \sqrt{\frac{(3.21 \times 10^{-5})5}{4\,380.5}} = 0.000\,191$$

$$s_b = \sqrt{\frac{s_y^2 \Sigma(x_i^2)}{D}} = \sqrt{\frac{(3.21 \times 10^{-5})(2\,628.29)}{4\,380.5}}$$
$$= 0.004\,39$$

Equation of the best line:

$$y = [0.021\,8\,(\pm 0.000\,2)]x + [0.471\,(\pm 0.004)]$$

**(c)** $x = \dfrac{y - b}{m} = \dfrac{0.973 - 0.471}{0.021\,8} = 23.0\;\mu\text{g}$

If you keep more digits for $m$ and $b$, $x = 23.07\;\mu\text{g}$.

Uncertainty in $x$ ($s_x$)

$$= \frac{s_y}{|m|} \sqrt{\frac{1}{k} + \frac{1}{n} + \frac{(y - \bar{y})^2}{m^2 \Sigma(x_i - \bar{x})^2}}$$

$$= \frac{0.005\,67}{|0.021\,77|} \sqrt{\frac{1}{1} + \frac{1}{5} + \frac{(0.973 - 0.878\,2)^2}{(0.021\,77)^2\,(876.1)}} = 0.29\;\mu\text{g}$$

Final answer is $23.1 \pm 0.3\;\mu\text{g}$.

## Chapter 5

**5-A.** **(a)** Standard deviation of 9 samples = $s = 0.000\,6_{44}$

$$\text{Mean blank} = y_{\text{blank}} = 0.001\,1_{89}$$
$$y_{\text{dl}} = y_{\text{blank}} + 3s = 0.001\,1_8 + (3)(0.000\,6_{44}) = 0.003\,1_{12}$$

**(b)** Minimum detectable concentration $= \dfrac{3s}{m} = \dfrac{(3)(0.000\,6_{44})}{2.24 \times 10^4\,\text{M}^{-1}}$
$$= 8._6 \times 10^{-8}\,\text{M}$$

| | A | B | C | D | E | F | G | H | I |
|---|---|---|---|---|---|---|---|---|---|
| **1** | Least-Squares Spreadsheet | | | | | | | | |
| **2** | | | | | | | | | |
| **3** | | x | y | | | | | | |
| **4** | | 0 | 0.466 | | | | | | |
| **5** | | 9.36 | 0.676 | | | | | | |
| **6** | | 18.72 | 0.883 | | | | | | |
| **7** | | 28.08 | 1.086 | | | | | | |
| **8** | | 37.44 | 1.280 | | | | | | |
| **9** | | | | | | | | | |
| **10** | | LINEST output: | | | | | | | |
| **11** | m | 0.02177 | 0.47060 | b | | | | | |
| **12** | $s_m$ | 0.00019 | 0.00439 | $s_b$ | | | | | |
| **13** | $R^2$ | 0.99977 | 0.00567 | $s_y$ | | | | | |
| **14** | | | | | | | | | |
| **15** | n = | 5 | B15 = COUNT(B4:B8) | | | | | | |
| **16** | Mean y = | 0.878 | B16 = AVERAGE(C4:C8) | | | | | | |
| **17** | $\Sigma(x_i\text{-mean }x)^2$ = | 876.096 | B17 = DEVSQ(B4:B8) | | | | | | |
| **18** | | | | | | | | | |
| **19** | Measured y = | 0.973 | Input | | | | | | |
| **20** | Number of replicate measurements of y (k) = | 1 | Input | | | | | | |
| **21** | Derived x | 23.0739 | B21 = (B19-C11)/B11 | | | | | | |
| **22** | $s_x$ = | 0.2878 | B22 = (C13/B11)*SQRT((1/B20)+(1/B15)+((B19-B16)^2)/(B11^2*B17)) | | | | | | |

Graph: Absorbance at 595 nm vs Protein (µg), with line $y = 0.0218x + 0.4706$.

Spreadsheet for Exercise 4-G.

**(c)** Lower limit of quantitation $= \dfrac{10s}{m} = \dfrac{(10)(0.000\ 6_{44})}{2.24 \times 10^4\ M^{-1}} = 2._9 \times 10^{-7}$ M

**5-B.** **(a)** $[Ni^{2+}]_f = [Ni^{2+}]_i \dfrac{V_i}{V_f} = [Ni^{2+}]_i \left(\dfrac{25.0}{25.5}\right) = 0.980_4[Ni^{2+}]_i$

**(b)** $[S]_f = (0.028\ 7\ M)\left(\dfrac{0.500}{25.5}\right) = 0.000\ 562_7$ M

**(c)** $\dfrac{[Ni^{2+}]_i}{0.000\ 562\ 7 + 0.980\ 4[Ni^{2+}]_i} = \dfrac{2.36\ \mu A}{3.79\ \mu A}$

$\Rightarrow [Ni^{2+}]_i = 9.00 \times 10^{-4}$ M

**5-C.** Use the standard mixture to find the response factor. We know that, when [X] = [S], the ratio of signals $A_X/A_S$ is 1.31.

$$\dfrac{A_X}{[X]} = F\left(\dfrac{A_S}{[S]}\right) \Rightarrow F = \dfrac{A_X/A_S}{[X]/[S]} = \dfrac{1.31}{1} = 1.31$$

In the mixture of unknown plus standard, the concentration of S is

$$[S] = (4.13\ \mu g/mL)\underbrace{\left(\dfrac{2.00}{10.0}\right)}_{} = 0.826\ \mu g/mL$$

$\quad\quad\quad\quad\quad\ $ Initial $\quad$ Dilution
$\quad\quad\quad\quad$ concentration $\ $ factor

For the unknown mixture: $F = \dfrac{A_X/A_S}{[X]/[S]}$

$$1.31 = \dfrac{0.808}{[X]/[0.826\ \mu g/mL]} \Rightarrow [X] = 0.509\ \mu g/mL$$

Because X was diluted from 5.00 to 10.0 mL in the mixture with S, the original concentration of X was $(10.0/5.00)(0.509\ \mu g/mL) = 1.02\ \mu g/mL$.

**5-D.** There are 9 points, so there are $9 - 2 = 7$ degrees of freedom. For 90% confidence, $t = 1.895$, so the 90% confidence interval is $\pm(1.895)(0.098\ mM) = \pm0.19\ mM$. For 99% confidence, $t = 3.500$, and the 99% confidence interval $= \pm(3.500)(0.098\ mM) = \pm0.34\ mM$.

**5-E.** For the data in this problem, mean $= 0.84_{11}$ and standard deviation $= 0.18_{88}$. Stability criteria are

• There should be no observations outside the action lines—One observation (day 101) lies above the upper action line.

• There are not 2 out of 3 consecutive measurements between warning and action lines—OK.

• There are not 7 consecutive measurements all above or all below the center line—OK.

• There are not 6 consecutive measurements all steadily increasing or all steadily decreasing, wherever they are located—OK.

• There are not 14 consecutive points alternating up and down, regardless of where they are located—OK.

• There is no obvious nonrandom pattern—OK.

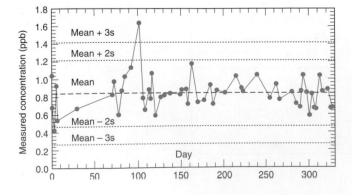

## Chapter 6

**6-A.** **(a)** $Ag^+ + Cl^- \rightleftharpoons AgCl(aq)$

$\underline{AgCl(s) \rightleftharpoons Ag^+ + Cl^-}$

$AgCl(s) \rightleftharpoons AgCl(aq)$

$K_1 = 2.0 \times 10^3$

$\underline{K_2 = 1.8 \times 10^{-10}}$

$K_3 = K_1K_2 = 3.6 \times 10^{-7}$

**(b)** The answer to (a) tells us $[AgCl(aq)] = 3.6 \times 10^{-7}$ M.

**(c)** $AgCl_2^- \rightleftharpoons AgCl(aq) + Cl^-$

$Ag^+ + Cl^- \rightleftharpoons AgCl(s)$

$\underline{AgCl(aq) \rightleftharpoons Ag^+ + Cl^-}$

$AgCl_2^- \rightleftharpoons AgCl(s) + Cl^-$

$K_1 = 1/(9.3 \times 10^1)$

$K_2 = 1/(1.8 \times 10^{-10})$

$\underline{K_3 = 1/(2.0 \times 10^3)}$

$K_4 = K_1K_2K_3 = 3.0 \times 10^4$

**6-B.** **(a)**

$$\dfrac{(x)(x)(1.00 + 8x)^8}{(0.010\ 0 - x)(0.010\ 0 - 2x)^2} = 1 \times 10^{11}$$

**(b)** $[Br^-]$ and $[Cr_2O_7^{2-}]$ will both be 0.005 00 M because $Cr^{3+}$ is the *limiting reagent*. The reaction requires 2 moles of $Cr^{3+}$ per mole of $BrO_3^-$. The $Cr^{3+}$ will be used up first, making 1 mole of $Br^-$ and 1 mole of $Cr_2O_7^{2-}$ per 2 moles of $Cr^{3+}$ consumed. To solve the preceding equation, we set $x = 0.005\ 00$ M in all terms except $[Cr^{3+}]$. The concentration of $Cr^{3+}$ will be a small, unknown quantity.

$$\dfrac{(0.005\ 00)(0.005\ 00)[1.00 + 8(0.005\ 00)]^8}{(0.010\ 0 - 0.005\ 00)[Cr^{3+}]^2} = 1 \times 10^{11}$$

$[Cr^{3+}] = 2._6 \times 10^{-7}$ M

$[BrO_3^-] = 0.010\ 0 - 0.005\ 00 = 0.005\ 00$ M

**6-C.** $K_{sp}$ for $La(IO_3)_3$ is small enough $(1.0 \times 10^{-11})$, so we presume that the concentration of iodate will not be altered by the small amount of $La(IO_3)_3$ that dissolves.

$$[La^{3+}] = \dfrac{K_{sp}}{[IO_3^-]^3} = \dfrac{1.0 \times 10^{-11}}{(0.050)^3} = 8.0 \times 10^{-8}\ M$$

The answer agrees with the assumption that iodate from $La(IO_3)_3 \ll$ 0.050 M.

**6-D.** We expect $Ca(IO_3)_2$ to be more soluble because its $K_{sp}$ is larger and the two salts have the same stoichiometry. If the stoichiometry were not the same, we could not directly compare values of $K_{sp}$. Our prediction could be wrong if, for example, the barium salt formed a great deal of the ion pairs $Ba(IO_3)^+$ or $Ba(IO_3)_2(aq)$ and the calcium salt did not form ion pairs.

**6-E.** $[Fe^{3+}][OH^-]^3 = (10^{-10})[OH^-]^3 = 1.6 \times 10^{-39}$

$\Rightarrow [OH^-] = 2.5 \times 10^{-10}$ M

$[Fe^{2+}][OH^-]^2 = (10^{-10})[OH^-]^2 = 7.9 \times 10^{-16}$

$\Rightarrow [OH^-] = 2.8 \times 10^{-3}$ M

**6-F.** We want to reduce $[Ce^{3+}]$ to 1.0% of 0.010 M = 0.000 10 M. The concentration of oxalate in equilibrium with 0.000 10 M $Ce^{3+}$ is computed as follows:

$$[Ce^{3+}]^2[C_2O_4^{2-}]^3 = K_{sp} = 3 \times 10^{-29}$$

$$(0.000\ 10)^2[C_2O_4^{2-}]^3 = 3 \times 10^{-29}$$

$$[C_2O_4^{2-}] = \left(\dfrac{3 \times 10^{-29}}{(0.000\ 10)^2}\right)^{1/3} = 1.4 \times 10^{-7}\ M$$

To see if $1.4 \times 10^{-7}$ M $C_2O_4^{2-}$ will precipitate 0.010 M $Ca^{2+}$, evaluate $Q$ for $CaC_2O_4$:

$$Q = [Ca^{2+}][C_2O_4^{2-}] = (0.010)(1.4 \times 10^{-7}) = 1.4 \times 10^{-9}$$

Because $Q < K_{sp}$ for $CaC_2O_4 (= 1.3 \times 10^{-8})$, $Ca^{2+}$ will not precipitate.

**6-G.** Assuming that all of the Ni is in the form $Ni(en)_3^{2+}$, $[Ni(en)_3^{2+}] = 1.00 \times 10^{-5}$ M. This uses up just $3 \times 10^{-5}$ mol of en, which leaves the en concentration at 0.100 M. Adding the three equations gives

$$Ni^{2+} + 3en \rightleftharpoons Ni(en)_3^{2+}$$

$$K = K_1K_2K_3 = 2.1_4 \times 10^{18}$$

$$[Ni^{2+}] = \frac{[Ni(en)_3^{2+}]}{K[en]^3}$$

$$= \frac{(1.00 \times 10^{-5})}{(2.1_4 \times 10^{18})(0.100)^3} = 4.7 \times 10^{-21} \text{ M}$$

Now we verify that $[Ni(en)^{2+}]$ and $[Ni(en)_2^{2+}] \ll 10^{-5}$ M:

$$[Ni(en)^{2+}] = K_1[Ni^{2+}][en] = 1.6 \times 10^{-14} \text{ M}$$

$$[Ni(en)_2^{2+}] = K_2[Ni(en)^{2+}][en] = 3.3 \times 10^{-9} \text{ M}$$

**6-H.** **(a)** Neutral—Neither $Na^+$ nor $Br^-$ has any acidic or basic properties.
**(b)** Basic—$CH_3CO_2^-$ is the conjugate base of acetic acid, and $Na^+$ is neither acidic nor basic.
**(c)** Acidic—$NH_4^+$ is the conjugate acid of $NH_3$, and $Cl^-$ is neither acidic nor basic.
**(d)** Basic—$PO_4^{3-}$ is a base, and $K^+$ is neither acidic nor basic.
**(e)** Neutral—Neither ion is acidic or basic.
**(f)** Basic—The quaternary ammonium ion is neither acidic nor basic, and the $C_6H_5CO_2^-$ anion is the conjugate base of benzoic acid.
**(g)** Acidic—$Fe^{3+}$ is acidic, and nitrate is neither acidic nor basic.

**6-I.** $K_{b1} = K_w/K_{a2} = 4.4 \times 10^{-9}$
$K_{b2} = K_w/K_{a1} = 1.6 \times 10^{-10}$

**6-J.** $K \equiv K_{b2} = K_w/K_{a2} = 1.2 \times 10^{-8}$

**6-K.** **(a)** $[H^+][OH^-] = x^2 = K_w \Rightarrow x = \sqrt{K_w} \Rightarrow pH = -\log \sqrt{K_w} = 7.469$ at 0°C, 7.082 at 20°C, and 6.770 at 40°C.
**(b)** Because $[D^+] = [OD^-]$ in pure $D_2O$, $K = 1.35 \times 10^{-15} = [D^+][OD^-] = [D^+]^2 \Rightarrow [D^+] = 3.67 \times 10^{-8}$ M $\Rightarrow pD = 7.435$.

## Chapter 7

**7-A.** **(a)** Formula mass of ascorbic acid = 176.124
0.197 0 g of ascorbic acid = 1.118 5 mmol

$$\text{Molarity of } I_3^- = 1.118\ 5 \text{ mmol}/29.41 \text{ mL}$$
$$= 0.038\ 03 \text{ M}$$

**(b)** 31.63 mL of $I_3^-$ = 1.203 mmol of $I_3^-$
= 1.203 mmol of ascorbic acid
= 0.211 9 g = 49.94% of the tablet

**7-B.** **(a)** $\dfrac{0.824 \text{ g acid}}{204.221 \text{ g/mol}} = 4.03$ mmol. This many mmol of NaOH is contained in 0.038 314 kg of NaOH solution

$$\Rightarrow \text{concentration} = \frac{4.03 \times 10^{-3} \text{ mol NaOH}}{0.038\ 314 \text{ kg solution}}$$
$$= 0.105_3 \text{ mol/kg solution}$$

**(b)** mol NaOH = $(0.057\ 911 \text{ kg})(0.105\ 3 \text{ mol/kg})$ = 6.10 mmol. Because 2 mol NaOH react with 1 mol $H_2SO_4$,

$$[H_2SO_4] = \frac{3.05 \text{ mmol}}{10.00 \text{ mL}} = 0.305 \text{ M}$$

**7-C.** 34.02 mL of 0.087 71 M NaOH = 2.983 9 mmol of $OH^-$. Let $x$ be the mass of malonic acid and $y$ be the mass of anilinium chloride. Then $x + y = 0.237\ 6$ g and

(Moles of anilinium chloride) + 2(moles of malonic acid) = 0.002 983 9

$$\frac{y}{129.59} + 2\left(\frac{x}{104.06}\right) = 0.002\ 983\ 9$$

Substituting $y = 0.237\ 6 - x$ gives $x = 0.099\ 97$ g = 42.07% malonic acid. Anilinium chloride = 57.93%.

**7-D.** The absorbance must be corrected by multiplying each observed absorbance by (total volume/initial volume). For example, at 36.0 μL, $A$ (corrected) = $(0.399)[(2\ 025 + 36)/2\ 025] = 0.406$. A graph of

corrected absorbance versus volume of $Pb^{2+}$ (μL) is similar to Figure 7-5, with the end point at 46.7 μL. The moles of $Pb^{2+}$ in this volume are $(46.7 \times 10^{-6} \text{ L})(7.515 \times 10^{-4} \text{ M}) = 3.510 \times 10^{-8}$ mol. The concentration of semi-xylenol orange is $(3.510 \times 10^{-8} \text{ mol})/(2.025 \times 10^{-3} \text{ L}) = 1.73 \times 10^{-5}$ M.

**7-E.** The reaction is $SCN^- + Cu^+ \rightarrow CuSCN(s)$. The equivalence point occurs when moles of $Cu^+$ = moles of $SCN^- \Rightarrow V_e = 100.0$ mL. Before the equivalence point, there is excess $SCN^-$ remaining in the solution. We calculate the molarity of $SCN^-$ and then find $[Cu^+]$ from the relation $[Cu^+] = K_{sp}/[SCN^-]$. For example, when 0.10 mL of $Cu^+$ has been added,

$$[SCN^-] = \left(\frac{100.0 - 0.10}{100.0}\right)(0.080\ 0)\left(\frac{50.0}{50.1}\right)$$
$$= 7.98 \times 10^{-2} \text{ M}$$

$$[Cu^+] = 4.8 \times 10^{-15}/7.98 \times 10^{-2}$$
$$= 6.0 \times 10^{-14}$$

$$pCu^+ = 13.22$$

At the equivalence point, $[Cu^+][SCN^-] = x^2 = K_{sp} \Rightarrow x = [Cu^+] = 6.9 \times 10^{-8} \Rightarrow pCu^+ = 7.16$.

Past the equivalence point, there is excess $Cu^+$. For example, when $V = 101.0$ mL,

$$[Cu^+] = (0.040\ 0)\left(\frac{101.0 - 100.0}{151.0}\right) = 2.6 \times 10^{-4} \text{ M}$$

$$pCu^+ = 3.58$$

| mL | pCu | mL | pCu | mL | pCu |
|---|---|---|---|---|---|
| 0.10 | 13.22 | 75.0 | 12.22 | 100.0 | 7.16 |
| 10.0 | 13.10 | 95.0 | 11.46 | 100.1 | 4.57 |
| 25.0 | 12.92 | 99.0 | 10.75 | 101.0 | 3.58 |
| 50.0 | 12.62 | 99.9 | 9.75 | 110.0 | 2.60 |

**7-F.** $V_e = 23.66$ mL for AgBr. At 2.00, 10.00, 22.00, and 23.00 mL, AgBr is partially precipitated and excess $Br^-$ remains.

At 2.00 mL:

$$[Ag^+] = \frac{K_{sp}(\text{for AgBr})}{[Br^-]}$$

$$= \frac{5.0 \times 10^{-13}}{\underbrace{\left(\frac{23.66 - 2.00}{23.66}\right)}_{\substack{\text{Fraction} \\ \text{remaining}}} \underbrace{(0.050\ 00)}_{\substack{\text{Original} \\ \text{molarity} \\ \text{of Br}^-}} \underbrace{\left(\frac{40.00}{42.00}\right)}_{\substack{\text{Dilution} \\ \text{factor}}}}$$

$$= 1.15 \times 10^{-11} \text{ M} \Rightarrow pAg^+ = 10.94$$

By similar reasoning, we find

at 10.00 mL: $pAg^+ = 10.66$
at 22.00 mL: $pAg^+ = 9.66$
at 23.00 mL: $pAg^+ = 9.25$

At 24.00, 30.00, and 40.00 mL, AgCl is precipitating and excess $Cl^-$ remains in solution.

At 24.00 mL:

$$[Ag^+] = \frac{K_{sp}(\text{for AgCl})}{[Cl^-]}$$

$$= \frac{1.8 \times 10^{-10}}{\left(\frac{47.32 - 24.00}{23.66}\right)(0.050\ 00)\left(\frac{40.00}{64.00}\right)}$$

$$= 5.8 \times 10^{-9} \text{ M} \Rightarrow pAg^+ = 8.23$$

By similar reasoning, we find

at 30.00 mL: $pAg^+ = 8.07$
at 40.00 mL: $pAg^+ = 7.63$

At the second equivalence point (47.32 mL), $[Ag^+] = [Cl^-]$, and we can write

$$[Ag^+][Cl^-] = x^2 = K_{sp}(\text{for AgCl})$$
$$\Rightarrow [Ag^+] = 1.34 \times 10^{-5}\,M$$
$$pAg^+ = 4.87$$

At 50.00 mL, there is an excess of $(50.00 - 47.32) = 2.68$ mL of $Ag^+$.

$$[Ag^+] = \left(\frac{2.68}{90.00}\right)(0.084\ 54\ M) = 2.5 \times 10^{-3}\,M$$
$$pAg^+ = 2.60$$

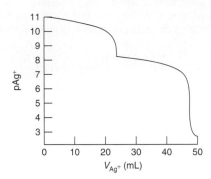

**7-G.** **(a)** 12.6 mL of $Ag^+$ are required to precipitate $I^-$.
$(27.7 - 12.6) = 15.1$ mL are required to precipitate $SCN^-$.

$$[SCN^-] = \frac{\text{moles of } Ag^+ \text{ needed to react with } SCN^-}{\text{original volume of } SCN^-}$$

$$= \frac{[27.7\ (\pm 0.3) - 12.6\ (\pm 0.4)][0.068\ 3\ (\pm 0.000\ 1)]}{50.00\ (\pm 0.05)}$$

$$= \frac{[15.1\ (\pm 0.5)][0.068\ 3\ (\pm 0.000\ 1)]}{50.00\ (\pm 0.05)}$$

$$= \frac{[15.1\ (\pm 3.31\%)][0.068\ 3\ (\pm 0.146\%)]}{50.00\ (\pm 0.100\%)}$$

$$= 0.020\ 6\ (\pm 0.000\ 7)\ M$$

**(b)** $[SCN^-](\pm 4.0\%) = \dfrac{[27.7\ (\pm 0.3) - 12.6\ (\pm ?)][0.068\ 3\ (\pm 0.000\ 1)]}{50.00\ (\pm 0.05)}$

Let the error in 15.1 mL be $y\%$:

$$(4.0\%)^2 = (y\%)^2 + (0.146\%)^2 + (0.100\%)^2$$
$$\Rightarrow y = 4.00\% = 0.603\ mL$$
$$27.7\ (\pm 0.3) - 12.6\ (\pm ?) = 15.1\ (\pm 0.603)$$
$$\Rightarrow 0.3^2 + ?^2 = 0.603^2 \Rightarrow ? = 0.5\ mL$$

## Chapter 8

**8-A.** **(a)** $\mu = \frac{1}{2}([K^+] \cdot 1^2 + [NO_3^-] \cdot (-1)^2) = 0.2$ mM

**(b)** $\mu = \frac{1}{2}([Cs^+] \cdot 1^2 + [CrO_4^{2-}] \cdot (-2)^2)$
$= \frac{1}{2}([0.4] \cdot 1 + [0.2] \cdot 4) = 0.6$ mM

**(c)** $\mu = \frac{1}{2}([Mg^{2+}] \cdot 2^2 + [Cl^-] \cdot (-1)^2 + [Al^{3+}] \cdot 3^2)$
$= \frac{1}{2}([0.2] \cdot 4 + [\ \ 0.4\ \ +\ \ 0.9\ \ ] \cdot 1 + [0.3] \cdot 9)$
              ↑       ↑
       From MgCl$_2$ From AlCl$_3$

$= 2.4$ mM

**8-B.** For 0.005 0 M $(CH_3CH_2CH_2)_4N^+Br^-$ plus 0.005 0 M $(CH_3)_4N^+Cl^-$, $\mu = 0.010$ M. The size of the ion $(CH_3CH_2CH_2)_4N^+$ is 800 pm. At $\mu = 0.01$ M, $\gamma = 0.912$ for an ion of charge $\pm 1$ with $\alpha = 800$ pm. $\mathcal{A} = (0.005\ 0)(0.912) = 0.004\ 6$.

**8-C.** $\mu = 0.060$ M from KSCN, assuming negligible solubility of AgSCN.

$$K_{sp} = [Ag^+]\gamma_{Ag^+}[SCN^-]\gamma_{SCN^-} = 1.1 \times 10^{-12}$$

The activity coefficients at $\mu = 0.060$ M are $\gamma_{Ag^+} = 0.79$ and $\gamma_{SCN^-} = 0.80$.

$$K_{sp} = [Ag^+](0.79)[0.060](0.80) = 1.1 \times 10^{-12}$$
$$\Rightarrow [Ag^+] = 2.9 \times 10^{-11}\,M.$$

**8-D.** At an ionic strength of 0.050 M, $\gamma_{H^+} = 0.86$ and $\gamma_{OH^-} = 0.81$.
$[H^+]\gamma_{H^+}[OH^-]\gamma_{OH^-} = (x)(0.86)(x)(0.81) = 1.0 \times 10^{-14} \Rightarrow x = [H^+] = 1.2 \times 10^{-7}\,M$. $pH = -\log[(1.2 \times 10^{-7})(0.86)] = 6.99$.

**8-E.** **(a)** Moles of $I^- = 2(\text{moles of } Hg_2^{2+})$

$$(V_e)(0.100\ M) = 2(40.0\ mL)(0.040\ 0\ M)$$
$$\Rightarrow V_e = 32.0\ mL$$

**(b)** Virtually all the $Hg_2^{2+}$ has precipitated, along with 3.20 mmol of $I^-$. The ions remaining in solution are

$$[NO_3^-] = \frac{3.20\ mmol}{100.0\ mL} = 0.032\ 0\ M$$

$$[I^-] = \frac{2.80\ mmol}{100.0\ mL} = 0.028\ 0\ M$$

$$[K^+] = \frac{6.00\ mmol}{100.0\ mL} = 0.060\ 0\ M$$

$$\mu = \tfrac{1}{2}\Sigma c_i z_i^2 = 0.060\ 0\ M$$

**(c)** $\mathcal{A}_{Hg_2^{2+}} = K_{sp}/\mathcal{A}_{I^-}^2$
$= 4.6 \times 10^{-29}/(0.028\ 0)^2(0.795)^2 = 9.3 \times 10^{-26}$
$\Rightarrow pHg_2^{2+} = -\log \mathcal{A}_{Hg_2^{2+}} = 25.03$

**8-F.** **(a)** $[Cl^-] = 2[Ca^{2+}]$

**(b)** $\underbrace{[Cl^-] + [CaCl^+]}_{\text{Species containing Cl}^-} = 2\underbrace{\{[Ca^{2+}] + [CaCl^+] + [CaOH^+]\}}_{\text{Species containing Ca}^{2+}}$

**(c)** $[Cl^-] + [OH^-] = 2[Ca^{2+}] + [CaCl^+] + [CaOH^+] + [H^+]$

**8-G.** Charge balance:
$$[F^-] + [HF_2^-] + [OH^-] = 2[Ca^{2+}] + [CaOH^+] + [CaF^+] + [H^+]$$
Mass balance: CaF$_2$ gives 2 mol F for each mol Ca.

$\underbrace{[F^-] + [CaF^+] + 2[CaF_2(aq)] + [HF] + 2[HF_2^-]}_{\text{Species containing F}^-}$

$= 2\underbrace{\{[Ca^{2+}] + [CaOH^+] + [CaF^+] + [CaF_2(aq)]\}}_{\text{Species containing Ca}^{2+}}$

**8-H.** Charge balance:
$$2[Ca^{2+}] + [CaOH^+] + [H^+] = [CaPO_4^-] + 3[PO_4^{3-}] + 2[HPO_4^{2-}] + [H_2PO_4^-] + [OH^-]$$

Mass balance: Equate 2(calcium species) = 3(phosphate species).
$2\underbrace{\{[Ca^{2+}] + [CaOH^+] + [CaPO_4^-]\}}_{\text{Species containing calcium}}$

$= 3\underbrace{\{[CaPO_4^-] + [PO_4^{3-}] + [HPO_4^{2-}] + [H_2PO_4^-] + [H_3PO_4]\}}_{\text{Species containing phosphate}}$

**8-I.** Pertinent reactions:

$$Mn(OH)_2(s) \overset{K_{sp}}{\rightleftharpoons} Mn^{2+} + 2OH^- \qquad K_{sp} = 1.6 \times 10^{-13}$$

$$Mn^{2+} + 2OH^- \overset{K_1}{\rightleftharpoons} MnOH^+ \qquad K_1 = 2.5 \times 10^3$$

$$H_2O \overset{K_w}{\rightleftharpoons} H^+ + OH^- \qquad K_w = 1.0 \times 10^{-14}$$

Charge balance: $2[Mn^{2+}] + [MnOH^+] + [H^+] = [OH^-]$

Mass balance: $\underbrace{[OH^-] + [MnOH^+]}_{\text{Species containing OH}^-} = 2\underbrace{\{[Mn^{2+}] + [MnOH^+]\}}_{\text{Species containing Mn}^{2+}} + [H^+]$

(Mass balance gives the same result as charge balance.)

Equilibrium constant expressions:

$$K_{sp} = [Mn^{2+}]\gamma_{Mn^{2+}}[OH^-]^2\gamma_{OH^-}^2$$

$$K_1 = \frac{[MnOH^+]\gamma_{MnOH^+}}{[Mn^{2+}]\gamma_{Mn^{2+}}[OH^-]\gamma_{OH^-}}$$

$$K_w = [H^+]\gamma_{H^+}[OH^-]\gamma_{OH^-}$$

From $K_1$, we write $[MnOH^+] = (K_1/\gamma_{MnOH^+})[Mn^{2+}]\gamma_{Mn^{2+}}[OH^-]\gamma_{OH^-}$.
Substitute for $[MnOH^+]$ in the charge balance:

$$2[Mn^{2+}] + (K_1/\gamma_{MnOH^+})[Mn^{2+}]\gamma_{Mn^{2+}}[OH^-]\gamma_{OH^-} + [H^+] = [OH^-] \quad (A)$$

In a basic solution, neglect $[H^+]$ in comparison with $[OH^-]$ and solve (A) for $[Mn^{2+}]$:

$$[Mn^{2+}] = \frac{[OH^-]}{2 + (K_i\gamma_{Mn^{2+}}/\gamma_{MnOH^+})[OH^-]\gamma_{OH^-}} \qquad (B)$$

Substitute (B) into $K_{sp}$:

$$K_{sp} = [Mn^{2+}]\gamma_{Mn^{2+}}[OH^-]^2\gamma_{OH^-}^2$$

$$K_{sp} = \left(\frac{\gamma_{Mn^{2+}}[OH^-]^3\gamma_{OH^-}^2}{2 + (K_i\gamma_{Mn^{2+}}/\gamma_{MnOH^+})[OH^-]\gamma_{OH^-}}\right) \qquad (C)$$

We take the ion size of $MnOH^+$ to be the same as $Mn^{2+}$, which is 600 pm. (But the charge of $MnOH^+$ is +1, not +2.) For $\mu = 0.10$ M, the activity coefficients are $\gamma_{Mn^{2+}} = 0.405$, $\gamma_{MnOH^+} = 0.80$, and $\gamma_{OH^-} = 0.76$. In the spreadsheet, we use GOAL SEEK to find the value $[OH^-] = 1.13 \times 10^{-4}$ M in cell C4 that makes the right side of Equation (C) equal to $1.6 \times 10^{-13}$ ($= K_{sp}$) in cell D4. With $[OH^-]$ known, we use $K_{sp}$ and $K_1$ to find:

$$[Mn^{2+}] = K_{sp}/(\gamma_{Mn^{2+}}[OH^-]^2\gamma_{OH^-}^2) = 5.36 \times 10^{-5}\ \text{M}$$

$$[MnOH^+] = (K_1/\gamma_{MnOH^+})([Mn^{2+}]\gamma_{Mn^{2+}}[OH^-]\gamma_{OH^-}) = 5.82 \times 10^{-6}\ \text{M}$$

You can check that the charge balance $2[Mn^{2+}] + [MnOH^+] \approx [OH^-]$ is satisfied.

|    | A | B | C | D |
|----|---|---|---|---|
| 1  | $Mn(OH)_2$ solubility with activity coefficients | | | |
| 2  | | | | |
| 3  | $K_{sp} =$ | | $[OH]_{guess} =$ | A10*[OH]$^3$*D10$^2$/(2+(K$_1$*A10/C10)[OH]*D10) = |
| 4  | 1.6E-13 | | 1.130E-04 | 1.6000E-13 |
| 5  | $K_1 =$ | | | |
| 6  | 2.5E+03 | | $[Mn^{2+}] =$ | $[MnOH^+] =$ |
| 7  | | | 5.358E-05 | 5.823E-06 |
| 8  | Activity coefficients: | | | |
| 9  | $Mn^{2+}$ | | $MnOH^+$ | $OH^-$ |
| 10 | 0.405 | | 0.80 | 0.76 |
| 11 | | | | |
| 12 | D4 = (A10*C4^3*D10^2)/(2+(A6*A10/C10)*C4*D10) | | | |
| 13 | C7 = A4/(A10*C4^2*D10^2) | | | |
| 14 | D7 = (A6/C10)*C7*A10*C4*D10 | | | |

Spreadsheet for Exercise 8-I.

# Chapter 9

**9-A.** $pH = -\log \mathcal{A}_{H^+}$. But $\mathcal{A}_{H^+} \mathcal{A}_{OH^-} = K_w \Rightarrow \mathcal{A}_{H^+} = K_w/\mathcal{A}_{OH^-}$. For $1.0 \times 10^{-2}$ M NaOH, $[OH^-] = 1.0 \times 10^{-2}$ M and $\gamma_{OH^-} = 0.900$ (using Table 8-1, with ionic strength $= 0.010$ M).

$$\begin{aligned}
\mathcal{A}_{H^+} &= \frac{K_w}{[OH^-]\gamma_{OH^-}} \\
&= \frac{1.0 \times 10^{-14}}{(1.0 \times 10^{-2})(0.900)} \\
&= 1.11 \times 10^{-12} \\
&\Rightarrow pH = -\log \mathcal{A}_{H^+} = 11.95
\end{aligned}$$

**9-B.** **(a)** Charge balance: $[H^+] = [OH^-] + [Br^-]$

Mass balance: $[Br^-] = 1.0 \times 10^{-8}$ M

Equilibrium: $[H^+][OH^-] = K_w$

Setting $[H^+] = x$ and $[Br^-] = 1.0 \times 10^{-8}$ M, the charge balance tells us that $[OH^-] = x - 1.0 \times 10^{-8}$. Putting this into the $K_w$ equilibrium gives

$$\begin{aligned}
(x)(x - 1.0 \times 10^{-8}) &= 1.0 \times 10^{-14} \\
&\Rightarrow x = 1.0_5 \times 10^{-7}\ \text{M} \\
&\Rightarrow pH = 6.98
\end{aligned}$$

**(b)** Charge balance: $[H^+] = [OH^-] + 2[SO_4^{2-}]$

Mass balance: $[SO_4^{2-}] = 1.0 \times 10^{-8}$ M

Equilibrium: $[H^+][OH^-] = K_w$

As before, writing $[H^+] = x$ and $[SO_4^{2-}] = 1.0 \times 10^{-8}$ M gives $[OH^-] = x - 2.0 \times 10^{-8}$ and $[H^+][OH^-] = (x)[x - (2.0 \times 10^{-8})] = 1.0 \times 10^{-14} \Rightarrow x = 1.10 \times 10^{-7}$ M $\Rightarrow pH = 6.96$.

**9-C.**

2-Nitrophenol  FM = 139.11
$C_6H_5NO_3$  $K_a = 5.89 \times 10^{-8}$

$$F_{HA}\ (\text{formal concentration}) = \frac{1.23\ \text{g}/(139.11\ \text{g/mol})}{0.250\ \text{L}}$$

$$= 0.035\ 4\ \text{M}$$

$$\begin{array}{ccc}
HA & \rightleftharpoons\ H^+ & +\quad A^- \\
F - x & x & x
\end{array}$$

$$\frac{x^2}{0.035\ 4 - x} = 5.89 \times 10^{-8}$$

$$\Rightarrow x = 4.56 \times 10^{-5}\ \text{M}$$

$$\Rightarrow pH = -\log x = 4.34$$

**9-D.**

$$\begin{array}{ccc}
& \rightleftharpoons\quad H^+ & + \\
F - x & x & x
\end{array}$$

But $[H^+] = 10^{-pH} = 6.9 \times 10^{-7}$ M $\Rightarrow [A^-] = 6.9 \times 10^{-7}$ M and $[HA] = 0.010 - [H^+] = 0.010$.

$$K_a = \frac{[H^+][A^-]}{[HA]} = \frac{(6.9 \times 10^{-7})^2}{0.010}$$

$$= 4.8 \times 10^{-11} \Rightarrow pK_a = 10.32$$

**9-E.** As $[HA] \rightarrow 0$, $pH \rightarrow 7$. If $pH = 7$,

$$\frac{[H^+][A^-]}{[HA]} = K_a \Rightarrow [A^-] = \frac{K_a}{[H^+]}[HA]$$

$$= \frac{10^{-5.00}}{10^{-7.00}}[HA] = 100[HA]$$

$$\alpha = \frac{[A^-]}{[HA] + [A^-]} = \frac{100[HA]}{[HA] + 100[HA]}$$

$$= \frac{100}{101} = 99\%$$

If $pK_a = 9.00$, we find $\alpha = 0.99\%$.

**9-F.** $CH_3CH_2CH_2CO_2^- + H_2O \rightleftharpoons CH_3CH_2CH_2CO_2H + OH^-$
        $F - x$                                         $x$           $x$

$$K_b = \frac{K_w}{K_a} = 6.58 \times 10^{-10}$$

$$\frac{x^2}{F - x} = K_b \Rightarrow x = 5.7_4 \times 10^{-6} \, M$$

$$pH = -\log\left(\frac{K_w}{x}\right) = 8.76$$

**9-G.** **(a)** $CH_3CH_2NH_2 + H_2O \rightleftharpoons CH_3CH_2NH_3^+ + OH^-$
            $F - x$                                       $x$         $x$

Because $pH = 11.82$, $[OH^-] = K_w/10^{-pH} = 6.6 \times 10^{-3} \, M = [BH^+]$.
$[B] = F - x = 0.093 \, M$.

$$K_b = \frac{[BH^+][OH^-]}{[B]} = \frac{(6.6 \times 10^{-3})^2}{0.093}$$

$$= 4.7 \times 10^{-4}$$

**(b)** $CH_3CH_2NH_3^+ \rightleftharpoons CH_3CH_2NH_2 + H^+$
      $F - x$                           $x$         $x$

$$K_a = \frac{K_w}{K_b} = 2.1 \times 10^{-11}$$

$$\frac{x^2}{F - x} = K_a \Rightarrow x = 1.4_5 \times 10^{-6} \, M$$

$$\Rightarrow pH = 5.84$$

**9-H.**

| Compound | $pK_a$ (for conjugate acid) |
|---|---|
| Ammonia | 9.24 $\leftarrow$ Most suitable, because $pK_a$ |
| Aniline | 4.60      is closest to pH |
| Hydrazine | 8.02 |
| Pyridine | 5.20 |

**9-I.** $pH = 4.25 + \log 0.75 = 4.13$

**9-J.** **(a)** $pH = pK_a + \log\dfrac{[B]}{[BH^+]}$

$$= 8.04 + \log\frac{[(1.00 \, g)/(74.08 \, g/mol)]}{[(1.00 \, g)/(110.54 \, g/mol)]} = 8.21$$

**(b)** $pH = pK_a + \log\dfrac{mol \, B}{mol \, BH^+}$

$$8.00 = 8.04 + \log\frac{mol \, B}{(1.00 \, g)/(110.54 \, g/mol)}$$

$$\Rightarrow mol \, B = 0.008\,25 = 0.611 \, g \text{ of glycine amide}$$

**(c)**

| | B | + | H$^+$ | $\rightarrow$ | BH$^+$ |
|---|---|---|---|---|---|
| Initial moles: | 0.013 499 | | 0.000 500 | | 0.009 046 |
| Final moles: | 0.012 999 | | — | | 0.009 546 |

$$pH = 8.04 + \log\left(\frac{0.012\,999}{0.009\,546}\right) = 8.17$$

**(d)**

| | BH$^+$ | + | OH$^-$ | $\rightarrow$ | B |
|---|---|---|---|---|---|
| Initial moles: | 0.009 546 | | 0.001 000 | | 0.012 999 |
| Final moles: | 0.008 546 | | — | | 0.013 999 |

$$pH = 8.04 + \log\left(\frac{0.013\,999}{0.008\,546}\right) = 8.25$$

**(e)** The solution in **(a)** contains 9.046 mmol glycine amide hydrochloride and 13.499 mmol glycine amide. Now we are adding 9.046 mmol of $OH^-$, which will convert all of the glycine amide hydrochloride into glycine amide. The new solution contains $9.046 + 13.499 = 22.545$ mmol of glycine amide in 190.46 mL. The concentration of glycine amide is $(22.545 \, mmol)/(190.46 \, mL) = 0.118_4 \, M$. The pH is determined by hydrolysis of glycine amide:

$$\frac{x^2}{0.118_4 - x} = K_b = \frac{K_w}{K_a} = \frac{10^{-14.00}}{10^{-8.04}} = 1.10 \times 10^{-6}$$

$$\Rightarrow x = 3.60 \times 10^{-4} \, M$$

$$pH = -\log\,(K_w/x) = 10.56$$

**9-K.** The reaction of phenylhydrazine with water is

$$B + H_2O \rightleftharpoons BH^+ + OH^- \quad K_b$$

We know that $pH = 8.13$, so we can find $[OH^-]$.

$$[OH^-] = \frac{\mathcal{A}_{OH^-}}{\gamma_{OH^-}} = \frac{K_w/10^{-pH}}{\gamma_{OH^-}} = 1.78 \times 10^{-6} \, M$$

(using $\gamma_{OH^-} = 0.76$ for $\mu = 0.10$ M)

$$K_b = \frac{[BH^+]\gamma_{BH^+}[OH^-]\gamma_{OH^-}}{[B]\gamma_B}$$

$$= \frac{(1.78 \times 10^{-6})(0.80)(1.78 \times 10^{-6})(0.76)}{[0.010 - (1.78 \times 10^{-6})](1.00)}$$

$$= 1.93 \times 10^{-10}$$

$$K_a = \frac{K_w}{K_b} = 5.19 \times 10^{-5} \rightleftharpoons pK_a = 4.28$$

To find $K_b$, we made use of the equality $[BH^+] = [OH^-]$.

**9-L.** We use GOAL SEEK to vary cell B5 until cell D4 is equal to $K_a$. The spreadsheet shows that $[H^+] = 4.236 \times 10^{-3}$ (cell B5) and $pH = 2.37$ in cell B8.

| | A | B | C | D | E |
|---|---|---|---|---|---|
| 1 | Ka = | 0.0031623 | | Reaction quotient | |
| 2 | Kw = | 1.00E-14 | | for Ka = | |
| 3 | FHA = | 0.03 | | [H+][A-]/[HA] = | |
| 4 | FA = | 0.015 | | 0.0031623 | |
| 5 | [H+] = | 4.236E-03 | | $\leftarrow$ Goal Seek solution | |
| 6 | [OH-] = | 2.361E-12 | | D4 = H*(FA+H-OH)/(FHA-H+OH) | |
| 7 | | | | B1 = 10^-2.5 | |
| 8 | pH = | 2.3730833 | | B6 = Kw/H | B8 = -log(H) |

If we were doing this problem by hand with the approximation that what we mix is what we get, $[H^+] = K_a[HA]/[A^-] = 10^{-2.50}[0.030]/[0.015] = 0.006\,32 \, M \Rightarrow pH = 2.20$.

## Chapter 10

**10-A.** (a) $H_2SO_3 \rightleftharpoons HSO_3^- + H^+$

$\qquad \quad 0.050 - x \quad\quad x \quad\quad\quad x$

$$\frac{x^2}{0.050 - x} = K_1 = 1.39 \times 10^{-2}$$

$$\Rightarrow x = 2.03 \times 10^{-2}$$

$$[HSO_3^-] = [H^+] = 2.03 \times 10^{-2}\,M$$

$$\Rightarrow pH = 1.69$$

$$[H_2SO_3] = 0.050 - x = 0.030\,M$$

$$[SO_3^{2-}] = \frac{K_2[HSO_3^-]}{[H^+]} = K_2$$

$$= 6.7 \times 10^{-8}\,M$$

(b) $\quad [H^+] = \sqrt{\dfrac{K_1K_2(0.050) + K_1K_w}{K_1 + (0.050)}}$

$$= 2.70 \times 10^{-5} \Rightarrow pH = 4.57$$

$$[H_2SO_3] = \frac{[H^+][HSO_3^-]}{K_1}$$

$$= \frac{(2.70 \times 10^{-5})(0.050)}{1.39 \times 10^{-2}}$$

$$= 9.7 \times 10^{-5}\,M$$

$$[SO_3^{2-}] = \frac{K_2[HSO_3^-]}{[H^+]} = 1.2 \times 10^{-4}\,M$$

$$[HSO_3^-] = 0.050\,M$$

(c) $\quad SO_3^{2-} + H_2O \rightleftharpoons HSO_3^- + OH^-$

$\qquad 0.050 - x \qquad\quad x \qquad\quad x$

$$\frac{x^2}{0.050 - x} = K_{b1} = \frac{K_w}{K_{a2}} = 1.49 \times 10^{-7}$$

$$[HSO_3^-] = x = 8.6 \times 10^{-5}\,M$$

$$[H^+] = \frac{K_w}{x} = 1.16 \times 10^{-10}\,M$$

$$\Rightarrow pH = 9.94$$

$$[SO_3^{2-}] = 0.050 - x = 0.050\,M$$

$$[H_2SO_3] = \frac{[H^+][HSO_3^-]}{K_1} = 7.2 \times 10^{-13}$$

**10-B.** (a) $\qquad pH = pK_2 \text{ (for } H_2CO_3) + \log\dfrac{[CO_3^{2-}]}{[HCO_3^-]}$

$$10.80 = 10.329 + \log\frac{(4.00/138.206)}{(x/84.007)}$$

$$\Rightarrow x = 0.822\,g$$

(b)

| | $CO_3^{2-}$ | $+$ | $H^+$ | $\rightarrow$ | $HCO_3^-$ |
|---|---|---|---|---|---|
| Initial moles: | $0.028\,9_4$ | | $0.010\,0$ | | $0.009\,78$ |
| Final moles: | $0.018\,9_4$ | | — | | $0.019\,7_8$ |

$$pH = 10.329 + \log\frac{0.018\,9_4}{0.019\,7_8} = 10.31$$

(c)

| | $CO_3^{2-}$ | $+$ | $H^+$ | $\rightarrow$ | $HCO_3^-$ |
|---|---|---|---|---|---|
| Initial moles: | $0.028\,9_4$ | | $x$ | | — |
| Final moles: | $0.028\,9_4 - x$ | | — | | $x$ |

$$10.00 = 10.329 + \log\frac{0.028\,9_4 - x}{x}$$

$$\Rightarrow x = 0.019\,7\,mol$$

$$\Rightarrow volume = \frac{0.019\,7\,mol}{0.320\,M} = 61.6\,mL$$

**10-C.**

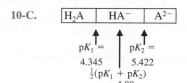

pH 4.40 is above $pK_1$. At pH $= pK_1$, there would be a 1:1 mixture of $H_2A$ and $HA^-$. We must add enough KOH to convert some $H_2A$ into $HA^-$ to create a mixture with pH $= 4.40$. 5.02 g $H_2A/(132.11$ g/mol$) = 0.038\,0$ mol $H_2A$.

| | $H_2A$ | $+$ | $OH^-$ | $\rightarrow$ | $HA^-$ |
|---|---|---|---|---|---|
| Initial moles: | $0.038\,0$ | | $x$ | | — |
| Final moles: | $0.038\,0 - x$ | | — | | $x$ |

$$pH = pK_1 + \log\frac{[HA^-]}{[H_2A]}$$

$$4.40 = 4.345 + \log\frac{x}{0.038\,0 - x} \Rightarrow x = 0.020\,2\,mol$$

Volume of KOH $= (0.020\,2$ mol$)/(0.800$ M$) = 25.2$ mL.

**10-D.** (a) Call the three forms of glutamine $H_2G^+$, HG, and $G^-$. The form shown is HG.

$$[H^+] = \sqrt{\frac{K_1K_2(0.010) + K_1K_w}{K_1 + 0.010}}$$

$$= 1.9_8 \times 10^{-6} \Rightarrow pH = 5.70$$

(b) Call the four forms of cysteine $H_3C^+$, $H_2C$, $HC^-$, and $C^{2-}$. The form shown is $HC^-$.

$$[H^+] = \sqrt{\frac{K_2K_3(0.010) + K_2K_w}{K_2 + 0.010}}$$

$$= 2.8_9 \times 10^{-10} \Rightarrow pH = 9.54$$

(c) Call the four forms of arginine $H_3A^{2+}$, $H_2A^+$, HA, and $A^-$. The form shown is HA.

$$[H^+] = \sqrt{\frac{K_2K_3(0.010) + K_2K_w}{K_2 + 0.010}}$$

$$= 4.2_8 \times 10^{-11} \Rightarrow pH = 10.37$$

**10-E.**

| | pH 9.00 | pH 11.00 |
|---|---|---|
| Principal species: | OH ... OH | O⁻ ... OH |
| Secondary species: | O⁻ ... OH | O⁻ ... O⁻ |
| Percentage in major form: | 66.5% | 52.9% |

The percentage in the major form was calculated with the formulas for $\alpha_{H_2A}$ (Equation 10-19 at pH 9.00) and $\alpha_{HA^-}$ (Equation 10-20 at pH 11.00).

**10-F.**

| | pH 9.0 | pH 10.0 |
|---|---|---|
| Predominant form: | $NH_3^+$<br>\|<br>$CHCH_2CH_2CO_2^-$<br>\|<br>$CO_2^-$ | $NH_2$<br>\|<br>$CHCH_2CH_2CO_2^-$<br>\|<br>$CO_2^-$ |
| Secondary form: | $NH_2$<br>\|<br>$CHCH_2CH_2CO_2^-$<br>\|<br>$CO_2^-$ | $NH_3^+$<br>\|<br>$CHCH_2CH_2CO_2^-$<br>\|<br>$CO_2^-$ |

| Predominant form: | $NH_2$ <br> $\vert$ <br> $CHCH_2$—⬡—$OH$ <br> $\vert$ <br> $CO_2^-$ | $NH_2$ <br> $\vert$ <br> $CHCH_2$—⬡—$OH$ <br> $\vert$ <br> $CO_2^-$ |
|---|---|---|
| Secondary form: | $NH_3^+$ <br> $\vert$ <br> $CHCH_2$—⬡—$O^-$ <br> $\vert$ <br> $CO_2^-$ | $NH_2$ <br> $\vert$ <br> $CHCH_2$—⬡—$O^-$ <br> $\vert$ <br> $CO_2^-$ |

**10-G.** The isoionic pH is the pH of a solution of pure neutral lysine, which is

$$NH_2$$
$$\vert$$
$$CHCH_2CH_2CH_2CH_2NH_3^+$$
$$\vert$$
$$CO_2^-$$

$$[H^+] = \sqrt{\frac{K_2K_3F + K_2K_w}{K_2 + F}} \Rightarrow pH = 9.93$$

**10-H.** We know that the isoelectric point will be near $\frac{1}{2}(pK_2 + pK_3) \approx 9.95$. At this pH, the fraction of lysine in the form $H_3L^{2+}$ is negligible. Therefore, the electroneutrality condition reduces to $[H_2L^+] = [L^-]$, for which the expression isoelectric pH $= \frac{1}{2}(pK_2 + pK_3) = 9.95$ applies.

## Chapter 11

**11-A.** The titration reaction is $H^+ + OH^- \rightarrow H_2O$ and $V_e = 5.00$ mL. Three representative calculations are given:

At 1.00 mL: $[OH^-] = \left(\frac{4.00}{5.00}\right)(0.010\ 0)\left(\frac{50.00}{51.00}\right)$

$$= 0.007\ 84\ M$$

$$pH = -\log\left(\frac{K_w}{[OH^-]}\right) = 11.89$$

At 5.00 mL: $H_2O \rightleftharpoons H^+ + OH^-$
$$\qquad\qquad x \qquad x$$

$$x^2 = K_w \Rightarrow x = 1.0 \times 10^{-7}\ M$$

$$pH = -\log x = 7.00$$

At 5.01 mL: $[H^+] = \left(\frac{0.01}{55.01}\right)(0.100)$

$$= 1.82 \times 10^{-5}\ M \Rightarrow pH = 4.74$$

| $V_a$ (mL) | pH | $V_a$ | pH | $V_a$ | pH |
|---|---|---|---|---|---|
| 0.00 | 12.00 | 4.50 | 10.96 | 5.10 | 3.74 |
| 1.00 | 11.89 | 4.90 | 10.26 | 5.50 | 3.05 |
| 2.00 | 11.76 | 4.99 | 9.26 | 6.00 | 2.75 |
| 3.00 | 11.58 | 5.00 | 7.00 | 8.00 | 2.29 |
| 4.00 | 11.27 | 5.01 | 4.74 | 10.00 | 2.08 |

**11-B.** The titration reaction is $HCO_2H + OH^- \rightarrow HCO_2^- + H_2O$ and $V_e = 50.0$ mL. For formic acid, $K_a = 1.80 \times 10^{-4}$. Four representative calculations are given:

At 0.0 mL: $\quad HA \rightleftharpoons H^+ + A^-$
$$\qquad\quad 0.050\ 0 - x \ \ x \qquad x$$

$$\frac{x^2}{0.050\ 0 - x} = K_a \Rightarrow x = 2.91 \times 10^{-3}$$

$$\Rightarrow pH = 2.54$$

At 48.0 mL:

| | HA | + | $OH^-$ | $\rightarrow$ | $A^-$ | + | $H_2O$ |
|---|---|---|---|---|---|---|---|
| Initial: | 50 | | 48 | | — | | — |
| Final: | 2 | | — | | 48 | | 48 |

$$pH = pK_a + \log\frac{[A^-]}{[HA]} = 3.744 + \log\frac{48.0}{2.0} = 5.12$$

At 50.0 mL: $A^- + H_2O \overset{K_b}{\rightleftharpoons} HA + OH^-$
$$\qquad\qquad F - x \qquad\qquad x \quad\ x$$

$$K_b = \frac{K_w}{K_a} \quad \text{and} \quad F = \left(\frac{50}{100}\right)(0.05)$$

$$\frac{x^2}{0.025\ 0 - x} = 5.56 \times 10^{-11} \Rightarrow x = 1.18 \times 10^{-6}\ M$$

$$pH = -\log\left(\frac{K_w}{x}\right) = 8.07$$

At 60.0 mL: $[OH^-] = \left(\frac{10.0}{110.0}\right)(0.050\ 0)$

$$= 4.55 \times 10^{-3}\ M \Rightarrow pH = 11.66$$

| $V_b$ (mL) | pH | $V_b$ | pH | $V_b$ | pH |
|---|---|---|---|---|---|
| 0.0 | 2.54 | 45.0 | 4.70 | 50.5 | 10.40 |
| 10.0 | 3.14 | 48.0 | 5.12 | 51.0 | 10.69 |
| 20.0 | 3.57 | 49.0 | 5.43 | 52.0 | 10.99 |
| 25.0 | 3.74 | 49.5 | 5.74 | 55.0 | 11.38 |
| 30.0 | 3.92 | 50.0 | 8.07 | 60.0 | 11.66 |
| 40.0 | 4.35 | | | | |

**11-C.** The titration reaction is $B + H^+ \rightarrow BH^+$ and $V_e = 50.0$ mL. Representative calculations:

At $V_a = 0.0$ mL: $\quad B + H_2O \rightleftharpoons BH^+ + OH^-$
$$\qquad\qquad\qquad 0.100 - x \qquad x \qquad x$$

$$\frac{x^2}{0.100 - x} = 2.6 \times 10^{-6} \Rightarrow x = 5.09 \times 10^{-4}$$

$$pH = -\log\left(\frac{K_w}{x}\right) = 10.71$$

At $V_a = 20.0$ mL:

| | B | + | $H^+$ | $\rightarrow$ | $BH^+$ |
|---|---|---|---|---|---|
| Initial: | 50.0 | | 20.0 | | — |
| Final: | 30.0 | | — | | 20.0 |

$$pH = pK_a \text{ (for } BH^+) + \log\frac{[B]}{[BH^+]}$$

$$= 8.41 + \log\frac{30.0}{20.0} = 8.59$$

At $V_a = V_e = 50.0$ mL: All B has been converted into the conjugate acid, $BH^+$. The formal concentration of $BH^+$ is $\left(\frac{100}{150}\right)(0.100) = 0.066\ 7$ M. The pH is determined by the reaction

$$BH^+ \rightleftharpoons B + H^+$$
$$0.066\ 7 - x \quad x \qquad x$$

$$\frac{x^2}{0.066\ 7 - x} = K_a = \frac{K_w}{K_b} \Rightarrow x = 1.60 \times 10^{-5} \Rightarrow pH = 4.80$$

At $V_a = 51.0$ mL: There is excess $H^+$:

$$[H^+] = \left(\frac{1.0}{151.0}\right)(0.200) = 1.32 \times 10^{-3}$$

$$\Rightarrow pH = 2.88$$

| $V_a$ (mL) | pH | $V_a$ | pH | $V_a$ | pH |
|---|---|---|---|---|---|
| 0.0 | 10.71 | 30.0 | 8.23 | 50.0 | 4.80 |
| 10.0 | 9.01 | 40.0 | 7.81 | 50.1 | 3.88 |
| 20.0 | 8.59 | 49.0 | 6.72 | 51.0 | 2.88 |
| 25.0 | 8.41 | 49.9 | 5.71 | 60.0 | 1.90 |

**11-D.** The titration reactions are

$$HO_2CCH_2CO_2H + OH^- \rightarrow {}^-O_2CCH_2CO_2H + H_2O$$
$${}^-O_2CCH_2CO_2H + OH^- \rightarrow {}^-O_2CCH_2CO_2^- + H_2O$$

and the equivalence points occur at 25.0 and 50.0 mL. We will designate malonic acid as $H_2M$.

At 0.0 mL: $H_2M \rightleftharpoons H^+ + HM^-$
$$0.050\,0 - x \quad x \qquad x$$

$$\frac{x^2}{0.050\,0 - x} = K_1 \Rightarrow x = 7.75 \times 10^{-3}$$
$$\Rightarrow pH = 2.11$$

At 8.0 mL:

|          | $H_2M$ | + | $OH^-$ | $\rightarrow$ | $HM^-$ | + | $H_2O$ |
|----------|--------|---|--------|---------------|--------|---|--------|
| Initial: | 25     |   | 8      |               | —      |   | —      |
| Final:   | 17     |   | —      |               | 8      |   | —      |

$$pH = pK_1 + \log\frac{[HM^-]}{[H_2M]} = 2.847 + \log\frac{8}{17}$$
$$= 2.52$$

At 12.5 mL: $V_b = \frac{1}{2}V_e \Rightarrow pH = pK_1 = 2.85$

At 19.3 mL:

|          | $H_2M$ | + | $OH^-$ | $\rightarrow$ | $HM^-$ | + | $H_2O$ |
|----------|--------|---|--------|---------------|--------|---|--------|
| Initial: | 25     |   | 19.3   |               | —      |   | —      |
| Final:   | 5.7    |   | —      |               | 19.3   |   | —      |

$$pH = pK_1 + \log\frac{19.3}{5.7} = 3.38$$

At 25.0 mL: At the first equivalence point, $H_2M$ has been converted into $HM^-$.

$$[H^+] = \sqrt{\frac{K_1K_2F + K_1K_w}{K_1 + F}}$$

where $F = \left(\frac{50}{75}\right)(0.050\,0) = 0.033\,3$ M.

$$[H^+] = 5.23 \times 10^{-5} \text{ M} \Rightarrow pH = 4.28$$

At 37.5 mL: $V_b = \frac{3}{2}V_e \Rightarrow pH = pK_2 = 5.70$

At 50.0 mL: At the second equivalence point, $H_2M$ has been converted into $M^{2-}$:

$$M^{2-} + H_2O \rightleftharpoons HM^- + OH^-$$
$$\left(\frac{50}{100}\right)(0.050\,0) - x \qquad x \qquad x$$

$$\frac{x^2}{0.025\,0 - x} = K_{b1} = \frac{K_w}{K_{a2}}$$
$$\Rightarrow x = 1.12 \times 10^{-5} \text{ M}$$
$$\Rightarrow pH = -\log\left(\frac{K_w}{x}\right) = 9.05$$

At 56.3 mL: There are 6.3 mL of excess NaOH.

$$[OH^-] = \left(\frac{6.3}{106.3}\right)(0.100) = 5.93 \times 10^{-3} \text{ M}$$
$$\Rightarrow pH = 11.77$$

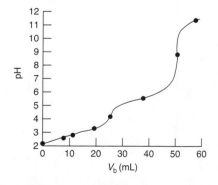

**11-E.**

$$\begin{array}{c}
\text{NH}_3^+ \\
| \\
\text{CHCH}_2 \\
| \\
\text{CO}_2^-
\end{array}
\quad
\text{(imidazole ring)} \quad \xrightarrow{H^+}$$

HHis

$$\begin{array}{c}
\text{NH}_3^+ \\
| \\
\text{CHCH}_2 \\
| \\
\text{CO}_2^-
\end{array}
\quad \xrightarrow{H^+} \quad
\begin{array}{c}
\text{NH}_3^+ \\
| \\
\text{CHCH}_2 \\
| \\
\text{CO}_2H
\end{array}$$

$$\text{H}_2\text{His}^+ \qquad\qquad \text{H}_3\text{His}^{2+}$$

The equivalence points occur at 25.0 and 50.0 mL.

At 0 mL: HHis is the second intermediate form derived from the triprotic acid $H_3His^{2+}$.

$$[H^+] = \sqrt{\frac{K_2K_3(0.050\,0) + K_2K_w}{K_2 + (0.050\,0)}}$$
$$= 2.37 \times 10^{-8} \text{ M} \Rightarrow pH = 7.62$$

At 4.0 mL:

|          | HHis | + | $H^+$ | $\rightarrow$ | $H_2His^+$ |
|----------|------|---|-------|---------------|------------|
| Initial: | 25   |   | 4     |               | —          |
| Final:   | 21   |   | —     |               | 4          |

$$pH = pK_2 + \log\frac{21}{4} = 6.69$$

At 12.5 mL: $pH = pK_2 = 5.97$

At 25.0 mL: The histidine has been converted into $H_2His^+$ at the formal concentration $F = \left(\frac{25}{50}\right)(0.050\,0) = 0.025\,0$ M.

$$[H^+] = \sqrt{\frac{K_1K_2F + K_1K_w}{K_1 + F}}$$
$$= 1.16 \times 10^{-4} \Rightarrow pH = 3.94$$

At 26.0 mL:

|          | $H_2His^+$ | + | $H^+$ | $\rightarrow$ | $H_3His^{2+}$ |
|----------|------------|---|-------|---------------|---------------|
| Initial: | 25         |   | 1     |               | —             |
| Final:   | 24         |   | —     |               | 1             |

$$pH = pK_1 + \log\frac{24}{1} = 2.98$$

The approximation that histidine reacts completely with HCl breaks down between 25 and 50 mL. If you used the titration equations in Table 11-6, you would find $pH = 3.28$, instead of 2.98, at $V_a = 26.0$ mL.

At 50.0 mL: The histidine has been converted into $H_3His$ at the formal concentration $F = \left(\frac{25}{75}\right)(0.050\,0) = 0.016\,7$ M.

$$H_3His^{2+} \rightleftharpoons H_2His^+ + H^+$$
$$0.016\,7 - x \qquad x \qquad x$$

$$\frac{x^2}{0.016\,7 - x} = K_1 \Rightarrow x = 0.011\,5 \text{ M} \Rightarrow pH = 1.94$$

**11-F.** Figure 11-1: bromothymol blue: blue $\rightarrow$ yellow
Figure 11-2: thymol blue: yellow $\rightarrow$ blue
Figure 11-3: thymolphthalein: colorless $\rightarrow$ blue

**11-G.** The titration reaction is $HA + OH^- \rightarrow A^- + H_2O$. It requires 1 mole of NaOH to react with 1 mole of HA. Therefore, the formal concentration of $A^-$ at the equivalence point is

$$\underbrace{\left(\frac{27.63}{127.63}\right)}_{\substack{\text{Dilution factor} \\ \text{for NaOH}}} \times \underbrace{(0.093\,81)}_{\substack{\text{Initial concentration} \\ \text{of NaOH}}} = 0.020\,31 \text{ M}$$

Because the pH is 10.99, $[OH^-] = 9.77 \times 10^{-4}$ and we can write

$$A^- + H_2O \rightleftharpoons HA + OH^-$$

$$K_b = \frac{[HA][OH^-]}{[A^-]} = \frac{(9.77 \times 10^{-4})^2}{0.020\,31 - (9.77 \times 10^{-4})}$$

$$= 4.94 \times 10^{-5}$$

$$K_a = \frac{K_w}{K_b} = 2.03 \times 10^{-10} \Rightarrow pK_a = 9.69$$

For the 19.47-mL point, we have

| | HA | + | OH⁻ | → | A⁻ | + | H₂O |
|---|---|---|---|---|---|---|---|
| Initial: | 27.63 | | 19.47 | | — | | — |
| Final: | 8.16 | | — | | 19.47 | | — |

$$pH = pK_a + \log\frac{[A^-]}{[HA]} = 9.69 + \log\frac{19.47}{8.16}$$

$$= 10.07$$

**11-H.** When $V_b = \frac{1}{2}V_e$, $[HA] = [A^-] = 0.033\,3$ M (using a correction for dilution by NaOH). $[Na^+] = 0.033\,3$ M as well. Ionic strength $= 0.033\,3$ M.

$$pK_a = pH - \log\frac{[A^-]\gamma_{A^-}}{[HA]\gamma_{HA}} \left(\begin{array}{c}\text{from}\\ \text{Equation 9-18}\end{array}\right)$$

$$= 4.62 - \log\frac{(0.033\,3)(0.854)}{(0.033\,3)(1.00)} = 4.69$$

The activity coefficient of A⁻ was found by interpolation in Table 8-1.

**11-I. (a)** The derivatives are shown in the spreadsheet below. In the first derivative graph, the maximum is near 119 mL. In Figure 11-7, the second derivative graph gives an end point of 118.9 μL.
**(b)** Column G in the spreadsheet gives $V_b(10^{-pH})$. In a graph of $V_b(10^{-pH})$ vs $V_b$, the points from 113 to 117 μL give a straight line whose slope is $-1.178 \times 10^6$ and whose intercept (end point) is 118.7 μL.

**11-J. (a)** pH 9.6 is past the equivalence point, so excess volume, $V$, is given by

$$[OH^-] = 10^{-4.4} = (0.100\,0\ M)\frac{V}{50.00 + 10.00 + V}$$

$$\Rightarrow V = 0.024\ mL$$

**(b)** pH 8.8 is before the equivalence point:

$$8.8 = 6.27 + \log\frac{[A^-]}{[HA]} \Rightarrow \frac{[A^-]}{[HA]} = 339$$

---

| Titration reaction: | | HA | + | OH⁻ | → | A⁻ | + | H₂O |
|---|---|---|---|---|---|---|---|---|
| Relative initial quantities: | | 10 | | $V$ | | — | | — |
| Relative final quantities: | | $10 - V$ | | — | | $V$ | | — |

To attain a ratio $[A^-]/[HA] = 339$, we need $V/(10 - V) = 339 \Rightarrow V = 9.97$ mL. The indicator error is $10 - 9.97 = 0.03$ mL.

**11-K. (a)** $A = 2\,080[HIn] + 14\,200[In^-]$
**(b)** $[HIn] = x$; $[In^-] = 1.84 \times 10^{-4} - x$

$$A = 0.868$$
$$= 2\,080x + 14\,200(1.84 \times 10^{-4} - x)$$
$$\Rightarrow x = 1.44 \times 10^{-4}\ M$$

$$pK_a = pH - \log\frac{[In^-]}{[HIn]}$$

$$= 6.23 - \log\frac{(1.84 \times 10^{-4}) - (1.44 \times 10^{-4})}{1.44 \times 10^{-4}}$$

$$= 6.79$$

## Chapter 12

**12-A.** For every mole of $K^+$ entering the first reaction, 4 moles of EDTA are produced in the second reaction.

Moles of EDTA = moles of $Zn^{2+}$ used in titration

$$[K^+] = \frac{\frac{1}{4}(\text{moles of } Zn^{2+})}{\text{volume of original sample}}$$

$$= \frac{\frac{1}{4}[28.73\ (\pm 0.03)][0.043\,7\ (\pm 0.000\,1)]}{250.0\ (\pm 0.1)}$$

$$= \frac{[\frac{1}{4}(\pm 0\%)][28.73\ (\pm 0.104\%)][0.043\,7\ (\pm 0.229\%)]}{250.0\ (\pm 0.040\,0\%)}$$

$$= 1.256\ (\pm 0.255\%) \times 10^{-3}\ M$$

$$= 1.256\ (\pm 0.003)\ mM$$

**12-B.** Total $Fe^{3+} + Cu^{2+}$ in 25.00 mL = (16.06 mL) × (0.050 83 M) = 0.816 3 mmol.

Second titration:

millimoles EDTA used: (25.00 mL)(0.050 83 M) = 1.270 8
millimoles $Pb^{2+}$ needed: (19.77 mL)(0.018 83 M) = 0.372 3
millimoles $Fe^{3+}$ present: (difference)    0.898 5

---

| | A | B | C | D | E | F | G |
|---|---|---|---|---|---|---|---|
| 1 | Derivatives of titration curve | | | | | | |
| 2 | | | | | | | |
| 3 | | | First Derivative | | Second Derivative | | |
| 4 | μL NaOH | pH | μL | Derivative | μL | Derivative | Vb*10^-pH |
| 5 | 107 | 6.921 | | | | | |
| 6 | 110 | 7.117 | 108.5 | 6.533E-02 | | | |
| 7 | 113 | 7.359 | 111.5 | 8.067E-02 | 110 | 5.11E-03 | 4.94E-06 |
| 8 | 114 | 7.457 | 113.5 | 9.800E-02 | 112.5 | 8.67E-03 | 3.98E-06 |
| 9 | 115 | 7.569 | 114.5 | 1.120E-01 | 114 | 1.40E-02 | 3.10E-06 |
| 10 | 116 | 7.705 | 115.5 | 1.360E-01 | 115 | 2.40E-02 | 2.29E-06 |
| 11 | 117 | 7.878 | 116.5 | 1.730E-01 | 116 | 3.70E-02 | 1.55E-06 |
| 12 | 118 | 8.090 | 117.5 | 2.120E-01 | 117 | 3.90E-02 | 9.59E-07 |
| 13 | 119 | 8.343 | 118.5 | 2.530E-01 | 118 | 4.10E-02 | 5.40E-07 |
| 14 | 120 | 8.591 | 119.5 | 2.480E-01 | 119 | -5.00E-03 | 3.08E-07 |
| 15 | 121 | 8.794 | 120.5 | 2.030E-01 | 120 | -4.50E-02 | 1.94E-07 |
| 16 | 122 | 8.952 | 121.5 | 1.580E-01 | 121 | -4.50E-02 | 1.36E-07 |
| 17 | | | | | | | |
| 18 | C6 = (A6+A5)/2 | | | E7 = (C7+C6)/2 | | G7 = A7*10^-B7 | |
| 19 | D6 = (B6-B5)/(A6-A5) | | | F7 = (D7-D6)/(C7-C6) | | | |

Spreadsheet for Exercise 11-I.

Because 50.00 mL of unknown were used in the second titration, the millimoles of $Fe^{3+}$ in 25.00 mL are 0.449 2. The millimoles of $Cu^{2+}$ in 25.00 mL are $0.816\ 3 - 0.449\ 2 = 0.367\ 1$ mmol/25.00 mL = 0.014 68 M.

**12-C.** Designating the total concentration of free EDTA as [EDTA], we can write

$$K_f' = \frac{[CuY^{2-}]}{[Cu^{2+}][EDTA]} = \alpha_{Y^{4-}}K_f = (2.9 \times 10^{-7})(10^{18.78})$$

$$= 1.7_4 \times 10^{12}$$

Representative calculations are shown here:

At 0.1 mL:

$$[EDTA] = \left(\frac{25.0 - 0.1}{25.0}\right)(0.040\ 0\ M)\left(\frac{50.0}{50.1}\right)$$

$$= 0.039\ 8\ M$$

$$[CuY^{2-}] = \left(\frac{0.1}{50.1}\right)(0.080\ 0\ M)$$

$$= 1.60 \times 10^{-4}\ M$$

$$[Cu^{2+}] = \frac{[CuY^{2-}]}{K_f'[EDTA]} = \frac{(1.60 \times 10^{-4}\ M)}{(1.7_4 \times 10^{12})(0.039\ 8\ M)}$$

$$= 2.3 \times 10^{-15}\ M$$

$$\Rightarrow pCu^{2+} = 14.64$$

At 25.0 mL: formal concentration of $CuY^{2-} = \left(\frac{25.0}{75.0}\right)(0.080\ 0\ M)$

$$= 0.026\ 7\ M$$

| | $Cu^{2+}$ | + | EDTA | $\rightleftharpoons$ | $CuY^{2-}$ |
|---|---|---|---|---|---|
| Initial concentration: | — | | — | | 0.026 7 |
| Final concentration: | $x$ | | $x$ | | 0.026 7 − $x$ |

$$\frac{0.026\ 7 - x}{x^2} = 1.7_4 \times 10^{12}$$

$$\Rightarrow [Cu^{2+}] = 1.2_4 \times 10^{-7}\ M$$

$$\Rightarrow pCu^{2+} = 6.91$$

At 26.0 mL: $\quad [Cu^{2+}] = \left(\frac{1.0}{76.0}\right)(0.080\ 0\ M)$

$$= 1.05 \times 10^{-3}\ M$$

$$\Rightarrow pCu^{2+} = 2.98$$

Summary:

| Volume (mL) | $pCu^{2+}$ | Volume | $pCu^{2+}$ |
|---|---|---|---|
| 0.1 | 14.64 | 24.0 | 10.86 |
| 5.0 | 12.84 | 25.0 | 6.91 |
| 10.0 | 12.42 | 26.0 | 2.98 |
| 15.0 | 12.07 | 30.0 | 2.30 |
| 20.0 | 11.64 | | |

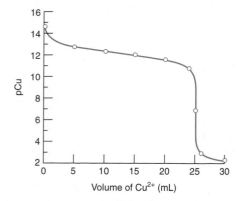

pCu (y-axis), Volume of $Cu^{2+}$ (mL) (x-axis)

**12-D.**
$$HY^{3-} \rightleftharpoons H^+ + Y^{4-} \quad K_6$$
$$H_2Y^{2-} \rightleftharpoons H^+ + HY^{3-} \quad K_5$$
$$\overline{H_2Y^{2-} \rightleftharpoons 2H^+ + Y^{4-} \quad K = K_5K_6}$$

$$= \frac{[H^+]^2[Y^{4-}]}{[H_2Y^{2-}]}$$

$$[H_2Y^{2-}] = \frac{[H^+]^2[Y^{4-}]}{K_5K_6} = \frac{[H^+]^2\alpha_{Y^{4-}}[EDTA]}{K_5K_6}$$

Using the values $[H^+] = 10^{-5.00}$, $\alpha_{Y^{4-}} = 2.9 \times 10^{-7}$, and $[EDTA] = 1.2_4 \times 10^{-7}$ gives $[H_2Y^{2-}] = 1.1 \times 10^{-7}\ M$.

**12-E. (a)** One volume of $Mn^{2+}$ will require two volumes of EDTA to reach the equivalence point. The formal concentration of $MnY^{2-}$ at the equivalence point is $\left(\frac{1}{3}\right)(0.010\ 0) = 0.003\ 33\ M$.

$$Mn^{2+} + EDTA \rightleftharpoons MnY^{2-}$$
$$x \qquad\quad x \qquad 0.003\ 33 - x$$

$$\frac{0.003\ 33 - x}{x^2} = \alpha_{Y^{4-}}K_f = (3.8 \times 10^{-4})10^{13.89} = 2.9 \times 10^{10}$$

$$\Rightarrow x = [Mn^{2+}] = 3.4 \times 10^{-7}\ M$$

**(b)** Because the pH is 7.00, the *ratio* $[H_3Y^-]/[H_2Y^{2-}]$ is constant throughout the *entire* titration.

$$\frac{[H_2Y^{2-}][H^+]}{[H_3Y^-]} = K_4$$

$$\Rightarrow \frac{[H_3Y^-]}{[H_2Y^{2-}]} = \frac{[H^+]}{K_4} = \frac{10^{-7.00}}{10^{-2.69}} = 4.9 \times 10^{-5}$$

**12-F.** $K_f$ for $CoY^{2-} = 10^{16.45} = 2.8 \times 10^{16}$

$$\alpha_{Y^{4-}} = 0.041\ \text{at pH } 9.00$$

$$\alpha_{Co^{2+}} = \frac{1}{1 + \beta_1[C_2O_4^{2-}] + \beta_2[C_2O_4^{2-}]^2}$$

$$= 6.8 \times 10^{-6}$$

(using $\beta_1 = K_1 = 10^{4.69}$ and $\beta_2 = K_1K_2 = 10^{7.15}$)

$$K_f' = \alpha_{Y^{4-}}K_f = 1.1_6 \times 10^{15}$$

$$K_f'' = \alpha_{Co^{2+}}\alpha_{Y^{4-}}K_f = 7.9 \times 10^9$$

At 0 mL: $[Co^{2+}] = \alpha_{Co^{2+}}(1.00 \times 10^{-3})$

$$= 6.8 \times 10^{-9}\ M \Rightarrow pCo^{2+} = 8.17$$

At 1.00 mL: $C_{Co^{2+}} = \left(\frac{1.00}{2.00}\right)(1.00 \times 10^{-3}\ M)\left(\frac{20.00}{21.00}\right)$

Fraction remaining, Initial concentration, Dilution factor

$$= 4.76 \times 10^{-4}\ M$$

$$[Co^{2+}] = \alpha_{Co^{2+}}C_{Co^{2+}}$$

$$= 3.2 \times 10^{-9}\ M \Rightarrow pCo^{2+} = 8.49$$

At 2.00 mL: This is the equivalence point.

$$C_{Co^{2+}} + EDTA \overset{K_f''}{\rightleftharpoons} CoY^{2-}$$
$$x \qquad\quad x \qquad \left(\frac{20.00}{22.00}\right)(1.00 \times 10^{-3}) - x$$

$$K_f'' = \frac{9.09 \times 10^{-4} - x}{x^2}$$

$$\Rightarrow x = 3.4 \times 10^{-7}\ M = C_{Co^{2+}}$$

$$[Co^{2+}] = \alpha_{Co^{2+}}C_{Co^{2+}}$$

$$= 2.3 \times 10^{-12}\ M \Rightarrow pCo^{2+} = 11.64$$

At 3.00 mL:

Concentration of excess EDTA $= \frac{1.00}{23.00}(1.00 \times 10^{-2}\ M)$

$$= 4.35 \times 10^{-4}\ M$$

Concentration of $CoY^{2-} = \frac{20.00}{23.00}(1.00 \times 10^{-3}\ M)$

$$= 8.70 \times 10^{-4}\ M$$

Knowing [EDTA] and [CoY²⁻], we can use the $K'_f$ equilibrium to find [Co²⁺]:

$$K'_f = \frac{[CoY^{2-}]}{[Co^{2+}][EDTA]} = \frac{[8.70 \times 10^{-4}\ M]}{[Co^{2+}][4.35 \times 10^{-4}\ M]}$$

$$\Rightarrow [Co^{2+}] = 1.7 \times 10^{-15}\ M \Rightarrow pCo^{2+} = 14.76$$

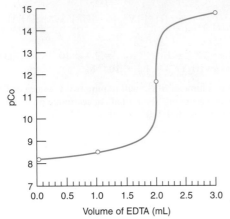

**12-G.** 25.0 mL of 0.120 M iminodiacetic acid = 3.00 mmol
25.0 mL of 0.050 0 M Cu²⁺ = 1.25 mmol

|                | Cu²⁺ | + | 2 iminodiacetic acid | ⇌ | CuX₂²⁻ |
|----------------|------|---|----------------------|---|--------|
| Initial mmol   | 1.25 |   | 3.00                 |   | —      |
| Final mmol     | —    |   | 0.50                 |   | 1.25   |

$$\frac{[CuX_2^{2-}]}{[Cu^{2+}][X^{2-}]^2} = K_f$$

$$\frac{[1.25/50.0]}{[Cu^{2+}][(0.50/50.0)(4.6 \times 10^{-3})]^2} = 3.5 \times 10^{16}$$

$$\Rightarrow [Cu^{2+}] = 3.4 \times 10^{-10}\ M$$

## Chapter 13

**13-A.** Hydroxybenzene = HA with $pK_{HA} = 9.997$

Dimethylamine = B from monoprotic BH⁺ with $pK_{BH^+} = 10.774$

Mixture contains 0.010 mol HA, 0.030 mol B, and 0.015 mol HCl in 1.00 L

Chemical reactions:
$HA \rightleftharpoons A^- + H^+$     $K_{HA} = 10^{-9.997}$
$BH^+ \rightleftharpoons B + H^+$     $K_{BH^+} = 10^{-10.774}$
$H_2O \rightleftharpoons H^+ + OH^-$     $K_w = 10^{-14.00}$

Charge balance: $[H^+] + [BH^+] = [OH^-] + [A^-] + [Cl^-]$

Mass balances: $[Cl^-] = 0.015\ M$
$[BH^+] + [B] = 0.030\ M \equiv F_B$
$[HA] + [A^-] = 0.010\ M = F_A$

We have seven equations and seven chemical species.

Fractional composition equations:

$$[BH^+] = \alpha_{BH^+}F_B = \frac{[H^+]F_B}{[H^+] + K_{BH^+}}$$

$$[B] = \alpha_B F_B = \frac{K_{BH^+}F_B}{[H^+] + K_{BH^+}}$$

$$[HA] = \alpha_{HA}F_A = \frac{[H^+]F_A}{[H^+] + K_{HA}}$$

$$[A^-] = \alpha_{A^-}F_A = \frac{K_{HA}F_A}{[H^+] + K_{HA}}$$

Substitute into charge balance:

$$[H^+] + \alpha_{BH^+}F_B = K_w/[H^+] + \alpha_{A^-}F_A + [0.015\ M] \quad \text{(A)}$$

We solve Equation A for [H⁺] by using SOLVER in the spreadsheet, with an initial guess of pH = 10 in cell H10. In the TOOLS menu, select SOLVER and choose Options. Set Precision to 1e-16 and click OK. In the SOLVER window, Set Target Cell E12 Equal To Value of 0 By Changing Cells H10. Click Solve and SOLVER finds pH = 10.33 in cell H10, giving a net charge near 0 in cell E12.

**13-B.** We use effective equilibrium constants, $K'$, as follows:

$$K_{HA} = \frac{[A^-]\gamma_{A^-}[H^+]\gamma_{H^+}}{[HA]\gamma_{HA}} = 10^{-9.997}$$

$$K'_{HA} = K_{HA}\left(\frac{\gamma_{HA}}{\gamma_{A^-}\gamma_{H^+}}\right) = \frac{[A^-][H^+]}{[HA]}$$

$$[HA] = \alpha_{HA}F_A = \frac{[H^+]F_A}{[H^+] + K'_{HA}}$$

$$[A^-] = \alpha_{A^-}F_A = \frac{K'_{HA}F_A}{[H^+] + K'_{HA}}$$

$$K_{BH^+} = \frac{[B]\gamma_B[H^+]\gamma_{H^+}}{[BH^+]\gamma_{BH^+}} = 10^{-10.774}$$

|    | A | B | C | D | E | F | G | H | I |
|----|---|---|---|---|---|---|---|---|---|
| 1  | Mixture of 0.010 M HA, 0.030 M B, and 0.015 M HCl | | | | | | | | |
| 2  |   |   |   |   |   |   |   |   |   |
| 3  | $F_A$ = | 0.010 | | $F_B$ = | 0.030 | | [Cl⁻] = | 0.015 | |
| 4  | $pK_{HA}$ = | 9.997 | | $pK_{BH^+}$ = | 10.774 | | $pK_w$ = | 14.000 | |
| 5  | $K_{HA}$ = | 1.01E-10 | | $K_{BH^+}$ = | 1.68E-11 | | $K_w$ = | 1.00E-14 | |
| 6  |   |   |   |   |   |   |   |   |   |
| 7  | Species in charge balance: | | | | | | Other concentrations: | | |
| 8  | [H⁺] = | 4.67E-11 | | [A⁻] = | 6.83E-03 | | [HA] = | 3.17E-03 | |
| 9  | [BH⁺] = | 2.20E-02 | | [Cl⁻] = | 0.015 | | [B] = | 7.95E-03 | |
| 10 |   |   |   | [OH⁻] = | 2.14E-04 | | pH = | 10.331 | |
| 11 |   |   |   |   |   |   | ↑ initial value is a guess | | |
| 12 | Positive charge minus negative charge | | | | -4.92E-17 | | = B8+B9-E8-E9-E10 | | |
| 13 | Formulas: | | | | | | | | |
| 14 | B5 = 10^-B4 | | B8 = 10^-H10 | | | | H5 = 10^-H4 | | |
| 15 | E5 = 10^-E4 | | E10 = H5/B8 | | | | E9 = H3 | | |
| 16 | B9 = B8*E3/(B8+E5) | | | | | | E8 = B5*B3/(B8+B5) | | |
| 17 | H9 = E5*E3/(B8+E5) | | | | | | H8 = B8*B3/(B8+B5) | | |

Spreadsheet for Exercise 13-A.

$$K'_{BH^+} = K_{BH^+}\left(\frac{\gamma_{BH^+}}{\gamma_B\gamma_{H^+}}\right) = \frac{[B][H^+]}{[BH^+]}$$

$$[BH^+] = \alpha_{BH^+}F_B = \frac{[H^+]F_B}{[H^+] + K'_{BH^+}}$$

$$[B] = \alpha_B F_B = \frac{K'_{BH^+}F_B}{[H^+] + K'_{BH^+}}$$

$$K_w = [H^+]\gamma_{H^+}[OH^-]\gamma_{OH^-} = 10^{-13.995}$$

$$K'_w = \frac{K_w}{\gamma_{H^+}\gamma_{OH^-}} = [H^+][OH^-]$$

$$[OH^-] = K'_w/[H^+]$$

$$pH = -\log([H^+]\gamma_{H^+})$$

$$[H^+] = (10^{-pH})/\gamma_{H^+}$$

The spreadsheet in Exercise 13-A is modified to add activity coefficients in cells A8:H9. Effective equilibrium constants are computed in row 5. Because we are going to the trouble of using activity coefficients, we use $pK_w = 13.995$ in cell H4 instead of the less accurate value of 14.00. With an initial ionic strength of 0 in cell C17 and a guess of pH = 10 in cell H14, the net charge in cell E16 is 0.005 56 $m$. Executing SOLVER to find the pH that reduces the net charge to near 0 gives the results shown in the spreadsheet.

The calculated ionic strength in cell C18 is 0.022 0 $m$. We write this number in cell C17 and use SOLVER again to find pH. Here are the results:

| Iteration | Ionic strength | pH |
|---|---|---|
| 1 | 0 | 10.331 |
| 2 | 0.022 0 | 10.357 |
| 3 | 0.022 5 | 10.358 |
| 4 | 0.022 5 | 10.358 |

| | A | B | C | D | E | F | G | H |
|---|---|---|---|---|---|---|---|---|
| 1 | Mixture of 0.010 M HA, 0.030 M B, and 0.015 M HCl | | | | | | | |
| 2 | With activities | | | | | | | |
| 3 | $F_A =$ | 0.010 | | $F_B =$ | 0.030 | | $[Cl^-] =$ | 0.015 |
| 4 | $pK_{HA} =$ | 9.997 | | $pK_{BH^+} =$ | 10.774 | | $pK_w =$ | 13.995 |
| 5 | $K_{HA}' =$ | 1.01E-10 | | $K_{BH^+}' =$ | 1.68E-11 | | $K_w' =$ | 1.01E-14 |
| 6 | | | | | | | | |
| 7 | Activity coefficients: | | | | | | | |
| 8 | $H^+ =$ | 1.00 | | $A^-$ | 1.00 | | HA | 1.00 |
| 9 | $OH^- =$ | 1.00 | | $BH^+$ | 1.00 | | B | 1.00 |
| 10 | | | | | | | | |
| 11 | Species in charge balance: | | | | | | Other concentrations: | |
| 12 | $[H^+] =$ | 4.67E-11 | | $[A^-] =$ | 6.83E-03 | | $[HA] =$ | 3.17E-03 |
| 13 | $[BH^+] =$ | 2.20E-02 | | $[Cl^-] =$ | 0.015 | | $[B] =$ | 7.95E-03 |
| 14 | | | | $[OH^-] =$ | 2.17E-04 | | pH = | 10.331 |
| 15 | | | | | | | ↑ initial value is a guess | |
| 16 | Positive charge minus negative charge = | | | | -3.24E-17 | = B12+B13-E12-E13-E14 | | |
| 17 | Ionic strength = | | 0.0000 | ← initial value is 0 | | | | |
| 18 | New ionic strength = | | 0.0220 | ← substitute this value into cell C17 for next iteration | | | | |
| 19 | | | | | | | | |
| 20 | Formulas: | | | | | | | |
| 21 | B5 = (10^-B4)*H8/(E8*B8) | | | | H8 = H9 = 1 | | | |
| 22 | E5 = (10^-E4)*E9/(H9*B8) | | | | E13 = H3 | | | |
| 23 | H5 = (10^-H4)/(B8*B9) | | | | | | | |
| 24 | B8 = B9 = E8 = E9 = 10^(-0.51*1^2*(SQRT($C$17)/(1+SQRT($C$17))-0.3*$C$17)) | | | | | | | |
| 25 | B12 = (10^-H14)/B8 | | | | E14 = H5/B12 | | | |
| 26 | B13 = B12*E3/(B12+E5) | | | | H13 = E5*E3/(B12+E5) | | | |
| 27 | E12 = B5*B3/(B12+B5) | | | | H12 = B12*B3/(B12+B5) | | | |
| 28 | C18 = 0.5*(B12+B13+E12+E13+E14) | | | | | | | |

Spreadsheet for Exercise 13-B.

**13-C.** **(a)** 2-Aminobenzoic acid = HA from diprotic $H_2A^+$

$pK_1 = 2.08$ $\quad$ $pK_2 = 4.96$

Dimethylamine = B from monoprotic $BH^+$ $\quad pK_a = 10.774$

Mixture contains 0.040 mol HA, 0.020 mol B, and 0.015 mol HCl in 1.00 L.

Chemical reactions:
$$H_2A^+ \rightleftharpoons HA + H^+ \quad K_1 = 10^{-2.08}$$
$$HA \rightleftharpoons A^- + H^+ \quad K_2 = 10^{-4.96}$$
$$BH^+ \rightleftharpoons B + H^+ \quad K_a = 10^{-10.774}$$
$$H_2O \rightleftharpoons H^+ + OH^- \quad K_w = 10^{-14.00}$$

Charge balance: $[H^+] + [H_2A^+] + [BH^+] = [OH^-] + [A^-] + [Cl^-]$

Mass balances: $[Cl^-] = 0.015$ M
$$[BH^+] + [B] = 0.020 \text{ M} \equiv F_B$$
$$[H_2A^+] + [HA] + [A^-] = 0.040 \text{ M} = F_A$$

We have eight equations and eight chemical species.

Fractional composition equations:

$$[BH^+] = \alpha_{BH^+}F_B = \frac{[H^+]F_B}{[H^+] + K_a}$$

$$[B] = \alpha_B F_B = \frac{K_a F_B}{[H^+] + K_a}$$

| | A | B | C | D | E | F | G | H | I |
|---|---|---|---|---|---|---|---|---|---|
| 1 | Mixture of 0.040 M HA, 0.020 M B, and 0.015 M HCl | | | | | | | | |
| 2 | | | | | | | | | |
| 3 | $F_A =$ | 0.040 | | $F_B =$ | 0.020 | | $[Cl^-] =$ | 0.015 | |
| 4 | $pK_1 =$ | 2.080 | | $pK_a =$ | 10.774 | | $K_w =$ | 1.00E-14 | |
| 5 | $pK_2 =$ | 4.960 | | $K_a =$ | 1.68E-11 | | | | |
| 6 | $K_1 =$ | 8.32E-03 | | | | | | | |
| 7 | $K_2 =$ | 1.10E-05 | | | | | | | |
| 8 | | | | | | | | | |
| 9 | Species in charge balance: | | | | | | Other concentrations: | | |
| 10 | $[H^+] =$ | 7.03E-05 | | $[H_2A^+] =$ | 2.90E-04 | | $[HA] =$ | 3.43E-02 | |
| 11 | $[BH^+] =$ | 2.00E-02 | | $[A^-] =$ | 5.36E-03 | | $[B] =$ | 4.79E-09 | |
| 12 | $[OH^-] =$ | 1.42E-10 | | $[Cl^-] =$ | 0.015 | | pH = | 4.153 | |
| 13 | | | | | | | ↑ initial value is a guess | | |
| 14 | Positive charge minus negative charge | | | | 0.00E+00 | | = B10+B11+E10-B12-E11-E12 | | |
| 15 | Formulas: | | | | | | | | |
| 16 | B16 = 10^-B4 | | B7 = 10^-B5 | | | | E5 = 10^-E4 | | |
| 17 | B10 =10^-H12 | | B12 =H4/B10 | | | | E12 = H3 | | |
| 18 | B11 = B10*E3/(B10+E5) | | | | | | | | |
| 19 | E10 = B10^2*B3/(B10^2+B10*B6+B6*B7) | | | | | | | | |
| 20 | E11 = B6*B7*B3/(B10^2+B10*B6+B6*B7) | | | | | | | | |
| 21 | H10 = B10*B6*B3/(B10^2+B10*B6+B6*B7) | | | | | | | | |
| 22 | H11 = E5*E3/(B10+E5) | | | | | | | | |

Spreadsheet for Exercise 13-C.

$$[H_2A^+] = \alpha_{H_2A^+}F_A = \frac{[H^+]^2F_A}{[H^+]^2 + [H^+]K_1 + K_1K_2}$$

$$[HA] = \alpha_{HA}F_A = \frac{K_1[H^+]F_A}{[H^+]^2 + [H^+]K_1 + K_1K_2}$$

$$[A^-] = \alpha_{A^-}F_A = \frac{K_1K_2F_A}{[H^+]^2 + [H^+]K_1 + K_1K_2}$$

Substitute into charge balance:

$$[H^+] + \alpha_{H_2A^+}F_A + \alpha_{BH^+}F_B = K_w/[H^+] + \alpha_{A^-}F_A + [0.015 \text{ M}] \quad \text{(A)}$$

Solve Equation A for $[H^+]$ by using SOLVER in the spreadsheet, with an initial guess of pH = 7 in cell H12. In the TOOLS menu, select SOLVER and choose Options. Set Precision to 1e-16 and click OK. In the SOLVER window, Set Target Cell E14 Equal To Value of 0 By Changing Cells H12. Click Solve and SOLVER finds pH = 4.15 in cell H12, giving a net charge near 0 in cell E14.

(b) From the concentrations in the spreadsheet, we find the following fractions of 2-aminobenzoic acid: $H_2A^+ = 0.7\%$, HA = 85.9%, and $A^- = 13.4\%$. The fractions of dimethylamine are $BH^+ = 100.0\%$ and B = 0.0%. The simple prediction is that HCl would consume B, giving 100% $BH^+$. The remaining 5 mmol B consumes 5 mmol HA, making 5 mmol $A^-$ and leaving 35 mmol HA.

Predicted fractions: $A^- = 5/40 = 12.5\%$, HA = $35/40 = 87.5\%$

Estimated pH = $pK_2 + \log([A^-]/[HA]) = 4.96 + \log(5/35) = 4.12$

13-D. Effective equilibrium constants:

$$H_2T \xrightleftharpoons{pK_1=3.036} HT^- + H^+ \qquad K_1' = K_1\left(\frac{\gamma_{H_2T}}{\gamma_{HT^-}\gamma_{H^+}}\right) = \frac{[HT^-][H^+]}{[H_2T]}$$

$$HT^- \xrightleftharpoons{pK_2=4.366} T^{2-} + H^+ \qquad K_2' = K_2\left(\frac{\gamma_{HT^-}}{\gamma_{T^{2-}}\gamma_{H^+}}\right) = \frac{[T^{2-}][H^+]}{[HT^-]}$$

$$PyH^+ \xrightleftharpoons{pK_a=5.20} Py + H^+ \qquad K_a' = K_a\left(\frac{\gamma_{PyH^+}}{\gamma_{Py}\gamma_{H^+}}\right) = \frac{[Py][H^+]}{[PyH^+]}$$

$$H_2O \xrightleftharpoons{pK_w=13.995} H^+ + OH^- \qquad K_w' = \frac{K_w}{\gamma_{H^+}\gamma_{OH^-}} = [H^+][OH^-]$$

$$[OH^-] = K_w'/[H^+] \qquad pH = -\log([H^+]\gamma_{H^+})$$

The spreadsheet on the next page is modified from the one in the text by the addition of activity coefficients in cells A10:H11. The ionic strength for computing activity coefficients with the Davies equation is in cell C20. The initial ionic strength is set to 0. The first guess for pH in cell H17 is 5. Then use SOLVER to find the pH in cell H17 that makes the net charge in cell E19 nearly 0. From this pH, all concentrations are computed and the new ionic strength is found in cell C21. This new ionic strength is typed into cell C20 for the next iteration. From the new ionic strength, new activity coefficients are computed and new values of $K'$ are computed in cells B6, B7, E5, and H5. The pH that satisfies the charge balance in cell E19 is then found again with SOLVER and the whole process is repeated until the ionic strength reaches a constant value.

| Iteration | Ionic strength | pH |
|---|---|---|
| 1 | 0 | 4.298 |
| 2 | 0.0523 | 4.116 |
| 3 | 0.0539 | 4.114 |
| 4 | 0.0540 | 4.114 |

13-E. $Mg^{2+} + SO_4^{2-} \rightleftharpoons MgSO_4(aq)$ $K_{ip} = 10^{2.23}$

If hydrolysis is neglected, $[Mg^{2+}] = [SO_4^{2-}]$ and $[MgSO_4] = F - [Mg^{2+}]$, where F is the formal concentration (0.025 M). We will solve the equation

$$K_{ip}' = K_{ip}\left(\frac{\gamma_{Mg^{2+}}\gamma_{SO_4^{2-}}}{\gamma_{MgSO_4}}\right) = \frac{[MgSO_4(aq)]}{[Mg^{2+}][SO_4^{2-}]} = \frac{F - [Mg^{2+}]}{[Mg^{2+}]^2} \quad \text{(A)}$$

with $\gamma_{MgSO_4} = 1$ because it is neutral.

(a) First, set $\gamma_{Mg^{2+}} = \gamma_{SO_4^{2-}} = 1.00$ and use SOLVER (or the quadratic equation) to find $[Mg^{2+}]$ to satisfy Equation A. In the spreadsheet on the next page, guess an initial value of $[Mg^{2+}] = 0.012\ 5$ M in cell B11. Open SOLVER from the TOOLS menu. Select SOLVER Options and set Precision = 1e-6. Click OK. In the SOLVER window, Set Target Cell F14 Equal To Value of 169.8 By Changing Cells B11. SOLVER finds $[Mg^{2+}] = 0.009\ 54$ M and $\mu = 0.038\ 2$ M.

(b) Write the value $\mu = 0.038\ 2$ M in cell F3. This ionic strength changes $K_{ip}'$ from 169.8 to 40.7. Execute SOLVER again, and this time Set Target Cell F14 Equal To Value of 40.7 By Changing Cells B11. SOLVER finds $[Mg^{2+}] = 0.015\ 4$ M and $\mu = 0.061\ 5$ M. Write 0.061 5 in cell F3 and repeat the process again. The succession of results is shown in the table. The fraction of ion pairing is $[MgSO_4(aq)]/F = 33\%$

| | A | B | C | D | E | F | G | H |
|---|---|---|---|---|---|---|---|---|
| 1 | Mixture of 0.020 M $Na^+HT^-$, 0.015 M $PyH^+Cl^-$, and 0.010 M KOH - with activity | | | | | | | |
| 2 | | | | | | | | |
| 3 | $F_{H2T} =$ | 0.020 | | $F_{PyH^+} =$ | 0.015 | | $[K^+] =$ | 0.010 |
| 4 | $pK_1 =$ | 3.036 | | $pK_a =$ | 5.20 | | $pK_w =$ | 13.995 |
| 5 | $pK_2 =$ | 4.366 | | $K_a' =$ | 6.31E-06 | | $K_w' =$ | 1.52E-14 |
| 6 | $K_1' =$ | 1.38E-03 | | | | | | |
| 7 | $K_2' =$ | 9.67E-05 | | | | | | |
| 8 | | | | | | | | |
| 9 | Activity coefficients from Davies equation: | | | | | | | |
| 10 | $H^+ =$ | 0.82 | | $HT^- =$ | 0.82 | | $OH^- =$ | 0.82 |
| 11 | $PyH^+ =$ | 0.82 | | $T^{2-} =$ | 0.45 | | | |
| 12 | | | | | | | | |
| 13 | Species in charge balance: | | | | | | Other concentrations: | |
| 14 | $[H^+] =$ | 9.42E-05 | | $[OH^-] =$ | 1.61E-10 | | $[H_2T] =$ | 6.52E-04 |
| 15 | $[PyH^+] =$ | 1.41E-02 | | $[HT^-] =$ | 9.54E-03 | | $[Py] =$ | 9.42E-04 |
| 16 | $[Na^+] =$ | 0.020 | | $[T^{2-}] =$ | 9.80E-03 | | | |
| 17 | $[K^+] =$ | 0.010 | | $[Cl^-] =$ | 0.015 | | pH = | 4.114 |
| 18 | | | | | | | ↑ initial value is a guess | |
| 19 | Positive charge minus negative charge = | | | | 2.78E-17 | | =B14+B15+B16+B17 | |
| 20 | Ionic strength = | 0.0540 | | ← initial value is 0 | | | -E14-E15-2*E16-E17 | |
| 21 | New ionic strength = | 0.0540 | | ← substitute this value into cell C17 for next iteration | | | | |
| 22 | | | | | | | | |
| 23 | Formulas: | | | | | | | |
| 24 | B6 = 10^-B4*(1/(E10*B10)) | | | B14 = (10^-H17)/B10 | | | E14 = H5/B14 | |
| 25 | B7 = 10^-B5*(E10/(E11*B10)) | | | B16 = B3 | | | E17 = E3 | |
| 26 | E5 = 10^-E4*(B11/B10) | | | B17 = H3 | | | | |
| 27 | H5 = (10^-H4)/(B10*H10) | | | | | | | |
| 28 | B10=B11=E10=H10 = 10^(-0.51*1^2*(SQRT($C$20)/(1+SQRT($C$20))-0.3*$C$20)) | | | | | | | |
| 29 | E11 = 10^(-0.51*2^2*(SQRT($C$20)/(1+SQRT($C$20))-0.3*$C$20)) | | | | | | | |
| 30 | B15 = B14*E3/(B14+E5) | | | | | | | |
| 31 | E15 = B14*B6*B3/(B14^2+B14*B6+B6*B7) | | | | | | | |
| 32 | E16 = B6*B7*B3/(B14^2+B14*B6+B6*B7) | | | | | | | |
| 33 | H14 = B14^2*B3/(B14^2+B14*B6+B6*B7) | | | | | | | |
| 34 | H15 = E5*E3/(B14+E5) | | | | | | | |
| 35 | C21 = 0.5*(B14+B15+B16+B17+E14+E15+4*E16+E17) | | | | | | | |

Spreadsheet for Exercise 13-D.

| | A | B | C | D | E | F |
|---|---|---|---|---|---|---|
| 1 | Ion pairing of $MgSO_4$ with Davies activity | | | | | |
| 2 | | | | | | |
| 3 | F = | 0.025 | | Ionic strength = | | 0.0000 |
| 4 | log $K_{ip}$ = | 2.23 | | New ionic strength = | | 0.0382 |
| 5 | $K_{ip}'$ = | 169.8 | | | | |
| 6 | | | | | | |
| 7 | Activity coefficients from Davies equation: | | | | | |
| 8 | $\gamma_{Mg}$ = | 1.00 | | $\gamma_{SO4}$ = | 1.00 | |
| 9 | | | | | | |
| 10 | Concentrations: | | | | | |
| 11 | $[Mg^{2+}]$ = | 0.00954 | | ← initial value is a guess | | |
| 12 | $[MgSO_4(aq)]$ = | 0.01546 | | $[SO_4^{2-}]$ = | 0.00954 | |
| 13 | | | | | | |
| 14 | | | Reaction quotient = $[MgSO_4(aq)]/([Mg^{2+}][SO_4^{2-}])$ = | | | 169.8 |
| 15 | % ion pair = | 61.8 | | = 100*B12/B3 | | |
| 16 | | | | | | |
| 17 | Formulas: | | | | | |
| 18 | B5 = (10^B4)*B8*E8 | | B12 = B3-B11 | | E12 = B11 | |
| 19 | B8 = E8 = 10^(-0.51*2^2*(SQRT($F$3)/(1+SQRT($F$3))-0.3*$F$3)) | | | | | |
| 20 | F14 = B12/(B11*E12) | | | F4 = 0.5*(B11*4+E12*4) | | |

Spreadsheet for Exercise 13-E.

| Iteration | Ionic strength (mM) | $[Mg^{2+}]$ (mM) |
|---|---|---|
| 1 | 0 | 9.54 |
| 2 | 38.2 | 15.4 |
| 3 | 61.5 | 16.5 |
| 4 | 66.0 | 16.7 |
| 5 | 66.7 | 16.7 |
| 6 | 66.8 | 16.7 |

(c) The ionic strength is 0.067 M, not 0.10 M.

**13-F.** (a) In addition to Reactions 13-32 through 13-36, we add the reaction

$$HF(aq) + F^- \rightleftharpoons HF_2^- \qquad K_{HF_2} = 10^{0.58}$$
$$\Rightarrow [HF_2^-] = K_{HF_2}[F^-][HF] = K_{HF_2}[F^-]([H^+][F^-]/K_{HF})$$
$$= K_{HF_2}[F^-]^2[H^+]/K_{HF}$$

The charge balance becomes

$$[H^+] + 2[Ca^{2+}] + [CaOH^+] + [CaF^+] - [OH^-] - [F^-] - [HF_2^-] = 0 \qquad (A)$$

and the mass balance is

$$2\{[Ca^{2+}] + [CaOH^+] + [CaF^+]\} = [F^-] + [HF] + [CaF^+] + 2[HF_2^-]$$
$$2[Ca^{2+}] + 2[CaOH^+] + [CaF^+] - [F^-] - [HF] - 2[HF_2^-] = 0$$

Substitute equilibrium expressions for the various species into the mass balance:

$$\frac{2K_{sp}}{[F^-]^2} + \frac{2K_aK_{sp}}{[H^+][F^-]^2} + \frac{K_{ip}K_{sp}}{[F^-]} - [F^-] - \frac{[H^+][F^-]}{K_{HF}} - \frac{2K_{HF_2}[F^-]^2[H^+]}{K_{HF}} = 0 \qquad (B)$$

The procedure at this point is the one we used in the text. For a given value of $[H^+]$, use SOLVER to find the value of $[F^-]$ that makes the left side of Equation B equal to 0. From $[H^+]$ and $[F^-]$, compute the remaining concentrations with the equilibrium expressions. Results are shown in the graph at the right.

(b) To find the pH of unbuffered solution, find $[H^+]$ at which the charge balance (A) is also satisfied. Because $[HF_2^-]$ is negligible near neutral pH, the pH is unchanged from the case worked in the text. The pH of unbuffered solution is 7.10 and concentrations are:

$$[F^-] = 4.00 \times 10^{-4} \text{ M} \qquad [Ca^{2+}] = 2.00 \times 10^{-4} \text{ M}$$
$$[HF] = 4.67 \times 10^{-8} \text{ M} \qquad [CaOH^+] = 5.03 \times 10^{-10} \text{ M}$$
$$[HF_2^-] = 7.11 \times 10^{-11} \text{ M} \qquad [CaF^+] = 3.44 \times 10^{-7} \text{ M}$$

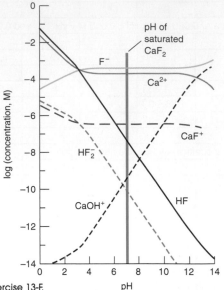

Graph for Exercise 13-F.

**13-G.**

| | | |
|---|---|---|
| $AgCN(s) \rightleftharpoons Ag^+ + CN^-$ | | $pK_{sp} = 15.66$ |
| $HCN(aq) \rightleftharpoons CN^- + H^+$ | | $pK_{HCN} = 9.21$ |
| $Ag^+ + H_2O \rightleftharpoons AgOH(aq) + H^+$ | | $pK_{Ag} = 12.0$ |

Express all concentrations in terms of $[Ag^+]$ and $[H^+]$:

$$[CN^-] = K_{sp}/[Ag^+]$$
$$[HCN(aq)] = \frac{[H^+][CN^-]}{K_{HCN}} = \frac{[H^+]K_{sp}}{K_{HCN}[Ag^+]}$$
$$[AgOH(aq)] = \frac{K_{Ag}[Ag^+]}{[H^+]}$$

Mass balance: total silver = total cyanide

$$[Ag^+] + [AgOH(aq)] = [CN^-] + [HCN(aq)]$$

Substitute expressions for concentrations into the mass balance:

$$[Ag^+] + \frac{K_{Ag}[Ag^+]}{[H^+]} = \frac{K_{sp}}{[Ag^+]} + \frac{[H^+]K_{sp}}{K_{HCN}[Ag^+]} \qquad (A)$$

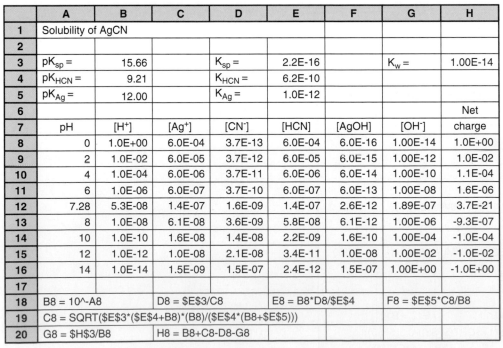

|  | A | B | C | D | E | F | G | H |
|---|---|---|---|---|---|---|---|---|
| 1 | Solubility of AgCN | | | | | | | |
| 2 | | | | | | | | |
| 3 | $pK_{sp} =$ | 15.66 | | $K_{sp} =$ | 2.2E-16 | | $K_w =$ | 1.00E-14 |
| 4 | $pK_{HCN} =$ | 9.21 | | $K_{HCN} =$ | 6.2E-10 | | | |
| 5 | $pK_{Ag} =$ | 12.00 | | $K_{Ag} =$ | 1.0E-12 | | | |
| 6 | | | | | | | | Net |
| 7 | pH | $[H^+]$ | $[Ag^+]$ | $[CN^-]$ | $[HCN]$ | $[AgOH]$ | $[OH^-]$ | charge |
| 8 | 0 | 1.0E+00 | 6.0E-04 | 3.7E-13 | 6.0E-04 | 6.0E-16 | 1.00E-14 | 1.0E+00 |
| 9 | 2 | 1.0E-02 | 6.0E-05 | 3.7E-12 | 6.0E-05 | 6.0E-15 | 1.00E-12 | 1.0E-02 |
| 10 | 4 | 1.0E-04 | 6.0E-06 | 3.7E-11 | 6.0E-06 | 6.0E-14 | 1.00E-10 | 1.1E-04 |
| 11 | 6 | 1.0E-06 | 6.0E-07 | 3.7E-10 | 6.0E-07 | 6.0E-13 | 1.00E-08 | 1.6E-06 |
| 12 | 7.28 | 5.3E-08 | 1.4E-07 | 1.6E-09 | 1.4E-07 | 2.6E-12 | 1.89E-07 | 3.7E-21 |
| 13 | 8 | 1.0E-08 | 6.1E-08 | 3.6E-09 | 5.8E-08 | 6.1E-12 | 1.00E-06 | -9.3E-07 |
| 14 | 10 | 1.0E-10 | 1.6E-08 | 1.4E-08 | 2.2E-09 | 1.6E-10 | 1.00E-04 | -1.0E-04 |
| 15 | 12 | 1.0E-12 | 1.0E-08 | 2.1E-08 | 3.4E-11 | 1.0E-08 | 1.00E-02 | -1.0E-02 |
| 16 | 14 | 1.0E-14 | 1.5E-09 | 1.5E-07 | 2.4E-12 | 1.5E-07 | 1.00E+00 | -1.0E+00 |
| 17 | | | | | | | | |
| 18 | B8 = 10^-A8 | | D8 = \$E\$3/C8 | | E8 = B8*D8/\$E\$4 | | F8 = \$E\$5*C8/B8 | |
| 19 | C8 = SQRT(\$E\$3*(\$E\$4+B8)*(B8)/(\$E\$4*(B8+\$E\$5))) | | | | | | | |
| 20 | G8 = \$H\$3/B8 | | H8 = B8+C8-D8-G8 | | | | | |

Spreadsheet for Exercise 13-G.

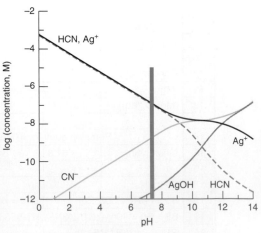

Graph for Exercise 13-G.

Rearrange Equation A to solve for $[Ag^+]$ as a function of $[H^+]$ or use SOLVER to find $[Ag^+]$ as a function of $[H^+]$. We will use the algebraic solution, which is easy for this exercise. Multiply both sides by $[Ag^+]$ and solve:

$$[Ag^+]^2 + \frac{K_{Ag}[Ag^+]^2}{[H^+]} = K_{sp} + \frac{[H^+]K_{sp}}{K_{HCN}}$$

$$[Ag^+]^2\left(\frac{[H^+] + K_{Ag}}{[H^+]}\right) = K_{sp}\left(\frac{K_{HCN} + [H^+]}{K_{HCN}}\right)$$

$$[Ag^+] = \sqrt{\frac{K_{sp}(K_{HCN} + [H^+])[H^+]}{K_{HCN}([H^+] + K_{Ag})}} \qquad (B)$$

The spreadsheet uses Equation B to find $[Ag^+]$ in column C. pH is input in column A. To find the pH of unbuffered solution, we find the pH at which the net charge in column H is zero. We used SOLVER to find that pH = 7.28 in cell A12 makes the net charge in cell H12 equal to 0.

**13-H. (a)** $\bar{n}_H = \dfrac{\text{moles of bound } H^+}{\text{total moles of weak acid}} = \dfrac{[HA]}{[HA] + [A^-]}$

$$= \frac{[HA]}{F_{HA}} = \frac{F_{HA} - [A^-]}{F_{HA}} \qquad (A)$$

Charge balance: $[H^+] + [Na^+] = [OH^-] + [Cl^-]_{HCl} + [A^-]$

or

$$-[A^-] = [OH^-] + [Cl^-]_{HCl} - [H^+] - [Na^+]$$

Put this expression for $-[A^-]$ into numerator of (A):

$$\bar{n}_H(\text{measured}) = \frac{F_{HA} + [OH^-] + [Cl^-]_{HCl} - [H^+] - [Na^+]}{F_{HA}}$$

$$= 1 + \frac{[OH^-] + [Cl^-]_{HCl} - [H^+] - [Na^+]}{F_{H_2A}}$$

Making the same substitutions used in Section 13-4 produces Equation 13-59 with $n = 1$. The expression for $\bar{n}_H(\text{theoretical})$ is just $\bar{n}_H(\text{theoretical}) = \alpha_{HA} = [H^+]/([H^+] + K_a)$.
**(b)** Optimized values of $pK_w'$ and $pK_a$ in cells B9 and B10 in the spreadsheet are 13.869 and 4.726. They were obtained from initial guesses of $pK_w' = 14$ and $pK_a = 5$ after executing SOLVER to minimize the sum of squared residuals in cell B11. The NIST database lists $pK_a = 4.757$ at $\mu = 0$ and $pK_a = 4.56$ at $\mu = 0.1$ M. Our observed

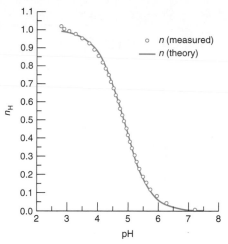

Graph for Exercise 13-H.

value of 4.726 at $\mu = 0.1$ M suggests that the titration experiment was not very accurate.

## Chapter 14

**14-A.** The cell voltage will be 1.35 V because all activities are unity.

$$I = P/E = 0.010\ 0\ W/1.35\ V = 7.41 \times 10^{-3}\ C/s$$
$$(7.41 \times 10^{-3}\ C/s)/(9.649 \times 10^4\ C/mol)$$
$$= 7.68 \times 10^{-8}\ mol\ e^-/s = 2.42\ mol\ e^-/365\ days$$
$$= 1.21\ mol\ HgO/365\ days = 0.262\ kg\ HgO$$
$$= 0.578\ lb$$

**14-B. (a)**
$$5Br_2(aq) + \cancel{10e^-} \rightleftharpoons 10Br^-$$
$$\underline{2IO_3^- + 12H^+ + \cancel{10e^-} \rightleftharpoons I_2(s) + 6H_2O}$$
$$I_2(s) + 5Br_2(aq) + 6H_2O \rightleftharpoons 2IO_3^- + 10Br^- + 12H^+$$

$E_+^\circ = 1.098\ V$

$\underline{E_-^\circ = 1.210\ V}$

$E^\circ = 1.098 - 1.210 = -0.112\ V$

$K = 10^{10(-0.112)/0.059\ 16} = 1 \times 10^{-19}$

|     | A | B | C | D | E | F | G |
|-----|---|---|---|---|---|---|---|
| 1 | Difference plot for acetic acid | | | | | | |
| 2 | | | | C15 = 10^-B15/$B$8 | | | |
| 3 | Titrant NaOH = | 0.4905 | $C_b$ (M) | D15 = 10^-$B$9/C15 | | | |
| 4 | Initial volume = | 200 | $V_o$ (mL) | E15 = $B$7+($B$6-$B$3*A15) | | | |
| 5 | Acetic acid = | 3.96 | L (mmol) | | -(C15-D15)*($B$4+A15))/$B$5 | | |
| 6 | HCl added = | 0.484 | A (mmol) | F15 = $C15/($C15+$E$10) | | | |
| 7 | Number of H⁺ = | 1 | n | G15 = (E15-F15)^2 | | | |
| 8 | Activity coeff = | 0.78 | $\gamma_H$ | | | | |
| 9 | $pK_w'$ = | 13.869 | | | | | |
| 10 | $pK_a$ = | 4.726 | | $K_a$ = | 1.881E-05 | = 10^-B10 | |
| 11 | $\Sigma(\text{resid})^2$ = | 0.0045 | = sum of column G | | | | |
| 12 | | | | | | | |
| 13 | v | pH | [H⁺] = | [OH⁻] = | Measured | Theoretical | (residuals)² = |
| 14 | mL NaOH | | (10^-pH)/$\gamma_H$ | (10^-pKw)/[H⁺] | $n_H$ | $n_H = \alpha_{HA}$ | $(n_{meas} - n_{theor})^2$ |
| 15 | 0.00 | 2.79 | 2.08E-03 | 6.50E-12 | 1.017 | 0.991 | 0.000685 |
| 16 | 0.30 | 2.89 | 1.65E-03 | 8.19E-12 | 1.002 | 0.989 | 0.000163 |
| 17 | : | | | | | | |
| 18 | 4.80 | 4.78 | 2.13E-05 | 6.36E-10 | 0.527 | 0.531 | 0.000018 |
| 19 | 5.10 | 4.85 | 1.81E-05 | 7.47E-10 | 0.490 | 0.491 | 0.000001 |
| 20 | : | | | | | | |
| 21 | 10.20 | 11.39 | 5.22E-12 | 2.59E-03 | -0.004 | 0.000 | 0.000014 |
| 22 | 10.50 | 11.54 | 3.70E-12 | 3.66E-03 | 0.016 | 0.000 | 0.000259 |

Spreadsheet for Exercise 13-H.

**(b)**
$$\underline{\begin{array}{l} Cr^{2+} + 2e^- \rightleftharpoons Cr(s) \\ Fe^{2+} + 2e^- \rightleftharpoons Fe(s) \end{array}}$$
$$Cr^{2+} + Fe(s) \rightleftharpoons Cr(s) + Fe^{2+}$$

$E°_+ = -0.89$ V

$E°_- = -0.44$ V

$E° = -0.89 - (-0.44) = -0.45$ V

$K = 10^{2(-0.45)/0.059\ 16} = 10^{-15}$

**(c)**
$$\underline{\begin{array}{l} Cl_2(g) + 2e^- \rightleftharpoons 2Cl^- \\ Mg^{2+} + 2e^- \rightleftharpoons Mg(s) \end{array}}$$
$$Mg(s) + Cl_2(g) \rightleftharpoons Mg^{2+} + 2Cl^-$$

$E°_+ = 1.360$ V

$E°_- = -2.360$ V

$E° = 1.360 - (-2.360) = 3.720$ V

$K = 10^{2(3.720)/0.059\ 16} = 6 \times 10^{125}$

**(d)**
$$\underline{\begin{array}{l} 3[MnO_2(s) + 4H^+ + 2e^- \rightleftharpoons Mn^{2+} + 2H_2O] \\ 2[MnO_4^- + 4H^+ + 3e^- \rightleftharpoons MnO_2(s) + 2H_2O] \end{array}}$$
$$5MnO_2(s) + 4H^+ \rightleftharpoons 2MnO_4^- + 3Mn^{2+} + 2H_2O$$

$E°_+ = 1.230$ V

$E°_- = 1.692$ V

$E° = 1.230 - 1.692 = -0.462$ V

$K = 10^{6(-0.462)/0.059\ 16} = 1 \times 10^{-47}$

An alternate way to answer **(d)** is:
$$\underline{\begin{array}{l} 5[MnO_2(s) + 4H^+ + 2e^- \rightleftharpoons Mn^{2+} + 2H_2O] \\ 2[MnO_4^- + 8H^+ + 5e^- \rightleftharpoons Mn^{2+} + 4H_2O] \end{array}}$$
$$5MnO_2(s) + 4H^+ \rightleftharpoons 2MnO_4^- + 3Mn^{2+} + 2H_2O$$

$E°_+ = 1.230$ V

$E°_- = 1.507$ V

$E° = 1.230 - 1.507 = -0.277$ V

$K = 10^{10(-0.277)/0.059\ 16} = 2 \times 10^{-47}$

**(e)**
$$\underline{\begin{array}{l} Ag^+ + e^- \rightleftharpoons Ag(s) \\ Ag(S_2O_3)_2^{3-} + e^- \rightleftharpoons Ag(s) + 2S_2O_3^{2-} \end{array}}$$
$$Ag^+ + 2S_2O_3^{2-} \rightleftharpoons Ag(S_2O_3)_2^{3-}$$

$E°_+ = 0.799$ V

$E°_- = 0.017$ V

$E° = 0.799 - 0.017 = 0.782$ V

$K = 10^{0.782/0.059\ 16} = 2 \times 10^{13}$

**(f)**
$$\underline{\begin{array}{l} CuI(s) + e^- \rightleftharpoons Cu(s) + I^- \\ Cu^+ + e^- \rightleftharpoons Cu(s) \end{array}}$$
$$CuI(s) \rightleftharpoons Cu^+ + I^-$$

$E°_+ = -0.185$ V

$E°_- = 0.518$ V

$E° = -0.185 - 0.518 = -0.703$ V

$K = 10^{-0.703/0.059\ 16} = 1 \times 10^{-12}$

**14-C. (a)**
$$\underline{\begin{array}{ll} Br_2(l) + 2e^- \rightleftharpoons 2Br^- & E°_+ = 1.078\ V \\ Fe^{2+} + 2e^- \rightleftharpoons Fe(s) & E°_- = -0.44\ V \end{array}}$$
$$Br_2(l) + Fe(s) \rightleftharpoons 2Br^- + Fe^{2+}$$

$$E = \left\{1.078 - \frac{0.059\ 16}{2} \log (0.050)^2\right\} -$$
$$\left\{-0.44 - \frac{0.059\ 16}{2} \log \frac{1}{0.010}\right\} = 1.155 - (-0.50) = 1.65\ V$$

Electrons flow from the more negative Fe electrode ($-0.50$ V) to the more positive Pt electrode (1.155 V).

**(b)**
$$\underline{\begin{array}{ll} Fe^{2+} + 2e^- \rightleftharpoons Fe(s) & E°_+ = -0.44\ V \\ Cu^{2+} + 2e^- \rightleftharpoons Cu(s) & E°_- = 0.339\ V \end{array}}$$
$$Fe^{2+} + Cu(s) \rightleftharpoons Fe(s) + Cu^{2+}$$

$$E = \left\{-0.44 - \frac{0.059\ 16}{2} \log \frac{1}{0.050}\right\} -$$
$$\left\{0.339 - \frac{0.059\ 16}{2} \log \frac{1}{0.020}\right\} = -0.48 - (0.289) = -0.77\ V$$

Electrons flow from the more negative Fe electrode ($-0.48$ V) to the more positive Cu electrode (0.289 V).

**(c)**
$$\underline{\begin{array}{ll} Cl_2(g) + 2e^- \rightleftharpoons 2Cl^- & E°_+ = 1.360\ V \\ Hg_2Cl_2(s) + 2e^- \rightleftharpoons 2Hg(l) + 2Cl^- & E°_- = 0.268\ V \end{array}}$$
$$Cl_2(g) + 2Hg(l) \rightleftharpoons Hg_2Cl_2(s)$$

$$E = \left\{1.360 - \frac{0.059\ 16}{2} \log \frac{(0.040)^2}{0.50}\right\} -$$
$$\left\{0.268 - \frac{0.059\ 16}{2} \log (0.060)^2\right\} = 1.434 - (0.340) = 1.094\ V$$

Electrons flow from the more negative Hg electrode (0.340 V) to the more positive Pt electrode (1.434 V).

**14-D. (a)** $H^+ + e^- \rightleftharpoons \frac{1}{2}H_2(g)$   $E°_+ = 0$ V
$$Ag^+ + e^- \rightleftharpoons Ag(s) \quad E°_- = 0.799\ V$$
$$E° = E°_+ - E°_- = -0.799\ V$$

$$E = \left\{0 - 0.059\ 16 \log \frac{P_{H_2}^{1/2}}{[H^+]}\right\} -$$
$$\left\{0.799 - 0.059\ 16 \log \frac{1}{[Ag^+]}\right\}$$

**(b)** $[Ag^+] = \dfrac{K_{sp}}{[I^-]} = \dfrac{8.3 \times 10^{-17}}{0.10} = 8.3 \times 10^{-16}$ M

$$E = \left\{0 - 0.059\ 16 \log \frac{\sqrt{0.20}}{0.10}\right\} -$$
$$\left\{0.799 - 0.059\ 16 \log \frac{1}{8.3 \times 10^{-16}}\right\} = -0.038 - (-0.093) = 0.055\ V$$

Electrons flow from the more negative Ag electrode ($-0.093$ V) to the more positive Pt electrode ($-0.038$ V).

**(c)** $H^+ + e^- \rightleftharpoons \frac{1}{2}H_2(g)$       $E°_+ = 0$ V
$$AgI(s) + e^- \rightleftharpoons Ag(s) + I^- \quad E°_- = ?$$

$$0.055 = \left\{0 - 0.059\ 16 \log \frac{\sqrt{0.20}}{0.10}\right\} - \{E°_- - 0.059\ 16 \log (0.10)\}$$
$$\Rightarrow E°_- = -0.153\ V$$

(Appendix H gives $E°_- = -0.152$ V.)

**14-E.** $Ag(CN)_2^- + e^- \rightleftharpoons Ag(s) + 2CN^-$   $E°_+ = -0.310$ V
$$Cu^{2+} + 2e^- \rightleftharpoons Cu(s) \qquad\qquad E°_- = 0.339\ V$$

$$E = \left\{-0.310 - 0.059\ 16 \log \frac{[CN^-]^2}{[Ag(CN)_2^-]}\right\} -$$
$$\left\{0.339 - \frac{0.059\ 16}{2} \log \frac{1}{[Cu^{2+}]}\right\}$$

We know that $[Ag(CN)_2^-] = 0.010$ M and $[Cu^{2+}] = 0.030$ M. To find $[CN^-]$ at pH 8.21, we write

$$\frac{[CN^-]}{[HCN]} = \frac{K_a}{[H^+]} \Rightarrow [CN^-] = 0.10\ [HCN]$$

But, because $[CN^-] + [HCN] = 0.10$ M, $[CN^-] = 0.009\ 1$ M.

Putting this concentration into the Nernst equation gives $E = -0.187 - (0.294) = -0.481$ V.

Electrons flow from the more negative Ag electrode ($-0.187$ V) to the more positive Cu electrode (0.294 V).

**14-F. (a)** $PuO_2^+ + e^- + 4H^+ \rightleftharpoons Pu^{4+} + 2H_2O$

$$\underline{\begin{array}{ll} PuO_2^{2+} \rightarrow PuO_2^+ & \Delta G = -1F\ (0.966) \\ PuO_2^+ \rightarrow Pu^{4+} & \Delta G = -1F\ E° \\ Pu^{4+} \rightarrow Pu^{3+} & \Delta G = -1F(1.006) \end{array}}$$
$$PuO_2^{2+} \rightarrow Pu^{3+} \qquad \Delta G = -3F\ (1.021)$$

$$-3F(1.021) = -1F(0.966) - 1FE° - 1F(1.006)$$
$$\Rightarrow E° = 1.091 \text{ V}$$

**(b)**
$$\begin{aligned}
&2PuO_2^{2+} + 2e^- \rightleftharpoons 2PuO_2^+ &&E_+° = 0.966 \text{ V}\\
&\tfrac{1}{2}O_2(g) + 2H^+ + 2e^- \rightleftharpoons H_2O &&E_-° = 1.229 \text{ V}\\
\hline
&2PuO_2^{2+} + H_2O \rightleftharpoons 2PuO_2^+ + \tfrac{1}{2}O_2(g) + 2H^+
\end{aligned}$$

$$E = \left\{ 0.966 - \frac{0.059\ 16}{2} \log \frac{[PuO_2^+]^2}{[PuO_2^{2+}]^2} \right\} - $$
$$\left\{ 1.229 - \frac{0.059\ 16}{2} \log \frac{1}{P_{O_2}^{1/2}[H^+]^2} \right\}$$

$[PuO_2^+]$ cancels $[PuO_2^{2+}]$ because they are equal. At pH 2.00, we insert $[H^+] = 10^{-2.00}$ and $P_{O_2} = 0.20$ to find $E = -0.134$ V. Because $E < 0$, the reaction is not spontaneous and water is not oxidized. At pH 7.00, we find $E = +0.161$ V, so water will be oxidized.

**14-G.**
$$\begin{aligned}
&2H^+ + 2e^- \rightleftharpoons H_2(g) &&E_+° = 0 \text{ V}\\
&Hg_2Cl_2(s) + 2e^- \rightleftharpoons 2Hg(l) + 2Cl^- &&E_-° = 0.268 \text{ V}
\end{aligned}$$

$$E = \left\{ -\frac{0.059\ 16}{2} \log \frac{P_{H_2}}{[H^+]^2} \right\} - \left\{ 0.268 - \frac{0.059\ 16}{2} \log[Cl^-]^2 \right\}$$

We find $[H^+]$ in the right half-cell by considering the acid-base chemistry of KHP, the intermediate form of a diprotic acid:

$$[H^+] = \sqrt{\frac{K_1 K_2(0.050) + K_1 K_w}{K_1 + 0.050}} = 6.5 \times 10^{-5} \text{ M}$$

$$E = \left\{ \frac{-0.059\ 16}{2} \log \frac{1}{(6.5 \times 10^{-5})^2} \right\} - $$
$$\left\{ 0.268 - \frac{0.059\ 16}{2} \log (0.10)^2 \right\} = -0.575 \text{ V}$$

**14-H.**
$$\begin{aligned}
&2H^+ + 2e^- \rightleftharpoons H_2(g) &&E_+° = 0 \text{ V}\\
&Hg^{2+} + 2e^- \rightleftharpoons Hg(l) &&E_-° = 0.852 \text{ V}
\end{aligned}$$

$$E = \left\{ \frac{-0.059\ 16}{2} \log \frac{P_{H_2}}{[H^+]^2} \right\} - \left\{ 0.852 - \frac{0.059\ 16}{2} \log \frac{1}{[Hg^{2+}]} \right\}$$

$$0.083 = \left\{ \frac{-0.059\ 16}{2} \log \frac{1}{1^2} \right\} - \left\{ 0.852 - \frac{0.059\ 16}{2} \log \frac{1}{[Hg^{2+}]} \right\}$$

$$\Rightarrow [Hg^{2+}] = 2.5 \times 10^{-32} \text{ M}$$

$[HgI_4^{2-}] = 0.001\ 0$ M. To make this much $HgI_4^{2-}$, the concentration of $I^-$ must have been reduced from 0.500 M to 0.496 M, because one $Hg^{2+}$ ion reacts with four $I^-$ ions.

$$K = \frac{[HgI_4^{2-}]}{[Hg^{2+}][I^-]^4} = \frac{(0.001\ 0)}{(2.5 \times 10^{-32})(0.496)^4}$$
$$= 7 \times 10^{29}$$

**14-I.**
$$\begin{aligned}
&CuY^{2-} + 2e^- \rightleftharpoons Cu(s) + Y^{4-} &&E_+° = ?\\
&Cu^{2+} + 2e^- \rightleftharpoons Cu(s) &&E_-° = 0.339 \text{ V}\\
\hline
&CuY^{2-} \xrightleftharpoons{1/K_f} Cu^{2+} + Y^{4-} &&E°
\end{aligned}$$

$$E° = \frac{0.059\ 16}{2} \log \frac{1}{K_f} = -0.556 \text{ V}$$

$$E_+° = E° + E_-° = -0.556 + 0.339 = -0.217 \text{ V}$$

**14-J.** To compare glucose and $H_2$ at pH = 0, we need to know $E°$ for each. For $H_2$, $E° = 0$ V. For glucose, we find $E°$ from $E°'$:

$$\underset{\text{Gluconic acid}}{HA} + 2H^+ + 2e^- \rightleftharpoons \underset{\text{Glucose}}{G} + H_2O$$

$$E = E° - \frac{0.059\ 16}{2} \log \frac{[G]}{[HA][H^+]^2}$$

But $F_G = [G]$ and $[HA] = \dfrac{[H^+]F_{HA}}{[H^+] + K_a}$. Putting these into the Nernst equation gives

$$E = E° - \frac{0.059\ 16}{2} \log \frac{F_G}{\left( \dfrac{[H^+]F_{HA}}{[H^+] + K_a} \right)[H^+]^2}$$

$$= E° - \frac{0.059\ 16}{2} \log \frac{[H^+] + K_a}{[H^+]^3} - \frac{0.059\ 16}{2} \log \frac{F_G}{F_{HA}}$$

This is $E°' = -0.45$ V when $[H^+] = 10^{-7}$

$$-0.45 \text{ V} = E° - \frac{0.059\ 16}{2} \log \frac{10^{-7.00} + 10^{-3.56}}{(10^{-7.00})^3}$$

$$\Rightarrow E° = +0.066 \text{ V for glucose}$$

Because $E°$ for $H_2$ is more negative than $E°$ for glucose, $H_2$ is the stronger reducing agent at pH 0.00.

**14-K. (a)** Each $H^+$ must provide $\tfrac{1}{2}(34.5 \text{ kJ})$ when it passes from outside to inside.

$$\Delta G = -\tfrac{1}{2}(34.5 \times 10^3 \text{ J}) = -RT \ln \frac{\mathcal{A}_{\text{high}}}{\mathcal{A}_{\text{low}}}$$

$$\frac{\mathcal{A}_{\text{high}}}{\mathcal{A}_{\text{low}}} = 1.05 \times 10^3$$

$$\Rightarrow \Delta pH = \log(1.05 \times 10^3)$$
$$= 3.02 \text{ pH units}$$

**(b)** $\Delta G = -nFE$ (where $n$ = charge of $H^+$ = 1)
$$-\tfrac{1}{2}(34.5 \times 10^3 \text{ J}) = -1FE \Rightarrow E = 0.179 \text{ V}$$

**(c)** If $\Delta pH = 1.00$, $\mathcal{A}_{\text{high}}/\mathcal{A}_{\text{low}} = 10$.

$$\Delta G(pH) = -RT \ln 10 = -5.7 \times 10^3 \text{ J}$$
$$\Delta G(\text{electric}) = [\tfrac{1}{2}(34.5) - 5.7] \text{ kJ} = 11.5 \text{ kJ}$$

$$E = \frac{\Delta G(\text{electric})}{F} = 0.120 \text{ V}$$

# Chapter 15

**15-A.** The reaction at the silver electrode (written as a reduction) is $Ag^+ + e^- \rightleftharpoons Ag(s)$, and the cell voltage is written as
$$E = E_+ - E_-$$

$$= (+0.200) - \left( 0.799 - 0.059\ 16 \log \frac{1}{[Ag^+]} \right)$$

$$= -0.599 - 0.059\ 16 \log [Ag^+]$$

Titration reactions:
$$\begin{aligned}
&Br^- + Ag^+ \rightarrow AgBr(s) &&(K_{sp} = 5.0 \times 10^{-13})\\
&Br^- + Tl^+ \rightarrow TlBr(s) &&(K_{sp} = 3.6 \times 10^{-6})
\end{aligned}$$

The two equivalence points are at 25.0 and 50.0 mL. Between 0 and 25 mL, there is unreacted $Ag^+$ in the solution.

At 1.0 mL: $[Ag^+] = \underbrace{\left(\dfrac{24.0}{25.0}\right)(0.050\ 0 \text{ M})}_{\substack{\text{Initial concentration} \\ \text{of } Ag^+}}\left(\dfrac{100.0}{101.0}\right)$

$$= 0.047\ 5 \text{ M} \Rightarrow E = -0.521 \text{ V}$$

At 15.0 mL: $[Ag^+] = \left(\dfrac{10.0}{25.0}\right)(0.050\ 0 \text{ M})\left(\dfrac{100.0}{115.0}\right)$

$$= 0.017\ 4 \text{ M} \Rightarrow E = -0.495 \text{ V}$$

At 24.0 mL: $[Ag^+] = \left(\dfrac{1.0}{25.0}\right)(0.050\ 0 \text{ M})\left(\dfrac{100.0}{124.0}\right)$

$$= 0.001\ 61 \text{ M} \Rightarrow E = -0.434 \text{ V}$$

At 24.9 mL: $[Ag^+] = \left(\dfrac{0.10}{25.0}\right)(0.050\ 0 \text{ M})\left(\dfrac{100.0}{124.9}\right)$

$$= 1.60 \times 10^{-4} \text{ M}$$
$$\Rightarrow E = -0.374 \text{ V}$$

Between 25 mL and 50 mL, all AgBr has precipitated and TlBr is in the process of precipitating. There is some unreacted $Tl^+$ left in solution in this region.

At 25.2 mL:

$$[Tl^+] = \left(\dfrac{24.8}{25.0}\right)(0.050\ 0 \text{ M})\left(\dfrac{100.0}{125.2}\right)$$
$$= 3.96 \times 10^{-2} \text{ M}$$

$$[Br^-] = K_{sp} \text{ (for TlBr)}/[Tl^+] = 9.0_9 \times 10^{-5} \text{ M}$$
$$[Ag^+] = K_{sp} \text{ (for AgBr)}/[Br^-] = 5.5 \times 10^{-9} \text{ M}$$
$$E = -0.599 - 0.059\ 16 \log [Ag^+]$$
$$= -0.110 \text{ V}$$

At 35.0 mL: $[Tl^+] = \left(\dfrac{15.0}{25.0}\right)(0.050\ 0 \text{ M})\left(\dfrac{100.0}{135.0}\right)$
$$= 0.022\ 2 \text{ M}$$
$$\Rightarrow [Br^-] = 1.62 \times 10^{-4} \text{ M}$$
$$\Rightarrow [Ag^+] = 3.08 \times 10^{-9} \text{ M}$$
$$\Rightarrow E = -0.095 \text{ V}$$

50.0 mL is the second equivalence point, at which $[Tl^+] = [Br^-]$.
At 50.0 mL: $[Tl^+][Br^-] = K_{sp}$ (for TlBr)
$$\Rightarrow [Tl^+] = \sqrt{K_{sp}} = 1.90 \times 10^{-3} \text{ M}$$
$$\Rightarrow [Br^-] = 1.90 \times 10^{-3} \text{ M}$$
$$\Rightarrow [Ag^+] = 2.64 \times 10^{-10} \text{ M} \Rightarrow E = -0.032 \text{ V}$$

At 60.0 mL, there is excess $Br^-$ in the solution:
$$[Br^-] = \left(\dfrac{10.0}{160.0}\right)(0.200 \text{ M}) = 0.012\ 5 \text{ M}$$
$$\Rightarrow [Ag^+] = 4.00 \times 10^{-11} \text{ M} \Rightarrow E = +0.016 \text{ V}$$

**15-B.** The cell voltage is given by Equation C, in which $K_f$ is the formation constant for $Hg(EDTA)^{2-}$ $(= 10^{21.5})$. To find the voltage, we must calculate $[HgY^{2-}]$ and $[Y^{4-}]$ at each point. The concentration of $HgY^{2-}$ is $1.0 \times 10^{-4}$ M when $V = 0$ and is thereafter affected only by dilution because $K_f(HgY^{2-}) \gg K_f(MgY^{2-})$. The concentration of $Y^{4-}$ is found from the Mg-EDTA equilibrium at all but the first point. At $V = 0$ mL, the Hg-EDTA equilibrium determines $[Y^{4-}]$.

At 0 mL:
$$\dfrac{[HgY^{2-}]}{[Hg^{2+}][EDTA]} = \alpha_{Y^{4-}} K_f \text{ (for } HgY^{4-}) = (0.30)(10^{21.5})$$
$$\dfrac{1.0 \times 10^{-4} - x}{(x)(x)} = 9.5 \times 10^{20}$$
$$\Rightarrow x = [EDTA] = 3.2 \times 10^{-13} \text{ M}$$
$$[Y^{4-}] = \alpha_{Y^{4-}} [EDTA]$$
$$= 9.7 \times 10^{-14} \text{ M}$$

Using Equation C, we write
$$E = 0.852 - 0.241$$
$$- \dfrac{0.059\ 16}{2} \log \dfrac{10^{21.5}}{1.0 \times 10^{-4}}$$
$$- \dfrac{0.059\ 16}{2} \log (9.7 \times 10^{-14})$$
$$= 0.242 \text{ V}$$

At 10.0 mL: Because $V_e = 25.0$ mL, $\frac{10}{25}$ of the $Mg^{2+}$ is in the form $MgY^{2-}$, and $\frac{15}{25}$ is in the form $Mg^{2+}$.
$$[Y^{4-}] = \dfrac{[MgY^{2-}]}{[Mg^{2+}]} \Big/ K_f \text{ (for } MgY^{2-})$$
$$= \left(\dfrac{10}{15}\right) \Big/ 6.2 \times 10^8 = 1.08 \times 10^{-9} \text{ M}$$
$$[HgY^{2-}] = \underbrace{\left(\dfrac{50.0}{60.0}\right)}_{\substack{\text{Dilution} \\ \text{factor}}}(1.0 \times 10^{-4} \text{ M}) = 8.33 \times 10^{-5} \text{ M}$$
$$E = 0.852 - 0.241 - \dfrac{0.059\ 16}{2} \log \dfrac{10^{21.5}}{8.33 \times 10^{-5}}$$
$$- \dfrac{0.059\ 16}{2} \log (1.08 \times 10^{-9})$$
$$= 0.120 \text{ V}$$

At 20.0 mL:
$$[Y^{4-}] = \left(\dfrac{20}{5}\right) \Big/ 6.2 \times 10^8 = 6.45 \times 10^{-9} \text{ M}$$
$$[HgY^{2-}] = \left(\dfrac{50.0}{70.0}\right)(1.0 \times 10^{-4} \text{ M}) = 7.14 \times 10^{-5} \text{ M}$$
$$\Rightarrow E = 0.095 \text{ V}$$

At 24.9 mL:
$$[Y^{4-}] = \left(\dfrac{24.9}{0.1}\right) \Big/ 6.2 \times 10^8$$
$$= 4.02 \times 10^{-7} \text{ M}$$
$$[HgY^{2-}] = \left(\dfrac{50.0}{74.9}\right)(1.0 \times 10^{-4} \text{ M})$$
$$= 6.68 \times 10^{-5} \text{ M}$$
$$\Rightarrow E = 0.041 \text{ V}$$

At 25.0 mL: This is the equivalence point, at which $[Mg^{2+}] = [EDTA]$.
$$\dfrac{[MgY^{2-}]}{[Mg^{2+}][EDTA]} = \alpha_{Y^{4-}} K_f \text{ (for } MgY^{2-})$$
$$\dfrac{\left(\dfrac{50.0}{75.0}\right)(0.010\ 0) - x}{x^2} = 1.85 \times 10^8$$
$$\Rightarrow x = 5.48 \times 10^{-6} \text{ M}$$
$$[Y^{4-}] = \alpha_{Y^{4-}} (6.0 \times 10^{-6} \text{ M})$$
$$= 1.80 \times 10^{-6} \text{ M}$$
$$[HgY^{2-}] = \left(\dfrac{50.0}{75.0}\right)(1.0 \times 10^{-4} \text{ M})$$
$$= 6.67 \times 10^{-5} \text{ M}$$
$$\Rightarrow E = 0.021 \text{ V}$$

At 26.0 mL: Now there is excess EDTA in the solution:
$$[Y^{4-}] = \alpha_{Y^{4-}} [EDTA] = (0.36)\left[\left(\dfrac{1.0}{76.0}\right)(0.020\ 0 \text{ M})\right]$$
$$= 9.47 \times 10^{-5} \text{ M}$$
$$[HgY^{2-}] = \left(\dfrac{50.0}{76.0}\right)(1.0 \times 10^{-4} \text{ M}) = 6.58 \times 10^{-5} \text{ M}$$
$$\Rightarrow E = -0.030 \text{ V}$$

**15-C.** At intermediate pH, the voltage will be constant at 100 mV. When $[OH^-] \approx [F^-]/10 = 10^{-6}$ M (pH = 8), the electrode begins to respond to $OH^-$ and the voltage will decrease (i.e., the electrode potential will change in the same direction as if more $F^-$ were being added). Near pH = 3.17 $(= pK_a$ for HF), $F^-$ reacts with $H^+$ and the concentration of free $F^-$ decreases. At pH = 1.17, $[F^-] \approx 1\%$ of $10^{-5}$ M $= 10^{-7}$ M, and $E \approx 100 + 2(59) = 218$ mV. A qualitative sketch of this behavior is shown here. The slope at high pH is less than 59 mV/pH unit, because the response of the electrode to $OH^-$ is less than the response to $F^-$.

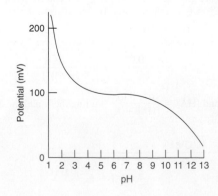

**15-D.** **(a)** For 1.00 mM $Na^+$ at pH 8.00, we can write

$$E = \text{constant} + 0.059\,16\,\log([Na^+] + 36[H^+]) - 0.038$$
$$= \text{constant} + 0.059\,16\,\log\,[(1.00 \times 10^{-3}) + (36 \times 10^{-8})]$$
$$\Rightarrow \text{constant} = +0.139\text{ V}$$

For 5.00 mM $Na^+$ at pH 8.00, we have

$$E = +0.139 + 0.059\,16\,\log\,[(5.00 \times 10^{-3}) + (36 \times 10^{-8})]$$
$$= 0.003\text{ V}$$

**(b)** For 1.00 mM $Na^+$ at pH 3.87, we have

$$E = +0.139 + 0.059\,16\,\log[(1.00 \times 10^{-3}) + (36 \times 10^{-3.87})]$$
$$= 0.007\text{ V}$$

**15-E.** A graph of $E(\text{mV})$ versus $\log[NH_3(M)]$ gives a straight line whose equation is $E = 563.4 + 59.05 \times \log[NH_3]$. For $E = 339.3$ mV, $[NH_3] = 1.60 \times 10^{-4}$ M. The sample analyzed contains $(100\text{ mL}) \times (1.60 \times 10^{-4}\text{ M}) = 0.016\,0$ mmol of nitrogen. But this sample represents just 2.00% (20.0 mL/1.00 L) of the food sample. Therefore, the food contains $0.016/0.020\,0 = 0.800$ mmol of nitrogen $= 11.2$ mg of N $= 3.59$ wt% nitrogen.

**15-F.** The function to plot on the $y$-axis is $(V_0 + V_S)\,10^{E/S}$, where $S = \beta RT \ln 10/nF$. $\beta$ is 0.985. Putting in $R = 8.314\,472$ J/(mol·K), $F = 96\,485.3$ C/mol, $T = 298.15$ K, and $n = -2$ gives $S = -0.029\,136$ J/C $= -0.029\,136$ V. [You can get the relation J/C = V from the equation $\Delta G = -nFE$, in which the units are J = (mol)(C/mol)(V).]

| $V_S$ (mL) | $E$ (V) | $y$ |
|---|---|---|
| 0 | 0.046 5 | 0.633 8 |
| 1.00 | 0.040 7 | 1.042 5 |
| 2.00 | 0.034 4 | 1.781 1 |
| 3.00 | 0.030 0 | 2.615 2 |
| 4.00 | 0.026 5 | 3.571 7 |

The data are plotted in Figure 15-23, which has a slope of $m = 0.744\,84$ and an intercept of $b = 0.439\,19$, giving an $x$-intercept of $-b/m = -0.58_{9\,65}$ mL. The concentration of original unknown is

$$c_X = -\frac{(x\text{-intercept})c_S}{V_0} = -\frac{(-0.58_{9\,65}\text{ mL})(1.78\text{ mM})}{25.0\text{ mL}}$$
$$= 4.2 \times 10^{-5}\text{ M}$$

(We decided that the last significant digit in the $x$-intercept was the 0.01 decimal place because the original data were only measured to the 0.01 decimal place.)

## Chapter 16

**16-A.** Titration reaction: $Sn^{2+} + 2Ce^{4+} \rightarrow Sn^{4+} + 2Ce^{3+}$

$$V_e = 10.0\text{ mL}$$

Representative calculations:

At 0.100 mL:

$$E_+ = 0.139 - \frac{0.059\,16}{2}\log\frac{[Sn^{2+}]}{[Sn^{4+}]}$$
$$= 0.139 - \frac{0.059\,16}{2}\log\frac{9.90}{0.100} = 0.080\text{ V}$$
$$E = E_+ - E_- = 0.080 - 0.241 = -0.161\text{ V}$$

At 10.00 mL:

$$2E_+ = 2(0.139) - 0.059\,16\,\log\frac{[Sn^{2+}]}{[Sn^{4+}]}$$

$$E_+ = 1.47 - 0.059\,16\,\log\frac{[Ce^{3+}]}{[Ce^{4+}]}$$
_____
$$3E_+ = 1.748 - 0.059\,16\,\log\frac{[Sn^{2+}][Ce^{3+}]}{[Sn^{4+}][Ce^{4+}]}$$

At the equivalence point, $[Sn^{4+}] = \frac{1}{2}[Ce^{3+}]$ and $[Sn^{2+}] = \frac{1}{2}[Ce^{4+}]$, which makes the log term 0. Therefore $3E_+ = 1.748$ and $E_+ = 0.583$ V.

$$E = E_+ - E_- = 0.583 - 0.241 = 0.342\text{ V}$$

At 10.10 mL:

$$E_+ = 1.47 - 0.059\,16\,\log\frac{[Ce^{3+}]}{[Ce^{4+}]}$$
$$= 1.47 - 0.059\,16\,\log\frac{10.0}{0.10} = 1.35_2\text{ V}$$
$$E = E_+ - E_- = 1.35_2 - 0.241 = 1.11\text{ V}$$

| mL | $E$ (V) | mL | $E$ (V) |
|---|---|---|---|
| 0.100 | −0.161 | 10.00 | 0.342 |
| 1.00 | −0.130 | 10.10 | 1.11 |
| 5.00 | −0.102 | 12.00 | 1.19 |
| 9.50 | −0.064 | | |

**16-B.** Standard potentials: indigo tetrasulfonate, 0.36 V; $Fe[CN]_6^{3-}\,|\,Fe[CN]_6^{4-}$, 0.356 V; $Tl^{3+}\,|\,Tl^{3+}$, 0.77 V. The end-point potential will be between 0.356 and 0.77 V. Indigo tetrasulfonate changes color near 0.36 V. Therefore it will not be a useful indicator for this titration.

**16-C.** Titration: $MnO_4^- + 5Fe^{2+} + 8H^+ \rightarrow Mn^{2+} + 5Fe^{3+} + 4H_2O$

$$Fe^{3+} + e^- \rightleftharpoons Fe^{2+} \qquad\qquad E° = 0.68\text{ V in 1 M }H_2SO_4$$
$$MnO_4^- + 8H^+ + 5e^- \rightarrow Mn^{2+} + 4H_2O \quad E° = 1.507\text{ V}$$

The equivalence point comes at 15.0 mL. Before the equivalence point:

$$E = E_+ - E_- = \left(0.68 - 0.059\,16\,\log\frac{[Fe^{2+}]}{[Fe^{3+}]}\right) - 0.241$$

1.0 mL: $[Fe^{2+}]/[Fe^{3+}] = 14.0/1.0 \Rightarrow E = 0.371$ V
7.5 mL: $[Fe^{2+}]/[Fe^{3+}] = 7.5/7.5 \Rightarrow E = 0.439$ V
14.0 mL: $[Fe^{2+}]/[Fe^{3+}] = 1.0/14.0 \Rightarrow E = 0.507$ V

At the equivalence point, use Equation E of Demonstration 16-1:

$$6E_+ = 8.215 - 0.059\,16\,\log\frac{1}{[H^+]^8}\overset{pH\,=\,0}{\Rightarrow} E_+ = 1.369\text{ V}$$

$$E = E_+ - E_- = 1.369 - 0.241 = 1.128\text{ V}$$

After the equivalence point:

$$E = E_+ - E_- = \left(1.507 - \frac{0.059\,16}{5}\log\frac{[Mn^{2+}]}{[MnO_4^-][H^+]^8}\right) - 0.241$$

16.0 mL: $[Mn^{2+}]/[MnO_4^-] = 15.0/1.0$ and $[H^+] = 1$ M
$$\Rightarrow E = 1.252\text{ V}$$

30.0 mL: $[Mn^{2+}]/[MnO_4^-] = 15.0/15.0$ and $[H^+] = 1$ M
$$\Rightarrow E = 1.266\text{ V}$$

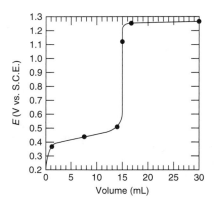

**16-D.** $I_3^- + 2S_2O_3^{2-} \rightarrow 3I^- + S_4O_6^{2-}$
823 μL of 0.098 8 M $S_2O_3^{2-} = 81.3_1$ μmol of $S_2O_3^{2-} = 40.6_6$ μmol of $I_3^-$. But 1 mole of unreacted $IO_4^-$ gives 1 mole of $I_3^-$. Therefore, $40.6_6$ μmol of $IO_4^-$ was left from the periodate oxidation of the amino acids. The original amount of $IO_4^-$ was 2.000 mL of 0.048 7 M $IO_4^- = 97.4_0$ μmol. The difference $(97.40 - 40.66 = 56.74)$ is the number of micromoles of serine + threonine in 128.6 mg of protein. But with FM = 58 600, 128.6 mg of protein = 2.195 μmol. (Serine + threonine)/protein = 56.74 μmol/ 2.195 μmol = $25.85 \approx 26$ residues/molecule.

**16-E.** The Gran plot of $V*10^{-E/0.059\ 16}$ versus $V$ is shown in Figure 16-4. The data from 8.5–12.5 mL appear to be on a straight line. The least-squares line through these four points has a slope of $m = -1.567\ 3 \times 10^{-11}$ and an intercept of $b = 2.170\ 2 \times 10^{-10}$. The x-intercept is $-b/m = 13.85$ mL. The amount of $Ce^{4+}$ required to reach the equivalence point is $(0.100\ \text{mmol/mL})(13.85\ \text{mL}) = 1.385$ mmol, and the concentration of unknown $Fe^{2+}$ is $1.385\ \text{mmol}/50.0\ \text{mL} = 0.027\ 7$ M.

| Titrant volume, $V$ (mL) | $E$ (volts) | $V*10^{-E/0.059\ 16}$ |
|---|---|---|
| 6.50 | 0.635 | $1.200\ 3 \times 10^{-10}$ |
| 8.50 | 0.651 | $8.421\ 0 \times 10^{-11}$ |
| 10.50 | 0.669 | $5.162\ 6 \times 10^{-11}$ |
| 11.50 | 0.680 | $3.685\ 1 \times 10^{-11}$ |
| 12.50 | 0.696 | $2.148\ 8 \times 10^{-11}$ |

## Chapter 17

**17-A.** Cathode: $2H^+ + 2e^- \rightleftharpoons H_2(g)$   $E° = 0$ V

Anode (written as a reduction):

$$\tfrac{1}{2}O_2(g) + 2H^+ + 2e^- \rightleftharpoons H_2O \quad E° = 1.229\ \text{V}$$

$$E(\text{cathode}) = 0 - \frac{0.059\ 16}{2}\log\frac{P_{H_2}}{[H^+]^2}$$

$$E(\text{anode}) = 1.229 - \frac{0.059\ 16}{2}\log\frac{1}{[H^+]^2 P_{O_2}^{1/2}}$$

$$E(\text{cell}) = E(\text{cathode}) - E(\text{anode})$$

$$= -1.229 - \frac{0.059\ 16}{2}\log P_{H_2} P_{O_2}^{1/2} = -1.229\ \text{V}$$

$$E = E(\text{cell}) - I\cdot R - \text{overpotentials}$$
$$= -1.229 - (0.100\ \text{A})(2.00\ \Omega)$$
$$\underbrace{-\ 0.85\ \text{V}}_{\substack{\text{Anode}\\\text{overpotential}}} \underbrace{-\ 0.068\ \text{V}}_{\substack{\text{Cathode}\\\text{overpotential}}} = -2.35\ \text{V}$$

From Table 17-1

For Au electrodes, the overpotentials are 0.963 and 0.390 V, giving $E = -2.78$ V.

**17-B.** **(a)** To electrolyze 0.010 M $SbO^+$ requires a potential of

$$E(\text{cathode}) = 0.208 - \frac{0.059\ 16}{3}\log\frac{1}{[SbO^+][H^+]^2}$$

$$= 0.208 - \frac{0.059\ 16}{3}\log\frac{1}{(0.010)(1.0)^2}$$

$$= 0.169\ \text{V}$$

$$E(\text{cathode versus Ag}\,|\,\text{AgCl}) = E(\text{versus S.H.E.}) - E(\text{Ag}\,|\,\text{AgCl})$$
$$= 0.169 - 0.197 = -0.028\ \text{V}$$

**(b)** The concentration of $[Cu^{2+}]$ that would be in equilibrium with $Cu(s)$ at 0.169 V is found as follows:

$$Cu^{2+} + 2e^- \rightleftharpoons Cu(s) \quad E° = 0.339$$

$$E(\text{cathode}) = 0.339 - \frac{0.059\ 16}{2}\log\frac{1}{[Cu^{2+}]}$$

$$0.169 = 0.339 - \frac{0.059\ 16}{2}\log\frac{1}{[Cu^{2+}]}$$

$$\Rightarrow [Cu^{2+}] = 1.8 \times 10^{-6}\ \text{M}$$

$$\text{Percentage of } Cu^{2+} \text{ not reduced} = \frac{1.8 \times 10^{-6}}{0.10} \times 100 = 1.8 \times 10^{-3}\%$$

$$\text{Percentage of } Cu^{2+} \text{ reduced} = 99.998\%$$

**17-C.** **(a)** $Co^{2+} + 2e^- \rightleftharpoons Co(s)$   $E° = -0.282$ V

$$E(\text{cathode versus S.H.E.}) = -0.282 - \frac{0.059\ 16}{2}\log\frac{1}{[Co^{2+}]}$$

Putting in $[Co^{2+}] = 1.0 \times 10^{-6}$ M gives $E = -0.459$ V and

$$E(\text{cathode versus S.C.E.}) = -0.459 - \underbrace{0.241}_{E(\text{S.C.E.})} = -0.700\ \text{V}$$

**(b)** $Co(C_2O_4)_2^{2-} + 2e^- \rightleftharpoons Co(s) + 2C_2O_4^{2-}$   $E° = -0.474$ V

$E(\text{cathode versus S.C.E.})$

$$= -0.474 - \frac{0.059\ 16}{2}\log\frac{[C_2O_4^{2-}]^2}{[Co(C_2O_4)_2^{2-}]} - 0.241$$

Putting in $[C_2O_4^{2-}] = 0.10$ M and $[Co(C_2O_4)_2^{2-}] = 1.0 \times 10^{-6}$ M gives $E = -0.833$ V.

**(c)** We can think of the reduction as $Co^{2+} + 2e^- \rightleftharpoons Co(s)$, for which $E° = -0.282$ V. But the concentration of $Co^{2+}$ is that tiny amount in equilibrium with 0.10 M EDTA plus $1.0 \times 10^{-6}$ M $Co(EDTA)^{2-}$. In Table 12-2, we find that the formation constant for $Co(EDTA)^{2-}$ is $10^{16.45} = 2.8 \times 10^{16}$.

$$K_f = \frac{[Co(EDTA)^{2-}]}{[Co^{2+}][EDTA^{4-}]} = \frac{[Co(EDTA)^{2-}]}{[Co^{2+}]\alpha_{Y^{4-}}\cdot F}$$

where F is the formal concentration of EDTA ($= 0.10$ M) and $\alpha_{Y^{4-}} = 3.8 \times 10^{-4}$ at pH 7.00 (Table 12-1). Putting in $[Co(EDTA)^{2-}] = 1.0 \times 10^{-6}$ M and solving for $[Co^{2+}]$ gives $[Co^{2+}] = 9.4 \times 10^{-19}$ M.

$$E = -0.282 - \frac{0.059\ 16}{2}\log\frac{1}{9.4 \times 10^{-19}} - 0.241$$

$$= -1.056\ \text{V}$$

**17-D.** **(a)** 75.00 mL of 0.023 80 M KSCN = 1.785 mmol of $SCN^-$, which gives 1.785 mmol of AgSCN, containing 0.103 7 g of SCN. Final mass = $12.463\ 8 + 0.103\ 7 = 12.567\ 5$ g.

**(b)** Anode: $AgBr(s) + e^- \rightleftharpoons Ag(s) + Br^-$   $E° = 0.071$ V

$$E(\text{anode}) = 0.071 - 0.059\ 16\ \log[Br^-]$$
$$= 0.071 - 0.059\ 16\ \log[0.10] = 0.130\ \text{V}$$

$$E(\text{cathode}) = E(\text{S.C.E.}) = 0.241\ \text{V}$$
$$E = E(\text{cathode}) - E(\text{anode}) = 0.111\ \text{V}$$

**(c)** To remove 99.99% of 0.10 M KI will leave $[I^-] = 1.0 \times 10^{-5}$ M. The concentration of $Ag^+$ in equilibrium with this much $I^-$ is $[Ag^+] = K_{sp}/[I^-] = (8.3 \times 10^{-17})/(1.0 \times 10^{-5}) = 8.3 \times 10^{-12}$ M. The concentration of $Ag^+$ in equilibrium with 0.10 M $Br^-$ is $[Ag^+] = K_{sp}/[I^-] = (5.0 \times 10^{-13})/(0.10) = 5.0 \times 10^{-12}$ M. Therefore $8.3 \times 10^{-12}$ M $Ag^+$ will begin to precipitate 0.10 M $Br^-$. The separation is not possible.

**17-E.** The corrected coulometric titration time is $387 - 6 = 381$ s. $q = It/F = (4.23\ \text{mA})(381\ \text{s})/(96\ 485\ \text{C/mol}) = 16.7\ \mu\text{mol}\ e^-$. Because $1e^-$ is equivalent to one $X^-$, the concentration of organohalide is 16.7 $\mu$M. If all halogen is Cl, this corresponds to 592 $\mu$g Cl/L.

**17-F.** **(a)** Use the internal standard equation with $X = Pb^{2+}$ and $S = Cd^{2+}$. From the standard mixture, we find the response factor, $F$:

$$\frac{\text{Signal}_X}{[X]} = F\left(\frac{\text{signal}_S}{[S]}\right)$$

$$\frac{1.58\ \mu\text{A}}{[41.8\ \mu\text{M}]} = F\left(\frac{1.64\ \mu\text{A}}{[32.3\ \mu\text{M}]}\right) \Rightarrow F = 0.744_5$$

$[Cd^{2+}]$ standard added to unknown

$$= \left(\frac{10.00}{50.00}\right)(3.23 \times 10^{-4}\ \text{M}) = 6.46 \times 10^{-5}\ \text{M}$$

For the unknown mixture, we can now say

$$\frac{\text{Signal}_X}{[X]} = F\left(\frac{\text{signal}_S}{[S]}\right)$$

$$\frac{3.00\ \mu\text{A}}{[Pb^{2+}]} = 0.744_5\left(\frac{2.00\ \mu\text{A}}{[64.6\ \mu\text{M}]}\right) \Rightarrow [Pb^{2+}] = 130._2\ \mu\text{M}$$

The concentration of $Pb^{2+}$ in diluted unknown is $130._2$ $\mu$M. In the undiluted unknown, the concentration is $\left(\frac{50.00}{25.00}\right)(130._2\ \mu\text{M}) = 2.60 \times 10^{-4}$ M.

**(b)** First find the relative uncertainty in the response factor:

$$F = \frac{(1.58 \pm 0.03)(32.3 \pm 0.1)}{(1.64 \pm 0.03)(41.8 \pm 0.1)}$$

$$\Rightarrow F = 0.744\,5 \pm 0.019\,9\ (\pm 2.67\%)$$

Then find the uncertainty in $[Pb^{2+}]$:

$$[Pb^{2+}] = \frac{(3.00 \pm 0.03)\dfrac{(10.00 \pm 0.05)}{(50.00 \pm 0.05)}(3.23\ (\pm 0.01) \times 10^{-4})}{(2.00 \pm 0.03)(0.744\,5 \pm 0.019\,9)}$$

$$\Rightarrow [Pb^{2+}] = 2.60\ (\pm 0.09) \times 10^{-4}\ M$$

**17-G.** We see two consecutive reductions. From the value of $E_{pa} - E_{pc}$, we find that one electron is involved in each reduction (using Equation 17-16). A possible sequence of reaction is

$$Co(III)(B_9C_2H_{11})_2^- \rightarrow Co(II)(B_9C_2H_{11})_2^{2-} \rightarrow Co(I)(B_9C_2H_{11})_2^{3-}$$

The equality of the anodic and cathodic peak heights suggests that the reactions are reversible. The expected sampled current (panel *a*) and square wave (panel *b*) polarograms are sketched below.

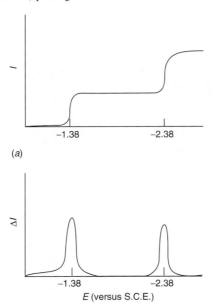

(a)

(b)

*E* (versus S.C.E.)

**17-H.** Electricity required for $H_2O$ in 0.847 6 g of polymer = $(63.16 - 4.23) = 58.93$ C.

$$\frac{58.93\ C}{96\,485\ C/mol} = 0.610\,8\ mmol\ of\ e^-$$

which corresponds to $\tfrac{1}{2}(0.610\,8) = 0.305\,4$ mmol of $I_2 = 0.305\,4$ mmol of $H_2O = 5.502$ mg $H_2O$

$$Water\ content = 100 \times \frac{5.502\ mg\ H_2O}{847.6\ mg\ polymer} = 0.649\,1\ wt\%$$

## Chapter 18

**18-A. (a)** $A = -\log P/P_0 = -\log T = -\log(0.45) = 0.347$
**(b)** Absorbance is proportional to concentration, so the absorbance will double to 0.694, giving $T = 10^{-A} = 10^{-0.694} = 0.202 \Rightarrow \%T = 20.2\%$.

**18-B. (a)** $\epsilon = \dfrac{A}{cb} = \dfrac{0.624 - 0.029}{(3.96 \times 10^{-4}\ M)(1.000\ cm)}$

$$= 1.50 \times 10^3\ M^{-1}\ cm^{-1}$$

**(b)** $c = \dfrac{A}{\epsilon b} = \dfrac{0.375 - 0.029}{(1.50 \times 10^3\ M^{-1}\ cm^{-1})(1.000\ cm)} = 2.31 \times 10^{-4}\ M$

**(c)** $c = \underbrace{\left(\dfrac{25.00\ mL}{2.00\ mL}\right)}_{Dilution\ factor} \dfrac{0.733 - 0.029}{(1.50 \times 10^3\ M^{-1}\ cm^{-1})(1.000\ cm)}$

$$= 5.87 \times 10^{-3}\ M$$

**18-C. (a)** $1.00 \times 10^{-2}$ g of $NH_4Cl$ in $1.00\ L = 1.869 \times 10^{-4}$ M. In the colored solution, the concentration is $(\tfrac{10}{50})(1.869 \times 10^{-4}\ M) = 3.739 \times 10^{-5}$ M. $\epsilon = A/bc = (0.308 - 0.140)/[(1.00)(3.739\times 10^{-5})] = 4.49_3 \times 10^3\ M^{-1}\ cm^{-1}$.

**(b)** $\dfrac{Absorbance\ of\ unknown}{Absorbance\ of\ reference}$

$$= \frac{0.592 - 0.140}{0.308 - 0.140}$$

$$= \frac{concentration\ of\ unknown}{concentration\ of\ reference}$$

$\Rightarrow$ concentration of $NH_3$ in unknown

$$= \left(\frac{0.452}{0.168}\right)(1.869 \times 10^{-4})$$

$$= 5.028 \times 10^{-4}\ M$$

100.00 mL of unknown

$$= 5.028 \times 10^{-5}\ mol\ of\ N$$

$$= 7.043 \times 10^{-4}\ g\ of\ N$$

$\Rightarrow$ weight % of N

$$= (7.043 \times 10^{-4}\ g)/(4.37 \times 10^{-3}\ g)$$

$$= 16.1\%$$

**18-D. (a)** Milligrams of Cu in flask C = $(1.00)(\tfrac{10}{250})(\tfrac{15}{30}) = 0.020\,0$ mg. This entire quantity is in the isoamyl alcohol (20.00 mL), so the concentration is $(2.00 \times 10^{-5}\ g)/[(0.020\,0\ L)(63.546\ g/mol)] = 1.57 \times 10^{-5}$ M.
**(b)** Observed absorbance

$$= absorbance\ due\ to\ Cu\ in\ rock + blank\ absorbance$$

$$= \epsilon bc + 0.056$$

$$= (7.90 \times 10^3)(1.00)(1.574 \times 10^{-5}) + 0.056$$

$$= 0.180$$

Note that the observed absorbance is equal to the absorbance from Cu in the rock *plus* the blank absorbance. In the lab we measure the observed absorbance and subtract the blank absorbance from it to find the absorbance due to copper.

**(c)** $\dfrac{Cu\ in\ unknown}{Cu\ in\ known} = \dfrac{A\ of\ unknown}{A\ of\ known}$

$$\frac{x\ mg}{1.00\ mg} = \frac{0.874 - 0.056}{0.180 - 0.056} \Rightarrow x = 6.60\ mg\ Cu$$

**18-E. (c)**

| Vol% acetone | Corrected absorbance |
|---|---|
| 10 | 0.215 |
| 20 | 0.411 |
| 30 | 0.578 |
| 40 | 0.745 |
| 50 | 0.913 |

(Your answers will be somewhat different from Dan's, depending on where you draw the baselines and how you measure the peaks.)
**(e)** $y = 0.017\,3_0\ (\pm 0.000\,3_3)x + 0.05_3\ (\pm 0.01_1)(s_y = 0.010_4)$

**(f)** $vol\% = \dfrac{0.611 - 0.053}{0.017\,30} = 32.3\%$

When a spreadsheet with all its extra digits is used, the answer is 32.2 vol%.

Uncertainty in $x\ (= s_X)$

$$= \frac{s_y}{|m|}\sqrt{\frac{1}{k} + \frac{1}{n} + \frac{(y - \bar{y})^2}{m^2 \Sigma(x_1 - \bar{x})^2}}$$

$$= \frac{0.010\,4}{0.017\,3}\sqrt{\frac{1}{3} + \frac{1}{5} + \frac{(0.611 - 0.572_4)^2}{0.017\,3^2\,(1\,000)}} = 0.4_4$$

Answer: $32.2 \pm 0.4$ vol%.

## Chapter 19

**19-A. (a)** $c = A/\epsilon b = 0.463/[(4\,170)(1.00)] = 1.110 \times 10^{-4}$ M $= 8.99$ g/L $= 8.99$ mg of transferrin/mL. The Fe concentration is $2.220 \times 10^{-4}$ M $= 0.012\,4$ g/L $= 12.4$ µg/mL.

**(b)**
$$A_\lambda = \Sigma \epsilon bc$$
$$0.424 = 4\,170[T] + 2\,290[D]$$

$$0.401 = 3\,540[T] + 2\,730[D]$$

where [T] and [D] are the concentrations of transferrin and desferrioxamine, respectively. Solving for [T] and [D] gives [T] $= 7.30 \times 10^{-5}$ M and [D] $= 5.22 \times 10^{-5}$ M. The fraction of iron in transferrin (which binds two ferric ions) is $2[T]/(2[T] + [D]) = 73.7\%$.

The spreadsheet solution looks like this:

| | A | B | C | D | E | F | G |
|---|---|---|---|---|---|---|---|
| 1 | Transferrin/Desferrioxamine mixture | | | | | | |
| 2 | | | | | | | |
| 3 | Wavelength | Coefficient matrix | | Absorbance | | Concentrations | |
| 4 | | | | of unknown | | in mixture | |
| 5 | 428 | 3540 | 2730 | 0.401 | | 7.2992E-05 | ← [TRF] |
| 6 | 470 | 4170 | 2290 | 0.424 | | 5.2238E-05 | ← [DFO] |
| 7 | | | K | A | | C | |

Spreadsheet for Exercise 19-A.

**19-B.**

| | A | B | C | D | E | F | G |
|---|---|---|---|---|---|---|---|
| 2 | | | | | Absorbance | Calculated | |
| 3 | Wavelength | Molar absorptivity | | | of mixture | absorbance | |
| 4 | (nm) | Tartrazine | Sunset yellow | Ponceau 4R | Am | Acalc | (Acalc-Am)^2 |
| 5 | 350 | 6.229E+03 | 2.019E+03 | 4.172E+03 | 0.557 | 0.536 | 0.0004 |
| 6 | 375 | 1.324E+04 | 4.474E+03 | 2.313E+03 | 0.853 | 0.837 | 0.0002 |
| 7 | 400 | 2.144E+04 | 7.403E+03 | 3.310E+03 | 1.332 | 1.343 | 0.0001 |
| 8 | 425 | 2.514E+04 | 8.551E+03 | 4.534E+03 | 1.603 | 1.600 | 0.0000 |
| 9 | 450 | 2.200E+04 | 1.275E+04 | 6.575E+03 | 1.792 | 1.801 | 0.0001 |
| 10 | 475 | 1.055E+04 | 1.940E+04 | 1.229E+04 | 2.006 | 1.999 | 0.0000 |
| 11 | 500 | 1.403E+03 | 1.869E+04 | 1.673E+04 | 1.821 | 1.834 | 0.0002 |
| 12 | 525 | 0.000E+00 | 7.641E+03 | 1.528E+04 | 1.155 | 1.130 | 0.0006 |
| 13 | 550 | 0.000E+00 | 3.959E+02 | 9.522E+03 | 0.445 | 0.474 | 0.0008 |
| 14 | 575 | 0.000E+00 | 0.000E+00 | 1.814E+03 | 0.084 | 0.086 | 0.0000 |
| 15 | | Least-squares | | | | Sum = | 0.0026 |
| 16 | | guessed concentrations | | | | | |
| 17 | Tartrazine | 3.71E-05 | | | | | |
| 18 | Sunset y | 5.27E-05 | | | | | |
| 19 | Ponceau 4R | 4.76E-05 | | | | | |

**19-C.** The appropriate Scatchard plot is a graph of $\Delta A/[X]$ versus $\Delta A$ (Equation 19-16).

| Experiment | $\Delta A$ | $\Delta A/[X]$ |
|---|---|---|
| 1 | 0.090 | 20 360 |
| 2 | 0.181 | 19 890 |
| 3 | 0.271 | 16 940 |
| 4 | 0.361 | 14 620 |
| 5 | 0.450 | 12 610 |
| 6 | 0.539 | 9 764 |
| 7 | 0.627 | 7 646 |
| 8 | 0.713 | 5 021 |
| 9 | 0.793 | 2 948 |
| 10 | 0.853 | 1 453 |
| 11 | 0.904 | 93.6 |

Points 2–11 lie on a reasonably straight line whose slope is $-2.72 \times 10^4$ M$^{-1}$, giving $K = 2.72 \times 10^4$ M$^{-1}$.

**19-D.**

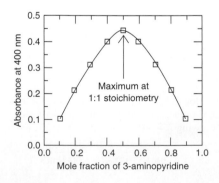

## Chapter 20

**20-A. (a)** For $\lambda = 10.00$ µm and $\Delta\lambda = 0.01$ µm, $\lambda/\Delta\lambda = 10.00/0.01 = 10^3$. These lines will be resolved.

**(b)** $\lambda = \dfrac{1}{\tilde{\nu}} = \dfrac{1}{(1\,000 \text{ cm}^{-1})(10^{-4} \text{ cm/µm})} = 10$ µm

$$\Delta\lambda = \frac{\lambda}{10^4} = 10^{-3}\ \mu m$$

$$\Rightarrow 10.001\ \mu m \text{ could be resolved from } 10.000\ \mu m.$$

$$\left.\begin{array}{l} 10.000\ \mu m = 1\ 000.0\ cm^{-1} \\ 10.001\ \mu m = 999.9\ cm^{-1} \end{array}\right\}\ \text{Difference} = 0.1\ cm^{-1}$$

(c) $5.0\ cm \times 2\ 500\ \text{lines/cm} = 12\ 500\ \text{lines}$

$$\begin{aligned} \text{Resolution} &= 1 \cdot 12\ 500 = 12\ 500 \text{ for } n = 1 \\ &= 10 \cdot 12\ 500 = 125\ 000 \text{ for } n = 10 \end{aligned}$$

(d) $\dfrac{\Delta\phi}{\Delta\lambda} = \dfrac{n}{d\cos\phi} = \dfrac{2}{\left(\dfrac{1\ mm}{250}\right)\cos 30°}$

$$= 577\ \text{radians/mm} = 0.577\ \text{radian/}\mu m$$

$$\text{Degrees} = \frac{\text{radians}}{\pi} \times 180$$

$$\Rightarrow \frac{\Delta\phi}{\Delta\lambda} = 33.1\ \text{degrees/}\mu m$$

The two wavelengths are $1\ 000\ cm^{-1} = 10.00\ \mu m$ and $1\ 001\ cm^{-1} = 9.99\ \mu m \Rightarrow \Delta\lambda = 0.01\ \mu m$.

$$\Delta\phi = 0.577\frac{\text{radian}}{\mu m} \times 0.01\ \mu m$$

$$= 6 \times 10^{-3}\ \text{radian} = 0.3°$$

**20-B.** True transmittance $= 10^{-1.000} = 0.100$. With 1.0% stray light, the apparent transmittance is

$$\text{Apparent transmittance} = \frac{P + S}{P_0 + S} = \frac{0.100 + 0.010}{1 + 0.010} = 0.109$$

The apparent absorbance is $-\log T = -\log 0.109 = 0.963$.

**20-C. (a)** $\Delta\tilde{\nu} = 1/2\delta = 1/(2 \cdot 1.266\ 0 \times 10^{-4}\ cm)$

$$= 3\ 949\ cm^{-1}$$

**(b)** Each interval is $1.266\ 0 \times 10^{-4}$ cm. 4 096 intervals $= (4\ 096)(1.266\ 0 \times 10^{-4}\ cm) = 0.518\ 6$ cm. This is a range of $\pm\Delta$, so $\Delta = 0.259\ 3$ cm.

**(c)** Resolution $\approx 1/\Delta = 1/(0.259\ 3\ cm)$

$$= 3.86\ cm^{-1}$$

**(d)** Mirror velocity $= 0.693$ cm/s

$$\text{Interval} = \frac{1.266\ 0 \times 10^{-4}\ cm}{0.693\ cm/s} = 183\ \mu s$$

**(e)** $(4\ 096\ \text{points})(183\ \mu s/\text{point}) = 0.748$ s

**(f)** The beamsplitter is germanium on KBr. The KBr absorbs light below $400\ cm^{-1}$, which the background transform shows clearly.

**20-D.** The graphs show that signal-to-noise ratio is proportional to $\sqrt{n}$.

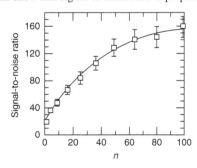

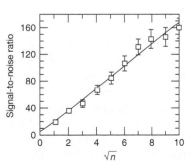

The confidence interval is $\pm ts/\sqrt{n}$, where $s$ is the standard deviation, $n$ is the number of experiments, and $t$ is Student's $t$ from Table 4-2 for 95% confidence and $n - 1$ degrees of freedom. For the first line of the table, $n = 8$, $s = 1.9$, and $t = 2.365$ for 7 degrees of freedom. 95% confidence interval $= \pm(2.365)(1.9)/\sqrt{8} = \pm 1.6$. For the remaining rows, 95% confidence interval $= 3.9, 5.1, 5.9, 9.0, 11.2, 14.0, 24.0, 23.9,$ and $27.2$.

## Chapter 21

**21-A.** A graph of intensity versus concentration of added standard has an $x$-intercept of $-0.164 \pm 0.005\ \mu g/mL$. Because the sample was diluted by a factor of 10, the original concentration is $1.64 \pm 0.05\ \mu g/mL$.

| | A | B | C | D |
|---|---|---|---|---|
| 1 | Standard Addition Constant Volume Least-Squares | | | |
| 2 | x = Added Sr | y | | |
| 3 | (µg/mL) | Signal | | |
| 4 | 0.000 | 309 | | |
| 5 | 0.081 | 452 | | |
| 6 | 0.162 | 600 | | |
| 7 | 0.243 | 765 | | |
| 8 | 0.324 | 906 | | |
| 9 | B11:C13 = LINEST(B4:B8,A4:A8,TRUE,TRUE) | | | |
| 10 | | LINEST output: | | |
| 11 | m | 1860.5 | 305.0 | b |
| 12 | $s_m$ | 26.3 | 5.2 | $s_b$ |
| 13 | $R^2$ | 0.9994 | 6.7 | $s_y$ |
| 14 | x-intercept = -b/m = | -0.164 | | |
| 15 | n = | 5 | = COUNT(A4:A8) | |
| 16 | Mean y = | 606.400 | = AVERAGE(B4:B8) | |
| 17 | $\sum(x_i - \text{mean } x)^2 =$ | 0.06561 | = DEVSQ(A4:A8) | |
| 18 | Std deviation of | | | |
| 19 | x-intercept = | 0.0049 | | |
| 20 | B19=(C13/ABS(B11))*SQRT((1/B15) + B16^2/(B11^2*B17)) | | | |

**21-B.** The concentration of Mn in the unknown mixture is $(13.5\ \mu g/mL)(1.00/6.00) = 2.25\ \mu g/mL$.

Standard mixture:

$$\frac{A_X}{[X]} = F\left(\frac{A_S}{[S]}\right)$$

$$\frac{1.05}{[2.50\ \mu g/mL]} = F\left(\frac{1.00}{[2.00\ \mu g/mL]}\right) \Rightarrow F = 0.840$$

Unknown mixture:

$$\frac{A_X}{[X]} = F\left(\frac{A_S}{[S]}\right)$$

$$\frac{0.185}{[Fe]} = 0.840\left(\frac{0.128}{[2.25\ \mu g/mL]}\right) \Rightarrow [Fe] = 3.87\ \mu g/mL$$

The original concentration of Fe must have been

$$\frac{6.00}{5.00}(3.87) = 4.65\ \mu g/mL = 8.33 \times 10^{-5}\ M$$

**21-C.** The ratio of signal to peak-to-peak noise level is measured to be 17 in the figure. The concentration of Fe needed to give a signal-to-noise ratio of 2 is $\left(\frac{2}{17}\right)(0.048\ 5\ \mu g/mL) = 0.005\ 7\ \mu g/mL\ (= 5.7\ \text{ppb})$.

**21-D. (a)** The higher result in experiment 2 compared with experiment 1 is probably the effect of diluting interfering species, so they do not interfere as much in experiment 2 as in experiment 1. Dilution lowers the concentration of species that might react with Li or make smoke that scatters light. In experiment 3, interference is present to the same extent as in experiment 2, but the standard addition procedure corrects for the interference. The whole point of standard addition is to measure the effect of the complex interfering matrix on the response to known quantities of analyte.

(b) Experiments 4–6 use a hotter flame than experiments 1–3. High temperature appears to eliminate most of the interference observed at lower temperature. Dilution has only a tiny effect on the results.

(c) Because it appears from experiments 1–3 that standard addition gives a true result, we surmise that experiments 3 and 6, and possibly 5, are within experimental error of each other. I would probably report the "true" value as the mean of experiments 3 and 6 (81.4 ppm). It might also be reasonable to take an average of experiments 3, 5, and 6 (80.8 ppm).

## Chapter 22

**22-A.** Resolving power $= \dfrac{m}{m_{1/2}} = \dfrac{53}{0.6_0} \approx 88$

We should be able to barely distinguish two peaks differing by 1 Da at a mass of 88 Da. We will probably not be able to distinguish two peaks at 100 and 101 Da.

**22-B.** $C_2H_5^+$:

$$\begin{array}{rr} & 2 \times 12.000\ 00 \\ & +5 \times \ \ 1.007\ 825 \\ -e^- \text{ mass} & -1 \times \ \ 0.000\ 55 \\ \hline & 29.038\ 58 \end{array}$$

$HCO^+$:

$$\begin{array}{rr} & 1 \times 12.000\ 00 \\ & 1 \times \ \ 1.007\ 825 \\ & +1 \times 15.994\ 91 \\ -e^- \text{ mass} & -1 \times \ \ 0.000\ 55 \\ \hline & 29.002\ 18 \end{array}$$

We need to distinguish a mass difference of $29.038\ 58 - 29.002\ 18 = 0.036\ 40$. The required resolving power is $m/\Delta m = 29.0/(0.036\ 40) = 7.97 \times 10^2 \approx 800$.

**22-C.** $^{35}Cl$ abundance $\equiv a = 0.757\ 8$    $^{37}Cl$ abundance $\equiv b = 0.242\ 2$

Relative abundance of $C_6H_4^{35}Cl_2 = a^2 = 0.574\ 2_6$

Relative abundance of $C_6H_4^{35}Cl^{37}Cl = 2ab = 0.367\ 0_8$

Relative abundance of $C_6H_4^{37}Cl_2 = b^2 = 0.058\ 66_1$

Relative abundances: $^{35}Cl_2 : {}^{35}Cl^{37}Cl : {}^{37}Cl_2 = 1 : 0.639\ 2 : 0.102\ 2$

Figure 22-7 shows the stick diagram.

**22-D. (a)** $C_{14}H_{12}$

$R + DB = c - h/2 + n/2 + 1 = 14 - 12/2 + 0/2 + 1 = 9$

A molecule with two rings + seven double bonds is *trans*-stilbene:

**(b)** $C_4H_{10}NO^+$

$R + DB = c - h/2 + n/2 + 1 = 4 - 10/2 + 1/2 + 1 = \frac{1}{2}$  Huh?

We came out with a fraction because the species is an ion in which at least one atom does not make its usual number of bonds. In the following structure, N makes four bonds instead of three:

A fragment in a mass spectrum

**22-E. (a)** The principal difference between the two spectra is the appearance of a significant peak at $m/z$ 72 in A that is missing in B. This even mass represents loss of a neutral molecule with a mass of 28 Da from the molecular ion. The McLafferty rearrangement can split $C_2H_4$ from 3-methyl-2 pentanone, but not from 3,3-dimethyl-2- butanone, which lacks a $\gamma$-CH group.

3-Methyl-2-pentanone          Ethylene          $m/z = 72$
28 Da

Spectrum A must be from 3-methyl-2-pentanone and spectrum B is from 3,3-dimethyl-2-butanone.

**(b)** Expected intensity of M+1 relative to $M^{+\cdot}$ for $C_6H_{12}O$:

Intensity $= \underbrace{6 \times 1.08\%}_{^{13}C} + \underbrace{12 \times 0.012\%}_{^{2}H} + \underbrace{1 \times 0.038\%}_{^{17}O} = 6.7\%$ of $M^{+\cdot}$

**22-F. (a)** —OH          $C_6H_6O$: $M^{+\cdot} = 94$

Rings + double bonds $= c - h/2 + n/2 + 1 = 6 - 6/2 + 0/2 + 1 = 4$

Expected intensity of M+1 from Table 22-2:

$$1.08(6) + 0.012(6) + 0.038(1) = 6.59\%$$
  Carbon    Hydrogen    Oxygen

Observed intensity of M+1 = 68/999 = 6.8%

Expected intensity of M+2 = 0.005 8(6)(5) + 0.205(1) = 0.38%

Observed intensity of M+2 = 0.3%

**(b)** —Br          $C_6H_5Br$: $M^{+\cdot} = 156$

The two nearly equal peaks at $m/z$ 156 and 158 scream out "bromine!"

Rings + double bonds $= c - h/2 + n/2 + 1 = 6 - 6/2 + 0/2 + 1 = 4$
                                             ↑
                          *h* includes H + Br

Expected intensity of M+1 = 1.08(6) + 0.012(5) = 6.54%
                   Carbon    Hydrogen

Observed intensity of M+1 = 46/566 = 8.1%

Expected intensity of M+2 = 0.005 8(6)(5) + 97.3(1) = 97.5%
                     Carbon              Bromine

Observed intensity of M+2 = 520/566 = 91.9%

The M+3 peak is the isotopic partner of the M+2 peak ($C_6H_5^{81}Br$). M+3 contains $^{81}Br$ plus either 1 $^{13}C$ or 1 $^{2}H$. Therefore, the expected intensity of M+3 (relative to $C_6H_5^{81}Br$ at M+2) is 1.08(6) + 0.012(5) = 6.54% of predicted intensity of $C_6H_5^{81}Br$ at M+2 = (0.065 4)(97.3) = 6.4% of $M^{+\cdot}$. Observed intensity of M+3 is 35/566 = 6.2%.

**(c)** —$CO_2H$          $C_7H_3O_2Cl_3$: $M^{+\cdot} = 224$

From Figure 22-7, the M : M+2: M+4 pattern looks like a molecule containing three chlorine atoms. The correct structure is shown here, but there is no way you could assign the composition to one isomer from the given data.

Rings + double bonds $= c - h/2 + n/2 + 1 = 7 - 6/2 + 0/2 + 1 = 5$

Expected intensity of M+1 from Table 22-2:

$$1.08(7) + 0.012(3) + 0.038(2) = 7.67\%$$
  Carbon    Hydrogen    Oxygen

Observed intensity of M+1 = 63/791 = 8.0%.

Expected intensity of M+2 = 0.005 8(7)(6) + 0.205(2) + 32.0(3) = 96.7%. Observed intensity of M+2 = 754/791 = 95.4%.

The M+3 peak is the isotopic partner of $C_7H_3O_2^{35}Cl_2^{37}Cl$ at M+2. M+3 contains one $^{37}Cl$ plus either 1 $^{13}C$ or 1 $^{2}H$ or 1 $^{17}O$. Expected intensity of

M+3 (relative to $C_7H_3O_2{}^{35}Cl_2{}^{37}Cl$ at M+2) = 1.08(7) + 0.012(3) + 0.038(2) = 7.67% of predicted intensity of $C_7H_3O_2{}^{35}Cl_2{}^{37}Cl$ at M+2. The predicted intensity of $C_7H_3O_2{}^{35}Cl_2{}^{37}Cl$ is 32.0(3) = 96.0% of $M^{+\cdot}$. Expected intensity of M+3 = 7.67% of 96.0% = 7.4% of $M^{+\cdot}$. Observed intensity = 60/791 = 7.6%.

M+4 is composed mainly of $C_7H_3O_2{}^{35}Cl{}^{37}Cl_2$ plus a small amount of $C_7H_3{}^{16}O{}^{18}O{}^{35}Cl_2{}^{37}Cl$. Other formulas such as $^{12}C_6{}^{13}CH_2{}^{16}O{}^{17}O{}^{35}Cl_2{}^{37}Cl$ also add up to M+4, but they are even less likely to occur because they have two minor isotopes ($^{13}C$ and $^{17}O$). Expected intensity of M+4 from $C_7H_3O_2{}^{35}Cl{}^{37}Cl_2$ is 5.11(3)(2) = 30.7% of $M^{+\cdot}$. The contribution from $C_7H_3{}^{16}O{}^{18}O{}^{35}Cl_2{}^{37}Cl$ is based on the predicted intensity of $C_7H_3O_2{}^{35}Cl_2{}^{37}Cl$ at M+2. The predicted intensity of $C_7H_3O_2{}^{35}Cl_2{}^{37}Cl$ is 32.0(3) = 96.0% of $M^{+\cdot}$. The predicted intensity from $C_7H_3{}^{16}O{}^{18}O{}^{35}Cl_2{}^{37}Cl$ at M+4 is 0.205(2) = 0.410% of 96.0% = 0.4%. Total expected intensity of M+4 is 30.7% + 0.4% = 31.1% of $M^{+\cdot}$. Observed intensity = 264/791 = 33.4%.

Expected intensity of M+5 from $^{12}C_6{}^{13}CH_3O_2{}^{35}Cl{}^{37}Cl_2$ and $^{12}C_7H_2{}^2HO_2{}^{35}Cl{}^{37}Cl_2$ and $C_7H_3{}^{16}O{}^{17}O{}^{35}Cl{}^{37}Cl_2$ is based on the predicted intensity of $C_7H_3O_2{}^{35}Cl{}^{37}Cl_2$ at M+4. M+5 should have 1.08(7) + 0.012(3) + 0.038(2) = 7.7% of $C_7H_3O_2{}^{35}Cl{}^{37}Cl_2$ at M+4 = 7.7% of 30.7% = 2.4%. Observed intensity = 19/791 = 2.4%.

Expected intensity of M+6 from $C_7H_3O_2{}^{37}Cl_3$ is 0.544(3)(2)(1) = 3.26% of $M^{+\cdot}$. There will also be a small contribution from $C_7H_3{}^{16}O{}^{18}O{}^{35}Cl{}^{37}Cl_2$, which will be 0.205(2) = 0.410% of predicted intensity of

$C_7H_3{}^{16}O{}^{35}Cl{}^{37}Cl_2 = 0.410\%$ of 30.7% of $M^{+\cdot}$ = 0.13% of $M^{+\cdot}$. The total expected intensity at M+6 is therefore 3.26 + 0.13 = 3.4% of $M^{+\cdot}$. Observed intensity = 29/791 = 3.7%.

(d)   OH   $C_4H_{10}O_2S_2$: $M^{+\cdot}$ = 154

HS —— / OH / SH

Sulfur gives a significant M+2 peak (4.52% of M per sulfur). The observed M+2 is 12/122 = 9.8%, which could represent two sulfur atoms. The composition $C_4H_{10}O_2S_2$ has two sulfur atoms and has a molecular mass of 154. The known structure is shown here, but you could not deduce the structure from the composition.

$$\text{Rings + double bonds} = c - h/2 + n/2 + 1$$
$$= 4 - 10/2 + 0/2 + 1 = 0$$

Expected intensity of M+1:

$$\underset{\text{Carbon}}{1.08(4)} + \underset{\text{Hydrogen}}{0.012(10)} + \underset{\text{Oxygen}}{0.038(2)} + \underset{\text{Sulfur}}{0.801(2)} = 6.12\%$$

Observed intensity of M+1 = 9/122 = 7.4%

Expected intensity of M+2 = 0.005 8(4)(3) + 0.205(2) + 4.52(2) = 9.52%. Observed intensity of M+2 = 12/122 = 9.8%.

**22-G.**

Analysis of electrospray mass spectrum of lysozyme

| Observed $m/z$ $\equiv m_n$ | $m_{n+1} - 1.008$ | $m_n - m_{n+1}$ | Charge $= n =$ $\dfrac{m_{n+1} - 1.008}{m_n - m_{n+1}}$ | Molecular mass $= n \times (m_n - 1.008)$ |
|---|---|---|---|---|
| 1 789.1 | 1 589.39 | 198.7 | 7.99 ≈ 8 | 14 304.7 |
| 1 590.4 | 1 430.49 | 158.9 | 9.00 ≈ 9 | 14 304.5 |
| 1 431.5 | 1 300.49 | 130.0 | 10.00 ≈ 10 | 14 304.9 |
| 1 301.5 | 1 192.09 | 108.4 | 11.00 ≈ 11 | 14 305.4 |
| 1 193.1 | — | — | 12 | 14 305.1 |
| | | | | mean = 14 304.9 (±0.3) |

**22-H. (a)** To find the response factor, we insert values from the first line of the table into the equation:

$$\frac{\text{Area of analyte signal}}{\text{Area of standard signal}} = F\left(\frac{\text{concentration of analyte}}{\text{concentration of standard}}\right)$$

$$\frac{11\ 438}{2\ 992} = F\left(\frac{13.60 \times 10^2}{3.70 \times 10^2}\right) = F = 1.04_0$$

For the next two sets of data, we find $F = 1.02_0$ and $1.06_4$, giving a mean value $F = 1.04_1$.

**(b)** The concentration of internal standard in the mixture of caffeine-$D_3$ plus cola is

$$(1.11 \text{ g/L}) \times \frac{0.050\ 0 \text{ mL}}{1.050 \text{ mL}} = 52.8_6 \text{ mg/L}$$

The concentration of caffeine in the chromatographed solution is

$$\frac{\text{Area of analyte signal}}{\text{Area of standard signal}} = F\left(\frac{\text{concentration of analyte}}{\text{concentration of standard}}\right)$$

$$\frac{1\ 733}{1\ 144} = 1.04_1\left(\frac{[\text{caffeine}]}{52.8_6 \text{ mg/L}}\right) \Rightarrow [\text{caffeine}] = 76.9 \text{ mg/L}$$

The unknown beverage had been diluted from 1.000 to 1.050 mL when the standard was added, so the concentration of caffeine in the original beverage was (1.050)(76.9 mg/L) = 80.8 mg/L.

## Chapter 23

**23-A. (a)** $k_1' = \dfrac{t_{r1} - t_m}{t_m}$

$$\Rightarrow t_m = \frac{t_{r1}}{k_1' + 1} = \frac{10.0}{5.00} = 2.00 \text{ min}$$

$$t_{r2} = t_m(k_2' + 1) = 2.00\ (5.00 + 1) = 12.0 \text{ min}$$

$$\sigma_1 = \frac{t_{r1}}{\sqrt{N}} = \frac{10.0}{\sqrt{1\ 000}} = 0.316 \text{ min}$$

$$\Rightarrow = w_{1/2} \text{ (peak 1)} = 2.35\sigma_1 = 0.74 \text{ min}$$

$$w_1 = 4\sigma_1 = 1.26 \text{ min}$$

$$\sigma_2 = \frac{t_{r2}}{\sqrt{N}} = \frac{12.0}{\sqrt{1\ 000}} = 0.379 \text{ min}$$

$$\Rightarrow w_{1/2} \text{ (peak 2)} = 2.35\sigma_2 = 0.89 \text{ min}$$

$$w_2 = 4\sigma_2 = 1.52 \text{ min}$$

**(b)**

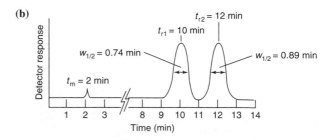

**(c)** Resolution $= \dfrac{\Delta t_r}{w_{av}} = \dfrac{2}{[(1.26 + 1.52)/2]} = 1.44$

**23-B.** **(a)** Fraction remaining $= q = \dfrac{V_1}{V_1 + KV_2}$

$$0.01 = \frac{10}{10 + 4.0V_2} \Rightarrow V_2 = 248 \text{ mL}$$

**(b)** $q^3 = 0.01 = \left(\dfrac{10}{10 + 4.0V_2}\right)^3 \Rightarrow V_2 = 9.1 \text{ mL}$, and total volume $=$ 27.3 mL.

**23-C.** **(a)** Relative distances measured from Figure 23-7:

$t_m = 10._4$

$t'_r = 39._8$ for octane

$t'_r = 76._0$ for nonane

$k' = t'_r/t_m = 3.8_3$ for octane and $7.3_1$ for nonane

**(b)** Let $t_s$ = time in stationary phase, $t_m$ = time in mobile phase, and $t$ be total time on column. We know that $k' = t_s/t_m$. But

$$t = t_s + t_m = t_s + \frac{t_s}{k'}$$

$$= t_s\left(1 + \frac{1}{k'}\right) = t_s\left(\frac{k'+1}{k'}\right)$$

Therefore,

$$t_s/t = \frac{k'}{k'+1} = 3.8_3/4.8_3 = 0.79_3$$

**(c)** $\alpha = t'_r(\text{nonane})/t'_r(\text{octane}) = 76._0/39._8 = 1.9_1$.

**(d)** $K = k'V_m/V_s = 3.8_3(V_m/\frac{1}{2}V_m) = 7.6_6$.

**23-D.** **(a)** For ethyl acetate, we measure $t_r = 11.3$ and $w = 1.5$ mm. Therefore, $N = 16t_r^2/w^2 = 910$ plates. For toluene, the values are $t_r = 36.2$, $w = 4.2$, and $N = 1\,200$ plates.

**(b)** We expect $w_{1/2} = (2.35/4)w$. The measured value of $w_{1/2}$ is in good agreement with the calculated value.

**23-E.** The column is overloaded, causing a gradual rise and an abrupt fall of the peak. As the sample size is decreased, the overloading decreases and the peak becomes more symmetric.

**23-F.** **(a)** We know that $\alpha = 1.068$ and $k'_1 = 5.16$. For resolution, we need to find the separation factor $\gamma = t_2/t_1$, where $t$ is retention time. The relation between capacity factor and retention time is $k'_1 = (t_1 - t_m)/t_m = t_1/t_m - 1 \Rightarrow t_1 = t_m(k'_1 + 1) = t_m(6.16)$. Also, $k'_2 = \alpha k'_1 = (1.068)(5.16) = 5.51_1$. Therefore, $t_2 = t_m(k'_2 + 1) = t_m(6.51_1)$.

$$\frac{t_2}{t_1} = \frac{t_m(6.51_1)}{t_m(6.16)} = 1.057_0$$

**(b)** Resolution $= \dfrac{\sqrt{N}}{4}(\gamma - 1)$

$$1.00 = \frac{\sqrt{N}}{4}(1.057_0 - 1) \Rightarrow N = 4.92 \times 10^3 \text{ plates}$$

Required length $= (4.92 \times 10^3 \text{ plates})(0.520 \text{ mm/plate}) = 2.56 \text{ m}$

**(c)** From (a), $t_1 = t_m(6.16) = (2.00 \text{ min})(6.16) = 12.32 \text{ min}$

$t_2 = t_m(6.51_1) = (2.00 \text{ min})(6.51_1) = 13.02 \text{ min}$

$w_{1/2} = \sqrt{\dfrac{5.55}{N}} t_r = \sqrt{\dfrac{5.55}{4.92 \times 10^3}}(12.32 \text{ min}) = 0.41 \text{ min for component 1}$

$w_{1/2} = \sqrt{\dfrac{5.55}{4.92 \times 10^3}}(13.02 \text{ min}) = 0.44 \text{ min for component 2}$

**(d)** $k' = KV_s/V_m$

$5.16 = K(0.30) \Rightarrow K = 17._2$

## Chapter 24

**24-A.** **(a)** $S = [\text{butanol}] = \dfrac{234 \text{ mg}/74.12 \text{ g/mol}}{10.0 \text{ mL}} = 0.315_7 \text{ M}$

$X = [\text{hexanol}] = \dfrac{312 \text{ mg}/102.17 \text{ g/mol}}{10.0 \text{ mL}} = 0.305_4 \text{ M}$

$$\frac{A_X}{[X]} = F\left(\frac{A_S}{[S]}\right) \Rightarrow \frac{1.45}{[0.305_4 \text{ M}]} = F\left(\frac{1.00}{[0.315_7 \text{ M}]}\right)$$
$$\Rightarrow F = 1.49_9$$

**(b)** I estimate the areas by measuring the height and $w_{1/2}$ in millimeters. Your answer will be different from mine if the figure size in your book is different from that in my manuscript. However, the relative peak areas should be the same.

Butanol: Height $= 41.3$ mm; $w_{1/2} = 2.2$ mm;

area $= 1.064 \times$ peak height $\times w_{1/2}$

$= 96._7 \text{ mm}^2$

Hexanol: Height $= 21.9$ mm; $w_{1/2} = 6.9$ mm;

area $= 161 \text{ mm}^2$

**(c)** The volume of solution is not stated, but concentration is directly proportional to the number of moles. We can substitute moles for concentrations in the internal standard equation:

$$\frac{A_X}{[X]} = F\left(\frac{A_S}{[S]}\right) \Rightarrow \frac{161 \text{ mm}^2}{\text{mg hexanol}/102.17 \text{ mg/mmol}}$$

$$= 1.49_9\left(\frac{96.7 \text{ mm}^2}{112 \text{ mg}/74.12 \text{ mg/mmol}}\right)$$

$$\Rightarrow \text{hexanol} = 171 \text{ mg}$$

**(d)** The greatest uncertainty is in the width of the fairly narrow butanol peak. The uncertainty in width is ~5–10%.

**24-B.** $S = [\text{pentanol}]$; $X = [\text{hexanol}]$. We will substitute mmol for concentrations, because the volume is unknown and concentrations are proportional to mmol.

For the standard mixture, we can write

$$\frac{A_X}{[X]} = F\left(\frac{A_S}{[S]}\right) \Rightarrow \frac{1\,570}{[1.53]} = F\left(\frac{922}{[1.06]}\right) \Rightarrow F = 1.18_0$$

For the unknown mixture,

$$\frac{816}{[X]} = 1.18_0\left(\frac{843}{[0.57]}\right) \Rightarrow [X] = 0.47 \text{ mmol}$$

**24-C.** **(a)** Between nonane ($C_9H_{20}$) and decane ($C_{10}H_{22}$).

**(b)** Adjusted retention times are 13.83 ($C_{10}$) and 15.42 min ($C_{11}$)

$$1\,050 = 100\left[10 + (11 - 10)\frac{\log t'_r(\text{unknown}) - \log 13.83}{\log 15.42 - \log 13.83}\right]$$

$$\Rightarrow t'_r(\text{unknown}) = 14.60 \text{ min} \Rightarrow t_r(\text{unknown}) = 16.40 \text{ min}$$

**24-D.** **(a)** A plot of $\log t'_r$ versus (number of carbon atoms) should be a fairly straight line for a homologous series of compounds.

| Peak | $t'_r$ | $\log t'_r$ |
|---|---|---|
| $n = 7$ | 2.9 | 0.46 |
| $n = 8$ | 5.4 | 0.73 |
| $n = 14$ | 85.8 | 1.93 |
| Unknown | 41.4 | 1.62 |

From a graph of $\log t'_r$ versus $n$, it appears that $n = 12$ for the unknown.

**(b)** $k' = t'_r/t_m = 41.4/1.1 = 38$

**24-E.** **(a)** Plate number is proportional to column length. If everything is the same except for length, we can say from Equation 23-30 that

$$\frac{1.5}{1.0} = \frac{R_2}{R_1} = \frac{\sqrt{N_2}}{\sqrt{N_1}} \Rightarrow N_2 = 2.25N_1$$

The column must be 2.25 times longer to achieve the desired resolution and the elution time will be 2.25 times longer.

**(b)** If everything is the same except for $\gamma$, we can say from Equation 23-30 that

$$\frac{1.5}{1.0} = \frac{R_2}{R_1} = \frac{\gamma_2 - 1}{\gamma_1 - 1} = \frac{\gamma_2 - 1}{1.013 - 1} \Rightarrow \gamma_2 = 1.020$$

Alcohols are polar, so we could probably increase the relative retention by choosing a more polar stationary phase. (Diphenyl)$_{0.05}$(dimethyl)$_{0.95}$

polysiloxane is listed as nonpolar. We could try an intermediate polarity phase such as $(\text{diphenyl})_{0.35}(\text{dimethyl})_{0.65}$ polysiloxane. The more polar phase will probably retain alcohols more strongly and increase the retention time. We have no way to predict how much the retention time will increase.

## Chapter 25

**25-A.** $\dfrac{\text{Area}_A}{[A]} = F\left(\dfrac{\text{Area}_B}{[B]}\right) \Rightarrow \dfrac{10.86}{[1.03]} = F\left(\dfrac{4.37}{[1.16]}\right) \Rightarrow F = 2.79_9$

The concentration of internal standard (B) mixed with unknown (A) is

$$12.49 \text{ mg}/25.00 \text{ mL} = 0.499\,6 \text{ mg/mL}$$

$$\dfrac{5.97}{[A]} = 2.79_9\left(\dfrac{6.38}{[0.499\,6]}\right) \Rightarrow [A] = 0.167_0 \text{ mg/mL}$$

[A] in original unknown $= \dfrac{25.00}{10.00}(0.167_0 \text{ mg/mL})$

$$= 0.418 \text{ mg/mL}$$

**25-B.** **(a)** Equation 23-16: $k' = \dfrac{t_r - t_m}{t_m} \Rightarrow \dfrac{t_1 - 1.00}{1.00} = 1.35$

$$\Rightarrow t_1 = 2.35 \text{ min}$$

Equation 23-15: $= \alpha = \dfrac{t'_{r2}}{t'_{r1}} \Rightarrow 4.53 = \dfrac{t_2 - 1.00}{t_1 - 1.00} \Rightarrow t_2 = 7.12 \text{ min}$

Equation 23-23: Resolution $= \dfrac{\Delta t_r}{w_{av}}$

$$\Rightarrow 7.7 = \dfrac{7.12 - 2.35}{w_{av}} \Rightarrow w_{av} = 0.62 \text{ min}$$

**(b)** From Equation 23-28b, we know that $w_{1/2}$ is proportional to $t_r$ if $N$ is constant. Therefore, $\dfrac{w_{1/2} \text{ (peak 1)}}{w_{1/2} \text{ (peak 2)}} = \dfrac{t_1}{t_2} = \dfrac{2.35}{7.12} = 0.330$. We know that $w_{av}$, the average width at the base, is 0.62 min. For each peak, $w = 4\sigma$ and $w_{1/2} = 2.35\sigma$, so $w = 1.70w_{1/2}$.

$$w_{av} = 0.62 = \tfrac{1}{2}(w_1 + w_2)$$
$$= \tfrac{1}{2}[1.70w_{1/2} \text{ (peak 1)} + 1.70w_{1/2} \text{ (peak 2)}]$$

Substituting $w_{1/2}$ (peak 1) $= 0.330w_{1/2}$ (peak 2) into the previous equation gives $w_{1/2}$ (peak 2) $= 0.54_8$ min. Then $w_{1/2}$ (peak 1) $= 0.330w_{1/2}$ (peak 2) $= 0.18_1$ min.

**(c)** Because the areas are equal, we can say

$$\text{Height}_R \times w_R = \text{height}_S \times w_S$$
$$\Rightarrow \dfrac{\text{height}_R}{\text{height}_S} = \dfrac{w_S}{w_R} = \dfrac{0.54_8}{0.18_1} = 3.0$$

**25-C. (a)**

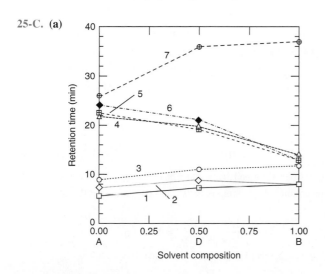

| Solvent composition: | Peaks | | | | | | |
|---|---|---|---|---|---|---|---|
| | 1 | 2 | 3 | 4 | 5 | 6 | 7 |
| 0.0 | 5.6 | 7.2 | 8.7 | 21.6 | 22.3 | 24.0 | 25.5 |
| 0.5 | 7.1 | 8.5 | 10.8 | 19.5 | 19.0 | 20.9 | 35.7 |
| 1.0 | 8.0 | 8.0 | 11.5 | 13.8 | 12.8 | 12.8 | 37.0 |
| Predicted positions (by linear interpolation): | | | | | | | |
| 0.25 | 6.35 | 7.85 | 9.75 | 21.05 | 20.65 | 22.45 | 30.60 |
| 0.75 | 7.55 | 8.25 | 11.15 | 16.65 | 15.90 | 16.85 | 36.35 |

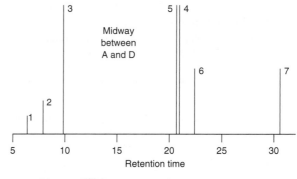

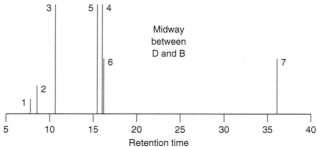

**(b)** A: 30% acetonitrile/70% buffer
B: 40% methanol/60% buffer
D: 15% acetonitrile/20% methanol/65% buffer
Between A and D:
  22.5% acetonitrile/10% methanol/67.5% buffer
Between D and B:
  7.5% acetonitrile/30% methanol/62.5% buffer

**25-D.**

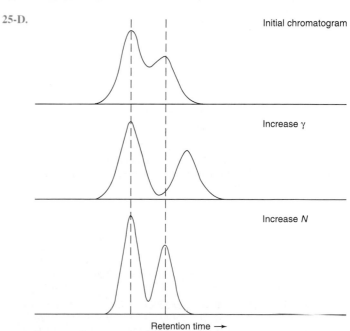

[Adapted from L. R. Snyder, J. J. Kirkland, and J. L. Glajch, *Practical HPLC Method Development* (New York: Wiley, 1997).]

# Chapter 26

**26-A.** 13.03 mL of 0.022 74 M NaOH = 0.296 3 mmol of $OH^-$, which must equal the total cation charge (= $2[VO^{2+}] + 2[H_2SO_4]$) in the 5.00-mL aliquot. 50.0 mL therefore contains 2.963 mmol of cation charge. The $VO^{2+}$ content is (50.0 mL) (0.024 3 M) = 1.215 mmol = 2.43 mmol of charge. The $H_2SO_4$ must therefore be $(2.963 - 2.43)/2 = 0.266_5$ mmol.

$$1.215 \text{ mmol } VOSO_4 = 0.198 \text{ g } VOSO_4 \text{ in}$$
$$0.244\ 7\text{-g sample} = 80.9\%$$

$$0.266_5 \text{ mmol } H_2SO_4 = 0.026\ 1 \text{ g } H_2SO_4 \text{ in}$$
$$0.244\ 7\text{-g sample} = 10.7\%$$

$$H_2O \text{ (by difference)} = 8.4\%$$

**26-B.** **(a)** Because the fractionation range of Sephadex G-50 is 1 500–30 000, hemoglobin should not be retained and ought to be eluted in a volume of 36.4 mL.
**(b)** The elution volume of $^{22}NaCl$ is $V_m$. Inserting $V_m = 36.4$ mL into the elution equation gives

$$K_{av} = \frac{V_r - V_0}{V_m - V_0} \Rightarrow 0.65 = \frac{V_r - 36.4}{109.8 - 36.4} \Rightarrow V_r = 84.1 \text{ mL}$$

**26-C.** **(a)**

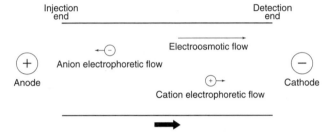

**(b)** $I^-$ has a greater mobility than $Cl^-$. Therefore, $I^-$ swims upstream faster than $Cl^-$ (because electrophoresis opposes electroosmosis) and is eluted later than $Cl^-$. The mobility of $Br^-$ is greater than that of $I^-$ in Table 15-1. Therefore, $Br^-$ will have a longer migration time than $I^-$.
**(c)** Bare $I^-$ is a larger ion than bare $Cl^-$, so the charge density in $I^-$ is lower than the charge density in $Cl^-$. Therefore, $I^-$ should have a smaller hydrated radius than $Cl^-$. This means that $I^-$ has less friction than $Cl^-$ and a greater mobility than $Cl^-$.

# Chapter 27

**27-A.** 1 mole of ethoxyl groups produces 1 mole of AgI. 29.03 mg of AgI = 0.123 65 mmol. The amount of compound analyzed is 25.42 mg/(417 g/mol) = 0.060 96 mmol. There are

$$\frac{0.123\ 65 \text{ mmol ethoxyl groups}}{0.060\ 96 \text{ mmol compound}}$$
$$= 2.03 \ (= 2) \text{ ethoxyl groups/molecule}$$

**27-B.** There is 1 mole of $SO_4^{2-}$ in each mole of each reactant and of the product. Let $x$ = g of $K_2SO_4$ and $y$ = g of $(NH_4)_2SO_4$.

$$x + y = 0.649 \text{ g} \tag{1}$$

$$\underbrace{\frac{x}{174.27}}_{\substack{\text{Moles of} \\ K_2SO_4}} + \underbrace{\frac{y}{132.14}}_{\substack{\text{Moles of} \\ (NH_4)_2SO_4}} = \underbrace{\frac{0.977}{233.39}}_{\substack{\text{Moles of} \\ BaSO_4}} \tag{2}$$

Making the substitution $y = 0.649 - x$ in Equation 2 gives $x = 0.397$ g = 61.1% of the sample.

**27-C.** Formula and atomic masses: Ba(137.327), Cl(35.453), K(39.098), $H_2O$(18.015), KCl(74.551), $BaCl_2 \cdot 2H_2O$ (244.26). $H_2O$ lost = 1.783 9 − 1.562 3 = 0.221 6 g = $1.230\ 1 \times 10^{-2}$ mol of $H_2O$. For every 2 moles of $H_2O$ lost, 1 mole of $BaCl_2 \cdot 2H_2O$ must have been present. $1.230\ 1 \times 10^{-2}$ mol of $H_2O$ implies that $6.150\ 4 \times 10^{-3}$ mol of $BaCl_2 \cdot 2H_2O$ must have

been present. This much $BaCl_2 \cdot 2H_2O$ equals 1.502 4 g. The Ba and Cl contents of the $BaCl_2 \cdot 2H_2O$ are

$$Ba = \left(\frac{137.33}{244.26}\right)(1.502\ 4 \text{ g}) = 0.844\ 69 \text{ g}$$

$$Cl = \left(\frac{2(35.453)}{244.26}\right)(1.502\ 4 \text{ g}) = 0.436\ 13 \text{ g}$$

Because the total sample weighs 1.783 9 g and contains 1.502 4 g of $BaCl_2 \cdot 2H_2O$, the sample must contain 1.783 9 − 1.502 4 = 0.281 5 g of KCl, which contains

$$K = \left(\frac{39.098}{74.551}\right)(0.281\ 5) = 0.147\ 63 \text{ g}$$

$$Cl = \left(\frac{35.453}{74.551}\right)(0.281\ 5) = 0.133\ 87 \text{ g}$$

Weight percent of each element:

$$Ba = \frac{0.844\ 69}{1.783\ 9} = 47.35\%$$

$$K = \frac{0.147\ 63}{1.783\ 9} = 8.276\%$$

$$Cl = \frac{0.436\ 13 + 0.133\ 87}{1.783\ 9} = 31.95\%$$

**27-D.** Let $x$ = mass of $Al(BF_4)_3$ and $y$ = mass of $Mg(NO_3)_3$. We can say that $x + y = 0.282\ 8$ g. We also know that

$$\begin{array}{l}\text{Moles of nitron} \\ \text{tetrafluoroborate}\end{array} = 3(\text{moles of } Al(BF_4)_3) = \frac{3x}{287.39}$$

$$\begin{array}{l}\text{Moles of} \\ \text{nitron nitrate}\end{array} = 2(\text{moles of } Mg(NO_3)_2) = \frac{2y}{148.31}$$

Equating the mass of product to the mass of nitron tetrafluoroborate plus the mass of nitron nitrate, we can write

$$\underbrace{1.322}_{\substack{\text{Mass of} \\ \text{product}}} = \underbrace{\left(\frac{3x}{287.39}\right)(400.18)}_{\substack{\text{Mass of nitron} \\ \text{tetrafluoroborate}}} + \underbrace{\left(\frac{2y}{148.31}\right)(375.39)}_{\substack{\text{Mass of nitron} \\ \text{nitrate}}}$$

Making the substitution $x = 0.282\ 8 - y$ allows us to find $y = 0.158\ 9$ g of $Mg(NO_3)_2 = 1.072$ mmol of Mg = 0.026 05 g of Mg = 9.210% of the original solid sample.

# Chapter 28

**28-A.** **(a)** Expected number of red marbles = $np_{red}$ = (1 000)(0.12) = 120. Expected number of yellow = $nq_{yellow}$ = (1 000)(0.88) = 880.
**(b)** Absolute: $\sigma_{red} = \sigma_{yellow} = \sqrt{npq}$
$$= \sqrt{(1\ 000)(0.12)(0.88)} = 10.28$$

Relative: $\sigma_{red}/n_{red}$ = 10.28/120 = 8.56%
$$\sigma_{yellow}/n_{yellow} = 10.28/880 = 1.17\%$$

**(c)** For 4 000 marbles, $n_{red}$ = 480 and $n_{yellow}$ = 3 520.
$$\sigma_{red} = \sigma_{yellow} = \sqrt{npq} = \sqrt{(4\ 000)(0.12)(0.88)} = 20.55.$$
$$\sigma_{red}/n_{red} = 4.28\%. \quad \sigma_{yellow}/n_{yellow} = 0.58\%$$

**(d)** 2, $\sqrt{n}$

**(e)** $\dfrac{\sigma_{red}}{n_{red}} = 0.02 = \dfrac{\sqrt{n(0.12)(0.88)}}{(0.12)n}$
$$\Rightarrow n = 1.83 \times 10^4$$

**28-B.** **(a)** $mR^2 = K_s \Rightarrow m(10)^2 = 36 \Rightarrow m = 0.36$ g
**(b)** An uncertainty of ±20 counts per second per gram is $100 \times 20/237$ = 8.4%.

$$n = \frac{t^2 s_s^2}{e^2} \approx \frac{(1.96)^2(0.10)^2}{(0.084)^2} = 5.4 \approx 5$$
$$\Rightarrow t = 2.776 \text{ (4 degrees of freedom)}$$

$$n \approx \frac{(2.776)^2(0.10)^2}{(0.084)^2} = 10.9 \approx 11 \Rightarrow t = 2.228$$

$$n \approx \frac{(2.228)^2(0.10)^2}{(0.084)^2} = 7.0 \approx 7 \Rightarrow t = 2.447$$

$$n \approx \frac{(2.447)^2(0.10)^2}{(0.084)^2} = 8.5 \approx 8 \Rightarrow t = 2.365$$

$$n \approx \frac{(2.365)^2(0.10)^2}{(0.084)^2} = 7.9 \approx 8$$

**28-C.** 1. The figure shows that coprecipitation occurs in distilled water, so it should first be shown that coprecipitation occurs in seawater, which has a high salt concentration. This can be done by adding standard $Pb^{2+}$ to real seawater and to artificial seawater (NaCl solution) and repeating the experiment in the figure.

2. To demonstrate that coprecipitation of $Pb^{2+}$ does not decrease the concentration of alkyl lead compounds, artificial seawater samples containing known quantities of alkyl lead compounds (but no $Pb^{2+}$) should be prepared and stripping analysis performed. Then a desired excess of $Pb^{2+}$ would be added to a similar solution containing an alkyl lead compound and coprecipitation with $BaSO_4$ would be carried out. The remaining solution would then be analyzed by stripping voltammetry. If the signal is the same as it was without $Pb^{2+}$ coprecipitation, then it is safe to say that coprecipitation does not reduce the concentration of alkyl lead in the seawater.

**28-D.** The acid-soluble inorganic matter and the organic material can probably be dissolved (and oxidized) together by wet ashing with $HNO_3 + H_2SO_4$ in a Teflon-lined bomb in a microwave oven. The insoluble residue should be washed well with water and the washings combined with the acid solution. After the residue has been dried, it can be fused with one of the fluxes in Table 28-6, dissolved in dilute acid, and combined with the previous solution.

# Answers to Problems

## Chapter 1

**3.** (a) milliwatt $= 10^{-3}$ watt  (b) picometer $= 10^{-12}$ meter
(c) kiloohm $= 10^3$ ohm  (d) microfarad $= 10^{-6}$ farad
(e) terajoule $= 10^{12}$ joule  (f) nanosecond $= 10^{-9}$ second
(g) femtogram $= 10^{-15}$ gram  (h) decipascal $= 10^{-1}$ pascal

**4.** (a) 100 fJ or 0.1 pJ  (b) 43.172 8 nF  (c) 299.79 THz
or 0.299 79 PHz  (d) 0.1 nm or 100 pm  (e) 21 TW  (f) 0.483 amol
or 483 zmol

**5.** (a) $5.4 \times 10^{12}$ kg of C  (b) $2.0 \times 10^{13}$ kg $CO_2$
(c) $2.0 \times 10^{10}$ ton $CO_2 = 4$ tons per person

**6.** $7.457 \times 10^4$ J/s, $6.416 \times 10^7$ cal/h

**7.** (a) 2.0 W/kg and 3.0 W/kg  (b) The person consumes $1.1 \times 10^2$ W.

**8.** (a) $\dfrac{0.025\ 4\ m}{1\ inch}$, 39.37 inches  (b) $0.214\ \dfrac{mile}{s}$, 770 mile/h
(c) $1.04 \times 10^3$ m, 1.04 km, 0.643 mile

**9.** $1.47 \times 10^3\ \dfrac{J}{s}$, $1.47 \times 10^3$ W

**11.** 6 tons/year

**14.** 1.10 M

**15.** 5.48 g

**16.** (a) $1.9 \times 10^{-7}$ bar  (b) 11 nM

**17.** $10^{-3}$ g/L, $10^3$ µg/L, 1 µg/mL, 1 mg/L

**18.** $7 \times 10^{-10}$ M

**19.** 26.5 g $HClO_4$, 11.1 g $H_2O$

**20.** (a) 1 670 g solution  (b) $1.18 \times 10^3$ g $HClO_4$  (c) 11.7 mol

**21.** 1.51 $m$

**22.** (a) 6.0 amol/vesicle  (b) $3.6 \times 10^6$ molecules
(c) $3.35 \times 10^{-20}$ m³, $3.35 \times 10^{-17}$ L  (d) 0.30 M

**23.** $4.4 \times 10^{-3}$ M, $6.7 \times 10^{-3}$ M

**24.** (a) 1 046 g, 376.6 g/L  (b) 9.07 $m$

**25.** Cal/g, Cal/ounce: shredded wheat (3.6, 102); doughnut (3.9, 111);
hamburger (2.8, 79); apple (0.48, 14)

**26.** $2.5 \times 10^6$ g $F^-$, $5.6 \times 10^6$ g NaF

**27.** (a) $2.11 \times 10^{-7}$ M  (b) Ar: $3.77 \times 10^{-4}$ M, Kr: $4.60 \times 10^{-8}$ M,
Xe: $3.5 \times 10^{-9}$ M

**28.** 6.18 g in a 2-L volumetric flask

**29.** Dissolve 6.18 g $B(OH)_3$ in 2.00 kg $H_2O$.

**30.** 3.2 L

**31.** 8.0 g

**32.** (a) 55.6 mL  (b) 1.80 g/mL

**33.** 1.51 g/mL

**34.** 1.29 mL

**35.** 14.4 g

## Chapter 2

**3.** $PbSiO_3$ is insoluble and will not leach into groundwater.

**4.** The upper "0" means that the reagent has no fire hazard. The
right-hand "0" indicates that the reagent is stable. The "3" tells us that the
reagent is corrosive or toxic and we should avoid skin contact or inhalation.

**5.** The lab notebook must: (1) state what was done; (2) state what was
observed; and (3) be understandable to a stranger.

**7.** The buoyancy correction is 1 when the substance being weighed has
the same density as the weight used to calibrate the balance.

**8.** 14.85 g

**9.** smallest: $PbO_2$; largest: lithium

**10.** 4.239 1 g; lower by 0.06%

**11.** (a) 0.000 164 g/mL  (b) 0.823 g

**12.** (a) 979 Pa  (b) 0.001 1 g/mL  (c) 1.001 0 g

**13.** 99.999 1 g

**20.** phosphorus pentoxide

**21.** 9.979 9 mL

**22.** 0.2%; 0.499 0 M

**23.** 49.947 g in vacuum; 49.892 g in air

**24.** true mass $= 50.506$ g; mass in air $= 50.484$ g

**25.** 0.70%

## Chapter 3

**1.** (a) 5  (b) 4  (c) 3

**2.** (a) 1.237  (b) 1.238  (c) 0.135  (d) 2.1  (e) 2.00

**3.** (a) 0.217  (b) 0.216  (c) 0.217

**4.** (b) 1.18 (three significant figures)  (c) 0.71 (two significant figures)

**5.** (a) 3.71  (b) 10.7  (c) $4.0 \times 10^1$  (d) $2.85 \times 10^{-6}$
(e) 12.625 1  (f) $6.0 \times 10^{-4}$  (g) 242

**6.** (a) 175.324  (b) 140.093 5 or 140.094

**7.** (a) 12.3  (b) 75.5  (c) $5.520 \times 10^3$  (d) 3.04
(e) $3.04 \times 10^{-10}$  (f) 11.9  (g) 4.600  (h) $4.9 \times 10^{-7}$

**11.** low; systematic

**12.** (a) 25.031 mL—systematic error; $\pm 0.009$ mL—random error
(b) 1.98 and 2.03 mL—systematic error; $\pm 0.01$ and $\pm 0.02$ mL—random
error  (c) random error  (d) random error

**13.** (a) Carmen  (b) Cynthia  (c) Chastity  (d) Cheryl

**14.** 3.124 ($\pm 0.005$), 3.124 ($\pm 0.2\%$)

**15.** (a) 2.1 ($\pm 0.2$ or $\pm 11\%$)  (b) 0.151 ($\pm 0.009$ or $\pm 6\%$)
(c) $0.22_3 (\pm 0.02_4$ or $\pm 11\%)$  (d) $0.097_1 (\pm 0.002_2$ or $\pm 2._2\%)$

**16.** (a) 10.18 ($\pm 0.07$ or $\pm 0.7\%$)  (b) 174 ($\pm 3$ or $\pm 2\%$)
(c) 0.147 ($\pm 0.003$ or $\pm 2\%$)  (d) 7.86 ($\pm 0.01$ or $\pm 0.1\%$)
(e) 2 185.8 ($\pm 0.8$ or $\pm 0.04\%$)  (f) $1.464_3 (\pm 0.007_8$ or $\pm 0.5_3\%)$
(g) $0.496_9 (\pm 0.006_9$ or $\pm 1.3_9\%)$

**18.** 255.184 $\pm$ 0.007

**19.** (b) 0.450 7 ($\pm 0.000$ 5) M

**20.** 1.035 7 ($\pm 0.000$ 2) g

**21.** 16.2 $\pm$ 0.1 mg

**22.** 0.667 $\pm$ 0.001 M

**23.** 6.022 136 9 (48) $\times 10^{23}$

## Chapter 4

**2.** (a) 0.682 6  (b) 0.954 6  (c) 0.341 3  (d) 0.191 5  (e) 0.149 8

**3.** (a) 1.527 67  (b) 0.001 26  (c) $1.59 \times 10^{-6}$

**4.** (a) 0.044 6  (b) 0.417 3  (c) 0.404 0

**5.** (a) 0.5  (b) 0.8%  (c) 8.7%

**11.** 90%: $0.14_8 \pm 0.02_8$; 99%: $0.14_8 \pm 0.05_6$

**12.** $\bar{x} \pm 0.000\ 10$ (1.527 83 to 1.528 03)

**13.** (a) deciliter $= 0.1$ L  (b) yes

**14.** Difference is *not* significant ($t_{calculated} = 2.4_3 < t_{table} = 2.776$).

**15.** Difference is *not* significant ($t_{calculated} = 2.429\ 7 < t_{table} = 2.776\ 5$).

16. Difference *is* significant ($t_{\text{calculated}} = 11.3 > t_{\text{table}} = 2.57$).

17. yes

18. 1–2 difference *is* significant; 2–3 difference is *not* significant.

19. Difference *is* significant at 95% and 99% levels.

20. Difference *is* significant in both cases.

21. (a) Differences are *not* significant.  (b) yes

22. Retain 216.

23. $m = -1.299\ (\pm0.001) \times 10^4$; $b = 3\ (\pm3) \times 10^2$

24. $m = 0.6_4 \pm 0.1_2$; $b = 0.9_3 \pm 0.2_6$

29. 10.1 μg

30. (a) $2.0_0 \pm 0.3_8$  (c) $0.2_6$

31. $10.1 \pm 0.2$ μg

32. (a) $m = 869 \pm 11$, $b = -22._1 \pm 8._9$  (b) 145.0 mV
(c) $0.19_2\ (\pm0.01_4)$ vol%

33. (b) $m = 15.12\ (\pm0.10)$, $b = 18.9\ (\pm3.8)$
(c) $31.9\ (\pm0.2)$ μM

34. 21.9 μg

35. (a) Entire range is linear.  (b) log (current) =
0.969 2 log (concentration) + 1.339  (c) 4.80 μg/mL

36. $15.2_2 \pm 0.8_6$ μg, $15._2 \pm 1._5$ μg

## Chapter 5

11. statement c

14. 1% noise data: $y = 26.075x + 12.455$    $R^2 = 0.999\ 3$
10% noise data: $y = 23.336x + 141.27$    $R^2 = 0.973\ 1$

15. (a) 22.2 ng/mL: precision = 23.8%, accuracy = 6.6%
88.2 ng/mL: precision = 13.9%, accuracy = −6.5%
314 ng/mL: precision = 7.8%, accuracy = −3.6%
(b) signal detection limit = $129._6$; detection limit = $4.8 \times 10^{-8}$ M;
quantitation limit = $1.6 \times 10^{-7}$ M

16. (a) 4%, 128%  (b) 1.4%

18. recovery = 96%; concentration detection limit = 0.064 μg/L

19. detection limit = 130 counts;
minimum detectable concentration = $4.8 \times 10^{-8}$ M

20. detection limits: 0.086, 0.102, 0.096, and 0.114 μg/mL;
mean = 0.10 μg/mL

23. (c) 1.04 ppm

24. (a) $8.72 \pm 0.43$ ppb  (b) 116 ppm  (c) ±6 ppm
(d) ±18 ppm

25. (a) tap water, 0.091 ng/mL; pond water, 22.2 ng/mL  (b) This is a
matrix effect. Something in pond water decreases the Eu(III) emission.

26. (a) 0.140  (b) standard deviation = ±0.005 M;
95% confidence = ±0.015 M

27. 1.56 (±0.08) mM; 95% confidence: 1.56 (±0.24) mM

29. (a) $0.168_4$  (b) 0.847 mM  (c) 6.16 mM  (d) 12.3 mM

30. 9.09 mM

31. response factor = slope of graph = $1.07_6$; standard
deviation = $0.06_7 = 6._2\%$

## Chapter 6

4. (a) $K = 1/[\text{Ag}^+]^3[\text{PO}_4^{3-}]$  (b) $K = P_{\text{CO}_2}^6/P_{\text{O}_2}^{15/2}$

5. $1.2 \times 10^{10}$

6. $2.0 \times 10^{-9}$

7. (a) decrease  (b) give off  (c) negative

8. $5 \times 10^{-11}$

9. (a) right  (b) right  (c) neither  (d) right  (e) smaller

10. (a) $4.7 \times 10^{-4}$ bar  (b) 153°C

11. (a) 7.82 kJ/mol  (b) A graph of ln $K$ vs $1/T$ will have a slope of
$-\Delta H^\circ/R$.

12. (a) right  (b) $P_{\text{H}_2} = 1\ 366$ Pa, $P_{\text{Br}_2} = 3\ 306$ Pa, $P_{\text{HBr}} = 57.0$ Pa
(c) neither  (d) formed

13. 0.663 mbar

14. $5 \times 10^{-8}$ M

15. 8.5 zM

16. $3.9 \times 10^{-7}$ M

17. (a) $2.1 \times 10^{-8}$ M  (b) $8.4 \times 10^{-4}$ M

19. no, 0.001 4 M

20. no

21. $\text{I}^- < \text{Br}^- < \text{Cl}^- < \text{CrO}_4^{2-}$

23. (a) $\text{BF}_3$  (b) $\text{AsF}_5$

24. 0.096 M

25. $[\text{Zn}^{2+}] = 2.9 \times 10^{-3}$ M, $[\text{ZnOH}^+] = 2.3 \times 10^{-5}$ M,
$[\text{Zn(OH)}_3^-] = 6.9 \times 10^{-7}$ M, $[\text{Zn(OH)}_4^{2-}] = 8.6 \times 10^{-14}$ M

26. 15%

27. $1.1 \times 10^{-5}$ M

30. (a) an adduct  (b) dative or coordinate covalent  (c) conjugate
(d) $[\text{H}^+] > [\text{OH}^-]$, $[\text{H}^+] < [\text{OH}^-]$

34. (a) HI  (b) $\text{H}_2\text{O}$

35. $2\text{H}_2\text{SO}_4 \rightleftharpoons \text{HSO}_4^- + \text{H}_3\text{SO}_4^+$

36. (a) $(\text{H}_3\text{O}^+, \text{H}_2\text{O})$; $(\text{H}_3\overset{+}{\text{N}}\text{CH}_2\text{CH}_2\text{NH}_3, \text{H}_3\overset{+}{\text{N}}\text{CH}_2\text{CH}_2\text{NH}_2)$
(b) $(\text{C}_6\text{H}_5\text{CO}_2\text{H}, \text{C}_6\text{H}_5\text{CO}_2^-)$; $(\text{C}_5\text{H}_5\text{NH}^+, \text{C}_5\text{H}_5\text{N})$

37. (a) 2.00  (b) 12.54  (c) 1.52  (d) −0.48  (e) 12.00

38. (a) 6.998  (b) 6.132

39. $1.0 \times 10^{-56}$

40. 7.8

41. (a) endothermic  (b) endothermic  (c) exothermic

44. $\text{Cl}_3\text{CCO}_2\text{H} \rightleftharpoons \text{Cl}_3\text{CCO}_2^- + \text{H}^+$

$\text{La}^{3+} + \text{H}_2\text{O} \rightleftharpoons \text{LaOH}^{2+} + \text{H}^+$

45.

$\text{HOCH}_2\text{CH}_2\text{S}^- + \text{H}_2\text{O} \rightleftharpoons \text{HOCH}_2\text{CH}_2\text{SH} + \text{OH}^-$

46. $K_a$: $\text{HCO}_3^- \rightleftharpoons \text{H}^+ + \text{CO}_3^{2-}$
$K_b$: $\text{HCO}_3^- + \text{H}_2\text{O} \rightleftharpoons \text{H}_2\text{CO}_3 + \text{OH}^-$

47. (a) $\text{H}_3\overset{+}{\text{N}}\text{CH}_2\text{CH}_2\overset{+}{\text{N}}\text{H}_3 \xrightleftharpoons{K_{a1}} \text{H}_2\text{NCH}_2\text{CH}_2\overset{+}{\text{N}}\text{H}_3 + \text{H}^+$

$\text{H}_2\text{NCH}_2\text{CH}_2\overset{+}{\text{N}}\text{H}_3 \xrightleftharpoons{K_{a2}} \text{H}_2\text{NCH}_2\text{CH}_2\text{NH}_2 + \text{H}^+$

(b) $^-\text{O}_2\text{CCH}_2\text{CO}_2^- + \text{H}_2\text{O} \xrightleftharpoons{K_{b1}} \text{HO}_2\text{CCH}_2\text{CO}_2^- + \text{OH}^-$

$\text{HO}_2\text{CCH}_2\text{CO}_2^- + \text{H}_2\text{O} \xrightleftharpoons{K_{b2}} \text{HO}_2\text{CCH}_2\text{CO}_2\text{H} + \text{OH}^-$

48. a, c

49. $\text{CN}^- + \text{H}_2\text{O} \rightleftharpoons \text{HCN} + \text{OH}^-$; $K_b = 1.6 \times 10^{-5}$

50. $\text{H}_2\text{PO}_4^- \xrightleftharpoons{K_{a2}} \text{HPO}_4^{2-} + \text{H}^+$

$\text{HC}_2\text{O}_4^- + \text{H}_2\text{O} \xrightleftharpoons{K_{b2}} \text{H}_2\text{C}_2\text{O}_4 + \text{OH}^-$

51. $K_{a1} = 7.04 \times 10^{-3}$, $K_{a2} = 6.29 \times 10^{-8}$, $K_{a3} = 7.1 \times 10^{-13}$

52. $2.9 \times 10^{-6}$

53. (a) $1.2 \times 10^{-2}$ M  (b) Solubility will be greater.

54. 0.22 g

55. right; $[\text{IO}_3^-] = 7.00$ mM, $[\text{Br}^-] = 15.0$ mM, $[\text{H}^+] = 17.0$ mM,
$[\text{Br}_2(aq)] = 11.0$ μM

**56.** $[IO_3^-] = 0.834$ mM, $[I^-] = 0.167$ mM, $[H^+] = 8.51 \times 10^{-7}$ M, $[I_2(aq)] = 0.500$ mM

## Chapter 7

**7.** 32.0 mL

**8.** 43.20 mL $KMnO_4$; 270.0 mL $H_2C_2O_4$

**9.** 0.149 M

**10.** 0.100 3 M

**11.** 92.0 wt%

**12.** 15.1%

**13.** **(a)** 0.020 34 M  **(b)** 0.125 7 g  **(c)** 0.019 83 M

**14.** 56.28 wt%

**15.** 8.17 wt%

**16.** 0.092 54 M

**17.** **(a)** 17 L  **(b)** 793 L  **(c)** $1.05 \times 10^3$ L

**19.** **(a)** $2.33 \times 10^{-7}$ mol Fe(III)  **(b)** $5.83 \times 10^{-5}$ M

**20.** theoretical equivalence point = 13.3 µL; observed end point = 12.2 µL = 1.83 Ga/transferrin. Ga does not appear to bind in the absence of oxalate.

**22.** **(a)** 13.08  **(b)** 8.04  **(c)** 2.53

**23.** $[Ag^+] = 9.1 \times 10^{-9}$ M; $Q = [Ag^+][Cl^-] = 2.8 \times 10^{-10} > K_{sp}$ for AgCl

**24.** **(a)** 6.06  **(b)** 3.94  **(c)** 2.69

**25.** $[AgCl(aq)] = 370$ nM, $[AgBr(aq)] = 20$ nM, $[AgI(aq)] = 0.32$ nM

**26.** $V_{e1} = 18.76$ mL, $V_{e2} = 37.52$ mL

**27.** **(a)** 14.45  **(b)** 13.80  **(c)** 8.07  **(d)** 4.87  **(e)** 2.61

**28.** **(a)** 19.00, 18.85, 18.65, 17.76, 14.17, 13.81, 7.83, 1.95  **(b)** no

**29.** $V_X = V_M^0(C_M^0 - [M^+] + [X^-])/(C_X^0 + [M^+] - [X^-])$

**36.** 947 mg

**37.** $[A] = 0.83_8 \pm 0.02_7$ M, $[B] = 0.44_0 \pm 0.02_7$ M, $[C] = 0.25_6 \pm 0.02_7$ M

## Chapter 8

**2.** **(a)** true  **(b)** true  **(c)** true

**3.** **(a)** 0.008 7 M  **(b)** $0.001_2$ M

**4.** **(a)** 0.660  **(b)** 0.54  **(c)** 0.18  **(d)** 0.83

**5.** $0.88_7$

**6.** **(a)** $0.42_2$  **(b)** $0.43_2$

**7.** $0.20_2$

**8.** increase

**9.** $7.0 \times 10^{-17}$ M

**10.** $6.6 \times 10^{-7}$ M

**11.** $\gamma_{H^+} = 0.86$, pH = 2.07

**12.** 11.94, 12.00

**13.** 0.329

**17.** $[H^+] + 2[Ca^{2+}] + [Ca(HCO_3)^+] + [Ca(OH)^+] + [K^+] = [OH^-] + [HCO_3^-] + 2[CO_3^{2-}] + [ClO_4^-]$

**18.** $[H^+] = [OH^-] + [HSO_4^-] + 2[SO_4^{2-}]$

**19.** $[H^+] = [OH^-] + [H_2AsO_4^-] + 2[HAsO_4^{2-}] + 3[AsO_4^{3-}]$

**20.** **(a)** charge: $2[Mg^{2+}] + [H^+] + [MgBr^+] + [MgOH^+] = [Br^-] + [OH^-]$
mass: $[MgBr^+] + [Br^-] = 2\{[Mg^{2+}] + [MgBr^+] + [MgOH^+]\}$
**(b)** $[Mg^{2+}] + [MgBr^+] + [MgOH^+] = 0.2$ M; $[MgBr^+] + [Br^-] = 0.4$ M

**21.** $2.3 \times 10^6$ N, $5.2 \times 10^5$ pounds

**22.** $[CH_3CO_2^-] + [CH_3CO_2H] = 0.1$ M

**23.** $[Y^{2-}] = [X_2Y_2^{2+}] + 2[X_2Y^{4+}]$

**24.** $3\{Fe^{3+}\} + [Fe(OH)^{2+}] + [Fe(OH)_2^+] + 2[Fe_2(OH)_2^{4+}] + [FeSO_4^+]\} = 2\{[FeSO_4^+] + [SO_4^{2-}] + [HSO_4^-]\}$

**25.** **(b)** $[Ca^{2+}] = 0.010$ 1 M, $[CaOH^+] = 0.005$ 1 M, $[OH^-] = 0.025$ 4 M, $[H^+] = 4 \times 10^{-13}$ M; solubility = 1.1 g/L

**26.** **(a)** 0.004 8 M  **(b)** $[Zn^{2+}] = 0.006$ 7 M, 33% is ion paired, $\mu = 0.027$ M

**27.** $[Li^+] = [F^-] = 0.050$ M, $[LiF(aq)] = 0.002$ 9 M

**28.** **(a)** $4.3 \times 10^{-5}$  **(b)** $4.9 \times 10^{-4}$ M = 20 mg/L  **(c)** 0.023 bar

**29.** 70.7, 85.2, 150.4, 243.6, 280.1 mV

## Chapter 9

**2.** **(a)** 3.00  **(b)** 12.00

**3.** 6.89, 0.61

**4.** **(a)** 0.809  **(b)** 0.791  **(c)** Activity coefficient depends slightly on counterion.

**5.** **(a)** ⬡—$CO_2H$ ⇌ ⬡—$CO_2^-$ + $H^+$  $K_a$

**(b)** ⬡—$CO_2^-$ + $H_2O$ ⇌ ⬡—$CO_2H$ + $OH^-$  $K_b$

**(c)** ⬡—$NH_2$ + $H_2O$ ⇌ ⬡—$\overset{+}{N}H_3$ + $OH^-$  $K_b$

**(d)** ⬡—$\overset{+}{N}H_3$ ⇌ ⬡—$NH_2$ + $H^+$  $K_a$

**6.** pH = 3.00, $\alpha = 0.995\%$

**7.** 5.50

**8.** 5.51, $3.1 \times 10^{-6}$ M, 0.060 M

**10.** 99% dissociation when F = $(0.010\ 2)K_a$

**11.** 4.20

**12.** 5.79

**13.** **(a)** 3.03, 9.4%  **(b)** 7.00, 99.9%

**14.** 5.64, 0.005 3%

**15.** 2.86, 14%

**16.** 99.6%, 96.5%

**19.** 11.00, 0.995%

**20.** 11.28, [B] = 0.058 M, $[BH^+] = 1.9 \times 10^{-3}$ M

**21.** 10.95

**22.** 0.007 6%, 0.024%, 0.57%

**23.** $3.6 \times 10^{-9}$

**24.** $4.1 \times 10^{-5}$

**25.** 0.999, 0.000 999

**32.** 4-aminobenzenesulfonic acid

**33.** 4.70

**34.** **(a)** 0.180  **(b)** 1.00  **(c)** 1.80

**35.** **(a)** 14  **(b)** $1.4 \times 10^{-7}$

**36.** **(a)** NaOH  **(b)** 1. Weigh out $(0.250\ L)(0.050\ 0\ M) = 0.012$ 5 mol of HEPES and dissolve in ~200 mL. 2. Adjust the pH to 7.45 with NaOH. 3. Dilute to 250 mL.

**37.** 3.38 mL

**38.** **(b)** 7.18  **(c)** 7.00  **(d)** 6.86 mL

**39.** **(a)** 2.56  **(b)** 2.61  **(c)** 2.86

**40.** 16.2 mL

**41.** **(a)** pH = 5.06, [HA] = 0.001 99 M, $[A^-] = 0.004$ 01 M

**42.** **(a)** Approximate pH = 11.70, more accurate pH = 11.48

**43.** 6.86

# Chapter 10

2. $H_3\overset{+}{N}-CH-CO_2^-$ (with R substituent above the CH); $pK$ values apply to $-\overset{+}{N}H_3$, $-CO_2H$, and, in some cases, R.

3. $4.37 \times 10^{-4}$, $8.93 \times 10^{-13}$

4. **(a)** pH = 2.51, $[H_2A]$ = 0.096 9 M, $[HA^-]$ = $3.11 \times 10^{-3}$ M, $[A^{2-}]$ = $1.00 \times 10^{-8}$ M **(b)** 6.00, $1.00 \times 10^{-3}$ M, $1.00 \times 10^{-1}$ M, $1.00 \times 10^{-3}$ M **(c)** 10.50, $1.00 \times 10^{-10}$ M, $3.16 \times 10^{-4}$ M, $9.97 \times 10^{-2}$ M

5. **(a)** pH = 1.95, $[H_2M]$ = 0.089 M, $[HM^-]$ = $1.12 \times 10^{-2}$ M, $[M^{2-}]$ = $2.01 \times 10^{-6}$ M **(b)** pH = 4.28, $[H_2M]$ = $3.7 \times 10^{-3}$ M, $[HM^-] \approx 0.100$ M, $[M^{2-}]$ = $3.8 \times 10^{-3}$ M **(c)** pH = 9.35, $[H_2M]$ = $7.04 \times 10^{-12}$ M, $[HM^-]$ = $2.23 \times 10^{-5}$ M, $[M^{2-}]$ = 0.100 M

6. pH = 11.60, $[B]$ = 0.296 M, $[BH^+]$ = $3.99 \times 10^{-3}$ M, $[BH_2^{2+}]$ = $2.15 \times 10^{-9}$ M

7. pH = 3.69, $[H_2A]$ = $3.0 \times 10^{-6}$ M, $[HA^-]$ = $7.9 \times 10^{-4}$ M, $[A^{2-}]$ = $2.1 \times 10^{-4}$ M

8. 4.03

9. **(a)** pH = 6.002, $[HA^-]$ = 0.009 8 M **(b)** pH = 4.50, $[HA^-]$ = 0.006 1 M

10. $[CO_2(aq)]$ = $10^{-4.9}$ M, pH = 5.67

11. 2.96 g

12. 2.22 mL

13. Procedure: Dissolve 10.0 mmol (1.23 g) picolinic acid in ~75 mL $H_2O$ in a beaker. Add NaOH (~5.63 mL) until the measured pH is 5.50. Transfer to a 100-mL volumetric flask and use small portions of $H_2O$ to rinse the beaker into the flask. Dilute to 100.0 mL and mix well.

14. 26.5 g $Na_2SO_4$ + 1.31 g $H_2SO_4$

15. no

16.

$$\overset{+}{N}H_3-CHCH_2CH_2CO_2H \overset{K_1}{\rightleftharpoons} \overset{+}{N}H_3-CHCH_2CH_2CO_2H$$
(with $CO_2H$ / $CO_2^-$ below respectively)

Glutamic acid

$$\overset{K_2}{\rightleftharpoons} \overset{+}{N}H_3-CHCH_2CH_2CO_2^- \overset{K_3}{\rightleftharpoons} NH_2-CHCH_2CH_2CO_2^-$$
(with $CO_2^-$ below each)

$$\overset{+}{N}H_3-CHCH_2-\bigcirc-OH \overset{K_1}{\rightleftharpoons} \overset{+}{N}H_3-CHCH_2-\bigcirc-OH$$
(with $CO_2H$ / $CO_2^-$ below)  Tyrosine

$$\overset{K_2}{\rightleftharpoons} NH_2-CHCH_2-\bigcirc-OH \overset{K_3}{\rightleftharpoons} NH_2-CHCH_2-\bigcirc-O^-$$
(with $CO_2^-$ below each)

17. **(a)** $2.8 \times 10^{-3}$ **(b)** $2.9 \times 10^{-8}$

18. **(a)** $NaH_2PO_4$ and $Na_2HPO_4$ would be simplest, but other combinations such as $H_3PO_4$ and $Na_3PO_4$ or $H_3PO_4$ and $Na_2HPO_4$ would work just fine. **(b)** 4.55 g $Na_2HPO_4$ + 2.15 g $NaH_2PO_4$ **(c)** One of several ways: Weigh out 0.050 0 mol $Na_2HPO_4$ and dissolve it in 900 mL of water. Add HCl while monitoring the pH with a pH electrode. When the pH is 7.45, stop adding HCl and dilute to exactly 1 L with $H_2O$.

19. pH = 5.64, $[H_2L^+]$ = 0.010 0 M, $[H_3L^{2+}]$ = $1.36 \times 10^{-6}$ M, $[HL]$ = $3.68 \times 10^{-6}$ M, $[L^-]$ = $2.40 \times 10^{-11}$ M

20. 78.9 mL

21. **(a)** 5.88 **(b)** 5.59

22. **(a)** HA **(b)** $A^-$ **(c)** 1.0, 0.10

23. **(a)** 4.00 **(b)** 8.00 **(c)** $H_2A$ **(d)** $HA^-$ **(e)** $A^{2-}$

24. **(a)** 9.00 **(b)** 9.00 **(c)** $BH^+$ **(d)** $1.0 \times 10^3$

25.

26. $\alpha_{HA}$ = 0.091, $\alpha_{A^-}$ = 0.909, $[A^-]/[HA]$ = 10

27. 0.91

28. $\alpha_{HA^-}$ = 0.123, 0.694

29. $\alpha_{H_2A^-}$ = 0.893, 0.500, $5.4 \times 10^{-5}$, $2.2 \times 10^{-5}$, $1.55 \times 10^{-12}$
$\alpha_{HA^-}$ = 0.107, 0.500, 0.651, 0.500, $1.86 \times 10^{-4}$
$\alpha_{A^{2-}}$ = $5.8 \times 10^{-7}$, $2.2 \times 10^{-5}$, 0.349, 0.500, 0.999 8

30. **(b)** $8.6 \times 10^{-6}$, 0.61, 0.39, $1.6 \times 10^{-6}$

31. 0.36

32. 96%

34. At pH 10: $\alpha_{H_3A}$ = $1.05 \times 10^{-9}$, $\alpha_{H_2A^-}$ = 0.040 9, $\alpha_{HA^{2-}}$ = 0.874, $\alpha_{A^{3-}}$ = 0.085 4

35. **(b)** $[Cr(OH)_3(aq)]$ = $10^{-6.84}$ M **(c)** $[Cr(OH)_2^+]$ = $10^{-4.44}$ M, $[Cr(OH)^{2+}]$ = $10^{-2.04}$ M

37. The *average* charge is 0. There is no pH at which *all* molecules have zero charge.

38. isoelectric pH 5.59, isoionic pH 5.72

40. pH of zero charge = 5.95

# Chapter 11

2. 13.00, 12.95, 12.68, 11.96, 10.96, 7.00, 3.04, 1.75

6. 3.00, 4.05, 5.00, 5.95, 7.00, 8.98, 10.96, 12.25

7. $V_e/11$; $10V_e/11$; $V_e$ = 0, pH = 2.80; $V_e/11$, pH = 3.60; $V_e/2$, pH = 4.60; $10V_e/11$, pH = 5.60; $V_e$, pH = 8.65; $1.2V_e$, pH = 11.96

8. 8.18

9. $5.4 \times 10^7$

10. 0.107 M

11. 9.72

14. 11.00, 9.95, 9.00, 8.05, 7.00, 5.02, 3.04, 1.75

15. $\frac{1}{2}V_e$

16. $2.2 \times 10^9$

17. 10.92, 9.57, 9.35, 8.15, 5.53, 2.74

18. **(a)** 9.45 **(b)** 2.55 **(c)** 5.15

20. positive

21. isoionic

23. 11.49, 10.95, 10.00, 9.05, 8.00, 6.95, 6.00, 5.05, 3.54, 1.79

24. 2.51, 3.05, 4.00, 4.95, 6.00, 7.05, 8.00, 8.95, 10.46, 12.21

25. 11.36, 10.21, 9.73, 9.25, 7.53, 5.81, 5.33, 4.86, 3.41, 2.11, 1.85

26. 5.01

27. **(a)** 1.99

28. **(b)** 7.13

29. 2.72

30. **(a)** 9.54 **(b)** $7.9 \times 10^{-10}$

31. 6.28 g

32. $pK_2$ = 9.84

34. end point = 23.39 mL

35. end point = 10.727 mL

39. $H_2SO_4$, HCl, $HNO_3$, or $HClO_4$

40. yellow, green, blue

41. (a) red  (b) orange  (c) yellow

42. (a) red  (b) orange  (c) yellow  (d) red

43. no (end point pH must be > 7)

44. (a) 2.47

45. (a) violet  (b) blue  (c) yellow

46. (a) 5.62

47. 2.859%

51. 0.079 34 mol/kg

52. $1.023_8$ g, systematic error = 0.08%, calculated HCl molarity is low

53. 0.100 0 M

54. 0.31 g

55. (a) 20.254 wt%  (b) 17.985 g

58. (a) acetic acid  (b) pyridine

65. (b) $K = 0.279$, pH = 4.16

70. 0.139 M

71. 0.815

## Chapter 12

2. (a) $2.7 \times 10^{-10}$  (b) 0.57

3. (a) $2.5 \times 10^7$  (b) $4.5 \times 10^{-5}$ M

4. 5.60 g

5. (a) 100.0 mL  (b) 0.016 7 M  (c) 0.041  (d) $4.1 \times 10^{10}$
(e) $7.8 \times 10^{-7}$ M  (f) $2.49 \times 10^{-10}$ M

6. (a) 2.93  (b) 6.79  (c) 10.52

7. 1.70, 2.18, 2.81, 3.87, 4.87, 6.85, 8.82, 10.51, 10.82

8. $\infty$, 10.30, 9.52, 8.44, 7.43, 6.15, 4.88, 3.20, 2.93

9. $4.6 \times 10^{-11}$ M

14. (a) 25  (b) 0.017

15. (a) 11.08  (b) 11.09  (c) 12.35  (d) 15.03  (e) 17.69

16. (b) $\alpha_{ML} = 0.28$, $\alpha_{ML_2} = 0.70$

17. (d) $[T] = 0.27_7$, $[Fe_aT] = 0.55_3$, $[Fe_bT] = 0.09_2$, $[Fe_2T] = 0.07_7$

19. (b) 1.34 mL, pNi = 6.00; 21.70 mL, pNi = 7.00; 26.23 mL, pNi = 17.00

23. 1. with metal ion indicators;  2. with a mercury electrode;
3. with an ion-selective electrode;  4. with a glass electrode

24. $HIn^{2-}$, wine-red, blue

25. Buffer (a): yellow $\rightarrow$ blue; other buffers: violet $\rightarrow$ blue, which is harder to see

29. Temporary hardness, due to $Ca(HCO_3)_2$, is lost by heating. Permanent hardness is derived from other salts, such as $CaSO_4$, and is not affected by heat.

30. 10.0 mL, 10.0 mL

31. 0.020 0 M

32. 0.995 mg

33. 21.45 mL

34. $[Ni^{2+}] = 0.012 4$ M, $[Zn^{2+}] = 0.007 18$ M

35. 0.024 30 M

36. 0.092 28 M

37. observed: 32.7 wt%; theoretical: 32.90 wt%

38. (a) Dominant form of EDTA at pH 4 is $H_2Y^{2-}$. Before $V_e$,
$Cu^{2+} + H_2Y^{2-} \rightarrow CuY^{2-} + 2H^+$ releases $2H^+$/EDTA. After $V_e$, no reaction consumes or releases $H^+$.

(b) Dominant form of EDTA at pH 8 is $HY^{3-}$. Before $V_e$,
$Mg^{2+} + H_2Y^{2-} \rightarrow MgY^{2-} + 2H^+$ releases $2H^+$/EDTA. After $V_e$, the reaction $OH^- + H_2Y^{2-} \rightarrow HY^{3-} + H_2O$ consumes 1 $OH^-$ per EDTA.

## Chapter 13

1. $PbS(s) + H^+ \rightleftharpoons Pb^{2+} + HS^-$

$PbCO_3(s) + H^+ \rightleftharpoons Pb^{2+} + HCO_3^-$

2. (a) pH = 9.98  (b) pH = 10.00  (c) pH = 9.45

3. pH = 9.95

4. predicted values: $pK_1' = 2.350$, $pK_2' = 9.562$

5. pH = 10.194 from spreadsheet and 10.197 by hand method

6. pH = 4.52

7. pH = 5.00

8. ionic strength = 0.025 M, pH = 4.94

9. (a) pH = 7.420  (b) pH = 7.403

10. pH = 4.44

11. (d) $[Fe^{3+}] = \dfrac{F_{SCN} - [SCN^-]}{\beta_1'[SCN^-] + 2\beta_2'[SCN^-]^2}$

(e) $[H^+] = \dfrac{K_a'[Fe^{3+}]}{F_{Fe} - [Fe^{3+}] - \beta_1'[Fe^{3+}][SCN^-] - \beta_2'[Fe^{3+}][SCN^-]^2}$

(f, g, h, i) $[SCN^-] = 2.03$ μM, $[Fe^{3+}] = 4.20$ mM, $[H^+] = 15.8$ mM, $[Fe(SCN)^{2+}] = 2.97$ μM, $[Fe(SCN)_2^+] = 0.106$ nM, $[FeOH^{2+}] = 0.802$ mM, ionic strength = 0.043 4 M, pH = 1.88,

$\dfrac{[Fe(SCN)^{2+}]}{\{[Fe^{3+}] + [FeOH^{2+}]\}[SCN^-]} = 293$

(j) $[SCN^-] = 2.81$ μM, $[Fe^{3+}] = 4.45$ mM, $[H^+] = 15.5$ mM, $[Fe(SCN)^{2+}] = 2.19$ μM, $[Fe(SCN)_2^+] = 0.068$ nM, $[FeOH^{2+}] = 0.546$ mM, ionic strength = 0.243 9 M, pH = 1.94,

$\dfrac{[Fe(SCN)^{2+}]}{\{[Fe^{3+}] + [FeOH^{2+}]\}[SCN^-]} = 156$

12. (a) $[SO_4^{2-}] = 1.50$ mM, $[La^{3+}] = 0.57$ mM, $[H^+] = 1.14$ μM, $[La(SO_4)^+] = 1.36$ mM, $[La(SO_4)_2^-] = 67$ μM, $[LaOH^{2+}] = 1.13$ μM, ionic strength = 0.006 29 M, pH = 5.98  (b) ionic strength of strong electrolyte = 15.0 mM; actual ionic strength = 6.3 mM  (c) 28.5%
(d) $pK_a$ for $HSO_4^-$ is 1.99 and we expect the solution to be near neutral pH.  (e) no; $[La^{3+}][OH^-]^3\gamma_{La^{3+}}\gamma_{OH^-}^3 = 2.3 \times 10^{-28} < K_{sp}$ for $La(OH)_3 = 2 \times 10^{-21}$

13. $[CN^-] = 3.26$ μM, $[H^+] = 1.29 \times 10^{-12}$ M, $[OH^-] = 0.012 9$ M, $[Ag^+] = 1.86$ nM, $[AgOH] = 0.187$ nM, $[Ag(OH)(CN)^-] = 101$ μM, $[Ag(CN)_2^-] = 0.100$ M, $[Ag(CN)_3^{2-}] = 1.94$ μM, $[HCN] = 1.90$ nM, $[Na^+] = 0.013 0$ M, $[K^+] = 0.100$ M, ionic strength = 0.113 M

14. fraction of Fe in each form: $[Fe^{2+}]$, 2.4%; $[FeG^+]$, 58.4%; $[FeG_2]$, 38.9%; $[FeG_3^-]$, 0.14%; $[FeOH^+]$, 0.11%.
fraction of glycine in each form: $[G^-]$, 0.50%; $[HG]$, 31.2%; $[H_2G^+]$, 0.000 1%; $[FeG^+]$, 29.2%; $2[FeG_2]$, 38.9%; $3[FeG_3^-]$, 0.21%

HCl added = 31.1 mmol; ionic strength = 32.9 mM

Chemistry: $FeG_2 \rightleftharpoons FeG^+ + G^-$ followed by $G^- + H^+ \rightleftharpoons HG$. $G^-$ released when $FeG_2$ dissolves requires HCl to lower the pH to 8.50.

15. (b) Fixing $pK_w'$ at 13.797 causes $\bar{n}_H$(measured) to deviate systematically above $\bar{n}_H$(theoretical) at the end of the titration when $\bar{n}_H$ should approach 0.

16. (a) $\bar{n}_H$(experimental) $= 3 + \dfrac{[OH^-] + [Cl^-]_{HCl} - [H^+] - [Na^+]}{F_{H_3A}}$

$\bar{n}_H$(theoretical) $= 3\alpha_{H_3A} + 2\alpha_{H_2A} + \alpha_{HA}$
(b) $pK_w' = 13.819$, $pK_1 = 8.33$, $pK_2 = 9.48$, $pK_3 = 10.19$

17. (b) pH = 4.61. In the absence of precipitation, at pH 4.61, the concentrations are $[Cu^{2+}] = 18.6$ mM, $[SO_4^{2-}] = 18.7$ mM, $[CuSO_4(aq)] = 6.3$ mM, $[HSO_4^-] = 20$ μM, $[CuOH^+] = 12$ μM, $[Cu_2(OH)^{3+}] = 28$ μM, and $[Cu_2(OH)_2^{2+}] = 5.8$ μM; the other species have lower concentrations.

**(c)** The solubility of $Cu(OH)_{1.5}(SO_4)_{0.25}(s)$ is exceeded above pH $\approx 4.5$. $CuO(s)$ solubility is exceeded above pH $\approx 5$. $Cu(OH)_2(s)$ solubility is exceeded above pH $\approx 5.5$. At pH 4.61, $Cu(OH)_{1.5}(SO_4)_{0.25}(s)$ will precipitate from 0.025 M $CuSO_4$.

18. **(b)** $[T^{2-}] = \dfrac{F_{H_2T}}{\dfrac{[H^+]^2}{K_1K_2} + \dfrac{[H^+]}{K_2} + 1 + K_{NaT^-}[Na^+] + K_{NaHT}[Na^+]\dfrac{[H^+]}{K_2}}$

**(c)** $[HT^-] = \dfrac{F_{H_2T}}{\dfrac{[H^+]}{K_1} + 1 + \dfrac{K_2}{[H^+]} + K_{NaT^-}[Na^+]\dfrac{K_2}{[H^+]} + K_{NaHT}[Na^+]}$

$[H_2T] = \dfrac{F_{H_2T}}{1 + \dfrac{K_1}{[H^+]} + \dfrac{K_1K_2}{[H^+]^2} + K_{NaT^-}[Na^+]\dfrac{K_1K_2}{[H^+]^2} + K_{NaHT}[Na^+]\dfrac{K_1}{[H^+]}}$

**(d)** pH = 4.264, $[PyH^+]$ = 0.013 4, $[Na^+]$ = 0.018 5, $[K^+]$ = 0.010 0, $[OH^-] = 1.84 \times 10^{-10}$, $[HT^-]$ = 0.010 0, $[T^{2-}]$ = 0.007 92, $[Cl^-]$ = 0.015 0, $[NaT^-]$ = 0.001 17, $[H_2T] = 5.93 \times 10^{-4}$, $[Py]$ = 0.001 56, $[NaHT] = 2.97 \times 10^{-4}$ M

## Chapter 14

2. **(a)** $6.241\ 509\ 48 \times 10^{18}$  **(b)** 96 485.338 3

3. **(a)** $71._5$ A  **(b)** 4.35 A  **(c)** 79 W

4. **(a)** $1.87 \times 10^{16}$ $e^-$/s  **(b)** $9.63 \times 10^{-19}$ J/$e^-$  **(c)** $5.60 \times 10^{-5}$ mol
**(d)** 447 V

5. **(a)** $I_2$  **(b)** $S_2O_3^{2-}$  **(c)** 861 C  **(d)** 14.3 A

6. **(a)** $NH_4^+$ and Al, reducing agents; $ClO_4^-$, oxidizing agent
**(b)** 9.576 kJ/g

8. **(a)** $Fe(s)\,|\,FeO(s)\,|\,KOH(aq)\,|\,Ag_2O(s)\,|\,Ag(s)$
$FeO(s) + H_2O + 2e^- \rightleftharpoons Fe(s) + 2OH^-$
$Ag_2O(s) + H_2O + 2e^- \rightleftharpoons 2Ag(s) + 2OH^-$
**(b)** $Pb(s)\,|\,PbSO_4(s)\,|\,K_2SO_4(aq)\,\|\,H_2SO_4(aq)\,|\,PbSO_4(s)\,|\,PbO_2(s)\,|\,Pb(s)$
$PbSO_4(s) + 2e^- \rightleftharpoons Pb(s) + SO_4^{2-}$
$PbO_2(s) + 4H^+ + SO_4^{2-} + 2e^- \rightleftharpoons PbSO_4(s) + 2H_2O$

9. $Fe^{3+} + e^- \rightleftharpoons Fe^{2+}$; $Cr_2O_7^{2-} + 14H^+ + 6e^- \rightleftharpoons 2Cr^{3+} + 7H_2O$

10. **(a)** Electrons flow from Zn to C.  **(b)** 1.32 kg

11. **(a)** $O_2 + 4H^+ + 4e^- \rightleftharpoons 2H_2O$ and $CH_2O + H_2O \rightleftharpoons$
$CO_2 + 4H^+ + 4e^-$  **(b)** 24–120 mA  **(c)** 7–36 mW
**(d)** $O_2 + 4H^+ + 4e^- \rightleftharpoons 2H_2O$ and $HS^- \rightleftharpoons S + H^+ + e^-$

12. $Cl_2$ has the most positive $E°$.

13. **(a)** Fe(III)  **(b)** Fe(II)

15. **(a)** $Cu^{2+} + Zn(s) \rightleftharpoons Cu(s) + Zn^{2+}$  **(b)** $Zn^{2+}$

16. **(b)** $-0.356$ V

17. **(a)** $Pt(s)\,|\,Br_2(l)\,|\,HBr(aq, 0.10\ M)\,\|\,Al(NO_3)_3(aq, 0.010\ M)\,|\,Al(s)$
**(b)** $E = -2.854$ V, $e^-$ flow from Al to Pt  **(c)** $Br_2$  **(d)** 1.31 kJ
**(e)** $2.69 \times 10^{-8}$ g/s

19. **(a)** 0.572 V  **(b)** 0.568 V

20. 0.799 2 V

21. $HOBr + 2e^- + H^+ \rightleftharpoons Br^- + H_2O$; 1.341 V

22. $3X^+ \rightleftharpoons X^{3+} + 2X(s)$; $E_2° > E_1°$

23. 0.580 V

24. **(a)** 1.33 V  **(b)** $1 \times 10^{45}$

25. **(a)** $K = 10^{47}$  **(b)** $K = 1.9 \times 10^{-6}$

26. **(b)** $K = 2 \times 10^{16}$  **(c)** $-0.02_0$ V  **(d)** 10 kJ  **(e)** 0.21

27. $K = 1.0 \times 10^{-9}$

28. 0.101 V

29. 34 g/L

30. 0.116 V

31. $-1.664$ V

32. $K = 3 \times 10^5$

33. **(a)** $Al_2O_3(s) + MgO(s) \rightleftharpoons MgAl_2O_4(s)$
**(b)** $-29.51$ kJ/mol  **(c)** $\Delta H° = -23.60$ kJ/mol, $\Delta S° = 5.90$ J/(K·mol)

35. **(b)** $0.14_3$ M

36. **(b)** A = $-0.414$ V, B = 0.059 16 V  **(c)** Hg → Pt

37. $9.6 \times 10^{-7}$

38. $5.7 \times 10^{14}$

39. 0.76

40. $7.5 \times 10^{-8}$

42. **(c)** 0.317 V

43. $-0.041$ V

44. $-0.268$ V

45. $-0.036$ V

46. $7.2 \times 10^{-4}$

47. $-0.447$ V

48. **(a)** $[Ox] = 3.82 \times 10^{-5}$ M, $[Red] = 1.88 \times 10^{-5}$ M
**(b)** $[S^-] = [Ox]$, $[S] = [Red]$  **(c)** $-0.092$ V

## Chapter 15

1. **(b)** 0.044 V

2. **(a)** 0.326 V  **(b)** 0.086 V  **(c)** 0.019 V  **(d)** $-0.021$ V  **(e)** 0.021 V

3. 0.684 V

4. 0.243 V

5. 0.627

6. **(c)** 0.068 V

8. 0.481 V; 0.445 V; 0.194 V; $-0.039$ V

9. **(a)** $K_I/0.033\ 3$  **(b)** $K_{Cl}/0.020\ 0$  **(d)** $2.2 \times 10^6$

10. $3 \times 10^{21}$

11. **(b)** $1 \times 10^{11}$

12. $0.29_6$ M

15. left

16. **(a)** 42.4 s  **(b)** 208 s

17. **(a)** $3._2 \times 10^{13}$  **(b)** 8%  **(c)** 49.0, 8%

19. **(c)** 0.1 M HCl | 1 mM KCl, 93.6 mV; 0.1 M HCl | 4 M KCl, 4.7 mV

22. 10.67

23. potassium hydrogen tartrate and potassium hydrogen phthalate

25. $+0.10$ pH unit

26. **(a)** 274 mV  **(b)** 285 mV

27. pH = 5.686; slope = $-57.17_3$ mV/pH unit;
theoretical slope = $-58.17$ mV/pH unit; $\beta$ = 0.983

28. **(b)** 0.465  **(c)** $Na_2HPO_4$ = 0.026 8 $m$ and $KH_2PO_4$ = 0.019 6 $m$

30. Smaller $k_{A,X}$ is better.

34. **(a)** $-0.407$ V  **(b)** $1.5_5 \times 10^{-2}$ M  **(c)** $1.5_2 \times 10^{-2}$ M

35. $+0.029\ 6$ V

36. 0.211 mg/L

37. Group 1: $K^+$; Group 2: $Sr^{2+}$ and $Ba^{2+}$; $[K^+] \approx 100[Li^+]$

38. $3.8 \times 10^{-9}$ M

39. **(a)** $E = 51.10\ (\pm0.24) + 28.14\ (\pm0.08_5)$ log $[Ca^{2+}]$ $(s_y = 0.2_7)$
**(b)** 0.951  **(c)** $2.43\ (\pm0.04) \times 10^{-3}$ M

40. $-0.331$ V

41. $3.0 \times 10^{-5}$ M

42. $E = 120.2 + 28.80$ log $([Ca^{2+}] + 6.0 \times 10^{-4} [Mg^{2+}])$

44. **(a)** $8.9 \times 10^{-8}$ M  **(b)** 1.90 mmol

46. **(a)** $1.13 \times 10^{-4}$  **(b)** $4.8 \times 10^4$

# Chapter 16

1. **(d)** 0.490, 0.526, 0.626, 0.99, 1.36, 1.42, 1.46 V

2. **(d)** 1.58, 1.50, 1.40, 0.733, 0.065, 0.005, $-0.036$ V

3. **(d)** $-0.120$, $-0.102$, $-0.052$, 0.21, 0.48, 0.53 V

4. **(b)** 0.570, 0.307, 0.184 V

5. **(d)** $-0.143$, $-0.102$, $-0.061$, 0.096, 0.408, 0.450

6. diphenylamine sulfonic acid: colorless $\rightarrow$ red-violet; diphenylbenzidine sulfonic acid: colorless $\rightarrow$ violet; *tris*-(2,2$'$-bipyridine)iron: red $\rightarrow$ pale blue; ferroin: red $\rightarrow$ pale blue

7. no

13. **(a)** $MnO_4^- + 8H^+ + 5e^- \rightleftharpoons Mn^{2+} + 4H_2O$
**(b)** $MnO_4^- + 4H^+ + 3e^- \rightleftharpoons MnO_2(s) + 2H_2O$
**(c)** $MnO_4^- + e^- \rightleftharpoons MnO_4^{2-}$

14. 0.011 29 M

15. 0.586 4 M

16. **(a)** Scheme 1: $6H^+ + 2MnO_4^- + 5H_2O_2 \rightarrow 2Mn^{2+} + 5O_2 + 8H_2O$
Scheme 2: $6H^+ + 2MnO_4^- + 3H_2O_2 \rightarrow 2Mn^{2+} + 4O_2 + 6H_2O$
**(b)** Scheme 1: 25.43 mL; Scheme 2: 42.38 mL

17. 3.826 mM

18. 41.9 wt%

19. 78.67 wt%

20. oxidation number = 3.761; 217 $\mu$g/g

23. iodometry

24. **(a)** 1.433 mmol **(b)** 0.076 09 M **(c)** 12.8 wt%
**(d)** Do not add starch until right before the end point.

25. 11.43 wt%; just before the end point

26. 0.007 744 M; just before the end point

27. **(a)** $7 \times 10^2$ **(b)** 1.0 **(c)** 0.34 g/L

28. mol $NH_3$ = 2 (initial mol $H_2SO_4$) $-$ mol thiosulfate

29. **(a)** no, no **(b)** $I_3^- + SO_3^{2-} + H_2O \rightarrow 3I^- + SO_4^{2-} + 2H^+$
**(c)** $5.079 \times 10^{-3}$ M, 406.6 mg/L **(d)** no: $t_{calculated} = 2.56 < t_{table} = 2.776$

30. 5.730 mg

31. **(a)** 0.125 **(b)** $6.875 \pm 0.038$

33. **(a)** 0.191 5 mmol **(b)** 2.80 **(c)** 0.20 **(d)** 0.141 3

34. Bi oxidation state = $+3.200\,0$ ($\pm 0.003\,3$) Cu oxidation state = $+2.200\,1$ ($\pm 0.004\,6$) Formula = $Bi_2Sr_2CaCu_2O_{8.400\,1\ (\pm 0.005\,7)}$

# Chapter 17

1. Difference is due to overpotential.

2. 2.68 h

3. $-1.228\,8$ V

4. **(a)** $-1.906$ V **(b)** 0.20 V **(c)** $-2.71$ V **(d)** $-2.82$ V

5. $V_2$

6. **(a)** $6.64 \times 10^3$ J **(b)** 0.012 4 g/h

9. anode, 54.77 wt%

10. $-0.619$ V, negative

11. $-0.744$ V

14. **(a)** $5._2 \times 10^{-9}$ mol **(b)** $0.000\,2_6$ mL

15. **(a)** $5.32 \times 10^{-5}$ mol **(b)** $2.66 \times 10^{-5}$ mol **(c)** $5.32 \times 10^{-3}$ M

16. 151 $\mu$g/mL

17. **(a)** current density = $1.00 \times 10^2$ A/m$^2$, overpotential = 0.85 V
**(b)** $-0.036$ V **(c)** 1.160 V **(d)** $-2.57$ V

18. $96\,486.6_7 \pm 0.2_8$ C/mol

19. **(a)** $H_2SO_3 <$ pH 1.86; pH 1.86 $< HSO_3^- <$ pH 7.17; $SO_3^{2-} >$ pH 7.17 **(b)** cathode: $H_2O + e^- \rightarrow \frac{1}{2}H_2(g) + OH^-$
anode: $3I^- \rightarrow I_3^- + 2e^-$

---

**(c)** $I_3^- + HSO_3^- + H_2O \rightarrow 3I^- + SO_4^{2-} + 3H^+$
$I_3^- + 2S_2O_3^{2-} \rightleftharpoons 3I^- + S_4O_6^{2-}$ **(d)** 3.64 mM

20. **(a)** $B = c$; $C = x$; $D = n$
$o + A = 2B \Rightarrow o + A = 2c \Rightarrow A = 2c - o$
$h + 2A = 3D + E \Rightarrow h + 2(2c - o) = 3n + E \Rightarrow E = h + 4c - 2o - 3n$
$F = E - C = h + 4c - 2o - 3n - c = h - c/2 + o - 3n$
**(b)** $F/4$ **(c)** $2.22_3 \times 10^{-8}$ mol **(d)** 52.7 mg $O_2$/L **(e)** $2.26 \times 10^{-4}$ M

23. 15 $\mu$m, $7.8 \times 10^2$ A/m$^2$

26. 0.12%

27. 0.096 mM

29. **(a)** $Cu^{2+} + 2e^- \rightarrow Cu(s)$ **(b)** $Cu(s) \rightarrow Cu^{2+} + 2e^-$ **(c)** 313 ppb

30. Estimated relative peak heights are 1, $1.5_6$, and $1.9_8$.
Fe(III) in seawater = $1.0 \times 10^2$ pM

31. peak B: $RNHOH \rightarrow RNO + 2H^+ + 2e^-$
peak C: $RNO + 2H^+ + 2e^- \rightarrow RNHOH$
There was no RNO present before the initial scan.

32. $7.8 \times 10^{-10}$ m$^2$/s

# Chapter 18

1. **(a)** double, **(b)** halve, **(c)** double

2. **(a)** 184 kJ/mol **(b)** 299 kJ/mol

3. $5.33 \times 10^{14}$ Hz, $1.78 \times 10^4$ cm$^{-1}$, $3.53 \times 10^{-19}$ J/photon, 213 kJ/mol

5. $\nu = 5.088\,491\,0$ and $5.083\,335\,8 \times 10^{14}$ Hz, $\lambda = 588.985\,54$ and 589.582 86 nm, $\tilde{\nu} = 1.697\,834\,5$ and $1.696\,114\,4 \times 10^4$ cm$^{-1}$

10. $3.56 \times 10^4$ M$^{-1}$ cm$^{-1}$

11. violet-blue

12. $2.19 \times 10^{-4}$ M

13. **(a)** 325 nm: $T = 0.90$, $A = 0.045$; 300 nm: $T = 0.061$, $A = 1.22$
**(b)** 2.0% **(c)** $T_{winter} = 0.142$; $T_{summer} = 0.095$; 49%

14. **(a)** 40.0 kJ/mol **(b)** $1.97 \times 10^2$ M$^{-1}$ cm$^{-1}$

16. **(a)** $6.97 \times 10^{-5}$ M **(b)** $6.97 \times 10^{-4}$ M
**(c)** 1.02 mg

17. yes

18. **(a)** $7.87 \times 10^4$ M$^{-1}$ cm$^{-1}$ **(b)** $1.98 \times 10^{-6}$ M

19. **(a)** $4.97 \times 10^4$ M$^{-1}$ cm$^{-1}$ **(b)** 4.69 mg

20. $\Delta E(S_1 - T_1) = 36$ kJ/mol

23. wavelength: absorption $<$ fluorescence $<$ phosphorescence

25. Fluorescence is proportional to concentration up to 5 $\mu$M (with 5%).

26. $3.56\ (\pm 0.07) \times 10^{-4}$ wt%; 95% confidence interval: $3.56\ (\pm 0.22) \times 10^{-4}$ wt%

# Chapter 19

1. $[X] = 8.03 \times 10^{-5}$ M, $[Y] = 2.62 \times 10^{-4}$ M

2. $[Cr_2O_7^{2-}] = 1.78 \times 10^{-4}$ M, $[MnO_4^-] = 8.36 \times 10^{-5}$ M

5. $[A] = 9.11 \times 10^{-3}$ M, $[B] = 4.68 \times 10^{-3}$ M

6. $[TB] = 1.22 \times 10^{-5}$ M, $[STB] = 9.30 \times 10^{-6}$ M,
$[MTB] = 1.32 \times 10^{-5}$ M

7. $[p$-xylene$] = 0.062\,7$ M, $[m$-xylene$] = 0.079\,5$ M,
$[o$-xylene$] = 0.075\,9$ M, $[$ethylbenzene$] = 0.076\,1$ M

8. $pK_a = 4.00$

9. **(f)** $[CO_2(aq)] = 3.0\ \mu$M **(g)** $\mu \approx 10^{-4}$ M, yes

10. $K = 4.0 \times 10^9$ M$^{-1}$, $S = 0.29$ to 0.84

11. **(b)** $K = 88$

12. **(b)** $K = 0.464$, $\varepsilon = 1.074 \times 10^4$ M$^{-1}$ cm$^{-1}$

13. **(b)** $K = 0.464$, $\varepsilon = 1.073 \times 10^4$ M$^{-1}$ cm$^{-1}$

14. **(a)** 1:1 **(b)** $K$ must not be very large. **(c)** to maintain constant ionic strength

16. yes

20. $pK_a \approx 10.8$

21. **(b)** $N_{av} = 55.9$ **(c)** $[M] = 0.227$ mM; $\overline{Q} = 0.881$ molecules per micelle **(d)** $P_0 = 0.414$; $P_1 = 0.365$; $P_2 = 0.161$

## Chapter 20

3. $D_2$

8. **(a)** $2.38 \times 10^3$ lines/cm **(b)** 143

10. **(a)** $1.7 \times 10^4$ **(b)** 0.05 nm **(c)** $5.9 \times 10^4$ **(d)** $0.000\,43°$, $0.013°$

11. $T = 0.036\,4$, $A = 1.439$

12. $0.124\,2$ mm

13. $77$ K: $1.99$ W/m²; $298$ K: $447$ W/m²

14. **(a)** $M_\lambda = 8.79 \times 10^9$ W/m³ at $2.00\ \mu$m; $M_\lambda = 1.164 \times 10^9$ W/m³ at $10.00\ \mu$m **(b)** $1.8 \times 10^2$ W/m² **(c)** $2.3 \times 10^1$ W/m²
**(d)** $\dfrac{M_{2.00\ \mu m}}{M_{10.00\ \mu m}} = 7.55$ at $1\,000$ K,
$\dfrac{M_{2.00\ \mu m}}{M_{10.00\ \mu m}} = 3.17 \times 10^{-22}$ at $100$ K

15. $2.51 \times 10^{-6}$

16. **(a)** $34°$ **(b)** $0°$

19. **(a)** $80.7°$ **(b)** $0.955$

20. **(a)** $61.04°$ **(b)** $51.06°$

21. $n_{prism} > \sqrt{2}$

22. **(b)** $76°$ **(b)** $0.20$

23. **(a)** $0.964$ **(b)** $343$ nm, $5.83 \times 10^{14}$ Hz

24. **(b)** blue

25. **(a)** $\pm 2$ cm **(c)** $0.5$ cm$^{-1}$ **(d)** $2.5$ mm

27. 7

29. [at cycles, predicted $S/N$] at $1\,000$, $60.0$ (observed); at $300$, $32.9$; at $100$, $19.0$; at $1$, $1.90$

## Chapter 21

10. Pb: $1.2 \pm 0.2$; Tl: $0.005 \pm 0.001$; Cd: $0.04 \pm 0.01$; Zn: $2.0 \pm 0.3$; Al: $7\,(\pm 2) \times 10^1$ ng/cm²

11. $589.3$ nm

12. $0.025$

13. Na: $0.003_8$ nm; Hg: $0.000\,5_6$ nm

14. **(a)** $283.0$ kJ/mol **(b)** $3.67 \times 10^{-6}$ **(c)** $+8.4\%$ **(d)** $1.03 \times 10^{-2}$

15. wavelength (nm):     591     328     154
$N^*/N_0$ at $2\,600$ K in flame:  $2.6 \times 10^{-4}$  $1.4 \times 10^{-7}$  $1.8 \times 10^{-16}$
$N^*/N_0$ at $6\,000$ K in plasma:  $5.2 \times 10^{-2}$  $2.0 \times 10^{-3}$  $1.2 \times 10^{-7}$

18. $0.429 \pm 0.012$ wt%

19. **(a)** $7.49\ \mu$g/mL **(b)** $25.6\ \mu$g/mL

20. $17.4 \pm 0.3\ \mu$g/mL

21. $4.54\ \mu$M

22. **(a)** CsCl inhibits ionization of Sn. **(b)** $m = 0.782 \pm 0.019$; $b = 0.86 \pm 1.56$; $R^2 = 0.997$ **(c)** Little interference at $189.927$ nm, which is the better choice of wavelengths. At $235.485$ nm, there is interference from Fe, Cu, Mn, Zn, Cr, and, perhaps, Mg.
**(d)** limit of detection $= 9\ \mu$g/L; limit of quantitation $= 31\ \mu$g/L
**(e)** $0.8$ mg/kg

## Chapter 22

3. $1$ Da $= 1.660\,54 \times 10^{-24}$ g, $83.5\ (\pm 2.3)$ fg

4. $58.5_1$ from the spectrum in the text

5. $1.5 \times 10^4$, yes

6. $\sim 3\,100$

7. $2.0 \times 10^6$, $3.4 \times 10^6$

8. $C_6H_{11}^+$

9. $^{31}P^+ = 30.973\,21$, $^{15}N^{16}O^+ = 30.994\,47$, $^{14}N^{16}OH^+ = 31.005\,25$

10. $1 : 1.946 : 0.946\,3$

11. $1 : 8.05 : 16.20$

12. **(a)** 4 **(b)** 6 **(c)** $1\frac{1}{2}$

13. **(a)** ⬡—Cl, $C_6H_5Cl$: $M^{+\cdot} = 112$

**(b)** Cl—⬡—Cl, $C_6H_4Cl_2$: $M^{+\cdot} = 146$

**(c)** ⬡—NH₂, $C_6H_7N$: $M^{+\cdot} = 93$

**(d)** $(CH_3)_2Hg$: $M^{+\cdot} = 228$ **(e)** $CH_2Br_2$: $M^{+\cdot} = 172$

**(f)** , 1,10-phenanthroline, $C_{12}H_8N_2$: $M^{+\cdot} = 180$

**(g)** ⬡—Fe—⬡, ferrocene, $C_{10}H_{10}Fe$: $M^{+\cdot} = 186$

14. $M^{+\cdot} = 206$, $CH^{79}Br_2{}^{35}Cl$

16. **(a)** mass of $p^+ + e^- =$ mass of $^1H$ **(b)** mass of $p^+ + n + e^- = 2.016\,489\,963$ Da; mass of $^2H = 2.014\,10$ Da
**(c)** $2.15 \times 10^8$ kJ/mol **(d)** $1.31 \times 10^3$ kJ/mol **(e)** $5 \times 10^5$

17. $0.342\,7 : 1 : 0.972\,8 : 0.315\,4$

18. (mass, intensity): $(84, 1)$ $(85, 0.152)$ $(86, 0.108)$ $(87, 0.010\,3)$ $(88, 0.003\,62)$ $(89, 0.000\,171)$ $(90, 0.000\,037)$

20. $4.39 \times 10^4$ m/s; $45.6\ \mu$s; $2.20 \times 10^4$ spectra/s; $1.56 \times 10^4$ spectra/s

22. $93$ m

27. **(a)** negative ion mode, neutral solution **(b)** $14.32$

28. $n_A = 12$ and $n_I = 20$; mean molecular mass (disregarding peak G) is $15\,126$ Da

29. charge $= 4$; molecular mass $= 7\,848.48$ Da

30.

| | | $[MNH_4]^+ = C_{37}H_{72}ON$ | $[MH]^+ = C_{37}H_{69}O$ |
|---|---|---|---|
| 37:3: | X + 1: | 41.2% predicted | 40.8% predicted |
| | | 35.8% observed | 23.0% observed |
| | X + 2: | 7.9% predicted | 7.9% predicted |
| | | 7.0% observed | 8.0% observed |
| 37:2: | | $[MNH_4]^+ = C_{37}H_{74}ON$ | $[MH]^+ = C_{37}H_{71}O$ |
| | X + 1: | 41.3% predicted | 40.8% predicted |
| | | 40.8% observed | 33.4% observed |
| | X + 2: | 7.9% predicted | 7.9% predicted |
| | | 3.7% observed | 8.4% observed |

32. **(d)** $7.63_9\ \mu$mol V/g

## Chapter 23

2. 3

7. **(a)** $0.080$ M **(b)** $0.50$

8. $0.088$

9. **(c)** $4.5$ **(d)** greater

11. **(a)** $0.16$ M in benzene **(b)** $2 \times 10^{-6}$ M in benzene

12. $2$ pH units

13. **(a)** $2.6 \times 10^4$ at pH 1 and $2.6 \times 10^{10}$ at pH 4 **(b)** $3.8 \times 10^{-4}$

15. 1-C, 2-D, 3-A, 4-E, 5-B

18. **(a)** $17.4$ cm/min **(b)** $0.592$ min **(c)** $6.51$ min

19. **(a)** $13.9$ m/min, $3.00$ mL/min **(b)** $k' = 7.02$, fraction of time $= 0.875$ **(c)** $295$

20. **(a)** $40$ cm long $\times$ $4.25$ cm diameter **(b)** $5.5$ mL/min
**(c)** $1.11$ cm/min for both

21. **(a)** $2.0$ **(b)** $0.33$ **(c)** $20$

22. 19 cm/min

23. 0.6, 6

24. $k' = 3.59$, $K = 4.69$

25. 603, 0.854

26. (a) $1.2_6 \times 10^4$  (b) 40 μm  (c) 0.72

27. (a) 1  (b) 2  (c) 1  (d) neither  (e) B  (f) B  (g) 1.25

29. 0.1 mm

32. 33 mL/min

36. 2.65 mm

37. (a) $1.1 \times 10^2$  (b) 0.89 mm

38. (a) $4.1 \times 10^3$  (b) Equation 23-28: $N = 7.72 \times 10^3$, Equation 23-29: $N = 7.75 \times 10^3$

39. resolution = 0.83

40. 10.4 mL

41. $110 \text{ s}^2$, $43 \text{ s}^2$, 26.9 s

42. $1.3 \times 10^3$

43. (a) $\gamma = 1.01$, $N = 640\,000$; $\gamma = 1.05$, $N = 25\,600$; $\gamma = 1.10$, $N = 6\,400$

44. (a) $k' = 11.25$, 11.45  (b) 1.018  (c) 1.017  (d) $C_6HF_5$: 60 800 plates, height = 0.493 mm; $C_6H_6$: 66 000 plates, height = 0.455 mm  (e) $C_6HF_5$: 55 700 plates; $C_6H_6$: 48 800 plates  (f) 0.96  (g) 0.97

## Chapter 24

8. (a) hexane < butanol < benzene < 2-pentanone < heptane < octane  (b) hexane < heptane < butanol < benzene < 2-pentanone < octane  (c) hexane < heptane < octane < benzene < 2-pentanone < butanol

9. (a) 3, 1, 2, 4, 5, 6  (b) 3, 4, 1, 2, 5, 6  (c) 3, 4, 5, 6, 2, 1

10. (a) 4.7 min, 1.3  (b) 1.8

11. 836

12. $a = 1.69 \times 10^3$ K, $b = -3.36$, 27.1 min

17. (b) 0.16 ng (narrow bore), 56 ng (wide bore)

19. (a) $1.25_3$  (c) 77.6 mM

20. 0.41 μM

21. 932

23. (d) nonsmoker: $78 \pm 5$ μg/L; nonsmoker whose parents smoke: $192 \pm 6$ μg/L

24. $[^{14}NO_2^-] = 1.8$ μM; $[^{14}NO_3^-] = 384$ μM

25. (a) $A = 0$  (b) $B = 2D_m$
(c) $C = C_s + C_m = \dfrac{2k'}{3(k'+1)^2}\dfrac{d^2}{D_s} + \dfrac{1 + 6k' + 11k'^2}{24(k'+1)^2}\dfrac{r^2}{D_m}$
(d) $u_x$ (optimum) $= \sqrt{\dfrac{B}{C}}$; $H_{min} = \sqrt{2B(C_s + C_m)}$

26. (a) 0.58, 1.9  (b) 0.058 mm, 0.19 mm  (c) $3.0 \times 10^5$  (d) 4.0

27. $D_m = 3.0 \times 10^{-5} \text{ m}^2/\text{s}$, $D_s = 5.0 \times 10^{-10} \text{ m}^2/\text{s}$

28. (b) limiting $m = KV_fC_0$  (c) 0.69%, 41%

29. (a) 6  (b) 350  (c) 350, 315, 280, 245, 210 = M⁺, (M-Cl)⁺, (M-2Cl)⁺, (M-3Cl)⁺, (M-4Cl)⁺, (M-5Cl)⁺  (d) For 5Cl, predicted relative abundances = 62.5 : 100 : 64.0 : 20.4 : 3.3 : 0.2. For 4Cl, abundances = 78.2 : 100 : 48.0 : 10.2 : 0.8. For 3Cl, abundances = 100 : 95.9 : 30.7 : 3.3. For 2Cl, abundances = 100 : 64.0 : 10.2.

## Chapter 25

1. (b) pressure gradient (= density gradient)

4. (a) $L = 33, 17, 10, 5$ cm

5. 0.14 min; 0.30 min

10. (a) 1 560 for L enantiomer and 1 310 for D enantiomer  (b) 1.25  (c) 1.35

11. (a) 18  (b) 10

12. (a) bonded reversed-phase chromatography  (b) bonded normal-phase chromatography  (c) ion-exchange or ion chromatography  (d) molecular exclusion chromatography  (e) ion-exchange chromatography  (f) molecular exclusion chromatography

14. $0.27 \text{ m}^2$

15. (a) shorter  (b) amine

16. 126 mm

18. (a) $t_m = 0.38$ min for column A and 0.26 min for column B  (d) $k' = 1.3$ for morphine 3-β-D-glucuronide and 3.3 for morphine  (e) 6.2

19. (a) m/z 304 is BH⁺ (cocaine protonated at N)  (b) loss of $C_6H_5CO_2H$  (e) phenyl group

20. (a) m/z 234 is MH⁺, m/z 84 is $C_5H_{10}N^+$  (b) 237 → 84

21. 1.1 min for both

29. 27.8 min predicted, 20.2 min observed

30. ~36 min

31. (b) between B and F: 30% methanol/8% tetrahydrofuran/62% buffer; between F and C: 10% methanol/24% tetrahydrofuran/66% buffer

32. D: 25% acetonitrile/30% methanol/45% buffer
E: 25% acetonitrile/20% tetrahydrofuran/55% buffer
F: 30% methanol/20% tetrahydrofuran/50% buffer
G: 16.7% acetonitrile/20% methanol/13.3% tetrahydrofuran/50% buffer

33. 38%

34. (a) lower  (b) higher

35. 40 to 70% acetonitrile in 60 min

36. (a) Change solvent strength, temperature, or pH. Use a different solvent or a different kind of stationary phase.  (b) slower flow rate, different temperature, longer column, smaller particle size

37. (a) ~29 min  (b) $k* = 12.9$, $F = 4.7$ mL/min, $m = 4.7$ mg, $t_G = 11.5$ min

## Chapter 26

6. 38–75 μm; 200/400 mesh

8. (a) 30  (b) 3.3  (c) increase

9. cation charge = 0.002 02 M, anion charge = −0.001 59 M; either some concentrations are inaccurate or some ionic material was not detected.

10. 38.0%

12. (b) 29 ng/mL

14. (a) $3.8 \times 10^2$ h  (b) 8.0 mA, 0.16 A

16. (b) 10 μm

17. (a) 40.2 mL  (b) 0.53

18. 0.16

19. (a) 2 000 Da  (b) 300 Da

20. (a) 5.7 mL  (b) 11.5 mL  (c) Solutes must be adsorbed.

21. 320 000

26. (a) 0.167 mm  (b) 0.016 s  (c) 0.000 40 s  (d) 0.064 s

29. (a) $1.15 \times 10^4$ Pa  (b) 1.17 m

30. (a) 29.5 fmol  (b) $3.00 \times 10^3$ V

31. $9.2 \times 10^4$ plates, $4.1 \times 10^3$ plates (My measurements are about 1/3 lower than the values labeled in the figure from the original source.)

32. (a) maleate  (b) Fumarate is eluted first.  (c) Maleate is eluted first.

33. (a) pH 2: 920 s; pH 12: 150 s  (b) pH 2: never; pH 12: 180 s

34. (a) $t_{120\,kV}/t_{28\,kV} = 4.3$ (observed ratio = 3.9)  (b) $N_{120\,kV}/N_{28\,kV} = 4.3$  (c) $\sigma_{120\,kV}/\sigma_{28\,kV} = 0.48$  (d) Increasing voltage decreases migration time, giving bands less time to spread apart by diffusion.

36. $1.3_5 \times 10^4$ plates, 30 μm

37. 20.5 min

38. light chain = 17 300 Da, heavy chain = 23 500 Da

39. $z_0 = -3.28$

40. $2.0 \times 10^5$ plates

41. thiamine < (niacinamide + riboflavin) < niacin. Thiamine is most soluble.

42. Cyclobarbital and thiopental each separate into two peaks because each has a chiral carbon atom.

44. 5.55

41. **(c)** $n_{obs} = 1.45 \times 10^5$, $n_{theory} = 2.06 \times 10^5$ **(d)** 1.6

## Chapter 27

9. 38 Hz/ag; 1 ag = 6 000 molecules

10. 0.022 86 M

11. 1.94 wt%

12. 225.3 g/mol

13. 0.085 38 g

14. 50.79 wt%

15. 0.191 4 g calcium carbonate, 0.107 3 g calcium oxide

16. 104.1 ppm

17. 7.22 mL

18. 0.339 g

19. **(a)** 19.98%

20. 14.5 wt% $K_2CO_3$, 14.6 wt% $NH_4Cl$

21. 40.4 wt%

22. 22.65 wt%

23. **(a)** 40.05 wt% **(b)** 39%

24. **(a)** 1.82 **(b)** $Y_2O_2(OH)Cl$ or $Y_2O(OH)_4$

25. **(b)** 0.204 ($\pm 0.004$)

26. **(a)** 5.5 mg/100 mL **(b)** 5.834 mg, yes

27. **(a)** Fe (98.3%), Pb (104.0%), Cd (98.6%), In (97.6%), Cr (<0.3%), Mn (36.5%), Co (6.4%), Ni (<0.3%), and Cu (44.2%) **(b)** $Fe^{3+}$, $Pb^{2+}$, $Cd^{2+}$, and $In^{3+}$ **(c)** 10

32. 11.69 mg $CO_2$, 2.051 mg $H_2O$

33. $C_8H_7NO_2SBrCl + 9\frac{1}{4}O_2 \rightarrow 8CO_2 + \frac{5}{2}H_2O + \frac{1}{2}N_2 + SO_2 + HBr + HCl$

34. $C_4H_9NO_2$

35. 10.5 wt%

36. $C_8H_{9.06 \pm 0.17}N_{0.997 \pm 0.010}$

37. 12.4 wt%

38. **(a)** 95% confidence: $10.16_0 \pm 1.93_6$ μmol $Cl^-$ (Experiment 1), $10.77_0 \pm 2.29_3$ μmol $Cl^-$ (Experiment 2) **(b)** difference is *not* significant **(c)** $24.2_{95}$ mg $BaSO_4$ **(d)** 4.35%

## Chapter 28

3. **(a)** 5% **(b)** 2.6%

4. 1.0 g

5. 120/170 mesh

6. $10^4 \pm 0.99\%$

7. **(a)** 15.8 **(b)** 1.647 **(c)** 474–526

8. 95%: 8; 90%: 6

9. **(a)** 5.0 g **(b)** 7

10. $0.34 \pm 0.14$ pg

11. **(a)** $Na_2CO_3$: 4.47 μg, $8.94 \times 10^5$ particles; $K_2CO_3$: 4.29 μg, $2.24 \times 10^7$ particles **(b)** $2.33 \times 10^4$ **(c)** $Na_2CO_3$: 3.28%; $K_2CO_3$: 0.131%

12. Zn, Fe, Co, Al

13. prevents possible explosion

17. **(a)** 53

18. 64.90 wt%

# Index

## Abbreviations

App. = Appendix   f = footnote   NR = Notes and References
b = box           i = illustration   p = problem
d = demonstration  m = marginal note   t = table

Triiodide, 340
Trimethylammonium chloride, 165
Trimethylphosphate, 633t
Trimethylsilyl derivative, 528i
Trinitrotoluene, 422p
Trioctylphosphine, 602i
Triple point, 9t, 568b
Triplet state, 388, 416
Triprotic acid, 114, 188–189,
    191i, 221t
TRIS, 169, 174t, 175t, 217t
Tris (2-aminoethyl) amine, 268p
Tris(2,2'-bipyridine)iron, 333t
Tris(2,2'-bipyridine)ruthenium(II),
    333t, 423p
Tris(hydroxymethyl aminomethane),
    169, 174t, 175t, 217t
$N$-Tris(hydroxymethyl)methyl-2-
    aminoethanesulfonic acid, 175t
$N$-Tris(hydroxymethyl)methylglycine,
    174t, 175t
Tris(5-nitro-1,10-phenanthroline)iron,
    333t
Tris(1,10-phenanthroline)iron, 333t
Trititrotoluene, 402
Tritium, 307
Triton X-100, 504d
Tropaeolin O, 215t
Tryptophan, 182t
TSK silica, 600t
Tswett, M. S., 506m
$t$ Test, 59–62, 64–65, NR2 (4.5)
Tumor, 74p, 229–230
Tungsten, 337t, 459
Tungsten carbide, 650, 651
Tungsten lamp, 426, 426i, 571
Tungsten trioxide, 467, 638
Turbidity, 131m
Two-tail $t$-test, 64, 65m
Tylenol, 360
Type A silica, 560
Type B silica, 560
Type C silica, 560
Tyrosine, 182t, 199

Ulcer, 424
Ultrapure acid, 123b
Ultrasonic mixing, 622
Ultrasonic nebulizer, 460
Ultrasonic vibration, 366
Ultraviolet:
    absorbance standard, 432t
    cutoff, 563t
    detector, 571, 571t, 613
    radiation, 10, 338b, 378, 380i
Uncertainty, 44–50, 58, App. C
    calibration curve, 71
    confidence interval, 58
    exponents, 47–48
    least-squares parameters, 68–69
    logarithms, 47–48
    molecular mass, 49
    pipet, 49–50
    principle, 462
    propagation, 49t
    sampling, 646–650
    standard addition, 90
    $x$-intercept, 90
Uncle Wilbur, 400p
Unified atomic mass unit, 476b
Unit matrix, 406
Units, 11
Units of measurement, 9–11
Unmentionables, 391b
Uracil, 576
Uranium, 337t, 628
Urea, 120p, 218, 424, 631, 633t
Uric acid, 360

Urine, 108i, 123, 553p, 655, 658
Use objective, 79

V (volt), 10t
Vacancy, 313i
Vacuum desiccator, 31i
Vacuum pump, 537
Vacuum tube, 307m
Vacuum ultraviolet, 462
Valine, 182t
Valinomycin, 304
Vanadium(V) oxide, 651
Vanadium(V) spectrum, 403i
Vanadate, 630t
Vanadyl sulfate, 623p
van Deemter equation, 517, 538
    electrophoresis, 604–605
    experimental graph, 518i, 537i,
        558i, 627p
    micellar electrokinetic
        chromatography, 627p
Variable voltage source, 349i
Variamine blue B base, 243i
Varian Cary E3 spectrophotometer,
    425i
Variance, 54
    additivity, 517, 646
    $F$ test, 63–64
Vascular disease, 1
Vasodilator, 2
Vector, 406
Velocity, electroosmotic, 607m, 609
Ventilation blood test, 312t
Vernier scale, 51p
Vesicle, 18p, 604
Vessel:
    digestion, 651, 655
    extraction, 657i
Vibration, molecular, 389
Vibrational frequency, 20
Vibrational relaxation, 390
Vibrational states, 390–391
Vibrational structure, 393
Vibrational transition, 390
Vicinal silanol groups, 560i
Vinegar pH, 108i
Virginia Polytechnic Institute, 277d
Viscosity, 558, 605, 607
Visible light, 380i
Vision, 435b
Vitamin A, 328t, 435b
Vitamin C, 232b, 289t, 290i, 343, 360
    formal potential, 289–291
Vitamin E, 328t
Void volume, 600
Vol%, 13
Volatile buffers, 574
Volatile electrolyte, 634
Volhard titration, 133, 134t
Volt, 10t, 272
Voltage, 272
    breakdown, 319m
    capillary electrophoresis, 609, 610
    open circuit, 350m
    ramp, 363
    scale, 300
    source, 349i
    unit, 10t
Voltaic cell, 274
Voltammetry, 362–370
Voltammogram, 362
Voltmeter, 277
Volume:
    column, 600
    flow rate, 508
    gravimetric analysis, 631
    units, 11t
Volume percent, 13

Volumetric analysis, 121
Volumetric flask, 14, 14i, 26–27

W (watt), 10t
Waage, P., 97m
Walden reductor, 336
Wall-coated open tubular column, 529
Warning line, 81b
Waste:
    disposal, 21b, 576m, NR1 (2.5)
    reduction in HPLC, 568m
    remediation, 327m
Wastewater, 228r, 338b
Watchglass, 31i
Water:
    acid-base effect on dissociation,
        161
    analysis, 1, 596i
    arsenic removal, 634m
    carbon dioxide content, 108
    coordination, 140
    critical constants, 568t
    density, 32t, 33
    diffusion coefficient, 513t
    distillation, 645t
    electrolysis, 350d
    expansion, 31–33
    hardness, 209b, 245b, 594m
    infrared absorption, 394
    ion chromatography, 596i
    ionization, 150–151, 161
    irrigation, 209b, 245b
    isotopic molecules, 394
    Karl Fischer titration, 370–372
    lead, 466i
    metal content impurities, 504d
    pH, 108
    "polishing," 594
    potential range, 563t
    purity, 564m
    refractive index, 438m
    resistivity, 594
    softener, 594m
    solvation of ions, 140
    temperature dependence of $K_w$, 107t
    thermal expansion, 31–33
    triple point, 9t
    water of hydration, 140
Watt, 10t, 273
Waveform, 367i
Waveguide, 439–440
Wavelength, 379
    color, 380i
    selection for analysis, 384
Wavenumber, 379
WCOT, 529
Weak acid, 110–114
    conjugate, 167
    diprotic, 181–188
    fraction of dissociation, 164,
        191–192
    intermediate form, 184
    pH calculation, 162–166
    polyprotic, 188–189
    reaction with strong base, 170b
    titration with strong base, 202–204
    titration with weak base, 218–219
Weak base, 110–114
    pH calculation, 166–167
    reaction with strong acid, 170b
    titration with weak acid, 218–219
Weak electrolyte, 12, 164, 165d
Weakly acidic ion exchanger, 589,
    590t
Weakly basic ion exchanger, 589, 590t
Weighing bottle, 23, 31i
Weighing errors, 23–24
Weighing procedure, 22–24

Weight:
    effect of altitude, 37p
    tolerance of lab standards, 24t
Weight percent, 13
Wenzel, T., 2
Weston cell, 374p
Wet ashing, 650, 654
Wetterhahn, K., 21m
Whistle noise, 449
White blood cell, 158
White chocolate, 6t
White light, 435
Whitener, 391b
White noise, 448
Wide-bore column, 550t
Wig-L-Bug, 650, 651i
Wind power, 415b
Wine, 346p, 375p
Work, 10t, 272–273
Working electrode, 349, 353
    potential range, 363m
    rotating disk, 361, 362i
Wright, W.H., 307m
Wt%, 13

$\bar{x}$ (mean value), 54
Xanthine, 289t
Xenon lamp, 571
X-Gal, 1
$x$-intercept, 90
X-ray, 380i
$p$-Xylene in iodine titrations, 340m
Xylenol orange, 210i, 242t, 243i, 419p

y (yocto), 10t
Y (yotta), 10t
Yalow, R., 411m
Yeast, 194b, 439
YO, 467i
Yocto, 10t
Yoctomole, 9
Yotta, 10t
Yttrium-90, 230
Yttrium aluminum garnet, 431i
Yttrium barium copper oxide, 327,
    342b, 467i, 642p
Yttrium hydroxychloride, 642p
Yttrium oxide, 642p

$z$ (multiple of standard deviation), 55
z (zepto), 10t
Z (Zetta), 10t
Zarontin, 43b
Zeeman effect, 466
Zeolite, 534i
Zepto, 10t
Zeptomole, 9
Zeros, 40
Zetta, 10t
Zinc:
    ammonia complexes, 239
    EDTA titrations, 244
    electrode, 302
    gravimetric analysis, 630t
    hydroxide complexes, 119p
    impurity in tungsten, 459
    iodimetric titration, 341t
    Jones reductor, 336
    masking, 245
    permanganate titration, 337t
    precipitation titration, 134t
    spectrophotometric analysis, 504d
Zinc sulfate reagent, 123b
Zirconia, 519i, 650
Zirconia oxygen sensor, 359b
Zone broadening, electrophoresis, 609
Zorbax silica, 560i
Zwitterion, 181
Zwitterionic bonded phase, 567i